SPACE SCIENCE SERIES

Tom Gehrels, General Editor

Planets, Stars and Nebulae Studied with Photopolarimetry
Tom Gehrels, editor, 1974, 1133 pages

Jupiter
Tom Gehrels, editor, 1976, 1254 pages

Planetary Satellites
Joseph A. Burns, editor, 1977, 598 pages

Protostars and Planets
Tom Gehrels, editor, 1978, 756 pages

Asteroids
Tom Gehrels, editor, 1979, 1181 pages

Comets
Laurel L. Wilkening, editor, 1982, 766 pages

Satellites of Jupiter
David Morrison, editor, 1982, 972 pages

Venus
D.M. Hunten, L. Colin, T.M. Donahue and V.I. Moroz, editors, 1983, 1143 pages

Saturn
Tom Gehrels and Mildred S. Matthews, editors, 1984, 968 pages

Planetary Rings
Richard Greenberg and André Brahic, editors, 1984, 784 pages

Protostars and Planets II
David C. Black and Mildred S. Matthews, editors, 1985, 1293 pages

Satellites
Joseph A. Burns and Mildred S. Matthews, editors, 1986, 1021 pages

The Galaxy and the Solar System
Roman Smoluchowski, John N. Bahcall and Mildred S. Matthews, editors, 1986, 485 pages

Meteorites and the Early Solar System
John F. Kerridge and Mildred S. Matthews, editors, 1988, 1269 pages

Mercury
Faith Vilas, Clark R. Chapman and Mildred S. Matthews, editors, 1988, 794 pages

Origin and Evolution of Planetary and Satellite Atmospheres
S.K. Atreya, J.B. Pollack and M.S. Matthews, editors, 1989, 881 pages

Asteroids II
Richard P. Binzel, Tom Gehrels and Mildred S. Matthews, editors, 1989, 1258 pages

Uranus
Jay T. Bergstralh, Ellis D. Miner and Mildred S. Matthews, editors, 1991, 1076 pages

The Sun in Time
C.P. Sonett, M.S. Giampapa and M.S. Matthews, editors, 1991, 996 pages

Solar Interior and Atmosphere
A.N. Cox, W.C. Livingston and M.S. Matthews, editors, 1991, 1414 pages

Mars
H.H. Kieffer, B.M. Jakosky, C.W. Snyder, and M.S. Matthews, editors, 1992, 1536 pages

Protostars and Planets III
E.H. Levy and J.I. Lunine, editors, 1993, 1596 pages

Resources of Near-Earth Space
J. Lewis, M.S. Matthews and M. Guerrieri, editors, 1993, in press

PROTOSTARS
AND
PLANETS III

Eugene H. Levy
Jonathan I. Lunine

Editors

With the editorial assistance of
M. Guerrieri and M. S. Matthews

With 91 collaborating authors

THE UNIVERSITY OF ARIZONA PRESS
TUCSON & LONDON

About the cover:

Within a star-forming region. This view is based in part on discoveries since the Protostars and Planets II meeting, including new infrared imaging. Approximately visual colors are used, except that deep red color sensitivity is extended as far as 2 μm into the infrared. A trapezium-like cluster of O and B stars has formed and cleared a cavity in the interstellar medium. Expansion has produced elephant trunk nebulae and globules. Infrared stars, both single and in groupings, can be seen within the densest portions of some of the globules. Several disk-shaped cocoon nebulae can be found, including one interacting pair. (Painting by William K. Hartmann)

The University of Arizona Press

Copyright © 1993
The Arizona Board of Regents
All Rights Reserved

⊗ This book is printed on acid-free, archival-quality paper.
Manufactured in the United States of America.

98 97 96 95 94 93   6 5 4 3 2 1

Library of Congress Cataloging-in-Publication Data

Protostars and planets III / Eugene H. Levy, Jonathan I. Lunine, editors ;
    with the assistance of Mary Guerrieri and Mildred S. Matthews ;
    with 91 collaborating authors.
        p.   cm. — (Space science series)
    Includes bibliographical references and index.
    ISBN 0-8165-1334-1
    1. Protostars. 2. Planetology. 3. Interstellar clouds.
    4. Stars—Formation. 5. Disks (Astrophysics) I. Levy, Eugene H.
(Eugene Howard), 1944– .   II. Lunine, Jonathan Irving.
III. Title: Protostars and planets III. IV. Title: Protostars and
planets III. V. Series.
QB806.P78   1993
523.8—dc20   92-44841
                                                       CIP

British Library Cataloguing in Publication data are available.

CONTENTS

Part III—DISKS AND OUTFLOWS

Part IV—DISK PROCESSES AND PLANETARY MATTER

COLLABORATING AUTHORS

Stewart, G. R., 1061
Strom, K. M., 245
Strom, S. E., 837
Swindle, T. D., 867
Tittemore, W. C., 1149
Tonks, W. B., 1339
Truran, J. W., 75
Tscharnuter, W. M., 921
van Dishoeck, E. F., 163
Vickery, A. M., 1339

Walter, F. M., 405
Weidenschilling, S. J., 1031
Weissman, P. R., 1177
Wilking, B. A., 429
Williams, R. E., 75
Woolum, D. S., 903
Zahnle, K., 1305
Zinnecker, H., 429
Zweibel, E. G., 279, 327

PREFACE

Modern astrophysics began near the turn of the century with the study of stars grounded in fundamental understanding of the structure and behavior of matter. Since then, astrophysical frontiers have been pushed simultaneously in two opposite directions: toward larger and larger scales to understand what we know as our universe, and toward smaller and smaller scales to understand the formation of stars and planets. In some sense, astrophysics becomes a particularly humanistic endeavor as it pushes toward planets. Human beings are peculiarly a planetary phenomenon; life itself is likely to be a planetary phenomenon throughout the universe. Attempts to understand the mechanisms of planet formation appeal to several deep human motivations: to understand ourselves in relation to the universe of which we are part, and to grasp the possibilities that the universe offers for planets and life elsewhere.

During the past several decades, the study of planetary-system formation has occupied increasing attention and has advanced rapidly. A number of developments have contributed to the florescence of this investigation. The empirical basis of our understanding has expanded greatly over the past three decades as a result of complementary advances made in planetary science and in traditional astrophysics. Exploration of the solar system by spacecraft has brought previously remote planets within the reach of direct study with instruments capable of measurements not possible with astronomical methods alone. Technological advances in measurement capabilities have provided microscopically detailed information about solar system matter not available in previous decades. Advances in the techniques and spectral ranges of astronomical observations have vastly expanded our knowledge of the interstellar medium, of stars and stellar birth, and of probable protoplanetary disks [see Plate 7 for a possible example], as well of the planets in our own solar system. Astronomical measurement capabilities are expanding at a rate that allows us seriously to contemplate the possibility of discovering and studying numerous other planetary systems in the next few decades. Finally, advances in theoretical understanding facilitated in part by calculations feasible only with modern high-speed computers have crystallized a conceptual basis for thinking about the formation of planetary systems in relation to the formation of stars.

Protostars & Planets III is the third in a series of books begun in 1978. This series has tracked the development of our understanding of star and planet-system formation, reflecting both impressive advances and continuing

[xi]

ignorance over the past two decades. The endeavor to understand planetary systems falls into the most modern of complex scientific categories. Planetary systems are too complex to have been predicted *a priori* from the basic equations of physics. It is unlikely that the formation of a star and planetary system will ever be modeled completely from first principles, tracking the evolution of some 10^{57} atoms from a diffuse interstellar assemblage to the highly structured state that is a star and planetary system. Instead, as in the study of all complex systems, we can expect that it will be the close interaction between observations, measurements, and theoretical investigations that fuels the continuing advance of our understanding. In recent years, observations have rendered the very cores of star-forming complexes open to astronomical scrutiny, revealing disk-shaped assemblages of dust and gas, which look for all the world like the precursors of planetary systems. New insights about distinct mechanisms of star formation at the high and low ends of the stellar-mass distribution have been gained through observations of molecular clouds. Seven years ago, when *Protostars & Planets II* was published, the nature of T-Tauri objects as forming stars with large amounts of surrounding gas and dust was understood, but little else was known about these curious systems. Today there is a whole taxonomy of T-Tauri stars (classical, weak, and naked), the various members of which may represent different stages of accretion and dissipation of the surrounding material. The "T-Tauri wind," assumed in the past to selectively sweep gas from early planetary atmospheres, now appears to be part and parcel of the disk-accretion process itself; the wind's role in setting the properties of the star and the environment of the surrounding planets remains unclear.

At the time of *Protostars & Planets II*, significant chemical and isotopic data on the solar system had already been accumulated, but little was yet known chemically about objects in the outer reaches of the solar system. Studies over the intervening seven years have produced inventories of molecular species in comets, Pluto, and the atmospheres and moons of Uranus and Neptune. Complementing these are increasingly sensitive observations of abundances in interstellar clouds including, for example, the first, and long-awaited, identification of methane in molecular clouds. Continuing accumulation of evidence about isotopic abundances and anomalies in meteorites (and, tentatively, in comets) has constrained further the timing of solar system formation against the tapestry of other events in our parent molecular cloud.

Multi-dimensional hydrodynamic and magnetohydrodynamic computer codes have advanced rapidly since *Protostars & Planets II*, facilitating calculations of molecular cloud dynamics, clump formation, cloud collapse, disk or binary star formation, disk evolution and dissipation, wind mechanics, giant planet formation through nucleated collapse of gas, grain and planetesimal accretional dynamics, tidal effects on disks, energetic plasma processes in disks, and impact-driven escape and evolution of atmospheres.

Close interaction between observations, measurements, theory, and numerical modeling is playing an important role in advancing our understanding. Maps of temperature and density in molecular clouds constrain models of the collapse of cloud clumps. Disk-temperature profiles, inferred spectroscopically, challenge complex numerical simulations of the evolution of these planet-forming structures. Of particular interest is how protoplanetary disks transport mass and angular momentum; only recently have theorists tried to face this issue from the point of view of two- and three-dimensional hydrodynamical codes. Newly refined and detailed models of giant-planet interiors, coupled with improved knowledge of atmospheric abundances, pose fresh complications in the physics of planet formation. Models of molecular cloud grain evolution now possess a rich complexity, driven by the need to explain an impressive bulk of chemical and physical evidence. At the same time, icy grains which became part of comets, and perhaps Pluto, may have been spared some of the substantial processing associated with the grains captured by objects closer to the center of our solar system; models explaining the current composition of the outer-most objects in the solar system must therefore account for a combination of molecular cloud and planetary disk processes. Energetic processes in disks, ranging from those responsible for FU Orionis phenomena to the elusive nebular lightning, have only recently been subjected to theoretical study, and their role in planet formation remains unclear.

This volume summarizes a field in which progress has been substantial since the publication of its predecessor. The book was generated in the usual way for the Space Science Series, with a lively conference followed by a special issue of *Icarus* and writing of the chapters which follow. The chapters were formatted by the Space Science Series editorial staff, with the TEXnical assistance of Mark McCaughrean. *Protostars and Planets III* is the first Space Science Series volume to be produced using TEX.

It is now almost unnecessary to point out the tight coupling between the fields of astrophysics, observational astronomy, planetology and plasma physics which characterizes the search to understand stellar and planetary origins. The careful reader should come away from this book with an appreciation for the variety and intricacy of processes which, in creating stars and planets, drive much of the physical and chemical evolution of galaxies.

E. H. Levy
J. I. Lunine

PART I
Clouds

THE COLLAPSE OF CLOUDS AND THE FORMATION AND EVOLUTION OF STARS AND DISKS

FRANK SHU, JOAN NAJITA, DANIELE GALLI, EVE OSTRIKER
University of California, Berkeley

and

SUSANA LIZANO
Universidad Nacional Autonoma de Mexico

We consider the interrelationships among the structure of molecular clouds; the collapse of rotating cloud cores; the formation of stars and disks; the origin of molecular outflows, protostellar winds, and highly collimated jets; the birth of planetary and binary systems; and the dynamics of star/disk/satellite interactions. Our discussion interweaves theory with the results of observations that span from millimeter wavelengths to X-rays.

I. OVERVIEW

This chapter gives a status report on the current astrophysical problems that confront the theory and observation of the collapse of clouds and the formation and evolution of stars and disks. The assignment contains too vast a topic to review in any detail in the allotted pages, and we have interpreted our task accordingly as setting the context for some of the following chapters by others. Parts of this review in essence can also be found in Shu (1991).

We start in Sec. II with the notion that two different modes seem to account for the birth of most stars in the Galaxy: a "closely packed" mode characterized by the more or less simultaneous formation of a tight group of many stars from large dense clumps of molecular gas and dust (see Chapters by Blitz and by Lada et al.), and a "loosely-aggregated" mode in which an unbound association of individual systems (some of which may be binaries) forms sporadically from well-separated, small, dense, cloud cores embedded within a more rarefied common envelope. We introduce the working hypothesis, adopted currently by many workers in the field, that the formation of sunlike stars by the second mode occurs in nearby dark clouds like the Taurus region in four conceptually distinct stages (Fig. 1).

Of the four stages *a–d* outlined in Fig. 1, the most surprising—the one totally unanticipated by prior theoretical developments—is *c*, the *bipolar outflow phase*. In Sec. III, we review the observational discoveries and

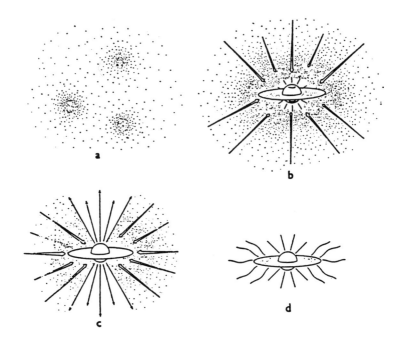

Figure 1.   The four stages of star formation. (a) Cores form within molecular cloud
    envelopes as magnetic and turbulent support is lost through ambipolar diffusion.
    (b) Protostar with a surrounding nebular disk forms at the center of a cloud core
    collapsing from inside-out. (c) A stellar wind breaks out along the rotational axis
    of the system, creating a bipolar flow. (d) The infall terminates, revealing a newly
    formed star with a circumstellar disk (figure from Shu et al. 1987*a*).

interpretations that have led up to our current empirical understanding of this
fascinating phenomenon. We pay particular attention to those aspects that
set tight constraints on possible theories. In Sec. IV, we begin a theoretical
discussion that indicates why, in retrospect, we should perhaps have known
all along that the formation of stars would necessarily involve the heavy *loss*
of mass. Given the angular momentum difficulty likely to be faced by any
object which forms by contraction through many orders of magnitude from
an initially extended, rotating state, the birth of stars (and, perhaps, galactic
nuclei) through the accretion of matter from the surroundings may simply not
be possible without the *simultaneous* accompaniment of powerful outflows.

    Under the working hypothesis, gas and dust in the first stage (Fig. 1*a*)
slowly contract under their own gravitation against the frictional support
provided by a background of ions and magnetic fields via the process of
ambipolar diffusion (see also Chapters by Heiles et al. and McKee et al.).
The principal feature in this stage involves a *quasi-static evolution toward a*
$1/r^2$ *density configuration* appropriate for a singular isothermal sphere. The

relative statistics of cores with and without embedded infrared sources, as well as the theory of ambipolar diffusion, suggests that the time scale over which the configuration is observable as a quiescent ammonia core before it enters dynamical collapse (i.e., before it contains an embedded infrared source) should span about 10^6 yr.

When the contracting configuration becomes sufficiently centrally concentrated, it enters stage 2 (Sec. V; cf. Chapter by Tscharnuter and Boss), wherein the cloud core gravitationally collapses from *inside out*. In such a situation, the inner regions form an accreting but otherwise secularly evolving protostar plus nebular disk. An infalling envelope of gas and dust that rains down from the overlying (slowly rotating) molecular cloud core covers the growing star plus disk. The visual extinction to the central star measures from several tens to a thousand or more, so the embedded source during this stage does not appear as an optically visible object, but must be studied principally by means of the infrared, submillimeter, and millimeter radiation produced by dust reprocessing in the surrounding envelope (see Chapter by Zinnecker et al.).

At some point during this phase of the evolution (see Sec. VI), a powerful wind breaks out along the rotational poles of the system, reversing the infall and sweeping up the material over the poles into two outwardly expanding shells of gas and dust. This stage (stage c) corresponds to the *bipolar outflow* phase observed spectroscopically at radio wavelengths by CO observers (cf. Chapter by Fukui et al.). Theory suggests that this stage features *combined* inflow (in the equatorial regions) and outflow (over the poles). Current consensus in the field holds that magnetohydrodynamic forces drive the wind sweeping up the molecular outflow. The main debate concerns whether the wind originates from the *star*, or from the *disk*, or from their *interface* (see Chapter by Königl and Ruden). From the mass infall rate as well as the statistics of numbers of embedded sources compared to revealed ones (T Tauri stars), we can estimate the combined time spent in stages *b* and *c* (when the system still gains mass in net) as roughly 10^5 yr, almost independent of mass.

As time proceeds, we envisage the angle occupied by the outflow to open up (like an umbrella) from the rotation axes and to spread, halting even the rain of infalling matter over the equator. At this point, stage d, the system becomes visible, even at ultraviolet, optical, and near-infrared wavelengths as a star plus disk to all outside observers (see Sec. VII; Chapter by Basri and Bertout). The location of the star in the Hertzsprung-Russell diagram yields a constraint on the accretion time scale (see Chapter by Stahler and Walter), and the numbers estimated by this method agree well with the dynamical calculations based on the concept of inside-out collapse from a molecular cloud core modeled as a (rotating) singular isothermal sphere. The principal questions during this stage then concern the mechanisms by which mass, angular momentum, and energy transport take place within the disk (see Chapter by Adams and Lin), and the nature of any companions (stellar or planetary) that may condense from such a disk (see Sec. VIII; Chapters by Bodenheimer et

al., Weidenschilling and Cuzzi, Lissauer and Stewart). Intimately tied to these
issues are the estimates of disk masses that can be obtained from observational
measurements at submillimeter and millimeter wavelengths (see Chapter by
Beckwith and Sargent). Another important question concerns whether disk
accretion mainly occurs episodically during FU Orionis outbursts or during
the relatively quiescent (normal) states as well (see Chapter by Hartmann et
al.). FU Orionis outbursts may have important connections with the meteoritic
evidence for transiently high nebular temperatures (see Chapter by Palme and
Boynton); however, the empirical need to appeal to such outbursts for geochemical peculiarities becomes less clear if nebular lightning truly offers a
solution for the origin of chondrules (see Chapter by Morfill et al.).

For the purposes of this book, we may append a fifth stage to the above
four: an epoch of disk clearing. The chapter by Strom et al. identifies this
phase with planet making and sets disk lifetimes at 10^6 to 10^7 yr. We can
usefully compare this astrophysical constraint with the models of giant planet
formation that do not begin to accrete large amounts of gas until a critical core
mass of solids has been reached (see Chapter by Podolak et al.). Edwards
et al. in their chapter argue persuasively that the presence of an inner disk
constitutes the crucial distinction between those T Tauri systems that drive
extraordinary winds (classical T Tauri stars) and those that do not (weak-line
or "naked" T Tauri stars). The absence of an absorbing wind (and inner disk)
makes the magnetic activity on the surfaces of the latter objects observable as
X-ray sources (cf. Chapter by Montmerle et al.). Spallation reactions driven by
energetic particle fluxes associated with the enhanced flaring activity observed
for these relatively gas-poor systems may help to explain some of the isotopic
anomalies seen in the meteoritic record. Surviving dust grains from the
protosolar core may have brought other anomalies intact into the solar nebula
(cf. Chapters by Cameron, Swindle, and Ott).

In Sec. IX, we recapitulate by discussing the question of the origin of
stellar and planetary masses. In particular, we emphasize the emerging view
that in environments where there exist more than enough material to form the
final condensed objects, stars and giant planets help, in part, to determine their
own masses: stars, by blowing powerful winds that shut off the continuing
infall from a molecular cloud; and giant planets, by opening up gaps that
turn off the continuing accumulation from a nebular disk (see also Chapter
by Lin and Papaloizou). A worrisome issue in the latter context concerns
how the solar system managed to clear the debris left over in the disk from
making the Sun and the planets. The Chapter by Duncan and Quinn makes
clear that within the region of the giant planets the solid debris could have
been cleared by gravitational interactions with the planetary bodies. The
observations of remnant particulate disks around main-sequence stars (see
Chapter by Backman and Paresce) demonstrate empirically, however, that
planetesimal dispersal mechanisms become inefficient at large distances from
the central star; such regions probably become the reservoir for (an inner belt
of) comets (see Chapter by Mumma et al.). The gas appears to pose a greater

difficulty. The old view envisages nebular gas left over from planet building to be dispersed by a T Tauri wind. The new view suggests that T Tauri winds do not exist in the absence of (the inner portions of) nebular disks; thus, Edwards et al. argue that the former cannot get rid of the latter (however, see Sec. IX).

The above summary gives a sample of the rich diversity of physical processes that confronts the research worker in the field of star and planet formation. This variety arises for a simple and basic reason: the problem spans physical conditions ranging from the depths of interstellar space to the interiors of stars and planets, involving all the known states of matter and forces of nature, with observational diagnostics available across practically the entire electromagnetic spectrum, and experimental access to relevant primitive materials that has no parallel in any other branch of astronomy and astrophysics. To attack its problems, the field has developed a broad range of technical tools—observational, theoretical, and experimental; and the full exploitation of this range of tools during the past decade has led to rapid and impressive progress.

II. BIMODAL STAR FORMATION

The empirical notion that the birth of low- and high-mass stars may involve separate mechanisms has a long and controversial history (Herbig 1962b; Mezger and Smith 1977; Elmegreen and Lada 1977; Gusten and Mezger 1982; Larson 1986; Scalo 1986; Walter and Boyd 1991). The name "bimodal star formation" usually attaches to this concept, but, more recently, the emphasis has changed from "low mass versus high mass" to "loosely aggregated versus closely packed" (see Chapter by Lada et al.). The latter phrasing roughly coincides with the older one of "associations versus clusters," except that it need not carry the connotation of "gravitationally unbound versus gravitationally bound."

The theoretical distinction between "loosely aggregated" and "closely packed" refers to whether gravitational collapse occurs independently for individual small cores to form single stars (or binaries); or whether it involves a large piece of a giant molecular cloud to produce a tight group of stars created more or less simultaneously. This tight group need not form a bound cluster if the winds or other violent events that accompany the formation of stars expel a large fraction of the gas not directly incorporated into stars (Lada et al. 1984; Elmegreen and Clemens 1985).

The occurrence of two separate modes of star formation has a natural theoretical explanation (Mestel 1985; Shu et al. 1987b) if we adopt the point of view that magnetic fields provide the primary agent of support of molecular clouds against their self-gravity (Mestel 1965a; Mouschovias 1976b; Nakano 1979; see also Chapters by Heiles et al. and McKee et al.). The inclusion of a conserved magnetic flux Φ threaded by an electrically conducting cloud introduces a natural mass scale (cf. Mouschovias and Spitzer 1976; Tomisaka et al. 1988*a*,1989),

$$M_\Phi \equiv 0.13\, G^{-1/2}\Phi \tag{1}$$

that is analogous to Chandrasekhar's limit M_{Ch} in the theory of white dwarfs. A critical mass arises whenever we try to balance Newtonian self-gravity by the internal pressure of a fluid which varies as the 4/3 power of the density (because of ultrarelativistic electron degeneracy pressure in the case of a white dwarf; because of the pressure of a frozen-in magnetic field in the case of a molecular cloud). White dwarfs with masses $M > M_{Ch}$ cannot be held up by electron degeneracy pressure alone, but must suffer overall collapse to neutron stars or black holes. Molecular clouds with masses $M > M_\Phi$ (supercritical case) cannot be held up by magnetic fields alone (even if perfectly frozen into the matter), but, in the absence of other substantial means of support, must collapse as a whole to form a closely packed group of stars. The only trick in this case concerns how to get a supercritical cloud or clump from an initially subcritical assemblage (or they would have all collapsed by now). Theory and observation both suggest a natural evolutionary course: the agglomeration, with an increase of the mass-to-flux ratio, of the discrete cloud clumps that comprise giant molecular complexes (Blitz and Shu 1980; Blitz 1987*b*,1990; Shu 1987; see also Chapters by Blitz and by Lada et al.).

The analogy between white dwarfs and molecular clumps breaks down for the subcritical case. White dwarfs with $M < M_{Ch}$ can last forever as uncollapsed degenerate objects (if nucleons and electrons are stable forms of matter) because quantum principles never weaken. Magnetic clouds with $M < M_\Phi$ initially cannot last forever because magnetic fields do weaken, and the local loss of magnetic flux from unrelated dense regions allows loosely aggregated star formation. In particular, in a lightly ionized gas, the fields can leak out of the neutral fraction by the process of *ambipolar diffusion* (Mestel and Spitzer 1956; Nakano 1979; Mouschovias 1978; Shu 1983). In Sec. IV, we shall examine in more detail the production of individual molecular cloud cores by this process. For the present, we merely note that theory predicts that self-gravitating cores sustained by magnetic fields will inevitably evolve to a state of spontaneous gravitational collapse. This prediction seemingly flies in the face of observations, which almost always find young stellar objects associated with outflows rather than inflows.

III. THE BIPOLAR OUTFLOW PHASE: OBSERVATIONS

At the heart of the central paradox concerning star formation lies the problem of *bipolar outflows* (for reviews, see Lada 1985; Welch et al. 1985; Bally 1987; Snell 1987; Fukui 1989; Rodriguez 1990). Astronomical visionaries (see, e.g., the reminiscences of Ambartsumian 1980) have long worried that forming astronomical systems frequently exhibit *expansion*, rather than the *contraction* that would be naively predicted by gravitational theories. Astronomical conservatives have long persisted in ignoring such warnings, doggedly pursuing the theoretical holy grail that bound objects, such as stars and planetary systems, should form from more rarefied precursors by a process of gravitational contraction. The reconciliation of this fundamental dichotomy remains the central challenge facing theorists and observers alike (see the review by Lada and Shu 1990).

In 1979, Cudworth and Herbig discovered that two H-H objects in a nearby dark cloud L1551 (Lynds 1962) exhibit very high proper motions (corresponding to $\sim$150 km s^{-1}) that trace back to an apparent point of origin near the location of an embedded infrared source IRS 5 found by Strom et al. (1976). Knapp et al. (1976) had earlier found that CO millimeter-wave spectra in this region possess linewidths of the order 10 km s^{-1}, too large according to Strom et al. to correspond to gravitational collapse. The latter authors suggested instead that the disturbance in the ambient molecular cloud material might be produced by a powerful outflow from IRS 5. Snell et al. (1980) mapped the CO emission in the L1551 region and verified this conjecture in a dramatic fashion. They found that the high-velocity CO surrounds the tracks of the fast H-H objects, taking the form of two lobes of gas moving in diametrically opposed directions from IRS 5 (see Fig. 2). Putting together all of the empirical clues, Snell et al. made the prescient proposal (Fig. 3) that a stellar wind must blow at 100 to 200 km s^{-1} from IRS 5, in directions parallel and anti-parallel to the rotation axis of a surrounding accretion disk, and that this collimated wind sweeps up ambient molecular cloud material into two thin shells, which manifest themselves as the observed bipolar lobes of CO emission. The thin-shell nature of the CO lobes of L1551 has since received empirical validation in the investigations of Snell and Schloerb (1985) and Moriarty-Schieven and Snell (1988). Except for the further interpretation that IRS 5 represents a *protostar*, which, apart from suffering mass loss in the polar directions, accretes matter from *infall* occurring in the equatorial regions (see panel *c* in Fig. 1), the proposal of Snell et al. (1980) corresponds in every detail to the theoretical model that we shall pursue in Sec. VI.

Shortly after the original suggestion, however, events moved to a different interim conclusion. In common with the central source of several other bipolar outflows (see reviews of Cohen 1984; Schwartz 1983), IRS 5 lies along a chain of HH objects (Strom et al. 1974; Mundt and Fried 1983) that may just constitute the brighter spots of a more or less continuous optical jet (Mundt 1985; Reipurth 1989c).

The emission knots probably arise as a result of the interactions of a highly collimated ionized stellar wind with the surrounding medium (possibly another [neutral] wind; see Stocke et al. [1988] and Shu et al. [1988]). The free-free thermal emission from the ionized stellar wind resolves itself in VLA radio-continuum measurements as a two-sided jet centered on IRS 5 (Bieging et al. 1984). It was natural to suppose that this ionized wind represents the driver for the bipolar molecular outflow. Unfortunately, the momentum input provided by the ionized wind, integrated over the likely lifetime of the system, falls below the value required to explain the moving CO lobes by one to two

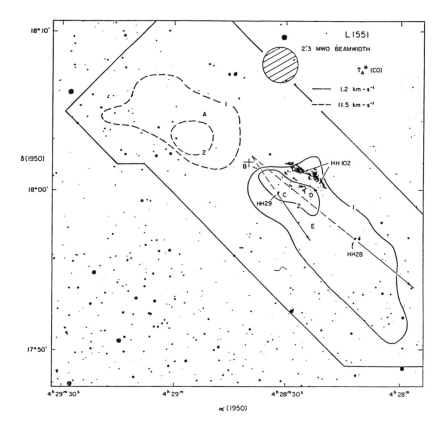

Figure 2.   Contour map of the antenna temperature of the $J = 1 - 0$ transition of ^{12}CO at high velocities, superposed on an optical photograph of the L1551 dark cloud. The cross indicates the position of IRS 5; also shown are the directions of the proper motions of two Herbig-Haro objects, HH 28 and HH 29 (figure from Snell et al. 1980).

orders of magnitude (for the likely case of momentum-driven rather than energy-driven flows), a conclusion that holds as well for all other well-studied bipolar flow sources (Bally and Lada 1983; Levreault 1985). By comparing the required momentum input to the photon luminosity divided by the speed of light, the same authors argued persuasively against the possible importance of radiation pressure in the outflow dynamics.

    The claim by Kaifu et al. (1984) of the detection of an extended, rotating, molecular disk encircling IRS 5, contributed to the puzzle. This result, and the inability of ionized stellar winds to drive the observed molecular flows, motivated Pudritz and Norman (1983,1986) and Uchida and Shibata (1985) to suggest that bipolar outflows originate, not as shells of molecular cloud gas

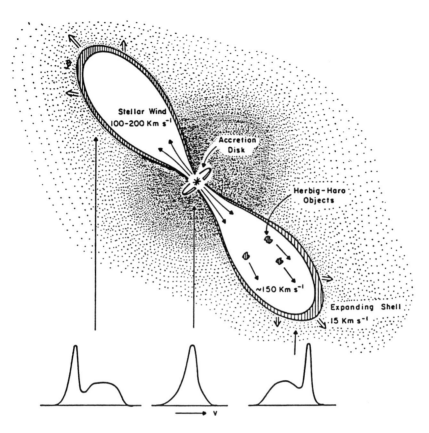

Figure 3. Schematic model for the bipolar flow in L1551 driven by a stellar wind emanating from IRS 5. At the bottom is depicted expected CO line profiles for different line of sights across the source (figure from Snell et al. 1980).

swept up by a stellar wind, but, following the work of Blandford (1976) and Lovelace (1976), as gaseous material driven magneto-centrifugally directly off large ($\sim 10^{17}$ cm) and massive ($\sim 10^2$ $M_\odot$) circumstellar disks. This suggestion seemingly received support from reports that one limb of the blueshifted lobe in L1551 exhibits rotation (Uchida et al. 1987a). However, subsequent observational studies have failed to confirm the presence of large and massive disks around young stellar objects with the properties required by the theoretical models (see, e.g., Batrla and Menten 1985; Menten and Walmsley 1985; Moriarty-Schieven and Snell 1988). Moreover, the claim of rotation in the blueshifted lobe may have been contaminated by confusion with another bipolar flow source in the same field of view (Moriarty-Schieven and Wannier 1991). Finally, from a theoretical point of view, the original large-disk models have severe energetic problems (Shu et al. 1987a; Pringle 1989a).

Recently, a number of different groups have attempted to salvage the disk model by postulating smaller and less massive disks, with the bulk of the wind emerging from radii much closer to the central star. This revised picture, however, suffers from the lack of a plausible injection mechanism. The paper by Blandford and Payne (1982) illuminates the basic difficulty.

Blandford and Payne consider a thin disk threaded by a poloidal magnetic field **B** in which rotation balances the radial component of gravity. If **B** has sufficient strength and a proper orientation, it can fling to infinity electrically conducting gas from the top and bottom surfaces of the disk. A freely sliding bead on a rigid wire anchored at one end to a point ϖ in a disk rotating at the angular velocity $\Omega_{\text{disk}}(\varpi)$ provides a useful analogy (Henriksen and Rayburn 1971). In such an analogy, the termination of the wire at the free end corresponds to the Alfvén surface, beyond which the magnetic field can no longer enforce corotation, even approximately, and the super-Alfvénic fluid motion becomes essentially ballistic. For the flinging effect to take place in a Keplerian disk (one in which all of the gravitational attraction comes from a central mass point), the poloidal field must enter or leave the disk at an angle larger than 30° from the normal. (The component of the centrifugal force parallel to the field [wire] in the meridional plane has no excess compared to gravity if the field [wire] makes too small an angle with respect to the rotation axis, e.g., if it points vertically through the disk.) However, the magnetic field on the two sides of a disk cannot *both* bend outward by a nonzero angle without generating a large kink across the midplane of the disk, one that would result in a very large $\nabla \times \mathbf{B}$, and which would yield, by Ampere's law, a very large current (infinitely large in a disk of infinitesimal thickness). The Lorentz force per unit volume,

$$\mathbf{f}_{\text{L}} = \frac{1}{4\pi}(\nabla \times \mathbf{B}) \times \mathbf{B} \qquad (2)$$

needed to produce order-unity departures from Keplerian motion above the disk (in order to drive a wind) must then have even bigger values inside the disk. The tendency for the field lines to want to straighten vertically (thereby shutting off the magneto-centrifugal acceleration) can be offset only by a heavy radial inflow through the disk, maintained, for example, by ambipolar diffusion in the presence of sub-Keplerian rotation inside the disk (Konigl 1989). Sub-Keplerian rotation everywhere in the disk necessitates the continuous removal of angular momentum from it (the role of the wind), but the postulated deficit of centrifugal support in the disk then works against lift-off of the gas in the first place. Indeed, no one has yet succeeded in constructing a self-consistent cool model of this type, demonstrating a smooth connection from the region of the sonic transition (near the disk surface) to the region of the Alfvénic transition (far from the disk surface).

Blandford and Payne themselves adopt another strategy. They accept the conclusion that the symmetry of a disk geometry naturally forces any poloidal magnetic field to thread vertically through the midplane of the disk, from where it *gradually* bows outward (see Fig. 4). To lift the material off the disk

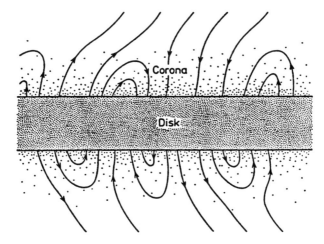

Figure 4. A schematic representation of a possible field geometry close to the disk (figure from Blandford and Payne 1982).

to heights where the magnetic field bends over sufficiently for the magneto-centrifugal mechanism to operate, they invoke the thermal pressure of a hot corona above (and below) the disk. In other words, their final model starts off as a *thermally driven wind*. A hot gas with thermal speeds comparable to the virial speeds characteristic of the depth of the gravitational potential in the inner parts of a protostellar disk, however, suffers tremendous radiative losses at the injection densities needed to supply the observed mass outflows. While astrophysical systems with highly luminous compact objects at their centers, such as active galactic nuclei or binary X-ray sources, might well provide sufficient energy to sustain analogous radiative losses, we cannot expect the same bounty from the central machine of protostellar systems (cf. the discussion of DeCampli [1981] that even the relatively low-powered T Tauri winds cannot be thermally driven).

In summary, at radii $\gtrsim 10^{17}$ cm, disks *can* maintain thermal speeds that compete with the local virial speeds, but such regions lack enough rotational energy to drive the observed flows. Greater centrifugal power exists at smaller radii, but lift-off from the disk then requires an unidentified nonthermal injection mechanism. Compounding these difficulties is the unlikelihood of strong magnetic fields in thin disks. Recall that the energy density of the field at the Alfvén surface must equal the kinetic energy density of a wind flowing close to its terminal velocity. But magnetic buoyancy effects probably limit the generation of fields by dynamo mechanisms acting within disks to energy densities comparable to or less than the prevailing thermal values (Stella and Rosner 1984). The inability of disks to retain large *toroidal* fields (Parker 1966; Matsumoto et al. 1988) limits the viability of mechanisms like those of Lovelace et al. (1991) that rely on the pressure gradient of such fields in the z-direction for the initial vertical acceleration of the gas.

Other theoretical arguments suggest that disk fields must be weak. Umebayashi and Nakano (1988) have computed the ionization state of a minimum-mass solar nebula. The large nebular densities encountered even in a minimum-mass model lead to very low ionization fractions, so low that magnetic fields would be safely decoupled by several orders of magnitude from the nebular gas throughout the disk, except for the innermost and outermost regions. Even in disks where Ohmic dissipation and ambipolar diffusion do not present severe obstacles, difficulties exist in supposing that magnetic fields acquire strengths sufficient to drive the observed outflows. For example, if accretion disks owe their viscosity to the powerful MHD instability recently discussed by Balbus and Hawley (1991) and Hawley and Balbus (1991), the *poloidal* component of the field cannot have energy densities much higher than thermal gas values. Indeed, the instability itself may serve as one of the ingredients of a self-consistent dynamo mechanism responsible for generating the poloidal component of the disk field in the first place. In addition, dynamically strong fields (energy densities comparable to that contained in the *rotation* of the part of the disk supplying the matter for the outflow) cannot have been dragged in and amplified by lightly ionized gas that collapsed from interstellar dimensions by ambipolar diffusion (see Sec. IV below), because mass infall at rates and speeds (free-fall) comparable to the inferred outflow values had to overcome the interstellar field to form the disk in the first place. Finally, even if we could surmount the above problems and drive a disk wind, we would *still* be left with an angular-momentum problem for the accreting central star (cf. the discussion in Sec. V). For all of these reasons, we believe that Snell et al. (1980) intrinsically had the right idea when they speculated that *stellar winds* must constitute the basic driver for bipolar outflows in young stellar objects.

The issue then reverts to the original question: how can stellar winds drive the observed molecular outflows when the *ionized* component lacks the requisite power? Many researchers arrived independently at the obvious answer: perhaps young stellar objects possess *neutral* winds that supply the missing momentum input. Detection of fast neutral winds would require the spectroscopic observation of broad and shallow radio emission lines, a difficult task because so little mass resides within the telescope beam at any particular time. Nevertheless, the search for *atomic* hydrogen winds has now succeeded in at least three sources: HH 7–11, L1551, and T Tau (Lizano et al. 1988; Giovanardi et al., in preparation; Ruiz et al. 1991). Fast neutral winds are also indicated through the detection of CO moving at velocities of ± 100 to 200 km s^{-1} in HH 7–11 and several other sources (Koo 1989; Margulis and Snell 1989; Masson et al. 1990; Bachiller and Cernicharo 1990; J. E. Carlstrom, personal communication). The mass-loss rate measured in HH 7–11 is especially impressive, $\dot{M}_w \approx 3 \times 10^{-6}$ M$_\odot$ yr^{-1}, more than sufficient to drive the known bipolar molecular outflow associated with this source. Thus, the discovery of massive neutral winds from (low-mass) protostars solves one of the principal remaining mysteries concerning bipolar outflows, namely, the ultimate source for their power. However, this empirical discovery pushes to the fore another puzzling question: how and why are stars born by *losing* mass at such tremendous rates?

IV. ROTATING, MAGNETIZED, MOLECULAR CLOUD CORES

A major breakthrough occurred in the field when Myers and Benson (1983) identified the sites for the birth of individual sunlike stars in the Taurus cloud as small dense cores of dust and molecular gas that appear as especially dark regions in the Palomar sky survey or as regions of NH_3 and CS emission (cf. the references in Evans 1991). One theory for the formation of such cores begins with the view that large molecular clouds are supported by a combination of magnetic fields and turbulent motions (see Shu et al. 1987*a*; Myers and Goodman 1988*a*, *b*). Because magnetic fields directly affect only the motions of the charged particles, magnetic support of the neutral component against its self-gravity arises only through the friction generated when neutrals slip relative to the ions and field via the process of ambipolar diffusion (Mestel and Spitzer 1956). This slippage occurs at enhanced rates in localized dense pockets where the ionization fraction is especially low. As the field diffuses out relative to the neutral gas, the Alfvén velocity $B/(4\pi\rho)^{1/2}$ drops, and the turbulence must also decay if fluctuating fluid motions are to remain sub-Alfvénic. Thus, both processes—field slippage relative to neutrals and the lowering of turbulent line widths in regions of increasing mass-to-flux— reinforce the tendency for quiet dense cores to separate from a more diffuse common envelope.

Assuming axial symmetry (no φ dependence) in spherical polar coordinates (r, θ, φ) as well as reflection symmetry about the midplane $\theta = \pi/2$, Lizano and Shu (1989) performed detailed ambipolar diffusion computations, using the ionization balance calculations of Elmegreen (1979*b*) and including empirically the effects of turbulence with the scaling laws found by radio observers (cf. Chapters by Blitz, by Lada et al. and by McKee at al.). They find that the time to form NH_3 cores, $\sim 10^6$ to 10^7 yr, and the additional time to proceed to gravitational collapse, a few times 10^5 yr, agree roughly with the spread in ages of T Tauri stars in the Taurus dark cloud (Cohen and Kuhi 1979) and with the statistics of cores with and without embedded infrared sources (Fuller and Myers 1987). Toward the end of its quiescent life, a molecular cloud core tends to acquire the density configuration of a (modified) singular isothermal sphere:

$$\rho(r, \theta) = \frac{a_{\text{eff}}^2}{2\pi G r^2} Q(\theta) \tag{3}$$

where a_{eff} is the effective isothermal sound speed, including the effects of a quasi-static magnetic field $\mathbf{B}_0$ and (as modeled) an isotropic turbulent pressure. The dimensionless function $Q(\theta)$, >1 toward the magnetic equator and <1 toward the poles, yields the flattening that occurs because the magnetic field contributes no support in the direction along $\mathbf{B}_0$. We define $Q(\theta)$ so that it has a value equal to unity when averaged over all solid angles, i.e.,

$$\int_0^{\pi/2} Q(\theta) \sin \theta \, d\theta = 1 \qquad (4)$$

hence, a_{eff}^2 in Eq. (3) gives the angle-averaged contributions of thermal, turbulent and magnetic support.

Myers et al. (1991a) have used statistical arguments to deduce that many cloud cores have prolate, rather than oblate, shapes. They note that this result poses difficulty for the above view that cloud cores represent quasi-equilibrium structures in which quasi-static (poloidal) magnetic fields play an important part in the support against self-gravitation (see also Bonnell and Bastien 1991). Tomisaka (1991) points out that equilibrium clouds with prolate shapes become possible if they possess toroidal magnetic fields of substantial strength. Even without toroidal fields, oblate shapes represent a natural theoretical expectation only if we model the effects of molecular-cloud turbulence as an *isotropic* pressure. If "turbulent" support arises, instead, from the propagation and dissipation of nonlinear Alfvén waves in an inhomogeneous medium (see Arons and Max [1975], as well as the discussion on pp. 36–37 of Shu et al. [1987a]), then the largest amount of momentum transfer might well occur *parallel* to the direction of the mean field $\mathbf{B}_0$, rather than perpendicular to it. For example, if ambipolar diffusion causes $\mathbf{B}_0$ to become nearly straight and uniform, as occurs in the detailed calculations before collapse ensues, then the mean field exerts little stress on average, and the anisotropic forces associated with the fluctuating component $\delta\mathbf{B}$ in total $\mathbf{B} = \mathbf{B}_0 + \delta\mathbf{B}$ may actually cause cloud cores to assume prolate instead of oblate shapes. This tendency for condensations to stretch out along the direction of the mean field would only be enhanced by tidal forces if cloud cores tend to form as individual links on a magnetic "sausage" (cf. Elmegreen 1985a; Lizano and Shu 1989). Bastien et al.'s (1991) study of the fragmentation of elongated cylindrical clouds into such individual links differs from that of Lizano and Shu (1989) in that the former authors ignore magnetic effects and they assume that the core-formation process occurs by dynamical collapse rather than by quasi-static contraction.

Apart from the delicate question of the sense of elongation of the shape function $Q(\theta)$, the prediction (Eq. 3) with regard to the *radial* variation of ρ agrees reasonably well with observations of isolated cores and Bok globules (see, e.g., Zhou et al. [1990b], who comment particularly that the *outer* velocity fields are more representative of quasi-static contraction than dynamical collapse). The solution as written represents an asymptotic result; actual configurations become gravitationally unstable in their central regions before a singular density cusp develops there. However, because the phase of quasi-static contraction produces very high central concentrations in the numerical models, the resulting dynamical behavior would probably closely resemble the inside-out collapse known analytically for the (exact) singular isothermal sphere without rotation (Shu 1977). Such a collapse solution builds up a central protostar at a rate,

$$\dot{M}_{\text{infall}} = 0.975\, a_{\text{eff}}^3 / G \tag{5}$$

that remains constant in time as long as the external reservoir of molecular cloud gas (with a r^{-2} density distribution) continues to last. For L1551, where $a_{\text{eff}} \sim 0.35$ km s^{-1}, $\dot{M}_{\text{infall}} = 1 \times 10^{-5}$ M$_\odot$ yr^{-1}.

The solution requires modification in the presence of rotation. Because ambipolar diffusion occurs relatively slowly, one might imagine that magnetic braking has time to torque down the cloud core sufficiently during the quasi-static condensation stage to allow even the direct formation of single stars, without subsequent angular momentum difficulties (see, e.g., Mouschovias 1978). Realistic estimates yield a somewhat different picture. Before the onset of dynamical collapse, magnetic braking can enforce more or less rigid rotation of the core at the angular velocity Ω of its surroundings (Mestel 1965a,1985; Mouschovias and Palelogou 1981). Once dynamical collapse starts, however, the infall velocities quickly become super-magnetosonic, and any initial angular momentum possessed by a fluid element carries into the interior. In the inside-out scenario presented above, dynamical collapse is initiated while much of the mass still has an *extended configuration*. For example, to contain 1 M$_\odot$, the wave of falling must typically engulf material out to $\sim 10^{17}$ cm. Even if such a core were perfectly magnetically coupled to its envelope before collapse, it would typically rotate at too large an initial angular speed Ω to allow the formation of just a single star with a dimension of $\sim 10^{11}$ cm. The simplest solution to the core's residual angular momentum problem is to form a star *plus a disk*.

Probably as a consequence of cloud magnetic braking, Arquilla and Goldsmith (1986) and Fuller and Myers (1987) find empirically that rotation does not usually play a dynamically important role on scales of a molecular cloud core and larger. This fact suggests the possibility of a perturbational analysis for the dynamical collapse problem (Terebey et al. 1984), with core rotation at an initially uniform rate Ω treatable as a small correction, in the outer parts, to the spherical self-similar solution for the collapse of a singular isothermal configuration. Large departures from sphericity occur only at radii comparable to or smaller than a centrifugal radius,

$$R_{\text{C}} \equiv \frac{G^3 M^3 \Omega^2}{16 a_{\text{eff}}^8} \tag{6}$$

where $M \equiv \dot{M}_{\text{infall}} t$ equals the total mass that has fallen in at time t. Typical combinations of a_{eff}, Ω, and M yield values for $R_{\text{C}} \sim 10^{15}$ cm.

For $R_* \sim 10^{11}$ cm $<< R_{\text{C}} \sim 10^{15}$ cm, the bulk of the freely falling matter (on parabolic streamlines because of the nonzero values of the specific angular momentum) does not strike the star directly, but forms a disk of size

$\sim R_C$ that swirls around the protostar. Radiative transfer calculations for the emergent spectral energy distribution in an infall model of this type give quite good fits to the data for IRS 5 L1551, if we choose $a_{eff} = 0.35$ km s^{-1}, $\Omega = 1 \times 10^{-13}$ rad s^{-1}, and $M = 1$ M$_\odot$ (Fig. 5).

Fermi is supposed to have remarked that with three free parameters, he could fit an elephant. Figure 5 gives an empirical proof of this claim. Indeed, one can readily discern a tail at millimeter frequencies, attached to a broad back at far- and mid-infrared frequencies, with a 10 μm silicate absorption feature separating the shoulder from the head and a 3.1 μm water-ice feature defining a tusk. The trunk of the elephant emerges from the droop to optical frequencies, although the models do not reproduce the slight raising of the trunk seen in the data points because they do not properly account for the presence of scattered near-infrared and optical light. By Fermi's standards, then, our failure to catch the trumpet call of the elephant, without the need to introduce additional free parameters, prevents us from claiming a complete victory.

However, we note that the three independent parameters that go into the fits, a_{eff}, Ω, and M, represent an irreducible set from an *a priori* theory, which have fundamental dynamical implications apart from the spectral energy fits. For example, the value $a_{eff} = 0.35$ km s^{-1} well describes the properties of the core density distribution around L1551 (cf. Eq. [3] and the review of Evans [1991]). The agreement between the size of the spatial maps at millimeter and submillimeter wavelengths (Walker et al. 1990) and at infrared wavelengths (Butner et al. 1991) with the predictions of the radiative-transfer calculations of our model also bodes well for the values of a_{eff} and M, derived in the spectral energy fits on the basis of providing the correct overall optical depth to the central star and the absolute luminosity scale (from protostar theory). Finally, the numerical value of Ω, chosen to fit the depth of the silicate absorption feature, yields a predicted size for the circumstellar disk, $R_C = 42$ AU, which compares well with the limits 45 ± 20 AU deduced for this source by Keene and Masson (1990) from radio interferometric measurements of the thermal emission from dust grains in the putative disk. More recently, Goodman et al. (1991; see also Menten and Walmsley 1985) report the detection of a velocity gradient ~ 4 km s^{-1} pc^{-1}, localized to the core region of L1551, consistent in magnitude with the model value $\Omega = 1 \times 10^{-13}$ s^{-1}, although the result may have been contaminated by the velocity shear introduced by the outflow.

A similar modeling of the spectral energy distribution of HH 7–11 suggests that this source also possesses infall at a rate $\dot{M}_{infall} \gtrsim 1 \times 10^{-5}$ M$_\odot$ yr^{-1} (F. C. Adams, personal communication). Our ability to fit the spectral energy distribution of famous *outflow* sources like L1551 and HH 7–11 with pure *inflow* models bears directly on the fundamental paradox of Sec. III. Many of the observed embedded systems (those in stage c of Fig. 1) must possess *both* outflow and inflow, with the blowing off of a small polar cap in deeply embedded sources making little difference for the problem of infrared reprocessing in the rest of the infalling envelope. The challenge to observers

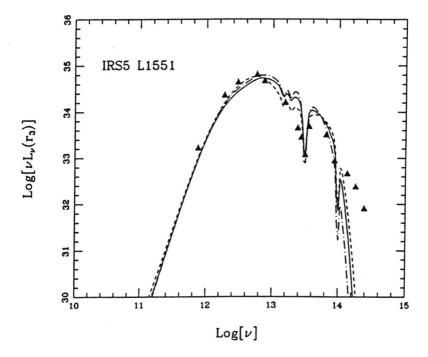

Figure 5. Spectral energy distribution for the bipolar outflow source L1551. The best fit for the data points (solid curve) results from choosing $a_{\text{eff}} = 0.35$ km s^{-1}, $\Omega = 10^{-13}$ s^{-1}, and $M = 0.975\, a_{\text{eff}}^3 t/G = 1$ M$_\odot$ (figure from Adams et al. 1987).

is to find spectroscopic evidence, in the low-luminosity sources, for the (rotating) collapse along the equatorial directions in an analogous manner that this task has been accomplished for high-luminosity protostars (see, e.g., Ho and Haschick 1986; Keto et al. 1987; Welch et al. 1987; Rudolph et al. 1990). In any case, here then lies a possible answer to the question: how can a star (SVS 13, the driving source in HH 7–11) form if it loses mass (through a neutral wind emerging from the poles) at the enormous rate $\dot{M}_{\text{w}} = 3 \times 10^{-6}$ M$_\odot$ yr^{-1}? Clearly, a net gain can still take place if the star is being "force-fed" (through the equatorial regions) at a rate $\dot{M}_{\text{infall}}$ which is $\gtrsim 3$ times larger yet. This empirical finding answers the "how" of our original question, but it does not address the question "why" a new-born star should be losing mass. A clue to the resolution of the latter issue lies in an examination of the very different way in which a star, in contrast to a disk, may deal with its angular momentum legacy.

V. PROTOSTAR FORMATION BY DISK ACCRETION

Notice that no mass scale that we can identify with stars emerges naturally in the theoretical picture of infall summarized in the previous section. To obtain a stellar mass $M_* \sim \dot{M}_{\mathrm{infall}} t_{\mathrm{infall}}$ from an object (a giant molecular cloud typically), whose characteristic mass scale much exceeds anything that we normally associate with stars, requires us to choose a small total duration t_{infall} (say, 10^5 yr) over which infall occurs. This duration might be set, for example, by the time required for an incipient outflow from the star to completely reverse the inflow. In spherical simulations of protostar formation and evolution (Stahler et al. 1980; Palla and Stahler 1990), the only event of significance to take place on such a time scale concerns the onset of deuterium burning and the establishment of an outer convection zone at the bottom of which the deuterium burns at a rate equal to that which accretion brings in a fresh supply. In low-mass protostars ($<2\ M_\odot$, i.e., precursors to T Tauri stars), deuterium burning occurs only slightly off-center; in intermediate-mass stars (2 to 8 $M_\odot$, i.e., precursors to Herbig Ae and Be stars), it occurs in a thin shell. In either case, the important feature is that the process induces an outer convection zone—a condition, when combined with rapid stellar rotation, that Shu and Terebey (1984) speculated would be conducive to dynamo action and to the appearance of strong magnetic fields on the surface of the star.

Calculations by Picklum and Shu (in preparation) demonstrate that accretion through a disk rather than by spherical infall does not modify the above conclusions appreciably, except to lower by a significant factor the emergent photon luminosity for a given rate of mass accretion. The latter effect may alleviate the discrepancies in time scales inferred for embedded protostars and revealed pre-main-sequence stars noted by Kenyon et al. (1990; see also Hartmann et al. 1991).

For star formation via the collapse of a cloud core modeled initially as a uniformly rotating singular isothermal sphere, the fraction of mass that suffers direct infall onto the star compared to that which first lands in the disk equals $1.29\ (R_*/R_C)^{1/3}$ (Adams and Shu 1986). This expression yields a small number (several percent) for stars with radii $\sim$ a few $R_\odot$ and disks with radii $\sim$10 to 100 AU. The small cross-sectional area of stars in comparison with their disks leads us to expect generically that most of the mass from a collapsing molecular cloud core, which eventually ends up inside the star, must first make its way through the disk (an original point of view argued first by Cameron [1962]; see also Mercer-Smith et al. [1984]).

The process of disk accretion is not well understood in any astrophysical system. The most frequently invoked mechanism involves the inward transport of mass and the outward transport of angular momentum by the friction associated with some form of anomalous viscosity in a differentially rotating disk (see, e.g., Lynden-Bell and Pringle 1974). In the case of the nebular disks that surround young stellar objects, Lin and Papaloizou (1985; see also Ruden and Lin 1986) identified thermal convection in the disk as a

plausible source of the turbulent viscosity, and this may well hold for the optically thick dusty disks believed to surround all classical T Tauri stars. More
recently, Balbus and Hawley (1991; see also Hawley and Balbus 1991) have
rediscovered a powerful magnetohydrodynamic instability (Chandrasekhar
1960,1965; Fricke 1969) that afflicts weakly magnetized differentially rotating systems in which the angular velocity of rotation decreases outwards
(as occurs in all known astrophysical disks). These developments hold great
promise for providing a physical foundation for the central assumption of
standard accretion-disk theory that the viscosity has an anomalous magnitude
much higher than molecular values (for a review, see Pringle 1981).

However, in the most extreme cases (so-called "flat spectrum sources";
see Sec. VII; the Chapter by Beckwith and Sargent), the far-infrared radiation
(relative to the near-infrared values) exceeds by a factor approaching 10^3, the
value one would have naively predicted by the steady-state viscous model
(Adams et al. 1988). The observed disks also have masses, as estimated from
their submillimeter and millimeter emission, that compare favorably with
those contained in the central stars (Adams et al. 1990; Beckwith et al. 1990;
Keene and Masson 1990).

The above considerations suggest an interesting scenario for protostellar
disk accretion (see also Chapter by Adams and Lin). The overall time scale
for viscous transport of mass and angular momentum scales roughly as

$$t_{vis} \sim \alpha^{-1}(R/H)^2 t_{rot} \qquad (7)$$

where (R/H) equals the aspect ratio of the disk (radius to vertical scale
height), t_{rot} is the rotation period at the outer edge of the disk, and α is a dimensionless parameter in the so-called "alpha" prescription for the anomalous
viscosity. For application to nebular disks, typically, $R/H \gtrsim 10$, $t_{rot} \sim 10^3$
yr, and $\alpha < 1$ (e.g., $\alpha \sim 10^{-2}$ in the convection models of Ruden and Lin
[1986]); thus, $t_{vis} > 10^5$ yr. The statistics of disks around classical T Tauri
stars sets a lifetime of $\sim 10^6$ to 10^7 yr (see Chapter by Strom et al.). This represents only a lower limit for t_{vis} because the disappearance of near-infrared
radiation only implies the disappearance of small dust particles surrounding
the central star, and not necessarily the viscous transport of all of the material,
gas and dust, from the entire disk.

Suppose rotating infall piles matter into the disk faster than viscous
accretion can transfer it to the central star. (Models for the main accretion
phase of low-mass protostars suggest that $M/\dot{M} \sim 10^5$ yr.) The disk would
then build up mass relative to the star until the disk becomes comparably
massive. At this point, strong gravitational instabilities (nonaxisymmetric
density waves) may develop in the disk that result, in the nonlinear regime, in
the inward transport of mass and the outward transport of angular momentum.
The energy dissipation associated with this "wave dredging" differs, however,
from the viscous mechanism, and we might expect to see a different resultant
radial distribution of temperature in the disk and a different emergent spectrum

(see Sec. VII). In particular, for the transport to occur *globally* (as indicated by the observations of flat-spectrum sources), from the star-disk interface to the outer edge of the disk $\sim 10^4$ times larger, we can restrict our search of unstable modes in nearly Keplerian disks to one-sided disturbances, in which circular streamlines distort to elliptical ones with foci at the star (Adams et al. 1989).

Unless such instabilities run away catastrophically to form a companion star (a possible origin for binary systems), we may speculate that they self-regulate the amount of mass in the disk during the infall phases (Fig. 1b, and c) so that it never consists of a large fraction of the total. Thus, we expect that some appreciable fraction f_1 of the total infall rate onto the system (mostly onto the disk), $\dot{M}_{infall}$, must make its way through the disk at an accretion rate $\dot{M}_{acc}$, eventually to be deposited onto the star,

$$\dot{M}_{acc} = f_1 \dot{M}_{infall}. \qquad (8)$$

As the detailed calculations of Adams et al. (1989) demonstrate, disks with temperature profiles consistent with the observational requirements suffer a strong one-armed spiral instability (see Fig. 6) if they have masses equal, say, to one-half or one-fourth of the total; thus a reasonable estimate for f_1 ranges from 1/2 to 3/4. We emphasize that this estimate for f_1 represents a *time-averaged* value; the instantaneous rate at which mass from the disk actually empties onto the star may suffer large fluctuations about an average value. In particular, the arguments of Sec. VII suggest that disk accretion induced by global gravitational instabilities may have intrinsic tendencies to occur in a nonsteady manner.

VI. STELLAR WINDS AND BIPOLAR FLOWS: THEORY

The matter entering the protostar from the disk at the rate $\dot{M}_{acc}$ carries a relatively large specific angular momentum. For example, compared to the specific angular momentum of disk matter circling in orbit just outside of a star's equator, the angular momentum per unit mass of a uniformly rotating polytrope of index 1.5 (a good model for a fully convective low-mass proto-star) is a small number $b = 0.136$, even if the star were to rotate at the brink of rupture (James 1964). Thus, the disk feeds material to the star that contains, per gram, about $b^{-1} \sim 7$ times more angular momentum than can be absorbed dynamically by the star. Hence, once the star begins to accrete matter from a centrifugally supported disk, it can increase its mass M_*, at most, only by an additional small fraction b before it reaches breakup. In practice, the star may be able to accept even less matter from the disk because even the matter accumulated by the star through direct infall carries nonzero values of specific angular momentum (see, e.g., Durisen et al. 1989b).

As the mass that can be accumulated by a star through direct infall from a realistically rotating molecular cloud core amounts to only several percent

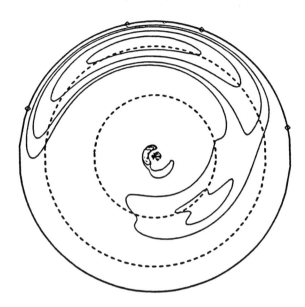

Figure 6. Contour plot of the lowest-order, growing, $m = 1$, mode in a star/disk system where the disk's mass equals the star's mass, but the disk's radius is 10^4 times larger than the star's radius (figure from Adams et al. 1989).

of a solar mass, and as the additional amount that can be accreted from a disk amounts to a small fraction of this tiny value, how do stars of sunlike masses and larger ever form?

Paczynski (1991) and Popham and Narayan (1991) calculated star plus disk models where the star, spun to (slightly faster than) breakup, can viscously transport angular momentum to an adjoining disk, and thereby continue to accrete an indefinite amount of mass. This model presumes the existence of a sufficiently efficient viscous coupling between star and disk as to transport away the excess specific angular momentum (above the fraction b) brought in by disk accretion, an assumption that may not hold if the mechanism of disk accretion is gravitational rather than viscous. In any case, their mechanism of continuous outward angular-momentum transport cannot work for T Tauri stars, which are observed to rotate at speeds significantly less than breakup. Furthermore, the almost ubiquitous presence of energetic bipolar flows among deeply embedded sources suggests that quiescent viscous torques do not provide the primary source of relief for the angular momentum difficulty of protostars in their main accretion phase.

Such objects evidently find a more spectacular way of shedding the excess angular momentum brought in by disk accretion. Shu et al. (1988) proposed that the protostar could fling off the excess in a powerful wind once the

protostar gets spun to breakup by the disk, if the star possesses sufficiently
strong surface magnetic fields (see also Hartmann and MacGregor 1982).

Schematically, the process works as follows. Like other rotating stars
with outer convection zones, a protostar probably has a network of open and
closed magnetic field lines that protrude from its surface (Fig. 7). An ordinary
(ionized) stellar wind blows out along the open field lines; this O-wind may
have a higher intensity than that characteristic of a normal star of the same
spectral type, luminosity class, and rotation rate, because of the enhancement
of stellar dynamo action associated with circulation currents induced by the
disk through Ekman pumping (see Sec. VII).

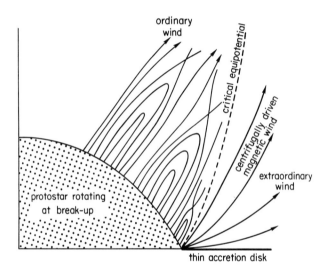

Figure 7.   Schematic model for heavy mass-loss driven magnetocentrifugally from
the equator of a protostar which is forced to rotate at breakup because of accretion
through a Keplerian disk. All streamlines to the right of the critical equipotential
(marked as a dashed curve) correspond to possible "downhill" paths for electrically
conducting gas attached to magnetic lines of force that rotate rigidly with the star.
The large slip velocities present in the interfaces between the X-wind and the O-
wind and between the X-wind and the disk may lead to turbulent mixing layers that
have observable radiative signatures (figure adapted from Shu et al. 1988).

If open field lines circulate into the equatorial regions of the protostar,
which spins at breakup by assumption, the O-wind may intensify even more
via the X-celerator mechanism (Shu et al. 1988) and become an extraordinary
wind. The magneto-centrifugal acceleration in the X-wind takes place much
as described in Sec. III on the Blandford-Payne mechanism for disk winds,
except that the geometry here no longer selects against field lines (equivalent

to streamlines for a conducting gas) that emerge in the "downhill" directions of the effective (corotating) potential. To carry away the requisite mass-loss rate $\dot{M}_w$, the sonic transition of the X-wind must occur typically in the photosphere of the star, where the material, being cool, has a relatively low ionization fraction. Thus, the X-wind is predominantly a *neutral* or *lightly ionized* wind (Natta et al. 1988; Ruden et al. 1990).

Consider the quasi-steady state wherein spinup by disk accretion at a rate $\dot{M}_{acc}$ is balanced by spin-down via a magnetized stellar wind at a rate $\dot{M}_w$ in such a way that at each stage of the process the star continues to rotate exactly at breakup. Suppose the specific angular momentum carried away in the X-wind, measured as an average over all mass-carrying streamlines, equals some multiple $\bar{J}$ of the specific angular momentum of the material in circular orbit at the equator of the star. If we further assume the equatorial radius to be proportional to the stellar mass, as roughly true for a star actively burning deuterium at its center, we find that the requisite mass-loss rate $\dot{M}_w$ equals some fraction f_2 of the disk accretion rate $\dot{M}_{acc}$ (cf. Eq. 1 of Shu et al. 1988):

$$\dot{M}_w = f_2 \dot{M}_{acc} \qquad \text{with} \qquad f_2 = \frac{1 - 2b}{\bar{J} - 2b} \qquad (9)$$

where $b = 0.136$ is the pure number defined earlier. (If only the equator of the protostar rotates near breakup, the effective value of b that appears in Eq. [9] could be [much] smaller.) To derive the above expression for f_2, we have assumed that all of the excess angular momentum brought in by accretion above that needed to keep the star exactly at breakup is carried away by the wind, i.e., that none of it is viscously transported to the disk.

The extent by which $\bar{J}$ exceeds unity depends on the strength of the stellar magnetic field in the neighborhood of the X-point of the equipotential. For escaping streamlines to have a finite terminal velocity v_w at infinity, the (rotating) magnetic field lines must exert sufficient torque as to make $\bar{J}$ exceed $3/2$. For example, in order for the terminal velocity of the X-wind to have a measured value equal to the breakup speed at the equator of the star (typically $\sim$150 km s^{-1} for low-mass protostars), $\bar{J}$ must have a value at least equal to 2. In fact, because some of the angular momentum is asymptotically carried away by magnetic stresses rather than all by the fluid, detailed numerical calculations suggest that the actual needed value of $\bar{J} \sim 3.6$, implying a fiducial value for $f_2 \sim 0.2$. Combining Eqs. (8) and (9), we now obtain a relationship between inflow and outflow rates,

$$\dot{M}_w = f \dot{M}_{infall} \qquad (10)$$

with $f \equiv f_1 f_2 \sim 0.1$ to 0.2 typically, in very rough agreement with the observations.

On the other hand, if the magnetic field emerging from the neighborhood of the X-point of the effective potential had infinite strength, X-wind gas would be flung to infinity in the entire sector from the equator to the critical equipotential surface (the dotted curve in Fig. 7). Because this equipotential surface (the last rigid "wire" to which an attached bead could still slide "downhill") bends up to turn asymptotically parallel to the rotation axis of the star (at a radius $= \sqrt{3}R_e$), the strong X-wind could act to focus and collimate an otherwise isotropic and more mild O-wind into an (ionized) optical jet (one on each side of the equator).

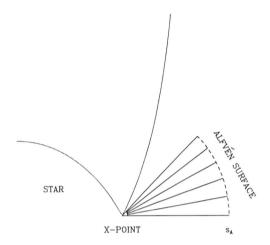

Figure 8.   Spreading of streamlines for an X-celerator driven wind in the meridional plane, computed in a frame that corotates with the star. The bottom-most streamline has been constrained to flow horizontally across the surface of a perfectly flat disk; the uppermost streamline has also been taken to be straight for sake of simplicity. In a more realistic model, the slip interfaces between the X-wind and the O-wind and between the X-wind and the disk surface are likely to be turbulent. A dashed curve marks the Alfvén surface, where the magnetic stresses no longer dominate over the gas inertia. The dotted curve marks the location of the critical equipotential surface that, together with its counterpart in the lower half of the diagram (not drawn), crosses the stellar surface at an X-point. (figure from Shu et al., in preparation).

For stellar magnetic fields of finite strength, the situation becomes more complicated. The method of matched asymptotic expansions provides a tractable approach to the resultant magnetohydrodynamic problem if we adopt the assumption of axial symmetry (Shu et al. 1991). Figure 8 gives a sample solution for a case with a dimensionless stellar field strength that produces Alfvén crossing at a distance equal to one additional stellar radius from the equator of the star. In this calculation, we have used upper and lower boundary conditions that artificially constrain the limiting streamlines, $\psi = 0$ and $\psi = 1$, to be perfectly straight (see below). Interior to the Alfvén surface defined by $\mathcal{A} = 0$, where $\mathcal{A}$ is an Alfvén discriminant, the stream function ψ satisfies an elliptic partial differential equation:

$$\nabla \cdot (\mathcal{A} \nabla \psi) = \mathcal{Q}. \tag{11}$$

This equation has the formal structure of steady-state heat conduction; here, however, we are concerned with the spreading of streamlines, and not the spreading of heat. The source term Q responsible for streamline spreading in the meridional plane equals a sum of three terms that are each proportional to the derivatives (across streamlines) of quantities that are conserved along streamlines: Bernoulli's constant $H(\psi)$, the ratio of magnetic and mass fluxes $\beta(\psi)$, and the specific angular momentum carried in both matter and field $J(\psi)$. Smooth passage through the sonic transition near the X-point of the effective potential yields $H(\psi)$; for the part of the problem much beyond the immediate neighborhood of the X-point of the critical equipotential, $H(\psi)$ goes to 0 in the limit of a cold flow. The requirement that the flow accelerates smoothly through the Alfvén surface, $\mathcal{A} = 0$, places a constraint on the normal derivative of ψ, $\nabla \mathcal{A} \cdot \nabla \psi = Q$, that determines the spatial location and shape of the Alfvén surface and the functional form of $J(\psi)$ if we are given $\beta(\psi)$. In other words, the geometry of the problem is entirely determined if we know the strength of the stellar magnetic field and the way that mass is loaded onto the open field lines at the equatorial belt of the protostar.

For the dimensional parameters that approximately apply to HH 7–11 or L1551, we find that a poloidal speed of $\sim$150 km s^{-1} (corresponding to $\bar{J} \sim 3.6$) can be achieved at the Alfvén surface when the equator of the protostar has open photospheric magnetic fields of the strength $\sim$4 kilogauss. Such fields appear reasonable if we extrapolate from the product of field strength and filling factor, 1 to 2 kilogauss, inferred for weak-line T Tauri stars by Basri and Marcy (1991) from the Zeeman broadening of spectral lines with high Landé-g factors.

It may be informative to elucidate why we believe the X-celerator mechanism to work for a star, but not for a disk (Sec. III). To carry the observed mass-loss rate, the density of the wind at the sonic surface must be large, close to photospheric values. Thus, the sound speed is low, and the sonic transition can occur only near X-points where the effective gravity vanishes, because in a frame which corotates with the footpoint of the magnetic field, the magnetic stresses cannot help the gas make a sonic transition (see Eq. [3] of Shu et al. 1988). On the other hand, if the gas is to continue to accelerate to higher speeds by magneto-centrifugal flinging after having made the sonic transition, the Lorentz force must be able to overcome both inertia and gravity. Because the densities are very high at the sonic surface, a smooth transition from a gas-pressure driven flow to a magnetohydrodynamically driven flow requires either the magnetic field $\mathbf{B}$, or its curl, $\nabla \times \mathbf{B}$, to be large, or both (cf. Eq. [2]). The geometry of a thin disk requires $\nabla \times \mathbf{B}$ to be large, but to keep the gas sub-Alfvénic in the post-sonic-transition region, $\mathbf{B}$ also has to be large. This combination proves fatal, we believe, to disk-wind models in the protostellar context.

An X-celerator driven wind from a protostar (or star-disk interface) fares better, because we require only $\mathbf{B}$ to be large, and we can relatively easily believe that magnetic fields *rooted in the deep convection zones of a rapidly rotating star* could achieve the requisite values. High-mass stars that lack outer convection zones have greater theoretical difficulty generating surface magnetic fields, although the instability mechanism discussed by Balbus and Hawley (1991; see also Fricke [1969] for specific application to the case of stars) offers a promising mechanism for such objects to generate moderate-to-small magnetic fields provided only that they suffer differential rotation of dynamical significance for the structure of the star (with Ω decreasing radially outward). Indeed, X-celerator driven winds in such a situation may offer an explanation why the terminal velocities of winds from high-mass protostars have such low values (100 to 200 km s^{-1}) in comparison with the high values (1000 to 2000 km s^{-1}) that characterize (probably radiation-driven) winds from main-sequence and evolved stars of early spectral type (see, e.g., Chiosi and Maeder 1986).

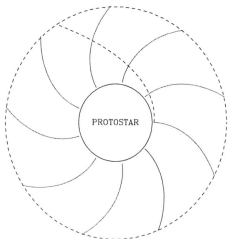

Figure 9. Field lines (dotted curves) and trajectory of a fluid element (dashed curve) for an X-celerator driven wind in the equatorial plane of an inertial frame of reference. The protostar rotates in a counter-clockwise sense, so that open field lines form spirals that trail in the sense of rotation. The dashed circle marks the location of the Alfvén surface in a calculation where we have explicitly assumed axial symmetry (figure from Najita et al., in preparation).

In Fig. 9, we plot magnetic field patterns and fluid trajectories of the flow in the equatorial plane of our X-celerator model. In a frame that rotates with the same angular speed as the star, streamlines and field lines coincide (dotted curves). In a stationary frame, however, the trajectory of an element of electrically conducting gas takes the form indicated by the dashed curve. The fluid element starts by making a sonic crossing close to the photosphere of the star. The acceleration to sonic speeds (7 or 8 km s^{-1} typically) can occur purely by the outward push of photospheric gas pressure in a small equatorial belt where the effective gravity nearly vanishes because we have assumed that the protostar rotates at breakup speeds. As the open magnetic line of force to which the gas is tied rotates in a counter-clockwise direction with the star, the excess rotation (above local Keplerian values) enforced by the field accelerates the gas along the field line (with negligible slip for the typical degrees of ionization computed by Ruden et al. [1990]), until the gas reaches the Alfvén velocity at the position of the dashed circle. At this point, if the gas speed exceeds the local escape velocity by a significant margin, the Alfvén speed will also typically equal a healthy fraction (say, ~90%) of the final terminal speed. Because the magnetic field does not have infinite strength, it cannot keep the gas corotating indefinitely with the star, and the pattern of field lines (in either the rotating or inertial frame) form trailing spirals. The field lines and streamlines form a similar lagging pattern at other latitudes, except that they climb in the vertical direction (toward the rotational poles) as the gas flows away from the protostar.

As already mentioned, the boundary conditions in Fig. 8 on the uppermost and bottom-most streamlines have been taken to be perfectly straight so as not to bias the results for the interior. When we do this, the intermediate streamlines have a tendency also to remain straight, with the gas at higher latitudes having larger densities and slower terminal speeds (for a magnetic field distribution that starts off approximately uniform at the sonic surface). We are currently trying to compute the amount of bending which takes place when the shape of the upper streamline is left more realistically as a free boundary (pressure balance between an X-wind and an O-wind). When such a calculation becomes available, we will be able to estimate the degree of collimation of the X-wind; in particular, we will be able to compute the angle dependence $P(\theta)$ of the rate of momentum injection (needed for any *a priori* calculation of the shapes of the swept-up shells of molecular gas):

$$\frac{\dot{M}_w v_w}{4\pi} P(\theta). \tag{12}$$

Even without detailed knowledge of $P(\theta)$, a simple order of magnitude estimate using the theory developed so far yields the velocity of the swept-up bipolar lobes of molecular gas roughly as the geometrical mean between the terminal speed of the wind v_w and the effective sound speed of the ambient molecular cloud core a_{eff} (Shu et al. 1991; see also Chapter by Königl and Ruden). For v_w measuring hundreds of km s^{-1}, and a_{eff} ranging from a fraction of a km s^{-1} (low-mass cores) to greater than 1 km s^{-1} (high-mass cores), we would then predict typical lobe speeds $\sim (v_w a_{eff})^{1/2} \sim 10$ km s^{-1}, in rough agreement with the observations.

For well-collimated flows, $P(\theta)$ will have values larger than unity for a small range of angles θ near the two poles, 0 and π. Our previous discussion leads us to suspect that younger sources, which may have stronger magnetic fields, possess more highly collimated outflows. We speculate that as the sources age, their rotation speeds progressively fall below the critical rate (see Sec. VII), their magnetic fields weaken, and their outflows fan out more in polar angle and sweep out an increasingly greater solid angle of the overlying envelope of gas and dust. In this fashion may the protostars in stage c of Fig. 1 become optically revealed as the T Tauri stars of stage d.

VII. REVEALED T TAURI STARS

The ability to study T Tauri stars at optical and near-infrared wavelengths has yielded tremendous observational dividends. Foremost in importance among the discoveries of the past decade has been the realization that many of the photometric and spectroscopic peculiarities (variability, strong emission lines, infrared and ultraviolet excesses, strong outflows) long known to characterize these systems (cf. the reviews of Herbig 1962b; Kuhi 1978; Cohen 1984; Imhoff and Appenzeller 1987; Bertout 1986,1989; Appenzeller and Mundt 1991) may owe their explanation to the presence of circumstellar disks. Several independent lines of evidence lead to the conclusion that circumstellar disks surround many young stellar objects, e.g., the large polarizations often seen in these sources (Elsasser and Staude 1978; Bastien and Menard 1988; Bastien et al. 1989), or the asymmetry of forbidden O I line profiles (Appenzeller et al. 1984; Edwards et al. 1987), or the motions of SiO masers (Plambeck et al. 1990). Here, however, we shall concentrate on the nature and implications of the infrared and ultraviolet excesses in T Tauri stars.

We begin with the ultraviolet excess (Kuhi 1974; Herbig and Goodrich 1986), which has been plausibly linked by Hartmann and Kenyon (1987a), and Bertout et al. (1988) to the action of a boundary layer between the star and the disk (see also Chapter by Basri and Bertout). For a conventional boundary layer to arise (gas heated to high temperatures by the frictional rubbing of a rapidly rotating disk against a slowly rotating star), a disk must abut the star.

Cabrit et al. (1990; see also Chapter by Edwards et al.) find an interesting correlation between the strength of the near-infrared excess (a measure of the amount of disk just beyond the boundary layer) and the strength of Hα emission (a measure of T Tauri wind power). Systems inferred to be missing inner disks (radii equal to a few stellar radii) exhibit relatively little wind power (weak-line T Tauri stars), whereas systems with appreciable inner disks have relatively strong winds (classical T Tauri stars). As millimeter and submillimeter investigations reveal no strong biases with respect to the question of whether weak-line and classical T Tauri stars possess outer disks (radii equal to thousands of stellar radii), Edwards and her colleagues interpret their finding to imply that pre-main-sequence and protostellar winds are driven by disk accretion (one that extends virtually right up to the surface of the star). At first sight, this interpretation seems to bode well for the X-celerator theory, which pinpoints the interface between star and disk as the source of the wind. In particular, theories that use self-similar solutions for disk winds (Blandford and Payne 1982; Konigl 1989) have a difficult time explaining why missing just a small portion (the innermost part) of a disk should completely change the nature of the solution. However, a deeper probing of the T Tauri results shows that they also pose a severe problem, potentially fatal, for X-celerator models.

For the X-celerator model of Sec. VI to work, we need to posit that the equator of the star, to which the open magnetic field lines are tied, rotates at breakup. The very fact that T Tauri stars possess boundary layers (if this is the correct interpretation for the ultraviolet excess) demonstrates, however, that their equatorial regions are not rotating at breakup. Even more damaging, direct spectroscopic investigation shows that most T Tauri stars rotate at speeds about an order of magnitude *slower* than breakup (Vogel and Kuhi 1981; Bouvier et al. 1986a; Hartmann et al. 1987; see the review of Bouvier 1991). Yet, Edwards et al. (see also Chapter by Hartmann et al.) deduce that T Tauri stars also satisfy an inflow-outflow relationship of the form of Eq. (9), $\dot{M}_w = f_2 \dot{M}_{acc}$, with f_2 numerically not very different from the value that we have deduced from X-celerator theory ($f_2 \sim 0.2$).

How can we resolve the discrepancy? Galli and Shu (in preparation) propose that a partial solution may be found by asking how an accretion disk actually tries to spin up a slowly rotating star. Their answer involves, not the inward diffusion of vorticity as in the models of Paczynski (1991) and Popham and Narayan (1991), but *Ekman pumping*. The mathematics for Ekman pumping is fairly involved; however, the basic physics is simple, and easily explained by analogy to spinup in a cup of tea.

Suppose we place a cup of tea on a turntable spinning at angular speed Ω, and we ask how long it takes for the liquid, initially at rest, to spin up to the same angular speed as the boundaries of the cup. If spinup occurred by the diffusion of vorticity from the boundary layer to the interior, the theoretical answer, in order of magnitude, would be given by the familiar formula:

$$t_{diff} \sim L^2/\nu \qquad (13)$$

where L represents the typical dimension of the cup, and ν gives the kinematic viscosity of tea. Putting in characteristic numbers, we would deduce $t_{diff} \sim$ tens of minutes. In fact, the actual spinup time empirically measures more in the neighborhood of tens of seconds, and is theoretically given as the geometric mean between the viscous diffusive time scale Eq. (13) and the overturn time scale Ω^{-1}:

$$t_{spinup} \sim (t_{diff}/\Omega)^{1/2}. \qquad (14)$$

The formula (14) represents essentially the time that it takes tea to circulate from the interior of the cup to the boundaries, where it can quickly match its rotational speed to that of the cup (Fig. 10a). In other words, Ekman pumping sets up a secondary circulation that brings the tea to the boundary layer rather than waiting for the effects of the accelerating agent to diffuse to the interior. Because the bottom of the cup speeds up the tea in contact with it relative to the tea in the interior, the pressure gradient needed to balance

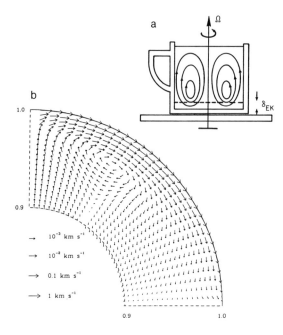

Figure 10.   (a) Ekman pumping in a teacup (schematic). A thin Ekman layer δ_{EK}
forms above the bounding surface perpendicular to the axis of rotation. (b) Ekman
pumping in a T Tauri star modeled via the spinup of a (uniformly rotating) polytrope
of index 1.5 with turbulent viscosity estimated by convective mixing-length theory.
A secondary circulation pattern sets up in the meridional plane when the star is
subjected to a shear stress at its surface that is highly concentrated toward the
equatorial plane. The velocity scale corresponds to a model in which $M_* = 0.8$
$M_\odot$, $R_* = 3\,R_\odot$, $\Omega_* R_* = 20$ km s^{-1}, and $\nu = 10^{14}$ cm^2 s^{-1}. The radial scale from
0.9 R_* to 1.0 R_* has been expanded for clarity of display (figure from Galli 1990).

the centrifugal force of rotation differs in the two regions. The difference in
pressure distribution outside and inside the boundary layer then pumps the
fluid from the former to the latter in such a way as to set up the secondary
circulation depicted in Fig. 10a.

We believe an analogous situation to hold for the case of T Tauri stars,
except that gravity, in addition to pressure and inertial terms, enters the equation for force balance. Schematically, the presence of an abutting disk tries
to spin up the equatorial regions of a star faster than its poles. The excess of
centrifugal support in the equatorial regions necessitates less vertical pressure
support against gravity there than elsewhere. As a consequence, the unbal-
anced horizontal pressure gradients will push gas from the surface layers of
the rest of the star toward the equator. Continuity requires those streamlines
converging onto the equator of the star that do not blow outwards in a wind to

resubmerge into the interior of the star, establishing a circulation pattern that looks as depicted in Fig. 10b.

Figure 10b represents a detailed calculation from Galli's (1990) doctoral thesis of Ekman pumping in a polytrope of index 1.5 (to represent a fully convective star). The calculation, however, contains only one of two possible effects that can lead to a secondary circulation: the *barotropic* response to the spinup induced by the frictional torque of an adjacent disk highly concentrated toward the equatorial regions. The rotational evolution, with $d\Omega/dt > 0$, is assumed to proceed keeping the zeroth-order rotation spatially uniform, which leads to perturbations in the angular speed of the stellar interior that are stratified across cylinders. If the true surfaces of constant Ω in T Tauri stars correspond—as they do in the case of the convection zone of the Sun (Dziembowski et al. 1989)—not to cylinders, but to radial cones, a second effect can arise because the surfaces of constant pressure do not then correspond to surfaces of constant density. In this situation, the *baroclinic* generation and maintenance of meridional mass flow may compete with or even dominate over the barotropic part. In other words, the meridional flows induced by a quasi-steady balance between spinup through a disk and spin-down via a wind may make the application of Fig. 10b to actual T Tauri stars somewhat oversimplified.

Nevertheless, Fig. 10b shows a robust qualitative feature of great importance to our current discussion: the circulation of matter in the photospheric and subphotospheric layers toward the equator. If such layers carry stellar magnetic field lines along with them, then we can easily visualize how open field lines of kilogauss strength might be brought from the rest of the star that rotates slowly (as required to satisfy the spectroscopic observations of T Tauri stars) to an equatorial band that rotates quite rapidly (by definition, at "breakup" when we enter the disk proper). Conceivably, the Doppler imaging (Vogt 1981) of classical T Tauri stars could test whether circulation currents of this geometry cause starspots to migrate toward the equator on time scales (radius of star divided by photospheric circulation speed) of weeks to months. If so, the resulting ejection of matter in a powerful magnetocentrifugal wind along open field lines might then qualitatively proceed, more or less, as we described in Sec. VI for an X-celerator-driven flow. As it takes on the order of a day for circulation-driven, inhomogeneous, magnetic structures to migrate through the sonic surface of a wind originating from an equatorial belt equal to a few percent of the radius of the T Tauri star, we can now understand why optical spectral lines believed to be wind diagnostics (in contrast to general photospheric diagnostics) should show profiles that vary wildly on that sort of time scale (G. Basri, personal communication).

The above speculation, that T Tauri winds originate basically in a highly magnetized and rapidly rotating boundary layer, amounts essentially to a compromise solution between a stellar-driven wind and a disk-driven wind. In this compromise, we make use of the known ability of a star to hold onto very strong magnetic fields, and we take advantage of the natural tendency for disks to possess very rapid rotation. More work, however, needs to be done on this problem before we can claim that we can safely import to this complex situation the main features of the solutions for more simple models.

We should note, however, that *closed* field lines may play a competing role to open ones. If stellar circulation currents force closed field lines of sufficient strength to thread through the disk (a process made easier at a late stage when the disk accretion rate onto the star drops to small values), and if the star becomes magnetically coupled largely to those parts of the disk that rotate fairly slowly, then we might have another explanation as to why T Tauri stars have projected rotational speeds $v\sin i \lesssim 15$ km s^{-1} (see Konigl [1991]). For this braking mechanism to work, however, the same magnetic torques associated with fields threading the gaseous disk need to empty out much of the rapidly rotating inner portions of the disk, which would otherwise act to spin up the star, rather than spin it down. Weak-line T Tauri stars may represent (evolved) systems missing such inner disks, and it is interesting to note that they do not possess the powerful (X-celerator-driven) winds that classical T Tauri stars do.

We turn now to the dynamics of the disk proper, and the origin of the infrared excesses of T Tauri stars. Since the pioneering work of Mendoza (1966,1968), astronomers have known that T Tauri stars emit significantly more infrared radiation than other stars of their spectral type (typically K subgiants). From the start, thermal emission from circumstellar dust has been suspected as the culprit; however, the geometry of the dust distribution remained obscure until Lynden-Bell and Pringle (1974) suggested that it might lie within a viscous accretion disk. In particular, Lynden-Bell and Pringle pointed out that an optically thick disk, vertically thin but spatially extended in distance ϖ from the rotation axis over several orders of magnitude, and possessing a power-law distribution of temperature,

$$T_D \propto \varpi^{-p} \tag{15}$$

would exhibit a power-law infrared energy distribution,

$$\nu F_\nu \propto \nu^n \qquad \text{where} \qquad n \equiv 4 - \frac{2}{p}. \tag{16}$$

Viscous accretion-disk theory for a system in quasi-steady state predicts $p = 3/4$, i.e., $n = 4/3$.

Rucinski (1985) examined the observational evidence and he concluded that, while the infrared excesses of T Tauri stars did well approximate a power-law distribution, the mean value of n is appreciably smaller than 4/3 (see also Rydgren and Zak 1987). Adams et al. (1987) noted that the presence of considerable infrared emission without a correspondingly large visual extinction of the central star implies that the circumstellar distribution of dust must exist in orbit in virtually a single plane about the star. In such a situation, a *non-accreting*, optically thick, flat, and radially extended disk would intercept 25% of the starlight and reprocess it to near- to far-infrared wavelengths, with disk-temperature and spectral-energy distributions also very nearly satisfying Eqs. (15) and (16), except that the coefficient of proportionality would be exactly known. Unshielded large dust particles at a distance r from a star would, of course, have a temperature distribution that satisfies $T_D \propto r^{-1/2}$. However, astronomers believe that young nebular disks are completely thick at optical wavelengths even vertically through the disk, so that starlight cannot shine directly on dust particles by propagating through the midplane, but must come at oblique angles from near the limbs of a star of size R_*. This effect introduces an extra geometrical factor of essentially R_*/r into the relation between absorbed starlight ($\propto [R_*/r]r^{-2}$) and emitted infrared radiation ($\propto T_D^4$), which makes the reprocessing temperature law $T_D \propto r^{-3/4}$ rather than $r^{-1/2}$. To remind ourselves of this important difference between the reprocessing in an opaque flat disk with faces perpendicular to the z direction and in an airless planetary surface with projected area perpendicular to r, we write the former law as $T_D \propto \varpi^{-3/4}$.

Adams et al. (1987) also demonstrated that the few cases of T Tauri systems that did show the canonical value $n = 4/3$ could indeed be explained in terms of *passive* or *reprocessing* disks; whereas, those disks that have the most extreme infrared excesses and, therefore, exhibit the most quantitative evidence for a nonstellar contribution to the intrinsic luminosity, correspond, not to $n = 4/3$, but to $n = 0$ (Adams et al. 1988; see Fig. 11a,b). Flat-spectrum sources with $n = 0$ require $p = 1/2$, i.e., $T_D \propto \varpi^{-1/2}$. A comprehensive survey by Beckwith et al. (1990; see also Chapter by Beckwith and Sargent) fitted temperature power-laws to the infrared excesses of classical T Tauri stars, and found that the derived values for the exponent p range from the extremes $p = 3/4$ (steep-spectrum sources) to $p = 1/2$ (flat-spectrum sources), with more sources having the latter value than the former. Strom claims, however (see his chapter in this volume), that the "flatness" of the spectra of many of the sources disappears if one uses long-wavelength measurements with smaller beams than the IRAS measurements.

Kenyon and Hartmann (1987) proposed that *geometric flaring* in purely reprocessing disks could flatten out the spectral energy distributions of the spatially flat models that predict $n = 4/3$, and thereby remove much of the discrepancy between theory and observation. A greater interception and scattering of stellar photons would also better account for the relatively large fractional polarization detected for some T Tauri stars (Bastien et al. 1989). In contrast, Hartmann and Kenyon (1988) note that the source FU Orionis, which may be undergoing enhanced accretion through a disk (for a dissenting view, see Herbig 1989*a*), *does* show the canonical value $n = 4/3$ (see also Adams et al. 1987), and they conclude that nebular disks may transfer much, if not most, of their mass onto the central stars through such recurrent episodic outbursts. Finally, Kenyon and Hartmann (1991) conclude that, as a class, many of the FU Orionis variables may still be surrounded, not by flared disks,

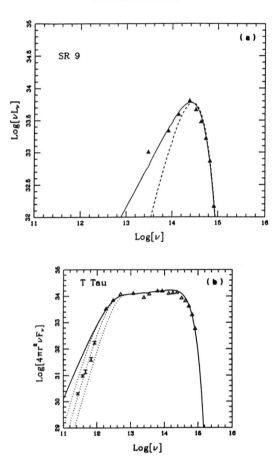

Figure 11.   (a) A steep-spectrum T Tauri star, SR 9. The dashed curve shows the expected spectrum for a star alone; the solid, for a star plus a spatially flat, but radially extended, disk that locally reprocesses the starlight that falls on it (figure from Adams et al. 1987) . (b) A flat-spectrum T Tauri star, T Tau itself. The curves show models (from Adams et al. 1988) for a star plus a spatially flat disk, with the latter heated by stellar photons as well as possessing an intrinsic source of luminosity with a temperature distribution, $T_D \propto \varpi^{-p}$, where $p = 0.515$. The disk extends from the stellar surface to a radius $R_D = 120$ AU, the minimum value needed to fit the extent of the flat portion of the infrared data points (triangles). The actual size also cannot be much larger before suffering truncation by the companion. (The system corresponds to a binary, but the model is insensitive to which star has the flat-spectrum disk, or even if both stars have disks.) The disk model in the solid curve has infinite mass and is optically thick at all frequencies; the dotted curves that turn over progressively toward higher frequencies because of declining optical depth have disk masses, respectively, of 1, 0.1, and 0.01 $M_\odot$. Measurements at submillimeter and millimeter frequencies (data points with error bars) can establish the disk mass, provided the dust opacities in the disk correspond with those of the model (figure from Adams et al. 1990).

but by dusty (infalling?) envelopes that contribute significant amounts of far-infrared emission through the reprocessing of starlight.

There exists a difficulty with explaining the most extreme flat-spectrum T Tauri stars (which includes some of the most famous examples in the class: T Tauri itself, HL Tau, DG Tau, etc.) as passive, geometrically flared disks. To reprocess the starlight into the observed amounts of infrared excess would require unrealistic ratios of the dust photospheric height H to disk radius ϖ. Appendix B of Ruden and Pollack (1991; see also Kusaka et al. 1970) gives the temperature of a flared disk illuminated by a central star of radius R_* and effective temperature T_* (at a radius $\varpi >> R_*$) as

$$T_D(\varpi) = T_* \left[\frac{2}{3\pi} \left(\frac{R_*}{\varpi} \right)^3 + \frac{1}{2} \left(\frac{R_*}{\varpi} \right)^2 \left(\frac{H}{\varpi} \right) \left(\frac{d \ln H}{d \ln \varpi} - 1 \right) \right]^{1/4}. \quad (17)$$

Notice that the canonical law $T_D \propto \varpi^{-3/4}$ dominates at small radii ϖ (where $H << R_*$), while shallower temperature gradients can arise at larger radii if H varies as a power of ϖ with an exponent s greater than unity (a geometrically flared disk). If the dust does not settle to the midplane but remains suspended in the gas (an inconsistent assumption if reprocessing disks are not turbulent), vertical hydrostatic equilibrium requires $H \propto \Omega^{-1} T_D^{1/2}$, so that $s = 9/7$ and $p = 3/7$ in the outer parts of a self-consistent, Keplerian, reprocessing disk. Such a disk typically requires $H/\varpi \gtrsim 1/3$ at large ϖ to produce an appreciable departure from the canonical case (Kenyon and Hartmann 1987). A circumstellar layer of dust with such an aspect ratio would conflict, however, with the observed fact that unobscured T Tauri stars are an order of magnitude more numerous than heavily extincted sources (in the Taurus molecular cloud). Moreover, for sources not seen *through* the disk, since $n = +4/3$ for $p = 3/4$ but $n = -2/3$ for $p = 3/7$, the spectrum would *fall* from the near- to mid-infrared but then *rise* toward the far infrared. Reprocessing of starlight in either flared disks or warped disks cannot generally reproduce the *single* power-law index, $n = 0$ from the near infrared to the far infrared, observed for the most extreme flat-spectrum sources.

Binarity is known to add complications in the case of T Tauri, and other sources of confusion may contribute to the observed departure from the expected spectral slope of $n = 4/3$. Taking the point of view, however, that physical significance could be assigned to the frequent occurrence of the single exponent $n = 0$, which neither viscous accretion nor passive reprocessing adequately explains, Adams et al. (1988) speculated that the flat-spectrum T Tauri stars must have *active* disks where the processes of transport of mass (inward) and angular momentum (outward) involve *nonlocal* mechanisms. Adams et al. (1989; see also Chapter by Adams and Lin) worked out one such mechanism: the generation of one-armed spiral density waves ($m = 1$ disturbances) by eccentric gravitational instabilities that involve the displacement of both star and disk from the center of mass of the system. Shu et al. (1990) gave an analytic demonstration, under a restrictive set of circumstances, that such modal eccentric instabilities would arise whenever the disk mass as a fraction of the whole (star plus disk) exceeds $3/4\pi \sim 0.24$. They also pointed out that the temperature law $T_D \propto \varpi^{-1/2}$ plays a special role in the linear theory of $m = 1$ disturbances in that this distribution keeps the fractional amplitude of the wave (ratio of disturbance to unperturbed surface densities) constant throughout much of the disk. As the fractional amplitude in a full theory provides a measure of the nonlinearity of the wave, Shu et al. (1990) speculated that nonlinear propagation and dissipation of excited $m = 1$ waves would provide the feedback that accounts for the temperature law $T_D \propto \varpi^{-1/2}$ in the first place. Ostriker and Shu (work in progress) are pursuing the correct nonlinear derivation of this result, and they are also investigating the possibility that stellar companions in binary systems may provide an alternative source for the excitation of $m = 1$ density waves.

We find it informative to comment on the basic reason why either low-freqeuncy unstable normal modes or fairly distant companions can provide, in principle, a redistribution of mass and angular momentum of a different qualitative nature than viscous accretion. A disturbance which rotates at a uniform angular speed Ω_p (see Sec. VIII) contains excess energy E_{wave} and angular momentum J_{wave} above the unperturbed values. These are related to each other by the formula (cf. Lynden-Bell and Kalnajs 1972),

$$E_{\text{wave}} = \Omega_p J_{\text{wave}}. \tag{18}$$

The same proportionality applies *locally* to the wave energy densities $\mathcal{E}_{\text{wave}}$ and angular-momentum densities $\mathcal{J}_{\text{wave}}$ (per unit area),

$$\mathcal{E}_{\text{wave}} = \Omega_p \mathcal{J}_{\text{wave}} \tag{19}$$

if the spiral waves are tightly wrapped (Lin and Shu 1964; Toomre 1969; Shu 1970), as they inevitably are for disturbances in most of a radially extended nebular disk (cf. Adams et al. 1989; Shu et al. 1990). Moreover, $\mathcal{E}_{\text{wave}}$ and $\mathcal{J}_{\text{wave}}$ are both positive or both negative depending on whether the wave rotates faster or slower than the matter; i.e., $\mathcal{E}_{\text{wave}}$ and $\mathcal{J}_{\text{wave}} > 0$ outside corotation, $\Omega_p > \Omega$, whereas $\mathcal{E}_{\text{wave}}$ and $\mathcal{J}_{\text{wave}} < 0$ inside corotation, $\Omega_p < \Omega$, where $\Omega(\varpi)$ represents the local angular speed of the disk.

On the other hand, in order for a gas element to remain in *circular* orbits when some external torque changes its angular momentum at a rate $\dot{J}$, its energy must change at a rate

$$\dot{E}_{\text{circ}} = \Omega \dot{J}. \tag{20}$$

When spiral density waves inside (outside) corotation dissipate they give a negative (positive) amount of angular momentum to the local gas, causing that gas to move to a radially smaller (larger) orbit. However, the energy lost (gained) by the gas element (a factor of Ω_p times the angular momentum lost [gained]) in the process of damping the wave will generally be inappropriate for the gas to remain in a circular orbit with the modified angular momentum. The excess orbital energy will be converted to heat at a rate

$$\dot{H} = (\Omega_p - \Omega)\dot{J} \qquad (21)$$

when the orbit of the gas element circularizes by hydrodynamic dissipative mechanisms.

Note that as the gas loses (gains) angular momentum inside (outside) corotation, the rate of conversion of energy into heat, $\dot{H}$, is everywhere positive. Note also that deep inside the disk where $\Omega >> \Omega_p$, the rate of heat gain is very large per unit rate of angular momentum transferred. In other words, although a slowly rotating disturbance in the outer part of the disk (either a normal mode or a companion star) cannot directly supply much energy to the inner disk (only a factor $\Omega_p \times \dot{J}$), it potentially comprises a huge reservoir for dumping the angular momentum of the system. Removing angular momentum from the part of the disk inside of corotation (by direct near-resonant coupling in $m = 1$ density waves) results in accretion. When the basic orbits cannot become very noncircular (because tightly wrapped waves cannot yield highly eccentric orbits without producing either radiative damping or streamline crossing and dissipative shocks), the heat released by this nonviscous form of accretion will provide, we believe, the luminosity excess that one sees in the flat-spectrum T Tauri stars.

In the inner portions of the disk, where $\Omega >> \Omega_p$, but still much beyond the star-disk boundary layer, we note that Eq. (21) for the disk heat input differs only by a factor of 3 from the corresponding formula given by viscous accretion-disk theory. (A much bigger difference exists beyond the corotation circle, where $\Omega_p >> \Omega$, but the physical dimensions of realistic disks may not extend to regions where the wave energy much dominates over the accretion energy.) The ability of wave transport (perhaps) to produce flat-spectrum sources may then depend critically on the fact that $m = 1$ wave dissipation yields a nonuniform rate of local disk accretion $\dot{M}_{acc}$, with $\dot{M}_{acc} \propto \varpi$ needed to produce a flat-spectrum source.

It might be argued that viscous-accretion disk models could also accomodate flat-spectrum sources if we allow ourselves the luxury of assuming $\dot{M}_{acc} \propto \varpi$. Hartmann and Kenyon (1988) dismissed this possibility on the grounds that it would require mass accretion rates in the outermost parts of a disk three or more orders of magnitude higher than the innermost regions. Compare the required behavior with theoretical models of viscous accretion. Independent of the details for the specification of the viscosity, the diffusive nature of the viscous process (if stable) inevitably yields in the relevant portions of the disk, a long-term evolution that satisfies the approximate condition, $\dot{M}_{acc} = a$ spatial constant (cf. Ruden and Lin 1986; Ruden and Pollack 1991). The same result need not apply to accretion disks driven by global instabilities. Indeed, the dynamics of quasi-steady wave generation, propagation, and dissipation will generally *not* correspond to the condition of steady-state accretion.

The implication that the (self-gravitating) disks in flat-spectrum T Tauri stars have much larger accretion rates in their outermost parts (say, $\sim 10^{-5}$ $M_\odot$ yr^{-1}) than their innermost parts (say, 10^{-8} to 10^{-7} $M_\odot$ yr^{-1}) may have important observational consequences. In particular, the tendency to pile up matter at a bottle-neck at the smallest radii leads, for a temperature distribution fixed by wave dissipation independent of the surface density distribution (see Shu et al. 1990), to a local lowering of Toomre's (1964) Q parameter. If Q falls sufficiently, violent $m > 1$ gravitational instabilities may make an appearance (see, e.g., Papaloizou and Savonije 1991; Tohline and Hachisu 1990). Such modes act over a much more restricted radial range than $m = 1$ modes, but they may induce a sudden release of the piled-up material onto the central star. Two very different gravitational modes of mass transport may therefore act: $m = 1$ waves, driven by distant orbiting companions or mild instabilities characteristic of the outer disk; and $m > 1$ instabilities with high growth rates characteristic of the inner disk. The relative ineffectiveness of the first mechanism for inducing mass transport at small ϖ leads to large amounts of matter caught in the "throat"; the clearing of this material by the second mechanism may then yield the sporadic "coughs" that Hartmann et al. identify in their chapter as FU Orionis outbursts. Much more work would have to be performed on this scenario, however, before we could characterize it as more than an interesting speculation.

VIII. BINARY STARS AND PLANETARY SYSTEMS

In their studies, Adams et al. (1989) and Shu et al. (1990) suggested that the existence of binary systems among the T Tauri stars may itself represent a manifestation of the prior action of $m = 1$ gravitational instabilities (see Chapters by Bodenheimer et al., and by Adams and Lin). Recent numerical simulations using smooth-particle-hydrodynamics codes (Lattanzio and Monaghan 1991; Bastien et al. 1991; F. C. Adams and W. Benz, personal communication) indicate that the formation of companion objects by non-axisymmetric disturbances (in particular, $m = 1$) in massive nebular disks may indeed be a viable physical process. Nevertheless, because the actual accumulation of mass occurs, at least in some systems, via *gradual* infall of a single molecular cloud core (cf. Secs. IV and V), we should not directly apply the results of simulations of fully accumulated stars plus disks to the actual binary-formation process. We believe it much more likely that *both* stars start off small (perhaps as the result of a runaway $m = 1$ gravitational instability), that the two protostars separately acquire inner accretion disks, that they may also jointly acquire a circumbinary disk, and that buildup of the entire system occurs substantially while the system lies deeply embedded within a common infalling envelope. Haro 6–10 and IRAS 16293–2422 may represent examples of such systems (see Zinnecker 1989a; G. A. Blake and J. E. Carlstrom, personal communication; for a review of alternative formation scenarios, see Pringle 1991).

An interesting aspect of the realization that binary-star formation may occur in the presence of circumstellar, or even circumbinary, disks concerns the eccentricity of the resulting orbits. Duquennoy and Mayor (1991) and, especially, Mathieu et al. (1989) have emphasized the important observational point that most binaries are probably born with eccentric orbits, and that the shorter-period ones only later become circularized by stellar tidal processes. This state of affairs should be contrasted with that applicable to planetary systems, where current belief holds that most planets are born with nearly circular orbits. How do we reconcile this difference if we believe that companion stars and planets are both born from disks?

A qualitative answer to this question may come from the work of Goldreich and Tremaine (1980) on the interactions of satellites and disks (see also Ward 1986; Chapter by Lin and Papaloizou). For satellite orbits of small eccentricity e and inclination i, a Fourier expansion of the potential associated with the satellite's gravity introduces a series of disturbances that have wave frequency ω given by (cf., e.g., Shu 1984)

$$\omega = m\Omega_s \pm n\kappa_s \pm p\mu_s \qquad (22)$$

where m, n, and p are positive integers, with κ_s and μ_s being the epicyclic and vertical oscillation frequencies, respectively, when we expand the satellite's eccentric and inclined orbit about the circular orbit with angular rotation speed Ω_s. The integer m yields the angular dependence $\propto e^{-im\varphi}$ of the disturbance potential, whereas n and p represent the number of powers of e and $\sin i$ that enter the coefficient of the specific term in the series expansion. We refer to the terms corresponding to $n = p = 0$, which can arise even if the satellite orbit is circular, as being of *lowest order*; higher-order terms depend on the orbit having nonzero eccentricity e or inclination i.

As the time dependence of the disturbance potential $\propto e^{i\omega t}$ and the angular dependence $\propto e^{-im\varphi}$, the pattern speed of the disturbance equals

$$\Omega_p = \omega/m. \qquad (23)$$

In a steady state (including dissipation), the satellite potential excites a density response in the disk that remains stationary in a frame that rotates at the pattern speed Ω_p. This response achieves its highest values at various resonance locations. *Corotation resonances* occur at radii ϖ where

$$\Omega(\varpi) = \Omega_p \qquad (24)$$

Lindblad resonances occur where

$$\Omega(\varpi) \pm \frac{1}{m}\kappa(\varpi) = \Omega_p \qquad (25)$$

and *vertical resonances* occur where

$$\Omega(\varpi) \pm \frac{1}{m}\mu(\varpi) = \Omega_p \qquad (26)$$

with $\Omega(\varpi)$, $\kappa(\varpi)$, and $\mu(\varpi)$ representing, respectively, the circular, epicyclic, and vertical oscillation frequencies of the local gas. *Inner* resonances occur where the minus sign is chosen in Eqs. (25) and (26); *outer* resonances, where the plus sign is chosen. For a nonaxisymmetric disturbance with given pattern speed Ω_p, we generally encounter in succession an inner resonance, a corotation circle, and an outer resonance, as we move radially outwards through the disk. For a Keplerian disk, all three frequencies are equal, $\Omega = \kappa = \mu$, and special significance attaches to $m = 1$ disturbances because the entire disk can be in near-inner Lindblad or vertical resonance with a distant companion (where $\Omega_p \sim 0$).

Consider now the gravitational back-reaction of the density response in the disk on the exciting satellite. The gravitational potential associated with the density response, being nonaxisymmetric and time dependent in an inertial frame (but time independent in a frame that rotates at Ω_p), conserves neither the orbital angular momentum J_s nor the orbital energy E_s of the satellite; however, the combination $E_s - \Omega_p J_s$ = Jacobi's constant *is* conserved. As a consequence, the back-reaction from the disk pumps energy and angular momentum into the orbit of the satellite in the ratio Ω_p, $\dot{E}_s = \Omega_p \dot{J}_s$, with $\dot{J}_s$ equal to minus the torque on the disk. For the strongest resonances (corotation or Lindblad), where $n = p = 0$, Eqs. (22) and (23) state that $\Omega_p = \Omega_s$; consequently to zeroth order in e (and $\sin i$), these strong resonances contribute energy and angular momentum to the satellite in a ratio as to keep circular orbits circular. To linear order in e^2, however, the strongest resonances change the satellite's eccentricity at rates comparable to the weaker resonances that have $n = 1$ and $p = 0$. The former always lead to eccentricity damping (growth) for inner (outer) Lindblad resonances; the latter contribute to eccentricity changes depending on the sign in $\Omega_p = \Omega_s \pm \kappa_s / m$ and on whether the Lindblad resonances are inner or outer ones. The torque exerted by corotation resonances (and therefore their contribution to eccentricity changes) depends on the sign of the gradient of the vorticity divided by the surface density.

In an untruncated Keplerian disk of uniform surface density, Goldreich and Tremaine (1980) demonstrate that the sum of the effects of all resonances has a slight dominance of eccentricity-damping resonances over eccentricity-exciting ones (see also Ward 1986 for estimates of the effects of surface-density gradients). This presumably explains why planets embedded in nebular disks acquire nearly circular orbits, although gap opening adds to uncertainties (see Lin and Papaloizou 1986a, b). The ability of strong resonances to open gaps is observed in the Saturnian ring-moonlet system (see the discussion of Cuzzi et al. 1984). For a massive companion such as another star, however, not only may the companion be born with an initial eccentricity, but wide gaps opened up on either side of the secondary star may lead to the disappearance of the strongest $m \neq 1$ resonances. The remaining resonances may amplify orbital eccentricities under some circumstances; for example, Artymowicz et al. (1991) find eccentricity growth in a numerical simulation of a binary surrounded by a circumbinary disk that is dominated by a $m = 2$ outer-Lindblad resonance with $\Omega_p = \Omega_s - \kappa_s/2$. On the other hand, the system GW Ori appears to be a binary with a circular orbit and with circumstellar and circumbinary disks (Mathieu et al. 1991). The inferred gap surrounding the companion eliminates the stronger $m \neq 1$ resonances; if true, the circumstellar disk may provide a circularizing influence on the binary orbit through the action of $m = 1$ inner-Lindblad near-resonances (Ostriker et al. 1991). In any case, if a companion star acquires large eccentricities, it may have a high efficiency for sweeping up the matter of the disk in which it is embedded.

In summary, the orbital characteristics of newly formed planets and companion stars might qualitatively differ from one another because the two types of objects originate by completely different processes (cf. Pringle [1991] for this point of view). On the other hand, even if they both descend from disks, we see that good mechanistic reasons exist why binary stars might acquire appreciably eccentric orbits, whereas planets usually do not. The difference in outcomes—binary stars or planets—might then depend on differences in initial conditions, namely, the total amount of angular momentum J_{tot} contained in the collapsing system. If J_{tot} is small, a substantial central mass could accumulate by direct infall before disk formation occurs. The stabilizing influence of a massive center of attraction may then prevent any runaway gravitational instabilties in the disk, leaving only the possibility of planetary condensation by chemical or physical means. If J_{tot} is large, the ratio of disk to star mass may reach order unity before the central object has grown sufficiently to prevent the gravitational fragmentation of the disk to form another body on an eccentric orbit. In this case, continued infall builds up a binary-star system.

IX. THE ORIGIN OF STELLAR AND PLANETARY MASSES

As the primary transformers of the simple elements that emerge from the big bang, stars constitute the fundamental agents of change in the visible universe. Of all its properties, a star's mass most affects its main-sequence appearance and later evolution. Given that giant molecular clouds, the raw material for forming stars, contain much more mass than required to make the final products, the most basic problem posed for theories of stellar origin concerns, then, how these objects acquire the masses that they do. At one time, astronomers sought for an answer to this important question in terms of a picture of hierarchical fragmentation, but severe doubt about this possibility has been cast by the realization that giant molecular clouds are not collapsing dynamically and have, in fact, generally a very low efficiency for stellar genesis (Zuckerman and Evans 1974).

The discovery that massive outflows ubiquitously accompany the birth pangs of stars has led to an alternative explanation, that forming stars might help to define their own masses. From this point of view, star formation represents basically an accretion process (a theoretical point of view with modern origins in the work of Larson [1969]), with the mass of the star continuing to build up until a powerful enough wind turns on to reverse the infall (first, over the poles, and later, even in the equatorial directions) and shut off the accretion. Indeed, given the basic angular-momentum difficulty faced by any condensing and rotating object which needs to contract by several orders of magnitude in linear scale, perhaps theorists should have anticipated all along that stars would not be able to form without simultaneously suffering tremendous mass loss. As it is, however, theorists from Laplace onwards managed to predict, in one form or another, only three of the four stages of star formation depicted schematically in Fig. 1.

A parallel situation holds for the issue of planetary masses, which we know today probably also results from a process of accumulation. At one time theorists sought to skirt the issue by postulating that the primitive solar nebula had only the amount of mass, when reconstituted to solar composition, needed to build the known planets. The planets then incorporated the masses that they did, essentially because that was all the material there was. (We exaggerate the degree to which "input equals output" in the so-called "mini-mum-mass model" because theorists still had the chore of explaining why the solar system formed with *nine* major planets, and not one, or a hundred.)

The modern view suggests that the birth of a star must indeed almost always proceed with the accompanying appearance of a disk. However, it also claims that the masses of the growing star and disk are unlikely to have a ratio of anything like 100 to 1. In the early stages, at least, the solar nebula may have had a comparable mass to the proto-Sun. Such a situation places sharp focus on the question how planets acquire the masses that they do if most disks have available more mass than needed to make the known planets.

A partial answer may have been supplied by the suggestion by Lin and Papaloizou (1985,1986a; see also their Chapter) that *tidal truncation* limits the maximum amount of mass acquirable by a *giant* planet. Tidal truncation works because a circularly orbiting body has a basic tendency to want to speed up more slowly rotating material exterior to its own orbit, and to slow down more rapidly rotating material interior to it. This tendency to give angular momentum to the matter on the outside and to remove it from matter on the inside (as already discussed for resonant interactions in Sec. VIII) secularly pushes gas on either side of a protoplanet away from its radial position, opening up a gap centered on the body's orbit. Viscous diffusion tends to fill in this gap, and for a true gap to form, the tidal torques have to dominate, leading, for a given level of turbulent viscosity, to a minimum protoplanetary mass that can open up a gap larger than its own physical size. (See Chapter by Lin and Papaloizou on the additional criterion that arises in a *gaseous* disk because the gap width must also exceed the vertical pressure scale-height, if gas pressure is not to close the gap.)

The observational knowledge that the tidal force of moonlets can clear analogous gaps in planetary rings (see, e.g., Showalter 1991) shows that this mechanism does have empirical validity; however, the simplest formulations of gap opening and shepherding theories do not yet yield good quantitative comparisons with observations of planetary rings and their moonlets (see, e.g., Goldreich and Porco 1987). Moreover, this mechanism applied to the solar nebula will not work to define the masses of the *terrestrial* planets, nor does it explain what happens to the excess disk material that does not go into planets. As already discussed in Sec. I, the fate of the solid debris may not pose a great problem given the demonstration that most of the space between the major planets yields secularly unstable orbits (see Chapter by Duncan and Quinn); however, the dispersal of the remnant gaseous component remains a serious problem, especially if massive protostellar and pre-main-sequence winds are driven by disk accretion.

A possible resolution of the problem may rest with Elmegreen's (1978b,-1979c) proposal that stellar winds erode disks, not by blowing the gas away (Cameron 1973), but by inducing turbulence in the wind-disk interface that enhances the *inward* viscous transport of matter. If, in turn, disk accretion drives powerful stellar winds, the process could continue to operate, in principle, until the disk has completely emptied onto the star. Indeed, the existence of a feedback loop—disk accretion drives wind → wind drives disk accretion, etc.—together with time delays, leads to the possibility of instabilities and limit cycles (see Bell et al. [1991] for an example of a negative feedback cycle leading to chaotic oscillations). Such feedback loops—modulated perhaps by magnetic cycles on the star itself—yield yet another possible mechanism for FU Orionis outbursts.

Finally, we note that disk accretion can be dammed by the formation of giant planets and the appearance of gaps across which matter does not flow. The removal of the gas from the outer regions of dammed disks may then require slow evaporation by ultraviolet radiation (Sekiya et al. 1980a; Hayashi et al. 1985; see review of Ohtsuki and Nakagawa 1988). The abundance of possibilities regarding the dispersal of the remnant gas in nebular disks yields a specific demonstration that we still possess ample problems to wrestle with before *Protostars and Planets IV* is written.

Acknowledgments. This work was supported in part by a NSF grant and in part under the auspices of a special NASA astrophysics theory program which funds a joint Center for Star Formation Studies at NASA-Ames Research Center, U. C. Berkeley, and U. C. Santa Cruz.

NUCLEOSYNTHESIS AND STAR FORMATION

A. G. W. CAMERON
Harvard-Smithsonian Center for Astrophysics

Discussions are given of the stellar nucleosynthesis sites of several extinct nuclides: ^{26}Al, ^{60}Fe, ^{135}Cs, ^{53}Mn, ^{92}Nb, ^{107}Pd, ^{129}I, ^{146}Sm, ^{182}Hf and ^{244}Pu. The mean lives of these nuclides range over 2 orders of magnitude, from 1.07 to 149 Myr. These are interpreted to include 1 equilibrium process product, 2 r-process products, 2 p-process products, 1 product of hot hydrogen burning, and 4 s-process products. It is shown that their abundances in the early solar system are consistent with the following postulated local history of the Galaxy. About 130 Myr before the formation of the solar system an OB Association was formed in a nearby molecular cloud, and the ultraviolet radiation from the more massive O stars destroyed the cloud, caused very large-scale mixing in the interstellar medium, and created new molecular cloud complexes. In the nearest of these, roughly 25 Myr before the formation of the solar system, a new OB Association again caused large-scale mixing and formed new molecular cloud complexes, including the one in which the solar system would be born. An AGB star in the last stages of its evolution wandered into our cloud shortly after its formation; its planetary nebula stage triggered a chain of formation of new stars that propagated through strong stellar wind and bipolar ejection processes occurring in the new stars themselves.

I. INTRODUCTION

The solar system, when formed, possessed interesting amounts of now-extinct radioactivities whose abundances have been preserved as isotopic anomalies in some phases of meteorites. If we know how a nuclide was made and what kind of stellar scenario was involved, then in principle we can date the interval between the last event or few events making the nuclide and the formation of the solar system. This chapter constitutes an attempt to do that, within a rather generalized model of the interstellar medium and of the molecular cloud complexes formed within it.

There is no significant body of literature on this subject. Papers announcing the discovery of various extinct nuclides have usually included an estimate of the production rate of the nuclide relative to some reference nuclide during the course of stellar nucleosynthesis, and have noted the time implied by the radioactive decay interval needed to reduce the isotope production ratio down to the observed value of the isotope ratio. But such estimates rarely include a serious attempt to deal with an appropriate astrophysical scenario in which the decay could have occurred. Thus, when the editors of this book asked me to write on this subject, in effect they challenged me to bring together information about the interstellar medium, star formation, stellar evolution, and nucleosynthesis to create a general scenario in which all of these ages would become meaningful. This chapter is my attempt to do this; my effort has extended over the past two years as interesting new information and new ideas have come to the fore. The relevant fields of research are very active, and it is hoped that the general scenario derived here will be useful to others as they test various parts of it and make appropriate modifications.

Table I shows the extinct radioactivities that have been measured in primitive solar system material, the comparison isotope (normally stable) used as a measure of the abundance of the extinct nuclide, and the ratios of the decay products measured in primitive material to their comparison isotopes. It also shows the mechanism of stellar nucleosynthesis which appears to have been primarily responsible for the production of the nuclide, at least in the general scenario adopted in this chapter. Most of the data has been taken from the review by Wasserburg (1985), but the data for ^{146}Sm is from Prinzhofer et al. (1989), the data for ^{53}Mn is from Birck and Allègre (1987), and the data for ^{60}Fe is a tentative detection by Birck and Lugmair (1988). Very new data from Harper et al. (1991a, b, c, d) on the radioactivities ^{182}Hf, ^{135}Cs and ^{92}Nb has contributed in an important way to the analysis given in this chapter. There is a considerable range of uncertainty in the observed abundance of ^{182}Hf, mostly due to uncertainties in the corrections that must be made; Harper prefers an abundance near the lower end of this range; the upper limit is from Ireland (1991) as interpreted by Harper (personal communication). The detections of ^{135}Cs and ^{92}Nb are also tentative and require confirmation. See also the Chapter by Swindle.

During the last few years there has been an intensive investigation of star formation taking place in dense molecular cloud complexes, primarily by collapse of denser cores which form in such clouds. The shorter-lived extinct radioactivities give some information about these events. However, typical lifetimes that have been estimated for these cloud complexes are usually in the range between 10 and 200 Myr (see, e.g., Myers 1990). The longer-lived extinct radioactivities have mean lives in this general range, and they thus tend to give information about the state of the interstellar medium prior to the formation of a molecular cloud complex in which it is hypothesized that the solar system was born. If the picture that we attempt to put together does not produce a self-consistent scenario, then we would be forced to conclude that the solar system was formed in some other, much rarer, way. It is likely that several of these alternative paths exist and are responsible for a small part of the star formation in the Galaxy.

II. FORMATION AND DESTRUCTION OF CLOUDS

It is generally accepted that a molecular cloud complex is formed through an extensive flow of interstellar gas along the magnetic field lines. The

TABLE I

Data Concerning Extinct Radioactivities

Nuclide	Mean Life[a]	Ref. Nuc.	Obs. Ratio	Abundance[b]	Process
^{26}Al	1.07	^{27}Al	5.0×10^{-5}	4.3	hot H-burning
^{60}Fe	2.2	^{56}Fe	1.6×10^{-6}	1.3	s-process
^{135}Cs	3.3	^{133}Cs	3	1.2×10^{-5}	s-process
^{53}Mn	5.3	^{55}Mn	4.4×10^{-5}	0.41	equilibrium
^{107}Pd	9.4	^{108}Pd	2.0×10^{-5}	7×10^{-6}	s-process
^{182}Hf	5.3	^{180}Hf	$8 - 80 \times 10^{-6}$	$4 - 40 \times 10^{-7}$	s-process
^{129}I	23.1	^{127}I	1.0×10^{-4}	1.3×10^{-4}	r-process
^{92}Nb	50	^{93}Nb	2.0×10^{-5}	1.8×10^{-5}	p-process
^{244}Pu	118	^{238}U	7×10^{-3}	1.4×10^{-4}	r-process
^{146}Sm	149	^{144}Sm	7×10^{-3}	5×10^{-5}	p-process

[a] In millions of years (Myr).
[b] Initial solar system abundance on the scale silicon = 10^6.

classical mechanism proposed for this process is known as the Parker (1968) mechanism; while the details of this mechanism are not considered today to be correct (see Chapter by Elmegreen), it makes a good conceptual starting point for our discussion. In this mechanism it is assumed that a section of the magnetic field lines lying in the galactic disk becomes elevated above the plane. The matter on the field lines is tied to them, at least on dynamic time scales, and feels a force tending to pull it toward the galactic plane. It responds by sliding down the field lines toward the plane. In this way, it is expected that molecular cloud complexes will be formed in the areas where the field lines rejoin the plane. The time required to form a molecular cloud complex is of the order of 20 Myr (Parker 1968). The mean number density of matter in the interstellar medium is about 1 cm^{-3}, and in a molecular cloud the mean density is usually in the range 10^2 to 10^3 cm^{-3}; it can be smaller in diffuse clouds and parts of giant molecular clouds (Myers 1990). Thus the characteristic distance through a cloud complex with a mass in the range 10^4 to 10^6 M$_\odot$ is from 5 to 70 pc.

It requires an energetic event to dissipate a molecular cloud. This event probably involves the formation of a number of O and B stars, which will flood their surroundings with ultraviolet radiation, thus ionizing the nearby gas. This will raise the temperature of the gas from 10 K to about 10^4 K, thus raising the thermal pressure by a similar factor above the mean in the interstellar medium. Under steady-state conditions around a typical O star, a Strömgren sphere of ionized hydrogen would be formed with a radius of 0.1 to 1 pc (Strömgren 1939). But the major pressure imbalance produced by the heating will cause the region to expand. Under steady-state conditions at a density of 1 cm^{-3}, the radius of the Strömgren sphere would become 100 times larger and it would ionize 1000 times as much material. Thus as the overheated gas expands more material becomes ionized. The dynamics of the process involves the propagation of an ionization front (Vandervoort 1963). Thus O stars with a range of masses would ionize anywhere from 10^2 to 10^6 M$_\odot$ of material in a cloud. The lower-mass O stars would disrupt star formation in a part of the cloud, and the higher mass ones would disrupt the cloud completely. Strong winds from O and B stars are also locally quite disruptive.

The disruption of the cloud would certainly be complete if star formation in it has produced an OB Association, as this would contain at least a few stars in the higher-mass range. In such a case, the cloud must respond by exploding. The expansion velocity will be comparable to sound speed at 10^4 K, or 10 km s^{-1}. Assuming momentum conservation, the expanding gases will slow to 1 km s^{-1} or less, thus mixing with at least ten times the mass of the molecular cloud. In the later stages of the expansion, when the overpressure is not too great, it should be expected that the galactic magnetic field will resist compression perpendicular to the field lines, and so there will be a preferential expansion along the direction of the field lines. The field lines also tend to inhibit actual mixing in the transverse direction, but the field lines can intermingle with one another and then mixing can occur by relatively short-range diffusion.

If the cloud mass was as much as 10^6 $M_\odot$, and if the total mass of the interstellar medium is about 4×10^9 $M_\odot$, then this one energetic event will have mixed 0.025% of the interstellar medium. Such events are undoubtedly the primary causes of large-scale mixing in the interstellar medium. The time interval between these events is probably comparable to the lifetimes of cloud complexes themselves, or one to two hundred Myr, or possibly less.

The most massive stars have main sequence lifetimes of about 3 Myr; in these stars the luminosity is proportional to the stellar mass and the mass fraction that burns hydrogen into helium on the main sequence is approximately constant (see, e.g., Shapiro and Teukolsky 1983). For stellar masses of about 10 $M_\odot$, the main sequence lifetime has increased to about 10^7 yr. But by that time the massive stars will have evolved to the end-points of their evolution; this means that most of them will have become Type II supernovae. To the high pressures generated by ionizing hydrogen will be added the "mechanical luminosity" of the supernovae. This causes a rapid inflation of the volume containing the supernovae ejecta and the embedded magnetic field lines. Hydrodynamic studies of this event have been carried out by Tomisaka (1990), and more recently by M. Norman and his colleagues (personal communication), who describe the inflated region as a superbubble in the interstellar medium. Tomisaka (1990) estimates the duration of the event as about 25 Myr, including the time for the beginning of the inflation.

The general expansion accompanying this energetic event will not only lift many field lines above the plane, but will also cause a substantial sidewards displacement of the magnetic field lines, thus causing the pinched field to squirt the embedded gas along the field lines, unless these field lines are quite tangled, in which case star formation may be forced in the small cloud that is thus formed. The gas on the elevated field lines will fall down the lines due to the gravitational attraction of the plane. This scenario is a substantial modification of the original Parker instability scenario. There is thus a variety of ways in which the gas flow along the field lines can lead to the formation of molecular cloud complexes. Because the field lines act as individuals, the actual processes involved are bound to be very complex, and twisting and crossing of the field lines can lead to mixing in the resulting clouds (Chapter by Elmegreen).

Because stars in OB Associations are generally observed to be expanding from a common point in space, it is likely that the stars in an Association were formed within a short period of time well after the formation of their cloud complex. Because some giant clouds have cores as massive as 10^4 $M_\odot$ (Myers 1990), the collapse of these cores is likely to form an unbound cluster, many members of which are quite massive. In such a case the stars in the cluster are born at essentially the same time, and we shall so assume. Therefore the formation of the superbubble, the large-scale interstellar mixing, and the formation of new molecular cloud complexes all should be well under way before stars in the OB Association of 10 $M_\odot$ and below have evolved off the main sequence. This has profound consequences for nucleosynthesis. Studies of the r-process and related nucleochronology (see, e.g., Cowan et al. 1991) have found that r-process elements do not appear in the most ancient stars until the abundance ratio of iron to hydrogen has risen to 10^{-3} of the present value, and s-process elements do not appear until the ratio has increased by a further factor of 10. Mathews and Cowan (1990) have attempted to compare these trends with a variety of galactic nucleosynthesis models, and they find that they cannot fit the trends by assuming that all supernovae contribute to the r-process, but they can get a reasonable fit if they assume that only the lower mass supernovae contribute to the r-process. The upper limit to the mass range producing the r-process is roughly 10 $M_\odot$. The shortest evolutionary time in which the r-process can be produced is thus about 10 Myr. It follows that the creation of the superbubble and the formation of new molecular clouds incorporates the products of the massive supernovae, including the products of the equilibrium process, but it does not incorporate the products of the r-process. The latter products will be blown off from lower mass supernovae in due course, but there will be only relatively local mixing processes available to distribute these products until the next large-scale mixing event occurs, probably 100 to 200 Myr later.

The most detailed recent estimate (Kennicutt 1984) is that main-sequence stars with masses below 8 $M_\odot$ lose enough mass so that they will end their lives by forming a white dwarf remnant (see discussion of AGB stars in Secs. VI and VII). Thus the spread in stellar mass that appears to contribute to the r-process is quite small. From the point of view of time scales, an 8 $M_\odot$ star has a main sequence lifetime of about 25 Myr (Shapiro and Teukolsky 1983), with subsequent evolution taking only a relatively short time, and thus this would be the longest period of time over which the r-process would take place following the formation of an OB Association.

III. ^{53}Mn

With a mean life of 5.3 Myr, ^{53}Mn is the fourth shortest lived of the extinct radioactivities. It is a product of what is called the equilibrium process, in which stellar material is processed at temperatures of 4×10^9 K or higher to a condition of nuclear statistical equilibrium. (Woosley et al. 1973). This processing occurs explosively, which means that the ^{53}Mn is a product of a supernova explosion. Such stellar material starts with a near equality of the numbers of neutrons and protons in the nuclei, and in the thermonuclear runaway there is not sufficient time for any significant number of beta decays to take place. Under these circumstances, the most abundant nucleus is ^{56}Ni, which subsequently decays to ^{56}Co and ^{56}Fe, thereby releasing the energy that appears in the supernova lightcurve. Mass number 53 is produced as ^{53}Fe and ^{53}Mn, and the mass number 55, leading to the reference nuclide ^{55}Mn is produced as ^{55}Co and ^{55}Fe.

Because of the equilibrium character of the resulting abundances, the relative production ratios (after beta decay) of ^{53}Mn, ^{55}Mn and ^{56}Fe are determined by their abundances in solar matter. The ^{53}Mn abundance comes from the abundance of its decay product, ^{53}Cr. These give a production ratio of ^{53}Mn/^{55}Mn of 0.13. The latest abundance compilation of Anders and Grevesse (1989) gives two values for the iron abundance, one derived from solar spectrum lines, and the other lower value derived from carbonaceous meteorites. I prefer to use the meteoritic value, giving a ^{53}Mn/^{56}Fe ratio of 0.00147 and a ^{55}Mn/^{56}Fe ratio of 0.0113.

Because of the short lifetime of ^{53}Mn, we are only interested in processes that transfer it quickly from production sites to the site of formation of the solar system. The prompt supernovae resulting from the massive stars in the OB Association that inflated the superbubble and formed the molecular cloud in which the solar system would be born are thus suitable production sites. The supernova explosions contributing to the interstellar superbubble should all take place within about a mean life of ^{53}Mn. These massive stars are each expected to produce about 0.1 $M_\odot$ of iron-peak elements (for a discussion of these iron masses see Thielemann et al. [1992]). If we assume that there were 40 such supernova explosions contributing to the superbubble (the number assumed by Norman and his colleagues in their simulations), then about 4 $M_\odot$ of fresh iron peak material would be contained in the superbubble, along with 0.045 $M_\odot$ of ^{55}Mn.

The ^{53}Mn/^{55}Mn abundance ratio observed to have been present in meteorites in the early solar system is 4.4×10^{-5}. This is but one of several quantities that we need to know in order to characterize the scenario unambiguously, so we must assume some of them in order to derive the remainder. Assume that there was a time delay of 25 Myr between the Type II supernovae formed by the massive stars from the OB Association and the formation of the solar system, utilizing the time scales associated with a Parker instability and found for the hydrodynamics of an interstellar superbubble by Norman, as discussed in the previous section. During this interval, the ^{53}Mn would have decayed by a factor 112, meaning that the initial ^{53}Mn/^{55}Mn ratio would be 0.0049 as geometrically diluted by mixing into cloud material but without any radioactive decay. The first geometric dilution factor is then $0.13/0.0049 = 26$.

The second geometric dilution factor consists of the ratio of the amount of ^{55}Mn produced by the supernovae to the amount of this nuclide in 1 $M_\odot$, or $0.045/1.92 \times 10^{-5} = 2350$. The product of these two factors gives 6.2×10^4 $M_\odot$ for the amount of material into which the supernovae ejecta is mixed. If the interstellar superbubble formed two molecular cloud complexes at the two ends of the lines of force of the magnetic field threading the superbubble, then each cloud complex would have a mass of at least 3×10^4 $M_\odot$. Because the radioactive debris can mix across lines of force only with difficulty, the actual masses of the cloud complexes could be considerably larger. What we get from this calculation is only the radioactively contaminated mass in the clouds. The calculation assumed that the solar system formed very soon after the cloud complex in which it was situated was formed. If there was a significant delay in this event, then further decay of the ^{53}Mn would have occurred, and the early solar system abundance could then only be obtained by reducing the amount of mass subject to radioactive contamination.

IV. r-PROCESS NUCLIDES

^{244}Pu is a product of the r-process. Because it is an actinide element, it cannot be produced by some relatively weak source of neutrons; a large source of neutrons is required. Studies by the author (unpublished) have shown that in order to produce many tens to more than one hundred neutrons per seed nucleus in the r-process production region, the source material must be compressed to about 10^{12} g cm^{-3}, where the neutrons are produced by electron capture, or to about 10^{10} g cm^{-3} at much higher temperatures. But studies of the supernova explosion mechanism have indicated that the "mass cut," which marks the point of separation between ejected matter and that which falls back onto the forming neutron star, has been compressed only to about 10^{10} g cm^{-3} (Mayle and Wilson 1988). The implication of this is that nuclei like ^{244}Pu must be formed in a small amount of matter asymmetrically ejected in the form of a jet if from high densities below the normal mass cut, or in a small amount of matter in a shell at the mass cut for very high temperatures; this is consistent with the r-process material requirements in the Galaxy which are satisfied if the average supernova ejects about 10^{-5} M$_\odot$ of r-process material. The mechanism by which either of these events might happen are not known, but the empirical fact that the r-process occurs only in stars of mass less than about 10 M$_\odot$, mentioned above, should be an important clue.

The following section on p-process nuclides describes a continuous production method for estimating the yields of the radioactivities that are long enough lived so that averaging procedures can be applied. But we have seen above that r-process nuclei are produced in a narrow range of stellar masses, 8 to 10 M$_\odot$, during a period of 10 to 25 Myr after formation of the OB Association. Then there will be a delay of order 10^8 yr before the next major mixing event occurs driven by a new OB Association. Thus in a local region of the Galaxy the r-process production is better represented by a series of spikes of production. Such a spike model for making abundance estimates will be used here.

There is no stable isotope of plutonium relative to which the abundance of ^{244}Pu can be measured. Therefore the usual reference isotope is ^{238}U, which, while not stable, is at least long-lived, with a mean life a little longer than the age of the solar system (6.5 Gyr). Wasserburg (1985) recommends a value for the ratio of ^{244}Pu/^{238}U of 7×10^{-3} at the time of formation of the solar system. This is close to the expected average value at that time in the interstellar medium. However, the relative formation rates of ^{244}Pu and ^{238}U are not very well known, except that they must be within a factor of a few of unity. Each of these nuclides has several progenitors which decay to it by emitting α-particles. If the r-process yields are strongly peaked at or beyond mass 244, then the ^{244}Pu and ^{238}U yields would be similar, but if the peak is well below mass 244, or if the yield curve is quite uniform, then ^{244}Pu would have a lower abundance. The value obtained in the author's unpublished r-process calculations for the ^{244}Pu/^{238}U production ratio is 0.7, and this value will be used here.

The mean life of ^{244}Pu is 1.18×10^8 yr. A simple spike model would assume a constant rate of production of the nuclide (and also of its reference nuclide ^{238}U) for a period of 10^{10} yr; this has been spike approximated by taking 100 spikes at 10^8 yr intervals and determining how much of the ^{244}Pu and ^{238}U that have ever been produced remain after the last spike. In a second modified spike model, the production rate has been multiplied by the weighting factor $\exp(-t/10^{10})$, where t is the time since the start of production, and a third modified model used the square of this weighting factor. Using the relative production ratio given above, at the time of the last spike the ^{244}Pu/^{238}U ratio is found to lie in the range 0.0166 to 0.0238 due to variations in the weighting factors. The observed ^{244}Pu/^{238}U ratio for the solar sysem is 0.007. To reach this value primarily by ^{244}Pu decay thus requires some 99 to 122 Myr, depending on which of the above two values of the post-spike abundance ratio is chosen. This decay interval has some further uncertainty owing to uncertainties in the production ratio of the two nuclides.

Four of the extinct radioactivities, ^{107}Pd, ^{129}I, ^{135}Cs and ^{182}Hf, are candidates to be made by the r-process, because they lie on the neutron-rich side of the s-process capture path and would not be traversed by an s-process event taking place with a relatively small neutron number density. However, the s-process at a high neutron number density would make large amounts of these nuclides. Here we first consider whether they can be r-process products.

Let us first make the hypothesis that the ^{129}I present in the early solar system was primarily made by the galactic r-process. This nuclide sits on top of the large r-process abundance peak associated with the 82 neutron closed shell, as does its comparison isotope ^{127}I (Cameron 1982). Thus we can consider r-process production of ^{129}I on the same basis as ^{244}Pu. The 23.1 Myr mean life of ^{129}I is shorter than the expected interval between large-scale mixing events in the interstellar medium that we have discussed. Because both the mixing and the formation of new cloud complexes appear to be triggered by the formation of the same OB Association, it appears that the mixing event will sweep up the r-process products formed after the previous large-scale mixing event.

Again we use the spike method, but owing to the shortness of the mean life of ^{129}I, we need to consider only the ^{129}I made in the last spike. Because we take T to be 10^{10} yr with spike intervals of 10^8 yr, the last spike would produce 0.01 of all the ^{129}I ever made. Estimating the production ratio of ^{129}I/^{127}I to be

1.27 from observed r-process abundances, we obtain a $^{129}I/^{127}I$ ratio of 0.0127 after the last spike. This would decay to the early solar system observed level of 10^{-4} in 4.8 mean lives of ^{129}I, or 110 Myr. This is remarkably consistent with the decay interval for ^{244}Pu, which was independently estimated above.

Next consider ^{182}Hf. Here the mean life is 13 Myr. This radioactivity has only recently been discovered (Harper et al. 1991a), and with the abundance known only to be in the range 8 to 80 $\times$ 10^{-6} relative to ^{180}Hf (these numbers include an uncertain correction factor of 3), it might seem too uncertain to be of much use. However, despite this uncertainty, this nuclide can be used to make a decisive test of the r-process origin of both itself and of ^{129}I.

The r-process is comparable in abundance to the s-process at mass 182. Using the periodic spike method mentioned above, some 110 Myr prior to the solar system 0.01 of all the ^{182}Hf ever made was present, making the $^{182}Hf/^{180}Hf$ ratio at that time 2.1 $\times$ 10^{-3} (using the r-process abundance estimates of Käppeler et al. [1989]). If, in parallel to the considerations for ^{129}I, the ^{182}Hf is now allowed to decay for 110 Myr, the $^{182}Hf/^{180}Hf$ ratio is decreased to 4.4 $\times$ 10^{-7}. The lower limit on the $^{182}Hf/^{180}Hf$ is a factor of 20 higher. Therefore it appears that the r-process can only make a minor contribution to the ^{182}Hf in the early solar system.

It is clear from this calculation that the radioactivities which are shorter-lived than ^{182}Hf cannot have received significant contributions from the r-process.

As we have noted above, the r-process starts about 10 Myr after the formation of an OB Association and continues until about 25 Myr after the formation of the Association. ^{129}I decayed for an additional period of 110 Myr before formation of the solar system. This would then place the time interval between formation of the two OB Associations also to be 110 Myr, with the solar system being formed about 140 Myr after formation of the first OB Association.

The tentative conclusion to be reached at this point in the discussion is that ^{129}I may be formed primarily as an r-process product, but that the shorter-lived radioactivities can receive only a minor contribution from the r-process at best.

V. THE p-PROCESS RADIOACTIVITIES

There are two p-process nuclides among the extinct radioactivities, ^{146}Sm and ^{92}Nb, the latter newly discovered by Harper et al. (1991b). We consider these in turn.

The p-process products are formed by the photodisintegration of heavy elements that have previously been built up by neutron capture. The term "p-process" was originally coined to denote an unspecified mechanism that would make nuclei on the proton-rich side of the valley of beta stability, but an alternative term "gamma-process" has recently been used to indicate the photodisintegrations specifically as distinct from proton capture. However, throughout this chapter we use the original term.

Because of the extreme environmental conditions required for nuclear photodisintegrations, the p-process nuclei must be products of supernova explosions (Woosley and Howard 1978). The photodisintegrations take place on heavy nuclei that have been built up to substantial abundances by the prior operation of the s-process in the presupernova star. A heavy element like samarium (the same would be marginally true of niobium) will only be built up significantly in an s-process using a ^{13}C neutron source in an asymptotic giant branch star (AGB star, to be described in more detail in Secs. VI and VII) (Hollowell and Iben 1989). These stars have relatively low masses (up to 2 $M_\odot$ or perhaps a little more) which do not ordinarily undergo supernova explosions unless (in some cases) they are in binary stellar systems.

The mean lives of the p-process nuclides are long enough so that we can use an averaging method to estimate their abundances in the interstellar medium. As long as the medium has not been isolated from the mixing currents in the interstellar medium (as in a molecular cloud complex), the average amount present in the medium relative to the total amount ever made is the ratio of the mean life τ to the total period T in which the radionuclide has been manufactured (roughly the age of the Galaxy) (Cameron 1962). Such an estimate assumes continuous nucleosynthesis at a constant rate, whereas it is evident that nucleosynthesis occurred at a more rapid rate near the birth of the Galaxy than at the time of formation of the solar system. To compensate for this, we take T to be a little too high, about 10 Gyr at the formation of the solar system. Because a supernova explosion would disrupt a molecular cloud, we must consider p-process products to be injected into the general interstellar medium and later introduced into a molecular cloud when it is formed.

Howard et al. (1991) have suggested that the p-process products in the vicinity of mass number 90 may also be built up by proton-capture reactions using protons released by carbon-burning at temperatures of about 3×10^9 K in Type Ia supernovae. These reactions would compete comparably with photodisintegrations under such conditions in Type Ia supernovae.

^{146}Sm is the longest lived of the extinct radioactivities discussed here, with a mean life of 149 Myr. Its reference isotope is ^{144}Sm, and Prinzhofer et al. (1989) have estimated the ^{146}Sm/^{144}Sm ratio at the time of formation of the solar system as about 0.015; Swindle (see his Chapter) estimates that this could be as much as a factor of three smaller. The latter estimate is in agreement with that of G. Lugmair (personal communication), who gives the ratio as 0.006 to 0.007 with an error of ± 0.0015. Woosley and Howard (1990) have estimated the production ratio of ^{146}Sm/^{144}Sm to lie in the range 0.01 to 0.4. F.-K. Thielemann (personal communication) has also made a preliminary estimate of this ratio which is about 20 times as large as that obtained by Woosley and Howard with the same range of uncertainties.

If we adopt the continuous nucleosynthesis procedure of estimating the abundance of ^{146}Sm in the interstellar medium, with the Woosley and Howard production ratio, we obtain a value in the range 1.5×10^{-4} to 0.006 relative to ^{144}Sm, which slightly overlaps the observed value range of 0.005 to 0.015. If we apply Thielemann's correction, the range becomes 3×10^{-3} to 0.12. The observed values lie entirely within this range. The uncertainties emphasize the need for further work to clarify the nuclear physics involved as well as the need for more experimental measurements.

For ^{92}Nb the mean life is about 50 Myr. The continuous nucleosynthesis model predicts that 0.005 of all the ^{92}Nb ever made would be present at the time of formation of the solar system. However, we can only guess at the ^{92}Nb/^{93}Nb production ratio. An extremely crude estimate can be made from systematics of p-process abundances. In the neighborhood of mass 92, the normal p-process abundances (of nuclei having even numbers of protons and neutrons) lie near 10 % of the even-even s-process products (however, they are considerably higher than this very close to ^{92}Nb because of the vicinity to the closed shell at 50 neutrons). In the heavier mass number range, the p-process abundances are roughly 3% of the s-process even-even products. There are only two stable odd-odd p-process nuclei, ^{138}La (very near a closed neutron shell) and ^{180}Ta, and these have abundances of 10% and 1%, respectively, of neighboring even-even p-process nuclides. However, the observed long-lived ^{180}Ta is only an isomeric state and its abundance would be abnormally low. Therefore we take the abundances of odd-odd p-process nuclei to be about 10% of the even-even ones. Thus from systematics alone, with very large errors, we estimate that ^{92}Nb/^{93}Nb production ratio would be about 0.01.

The continuous nucleosynthesis model predicts that the observed ^{92}Nb/^{93}Nb ratio should be about 5×10^{-5}, compared to the observed value of 2×10^{-5}. In view of the large errors involved, and the tendency of the continuous nucleosynthesis model to predict somewhat high abundances, these two numbers are in essential agreement.

VI. THE MOLECULAR CLOUD ENVIRONMENT

To set the stage for the following discussion, it is useful to write down a few numbers characteristic of the molecular cloud environment. As noted previously, the average density of particles in the interstellar medium is about one particle per cubic centimeter. The average density in a molecular cloud is about 10^2 to 10^3 cm^{-3}. Cores form inside the cloud complex and have characteristic densities of in the range 10^4 to 10^5 cm^{-3} (Myers 1990). Small cores in massive clouds can have densities in the range 10^6 to 10^7 cm^{-3}. It is difficult to define the mass of a core because there is a smooth transition from the background density of the cloud to that in the interior of the core. The cores themselves have a fairly large range of masses, although there seems to be a smaller range within a given part of a cloud complex than the total range in general. We shall consider a nominal core mass of about 1 M$_\odot$ for the central part of a core. The temperature within the cloud complex appears to be a surprisingly uniform 10 K. The Doppler broadening of spectral lines emitted from core material corresponds to about Mach 0.7, or about 0.2 km s^{-1}.

Nevertheless, Doppler shifts from material outside the cores typically reveal radial velocities of about 0.5 to 1 km s^{-1} and therefore the velocities are characteristically supersonic (Myers 1990; Myers et al. 1991b). Such supersonic motions are strongly dissipative through shocks, and therefore they cannot persist for very long without being frequently and strongly driven by disturbances in the gas which effectively act like supersonic pistons. Because the effect is so widespread, the source of the supersonic turbulence must also be widespread, and in order for the turbulence to be persistent, very large amounts of mass must be set into motion. Possible sources for this include large-scale generation of hydromagnetic waves by gas motions in the cloud complex which are rapidly dissipated into the sub-Alfvénic velocity range, and the motion of the background stars that are embedded in the cloud and move with random velocities of 10 to 20 km s^{-1} relative to it. The accelerations impulsively given to the cloud gas by the passages of these stars (like the stellar acceleration of comets in the Oort Cloud) appear to be of the right order. The nonthermal behavior is some superposition of supersonic turbulence from these and other pervasive and widespread sources. We shall discuss several potential sources of much larger disturbances in the following paragraphs.

A major question of great relevance to us is the mechanism of formation of the cores. One possibility, discussed extensively by Shu (1983), is that the cores arise from density fluctuations within the cloud that attract surrounding material, which flows very slowly inwards to form the core by ambipolar diffusion, in which the matter slowly diffuses across the magnetic field lines. This generally requires the conditions to be rather quiescent. The characteristic radius of a core is about 0.1 pc. Thus the ambipolar diffusion needs to concentrate the gas from a sphere of radius about 1 pc. The ambipolar diffusion time is probably a few Myr (Shu 1983).

Cores can be formed much faster if the motion of the gas can be primarily along the field lines rather than across them. This would require the action of some disturbing event within the cloud that would accelerate the gas to a substantial velocity, followed by a channeling effect due to the surrounding magnetic pressure perpendicular to the field lines in the local flux tube. The radius of the affected part of the flux tube would be larger than 0.1 pc and smaller than 1 pc. As mass moves along the flux tube away from the point of the disturbance, the mass it encounters becomes swept up into a density enhancement traveling along the tube, but progressively more slowly as the amount of mass increases. Also, as the mass increases, some gravitational compression of it becomes likely, leading to a pinching together of the local field lines. In this picture there may well be two cores formed, one in each of the two directions along the flux tube from the point of the disturbance. The time taken for this longitudinal compression may typically be < 1 Myr (Cameron 1984).

These two pictures may well be extremes, with the truth most likely lying in between them. Thus, longitudinal compression may produce a substantial density enhancement, followed by ambipolar diffusion in the final stage of the core formation. Certainly, when a core becomes sufficiently concentrated to go into gravitational collapse, the field lines will be diffusing out all the time, although the net effect of the collapse will be a substantial enhancement in the magnetic field strength.

There can be a variety of disturbing events within the interior of a cloud complex. A star may come to the end of its active lifetime and eject material. Such events depend upon the accidental presence of a star within the cloud complex when its eruption takes place. But because star formation is an intrinsic process of the cloud complex itself, by far the most abundant disturbing events are likely to be those associated with star formation.

A typical star in the galactic plane moves relative to a molecular cloud with a velocity of 10 to 20 km s^{-1}. A typical path through such a cloud has a length of order 15 pc. Therefore the typical star takes about 4 to 8 $\times$ 10^4 yr to traverse the cloud. However, stars formed within the cloud will normally have only a small velocity relative to it, and their residence time within it will be much longer.

Ordinary main-sequence stars have rather little effect on a molecular cloud when they pass through, unless they have a high mass. A star like the Sun emits sufficiently little ultraviolet radiation that only that part of the cloud within a small fraction of a parsec becomes ionized, too little to produce a significant perturbation to the cloud parameters. The most massive stars will destroy the cloud, as noted previously. Stars of intermediate mass may not destroy the cloud, but they are certainly likely to disrupt it locally, probably interfering with star-formation processes close to the intermediate mass star, but possibly enhancing it farther away due to induced compression within the cloud gas. Most observations of star formation in molecular clouds are associated with environments not containing these intermediate and massive stars, so we shall concentrate our discussion on them.

Stars of one up to a few solar masses produce the most interesting effects within a molecular cloud when they approach the ends of their post-main-sequence lifetimes. These stars are called asymptotic giant branch (AGB) stars. They have evolved through their hydrogen and helium burning stages, they have ascended the giant branch in their color-magnitude diagrams for the second time, and they end their active lifetimes near the tip of the giant branch where they eject typically a few tenths of a solar mass of envelope material as a planetary nebula, leaving a white dwarf remnant (Hollowell and Iben 1989). Prior to this, they will be the sources of strong stellar winds.

Thus the AGB star creates quite a disturbance within the molecular cloud during the final stage of its evolutionary lifetime. Consider only the final episode in which 0.1 or 0.2 M$_\odot$ is expelled to form a planetary nebula. This material is typically ejected with a velocity of several tens of km s^{-1}. This acts like a point explosion within the molecular cloud. An oversimplified picture of the resulting events is the following one. The ejected material can move freely along the direction of the lines of force of the local magnetic field, but the motion perpendicular to the field lines compresses the field, which rebounds to redirect the ejecta along the field lines. In this way, the point explosion generates two pistons moving in opposite directions along the field lines, sweeping mass before them as they slow down. By the time the pistons have slowed to local sound speed, they will have swept up roughly 100 times their original mass, or of order 5 to 10 $M_\odot$ for each piston (considering only momentum conservation). This is precisely the picture described above as required to form molecular cloud cores that will become unstable to gravitational collapse. Thus the final eruption of the AGB star is expected to act as a trigger for star formation within the cloud (Cameron 1984). With magnetic field entanglements and other variations in the field structure, a more realistic picture of what happens may differ significantly from this description.

Now consider the disturbances that originate from newly formed stars. These are observed to be of two kinds. Such newly formed stars are known by a variety of names; we shall call them young stellar objects (YSO) when they are very new and usually at least partly obscured, and at a later stage we shall call them T Tauri stars. The YSO are observed to emit very strong stellar winds. They are also observed to emit jets of material in opposite directions, a process called bipolar ejection. Within broad observational uncertainties, the amount of momentum in these jets is comparable to the amount of momentum in the ejection of a planetary nebula from an AGB star. Similar considerations apply, and thus it appears probable that newly formed stars can act as triggers to form yet another pair of new stars. If conditions were ideal, a cascade would develop from the original AGB trigger that would double the rate of star formation in each generation (Cameron 1984). Conditions are unlikely to be sufficiently ideal for that, but nevertheless a great deal of star formation can ensue.

Self-propagating star formation has been suggested before in a variety of contexts (see Elmegreen [1987b] for a list of these proposals). Most of the proposals involve considerably more energetic events than those discussed here, and the effects are expected to act over greater distances than the rather local processes with which we are concerned.

Let us make a crude estimate of the mixing induced by star formation within the molecular cloud. There is of order 10^9 $M_\odot$ of gas in molecular clouds in the Galaxy, and in this gas about one star is formed per year, or about 10^{-9} stars per solar mass of gas per year. Because the mass outflow from a young star has an ejection velocity of order 100 km s^{-1}, then it follows that this ejecta will mix with about 100 times as much material in the cloud when the material is slowed to typical turbulent velocities of order 1 km s^{-1} in the cloud, again considering only momentum conservation. Thus something like 10^{-7} of the gas is mixed per year, for an interval between mixings of 10 Myr. However, since there are regions in which the star-formation activity is concentrated, clearly the deduced mixing interval has a large dispersion. The turbulent velocities themselves will produce a diffusive mixing, but this will not rapidly spread the boundaries of a region that has been seeded with some radioactivity.

Another form of cloud disruption can occur because of a nova eruption. In this case hydrogen accretes onto a white dwarf component of a binary system. Thermonuclear ignition occurs in a thin shell near the surface of the white dwarf, heating and greatly expanding the upper layers of the envelope, which then initiates a substantial blowoff of mass. However, only a small fraction of the total stellar mass is ejected in the eruption, of order 10^{-4} to 10^{-5} of a solar mass. However, these eruptions are observed to repeat for the less energetic novae, and it is likely that all novae repeat as long as the material transferred onto them from their binary companions remains available. In a given eruption, the fraction of the ejected mass containing ^{26}Al is likely to be small. According to Weiss and Truran (1990), a nova in which hydrogen burning occurs only in solar composition material should eject material with a contaminated mass fraction of 10^{-6} to 10^{-4}, as is also true of a star with carbon and oxygen mixed into the burning material, but a star whose burning layer is contaminated with underlying oxygen-neon-magnesium may eject a mass fraction of 10^{-4} to 5×10^{-3}. This last amount probably occurs in about 25% of all novae.

When the nova eruption occurs, about 10^{45} erg of ultraviolet radiation is produced, a fair fraction of it above the hydrogen ionization threshold, so that something approaching 10^{56} hydrogen atoms will be ionized within roughly a parsec. At molecular cloud densities this hydrogen will recombine in about 100 yr. The speed of sound in the ionized hydrogen (at 10^4 K) is 10 km s^{-1}, and a sound wave will traverse 10^{-3} pc during the recombination time. Thus an organized spherical expansion does not have time to become established during the recombination time. Following recombination, hydrogen molecules can form again on grain surfaces. The time to do this is roughly the same as the time for a random hydrogen atom to collide with a grain, or 3×10^4 yr. In this time, ordinary gas cooling processes will bring the temperature down to 100 K or somewhat less. But some hydrogen molecules will form relatively quickly and provide a more efficient cooling. Thus the net effect of the nova explosion is to heat a region of order 1 pc in radius which will undergo significant expansion in a time of order 10^5 yr.

As the nova eruptions recur, a series of these heating episodes will take place along the track of the binary system in which the nova episodes are taking place. If the random velocity of the binary relative to the cloud is 10 km s^{-1}, then a cylindrical length of order 10 pc will be heated and will expand. This is substantially disruptive to the molecular cloud, and it is not consistent with the conditions observed in which the core collapses take place. Whether it can lead to star formation by a different mechanism is a separate question that will not be considered here.

VII. ^{26}Al

The mean life of ^{26}Al is 1.07 Myr, the shortest lived of the extinct radioactivities. We therefore expect that we should look for a stellar source within a cloud complex. There are a number of ways in which ^{26}Al can be made in stars, all of them involving the capture of a proton on ^{25}Mg, in the reaction ^{25}Mg$(p,\gamma)^{26}$Al. This reaction occurs at slow stellar evolution rates at a temperature of 5×10^7 K, and it occurs very rapidly at temperatures higher by a factor of 4 or more (Champagne et al. 1984).

A supernova is expected to produce a ^{26}Al/^{27}Al ratio of order 10^{-3} (Truran and Cameron 1978; Arnett and Wefel 1978; Woosley and Weaver 1980; Arnould et al. 1980). This ratio will be reduced to the observed meteoritic value in not more than 5 mean lives, or about 5 Myr. Because a supernova explosion will eject a good fraction of 1 M$_\odot$ (or more) of material at velocities of several thousand km s^{-1}, it would be highly destructive to a molecular cloud environment, and only a different scenario which utilizes the violence of the explosion, such as the supernova trigger hypothesis (Cameron and Truran 1977), might be consistent with solar system ^{26}Al. However, this scenario does not quantitatively support the production of the other extinct radioactivities.

^{26}Al is also expected to be formed in a nova eruption (Truran 1982). As is discussed above, relatively little material is ejected in such an eruption, and this material is not mixed efficiently into the surrounding mass that is briefly ionized following the eruption. Thus novae are not expected to be important sources of any extinct radioactivity in the solar system. Yet another class of objects in which ^{26}Al can be produced is that of the AGB stars. In the giant phase of their evolution they will have (periodically) a hydrogen-burning shell in which the temperature is at least 5×10^7 K, so that ^{26}Al will be produced in this shell region (Cameron 1984). In such giant stars, the outer convective region extends down from the surface to great depths. As the star approaches the planetary nebula ejection stage, it is expected that the outer convection zone will extend all the way down to the hydrogen-burning shell (Norgaard 1980), at which point it has been called a "hot bottom" configuration. Then the ^{26}Al produced in the shell can be mixed efficiently to the surface and hence expelled in a stellar wind. Important new insights are developing from the examination of meteoritic graphite and silicon carbide grains which are believed to have been formed in AGB stars (Zinner et al. 1991a). Large numbers of these grains show evidence for the one-time presence of ^{26}Al with ^{26}Al/^{27}Al ratios of order 10^{-3} or 10^{-4}. However, a few of the grains show much larger amounts of ^{26}Al; in one SiC grain the ^{26}Al/^{27}Al ratio is 0.2; this same grain is depleted in ^{13}C relative to ^{12}C, enriched in ^{15}N relative to ^{14}N, and depleted in ^{29}Si and ^{30}Si relative to ^{28}Si. In a second, larger, grain the ^{26}Al/^{27}Al ratio is 0.23 (Zinner et al. 1991b), and the other depletion and enhancement patterns are similar. In addition, the isotope ^{44}Ca is anomalously large. In the following paragraph, the suggestion is made that these are signatures of the hot bottom stage of the AGB star.

Before the hot bottom stage, there will have been a number of small mass exchanges between the hydrogen envelope and the helium layer; in the course of these, the hydrogen envelope will have been mixed with the products formed in the hydrogen-burning shell, including ^{26}Al, and most grains formed and expelled at this time will have small ^{26}Al/^{27}Al ratios. In the hot bottom stage, the entire hydrogen envelope circulates through the hydrogen-burning shell, so that much of the ^{25}Mg will be converted to ^{26}Al by proton capture. Because this shell is relatively close to the surface in the late evolutionary stage, its temperature will be quite high, probably about 7×10^7 K or more. ^{29}Si and ^{30}Si are much more rapidly destroyed by proton capture than ^{28}Si (Woosley and Hoffman 1986). Also at such a high temperature ^{15}N is destroyed less rapidly relative to ^{14}N than is true at lower temperatures, and so ^{15}N will become enriched. From time to time, material will be mixed in from below as the hydrogen-burning shell is shut down, and if the grains are ejected at such times, then, as discussed later, ^{12}C formed in the helium layer will be enriched, and it is possible that enough neutron-produced ^{44}Ca can be brought up from below to become visible as an anomaly (this is further discussed in the following section). The ^{26}Al/^{27}Al ratio may become significantly higher than 0.23, but we must await more experimental evidence to be sure of this. Zinner's two highly anomalous grains are unusually large, and they are thus candidates to have been made during the final AGB star stage in which the planetary nebula was expelled; under such circumstances the relative gas expansion rate is smaller than in a wind, and grains would have an opportunity to grow for a longer period of time after nucleation.

The winds are strong in high-luminosity red giant stars. A rough estimate is that 0.5 $M_\odot$ will be expelled in the 1 Myr prior to the planetary nebula ejection from an average star operating with a ^{13}C neutron source (discussed below), which assumes that the average such star has an initial mass of order 1.5 $M_\odot$. In the planetary nebula itself, another 0.1 or 0.2 $M_\odot$ is expelled which will also contain ^{26}Al. If such a star also has a velocity relative to the cloud of about 10 km s^{-1}, then it will travel 10 pc in this last 1 Myr, a good fraction of typical cloud dimensions. If the wind has a velocity of order 100 km s^{-1}, then the 0.5 $M_\odot$ of material ejected in the wind will mix into about 50 $M_\odot$ of material (considering only momentum conservation) before being slowed down to typical cloud turbulent velocities. If the ^{26}Al abundance is about equal to 0.25 of the abundance of ^{27}Al (from Zinner's measurements), then the ^{26}Al/^{27}Al ratio in this mixed material will probably be about 2.5×10^{-3}. This is a factor of 50 higher than the ratio observed in the solar system. Thus some 3.9 mean lives of ^{26}Al would be necessary to bring the ^{26}Al abundance down to the solar system level. This AGB source of ^{26}Al seems the most plausible of the possible sources of this radionuclide in the early solar system. We now consider the frequency of such contamination in a molecular cloud.

Under steady-state conditions in the Galaxy, about one star is formed per year, primarily in molecular clouds, and therefore one star comes to the end of its life every year, most of these being AGB stars in the lower-mass range. The end points of stellar evolution can be reached at arbitrary points in space. If we assume that one-quarter of the galactic mass is in molecular clouds, then the fractional volume within the galactic disk occupied by molecular clouds is about 2.5×10^{-4}, and this number is also the number of AGB stars that end their lives within a molecular cloud per year. If the ejecta from the terminal stage of an AGB star contaminates 50 $M_\odot$ of material in its vicinity, as discussed above, then 0.012 $M_\odot$ of molecular gas is contaminated per year in the Galaxy. In a mean life of ^{26}Al, 1.3×10^4 $M_\odot$ of material will be directly contaminated by AGB ejecta. This is small compared to the amount of mass that goes into new stars.

The net effect is that only a small part of the material within molecular clouds would be contaminated with ^{26}Al. If this were the entire story, then one would conclude that the probability of having the solar system emerge from average star formation within a molecular cloud complex is very small. But there is an escape clause. The possibility was mentioned above that events like the AGB stars and the bipolar ejecta from young stars can themselves trigger the formation of additional stars, leading to the formation of the close groups commonly seen in these clouds (Myers and Benson 1983). For purposes of illustration consider an oversimplified and ideal model. If the star formation events triggered by AGB stars should go through 12 generations, each generation doubling the number of star formation sites, then 4096 stars would be formed. Because the average stellar mass is considerably less than one $M_\odot$, these 4096 stars would have something like 10^3 $M_\odot$ of material in them, which is an order of magnitude larger than the estimated amount of material that would be contaminated by a single AGB star event. The postulated chain of star formation would clearly induce some further mixing in the cloud. With this additional geometric dilution, the ^{26}Al abundance in the contaminated material would start out up to an order of magnitude higher than the solar system level, but this can be in error by a substantial factor. In Cameron (1984) it was estimated that the time required to complete a step in the above ideal star-formation chain would be about 5×10^5 yr. The full chain would thus take about 6 Myr. The early steps in the chain would produce stars with an ^{26}Al abundance higher than that in the solar system, but in the majority of them the ^{26}Al abundance level would be lower than in the solar system by a modest factor.

In summary, it appears that the local molecular cloud in the vicinity of the path of the AGB star would become contaminated with the ^{26}Al/^{27}Al ratio initially at a level about 1 order of magnitude above that found in the early solar system. Some combination of mixing and decay would reduce this ratio to the solar system level. If the reduction were entirely due to decay, then about 1 or 2 Myr would pass between the time the material was ejected from the AGB star and the formation of the solar system.

It is necessary to caution that many quite isolated new stars appear in molecular clouds, and these may have been formed by some other kind of triggering process or even by a spontaneous core formation and collapse, so that they need not have any contamination by ejecta from an AGB star in its final phases. For the solar system, the evidence is that this contamination occurred.

VIII. s-PROCESS RADIOACTIVITIES

Previously (Sec. IV) we considered whether the observed abundances of the nuclides ^{107}Pd, ^{129}I, ^{135}Cs and ^{182}Hf could have been produced by the galactic r-process, and we concluded that only ^{129}I could have been obtained in this way. The other nuclides have shorter mean lives and hence would have decayed to too low a level to fit the solar system abundances. The ^{129}I identification was only tentative, however, and it remains to be determined whether all four of these nuclides can have been provided to the solar system as a result of the s-process taking place in the helium layer of the AGB star which provided the ^{26}Al from the hydrogen envelope.

It appears that the main contributions to the s-process in nature are made in an AGB star of 1 to 2 $M_\odot$ (Hollowell and Iben 1989). Such a star has been through both hydrogen and helium burning in its central regions, and it therefore possesses hydrogen and helium shell burning sources. However, these shell sources are not active simultaneously, owing to the high-temperature sensitivity of the helium burning shell. After a period of hydrogen shell burning, the helium shell will undergo a thermonuclear runaway, leading to an extensive convection zone in the helium region that extends for a time to the lower boundary of the hydrogen envelope. After this outburst, the helium shell burning gradually dies away and the helium shell becomes dormant. Shortly thereafter the hydrogen burning shell is regenerated; the outer convection zone extends all the way down to the hydrogen burning shell, and some penetrative mixing into the top of the underlying helium zone occurs. This has two important effects: it mixes the products of the neutron capture in the helium layer out to the surface of the star, and it provides via the admixed hydrogen additional fuel for neutron production. The mixing brings together ^{12}C made in helium burning with the hydrogen, and the sequence ^{12}C(p,γ) ^{13}N$(\beta^+\nu)$^{13}C(α,n)^{16}O then creates the neutrons as the mixture of hydrogen and carbon is swept down to the base of the helium layer during the next helium flash. The optimum mixing ratio is 1 proton to 10 carbons, so that very little ^{14}N neutron absorbing poison is produced by the proton capture, and this mixing ratio appears close to that achieved in the AGB star.

The AGB star evolves through a series of these helium shell flashes, during which time carbon and oxygen formed in the helium region are deposited on the core below the helium-carbon boundary, and fresh helium is added to the helium region at the top as a result of the hydrogen burning. Thus in each flash, the helium region has been somewhat diluted by fresh material that has not been exposed to neutrons before, and which thus maintains a supply of seed nuclei in the vicinity of the iron abundance peak. It has been found that this type of behavior is required so that the material in the helium zone can have been exposed to a roughly exponential distribution of neutron fluences, in accord with the observational inference.

More massive AGB stars can produce neutrons by purely helium reactions: as in $^{14}N(\alpha,\gamma)^{18}F(\beta^+\nu)^{18}O(\alpha,\gamma)^{22}Ne(\alpha,n)^{25}Mg$. However, the total number of neutrons generated per iron seed nucleus in this way is smaller than in the lower-mass AGB star case, because the neutron source material is not renewable. Thus it is unlikely that the more massive stars can be as effective in producing ^{107}Pd as the lower-mass ones.

Analyses of the s-process abundance yields have shown that a neutron number density of 3×10^8 cm^{-3} characterizes the final abundances produced by the process (Käppeler et al. 1990). It should be noted that this conclusion is particularly sensitive to the lower neutron densities in a distribution of such densities, because lower densities allow the formation of nuclides in "protected" positions close to the valley of beta stability, whereas the neutron capture path does not traverse such nuclei if the neutron density is high. Bazan (1991) has concluded that this "freezing neutron density" is 3×10^7 cm^{-3}. Bazan has also found that in the course of a helium shell flash, under a wide range of conditions, the neutron number density reaches a peak value close to 1×10^{10} cm^{-3}. Also, during the course of a flash, the amount of material in the helium region convective zone increases while the neutron density is increasing, but it reaches its maximum extent after the neutron number density has abruptly dropped to a very low value. While the convection zone is near its maximum extent, the low neutron number density can regenerate nuclear abundances in protected positions near the valley of beta stability, but there is no substantial destruction of the abundance patterns established at maximum neutron number density. This means that the material left closest to the hydrogen boundary, which will be dredged up later, will be that characteristic of the highest neutron densities, or 3×10^9 to 1×10^{10} cm^{-3}, even though an analysis of the Käppeler et al. (1990) type would indicate a much lower density (Hollowell and Iben 1990; Iben, personal communication).

The importance of this consideration may be seen from the following. I have calculated a steady-flow s-process through nuclei near the valley of beta stability, using an r-process network in which these nuclei were included (a similar r-process steady-flow calculation was reported by Cameron et al. [1983]). These calculations used nuclear data assembled by F.-K. Thielemann and J. J. Cowan (unpublished). The ratios of the four potential s-process extinct radioactivities relative to their reference nuclei are shown for several values of the neutron number density in Table II. Also shown in Table II are the effective production ratios in which the reference nuclide abundances are multiplied by the normal s-process abundance and divided by the sum of the s- and r-process abundances. This correction factor used the separation of nuclide abundances into s- and r-process components by Käppeler et al. (1989). Finally, Table II also shows the ratios of the corrected production ratios to the observed ones.

TABLE II

s-Process Ratios as a Function of Neutron Number Density[a]

Nuc. Ratio	3×10^8	3×10^9	1×10^{10}	1×10^{11}
^{107}Pd/^{108}Pd	0.15	0.15	0.15	0.15
^{129}I/^{127}I	0.0037	0.033	0.102	0.55
^{135}Cs/^{133}Cs	0.095	0.72	1.38	1.72
^{182}Hf/^{180}Hf	0.0040	0.043	0.131	0.60
^{107}Pd/^{108}Pd	0.086	0.086	0.086	0.086
^{129}I/^{127}I	0.00020	0.0018	0.0057	0.031
^{135}Cs/^{133}Cs	0.014	0.11	0.21	0.26
^{182}Hf/^{180}Hf	0.0026	0.027	0.084	0.38
^{107}Pd/^{108}Pd	4,300	4,300	4,300	4,300
^{129}I/^{127}I	2.0	18	57	310
^{182}Hf/^{180}Hf	< 325	< 3,400	< 10,500	< 47,500
^{182}Hf/^{180}Hf	> 32.5	> 340	> 1,050	> 4,750

[a] Nuc. ratio is the ratio of abundances of indicated isotopes. Neutron number densities are in cm^{-3}. The first set of four rows shows the relative production ratios in the s-process. The second set of four rows shows the production ratios after multiplying the reference s-process abundance by the normal fractional production of the reference nucleus in the combined s- and r-processes. The third set of five rows shows the corrected production ratios relative to the observed ratios; the last two of these rows correspond to the lower and upper limits on the ^{182}Hf/^{180}Hf observed ratio.

It should be noted that we had concluded from the ^{26}Al discussion that the interval between ejection of material from this unique AGB star, that has contributed very short-lived radioactivities to the solar system, and the formation of the primitive solar nebula, is very short, probably not more than 2 Myr. Hence the correction to the abundances in Table II for decay during this interval is small and no correction is made here. Therefore the production ratios shown in the last set of rows of Table II should be proportional to the abundances of the three extinct radioactivities, all other factors being the same.

The numbers obtained in the last set of rows of Table II should correspond to a common factor which is the geometric dilution needed to bring the production ratios of the extinct radioactivities down to the level observed in the early solar system, provided that the assumption is correct that these four nuclides are products of the s-process in the AGB star. Another hidden assumption here is that the shape of the s-process abundance curve for the AGB star is the same as that in the solar system abundance distribution generally, which is probably a quite good assumption. From the earlier discussion of the AGB star neutron production environment, it should be expected that the best fit to a common geometric dilution factor should be for a neutron number density of 3×10^9 to 10^{10} cm^{-3}.

First, it should be noted that the s-process production of ^{129}I is much too small relative to the other nuclides. This confirms the previous tentative assignment of this nuclide to the galactic r-process.

Considering the substantial errors in making all of these estimates, the other three nuclides are consistent with a common production by the s-process with a neutron number density in the expected range. They could not be made at the lowest or highest neutron number densities shown in Table II. The common geometric dilution factor is roughly 4000. This favors an abundance of ^{182}Hf near the upper end of its range of uncertainty.

In the preceding section, it was estimated that the geometric dilution that occurred after ^{26}Al was ejected from the AGB star was about a factor of 100. If this is correct, the total geometric dilution of a factor 4,000 implies that the material from the helium region of the AGB star is geometrically diluted by a factor of 40 by mixing into the hydrogen envelope as a result of dredge-up toward the end of the AGB active lifetime. This 2.5% percent admixture from the helium layer is consistent with estimates that have been made from other considerations (Bazan 1991).

Thus we confirm that ^{107}Pd, ^{135}Cs and ^{182}Hf are primarily the products of the s-process operating in the AGB star that we postulate to have triggered a chain of star formation that included formation of the solar system.

Although ^{60}Fe has been listed in Table I as an extinct radionuclide, it should be kept in mind that the reality of its existence in the early solar system is not firmly established. With its mean life of 2.2 Myr, it is the second shortest lived of the extinct radioactivities. It can be made by both s- and r-processes, but from our earlier discussions it is evident that any r-process component of the abundance would long since have decayed before the solar system was formed. However, it can be formed by the s-process in the same AGB star that we have been discussing here.

The abundance of ^{60}Fe relative to ^{56}Fe cannot be estimated by the method of steady-flow neutron capture in the manner done above. ^{56}Fe is the principal seed nucleus for the s-process, and in any one pulse of neutron capture in the AGB star, the ^{56}Fe will capture enough neutrons to become quite depleted; when hydrogen burns to helium and deposits material on the top of the helium layer, the ^{56}Fe that accompanies the converted hydrogen provides a fresh source of seed nuclei for the next pulse. Thus the amount of ^{60}Fe remaining after a pulse depends upon the detailed history of the pulse. In the steady-flow calculations for neutron capture that were described above, the neutron capture path is centered at ^{60}Fe for neutron number densities of 3×10^9 and 10^{10} cm^{-3}. Thus neutron capture on ^{56}Fe can be expected to lead straight to ^{60}Fe. As the neutrons are captured, the material initially present as ^{56}Fe forms an abundance peak that moves upward in mass number and spreads out as it goes. An initial guess is that the ^{60}Fe/^{56}Fe abundance ratio is roughly of order 10^{-2} during the neutron capture phases. If the material is now subjected to a geometric dilution by a factor of 4000, as indicated above, the order of magnitude of the ^{60}Fe/^{56}Fe ratio at the time of formation of the solar system is about 2.5×10^{-6}. This will be further diminished by a little less than a mean life of radioactive decay of the ^{60}Fe, making the abundance ratio compatible with the observed value of 1.6×10^{-6}, within large uncertainties.

Let us return briefly to the question of the abundances in the two large SiC grains analyzed by Zinner and his colleagues, described in the preceding section. The excesses of ^{12}C and ^{44}Ca were postulated to be introduced into the hydrogen envelope by mixing from the helium layer. Of course, the ^{12}C is a main product of the helium burning that takes place at the base of the helium layer. In the author's steady-flow neutron capture calculations at neutron number densities of 3×10^9 and 10^{10} cm^{-3}, the neutron capture path is centered on the calcium isotopes with mass numbers 42 through 46, including 44. The calculated cross section for capture on ^{44}Ca is lower than those for the other calcium isotopes, but not by a large enough factor to account for the ^{44}Ca peak observed by Zinner and his colleagues. However, the level spacing at the neutron binding energy in the ^{45}Ca compound nucleus is comparable to the mean thermal energy at which the capture takes place, indicating that the usual calculational method of estimating neutron capture cross sections, which averages over the expected resonances, is highly unreliable here; therefore the cross section for capture on ^{44}Ca could be very much less than estimated. It will be so postulated here. Then the situation would be similar to that of ^{60}Fe discussed above; the highly abundant seed nucleus ^{40}Ca would feed into the neutron capture path at ^{42}Ca, and its abundance would be periodically refreshed as hydrogen is converted into helium and added to the helium layer.

IX. UNOBSERVED EXTINCT RADIOACTIVITIES

The above analyses provide a crude basis for making some predictions about the expected abundance levels in the early solar system for several extinct radionuclides with reasonably long mean lives, some of which have been looked for in meteorites. In the above analyses, an estimate has been made of the production and dilution of the radionuclide into the surrounding medium, followed by a period of decay which would be required in order to bring the estimated diluted abundance down to observed solar system levels. There can be substantial errors in these estimates. In applying the resulting methods to make predictions of early solar system abundances for additional radionuclides, one can get led astray by these errors, and so the derived numbers for the abundances given below should be taken as just crude indications of the order of magnitude to be expected. The radionuclides are discussed in order of increasing mean life.

^{41}Ca: $\tau \approx 1.9 \times 10^5$ yr. This nuclide is normally made as part of the oxygen or silicon burning processes, so it is primarily a product of supernova explosions. The closest supernova events would be those producing ^{53}Mn, with a decay time of roughly 25 Myr. After this decay time, the ^{41}Ca abundance would be completely unobservable. It can also be made by neutron capture in ^{40}Ca in the neutron production stage of an AGB star; here the above discussion concerning ^{44}Ca is relevant. If we estimate the abundance of ^{41}Ca to be very roughly maintained at 0.01 of the ^{40}Ca, then the helium layer ^{41}Ca/^{41}K abundance ratio would become about 2.5. After dilution by a factor of 4000, this ratio becomes 6×10^{-4}. For every million years of subsequent decay, the abundance ratio decreases by a factor of 200. The observed upper limit is 8×10^{-9} (Hutcheon et al. 1984). The estimated abundance would decay to the observed upper limit in 2.1 Myr. As the initial abundance was just a guess, this seems roughly compatible with the decay period found for ^{26}Al, and it would appear that searching for this extinct nuclide remains a worthwhile goal.

^{247}Cm: $\tau \approx 22.5$ Myr. This is an r-process product which decays to ^{235}U, which should thus be regarded as the reference isotope. The fraction of all ^{247}Cm ever made that was present at the time of origin of the solar system was about 7×10^{-5}, making this estimate by the method of periodic spikes and allowing 110 Myr of decay after the last spike. The ^{235}U has a mean life of 1 Gyr, so the fraction of it that was ever made at the time of origin of the solar system is 0.1, with negligible subsequent decay. This gives a prediction for ^{247}Cm/^{235}U of about 7×10^{-4}, to be multiplied by a relative production factor. If the r-process yields peak around mass number 250, then the production factor is nearly unity, but if it peaks around mass number 235 or is fairly flat, then the production factor found in the author's could be as low as 0.1. The production factor that I find in my r-process calculations varies in the range between 0.4 and 0.7. Thus is estimated the probable solar system ^{247}Cm/^{235}U abundance ratio to be about 4×10^{-4}. The measured upper limit to ^{247}Cm/^{235}U is 4×10^{-3} (Chen and Wasserburg 1981a, b).

X. DISCUSSION

In this survey of the ten extinct radionuclides, we have found a variety of responsible nucleosynthesis mechanisms. These include high temperature hydrogen-burning, the s-process (four times), the equilibrium process, the r-process (twice), and the p-process (twice). We concluded that five of the radioactivities were likely to be due to an AGB star reaching the final stages of its active evolutionary lifetime within the molecular cloud in which the solar system would form, and that the other five were due to supernova explosions taking place in the interstellar medium prior to the formation of the molecular cloud complex (in some cases, long before).

We also concluded that the probability was low of inheriting nucleosynthesis products from an AGB star contaminating a rather small part of the molecular cloud complex unless that star also triggered a chain of star formation that would be propagated by energetic events accompanying the star formation process itself. We cited some physical reasons for expecting that such triggering should occur.

The great majority of star formation currently observed to occur in the Galaxy takes place in molecular cloud complexes. One of the major objectives of this survey has been to see whether the extinct radioactivities found to have been present in primitive solar system matter could have been introduced in a timely manner into a core within such a molecular cloud complex. Our analyses were necessarily very crude due to some poorly known numbers, but we did conclude that the observed solar system extinct radionuclides were consistent with the formation of the Sun in such a molecular cloud environment.

The discussion given here of the abundances of now-extinct radioactivities in the early solar system was developed with the following postulated local history of the Galaxy, with which the abundances appear to be consistent. We consider the times to be relative to the time of formation of the solar system, and all of them will be before the solar system (BSS).

1. 130 or 140 Myr BSS. An OB Association formed in a local "great grandparent" molecular cloud. The cloud was destroyed and a large-scale mixing event took place in nearby space, leading to the formation of a new local "grandparent" cloud.

2. 110 to 120 Myr BSS. The OB Association stars in the range 8 to 10 $M_\odot$ produced r-process nuclides. These remained rather localized to the sites of production. We obtained ^{129}I and ^{244}Pu from this event.

3. Prior to about 25 Myr BSS. There was a continuing production of p-process nuclides from imploding supernovae in suitable binary systems. These nuclides were also rather localized to the sites of production. We

obtained ^{92}Nb and ^{146}Sm from these events.

4. 25 Myr BSS. An OB Association formed in the grandparent cloud. This was destroyed with large-scale mixing and a new "parent" cloud was formed. The mixing brought in all locally produced r-process and p-process nuclides. ^{53}Mn produced in the highest mass O stars in the OB Association was also mixed into the parent cloud.

5. About 2 Myr BSS. A relatively low-mass AGB star, accidentally located in the parent cloud, came to the end of its active lifetime and triggered a chain of star formation that led to the formation of the solar system. We obtained ^{26}Al, ^{60}Fe, ^{107}Pd, ^{135}Cs and ^{182}Hf from this event.

There is rather little redundancy in this set of interpretations. It has been necessary to draw extensively on information from the fields of nucleosynthesis, stellar evolution and supernova explosions, and the theory of the interstellar medium and its structure, as well as meteoritics and planetary sciences. Further work can and should test several aspects of this set of scenarios.

Acknowledgment. I am especially indebted to my colleague F. Thielemann, who has been a patient and helpful source of information on quantities needed for nucleosynthesis estimates. I am also indebted to J. Wasserburg, T. Lee, J. Truran, J. Cowan, P. Myers, G. Lugmair, C. Harper, I. Iben, B. Elmegreen and T. Swindle for helpful discussions of some of these issues.

CLASSICAL NOVAE: CONTRIBUTIONS TO THE INTERSTELLAR MEDIUM

ROBERT D. GEHRZ
University of Minnesota

JAMES W. TRURAN
University of Illinois

and

ROBERT E. WILLIAMS
Cerro Tololo Inter-American Observatory

Classical novae, explosions that result from thermonuclear runaways on the surfaces of white dwarfs accreting matter in close binary systems, are sporadically injecting material processed by explosive nucleosynthesis into the interstellar medium. Although novae probably have processed less than ~0.3% of the interstellar matter in the Galaxy, there is theoretical and observational evidence suggesting that they may be important sources of the nuclides ^{7}Li, ^{13}C, ^{15}N, ^{17}O, as well as of ^{22}Na and ^{26}Al. The latter nuclides are astrophysically important radioactive isotopes that could have been involved in the production of the ^{22}Ne (Ne-E) and ^{26}Mg enrichments in meteoritic inclusions believed to contain important clues about the chemical and mineral contents of the primitive solar nebula. These inclusions may be partially composed of dust condensed in nova outbursts. We review theoretical results predicting yields of these various isotopes in nova outbursts, and conclude that most of the heavy isotope anomalies are produced by the approximately 25% of novae that occur in systems containing massive (≥ 1.2 M$_\odot$) O-Ne-Mg white dwarfs. Ultraviolet, optical and infrared emission line spectra of classical novae reveal the abundances of some of the gas-phase elements present in the ejecta. Recent studies show that the ejecta in some novae can be strongly cooled by near- and mid-infrared radiation from forbidden-line radiation from highly ionized atomic states. We compare the abundances deduced from observations with the theoretical predictions, and suggest that future studies of infrared coronal emission lines may provide additional key information. Novae produce only ~0.1% of the galactic "stardust" (dust condensed in stellar outflows), but it may be some of the most interesting dust. Novae appear capable of producing astrophysical dust of every known chemical and mineral composition. We summarize the dust-production scenario for novae, and argue that explosions on O-Ne-Mg white dwarfs may lead to the formation of dust grains that carry the Ne-E and ^{26}Mg anomalies. Finally, we attempt to place quantitative constraints on the degree to which classical novae participate in the production of chemical anomalies, both in the primitive solar system and on a galactic scale. Diffuse galactic gamma-ray fluxes are especially useful for assessing the ^{22}Na and ^{26}Al yields from novae.

I. INTRODUCTION

Infrared observations support a picture of galactic chemical evolution in which grains condensed in the ejecta of stars may be a significant reservoir for the transportation of metals in the galactic ecosystem. Metals produced during the advanced stages of nucleosynthesis can be condensed into grains in circumstellar shells during late stages of stellar evolution, and injected into the interstellar medium (ISM) where they are processed and eventually incorporated into new stellar and planetary systems. Once in the ISM, these grains may be further processed by supernova shocks and in molecular clouds, before being incorporated into young stars and planetary systems during star formation in the clouds. Solids that may be remnants of the formation phase are found in our own solar system and around some other main sequence stars. Wolf-Rayet stars, novae and supernovae may also be capable of injecting copious amounts of gas-phase condensibles into the ISM, where they are later processed onto grains in the cores of dense molecular clouds.

Grains in circumstellar shells, the ISM and the solar system are similar in general mineral content. The same four basic mineral types—amorphous carbon (iron?), silicates, silicon carbide and hydrocarbons—are present in all three environments. There are important differences between the grains in these diverse environments (see Gehrz 1989b). ISM grains appear to be smaller than their circumstellar counterparts, and grains in some comets were apparently annealed at high temperature. Nonetheless, several properties of cometary dust and meteorite inclusions suggest that dust made in circumstellar shells (stardust) was incorporated directly into solid bodies in the primitive solar nebula. For example, some meteorites contain small dust inclusions with chemical abundance anomalies that strongly suggest that these grains condensed in the immediate ejecta of nova and supernova eruptions. This would imply that some grains may have survived from the circumstellar to the solar environments relatively unchanged. On the other hand, it has been argued that it is difficult for stardust to survive destruction by supernova shocks in the ISM and by high temperatures during the early evolution of young stellar objects.

Nucleation and growth of stardust in a circumstellar environment has been confirmed by infrared studies of classical novae. Although they contribute only a small amount of stardust to the ISM of a galaxy, classical novae are the only objects in which it has been possible to observe the process of circumstellar dust formation directly. Observations have shown that the infrared development of a nova outburst is governed by the evolution of the grains in the outflow, and have defined the conditions under which grains nucleate and grow as they are expelled into the ISM.

In this review, we concentrate specifically on the stardust and gas-phase chemical anomalies that may be contributed to the galactic ISM by classical novae. Novae are concentrated toward the center of the Galaxy, where they suffer obscuration. Therefore, some of their global properties have only recently been estimated from studies of nearby galaxies. It is clear that they are associated with the old stellar population in galaxies, in that both the spatial distribution and the relative frequency of novae are well correlated with the infrared K-band luminosity (Ciardullo et al. 1990). Although the distribution of novae near the Sun suggests a concentration toward the disk of the Galaxy, the majority of novae belong to the halo or at least to a thick disk population (Warner 1989). The influence of nova ejecta on the ISM is therefore global. Roughly 30 novae per year occur in the Galaxy (Arp 1956; Capaccioli et al. 1989), and they all have in common the fact that they eject material at high velocities during outburst, with this material usually enriched in elements heavier than He.

II. NATURE OF THE OUTBURST

The theory of classical nova outbursts has advanced at a significant pace over the past two decades. Two factors contributing to these rapid developments are progress in theoretical modeling of the outbursts and, particularly, a growing body of observational data describing novae in outburst. Indeed, observations of novae are now available in virtually all interesting wavelength ranges (Starrfield 1989a). Recent observations have served both to constrain the theoretical models and to provide a quantitative measure of the influence of classical nova explosions on the surrounding ISM.

It is now commonly understood that a nova outburst results from a thermonuclear runaway on the surface of a white dwarf accreting matter in a close binary system. The companion stars in these "cataclysmic variable" systems are observed to be late-type stars that are filling their Roche lobes and passing matter through the inner Lagrangian point, via a disk, onto the surface of the white dwarf. Accumulation of a hydrogen-rich envelope on the white dwarf continues until a critical pressure is achieved at the base of the accreted envelope, and a thermonuclear runaway ensues. There follows: (1) a rapid rise of the luminosity to a maximum value that approaches (and sometimes exceeds) the Eddington limit; (2) a period over which the rekindled hydrogen burning shell powers the system at approximately constant bolometric luminosity; and (3) a gradual return of the system to its pre-outburst state. The detailed features of such outbursts have been reviewed by a number of authors (Truran 1982; Gehrz 1988; Starrfield 1989b).

Hot gas expelled in the explosion is initially seen as an expanding pseudophotosphere, but free-free and line emission are observed when the expanding fireball becomes optically thin. A dust condensation phase, characterized by rising infrared emission and sudden visible absorption occurs in many novae within 30 to 80 days following the eruption. Novae produce silicates, silicon carbide, carbon and hydrocarbons. Several have condensed all four types of grains in a single eruption, suggesting that novae may have a complex temperature, density and abundance structure in their ejecta.

Spectroscopy of infrared coronal emission lines in classical novae has provided new insight into the physics of the nova phenomenon. Abundance information implies that some novae may condense anomalous grains similar to those found in meteorite inclusions and has provided constraints on models of the progenitor stars of nova systems. One recent suggestion to explain these observations is that a substantial fraction of observed nova binary systems may contain O-Ne-Mg white dwarfs that are the predicted end product of the evolution of some intermediate-mass stars.

We are specifically concerned in these proceedings with the contributions of novae to the ISM. These contributions can be examined by considering the nature of the contributions from an individual "average" nova event. The critical parameter is the white-dwarf mass. Truran and Livio (1986,1989) recently pointed out that the typical masses of the white dwarfs in nova systems observed in outburst are significantly higher than the 0.6 to 0.7 $M_\odot$ range characteristic of single white dwarfs. This is understood to be a consequence of selection effects, favoring massive white dwarfs that require significantly less accreted matter to trigger a thermonuclear runaway and therefore experience recurring outbursts on much shorter time scales. Truran and Livio estimate the average white dwarf mass in observed classical nova systems to be approximately 1.1 to 1.2 $M_\odot$; a comparable value has recently been determined by Politano et al. (1990). The relative frequencies of occurrence of white-dwarf masses $f(WD)$ from these two references are presented in Table I, together with an estimate of the (accreted) envelope mass required to initiate runaway (column 3), and an approximate determination of the recurrence time scale (column 4) based on the assumption that all systems have an accretion rate of $\sim 10^{-9}$ $M_\odot$ yr^{-1} (King 1989).

The mean envelope mass required to initiate a runaway that follows from the frequency estimates is approximately 2×10^{-4} $M_\odot$ (note that this is slightly larger than the envelope mass of 5×10^{-5} $M_\odot$ required for a white dwarf of the mean mass 1.1 to 1.2 $M_\odot$). The theoretical models and general energy considerations indicate that most of the accreted matter is ultimately ejected in the outburst and thus returned to the ISM . The mean mass returned by a nova outburst to the ISM is thus of the order of 2×10^{-4} $M_\odot$. If we assume that the nova rate of ~ 30 per year for M31 in Andromeda (Arp 1956; Capaccioli et al. 1989) can be applied to our Galaxy, it follows that novae introduce approximately 6×10^{-3} $M_\odot$ yr^{-1} of processed matter into the ISM. Supernovae, occurring at a rate of 1 per 50 yr and ejecting approximately 3 $M_\odot$ per event, introduce roughly 0.06 $M_\odot$ yr^{-1} of processed matter into the ISM. Therefore, if novae are to contribute significantly to nucleosynthesis, the levels of overproduction of heavy elements in novae must be on the order of 10 times the levels characteristic of supernova ejecta. This crucial question of the global contribution of novae will be emphasized several times in the discussions that follow.

The mass estimate provided here also allows us to determine the energy contributions of individual nova events to the ISM. Velocity determinations

TABLE I

Mean Properties of Classical Nova White Dwarfs

Mass ($M_\odot$)	$f_{TL}(WD)$	ΔM_H ($M_\odot$)	τ_{REC} (yr)	$f_{PLTW}(WD)$
0.6	0.103	1.3×10^{-3}	1.3×10^{6}	0.035
0.7	0.053	7.3×10^{-4}	7.3×10^{5}	0.087
0.8	0.042	4.2×10^{-4}	4.2×10^{5}	0.076
0.9	0.040	2.4×10^{-4}	2.4×10^{5}	0.060
1.0	0.046	1.3×10^{-4}	1.3×10^{5}	0.058
1.2	0.100	2.8×10^{-5}	6.4×10^{4}	0.098
1.3	0.232	9.0×10^{-6}	9.0×10^{3}	0.212
1.35	0.322	4.0×10^{-6}	4.0×10^{3}	0.306

for the principal ejecta of novae lie in the range 400 to 2000 km s^{-1}, although some of the material appears to be ejected at velocities as high as 10^4 km s^{-1}. This implies that the typical output kinetic energies will fall in the range 10^{43} to 10^{45} erg, well below values typical of supernovae, and that novae do not constitute a significant source of energy for the heating of the ISM. Spectroscopic studies of classical novae have revealed two very important features:

(1) The presence of extremely large concentrations of C, N and O nuclei and, more recently, of O, Ne and Mg nuclei in the ejecta;
(2) The formation of grains in a substantial number of the slower novae.

The issue of grain formation and its implications is addressed in Sec. V. It is appropriate at this stage, however, to examine briefly the theoretical implications of the heavy-element enrichments.

III THERMONUCLEAR PRODUCTION OF HEAVY NUCLIDES

Nucleosynthesis associated with nova events is limited to the effects of the proton-induced reactions that can proceed at the temperatures of 1.8×10^8 to 3×10^8 K typically achieved in their hydrogen burning shells. CNO cycle hydrogen burning will, of course, act to redistribute these nuclei and thereby alter the concentrations of the isotopes of carbon, nitrogen and oxygen. The observational situation regarding the abundances of these and heavier nuclei in nova ejecta is reviewed in the next section, and the implications of the observed enrichments of heavy elements is addressed. In this section, we review the possible role of novae in the synthesis of ^{7}Li and in the generation of significant abundances of the radioactive nuclei ^{26}Al and ^{22}Na.

It is important to recognize in advance that the relative contribution of novae to nucleosynthesis is small, and that detailed predictions regarding their contributions are quite uncertain. We do not know with confidence either the rate of nova outbursts as a function of time over the history of our Galaxy or the average mass ejected per outburst. Nonetheless, the total amount of processed matter contributed by novae to the ISM may be roughly estimated by assuming that (1) the current rate of nova events is 30 yr^{-1} (Arp 1956; Capaccioli et al. 1989); (2) the rate has remained constant over the lifetime $\sim 10^{10}$ yr of the galactic disk; and (3) the average mass ejected per nova event is $\sim 2 \times 10^{-4}$ M$_\odot$. This implies a total mass processed by novae of $\sim 6 \times 10^7$ M$_\odot$. By comparison, supernovae occurring at a rate of 1 per 50 yr and ejecting ~ 3 M$_\odot$ per event would have processed 6×10^8 M$_\odot$ during the same period. As the mass processed through novae represents only a fraction $<1/150$ of the mass of interstellar matter ($\sim 10^{10}$ M$_\odot$) in our Galaxy, it follows that significant contributions require enrichments, relative to solar system abundances, in nova ejecta of factors of >150. This would apparently rule out the possibility that novae are important contributors to galactic abundances of the dominant isotopes of carbon and oxygen, and probably of nitrogen as well. Novae may, however, produce important amounts of the rarer isotopes ^{13}C, ^{15}N and ^{17}O.

We have seen that classical nova outbursts provide an environment in which hydrogen-burning reactions proceed on carbon, nitrogen and oxygen (CNO) nuclei, and heavier nuclei, at high temperatures and densities on a dynamic time scale. For such conditions, the production of potentially interesting and detectable abundance levels of the radioactive isotopes ^{22}Na and ^{26}Al can also occur, and novae are believed by many to be potentially important nucleosynthesis sites for these elements. Weiss and Truran (1990) recently modeled the nucleosynthesis accompanying nova explosions for representative temperature histories extracted from hydrodynamic models, utilizing a significantly expanded nuclear reaction network (see Nofar et al. 1991). Their results confirm the earlier findings of Hillebrandt and Thielemann (1982) and Wiescher et al. (1986) that extremely low levels of both ^{22}Na and ^{26}Al are expected to form in nova envelopes of initial heavy-element composition comparable to that of solar system matter. This implies both that we should not expect to find detectable levels of ^{22}Na in the ejecta of slow novae (e.g., novae with relatively low heavy-element concentrations) and that such novae do not contribute significantly to the abundance of ^{26}Al in the Galaxy. Moreover, enriched CNO concentrations alone do not guarantee any significantly increased ^{22}Na and ^{26}Al production. The calculations by Weiss and Truran (1990) predict low concentrations of ^{22}Na and ^{26}Al even with the assumption of matter enriched to a level $Z_{CNO} = 0.25$. The conclusions drawn above to the effect that these systems characterized by solar abundances of CNO elements can neither produce detectable levels of ^{22}Na nor contribute significantly to the abundance of ^{26}Al in galactic matter are equally appropriate to the case of novae with moderate CNO enrichments.

As anticipated, greatly increased ^{22}Na and ^{26}Al production results for envelopes characterized by substantial initial enrichments of elements in the range from neon to aluminum. Because these abundance enrichments are presumed to arise as a consequence of the dredge-up of core matter (see, e.g., Livio and Truran 1990), it follows that such systems involve more massive O-Ne-Mg white dwarfs. Theoretical estimates (Truran and Livio 1986; Truran 1990) and spectroscopic analyses of the ejecta of recent novae (Table I) indicate that such neon rich novae comprise approximately 1/4 to 1/3 of all observed classical nova outbursts. For these cases, burning by means of the neon-sodium and magnesium-aluminum hydrogen burning sequences can yield substantial concentrations of ^{22}Na and ^{26}Al. Specifically, for the choice of an initial composition consisting of matter enriched to a level Z = 0.25 in the products of stellar carbon burning (Arnett and Truran 1969), the calculations of Weiss and Truran (1990) predict that the abundance levels of ^{22}Na and ^{26}Al formed can be 1 to 2 orders of magnitude larger than for the case of matter of initial solar composition. They conclude that such neon-rich novae may represent an important source of ^{26}Al in our Milky Way Galaxy. Future observations of the distribution of ^{26}Al in the Galaxy may provide constraints upon the progenitors and, perhaps, indicate whether red giants, novae, or supernovae represent the dominant source.

The concentrations of ^{22}Na predicted for the ejecta of novae involving O-Ne-Mg white dwarfs are sufficiently high that we may also expect relatively nearby novae to produce detectable flux levels of ^{22}Na decay gamma rays. Weiss and Truran (1990) estimated that a nova known to be both enriched in neon and magnesium and at a distance of 500 pc may yield a gamma ray flux at a level of $\sim 6 \times 10^{-5}$ cm^{-2} s^{-1}. Such novae should be considered as promising targets for the Gamma Ray Observatory. (Note that, in contrast, neither slow novae nor fast novae enriched only in carbon, nitrogen and oxygen are likely to produce detectable fluxes.) Livio et al. (1990) also argued that the Compton degraded gamma rays from ^{22}Na decay might contribute to the observed X-ray fluxes from several recent novae (Ögelman et al. 1987).

We have emphasized the importance of the detection of ^{22}Na gamma rays from novae as a means of imposing important constraints on theoretical models of nova outbursts and their role in galactic nucleosynthesis. In this context, however, it is important to recognize that the ^{22}Na concentrations produced in novae are quite generally insufficient to account for the observed abundance of its decay product ^{22}Ne in the Galaxy. Truran and Hillebrandt (1986) showed that the observational upper limit on the gamma-ray flux (Mahoney et al. 1982,1984) is such that the total amount of ^{22}Na ever formed constitutes less than $\sim 1\%$ of the mass of ^{22}Ne in the Galaxy today. While novae and other astrophysical events may make interesting amounts of ^{22}Na with respect, for example, to the generation of Ne-E anomalies in meteorites, most of the mass of ^{22}Ne in the Galaxy is actually incorporated into the ISM directly as ^{22}Ne. The relative levels of overproduction of ^{22}Na and ^{26}Al in novae, as predicted by recent calculations, are consistent with the conclusion that novae may contribute to galactic nucleosynthesis of ^{26}Al but not to that of ^{22}Ne.

Significant production of ^{7}Li can also occur in the thermonuclear runaways that define the outbursts of classical novae (Arnould and Norgaard 1975; Starrfield et al. 1978). The mode of ^{7}Li production here is similar to that in red giants: ^{7}Be is carried outward by convection to cooler regions of the envelope on a sufficiently rapid time scale to ensure that it will not be destroyed via ^{7}Be (p,γ) ^{8}B. The total mass of ^{7}Li ejected is a sensitive function of the conditions achieved in the outburst and may therefore be expected to vary from event to event. Model calculations (Starrfield et al. 1978) predict that ^{7}Li enrichments of factors of up to $\sim 10^2$ to 10^3 may characterize the ejecta of novae under some circumstances. The critical dependencies with regard to ^{7}Li production are on the temperature history of the matter and the convective character of the envelope over the course of the runaway (both of which may be a function of the speed class of the nova); the initial ^{3}He concentration of the envelope matter is also of interest, as much of the ^{3}He may be converted readily to ^{7}Be.

The integrated contributions of classical novae to the abundance of ^{7}Li in galactic matter cannot be estimated very reliably because of uncertainties in our knowledge of the properties of the progenitors of classical nova systems and of their rate of formation over the past history of the Galaxy. Nevertheless, a statement regarding production of ^{7}Li in novae is possible. The theoretical models by Starrfield et al. (1978) predict average enrichment factors for the rare isotopes ^{13}C and ^{15}N that are ~ 10 times that of ^{7}Li. If these relative production ratios are representative of classical nova systems, it must follow that novae can account at best for only $\sim 10\%$ of the ^{7}Li in the Galaxy (when one assumes that ^{13}C and ^{15}N have their origin entirely in nova events). Further detailed calculations of nucleosynthesis coupled to hydrodynamic models of novae are required to clarify the situation for ^{7}Li.

IV. ULTRAVIOLET/INFRARED SPECTRA OBSERVATIONS OF HEAVY NUCLIDES

A. Gas-Phase Abundances

The study of nova ejecta is central to the understanding of the nature of the outburst. Gaseous shells having a range of mass from $M_{ej} \sim 10^{-7}$ to 10^{-3} $M_\odot$ are ejected with velocities of 4×10^2 to 10^4 km s^{-1} from each outburst, and this material may be an important nucleosynthesis source for certain elements and/or isotopes in the ISM. The fact that some novae form considerable quantities of dust soon after outburst may also have consequences for star formation in the galactic disk. The relative fraction of the ejected mass that precipitates out of the gas into grains is difficult to ascertain but, judging from the sudden decrease in the line emission from certain elements at the time that dust condensation occurs, it is substantial for some novae.

Elemental abundances in nova ejecta can, in principal, be determined by two entirely different procedures. For several days following the outburst, some nova envelopes exhibit an absorption spectrum from an expanding pseudophotosphere, that can be analyzed in the same manner as a stellar atmosphere. Because of the high lack of equilibrium and rapidly changing environment of the envelope, most of the normal atmospheric techniques are not valid, and they have produced inconsistent results (Williams 1977). Consequently, few such studies were attempted in the last decade.

A more effective approach to determining element abundances is to analyze the emission spectrum of the ejecta that typically appears within a week of the outburst. This spectrum shows many similarities to the spectra of nebulae and emission line galaxies. Although the conditions giving rise to the emission lines are very complicated immediately after outburst, in a short time a statistical equilibrium is established and the lines become optically thin, greatly simplifying the radiative transfer.

All emission line fluxes depend directly on the ion abundance and another parameter: either N_e, T_e, or the continuum radiation intensity, depending upon the particular excitation mechanism. Relatively few lines are excited by radiative excitation, especially when the radiation field weakens more than a few weeks after outburst, and if one considers the relative strengths of other lines; i.e., for those formed by collisional excitation and recombination, the dependence on N_e cancels, leaving T_e as the only remaining parameter. Thus, the line ratios of most lines will directly yield ion abundances if the electron temperature is known. One can even avoid a dependence of line ratios on T_e if one selects lines originating from similar ionization potentials. Fairly reliable abundances can therefore be determined for some ions. Reliable element abundance determinations require that one account for all of the ions of each element, even though many of the ions do not have observable lines in accessible regions of the spectrum. One is therefore forced to compare relative abundances of ions having similar ionization potentials, and to assume that these ion ratios reflect the relative element abundances. An alternative to this is to try to build a complete ionization model for the ejecta and to make theoretical corrections for unobserved ionization stages. As the post-outburst geometry of the ejecta is poorly known, but obviously complicated, reliance on models for these corrections is inadvisable. Thus, the operative procedure has been to take several ionization stages in elements in which the ionization potentials are similar and to equate the element abundances to those of the ions, i.e., to assume that N/O = (N III + N IV)/(O III + O IV), for example. If the above procedure can be applied to collisionally excited lines with similar excitation potentials, there is little dependence on T_e, and the resulting abundances are less uncertain.

B. Infrared Fine-Structure Lines

Many novae develop infrared emission lines from transitions among the fine-structure levels of the ground states of ions of the heavy elements. Although lines such as [Ne II] λ12.8 μm do occur, and can be dominant sources of cooling of the ejecta in some novae (see Gehrz et al. 1985,1986), it is more common that higher ionization stages predominate in the infrared spectra, e.g., [Mg VIII]λ3.02 μm and [Si IX]λ3.92 μm (Grasdalen and Joyce 1976; Greenhouse et al. 1988,1990).

Because of the very low excitation potentials of the infrared lines, $kT_e >$ χ_{12} for all such transitions, and therefore the line intensities do not have a dependence upon T_e. Relative strengths of the infrared lines yield reliable ion abundances, since they are probably all collisionally dominated, leading to the same dependence upon density. Intercomparison of these lines produces relative abundances of Mg, Na, Al and Si.

C. Physics of the Infrared Coronal Phase

Infrared spectroscopic observations of infrared forbidden-line-structure lines were identified in the six recent novae V1500 Cyg 1975, Cyg 1986, QU Vul 1984, OS And 1986, V827 Her 1987 and V2214 Oph (Grasdalen and Joyce 1976; Gehrz et al. 1985; Greenhouse et al. 1988,1990; Benjamin and Dinerstein 1990) confirming a theoretical prediction by Ferland and Shields (1978a, b) that such emission could be an important source of cooling in nova shells. The strongest of these lines was 12.8μm [Ne II] emission from QU Vul (Gehrz et al. 1985). These data showed that infrared spectroscopic information can be important for the identification of, and the estimation of abundances of atomic species in nova ejecta. Greenhouse et al. (1988,1990) established the importance of the infrared coronal emission phase in classical novae for estimating physical conditions in the ejecta, and for determining important chemical abundances related both to nucleosynthesis associated with the nova progenitor and to nucleosynthesis occurring in the thermonuclear runaway during the nova eruption. While the infrared transitions themselves are collisionally excited, the ionization states responsible for the transitions must be largely radiatively excited (Benjamin and Dinerstein 1990; Williams 1991). In some cases, it seems possible that a collisional excitation component may be required to explain the ionization structure (see Greenhouse et al. 1990). Temperature conditions associated with the coronal emission are not fully understood. On one hand, excitation temperatures inferred for the photoionization of the coronal states are as high as 5×10^5 to 10^6 K. On the other hand, the absence of certain lines that would be expected to occur suggests temperatures well below this. Infrared speckle interferometric measurements show that the coronal emission occurs in the principal ejecta rather than close to the central engine. High-resolution near-infrared spectroscopy shows that the velocity widths of coronal lines are likewise consistent with their production in the principal ejecta.

Analyses of the abundances of the coronal species show that the ejecta of many novae are rich in O, Ne and Mg. For example, the infrared coronal emission lines in QU Vul (Gehrz et al. 1985; Greenhouse et al. 1988) lead to the conclusion that some novae may be overabundant in neon and silicon

(Gehrz et al. 1984,1986; Gehrz et al. 1985; Williams et al. 1985; Gehrz et al. 1988). Starrfield et al. (1985*b*) and Truran and Livio (1986) argued that these novae result from accretion of matter onto O-Ne-Mg white dwarfs that are the evolutionary endproduct of intermediate-mass (8 to 12 $M_\odot$) stars. The nucleosynthesis associated with such nova explosions can produce significant quantities of the radioactive isotopes ^{22}Na and ^{26}Al that positron decay to ^{22}Ne (2.7 yr) and ^{26}Mg (3×10^5 yr). These materials could be trapped in grains during the dust formation phase of novae. Grain inclusions with significant overabundances of ^{22}Ne and ^{26}Mg have been discovered in solar system meteorites. The inference is that these meteoritic inclusions could have been produced by the condensation of grain materials in the immediate vicinity of sources of explosive nucleosynthesis, such as novae or supernovae (see Clayton 1982; Truran 1985,1990; Gehrz 1988; Weiss and Truran 1990). Theoretical studies suggest that these abundances may reveal important information about nucleosynthesis during both the outburst and or the evolution of the progenitor.

D. Gas Abundances in Nova Ejecta

There are now a number of novae that have experienced outbursts in the last decade, for which post-outburst abundances were determined by the methods described above (Snijders et al. 1987; Williams 1985; Snijders 1990). Because such studies are based on emission-line fluxes, they sample only the gas, and not the dust. The relative dust/gas ratio is difficult to determine, and therefore the gaseous depletion of certain elements from the gas phase by condensation into dust is unknown. In some novae that form optically thick dust shells, the evolution of the emission spectra of certain elements suggests that a substantial fraction of some elements may precipitate into dust in the months after outburst. The gas-phase abundances may therefore not reflect the actual distribution produced by the outburst and injected into the ISM. It is usually the case, however, that thermal emission from dust is absent from the infrared spectra of novae when infrared forbidden-line emission is important, so that one can presume that the emission line phase refers to an epoch when the condensible elements are still mainly in the gas phase. Nonetheless, because evidence for extensive dust formation occurs in only ∼1/3 of all novae, the gaseous abundances of a sample of many novae should, in fact, characterize the true ejecta abundances rather well. A compilation is given in Tables II and III of gas-phase abundances for those novae whose emission spectra were analyzed in the past 15 yr. In spite of the uncertainties associated with the results for individual objects, certain features are clear that may be taken to be general characteristics of nova outbursts:

1. He is generally enhanced with respect to H in the ejecta; i.e., H is depleted.
2. The CNO nuclei are highly enriched, with N typically being the most abundant of these elements.

TABLE II

Heavy-Element Abundances in Novae

Object	Year	Ref[a]	Mass Fractions											
			H	He	C	N	O	Ne	Na	Mg	Al	Si	S	Fe
RR Pic	1925	1	0.53	0.43	0.0039	0.022	0.0058	0.011	—	—	—	—	—	—
HR Del	1967	2	0.45	0.48	—	0.027	0.047	0.0030	—	—	—	—	—	—
T Aur	1891	3	0.47	0.40	—	0.079	0.051	—	—	—	—	—	—	—
PW Vul	1984	4	0.69	0.25	0.0033	0.049	0.014	0.023	—	—	—	—	—	—
V1500 Cyg	1975	5	0.49	0.21	0.070	0.075	0.13	0.023	—	—	—	—	—	—
V1668 Cyg	1978	6	0.45	0.23	0.047	0.14	0.13	0.0068	—	—	—	—	—	—
V693 CrA	1981	7	0.29	0.32	0.046	0.080	0.12	0.17	0.0016	0.0076	0.0043	0.0022	—	—
GQ Mus	1983	4	0.27	0.32	0.016	0.19	0.19	0.0034	—	0.0014	0.00056	0.0028	0.0016	0.00047
DQ Her	1934	8	0.34	0.095	0.045	0.23	0.29	—	—	—	—	—	—	—
V1370 Aql	1982	9	0.053	0.088	0.035	0.14	0.051	0.52	—	0.0067	—	0.0018	0.10	0.0045

[a] 1: Williams and Gallagher 1979; 2: Tylenda 1978; 3. Gallagher et al. 1980; 4: Salzar et al. 1990; 5: Stickland et al. 1981; 6: Williams et al. 1985; 7: Williams et al. 1978; 8: Snijders et al. 1987.

3. Perhaps as many as 33% of the novae show a strong enrichment of neon. These "neon" novae usually also show enhancements of elements with $Z > 10$, e.g., Mg and Si.

The first two of these characteristics can, in principle, derive in part (e.g., H depletion and high N) from the outburst phenomenon, as these features can follow naturally from the proton-capture reactions that occur in the thermonuclear runaway. The high-neon abundance is difficult to produce in normal outburst models on CO white dwarfs, and is believed to result from material mixed from O-Ne-Mg white dwarfs in those novae (Starrfield et al. 1986; Truran and Livio 1986). Moreover, the high-CNO abundance enrichments also demand outward mixing of core matter (Truran 1990; Livio and Truran 1990).

Spectroscopy of the emission lines from ions does not enable isotopic abundances to be determined; the isotope shifts are too small to be detected. Thus, the neon could be ^{20}Ne or ^{22}Ne, and the aluminum could be ^{26}Al or ^{27}Al. Any isotopic information must be deduced from theory or, perhaps, as in the case of ^{22}Na, from gamma-ray observations of novae after outburst. The direct evidence for the gaseous contribution of novae to the element abundances of the ISM is limited to the information in Tables II and III. Gas of this composition is ejected into the Galaxy by roughly 30 novae per year, each of which contributes about 2×10^{-5} $M_\odot$ of gas at velocities (see Table IV) usually exceeding the escape velocity of the Galaxy ($\sim$300 km s^{-1} for a galactocentric distance of 10 kpc). Thus, the ejecta must interact with a greater mass of ambient gas if they are to be confined to the Galaxy, eventually settling into the ISM.

V. INFRARED OBSERVATIONS OF DUST FORMED FROM HEAVY NUCLIDES

The most convincing cases for transient circumstellar dust formation are those based upon infrared observations of the temporal development of classical nova systems; see Bode 1988,1989; Bode and Evans 1989; Gehrz 1988; Starrfield 1989a for reviews of these observations. Their infrared temporal development progresses in several identifiable stages (Gehrz 1988). The initial eruption results from a thermonuclear runaway on the surface of a white dwarf that has been accreting matter from a companion star through the inner Lagrangian point in a close binary system. The hot gas expelled in the explosion is initially seen as an expanding pseudophotosphere, or "fireball." Free-free and line emission are observed when the expanding fireball becomes optically thin. A dust condensation phase, characterized by declining visual light and rising infrared emission, occurs in many novae within 50 to 200 days following the eruption. The infrared emission continues to rise as the grains grow to a maximum radius of 0.2 to 0.3 μm within a few hundred days after their condensation, and then falls as the mature grains are dispersed by the

TABLE III
Helium and Heavy-Element Abundances in Novae

Nova		He/H	Z	Ref.[a]	Enriched Fraction
T Aur	1891	0.21	0.13	2	0.36
R Pic	1925	0.20	0.039	1	0.28
DQ Her	1934	0.08	0.56	1	0.55
CP Lac	1936	0.11 ± 0.02	—	1	0.08
RR Tel	1946	0.19	—	1	0.24
DK Lac	1950	0.22 ± 0.04	—	1	0.30
V446 Her	1960	0.19 ± 0.03	—	1	0.24
V553 Her	1963	0.18 ± 0.03	—	1	0.23
HR Del	1967	0.23 ± 0.05	0.077	1	0.35
V1500 Cyg	1975	0.11 ± 0.01	0.30	1	0.34
V1668 Cyg	1978	0.12	0.32	3	0.38
V693 Cr A	1981	0.28	0.38	4	0.61
V1370 Aql	1982	0.40	0.86	5	0.93
GQ Mus	1983	0.29	0.42	6	0.64
PW Vul	1984	0.09	0.067	7	
V1819 Cyg	1986	0.19	—	8	0.24

[a] 1: Ferland 1979; 2: Gallagher et al. 1980; 3: Stickland et al. 1981; 4: Williams 1985; Williams et al. 1985; 5: Snijders et al. 1987; 6: Hassal et al. 1990; 7: Salzar et al. 1990; 8: Whitney and Clayton 1989.

outflow into the ISM (Gehrz et al. 1980a, b). Thus, nova grains initially grow much larger than interstellar grains, which typically have radii in the range from 0.01 to 0.12 μm (see Greenberg 1989a). However, the rate of decline of the infrared radiation and the relatively slow cooling rate of the grains in the outflow suggest that the grains begin to decrease in radius again shortly after having grown to their maximum size. The observations are consistent

TABLE IV

Summary of Dust Formation in 21 Recent Classical Novae[a]

Name	Year	V_o (km s⁻¹)	Dust Optical Depth $\tau_d = \dfrac{(L_{IR})_{max}}{(L_o)_{max}}$	t_d for IR max (days)	Shell Density at R_c (g cm⁻²)	Dust Types Formed[c]	t_3[e] (days)	M_{gas} (M⊙)	M_{dust} (M⊙)	Gas to Dust Ratio M_{gas}/M_{dust}
FH Ser	1970	560	0.50	60	—	C	62	*	*	*
V1229 Aql	1970	575	0.55	—	—	C	37	*	*	*
V1301 Aql	1975	—	—	—	—	C	35	*	*	*
V1500 Cyg[b]	1975	1180	?	100/100	3×10^{-17}	—	3.6	5×10^{-6}–8×10^{-4}	*	*
NQ Vul	1976	750	1.00	32/80	2.3×10^{-15}	C	65	10^{-4}	2×10^{-7}	520
V4021 Sgr	1977	—	≤0.60	—	—	C	70	*	*	*
LW Ser	1978	1250	0.70	23/75	3.0×10^{-16}	C	50	2×10^{-5}	3.6×10^{-7}	56
V1668 Cyg	1978	1300	0.08	33/57	7.6×10^{-17}	C	23	2×10^{-5}	2×10^{-8}	800
V1370 Aql[d]	1982	2800	0.50	≤16/≤37	—	C,SiC,SiO₂		*	*	*
GQ Mus	1983	600	no dust	—	≤10^{-17}	no dust	45	≤2.6×10^{-6}	*	*
PW Vul	1984 #1	285	3×10^{-3}	152/≤280	2.0×10^{-17}	C	97	≤3.2×10^{-6}	5.1×10^{-10}	6.3×10^{-3}
QU Vul[b]	1984 #2	1–5000	3×10^{-3}	40–200/240	—	SiO₂	40	*	10^{-8}	*
OS And	1986	900	no dust	—	1.1×10^{-17}	no dust	22	*	*	*
V1819 Cyg[b]	1986	1000	no dust	—	—	no dust	87–104	*	*	*
V842 Cen	1986	1200	1.00	36/87	—	C,SiO₂,HC	48	*	*	*
V827 Her[b]	1987	1000	0.10	43/83	—	C	55	*	*	*
V4135 Sgr	1987	500	—	—	—			*	*	*
QV Vul	1987	700	1.00	56/115	2.0×10^{-16}	C,SiO₂,HC	30	3×10^{-5}	3.4×10^{-8}	890
LMC 1988 #1	1988	800	0.06	59/57	—	C?	43	*	*	*
LMC 1988 #2	1988	1500	—	—	—	—	15	*	*	*
V2214 Oph	1988	500	—	—	—	—	73	*	*	*

* Insufficient data to provide a definitive estimate. [a] Data from Gehrz (1988, references in his Figs. 2, 3), Gehrz (1990), and IAU Circulars. [b] Infrared coronal emission lines recorded. [c] HC = Hydrocarbons. [d] The 7–14 μm emission may have been caused by a combination of SiC and SiO₂. [e] t_3 from Duerbeck (1987, and personal communication [1990]).

with the hypothesis that the nova grains could be processed to interstellar grain sizes by evaporation or sputtering before they eventually reach the ISM. About 10^{-8} to 10^{-6} $M_\odot$ of dust forms in each episode.

Although most novae produce carbon (possibly iron?) dust, recent observations suggest that nova explosions can produce every other type of astrophysical grain, including silicates, silicon carbide (SiC), and hydrocarbons (see Gehrz 1988,1989a, b; Hyland and MacGregor 1989). Recently, it has been shown that nova shells can be strongly cooled by infrared fine structure forbidden-line emission (Gehrz et. al. 1985).

A. Abundances from Observations of Dust

Table IV summarizes our knowledge of outburst and shell parameters for 21 recent novae from optical/infrared observations. The mass of gas in the ejecta can be measured using the energy distribution, as determined by infrared techniques. As the pseudophotosphere begins to become optically thin, the energy distribution deviates slightly from that of a blackbody, but the ejecta are still hot enough that the Thomson (electron) scattering opacity κ_t dominates radiative transfer in the shell. In this case, the shell mass is given by $M = \pi R^2 \kappa_T^{-1}$ where the shell radius $R = tV_o$ and t is the time since outburst. Later, when the ejecta cool sufficiently that thermal Bremsstrahlung dominates the shell opacity, the shell mass of a shell of depth ℓ pc and electron density n_e cm^{-3} can be determined through the relationships $n_e^2 \lambda_c^2 l = 10^{18}$ and $M \sim \pi R^2 l n_e H$ where λ_c is the cutoff wavelength in μm at which free-free self-absorption becomes important, and H is the mass of the nucleon. The mass of dust can be inferred from the dust shell opacity required to produce the observed visual extinction for novae that produce optically thick shells. Opacities for carbon, SiC, silicates and iron are given by Gilman (1974a, b) and Draine (1985b). The gas to dust ratios calculated in this way imply overabundances of carbon and silicate materials in a few cases.

B. Mineral Composition of the Grains

Although most novae condense carbon or iron dust, there is evidence that astrophysical dust of every known chemical and mineral composition can condense in nova ejecta. QU Vul 1984, V842 Cen 1986 and QV Vul 1987 (Gehrz 1990) formed silicate grains; V1301 Aql 1975 may have formed SiC or a combination of SiC and silicates (Gehrz et al. 1984), and V842 Cen 1986 (Hyland and Macgregor 1989) and QV Vul 1987 (Greenhouse et al. 1988,1990) apparently formed hydrocarbons. The hard radiation from the nova remnant may provide the ultraviolet radiation that is believed to be important in creating hydrocarbons from amorphous carbon (Allamandola 1984; Allamandola et al. 1987).

The formation of more than one type of grain has now been recorded for at least three novae (Gehrz 1990). Both V842 Cen 1986 and QV Vul 1987 apparently formed carbon and silicates dust, and hydrocarbons. It is difficult to understand how both silicate grains and carbon grains can form co-spatially in the ejecta if silicate grain condensation requires an environment where C < O and carbon grain condensation requires an environment with C > O (Hackwell 1971,1972). One possibility is that there are significant abundance, temperature and density gradients within the ejecta. Perhaps the polar plumes and equatorial ring that are believed to be produced in the explosion (Gallagher and Starrfield 1978) have different properties.

C. Continuum Emission from Dust

Most dust-producing novae condense dust that emits a gray or black, smooth 1 to 25 μm continuum that is devoid of emission features. Both carbon and iron grains can produce this spectral signature (Gilman 1974a, b). Because a similar continuum is characteristic of the circumstellar emission from carbon rich stars (see, e.g., Gehrz and Hackwell 1976), it has generally been assumed that this continuum in classical nova systems is produced by amorphous carbon grains (see Gehrz 1989a). The genesis of the argument that smooth infrared continua result from circumstellar carbon dust in carbon stars is an equilibrium condensation calculation showing that carbon grains condense if C > O and silicate grains condense if O > C (Gaustad 1963, Hackwell 1971,1972). A fundamental problem with the carbon grain hypothesis for novae is that it is unlikely that C > O can occur in nova ejecta that result from thermonuclear reactions on CO and O-NE-Mg white dwarfs (Truran 1990); theoretical models generally predict that C < O for runaways on both CO and O-Ne-Mg white dwarfs. If this is the case, the formation of carbon grains seems improbable, unless the condensation of carbon proceeds in a rather different manner than is predicted by the equilibrium calculation.

An alternative explanation for the continuum dust emission in novae is that the grains are iron. We can use the opacities and Planck mean cross sections given by Gilman (1974a, b) to estimate the mass of iron dust that would be required to produce the observed visual extinction in a typical optically thick nova shell. As an example, we consider NQ Vul 1976, a nova that formed a shell with a visual optical depth of $\tau_v = 4.6$ about 80 days after the eruption; the shell involved 10^{-4} $M_\odot$ of gas expanding at 750 km s^{-1}. In this case, the observational parameters require a dust mass of 10^{-6} $M_\odot$ of iron or 1.4×10^{-7} $M_\odot$ of carbon, leading to gas-to-dust ratios of 100 and 700 for iron and carbon, respectively. The gas-to-dust ratios for iron and carbon in solar material are 670 and 250, respectively. We conclude that NQ Vul's ejecta would have had to have been enhanced in iron by at least a factor of 7, if the dust were all iron grains. The ejecta of one dust forming nova, V1370 Aql 1982, may have been enhanced in iron by a factor of 4 (see Tables II and III), although there is no obvious way to account for such an iron enrichment in the context of the standard thermonuclear runaway model. No enhancement of carbon is implied by the visual extinction in NQ Vul, but at least one dust forming nova, LW Ser 1978, would require a carbon enhancement of a factor of ≥ 5 over solar. We conclude that the existing evidence suggests that the dust responsible for the continuum radiation in novae is most likely carbon.

D. THE CONDENSATION PROCESS

A significant question is how can grains nucleate and grow in the hard radiation field of the central engine. It is generally accepted that the dust must form in clumps of material that shield the grains from extreme-ultraviolet radiation (see Rawlings 1988; Martin 1989; Bode and Evans 1989). The visual lightcurve of V842 Cen 1986 provided some interesting circumstantial evidence for clumps in the ejecta. The very deep and short-lived minimum during the transition phase of that nova could have been caused by a dense dust knot that obscured the line of sight for a very short time (Gehrz 1990). If clumps are the major sites of dust production, the abundances derived above may be significantly in error. Albinson and Evans (1987) argued that carbine whiskers can form under some conditions. Perhaps iron is at least an important factor in producing seed nuclei, as suggested by Lewis and Ney (1979).

A novel suggestion by Matese et al. (1989) is that the outburst can release a significant amount of pre-existing dust from comet nuclei in a cometesimal disk and that this material provides seed nuclei that then accrete material in the ejecta. Stern and Shull (1990) determined that up to 10^{-3} $M_\odot$ of dust may return to the ISM per year from this source. Dust released from planetesimals in sufficient quantities could cause a sudden obscuration of the visible light from the remnant; it can be shown that the opacity of a given circumstellar mass of solid planetesimals increases as a^{-1} where a is the radius of the average planetesimal (Gehrz 1989b).

VI. ABUNDANCES OF HEAVY NUCLIDES AND COMPARISON WITH THEORY

Theoretical studies show that Li, CNO and Ne-Na-Mg-Al are expected to be enhanced by substantial factors over solar abundances in runaways on CNO and O-Mg-Ne white dwarfs, as discussed in Sec. III. Elemental abundance data are now becoming increasingly available for a number of classical nova systems (for reviews, see Truran 1985; Truran and Livio 1986; Williams 1985; Truran 1990). The abundance data concerning hydrogen, helium and some heavy elements are summarized in Tables II and III. Table II presents, specifically, the mass fractions (where known) in the form of the elements hydrogen, helium, carbon, nitrogen, oxygen, neon, sodium, magnesium, aluminum, silicon, sulphur and iron, adapted from the indicated references. Table III provides helium-to-hydrogen ratios for a somewhat larger sample of novae. Here again, where known, the total mass fractions Z in the form of heavy elements are also tabulated. For purposes of comparison, we note that solar system matter is characterized by a helium-to-hydrogen ratio by number He/H = 0.08 and a heavy-element mass fraction Z solar = 0.019 (see

Cameron 1982,1992). The column labeled enriched fraction in Table III gives the total mass fraction of enriched matter in the form of both helium and heavy elements.

It is apparent that all classical novae with relatively reliable abundance determinations show enrichments relative to solar composition in helium and/or heavy elements. This statement is fairly secure even when the large uncertainties in the abundance determinations are taken into account. We conclude that envelope enrichment in classical novae is a very general phenomenon. Truran and Livio (1986) argued that it is extremely unlikely that the source of these enrichments is either the mass transfer from the secondary star or nuclear transformations accompanying the outburst, and concluded that the abundance patterns suggest that some fraction of the envelope is dredged up from the underlying white dwarf. Given the presence of large concentrations of neon and heavier elements in the ejecta of several recent novae (e.g., V693 CrA 1981 and V1370 Aql 1982), it is interesting to note that this conclusion has the consequence that massive O-Ne-Mg white dwarfs (as opposed to CO white dwarfs) occur quite frequently in classical nova systems.

The important question that immediately arises is what is the mechanism that is responsible for the mixing between the accreted envelope and the white dwarf core? This was addressed in the recent paper by Livio and Truran (1990). Four possible mechanisms were suggested: (1) diffusion induced convection; (2) shear mixing; (3) convective overshoot-induced flame propagation; and (4) convection-induced shear mixing. Livio and Truran (1990) discuss each of these mechanisms, pointing out their strengths and weaknesses, and attempt to identify critical observations that will help to determine which mixing mechanism is most likely to dominate.

VII. INTEGRATED CONTRIBUTIONS ON LOCAL AND GALACTIC SCALES

Estimates of the enrichment of ISM material by classical nova ejecta are important for understanding possible inputs to the primitive solar nebula. We discuss below the general concept of ISM enhancement by ejections from novae, and review specific theoretical predictions for the capability of novae to produce enhancements of specific nuclides.

A. Global Processing of ISM Material by Classical Novae

Recent optical/infrared spectral observations and theoretical studies support the proposition that thermonuclear runaways on massive (1.1 to 1.3 $M_\odot$) O-Ne-Mg white dwarfs accreting matter in close binary systems account for $\sim$25% of the 30 classical novae that occur each year in our Galaxy. Assuming that each of these novae ejects $\sim 3 \times 10^{-5}$ $M_\odot$ per eruption, and that the Galaxy has an age of $\sim 10^{10}$ yr, we find that $\sim 10^7$ $M_\odot$ has been processed through nova eruptions of these massive novae during the lifetime of the Galaxy which is only $\sim$0.3% of the 3×10^9 $M_\odot$ of gas estimated to be in the present ISM.

We conclude that nova eruptions must give a yield of 100 to 300 times solar abundance of any given nuclide if they are to be a significant source of that nuclide on a galactic scale. It is, of course, possible that nova eruptions could be significant sources of overabundances of certain nuclides associated with star formation on a local scale where the ejecta could be concentrated into a molecular cloud shortly before a star formation event within the cloud (see, e.g., Cameron 1984). We show below that theoretical predictions suggest that novae can produce significant abundance anomalies if the eruptions occur on O-Ne-Mg white dwarfs.

B. Contributions to Specific Anomalies

The Ne-E and ^{26}Mg anomalies found in meteorites are especially interesting in the context of classical novae in view of the large amounts of ^{22}Na, and ^{26}Al that can be produced in runaways on O-Ne-Mg white dwarfs (Weiss and Truran 1990). *1. Ne-E.* ^{22}Na positron decays to ^{22}Ne via the reaction ^{22}Na $\rightarrow$ ^{22}Ne $+ e^+ + \gamma + \nu$, leading to the emission of 1275 KeV gamma rays that can be detected from Earth. The mean life of the reaction is only 3.9 yr, so that novae will be short-lived sources of gamma rays from ^{22}Na decay, and the galactic background should be very anisotropic, with strong peaks in the direction of the most recent novae. This short mean life also implies that ^{22}Ne trapped in grains probably requires that the parent ^{22}Na was trapped in the grains very shortly after their formation, probably at the site of formation. No detections of 1275 KeV gamma rays have been made to this date, but several upper limits were placed on the diffuse galactic emission and specific nova events. Leventhal et al. (1977) found that V1500 Cyg 1975 must have produced $<10^{-6}$ M$_\odot$ of ^{22}Na. Truran and Hillebrandt (1986) argued that the observational upper limit on the gamma-ray flux (Mahoney et al. 1982, 1984) is such that the total amount of ^{22}Na ever formed constitutes less than approximately 1 % of the mass of ^{22}Ne in the Galaxy today. Higdon and Fowler (1987) concluded that limits to the diffuse galactic 1275 KeV flux suggested a production of less than 6×10^{-7} M$_\odot$ of ^{22}Na per nova event. The calculations of Weiss and Truran (1990) discussed in Sec. III predict that the average yield of ^{22}Na per O-Ne-Mg outburst should be between 10^{-8} and 10^{-7} M$_\odot$. Under these conditions, the galactic background will be difficult to detect with current technologies, but NASA's Gamma Ray Observatory may be able to detect individual novae within 0.5 to 1 kpc of Earth.

   2. ^{26}Al. Classical novae appear to be a promising source of ^{26}Al, at least on local scales. There is chemical evidence that the ^{26}Al that produced the ^{26}Mg anomaly in meteorites was alive in the primitive solar system, implying that the parent nova (if a nova and not a red giant was, indeed, the source) was relatively nearby (see, e.g., the discussion by Cameron [1984]). Gamma radiation at 1809 KeV will be produced via the decay of ^{26}Al to ^{26}Mg by the reaction ^{26}Al $\rightarrow$ ^{26}Mg $+ e^+ + \gamma + \nu$ that has a mean life of 10^6 yr. This long mean life would enable ejecta from many eruptions of a single nova system to inject live ^{26}Al into the primitive solar system during its formation. The long mean life also implies that the galactic background at 1809 KeV from novae should be fairly isotropic, with components from disk and halo nova populations. Assuming that approximately 10 novae per year are O-Ne-Mg novae capable of injecting into the ISM 3×10^{-5} $M_\odot$ of material with a fraction by mass of ^{22}Al of 2×10^{-3}, the steady-state galactic mass of ^{26}Al from novae is about 0.4 $M_\odot$. The measured 1809 KeV background implies that there is $\sim$3.3 $M_\odot$ of ^{26}Al in the galactic ISM (see Higdon and Fowler 1989). Although novae therefore do not appear to be a dominant source of ^{26}Al on a galactic scale, they may contribute significantly to the ^{26}Al concentration on a local scale in the ISM.

Acknowledgments. RDG is supported by NASA, the United States Air Force, the National Science Foundation and the University of Minnesota Graduate School. JWT acknowledges support under a grant from the National Science Foundation and the hospitality of the Aspen Center for Physics where work on various portions of this chapter was completed.

FORMATION OF INTERSTELLAR CLOUDS AND STRUCTURE

BRUCE G. ELMEGREEN
T. J. Watson Research Center

Cloud formation is discussed for an interstellar medium that is clumpy, supersonic and magnetic. Two types of clouds, diffuse and self-gravitating, are distinguished by a dimensionless ratio of pressure to self-gravity. Diffuse clouds probably form by thermal instabilities and pressurized accumulation of low-density gas, and self-gravitating clouds probably form by condensation in dense regions, such as shells and spiral arms. Unbound clouds are a third structural type. Superclouds and cloud associations are the largest part of the hierarchy of cloud structures. They apparently form by coalescence, condensation and fragmentation in spiral arms, and then follow a standard sequence of events in which dissipation and collapse lead to star formation, disruption forms shells and giant bubbles, and then shell collapse forms more molecular clouds and stars. Eventually the gas disperses somewhere in the interarm region, after ~ 100 Myr of activity. The generally supersonic and supervirial velocities between clouds imply that most collisions are shattering if they occur at the full random speed. In this case the coagulation of small clouds into large complexes will be driven by collision-induced cooling instabilities and self-gravity in the cloud fluid rather than sticky collisions.If the mean collision speed scales with cloud mass, then random coalescence could be more important. Magnetic fields affect cloud formation by limiting the random motions of individual clouds, by transferring linear momentum between clouds without physical contact, by removing angular momentum from growing condensations, and by directly pushing, shocking and twisting the gas in nonlinear waves. Differences between astrophysical and laboratory turbulence are summarized.

I. DEFINING THE PROBLEM OF CLOUD FORMATION

Theories of interstellar cloud formation generally follow the available observations. When optical absorption lines in the 1930s to 1950s showed an average of eight interstellar features per kiloparsec (Blaauw 1952), the gas that caused this absorption was described as originating in well-separated clouds, sometimes assumed to be spherical, moving in a more or less ballistic fashion until they collided with other clouds (see, e.g., Oort 1954). Cloud formation was then viewed as a random coagulation process, with star formation disrupting the largest clouds to reform the smallest clouds. When the highest velocity features were found to have anomalous metallicities (Routley and Spitzer 1952), and to exceed in abundance the extrapolation from a gaussian velocity distribution (Munch 1957), as well as the extrapolation to large heights above the galactic plane based on an isotropic velocity distribution (Munch and Zirin 1961), a second model for their origin was proposed, involving expansion and shock fronts around early-type stars and supernova remnants (Spitzer 1968*b*; Siluk and Silk 1974).

These early theories became increasingly refined as the observations grew more detailed, first with the discovery of a warm, neutral intercloud medium (Clark 1965) leading to two-phase models based on thermal instabilities (Field et al. 1969) and cloud coagulation (Field and Saslaw 1965), and then with the discovery of the ionized (Reynolds et al. 1974) and hot (Jenkins and Meloy 1974; Burstein et al. 1976) intercloud phases, leading to frothy models (Brand and Zealey 1975) in which clouds form in swept-up shells around H II regions (Hills 1972; Bania and Lyon 1980) and supernovae (Cox and Smith 1974; Salpeter 1976; McKee and Ostriker 1977). At the same time, various large-scale instabilities in the interstellar gas (Goldreich and Lynden Bell 1965*a*; Parker 1966) were offered as explanations for the largest clouds and spiral-like features observed locally and in other galaxies. Similar formation theories involving random coagulation (Kwan 1979; Norman and Silk 1980*b*), pressurized shells (Tenorio-Tagle 1981; Olano 1982; Elmegreen 1982*a*; Franco et al. 1988*a*), and various instabilities (Elmegreen 1979; Blitz and Shu 1980; Cowie 1981) followed the discovery of giant molecular clouds in the mid-1970s. Reviews of the history of these theories, and of the many observations that stimulated them, can be found in Elmegreen (1987*a*,1990*a*), Kwan (1988), Larson (1988), Tenorio-Tagle and Bodenheimer (1989), Balbus (1990), and Franco (1990).

Observations of the interstellar gas now present a very different problem to solve. We see that the gas is highly textured (Low et al. 1984; Perault et al. 1985; Bally et al. 1987; Loren 1989*a, b*; Stutzki and Gusten 1990; Dickman et al. 1990; Falgarone et al. 1991) and although most of the mass is in the form of amorphous density concentrations that can still be named clouds or clumps, the geometries and juxtapositions of these concentrations defy mathematical description. According to Scalo (1990), the gas is not just a collection of isolated spheres, filaments and shells, as we formerly thought, but an interconnected network that has only been made to look like spheres, filaments and shells by inadequate resolution, undersampling, and line-of-sight overlaps (see also Deul and Burton 1990). This revision implies that the coagulation theories no longer have their isolated objects to coagulate, the shell theories have lost their uniform preshock media to compress, and the instability theories have been using an oversimplified equation of state.

We also know now that thermal motions are usually negligible compared to cloud line widths and the cloud-to-cloud dispersion (Blaauw 1952; Barrett et al. 1964; Radhakrishnan and Goss 1972; Myers and Benson 1983; Stark 1984), in which case the thermal instability, which requires thermal pressures to be important, may be relevant on small scales only in cold gas (Yoshii and Sabano 1980; Gilden 1984) and in the cooling regions behind shock fronts (Avedisova 1974; Mufson 1975; McCray and Stein 1975), and perhaps on large scales in the tenuous warm (Parravano 1987; Heiles 1989; Lioure and Chieze 1990) and hot (Bregman 1980; Lepp et al. 1985) phases of the intercloud medium (see review in Begelman 1990). Modifications of the thermal instability may have other applications, however, if stirring from supernovae and other sources is considered as a type of heating mechanism for bulk cloud motions, and if cloud collisions, shocks and other nonlinear magnetohydrodynamic processes are considered as macroscopic cooling mechanisms for these motions (Struck-Marcell and Scalo 1984; Tomisaka 1987; Elmegreen 1989*d*).

Evidently the first problems to solve, before detailed studies can be devoted to cloud formation, are to find mathematical descriptions for the distribution and motion of the interstellar gas, and to assess the relative importance of the various forces that regulate these motions, such as magnetic, gravitational, thermal pressure and inertial forces (i.e, the ram pressure from supersonic collisions or the $v \cdot \nabla v$ term in turbulence theory). Then models using the proper balance of forces can be tuned to match the descriptions of gas in various environments. Unfortunately, finding a description for the structure does not appear to be very simple, and finding an explanation for the origin of what is eventually described may be more difficult than describing it.

This chapter takes the point of view that the study of cloud formation can be divided into two parts, and that we have made progress in only one of these parts. These two parts are (1) the identification of physical processes that cause rarefied, generally clumpy gas to condense into big cloudy objects, and (2) the mathematical description of the structure of the condensations that form by these processes. The first of these parts has several reasonable solutions, but the second has hardly been approached.

A good example of how cloud formation can be understood in terms of a condensation theory but not in terms of a theory for the origin of structure arises for terrestrial clouds. There the formation process is known to involve the condensation of water droplets in convection cells that rise to where the temperature falls below the dew point. But the origin of the structure inside these clouds, the convoluted boundaries and wisps, is not understood in detail, and may be more a result of droplet segregation in a turbulent medium than droplet formation (Lovejoy 1982). Similarly, in the interstellar medium, the growth of various instabilities and other condensation processes can be described using reasonable assumptions about heating, cooling and magnetic effects, but these theories alone do not predict the structures of the resulting density enhancements, even if the statistical mass distribution function for the condensations and their idealized shapes (e.g., clumps, filaments, etc.) can be determined from first principles. Understanding the structure and dynamics together seems to require computer models in which the instabilities are followed in time with a resolution comparable to that of the observations. Such models are in progress for cosmological studies (see, e.g., Park 1990), but not yet for interstellar theory.

The purpose of this review is to summarize the condensation and coagulation theories for interstellar cloud formation, which is the first part of the division suggested above, and, for the second part of the division, to discuss briefly the connection between interstellar structure and laboratory turbulence. One hopes that these two parts are eventually made indistinguishable by one comprehensive theory.

Section II begins by highlighting the physical differences between diffuse and self-gravitating interstellar clouds, based on the dimensionless ratio of external pressure to internal gravitational binding-energy density, and suggests a difference in their formation routes. This section also notes that diffuse clouds are not always atomic and self-gravitating clouds are not always molecular, even though this correspondence usually appears in the solar neighborhood.

The subsequent sections summarize various implications for cloud formation theories that follow from three observed features: the apparent hierarchical structure of the gas (Sec. III), the prevalence of supersonic motions (Sec. IV), and the ubiquitous magnetic field (Sec. V). Section VI then discusses scenarios for cloud and star formation. All of these scenarios are made very uncertain by the supersonic and magnetic nature of the interstellar gas and by the role of spiral density waves.

The exact nature of the random motions, or "astrophysical" turbulence, that is observed, is also unknown, as discussed in Sec. VII. It seems likely that the bulk of the interstellar medium is turbulent, but not in the same way as a laboratory fluid because strong magnetic fields limit vorticity and the random diffusion of clumps, because there are no hard boundaries as in a laboratory, and because the kinetic energy is applied and dissipated on all scales, without a clear cascade from large to small scales as in Kolmogorov turbulence. Gravity and embedded energy sources such as stars are also much more influential in astrophysical turbulence than in the laboratory.

II. DEFINING THE STRUCTURE OF CLOUDS

A. Three Structural Types

For the purposesof this discussion, interstellar clouds will be categorized into three structural types: diffuse, self-gravitating and unbound. The difference between the first two depends on the relative importance of pressure and self-gravity for confinement. Unbound clouds are not confined by either pressure or gravity, and are presumably in the process of dispersing. This distinction between structural types is convenient for studies of cloud formation because the diffuse and self-gravitating types probably have different formation mechanisms. Structurally diffuse clouds should form by processes in which pressure is important, such as shock accumulation and thermal instabilities. Self-gravitating clouds should form where gravity is important, as in galactic spiral arms, dense shells and diffuse regions that have become cold or quiescent as a result of dissipation. Unbound clouds may form in the same way as diffuse clouds, by the pressurized accumulation of gas, for example, but the high pressure that formed them must have disappeared very recently. All of these cloud types can presumably grow by coagulation of other clouds too (Sec. IV).

A single dimensionless parameter is used to distinguish between diffuse and self-gravitating clouds. This parameter involves the pressure P external to the cloud, and the average mass column density inside the cloud σ_c. The origin of the external pressure is unknown, but it is probably a combination of thermal pressure and ram pressure from external flows; it may be highly variable with time and position. The column density is also poorly defined because it depends on an arbitrary choice of cloud boundaries. Even with these uncertainties, the dimensionless ratio $P/G\sigma_c^2$ is a convenient measure of the relative importance of gravity and external pressure in binding a cloud together. If $P/G\sigma_c^2$ is much larger than 1, then self-gravity is relatively unimportant for a cloud and the shape and size should be governed primarily by external flows and pressure gradients. Such clouds should also have a more or less uniform average density inside their boundaries (although with clumps and filaments), and a structureless, or diffuse shape. A "standard cloud" for example (Spitzer 1978) has a value of $P/G\sigma_c^2 \sim 30$. Interstellar condensations with small values of $P/G\sigma_c^2$ may be significantly self-gravitating (if they are not unbound), and consequently stratified by internal pressure and average-density gradients. Most molecular clouds in our Galaxy are probably of this latter type, having values of $P/G\sigma_c^2$ between 0.01 and 0.1. Because of the importance of self-gravity, these are the clouds where stars form. We refer to them here as self-gravitating clouds (not molecular clouds) to emphasize their physical (not chemical) state.

Most likely, real clouds span a wide range of $P/G\sigma_c^2$ with no bimodal separation between diffuse and self-gravitating types. Indeed, Dickey and Garwood (1989) find a continuous mass spectrum for clouds, with only a slight change in slope to indicate the transition from diffuse to self-gravitating types.

Our definition of diffuse clouds differs from the operational one in which they are taken to be any cool objects observed in absorption. For example, the CS absorbing clouds on the line of sight to W49A were labeled diffuse by Miyawaki et al. 1988, even though these clouds are rather dense (100 cm^{-3}) and may be self-gravitating. Similar dense clouds were studied in absorption by Cardelli et al. (1990). Diffuse clouds that are more opaque than those commonly observed at 21 cm and in the ultraviolet are sometimes called translucent (van Dishoeck and Black 1989). If these clouds are also influenced more by external pressures than internal gravity, then our structural classification system still places them in the diffuse category, although they may be transition cases ($P/G\sigma_c^2 \sim 1$). The distinction between diffuse and dark molecular clouds made by Myers (1989) is analogous to our distinction between diffuse and self-gravitating clouds, although Myers considers only low-mass clouds ($<10^4$ M$_\odot$), which generally look dark when they are self-gravitating. Extremely high mass clouds (10^7 M$_\odot$) can be self-gravitating too, but virtually invisible in extinction studies and also primarily atomic

(Sec. II.B). Thus the structural distinction based on $P/G\sigma_c^2$ is more useful for clouds in general, and also more illustrative of possible formation routes, than distinctions based on opacity or molecular abundance alone.

Unbound clouds are difficult to recognize unless the recent history of external pressure is known. This pressure includes both the uniform thermal pressure from surrounding gases and the nonuniform, time-averaged pressure from external flows, cloud collisions, magnetic waves and other variable sources. The cloud is unbound if two conditions are satisfied: $P_{int}/G\sigma_c^2 \gg 1$ and $P_{int} \gg P$ for internal pressure P_{int}. Note that a non-self-gravitating cloud with a low external pressure at the present time can still be pressure bound if the time-averaged external pressure, including pressure fluctuations in the past and future, is large enough to bind it on average. Such stochastic binding requires that the frequency of pressure fluctuations exceed the cloud expansion rate. Stochastic pressure binding may be applicable to most diffuse clouds because the external pressure environment of all clouds is likely to be highly variable with random supernovae, passing bright stars, and magnetic and physical cloud collisions. Clumpy regions that are swept up by a shock front should also have a variable post-shock pressure (including clump motions) that satisfies the global shock conditions only on average; large transients should occur as each new clump joins the front (Elmegreen 1988a).

The dimensionless pressure $P/G\sigma_c^2$ may also be useful as a tracer for diffuse and self-gravitating regions in large maps of interstellar gas. The boundaries of these regions are usually ambiguous if only the molecular emission line contours are used to delineate them, but if contours of $P/G\sigma_c^2$ are used instead, then some of the complicated structures that are likely to be present could become more clear. For example, it is possible that a cloud is structurally diffuse on one scale, where P and σ_c locally satisfy $P/G\sigma_c^2 \gg 1$, but that it contains self-gravitating pieces where $P/G\sigma_c^2 < 1$ on a smaller scale, and at the same time is part of a larger cloud that is also self-gravitating ($P/G\sigma_c^2 < 1$) at a larger boundary. This would be the case for a giant self-gravitating cloud complex that has diffuse cloud pieces on its periphery (see, e.g., Federman and Willson 1982), where the pressure may be high because of a history of supernova explosions, and which also has dense, self-gravitating clumps inside of the diffuse pieces, as observed, for example, by Stacy et al. (1989). Other combinations are possible too. In practice, one might try to measure the density ρ, density gradient $k\rho$ and velocity dispersion c everywhere in a three-dimensional region, and then determine the local value of $P/G\sigma_c^2 \sim k^2c^2/G\rho$ as an indicator of the relative importance of pressure and self-gravity at each location. A contour map of this quantity should distinguish between self-gravitating "objects," which have some structural integrity, and diffuse material, which is amorphous and poorly defined by other means. In such a map, the contours enclosing low values of $k^2c^2/G\rho$ would designate the self-gravitating objects. The actual threshold for "strong" self-gravity could be adjusted, but once chosen, the mapping procedure becomes objective and potentially informative.

B. Two Chemical Subtypes

The distinction between molecular and atomic clouds is based on a different combination of cloud properties than the distinction between diffuse and self-gravitating clouds. This difference took a while to recognize. The first decade of CO observations taught us that most dark clouds are also molecular, and that most molecular clouds are strongly self-gravitating. We also learned that most atomic clouds, like the diffuse clouds studied with optical and ultraviolet absorption, are not strongly self-gravitating. A change occurred when giant dust complexes and star-forming regions in irregular galaxies were found to emit very little CO radiation (Elmegreen et al. 1980; see review in Thronson 1988), when giant H I clouds in our Galaxy and other galaxies were discovered to be virialized (Elmegreen and Elmegreen 1987), and when diffuse-looking, non-self-gravitating, possibly unbound clouds at high galactic latitude were observed to emit CO (Magnani et al. 1985). We also have examples now of clouds that are primarily composed of H_2 molecules but which have very low column densities of CO (Blitz et al. 1990); this may also be the case for some of the giant cloud complexes in irregular (Israel et al. 1986) and other galaxies (Israel et al. 1990).

Evidently both diffuse and self-gravitating clouds can be either atomic or molecular. We refer here to four basic types of bound clouds: molecular self-gravitating, molecular diffuse, atomic self-gravitating, and atomic diffuse, and give below the expected conditions for each type and observed examples. The molecular nature of a cloud is assumed for this discussion to be based on the relative fraction of H_2, but the discussion could refer to CO instead if the CO formation criterion were used (see, e.g., van Dishoeck and Black 1988b; Federman et al. 1990).

Molecular hydrogen begins to appear in great abundance when the H_2 formation rate on dust balances the photodissociation rate from line absorption of background starlight. Jura (1974) and Federman et al. (1979) calibrated this condition from ultraviolet observations. Their results for diffuse clouds can be approximated by the criterion that molecular hydrogen appears when the path integral of the hydrogen-grain collision rate, which is proportional to $n^2DZ \sim nNZ$, exceeds the absorbed incident ultraviolet flux, which is proportional to $\sim \phi N^{1/3}$. Here, n is the rms density of H nucleons, D is the cloud size, $N = \langle n^2 \rangle D / \langle n \rangle$ is the effective column density through the cloud, considering clumpiness, Z is the grain surface area per atom, which is written here as a metallicity, and ϕ is the radiation field. The factor $N^{1/3}$ is an approximation (Elmegreen 1989a) to the column density dependence in Federman et al. (1979), which arises because clouds with higher column densities absorb more dissociative radiation in the line wings; one could also use a Voigt function. Under most conditions, the absorption of photodissociative radiation arises in the molecular line transitions, not in dust, although extinction is still important in the ultraviolet. This implies that the molecular fraction depends on ambient pressure (through n) and not just extinction (through N), i.e., that clouds with low extinction can be molecular if the pressure is high enough.

The numerical calibration suggested by Federman et al. (1979) gives a condition for a high H_2 fraction in diffuse clouds:

$$S \equiv \left(\frac{n}{60\text{cm}^{-3}}\right) \left(\frac{N}{5 \times 10^{20}\text{cm}^{-2}}\right)^{2/3} \left(\frac{Z/Z_\odot}{\phi/\phi_\odot}\right) \gg 1 \qquad (1)$$

where S is a shielding function and $Z_\odot$ and $\phi_\odot$ are values in the solar neighborhood. Small values of S presumably correspond to predominantly atomic clouds. If we also write the dimensionless pressure $P/G\sigma_c^2$ in terms of the same parameter values, using, for convenience, the cloud-averaged (instead of surface) pressure $P = \rho c^2$ and the column density $\sigma_c = m(H)N$, then

$$\frac{P}{G\sigma_c^2} = 15 \left(\frac{n}{60\text{cm}^{-3}}\right) \left(\frac{c}{1\text{km s}^{-1}}\right)^2 \left(\frac{N}{5 \times 10^{20}\text{cm}^{-2}}\right)^{-2}. \qquad (2)$$

Now the conditions for each of the four cloud types can be determined. Generally, when N is low and all of the other parameters are typical for the solar neighborhood, S will be small and $P/G\sigma_c^2$ will be large, so the cloud is an atomic diffuse cloud. When N is large and the other parameters are typical, S will be large and $P/G\sigma_c^2$ will be small, so the cloud is molecular and strongly self-gravitating. These two cases include most of the clouds in the solar neighborhood, which span the range from atomic to molecular as the column density increases. A less typical case arises when n is large and N is small, as in a small region or thin shell that is exposed to a high pressure. Then both S and $P/G\sigma_c^2$ can be large and the cloud can be both molecular and diffuse (with the exception of some shock fronts and other regions too hot or young to form molecules). The other unusual case is when n is small and N is large, which implies that the radius and mass are large (10^7 M$_\odot$). Then S and $P/G\sigma_c^2$ are both small and the cloud can be mostly atomic but still self-gravitating (with the exception of extremely low-density objects, which can become unbound by galactic tidal forces). Variations in ϕ and Z give odd cases too. Regions with large ϕ or low Z, for example, can have mostly atomic self-gravitating clouds, even with a relatively large N. This latter situation presumably occurs in irregular galaxies (see, e.g., Skillman 1987).

Real clouds are more complicated than this four-part classification scheme would suggest because they can have subparts that are a different type than the main clouds, all depending on local values of n, c, N, ϕ and Z. For example, atomic self-gravitating clouds often have molecular self-gravitating cores (Sec. III), and molecular self-gravitating clouds can have molecular diffuse wisps and clumps inside of them (see, e.g., Falgarone and Perault 1988). Molecular self-gravitating clouds can also have atomic clumps (diffuse or self-gravitating) inside if ϕ is large there from an embedded star (van der Werf and Goss 1990).

Specific examples of diffuse molecular clouds, some of which may be unbound, include (1) local high latitude clouds and their pieces (Magnani et al. 1985,1989,1990a; Lada and Blitz 1988; Mebold 1989), which may have a large $P / G\sigma_c^2$ because of high pressures from the Sco-Cen OB association; (2) uniform-density molecular clouds in various stages of disruption around star clusters (Leisawitz et al. 1989,1990), which may have large P from the clusters, and (3) non-self-gravitating molecular clumps along the peripheries of dense H II regions, such as NGC 7538 (Pratap et al. 1990), which are presumably pressurized by the nebulae. Maloney (1990b) suggests that essentially all molecular clouds are of this type, i.e., bound more by external pressure than self-gravity, and Issa et al. (1990) suggest that the largest molecular clouds are not virialized but defined instead by artificial boundaries in crowded regions. Evidently, it is difficult to know if some clouds are virialized or not, but in general one expects true diffuse clouds to contain very little star formation (see, e.g., Magnani et al. 1990b).

Diffuse molecular emission is also evident on a galactic scale. Parts of the molecular dust lanes in galaxies (Kaufman et al. 1989) could be non-self-gravitating because of the high ram pressure from the convergent flow. Diffuse molecular clouds in the inner part of our Galaxy were discussed by Polk et al. (1988), in the interarm regions of M83 by Wiklind et al. (1990), and in NGC 6946 by Casoli et al (1990). Warm CO emission from optically thin gas in IC 342 (Eckart et al. 1990; Wall and Jaffee 1990) may be diffuse also.

Examples of atomic self-gravitating clouds are the giant H I emission features in star-forming regions of irregular galaxies, where the metallicity is low (Skillman 1987; Viallefond 1988), and the largest self-gravitating clouds in our Galaxy (Elmegreen and Elmegreen 1987). Some of the atomic gas downstream from the dust lanes in M83 (Allen et al. 1986) and M51 (Tilanus and Allen 1989) could be in this category too, because of the high radiation fields.

Examples of truly unbound clouds are difficult to recognize because of the possibility that continuous pressure fluctuations occur, and that current pressures are only temporarily low. Nevertheless, high-latitude clouds that appear to be unbound have been studied by Pound et al. (1990) and Meyerdierks et al. (1990), who suggest that supernovae shocks compressed them recently and then cooled, bringing their environments to a low pressure. Support for this idea comes from Verschuur (1990), who suggests that two intermediate velocity H I clouds were recently stripped and displaced from their former molecular cores (which are still observed) by passing supernova shocks. The entire Ophiuchus cloud complex could be unbound as well, as a result of disruptive pressures from the nearby Sco-Cen OB association (de Geus et al. 1990), although many of the cores and clumps could still be bound internally.

In addition to this simple distinction between clouds that are primarily atomic or primarily molecular, all molecular clouds should also have photodissociated surfaces that are largely atomic. Examples of atomic surface layers were discussed by Chromey et al. (1989), Anderson et al. (1990) and Pound et al. (1990). Such atomic shielding layers may, in some cases, be difficult to distinguish, both morphologically and physically, from self-gravitating atomic cloud entities that happen to have dense molecular cores. Presumably the difference between these two types of atomic regions can be determined from the self-gravitational binding of the atomic part alone. A simple shielding layer should be bound to the whole cloud but not to itself, unlike a giant H I cloud complex. Thus the line width of the H I envelope in the case of simple shielding should not satisfy the virial theorem given the total H I mass and the radius of the cloud, but the line width of a giant H I complex should satisfy the virial theorem. Using this criterion, the H I around the Orion cloud, which is bound to the whole complex but not to itself (Chromey et al. 1989), is a simple shielding layer for the enclosed molecular gas, and the giant H I complexes in the inner Galaxy, which are wholly virialized (Elmegreen and Elmegreen 1987), are distinct entities that happen to have dense molecular cores (e.g., M17).

III. CLOUD FORMATION AND HIERARCHICAL STRUCTURE

Numerous observations suggest that most of the gas in the interstellar medium is a combination of clumps, filaments, holes, tunnels and various other structures in a more or less connected magnetic network. This structure is evident in projection against the sky on maps that have a sufficiently large ratio of survey area to resolution size (Low et al. 1984; Bally et al. 1987; Scalo 1990). What is remarkable about the structure is that in some regions, it appears to extend in an approximately self-similar fashion, with no obvious break in structural properties, from giant spiral arm clouds and shells, hundreds of parsecs in size, to tiny clouds, clumps, filaments and holes that are less than 0.1 pc in size (Scalo 1985; Falgarone and Perault 1987,1988; Efremov 1989). Such self-similarity is very uncertain, however, and it may not be ubiquitous. For example, Perault et al. (1985) suggested that the substructure in one particular cloud has a lower length scale of $\sim$1 pc, below which the gas is more uniform. Also, on a larger scale, many of the giant clouds and holes seen in high-resolution maps of other galaxies (Casoli et al. 1990; Brinks and Bajaja 1986) have a structure distinctly different than the clumps, wisps and holes observed in nearby clouds, presumably because of the influence of stellar density waves and galactic shear on the large scale.

The origin of the apparent hierarchy is an important unsolved problem, not unrelated to cloud formation. Observations on the largest scale may offer a partial solution. The largest clouds in our Galaxy appear in the spiral arms as H I emission regions containing 10^7 M$_\odot$ of neutral hydrogen at an average density of $\sim$10 cm^{-3} (McGee and Milton 1964). These clouds, along with similar clouds in other galaxies (Viallefond et al. 1982), are apparently dense enough to be bound against galactic tidal forces (Elmegreen 1987b; Waller and Hodge 1991), which are particularly low in the arms (Elmegreen 1987d), and their internal 21-cm velocity dispersions make them appear virialized (Elmegreen and Elmegreen 1987). Thus, many are well-defined objects, sometimes referred to as "superclouds," in analogy with the "superassociations" of Ambartsumian et al. (1963). Lo et al. (1987) and Rand and Kulkarni (1990) find similar clouds in the spiral arms of M51, but because M51 is generally more molecular than other galaxies (perhaps it has a higher pressure; see Sec. II), these largest clouds are mostly molecular. Rand and Kulkarni (1990) refer to them as "giant molecular associations." The common appearance of such giant clouds in the main spiral arms of our Galaxy and other galaxies makes it likely that they were formed by processes related to compression in a density wave. They may be the first generation of clouds that form in a region after such a wave passes by. Detailed studies of their properties may elucidate the mechanisms by which spiral arms trigger cloud and star formation.

The denser regions of superclouds, which contribute to the impression that there is hierarchical structure in the interstellar medium (Scalo 1985; Efremov 1989), often show up as giant molecular clouds. Such a hierarchy is illustrated for the Carina arm by Grabelsky et al. (1987), for the Sagittarius arm by Elmegreen and Elmegreen (1987), and for a 3-kpc region around the solar neighborhood by Efremov and Sitnik (1988). Similar structures have been resolved in other galaxies too, including spiral arm complexes in M31 (Lada et al. 1988a) and NGC 6946 (Casoli et al. 1990). In fact, many of the star-forming clouds in spiral arms could be the cores of much larger, low-density complexes. In the Carina arm, for example, nearly every observed molecular cloud (Grabelsky et al. 1987) is associated with one of the 10^7 M$_\odot$ H I clouds found by McGee and Milton (1964). In M101 (Viallefond et al. 1982) and M33 (Viallefond et al. 1986) the bright H II regions also tend to be associated with giant H I clouds. It follows that most giant molecular clouds in the main spiral arms of galaxies could be the dense, strongly self-gravitating fragments of larger cloud complexes that were directly triggered by the stellar wave (Elmegreen and Elmegreen 1983).

Not all giant molecular clouds are parts of obvious superclouds, however, and this may be an important clue to the mechanisms of cloud formation. Several other well-studied molecular clouds in our Galaxy, such as those associated with giant shells or those in the interarm regions (Solomon et al. 1985) apparently have no well-defined superclouds associated with them. For example, the local clouds in Orion, Sco-Cen, Monoceros and Perseus are relatively bare molecular clouds. They all have H I envelopes (Chromey et al. 1989; Strauss et al. 1979; Puchalsky et al., in preparation; Sancisi et al. 1974), but unlike the main spiral arm clouds that are inside superclouds (e.g., M17), the envelopes of the local clouds are relatively small, having an H I mass only comparable to or less than the molecular mass (see, e.g., Efremov and Sitnik 1988). Thus the H I around bare molecular clouds is probably only a photodissociated surface layer and not a distinct self-bound entity (although it is probably bound to the molecular cloud; cf. Sec. II.B).

A second clue along these lines is that several of the local bare molecular clouds appear to be part of Lindblad's expanding ring (Olano 1982; Elmegreen 1982*a*), in which case they probably formed in a giant pressurized shell (e.g., see also Franco et al. 1988). Moreover, the region is not young, as if it were a first generation of star formation, because most of the star formation in the solar neighborhood, which is in Gould's Belt, began its present activity some 60 Myr ago (Stothers and Frogel 1974; Palous 1986). In addition, the local clouds are not in a main spiral arm (they appear to be in only a spur; Blaauw 1985), but if we extrapolate the position of the solar neighborhood back for this 60 Myr, using the spiral pattern speed for our Galaxy given by Yuan (1969), then the gas would be at the position of the Carina arm, which is a main arm. The Sirius supercluster of stars in the solar neighborhood has also been traced back to an origin in the Carina arm (Palous and Houck 1986).

These observations suggest a scenario for cloud and star formation in the solar neighborhood that may be representative of other regions too. The local star-forming activity began 60 Myr ago when the Carina arm passed through the local gas. The triggering process was probably one of collection of this local gas into one or more superclouds, which fragmented into a first generation of molecular clouds and star formation. One of the associations that formed at that time, the now dispersed Cas-Tau association (Blaauw 1984), blew a giant bubble that made Lindblad's ring today (this association is in the center of the ring and it has the right age; for an even larger-scale expansion, see Palous [1987]). When the ring fragmented some 20 Myr ago, the nearest giant molecular clouds in Orion, Sco-Cen and Perseus formed as a second generation, and more recently, the Ophiuchus and Taurus (Olano and Pöppel 1987) clouds may have formed or been compressed as a third generation. The original low-density supercloud probably was dispersed by this time.

This scenario is largely speculative, but it fits the observed distribution and ages of the nearest clouds and star clusters in the solar neighborhood, and it is consistent with cloud- and star-formation processes observed elsewhere (Elmegreen and Elmegreen 1983; Efremov 1989). If true, it suggests that the largest scales in the hierarchy of gas structure, the superclouds and the embedded giant molecular clouds, form by pressurization, condensation and fragmentation of gas in spiral density waves. The densest cores of the giant molecular clouds, where the stars actually form, and other smaller steps in the structural hierarchy, could be a continuation of this fragmentation process. But the scenario also suggests that not all molecular clouds form this way, that they are not all part of a self-similar hierarchical structure extending to larger scales. Some molecular clouds are parts of rings and shells, which are not structurally similar to superclouds, and other molecular clouds are probably bare, the largest objects in their hierarchies. In this latter case, the extended gas that was formerly involved in the formation of the molecular clouds, and which may have once connected them to other dense clouds, has been dispersed or ionized so that it is currently unrecognizable. The associated H I envelopes are then simple shielding layers.

Hierarchical structure on smaller scales, as in low-mass molecular clouds and diffuse clouds, could also result from fragmentation and cooling after the main cloud forms. This might involve thermal instabilities if gravity is not important. But there are alternative possibilities (which also apply to larger clouds although they were not discussed above). For example, the pieces of a cloud could have existed prior to the cloud's formation and been brought into the cloud during the mass accumulation process, which may include shock accumulation of the ambient clumpy medium, random coagulation of formerly independent clouds, or large-scale instabilities. Some of the substructure could also grow behind shocks inside the cloud, as a result of spontaneous steepening of magnetosonic waves (Zweibel and Josafatsson 1983; Elmegreen 1990b), or in shells around embedded stars (Norman and Silk 1980a). Hierarchical and fractal structures could also result from collisional fragmentation of clumps inside the cloud (see, e.g., Nozakura 1990). These processes are discussed in more detail in Sec. VI.

There is no reason to think that the hierarchy of density structures is not accompanied by an analogous hierarchy of clump and cloud formation processes, which structure the interiors of clouds in the same way as they structure the general interstellar medium, but scaled down in size. Bally (1989), for example, suggests that much of the substructure in molecular clouds results from agitation by local stars, which is a structuring process similar to that in the general interstellar medium.

IV. CLOUD FORMATION AND SUPERSONIC MOTIONS

The independent motions of individual clouds and clumps appear to be somewhat random, giving broad and nearly gaussian spectral lines whenever several clouds or clumps are unresolved in a telescope beam (Perault et al. 1985; Kwan and Sanders 1986; Tauber and Goldsmith 1990). This randomness may also persist on larger scales in the hierarchy of cloud structures, as in the correlated motions of clusters of clouds which may be random with respect to each other. Such random motions are the likely result of cloud or cloud-cluster interactions of a gravitational or magnetic nature, analogous to two-body scattering or turbulence, just as the random motions of atoms in a gas result from collisions.

Deviations from purely gaussian motions appear in the form of broad and faint-line wings. These line wings sometimes occur in regions that are far from obvious embedded stars, and so presumably originate with purely hydrodynamic processes. Proposed explanations are that the high-velocity gas is an unbound interclump medium (Blitz and Stark 1986), that the wings result from high-speed clump collisions (Keto and Lattanzio 1989) or turbulent intermittency (Falgarone and Phillips 1990), or that they arise in nonlinear Alfvén waves (a form of turbulence) on the periphery of the dense cores, where the Alfvén velocity is expected to exceed the cloud's virial velocity because of the lower average density (Elmegreen 1990b). Observations of the distribution and average density of the wing-emitting regions could presumably determine the origin of this gas. Magnani et al. (1990*a*), for example, suggest that in some sources this emission comes from the periphery of the cores, where the gas has a density of ~0.1 times the peak value, and Falgarone et al. (1991) find that the excitation density of the wing-emitting gas is the same as the excitation density in the line-core. These observations can be mutually consistent if both the line-core and the line-wing emission come from equally dense clumps which have a filling factor 10 times higher for the line-core emission than for the wing emission. This contrast in average density is consistent with the predictions of the wave theory, and the observed relative positions of the core and wing-emitting regions are consistent with the collision theory (the cores are then the shocked gas that is at rest between the colliding clumps). Unfortunately, the line wings in interstellar clouds are very weak and at the present time can be observed only by adding together many spectra. Then, there is a possibility that some of the wing emission is associated with stellar winds and other forced motions instead of turbulence (see, e.g., Margulis et al. 1986,1988).

The motion of interstellar gas also has a systematic component wherever gravity or magnetism exert significant long range forces, acting on many clumps simultaneously, or where the pressures from stars or cloud-cloud collisions sweep up large regions in a type of shock, or cloud-collision front. Such systematic motions occur on a variety of scales, ranging from contracting envelopes, rotating disks and bipolar expansions around protostars, to the formation, condensation and disruption of whole clouds, to the rotation, shear and spiral flows of a galaxy. The importance of systematic flows for cloud formation arises when large-scale shock fronts compress the gas in bulk, forming shells and other cloudy structures, and when Kelvin-Helmholtz instabilities form clouds in shearing regions. Hunter and Whitaker (1989), for example, suggest that some of the clumps and small-scale structures observed in clouds result from Kelvin-Helmholtz instabilities, and Fleck (1989) attributes cloud rotation to them. Other systematic motions, such as linear and torsional waves that are probably magnetic in nature were discussed by Shuter et al. (1987), Heiles (1988) and Uchida et al. (1990), and wave-like motions, in general, were fitted to molecular line profiles by Stenholm (1990).

One of the most important characteristics of the random motion of clumps is their high speed relative to both their internal thermal speed and their internal velocity dispersion. These high speeds imply that most clump collisions are supersonic, and if one identifies the internal velocity dispersion with an Alfvén velocity for magnetic waves, then the motions are probably also super-Alfvénic (see, e.g., Heiles 1989). In that case, clump collisions lead to shocks and significant energy dissipation. Such dissipation makes the large-scale properties of the interstellar gas, taken as an interconnected network of clumps or tiny shocks, for example, qualitatively different from the large-scale properties of a conventional fluid, taken as a system of atoms or molecules. Because of the dissipation, a better analogy may be with a molecular fluid where the thermal energy greatly exceeds the binding energy of the molecules. Such a fluid is not in equilibrium, but will quickly dissociate its molecules and cool. Similarly, the interstellar fluid may rarely find a stable equilibrium, but may always be in a state of rapid dissipation following constant or sporadic agitation by stellar energy sources, gravitational interactions, inertial forces from collisions or ram pressure, and various types of sonic and magnetic waves and shocks.

It follows that the adiabatic equation of state, $P \propto \rho^\gamma$, which has often been used for interstellar gas dynamics, is largely irrelevant for the interstellar fluid, even if γ is taken to be much less than 1. In fact, the effective value of γ, if one is to be defined, should depend on the rate of change of the density, in addition to the density and the relevant cooling and heating rates. An effective value of γ was given by Elmegreen (1991a) for a particular model of large-scale cloud formation by combined instabilities:

$$\gamma_{\text{eff}} = \frac{\gamma \omega - \omega_c (2l + s - m - 2r)}{\omega + \omega_c (m - s)} \tag{3}$$

where $\gamma = 5/3$ for purely translational cloud motions, ω is the rate of condensation into a cloud, $\omega_c = (\gamma - 1)\Lambda_0/(2P)$ is the cooling rate, and l, s, m, r are powers in the cooling and heating rates, given by $\Lambda \propto \rho^l c^m$ and $\Gamma \propto \rho^r c^s$ for density ρ and interstellar velocity dispersion c; Λ_0 is the value of Λ in equilibrium. For typical large-scale instabilities with cloud collisional cooling ($l = 2$, $m = 3$), $\gamma_{\text{eff}} \sim 0.3$, which is comparable to the average value of $\mathrm{d}\ln P / \mathrm{d}\ln \rho \sim 0.25$ obtained from Myers' (1978) compilation of cloud properties. If $r < 0.5(1 + s)$, then $\gamma_{\text{eff}} < 0$ for some values of ω and there is an instability analogous to the thermal instability at constant pressure.

Supersonic motions also imply that sticking collisions between separate clouds should be rare. The individual velocities of diffuse clouds, for example, are so large compared with their internal dispersions and temperatures that direct hits between them should be destructive or shredding (Hausman 1981). Self-gravitating clouds should also disperse upon collision, unless they are either very massive or the relative velocity decreases with decreasing mass. Calculations suggest that if the relative cloud velocity exceeds approximately 2 or 3 times the escape velocity of the largest cloud, then the nonoverlapping and overlapping pieces will not be gravitationally bound to each other after the collision ends, and the clouds will fragment or explode upon impact instead of coalesce (Vazquez and Scalo 1989; Elmegreen 1990a; Lattanzio and Elmegreen 1991). Such high velocities may be common, considering that the relative velocity between two clouds is, on average, $6^{0.5} = 2.4$ times the observed one-dimensional dispersion. This dispersion has a random component between 3 km s^{-1} (Clemens 1985; Alvarez et al. 1990) and $\sim$7 km s^{-1} (Stark and Brand 1989), so the relative cloud velocity is between 7 km s^{-1} and 17 km s^{-1}, nearly independent of cloud mass for all but the largest clouds. If the coalescence limit is 2 times the escape speed for a typical collision, then only clouds larger than 0.1×10^5 M$_\odot$ to 4×10^5 M$_\odot$ can stick to each other after a random collision, considering the two dispersions, respectively. These numbers assume that the escape velocity for a cloud of mass M in solar masses equals $0.34\,M^{0.25}$ km s^{-1}, from the scaling relations in Solomon et al. (1987).

These limiting masses for coalescence allow giant molecular clouds such as the Orion cloud to coalesce after a low speed impact, and this includes at least half of the molecular mass in the Galaxy considering the $M^{-1.5}$ cloud spectrum, but the limiting mass is too large for a complete Oort cycle of cloud coalescence from the smallest to the largest molecular clouds. Coalescence is possible above these limits if the clouds are self-gravitating so they stick, but then the geometric cross section for collisions should scale with the cloud-mass, as observed for molecular clouds (Solomon et al. 1987), and the resulting cloud mass spectrum should vary as M^{-2} in the Oort model (Elmegreen 1989e) instead of the observed $M^{-1.5}$ (Dickey and Garwood 1989).

The coalescence model of cloud formation works better when the velocities, masses and separations between clouds are correlated in the sense that low-mass clouds move slower relative to each other than high-mass clouds. This might be the case in a turbulent, fractal fluid where small neighborhoods have small-scale structure and low relative velocities (see, e.g., Falgarone and Phillips 1991). Then both high and low-mass clouds can hit each other at low enough velocities relative to their escape speeds to allow coalescence as often as fragmentation. The turbulent nature of the interstellar fluid might even be based on this condition, that cloud masses and positions take on a distribution such that the fragmentation probability equals the coalescence probability for the given amount of kinetic energy density. Indeed, fractal structures have been shown to result from cloud interactions (Nozakura 1990).

Can magnetic fields make cloud collisions more sticky? A naive answer is no: during a collision between two clouds on different magnetic flux tubes, the magnetic field lines that connect each cloud to the ambient medium should be distorted by the other cloud, producing a magnetic force that is generally repulsive. But a more subtle answer to this question may be affirmative if the magnetic field diffuses or reconnects significantly during the interaction. The outcome of such a magnetic collision has never been modeled, but it seems possible that the field can diffuse or reconnect to a geometry that eventually helps bind the clouds together gravitationally in the direction parallel to the field. Obviously, if two clouds start on the same flux tube, their collision should be somewhat sticky, but such collisions probably represent only a small fraction of cloud interactions. However, diffusion or reconnection could mix parts of their flux tubes and have the same effect.

The sticking problem with the random coagulation model of cloud growth calls into question some of the assumptions made for computer simulations of cloud dynamics, which show substantial cloud coagulation in spiral arms even though the collision velocities in the spiral arm streams are large. Most of these simulations assume that the cloud particles live forever, drifting freely between collisions, but magnetic fields and internal star formation would seem to prevent such prolonged drifting in real galaxies. Many models also assume that some of the clouds stick upon impact, or that they lose relative energy after a bounce but do not destroy each other. Then coagulation is more likely. But the real situation may be very different, with the interarm clouds getting shattered and destroyed after colliding with clouds in the arms, producing a more diffuse type of gas (dust lanes?) instead of discrete cloud complexes. The statistical nature of the cloud distribution can also give the appearance of groupings and complexes for those interarm clouds that do not disperse in the arms (Adler and Roberts 1988), but such random groupings are transient and do not generally produce virialized objects like the molecular clouds (Lee et al. 1990) or superclouds in our Galaxy. Computer models also do not indicate why the cloud particles coalesce into complexes. If the cloud particles bounce off of each other with a reduced velocity after the collision, as is sometimes assumed for cloud cooling, then the simulated condensation into giant complexes cannot result directly from sticky coagulation, but probably results instead from cooling instabilities, in which the internal pressure in an incipient complex decreases rapidly because the component clouds dissipate their relative energy, and then continued collisions with fast-moving external clouds compress the complex to a high density. Such a cloud-formation process may look like sticky coalescence in a computer simulation, but it is really an instability related to the classical thermal instability.

Supersonic motions also imply that the length scale at which random kinetic energy is dissipated spans the entire range over which these motions are observed, occurring at shock fronts between any two colliding eddies or clouds, and in pervasive nonlinear magnetic waves. This differs from subsonic turbulence in laboratory fluids, for which the dissipation is thought to occur at the end of a swirling energy cascade to small scales. The kinematic energy may also be put into the interstellar fluid on all scales, in the form of wind or radiation pressure-driven bubbles or explosions from stars with a variety of luminosities, or in gravitational scatterings between clumps that are magnetically or viscously tied to the rest of the gas.

V. CLOUD FORMATION AND MAGNETIC FIELDS

Another important characteristic of the interstellar gas is the pervasive magnetic field. In the general interstellar medium, this field has a uniform component of $\sim 1.6 \mu$gauss and a seemingly random component of $\sim 5 \mu$gauss (Rand and Kulkarni 1989; see also Beck et al. 1989; Beck 1991). The relatively large value of this random component suggests that the random and expansive motions of the gas push the field around significantly, with a short time scale, but the general uniformity of the average field direction (Mathewson and Ford 1970) implies that such pushing does not completely control the field line structure. The field should respond with a restoring force from $B \cdot \nabla B$ tension that ultimately controls the gas on large scales.

In dense regions, the field is often so strong that the magnetic pressure is comparable to or larger than the kinematic pressure from random motions (Heiles 1989; Myers and Goodman 1988a). This implies that the field should influence these motions, which is probably the case in many regions, especially at low densities (see, e.g., Heyer et al. 1987). However, the orientation of the field in general seems to be independent of detailed structural features (Goodman et al. 1990), as if the field has a relatively minor role in cloud formation.

How can this discrepancy be resolved? A simple explanation is that the field lines are helical (Bally 1989), in which case the lines on the near side have some angle with respect to the cloud axis. But then the average orientation of the field, for the front and back sides combined, should still be parallel to the filament, and polarization observations using background stars should show this parallel direction, on average.

Another explanation is that a dense cloud's orientation is influenced by pressure gradients on a small scale, as would be the case if the cloud were swept up by some pressure source, but the field's orientation is determined by larger-scale processes, such as the positions and motions of other more remote clouds to which the field is also attached. The consequences of this idea are easily imagined. For example, a parcel of gas in a magnetic field can be accumulated into a cloud by a neighboring high pressure event, and the cloud can have any orientation relative to the initial field if the pressure is high enough. The field lines inside the cloud should follow the gas at first, in which case the component parallel to the pressure front should increase linearly with density; this orients the internal field so that it is nearly parallel to the cloud axis. But after a diffusion time (and several oscillations), the field lines inside the cloud should return to their initial orientation without much of a change in the cloud's overall orientation. Then the internal field becomes parallel to the external field again, and it will, in general, have no relation to the final orientation of the cloud. A condition for such random orientations between magnetic fields and cloud structure in this model is that the diffusion or straightening time should be less than or comparable to the time scale for cloud accumulation.

Consideration of the magnetic field is essential for theories of cloud formation: (1) It limits the random motions of individual clouds by dragging along low-density gas. (2) It transfers linear momentum between clouds without physical contact. (3) It removes angular momentum from a cloud during condensation. (4) Nonlinear waves also act as a source of pressure for pushing, twisting and shocking the gas into dense cloudy regions.

The first of these four processes makes it unlikely that a dense molecular cloud will drift freely for a geometric mean free path (Elmegreen 1981b). The characteristic confinement length corresponds to a distorted volume of the cloud's magnetic flux tube that contains a mass of other clouds equal to the cloud's mass. For dense molecular clouds with virial equilibrium field strengths, this length implies a maximum excursion of only several hundred

(<500) parsecs from an equilibrium position, which is much less than the geometric mean free path (~ 2000 pc). Theories of cloud formation involving random linear drifts followed by physical coagulations are therefore oversimplified, as are N-body computer models of cloud fluids with no intercloud medium or interlinking magnetic field structure. The relatively uniform field structure in nearby clouds also implies that these clouds are not composed of numerous pieces which drifted from random directions carrying their field lines with them.

Strong magnetic confinement also implies that one of the characteristics of a terrestrial turbulent fluid, i.e., the random diffusion of every pair of points away from each other, may not apply to interstellar fluids. The clumps may just oscillate on their interconnecting field lines (see, e.g., Elmegreen 1988b; Zweibel 1990a), as if they were trapped in a net.

The second of the four magnetic processes mentioned above implies that both diffuse clouds and the small clumps inside dense molecular clouds are likely to interact more by purely magnetic linkages than by direct physical collisions (Clifford and Elmegreen 1983; Falgarone and Puget 1986). The relative importance of these two interactions depends on the ratio $(\rho_c v_A / \rho v)^{2/3}$ for average and clump densities ρ and ρ_c, clump Alfvén velocity v_A, and relative clump velocity v. Large values of this ratio make magnetic linking interactions more important than physical collisions as a means of momentum exchange. When the interclump mass is low, the ratio ρ / ρ_c is the clump filling factor f. For self-gravitating clumps inside a self-gravitating cloud, the ratio v_A / v is approximately $(R_c / R)^{1/2}$ for size-line width relation $v \propto R^{1/2}$ and cloud and clump sizes R and R_c (see, e.g., Perault et al. 1986). Then the quantity $(\rho_c v_A / \rho v)^{2/3}$ equals approximately $(f^5 N)^{-1/9}$ for number of clumps N. Clumps typically have $f < 0.1$ (Falgarone and Phillips 1991), so this ratio exceeds unity for $N < 10^5$; in that case, magnetic linking interactions should dominate the collisions. For non-self-gravitating pieces in a cloud, the magnetic field could be more uniform in the clump and interclump regions, and then $v_A / v \sim (\rho_c / \rho)^{-0.5}$ so that $(\rho_c v_A / \rho v)^{2/3} \sim f^{-1/3}$, which also exceeds 1. Magnetic linking interactions give the interstellar fluid a cohesion and resilience that is not possible with purely physical cloud interactions.

In the third process mentioned above, magnetic fields exchange angular momentum between clouds and their environment, with a spin-down time approximately equal to the time for an Alfvén wave to traverse an external mass having a moment of inertia equal to the cloud's moment of inertia (see Chapter by McKee et al.). This implies that clouds and cloud complexes should not be rotating significantly, and it removes some of the stability against self-gravitational collapse given to the interstellar fluid by galactic rotation (Elmegreen 1987c,1991).

Angular momentum exchanges by a magnetic field should also limit the vorticity of the interstellar fluid, strongly in two dimensions and weakly along the field (see, e.g., Dorfi 1990). This could remove one of the primary mechanisms for kinetic energy cascade from large to small scales in a terrestrial turbulent fluid, i.e., vorticity stretching (cf. Sec. VII), although other mechanisms for energy cascade could replace it. The transfer of energy between different scales in the interstellar medium could be driven by magnetic stresses and torques, i.e., the nonlinear $B \cdot \nabla B$ term in addition to, or instead of, the nonlinear $v \cdot \nabla v$ term, which is more important for weakly magnetic turbulence. This energy transfer may proceed from small to large scales, as when a moving, spinning or expanding region magnetically drags along the surrounding gas (Uchida et al. 1990), as well as from large to small scales, as when two colliding clouds pump energy into internal Alfvén waves which steepen into shocks (see, e.g., Falgarone and Puget 1986).

Magnetic fields can also form clouds directly, by the parallel-pushing action of nonlinear waves, or by twisting the gas into ropes. In the first case, strong plane-polarized waves will couple the transverse and parallel motions of the field and gas, so that strong transverse field-line deformations, as might result from explosions or random field-line entanglements, can lead to magnetic shock fronts and other pulses of dense gas traveling along the field at or slightly below the Alfvén speed. Because such deformations should be common in the interstellar medium, collisions between these parallel-moving pulses can form very dense, transient clouds and small-scale structure. The shock fronts can also clear out much of the gas between the deformations, producing a low-density intercloud medium. Numerical simulations of such collisions are in Elmegreen (1990b), and an illustration of the long-time evolution of numerous interacting magnetic waves is in Elmegreen (1991b). These simulations suggest that some cloud clumps could be transient shock-like fronts between regions of converging flows; the clumps last for the convergence time, which is several internal wave-crossing times. In the second case, the angular momentum drain from one part of a cloud can cause another part to spin up, and then the field lines can become helical and exert a tension that compresses the gas into a dense filament (Fukui and Mizuno 1991). Evidence for such twisting compression in Orion is discussed by Uchida et al. (1991).

Magnetic fields influence almost every aspect of cloud formation and gas dynamics. In some situations, this influence is probably so large that models which do not consider magnetic effects may have limited application. Some cloud formation models are based entirely on magnetic effects, such as the Parker (1966) instability and nonlinear wave-pushing models. Other models are greatly helped by the fields, such as the gravitational instability model and the coagulation model where pre-existing clouds and their field lines link together in a nondestructive fashion behind moving pressure fronts. Still other models, such as the random coalescence model in which dense clouds move independently for over a mean free path, are probably oversimplified because they do not consider the confining influence of magnetic fields.

General models of interstellar gas dynamics may also need revision when the magnetic field is included. For example, the three-phase model by McKee and Ostriker (1977) proposes that random supernova explosions become so large that they overlap in space. Their primary assumption is that a supernova shock front passes over individual clouds without imparting much momentum to them (McKee and Cowie 1975); then most of the shock goes around the cloud and grows to a large size. But this assumption may not be true if all of the clouds are strongly connected to each other and to the intercloud medium by magnetic fields. Then the clouds could move together with the intercloud medium, and most of the enclosed mass should stay with the front, especially at late times (Elmegreen 1981a,1988a). As a result, a strongly magnetic front should slow down faster and stall in a shorter distance than a nonmagnetic front that sweeps up only the intercloud medium. One of the observational consequences of this effect is that large supernova shells should have low-density interiors because the clouds that were there before the explosion were removed and not evaporated in place. Of course, such magnetic clearing should occur primarily at late times when the supernova pressure is not so large that it severely distorts and then detaches (by reconnection) the field lines that emerge from an impacted cloud. Another magnetic effect is that the expansion of a supernova shock in even a homogeneous medium should be resisted by magnetic tension in the swept-up shell (Tomisaka 1990; Ferriere et al. 1991). Both of these magnetic effects, the added shell mass and the surface tension, should decrease the porosity of the hot intercloud phase by limiting the expansion of individual supernovae remnants (see also Cox 1990).

Another potentially important effect of magnetic fields is the cushioning of collisions between clouds. Magnetic linking collisions, for example, may allow small clumps to be gently collected into moving pressure fronts without disruptive collisions. This implies that some of the clumpy structure observed in clouds could have been present in the ambient gas before the cloud formed, and not been disrupted by the (supersonic) violence of the formation process. Whether such soft coalescence can preserve the observed fractal structure of the clumps is unknown. Perhaps many of these effects will be observed in computer simulations when magnetic fields are generally included.

VI. CLOUD FORMATION SCENARIOS

The various theories of cloud formation can be divided into three categories: cloud formation by spontaneous instabilities in the ambient interstellar gas, including gravitational, thermal, Kelvin-Helmholtz and Rayleigh-Taylor instabilities; cloud formation by pressurized accumulation of ambient gas into shells, sheets, cometary globules and chimney walls; and cloud formation by random coalescence of existing clouds. Most practical theories today are combinations of these. Many of these theories were mentioned in the preceding sections. More detailed and specific comments are given here.

Lioure and Chieze (1990) suggest that cloud formation occurs in two steps, first with a thermal instability that condenses the warm H I phase into cool diffuse clouds, and then with a coagulation of the diffuse clouds into self-gravitating cloud complexes (see also Field and Saslaw 1965). Cloud

formation by thermal instabilities in the warm H I was also investigated by
Parravano (1987,1989), who suggested that the interstellar pressure can regulate itself to remain close to the maximum value for two stable phases
(Parravano et al. 1988,1990). Thermal instabilities in the warm H I phase
seem very likely, in fact (Heiles 1989), but whether these instabilities lead
to the observed diffuse clouds (of mass 10^2 $M_\odot$ or larger) as assumed by
these authors, or only to very small-scale (< 1 pc) cool structures inside the
warm H I regions is uncertain. In fact, thermal instabilities tend to favor small
scales (Field 1965; Smith 1989) and may even generate a fractal structure
(Elphick et al. 1991), and the mobility of the small pieces that condense out
of the warm gas may be limited by magnetic effects (cf. Sec. V). Moreover,
alternate scenarios for the formation of the observed diffuse clouds, such as
accretion in pervasive shocks from supernovae and other sources (see, e.g.,
McKee and Ostriker 1977; Chiang and Prendergast 1985), are difficult to rule
out. Most likely, some diffuse clouds form by thermal instabilities and others
form in shocks of various types (Sec. II.A), but whether these clouds move
freely enough to collide and slowly enough to coalesce is difficult to know
without further modeling.

The coalescence of small interarm clouds into giant spiral-arm cloud
complexes was discussed and modeled by Casoli and Combes (1982), Kwan
and Valdez (1983), Tomisaka (1984,1986), Combes and Gerin (1985), Johns
and Nelson (1986), Roberts and Steward (1987) and others. Such coalescence
seems inevitable, but the details of the coalescence process and of the gas
flow in a density wave are not understood. Diffuse clouds may be destroyed
when they enter a spiral arm, splattered by mutual collisions and strongly
pushed around by pressures from stellar radiation and giant H II regions.
Their remnants could then disperse, creating the somewhat uniform gas that
is observed as a long and continuous dust lane. The dense, self-gravitating
clouds that enter a spiral arm should move differently than the diffuse clouds,
colliding infrequently with other self-gravitating clouds (because of their long
mean free path) and responding more freely to the spiral potential—perhaps
in a somewhat ballistic fashion (because external pressures are less important
for these clouds than for diffuse clouds; see Sec. II). Whether or not they
coalesce depends on their relative velocity (Sec. IV), and this depends on the
strength of the arm and on viscous effects such as magnetic drag (Sec. V).
Perhaps the largest interarm clouds coalesce upon impact in an arm, with the
considerable amount of mutual destruction that is expected in such a collision
leading to the formation of the high-dispersion H I component that is seen as
a supercloud (Sec. III). Subsequent cooling then reforms the molecular cores
and ultimately leads to star formation.

Models which emphasize the destructive power of cloud collisions but
still produce coalescence and giant complexes utilize a different, indirect route
for cloud assembly. Struck-Marcel and Scalo (1984), for example, suggest
that collisions between pre-existing diffuse clouds can be largely destructive
and shredding, rather than directly coalescing, but that these collisions are

still important as a coolant for the diffuse cloud fluid (i.e., they lower the velocity dispersion). This cooling may then lead to a thermal-like insta- bility in which slightly overdense regions diminish their dispersions faster than they are increased by random supernovae and other discrete pressure sources. When combined with gravitational forces between the clouds, the resultant gravitational-thermal instability (Elmegreen 1989c) can form cloud complexes with a much wider mass range than the pure Jeans instability (for an adiabatic fluid), and at a rate comparable to both the diffuse-cloud collision rate and the Jeans rate, which is ~20 Myr, depending on density. Tomisaka (1987) considered this combined instability in the gas that flows through a spiral density wave, and Cowie (1981) considered cloud collisional cooling and gravitational instabilities too, but without the intermediate stage of a thermal-like instability.

The existence of such macroscopic cooling instabilities is uncertain be- cause it depends on unknown details of the stirring and energy dissipation processes for the interstellar medium. Many of these processes depend on length scale, so the velocity dispersion should depend on scale also. This im- plies that the equilibrium or initial state could be turbulent, in which case the theory of gravitational instabilities should be revised (Chandrasekhar 1951a; Bonazzola et al. 1987; Leorat et al. 1990). Nevertheless, applications of the conventional Jeans analysis with cooling have been made to rotating and shearing galaxy disks that are infinitesimally thin (Elmegreen 1989c) or thick and also subject to the magnetic Rayleigh-Taylor (Parker 1966) instability in the third dimension (Elmegreen 1991a). In this latter study, the shearing wavelets that grow in the gas by the so-called swing amplifier were found to be gravitationally unstable to fragmentation along their length, producing discrete, self-gravitating clouds out of a continuous (clumpy, magnetic) fluid. Superclouds follow naturally from the instability model, because they have the expected characteristic mass (see Skillman 1987). Smaller clouds can form by the same mechanism also, but this depends on the rate of energy dissipation in the ambient gas. If the kinematic energy dissipates too slowly, then the smaller self-gravitating clouds would have to form inside the larger clouds, where the density is already high (or they form by other mechanisms).

A limitation to this instability model is that the collapse of gas into dis- crete clouds has never been followed for a realistic flow in a density wave, considering the simultaneous triggering of magnetic Rayleigh-Taylor, gravi- tational and thermal instabilities in three dimensions, with shear and density variations appropriate for the wave. Balbus and Cowie (1985) calculated a criterion for the collapse of whole spirals parallel to the flow direction, which is like the criterion for ring instabilities in a disk, but this collapse does not produce discrete clouds, and shear variations parallel to the spiral were not considered (nor were magnetic fields). Balbus (1988) considered such shear variations and got more realistic effects such as spurs, but again without mag- netic fields. Elmegreen (1987c) considered shear instabilities in magnetic gas with various shear rates appropriate to spiral arms and interarms, but the time dependence of this shear and of other flow properties in the arm were not considered, nor was the fragmentation of the shearing wavelets into discrete clouds. This latter effect was addressed in Elmegreen (1991b), but with no density wave variations at all. All of these analyses point to an increase in the instability rate in spiral arms because of the high density and low shear rates there, with the resultant formation of giant cloud complexes resembling the observed superclouds, but the conditions for when this should occur are vague (e.g., is there a minimum spiral arm strength and does the instability operate faster than the arm flow-through time?). The application of this instability model to the formation of low-mass self-gravitating clouds is also unknown, because of uncertainties involved with the thermal instability part of the model (i.e., with cloud stirring and various cooling effects). The Parker instability part of the model for cloud growth has been studied in more detail by Matsumoto et al. (1988), Shibata et al. (1989) and others.

Another complication is that spiral-arm cloud formation could differ in strong and weak spiral arms (Elmegreen 1987b,1988a), with pervasive cloud destruction and the formation of a homogeneously dense compression front (dust lane) followed by large-scale instabilities and supercloud formation occurring only in the case of strong arms. Weak-arm galaxies may scatter the existing dense clouds without severe destruction and create a dust lane by the dispersal of only the diffuse clouds. Examples of these two cases seem to be M51 (Rand and Kulkarni 1990) and M83 (Lord and Kenney 1991), respectively.

Observational evidence for cloud formation in swept-up shells was reviewed by Elmegreen (1987b) for small scales, and by Tenorio-Tagle and Bodenheimer (1989) for large scales (see also Deul and Hartog 1990). Both diffuse and self-gravitating clouds can form this way, but in the latter case a second step is required, in which the ring or shell that forms directly by the high-pressure event collapses gravitationally along its perimeter. The theory for the expansion has been progressing steadily (see, e.g., Tenorio-Tagle and Palous 1987; Norman and Ikeuchi 1989; Palous et al. 1990), but the theory for the collapse is in a poor state. McCray and Kafatos (1987) suggest that the time scale for the collapse of a large shell is the gravitational free-fall time at the compressed density, but transverse flows behind the shock may destroy or modify incipient condensations on these short time scales (unless the geometry is special; Kimura and Tosa 1988) giving a longer time for monotonic collapse (Elmegreen 1989b). Transverse flows and magnetic forces also redistribute mass inside the shells (Ferriere et al. 1990; Tenorio-Tagle et al. 1990). Magnetic effects during the collapse are generally unknown, and the possibility of blow out perpendicular to the disk (MacLow et al. 1989; Igumentshchev et al. 1990; Tomisaka 1990; Tenorio-Tagle et al. 1990), which would limit the total accumulation of material in the midplane, is a concern (Cox 1990).

VII. CLOUD FORMATION AND TURBULENCE

The above discussion summarizes some of the constraints on theories of cloud formation that are imposed by the hierarchically clumpy, supersonic, and magnetic nature of the interstellar gas. Obviously there is no complete theory of cloud formation yet, but the various parts of a complete theory seem to be evident, and a quantitative understanding of them seems to require only more calculations. But is this the way to proceed?

Most of the theories assume that the various cloud formation processes can be studied separately, as if in isolation, and that they combine without much change into a complete theory for the whole interstellar medium. This is basically a microscopic point of view, in which the local physics is probed to whatever depth is desired. However, the microscopic view currently fails to explain the self-similar (fractal) structure of clouds, and it may also fail to explain, for the same reason, the power-law correlations between cloud properties and the observed power-law mass spectrum (in spite of several theories in the literature which contain microscopic-type explanations for these power laws). How does one derive a cloud mass spectrum when some of the clouds form as shells around high-pressure OB associations, some form by fragmentation or collisional fracturing of larger objects, and some form by coalescence of smaller objects—all equally plausible and likely scenarios for cloud formation?

A different point of view is to ignore the microscopic details of interstellar gas processes entirely and to concentrate instead on the macroscopic properties of the fluid, taken as a complete system. This would presumably involve various conservation laws, assumptions of detailed balancing, equilibrium states and so on, and all of the details such as supernovae explosions, instabilities, etc. could enter the system as stochastic noise with a scale-dependent power. This is the approach often applied to laboratory systems that exhibit fractal structures and power laws, such as systems with second-order phase transitions, which have infinite correlation lengths (see, e.g., Ma 1976). Indeed, many of the details of laboratory phase transitions, such as the power-law indices, are independent of the microscopic details of the interactions between atoms and molecules, but depend only on gross properties of the system, such as the number of dimensions. The macroscopic approach is also generally applied to laboratory turbulence as an explanation for the power index of the velocity-position correlation (Kolmogorov 1941), without detailed consideration of the exact fluid motions or viscous coefficient.

Perhaps a future generation of theories will combine these two points of view, or perhaps detailed computer simulations, with all of the microscopic physics included, will show the scale invariant properties of the macroscopic system whether or not it makes their origin transparent. A problem with combined theories, however, is that if the power law and fractal behavior of the macroscopic system is independent of microscopic details, then the detailed physics that is put into the theory could be wrong and still the right general characteristics of the gas could be derived. This is often a problem with computer simulations, which can be made to look realistic with very simple, and in some cases, inappropriate assumptions about the detailed physics. Obviously both points of view, the microscopic and the macroscopic, should be investigated simultaneously.

The discussions in the preceding sections questioned the application of what we know about laboratory turbulence to interstellar gas processes. Such cautions are useful at this time, and should not be mistaken as an implication that the interstellar medium is not turbulent at all. Indeed, most of the interstellar gas is probably very turbulent: it has a random element, as does turbulence, nonlinearities in the fluid equations are important, and both laboratory and astronomical systems seem to have fractal structures (Beech 1987; Bazell and Desert 1988; Scalo 1990; Dickman et al 1990; Falgarone et al. 1991). Falgarone and Phillips (1990) suggest further that the broad line wings in interstellar clouds are another manifestation of turbulence; in their interpretation, the line profile is not a probability distribution for velocities, which would be nearly gaussian for turbulence (Batchelor 1967; Monin and Yaglom 1981), but a distribution of velocity differences inside a mapped region or telescope beam. Yet even with these similarities between astronomical and laboratory motions, the interstellar gas is still very different from a laboratory fluid. To emphasize this difference here, we use the terms "astrophysical turbulence" and "interstellar turbulence" (Scalo 1987) instead of just "turbulence." Indeed, it seems possible that some of the self-similar, nonlinear motion and structure currently ascribed to turbulence may later prove to be caused by other processes. An example might be the structure and velocities that result from overlapping wind-swept shells with a power-law distribution of stellar wind luminosities. Another example was given by Dickman et al. (1990), who noted that the area-perimeter correlation for molecular clouds, which suggests that the clouds have a fractal structure reminiscent of turbulence, could instead result from a random positioning of unresolved clumps with no required connection to turbulence. Fractal structures may arise from thermal instabilities also (Elphick et al. 1991).

How does the interstellar gas differ from turbulent fluids usually studied in the laboratory? There seem to be at least four characteristics of astrophysical gas dynamics that distinguish it from laboratory turbulence: highly supersonic motions, strong magnetism, strong self-gravity, and the presence of numerous embedded (stellar) energy sources.

Section IV suggested that the supersonic nature of the interstellar motions should cause much of the dissipation of energy to occur in shock fronts between colliding clumps or eddies. This shocking process short-circuits the usual cascade of energy from large to small scales and shunts the energy directly from every scale to the smallest scale (which is the shock thickness, equal to the molecular mean free path, or the ion-neutral collision scale in the case of a magnetic shock). Shocks also make the fluid highly compressible. Theory suggests that such compressibility makes the velocity-distance relation steeper than in an incompressible fluid (Fleck 1983; Biglari and Diamond 1988), in which case the observed relation $v \propto l^{1/3}$ for astrophysical gas (Falgarone and Phillips 1990) may be uncomfortably close to the relation observed in the laboratory for incompressible fluids.

The shocks in supersonic turbulence might also be expected to take the form of thin layers that appear as sheets or filaments in projection (Passot et al. 1988). Such sheets and filaments, unless they are unresolved, appear different from most of the interstellar gas, which looks more clumpy. In addition, the observed fractal structure of interstellar clouds, which is a property of the distribution of mass, differs significantly from the fractal structure discussed for laboratory turbulence, which is a property of the space in which energy is dissipated, the density being uniform. The dissipating regions in subsonic laboratory turbulence seem to consist of layered vortex sheet-like structures (Schwarz 1990), which have not yet been observed in astronomical sources.

Magnetic fields should also limit the motion of interstellar gas much more than the gas in laboratory fluids (see, e.g., Chaboyer and Henriksen 1990; Kahn and Breitschwerdt 1990). It may prevent or severely constrain vortical motions altogether and prevent the random diffusion of eddies and subsequent vortex stretching, which helps drive the energy cascade in laboratory fluids. Nonlinear magnetic wave interactions could drive turbulence in very different ways, replacing the vortex stretching in laboratory fluids with other nonlinear processes that cause energy to cascade. Alternatively, the most massive parts of interstellar clouds may just oscillate regularly on their magnetic field lines, driven by a balance between gravitational forces and magnetic tension. Such oscillations could drive nonlinear magnetic waves and turbulence elsewhere, but not be turbulent themselves.

Self-gravity is also likely to be important in structuring the interstellar gas. Gravitational collapse can increase the eccentricity of prolate or oblate objects (Smith 1989), exaggerating an already filamentary or sheet-like structure, and it can change the flat shocked layer between two colliding eddies or clumps into a globular feature. Note that the collapse time is comparable to the crossing time for a virialized object, and this is the approximate duration of a shock between two large eddies or clumps.

Embedded stars should also give structure and motions to the gas. Both field stars and newborn stars in a cloud probably have a power-law distribution of luminosities and winds, and they should push the gas around into structures that have a power law distribution too (see, e.g., Norman and Silk 1980a). Whether this structure resembles interstellar clouds cannot be determined without detailed modeling.

These considerations imply that many of the characteristics of interstellar gas that are currently attributed to turbulence may not really be the exclusive result of nonlinear terms in the equation of motion for isolated parcels of gas, but could instead be the result of power-law energy sources, gravitationally induced motions and waves, and mutual interactions between mass elements that have a random but nonturbulent nature. The net result of all of these processes could differ significantly from terrestrial turbulence, thereby limiting direct analogies, but the interstellar fluid could still be turbulent in a broader sense.

Perhaps something more analogous to laboratory turbulence is realized on very small scales in the interstellar medium (see, e.g., Spangler and Gwinn 1990), where thermal pressure is important, gravity is unimportant, magnetic diffusion is so fast that the field lines are relatively straight, and the nearest star is relatively far away. This smallest scale for astrophysical turbulence might be the largest scale for conventional turbulence, i.e., for nonlinear fluid processes that resemble those seen in the laboratory. Such small-scale turbulence may then cascade down in the usual fashion until it reaches the microscopic scale of the collision mean free path. As long as the nonlinear terms in the equations of motion dominate the dynamics, the interstellar fluid can presumably be self-similar over a wide range of scales, from 0.01 pc to 1000 pc for example, but if other terms are important too, such as the pressure from stars, self-gravity, galactic shear and so on, then such a wide range for self-similarity might be difficult to explain.

Acknowledgments. This review benefited from comments by J. Holliman and C. McKee, and from extensive discussions, particularly about turbulence, with E. Falgarone, J. Scalo and K. Schwarz.

GIANT MOLECULAR CLOUDS

LEO BLITZ
University of Maryland

The properties of galactic giant molecular clouds (GMCs) in the solar vicinity and in the inner Galaxy are reviewed. Special attention is given to the role of the clouds in forming stars. The question of whether all GMCs form stars is raised and it is shown that there is little evidence that GMCs anywhere in the Galaxy are devoid of star formation, even O star formation. The angular momentum of local GMCs is then discussed in relation to the angular momentum of the diffuse interstellar medium. At least three GMCs in the solar neighborhood have retrograde rotation in an inertial frame of reference if their large-scale velocity gradients are due to rotation. Three GMCs in different evolutionary states are identified, and some of the differences in their properties are identified. The internal structure of GMCs is then discussed in some detail. It is argued that clumps are the fundamental units in which the GMC mass is distributed. Clump properties are discussed quantitatively, and the mass spectrum of five GMCs are compared and shown to be remarkably similar. Although the majority of clumps do not appear to be gravitationally bound, the clumps can be confined by the pressure of an atomic intercloud medium. The observations presented in this review are used to suggest a tentative outline for the evolution of GMCs.

I. INTRODUCTION

The fundamental goal of the study of molecular clouds is to understand how they form stars. All present-day star formation takes place in molecular clouds, so we may think of them as providing the initial conditions for the process of star formation. Yet, while all stars may form in molecular clouds, there are at least some small molecular clouds that do not form stars. Furthermore, even within those clouds that do form stars, it appears that only a small fraction of the mass of a cloud actively takes part in the star-formation process. What controls both the presence and the absence of star formation in molecular clouds? There seems to be an inevitability to the star-formation process. Observational evidence from surveys of stars and gas indicates that almost all of the giant molecular clouds (GMCs) in the solar vicinity are currently forming stars. How is this inevitability to be reconciled with the absence of star formation in most of the molecular mass of the Galaxy?

We begin by asking a few broad questions about molecular clouds. (1) How do molecular clouds, especially the GMCs, form? (2) Once a molecular cloud forms, how does it evolve to generate the entities that eventually become stars? (3) What is the basic unit of a molecular cloud that produces stars, and how does it evolve to form a star? None of these questions has been answered as yet, but partial answers are beginning to emerge. This chapter will review some of the observational material relevant to these questions and will concentrate on the larger-scale aspects of the structure and evolution of GMCs. It will be argued that an understanding of the evolution of the interstellar medium into stars requires a detailed knowledge of the clumpy structure of molecular clouds. It will be shown that it is possible to identify molecular clouds in different evolutionary states, and that evolutionary effects within a particular molecular cloud, the Rosette Molecular Cloud, are observable in the dispersion of relative velocities of the identifiable clumps. The observations of a number of star-forming molecular clouds are reviewed with the aim of providing generalizations that are useful for future work. Among the topics not covered is the energy balance within molecular clouds. A good review of this subject is given by Genzel (1991b). We begin with a review of the large-scale properties of GMCs.

II. GLOBAL PROPERTIES OF GMCs

It is rather amazing that 15 yr since the identification of giant molecular clouds, there is no generally accepted definition of what a GMC is. There seems to be little disagreement about the classification of the largest clouds as GMCs, but an all inclusive definition of what a GMC is has proven elusive. A large part of the problem is that the various studies of the mass spectrum of molecular clouds indicate that the spectrum is well fit by a power law (see below) and there is consequently no natural size or mass scale for molecular clouds. What we call a GMC is therefore largely a question of taste. For the purposes of this chapter, any molecular cloud that has a mass $\gtrsim 10^5$ $M_\odot$ will be defined to definitely be a GMC, a cloud with a mass 10^5 $M_\odot \gtrsim M(\text{cloud}) \gtrsim 10^4$ $M_\odot$ probably to be a GMC, and a cloud with a mass $\lesssim 10^4$ $M_\odot$ probably not to be a GMC. In most cases, this fuzziness does not cause serious problems.

Giant molecular clouds have been studied as a whole largely through their CO emission in the radio portion of the spectrum. The studies have been mainly of two kinds. (1) Observations of individual objects, where a cloud is identified, often by its association with a visible H II region, and then mapped to its outer boundaries. (2) Surveys of the galactic plane in CO, where GMCs are identified through some objective criterion. In the second case, GMCs are generally defined down to some contour level because of the possibility of confusion with other gas along the line of sight. Surprisingly, the general properties of GMCs defined in this way are not significantly different from those identified by the first method. Table I gives the properties derived for clouds in the solar vicinity (see, e.g., Blitz 1987b). Inner Galaxy molecular clouds may be somewhat denser and more opaque (see McKee 1989; Solomon et al. 1987), but there is no evidence that they form a separate population from the local clouds. This important conclusion suggests that detailed studies of the molecular clouds near the Sun can tell us about the ensemble properties of GMCs everywhere in the disk. An exception is likely to be in the innermost regions of the Galaxy such as the molecular disk within 400 pc of the center.

TABLE I

Global Properties of Solar Neighborhood
Giant Molecular Clouds

Mass	$1\text{--}2 \times 10^5$ $M_\odot$
Mean diameter	45 pc
Projected surface area	2.1×10^3 pc^2
Volume	9.6×10^4 pc^3
Volume averaged $N(H_2)$	$\sim$50 cm^{-3}
Mean $N(H_2)$	$3\text{--}6 \times 10^{21}$ cm^{-2}
Local surface density	$\sim$4 kpc^{-2}
Mean separation	$\sim$500 pc

An example of a local GMC is shown in Fig. 1, the L1641 cloud in Orion, in which the Orion Nebula is located. The figure shows the emission from the ^{13}CO $J=1-0$ transition in various velocity bins, and in the last panel, the emission integrated over the cloud as a whole is shown. The figure shows several important features that are common to many local GMCs. (a) The ridge line of the final panel closely parallels the galactic plane (it is located at $b = -19°4$). Most local GMCs are similarly elongated (Blitz 1978; Stark and Blitz 1978). (b) The ^{13}CO emission shows a strong velocity gradient along the length of the cloud which is generally interpreted as rotation. The subject of angular momentum in GMCs is discussed in Sec. IV. (c) The ^{13}CO emission is seen to break up into discrete clumps. In the L1641 cloud, the clumpiness is apparently quite filamentary. The clumpiness of GMCs is discussed in detail in Secs. VII, VIII and IX.

From the study of local GMCs, the following general conclusions may be drawn:

1. GMCs are discrete objects with well defined boundaries (see, e.g., Blitz and Thaddeus [1980] for a quantitative discussion of this point—see especially their Appendix B). The well-defined boundaries suggest that there is a phase transition at the edges of a molecular cloud. This point is discussed in greater detail below.

2. GMCs are not uniform entities, but are always composed of numerous dense clumps and have small-volume filling fractions (see Sec. VII for a quantitative discussion). These clumps appear to have a range of geometries from spherical to highly filamentary (see the maps in the references to Table IV below as well as Bally et al. [1987]). Nevertheless, maps of the CO emission from the complexes suggest that the surface filling fraction of the clumps is almost always near unity, where unity is defined as one or more clumps along the line of sight (Blitz 1980, and references therein).

3. The CO and ^{13}CO line widths of the clumps are *always* wider than

L. BLITZ

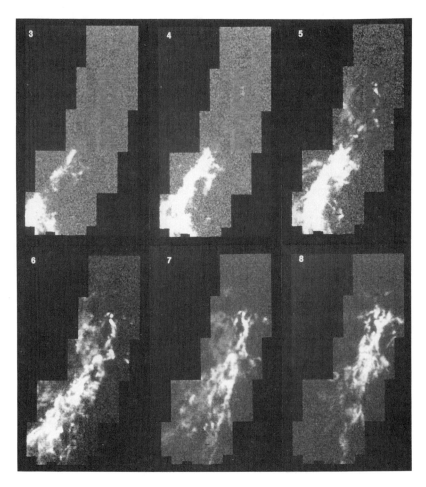

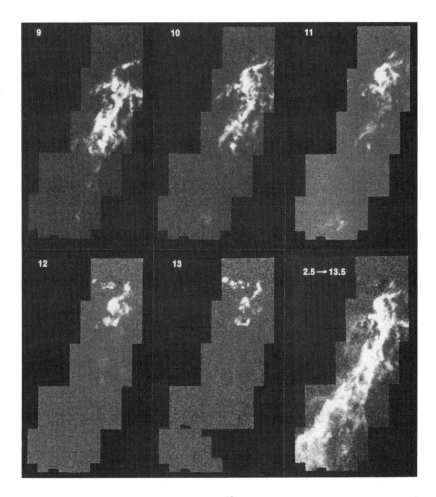

Figure 1.   A series of maps, each showing ^{13}CO emission integrated over a 1 km s^{-1} wide velocity range illustrating the internal structure of the Orion A cloud. The number in each subpanel is the LSR velocity range shown. Each panel shows a region roughly $2° \times 5°$ in extent with an angular resolution of about 100″. The galactic plane closely parallels the ridge line in the final panel. Figure from Bally et al. (1987).

the thermal widths implied by the excitation temperature. The linewidths are usually interpreted as the result of bulk motions associated with turbulence and/or magnetic fields.

4. GMCs are gravitationally bound (see, e.g., Kutner et al. 1977; Elmegreen et al. 1979; Blitz 1980). GMC masses are orders of magnitude larger than the Jeans mass computed from the excitation temperature of the CO. An indirect argument for the boundedness of local GMCs comes from considerations of their internal pressures (see Sec. VII). Typical internal pressures, P/k, are $\sim 1 \times 10^5$ K cm^{-3}, much larger than the mean local interstellar pressure of $\sim 1 \times 10^4$ K cm^{-3} (Bloemen 1987). Because these clouds are known to exist long enough to give rise to several generations of stars, or ~ 20 Myr (Blaauw 1964), the GMCs would have dispersed if they were not held together by gravity.

5. All OB associations form from GMCs; thus GMCs are the nucleation sites for nearly all star formation in the Milky Way (Zuckerman and Palmer 1974; Blitz 1978,1980, and references therein).

6. Within the uncertainties, the cloud-to-cloud velocity dispersion of local GMCs (with typical masses of $\sim 2 \times 10^5$ M$_\odot$) is the same as that of the small molecular clouds (with typical masses of ~ 50 M$_\odot$) found at high galactic latitude (Blitz 1978; Stark 1984; Magnani et al. 1985). The velocity dispersion of local molecular clouds therefore appears to be independent of mass over 3 or 4 orders of magnitude. If clouds are formed by collisional agglomeration of smaller independently moving clouds, then cloud velocity dispersions would be proportional to $M^{-0.5}$, unless there were some way to selectively re-energize the more massive clouds, an unlikely prospect. Furthermore, the velocity dispersion of molecular clouds seems to vary only weakly with galactocentric distance (Liszt and Burton 1983; Stark 1984; Clemens 1985). The constancy of the velocity dispersion with radius can also be inferred from the radial variation of CO scale height (see, e.g., Sanders et al. 1984). The half thickness of the CO layer increases from 50 to 70 pc when R increases from 0.4 to 0.8 R_0. The scale length for the stars in the disk of the Milky Way is about 5 kpc (van der Kruit 1987). Because $(v_g)^2 \propto (h_g)^2 \rho_*$, the increase in the scale height h_g just makes up for the decrease in the midplane stellar density ρ_*, and the velocity dispersion of the gas, v_g, is nearly constant.

For the most massive clouds, Stark (1983) has shown that the scale height, and by implication the vertical velocity dispersion of GMCs, is smallest for the largest clouds at a given radius. Cloud-cloud collisions may therefore be important for the largest, most massive clouds. All of these results taken together suggest that collisional processes are not important in the formation of GMCs except possibly in the inner Galaxy at the highest masses ($\sim 10^6$ M$_\odot$).

It is noteworthy, however, that within a GMC, collisional processes do seem to be important. This point is discussed in more detail in Sec. VII.

7. The survey of local molecular material within 1 kpc of the Sun (Dame et al. 1986) finds no GMCs without star formation. In fact, within 3 kpc of the

Sun, only one GMC is found without evidence of star formation (Maddalena and Thaddeus 1985). This cloud is discussed in more detail below. In the solar vicinity at least, molecular clouds without star formation are quite rare. It therefore appears that star formation is quite rapid after the formation of a GMC.

8. The above result, when combined with calculations of the destructive processes associated with the formation of massive stars, implies that the GMCs are quite young, $\sim$30 Myr (Blitz and Shu 1980). That is, because all GMCs appear to be sites of star formation, especially massive star formation, the clouds do not appear to be able to survive the birth of more than a few generations of massive stars, an argument that is consistent with various other lines of reasoning (Blitz and Shu 1980). Direct observational evidence for lifetimes of this order was first presented by Bash et al. (1977), and subsequently by Leisawitz et al. (1986) from the association (or lack thereof) of molecular material with young clusters of varying ages. Cohen et al. (1980) have come to the same conclusion from the confinement of the molecular arms in the outer Galaxy to narrow velocity ranges.

The primary argument for long-lived clouds in the Milky Way had come from continuity arguments based on the ratio of atomic to molecular gas in the molecular ring of the Milky Way (see, e.g., Solomon and Sanders 1980). That is, if the mass in a ring at some radius R is overwhelmingly molecular, and if GMC formation is initiated primarily by compression of diffuse gas in spiral arms, then in a steady state the mean ages of molecular clouds should be much greater than the mean time between spiral arm passages, $\sim$100 Myr at $0.5\,R_0$. However, the Bloemen et al. (1986) COS-B gamma-ray analysis shows that at virtually all radii, the atomic gas dominates the surface density, and even at the peak of the molecular ring, the excess of molecular gas is small. Some authors (see, e.g., Wolfendale 1991) prefer even lower molecular abundances. The data therefore suggest that the primary argument for long-lived clouds in the disk of the Milky Way is not tenable. A corollary of this conclusion is that clouds found in the interarm regions may have formed there; throughout most of the Galaxy, there is sufficient atomic gas to condense into clouds, even between spiral arms. It should be pointed out, however, that in the molecular disk within 400 pc of the galactic center and in some extreme molecule-rich galaxies, the molecular gas may be so abundant, that clouds may be quite long lived, especially if star formation can be inhibited in them.

9. A large fraction of the stars formed in GMCs do not return their material to the interstellar medium in a Hubble time. Therefore, either the GMCs were more numerous in the past, or the interstellar medium is replenished by infall or inflow from the outer reaches of the Galaxy (see, e.g., Lacey and Fall 1985).

Studies of GMCs based on inner Galaxy CO surveys run into the problem of finding an objective way to identify the molecular clouds. The degree of blending of the spectral lines, the large amount of foreground and background emission, and the necessity of using kinematic criteria for identifying the clouds add a significant amount of uncertainty to the identification of

GMCs (see, e.g., Adler 1988; Blitz 1987a). Nevertheless, a number of important properties of GMCs are inferred from the galactic CO surveys, and the searches for molecular clouds in the outer Galaxy. Some of the more important of these are the following:

(a) There appears to be a linewidth-size relation for GMCs in the inner Galaxy. Larson (1981) was the first to point out the existence of a power-law relationship between the sizes of molecular clouds and their intrinsic velocity dispersions. From various recent studies, the results are given in Table II. In this table, the quantity on the left is either the full width at half maximum or the one-dimensional velocity dispersion in km s^{-1}, and the quantity on the right is some measure of the mean linear size of a GMC in pc. The scatter in all of the studies is not terribly large, but note that the second and the third determinations are obtained from the same data set. The close agreement between the various observers in the value of the power-law exponent of the linewidth-size relation is so striking, that unless there is a selection effect common to all of the studies (see, e.g., Blitz 1987a; Kegel 1989), it suggests that the relation underlies some fundamental property of GMCs.

TABLE II
Linewidth-Size Relation for GMCs

Relation[a]	References
$\Delta V = 1.20R^{0.50}$	Dame et al. (1986)
$\sigma_V = 1.0S^{0.50}$	Solomon et al. (1987)
$\sigma_V = 0.31D^{0.55}$	Scoville et al. (1987)
$\Delta V = 0.88D^{0.62}$	Sanders et al. (1985)
$\Delta V = 0.85D^{0.63}$	Leisawitz (1990)

[a] ΔV = full-width at half maximum in km s^{-1}; σ_V = one-dimensional velocity dispersion in km s^{-1}; R, S and D are measures of the mean linear size of a GMC in parsecs.

Nevertheless, both the reality and the implications of the linewidth-size relation have been challenged by several observers. Adler and Roberts (1992) have argued on the basis of N-body simulations that many of the clouds identified in galactic surveys are likely to be chance superpositions of unrelated clouds. Nevertheless, these spurious clouds produce a linewidth-size relation with the observed slope. The reality of the relation has been questioned by Issa et al. (1990) from an analysis of published survey data. These authors showed that if the Solomon et al. (1987) method of analysis is used on their own survey data, but at (a) positions not centered on their clouds, and (b) at random locations in the survey, one obtains a linewidth-size relation similar to that of the original analysis. This result is puzzling unless the linewidth-size relation is an artifact of the dataset itself, rather than of the dynamical state of the clouds. In addition, Issa et al. also lowered the threshold of their analysis to a point where one expects clouds to be blended together into apparent clouds that are unrelated to one another. Again, Issa et al. obtained a linewidth-size relation indistinguishable from that of Solomon et al. suggesting that there is no physical significance to the linewidth-size relation. Its application to clouds to obtain, for example, their distances, may be flawed. Analysis of the Bell Labs ^{13}CO galactic plane survey, should be able to resolve this issue because of the higher contrast in the ^{13}CO data compared to that of CO.

(b) Solomon et al. (1987) have shown that for GMCs in the inner Galaxy there is a correlation between CO line luminosities and virial masses, and thus that (1) the CO line is a good tracer of the mass of a GMC, and (2) GMCs are in approximate virial equilibrium. Maloney (1988,1990a) has challenged this interpretation on two counts. First, he argues that the CO luminosity-virial mass correlation can be an artifact of the linewidth-size relation alone. Second, he argues that one obtains a linewidth-size relation indistinguishable from the data if one models an ensemble of *pressure* confined clouds. Because the CO luminosity-virial mass correlation may be the result of pressure confined clouds and not gravity, it need not imply that GMCs are virialized. However, it seems to this author that although pressure may confine the relatively small clouds, the evidence for gravitationally bound GMCs is overwhelming; see point (4) above.

(c) Independent estimates of GMC masses from gamma-ray observations (Bloemen et al. 1984,1986) and from IRAS observations (see, e.g., Boulanger and Perault 1988), suggest values for the CO/H_2 conversion ratio which imply that GMCs are gravitationally bound. If the clouds are in virial equilibrium, then

$$(\Delta V)^2 = \alpha GM/R \qquad (1)$$

where α is a constant near unity.

If $(\Delta V) \sim R^{0.5}$, then $M \sim R^2$ which in turn implies that the mean H_2 column density of GMCs is constant. This simple conclusion is presumably telling us something important about either how GMCs form or how they regulate themselves. McKee (1989) has assumed the latter in his recent theory of photo-regulated star formation in GMCs. Kegel (1989) however has criticized the constancy of the H_2 column density of GMCs as an artifact of observational selection. This issue can be resolved by aperture synthesis observations of GMCs in other galaxies.

(d) GMCs in the outer Galaxy appear to have lower excitation temperatures than GMCs in the inner Galaxy (Mead and Kutner 1988). It is unclear at present whether this is due to a lower external heating rate, decreased star-formation rate or both. In any event, the outer Galaxy GMCs have lower CO luminosities than those in the solar vicinity (Digel et al. 1990), which suggests that the CO/H_2 conversion ratio is a function of galactocentric distance.

(e) The distribution of masses for GMCs in the inner Galaxy is found to be (for masses in excess of 10^5 $M_\odot$), $dN(M)/dM \propto M^{-1.5}$ (Solomon et al. 1987). The power-law exponent is the same as that found for the distribution of clump masses in a GMC (see Sec. VII). If one tries to determine at what cloud mass half of the H_2 mass in the Galaxy resides, the various analyses of the inner Galaxy CO give results that fall within a factor of 2 of the range 1 to 2×10^5 M$_\odot$ when account is made of the different assumptions used and assuming that the power law doesn't steepen significantly at low masses. This is very close to the mean value in the solar vicinity (Stark and Blitz 1978).

One of the glaring deficiencies in galactic studies of GMCs is a quantitative study of how the properties of GMCs vary with galactic radius. For example, we might expect that to be stable against the larger tidal forces and the larger energy density of dissociating radiation, that inner Galaxy clouds would be denser than the clouds at larger galactocentric distance. Such a conclusion was reached from an indirect analysis of Liszt et al. (1981); however, no direct confirmation has been made. Such a study would be particularly useful in trying to understand how different galactic environments affect the formation and evolution of GMCs. It has been known since the Altenhoff et al. (1970,1978) 5 GHz surveys of the galactic plane, for example, that giant H II regions are much more common in what later became known as the molecular ring than they are at larger galactic radii. Surely the greater efficiency with which the inner Galaxy GMCs converts molecular gas into O stars must be a reflection of differing molecular cloud properties, but which properties? There have been a few studies of molecular clouds in the outer Galaxy that suggest that there are differences between them and the inner Galaxy clouds (see, e.g., Mead and Kutner 1988; Digel et al. 1990), but other than these, systematic studies of the gradients of the large-scale properties of GMCs as a function of galactic radius are almost totally lacking. It would seem imperative that even a rudimentary understanding of extragalactic CO emission would require an understanding of how GMC properties vary within the Milky Way.

III. DO ALL GMCs FORM STARS?

As mentioned above, various surveys of the molecular gas in the vicinity of the Sun have turned up only one GMC within 3 kpc that is devoid of star formation. On the other hand, Mooney and Solomon (1988) argue that at least 25% of the GMCs in the inner Galaxy are devoid of massive star formation. Therefore, on the basis of local observations, the answer to the question posed in the heading is essentially yes, and on the basis of inner Galaxy observations, the answer would be no. We must keep in mind that clouds without O star formation may still be forming stars of lower mass in great abundance, so strictly speaking, the solar vicinity and inner Galaxy observations are not necessarily incommensurate. Care must therefore be taken in defining the question so that answers are not biased by studies with different sensitivity limits.

Mooney and Solomon (1988) approach the question by looking at the 100 μm surface brightness of inner Galaxy molecular clouds and comparing the far-infrared luminosities, L_{IR} to the virial masses of the clouds, M_{VT}. They find that the ratio L_{IR}/M_{VT} varies by 2 orders of magnitude for a given mass, but the mean is independent of mass. Molecular clouds with masses greater than 10^4 $M_\odot$ (i.e., all of which are probably GMCs) are equally efficient, on average, at forming stars. On the other hand, they find a number of GMCs with low infrared surface brightnesses. They argue that these clouds are not rare and that therefore, about 25% of GMCs have no massive stars forming within them. Let us examine this latter conclusion with reference to what we know about nearby molcular clouds with known stellar contents.

Consider, for example, the Taurus and Ophiuchus clouds. Each has a mass of about 10^4 $M_\odot$ (Ungerechts and Thaddeus 1987; de Geus et al. 1990), and from the definitions given in Sec. II, these are probably not GMCs. Nevertheless, both clouds have substantial associated star formation, but no associated H II regions. The star-formation efficiency in the core of Ophiuchus, for example, may be as high as 25% (Lada and Wilking 1984). Although these cloud complexes both have stars as early as B associated with them, it would be very difficult to determine whether clouds in the inner part of the Milky Way that otherwise appear to be devoid of signs of star formation are like these clouds. Ophiuchus, for example, has a 100 μm infrared luminosity of about 7000 $L_\odot$, and the infrared luminosity is consistent with it being reradiated starlight from embedded sources (Greene and Young 1989). The overall star-formation efficiency of the Ophiuchus complex (zero age main sequence and pre-main-sequence stellar mass divided by the mass of H_2) is about 2 to 3% (Wilking et al. 1989a; E. Lada, personal communication), equal to the mean in the plane of the Galaxy (Myers et al. 1986). This value must be typical of GMCs in the Milky Way if most star formation takes place in GMCs. The level of activity found in Ophiuchus, even if it were scaled up for clouds with masses which would here be considered definitely to be GMCs, that is, masses greater than 10^5 $M_\odot$, would not be recognizable from studies such as those of Mooney and Solomon (1988).

However, we run into difficulties even if we attempt to determine whether all OB star formation would be detectable in galactic surveys, where we define OB stars as those that give rise to H II regions. To see this, it is useful to refer ahead to Fig. 3a which shows the 100 μm surface brightness of the cloud complex associated with the Rosette Nebula as observed by IRAS. The peak brightness of the map, which has had the background removed, is 720 MJy sr^{-1}, and the infrared luminosity of the cloud is $\sim 5 \times 10^5$ $L_\odot$.

Mooney and Solomon find that there is a class of infrared-quiet molecular clouds, many with masses $> 10^5$ $M_\odot$, which have peak 100 μm surface brightnesses only 1/20 that of the infrared-strong clouds. The infrared-strong clouds are defined by strong localized infrared emission which, in most cases, seems to result from the action of an H II region. If it were located in the inner Galaxy, the Rosette Molecular Cloud would probably be classified as infrared quiet based on the definition for infrared-strong clouds given by Mooney and Solomon. That is, an infrared-strong cloud must have a peak 100 μm surface brightness more than a factor of 2 above the local backgrond. In the longitude range $40° > l > 0°$, the mean surface brightness of the galactic plane between

$0.75° < b < −0.75°$ is never less than 1000 MJy sr⁻¹ (Sodrowski et al. 1987), thus in a typical location in the inner Galaxy, the Rosette Molecular Cloud would not meet the brightness criterion. Incidentally, the Rosette H II region itself is actually a minimum on the infrared maps. Furthermore, its peak 100 μm surface brightness is only 3.5 times that of the mean of the infrared-quiet clouds, ∼200 MJy sr⁻¹. Nevertheless, the Rosette Molecular Cloud is particularly rich in O stars even though its size and mass are typical of GMCs in the solar vicinity.

The nebula itself is illuminated by 7 stars of spectral type O4 to B0 stars (Pérez et al. 1987) and the stellar association has 13 stars of spectral type O 9.5 or earlier (Morgan et al. 1965; Turner 1976). The association is about as rich as the Orion OB1 association, the most spectacular star-forming region within 1 kpc of the Sun. However, in spite of the richness of its OB star formation, the RMC would probably have been classified by Mooney and Solomon as being devoid of OB star formation on the basis of its surface brightness, probably because there are no *embedded* O or early B stars. Furthermore, if the Rosette OB association had fewer O and B stars, its far-infrared surface brightness and luminosity would likely be even smaller. It is quite possible, therefore, that all of the clouds classified as infrared-quiet are Rosette-like objects or similar sources with fewer O and B stars.

A similar problem exists if one tries to use H II regions to trace the star formation in inner Galaxy GMCs as was done by Myers et al. (1986). They used the Altenhoff et al. (1970,1978) 5 GHz surveys to find H II regions associated with molecular clouds, and found a few GMCs (a smaller percentage than Mooney and Solomon) with very low star-formation efficiencies based on the absence of detectable 5 GHz flux. However, the Rosette Nebula is known to have an emission measure of about 3000 cm⁻⁶ pc (Bottinelli and Gouguenheim 1964), well below the detection limit at which H II regions are unambiguously identified in the Altenhoff survey. Other solar vicinity OB associations (e.g., the Mon OB1 association which contains the well known young cluster NGC 2264) have H II regions with even lower emission measures. Low-level infrared or radiocontinuum emission may simply signal that the conditions in inner Galaxy molecular clouds without giant H II regions are like those in the Rosette Nebula, and not due to the absence of O stars, or star formation in general. Such situations can arise when the action of O stars destroys the dust, when the angle subtended by the molecular cloud at the ionizing stars is small, or both.

Lockman (1990) has considered the detectability of the Rosette Nebula in the inner Galaxy in a recent review. He concludes that the Rosette Nebula would not have been detected in *any* discrete source survey of the inner Galaxy to date, including his own sensitive recombination line survey (1989). Consequently, the abundance of low surface brightness H II region/molecular cloud complexes in the inner Galaxy is unknown, and may account for the clouds identified by Mooney and Solomon and by Myers et al. that have little apparent star formation.

There may therefore be no discrepancy between the fraction of GMCs that form stars in the solar vicinity (virtually all) and that found in the inner Galaxy. The Mooney and Solomon clouds not only may have star-formation efficiencies similar to that of other GMCs in the Galaxy, but they could contain substantial OB star formation and still be infrared quiet according to their definition. On the other hand, some of the clouds in their sample may have low star-formation rates, but much more sensitive observations are necessary to determine this. The Maddalena-Thaddeus cloud remains the only known molecular cloud where the star-formation rate within the cloud is known to be a small fraction of the rate observed in local OB associations. It is possible with directed observations to set better limits on the stellar content of individual inner Galaxy molecular clouds than are currently available, and it may be possible to find clouds with star-formation rates significantly smaller than those found in local GMCs. On the other hand, if the inner Galaxy clouds form stars with the same efficiency as local GMCs (Mooney and Solomon 1988), we might expect that finding GMCs with significantly depressed star formation efficiencies inside the solar circle will prove difficult.

We conclude, on the basis of both local and Galaxy-wide studies, that the fraction of GMCs devoid of star formation is very small, probably less than 10%, but no more than 25%. This fraction suggests that the onset of star formation in a GMC is only a small fraction of its age, and that a GMC which is found to be truly without young stars must be very young, possibly 3 Myr, if the mean ages of GMCs are 3 Myr, as is the prevailing view. The alternative view requires that the small fraction of GMCs without star formation comprise a separate population of clouds in which the formation of stars is strongly inhibited for some reason. Detailed comparison of objects like the Maddalena-Thaddeus cloud with other GMCs should help to determine whether the latter viewpoint has any validity.

IV. ANGULAR MOMENTUM

The angular momentum of a GMC should reflect the angular momentum of the interstellar medium from which the cloud formed, thus the specific angular momentum of a GMC (that is, the total angular momentum per unit mass of a cloud complex) is an important parameter for understanding its history. Although much has been written on the evolution of angular momentum in connection with the formation of individual stars (see, e.g., Mouschovias 1991a; Chapter by McKee et al., and references therein), little is known about the angular momentum of GMCs. Mestel (1966b) considered the problem of the angular momentum of a planet condensing from a protoplanetary disk; this work is applicable in part to the problem of GMC formation from the disk of the Galaxy. In the case of a GMC that condenses from the general interstellar medium, however, it is possible, perhaps even likely that magnetic fields play a role (see, e.g., Mouschovias et al. 1974), a possibility not considered originally by Mestel, but which is discussed in his chapter in *Protostars and*

Planets II (Black and Matthews 1985). In what follows below we will first evaluate the evidence for the overall rotation of GMCs. We will then consider how the specific angular momentum of a GMC compares to that of the material from which a cloud formed.

A. Rotation vs Shear

That GMCs rotate has long been inferred from the velocity gradients observed across them (see, e.g., Kutner et al. 1977). The Kutner et al. observations of the Orion cloud show a continuous velocity gradient suggestive of solid body rotation (see Fig. 1 above). However, the velocity gradients may also be due to shear resulting from the momentum deposited by energetic stellar winds. Bally (1989) has argued, for example, that the large-scale velocity gradient in the L1641 cloud in Orion results from the interaction of the molecular gas with the energetic stellar winds from the O stars in the association, and that the gradient is due to a wind induced differential linear acceleration of the molecular material rather than rotation. Could it be that the large-scale gradients of all GMCs in which they are found are the result of similar stellar wind activity?

Table III gives the values of the largest known velocity gradients for entire GMCs. The values are uncertain by about 20%. The negative sign indicates that the inferred angular velocity is antiparallel to the angular velocity vector of the galactic disk.

TABLE III
Measurable Velocity Gradients for Giant Molecular Clouds

Cloud	Gradient km s^{-1} pc^{-1}	$1/2\,R^2\Omega$ km s^{-1} pc	Reference
Rosette	−0.18	−47	Blitz and Thaddeus (1980)
Mon R1	∼ −0.20	−45	Blitz (1978)
W3	0.30	27	Thronson et al. (1985)
Orion	−0.10	−74	Kutner et al. (1977)
Typical GMC	< ±0.05	< ±15	various; Blitz (1980)

Other velocity gradients published for GMCs are significantly smaller in absolute value than the values quoted above. Note that the observed gradients listed in Table III are much larger than the galactic shear due to differential rotation. In each case, the most recent burst of star formation has taken place at or near the edge of a prolate molecular cloud, and, with the exception of the W3 cloud, the gradient is in a direction roughly parallel to the galactic plane along the long axis of the cloud. In the W3 cloud, the gradient is nearly perpendicular to the plane of the Galaxy in a layer argued by Lada et al. (1978*b*) and Thronson et al. (1985) to be swept up by the Cas OB6 association. It appears likely that the velocity gradient in this cloud is related to the sweeping action of the stellar association. Can an OB association accelerate the molecular gas in a GMC along the *long* axis of the cloud in such a way as to induce the observed velocity gradients if the stars are situated at one end? The linear shear thus induced can in principle produce a positive or negative velocity gradient depending on the orientation of the cloud with respect to the line of sight. In what follows, we consider only the three clouds with gradients roughly parallel to the plane; we omit W3.

If GMC velocity gradients were due to the acceleration of cloud material by stellar winds, then the geometry of the complexes must have undergone a considerable rearrangement. Consider the last panel shown in Fig. 1. The stars capable of producing stellar winds are located just beyond the upper edge of the cloud along the line defined by the long axis of the cloud. The material most distant from the accelerating stars must have at one time been closest to them. The time scale required to reach the present state is $\sim$10 to 20 Myr. The material must have remained at all times molecular, because there is no evidence in any of the clouds that the gas has condensed from a previously atomic or ionized state. Nevertheless, we know that H II regions tend to ionize and evaporate the molecular gas rather than simply to accelerate it. A good example is the molecular gas most closely associated with the Rosette Nebula. The mean density of the optical nebula is nearly equal to the mean density of the molecular cloud suggesting that the stars have simply caused a phase change in the pre-existing gas. Also, there is remnant molecular gas seen in projection against the optical nebula, the well-known optical globules, and these have been shown to be evaporating (Herbig 1974; Schneps et al. 1980). In the three GMCs of interest here, the clouds are prolate with the long axis pointing away from the direction of the accelerating stars. It is hard to understand how it would be possible to accelerate such a large mass of neutral material so that the acceleration is linear rather than shell-like.

Perhaps the most serious objection to a wind induced gradient for the GMCs comes from the conservation of linear momentum. If the clouds in Table III have a mean mass of $\sim 10^5$ $M_\odot$, the linear momentum imparted to the clouds has a typical value of $\sim 5 \times 10^5$ $M_\odot$ km s^{-1}. This implies, from Newton's third law, that gas must have been accelerated in the opposite direction with the same momentum. Orion has a bubble of ionized gas expanding in a direction opposite to the molecular cloud (Reynolds and Ogden 1979), but the mean radial velocity of the ionized gas has the wrong sign to conserve the linear momentum. The ionized gas, therefore, should have no relation to the velocity gradient in the cloud. Furthermore, no such "counterjets" of material are evident in the vicinity of any of the other two GMCs in either atomic, molecular or ionized gas. Moreover, if the mass of any such gas is small compared to that of the GMC itself, it would be accelerated to velocities that would stand out clearly in spectral line surveys of such gas in the galactic plane.

It therefore seems reasonable to conclude that the velocity gradients observed in CO are due to rotation rather than shear, and the remainder of this section proceeds from this conclusion.

B. Conservation of Angular Momentum

1. Initial Angular-Momentum States. The specific angular momentum, J/M, of the general ISM with which one wants to compare that of a GMC depends in detail on how a cloud forms and the angular momentum distribution of the disk of the galaxy in which it is located. We might consider, for example, that a cloud forms when a disk-like region becomes gravitationally unstable and collapses. In this case, we expect a result similar to that considered analytically by Mestel (1966*b*) who showed that for a cloud that has condensed from a disk with a flat or rising rotation curve, the angular momentum is always prograde in an inertial frame of reference. If the rotation curve is falling, then the rotation can be either prograde or retrograde depending on the details of the cloud formation. Furthermore, it is the form of the rotation law *locally* that determines the initial angular momentum of the cloud.

We wish to calculate $1/2R^2\Omega$, the specific angular momentum for the interstellar medium from which a local GMC forms. The quantity R is the size of the region that has collapsed to form a GMC. Let us assume first that the cloud is formed from the condensation of atomic gas only. We now ask from what radius must a cloud contract to form a cloud with a mass of 2×10^5 $M_\odot$? Let us assume that the geometry of the initial configuration of the condensing gas is that of a cylinder with a diameter equal to that of its height. This assumption will not have a large effect on the final results unless the initial geometry is very different in one direction from the others. The surface density of atomic gas in the vicinity of the Sun is 5 $M_\odot$ pc^{-2} (Henderson et al. 1982). The local effective scale height of the atomic gas is 200 pc (Falgarone and Lequeux 1973); thus the midplane density of atomic gas is 0.5 cm^{-3}. The radius of a cylinder that would contract to form a typical GMC in the solar vicinity is therefore 140 pc.

If the GMCs formed from initially *molecular* gas, R would increase to 220 pc because the surface density of molecular gas locally is 1.3 $M_\odot$ pc^{-2} (Dame et al. 1987). However, it is difficult to see how a GMC could form from molecular gas without dragging some atomic gas along with it. If the GMCs form from a combination of atomic and molecular gas, R will of course have some intermediate value. The dependence of R on the initial geometry is weak because it scales as the 1/3 power of the initial mass. We may then take the value of 140 pc for R to be a reasonable lower limit, but even considering the uncertainties of geometry and the inclusion of molecular gas, it is probably not in error by more than about 50%.

The initial value of Ω depends on the details of the collapse (or equivalently accretion) of the gas that forms the GMC. Consider what happens for collapse along a line of constant radius in the limit that all of the collapsing gas lies at a single distance from the galactic center. This needle-like collapse has zero angular velocity in the rotating frame centered on the local standard of rest (LSR) (which I will call the LSR frame), and is +0.025 km s^{-1} pc^{-1} in the inertial frame. Consider now a similar collapse which takes place in the perpendicular direction, along a line of galactic radius. In the LSR frame, Ω is determined by the large scale shear in the Galaxy. Assuming a flat rotation curve, the value of Ω in the solar neighborhood is -0.025 km s^{-1} pc^{-1}; in the inertial frame, Ω is zero. We find therefore that the value of the initial angular momentum depends on the details of the collapse. The angular momentum of the gas that initially forms a GMC can have a range of values, therefore, from 0 to 250 km s^{-1} pc (or 0 to 7.7×10^{25} cm^2 s^{-1}) assuming that the GMC formed in a location where the rotation curve is flat. Allowing for the uncertainty in R, this value can be higher by about a factor of 2.

   2. *Final Angular-Momentum States.* Consider a typical GMC in the solar vicinity with a mass of 2×10^5 M$_\odot$. Such clouds have a typical dimension of 45 pc (Blitz 1978), a maximum dimension of about 100 pc and tend to be elongated along the galactic plane (Stark and Blitz 1978; Blitz 1980). If the true shape of the clouds is that of a cigar, the specific angular momentum is $1/3R^2\Omega$; if it is a disk, then the specific angular momentum is $1/2R^2\Omega$. As stated above, the values given in Table III are the largest published velocity gradients for entire GMCs. The velocity gradient of a *typical* GMC is less than half that of the L1641 cloud, and published maps make it difficult to discern gradients smaller than about 0.02 km s^{-1} pc^{-1}. In a survey of the literature, Blitz (1980) found that only half of the GMCs known at that time had a measurable velocity gradient. Thus, if the velocity gradient is entirely due to rotation, Ω has an extreme value for GMCs of 0.3 km s^{-1} pc^{-1} in the solar vicinity, but has a typical value somewhat less than 0.05 km s^{-1} pc^{-1}. The specific angular momentum of a typical GMC is therefore somewhat less than about ± 15 pc km s^{-1} or 5×10^{24} cm^2 s^{-1}, for clouds with a typical diameter of 45 pc. The specific angular momentum for the four clouds considered in Table III is listed in the table.

   For the clouds with measurable gradients, the sense of the rotation is *retrograde* with respect to the galactic rotation in a rotating coordinate frame of reference centered on the LSR; that is, the three clouds (other than W3) have higher velocities on the low longitude side of the cloud. To transform to an inertial frame of reference, we must add the angular velocity of the LSR, Ω_0, to the observed angular velocities; about 0.025 km s^{-1} pc^{-1}. This value is small compared to the measured angular velocites for the GMCs in the table above. Thus, the sense of the angular momenta of the three clouds with negative gradients is retrograde even in an inertial frame of reference. The GMCs without measurable gradients have angular momenta consistent with zero in the inertial reference frame. It was noted above that most of the mass in GMCs typically resides in dense clumps which may themselves be spinning. However, because of the R^2 dependence of the angular momentum, and because clumps do not, as a general rule, show large velocity gradients (although there are many specific counterexamples), the angular momenta of the clumps are not likely to contribute significantly to the total angular momentum of a GMC.

   3. *Comparison.* Because the clouds listed in Table III are strongly

counter-rotating, that is, rotating in a sense opposite to the Galaxy, in the inertial frame, these clouds cannot have conserved angular momentum from the initial states calculated above. Mestel (1966*b*) pointed out that retrograde rotation can occur if the disk from which the clouds form has a falling rotation curve. The Oort A constant expected in the solar vicinity for a flat rotation curve is 12.5 km s^{-1} kpc^{-1}. However, almost all measurements of this quantity suggest that the value is 15 km s^{-1} kpc^{-1} (Kerr and Lynden-Bell 1986), a value which implies that locally the rotation curve is falling. Such local variations are expected in the vicinity of a spiral arm. The value of the Oort A constant does not affect the collapse of a cloud along a line of constant radius. However, for the case of radial collapse, Ω has a value of -0.005 km s^{-1} pc^{-1} in the inertial frame. The corresponding specific angular momentum is ~50 km s^{-1} pc, but may be somewhat higher if the initial geometry is different from the one considered above.

Thus all of the clouds with large gradients have angular momenta within the values expected from collapse from the diffuse interstellar medium. However, the counter-rotating clouds put a severe constraint on how the clouds can have formed. That is, unless they did not form by condensation, they must have formed where the rotation curve is locally falling, and they must have collapsed very nearly along a line of galactic radius. Because this constraint is so severe, it is important to establish unequivocally that the large velocity gradients are due to rotation and not to shear. In this way, statistical studies of the angular velocities of many GMCs can provide important information about how the clouds formed.

Now, consider the *typical* GMCs. These have angular momenta which are only a small fraction of the allowable range. Does this mean that clouds only form from low angular-momentum initial states, or does it tell us that the initial angular momentum is frequently large, but in most cases is shed at an early stage as might be the case for magnetic braking? With the current data there is no way of deciding between these alternatives. Nevertheless, finding the answer to this question will likely tell us much about how GMCs form and evolve.

It is important to recognize that the observed values of J/M quoted above are reasonably well determined. For a molecular cloud, the velocity differences across a cloud can be measured to an accuracy of about 1 km s^{-1} (measurement errors are considerably smaller). Large velocity gradients are therefore measurable to an accuracy of 10 to 20%. The major source of uncertainty is probably the distance to a cloud, which, for clouds like Orion and the Rosette are probably as low as about 20%. For the ISM, the largest uncertainties are the value of the Oort A constant, which is probably known to an accuracy of 20%, and the midplane density of the atomic hydrogen gas, which translates into the distance R out to which the ISM must be collected to form a GMC. The uncertainty in the density is probably less than 50%, but in any event, the distance R depends only on the 1/3 power of the midplane density, and thus J/M on the the 2/3 power. The major uncertainty for the are formed. For example, if the efficiency of transformation of the ISM into molecular material is low, then R may be bigger, and the resulting J/M may be larger.

Studies of the angular momentum distribution among the GMCs simply do not exist at present in spite of the importance that they have for understanding how the molecular clouds form from the interstellar medium. It is unknown, for example, whether there are any examples of clouds with measurable prograde rotation in the LSR frame. Even the results quoted above come partly from unpublished analyses of molecular clouds and from personal communications. There is a wealth of information, however, in the large-scale surveys of the CO distribution in the Galaxy. For example, there have been hundreds of GMCs that have been catalogued (Solomon et al. 1987; Scoville et al. 1987) in the inner Galaxy. It would be an easy task to determine to what degree GMCs which exhibit measurable velocity gradients will have the sense of the angular momentum antiparallel to the angular momentum of the Galaxy, yet this has never been done. Nor has a detailed study of the angular momentum of GMCs as a function of galactic radius been done. Many other questions about the distribution of angular momentum are answerable with existing data.

V. CLOUDS IN DIFFERENT EVOLUTIONARY STATES

Within have at least some traces of star formation. However, in 1985, Maddalena and Thaddeus found a cloud in the outer Galaxy with a mass of $\sim 10^6$ $M_\odot$, a longest diameter of about 150 pc (both of these numbers assume a kinematic distance of 3 kpc), without any obvious traces of star formation activity. Moreover, the cloud has relatively weak, broad CO lines which appear to be different from those seen in most of the clouds observed in the local solar neighborhood. Maddalena and Thaddeus speculated that the cloud is so young that it has not yet had time to form stars.

If this hypothesis is correct, there should not be any buried or embedded population of stars within the cloud. It is possible, for example, that there is a large H II region on the far side of the cloud which is obscured by the intervening dust (even this hypothesis is unlikely because H II regions are almost always accompanied by strong CO peaks). In order to look for the effects of heating from an embedded population of stars, maps of the cloud have been made using the IRAS 100 μm database, and the result is shown in Fig. 2a (Puchalsky and Blitz, in preparation). Figure 2b shows the CO emission associated with the Maddalena-Thaddeus cloud taken from their paper. For comparison, a similar map of the 100 μm emission from the Rosette Molecular Cloud is shown in Fig. 3a, and the molecular cloud is shown in Fig. 3b. Both maps have the zodiacal emission removed and have had a background subtracted. Remarkably, the highest contour in Fig. 2a is lower than the *lowest* contour in Fig. 3a. The average 100 μm emissivity for the Maddalena-Thaddeus cloud is more than *2 orders of magnitude* lower

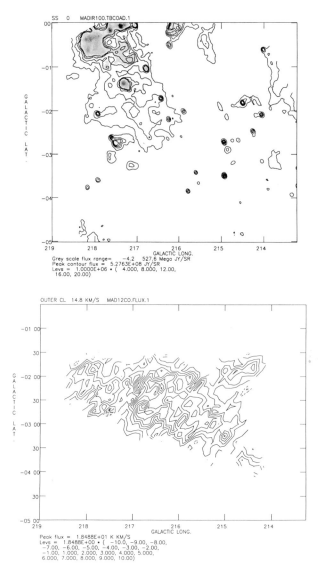

Figure 2.   Figures 2 and 3 show IRAS 100 μm emission and CO emission from two GMCs in apparently different evolutionary states. (a) IRAS 100 μm emission from the region of the Maddalena-Thaddeus GMC. The molecular cloud is located in the range $-1°45' > b > -3°30'$ and $214° > l > 219°$. There is no IRAS flux detectable in the figure that can be definitely associated with the molecular cloud. The largest flux detected in the direction of the cloud is about 20 MJy sr^{-1}. (b) CO emission integrated over all velocities associated with the Maddalena-Thaddeus Cloud. Both (a) and (b) are on very nearly the same angular scale. Note the almost complete absence of infrared emission from the region where the CO emission is detected.

for the Maddalena-Thaddeus cloud is more than *2 orders of magnitude* lower than that of the Rosette Molecular Cloud. Because the far-infrared emission is generally thought to come from reradiated starlight, by comparison with the Rosette, which has an infrared emissivity typical of GMCs in the solar vicinity (Boulanger and Perault 1988), the Maddalena-Thaddeus cloud is extremely deficient in embedded stars.

This evidence supports the hypothesis that the Maddalena-Thaddeus cloud is so young that it has not yet had time to form stars. It is therefore an excellent candidate to examine the differences in structure between it and a more evolved cloud like the Rosette Cloud, or the Orion Molecular Cloud. But if the Maddalena-Thaddeus cloud is too young to have yet formed stars, and the Rosette and Orion clouds are middle aged clouds still in the throes of star formation, is it possible to find a demonstrably old cloud, one which is now showing only the last vestiges of star formation? The answer appears to be yes. Probably the best candidate for a remnant molecular cloud is the small cloud associated with the Lac OB1 association. Although no systematic study of the molecular gas in this region has yet been undertaken, the overall morphology can be seen in Fig. 4 which shows an IRAS map of the region at a wavelength of 100 μm.

The Lac OB1 association is one of the oldest OB associations in the solar vicinity with an age of $\sim$20 Myr (Blaauw 1964). There are two subassociations, the younger of which has an age of $\sim$6 Myr. Very sparsely sampled observations (Blitz, unpublished) indicate that there is almost no molecular gas remaining in the region; what there is appears to be concentrated in the knot centered at $\alpha = 22°30'$, $\delta = 40°.5$. This knot is the location of the reflection nebula DG187. The stars, on the other hand, are spread throughout the area of the map. At the 500 pc distance of the OB association, the remnant cloud appears to be no more than 10 pc in extent. The large loop seen in the figure is probably dust associated with a shell of atomic hydrogen, and is evidently the result of the stellar winds, and possibly supernovae from the stellar association, which, together with the ionizing radiation from the O stars, appear to have effectively destroyed the molecular cloud. Very little is known about the overall gas and dust content of the region, but it presents a good opportunity to study what appears to be the last gasp of the star-formation process. Studies of such regions promise to explain how the gas and dust in a molecular cloud that has not been converted into stars is ultimately returned to the interstellar medium.

It is also possible to find clouds in different evolutionary states statistically, as has been done by Leisawitz (1990). He has surveyed the molecular gas near numerous open clusters of varing ages and finds that molecular clouds associated with open clusters older than 10 Myr are significantly smaller than the clouds associated with younger clusters. He argues that this result is not due to observational selection and suggests that molecular clouds are destroyed or dispersed on a time scale of $\sim$10 Myr.

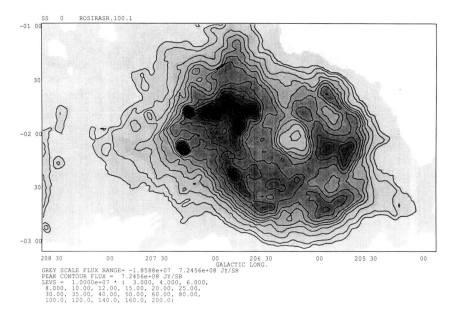

GREY SCALE FLUX RANGE= -1.8588e+07 7.2456e+08 JY/SR
PEAK CONTOUR FLUX = 7.2456e+08 JY/SR
LEVS =   1.0000e+07 * (3.000, 4.000, 6.000,
   8.000, 10.00, 12.00, 15.00, 20.00, 25.00,
   30.00, 35.00, 40.00, 50.00, 60.00, 80.00,
   100.0, 120.0, 140.0, 160.0, 200.0)

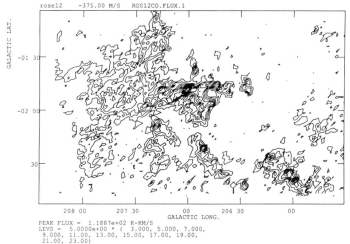

rose12   -375.00 M/S   ROS12CO.FLUX.1

PEAK FLUX = 1.1887e+02 K-KM/S
LEVS =   5.0000e+00 * (3.000, 5.000, 7.000,
   9.000, 11.00, 13.00, 15.00, 17.00, 19.00,
   21.00, 23.00)

Figure 3.   (a) IRAS μm emission associated with the Rosette Molecular Cloud. Note that the *lowest* contour in the emission is 50% higher than the *highest* contour in Fig. 2a. The difference in the mean emission from the two clouds is at least 2 orders of magnitude. The optical nebula is centered on the hole in the infrared emission at $l = 206°\!.2 b = -2°$. (b) Map of the CO emission integrated over all velocities associated with the Rosette Molecular Cloud from data taken with the Bell Labs' 7-m antenna. The angular scale is very nearly the same as that shown in (a) (figure from Blitz et al., in preparation).

2240+43_ LACZERO.B4.1

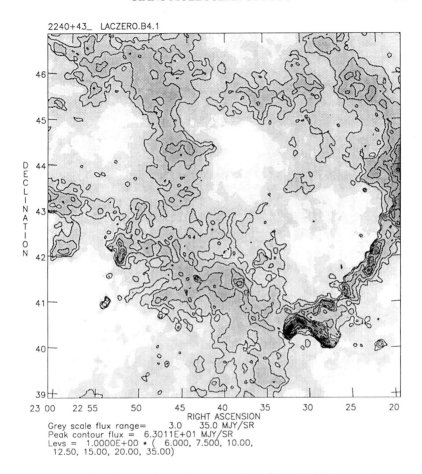

Grey scale flux range=   3.0   35.0 MJY/SR
Peak contour flux = 6.3011E+01 MJY/SR
Levs =   1.0000E+00 * (6.000, 7.500, 10.00,
   12.50, 15.00, 20.00, 35.00)

Figure 4.   IRAS 100 μm emission from the region of Lac OB1. The only location from which CO has been detected to date is the small knot at $\alpha = 22^h 30^m \delta = 40°5$ which is associated with the reflection nebula DG187. This figure was produced by E. de Geus.

VI. RELATIONSHIP TO ATOMIC HYDROGEN

The relatively sharp boundaries of GMCs are inferred from the quantitative analysis of Blitz and Thaddeus (1980), and the appearance of many GMCs on the Palomar Observatory Sky Survey prints. In the latter case, a well-defined region of dust obscuration is observed for many GMCs, and these follow the outermost contours of the CO emission quite closely. This is especially true for GMCs within 1 kpc of the Sun where the contrast between the foreground dust obscuration and the background stars is the highest. Good examples are the Orion Molecular Cloud (especially the boundary at the lowest galactic

latitudes (see the CO maps of Kutner et al. [1977] and the ^{13}CO maps of
Bally et al. [1987]), the Mon OB1 molecular clouds (see Blitz 1980), and the
Ophiuchus molecular clouds (see the CO maps made by de Geus et al. [1990]
and Loren [1989a]). There are many other examples that illustrate this point.

These sharp boundaries suggest that there is a phase transition that takes
place at the boundaries of the clouds, but what then is the state of the gas at
the low-density side of the phase transition of the GMC? Wannier et al. (1983)
have shown that for a few GMCs, there is a thin layer of atomic hydrogen in
a transition zone probably associated with a photodissociation region. In a
larger-scale unpublished analysis of the atomic clouds associated with local
GMCs, Blitz and Terndrup (unpublished) have analyzed the HI emission
from the velocity range detected in CO for a number of GMCs in the solar
vicinity. One such map was shown in Blitz (1987a). Figure 5 is another such
map; it shows the HI column density associated with the molecular cloud
accompanying the Per OB2 association (this is the cloud that contains the
NGC 1333 star-forming region). The map contours are in units of 10^{21} cm^{-2};
the contour marked 10 therefore is associated with an extinction (A_v) of about
0.5 mag. The map itself shows the HI emission integrated over the velocity
range 4.2 to 12.7 km s^{-1} , the velocity range associated with the Per OB2
cloud, which is shown as the shaded area in the figure. The association of
the atomic gas with the molecular cloud is quite obvious. All of the GMCs
surveyed to date show atomic envelopes similar to that shown in Fig. 5.

In order to quantify the relationship between the atomic and molecular
gas, Blitz and Terndrup estimated the mass of atomic gas associated with the
molecular clouds they studied by estimating the angular extent to which the
atomic gas shows an excess over the background in the relevant velocity range.
These masses are then plotted as a function of $M(H_2)$ derived from the CO
maps using the CO/H_2 conversion ratio of Bloemen et al. (1986). The results
are plotted in Fig. 6. The error bars are from estimates of the uncertainty in
defining the background level for the atomic clouds, and in determining the
CO/H_2 conversion ratio for an individual molecular cloud. What the figure
clearly shows is that the molecular clouds have atomic envelopes that are as
massive as the molecular clouds in most cases. Note, however, that the atomic
clouds are much more extended; the HI masses pertain to the entire region in
which HI is seen above the background. The individual maps show a very
small range in the peak column density of the atomic gas associated with the
molecular clouds; that is, only 1 to 2×10^{21} cm^{-2}, or an A_v of 0.5 to 1.0 mag.
A similar result has been found for diffuse molecular clouds in the Milky Way
(Savage et al. 1977), and for a GMC in M31 (Lada et al. 1988a).

VII. THE CLUMPY STRUCTURE OF GMCs

As early as 1980, Blitz and Shu noted that all GMCs that have been observed
up until that time exhibit clumpy structure. The evidence at the time suggested
that the density contrast between the clumps and the interclump medium is

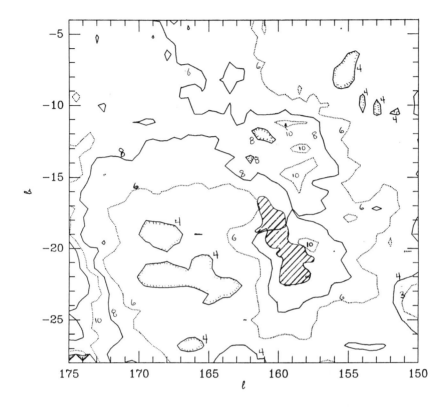

Figure 5.   Map of the atomic hydrogen emission associated with the Per OB2 molecular cloud. The emission is integrated over the velocity range 4.2 to 12.7 km s^{-1}, and the contours are in units of 10^{20} cm^{-2}. The molecular cloud is shown as the shaded region. There is clearly enhanced emission in the vicinity of the molecular cloud. This map is typical of all of the GMCs in the solar vicinity.

large because the densities required to detect CO at the observed antenna temperatures are an order of magnitude greater than the mean densities inferred if the clouds have dimensions along the line of sight comparable to their dimensions in the plane of the sky and the CO is uniformly distributed within it. Thus, if the H_2 is clumped so that the volume filling fraction is ~0.1 or less, implying a clump/interclump density contrast $\gtrsim 10$, the discrepancy disappears. This conclusion was qualitatively confirmed from the velocity structure observed in the ^{13}CO maps of the dense ridge of the Rosette Molecular Cloud (Blitz and Thaddeus 1980).

    Recently, Scalo (1988,1990) has attacked much of the standard picture of the description of the interstellar medium. He has argued that concepts such as clouds and clumps are frequently artifacts of the limited dynamic range inherent in many observations, selection biases, etc. However, high-quality three-dimensional data cubes have recently become available from

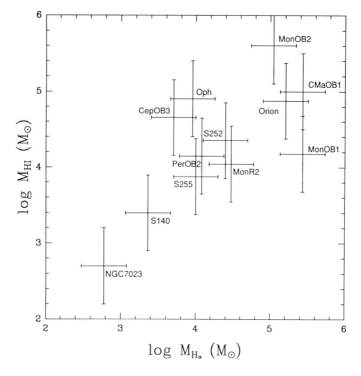

Figure 6.   Plot of the mass of atomic hydrogen emission associated with a number
of GMCs in the solar vicinity. The atomic hydrogen mass associated with the GMC
is taken over a narrow velocity range centered on the molecular cloud velocity and
the molecular mass is taken from the CO emission integrated over the area of the
molecular cloud.

observations of the generally optically thin ^{13}CO molecule. These cubes have
high signal-to-noise ratios, and have many resolution elements across the
projected cloud areas. Furthermore, the inclusion of the velocity information
can circumvent many of the objections raised by Scalo in his provocative
study. It is the inferences made from the ^{13}CO observations that are discussed
below.

      In principle, an understanding of the clumpy structure of molecular clouds
holds a great deal of information about their evolution. For example, is the
clumpy structure a result of the process of star formation, or is the clumpiness
primordial? If the clumpiness is primordial, one should be able to see the
process of clump formation taking place in the diffuse interstellar medium.
In either case, detailed observations of the kinematics of the clumps should
be able to resolve the issue. Does the interclump medium play a role in
the structure and evolution of the GMCs? In what form is the interclump
medium? How do the clumps eventually form stars? OB associations and
massive star clusters must eventually form from large massive clumps. Can

we find these clumps? Because collisions between clumps should be highly inelastic, the kinematics of the clump ensemble should provide a great deal of information about the history of a GMC.

The density structure of GMCs is best inferred from a pervasive optically thin molecule. The best current candidate is ^{13}CO, but receivers sensitive enough to observe ^{13}CO over the large angular extents needed to map a GMC fully have been available only since the early 1980s. Thus, fully quantitative information on the density structure of GMCs has not become available until fairly recently. Large scale ^{13}CO maps which delineate the clumpiness of GMCs are now available for the Rosette Molecular Cloud (Blitz and Stark 1986), the Orion Molecular Cloud (Bally et al. 1987a), parts of the Cep OB3 molecular cloud (Carr 1987), parts of the Ophiuchus molecular cloud (Loren 1989a; Nozawa et al. 1991), and some anonymous CO clouds toward l=90°, b=3° (Perault et al. 1985). Extensive observations of the clumpiness of molecular clouds also have been made in the Orion B molecular cloud using CS (Lada et al. 1991a) and in part of the M17 molecular cloud using $C^{18}O$ (Stutzki and Güsten 1990).

We review here some of the results from the ^{13}CO observations of the Rosette Molecular Cloud (Blitz et al., in preparation) and compare them to other published work on clumpiness. Blitz and Stark (1986) showed that the clumpy structure of the Rosette Molecular Cloud becomes especially clear when analyzed using position-velocity diagrams. Such plots are superior to channel maps (maps made in two spatial dimensions at a particular velocity interval) because small differences in clump velocities are not as apparent in the channel maps. Because the velocity differences between clumps are frequently smaller than the line-width of the CO emission from an individual clump, channel maps tend to blend the emission from several adjacent clumps into one large entity. On the other hand, because the velocity resolution of the position-velocity maps is generally high enough that the individual line profiles and the velocity differences between clumps is easily seen, the identification of separate kinematic units is easier to accomplish.

For the Rosette Molecular Cloud, analysis by means of position-velocity maps has made it possible to produce a catalog of individual clumps because in that cloud the separation between the clumps is large enough to make the identifications of individual objects possible by eye. A similar procedure has been carried out by Loren (1989a) for the streamers in Ophiuchus, and by Carr (1987) for a piece of the Cep OB3 molecular cloud. What makes the Rosette study different is that enough of the molecular cloud has been observed to make conclusions about the structure of the cloud as a whole. It is unclear, however, to what degree the procedure of identifying clumps by eye will work for other GMCs. This is simply another way of saying that it is unclear at present to what degree the structure of the Rosette Molecular Cloud is typical of other GMCs. It should be pointed out, however, that in terms of its size and mass, the Rosette Molecular Cloud is quite typical of other GMCs in the solar vicinity.

Blitz, Stark and Long reach the following conclusions from an analysis of their data:

1. Between 60 and 90% of the H_2 mass resides in the clumps they have identified in their catalog. Therefore, most of the molecular mass is in the clumps and is not in a distributed component.

2. The mean H_2 density for all of the clumps is $\sim 1 \times 10^3$ cm^{-3}. The volume averaged density for the entire complex is ~ 25 cm^{-3} (Blitz and Thaddeus 1980); therefore the volume filling fraction of the clumps is $\sim 2.5\%$. A similarly low-volume filling fraction of 3% is derived for the G90+03 molecular cloud complex observed by Perault et al. (1985). If we assume that 10 to 50% of the H_2 mass is in a diffuse component not related to the clumps, then the interclump density is ~ 2.5 to 12.5 cm^{-3}. The existence of the H I envelopes suggests that the interclump medium in GMCs also contains an atomic component, but because the envelopes are much more extended than the molecular clouds, the average H I density within the volume of the molecular cloud is unlikely to be more than about ~ 15 cm^{-3}. Thus, the clumps are not small density enhancements in a relatively smooth substrate, but are dense blobs held together by their mutual gravitational attraction which are moving through a tenuous interclump medium.

It should be noted that although the ^{13}CO observations provide the best quantitative estimates of the density contrast of the clumps, important confirmation of this contrast comes from the large spatial extent of the C II/C I regions around O stars (Stutzki et al. 1988; Genzel 1991b). That is, the large penetration of the C II emission regions into molecular clouds requires a clump/interclump density ratio of 10 to 100, similar to what is found from ^{13}CO mapping of the Rosette and G90+03 clouds. Thus, very different kinds of observations of GMCs in different parts of the sky reach similar conclusions about the density contrast of their clumps.

An interesting question arises as to whether the clumps themselves have internal structure. Because the clumps have a mass spectrum different from the Salpeter or Miller-Scalo Initial Mass Function (see below), the implication is that the clumps in which the star formation takes place must be fragmented on still smaller scales. Direct evidence for this comes from the C^{18}O mapping of two clumps in G90+03 by Perault et al. (1985), and from unpublished maps of ^{13}CO, C^{18}O, and CS of a large clump in the Rosette Molecular Cloud (Blitz and Puchalsky, unpublished). However, to date, this subclumping has not been well characterized quantitatively. Presumably it is the subclumping within a GMC, that is, the fragmentation of the individual clumps, that is the direct precursor of individual stars (see Sec. IX).

3. The distribution of clump mass follows a power law such that

$$dN(m) = N_0 M^{-1.54} dM \qquad (2)$$

where $dN(m)/dM$ is the number of clouds per solar mass interval, and N_0 is 460 $M_\odot^{-1}$.

There are a number of studies that have been done to date which indicate that the clump mass spectrum is a power law and that the power-law exponent is similar to that obtained for the Rosette Molecular Cloud. The observations are summarized in Table IV below. The Cep OB3 spectrum was calculated from the original unpublished list kindly provided by J. Carr and does not include the lowest-mass bin, which is likely to suffer from incompleteness. The Loren value, quoted in the original paper as -1.1 is incorrect, the value given in Table IV was recomputed from the data presented in his paper.

TABLE IV
Clump Mass Spectrum Power Law

Exponent	Tracer	Source	Reference
-1.5	^{13}CO	Rosette	Blitz et al. (1991a)
-1.6	CS	Orion B	Lada et al. (1991)
-1.4	^{13}CO	Cep OB3	Carr (1987)
-1.7	$C^{18}O$	M17SW	Stutzki and Güsten (1990)
-1.7	^{13}CO	Opiuchus	Loren (1989a)
-1.7	^{13}CO	Ophiuchus	Nozawa et al. (1991)

That all five studies should obtain such a similar power law is even more remarkable than the similarity of the power-law exponent in the linewidth-size relation for GMCs given in Sec. II. The data for the different clouds are sensitive to different ranges of mass and density, and use different reduction and analysis techniques. Furthermore, unlike the data used to obtain the molecular cloud distribution, all of the measurements are determined from fundamentally different sources. Given the uncertainties in the determinations of the slopes, it is reasonable to conclude that for the molecular clouds that are well studied to date, there is a universal mass spectrum for the clumps within a GMC, and that the spectrum is a power law with an exponent of -1.6 with an uncertainty of $\sim 10\%$ and that the power law seems to hold over three orders of magnitude in mass, i.e., from ~ 1 $M_\odot$ to about 3000 $M_\odot$. A reasonable inference is that the processes that determine the distribution of clump masses are rather similar from cloud to cloud. Clouds that show significantly different values of the mass spectrum are then likely to have had different dynamical histories.

It is noteworthy that none of the power-law exponents resembles the initial mass function, however. This suggests that the clumps that are observed in these studies are not the ones that form individual stars. Unless the clumps somehow have a star formation efficiency that varies with the mass of the clump (Zinnecker 1989b), the formation of individual stars lies deeper within the clumps that have so far been identified. Some of the larger clumps have sufficient mass to form entire clusters or OB associations, so this result is not entirely surprising. At some level, one expects to find the IMF mirrored in the spectrum of condensations within a GMC, but it appears that these condensations will be identified as the substructure within individual clumps.

This concept has meaning only if it is possible to identify a scale or a parameter that can differentiate the clumps within a complex from the structure within the clumps, but this has not yet been accomplished. It is likely that high-resolution interferometric observations will be needed to observe the true star-forming condensations in most GMCs.

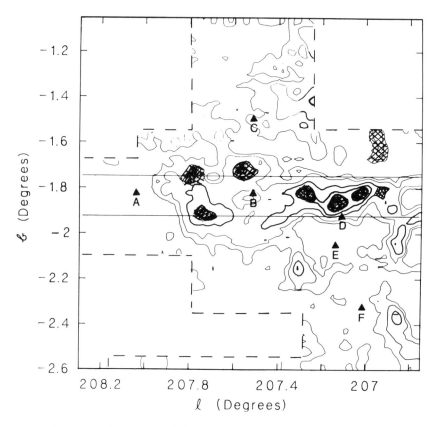

Figure 7.   Map of the ^{13}CO emission from the Rosette Molecular Cloud taken from Blitz and Stark (1986). The shaded areas are the locations of the eight most massive clumps associated with the molecular cloud complex.

4. Although most of the mass in the Rosette Molecular Cloud is in clumps that are at or near the point of being gravitationally bound, most of the clumps are not gravitationally bound. That is, half of the mass is in the 10 most massive clumps of the 86 catalogued; however, the 40 or so lowest mass clumps do not appear to be gravitationally bound. A similar conclusion was reached by Carr (1987) and Loren (1989a) who found that virtually none of the clumps in the clouds they observed are gravitationally bound. This point is discussed in greater detail in the following section.

5. The most massive clumps in the Rosette Molecular Cloud lie close to the midplane. This can be seen from Fig. 7 which shows the eight most massive clumps shaded on the map of the velocity integrated ^{13}CO emission published by Blitz and Stark. These clumps as a group are the most tightly bound of all of the clumps as measured by the ratio of their gravitational to internal kinetic energy. Thus, the next cluster of stars to form in the cloud complex is most likely to form from one of these clumps.

6. The clumps show a strong velocity segregation with mass. Figure 8 is a plot of the clump-to-clump velocity dispersion as a function of mass for the 86 catalogued clumps. The most massive clumps, the ones that lie closest to the midplane of the complex, also show the smallest velocity dispersion. *These results strongly suggest that the clumps have undergone dynamical evolution in the time since the complex has formed.* That is, the clump maps and the clump kinematics show a clear signature of inelastic collisions that indicate that the cloud has evolved considerably since it first formed. Because the observations were made in a part of the complex that has not been affected by the Rosette Nebula itself, and because there is little evidence for star formation in the mapped area, it appears that the observed structure and kinematics are the result of the dynamics of the clumps themselves without the intervention of energetic phenomena associated with star formation (e.g., H II regions, stellar winds, protostellar outflows, supernova explosions). Furthermore, the young stars that are found within the mapped boundaries (Cox et al. 1990) all appear to be embedded in dense clumps and would not therefore affect the overall dynamics. If the young stars and molecular clumps in this region are the result of either photoionization-regulated star formation such as that described by McKee (1989) or by the mechanism of Norman and Silk (1980), then substantial numbers of T Tauri stars which have not yet been identified would have to be found projected on the cold part of the molecular cloud.

7. There is a strong correlation between the embedded IRAS point sources and the clumps identified in ^{13}CO (Cox et al. 1990). The brightest infrared sources are associated with the most massive clumps. In the Rosette Molecular Cloud, most of the sources have flux densities that are strongly increasing functions of wavelength suggesting that they are very young. Except for GL-961, none of the sources has been investigated in any detail.

VIII. BOUNDEDNESS OF THE CLUMPS

In the Blitz et al. (in preparation) study of the clumps associated with the Rosette Molecular Cloud, at least half of the clumps are not gravitationally bound. The precise fraction is difficult to assess because of the uncertainties in the luminous masses derived from the ^{13}CO line strengths. Carr (1987) and Loren (1989a) find that *none* of the clumps in the clouds they observed are gravitationally bound. The latter two authors conclude that because the clumps are so far from being gravitationally bound that the clumps must be expanding. This conclusion is, however, difficult to accept. First, the clumps

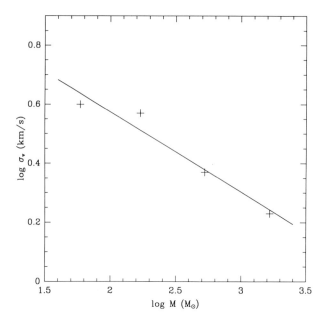

Figure 8. Plot of the clump-to-clump velocity dispersion as a function of mass in the Rosette Molecular Cloud. The straight line is a least squares fit to the data and has a power law index of −0.27.

would be expanding on the scale of a crossing time which, for the clumps they measured is less than 1 Myr. The clumps would all be dissolving unless there were also clumps that were forming at the same rate, but such clumps have not been identified. Second, unless all three clouds were observed at some special epoch, then mass conservation would require that

$$\frac{n(\text{clump})}{n(\text{interclump})} = \frac{t(\text{dissolve})}{t(\text{form})}. \tag{3}$$

Direct evidence from the Rosette Molecular Cloud indicates that the left-hand side of the equation is of the order of 100 or more, which requires that the formation time scale is no more than ∼0.01 Myr if the structure of the Rosette Molecular Cloud and the Ophiuchus clouds is similar. This improbably short time would seem to require that the interclump medium would be filled with stars that could sweep the interclump gas together quickly to form the clumps as quickly as they dissolve, a requirement which does not seem to be supported by the observations.

On the other hand, the clumps could be bound together by the pressure of the interclump medium. Such a medium would be in hydrostatic equilibrium with the gravitational potential of the GMC. Let us calculate what is required and compare it to the observations of the Rosette GMC. The pressure inside a clump is given by:

$$\frac{P\,(\text{clump})}{k} = \frac{nm(H_2) < v_{1D}^2 >}{k} \tag{4}$$

where we obtain from observations of the Rosette Molecular Cloud the following mean values: $n = 500 - 1000$ cm^{-3}; $<v_{1D}^2>^{0.5}=0.70$ km s^{-1}; and $P/k = 6 - 12 \times 10^4$ K cm^{-3}.

This pressure is a kinetic pressure determined from the linewidth of a typical clump. The thermal pressure, given by nT, is frequently less than the value quoted above away from regions of active star formation; typical excitation temperatures are 5 to 15 K as derived from the optically thick CO line. Therefore, the linewidth and pressures within a clump are generally dominated by nonthermal bulk motions of the gas.

The mean pressure within a GMC due to its self-gravity is given by:

$$\frac{P_{\text{grav}}}{k} = \frac{\alpha GM <n>m(H_2)}{Rk} \tag{5}$$

where M and R are the mass and radius of the GMC and $<n>$ is the volume averaged density of the complex. The constant α is of order unity and depends on geometry and viewing angle. It has recently been computed by Bertoldi and McKee (1992) for a variety of interesting geometries. Again for the RMC we have: $<n> = 23$ cm^{-3}; $M = 1.1 \times 10^5$ M$_\odot$; $R = 23$ pc; and $P_{\text{grav}}/k = 12 \times 10^4$ K cm^{-3}. Here R is found by taking the square root of the projected surface area, and α is taken to be 1.

Thus, the pressure that the interclump gas would have if it were in pressure equilibrium with the self-gravity of the complex is equal, within the uncertainties, to the pressure needed to confine the clumps. It was argued in Sec. VI above that the atomic hydrogen is very likely to be the interclump gas. Indeed, detailed observations of the atomic hydrogen associated with the Rosette (Kuchar et al., in preparation) made with the Arecibo telescope suggest that there is a detailed anticorrelation between CO and HI maxima in the cloud, lending support to the idea that warm HI is likely to be the interclump gas. The pressure of this gas can be obtained from published observations of the HI associated with the Rosette Molecular Cloud at low resolution made by Raimond (1966). These are as follows:

$$\frac{P\,(\text{HI})}{k} = \frac{nm(\text{HI})<v_{1D}^2>}{k} \tag{6}$$

where n in this case is the mean HI interclump density estimated from both the Raimond (1966) and the Kuchar et al. (in preparation) observations, and the other values are as follows: $<n> = 7$ cm^{-3}; $<v_{1D}^2>^{0.5}=10.6$ km s^{-1}; and hence $P\,(\text{HI})/k = 10 \times 10^4$ K cm^{-3}. Thus it seems quite reasonable that the clumps are confined by the pressure of the interstellar gas.

But why is it then, that *all* of the clumps in the Ophiuchus and Cep OB3 observations are found not to be bound by gravity, whereas many of the clumps in the Rosette Molecular Cloud are self-gravitating? The answer is the differing linear resolutions of the observations. In both the Loren and Carr observations, the clouds are considerably nearer than the Rosette; for the Ophiuchus observations, the difference is an order of magnitude. With one exception, the most massive clumps in the Loren and Carr samples are all within the mass range for which only unbound clumps are found in the Rosette. Their observations are therefore quite compatible with the Rosette observations. In fact, the three sets of observations can be combined to conclude that there seems to be a mass below which the clumps at the densities sampled by the ^{13}CO molecule are not gravitationally bound, and above which the clumps are bound. This mass appears to be $\sim$300 to 500 M$_\odot$. Apparently, then, low-mass clumps such as the cores investigated by Myers and Benson 1983 must be much denser than the clumps observed in ^{13}CO to be gravitationally bound.

Bertoldi and McKee (1992) have extended the arguments presented here to include the effects of magnetic fields. In an extensive work on the gravitational stability of the clumps, they show, among other things, that the magnetic pressure is insufficient to support the most massive clumps, thus internally generated turbulence is necessary to stabilize them. They suggest that low-mass star formation, which is observed in at least some of these clumps, is the source of this turbulence.

IX. STAR FORMATION IN THE CLUMPS

If the clump mass spectrum suggests that the clumps (at least those that have been identified in studies made to date), are not the progenitors of individual stars, then what is the role of the clumps in the star formation process? Since Roberts (1957) first showed that most stars form in clusters and associations, a result that was later confirmed by Miller and Scalo (1978), it seems natural to suppose that the more massive clumps give rise to these stellar groups. The masses of the largest clumps in the Rosette Molecular Cloud, for example, are comparable to the masses of OB associations (assuming that they form with a normal IMF) and open clusters.

A significant advance in understanding the star formation in GMCs was recently made by Lada et al. (1991*a*) who examined the Orion B molecular cloud for CS clumps and also did a near-infrared survey to look for embedded stellar objects. She was able to identify 39 dense clumps of molecular gas in her survey; of these nearly all the star formation is limited to the three most massive clumps. These three clumps account for only 30% of the mass of dense gas and $\leq$8% of the total gas mass in the area she surveyed. Yet when account is taken of background sources, she finds that $\sim$96% of the of the star formation in the surveyed region is associated with the dense clumps. This work is discussed in more detail in the Chapter by Lada et al.

The conclusion seems to be that the clumps, especially massive, dense clumps are the units of star formation in GMCs. It will be important to extend Lada's work to other clouds to see to what degree her results can be generalized. But the evidence seems to indicate that regions of incipient star formation can be identified from the molecular observations. The key to understanding how molecular clouds form stars now seems to require detailed investigations into the structure of the dense clumps with an eye to understanding how density inhomogeneities develop in these objects and what is the cause of their internal motions.

X. THE EVOLUTION OF GIANT MOLECULAR CLOUDS

The observations cited in this review, taken together, suggest a tentative picture of the evolution of a GMC. We imagine first, that by some process, the interstellar medium collects enough mass together to form a GMC. Methods by which this can in principle be done include spiral arm shocks (Roberts 1957; Cowie 1981; Elmegreen 1982b), shear instabilities (see, e.g., Toomre 1964; see also Elmegreen 1991a and references therein), Rayleigh-Taylor instabilities (see, e.g., Mouschovias et al. 1974), and through "supershells" driven by multiple supernovae (Öpik 1953; Heiles 1979; McCray and Kafatos 1987). Evidence that some other method is also at work comes from observations of other galaxies which suggest that the star-formation rate is not strongly dependent on spiral arm morphology; even galaxies without coherent spiral arms form stars quite efficiently (Stark et al. 1987). In any event, the dominant process by which the diffuse interstellar medium is collected into clouds in the solar vicinity has not yet been identified.

As enough material is collected together to form a cloud, when the column density of atomic hydrogen exceeds a threshold of about 10^{21} cm^{-2}, the associated visual extinction of about 0.5 mag is sufficient to shield the gas from the ultraviolet radiation that dissociates the molecules that form, and the cloud begins to turn molecular. It is presumably early in the process of turning molecular that the clumps have begun to form. Evidence for this comes not only from the Rosette Molecular Cloud, where the kinematics suggest that clumpiness precedes the star-formation process and then evolves, but from observations of high-latitude molecular clouds very close to the Sun. Many of these clouds appear to be very young (≤ 2 Myr), and all show evidence of clumpiness that must surely be primordial (Magnani 1986). Furthermore, the properties of these clouds are quite similar to the gravitationally unbound clumps in the Rosette Molecular Cloud, Cep OB3, Ophiuchus and M17.

It is not necessary for the interstellar medium to produce proto-GMCs that are gravitationally bound. A gravitationally neutral cloud, that is, one in which the internal kinetic energy is just equal to its potential energy, can become gravitationally bound, because the clumps that have formed will collide, and the kinetic energy of the ensemble of clumps can be radiated away because the clump collisions are so inelastic. The details of such a process are amenable to numerical modeling. If a GMC forms in this way, then it is natural that the atomic gas from which it formed remains as the interclump gas. The H I envelope that remains in this case is largely the remnant of the primordial GMC. All of the gas will remain in pressure equilibrium, and as the cloud becomes gravitationally bound, the interclump medium will respond to the gravitational potential well of the GMC. The pressure of the interclump medium will therefore increase above the nominal interstellar value, putting the clumps within the cloud under higher pressure than they had initially.

The magnetic field can support a clump against gravity up to a certain critical mass (see, e.g., Mestel and Spitzer 1956; Mouschovias 1987a; McKee 1989). Once a clump becomes magnetically supercritical, then even the magnetic field cannot prevent collapse. Thus star formation will take place in the clumps that have grown to be the largest and densest through collisions. Some support of the clumps themselves can also come from the energy input from protostellar and neostellar winds (see, e.g., Margulis et al. 1988). It may very well be that a process of self-regulating star formation similar to that described by McKee (1989) may then take place, until the cloud is ultimately consumed from the dissociating effects of the H II regions, stellar winds, and supernova remnants of the stars that formed within it.

Note, however, that the process of molecular cloud evolution described above has no need for sequential star formation. This is not to say that sequential star formation is irrelevant, but that some of the difficulties with the original theory of Elmegreen and Lada (1977) can be avoided with the evolutionary scheme outlined above. For example, Elmegreen and Lada noted that there is a need for an initial trigger to generate the first subgroup of O stars in their theory. Furthermore, the theory is based on the assumption that an ionization front propagates into an initially *homogeneous* molecular cloud; as discussed above, such clouds are not observed. Furthermore, in some cases, young OB subgroups have been identified that lie *between* older subgroups; as for example in Cep OB3 (Sargent 1977), and in the Sco-Cen OB association (Blaauw 1964; de Geus 1988). In the outline of GMC evolution described here, star formation proceeds in a quasi-random way within a molecular cloud. That is, the collisions between clumps in a molecular cloud proceed until one of the clumps becomes unstable to the process of formation of an OB subgroup. The stellar activity will ionize the molecular gas in its vicinity, and in a time probably less than 1 Myr (depending on a number of variables), this subgroup and its attendant H II region will appear to be at the edge of the molecular cloud. The next large clump to form an OB association will then be determined by the collisional processes between the remaining clumps in the complex, and will be independent of the previous star formation history of the cloud. In this way, there is no need to postulate a special event to produce the first generation of stars, and the subsequent generations of stars do not necessarily have a causal connection to previous generations. The apparently random orientation of the subgroups associated with many GMCs occurs quite naturally. Cases like the ordered orientation of the subgroups of the Orion complex therefore would be the result of chance (and are not very unlikely).

The degree to which this evolutionary picture is true will depend on the results obtained from the ^{13}CO mapping of a number of GMC complexes, the identification of more complexes in different evolutionary states and the identification of embedded cluster of stars within the complexes. Once the process of the formation and evolution of GMCs is better understood in the Milky Way, we can then apply our understanding to other galaxies where the environmental conditions are often markedly different.

Acknowledgements. This work was partially supported by an NSF grant and funding from the State of Maryland to the Laboratory for Millimeter-wave Astronomy. I wish to thank C. Lada, E. Lada, C. McKee, J. Pringle, L. Spitzer and J. Williams for a critical reading of the manuscript; as well as J. Bally, C. Gammie, C. Lada and D. Spergel for stimulating conversations on various subjects included in this review.

THE CHEMICAL EVOLUTION OF PROTOSTELLAR AND PROTOPLANETARY MATTER

EWINE F. VAN DISHOECK
California Institute of Technology

GEOFFREY A. BLAKE
California Institute of Technology

B. T. DRAINE
Princeton University

and

J. I. LUNINE
University of Arizona

The different processes that can affect the chemical composition of matter as it evolves from quiescent molecular clouds into protostellar and protoplanetary regions are discussed. Millimeter observations show that the chemical state of dense interstellar clouds prior to star formation is highly inhomogeneous: cold, dark clouds such as TMC-1 and L 134N show large chemical gradients on scales of a few tenths of a parsec, which are not well understood. Chemical models based on gas-phase ion-molecule reactions are moderately successful in reproducing the observed molecular abundances, but their predictive power is limited by unknown rates for several crucial reactions. Also, observational data for several important molecules are lacking. As a result, the abundances of the dominant oxygen- and nitrogen-bearing molecules prior to star formation are poorly determined. Some atoms and molecules are found to be depleted significantly onto grains, but the mechanisms for returning species to the gas phase in cold clouds are still uncertain. In star-forming regions, the high temperatures induced by radiation from newly formed stars can evaporate volatile grain mantles, and can open up additional gas-phase reaction channels. Powerful shocks associated with the outflows can also return more refractory material such as silicon and sulfur to the gas phase. These processes are operative in the best studied high-mass star-formation region Orion/KL, where interferometric observations reveal a complex chemistry with variations on scales of <2000 AU. For low-mass stars, observations of the chemistry in circumstellar disks on scales of 500 to 10,000 AU are only just becoming available. Initial studies of the young stellar object IRAS 16293–2422 also show large chemical gradients on scales of less than 1000 AU, but the chemical abundances appear less affected by the star-formation process than in Orion/KL. Systematic studies of the chemistry in star-forming regions are still lacking. On smaller scales of 50 to 500 AU corresponding to the outer solar nebula, observations of only the nearest and brightest objects are now possible, and detailed models of the chemistry have yet to be made. On scales appropriate to the solar nebula, 5 to 50 AU, material is thought to be physically and chemically modified in a number of ways: accretional shock heating, gas-drag heating on grains, and lightning are possible sources of energy. In the inner part of the nebula, temperatures are high enough for gas phase reactions to alter the molecular composition significantly; however, the amount of radial mixing between the inner and outer parts of the nebula may be limited. The extent to which outer solar nebula material retained its interstellar composition is therefore a highly complex issue, dependent on the body and the molecular/isotopic signature being considered. Comets almost certainly contain the most pristine matter, and a detailed comparison between the observed abundances in comet Halley and those found in interstellar clouds is made.

I. INTRODUCTION

The chemical evolution of protostellar nebulae is intimately related to the chemical state of the molecular clouds from which the stars form. It is well known that molecular clouds come in different varieties, ranging from the diffuse clouds to the giant molecular clouds, with distinctive chemical characteristics. The major question addressed in this chapter is the extent to which this chemical state is retained during the formation of stars and planets.

That the chemical state of an interstellar cloud will be modified by the star-formation process is evident. Young embedded or neighboring stars can raise the kinetic temperature from 10 K to a few 100 K. In addition, energetic protostellar outflows create shocks which can raise the temperature locally to more than 2000 K. Such high temperatures not only drive the chemistry from the cold, ion-molecule gas-phase scheme toward an equilibrium chemistry, but they can also evaporate the icy organic mantles of the grains and return material to the gas phase. A major goal of astrochemistry is to understand this chemical evolution well enough that the various molecules can be used as tracers of specific physical activity in the protostellar environment. For example, can the abundances and distribution of some molecules be used to probe the outflow activity of a protostellar system? If so, what is driving the chemical selectivity, and how does it depend on stellar type or evolutionary state? Another important question concerns the amount of gas-phase depletion of molecules onto the grains in a collapsing molecular cloud. When does it occur and how? To what extent are the molecular abundances modified by grain chemistry, and what are the critical processes controlling the return to the gas phase?

The reverse question of how the initial chemical state of a cloud affects the star-formation process is equally important. On theoretical grounds, the rates of collapse, and ultimately the efficiency of star formation in the cloud, are expected to depend on the cooling rates, which, in turn, are governed by the chemical composition of the gas. The efficiency of forming stars also depends sensitively on parameters such as the fractional ionization in the cloud, which is controlled by the cosmic-ray ionization rate and subsequent chemical processes. Similarly, the evolution of the grain size distribution and composition in a collapsing molecular cloud may influence profoundly the chemical and physical state of the early solar system. Unfortunately, few observational or theoretical facts regarding these points are yet available, so that they will receive relatively little attention throughout the rest of the chapter.

Once the interstellar matter enters a protostellar nebula, it will be subjected to numerous physical processes which can further alter its composition. The amount of transformation will depend on the exact position in the solar nebula, with only minor changes occurring in the cool outer part, but nearly complete reprocessing of grains and molecules taking place in the hot inner part. One of the major uncertainties here is the distance at which the transformation from kinetically controlled interstellar chemistry to "nebular" equilibrium chemistry occurs, and the amount of mixing between the two phases. The chemical composition of the various objects in our own outer solar system, such as the outer planets, comets and meteorites, may provide important clues in this respect.

While the above areas of research cover a wide range of physical and chemical problems in all phases of the life cycle of molecular clouds, they share the common thread that a combination of observational and experimental data with detailed theoretical modeling is required to make significant headway. Indeed, vigorous complementary laboratory and theoretical programs addressing fundamental chemical physical problems are essential to unravel the myriad of chemical processes occurring during the star-formation process.

This chapter summarizes systematically our current understanding of the chemical state of each of the phases of the star-formation process on scales ranging from interstellar clouds to that appropriate to the solar nebula. Sections II and III contain a critical discussion of the chemistry in interstellar clouds prior to star formation on scales >10,000 AU, both from a theoretical and an observational point of view. This is followed (Sec. IV) by a summary of the observational data on the chemistry in regions of recent high-mass and low-mass star formation on scales of 500 to 10,000 AU. The availability of a new generation of large aperture single dishes and interferometers at millimeter and submillimeter wavelengths, combined with boosts in receiver sensitivities, have increased our information in this area significantly in recent years. However, the molecules have so far been used mostly as diagnostics of physical conditions such as temperature, density and velocity structure, with little regard to the chemistry. Detailed chemical studies have been performed for a few specific objects, but systematic observations of molecular abundances as functions of, for example, luminosity of the central star or evolutionary state have yet to be carried out. As a consequence, the discussion in Sec. IV is highly fragmentary and incomplete, and focuses only on those molecules whose abundances are thought to be most affected by the star-formation process. Section V summarizes more quantitatively our current understanding of the abundances of the major carbon- , oxygen- and nitrogen-bearing species, both in the gas and solid phase.

TABLE I

Angular Sizes of Protostellar Objects and Capabilities of Telescopes[a]

Linear size	Angular Size				Telescope[b]	Angular Resolution			
	Taurus 140 pc	Orion 450 pc	M17 2.2 kpc	Galactic Center 8.5 kpc		115 GHz	230 GHz	345 GHz	810 GHz
5 AU	0."04	0."01	—	—	45 m (Nobeyama)	15"	—	—	—
Inner solar nebula									
100 AU	0."7	0."2	0."05	—	30 m (IRAM)	22"	12"	7"	—
Outer solar nebula									
1000 AU	7"	2"	0."5	0."1	15 m (JCMT/SEST)	44"	20"	15"	6"
Presolar nebula									
0.05 pc	74"	23"	5"	1"	10 m (CSO)	—	30"	20"	9"
Cloud core									
0.5 pc	12'	4'	50"	12"	Interferometer (OVRO/ BIMA/Nobeyama/IRAM)	4–7"	1–2"	—	—
Cloud									

[a] The table only lists the capabilities of currently operating telescopes, not those of future projects.
[b] IRAM = Institute de Radio Astronomie Millimetrique; JCMT = James Clerk Maxwell Telescope; CSO = Caltech Submillimeter Observatory; SEST = Swedish-ESO Submillimetre Telescope; OVRO = Owens Valley Radio Observatory; BIMA = Berkeley-Illinois-Maryland Array.

Although several millimeter interferometers are now operative, these facilities still lack the sensitivity to achieve spatial resolutions which directly probe the chemistry in the circumstellar disks on scales of 50 to 500 AU in any but the nearest sources. This problem is illustrated in Table I, which summarizes on the right-hand side the capabilities of current millimeter telescopes in terms of spatial resolution, and compares them with typical sizes of the (pre-) solar nebula and parental cloud core on the left-hand side. It is seen that cloud cores are readily resolved up to distances comparable to Orion with large single-dish telescopes, but that the outer pre-solar nebula can only be probed by interferometers in the nearest molecular clouds such as Taurus and Ophiuchus. In spite of this lack of observational data for stars other than our Sun, Sec. VI discusses the processes which may affect the chemical state during formation of protosolar and protoplanetary nebulae. On scales of 1 to 30 AU, the recent appearances of a number of bright comets, as well as the Voyager encounters with the outer planets, have provided a wealth of new observational data on the chemical composition of bodies in the outer solar nebula, which can be used to test the various formation and evolutionary scenarios, at least for our own solar system.

II. INTERSTELLAR CHEMISTRY

Any study of the chemical evolution of protostellar regions must start with a description of the physical and chemical state of the natal interstellar molecular cloud. It has become apparent in recent years that this initial chemical state is still poorly constrained, both observationally and theoretically. At the time when *Protostars and Planets II* was published (Black and Matthews 1985), the more optimistic view was that the chemistry in molecular clouds is reasonably well described by gas-phase ion-molecule chemistry, apart from some minor details (Irvine et al. 1985; Herbst 1985). Although in a broad sense these models are still valid, the large amount of new observational data poses significant problems to the "standard" description, and no coherent picture has yet emerged to account for the sometimes bewildering array of observed chemical variations from cloud to cloud. In the following, we will first give a general description of the physical characteristics of the different kinds of molecular clouds. This is followed by a general discussion of interstellar molecules and observational techniques. Finally, a brief summary of the various modeling efforts is given. There has been no shortage of reviews on interstellar chemistry recently (for more details, see, e.g., Irvine et al. 1985,1987; Dalgarno 1987,1991; Herbst and Winnewisser 1987; van Dishoeck 1988a; Guélin 1988; Turner and Ziurys 1988; Turner 1989c; Irvine and Knacke 1989; Millar 1990; Genzel 1991a; Walmsley 1991). Books on the subject in general include *Interstellar Chemistry* by Duley and Williams (1984), *Astrochemistry, IAU Symposium 120*, edited by Vardya and Tarafdar (1987), *Molecular Astrophysics*, edited by Hartquist (1990), and *Astrochemistry of Cosmic Phenomena, IAU Symposium 150*, edited by Singh (1992).

A. Molecular Clouds

Table II summarizes the different kinds of molecular clouds that can be distinguished on the basis of their physical characteristics (see, e.g., Turner
1988,1989c; Wilson and Walmsley 1989; Friberg and Hjalmarson 1990;
Chapter by Blitz). The most tenuous are the diffuse molecular clouds, which
are characterized by their low total visual extinction, $A_V \approx 1$ mag. These
clouds are studied observationally primarily by the electronic absorption lines
of their constituents against bright background stars. Ultraviolet photons
from the interstellar radiation field can penetrate diffuse clouds and rapidly
destroy most molecules, so that the atomic fraction is high and only the very
simplest diatomic species are found. The translucent clouds with $A_V \approx 1$ to
5 mag form the bridge between the diffuse and the dark molecular clouds.
Their densities are somewhat higher and the temperatures lower than those in
diffuse clouds. Photoprocesses still play a significant role in the outer part of
translucent clouds, but their importance diminishes rapidly toward the center.
The high-latitude cirrus molecular clouds, seen by their CO millimeter and
IRAS 100 μm emission, fall into the same category, as do the outer envelopes
of dark molecular clouds. No star formation has yet been observed to occur in
diffuse or translucent clouds, probably because of the high fractional ionization. A detailed overview of their physical and chemical structure has been
given by van Dishoeck and Black (1988a) and van Dishoeck (1990).

The cold, dark molecular clouds are visible as dark patches on the sky,
and show a complex morphology. In Fig. 1, the outlines of the CO $J=1\rightarrow0$
emission from the well-studied dark cloud complex in Taurus are presented
(Ungerechts and Thaddeus 1987). It consists of a large number of clouds with
average densities of 10^3 to 10^4 cm^{-3} which are part of one or more cloud
complexes with average densities of 10^2 to 10^3 cm^{-3}. These complexes can
extend over several tens of parsecs, corresponding to several square degrees on
the sky. Their *average* densities are simply obtained by dividing the total mass
by the observed dimensions. Further mapping at higher angular resolution
reveals small individual cores or clumps in the clouds, with densities of 10^4
to 10^5 cm^{-3} or more and temperatures as low as 10 K. It is in these cores that
low-mass (<2 M$_\odot$) star formation has been observed to occur. In addition,
they are the sites where many complex molecules have been detected. It is
clear that the distinction between cloud complexes, clouds and cores or clumps
is rather arbitrary and ill defined (Goldsmith 1987), and the term cloud will
be used rather loosely throughout this chapter to refer also to the dense cores
or clumps, which are the main topic.

The giant molecular clouds (GMC) have a similarly complex morphology,
but differ from the dark clouds by being more massive and warmer. The
average density in the GMC complexes and clouds is similar to that of their
dark cloud counterparts, but densities as high as 10^8 cm^{-3} have been found
in GMC cores or clumps. These GMC cores are the sites of massive star
formation, and are heated significantly by young embedded or neighboring

TABLE II
Physical Characteristics of Molecular Regions in the Interstellar Medium[a]

	Density (cm^{-3})	T (K)	Mass $M_\odot$	A_V (mag)	Size (pc)	ΔV (km s^{-1})	Examples
Diffuse Clouds	$100-800$	$30-80$	$1-100$	$\lesssim 1$	$1-5$	$0.5-3$	ζ Oph
Translucent Clouds	$500-5000$	$15-50$	$3-100$	$1-5$	$0.5-5$	$0.5-3$	HD 169454; High-latitude clouds
Cold, Dark Clouds							
complex	10^2-10^3	$\gtrsim 10$	10^3-10^4	$1-2$	$6-20$	$1-3$	Taurus-Auriga
clouds	10^2-10^4	$\gtrsim 10$	$10-10^3$	$2-5$	$0.2-4$	$0.5-1.5$	B1, B5
cores/clumps	10^4-10^5	≈ 10	$0.3-10$	$5-25$	$0.05-0.4$	$0.2-0.4$	TMC-1, B335
Giant Molecular Clouds							
complex	$100-300$	$15-20$	$10^5-3\times10^6$	$1-2$	$20-80$	$6-15$	M 17, Orion
clouds	10^2-10^4	$\gtrsim 20$	10^3-10^5	$\gtrsim 2$	$3-20$	$3-12$	Orion OMC-1, W3 A
warm clumps	10^4-10^7	$25-70$	$1-10^3$	$5-1000$	$0.05-3$	$1-3$	M 17 clumps, Orion 1′.5 S
hot cores	10^7-10^9	$100-200$	$10-10^3$	$50-1000$	$0.05-1$	$1-10$	Orion hot core

[a] Table adapted from Goldsmith (1987), Turner (1989a) and Friberg and Hjalmarson (1990).

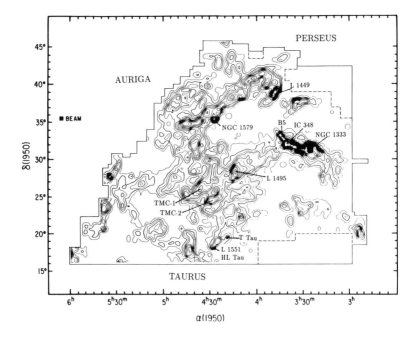

Figure 1. Velocity integrated intensity of CO emission in the Taurus molecular cloud
   complex. The lowest contour is 0.5 K km s^{-1}, and the separation between contours
   is 1.5 K km s^{-1}. The border of the surveyed region is indicated by the outer solid
   line. Various clouds such as B5 and cloud cores such as TMC-1 discussed in the
   text are indicated (figure adapted from Ungerechts and Thaddeus 1987).

stars, which may raise the kinetic temperatures as high as 200 K. The chemical
signatures of clouds in which massive star formation occurs are therefore likely
to differ from those in cold, dark clouds. Low-mass star formation also takes
place in the GMC; the question whether our own Sun formed in a Taurus-like
cloud or in a GMC is therefore not settled, and both scenarios need to be
investigated.

   Analysis of masses and line widths suggests that the dark and GMC
complexes, clouds and cores are probably close to virial equilibrium, so that
self-gravity is important in controlling their structure and evolution. The
diffuse and translucent clouds are not in virial equilibrium, but are probably
bounded by some external pressure.

   In discussing the chemistry of the various types of clouds, it is instructive
to have some indication of their ages and expected lifetimes. Unfortunately,
there are very few observational constraints on these parameters (Myers 1990).
The age of a cloud in virial equilibrium is probably at least the time needed to
reach equilibrium, which is a few free-fall times. This leads to lower limits on
the ages of ~10 Myr for the dark and giant molecular cloud complexes, which

are consistent with those derived from the ages of the oldest nearby stars which presumably formed from the same cloud. Dark and giant molecular cloud complexes can be disrupted by a variety of processes such as energetic winds and radiation from embedded stars, supernovae, and cloud-cloud collisions, on time scales of 10^7 to 10^8 yr.

More relevant for the discussion in this chapter are the ages and expected lifetimes of the cores in the clouds, which may be significantly shorter. If star formation has already occurred, the lifetime of the core may be limited by the dynamical time scale of the associated molecular outflows, which is only 10^4 to 10^5 yr. Cores without stars may be disrupted by the onset of star formation, which has a time scale of $\sim 10^5$ to 10^6 yr (Myers 1990). On the other hand, observations suggest that cores are supported against collapse by turbulence or other mechanisms for longer periods of time. Also, gaseous disks around young stellar objects appear to persist as long as 10^7 yr (see Chapter by Beckwith and Sargent). Therefore, the ages and expected lifetimes for cores and dense clumps are highly uncertain, and may range from 0.1 to 10 Myr.

B. Interstellar Molecules

1. Gas Phase. Table III summarizes the nearly 80 different molecules that have been detected in interstellar clouds. This table does not include the many isotopic varieties studied, and also lacks the many molecules that have been searched for but not detected, such as molecular oxygen (O_2), metal hydrides (MgH, NaH), and even amino acids like glycine. Often, stringent upper limits on these species are available (see, e.g., Irvine et al. 1987,1991 for overviews), which can be as useful in constraining the chemistry as the actual detections. Molecular hydrogen, H_2, is by far the most abundant molecule in the clouds, with other molecules present only in trace amounts. Their abundances with respect to H_2 range from 10^{-4} (CO) to 10^{-10} or less (e.g., HCS^+, PN).

The first conclusion to be drawn from Table III is that the chemical composition of interstellar clouds is far from thermodynamic equilibrium. This is evidenced, for example, by the fact that most of the detected organic molecules are linear and highly unsaturated, in spite of the reducing environment. Although the lack of detected saturated, branched hydrocarbons is partly an observational selection effect owing to their more unfavorable partition function, the available detections and upper limits clearly suggest that they are less abundant than the unsaturated linear ones. Also, the relatively large abundances of molecular ions such as HCO^+ attest to a kinetically controlled, rather than an equilibrium chemistry.

In spite of the wealth of chemical information contained in Table III, there are several important limitations to the data. First, most of the molecules have been discovered by their rotational emission lines at millimeter and submillimeter wavelengths. Thus, only molecules with a permanent dipole moment can be detected. Symmetric molecules such as H_2, CH_4, C_2H_2 and N_2 cannot be observed in this way. Molecules like H_2 and C_2 have been seen by their electronic absorption lines at optical wavelengths against bright

TABLE III
Identified Interstellar and Circumstellar Molecules[a]

Species	Name	Species	Name	Species	Name
H_2	molecular hydrogen	C_2H_2	acetylene	C_6H	
C_2	diatomic carbon	C_3H	propynylidyne (l and c)	CH_2CHCN	vinyl cyanide
CH	methylidyne	H_2CO	formaldehyde	CH_3C_2H	methylacetylene
CH^+	methylidyne ion	NH_3	ammonia	CH_3CHO	acetaldehyde
CN	cyanogen	$HNCO$	isocyanic acid	CH_3NH_2	methylamine
CO	carbon monoxide	$HOCO^+$	protonated carbon dioxide	HC_5N	cyanodiacetylene
CS	carbon monosulfide	$HCNH^+$	protonated hydrogen cyanide		
OH	hydroxyl	$HNCS$	isothiocyanic acid		
HCl	hydrogen chloride	C_3N	cyanoethynyl	$HCOOCH_3$	methyl formate
NO	nitric oxide	C_3O	tricarbon monoxide	CH_3C_3N	methylcyanoacetylene
NS	nitrogen sulfide	H_2CS	thioformaldehyde	CH_3C_4H	methyldiacetylene
SiC	silicon carbide*	H_3O^+	hydronium ion	CH_3CH_3O	dimethyl ether
SiO	silicon monoxide	C_3S		CH_3CH_2CN	ethyl cyanide
SiS	silicon sulfide	HC_2N		CH_3CH_2OH	ethanol
SO	sulfur monoxide			HC_7N	cyanohexatriyne
PN		C_4H	butadiynyl		
CP	*	C_3H_2	cyclopropenylidene	CH_3C_4CN	
SO^+	sulfoxide ion	H_2CCC	propadienylidene	CH_3CH_3CO	acetone†
$NaCl$	sodium chloride*	$HCOOH$	formic acid		
$AlCl$	aluminum chloride*	CH_2CO	ketene	HC_9N	cyano-octa-tetra-yne
KCl	potassium chloride*	HC_3N	cyanoacetylene		
AlF	aluminum fluoride*†	CH_2CN	cyanomethyl		
NH	nitrogen hydride				

Formula	Name	Formula	Name	Formula	Name
SiN	*	NH$_2$CN	cyanamide	HC$_{11}$N	cyano-deca-penta-yne
	†	CH$_2$NH	methanimine	CH$_4$	methane
H$_2$D$^+$				SiH$_4$	silane*
C$_2$H	ethynyl			C$_4$Si	*
CH$_2$	methylene†	C$_5$H	pentynylidyne	C$_5$	pentatomic carbon*
HCN	hydrogen cyanide	C$_2$H$_4$	ethylene*	HCCNC	isocyanoacetylene
HNC	hydrogen isocyanide	H$_2$CCCC	butatrienylidene		
HCO	formyl	CH$_3$OH	methanol		
HCO$^+$	formyl ion	CH$_3$CN	methyl cyanide		
HOC$^+$	isoformyl ion†	CH$_3$NC	methyl isocyanide		
N$_2$H$^+$	protonated nitrogen	CH$_3$SH	methyl mercaptan		
HNO	nitroxyl	NH$_2$CHO	formamide		
H$_2$O	water	HC$_3$HO	propynal		
HCS$^+$	thioformyl ion				
H$_2$S	hydrogen sulfide				
OCS	carbonyl sulfide				
SO$_2$	sulfur dioxide				
SiC$_2$	silicon dicarbide*				
C$_2$O	dicarbon monoxide				
C$_3$	triatomic carbon*				
C$_2$S					

a As of July 1992. Reported but doubtful, unconfirmed, or rejected: CO$^+$, CS$^+$, NaOH, NH$_2$CH$_2$COOH, CH$_2$CH$_2$O.

* Detected in circumstellar envelopes only.

† Tentative.

background stars, but this technique is limited to relatively diffuse interstellar clouds. Other molecules, such as H_3^+, CH_4 and C_2H_2 have been searched for through their vibration-rotation absorption lines at infrared wavelengths against embedded or background sources. The drawback of these studies is that they can only be applied to a specific line of sight in a dense cloud in which star formation has often already occurred. Finally, the Earth's atmosphere prevents the detection of the principal lines of molecules like H_2O, O_2 and CO_2. As a result, our information on the abundances of some of the *dominant* carbon-, oxygen- and nitrogen-bearing molecules in interstellar clouds is very indirect and incomplete, in spite of the fact that the abundances of some trace molecules are known quite accurately.

Second, the column densities and abundances of molecules that have been detected are often uncertain by factors of at least 2 or 3, and, in some cases, up to an order of magnitude. The difficulties in deriving molecular abundances from millimeter observations have been clearly outlined, for example, by Irvine et al. (1985,1987). One of the major problems is that H_2 cannot yet be observed directly in dense quiescent clouds, so that its column density needs to be inferred indirectly from some trace molecule like CO (van Dishoeck and Black 1987).

Third, most of the molecules listed in Table III have been detected in only two or three well-studied sources: the Orion and Sgr B2 giant molecular clouds, and the cold, dark Taurus molecular cloud, TMC-1. To what extent these regions are representative of other molecular clouds is still an open question. In fact, it will be argued in Secs. III and IV that the observed positions are probably chemically quite distinct from most other clouds, or even from other positions in the same cloud, and that their chemical richness may be due to a very specific time in their chemical evolution, a time at which we are observing them fortuitously. This is evidenced by some striking differences between the observed molecular abundances in the Orion, TMC-1 and Sgr B2 clouds. For example, saturated molecules such as CH_3CH_2OH are much more abundant in Sgr B2 than in the specific part of the Orion cloud called the extended ridge or in TMC-1, whereas molecules like SO_2 and SO are much more prominent in Orion than in the Taurus clouds. On the other hand, TMC-1 is one of the few interstellar regions with high abundances of carbon-chain molecules like HC_7N and C_4H. In spite of their different physical characteristics, however, it appears that the abundances of some simple molecules like CO, HCN and CS are fairly similar in the three regions.

The rate at which new discoveries are made is unlikely to decrease over the next decade owing to the continuing improvements in detector sensitivities and the availability of new large observational facilities at millimeter and submillimeter wavelengths (see Table I). Some highlights of the discoveries in the last few years are listed below (see van Dishoeck [1988a] for references to the original observations):

1. *Carbon-chain molecules*:   New carbon-chain molecules continue to be found in cold dark clouds, in particular in TMC-1. The sequence:

$$HCN, HC_3N, HC_5N, \ldots, HC_{11}N$$

has been known to exist for more than a decade (Avery 1987), although even-numbered varieties such as HC_2N and isomers such as HCCNC are much less abundant. In recent years, several new members of the series:

$$CH, C_2H, C_3H, C_4H, C_5H, C_6H$$

$$CS, C_2S, C_3S$$

$$CO, C_2O, C_3O, C_5O$$

have been discovered. In addition, the detection of the first interstellar carbon carbenes, H_2CCC and H_2CCCC has just been reported (Cernicharo et al. 1991*a*, *b*).

2. *Ring molecules*: In contrast with the wealth of carbon-chain molecules, only two ring molecules have so far been discovered in interstellar clouds: c-C_3H_2 and c-C_3H. However, c-C_3H_2 appears to be more widespread throughout the interstellar medium than the carbon-chain molecules, as it has been detected not only in cold dark clouds and warm giant molecular clouds, but also in relatively diffuse clouds (Madden et al. 1989). The cyclic form C_3H_2 appears to be 2 orders of magnitude more abundant than the linear form H_2CCC.

3. *Protonated molecules*: Molecules like HCO^+ and N_2H^+, i.e., protonated CO and N_2, have been known for some time. More recently, protonated CS and CO_2-HCS^+ and $HOCO^+$-, as well as protonated HCN and H_2O-$HCNH^+$ and H_3O^+ (Wootten et al. 1991; Phillips et al. 1992) have been added to the list. Observations of these ions are important, because they provide indirect information on the abundances of molecules like N_2 and CO_2, which cannot be measured directly.

4. *Deuterated molecules*: Observations of deuterated molecules continue to show significant fractionation effects, not only in cold clouds, but also, quite surprisingly, in some warm regions. Table IV summarizes the observed abundance ratios. Recently, the first discovery of a doubly deuterated molecule, D_2CO, has been reported (Turner 1990).

5. *Second-row molecules*: Sensitive searches have been made for molecules containing second-row atoms, as relatively little is yet known about the chemistry of these species (Ziurys 1990; Turner 1991). Of particular interest are searches for metal-hydrides like MgH, which may provide information on the depletion of the metals.

6. *Silicon carbides*: The molecules SiC, c-SiC_2 and SiC_4 have been detected in the gas phase in the carbon-rich circumstellar envelope IRC+10216 (Thaddeus et al. 1984; Cernicharo et al. 1989; Ohishi et al. 1989). This

raises the question whether a large fraction of the metals in general could be locked up in metal carbides such as FeC and MgC. Laboratory spectra of most of these species are still lacking, however.

More details on the abundances of the various molecules will be presented in Sec. V and the Appendix.

2. *Solid Phase.* Although there is still considerable discussion concerning the composition and origin of interstellar grains, some basic features appear well established. The central element in all grain models is a population of "refractory" grains, which are presumed to provide the extinction in diffuse clouds. These refractory particles are composed of silicates plus some carbonaceous material, possibly graphite, and range in size from $\gtrsim 0.25\,\mu$m down to $\sim 0.01\,\mu$m or smaller. They "lock up" a substantial fraction of the chemically active elements, such as C, Si and Fe. In dense clouds, there is direct evidence from infrared spectroscopic observations that these refractory cores are surrounded by icy grain mantles consisting of molecules like H_2O, CO and CH_3OH. In more diffuse clouds, the cores may have an organic refractory layer containing complex molecules resulting from ultraviolet processing of mantles. In addition, large molecules or ultra-small grains consisting of 25 to 200 carbon atoms are thought to be present in interstellar clouds on the basis of infrared emission features observed in the 3.3 to 11 μm range toward reflection nebulae and the 12 to 60 μm IRAS emission. The most popular identification of the species responsible for the infrared features is some type of polycyclic aromatic hydrocarbons (PAH). More details about the grain composition can be found in Sec. V and the Appendix.

C. Chemical Models

1. Types of Models. The observed composition in interstellar clouds discussed above indicates that the chemistry is not in thermodynamic equilibrium, but that it is kinetically controlled by two-body processes in a low-density, low-temperature environment. About 20 years ago, it was realized that reactions between ions and molecules are much more rapid at low temperatures than reactions between two neutral species (Herbst and Klemperer 1973). The latter reactions often have barriers which are much higher than the available energies in the clouds. This realization led to a network of gas-phase ion-molecule reactions, which was very successful at explaining at least qualitatively many of the observed characteristics, and which has since become the basis of nearly all modeling efforts. A brief introduction to this chemistry is given in the Appendix. That gas-phase processes continue to play a dominant role even in the densest interstellar clouds is evidenced by the large observed abundances of ions and the high deuterium fractionation in organic molecules in cold clouds. An important question is to what extent the *pure* gas-phase networks can reproduce the measured abundances, and whether the grains play an active or passive role in the chemistry in the cloud.

The importance of grains in various aspects of the chemistry is evident.

TABLE IV

Observed Deuterium Fractionation Ratios*

Molecule	TMC-1	Warm Clouds	Hot Cores[a]	Halley	Outer
DCO^+/HCO^+	0.015	$\lesssim 0.002$	—		
DCN/HCN	0.023	0.002–0.006	0.005–0.02[b]		
DNC/HNC	0.015	0.01	—		
DC_3N/HC_3N	0.015	—	—		
DC_5N/HC_5N	0.013	—	—		
N_2D^+/N_2H^+	< 0.04	—	—		
C_2D/C_2H	0.01	0.045	—		
C_4D/C_4H	0.004[c]	—	—		
C_3HD/C_3H_2	0.08–0.16	—	—		
CH_2DC_2H/CH_3C_2H	0.06	—	< 0.04[j]		
CH_2DCN/CH_3CN	—	—	0.01[k]		
$HDCO/H_2CO$	0.015	—	0.14[d]		
$D_2CO/HDCO$	—	—	0.02[d]		
NH_2D/NH_3	< 0.02	—	0.06[e]		
HDO/H_2O	—	—	0.001–0.004[f]	$(0.6–5)\times10^{-4g}$	
CH_3OD/CH_3OH	—	—	0.01–0.06[h]	—	—
CH_3D/CH_4	—	—	—	—	$(0.4–2)\times10^{-4i}$

* Table adapted from Millar et al. (1989) and Irvine and Knacke (1989). [a] Tabulated values refer to the Orion hot core or compact ridge, which both have $T > 50$ K. [b] Mangum et al. (1991). [c] Turner (1989a). [d] Turner (1990). [e] Re-assessment of Turner (1990) of data of Walmsley et al. (1987). [f] Jacq et al. (1990). [g] Eberhardt et al. (1987a). [h] Mauersberger et al. (1988). [i] Lutz et al. (1990). [j] Gerin et al. (1992a). [k] Gerin et al. (1992b).

Clearly, they are crucial in shielding the molecules from the dissociating ultraviolet radiation. Also, the large abundance of H_2 even in diffuse clouds can only be understood if the formation of this molecule takes place on the surfaces of grains. Moreover, solid H_2O and CO have been observed directly in dense interstellar clouds. This is not surprising, since the time scale for a species to be deposited on a grain surface is only

$$\tau \approx 2 \times 10^9/(\alpha \, n_H) \text{ yr} \qquad (1)$$

where α is the probability that the species will stick to the grain, which is thought to lie between 0.1 and 1 (see Appendix). As Fig. 2 shows, this time scale is short compared with, e.g., the free-fall time scale or the expected lifetime of the cloud for densities $n_H > 10^4$ cm^{-3}. In fact, the grain accretion process by itself removes heavy species from the gas phase so fast that one would expect many opaque dense clouds devoid of gas-phase molecules like CO. As no such cloud has yet been found, there must be efficient mechanisms which return the adsorbed molecules back to the gas phase. Several possible mechanisms in cold, dark clouds are discussed in the Appendix, but no consensus has yet been reached as to their relative importance. Of course, in regions of star formation, the higher temperatures can sublime the grain mantles. The temperature required for thermal sublimation of pure CO ice is only 17 K, whereas that for pure H_2O ice is 90 K. "Dirty ice" mixtures probably have sublimation temperatures in between these values.

On the basis of these considerations, the models can be divided into several categories. First, there are the pure gas-phase models which adopt the basic ion-molecule chemistry scheme. Recent examples can be found in Herbst and Leung (1989), Langer and Graedel (1989) and Millar et al. (1991b). In the simplest "standard" form, these models are homogeneous with a fixed temperature and density. They solve the set of rate equations either at steady-state or as a function of time. Several recent improvements to this standard model are discussed in the Appendix, such as efforts to include the clumpy structure of molecular clouds; calculations of the effects of cosmic-ray-induced photodissociation processes inside dense clouds; studies of the influence of large molecules such as the PAH on the chemistry and ionization balance; considerations of the depth-dependent structure of clouds located close to young, hot stars; and treatments of evolutionary processes.

Second, several models have been developed which include the effects of grains on the chemistry. The simplest models of, e.g., Iglesias (1977) and Millar and Nejad (1985) consider only the time-dependent accretion of molecules onto grains in a dense cloud, but ignore any surface processing. The models of, e.g., Tielens and Hagen (1982) and Brown and Charnley (1990) include subsequent reactions on the surfaces to form other molecules, but have no mechanism for returning the molecules to the gas phase. Nevertheless, they make important predictions about the composition of the grain mantles. The most sophisticated models are those of d'Hendecourt et al. (1985) and

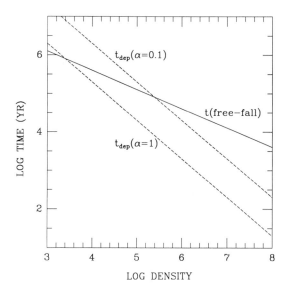

Figure 2. Comparison of free-fall and depletion time scales. The dashed lines represent the time scale for molecules to condense out on grain surfaces as a function of hydrogen density for two values of the sticking coefficient α. The figure shows that even for $\alpha = 0.1$, a large fraction of the molecules is expected to "freeze-out" in one free-fall time at a density of 10^5 cm^{-3}. For comparison, the time scale to reach chemical equilibrium is typically 10^6 to 10^7 yr, the expected lifetime of a cloud is 10^6 to 10^8 yr, and the dynamical time scale of molecular outflows is typically 10^4 to 10^5 yr (figure adapted from Walmsley 1991).

Breukers (1991), which include desorption mechanisms and investigate the influence of the returned molecules on the gas-phase composition as a function of time. As is discussed in more detail in the Appendix, different compositions of grain mantles can occur under different conditions. In an atomic hydrogen-rich atmosphere, which is the situation that most likely prevails in the clouds, molecules like H_2O, H_2CO, CH_3OH, and possibly NH_3 and CH_4 are expected to dominate (see Fig. 3). The major effect on the gas-phase chemistry is an increase in the abundances of these molecules when they are released from the surface by some mechanism.

   Third, detailed models have been developed to investigate the chemistry at high temperatures, $T \gtrsim 2000$ K, such as encountered in shocks. As described in the Appendix, two different kinds of shocks can be distinguished. The so-called J shocks are usually powerful enough to dissociate molecules completely. Only when the post-shock gas cools can the molecules reform, although often in different configurations compared with the pre-shock gas. Recent examples have been given by Neufeld and Dalgarno (1989) and Hollenbach and McKee (1989). Such powerful shocks can also destroy the

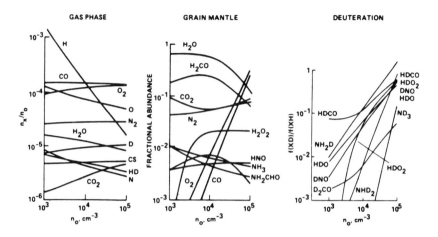

Figure 3. A comparison of the calculated gas-phase and solid-state composition of a molecular cloud. When atomic H dominates over heavy species in the gas phase, the accreting species will be hydrogenated. Otherwise grain mantles will reflect the gas phase more directly. Large deuterium enrichments are expected to be a characteristic of grain mantles as a result of the high atomic D/H ratio in the gas phase (Tielens 1989).

refractory grain cores and return Si and Fe to the gas phase. In C-type shocks, the peak temperatures are lower, so that most molecules survive the passage of the shock wave. The chemistry is dominated by neutral-neutral rather than ion-molecule reactions, although subtle effects due to differential streaming of the ions and neutrals can occur (Draine et al. 1983). Compared with the low-temperature cases, the shock models are characterized by large abundances of H_2O and OH, of sulphur-bearing molecules such as H_2S, and silicon-containing molecules such as SiO.

The adopted gas-phase elemental abundances are crucial parameters in the models. The fraction of species X in the solid phase with respect to the solar abundance is denoted by the depletion factor δ_X. [Note that in some papers δ_X refers to the abundance fraction in the gas phase.] We adopt as our reference the abundances of the solar system at its time of formation ~ 4.5 Gyr ago as summarized by Anders and Grevesse (1989): $O/H = 8.5 \times 10^{-4}$, $N/H = 1.1 \times 10^{-4}$ and $Si/H = 3.6 \times 10^{-5}$. The only exception is carbon, for which we use $C/H = 4.0 \times 10^{-4}$ as derived by Grevesse et al. (1991).

Oxygen and nitrogen are usually assumed to be relatively undepleted ($\delta_X \lesssim 0.25$), but carbon can be depleted by 60% or more in dense clouds and silicon is often taken to be nearly completely removed (see Sec. V). It should be noted that most molecular abundances do not scale linearly with the depletion factors, because of the complicated structure of the chemistry networks. Of particular interest is the C/O ratio in the gas phase. For any reasonable grain model (see Sec. V), this ratio is less than unity so that virtually all gas-phase carbon is locked up in CO. The remainder of the gas-phase oxygen is then thought to be tied up in molecules such as H_2O and O_2. If C/O were greater than unity in the gas phase, all oxygen would be locked up in CO and the remaining carbon would be contained in more complex carbon-containing molecules.

It is important to recognize that elemental abundances in the interstellar medium today may differ appreciably from solar abundances, which represent a particular ~ 1 $M_\odot$ sample from the interstellar medium 4.5 Gyr ago. In fact, detailed modeling of the Orion H II region (Baldwin et al. 1991) indicates a gas-phase O/H ratio only 30% of the solar value. Isotopic studies of mete-orites, revealing remarkably high abundances of radioactive ^{26}Al at the time of chondrite formation, may indicate contamination of the primordial solar nebula by fresh nucleosynthetic material from a nearby supernova explosion, nova, or wind from an asymptotic giant branch star (Chapter by Cameron). The hypothesized sources of ^{26}Al may also have affected the abundances of C, N and O.

2. Comparison with Observations. As an illustration of the reliability of the model results, we summarize in Table V the computed abundances in the latest gas-phase models of Herbst and Leung (1989), Langer and Graedel (1989) and Gredel et al. (1989), and compare them with observations in TMC-1. The first conclusion to be drawn from Table V is that at steady state, the abundances in the different models disagree in some cases by more than an order of magnitude, even for simple molecules. Although some of these discrepancies can be traced to slight differences in adopted physical conditions and elemental abundances, they mostly result from differences in the adopted reaction schemes. Indeed, recent comparisons between the groups have been very valuable in establishing the sensitivity of various aspects of the models to differing assumptions about one or two key reactions (Dalgarno 1986; Millar et al. 1987). They also provide a warning that, in spite of their relative agreement in some cases, the computed abundances should not be trusted to more than an order of magnitude on an absolute scale, and that special care must be taken in quoting older model results which use outdated chemical reaction sets.

When compared with observations, it appears that the pure gas-phase steady-state models can reproduce the observed abundances of simple molecules like CO, CS and HCO^+, but that they fail to produce enough complex molecules like C_3H_2, C_4H and HC_5N by several orders of magnitude. They also result in too little NH_3 if the latest experimental information on its formation processes is taken into account, and in many cases too much O_2 and H_2O compared with the measured values or upper limits for other clouds. One possible explanation for the discrepancy with the complex molecules is that the cloud chemistry has not yet reached steady state. In particular, the complete conversion of C to CO in a cloud that is initially mostly atomic takes approximately 1 Myr. Thus, at earlier times, there is a significant amount of atomic carbon available in the cloud which can be inserted into the carbon

E. F. VAN DISHOECK ET AL.

TABLE V

Selected Model Abundances Relative to H_2

Species	HL89[a]		LG89[b]	GLDH89[c]	TMC-1[d]
	Early	Steady State	Steady State	Steady State	South
C	6(−5)	7(−9)	3(−8)	6(−7)	—
C^+	2(−9)	1(−9)	3(−9)	4(−10)	—
CO	8(−5)	1(−4)	7(−5)	1(−4)	8(−5)
CH	4(−8)	2(−10)	1(−9)	—	2(−8)
C_2	3(−8)	2(−10)	1(−9)	—	5(−8)
C_2H	7(−8)	3(−10)	2(−9)	4(−10)	8(−8)
C_4H	8(−8)	6(−12)	—	—	2(−8)
CH_4	5(−6)	2(−7)	2(−9)	1(−7)	—
C_2H_2	9(−7)	2(−8)	—	8(−10)	—
C_3H_2	2(−8)	1(−10)	—	9(−12)	2(−8)
O	3(−4)	8(−5)	2(−5)	1(−4)	—
OH	1(−7)	2(−7)	7(−9)	1(−8)	3(−7)
O_2	5(−7)	6(−5)	6(−6)	3(−5)	—
H_2O	3(−6)	1(−6)	9(−8)	2(−5)	—
H_2CO	3(−7)	3(−9)	1(−9)	—	2(−8)
CO_2	4(−7)	2(−7)	—	—	—
CH_3OH	2(−9)	5(−12)	—	—	2(−9)[e]
HCO^+	5(−9)	1(−8)	4(−9)	—	8(−9)
N	4(−5)	3(−6)	3(−5)	—	—
N_2	1(−7)	2(−5)	6(−6)	—	—
NO	3(−8)	2(−7)	7(−8)	—	≤ 3(−8)[f]
NH_3	3(−8)	5(−8)	6(−9)	4(−8)	2(−8)
HCN	2(−7)	2(−9)	1(−9)	1(−9)	2(−8)
HC_3N	5(−9)	3(−12)	—	—	6(−9)
HC_5N	1(−9)	7(−15)	—	—	3(−9)
CS	1(−8)	1(−8)	4(−9)	—	1(−8)
SO	2(−9)	2(−8)	—	—	5(−9)
SO_2	8(−10)	1(−8)	—	—	< 1(−9)
H_2S	5(−11)	2(−10)	—	—	< 5(−10)[g]
SiO	7(−10)	7(−10)	—	—	< 2(−12)[h]
e	3(−8)	3(−8)	8(−8)	—	—

[a] From Herbst and Leung (1989) for $n(H_2)= 2 \times 10^4$ cm^{-3} and $T = 10$ K with low metal abundances and standard rates. The early results refer to $t \approx 10^5$ yr. [b] From Langer and Graedel (1989), their run 8 with $n(H_2)= 5 \times 10^4$ cm^{-3} and $T = 20$ K. These models employ low metal abundances, and take cosmic-ray-induced photodissociation of CO into account. [c] From Gredel et al. (1989), their case C with $n(H_2)= 10^4$ cm^{-3} and $T = 50$ K. These models employ low metal abundances, and take cosmic-ray-induced photodissociation of all molecules into account. [d] Observed abundances from Irvine et al. (1987), unless otherwise indicated. [e] Friberg et al. (1988). [f] McGonagle et al. (1990). [g] Minh et al. (1989). [h] Ziurys et al. (1989).

chains to produce more complex hydrocarbons. As Fig. 4 illustrates, their abundances reach maximum values at early times that are 3 to 4 orders of magnitude larger than the steady-state results. On the other hand, the concentrations of molecules like CO, NH_3 and O_2 increase steadily with time. For the carbon- and oxygen-bearing species, the equilibrium abundances are reached in a few Myr. Although this time scale is uncertain by a factor of a few due to uncertainties in crucial reaction rates, it is comparable to the estimated lifetime of a clump in a molecular cloud complex (see Sec. II.A). The abundances of nitrogen-containing molecules such as N_2 and NO do not attain steady state until 10 Myr, because they are produced by slow neutral-neutral reactions.

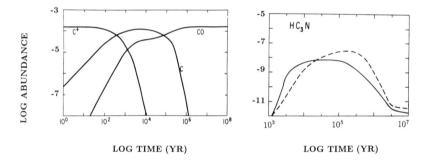

Figure 4. *Left*: Abundances of C^+, C and CO with respect to H_2 as functions of time for a cloud with $n_H = 2 \times 10^4$ cm^{-3} (Leung et al. 1984). *Right*: Abundance of a typical complex carbon-bearing molecule as a function of time. The dashed line is for a similar model as in the left figure. The full line refers to a model in which more rapid ion-polar molecule reaction rates are employed (Herbst and Leung 1986).

It should be noted that the scenario of a young age for TMC-1 is not the only possible explanation for the large observed abundances of carbon-chain molecules: in fact, only a large fraction of atomic or ionized carbon in the gas phase is required in the models, which may result from incomplete conversion of carbon to CO due to any other mechanism (Suzuki 1983).

The above example reinforces the general conclusion about interstellar cloud models, namely that they are only moderately successful in reproducing the observed abundances. A large source of uncertainty still lies in the adopted reaction rates, which can introduce orders of magnitude uncertainty in the predicted abundances, but which are not always widely recognized. Nevertheless, the gas phase models have at least some predictive, quantitative power. Our understanding of processes involving grains is still much more limited and these models can be used only for qualitative statements at best. Much more basic laboratory study of reactions on surfaces representative of interstellar grains needs to be performed in order to remedy this situation.

III. CHEMICAL STATE OF DENSE CLOUDS PRIOR TO STAR FORMATION

The discussion in Sec. II focused on interstellar molecules and chemical models in general, and compared the models with observed molecular abundances at one specific position in the Taurus clouds. In this section, we investigate from an observational point of view how representative the chemistry at this position is of that of other clouds or clumps which currently show no indication of ongoing star formation. It appears that these dark, dense clouds are very inhomogeneous on scales of 20 to 60″, and that as a result their chemical state is poorly constrained observationally.

A. TMC-1 Cloud

The TMC-1 core is a small elongated, dense ridge of approximate dimensions 0.6×0.06 pc in the Taurus clouds, with a total mass <40 M$_\odot$ (see Fig. 1). Although low-mass star formation occurs throughout the Taurus cloud complex, including sites close to TMC-1, no direct evidence for recent star formation has been found within the TMC-1 ridge. It appears to be a quiescent, self-gravitating clump of gas with a minor axis of the order of a Jeans length. The narrow observed molecular line widths (<1 km s^{-1}) show no signs of collapse. Thus, TMC-1 is thought to be a cloud close to hydrostatic equilibrium, which may be the site for the next generation of pre-main-sequence stars.

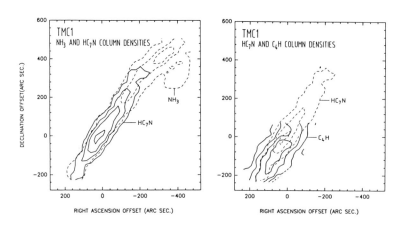

Figure 5. (a) Column density distribution of HC$_7$N (full lines) in TMC-1 compared with that of NH$_3$ (dashed lines). (b) Column density distribution of C$_4$H (full lines) compared with that of HC$_7$N (dashed lines). Possible gradients in excitation temperature across the TMC-1 core have been neglected in making these maps (see text) (Olano et al. 1988).

Figure 5 shows the NH_3 and HC_7N column density distribution along the TMC-1 ridge at about $40''$ (0.03 pc at the adopted distance of 140 pc) resolution (Olano et al. 1988). The (0,0) position is the peak in the cyanopolyyne emission to which most of the observations of the molecules listed in Table V refer. It is apparent that these two molecules have very different distributions along the ridge, as was also found in earlier studies at lower angular resolution (see, e.g., Little et al. 1979; Tölle et al. 1981). The ammonia column density peaks about $7'$ (0.3 pc) to the NW of the maximum of the complex carbon-bearing molecules. Care has to be taken in interpreting these maps, however, because variations in the emission can be caused either by variations in the excitation conditions, or by true chemical abundance variations along the ridge (see, e.g., Bujarrabal et al. 1981). For TMC-1, these two possibilities have been considered carefully, and it was concluded that the HC_7N/NH_3 abundance ratio varies genuinely by more than an order of magnitude over the scale of a few tenths of a pc. The physical conditions have been constrained from a variety of diagnostics. The temperature, as determined from the NH_3 (1,1) and (2,2) excitation, is about 10 K, whereas estimates of the densities range from $\sim 3 \times 10^4$ to 10^5 cm^{-3}. There is no consensus yet whether there is any gradient in density along the ridge. In any case, it is insufficient to explain the different distributions shown in Fig. 5 in terms of different excitation effects. On the other hand, small differences of order $1'$ (0.04 pc) have been found in the distribution of, e.g., HC_7N and C_4H in the southern part of the ridge, which may well be caused by excitation effects if the H_2 density falls off rapidly to the south (Fig. 5b). Maps of HC_3N and C_3H_2 in the southern part at even higher angular resolution of $20''$ have been presented by Guélin and Cernicharo (1988). It is seen that the ridge breaks up into a chain of tiny clumps with a characteristic size of $1'$ (0.04 pc or 8000 AU), which is comparable to the size of the gas disks surrounding young stellar objects such as HL Tauri (see Chapter by Beckwith and Sargent).

What can cause the striking chemical differences between the northern and southern peak? As discussed in Sec. II, most chemical models have attempted to reproduce the observed abundances in the southern peak, but have succeeded only in a time-dependent calculation at an early age of a few times 10^5 yr. These models are somewhat artificial in that they presuppose that the carbon is initially mostly atomic, but that hydrogen is in molecular form, and that the physical conditions are fixed with time. Nevertheless, this chemical time scale is comparable to the probable age of the TMC-1 ridge (Olano et al. 1988). However, in order to explain the abundance variation along the ridge, the northern part would have to be significantly older than the southern part. This would be consistent with the larger abundances of nitrogen- and sulfur-bearing molecules such as NH_3, NO and H_2S in the north, because the chemical time scales for these species are longer. TMC-1 may be an unusual cloud because of its elongated structure, which has been speculated to arise through a shock caused by cloud-cloud collisions. On the other hand, the observed small line widths and the absence of any significant velocity gradient indicate that all the energy of a possible shock has already been dissipated, so that we are probably seeing the "fossilized remnant" of this collision, which is being preserved by self-gravitation (Tölle et al. 1981). In this picture, an age gradient across the ridge is not easily accounted for. Variations in density along the ridge from 10^4 to 10^5 cm^{-3} have little effect on the computed abundances in a pure gas-phase model, but do significantly affect the time scale for depletion onto the grains (cf. Fig. 2). If the density were higher in the northern part, more carbon-chain molecules could be removed from the gas phase. However, there appears little evidence for differential depletion of other molecules between the two peaks, and NH$_3$, which may form on grains, is even more abundant in the north. Better observational constraints on possible density gradients along the ridge, combined with more detailed modeling of the grain depletion processes are needed to elucidate the cause for the observed variations.

B. L 134N Cloud

A second example of a cold, dark cloud which shows no obvious evidence of star formation is the L 134N cloud, also known as L 183. The chemical structure of the densest part of this cloud has been studied in great detail by Swade (1989a, b). Figure 6 reproduces his observed contour maps of integrated emission of a number of representative molecules.

The core region of L 134N, as seen in C^{18}O, extends over an area of about $8' \times 8'$, which, at a distance of 160 pc, corresponds to about 0.3×0.3 pc. Its estimated total mass is about 20 M$_\odot$. It is apparent that the emission of some molecules is fairly similar (e.g., C^{18}O and CS), but that other molecules (e.g., HCO$^+$, SO and NH$_3$) show widely different distributions. For example, the CO and CS maps both have a NE-SW extension, whereas the HCO$^+$ and NH$_3$ maps are much more elongated in the N-S direction. C$_3$H$_2$ also peaks in the northern part, but does not show a corresponding peak in the south at the same position as NH$_3$ or HCO$^+$. The SO emission, on the other hand, peaks more to the west than that of any other molecule. To what extent are these differences due to varying excitation conditions or to actual abundance variations? The physical conditions in L 134N have been derived from a number of diagnostic species, and a kinetic temperature $T \approx 12$ K and peak density $n(\mathrm{H_2}) \approx 3 \times 10^4$ cm^{-3} have been found, very similar to the conditions in the TMC-1 ridge. Some variations are seen in density across the L 134N core, but the different map morphologies are mostly due to chemical abundance variations on scales of a few tenths of a pc. Swade and Schloerb (1992) have suggested that the different distributions of CS and SO indicate an oxygen abundance gradient across the L 134N core, with oxygen being depleted in the highest density core around the (0,0) position. Atomic carbon has been detected near the (0,0) position in L 134N (Phillips and Huggins 1981), but no data are available on other positions in this cloud or TMC-1 to test the hypothesis. Also, no detailed chemical modeling has yet been done to investigate how large such a gradient would have to be to explain the different

CS and SO distributions, and how it could account for the observations of other molecules. The age of the L 134N core is estimated to be at least 1 Myr, so that time-dependent effects probably play a smaller role in the chemistry of this cloud than in TMC-1.

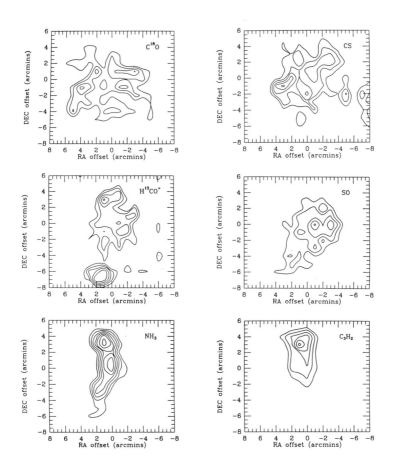

Figure 6.   Contour maps of the integrated intensity of various molecules in the L 134N molecular core. The different map morphologies are thought to reflect true chemical abundance variations on scales of a few arcmin (0.2 pc) (Swade 1989a).

C. Discussion

The cases of TMC-1 and L 134N illustrate that significant chemical abundance gradients occur over scales of only a few tenths of a pc within a single, supposedly quiescent, cloud. Comparison of the abundances in the TMC-1 and L 134N cores shows some interesting differences and similarities between the two clouds as well, which are illustrated in Table VI of Sec. IV. For example, the long-carbon chain molecules are 1 to 2 orders of magnitude less abundant anywhere in L 134N than in the southern part of TMC-1. On the other hand, SO and SO_2 are more abundant in the western part of L 134N compared with the southern part of TMC-1. If anything, the chemistry in L 134N resembles that found in the northern part of TMC-1. In spite of these differences, some molecules have similar abundances in the two clouds, such as the fairly complex molecule methanol (Friberg et al. 1988). The fact that methanol does not peak at the position of that of the long-carbon chain molecules suggests that it has a different origin.

Limited surveys of the chemistry have been made for a number of other dark cores, both with and without star formation. None of these cores has as rich a chemistry as TMC-1, although carbon-chain molecules have been detected in nearly half of them (Cernicharo et al. 1984; Suzuki et al. 1992). One striking result is that the complex carbon molecules are not found in any region in which star formation has already occurred, such as L 1551. On the other hand, more than 2/3 of the cores seen in NH_3 have associated IRAS sources (Benson and Myers 1989). These surveys appear to confirm the picture that regions in which NH_3 is abundant represent a later stage in the evolution of the cloud than regions in which the complex carbon molecules are present. Whether the larger NH_3 abundance in regions of star formation is solely the result of the longer time scales or different physical conditions, or whether, e.g., release from grain mantles plays a role, is still an open question. More detailed comparisons of the abundances of other molecules in such surveys will be very valuable.

The observed correlation between NH_3 cores and star formation also raises the question whether there may yet be alternative explanations for the observed chemical gradients in the TMC-1 and L 134N clouds. If cold, low-luminosity protostars in a very early stage of evolution were present, they would not show up as strong IRAS point sources, but may still affect the chemistry if they are already in the outflow stage. Such outflows could, for example, evaporate grain mantles and create local shocks in which sulfur-bearing molecules could be formed. In fact, a low-luminosity IRAS source with associated high-velocity CO outflow emission has been detected only 2' W of the northern TMC-1 NH_3 peak (Terebey et al. 1989). More information on the distance over which such outflows could affect the chemistry would be very useful. Deeper searches for protostellar activity in the L 134 N core are warranted.

In summary, the question "what is the chemical composition of a dark cloud prior to star formation" cannot be answered in detail at present without an understanding of the physical and chemical processes that give rise to the observed variations. Several possible causes have been identified, such as age gradients, elemental abundance gradients, and maybe even star formation in its earliest stages, but detailed modeling and systematic observations are still lacking.

IV. CHEMICAL STATE OF STAR-FORMING REGIONS

The preceeding sections outline the chemical composition and evolution of molecular clouds prior to and during the incipient stages of star formation. As is outlined below, observationally we are now in a position to examine *directly* the chemical evolution which accompanies the star-formation process, at least on scales of 500 to 10,000 AU. It is generally accepted that, as stars condense out of molecular clouds, the circumstellar gas will gradually become flattened so that eventually the remnant cloud around the protostellar object will have a disk-like morphology. Until very recently, the structure of this circumstellar gas could be inferred only from indirect clues. However, with the advent of interferometric techniques at millimeter and infrared wavelengths, it has become possible to identify such disks directly. Around the low mass objects HL Tauri, L1551 IRS5 and IRAS 16293–2422, for example, are extended, flattened structures with radii of order a few 100 AU (Mundy et al. 1986*b*; Sargent and Beckwith 1987; Sargent 1989; Keene and Masson 1990). Thus far, observations have concentrated on the dust continuum and isotopic CO emission from these objects with the intent of obtaining an overview of the spatial structure, mass and kinematics of the system, but little work has been reported on their chemical structure. While dynamical studies ultimately require the resources of instruments capable of higher resolution, the large-aperture millimeter and sub-millimeter telescopes which have recently come on-line (IRAM, JCMT, CSO, Nobeyama) are well suited for initial exploration of the chemical nature of these disks and their surroundings (see Table I). For example, broadband molecular line surveys can delineate the "initial conditions" accompanying the collapse of dense clouds into protostellar objects; while deep integrations on selected species and interferometric mapping can examine the chemical heterogeneity of the circumstellar disks or cores themselves.

As a result of recent surveys of young stellar systems at millimeter (Wilking et al. 1989*b*) and sub-millimeter (Beckwith et al. 1990) wavelengths, over 100 cold compact dust regions associated with young stellar objects have been identified. In addition, new mid-infrared cameras and high resolution far-infrared data are beginning to reveal more detailed information about the warm dust. This list contains sources which span stellar luminosities from 1 to 10^5 $L_\odot$ and a range of obscuration from visible stars to objects so deeply embedded that no 2-μm source is detected. Estimates of the masses of circumstellar material range widely from <0.1 to 10 $M_\odot$ or more. As two of the nearest and brightest star-forming molecular clouds, the core of the Orion nebula and the IRAS 16293–2422 core in the ρ Ophiuchi cloud present unique opportunities to examine the complexity and evolution of interstellar chemistry in two very dissimilar environments, i.e., high- vs low-mass star formation. These two regions have been studied chemically in so much more detail than any other regions that it is instructive to discuss them first. Emphasis will be placed on the alteration of the chemical composition of the parent molecular cloud by the star formation process.

A. High-Mass Star-Forming Regions

1. Orion/KL: Observations. The Orion/KL or OMC-1 molecular cloud core, centered behind the Trapezium stars in the sword of the Orion constellation and including the luminous infrared sources BN and IRc2, is perhaps the most widely known and best studied example of a young stellar nebula in which high-mass star formation occurs. Indeed, because of its proximity, large intrinsic luminosity (2 to 10×10^4 L$_\odot$), and enormous quantity of surrounding material, the IRc2 region is one of the brightest compact objects in the sky from mid-infrared to millimeter wavelengths. Its unique strength in molecular line emission is obvious from its central role in molecular line searches since the early days of molecular radio astronomy. In more recent years, these strong lines have made IRc2 an irresistible source for millimeter interferometer studies of molecular spatial distributions (Plambeck et al. 1985; Vogel et al. 1985; Masson and Mundy 1988; Plambeck and Wright 1987,1988; Mangum et al. 1990).

The connection of the Orion/KL core with the Orion-Monoceros giant molecular cloud complex is visible in maps of many molecular species as a strip, or ridge, of material running roughly N-S near the Trapezium. The linewidths in the ridge material are at most 4 to 5 km s^{-1}, and decrease with increasing spatial resolution. Condensations in the Orion ridge are nicely sampled by high-density tracers such as CS, as is demonstrated in the left-hand side of Fig. 7. Three condensations are clearly visible, with dimensions from 0.03 to 0.11 pc and virial masses from 30 to 80 M$_\odot$. The largest condensation is centered on the main source in the region, IRc2, a roughly 25 M$_\odot$ star deeply embedded in the local material (Genzel and Stutzki 1989). The 1.5'S source is the only other clump in Fig. 7 which also shows obvious molecular signposts of star formation activity (Ziurys et al. 1989).

In addition to the "spike" or ridge gas, a number of other distinct kinematic components are detected near IRc2. Centered approximately 2″ south of IRc2 is the "hot core" component which is a $\sim$10 M$_\odot$ clumping of gas and dust about 15″ (0.03 pc or 7000 AU) in extent. This material is associated with and strongly heated by IRc2 to at least 100 to 200 K, but does not appear to be a protostellar disk itself (Plambeck and Wright 1987; Genzel and Stutzki 1989), although such a disk may exist within 80 AU of IRc2 (Plambeck et al. 1990). A strong bipolar outflow from IRc2 has interacted with the structure in the local molecular cloud to create a "doughnut" of slowly expanding shocked cloud material (the "low-velocity plateau" component characterized by line widths $\Delta V \leq 25$ km s^{-1}) roughly normal to the flow axis. Along the flow axis the velocities are correspondingly higher due to the faster density decrease (the "high-velocity plateau"), as is evident in the emission from CO, HCO$^+$ and H$_2$ (Vogel et al. 1984; Masson and Mundy 1988). Finally, a compressed clump of cloud material called the "compact ridge" is visible some 10 to

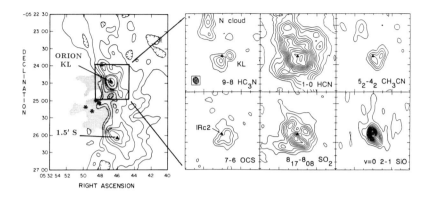

Figure 7. *Left*: OVRO/Onsala image of the integrated CS $J=2\rightarrow1$ emission from the Orion ridge. The asterisks denote the positions of the Trapezium stars, while the triangles mark the positions of the millimeter continuum sources (Mundy et al. 1988). *Right*: Hat Creek aperture synthesis maps of molecular emission towards Orion/KL. The contour levels for the first four $2'$ square maps step by 5 K, while for the last two the interval is 10 K. The tickmarks are spaced by $10''$ (Plambeck and Wright 1988).

$15''$ to the southwest of IRc2. Figure 8 summarizes the various components surrounding IRc2 graphically in more detail.

Chemistry adds considerably to the complexity of the region as variations in temperature, atomic abundances, and chemical history drive variations in the molecular composition of each component. For example, interferometer maps of species are often very different in appearance, as the right-hand side of Fig. 7 demonstrates. This figure, obtained by the Hat Creek millimeter-wave interferometer (now called BIMA), outlines dramatic differences in molecular emission on the 3 to $4''$ scale. When combined with the temperature and overall abundance constraints provided by the spectral line surveys outlined below, these maps demand that true chemical heterogeneity exists on scales of 2000 AU or less.

Due to the opaque nature of the atmosphere in many wavelength regions and the relatively insensitive nature of early receiver/telescope combinations, much of the initial work on the chemistry of dense molecular clouds pro- ceeded through a biased selection of molecules on a line-by-line basis, such as discussed in Sec. III. However, *unbiased* surveys of small regions in the millimeter-wave spectrum have been recorded for the brightest objects in the sky like Orion/KL (Lovas et al. 1976,1979; Johansson et al. 1984; Sutton et al. 1985; Blake et al. 1986; Turner 1989*b*; Jewell et al. 1989). More recently, the new submillimeter facilities have allowed the surveys to be extended to high-frequency atmospheric windows. The main advantage of line surveys

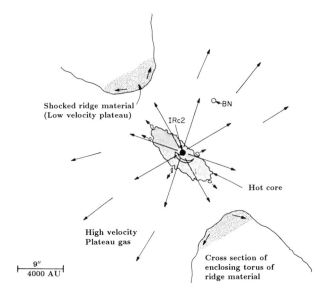

Figure 8. Schematic model for the Orion/KL outflow region around IRc2. The hot core material (shaded region) lies mainly in front of IRc2. It is being ablated by a wind from IRc2; the resulting material forms the high-velocity plateau. The volume around IRc2 has been largely cleared of material except where the outflow has been stopped by the dense ridge gas. The resulting shocked region (dotted) is responsible for the low-velocity plateau (figure adapted from Masson and Mundy 1988).

vs selected observations is completeness, but they also result in better calibration, certainty of identification, and in a measurement of the total integrated line flux.

These points are illustrated by a recently completed CSO 345 GHz survey of the Orion/KL region, which yields dramatically more intense emission than is prevalent at longer wavelengths due to the high temperature and large mass of the central source (Groesbeck et al. 1993). Two positions have been observed to examine the nature of chemical abundance variations on small size scales, in this case the classical Orion/KL "plateau"/hot core source and the warm core $1\rlap{.}''5$ S (Fig. 7). The differences in the spectra between the two sources are striking, as is demonstrated by Fig. 9.

Indeed, an incredible complexity is revealed for Orion/KL, but only few lines are visible at the $1\rlap{.}''5$ S position in spite of the fact that the total H_2 column density is very similar. The myriad lines in the Orion/KL spectra arise mostly from high-energy levels (>300 K) of asymmetric top molecules such as methyl formate, sulfur dioxide, or methanol, and result in a quasi-continuum. Only in a few selected bands can the true continuum level of

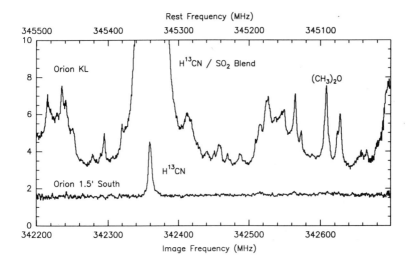

Figure 9. CSO 500 MHz double sideband spectrum of the Orion/KL and Orion 1.5 S sources near the H^{13}CN $J=4\rightarrow3$ transition at 345.34 GHz. Both spectra have the same vertical scale, and the total integration time was <10 minutes. Note the enormous difference in the number of lines between the two sources (Groesbeck et al. 1993).

$T_A^* \approx 1.7$ K be reached. The most striking chemical difference between the 1.5 S source and the Orion/KL region is the nearly complete absence of SO$_2$ in the former, despite easily observable SO and SiO spectral lines. In addition, several of the species more traditionally associated with cooler molecular clouds, such as C$_2$H, are observed to have greater abundances in the southern source. The difference in line density between the two sources also illustrates that care has to be taken in the interpretation of continuum measurements of distant unresolved sources: if the source is like the 1.5'S source, the continuum is mostly due to dust emission, whereas if it is like Orion/KL, a significant fraction of the continuum may actually be due to lines. This could lead to erroneous conclusions, e.g., about the gas-to-dust ratio.

  Orion/KL: Chemistry. In the following, the different chemistries of the Orion core region (the extended ridge, the plateau, the hot core, and the compact ridge) will be discussed in more detail. Results for selected molecules are summarized in Table VI in Sec. IV.B. In spite of the tremendously heterogeneous composition of this region, it is heartening to note that it may be interpreted at least qualitatively in the framework of the interaction between a quiescent molecular cloud chemistry and that induced by massive star formation (Blake et al. 1987).

  The chemistry of the extended ridge material is similar to that observed in a number of other cool, quiescent clouds described in Sec. III: a temperature increase from 10 to 50 K and an increase in density by an order of magnitude seem to have little influence on the chemistry. In contrast, the chemical composition of the Orion/KL plateau source is dominated by high-temperature chemistry, including shocks and grain-mantle evaporation. The most striking result for the plateau source is the large drop in the abundances of carbon-rich species like C, C_2H and CN and the concomitant rise in the abundances of oxygen- and sulfur-containing species such as H_2O, SiO, H_2S and SO_2. The observed ratio of carbon and oxygen contained in molecules other than CO and water is $C/O \leq 0.5$ in the plateau source. In contrast, the same ratio is greater than unity in the ridge material. The higher oxygen abundance and the very high temperatures behind shock fronts are expected to drive most of the oxygen into H_2O and OH and carbon into stable forms such as CO. Also, they will enhance the abundances of more refractory species such as sulfur and silicon via grain disruption or evaporation, as is evidenced by large abundances of gas-phase atomic silicon (Haas et al. 1991). The considerable abundance of "fragile" species such as H_2CO and the large residual deuterium enhancement in water imply an extremely clumpy density structure in the outflowing wind.

High-temperature chemistry is also evident in observations of the hot core, but with a considerably different signature. Again, the abundances of reactive, carbon-rich species are significantly reduced, with the observed C/O ratio in molecular form less than unity. Because the heating is caused by radiation rather than by shocks, grain vaporization does not occur and the abundances of the refractory sulfur and silicon species remain low. The high densities and temperatures in the hot core reduce the effectiveness of ion-molecule networks, and are sufficient to release much of the grain-mantle material stored over the cloud lifetime. Indeed, the composition of the hot core is most consistent with a chemistry that was dominated by ion-molecule reactions in a cold, quiescent dense cloud with accretion onto grains, followed by evaporation of the molecules from the grains once star formation has occurred (Brown et al. 1988). Because of the relatively slow gas-phase reactions, the ejected mantle material has not yet been chemically modified. It is worth recalling, however, the considerable uncertainties in quiescent cloud gas-phase and grain-mantle chemistries outlined in Secs. II and III, so that it is difficult to test this scenario in detail. Nevertheless, support for this picture comes from the unexpectedly large observed abundances of HDO (Jacq et al. 1990), NH_2D (Walmsley et al. 1987), CH_3OD (Mauersberger et al. 1988), DCN (Mangum et al. 1991), and D_2CO (Turner 1990) in the hot core/compact ridge region, which indicate that significant fractionation is retained even though the kinetic temperature is well above 100 K (see Table IV). The most likely explanation is that the observed deuterated molecules are "fossil" water, ammonia, methanol etc. resulting from efficient deuterium fractionation at low temperatures both in the gas and/or the grains, trapped onto grains throughout the cloud lifetime, and released only recently because of heating by a newly formed star (Plambeck and Wright 1987; Brown and

Millar 1989*a, b*; Walmsley 1992; see also the chapter Appendix and Fig. 3). If IRc2 is responsible for this heating, it cannot have existed for more than $\sim 10^4$ yr, because otherwise the deuterium enhancements would have been reset by chemical reactions to their high-temperature values. This time scale is comparable to that derived from the SiO and H_2O maser flows. Grain chemistry may also be responsible for the enhanced abundance of H_2CO, and for the conversion of unsaturated carbon-chain molecules into more hydrogenated species like ethyl cyanide, because the relevant gas phase routes are highly endothermic.

Finally, the importance of chemical mixing between active cores like Orion/KL and surrounding cloud material is demonstrated by the composition of the compact ridge source in the southwest. It is here that the abundances of complex oxygen containing organics such as CH_3OCH_3 and $HCOOCH_3$ peak. These species could either be produced by grain-mantle hydrogenation of CO, etc., or by gas-phase routes. An earlier gas-phase suggestion involving mixing of water from the plateau source with ion-rich gas from the ridge (Blake et al. 1987) does not appear to explain the observed abundances via an ion-molecule route when subjected to detailed numerical tests, although the direct injection of methanol from grain surfaces followed by an ion-molecule chemistry appears to be more promising (Millar et al. 1991*a*). These mixed gas/grain models could most likely account for the abundances, but do not currently explain why the chemical composition in the compact ridge is different from that in the hot core. Also, it is difficult with such models to account quantitatively for the observed selectivity, namely that dimethyl ether (CH_3OCH_3) and methyl formate ($HCOOCH_3$) are found to be much more abundant than their isomers ethanol (CH_3CH_2OH) and acetic acid (CH_3COOH). Another point in favor of a different chemistry in the compact ridge than that proposed for the hot core is provided by recent high angular resolution observations of methanol (Plambeck and Wright 1988; Wilson et al. 1989). If methanol and the other related species were produced by grain-mantle catalysis, then they should show observational correlation with other grain-mantle products. For example, ammonia, ethyl cyanide and water (as traced by HDO) show striking similarities by peaking at the hot core position. Methanol, however, appears to be *anti*correlated with these species by peaking at the compact ridge; and thus any model which accounts for its chemistry cannot be similar in its details to the hot core. Specifically, a mechanism must be found to alter grain mantle chemistry on small size scales (cf. Tielens et al. 1991).

2. Other Regions. The Orion/KL observations show that several chemically distinct regions occur on scales of 2000 to 10,000 AU, corresponding to 5 to 25″ at the distance of Orion. Thus, for regions at greater distances observed with comparable beam sizes, we may expect enormous unresolved complexity if the region is at a similar stage in its evolution as Orion/KL. The Sgr B2 giant molecular cloud complex has been surveyed at various wavelengths and shows, not surprisingly, significant chemical heterogeneity on small angular scales (Goldsmith et al. 1987; Sutton et al. 1991). However, because these clouds are very far removed, the origin of the chemical variations is even less clear than in the case of Orion. Some regions with physical and chemical properties similar to the Orion hot core have recently been found, such as G 34.3+0.15 (Henkel et al. 1987; Heaton et al. 1989), but these are also at much larger distances. Distant massive GMC such as M 17 (see, e.g., Stutzki et al. 1988), W3 (see, e.g., Wright et al. 1984; Dickel and Goss 1987; Hayashi et al. 1989), NGC 7538 (Pratap et al. 1990), and NGC 6334 (Bachiller and Cernicharo 1990) show strong atomic and molecular emission lines, but have been studied in much less detail chemically and are difficult to interpret, owing to the poor spatial resolution.

The only other nearby molecular clouds with high-mass star formation which have been studied in some detail are the NGC 2024 region, also known as Orion B (see, e.g., Mundy et al. 1987; Barnes and Crutcher 1990), the region surrounding NGC 2071 IRS 1 (Yamashita et al. 1989; Zhou et al. 1990a) and the ρ Ophiuchi A and B cloud cores (see, e.g., Martin-Pintado et al. 1983; Zeng et al. 1984; Loren and Wootten 1986; Wootten and Loren 1988; Sasselov and Rucinski 1990; Loren et al. 1990), although the latter cases are sometimes classified under low-mass star formation. All these cores are known to have energetic molecular outflows, and molecules such as SO_2, HCO^+ and even DCO^+ are readily detected. In addition, multitransition observations of H_2CO, NH_3 and CS have been used to constrain the density structure of the clumps. True molecular abundance gradients may be present across the cores, and molecules appear to be depleted from the gas phase in the densest parts (Mezger et al. 1992; Mauersberger et al. 1992). Another interesting study in this respect is that of Loren and Wootten (1986), who systematically observed a number of molecular lines at 3 positions in the ρ Ophiuchi A and B cores.

Because the observational data on these other regions are so fragmentary, it is difficult to establish any possible evolutionary sequence for the chemistry in high-mass star-forming regions. For example, is the 1.5 S source in Orion indeed at an earlier evolutionary stage than Orion/KL as argued by McMullin et al. (1992)? How long does the Orion/KL phase in which molecules such as SO_2 are so much more abundant last, and what is the underlying cause? What is the fate of the so-called photodissociation regions in, e.g., M 17, which are close to young hot stars? Is their fractional ionization too high for star formation to occur? More systematic studies on the chemical abundances in warm and cold dense clumps, such as those initiated by Loren and Wootten (1986) for ρ Ophiuchi, are needed to settle these issues. Clearly, only the warmest of them resemble Orion/KL in terms of chemical complexity, emphasizing the unique evolutionary state of this object.

Another important method to probe both the cold gas of the molecular core and the warm, dense gas close to the massive protostar is through infrared absorption lines toward embedded sources such as Orion BN/KL, AFGL 2591 and NGC 2264 (Scoville et al. 1983; Mitchell et al. 1989,1990). Such studies are valuable not only to constrain the physical structure of the circumstellar gas environment on much smaller scales than is currently possible with millimeter techniques, but may also provide information on the chemistry in this inner part, if extended to other molecules (Evans et al. 1991). Such high-resolution infrared observations are expected to become more important in the near future due to rapid technological advances in this area, and planned space-based missions such as the Infrared Space Observatory.

B. Low-Mass Star-Forming Regions

As noted above, by combining high angular resolution interferometric observations to image molecular distributions with broadband single-dish spectroscopic surveys to accurately constrain excitation and abundances, it has become possible to examine in detail the nature of the chemistry of bright sources such as Orion. Observations of fainter, low-mass star-forming regions have become possible only recently. We focus the discussion here on IRAS 16293–2422, one of the better studied examples of local low-mass star formation.

1. IRAS 16293–2422 Object. IRAS 16293–2422 is a young, far-infrared object of 27 $L_\odot$ located in the nearby (160 pc) Ophiuchus cloud complex. It was originally identified as a possible protostar with a molecular outflow (Walker et al. 1986; Wootten and Loren 1987) and was suggested to have a spherically symmetric infall (Walker et al. 1986). While the claim for infall is still controversial (Menten et al. 1987), the central object associated with the far-infrared emission is a rich source of molecular line and long-wavelength dust emission.

Dust continuum emission from IRAS 16293–2422 has been observed at a wide range of wavelengths with the most revealing data being recent $\lambda = 2.7$ mm interferometer maps at high resolution ($4\rlap{.}''3 \times 2\rlap{.}''4$) (see Fig. 10d), which resolve this region into two components separated by 830 AU ($5\rlap{.}''2$) along a northwest-southeast axis (Mundy et al. 1986b,1992). The majority of the dust in the system is contained in these two components which are coincident with weak centimeter emission from ionized gas (Wootten 1989). The southeastern source is marginally resolved with a source size of 640 AU ($4''$); the northern source is unresolved (FWHM <200 AU; $1\rlap{.}''5$). Observations with the Kuiper Airborne Observatory at 50 and 100 μm have confirmed that the majority of the far-infrared emission emanates from these sources with a maximum allowable source size of $15''$ (Butner et al., in preparation). The overall spectral energy distribution of the region is well-fitted by 40 K dust with a λ^{-1} to λ^{-2} emissivity law. Recent groundbased 10 and 20 μm measurements have failed to detect emission, consistent with the cool dust temperature.

Interferometer maps of molecular line emission in the $C^{18}O$ $J=1\rightarrow0$, NH_3 (1,1) and CS $J=2\rightarrow1$ transition have revealed elongated gas structures coincident with the dust distribution (Mundy et al. 1990; Walker et al. 1990). The $C^{18}O$ emission, seen in Fig. 10a, shows a velocity shift from northwest to southeast of 2.4 km s^{-1} consistent with orbital motion. Assuming point masses, the mass of the system is around $1.5/\sin i$ $M_\odot$, where i, the orbital

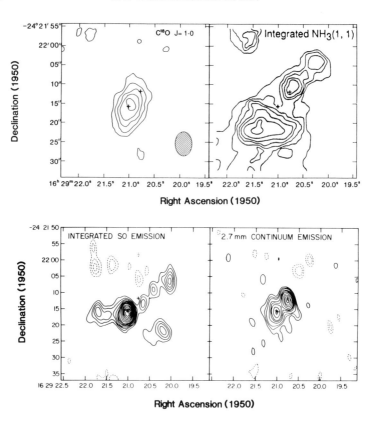

Figure 10. Composite of emission maps of IRAS 16293–2422. (a) Velocity integrated
$C^{18}O$ $J=1\rightarrow0$ emission obtained with the Owens Valley Interferometer at $6\rlap{.}''3 \times 4\rlap{.}''5$
resolution; (b) NH$_3$ (1,1) emission obtained with the VLA at $6'' \times 6''$ resolution;
(c) and (d) Integrated SO 2_3–1_2 and 2.7 mm continuum emission mapped at $2\rlap{.}''4 \times$
$4\rlap{.}''2$ with the Owens Valley Interferometer. The pluses indicate the positions of the
2 cm radio sources. Note that the object has broken into two sources separated by
830 AU, but that the SO emission is mainly associated with only one of the two.

inclination, is likely to be near 90°. The CS $J=2\rightarrow1$ emission reveals a
similar structure. On the other hand, the NH$_3$ emission, shown in Fig. 10b,
arises from a larger (8000 AU diameter) region with a kinetic temperature of
15 to 20 K and a total mass of $\sim$0.2 M$_\odot$ in the outer region. Yet a different
distribution is seen for the SO molecule, as Fig. 10c demonstrates: very strong
SO 2_3–1_2 emission arises from the southeastern component, but no detectable
emission comes from the other component (Mundy et al. 1992). Finally, the
HCO$^+$ 1–0 emission appears more associated with the outflow, being extended
considerably NE-SW (Mundy 1990). Thus, at 500 to 800 AU resolution, the
various molecular species exhibit significantly different spatial distribution,
much as is the case in the Orion/KL region.

IRAS 16293–2422 therefore appears to be a very young binary system in which both components are still surrounded by significant amounts of gas and dust. The southeastern source is the more active of the two with the double-lobed radio emission, strong SO emission, as well as H_2O masers tracing its outflow activity (Wilking and Claussen 1987; Menten et al. 1990). The NH_3 emission appears to be distributed in a ring with radius 3000 to 4000 AU surrounding the central sources. The dearth of ammonia emission from the central components is likely to be due to an actual decrease in NH_3 abundance. These points are summarized graphically in Fig. 11.

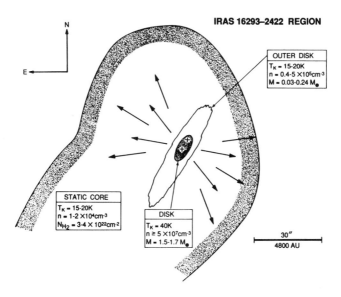

Figure 11. Graphical depiction of the IRAS 16293–2422 molecular cloud core and circumstellar disk(s) (Mundy et al. 1990).

Single-dish spectra at submillimeter wavelengths cannot spatially resolve such distributions (see Table I), but can provide valuable information on the overall chemical composition of the region. Figure 12 demonstrates the quality of data that is now obtainable in only a few minutes of integration time (Blake et al. 1993). Many different molecules are readily detected, and in several cases more than one line is visible, so that rotation excitation diagrams can be constructed. The rotation diagram temperatures of SO_2 and CH_3OH are ~80 K, significantly higher than the dust temperature of 40 K.

The rich molecular spectrum observed for IRAS 16293–2422 does not, however, imply much higher abundances compared with, e.g., cold clouds. The main reason that the lines in this source are so strong is simply that the total

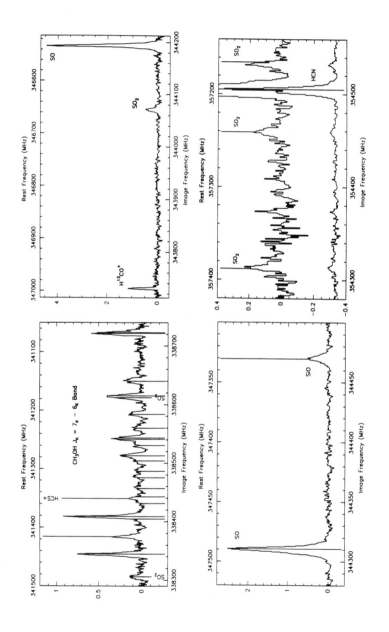

Figure 12.   Composite of sub-millimeter spectra of IRAS 16293–2422 obtained with the CSO: (a) CH_3OH $J_K = 7_K \rightarrow 6_K$ band; (b) SO and SiO emission; (c) the HCN $J = 4 \rightarrow 3$ transition with various SO_2 lines nearby; and (d) the HCO^+ $J = 4 \rightarrow 3$ transition. All spectra are double sideband, with the frequency ranges observed labeled on the top and bottom axes.

column density of matter is at least an order of magnitude higher compared with, e.g., TMC-1. In addition, the higher temperature and density facilitate observation of the high-lying transitions. The inferred abundances for IRAS 16293–2422 averaged over a $20''$ beam are presented in Table VI, and are surprisingly similar to those found in the cold clouds, with the exception of SiO. Silicon monoxide is suspected to have at least moderate activation barriers which affect its formation, consistent with its larger abundance in IRAS 16293–2422 (Ziurys et al. 1989; Langer and Glassgold 1990), and is also enhanced by shock destruction of grains (Martin-Pintado et al. 1992).

One might therefore draw the conclusion from Table VI that the abundances of molecules like SO in low-mass regions such as IRAS 16293–2422 are not affected by the star-formation process. However, the interferometer observations of SO clearly show that this is an oversimplification, because the molecule is only concentrated around one of the two sources. The SO abundance in the southern source is close to the value listed in Table VI, but that in the northern source is roughly a factor 10 below even the dark-cloud value. As most of the protostellar activity arises from the southern source, it is tempting to speculate that the SO is in that case an interaction product of the outflow with the surrounding molecular material, which probably releases sulfur and oxygen back into the gas phase, much the same as in the Orion plateau source. However, the similarity of the SO abundance in the southern source and that in cold clouds suggests that this process is not very effective, presumably because the outflow is less powerful than in Orion. The difference with the northern source may result from larger depletion of SO onto grains. The northern source is presumably younger than the southern one, given its unusual continuum flux distribution and the lack of an outflow (Mundy et al. 1992).

The disparity in relative distributions of $C^{18}O$ and NH_3, seen in Fig. 10, may shed additional light on this question. It is clear that gas-phase NH_3 is down by at least an order of magnitude in the inner core. This behavior is also seen in ammonia relative to CS and dust in the more massive NGC 2071 IRS1 region (Zhou et al. 1990a). The question is whether the NH_3 surrounding IRAS 16293–2422 is remnant interstellar cloud material, or whether it has been depleted and/or formed on grains in the inner part and subsequently returned back to the gas phase. If anything, the NH_3 abundance around the northern source appears larger than that in the southern part, opposite to the case of SO. It should be recalled that variations in the NH_3 abundance on small scales have also been seen in cold clouds such as TMC-1 and L 134N, where freeze-out is probably not the explanation. On the other hand, NH_2D is also prominent in IRAS 16293–2422 with $NH_2D/NH_3 \approx 3 \times 10^{-3}$, as are other deuterated molecules such as HDCO. This amount of deuterium fractionation is similar to that found for the Orion hot core, where it is thought to result from grain-surface processes. Water is predicted to have the largest abundance and excitation contrast between the protostellar source and the surrounding molecular cloud, but is presently unobservable due to atmospheric attenuation

TABLE VI

Abundances of Selected Molecules in Quiescent and Star-Forming Clouds[a]

| Species | Low-Mass Star Formation | | | | | | High-Mass Star Formation | | | |
	TMC-1 S	Ref.[b]	L 134N	Ref.[b]	IRAS 16293–2422	Ref.[b]	Orion/KL Plateau	Ref.[b]	Orion Hot core	Ref.[b]
CO	8(-5)	1	8(-5)	1	1(-4)	18	1(-4)	11	1(-4)	11
HCO⁺	6(-9)	2	8(-9)	1	1(-9)	17	>1(-8)	16	<1(-9)	16
H₂CO	2(-8)	1	2(-8)	1	1(-9)		3(-8)	11	>6(-8)	9
CH₃OH	2(-9)	8	3(-9)	8	3(-9)	17	<1(-8)	19	(0.1 – 1)(-6)	13
NH₃	3(-8)	4	1(-7)	5	2(-7)	18	<1(-8)	19	1(-7)	14
HCN	1(-8)	3	3(-9)	3	7(-10)	17	3(-7)	11	2(-7)	15
HC₅N	3(-9)	1	1(-10)	1	—		<1(-11)	19	<1(-12)	19
CH₃CN	7(-10)	7	<4(-10)	7	1(-10)		<5(-10)	19	5(-9)	19
C₄H	1(-7)	4	<6(-9)	5	—		<1(-11)	19	<1(-12)	19
c-C₃H₂	1(-8)	6	2(-9)	5	5(-11)	17	<4(-9)	19	<4(-10)	19
CS	1(-8)	1	7(-10)	5	1(-9)	17	2(-8)	11	>1(-8)	11
SO	5(-9)	1	2(-8)	5	2(-9)	17	5(-7)	11	≲(-8)	11
SO₂	<1(-9)	1	2(-9)	5	1(-9)	17	5(-7)	11	≲2(-8)	11
H₂S	<5(-10)	10	8(-10)	10	1(-9)		(0.1 – 4)(-6)	11,12	5(-6)	11
SiO	<2(-12)	3	<4(-12)	3	1(-10)	17	(0.3 – 2)(-7)	11,3	—	12

[a] Warning: This table lists the abundances quoted in the references; no attempt has been made to systematically re-evaluate the data. Abundances for a given source derived by different authors may vary by more than an order of magnitude.

[b] References: 1. Irvine et al. 1987; 2. Guélin et al. 1982; 3. Ziurys et al. 1989; 4. Olano et al. 1988; 5. Madden et al. 1989; 6. Swade 1989b; 7. Matthews and Sears 1983; 8. Friberg et al. 1988; 9. Mangum et al. 1990; 10. Minh et al. 1989; 11. Blake et al. 1987; 12. Minh et al. 1990; 13. Menten et al. 1988; 14. See footnote 10, Table IX; 15. Mangum et al. 1991; 16. Vogel et al. 1984; 17. Blake et al. 1993; 18. Mundy et al. 1990; 19. Blake, personal communication.

even at airborne altitudes, except for the masing lines. However, the HDO abundances may exhibit similar characteristics, so that further searches for this species in star-forming regions will be very valuable.

2. Other Regions. Many other dark cloud cores in which low-mass star formation is currently taking place have been observed in a variety of molecules. Most of these cores, such as B1, B5, B335 and L1551, contain extended bipolar outflows, and have associated IRAS or far-infrared sources, with luminosities ranging from <5 to 30 $L_\odot$. As for the high-mass regions, they have been studied principally in lines of the CO, NH_3, CS and H_2CO molecules with the aim of deriving the physical structure of the clouds (Fuller and Myers 1987; Benson and Myers 1989; Zhou et al. 1989; Menten et al. 1989; Zhou et al. 1990b). Little coherent information is available on the chemistry in these cores. Single-dish data are available for a number of molecules scattered throughout the literature, but they have often been obtained with widely different beam sizes ranging from 20″ to 5′, which complicates the determination of relative abundances. Nevertheless, as mentioned in Sec. III.C, it appears that long carbon-chain molecules are much less abundant in star-forming regions than, e.g., in the southern part of TMC-1. On the other hand, molecules such as NH_3 and CH_3OH seem to have abundances similar to those found in, e.g., L 134N. A recent study of the B1 cloud indicates that its SiO abundance is enhanced compared with TMC-1, especially in the direction of the IRAS source located near the center of the NH_3 core where high-velocity CO gas has also been found (Bachiller et al. 1990b). The same trend as observed for IRAS 16293–2422, namely that most molecular abundances are similar to those in cold clouds, but that selected species such as SiO are enhanced, thus appears to occur in other star-forming clouds. On much smaller scales, however, significant chemical gradients may be present.

An intriguing case for further study is provided by the B5 cloud, which has been mapped in detail in CO by Goldsmith et al. (1986) and Fuller et al. (1991). They suggest that in B5 (and probably in most other low-mass star-forming regions) material cycles continuously between the clumps and a tenuous interclump medium, on a time scale of a few Myr (Charnley et al. 1990). The clumps are formed in the collapse of the interclump gas, and are destroyed by T Tauri winds from stars forming at the centers of the clumps. Dynamical models for this scenario have been constructed by Charnley et al. (1988,1990), but not enough observations of other species in B5 have yet been reported to test them in detail. Nevertheless, such models are an important first step to include the star-formation process into the chemistry of a cloud as a whole.

IRAS 16293–2422 is the only low-mass protostellar disk in which the chemistry has been studied in some detail using interferometer techniques. Observations of the distribution of molecules other than CO on scales of 500 to 1000 AU in well-established circumstellar disks such as those of HL Tau and T Tau are becoming feasible with current interferometer facilities, and significant advances in our understanding are expected in the coming years

(Ohashi et al. 1991). These young stellar objects are clearly at a further stage in their evolution, as most of the associated molecular cloud material has been dispersed already. As a result, the molecular lines are much weaker than those seen in, e.g., IRAS 16293–2422, when observed with a single-dish telescope at 20 to 30″ resolution because of the smaller *averaged* column densities. Initial studies by Blake et al. (1992) suggest that the abundance and excitation of molecules like CS may be different in the circumstellar disk of HL Tau compared with, e.g., IRAS 16293–2422. In particular, polar molecules like CS appear significantly depleted from the gas phase. The observed molecular abundances in disks may eventually help to place limits on the distance at which the transition from LTE chemistry to interstellar chemistry occurs, and to probe the chemistry which occurs across the accretion shock.

Of particular chemical interest is the recent detection of extremely high-velocity molecular flows associated with young stellar objects (Lizano et al. 1988; Koo 1989,1990; Masson et al. 1990; Bachiller et al. 1990*b*). The best-studied case is that of HH 7–11, where CO with velocities extending up to 160 km s^{-1} from line center has been found. The major question is whether the CO is present in the jet itself, or whether it is ambient material which has recently been swept up by the jet. The first possibility was suggested by Lizano et al. (1988), and detailed chemical models of molecule formation in fast winds were developed by Glassgold et al. (1989,1991). They find that significant quantities of CO can indeed be formed in the winds, provided that the temperature decreases rapidly with distance from the protostar. However, Masson et al. (1990) argue on the basis of higher spatial resolution data that the high-velocity CO is mostly swept-up ambient material accelerated by the interaction with the jet. The high-velocity CO is found to be highly localized and coincides with the strong HCO^{+} (Rudolph and Welch 1988) and shocked H$_2$ emission (Garden et al. 1990). The main difficulty with this model is that the CO is unlikely to survive such a powerful shock. Searches for atomic species such as C I should indicate whether some of the CO has indeed dissociated downstream. Detailed chemical models for this scenario are being developed by Wolfire and Königl (1991). It would be interesting to determine how ubiquitous these extremely high-velocity flows are, and whether they represent a specific phase in the evolution of young stellar objects.

3. An Evolutionary Chemical Sequence? Although the discussion in the previous sections has been highly fragmentary, an attempt has been made in Table VI to summarize the abundances of a few characteristic species in an evolutionary sequence. The earliest stage is thought to be formed by the TMC-1 S position, which is very rich in large carbon-chain molecules, and which is presumably very young, <1 Myr. The next phase is given by cold cores such as L 134N, in which no star formation has yet occurred or in which nearby star formation is in its very earliest stages. These clouds are characterized by an absence of complex carbon-bearing molecules, and by larger abundances of NH$_3$ and SO. The related stage, in which the molecular cloud core has collapsed and in which a young star is being formed, has not yet been clearly identified, but most molecules are expected to be condensed onto grains at this phase, complicating the detection (Rawlings et al. 1992). These stages are followed by that of the IRAS 16293–2422 core, in which low-mass star formation is obviously taking place, but in which the young star is still surrounded by a significant amount of the original cloud material. A powerful outflow has already developed, although accretion may still be occurring. At this point, the source has a rich molecular spectrum, mostly because of the large mass associated with it and because of the higher temperatures and densities. Many molecular abundances are still similar to those found in L 134N, but interferometric images show dramatic variations of at least a factor 10 on the 500 to 600 AU scale. Thus, the low-mass star-formation process may alter molecular abundances selectively, but the degree of alteration is not as severe as for the high-mass case, at least not at the point of evolution of IRAS 16293–2422. Finally, the young stellar object reaches the stage in which most of the surrounding cloud material has been blown away and a well-established circumstellar disk remains, such as seen for HL Tau and T Tau. When viewed with 20 to 30″ beams, these objects do not contain strong molecular lines, although species such as HCN and HCO^+ are readily detectable in T Tau. Much higher spatial resolution is necessary to probe the chemistry in the circumstellar disks.

V. ABUNDANCES OVERVIEW

In Secs. III and IV, a discussion of abundances and small-scale chemical heterogeneity has been given at various stages of the star formation process. These sections focused mostly on those molecules whose abundances are most affected by the star-formation process, but which are often only minor species in terms of overall composition. Here we investigate more quantitatively the actual composition of the major species in the various regions and the fraction of elemental abundances locked up in the different molecules in the gas or solid phase. All fractional abundances have as reference the solar abundances quoted in Sec. II.

A. GENERAL DISCUSSION

1. Gas-Phase Overview. In Tables V and VI, the observed molecular abundances in the various clouds have been summarized. Inspection shows that this list contains mostly molecules with very small abundances of $<10^{-7}$, which lock up only a small fraction of the available elements. Only carbon monoxide may contain up to 25% of the carbon budget. As noted in Sec. II, virtually no observational constraints on the principal oxygen and nitrogen species are available, because neither H_2O, nor O_2, O, N_2 or N can be observed in cold clouds directly from the ground. In what form are these elements? Model results are unfortunately not very reliable. The models of Herbst and Leung (1989), which use $f \approx 0.5$ for the branching ratio to H_2O and OH in the H_3O^+ dissociative recombination, predict that most of the gas-phase oxygen is taken up in steady state by O_2, followed by O and H_2O in proportions 1:0.46:0.04. If cosmic-ray-induced photodissociation is included, Gredel et al. (1989) predict oxygen to be mostly in atomic form, with $O:O_2:H_2O = 1:0.25:0.20$. It appears that there is no consensus yet about the dominant form of gas-phase oxygen, which may affect the model predictions for other molecules in the clouds as well. In warm clouds, indirect limits on the gas-phase O_2 abundance have been obtained from searches for the $^{16}O^{18}O$ transition at 234 GHz (Black and Smith 1984; Goldsmith et al. 1985; Liszt and Vanden Bout 1985), which indicate $O_2/CO < 0.1$–0.5. In such clouds, the higher-lying transitions of H_2O can also be observed, but lead to abundances of gas-phase H_2O with respect to CO of only 1 to 10%. Thus, neither O_2 nor H_2O appear to tie up a significant fraction of the oxygen budget in the gas phase.

Some indirect limits on the N_2 abundance are available from observations of the N_2H^+ ion. For the well-studied case of L 134N, $N_2/CO \approx 0.08$ is found, corresponding to $N_2/H_2 \approx 8 \times 10^{-6}$, if $CO/H_2 \approx 10^{-4}$. For comparison, $NO/CO \approx 7 \times 10^{-4}$ and $NH_3/CO \approx 10^{-3}$, so that N_2 is apparently the dominant nitrogen-bearing molecule, consistent with the steady-state models (McGonagle et al. 1990; Womack et al. 1992; van Dishoeck et al. 1992). However, the amount of nitrogen still unaccounted for may be as large as 90%. Part of this nitrogen is probably in the form of atomic nitrogen, which is predicted to be the dominant form at early times, and part of it may be depleted onto grains.

2. *Solid-Phase Overview: Refractory Components.* The refractory grain cores are thought to lock up a substantial fraction of the chemically active elements Si, Fe and C. Several types of refractory material appear to be present, based on the spectral features produced by dust in diffuse clouds: these spectral features are listed in Table VII. For each spectral feature, this Table gives: (1) the central wavelength λ; (2) $\Delta\tau_\lambda/N_H$, the maximum optical depth per H atom; (3) $\Delta\tau/N_H$ integrated over $d\tilde{\nu}$ (where $\tilde{\nu} \equiv \lambda^{-1}$); (4) proposed identification(s); (5) for each identification X, the expected value of σ_λ, the cross section per X at band center; (6) the "band strength" $\int \sigma \, d\tilde{\nu}$; and (7) the abundance of X (relative to H) required to account for the observed feature.

The strong interstellar 9.7 μm feature is identified as the Si-O stretching mode in amorphous silicate material. The identification as silicate appears to be secure. First, silicates are theoretically expected as major condensates in cooling gas with cosmic abundances; second, the feature is seen in *emission* in dusty outflows from oxygen-rich red giants; and third, the expected feature due to O-Si-O bending is seen at $\sim 18 \, \mu$m. The strength of the 9.7 μm feature demands that a substantial fraction of the refractory material be in some form of silicate, perhaps amorphous olivine ($Mg_x Fe_{1-x} SiO_4$), containing practically 100% of the interstellar Si and Fe (see, e.g., Draine and Lee [1984] for a discussion of the silicate band strength and the depletion of interstellar Si). Assuming 4 O per Si, and $\delta_{Si} \approx 1$, these silicate grains probably account for $\delta_O \approx 0.17$ of the oxygen.

TABLE VII
Solid-State Extinction Features in Diffuse Clouds[a]

λ (μm)	$\dfrac{\Delta\tau_\lambda}{N_H}$ (10^{-23} cm^2)	$\dfrac{\int\Delta\tau\,d\tilde{v}}{N_H}$ (10^{-20} cm)	X	σ_λ (10^{-19} cm^2)	$\int\sigma\,d\tilde{v}$ (10^{-16} cm)	$\dfrac{N(X)}{N_H}$ (10^{-5})	Notes[b]
0.22	53	820	graphite?	100	1600	5.3	1,2
0.443	1.6	1.8	?	?	$8850f_{.443}$	$2\times10^{-4}f_{.443}^{-1}$	3
1.32	0.18	0.004	?	?	$8850f_{1.32}$	$5\times10^{-6}f_{1.32}^{-1}$	4
3.4	0.23	0.023	C in HAC?	0.11?	.009?	26?	5,6,7
"	"	"	C in hydroc.?	?	0.1?	2?	8
9.7	2.5	0.62	silicates	6	1.6	4	9,10,11
19.	1.0	0.2	silicates	2	0.5	4	11,12

[a] For mean extinction law with $N_H/E(B-V) = 5.8\times10^{21}$ cm^{-2} (Bohlin et al. 1978), $E(B-V)/E(J-K)$=1.91 (Rieke and Lebofsky 1985), $A_{9.7} = 0.35E(J-K)$ (Roche and Aitken 1984; Draine 1989a), $\Delta A_{9.7}\approx0.30\ E(J-K)$ (cf. Draine 1989a).

[b] Notes: 1. $\Delta\tau_\lambda/N_H$ is based on mean Drude fit parameters of Fitzpatrick and Massa (1986); 2. σ_λ is for 2:1 oblate spheroids with $a_{\rm eff}<0.01\ \mu$m (Draine 1989b), and assumed density $\rho = 2$ g cm^{-3}; 3. From $W_{.433}/E(B-V)$=2.04 Å for 25 stars, and $\Delta\tau/W = 0.51/$Å toward HD183143 (Herbig 1975). Here $f_{.433}$ is the (unknown) oscillator strength for the 0.443 μm band; 4. From $W_{1.32}/E(B-V)$=0.42 Å and FWHM=4.0 Å (Joblin et al. 1990). Here $f_{1.32}$ is the (unknown) oscillator strength for the 1.32 μm band; 5. $\Delta\tau/N_H$ is estimated from $\Delta\tau_{3,4}$ toward Galactic center (McFadzean et al. 1989) and VI Cyg 12 (Adamson et al. 1990), with N_H estimated from $\Delta\tau_{9.7}\approx3.6$ toward GC (Roche and Aitken 1985) and 0.6 toward VI Cyg 12 (Roche and Aitken 1984); 6. $\int\Delta\tau\,d\tilde{v}/\Delta\tau_{3,4} = 100$ cm^{-1} from GC-IRS7 (Butchart et al. 1986; Adamson et al. 1990); 7. σ_λ and $\int\sigma\,d\tilde{v}$ for hydrogenated amorphous carbon with H/C$\approx$0.5 (Dischler et al. 1983); 8. $\int\sigma\,d\tilde{v} =$ A3.82 $\times$ 10^{-17} cm, where $A\approx0.1$–0.5 per C in saturated hydrocarbons (Wexler 1967); we take $A\approx0.25$; 9. $\Delta A_{9.7}/E(J-K)\approx0.30$ (see footnote a). Note, however, that Cohen et al. (1989a) argue for a significantly smaller value; 10. $\int\Delta\tau\,d\tilde{v}/\Delta\tau_{9.7}\approx250$ cm^{-1} based on μ Cep emissivity of Roche and Aitken (1984); 11. σ_λ and $\int\sigma\,d\tilde{v}$ are estimates based on assumption that features are produced by silicates with approximately cosmic silicon abundance $\sim$4 $\times$ 10^{-5}; 12. $A_{19}/A_{9.7}\approx0.4$ (cf. Draine 1989a). The 19 μm feature is broad and blended with long-wavelength tail of 8.7 μm feature; choice of baseline is therefore difficult. We take $\Delta A_{19}\approx\Delta A_{19}$. $\int\tau\,d\tilde{v}$ is integrated from 15–23 μm.

There are several reasons for believing carbon to be an important component of diffuse-cloud dust. The most prominent interstellar extinction feature
is the 2175 Å "bump" (see the recent review by Draine 1989b). While noncar-
bonaceous "carriers" such as OH$^-$ on small silicate grains have been proposed
(Steel and Duley 1987), the λ 2175 Å feature is generally attributed to $\pi \rightarrow \pi^*$
transitions in small ($a \lesssim 100$ Å) grains of graphite or some other carbon-rich
material. If due to graphite, $\delta_C \gtrsim 0.15$ is required to account for the 2175 Å
feature.

A broad extinction feature at 3.4 μm has been seen toward the Galactic
center and recently detected toward VI Cyg No. 12 (Adamson et al. 1990)
and other sources (Sandford et al. 1991). Existing upper limits do not rule
out the presence of this feature on all diffuse-cloud lines of sight. The feature
is generally considered to be due to C-H stretching modes, but the specific
material is uncertain. If due to hydrogenated amorphous carbon (HAC), as
proposed by Duley et al. (1989), the HAC must lock up $\delta_C \approx 0.45$, based on
the bandstrength measured by Dischler et al. (1983). If, on the other hand, the
feature is due to some mixture of saturated hydrocarbon molecules, perhaps
only $\delta_C \approx 0.05$–0.10 may be required (cf. note 8, Table VII; Sandford et al.
1991).

In addition to the above features, there is a large number of so-called
"diffuse interstellar bands" which remain unidentified. Two diffuse interstellar
bands have been selected for inclusion in Table VII: the strongest at 4430 Å and
the longest-wavelength DIB at 1.32 μm. If these features are due to "allowed"
electronic transitions with oscillator strengths $f \gtrsim 0.1$, then the features could
be ascribed to minor trace constituents of the grains. On the other hand,
if the 4430 Å feature is caused by a "forbidden" electronic transition with
$f \lesssim 10^{-3}$, then the substance responsible for the feature could contribute a
significant fraction of the interstellar grain mass. The 1.32 μm feature is
much weaker, but might be due to a vibrational mode, perhaps an overtone of
a "fundamental" band at 2.6 or 4.0 μm, in which case an oscillator strength
$f \lesssim 10^{-4}$ would be expected, and the material responsible could possibly be a
significant grain constituent.

Finally, recent models to explain either the infrared emission features
observed in the 3.3 to 11 μm range toward reflection nebulae (see, e.g.,
Léger and Puget 1984; Léger et al. 1989; Allamandola 1989) or the 12 to
60 μm emission observed by IRAS (Draine and Anderson 1985) postulate the
existence of large numbers of clusters of 25 to 200 carbon atoms, which may
be considered either as large molecules such as the PAH or ultra-small grains.
Models which attempt to quantitatively account for both the near-infrared
emission features and the continuous emission from 1 to 20 μm appear to
require $\delta_C \approx 0.1$ in these ultra-small grains (Léger et al. 1989; Désert et al.
1990).

What overall depletions are therefore expected for C in the different
refractory grain models? The "graphite-silicate" model (Draine and Lee
1984) has $\delta_C \approx 0.62$ in graphite grains, assuming a density $\rho = 2\,\mathrm{g\,cm}^{-3}$ for the

graphite grains; the model of Duley et al. (1989) has $\delta_C \approx 0.45$ in hydrogenated amorphous carbon coatings on the silicate grains; and Greenberg's model (Greenberg 1989a, b; Hong and Greenberg 1980) uses $\delta_C \approx 0.64$, of which ~ 0.39 is in a photoprocessed refractory organic residue coating the silicates, and ~ 0.25 in small graphitic grains. A reasonable estimate would appear to be $\delta_C \approx 0.6 \pm 0.2$.

   3. Solid-Phase Overview: Volatile Components. In dark clouds, new spectral features make their appearance, which are attributed to the presence of icy mantles deposited on top of the refractory grain cores. The strongest feature is the 3.08 μm feature which is securely identified as amorphous H_2O ice. A list of well-established absorption features and their proposed identifications is given in Table VIII; the reader is referred to several recent review articles (Tielens and Allamandola 1987a, b; Whittet 1988,1992; Tielens 1989) in which the observations and identifications are discussed. Table VIII is organized in a manner similar to Table VII, except that: (1) the observed feature strengths in column 2 are given relative to the strength of the 3.08 μm feature; (2) column 3 gives the observed "width" of the feature in cm^{-1}; and (3) proposed abundances are given relative to H_2O. Many of these absorption features have been discovered only recently due to rapid improvements in the sensitivity and resolution of infrared spectrometers.

   Only for H_2O, CO, CH_3OH, and possibly CO_2 are the identifications secure. Basic laboratory spectroscopy of icy mixtures is needed to remove the many remaining question marks in Table VIII. A 2.97 μm feature in the spectrum of BN, originally attributed to NH_3 mixed with H_2O (Knacke et al. 1982), has recently been ascribed to scattering into a large beam by H_2O grains (Knacke and McCorkle 1987), although this explanation has been questioned by Smith et al. (1988,1989), who argue that the 2.97 μm feature is a true absorption feature. Of particular interest are the recent possible identifications of solid CO_2 and solid CH_4. The presence of solid CO_2 has been proposed by d'Hendecourt and de Muizon (1989) on the basis of IRAS LRS spectra toward two stars. The inferred abundance of solid CO_2 is comparable to that of solid CO. Unfortunately, no solid H_2O data are available for these same lines of sight. Solid CH_4 has tentatively been identified by Lacy et al. (1991) from high-resolution spectra toward W33A and NGC 7538, with an abundance comparable to that of solid CO.

   It is important to realize that Table VIII attempts to summarize a highly inhomogeneous set of data—some of the weaker features have only been detected on a single line of sight (toward W33A), and the possible variability relative to H_2O for these features is therefore unknown. Even along a single line of sight, different grain components may be present (Tielens et al. 1991). Note also that our depletion estimates, based on the ratio $\Delta\tau_{3.08}/\Delta\tau_{9.7}$ (because the total column density N_H cannot be measured directly) are somewhat uncertain because of the need to correct for silicate emission features in the spectra of the sources. Finally, most of the data in Table VIII refer to clouds

TABLE VIII
Solid-State Extinction Features Unique to Dense Clouds

λ (μm)	$\dfrac{\int \Delta\tau d\bar{\nu}}{\Delta\tau_{3.08}}$ (cm^{-1})	X	σ_λ (10^{-19} cm^2)	$\int \sigma d\bar{\nu}$ (10^{-16}cm)	$\dfrac{N(X)}{N(H_2O)}$	Notes[a]
2.85	100	OH stretch?	?	?	?	1
2.95	80	?	?	?	?	2
3.08	300	H_2O ice	5.7	1.9	1.	3,4,5
3.3	200	?	?	?	?	6
3.53	43	CH_3OH	2.6	0.076	0.07	7,8
3.95	20	H_2S?	?	0.29?	< 0.01	9,28
4.62	30	XCN?	?	?	?	10
"	"	OCN^-?	?	?	?	11
"	"	SiH?	?	> 0.7	< 0.02	12
4.67	5	CO	19.	0.17	0–0.13	13–17
4.9	14	OCS?	?	1.?	0.001?	7,18
6.0	150	H_2O	0.5	0.084	1.	8,19
6.85	100	CH_3OH?	1.	0.1	0.4?	8,19,20
"	"	NH_4^+?	20.	2.	0.02–0.1?	21
7.65	40	CH_4?	?	0.06	0.04?	22,23
13.5	250	H_2O	1	0.26	1.	8,24,25
15.2	50	CO_2	30.	0.41	0–0.03	24,25,26
45.	30	H_2O	0.5	0.04	1.	27

[a] Notes: 1. Smith et al. 1989; 2. In spectrum of OH-IR source OH 231.8+4.2 (Smith et al. 1988); hint of feature in spectra of several protostars (Smith et al. 1989); 3. Profile varies from source to source (Smith et al. 1989); 4. $\Delta\sigma_\lambda = 5.7 \times 10^{-19}$ cm^2 and $\int \Delta\sigma d\bar{\nu} = 1.9 \times 10^{-16}$ cm^2 calculated for $a = 0.3$ μm spherical core with 0.1 μm coating of H_2O ice, using refractive index $m = 1.55 + 0.05i$ of amorphous H_2O measured by Kitta and Kratschmer (1983). Peak σ_λ occurs at 3.05 μm for this case. σ_λ and $\int \sigma d\bar{\nu}$ agree well with film transmission results of d'Hendecourt and Allamandola (1986); 5. Excluding 3 stars with questionable $\Delta\tau_{9.7}$, $\Delta\tau_{3.1}/\Delta\tau_{9.7}$ reaches values as large as 1.20 in the Taurus dark cloud (Whittet et al. 1988). Assuming $\Delta\tau_{9.7}/N_H = 2.5 \times 10^{-23}$ cm^2 we infer that up to ~6% of the O may be in H_2O; 6. In spectrum of OH-IR star OH 231.8+4.2 (Smith et al. 1988); 7. Toward W33A; Baas et al. (1988); 8. σ_λ and $\int \sigma d\bar{\nu}$ from d'Hendecourt and Allamandola (1986); 9. Toward W33A; Geballe et al. (1985); 10. Toward W33A and NGC7538/IRS9 (Lacy et al. 1984); 11. Grim and Greenberg 1987a; 12. Nuth and Moore 1988; 13. Lacy et al. 1984; 14. Geballe 1986; 15. Mitchell et al. 1988; 16. Sandford et al. 1988; 17. Whittet et al. 1989; 18. See discussion by Tielens and Allamandola 1987a; 19. See 5–8 μm spectra in Tielens (1989); 20. Note that N(CH$_3$OH) from 3.53 μm feature is smaller by factor of 5; 21. Grim et al. 1989; critical discussion in Tielens 1989; 22. Tentative detection of CH$_4$ toward W33A, NGC7538/1, and NGC7538/9 (Lacy et al. 1991); 23. $\int \sigma d\bar{\nu}$ from d'Hendecourt and Allamandola 1986; 24. Cox 1989; 25. d'Hendecourt and de Muizon 1989; 26. Sandford and Allamandola 1990a; 27. Erickson et al. 1981; Papoular 1981; Drapatz et al. 1983; 28. Smith 1991.

in which star formation has already occurred, so that part of the mantles may have evaporated.

For each identification in Table VIII, the inferred abundance relative to H_2O is listed. It is evident that the abundances are in general small compared to H_2O. The only possible exception is CH_3OH, for which the abundance (and therefore the identification) is controversial: the 3.53 μm band gives a much lower abundance than the 6.85 μm band, which may indicate that the 6.85 μm band is primarily due to some other substance, e.g., NH_4^+ (see also Schutte et al. 1992). Using $\Delta\tau_{9.7}/N_H$ from Table VII, and $\sigma_{3.08}$ from Table VIII, we obtain

$$\frac{N(H_2O, \text{ice})}{N_H} \approx 4.4 \times 10^{-5} \frac{\Delta\tau_{3.08}}{\Delta\tau_{9.7}} \tag{2}.$$

Excluding the stars Elias 3, 16 and 18, for which the measured silicate absorptions may be suspect, the largest value of $\Delta\tau_{3.08}/\Delta\tau_{9.7}$ found in the Taurus clouds is 1.2 (Whittet et al. 1988), indicating that H_2O contains up to $\delta_O \approx 0.06$ of the total oxygen. The largest value of $\Delta\tau_{3.08}/\Delta\tau_{9.7}$ reported for dense clouds is 0.8 toward AFGL 961 (Willner et al. 1982), implying that $\sim$4% of the oxygen is in the form of solid H_2O in this cloud. In no case has the bulk of the oxygen been found to be condensed into solid H_2O. The amount of carbon taken up in ices is even less: at most 1% is found in solid CO, CO_2 or CH_4. Up to 5% of the nitrogen may be in the form of solid NH_3.

In summary, the refractory and volatile-grain components together contain up to 23% of the available oxygen in the cloud in known species, about 60$\pm$20% of the disposable carbon, close to 100% of the available silicon and iron, but only a few % of the nitrogen. Substantial amounts of oxygen and nitrogen may still be tied up in unobservable species such as solid O_2 and N_2.

B. Orion/KL Abundances

Because Orion/KL, and in particular the hot core, is one of the best chemically characterized sources, it is instructive to summarize the observed gas-phase and grain-surface abundances for this case in detail. Orion has several observational advantages over other regions. First, it is warm so that lines of molecules like H_2O which are not excited in cold clouds become observable. Second, at least two bright embedded infrared sources, BN and IRc2, are available against which absorption line observations of both solid dust features and molecules can be performed. Absorption lines toward IRc2 sample mostly hot core material, whereas those toward BN probe mainly extended ridge material and a little bit of the high-velocity plateau gas, in addition to the immediate circumstellar gas and dust.

The results are summarized in Table IX. It is seen that CO is by far the most abundant gas-phase molecule in the hot core with $CO/H_2 \approx 1.2 \times 10^{-4}$, accounting for $\sim$15% of the carbon in the cloud. The amount of atomic C in the gas phase is uncertain, but is unlikely to be more than 10% of that of CO, and may be as low as 1% in the hot core. Limits on the CH_4 abundance of $CH_4/CO < 0.01$ have been obtained from searches for the millimeter CH_3D

lines (Blake et al. 1987), and from limits on the infrared absorption lines of CH_4 toward BN (Knacke et al. 1985,1988b). These limits are consistent with the probable discovery of CH_4 by Lacy et al. (1991) toward IRc2 at a level $CH_4/CO \approx 10^{-3}$. Acetylene, C_2H_2, has also been detected toward IRc2 with an abundance $C_2H_2/CO \approx 10^{-3}$ (Lacy et al. 1989b). Thus, neither gas-phase methane nor acetylene are significant, and the total amount of carbon accounted for in the gas phase is about 15%. The amount of carbon in the solid phase has been estimated to be $60 \pm 20\%$, so that at most a small fraction of the carbon could be hidden in some as yet unidentified species.

In contrast, the oxygen budget in the cloud is much more uncertain. H_2O has been observed directly in Orion by its transitions at 183 GHz (Waters et al. 1980) and 380 GHz (Phillips et al. 1980) from the Kuiper Airborne Observatory, but because these lines arise from high-lying levels, the inferred abundances are quite uncertain. Very recently, the 183 GHz line has also been detected by groundbased observations at much higher angular resolution (Cernicharo et al. 1990). More reliable estimates can be made from observations of the optically thin $H_2^{18}O$ 203 GHz line, which lead to $H_2O/H_2 \approx 10^{-5}$ in the hot core (Phillips et al. 1978; Jacq et al. 1988,1990). An unsuccessful search for the lowest 547 GHz transition of $H_2^{18}O$ gives $H_2O/CO < 0.01$, but this limit probably applies to the extended ridge rather than the hot core (Wannier et al. 1991). Infrared absorption lines of gas-phase H_2O have probably been seen toward BN, leading to $H_2O/CO \approx 0.03 \pm 0.02$ (Knacke et al. 1988b; Knacke and Larson 1991). As mentioned in Sec. IV, HDO is readily detected in the hot core with an abundance $HDO/H_2 \approx 5 \times 10^{-8}$ (Blake et al. 1987; Plambeck and Wright 1987; Jacq et al. 1990), suggesting $HDO/H_2O \approx 0.004$. Thus, many lines of evidence indicate that the gas-phase H_2O abundance in Orion/KL is significantly less than that of CO, and accounts for at most a few % of the total oxygen budget. Indirect limits on the gas-phase O_2 abundance from searches for the $^{16}O^{18}O$ 234 GHz transition give $O_2/CO < 1$ for the Orion hot core (Blake et al. 1987), so that only a few % of the oxygen can be locked up in O_2. Atomic oxygen can be observed by its fine-structure transitions at 63 and 145 μm from airborne platforms. Strong O I 63 μm emission in Orion has been observed by Werner et al. (1984), but the interpretation is complicated by the fact that it arises partly in the shocked plateau gas, and partly in the photodissociation region close to the Trapezium stars. Nevertheless, conservative estimates of the atomic oxygen abundance in the hot core region indicate $O/CO < 1$. Gas-phase CO itself takes up $< 10\%$ of the available oxygen. CO_2 could conceivably be an important oxygen-bearing molecule, but cannot be observed directly in the gas phase by millimeter transitions. The protonated version $HOCO^+$ has not been detected in Orion. In other molecular clouds, limits of $CO_2/CO < 0.01$ have been inferred, except for the galactic center where $CO_2/CO \approx 1$ (Minh et al. 1988).

In summary, at most 25% of the oxygen in the Orion hot core is detected in gas-phase species. As discussed above, another 20 to 25% may be locked up in known species in grains. This leaves more than 50% of the oxygen

TABLE IX

Observed Abundances X/H_2 in Orion

	Species	Hot Core	Ref.[a]	Ext. Ridge/BN	Ref.[a]	Other Clouds	Ref[a]	Abundance
Gas phase:	CO	$1(-4)$	1	$5(-5)$	1	$(0.5-2)(-4)$	2	15–40% of C; few % of O
	C	—		$3(-6)$	1	$1(-5)$	2	few % of C
	CH_4	$1(-7)$	3	$<1(-6)$	4	$1(-7)$	3	negligible
	C_2H_2	$1(-7)$	5	—		$1(-7)$	5	negligible
	H_2O	$1(-5)$	6	$\lesssim 1(-6)$	6	$\lesssim 1(-5)$	6	few % of O
	O_2	$<1(-4)$	1	—		$<(1-5)(-5)$	6	< few % of O
	O	$<1(-4)$	6	$<1(-4)$	6	—		?
	CO_2	—		$<1(-6)$	7	$<1(-6)$	7	<1% of O
	H_2CO	$>6(-8)$	8	—		$(1-3)(-8)$	8	negligible
	CH_3OH	$(0.1-1)(-6)$	11	$1(-8)$	11	$(0.1-1)(-8)$	12	negligible
	N_2	—		—		$1(-6)-1(-4)$	9,17	dominant?
	NH_3	$1(-7)$	10	—		$(0.2-2)(-7)$	2	< 1% of N
	N	—		—		—		?
Solid phase:	H_2O	—		$3(-5)$	13	$(0-1)(-4)$	14	up to 6% of O
	CO	—		$<3(-7)$	15	$(0-1)(-5)$	14	up to 1% of C; 0.5% of O
	CO_2	—		—		$(0-3)(-6)$	16	up to 0.3% of C, O
	NH_3	—		—		$<1(-5)$	6	less than 5% of N
	CH_4	—		—		$(0.1-1)(-5)$	3	up to 1% of C
	CH_3OH	—		—		$(0.1-1)(-6)$	14	negligible

[a] References: 1. Blake et al. 1987; 2. Irvine et al. 1987; 3. Lacy et al. 1991; 4. Knacke et al. 1988a; 5. Lacy et al. 1989b; 6. see text; 7. Minh et al. 1988; 8. Mangum et al. 1990; 9. McGonagle et al. 1990; 10. Based on $N(NH_3) \approx 2 \times 10^{17}$ cm^{-2} (Jacq et al. 1990) and $N(H_2) = 2 \times 10^{24}$ cm^{-2} (Mundy et al. 1986a); Genzel et al. (1982) suggest a higher NH_3 abundance in the hot core of $1(-6) - 1(-5)$; 11. Menten et al. 1988; 12. Friberg et al. 1988; 13. Knacke et al. 1988b; Knacke and Larson 1991; 14. See Table VIII; 15. Geballe 1986; 16. d'Hendecourt and de Muizon 1989; 17. Womack et al. 1992; van Dishoeck et al. 1992.

budget unaccounted for, assuming solar abundances. If, however, the oxygen abundance in Orion is indeed significantly less than solar (cf. Sec. II.C), the major oxygen species may have been identified.

On the basis of the models, most of the gas-phase nitrogen is thought to reside in the form of N_2, but observations of the N_2H^+ ion suggest that it takes up at most a few %. The ammonia abundance in the hot core is controversial, but may be as low as 10^{-7} (see Table IX), i.e., similar to that found in cold clouds. A few % of the nitrogen may be in the form of solid NH_3. This leaves more than 90% of the nitrogen in unobserved species. It is likely that a significant fraction of this is in the form of gas-phase N and N_2.

It is thus concluded that in spite of the fact that Orion/KL is one of the best studied sources in the sky, major uncertainties remain in the principal oxygen and nitrogen reservoirs in these clouds. Even less is known about the extent to which the abundances of these major species are influenced by the star formation process, and about the form in which they enter a solar nebula.

VI. CHEMICAL EVOLUTION OF THE SOLAR NEBULA

A. Chemical Processes

Gases and solids processed in the star-forming region in which the Sun was born eventually fall into the solar nebula where they undergo further chemical and physical transformations before incorporation in solid bodies. A general discussion of chemistry in protoplanetary nebulae is provided in the chapter by Prinn and Chang. Here, we outline the driving physical processes behind the chemistry and the implications for the fate of infalling cloud material, with particular emphasis on how the original chemical and physical mix has been altered. Gases and grains may experience, in rough order, the nebular accretion shock, drag heating of grains, transport of grains radially and consequent thermal chemistry, and energetic chemistry associated with lightning, nebular flares, and ultraviolet irradiation. Additionally, some material may find its way into the denser nebulae surrounding the forming giant planets, and be further altered there (see Chapter by Lunine and Tittemore). Condensation of volatiles and/or trapping in a more refractory matrix are the final processing steps considered here. Figure 13 summarizes these processes graphically.

A wide range of solar nebula models exist in the literature (see Chapters by Morfill et al. and by Palme and Boynton). For our purposes, the following, somewhat generalized properties are invoked: (1) the nebula has a radial temperature which decreases with increasing distance; (2) the nebular gas is crudely of solar composition; (3) the nebula gains material in the form of gas and grains from the surrounding interstellar cloud which has a composition described in Secs. II–IV, and summarized in Tables V–IX. Thus, while specific models of chemical modification of infalling material require a very specific set of nebular physical processes, the following discussion is framed in a more general way to illustrate more clearly the principles involved. The inner nebula is crudely defined as the region up to 5 AU, whereas the outer

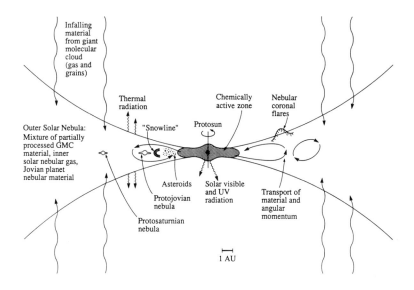

Figure 13. Schematic illustration of the physical and chemical processes in the solar nebula, illustrating the complex interplay between different parts of the nebula, between the nebula and the infalling interstellar cloud material, and between the solar nebula and sub–nebulae surrounding the giant planets. Two processes which further affect the material but which are not shown are an accretion shock, and drag heating of infalling grains (Lunine 1989c).

nebula extends to about 100 AU. In such a model, the density is roughly 10^{14} cm^{-3} and the temperature $T \approx 800$ K at 1 AU, decreasing to 10^{11} cm^{-3} and 80 K, respectively, at 15 AU. Thus, the densities are much higher than those found in the molecular cloud cores from which stars like the Sun formed, and three-body processes are important in the chemistry.

    1. Thermal Chemistry in the Nebula. Prinn and Fegley (1989) computed the energy available for chemical processing of material by various nebular mechanisms, and concluded that, because of the high densities, thermal equilibrium chemistry (i.e., that driven at a sufficiently high rate by virtue of the gas temperature) was by far the most important. They also reviewed the important reaction schemes associated with interchange of carbon-, oxygen- and nitrogen-bearing species. As one proceeds radially outward in the nebula, the direction of the reactions is generally such as to increase the abundances of the reduced species in the gas. Thus, in thermodynamic equilibrium, methane would increase at the expense of carbon monoxide in the outer nebula, and ammonia at the expense of molecular nitrogen (Lewis and Prinn 1980). At lower temperatures, carbon dioxide is also produced at the expense of carbon monoxide, though the compositional gradient is less steep than for methane. However, the reaction rates are strongly temperature dependent, and can be represented as an Arrhenius relationship for the rate-limiting step, and therefore the molecular abundances in a parcel of nebular gas depend rather sensitively on the length of time that parcel retains its identity and radial location. In other words, physical mixing time scales, and in their absence, the dynamic lifetime of the nebula, determine the gas-phase abundances radially outward from the chemically-active zone. Some of the key reaction rates, particularly those of the carbon species, are poorly known (Yung et al. 1988). A final complication is that the chemical-reaction rates can be greatly accelerated by catalyzing the processes on metal grains (Vannice 1975).

As a result of all these processes, the outer nebula is expected to contain less CH_4 and NH_3 than would be expected on the basis of equilibrium considerations, if the carbon arrives mostly in the form of CO and the nitrogen in the form of N_2 from the interstellar cloud. For example, using the graphs in Prinn and Fegley (1989), the recently determined interstellar value of CH_4/CO (gas plus grains) of 0.01 (Lacy et al. 1989b) would be reproduced thermochemically in the nebula at a temperature of roughly 750 K. However, this high a ratio could be achieved within reasonable nebular mixing times only if grain catalysis were efficient; in the absence of such catalysis, quenching would occur at a CH_4/CO ratio 5 orders of magnitude smaller. Conversely, significantly larger mixing ratios of methane in the outer nebula would require lower quench temperatures (i.e., the temperature at the nebular radius at which the chemical destruction time equals the radial mixing time), which are likely to have been unachievable during the astrophysical lifetime of the nebula (of order 1 Myr).

2. Mixing of Inner and Outer Nebular Material. In the absence of any radial mixing of nebular gases, two chemically distinct zones would occur: an inner zone in which material comes to chemical equilibrium, and an outer region whose composition reflects that of the "last" radial zone at which equilibrium could be achieved over the duration of the nebula. This is an oversimplification for several reasons, including the fact that different molecular species are interconverted by reaction networks with distinct limiting rates and hence quench temperatures. Thus, one must define quenched zones for each of the CO-CH_4, CO-CO_2 and N_2-NH_3 interconversions. Radial mixing of nebular material acts to bring chemically "unaltered" material inward to chemically active zones, and return partially processed material back to the inert, outer nebula. (Here again we are oversimplifying the picture for heuristic purposes; as discussed below, other chemical processes may be active in the outer parts of the nebular disk.) The degree to which material is mixed radially in the nebula is crucial to an understanding of the gas-phase composition of the nebula, yet current nebular physical models do a poor job of addressing this question. Stevenson (1990) and Prinn (1990) examined semi-analytic models of a viscously evolving solar nebula to set limits on the extent of radial mixing. While a viscously evolving disk will inevitably transport some material radially, the amount cannot be determined based on a linearized treatment of momentum and mass transport terms, as such a treatment leaves the relative magnitudes of angular momentum and mass transport undetermined (Prinn 1990). More complex models of coupled momentum and mass transport in the solar nebula are required, which unavoidably will involve additional parameters and ambiguities. The implications of nebular transport for carbon- and nitrogen-bearing species are described below in Sec. VI.B.1. on comets.

Solar nebula thermochemistry will also act to reduce the enhancements of deuterated species in molecules heavier than hydrogen created in the interstellar medium. The higher temperatures and preponderance of neutral-neutral reactions in the solar nebula shift the abundances back toward higher-temperature equilibrium, in which HD is the preferred deuterated carrier. Constraints on D-to-H ratios in the outer solar nebula may therefore set limits on the amount of chemical processing in the inner nebula and subsequent outward mixing, although to date this has not been attempted quantitatively.

3. Processing of Material in Giant Planet Subnebulae. The formation of the giant planets apparently resulted in bound, gaseous disks which produced the regular satellite systems (see Chapters by Podolak et al., Bodenheimer et al. and Lunine and Tittemore). Such disks were not necessarily of similar composition to the solar nebula, as they could have been spun-out from the atmospheres of the forming giant planets. In such a case, the rock and ice abundances may have been greatly enhanced relative to solar, with highly uncertain ratios. Quantitative chemical modeling of such disks, for an assumed solar abundance, has been conducted for Jupiter and Saturn by Prinn and Fegley (1989). By setting these disks to a minimum mass required to account for the regular satellites, and constructing simple models of disk energy balance, one can show that the pressures in the satellite-forming zone are as much as 6 orders of magnitude higher than in the surrounding solar nebula (Lunine and Stevenson 1982). Consequently, for a given temperature, the equilibria are strongly shifted toward the reduced nitrogen- and carbon-bearing species. Such nebulae are potentially significant sources for ammonia and methane, reprocessing solar nebula carbon monoxide and nitrogen and returning the products (Prinn and Fegley 1989). The difficulty is in estimating the amount of material which may plausibly escape these gravitationally bound disks into the solar nebula. Prinn and Fegley propose that ejection of circumplanetary planetesimals into solar orbit may be the most plausible process, but to date no detailed work on this or other dynamical mechanisms has been undertaken.

4. Chemistry at the Nebular Accretion Shock. Gas and grains, undergoing free-fall into a medium of increasing gas density, eventually encounter a shock front referred to as the accretion shock. Here material is heated to varying degrees permitting evaporation of volatiles and acceleration of chemical reaction rates. Astrophysical shocks and resulting chemistry have been extensively examined in the literature (see, e.g., Mitchell 1984; Shull and Draine 1987; Hollenbach and McKee 1989; Neufeld and Dalgarno 1989), but

generally have not been specialized to the case of the solar nebula. The chemical modification of the shocked material depends on (1) velocities through the
shock; (2) grain size (and hence coupling to the gas); and (3) initial chemical
state of the material. Mitchell (1984) shows that for a shocked interstellar
cloud, the initial abundance of CO vs elemental carbon as well as the shock
velocity critically affect the post-shock abundances. Substantial quantities
of methane could only be produced in a high-velocity shock (>10 km s^{-1}),
again dependent upon the initial state of the carbon. One key difference for
the case of the nebular shock is that much of the infalling material may be in
the form of grains, which depending on their size may largely decouple from
the shocked gas and be subject to relatively little processing. Initial results of
models by Neufeld and Hollenbach (in preparation) indicate that most of the
oxygen is converted into water, and that grains most likely survive, except at
distances <3 AU.

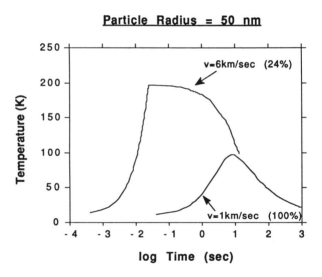

Figure 14.   Temperature vs time for a water ice grain falling into the outer part of the
solar nebula at two initial velocities. A gas density typical of the outer solar nebula,
10^{-11} g cm^{-3}, is used. The number in parentheses is the amount of mass remining
after the heating/sublimation episode (figure adapted from Engel et al. 1990).

5. Drag Heating of Grains.   Grains which have undergone free-fall
into the solar nebula and survived passage through the accretion shock are
subjected to gas drag forces which bring the particles toward a state of Keplerian orbit. Interstellar grains, being small submicron-sized particles, will
continue to be largely borne by the gas and hence not in true Keplerian rotation; however in terms of chemical modification, it is the decoupling of the
grains from the gas during infall (probably not in the accretion shock) which

concerns us most. Wood (1984) postulated that meteorite chondrules might have been formed by drag heating, but required rather extreme conditions to melt refractory grains. Lunine (1989*b*) and Engel et al. (1990) showed that drag heating could raise the temperature of icy grains entering the outer part of the solar system to sufficient temperatures to cause significant mass loss. Figure 14 shows an example of the results, for a small (0.05 μm) water ice interstellar grain falling in with two initial velocities. The percentages show the amount of grain material which survives infall. The results turn out to be highly sensitive to the ability of the grain to effectively radiate in the infrared, i.e., to the choice of emissivity, and only moderately sensitive to the nebular gas density. Lunine et al. (1991), in a follow-on study, used Mie-scattering theory to explicitly compute the radiating properties of the subliming grains, and examined the effects of other volatile species on the grain sublimation. While the more precise computation of thermal radiation results in higher temperatures and hence more mass loss, the presence of species more volatile than water ice acts as a thermostat and keeps temperatures lower, resulting in loss of less water. Temperatures become large enough to cause reactions in any unstable radical species which may be trapped in the grains, but little else occurs chemically during the short period of heating. Nonetheless, the results indicate that perhaps half of the icy grain mass entering the solar nebula is returned to the gas phase; for this material to be incorporated in icy outer solar system bodies, condensation and/or physical trapping must subsequently occur within the nebular environment. These processes are quantified in Lunine et al. (1991).

   6. Trapping of Volatiles in Ices. Return of volatile molecular species to condensed phases requires that they be either condensed out or physically trapped in water ice. Examination of theoretical nebular temperature profiles indicate that water ice should be fully condensed beyond 5 AU; condensation of ammonia (either as a pure ice or stoichiometric ammonia hydrate) is uncertain because it is predicted to be highly underabundant based on nebular thermochemical models. However, one can argue that interstellar ammonia, delivered to the nebular gas by grain-drag evaporation, may not be destroyed by nebular thermochemistry if radial mixing is limited. We provide supporting evidence for this below. Under such circumstances ammonia condensation is assured at or beyond roughly 10 AU. Carbon dioxide may also condense in the outer solar system if it represents a significant sink of carbon. Beyond carbon dioxide, however, direct condensation of species such as methane, carbon monoxide, nitrogen are problematic, depending on the temperature-pressure dependence in the outer parts of the nebula which is poorly defined. The more likely mechanism by which such species find their way (back) into the condensed phase is by physico-chemical trapping in water ice. Trapping in other phases is possible too (e.g., in graphite), but the large abundance of water ice coupled with its high specific adsorption area make it a compelling reservoir. Two trapping mechanisms have been explored in detail: adsorption onto surface sites, and clathration, in which a structural phase change in the water ice creates three-dimensional voids into which the volatile species are inserted (Stevenson and Lunine 1988).

The amount of material trapped by these processes is a strong function of temperature. Clathration, being a phase change, exhibits a well-defined onset for a given set of gas abundances, and can accommodate up to 13% by number of non-water-ice volatile species. However, efficient clathration requires good exposure of the gas and ice; clathration in the interior of ice grains is strongly kinetically inhibited (Lunine and Stevenson 1985). Adsorption into crystalline ice generally traps much less volatile material than does clathration, but this is not true of amorphous ice which can trap as much as clathrate at very low temperatures (Mayer and Pletzer 1986; Bar-Nun et al. 1987). Again, however, this efficient trapping requires that the trapped gas be available at the time of amorphous ice condensation, which has traditionally been considered unlikely for the solar nebula. However, more recent experiments (Bar-Nun et al. 1988) and theoretical models (Lunine et al. 1991) suggest mechanisms by which such efficient trapping may have taken place.

Of most interest to us here is the chemical fractionation effects associated with water ice trapping of more volatile species. Direct condensation, of course, is potentially the strongest form of what is essentially fractional distillation: if methane were to condense out but not carbon monoxide, icy objects in the outer solar system would exhibit large amounts of methane but little or no carbon monoxide. However, at temperatures above the condensation points of the pure ices, selective clathration or adsorption could produce fractionation effects with the same trends, though not of the same severity.

Figure 15, modified from Lunine (1989c), illustrates this effect for methane and carbon monoxide trapped in clathrate, and demonstrates that the degree of fractionation is a function of total gas uptake. Imagine having a cometary water-ice sample in the laboratory in which the ratio of CO to CH_4 were measured to be unity. Figure 15 then shows the ratio CO/CH_4 in the original nebular gas as a function of the total amount of gas trapped in the ice. An abscissa value of unity corresponds to the maximum clathration ratio of 1:6 volatile gas to ice. Smaller values physically represent kinetic inhibition of gas uptake, due to poor contact between the gas and ice, or limited time available for uptake. The figure is not general, for it assumes that the ice was present in a region of the nebula cold enough that both methane and carbon monoxide clathrates would be separately stable (Lunine and Stevenson 1985). Hence, for abscissa values toward the right-hand end, after all of the methane has been trapped, cage sites are available for carbon monoxide and will be readily filled by that gas. Had we picked a nebular region at which methane but not carbon monoxide clathrate hydrate was stable, the curve would not decrease so steeply to the right.

Based on volatility considerations (Miller 1961) and detailed models (Lunine and Stevenson 1985), CO_2, CH_4, CO and N_2 will incorporate in clathrate or adsorption sites in decreasing propensity. Ammonia strongly hydrogen bonds to water, hence forming a stoichiometric crystal much more

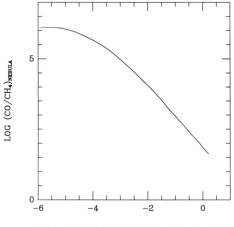

LOG (FRACTION OF H₂O ICE EXPOSED TO GAS)

Figure 15. Carbon monoxide to methane ratio in the initial nebular gas from which a grain of clathrate has been recovered. The clathrate grain is assumed to have a measured CO/CH_4 ratio of unity. The ordinate is plotted as a function of the amount of gas taken up in the clathrate, where a value of $0 = \log(1)$ corresponds to the maximum possible gas uptake, one volatile molecule for every six molecules of water (figure based on Lunine 1989*b*).

readily than clathrate. Carbon dioxide, at low temperatures, tends to condense out as dry ice rather than be trapped in clathrate or adsorption sites. Finally, no detailed work has been done to date on the isotopic effects of nebular gas trapping, although vapor pressure effects on condensation can alter the mole fraction of deuterated species relative to the gas phase. Such alterations are not strong compared to interstellar enhancements but can be in either direction depending on the temperature (Bigeleisen 1961). In any event, it is clear that the composition of the condensed phases in nebular planetesimals is not identical to that of the gas phase, and requires an understanding of the mechanisms of trapping in order for the original gaseous mix to be inferred.

In these discussions, we assume that direct gravitational capture of nebular gas by objects up to the size of the largest satellites is not a plausible source of observed volatiles; this is certainly the case dynamically because plausible subsequent processes of atmospheric escape could not leave behind the small solar abundances of heavier species while removing all of the molecular hydrogen (see, e.g., Hunten et al. 1989).

7. Energetic Chemistry in the Nebula. In addition to the processes described above, special effects may alter the abundances of species in planet-forming bodies from that in the original interstellar cloud. These include photolysis driven by solar ultraviolet, radiochemistry powered by [26]Al, light-

ning and nebular flares. Prinn and Fegley (1989) have assessed the importance
of the first three of these processes for solar nebula chemistry. They compared
the energy flux from these processes, usable for chemistry, to the energy flux
available for inner nebula thermochemistry. Assuming live ^{26}Al was available, they show that the energy flux available for radiochemistry was 7 orders
of magnitude less than that from thermochemistry. As at least 10^{-3} of the
thermochemically processed gas could have been transported to the outer
nebula ($\sim$100 AU) by both the Prinn (1990) and Stevenson (1990) transport
models, it is clear that the radiochemistry is unimportant relative to thermochemistry even if it acted locally in the outer part of the nebula. Solar and
stellar ultraviolet radiation cannot penetrate very far into the gas and dust
disk of the solar nebula; stellar photons are more efficient than solar because
of the geometry. Prinn and Fegley estimate, based on the current observed
stellar ultraviolet flux, that the usable energy flux for nebular chemistry is $\sim$4
orders of magnitude lower than that due to thermochemistry. However, if the
molecular cloud ultraviolet flux were enhanced, such photochemistry could
have been predominant in parts of the outer nebula if surrounding gas and dust
were patchy and radial mixing inhibited so that inner nebula thermochemistry
were less important. It is difficult to estimate the energy flux available from
lightning because the nebular mechanism is not well understood. Prinn and
Fegley estimate a usable energy flux from lightning and thundershocks of 0.1
that of thermochemistry. Clearly this could be an important source of chemical reactions in the outer part of the nebula. Further work on the mechanisms,
and radial dependence thereof, of lightning generation may alter this estimate
(see Chapter by Morfill et al.).

The Levy and Araki (1989) have proposed a mechanism for making meteorite
chondrules by generation of nebular flares. The physics of the flares is entirely
analogous to solar flares. Such flares will engender heating of the surrounding
gas and are a potential source of energy. The minimum energy flux from such
flares can be estimated by assuming that they are the primary means by which
chondrules were created. Using the fraction of primitive meteorites in the form
of chondrules, the required energy release of 10^{34} ergs/nebula corresponds to
an average flux roughly 8 orders of magnitude lower than the energy available
from thermochemistry estimated by Prinn and Fegley (1989). Although this
is a lower limit, two other issues mitigate against the importance of nebular
flares for altering the nebular chemistry: (1) flares occur high above the disk
midplane, and (2) a value many orders of magnitude larger than the lower
limit given would require large magnetic fields which may be incompatible
with other aspects of nebular models. There is sufficient uncertainty here that
a careful look at flares as a nebular chemistry source is required.

The net effect of the lightning and thundershock-induced chemistry described by Prinn and Fegley (1989) is to produce significant quantities of
hydrogen cyanide. Maximum yields in the solar nebula correspond to 0.3%
of the nitrogen locked up as HCN; for the Jovian nebula Prinn and Fegley
estimate as much as 7%. Shock chemistry may also partially reset the D to

H ratio in molecules heavier than hydrogen, as well as have other isotopic effects which have not been investigated.

B. SOLAR SYSTEM OBSERVATIONAL DATA

1. Comets. Because comets are generally agreed to be the least evolved samples of water ice and more volatile species in the solar system, they may be the most valuable link between solar nebula and interstellar chemistry. A proper chemical inventory of cometary ices and volatiles requires either a laboratory or *in-situ* analysis of cometary nucleus samples, in which the molecular abundances, ice phases and mechanisms of trapping are assessed directly. A big step forward toward this goal will be the laboratory analysis of a comet sample returned to Earth, such as is planned for the Rosetta mission. The best currently available assay of cometary volatile species comes from the combined ground- and space-based measurements of comet Halley in 1986. It should be remembered that these analyses involved measurement of molecular fragments, or daughter products, in the comet coma; nucleus abundances were inferred from chemical models of the coma. No information on the physical modes of trapping of the molecules in the nucleus could be obtained. Nonetheless, consideration of these data sets along with those of other comets is illuminating for the issue of nebular modification of the original mix of gas and dust coming from the giant molecular cloud.

Table X is a current interpretation of the 1986 results for Halley abundances. Much of it is the same as that reviewed in Lunine (1989*b*) and Weaver (1989), and the discussion in those papers should be consulted for the uncertainties (see also Chapter by Mumma et al.). Changes include a decreased ammonia abundance based on the telescopic studies of Wyckoff et al. (1988;1991*a*). The methane abundance is particularly controversial. The value quoted here derives from the analysis of the spatial distribution of CH before perihelion, assuming that CH is a granddaughter product of CH_4 (Allen et al. 1989). Subsequent searches for the infrared lines of CH_4 at postperihelion were unsuccessful, yielding upper limits of $CH_4/H_2O<0.01$ (Kawara et al. 1988). Also, Boice et al. (1990) argue for an upper limit $CH_4/CO<0.005$, based on a re-interpretation of the Giotto Ion Mass Spectrometer data, although their arguments are not compelling. If the values listed in Table X apply, the ratio of methane to carbon monoxide in Halley is comparable to that found in some interstellar clouds (Lacy et al. 1991), and much higher than that predicted by standard solar nebula models. Similarly, the ammonia abundance relative to water is much higher than that given by the nebular models. Lunine (1989*b*) proposed that the methane abundance in Halley was an indication of more effective catalysis of nebular thermochemistry than had been previously assumed in published models, along with the fractionation effects associated with trapping of these species in water ice. Engel et al. (1990) quantified this, arguing that laboratory studies did not rule out the synthesis of methane in Fischer-Tropsch-type reactions on grains, and that therefore low nebula quench temperatures for the CO-CH_4 interconversions

TABLE X

Comparison of Abundances in Comets with Interstellar Clouds

Species	Halley[a]	Other Comets[a]	ISM (gas)[b]	ISM (grains)[b]
H_2O	1	1	1	1
CO	$0.05 - 0.1$	$0.02 - 0.07$	10	$0 - 0.13$
N_2	<0.05	<0.001	$0.1 - 10$	—
CH_4	0.05^c	—	0.01	$0 - 0.05$
NH_3	$0.001 - 0.01$	$0.001 - 0.02$	0.01	< 0.1
CO_2	$0.02 - 0.04$	—		$0 - 0.03$
H_2CO	$0.01 - 0.1$	0.005^d	< 0.1	—
CH_3OH	0.01	0.01^d	$0.002 - 0.01$	0.01
HCN	0.001	$0.0002 - 0.002^d$	$0.01 - 0.1$	0.01
H_2S	$?^f$	0.002^d	0.02^e	—
$^{12}CN/^{13}CN$	65 ± 9^h	—	0.5^g	$< 0.01^i$
			$45 - 90$	

[a] Based on review of Weaver(1989), unless otherwise indicated.
[b] Based on Table VIII; the gas-phase abundances refer mostly to the Orion hot core.
[c] Boice et al. 1990 suggest an abundance less than 0.0005.
[d] Bockelée-Morvan et al. 1990.
[e] Mangum et al. 1991.
[f] Probably detected, but abundance not known.
[g] Minh et al. 1990.
[h] See text.
[i] Smith 1991.

could not be ruled out. Under these conditions, 0.1 to 1% of the carbon might be in the form of methane at the edge of the chemically active region. This abundance is then diluted in the outer nebula because of inefficient radial mixing (and their low assumed methane abundance in the native gas of interstellar composition), but the methane fraction relative to carbon monoxide is then boosted in water ice by the effects of clathration or adsorption.

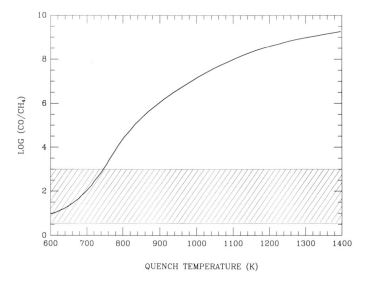

Figure 16. Ratio of carbon monoxide to methane in the solar nebula as a function of quench temperature for the grain catalysis process discussed in the text. The rough acceptable range based on the Halley data, corrected for the fractionation effects of ice trapping, is given by the shaded region. Thus, to be consistent with the Halley data, one must choose a quench temperature that puts the curve in the shaded region. It is assumed that nebular mixing is only moderately efficient, as discussed in the text (figure based on Engel et al. 1990).

Figure 16 shows the results from Engel et al. (1990) for the CO-CH_4 ratio. Quench temperatures from 600 to 1400 K are chosen, and the assumed amount of nebular radial mixing corresponds to a conservative case from Stevenson (1990), wherein 1% of the material in the outer solar nebula has been processed in the inner disk and remixed outwards. The figure assumes a C/O ratio of 0.6; while Anders and Grevesse (1989) and Grevesse et al. (1991) recently redetermined this ratio to be 0.45–0.5, consideration of cold trapping of water by condensation (Lunine and Stevenson 1988) argues in favor of a ratio closer to unity. In any event, the results are mostly sensitive to the quench temperature. All values correspond to some significant degree of catalysis of the interchange reactions on grains; if such catalysis is inhibited, as argued by Prinn and Fegley (1989), then the CH_4 in comet Halley cannot be a product of solar nebula processing of CO. If, on the other hand, grain catalysis can be argued to have been operative in the nebula, Halley's methane abundance could indeed be a solar nebula product. Likewise, the carbon dioxide abundance measured in Halley could have been manufactured from CO in the solar nebula, though somewhat lower quench temperatures or higher efficiency of radial mixing are required. HCN, on the other hand, if a solar nebula origin were to be invoked, would best be explained as a product of energetic chemistry associated with lightning and thundershocks as described above.

Other species measured in Halley cannot be explained by thermochemistry in the solar nebula. Engel et al. (1990) concluded that no plausible model of solar nebula thermochemistry could manufacture the abundance of ammonia, relative to water, seen in comet Halley. They concluded that ammonia was evidence of preserved volatile species from the molecular cloud. Prinn and Fegley (1989) argue for an alternative explanation, that both methane and ammonia were manufactured in the high pressure nebulae surrounding the giant planets, and then transported outwards to form part of Halley. They propose that the most plausible delivery mechanism is in the form of solid bodies collisionally scattered from orbits around the giant planets into the solar nebula. Formaldehyde, on the other hand, is difficult to manufacture within solar or giant planet nebular environments, and if the abundance determined by Mumma and Reuter (1989) is correct, H_2CO may be of interstellar origin. Fegley (1992) offers an alternative interpretation: that formaldehyde is indicative of active photochemical processes in the solar nebula. This, and the possibility that nebular Fischer-Tropsch reactions provided large amounts of formaldehyde (Fegley 1992), require detailed modeling and laboratory studies for their validity to be evaluated.

Two isotopic ratios in Halley have been reasonably well-constrained: the D/H value and $^{12}C/^{13}C$, listed in the table (see also Table IV). When interpreting D/H ratios, it is essential to distinguish between the various deuterated species and the different histories they may have experienced. Halley's ratio is for water, and while it compares well with enhanced deuterium in methane in the outer solar system, methane and water had very different histories. Lutz et al. (1990) argue that the high D/H in Uranus, Neptune, Titan and Halley are the result of the presence of an interstellar component in each of these bodies. Alternatively, Yung et al. (1988) conclude that photochemical enhancement of deuterium in the solar nebula may have occurred for both methane and water. However, the fact that water condenses readily in most of the nebula, coupled with the need to bring these materials well above the midplane to enable exposure to stellar ultraviolet, suggests that these two molecules may have had very different photochemical histories and the respective deuterated fractions ought not necessarily to be similar. Also, Titan's deuterated methane component may have been further enhanced by atmospheric photochemistry after formation (Pinto et al. 1986), further decoupling the history of the two molecules. In summary, the relationship of deuterated water in Halley to deuterated methane in the rest of the outer solar system is by no means clear, nor is the origin of the enrichment necessarily interstellar. A sample return mission to a comet would clearly be of great importance to settle this issue.

The carbon isotopic ratio $^{12}C/^{13}C$ of roughly 65 in Halley appears different from that measured in meteorites and other comets (Wyckoff et al. 1989). This might be used to argue for an interstellar origin for Halley, and perhaps even a special one for this comet. Such an invocation appears premature, on several grounds. First, only Halley has had a carbon isotope ratio reported in CN, published values for other comets are from C_2, and chemical processes which produce the isotopic enhancement are different for the two. In fact, an analysis of the available $^{12}C^{13}C$ spectra of Halley yields $^{12}C/^{13}C$ around 90 (Gredel 1987). Second, Fegley (1992) points out that carbon isotopic measurements for the outer solar system range from roughly 20 to 160, illustrating not only the observational difficulties but the possibility that the history of this isotopic ratio is complex. The inner solar system value of 89 is derived from meteorites and may represent yet a different history and set of processes. Finally, the interstellar $^{12}C/^{13}C$ ratio has been the subject of considerable discussion, with values ranging from 43±6 (Hawkins and Jura 1987) to 77±3 (Stahler et al. 1989), although the most recent data appear to favor a value around 65 (Langer and Penzias 1992; Crane et al. 1991; Centurión and Vladilo 1991). It is therefore premature to argue that the Halley value must represent an interstellar origin distinct from the solar system.

Another clue to their origin may be provided by the ortho/para ratio of the cometary molecules. Mumma et al. (1988b) find an ortho/para ratio of H_2O in comet Halley that can be characterized by a nuclear spin temperature of ~25 K. Such a spin temperature is similar to that derived from the observed ortho/para ratios of H_2CO and H_2CS in cold interstellar clouds such as TMC-1, which in turn is thought to reflect the temperature of the grains on which the molecules formed (Kahane et al. 1984; Minh et al. 1991). However, gas-phase exchange reactions with ions can also establish ortho/para ratios which are characterized by a low spin temperature.

In summary, a purely interstellar origin for Halley cannot be ruled out; nonetheless, we do not know enough to dismiss substantial solar nebula processing either. The heating calculations of Engel et al. (1990) demonstrate that the fate of grains falling into the solar nebula is to be partially preserved in solid form and partially evaporated. The subsequent history of such grains is that they will later condense or adsorb partially processed solar nebula gases as a layer surrounding a core of relatively unaltered material. The result would be a comet comprised of an amalgam of chemistry reflecting the whole history of planetary system formation. Ultimately, detailed study of cometary samples is required to elucidate the rich history locked in the icy grains.

   2. *Outer Planets and Satellites.* The origin and evolution of outer planet atmospheres is reviewed by Gautier and Owen (1989) and Lunine (1989c), and will not be repeated here. Because of the re-equilibration of gases throughout

their massive, hot envelopes, these bodies are not good indicators of early solar nebula processing of interstellar material. The exception, isotopic ratios involving deuterium and carbon, have been obliquely referred to above. The state of observations of the carbon isotope ratios is too primitive to be of help in the present discussion. The challenge with regard to deuterium is to understand the relationship between HD and CH_3D at present, as the deuterium re-equilibrates between these species. Progress is being made in this area (Fegley and Prinn 1988), to the point that Lutz et al. (1990) have begun constructing a history of deuterium reservoirs in the outer solar system. A stubborn and outstanding problem in ascribing enriched deuterium to a preserved, interstellar component is current inability to characterize and constrain the ultraviolet flux in the early solar nebula. How much material was transported high above the nebular disk, exposed to ultraviolet and chemically altered, must be quantified before one can evaluate the alternative that the enriched deuterium in methane was photochemically created (Yung et al. 1988). In Titan's atmosphere, which contains a deuterated methane enrichment nearly identical to the deuterated water enrichment on Earth (de Bergh et al. 1988), it is possible to show that atmospheric photochemistry could be responsible for a significant fraction of the enhancement (Pinto et al. 1986). It cannot be totally ruled out that an early epoch of enhanced photolysis or even impact-shock chemistry on Titan produced today's observed enrichment.

The large icy bodies of the outer solar system tell an interesting story with regard to carbon and nitrogen species, as covered in the Chapter by Lunine and Tittemore. Carbon monoxide is apparently a minor component of Triton's surface, based on Voyager UVS data (Broadfoot et al. 1989), and preliminary reduction of near-infrared spectroscopy (Cruikshank et al. 1991). This suggests chemical evolution of the component grains after contact was lost with the gas of the solar nebula or parent molecular cloud (Lunine et al. 1991), or processing after the formation of Triton itself (Shock and McKinnon 1992). Further refinement of the CO upper limit for Triton from Voyager data is critical, as is a more precise determination of Pluto's atmospheric composition. The surface has methane, nitrogen and carbon monoxide, based on spectroscopic detection of frosts (Owen et al. 1992). The relative abundance of these last two in the atmosphere is undetermined (Yelle and Lunine 1989), but obviously is of high interest. Finally, the inability to find ammonia on any surface in the outer solar system is curious, but should not be taken as a signal that it was absent from the solar nebula. Surface ultraviolet or charged-particle chemistry has undoubtedly transformed ammonia to hydrazine and perhaps even molecular nitrogen in the optical surfaces of icy satellites. Searches by Cassini for polar volatiles on the icy satellite Enceladus, and others, would be illuminating in this regard.

3. Isotopic Effects in Meteorites. The reader is referred here to the Chapters by Ott, and by Palme and Boynton, for a discussion of isotopes in meteorites. We mention here only the deuterium. Clearly the deuterium record in meteorites indicates processes for enhancing deuterium in interstellar clouds (Yang and Epstein 1983). The relationship between the highly fractionated organic component and enhanced, deuterated methane awaits an understanding of the origin of methane in the outer solar system, but it is tempting to ask whether methane might not be the product of heavy organics in the hydrogen-rich environment of the solar nebula (Walker et al. 1970). While this is distinct from the origins of solar system methane described above, an outstanding uncertainty remains in how Fischer-Tropsch-type reactions operate in a system like the solar nebula, for which no real laboratory analog has been constructed. Clearly a severe test would be the measurement of the CH_3D/CH_4 ratio in a comet. While some might regard it as ironic if solar system methane turned out to be a product of heavier, interstellar organics, irony appears to be one of nature's strong suits.

VII. CONCLUDING REMARKS

The preceding sections have amply demonstrated that many different processes can modify the chemical composition of a cloud as it evolves from its initial quiescent state to protostellar and protoplanetary regions. Although systematic data on molecular abundances at the various stages of the star-formation process are still lacking, the observed chemical gradients within actively star-forming regions such as Orion/KL and IRAS 16293–2422 stress more and more the close interdependence between source dynamics, energy balance and chemical composition. No longer can the initial stages of star formation be studied ignoring these chemical changes, and vice versa.

In spite of the wealth of new observational data, many questions remain. For example, what are the primary oxygen- and nitrogen-bearing species in cold, dark clouds prior to star formation? How are they affected by the star formation process? Unfortunately, observational searches for H_2O, O_2, O, N_2 and N in cold clouds will remain difficult to impossible for some time to come. Even in warm clouds, where H_2O can be observed, up to 50% of the oxygen may be unaccounted for, if the elemental abundances in these clouds are close to solar. Finding the missing source of oxygen, whether in the gas phase or on grains, will be one of the major challenges for the next decade. Another challenge will be to find a truly quiescent dark cloud core in which the chemistry has not yet been affected by star formation, to delineate the initial conditions and to test the "standard" interstellar chemistry models. Even in the classical test cases like TMC-1, the chemistry may be affected to some extent by the earliest stages of star formation, which may at least partly explain the chemical gradients observed in these clouds on scales of only a few tenths of a pc.

The example of Orion/KL demonstrates that star formation modifies the chemistry in at least two ways: by changing the molecular depletions and by creating energetic events. The observed high abundances of deuterated and even doubly deuterated molecules in the hot core in Orion/KL form perhaps the best evidence that molecular mantles are released from grains when heated to temperatures in excess of 100 K. The distribution of NH_3 in IRAS 16293–2422 may be one of the first case studies of the depletion of molecules in the inner core, and return to the gas phase in the outer part. The extent to which grains actively form new molecules is still an open question, however, and the actual chemistry of NH_3 in interstellar clouds is still poorly understood. Observations of CS toward HL Tau also suggest large depletions of polar molecules in its circumstellar disk.

The two most obvious signposts of energetic phenomena associated with star formation are infrared emission from warm dust and large scale molecular outflows. The chemistry is expected to respond to these energetic events. Physically, the increased temperature opens up normally closed chemical channels for the destruction and creation of species such as H_2O. If strong enough, the shocks not only can vaporize the mantles, but also disrupt grain cores and return sulfur, silicon and perhaps oxygen to the gas phase. The large amounts of SO, SO_2 and SiO observed in star-forming regions like Orion/KL most likely result from the interaction of the surrounding molecular material with the outflow, although the details of the chemistry still need to be worked out. The enhancements may depend on the luminosity of the source, however, and on the evolutionary state, as demonstrated by the fact that the abundances in IRAS 16293–2422 are only little affected compared with cold clouds.

On scales appropriate to the solar nebula (< 100 AU), many processes have been identified that can change the chemical composition of material entering even the outer part of the solar nebula. The relative importance of these processes as a function of position in the nebula is still poorly determined observationally. It must be understood that different material will be altered to different degrees, and the *whole* record obtained from refractory phases, volatiles, and isotopic enhancements must be considered in quantifying the history of interstellar material which became solar nebula and then planetary material. The clues that we currently have suggest that in the outer part of the nebula some original molecular cloud material was preserved to a significant degree. Desperately needed are *in situ* or sample return analyses of comet nucleus material, and further inventory of the isotopic and molecular record on other outer solar system bodies.

Observations of the distribution of molecules in circumstellar disks around other stars on scales of 5 to 50 AU would be very valuable as tests of the theories of our own solar system formation, but must await the development of higher-resolution interferometric techniques, such as might be provided by the millimeter array. Nevertheless, some insight into the extent to which interstellar material is modified can be obtained from the single dish molecular surveys on scales of a few 1000 AU combined with interferometric imaging on scales of 500 to 1000 AU. High-resolution infrared absorption line observations toward embedded sources will also be very valuable for the determination of abundances in both gas and solid phases close to the young stellar object, but additional laboratory measurements on infrared absorption in synthetic grain mantle analogs will be required to fully interpret the astronomical spectra. By applying this array of observational techniques to a much wider range of objects with differing luminosities, we are hopefully in a position to unravel, for the first time, the driving chemical reactions and physical processes that steer the formation of stars and planets.

APPENDIX

INTERSTELLAR CHEMISTRY AND MODELS

This Appendix contains a more detailed description of the interstellar chemistry and models discussed in Sec. II. First, a brief introduction into the basic ion-molecule gas-phase chemistry scheme is given, and some recent developments regarding the models are discussed. This is followed by a description of chemical processes involving grains, and a summary of model results regarding grain mantles. Finally, a few remarks concerning shock chemistry are made. More details can be found in the reviews mentioned at the beginning of Sec. II.

A. Basic Processes

The thousands of reactions that enter all interstellar and protostellar chemical networks can be categorized in only a few different classes, as summarized in Table A1 (Dalgarno 1987; van Dishoeck 1988*a*). The two basic processes by which molecular bonds can be formed from atomic precursors in the low-density interstellar environment are radiative association, which is accompanied by the emission of a photon, and grain-surface formation, in which the grain carries off the energy released upon formation of the molecular bond. The major destruction process of neutral molecules in any environment in which ultraviolet photons are present is photodissociation, whereas molecular ions are usually most rapidly removed by dissociative recombination with electrons. In regions of very high temperature (>3000 K) and density, collisional dissociation can play a role. Chemical reactions can rearrange bonds, once diatomic species have been formed. In an ion-molecule exchange reaction, the bonds in the new molecule are formed at the expense of those of the reactant species. Another possible outcome of an ion-molecule reaction is charge-exchange. Ion molecule reactions are usually very fast at low temperatures if the reaction is exothermic, with rate coefficients close to those given by the simple Langevin theory ($\sim 10^{-9}$ cm^3 s^{-1}). Neutral-neutral reactions are much slower than the ion-molecule reactions at low temperatures due to the weak long-range attraction and potential barriers along the reaction path. They become significant, however, at higher temperatures.

A key aspect of interstellar chemistry is the fact that H$_2$ is by far the most important constituent of the cloud. Thus, if the reaction of an ion or molecule with H$_2$ is exothermic, that process will be the dominant reaction path. Only if the ion does not react with H$_2$ can reactions with less abundant species effectively compete to produce other molecules. A second essential

TABLE A1

Classes of Chemical Reactions

Type	Process	Rate Coefficient
Formation Processes		
Radiative association	$X + Y \rightarrow XY + h\nu$	10^{-16}–10^{-9}
Grain surface formation	$X + Y{:}g \rightarrow XY + g$	$\sim 10^{-18}$
Destruction Processes		
Photodissociation	$XY + h\nu \rightarrow X + Y$	$\sim 10^{-10}$–10^{-8} s^{-1}
Dissociative recombination	$XY^+ + e \rightarrow X + Y$	$\sim 10^{-6}$
Collisional dissociation	$XY + M \rightarrow X + Y + M$	—
Chemical Processes		
Ion-molecule exchange	$X^+ + YZ \rightarrow XY^+ + Z$	$\sim 10^{-9}$
Charge-transfer	$X^+ + YZ \rightarrow X + YZ^+$	$\sim 10^{-9}$
Neutral-neutral	$X + YZ \rightarrow XY + Z$	$\sim 10^{-12}$

[a] Approximate rate coefficients appropriate for cold dark clouds. All rate coefficients are sensitive to temperature. For photodissociation, the rates in s^{-1} in the unattenuated interstellar radiation field are listed.

assumption in the chemistry networks is the presence of sufficient amounts of reactive ions to drive the ion-molecule reaction network. In diffuse and translucent interstellar clouds, these ions are created by the photoionization of elements with ionization potentials less than 13.6 eV, such as C, S and Si, but not O and N. Inside dense clouds, where the interstellar radiation field no longer penetrates, cosmic-ray ionization produces He^+ and H_2^+ ions, and H_2^+ subsequently reacts rapidly with H_2 to form H_3^+. The rate at which cosmic rays ionize hydrogen is $\sim 10^{-17}$ to 10^{-16} s^{-1}, and it is usually assumed that they can penetrate dense clouds at least up to column densities of 10^{24} cm^{-2}, although their penetration deep into star-forming regions is still uncertain.

Consider as an example the first part of the oxygen network, illustrated in Fig. A1. The H_3^+ resulting from cosmic-ray ionization reacts with atomic oxygen to form OH^+. Once OH^+ is formed, a sequence of rapid hydrogen abstraction reactions with H_2 follows, in which H_2O^+ and then H_3O^+ are made. This sequence stops when an ion is formed (H_3O^+ in this case) for which the reaction with H_2 is endothermic. H_3O^+ can thus react with other species in the cloud, and will most likely dissociatively recombine to form OH and H_2O, unless the electron abundance is very low, $n(e)/n(H_2) < 10^{-8}$. OH can subsequently react with neutral or ionized carbon to form CO, and with neutral atomic oxygen to form O_2. Only at high temperatures can it react rapidly with H_2 to produce H_2O.

The carbon chemistry in dense clouds is initiated by a similar reaction

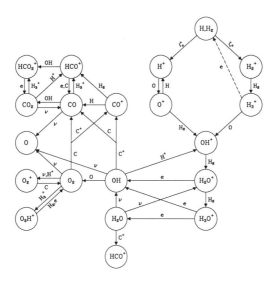

Figure A1. The most important interstellar chemical reactions involving oxygen-bearing molecules (figure from van Dishoeck 1988*a*).

between C and H_3^+ to form CH^+. In more diffuse regions, C can be pho-toionized directly and the radiative association reaction between C^+ and H_2 then starts the carbon network. By contrast, the sequence of reactions that launches the nitrogen chemistry is more uncertain, because the reaction of N with H_3^+ to form NH^+ is endothermic, while that to form NH_2^+ has significant barriers. The only other possible initiating reaction is that between N^+ and H_2 to form NH^+, which is slightly endothermic (by 19 meV), but which can proceed if either H_2 is rotationally excited, or if the N^+ atom has excess trans-lational energy, e.g., as a result of its production by dissociative ionization of nitrogen-bearing molecules such as N_2 or CN. The latter scheme requires, of course, the prior presence of nitrogen molecules in the cloud. These can be formed through coupling of the oxygen, carbon and nitrogen chemistries. For example, the NO molecule can result from the neutral-neutral reaction of N with OH. Much more efficient are the reactions involving carbon, because C^+ is easily inserted into carbon chains, and relatively rapid condensation (e.g., $CH_3^+ + C_4H_2 \rightarrow C_5H_3^+ + H_2$) and radiative association (e.g., $C_3H^+ + H_2 \rightarrow C_3H_3^+ + h\nu$) reactions can occur, leading to the more complex hydrocarbon molecules.

An essential element of the chemical models is the availability of reli-able reaction rates for these processes (see Millar et al. 1991*b* for a recent summary). Because the physical conditions in interstellar clouds are quite different from those normally found in a laboratory on Earth, relatively few reaction rates have been measured at the relevant temperatures and densities. This often leads to major uncertainties in the predicted abundances, which need to be taken into account in comparison with observations. Although educated guesses can be made for the majority of the unmeasured cases, important gaps remain for some crucial reactions. In fact, one of the most important aspects of the modeling efforts is to identify the key reactions in the networks for which accurate rates are essential. This, in turn, stimulates complementary laboratory or theoretical efforts to determine these rates, and often unexpected results are obtained, not only for individual reactions, but even for whole categories of reactions. Some recent developments in this area are summarized below:

1. *Ion-polar molecule reactions*: Until a few years ago, most ion-molecule reactions were measured either at room temperature or at 80 K, and it was assumed that if the reaction is fast at those temperatures, it remains equally fast at lower temperatures. It is now possible to measure ion-molecule reactions at temperatures as low as 10 K (Rowe 1988), and several exceptions to this simple rule have been found. In particular, reactions of ions with molecules that have a sizable permanent dipole moment (e.g., C^+ with H_2O) have been found to proceed 1 to 2 orders of magnitude more rapidly than the Langevin rate at the lowest temperatures, owing to the enhanced long-range attraction.

2. *Dissociative recombination branching ratios*: There has been considerable theoretical debate concerning the products that arise from the dissociative recombination of polyatomic ions (Bates and Herbst 1988b; Adams and Smith 1988). For example, does the dissociative recombination of H_3O^+ produce mainly H_2O or OH?

$$H_3O^+ + e \xrightarrow{1-f} H_2O + H$$

$$\xrightarrow{f} OH + 2H \text{ or } H_2. \tag{A1}$$

The theoretical values for f have fluctuated between 0 and 0.85 over the last few years. Such variations directly affect the amounts of H_2O and OH, and consequently O_2, in the models. Recently, the first experimental determination of this branching ratio has been reported, $f \approx 0.65$ (Herd et al. 1990), which results in less H_2O and more O_2 in the models than if f were 0. Note also that dissociative recombination is usually a very fast process, but that some exceptionally slow cases have been found. In particular, the dissociative recombination of H_3^+ may be extremely slow under interstellar conditions, although this issue has not yet been settled (Mitchell 1990).

3. *Radiative association*: Rates for radiative association reactions continue to be uncertain by 1 to 2 orders of magnitude (Bates and Herbst 1988a).

Because a photon has to be emitted during the formation of the molecular bond, the process is extremely slow and difficult to measure in the laboratory. The theoretical rates have increased considerably over the last few years, when it was realized that excited electronic states may play a significant role in the process. However, for the one reaction which has been studied in the laboratory,

$$CH_3^+ + H_2 \rightarrow CH_5^+ + h\nu \qquad (A2)$$

the experiments give a rate that is a factor of 20 lower than the theoretical estimates (Gerlich and Kaefer 1989).

4. *Photodissociation*:   Most photodissociation rates have increased up to an order of magnitude over the last decade, as more laboratory data and better calculations become available (van Dishoeck 1988b).

B. Gas-Phase Models

The models presented in Table V of Sec. II are the simplest "standard" models, in the sense that they are homogeneous and consider only gas-phase reactions at fixed temperature and density. They also compute only relative abundances at one position in the cloud, and equate abundance ratios with column density ratios. Several recent improvements to this standard model are listed below. Most of them have been stimulated by the large observed abundances of atomic carbon and complex molecules in the clouds, which are not readily explained in the standard models.

1. Clumpy Models. There is growing observational evidence that interstellar clouds are far from homogeneous, but that they have a very clumpy structure (Falgarone and Pérault 1987).   In particular, the large measured column densities of C and C^+ and their widespread abundance throughout clouds such as M 17 suggest an inhomogeneous structure through which the ultraviolet radiation can penetrate deeper into the cloud and dissociate CO (Keene et al. 1985; Stutzki et al. 1988; White and Padman 1991). Detailed models of such clumpy clouds are in the process of being developed. They can no longer be restricted to relative abundances, but need to compute actual column densities and line intensities which include the contributions from the edges of the clumps.

2. Cosmic-Ray-Induced Photons.   A dilute flux of ultraviolet photons can also be maintained inside the densest clumps through the interaction of cosmic rays with hydrogen (Prasad and Tarafdar 1983):

$$H \text{ or } H_2 + \text{cosmic ray} \rightarrow H^+ \text{ or } H_2^+ + e^* + \text{cosmic ray}$$
$$e^* + H_2 \rightarrow H_2^* \qquad \qquad . \qquad (A3)$$
$$H_2^* \rightarrow H_2 + h\nu$$

The electrons e^* produced in the cosmic-ray ionization of H and H_2 have sufficient energy to excite H_2 to H_2^* electronically. The excitations are followed by spontaneous emission of photons in the wavelength range between 800 and 2000 Å. The resulting photodissociation rates of molecules due to the photons produced by this process have been computed by Gredel et al. (1989), and are, in the case of CO, sufficient to maintain a significant fraction of atomic carbon inside the cloud. Note that ionizations by X-ray emission from embedded young stars can have similar effects at comparable rates (Krolik and Kallman 1983).

3. Large Molecules. The presence of very small grains or large molecules such as the PAH can influence the chemistry as well as the ionization balance (Omont 1986; Lepp and Dalgarno 1988; Draine and Sutin 1987). Their most important effect is to neutralize atomic and molecular ions without destroying them, so that the abundances of complex molecules are enhanced. Also, the free electrons can attach to the PAH to form PAH^- ions, which may carry most of the negative charge in the cloud. In that case, the limits on the fractional ionization $x(e)$ are lowered substantially. Finally, the inclusion of large molecules in the chemistry removes the dependence of cloud-model results on metal abundance, which was found in earlier studies.

4. Photodissociation Models. Models of clouds exposed to intense ultraviolet radiation due to a nearby young star have been developed, e.g., by Tielens and Hollenbach (1985) and van Dishoeck and Black (1988*b*). Such models treat the depth dependence of the concentrations and photodissociation rates in detail and calculate column densities, which can be compared directly with observations. Large amounts of atomic carbon are readily formed in these models, although those of complex molecules have not yet been investigated in detail. Attempts to include the observed clumpy structure into the models have been made, e.g., by Burton et al. (1990).

5. Evolutionary Models. Models in which the physical parameters are no longer fixed, but in which density and temperature evolve with time from a diffuse to a dense state according to the equations of motion for gravitational collapse have been developed by Tarafdar et al. (1985) and Prasad et al. (1987). Because the clouds spend most of their lifetime as diffuse clouds where the amount of atomic carbon relative to CO is high, they produce significant amounts of complex molecules. One difficulty with these models is that they may lead to star-formation rates that are too large compared with observations, unless some arbitrary mechanism is introduced that slows down the evolution or prevents stars from forming. Dynamical models in which material is repeatedly cycled between clumps and the tenuous interclump medium by ablation of material from the clumps by newly formed stars have been investigated by Charnley et al. (1988,1990).

One problem with the new models is that they often include only one or two of these new developments, but ignore others. Such studies are useful to investigate their influence on the computed abundances compared with the standard model (see, e.g., Millar 1990). However, the effects sometimes go in opposite directions, so that the overall result is not clear, and comparison with observations less meaningful.

An important parameter in the chemistry, and also in the star-formation process, is the fractional ionization in the cloud, $x(e) = n(e)/n(H_2)$, which determines the efficiency with which the magnetic field couples to the cloud. In diffuse and translucent clouds, the electrons are produced primarily by photoionization of atomic carbon, and the resulting ionization is high, $x(e) \approx 10^{-4}$. Such a high fractional ionization probably prevents star formation in these clouds (McKee 1989). Inside dense clouds, the ionization fraction is determined by the cosmic-ray ionization rate ζ, and the rate at which ions recombine either with electrons or with negatively charged grains: $x(e) \propto (\zeta/n)^{1/2}$. Upper limits on the fractional ionization in dense clouds can be derived from the observed DCO^+/HCO^+ abundance ratio, and values of $x(e) \lesssim 10^{-6}$ are found if $\zeta \approx 10^{-17}$ to 10^{-16} s^{-1} (Dalgarno and Lepp 1984).

C. Grain Chemistry

Quite aside from being the largest "molecules" present, dust grains also play an intimate role in interstellar chemistry. Their importance is based on three distinct functions: they serve to shield molecular regions from the dissociating interstellar radiation; they catalyze the formation of molecules, particularly the most important molecule, H_2; and they remove (perhaps selectively) molecules from the gas phase. Good summaries of the various chemical processes involving grains have been given by Tielens and Allamandola (1987a, b) and Turner (1989c).

1. Ultraviolet Opacity. The ultraviolet extinction law is well determined for diffuse interstellar gas in the solar neighborhood to wavelengths as short as 1000 Å. Considerable variations in the ultraviolet extinction are seen from one line of sight to another (see, e.g., Cardelli, et al. 1989), but on average the dust provides an extinction cross section per H nucleon which is given to within ±20% by $\sigma \approx 2.5 \times 10^{-21}(1000 \text{ Å}/\lambda)$ cm^2 for $2500 > \lambda > 1000$ Å. In dense regions the extinction per H nucleon may be smaller than in diffuse clouds by a factor of 2 or more. This reduction in extinction per H nucleon is presumed to be due to coagulation of small grains (Jura 1980).

The ultraviolet albedo ω and scattering asymmetry parameter g remain somewhat uncertain (Witt 1989), but for diffuse regions both observations and models (see, e.g., Draine and Lee 1984) are consistent with $\omega \approx 0.5 \pm 0.1$ and $g = \langle \cos\theta \rangle \approx 0.5 \pm 0.1$ over the 1000 to 2000 Å region. Unfortunately, the ultraviolet scattering parameters g and ω are essentially unknown in dense clouds. Because photodissociation and photoionization play a major role in gas-phase chemistry in diffuse and translucent regions with $A_V < 2$ mag, models of such clouds will remain uncertain until the ultraviolet scattering properties of the dust therein are better characterized.

2. Gas-Grain Interactions. Dust grains play a primary role in interstellar chemistry by catalyzing the formation of H_2. If Σ is the projected grain surface area per H nucleon, and $n_H = n(H) + 2n(H_2)$ the density of H nucleons, then the time scale for a neutral X to collide with a dust grain is $\tau^{-1} = R_{coll}n_H$, where the "rate coefficient" for collision with a grain surface is given in

$cm^3 s^{-1}$ by

$$R_{coll} = (8kT/\pi m_X)^{1/2}\Sigma = 6.5 \times 10^{-17}(T/20 \text{ K})^{1/2}A_X^{-1/2}\Sigma_{21} \qquad (A4)$$

where A_X is the molecular weight, and $\Sigma_{21} \equiv \Sigma/10^{-21}$ cm^2. For a graphite-silicate grain model designed to reproduce the observed interstellar extinction, $\Sigma_{21} \approx 1.1$ (Mathis et al. 1977; Draine and Lee 1984).

The rate of formation of H_2 on dust grains is given by $R_{H_2} = 0.5\epsilon R_{coll} \approx 3.2 \times 10^{-17}\epsilon(T/20 \text{ K})^{1/2}\Sigma_{21}$ $cm^3 s^{-1}$, where ϵ is the fraction of the H atoms arriving at a grain surface that are converted to H_2. In $T \approx 80$ K diffuse clouds, ultraviolet absorption-line observations have been used to estimate $R \approx 3 \times 10^{-17}$ cm^3 s^{-1} (Jura 1975), seemingly confirming the original theoretical analyses of H_2 catalysis by grains (Hollenbach and Salpeter 1971) which concluded that the efficiency ϵ should be of order unity. The situation is, however, not so simple, because of the presence of the ultra-small grains. The surface area provided by these particles would be considerable: e.g., $\Sigma_{21} \approx 9$ for a size distribution favored by Draine and Anderson (1985). Such an enhanced population of ultra-small grains could play a major role in both surface chemistry and recombination of free charges (Draine and Sutin 1987). The fact that the H_2 formation rate coefficient deduced from observations is $R_{H_2} \approx 3 \times 10^{-17}$ cm^3 s^{-1} indicates that these particles, if actually present, do *not* efficiently catalyze the formation of H_2 in diffuse clouds. This reduced efficiency may be a consequence of frequent thermal "spikes" which may result in desorption of adsorbed H atoms before another H arrives with which to recombine.

Unfortunately, we have little direct information on the values of R appropriate in denser regions; the observed decrease in the extinction per H nucleon in going from diffuse to dense regions, and the corresponding decrease in inferred grain surface area, would suggest that R will follow a similar reduction. However, in regions where the atomic density is increased and the ultraviolet radiation field is attenuated, thermal spikes may be less effective at preventing H_2 formation on the ultra-small grains, if they are in fact present in dense regions, so that the effective R may remain similar.

In addition to acting as a source of H_2, grains may catalyze the formation of other species. The time scale in years for a species X to be deposited on a grain surface is only

$$\tau = (Rn_H)^{-1} =$$

$$2 \times 10^6 \alpha^{-1}(10^3 cm^{-3}/n_H)(A_X/20)^{1/2}(T/20K)^{-1/2}\Sigma_{21}^{-1} \qquad (A5)$$

where α is the probability that species X will stick to the grain, which is thought to lie between 0.1 and 1 for most species. Thus, the grains may act as very effective cold traps, especially if numerous small grains are present. A major question in grain chemistry is the extent to which the grains act only as *passive* sinks, and the extent to which surface catalytic reactions can *actively* produce new molecules. Jones and Williams (1984), e.g., have argued that the observed H_2O ice mantles in the Taurus molecular cloud cannot have been formed by accretion of H_2O, and therefore must be due to formation of H_2O on the grain surface itself. Additional observational evidence for active grain chemistry has been presented in Sec. IV.A. Note that trapping of H_2 itself is not possible; while there may be a surface monolayer of H_2 adsorbed on the grain substrate, this is a negligible fraction of the total H_2 reservoir. The vapor pressure of solid H_2 is too high for more than a monolayer to accumulate.

As noted in Sec. II and Fig. 2, the accretion process by itself would quickly remove heavy species such as C and CO from the gas phase on a time scale that is short compared with the estimated lifetime of the clouds, or even the time scale to reach chemical equilibrium in the gas phase. As no opaque clouds void of gas-phase molecules like CO have yet been found, this raises the question of how the trapping process is circumvented. Various transient processes which may act to return adsorbed species to the gas have been discussed by Leger et al. (1985), including grain-grain collisions triggering "chemical explosions" (d'Hendecourt et al. 1982), and thermal spikes due to ultraviolet and X-ray photons and cosmic rays. Transient increases in the radiation field due to, e.g., FU Ori outbursts could conceivably be important (Draine 1985a). Duley et al. (1989) have recently argued that "spot heating" of amorphous carbon grains by photons can result in effective desorption of volatiles, whereas Boland and de Jong (1982) suggested that material is continuously recycled from the dense part to the more diffuse edge of the cloud by turbulent mixing. Some combination of these processes is apparently effective in quiescent clouds, but it is not yet clear which are the most important.

In regions of star formation, the higher temperatures resulting from the radiation of embedded stars or shocks can evaporate the grain mantles and return species to the gas phase. The temperature required for thermal evaporation of pure CO ice is only 17 K, whereas that for pure H_2O ice is 90 K. "Dirty ice" mixtures probably have sublimation temperatures in between these values. If the clouds are subjected to strong shocks ($v_s > 200$ km s^{-1}), not only the volatile mantles, but also the refractory cores can be destroyed, returning silicon, iron, carbon and other species to the gas phase (Seab 1987).

3. Grain Mantle Models. Detailed theoretical models of the composition of icy grain mantles have been developed, e.g., by Tielens and Hagen (1982) and d'Hendecourt et al. (1985). The results depend sensitively on the physical conditions in the cloud, and in particular on the corresponding gas-phase composition. This can be understood by realizing that at low temperatures ($T \approx 10$ K), only lightly bound species such as H, H_2, C, N, O and S can migrate over the surface, whereas molecules such as CO, N_2, O_2, NH_3 and H_2O are trapped in both physi- and chemi-sorbed sites. Thus, only reactions among the mobile species themselves, or between a mobile atom and a trapped molecule can occur, with the result that the mantle composition depends sensitively on the relative amounts of H, C, O and N in the gas phase. Three different chemical regimes can be distinguished, which have been illustrated in Fig. 3 in Sec. II (Tielens 1989): at low densities ($n < 10^4$ cm^{-3}), atomic H is more abundant in the gas phase than heavy species, and dominates the grain surface reactions. Under such reducing conditions, most of the accreted gas-phase species such as CO and O_2 are transformed into hydrides like H_2O and H_2CO. Subsequent reactions of H with H_2CO may form CH_3OH. Note, however, that the predicted mantle abundance of NH_3 is quite low in these particular models, in which the reaction of N_2 with H is inhibited by a large barrier. Only a large fraction of gas-phase atomic N can lead to a substantial amount of NH_3. CH_4 is also not predicted to be significant in these steady-state models, although it reaches a fractional abundance of more than 10% at early times.

At higher densities ($n > 5 \times 10^4$ cm^{-3}), the gas-phase abundances of CO, O, O_2 and N_2 are larger than those of H, so that O becomes the most mobile atom on the surfaces. If the O/O_2 ratio in the gas phase is low, few surface reactions occur, resulting in an inert grain mantle whose composition reflects that of the gas phase. For high O/O_2 abundance ratios, the reaction O + CO $\rightarrow$ CO_2 makes CO_2 a major constituent.

The grain surface processes can also result in large deuteration effects, because the gas-phase atomic D to H ratio ($\approx 10^{-2}$) is much larger than the overall deuterium abundance ratio ($\approx 10^{-5}$). Most of the gas-phase deuterium is locked up in HD, but the sequence of reactions HD + H_3^+ $\rightarrow$ H_2D^+ + H_2; H_2D^+ + CO $\rightarrow$ DCO^+ + H_2; DCO^+ + e $\rightarrow$ D + CO can enhance D relative to H (Tielens 1983; Dalgarno and Lepp 1984). Because D is nearly as mobile as H on the surface, molecules like HDO, NH_2D and CH_3OD are readily formed in the grain mantles. Doubly deuterated molecules such as NHD_2 and D_2CO probably arise exclusively from grain mantles (Brown and Millar 1989b; Turner 1990).

In summary, different types of grain mantles can occur under different conditions in molecular clouds. Because most grain mantles observed so far are dominated by H_2O and possibly CH_3OH (see Table VIII), it appears that the first scenario, in which grain mantles are formed at relatively low densities in a hydrogen-rich atmosphere, prevails. However, it should be stressed again that the model grain mantles primarily reflect the gas–phase composition, which, as discussed in Sec. II.B (see also Sec. III.C and Sec. V), is highly uncertain. In particular, the recent controversies surrounding some crucial gas-phase reaction rates result in ill-determined O/O_2, N/N_2 and C/CO gas-phase abundance ratios. Also, the actual amount of atomic H inside dense clouds is not well known. The effects of these uncertainties on the grain mantle composition have not yet been explored in detail. In addition, subsequent processing of mantle material by ultraviolet photons and/or cosmic rays can occur. More laboratory work is needed on the basic chemical processes that can take place on the grains, as well as on basic solid-state spectroscopy.

4. Shock Chemistry. Models of shocked regions may be divided into several classes (see Dalgarno 1984; Shull and Draine 1987; Hollenbach et al. 1989 for reviews). First, there are the classical J-type shocks with or without magnetic precursors. J-type shocks are characterized by the presence of an abrupt transition layer where the hydrodynamic variables jump discontinuously. The chemistry in J-type shocks has been discussed, e.g., by Hollenbach and McKee (1979), Hartquist et al. (1980), Mitchell (1984) and Graff and Dalgarno (1987). If the shock is fast enough, the molecules are dissociated and they reform in different configurations as the post-shock gas cools. Sophisticated dissociative shock models have recently been presented by Neufeld and Dalgarno (1989) and Hollenbach and McKee (1989).

The second class of shocks, called C-type, occurs if the fractional ionization is low and the magnetic field sufficiently strong. Here the hydrodynamic parameters vary smoothly from pre-shock to post-shock values, and the peak temperature is generally lower than in J shocks. Detailed calculations of the physical structure of such a shock in a dense cloud have been performed by Draine et al. (1983), but these models include only a limited chemistry.

The results of the models depend, of course, on whether the shock is of J or C type, and whether it passes through a diffuse or dense cloud. Nevertheless, some general features emerge. In particular, the formation of molecules no longer occurs by ion-molecule reactions, but proceeds mostly through neutral-neutral reactions of the type

$$H_nX + H_2 \rightarrow H_{n+1}X + H. \tag{A6}$$

A good example is provided by the oxygen chemistry, where both OH and H_2O are formed efficiently through the reactions

$$\begin{aligned} O + H_2 &\rightarrow OH + H \\ OH + H_2 &\rightarrow H_2O + H. \end{aligned} \tag{A7}$$

Compared with cold quiescent clouds, the shock models thus predict enhancements of OH and H_2O, of sulfur-bearing molecules like H_2S for which the reactions are endothermic, and of molecules such as SiO and HCN.

PART II
Star Formation

ENVIRONMENTS OF STAR FORMATION: RELATIONSHIP BETWEEN MOLECULAR CLOUDS, DENSE CORES AND YOUNG STARS

ELIZABETH A. LADA
Harvard-Smithsonian Center for Astrophysics

KAREN M. STROM
University of Massachusetts

and

PHILIP C. MYERS
Harvard-Smithsonian Center for Astrophysics

In this chapter, the star-forming properties of three molecular cloud complexes, Taurus-Auriga, Ophiuchus and Orion, are discussed. In particular, recent observational knowledge of the distribution and properties of dense molecular gas and young stellar objects within each complex is reviewed. Studies of the two nearest molecular cloud complexes, Taurus and Ophiuchus, have shown that at least two modes or environments of star formation exist: isolated and clustered. Studies of the nearest giant molecular clouds in the Orion complex also show evidence for both modes. However, the recent and extensive census for dense cores and young stellar objects in the L1630 (Orion B) cloud has clearly shown that the clustered mode of star formation is dominant in this region. Existing observations of the L1641 GMC (Orion A) suggest that the clustered mode of star formation may also dominate in this cloud. If star formation in L1630 is representative of star formation in other giant molecular clouds, then the clustered mode of star formation may be the dominant mode of star formation in our Galaxy.

I. INTRODUCTION

Understanding how stars form is one of the major goals of modern astrophysics. In order to fully understand star formation, one must study both the prenatal material and the recent products of star formation. Over the last two decades, radio and infrared observations have revealed that stars in our Galaxy form from the densest, dust-enshrouded regions of molecular clouds. It is therefore important for star formation studies to acquire a knowledge of the distribution and properties of dense gas within molecular clouds and the distribution and properties of the young stellar objects (YSOs) formed within molecular clouds. Such knowledge not only helps to determine the conditions and mechanisms necessary to produce stars but also provides valuable insights into the relationship between young stars and their environments.

In this chapter, we will discuss recent observations of the dense gas and associated YSOs in three star forming complexes: Taurus-Auriga, Ophiuchus, and Orion. These are regions for which both the molecular gas content and young stellar content have been the most extensively studied.

II. STAR FORMATION IN NEARBY CLOUDS

A. Taurus-Auriga Molecular Complex

Nearby molecular cloud complexes provide excellent laboratories for investigations of the star forming process because they have the advantage of being studied with high angular resolution and high sensitivity. One of the best-studied examples of a nearby region forming predominantly low-mass stars is the Taurus-Auriga complex. It is located at $\sim$140 pc (Elias 1978b), towards $l = 170°$ and $b = -15°$, and extends over several tens of parsecs. The complex contains $\sim 10^4$ $M_\odot$ of material (Wouterloot and Habing 1985; Cernicharo et al. 1985; Ungerechts and Thaddeus 1987) and is composed of several well known dark clouds (Barnard 1919; Lynds 1962). Figure 1 presents a visual photograph of the central region of the Taurus-Auriga complex. The dark clouds L1495 (right center) and L1534 or TMC1 (lower left corner) are easily seen in this picture as regions of visual obscuration. The large-scale structure of Taurus-Auriga has been studied using ^{13}CO (Kleiner and Dickman 1984,1985), star counts (Cernicharo et al. 1985), OH (Wouterloot and Habing 1985), CO (Baran 1982; Ungerechts and Thaddeus 1987) and IRAS 100 μm column density measurements (Scalo 1990). These studies have all revealed very complex, irregular and filamentary structures, typical of molecular cloud complexes in our Galaxy (see Chapter by Blitz).

In addition to the diffuse, filamentary structures, small dense cores have been found within Taurus-Auriga by molecular line surveys of regions of optical obscuration (Myers et al. 1983; Myers and Benson 1983). The cores are well-defined structures and their physical properties are determined from molecular line maps of molecules sensitive to gas densities $\geq 10^4$ cm^{-3}, such as NH$_3$ (Myers and Benson 1983; Benson and Myers 1989), CS (Zhou et al. 1989), HC$_3$N (Sorochenko et al. 1986) and DCO$^+$ (Butner et al. 1992). The most extensive studies of the properties of cores in Taurus have been carried out using NH$_3$ (Myers and Benson 1983; Benson and Myers 1989). Examples of NH$_3$ cores are shown in Figs. 2 and 3. These cores are found in regions of high visual obscuration as is evident from the Palomar Sky Survey plates. The NH$_3$ observations reveal that the cores are small, having sizes on the order of 0.1 pc and masses on the order of a few solar masses. In addition, the cores exhibit kinetic temperatures of $\sim$10 K and densities $\sim 10^4$ cm^{-3}. One of the most notable characteristics of the cores is their very narrow NH$_3$ line widths $\sim$0.3 km s^{-1} (FWHM). Analysis of the NH$_3$ line profiles have revealed that thermal motions dominate the line broadening and that nonthermal motions

Figure 1.   A visual photograph of the central regions on the Taurus complex (figure
from Barnard 1927).

may be due to subsonic microturbulence. Similarly narrow line widths are
also evident in lines of HC_3N (Sorochenko et al. 1986), C_3H_2 (Cox et al.
1989).

The core properties described above are not restricted to Taurus cores.
They also apply to most nearby dark cloud cores such as those found in Ophiuchus and Cygnus, but the Taurus complex is the best studied complex: 14 of
the 41 dense core maps in Benson and Myers (1989) are in Taurus. It is important to note however, that the properties derived for the cores can vary with
the molecular line observed (Myers 1985). For example, CS observations of
a sample of nearby "NH_3 cores" in Taurus and other nearby clouds (Zhou
et al. 1989; Fuller 1989) have revealed that the core properties determined
from CS differ with those determined from previous NH_3 observations. Core
sizes measured in CS are on average 1.5 times larger than sizes measured
in NH_3 and the CS line widths are $\sim$ a factor of 2 wider than the NH_3 line
widths. In addition, estimates of core mass and density are also larger by 1 to

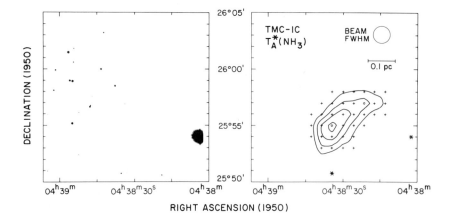

Figure 2.   An example of a typical NH₃ core in the Taurus molecular cloud complex.
An enlargement of the PSS Print is compared with an NH₃ map for TMC1-1C. The
NH₃ contour levels are 0.10, 0.30, 0.50 and 0.70 K (figure from Benson and Myers
1989).

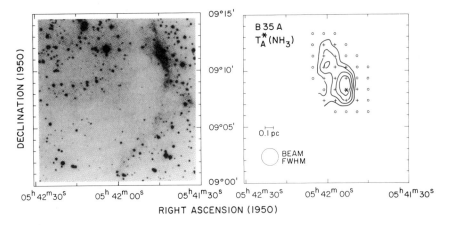

Figure 3.   An example of a nearby dense core and its associated IRAS source. A
reproduction of the PSS Print is compared with NH₃ maps for B35. The NH₃
contour levels equal 0.15, 0.20, 0.25 and 0.30. IRAS sources associated with the
cores are represented by asterisks. (figure from Benson and Myers 1989).

2 orders of magnitude. The differences in size and line width are perplexing
because the CS transitions are expected to trace densities comparable to or
higher than NH₃ (Zhou et al. 1989). In particular, if the NH₃ and CS emission originate from similar regions and thermal motions dominate, then one
would expect the CS lines to be narrower than the NH₃ lines, because CS is a
heavier molecule. Yet, the widths of the CS lines are substantially larger than
those of the NH₃ lines suggesting that nonthermal (e.g., turbulent) motions
are significant. Fuller (1989) has suggested that the CS observations trace

somewhat lower densities than the NH_3 due to high optical depth and radiative trapping in the CS line. However, Zhou et al. (1989), have ruled out optical depth effects as the sole cause of CS line broadening in at least two sources based on $C^{34}S$ observations. They suggest that chemical differentiation in the cores (e.g., underabundance of NH_3 relative to CS) may be responsible for the differences in the CS and NH_3 properties.

Butner et al. (1992) have surveyed the Myers and Benson (1983) sample of 27 NH_3 cores for $DCO^+(J=1{\to}0, J=2{\to}1$ and $J=3{\to}2)$ emission and have mapped 9 cores in $DCO^+(J=1{\to}0)$. The DCO^+ molecule is expected to be abundant in cold, dense gas. The critical densities of the DCO^+ transitions are $\sim 10^5$ cm^{-3}, similar to the critical densities for the CS transitions. Therefore the DCO^+ observations provide an independent measurement of the properties of the dense cores. Core sizes and shapes measured in $DCO^+(J=1{\to}0)$ are similar to those measured in NH_3. However, core densities derived from the DCO^+ multi-transition observations are 3 to 10 times higher than densities estimated from NH_3 observations. In addition, the DCO^+ line widths are found to be significantly broader than the NH_3 line widths, although somewhat narrower than the CS line widths. Butner et al. (1992) conclude that many of the broad DCO^+ line widths do not arise from optical depth effects, scattering or foreground absorption. On the other hand, it is difficult to identify any mechanism, other than maser emission, which would make the widths of NH_3, C_3H_2, and HC_3N significantly narrower than the true distribution of molecular velocities. Furthermore, the relative intensities of NH_3 lines in dark clouds are inconsistent with maser emission. The reason why line width differences among various molecular species exist is still a puzzle and may be due to effects of interstellar chemistry (Zhou et al. 1989; Butner et al. 1992). However, it appears that the dynamical states of dense cores can not be simply characterized as either thermal or turbulent. Clearly further studies are needed to resolve the apparent inconsistencies among dense core line widths.

One core property that appears to be independent of the choice of molecular line tracer is core shape. An NH_3 survey for nearby dense cores has shown that cores are significantly elongated, having aspect ratios between 1.4 to 2.3. (Benson and Myers 1989). Butner et al. (1992) compare DCO^+ and NH_3 maps for 9 cores and find that the maps have similar shapes. Myers et al. (1991a) compare NH_3 (Benson and Myers 1989), $C^{18}O$ and CS (Fuller 1989) maps of 16 nearby cores. They find that the maps for each core are similar in position, elongation and orientation and therefore suggest that elongation is a common characteristic of dense cores. This elongation probably corresponds to a prolate, rather than oblate shape in at least six cores (Myers et al. 1991a). This has important consequences for theories of star formation, for it implies that cores cannot be simply modeled as isolated, self-gravitating regions of quasi-stable equilibrium.

Dense cores, such as those found in Taurus, are sites of star formation. This is demonstrated by the observations that many dense cores are associated with known T Tauri stars (Myers and Benson 1983), bipolar outflows (Fuller and Myers 1987; Myers et al. 1988) and low luminosity IRAS sources (Beichman et al. 1986; Myers et al. 1987). In fact, IRAS sources were detected in approximately half of the known $\sim$100 nearby cores (Benson and Myers 1989). Furthermore, IRAS sources having spectral energy distributions with positive slopes (Class I or steep spectrum sources) are found in closest proximity to dense core centers. Typically one to a few stars are found per core. Figure 3 compares the position of the dense core, B35 A with its associated IRAS source. In this example, the IRAS source is located near the peak of the NH_3 emission.

To fully characterize the star-forming activity within a given complex requires knowledge of both the dense gas and the population of YSOs within the cloud. Knowledge of the distribution of YSOs is of particular interest. The first large-scale survey of the embedded stellar population in the Taurus complex was obtained at 2μm using a single detector (Elias 1978c). The spatial resolution ($1'-2'$) and the sensitivity ($m_K \geq 7.5$) of this survey were poor by current standards but nonetheless, Elias was able to detect $\sim$200 sources. Most of these sources were found to be field stars. The sources associated with the cloud appeared to be distributed in small groups, scattered throughout the cloud. Jones and Herbig (1979) and Larson (1982) arrived at similar conclusions based on optical studies of the distribution of young stars in this complex.

Recently, Kenyon et al. (1990) have completed a survey of the young stellar population in the Taurus-Auriga region, using the IRAS data base. Their survey is complete for luminosities $L > 0.5$ $L_\odot$, making it one of the most sensitive surveys for YSOs. They have identified $\sim$105 objects clearly associated with the Taurus complex. Only 7 of these objects are newly identified pre-main-sequence stars (i.e., optically visible T Tauri stars). Six are new deeply embedded objects (i.e., optically invisible IRAS sources with red spectra). The distribution of YSOs in the Taurus complex is shown in Figs. 4 and 5 superimposed on maps of CO emission (Ungerechts and Thaddeus 1987; Kenyon et al. 1990) and IRAS 100μm column density, respectively. Examining the distribution of YSOs, we find that the YSOs in Taurus are located over a large area of the molecular cloud. Roughly 100 YSOs are found in a 15 pc × 20 pc region. Although the sources for the most part are scattered throughout the cloud in agreement with earlier studies, some weak clustering of sources is also apparent in Figs. 4 and 5.

Although studies of dense cores in Taurus and other nearby regions suggest that cores are frequently associated with star formation, they do not address the question of whether star formation occurs exclusively in dense molecular cores. Answering this question requires comparing the surveys for embedded YSOs with surveys of the molecular gas content in a given cloud complex. Comparing the distribution of YSOs with the distribution of CO in Taurus (Fig. 4), we find that the YSOs are all located within the molecular cloud boundaries, illustrating that stars do form in molecular clouds.

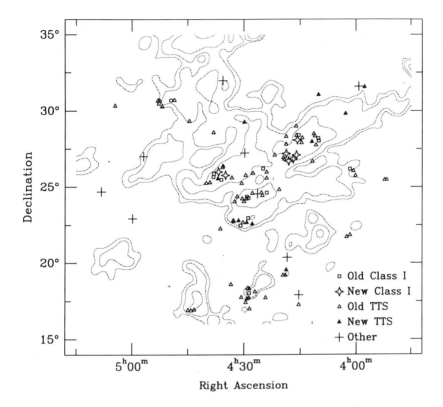

Figure 4. The distribution of young stellar objects in Taurus-Auriga is shown superimposed on a map of CO integrated intensity (figure from Kenyon et al. 1990). The CO contours are presented for integrated intensities equal to 5, 10, 25, and 40 K km s^{-1}.

Furthermore, most of the YSOs in Taurus are located in regions of relatively high column density (Fig. 5). While high column densities do not necessarily imply high volume density, recent work on clumping in molecular clouds has found that regions of high column density are usually also regions of high volume density (Stutzki and Gusten 1990). Finally, since IRAS surveys for embedded objects have not found a significantly large number of new sources (Kenyon et al. 1990), and since previously identified embedded sources in Taurus are known to be associated with dense molecular gas (Beichman et al. 1986; Myers et al. 1987), it appears that star formation in Taurus is restricted to the dense gas. However, unbiased studies of the distribution of the dense gas in Taurus are needed to quantify this observation.

Determining the natures and evolutionary states of YSOs is essential for investigating the star-forming history of a molecular cloud complex. Because stars appear to form from the densest regions of molecular clouds, it is expected that YSOs will be associated with large amounts of gas and dust.

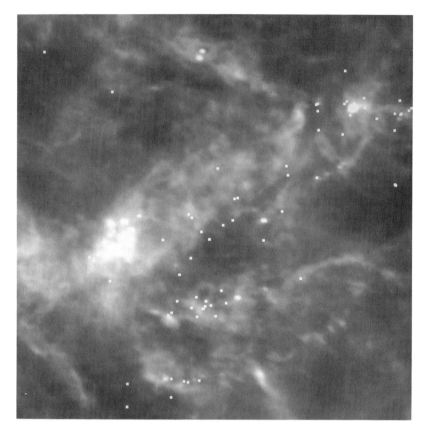

Figure 5.   The distribution of young stellar objects in Taurus-Auriga is shown superimposed on a map of IRAS 100 μm column density (figure courtesy of S. Kenyon).

Consequently, they will radiate a significant fraction of their energy at infrared
wavelengths and their infrared spectral energy distributions should reflect the
nature and distribution of the material surrounding them and their evolutionary state (Lada 1987; Adams et al. 1987). In fact, YSOs can be meaningfully
classified into three distinct morphological classes based on the shapes of their
broad band (1–100 μm) energy distributions (Lada and Wilking 1984; Adams
et al. 1987; Myers et al. 1987). These are illustrated in Fig. 6. Class I sources
have strong "infrared excesses" (i.e., infrared energy distributions that are
broader than a blackbody distribution) and their spectral energy distributions
are characterized by positive slopes at wavelengths longer than 2 μm.   In
addition, many of these sources are difficult to detect at wavelengths short-

ward of 12 μm. Class I sources are likely protostellar in nature (Adams and Shu 1986; Adams et al. 1987). Class II sources also have infrared excesses, but their energy distributions are characterized by flat or decreasing slopes at wavelengths longer than 2 μm. They are surrounded by considerably less circumstellar material than protostellar objects. Class II sources are optically visible and have emission-line characteristics of classical T Tauri stars. The infrared excess in Class II sources is thought to arise in circumstellar disks (Rucinski 1985; Adams et al. 1987; Kenyon and Hartmann 1987; Rydgren and Zak 1987; Strom et al. 1988). Class III objects have little or no infrared excess emission and their energy distributions are characterized by reddened blackbody functions. These sources are also optically visible and include young main-sequence stars and also younger pre-main-sequence stars that no longer have optically thick circumstellar disks such as the weak-lined or naked T Tauri stars (see, e.g., Walter 1987a). The classifications described above are thought to represent phases in an evolutionary sequence of YSOs from protostar to young main-sequence star (Lada 1987; Adams et al. 1987).

Broadband infrared energy distributions have been constructed for the IRAS population in Taurus (Beichman et al. 1986; Myers et al. 1987; Kenyon et al. 1990). As a result, 26 IRAS sources known to be associated with the cloud have been classified as Class I sources and 59 as Class II sources (Kenyon et al. 1990). In addition, there are about 20 Class III sources in Taurus (Walter et al. 1988; Strom et al. 1989b). One should note that the sample of Class III sources is most likely incomplete and the total number of these sources may be comparable to the number of Class II sources (Walter et al. 1988; Kenyon and Hartmann 1990; Kenyon et al. 1990; Strom et al. 1990). Kenyon et al. (1990) estimate that Taurus-Auriga contains $\sim$100 to 140 pre-main-sequence stars and consequently the ratio of embedded sources to pre-main-sequence sources is small. These statistics are consistent with the idea that most sources have an optically invisible "embedded" phase of a $(1-2)\times10^5$ yr, comparable to the free-fall time of a dense core, followed by a visible T Tauri phase of a few $\times$ 10^6 yr.

Using the infrared energy distributions, Kenyon et al. (1990) have constructed a luminosity function for the Taurus complex (see Chapter by Zinnecker et al.). Surprisingly, the luminosities of the embedded (Class I) sources do not obviously differ from the luminosities of the optically visible T Tauri stars. This fact is inconsistent with the protostellar collapse calculations, where the protostellar luminosity must be considerably greater than that of the resulting star if the collapse time is short relative to the T Tauri star lifetime. Kenyon et al. (1990) construct models for the luminosity functions of accreting protostars, examining the sensitivity of the resultant luminosity functions to the input conditions, in order to place constraints on the model parameters from the observations. They find that, if the assumed age of the T Tauri population is correct and the time scale for the embedded phase $\sim(1-2) \times 10^5$ yr, the mass accretion rate onto the stellar photosphere for the embedded objects cannot be steady; it must be much lower than previously

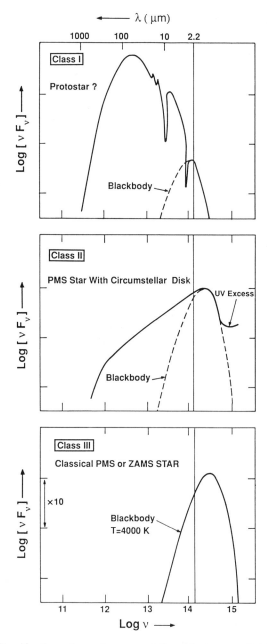

Figure 6.   Classification scheme for young stellar object energy distributions (figure from Lada 1987a).

assumed for most of the time in order that the luminosity of the embedded sources be as low as observed. This might indicate that most of the infalling material is stored in a circumstellar disk to be accreted over a longer period of time or in bursts such as FU Ori outbursts. Hartmann and Kenyon (1990) have estimated that as much as 0.1 of the final stellar mass is accreted from the disk during the T Tauri phase of evolution. If the true ages of the T Tauri stars have been underestimated by a factor of ~3 and material is accreted slowly from the disk during that time, the models agree within a factor of 2 with the observed luminosity function.

B. Ophiuchus Molecular Complex

Another example of a nearby star forming region is the Ophiuchus molecular cloud complex (see Wilking 1992 for a review). The Ophiuchus complex is located near the Scorpius-Centaurus OB association. Distance estimates to the molecular complex range from the traditional value of 160 pc (Bertiau 1958; Whittet 1974; Chini 1981) to a more recent value of 125 ± 25 pc (de Geus et al. 1990). Many similarities exist between the Ophiuchus and Taurus complexes. Like Taurus, the Ophiuchus complex is a region of low-mass star formation. It contains a number of dark clouds (Lynds 1962), and the total mass of the complex, measured in CO, is $\sim 10^4$ $M_\odot$ (de Geus et al. 1990). Figure 7 presents an optical photograph of the L1688 cloud, located in the western part of the Ophiuchus complex (Barnard 1927). Large-scale molecular line surveys of Ophiuchus using CO (de Geus et al. 1990) and ^{13}CO (Loren 1989a, b) have revealed filamentary or clumpy structures. In fact 89 distinct ^{13}CO structures have been identified by Loren (1989a, b). In addition, several dense NH_3 cores, having properties similar to cores found in Taurus, have been identified in this complex (Myers and Benson 1983; Benson and Myers 1989).

The Ophiuchus star-forming region is distinguished by the presence of a large centrally condensed core with active star formation (Wilking and Lada 1983). This core is located near the star Rho Oph, in the dark cloud L1688 (Rho Ophiuchi), the westernmost cloud of the complex. It is a region of high visual extinction, with A_v as high as ~50–100 mag (Vrba et al. 1975,1976; Chini et al. 1977,1981; Wilking and Lada 1983). Wilking and Lada (1983) have mapped the core in $C^{18}O$, which is expected to be a good indicator of column density. They find the core to be 1 pc × 2 pc in size and to contain ~ 600 $M_\odot$.

In addition to being a region of high visual extinction and high column density, high-density gas is also concentrated in the Rho Ophiuchi cloud core. Early observations of the density sensitive molecules, SO (Gottlieb et al. 1978) and H_2CO (Loren et al. 1983) identified two concentrations of dense gas, Rho Oph A and B within the cloud core. Density estimates of these concentrations indicate peak densities of $\sim 10^5$ cm^{-3} for Rho Oph A and $\sim 10^6$ cm^{-3} for Rho Oph B (Loren et al. 1983). Recently, an extensive survey for DCO$^+$ emission has been completed in the Ophiuchus complex (Loren et al. 1990). As a result, 12 cold, dense condensations were identified. These DCO$^+$ cores have masses ranging from 8 to 44 $M_\odot$ and densities ranging from $10^{4.5}$ to 10^5 cm^{-3}. In addition, DCO$^+$ line widths range from 0.4 to

Figure 7. A visual photograph of the dark cloud of ρ Ophiuchi (figure from Barnard 1927).

1.1 km s^{-1}. Of the 12 dense condensations identified, 8 are located within the Rho Ophiuchi cloud core as defined by C^{18}O. Moreover, the distribution of these 8 DCO^{+} cores appears to be highly clustered (Loren et al. 1990), indicating that the dense gas is localized in the Rho Ophiuchi cloud core.

The Ophiuchus complex is further distinguished by the localized concentration of YSOs in the Rho Ophiuchi cloud. Hα objective prism surveys of the Ophiuchus complex (Haro 1949; Dolidze and Arakelyan 1959; Wilking et al. 1987) have shown that the majority of young T-Tauri like objects are associated with the western half of Rho Ophiuchi (Fig. 8). IRAS based studies of the embedded stellar population (Ichikawa and Nishida 1989) have also found a concentration of sources in this cloud. Furthermore, near-infrared observations of the Rho Ophiuchi cloud core have revealed the presence of a dense stellar cluster (Grasdalen et al. 1973; Vrba et al. 1975; Elias 1978a; Wilking and Lada 1983).

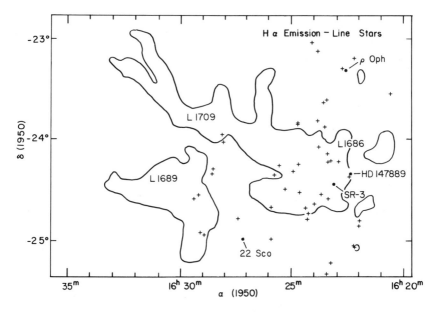

Figure 8.   The distribution of Hα stars is shown (figure from Wilking et al. 1987a). The molecular clouds are outlined by contours of ^{13}CO emission. The contour level represents $T_A^* = 3$ K (Loren and Wootten 1986).

The young stellar population in the Rho Ophiuchi cloud core has been carefully studied using both near-infrared and IRAS observations (Wilking and Lada 1983; Lada and Wilking 1984; Young et al. 1986; Wilking et al. 1989a). The distribution of YSOs is shown in Fig. 9. A total of 78 YSOs have been identified as members of the cluster and 39 additional infrared sources lie within the core boundary, but remain unclassified (Wilking et al. 1989a). Spectral energy distributions have been used to classify the majority of these sources. As a result, 24 YSOs have been identified as Class I sources and 27 as Class II sources. The nearly equal number of Class I and II sources suggests that the lifetimes of the stars in the embedded state (Class I) may be comparable to the average lifetime of the T Tauri (Class II) stars in the cloud, approximately 4×10^5 years (Wilking et al. 1989a). This would imply that the lifetime of the embedded phase in Rho Ophiuchi is somewhat longer than that in Taurus. However, one should note that the survey for Class II sources in the Rho Ophiuchi cloud is incomplete due to IRAS insensitivity and that Class III sources are not considered in this estimate. Accounting for the incompleteness of Class II and Class III sources, Wilking et al. (1989a) estimate the lifetime of the embedded phase to be $(1–4) \times 10^5$ yr, consistent with the estimates for Taurus.

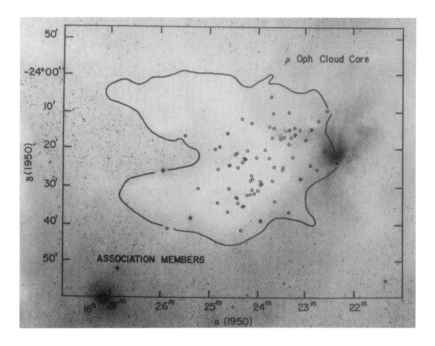

Figure 9.   The distribution of young stellar objects associated with the ρ Ophiuchi molecular cloud (figure from Wilking et al. 1989a). The YSOs associated with the molecular cloud core are displayed as open circles, superposed on the red photograph of the Palomar Sky Survey. The boundaries of the molecular gas (^{13}CO; Loren 1989a) are presented by a solid contour).

Wilking et al. (1989a) construct a luminosity function for the Rho Oph core and find that the YSO luminosities are segregated according to the shapes of their spectral energy distributions. Class I sources appear to dominate at intermediate luminosities while Class II sources dominate at lower luminosities. For example, 82% of the YSOs having luminosities in the range of 5.6 $L_\odot$ to 56 $L_\odot$, are Class I sources whereas 67% of the sources with $0.5 L_\odot < L < 5.6 L_\odot$, are Class II objects. This result implies that stars in Rho Oph undergo luminosity evolution as they progress from Class I to Class II objects, consistent with theoretical expectations (see, e.g., Stahler et al. 1980; Adams and Shu 1986; Adams et al. 1987). However this is quite different

TABLE I
Star-Forming Properties of Nearby Clouds

Regions containing 100 stars	Area (pc^2)	Stellar Density (number/pc^2)	Gas Mass (M$_\odot$)	Mode of Star Formation
Taurus	300	0.3	10^4	isolated
ρ Oph Core	2	50	600	clustered

than the situation in Taurus as described earlier in this review, where Class I sources have comparable and even lower luminosities than Class II sources. An alternate interpretation of the luminosity segregation is that stars in the Rho Oph cloud core are being produced sequentially in mass, with the most recent episode of star formation producing mostly intermediate mass stars. In addition, the observations suggest that a deficiency of intermediate mass stars relative to the initial luminosity function exists in the Rho Oph cloud.

Finally, estimates of the star formation efficiency (SFE) of the Rho Oph core reveal a SFE>20% (Wilking and Lada 1983; Wilking et al. 1989*a*). Such high star formation efficiencies suggest that the embedded cluster may survive as an gravitationally bound open cluster when it emerges from its core (Wilking and Lada 1983; Lada et al. 1984).

III. CLUSTERED VERSUS ISOLATED STAR FORMATION

From studies of nearby molecular cloud complexes, two different pictures of the star-forming process have emerged (see, e.g., Lada 1987; Shu et al. 1987*a*). Observations of low-mass cores in Taurus and Ophiuchus (Myers 1985) have presented us with one picture in which individual stars form from small, isolated dense cores distributed throughout a molecular cloud. These observations have led to the development of the first detailed theory of star formation (Shu et al. 1987; Lada and Shu 1990). In this theory, a dense core, supported primarily by thermal and magnetic pressures, slowly contracts, via ambipolar diffusion, to form a density structure similar to that of an isothermal sphere. At this point, gravitational collapse begins, first in the center of the core and then propagating outward (inside-out collapse). The inner regions form a protostar and if the dense core is slowly rotating initially, a disk will also form within the centrifugal radius, R_c. Material continues to fall onto the star-disk system from the surrounding envelope of dust and gas. At some point during the evolution of this system, a strong wind or outflow develops. The outflow reverses the infall of the outer envelope and eventually reveals the young star and its circumstellar disk.

A very different picture of star formation is presented by observations of the Rho Ophiuchi molecular cloud core. Here, a rich cluster of stars is forming from a single massive concentration of dense gas. In this case, star formation is not occurring in an isolated environment but rather in a densely packed one. To illustrate this point, we can compare the surface density of

YSOs in the Rho Ophiuchus cloud core with that of YSOs in regions of Taurus (Table I). We have learned from the previous discussions that approximately 100 YSOs are located in a 2-pc^2 area of the Rho Ophiuchi cloud core. In Taurus, an equal number of YSOs are found in a much larger area, covering roughly 300 pc^2. Consequently, the stellar surface density in Rho Ophiuchi (50 stars/pc^2) is much higher than the corresponding density (0.3 stars/pc^2) in Taurus. Although some weak clustering is also apparent in Taurus (see Figs. 4 and 5), the surface density of stars, even in the clustered regions is still an order of magnitude lower than in the Rho Ophiuchi cluster (Lada and Myers, in preparation).

The physical conditions within massive cores producing rich clusters of stars are expected to be quite different from those of isolated cores forming single stars. Consequently, the physics governing star-formation in these two environments may differ in some fundamental way not accounted for in present theoretical models. For example, the star-formation efficiency of the Rho Ophiuchi cloud core is higher than the efficiency of regions in Taurus which are producing an equal number of YSOs. In the Rho Oph core, $\sim$100 stars are forming from $\sim$600 $M_\odot$ of gas, whereas $\sim$100 stars are forming from $\sim$10^4 $M_\odot$ of gas in Taurus.

The differences observed in star-forming environments suggest that two distinct modes of star formation may exist, (1) an isolated or distributed mode and (2) a clustered mode. Isolated star formation can be characterized by low stellar density and low overall star-formation efficiency. In this mode, 1 to a few stars form from small, well-defined dense cores *distributed* through a molecular cloud. Clustered star formation can be characterized by high stellar densities and relatively high star-formation efficiency. In this case, groups of many stars form from single massive concentrations of gas. At the present time, it is unclear whether current theoretical models can account for the differences in these modes of star formation.

IV. STAR FORMATION IN GIANT MOLECULAR CLOUDS: ORION COMPLEX

While studies of nearby clouds reveal much about the star-formation process, they do not necessarily tell us how the majority of the stars in the Galaxy form. Most stars in our Galaxy form in giant molecular clouds (GMCs) (Roberts 1957; Miller and Scalo 1978; Blitz 1980). GMCs are large, massive collections of gas and dust, extending over 100s of parsecs and having masses on the order of 10^5 $M_\odot$ (see Chapter by Blitz). They are notable for being the sites of OB star formation but produce stars of both high and low mass (Lada et al. 1991*b*). It is important to study the properties of these clouds and to determine how GMCs produce stars. For example, how do the properties of the star-forming cores in GMCs compare to those in nearby dark clouds, and does star formation in GMCs occur in the clustered or isolated mode or by some combination of both modes?

The Orion molecular cloud complex is the nearest giant star-forming complex to the Sun. Distance estimates to the Orion clouds have traditionally ranged from ∼400 to 500 pc. The best estimates, ranging from 390 to 415 pc, result from a study of the distances to B stars in the Orion association (Anthony-Twarog 1982). A recent review of the star-forming properties of this region is given by Genzel and Stutzki (1989). The large-scale gas distribution of the complex has been mapped in CO (Maddalena et al. 1986). Most of the molecular material is concentrated in two large clouds, Orion A and Orion B . Both are elongated structures. Orion A or L1641 extends southward from the Orion Nebula for about 6 deg. The Orion B or L1630 cloud extends northward from the Horsehead Nebula for about 4 degrees. Both clouds contain ∼10^5 $M_\odot$ of material, respectively, and are associated with regions of active star formation such as M42, NGC 1977, NGC 2023, NGC 2024, NGC 2068 and NGC 2071. The L1641 and L1630 clouds have also been extensively mapped in ^{13}CO (Bally et al. 1987,1992; see Chapter by Blitz). Both clouds have very clumpy and filamentary structures. In this sense, these massive clouds are very similar to the nearby Taurus and Ophiuchus clouds.

A. The L1641 Molecular Cloud

Recently, the properties of dense cores and young stellar objects in the L1641 molecular cloud have been examined. A survey of the NH_3 emission toward IRAS sources in the Orion region (Wouterloot et al. 1988) has revealed the presence of dense gas. This survey detected NH_3 emission towards 18 IRAS sources. Subsequently many of these sources have been mapped in NH_3 (Harju et al. 1991) with a linear resolution comparable to that used to study NH_3 cores in Taurus (Myers and Benson 1983). As a result, 24 dense cores have been found in the Orion complex, 18 of these located in L1641. The cores have significantly larger NH_3 sizes, having median core diameters twice as large as the Taurus cores (Fig. 10). NH_3 line widths are also substantially larger in Orion (Fig. 10). However, the Orion cores are not as large as would be expected from the correlation between line width and size for other cores (Myers 1983). Harju et al. (1991) suggest that CO outflows associated with many of these IRAS sources (Wouterloot et al. 1989) may cause the observed line widths to increase, leading to the observed deviations from the predicted relationships. Finally, the L1641 cores also have higher kinetic temperatures and larger masses and are associated with IRAS sources that are much more luminous than those in the cores found in Taurus. It is important to note, however, that the comparison between Orion and Taurus cores may be somewhat misleading. The NH_3 cores studied in Orion were IRAS selected and consequently the sensitivity to sources in Orion is lower than that in Taurus (see discussion that follows). It is therefore possible that the L1641 molecular cloud also contains cores similar to those found in Taurus, but that these cores were not detected.

Studies of the IRAS selected young stellar population have also been carried out in the L1641 molecular cloud (Strom et al. 1989a, b, c). Strom

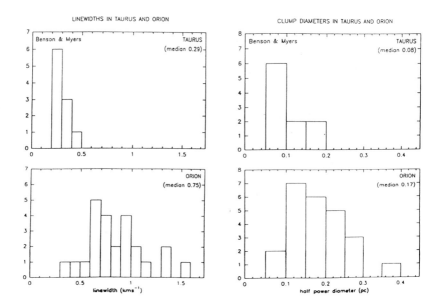

Figure 10.   A comparison of NH_3 core (a) line widths and (b) diameters in Taurus
and Orion (figure from Harju et al. 1991).

et al. (1989c) found 93 sources from the IRAS Point Source Catalog (PSC)
to be associated with this cloud. Figure 11 shows the distribution of these
sources superimposed on the ^{13}CO map of Bally et al. (1987). The IRAS
sources are generally found in regions of high column density. In addition,
the sources appear to be scattered throughout the cloud. This is reminiscent
of the distribution of sources seen in Taurus.

In the Taurus-Auriga clouds, due to their proximity, we can assemble
virtually the entire stellar population of the clouds, while in L1641 we are
much more restricted. The IRAS selected sample in L1641 is only complete
to ~ 6 $L_{\odot}$(Strom et al. 1989c). This completeness limit is much higher than
the corresponding limit (0.5 $L_{\odot}$) for the Taurus-Auriga clouds. It is therefore
difficult to compare the stellar populations of these two regions directly. Some
useful comparisons can be made however, but one should keep in mind the
completeness differences of the two samples. In Taurus-Auriga over 80% of
the IRAS PSC sources have optical identifications while, in L1641, only 53%
of the IRAS sources have such identifications even though we are observing
sources that are intrinsically more luminous sources. The reason for this
difference can be found in the fact that L1641 contains a larger percentage
of sources with spectra which are flat or rising into the far-infrared (Strom et
al. 1989a) than does Taurus (Fig. 12). Specifically, only 29% of the IRAS
sources in Taurus have Class I type spectral energy distributions and only 2%
of the sample is detected only at wavelengths longward of 12 μm. In contrast,

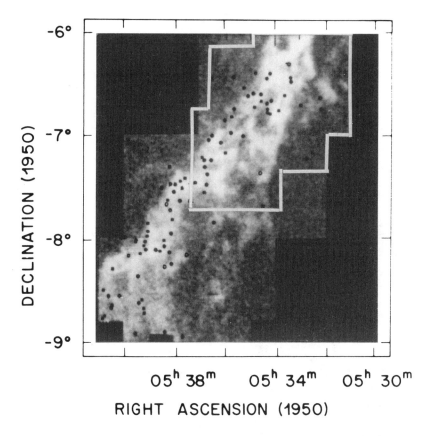

Figure 11.   The distribution of IRAS sources associated with the L1641 molecular
cloud (figure from Strom et al. 1989*a*). The associated IRAS sources are shown
superposed on a ^{13}CO map (Bally et al. 1987).

76% of the sources in L1641 are classified as Class I sources and 32% are
detected only at $\lambda > 12\mu$m.

Comparison of dense cores and YSOs in L1641 reveals that 8 of 12 steep
spectrum sources (i.e., with slope of their spectral energy distribution $n>1$)
are associated with NH$_3$ emission while only 1 of 8 sources with $n \leq 1$ has
detectable NH$_3$ emission (note that here, $n = $ d(log$\lambda F \lambda$)/d(logλ) evaluated at
2.2 μm and 25 μm) (Strom et al. 1989*b*). Thus, in L1641 the steep spectrum or
Class I IRAS sources are preferentially associated with the dense molecular
cores as they are in Taurus, Ophiuchus and other nearby molecular clouds
(Myers et al. 1987).

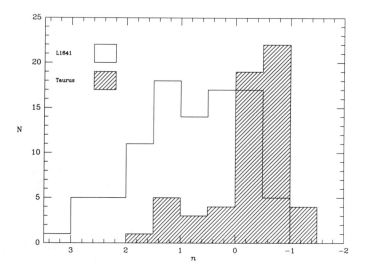

Figure 12.   A comparison of the frequency distribution of the slope of the spectral
energy distribution n for sources in L1641 and Taurus (figure from Strom et al.
1989b). For L1641, the distribution is incomplete below $n = 1$.

Myers et al. (1987), Strom et al. (1989c) and Hughes (1989) find that
the slope of the spectral energy distribution n is correlated with A_V, the visual extinction of the circumstellar dust. Sources located within cores with
high extinction show large values of n because: (a) extinction at 2.2 μm de-
presses the near-infrared flux, and (b) the core material reprocesses the optical
and near-infrared radiation from the embedded sources and re-emits these
absorbed photons in the mid- and far-infrared, thus increasing the observed
flux at $\lambda \geq 25$ μm. The higher proportion of IRAS sources with steep spectral energy distributions in L1641, compared to Taurus-Auriga, suggests that
star-forming dense cores have greater optical depths in L1641 than in Taurus.
Strom et al. (1989a) have used deep 2.2 μm searches to locate and image 10
of 11 IRAS selected cores and have obtained near-infrared fluxes for 10 and
colors for 6 of the associated YSOs. Estimates of A_V were obtained for these
core sources from the (H–K) color index, assuming that the spectral energy
distributions of the YSOs were similar to that of a typical T Tauri star in
Taurus. The derived values range from 40 to 80 mag.
    The studies discussed above have mainly emphasized the differences in
the star-forming properties of the Taurus and L1641 regions, namely that the
cores in L1641 appear to be larger and more massive and are producing more
luminous, and hence presumably more massive stars than those in Taurus.
This is consistent with the observation that the L1641 cloud as a whole
contains about ten times more molecular mass than do the Taurus clouds.
If we compare the distribution of IRAS sources, however, we see that the

two regions are morphologically similar. As in Taurus, the IRAS sources in L1641 are spread over a large region of the cloud. Star formation is also occurring in well-defined dense cores. Sensitive $2.2\,\mu$m observations of 9 Class I IRAS sources reveal that the majority of the sources are single and pointlike (Strom et al. 1989a), indicating that only small number of stars form per core. (However, the magnitudes of many of these objects were not far above the detection threshold and it is possible that many fainter objects in these fields remain undetected.) These characteristics suggest that the isolated or distributed mode of star formation is occurring in L1641. However, we know from optical studies that clusters are also present in this cloud. One well-known cluster is the Trapezium (M42, Orion A) cluster (Trumpler 1931; Herbig and Terndrup 1986). Recently, McCaughrean et al. (1989) has surveyed this cluster in the near infrared and has found a total of 480 sources. The properties of this cluster are discussed in the Chapter by Zinnecker et al. In addition to this cluster, Strom et al. (1989b) have discovered a possible dense, but much smaller cluster associated with the IRAS source, 05338-0624 (Fig. 13). This IRAS source has the highest luminosity of the steep spectrum sources in L1641 and is associated with the bipolar outflow, L1641-N (Fukui et al. 1986; Wilking et al. 1990). Harju et al. (1991) estimate a clump mass of $\sim$64 $M_\odot$ from their NH_3 map of the region. The cluster contains at least 20 stellar or semi-stellar objects located within $35''$ (0.08 pc.) radius. The resulting stellar density is therefore very high. The star-formation efficiency must lie between 4% (if all objects are assumed to be 0.1 $M_\odot$, a least likely case) and 40% (if all objects are assumed to be 1 $M_\odot$, a more likely case) (Strom et al. 1989b).

Both the clustered and isolated modes of star formation seem to occur in the L1641 molecular cloud. But what fraction of stars form in clusters vs isolated regions? If we add up all the embedded sources associated with clusters, we find that $\sim$85% of the YSOs known to be associated with L1641 are located in clusters. This number is uncertain and may be an overestimate, because the IRAS survey is not sensitive to the entire stellar mass spectrum and hence more low-mass stars may be distributed throughout the cloud.

B. The L1630 Molecular Cloud

In order to more confidently determine the nature of star formation within a GMC, an accurate census of the dense cores and embedded YSOs within the cloud is required. To date only one GMC, the L1630 molecular cloud, has been systematically and extensively surveyed for both dense gas and embedded infrared sources (Lada 1990; Lada et al. 1991a, b; Lada et al. 1992). The goals of these surveys were to determine the distribution and properties of the dense cores and young stellar objects within L1630 and most importantly to systematically investigate the relationship between the dense cores and the YSOs.

To identify the dense cores the L1630 molecular cloud was surveyed in the $J=2\rightarrow1$ transition of CS (Lada et al. 1991a). The total area covered

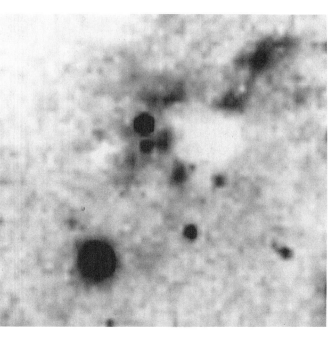

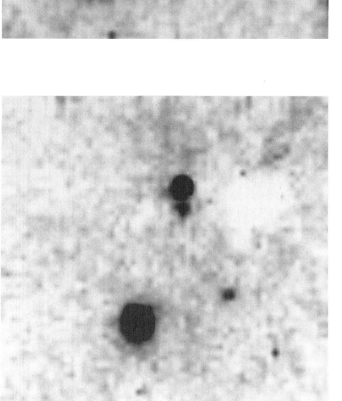

H

K

Figure 13. Near-infrared images of a small, dense cluster associated with IRAS source 05338-0624 (figure from Strom et al. 1989c). The left panel presents the H (1.65 μm) image, while the right panel shows the K (2.2 μm) image of the same region.

by the survey was ~3.6 square degrees or ~20% of the molecular cloud as measured in CO (Maddalena et al. 1986). The results of the CS(2→1) survey are presented in Fig. 14 in the form of an integrated intensity map. Emission was detected, at a 3σ level over approximately 10% of the area surveyed, revealing very clumpy structures. The CS emission was found to be bright in all previously known star-forming regions (e.g., NGC 2071, NGC 2068, M78, NGC 2024, NGC 2023 and the Horsehead Nebula). In addition to these well known sources, many previously unknown dense condensations were found. In fact, 42 individual dense clumps or cores were identified at a 5σ level above the noise. The spatial distribution of these cores is shown in Fig. 14. Most of the cores appear to be distributed about a line extending southwest to northeast, corresponding to a ridge of strong CO emission (Maddalena et al. 1986).

The dense cores identified by the CS survey exhibit a range of properties. Core sizes range from <0.10 pc to 0.53 pc and the distribution of sizes is such that there exist many more small cores than large ones. Most cores were found to be elongated, having aspect ratios between 1.0 and 3.3. The average aspect ratio for all the cores equals 1.8, similar to the ratios found for cores in Taurus (Benson and Myers 1989), Ophiuchus (Loren 1989a) and other nearby clouds (Clemens and Barvanis 1988). CS line widths, measured from composite core spectra, range from 0.73 to 2.24 km s^{-1}. In addition, virial masses range from ≤ 8 to 500 M$_\odot$. Most cores have masses < than 100 M$_\odot$ and only 5 cores have masses >200 M$_\odot$. The spectrum of core masses is presented in Fig. 15. For $M > 20$M$_\odot$, the distribution can be described by the power law, $dN/dM \propto M^{-1.6}$, where N equals the number of cores per solar mass interval. It is interesting to note that all clump mass spectra determined to date (for several different molecular clouds) have a similar spectral index (see Chapter by Blitz). Furthermore, this spectral index, $\alpha = -1.6$, implies that *a significant amount of the mass of the dense gas within a GMC is contained in the most massive cores*. Indeed ~50% of the total core mass in L1630 is contained within the 5 most massive cores, having masses >200 M$_\odot$. These 5 massive cores cover approximately 1% of the total area surveyed, indicating that the dense gas in the surveyed region is confined to a small area. Moreover, the cores appear to be spatially distributed in groups with the small cores clustered around the most massive ones (Fig. 14). This further indicates that the dense gas is highly localized or clustered within this cloud.

In order to study the population of YSOs, a significant portion of the L1630 molecular cloud was surveyed at near infrared wavelengths using an infrared array camera (Lada et al. 1991b). Three thousand 1' x 1' fields were surveyed at 2.2 μm covering an area of approximately 0.8 square degrees. Fifty percent of the area surveyed contained CS(2→1) emission above a 3σ level. The survey is estimated to be complete to a K magnitude of 13 which at the distance to L1630, corresponds to a 0.6 M$_\odot$ main-sequence dwarf. Therefore the observations were sensitive enough to detect both high- and low-mass YSOs in the cloud.

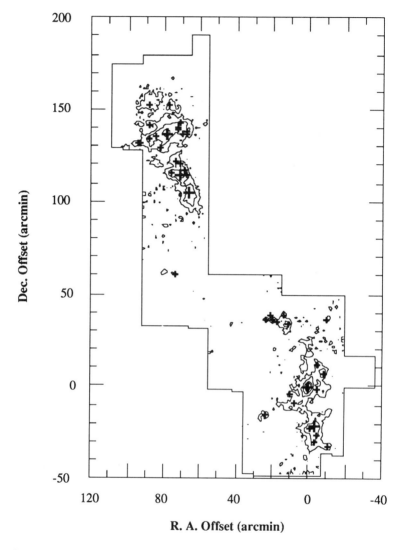

Figure 14.   The distribution of dense gas in the L1630 molecular cloud (figure from
Lada 1992). The contours represent CS $(2 \rightarrow 1)$ integrated intensity emission. The
lowest contour level equals a 3σ detection above the noise. In addition, dense cores,
identified by Lada et al. (1991a) are shown as crosses with the most massive cores
represented by large crosses.

As a result of this survey, 912 sources having $m_K < 13$ were identified. The
distribution of these sources is shown in Fig. 16. Using statistical arguments,
Lada et al. (1991b) estimate that $\sim$50% of the sources having $m_K < 13$ are
associated with the molecular cloud. Even with this large contribution from

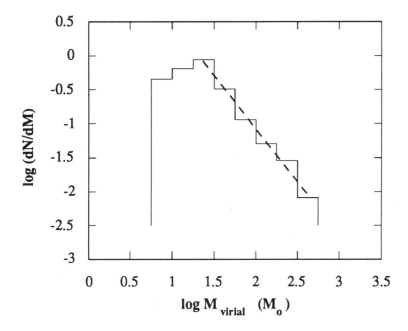

Figure 15.   The mass spectrum for dense cores in the L1630 cloud is plotted (figure from Lada et al. 1991a). For $M > 20$ M$_\odot$, the mass spectrum is fit by the power law: $dN/dM \propto M^{-1.6}$. The turnover in the mass spectrum may be due to incomplete sampling.

a uniform background distribution, the observed sources appear grouped or clustered. In fact, four spatially distinct, embedded clusters were identified, where an embedded cluster is defined as a region in the sky where the source density significantly increases over the background star density (see, e.g., Lada and Lada 1991). Figure 17 presents the surface density distribution of the 2.2 μm sources, showing the locations of the four embedded clusters. These clusters are associated with the well-known star-formation regions, NGC 2071, 2068, 2024 and 2023.

The basic physical properties of the these and other young embedded clusters are reviewed by Lada and Lada (1991). The L1630 embedded clusters exhibit a range in size ($r = 0.3$ pc to 0.9 pc) and associated number of sources ($\sim$20 to 300). The smallest cluster is associated with the reflection nebula NGC 2023. The largest and most spectacular of the four clusters is associated with the H II region NGC 2024. Optical images of this region are noted for the presence of a dark, obscuring dust lane (Fig. 18). Early infrared studies of the H II region (Grasdalen 1974) identified a bright embedded infrared source in the middle of the dust lane. More recently, studies using single beam photometers have discovered $\sim$30 near-infrared sources in this region (Barnes et al. 1989). Figure 19 presents K band image obtained by Lada et

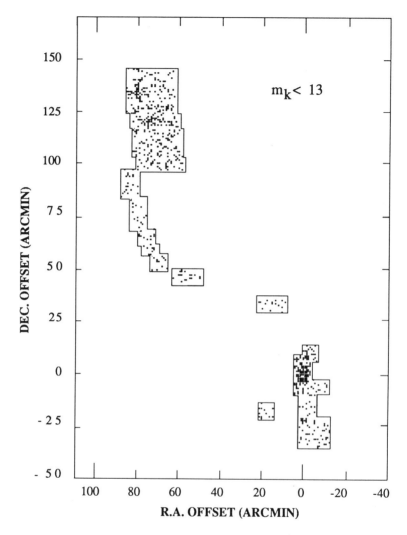

Figure 16.   The distribution of 2.2 μm sources ($m_K < 13$) towards the L1630 cloud. The solid lines correspond to the regions surveyed (figure from Lada et al. 1991b).

al. (1991b). This image illustrates the advantage of high-resolution imaging cameras.

The most striking result of the 2.2 μm survey is that the vast majority of the sources detected are concentrated in the four embedded clusters. Lada et al. (1991b) find that 58% of the sources detected by the survey are contained within the four clusters. After correction for the presence of background/foreground field stars, Lada et al. (1991b) estimate that ∼96% of the sources *associated* with the molecular cloud are contained within the four

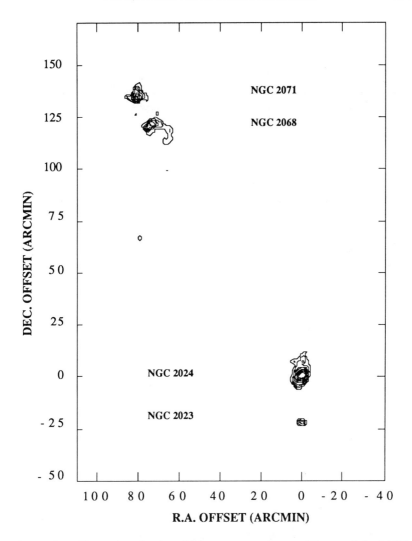

Figure 17.   The surface density of 2.2 μm sources (m_K < 12) toward the L1630 molecular cloud is shown in the form of a contour map. The lowest contour level corresponds to 5 times the background/foreground density of stars and subsequent levels are multiples of this value. Four embedded clusters were identified by Lada et al. (1991*b*) from this map at 10 times the background density. These clusters are associated with the well-known star formation regions NGC 2071, NGC 2068, NGC 2024 and NGC 2023. (figure from Lada et al. 1991*b*).

clusters. Furthermore, the total area covered by the four embedded clusters equals only 18% of the total region surveyed. From these results, Lada et al. (1991*b*) conclude that star formation in L1630 is a highly localized process even for stars whose masses are as low as the mass of the Sun.

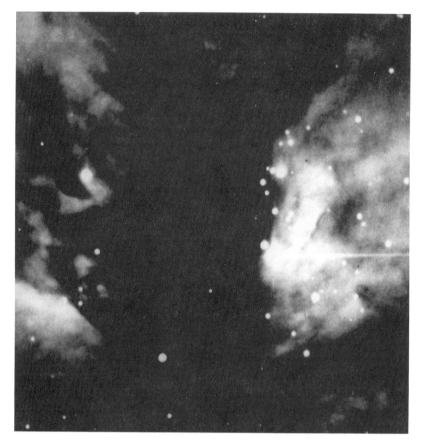

Figure 18.   CCD I-band image of NGC 2024 (figure courtesy of R. Probst). The well-known dust lane is prominent in this figure.

The CS and 2.2 μm surveys have revealed that both the distribution of the dense gas and the distribution of embedded infrared sources are highly localized within the L1630 molecular cloud. Comparison of these two surveys shows that the embedded clusters are coincident or nearly coincident with 4 of the 5 most massive CS cores (Lada 1992). These findings are summarized in Fig. 20 which displays the locations and extents of the embedded clusters and the dense gas in the L1630 molecular cloud. As the embedded clusters contain the majority of YSOs associated with L1630, these results clearly show that star formation in L1630 occurs almost exclusively in dense gas. Furthermore, it appears that star formation in L1630 not only requires high densities but is also strongly favored in regions having substantial amounts of mass.

Surprisingly, even though star formation is confined to the dense regions of the L1630 molecular cloud, it is not occurring uniformly throughout the

Figure 19.   Distribution of 2.2 μm sources in NGC 2024. The K-band image (figure from Lada et al. 1991b) reveals the presence of a rich embedded cluster. The image is a mosaic of 64 1′ × 1′ fields. The sensitivity limit of this image is estimated to be $m_K < 14$.

dense gas (Lada 1992). In fact, ~97% of the embedded sources found in clusters are associated with only 3 massive CS cores. These 3 cores have a combined mass of ~1200 $M_\odot$ which corresponds to only 30% of the total core mass of the molecular cloud. This result is quite striking for it implies that the 3 massive cores may be distinct in their star-forming properties from the remainder of the dense gas. If all the cores had the same star-formation efficiency, one would expect the three most massive cores to have only 30% of the embedded sources rather than the 97 % observed. Furthermore, the L1630 cloud was found to contain a massive core (LBS 23) that is not associated with a recognizable cluster and another massive core (NGC 2023) which is associated with only a very poor cluster. The level of star-formation activity in these two massive cores is considerably lower than that in the three other comparable mass cores which are producing rich clusters. This is reflected in

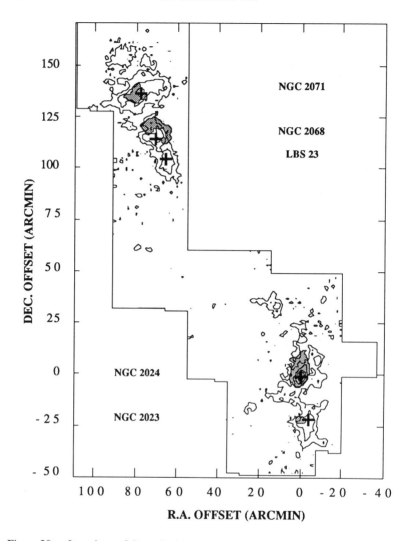

Figure 20.   Locations of the embedded stellar clusters and dense cores in the L1630 molecular cloud (figure from Lada 1992). The shaded regions represent the location and extent of the embedded clusters. The distribution of dense gas is presented as intensity contours of CS(2→1) emission. In addition, the peak intensity positions of the 5 most massive CS cores ($M > 200$ $M_\odot$) are represented by crosses.

the derived star-formation efficiencies (Table II). The cores NGC 2024 and NGC 2071 contain 0.7 stars per $M_\odot$ of dense gas, while the cores, NGC 2023 and LBS 23 contain only 0.07 stars per $M_\odot$ of dense gas (Lada 1992). It therefore appears that high gas density and high gas mass may be necessary, but not sufficient conditions for formation of star clusters.

TABLE II
Star-Formation Efficiencies in L1630

Core	Number of Cluster Sources	Core Mass ($M_\odot$)	Number of YSOs per Core Mass ($M_\odot^{-1}$)
LBS 23[a]	18	230	0.08
NGC 2023	21	294	0.07
NGC 2071	105	456	0.23
NGC 2068	192	266	0.72
NGC 2024	309	430	0.72

[a] Not identified as associated with a cluster by the criteria of Lada et al. (1991b) due to low surface density of sources.

The surveys for dense cores and embedded infrared sources in L1630 have revealed that the vast majority of the stars, located in the surveyed regions, were formed in three rich embedded clusters which in turn, formed from three of the largest and most massive dense cores in the cloud, at relatively high star-formation efficiency. It therefore appears that star formation in these regions of L1630 is occurring in the clustered mode of star formation, resembling the star-forming process in the Ophiuchus molecular cloud core and L1641 clusters. In addition, there is no evidence for any significant star-formation activity of either high- or low-mass stars occurring outside these clusters. This implies that isolated or distributed star formation, similar to what is occurring in regions of the Taurus, Ophiuchus and L1641 molecular clouds, does not *significantly* contribute to the star-forming process in the L1630 cloud. Apparently, clustered star formation is the dominant mode of star formation in the L1630 molecular cloud (Lada et al. 1991b; Lada 1992). This leads to the astonishing conclusion that if L1630 is a typical GMC, most star formation in GMCs and therefore in the Galaxy may occur in the environment of dense clusters and not in isolated protostellar systems.

In order to understand how the vast majority of stars in L1630 (and possibly the Galaxy) formed, we must understand how a massive dense core with high star-forming efficiency is produced. In this regard, it is interesting to consider the five massive dense cores found in L1630. As we have already learned, these cores exhibit a range of star-forming efficiencies with three cores having high efficiencies compared to the remaining two. The origin of these differences could be due to differences in the fundamental physical properties of the cores such as density or magnetic fields or could be due to evolutionary effects. Comparison of the physical conditions in cores with different star-formation efficiencies should provide important insights into the nature of the star-formation processes in molecular clouds.

Clouds with embedded clusters appear to differ from clouds without embedded clusters in one respect which may be significant for understanding the role of magnetic fields in their evolution. The position angles of optical polarization, measured toward background stars near in projection to a cloud, generally have a number distribution with a single local maximum, and a width greater than that due to measurement uncertainty. The width of the distribution is significantly broader in regions with embedded clusters in Ophiuchus, Perseus, Corona Australis and Lupus than in associated clouds with less mass and fewer stars (Myers and Goodman 1991). This increased dispersion in polarization direction might arise if the cluster gas has accumulated with infall speed greater than the Alfven speed so that the associated field lines bend due to gravity faster than they can straighten by radiating Alfvén waves.

Finally, observations of GMCs indicate that their overall star-formation efficiency (SFE) is low, on the order of a few percent (Duerr et al. 1982). The question arises, why is the SFE so small? A key to answering this question may be found by considering the volumes within which the SFEs are measured (Lada 1987). The overall SFE for the entire region surveyed in the L1630 molecular cloud is only 3 to 4%. In contrast, the SFE of the 3 massive cores responsible for the bulk of the star-forming activity in this cloud is ~30 to 40% (Lada 1992) . It therefore appears that only a small fraction of the mass of a GMC is suitable for star formation.

V. CONCLUSION

Observations of dense gas and associated YSOs are powerful probes of the star-formation processes in molecular clouds. Such studies in the nearest molecular cloud complexes have revealed that at least two modes or environments of star formation exist. For example, in Taurus, stars appear to be forming in an isolated or distributed mode of star formation, characterized by low stellar densities and low overall star-formation efficiency. In this mode, one to a few stars form in small, isolated dense cores distributed throughout the complex. In contrast, clustered star formation, characterized by high stellar densities and high star formation efficiency is occurring in the Rho Ophiuchus cloud. Here a relatively rich cluster of stars (~100) is forming from a single massive concentration of dense gas. The relative importance of these two modes in the production of stars in our Galaxy is still unclear. Indeed, these two examples may represent the extremes in a spectrum of a cluster forming process which ranges from the formation of isolated protostars, to small protostellar groups, to rich clusters. However, recent unbiased studies of the dense gas and associated young stellar content of the L1630 molecular cloud have revealed that the clustered mode of star formation is clearly dominant in this GMC. In addition, existing observations suggest that the clustered mode of star formation may also dominate in the L1641 cloud. If star formation in L1630 is representative of star formation in other giant molecular clouds, then clustered star formation may be the dominant mode of star formation in our Galaxy.

Most present day theories of star formation deal with the formation of stars within individual cores. While this is important, the results discussed in this chapter suggest that in order to understand how most stars form in the L1630 molecular cloud and possibly throughout the Galaxy, one must understand how massive cores form and produce rich clusters of stars with high efficiencies.

Acknowledgments.   We would like to thank L. Blitz, H. Butner, C. Lada and B. Wilking for their thoughtful comments on earlier versions of this chapter. We are also grateful to H. Butner, G. Fuller and C. Lada for many enlightening discussions on the subject matter presented here. Finally, we thank S. Kenyon, C. Lada and B. Wilking for assistance in preparation of the figures.

MAGNETIC FIELDS IN STAR-FORMING REGIONS: OBSERVATIONS

CARL HEILES, ALYSSA A. GOODMAN, CHRISTOPHER F. McKEE
University of California, Berkeley

and

ELLEN G. ZWEIBEL
University of Colorado, Boulder

We review the observational aspects of magnetic fields in dense, star-forming regions. First we discuss ways to observe the field. These include direct methods, which consist of the measurement of both linear and circular polarization of spectral line and continuum radiation; and indirect methods, consisting of the angular distribution of H_2O masers on the sky and the measurement of ambipolar diffusion. Next we discuss selected observational results, focusing on detailed discussions of a small number of points rather than a generalized discussion that covers the waterfront. First we discuss the Orion/BN-KL region in detail, both on the small and large scales. Next we discuss the derivation of the complete magnetic vector, including both the systematic and fluctuating component, from a large sample of Zeeman and linear polarization measurements for the L204 dark cloud. Third we discuss the virial theorem as it applies to dark clouds in general and one dark cloud, Barnard 1, in particular. Finally we critically discuss the numerous claims for alignment of cloud structural features with the plane-of-the-sky component of the magnetic field, and find that many of these have not been definitively established.

This is the first of two chapters in this book on magnetic fields in star-forming regions, and deals with the observations; the second chapter by McKee et al. deals with theory.

Magnetic fields reveal themselves by polarizing radiation. The first observational indication of interstellar magnetism was linear polarization of optical starlight, produced by extinction of aligned interstellar dust grains. During the past few years, such polarization has also been observed in the infrared, and even in the thermal radiation emitted by the grains. A second indication, the Zeeman effect, produces circular polarization, and the first detections in dense regions have occurred during the past few years. In addition, magnetic fields should reveal themselves in ways not yet observed, and we expect that people will soon attempt to detect these different effects.

Here we review both the observational manifestations of magnetic fields and some aspects of the current observational knowledge. Previous reviews (Crutcher 1988; Troland 1990) summarize all existing observational results;

here we cover results for only a few objects, but discuss them extensively. Section I covers the observational manifestations of magnetic fields. These include the direct mechanisms that produce both linear and circular polarization of spectral lines and radiation emitted or attenuated by dust, and indirect indications such as ambipolar diffusion and the distribution of masers in the sky. The remaining sections cover the observations and their interpretation, usually employing particular examples. Section II covers the Orion region, the nearest and best-studied hotbed of young, massive stars. Section III discusses a technique by which the three-dimensional magnetic field can be described, and applies this technique to the dark cloud L204. Section IV discusses the virial theorem and its application to, first, the dark cloud Barnard 1, and second, to statistical ensembles of clouds. Section V discusses alignments of cloud structure, rotation, and magnetic field. Finally, Sec. VI presents our perspective and summary discussion.

To begin with, a word on notation is in order. We use **B** to designate the full vector (magnitude and direction) magnetic field, $\hat{B}$ to designate the direction alone (within the two-fold ambiguity of "which way the vector points"), and B to designate the magnitude alone. The subscripts $\parallel$ and $\perp$ designate the line-of-sight and the plane-of-the-sky components, respectively.

I. MANIFESTATIONS OF MAGNETIC FIELDS IN DENSE CLOUDS

A. Zeeman Splitting

The Zeeman effect arises from the coupling of an atom's or a molecule's magnetic moment with an external magnetic field. The magnetic field removes the degeneracy in states with nonzero angular momentum, which splits a transition into a number of π and σ components. The σ components are elliptically polarized, and the circularly-polarized portions are used to detect the splitting. Individual σ transitions appear at $\pm \Delta \nu_Z$ where ν is frequency, and the circularly polarized portion of their amplitudes $\propto \cos^2 \gamma$, where γ is the angle between $\hat{B}$ and the line of sight.

For any species with an electronic orbital angular momentum or spin

$$\Delta \nu_Z = \frac{g \mu_\circ}{h} B = \frac{b}{2} B \qquad (1)$$

where g is the Landé g-factor appropriate to the transition, $\mu_\circ$ is the Bohr magneton, h is Planck's constant, and b is a convenient coefficient for calculating the splitting. For species with no unpaired electron spin, but nonzero nuclear magnetic moment, such as H_2O, the Bohr magneton is replaced with the nuclear magneton in calculating the Zeeman splitting, which reduces the order of magnitude of the effect by a factor $m_e/m_p = 1/1836$. Table I provides b-values for many astrophysically important species; these values are either weighted over sublevels by opacity or are for the strongest subcomponent, which is proper in the optically thin case but not necessarily in any other case. Calculation of a g-factor is often not a trivial undertaking, and some of the b-values in Table I may yet be revised. Often, a g-factor is not considered known until it has been measured in the laboratory.

Under essentially all interstellar conditions except in OH masers, $\Delta v_Z <<$ Δv, where Δv is the full line width at half-intensity (FWHM). This makes Zeeman splitting very difficult to detect. Success requires either a species with a large magnetic moment and/or a region of very high field strength. To date, detections of Zeeman splitting have been reported for only four species in the interstellar medium: H I (emission, absorption and self-absorption); OH (emission, absorption and stimulated emission); C_2S (emission, but the result is probably spurious [see, e.g., Güsten and Fiebig 1990 and personal communication]); and H_2O (stimulated emission [see Fiebig and Güsten 1989], but this may be "fake Zeeman splitting" [see Sec. I.B.1c]). H I traces low-density atomic regions and the outer portions of molecular clouds. OH traces molecular clouds of moderate density, $n_H \lesssim 2500$ cm^{-3} (or perhaps greater; Sec. I.A.2) and C_2S probably traces somewhat higher densities than OH, perhaps $n_H \sim 10^3$ to 10^5 cm^{-3}. OH masers trace $n \sim 10^7$ cm^{-3}. H_2O masers trace $n \sim 10^9$ cm^{-3}; H_2O has no unpaired electron spin, and Zeeman splitting is detectable only because the field strength increases with density. B ranges from a few μG in H I to perhaps over 40 mG in H_2O masers (see Sec. I.B.4a).

In the usual case in which $\Delta v_Z << \Delta v$, Zeeman splitting is best observed by subtracting the left from the right circular polarization to produce the Stokes V spectrum. This exhibits a scaled derivative of the line profile, known as a "Zeeman pattern." (See Fig. 2 in Sec. IV, for an example.) The amplitude of this pattern depends not only on splitting itself, but also the amplitude of the circularly polarized portion of the σ component; the net result is that the amplitude depends on the line-of-sight component $B_\parallel = B \cos \gamma$. Unfortunately, there is no way to separate $\cos \gamma$ from B. In contrast, if $\Delta v_Z > \Delta v$, the splitting is directly measurable and provides B itself, not $B_\parallel$; this is the case for OH masers.

For the usual case $\Delta v_Z << \Delta v$,

$$V(v) = \frac{dT(v)}{dv} b B_\parallel \tag{2}$$

where $T(v)$ is the total intensity. For a Gaussian line profile having central intensity T_A and FWHM Δv, $|V(max)| = 1.43 T_A b |B_\parallel| / \Delta v$. Detectability increases with the ratio of splitting to line width $bB_\parallel / \Delta v^{1/2}$; this favors low frequencies because Δv arises from Doppler broadening and is proportional to the line frequency. Compared to the splitting factor and the field strength, velocity width is relatively unimportant, both because the sensitivity varies only as $\Delta v^{1/2}$ and because widths usually lie within the fairly narrow range of 1 to 10 km s^{-1}. Table I provides estimates of $|V(max)|/T_A$ for a field strength derived from the empirical relation $B_{\mu G} = 0.51 \Delta v_{km\ s^{-1}} n_{H,cm^{-3}}^{1/2}$ (Myers and Goodman 1988a); here n_H is the volume density of hydrogen nuclei. As Table I illustrates, detecting the Zeeman effect is difficult. As

TABLE I

Candidates for Zeeman Observations

	ν [GHz] (a)	(b)	b [Hz μG^{-1}] (c)	n [cm^{-3}] (d)	Δv [km s^{-1}] (e)	B [μG] (f)	$|V_{max}|/T_A$ (g)
Atomic Transitions:							
HI $^2S_{\frac{1}{2}}$, F=1–0	1.420	1, 1	2.80	1×10^2	2.0	10	2×10^{-3}
HI recombination lines[h]	1–400	16, 17	2.80	10^2–10^7	2.0	100?	$\lesssim 5 \times 10^{-3}$
Molecular Transitions, Splitting Determined by Electronic Magnetic Moment:							
CH $^2\Pi_{3/2}$, J=3/2, F=2–2	0.7017	2, 12	1.96	1×10^6	2.0	1020	3×10^{-1}
CH $^2\Pi_{3/2}$, J=3/2, F=1–1	0.7248	2, 12	3.27	1×10^6	2.0	1020	5×10^{-1}
OH $^2\Pi_{3/2}$, J=3/2, F=1–2	1.612	3, 3	1.31	5×10^3	1.0	36	6×10^{-3}
OH $^2\Pi_{3/2}$, J=3/2, F=1–1	1.665	3, 3	3.27	5×10^3	1.0	36	2×10^{-2}
OH $^2\Pi_{3/2}$, J=3/2, F=2–2	1.667	3, 2	1.96	5×10^3	1.0	36	9×10^{-3}
OH $^2\Pi_{3/2}$, J=5/2, F=2–1	1.720	3, 3	1.31	5×10^3	1.0	36	6×10^{-3}
OH $^2\Pi_{3/2}$, J=5/2, F=2–3	6.016	3, 3	0.68	5×10^3	1.0	36	9×10^{-4} (p)
OH $^2\Pi_{3/2}$, J=5/2, F=2–2	6.031	3, 3	1.58	5×10^3	1.0	36	2×10^{-3} (p)
OH $^2\Pi_{3/2}$, J=5/2, F=3–3	6.035	3, 3	1.13	5×10^3	1.0	36	1×10^{-3} (p)
OH $^2\Pi_{3/2}$, J=5/2, F=3–2	6.049	3, 3	0.68	5×10^3	1.0	36	9×10^{-4} (p)
C$_4$H N=1–0, J=3/2–1/2, F=2–1	9.4976	4, 13	1.40	3×10^4	0.5	44	3×10^{-3}
C$_4$H N=2–1, J=5/2–3/2, F=2–1	19.0147	4, 13	1.30	3×10^4	0.5	44	1×10^{-3}
C$_4$H N=2–1, J=5/2–3/2, F=3–2	19.0151	4, 13	0.93	3×10^4	0.5	44	9×10^{-4}
C$_2$S $J_N = 1_0$–0_1	11.12	5, 14	$0.84^{(l)}$	1×10^5	0.5	81	3×10^{-3}
SO $J_N = 2_2 - 1_1$	86.094	6, 6	$0.47^{(m)}$	5×10^4	1.0	114	1×10^{-4}

Species	Transition	(a)	(b)	(c)	(d)	(e)	(f)	(g/h)
O_2	N=1, J=1-0 (above atmosphere)	56.264	7, 8	2.80	5×10^4	0.5	57	1×10^{-3} (q)
O_2	N=1, J=2-1 (above atmosphere)	118.75	7, 8	2.80	5×10^4	0.5	57	6×10^{-4} (q)
CN	N=1-0, J=3/2-1/2, F=3/2-1/2	113.49	8, 8	2.20	1×10^4	1.0	51	2×10^{-4}
CN	N=2-1, J=3/2, F=3/2-5/2	226.33	8, 8	2.60	1×10^4	1.0	51	1×10^{-4}
C_2H	N=1-0, J=3/2-1/2, F=2-1	87.317	9, 13	1.40	1×10^5	2.0	323	6×10^{-4}

Molecular Transitions, Splitting Determined by Nuclear Magnetic Moment:

Species	Transition	(a)	(b)	(c)	(d)	(e)	(f)	(g/h)
OH	$^2\pi_{1/2}$, J=1/2, F=0-1	4.66	3, 3	~0.001	1×10^7	2.0	3×10^3	7×10^{-5}
OH	$^2\pi_{1/2}$, J=1/2, F=1-1	4.751	3, 3	~0.001	1×10^7	2.0	3×10^3	7×10^{-5}
OH	$^2\pi_{1/2}$, J=1/2, F=1-0	4.766	3, 3	~0.001	1×10^7	2.0	3×10^3	7×10^{-5}
H_2O	Hyperfines of $(6_{16} - 5_{23})$	22.235	10, 10	0.0029	1×10^9	2.0	3×10^4	5×10^{-4}
NH_3	Inversion transitions, e.g. JK=33[n]	~78	11, 15	0.00072	1×10^7	2.0	3×10^3	3×10^{-6}

References: 1. Kulkarni and Heiles 1988; 2. Ziurys and Turner 1985; 3. Davies 1974; 4. Gottleib et al. 1983a; 5. Saito et al. 1987; 6. Clark and Johnson 1974; 7. Hill and Gordy 1954; 8. Bel and Leroy 1989; 9. Gottleib et al. 1983b; 10. Fiebig and Güsten 1989; 11. Extrapolated from Poynter and Kakar 1975; 12. Assumed equal to OH values for same transition designation; 13. Estimated based on first-order calculation; 14. Güsten and Fiebig 1990; 15. Jen 1948; 16. Lilley and Palmer 1968; 17. Greve and Pauls 1980.

Notes: (a) approximate transition frequency; (b) references for columns (a) and (c); (c) b, the Zeeman splitting factor in Hz μG^{-1}; (d) approximate highest density traced by this transition, in LTE for the species with splitting determined by an electronic magnetic moment, or maser emission for the species with splitting is determined by a nuclear magnetic moment; (e) typical observed line width of this transition for LTE cases, or of an LTE transition tracing the same density in the case of stimulated emission; (f) estimated $B[\mu G] = 0.51(\Delta v[\mathrm{km\ s^{-1}}])(n[\mathrm{cm^{-3}}])^{1/2}$ (Myers and Goodman 1988a), and we assumed $B_{\parallel} = B/2$; (g) see Eq. (2); (h) recombination lines occur in a wide variety of environments and can even show maser emission (Martin-Pintado et al. 1990); (l) the b-value is quoted by the authors as uncertain to within a factor of 2; (m) the b-value listed is an average of the values listed by the authors for the transitions all very close to this frequency—it should be noted that Bel and Leroy 1989 present a different value; (n) although stimulated emission has been observed in NH_3 transitions, (Madden, personal communication) the transition listed is hypothetical; (p) these transitions are mostly observed as masers; (q) it is not feasible to observe these transitions from the ground. (r) these are LTE line strengths, presumably the transition would be much stronger in stimulated emission.

Other species: Zeeman candidates suggested to date, but not included in the table above, are transitions of: NS, CP, HCO and CH_2CN which have been detected in the interstellar medium in states with nonzero electronic magnetic moment (Turner, personal communication); and CS, SiO, and SiS, which have been detected in states with no electronic magnetic moment, but finite nuclear magnetic moment.

expected, H I and OH appear among the best candidates. Less expected, perhaps, is the predicted feasibility of CH Zeeman experiments at 0.7 GHz, which occurs both because the transition is in an excited rotation level, which requires high density for its excitation, and because of the low frequency of the line.

Astronomical convention defines a magnetic field that points toward the observer as negative in sign. In Zeeman experiments, this means that if the right circularly polarized (RCP) component (defined according to the IEEE convention as a wave that rotates clockwise as seen by the source) is observed at higher frequency than the LCP component, then the field points toward the observer and is given a negative sign.

1. Zeeman Splitting of H I. H_2 dominates H I in molecular clouds by definition. Nevertheless, sometimes H I is observed to be associated with dark clouds. This takes two forms: one, cold H I in the cloud produces "self-absorption" of the 21-cm line against warmer background 21-cm line emission (see, e.g., Knapp 1974); and two, warm H I on the outside of a cloud tends to produce excess 21-cm line emission (Andersson et al. 1991). Zeeman splitting has been seen from both forms (near Orion in one single spectrum; Heiles and Troland 1982).

Astrochemical theory indicates that in the ultraviolet-free centers of dense clouds the volume density $n_{HI} \sim 1.2$ cm^{-3}, almost independent of n_{H_2} (Leung et al. 1984). As a representative example, consider the "medium" size of cloud Barnard 1 (B1), as defined in Table III, and discussed in Sec. IV.B. Although the B1 H I profile as observed at Arecibo does not exhibit self-absorption, the cloud should be representative in its physical properties. The central portion should have an H I column density $N_{HI} \sim 4 \times 10^{18}$ cm^{-2} at temperature ~ 15 K, which would provide a 21-cm line opacity of ~ 0.15. This is too small to produce the self-absorption that is observed in some clouds.

This indicates that most of the H I responsible for self-absorption lies near the edge of the cloud where UV is more active in dissociating H_2 (van Dishoeck and Black 1986). Thus H I should probe the outer portions of molecular clouds, although this has yet to be observationally verified. In contrast, OH is a better high-density probe, so comparing H I with OH should provide information on magnetic structure within the cloud. However, if a cloud is sufficiently patchy then UV can dissociate H_2 throughout the cloud volume, and H I results could be equivalent to OH.

Zeeman splitting of H I in self-absorption has been measured for two clouds, L204 (Heiles 1988) and Ophiuchus (Troland et al. 1991). Only for Ophiuchus do there exist OH Zeeman splitting data for comparison. The H I and OH lines differ in velocity by ~ 1 km s^{-1}, and the measured field strengths $B_\parallel$ are nearly identical. This single result does not permit much interpretation, so we are currently surveying a large portion of the Ophiuchus dark cloud complex and several other known self-absorption regions for Zeeman splitting.

2. Zeeman Splitting of OH in Emission. The OH molecule is detected in most molecular clouds that have extinction $\gtrsim 1$ mag. The lowest rotational level, $^2\Pi_{\frac{3}{2}}$, $J = 3/2$, is always populated and exhibits Λ-doubling and hyperfine splitting, which produces transitions near 1612, 1665, 1667 and 1720 MHz. In addition to the weak emission from molecular clouds, OH often produces maser emission from very high-density regions located near H II regions and from circumstellar ejecta from cool stars (Sec. I.B.4). The unpaired electron and low transition frequencies make OH the most sensitive tracer of magnetic fields in molecular regions.

The transition probabilities of the 1665 and 1667 MHz lines are 5 and 9 times, respectively, those of 1612 and 1720 MHz lines, and their g-factors are also larger (Table I). For these reasons, it is the 1665 and 1667 MHz lines which are observed, usually simultaneously, in observations of molecular clouds. However, if the 1665 and 1667 lines are opaque, it may be wise to observe the 1612 and 1720 MHz lines in order to see into the more dense portion of a cloud.

Although OH is the best tracer of magnetic fields in molecular regions, it may not be a good tracer of H_2. In clouds that are not too dense, both observations and theory imply that OH is a good tracer of H_2. However, in cold dense clouds there are no observational indications, and theory implies that OH is *not* a good tracer of H_2. The primary question is, up to how large a density does OH remain a good tracer of H_2?

Observationally, for moderate extinctions of <7 mag the fractional abundance of OH with respect to H (total H, H I + $2H_2$) X_{OH} is constant and $\sim 4 \times 10^{-8}$ (Crutcher 1979). The major observational work that extends the range of *column* density as high as 7 mag is the detailed study of the ρ Oph dark cloud by Myers et al. (1978); this study finds the corresponding *volume* density to be $n_H \sim 2500$ cm^{-3}. Crutcher (1988) argues that OH traces n_H up to $\sim 3 \times 10^4$ cm^{-3} in TMC-1, but in our opinion this argument needs to be better substantiated and to be made quantitative.

Theoretically, in clouds that are not too dense, starlight dominates the ionization and is involved both in the production and destruction of many molecules. Under these conditions X_{OH} is predicted to be roughly independent of n_H for the range $250 < n_H < 1000$ cm^{-3}, although X_{OH} does depend on temperature, particularly for $T \gtrsim 50$ K (van Dishoeck and Black 1986). However, the theoretical estimate of these density limits is inaccurate because of uncertainty regarding the exact process responsible for the formation of OH (Lepp et al. 1987; Gredel et al. 1989; Herd et al. 1990). Thus it seems reasonable to use the observations as a guide and extend the upper limit of the range of n_H from 1000 to at least 2500 cm^{-3}.

At higher densities, we must rely exclusively on theoretical astrochemical calculations. The question of what constitutes a "high" density depends on the degree to which starlight is excluded, which in turn depends on the degree of nonuniformity of a cloud, and in particular its porosity. The absence of starlight makes a big difference because the sequence of reactions that form OH involves ion-molecule reactions, whose rate is limited by the ionization rate. Without starlight, ionization results from cosmic rays and OH destruction from reactions with neutral atoms, which to first order makes X_{OH} far from constant; instead, n_{OH} is almost independent of n_{H_2} (Herbst and Klemperer 1973). Calculations that include a more complete set of reactions (see, e.g., Leung et al. 1984) indicate X_{OH} approximately $\propto n_{H_2}^{-0.8}$.

This near-independence of n_{OH} with respect to n_{H_2} is definitely not universally true at very high densities. Theoretically, high-temperature regions should produce copious amounts of OH. High temperatures can be produced by shocks, either with magnetic fields (Draine and Katz 1986) or without (Mitchell and Watt 1985); X_{OH} can be much higher than usual with shock velocities of somewhat less than 10 km s^{-1}. Such production has presumably occurred in OH masers, where X_{OH} is high and $n_H \sim 10^7$ cm^{-3}; the high OH abundance can be understood only if these regions have been subjected to high temperatures.

3. Aperture-Synthesis Techniques. Observationally, understanding the role of magnetic fields in star formation requires obtaining information on small angular scales. This, in turn, requires use of aperture synthesis arrays such as the Very Large Array (VLA). However, as the telescopes are spread farther apart to obtain better angular resolution, the sensitivity to surface brightness is decreased. The major limitation of most Zeeman measurements on emission lines is sensitivity to brightness temperature. To obtain sufficient sensitivity with an array, the telescopes must be placed as close together as possible. At no existing array can the telescopes be clustered tightly enough to obtain useful surface-brightness sensitivity for most Zeeman measurements of emission lines.

This sensitivity problem can be circumvented by observing opaque lines in absorption against a continuum source having high surface brightness. Compact H II regions are the prime sources, both because they are bright and because they are the very places where star formation has recently occurred. The only other type of bright continuum source is supernova remnants such as Cas A, which can provide information on molecular regions, but these regions are not necessarily characterized by star formation (Schwarz et al. 1986; Heiles and Stevens 1986).

Zeeman observations of OH lines in absorption against compact H II regions are usually (but not always) contaminated by highly polarized OH maser emission. This leaves the 21-cm H I line, which is unfortunate because it does not highlight the dense molecular gas.

H II regions are special: part of the photodissociation region, which is the interface between the ionized and molecular gas, contains primarily H I, not H$_2$ (Tielens and Hollenbach 1985). If the exciting stars are early O stars then the H I is predicted to be warm, ~ 1000 K, but to have high column density, $\sim 4 \times 10^{21}$ cm^{-2}, so its 21-cm line opacity should be reasonably high, ~ 1; if the exciting stars are cooler, then the H I should be cooler. Unfortunately, no H I absorption line has ever been interpreted as being associated with a photodissociation region. To be in such a region, the H I velocity should be more negative than the H II velocity by amounts ranging up to ~10 km s^{-1} for a young H II region. Instead, all currently observed H I absorption lines seem to result from foreground H I that may lie close to the H II region but has not yet been dynamically perturbed by the ionization front. Much of this H I should lie preferentially in the outer layers of the parent molecular cloud. It should reflect the magnetic field in the outer, less dense portions of the cloud (Sec. I.A.1).

There are currently only two H II regions, W3 and Orion A, for which aperture synthesis Zeeman observations exist; they are discussed in Sec. II below. Orion A has a strong, easily detected field; for W3 there are two independent studies, and they do not agree.

B. Polarization of Radio Spectral Lines Arising From Radiative Transfer Effects

Both linear and circular polarization resulting from radiative transfer effects of radio lines have been predicted in many environments, but observed only in masers. In our discussion below, we will conclude that, apart from masers, both linear and circular polarization will be difficult to detect. In both masers and nonmasers, circular polarization can falsely mimic the Zeeman effect in some cases.

1. Physical Principles

a. Linear polarization. If an external magnetic field orients the quantization axis of a molecule, then anisotropic radiative excitation can produce differing populations of the magnetic substates, which leads to different optical depths for the π and σ components. These components are orthogonally polarized. Thus, if a background source suffers absorption that differs in the two polarizations, the observed absorption line will be polarized. This phenomenon is called "linear dichroism" (Kylafis and Shapiro 1983). Emission lines from the cloud can be similarly lineary polarized.

b. Circular polarization. Circular polarization can also occur. The production of linear polarization relies on differing opacities for the polarizations that are parallel and perpendicular to $\hat{B}_\perp$ (the optical axes). These opacities are the imaginary part of the index of refraction. According to the Kramers-Kronig dispersion relation, this inevitably leads to different phase velocities for orthogonal linear polarizations that are aligned along the optical axes. This turns linear polarization into circular polarization by a process that is similar to Faraday rotation, which occurs when the phase velocities are different for the orthogonal circular polarizations.

First, consider absorption of a linearly polarized background source. If the polarization direction is *not* parallel to either optical axis, then the wave can be decomposed into two waves in both axes. The differing phase velocities then produce elliptical polarization; the intensity of the circular component varies sinusoidally with distance along the direction of propagation. The polarization always changes sign across the line center, and looks similar to signature of the Zeeman effect. This is known as "linear birefringence." The fractional circular polarization p_c is maximum when the polarization of the background is oriented at 45° with respect to the optical axes. In this case, for a Maxwellian velocity distribution and to first order in $\delta\tau_0$, the difference in line center opacities for the two optical axes, the maximum Stokes parameter V occurs at the 0.43 intensity point and is about 0.61 $\delta\tau_0$. Thus, V can be comparable to the change in linear polarization.

Next, consider either absorption of an unpolarized background source or the line emission itself. Here, the differing opacities produce linear polarization within the cloud. This polarization lies along one of the optical axes, so there is no possibility for circular polarization. However, if the direction of $\hat{B}_\perp$ changes along the line of sight (the field twists along the line of sight), then the optical axes follow the twist and linear birefringence can occur. The circular polarization depends on the amount of twist. For typical cases, we expect $p_c \lesssim p_l^2$ because the circular polarization is a second-order effect. Two illustrative examples, in which the field twists uniformly with distance along the line of sight, are given by Deguchi and Watson (1985). Our expectation is borne out in the first example in their paper, but not in the second. The explanation is simple: in the second example, the twist is so severe (5 radians) that the linear polarization "interferes with itself" and is thereby decreased in intensity. In the real Universe such a large twist might be produced by a large-amplitude circularly polarized Alfvén wave. Such severe twists in a static field may be rare.

c. *Fake Zeeman splitting.* In the latter "unpolarized background source" or line emission case, p_c should be typically so small that it would be very difficult to detect, and we would thereby recommend that observers concentrate on other, less difficult observations. Nevertheless, the effect may loom large in importance because it mimics—and can easily be mistaken for—the Zeeman effect (Deguchi and Watson 1985). The Zeeman effect produces small p_c, about $1.5\Delta\nu_Z/\Delta\nu$, where $\Delta\nu_Z$ is the Zeeman splitting and $\Delta\nu$ the line width. This amounts to one percent or less in many cases, which is small enough to be produced by linear birefringence. If so, linear polarization near the level $p_l \sim p_c^{1/2}$ should also be present.

d. *Excitation considerations.* The physical mechanism that produces differing populations of the magnetic substates rests on the fact that π and σ components, which are orthogonally polarized, have different directional dependencies on their interaction with radiation. If an external magnetic field orients the quantization axis, and if the molecule is subject to an anisotropic radiation field, then the populations of the associated states are affected differently by absorption and stimulated emission of the radiation, and the π and σ components have different intensities. To obtain a significant population difference, the collisional rate must not be too high. In addition, the Zeeman splitting must be larger than the collisional and the spontaneous emission rates, but small compared to the line width. These conditions are easily satisfied for many transitions.

For a two-level system, only the line radiation itself can produce an effect, and we can easily get a feel for the direction of polarization. The cross section for the π component $\propto \sin^2 \gamma$ and that for the σ components $\propto 1 + \cos^2 \gamma$, where γ is the angle between the propagation direction and the magnetic field. If the intensity of radiation I is larger along the axis defined by the magnetic field ($\gamma = 0°$), then the π state does not interact as strongly with the external radiation field. Photon trapping is less effective for the π component, so it has a smaller excitation temperature than the σ component. The π component has a smaller intensity than the σ component, so the emission is polarized perpendicular to the magnetic field.

A real molecule has multiple rotational levels and, in addition, vibrational levels that can be excited by an external source of infrared radiation. Infrared radiation is likely to be directional. This anisotropic radiation can affect the level populations to a greater degree than other local physical conditions if the source is strong, differentially populating the magnetic substates and leading to polarization. An extreme example is circumstellar material, in particular SiO masers; even in interstellar masers the radiative excitation is almost inherently anisotropic (see, e.g., Western and Watson 1984). In molecular clouds, infrared significantly affects the excitation of levels involved in the 18- and 5-cm transitions of OH (Burdyuzha and Varshalovich 1972; Crutcher 1977; Sec. I.B.3 below).

Calculating the polarization requires the full solution of radiative transfer and level population equations, and as such depends on the physical geometry and the velocity field. It is far from a trivial exercise, and solutions have only been calculated for simple geometries or using the large velocity gradient approximation for simple velocity fields. These have large anisotropies and produce quite large polarizations. Realistic velocity fields should be more complicated, resulting in significantly smaller polarization.

2. Emission Lines: Model Calculations. To discuss the predicted observables we consider two illustrative examples introduced by Kylafis (1983a) in which the anisotropy is provided by the velocity field. Even though these examples use simple geometries and velocity fields, their solution is complicated and requires numerical techniques. The symmetry axis (z axis) is perpendicular to the line of sight at position angle $\theta = 0°$. The magnetic field is also perpendicular to the line of sight, but can have arbitrary position angle θ_B. The velocity field is symmetric with respect to the z axis. In the one-dimensional example, the cloud expands in the z direction with uniform velocity gradient. In the two-dimensional example, the cloud expands axisymmetrically with no motion in the z direction.

a. Intensity. Kylafis (1983a) provided solutions over the full range of physical parameters for his two examples, but restricted the treatment to a two-level system. Deguchi and Watson (1984) extended the treatment to the full rotational ladder for CO and CS, and found that the polarization was smaller than for the two-level case by a factor ~ 2. We use results from their figures here, for which the magnetic field is parallel to the symmetry (z) axis ($\theta_B = 0°$).

The *fractional* linear polarization, denoted as $p_l = (U^2 + V^2)^{1/2}/I$, depends sensitively on C/A, the ratio of collisional rate to Einstein A, and also on the optical depth. As $C/A \rightarrow 0$, p_l becomes large (~ 0.10)—but as $C/A \rightarrow 0$ the *line intensity* also $\rightarrow 0$, so the high fractional polarization is unobservable. As C/A increases, the line intensity increases, but p_l decreases significantly. The linearly *polarized intensity* $P_l = (U^2 + V^2)^{1/2}$ peaks at $C/A \sim 1$, at which point p_l is only ~ 0.01 for the two-dimensional and 0.03 to 0.12 (depending on the viewing angle) for the one-dimensional case. For gas temperature 30 K, the maximum polarized brightness temperature is about 0.3 K in the two-dimensional and 3 K in the one-dimensional case.

These two simple examples of Kylafis show that p_l depends extremely sensitively on the velocity field. Simply going from one dimensional to two dimensional reduces the polarization to the point of being observable only with difficulty. It seems reasonable to conclude that for more realistic velocity fields the polarization will be very small indeed.

b. Direction. The polarization is either parallel or perpendicular to $\hat{B}_\perp$, but it is not so easy to determine which. Again we consider the two examples of Kylafis (1983a).

First consider the behavior of the position angle of the linear polarization θ_{pol} as a function of the position angle of the magnetic field θ_B. In the one-dimensional case, the polarization is perpendicular to the magnetic field for $54° < \theta_B < 125°.3$, and parallel otherwise. In contrast, the two-dimensional case is *precisely opposite*. This behavior can be understood from the basic description given at the beginning of Sec. I.B.1d because the radiation intensity is larger along directions having no velocity gradient. Surprisingly, as a function of θ_{pol}, this dependence translates into the field being *either* parallel *or* perpendicular to the observed polarization (i.e., $\theta_B = \theta_{pol}$ or $\theta_B = \theta_{pol} + 90°$) for a large range of θ_{pol}.

Next consider a three-dimensional case with isotropic velocity field; this has not been treated in the literature, so we must extrapolate the above one- and two-dimensional treatments to this case. Suppose that the magnetic field is either uniform, along the z axis, or a dipole field oriented along the z axis; in either case, the field direction $\theta_B = 0°$ at both the "poles" and the "equator." It would seem that the one-dimensional example should apply at the poles, while the two-dimensional example should apply at the equator. If so, then at the poles the observed polarization would be parallel to the field, i.e., $\delta = 0°$, while at the equator it should be perpendicular with $\delta = 90°$. Thus, as we move around a circle in the sky centered on the source, the observed polarization direction would change four times, from $\delta = 0°$ to $90°$ to $0°$ to $90°$. If one were observing with a single dish that was unable to resolve these changes the net polarization would be very small. If one used an array with adequate angular resolution, how would one interpret the observations?

Lis et al. (1988) relaxed the large velocity gradient assumption and considered a spherically symmetric cloud with a dense core and a radial magnetic field. They calculated $p_l \lesssim 0.02$, with 0.02 occurring only in the most optimistic case. Polarization increases with decreasing C/A, which favors molecules with high dipole moments; increasing complexity of hyperfine structure also decreases polarization.

In the absence of calculations for other, more realistic situations, we can only speculate. The direction of polarization is a sensitive function of the details of the velocity field. It seems likely that in more complicated cases the polarization direction might change direction rapidly with position, both on the sky and along the line of sight, and this should lead to much smaller p_l than is predicted for the simple situations. In this spirit, the negative results of attempts to observe polarization by Wannier et al. (1983) and by Lis et al. (1988) come as no surprise.

3. Absorption Lines. To discuss radiative excitation in the absorption-line case we first consider anisotropic radiation at the frequency of the absorption line. The source can be either a spectral line emitter, such as a warm CO cloud affecting a colder foreground CO cloud, or a continuum source such as a bright H II region. An example of each interstellar case was given by Kylafis (1983*b*). His examples used a three-dimensional isotropic velocity field, which presumably is less efficient than one- or two-dimensional cases in producing large polarization, and were geometrically optimized to produce large polarizations. Apart from these geometrical attributes, the three most important parameters are C/A, the solid angle of the source Ω_0, and the brightness temperature of the source T_s.

The former example, with a warm CO cloud ($T_s = 20$ K) having angular diameter $\sim 45°$ as seen by the colder foreground ($T = 10$ K) CO cloud, produces $p_l \sim 0.06$ for $n_{H_2} = 3000$ cm^{-3}. The latter example applies mainly at cm wavelengths, where continuum sources are bright enough to matter. For the H_2CO 6-cm transition absorbing a bright H II region ($T_s = 1000$ K) with angular diameter $20°$ as seen by the cloud, $p_l \sim 0.09$ for $n_{H_2} = 3000$ cm^{-3}. In both cases, 3000 cm^{-3} is the density for which the induced radiative and the collisional rates are roughly equal; p_l increases with smaller C/A (smaller n_{H_2}).

These examples seem reasonably realistic, and it seems worth making a serious observational effort to detect the linear polarization of absorption lines. Of course, anisotropic excitation at frequencies other than the line frequency, for example infrared radiation, can indirectly excite the line. Interpretation of results will not be without ambiguity, because as in the case of polarized emission lines, the polarization is either parallel or perpendicular to $\hat{B}_\perp$, depending on geometry. The requirement that the background source subtend a large solid angle is crucial to the mechanism, and observers should select clouds for which other evidence favors this geometry.

In particular, we believe that the 1665 and 1667 MHz lines of OH provide an interesting test case, although one for which the excitation arises from infrared radiation instead of from radiation at the line frequency itself. The *b*-factors from Table I have the ratio 1.67. If the polarization results from Zeeman splitting, then the ratio $p_{c,1665}/p_{c,1667}$ should equal 1.67, while if the polarization results from linear birefringence the ratio should equal unity. For all observed cases (except one) the ratio is close to 1.67. For the lines in absorption against W22, a ratio of 1.67 lies quite far from the limits allowed by observational uncertainty; instead, the ratio is nearly unity (Kazès and Crutcher 1986), with both lines having $p_c \sim 0.015$. W22 is an H II region, so it is unpolarized. Therefore, if the circular polarization does indeed result from linear birefringence, then linear polarization should exist at the level of $\gtrsim 0.12$. This would require an intense source of anisotropic far-infrared radiation; independent evidence for this exists in the form of a strongly anomalous 1665/1667 intensity ratio (Crutcher 1979). W22 itself cannot be responsible, because the absorbing cloud lies far from W22.

4. Polarization in Masers. Both the circular and linear polarization of masers can provide information on the magnetic field.

a. Zeeman splitting, fake Zeeman splitting. Zeeman splitting of OH masers is often very easy to observe. First, the maser regions have high densities, which usually means high field strengths. Second, the OH molecule has a high b resulting from electronic angular momentum. As a result, the splitting is often larger than the line width, and OH masers often exhibit a classical, well-separated Zeeman pair, with each component almost 100% circularly polarized, in which case the splitting provides the total field strength B instead of line-of-sight component $B_\parallel$. A sufficiently large velocity gradient can cause selective maser amplification of only one of these components, leading to a single component with high circular polarization; in this case, the field strength cannot be derived. The position angle of linear polarization of OH masers provides no information on $\hat{B}_\perp$ (Nedoluha and Watson 1990a).

An important selection effect is implied for OH masers. Field strengths can only be derived when both σ components are observed, which requires that the Zeeman splitting be larger than about 3 times the total velocity difference through the maser (Nedoluha and Watson 1990a). Thus, magnetic fields derived from OH masers emphasize regions that have either small velocity gradients or large magnetic fields. Thus, in a region with large velocity gradients (i.e., any region in which some of the OH masers exhibit only a single component with high circular polarization), the field strengths sampled by those OH masers that exhibit both σ components are larger than average for the region.

In masers, there is a problem for lines having more than one subtransition with different b-values. For example, the 1612 MHz line of OH has three σ components, and the 22-GHz line of H_2O has six hyperfine components (each of which has many σ components). The maser process can amplify transitions from some sublevels at the expense of others, so it is not clear which value of b to adopt; Nedoluha and Watson (1991) find that a single hyperfine component is rarely dominant for H_2O masers. Furthermore, observers of OH masers have not consistently used proper values of b when interpreting their data, and some of the quoted magnetic fields are simply wrong.

Fiebig and Güsten (1989) have detected weak circular polarization in H_2O

masers for one maser in each of four H II regions, with typical $B_\parallel \sim 35$ mG; H_2O masers sample densities of 10^8 to 10^{10} cm^{-3} (Elitzur et al. 1989). The circular polarization $p_c \sim 0.001$, which is extremely small. If linear polarization is present at the level of a few percent, then the circular polarization could be a result of linear birefringence (Sec. I.B.1b,c). Nedoluha and Watson (1990b) have made calculations of this effect for H_2O masers and find that this fake Zeeman splitting (which results only from radiative transfer effects) is in fact likely to occur. Future observations should include all Stokes parameters, which makes the task more difficult.

For OH masers associated with H II regions, Reid and Silverstein (1990) have compiled and assessed many measurements. Field strengths range up to 7 mG and average 3.6 mG. OH masers sample densities of 10^6 to 10^8 cm^{-3} (Reid and Moran 1981). If we ignore the possibility that the H_2O and OH field strengths are biased towards high values because of observational selection effects, then comparison of their field strengths and probable densities suggests that very roughly, $B \propto n^{1/2}$ in this regime. Unfortunately, no OH and H_2O masers for which Zeeman measurements exist have identical velocities and positions, so we cannot as yet use maser observations to study the detailed dependence of B on n in a given region.

Reid and Silverstein come to a remarkable conclusion: the *directions* of the fields in OH masers are systematically aligned over large segments of the Galaxy. This indicates that the field direction in a maser is related to the direction of the galactic field in its immediate vicinity. This, in turn, indicates that the field direction is largely preserved during the contraction of clouds to high volume densities. This is surprising, because during this contraction process one would expect the competing effects of gravitation, angular momentum and shocks to randomize the field in the dense cloud with respect to the surroundings. We discuss a particular example, Orion, below in Sec. II.A, and conclude that Orion seems, on the surface, to be preserving field direction over a large range of density, but that a deeper look brings cause for skepticism.

b. Linear polarization. Linear polarization can provide useful information on $\hat{B}_\perp$ under some circumstances for masers in which the Zeeman splitting is much smaller than the line width. Most discussions are based on the theoretical analysis of $J = 1 - 0$ transitions, for which the production of linear polarization requires that several constraints on ratios of pumping rates, decay rates and Zeeman splitting be satisfied (Goldreich et al. 1973).

However, recent theoretical calculations for higher-J transitions and realistic ratios (Deguchi and Watson 1990; Nedoluha and Watson 1990c) show that observations of H_2O masers have been misinterpreted by using the earlier theory. Garay et al. (1989) provide a convenient observationally-oriented summary of the results for $J = 1 - 0$ transitions, which have been generally assumed to apply also to higher-J transitions; this summary must now be modified. The details are beyond the scope of this chapter, but we present some brief highlights.

Two particularly important parameters (among others) are (1) $g\Omega/R$, the ratio of Zeeman splitting to the stimulated emission rate of the maser, and (2) R/Γ, the degree of maser saturation. In previous treatments, $g\Omega/R$ was usually taken to be very large, but in reality it is expected to be of order unity. When $g\Omega/R >> 1$ and $(g\Omega)^2/R\Gamma >> 1$, the linear polarization is either parallel or perpendicular to $\hat{B}_\perp$, depending on γ, the angle between the magnetic field and the line of sight. However, the highest linear polarization occurs when the two parameters are roughly equal, i.e., $g\Omega/R \sim R/\Gamma$, and in this case the position angle of polarization is *not* either parallel or perpendicular to $B_\perp$ (see Nedoluha and Watson 1990c, Sec. III). Thus, when the polarization is largest (and thus easiest to observe), its direction bears no simple relation to $\hat{B}_\perp$.

The existence of large linear polarization in H_2O masers has been used to derive magnetic field strengths in the masers. For example, Garay et al. (1989) suggested that the $\sim 60\%$ polarization of the Orion "supermaser" implies that it has $B \sim 30$ mG. However, the new theory indicates that anisotropic pumping can produce large polarizations even without strong fields, and in particular, it is required for very strong fractional polarization ($\gtrsim 1/3$). Therefore, one should be skeptical of all field strengths derived from nothing more than the existence of large linear polarization.

C. Polarization Caused by Aligned Grains

Optical starlight is almost universally linearly polarized, to a degree that increases with extinction. The polarization arises from extinction by systematically oriented dust grains. As with spectral lines (Sec. I.B.1b), weak circular polarization can result from a twisted field.

Only magnetic orientation is sufficiently general and powerful to provide the universality, although in individual circumstances other agents, such as gas streaming or photons, may dominate. But *even if these other agents operate*, the polarization must be either parallel or perperpendicular to the magnetic field under all reasonable interstellar conditions. The reason is that the Barnett effect, combined with suprathermal grain rotation, causes the grains to precess about the magnetic field on a very short time scale (Purcell 1979). (Such precession is also produced, albeit less strongly, by purely electromagnetic forces on a charged grain.) Thus the time-averaged image of the grain, seen in projection against the sky by the observer, is always symmetric with respect to the magnetic field. If other, stronger agents compete with magnetic alignment, then they can either reverse the sense of or reduce the magnitude of the net polarization compared to what would be obtained with pure magnetic alignment. However, they cannot cause the polarization to depart from being either parallel or perpendicular to $\hat{B}_\perp$.

In magnetic alignment, a needle-like grain spins primarily end-over-end or an oblate grain spins as tossed pizza dough (Lee and Draine 1985) around axes that are parallel to $\hat{B}$. The polarization that is most strongly attenuated by this field of systematically aligned grains lies along the longest projected grain area, which is perpendicular to $\hat{B}_\perp$. Therefore, the observed linear polarization of absorbed light is parallel to $\hat{B}_\perp$. The alignment also produces linear polarization of far-infrared and sub-mm emission from the grains (Sec. I.C.2), which is perpendicular to $\hat{B}_\perp$. These relations between field direction and polarization often provide a satisfying interpretive picture concerning the *direction* of the field. For example, the many-thousand sample of stars of Mathewson and Ford (1970) shows that the large-scale galactic field lies predominantly in the galactic plane.

The *percentage polarization* provides no useful information on field strength. The polarization/extinction ratio depends on the degree of alignment. Suprathermal rotation increases the effectiveness of magnetic alignment against the randomization caused by gas-grain collisions. This allows good alignment even in weak fields, in contrast to the pure thermal rotation case treated by Davis and Greenstein (1951). However, the alignment occurs on a non-negligible time scale. The predicted time required to align the grains' spin axes τ_g depends on grain properties and, roughly, $\tau_g \sim 4 \times 10^5$ yr/$(B/30\,\mu G)^2$; it can be of order the lifetime of the cloud under study (Purcell 1979; Goodman 1989). Thus the grains may not be very well aligned and the polarization may be small, even with a strong field.

1. Interpretive Aspects of Polarization of Starlight by Extinction. The anisotropy of magnetic forces creates the potential for magnetically related asymmetries of interstellar structures. But one cannot automatically assume that an alignment between structures and magnetic field means that the field is responsible, because there exist several other anisotropic influences. For example, a shock sweeps up everything and tends to align everything perpendicular to the shock velocity. Furthermore, the magnetic field can only directly affect the ionized component of the interstellar gas, and all effects on the neutral component are due to collisions, causing a time lag (Sec. I.D.3; Chapter by McKee et al., Sec. V). Therefore, in cold regions with very low fractional ionization, even an "energetically significant" magnetic field may not dominate the dynamics on short time scales.

The literature is replete with reports of alignments of objects, such as elongated clouds and bipolar flows, with $\hat{B}_\perp$ (Sec. V contains an incomplete review). The emerging bank of observational data and theoretical results has made magnetic fields astrophysically fashionable, and the temptation to assign observed effects to magnetism is strong. We suspect that some of these assignments are misplaced. There are several cautions that must be considered before such assignments can reliably be made:

1. Correlations or alignments found within a cloud might not be peculiar to the cloud itself. Instead, they might characterize not only the cloud but also the region within which it is imbedded. Observers must map regions on both small and large scales so that the conditions within an individual object can be separated from those of the ambient medium.

2. The direction of $\hat{B}_\perp$ is difficult to measure *within* a dense cloud because of

the large extinction. Optical polarization probes the $A_V \lesssim 3$ mag gas at the periphery of the clouds. In contrast, more difficult infrared polarization probes the denser portions. For example, 2.2 μm can probe regions where $A_V = 20$ mag ($n \sim 10^5$ cm^{-3} for size 0.05 pc). In compensation for the more difficult observations, the interpretation is simpler; virtually all of the dust that polarizes the infrared polarization must be associated with dense gas in the cloud of interest.

3. Sources observed in near-infrared polarization studies are often not true background stars, but are instead embedded deeply within the clouds. Such objects are often intrinsically polarized, and must be dealt with appropriately; they include sources associated with extended infrared emission and young (e.g., T Tauri) stars with disks. They can be identified by obtaining appropriate spectrophotometry.

4. Alignments between objects located within a dense cloud, such as bipolar flows, and the ambient magnetic field as measured by optical polarization of starlight near the cloud have been reported and seem to be common. However, the polarization near the cloud is produced by dust grains along the *entire* line of sight and may not reflect $\hat{B}_\perp$ in the ambient medium near the cloud. This can be investigated only by the tedious procedure of observing a large sample of stars at different distances and in a large area surrounding the cloud.

5. It is imperative to make the distinction between cloud features aligned with each other from those that are aligned with the field. The mere existence of aligned features does not imply magnetic dominance.

6. "Alignments" can only be observed on the plane of the sky. The distribution of three-dimensional angles that produce a given two-dimensional angle in projection is only very weakly peaked at the observed value. Thus, large samples are required to relate three-dimensional orientations to the observed two-dimensional orientations.

2. Polarization of Grain Thermal Emission. According to Kirchoff's law, aligned grains that produce polarization by absorption of background radiation must also produce polarization of their own emitted thermal radiation. Thus the direction of polarization of the emitted radiation is the same as that in which the absorption is more effective. Normally, the magnetic field is perpendicular to the polarization of the emitted radiation. For details, we refer the reader to the excellent review by Hildebrand (1988), and here emphasize two of his most important points.

First, the ability to trace the magnetic field by making observations in emission is extremely important, because it allows one to map the magnetic field at will, without requiring a background source of radiation. The emission is detectable both at far-infrared and mm wavelengths. mm-wavelength observations have the advantage of higher attainable angular resolution by using interferometric techniques.

Second, if one is to interpret the polarization of diffuse radiation as resulting from aligned grains, then one must ensure that scattered radiation, which is intrinsically highly polarized, is negligible. For realistic grains, the ratio of scattering to absorption decreases as λ^{-3} and is about 4×10^{-6} at $\lambda = 100\ \mu$m (Novak et al. 1989). Thus, at far-infrared and sub-mm wavelengths, polarized diffuse radiation can reliably be assumed to be produced by thermal emission from aligned grains. At such long wavelengths, grains have no resonant absorption features, and if the Rayleigh-Jeans limit also applies, then the fractional polarization of the thermal radiation (but not that of the extinction) should be independent of wavelength. For well-aligned, highly elongated grains the fractional polarization can exceed 10%, but with realistic grain models will be $\lesssim 4\%$ (Novak 1991, personal communication).

However, two simple circumstances can produce a wavelength dependence. If there is a mixture of cool and sufficiently warm grains along the line of sight, as may easily happen in molecular clouds containing protostars, then far-infrared wavelengths might lie on the Wien side of the thermal peak for the cool grains and on the Rayleigh-Jeans side for the warm grains, so the infrared observations would emphasize the warm grains. And far-infrared observations taken from the Kuiper Airborne Observatory have worse angular resolution than mm-wave observations, which may have produced the differences in angle of linear polarization that are probably observed in the BN-KL region of Orion (Novak et al. 1990; Sec. II.A.1 below).

In the mid-infrared ($\sim 10\ \mu$m), scattering may contribute to producing some of the observed polarization. The spectra of many deeply embedded mid-infrared sources are best fit by models that include both emission and absorption by the same grain material. These embedded sources are often seen at least partly in scattered light, which is itself highly polarized. The polarization of thermal emission varies rapidly with wavelength in the vicinity of grain absorption resonances, which can be used to infer the properties of both the grains and the magnetic field (see, e.g. Lee and Draine 1985; Sec. II.A.1 below). If spectropolarimetry shows that the mid-infrared polarization angle is independent of wavelength (although in the vicinity of grain absorption features, the position angle can flip by 90° [Draine and Lee 1984]), and if the fractional polarization depends properly on wavelength, then it is reasonable to conclude that the polarization results not from scattering but from aligned grains. Even in this case, however, it is difficult to interpret the observations, because emission and absorption by the same grains give orthogonal positional angles.

D. Some Indirect Indications of Magnetic Fields

1. The Angular Distribution of H_2O Masers. The angular distribution on the sky of H_2O masers may indicate the direction of the plane-of-the-sky field $\hat{B}_\perp$ and conceivably even the ratio of field strengths $B_\perp/B_\parallel$. Elitzur et al. (1989) have presented a comprehensive model for the H_2O masers observed in star-forming regions in which the masers occur behind shocks. The model is in reasonable agreement with observations and is reasonably efficient from the energy standpoint. Magnetic fields dominate the post-shock pressure, leading to the expectation that $B \propto n_H^{1/2}$, which is reasonably accurate, at least in Orion (Sec. II.A).

The model makes the very important prediction that the maser luminosity $\propto B_{\perp,shock}^4$, where $B_{\perp,shock}$ is the strength of the preshock magnetic field component that is perpendicular to the shock velocity. Suppose a protostar emits an isotropic wind that produces a spherical shock in the ambient medium, which is threaded by a uniform magnetic field. Then the theory predicts the masers located in a ring around the "magnetic equator" to be very much brighter than those located toward the "magnetic poles." Thus, an observer should see this ring in projection as an ellipse on the plane of the sky. The major axis would be perpendicular to $B_\perp$ and the ratio of minor to major axes would be equal to $B_\parallel/B_\perp$. The ring should exhibit the expansion of the shock, and may also exhibit rotation of the ambient pre-shocked gas. The masers are velocity-coherent filaments and emit along their axes, which could conceivably produce a geometrical preference for observability that varies systematically around the ring. In Sec. II.A.3 below we find that the positional distribution of masers in Orion is consistent with the Elitzur et al. theory.

2. *Faraday Rotation.* Faraday rotation of the direction of linear polarization is $\propto \lambda^2 \int n_e B_\parallel \, dl$ and is usually considered to be the domain of galactic structure (long path lengths) and of low-frequency radio astronomy. However, with the possibility for high volume densities and magnetic fields in star forming regions, measurable Faraday rotation can occur.

Simonetti and Cordes (1986) used the VLA at dm wavelengths to observe small sets of extragalactic sources fortuitously located behind the bipolar flows Cep A and L1551. They detected significant changes of Faraday rotation measures amounting to 150 and 30 rad m^{-2}, respectively, and ascribed these changes to conditions within the outflows. For a path length of 0.1 pc, 30 rad m^{-2} corresponds to $n_e B_\parallel = 400$ cm^{-3} μG. Unfortunately, there are no independent estimates of n_e, but the absence of radio bremsstrahlung emission sets an upper limit to the emission measure $n_e^2 l$, which in turn specifies a minimum field strength of ~ 1 μG.

Novak et al. (1990) detected a $18°$ difference for the position angle of linear polarization between the far-infrared and 1.3-mm wavelengths in the BN-KL region of Orion (Sec. II.A.1 below); the difference was confirmed by Flett and Murray (1991). The most likely explanation involves the differing angular resolutions, but Faraday rotation is not ruled out. The required rotation measure of $\sim 2 \times 10^5$ rad m^{-2} could be produced by $n_e \sim 2000$ cm^{-3} and $B_\parallel \sim 300$ μG, parameters which could conceivably describe the ionized gas that lies in front of BN. If Faraday rotation is responsible, the definitive observation would be a measurement at 3-mm wavelength.

3. *Ambipolar Diffusion.* Ambipolar diffusion is a motion of charged particles, which are directly tied to the magnetic field, with respect to the neutrals, which are not (Chapter by McKee et al., Sec. V). If the density of charged

particles is high enough, the collisional drag makes the relative velocity very small. However, in the ultraviolet-free interiors of molecular clouds, charged particles can be sufficiently rare that the velocity is measurable. Thus, if charged species such as HCO^+ have systematically different velocities from neutral species, one might conclude that one has measured the drift velocity v_D (see Chapter by McKee et al. [hereafter McK], their Eq. 55):

$$v_D \sim 3 \frac{B^2_{\mu G}}{n^{3/2}_H \ell_{pc}} \text{ km s}^{-1} \tag{4}$$

where n_H is the volume density of H-nuclei ($n_{HI} + 2n_{H_2}$) in units of cm^{-3} and ℓ_{pc} is the length scale for variation in **B** in pc. If ℓ is 0.1 of the subcore diameter in cloud B1 (Sec. IV.B), $v_D \sim 0.05$ km s^{-1}, which might be measurable. However, if the ambipolar diffusion is similar on both the front and back side of the cloud, then the observable effect would be a small increase in line width instead of a velocity difference, which would be undetectable.

II. OBSERVATIONAL EXAMPLES: H II REGIONS

A. Orion: Zeeman Studies, Masers, Linear Polarization and Structural Alignments

The Orion region occupies a place of central importance because it is the nearest region of massive star formation. Genzel and Stutzki (1989) present a comprehensive review of the morphology and physical conditions, and here we extract the major points. These authors also review the large-scale properties of the magnetic field in the L1641 dark cloud. We concentrate here on the properties of the field in the Orion A region, the most massive concentration in L1641.

Orion A is a "blister" H II region protruding from the near side of a small (~6 arcmin), very dense north-south ridge about 1 pc in length known as Orion Molecular Cloud 1, or OMC-1. These are imbedded within a much larger (several degrees) elongated (~40 × 2 pc) giant molecular cloud known as the Orion A Molecular Ridge, which is in turn imbedded in a much larger and lower-density H I cloud (Chromey et al. 1989). There are other large molecular clouds immediately adjacent.

Within OMC-1 exist the famous Becklin-Neugebauer point source (the BN object) and the Kleinmann-Low nebula (the KL nebula); this smaller region is known simply as BN-KL. BN-KL contains a cornucopia of interesting objects. The two most luminous sources are IR compact source number 2 (IRc2) and BN itself. IRc2 is the brightest, with a luminosity of 2 to 10 × 10^4 L$_\odot$, and is also the center of a powerful bipolar outflow which is highlighted by esoteric and energetic phenomena such as OH masers, SiO masers, H_2O masers, Herbig-Haro objects, and vibrationally excited H_2 lines.

Information on magnetic fields in the Orion region has been obtained from linear polarization of optical, mid-infrared, far-infrared, and mm-wave radiation, from Zeeman splitting of H I and OH in absorption against the H II region, from Zeeman splitting of OH and H_2O masers, and from linear polarization of H_2O masers. The OH and H I absorption measurements refer to gas on the near side of the H II region, and the infrared sources and masers are on the far side.

1. The Becklin-Neugebauer (BN) Point Source: A Magnetic Gold Mine. The BN object in Orion is a very bright, easily identifiable (nowadays) infrared point source in the Orion region. It is embedded in the dense molecular cloud OMC-1, heavily attenuated by dust, and has been the subject of extensive spectrophotometric and spectropolarimetric observations in the infrared. The analysis is complicated by the fact that the intrinsic spectrum of the BN object is unknown. However, by assuming that BN is intrinsically unpolarized, the spectropolarimetric data allowed Lee and Draine (1985) to deduce the polarization caused by aligned dust grains and, consequently, to infer a great deal about the magnetic field and the details of grain composition and shape in front of BN, and even the intrinsic spectrum of BN.

High linear polarization ($\sim15\%$) that varies rapidly with wavelength is observed in the vicinities of the 3.1 μm "ice" (H_2O and NH_3) and 9.7 μm "silicate" absorption features, and weak ($\sim1\%$) circular polarization is observed near 3 μm. In Lee and Draine's model, these data are best fit with a combination of graphite and silicate grains with ice mantles. The graphite grains are large enough so that scattering is negligible but magnetic dipole absorption is not. The graphite grains are oblate with the c-axis of the graphite parallel to the symmetry axis ("pancakes" of graphite flakes). The silicate grains are also oblate and have poorly constrained sizes, but they must have mantle volume larger than half the core volume or else they produce too much polarization near 10 μm.

Of more interest for the current review is the magnetic field. The position angle of the linear polarization is $118° \pm 0.4°$ in the mid-infrared (Aitken et al. 1985), which gives the direction of $\hat{B}_\perp$ because the polarization results from extinction, not thermal emission. Lee and Draine cannot reproduce the observed high linear polarizations unless the grain alignment is nearly perfect. Even so, the field must be not only well ordered, but also lie nearly in the plane of the sky. For a uniform field, the observed polarization varies as $\sin^2 \gamma$, where γ is the angle between the magnetic field and the line of sight (our definition of γ is orthogonal to that of the Lee and Draine model). Formally, Lee and Draine require $\gamma \gtrsim 40°$, but in reality other effects (imperfect alignment and field nonuniformities) force an even higher limit on γ. The corollary is, of course, that the high polarization also implies a highly ordered field near BN. However, the field is not perfectly uniform: the observed circular polarization near 3 μm requires that the field "twist" by $\sim30°$ along the line of sight near BN. A similar twist is seen directly in $\hat{B}_\perp$ by Burton et al. (1991; also Sec. II.A.3 below).

Chernoff et al. (1982) use shock models for BN-KL to estimate theoretically that $n_H \approx 2 \times 10^5$ cm^{-3} and $B \approx 0.45$ mG in this region, similar to values derived in a similar model by Draine and Roberge (1982). Such field strengths are also required by the Lee and Draine model to align the grains. We regard these estimates as sufficiently secure to insert in Table II, which summarizes field strengths in the Orion region.

   2. The Orion Region: The Field Strength B. The field strength is most reliably derived from Zeeman splitting. This has been measured on several size and density scales: for the large-scale environment with H I and OH in absorption against the H II region, and for the BN-KL region with OH masers and one H_2O maser (we assume here—possibly erroneously [Sec. I.B.4*a*]—that the circular polarization measured in the H_2O maser is a result of Zeeman splitting). The field strength can also be derived from theoretical considerations as applied to observations. This has been done for the BN object (Sec. II.A.1 above) and for the time-variable H_2O "supermaser" (but possibly spuriously; see Sec. I.B.4*b*).

   The known field strengths are summarized in Table II. Volume densities sampled by the different techniques range from $\gtrsim 400$ to $\sim 10^9$ cm^{-3}, a range of about 6 orders of magnitude, and the total magnetic field strengths B range from ~ 0.05 to ~ 40 mG, a range of about 3 orders of magnitude. A reasonable fit to Table II, particularly if the lowest-density H I data are excluded, is $B = 1.0(n_H/10^6 \text{ cm}^{-3})^{1/2}$ mG. This corresponds to a density-independent Alfvén velocity $V_A = 1.8$ km s^{-1}; it may be significant that molecular line widths tend to be comparable to this value.

   a. The large-scale environment. The H I absorption line comprises two velocity components. In our opinion, this H I is not in the photodissociation region and does not directly abut the H II region. We support our opinion by two facts. One, the H I velocities of ~ 0 and 6 km s^{-1} are 3 and 9 km s^{-1} more *positive* than the H II velocity, while the absorption gas is in front of the H II region and, if it were gas from the photodissociation region, should be moving at *negative* velocity with respect to the H II region. Two, the H I is cold, but given the exciting stars for Orion, the H I in the photodissociation region should be warm (Tielens and Hollenbach 1985).

   In the H I absorption lines, Troland et al. (1989) mapped the field strength with 25″ resolution over most of the H II region, which occupies an area about 5′ in diameter. They found $B_{\parallel}$ to range from –43 to –107 μG. In both OH and H I absorption lines, Troland et al. (1986) measured Zeeman splitting for a single spot on the eastern edge of the H II region; they obtained –125 and –49 μG, respectively. We adopt $B = -100$ μG, and if this uniformly fills a 5′-diameter circle, the magnetic flux is –2.0 mG arcmin^2.

   b. The masers. The molecular "doughnut" (Plambeck et al. 1982) is a disk of dense (up to $\sim 10^7$ cm^{-3}) gas with outer diameter ~ 44″ and a hole in the middle. It surrounds IRc2 and expands at ~ 20 km s^{-1}. Its total mass is ~ 15 M$_\odot$, which seems to be too massive to have been ejected directly from IRc2. The expansion time scale (radius/velocity) ~ 3000 yr and the kinetic energy $\sim 5 \times 10^{46}$ erg, which amounts to $\sim 0.5\%$ of the luminosity of IRc2 for 3000 yr. Thus it is reasonable to consider the doughnut as being composed

TABLE II
Magnetic Field Strengths in the Orion Region

REGION	ΔRA[a]	ΔDEC[a]	V[b]	B[c]	n_H	ref
H$_2$O maser	−23	−9	−49.7	−38[g]	~10^9	FG, GD
H$_2$O supermaser	−5	−7	7.7	~$|30|$[g]	~10^9	GMH
OH maser	−9	4	6.0	−1.4	~10^7	JMN[d]
OH maser	4	−13	7.7	−1.3	~10^7	JMN[d]
OH maser	−3	−1	8.3	−2.0	~10^7	JMN[d]
OH maser	1	1	14.8	−1.3	~10^7	JMN[d]
OH maser	1	1	23.7	−0.8	~10^7	JMN[d]
BN dust	−6	6	—	$|0.45|$[c]	~2×10^5	—
OH abs[f]	~10^4	~50	6.0	−0.12	~10^4	TCK
HI abs[f]	~10^4	~50	~5.0	−0.05	$\gtrsim$400	TCK
HI abs[e]	~0	~0	~6	~−0.10	$\gtrsim$400	THG
HI abs[e]	~0	~0	~0	~−0.10	$\gtrsim$400	THG

[a] Position offsets in arcsec from IRc2.

[b] Component velocity, km s^{-1}.

[c] Measured B, mG. Values are from Zeeman splitting and the sign indicates the direction of the line-of-sight component unless surrounded by | |, in which case the sign is undetermined. For OH masers, total B; otherwise, $B_{\parallel}$. The H$_2$O maser results may be spurious (see footnote g). The BN result is inferred from dust alignment and from theoretical arguments (Sec. II.A.1).

[d] JMN neglected to quote the sign of B, and for the 1665 MHz line used an incorrect Landé g-factor (our factor b in Table I). For the 1612 MHz line, we assumed $b = 1.31$, the intensity-weighted factor quoted in Table I.

[e] THG mapped an area about 4′ in diameter. $B_{\parallel}$ varied from −0.04 to −0.17 mG and may have been correlated with total column density. There was no apparent correlation between the two velocity components.

[f] Data were taken with Nancay telescope, which has a narrow beam in the ew direction and a broad beam in the ns direction. The ns position is determined by the product of the OH opacity and the continuum intensity averaged over the telescope beam; the quoted ns offset is an estimate.

[g] The H$_2$O maser result is from Zeeman splitting, and may be 'fake Zeeman splitting' (Sec. I.B.4a). The H$_2$O supermaser result is inferred from the existence of strong linear polarization using the older theories and may be spurious (Sec. I.B.4b). References: GD, Genzel and Downes (1977); FG, Fiebig and Güsten (1989); GMH, Garay et al. (1989); JMN, Johnston et al. (1989); TCK, Troland et al. (1986); THG, Troland et al. (1989).

of ambient material that has been swept up (i.e., shocked) by less massive, faster-moving or high-temperature ejecta from IRc2.

Such shocks bring us to the magnetic field/H_2O maser discussion of Sec. I.D.1. There are two classes of H_2O maser, the 18 km s^{-1} (low-velocity) flow and the 30 to 100 km s^{-1} (high-velocity) flow (Genzel et al. 1981). The low-velocity masers lie in the plane of the doughnut and trace its *outside* with diameter ~44″; they have $n_H \sim 10^9$ cm^{-3} and $B \sim 30$ mG, corresponding to an Alfvén velocity $V_A \simeq 1.8$ km s^{-1}. From the discussion of Elitzur et al. (1989) in Sec. I.D.1, this implies that the ambient magnetic field lies along the axis of the doughnut. The high-velocity masers lie on the axis of the doughnut and are probably the equivalent of Herbig-Haro objects.

The calculations of Elitzur et al. were for the case of fast, dissociative shocks, but 18 km s^{-1} H_2O masers are not fast enough. However, the conditions in slower, nondissociative shocks can be similar, so it is likely that the low-velocity H_2O masers arise in shocks. Much of the emission in nondissociative shocks is predicted to occur near the velocity of the *unshocked* gas, where the heating due to ambipolar diffusion is greatest. However, in order to get saturated maser emission in masers of the observed size (which is comparable to the shock thickness), the density must be of order 10^9 cm^{-3} (Elitzur et al. 1989); this suggests that the maser emission occurs at the density, and hence the velocity, of the *shocked* gas. The location of the H_2O masers at the outer perimeter of the doughnut is also consistent with the idea that they are in recently shocked gas.

Most of the OH masers lie within 8″ of IRc2 and have velocities within 18 km s^{-1} of IRc2 (Johnston et al. 1989). Thus it seems that the OH masers trace the *inside* of the doughnut, because both the velocities and gas densities match. The OH masers have $n_H \sim 10^7$ cm^{-3} and $B \sim 1.4$ mG, which gives $V_A \simeq 0.8$ km s^{-1}, about 2.2 times smaller than V_A for the H_2O masers. The OH maser is a ground-state transition, so exciting it does not require the temperature to be above several hundred degrees as does the water maser. (But the *production* of OH requires either high temperatures [Sec. I.A.2] or photodissociation of H_2O.) Thus the location of the OH masers at the inner edge of the doughnut is consistent with either radiative pumping or with weak shocks due to variations in the wind luminosity of IRc2.

Placing both the OH and the H_2O masers in the shocked gas leaves us with the apparently inexplicable puzzle as to why the pressure in the H_2O masers is so much greater than that in the OH masers, even though their radial velocities are the same. We cannot offer an explanation.

c. Equal magnetic fluxes? First, we estimate the magnetic flux through the doughnut. The OH masers see the magnetic field near the inside of the doughnut, about -1.4 mG, and the H_2O masers at the outside, about -30 mG. We do not know how the field strength varies within the doughnut, but if we brazenly assume the average field strength to be -5.4 mG then the doughnut, which we take as a circle 44″ in diameter with a 16″ diameter hole, contains the same magnetic flux as the ambient gas, about -2.0 mG arcmin^2. This is equal to the flux obtained above for the diffuse ambient gas.

There is, in fact, absolutely no reason for these magnetic fluxes to be equal. Most importantly, the doughnut contains only a small fraction of the dense molecular gas, and presumably most of the magnetic field lines permeate the majority of the molecular gas. In addition, the ambient-medium flux was obtained from the H I absorption observations and thereby depends on the size of the H II region, which is determined by completely different considerations.

The approximate equality of the magnetic fluxes is probably just a coincidence, but we cannot be sure. This matter is discussed further in Sec. II.A.4 below.

3. The Orion Region and BN-KL in Particular: $\hat{B}_\perp$ *from Linear Polarization.* In the vicinity of BN-KL, polarization of the thermal *emission* from aligned dust grains is observed. Far-infrared emission has been measured with $40''$ angular resolution by Novak et al. (1989) and Gonatas et al. (1990), sub-mm-wave emission with $\sim 16''$ by Flett and Murray (1991), and mm-wave emission with $\sim 22''$ resolution by Barvainis et al. (1988) and Novak et al. (1990). Position angles derived by the various mm-wave measurements differ by about $24°$, which is unimportant for the current discussion. Position angles for the far-infrared and mm wavelengths are similar, as expected (Sec. I.C.2). The position angle of the field as derived from the emitting grains, $\theta_{B,dust}$, is about $110°$.

Higher angular resolution is required to establish whether small-scale variations of $\theta_{B,dust}$ exist. Unless one reverts to interferometric techniques, this requires going to shorter infrared wavelengths. Within $\sim 1'$ of IRc2, Burton et al. (1991) have achieved $\sim 2''$ resolution by mapping the linear polarization of the emission from the v=1–0 $S(1)$ line of H_2 at $2.1218 \,\mu$m. The emission is polarized by aligned *absorbing* grains that lie in front of the excited H_2. The line emission provides an excellent background source because it is extended; this not only allows mapping, but also should reduce the effects of scattering. Outside $\sim 1'$ from the BN object, the polarization is very large and its direction is roughly perpendicular to the radial line to BN, which implies that it results from scattering.

But closer to BN, the polarization is smaller and systematically oriented with position angle $\theta_{pol} \sim 130°$. Near ($\sim 10''$ from) BN, θ_{pol} "twists" by $\sim \pm 15°$. A similar twist, but for the direction along the line of sight, is inferred from the existence of circular polarization of the absorbed BN radiation (Sec. II.A.1 above). Perhaps this second twist is *directly* indicated by the $\sim 20°$-difference between $\theta_{B,dust}$ and θ_{pol} (i.e., between the *emitting* dust seen at far-infrared and sub-mm wavelengths and the *absorbing* dust seen by Burton et al.).

Aitken et al. (1985) made $4''$ resolution spectropolarimetric observations of the mid-infrared (8–13 μm) *continuum* radiation from the six infrared sources within $\sim 12''$ of IRc2. In contrast to Burton et al. (1991), they found the position angle to vary considerably among the six positions. For all positions except IRc2, they concluded that the polarization is produced by aligned grains, in contrast to the situation at 3.8 μm (Werner et al. 1983). However, the discrepancy with the results of Burton et al. suggests that Aitken et al. may have underestimated the scattering.

$\theta_{B,dust} \sim 110°$ is in very good agreement with the direction ($118° \pm 4°$) obtained for the point source BN by Aitken et al. (1985; Sec. II.A.1) and, of course, is close to the average angle $\theta_{pol} \approx 130°$ obtained by Burton et al. It is also close to the position angle of $\sim 100°$ for the huge surrounding area $\sim 1°$ in extent obtained from optical polarization of background stars, which indicates the field direction in the ambient matter surrounding the dense molecular cloud. Thus we conclude that the field direction θ_B is about $110°$ in the vicinity of BN-KL where the dust is warm and it has small twists in the surrounding region outside a distance $\sim 10''$ (~ 5000 AU ~ 0.02 pc) from BN-KL.

Position angles describe the orientations of various structural features in the interior of Orion, and we might expect them to be related to the local direction of the field. We exhibit them in Fig. 1 and discuss them in the following paragraphs:

(1) The low-velocity H_2O masers that are located within $\sim 10''$ of IRc2 tend to lie near a line having position angle $\theta_{line} \sim 30°$ (Knowles and Batchelor 1978; Genzel et al. 1981). From our discussion in Sec. I.D.1., $\hat{B}_\perp$ should be perpendicular to the line, at a position angle $120°$, which is only $10°$ different from $\theta_{B,dust}$. We expect the masers to lie in an ellipse; the thickness of their positional distribution about the line implies that γ, the angle between the magnetic field and the line of sight, is $\sim 60°$.

(2) The group of most intense OH masers, populations 1 and 2 of Norris (1984), lie on a line having position angle $\sim 60°$. This differs by $30°$ from that of the H_2O masers. The OH masers lie toward the inside of the doughnut, while the H_2O masers lie on the outside; the field directions may differ in these regions. Alternatively, the prediction of the Elitzur et al. theory, that maser luminosity $\propto B_{\perp,shock}^4$, may not apply to OH masers.

(3) On the very smallest scale, the position angle of the bipolar flow from IRc2 is $\sim 130°$; this is our estimate using the SO map of Erickson et al. (1982) and the SiO map of Wright et al. (1983). This is only about $20°$ from $\theta_{B,dust}$, and is equal to the position angle of the absorbing dust θ_{pol} seen by Burton et al. (1991); this implies that the flow is nearly aligned with the field directions in both the immediate vicinity and in the larger-scale ambient medium.

(4) Judging from the map of Vogel et al. (1984), the axis of the molecular doughnut has position angle $140°$, which is nearly perfectly aligned with the bipolar flow from IRc2 and is only $\sim 30°$ from $\theta_{B,dust}$.

(5) On the much larger scale of $\sim 0.5°$ (~ 4 pc), the molecular ridge in which the Orion complex is embedded has a position angle of $\sim 10°$, and it contains filaments running along its length (Bally et al. 1987). This direction is within $\sim 10°$ of the field direction implied by $\theta_{B,dust}$.

(6) The direction of linear polarization of H_2O masers is ambiguous. H_2O

masers in Orion tend to be linearly polarized and are time-variable on a scale
as short as weeks. Unfortunately, however, there exist few published reports
of the position angle of linear polarization, only for 5 maser components
by Knowles and Batchelor (1978) and for the "supermaser" by Garay et
al. (1989). Furthermore, the polarization direction can depend on maser
parameters (Sec. I.B.4*b*). For these reasons we exclude maser polarization
from the present discussion.

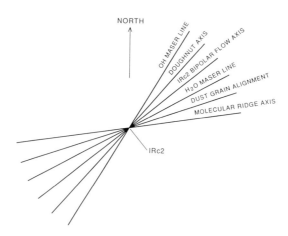

Figure 1.   Directions of $\hat{B}_\perp$ near IRc2 in Orion derived from the possible indicators
discussed in the text. The direction for dust-grain alignment indicates the polarization of both the point source BN and the dense molecular cloud OMC-1. Aside
from observational uncertainties, differences might be caused by changes in $\hat{B}_\perp$
with position.

Figure 1 shows that all of the above angles lie within 20° of 130°. The
above structural features represent anisotropies that could easily result from
magnetic (or other anisotropic) forces. We conclude that position angles of
various structural features may provide significant information on the magnetic field. However, we need much more than a statistical sample of one H II
region. There is much to be done in the way of further observations, both on
Orion itself and on other H II regions.

   4. From the Smallest to the Largest Scales. We remarked above in Sec.
II.A.2 that the line-of-sight components of magnetic fluxes in the ambient
gas and in the doughnut are approximately equal, the possible implication
being that most of the magnetic lines that thread the doughnut also thread the
ambient gas. Furthermore, they have the same sign. This implies that the
fields are not too highly inclined to the line of sight. On the other hand, we
also discussed how both the BN dust alignment model and the distribution of
H_2O masers on the sky imply that the field is not too closely aligned with the

line of sight. These somewhat contradictory requirements imply that γ, the angle between the magnetic field and the line of sight, is $\sim 45°$.

With $\gamma \sim 45°$, $B_{\parallel}$ can be negative throughout the region while the field lines thread both the ambient gas and the dense molecular gas, satisfy the constraint $\nabla \cdot \mathbf{B} = 0$, and do not bend unduly sharply. This is because the H II region is roughly spherical (Balick et al. 1974), so that as the field lines diverge from the doughnut, which is located behind the H II region, to the ambient gas, which is located in front, they need not bend by more than $\sim 30°$.

This all points to Orion/BN-KL as being an excellent example of the general (and surprising) trend found by Reid and Silverstein (1990) for the OH maser field directions to reflect the galactic field directions in their vicinity (Sec. I.B.4a). However, on slightly larger angular scales, the situation is not so clear-cut. Zeeman observations of OH in absorption against the H II region Orion B (NGC 2024), only 4° away from Orion A, show a *positive* field (Heiles and Stevens 1986). Thus, while Orion A is in accord with Reid and Silverstein (1990), Orion B is not.

This less than ideal state of affairs elicits two comments. First, Table II reveals the signs of $B_{\parallel}$ for many measurements, but these can be considered to sample only two independent regions: the ambient gas overlying the H II region and the masers. The probability that the signs are the same in these two regions is equal to 1/2, so the result of sign equality is hardly a surprise. Second, concerning the results of Reid and Silverstein, we suspect that the last word has not yet been written concerning the general trend for OH maser fields to reflect the ambient galactic field.

B. W3: Discrepant Aperture Synthesis Results

For W3 there are two independent studies of Zeeman splitting of the H I line in absorption against the H II region. They do not agree. Troland et al. (1989) used the VLA with 59″ resolution and van der Werf and Goss (1990) used the WSRT (Westerbrook Synthesis Radio Telescope) with 16″ and 30″ resolution. Troland et al. detected significant polarization that should have been seen by van der Werf and Goss, but was not; the latter detected polarization that was too weak to have been seen by Troland et al. We cannot imagine a cause for the discrepancy except that one of the observational results is simply wrong. We conclude that discussions of physical interpretation are better left until reliable data are obtained. The sensitivity must be excellent. The angular resolution of Troland et al. is clearly inadequate, and future studies should use even better resolution than that of van der Werf and Goss.

III. OBSERVATIONAL EXAMPLES: DERIVING B

A. The Method: Statistical Treatment of Zeeman and Linear Polarization Measurements

Myers and Goodman (1991) have devised a method to specify almost completely the uniform component of $\mathbf{B}$; here we present a brief summary and the application to the dark cloud L204. The method involves relating measured fluctuations in $B_\parallel$ to both measured fluctuations in the position angle of linear polarization and to fluctuations in $B_\perp$ with a model. It requires (1) good statistical samples of $B_\parallel$ from Zeeman splitting and of θ_B, the position angle of $\hat{B}_\perp$, from optical polarization; and (2) a model for the behavior of the fluctuations in **B**. A related model by Jones (1989) uses only the optical polarization data, and is able to derive the direction of the mean field and the relative strengths of the random and uniform field components, but any model that does not include Zeeman measurements cannot derive the absolute field strength.

Consider the total magnetic field **B** at any point in space to be the vector sum of a straight "uniform" field, $\mathbf{B_0}$, and a spatially varying, nonuniform field, $\mathbf{B_r}$. If we average many Zeeman observations within a given region, the mean value is a good estimate of $B_{\parallel,0}$, and the dispersion about the mean approximates $\sigma_{B_\parallel}$, the dispersion in the line-of-sight field. Similarly, optical polarization maps provide the mean position angle of $\hat{B}_\perp$, $\langle \theta_{B_\perp} \rangle$, and its dispersion σ_{θ_B}; clearly, σ_{θ_B} is related to $\sigma_{B_\perp}/B_{\perp,0}$.

Now introduce a model that relates $\sigma_{B_\parallel}/B_{\parallel,0}$ to $\sigma_{B_\perp}/B_{\perp,0}$ and also relates $\sigma_{B_\parallel}$ to $\sigma_{B_\perp}$. Then given the quantities derived from observations, namely $B_{\parallel,0}$, $\sigma_{B_\parallel}$, $\hat{B}_\perp$, and σ_{θ_B}, the magnitude of $B_{\perp,0}$ can be derived. This, together with $B_{\parallel,0}$, $\sigma_{B_\parallel}$, $\langle \theta_{B_\perp} \rangle$, and an estimate of the number of correlation lengths (N) of the field along the line-of-sight, provides $\mathbf{B_0}$ (with a two-fold directional ambiguity), and the average total field strength, $|\mathbf{B}|$.

Models include one for which $\mathbf{B_r}$ is essentially two-dimensional, resulting from Alfvén waves (Zweibel 1990a; Myers and Goodman 1991), and a "turbulent" field model, in which $\mathbf{B_r}$ is distributed isotropically in three dimensions (Myers and Goodman 1991). The isotropic model is more directly applicable in practice, but may be unrealistic because transverse fluctuations undergo weaker nonlinear damping than do aligned or longitudinal fluctuations. An ideal model would incorporate both density and magnetic field fluctuations using the concepts first enunciated by Chandrasekhar and Fermi (1953). In any model, there is the practical problem of separating fluctuations from systematic gradients.

There are some complications and uncertainties. First, measurement uncertainties increase $\sigma_{B_\parallel}$ above its true value, thus increasing the derived value of $B_{\perp,0}$. Second, both Zeeman and linear polarization measurements average along the line of sight. If there are N independent elements along the line of sight, then the fluctuations are reduced by $N^{1/2}$. It is impossible to determine N definitively, so one must make an educated guess and live with the resulting uncertainties. Finally, because polarization measurements do not give the sign of $\hat{B}_\perp$, one cannot derive the full vector $\mathbf{B_{\perp,0}}$ without a two-fold directional ambiguity, although in some cases, as in the L204 example given below (Sec. III.B), the ambiguity can be resolved by additional information.

We mention two important caveats. First, the Zeeman and polarization observations highlight different portions of the cloud unless care is exercised in selecting the statistical samples. Zeeman observations emphasize the regions of high $B_\parallel$, which tend to be those of high n; polarization observations are most easily obtained in regions of low extinction, which tend to have low n. Second, the result depends on the model used to describe the fluctuations in **B**. Magnetic fluctuations are not inherently isotropic, so the relation between $\sigma_{B_\parallel}$ and $\sigma_{B_\perp}$ depends on γ, the angle between $\hat{B}$ and the line of sight, which is itself derived from the model. Finally, in the best of worlds—but not necessarily in the real world—the model is consistent both with basic physical principles and with their application to the specific physical conditions in the region.

B. An Example: L204

In the dark cloud Lynds 204 (L204), Heiles (1988) has made H I Zeeman measurements on a grid of 27 positions covering approximately a 6 by 15 pc area. McCutcheon et al. (1986) have made well-sampled optical polarization measurements which cover a size scale equivalent to the Zeeman measurements and show a nonrandom plane-of-the-sky field. This is currently the only region where the method described in Sec. III.A could be applied.

The Zeeman spectra in L204 are best fit by a two-component model, comprised of a wide emission profile plus a narrower self-absorption feature. In this chapter, we consider only the self-absorption data, and we assume $N = 1$; Myers and Goodman (1991) also quote results for the main emission component and for various assumed values of N. The field values exhibit a tendency to correlate with velocity, so in calculating the mean and standard deviation for the measurements, Myers and Goodman (1991) must first subtract out this velocity dependence. For the $N = 1$ case, they find $B_{\parallel,0} = 6.4\,\mu G$ and $\sigma_{B_\parallel} = 4.5\,\mu G$. In analyzing the distribution of θ_B, they use the isotropic model and find $B_{\perp,0}/\sigma_{B_\perp} = 2.5$. Thus $B_{\perp,0} = 11\,\mu G$. Adding $B_{\parallel,0}$ and $B_{\perp,0}$ in quadrature gives $B_0 = 13\,\mu G$, so that $\gamma = \pm 60°$. Based on two correlations, one between the cloud shape and line-of-sight velocity and one between $B_\parallel$ and line-of-sight velocity, Heiles (1988) infers that $\gamma > 0$. The optical polarization data give $\langle \theta_{B_\perp} \rangle = 71°$, so the absolute direction and magnitude of the uniform component, i.e., $\mathbf{B_0}$, can be specified. Similarly, the magnitude of the total field, obtained by adding the uniform and fluctuating components in quadrature, is $|\mathbf{B}| = 15\,\mu G$; for comparison, the $N = 10$ assumption yields $28\,\mu G$. The actual value of N is thought to be between 1 and 10.

With a value of $N \sim 3$, the energy in the nonuniform magnetic field is about equal to the energy in the uniform field, and L204 is in approximate virial equilibrium. The correlations among cloud shape, line-of-sight velocity, and line-of-sight field found by Heiles (1988) also imply the presence of large-amplitude Alfvén waves, which may make the three-dimensional isotropic model inapplicable. Results for the (perhaps more relevant) case where $\mathbf{B_r}$ is wave-like (i.e., isotropic in two-dimensions) are given in Myers and Goodman (1991).

IV. OBSERVATIONAL EXAMPLES: THE VIRIAL THEOREM AND CLOUDS

A. The Virial Theorem: Observational Considerations

Many observational treatments of molecular clouds rely on the virial theorem, written for a spherical cloud of uniform density and mass M (McK Eq. 9):

$$|W| + 3P_0V = 2T + \mathcal{M} \tag{5}$$

where W is the gravitational potential energy $-3GM^2/5R$ for a uniform cloud; $3P_0V$ is the external pressure term, equal to $2T_0$ in McK Eq. (3); T (McK Eq. 2) is the total kinetic energy $0.27M \, \Delta v^2$ ($2T$ is $3\bar{P}V_{cl}$ in McK Eq. 9); and $\mathcal{M}$ is the total *net* magnetic energy $0.1B^2R^3$, where the factor 0.1 comes from setting $b = 0.3$ in McK Eq. (16), which is valid for self-gravitating clouds but not pressure-confined clouds. The external pressure P_0 includes both thermal and turbulent pressure, and for the case of a dark cloud surrounded by low-density gas should also include the interstellar radiation pressure. Here we use the observationally oriented Δv, which is the line width at half-peak intensity; McKee et al. use in their chapter the theoretically oriented one-dimensional velocity dispersion σ. For a Gaussian velocity distribution the two are related by

$$\sigma^2 = 0.18 \, \Delta v^2. \tag{6}$$

The McKee et al. chapter defines several additional virial-related parameters. Observers often neglect the surface terms and the magnetic field. Under this assumption, equilibrium is obtained when the parameter $\alpha \equiv 2T/|W|$ (McK Eq. 33) is equal to unity. If virial equilibrium prevails, then the question remains as to the stability of the equilibrium. The stability can be expressed in terms of various *critical masses*. In the absence of a magnetic field, the critical mass is M_J, the familiar Jeans mass: if $M/M_J > 1$, then the cloud is unstable to collapse. In the absence of gas pressure, the corresponding magnetic critical mass is M_Φ: if $M/M_\Phi > 1$, the cloud is "magnetically supercritical" and the equilibrium is unstable to collapse. The distinction between magnetically subcritical and supercritical clouds has been invoked to explain the possible bimodal character of star formation ("low-mass, low-efficiency" vs "high-mass, high-efficiency," respectively [Shu et al. 1987a]). More generally, the critical mass is M_{cr}, which is approximately equal to $M_J + M_\Phi$.

The virial theorem allows the overall equilibrium to be described in terms of global cloud parameters, which can be derived directly from observational data. This is a minimalist approach, because it avoids dealing with the internal structure of clouds. The problem with this approach is the uncertainties in the virial terms. This problem is well exemplified by our discussion of the B1 cloud below in Sec. IV.B and, apart from matters such as surface terms and nonuniform cloud structure, involves three specific observational quantities: the assumed distance D, the conversion of $B_\parallel$ to B, and the determination of the true H_2 content.

The distance uncertainty can be very serious and should never be ignored or minimized. Its seriousness depends on the method by which density and mass are determined. First, we have $\mathcal{M} \propto D^3$. If masses and volume densities are inferred from observations of *column* density, then $|W| \propto D^3$ while $P_0 V$ and $\mathcal{T} \propto D^2$. Thus, virial equilibrium depends on the *first* power of the distance. However, if masses are determined from observations of *volume* density, then $|W| \propto D^5$ and $P_0 V$ and $\mathcal{T} \propto D^3$; virial equilibrium depends on the *square* of the distance.

The second problem is a matter of projection angle. We usually measure $B_{\parallel}$ instead of B. For a large sample $\langle B \rangle = 2\langle B_{\parallel} \rangle$ and $\langle B^2 \rangle = 3\langle B_{\parallel}^2 \rangle$. However, an individual case is generally different. The matter is important, because the magnetic term in the virial theorem $\mathcal{M} \propto B^2$ so the arbitrary use of the large-sample relations increases $\mathcal{M}$ by a factor of 3 over its minimum value.

The third problem involves determination of the volume density n_{H_2} or, equivalently, the column density N_{H_2} or cloud mass M. Many clouds have been observed only in the (very) optically thick ^{12}CO line, and for these one usually assumes that M or $N_{H_2} \propto$ either the CO line luminosity or the line width (Solomon et al. 1987). The basis for this procedure involves assuming that the cloud is in virial equilibrium and has standard properties (e.g. $\langle n_H \rangle \sim 200 \, cm^{-3}$, $T \sim 10 \, K$; see the succinct review by Maloney [1990b]). Clearly, this procedure is undesirable in the present context, where ^{12}CO optical depths are high, and derived values of N_{H_2} are very uncertain. Density estimates using $C^{18}O$ or ^{13}CO observations are more reliable, but only up to moderate density ($n \sim 10^4$) where optical depth again begins to become a problem.

In principle, N_{H_2} can be determined by observing dust associated with molecular clouds. Observations of extinction due to dust are limited by the dust optical depth at any given wavelength, and therefore cannot be used to measure column density when the cloud becomes opaque. Observations of far-infrared or sub-mm emission provide the temperature-weighted (relative to the cosmic background) dust column density if the dust opacity and dust-to-gas ratio are known (Draine 1990). But in opaque clouds, dust inside the cloud cannot be heated by starlight and will not be revealed in thermal emission unless there is an imbedded heating source such as a protostar. In practice, neither the dust opacity at long wavelengths (e.g., $600 \, \mu m$) nor the dust temperature distribution along the line of sight are well understood, so this technique cannot yet be reliably applied.

In summary, it is very difficult to obtain accurate measurements of *column* density. The *volume* density n_{H_2} can be determined directly from observations of molecular transitions, because these transitions are excited by collisions with H_2 molecules. However, in practice, there is rarely sufficient knowledge to obtain accurate results, because the derived densities also depend on the details of radiative transfer, which in turn depend on the velocity field. Models are usually employed, and the resulting volume densities are uncertain by factors of order 2 or 3 (White 1977). Turner and Ziurys (1988) provide a contemporary summary of this problem.

In applying the virial theorem, even with distance uncertainties aside, it seems unlikely that the terms can be specified relative to each other more accurately than a factor of 2. Factors of 2 are good from the standpoint of astronomical accuracy but inadequate for a definitive discussion involving the virial theorem, as we shall see in our discussion of B1 below. If $\mathcal{M}$ is comparable to other terms in the virial theorem which are, in turn, comparable to one another, then the uncertainties allow virial equilibrium to exist with or without $\mathcal{M}$.

B. Barnard 1 (B1): The Magnetic Field, Virial Equilibrium and Stability

B1 is, to our knowledge, the dark cloud about which most is known from the combined standpoints of structure and magnetic field, and represents the best example for application of the virial theorem. B1 is a well-defined cloud in the Perseus region. It was selected for Zeeman observations because it has the largest value of $\Delta v/R$, which implies a large magnetic field strength if it is in virial equilibrium with $\alpha \approx 1$ (Myers and Goodman 1988a); this may mean that it is not a representative cloud. B1 has been mapped in several molecular transitions, including ^{13}CO with 4.4 arcmin resolution by Bachiller and Cernicharo (1984); OH with 3 arcmin resolution by Goodman et al. (1989); and NH$_3$ with 40 arcsec angular resolution by Bachiller et al. (1990b). For these molecules—^{13}CO, OH and NH$_3$—the sizes decrease and the line widths decrease monotonically. This is a reasonable result, because the three molecules trace increasingly dense regimes of H$_2$ volume density, and according to the empirical "Larson's laws" (Larson 1981; Solomon et al. 1987), size $\propto n_{H_2}^{-1}$ and line width $\propto n_{H_2}^{-1/2}$.

The appearance of B1 in these three molecules is depicted in Fig. 2. As the angular scale gets smaller B1 gets more complicated. We define the cloud *core* as the $\sim 3'$-diameter portion of the cloud highlighted by the NH$_3$ lines. The core contains an IRAS source that exhibits an apparent outflow (E. Lada, personal communication), and thus is a protostar or newly formed star. The cloud core consists of two or three condensations, which we define as subcores, separated by about 1.2 arcmin, in which $n_{H_2} \sim 4 \times 10^4$ and 8×10^4 cm^{-3}. If these subcores themselves have smaller substructure, it would not have been resolved by existing observations.

Zeeman splitting of the OH 1665 and 1667 MHz lines was measured by Goodman et al. and Goodman (1989). On the core, $B_{\parallel} \sim -27\ \mu$G, while 4' to the southwest $B_{\parallel} \sim -12\ \mu$G (Fig. 2). On the core of B1, the brightness temperature of the OH 1667 MHz line peaks at ~ 1.3 K, while those of the optically thick ^{13}CO and NH$_3$ lines peak at ~ 6 K. This indicates that the OH line is optically thin. However, direct comparison of the 1667 and 1665 MHz lines indicates a 1667 MHz optical depth of 1.22. This is a potentially serious discrepancy. Its resolution probably lies in the 1667 and 1665 MHz lines having unequal excitation temperatures, as has been observed in other dark

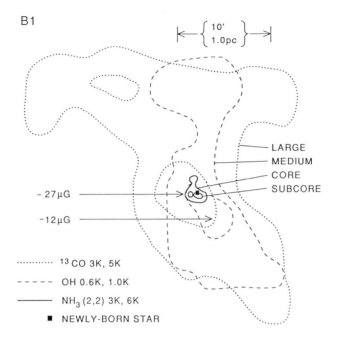

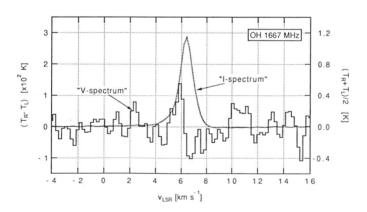

Figure 2.   *Upper panel*: B1 as observed in ^{13}CO, OH and NH$_3$. The three molecules highlight increasingly smaller size scales (see text). Only two contours are shown for each molecule, one near the peak and one about half the peak intensity. The four size ranges, ranging from large to subcore, are indicated, as are the two $B_\parallel$'s derived from OH Zeeman splitting. There are several young stars in the region; we show only the one located in the core. *Lower panel*: The Zeeman effect in the B1 core region, as observed at Arecibo (Goodman et al. 1989). Fits of the derivative of the Stokes I spectrum to the Stokes V spectrum (at 1665 and 1667 MHz) give $B_\parallel = -27 \pm 4\ \mu$G.

clouds by Crutcher (1977). This is probably caused by excitation of the OH rotational ladders by far-infrared radiation. In the case of B1, a difference in excitation temperature of ~3 degrees would explain the discrepancy; this is only somewhat larger than Crutcher obtained for OH along the line of sight to 3C123, which is not far from B1 on the plane of the sky.

1. The Field Strength in the Densest Portions of the Core. Goodman's (1989) off-core result of only –12 μG is less than half the field strength measured on the core, which implies that the field strength might increase in the core. Further, if we were able to sample exclusively the core gas, then the measured field strength might be even greater than the –27 μG obtained from the OH, because OH may not sample the densest portions of the core. This is indicated by both an observational and a theoretical argument.

(1) Observationally, the OH line width is nearly twice as wide as the NH_3 line. According to Larson's empirical laws, line width tends to decrease with decreasing size and increasing volume density; this implies that the NH_3 samples a denser portion than the OH. However, we cannot unequivocally conclude that this is the case because the NH_3 maps show that considerable macroscopic velocity structure exists within the larger OH telescope beam, and there may be enough beam smearing to produce the observed OH line width. Furthermore, Crutcher et al. (in preparation) are in the process of determining whether the observed line width is wide because of the contribution of the "large" and "medium" sizes to the OH emission. At the time of this review, this observational question has not yet been resolved.

(2) At the high densities in the core, astrochemistry predicts that the OH volume density is roughly independent of the H_2 volume density, and therefore not a good tracer of H_2 (Sec. I.A.2).

The question of just how well OH traces high-density regions is a major one and deserves intense observational and theoretical attention.

2. The Virial Theorem Applied to B1. Figure 2 depicts four size ranges of B1: large, medium, core and subcore. We have reasonably complete information on physical conditions for each size range, which allows us to evaluate the virial terms in Eq. (5) and other quantities. Unfortunately, we do not have a complete sample of $B_\parallel$ in the four size ranges, and there is always the annoying matter of converting $B_\parallel$ to B. We assume $B = 2B_\parallel$ (and $B^2 = 4B_\parallel^2$ instead of the large-sample correction $B^2 = 3B_\parallel^2$). For the medium size and core, the observations provide $B_\parallel = -12$ and -27 μG, respectively. For the subcore, we assume that $B_\parallel$ is the same -27 μG that it is for the core. This may well be incorrect, because $B_\parallel$ increases as we go from the medium size to the core, but it serves as a reasonably interesting example.

The external pressure term is not usually included in observational discussions, but it is important for the three smallest size ranges. Let P_0 be the external pressure at the boundary of each size range. At the boundary of the largest size range, P_0 should equal the ambient interstellar pressure, for which we adopt $P_0 = 1.6 \times 10^4 k_B$ dyne cm^{-2} (Chapter by McKee et al.); we neglect the contribution from radiation pressure, which would add about

10% (Spitzer 1978). For the smaller size ranges, we set P_0 equal to the total pressure of the gas in the next larger size, calculated from the line width and density. That is, we take

$$P_0 = \rho \sigma^2 = 0.18 \times 10^{10} \rho \, \Delta v_5^2 \tag{7}$$

which amounts to assuming that the velocity field is "microturbulent." There is some theoretical justification for this assumption, because in the subcore the damping length of MHD turbulence is about 0.005 pc (McK Eq. 38), which is 20 times smaller than the size of the subcore.

Table III gives numerical values for the virial terms. We emphasize that the accuracies are low. Virial equilibrium is easily achieved for any size by changing the mass by a factor of less than 2, which is within the observational uncertainties. For all size ranges, the virial theorem is satisfied as well by excluding $\mathcal{M}$ as by including it, which is the inevitable consequence of the uncertainties and having all terms in the virial theorem comparable.

Table III also lists $\alpha (\equiv 2\mathcal{T}/|W|$, McK Eq. (33); $\alpha = 1$ implies gravitational virial equilibrium in the absence of magnetism and surface terms). α varies from 1.2 to 2.0, which is commensurate with approximate gravitational virial equilibrium. As mentioned in McKee et al.'s chapter, observers often estimate cloud masses by taking $\alpha = 1\text{-}2$. This appears to be a valid procedure. However, α cannot be measured accurately enough to provide independent information on M/M_Φ or $P_0/\bar{P}$. If we had computed α from McK Eq. (34) and set $b' = b$, we would have obtained $\alpha \sim 1.2$ for all sizes but the medium size; the differences between this and the values in Table III presumably reflect the departure of the real world from the ideal theoretical model. Observational uncertainties render McK Eq. (34) less useful than McK Eq. (33) in practical applications.

We now forge ahead and discuss the stability of the virial equilibrium, temporarily assuming absolute accuracy. The magnetic field is crucial for the stability. Table III lists the cloud mass M in terms of the magnetic critical mass M_Φ (McK Eq. 17b), the Jeans mass M_J (McK Eq. 12), and the critical mass M_{cr} (McK Eq. 22). All sizes have $M/M_\Phi > 1$ and are magnetically supercritical, which means that without gas pressure, the cloud would collapse. The medium size and core have $M/M_J \sim 2.2$, which means that in the absence of the magnetic field, they would also be unstable to collapse. But the *combination* of magnetic and gas pressure prevents collapse: M_{cr} includes both, and these two sizes have $M/M_{cr} \sim 1$, which either removes the instability or makes it much less pronounced.

We emphasize again that the statements in the above paragraph are based on uncertain parameters. If we had used a more conventional definition of cloud radius, or different values for n_{H_2}, we would have reached different conclusions. Nevertheless, we are encouraged by the result $M/M_{cr} \sim 1$ for the core. The core has split into three subcores, and it also contains a protostar or new star. These are *observational* indications that the core of B1 is on the

TABLE III
B1 at Various Size Scales

Size	$n(H_2)$ cm^{-3}	$N(H_2)^a$ cm^{-2}	Diam. pc	ΔV_5 km/s	B μG	Mass $M_\odot$	$-W$ erg	$3P_0V^e$ erg	$2T$ erg	$\mathcal{M}$ erg	M/M_Φ^f	M/M_J^g	M/M_{cr}^h	α^i
Large^b	900	5(21)	2.7	2.1	-10^d	630	1.5(46)	.20(46)	3.0(46)	.07(46)	4.5	0.43	0.34	2.0
Medium^b	2600	6(21)	1.1	1.3	-24^c	120	1.5(45)	2.0(45)	2.3(45)	.30(45)	2.2	2.2	1.13	1.6
Core^b	20000	10(21)	0.24	0.7	-54^c	10	4.3(43)	2.4(43)	5.3(43)	1.6(43)	1.64	2.3	0.99	1.2
Subcore^j	70000	14(21)	0.10	0.7	-54^d	2.5	6.4(42)	3.8(42)	13.0(42)	1.1(42)	2.4	0.85	0.65	2.0

^{a} $N(H_2)$ is defined as the total number of H_2 molecules in the cloud divided by the cloud area.

^{b} From Goodman et al. (1989), taking R in their Table 2 as the cloud diameter.

^{c} Twice the measured value, to account for orientation to line of sight.

^{d} Assumed. For the core, where B probably increases, this value is probably a lower limit.

^{e} From Eq. (7).

^{f} M_Φ is the magnetic critical mass from McK Eq. (17b) and McK Eq. (19).

^{g} M_J is the Jeans mass from McK Eq. (12) for the isothermal case with $c_J = 1.18$.

^{i} α is the ratio of turbulent plus thermal energy to the gravitational energy (McK Eq. 33).

^{j} From Bachiller et al. (1990); these parameters apply to a single NH_3 subcore, and differ from the parameters given by those authors, which apply to the whole NH_3 core.

verge of instability. With $M/M_{cr} \sim 1$, we have a *theoretical* indication that it is close to instability. This argues that our numerical estimates in Table III are reasonably accurate.

As explained in McKee et al.'s Chapter, Sec. VI.A, low-mass stars form after magnetically subcritical clouds evolve via ambipolar diffusion until their cores become magnetically supercritical (Mouschovias 1977; Nakano 1979; Shu et al. 1987*a*). If this scenario were true for the B1 subcore (in which a protostar has recently formed), then the subcore should be magnetically supercritical ($M/M_\phi > 1$) and the surroundings magnetically subcritical ($M/M_\phi < 1$). However, ($M/M_\phi > 1$) for all four size scales in B1. This may be a result of either observational uncertainties or inapplicablity of the theory.

C. Statistical Studies of Clouds for which Zeeman Measurements Exist

Myers and Goodman 1988*a* applied the virial theorem without surface terms systematically to all appropriate regions for which cloud maps *and* Zeeman splitting measurements of field strength existed. Their sample includes 2 atomic clouds, 6 molecular clouds (one of which, Orion, contains measurements in three different density regimes) and 6 H II regions associated with dense gas containing OH masers. They showed that all the regions roughly satisfy the virial theorem (Eq. 5) with approximate equality of $|W|$, $2\mathcal{T}$ and $\mathcal{M}$. They used this conclusion to support the idea that $(B/\Delta v) \propto n_{H_2}^{1/2}$, i.e., that Δv is related to the Alfvén velocity.

Nevertheless, we must be a bit cautious. Virial terms for OH masers are really virial terms for clouds in which the masers are imbedded, having been calculated using measurements of cloud properties obtained from other molecules that highlight the clouds, such as NH_3. In W3(OH), the OH masers trace the same gas as the NH_3 (Reid et al. 1987), and this conclusion may well apply elsewhere, but we cannot be absolutely certain of this. In addition, all the points in the sample are significantly biased towards having high magnetic fields because observers try to obtain measurements instead of upper limits.

The statistical reliability of the Myers and Goodman (1988*a*) conclusions is much increased by the large range of n_{H_2} and B over which approximate equilibrium is found. If, however, we consider the 6 dark clouds and 6 maser regions as separate groups, the ranges in n_{H_2} and B are much smaller, the intrinsic scatter in the data becomes more important, and the conclusions less certain. This is particularly the case for the conclusion that $(B/\Delta v) \propto n_{H_2}^{1/2}$: this is not a very good fit for the dark clouds alone, and the masers alone are better fit by a constant value for $(B/\Delta v)$.

Nevertheless, their primary conclusion, that the virial theorem is satisfied with rough equipartition between the kinetic and magnetic terms, is approximately valid for the non-maser clouds, and this statement can also be made for the masers if their conditions are the same as those in the larger clouds in which they are embedded.

D. Statistical Studies of Clouds for which Zeeman Measurements Do Not Exist

Myers and Goodman (1988b) have also applied the virial theorem to a large sample of 120 clouds for which observational data exist on size, velocity dispersion, and volume density, but not on magnetic field. They concluded that this sample is consistent with the smaller sample for which Zeeman measurements do exist if there is rough equipartition among $|W|$, $2T$, and $\mathcal{M}$ in Eq. (5). Their conclusions are independent of cloud mass and volume density, because they depend only on the directly observed quantities Δv and cloud radius R. This is fortunate, because the references from which cloud parameters were obtained use methods for deriving cloud mass that, for an individual cloud, can differ by an order of magnitude; this illustrates the difficulty in deriving accurate masses.

Nevertheless, we consider that their conclusions are not much more than a consistency argument. The assumption of equipartition implies that $B \propto \Delta v^2/R$. Their basic result is that $\Delta v^2 \propto R$, which implies that clouds typically have $B \sim 30\,\mu G$, $\pm$ a factor of 3 or so. However, the result $\Delta v^2 \propto R$ is also a restatement of Larson's laws, and given our incomplete understanding of the interstellar medium, we have no special justification for assuming that magnetism is responsible for Larson's laws (except, of course, the results of Myers and Goodman (1988a; Sec. IV.C).

Equipartition between $2T$ and $\mathcal{M}$ seems to neatly avoid the serious problem associated with highly supersonic line widths: it makes the line width comparable to the Alfvén velocity (Arons and Max 1975), eliminating the necessity for the observed macroscopic motions to produce strong shocks, which have a very high rate of energy dissipation. Zweibel and Josafatsson (1983) show that Alfvén waves are also subject to dissipation; this dissipation is not excessive and may contribute to the heating of the cloud (McKee et al. Chapter, Sec. V.B). Thus the dissipation problem is ameliorated.

In our opinion, discussions of the magnetic properties of clouds that are based on statistical analyses of the virial theorem alone, without actual data on magnetic field strength, should be interpreted with care.

V. ORIENTATION OF CLOUD FEATURE WITH RESPECT TO THE FIELD

Goodman (1991) has reviewed existing optical polarization maps of background starlight. Such maps cover portions of several molecular cloud complexes, including Taurus, Perseus, Ophiuchus, Orion and L204. The polarization maps typically exhibit a well-defined mean polarization position angle, with fluctuations about the mean which differ in magnitude from region to region. These regions also contain well-defined molecular clouds, many of which are elongated and some of which, such as Orion, are extremely so. Are these features aligned with the magnetic field? Section I.C.1 presented some basic cautions required to establish alignment reliably; here we illustrate some of those with observational examples, and also summarize the recent literature concerning alignments.

We reiterate two important cautions that we listed in Sec. I.C.1. One, we must distinguish alignment *between cloud features* from that with the *magnetic field*. The evidence to date does not unequivocally show the magnetic field direction to be correlated with the orientations of cloud features. Two, one must take into account the statistics of the projection of three-dimensional angles between features into the two-dimensional plane of the sky. This point is almost always ignored in interpretive discussions of all types of alignment.

A. Basic Cautions: Observational Examples

Polarization maps of individual clouds, presented in isolation and removed from their environments, may be misleading. [Note: Also see Sec. I.C.1] For example, consider the elongated dark clouds B216/217 in Taurus (Fig. 3) and L1755 in Ophiuchus (Fig. 4). Analyzed individually, it would appear that $\hat{B}_\perp$ in the B216/217 cloud is perpendicular to the cloud's long axis, while in L1755 $\hat{B}_\perp$ is parallel to the cloud's long axis. Several discussions in past literature have emphasized such apparent alignments. If, however, we look at these clouds in the context of their respective complex-wide polarization maps (Figs. 3 and 4) we see, in both cases, that the polarization observed fits into a smooth, larger-scale field, and that the cloud orientation could be fortuitous. These figures also illustrate the tendency of optical polarization to probe only the edge, not the center, of a cloud.

In both infrared and optical polarization studies, there is a danger of examining maps at too small a scale. For example, Turnshek et al. (1980), in their study of NGC 1333, found a bimodal distribution in polarization position angle, which is also apparent at 2.2 μm (Tamura et al. 1988). They further deduced that there is a significant spatial arrangement of the two different distributions within their map. However, Goodman et al. (1990) extended the map to include all of the Perseus complex and found that the bimodal distribution persists throughout the entire complex, so that the spatial correlation reported by Turnshek et al. may have been an artifact.

This narrow-view selection effect might also be at work in the analysis of Bok globule 2 discussed below.

B. Relation of Fields Within Clouds to the Ambient Field

The relatively small number of near-infrared polarization maps of dense clouds indicate that the magnetic field within the denser portions of a cloud does not deviate in direction very much from the "ambient" field indicated by optical polarization observations. In Orion (Sec. II.A), the directions are similar. In the densest part of the ρ-Ophiuchus molecular cloud, Wilking et al. (1979) found that the near-infrared polarization is similar in direction and dispersion to the surrounding optical polarization. The same situation exists in Taurus, where Tamura and Sato (1989) (with large-scale, but very sparse coverage) find no significant difference between optical and 2.2 μm position angles.

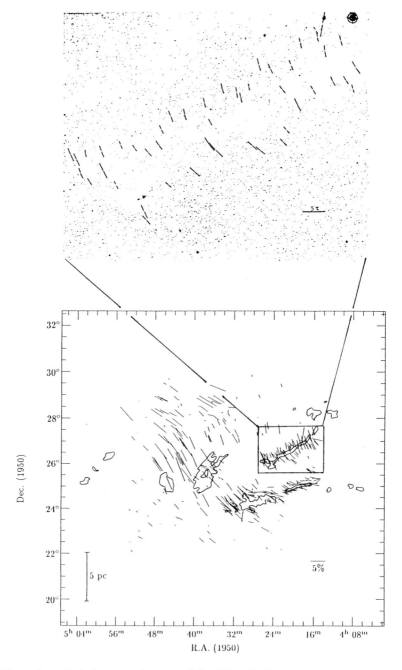

Figure 3.   Optical polarization map of the B216–B217 region (top) of the Taurus
Molecular Cloud Complex (bottom). Polarization data is from Moneti et al. (1984),
Heyer et al. (1987) and Goodman et al. (1990). Photograph from Palomar Sky
Survey. Contours superposed on the lower panel indicate the extent of the dark
clouds in the region (from Tamura and Sato 1989).

In three Bok globules, polarization vectors appear to be unaffected by the presence of the globule (Klebe and Jones 1990). In the dark cloud, Heiles Cloud 2, Tamura et al. (1987) possibly find weak evidence in their $2.2\,\mu$m polarization maps for compression of field lines, although the mean orientation of the infrared polarization vectors is virtually identical to the well-determined mean optical polarization direction in the region.

In contrast, the interior of Bok globule 2 in the Southern Coalsack exhibits a definite change in the infrared polarization position angle (Jones et al. 1984). The total number of cases where comparisons can be made is small, but most do not show an easily discernable difference between the optical and infrared results.

C. Relation of Cloud Structure to the Ambient Magnetic Field

At the several-pc scale traced by extinction, ^{13}CO, and OH, many dark clouds are highly elongated. Theoretical models predict an orientation of the long axis either perpendicular or parallel to the field lines as indicated by optical polarization observations, depending on whether the cloud has collapsed or expanded along the field lines. However, in a comprehensive study of many clouds, Goodman et al. (1990) show that neither orientation is generally observed. Instead, it appears that the magnetic field in a complex is ordered on the several tens of pc scale, and that individual dark clouds within the complex are not oriented in a special way with respect to the magnetic field, although they are perhaps aligned with each other.

On a smaller ($\lesssim 1$ pc) scale, observations of bipolar outflows have been interpreted as being aligned with the local large-scale magnetic field. However, such alignments are not always observed and may even be rare. Cohen et al. (1984) reported that half of a sample of 10 outflows (located in several different clouds) are oriented within $\pm 20°$ of the local field direction, which in turn is determined from previously existing polarization measurements of background stars relatively near the outflow in question. In B335 and L723, Vrba et al. (1986b) constructed optical polarization maps on an appropriate size scale and found good field/outflow alignment. Then, however, Langer et al. (1986) detected a new component to the CO outflow in B335, and Hodapp (1987) constructed new near-infrared polarization maps of B335, and the situation became much more complicated. It now appears that there is more than one preferred direction in the polarization map and more than one outflow, and it is not completely clear how these features are related. Several other sources do not show alignments. Wooten and Loren (1987) have written a paper summarily entitled "L1689N: Misalignment Between a Bipolar Outflow and a Magnetic Field." In a recent study of outflows by Hodapp (1990), in which polarization was explicitly mapped near outflows, one of the three cases shows an outflow roughly parallel to $\hat{B}_\perp$, one is roughly perpendicular, and the third is neither parallel nor perpendicular.

On the even smaller scale of optical jets, Heyer et al. (1987) searched for alignments between the jets and the local field at an appropriately small

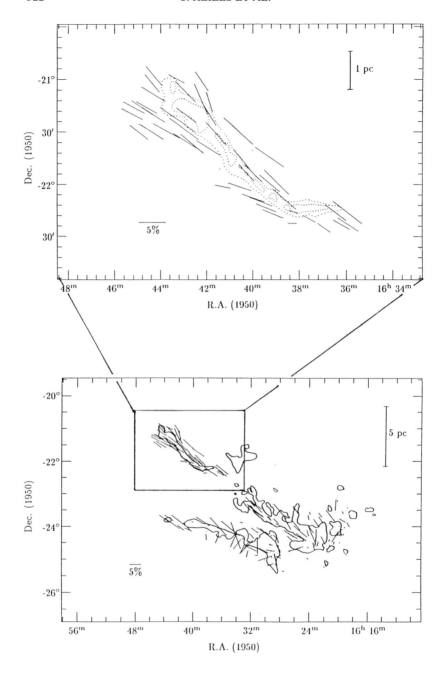

Figure 4.   Optical polarization map of the L1755 region (top) of the Ophiuchus
Molecular Cloud Complex (bottom). Polarization data is from Vrba et al. (1976)
and Goodman et al. (1990). Contours superposed on both panels indicate the extent
of dark clouds in the region, as traced by ^{13}CO emission (Loren 1989b).

scale by constructing CCD polarization maps of background starlight. These give a high-resolution view of the local, rather than the global $\hat{B}_\perp$. In the three cases they studied (SVS 13/HH 7–11, HH 33/HH 40, and HH 12), two jets were well aligned ($\lesssim 15°$) with the local magnetic field, but the third (HH 12) was oriented 60° from the field direction and is located only 2′ to the north of HH 7–11. The authors conclude that the relation between a jet axis and the magnetic field direction may depend on the properties of the dense gas surrounding the source powering the jet, and they favor ionization as the important property. Numerous observations indicate the existence of large density gradients within clouds, and we believe that density gradients should also be included in the list of possibilities.

D. Relation of Cloud Structures with Each Other

There is no good evidence for the global alignment of CO outflows in a cloud complex. For example, in L1641 the five outflows for which a position angle can be defined to better than ±40° span a 135° range in orientation (Goodman and Myers 1991)—despite the seemingly contradictory fact that alignment is observed among the much smaller-scale ($\sim$0.05 pc) optical jets associated with young stars in the cloud.

These smaller jets are identified as fast-moving jets from optical observations of regions around young stars associated with HH objects. Strom et al. (1986) and Reipurth (1989a, b) have found more than ten such jets within L1641. Eleven of the 13 jets have position angles between 130° and 170°, with the other two being very far outside this range. This is good evidence for alignment of the 11 jets with each other at a position angle near 150°. However, $\hat{B}_\perp$ as derived from optical polarization has position angle 110° (Vrba et al. 1988a), 40° away from the 150° which seems to characterize most of the flows. Thus the case for *magnetic* alignment has not been made. In contrast, density gradients present an attractive hypothesis for the alignment, because of the filamentary nature of this cloud (see Bally et al. 1987).

In Taurus, we used the results of Mundt et al. (1987) to find that the position angles of the 7 known optical jets have standard deviation 119° and are therefore uncorrelated on the complex-wide scale. These authors also emphasize that the jets are not linear, and conclude that asymmetries in optical jets are probably caused by asymmetric density distributions in the surrounding molecular cloud, and are quite possibly important for L1641.

E. Relation of Cloud Structure to Rotation

There appears to be no correlation between structure and rotation for cloud cores, which are elongated (Myers et al. 1991a) and often exhibit velocity gradients which can be interpreted as solid-body rotation (Fuller 1989; Goodman et al. 1991b). Rotation, as deduced from fits to observed velocity gradients, is not found to be energetically significant in the support of the cores (Goodman et al. 1991b). In a preliminary analysis of NH_3 data (Benson and Myers 1989) on cores in Taurus, no obvious structural correlations are found (Goodman and Myers 1991). In a set of CS $J = 2 - 1$ observations of seven dense cores, Heyer et al. (1986) also find that the distribution of dense gas is often elongated, but that it shows no preferred orientation with respect to the field, to the rotational axis of the core, or to the outflow direction.

Molecular clouds seem to exhibit little rotation. This is in accord with theoretical prediction. For clouds that are dense relative to their surroundings, McK Eq. (45) applies; it predicts that the magnetic braking time is fairly short.

VI. PERSPECTIVE AND COMMENTARY

A. Techniques

The line-of-sight component of the magnetic field $B_\parallel$ can be measured using Zeeman splitting (Sec. I.A). With molecules, there is often a danger that Zeeman splitting is a result of linear birefringence, and this possibility should be checked, particularly for H_2O masers. OH is the most widely used probe of molecular clouds because its Zeeman splitting is easiest to detect. However, the degree to which OH traces really *dense* regions is unclear (Sec. I.A.2). This is a major and fundamental question whose answer requires extensive observational and theoretical effort. Magnetic field strengths derived from OH masers in regions containing large velocity gradients are biased towards large values (Sec. I.B.4*a*). The few other suitable candidates for Zeeman observations are listed in Table I.

The direction of $\hat{B}_\perp$ can be obtained from linear polarization of either attenuated background radiation or grain thermal emission, which is always either parallel or perpendicular to $\hat{B}_\perp$ (Sec. I.C). We expect linear polarization of spectral lines to be weak in most cases, with ambiguous interpretation even if it is detected (Sec. I.B). The structure of **B** within clouds, as revealed by the structure of $\hat{B}_\perp$, seems to be the best way to determine the importance of **B** on cloud dynamics and evolution (see below). This cannot be obtained only from polarization of optical starlight, which can only probe the outer, less-dense portions of clouds. Infrared measurements, both of attenuated starlight and grain thermal emission, are required to probe the dense portions of clouds. Good angular sampling and resolution are required for reliable interpretation.

The goal of deriving $\langle \mathbf{B} \rangle$ (the uniform component) within an object, both in magnitude and direction, can be realized by obtaining a large sample of measurements of Zeeman splitting and linear polarization and, in addition, by applying a model of the field fluctuations (Sec. III). Statistical information on the fluctuations can also be obtained.

There are several possible indirect probes of the magnetic field (Sec. I.D). Shock-produced H_2O masers are predicted to be most luminous when the shock velocity is perpendicular to $\hat{B}$, and this should make the masers lie in an ellipse on the sky as is observed in Orion (Sec. II.A.3). Faraday rotation in small star-forming regions may be observable. And ambipolar diffusion, which affects ions differently from neutrals, may be directly observable.

In our opinion, the most rewarding observational paths will involve Zee-man splitting and linear polarization caused by aligned grains. For one desir-ing reliably useful results, such as a Ph. D. student, these are the recommended paths. For one who can afford to gamble on wasted effort, such as an inde-pendently wealthy middle-aged full professor, we suggest the indirect probes mentioned in the above paragraph.

B. Results and Interpretation

Combining all information on $B_\parallel$ and $\hat{B}_\perp$ can provide a reasonably complete picture of **B** within a region. In this way, we described a crude picture of **B** in Orion (Sec. II) and a surprisingly complete picture of L204 derived by Myers and Goodman (1991) (Sec. III). This shows the value of multi-pronged attacks: observers who employ different specialties should cooperate in selecting objects for study.

When results for all the regions where magnetic fields strengths have been measured are considered, over the sampled density range from 10 to 10^{10} cm^{-3}, B runs from just a few μG up to several tens of mG. The existing data are well fit by the relation $B = An^\kappa$, where $\kappa = 0.5$ and $A \sim 1.5$, with a fair amount of scatter about this trend. (See Troland and Heiles [1986], Fiebig and Güsten [1989], and Myers and Goodman [1988a] for summary plots of B vs n_H.) (Here B is expressed in μG, n in cm^{-3}, and velocity in km s^{-1}.) With $\kappa = 0.5$, A can be thought of as giving a "typical" Alfvén speed; the "fit" to the existing data gives 2.1 km s^{-1}, with a scatter corresponding to about a factor of 3.

Studies using the virial theorem are interesting, but the usefulness is limited because of the inherent inaccuracies. Furthermore, it must be applied correctly. If cloud B1 is representative (Sec. IV.B), the magnetic term $\mathcal{M}$ is important, but less so than the external pressure term $3P_0V$ (Eq. 5). However, the external pressure term is rarely included in observational analyses. The inherent inaccuracies in the virial terms suggest that the direct measurement of magnetic topology using the direction of $\hat{B}_\perp$, together with theoretical models, may provide a more useful indication of the influence of magnetic forces on cloud structure and dynamics.

Past published papers aside, the jury is still out concerning the align-ments of clouds and their internal structures with either $\hat{B}$ or with each other (Sec. V). Conclusions based on qualitative appearance are not sufficient. In many individual cases, including Orion (Sec. I.B), alignments are observed. However, a definitive statistical study is required to establish the generality and validity of the alignments. Reliable conclusions require an enormous amount of work (Sec. I.C.1) in obtaining sufficient statistical data, and this has been done only recently for a small number of regions.

Acknowledgment. We acknowledge, with extreme pleasure, the dedicated efforts of the many observers and theorists who have worked so hard to bring the state of knowledge to its present level. If we had written this Chapter ten years ago, it would have been so short as to *easily* fit into the page limits prescribed for this book. As it is, we have unavoidably exceeded those limits by approximately a factor of two. CH acknowledges with pleasure the hospitality of the Joint Institute for Laboratory Astrophysics, which provided the ideal environment for writing his portion of this Chapter. We are grateful to R. Bachiller, V. Burdyuzha, R. Crutcher, R. Güsten, R. Hildebrand, T. Jones, K. Menten, M. Reid and T. Troland for providing unpublished material; and to J. Bally, G. Fuller, M. Goss, R. Hildebrand, T. Jones, N. Ladd, N. Kylafis, P. Myers, B. Turner and P. van der Werf for instructive discussions and comments. We especially thank B. Draine, S. Spangler and W. Watson for careful reading of an early version of the manuscript and for a number of detailed and constructively critical comments. CH was supported in part by an NSF grant. AG is supported in part by a President's Fellowship at the University of California, Berkeley. CFM's research is supported by an NSF grant; his research on star formation is supported in part by a NASA grant to the Center for Star Formation Studies. EAZ was supported in part by an NSF grant, and by a grant of the Institute for Theoretical Physics.

MAGNETIC FIELDS IN STAR-FORMING REGIONS: THEORY

CHRISTOPHER F. McKEE
University of California, Berkeley

ELLEN G. ZWEIBEL
University of Colorado

and

ALYSSA A. GOODMAN, CARL HEILES
University of California, Berkeley

Magnetic fields are essential in supporting molecular clouds and the clumps within them from gravitational collapse. The virial theorem, including the effects of fluctuating magnetic fields and a turbulent ambient medium, can be used to show that observations imply that the magnetic flux threading a cloud and the internal motions within it are of comparable importance in cloud support. For clouds and clumps within clouds in which the magnetic energy exceeds the gravitational energy, star formation occurs on the time scale for the field to leak out of the cloud by ambipolar diffusion; the field is effective in removing the angular momentum from the cloud in this case. Reduction of the magnetic flux to the small values found in stars occurs only after the collapsing cloud reaches very high densities. Massive stars may well form in regions in which the magnetic energy is less than the gravitational energy, and this process is less well understood. It appears that molecular clouds form as a result of gravitational instabilities in the interstellar medium, but a full understanding of the role of the magnetic field in the formation and subsequent evolution of molecular clouds remains a challenge for the future.

I. INTRODUCTION

The theory of magnetic fields in star formation has a long and distinguished history, with major contributions by Mestel, Spitzer, Mouschovias, Elmegreen, Nakano and many others. Attention was drawn to magnetic fields early on by a fundamental question: how do stars form with so little magnetic flux? The observation of magnetic fields in star-forming regions, on the other hand, has a short, but also distinguished history which is reviewed in a companion Chapter by Heiles et al. Indeed, the major development in our understanding of magnetic fields in star-forming regions since *Protostars and Planets II* (Black and Matthews 1985) was published has come from observations which show that these fields are about as strong as theoretically expected.

This review is an attempt to provide a comprehensive view of our understanding (and our ignorance) of the role of magnetic fields in the structure and evolution of the molecular clouds and cloud cores out of which stars are born. We do not address the effects of magnetic fields in the formation and evolution of protostars and protostellar disks; this topic is covered in the Chapter by Königl and Ruden. We have drawn on several excellent previous reviews by Nakano (1984), Mestel (1985), Mouschovias (1987a, b), and Shu et al. (1987b). Literature up to mid 1990 is thoroughly reviewed, whereas subsequent literature is unevenly reviewed. We begin with a discussion of the virial theorem, which provides a framework in which to treat the competing effects of magnetic fields and gravity on the structure of molecular clouds (Sec. II). The utility of the virial theorem in estimating the masses of molecular clouds is assessed. Models of molecular clouds, both static and dynamic (i.e., explicitly allowing for turbulence) are reviewed in Sec. III. The rotation of molecular clouds and the effectiveness in magnetic fields in braking the rotation are discussed in Sec. IV. To this point, the discussion has assumed that the flux is frozen to the matter. In Sec. V, we consider violation of this assumption by ambipolar diffusion, a process of fundamental importance in star formation, and by magnetic reconnection, which is less well understood. The effects of ambipolar diffusion and flux loss on star formation (subcritical vs supercritical star formation, contraction of magnetized clouds, and the rate of star formation) are described in Sec. VI. A somewhat speculative discussion of the magnetic evolution of molecular clouds is given in Sec. VII. Our conclusions are briefly summarized in Sec. VIII.

II. MAGNETIC FIELDS VERSUS GRAVITY

A. The Virial Theorem

The effect of magnetic fields on the structure and evolution of molecular clouds can be estimated by the virial theorem (Chandrasekhar and Fermi 1953; Mestel and Spitzer 1956; Strittmatter 1966), which can be written as

$$\frac{1}{2}\ddot{I} = 2(\mathcal{T} - \mathcal{T}_0) + \mathcal{M} + \mathcal{W}. \tag{1}$$

The left-hand side of this equation expresses the effects of changes in the shape or volume of the cloud through variations in $I = \int \rho r^2 dV$, which is like the moment of inertia. The kinetic energy of the cloud is

$$\mathcal{T} = \int_{V_{cl}} \left(\frac{3}{2}P_{th} + \frac{1}{2}\rho v^2\right) dV = \frac{3}{2}\bar{P}V_{cl} \tag{2}$$

where v is the fluid velocity relative to the center of mass of the cloud, $\bar{P}$ is the mean total pressure in the cloud, thermal plus turbulent, and V_{cl} is the volume of the cloud. We are defining the turbulent pressure broadly to include all turbulent motions, such as large amplitude Alfvén waves, as well as the motions associated with large-scale flows, such as rotation, expansion, contraction, or changes in shape. We shall focus on nonrotating clouds in this section for simplicity; in any case, rotation is observed to be not very important in molecular clouds (Goldsmith and Arquilla 1985; see Sec. IV.A). The term in the virial theorem that represents the surface pressure may be expressed as

$$\mathcal{T}_0 = \int_{S_{cl}} P_{th}\, \mathbf{r} \cdot d\mathbf{S} \simeq \frac{3}{2} P_0 V_{cl} \tag{3}$$

where P_0 is the total pressure in the ambient medium. The demonstration that the integral of the thermal pressure over the cloud surface is proportional to the total pressure in the ambient medium P_0, including the turbulent pressure, is given by McKee and Zweibel (1992). They show that for purely hydrodynamic turbulence ($B = 0$), the surface term $\mathcal{T}_0$ is actually somewhat less than $(3/2)P_0 V_{cl}$ because the turbulence in the intercloud medium is anisotropic near the cloud; we shall ignore this complication here. The premise of the demonstration is that the intercloud medium has a much lower density than the cloud, which should be valid for star-forming molecular clouds in the interstellar medium.

The magnetic term $\mathcal{M}$ is the net magnetic energy:

$$\mathcal{M} = \frac{1}{8\pi} \int_{V_a} B^2 dV + \frac{1}{4\pi} \int_{S_a} (\mathbf{r} \cdot \mathbf{B})\mathbf{B} \cdot d\mathbf{S} - \frac{1}{8\pi} \int_{S_a} B^2 \mathbf{r} \cdot d\mathbf{S} \tag{4}$$

where V_a is the volume of integration, and S_a is the bounding surface. This has a simple form if the intercloud medium is of low density and in a steady state: V_a then can be taken to be large enough that the field has dropped to its asymptotic value in the ambient medium B_0 (which we shall term the ambient field), and $\mathcal{M}$ becomes (McKee and Zweibel 1992)

$$\mathcal{M} = \frac{1}{8\pi} \int_{V_a} (B^2 - B_0^2) dV. \tag{5}$$

Fluctuations $\delta \mathbf{B}$ of the magnetic field will in general accompany a turbulent velocity field. The contribution of these fluctuations to $\mathcal{M}$ can be shown to be of the same form as Eq. (4), with B^2 and B_0^2 replaced by time averages of δB^2 and δB_0^2. Hence, the magnetic term is just the total magnetic energy associated with the cloud, including the fluctuating field; because the presence of a field inside the cloud distorts the field outside, the energy is not confined to the volume of the cloud.

Finally, if the gravitational field is dominated by the self-gravity of the cloud, then the gravitational term W is simply the binding energy of the cloud,

$$W = -\frac{3}{5} a \left(\frac{GM^2}{R} \right) \tag{6}$$

where the coefficient a allows for deviations from the simplest case of a uniform, spherical cloud. For example, a spherical cloud with a density distribution $\rho(r) \propto r^{-k_\rho}$ has $a = (1 - k_\rho/3)/(1 - 2k_\rho/5)$; a thin, uniform disk has $a = \pi/2$. The case of arbitrary oblate and prolate clouds has been considered by Bertoldi and McKee (1992).

We shall assume that the cloud is approximately in virial equilibrium, so that the left-hand side of Eq. (1) is negligible. To see when this is valid, consider a uniform sphere of mass M and radius R, which has $I = \frac{2}{5}MR^2$. Define the characteristic time scale for I to vary, t_I, by the relation $|d^2I/dt^2| \equiv I/t_I^2$. We shall compare t_I to the gravitational free-fall time for a sphere of mean density $\bar{\rho}$,

$$t_{ff} = \left(\frac{3\pi}{32G\bar{\rho}} \right)^{1/2} = \frac{4.35 \times 10^7}{\bar{n}_H^{1/2}} \text{yr} \qquad (7)$$

where $\bar{n}_H$ is the mean density of hydrogen nuclei in the cloud; if the gas is entirely molecular, then $n_H = 2n(H_2)$. The ratio of the of the left-hand side of Eq. (1) to the gravitational energy is then

$$\frac{|\frac{1}{2}\ddot{I}|}{|W|} \simeq 0.5 \left(\frac{t_{ff}}{t_I} \right)^2. \qquad (8)$$

If the volume or shape of a self-gravitating cloud is changing on a time scale t_I exceeding about two free-fall times, the cloud may be considered to be in approximate gravitational virial equilibrium. Clouds that are confined by the pressure of the surrounding medium rather than by their own self-gravity can also be in virial equilibrium; in this case, it is necessary for t_I to exceed the sound crossing time of the cloud. Even if the cloud is not in equilibrium, so that the variation in the moment of inertia is important, the average value of $\ddot{I}$ for an ensemble of clouds will tend to vanish: the equilibrium virial theorem will be more accurate for a sample of observed clouds than for an individual one.

The virial theorem then simplifies to

$$3(\bar{P} - P_0)V_{cl} - \frac{3}{5}a\frac{GM^2}{R}\left(1 - \frac{M}{|W|} \right) = 0 \qquad (9)$$

for equilibrium clouds. The rationale for taking the ratio of the magnetic and gravitational terms will become apparent in Sec. II.C below.

B. Pressure-Supported Clouds ($B = 0$)

In order to understand the effects of magnetic fields on molecular clouds, it is first necessary to consider the field-free case. In this case, the cloud is spherical, and the virial equilibrium Eq. (9) with $\mathcal{M} = 0$ implies

$$\bar{P} - P_0 = \left(\frac{3\pi a}{20} \right) G\Sigma^2 \qquad (10)$$

where $\Sigma \equiv M/\pi R^2$ is the mean surface density of the cloud. This result is relatively independent of the shape of the cloud, so it provides a useful estimate of the gas pressure even when a field is present. For example, for a planar cloud, integration of the equation of hydrostatic equilibrium shows that the pressure in the midplane is $(\pi/2)G\Sigma^2$, so that, independent of the shape of the cloud, the pressure due to self-gravity is of order $G\Sigma^2$. Numerically, this can be expressed in terms of the mean column of hydrogen, $N_H \equiv 10^{21}N_{21}$ cm^{-2}, or the mean extinction, $\bar{A}_V = 0.5N_{21}$ mag. For example, for a spherical cloud with a r^{-1} density profile, we have

$$\frac{\bar{P} - P_0}{k_B} = 1340\, N_{21}^2 \quad \text{cm}^{-3}\, \text{K} = 5540\, \bar{A}_V^2 \quad \text{cm}^{-3}\, \text{K} \tag{11}$$

where k_B is Boltzmann's constant. Because the interstellar kinetic pressure is about $(1$ to $2)\times 10^4$ cm^{-3} K (see, e.g., Cox 1988), this means that any cloud with a mean extinction exceeding 2 mag is strongly self-gravitating ($\bar{P} \gtrsim 2P_0$). Equation (11) is satisfied by the cloud B1 discussed in the Chapter by Heiles et al. to within a factor 2 for the four different size scales considered there, consistent with the observation that the magnetic field is not dominant in that cloud.

Calculations of the structure of nonmagnetic, isothermal clouds were carried out a number of years ago by Bonnor (1956) and Ebert (1955). For a given external pressure, equilibrium is possible only up to a maximum mass, sometimes termed the critical mass, M_{cr}. The critical mass also represents the threshold for gravitational instability. Below the critical mass, there are additional equilibria with higher central densities; indeed, if the mass equals that of a critical isothermal sphere with $\rho \propto r^{-2}$, there are an infinite number of equilibria. Only the one with the lowest central concentration is stable (see, e.g., Stahler 1983). Bonnor and Ebert showed that the maximum stable mass for a given external pressure is

$$M_J = 1.18\frac{\sigma^4}{(G^3P_0)^{1/2}} \tag{12}$$

where $\sigma^2 = \bar{P}/\bar{\rho}$ is the one-dimensional velocity dispersion; for purely thermal motions, $\sigma^2 = k_BT/\mu$, where μ is the mean mass per particle. This mass is often called the Bonnor-Ebert mass. However, we prefer to term it the Jeans mass, because it corresponds to the widespread intuitive concept of the Jeans mass as the maximum mass that is stable against gravitational collapse. The surface density corresponding to the Jeans mass depends only on the pressure,

$$\Sigma_J = \frac{M_J}{\pi R^2} = 1.60\left(\frac{P_0}{G}\right)^{1/2}. \tag{13}$$

Physically, the Jeans mass corresponds to the mass such that the gravitational binding energy is comparable to the internal energy, and as a result it can be inferred from the virial theorem with reasonable accuracy (McCrea 1957; Spitzer 1968a). Solving the virial equation for the ambient pressure yields

$$P_0 = \frac{3M\sigma^2}{4\pi R^3} - \frac{3aGM^2}{20\pi R^4}. \tag{14}$$

As R is varied, P_0 reaches a maximum value for a given mass. Correspondingly, this maximum value of P_0 determines the maximum mass which can be supported by gas pressure, M_J. This treatment is approximate because the parameter a, which is a weak function of R, is held fixed during the variation; the effective value of a in Eq. (14) corresponding to the Jeans mass is 1.31. Molecular clouds typically have $M \sim (1-2)M_J$ (see Sec. II.E below), where the velocity dispersion σ is due primarily to turbulence. Low-mass clumps within a cloud often have smaller line widths, so that thermal broadening is more important (Myers 1985). In the limit in which the clump is entirely supported by thermal pressure, we have $\sigma = 0.188(T/10 \text{ K})^{1/2} \text{ km s}^{-1}$, and the Jeans mass is

$$M_J = 11.5 \frac{(T/10 \text{ K})^2}{(P_{0,\text{clump}}/10^3 \text{ cm}^{-3} \text{ K})^{1/2}} \text{M}_\odot = 5.7 \frac{(T/10 \text{ K})^2}{\bar{A}_V} \text{M}_\odot. \tag{15}$$

Here $P_{0,\text{clump}}$ is the total gas pressure at the surface of the clump; the clump itself has been assumed to be supported entirely by thermal pressure. We have assumed that $P_{0,\text{clump}}$ is of order the mean pressure in the cloud as a whole so that it can be estimated in terms of $\bar{A}_V$, the mean extinction of the entire cloud, from Eq. (11); we have also assumed that it is large compared to P_0, the pressure at the surface of the entire cloud. Equation (15) shows that the scale of a solar mass is naturally selected in observed molecular clouds, which have $\bar{A}_V \sim 3 - 8$. Low-mass stars ($M \lesssim \frac{1}{2}\text{M}_\odot$) that form in self-gravitating molecular clouds must do so either in regions of very high extinction ($\bar{A}_V \gtrsim 10$) or as fragments of larger clumps.

C. Effects of a Steady, Poloidal Magnetic Field

It was recognized early on that the magnetic and gravitational terms in the virial theorem scale in the same same way with R: for a poloidal field (one with no azimuthal component) that is time independent, the magnetic energy scales as $\mathcal{M} \propto B^2 R^3 \propto \Phi^2/R$, where Φ is the magnetic flux threading the cloud, whereas the gravitational energy scales as $W \propto M^2/R$. This argument is quite crude because, as we have seen above, the magnetic energy actually scales as $(B^2 - B_0^2)R^3$, and the difference $B - B_0$ also depends on the size of the cloud. Nonetheless, this argument reveals a basic result that is borne out by more complete calculation; if the mass-to-flux ratio exceeds a certain value (magnetically supercritical clouds), the magnetic field cannot prevent gravitational collapse; if it is less than this value (magnetically subcritical clouds), the magnetic energy is large enough that collapse is impossible, so long as flux freezing holds.

First consider the case in which the gas pressure vanishes ($M_J = 0$); the gas then settles into a thin disk normal to the field (Parker 1974; Mestel and Ray 1985; Barker and Mestel 1990). Assume that the distribution of the mass per flux is specified—e.g., that corresponding to a uniform field threading a spheroidal cloud. If the ambient field B_0 also vanishes, then there is a unique equilibrium mass for a given flux. On the other hand, if the field in the cloud is connected to an ambient field, then there is a sequence of equilibria of increasing mass for a given flux (magnetically subcritical clouds) up to the magnetic critical mass M_Φ, where the magnetic and gravitational energies are equal. To determine M_Φ, we define $\bar{B} \equiv \Phi/(\pi R^2)$ to be the mean field threading the cloud, and we define the parameter b, which is of order unity, to be proportional to the net magnetic energy of the critical cloud (cf. Nakano 1984):

$$\mathcal{M}_{cr} \equiv \frac{b}{3}\bar{B}^2 R^3 = \left(\frac{b}{3\pi^2}\right)\frac{\Phi^2}{R}. \qquad (16)$$

Note that the dependence on the ambient field has been absorbed into b. In general, the cloud will be flattened along the field, and in this case R is the semimajor axis. For $B_0 = 0$ (completely closed field topology), Strittmatter (1966) found that $b = 3/4$ for a sphere and $b = \pi^{-1}$ for a thin disk; for a completely open field topology, the numerical calculations cited below give $b \simeq 0.3$ for clouds on the verge of gravitational collapse. Then the magnetic critical mass M_Φ is determined by the condition $|\mathcal{W}_{cr}| = \mathcal{M}_{cr}$:

$$\frac{3}{5}a\frac{GM_\Phi^2}{R} \equiv \frac{b}{3\pi^2}\left(\frac{\Phi^2}{R}\right) \qquad (17a)$$

so that

$$M_\Phi^2 = \left(\frac{5b}{9\pi^2 a}\right)\frac{\Phi^2}{G} \equiv c_\Phi^2 \frac{\Phi^2}{G}. \qquad (17b)$$

For axisymmetric modes, M_Φ is also the maximum stable mass (Tomisaka et al. 1988b). It remains to be demonstrated that this is also true for non-axisymmetric modes.

Equation (16) for the magnetic energy can be extended to subcritical masses ($M < M_\Phi$) by noting that $\mathcal{M}$ remains equal to $|\mathcal{W}|$ in this case because there is no thermal or turbulent energy ($M_J = 0$). The condition $\mathcal{M} = |\mathcal{W}|$ implies

$$\mathcal{M} = \frac{b}{3}\bar{B}^2 R^3 \left(\frac{M}{M_\Phi}\right)^2 \qquad (M_J = 0). \qquad (18)$$

The magnetic energy in the subcritical case is less than that in the critical case by a factor $(M/M_\Phi)^2$ because the field is less distorted.

Next consider the more general case in which thermal gas pressure (turbulent pressure involves time-dependent fields and is discussed in Sec. II.D below) also contributes to the support of the cloud ($M_J > 0$). The critical mass M_{cr}, which is the maximum equilibrium mass, now exceeds M_Φ because the gas pressure adds to the support against gravitational collapse. For clouds which are strongly self-gravitating (i.e., close to the critical mass), the magnetic energy is proportional to M_Φ^2 [Eqs. (16) and (17b)] and the gravitational energy is proportional to $M^2 \simeq M_{cr}^2$; as a result, the ratio of magnetic to gravitational energies is

$$\frac{\mathcal{M}}{|W|} \simeq \left(\frac{M_\Phi}{M_{cr}}\right)^2. \tag{19}$$

This same relation should continue to hold for subcritical clouds. Consider a cold cloud with mass $M = 0.1 M_{cr}$, for example; because it is cold, $M_\Phi = M_{cr}$. The energy ratio is unity in this case. Now heat the cloud, while at the same time increasing its mass so that it remains at $0.1 M_{cr}$. Under these conditions, the radius of the cloud will not change significantly, and its magnetic energy will remain about constant [$\sim G(0.1 M_\Phi)^2/R$]. On the other hand, the gravitational energy will increase as $G(0.1M)^2/R = G(0.1 M_{cr})^2/R$, and Eq. (19) remains valid. As a consequence, the magnetic energy may be expressed in general as

$$\mathcal{M} \simeq \frac{b}{3}\bar{B}^2 R^3 \left(\frac{M}{M_{cr}}\right)^2. \tag{20}$$

The ratio of the *net* magnetic energy $\mathcal{M}$ to the magnetic energy in the cloud is then

$$\frac{\mathcal{M}}{(B^2/8\pi)V_{cl}} \simeq 2b \left(\frac{M}{M_{cr}}\right)^2. \tag{21}$$

This ratio becomes small for very subcritical clouds because the field becomes almost uniform, making the net magnetic energy small. The accuracy of Eqs. (19)–(21) remains to be checked for such clouds.

With the aid of the virial theorem, it is straightforward to estimate the maximum mass M_{cr} for a magnetized cloud using a procedure similar to that used for the nonmagnetic cloud above (Mouschovias and Spitzer 1976; cf. Nakano 1984):

$$M_{cr} = c_J M_J \left[1 - \left(\frac{M_\Phi}{M_{cr}}\right)^2\right]^{-3/2} \tag{22}$$

where $c_J = (3.15/1.40a^3)^{1/2}$ is a constant to be determined; it allows for the value of a to differ from that appropriate for the case of no magnetic field. The results of Mouschovias and Spitzer (1976) correspond to $c_J = 1.16$ and $c_\Phi = 0.126$. The more extensive numerical calculations of Tomisaka et al. (1988b) give similar results, $c_J = 1.18$ and

$$c_\Phi \equiv \frac{M_\Phi}{\Phi/G^{1/2}} = 0.12. \tag{23}$$

We shall adopt the latter values for numerical evaluation; they correspond to $a = 1.17$ and $b = 0.30$. Tomisaka et al. (1988b) show that there is one stable equilibrium for $M < M_\Phi$; two equilibria, of which only the less dense one is stable, for $M_\Phi < M < M_{cr}$; and no equilibria for $M_{cr} < M$.

This numerical value for M_Φ, the maximum equilibrium mass in the absence of gas pressure, is based on a particular distribution of flux within the cloud. It is convenient to define the mass to flux distribution in terms of a "parent cloud" in which the field and density are both uniform, as they would be in the absence of gravity. The value of c_Φ given above corresponds to a spheroidal cloud; the ratio of the major and minor axes is arbitrary. Tomisaka et al. (1988b) have shown that the stability of clouds with axisymmetric flux distributions is governed primarily by the flux to mass ratio on the axis of the cloud, $d\Phi/dM$ evaluated at $r = 0$. Eq. (22) can be generalized to other flux distributions by the replacement

$$\frac{M_\Phi}{M_{cr}} \rightarrow \frac{0.17}{G^{1/2}} \left(\frac{d\Phi}{dM} \right)_{r=0}. \qquad (24)$$

For the particular case of a spheroidal parent cloud, a more accurate expression is obtained by replacing 0.17 by $0.18 = (3/2)c_\Phi$, as the column density along the axis of a spheroid exceeds the mean value of the column density by a factor 3/2. The value of the central column density of a critical cloud is $0.17\bar{B}/G^{1/2}$, which is quite close to the maximum stable column density of a slab of gas with a uniform field perpendicular to the slab, $B/(2\pi G^{1/2}) \simeq 0.16B/G^{1/2}$ (Nakano 1984).

The numerical value of the field in the cloud can be related to its column density by noting that $M/M_\Phi \propto \Sigma/\bar{B} \propto N_H/\bar{B}$, so that

$$\bar{B} = 5.05 \left(\frac{N_{21}}{M/M_\Phi} \right) \ \mu G = 10.1 \left(\frac{\bar{A}_V}{M/M_\Phi} \right) \ \mu G. \qquad (25)$$

For example, a cloud with $\bar{A}_V \simeq 6$ and $M \simeq 2M_\Phi$, which we shall see is typical (Sec. II.E), has a mean field of 30 μG. To evaluate M_Φ numerically, it is convenient to define the characteristic mass M_B, which is independent of the size of the cloud, by the relation (Mouschovias and Spitzer 1976):

$$\frac{M_B}{M} \equiv \left(\frac{M_\Phi}{M} \right)^3. \qquad (26)$$

The ratio M_B/M determines whether a cloud is magnetically supercritical, just as does M_Φ/M. For a spheroidal cloud of volume $4\pi R^2 Z/3$, we have

$$M_B = 512 \left(\frac{R}{Z} \right)^2 \frac{\bar{B}_{1.5}^3}{(\bar{n}_3)^2} \ M_\odot \qquad (27)$$

which is independent of the cloud size, as stated above; here $B_{1.5}$ is the field in units of $10^{1.5} \ \mu G$ and $\bar{n}_3 \equiv (\bar{n}_H/10^3 \ cm^{-3})$ is the density of hydrogen nuclei in units of 10^3 hydrogen nuclei cm^{-3}. Thus, for example, stellar mass clumps of density 10^3 cm^{-3} in a giant molecular cloud (GMC) with a mean field of 30 μG are very magnetically subcritical. Note that M_B, like M_Φ, remains constant if the mass and flux of the cloud are constant. Spherical atomic clouds in the diffuse interstellar medium (ISM) have $B \sim 3~\mu$G and $\bar{n}_H \sim 30$ cm^{-3}, so that $M_B \sim 500~M_\odot$, independent of size; if such clouds were subsequently incorporated as clumps into a molecular cloud, they would all retain the same value of M_B. The observed values of M_B for clumps in GMCs are uncertain, but appear to be somewhat smaller than this (Bertoldi and McKee 1992).

An approximate solution of Eq. (22) which is accurate to within 5% for $M_{cr} < 8M_\Phi$ is (McKee 1989)

$$M_{cr} \simeq M_J + M_\Phi. \qquad (28)$$

For larger values of M_{cr}/M_Φ the critical mass approaches the Jeans mass and Eq. (28) becomes more accurate than Eq. (22). Equation (28) clearly demonstrates that the effects of gas pressure and magnetic pressure in supporting the cloud against gravitational collapse are approximately additive.

D. Time-Dependent Magnetic Fields

The primary case considered to date in the literature is that of a steady axisymmetric field. Observed fields are more complex; because the observed turbulent velocity fields will induce turbulent magnetic fields, the latter must be time dependent, thereby increasing the strength of the field while leaving the true flux constant. The role of time-dependent motions (primarily Alfvén waves) in the structure of molecular clouds is discussed in Sec. III.B below. Here we give a brief discussion of how wave motions may be included in the virial theorem, following the treatment of McKee and Zweibel (1992).

If we divide the energies into steady and wave components, $\mathcal{T} = \mathcal{T}_s + \mathcal{T}_w$, etc., we can express the virial theorem for a cloud in a steady state as

$$2(\mathcal{T}_s - \mathcal{T}_{0s}) + \mathcal{M}_s + \mathcal{W}_s + 2(\mathcal{T}_w - \mathcal{T}_{0w}) + \mathcal{M}_w + \mathcal{W}_w = 0. \qquad (29)$$

This division is precise only if the cloud is very nearly in a steady state, so that there is a clear distinction between the internal wave motions and the dynamical evolution of the cloud as a whole. The steady kinetic energy $\mathcal{T}_s$ includes thermal energy and the energy associated with steady rotation; all the turbulent motions are contained in $\mathcal{T}_w$ for the cloud and $\mathcal{T}_{0w}$ for the intercloud medium. Now, the level of magnetic fluctuations is related to the associated fluctuations in the kinetic energy by the principle of equipartition (Whitham 1974; Zweibel and McKee, in preparation). This states that the kinetic energy of the wave equals the total potential energy associated with the wave, including the magnetic, gravitational and thermal energies. The thermal term is negligible for Alfvén waves, so we ignore it here. The equipartition principle then becomes

$$\mathcal{M}_w + \mathcal{W}_w = \mathcal{T}_w - \mathcal{T}_{0w}. \tag{30}$$

Defining the coefficient

$$c_w \equiv 1 + \frac{\mathcal{T}_w - \mathcal{T}_{0w}}{2(\mathcal{T} - \mathcal{T}_0)} \tag{31}$$

we can express the virial theorem as

$$2c_w(\mathcal{T} - \mathcal{T}_0) + \mathcal{M}_s + \mathcal{W}_s = 0. \tag{32}$$

This is of the same form as Eq. (9), except that the kinetic energy term is multiplied by a factor c_w and the magnetic and gravitational terms are based on the time-averaged field and density, respectively.

Molecular clouds and the CO clumps within the clouds are generally highly turbulent, with $\mathcal{T} \simeq \mathcal{T}_w$; in this case, $c_w = 3/2$. On the other hand, thermal motions are often dominant in cloud cores (Myers 1985), and in such cores, c_w is closer to unity. In general then, we expect $1 \leq c_w \leq 3/2$.

The effect of the wave energy on the critical masses described in Sec. II.B and C above is straightforward: the Jeans mass is increased by a factor $c_w^{3/2} <$ 1.84, whereas the magnetic critical masses M_Φ and M_B are unchanged so long as time-averaged field B_s is used. As the flux is determined by B_s, not by the rms field $B_{\text{rms}} = (B_s^2 + B_w^2)^{1/2}$, the critical mass M_Φ is completely unaffected by the inclusion of time-dependent fields. As discussed in the Chapter by Heiles et al., the strength of the observed field is determined from Zeeman observations that average along the line of sight. This spatially averaged field should be closer to B_s than to the rms field, making this form of the virial theorem particularly suitable for analyzing observations.

E. Implications

The mass of a molecular cloud is often estimated from the virial theorem, as direct measurements of the the mass from ^{13}CO or extinction observations show that this is generally valid, though the scatter is considerable (Larson 1981). An approximate form of the virial theorem Eq. (9), ignoring the complications associated with internal density variations (a), surface pressure (P_0), and magnetic fields ($\mathcal{M}$), gives a mass $M = 5\sigma^2 R/G$. We introduce the *virial parameter* α, which measures the ratio of the thermal plus turbulent energy to the gravitational energy, to make this relation exact:

$$M \equiv \frac{1}{\alpha} \left(\frac{5\sigma^2 R}{G} \right). \tag{33}$$

For a cloud in approximate virial equilibrium ($t_l \gtrsim 2t_{ff}$—see above), the value of α can be evaluated from Eq. (9),

$$\alpha = \frac{a}{(1 - P_0/\bar{P})} \left(1 - \frac{\mathcal{M}}{|\mathcal{W}|}\right) \simeq \frac{1 - (M_\Phi/M_{cr})^2}{1 - (P_0/\bar{P})}. \qquad (34)$$

The second step follows from Eq. (19) and from the fact that the parameter a is of order unity.

In applications of Eq. (33) to estimate cloud masses, the surface pressure and magnetic energy are generally ignored, and α is taken to be of order unity: for example, Larson took $\alpha = 2.0$, Solomon et al. (1987) adopted $\alpha = 1.11$, and Myers and Goodman (1988a, b) set $\alpha = 1.0$. Barring a fortuitous cancellation, it is clear from Eq. (34) that this procedure is valid only if two conditions are satisfied. First, the mean pressure in the cloud must significantly exceed the surface pressure ($\bar{P} >> P_0$); otherwise the observed line width reflects the confinement by external pressure and the mass would be overestimated. Second, the cloud must not be too highly magnetized (e.g., a cloud at the critical mass should have $M = M_{cr} \gtrsim 2M_\Phi$); otherwise the observed line width reflects the small role played by the gas pressure in supporting the cloud and the mass would be underestimated. In the absence of a magnetic field, an isothermal cloud at the Jeans mass has $\alpha = 2.04$. Magnetized clouds have smaller values of α: estimating the cloud radius for a flattened cloud as $(R^2 Z_{max})^{1/3}$, where Z_{max} is the maximum height of the cloud above the plane of symmetry, we find that Mouschovias's (1976) models have $M \simeq 2M_\Phi$ and $1.1 \lesssim \alpha \lesssim 1.33$, consistent with the Tomiska et al. (1988b) result $\alpha = 1.19$ for their model with $M \simeq 2M_\Phi$; on the other hand, their most strongly magnetized model (their $\beta_0 = 0.02$) has $M \simeq M_\Phi$ and $\alpha \simeq 1/3$. Elmegreen (1989a) has estimated the effect of the magnetic field on virial mass estimates using the Nakano (1984) model for the field, and finds that α is typically close to unity. It is clear that in the absence of some knowledge of the magnetization of the cloud, the virial estimate of the cloud mass is subject to large uncertainties. For the case in which turbulent motions and the magnetic field are of comparable importance in supporting the cloud and in which the cloud is close to the critical mass ($M_J \sim M_\Phi \sim 0.5M_{cr}$), the virial estimate with $\alpha \simeq 1$ can in principle provide a reasonably accurate estimate of the mass.

These conditions are just those found by Myers and Goodman (1988a, b). Figure 3 in their 1988b paper plots the quantity $15\sigma^2/(4\pi\bar{\rho}GR^2)$, which (assuming a spherical cloud) is just α. Their data show that $\alpha \simeq 1$, albeit with considerable scatter; the mean value is 1.2. In their 1988a paper, they compare a theoretical field strength B_{eq} with measured values. Equations (17) and (33) give the relation between B_{eq} and the actual mean field $\bar{B}$ as

$$B_{eq} \equiv \frac{3\sigma^2}{R} \left(\frac{5}{G}\right)^{1/2} = \frac{3\pi c_\Phi \alpha}{\sqrt{5}} \left(\frac{M}{M_\Phi}\right)\bar{B} = \frac{\alpha}{1.98} \left(\frac{M}{M_\Phi}\right)\bar{B}. \qquad (35)$$

They find good agreement between B_{eq} and observation, on average; hence, as $\alpha \simeq 1$, it follows that typically $M \simeq 2M_\Phi$. In particular, this result holds, within the observational uncertainties, for the dark cloud B1, which is discussed at some length in the Chapter by Heiles et al. (see their Table III). Myers and Goodman assume that the observed line width (excluding any thermal contribution) satisfies $\sigma^2 = v_A^2/3$, where v_A is the Alfvén velocity. Direct evaluation shows that

$$\frac{\sigma}{v_A/\sqrt{3}} = \frac{\sqrt{\alpha}}{1.98} \left(\frac{M}{M_\Phi} \right) \tag{36}$$

for a spherical cloud, so that this assumption is satisfied for the typical observed values of α and M/M_Φ. Thus, Myers and Goodman's approximate analysis, which ignored surface terms in the virial theorem and did not treat the magnetic field self-consistently, remains reasonably accurate when these effects are included properly, as we have done here.

Molecular clouds are observed to be at substantially higher pressures than the ambient medium: for example, a cloud with an extinction of 6 magnitudes has a mean pressure of $\sim 2 \times 10^5$ cm^{-3} K according to Eq. (11), far greater than the typical interstellar pressure of 1 to 2×10^4 cm^{-3} K. This is possible only for clouds which are close to the critical mass. For isothermal clouds, $\bar{P}/P_0$ drops from its maximum value of ~ 2.5 at $M/M_J = 1$ to only ~ 1.36 at $M/M_J = 0.5$. Mestel (1985) has emphasized that the same result holds for magnetized clouds: having $\bar{P}$ significantly greater than P_0 requires that M be close to M_{cr}. Reference to the numerical calculations of Tomisaka et al. (1988b) shows that, for the case in which the magnetic and gas pressures are comparable in the ambient medium, the central pressure drops from ~ 30 times the ambient value at $M = M_{cr}$ to only about twice the ambient value at $M = 1/2M_{cr}$. Hence, molecular clouds are almost critical, $M \simeq M_{cr}$. As GMCs are generally not highly flattened, random motions in the gas must also be important in supporting the cloud against gravity. It follows that $M_J \gtrsim M_\Phi$, and hence that $M \simeq M_{cr} \simeq M_J + M_\Phi \gtrsim 2M_\Phi$: GMCs are magnetically supercritical (McKee 1989). We do not expect M to be too much greater than M_Φ, both because the turbulence would then be highly super-Alfvénic (Eq. 36) and subject to rapid dissipation, and because the large amplitude motions would tend to amplify the field. This is consistent with the observational result that $M \simeq 2M_\Phi$ discussed above. Then, as $M \simeq M_{cr}$, we conclude that $M_\Phi \simeq M_J$: molecular clouds appear to be supported about equally by magnetic flux threading the cloud and by MHD turbulence, a result which appears to have been first suggested by Elmegreen (1978a). The existing data, which is discussed in the Chapter by Heiles et al., is consistent with these conclusions, but at present observational uncertainties make it impossible to determine the mass of a given cloud, or its critical mass, to better than a factor 2.

These conclusions need not apply to clumps within GMCs, however; because the clumps are not necessarily highly pressured with respect to their surroundings, they are not necessarily close to critical. Observational data which suggest that clumps obey the same scaling laws as entire molecular clouds (Larson 1981; Myers and Goodman 1988b) are subject to observational selection for small clumps. More complete surveys of low-mass clumps in ^{13}CO show that such clumps in molecular clouds are typically subcritical (Carr 1987; Loren 1989). These subcritical clumps are confined by the pressure in the molecular cloud, not by their self-gravity; they apparently continue to obey the equipartition relation $\sigma \simeq v_A/\sqrt{3}$, however (Bertoldi and McKee 1992). On the other hand, the most massive clumps in GMCs ($M \gtrsim 10^2$ to 10^3 $M_\odot$, depending on the GMC), which contain a significant fraction of the mass, are magnetically supercritical and in gravitational virial equilibrium, like GMCs as a whole (Bertoldi and McKee 1992).

III. MAGNETIC STRUCTURE OF MOLECULAR CLOUDS

A. Static Cloud Models

The usual idealized theoretical picture of a nonrotating, magnetically supported cloud is an oblate spheroid with its minor axis parallel to the magnetic field. The field-line configuration resembles an hourglass, compressed by the cloud and becoming uniform at large distances. Such a model can be imagined as the outcome of an evolutionary process, in which some volume of gas in an initially uniform medium with infinite electrical conductivity is designated the "parent cloud," and then cooled or otherwise driven out of equilibrium so that it contracts under the influence of gravity to another equilibrium state. As indicated in Sec. II, this picture is relevant primarily to clouds or clumps within clouds that are smaller than a few tens to a few hundred solar masses; larger clouds are magnetically supercritical, and turbulence is essential for their support.

If we measure density and magnetic field in units of their initial values, then three parameters and one function characterize this problem. The three parameters are the cloud mass normalized to the Jeans mass (M/M_J), the ratio of gas to magnetic pressure at infinity ($\beta_0 = 8\pi P_0/B_0^2$), and the initial ratio of external to cloud gas pressure. Alternatively, the parent cloud may be considered to be in pressure equilibrium with its surroundings; the cloud would then be spheroidal in shape, and the third parameter is the aspect ratio of the spheroid. The effects of the second and third parameters are contained primarily in the ratio M/M_Φ (see Sec. II). The equilibrium also depends on the function describing the distribution of the mass to flux in the cloud; this is often taken to be that of a uniform field threading a sphere. Although it is simple to sketch this picture, it is rather difficult to quantify it. Some relationships between the magnetic geometry and the degree of flattening can be derived analytically from the tensor virial theorem (Zweibel 1990b). Parker (1973) was the first to construct self-gravitating, magnetized equilibrium models. His method, although self-consistent, did not control the magnetic topology, and the magnetic fields in some of his models were only partially connected to the ambient field. Such clouds could not condense from a medium with a uniform magnetic field if the flux were frozen.

The calculations by Mouschovias (1976a, b) do specify the magnetic topology, and are the realization of the initial value problem described above. Mouschovias found the critical states already described in Sec. II—clouds which are at the maximum possible equilibrium mass given their temperature, magnetic flux, and ambient pressure. His published results include clouds with central to surface density ratio ρ_c/ρ_s up to ~23, and he found the important result that the value of B at cloud center B_c scales as $\rho_c^{1/2}$. Isotropic contraction would have given an exponent of 2/3.

Tomisaka et al. (1988a, b) studied a problem similar to Mouschovias', but they departed somewhat from the initial value formulation by specifying ρ_c/ρ_s instead of the initial pressure contrast as a third parameter. For $M > M_\Phi$, they found second, more centrally condensed models as well as those found by Mouschovias (as we mentioned in Sec. II, the unmagnetized problem has an infinite number of solutions); they showed that the more centrally condensed solutions are almost certainly unstable to collapse. Their published models include both stable and unstable clouds, and have ρ_c/ρ_s as high as 10^4; stable clouds are restricted to $\rho_c/\rho_s < 30$ for $\beta_0 = 1$ and $\rho_c/\rho_s < 10^3$ for $\beta_0 = 0.02$. For large values of ρ_c/ρ_s the density distribution in the cloud approaches $1/r^2$, just as in the case of unmagnetized clouds (Bodenheimer and Sweigert 1968). These calculations have been extended to include toroidal fields by Tomisaka (1991).

Tomisaka et al. (1988b) proposed a modification to Mouschovias' $B_c \propto \rho_c^{1/2}$ scaling law:

$$\frac{B_c}{B_0} \sim \left[1 + \frac{4\pi\rho_c\sigma^2}{B_0^2}\left(1 - \frac{\rho_s}{\rho_c}\right)\right]^{1/2}. \tag{37}$$

If the ambient field B_0 is large enough, the cloud collapse can be highly aligned with the magnetic field, which is then hardly compressed at all. On the other hand, if the magnetic pressure at infinity is much less than the central gas pressure, B_c scales as $\rho_c^{1/2}$, with the constant of proportionality such that the Alfvén and acoustic speeds at the center of the cloud are equal. It should be noted that this scaling applies only to clouds that are approximately magnetically critical ($M \sim M_\Phi$), because it was obtained for magnetized clouds ($M_\Phi \gtrsim M_J$) that must be close to the critical mass ($M \sim M_{cr} \sim M_\Phi$) in order that there be a large pressure drop from the center to the edge of the cloud. The approximate equality of the velocity dispersion σ and the Alfvén velocity v_A then follows from the fact that a cloud with $M \sim M_\Phi$ is supported in the radial direction by the magnetic pressure and in the vertical direction by the thermal pressure; near the center, these pressures must be approximately equal (cf. Umebayashi and Nakano 1990). As a consequence, the central regions of highly condensed clouds are not magnetically dominated, even though the field may be quite strong there. In fact, the models of Tomisaka et al. (1988b) show that for $\beta_0 \lesssim 1$, the field compression is not large: $B_c/B_0 \lesssim 5$ and $\bar{B}/B_0 \lesssim 1.1$.

Cloud models based on the tensor virial theorem, which do not constrain the magnetic topology, show that when $\bar{B}/B_0$ is large, the field lines are not all connected with the asymptotic field (Strittmatter 1966; Zweibel 1990b). This may be understood intuitively by noting that when $\bar{B}/B_0$ is very large, the field lines passing through the cloud must spread out by a large amount in order to connect properly with the field at infinity. The requirement $\nabla \cdot B = 0$ places a geometrical constraint on the degree of spreading, and most of the field lines must then return to the cloud. A distinctive feature of a closed topology is that the mean field would reverse direction, which could be detected through observations of Zeeman splitting.

Observed clouds appear to have values of $\bar{B}/B_0 \sim 10$, which greatly exceed those in any of the numerical models to date. Including the H I envelopes surrounding the molecular clouds cannot significantly reduce the discrepancy because the envelopes do not have a large fraction of the mass for typical star-forming clouds with $\bar{A}_V \gtrsim 6$ (Elmegreen 1985a). For example, the atomic envelope around the Orion molecular cloud has $\sim 30\%$ of the mass of the entire cloud (Chromey et al. 1989). There are several possible explanations for the large observed values of $\bar{B}/B_0$. One possibility is that molecular clouds occur in regions of high interstellar pressure and high magnetic field; 21 cm Zeeman observations could determine if this is the case. Another possibility is that the clouds are not isothermal, but rather have an effective polytropic index corresponding to a specific heat ratio exceeding unity so that the clouds are cooler at the surface, like stars (McKee and Lin 1988); Elmegreen (1989a) has shown that this may indeed give greater compressions. It should be noted that the "specific heat" here refers to the turbulence in the cloud, not to the thermal properties, and as yet there is no theory for its value. Finally, there is the possibility that the field topology is partially closed, because, as just discussed, this can lead to higher internal fields. One way this can occur is by reconnection, which is discussed in Sec. V.C below.

B. Dynamical Effects on Cloud Structure

One of the fundamental questions concerning molecular clouds is the nature of the line widths, which greatly exceed the sound velocity in the gas. Arons and Max (1975) first suggested that these motions are large amplitude Alfvén waves, and this idea has been the basis of much of the subsequent theoretical work (see, e.g., Zweibel and Josafatsson 1983; Falgarone and Puget 1986; Carlberg and Pudritz 1991; Elmegreen 1990b). As described in the Chapter by Heiles et al., observed fields in molecular clouds are strong enough that the Alfvén velocity is comparable to the observed velocity dispersion, thereby confirming Arons and Max's hypothesis. As described in Sec. II, it is possible to include the effects of fluctuating fields on the structure of the cloud in terms of a turbulent pressure (see, e.g., Lizano and Shu 1989). Here we consider hydromagnetic fluctuations explicitly, on both the large scale of individual

GMC's and the small scale corresponding to individual clumps.

On the large scales, one of the most basic questions is the role of the magnetic field in the dissipation of internal cloud motions. The mechanism by which internal random velocities are apparently maintained over several free-fall times remains problematic, and we can ask whether a large-scale magnetic field enhances or reduces the dissipation. Many of the relevant processes have been discussed by Elmegreen (1985b). He considered radiation of the energy of internal cloud motions as Alfvén waves which propagate along the field lines away from the GMC, as well as remote cloud collisions, in which clouds interact via forces transmitted by the magnetic fields which thread them without direct physical contact. Both of these mechanisms increase the dissipation. Two other factors mitigate the dissipation, however. These are the increased elasticity of direct collisions between magnetized clouds and the inhibition of collisions by magnetic tension forces (the latter effect was estimated by Zweibel 1990a). At the present time, the conditions under which magnetic fields either increase or decrease the rate of bulk kinetic energy dissipation are not fully clear and more work is needed on the problem.

The small-scale fluctuation spectrum is coupled to the large-scale spectrum, as was recognized by Falgarone and Puget (1986). These authors computed the internal Alfvénic disturbance induced by grazing clump-clump collisions transverse to the ambient magnetic field. Relatively long wavelength modes (with wavelength comparable to the size of the clump) are generated by this process. These waves either decay directly or steepen into short-wavelength fluctuations. It is not firmly established that the rate of clump-clump collisions is sufficient to explain the nonthermal velocities seen in clumps, and internal sources or transmittal of other disturbances from the interclump medium may also be required. As pointed out by Shu et al. (1987b), the existence of waves in a clump is not sufficient to support it against gravitational collapse: in addition, the energy density of the waves must decrease outward, which could indicate an internal energy source.

Clumps of sufficiently small size cannot sustain a hydromagnetic wave population, and thus cannot be turbulently supported (Mouschovias 1987a; Carlberg and Pudritz 1991). The critical size results from the cutoff wavenumber (or frequency) given by Eq. (52) below, which gives a cutoff clump radius,

$$R_{\text{cut}} = \frac{\pi}{k_{\text{cut}}} = \frac{\pi V_A}{2\nu_{ni}} \simeq 0.20 \left(\frac{B_{1.5}}{K_{i-5}\,n_3} \right) \text{pc} \tag{38}$$

corresponding to a mass

$$M_{\text{cut}} \simeq 1.1 \left(\frac{B_{1.5}^3}{K_{i-5}^3 n_3^2} \right) \, \text{M}_\odot = \left(\frac{2.2 \times 10^{-3}}{K_{i-5}^3} \right) M_B \tag{39}$$

where the magnetic mass M_B is given by Eq. (27), the neutral-ion collision frequency ν_{ni} is given in Eq. (49), and the ionization parameter K_i is defined in Eq. (48). Clumps of lower mass must have essentially negligible pressure associated with turbulent magnetic fields; however, they are well supported by the static magnetic field since $M << M_B$. Cloud cores are observed to have $M \gtrsim M_B$ (e.g., Table III in the Chapber by Heiles et al.) and so are large compared to M_{cut}; nonetheless, since the wavelength $\propto M^{1/3}$, the range of wavelengths which can propagate in such cores is severely restricted (Carlberg and Pudritz 1991).

On length scales less than R_{cut} and less than a gravitational scale height, the thermal pressure along field lines is constant. If the boundaries of the clumps within clouds are sharper than this, then the clump and the interclump medium would have the same thermal pressure. The interclump medium could have a substantial atomic component (Blitz et al., in preparation). If this component of the interclump medium had a temperature similar to that of the intercloud medium of the diffuse interstellar medium ($T \sim 10^4$ K), then its thermal velocity would be larger than the turbulent velocity in the cloud, enabling the thermal pressure of the interclump medium to be constant throughout the molecular cloud and about the same as that in the ambient medium (McKee 1989); the surface layers of the clumps in the cloud would be at the same thermal pressure. The thermal pressure in a typical clump ($n \sim 10^3$ cm^{-3}, $T \sim 10$ K) is indeed comparable to that in the diffuse interstellar medium; cloud cores, being self-gravitating, have higher pressures.

If turbulence is the primary source of support against gravitational collapse, the Jeans criterion can be modified substantially. It was first pointed out by Chandrasekhar (1951b) that the stability of a region of some given size L will be determined by the turbulent energy on scales less than L. This raises the intriguing possibility that shorter scales can be less stable than longer scales. Bonazzola et al. (1987) and Pudritz (1990) have elaborated upon this basic idea. (However, note that these treatments of the gravitational stability of turbulent media have not allowed for the clumpiness of molecular clouds, which renders the mean density, and hence the effect of gravity, substantially less on large scales than on the scale of individual clumps.) The gravitational stability properties of the medium, in this picture, depend upon the turbulence spectrum, i.e., the relative distribution of power at different scales. At present, there is no *ab initio* calculation of this spectrum, but the cutoff at short wavelength clearly depends upon the damping, just as for the equilibria discussed above. In the first numerical calculation of the effects of a given spectrum of hydromagnetic waves on an initially isothermal cloud, Carlberg and Pudritz (1991) have found that the cloud undergoes a global, isotropic collapse at about 1/4 the free-fall rate.

IV. ROTATION AND MAGNETIC BRAKING

A. Rotation of Molecular Clouds

Magnetic fields couple molecular clouds to the rest of the interstellar medium and thereby regulate the rotation of the clouds. How important is rotation to the structure of the clouds? Stahler (1983b) studied the equilibria of rotating, isothermal clouds with an angular momentum distribution corresponding to that of a uniform sphere. Kiguchi et al. (1987) studied a wider range of conditions than Stahler, but, as they pointed out, nonaxisymmetric instabilities are likely to disrupt the clouds rotating more rapidly than those considered by Stahler. Tomisaka et al. (1989) have provided a simple interpretation of the results of Kiguchi et al.: just as M_J represents the maximum mass that can be supported by gas pressure and M_Φ represents the maximum mass that can be supported by magnetic fields, so

$$M_{\rm rot} = 5.1 \left(\frac{\sigma J}{GM} \right) \tag{40}$$

where J is the angular momentum of the cloud, represents the maximum mass that can be supported by rotation. In contrast to M_Φ, which can take on arbitrary values relative to M_J, there is a limit $M_{\rm rot}/M_J \lesssim 3.6$ in Stahler's models; in other words, some thermal support is essential for nonmagnetized rotating clouds in order to avoid instability. Tomisaka et al. have shown how to include the effect of magnetic fields on the critical mass: with the approximation (Eq. 28), their result for the critical mass for magnetized, rotating clouds becomes

$$M_{\rm cr} \simeq \left[(M_J + M_\Phi)^2 + M_{\rm rot}^2 \right]^{1/2} . \tag{41}$$

This simple approximation agrees with the numerical calculations of Tomisaka et al. (1989, Table 4) to within ~5%.

We can characterize the rotation of a cloud by the parameter

$$\alpha_\Omega \equiv \frac{\Omega^2 R^3}{GM} = 231 \frac{\Omega_{\rm ksp}^2 R_{\rm pc}^3}{M/M_\odot} = 1.61 \frac{\Omega_{\rm ksp}^2}{\bar{n}_3} \tag{42}$$

where $\Omega = J/I$ is the mean angular velocity of the cloud and $\Omega_{\rm ksp}$ is in units of km s^{-1} pc^{-1}. This parameter measures the ratio of the centrifugal force to the gravitational force, and is similar to the rotation parameter α used by Mestel and Paris (1984); for an oblate spheroid with an eccentricity <0.7, the two quantities differ by $<20\%$. The rotational mass can be expressed in terms of α_Ω by $M_{\rm rot} \simeq 0.64(\alpha_\Omega\alpha)^{1/2}M$, where we have used Stahler's (1983b) results to evaluate the moment of inertia at the maximum stable mass. Based on Stahler's models, we find that a sufficient condition for clouds to be primarily supported by gas pressure rather than rotation (i.e., $M_{\rm rot} < M_J$) is $\alpha_\Omega < 0.5$. The rotational kinetic energy $\mathcal{T}_{\rm rot}$ is directly related to α_Ω: Stahler's (1983b) results show that $\alpha_\Omega \simeq (3 \sim 6)\mathcal{T}_{\rm rot}/|\mathcal{W}|$, where the factor 3 is appropriate for a rigidly rotating sphere. The parameter α_Ω is also useful for characterizing the disk formed by a cloud which collapses while conserving its angular momentum: if the cloud begins at a radius R_0 with a rotation parameter $\alpha_{\Omega 0} << 1$, then, as $\alpha_{\Omega} \propto J^2/R$, it will form a disk with a radius $R_c \simeq \alpha_{\Omega 0} R_0$ (cf. Terebey et al. 1984).

Remarkably enough, the importance of rotation in observed molecular clouds remains somewhat unclear. Based on CO observations, Goldsmith and Arquilla (1985) have reported that the line widths of rotating clouds do not differ significantly from those of non-rotating clouds; we interpret this as implying $\Omega D < \Delta v$, where $\Delta v = (8 \ln 2)^{1/2} \sigma$ is the FWHM of the line and $D = 2R$ is the cloud diameter. As

$$\frac{\alpha_{\Omega}}{\alpha} = \frac{\Omega^2 R^2}{5\sigma^2} = 0.28 \left(\frac{\Omega D}{\Delta v} \right)^2 \tag{43}$$

and as $\alpha \sim 1$ (Sec. II), we infer that $\alpha_{\Omega} \lesssim 1/4$, so that rotation is not that important in supporting the clouds in their sample. This conclusion is consistent with that of Goldsmith and Arquilla. In fact, they point out that their sample is biased towards clouds which show rotation, so the mean rotation of all clouds could be smaller yet. On the other hand, evaluation of α_{Ω} from their mean value, $\Omega_{\text{ksp}} = 0.64 \bar{n}_3^{0.4}$, gives $\alpha_{\Omega} \simeq 0.65 \bar{n}_3^{-0.2} \sim 1$, which would suggest that rotation is essential. Arquilla and Goldsmith (1986) made a detailed study of eight clouds, six with reported rotation. Five of the eight clouds in this biased sample showed $T_{\text{rot}}/|\mathcal{W}| \gtrsim 0.3$, corresponding to $\alpha_{\Omega} \gtrsim 1$, so that rotation is important in the structure of these clouds. Although the numerical evaluation of α_{Ω} is uncertain, it appears that rotation is significant in at least some clouds. Rotation is observed to be unimportant in the cloud B1, which is extensively discussed in the Chapter by Heiles et al. Furthermore, recent observations of cloud cores in NH_3 by Goodman et al. (1991a) confirm that rotation is not important ($\alpha_{\Omega} << 1$) for these objects, at least. Observations are needed to determine the fraction of clouds that are rotationally supported, and the conditions under which this occurs.

B. Magnetic Braking

The role of magnetic fields in regulating the rotation of clouds seems to have first been studied by Ebert et al. (1960), and subsequently by Mestel (1965b); it has been reviewed recently by Mestel and Paris (1984), Mestel (1985), and Mouschovias (1987a, b). The basic process is quite simple: if the rotation of the cloud differs from that of the ambient medium, torsional Alfvén waves are launched into the ambient medium (and into the interior of the cloud) which transport angular momentum and tend to bring the cloud into corotation with the ambient medium. The characteristic time scale for this magnetic braking, t_b, is basically the time for the outwardly propagating Alfvén waves to engulf an amount of ambient material with a moment of inertia equal to that of the cloud. The braking time thus depends on the geometry of the field, because the rate at which the field spreads out determines the rate of increase of the moment of inertia inside the wavefront. The simplest case is that of a cloud embedded in a uniform magnetic field $\mathbf{B_0}$ with its angular momentum parallel to $\mathbf{B_0}$ (Ebert et al. 1960; Mouschovias and Paleologou 1980; Mestel and Paris 1984). For a spherical cloud, the braking time is

$$t_b = \frac{8}{15} \left(\frac{\bar{\rho}}{\rho_0} \right) \frac{R}{v_{A0}} \simeq 0.3 \frac{M}{M_\Phi} \left(\frac{\bar{\rho}}{\rho_0} \right)^{1/2} t_{ff} \qquad (44)$$

where v_{A0} is the Alfvén velocity of the ambient medium, etc. If the cloud is cylindrical rather than spherical, the result is the same, except that the factor $(8/15)R$ is replaced by Z, the half-height of the cylinder (Mouschovias and Paleologou 1980). This result shows that magnetically subcritical clouds ($M/M_\Phi < 1$) which are not very dense are braked on an extremely short time scale. Magnetically supercritical clouds that are substantially denser than their surroundings ($\bar{\rho} \gtrsim 10\rho_0$) would be braked slowly enough that rotation could inhibit collapse, but the assumption of a uniform field is less likely to be valid in this case.

Clouds are observed to have fields substantially larger than the ambient one, suggesting that (if the cloud field is connected to the ambient field) a radial field in the ambient medium is a better approximation than a uniform one. Gillis et al. (1974,1979) and Mestel and Paris (1979,1984) have shown that in this case the cloud is coupled to a larger volume of ambient gas and the braking time is somewhat shorter; in our notation, their result is

$$t_b \simeq 0.2 \frac{M}{M_\Phi} \left(\frac{\bar{\rho}}{\rho_0} \right)^{1/10} t_{ff}. \qquad (45)$$

Thus, if the field in the cloud substantially exceeds the ambient one, the braking time is almost entirely dependent on the ratio of the mass to the magnetic critical mass, M/M_Φ, and almost independent of the density contrast. The numerical coefficient in Eq. (45) is several times smaller than that found by Mestel and Paris (1984) because we have used a more accurate value for M_Φ. This result implies that prior to the onset of collapse, rotation should be unimportant in clouds. If the cloud undergoes gravitational contraction at a velocity of order of or faster than the Alfvén velocity, this expression breaks down because the Alfvén waves responsible for braking the cloud become trapped (Mouschovias 1989). The cloud will then evolve to a state in which centrifugal support is important and the evolution is governed by magnetic braking (see, e.g., Mestel and Paris 1984).

The results in Eqs. (44) and (45) are both based on the assumption that the angular momentum is parallel to the field; in the opposite case in which the two vectors are orthogonal, the braking time is shorter (Mouschovias and Paleogolou 1979), although the difference from the parallel case is small in the case of a spherical cloud with a compressed field (Mestel and Paris 1984). All the calculations are based on flux freezing (i.e., $t_b \ll t_{AD}$), an assumption verified by Mouschovias and Paleologou (1986).

The numerical calculations carried out to date are consistent with the analytic results just described. Dorfi (1982) followed the evolution of a magnetized, rotating cloud with $M \simeq M_\Phi$ in three dimensions, and he found that the braking time for the case with **J** parallel to $\mathbf{B_0}$ is $t_b \simeq 0.5 t_{ff}$, consistent with Eq. (44). For the case in which **J** is perpendicular to $\mathbf{B_0}$, he found that the braking time is a good deal shorter, $t_b \lesssim 0.1 t_{ff}$, consistent with the estimate of Mouschovias and Paleologou (1979). He assumed flux freezing, but found substantial numerical diffusion of the field in his calculations. Tomisaka et al. (1990) have carried out calculations which include ambipolar diffusion, but are restricted to quasi-static flows with axisymmetry; because of these restrictions, their calculations are numerically more robust than those of Dorfi (1982). They considered the case in which the ambient medium had a density small compared to that of the cloud, but a field of the same magnitude. The assumption of an approximately uniform field implies that their calculation corresponds to Eq. (44); they confirmed the validity of this equation, as did Dorfi. On the other hand, the assumption of a low external density made the braking time quite long, comparable to the ambipolar diffusion time. It would be worthwhile to carry out such a calculation for the more realistic case in which there is a contrast in field strength as well as density between the cloud and intercloud medium.

How important then is rotation in molecular clouds? Observations are ambiguous, although they suggest that rotation is generally not very important. Theory is less ambiguous provided that the field topology is an open one; it predicts a braking time that is comparable to or less than the free-fall time t_{ff}, depending on the rate at which the field diverges from the cloud and on the density contrast with the ambient medium (Eqs. 44,45). Magnetically subcritical clouds evolve on the ambipolar diffusion time, which is much greater than t_{ff} (cf. Eq. 51 below), so such clouds should be slow rotators. For a dense, magnetically supercritical cloud on a cylindrical flux tube, Eq. (44) predicts that the contraction of the cloud would be limited by the rate of braking, so the cloud would be in approximate virial equilibrium. On the other hand, the field in molecular clouds is observed to be about an order of magnitude greater than that in the diffuse ISM, and the field in the gas collapsing onto a protostar is predicted to be greater still. In these cases, the magnetic braking should be sufficiently efficient that the angular momentum is transported outward at about the same rate that collapse occurs. Hence, a dynamical calculation is needed to determine the final angular momentum, which in turn determines the size of the protostellar accretion disk (Terebey et al. 1984) and whether a binary star forms (Mouschovias 1977). In principle, such a calculation should allow for the possibility of magnetic reconnection (Sec. V.C); existing calculations assume that the field in the cloud is fully connected with the ambient field, but the braking time probably would be greater if the field were partially closed (see Kulsrud 1971). Observationally, it would be worthwhile to study the topology of the field of clouds which appear to be relatively rapid rotators to see if there is any evidence for closed field lines.

V. THE PHYSICS OF AMBIPOLAR DIFFUSION AND FLUX REDISTRIBUTION

A. Time Scales for Ambipolar Diffusion

To this point, we have assumed that the magnetic field is frozen to the gas, which is generally a good approximation. In fact, under interstellar conditions the field is frozen only to the small number of ionized particles in the gas, and it can act on neutral particles only indirectly through collisions. As a result, the field slips through the neutrals at a velocity v_D, exerting a force per unit volume $\rho_n \nu_{ni} v_D$ on the neutrals, where ρ_n is the density of the neutral particles and ν_{ni} is the neutral-ion collision frequency. This process is often referred to as "ambipolar diffusion" (Spitzer 1978). Neglecting the gas pressure for simplicity, we have for the equations of motion for the neutrals and charged particles

$$\rho_n \frac{d\mathbf{v}_n}{dt} = \rho_n \nu_{ni} \mathbf{v}_D + \rho_n \mathbf{g} \tag{46}$$

$$\rho_i \frac{d\mathbf{v}_i}{dt} = -\rho_n \nu_{ni} \mathbf{v}_D + \frac{(\nabla \times \mathbf{B}) \times \mathbf{B}}{4\pi} \simeq 0. \tag{47}$$

The low ion density renders the charged particle inertia negligible unless the frequency of the motion is so high that the charged particles are uncoupled from the neutrals; we do not consider such high frequencies here. In our applications, the ionization is sufficiently low that we can set the neutral density equal to the total density, $\rho_n \simeq \rho$.

Equations (46) and (47) imply that the drift velocity v_D varies inversely with the collision frequency $\nu_{ni} = n_i \langle \sigma v \rangle$. Numerically, the collision rate coefficient $\langle \sigma v \rangle \simeq 1.5 \times 10^{-9}$ cm^3 s^{-1} for molecular gas of cosmic abundances (Nakano 1984). At high densities ($n \gtrsim 10^8$ cm^{-3}), grain drag increases this somewhat (Elmegreen 1979b; Nakano 1984), and at high velocities ($v_D \gtrsim 10$ km s^{-1}), the cross section becomes geometric and $\nu_{ni} \propto v_D$ (Mouschovias and Paleologou 1981). The ionization of the gas is determined by balancing the rate at which the ions are formed, $\zeta_H n_H$, with the rate at which they are destroyed by recombination, which is generally proportional to n_i^2. As a result, the ionization can be expressed as

$$x_i \equiv \frac{n_i}{n_H} \equiv \frac{K_i}{n_H^{1/2}} \tag{48}$$

where K_i is a constant. If the ionization is due to cosmic rays, then $\zeta_H \simeq 10^{-17}$ s^{-1} (Spitzer 1978). For this value of the cosmic-ray ionization rate, and assuming that all grains have a size 0.15 μm, Elmegreen (1979b) found $K_i \simeq 10^{-5}$ cm$^{-3/2}$; in general, $K_i \propto \zeta_H^{1/2}$. More recently, McKee (1989) found $K_{i-5} \equiv K_i/(10^{-5}$ cm$^{-3/2}) \simeq 0.5$ for an MRN grain distribution (Mathis et al. 1977) and for densities high enough that an ion such as HCO$^+$, rather than H$_3^+$, is the dominant molecular ion ($n_H \gtrsim 10^3$ cm^{-3}). Lepp and Dalgarno (1988) find that the inclusion of a substantial concentration of large molecules such as polycyclic aromatic hydrocarbons (PAHs) increases the ionization by up to a factor of ~ 2. The presence of large molecules or small grains implies that the $n_H^{-1/2}$ scaling for the ionization will persist to higher densities than found by Elmegreen (1979b). Altogether, the value $K_{i-5} \simeq 1$ appears to be a fairly robust estimate for the ionization in dense molecular gas ionized by cosmic rays. The collision time is then

$$\tau_{ni} \equiv \frac{1}{\nu_{ni}} = \frac{6.7 \times 10^4}{K_{i-5} n_3^{1/2}} \text{ yr} \tag{49}$$

provided $v_D \lesssim 10$ km s^{-1}. It should be noted, however, that much of the molecular gas in the Galaxy is photoionized by far-ultraviolet radiation, and for such gas the ionization is higher and K_i is no longer constant (McKee 1989). Models of photodissociation regions (Tielens and Hollenbach 1985) show that the ionization of gas in which the carbon is locked up in CO is due primarily to S$^+$ in regions of relatively low extinction, with $x_i \sim 10^{-5}$; for typical densities $n_H \sim 10^3$ cm^{-3}, this gives $K_{i-5} \simeq 30$. At somewhat higher extinctions, the ionization is due to Mg$^+$ and Fe$^+$, with $x_i \sim 10^{-6}$; this corresponds to $K_{i-5} \simeq 3$ at the typical density. Cosmic-ray ionization takes over for $A_V \gtrsim 4$ in the typical interstellar radiation field.

1. Ambipolar Diffusion Under the Influence of Gravity. Consider a magnetized cloud held together by its self-gravity. Because the neutrals do not directly feel the effects of the magnetic field, they fall at a rate limited by the friction with the charged particles, which are tied to the field. Under the assumption that the contraction is slow, the inertia is negligible and the drift velocity is simply $v_D = g\tau_{ni}$ from Eq. (46). For an oblate spheroid of semiminor axis Z parallel to the ambient field, the gravitational acceleration is $g \simeq GM/R^2 = (4\pi/3)G\rho Z$. The ambipolar diffusion time for the core of the cloud,

$$t_{AD} = \frac{Z}{v_D} = \frac{3}{4\pi G\rho\tau_{ni}} = \frac{3\langle\sigma v\rangle}{4\pi G\mu_H} x_i \tag{50}$$

is directly proportional to the ionization; here $\mu_H \equiv \rho/n_H = 2.34 \times 10^{-24}$ g is the mass per hydrogen nucleus for a gas of cosmic abundances. Evaluation of Eq. (50) gives $t_{AD} = 7.3 \times 10^{13} x_i$ yr. This is quite close to the typical value for the time to core collapse found by Nakano (1984) in his numerical calculations, $\sim 8 \times 10^{13} x_i$ yr. Because the time for the entire cloud to undergo ambipolar diffusion is longer by a factor R/Z, which is typically of order 2 for Mouschovias's (1976b) clouds, a more characteristic value for the ambipolar diffusion time for the entire cloud, not just the core, is $1.6 \times 10^{14} x_i$ yr (McKee 1989).

Because the ionization is proportional to $n^{-1/2}$, Eq. (50) implies that the ratio of the ambipolar diffusion time to the free-fall time t_{ff} is a constant (see, e.g., Mouschovias 1987a):

$$\frac{t_{AD}}{t_{ff}} = \left(\frac{6}{\pi^3 G\mu_H}\right)^{1/2} \langle\sigma v\rangle K_i \simeq 17 K_{i-5}. \tag{51}$$

The fact that for typical interstellar conditions the ambipolar diffusion time is long compared to the free-fall time is extremely important, because it ensures that magnetically subcritical clumps do indeed evolve quasi-statically. Had we adopted a cylindrical or slab geometry, we would have obtained the same result but with the coefficient of K_{i-5} reduced by a factor of ~ 1.5 and 2, respectively.

2. *Ambipolar Diffusion in Hydromagnetic Waves.* Ion-neutral friction is an important damping mechanism for hydromagnetic waves. The propagation of these waves in the absence of thermal pressure or gravity has been described by Kulsrud and Pearce (1969), while Ferriere et al. (1988) have included the effects of pressure, and Langer (1978) and Pudritz (1990) have considered the effects of gravity. The basic physics can be extracted from Eqs. (46) and (47) (with the gravitational term omitted). At sufficiently low frequencies or low wavenumbers, the charged and neutral components are well coupled and the wave speed (the Alfvén velocity), $v_A = B/(4\pi\rho)^{1/2}$, is determined by the mass density of the entire medium. Equation (46) then yields the drift velocity $v_D = \omega\delta v/v_{ni}$ in terms of the wave amplitude δv. Now the energy density of the wave, including both magnetic and kinetic energy, is $\epsilon = \rho\delta v^2$, and the rate of energy dissipation is $\dot\epsilon = \rho v_{ni}v_D^2 = \rho\omega^2\delta v^2/v_{ni}$. In terms of the damping rate for the amplitude of the wave Γ, we have $2\Gamma\epsilon = \dot\epsilon$, which implies

$$\Gamma = \frac{\omega^2}{2v_{ni}}.$$

(52)

This heuristic argument tacitly assumes that $\Gamma << \omega$. According to Eq. (52), the waves are critically damped (i.e., $\Gamma = \omega$) at $\omega = 2v_{ni}$. This frequency corresponds to a wavenumber $k = 2v_{ni}/v_A \equiv k_{cut}$, above which we expect very little power. A more precise calculation (Kulsrud and Pearce 1969) shows that the real part of the frequency actually vanishes at $k = k_{cut}$. On the other hand, if the waves are emitted from a source, such as a collision between two clumps (Sec. III.B), then the frequency is real and the wavenumber complex; the damping length is about a wavelength at $k = k_{cut}$.

Hydromagnetic waves in which δB^2 varies on the scale of the wavelength (i.e., linearly polarized or unpolarized waves, not circularly polarized waves) steepen as they propagate, and this leads to nonlinear damping at a rate

$$\Gamma_{nl} \simeq k v_A \left(\frac{\delta B}{B}\right)^{q_d}$$

(53)

where $q_d = 1$ for magnetosonic waves propagating perpendicular to the magnetic field, and $q_d = 2$ for Alfvén waves propagating along the field (Cohen and Kulsrud 1974; Zweibel and Josafatsson 1983). Nonlinear damping exceeds ion-neutral damping for

$$k < k_{nl} \equiv k_{cut}\left(\frac{\delta B}{B}\right)^{q_d} = \frac{2v_{ni}}{v_A}\left(\frac{\delta B}{B}\right)^{q_d}.$$

(54)

Thus, large amplitude waves evolve by steepening until ion-neutral collisions damp them when the wavenumber has increased to of order the cutoff
wavenumber k_{cut}.

B. Ambipolar Drift as a Heating Process

Frictional heating is an inevitable consequence of ambipolar diffusion, and
occurs in both steady and fluctuating magnetic fields. Recall that the magnetic
force density $\mathbf{J} \times \mathbf{B}/c$ is the sum of the magnetic pressure gradient $-\nabla B^2/8\pi$
and the magnetic tension $\mathbf{B} \cdot \nabla \mathbf{B}/4\pi$. If we define the characteristic length
scale for variations in the magnetic field, ℓ_B, by $|\mathbf{J} \times \mathbf{B}|/c \equiv B^2/(4\pi\ell_B)$,
then the drift velocity becomes

$$v_D = \frac{B^2}{4\pi\ell_B\rho v_{ni}} = \frac{v_A^2}{\ell_B v_{ni}} \tag{55}$$

from Eq. (47), and the heating rate is

$$\dot{\epsilon} = \rho v_{ni} v_D^2 = \frac{B^4}{(4\pi\ell_B)^2 \rho v_{ni}}. \tag{56}$$

Equation (56) is provocative but difficult to evaluate, because of the sensitivity
to the length scale ℓ_B, which is very difficult to measure, and to B.

Scalo (1977) was the first to consider ambipolar diffusion heating of
clouds by the mean magnetic field. He took ℓ_B to be about equal to the cloud
size R, and he then calculated the equilibrium temperature as a function of
density assuming $R \propto n^{-1/3}$ and $B \simeq (1.5\mu G)(0.6n_{\text{H}})^y$. The resulting cloud
temperatures were too high to be consistent with $y = 2/3$ but marginally
consistent with $y = 1/2$ (corresponding to field strengths of 110 μG and
37 μG at $n_{\text{H}} = 10^3$ cm^{-3}, respectively). In our notation, the heating rate in a
static magnetic field under Scalo's assumption that $\ell_B = R$ is

$$\dot{\epsilon}_s = 2.6 \times 10^{-25} \left(\frac{B_{1.5}}{M/M_\Phi}\right)^2 \frac{n_3^{1/2}}{K_{i-5}} \quad \text{erg cm}^{-3} \text{ s}^{-1} \tag{57}$$

where we have used Eqs. (22) and (49). This estimate is quite crude: in
order that $\ell_B \sim R$, it is necessary for the mass M to be close to the critical
mass M_{cr}; on the other hand, calculations show that clouds become highly
centrally concentrated as $M \to M_{\text{cr}}$ (see, e.g., Tomisaka et al. 1988b), thereby
violating the assumption that the density and field are approximately uniform.
If the parameters entering Eq. (57) are set equal to unity, this heating rate is
comparable to the cosmic-ray heating rate estimated by Goldsmith and Langer
(1978), which is $3.2 \times 10^{-25} n_3$ erg cm^{-3} s^{-1}.

Lizano and Shu (1987) have gone beyond the crude gradient length scale
parameterization by calculating the heating rates in models of self-gravitating,
magnetically supported slabs (Shu 1983). In the case studied by Lizano and

Shu, the pressure is assumed to be provided by both a static magnetic field and by Alfvén waves with an energy density that is a fixed fraction of the mean magnetic energy density. Under these conditions, the field is largest in the midplane, whereas the field gradients are steepest in the outer parts of the slab. The results of Lizano and Shu show that ambipolar diffusion heating is most important in models with large field strengths, but that in any given model the heating increases with distance from the midplane, so much so as to produce a temperature inversion. It is not clear that this inversion would persist in a spheroidal cloud, however.

Hydromagnetic wave damping also heats the gas (Elmegreen et al. 1978; Zweibel and Josafatsson 1983). Noting that $v_D = \omega \delta v / v_{ni}$ from Eq. (46), we find that the heating due to waves of wavenumber k is

$$\dot{\epsilon}_w = \frac{2}{3} \rho \delta v^2 \left(\frac{k^2 v_A^2}{v_{ni}} \right) = \frac{2}{3} \left(\frac{\delta v}{v_A} \right)^2 (kR)^2 \dot{\epsilon}_s \qquad (58)$$

where we have assumed that the turbulence is isotropic and we have neglected gravity; only the components of $\delta \mathbf{v}$ perpendicular to $\mathbf{B}$ contribute to the heating. Wave heating is more effective than the static heating considered by Scalo if the wavelength is small and the amplitude of the waves large. As Elmegreen et al. (1978) point out, the heating rate will increase if the gas is compressed, as, for example, by a shock, because the compression increases both B and k. As an example of wave heating, consider the heating due to a spectrum of waves extending from k_{min} to k_{max} with $(\delta v^2)_k = \delta v^2 (k_{min}/k^2)$; this is consistent with Larson's (1981) relation that the line width on a scale of R is proportional to $R^{1/2}$. One can readily show that the heating rate is given by Eq. (58) with k^2 replaced by $k_{min} k_{max}$. For a minimum wavenumber given by $k_{min} R = 1$, and a maximum wavenumber corresponding to a wave with a damping length equal to the cloud diameter, $k_{max} = (v_{ni}/R v_A)^{1/2}$, one finds that the heating rate is a little more than twice the static value given in Eq. (57). As Goldsmith and Langer (1978) show, cosmic ray heating is of the right order of magnitude to account for the observed temperatures of dark clouds, and it appears that wave heating is of comparable importance. Carlberg and Pudritz (1991) have come to a similar conclusion. Ferriere et al. (1988) have shown that hydromagnetic wave damping is an important heating mechanism in the intercloud gas in the ISM, which has a temperature of order 10^4 K, and it is possible that this process can heat the interclump gas in molecular clouds to a similar temperature (McKee 1989).

C. Effects of Ohmic Dissipation and Magnetic Reconnection

Ambipolar diffusion preserves the topology of the magnetic field, because the field lines are frozen to the charged component of the medium. Yet, topological change is an essential aspect of star formation: it must occur at some point between the time a molecular cloud forms with a field that is well connected with that of the diffuse interstellar medium, and the time a star forms out of the cloud with a field that is substantially detached from the interstellar field. At present, we have no direct observational evidence as to when this reconnection occurs, which poses a substantial observational problem for the future.

Reconnection can occur only if the medium is resistive. A lower limit to the rate of reconnection is set by pure ohmic diffusion, which occurs on a characteristic time scale $t_{OD} \equiv 4\pi \ell_B^2/\eta c^2$, where ℓ_B is the characteristic length scale over which the field reverses direction and η is the resistivity. In molecular clouds, η is determined by electron collisions with neutral molecules, $\eta \sim \langle \sigma v \rangle_{en} m_e/e^2 x_i$. With the ionization fraction x_i given by Eq. (48), we have, in years

$$t_{OD} \sim \frac{5.9 \times 10^{20} \ell_{B,18}^2}{(n_H T)^{\frac{1}{2}}} \qquad (59)$$

which is extremely long at normal molecular cloud densities. In order for this process to be competitive with ambipolar diffusion, the density must be very high, $\gtrsim 10^{12}$ cm^{-3} (Umebayashi and Nakano 1990). In reality, dynamics plays a crucial role in reconnection, and the rate is intermediate between the diffusion rate t_{OD}^{-1} and the dynamical rate $t_{Alf}^{-1} \equiv (\ell_B/v_A)^{-1}$. Reconnection may occur spontaneously in a sheared magnetic field, due to instability, or may be externally driven by a flow. If reconnection is initiated by magnetic tearing instabilities (Furth et al. 1963) when the field becomes sufficiently sheared, then the growth time t_{rec} of these instabilities is roughly proportional to the geometric mean of the Alfvén crossing time t_{Alf} and the ohmic diffusion time t_{OD}, where in this case the length scale ℓ_B is the characteristic size of the shear layer. Rapid tearing requires very short gradient length scales for the magnetic field. Such scales may develop naturally through the interaction of the field and motions of the gas, even if the structure in the flow is on much larger scales (Moffatt 1978; Parker 1979). Hewitt et al. (1989) have shown that if ℓ_B becomes less than a mean free path, then the reconnection proceeds extremely efficiently.

It is possible that GMCs with large magnetic fields ($\bar{B} >> B_0$) already have undergone a significant amount of reconnection, as the concentration of the field $\bar{B}/B_0$ can be much greater in magnetically disconnected clumps (Strittmatter 1966; Zweibel 1990b). These authors modeled the cloud as a homogeneous, isothermal body of gas with a uniform internal field $\bar{B}$ (in our notation) immersed in a medium with a field $q\bar{B}$; for the case in which the temperature of the gas is negligible, so that $M_{cr} = M_\Phi$, their results imply that the coefficient in the magnetic critical mass (Eq. 23) is

$$c_\Phi = \frac{\sqrt{10}}{3\pi^2} \max \left[(1-q)(1+3q)\right]^{1/2} = 0.123 \qquad (60)$$

in close agreement with the result of Mouschovias and Spitzer (1976) and Tomisaka et al. (1988b). (The maximum occurs at $q = 1/3$, which is significantly smaller than the values of q in the numerical models, however.) They showed that for q less than some critical value q_c, the field develops a partially closed topology. For spherical clouds, $q_c = 1/3$; for a flattened disk, $q_c \rightarrow 1$, so that a flattened cloud with $q = 1/3$ would have a substantially closed topology. For fixed q, the coefficient c_Φ is given by Eq. (60) without the "max." This suggests that a cloud with a substantially closed field topology has a magnetic critical mass close to that for a connected topology, although this should be verified by accurate numerical calculation. This conclusion is consistent with the result of Tomisaka et al. (1988b) that it is primarily the central-flux-to-mass ratio that determines the stability of the cloud.

In a prescient paper, Mestel and Strittmatter (1967) considered the resistive relaxation of a magnetic field which had been drawn in by rapid collapse (Mestel 1966a). The field evolves from fully connected to partially connected, and eventually returns to its initial uniform configuration. As they pointed out, the resistivity in interstellar clouds is too low for this process to be of importance there (their work was done prior to the discovery of molecular clouds, so they considered atomic clouds). It is now known that in the later stages of gravitational collapse, the resistivity becomes sufficiently large that the ohmic diffusion time becomes shorter than the dynamical time (Umebayashi and Nakano 1990), making their work of greater relevance than they foresaw.

Reconnection may also play a role in the internal structure of a molecular cloud. If the magnetic field of a GMC is weaker than the equipartion value of $(12\pi \bar{\rho}\sigma^2)^{1/2}$, and if the clumps which dominate the inertia are not so small that ambipolar diffusion rapidly removes the flux, then it appears inevitable that the field becomes tangled by the motions of the gas and thereby subject to reconnection. For example, Clifford and Elmegreen (1983) show sketches of misaligned magnetic flux tubes brought into proximity by the motion of the clumps to which they are attached. Magnetic reconnection in the interclump medium would have several important consequences: the field could be kept fairly smooth instead of progressively more tangled; the field would gradually become detached from the mean interstellar field; once a clump had a disconnected field, it would interact with the field of the rest of the cloud much more weakly; and magnetic braking, in which angular momentum is carried away along the interclump field lines (see Sec. IV.B above), would proceed quite differently (Kulsrud 1971). Reconnection could also lead to heating and particle acceleration. It is clear that magnetic reconnection remains one of the major problems of plasma astrophysics, and is crucial to our understanding of the evolution of magnetic fields in star-forming regions.

VI. AMBIPOLAR DIFFUSION AND FLUX REDISTRIBUTION: ROLE IN STAR FORMATION

As discussed in the Sec. I, magnetic fields are intrinsic to two of the classical problems of star formation, the magnetic flux problem and the angular momentum problem. The discussion in the Chapter by Heiles et al. and in Secs.

II–V has demonstrated the importance of magnetic fields in the structure of the molecular clouds out of which stars are born. However, once the stars have formed, magnetic fields are no longer energetically significant. To see this quantitatively, note that Eq. (25) indicates that the ratio of the mass to the magnetic critical mass, M/M_Φ, is typically within an order of magnitude of unity for either diffuse atomic clouds with fields of order 3 μG, or molecular clouds with fields of order 30 μG. A star like the Sun, however, has a magnetic critical mass $M_\Phi = 0.12\pi R^2 \bar{B}/G^{1/2} \simeq 3 \times 10^{-6}$ M$_\odot$, even if the mean field is 10^3 G. Thus, for such a star $M/M_\Phi \sim 10^{5.5}$, virtually all the flux in the initial molecular cloud has been lost, and the magnetic energy is negligible. There is general agreement that ambipolar diffusion is a significant mechanism for flux loss (Mestel and Spitzer 1956); reconnection may also be important (Mestel 1965b; Mestel and Strittmatter 1967). The question is, when does the flux loss occur—in the interstellar or in the protostellar stage? The answer to this question is essential if one is to solve the angular momentum problem, as it determines the efficiency with which the fields can remove the angular momentum of the contracting gas through magnetic braking.

A. Magnetically Subcritical vs Magnetically Supercritical Star Formation

For a frozen-in field, magnetically subcritical clouds (those with $M < M_\Phi$) can never undergo gravitational collapse, whereas magnetically supercritical clouds ($M > M_\Phi$) can never be prevented from collapse by the magnetic field (Sec. II). The evolution of these types of clouds is therefore quite different (Mestel 1985): magnetically subcritical clouds evolve on an ambipolar diffusion time scale, which is generally quite long (Sec. V.A), whereas magnetically supercritical clouds evolve on a dynamical time scale set by the rate of dissipation of the internal turbulence and the rate of loss of angular momentum. This distinction has led Shu et al. (1987b) to suggest that magnetically subcritical clouds are associated with low-mass star formation, which is observed to occur at a relatively low efficiency, whereas magnetically supercritical clouds form stars with a relatively high efficiency, and may preferentially form high-mass stars.

This distinction can be refined somewhat. Magnetically subcritical clumps evolve via ambipolar diffusion until their cores become magnetically supercritical. The subsequent "inside-out" collapse (Shu et al. 1987b) is supercritical, but it occurs around a well-defined core. As the core will have some angular momentum, it will form a disk upon collapse (Terebey et al. 1984) at the same time the star is forming. However, the collapse of a supercritical clump is quite different, particularly if the angular momentum is low. In this case, there is no well-defined initial core, so the collapse is like that considered by Zel'dovich (1970) for the formation of large-scale structure in the Universe: the clump will first collapse into a sheet (a "pancake") and then fragment. In other words, disk formation precedes star formation in the supercritical case. Having the disk form before the star could resolve a difficulty in forming massive stars (Nakano 1989b), because spherically symmetric models indicate that the radiation pressure on dust grains (assuming the standard dust-to-gas ratio) is too great to allow massive stars to form by spherical collapse (Wolfire and Cassinelli 1987). The formation of a ring of massive stars in W49A (Welch et al. 1987) is a possible example of supercritical star formation on a large scale.

B. Magnetic Flux Loss in Magnetically Subcritical Star Formation

Consider an isolated, magnetically subcritical clump in a molecular cloud. In view of the efficiency of magnetic braking (Sec. IV), assume that the cloud is not rotating. Under the influence of ambipolar diffusion, the clump will evolve toward greater densities until either it settles into a stable pressure-supported configuration (a "failed core" in the terminology of Lizano and Shu [1989], which is possible for $M < M_J$), or it begins a supercritical collapse leading to the formation of a star. Most studies of this problem have concluded that only a modest amount of magnetic flux can be lost in this "interstellar" stage of star formation (Nakano 1979,1982,1983; Nakano and Umebayashi 1986a, b; Umebayashi and Nakano 1990; Black and Scott 1982; Mestel and Paris 1984; Mestel 1985; Lizano and Shu 1989; Tomisaka et al. 1990). These studies indicate that there are several distinct stages in the evolution of the clump:

1. Core Formation. So long as $M < M_{cr}$, the clump contracts quasi-statically due to ambipolar diffusion. The flux leakage occurs primarily from the central region of the clump, redistributing the flux into the outer parts of the clump (Mouschovias 1976b; Nakano 1979). As a result M_Φ, and therefore M_{cr}, drop, and a cloud core forms that is in approximate gravitational virial equilibrium. The mass of the core is approximately equal to the mass of the star (or stars) that will eventually form; the outer parts of the clump, being magnetically subcritical, cannot undergo collapse (Mouschovias 1976b; Nakano 1979). Because interstellar clouds typically have $M/M_\Phi \gtrsim 0.1$, only a small fraction of the magnetic flux problem can be solved in this stage. Nonetheless, the long time scale associated with this stage has a major effect on the evolution of the cloud (Mouschovias 1977) and on the rate of star formation (Sec. VI.C).

2. Dynamical Collapse. Once M_{cr} drops below M, gravity exceeds the combined effects of gas pressure and magnetic fields and collapse ensues. So long as the ambipolar drift velocity is small compared to the infall velocity, the flux-to-mass ratio remains approximately constant. Now the free-fall velocity is $v_{ff} = (2GM_t/R)^{1/2}$, where $M_t \equiv M + M_*$ is the sum of the mass of the infalling gas M and the mass of a central protostar/disk if one has formed. The ratio of the free-fall velocity to the ambipolar drift velocity v_D is essentially the same as the ratio t_{AD}/t_{ff} given in Eq. (51), without the assumption of hydrostatic equilibrium. Setting the magnetic scale length ℓ_B

in Eq. (55) equal to R, we find

$$\frac{v_{ff}}{v_D} \simeq 11 \left(\frac{M}{M_\Phi}\right)^2 \left(\frac{M_t}{M}\right)^{1/2} \left(\frac{R}{Z}\right)^{3/2} K_{i-5}. \qquad (61)$$

Insofar as any flux leaks out in the collapse stage, M/M_Φ increases and the field becomes more tightly coupled to the gas, reducing the rate of flux loss; in other words, a weaker field is less effective at driving the charged particles through the neutrals (see also Mouschovias 1989)

3. *Magnetic Decoupling.* Substantial flux loss can occur in the collapse only when the ionization parameter K_i drops below its normal interstellar value so that the drift velocity v_D becomes comparable to the free-fall velocity v_{ff}. Nakano and Umebayashi (1986a, b) find that this occurs only for densities $n_H \gtrsim 10^{12}$ cm^{-3}. For the cases they considered, such high densities are reached only after the gas is within a few AU of the protostar, i.e., the decoupling occurs at protostellar, rather than interstellar, densities. These densities are sufficiently high that ohmic dissipation is comparable to ambipolar diffusion: the field is no longer even well coupled to the charged particles. More recent calculations, including more accurate chemical reactions (Umebayashi and Nakano 1990) and allowing for the effect of a distribution of grain sizes (Nishi et al. 1991) show that the critical density at which the field decouples from the gas may be as low as 2×10^{10} cm^{-3}.

4. *Magnetic Recoupling.* When the contracting core becomes opaque, the temperature rises, eventually re-ionizing the gas and trapping the flux. When allowance for rotation is made, much of the flux will be in the accretion disk surrounding the protostar, as discussed in the Chapter by Königl and Ruden.

An alternative view of magnetic flux loss in low-mass star formation has been given by Mouschovias and his collaborators (Mouschovias 1979; Mouschovias and Paleologou 1981; Mouschovias et al. 1985; Mouschovias 1987a, b; Mouschovias and Morton 1991). Much of this work focused on ambipolar diffusion in a slab with a magnetic field in the plane of the slab, and led them to the conclusion that the flux loss occurs at interstellar ($n_H \lesssim 10^7$ cm^{-3}) rather than protostellar densities. That their solutions are basically correct has been verified analytically by Shu (1983). However, the relevance of their solution to the astrophysical problem has been questioned by Mestel (1985) and by Lizano and Shu (1989), who point out that gravity saturates in one dimension; i.e., the force on a given mass element is independent of its height above the midplane. Tomisaka et al. (1990) have also shown that the evolution of the mass-to-flux ratio in a planar geometry is quite distinct from that in a spherical geometry. More generally, the distinction between a spherical collapse on the one hand and a planar or cylindrical collapse on the other is that the latter two are always magnetically subcritical: in these cases, the magnetic forces (assuming a frozen-in field) will always exceed the gravitational ones under sufficient compression. To see this, recall that the pressure due to gravity at the center of any of these configurations is $P_{gr} \sim G\Sigma^2$ from the equation of hydrostatic equilibrium. (Note that Σ for the cylinder is measured for a surface containing the axis.) Comparing this with the magnetic pressure $P_B = B^2/8\pi$, we have

$$\frac{P_B}{P_{gr}} \sim \frac{B^2}{G\Sigma^2} \propto \begin{cases} L^{-2}/L^0 = L^{-2}, & \text{slab} \\ L^{-4}/L^{-2} = L^{-2}, & \text{cylinder} \\ L^{-4}/L^{-4} = L^0, & \text{sphere.} \end{cases} \quad (62)$$

Hence, for both the slab and cylindrical geometries, the magnetic forces dominate at sufficiently small scales, so that they can never be truly supercritical; in the spherical case, the magnetic and gravitational forces scale in the same manner, which is what enabled us to introduce the magnetic critical mass in Sec. II. A spheroid which is very oblate is similar to a slab with the field perpendicular to the plane; in contrast to the slab with the in-plane field, there is a critical surface density $\propto B/G^{1/2}$ in this case, and the cloud can be supercritical. On the other hand, a spheroid which is very prolate behaves like a cylinder provided the contraction velocity along the field is not much greater than that in the perpendicular direction; such a prolate spheroid cannot be magnetically supercritical. The distinction between the planar and cylindrical geometries on the one hand and the spherical geometry on the other can be succinctly summarized: a cloud with a uniform magnetic field can be magnetically supercritical only if flow is permitted along field lines; in the absence of such flow, the magnetic pressure will eventually dominate the pressure due to self gravity as the cloud contracts.

The consequence of this distinction is that gravitational collapse cannot occur in one or two dimensions so long as flux freezing holds; as Mouschovias and Morton (1991) state, collapse in these cases can occur only as a result of ambipolar diffusion. As flux-freezing collapse *can* occur in three dimensions, calculations in one or two dimensions are of limited relevance to resolving the magnetic flux problem in star formation, a point acknowledged by Mouschovias (1989). Theoretical modeling of collapse in three dimensions demonstrates that the magnetic flux problem must be resolved at high densities (at least $\gtrsim 10^{10}$ cm^{-3}, and probably $\gtrsim 10^{11}$ cm^{-3}).

There is a further difficulty with the slab and cylindrical geometries, which has been pointed out by Nakano (1984). If the thermal pressure is small, then the Jeans length will be small compared to L and the cloud will fragment, losing its cylindrical symmetry; on the other hand, if the gas pressure is large enough to overcome this difficulty, then the rate of ambipolar diffusion is substantially reduced.

Numerical calculations of three-dimensional, axisymmetric collapse with ambipolar diffusion have been carried out by Black and Scott (1982). The simple form, Eq. (48), for the ionization breaks down at high density, so they adopted an ionization that varies as a power of the density, $x_i \propto n^{-q_i}$. To represent this case, the parameter K_i must vary as $n^{-(q_i-1/2)}$; recall that

$q_i = 1/2$ for the case of cosmic-ray ionization discussed in Sec. V.A above. Once the dynamical collapse has begun, the flux remains frozen so long as $v_{ff} >> v_D$. If q_i exceeds 1/2, however, K_i drops and eventually $v_D \simeq v_{ff}$. Now note that in Eq. (61), $M/M_\Phi \simeq \Sigma/B$, where $\Sigma = M/\pi R^2$ is the surface density; then $v_{ff}/v_D \propto \Sigma^2 K_i/B^2$. If the drop in the ionization is slow enough that $\Sigma^2 K_i$ increases with density, then B will continue to increase as well, though perhaps more slowly than when the flux was frozen, and v_D will remain of order v_{ff}. On the other hand, if the ionization drops so rapidly that $\Sigma^2 K_i$ decreases with density, then the field is essentially completely decoupled from the neutrals and B will tend to remain constant as the density increases. (Equation [61] does not apply in this case because the magnetic scale length ℓ_B becomes much greater than R.) These three cases correspond qualitatively to the three types of solution found by Black and Scott (1982).

In summary, the controlling factors in magnetic flux loss are the geometry and ionization. For the idealized case of a planar contraction perpendicular to **B**, the magnetic field will tend to prevent gravitational collapse; as a result, the flux loss will occur on the ambipolar diffusion time scale at relatively low densities. For the more realistic case of a three-dimensional collapse, the flux remains approximately frozen to the gas so long as the ionization varies as $n^{-1/2}$; in this case, the flux loss must occur at high densities.

C. Magnetic Fields and the Rate of Star Formation

In order for a clump of gas to collapse, it must overcome the combined effects of gas pressure (represented by M_J), magnetic fields (M_Φ), and angular momentum (M_{rot}); in other words, its mass must exceed the critical mass M_{cr} given in Eq. (41). We have seen that angular momentum is generally rendered unimportant by magnetic braking, so we focus on M_J and M_Φ. For a molecular cloud which is at a substantially higher pressure than its surroundings, we have argued that the mass must be close to M_{cr} and therefore must exceed both M_J and M_Φ (see Sec. II). On the other hand, prior to ambipolar diffusion, stellar mass clumps inside molecular clouds are by and large magnetically subcritical—the magnetic mass M_B is typically of order $10^{2.5}$ $M_\odot$ according to Eq. (27) with $n_H \sim 10^3$ cm^{-3} and $B \sim 30$ μG. Magnetically supercritical clumps ($M > M_\Phi$, or, equivalently, $M > M_B$) may have subclumps and are therefore potential sites of both subcritical and supercritical star formation (Sec. VI.A).

Subcritical star formation is believed to be the dominant mode of star formation in the Galaxy (Shu et al. 1987a), and it proceeds via ambipolar diffusion. McKee (1989) has estimated the rate of star formation in molecular clouds on the basis that the rate is set by the ambipolar diffusion time in the typical molecular gas, which has a density $n_H \sim 10^3$ cm^{-3}; the higher-density gas seen in molecular cloud cores has presumably already undergone ambipolar diffusion and is now close to forming stars. Penetrating far-ultraviolet radiation can photo-ionize metals such as magnesium and iron until the extinction exceeds a few, and the ambipolar diffusion time in this photo-ionized gas is quite long. As a result, most of the star formation occurs in the shielded regions of the cloud where cosmic-ray ionization is dominant. For typical conditions in the Galaxy, McKee (1989) estimates that the star-formation time in a molecular cloud with a mean visual extinction $\bar{A}_V$ is,

$$t_{g*} \equiv \frac{M}{\dot{M}_*} \simeq \frac{2 \times 10^7}{n_3^{1/2}} \exp(16/\bar{A}_V) \text{ yr} \tag{63}$$

where $\dot{M}_*$ is the rate at which mass in the cloud is converted into stars. The star-formation time scale is exponentially sensitive to the mean extinction; cold molecular clouds with $\bar{A}_V \sim 2$ (Falgarone and Puget 1986) should have little star formation. On the other hand, typical warm molecular clouds with $\bar{A}_V \simeq 5$ to 7 should have $t_{g*} \sim 3 \times 10^8$ yr. If about half the molecular gas in the Galaxy is warm (Solomon and Rivolo 1989), corresponding to about 10^9 $M_\odot$, this gives a galactic star-formation rate of 3 $M_\odot$ yr^{-1}, consistent with observation.

Star formation can also be induced—triggered by a mechanism external to the clump—although this is believed to be less important than the spontaneous mechanism just described. Shocks, which can be due to supernovae or to cloud–cloud collisions, have been invoked frequently as a mechanism for inducing star formation. Elmegreen and Elmegreen (1989) have argued that, so long as flux freezing holds, magnetic fields can inhibit this process. For magnetically subcritical clumps, this conclusion is obvious, because compression, unless predominantly along the field, leaves M/M_Φ less than unity so that the clump remains stable. For magnetically supercritical clumps, their argument can be strengthened as follows. The condition for star formation to be induced is that the ratio of the free-fall time t_{ff} to the dynamical time t_d must decrease as a result of the shock. For cloud-cloud collisions and for shocks striking clouds which are not too small, the dynamical time is simply $t_d = L/\sigma$, where L is the characteristic size of the compressed region (e.g., the thickness of the shocked layer for a one-dimensional compression). Now the velocity dispersion is $\sigma = (P/\rho)^{1/2} \propto \rho^{(\gamma-1)/2}$; if the compression occurs in ν dimensions, so that $\rho \propto L^{-\nu}$, then we have

$$\frac{t_{ff}}{t_d} \propto \frac{\rho^{-1/2}}{L/\sigma} \propto \rho^{\frac{\gamma}{2}+\frac{1}{\nu}-1}. \tag{64}$$

For a magnetically supercritical clump, the gas pressure is dominant and $\gamma \sim 1$; three-dimensional contraction can lead to gravitational collapse, but in the more likely case in which the shock compression is primarily one dimensional, collapse usually will not occur. The only way that one-dimensional compression can lead to collapse is if the cloud is initially magnetically supported and if the compression along the field is sufficient to render the cloud magnetically supercritical while leaving $t_{ff}/t_d < 1$. If the cloud is initially supported by gas pressure ($M < M_J$ and $M > M_\Phi$), and if $\gamma \gtrsim 1$, then the compression must be approximately three dimensional to induce collapse. On the other hand, if the compression leads to a phase change (e.g., 100 K atomic gas to 10 K molecular gas), then γ is significantly less than unity and compression can lead to collapse over a wider range of conditions.

Under what conditions can induced star formation occur? In a classic paper, Elmegreen and Lada (1977) showed that ionization-shock fronts driven by associations of massive stars can drive gas along the mean field direction until it becomes magnetically supercritical and another generation of massive stars can form. This model fits the observations of Orion reasonably well, but the universality of the mechanism remains to be established. Scoville et al. (1987) have argued that the correlation of giant H II regions with molecular gas indicates that star formation can be induced by collisions between molecular clouds. The argument summarized in Eq. (64) suggests that the compression itself is not effective in accelerating the rate of star formation, because the hydrodynamic expansion time of two clouds which have collided is reduced more than the gravitational contraction time. What the collision does, however, is to increase dramatically the mass of molecular gas that is shielded from far-ultraviolet radiation; this can increase the rate of star formation by increasing the rate of ambipolar diffusion in the clumps comprising the combined clouds (McKee 1989). Pringle (1989c) has argued that the prevalence of binary stars can be best understood if most low-mass star formation is induced by collisions among weakly magnetized clumps that are close to gravitational collapse, in contrast to the canonical view of low-mass star formation summarized by Shu et al. (1987b).

Star formation can also be induced by the three-dimensional, radiation driven implosion of a clump of gas exposed to the ionizing radiation of one or more massive stars (Klein et al. 1983). Radiation-driven implosion is the first stage of cloud photoevaporation (Bertoldi 1989); if the clump is initially magnetically supercritical, as assumed by Klein et al. (1983), or if the core of the clump becomes supercritical as a result of ambipolar diffusion in the shocks compressing the clump (Bertoldi et al. 1991), then star formation can occur promptly. Possible evidence for the latter process has been found by Sugitani et al. (1989), who have observed molecular outflows in several bright-rimmed globules associated with H II regions.

In summary, low-mass star formation in the Galaxy appears to be due primarily to the gravitational collapse of clumps which are initially magnetically subcritical and become supercritical as a result of ambipolar diffusion. In principle, the formation of both low-mass and high-mass stars can be induced by shock compression, but only in certain cases. For a magnetically supercritical clump with $\gamma \sim 1$, the compression must be approximately three dimensional to induce collapse, as in the case of radiation-driven implosions. For a magnetically subcritical clump, induced star formation requires either compression along the field, as in the Elmegreen-Lada mechanism, or enhanced ambipolar diffusion as a result of increased density, reduced flux of ionizing radiation, or steeper gradients associated with shocks.

VII. MAGNETIC EVOLUTION OF MOLECULAR CLOUDS

We conclude our discussion of magnetic fields in molecular clouds by addressing a basic question: how did the fields in these clouds evolve from the much weaker field observed in the diffuse interstellar medium? As a first step, we may determine the volume of interstellar medium out of which the cloud formed, the accumulation volume (Mestel 1985; McKee and Lin 1988). For simplicity, we assume that the accumulation volume is an ellipsoid of radius R_0 normal to the field and length L_0 along the field. The mass $M = 2\pi R_0^2 L_0 n_0 \mu_H / 3$ and (assuming flux freezing) the flux $\Phi = \pi R_0^2 B_0$ are the same for the accumulation volume as for the actual cloud. Then the length of the accumulation volume is readily shown to be,

$$L_0 = 96 \left(\frac{B_{0\mu}}{n_0} \right) \frac{M}{M_\Phi} \text{ pc} \tag{65}$$

and the radius is

$$R_0 = 380 \left(\frac{M_6}{B_{0\mu}} \right)^{1/2} \left(\frac{M}{M_\Phi} \right)^{-1/2} \text{ pc} \tag{66}$$

where $B_{0\mu} = B_0/1 \ \mu G$ and $M_6 = M/10^6 \ M_\odot$. Our value for L_0 is smaller than Mestel's (1985) because our value of M_Φ is smaller than his. Note that L_0 is independent of the size of the molecular cloud: the assumption of flux freezing implies that M/M_Φ is the same for the accumulation volume as for the final cloud, so that $M/M_\Phi \propto \Sigma/\bar{B} \propto n_0 L_0/B_0$. For the typical case $n_0 = 1 \text{ cm}^{-3}, B_0 = 3 \ \mu G$, and $M = 2M_\Phi$, this length is about 600 pc. Under the same conditions, the radius is about 250 $M_6^{1/2}$ pc, so that the accumulation volume of a $10^6 \ M_\odot$ cloud is roughly spherical. For such large molecular clouds this radius exceeds the scale height of the gas in the galactic disk, so the accumulation volume is actually triaxial for such clouds, with R being the geometric mean of the radii normal to the field. The accumulation volume is large enough to contain many diffuse interstellar clouds, which subsequently become the initial clumps in the molecular cloud (Elmegreen 1985a). These same equations apply to the accumulation volume for a clump in a GMC which originates as a diffuse interstellar cloud, as suggested by Elmegreen (1985a). As the magnetic field in a diffuse interstellar cloud is comparable to the mean interstellar field, the accumulation length L_0 for a clump is smaller than that for a GMC by the ratio of the initial density of the diffuse cloud to the mean interstellar density, and the mass of the clump is correspondingly smaller as well.

What is the physical process that causes the gas in the accumulation volume to condense into a molecular cloud? Elmegreen (1990a) has reviewed the various cloud formation mechanisms that have been proposed—cloud coalescence due to cloud-cloud collisions, gravitational instability in the gas, including the Parker instability (Parker 1966), and shock compression—and he concludes that gravitational instability is the most promising explanation at present. The importance of the self-gravity of the gas in the process of cloud formation is demonstrated by the fact that the length scale for the Jeans instability is comparable to L_0. Define the effective scale height of the gas in the disk, H_{eff}, by $\Sigma_d = 2\rho_0 H_{eff}$, where Σ_d is the total surface density of gas in the disk and ρ_0 is the gas density in the midplane. This scale height is less than the value it would have in the absence of other matter in the disk by a factor f_g:

$$H_{eff} = f_g \left(\frac{\sigma^2}{2\pi G\rho_0} \right)^{1/2}. \tag{67}$$

Because the density of the gas in the solar neighborhood is about 1/4 of the total density (Binney and Tremaine 1987), and its scale height is less than that of the stars, f_g is of order 0.5 in the galactic disk. In terms of this scale height, half the minimum wavelength for the Jeans instability is $\pi(\sigma^2/4\pi G\rho_0)^{1/2} = \pi H_{eff}/(f_g\sqrt{2})$. Now consider the accumulation length L_0. Adapting Parker's notation, we write $B_0^2/8\pi = \alpha_B P_0$; it is then straightforward to show that

$$\frac{L_0}{H_{eff}} = 2.26\frac{\alpha_B^{1/2}}{f_g} \left(\frac{M}{M_\Phi} \right). \tag{68}$$

For typical values $\alpha_B \sim 0.1 - 0.2$, $f_g \sim 0.5$, and $M \sim 2M_\Phi$, this result leads to an accumulation length $L_0 \simeq (3 \sim 4)H_{eff}$, in agreement with the Jeans length. Thus self-gravity must play a role in the formation of giant molecular clouds. This could have been foreseen: because the ratio M/M_Φ is constant as the cloud contracts from its initial state in the diffuse ISM, the fact that gravity is important in the final cloud implies that it was in the initial cloud as well.

When the rotation of the Galaxy is included, the normal criterion for gravitational instability is that $Q \equiv \kappa\sigma/(\pi G\Sigma_d) \lesssim 1$ (Toomre 1964), where κ is the epicyclic frequency. This condition ensures that there is a range of wavelengths which exceed the Jeans length, but are small enough that they are not stabilized by rotation. Elmegreen (1987c,1989c), however, has suggested that magnetic fields render this criterion irrelevant, based on the fact that an arbitrarily weak magnetic field normal to the rotation axis eliminates rotational stabilization (Chandrasekhar 1961). Physically, this destabilization occurs because the field, given enough time, can prevent the gas from spinning up as it contracts. As we have seen in Sec. IV.B, this magnetic braking is highly efficient for observed field strengths. How then can one account for the observation that the criterion $Q \lesssim 1$ describes the onset of star formation in external galaxies extremely well (Quirk 1972; Kennicutt 1989)? In other words, why do not magnetic fields permit star formation to occur at lower surface densities, albeit more slowly? We conjecture that this is because galactic disks are differentially rotating: a perturbation with $Q > 1$ will be destroyed by shear unless the field is so strong that it can maintain rigid rotation. Recent calculations by Elmegreen (1991a) demonstrate that fragmentation of shearing perturbations does in fact require $Q \lesssim 1$.

It appears, then, that regions of a galactic disk which accumulate enough gas so that $Q \lesssim 1$ will be subject to gravitational instability, perhaps aided by the buckling of the magnetic field as envisioned by Parker. As the gas in the accumulation volume contracts, the mean magnetic field will grow as $B \simeq 5\bar{A}_V$ for $M \simeq 2M_\Phi$ (Eq. 25); the initial extinction of the accumulation volume is 0.6 mag for an initial field of 3 μG. The field begins to deviate significantly from its interstellar value for $\bar{A}_V \sim 1$, which is also the point at which clouds become substantially molecular in the solar neighborhood. So long as the mean extinction is relatively low, star formation is inhibited (McKee 1989) and the clouds remain cold; Perault et al. (1985) have studied a population of such clouds with a mean extinction $\bar{A}_V \sim 2$. The energy dissipation in a magnetized cloud is relatively inefficient (Elmegreen 1985b), so that molecular clouds should contract gradually, over several free-fall times (Carlberg and Pudritz 1991). The morphology of the clumps in the cloud will evolve as the cloud contracts: on the one hand, the gradual increase in the mean field will tend to squeeze non-self-gravitating clumps out into filamentary structures, such as those observed by Bally et al. (1987); on the other hand, this tendency to form structures along the field will be counteracted by the increase in the density as the gas converts from atomic to molecular form and gradually cools to 10 K. Isolated molecular clouds will evolve with M/M_Φ constant. If clouds coalesce (which should be relatively unimportant at typical cloud-cloud velocity dispersions according to Elmegreen [1990a]), M/M_Φ will increase unless the two clouds are on completely separate flux tubes. As the cloud contracts, it becomes increasingly shielded from the interstellar radiation field, reducing the ionization and permitting more rapid star formation. When the mean extinction of the cloud reaches $\bar{A}_V \sim 6$ to 8, whether through contraction alone or through contraction abetted by coalescence, the energy injection from the bipolar flows of newly formed stars can slow, or perhaps halt, the further contraction of the cloud (Norman and Silk 1989a; McKee 1989); such clouds correspond to the warm molecular clouds studied by Scoville et al. (1987) and Solomon et al. (1987). At this stage, the mean magnetic field is $\sim$30 to 40 μG, $\sim$10 times the interstellar value. Eventually, the cloud is disrupted by the formation of massive stars, and the cycle begins anew.

VIII. CONCLUSION

Substantial progress has been made toward the solution of the two classical problems of star formation, the angular momentum problem and the magnetic flux problem. Magnetic braking appears to be relatively efficient, although it would be worthwhile to have more extensive observations of cloud rotation in order to test the theory more precisely. For low-mass stars, the loss of magnetic flux is thought to occur in two stages. First, ambipolar diffusion permits a clump which was initially magnetically subcritical to evolve to the point that it is magnetically supercritical and to begin its collapse. The second stage, in which most of the flux is lost, must occur at relatively high densities, but the details remain obscure because of uncertainties in the ionization associated with tiny grains and large molecules; direct observation of the ionization in cloud cores would represent a major advance. The formation of massive stars is less well understood. They may form from clumps which are initially magnetically supercritical, and in this case disk formation may precede star formation. Star formation can also be induced by an external mechanism, such as a shock, which acts to reduce the flux-to-mass ratio; this may occur directly, by driving gas along the field, or indirectly, by accelerating the rate of ambipolar diffusion.

There is a third problem, however, on which less progress has been made. For want of a better name, we shall term it the magnetic topology problem: how does the topology of the field evolve as a molecular cloud forms out of the ISM and as clumps within the cloud contract to form stars? This problem has several aspects. One of the most basic is the determination of when the field in a protostellar clump becomes disconnected from the field in the ambient molecular cloud. The topology of the field of the cloud as a whole is also uncertain; the observations summarized in the Chapter by Heiles et al. suggest that the ratio of mean field in a cloud to the ambient field (up to a factor 10) is well beyond that of any published theoretical models. This poses the observational challenge of determining if and how the cloud field joins up with the ambient interstellar field, perhaps in the H I envelope surrounding the cloud; and it poses the theoretical challenge of producing models which can match these observations. Given the rate at which the field is advancing, we anticipate substantial progress on these issues will have been made by the time that the "Protostars and Planets IV" book is written.

Acknowledgments. We wish to thank F. Shu for suggesting that we undertake this review and for helpful comments. We are grateful to F. Bertoldi, B. Elmegreen, E. Falgarone, L. Mestel, T. Mouschovias and R. Pudritz for useful remarks. The research of CFM on star formation is supported by a NASA grant to the Center for Star-Formation Studies; his research on the interstellar medium is supported by a grant from the NSF. EGZ acknowledges support also through an NSF grant to the University of Colorado and to the Institute for Theoretical Physics, where a portion of this chapter was written. AG is supported in part by a President's Fellowship at UC Berkeley. CH acknowledges with pleasure the hospitality of the Joint Institute for Laboratory Astrophysics. His research is supported in part by a grant from NSF.

STELLAR MULTIPLE SYSTEMS: CONSTRAINTS ON THE MECHANISM OF ORIGIN

PETER BODENHEIMER
Lick Observatory

TAMARA RUZMAIKINA
O. Yu. Schmidt Institute of the Physics of the Earth

and

ROBERT D. MATHIEU
University of Wisconsin

The discovery rate of pre-main-sequence binary and multiple systems has increased dramatically over the last few years, providing us with our first insights into the nature of such systems at young ages. The observational data derived from spectroscopic, occultation, speckle, visual, and other techniques are reviewed. Where possible, the pre-main-sequence binary population is compared with main-sequence systems. These results are interpreted in terms of binary formation mechanisms. While wide binaries (P>100 yr) are ordinarily explained by a fragmentation process during the collapse of a rotating protostar, origin of some systems by capture is a possibility. The origin of close systems is not well understood: various processes, such as fragmentation during late stages of collapse, gravitational instabilities in disks, or orbital decay from a longer-period system are examined. Many important questions remain unanswered: (1) Can binary formation be explained by a single process for all separations? (2) What is the origin of the angular momentum of binary systems? (3) Under what circumstances do single stars rather than binaries form? (4) What is the relationship between disks and binaries? Speculations regarding these questions are discussed with reference to the evolution of the angular momentum distribution in collapsing protostars and circumstellar disks.

I. INTRODUCTION

The observational discovery of binary and multiple systems among young stars and the quest for clarification of the mode of origin of these systems is a complicated subject and one that has developed rapidly in the past few years. This review summarizes recent observational and theoretical results, concentrating on the solar mass range. Further information may be found in the reviews of observational material by Abt (1983), Reipurth (1988), Mathieu (1989) and Zinnecker (1989a) and in the reviews of theoretical material by

Tassoul (1978), Lucy (1981), Zinnecker (1984a), Boss (1988,1989), Miyama (1989a), Lebovitz (1989), Durisen et al. (1989a), Verbunt (1989) and Pringle (1991). The issue of the presence of low-mass companions is not dealt with at length, but recent discussions have been presented by Zuckerman (1989) and Latham (1989). In Sec. II we discuss the observational material. The main-sequence binary population in the solar mass range has been studied extensively, and the systematic behavior that is found must be explained by any theory of binary origin. Pre-main-sequence binaries provide information closer to the time of origin; observational searches are just now entering the phase of systematic surveys. Information has been obtained by several methods: spectroscopic, lunar occultation, speckle and slow slit scan techniques, and optical and infrared imaging. One purpose of the observations is to provide constraints on the mechanism or mechanisms of binary origin. Section III then considers the important question of the available angular momentum in the precursor material out of which the binaries form. The fate of a collapsing cloud depends strongly on the value and distribution of the angular momentum in the dense core of a molecular cloud where it originates. Observed rotation rates in molecular clouds and cloud cores are compared with the angular momenta in the orbits of binary systems and in the rotation of single stars. Physical mechanisms for the redistribution of angular momentum during protostellar evolution are discussed. Section IV then discusses formation theories, emphasizing the closely related processes of fragmentation during protostellar collapse, during disk formation, and after an equilibrium disk has been formed. Other processes, such as capture, orbital decay, and fission are also considered. Finally, Sec. V reviews the mostly speculative suggestions regarding the situations in which single stars rather than multiple systems are formed, and Sec. VI provides a summary with a view toward future work.

II. OBSERVATIONAL EVIDENCE

A. The Main-Sequence Solar-Mass Binary Population

Certainly some of the most valuable clues to the process of binary star formation lie in the nature of the product, that is, the main-sequence binary population. For more than a decade, the classic study of Abt and Levy (1976, henceforth AL; see also Abt 1983) has been the standard description of the main-sequence solar-mass binary population. Recently Morbey and Griffin (1987) have argued that a large fraction of the spectroscopic orbit solutions presented there are not statistically significant and that in most of these cases there is no evidence that the stars are binaries. Abt (1987) reanalyzed his results in light of these findings, noting that the description of the solar-mass binary population was nonetheless little changed. Recent technological advances (in particular in the realm of radial-velocity measurements and speckle interferometry) have permitted several extensive long-term surveys which will soon be ripe for analysis; we can expect an improved description of the main-sequence solar-mass binary population in the next few years. For example,

Duquennoy and Mayor (1991, henceforth DM) have reported on the Swiss study of 164 F7-G9, luminosity class IV–V, V and VI stars, selected from the Gliese catalog to be within a parallax-limited volume.

Given the rapidly changing state of the observations, here we only discuss several essential points of the main-sequence binary population, based on the revised AL sample and the DM study, unless otherwise noted.

1. In these surveys, approximately half of the stars are detected to be components of multiple systems. Due to detection biases, the true frequency of multiple systems is certainly substantially higher; for example, Abt (1983) argues that 78% of the AL sample has at least one stellar companion. Most (~85%) of the stars observed to be multiple are detected as binaries.

2. The orbital period distribution is continuous in the range $0 < \log$ period (days) <9, with a single maximum. DM argue for a median period of 180 yr, somewhat longer than that found by AL.

3. The mass-ratio q distribution has been the subject of much study with widely varying results. While it has generally been found that the secondary mass distribution of longer-period binaries rises continuously toward smaller mass (similar to the Van Rhijn distribution according to AL or to derived initial mass functions according to DM), an essential issue has been whether shorter-period binaries tend to have smaller primary-secondary mass ratios (closer to unity). AL and Abt (1987) have argued for such a trend, with the transition occurring at a period of ~100 yr. On the other hand, DM find no support for differing secondary mass functions between binaries with periods longer than or shorter than 100 yr. Historically, results derived from cataloged data have differed greatly, presumably due to selection biases. A recent careful analysis of cataloged data by Halbwachs (1987) showed a secondary mass distribution similar to that of DM with no evidence for a period dependence. This issue of the dependence of the secondary mass distribution on orbital period is critical; for example, a transition as suggested by AL may well imply separate formation mechanisms for short- vs long-period binaries, with the transition period perhaps related to the maximum size of circumstellar accretion disks. From the point of view of binary formation, resolution of this question is perhaps the most essential goal of the next generation of main-sequence binary surveys.

4. The orbital eccentricity distribution is clearly dependent on orbital period. The binaries with the shortest periods have circular orbits; DM find that the transition from circular to eccentric orbits occurs at a period between 10 and 13 days. Indeed, the transition is remarkably well defined, given that the sample includes stars of all ages. Longer-period binaries are found to have essentially all eccentricities. Heintz (1969) has argued that the mean orbital eccentricity is ~0.5, independent of orbital period between 1 yr and 500 yr. DM argue that for binaries with periods between 11 days and 1000 days, the eccentricity distribution is bell shaped with a maximum at ~0.3. For longer-period binaries, they find that the relative number of binaries having high eccentricity increases, and they suggest the distribution is more like the

dynamically relaxed distribution $f(e) = 2e$.

These main-sequence eccentricity distributions have significant implications for the processes of binary formation and evolution during the pre-main-sequence (PMS) phase. That the shortest period orbits tend to be circular has been recognized for some time; the long-standing explanation has been tidal circularization during the main-sequence phase of evolution (see, e.g., Zahn 1977; Koch and Hrivnak 1981). This interpretation was reinforced by the discovery of similar transitions in the Hyades and Praesepe open clusters (age ∼0.8 Gyr) at a period of 5.7 days, in M67 (age ∼5 Gyr, comparable to the expected mean age of the DM G-dwarf sample) at a period of 10 to 11 days, and in the halo of the Galaxy at a period between 13 and 18 days (Mayor and Mermilliod 1984; Mathieu and Mazeh 1988; Jasniewicz and Mayor 1988; Latham et al. 1988). Mathieu and Mazeh (1988) noted that this increase in transition period with binary age is consistent with effective tidal circularization on the main sequence.

Recently, however, Zahn (1989) has argued that viscous dissipation in stellar envelopes, and consequently tidal circularization, is less effective than previously thought. Subsequently, Zahn and Bouchet (1989) have done a theoretical analysis of the expected tidal circularization as binary components evolve from the birthline to the main sequence. They come to the important conclusion that the circularization of a binary orbit takes place almost entirely at the very beginning of the Hayashi phase, and that there is little further decrease of the eccentricity on the main sequence. For a variety of assumptions regarding initial conditions and binary component masses, they predict that transition periods between 7.2 and 8.5 days should be found for low-mass binaries of all ages. Recognizing that both shorter and longer transition periods have already been found in several systems of varying ages, they suggested alternative *ad hoc* explanations for those systems (such as mass transfer and evolution of the stellar interiors).

That the observed monotonic increase in transition period with main-sequence population age can be so easily explained away is not clear; the reader is referred to the review of Mazeh et al. (1990) for an alternative view. On the observational side, both the number of binary samples with detected transition periods and the number of stars defining the transition periods are small. In addition, in some samples counterexamples to the premise of a single transition period exist. The numbers of binary samples and orbits must be increased, particularly at ages of less than 10^9 yr. The theory of stellar tidal circularization, particularly in regard to the dissipation mechanism and consequently the circularization time scale, also remains uncertain (Tassoul 1987,1988,1990; Goldman and Mazeh 1991). Nonetheless, the work of Zahn and Bouchet is seminal in its attempt to quantify the evolution of binary orbits during the PMS phase. Future work must both clarify the role of tidal circularization and incorporate other relevant processes, such as associated disks and mass loss (see, e.g., Artymowicz et al. 1991).

Taking a broader view in orbital period, among main-sequence binaries with periods longer than ≈ 10 days, the vast majority have eccentric orbits. Equally important, the observed eccentricity distribution shows a significant frequency over essentially the entire range of eccentricity. Thus any theory of binary formation or early orbital evolution must not overly tend toward either circular or high-eccentricity orbits. The challenge to the observers is to provide more precise definition to the eccentricity distribution, particularly as a function of period.

B. The Pre-Main-Sequence Low-Mass Binary Population

While the main-sequence binary population provides a valuable boundary condition on the processes of binary formation and PMS evolution, clearly the more direct approach is to observe binaries in the process of forming and evolving. Substantial advances in this regard have been made since the publication of *Protostars and Planets II* (Black and Matthews 1985). The field is young and still largely in the stage of collecting objects for analysis. As yet, no comprehensive study of the young binary population comparable in magnitude to those for the main-sequence binary population has been done.

The following review of the observed PMS binary population is organized by discovery technique. One important generalization should be made, however. Binaries of all periods have been found among PMS stars of all ages, including the very youngest. DF Tau and SVS 20 are examples of binaries very near the birthline (ages $\sim 10^5$ yr). Binary formation evidently can occur very early in the star formation process.

1. Spectroscopic Binaries. The first orbit for a low-mass PMS spectroscopic binary was that derived for V826 Tau by Mundt et al. (1983). At present, the roster of identified low-mass PMS spectroscopic binaries stands at more than 20. Those published at the time of writing are listed in order of increasing period in Table I, where the orbital periods, eccentricities, and the relevant references are presented. Arguably, EK Cep might not be considered a PMS binary as the primary is a 2.0 $M_\odot$ star on the ZAMS. However, Popper (1987) argues that the 1.1 $M_\odot$ secondary is still contracting to the main sequence, and hence we include it here. The high frequency of short-period binaries in the list of Table I reflects the fact that systematic monitoring of large samples of PMS stars has only recently begun.

Only the binaries discovered by Mathieu, Walter and Myers (1989, henceforth MWM) are the product of a comprehensive radial-velocity survey. The MWM study included an X-ray selected sample of some 50 PMS stars in the Taurus-Auriga, Ophiuchus and Corona Australis star-forming regions. For binaries in this sample with periods of less than 100 days, they find the detected PMS binary frequency to be $9\% \pm 4\%$, indistinguishable from the corresponding detected binary frequency among main-sequence stars ($\sim 12\%$).

Spectroscopic techniques have the advantage that they are sensitive to the shortest period binaries and hence often provide not only evidence of binarity but also orbital solutions; the only orbits yet determined for PMS binaries have been derived from spectroscopic data. Knowledge of the orbital

P. BODENHEIMER ET AL.

TABLE I

Roster of PMS Spectroscopic Binaries

Object	P (d)	e	Reference[a]
155913-2233	2.4238	0.0	MWM
V4046 Sgr	2.43	0.0	BDLR
V826 Tau	3.8878	0.0	MR
EK Cep	4.4278	0.11	TPO
160905-1859	10.400	0.17	MWM
AK Sco	13.6093	0.47	AND
P1540	33.73	0.12	MM
162814−2427	35.95	0.48	MWM
162819−2423S	89.1	0.41	MWM
160814−1857	144.7	0.26	MWM
GW Ori	242	0.04 ±0.06	MAL

[a] AND = Andersen et al. (1989); BDLR = Byrne (1986), de la Reza et al. (1986); MAL = Mathieu et al. (1991); MM = Marschall and Mathieu (1988); MR = Mundt et al. (1983), Reipurth et al. (1990); MWM = Mathieu et al. (1989); TPO = Tomkin (1983), Popper (1987).

elements can provide important insights into the nature of the binary formation process itself. In the remainder of this section we highlight several of the more interesting discoveries to date.

i. Orbital Eccentricities. Examination of Table I shows that all binaries with periods of <4 days have circular orbits, while those with longer periods have eccentric orbits (with the one notable exception of GW Ori). In addition, Mathieu, Walter and Myers (in preparation) have found two additional PMS binaries of interest here, in particular a binary with a circular orbit having a period of 4.253 days and another with an eccentric orbit ($e = 0.26$) having a period of 7.46 days.

These data reflect directly on the issue of orbital evolution during the PMS phase. Taken straightforwardly, the binaries listed in Table I indicate a transition from circular to eccentric orbits occurring at a period of 4 days, the shortest transition period yet found. However, this conclusion rests on only one binary, EK Cep, so that a PMS transition period is not well established. Should enlargement of the PMS sample support such a short transition period, the indication would be that tidal circularization during the PMS phase is not as effective as Zahn and Bouchet (1989) suggest (or that other processes are also active in determining orbital eccentricity), and that significant tidal circularization occurs during the main-sequence phase. However, excepting EK Cep, the present sample is consistent with the predictions of Zahn and Bouchet (1989) for PMS tidal circularization. The existence of another PMS binary with an eccentric orbit at 7.46 days, when compared to transition periods of >10 days among both disk and halo solar-mass field stars as well as in M67, does suggest that tidal circularization progresses on the main sequence.

In any case, the complex evolution of internal stellar structure, circumbinary disk structure, and orbital separation prior to 1 Myr must ultimately produce the robust result of circular orbits for binaries with periods <4 days. Given the variety of influences on a binary orbit between formation and the main sequence (e.g., disks, mass accretion, mass outflows, encounters), some or all of which may drive orbital eccentricity (Goldreich and Tremaine 1980; Artymowicz et al. 1991), the uniformly circular orbits among the shortest-period binaries suggests that these orbits are established relatively late in the orbital evolution of a young binary. The apparent sensitivity of the circularization mechanism to orbital period may indicate that stellar tidal circularization has the final word in these short-period systems.

The case of GW Ori may be an important counterpoint to this simple picture of the PMS eccentricity distribution. The spectral energy distribution of GW Ori indicates the presence of a circumprimary disk and possibly a circumbinary disk as well (Mathieu, Adams and Latham 1991, henceforth MAL; see discussion below). Thus GW Ori provides the rare opportunity to examine a young, short-period binary in a disk environment. Notably, MAL find an orbital eccentricity of 0.04 ± 0.06, despite the rather long period of 242 days. Formally, the measured eccentricity is indistinguishable from zero, but the measurement error is large, so it would be premature to assert that the orbit is circular. Certainly, however, the eccentricity is small. Interestingly, Artymowicz et al. (1991) argue that disks should very rapidly drive a binary orbit to high eccentricities. Earlier work (Goldreich and Tremaine 1980; Ward 1988) also found that the orbital evolution would be rapid, but in the case of planetesimal companions the sense of the eccentricity evolution was very sensitive to the details of the disk structure. Thus, if GW Ori does presently have a circular orbit and will ultimately have an eccentric orbit, it may provide important insight into the details of disk evolution.

ii. Mass Determinations. To date the only double-lined eclipsing PMS binary discovered is EK Cep, which provides a dynamical mass determination for the secondary of $1.12\ M_\odot$. A preliminary analysis by Popper (1987) finds the position of the secondary in the theoretical HR diagram to be between the $1.0\ M_\odot$ and $1.25\ M_\odot$ evolutionary tracks of Iben at an age of 20 Myr. This result is important, but the secondary is sufficiently near the main sequence that it would be remarkable indeed if the theoretical mass calibration were greatly in error.

The double-lined binary $162814-2427$ is substantially younger with an age of 1 Myr (MWM). The system is notable in that the dynamical lower limit on the primary mass is $1\ M_\odot$. However, MWM have placed the primary on the theoretical HR diagram and found theoretical mass estimates of $0.7\ M_\odot$ and $0.8\ M_\odot$ (using models of VandenBerg [in preparation] and Cohen and Kuhi [1979], respectively). This inconsistency may reflect on the validity of using these theoretical tracks for estimating masses of young low-mass stars (at least in short-period binaries). However, the uncertainty in the effective temperature of the primary permits theoretical values consistent with the dynamical mass limit (although only marginally so in the case of the VandenBerg models).

Marschall and Mathieu (1988) have used the double-lined binary P1540 to test the relative mass calibrations of the theoretical tracks. They deconvolved the composite light, placing both the primary and secondary on the theoretical HR diagram. Using the PMS tracks of Cohen and Kuhi (1979) and assuming coeval formation, they could not find consistency between the theoretical and dynamical mass ratios. By relaxing the assumption of coevality, they could make the two mass ratios consistent, although only at the limits of the estimated errors for the secondary effective temperature and luminosity. Again, this binary raises doubts regarding either the mass and age calibrations of the theoretical tracks or the assumption of coevality. However, there is an interesting ambiguity with P1540, as the binary has a space velocity 3.5 times the velocity dispersion of the Trapezium cluster upon which it is projected. If the binary formed in the cluster, then it may have recently undergone a stellar encounter during which exchange of a binary component with the encountered star may have occurred. In this scenario, coevality of the binary components would no longer be expected.

Increasing the number of fundamental mass determinations, particularly for binary components with ages of 1 Myr or less, must be a primary goal of future spectroscopic binary work. While discovery of eclipsing systems is entirely a matter of fortune, analyses of double-lined PMS binaries similar to that for P1540 can provide critical tests of both relative mass calibrations and coevality of formation.

iii. Circumbinary Material. Spectroscopic binaries among PMS stars with significant infrared excesses or other diagnostics for the presence of substantial circumstellar disks represent important opportunities to study the evolution of disks in the binary environment. In fact, few spectroscopic binaries have been found among those T Tauri stars which show evidence for substantial circumstellar disks (see Mathieu [1989] and MWM for a more complete discussion of the observational situation). This paucity of detection is surprising (Herbig 1962*b*), given the amount of spectroscopic attention that has been given to these stars and the ease with which spectroscopic binaries have been detected among main-sequence stars and the naked T Tauri stars. It may be an important clue in understanding the interaction of binaries and circumstellar disks. Nonetheless, several PMS stars with substantial infrared excesses and line emission have been found to be spectroscopic binaries. Here we briefly discuss two of the more interesting cases, AK Sco and GW Ori.

The double-lined binary AK Sco has been studied extensively by Andersen et al. (1989), who found an orbit with a period of 13.6 days and an eccentricity of 0.47. Evidence for the existence of circumstellar material in the system rests largely on near- and far-infrared excesses. In addition, the star is variable at the 1 mag level, and Hα has been observed in emission with a central reversal (although no other emission lines are present in the spectrum). Remarkably, Andersen et al. (1989) found the relative luminosities of the stars to vary in the blue by as much as a factor 2 from one orbital cycle to the next. They ascribe both the infrared emission and the light variability to the presence of local dust clouds, in particular a cold (160 K) component with solar-system size scale and a hot (1600 K) component having a size scale comparable to the binary orbit. Alternatively, C. Lada (personal communication to Andersen) has suggested that the near infrared emission could as well be attributed to the power-law emission of a circumbinary disk. Perhaps the most exciting opportunity afforded by AK Sco is a relatively detailed map of the distribution of the circumbinary material; combining aperture photometry with the relative luminosities of each component throughout an orbit may permit determination of the extinction to each star at each point in the orbit.

GW Ori is a classic T Tauri star with strong Hα emission and with excess infrared luminosity comparable to the photospheric luminosity. The star is also a single-lined spectroscopic binary with a period of 242 days, corresponding to a separation of slightly greater than 1 AU (MAL). The spectral energy distribution of GW Ori is shown in Fig. 1, as derived from the literature. The pronounced dip in the energy distribution between 2 μm and 20 μm and the presence of a strong silicate emission feature at 10 μm are notable.

One intriguing explanation for the structure in the flux distribution is a nearly evacuated gap in an otherwise optically thick disk. Indeed, the presence of a companion is expected to create such gaps in circumstellar disks on size scales comparable to the orbital separation (Lin and Papaloizou 1986; see also their Chapter). In this picture, the near-infrared excess is provided by at least one circumstellar disk (likely circumprimary), while the far-infrared emission is attributed to a circumbinary disk. The silicate emission is attributed to optically thin hot dust between the two disks. The spectral energy distribution of such a configuration has been computed by MAL (Fig. 1). They find that a disk with a gap from 0.17 AU to 3.3 AU provides a reasonable match to the observed flux distribution; the model is consistent with the independent constraint that the gap location correspond to that of the companion orbit. However, the large disk luminosity, particularly in the far infrared, is difficult to produce through accretion mechanisms, as discussed in detail by MAL. This problem can be relieved somewhat if one attributes the far-infrared emission to re-processed photospheric light from a shell at distances greater than 200 AU from the binary (MAL). An appropriately flared disk is another possibility. In any case, the energy budget is such that the entire infrared luminosity cannot be explained as reprocessed photospheric light. Even a partial contribution from energy production in the disk through accretion raises fascinating questions regarding the dynamics of accretion in the binary environment.

2. Spatially Resolved Binaries. At present much of our knowledge garnered from spatially resolved pairs of PMS stars lies not in their binary properties *per se,* but rather derives from the fact that binaries consist of two stars which are presumed to have the same age and which typically have different masses. Spatially resolved binaries permit direct examination of the flux distributions and derived parameters (L_{bol}, T_{eff}, theoretical ages,

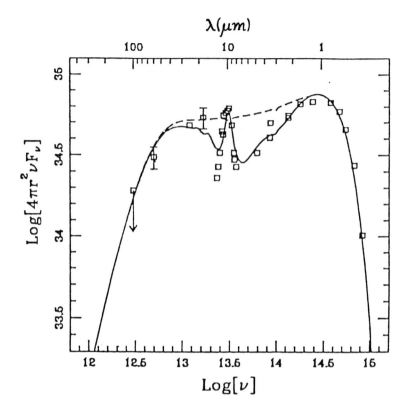

Figure 1.   The spectral energy distribution (open boxes) of the binary T Tauri star GW Ori, as compiled from the literature. Particularly notable are the dip in the flux distribution at mid-infrared wavelengths and the prominent silicate emission feature at 10 μm. The solid curve is the spectral energy distribution of the primary star and a disk having a gap between 0.17 AU and 3.3 AU. The dashed curve shows the spectral energy distribution of the same star-disk model without such a gap (figure from Mathieu et al. 1991).

infrared excesses, lithium abundances, etc.)   for each star independently. Thus, resolved binaries represent laboratories in which to study the mass dependence of PMS stellar and circumstellar evolution directly, always given the caveat that as circumstellar disk length scales are often greater than or comparable to the binary separations, evolution in the binary environment may not reflect evolution in isolation.

   i. Lunar Occultation. Occultation techniques are sensitive (at 2.2 μm) to binaries with separations from a few milli-arcseconds ($\sim$1 AU at 150 pc) to of order 1 arcsec, to be compared with the diffraction limit of a 4 m telescope of 140 milli-arcseconds. As such, lunar occultation observations are uniquely sensitive to binaries with periods at the peak of the main-sequence binary frequency distribution and overlap with both spectroscopic observations and techniques with lower spatial resolution. By good fortune, two of the nearest star-forming regions, Taurus and Ophiuchus, lie on the lunar path.

The most extensive lunar occultation studies of PMS stars have been done by Simon and his collaborators (Simon et al. 1987; Chen et al. 1990; Chen 1990). Combining the Taurus and Ophiuchus regions, they have observed occultation events for 31 objects and have found 9 double systems, several of which have projected separations of 10 AU or less (Table II). This detected binary frequency (29%) is in fact comparable to the frequency of Abt and Levy (1976) in the same regime of projected semimajor axis. Direct comparisons of the *detected* binary frequencies must be interpreted with caution, however, because the selection effects associated with the two studies are very different; as yet no attempt has been made to make a comparison after correcting for these biases.

TABLE II
Roster of PMS Lunar Occultation Binaries

Object	D^a		2μm Flux Ratio
	milliarcsec	AU	
HQ Tau	5	0.7	0.19
FF Tau	37	5.1	0.25
DI Tau	72	10.0	0.13
DF Tau	73	10.0	0.68
FW Tau	118	16.5	0.95
ROX 31	130	20.8	0.78
SR 12	186	29.8	0.85
FS Tau A	259	36.3	0.08
FV Tau	570	79.8	0.64

[a] The angular separations D are the projections of the true angular separation on the direction of occultation; as such they represent lower bounds. The one exception is DF Tau, where multiple occultation events permit a determination of the true angular separation (Chen 1990), which is given here.

There is no preferred value for the detected secondary-to-primary flux ratios. The distribution is reasonably uniform over the range 0.1 to 1. Because selection effects bias against systems that have large flux differences, the true distribution may be more concentrated toward smaller flux ratios (Chen 1990). This result must be interpreted with caution since the observations were made only at K, and the contributions to the light in many cases do not arise solely from the stellar photosphere (Moneti and Zinnecker 1991). Consequently, conversions from flux ratios to mass ratios are uncertain (Chen 1990).

Several of the detected binaries are active T Tauri stars. DF Tau in particular has large ultraviolet and infrared excesses, a veiled optical spectrum, and strong Balmer lines in emission. With a projected separation of only 10 AU the orbital period is likely to be only a few tens of years. Thus DF Tau, along with GW Ori and V4046 Sgr, show that the short-period binary environment does not necessarily preclude T Tauri activity. These

T Tauri diagnostics are often interpreted as deriving from the presence of active circumstellar accretion disks. For example, Bertout et al. (1988) have argued that the ultraviolet excess of DF Tau is indicative of an accretion rate of a few times 10^{-7} $M_\odot$ yr^{-1} into a boundary layer at the stellar surface. If this interpretation is correct, it is in contrast to the theoretical work of Artymowicz et al. (1991), who find that the formation of a binary in an accretion disk terminates accretion onto the binary components.

ii. Speckle/Slit Scan. Fittingly, the application of high-resolution techniques to PMS stars was stimulated by speckle observations of T Tau itself. Dyck et al. (1982) used one-dimensional near-infrared speckle interferometry to detect a companion at a separation of 0.61 arcsec (~100 AU). A positional ambiguity was resolved by Schwartz et al. (1984), the companion being south of T Tau. The companion is substantially cooler than T Tau, fainter at all wavelengths shorter than 4.8 μm and brighter at radio wavelengths, perhaps due to a mass outflow. Shortly thereafter, Nisenson et al. (1985) reported an optical companion 0.27 arcsec north of T Tau. Dynamically the discovery of this third component with a projected separation from T Tau comparable to that of the infrared companion is intriguing; if the true separations are similarly distributed, the triplet is unstable. However, Ghez et al. (1990) have done two-dimensional imaging at optical and near-infrared wavelengths and one-dimensional slit scans at 10 μm. They do not confirm the presence of the third (northern) component.

The discovery of the companion to T Tau augured the importance of using infrared techniques for the discovery of PMS binaries. In the last few years infrared speckle and slow slit-scan techniques have uncovered several infrared companions to low-mass young stellar objects (e.g., SVS20: Eiroa et al. 1987, Eiroa and Leinert 1987; DoAr24E [alias Elias 22] and Glass I: Chelli et al. 1988; Sz 30: see Reipurth 1988; Haro 6–10: Leinert and Haas 1989; DK Tau: Weintraub et al. 1989c; XZ Tau: Haas et al. 1990; MWC 863 [alias Elias 49]: see Zinnecker 1989a).

In some cases the companions are more luminous at infrared wavelengths than the optical component of the binary. One such binary, Glass I, is particularly intriguing. In Fig. 2 we show the flux distributions in the West-East direction at 1.25, 1.63, 2.19 and 3.61 μm; the infrared component is substantially brighter than the optical component at wavelengths greater than 2 μm (Chelli et al. 1988). More significantly, attributing the IRAS fluxes to the infrared companion, one finds that the infrared component has a higher bolometric luminosity (4 $L_\odot$) than the optical component (0.75 $L_\odot$). Haro 6–10 and XZ Tau show similar properties.

The interpretation of these three binaries is not clear. Chelli et al. (1988) linked the infrared luminosity of the more luminous component of Glass I to the reprocessing of a dust envelope, noting that it was unusual that the less luminous object had first dispersed its envelope. Nonetheless, given the youth indicated by the extreme T Tauri characteristics of the optical component, they find both components to be of comparable age. Zinnecker (1989a) suggests

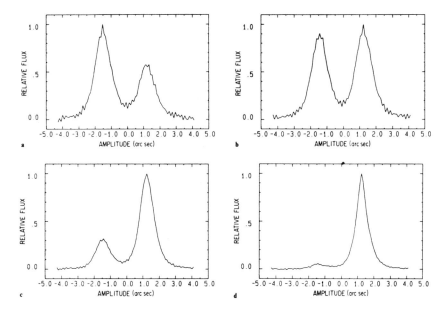

Figure 2.   Slit scans of the binary Glass 1 in the East-West direction at wavelengths of (a) 1.25 μm, (b) 1.63 μm (c) 2.19 μm and (d) 3.61 μm (figure from Chelli et al. 1988.) The bolometric luminosity of the infrared companion is greater than that of the optically visible star.

that higher accretion rates onto the infrared companions might lead to an over-sized, cooler star or substantial contribution to the luminosity from the disk accretion itself. He also notes that non-coplanarity of the disks might lead to differing extinctions, perhaps indicative of a capture origin. In any case, the discovery and detailed study of more binaries with such unexpected relative spectral distributions may provide substantial insight into the formation of binary companions and the evolution of disks in the binary environment.

The detection rate of infrared companions is accelerating; indeed at the time of writing the number of unpublished detections likely exceeds the number referenced above. While infrared observations have several important advantages, optical speckle work has also proven effective (see, e.g., Baier et al. 1985). Finally, the application of similar techniques to embedded massive stars has been equally fruitful in uncovering pairs of infrared sources (see, e.g., Dyck and Howell 1982; Dyck and Staude 1982; McCarthy 1982; Castelaz et al. 1985), although typically at much larger physical separations. In general, the application of high-resolution techniques to PMS stars is about to yield a wealth of important results.

iii.   Visual Binaries. By their very nature, "visual" pairs (including infrared pairs with separations similar to classical visual pairs) are often discovered serendipitously. In his classic review of T Tauri stars, Herbig

(1962*b*) tabulated 29 pairs in which at least one star was a T Tauri and commented that Joy had noted the presence of pairs in his discovery studies (Joy and van Biesbroeck 1944). Cohen and Kuhi (1979) listed 34 pairs serendipitously discovered with the Lick acquisition TV during the course of their spectroscopic survey. Reipurth (1988) reports the discovery of 28 pairs found during a systematic CCD survey of southern low-mass PMS stars, with an increase in frequency with smaller separation. Walter et al. (1988) find six visual pairs of naked T Tauri stars among 37 such systems in Taurus-Auriga. While some chance projections are certainly present in these samples, clearly visual companions to low-mass PMS stars are a common phenomenon. Time scale arguments show that systems with projected separations within tens of arcseconds at 150 pc are likely to be bound when considered in isolation, although the evolution of such systems in the molecular cloud environment remains to be investigated.

Despite this long history of visual pairs among young low-mass stars, little systematic analysis of the population has been done to date. However, at present array detectors (both optical and near-infrared) are being used at many observatories for imaging surveys of the PMS stellar population. Near-infrared imaging in particular holds great promise; it has proven effective in detecting companions to optically visible young stars, the companions often being of later spectral type and faint in the optical (see, e.g., Reipurth and Gee 1986; Sandell et al. 1987). Also, near-infrared imaging has permitted surveys for binaries in embedded regions (see, e.g., Rieke et al. 1989; Barsony et al. 1989*a*; Zinnecker 1989*a*). Finally, Wootten (1989) has found a candidate protobinary star (separation of 5 sec or 750 AU) with centimeter continuum maps of IRAS 16293−222 obtained with the VLA. The nature of the wide binary population should be substantially clarified in the very near future.

Wide pairs can be used to test the consistency of the assumption of coeval formation and the theoretical evolutionary tracks and to examine standard techniques for estimating PMS stellar ages. When analyzed independently, each component of a wide pair should fall on the same theoretical isochrone under these assumptions. Walter et al. (1988) have performed this test with five naked T Tauri pairs in the Taurus-Auriga region and found good agreement in the estimated ages for each pair of stars. Unfortunately, the youngest of the pairs has an age somewhat in excess of 1 Myr, so that none are young enough to be sensitive tests of the earliest stages of PMS evolution. In fact, there may be fundamental limitations to performing such a test upon the youngest stars, for the presence of disks and accretion produces substantial uncertainty in defining the photospheric parameters (Hartmann and Kenyon 1990; Kenyon and Hartmann 1990).

Recently, Moneti and Zinnecker (1991) have completed a near-infrared imaging study of nine PMS multiple systems. They find that the K brightness ratios range from 0.05 to 0.5 (median 0.1), somewhat smaller than the ratios found for the lunar occultation binaries. In nearly all cases the brighter component at K is the redder at J–K. Moneti and Zinnecker attribute the excess flux at K to disk luminosity and find that the ratio of disk to stellar luminosity increases with stellar luminosity. As they find a similar color-magnitude distribution for isolated T Tauri stars, they argue that in these wide binaries the two stars (and associated circumstellar disks) do not interact. This may suggest that the disks are smaller than about half the minimum separation of the binaries (∼200 AU in radius).

In a similar vein, Beckwith et al. (1990) and Beckwith and Sargent (see their chapter, especially their Fig. 5) have found an indication that the existence of a binary companion may act to limit the mass of associated disks. These investigators have observed at 1.3 mm a sample of 17 PMS binaries, most having projected separations between 1 AU and 1000 AU. Their observations permit estimates of disk masses; intriguingly, they find detectable disk masses only among those binaries with projected separations of $\gtrsim 100$ AU. This transition point is significant in that 100 AU is a typical value deduced for the outer radii of circumstellar disks; the result suggests that companions embedded within such disks act to reduce the total disk mass. As they note, the upper limits on the disk masses of those binaries with projected separations <100 AU are only slightly smaller than the detected masses. In addition, the expected number of detections among these closer binaries is only a few, so small-number statistics are a concern. Nonetheless, the result is reasonable theoretically and, if confirmed by future work, critical to our understanding of disk evolution in the young binary environment.

To conclude, we summarize those findings which most directly constrain the processes of binary formation and early evolution. Approximately half of main-sequence solar-mass stars are detected as binary systems; the actual binary frequency must be substantially higher (e.g., ∼80%; Abt 1983). Thus binary formation is the *primary* branch of the star formation process. The binary period distribution ranges over 10 orders of magnitude and is continuous with a single maximum. For longer-period binaries, the mass distribution of the secondary stars rises to smaller masses, similar to field initial mass functions. Whether the secondary mass distribution for the shortest-period binaries follows the same distribution is not yet settled and forms a critical gap in the observational picture. The vast majority of Population I main-sequence binaries with orbital periods longer than ∼10 days have eccentric orbits, while shorter-period systems have circular orbits. The transition period from circular to eccentric orbits depends on age.

Binaries have been found among stars very near the stellar birthline, so that binaries can form at ages of $<10^5$ yr. Observations to date suggest that binary components form coevally within an accuracy of ∼1 Myr. The first studies suggest that the PMS binary frequency and orbital period distribution are similar to those of main-sequence binaries (excepting a possible deficiency of short-period binaries among classical T Tauri stars); however, rigorous comparisons have yet to be done. PMS binaries have circular orbits for periods of less than 4 days and eccentric orbits for longer periods, with the one notable exception of GW Ori, a binary-disk system which has a circular or near-circular orbit at a period of 242 days. These orbital period and eccentricity distributions are likely the result of evolution prior to and during the PMS stage due to coupling with circumstellar disks, mass redistribution, and stellar tidal interaction. Tantalizing evidence for the evolution of disks in the binary environment may be found in the flux distribution and very low orbital eccentricity of GW Ori, in the lack of detection of 1.3 mm emission among binaries with separations of less than 100 AU, and in the lack of detection of the shortest-period binaries among classical T Tauri stars. These results suggest that the amount of disk material is substantially reduced at scale lengths comparable to the binary separation. Nonetheless, observations indicate that accretion near the stellar surface(s) can continue in the binary environment. Young binaries are observationally accessible analogs to planetary systems, and the study of binary-disk systems will be a particularly fruitful forefront for observers in the near future.

III. ANGULAR MOMENTUM IN PROTOSTELLAR MATERIAL

We inquire here whether the orbital angular momentum of a binary or multiple system can reasonably be associated with the remnant of the angular momentum of the molecular cloud core from which the stellar system formed. In this section we summarize observational data on rotation of molecular cloud material, discuss the turbulent nature of the rotation of cloud cores, and mention mechanisms of angular momentum redistribution at the different stages of star formation.

A. Rotation Rates in Molecular Clouds and Cloud Cores

Observations reveal that large molecular clouds are quite inhomogeneous and clumpy. Frequently they show inhomogeneities down to the smallest resolvable scales, which are smaller than the likely Jeans length in the clouds. Stars of roughly solar mass or less form in the small-scale (≤ 0.1 pc) densest region of molecular clouds, which are referred to as cloud cores. The chaotic internal motions suggest that it is reasonable to consider the clouds as turbulent, taking into account that turbulent flows actually consist of a hierarchy of small-scale irregularities superimposed on larger-scale more systematic motions such as rotation or expansion (Larson 1981; Dickman 1985; Kleiner and Dickman 1987).

The turbulence in the idealized incompressible fluid of classical laboratory hydrodynamics is characterized by a hierarchy of vortices on a range of length scales down to the scale where dissipation becomes important. However, the interstellar turbulence may be quite different and more complicated. The observed internal velocities could refer to the independent motions of clumps, filaments and other wispy condensations superimposed on a background of warmer less-dense matter. Such motions could, in principle, be vortex free, for example, if the motions result from fragmentation of an initially hydrostatic nonrotating cloud. The random motions of fragments in this case could describe motions of clumps under the action of mutual gravitation of the system of clumps. Such a system slowly contracts at the rate set by dissipation due to clump collisions and drag, but the kinetic energy of clumps is continually extracted from the gravitational field of their parent cloud rather than from shear motion on a larger scale as in incompressible turbulent flow (Scalo and Pumphrey 1982). Furthermore, the turbulence in the interstellar medium is affected by energy deposition and energy sinks on several scales at once. It is not surprising therefore that the real interstellar turbulence in molecular clouds has much smaller characteristic rotational velocities on large scales than would be expected for Kolmogoroff turbulence, based on the velocity dispersion on the small scales (Scalo 1984; Kleiner and Dickman 1985). Thus the velocity dispersion in the cloud is not a direct measure of rotation, which must be measured independently.

The rotation of interstellar clouds or their subregions is usually determined by observation of systematic shifts in the velocity of the centroid of a spectral line along a cut through the cloud (Goldsmith and Arquilla 1985). Naturally the shift must be not too small compared with the width of the spectral line. Therefore the lower limit on the rotational rate which can be measured depends on the dimension of the cloud and its temperature or turbulent velocity. If a cloud possessed a highly ordered velocity field, the observation of a gradient of the radial velocity across the cloud would be sufficient proof of the presence of rotation. However, in fact, different types of mass motion can produce gradients more or less similar to those resulting from rotation. Therefore additional information (such as detailed velocity field, mass distribution, structure of the magnetic field) is needed to assess the probable validity of the rotation model. Arquilla and Goldsmith (1986) have undertaken a detailed study of the velocity fields of eight clouds, six of which were previously referred to in the literature as fast rotators. They analyze the radial-velocity distribution using well-filled maps of ^{13}CO spectra, generally more coarsely sampled ^{12}CO maps, and extensive star counts. Kinematic models of the sources were developed from the examination of the relevant spatial velocity diagrams and ^{13}CO column-density maps. The data strongly supported the presence of rotation in only three program clouds (B361, L1253, L1257). One object (CRL 437) has been removed from the list of rotating clouds, and the rotational model is in difficulty for the globules B163 and B163 SW. As most of these clouds were known to have large velocity gradients across them, these results imply that rapid rotation in dark clouds may be less common than hitherto suspected, because in most other measurements of rotation, such a detailed analysis was not applied. In general, there is a range of deduced values of the rotational velocity at each size scale, and the mean specific angular momentum clearly decreases at smaller scales (Goldsmith and Arquilla 1985). Typical results indicate that clouds with radius 0.5 to 1.0 pc and a few hundred solar masses rotate with $\omega \sim 3$ to 11×10^{-14} rad s^{-1}. The typical specific angular momentum of the clouds that rotate is therefore $j \sim 10^{23}$ cm^2 s^{-1}. Most of the clouds seem to rotate rigidly, although the outer envelope of B361 shows evidence for differential rotation.

Similar difficulties occur in the measurement of rotation of cloud cores. Bipolar outflows from embedded low-mass young stars could create the observable shift across some cores; recent observations reveal a high fraction of outflows in cores with stars. Therefore the measured velocity gradient can be considered only as an upper limit on the rotational velocity projected onto the line of sight (Fuller and Myers 1987; Mathieu et al. 1988). In small cores, the velocity gradients are less than the observational limit in approximately half of studied cases (Myers and Benson 1983; Ungerechts et al. 1982; Heyer 1988). In Heyer's (1988) work on cores in Taurus the mass range is 0.3 to 38 $M_\odot$, the range in radius is 0.1 to 0.5 pc, the mean density range is 0.1 to 2.6×10^4 cm^{-3}, and the observed velocity gradients are 0.2 to 1.5 km s^{-1} pc^{-1}. Corresponding angular velocities, Ω are 0.6 to 5×10^{-14} rad s^{-1}, and values of j (at the outer edge) fall in the range 1 to 45×10^{21} cm^2 s^{-1}, increasing on the larger scales. The small (0.1 pc) cores, which represent initial conditions for star formation, have $j \sim 10^{21}$ cm^2 s^{-1}, if they are observed to rotate. Similar results were found in earlier work (Myers and Benson 1983; Ungerechts et al. 1982; Harris et al. 1983; Wadiak et al. 1985). It is worth noting, however, that the observational limit at this scale is $\Omega \sim 2 \times 10^{-14}$ s^{-1} or $j \sim 10^{21}$ cm^2 s^{-1}. Therefore the observations do not exclude the existence of cores with much slower rotation. The cores are usually centrally condensed in density: $\rho \sim r^{-p}$ with $1 \leq p \leq 2$ (Myers et al. 1987). It is not known whether differential rotation is present.

The typical angular momenta observed in cloud cores may be compared with those in binary systems and single stars (Table III). This table emphasizes a number of points. First, the angular momentum associated with orbital motion in binaries is considerably larger than that associated with rotation of individual stars. Second, the orbital angular momenta in long-period binaries have the same order of magnitude as those inferred from observations of cloud cores. However, it is not known whether some cloud cores have angular momenta consistent with those of the systems with the shorter periods. These close systems do have far too much angular momentum to be explained by fission of (now slowly) rotating T Tauri stars. Third, the value of j in the solar nebula, as deduced from Jupiter's orbit, is comparable to that of binary orbits of intermediate period, raising the question of whether the solar system formed from a very slowly rotating cloud ($j \leq 10^{19}$) with outward transport of angular momentum occurring into a small amount of mass at a later stage, or from one with $j \sim 10^{20}$. This question is further discussed in Sec. V.

B. Physical Nature of Rotation in the Cloud Cores

Interstellar turbulence (Cameron 1962) and galactic differential rotation (Hopper and Disney 1974) were considered to be the main mechanisms causing cloud rotation. The absence of a preferred orientation for the rotational axes of single stars or the orbital planes of binaries (Kraft 1970) or, in fact, the rotational axes of molecular clouds themselves (Goldsmith and Arquilla 1985),

TABLE III

Characteristic Values of Specific Angular Momentum

Object	J/M (cm^2 s^{-1})
Binary (10^4 yr period)	4×10^{20}–10^{21}
Binary (10 yr period)	4×10^{19}–10^{20}
Binary (3 day period)	4×10^{18}–10^{19}
T Tauri star (beginning of contraction phase)	5×10^{17}
Jupiter (orbit)	10^{20}
Present Sun	10^{15}
Molecular cloud (scale 1 pc)	10^{23}
Molecular cloud core (scale 0.1 pc)	10^{21}

means that the rotation of the cores in molecular clouds is probably caused by interstellar turbulence. Larson (1981) pointed out that in the scale range $0.05 \leq L \leq 60$ pc the turbulent velocity is v_t (km s^{-1}) $= 1.1 \, L^{0.38}$(pc). The Kolmogoroff spectrum, which describes a dissipationless cascade of vortices, has a power-law index of 1/3. Subsequent studies have shown, however, that there is appreciable scatter in the data and that the power-law index in different clouds ranges from 0.3 to 0.6, implying that the energy content of random motions does not depend on the length scale in some clouds (as in Kolmogoroff turbulence) but increases with length scale in others. Thus (see also Sec. III.A) the turbulence in molecular clouds probably does not represent the single dissipationless cascade from the scale determined by the cloud dimension to the scale in which the turbulence is dissipated due to molecular viscosity. The similarity with the Kolmogoroff spectrum, however, is more likely on the intermediate length scales where the turbulence is subsonic and where there is no significant dissipation. If so, the turbulence in molecular clouds could be considered as a superposition of cascades with randomly distributed energy sources on several length scales. One could expect that under such circumstances the interstellar turbulence would have an intermittent character (Ruzmaikina 1986,1988). The intermittency means that turbulent vortices are distributed nonuniformly in time and space, i.e., small regions of intense rotational motions, whose locations change with time, are separated by extended quiet regions.

Laboratory investigations of turbulence in an incompressible fluid show that intermittency is most marked in strong turbulence at high Reynolds numbers on the dissipative length scale $l_d = (v^3/\epsilon)^{1/4}$, where v is the kinematic viscosity and ϵ is the energy dissipation rate per unit mass (Stewart and Townsend 1951; Kuo and Corrsin 1971). If applied to the turbulence in molecular clouds, the intermittency suggests the existence of a number of regions with relatively small turbulent velocities on the length scales in which dissipation is important.

In magnetized molecular clouds there are two dissipative length scales: one is associated with the molecular viscosity and the second with magnetic viscosity. The magnetic viscosity v_{ad} is produced by ambipolar diffusion, which is the most effective mechanism of dissipation of the magnetic field and turbulence. It is much larger than the molecular viscosity as long as an equipartition exists between turbulent and magnetic energy for $l \geq l_d$, that is, $H^2 \sim 4\pi\rho v_t^2$, where H is the intensity of the magnetic field and ρ is the density. The degree of ionization is defined by n_i/n_n, where n_i and n_n are the particle densities of ions and neutrals, respectively. It is governed by ionization of gas by cosmic rays (with rate $\zeta = 10^{-17}$ to 10^{-18} s^{-1}) and by recombination of ions on the dust particles. The velocity of ions relative to neutrals $v_{in} = v_i - v_n$ is of the order of the turbulent velocity on the length scale of dissipation l_d. Then

$$v_{ad} = \frac{v_t^2 m}{2\mu_{in}\langle\sigma u\rangle_{in}n_i} \tag{1}$$

where μ_{in} is the reduced mass of ions and neutrals and $\langle\sigma u\rangle_{in}$ is the mean rate of collisions between ions and neutrals. The dissipation length scale l_{ad} associated with ambipolar diffusion is

$$l_{ad} = \frac{v_t m}{2\mu_{in}\langle\sigma u\rangle_{in}n_i} = 0.08 \left(\frac{\zeta}{10^{-17}s^{-1}}\right) \text{ pc.} \tag{2}$$

For parameters typical for the dense regions of molecular clouds, this formula gives values comparable to the dimension of the small cores containing only one or a few solar masses.

The laboratory intermittent turbulence obeys a log-normal distribution of energy dissipation rate ϵ. If the vorticity on the length scale l_d in molecular clouds follows the same distribution, then the probability for a cloud core to have an angular momentum per unit mass notably different from the average is much larger than for normal turbulence (Ruzmaikina 1986,1988). The log-normal distribution of the angular momentum J means that a probability to have a given $J = \psi <J>$ is

$$P(\psi) = \frac{1}{\sqrt{2\pi}D\psi}\exp\left[\frac{-\left(\ln\psi + \frac{D^2}{2}\right)^2}{D^2}\right] \tag{3}$$

where $D^2 = \langle(\ln J - \langle\ln J\rangle)^2\rangle$. Unfortunately, there are few measurements of the gradient of radial velocity across dense cores with dimensions ≤ 0.1 pc and mass of about 1 $M_\odot$. Assuming the measured velocity gradient is entirely caused by rotation and that a typical value for the cases when no rotation was observed is one-half of the upper limit, then $<J> = 7.8 \times 10^{53}$ g cm^2 s^{-1} and $D = 0.934$ according to the observational data. Then the fraction of slowly rotating dense cores could be large (Ruzmaikina 1989). This argument, although not strongly confirmed by observations, suggests that the turbulent motions of the interstellar gas could produce the angular momenta of the full range of binary orbital periods and possibly also of the single stars. However, the picture is undoubtedly more complicated than a simple one-to-one mapping between initial cloud rotation rate and final binary period, as both fragmentation and redistribution of angular momentum must be taken into account.

C. Mechanisms for Redistribution of Angular Momentum

The importance of the redistribution of the angular momentum at the stage of star formation is connected with a significant enhancement of density and decrease in the moment of inertia of the collapsing cloud. For example, if a sphere containing 1 $M_\odot$ contracts from an initial density of 10^{-24} g cm^{-3} to the average density of the Sun (1 g cm^{-3}), then the radius decreases by a factor 10^8 and the moment of inertia by a factor 10^{16}. It is clear, therefore, that if the sphere contracts with local conservation of angular momentum, then even negligible initial rotation could be dynamically important during contraction and could stop the contraction in the direction perpendicular to the rotation axis, resulting in fragmentation. The redistribution of angular momentum could be a crucial factor in determining what kind of system is produced.

The efficiency of redistribution depends strongly on the stage of protostellar evolution. It definitely could be high at the early stages at molecular cloud densities ($\sim 10^3$ cm^{-3}). If the typical magnetic field in a cloud is $\sim 10^{-5}$ gauss and if the magnetic field is frozen into the material, then the field is effective in braking the rotation of the cloud and in redistributing the angular momentum inside the cloud (Mouschovias 1977; Mestel 1985; Ruzmaikina 1985; Shu et al. 1987; Chapter by McKee et al.). The time scale for these effects is 10^6 to 10^7 yr, which is shorter than the evolution time of the cloud as long as the coupling between the field and gas is maintained. However, this coupling decreases because of ambipolar diffusion as the density approaches the values for the dense cores of molecular clouds (Mestel and Spitzer 1956; Lizano and Shu 1989; Ruzmaikina 1981a,1985). Magnetic braking becomes less effective, but ambipolar diffusion itself can still be effective in redistributing the angular momentum because of the relatively large effective viscosity v_{ad} associated with the drag of the ionized component moving through the neutral component (Sec. III.B).

During the protostellar collapse itself, the evolution time is comparable to the dynamic time, and there will not be appreciable angular momentum transport by any process, unless initial nonaxisymmetric perturbations are especially large. The magnetic field is dynamically unimportant on the time scale of collapse because of the effects of ambipolar diffusion, which restricts the enhancement of the magnetic field during collapse and decouples the field from the matter. The redistribution of angular momentum associated with v_{ad} has a time scale longer than that of collapse. Similarly, turbulent velocities are not likely to exceed the sound speed because of greatly increased dissipation at supersonic speeds. Numerical simulations show, however, that the collapse is supersonic at most stages. The time scale for angular momentum transport by turbulent viscosity is larger than the collapse time as long as the turbulent velocity $v_t < v_{collapse}$. Also, nonaxisymmetric perturbations will not produce appreciable transfer by gravitational torques because the perturbations do not grow appreciably with respect to the collapsing background until the collapse has been slowed considerably by rotational and pressure effects. After about one free-fall time, when the cloud has become rather disk-like and is approaching equilibrium, the perturbations can grow and fragmentation can occur; the process involves conversion of spin angular momentum of the cloud into orbital motion of the fragments (Sec. IV).

If the critical condition for fragmentation is not reached, however, then the outcome is likely to be a central object plus a disk. A number of transport processes now become possible—turbulent, magnetic and gravitational torques produced by spiral density waves (see review by Larson 1989). The question arises whether the disk can ever become unstable to the formation of a binary if it has not already fragmented during its formation. The answer will depend on the time scales of the various processes.

The disk is very likely to be turbulent because of the high Reynolds number $R_e \sim (GMR)^{1/2}/\nu \geq 10^{10}$, where ν refers to the molecular viscosity. Mechanisms which could, in principle, result in turbulence include shear motion due to differential rotation (Zeldovich 1981), relative motion of accreting material with respect to the disk (Cameron 1962), and thermal convection (Lin and Papaloizou 1980). At present the best developed theory is that of turbulent viscosity arising from convective instability in the vertical direction. The results of such calculations are approximately represented by the viscosity parameter α (Shakura and Sunyaev 1973) which is $\sim 10^{-2}$ (Lin and Papaloizou 1980) or 10^{-2} to 10^{-4} (Cabot et al. 1987a, b). The corresponding time scale t_r for redistribution of the angular momentum over a length scale of the order of the present-day size of the solar system ranges from 10^5 to 10^7 yr. Estimates of the turbulent viscosity in the disk arising from shear between accreting material and the disk give $\alpha \sim 10^{-2}$ and $t_r \sim 1$ to a few times 10^5 yr (Ruzmaikina 1982). The shear flow will definitely induce turbulence near the surface of the disk, but it is also possible that turbulence can be generated throughout the entire thickness of the disk. This statement is supported by measurements of turbulence in the ocean, where the turbulence is excited at the surface by wind. In spite of a stable density stratification, intermittent turbulence is measured down to depths of several km. Finally, turbulence associated with differential rotation can proceed through development and interaction of global nonaxisymmetric unstable modes (Sekiya and Miyama 1988b). The growth time is $\sim 10^3$ to 10^4 yr, but the evolution time t_r is uncertain.

Conditions for the occurrence of axisymmetric gravitational instability in a disk have been investigated by Safronov (1960) and Toomre (1964). Numerical calculations of a disk with a central point mass (Cassen et al. 1981) indicate that nonaxisymmetric effects in the form of trailing spiral waves develop, which result in transfer of angular momentum if $M_{disk} > M_{central}$. A disk of moderate mass $M_{disk} \sim M_{central}$ can be unstable to nonaxisymmetric disturbances having growth rates comparable to the orbital frequency at the outer edge (Adams et al. 1989). Lin and Pringle (1987) derived a time scale $t_r \sim 10\Omega^{-1}$. However, it is not clear whether angular momentum is transported over large radial distances. Vertical temperature stratification in the disk can refract waves with azimuthal wave number $m \geq 2$ to high altitudes, where they may steepen and dissipate (Lin et al. 1990). Adams et al. (1989) suggest that spiral waves with $m = 1$ may be most effective at transporting angular momentum over large radial distances. These are eccentric modes in which the star and the disturbance orbit around their common center of mass. The unstable modes can encompass the entire disk and be driven nonlinearly by these eccentric motions. In general, the gravitational instabilities result in transport of angular momentum through spiral waves on a near-dynamical time scale. However, it is possible that under special conditions the $m = 1$ instability could result in the formation of a binary (Sec. IV).

Magnetic fields amplified during the collapse of a cloud or generated by a hydromagnetic dynamo can result in angular momentum transport in the inner part of the disk (Lüst and Schlüter 1955; Hoyle 1960; Alfvén and Arrhenius 1976; Levy and Sonett 1978; Ruzmaikina 1981a,1985; Hayashi 1981; Stepinski and Levy 1988). The central region is hot enough for the evaporation of dust particles and thermal ionization of the alkaline metals ($T \geq 1600$ K). According to some models of protostellar evolution the radius of the region where the field is coupled sufficiently to the gas changes from a few tenths to about 1 AU (Makalkin 1987; Ruzmaikina and Maeva 1986). At larger distances it is generally thought that the degree of ionization near the central plane would be much lower because the thermal ionization is negligible and the disk is opaque to external sources of ionization such as cosmic rays.

However, non-negligible magnetic effects can occur also in those regions that are transparent to galactic cosmic rays—at high altitudes above the midplane and in the outer regions of the disk where the surface density is $< 10^2$ g cm^{-2}. From data on molecular clouds, one can expect that the cosmic ray ionization rate $\zeta \sim 10^{-17}$ s^{-1}, which makes possible a moderate enhancement of the magnetic field.

The maximum intensity of the azimuthal magnetic field B_ϕ that can be generated from an axially symmetric poloidal component B_p is limited by Ohmic dissipation to $B_\phi \leq R_m B_p$ (where R_m is the magnetic Reynolds number associated with the differential rotation), or by ambipolar diffusion (Levy et al. 1991). In a Keplerian disk the limiting value is

$$B_\phi^2 \leq 10^{11} \rho \left(\frac{\xi}{10^{-17}\text{s}^{-1}} \right) \sqrt{R/1\text{AU}} \sqrt{M/1\text{M}_\odot} \quad \text{gauss}^2. \qquad (4)$$

In the case when the solar nebula is turbulent, the hydromagnetic dynamo could generate both B_ϕ and B_p. Ambipolar diffusion restricts the maximum intensity of the field to (Levy et al. 1991)

$$B_\phi \sim 4\alpha^{1/4} \left(\frac{\xi}{10^{-17} \text{ s}^{-1}} \right) \sqrt{R/1 \text{ AU}} \ \sqrt{M/1 \text{ M}_\odot} \text{ gauss.} \qquad (5)$$

In general, B_p is less than B_ϕ by a factor of $\sqrt{D}$, where D is the dynamo number ($\propto \alpha^{-1}$; α is the turbulent viscosity parameter). The time scale of redistribution of angular momentum has the lower limit $t_r = \rho\sqrt{GMR}/(B_p B_\phi) \sim 10^8$ to 10^{10} s in the region beyond several AU where the matter is transparent to cosmic rays. Even if the field in the outer part of the nebula is just a remnant of the compressed interstellar field, t_r can be short compared with accretion times. Then $t_r \sim (\rho/10^{-11}) \times 10^9/(B_p B_\phi) \sim (\rho/10^{-11}) \times 10^9/B^2$ is less than the accretion time (10^5 yr) if $B \geq 1.7 \times 10^{-2}$. If $B \propto \rho^\gamma$ during collapse and the interstellar field in a cloud core is 10^{-5} gauss, then the required value for γ is 0.4; if the initial field is as high as 10^{-3} gauss, the value of γ is only 0.125. Thus only a modest enhancement of the interstellar field can have significant consequences once the disk stage is reached.

To summarize, the magnetic field could be an effective mechanism for redistribution of angular momentum both in the early stages of star formation and in a disk. Magnetic transport probably accounts for the difference of 2 orders of magnitude in the specific angular momentum of molecular clouds themselves and their cold, dense cores. In a disk, magnetic effects can be important except in the dense, cold, essentially un-ionized region at intermediate distances from the central object (e.g., 1 to 10 AU for central mass $M_c = 1$ M$_\odot$ and disk mass $M_d \sim 10^{-2}$ to 10^{-1} M$_\odot$). The time scale in the inner warm region can be as short as 10 times the orbital frequency (Hayashi 1981). In the outer regions the times could fall in the range 1 to 10^5 yr. Gravitational instability, generally in a relatively massive disk, can also result in rapid transport, on time scales again ~ 10 orbital periods. Turbulent shear flow during the infall phase is estimated to give transport times of $\sim 10^5$ yr. Turbulent convection is likely to be present in the 0.5 to 20 AU region where grain opacity dominates and where the nebula is optically thick. The resulting transport times fall in the range 10^5 to 10^7 yr. If a binary is to form in a disk, its formation time must not be much longer than that of the fastest applicable mechanism for angular momentum transport; the appropriate conditions have not been determined.

IV. THEORIES OF BINARY FORMATION

Theoretical work at the present time is still in the stage of identifying the primary mechanism for binary formation. Theory is not yet able to clarify the important questions concerning the binary frequency and the distributions of orbital periods and mass ratios. It is also not clear under what circumstances single stars form (see Sec. V). In fact, for a cloud of given mass and total angular momentum, it is energetically favorable to form a single star with a low-mass disk or a binary with a very large mass ratio (Lynden-Bell and Pringle 1974). One might expect that most stars would be found in such systems, but observations show otherwise. Presumably, the angular momentum redistribution required to produce them is in many cases not effective enough to prevent fragmentation at a relatively early phase of protostellar evolution. In this section we discuss, first, binary formation by fragmentation during the hydrodynamic collapse of a protostar; second, fragmentation after a relatively massive disk has formed; and third, other binary formation processes such as capture and fission.

A. Fragmentation During Protostellar Collapse

The theory of fragmentation was introduced by Hoyle (1953) but without the crucial element of rotation. Recent work on fragmentation during the protostar phase has concentrated on numerical calculations of hydrodynamical collapse with rotation in three space dimensions. The evolution of the protostar may be divided into the earlier optically thin isothermal phase and the later adiabatic phase, which sets in when the density ρ has increased to values above 10^{-13} g cm^{-3}. The three-dimensional calculations generally start out with uniform density, uniform angular velocity, and small nonaxisymmetric perturbations in the density. The importance of rotation is measured by the ratio β of rotational to gravitational energy. If angular momentum is conserved locally, then $\beta \propto \rho^{1/3}$ in the limit of slow rotation. During the first free-fall time the cloud collapses, becomes centrally condensed, and develops rotational flattening; however little growth of the perturbations occurs, relative to the collapsing background. The collapse then slows down because of pressure effects parallel to the rotation axis and rotational effects perpendicular to it. During the process of the formation of a disk-like structure, the system fragments, generally through a nonaxisymmetric pattern, into two or more orbiting subcondensations. The properties of the fragments depend on the initial perturbation. An $m = 2$ perturbation will generally result in fragments of roughly equal mass, with a total mass of about 15% of the original cloud mass over the time scale of the calculations, which are generally run only 1 to 1.5 initial free-fall times (Boss 1986a). A perturbation closer to $m = 1$ will result in highly unequal-mass fragments (Boss 1990). The typical fragment has a residual spin angular momentum per unit mass which is an order of magnitude or more smaller than that of the original cloud.

Extensive calculations have been carried out for the isothermal case. If rotational effects are to become important while the gas is still isothermal, the degree of collapse must be relatively mild, and therefore the initial angular momenta must be relatively large; wide systems are produced. Exploratory calculations with $m = 2$ perturbations of relatively large amplitude produced binaries (Bodenheimer et al. 1980; Boss 1980; Bodenheimer and Boss 1981); more recent calculations with small perturbations have produced larger numbers of fragments. If we define α_i and β_i to be the initial ratios of thermal and rotational energy, respectively, to the absolute value of the gravitational

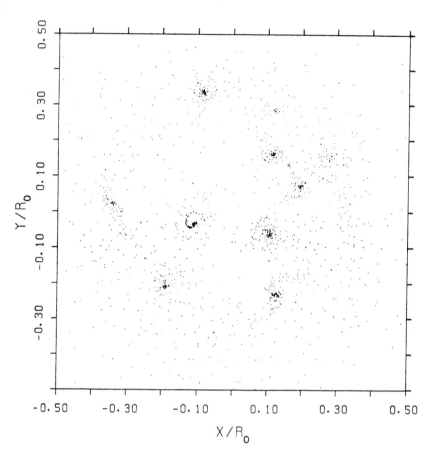

Figure 3.   Fragmentation of an isothermal cloud with $\alpha_i = 0.2$, $\beta_i = 0.3$ (Miyama et al. 1984). Projected particle positions are plotted in the equatorial (x,y) plane, after 2.19 initial free-fall times. R_o is the outer radius.

energy, then we find that the number of fragments increases as α_i decreases, as was demonstrated by Larson (1978) with a simple particle core. Figure 3 shows the result of Miyama et al. (1984) with $\alpha_i = 0.2$ and $\beta_i = 0.3$, starting with density fluctuations $\delta\rho/\rho \sim 0.05$, in which about 8 fragments formed. The numerical method is a modification of the smoothed-particle hydrodynamic scheme originally developed by Lucy (1977). The individual fragments, which have roughly equal mass, also have low enough α_i and β_i so that they can collapse and fragment again, forming hierarchical multiple systems (Larson 1972; Bodenheimer 1978). A recent calculation by Monaghan and Lattanzio (1991), including molecular cooling in the optically thin phase, shows similar results. Their smoothed-particle calculations were carried out with up to 30000 particles, starting with random initial density perturbations with typical initial amplitudes of 14%. With $\alpha_i = 0.47$ and $\beta_i = 0.3$, the system evolved, after cooling, into 6 fragments in a roughly ring-like configuration (Fig. 4).

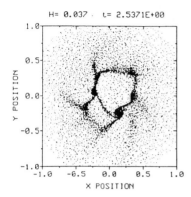 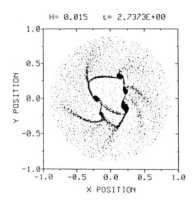

Figure 4.   Fragmentation of a cooling cloud with $\alpha_i = 0.25$, $\beta_i = 0.04$ (Monaghan and Lattanzio 1991). Projected particle positions are plotted in the equatorial (x,y) plane at two different times (given in units of $R^3 M^{-1} G^{-1}$, where R and M are the initial radius and mass of the cloud, respectively). The resolution length H is given in terms of R.

For values of the initial angular momentum that are small enough to produce a close binary, the central part of the cloud becomes opaque before rotational effects become significant in promoting fragmentation, and fragmentation will tend to be suppressed (Safronov and Ruzmaikina 1978; Ruzmaikina 1988). Calculations of fragmentation in the adiabatic phase are more difficult, because the cloud develops a large density contrast before the central density reaches 10^{-13} g cm^{-3}, and the inner regions, which contain only a small fraction of the total mass, must be adequately resolved. Calculations for purely adiabatic collapse (Boss 1980) and for the transition between the optically thin and thick regions (Boss 1986a) give the general result that if α_i is high at the beginning of adiabatic collapse, fragmentation is suppressed. For $\alpha_i \sim 0.05$ to 0.1, fragmentation can occur. Figure 5 shows a calculation of Boss (1986a) with $\alpha_i = 0.25$ and $\beta_i = 0.04$, starting in the isothermal phase and including radiative transfer. In this particular case, fragmentation into a binary occurs just before the adiabatic phase sets in. The distance between the orbiting fragments is 110 AU and the period is ~ 1000 yr, emphasizing the fact that fragmentation in the isothermal phase can explain only the longer-period systems. The reason is related to the fact that the central part of a protostar becomes optically thick on a scale of 100 AU because of grain opacity. In a low-angular-momentum cloud rotational effects become important in promoting fragmentation only in the optically thick region, where nearly adiabatic heating and the consequent increase in pressure tend to smooth out perturbations. However, when the central region reaches a temperature of

2000 K, molecular dissociation occurs, the adiabatic Γs of the gas decrease to values close to those appropriate for the isothermal phase, and conditions again become favorable for fragmentation (Larson 1972). The scale of this region is <1 AU, and the possibility of forming close binary systems at this stage exists but has not been adequately tested. Problems with this suggestion that must be considered include (1) the amount of mass initially present in the unstable region is very small, (2) later addition of material of higher angular momentum from the outer part of the cloud will tend to separate the system, and (3) generation of spiral waves in the cloud by the binary will tend to bring the components closer together.

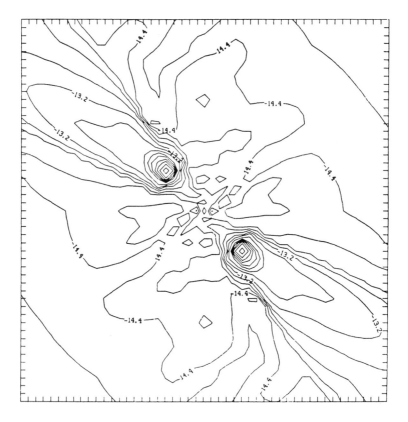

Figure 5. Fragmentation calculation including radiative transfer with $\alpha_i = 0.25$, $\beta_i = 0.04$ (Boss 1986a). Contours of equal density in the equatorial plane are plotted after one freefall time. Sample values of log ρ are indicated; the initial density was 2×10^{-18} g cm^{-3}. The length of one side of the box is 360 AU.

Alternate modes of fragmentation have been suggested, for example, in the collapse of cylindrical nonrotating clouds (Bastien 1983; Rouleau and Bastien 1990). A related mechanism involves elongated, prolate filamentary fragments rotating end over end (Zinnecker 1989a), in which, after fragmentation, the components first move almost directly toward each other until they reach the centrifugal barrier. Numerical calculations by Bonnell et al. (1991) indicate the formation of multiple systems by this process. Such elongated dark clouds are commonly observed in star-forming regions. Finally, it has been argued (Boss 1988) that fragmentation can occur on all scales and can explain the full range of binary orbital angular momenta. To explain the formation of close binaries by this process he relies on first, orbital decay of longer-period binaries by gravitational torques, and second, hierarchical fragmentation in which low-α initial conditions are provided in the adiabatic phase by previous fragmentation near the end of the isothermal phase. His numerical results (Boss 1986a) show that fragmentation in the adiabatic phase can produce systems with separations as small as 1 AU, given suitable initial conditions.

General criteria for the fragmentation of rotating collapsing clouds have been developed. For isothermal collapse, Miyama et al. (1984) have shown that fragmentation occurs if $\alpha_i \beta_i < 0.12$. Above this limit, nonfragmenting equilibria or disk systems are formed. For an adiabatic gas with $\gamma = 1.4$, fragmentation occurs if $\alpha_i < 0.09 \beta_i^{0.2}$ (Miyama 1989a; Hachisu et al. 1987; Tohline 1981; Boss 1981). This criterion is not necessarily useful for protostars, which begin their collapse in the isothermal phase and evolve into the adiabatic phase only in their central regions at a later time. Thus wide binaries (those with relatively high angular momentum) can be understood to have been formed by fragmentation in the isothermal phase. But a major problem remains concerning the origin of close binary systems. A further possibility, apart from fragmentation during dissociation or hierachical fragmentation during the adiabatic phase, is that clouds with relatively low angular momentum collapse through the adiabatic phase without fragmenting, then evolve to a central object plus a relatively massive equilibrium disk. Gravitational instability in such a disk could then result in a close binary.

B. Disk Fragmentation

The standard indicator for the occurrence of gravitational instability in a thin disk, which could, under the right conditions, lead to fragmentation, is the Toomre (1964) $Q = \kappa c/(\pi G \sigma)$, where κ is the epicyclic frequency, c is the isothermal sound speed, and σ is the local surface density. Axisymmetric perturbations are stabilized if $Q > Q_c$, where $Q_c \sim 1$. Larson (1985) generalized this condition to thin gaseous disks which are not necessarily isothermal, obtaining the same formula with c evaluated at the midplane. Values of Q_c range from 0.53 to 0.68, depending upon the equation of state. Hachisu et al. (1987,1988) develop a revised formulation of the criterion. They give convincing arguments that the condition for occurrence of an axisymmetric ring instability in the equilibrium configuration also gives the condition for instability to fragmentation. Two conditions must be met if the disk is to fragment. First, the ratio $t_{ff}/t_s < f_J$, where t_{ff} is the free-fall time at mid-plane, t_s is the sound travel time around the circumference of the disk, and $f_J \sim 0.1$. Second, $t_{ff}/t_{ep} < f_R$, where $t_{ep} = 2\pi/\kappa$, and $f_R \sim 0.15$. The second criterion is generally the governing one, and it can be expressed as $\frac{\kappa^2}{\pi G\rho} < 1$ for instability. The improvements in this criterion include the fact that it is not limited to thin disks, it is independent of the equation of state and the rotation law, and that it gives a criterion for *instability*, while Q is actually a stability indicator. Their results show that a collapsing cloud which conserves angular momentum is likely to fragment if the axisymmetric equilibrium configuration towards which it is collapsing is ring-like. This suggestion is consistent with three-dimensional numerical simulations. For example, in the isothermal case (Miyama et al. 1984) the criterion for fragmentation ($\alpha_i \beta_i < 0.12$) is also the criterion for ring formation in the equilibrium isothermal disk with the same angular momentum distribution.

Investigations of the nonaxisymmetric gravitational instability of thin disks have been performed by Adams et al. (1989) and Shu et al. (1990). This effect has been referred to as the "sling instability," meaning "stimulation by the long-range interaction of Newtonian gravity." They concentrate on the $m = 1$ (one-armed spiral) mode and find that eccentric distortions can grow on a time scale comparable to the orbital frequency at the outer edge of the disk. Growth times become shorter if the mass of the disk, relative to the central object, increases or if the temperature decreases. More precisely, growth times can be shorter than the 10^7 yr lifetimes of disks around T Tauri stars (Strom et al. 1989d) if $Q \leq 3$ at the corotation radius. Thus disks that are stable to axisymmetric modes (that is, those with $Q > 1$) according to the Toomre criterion can be unstable to nonaxisymmetric modes. Because of the eccentric nature of the unstable mode, they speculate that binary formation in the disk is a possible outcome. An important result of the linear analytic and numerical calculations is the determination of the so-called maximum-mass solar nebula. The disk is found to be stable to all gravitational disturbances below a critical value of its mass (which depends also on the temperature in the disk). The critical ratio of disk mass to total mass (disk plus central object) is found to be 0.24 if the Toomre Q value at the outer edge of the disk is 1. The speculation is that disks with lower mass can form planetary systems while those with greater mass may transport angular momentum through spiral waves or may develop a binary companion. Nonlinear calculations are necessary to verify this suggestion and to provide further insight into the question of the origin of close binaries. The first of such calculations (Yang et al. 1991), shows that star-disk systems resulting from the collapse of cloud cores become dynamically unstable to the generation of spiral modes right at the time of formation. It was not determined whether fragmentation of the disk occurred. Future work along these lines is necessary to predict, for example, the expected mass-ratio distribution from a disk instability.

C. Other Possibilities

The classical fission theory, involving the breakup of a rapidly rotating equilibrium object into two or more pieces in orbit as a result of dynamical instability to nonaxisymmetric perturbations, has often been suggested as a mechanism for the formation of close binary systems. However, there are numerous difficulties (reviewed by Tassoul 1978). For example, (1) pre-main-sequence single stars have too little spin angular momentum to be consistent with that of a typical close binary orbit (Table III); (2) angular momentum considerations (see Pringle 1989c) indicate that the mass ratio q of the binary would have to be small, ~ 0.1, contrary to observations; (3) the required initial condition may never be attained; and (4) numerical three-dimensional simulations of the process show that breakup does not occur. To consider the last two points, it is known that the classical Maclaurin spheroid becomes unstable to nonaxisymmetic perturbations when the ratio β of the rotational energy to the absolute value of the gravitational energy exceeds 0.27. Similar critical values of β hold for polytropes. However, in a rapidly rotating stellar-like object with polytropic index $n \geq 0.8$, the critical value of β will not be reached if a mechanism for angular momentum transport, such as the magnetic field, is efficient. The transport tends to result in uniform rotation in the interior and outflow in the equatorial plane to form a disk (Ruzmaikina 1981b). Even if the critical condition for instability can be reached (Fig. 6), numerical simulations (Durisen et al. 1986; Williams and Tohline 1988) for polytropes with $\beta = 0.3$ to 0.4 show that a bar-like structure forms in the center, then spiral arms form, angular momentum is transferred outward, and the end result is a ring surrounding a central bar that is stable to nonaxisymmetric perturbations. No fission occurs.

Related numerical studies have been performed for polytropic equilibrium toroidal configurations (Tohline and Hachisu 1990). These structures can be dynamically unstable for values of β as low as 0.16; however, in the range $0.17 \leq \beta \leq 0.27$, they do not fragment but instead develop ellipsoidal shapes and transfer angular momentum outward. It is not excluded, however, that fragmentation of tori could occur for other values of β or for other equations of state (see, e.g., Norman and Wilson 1978). Lebovitz (1989) has recently defended the fission theory, suggesting but not proving that other initial conditions from those chosen for the numerical simulations could lead to fission.

Capture (Stoney 1867) is often mentioned as a binary formation mechanism. The process requires either (1) the presence of a third object to carry off the excess energy; (2) tidal dissipation in a close two-body encounter; or (3) the presence of a dissipative medium in the system. The likelihood of the first two processes in the galactic disk or in associations and the effectiveness of the third process in protostellar envelopes has been shown to be negligibly small (Boss 1988). However, three-body encounters and tidal captures may

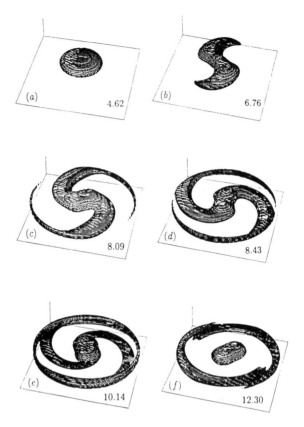

Figure 6. Fission calculation for an equilibrium polytrope with index $n = 0.8$ and $\beta_i = 0.31$. A time series of isodensity surfaces with $\rho/\rho_c = 10^{-3}$ is shown, where ρ_c is the central density. Times are given in terms of the central rotation period (Williams and Tohline 1988).

be of importance in the dense cores of globular clusters (Verbunt 1989; see Abt 1983). Larson (1990) has suggested that protostellar disks can provide the dissipation needed to promote capture in dense young clusters such as the Trapezium cluster associated with the Orion Nebula, although calculations by Clarke and Pringle (1991a) raise difficulties with this mechanism.

An important variant on the capture scenario involves the gravitational and dissipative interactions of a system of loosely bound fragments within a dense molecular core, in analogy to binary production in the disintegration of systems with a small number of stars (see, e.g., van Albada 1968a, b). The formation of the systems could either be induced by cloud collisions (Pringle 1989c) or occur by fragmentation during collapse, as illustrated in Figs. 3 and 4. Clarke and Pringle (1991b) have performed dynamical calculations for triple-star systems, including dissipation induced by the associated disks.

They find that a wide variety of binary orbital periods and eccentricities are possible outcomes, depending on only small changes in the initial conditions. A key concern here is that the orbital time scale of these interactions is comparable to the fragment collapse time scale. Thus, Boss (1986a) in a numerical study of fragmentation, shows that some of the incipient binaries have orbital decay times (by loss of angular momentum to trailing spiral arms) shorter than the contraction time of the components. He suggests that this process would lead to mergers, but it is also possible that close binaries would result. However, the numerical calculations have not been carried out long enough to prove this conjecture, and several physical processes have not been considered, such as H_2 dissociation and the accretion of material with higher specific angular momentum.

V. CRITERIA FOR THE FORMATION OF SINGLE STARS

Numerous suggestions have been made concerning the formation of single stars. For example, they could form as a result of fragment interactions in an incipient multiple system, resulting in the ejection of one member. Or, they could result from the spiralling together, as a result of gravitational torques, of binary components that have formed by fragmentation (Boss 1986a). This section concentrates, however, on direct formation of a single star as a result of the collapse of a rotating cloud, for various values of the specific angular momentum.

Note that the maximum possible angular momentum of a 1 $M_\odot$ star is about 10^{51} g cm^2 s^{-1} (Ruzmaikina 1981b), if it has the present solar structure and is rotating at breakup. Most T Tauri stars rotate significantly slower than the breakup velocity (Vogel and Kuhi 1981; Hartmann et al. 1986; Hartmann and Stauffer 1989). However, even a minimum-mass solar nebula adds a considerable amount of angular momentum. Reconstruction of the nebula by adding sufficient H and He to the present planets to recover the solar chemical composition (Weidenschilling 1977b), and adding a comparable mass to allow for ejection of material by the gravitational effects of the giant planets, gives a minimum mass M and angular momentum J of 0.02 $M_\odot$ and 5×10^{51} g cm^2 s^{-1}, respectively (for radius $R = 10$ AU). For this reconstituted disk model, M and J increase as $R^{1/2}$ and R, respectively, resulting in $M = 0.04$ $M_\odot$ and $J = 2 \times 10^{52}$ g cm^2 s^{-1} for the solar nebula with radius 40 AU. Therefore it is reasonable to consider a cloud core with specific angular momentum $j \sim 10^{19}$ cm^2 s^{-1} as an initial condition for the formation of the solar system.

If this core has uniform rotation and uniform density, then the centrifugal radius to which it will collapse is ~0.05 AU. Formation of a single star is plausible in this case because the rotation is slow enough so that a stellar core in hydrostatic equilibrium can form before fragmentation can occur (Ruzmaikina 1981a, b). The initial mass, radius and central density of the core will be similar to those in the nonrotating case (Larson 1969): $M_c \sim 10^{-2}$ $M_\odot$,

$R_c \sim 3 \times 10^{11}$ cm, $\rho_c \sim 10^{-2}$ g cm^{-3}. The temperature is high enough so that there is substantial ionization. This core accretes from the surrounding cloud and develops differential rotation, with angular velocity increasing inward. It is likely to possess a magnetic field of strength 10 to 10^3 gauss, resulting from intensification of the interstellar magnetic field (Ruzmaikina 1985). The outward transport of angular momentum by magnetic torque prevents the core from becoming unstable and produces an outflow of gas from the core equator to form an embryonic disk (Ruzmaikina 1981a,1985). Another mechanism that might inhibit core fragmentation and produce outward angular momentum transport is the generation of a spiral density wave in the envelope by a nonaxisymmetric core (Yuan and Cassen 1985). This model, therefore, involves accretion of most of the infalling material onto the star, and ejection of a disk during the later stages of infall.

Crude estimates of the mean radial density profiles of molecular cloud cores give $\rho \propto R^{-p}$ with $1 < p < 2$, in the radial range from 0.1 to 0.5 pc. More detailed information on density and velocity distributions within small cloud cores is still lacking (Myers et al. 1987). For the purpose of making estimates in the following discussion, we use two limiting descriptions of cloud cores:

$$\Omega = \text{const for } \rho = \text{const} \tag{6a}$$

$$\Omega = \text{const for } \rho \propto R^{-2}. \tag{6b}$$

Relation (6a) could result from the initiation of collapse by external sources, such as a fast contraction behind a shock front. On the other hand, a cloud which evolved through a slow nonhomogeneous contraction while being supported primarily by a magnetic field could have had effective redistribution of angular momentum to result in relation (6b), which corresponds to the uniformly rotating singular isothermal sphere (Shu 1977).

We estimate here the maximum angular momentum J_c consistent with the formation of a single star in this scenario. The absence of low-mass (10 to 80 Jupiter masses) binary companions to main-sequence stars (Campbell 1989; Marcy and Benitz 1989) is consistent with a relatively low (1 to 5 $\times$ 10^{52}) value for J_c (Ruzmaikina 1989). For $J < J_c$ one would expect the formation of a single star with a low-mass disk that is stable to fragmentation. For $J > J_c$, fragmentation during the stage of dynamical collapse is likely, resulting in roughly equal masses.

For the formation of a single star, we assume the following conditions must hold: first, a star-like core of $\geq 10^{-2}$ M$_\odot$ must form. Second, the mass of the disk cannot exceed, roughly, the mass of the star; otherwise the disk is likely to be unstable to binary formation. Third, the maximum centrifugal radius for the accreting envelope must be relatively small in radial extent (1 to a few AU) so that it will be hot, the magnetic Reynolds number will be $>> 1$, and the field can efficiently transport angular momentum.

In case (6a), the first condition is not well determined but falls in the range (Ruzmaikina 1981b)

$$4 \times 10^{51} \left(\frac{M}{M_\odot} \right)^{5/3} < J_c < 2 \times 10^{52} \left(\frac{M}{M_\odot} \right)^{5/3}. \qquad (7)$$

The relation between the core mass and the total mass is given by $M_c = 0.36M(R_c GM^3/J^2)^{1/3}$, where R_c is the core radius, in the limit $R_c << R_k$, where the centrifugal radius $R_k = j^2/GM$ (Levy et al. 1991). It follows from this relation that $M_{\text{disk}} \sim M_c$ at the end of accretion when $R_c GM^3/J^2 \sim 1$. Using $R_c = 3 \times 10^{11}$ cm (Stahler et al. 1980), we obtain $J < 1.3 \times 10^{52}$ g cm^2 s^{-1} for 1 $M_\odot$. This result is similar to that given by Eq. (7). Thus the first and second conditions give comparable values of J_c in this case, and they restrict J_c more strongly than does the third condition.

For case (6b), the first condition gives a high value of $J_c \sim 2 \times 10^{54}$ $(M/M_\odot)^3$ (Levy et al. 1991), so in fact the second and third conditions provide stricter limits. In this case, the ratio of M_c to M is given by $M_c = 0.5M(R_c GM^3/J^2)^3$ (Levy et al. 1991). It follows that $R_c GM^3/J^2 \sim 1$ as in Case 6a, so $J_c = 1.3 \times 10^{52}$ g cm^2 s^{-1}, from the second condition. On the other hand, the third condition states that the centrifugal radius $R_k < R_d$, where R_d is one to a few AU. At the end of accretion, $j = 9/2(J/M)$, from which we find that $J_c \sim 2 \times 10^{52}$ g cm^2 s^{-1} for $R_d = 1$ AU and $M = 1$ $M_\odot$. If $R_d > 1$ AU, then $J_c \propto (R_d/1AU)^{1/2}$. Thus the second and third conditions give comparable values of J_c. The argument indicates, therefore that $J/M \sim 10^{19}$ is the upper limit for formation of a single star.

For a cloud with $J = 2 \times 10^{52}$ and 1 $M_\odot$, $R_k = 0.3$ AU or 1 AU for distributions (6a) and (6b), respectively. The evolution of the system depends, in this case, on the relative time scales of infall (t_a) and angular momentum redistribution (t_r). If redistribution is inefficient, then the final radius of the disk $R_d \sim R_k$. But if $t_r < t_a$ then $R_d > R_k$. In that case, M_d will be small, $\sim 10^{-2}$ to 10^{-1} $M_\odot$. For example, with $J = 2 \times 10^{52}$ and $R_d = 40$ AU and with $\sigma \propto R^{-3/2}$, $M_d = 0.08$ $M_\odot$. This relatively low mass is consistent with angular momentum transport by magnetic fields for $R < 1$ AU and in the transparent outer regions. In the intermediate weakly ionized region some other transport mechanism must be effective, such as turbulence or gravitational torques. In summary, the low angular momentum case ($j \sim 10^{19}$) can result in single star formation as a consequence of angular momentum transport outward after most of the matter has fallen into the central star.

However, two-dimensional simulations of the formation of the solar nebula from collapse of a rotating cloud (Tscharnuter 1989; Bodenheimer et al. 1990) often assume higher values of j. For example, $j \sim 10^{20}$ seems to be required for agreement between calculated and observed spectra of a number of infrared sources (Adams et al. 1987). In this case, the central part of the collapsing protostar will become opaque before fragmentation occurs (Safronov and Ruzmaikina 1985), and it is not immediately clear whether or not fragmentation can occur at a later stage (for example, during dissociation of H_2; see above). According to Boss (1987) a cloud with $j \sim 10^{20}$ with distribution (6a) is likely to fragment during collapse, but a cloud with a centrally condensed distribution will evolve into a single star. The numerical calculations have probably not been carried out for sufficient time or with sufficient resolution in the central regions to prove this hypothesis. However, if true, the (6b) distribution with this value of j will evolve into a central star with a disk with $R_d \sim 100$ AU. Then, inward transport of mass from disk to star is required, rather than outward transport for disk formation in the case of $j \sim 10^{19}$. However, if there is no angular momentum transport during accretion, the disk is likely to be massive and gravitationally unstable. The instability could be an advantage if it promoted rapid angular momentum transport; on the other hand, it could prove to be a difficulty if binary formation resulted. A further problem lies in the difficulty of removing the excess mass in the disk after the presumed planetary system has formed. Further evolutionary calculations are necessary to determine the effectiveness of angular momentum transport and radiative cooling on the stability of the disk.

Even higher angular momenta are possible, as Miyama (1989b) proposes. Single stars could form from clouds with j near the observed value of $\sim 10^{21}$ cm^2 s^{-1} if α_i and β_i are in the range where the cloud core is stable to fragmentation ($\alpha_i \beta_i > 0.12$). Such a cloud will tend to evolve first into a rotationally supported isothermal equilibrium. With a proper choice of α_i and β_i the cloud will be unstable to nonaxisymmetric perturbations ($\beta > 0.27$) although stable to fragmentation. The instability would result in the formation of a bar-like structure and transport of angular momentum out of the central region by gravitational torques (Durisen et al. 1986). When sufficient angular momentum has been removed from the central regions, they will collapse from a centrally condensed initial condition, and the result will be similar to that of one of the previously discussed cases ($j \sim 10^{19}$ to 10^{20} cm^2 s^{-1}). The angular momentum is transferred to the outer parts of the core which then do not collapse.

VI. SUMMARY AND CONCLUSIONS

Observations show that a significant fraction of stars form as binary systems; the observational constraints on binary formation are summarized at the end of Sec. II. The origin of the orbital angular momentum in those systems is most likely associated with the turbulent motions in molecular clouds from which the stars form. Observational information on velocity gradients across many cores is consistent with rotation, and the specific angular momenta are comparable with those of the wider binary systems. However, in some cases, the angular momenta may have been overestimated because velocity gradients can also be produced by other dynamical effects, such as inclined bipolar outflows. Further detailed observations with improved spatial and velocity resolution and improved dynamical analysis are required. These observations would also help to determine the fraction of cores with relatively low angular momenta in the range necessary to form single stars or close binaries. They could also help to provide radial distributions of angular momentum within individual cores and thereby constrain initial conditions for protostar collapse.

Many different processes could account for the origin of binary systems; these can roughly be classified as capture, fission, and fragmentation mechanisms. Here we summarize the observational and theoretical arguments that apply to each of these processes.

Binary formation by capture of independently formed stars could occur by the three-body process, by tidal capture, or in the presence of a dissipative medium. However, current estimates indicate that such events are far too rare in the low stellar densities of the galactic field or even of associations to be solely responsible for the present field binary population. (Such captures may occur in very compact young clusters, such as the Trapezium Cluster, or in the cores of globular clusters.) Furthermore, three-body capture would tend to produce very wide orbits and tidal capture very close orbits; neither distribution (nor likely their union) would be in agreement with the observed period distribution. Capture processes do produce a wide range of orbital eccentricities, as observed. A more critical observational test would be the nonalignment in binaries of the stellar spin axes with each other and with the orbital plane. This test represents a very challenging but critical observational program. Finally, noncoevality of binary components would favor the capture theory. However, at present there is no evidence for such noncoevality on time scales >1 Myr. Thus if capture is a significant mechanism in binary formation, it must occur soon after the formation of each star.

Fission is traditionally associated with nonaxisymmetric instabilities in a centrally concentrated object, which is in equilibrium and is rotating more rapidly than a critical rate. The mechanism would produce close binaries; however, a stellar core with angular momentum low enough to form a close binary will probably transport angular momentum outward, resulting in a single star plus a disk, before it has a chance to reach the critical condition for fission. Even if the critical condition is reached, numerical calculations indicate that fission does not occur. Observationally, pre-main-sequence stars are not observed to rotate rapidly. Furthermore, fission theory predicts large mass ratios ($q \sim 0.1$) while the great majority of the observed systems have higher values of q.

Fragmentation during the isothermal stage of protostar evolution could serve as the initial step toward producing binaries with a wide range of periods, as observed. Calculations indicate that such fragmentation can straightforwardly produce systems with periods greater than a few hundred years. Earlier simulations of fragmentation tended to produce circular orbits. However, recent work suggests that fragmentation will typically produce a moderate number of fragments (Figs. 3 and 4), possibly followed by gravitational and dissipative disk interactions leading to captures and ejections. A key issue here is the time scale of these interactions compared with the fragment collapse time scale. For rapidly collapsing fragments, these processes can produce a wide distribution of orbital eccentricities as well as some short-period systems, although if there are many fragments an excessive number may be ejected as single stars (Clarke and Pringle 1991*b*). Another possibility for short-period systems is hierarchical fragmentation, involving subfragmentation during the adiabatic phase or during the collapse induced by molecular dissociation. An argument in favor of fragmentation as a formation mechanism is that it naturally produces hierarchical multiple systems, as commonly observed. Also, Zinnecker's (1984) hierarchical fragmentation theory is consistent with the observed gaussian-like distribution of periods. Finally, fragmentation may occur at a much later stage in a moderate-mass disk, perhaps through an $m = 1$ gravitational instability, which also would produce eccentric orbits.

In conclusion, we mention another important problem for future work. Recent observations indicate a deficiency of low-mass (0.01 to 0.08 $M_\odot$) companions to main-sequence stars of spectral types G, K and M. This fact may be significant for the question of single-star vs binary-star formation. One possible explanation could be the following. If the angular momentum of a cloud is below a relatively low J_c, a star-like core can form, and the result would be a single star with a relatively low-mass disk that could not fragment. But above J_c, fragmentation would occur during dynamic collapse, and the formation of comparable masses would be favored.

Acknowledgments. This work was supported in part through a National Science Foundation grant to the University of California, in part through a special NASA theory program which supports a joint Center for Star Formation Studies at NASA/Ames Research Center, University of California at Berkeley, and University of California at Santa Cruz, in part through a National Science Foundation grant to the University of Wisconsin, and in part through a Presidential Young Investigator award to R. Mathieu.

PRE-MAIN-SEQUENCE EVOLUTION AND THE BIRTH POPULATION

STEVEN W. STAHLER
Massachusetts Institute of Technology

and

FREDERICK M. WALTER
State University of New York at Stony Brook

We present an overview of what is known, both theoretically and observationally, concerning the evolution of stars prior to their ignition of hydrogen. We concentrate on the optically visible phase of pre-main-sequence contraction, when the star has already emerged from its parent cloud of gas and dust. This subject has a venerable history, and the classical model developed thirty years ago remains the basis of current theoretical treatments. Accordingly, we first review these historical developments, then discuss the modifications now in progress. There has been a wealth of observational data on pre-main-sequence stars obtained in recent years. In our treatment of the observational issues, we stress the morphological distinction between various classes of young stars, and the possible evolutionary link between these classes. The subject of pre-main-sequence evolution is a broad one, encompassing results from a large number of investigators over many years. We cannot hope in this limited space to present a comprehensive treatment. Our discussion is pedagogical, and our choice of topics is largely a reflection of individual taste and expertise.

I. THEORETICAL FOUNDATIONS

A. Classical Theory and Its Applications

1. Quasi-Static Contraction. The theory of pre-main-sequence (PMS) evolution attempts to provide a comprehensive physical account of optically visible stars prior to the onset of hydrogen burning. Over the last ten years, it has become clear that the classical theory first established by Hayashi and his contemporaries (Hayashi et al. 1962; Iben 1965; Ezer and Cameron 1967) is inadequate when confronted with the observations. While the shortcomings of this theory are well known to all workers in the field, it is nevertheless true that many of the basic features of the earlier models will continue to provide a framework for future endeavors. For this reason, and because the current ideas have not yet cohered into a unified account, we begin by reviewing the classical model of pre-main-sequence evolution.

The extraordinary longevity of stars is made possible by a delicate balance between the expansive effect of thermal pressure and the attractive force of gravity. In main-sequence stars, the pressure is derived from the heat supplied by nuclear reactions near the star's center. This heat diffuses slowly through the stellar interior and is ultimately lost by radiation at the surface. In a star too young to burn nuclear fuel, the heat loss leads to gravitational contraction. The use of the word "contraction" is deliberate, and is meant to describe a process slow compared to the dynamical, or "free-fall," time scale. That is, the shrinking of the star that is the primary characteristic of pre-main-sequence evolution is controlled by the rate at which the stellar surface radiates away internal thermal energy. In true free-fall collapse, on the other hand, gravity itself sets the time scale, which is exceedingly brief (about an hour) for an object of typical stellar density. To emphasize this difference, the contraction of young stars is also said to be "quasi-static." At any instant, the approximation of hydrostatic equilibrium is an excellent one, but the star nevertheless shrinks over the heat-leakage, or "Kelvin-Helmholtz," time scale.

A quantitative expression for the Kelvin-Helmholtz time is obtained by first estimating the total thermal energy content of the star. By the virial theorem, the energy contained in thermal kinetic energy is one-half the gravitational potential energy. The latter is given, within a factor of order unity, by GM_*^2/R_*, for a star of mass M_* and radius R_*. Thus, t_{KH}, the Kelvin-Helmholtz time, is given by

$$t_{KH} \equiv \frac{GM_*^2}{R_* L_*} \tag{1}$$

where L_* is the stellar luminosity. To get a feel for this quantity, we may substitute solar values for M_*, R_*, and L_*, obtaining 3×10^7 yr. This is the time it took for the Sun's radius to shrink by roughly a factor of 2 immediately prior to hydrogen ignition. Since the main-sequence lifetime of the Sun is about 10 Gyr, the entire phase of pre-main-sequence evolution is exceedingly brief in comparison, and the relevant candidate stars are correspondingly rare.

The virial theorem may be used also to estimate the internal temperatures of pre-main-sequence stars. Applying the perfect gas law, which is appropriate for the interiors of stars of ordinary mass, we find that the thermal energy can be written as $M_* \Re \langle T \rangle / \mu$, where $\langle T \rangle$ is the mass-averaged interior temperature and μ is the mean molecular weight. Since both the thermal and gravitational energies grow larger in magnitude as the radius shrinks, we see that $\langle T \rangle$ increases in the course of contraction. Thus pre-main-sequence stars are beautiful examples in Nature of objects with negative heat capacity—their temperatures rise as a result of the loss of heat. It is the relentless, compressive force of gravity which underlies this seemingly paradoxical effect.

2. The Role of Convection. Historically, the basic idea of quasi-static contraction was fully accepted by the beginning of this century. Nevertheless, it was not until the early 1960's that adequate interior models of pre-main-sequence stars were constructed. The difficulty lay in deciding the mode of energy transport through the stellar interior, radiation vs convection. This issue turns out to depend in a subtle manner on the properties of the star's subphotospheric layers.

The physical quantity of key importance is the entropy per unit mass. Figure 1 shows schematically the distribution of this quantity in the outer layers of a young star of roughly solar mass. As the entire region of interest comprises a negligible amount of mass, the specific entropy is shown as a function of pressure; the star's center lies to the right. The layers just inside the stellar photosphere have such low opacity that they are always radiatively stable against convection. For the rather low surface temperatures of pre-main-sequence stars, the opacity in this region is provided by H^- ions, which obtain their electrons from easily ionized metals. The supply of electrons, and hence the opacity, increases dramatically with temperature. At temperatures near 10^4 K, another opacity source comes into play—the photoionization of hydrogen. The combined opacity therefore climbs steeply as we move inward from the photosphere. Eventually, at the pressure marked P_{rc} in Fig. 1, the temperature gradient needed to push out the interior luminosity is so high that the material is convectively unstable.

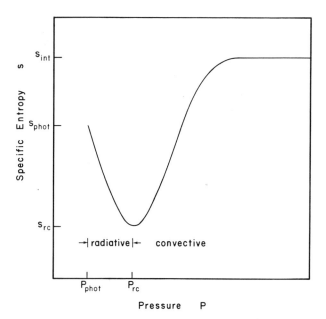

Figure 1. Specific entropy in the surface layers of a PMS star (schematic). Beginning at the photospheric pressure P_{phot}, the entropy falls from s_{phot} to s_{rc}, its value at the radiative-convective interface. The entropy then climbs the super-adiabatic gradient until it levels off at the interior value s_{int}.

Going from P_{phot} to P_{rc}, the specific entropy falls steeply because of the sharp increase in pressure. Inside this point, Schwarzschild's criterion dictates that the entropy must rise, so that s_{rc}, the entropy at P_{rc}, is a local minimum. The onset of hydrogen ionization leads to a very rapid entropy rise at first. Once ionization is complete, the rise is slowed to the characteristically tiny entropy gradient needed for convective transport in stellar interiors (see, e.g., Clayton 1968). Hence, the specific entropy quickly reaches the asymptotic value denoted s_{int} in the figure. This value must be matched by the outer limit of the star's interior entropy distribution. Finally, we note that the high-temperature sensitivity of the opacity implies that the entire entropy scale in the figure would be moved upward appreciably if the transported luminosity, or, equivalently, the value of effective temperature, were to increase by only a modest amount.

It was Hayashi (1961) who first realized that this outer entropy structure implies that the interiors of young stars can be *fully* convective. To show this, he first demonstrated the important result that any star of fixed mass and radius has a *minimum effective temperature*. Imagine a sequence of such stars with outer convection zones of varying depth. The requirement of fixed radius, together with the sensitivity of the radius to internal entropy, implies that the mass-averaged entropy in all models must be very similar. Thus, models with thin outer zones have correspondingly high values of s_{int} to compensate for the lower entropy values throughout their radiative interiors. It follows that the *minimum* value of s_{int} in this sequence is attained for that model with the fully convective interior. Since, as we have already noted, the value of s_{int} increases steeply as a function of T_{eff}, the existence of a minimum T_{eff} is established. The same temperature sensitivity of the opacity which plays a key role in this proof also implies that these minimum temperatures do not vary greatly as a function of stellar mass and radius. For solar-type stars, the "Hayashi temperature" is in the range from 4000 to 5000 K.

The fact that stars have a minimum surface temperature implies that sufficiently young ones will be fully convective, and hence will have surface temperatures equal to the minimum value. Young stars have large radii, so that their emitted surface luminosity, L_{surf}, is also large:

$$L_{\mathrm{surf}} = 4\pi R_*^2 \sigma T_{\mathrm{eff}}^4 . \tag{2}$$

On the other hand, L_{rad}, the luminosity which can be transported by radiation in a star, is given by

$$L_{\mathrm{rad}} = L_0 \left(\frac{M_*}{M_\odot}\right)^{11/2} \left(\frac{R_*}{R_\odot}\right)^{-1/2} \tag{3}$$

where L_0 is a luminosity of order 1 $L_\odot$ (Cox and Giuli 1968). Thus, if we consider progressively younger and larger stars of fixed mass, we inevitably reach the point where L_{surf} exceeds L_{rad}. For larger radii, the energy transport which the star needs to replenish the loss from its photosphere must be achieved by convection.

The solid curve in Fig. 2 shows schematically a typical pre-main-sequence track, as derived by Hayashi and his contemporaries. The shaded portion of the figure represents the region in which the star would have an effective temperature below Hayashi's minimum value; this region is thus "forbidden" to the evolving star. The heavy dotted line shows how the border of the forbidden region intersects the zero-age main sequence (ZAMS). It is seen that the true evolutionary track departs from this border and approaches the ZAMS along a more horizontal path. Along this portion of the track, the star is no longer fully convective, but has developed an inner core that is radiatively stable. For historical interest, the dashed curve labeled "HLL" is the older pre-main-sequence track of Henyey et al. (1955), which ignored convection entirely in the evolving young star.

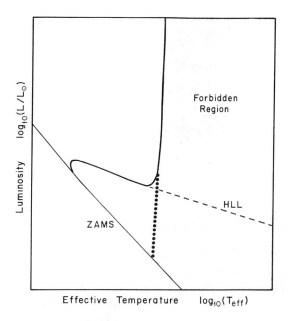

Figure 2.   A theoretical PMS track (schematic). A star of a fixed mass is first convective and descends a nearly vertical path, eventually turning onto the more horizontal radiative track discovered by Henyey et al. (1955). The vertical path and its downward extension (dotted line) form the border of the forbidden region of Hayashi (1961).

The fact that young stars tend to be convectively unstable is of key importance in the interpretation of their observed properties. Moreover, the interior structure and contraction of fully convective stars can be described in a mathematically simple and elegant fashion. Since the interior specific entropy is nearly uniform, the stars can be modeled as polytropes of index

$n = 3/2$ (Clayton 1968). The distribution of all physical variables in stars of arbitrary mass and radius can be obtained by proper rescaling of the solution to the dimensionless Lane-Emden equation. One finds that the star's contraction is "homologous," i.e., that the run of pressure, temperature, etc. at one instant can be related to the run at another time through multiplication by a spatially-independent scale factor. In physical terms, the rapid and efficient homogenization by convection of the interior entropy implies that the star looks the same at any time, but is compressed globally as its evolution proceeds.

3. Nuclear Burning. The classical model provides a simple account of the onset of main-sequence hydrogen burning. We have seen that the internal temperature of a young star paradoxically increases as it radiates heat into space. At early times, the central temperature is simply too low for hydrogen to ignite, but sufficient contraction inevitably raises the temperature to the requisite 10^7 K. As the nuclear burning in the central core gradually increases, its energy input begins to resupply the energy continually lost from the surface. Concurrently, the rate of gravitational contraction slows. For a brief time, in fact, the nuclear-generated luminosity is *greater* than the surface value, and the star temporarily expands. Soon, however, the star settles into a thermally stable state in which the nuclear and surface luminosities are essentially identical, and the rate of internal entropy change is extremely low. At this point, the star has joined the ZAMS. The details of this transition were followed numerically for a large range of stellar masses by a number of authors, most notably Iben (1965) and Ezer and Cameron (1967). The theoretical ZAMS found by these authors agreed very well with the observational results from color-magnitude diagrams (see the review by Iben 1967). It is hardly surprising, therefore, that the classical pre-mainsequence tracks in the HR diagram have formed a basis of research for almost three decades.

Stars that reach the point of hydrogen ignition all have nearly the same central temperature, and hence, similar values of the ratio $M_* R_*^{-1}$. Therefore, as we consider successively lower-mass stars along the ZAMS, the average internal density *rises* as M_*^{-2}. For stars of sufficiently low mass, the internal density is so high that electron degeneracy becomes the dominant mode of pressure support. One finds, for such stars, that T_c first rises during contraction, reaches a maximum value *below* 10^7K, then falls to zero as the contraction proceeds and degeneracy takes over. Kumar (1963) first established that, as a result of this effect, stars below 0.08 $M_\odot$ would never reach the ZAMS. Such objects are known as "brown dwarfs." Their possible presence in space has been the subject of intense inquiry and debate in recent years.

Other authors have repeated the contraction calculations, often obtaining noticeably different Hayashi temperatures (see, e.g, Cohen and Kuhi 1979; Vandenberg et al. 1983). Most of these differences can be attributed to the treatment of opacity in the subphotospheric layers. For example, Cohen and Kuhi were the first to display the effect of including molecular lines, which are important in lower-mass stars. Because the actual value of the surface temperature depends somewhat on the opacity (see Fig. 6-6 of Clayton 1968), the effect is to shift the tracks by modest amounts. There has also been interest, because of the possibility of disk accretion (see below), in including the effect of mass addition. Unfortunately, current knowledge of the disk accretion process is too poor to predict the physical condition of matter as it flows onto the stellar surface. As we have seen, it is the *specific entropy* of this gas which counts the most. Until the quantity can be calculated, the effect of disk accretion on PMS evolution cannot be gauged accurately.

Even before the completion of the first detailed evolutionary calculations, it was realized that other nuclear processes prior to main-sequence hydrogen burning could be of importance. In a brief but prescient contribution, Salpeter (1953) pointed out that appreciable concentrations of the light nuclei D, Li, Be and B would temporarily halt gravitational contraction, and that their ignition would occur at central temperatures not far above 10^6 K. Soon after the publication of the results of Iben, Bodenheimer (1966) performed a series of numerical calculations that followed this early burning. In the case of deuterium, which is by far the most abundant of these species in interstellar matter, the predicted slowdown of contraction indeed occurred. Grossman et al. (1974) followed Bodenheimer's study with a more detailed account of deuterium burning. By careful consideration of kinetic and thermodynamic effects at relatively high densities and low temperatures, they were able to describe accurately stars of a few tenths of a solar mass or less, for which deuterium ignition actually stalled contraction until the fuel was exhausted. They found that for M_* less than 0.01 $M_\odot$, the onset of electron degeneracy pressure during contraction prevented the central temperature from even rising to the point of deuterium ignition. In the HR diagram, they pointed to the existence of a "deuterium main sequence," found by joining the points on individual pre-main-sequence tracks where deuterium is able to halt the contraction temporarily.

The early investigators of deuterium burning all assumed that [D/H], the interstellar ratio by number of deuterium to hydrogen, was close to the terrestrial one of 2×10^{-4} (Hageman et al. 1970). Within a short time, however, direct observations of the interstellar medium were lowering this value by an order of magnitude (York and Rogerson 1976). Mazzitelli and Moretti (1980) redid the calculation of pre-main-sequence burning with the updated ratio of 2.5×10^{-5}. Not surprisingly, the decreased deuterium had less effect on contraction, so that the upper mass limit of the deuterium main sequence was effectively lowered. The curious fact was also noted that few, if any, observed young stars appeared to lie above this curve in the HR diagram. Although a smaller number of more luminous stars was naturally expected as a result of their faster contraction rates, there was no reason for the apparent cutoff. Mazzitelli and Moretti concluded that the appearance of the deuterium main sequence as the upper boundary in the distribution of young stars was a "coincidence" that was "probably accidental." As explained below, the coincidence has now found a satisfactory explanation in terms of the protostellar origin of pre-main-sequence stars.

Other studies of nuclear processes have focused on lithium as a possible diagnostic of stellar youth. The idea is that lithium should only be depleted in stars in which the base of the outer convection zone attains a temperature exceeding the ignition value of 3×10^6 K. This condition is never met in the relatively massive stars where the radiative-convective interface retreats to the surface before attaining the critical temperature. Indeed, it was observed that early F stars retained their primordial abundance ([Li/H] $\approx 10^{-9}$), while later G and K-type stars, with deeper convective envelopes, showed evidence of depletion (Wallerstein et al. 1965). For these cooler stars, Bodenheimer (1965) found a good correlation between the calculated lithium depletion during pre-main-sequence contraction for different masses and the observed abundances as a function of spectral type along the main sequence in the Hyades. More recent observations both in the Hyades and younger clusters, however, have clouded the picture, with an apparent spread in depletion among stars of similar mass and age (see, e.g., Strom et al. 1989*b*). It has been suggested that meridional mixing associated with stellar rotation could underly this puzzling result (Pinsonneault et al. 1989). Rotational mixing could also explain the remarkable *dip* in the lithium abundance as a function of spectral type discovered in Hyades F stars by Boesgaard and Tripico (1986).

   4. Rotation. The classical studies of Hayashi and his contemporaries avoided entirely the issue of angular momentum in their contracting stars. It was recognized, of course, that appreciable angular momentum in the youngest, most distended configurations implied severe spinup and possibly mass loss during the subsequent evolution, but lack of observational evidence deprived the theorists of any strong motivation for detailed calculations. Two significant exceptions are Bodenheimer and Ostriker (1970) and Moss (1973), who followed the contraction of rapidly rotating, centrifugally distorted configurations. Moss, in particular, advocated that the equatorial mass shed during contraction would orbit the star as a circumstellar disk. His study, now largely forgotten, foreshadowed the current high level of interest in pre-main-sequence disks.

   The angular momentum issue immediately came to the fore with the discovery by Vogel and Kuhi (1981) that solar-type pre-main-sequence stars (i.e., T Tauri stars) are *slow rotators*, in the sense that their surface speeds are a minor fraction (typically less than 0.1) of breakup (see also Hartmann et al. 1986). This finding, together with the earlier discovery of massive winds from T Tauri stars (Kuhi 1964), supplied the needed impetus to the theorists. Hartmann and McGregor (1982), following the seminal work on the solar wind torque by Weber and Davis (1967), suggested that centrifugally driven winds, coupled through ionization to a stellar magnetic field, could provide efficient braking in young stars. Endal and Sofia (1982) used the Weber-Davis mechanism to follow numerically the spindown of a solar-type star, assuming that the magnetic torque acted directly on a convective envelope in a state of rigid rotation. They also considered the effect of rotational instabilities on redistributing angular momentum in the radiative interior.

Other workers have offered similar parameterized treatments of the problem, but a self-consistent incorporation of rotation into pre-main-sequence theory has yet to be achieved. The original Weber-Davis picture of magnetic braking remains uncontested as a general framework, but its detailed implementation is formidable. In the slow rotation limit, the three-dimensional configuration of the magnetic field has been calculated by Suess and Nerney (1973), but determination of the field structure for rapid rotation, the stellar mass loss rate, and the influence on the field of the depth of convection are all unsolved problems.

Meanwhile, improvements in detector sensitivity and spectral analysis have resulted in a wealth of observational data. The original upper limits of Vogel and Kuhi have now been replaced by firm measurements of rotation velocities for dozens of stars (Hartmann et al. 1986, Bouvier et al. 1986a). To date, there is no convincing evidence for any systematic variation of rotational velocity along pre-main-sequence tracks. On the other hand, the new observations have revealed a population of rapid rotators (with velocities exceeding 100 km s^{-1}) in the Pleiades and Alpha Persei clusters, among stars thought to be just joining the ZAMS (Van Leeuwen et al. 1983; Stauffer et al. 1984). These stars appear coincident in the diagram with other slow rotators, and it is not clear if this fact indicates a corresponding age or angular momentum spread in the initial population. Nevertheless, it is encouraging that such a spinup, prior to final main-sequence spindown, is obtained naturally as the result of the formation of the relatively dense radiative core (Endal and Sofia 1982).

B. Recent Developments

1. Protostellar Origins. In classical pre-main-sequence theory, the initial configuration of a star was essentially arbitrary. As long as the star was assumed to be fully convective, it had to be a member of the unique sequence of homologous models corresponding to the given mass. Different choices of initial radius, therefore, merely shifted by a fixed amount the elapsed time to any subsequent state. Cameron (1962) and Hayashi (1966) argued that the relevant starting radii were very large, e.g., about 50 R$_\odot$ for a 1 M$_\odot$ star. If this were the case, then the corresponding Kelvin-Helmholtz time would be exceedingly brief (see Eq. 1). The contraction time to a state with much smaller radius would therefore be virtually independent of the initial value. In addition, Bodenheimer (1965) and Von Sengbusch (1968) relaxed the assumption of a fully convective interior. For solar-type stars with large initial radii, they showed that even fully radiative configurations quickly become convectively unstable and rejoin the standard vertical evolutionary tracks.

It had long been recognized that the quasistatic evolution appropriate for pre-main-sequence evolution could not describe the earlier, more diffuse configurations. Gaustad (1963) showed that the cloud fragment destined to become a star must collapse dynamically as long as it was optically thin to its own cooling radiation. In this "protostar" phase, the relevant time scale replacing Eq. (1) is the much shorter free-fall time, defined as

$$t_{ff} \equiv (\rho G)^{-1/2} \qquad (4)$$
$$= 7 \times 10^5 \left(\frac{n}{10^4}\right)^{-1/2} \text{yr}$$

where ρ and n are the cloud's mass and number density, respectively, and where the numerical estimate is for a typical cloud composed of molecular hydrogen. Although the details had not been worked out, it was assumed that this cloud collapse would end with the formation of a nearly hydrostatic object that became the pre-main-sequence star. The large initial radii advocated by Cameron and Hayashi were based on the belief that negligible energy was radiated away during the brief collapse phase.

The modern theory of protostars began when Larson (1969) first followed numerically the collapse of a protostellar cloud. Assuming for simplicity that the collapse was spherical, Larson integrated the equations of hydrodynamics and radiative transfer for a temporal sequence of cloud models, beginning with a diffuse, marginally unstable configuration. The most important result he obtained was that the collapse was distinctly *nonhomologous*. The region of highest density near the center collapsed first, while the outer layers remained temporarily static. The freely falling region spread outward at the speed of sound, reaching the cloud boundary at a time close to that in Eq. (4). [Note that the sound travel time and the free-fall collapse time are nearly equal in any cloud initially close to hydrostatic equilibrium.] As material within this region reached the center, it was arrested in a strongly radiating shock front. Inside the shock, this material settled slowly onto a growing *hydrostatic core*, which was the object destined to become a pre-main-sequence star. The photons emitted at the shock must diffuse through the dust in the infalling envelope before they can be detected; in the process, they are degraded into the infrared. Thus protostars, unlike their pre-main-sequence descendents, are optically invisible.

Larson's simulation demonstrated that considerable energy is lost at the shock during the protostar phase. Thus, the assumption of large initial radius was incorrect, and the early properties of pre-main-sequence stars were seen to depend critically on their prior history. Throughout the 1970s, many researchers attempted to refine Larson's calculation, but failed to arrive at consistent quantitative results for the hydrostatic outcome of the collapse. The issue was finally resolved by careful numerical treatment of the hydrostatic core and the bounding shock (Winkler and Newman 1980; Stahler et al. 1980). The final outcome was that the starting radius of a 1 $M_\odot$ star was a mere 5 $R_\odot$, a full factor of 10 smaller than the old result. This difference is a direct and dramatic manifestation of the severe energy loss associated with the accretion shock front.

2. Deuterium and the Birthline. The old issue of deuterium burning gained new attention when it was realized that protostars could readily attain the critical ignition temperature of 10^6 K (Stahler et al. 1980). Such high interior temperatures are a result of the relatively small radii reached during accretion. Prior to ignition, the protostar is built up by the low-entropy material that has cooled and settled inside of the shock. Once deuterium ignites, typically when the star has reached a few tenths of a solar mass, this intense heat source quickly turns the star convectively unstable.

As seen in Fig. 3, the burning progresses through a number of distinct phases. When the protostar is still fully convective, fresh deuterium that is continually landing on the surface can be rapidly swept into the center by turbulent eddies. In this manner, the star reaches a state of steady-state burning, in which the fuel is ignited at the same rate it is added. For a representative protostellar accretion rate of 1×10^{-5} $M_\odot$ yr^{-1}, steady-state burning generates a luminosity of 15 $L_\odot$.

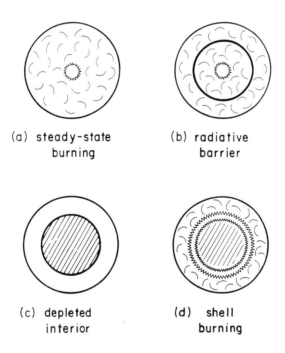

(a) steady-state burning

(b) radiative barrier

(c) depleted interior

(d) shell burning

Figure 3. Deuterium burning in protostars. (a) At low masses, accreted deuterium is fed in a steady-state fashion to the burning center by convective eddies. (b) Eventually, a radiative barrier appears, cutting off the supply to the center. (c) The central region soon exhausts its deuterium and reverts to a radiatively stable state. (d) At higher mass, deuterium ignites as a shell source just outside the depleted central region (figure from Palla and Stahler 1990).

As the mass of the star grows, its internal temperature climbs slowly, leading to a drop in the average opacity. At some point, the luminosity generated by deuterium is insufficient to drive convection. The entire star does not turn radiatively stable at once. Instead, a localized region, whose exact location depends on the detailed evolution, becomes stable and acts as a *radiative barrier*. Inside the barrier, the freshly accreted deuterium can no longer reach the center. The interior of the star quickly exhausts its fuel and reverts to radiative stability. Deuterium continues to pile onto the outside until the temperature at the original location of the barrier climbs to the ignition point. When this happens, a shell-burning phase ensues. With further increase in stellar mass, the shell gradually retreats toward the surface.

While the protostar is undergoing steady-state burning, deuterium acts effectively as a thermostat, keeping the star's central temperature close to the ignition value of 10^6 K. As the central temperature is proportional to the stellar mass divided by the radius, the thermostatic effect also implies a linear mass-radius relation for protostars. The physical basis for the deuterium thermostat can be easily understood. If the stellar radius increased so slowly that the central temperature rose appreciably, the deuterium burning would increase dramatically, because the energy generation rate of the fusion process varies roughly as T^{12} (Harris et al. 1983). The extra energy input would, of course, swell the radius back to its proper value.

The existence of a mass-radius relation for protostars implies that pre-main-sequence stars should first appear along a well-defined locus in the HR diagram (Stahler 1983a). Recall that, in the classical theory, the star's initial configuration was represented by an essentially arbitrary point somewhere on the vertical portion of its evolutionary track. The mass-radius relation can now be used to eliminate this arbitrariness and establish a stellar "birthline." For low-mass stars, the resulting curve corresponds essentially to the old deuterium main sequence. Along the birthline, stars of various masses first appear as optically visible objects. They then slowly contract to the ZAMS, each star following the remaining portion of its classical track. According to this theory, observed pre-main-sequence stars should be found on the birthline or below it, but never above. As shown in Fig. 4 for the case of the well-studied Taurus-Auriga molecular cloud complex, this prediction is in good agreement with current observations of T Tauri stars.

The deuterium thermostat is effective as long as the nuclear energy supply is comparable to the star's gravitational binding energy. This condition begins to fail for higher masses. However, with the ignition of the *deuterium shell source*, the protostellar radius swells once more (Palla and Stahler 1990). This swelling is only temporary, as the increasing influence of gravity soon creates rapid contraction. If the protostellar mass-radius relation is again used to predict the initial positions of visible stars in the HR diagram, the resulting birthline is found to intersect the ZAMS at a stellar mass of about 8 $M_\odot$ (see Fig. 5). *For stars with masses above this critical value, there is no optical pre-main-sequence phase*, because the rapid gravitational contraction in these

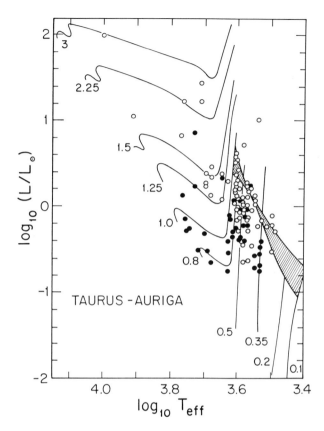

Figure 4. The birthline for low-mass stars (from Stahler 1988).The hatched region shows the variation in the birthline for protostellar mass accretion rates from 2×10^{-6} to 1×10^{-5} $M_{\odot}$ yr^{-1}. The lighter solid curves are the PMS tracks of Iben (1965), each track labeled by the corresponding mass in solar units. Open circles are observations of CTTS in Taurus-Auriga by Cohen and Kuhi (1979), while filled circles are NTTS from Walter et al. (1988).

objects leads to hydrogen ignition during the accretion phase. Figure 5 shows that the pre-main-sequence stars of intermediate mass, the Herbig Ae and Be stars, also have an upper H-R boundary consistent with the extended birthline.

3. Outflows. Observations of the molecular cloud clumps that appear to be forming low-mass stars show that they typically contain more gas than will be present in the final star (Myers and Benson 1983). Hence, it is generally believed that the protostar itself must somehow provide the means for ending its accretion of the infalling envelope. It is now known that a large fraction of embedded infrared sources are associated with strong outflows of molecular gas (Terebey et al. 1989) and that the cloud clumps in these cases have higher internal velocity dispersions (Myers et al. 1988). In light of these and similar findings, there is a growing consensus that the end of protostellar accretion and

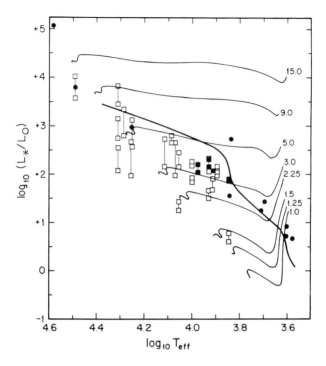

Figure 5.   The birthline for intermediate-mass stars (from Palla and Stahler 1990). The birthline is the heavy curve cutting across the lighter PMS tracks of Iben (1965). Each track is labeled by the corresponding mass in solar units. The squares are observed Herbig Ae and Be stars from Finkenzeller and Mundt (1984), and the filled symbols (circles and squares) are the outflow sources from Levreault (1988b).

the dispersal of the parent cloud are due to a powerful stellar wind, and that the molecular outflows are caused by the impact of these winds on surrounding cloud material. This view has recently been strengthened by the finding that those optically visible stars which have associated molecular outflows appear to lie along the stellar birthline (Levreault 1988; Palla and Stahler 1990; see also Fig. 5).

Within the context of spherical accretion, there is no likely source of energy for such winds. Deuterium does represent, as we have seen, a large energy reservoir. This energy is not available near the surface, because it is almost entirely absorbed by internal layers. Moreover, the thermostatic nature of deuterium burning places a severe constraint on any process dependent on it (Stahler 1988). For example, in the picture of Shu (1985), the wind energy is derived from differential rotation of the protostar. Once deuterium turns the star fully convective, a lower-energy state of rigid rotation can be attained, with the extra released in the form of internal magnetic fields that buoy upward to the surface. The problem with such a picture is that the released energy, if it is to be as large as needed, will swell the star. This swelling will lower the central temperature and immediately cut off the deuterium triggering mechanism.

Any models in which the star draws on the rotational motion of accreting matter for generating the wind must also face the fact that even the youngest pre-main-sequence stars, those on the birthline, are rotating relatively slowly. Hence, if the winds are created by centrifugal ejection during the protostar phase (Hartmann and McGregor 1982; Shu et al. 1988), there must follow a period of rapid braking in which the stellar mass loss is driven by a different process. This braking must be so efficient that it is completed by the time the star is fully revealed on the birthline, a period as brief as 10^4 yr (Fuller and Myers 1987). Wind models employing centrifugal ejection from circumstellar disks (Pudritz and Norman 1986; Konigl 1987) face a similar problem, because the extraction of angular momentum from the orbiting material allows it to spiral onto the star and spin it up. Once again, a brief period of efficent braking is required.

II. OBSERVATIONAL OVERVIEW

A. Basic Stellar Properties

1. Classical T Tauri Stars. Historically, the first stars identified as pre-main-sequence objects were those of the T Tauri class (Joy 1945). It is now known that these "classical T Tauri stars" (CTTS) are also of *low mass*, i.e., $\sim 2\,M_\odot$ or less. Mass determination has, in most cases, been made by comparison with theoretical pre-main-sequence tracks, but surface gravity measurements have also been used as confirmation (McNamara 1976). In this section, we offer an abbreviated summary of CTTS properties. The literature here is vast, and the reader interested in more detailed discussion is referred to recent, more extensive reviews, such as the Chapter by Basri and Bertout.

The general characteristics of the CTTS class are: (1) Infrared and ultraviolet excesses in the continuum spectrum; these excesses are measured with respect to a "normal" star of the same spectral type; (2) strong emission lines associated with low-excitation atomic species, chiefly the hydrogen Balmer lines, Ca II, and Fe II; (3) copious mass loss, as evidenced by P Cygni profiles and forbidden line emission (see Edwards and Strom 1987); (4) a stellar spectrum in which the absorption lines often appear partially filled in, or veiled; (5) irregular variability of the continuum and line fluxes, and of the line profiles. The T Tauri stars are generally found in loose clusters known as T associations, analogous to the higher-mass OB associations.

From their bolometric luminosities and effective temperatures, the CTTS are located above the main sequence, with ages ranging from 10^5 to 10^7 yr. These ages must, of course, be measured with respect to the birthline, but the reduction in age from the traditional figure using large initial radii is only appreciable for the youngest objects (Stahler 1983a). It is important to keep in mind that not all CTTS exhibit every characteristic listed here. Furthermore, both spectral and photometric properties of these stars are highly variable. Brightness variations of up to 5 magnitudes can occur on time scales of days to years, often in conjunction with changes of color (cf. Appenzeller and Mundt 1989). Emission line profiles can also change on similar time scales (e.g., Brown 1985).

In recent years, it has become popular to identify the continuum excess emission as arising from a circumstellar disk (see, e.g., Bertout et al. 1988). The infrared component is thought to arise in the dust of the cooler, outer regions of this disk, while the ultraviolet excess is attributed to a hot star-disk boundary layer, where disk matter is slowed from its rapid orbital motion to join the slowly rotating stellar surface. Thus, in the current paradigm, CTTS are characterized by optically thick disks, which can be either *passive* or *actively accreting*. In a passive disk, the accretion luminosity is dwarfed by the luminosity of starlight absorbed and reradiated at longer wavelengths by the disk. There exists, unfortunately, a significant gap between this phenomenological model and pre-main-sequence theory. Thus, there is no generally accepted account of the structure of the disks, for the mechanism of their accretion, or for the cause of the rotational braking in the boundary layer. Nevertheless, the accretion-disk picture has proved useful as a working hypothesis.

As CTTS eventually become normal main-sequence objects, it is natural to search for evolutionary trends. Thus, one might expect to see a decline in the infrared excess among progressively older stars. To date, quantitative evidence for such a trend is still lacking. Simon et al. (1985), Walter and Barry (1991), and Feigelson and Kriss (1989) show that the class of CTTS displays more surface activity than older low-mass stars, but there is no firm indication of variation with age within the CTTS class itself. A complicating factor here is that the ages of the most active CTTS, the so-called continuum stars, are difficult to determine. Their photospheric absorption lines are partially filled in ("veiled") by circumstellar emission, with the consequence that neither stellar luminosities nor effective temperatures can be determined accurately. Using recent techniques for separating the photospheric and circumstellar components of the luminosity, it should be possible to make better estimates of CTTS ages, and thus to re-examine this important issue.

2. Naked T Tauri Stars. From their location in the HR diagrams of populous T associations, such as Rho Ophiuchus and Taurus-Auriga, it is apparent that most T Tauri stars cluster in age between 10^5 and 3×10^6 yr. However, we have seen from Eq. (1) that the total contraction time to the ZAMS is of order 10^7 yr for solar-mass stars. Because of the steep decline of luminosity, this time is even longer for stars of lower mass. Herbig (1978) pointed out that these figures imply a population of "post T Tauri" objects which would easily outnumber the CTTS in cloud complexes which have been producing stars for a sufficiently long time.

Until about 1980, there was little evidence for another type of low-mass pre-main-sequence star. However, X-ray observations of regions of star formation by the Einstein Observatory revealed many stellar sources, only a few of which corresponded to catalogued CTTS (Ku and Chanan 1979; Montmerle et al. 1983*a*; Walter 1986; Feigelson et al. 1987). Further spectroscopic and photometric observations of the optical counterparts (see, e.g., Walter et al. 1988; Strom et al. 1990) confirmed the pre-main-sequence nature of these stars, but showed them to lack the defining characteristics of CTTS. From their positions in the HR diagram, the CTTS and a substantial fraction of the naked T Tauri stars (NTTS) have similar masses and ages (Fig. 4). However, the NTTS population, at least in Taurus-Auriga, also includes an older component (see Sec. II.B).

The general characteristics of the NTTS, in addition to their X-ray emission, are:

1. A normal stellar photospheric flux distribution, with no significant excesses in the optical and near-infrared. Ultraviolet excesses are similar to those in stars with very active chromospheres.
2. Modestly strong emission lines of Ca II H and K, and $H\alpha$. Except for late K and M stars, the $H\alpha$ emission is generally insufficient to fill in the photospheric absorption and appear as a distinct line above the continuum.
3. Symmetric emission line profiles, with no evidence for forbidden-line emission.
4. A normal photospheric absorption line spectrum, with no evidence for veiling.
5. Periodic photometric variability.

The physical basis for the differences between the CTTS and NTTS can be addressed within the disk paradigm. The lack of infrared excess and continuum veiling has been taken to imply the absence of an active accretion disk, and, in most cases, the absence of an optically thick passive disk, as well (Skrutskie et al. 1990). The emission line fluxes from NTTS are consistent with active solar-like chromospheres, and the X-ray emission is similarly indicative of a solar-like corona (Walter and Kuhi 1984; Walter and Barry 1991). The absence of forbidden-line emission and the symmetric line profiles suggest a low level of mass loss. Finally, the periodic variability most likely arises from the rotation of an asymmetric distribution of dark starspots (Rydgren and Vrba 1983), as is common among active, late-type stars, including the RS CVn and BY Dra systems (see, e.g., Vogt 1983).

Herbig and Bell (1988) use the term "weak-lined T Tauri stars" (WTTS) to indicate low-mass PMS stars with $H\alpha$ equivalent widths < 10 Å. All NTTS are WTTS, but the converse is not true. This fact has led to some confusion in the community. The equivalent width of $H\alpha$ is a useful observational characteristic, but not a meaningful physical parameter when discussing stars with a wide range of photospheric temperatures. The equivalent width in a line is proportional to the surface flux in that line, and inversely proportional to T_{eff}^4. Thus, G-type CTTS with active disks, such as SU Aur, are WTTS, but are very different from the NTTS. Similarly, there are cooler WTTS with large near-infrared excesses; these also are quite distinct from NTTS. The distinction between NTTS and WTTS is further illustrated in Fig. 2 of Walter and Barry (1991).

3. The Herbig Ae/Be stars. The Herbig Ae and Be stars (Herbig 1960) are the intermediate-mass analogs of the classical T Tauri stars. They are stars of spectral types A and B with strong T Tauri-like emission lines, and are found in regions of star formation. By definition, a Herbig Ae/Be star must illuminate a nebulosity. After Herbig's original identification of this class, their position above the main sequence was confirmed through measurements of their surface gravity (Strom et al. 1972). It has recently been shown that they also lie near, but below, the theoretical birthline (Palla and Stahler 1990; see also Fig. 5). Their inferred ages range from 10^5 to 10^6 yr.

A catalog of 57 Herbig Ae/Be stars has been compiled by Finkenzeller and Mundt (1984). These authors noted the common appearance of P Cygni profiles, indicative of strong stellar winds. The width of the Hα line indicates typical velocities of 500 km s^{-1}. These wind properties appear comparable to those of the high-mass loss CTTS (Mundt 1984). Stars earlier than spectral type A0 often have double-peaked emission, as is typical also of the classical Be stars. However, these stars are clearly distinguishable from classical Be stars in an infrared two-color diagram (see Fig. 5 of Finkenzeller and Mundt 1984). A catalog of line profiles of 26 Ae and Be stars was prepared by Finkenzeller and Jankovics (1984).

A majority of Herbig Ae/Be stars exhibit at least 5% variability of their continuum luminosities. This variability tends to be more pronounced among the later-type stars, but the data are insufficient to establish whether the fluctuations are periodic. Finkenzeller and Mundt suggest that the appearance of P Cygni profiles and significant variability may be indicative of vigorous surface convection. As is apparent from Fig. 5, however, these stars are on the *radiative* portion of their evolutionary tracks, and hence have no outer convection at all in the classical theory (Iben 1965). The resolution of this paradox may be provided by the shell-burning of deuterium left over from the protostar phase (Palla and Stahler 1990).

Garrison (1979) showed that the Herbig Ae/Be stars tend to exhibit near-infrared excesses and 3650 Å Balmer discontinuities that are smaller than expected for their spectral types. These shallow discontinuities could be created by emission from a star-disk boundary layer, the mechanism currently believed responsible for putting the Balmer jump into emission in the CTTS. The near-infrared excesses could, in principle, be caused by the same hydrogen continuum, but thermal emission from warm dust, as in the CTTS, is the preferred explanation. The light from these stars is often significantly polarized, as expected if the extinction arises in circumstellar material (Vrba et al. 1979). The 10-μm Si absorption feature has been observed in some cases, as have dust-related near-infrared emission features (see references in

Finkenzeller and Mundt 1984).

Praderie et al. (1982) and Felenbock et al. (1983) find that there are many similarities between the circumstellar emission of AB Aur, the brightest of the Herbig Ae stars, and that of the CTTS. In general, the Herbig Ae/Be stars share the following characteristics with the CTTS: circumstellar gas and line emission, variability, strong stellar winds, and association with dark clouds. As in the lower-mass case, individual objects exhibit a range of characteristics. Finally, there are at least a dozen X-ray bright late-B stars lacking Hα emission associated with the Orion Nebula (Caillault and Zoonematkermani 1989), and similar late-B stars associated with the Taurus, Corona Australis, and L1641 clouds (Walter et al. 1988; Strom et al. 1990). It is tempting to identify these stars as the intermediate-mass equivalents of the NTTS.

B. Observed Initial Stellar Populations

1. The Taurus-Auriga Complex. The low-masspopulation in Taurus-Auriga consists of the two observationally distinct groups described above: the CTTS and the NTTS. About 100 CTTS have been catalogued (Herbig and Bell 1988), and it is unlikely that significant numbers of CTTS associated with the Taurus clouds remain undiscovered. Any uncatalogued CTTS remaining in Tau-Aur are probably highly reddened. The number of NTTS is less well known. Walter et al. (1988) discussed the X-ray selected NTTS, and Herbig et al. (1986) discussed a sample of optically selected NTTS. In all, about 40 NTTS have been catalogued in the region (This figure excludes several stars previously catalogued as CTTS, such as V410 Tau or DI Tau). The difficulty in determining the number of NTTS is that, unlike the CTTS, there are no extreme characteristics which are easily detected in surveys. The characteristic most amenable to large-scale surveys appears to be the coronal X-ray emission (Walter et al. 1988), although the presence of Li I 6707 absorption must be used to confirm youth. Walter et al. (1988) estimated the true number of NTTS in Taurus-Auriga by extrapolating from the X-ray selected sample. Correcting for instrumental factors and accounting for the observed level of X-ray variability, they estimated an average 30% completeness. This result assumes that the observed X-ray luminosity function is the true luminosity function; a large population of less X-ray-luminous NTTS would decrease the assumed completeness. This 30% completeness corrects only those regions which have been observed in X-rays; about 90% of the Taurus-Auriga region was not observed by EINSTEIN. On the assumption of a uniform spatial distribution of NTTS across the entire region, Walter et al. extrapolated the total number of NTTS to be 35 × 3 × 10, or about 1000.

The ratio of NTTS to CTTS in Taurus-Auriga thus seems to be about 10. Such a result is in general agreement with expectations if (a) star formation has been proceeding in the region for at least 10^7 yr, and (b) the NTTS indeed represent the "post-T Tauri" population of Herbig (1978). From the HR diagram in Fig. 4, it is indeed apparent that the NTTS exist at lower luminosities, and hence greater ages, than the CTTS. However, the fact that both types of stars can be found close to the birthline is a puzzling complication. It is also significant that the CTTS and NTTS populations exhibit different spatial distributions. The CTTS tend to be concentrated near the dark clouds. The younger NTTS are co-spatial with the CTTS, but there is an older population of NTTS west of the main Taurus clouds (Walter et al. 1988), which may represent an earlier episode of star formation associated with the Cas-Tau OB association (Blaauw 1956). After correction for the X-ray incompleteness, the ratio of NTTS to CTTS, integrated over the whole of the complex, is unity at a stellar age of ~ 1 Myr. The oldest (least luminous) CTTS appear to have ages of order 10 Myr, while the old population of NTTS is on radiative tracks, close to the ZAMS, with typical ages of 30 Myr.

2. *Low-Mass Stars in OB Associations.* The Taurus-Auriga region is an example of a star-formation region dominated by the low-mass stars. A number of researchers have cited evidence that star formation is "bi-modal," i.e., that high- and low-mass stars form in different locations under different physical conditions (see, e.g., Larson 1986). If this view is correct, the initial mass function (IMF) inferred from counting field stars must represent the juxtaposition of many unrelated IMF. In any specific star-formation region, there should be no correlation between the numbers of high- and low-mass stars. One can examine this hypothesis by directly observing the stellar mass spectrum in regions of star formation, prior to merger with or dilution by the field. The result of such studies, conducted using X-ray data, is that large numbers of low-mass stars appear to be associated with nearby OB associations.

Walter (1987b) reported preliminary results from X-ray observations of 4 square degrees of the II Scorpio OB association. In this small area, 30 NTTS were discovered, along with one new CTTS. There are 9 B-star members of the OB association with ages of 5 to 8 Myr (de Geus 1988) within these fields. The properties of the NTTS are similar to those in Taurus-Auriga, as are the distances, so the completeness of the X-ray data should be comparable. Accounting for the completeness, there should be about 100 NTTS in these 4 square degrees, with a ratio of NTTS to CTTS of ~ 50. The projected space density (stars per square parsec) is three times that in Taurus-Auriga. The ratio of B stars to low-mass stars in this OB association is consistent with a Salpeter (1955) IMF from 0.3 to 10 $M_\odot$.

A similar result is found in the Orion OBI association, where Walter et al. (in preparation) have identified some 150 low-mass NTTS based on X-ray images of fields near Lambda Orionis, the B35 dark cloud, the belt of Orion, and the periphery of the Orion nebula. Furthermore, some 200 low-mass stars associated with the belt of Orion have been identified in the Hα survey of Kogure et al. (1989). Strom et al. (1990) find a 3:1 ratio of NTTS to CTTS in the L1641 dark cloud (presumably, this figure is lower than the 10:1 ratio in Taurus-Auriga because L1641 has not been producing stars as long). Throughout Orion, the numbers of NTTS greatly exceed those of CTTS. The Orion OBI association is 3 times more distant than either II Scorpio or Taurus-

Auriga, so the X-ray surveys are less complete. The observed space density of NTTS is about 3 times that in II Scorpio; correction for the distance and the larger incompleteness implies that the projected space density of NTTS is somewhat larger than in II Scorpio.

3. *High-Mass Stars in T Associations.* The results for low-mass stars in OB associations indicate the possible presence of high-mass stars in regions traditionally thought to be producing only low-mass objects. Walter and Boyd (1991) reasoned that, if the extrapolation of 1000 NTTS in Taurus-Auriga is correct, then there should be a significant number of B stars near this association. The exact number depends on the choice of IMF, ranging from ~30 for a Salpeter function to 60 for a Miller-Scalo (1979) IMF. That so many B stars could be hidden so close to the Sun would be surprising. However, the Taurus-Auriga association is projected on the Cas-Tau OB association (Blaauw 1956). The OB association has an age of ~30 Myr, and the nearest part is within 200 pc of the Sun. In fact, the kinematics of the Cas-Tau association's subgroups 6 and 7 are very similar to those of the T Tauri stars, and Cas-Tau and the T associations may well be physically related.

Walter and Boyd (1991) examined the space motions of the B stars in the Bright Star catalog, and found that 29 B stars in 21 systems are possible kinematic members of the T association, within the observational errors in the space motions. They conclude that the Taurus-Auriga complex is not devoid of high-mass stars. The estimated B star age of 30 Myr is similar to that of the oldest NTTS. There are a few younger high-mass stars associated with the 1 Myr old T Tauri stars, including the embedded B2 star HK Tau/G1 (Cohen and Kuhi 1979) and the B7V star 72 Tau (Jones and Herbig 1979). The T association may thus be the most recent episode of star formation in a larger complex which spawned the Cas-Tau OB association and the older NTTS 10 Myr ago.

C. Comparison of Observation with Theory

1. Distributions in the HR Diagram. Theoretical models of pre-main-sequence stars predict the stellar luminosity, effective temperature, and radius as a function of age. Because it is impossible to witness the evolution of these quantities in individual stars, the correctness of the models must be tested indirectly from the distribution of stars in the HR diagram.

The survey of Cohen and Kuhi (1979) showed that the more luminous CTTS lie in the region of the HR diagram predicted for young stars on convective tracks, while many of the higher-mass CTTS appeared to be on radiative tracks. Walter et al. (1988) placed the NTTS on the HR diagram of Taurus-Auriga. While some of the NTTS were coincident with the CTTS, many were found to lie on radiative tracks closer to the ZAMS, as seen in Fig. 5. Although the location of many CTTS and NTTS on convective tracks is consistent with their observed surface activity, direct evidence that the stars actually follow the theoretical tracks was first provided by the agreement of the theoretical birthline with the observed upper boundary of the stellar distribution (Stahler 1983a,1988). As the construction of the birthline uses both a protostar mass-radius relation and the pre-main-sequence tracks, this agreement provides strong support for both aspects of the theory. The mass-radius relation was constructed with a simplified treatment that ignored any rotation in the accretion flow or in the central protostar. The success of this relation in predicting the birthline can in part be attributed to the fact that the youngest optically visible stars are in fact slow rotators (Durisen et al. 1989b).

The recent discovery of a number of pre-main-sequence binaries has provided an opportunity for verifying the theoretical ages and masses of young stars. Walter et al. (1988) show that the NTTS in wide binary systems lie on similar isochrones, as would be expected if the stars form coevally. To date, few spectroscopic binaries, and no eclipsing binaries, are known among the low-mass PMS stars (Mathieu et al. 1989). In the cases of the two known double-lined spectroscopic binary NTTS, the inferred masses and ages are consistent with the evolutionary tracks, but only marginally so in the case of the system P1540 (Marschall and Mathieu 1988).

In comparing theory with observation, it is necessary to keep in mind that there are substantial systematic uncertainties that plague any empirical determination of the stellar luminosity. These include the errors associated with assigning a precise distance to an individual star, in separating the photospheric from the nonstellar components of the luminosity, and in evaluating the reddening corrections. In Taurus-Auriga, the depth of the clouds contributes an approximately 20% uncertainty in distance, which translates to a 40% uncertainty in luminosity. de Geus (1988) estimates a depth of 90 pc for the Ophiuchus dark clouds, which translates into a greater luminosity uncertainty. Corrections of the observed flux distribution for disk and boundary layer contributions, as well as for shadowing of the photosphere by the disk, are highly model dependent. The proper consideration of extinction in star-forming regions must be done with care, since R, the ratio of total to selective absorption, is frequently greater than the standard 3.1 (Mathis 1990) in star-forming regions. Moreover, visual extinctions can vary widely, even on close lines of sight (Cohen and Kuhi 1979). The net effect of these systematic uncertainties is to make the luminosity of any particular star rather uncertain, although the mean parameters of a large sample should be reasonably accurate.

  2. Observational Constraints on Star and Disk Evolution. It seems reasonable to assume that as stars first become visible at the birthline they exhibit the properties of the CTTS. In fact, one would like to go a step further and argue that the most extreme CTTS, the continuum stars, are the youngest and least evolved of the low-mass stars. Unfortunately, as we have emphasized, this simple picture is complicated by the presence of NTTS near the birthline.

The topic of how the evolution of the circumstellar material proceeds has attracted interest of late. In terms of the disk paradigm, the transition from CTTS to NTTS involves the dissipation of the circumstellar disk. Depending on the dissipation mechanism, one might expect to see some low-mass stars with characteristics of both classes. Such stars would yield important insights into time scales for disk dissipation, and perhaps provide constraints on how rapidly planets must form (if planetary formation is the rule).

Before one can attempt to quantify such considerations, a clear empirical distinction must be established between those stars which are believed to possess active disks and those which do not. Traditionally, the defining characteristic has been the presence or absence of excess emission at near-infrared wavelengths. Strom et al. (1989b) discussed the near-infrared colors of NTTS, and concluded that a significant fraction had excess emission from circumstellar material. Walter et al. (1989) analyzed the same data, and concluded that the near-infrared colors of NTTS were due to starspots with filling factors of up to 30%, similar to those observed in RS CVn systems and other active stars (see, e.g., Vogt 1983). The color temperature of the excesses is 2000 to 3000 K, far too high for circumstellar dust, but reasonable for starspots. The existence of starspots is confirmed by the periodic photometric variability (see, e.g., Rydgren and Vrba 1983). Among the stars first identified as NTTS in Tau-Aur, only V836 Tau and IP Tau show evidence for emission from warm dust at ~1000 K.

As a useful first step toward quantifying disk evolution, Strom et al. (1989b) have introduced the concepts of disk "survival" and "transition" times. The survival time measures the interval since the star's appearance at the birthline at which the disks disappear. Walter et al. (1988) have argued that this interval is difficult to pin down, at least in the Taurus-Auriga complex, because of the co-existence of NTTS with ages <1 Myr and CTTS with ages of nearly 10 Myr.

The transition time, defined as the interval over which the disk ceases to be detectable, may represent the time required for dust in the inner disk to clump into macroscopic agglomerations prior to forming terrestrial-type planets. This time can be estimated by counting the numbers of stars with intermediate characteristics. Walter et al. (1988) find that 5 to 10% of their observed sample actually have characteristics intermediate between the NTTS and CTTS. For example, V836 Tau exhibits a small near-infrared excess (Mundt et al. 1983), and a Hα equivalent width slightly larger than the other NTTS. If ~10% of the low-mass pre-main-sequence stars are in this transition state, and if the *mean* age of the low-mass population is a few Myr, then the implication is that the disks of the CTTS disappear on a time scale of a few times 10^5 yr (Walter et al. 1988). Observations of NTTS in the II Scorpio OB association (Walter et al., in preparation) generally confirm that the transition time is short. Of 30 NTTS, only 2 show significant near-infrared excesses. For a mean association age of ~5 Myr, transition time is again of order 10^5 yr. Skrutskie et al. (1990) arrived at a similar transition time.

It is crucial to understand that the observational evolutionary sequence from CTTS to NTTS as described here says little about the evolution of the star itself. The CTTS-to-NTTS transition is being viewed as a change in the *circumstellar environment* of the star, and likely has little effect on (and is probably little influenced by) the star's quasi-static contraction, unless a significant fraction of the stellar mass is accumulated from its disk. Recognition of this independence of star and disk evolution could help to explain the curious presence of very young NTTS. From a theoretical viewpoint, dissipation of the disk represents a major challange. Gravitational instability is unlikely to be effective long after protostellar accretion has ended, while mass transport due to convective processes seems to be far too slow (see, however, the Chapter by Adams and Lin).

The evolution from the NTTS to the ZAMS appears to be undramatic from an observational perspective. Chromospheric and coronal properties of the older pre-main-sequence stars appear very similar to those of their ZAMS counterparts (Walter and Barry 1991). The one identifiable change seems to be in the rotation of the star. As we have already mentioned, the studies of rotational speeds in open clusters indicate that low-mass stars spin up due to the growth of internal radiative cores prior to hydrogen ignition (Stauffer et al. 1984).

THE INITIAL STELLAR POPULATION

HANS ZINNECKER
Universität Würzburg

MARK J. McCAUGHREAN
University of Arizona

and

BRUCE A. WILKING
University of Missouri, St Louis

The collections of young stars with ages of 10 Myr or less found in association with molecular clouds are referred to as the initial stellar population. The majority of this population is comprised of low-mass ($<3\,M_\odot$) objects with long time scales for thermal contraction to hydrogen burning stars. As a result, they can display a wide variety of evolutionary states ranging from deeply embedded infrared sources to optically visible stars with enhanced magnetic surface activity. Hence, the study of the initial stellar population requires observational techniques spanning the entire electromagnetic spectrum. Two recent advances in infrared astronomy, the Infrared Astronomical Satellite all-sky survey, and the growing availability of infrared array detectors, have dramatically improved our sensitivity to heavily obscured solar- and sub-solar-mass objects, and greatly enhanced our ability to sample completely the young stellar population over an entire cloud. We begin this chapter with a brief discussion of current theories for the origin of stellar masses and the initial mass function, focusing on the formation of stars in clusters, and the possibility that several different formation mechanisms are at work in the Galaxy. Motivated by recent infrared camera observations of obscured clusters, models are presented for the evolution of the $2.2\,\mu$m pre-main-sequence luminosity function of a young coeval star cluster, with a discussion of the implications the results have for the conversion of an observed luminosity function into the underlying mass function. A description of theories which allow spectral energy distributions of young stellar objects to be used as diagnostics of their evolutionary states is given. The observational section of this chapter begins with a discussion of techniques employed to study the initial stellar population. Observations of young stellar populations over a wide range of distances from the Sun are reviewed, beginning with low-mass aggregates in nearby dark clouds and ending with clusters surrounding massive stars in distant giant molecular clouds. Particular attention is paid to populations within 500 pc of the Sun, including those associated with the ρ Ophiuchi, Taurus-Auriga and Orion molecular clouds, where a significant number of solar- and sub-solar-mass objects are observed. The question of the existence of sub-stellar objects (brown dwarfs) in nearby young clusters and regions of star formation is addressed. Finally, we briefly discuss current problems and future directions for studies of the initial stellar population.

[429]

I. DEFINITION, MOTIVATION, AND SCOPE

The term "initial stellar population" has been coined to describe the ensembles of young stellar objects (YSOs) with typical ages of 1 to 10 Myr or younger, found associated with molecular clouds. Because the objects cannot have moved far from their birthplaces, studying the distribution of their masses gives us insight into the origin of the stellar mass function at birth, i.e., the initial mass function (IMF), averaged over a molecular cloud. We may hope to find out whether or not the IMF is different in different types of molecular clouds, or in regions where star formation appears to be spontaneous rather than externally triggered. By studying the youngest observable populations in star-forming molecular clouds, we may ultimately be able to tackle some of the following questions:

1. Are there significant variations in the stellar mass spectrum from cloud to cloud? Is there a difference in the low-mass stellar content between regions that have or have not formed massive stars?
2. Is there mass segregation in a young cluster? Do massive stars tend to form near the cluster center? Are there halos of low-mass stars around the clusters?
3. Is the most massive star that can form related to the depth of the gravitational potential of the protocluster?
4. Is there a cutoff at some characteristic low mass or are there many brown dwarfs forming in clouds?
5. Do all stars in a cluster form at the same time? If not, what governs the age spread in a very young cluster? Do low-mass stars form first? Do stars form sequentially in mass?

If we can answer these questions, we will go a long way towards understanding the details of the star-formation process. We may be able to determine the rules of star formation, if any. By trying to observe collections of young stars as early as possible, we can hope to identify temporal and spatial correlations between stars of different masses (e.g., initial mass segregation and/or hierarchical stellar groupings) before they are erased by averaging over the field.

Studies of the initial stellar populations in molecular clouds face two problems—one is theoretical, the other observational. The theoretical problem is that it is not trivial to relate the luminosity of most YSOs to their mass. Only massive stars are born almost directly as hydrogen-burning main-sequence objects. Lower-mass objects exhibit luminosity evolution along pre-main-sequence (PMS) tracks, taking progressively longer times to reach the main sequence, e.g., a few$\times 10$ Myr for a $1\,M_\odot$ star. Thus the conversion of an observed luminosity function into a mass function by virtue of a time-independent main-sequence mass-luminosity relation is not possible: this step requires a theory that describes changes in the bolometric luminosity or absolute infrared brightness of PMS objects as a function of age and mass.

The observational problem is that the photospheres of YSOs are partially or totally obscured by dust at optical wavelengths. This dust can absorb much of the YSO luminosity and reprocess it to infrared wavelengths from 1 to $100\,\mu$m, depending on the proximity of the dust to the YSO. In addition, in clouds where massive stars have formed, the accompanying H II region nebulosity is often bright and can present a serious contrast problem in the optical. Therefore, in general, the earliest phases in the life of a YSO are best studied in the infrared, where YSOs are most luminous, where dust obscuration is vastly reduced, and where contrast against any nebular background is increased. Until recently, however, only single-element infrared detectors were available, making it impossible to carry out sensitive studies of large ensembles of sources. Instead, most studies concentrated on low-sensitivity, low-resolution surveys, or detailed studies of individual or small groups of objects. In addition, groundbased studies at thermal infrared wavelengths were severely hampered by the enormous background flux from the warm telescope and lower atmosphere.

With the introduction of infrared array detectors in the past five years, it is now possible to obtain high-resolution ($\sim$1 arcsec), high-sensitivity near-infrared (1–5 μm) images over relatively large areas of the sky. The same period saw detailed analyses of data from the Infrared Astronomical Satellite (IRAS) mid- and far-infrared (12–100 μm) all-sky survey. These two factors have led to a flood of new information about the embedded stellar populations of galactic star-forming regions. In this chapter, we review both the observational and theoretical progress in studying the initial stellar population since the publication of *Protostars and Planets II* in 1985. For a clear picture of our prior state of knowledge regarding fragmentation, star formation (including that in OB associations and young clusters), and the IMF, we refer the reader to reviews from the previous books in this series (e.g., Lada et al. 1978a; Scalo 1978; Silk 1978; Klein et al. 1985; Evans 1985; Scalo 1985; Strom 1985; Wilking and Lada 1985).

II. THEORETICAL CONCEPTS

We begin this section with some basic ideas regarding the origin of stellar masses and the IMF, as one of our prime goals is to understand which types of stars are born in the various kinds of molecular cloud. We continue with a simplified analysis of the time dependence of the mass-luminosity relation for low-mass PMS stars. Combining this relation with a given mass function, we examine the time-dependent behavior of the corresponding luminosity function. The model predictions suggest a new technique that may prove very useful for the interpretation of empirical 2.2 μm luminosity functions of embedded clusters, surely a very topical subject in the next few years. Finally, we discuss some of the more realistic aspects of YSOs, in particular the interpretation of the spectral energy distributions in terms of the evolution of YSOs and their circumstellar material. These considerations lead to a better delineation of the components of a YSO (star + disk + halo) that must be included in radiative transfer calculations to allow fits in observable parameter space.

A. The Origin of Stellar Masses and the IMF

An increasing body of observational data leads to a picture of star formation in which stars are born in individual dense clumps or protostellar fragments that appear to be more massive than the stars that eventually form from them (see Myers [1991], and references therein). This fact suggests that the origin of stellar masses and the IMF are inextricably linked to the mass distribution of dense clumps and the efficiency with which an individual dense clump forms a star or a group of stars. Through spatially- and velocity-resolved molecular line observations, it has recently become possible to identify dense clumps in molecular clouds and establish their mass distribution over a wide range of spatial and density scales (Blitz 1987b; Stutzki and Güsten 1990; Lada et al. 1991a; Chapter by Blitz). The mass distribution (dN/dM) appears to be very similar in each cloud studied so far and may be fairly universal: it scales as the clump mass M to roughly the -1.5 power. In order to obtain the IMF of stars born in these clumps, we require an "initial-final mass relation" to translate the (initial) mass distribution of the clumps into a (final) mass distribution of stars (Zinnecker 1989b,1990). If the final stellar mass m scales with the initial clump mass M as $m \propto M^p$ (with $p<1$; cf. the observations of Stacey et al. [1988]), one can easily show that the stellar mass spectrum is a power law with index $x = -(0.5 + p)/p$. For example, to produce the index of the Salpeter IMF ($dN/dm \propto m^x$, $x = -2.35$; Salpeter 1955) requires $p \sim 0.4$. The observed mass distribution of dense clumps may be a result of collisional fragmentation (see, e.g., Nozakura [1990], and references therein) and/or the hierarchical structure of interstellar clouds (see, e.g., Elmegreen's [1985c] model, and Zinnecker [1985]). The physical justification for the power p in the initial-final mass relation may come from the theory of protostellar winds dispersing the parent clumps (for low-mass stars: see Shu et al. [1991]) and the theory of radiation pressure acting on dust grains in protostellar envelopes to reverse their infall (for high-mass stars: see Yorke [1980] and Wolfire and Cassinelli [1986]). In addition, it would seem that magnetic field tension can help prevent complete accretion of the envelope mass, especially for low-mass stars (Nakano 1984; Lizano and Shu 1989; Mouschovias 1991b). All these ideas are still rather crude and the details have yet to be worked out, but the suggestion is that it will be progressively harder for the more massive clumps to accrete a substantial fraction of the original clump mass, while for the lower-mass clumps relatively more dense gas ends up in the star. This theory therefore suggests splitting any understanding of the IMF into two parts, namely the clump mass distribution and an initial-final mass relation. The advantage is that, at least in principle, each part can be tested separately by observations.

The theory as outlined above neglects various details of the star-forming process. For example, is there a characteristic mass for the clumps (con: Shu et al. 1987*a*; pro: Mouschovias 1991*b*)? Does every clump form a star (Larson 1991)? Do larger clumps subfragment into several yet denser clumps (Zinnecker 1990)? Current theory suggests that indeed there *is* a characteristic clump mass ($\sim 1\,M_\odot$), selected and preserved by self-gravity in a dense, turbulent, and magnetic interstellar medium (Mouschovias 1991*b*; Larson 1991). It is suspected that every clump above some critical density *does* form a star, and that subfragmentation *does* become increasingly important for progressively larger clump masses (Larson 1978,1991). Indeed, if the probability of subfragmentation increases with clump mass, this will steepen the slope of the original clump mass spectrum, as required to explain the observed IMF. Furthermore, this process can produce an IMF which does not have a single power-law slope, but a slope whose index changes gradually with mass, much like the Miller-Scalo IMF (Miller and Scalo 1979)—in this case, the variable p would have to be a function of the clump mass.

Implicit in these considerations is the route by which the IMF arises in a system of stars such as a protocluster (cf. Scalo 1986; Zinnecker 1986*a*). Larson (1990*a*) and Myers (1991) have discussed the possible evolution of protoclusters, with emphasis on dissipational processes. Larson (1991) further elaborated on the origin of the slope of the upper IMF in a protocluster, suggesting that the power-law extension of the IMF toward higher masses may result from accretional growth of some stars to much larger masses. However, there must be a limit to accretion on each star, and a key question on the way to a full understanding of the origin of the IMF is the process by which accretion is cut off. Does the protostar terminate the mass inflow by switching on a protostellar wind, thereby defining its own mass (self-limited accretion; Shu et al. 1987*a*)? Is the reservoir for accretion limited by the other stars competing for material (competitive accretion; Zinnecker 1982)? Or is the amount of accretable material mostly dictated by the initial conditions (e.g., angular momentum limited accretion; Larson 1978)? Hopefully, detailed observations will provide us with some clues.

Finally, in any discussion of the IMF, we cannot ignore the fact that most stars are binaries. In particular, with regards the observational determinations of stellar luminosity and mass functions presented later in this chapter, the effect of unresolved binaries may be quite important, and has recently been discussed in some detail (Kroupa et al. 1991; Piskunov and Malkov 1991; Reid 1991). However, these analyses, along with the origin of binary stars and the origin of their mass ratios, are beyond the scope of this chapter, and the reader is referred to the Chapter by Bodenheimer et al. for more information.

1. Isolated vs Collective Star Formation

There are several distinct effects that come into play regarding collective as opposed to isolated star formation, including tidal forces, mass segregation, and radiative implosion. These effects imply qualitatively observable consequences.

First we consider tidal effects. Accretion may be inhibited by the tidal forces generated by a star on the disk or protostellar envelope of a neighboring stellar core, or perhaps more importantly, by a protocluster cloud on smaller extended fragments that are randomly positioned off-center within it. Both effects limit the reservoir for accretion, unlike the single star case considered by Shu et al. (1987*a*) where only the star itself (i.e., via its wind) can limit the accretion. At the same time that tidal forces limit the reservoir for accretion, they may also increase the *rate* at which the remaining material is accreted, due to the tidal torques that are present. Thus overall, young stars in a protocluster may have less massive but more rapidly evolving circumstellar disks than isolated objects of the same age.

A second important effect in a protocluster is mass segregation. Are the more massive stars born in the denser, more central regions of the cloud? A naïve application of the Jeans criterion for a centrally condensed isothermal cloud would predict the opposite outcome, with low-mass stars in the center and high-mass stars at the periphery. However, in the center, the clump-clump collision time scale is shorter than the collapse time scale of an individual clump. This effect, along with accretion of residual gas sinking towards the center due to dissipational processes, leads to a prediction that the more massive stars will indeed form near the cloud center. More specifically, we predict that the mass of the most massive star should scale with the depth of the gravitational potential well at the cloud center, because the deeper the well, the more difficult it is for gas to escape. Clumps farther from the center will avoid collisions more readily and will remain small, forming lower-mass stars.

A third effect concerns star formation triggered by the radiative implosion of clumps, initiated by ionization shock fronts from newly formed OB stars in the cluster (Klein et al. 1980). Close to the massive stars the clumps should be completely obliterated, creating a zone of avoidance (R. Klein, personal communication), while at greater distances, dense enough clumps will survive to form other high-mass or intermediate-mass stars. Infrared imaging observations might be able to reveal evidence for gaps around the most massive members of a cluster.

In summary, the conditions that arise only in a cluster environment could play a dominant role in producing the wide range of masses observed in the stellar mass spectrum. In particular, perhaps massive stars can only form in a cluster. Furthermore and equally important, the majority of field stars may be former members of clusters and OB associations (see Chapter by Lada et al.), and therefore the form of the *field star* IMF may be determined more by the effects of the cluster environment than by any other factor.

2. Bimodal Star Formation

The concept of a decoupling of low- and high-mass star formation originated with Herbig (1962*b*). In his picture, low-mass stars form continuously over a long period, only to be interrupted by the relatively rapid formation of high-mass stars, the violence of which leads to dissipation of the remaining molecular material. This idea of temporal variations in the star-formation process with respect to stellar mass was further codified to refer to separate formation mechanisms for high- and low-mass stars, perhaps leading to distinct formation sites for OB stars and low-mass stars. This is the basis for so-called bimodal star formation (see, e.g., Eggen 1976; Mezger and Smith 1977; Güsten and Mezger 1982). However, as noted by Herbig (1962b), OB associations commonly have related associations of low-mass T Tauri stars, whereas the converse is not always true. Therefore, one continuous process could be at work, with an upper mass limit in regions of low-mass star formation set by, say, the initial reservoir of cloud material. The proof for separate mechanisms for high- and low-mass stars would be a region that has formed high-mass stars exclusively, implying a process that either prevented or biased against low-mass star formation.

Perhaps the strongest evidence for bimodal star formation has come from the study of starburst galaxies where low M/L ratios are best explained by the formation of predominantly high-mass stars in the nuclear regions (cf. Larson [1986], and references therein). Whether or not there are individual H II regions in our Galaxy whose exciting stars are *not* accompanied by low-mass stars is a very important question, with many implications for the origin of the IMF, both within a given cloud and throughout the field. However, until recently, finding firm observational evidence for this has been hindered by the difficulties encountered in detecting low-mass stars in regions where more massive stars are known to have formed.

In the spirit of bimodal star formation, Shu et al. (1987a) and Lizano and Shu (1989), following ideas by Mestel (1985), proposed a model whereby two distinct star-forming mechanisms can operate depending upon the ratio of magnetic support to cloud self-gravity. A small cloud dominated by magnetic support evolves slowly via ambipolar diffusion to form a low-mass core and subsequently a low-mass star. Conversely, if enough of these smaller clumps can agglomerate into larger clumps, for example in a giant molecular cloud via some external trigger, gravity can overcome the magnetic support, and these clumps will collapse rapidly, forming high-mass stars. In the latter scenario, fragmentation can also take place, perhaps forming low-mass stars. This model also explains the low star-formation efficiency seen in complexes with only low-mass cores and stars. The long time scale for low-mass core evolution leads to an incoherence across the cloud complex, and the collapse of some cores may be disrupted by the stellar winds of nearby newly formed stars.

Mouschovias (1987a) took another perspective on the difference between low- and high-mass star formation. In his scenario, low- and intermediate-mass stars form in quiescent clouds collapsing at sub-Alfvénic speeds, while a cloud initially supported by magnetic fields then imploded by an external trigger may collapse rapidly at Alfvénic or super-Alfvénic speeds, forming more massive stars.

It is important to note that neither of these models seems to preclude the formation of low-mass stars alongside the high-mass stars. Indeed, if the high-mass stars form first, they may trigger low-mass star formation in the clumpy medium by radiative implosion (Klein et al. 1980,1983; La Rosa 1983).

As our sensitivity to low-mass stars in the more distant massive star-forming regions improves, our ideas of bimodal star formation will become better developed. A possible example of bimodal star formation in our Galaxy, in the star-forming region S 255/257, is described below in Sec. III.C.2.

B. Evolution of a Model PMS Infrared Luminosity Function

From an observational standpoint, the near-infrared (1–2.5 μm) is perhaps the best wavelength range over which to find young low-mass stars in clusters. Using array detectors, it is now easy to detect and measure the near-infrared fluxes of large numbers of stars in such clusters, as we shall see in Sec. III. Unfortunately, it remains hard to make immediate astrophysical sense of these raw measurements. Because the stars are heavily reddened and may have large infrared excesses, it is not easy to determine the total stellar luminosity. Because the stars are young, a time-independent main-sequence mass-luminosity relation cannot be used to determine the stellar mass, and a time-dependent PMS relation must be used: for example, an empirical fit to the models presented below shows the near-infrared (K) luminosity for low-mass stars evolving down their Hayashi tracks changing as $L_K \propto M^{1.6} \times t^{-2/3}$.

The classical approach to these problems for PMS stars is to measure the underlying spectral type of each star and thence the reddening towards it. The luminosity is determined by integrating under the spectral energy distribution, and by combining this with the effective temperature, the age and mass of each star can be found by comparison with PMS evolutionary tracks in an HR diagram (Cohen and Kuhi 1979). However, it is difficult to obtain spectral types and reddening for all the members of a dense, heavily obscured PMS cluster, and some assumptions have to be made. For example, in their study of the Trapezium Cluster, Herbig and Terndrup (1986) measured a few stars spectroscopically and then used the mean derived extinction to deredden the rest of the cluster members. As an alternative, Straw et al. (1989) assumed all the stars they measured in NGC 6334 were on the zero-age main sequence (ZAMS), and moved them parallel to the reddening vector in the HR diagram until they intercepted the ZAMS locus. In the former case, it is highly unlikely that all stars in an embedded cluster have the same extinction. In the latter case, assuming the stars to be on the ZAMS when they are more likely to be PMS will overestimate their mass. However, if there was some other way of estimating the age of a PMS star or even a whole cluster, it should be possible to deredden the stars in the HR diagram until they intercept the appropriate PMS isochrone, thus determining their masses. The possibility of infrared excesses will complicate such simple approaches, but to some extent that can be dealt with by looking at a star's location in a near-infrared color-color diagram. In these diagrams, the reddening vector, isochrones, and isomass tracks all form a roughly parallel locus, whereas stars with infrared excesses fall below this line.

With this problem in mind, and as part of an on-going study of young clusters and how observations may be converted into physically useful parameters, we have performed some simple modeling which allows us to explore the phase space covered by young low-mass stars in near-infrared coordinates. One particular point of interest is the predicted PMS evolution of the near-infrared luminosity function. Some qualitative results relating to this are given here.

1. Modeling

We have taken a homogeneous grid of PMS tracks (0.1 to 10 Myr) for low-mass (0.08 to 2.5 $M_\odot$) stars, kindly provided to us by I. Mazzitelli, and converted them from $T_{\rm eff}$, $R/R_\odot$ coordinates into near-infrared magnitudes assuming the stars to be blackbodies, and using the Sun as a zero point, also assumed to be a blackbody. As the Sun is *not* accurately represented by a blackbody, and low-mass PMS stars even less so, this work must be repeated using synthetic spectra. However, for illustrative purposes here, the results obtained using blackbodies give a qualitative if not quantitative picture. Normally an observed luminosity function is converted to a mass function via some empirical or model mass-luminosity relation. Here we reverse the procedure. To derive a near-infrared 2.2 μm (K) luminosity function (i.e., dN/dK), we begin by assuming a given mass function (i.e., $dN/d\log M$) and then evaluate the following equation:

$$\frac{dN}{dK} = \frac{dN}{d\log M} \times \frac{d\log M}{dK} \qquad (1)$$

where $d\log M/dK$ is the slope of mass–K luminosity relation, and $K = -2.5\log(L_K/L_{\odot_K})$. For these models, we used the half-gaussian form of the Miller-Scalo IMF (Miller and Scalo 1979) and the slope of the mass–K luminosity relation as derived from the transformed PMS tracks at each age and mass. It has been pointed out many times (see, e.g., D'Antona 1986; Bahcall 1986; Kroupa et al. 1990) that it is the *slope* of the mass-luminosity relation that is involved here, and that simple inflections in that relation can lead to significant features in the luminosity function, as we shall see below.

2. Results

Figures 1 a,b,c,d show the model mass–K luminosity relations at four different ages (0.3, 0.7, 1 and 2 Myr), while Figs. 2 a,b,c,d show the corresponding model K luminosity functions (KLFs) for a coeval cluster of low-mass stars at the same four ages. Note that the distance to the cluster is assumed to be 10 pc, i.e., in these figures, $K \equiv M_K$. For reference, Fig. 3 shows an empirical

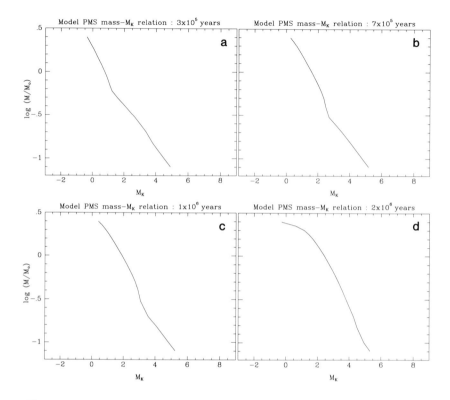

Figure 1 a,b,c,d.   Time-dependent mass–M_K luminosity relations, as derived from
homogeneous tracks of pre-main-sequence evolution calculated and kindly provided
by I. Mazzitelli. Four ages are shown: 0.3, 0.7, 1, and 2 Myr.

mass–M_K relation and corresponding spectral types for main-sequence stars.
Also note that 0.3 $M_\odot$ corresponds to $\log M = -0.5$ in these diagrams.

At 0.1 Myr (not shown), the shape of the KLF largely reflects the shape
of the Miller-Scalo IMF, i.e., turning down and flattening towards the lower
masses, corresponding to the almost straight mass–K luminosity relation at
this age.   Whether or not such a KLF would ever be seen in practice is
debatable however, as stars this young would still be deeply embedded in
their protostellar cocoons, making it unlikely that the "naked" PMS stars
would be seen at this stage. By 0.3 Myr, features become apparent in the
KLF, and by 0.7 Myr, there is a sharp peak in the KLF which can be traced
to a sharp inflection at the corresponding point in the mass–K luminosity
relation. This point of inflection is seen to move towards lower masses with
time, and by 2 Myr the peak in the KLF has moved down close to our chosen
mass cutoff at 0.08 $M_\odot$. By 10 Myr (not shown), the peak is long gone.

The inflection in the mass–K luminosity relation is a feature caused by

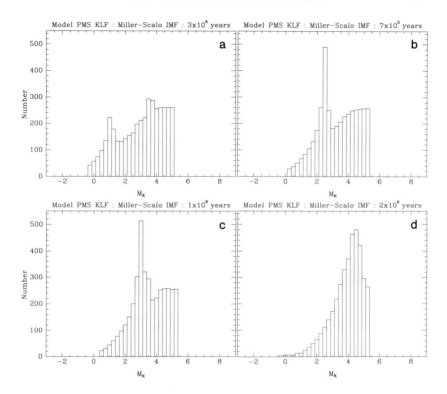

Figure 2 a,b,c,d.   Model 2.2 μm (M_K) luminosity functions for a coeval cluster of low-mass stars at the same four ages shown in Fig. 1. The luminosity functions were calculated by convolving the mass–M_K relations with a Miller-Scalo IMF. Note the strong peak due to deuterium burning in the last three luminosity functions.

deuterium burning in contracting PMS stars. D-burning occurs when the Kelvin-Helmholtz contraction has proceeded far enough to raise the central temperature to $\sim 10^6$ K. For a $0.3\,M_\odot$ PMS star, D-burning occurs at an age of ~ 1 Myr and lasts for ~ 1 Myr, until most of the central deuterium is consumed. These time scales are fairly independent of the precise value of the initial deuterium abundance, although for reference, the Mazzitelli models were calculated using a deuterium abundance of 2×10^{-5} by mass, as derived for the solar neighborhood (York and Rogerson 1976). While deuterium is burning, the corresponding nuclear energy generation provides substantial support against continuing PMS contraction, i.e., for a while the star's radius and luminosity remain almost constant (see Fig. 2 of Mazzitelli and Moretti 1980). While the deuterium is burning strongly in stars of a given mass, D-burning is ending in slightly higher-mass stars, and has yet to begin in lower-mass stars. Thus, at the mass where peak D-burning is occurring (i.e., the mass whose time scale for the onset of strong D-burning is equal to the

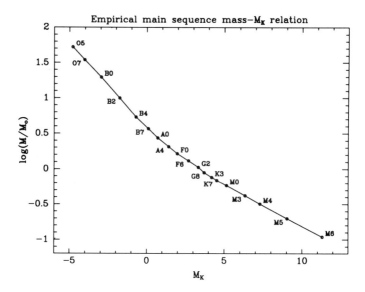

Figure 3.   An empirical mass–M_K luminosity relation for main-sequence stars, derived from the data of Miller and Scalo (1979), and Koornneef (1983*b*), following Straw et al. (1989). Corresponding spectral types are also marked. It should be noted that there is considerable uncertainty in this relation for spectral types K0 and later, as seen when compared to other empirical determinations (M. Simon, personal communication; Henry 1991).

present age of the coeval cluster in question), there is a kink in the mass-luminosity relation. This leads to an increase in the value of d log M/dK, and a boost in the luminosity function at the corresponding K magnitude, creating a peak and the subsequent turnover. A physical interpretation is that as the cluster members are decreasing in luminosity, evolving down their Hayashi tracks, stars in a certain mass range temporarily decrease less, and pile up in a slightly brighter than expected magnitude bin. It is, however, important to note that at no point in our completely coeval model cluster does a star of mass M appear more luminous than a star of mass $M + \delta M$, as the D-burning never completely halts the PMS contraction.

It is a general feature that the duration of the D-burning phase for a given star is roughly equal to its age at the onset of D-burning (cf. Mazzitelli and Moretti 1980). For example, D-burning starts in a PMS star of 1 $M_\odot$ at ~0.1 Myr and lasts approximately 0.1 Myr. These time scales are in accordance with the location of Stahler's (1983*a*) "birthline" in the HR diagram, the location below which optical T Tauri stars seem to appear for the first time, thought to be set by the D-burning "quasi-main sequence."

So we can see that for young clusters, features such as peaks and turnovers in the K luminosity function cannot be *assumed* to imply corresponding features in the IMF; rather, they may often reflect the complex process of PMS evolution. Therefore we caution against overly simplistic analyses of near-infrared luminosity functions and inferences for the IMF. On the other hand, such features might still be used to our advantage, namely as a possible age indicator. This somewhat surprising result might even hold when our simplifying assumptions are replaced in more realistic models, e.g., allowing for age spreads and infrared excesses due to circumstellar disks. We will also have to include intermediate-mass stars in this procedure, which we have omitted here. The PMS evolution of these stars is currently under investigation (Palla and Stahler 1990,1991; see also Chapter by Stahler and Walter).

C. Spectral Energy Distributions

The previous discussion was idealized by considering only naked PMS stars without any circumstellar material. But it is known from observations of infrared excesses and sub-millimeter continuum emission that YSOs can possess substantial amounts of circumstellar dust. The distribution and quantity of this dust determines the shape of the emergent energy distribution from the YSO. Low-mass YSOs gradually dissipate their circumstellar dust as they approach the main sequence and over their extended evolutionary time scales, the shapes of their spectral energy distributions (SEDs) change dramatically. Therefore, by constructing the energy distribution of a YSO over the 1 to $100\,\mu m$ spectral region, the evolutionary state can be inferred. Empirically, sources can be classified according to the slope of their 2 to $25\,\mu m$ SED in a $\log \lambda F_\lambda$ vs $\log \lambda$ plot (see, e.g., Lada 1988a). The most heavily obscured and presumably youngest YSOs are optically invisible and have SEDs which rise into the far-infrared with a slope greater than zero: these are Class I sources. Visible but reddened YSOs with strong mid-infrared excesses have SEDs with slopes between 0 and -2 and are classified as Class II sources: these objects are usually associated with classical T Tauri stars. Finally, YSOs which have weak or no infrared excesses have SEDs resembling reddened blackbodies (slope ~ -3). These objects, associated with weak emission T Tauri stars, are the Class III sources.

It has been suggested that the variations of shape in the SEDs observed for YSOs represent a quasi-continuous evolutionary sequence from protostars (Class I) to weak-emission T Tauri stars (Class III) (see, e.g., Adams and Shu 1986; Adams et al. 1987,1988; Myers et al. 1987; Rucinski 1985). As shown schematically in Fig. 4, Class I SEDs can be successfully modeled as accreting protostars comprised of three components (stellar core, circumstellar disk, and a quasi-spherical envelope or halo), with the far-infrared emission arising from the infalling envelope and the mid-infrared emission from the spatially thin disk. A stellar wind soon develops which breaks out at the rotational poles of the envelope, thereby reducing the extinction to the central YSO. Emission from the central object can now be observed at near-infrared wavelengths, producing a double-peaked, or Class IID, SED. As the wind dissipates the residual envelope, the central YSO plus disk are fully revealed resulting in

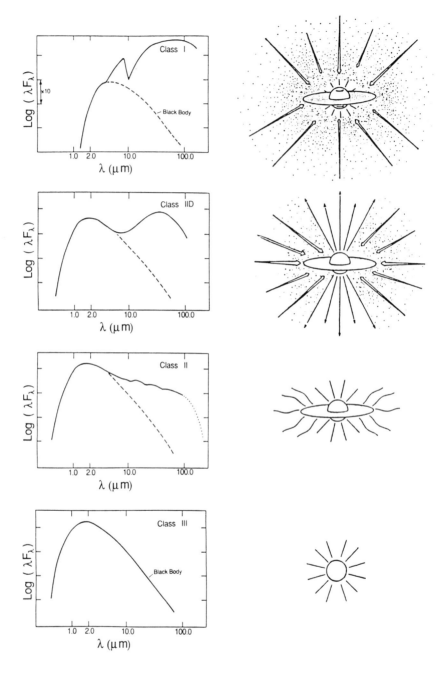

Figure 4.   A schematic of the sequence of protostellar evolution, along with the accompanying evolution of the spectral energy distribution (SED). This figure, taken from Wilking (1989), illustrates the ideas described by Lada (1988a) and in the review article by Shu et al. (1987a).

either a flat spectrum SED, indicative of $r^{-1/2}$ disk temperature gradient, or a Class II SED whose slope approaches $-4/3$, characteristic of a disk with an $r^{-3/4}$ temperature gradient. Finally, as the disk is dissipated, a Class III SED is observed. The time a given YSO spends in each of the Class I, II and III phases is believed to be longer for the later stages; e.g., for a $1\,M_\odot$ star, 1 to 0.2 Myr, 1 Myr, and 10 Myr, respectively. These are only rough estimates, and we should keep in mind that the discrete classes are just an approximation of a continuous evolution.

To further characterize the evolutionary state of a protostellar Class I (or IID) object, it is possible to to use its total luminosity L and the visual extinction to the stellar surface, A_V. Evolutionary L–A_V diagrams have recently been calculated from models of rotating low-mass YSOs in an attempt to provide the equivalent of an HR diagram for protostars (Adams 1990; see Chapter by Adams and Lin). As accretion proceeds for a given rotation rate, objects increase in luminosity while the visual extinction through the infalling envelope drops. This work extends that of Yorke and Shustov (1981) who had originally suggested the use of such diagrams for young massive stars embedded in spherical dust cocoons. Observationally, the total luminosity can be determined by integrating the SED; A_V is more difficult to determine but can be estimated from the depth of the $10\,\mu m$ silicate absorption features (Rieke and Lebofsky 1985), from a comparison of the intrinsic and the observed ratio of near-infrared hydrogen recombination lines (see, e.g., Alonso-Costa and Kwan 1989; Evans et al. 1987), or from near-infrared color indices which are free of excess emission (Myers et al. 1987). However, for very high extinctions ($A_V >> 100$ mag) none of these methods work, and even for lower extinctions each method suffers from specific problems, e.g., how to separate the intrinsic object extinction from foreground cloud extinction.

While Class I objects cannot be placed on a conventional HR diagram, Class II and III sources usually can. It must be noted however, that only the stellar luminosity should be used in placing the YSO on PMS tracks, not the total luminosity, which often also includes a significant contribution from an active or even a passive disk (Bertout et al. 1988; Kenyon and Hartmann 1987). Strom et al. (1989d) and Cohen et al. (1989b) have discussed methods for splitting the stellar and circumstellar contributions to the luminosities of T Tauri stars with disks, but the problem is far from trivial. Recently, Kenyon and Hartmann (1990) have investigated how a random distribution of disk inclination angles, coupled with a plausible range of accretion rates, can introduce a significant scatter in apparent luminosities for intrinsically identical stars. Age determinations for many PMS stars may therefore be uncertain by as much as factors of 2 to 3. Ultimately, PMS tracks need to be calculated where the luminosity arises not only from Kelvin-Helmholtz contraction, but also in part from disk accretion (see, e.g., Stringfellow 1989).

In summary, the current theory of YSO SEDs suggests that Class I objects consist of three distinct components, Class II of two, and Class III essentially of only one, i.e., the naked young star. However, this theory may be too simple. For example, Class I models may *not* require much of a disk (F. Adams, personal communication; Butner et al. 1991), while Class III models could include a disk with a large central hole (Beckwith et al. 1990). Also the time-dependent and possibly episodic nature of disk accretion should be considered in future models. For example, recent observations of YSOs in ρ Oph indicate a poor correlation between the mass of circumstellar material and the strength of the infrared excess (André et al. 1990*b*), suggesting variations in the compactness of disk material within SED classes. The origin of the $r^{-1/2}$ disk temperature gradient in flat-spectrum sources poses a particular problem, and suggests that future YSO evolutionary models should consider active as well as passive disks. Finally, the likelihood of binary and multiple YSOs sharing circumstellar material needs to be accounted for in analyzing SEDs.

III. OBSERVATIONS OF INITIAL STELLAR POPULATIONS

In this section, we will review the various observational techniques, from radio to X rays, employed to study YSOs over the full range of evolutionary states from protostar to weak-emission T Tauri star. The observational picture has changed dramatically since 1984 due to the success of IRAS and the growing availability of infrared array cameras. Examples of studies of the initial stellar population are given for diverse molecular environments ranging from nearby dark clouds, through giant molecular clouds, to regions of star formation in the Magellanic Clouds.

A. Techniques and Their Selection Effects

In order to find and study low-mass YSOs in all phases of PMS evolution, it is necessary to use a wide variety of observational techniques spanning the entire electromagnetic spectrum. At the same time, no single technique is likely to produce a complete sample of YSOs in any given star-formation region. The following are some of the techniques that have been used to reveal the whole range of YSOs, from Class I to Class III:

1. Far-infrared surveys (e.g., the IRAS all-sky 12–100 μm survey)
2. Near-infrared surveys (e.g., 1–2.5 μm)
3. 2.6 mm CO line surveys
4. 6 cm radio continuum surveys (e.g., with the VLA)
5. Optical objective prism surveys
6. X-ray surveys (e.g., with the EINSTEIN Observatory)
7. Optical proper motion surveys

Next, we shall discuss these techniques in more detail, paying particular attention to the inevitable selection effects and other shortcomings associated with each of them.

Since 1974, the Kuiper Airborne Observatory (KAO) has carried out studies of both massive and low-mass embedded YSOs at far-infrared wavelengths (see Thronson and Erickson [1984] for a review). Angular resolutions are typically 50 arcsec at 50 and 100 μm; for strong sources, diffraction limited resolution is possible, e.g., $\sim$23 arcsec at 100 μm (Lester et al. 1986). The KAO is somewhat limited in its spatial coverage and sensitivity, and therefore a major step forward was taken in 1983 when the cryogenically cooled IRAS carried out an all-sky survey at four mid- to far-infrared wavelengths (12, 25, 60, 100 μm) with a resolution of 0.75, 0.75, 1.4, and 3 arcmin, respectively. In addition to finding a number of deeply embedded YSOs (Beichman et al. 1986; Beichman 1987), IRAS was also sensitive to many of the previously known emission-line objects and objects associated with optical nebulosity. Thus IRAS detected several classes of YSOs and would have been superior to optical and near-infrared surveys were it not for a lack of sensitivity to low-luminosity objects (e.g., faint T Tauri stars in nearby dark clouds) and the low spatial resolution. Combined, these shortcomings implied that at the distance of Orion, IRAS was not only sensitivity limited but also confusion limited. In some cases, high-resolution pointed IRAS observations have overcome the source confusion problem in *nearby* dark clouds (e.g., in ρ Oph: Young et al. 1986; in Cha I: Prusti et al. 1991), as have follow-up KAO observations. However, at the distance of even the nearest H II regions, the IRAS and KAO views are blurred.

Groundbased mid-infrared observations with larger telescopes can deliver higher spatial resolution, but at a great expense in sensitivity, due to the enormous thermal background encountered at 10 to 20 μm. In contrast, the thermal background is negligible shortward of 2.5 μm, and near-infrared observations at 1 to 2.5 μm can result in high spatial resolution *and* high sensitivity. The advantages of near-infrared observations are that Class II and III YSOs emit most of their luminosity in that wavelength range; extinction due to foreground dust is greatly reduced compared to optical wavelengths; and competition from the bright ionized nebulosity associated with H II regions is considerably reduced. In general, only the very youngest and coolest, and/or most deeply embedded YSOs will escape detection at near-infrared wavelengths, thus requiring far-infrared and sub-millimeter observations. In the past, near-infrared surveys were limited by the need to scan a single detector over a region of star formation, resulting in a serious compromise between sensitivity, resolution, and areal coverage. Employing sensitive two-dimensional infrared detector arrays (with $\sim$64$\times$64 to 256$\times$256 pixels), near-infrared cameras are now being used to image large areas of star-forming regions. These observations result in simultaneous high spatial resolution (seeing limited, $\sim$1 arcsec FWHM) and sensitivity (point sources as faint as $K\sim$17 mag are visible in about two minutes on a 2 meter class telescope). It is important to note that for the majority of massive star-forming regions in the Galaxy (1–3 kpc), simultaneous high resolution and sensitivity are required to prevent a bias against detection of the lowest-mass stars, which would otherwise be lost in the "glare" of the ionizing massive stars. As we shall see below, the power of near-infrared cameras to reveal large clusters of low-mass stars associated with H II regions is proving to be a major step forward in YSO surveys. Perhaps the major complaint against these near-infrared surveys is that they can be *too* effective: in addition to revealing almost the entire YSO population associated with a star-forming region, these surveys can also detect large numbers of stars along the line of sight to and beyond the region, due to the effectiveness with which near-infrared photons can penetrate dust, and the general preponderance of the galactic field-star population towards low-mass, red objects. Surveys of this kind must employ reasonably sophisticated sorting criteria and/or statistical techniques to winnow out the field population, particularly in the more distant regions close to the galactic plane and within the inner Galaxy.

Mid-infrared (10 and 20 μm) cameras are also being built and becoming operational. They will complement observations at near- and far-infrared wavelengths and will serve as a high-resolution probe of deeply embedded YSOs, although again, the extremely high thermal background seen from the ground will limit them to objects that are sufficiently bright, on the order of a few hundred mJy at 10 μm.

Deeply embedded YSOs, as well as some Class II objects, are associated with energetic mass loss revealed as high-velocity molecular gas, Herbig-Haro objects, and/or free-free emission. Unbiased surveys for CO outflows in whole clouds such as NGC 2264 and the L 1641 cloud in Orion have revealed the crude positions of the sources that drive these outflows (Margulis and Lada 1988; Fukui et al. 1989; Fukui 1989). Sensitivity and resolution problems also plague this technique, and primarily higher luminosity sources have been found. If a source is close to the cloud surface, optical Herbig-Haro objects and jets may result, and infrared searches around them can be used to locate their obscured driving sources (Cohen and Schwartz 1983; Reipurth 1989c). There must be many cases, however, where the HH objects and jets are themselves deeply embedded in the molecular clouds, avoiding detection in the optical: few optical HH objects are seen near the ρ Oph cloud for example. Infrared cameras offer the potential for revealing HH objects and jets even in highly obscured regions: a recently discovered highly collimated structure in a dense core near IC 348 (Fig. 5; see also Color Plate 1) is invisible at wavelengths less than 1 μm, and may be a HH-like object, perhaps radiating strongly in the near-infrared lines of shocked molecular hydrogen and [Fe II].

Potentially promising for uncovering deeply embedded low-luminosity YSOs, as well as Class III objects, are large-scale radio continuum surveys carried out at the VLA (20 cm and 6 cm wavelengths). Typical maps can cover a few square degrees with synthesized beams of order 10 to 20 arcsec and source positions better than 5 arcsec. This technique is very sensitive to small amounts of ionized gas arising from stellar winds (Class I sources) or magnetic surface activity (Class III sources). However, it turns out that most low-mass YSOs have only low levels of radio continuum emission. For example, André et al. (1987) and Stine et al. (1988) conducted a VLA survey of 4 square degrees in the ρ Oph cloud. Only a few of the known YSOs were detected and none of the optical T Tauri stars: 90% of the YSOs remained

Figure 5.   A K (2.2 μm) image of the IC 348-IR region, covering ~3.2×3.2 arcmin (~0.3×0.3 pc at 300 pc). A previously unknown jet-like feature is revealed, possibly a Herbig-Haro object. The associated driving source, which must be a very young and highly obscured star, has not yet been identified. The data were obtained using a 256×256 pixel HgCdTe array on the University of Hawaii 2.2-m telescope.

undetected down to a limit of 1 mJy. The bulk of the sources detected at 20 cm at the VLA are unrelated to known YSOs, which calls for tedious identification procedures involving the determination of radio continuum spectral indices and thus multi-wavelength data. However, the radio continuum observations did discover a few new, very interesting embedded young sources which would otherwise have been overlooked, e.g., VLA 1 in the HL/XZ Tau region (Brown et al. 1985) and recently VLA 1623 in ρ Oph near GSS 30 (André et al. 1990a).

Objective prism surveys (Hα) have been very powerful in revealing

classical T Tauri stars characterized by strong emission lines at visible wavelengths (see, e.g., Herbig and Bell 1988). Also, surveys for Ca emission
have been undertaken to search for weak-emission T Tauri stars, but have
discovered few, adding only about 20 emission-line stars in Taurus, for example, (Herbig et al. 1986). One caveat in using Hα surveys is that the line
strength could be highly variable. For example, emission-line objects found
on prism plates taken at one epoch have shown less than 50% overlap with
those found on plates taken just a decade or so later: e.g., compare Dolidze
and Arakelyan (1959) with Wilking et al. (1987) who surveyed the same area
in the ρ Ophiuchi dark cloud. This variability is worrisome, undermining
the usefulness of single-epoch Hα surveys to assess the number of young
low-mass stars in even lightly obscured star-forming regions.

   X-ray surveys pick out young stars with magnetic surface activity and
have uncovered a population of YSOs which have no or only weak Hα
emission and thus were missed in the classical Hα surveys (Feigelson and
de Campli 1981; Montmerle et al. 1983a; Walter 1986; Walter et al. 1988;
Feigelson and Kriss 1989). Apparently, these weak-emission T Tauri stars
have dissipated most of their inner circumstellar gas and dust: however, they
still may have *outer* dust disks detectable at millimeter continuum wavelengths
(cf. Beckwith et al. 1990). Because of the limited nature of X-ray surveys with
the EINSTEIN Observatory and the highly variable nature of the emission, we
cannot say at present how many of these Class III objects have been missed.
With the ROSAT X-ray satellite, launched in June 1990, we should be able to
address this problem in the near future. However, even moderately embedded
Class III objects can escape detection due to the absorption of soft X rays by
small amounts of cloud material; an optical extinction of $A_V \sim 5$ ($\tau_{1.5\mathrm{keV}} \sim 1$)
may suffice to lose these young stars.

   One of the most unbiased techniques used to identify lightly-obscured
YSOs in nearby star-forming regions is an optical proper motion survey. For
the Taurus clouds, the pioneering work of Jones and Herbig (1979) was recently extended by Hartmann et al. (1991), who covered ∼9 square degrees to
an estimated completeness limit of $V=15$ mag. Optical spectra were obtained
for the subset of stars with proper motions consistent with Taurus membership,
leading to an increase of about 20% in the number of PMS objects in the main
survey region, essentially the value predicted by Jones and Herbig (1979).
The newly discovered PMS stars have ages around 1 Myr, even though much
older stars could have been detected. Thus the problem "where are all the
post T Tauri stars?" (Herbig et al. 1986; Walter et al. 1988) remains: they do
not seem to be there. It is unlikely that a large population of Class III objects
has been missed due to selection effects: the only apparent biasing effect is
caused by the movement of the photocenter of binaries due to their mutual
orbital motion rather than proper motion.

   Proper motion surveys for other regions are badly needed, but the effort
is huge, and only a few other populations have been examined, e.g., the Orion
region centered on the Trapezium Cluster (Parenago 1954; Jones and Walker

1988; van Altena et al. 1988; McNamara et al. 1989). As many members of these populations remain embedded in molecular cloud material, ultimately we will need proper motion surveys in the near-infrared.

B. Young Stars Associated with Dark Clouds

Low-mass star formation is best studied in the nearby (<200 pc) dark cloud complexes. There one can apply the greatest variety of observational techniques with maximum sensitivity to low-luminosity YSOs and with the highest linear spatial resolution. Additionally, most of these dark clouds lie well outside of the plane of the Galaxy, so that confusion from background stars is minimized.

1. Taurus-Auriga and ρ Ophiuchi

Among the most intensively studied star-forming regions are the Taurus-Auriga and ρ Ophiuchi dark cloud complexes (Figs. 6 a,b). These are intermediate-mass molecular complexes ($\sim 10^4 \, M_\odot$) which lie close to the Sun, at distances of about 140 pc and 160 pc respectively. At these distances, their sub-solar mass populations of YSOs are easily accessible to groundbased observers even at mid-infrared wavelengths. Indeed, of the ~ 100 YSOs identified with each cloud, almost all are low-luminosity (<5 $L_\odot$) objects. These YSOs cover the full range of expected evolutionary states, from heavily obscured Class I sources to optically visible weak-emission Class III stars. HR diagrams of the T Tauri populations in these clouds coupled with theoretical PMS tracks suggest that star formation has been going on for at least the past few million years (Cohen and Kuhi 1979; Stahler 1983a); yet as we discuss below, current evidence suggests that the mechanism for star formation may have been different in the two clouds.

While these molecular cloud complexes have similar masses and are formation sites for comparable numbers of low-mass stars, there are important differences between their young stellar populations. Some, if not all, of these differences can be traced to the different interstellar and molecular environments in which the stars have formed. First, the interstellar radiation field is less intense in Taurus-Auriga than in ρ Oph. In the Taurus complex, there are only 8 associated B stars within a 12 pc projected radius of the cloud center (Elias 1978c; Walter and Boyd 1991). In contrast, the ρ Oph cloud lies in the Upper Scorpius subgroup of the Sco-Cen OB association, and has about 20 B or A0 stars within a four times smaller area (radius ~ 6 pc), with the B2 V star (HD 147889) and a B3–B5 V star (Source 1) embedded in the western edge of the complex (Elias 1978a; de Geus et al. 1989). As a result, warmer dust and gas temperatures are observed in ρ Oph relative to Taurus-Auriga (see, e.g., Greene and Young 1989; Jarrett et al. 1989; Snell et al. 1989).

A second important difference between these young stellar populations relates to their distributions. Most of the YSOs in the Taurus-Auriga cloud are found in the vicinity of molecular cores with typical masses of 1 $M_\odot$, densities of 10^4 to 10^5 cm^{-3}, temperatures ~ 10 K, and visual extinctions of 5

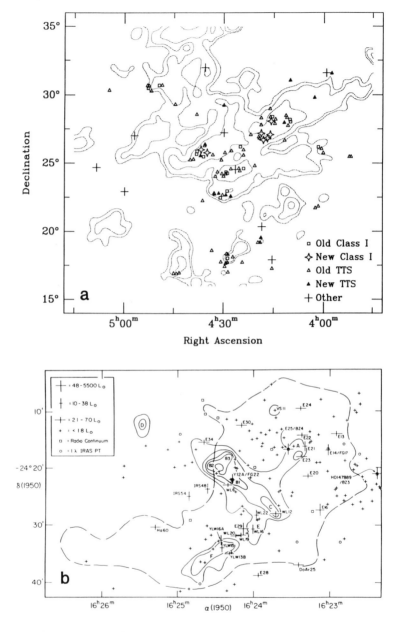

Figure 6 a,b.   The spatial distribution of YSOs in the Taurus-Auriga dark-cloud
complex (covering ~2400 pc²) from Kenyon et al. (1990), and the spatial distribution of YSOs in the ρ Ophiuchi dark cloud core covering ~4 pc² (figure from Loren
et al. 1990).

mag to 10 mag (see Myers [1985], and references therein; Myers et al. [1987]). The cores, and hence the YSOs, are distributed throughout a low-extinction, filamentary molecular complex covering about $2400 \, pc^2$ (Ungerechts and Thaddeus 1987). On the other hand, most YSOs in ρ Oph are concentrated toward a 1×2 pc ridge of high column density gas at the western edge of the complex, containing about $550 \, M_\odot$ of molecular gas (Vrba et al. 1975; Elias 1978a; Wilking and Lada 1983; Loren 1989a,b). Characteristic visual extinctions in this region are 50 mag to 100 mag. High-density molecular cores are found within the ridge, displaying higher masses and temperatures than their Taurus-Auriga counterparts (Loren et al. 1990). A lower density of YSOs is found in the filamentary streamers of Ophiuchus, forming in the vicinity of molecular cores in the manner observed in Taurus-Auriga.

The stellar density within the high column density ρ Oph cloud core is about 150 stars pc^{-3}, and the star-formation efficiency (SFE) is roughly 22% (Wilking et al. 1989a). Thus it has been suggested that given the time-dependent nature of the SFE and the slow release of gas from the cloud, a gravitationally bound stellar cluster of order 100 stars will ultimately emerge from the core (Wilking and Lada 1985). While it is true that the SFE in Taurus-Auriga is high in the vicinity of dense cores, the current distribution of YSOs does not suggest the formation of a large bound stellar group there.

Significant differences between these stellar populations also arise when a detailed comparison of the luminosity functions of their Class I and Class II sources is made (see, e.g., Wilking et al. 1989a; Kenyon et al. 1990). Spectral energy distributions for the YSOs and the resulting luminosity functions have been constructed by synthesizing visible-wavelength data with near- to far-infrared data, thus yielding the most reliable estimates for the YSOs' SED classes and bolometric luminosities. In regions of low source density, the completeness limit imposed by the IRAS sensitivity for the luminosity function in the Taurus-Auriga cloud is $>0.3 \, L_\odot$ for Class I sources and $>0.5 \, L_\odot$ for Class II stars. In ρ Oph, these limits are higher due to greater extinction in the cloud: Class I sources are completely sampled throughout the entire depth of the cloud core for $>1 \, L_\odot$, but $1 \, L_\odot$ Class II sources are completely sampled in only the outer $A_V \sim 35$ mag to 40 mag of the cloud. The luminosity functions for these two clouds are shown in Fig. 7, with the number of Class I and Class II sources in each bin indicated. In ρ Oph, the intermediate luminosity range (5.6–$56 \, L_\odot$) is dominated by Class I objects (80%) and their number drops with decreasing luminosity. This observation can be understood if these Class I sources form a population of heavily obscured 0.1 to 1 $M_\odot$ objects with an excess of luminosity relative to their Class II counterparts, i.e., they derive part or all of their luminosity from accretion. Alternatively, it is possible that these Class I sources are very young 2 to 3 $M_\odot$ objects, implying that the cloud is forming stars sequentially in mass. In Taurus-Auriga, there are fewer Class I objects relative to Class II at intermediate luminosities (45%), and their number *increases* with decreasing luminosity. There are several selection effects present in ρ Oph which could give rise to the observed differences

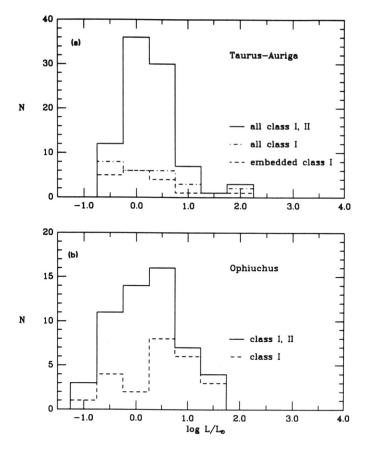

Figure 7.   A comparison of the bolometric luminosity functions for young stellar objects in the Taurus-Auriga dark cloud and in the ρ Oph molecular cloud core. The number of Class I and II sources in each bin is indicated. Significant differences are apparent between the two luminosity-limited samples. Figure taken from Kenyon et al. (1990, their Fig. 8).

in these luminosity functions; for example, confusion in the IRAS data and the high visual extinction in ρ Oph could account for the drop in the number of Class I sources with decreasing luminosity. However, the segregation of luminosities by SED class for $L > 5.6\,L_\odot$ appears to be a real effect in ρ Oph and cannot be accounted for by source confusion in the IRAS data (as determined by deep near-infrared imaging) or by reddening of Class II sources to produce Class I SEDs (as indicated by rising 25–100 μm SEDs). While the origin of the differences between these luminosity functions is unknown, several explanations are plausible. If intermediate-luminosity Class I sources in ρ Oph are 2 to 3 $M_\odot$ objects, then perhaps stars in Taurus-Auriga have not formed as recently or are not forming sequentially in mass. On the other hand, if these Class I sources are 0.1 to 1 $M_\odot$ objects, then perhaps the accretion rate, and hence luminosity, of protostars is lower in Taurus-Auriga relative to ρ Oph, as predicted by models for spherical accretion in cooler clouds (Adams et al. 1987).

Investigations into the low-mass end of the IMF of the ρ Oph Cluster are now possible using infrared array cameras, with high visual extinctions in the core minimizing confusion with background field stars. A moderately deep 2.2 μm survey (to $K \sim 15$ mag) of a 20 square arcmin area of the core by Rieke et al. (1989) revealed no new sources beyond those found by Wilking and Lada (1983), suggesting a deficiency of YSOs with masses below a few tenths of a solar mass. However, a subsequent survey by Barsony et al. (1989a) to $K \sim 14$ mag of a larger area (144 square arcmin) led to the discovery of 35 new sources, and most recently, Greene (1991) surveyed some 650 square arcmin on and off the densest parts of the cloud, finding almost 500 sources to a completeness limit of $K \sim 13$ mag. As such surveys grow in area coverage, it becomes increasingly important to determine what fraction of these new sources are indeed embedded in the cloud by using multi-color infrared photometry (cf. Greene 1991). However, one possible early conclusion is that the IMF of a cloud can vary dramatically over size scales less than 0.5 pc, and that large areas of a cloud must be considered before comparisons are made with other clouds.

The differences noted above between the populations of YSOs in Taurus-Auriga and ρ Oph have led to the idea that different mechanisms of star formation have operated in these clouds, but thus far the observational evidence is sparse. Vrba (1977) and Loren and Wootten (1986) have proposed that compression of the ρ Oph cloud by a shock wave propagating from the southwest has shaped the morphology of the cloud and dense cores, generated small velocity shifts in the molecular gas, and produced the high SFE in the core region. A more recent explanation for the high SFE in the ρ Oph core has been given by Uchida et al. (1990), based upon their discovery of rotational motions about the major axis of the filamentary streamer to the northeast. They suggest that these motions provide an angular momentum drain from the cloud core resulting in enhanced star formation. Consistent with these ideas that the ρ Oph core has lost its internal support, Shu et al. (1987a) suggest that ρ Oph is an example of star formation resulting from cloud contraction where self-gravity dominates magnetic support, thus producing a high SFE. In contrast, magnetic fields may dominate the evolution of molecular cores in Taurus-Auriga, resulting in longer contraction times and lower values for the SFE.

2. Other Dark Clouds and Globules

There are a number of star-forming regions within several hundred parsecs of the Sun whose young stellar populations have yet to be examined as extensively as those in the Taurus-Auriga cloud or ρ Oph core. These regions include the Lupus dark cloud (125 pc), the R Coronae Australis cloud

(130 pc), the filamentary streamers in Ophiuchus, L 1689/L 1709 (160 pc), the Chamaeleon dark clouds (115–220 pc), and globules selected from southern and northern hemisphere photographic surveys (≤500 pc). Future studies of these regions will provide important comparisons with Taurus-Auriga and ρ Oph, and give further insight into the process of low-mass star formation in dark clouds. Investigations of the YSO populations in several of these regions are briefly described below.

Like Taurus-Auriga, the Lupus dark cloud complex also has a filamentary appearance. It consists of 4 sub-regions, of which 3 are actively forming low-mass stars (see Krautter [1991] for a review). Preliminary indications are that the YSO population also resembles that in Taurus-Auriga, i.e., there appears to be a low ratio of embedded Class I sources to emission-line Class II stars. However, previous surveys for YSOs have selected against heavily obscured objects, and confirmation of the scarcity of Class I objects awaits a thorough analysis of the IRAS data for the region (see, e.g., Carballo et al. 1991) and near-infrared imaging of the cloud. Since Class I objects are most often found associated with mass loss, one might expect a low incidence of HH objects/optical jets in Lupus. This is in fact observed (Krautter 1986; Heyer and Graham 1989), although high visual extinction in the cloud can produce the same effect (e.g., as seen in the ρ Oph core). The early indications are that Lupus is an older star-forming region where activity has already died down. Consistent with this picture is the fact that the Lupus T Tauri population is dominated by M-type stars (Krautter and Kelemen 1987; see also Appenzeller et al. 1983). It will be interesting to see whether the ROSAT X-ray survey will reveal the large population of weak-emission T Tauri stars, particularly those with K spectral types, expected for a more mature star-forming region.

The R Coronae Australis dark cloud bears many similarities to the ρ Oph complex. The clouds are located about $17°$ out of the plane of the Galaxy, almost mirror images at galactic longitude $l \sim 355°$. The morphologies of their molecular gas are also similar, with lower-density streamers of gas and dust extending from centrally condensed cores (Loren et al. 1983), although different mechanisms could produce this morphology (Vrba 1977; Vrba et al. 1981; Fleck 1984). Like ρ Oph, the most active star formation is in the core, where 13 YSOs distributed over a $0.08 \, pc^2$ area centered on the star R Cr A comprise the Coronet Cluster (see, e.g., Taylor and Storey 1984; Wilking et al. 1986). In the absence of massive stars, the core could be the site of formation of a small gravitationally bound cluster, and yet overall, the density of YSOs in the R Cr A cloud appears lower than in ρ Oph. Hα studies have yielded only ≤17 bona-fide emission-line stars (Marraco and Rydgren 1981) while at least 65 have been identified in ρ Oph (Wilking et al. 1987). Preliminary analyses of IRAS coadded survey data of the cloud core and vicinity reveal less than half the number of mid- to far-infrared sources seen in ρ Oph (16 vs 44) and only three of these are invisible, Class I sources (Wilking et al. 1991). The difference between the number of young stars in the two molecular clouds appears to be in rough proportion to their masses, i.e., there is a smaller reservoir of molecular gas in the R Cr A cloud. Extensive near-infrared imaging of the cloud core is needed to characterize the embedded population in more detail.

The Chamaeleon dark clouds (Cha I and Cha II) have recently been extensively explored through optical, near-infrared, far-infrared, and X-ray observations (Hyland et al. 1982; Jones et al. 1985; Whittet et al. 1987,1991a,b; Assendorp et al. 1990; Prusti et al. 1992; Feigelson and Kriss 1989; Schwartz 1991; Gauvin and Strom 1991). The distance is a particularly controversial parameter, and is somewhere in the range 115 to 220 pc (Whittet et al. 1987; Gauvin and Strom 1991). The streaky, filamentary structure found in the complex is once again reminiscent of the Taurus-Auriga clouds (see, e.g., Mattila et al. 1989), although the star formation in Cha I is much more localized than in Taurus-Auriga: of the 81 YSOs identified over a 50 pc^2 area of the Cha I cloud, the great majority are concentrated in a $\sim$12 to 15 pc^2 region near the reflection nebulae Ced 110, 111 and 112 (Schwartz 1991; Prusti et al. 1992). The distribution of spectral types for the optically visible T Tauri stars in Cha I is heavily weighted towards K2–M0.5 stars, suggesting a higher characteristic stellar mass than in the Lupus cloud (Krautter 1991). Prusti et al. (1992) combined near-infrared photometry and IRAS fluxes to determine SEDs and bolometric luminosities for 60 of the YSOs, and thence derived a luminosity function for the association. When compared to the equivalent luminosity functions for ρ Oph and Taurus-Auriga, Cha I has the lowest average luminosity of all three, and fewer Class I objects (six) than either of the others. One possible interpretation is that the Cha I population is older. However, Gauvin and Strom (1991) compared the masses and ages of the optically visible YSOs in Cha I with those in Taurus-Auriga and found no significant differences. They suggest that the dearth of Class I objects in Cha I might indicate an episodic or variable star formation rate, with few stars formed recently. Finally, only $\sim$60% of the previously known T Tauri stars in Cha I were detected by IRAS, in either the main survey or the more sensitive pointed observations. As Whittet et al. (1991b) note, this implies that the IRAS data give an incomplete census of low-mass Class II or III YSOs, even in the nearby dark clouds: deep near-infrared surveys such as those carried out by Hyland et al. (1982) and Jones et al. (1985) for Cha I are probably more effective in identifying Class II and III YSOs.

The Cha II cloud contains no reflection nebulae and is less conspicuous than the more extensively studied Cha I. Schwartz (1977) found a small number ($\sim$19) of Hα emission-line stars along the line of sight towards Cha II, and these objects were recently confirmed as low-luminosity (0.1–1.3 L$_\odot$) young classical T Tauri stars based on their near- and and far-infrared SEDs (Whittet et al. 1991a). Apart from one luminous ($\sim$40 L$_\odot$) Class I source with an associated outflow, the Cha II cloud lacks objects with luminosities >2 L$_\odot$ compared to Cha I, and the lower-luminosity objects are not obviously clustered.

Recently, infrared studies have been made of more isolated star formation in small (<10 arcmin diameter), dark clouds within about 500 pc of the Sun, commonly referred to as Bok globules. Persi et al. (1990) have used the IRAS Point Source Catalog to analyze 482 southern hemisphere dark globules from a list compiled by Hartley et al. (1986). They have found only about 10% (53) are associated with point sources displaying a rising 12 to 60 μm flux distribution. Using near-infrared photometry, they have determined that 17 of these IRAS sources are PMS objects, with typical luminosities of 10 to $1000 \times (d/500 \text{ pc})^2 L_\odot$. Persi et al. (1990) conclude that star formation in small dark clouds may be very inefficient; however, as discussed earlier, observations over a broad range of wavelengths are necessary to completely sample a YSO population. A similar study of northern hemisphere ($-36° < \delta < 90°$) globules has been made by Yun and Clemens (1990). They used IRAS co-added images to search for evidence of star formation among the sample of 248 isolated clouds catalogued by Clemens and Barvainis (1988). They find 23% of the sample have associated far-infrared point sources, but argue, on the basis of the limiting sensitivity of their technique to PMS stars with $M > 0.7 M_\odot$, that *all* globules in their sample could harbor newly formed stars if the embedded stellar mass distribution follows the Miller-Scalo IMF. Clearly, the Persi et al. (1990) and Yun and Clemens (1990) samples provide useful target lists for follow-up studies with near-infrared cameras, ISO and SIRTF.

C. Young Stars Associated with Giant Molecular Clouds

In order to study the high-mass end of the IMF, we must look to the giant molecular clouds (GMCs) where young OB stars are found. However, GMCs are not only birthplaces of massive stars but also of low-mass stars. In fact, most of the integrated stellar mass in the IMF is in low-mass stars and because most of the molecular mass is in GMCs, it follows that most low-mass stars in the Galaxy should form in GMCs, rather than in smaller, less massive dark-cloud complexes like Taurus-Auriga and ρ Oph.

1. Orion (400–500 pc)

At 160 pc, the Sco-Cen OB association has the distinction of being the nearest to the Sun. However, there is no on-going high-mass star formation in Sco-Cen, although low-mass stars are still forming in the associated ρ Oph dark cloud. Thus it is the Orion OB association and GMC that present us with our closest view of on-going simultaneous low- and high-mass star formation.

We begin by delineating the young populations *near* the Orion cloud complex, namely the stellar content of the adjacent OB association, Ori OB1. This is followed by descriptions of the embedded clusters in the Orion A (L 1641) and Orion B (L 1630) cloud, starting with an in-depth discussion of the Trapezium Cluster, followed by briefer accounts of the clusters associated with OMC-2, NGC 2024, NGC 2023, NGC 2071, and NGC 2068. After discussing the clusters, we also discuss the underlying populations of their host clouds L 1641 and L 1630. Finally, we discuss the young stars in the nearby λ Ori region.

The Stellar Content of the OB Subgroups The various subgroups (1a, 1b, 1c, 1d) of the Ori OB1 association comprise a so-called "fossil record" (Blaauw 1991) of relatively recent star formation, while new stars are still being formed in the nearby Orion A and B molecular clouds (see Fig. 8). These various components are perhaps all linked via sequential star formation, i.e., the compressive action, via winds and supernovae, of one generation of O stars on adjacent molecular material leading to yet another burst of star formation (cf. Elmegreen and Lada 1977). A speculative sequence of events in Orion may have started with a first phase of star formation leading to the 1a subgroup some 7 to 12 Myr ago, perhaps triggered by the passage of the Orion GMC through the galactic disk (Franco et al. 1988) or by the effects of a runaway O star exploding inside the Orion molecular cloud (Blaauw 1991; A. Blaauw, personal communication). The 1a O stars may then have triggered the younger 1b subgroup (∼5 Myr), which includes the Orion Belt stars. The 1b subgroup is located between Orion B and Orion A, and it is tempting to speculate that in its turn, the 1b subgroup may have split the two clouds and triggered star formation in both directions at once (a bifurcation), i.e., leading to the NGC 2024 Cluster in Orion B to the north and the subgroups 1c (a few Myr) and 1d (the Trapezium Cluster; ∼1 Myr) in Orion A to the south.

The stellar content of 1a, 1b, and 1c subgroups was studied by Warren and Hesser (1977,1978), with the conclusion that only stars earlier than A0 (about 3 $M_\odot$) are completely known. The number of B stars in 1a, 1b and 1c are 121, 96, and 36 respectively, and the estimated number of supernovae thought to have occurred in 1a and 1b is 9 and 3, respectively (A. Blaauw, personal communication). Clearly, subgroup 1a must have been a very active site of star formation some 10 Myr ago, much more active than the Orion clouds today.

As mentioned above, little is known about the low-mass stellar content of the Ori OB1 association. However, recent Kiso Schmidt Hα surveys of the 1a and 1b subgroups have revealed some 250 and 150 new emission-line stars respectively, with magnitudes in the range V=13 mag to 17 mag, i.e., most likely low-mass T Tauri stars (Kogure et al. 1989; Wiramihardja et al. 1989). A sizeable fraction (about 30%) of these emission-line stars were found to be variable in Hα intensity over a time span of 18 months, again highlighting the limitations of single epoch searches for YSOs via Hα surveys. Future work will place the newly revealed Orion association members on the HR diagram and determine their masses, perhaps even their mass function (cf. Cohen and Kuhi 1979).

Proper motion surveys should be less biased, and the 1c subgroup (also known as the Orion Cluster, as distinct from the Trapezium Cluster, subgroup 1d) has long been a target for this technique, beginning with the classic work by Parenago (1954). He listed almost 3000 stars brighter than B∼17 mag in a 9

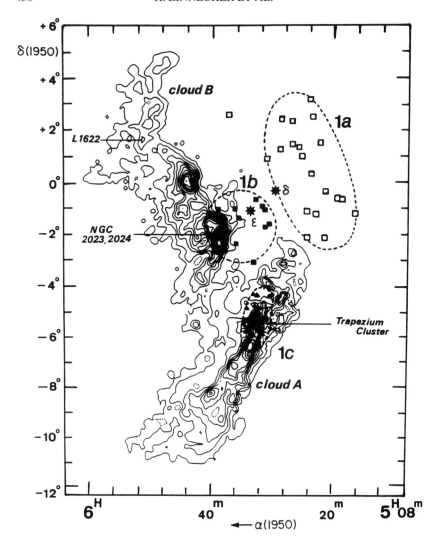

Figure 8.   The Ori OB1 association showing the subgroups 1a, 1b and 1c with respect to the molecular clouds Orion A (L 1641) and Orion B (L 1630) (Blaauw 1991). The 1a, 1b and 1c sub-groups cover roughly 45×25, 25×20, and 20×10 pc, and are approximately 7–12, 5, and 3 Myr old, respectively. The 1d subgroup (the Trapezium Cluster) lies at the center of the 1c subgroup.

square degree area centered on the Orion Nebula, but due to the relatively crude proper motions, only tentative membership probabilities could be derived. Recent studies have obtained significantly improved results from new and archival material spanning up to 77 years. McNamara et al. (1989) studied a somewhat smaller region (7 square degrees) excluding the central emission nebulosity, and analyzed 630 stars to a limit of $B\sim14$ mag, finding that approximately 40% of the stars are probable members. Restricting their study to the 73 stars brighter than $V\sim12.5$ mag within 30 arcmin of the Trapezium, van Altena et al. (1988) obtained very high precision proper motions, and assigned membership to about 67% of the stars. Lastly, using deep red plates (down to $I\sim13$ mag in the bright nebulosity, and $I\sim16$ mag in the outer region) to cover a smaller 15 arcmin radius around the Trapezium, Jones and Walker (1988) detected more than 1000 stars, and found $\sim90\%$ to be members of the Orion complex. From 693 low error and high membership probability stars, Jones and Walker (1988) derived a one-dimensional velocity dispersion of about 2.3 km s^{-1}, with a distribution that is more or less isotropic and independent of the distance from the Trapezium. Considering only the brighter stars, both Jones and Walker (1988) and van Altena et al. (1988) found a dispersion closer to 1.5 km s^{-1}, and the latter study revealed that the two brightest stars in the Trapezium Cluster (θ^1Ori C and θ^2Ori A) have very high proper motions, indicating that both are being ejected, and will be more than 5 pc from the cluster within 1 Myr.

Finally, it is worth noting that a special program being carried out with the HIPPARCOS satellite is beginning to provide parallaxes for hundreds of the brighter stars in the Orion region.

The Trapezium Cluster. At the core of the Orion 1c subgroup, near the apex of the L 1641 molecular cloud, is the famous Orion Nebula (M 42, NGC 1976). Much is known about this nearby H II region, the massive stars of the Trapezium OB association (i.e., the 1d subgroup) that ionize it, the background parent molecular cloud OMC-1, and the very young sources of the BN-KL complex embedded within it. Yet there has been remarkably little study of the *low-mass* stellar content of the region, the so-called Trapezium Cluster (Herbig 1982).

Despite the extremely bright H II region nebulosity, it has been known for a long time that the Orion Nebula is host to an unusually large number of faint red stars (Trumpler 1931; Baade and Minkowski 1937). Recently, Herbig and Terndrup (1986) used narrowband optical (at ~5500 and 8000Å) imaging photometry to reduce the competition from nebular line emission, detecting ~150 stars over a region approximately 3×3 arcmin in size. Excluding the OB stars, an extremely high stellar density in excess of 2200 pc^{-3} was derived, although the exact value depends on the volume the cluster is assumed to occupy. Some 68 cluster members were well enough measured to be placed on an HR diagram, allowing estimates of ages and masses by comparison with theoretical PMS tracks. A uniform dereddening, determined from spectroscopy of only 10 stars, was applied to all cluster members. The majority of the stars fell youngward of the 1 Myr isochrone, and including the OB stars, a mean mass density of 3000 $M_\odot$ pc^{-3} was derived, equivalent to 0.6×10^5 H$_2$ molecules cm^{-3}. This latter value is typical of the nearby molecular cloud material, implying that the cluster must have formed with

Figure 9 a,b.   A K (2.2 μm) mosaic covering the central 5×5 arcmin (~0.65×0.65 pc
at 450 pc) of the Trapezium Cluster. Some 500 stars are seen to a completeness
limit of K~15.5 mag. The data were obtained using a 62×58 pixel InSb array
on the United Kingdom 3.8-m Infrared Telescope (McCaughrean et al. 1992). For
comparison, an optical (I) photograph of the same region is also shown opposite
(photograph taken at the Anglo-Australian Telescope, and kindly provided by D.
Malin).

a high star-formation efficiency (> 10%) from the cloud (Genzel and Stutzki
1989).

However, optical studies tell less than half the story. As we have discussed
earlier in this chapter, the near-infrared is a much better wavelength range
over which to survey young clusters, and even relatively low resolution and
sensitivity raster-scan studies at 2.2 μm demonstrated the existence of a yet
higher stellar density than seen in the optical (Hyland et al. 1984). With the
introduction of infrared imaging cameras, the Trapezium Cluster was fully
revealed, with over 500 members in a 5×5 arcmin region centered on the
Trapezium OB stars (McCaughrean 1988; McCaughrean et al. 1992).

Figure 9 a shows a 2.2 μm image of the Trapezium Cluster: by comparison

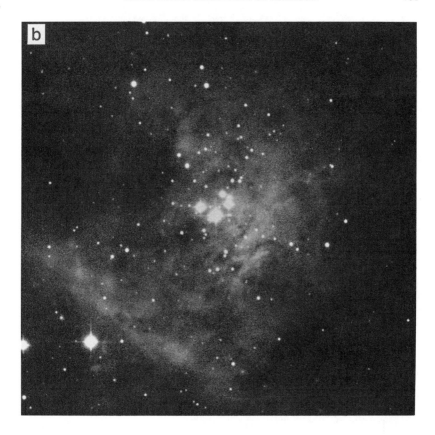

with an optical image of the same region (Fig. 9 b), the cluster appears more symmetric as the optically obscured stars, including those behind the eastern "dark bay," are revealed. (See Color Plates 2 and 3.) The massive Trapezium OB stars are located right at the center of the cluster, a feature common to many young clusters, as we shall see later (e.g., R 136a in 30 Doradus). The extraordinary stellar density of the Trapezium Cluster can be well illustrated by comparing it to the embedded cluster in the core of ρ Oph. Figure 10 a shows the stars in the central 5×5 arcmin of the Trapezium Cluster, while Fig. 10 b shows the core of ρ Oph as it would appear if projected to the distance of the Orion Nebula.

In the central 1 arcmin (0.14 pc) diameter core of the Trapezium Cluster, the stellar number density exceeds 10^4 pc^{-3}, although again the exact value depends somewhat on the three-dimensional geometry of the cluster. This density implies a mean separation between the cluster stars of $\sim$0.045 pc (1.5×10^{15} m or 10^4 AU). Assuming this length to be equal to the fragmentation scale length in the original molecular cloud, and that fragmentation occurs promptly (e.g., after shock compression) to create protostellar fragments with masses similar to those of the observed low-mass stars ($\sim$0.2–0.6 M$_\odot$), we

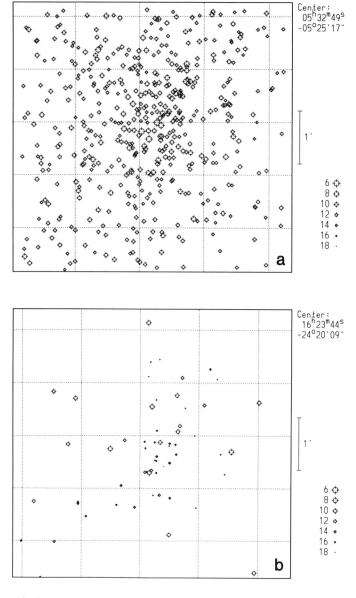

Figure 10 a,b.   A comparison between the stellar density of the Trapezium Cluster and that of the embedded ρ Oph Cluster. For this purpose the spatial separations and 2.2 μm luminosities of the ρ Oph Cluster stars have been adjusted to place the cluster at an apparent distance of 450 pc, the distance of the Trapezium Cluster. It is obvious that the Trapezium Cluster is denser and that its members have a greater mean 2.2 μm luminosity. Trapezium Cluster data from McCaughrean et al. (1992); ρ Oph data from Wilking et al. (1989a) and T. Morgan and G. Rieke (personal communication).

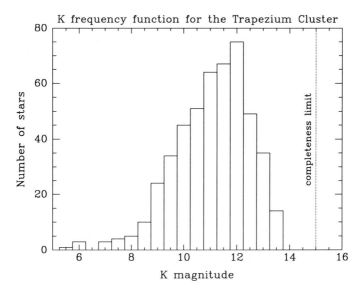

Figure 11.   The raw (uncorrected for extinction) K (2.2 μm) luminosity function for ~450 stars of the Trapezium Cluster, as derived from the infrared mosaic of the inner 5×5 arcmin of the cluster (McCaughrean et al. 1992). Note the very strong turnover in the luminosity function at $K \sim 12$ mag, well above the completeness limit of 15.5 mag.

can derive a gas density and temperature by requiring the separation and mass to be equal to the Jeans length and mass respectively. In this case, we derive reasonable values of $\sim 10^6$ cm^{-3} for the density and ~ 10 to 20 K for the temperature. This raises the possibility that the original protostars in the Trapezium Cluster may actually have been touching. The protostars probably had a core-envelope structure, with the cores competing gravitationally for material from the weakly bound envelopes, continuing until the stellar mass spectrum was frozen due to lack of accretable material.   Such a process can produce a reasonable IMF (Zinnecker 1982).   An alternative scenario for the origin and evolution of a protocluster has been discussed by Myers (1991), predicting noncoeval formation of cluster stars in a continually deeper potential well due to gradual accumulation of protocluster gas (cf. Larson 1982).

    There is an important point to be made about the Trapezium Cluster: it may be the only nearby star-forming region where we are dealing with a more or less complete and relatively uncontaminated sample. First, the infrared images go deep enough to almost completely sample the entire stellar population, with a strong peak in the K frequency function at 12 mag, well above the detection limit of 15.5 mag (see Fig. 11). The only substantial incompleteness in these data is due to unresolved binaries—very high-resolution imaging will be needed to probe binaries with separations ~ 0.1 arcsec (~ 45 AU, the peak in the distribution of semimajor axes for main-sequence binary stars), and indeed, recent HST imaging has revealed binaries at these separations in the Trapezium Cluster. As the Orion Nebula is nearby, well out of the galactic plane at $b \sim -19°$, and towards the galactic anticenter at $l \sim 209°$, there is minimal foreground or background field star contamination. The only major potential for contamination comes from sources embedded in the background molecular cloud OMC-1 and indeed, there are a relatively small number of very red sources, including the well-known BN-KL complex, which are projected almost exclusively along the ridge of dense gas seen in submillimeter maps of the region (McCaughrean 1988; Mezger et al. 1990). One difficulty does arise however, in determining spatial extent of the Trapezium Cluster: recent larger near-infrared mosaics covering 15×15 arcmin reveal a considerable number of stars over the wider area, albeit at a substantially lower density than in the cluster core. The problem is that many of the stars farther from the core are probably members of the 1c subgroup, rather than the Trapezium Cluster, i.e., the 1d subgroup. The question is, can any real distinction be drawn between the two? Are the Trapezium Cluster stars simply the youngest in a continuous phase of star formation that began with the 1c subgroup, or are the two subgroups well separated in time? Either way, with one projected on the other, care must be taken in drawing conclusions regarding coevality etc., as discussed below.

After applying a small extinction correction of $A_K \sim 0.5$ mag ($A_V \sim 5$ mag), the peak in the Trapezium Cluster K luminosity function at ~ 11.5 mag corresponds to a naked PMS star of $\sim 0.3\,M_\odot$, assuming an age of 1 Myr: erroneously assuming a main-sequence mass-K relation would give a much higher mass of $1\,M_\odot$. More subtly, it would also be wrong to claim that because the peak in the *luminosity* function corresponds to a $0.3\,M_\odot$ star, that there is also a peak in the *mass* function at $0.3\,M_\odot$: as we have seen in Sec. II.B, the problem is much more complicated than that, due to the time dependence of the mass-luminosity relation for PMS stars. In light of this, it is worth comparing the Trapezium Cluster KLF with the model KLFs shown in Fig. 2. The precise distance to the Trapezium Cluster still remains somewhat uncertain, but if we assume the widest possible range of 360 to 560 pc (corresponding to distance moduli of 7.8 and 8.7 mag: Genzel et al. 1981; Anthony-Twarog 1982), we can see that the measured peak in the cluster KLF lies in the range $M_K \sim 2.8$ to 3.7 mag. By comparing this to the position of the D-burning peak in the model KLFs, we would estimate an age of 1 to 2 Myr for the Trapezium Cluster, similar to the age determined by other methods (Herbig and Terndrup 1986). Despite the uncertainty in the distance to the Trapezium Cluster, that the observed Trapezium Cluster KLF has not been corrected for extinction, and that neither the observed or modeled KLFs account for infrared excess, this comparison nevertheless shows that the Trapezium Cluster is at least *qualitatively consistent* with a coeval, 1 to 2 Myr old cluster of stars drawn from a fully populated Miller-Scalo IMF.

A more rigorous analysis of the cluster mass function can be attempted using multi-color infrared data. McCaughrean et al. (1992) have placed the stars on infrared color-magnitude (i.e., K vs $J-K$), and color-color (i.e., $J-H$ vs $H-K$) diagrams, along with suitably transformed PMS tracks and isochrones. As discussed in Sec. II.B, it is almost impossible to make a unique determination of the mass of a PMS star if neither the reddening or age are known. Judging from the spread of stars along the reddening vector in the color-color diagram, there is a wide range of line-of-sight extinctions across the cluster ($A_V \sim 2$ to 20 mag), and therefore the Herbig and Terndrup (1986) assumption of constant reddening is not appropriate. McCaughrean et al. (1992) suggest the more physically realistic assumption of a constant age. Under this assumption, all the stars would be moved parallel to the reddening vector in the color-magnitude diagram until they intercept the chosen PMS isochrone, in this case 1 to 2 Myr. Counting the number of stars falling between any two PMS mass tracks, an approximate IMF would then be derived. As mentioned earlier, this procedure depends on there being little or no infrared excess from circumstellar disks. This question remains open for dense embedded clusters in general, although few of the "normal" Trapezium Cluster stars (i.e., those not associated with dense molecular cloud cores behind the cluster—see below) show much $H-K$ excess in the color-color diagram. Nevertheless, it is worth considering the effect that real excesses would have on this naïve dereddening technique. The foreground dust reddening would be overestimated, but the resulting error in the calculated mass would depend on the nature of the excess: if it was due to the reprocessing of stellar flux (a "passive" disk), the mass would be overestimated, whereas if it was due to heating of the disk by accretion of infalling matter (an "active" disk), the true stellar mass would be underestimated.

It is clear that the Trapezium Cluster warrants considerable future observational attention, as perhaps nowhere else is there a cluster near enough to us and relatively isolated from sources of contamination that we stand a fighting chance of understanding the birth, evolution, and ultimate fate of its members. In this way, it can also serve as an important prototype for other embedded clusters. With this in mind, we now discuss some specific questions about the Trapezium Cluster, questions which may also be applied to other young embedded clusters:

a. Are the Trapezium Cluster stars coeval? In their study of the optically visible population, Herbig and Terndrup (1986) found little evidence for a relationship between mass and age in the cluster, with most of the stars ≤ 1 Myr old. However, more recently, it has been suggested that the shape of the cumulative near-infrared (K) luminosity function indicates a wide spread in ages (Gatley et al. 1991). Based on the fact that the fainter section of their cluster LF parallels the field star LF, while the brighter part is shallower, and more like that expected for a Salpeter IMF, the claim is that the low-mass stars are older than the high-mass stars. However, in making these comparisons, Gatley et al. (1991) used a main-sequence field star LF, and a single power-law conversion from mass to K luminosity derived for massive main-sequence stars. As the Trapezium Cluster is composed of mostly low-mass pre-main-sequence stars, these comparisons with main-sequence derived LFs are inappropriate. As we saw earlier, the differential KLF for the Trapezium Cluster of McCaughrean et al. (1992) is consistent with model PMS luminosity functions for a completely coeval cluster. Again, while this does not imply that the Trapezium Cluster *is* coeval, any claim of noncoevality based on slope changes in the cumulative luminosity function is probably without basis.

Perhaps a more acceptable hypothesis given the lack of compelling evidence to the contrary, is that the Trapezium Cluster is more or less coeval, in the sense that the formation of the majority of its members was initiated at some "time zero" by the same external event. This is necessarily somewhat vague. It is possible for example that most of the cluster members formed at the same time, but that when the high-mass stars "turned on," they initiated a second burst of low-mass star formation via radiative implosion. Indeed, some stars seen in this region lie in molecular cloud OMC-1 behind the H II region, and *are* probably younger than average. Also, as we discussed above, care must be taken in defining the spatial extent of the Trapezium Cluster: at some radius, confusion with members of the older 1c subgroup will become a problem.

b. Why are the Trapezium OB stars at the center? With star formation initiated in a dense core, more stars would form towards the center where the most material was available, with the most massive stars forming at the core by gradual mass accumulation due to clump-clump collisions in a deep potential well. This would give rise to the high degree of apparent spherical symmetry observed in the Trapezium Cluster, with a higher stellar density towards the center, and the highest-mass (OB) stars right in the middle.

Another possibility is that the OB stars arrived at the center of the cluster by dynamical evolution. The one-dimensional RMS velocity dispersion of the optically visible stars near the cluster center is ~ 2.5 km s^{-1} (Jones and Walker 1988), enough for about three crossings of the central parsec, assuming an age of 1 Myr. However, given the apparent independence of the velocity dispersions with respect to mass and distance from the cluster center, it seems as though the mass segregation in the Trapezium Cluster is largely due to initial conditions, i.e., the more massive stars *did* form in the cluster center (Jones and Walker 1988).

At the same time, the Trapezium OB group itself is unlikely to be dynamically stable, with at least four massive components in close proximity: indeed the most massive star, θ^1Ori C, has a high proper motion and seems to be on the verge of being dynamically ejected (van Altena et al. 1988). Two of the other Trapezium OB stars (θ^1Ori B and θ^1Ori A) are eclipsing binaries (Lohsen 1976; Bossi et al. 1989; Popper and Plavec 1976), which suggests that their angular momentum vectors were closely aligned at the time of formation. One possible theory for the formation of the OB group that would result in such an alignment involves the fragmentation of a rotating ring (e.g., Bodenheimer 1978; Lucy 1981). However, also see Larson (1990) for alternate possibilities.

c. Is there evidence for substructure? In addition to the "normal" members of the Trapezium Cluster, there are a number ($\sim$10% of the cluster) of very red sources seen almost exclusively to the west of the OB stars, projected against a ridge of dense sub-millimeter peaks (Mezger et al. 1990; McCaughrean et al. 1992). These sources include the BN-KL complex and those associated with a secondary highly collimated molecular outflow, OMC-1S (McCaughrean 1988; Schmid-Burgk et al. 1990). How are these related to the Trapezium Cluster? Are they spatially distinct star-formation sites, or should they all be considered together? The BN-KL and OMC-1S sources are embedded in the molecular cloud OMC-1 behind the H II region M 42, and are probably younger than the normal Trapezium Cluster stars which are mainly in the H II region, in front of the cloud (Herbig and Terndrup 1986). However, it is possible that they will merge at some point, implying that some 1 Myr from now there will be an age spread on the order of several Myr in the Trapezium Cluster. Thus in principle, features in cluster luminosity functions may be due to age spreads of the same order as the cluster age, making interpretation of the luminosity function complicated as it would depend not only on the IMF, but also on the history of star formation in the cluster. However, in the case of the Trapezium Cluster at least, the younger ridge sources constitute only $\sim$10% of the total cluster population, and their effect on the current cluster luminosity function and any derived IMF is probably minimal.

d. Is the low-mass IMF of the Trapezium Cluster different from that in regions like Taurus-Auriga? Historically, the nebulosity of the Orion Nebula and other H II regions has created a bias against the detection of low-mass stars located near the massive stars which cause the nebular emission. Now that near-infrared imaging is revealing the cospatial low-mass clusters, the notion that the Orion 1d subgroup (i.e., the Trapezium OB stars) has fewer associated low-mass stars and therefore a different IMF for low-mass stars when compared to regions like Taurus-Auriga (cf. Larson 1982), becomes difficult to maintain. A more plausible suggestion is that the low-mass end of the IMF is the same for the two and that it is the high-mass end that is different, i.e., there are fewer massive stars in the Taurus-Auriga complex.

e. Do the Trapezium Cluster stars have circumstellar disks? In low-density star-forming regions, young stars are often seen to exhibit infrared and millimeter excesses attributed to dust in circumstellar disks. However, given the very high stellar density in the Trapezium Cluster, disk-disk interactions are very likely, and it is conceivable that the disks are stripped completely at an early stage, or perhaps that outer disks are removed, while inner disks survive. These ideas can be tested in the Trapezium Cluster by combining 3 to 5 μm imaging with the existing 2.2 μm data, to see if cluster members show a significant near-infrared excesses due to hot dust close to the sources. Also, 10 μm and millimeter continuum studies are needed to search for the progressively colder outer parts of disks. An unsuccessful attempt to detect 1.3 mm dust continuum from some cluster members was made by

Chini, McCaughrean and Zinnecker (unpublished data), using IRAM with an 11 arcsec beam. The main problem with this single beam experiment was confusion with extended and structured emission from the background molecular cloud: what is needed is an interferometric study to resolve out the background emission.

f. Could the Sun have formed in a region like the Trapezium Cluster? Related to the question of disks is the question of the formation of planets from such disks, and in particular, whether or not the Sun and its planetary system could have formed in an environment like the Trapezium Cluster. After all, even *bound* galactic clusters dissolve on a time scale of $\sim$100 Myr due to the tidal effects of passing molecular clouds, and thus the fact that the Sun is not *currently* in a cluster does not tell us whether or not it once was. The appeal of the Sun forming in an OB cluster is that supernova pollution of the protoplanetary disk is a natural consequence of the cluster environment (Reeves 1978). Also, the misalignment between the rotation axis of the Sun and the ecliptic might be explained by gravitational perturbations by other cluster members (Mottmann 1977; Herbig 1982; Tremaine 1991). At the very least, it is important not to assume *a priori* that the Sun was born in a low-density region like Taurus-Auriga, as the great majority of low-mass stars appear to be born in clusters, in close proximity to other stars and not in relative isolation.

g. What is the fate of the Trapezium Cluster? Will the Trapezium Cluster remain bound? Based on the cluster velocity dispersion and the mass of the cluster in stars as determined by Herbig and Terndrup (1986), Jones and Walker (1988) concluded that the cluster mass is an order of magnitude less than the mass required to be in virial equilibrium or bound. Although many more stars have been revealed in the near-infrared, they are mainly low-mass stars, and the total cluster mass is probably still too small to keep the cluster bound if the same velocity dispersion applies to those stars. However, combining stars with the ionized and molecular gas, the virial criterion may be satisfied (Genzel and Stutzki 1989). The amount of gas ionized by the OB stars and lost due to over-pressure in the H II region is between 10 to $100\,M_{\odot}$ (F. Yusef-Zadeh, personal communication), less than the cluster mass and therefore not enough to unbind the cluster. Observations of fine structure lines in the far-infrared might help to determine more accurately the amount of ionized gas in and around the Trapezium Cluster.

Clearly, any mass loss loosens the cluster, and it is likely that it has expanded already and will continue to do so. However, if the mass loss is slow enough, a cluster can revirialize with a lower velocity dispersion and remain bound (Lada et al. 1984; Verschueren and David 1989). Therefore, whether or not the Trapezium Cluster will become an unbound OB association with attendant low-mass stars is unknown. It is tempting to speculate that there may still be clumps of dense gas between the cluster stars, not yet evaporated by the ionizing radiation, and perhaps helping to bind the cluster temporarily. Such clumps would be subject to considerable ablation and even radiative implosion, perhaps leading to another generation of low-mass star formation in the Trapezium Cluster, triggered by the radiation of the OB stars. In this context, it is worth recalling the partially ionized globules seen close to the OB stars (Garay et al. 1987; Churchwell et al. 1987) and the warm ($\sim$300 K) dust arcs and shells seen in the same region, although the latter are notably *not* coincident with the former (McCaughrean and Gezari 1991). Also there are the small ($\sim$ few $M_\odot$) ammonia cores seen in the BN-KL complex (Migenes et al. 1989).

The OMC-2 Cluster. Some 12 arcmin north of the Trapezium Cluster there is a cluster of compact near-infrared sources associated with a part of the L 1641 molecular cloud known as OMC-2 (Gatley et al. 1974). Recently, Pendleton et al. (1986), Rayner et al. (1989), and Johnson et al. (1990) have studied this cluster using high-resolution infrared mapping, imaging, and polarimetry. Associated with a peak in molecular gas and dust emission and completely obscured in the visible, it is a small cluster with only a dozen sources to $K \sim 17$ mag within a 0.3 pc diameter field. Three of the sources (IRS 1, 2, 4) are deeply embedded ($A_V \sim 30$ to 40 mag), and are probably intermediate-mass YSOs, i.e., Herbig Ae/Be stars. Two sources illuminate bipolar reflection nebulae, implying that circumstellar disks may be present. Direct mapping of the sub-millimeter dust continuum emission could prove the presence of such disks; this group of very young stars might be an ideal place to study the effects of disk-disk interactions, if they have occurred. Most of the remaining sources are low-mass stars with relatively little extinction, possibly members of the wider Orion Cluster (1c subgroup) rather than directly associated with the OMC-2 core. At present it is not known whether a more extended cluster of low-luminosity PMS objects has formed in the same core as the bright embedded sources.

NGC 2024, NGC 2023, NGC 2071, and NGC 2068. The embedded clusters associated with star-forming regions NGC 2024, NGC 2023, NGC 2071, and NGC 2068 in the L 1630 molecular cloud are discussed in detail in the Chapter by by Lada et al. We give a brief description here to examine to what degree they may be viewed as analogs to the Trapezium and OMC-2 clusters in the nearby L 1641 cloud. It was known from infrared raster-scan maps that the ionizing star of the H II region NGC 2024 is associated with an embedded cluster of young, lower-mass stars (Barnes et al. 1989). But again, only with infrared cameras was the true extent revealed, with about 300 members brighter than $K \sim 14$ mag within a radius of 1 pc (Fowler et al. 1987; Lada et al. 1991*b*). Like the Trapezium Cluster, the spatial distribution of the young stars is centrally condensed, with the more luminous objects located toward the cluster center. Similarly, the near-infrared luminosity function rises to about $K \sim 12$ to 13 mag before falling again. Lada et al. (1991*b*) noted that up to the turnover at least, the cumulative KLF for NGC 2024 is steeper than the Salpeter initial mass function.

Some 25 arcmin south is the reflection nebula NGC 2023. Sellgren (1983) surveyed the associated cloud to $K\sim12$ mag, finding a small cluster of 16 infrared sources within a $\sim0.5\times0.5$ pc area, somewhat similar to OMC-2. The brightest source in NGC 2023 is an unreddened B1.5 V main-sequence star that illuminates the nebula, but interestingly is not at the center of the cluster. Surprisingly, a deeper imaging survey to $K\sim15$ mag by De Poy et al. (1990) found no additional cluster members; the flattening of their cumulative near-infrared luminosity function near $K\sim12$ mag was interpreted as either a dearth of low-mass stars or as preferential obscuration of the lowest luminosity sources.

Finally, some 2.5 degrees to the north-east of NGC 2024, there are two other embedded clusters associated with NGC 2068 (~200 members) and NGC 2071 (~100 members). Further details can be found elsewhere (Sellgren 1983; Lada et al. 1991b; Lada and Lada 1991), but relevant to the discussion here, it is worth noting that the KLF for NGC 2071 also shows evidence for a turnover at $K\sim12$ mag (Lada et al. 1991b).

The K luminosity functions of the L 1630 clusters were interpreted on the basis of a time-independent main-sequence mass-luminosity relation, instead of accounting for PMS luminosity evolution. In fact, the cumulative KLFs for all the L 1630 clusters are similar in shape to that of the Trapezium Cluster, and as we have seen earlier, the latter is at least qualitatively consistent with a ~1 Myr old coeval PMS population with a monotonically rising Miller-Scalo IMF. On this basis, there is probably no real justification for claiming a lack of low-mass stars in any of these regions. Even so, the alternative hypothesis of preferential obscuration of the lowest-mass stars can be tested by 10 μm imaging, as suggested by De Poy et al. (1990) in the case of NGC 2023.

The Underlying Stellar Populations in L 1641 and L 1630. It is clear that there is vigorous high- and low-mass star formation in the form of embedded clusters in both L 1641 and L 1630. These clusters occupy only a small fraction of the total mass and volume of the clouds, suggesting that star formation is very localized. However, both L 1641 (Bally et al. 1987; Bally et al. in preparation; see Chapter by Blitz) and L 1630 (Lada 1990; Lada et al. 1991a) are clumpy and filamentary like the Taurus-Auriga dark clouds, where isolated star formation is seen throughout the region. Therefore, it is necessary to examine the Orion clouds for evidence of any "pedestal population" in order to assess the relative importance of the cluster and isolated modes of star formation. A wider discussion of the clouds and their stellar content is given in the Chapter by Lada et al.

Nakajima et al. (1986) surveyed some 1.4 square degrees of the L 1641 cloud in the near-infrared to a limiting magnitude of $K\sim9.5$ mag. Of the ~100 sources found, 15 sources were classified as PMS stars based on their near-infrared color excesses, 6 of them newly identified. To the limits probed by Nakajima et al. (1986), the near-infrared luminosity function of L 1641 appears similar to that of L 1630 and the Trapezium Cluster.

Strom et al. (1989*c*) examined the properties of 123 IRAS sources over a ~9.5 square degree area in the field of L 1641, finding 93 of them to be young stars associated with the molecular cloud. Based on their SEDs, roughly 60% of the whole sample (123 objects) were classified as Class I sources, most of them without optical or near-infrared counterparts; the majority of the remainder are Class II sources with optical/near-infrared counterparts, and spectra typical of T Tauri stars and young emission-line stars; the rest are Class III sources, all but one of which lie either in front of or behind L 1641. When compared to Taurus-Auriga, L 1641 is found to have a much larger fraction of luminous Class I sources: ~76% of the 93 IRAS sources associated with L 1641 have Class I SEDs, compared to ~25% of the sources in Taurus-Auriga. This difference is attributed to the fact that typical molecular cloud cores in L 1641 have higher optical depths than those in Taurus-Auriga, which leads to greater extinction and reprocessing of optical and near-infrared light out to far-infrared wavelengths. Also, as Orion is farther away than Taurus-Auriga, we must beware of selection effects: the completeness limit for the L 1641 IRAS sources is ~6 $L_\odot$ (Strom et al. 1989*c*), compared to ~0.5 $L_\odot$ for Taurus-Auriga (Kenyon et al. 1990).

Thirty objects in the sample of Strom et al. (1989*c*) were detected by IRAS only, having no optical or near-infrared (5σ limit K~12.8 mag) counterparts. Nine of these IRAS-only sources were studied further through deeper near-infrared imaging (to K~16 mag) by Strom et al. (1989*a*). Most were found to be detected at K~15 mag, with all but one as single, isolated sources. This perhaps indicates that only a small number (~1) of stars form per core, with the caveat that these sources were only just detected in the deep imaging survey, and that there may be small clusters of lower-luminosity objects around each IRAS source. The obvious exception is IRAS 05338−0624, where Strom et al. (1989*b*) found a cluster of ~20 sources within a diameter of 0.16 pc. This is a remarkably dense clustering, although it is probably not fair to interpret all 20 sources as stars, as not all appear point-like (Strom et al. 1989*b*; K.-W. Hodapp, personal communication).

A study of the young stellar population of L 1641 was also made in X-rays using EINSTEIN IPC images (Strom et al. 1990). These images were centered near V 380 Ori (NGC 1999) and cover an area approximately half that of the IRAS study. Some 65 X-ray sources were found, 61 with stellar counterparts: HH 1 and HH 2 are two of the non-stellar X-ray sources. Five of the X-ray sources appear to coincide with optically identified IRAS sources: however, most of the IRAS sources suffer too much extinction to be seen in X-rays. Roughly two-thirds of the X-ray sources show no optical Hα emission, indicating once again that X-ray surveys can uncover substantial numbers of PMS stars, apparently somewhat older than classical T Tauris. Conversely, of the 25 Hα stars found in the objective prism survey of Parsamian and Chavira (1982), 15 were not detected in X-rays. Thus the ratio of the number of *all* PMS stars (i.e., X-ray stars plus Hα stars *not* detected in X-rays) to Hα stars in L 1641 is roughly 2. Although this number is somewhat dependent on limiting magnitude and completeness, it is nevertheless in marked contrast to the predicted ratio of ~10, based on an extrapolation of Taurus-Auriga sources (Walter et al. 1988; Strom et al. 1990, Appendix A).

Therefore, there *is* a pedestal population of YSOs in L 1641, although the cluster mode of star formation appears to dominate in this region. A rough comparison of the number of sources in the known clusters (Trapezium Cluster, OMC-2, IRAS 05338−0624) to the total number of sources in the cloud (i.e., the cluster sources plus the isolated IRAS, X-ray, near-infrared, and Hα sources), shows that about 80% of the stars in L 1641 are in clusters. However, most of the surveys of the pedestal population are likely to be quite incomplete, making this ratio an upper estimate: the X-ray and Hα surveys suffer from extinction, and the IRAS survey is complete only to ~6 L$_\odot$ (Strom et al. 1989c). A more comprehensive near-infrared imaging survey is required to make a better assessment of the L 1641 pedestal population.

Such a survey has been made for the L 1630 cloud (Lada 1990; Lada et al. 1991b). From a survey of the cloud for dense CS $J = 2 \rightarrow 1$ gas (Lada et al. 1991a), regions with and without dense gas totalling ~0.7 square degrees were then surveyed to K~13 mag. Some 900 sources were detected, of which almost 60% are associated with the previously discussed clusters (NGC 2024, NGC 2023, NGC 2068, NGC 2071). Indeed, after correcting for background and foreground field-star contamination of their survey, Lada et al. (1991b) estimated that the cluster members make up as much as ~96% of the total number of stars actually associated with L 1630. This indicates an overwhelming predominance of the cluster mode of star formation in L 1630, although considerably more work is needed to distinguish genuine cloud members from field stars between the clusters, and even deeper surveys may be needed to fully probe the pedestal population.

The dominance of the cluster mode in L 1630 may be at least in part due to the effects of external triggering. The L 1630 clusters tend to lie on the western edge of the cloud, close to the young Ori 1a and 1b subgroups. Also, there is a strikingly high concentration of Hα stars on the western edge of the L 1630 molecular ridge (Wiramihardja et al. 1989), and nearby, in the bright-rimmed globule Ori-I-2, the YSO IRAS 05335−0146 may have been triggered by an O star (Sugitani et al. 1989). Thus triggering by the Ori 1a,b stars may have played an important part in the formation of the L 1630 clusters. When discussing this sort of triggering effect however, it is unclear whether the "pressure" of the OB association actually produces protocluster condensations, or just pushes on pre-existing condensations: it may be both. Indeed, some form of two-step process does appear to be in operation in L 1630. While four of the five most massive (>200 M$_\odot$) cores in L 1630 are host to the four dense stellar clusters, the fifth core (LBS 23) does not appear to be forming stars at present (Lada 1990; Lada et al. 1991a,b).

In summary, the current star formation in both L 1630 and L 1641 appears to be dominated by the cluster mode, although more work is needed to obtain a more accurate census of the pedestal populations. One point to bear in mind when determining the relative importance of the cluster and isolated modes in a GMC, is that a part of a molecular cloud currently forming only a very low space density of stars may also become the site of a dense cluster in the future, perhaps due to material being piled up and/or triggered by massive stars of an earlier generation.

The λ Orionis Region. The λ Orionis region of star formation lies some 15 degrees north of the Orion Nebula (see Shevchenko [1979] for a POSS mosaic). The OB association consists of an O 8 III star (λ Ori itself) and 12 B stars at the center of a large spherical H II region, at the edge of which lies an H I shell coincident with a ring of dark molecular clouds, the latter seen to glow spectacularly in IRAS 100 μm images (Maddalena and Morris 1987). The whole complex has a diameter of ~10 degrees, ~70 pc at a distance of 400 pc. Some 83 Hα emission objects, i.e., low-mass PMS stars, have been found in the region. Most are concentrated on the bright-rimmed inside edges of the dark clouds B 30 and B 35, i.e., on the side of the clouds facing the OB association, with most of the remainder scattered in a more or less linear fashion between the two clouds (Duerr et al. 1982). Duerr et al. (1982) noted that the amount of material in the H I shell (~10^5 M$_\odot$) is comparable to the amount of molecular gas found in a GMC. Thus they proposed that the region around λ Ori is essentially a well-preserved fossil GMC, recently exposed by the action of the star. They find a star-formation efficiency in the region of only 0.2 to 0.3%.

Mathieu and Latham (unpublished data, as described by Mathieu [1986]) have made a radial velocity study of about 30 of the Hα stars in the λ Ori region, and found that the one-dimensional stellar velocity distribution is small, ~2 km s^{-1}, similar to the dispersions found for stars in other GMCs (e.g., the Ori OB1c subgroup: van Altena et al. 1988; Jones and Walker 1988). Locally, near B 30 and B 35, the stellar velocity dispersions are lower, at about 1 km s^{-1}, similar to the CO line widths in the clouds themselves (Maddalena et al. 1986; Maddalena and Morris 1987), and therefore these stars are probably bound to these clouds, much as the stars were bound to the original GMC. However, most of the original molecular material in the region has been swept into an expanding shell (Maddalena and Morris 1987), and the escape velocity for the global system is much lower than the current stellar velocities: with the removal of the gas, the stellar system has become unbound and will ultimately disperse.

2. Other Galactic H II Regions

The bulk of the Galaxy's giant molecular clouds and massive star-forming regions lie at distances greater than that of the Orion complex. Nevertheless, even low-mass stars can now be detected in many of these regions, and in the past few years, many dense embedded clusters have been revealed. Some of these are summarized here, roughly in order of increasing heliocentric distance.

Figure 12.   A K (2.2 μm) mosaic of the H II region S 106, covering $\sim$10×10 arcmin
($\sim$1.75×1.75 pc at 600 pc). The data were obtained using a 256×256 pixel HgCdTe
infrared array at the University of Hawaii 2.2-m (Rayner 1992). Some 200 stars are
probable members of a cluster centered near S 106 IRS 4.

S 106 (600 pc). In studies of S 106, the focus has been an embedded O9–B0 V
star (IRS 4) and the biconical H II region which it excites (see, e.g., Staude
et al. 1982; Bally et al. 1983; Hayashi et al. 1990). In particular, there is
some controversy over whether or not a dense molecular disk or ring structure
exists around IRS 4, constraining the ionization to form the biconical nebula
(Mezger et al. 1988; Barsony et al. 1989b; Loushin et al. 1990). Figure 12
shows a recent 2.2 μm image, revealing both a large number of embedded
point sources, and highly structured loops and arcs in the nebula (Rayner
1992; see also Color Plate 4). Analysis of an earlier 2.2 μm image has shown
that IRS 4 is near the center of a cluster of over 200 point sources, with

about 90% of them within a radius of 1.5 arcmin of the OB star (Rayner et al. 1991a; Hodapp and Rayner 1991). With a completeness limit of $K \sim 14$ mag, the faintest objects detected by Hodapp and Rayner (1991) are probably low-mass PMS stars between 0.2 and 0.5 $M_\odot$, based on a distance of 600 pc, a derived average extinction of $A_V \sim 13.5$ mag, and an assumed cluster age of 1 Myr. A direct age estimate was made by Hodapp and Rayner (1991), who found that only 4 out of 80 cluster members with masses above 1 $M_\odot$ show evidence for local nebulosity and high intrinsic polarization, and therefore ongoing mass loss. This, combined with a typical duration of 5×10^4 yr for the mass-outflow phase in YSOs (Fukui 1989), suggests to them that star formation has been going on in S 106 for 1 to 2 Myr, i.e., they infer an age spread of this order, if the star-formation rate has been constant. The latter is an implicit assumption: there is no obvious evidence to counter the possibility that most of the cluster formed in a single short burst, and that there is some secondary star formation occurring now, perhaps including IRS 4 itself.

The S 106 K luminosity function does not exhibit any hint of a turnover down to the completeness limit of 14 mag. In contrast, if the Trapezium Cluster were moved to a distance of 600 pc and an extra $A_K \sim 1$ mag of extinction added, its luminosity function turnover would still be detectable to that limit. Therefore at first sight, it seems as though the two clusters may be different. On one hand, if the S 106 cluster is *older* than the Trapezium Cluster, then any peak in the luminosity function due to deuterium burning (Sec. II.B) would have moved to lower masses and be undetected at the current completeness limit. On the other hand however, the angular spacing between the S 106 stars seems smaller than that for the Trapezium Cluster stars, which could indicate an even higher stellar density and perhaps a *younger* age for the S 106 cluster. Perhaps a simpler explanation that would fully reconcile these contradictory pieces of evidence is that S 106 is farther away than the nominal 600 pc. Indeed, uncertainty in the spectral classification of IRS 4 and the number of stars foreground to the nebula imply that it could be as far away as 2 kpc (Th. Neckel, personal communcation; Rayner 1992).

NGC 2264 (900 pc). The young ($\sim$10 Myr) open cluster NGC 2264 lies at a distance of 900 pc, in the Mon OB1 association between the O star S Mon and the famous Cone Nebula. From their optical color-magnitude diagram, Adams et al. (1983) found that, contrary to previous claims, the cluster does have low-mass stars, and also that there is an age spread in its members of $\sim$10 Myr. Stahler's (1985) analysis confirmed this age spread, but did not confirm the sequential build-up in mass space nor the exponential increase in star-formation rate in the cluster claimed by Adams et al. (1983). In addition, selection effects prevent us from trusting that the optical surveys have revealed the complete stellar population. IRAS data on the embedded cluster have been analysed by Margulis et al. (1989), who identified 18 Class I and 8 Class II sources, with the caveat that two-thirds of these sources are probably associated with two or more YSOs. All seven of the far-infrared sources with

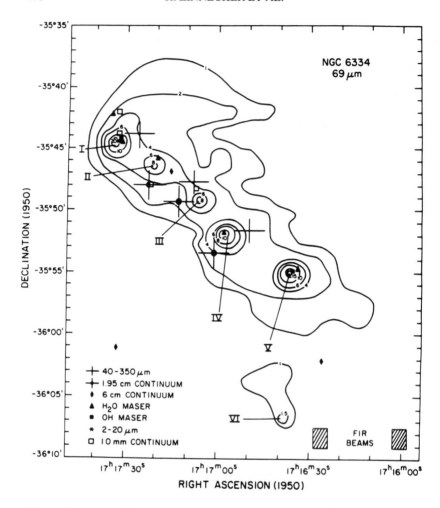

Figure 13. A KAO map (40–250 μm filter, effective wavelength 69 μm) of the NGC 6334 molecular cloud and star formation region, from McBreen et al. (1979). Note the characteristic spacing of the luminous far-infrared sources, indicated by the Roman numerals. The beam size is of order 1 to 2 arcmin.

$> 100\,L_\odot$ were found to be Class I objects. Near-infrared imaging is required to reveal the embedded population in this region, because part of the "optical" cluster is obscured by a molecular cloud (Crutcher et al. 1978), and may in fact be cut in two by it (Mathieu 1986).

NGC 6334 (1.7 kpc). Figure 13 shows a far-infrared map of the NGC 6334 molecular cloud in which several separate condensations are discerned, with a projected spacing of around 10 pc at a distance of 1.7 kpc. This region has recently been studied in detail by Straw and Hyland (1989) and Straw et al.

(1989). Straw et al. (1989) made a near-infrared raster-scan survey of the embedded populations in 11 regions of NGC 6334 that show indications of ongoing star formation, and found several hundred sources to a completeness limit of $K \sim 14$ mag, corresponding to a main-sequence stellar mass of $\sim 1 \, M_\odot$ if unobscured, or $\sim 4 \, M_\odot$ if embedded in the molecular cloud cores typically with $A_V \sim 30$ to 40 mag. In several of the regions, a large fraction of the sources were seen to have $H - K$ excesses, indicating PMS objects. In order to determine masses, the majority of the sources were plotted in an infrared color-magnitude diagram ($J - K$ vs K), and then projected back along a standard reddening vector until they intercepted the main sequence. A main-sequence absolute M_K magnitude versus mass relation (similar to that shown in Fig. 3) was then used to determine the individual masses, and thence the IMF. Sources showing some evidence (e.g., spectroscopic) for being PMS objects were dereddened back to suitably transformed PMS tracks, although as noted elsewhere in this chapter, masses derived by this procedure depend strongly on the chosen isochrone. The general conclusion from the work of Straw et al. (1989) is that the stellar mass distribution in the NGC 6334 complex is similar to that found in other young visible clusters and embedded clusters, i.e., roughly equal to the Salpeter IMF.

There are a number of problems with the general approach of Straw et al. (1989) which illustrate some of the difficulties encountered when converting measured fluxes into masses for embedded stars. First, many of the objects were red enough that they were only detected at K but not at J, leading to a complicated dereddening scheme in which the J magnitude was calculated via a number of *ad hoc* assumptions. Second, some of the objects may be background giants, which must be discriminated against spectroscopically and removed from the IMF determinations, although Straw et al. (1989) calculated that in most of their regions, contamination by background field sources was not that important. Third, Straw et al. (1989) did not discuss the effects of infrared excess emission at all. Finally, some of the sources assumed to be on the main sequence may actually be young, and Straw et al. (1989) assessed the error that can be made by assuming the stars to be main rather than pre-main sequence: it can be huge. For example, an object at $K = 11.0$ mag and $J - K = 3.0$ mag in the color-magnitude diagram could either be a $5 \, M_\odot$ ZAMS object obscured by $A_V \sim 4$ mag, or an unobscured young (10^5 yr) T Tauri star of mass $1 \, M_\odot$ at the top of its Hayashi track. Deep optical CCD imaging in the I-band (cf. Hunter and Massey 1990) may help separate heavily extincted sources ($A_V > 15$ mag) from less heavily extincted ones, which in turn will help to discriminate ZAMS from PMS stars.

M 17 (2.2 kpc). The M 17 H II region and the adjacent molecular cloud is a well-studied site of massive star formation and has a very high ratio of far-infrared luminosity to molecular gas mass (Mooney and Solomon 1988). A cluster of OB stars within an obscured portion of the H II region is responsible for most of the observed radio continuum and far-infrared emission (Beetz

Figure 14.   A near-infrared (1–2.5 μm) mosaic of M 17 covering ∼9×9 arcmin
(∼5.8×5.8 pc). Approximately 100 stars (B9 or earlier) towards the center of the
image have been identified as cluster members. The data were obtained using a
256×256 pixel PtSi array on the KPNO 2.1-m telescope (Lada et al. 1991).

et al. 1976; Gatley et al. 1979; Chini et al. 1980). Recent near-infrared imaging
at J, H, and K of a 9×9 arcmin area of the complex (Fig. 14) clearly defines
the true size and extent of the cluster, revealing about 100 stars of spectral type
B9 and earlier within a 2.8 arcmin (1.8 pc at 2.2 kpc) radius (Lada et al. 1991;
see also Color Plate 5). In addition, these observations indicate that most of
the cluster members possess infrared excesses, suggesting that circumstellar
disks may play a more important rôle in the early evolution of OB stars than
previously suspected. The cumulative K luminosity function for the cluster
stars, after subtraction of an unobscured background field, displays a slope
which is significantly shallower than that observed for field stars, but which

is consistent with that of the Salpeter IMF. Despite the fact that the detected massive cluster members are believed to be on the main sequence, Lada et al. (1991) point out that comparisons with the IMF may be complicated by the apparent ubiquity of excess emission at 2.2 μm.

Deeper infrared images should reveal the extent of the low-mass population of the cluster. However, it is worth noting that for regions such as M 17, within the inner Galaxy and close to the line of sight towards the galactic center, confusion will be extreme due to the very high density of field stars, particularly those behind the star-forming region which are revealed as the extinction in the molecular cloud is penetrated in the near-infrared. Separating cluster and field stars will be very difficult in such regions.

In a larger scale study, Elmegreen et al. (1988) searched for IRAS sources coincident with bright photographic near-infrared stars over a 25 square degree region of M 17 SW, the GMC to the south-west of M 17. Of the almost 2000 IRAS point sources in this region, some 360 were classified as embedded based on their positions in either the 12-25-60 μm or 25-60-100 μm IRAS color-color diagrams, with 22 sources classified as embedded in both. Some of these 360 sources were identified with previously known compact H II regions, three of them beyond M 17, while many of the remainder are consistent with embedded late O to late B type stars. By blinking B and I photographs, just over 2000 bright near-infrared stars were identified in the same region, of which 20 were found to lie within 1 arcmin of an embedded IRAS source. Elmegreen et al. (1988) suggest that at least 60% of these coincidences are likely to be physical associations. Some 13 of these stars were seen on both the B and I plates, and assuming these sources to be at the distance of M 17, their colors and magnitudes are also consistent with embedded ($A_V \sim 7$ mag) B stars. Thus these results lend support to the idea that there may be 20 to 50 B stars in the M 17 SW cloud (Stutzki et al. 1988), and therefore also suggest a large population of lower-mass PMS stars.

S 255/S 257 (2.5 kpc). S 255 and S 257 are diffuse, low-excitation H II regions associated with two early B stars, part of a small complex of H II regions (S 254–S 258) in the Gem OB1 association. They have diameters of $\sim$2.5 pc and their centroids have a projected separation of 3.8 pc, assuming a distance of 2.5 kpc to the complex. Between them is a dense molecular cloud (Evans et al. 1977; Mezger et al. 1988; Heyer et al. 1989) with a double, embedded near/mid-infrared source (S 255-IR) which radiates about 8×10^4 L$_\odot$ in the far-infrared (Beichman et al. 1979). Near-infrared images of the region show the bright embedded source S 255-IR surrounded by a dense cluster of $\sim$175 lower-luminosity objects within a $\sim$1 pc radius (McCaughrean et al. 1991; Tamura et al. 1991; see Fig. 15 and Color Plate 6), most of them probably cluster members judging from the local field-star density. The cluster appears to be sandwiched between two peaks of molecular emission (Snell and Bally 1986; Heyer et al. 1989), apparently at the position where two clouds are colliding, and it is possible that the cluster formation was triggered by the

Figure 15. A K (2.2 μm) mosaic of the S 255/S 257/S 255-IR region covering 8×5.2 arcmin (~5.8×3.8 pc at 2.5 kpc). At left center and right center are the two B0 stars which excite the H II regions S 255 and S 257: note that they have no cospatial clusters. Between the two H II regions, the infrared cluster associated with the source S 255-IR is seen. The data were obtained using a 128×128 HgCdTe infrared array at the Steward Observatory 2.3-m telescope (McCaughrean et al. 1991).

compression due to that collision. Perhaps the most remarkable feature of this region is that in contrast to the dense cluster between them, the two B stars exciting the H II regions S 255 and S 257 appear to be more or less isolated: they do not seem to have cospatial low-mass clusters, at least to a 5σ point-source sensitivity limit of $K \sim 16.5$ mag, corresponding to a ~ 1 Myr $0.2\,M_\odot$ star behind $A_V \sim 10$ mag. There are several possible explanations here. Firstly, the B stars may have been dynamically ejected from the embedded cluster, a possibility that can be tested by radial velocity measurements. Secondly, perhaps the high-mass stars formed first in S 255 and S 257, and low-mass stars have yet to form; this seems unlikely in light of the nearby dense cluster of high- and low-mass stars that shows every sign of being younger than the S 255 and S 257 H II regions. Lastly, perhaps a cloud *can* form high-mass stars without also forming low-mass stars: this would support the theory of separate modes of high-mass and low-mass star formation, i.e., bimodal star formation (cf. Güsten and Mezger 1982), in which case this complex appears to be our only concrete example of such so far.

W 3 (OH) (3 kpc). The ultracompact H II region W 3 (OH) coincides with a cold, luminous source of infrared, millimeter continuum, and OH maser emission, which points to the existence of an embedded massive star or group of stars (Wynn-Williams et al. 1972; Reid et al. 1980; Dreher and Welch 1981; Turner and Welch 1984). A 2.2 μm image of W 3 (OH) (Fig. 16) reveals an associated cluster of lower-luminosity embedded sources (Rayner et al. 1991*a*): more than 150 objects are found within a 1 pc radius, assuming a distance of 3 kpc. However, the most important feature of this image may be that which is unseen. High-resolution far-infrared observations by Campbell et al. (1989) have resolved the compact source (FWHM $\sim$ 13 arcsec at 100 μm), and their models allow us to estimate the dust column density in the volume over which the stars are distributed (see also Richardson et al. [1989]). With the assumption of a central cavity of radius 3.3×10^{14} m (the radius of the ultracompact H II region) and an $r^{-1.5}$ density profile outside this radius, they derive a visual extinction towards the central compact source of ~ 1000 mag, which implies an $A_V > 30$ mag even at a radius of 10 arcsec (0.15 pc) from the cluster center. Thus somewhat surprisingly, even near-infrared images of this region may give an incomplete picture of the young stellar population near the core, showing only the stars embedded in the outer skin or shell of the protocluster condensation.

Other Galactic Clusters. In the last few years, near-infrared cameras have revealed many other young embedded clusters in the Galaxy, many of which have so far only been discussed in conference proceedings, preprints, and unrefereed journals. There are too many to discuss in detail, but some are listed here for completeness. These include clusters associated with the Serpens dark cloud (Eiroa and Casali 1989), NGC 3603 (Moneti and Zinnecker in preparation, as shown by Melnick 1989), W 75 N (Moore et al. 1991), S 228

Figure 16.   A K (2.2 μm) image of the infrared cluster associated with the W 3 (OH) ultracompact H II region, covering $\sim$3.5$\times$3.5 arcmin ($\sim$3$\times$3 pc at 3 kpc). There is so much extinction towards the core of the cluster that the innermost stars are probably not detected in this image. The data were obtained using a 256$\times$256 pixel HgCdTe infrared array at the University of Hawaii 2.2-m telescope (Rayner et al. 1991a).

(Carpenter et al. 1991), LkHα 101 (Barsony et al. 1991), Mon R2 (Aspin and Walther 1991), GM 24 (Tapia et al. 1991), S 235, S 269, GL 437 (Rayner et al. 1991a), W 51 (Goldader and Wynn-Williams 1991), and NGC 7538 (Rayner et al. 1991b).

3. Clusters in the Magellanic Clouds (55 kpc)

Studies of the upper IMF in Large and Small Magellanic Cloud clusters are tractable using sensitive CCD detectors and sophisticated software for

crowded-field photometry. The advantage of these studies is that Magellanic
Cloud young clusters both extend to higher stellar masses (hence the name
"blue populous clusters") and have a larger number of stars of all masses
than open clusters in our Galaxy (Elson et al. 1989). Thus they can provide
more meaningful statistics overall,. and in particular at the high-mass end.
Furthermore, as the clusters are all at the same distance, an intercomparison
of their individual mass functions is easier.

30 Doradus. The core of the 30 Doradus or Tarantula Nebula in the Large
Magellanic Cloud (LMC) has long been a focus of attention. Weigelt and
Baier (1985) resolved the luminous central object R 136 into at least eight
individual stars using two-dimensional optical speckle interferometry, a result recently confirmed by direct ultraviolet imaging using the FOC of the
Hubble Space Telescope (Weigelt et al. 1991). On a slightly larger scale, a
deconvolved direct image taken with the ESO 2.2-m revealed some 30 stars in
a $\sim 4 \times 4$ arcsec ($\sim 1 \times 1$ pc at 55 kpc) region surrounding R 136 (Maaswinkel
et al. 1988). More recently this region was imaged with the Planetary Camera
of the WF/PC on the HST. In the raw image (i.e., not deconvolved to remove
the spherical aberration of the HST primary), at least 300 stars are seen within
a 2 arcsec radius at the R 136 core. For the 22+ visible B and earlier type stars
within a 1 arcsec diameter circle centered on R 136, a conservative minimum
stellar density of $5 \times 10^4 \, M_\odot \, pc^{-3}$ has been estimated (B. Campbell, personal
communication). See Fig. 17 a for the HST WF/PC image and Fig. 17 b for
a recent groundbased speckle image of a slightly smaller region (Pehlemann
et al. 1992).

While the 30 Dor cluster has many more high-mass stars and extends to
much higher individual stellar masses (up to $250 \, M_\odot$), there are nevertheless
many similarities with the prototypical galactic young cluster, the Trapezium
Cluster: the most luminous stars in 30 Dor are at the center of the cluster,
surrounded by lower-mass stars; the cluster age is ~ 1 Myr (Melnick 1985);
and in addition to the optical cluster, there is evidence for recent and ongoing
star formation in this region, in the form of four highly luminous (1 to $5 \times$
$10^4 \, L_\odot$) infrared sources found in a $\sim 80 \times 110$ pc survey centered on R 136
(Hyland and Jones 1991), and a number of red reflection nebulae interpreted
as clumps of gas with embedded young massive stars (Walborn and Blades
1987; Walborn 1990).

Based on the cospatial presence of red supergiants and OB stars, Mc-
Gregor and Hyland (1981) suggested that there had been at least two distinct
bursts of star formation in the 30 Dor region. They advanced the idea that a
wave of star formation had moved from the periphery to the center (R 136)
of the 30 Dor cluster and that new stars had formed at the intersection of the
prominent wind-blown bubbles or shells in the region, powered by older massive stars in the outer parts of the cluster. However, Melnick (1985) pointed
out that the red supergiants seem to be uniformly distributed in the region,
suggesting that relatively quiescent star formation may have been occurring

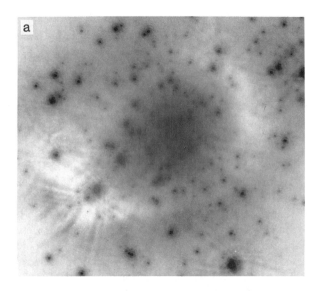

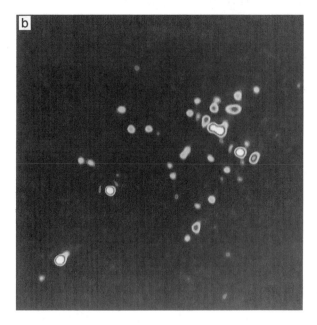

Figure 17 a,b.   Optical image ($\sim$5740 Å) of the central $\sim$11.5×10 arcsec ($\sim$3.1×2.7 pc at 55 kpc) region of the 30 Dor cluster, taken with the WF/PC of the HST (B. Campbell, personal communication). The apparent ring in the nebulosity surrounding the cluster center is an artifact of the photographic technique used to compress the dynamic range across the print, and is not present in the original data. Note that the massive stars are located near the cluster center. For comparison, a groundbased optical image of the inner $\sim$4.8×4.8 arcsec ($\sim$1.3×1.3 pc) region obtained using speckle techniques is also shown (Pehlemann et al. 1992).

throughout the region for 10 to 20 Myr, with a single violent burst of star formation leading to the nebula and massive OB cluster more recently, only some 2 Myr ago.

Another scenario for the triggering of the most recent phase of star formation is suggested by the impression that star clusters seem to form at the apex of a cloud exposed to external compression, combined with the striking fact that 30 Dor, the most luminous star-forming region in the LMC, is located on the leading edge of the motion of the LMC through the halo gas of our Galaxy. X-ray observations with ROSAT should reveal the hot gas associated with any bow shock preceding the LMC, and hydrodynamical calculations should then demonstrate whether or not this particular form of external compression could have triggered the formation of the 30 Dor cluster.

Finally, 30 Dor is near the end of the LMC bar, i.e., the bulk of the old stars, suggesting yet another possible source of triggering: in many irregular galaxies, vigorous star formation is seen to occur near the end of an underlying stellar bar, perhaps due to increased gas compression there (Elmegreen and Elmegreen 1980; Gallagher and Hunter 1984).

An all-important question is whether or not 30 Dor is home to any low-mass (i.e., subsolar) stars. As it is the nearest prototype of a starburst system, the current paradigm for starbursts would predict the supression of low-mass star formation in 30 Dor. There is currently no direct proof for a lack of low-mass stars in 30 Dor, and even though their detection would be extremely difficult, this very important experiment may be possible in the near future, by coupling adaptive optics, infrared arrays, and an 8-m class telescope. If the 30 Dor cluster is found to contain low-mass stars, there would be important implications for the notion of bimodal star formation in more distant starburst galaxies. Also, any such discovery would reinforce the hypothesis that the 30 Dor cluster may actually be a young globular cluster, i.e., similar to those that formed 10 to 15 Gyr ago in the haloes of most galaxies, including our own. Its mass of $\sim 10^5$ $M_\odot$ (Melnick 1985) is indeed similar to that of the old globular clusters.

Finally, because the 30 Dor cluster is so massive, the whole range of stellar masses should be populated in significant numbers according to any reasonable IMF. Current population synthesis models do not contain good templates for young metal-poor populations (cf. Bica and Alloin 1986a,b), and therefore the stellar content of 30 Dor might be considered a unique "template initial stellar population," useful for models of protogalactic evolution at times when the metallicity was far below solar.

Other LMC/SMC Clusters. In addition to the young 30 Dor cluster, the Magellanic Clouds are home to a number of somewhat older (10 to 100 Myr) clusters, the luminosity and mass functions of several of which have recently been studied in detail by Mateo (1988) and Sagar and Richtler (1991) using CCDs, and by Elson et al. (1989) using photographic plates.

As part of his study of LMC and SMC clusters across a wide age range,

Mateo (1988) determined the IMFs for the young clusters NGC 330 in the SMC ($\sim$10 Myr) and NGC 1711 in the LMC ($\sim$50 Myr), finding both to be characterized by a power law with slope $y = -2.4\pm0.6$ (cf. $y = -1.35$ for the $dN/d\log m \propto m^y$ form of the Salpeter IMF) for stars in the mass ranges $\sim$2 to 12 $M_\odot$ and $\sim$2 to 7 $M_\odot$ respectively. In contrast, Sagar and Richtler (1991) found a slope of $y = -1.5\pm0.3$ for NGC 1711: it is likely that this significant disagreement is due to the different incompleteness corrections applied, the technique of Mateo (1988) tending to overcorrect, resulting in a steeper IMF, and that of Sagar and Richtler (1991) tending to undercorrect, leading to a shallower IMF. It may well be that the correct IMF slope for NGC 1711 is somewhere between the two results, $y\sim-1.9$ (Mateo 1990; M. Mateo, personal communication).

For two other clusters in their sample, NGC 2164 and NGC 2214, Sagar and Richtler (1991) derived roughly the same slope of $y\sim-1.1$, close to the Salpeter and Miller-Scalo values for stars in the corresponding mass range of 2 to 14 $M_\odot$. From their photographic photometry of these clusters however, Elson et al. (1989) derived almost flat IMFs, with slopes $y = -0.8$ and $y = +0.2$ across the mass range $\sim$1.5 to 6 $M_\odot$, a result at least partially accounted for by their not using PSF fitting techniques: crowding tends to discriminate preferentially against fainter stars, thus flattening the derived IMF slope. Following this tack even further, Sagar and Richtler (1991) discussed the effect of unresolved optical and physical binaries. If plentiful ($>$50%), these can also flatten the slope of the *derived* IMF significantly, particularly if the slope of the *intrinsic* IMF is already rather flat. For example, in the extreme case of all observed stars actually being paired with another star of random mass, an intrinsic Salpeter slope of $y = -1.35$ would be flattened to a measured value of $y\sim-1.0$. Finally, looking at their study of five LMC/SMC clusters (NGC 1711, NGC 2004, NGC 2100, NGC 2614, NGC 2214) as a whole, Sagar and Richtler (1991) found no strong evidence for mass segregation (cf. Sagar et al. [1988], and references therein), although they did not study the crowded inner cores ($<$5 pc radius) of the clusters.

There remain significant uncertainties in IMF slope determinations for LMC/SMC clusters due to their relatively large distance, corresponding technique questions (e.g., PSF fitting, incompleteness corrections), and choice of comparison stellar models. However, there does seem to be some consensus emerging, with IMF slopes slightly steeper than Salpeter slopes, and evidence for cluster-to-cluster slope variations (M. Mateo, personal communication). Higher spatial resolution data may begin to settle some of these issues in the near future.

IV. THE QUESTION OF THE EXISTENCE OF BROWN DWARFS

Since publication of *Protostars and Planets II* (Black and Matthews, eds. 1984), there has been an increasing interest in finding observational evidence for the existence of brown dwarfs, i.e., substellar objects ($<$0.08 $M_\odot$) that

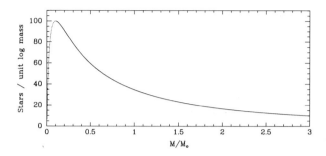

Figure 18.   Log-normal model IMF for low-mass stars according to Zinnecker (1984*b*). The IMF shown here is the relative number of newly born stars as a function of mass (in units of solar mass) per unit log mass interval. This curve is very similar to the well-known Miller-Scalo IMF, except for the (log-normally continued) extrapolation to substellar masses.

are not massive enough to ignite hydrogen burning (D'Antona and Mazzitelli 1985). There are many low-mass stars just above this critical mass, and it seems reasonable to expect many objects below it, as the minimum Jeans mass in a collapsing cloud is $<0.01 \, M_\odot$ (Low and Lynden-Bell 1976; Rees 1976). Also, in a more general sense, it seems unlikely that the physical processes of nuclear fusion that determine whether or not a star is massive enough to burn hydrogen have much to do with the processes that governed the fragmentation and initial phases of collapse that resulted in that star. If the empirically determined Miller-Scalo IMF for stars (a half-gaussian in log-mass, peaking at $0.1 \, M_\odot$) extrapolates as a descending gaussian below the peak, as some hierarchical fragmentation models would suggest (Zinnecker et al. 1980; Zinnecker 1984*b*; see Fig. 18), one would expect as many sub-stellar objects in the range 0.01 to $0.1 \, M_\odot$ as PMS stars in the range 0.1 to $1.0 \, M_\odot$. Liebert and Probst (1987) and Stevenson (1991) have presented extensive, thorough reviews of the theory and observations of these objects, but in the present context of young stellar populations, we shall give a brief account of searches for *young* brown dwarfs in regions such as Taurus-Auriga and the Pleiades, where low-mass stars are known to have recently formed or are still forming.

    There are major observational advantages in searching for young brown dwarfs. First, the search area is much better defined, because regions with young stars, i.e., clusters, are more coherent in space and time than the galactic disk in general. Second, young brown dwarfs are easier to detect than old brown dwarfs because they are still rather bright due to very slow quasi-static contraction. For example, at a distance of 125 pc, probable Pleiades brown dwarfs (age $\sim$100 Myr) are expected to be at $I \sim 17.5$ mag and $K \sim 14.5$ mag (Simons and Becklin [1991*b*], derived from Burrows et al. [1989]). In

Taurus-Auriga at 140 pc, probable brown dwarfs should be at $I \sim 14.0$ mag and $K \sim 11.0$ mag assuming an age of ~ 1 Myr, or $I \sim 15.7$ mag and $K \sim 12.7$ mag at ~ 10 Myr (derived from the PMS tracks discussed in Sec. II.B, provided by I. Mazzitelli). These magnitudes are for objects on the star : brown dwarf dividing line at $0.08\,M_\odot$, and therefore genuine brown dwarfs ($< 0.08\,M_\odot$) would be somewhat fainter. Even so, such objects should still be well within the capabilities of current optical and infrared imaging systems: it is their unambiguous identification that poses significant problems. This is mainly due to contamination by background field stars, but also partly due to a lack of knowledge about exactly where to draw the line between very low-mass stars and brown dwarfs in a young cluster. On the latter point however, recent theoretical models suggest that in the HR diagram of a young (≤ 100 Myr) cluster, brown dwarfs should be well segregated from very low mass stars, primarily in T_{eff} rather than L (Stringfellow 1991).

If the putative brown dwarf population *can* be kinematically and spectro-scopically identified for a given star-forming region of known distance and age, we should then be able to determine whether the luminosity function is rising or falling with decreasing luminosity. This question has obvious implications for the IMF and for the "missing mass" problem. In order to address these wider issues however, it is not enough to search for brown dwarfs occurring as binary companions (see, e.g., Zuckerman and Becklin 1987; Becklin and Zuckerman 1988; Henry and McCarthy 1990): it is also necessary to assess the existence of a possible "free-floating" population. In that context, we will discuss searches for this population in Taurus-Auriga, ρ Oph, the Pleiades, and the Hyades.

Forrest et al. (1989) carried out a $2.2\,\mu\text{m}$ (K) imaging survey in the im-mediate vicinity (within a few tens of arcsec) of 26 known PMS members of the Taurus-Auriga association. Of the 20 new objects detected, four were classified as Taurus-Auriga members based on their proper motions. These four have K magnitudes ranging from 14.2 to 15.8 mag, leading to estimated masses $\sim 0.01\,M_\odot$, and all are far enough separated from the nearby PMS star that they would more likely be free-floating than part of a bound mul-tiple system. However, these candidate Taurus-Auriga brown dwarfs seem strangely scattered in the near-infrared color-color diagram, and follow-up optical spectroscopy seemed to rule out the possibility that they are very cool objects (Stauffer et al. 1991). So at least for the time being, there appears to be no strong case for a large brown dwarf population in Taurus-Auriga.

The same seems to be true for ρ Oph, the subject of several recent near-infrared surveys. While Rieke et al. (1989) surveyed the ρ Oph dark cloud core to a limiting magnitude of $K \sim 14$ mag and found no new sources over and above those found in a previous shallower survey to $K \sim 12$ mag (Wilking and Lada 1983), another survey to $K \sim 14$ mag over a larger area did reveal some 35 new sources (Barsony et al. 1989*b*). However, there is no indication that these sources are necessarily substellar, and although an even deeper survey (to $K \sim 15.5$ mag) finally uncovered three, as yet unconfirmed, brown dwarf candidates (Rieke and Rieke 1990), the present conclusion is that there is no evidence for a *substantial* brown dwarf population in ρ Oph.

A limited start at the search for brown dwarfs in the Pleiades was made by Skrutskie et al. (1986), who looked at eight known cluster members and found no faint near-infrared companions. Searching for a free-floating population of brown dwarfs, CCD surveys in the central region of the Pleiades were made by Jameson and Skillen (1989) and Stauffer et al. (1989), covering 125 square arcmin at R and I, and 900 square arcmin at V and I, respectively. Although in both cases a small number ($\sim$5) of very faint, very red sources were found to fall in the part of the color-magnitude diagram expected for brown dwarfs, Stauffer et al. (1989) concluded that the mass distribution peaks near 0.2 $M_\odot$, and falls rapidly thereafter, so that brown dwarfs are not a major component of the Pleiades.

In contrast, Hambly and Jameson (1991) used R and I Schmidt photographic data to study the faint stellar and substellar content of a 3 degree diameter region centered on the Pleiades. After a statistical subtraction of field-star contamination as determined from the outer regions of their plates, they placed a lower limit of 30 on the potential number of brown dwarfs in the whole cluster, and concluded from their I luminosity function that the mass function is flat at the lowest masses. Finally, looking in the wavelength region where low-mass stars and brown dwarfs are most luminous, Simons and Becklin (1991a,b) surveyed some 200 square arcmin over several regions of the Pleiades in the near-infrared using the UKIRT infrared camera to $K\sim$17 mag, and in the optical using a CCD camera. Control fields 5 degrees away at the same galactic latitude were also surveyed to assess the field population. Plotting an I vs $I-K$ color-magnitude diagram overlaid by theoretical brown dwarf curves at the appropriate age, they isolated sources that lay close to the lower main sequence. After correcting for the background population, they were left with some 22$\pm$10 free-floating objects with nominal masses in the range $\sim$0.04 to 0.1 $M_\odot$, and a mass function that continues to rise steeply through the star : brown dwarf dividing line. When extrapolated to the whole cluster, these results imply a much larger population of substellar objects than that of Hambly and Jameson (1991), with objects in the 0.04 to 0.1 $M_\odot$ range contributing some 200 $M_\odot$ to the total Pleiades cluster mass of $\sim$1000 $M_\odot$, as determined from velocity dispersion measurements (van Leeuwen 1980,1983).

The results of Hambly and Jameson (1991) and Simons and Becklin (1991b) indicate that if most stars in the galactic disk are formed in clusters, and if the Pleiades is a typical cluster, then it is reasonable to suppose that brown dwarfs in the field are common. However, it must be noted that all these large-area surveys of the Pleiades used the temperature calibration for low-mass stars derived by Berriman and Reid (1987), which Stringfellow (1991) has recently found to be erroneous. This discovery places the results of Jameson and Skillen (1989) and Hambly and Jameson (1991) in serious doubt, and those of Stauffer et al. (1989) and Simons and Becklin (1991b) somewhat less so. This finding, along with the present degree of uncertainty in evolutionary models of very low-mass stars and brown dwarfs against which the observations are tested, suggest caution in interpreting the observational data. Also, as all the various authors point out, definite confirmation of any of these brown dwarf candidates as Pleiades members awaits accurate proper motion studies.

The older (600 Myr) but closer (~40 pc) Hyades cluster has also been surveyed for faint members (Leggett and Hawkins 1988,1989). Using Schmidt plates, they selected objects with large $R-I$ colors for follow-up near-infrared (J,H,K) photometry, in order to derive near-infrared luminosity functions. It is worth noting here that it probably makes more sense to talk in terms of the near-infrared LF for low-mass stars and brown dwarfs than the usual optical LF, since the former is much closer to the bolometric LF than the latter, due to the relatively cool temperature of these objects. Leggett and Hawkins (1988,1989) found a near-infrared LF for the Hyades which peaks at absolute magnitudes $M_J=7.6$ mag, $M_H=7.0$ mag, and $M_K=6.7$ mag, corresponding to ~$0.2\,M_\odot$. Henry and McCarthy (1990) dispute the significance of this peak, suggesting that unresolved binaries may be boosting the bright end of the luminosity function, while depressing the faint end. Hubbard et al. (1990a) calculated bolometric luminosity functions for stars and substellar objects spanning the range 0.03 to $0.2\,M_\odot$, and compared their model luminosity functions for the age of the Hyades with the observations of Leggett and Hawkins (1988,1989). They inferred that the mass distribution does not rise with decreasing mass over the mass interval studied. This is consistent with the negative result of a high-sensitivity 1 to $2\,\mu$m photometric search for cool, low-mass companions to 8 white dwarfs in the Hyades (Zuckerman and Becklin 1987).

It is interesting to note (cf. Hubbard et al. 1990a) that it is at a cluster age about equal to that of the Hyades that the luminosity function begins to separate into two components: that of the brown dwarfs and that of the lower main sequence. Therefore, deeper surveys of the Hyades may be able to differentiate between these two components. In older clusters, the brown dwarfs become harder to detect, because they will have evolved to even lower luminosity. On the other hand, Stringfellow (1991) presents isochrones spanning a mass range of 0.01 to $0.2\,M_\odot$ and an age range of 1 to 300 Myr. As mentioned above, he emphasizes that the best opportunity to discover brown dwarfs may be in even younger clusters (≤ 100 Myr), where the brown dwarfs should also be well segregated from the very low-mass stars in the HR diagram.

Finally, it is worth reporting the observational results on the low-mass stellar content of much older systems, globular clusters, as some of these may yet contain their low-mass initial stellar population. Drukier et al. (1988) and Richer et al. (1990) have both investigated the luminosity function of M 13 (metallicity [Fe/H] $= -1.4$) and concluded that the derived mass function rises steeply below $0.4\,M_\odot$ with no sign of flattening at the lowest observable masses (0.2 to 0.4 $M_\odot$). The more metal-rich M 71 and more metal-poor NGC 6397 exhibit flatter (but still rising) mass functions in the same mass range (Richer et al. 1990; Fahlman et al. 1989). Thus, contrary to earlier data (McClure et al. 1986), the most recent CCD photometry does not indicate a correlation between the slope of the mass distribution at the lowest masses and the metallicity of a cluster, but instead suggests that the steepest mass functions are found in the dynamically least-evolved clusters, such as M 13. Dynamically older systems, such as M 71 and NGC 6397, may have lost a much greater fraction of their low-mass stars due to evaporation, thus making for a shallower slope at the low end of their mass functions (Richer et al. 1990). This is now further confirmed for a larger sample of globular clusters, implying a very large initial population of low-mass stars and perhaps even brown dwarfs in globular clusters, and possibly in the galactic halo in general (Richer et al. 1991). Thus it seems that the conditions for forming very low-mass stars and brown dwarfs may be more favorable in Population II than in Population I regions (cf. Zinnecker 1986b).

V. SUMMARY AND OUTLOOK

In conclusion, we return to the questions posed in the introduction, summarize our present knowledge and ignorance about young stellar populations, and take a speculative look forward to some of the problems that may be addressed between now and the time of the *Protostars and Planets IV* book.

In most H II regions, massive stars do not appear alone but are usually accompanied by a cluster of low-mass PMS stars; i.e., low-mass stars and high-mass stars form in the same molecular cloud cores. Evidence for high-mass stars without accompanying low-mass stars (bimodal star formation) is very rare, while there are many examples of the converse, namely low-mass stars without high-mass stars. Taking the Galaxy as a whole however, it remains an open question whether the majority of low-mass stars form as cluster members near OB stars, or as a more widely dispersed background population in giant molecular clouds unrelated to OB stars. Recent results discussed in this and other chapters of this book seem to indicate a move toward the former hypothesis.

Clusters of PMS stars associated with OB stars form from dense gas concentrations in molecular clouds. It appears that some clusters owe their origin to the triggering push of adjacent active regions which compress pre-existing overdense clumps (obstacles); e.g., the ρ Oph cluster and those in L 1630. Thus there is evidence for sequential star formation (cf. Elmegreen 1989b). In some cases, two daughter sites of star formation may spring from one mother site: as discussed earlier, the Trapezium Cluster and the NGC 2024 cluster may both be offspring of the Orion 1b association, and a similar form of bifurcation may also have operated in the Sco-Cen association (de Geus et al. 1989). Other clusters may form as self-initiated independent condensations in molecular clouds, e.g., as in NGC 6334. If star formation in clusters is triggered, the age spread of the cluster stars should be small, while in self-initiated protocluster condensations, the individual clumps should have a larger age spread, probably governed by the time scale of the loss of magnetic support in the individual protostellar clumps.

The evolution of the IMF, star-formation rate, and star-formation efficiency in a protocluster are not yet well known. The IMF is thought to evolve from the clump mass spectrum, which in turn must be regulated by a balance of clump collision and fragmentation processes. The clump mass spectrum is relatively flat ($dN/dM \propto M^{-1.5}$), with most mass in the massive clumps, while the final stellar mass spectrum is steeper ($dN/dm \propto m^{-2.35}$), with most mass in the low-mass stars. Preferential fragmentation of the massive clumps and mass-dependent feedback processes could steepen the flat clump spectrum by the required amount to result in the final stellar mass spectrum. Clump collisions should be more prevalent near the protocluster center, in which case we would expect more massive clumps near the center, implying that the more massive stars are born predominantly near the cluster core. The more massive a protocluster cloud, the deeper its central potential well, which should lead to a correlation between total cluster mass and the most massive stars formed in the cluster, as seems to be observed.

At the time that *Protostars and Planets II* (Black and Matthews, eds. 1984) was written, it was generally felt that low-mass stars form first and over a long time scale (a few times 10 Myr) throughout a GMC. If and when high-mass stars form, they dissipate the reservoir of material, preventing further star formation (Strom 1985). This paradigm was based on observational evidence including the age spread of T Tauri stars (Cohen and Kuhi 1979) and of the low-mass members of the Pleiades (Herbig 1962*a*). The same "low-mass first, high-mass later" buildup in the IMF was found for the young clusters NGC 2264 and NGC 6530 (see, e.g., Iben and Talbot 1966; Adams et al. 1983), although Strom (1985) cautioned that these conclusions rested heavily on comparisons of the observational data with theoretical PMS tracks. Indeed, subsequent re-analyses have shown the same data to be more consistent with simultaneous formation of stars of all masses and a roughly constant star-formation rate in young clusters (Stahler 1985; Schroeder and Comins 1988).

There probably is a low level of low-mass star formation throughout a GMC to begin with, but when massive stars finally form, they are generally accompanied by vigorous low-mass star formation in the same molecular core, as indicated by the extensive clusters of low-mass stars seen surrounding OB stars in their H II regions. It is unknown to what degree these events are truly simultaneous: for example, the low-mass star formation may actually be triggered by radiative implosion or the powerful winds from the massive stars. In any case, the notion of "coevality" in a cluster may be hard to define both theoretically and observationally on time scales of <1 Myr. Even if the collapse of a dense molecular core is triggered at some "time zero" by an external event, the collapse and accretion phases for stars over a range of masses may be quite different. Low-mass stars may acquire their final mass first, but will then take a long time to reach the ZAMS; high-mass stars may take longer to accumulate their mass, but then evolve rapidly onto the ZAMS. By the time the cluster becomes visible, there may well be an *apparent* spread in stellar ages. Observationally, any test of coevality depends on comparing the positions of stars in an HR diagram with theoretical isochrones. The isochrones are difficult to model for both high- and low-mass stars (e.g., are the accretion rates constant?), and it is difficult to place photospheres in an HR diagram accurately when they are partially or totally obscured by dust. At this time, only the rather general conclusion can be made that in all clouds, stars have formed coevally to within 10 Myr, while in the cases of ρ Oph and the Trapezium Cluster, it appears that most stars have formed coevally to within a few Myr.

In more detail, complex star-formation histories in star clusters may be the rule rather than the exception. We cannot trust present PMS tracks enough to unravel the complex time sequences of star formation with confidence. It can be speculated that the time evolution of the total star-formation rate in a cluster may be described by $1/f$ noise (cf. Bak et al. 1988). This implies self-organized weak chaos, characterized by "memory effects": rare short periods of vigorous star formation alternate with more frequent periods of low-level activity.

Similarly, the star-formation efficiency in a protocluster is a highly time-dependent quantity, namely the integral of the star-formation rate up to a given time, if star-formation is still occurring. The asymptotic, overall star-formation efficiency with which a cluster will form probably depends on the initial conditions, most importantly on the mean gas density but probably also on the total mass of the protocluster (cf. Pandey et al. 1990). High protocluster densities apparently lead to high star-formation efficiencies: for example, in the Trapezium Cluster, the protostellar envelopes must have been touching each other at the time of formation. We speculate that triggered clusters will form with a higher star-formation efficiency, will be more compact, and also more coeval, than self-initiated clusters. Conversely, it is clear observationally that outside of massive cores of molecular clouds, the star-formation efficiency is always low: for example, the total stellar mass to total gas mass ratio is at most a few percent in the Taurus-Auriga complex (which lacks a massive core), as it is outside the core of the ρ Oph region.

The study of star formation in clusters as opposed to isolated star formation raises many interesting new problems: for example, it may be that a full distribution of stellar masses with a large spread in mass only comes about because the primary sites of star formation are protoclusters in which a rich variety of competing processes takes place. If so, future studies of the initial stellar population should focus even more on young star clusters.

Observationally at least, this task will be made easier by likely technological innovations and evolutions during the next decade. Infrared array detectors have emerged as perhaps the key component for studies of young stellar populations. These detectors have increased in size from 32×32 to

256×256 pixels since the production of *Protostars and Planets II* in 1984, and arrays with more than 10^6 pixels are anticipated within the next few years, allowing us to survey all galactic massive star-forming regions for all but the very youngest and most deeply embedded stars. By using these detectors at high spatial resolution, even the remotest galactic clusters will be disentangled, and binarity in the nearer regions probed. Active and adaptive optics techniques will be used on the new generation of large groundbased telescopes ($\sim$8 to 16-m effective aperture) to achieve extraordinary resolutions (between $\sim$0.02 and 0.2 arcsec) in the near-infrared, while a near-infrared camera under development for the Hubble Space Telescope will yield a unique combination of high spatial resolution and very low background at 1 to 2 μm. Near-infrared spectroscopy will also gain enormously from advanced arrays: the first generation of array spectrometers are now available, and promise much in the understanding of the physical nature of sources revealed by camera surveys.

Mid- and far-infrared array detectors may be somewhat behind the near-infrared arrays in terms of size, but major technological breakthroughs have been made in preparation for the next generation of cryogenic infrared space observatories, ISO and SIRTF. While IRAS used a small number of individual detectors for an all-sky survey, both ISO and SIRTF will use large array detectors for pointed observations, and the great increase in sensitivity should result in fundamental changes in our knowledge of the youngest, coolest, and most deeply embedded sources in star-formation regions. At far-infrared and sub-millimeter wavelengths, high sensitivity and spatial resolution will be made possible with SOFIA, an airborne 2.5-m telescope.

Complementary progress is being made at other wavelengths. The ROSAT X-ray observatory, already in operation, will detect many new weak-emission T Tauri stars in the nearer regions such as Taurus-Auriga; the larger AXAF will provide the spatial resolution and sensitivity needed to probe the X-ray emission from young stars in the clusters associated with GMCs. Studies of molecular clouds—the raw material of star formation—are benefitting from the new large millimeter and sub-millimeter telescopes already in operation or under construction, including the IRAM 30-m, the 15-m SEST, the 15-m JCMT, the 10-m CSO, the Nobeyama 30-m, and the 10-m SMT. Linking arrays of telescopes, millimeter interferometry will allow studies of the smallest scales of molecular cloud cores and clumps, in an effort to relate the cloud mass spectrum to the stellar mass spectrum. In addition, sub-millimeter interferometers should allow us to resolve cold disks around young stars where planets and/or binary stars are believed to form.

With the ever larger fields of view, higher spatial and spectral resolution, and greater sensitivity afforded by new instrumentation and techniques, we will be able to take a more global view of star formation within a cluster, a molecular cloud complex, or our whole Galaxy, rather than concentrating intently on the detailed physics of single objects. For the interpretation of these larger-scale observations, we require theoretical advances, such as improved PMS evolutionary tracks, including the effects of disk accretion. At the same time, the new observations can be expected to stimulate new theories.

An understanding of the problems of star formation and early stellar evolution is clearly an important part of our understanding of the origin of the solar system and planet formation. Also, as stars are the basic building blocks of galaxies, these very same problems bear heavily on our comprehension of galaxy formation and evolution, and many of the large-scale cosmological questions related to the early Universe. Therefore, if most stars form in dense clusters, then it is clear that the formation and early evolution of clusters and the stars that comprise them must be better understood.

Acknowledgments. Many people have contributed to making this work as inclusive and up to date as possible, a particularly difficult problem considering the rate at which new data are becoming available, especially from infrared arrays and space-based instruments. We would like to thank A. Blaauw, M. Haas, T. Henry, G. Herbig, K.-W. Hodapp, C. Lada, C. Leinert, M. Mateo, H. Richer, G. Rieke, R. Sagar, D. Simons, J. Stauffer, G. Stringfellow and F. Walter. We would particularly like to thank J. Rayner for allowing us to make extensive use of as yet unpublished images and information from the collaboration between him and two of us (HZ and MJM). We would also like to thank I. Mazzitelli for providing the homogeneous PMS tracks used in Sec. II.B; S. Kenyon for providing Fig. 7; the WF/PC Investigation Definition Team, B. Campbell, and T. Lauer for the HST image of 30 Doradus; and G. Weigelt for the speckle image of R 136. Thanks are also due to R. Genzel for challenging us to make some sense of the near-infrared images of star-forming regions and the modeling of 2.2 μm luminosity functions. This work was begun while HZ enjoyed a visiting astronomer position at the Institute for Astronomy at the University of Hawaii, and was later supported by the Deutsche Forschungsgemeinschaft. BW would like to thank C. Lada and J. Moran for their hospitality and generosity during his tenure as a Visiting Scientist at the Harvard-Smithsonian Center for Astrophysics. Part of this work was funded by a NASA ADP grant to BW. Finally, numerous U.S. and foreign government agencies are acknowledged for their continued funding of the national and international computer networks: without e-mail, this task would have been impossible.

YOUNG STARS: EPISODIC PHENOMENA, ACTIVITY AND VARIABILITY

LEE HARTMANN, SCOTT KENYON
Harvard-Smithsonian Center for Astrophysics

and

PATRICK HARTIGAN
University of Massachusetts

Much of the variability of pre-main-sequence stars is now attributed to changes in the mass accretion rates of circumstellar disks. The largest variations are observed in the FU Orionis outbursts, in which disk accretion rates increase by 2 orders of magnitude or more. At outburst maximum, accretion rates onto the central T Tauri star are $\gtrsim 10^{-4}$ $M_\odot$ yr^{-1}. Event statistics indicate that many, if not most, low-mass stars undergo multiple FU Orionis events, ultimately accreting at least 5% to 10% of the total stellar mass from the disk during these brief episodes. As much as 0.01 $M_\odot$ can be accreted in an individual event, demonstrating that FU Orionis disks are fairly massive. The rapid accretion causes the disk to become very hot during outbursts, which may have implications for solar nebula evolution. FU Orionis disk accretion produces very energetic winds, capable of driving bipolar molecular flows and Herbig-Haro objects. Studies of the interaction of pre-main-sequence winds with the interstellar medium suggest that these flows are not steady, but are composed of separate ejection events. The link between ejection and accretion may permit the inference of the accretion histories of individual objects from analyses of pre-main-sequence jets.

I. INTRODUCTION

Over the last few years several lines of research have suggested that disk accretion is an important phenomenon in early stellar evolution. Recent observational estimates suggest that at least 5% to 10% of the mass of an average low-mass star is accreted from its surrounding disk during optically visible stages of evolution (Hartmann and Kenyon 1990). The total amount of disk matter actually accreted onto a young star could be considerably larger than this estimate, because disk accretion probably also occurs during the earliest evolutionary phases, when young stars are completely shrouded from our view by their dusty envelopes.

Disk accretion appears to be highly time variable. The low-mass, pre-main-sequence T Tauri stars can vary in brightness by a factor of 10 or more on time scales of months to years (see, e.g., the review by Appenzeller and

Mundt [1989]), and fluctuations of the optical and ultraviolet excess emission within a 24-hr time interval seem quite common (Hartigan et al. 1991; Basri and Batalha 1990; Chapter by Basri and Bertout). Much of this variability is probably caused by changes in disk accretion rates (though not all; see Gahm et al. 1989; Herbig 1990). The most spectacular variations in T Tauri disk accretion are found in the classical FU Orionis (Fuor) outbursts (Herbig 1977,1989*a*), in which the mass accretion rate increases by up to 3 orders of magnitude on a time scale of a year or so (Hartmann and Kenyon 1985). Accretion rates of $10^{-4} \, M_\odot \, yr^{-1}$ or more appear to be characteristic of Fuor disks, and 10^{-3} to $10^{-2} \, M_\odot$ can be accumulated by the central T Tauri star during a single outburst lasting for decades or more. Such high accretion rates imply that the disk and central star evolve much more rapidly than previously thought, at least for short periods of time.

Fuor accretion disks are especially interesting for several reasons. First of all, Fuor outbursts tell us that protostellar disks can be quite massive—the masses typically are larger than the minimum mass solar nebula and perhaps are large enough for gravitational instabilities to be important. The high disk surface temperatures produce a wealth of atomic and molecular lines ideal for detailed optical and near-infrared spectroscopic studies. The time-variability of Fuor accretion yields important insights into the properties and physics of protostellar disks which are difficult or impossible to determine from first principles. The extreme variability of Fuors also implies a rapid cycling of disk temperatures and mass infall rates, which may be of some significance for certain aspects of planetary formation. Finally, Fuors are often associated with bipolar flows and jets, and they appear to be capable of rapidly ejecting material (roughly 10% of the mass accretion rate) from the surfaces of their inner disks. Their brightness makes it possible to study the wind acceleration region in some detail and constrain the mechanisms of mass loss.

In the following sections, we concentrate on the behavior of Fuors as the extreme example of pre-main-sequence accretion disk variability. We begin by outlining the standard picture of disk accretion for the typical solar-type, young (T Tauri) star in Sec. II. Trusting souls can skip Secs. III, IV and V, where we summarize the astronomical arguments supporting the standard picture, and proceed to Sec. VI, where we consider the implications of the observations for the physics of protostellar disks. Finally, in Sec. VII we describe observations of mass loss from Fuors, and consider the implications for the physics of disk winds and effects on the neighboring interstellar medium.

II. OVERVIEW

The early phases of low-mass star formation are thought to involve the collapse of a compact, dusty, opaque gas cloud into a star plus disk system (see Shu et al. 1987, and references therein). The best observational constraints suggest that this infall phase typically lasts $\gtrsim 2 \times 10^5$ yr (Myers et al. 1987; Kenyon et al. 1990), which is reasonably consistent with theoretical predictions (Adams et al. 1987). Mass loss from the central object eventually clears away some of the dusty circumstellar material to reveal a newly formed T Tauri star (Shu et al. 1987a; Chapter by Basri and Bertout).

Our picture of the accretion events following the onset of protostellar collapse is illustrated schematically in Fig. 1. Once the infalling envelope has been exhausted and/or expelled, material from the disk generally accretes onto the central star at a modest rate, $\sim 10^{-7}$ $M_\odot$ yr^{-1} (Chapter by Basri and Bertout). This "background" disk accretion is punctuated by Fuor eruptions in which the mass flow rate through the disk reaches $\sim 10^{-4}$ $M_\odot$ yr^{-1} for an uncertain length of time (~ 10 to 1000 yr). The large increase in the accretion rate causes the disk surface temperature to increase substantially (see Fig. 2a). The increase in the disk *midplane* temperature could be considerably larger than the surface temperature jump, because disks in the "T Tauri" state may be nearly isothermal (Kenyon and Hartmann 1987), while Fuor disks are likely to have relatively warm interiors (Clarke et al. 1990; Sec. IV).

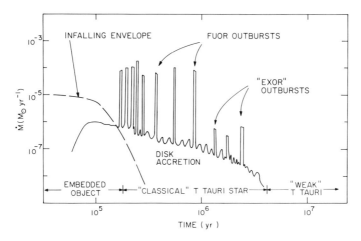

Figure 1.   Schematic illustration of the variation of accretion rate ($\dot{M}$ in $M_\odot$ yr^{-1}) with time for a young star. Mass from the circumstellar envelope initially falls onto the star quite rapidly, but material primarily lands on the disk at later times. Accretion from the disk onto the star increases from zero to some significant background rate and then slowly decreases with time during the T Tauri phase of evolution. The background accretion is supplemented by recurrent periods of very high $\dot{M}$ (Fuors) and moderately high $\dot{M}$ (Exors). Accretion from the disk ceases eventually, and the young star is then called a weak-emission T Tauri star.

The number of Fuor outbursts observed within 1 kpc of the Sun in the last 50 yr is approximately ten times larger than the estimated star-formation rate within this same region (Hartmann and Kenyon 1985; Herbig 1989a), so the "average" young star must undergo ~ 10 Fuor events during its lifetime. Fuor eruptions may be more common at earlier times, because all known Fuors are

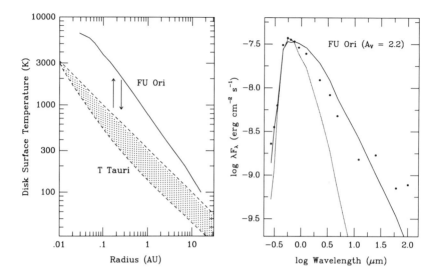

Figure 2.   Properties of protostellar accretion disks. (a) The variation of disk surface
temperature with radius in a Fuor disk (dotted line), determined from observations
of the emergent spectrum.   The inferred range of disk temperatures for typical
T Tauri stars is indicated by the dotted area. (b) The right panel shows the observed
spectral energy distribution of FU Ori (filled circles) from Kenyon et al. (1988). The
FU Ori temperature distribution shown in (a) produces the broad energy distribution
denoted by the solid line. The energy distribution of a normal G-type star (essentially
a one-temperature blackbody distribution) is plotted as a lighter line.

associated with reflection nebulae (Goodrich 1987; see Table I), whereas only
a fraction of T Tauri stars are associated with such nebulae.

Aside from Fuor outbursts, a T Tauri star may undergo more modest
bursts of accretion which can be associated with the "Exor" events identified
by Herbig (1977,1989a). Figure 1 places Exor eruptions near the end of the
lifetime of a classical T Tauri star when the background accretion rate is small,
but it is possible that Exor events are interspersed among the much larger Fuor
events. The Exors may also be more frequent than Fuors; the recurrence rate
is not known.

Present estimates suggest that the typical T Tauri star accretes roughly
the same amount of disk material in Fuor events as in relatively steady accretion: $\gtrsim 0.05$ $M_\odot$ in Fuor episodes (Hartmann and Kenyon 1987a) and
~ 0.1 $M_\odot$ in T Tauri disk accretion (Hartmann and Kenyon 1990). The classical T Tauri phase ends when the disk becomes less efficient at transporting
material radially inwards. Pre-main-sequence stars which no longer accrete
from their disks have a much smaller range of activity than classical T Tauri
stars and have become known as "weak-emission T Tauri stars" (see Chapter
by Montmerle et al.).

TABLE I

Fuor Properties

Object	Rise Time		Decay Time		d(Kpc)	$\dot{M}^a$	CO flow	Jet/ HH	References[b]
FU Ori	~ 1	yr	~ 100	yr	0.5	10^{-4}	no	no	1,2,4
V1057 Cyg	~ 1	yr	~ 10	yr	0.6	10^{-4}	yes	no	1,2,4,17
V1515 Cyg	~ 20	yr	~ 30	yr	1.0	10^{-4}	no	no	1,2,4
V1735 Cyg	< 8	yr	> 20	yr	0.9	?	yes	no	3,4,17
V346 Nor	< 5	yr	> 5	yr	0.7	?	yes	yes	6,11
L1551 IRS5	?		?		0.15	10^{-5}	yes	yes	7,10,12
Z CMa	?		> 100	yr	1.1	10^{-5}	yes	yes	4,5,13,14,16
BBW 76	?		~ 40	yr	1.7?	?	?	no	8,9,15
RNO 1B	?		?		0.8	?	yes?	no	18

[a] in $M_\odot$ yr^{-1}.
[b] References: 1,2 = Herbig 1977,1989a; 3 = Elias 1978a; 4 = Levreault 1983; 5 = Covino et al. 1984; 6 = Graham and Frogel 1985; 7 = Mundt et al. 1985; 8,9 = Reipurth 1985,1989c; 10 = Carr et al. 1987; 11 = Cohen and Schwartz 1987; 12 = Stocke et al. 1988; 13 = Hartmann et al. 1989; 14 = Poetzel et al. 1989; 15 = Eislöffel et al. 1990; 16 = Ray 1990, personal communication; 17 = Rodríguez et al. 1990; 18 = Staude and Neckel 1991.

We emphasize that the evolution outlined in Fig. 1 is highly schematic and that all stars may not follow the same evolutionary path. In particular, there is no guarantee that the solar system experienced Fuor events. Even so, Fuor eruptions must represent a fairly common phenomenon in the earliest phases of star formation, because they appear to be more frequent than individual star-forming events.

We adopt a spectroscopic definition for Fuors in this chapter and consider objects with similar spectra (similar surface temperature distributions). This choice groups together systems with qualitatively different outburst behavior and includes several objects with no recorded outburst. It is possible that some objects listed in Table I are extremely young and have continuously rapid disk accretion instead of recurrent eruptions. Others may be starting to run out of accretable disk material, so their "engine" starts to "cough" (outburst) as the "fuel" runs out.

III. WHY DISKS?

Fuors are relatively luminous objects found in regions of star formation (Herbig 1977,1989a). Attention was drawn to the first members of this class because of large increases in optical brightness (factors $\sim 10^2$) on time scales of about 1 yr. A single pre-outburst spectrum of one object, V1057 Cygni, indicated that the Fuor outburst was associated with a faint T Tauri star. This result is very important, because it demonstrates that the outburst of V1057 Cyg was not simply the result of disruption or dispersal of a dust cloud, but involved a change in the intrinsic spectrum (Herbig 1977).

The accretion disk model for Fuors (Hartmann and Kenyon 1985; Lin and Papaloizou 1985) initially was motivated by analogy to other astrophysical objects with similar outburst behavior and by the need to explain the observed broad spectral energy distributions (Fig. 2). Assuming optically thick emission, the $\sim$2000 K gas responsible for the $2\,\mu$m continuum of FU Ori must have an area $\sim$10 to 20 times larger than the $\sim$5000 K gas producing the optical continuum at a wavelength of $\sim$0.5 μm. Similarly, the $\sim$300 K material responsible for the 10 μm excess must have $\sim$100 times the emitting area of the optical continuum region, etc. The variation of temperature with radius expected from a disk naturally explains these observations (Lynden-Bell and Pringle 1974), and a simple steady accretion disk model explains the observed infrared excess emission of FU Ori (Fig. 2), V1057 Cyg, and V1515 Cyg reasonably well. Unfortunately, at the large distances of known Fuors, direct spatial resolution of the optical and near-infrared emitting disk regions (sizes $\lesssim$1 AU observed at distances of $\gtrsim$500 pc) is presently impossible.

The observed variation of spectral type with wavelength provides additional evidence that a Fuor possesses a significant range of emitting temperatures (Herbig 1977). Strong water vapor absorption features detected in FU Ori and V1057 Cyg at 1.6 μm are only seen in stars with surface temperatures $\lesssim$4000 K (Mould et al. 1978), while the optical and ultraviolet spectra of these objects imply surface temperatures closer to 6000 to 7000 K (cf. Kenyon et al. 1988,1989).

A fundamental, testable prediction of the disk model follows from the apparent variation of surface temperature with the wavelength of observation. Because the spectrum at progressively longer wavelengths is dominated by the cooler, outer disk regions, the observed velocity broadening of spectral features observed at long wavelengths is smaller than the broadening of lines at short wavelengths (assuming Keplerian rotation). This fundamental prediction of the disk model, pointed out by Hartmann and Kenyon (1985), was later verified for FU Ori and V1057 Cyg by Hartmann and Kenyon (1987b,1988), who showed that the rotational broadening observed at 2.2 μm was $\sim$2/3 of the rotational broadening observed at 0.6 μm (Fig. 3). Welty et al. (1991) later presented evidence for a modest, continuous variation of rotational velocity over the range of wavelengths between 0.5 μm and 0.9 μm in V1057 Cyg.

The disk hypothesis can be tested in quantitative detail by constructing a model disk spectrum and comparing it with observations. For simplicity, the disk is assumed to be in a steady state, which may be reasonable for objects on the slowly varying portions of their lightcurves. If this optically thick disk emits roughly as a blackbody, and if the gravitational energy released by the infalling material is dissipated and radiated locally, then the energy balance equation for a disk annulus at radius R with width ΔR is

$$\frac{GM_*\dot{M}}{R}\frac{\Delta R}{R} = 2 \times 2\pi R\,\Delta R\,\sigma T^4 \qquad (1)$$

where G is the gravitational constant, σ is the Stefan-Boltzmann constant, M_* is the mass of the central star, and T is the temperature of the annulus. This equation can be rearranged to yield

$$\sigma T^4 \sim \frac{GM_*\dot{M}}{4\pi R^3} \qquad (2)$$

(see Pringle [1981] for details and minor modifications which depend upon the inner boundary condition). The scaling of the temperature distribution is set by the maximum disk temperature, which occurs in the innermost disk regions, and this maximum temperature can be estimated from the observed spectra of Fuors at short (optical) wavelengths.

For the purposes of calculating the total disk spectrum, each annulus is assumed to radiate as a star of the appropriate effective temperature; we used supergiant stars, which have surface gravities similar to those expected at the disk surface (Kenyon et al. 1988). This assumption implies that essentially all of the accretion energy is released well below the disk photosphere, which may not be strictly correct. The spectrum of each annulus is convolved with the velocity broadening of Keplerian rotation, and the final synthesized model disk spectrum is the sum of the individual annular spectra.

A simple way to analyze the rotational broadening of the absorption lines is through cross-correlation analysis. Cross correlation of an object spectrum with a spectrum of a much more slowly rotating template star produces a correlation peak whose width is a measure of the rotational broadening, and whose shape indicates the line profile, averaged over many lines in the spectral bandpass adopted. We cross correlated both observed and synthesized spectra with the same template spectrum. The central stellar mass and the disk inclination are free parameters in the model; we scaled the rotational velocity to match the optical line broadening. The disk model then predicts the rotation that should be observed in the near infrared (and other spectral regions).

Figure 3 shows the comparison between theory and observation for FU Ori and V1057 Cyg. The agreement between the velocity broadening predicted for a Keplerian disk and the observed velocity broadening is reasonably good. The discrepancy between theory and observations, at the $\sim$20% level, can easily be attributed to uncertainties in the spectral modeling (Kenyon et al. 1988). In particular, the lack of adequate stellar spectral models for regions with surface temperatures $T < 2000$ K, which contribute 20% of the light at 2.2 μm, makes it difficult to model this portion of the spectrum with great accuracy.

The predicted cross-correlation peaks are "dimpled" or "doubled," and the observed cross-correlation peaks, which indicate the average line profile,

Normalized Correlation Amplitude

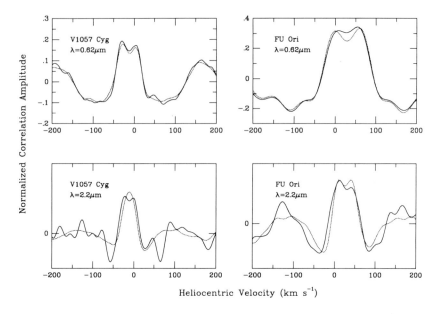

Heliocentric Velocity (km s^{-1})

Figure 3.   Comparison of observed (solid lines) and predicted (dotted lines) cross-correlation peaks for V1057 Cyg and FU Ori (Kenyon and Hartmann 1990). The velocity scales of the models have been adjusted to reproduce the observed optical correlation-peak widths. The model cross-correlation peak for V1057 Cyg at 2.2 μm is not doubled because the rotational velocity is not very large in comparison with the instrumental resolution.

show similar structure. High-quality spectra of the Fuors show double peaked behavior in individual line profiles in the red spectral region (Fig. 4). A spherical, rotating star of uniform surface brightness produces parabolic line profiles, unlike these observed profiles, but a disk model can account for this behavior quite naturally. A rotating, flat, narrow ring exhibits a line profile, ϕ, as a function of velocity shift from line center ΔV of the form

$$\phi(\Delta V) = \left(1 - \left(\frac{\Delta V}{V_{max}}\right)^2\right)^{-1/2} , \quad -V_{max} < \Delta V < V_{max} \qquad (3)$$

where V_{max} is the rotational velocity of the ring corrected for its inclination to our line of sight (e.g., v sin i; cf. Kenyon et al. 1988). For a disk model, the net profile at a given wavelength of observation is the sum of profiles from several neighboring annuli with different rotational velocities, and this differential rotation somewhat rounds or blurs the profile indicated by Eq. (3). The cross-correlation peak profiles in Fig. 3 indicate that the average line profile is reasonably well matched by the disk double-peaked profile. Direct modeling of the observed spectra also shows (Fig. 5) that the disk profiles work very well in explaining the peculiarities of the observed line profiles for several Fuors in many (but not all) spectral regions (see Sec. V).

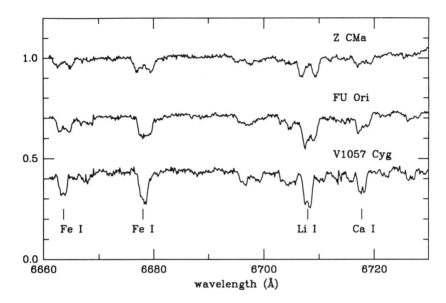

Figure 4.   Comparison of optical line profiles in three Fuors. Note the double-peaked structure characteristic of a rotating disk.

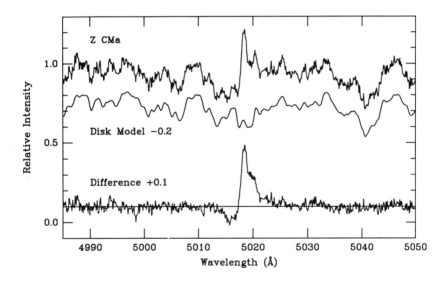

Figure 5.   A portion of the observed optical spectrum of Z CMa, compared with a disk model spectrum. The model spectrum (middle plot) has been subtracted from the observed spectrum (upper plot) to yield a difference spectrum (lower plot) which is essentially featureless except for a strong Fe II λ 5018 "P Cygni" line profile, formed in the strong wind of Z CMa (figure from Welty 1991).

In summary, the disk model is favored for Fuors because it naturally explains the observed broad spectral energy distribution and variation of spectral type with wavelength; it *quantitatively* accounts for the variation of rotation with wavelength; and it simply explains much of the peculiar line profile structure ("doubled" lines).

IV. WHY DISK ACCRETION?

The case for disk accretion is indirect, because it is not possible to observe the low-velocity drift of material from the outer disk to the inner disk and onto the central star. But given the spectroscopic and photometric evidence for disks reviewed in the previous section, disk accretion provides a plausible mechanism for the energy ($\sim 10^{45}$ to 10^{46} erg) released in a typical Fuor event. Accretion disks are known in other astrophysical contexts which exhibit qualitatively similar outburst behavior, and attempts have been made to scale physical parameters to Fuor disks (Lin and Papaloizou 1985; Clarke et al. 1989,1990).

Other mechanisms seem incapable of producing such a large amount of energy. Nuclear reactions probably are inadequate, because the only nuclear fuels which can be ignited in pre-main-sequence stars, deuterium and lithium (Stahler 1983,1988), are unlikely to supply enough energy. Furthermore, it is not obvious why nuclear events should be repetitive, as required by source statistics, or how energy produced by nuclear reactions deep in the stellar interior can account for the short rise times observed in most Fuors.

Stahler (1989a) has noted that spherically symmetric accretion produces an entropy distribution that is inconsistent with the structure of stars on radiative evolutionary tracks. He predicts that high-mass ($M > 1$ M$_\odot$) stars may undergo a large increase in their brightness over a short time scale when accretion ends and the internal structure of the star evolves to its proper radiative structure. This mechanism does not appear to apply to most Fuors, because the event statistics require that most stars, not just the massive ones, must undergo at least several eruptions.

The disk accretion model also explains why Fuors generally show strong CO absorption bands at 2.3 μm, while most T Tauri stars have featureless continua or comparatively weak CO emission or absorption bands even though their 2 μm continua also must be produced in disks (Kenyon and Hartmann 1987; Carr 1989). Simple accretion disk models for Fuors predict a temperature that decreases away from the midplane (cf. Clarke et al. 1990), which is exactly what is needed to produce strong absorption lines. However, the accretion rates in T Tauri disks are much lower than in Fuor disks, so disk heating by light from the central star is much more important. Absorption of stellar radiation tends to produce a more isothermal, or even inverted, disk temperature distribution, which results in weak CO absorption, or even CO emission, in T Tauri disks (Calvet et al. 1991a; see also Carr 1989).

In principle, the broad spectral energy distributions of Fuors can also be modeled by a nonaccreting or "passive" disk surrounding a central star, a model which can be applied to many T Tauri stars (Adams et al. 1987). The optical radiation in this picture comes from the central star, while the infrared excess emission is produced by stellar light absorbed by the disk and re-radiated at longer wavelengths. The passive disk model can account for the observed broad energy distributions, but it does not explain the source of outburst energy and also encounters difficulties with other aspects of the observations. For instance, it is difficult to interpret the differences between the CO spectra of Fuors and T Tauri stars, as described above, if both types of disks are purely passive. Welty et al. (1990) suggested that V1057 Cyg exhibits a *continuous* variation of rotation with wavelength in the optical region, difficult to explain with a stellar model but in quantitive agreement with disk model predictions. Finally, the star plus passive disk model also requires the central object to be rotating near breakup velocity, both to explain the ratio of infrared to optical rotational broadening (Kenyon et al. 1988), as well as to explain the doubled line profiles. It is far more straightforward simply to associate the optical spectrum with the inner disk.

IV. DISK MODEL PROBLEMS AND ALTERNATIVES

The case for Fuors as accretion disks becomes clearer when we consider criticisms of the disk model and alternative proposals. In particular, Herbig (1989) has urged caution in adopting the disk model for Fuors. Herbig argues that the profiles of many lines in FU Ori are more complex than those predicted by the simple disk model. While this is true, to the writers' knowledge no high-quality (S/N $\gtrsim$50–100) spectra of FU Ori at wavelengths exceeding 0.62 μm have failed to show the line doubling, at least since 1985. Lines at shorter wavelengths in FU Ori do not appear doubled but are more parabolic in shape (at the *same* time that the longer wavelength lines show doubling). These short-wavelength lines also exhibit a systematic *blueshift* relative to the center-of-mass velocity of the system (Hartmann and Kenyon 1985,1987a, c; Kenyon et al. 1988). Our explanation for this behavior, developed in more detail in Sec. VII, is that a powerful wind emanates from FU Ori, which blueshifts the profiles of strong lines and obscures the double-peaked structure. The weakest lines, found preferentially at longer wavelengths, are formed deep in the disk photosphere and thus show only the disk rotation.

Double-peaked profiles are a ubiquitous feature of Fuors; they have been observed in FU Ori and V1057 Cyg (Kenyon et al. 1988), in Z CMa (Hartmann et al. 1989), in BBW 76 (Eislöffel et al. 1990), and in RNO 1B (Staude and Neckel 1991). In fact, every Fuor where adequate spectral resolution and signal to noise can be obtained shows double-peaked absorption profiles (cf. Fig. 4). The disk model naturally explains this general observational result.

Herbig has also pointed out that V1057 Cyg was observed to be rotating faster at maximum light (1971) than at present epochs, whereas an approximation of the outburst by a sequence of steady disk models with differing accretion rates would predict the opposite effect. The $\sim$5000 K regions of the disk dominate the optical spectrum, and since $T \propto \dot{M}^{1/4} R^{-3/4}$ (Eq. 2), the optically emitting region is slightly larger at maximum light, and thus should exhibit a smaller rotational velocity. Recent observations indicate that Z CMa also may rotate somewhat more rapidly during its brighter states (Hessman et al. 1991), but the result is marginally significant. Although steady disk models cannot explain such behavior, the observations may simply mean that the disk does not maintain a steady-state temperature distribution when the accretion rate varies by factors of up to 100.

Herbig (1989a) has suggested that the spectral energy distributions of Fuors might arise from an extended, differentially rotating stellar atmosphere. This idea runs into practical difficulties because of the very large spatial extension needed to produce the infrared excess, based on blackbody arguments (Sec. III). For example, the 2 μm continuum must come from a region at least 3 to 4 times the optical radius. This region cannot be supported by thermal pressure. It is very difficult to construct an appropriate wind model to explain this atmospheric extension. Wind models naturally tend to accelerate to something comparable to the escape velocity over several stellar radii, but the CO lines show no evidence for high-velocity blueshifted absorption or profile asymmetries, limiting expansion velocities to $\leq$0.2 of the escape velocity (see Fig. 3). Rapid rotation is the simplest and most obvious mechanism to keep the infrared photosphere suspended at such a large distance from the star; however, a rapidly rotating star would produce differential rotation in the *opposite* sense of that observed, because infrared lines produced in the cool equatorial regions should have a much larger rotational velocity than the polar regions responsible for the optical emission. Binary models can also be ruled out for Fuors, because the large radial velocity variations expected from a close binary have not been observed (see Hartmann and Kenyon 1985).

Simon and Joyce (1988) also raised concerns that during the decline of V1057 Cyg, the spectral energy distribution was somewhat flatter, i.e., had more infrared excess, than is predicted by steady disk models. In a recent detailed study of V1057 Cyg (Kenyon and Hartmann 1991), it is shown that the evolution of the spectral energy distribution can be explained by a two-component model: an accretion disk, with slowly decreasing mass flow rate; and a dusty circumstellar envelope, which absorbs a fixed amount of light from the disk and re-radiates the energy at longer wavelengths. In our model, the distant dust envelope is responsible for the large infrared excesses at wavelengths $\gtrsim$10 μm which cannot be explained with a simple disk model.

It is important to remember the approximation of a time-varying accretion disk by a series of steady disk models with differing accretion rates cannot be strictly correct. The characteristic time scales of accretion must vary as a function of radius, so it is not possible for all regions of a varying disk to have exactly the same $\dot{M}$. It makes sense to use steady-state models well after outburst, when the Fuors generally exhibit a slow decay in light, but such models are likely to be inappropriate close to outburst.

We conclude that the accretion disk model currently provides a satisfactory explanation of Fuor properties. This model naturally explains the broad spectral energy distribution and variation of spectral type with wavelength; it quantitatively accounts for the variation of rotation with wavelength and much of the peculiar line profile structure ("doubled" lines) observed, and it provides a convenient explanation of outbursts, consistent with the behavior of other astrophysical disks.

VI. DISK PHYSICS AND PROPERTIES

Observations of Fuors yield several important physical parameters for protostellar disks. By matching an accretion disk model to the observed spectral energy distribution, the disk surface temperatures as a function of radius can be inferred fairly straightforwardly if the distance is known (Fig. 2). The mass-accretion rate through the disk can be determined if the mass and the radius of the central star are known. However, the central star is not observed directly, and so observations of the inner regions of the disk must be used to estimate these quantities.

The disk luminosity L_{disk} is related to the maximum disk surface temperature T_{max} and the inner disk radius R_{in} by

$$L_{disk} = \frac{GM_*\dot{M}}{2R_{in}} = \text{constant} \times \sigma T_{max}^4 R_{in}^2. \qquad (4)$$

The short-wavelength optical and ultraviolet spectra provide estimates for T_{max} (Kenyon et al. 1988,1989). After making the necessary extinction corrections, the observed spectral energy distribution yields $L_{disk} \cos i$, where i is the inclination of the disk axis to the line of sight; this quantity and the estimated T_{max} determine $R_{in}^2 \cos i$ using Eq. (4). The projected rotational velocity $v \sin i$ constrains the quantity $(M_*/R_{in})^{1/2} \sin i$.

The inclinations of the Fuors are not really known. Adopting values of $i \sim 45°$–$60°$, the resulting stellar parameters for most Fuors are $M_* \sim 0.1$–$0.5\ M_\odot$ and $R_{in} \sim 4$–$6\ R_\odot$, typical values for low-mass T Tauri stars. The derived parameters of Z CMa are $M_* \sim 1.0$–$2.0\ M_\odot$ and $R_{in} \sim 10$–$20\ R_\odot$, but these values are quite uncertain due to the difficulty in interpreting the spectral energy distribution of this object (Hartmann et al. 1989).

Once M_* and R_{in} are known, $\dot{M}$ is estimated from L_{disk} (Eq. 4). The total amount of mass accreted during an eruption ΔM then follows from $\dot{M}$ and the length of an eruption. Estimates made in this way indicate that Fuor disks are more massive than the minimum-mass solar nebula. The central star in FU Ori is estimated to have accreted $\Delta M \sim 10^{-2}\ M_\odot$ since the outburst began in 1939. Parameters are more uncertain for Z CMa because we do not know how long this object has remained bright. The estimated accretion rate $\dot{M} \sim 10^{-3}\ M_\odot\ yr^{-1}$ suggests that roughly $0.1\ M_\odot$ has been added to the central

central star since the middle 1800s. Presumably, the disk was initially more massive than the amount accreted.

A more general estimate for the amount of accreted material can be made from the local star formation rate ($\sim 10^{-2}$ kpc^{-2} yr^{-1}; Miller and Scalo 1979), assuming all stars go through Fuor episodes. The observed Fuor rate since 1939 suggests that each low-mass star accretes on average ~ 0.05 M$_\odot$. The total accreted mass per star increases to ~ 0.15 M$_\odot$ if Z CMa is included, demonstrating that large uncertainties are present due to small-number statistics.

Other (more model-dependent) constraints on disk properties can be made from an analysis of Fuor variability, in much the same way as in studies of dwarf nova accretion disks. As suggested by D. Lin, suppose that the rapid evolution from a T Tauri disk into a Fuor disk is caused by a sharp transition front which propagates inwards from outer disk regions. The optical emission arises from an area having a radius of roughly 10 R$_\odot$, so the radial infall velocity must be ~ 0.2 km s^{-1} to achieve the observed rise time of ~ 1 yr. The surface density of the disk follows from $\dot{M}$ and is roughly $\Sigma \sim 10^5$ g cm^{-3}. This estimate for Σ is at the high end of the range employed in standard models for the solar system disk (see, e.g., Ruden and Lin 1986).

The effective viscosity of the disk ν is related to the drift velocity, v_r, and the disk radius by

$$\nu\Sigma = \frac{\dot{M}}{3\pi}[1 - (R_*/R)^{1/2}] = \frac{2\pi R \Sigma v_r}{3\pi}[1 - (R_*/R)^{1/2}]. \qquad (5)$$

If we adopt the standard "α-model" for the viscosity, then $\nu = \alpha c_s H$, where c_s is the sound speed and H is the local vertical scale height. The properties of Fuor disks suggest $\alpha \sim 10^{-2}$ to 10^{-3} (Clarke et al. 1990).

If we assume that v_r is roughly constant in the disk, then the $\Delta M \sim 10^{-2}$ M$_\odot$ accreted by the central star in FU Ori in the past 50 yr originated in a region with outer radius ~ 2 AU. The actual outer radius for the accreted material is probably somewhat smaller than this estimate, because v_r is more likely to decrease with increasing radius than to remain constant with radius. Nevertheless, this simple calculation suggests that the FU Ori disk contains a mass comparable to the minimum mass solar nebula inside the orbits of the terrestrial planets.

Crude estimates for disk midplane temperatures also can be obtained from the surface density. If the disk behaves as a simple plane-parallel atmosphere, then the temperature should scale with optical depth τ as $T^4 = T_o^4(\tau + 2/3)$ (Mihalas 1978). Using the above estimates for Σ, models for the interior temperature distribution suggest $\tau \sim 10^4$ (Clarke et al. 1990), which would imply that the midplane temperature should be ~ 10 times larger than the surface temperature, at least in the inner disk ($R < 2$ AU; see Fig. 2).

The high midplane temperatures predicted by these disk models suggest a possible explanation for the apparent absence of boundary-layer emission in Fuors. The boundary layer is the region where inner disk material rotating in nearly Keplerian orbits comes to rest onto the (presumed slowly rotating) central star (Lynden-Bell and Pringle 1974). Simple boundary-layer theory predicts high-temperature emission, because the boundary-layer is assumed to have a small surface area and must radiate a total energy comparable to the rest of the accretion disk. FU Ori and Z CMa show no evidence for the hot (3×10^4 K) boundary-layer emission predicted by the simple theory (Kenyon et al. 1989). However, the boundary-layer theory also assumes that the disk is spatially thin ($H/R \sim 0.02$), whereas the models calculated by Clarke et al. (1989,1990) suggest that the disk may be much thicker, $H/R \sim 0.2$ (see also Lin and Papaloizou 1985). If the boundary layer has a scale height comparable to that of the inner disk, then any boundary-layer radiation might well be "swallowed" or absorbed by the thick disk.

Put another way, the 3×10^4 K boundary-layer temperature implied by the simple theory is *less* than the predicted inner disk midplane temperatures (at least 4×10^4 K, for $\alpha = 1$; Clarke et al. 1989). Thus it is not surprising that some modification to the simple boundary-layer picture is required. The basic boundary-layer theory should be much more applicable to the low-accretion T Tauri state, when the disk is much colder and thus much thinner. Indeed, simple boundary layer models appear to work quite well for T Tauri stars (Chapter by Basri and Bertout).

The nature of the mechanism(s) producing the accretion outbursts is (are) not clear, and will probably remain uncertain for some time to come, in view of our poor understanding of viscosity. One point of importance is that the relatively large masses inferred for Fuor disks

$$M(\text{disk}) > \text{total mass accreted} \sim 0.05 \ M_\odot \qquad (6)$$

suggest that gravitational instabilities might play an important role in driving Fuor accretion (see also Lin and Pringle 1987; Adams et al. 1989; Shu et al. 1990; Chapter by Adams and Lin). Rapid Fuor accretion might occur only when the disk is gravitationally unstable, which requires a minimum disk mass. If infalling material is currently landing on the outer disk, as suggested by Kenyon and Hartmann (1991), outbursts could be produced as long as the accretion rate though the inner disk is not closely tied to the rate at which mass is added from the infalling envelope. If the inner disk accretion rate is small enough, then the disk surface density could increase slowly until a critical mass is reached, at which point accretion becomes very rapid and the disk drains onto the central star. This process could repeat as long as material continues to fall onto the outer disk.

If the disks are gravitationally unstable, the instability almost certainly occurs in the outer disk, outside the region responsible for the optical emission. Many Fuor lightcurves show an initially rapid rise followed by a slow decay (Fig. 6).

By analogy with dwarf nova systems, this behavior suggests an "outside-in" transition wave, in which an instability in the outer portions of the disk

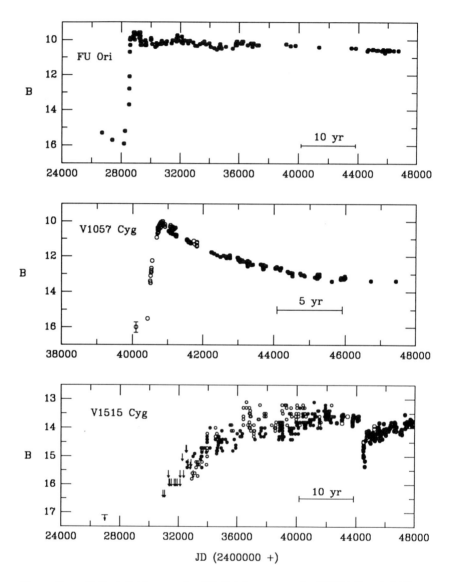

Figure 6.   Optical lightcurves for FU Ori (upper panel), V1057 Cyg (middle panel), and V1515 Cyg (lower panel). The data have been compiled from various sources: V1057 Cyg (Kenyon et al. 1988; Simon and Joyce 1988 and references therein); V1515 Cyg (Herbig 1977; Landolt 1977; Tsvetkova 1982; Kolotilov and Petrov 1983); FU Ori (Herbig 1977; Kolotilov and Petrov 1985; Kenyon et al. 1988). Five magnitudes corresponds to a factor of 100 in brightness.

moves inward. The rapid rise in optical light in this picture is produced by the propogation of a transition wave to the inner disk at (roughly) the local sound speed, while the slow decay follows from the length of time required for material to fall in from the outer disk. A gravitational instability could trigger such behavior (see, e.g., Adams et al. 1989). Alternatively, a close passage of a companion star might similarly perturb the outer disk (A. Toomre, personal communication). An important observational feature of an outside-in eruption is that the mid-infrared lightcurve should begin to rise prior to the optical lightcurve, as illustrated very schematically in Fig. 11 of Kenyon and Hartmann (1988). Perhaps some future Fuor outburst will be identified initally at infrared wavelengths.

Not all Fuors show the same type of light-curve. In particular, V1515 Cyg has exhibited a slow rise lasting many decades (Fig. 6). This evolution is similar to that expected from an "inside-out" eruption which begins near the inner edge of the disk and propogates to the outer disk. Aside from a slow rise in optical brightness, V1515 Cyg has also exhibited a pronounced minimum near JD 2444000 (1979–1981; see Fig. 6). The factor of 6 decrease in B light occured in ≤ 300 days, and the system returned to its maximum brightness after several years. The observed variation in the overall brightness, $\delta B \sim 1.75$ mag, and the color index, $\delta(B - V) \sim 0.4$ mag, are consistent with a simple increase in the reddening along the line of sight. Thus, this event may have resulted from an occultation of the source by an external dust cloud, rather than an intrinsic change in the structure of the disk.

Thermal disk instabilities, which result when the local heating rate is more temperature sensitive than the local cooling rate, also are a possible outburst mechanism (Lin and Papaloizou 1985; Clarke et al. 1989,1990). Unfortunately, thermal instabilities tend to occur in a narrow range of disk radii and result in relatively short-lived eruptions. These "standard" models break down if the disk scale height becomes large, because radial transport of energy then becomes an important feature of the calculation. Clarke et al. (1989) showed that radial transport can act to stabilize the disk against transitions to cooler (lower luminosity) states and produce eruptions which remain luminous for many decades.

Much theoretical work remains to be done before Fuor outbursts are understood. On the observational side, improved technology should yield simultaneous optical and infrared photometry of outbursts, which will enable us to study accretion variability over a much wider range of disk radii than before, providing additional clues to the mechanisms of outbursts.

VII. FUOR WINDS AND BIPOLAR FLOWS

A. Disk Winds

In addition to the phenomena we have directly associated with disk accretion, Fuors have very powerful winds (Herbig 1977; Bastian and Mundt 1985). The mass-loss rate in the best studied case, FU Ori, is $\dot{M}_{\mathrm{wind}} \sim 10^{-5} M_\odot \, \mathrm{yr}^{-1}$ (Croswell et al. 1987). This mass-loss rate is very much higher than that observed from T Tauri stars (see Chapter by Edwards et al.), suggesting that the enhanced outflow of Fuors is related to the accretion outburst. Currently, it is thought that the winds of T Tauri stars derive their energy from accretion (Cabrit et al. 1990; Hartigan et al. 1990*b*), and the Fuors may represent the extreme limit of mass loss generated by rapid disk accretion.

The amount of energy involved in Fuor winds is substantial. As the mass-loss rate of FU Ori is roughly 10% of the mass accretion rate, and the observed terminal velocity of the wind, V_{wind}, is comparable to the escape velocity at the stellar surface, the kinetic energy in the outflow is $\gtrsim 10\%$ of the accretion luminosity. Thus, the mass loss mechanism must be able to convert accretion energy into outflow kinetic energy very efficiently.

Radiation pressure cannot be responsible for Fuor mass loss, because the surface temperature of the disk is only $\sim$6000 K. The low temperature inferred for material in the wind (also roughly 6000 K) implies that thermal pressure gradients are not important in driving the flow (Croswell et al. 1987). Thus, the most likely wind acceleration mechanism is some type of action by magnetic fields.

Rapid rotation is an obvious consequence of disk accretion and may have a crucial effect in driving mass loss. If the magnetic fields are sufficiently strong, gas tied to rotating field lines can be flung outward with high efficiency (Blandford and Payne 1982; Pudritz and Norman 1983; Chapter by Königl and Ruden), even if the wind is very cold (Hartmann and MacGregor 1982). Two variations of this basic picture can be envisaged, as illustrated in Fig. 7.

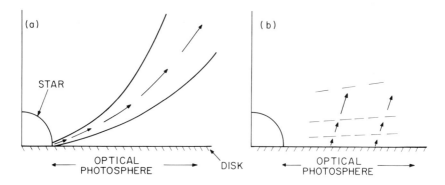

Figure 7.   Two alternative wind models for Fuors. In (a), the wind comes from the near-equatorial regions of the rapidly rotating central star. In (b), the wind originates directly from the surface of the (rapidly rotating) accretion disk. Only the meridional velocity vectors are indicated.

In the first picture, the central star rotates nearly at breakup velocity as a result of the accretion of high-angular momentum material from the disk (Shu et al. 1988). Material is then flung outward from regions near the stellar equator. In the second picture, material is accelerated directly from the surface of the disk along magnetic field lines (Blandford and Payne 1982; Pudritz and Norman 1983; Königl 1989; Lovelace et al. 1991). Note that Fig. 7 denotes only the poloidal or meridional velocity component; a large azimuthal velocity component also is present which is not shown.

Blueshifted absorption lines are formed in the outwardly accelerating wind material seen in projection against the inner disk, where the optical continuum emission originates. Although the very strong $H\alpha$ line in FU Ori exhibits large blueshifted absorption, indicating wind velocities of up to 300 km s^{-1}, lines of moderate strength show much smaller velocity shifts, $\sim$5 to 10 km s^{-1} (Hartmann and Kenyon 1985; Welty 1990). These small blueshifts are difficult to explain by a stellar wind seen against a disk (Fig. 7a). In most wind models, acceleration to a velocity of order the escape velocity generally occurs within a few stellar radii. Thus, low outflow velocities $\lesssim$0.1 of the escape velocity should occur only near the star, which is seen in projection against a very small fraction of the optical disk photosphere. Any low-velocity absorption should therefore be quite faint, greatly diluted by light from the extended optical disk. On the other hand, a disk wind model (Fig. 7b) can easily explain the small blueshifts, because the low-velocity wind regions near the disk surface can extend over the entire optical photosphere of the disk. In this model, the high-velocity absorption arises in strong lines formed many scale heights above the disk surface.

Observations of Fuor winds have several other important implications for understanding the mass-loss mechanisms:

1. The outflows are at least partly collimated *as they are accelerated.* This tentative conclusion is based on the lack of high-velocity, redshifted emission in any Fuor. An isotropic wind would produce much more redshifted emission than is observed, unless all Fuors are observed nearly pole-on (see the discussion in Croswell et al. 1987).

2. The mass-loss rate from disk regions at $r \sim 10\,R_*$ appears to be smaller than the mass-loss rate deduced from optical absorption lines, which are formed at $r \sim 2\,R_*$ (Kenyon and Hartmann 1990). (The outer disk mass-loss rate computed by Kenyon and Hartmann from the $2\,\mu$m CO lines is probably too small because the distribution over rotational states was neglected in the optical depth calculation.) The estimates of mass-loss rates from infrared studies should improve dramatically in the next few years as high-resolution infrared spectrometers with array detectors become available.

3. The wind probably must be mechanically heated to keep temperatures sufficiently high to produce $H\alpha$ absorption (Croswell et al. 1987). Mechanical heating is also suggested by the presence of strong Mg II emission lines in FU Ori and Z CMa (Ewald et al. 1986; Kenyon et al. 1989). MHD waves are an attractive possibility for producing this heating, and the momentum deposited by such waves near the disk might also assist in initiating the mass loss (cf. Blandford and Payne 1982).

B. Bipolar Flows and Jets

Many young stars are observed to have bipolar outflows and/or jets (see Chapter by Edwards et al.). Accretion-powered winds provide a natural energy source to power such bipolar outflows (see, e.g., Pudritz and Norman 1983; Chapter by Königl and Ruden). We suggest that Fuor disk accretion-driven winds are responsible for many energetic bipolar flows, for the following reasons:

1. The energetics of some Herbig-Haro (HH) objects and jets seem to require very dense outflows, and Fuor winds are the most massive directly observed outflows among pre-main-sequence objects. A causal connection between Fuor winds and HH objects depends upon the (unknown) duty cycle, but even a single "short" burst of mass loss at $\sim 10^{-5}$ $M_\odot$ yr^{-1} for 10^2 yr, as in FU Ori, can still produce a mass-loss rate of $10^{-7} M_\odot$ yr^{-1} averaged over a typical flow dynamical time of 10^4 yr.
2. Fuors are commonly associated with jets and HH objects (Reipurth 1989b). The most famous bipolar flow source, L1551 IRS5, is spectroscopically similar to other Fuors (Mundt et al. 1985; Carr et al. 1987; Stocke et al. 1988), although it may be a relatively low-luminosity member of the class (Table I). The HH57 emission nebula was known before the associated Fuor erupted (Graham and Frogel 1985), and a jet has recently been discovered in Z CMa (Poetzel et al. 1989).

Mass outflows from young stars should be episodic if accretion is highly variable, and is related to mass loss (Reipurth 1989c). The identification of time-dependent outflows can be complicated. Multiple emission knots (shocks) in an outflow do not necessarily require a time-dependent wind, because continuous jets can produce a variety of shocks within the jet and along its surface (see, e.g., Norman et al. 1985; Lind et al. 1989; Blondin et al. 1989; Cantó et al. 1989). However, continuous jet models have several difficulties. The spacing of knots along the HH 34 jet is too small to be explained with simple crossing shocks along a continuous jet (Cantó et al. 1990), and may arise from nonsteady flow (Cantó et al. 1990). Multiple ejections have been proposed to explain the large linewidths observed in HH 32 (Solf et al. 1986; Raga 1986; Hartigan et al. 1986) and in L1551 (Stocke et al. 1988).

Perhaps the clearest argument for multiple ejections arises from cases of multiple bow shocks (cf. Reipurth 1989b). In a simple jet, the bow shock is the strong, rounded outer shock, which compresses and heats the ambient medium. Bow shocks are relatively simple to identify both because of their distinctive shape and their well-known kinematic properties. A later ejection plowing into the wake of an earlier bow shock will have similar characteristics to the first, but its emission will be displaced from rest velocity of the ambient cloud by an amount equal to the radial velocity of the wake of the initial ejection.

Reipurth (1989*a*) found two bow-shaped structures in HH 111 and suggested that multiple ejections provide the best explanation for the flow. The HH 47 jet provides particularly strong evidence for the multiple ejection picture because detailed kinematic information is available. The outflow in HH 47 arises from a single, isolated globule and can be studied with excellent spatial resolution without confusion from overlapping material ejected from other sources (Fig. 8; Hartigan et al. 1990*b*).

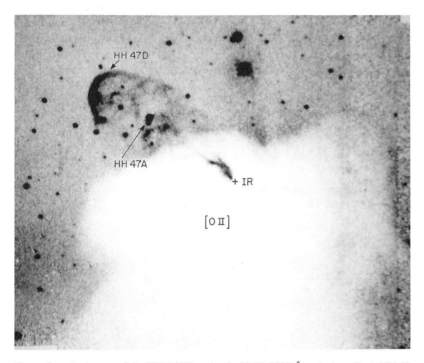

Figure 8.   An image of the HH 46/47 region in [O II] 3727 Å emission. Both HH 47 A and D have shapes and velocity shifts consistent with bow shocks, suggesting multiple ejection events from the embedded young stellar object (see text).

Spatially resolved Hα and [Sec. II] emission-line profiles of HH 47D show a hook-shaped position-velocity diagram which shows a striking similarity to the predictions of bow shock models. HH 47A has the shape and velocity characteristics of a second ejection; in particular, its line profiles are shifted from zero radial velocity by the amount expected for a bow-shock moving into the wake of HH 47D (Hartigan et al. 1990*b*); see also Meaburn and Dyson 1987; Reipurth 1989*c*). The elapsed time between the two ejections is roughly 2000 yr.

A combination of a disk wind and a stellar wind is another viable model for HH 147. In this model the region bounded by the outer bow shock (HH 47D) is filled with a poorly collimated disk wind, and a highly collimated jet from stellar wind produces the bow shock HH 47A. Although this model differs conceptually from a multiple ejection model, the two scenarios are similar kinematically in that both have a high-velocity collimated flow moving through a less collimated, slower flow. A similar model has been used by Stocke et al. (1988) to explain the Herbig-Haro objects in the L1 551 outflow.

Simple, continuous jet models are hard pressed to explain the kinematic data for HH 47 and are likely to fail in other regions as well (see the discussion in Hartigan et al. 1990*b*). As more examples of multiple ejections are identified, it should become feasible to determine the frequency of multiple ejections in the collimated outflows of pre-main-sequence stars. To the degree that accretion and outflows are connected, it may then be possible to trace the accretion history of a young star by studying shocks in the outflowing wind. Such analyses offer the opportunity for enhancing our understanding of the mechanisms responsible for episodic accretion in young stellar objects.

PART III
Disks and Outflows

THE OCCURRENCE AND PROPERTIES OF
DISKS AROUND YOUNG STARS

STEVEN V. W. BECKWITH
Cornell University

and

ANNEILA I. SARGENT
California Institute of Technology

Between 25 and 50% of pre-main-sequence stars in nearby dark clouds have detectable circumstellar disks. Characteristics such as radial temperature distribution and total mass are accurately defined by their spectral energy distributions. The outer disks extend as far as several hundred AU from the stars; some exhibit nearly Keplerian rotation curves at these large radii. The global properties of such disks often resemble those attributed to the primitive solar nebula, suggesting that conditions appropriate for planet formation commonly accompany the birth of low-mass stars. Disk masses, between 0.001 and 1 $M_\odot$, are generally lower than those of the stars, and may represent only a fraction ($\lesssim 10\%$) of the total system mass. The paucity of near-infrared radiation from some disks implies that in the inner regions there are gaps where the opacity from small particles becomes vanishingly small. Coagulation into large particles may have reduced the opacity coefficient without reducing the total amount of material. The temperatures in the outer parts of the disks (tens of AU from the stars) are too high to result from accretion and stellar radiation alone, although it is energetically possible for these processes to produce the observed luminosities. The total disk luminosity—dominated by far-infrared radiation—is, in fact, correlated with optically detectable inner disk activity, presumably the result of accretion and mass loss. For stars younger than 10 Myr, there is little indication of evolution of disk mass or luminosity. There is evidence that gaps in the inner disks develop preferentially in the oldest objects, suggesting that, with time, matter is lost or accumulates into large particles such as planetesimals, which cannot yet be detected. Studies encompassing a wider range of ages and environments are still required to establish unambiguously the time scales for planet formation around other stars.

I. INTRODUCTION

The title of this book presumes a link between protostars and planets. All plausible theories of the origins of the solar system require the existence of an extended disk of gas and dust around the young Sun before planets formed. Circumstellar disks are also expected as a natural byproduct of stellar birth in molecular clouds. Our planetary system most likely evolved from such a disk, the primitive solar nebula, with almost all the residual gas disappearing before the proto-Sun reached the main sequence over 4 Gyr ago. Disks, then, are the thread connecting protostars and planets; the detection and study of disks around very young stars is a necessary step to ascertaining if other planetary systems are born in the same manner as our own.

Insight into the potential number of planetary systems in the Galaxy can be obtained from the rate of occurrence of disks around other stars. More importantly, a knowledge of the properties of these disks may well provide details about the very early evolution of the solar system. Such details are essential to a complete understanding of planetary formation and are unlikely to be uncovered from the ancient record in meteors, comets, the planets themselves or their moons. For example, the rate at which the early solar nebula evolved depended on the density, distribution, temperature and viscosity of the gas. These properties can be observed directly only in disks around other stars. The origin of the outermost giant planets remains a mystery, and the role of magnetic fields, turbulence, convection and gravitational waves in planet formation is perhaps open to observational tests through an examination of nearby disks. Quite apart from the potential impact on planetary formation theory, the identification and study of disks around young stars should lead to significant advances in the field of early stellar evolution.

With modern observational techniques, it is possible to detect even tenuous circumstellar disks and characterize their gross properties—mass, luminosity, temperature structure and size. Comparing these properties in large samples allows insights into factors that affect the disk evolution. Since most of the techniques have become available only within the last ten years, this is a propitious time to review the observations. Here, we discuss the occurrence and properties of disks, emphasizing in particular how these may relate to planet formation and the evolution of the solar system.

II. IDENTIFICATION OF DISKS

While there was general enthusiasm for the *concept* of protostellar, possibly pre-planetary, disks at the the the time of the Protostars and Planets II book, there had been no direct identifications of such objects, let alone an understanding of their properties (Harvey 1985). Circumstantial evidence for their existence was, however, considerable. The presence of optical jets (Mundt 1985) and bipolar outflows of molecular gas (Snell et al. 1980) emanating from the vicinity of young stellar objects (YSO's) implied axially (*not* spherically) symmetric distributions of the circumstellar material. Polarization measurements showing that these jets and ouflows were generally aligned with the magnetic field direction (Bastien 1982; Hodapp 1984) were also interpreted in terms of a flattened circumstellar particle distribution (Elsässer and Staude 1978). The lack of correlation between interstellar reddening and infrared excess suggested that even visible pre-main-sequence objects, such as T Tauri stars, still have associated circumstellar material (Rydgren and Cohen 1985).

For these, the emission line profiles show preferentially blue-shifted wings, suggesting the presence of disks which occlude the red-shifted emission (Appenzeller 1983). Since T Tauri stars are of approximately solar mass, with pre-main sequence ages between 10^5 and 10^8 yr, they are ideal candidates to search for examples of the primitive solar nebula in its early evolution.

Such studies as those described above have been improved upon and extended over the last six years. Other chapters in this volume (see Strom et al.; Backman and Paresce; Edwards et al; and Basri and Bertout) affirm that the indirect evidence for disk-like structure around pre-main-sequence stars is now overwhelming. As we describe below, observational characteristics which *imply* the existence of disks can be exploited to provide statistical samples of possible pre-planetary nebulae, and to ascertain their gross properties. At the same time, more detailed information on a smaller number of individual disks can be obtained through *direct* observations.

A. Direct Images

To date, only a few disks have been detected directly in images which trace the density distributions of the gas or dust. These represent merely the tip of the iceberg; the next years should see an abundance of images of potential proto-planetary disks.

The most dramatic evidence for circumstellar disks orbiting pre-main-sequence stars comes from aperture synthesis mapping of mm-wave continuum and molecular line emission. The advent of mm-wave interferometers capable of imaging molecular species such as CO on seconds-of-arc scales has opened a new era for the direct study of disks. At these long wavelengths, the cold gas at distances of a few hundred AU or more from a star radiates strongly enough to be detected in the low-order rotation lines of a number of molecules. At shorter wavelengths accessible from the ground ($\lambda \lesssim 20$ μm), the transition energies require rather hot gas for significant excitation. Sufficiently hot gas exists only within a few AU of the stars, subtending angles of a few millisec of arc in the nearest star-forming regions, far too small to be resolved by the largest instruments currently available.

Figure 1 shows a particularly clear example of a circumstellar disk seen in ^{13}CO images of HL Tauri made with the Owens Valley Radio Observatory (OVRO) mm-wave interferometer (Sargent and Beckwith 1987,1991). The highly flattened structure beautifully illustrates the emission from a nearly edge-on disk. The velocities of the gas as a function of distance from the star fit a simple Keplerian rotation curve surprisingly well (Fig. 2), implying that the gravitational potential is dominated by the star. With a radius of $\sim$2000 AU and mass ≈ 0.1 $M_\odot$, there is no doubt that HL Tau is surrounded by a disk similar in its gross characteristics to the primitive solar nebula.

Several other low-mass stars show varying degrees of disk morphology (cf. Sargent 1989, and references therein), probably reflecting different inclinations to the line of sight. Among the more impressive is IRAS 16293–24, shown in Fig. 3, an embedded object in the nearby (160 pc) Ophiuchus dark

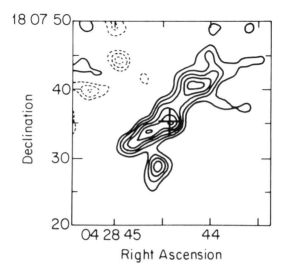

Figure 1.   The emission from the $J = 1 - 0$ rotational transition of ^{13}CO in a single 2.6 km s^{-1} channel at 2.7 mm from the disk around the T Tauri star HL Tauri. At the 160 pc distance to this star, the ~15″ extent of the disk corresponds to ~2000 AU. The highly flattened distribution is perpendicular to the axis of mass outflow from HL Tau and defines an edge-on disk.

cloud. OVRO and VLA aperture synthesis maps of IRAS 16293–24, in C^{18}O and NH$_3$, suggest a moderately massive disk bound by stellar gravity (Mundy et al. 1986b,1990). Higher-resolution images of the 2.7 mm continuum emission show that this structure is comprised of two separate components and is possibly a young binary system (Mundy et al. 1992). The stars L1551 IRS5, DG Tau, and T Tau also exhibit resolved, disk-like structures in OVRO molecular line maps (Sargent et al. 1988; Padin et al. 1989; Sargent and Beckwith 1989; Weintraub et al. 1989a). Owing to the large investments of time required for even one map, the number of stars observed with this technique is small. Nevertheless, these interferometric images make a strong case for the presence of disk-like structures extending several hundred AU or more around young stars of solar mass.

As the stars age, the disks become optically thin, first at infrared wavelengths and finally at visual wavelengths. Short wavelength observations are essential to detect these tenuous disks whose low optical depths may result either from an actual decrease in the amount of material or, more probably, from coagulation of the grains into large particles and planetesimals. The dramatic image of β Pictoris (Smith and Terrile 1984) demonstrates the power of optical images to uncover very small amounts of solid particles in orbit around older (~0.1 Gyr) stars. In addition, analyses of the optical spectra can provide much-needed information about the velocity fields of material around

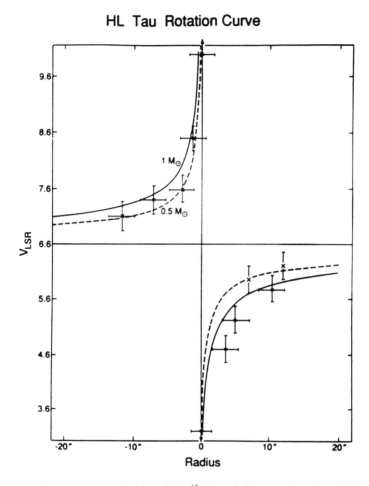

Figure 2. The measured velocities of the ^{13}CO emission as a function of distance from the star along a NW-SE line in the plane of the disk shown in Fig. 1. The solid lines are Keplerian orbital velocities expected at the projected distances around a 0.5 M$_\odot$ star.

such main-sequence objects (see, e.g., Chapter by Backman and Paresce). It is hoped that, even with degraded optical performance, the Hubble Space Telescope can observe the scattered optical light from the entrained particles around many young stars with resolution better than 10 AU.

B. Indirect Identification

Most disks around young stars are identified indirectly. Fortunately, these identifications are usually unambiguous, making it possible to carry out surveys. The essential points to bear in mind are (cf. Beckwith et al. 1990):

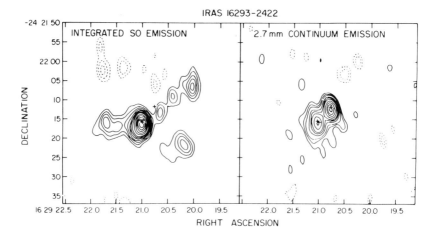

Figure 3. Integrated SO J = 2,3–2,1 line emission (left) and 2.7 mm continuum
emission (right) from IRAS 16293–2422 from Mundy et al. 1992. The beam for
both maps was 4.5 × 2.5 seconds of arc at a position angle of 13°6. The crosses
mark the position of the cm wavelength emission sources.

1. In the mixture of gas and dust prevalent in the interstellar medium
   (gas:dust = 100:1 by mass), the dust provides all the continuous opacity
   at temperatures below ~1000 K.
2. As cited below, many T Tauri stars emit strongly at far-infrared and mm
   wavelengths; the opacities imply total masses of order 0.01 $M_\odot$ (gas and
   dust) or more.
3. A quantity of 0.01 $M_\odot$ of gas and dust uniformly distributed in a sphere
   50 AU in radius will have a visual optical depth of ~500.
4. The visual optical depth of obscuration in front of T Tauri stars is almost
   always <3.

The first two points suggest that disks will emit at infrared wavelengths if
they are heated; radiation from the central stars is sufficient to heat the disks
in almost all cases. The last three statements are reconciled only if the stars
are surrounded by substantial quantities of matter which is distributed in
such a manner as to fill very little of the solid angle around a star. This is
most easily realized if the matter orbits the star in a thin disk. As planar
configurations result naturally in most scenarios for cloud collapse and star
formation, the observation of long wavelength emission from *visible* stars
provides compelling, albeit indirect, evidence for the presence of circumstellar
disks.

A standard mixture of gas (mostly hydrogen) and dust with a total mass
of 0.01 $M_\odot$ (minimum solar nebula), spread uniformly throughout a circular
disk, 50 AU in radius (solar system dimensions), will be opaque at all wave-

lengths shortwards of 100 μm. We expect the disk temperature to *decrease* at increasing distances from the star. Near the stellar surface ($\sim$0.01 AU), it will be close to the several thousand degree stellar photosphere. It drops below 1000 K within 10 stellar radii, approximately 0.1 AU from the star. At the radii where terrestrial planets might originate, 0.2 to 2 AU, the temperatures are a few hundred degrees, and emission is primarily at the thermal infrared wavelengths, 5 to 20 μm. At larger distances, 2 to 50 AU, where the gaseous planets form, the disk is quite cold, radiating in the far-infrared and submillimeter range. As a result, the wavelengths from the thermal infrared and beyond are most important for identifying radiating disks which are potential sites for planet formation.

Because very little matter will make a disk optically thick at infrared wavelengths, the spectral energy distribution shortwards of 100 μm depends entirely on the temperature distribution and is insensitive to the density structure and total mass (Lynden-Bell and Pringle 1974; Pringle 1981; Adams et al. 1987; Beckwith et al. 1990). In general, a power-law radial temperature dependence, $T(r) \propto r^{-q}$, leads to a power law spectral energy distribution, $\nu F_\nu \propto \nu^\alpha$, where $\alpha = 4 - \frac{2}{q}$; $T(r)$ is the temperature at distance r from the star, and F_ν is the flux density at frequency ν. Both the shape and normalization of νF_ν are uniquely determined by $T(r)$. The normalization depends on a high power of T; typically, $\nu F_\nu \sim T^4$. We have ignored any azimuthal or vertical temperature variations but will return to those points.

The *minimum* temperature that a disk will have is that established by heating from the star; a flat disk will be colder than a flared disk, for example, because it subtends a smaller solid angle. At large distances from the star, the disk temperature, $T(r)$, is given by

$$T(r) = \left(\frac{2}{3\pi}\right)^{\frac{1}{4}} T_* \left(\frac{r}{R_*}\right)^{-\frac{3}{4}} \tag{1}$$

where T_* and R_* are the stellar temperature and radius, respectively (Adams et al. 1988). A flat, opaque disk which extends from the stellar surface to at least a few AU will emit at the very minimum the radiation implied by this thermal distribution. According to the discussion above, this emission will have a frequency dependence $\nu F_\nu \propto \nu^{\frac{4}{3}}$ and a total luminosity one-fourth of the stellar luminosity. In theory, a flat disk heated solely by viscous accretion at the midplane would have the same radial temperature dependence, $T(r) \propto r^{-\frac{3}{4}}$, but the emission would, of course, only add to that stimulated by stellar radiation.

Skrutskie et al. (1990) have shown that for most nearby T Tauri stars, at wavelengths of 10 μm or less, this minimum emission is easily within reach of the largest telescopes. Likewise, Strom et al. (1989d) and Cohen et al. (1989b) have identified numerous disks from the IRAS database using measurements out to 100 μm. Our ability to detect circumstellar disks is therefore quite good, limited in most cases only by the amount of material in orbit around the stars.

It is useful to recall that the opacity for interstellar grains (typical size <0.1 μm) *decreases* as the wavelength *increases.* Even as particles grow, the opacity should be small at long enough wavelengths. Although it requires very little dust to make a disk optically thick in the near infrared ($\lambda \sim 1$ μm), rather a lot is needed to make the opacity greater than unity at 1 mm wavelength.

We expect, then, that at mm wavelengths the disks will be predominantly transparent. Owing to the low optical depth, the thermal emission depends equally on the material temperature and the column density (or, for a spatially unresolved disk, the total mass). This long wavelength emission is in the Rayleigh-Jeans limit. Where the surface density $\Sigma(r)$ is small enough to be optically thin, its intensity is given by the integral of the product of temperature and density over the surface of the disk

$$\nu F_\nu \propto \int \kappa_\nu(r)\Sigma(r)T(r)2\pi r\,\mathrm{d}r \qquad (2)$$

where the mass opacity, $\kappa_\nu(r)$, is the most important unknown factor. Crudely speaking, νF_ν is a product of the total mass and the characteristic temperature, modified by a weighting factor that depends only on the *form* of the density and temperature distributions. As this weighting factor remains relatively constant over a plausible range of distributions of $\Sigma(r)$ and $T(r)$, the disk mass, M_d, can be established from the mm-wave flux. Beckwith et al. (1990) and Adams et al. (1990) have identified disks and determined masses from such measurements. The major uncertainty in M_d arises in the value adopted for the mass opacity coefficient κ_ν (see also Beckwith and Sargent 1991).

The opacity coefficient is usually assumed to be independent of r. It need not be: coagulation in the inner, dense parts of the disk may greatly decrease κ_ν relative to its value in the outer regions. This effect is minimized by estimating M_d from long wavelength observations because the opacity coefficient should be independent of particle size as long as the particles are smaller than the observing wavelength (see, e.g., Hildebrand 1983). In fact, the expected change in κ_ν with grain size might be turned to advantage by using short wavelength observations to search for disks where coagulation or clearing has already taken place.

The temperatures described above are characteristic of an ordinary T Tauri star. During periods of strong heating by rapid accretion, or by FU Orionis-type eruptions (see Chapter by Hartmann et al.), the disks may become considerably warmer, and the matter distribution may not fall off smoothly with distance from the star. Given the paucity of observations of such phenomena, these circumstances are probably rare. Thus, as we have argued, it is relatively straightforward to identify disks around typical pre-main-sequence stars of approximately solar mass, provided M_d is sufficiently large, $\gtrsim 0.01$ $M_\odot$.

III. THE OCCURRENCE OF DISKS

As discussed above, surveys for disks around pre-main-sequence stars may involve searches for either infrared or mm-wave emission. In either case, the arguments presented above show that the detection of such emission implies planar matter distributions. Four such surveys include large enough populations ($\gtrsim 50$) to permit significant evaluation of the frequency with which disks occur. Although, in principle, optical searches can discover relatively tenuous disks, no such searches have yet been undertaken.

Cohen et al. (1989b), Strom et al. (1989d), and Kenyon et al. (1990) examined the IRAS database for far-infrared emission from stars in the Taurus-Auriga dark clouds. Between 25 and 50% of the objects are detected in the far-infrared, implying a rather high probability for disk formation around solar mass stars. Kenyon et al. (1990) searched for pre-main-sequence stars not previously known to be associated with the clouds and so were able to estimate the completeness of the earlier surveys. They conclude that surveys for pre-main-sequence stars in Taurus-Auriga are essentially complete for luminosities $L > 0.5\,L_\odot$. Almost all solar-mass stars younger than ~ 10 Myr fall into this category. The stars in these surveys are predominantly of the classical T Tauri type, with strong hydrogen emission lines, easily identifiable in objective prism plate surveys of dark clouds (Joy 1949; Herbig 1962b).

Beckwith et al. (1990) searched 86 stars in Taurus-Auriga for the 1.3 mm thermal radiation indicative of disks; André et al. (1991) made a similar survey of stars in Ophiuchus. Millimeter-wave observations have three distinct advantages over those in the far infrared; they may be made from the ground; the attainable spatial resolution is superior (11 arcsec vs ≥ 1 arcmin), and they are sensitive to the *mass* of emitting particles (as described in Sec. II). Disks were uncovered in approximately 40% of these samples, an almost identical detection rate to the far-infrared surveys. These samples are also biased toward the classical T Tauri stars.

Disks and potential pre-solar nebulae are evidently common around classical T Tauri stars. The best available evidence for completeness of the samples (Kenyon et al. 1990) indicates that the detection rate is unlikely to drop below 20%, even if few disks are found in more complete samples of pre-main-sequence stars. It is, in fact, likely that with improved observational sensitivity even more disks will be seen; many detections occur near the flux density limits of the extant surveys. Consequently, there is an ample supply of candidates to examine for possible planet formation. This is perhaps the biggest change since the Protostars and Planets II volume.

Another class of pre-main-sequence stars exhibits X-ray emission but *no* strong optical lines (Montmerle et al. 1983a; Walter 1986a; Feigelson et al. 1987). Initially called "naked" T Tauri stars (Walter 1986), and more recently referred to as weak-line T Tauris, these objects share the important characteristics of their classical sisters: ages <0.1 Gyr, masses near that of the Sun, locations in dark clouds such as Taurus-Auriga, but no strong emission-line signatures. Since T Tauri stars are identified most readily in objective prism surveys, a population of pre-main-sequence stars with very weak Hα emission could easily go unnoticed. Despite the initial claim that there may be as many as 10 weak-line stars for every classical T Tauri star (Walter et al. 1988), it appears that weak-line stars are actually in the minority (Kenyon et al. 1990; Hartmann et al. 1991).

The samples of Strom et al. (1989) and Kenyon et al. (1990) contain quite a few weak-line T Tauris. Many of these have far-infrared flux distributions indicative of circumstellar disks. For the limited number of weak-line T Tauris observed by Beckwith et al. (1990) at 1.3 mm, the detection rate was 30%. On this evidence, it is clear that the weak-line stars are *not* necessarily "naked." Indeed, they are quite likely to support disks; the masses of weak-line T Tauri disks are as large as those surrounding classical T Tauri stars (a few percent of a solar mass). Indeed, the disk densities may differ little between the classical and weak-line stars, although there is still insufficient evidence to reach strong conclusions.

On the other hand, none of the weak-line stars have very luminous disks. The infrared emission is in all cases relatively small. Part of the difficulty with searching for disks around weak-line stars is that they are faint at all wavelengths where the disk radiation dominates that from the star. There is certainly some connection between the strength of the emission lines and the properties of the disks, but "naked" appears to be a misnomer.

In our view, the major difference between the classical and weak-line stars is the level of energetic activity such as accretion, which occurs predominantly in the boundary layer between the disk and the star. There appears to be a good correlation between the disk luminosity, its accretion rate, and the strength of the emission lines (Chapter by Edwards et al.). Because the far infrared radiation depends on a high power of the local disk temperature, a lower level of heating will markedly decrease the infrared radiation emitted by the disk until it reaches the lower limit set by stellar heating. The visual emission lines originate in some fairly high temperature region, the boundary layer or the wind. Their intensity can have little bearing on the mass of the disk, which depends largely on the cool material extending many AU from the star.

Yet the accretion rate (or "energetic activity") could depend on the reservoir of material. A low accretion rate may well signal the depletion of at least the inner regions of a disk, and so the class of weak-line stars could be important to understanding the full range of behavior in circumstellar disks. To assess the actual rate of planet formation, complete samples of pre-main-sequence objects are essential to statistical surveys. Over the next few years, more extensive searches for disks around the weak-line stars using long wavelength ($\lesssim 1$ mm) emission as a diagnostic will help elicit the variety of circumstances leading to the formation of a circumstellar disk.

IV. THE PROPERTIES OF DISKS

A. Disk Temperatures

Because of the nearly one-to-one relationship between disk temperatures and infrared emission, the infrared spectra imply specific radial temperature distributions. The observed spectral energy distributions are consistent with power-laws in frequency, thus implying temperature distributions which are power-law in disk radius. It is convenient to write $T(r) = T_1 \left(\frac{r}{1\text{AU}}\right)^{-q}$ and discuss the results in terms of the power-law index q, and the temperature at 1 AU, T_1. For purely stellar heating of disks, $q = 0.75$ and T_1 is ~ 100 K for a typical T Tauri star.

For their sample, Beckwith et al. (1990) found disk temperatures, T_1, ranging from 50 to 400 K, the average being 120 K. The majority of the disks have total luminosities smaller than those of their central stars but larger than expected from a disk heated only by stellar radiation. The hottest, and therefore, most luminous disks radiate more energy than their central stars by factors of 3 or more. Some additional heat input is required to explain these large luminosities; the prevailing view attributes the difference to accretion, but this is by no means firmly established.

Power-law indices q range from ~ 0.5 to 0.75 among the Beckwith et al. (1990) sample of *visible* stars, where there can be little uncertainty about the extinction corrections. A large number of disks have $q \sim 0.5$; the number rapidly decreases with increasing q. Only a few objects show spectra consistent with pure accretion or radiative reprocessing, $q = 0.75$; the temperatures in most of the disks decline more slowly with radius. There is no generally accepted explanation for the high temperatures in the outer parts of the disks. At least four quite different theories have been put forward to explain the low values of q: heating by wave dissipation of $m = 1$ instability modes (Shu et al. 1990); heating by spiral density waves in the disks (Chapters by Adams and Lin and Shu et al.); heating by the central star of flared disks (Kenyon and Hartmann 1987); and heating by winds emanating from the central regions (Safier and Königl 1992). The small values of q which lead to flat spectral energy distributions do not yet have an obvious explanation from simple considerations of disk physics. There is a correlation between q and T_1 in that the hottest disks all have $q = 0.5$. Evidently, whatever provides the additional heat input for those disks preferentially heats the outer radii (Beckwith et al. 1990).

The *observed* temperatures at radii of interest for planet formation, 0.5 to 20 AU, are cool, well below melting temperatures of refractory particles and often low enough to condense water ice from the vapor phase. It should be noted that high disk opacities may require temperature gradients in the direction perpendicular to the plane of the disk, and the midplane temperatures might be higher than those observed at the disk "photosphere." Nevertheless, these surface temperatures are low. Although it is possible that the disks might experience brief periods of high temperature from, for example, rapid accretion "events" associated with FU Orionis eruptions (Chapter by Hartmann et al.), these events cannot occupy more than a tiny fraction of the disk lifetime, judging by the infrequency of FU Orionis stars among large samples. Therefore, conditions within the disks are favorable for the accumulation of solid planetesimals, consistent with theories of the early solar system.

It is noteworthy that the high temperatures in the inner disks, $T > 500$ K for $r \lesssim 0.3$ AU, are almost entirely responsible for near-infrared radiation between 2 and 10 μm. In some cases, this radiation is much weaker than expected from an optically thick disk heated by the central star (i.e., the *minimum* amount of radiation from a disk). The most straightforward conclusion is that the inner regions are optically thin and, by implication, virtually devoid of small particles (Skrutskie et al. 1990). Either gaps have developed (cf. Chapter by Strom et al.), or the particles have clumped into larger bodies whose opacity per unit mass is exceedingly low. Such would be the case during the early phases of planet formation.

There are other ways to explain the decreased near-infrared flux: flared disks viewed at high aspect angles can exhibit decreased emission from the inner regions (i.e., at near-infrared wavelengths) with little diminution of the outer disk radiation (S. J. Kenyon 1990, personal communication). Each concept merely adds another free parameter to the flat disk model. But there *is* economy of hypothesis in assuming that the inner disk opacity drops as the disk evolves. It is here that the densities are highest, promoting coagulation, and that dynamical time scales are shortest, implying a means to clear material relatively quickly. Gaps could be caused by the formation of small bodies whose gravitational perturbations make the orbits of particles unstable. The onset of coagulation into particles much larger than 10 μm. would serve equally to diminish the opacity and create the appearance of a gap. In principle, these cases can be distinguished by observations of the *gas* in the inner regions. The appearance of gaps may be the first vital signs of nascent planets, signaling an important leap in our ability to witness the birth of a planetary system.

B. Disk Masses

Derivations of disk mass depend on the relationship between the optically thin mm emission, the disk temperature, and the mass opacity. The long wavelength emission varies almost linearly with the disk temperatures; these can be determined rather accurately from the infrared spectra, as discussed above. The principal uncertainty in the mass estimates is introduced by the mass opacity κ_ν. Nevertheless, if we assume κ_ν does not vary much from disk to disk, the distribution of disk masses may be studied without precise knowledge of the normalization.

There have been several attempts to determine disk masses from mm-wave observations, the most extensive being that of Beckwith et al. (1990). More limited samples observed by Adams et al. (1990) and Weintraub et al. (1989b) are generally consistent with the larger survey results, although the derived disk masses are higher owing to their different choices for κ_ν. Beckwith et al. used an opacity coefficient approximately 5 times larger than that adopted by Adams et al. (1990), who had adopted the value proposed by Hildebrand (1983) and his co-workers. The range of κ_ν is reflected in the derived masses, with the Beckwith et al. results biased toward the low-mass side.

Within these uncertainties, disk masses range from ~ 0.001 $M_\odot$ to approximately 1 $M_\odot$, with an average value near 0.03 $M_\odot$. Thus, the vast majority of disks detected in the mm-wavelength surveys are more massive than the minimum solar nebula, 0.01 $M_\odot$ (see, e.g., Weidenschilling 1977b), and have sufficient material to form a planetary system like our own. At the lower end, the mass distribution is incomplete, limited by the sensitivity of the observations. A population of low-mass or low-temperature disks could have gone undetected. On the high end, there are very few disks which are comparable in mass to their central stars; in all cases, the uncertainties allow for disk masses much less than 1 $M_\odot$. In one case (HL Tauri), constraints from the observed rotation curve of the gas at a few hundred AU suggest somewhat less mass than the estimate from the mm-wave thermal emission.

There are, as yet, no clear correlations between disk mass and other stellar characteristics such as stellar age (up to ~ 10 Myr), stellar mass (range 0.2 to 2 $M_\odot$), Hα equivalent width, and stellar luminosity (Beckwith et al. 1990). There are modest uncertainties in the stellar characteristics which do not exclude subtle effects, but pronounced trends (e.g., disk mass proportional to stellar mass) should have been seen in the present samples. There *is* a small inverse correlation between disk mass and disk temperature: hotter disks tend to have less mass. This correlation is probably an artifact of the mass derivation procedure, in which the mm-wave flux density is divided by the product of the average temperature and a calibration factor involving the opacity.

We can imagine that significant coagulation will change κ_ν with time and will vary from disk to disk. The resulting scatter in derived M_d could mask evolutionary trends. The onset of coagulation *lowers* κ_ν, and hence *decreases* the derived disk mass. If coagulation occurs preferentially in older disks, we still expect a correlation between the derived M_d and stellar age. None is seen, showing that significant coagulation does not always occur within the first few Myr.

Only a few of the weak-line T Tauri stars have been observed adequately to determine disk masses. Even for these, there is no clear trend of decreasing disk mass with decreasing Hα emission. There is, as expected, no obvious correlation between the high-energy indicators of a disk (Hα emission, X-rays, near-infrared excess) and its mass, most of which is probably an AU or more from the star (see Sec. III).

C. Evolutionary Characteristics

Rather massive disks surround many pre-main-sequence stars. Around main sequence stars, such as Vega and β Pictoris, there are no known *massive* disks ($M_d \sim 0.01$ M$_\odot$.), or at least none comprising small ($\lesssim 1$ mm) particles. The particle masses *observed* around main-sequence stars are orders of magnitude smaller than those detected around T Tauri stars (Backman and Gillett 1987; Chapter by Backman and Paresce). The disks must therefore evolve, in the sense of losing their small particles and (perhaps) associated gas. The process which drives this evolution is central to the issue of planet formation, since the buildup of large bodies naturally causes the disks to become transparent. But planet formation is not the only way: accretion onto the star or expulsion by strong winds may deplete the material independently, *preventing* the formation of large, solid bodies.

As described above, there is no obvious evolution of disk mass among a sample of stars <10 Myr old. Examination of either the average disk mass versus age or the fraction of stars with detected disks as a function of stellar age fails to demonstrate a decrease of disk matter within the first few Myr of a disk's lifetime. This result is interesting for two reasons. First, some disks must persist long enough for accumulation into planetesimals to begin, according to our best current theories for planet formation. The evidence shows that planet formation can certainly occur around some of these stars, and it could emerge as the dominant way to lower the disk opacity. Second, winds are most vigorous early in the T Tauri stage and decrease markedly after a few Myr. If disks persist well through the phase of strong mass loss (and, presumably, accretion), it is difficult to see how a T Tauri wind can later destroy a disk (Beckwith et al. 1990; Chapter by Edwards et al.). Although such a wind is commonly included in theories of early solar evolution as a means of clearing light elements from the solar nebula after the formation of the giant planets (Horedt 1978; Elmegreen 1978b), we have no evidence that *observed* winds from T Tauri stars can effect the clearing.

This interpretation presupposes that the disk matter is not continuously depleted, through accretion onto the star and mass loss in a wind and replenished, via accretion from the surrounding dark cloud, throughout its lifetime. If the very high mass loss rates inferred for some stars ($\dot{M} \sim 10^{-7}$ M$_\odot$ yr^{-1}; Chapter by Edwards et al.) persist for several Myr *and* most of this mass originates from the circumstellar disks, the observed reservoir of material, $\lesssim 0.1$ M$_\odot$, is incompatible with the total mass lost, unless the disk mass is continually replenished. Of course, it is not clear if high mass loss rates are continuous or episodic, and the material could originate from the star itself. A greater understanding of the mass loss process is clearly needed. The high mass-loss rates are cited to buttress the case for high-mass disks (Shu et al. 1990).

Skrutskie et al. (1990) present good evidence for a decrease in the material in the *inner* regions of disks, the gaps discussed above, for stars older than 10 Myr. Old T Tauri stars generally display a paucity of near-infrared emission, characteristic of particulate matter within a few tenths AU of the stars. In future research, we might hope to examine these stars over a wide range of wavelengths to understand how the more distant disk material evolves. Apparently, 10 Myr is approximately the time scale for significant changes in disk structure, a time scale which is not seriously at odds with our theoretical notions of planet formation. By focussing attention on these older stars, there is a good chance that we can quantify the evolutionary changes in sufficient detail to constrain our picture of planetesimal growth.

We have concentrated so far on the *particles* in the disks because they are responsible for the strong infrared and mm-wave emission. We might well ask whether the *gas* evolves on the same time scales. Gas certainly dominates the mass of the youngest disks, such as HL Tau (Sargent and Beckwith 1987), but there are no statistical assessments of gas content in the older disks. In principle, gas content can be observed directly in rotational transitions of the trace molecules such as CO using mm-wave interferometers. A survey for gas disks is an excellent future project for arrays with many ($\gtrsim 5$) telescopes but is impractical for the three-element arrays presently available. Gas might also be detected in absorption in disk "photospheres" with the next generation of infrared space telescopes such as SIRTF. The essential point is that the expected cool temperatures at disk radii a few AU from the stars mean gas signatures will appear at wavelengths $> 10 \ \mu$m.

D. Chemical Composition: Particle Opacities

The mass opacity of particles, κ_ν, is of more than passing interest because it is primarily responsible for the assessment of disk mass and because compositional changes, mirrored by changes in κ_ν, might yield valuable clues to the processes taking place during the growth of large particles in planet-building nebulae. There is considerable debate about the mass opacity for interstellar dust at mm and sub-mm wavelengths (Mathis et al. 1977; Hildebrand 1983, Draine and Lee 1984; Wright 1987). Compounding the problem is the observation that grain size distributions certainly vary among common interstellar environments (diffuse ISM, molecular clouds, condensations within clouds), and it is not known how the disk environment will affect the size and nature of growth of solid particles. Perhaps the only consensus is that the opacity of grains will decrease when the wavelength exceeds the grain size, but the exact form and value of that decrease is poorly known and difficult to determine.

Hildebrand (1983) showed that the sub-mm opacity of small ($\lesssim 100 \ \mu$m.) particles depends only on the total *volume*, and therefore mass, of the material, regardless of size distribution, *if* the particles are convex—that is, the ratio of volume to surface area scales linearly with the size of the particles. All spheroidal shapes (including needles, cylinders, pancakes, etc.) fall into this category. It appears also to be true for particles of fractal dimension (ratio of volume to surface area increases as size to the power α, where α is < unity) so long as the wavelength exceeds the largest constituents (Wright 1987). As a first approximation, we can ignore the size distribution so long as most of the mass is in particles smaller than the wavelength of observation. This removes much of the freedom in choosing a bulk opacity coefficient.

The complex index of refraction is then directly related to bulk opacity through the Kramers-Kronig relationship. In principle, measurements of the indices of refraction for the kinds of materials which should constitute astrophysical particles lead to an accurate characterization of the bulk opacity. Unfortunately, some of the most important constituents are poorly understood (e.g., amorphous silicates) and often lack measurements at suitable wavelengths. In many cases, the index of refraction is temperature dependent. The bulk opacities for interstellar particles are most uncertain at the mm-wavelengths of interest here.

This uncertainty has generated debate about the masses for disks derived from observations of the thermal emission. Until very recently, the standard opacities were assumed to fall as the inverse square of the wavelength, conventionally characterized as a power law: $\kappa_\nu \propto \lambda^{-\beta}$, with $\beta = 2$. The constant of proportionality is estimated (observed is, perhaps, too strong a word) at a wavelength of 350 μm in dark clouds and extrapolated to longer wavelengths (Hildebrand 1983). The resulting curve fits reasonably well with theoretical estimates of the behavior for a mix of silicate and graphite material (Draine and Lee 1984), subject to the rather large uncertainties inherent in these estimates. The bulk opacity at 1.3 mm given by Draine and Lee (1984) is 0.002 cm^2 g^{-1}. Hildebrand (1983) indicates an opacity of 0.004 cm^2 g^{-1}, and most other authors give higher values. The value could be enhanced by factors as large as 30 by conducting particles of fractal dimension, that is if the particle volume increased more slowly than the cube of the size (Wright 1987). It is perhaps worth noting that Meakin and Donn (1988) favor the formation of fractal particles in their models of the early solar nebula.

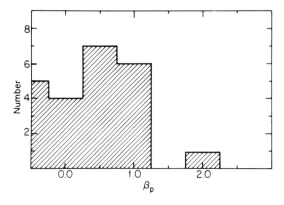

Figure 4. A histogram showing the distribution of sub-mm spectral indices ($F_\nu \propto \nu^{\beta+2}$) derived for 23 disk sources in Taurus-Auriga (Beckwith and Sargent 1991). The index β is the power-law assigned to the opacity coefficient κ_ν discussed in the text.

The wavelength dependence may be checked from sub-mm observations of disks, for example, but there is no straightforward means of establishing the constant of proportionality which is chiefly of interest. It is now apparent that most of these objects have solid particle emission with a weaker wavelength dependence than assumed in conventional models of interstellar particles (Beckwith et al. 1986; Sargent and Beckwith 1987; Woody et al. 1989; Adams et al. 1990; Weintraub et al. 1989b; Beckwith and Sargent 1991). Figure 4 is a histogram of the power-law β for a sample of 19 disks in Taurus-Auriga taken from Beckwith and Sargent (1991) showing the propensity toward $\beta \lesssim 1$. If we keep in mind that disks of high optical depth will appear to have $\beta = 0$, we might interpret the graph as showing a preference for $\beta = 1$, the distribution resulting from optical depth effects. This figure shows the standard opacity laws might need to be modified for the problem at hand, without indicating how the overall coefficient will change at 1.3 mm.

Pollack (personal communication) recently pointed out that long wavelength laboratory measurements have been made for a number of relevant materials, notably amorphous silicate, water ice, and graphite. At 1.3 mm, silicate will dominate the bulk opacity (assuming cosmic abundance) and will yield an index $\beta = 1$. By his estimate, the opacity coefficient at 1.3 mm is nearly identical to the Hildebrand value, but with a different power-law dependence. Different materials may dominate the opacity at different wavelengths, however, and water ice should become most important somewhere around 300 μm with a $\beta = 2$ dependence. Thus, it may be possible to assess the relative abundance of different materials through careful observations of the sub-mm spectrum, once good laboratory data are available.

If we adopt Pollack's value for κ_ν at 1.3 mm, the disk masses derived by Beckwith et al. (1990) should be raised by a factor of 5, making the average disk mass 0.15 $M_\odot$ for the objects in Taurus-Auriga. The larger disk masses alleviate the difficulty posed by inferred high accretion rates and lend credence to the ideas put forward by Shu and his collaborators regarding critical disk masses to make wave action in the disks important (Shu et al. 1990). The uncertainties in the opacity are not necessarily smaller, but the reasons for assigning a value to κ_ν are, perhaps, on firmer ground (Beckwith and Sargent 1991).

V. CHALLENGES

At the time of *Protostars and Planets II*, a circumstellar disk was a theoretical concept required to explain the primitive solar nebula, but almost nothing was known about the existence or properties of disks around young stars. It is safe to say that the observational evidence for disks around many stars is now overwhelming, and we have assessed the gross characteristics (mass, temperature, size, velocity field) for a relatively large sample. Some important refinements must await the availability of very high angular resolution from mm arrays and the Hubble Space Telescope. In the interim, there are several

areas in which observers and theorists might make progress, and we pose
these in the form of challenges to the respective camps.

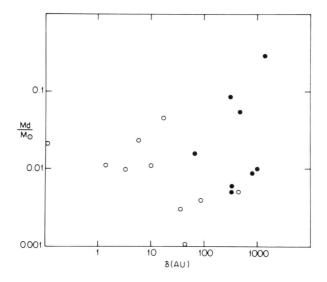

Figure 5.   The disk mass is plotted versus companion separation for 17 stars known
to be in binary systems. Filled circles denote those disks detected at 1.3 mm; open
circles indicate non detections, so the disk mass is an upper limit only. In many
cases, infrared radiation from these non detections indicates the presence of disk
material, but in smaller amounts than occur for the stars detected at 1.3 mm.

A. Observational Challenges

1. Binary Stars and Disks.   There is weak statistical evidence that close
companion stars do inhibit the formation of a disk. Zinnecker (1990*a*) presents
a review of known multiple star systems among low-mass, pre-main-sequence
stars with separations ranging from a few stellar radii to a few thousand AU.
Figure 5 is a plot of disk mass versus component separation for those systems
searched for circumstellar disks. The filled circles show stars detected at
1.3 mm, therefore having disk signatures; the open circles indicate upper
limits. Almost every system with component separations $\lesssim$100 AU has low
mm-wave flux, indicated by the upper limits to disk mass, whereas all the
large separation systems are detected. Unfortunately, the upper limits are
only slightly smaller, on average, than the detections, so it is too early to make
a strong statement about the dynamics. Yet the result is encouraging from a
theoretical point of view. Gas orbits will be unstable at circumstellar radii
of the same order as component separation in a multiple star system, and we
expect to see a decrease in disk mass when the separations become smaller
than the size of a typical disk, probably of order 100 AU.

The effect of binary separation on disk mass (or disk radius) is open to observational test. Speckle interferometry at near-infrared wavelengths has already demonstrated the binary nature of such pre-main-sequence objects as T Tauri (Dyck et al. 1982), Z Canis Majoris (Koresko et al. 1989), and XZ Tauri (Haas et al. 1990). The ever-continuing refinements to this technique which will enable systematic searches for pre-main sequence binaries at intermediate separations from a few to a few hundred AU should considerably enlarge the sample. The high resolution afforded by HST is an important part of these searches. Equally important, very deep observations of these systems at sub-mm wavelengths will detect low-mass disks, should they exist. The results will bear on the relative roles of disks and companions serving as reservoirs for the energy and angular momentum during the formation of a star.

2. *Interior Gaps.* Many of these stars have low 5- and 10-μm flux densities, yet detectable emission at 60 and 100 μm. It is very difficult to account for these spectral energy distributions unless the regions close to the stellar photospheres are devoid of small particles. The gaps apparently occur more frequently in stars older than 10 Myr, suggesting a clearing with age (Skrutskie et al. 1990). There are probably several ways to clear gaps in the inner disks; one of the more attractive and interesting is dynamical clearing through the gravitational influence of planet-sized bodies in the inner disks. Another is to deplete the small particles into larger bodies through coagulation, thus decreasing the opacity coefficient without changing the actual matter density.

Identification of gaps requires sensitive photometry in the thermal infrared, and this has already been carried out for limited samples of stars. Broadening these samples ought to quantify the stellar and disk characteristics which correlate well with clearing and firmly establish the evolutionary status of the gaps. Furthermore, it will be useful to affirm that gaps are cleared in the *gas* distribution as well as the distribution of small particles, as expected if the matter density is actually changed.

Theories of solar system formation usually have Jupiter forming first, clearing a narrow gap at 5 AU. At present, we lack the observational tools to distinguish a gap in the outer parts of disks, since the spectral energy distributions will change very little. It is conceivable that such gaps could be seen with long-baseline interferometers working in the thermal infrared regime. The development of aperture synthesis instruments of this sort could provide potentially powerful ways to examine planet formation in great detail.

3. *Disks Around "Older" Pre-Main-Sequence Stars.* We know the disks have largely disappeared by the time the stars reach the main sequence. The faint remnants seen around Vega and β Pictoris can be only tiny fractions of the original disks. The time scale for disk dissipation and the mechanism: stellar winds, accretion onto the stars; or creation of large, orbiting companions is open to observational test in the post-T Tauri stars. In all likelihood, it will be necessary to go beyond the proximate Taurus-Auriga and Ophiuchus clouds to identify significant numbers of these transition objects.

The Hubble Space Telescope is likely to be a potent instrument for study of tenuous disks around older pre-main-sequence stars, those between 10 and 100 Myr old. No longer opaque at visual wavelengths, the particles in these optically thin disks will scatter light from the central stars, enabling detection in high-dynamic range images. Although the amount of scattered light is small, the high optical mass opacity should make it possible to see very low-mass disks.

4. Particle Composition. At sub-mm wavelengths where most of these disks are optically thin, the spectral energy distributions reveal the wavelength dependence of the particle opacity. The large aperture sub-mm telescopes now in operation should provide a good sample of the distributions in the limited windows accessible from the ground. A large airborne telescope, such as SOFIA, or a cryogenic space telescope, such as SIRTF, might determine the continuous energy distributions for at least the brightest disks.

With continuous energy distributions, it may finally be possible to identify the substances which dominate the opacity in different wavelength regimes. Pollack's preliminary estimates suggest that silicates will dominate beyond a few $100 \, \mu$m and water ice at shorter wavelengths, the relative contributions seen through the turnover wavelength, where we expect a change in the slope of the opacity coefficient. It is conceivable that large particles in older disks will be detected through their signature in the sub-mm spectra.

B. Theoretical Challenges

1. Gaps and the Influence of Companions. Just as the observers must refine the empirical evidence for gaps in the inner disks, so must the theorists address the issue of how different gap-clearing processes can be differentiated. Clearing by small companions is a particularly interesting case, since it is a natural byproduct of planet formation. The challenge will be to understand if the influence of planetesimals has uniquely observable characteristics or if a single means of creating gaps is so superior to all others as to provide strong indirect evidence of its handiwork. An assessment of coagulation is similarly interesting and, in principle, subject to observational test.

2. Heating of Disks. The almost inescapable conclusion from the observed far-infrared flux distributions is that the outer disk temperatures are elevated above those expected from the simplest ideas about disk heating. If the opacities are very high, it is possible that vertical temperature gradients bias the observational results to reflect the midplane temperatures improperly, where internal heating presumably occurs. As described above, we are aware of at least four quite different ways to heat the outer disks. The strongest observational clue, that the heating is most effective when the accretion is vigorous, perhaps favors disk flaring or heating by the winds, but there are other reasons for rejecting such scenarios. At this writing, the luminosity redistribution (since the luminosity should be dominated by processes occuring in the *inner* disk) appears to be an important unsolved problem.

3. Persistence of Disks. Disks persist for at least several Myr. Yet the time scales to begin significant growth of planetesimals are relatively short, perhaps a few 10^5 yr or less. It is not clear whether this is a real problem or only reflects the uncertainty in modeling particle growth in the presence of complicating factors such as turbulence and electromagnetic forces. Persistence time scales are potentially important for assessing the means by which volatiles are lost in the latter stages of planet creation. Perhaps the ever-increasing observational constraints will allow a refinement of theory in ways useful to understanding our own origins.

   4. Disk Dissipation: Alternatives to Planet Formation. In the same vein, there is an apparent mismatch between the standard model—disruption by a "T Tauri" wind—and the observation that the strongest winds shut off before the disks lose very much of their mass; nor is there evidence for a recurrence of energetic winds which could destroy the rather robust disks persisting beyond a few Myr (Chapter by Edwards et al.). Are their other or better ways of destroying these older disks? Or is planet formation our best hope?

Acknowledgments. Research by the authors and the work going into this review is supported by the NSF and NASA; the authors are grateful to these agencies for their continued support of this burgeoning field.

T TAURI STARS AND THEIR ACCRETION DISKS

GIBOR BASRI
University of California, Berkeley

and

CLAUDE BERTOUT
Institut d'Astrophysique de Paris

The T Tauri stars are solar-type stars in their pre-main-sequence phase. The main distinguishing characteristics of the classical T Tauri stars are excess continuum emission from the ultraviolet to the far infrared and strong emission in selected lines, particularly Hα. These characteristics are thought to be due to both mass influx and mass loss in the systems. Such flows are now thought to be due to the presence of an accretion disk around the star—the final phase of star formation and possible precursor to planet formation. The stars themselves are both more rapidly rotating and more convective than their main-sequence counterparts, which leads to increased magnetic activity, a partner in producing the "T Tauri phenomena." This chapter is focused on the evidence for accretion through a disk as the source of the emission excesses, particularly those seen in the continuum between 0.1 and 10 μm and in the strong permitted emission lines. We present a brief overview of the classical T Tauri phenomena, and a detailed discussion of how accretion-disk models can explain them. We explore their relation to the weak-lined T Tauri stars, and mention some effects disk accretion could have on the evolutionary status of young stars. Finally, we suggest some directions that research in this area should take in the near future.

I. INTRODUCTION

Since the realization that the Hα emission stars first studied by Joy (1945) are really solar-type pre-main-sequence stars, we have been presented with the opportunity to directly observe conditions that may have prevailed at the beginning of our own solar system. The principal defining characteristics of T Tauri stars currently in use (besides their late spectral types) are kinematic association with molecular complexes and presence of strong ($W_\lambda \gtrsim 100$ mÅ) lithium absorption at 6707 Å. Both characteristics presumably reflect the youth of T Tauri stars. Hα emission, the historical defining criterion (Herbig 1962b), was shown over the last few years to signal only the most active fraction of young low-mass stars, now referred to as classical T Tauri stars (CTTS). Many late-type young stellar objects with W_{eq}(Hα) $\lesssim 10$ Å, discovered in X-ray surveys (see Walter et al. 1988; Feigelson et al. 1991), do not stand out at optical wavelengths but are relatively strong X-ray emitters

$(L_X \approx 10^{-3} L_{bol})$. They are called weak-emission line T Tauri stars (WTTS) or sometimes naked T Tauri stars (NTTS), and may be several times more numerous than CTTS. The dividing line between these subclasses is somewhat arbitrary and ill defined; we suggest a physical division in Sec. V. The basic properties of WTTS are reviewed in the Chapter by Montmerle et al. Here, we are concerned mainly with properties and models of CTTS. Extensive reviews of these stars have recently appeared by Bertout (1989) and Appenzeller and Mundt (1989).

CTTS exhibit a number of spectral anomalies. In addition to Hα and Ca II lines, the higher Balmer lines are often also seen in emission. In almost half of the CTTS, there is also forbidden line emission of [O I] and, less frequently, of [S II] (Cohen and Kuhi 1979). In extreme T Tauri stars, i.e., in the strongest emission-line cases, many other emission lines of iron, titanium, sodium, helium and calcium are present. Excess continuum emission (over the continuum flux level of main-sequence stars with comparable spectral types) is also quite common, particularly in the ultraviolet and infrared spectral ranges. The infrared excess is discussed in Sec. II. The ultraviolet excess often carries into the optical range, where it "veils" the photospheric absorption lines, sometimes so heavily that their spectral type cannot be determined from low-resolution spectrograms but only from high signal-to-noise, high-resolution data. Compared to their main-sequence counterparts, these stars rotate rapidly (5 to 30 km s^{-1} or higher), but their rotation is quite slow compared to their breakup velocities (typically 250 km s^{-1}).

CTTS are irregular photometric variables at all wavelengths. The amplitude of the optical variability can be from a few hundredths of a magnitude to several magnitudes, and the lightcurves vary from star to star. The T Tauri class, as defined from spectroscopic characteristics (Herbig 1962b), indeed includes stars belonging to all three classes of nebular variables (see Glasby 1974).

There is a remarkable variety of unusual phenomena in T Tauri stars, which all fall into three basic categories: magnetic (chromospheric) activity, outflow phenomena, and accretion phenomena.

A. Magnetic Activity

From the outset, the superficial similarity of the emission spectrum in T Tauri stars to the solar chromosphere has been noted (Joy 1945). The line profiles, however, are typically much broader and more asymmetric than solar lines. Strong surface magnetic activity is established through a number of different diagnostics. The presence of light modulation with a period of a few days on some stars is thought to be a rotational modulation by sunspot-like concentrations of magnetic field in the photosphere; this interpretation is supported by detailed light-curve models in a few cases which suggest substantial surface coverage (Bouvier and Bertout 1989). The traditional chromospheric diagnostic, Ca II emission, is quite strong in T Tauri stars, and in WTTS the Ca II lines appear very much as they do on magnetically active main-sequence stars. The amount and characteristics of the X-ray emission in WTTS are also entirely consistent with an origin in hot coronae such as those observed in RS CVn systems (see Chapter by Montmerle et al.). This conclusion is reinforced by evidence that the X-ray luminosity is a function of stellar rotation, accepted as a strong indicator of magnetic dynamo activity (Bouvier 1990). Finally, a technique for direct measurements of the total magnetic flux on T Tauri stars pioneered by Basri and Marcy (1991) indicates that it lies at the high extreme of the total magnetic flux seen on young main-sequence stars. Following a suggestion by Herbig (1970), it has been demonstrated that if the stellar chromosphere occurs deep in the atmosphere (at a continuum optical depth at 5000 Å of, say, 0.1 to 1), the fluxes of some emission lines might be reproduced and even the continuum jumps and veiling (Cram 1979; Calvet et al. 1984). Finkenzeller and Basri (1987) showed that differential filling-in of photospheric lines in WTTS is consistent with the deep-chromosphere explanation. The real difficulty with the hypothesis that the radiative excesses seen in T Tauri stars are caused by the presence of a deep chromosphere comes when one evaluates the total amount of excess energy present in the optical spectrum of CTTS with strong emission features. The true stellar photospheric contribution can be estimated by studying the absorption line spectrum (and the extent of veiling thereof). In some cases, the absorption lines are reduced substantially in depth across the spectrum, implying that major fractions or even multiples of the total photospheric luminosity appear as excess continuum light. These large radiative losses cannot be generated by solar-type magnetic dynamo processes alone even if one assumes complete coverage by active regions of the stellar surface (Calvet and Albarrán 1984).

B. Outflow Phenomena

The evidence for mass loss in young stellar objects is covered in the Chapter by Edwards et al. It suffices to recall here that mass-loss rates in the range 10^{-8} to $10^{-7} M_\odot$ yr^{-1} were derived from simple models for the Hα line of a few bright CTTS (Kuhi 1964). DeCampli (1981) was first to emphasize the difficulty for a late-type, low-luminosity star to eject mass at such high rates. He discussed the constraints on outflow models imposed by observations and concluded that although Alfvén-wave-driven winds could conceivably be driving mass losses of a few $10^{-8} M_\odot$ yr^{-1} in T Tauri stars, no known physical mechanism involving the star alone could account for the strongest CTTS winds. Further computations of Hα line profiles as produced by Alfvén-wave-driven winds by Hartmann et al. (1982) demonstrated that turbulent broadening was important in these winds, so that earlier models which neglected this effect systematically overestimated T Tauri mass-loss rates. The fact that only a few CTTS are detected at centimetric wavelengths (see, e.g., Bieging et al. 1984) apparently confirms that their typical mass loss is not greater than, say, a few $10^{-8} M_\odot$ yr^{-1} in the wind's ionized component. But recent analysis of optical forbidden and neutral lines such as those of sodium have led to the belief that T Tauri winds are mostly made up of atomic, rather than fully ionized gas, and the derived mass-loss rates soared again to much higher values (see Chapter by Edwards et al.). The exact mechanism by which the wind is driven out and collimated thus remains a major outstanding question in research on T Tauri stars and embedded young stellar objects.

C. Accretion Phenomena

First evidence for accretion in CTTS also comes from the shape of hydrogen line profiles. Walker (1972) identified a subclass of CTTS that he named YY Orionis stars after their prototype, which displayed (at times) inverse P Cygni profiles at their Balmer lines. Interestingly, another common property of these objects is their strong ultraviolet excess, which we now know is related to accretion (see Sec. III). Walker proposed that these stars were accreting matter, either through spherical infall or through disk accretion. Some of the work done in the framework of the spherical infall picture is summarized in the first *Protostars and Planets* book (Bertout and Yorke 1978). High-resolution spectroscopy later demonstrated that matter outflow was prevalent even in YY Orionis stars (see, e.g., Mundt 1984), and the spherical infall picture was abandoned. However, the evidence that mass accretion takes place in a number of CTTS at velocities in the 300 to 500 km s^{-1} range, i.e., comparable to free fall, is clearly established. Simultaneous accretion and outflow of matter is also hard evidence that a successful model must reproduce. These two constraints originate from studies of YY Orionis stars, which thus appear to be primary test objects for realistic models of T Tauri stars.

Plausible explanations for all the different facets of T Tauri activity have appeared, vanished, and been resurrected several times. Most of the early models that were briefly mentioned above focus on one or the other of the above categories. However, a unifying picture has begun to emerge over the last few years. It involves a magnetically active, solar-type star, surrounded by the circumstellar disk that is the natural outcome of the gravitational collapse of a dense molecular core with nonzero angular momentum. In this picture, both CTTS and WTTS display the characteristic properties of the magnetically active central star, and the specific properties of CTTS result from interaction between the disk and the star. WTTS, on the other hand, have either lost their disks or do not interact with one.

The disk hypothesis has proved fairly successful in explaining many properties of T Tauri stars. The next section provides a review of the current observational evidence for disks, and Sec. III discusses existing accretion disk models compared with observed spectral energy distributions. Section IV deals with spectroscopic tests of the accretion disk model. Finally, Sec. V discusses some of the implications of accretion disks to the evolutionary status of young stars, with Sec. VI giving a summary. The obvious successes of simple disk models should not, however, hide the fact that very little is known about the underlying physics of disks, about the nature of the interaction between disk and star, and about the generation of protostellar winds. The truly successful paradigm which self-consistently explains all observational facts, particularly the relationship between accretion and outflow, has yet to be worked out. Our aim in the following is merely to point out that the successful model is likely to involve an accretion disk. Whether accretion is driven by viscous torques, as envisioned in current models, or by another mechanism is, in our opinion, still debatable.

II. EVIDENCE FOR CIRCUMSTELLAR DISKS

A primary reason supporting a disk geometry rather than a more isotropic distribution of dust comes from the visual extinction towards T Tauri stars. If the line of sight passed through the amount of dust required to produce the observed infrared luminosity, it would produce much greater extinction than is measured by the optical reddening of the T Tauri stars (Myers et al. 1987). The line of sight therefore must miss most of the dust present, which instead absorbs a fraction of starlight not coming towards the observer and then re-emits it thermally at infrared wavelengths into the line of sight. While for a single system one could be looking through a gap or hole in the dust distribution, the fact that this condition applies to most CTTS leaves little choice but to conclude that the dust distribution is substantially flattened.

A. Direct Imaging

At the distance of the closest star-forming regions such as the Taurus molecular cloud, a solar-system-sized disk (say, 100 AU) subtends an angle of $1\rlap{.}''3$. The theoretical resolution of a 4 m telescope at $2\,\mu$m is about a tenth of that. Unfortunately, the portion of the disk which emits thermally in the near infrared is much smaller than 1 AU, so one must rely on detection of scattered light from the whole disk. In order to take full advantage of the increased scattering at shorter wavelengths, one should ultimately employ space instruments like HST. Reaching the diffraction limit on the ground is possible today only in the infrared because the coherence length of atmospheric turbulent elements at these wavelengths is large enough to be corrected for by current technologies (e.g., adaptive optics, interferometry).

Both infrared speckle interferometric and imaging techniques have already begun to provide direct images of circumstellar structures (see, e.g., Beckwith et al. 1985; Monin et al. 1990), but distinguishing between disk and other geometries from deconvolution of the data is difficult. Millimeter-wave interferometry currently provides a resolution of a few arcseconds and has the advantage of directly mapping thermal emission without interference from the star. Sargent and Beckwith (1987) mapped a flattened cool circumstellar structure around HL Tau, but it is much more extended than the solar system. The relationship between this structure and the warmer smaller flattened object "seen" at shorter wavelengths is still unclear. In any case, resolving protostellar disks has become a topic of hot international competition which should bear fascinating results in the next few years. This topic is also dealt with in the Chapter by Beckwith and Sargent.

B. Spectral Energy Distributions

The suggestion that the CTTS might be accretion disk systems was made by Lynden-Bell and Pringle (1974; hereafter LBP), who foresaw and also worked out the main details of the current paradigm. They developed the theory of flat optically thick viscous disks and showed that an infrared continuum distribution (λF_λ) flatter than a single temperature blackbody would result from the disk presence. It arises because the disk will be hotter nearer the star, due both to the increased absorption of stellar radiation by the disk and to the star's increasing gravity which increases the accretion power. The nearby dust contributes near-infrared light and the farther away cooler dust emits at increasingly longer wavelengths. The spectral slope is predicted by the LBP accretion disk model to vary as $\lambda^{-4/3}$. LBP also suggested that if the star were rotating slowly compared to breakup rotation (pure conjecture at the time), a strong ultraviolet excess would be generated in the boundary layer between disk and star. These ideas did not catch on at first, probably because the infrared and ultraviolet evidence that now makes them compelling was not then available. Thus, the previous edition of this volume (Black and Matthews 1984) shows the community feeling that disks may prove a very useful concept, but presents rather little hard evidence for them even at that time. An analysis of IRAS data by Rucinski (1985) then demonstrated that many CTTS had infrared spectral slopes that were more or less compatible with accretion disk models, and detailed analysis of CTTS spectral energy distributions at ultraviolet, optical, and infrared wavelengths confirmed the validity of the disk paradigm.

On the basis of a statistically significant study of the infrared sources in the ρ Ophiuchi star-forming region, Lada and Wilking (1984) suggested the existence of three main classes of young stellar objects according to their near and mid-infrared spectra: Class I stars are optically invisible and their spectra rise steeply to the infrared; Class II stars look basically stellar in the optical but have infrared excesses which decrease more slowly than a blackbody in the infrared; and Class III stars have very little near-infrared excess. Adams et al. (1987) then proposed an interpretation of these classes in evolutionary terms: Class I are protostellar, embedded objects still surrounded by infalling molecular gas; Class II are more evolved objects with circumstellar disks that have expelled their infalling envelopes with the help of their strong winds (i.e., they are the CTTS); and Class III objects are pre-main-sequence objects (the WTTS) that have lost much of their circumstellar material.

All three classes may actually have disks. The presence of a disk in Class I sources is difficult to assess because the infalling cloud is also still present, but spectral energy distributions of these objects have been shown to be consistent with the presence of disks by Adams and Shu (1986). Infrared sources with flat spectra (in the $\log\lambda F_\lambda$ vs $\log\lambda$ plane) are intermediate between Classes I and II. Because many of these are not well extinguished, it is likely that a disk is present. The flatness of the spectrum may come from a nonclassical temperature distribution in the disk (see Chapter by Adams and Lin), or reprocessing from dust out of the disk plane (Kenyon and Hartmann 1987). Alternatively, the overall spectral energy distributions of some of these objects might be composite spectra of several unresolved sources, as demonstrated by Leinert and Haas (1989) for Haro 6–10 and by Maihara and Kataza (1991) for T Tau itself. High-resolution infrared imaging of flat spectra sources will help settle this dispute. Most Class I and II objects, as well as a number of Class III stars (see Chapters by Strom et al. and by Montmerle et al.), are detected in the submillimetric and millimetric continua (see, e.g., Beckwith et al. 1990). This is again consistent with the presence of circumstellar matter but does not imply by itself any given geometry of the dust. Detailed models of spectral energy distributions of Class II sources based on the presence of a disk are presented in Sec. III.

C. Spectroscopic Evidence for Disks

Analysis of the optical and near-infrared line spectra of FU Orionis objects gives evidence for the expected differential rotation in the disk (see Chapter by Hartmann et al.). The cool, molecular disks surrounding CTTS do not allow similar analysis. In these objects, the strongest spectroscopic piece of evidence for the presence of a disk comes from a wind tracer, the [O I] and [S II] forbidden lines that form in low electronic density regions. The shape of these optical forbidden lines is typically asymmetric, with the red wing quite diminished or missing. This has been interpreted by Appenzeller et al. (1984) and Edwards et al. (1987) as the signature of mass loss from the star in which our view of the receding flow is cut off by the presence of a disk (see Chapter by Edwards et al.). With a naive analysis, the size of the disk implied by the emission measure in the forbidden lines is about 100 AU, although it might be substantially smaller if the electron density is smaller due to partial recombination.

D. Polarization Maps

Although the origin of linear polarization in young stellar objects has long been a matter of debate, scattering by dust grains recently emerged as the most likely polarization mechanism. Bastien and Ménard (1990) demonstrated that polarization maps of young stellar objects can be understood in the framework of multiple scattering in a disk/lobe geometry. Both the disk size and its view angle can be determined from a comparison of observed and synthetic maps (Ménard, in preparation). So far, this analysis is restricted to embedded sources surrounded by extensive disks. High-resolution near-infrared polarization maps are, however, becoming possible with the advent of 256×256 detectors and adaptive optics, and will soon make it possible to determine the main physical properties of T Tauri disks independently from other methods.

III. MODELS OF T TAURI STARS

A. Basic Assumptions of Accretion-Disk Models

The current model of a T Tauri star consists of (i) a central late-type star surrounded by (ii) a geometrically thin, dusty accretion disk that interacts with the star via (iii) a boundary layer. Model spectral energy distributions emitted by such systems can be computed and compared to observations. The accretion-disk model devised by LBP and Shakura and Sunyaev (1973) assumes that *local* processes induce a viscous coupling between neighboring disk annuli, thereby transporting angular momentum through the disk. Note that there could be other, more global ways of redistributing angular momentum, e.g., through density waves. These are not considered in this model. Once a physical mechanism for transporting angular momentum is specified, the equations governing the disk structure, as well as its evolution in time, can be written down.

The basic angular momentum transport mechanism considered by LBP is kinematic viscosity. In a disk where the gas is rotating differentially, any chaotic motions in the gas will give rise to viscous forces (shear viscosity). Gas particles moving along two neighboring streamlines at R and $R+dR$ with angular velocities, respectively, $\Omega(R)$ and $\Omega(R+dR)$ have different amounts of angular momentum. Random motions then lead to angular momentum transport resulting in a viscous torque exerted on the outer streamline by the inner streamline. This in turn will result in local dissipation of energy by the gas, and hence to mass accretion at a rate $\dot{M}$ that is assumed constant throughout the disk. This assumption is made for convenience only; there is no *a priori* reason why viscosity should adjust itself in such a way as to make a steady-state disk. If one assumes that the disk is not self-gravitating, then its rotation is quasi-Keplerian, i.e., $\Omega = \sqrt{GM_*/R^3}$, where M_* is the stellar mass and R the distance from the star's center. The hypothesis of a quasi-Keplerian disk can be shown to imply that the disk is geometrically thin. The expression for the energy dissipation rate $D(R)$ at $R >> R_*$ is easily derived from first principles (see Adams et al. 1987). It is

$$D(R) = \frac{\dot{M}}{4\pi} R\Omega \frac{d\Omega}{dR} = \frac{3GM_*\dot{M}}{8\pi R^3}. \tag{1}$$

No assumption regarding the physical properties of the kinematic viscosity is necessary to derive the above equation. If a rotation law other than Keplerian was valid as would be the case, e.g., if the viscous torques were of magnetic rather than kinematic origin, then $D(R)$ would not necessarily be $\propto R^{-3}$.

In order to determine the disk density, one must, however, assume something about the viscosity because its magnitude determines the angular momentum flow. LBP derive the disk density under the assumption that the kinematic viscosity is constant within the disk. Shakura and Sunyaev (1973) derive another analytical solution based on the same basic assumption but viscosity is assumed to be proportional to the local scale height times the local sound speed, with the proportionality constant (called α) being restricted to values ≤ 1. Underlying this formulation are the ideas that turbulent eddies cannot be larger than the disk height and that any supersonic turbulence should rapidly become subsonic because of the formation of internal shocks in the disk. Both viscosity prescriptions are *ad hoc* parameterizations that reflect our ignorance of the nature of kinematic viscosity in disks; they give qualitatively comparable results for the run of density with radius.

The disk's temperature structure determines the emitted spectrum. To compute it, one must make an assumption about the radiative transfer. LBP presented their hypothesis that viscous energy released in a given disk annulus is radiated away through both faces of that annulus and that the radial radiative flux is zero. While justified for the optically thick, geometrically infinitely thin disks envisioned by LBP, this assumption must be abandoned in the more complex radiative transfer approaches that are now being developed.

Also important for the disk temperature is the heating by stellar photons. Note that there are two ways in which the central star influences disk properties. Its mass and radius determine the potential well seen by the disk, i.e., the viscous energy dissipation rate, and hence the disk temperature. But the local disk temperature also depends upon the stellar effective temperature, which together with geometrical factors determines the local rate of heating by the central star. Adams and Shu (1986) and others computed the resulting disk temperature at the photospheric level. At large distance R from the star, and assuming that the disk is flat and infinitely thin, one finds that the local rate of heating due to reprocessing of photons originating from a star with radius R_* and effective temperature T_* is

$$F(R) = \frac{2\sigma T_*^4 R_*^3}{3\pi R^3}. \tag{2}$$

This equation must be modified when the disk is not flat (Kenyon and Hartmann 1987; Ruden and Pollack 1991), or when a finite disk atmospheric structure is taken into account (Malbet and Bertout 1991). At large distances from the central star, the above assumptions lead to the following equation for the disk effective temperature $T_D(R)$:

$$\sigma T_D^4(R) = \frac{3GM_*\dot{M}}{8\pi R^3} + \frac{2\sigma T_*^4 R_*^3}{3\pi R^3} \tag{3}$$

where the first term on the right-hand side represents the viscous energy dissipation rate and the second term takes into account the reprocessing of stellar photons. Note that both terms are proportional to R^{-3} in flat, Keplerian disks. This means that it is difficult to determine whether a disk is passive (purely reprocessing) or active (accreting), from the infrared alone. Only when the infrared luminosity is more that 50% of the stellar luminosity (25% for a flat disk) can one be certain that accretion is taking place (see Hartmann and Kenyon 1987a), but in either case, the presence of a disk is indicated.

Analyses by Strom et al. (1988) and Cohen et al. (1989b) show that at least the most active of the CTTS can be shown to be accreting on that basis.

Equation (3) describes the temperature far away from the central star. When computing the inner disk structure, one must assume something about the inner boundary of the disk, and more specifically about the way angular momentum is transferred from disk to star. LBP imposed the condition that the star exerts no torque on the inner edge of the disk, which also implies the existence of a boundary layer between the slowly rotating T Tauri stars (typically 20 km s^{-1} at the equator; see Hartmann et al. 1986 and Bouvier et al. 1986a) and the inner edge of the Keplerian disk, where matter is circling the star at about 250 km s^{-1}.

The LBP inner boundary condition assumes that the extent b of the boundary layer is much smaller than the stellar radius and that the angular velocity at the inner disk edge is comparable to the Keplerian velocity at the star, i.e., $\Omega(R_* + b) \sim \sqrt{GM_*/R_*^3}$. Energetic properties of the disk/boundary layer system directly follow from this inner boundary condition. One finds for the disk luminosity L_D and the boundary layer luminosity L_{bl}:

$$L_D = \frac{GM_* \dot{M}}{2R_*} = \frac{L_{acc}}{2} = L_{bl}. \tag{4}$$

One thus concludes that half of the total accretion luminosity L_{acc} is advected from the disk into the boundary layer; this corresponds to the Keplerian kinetic energy rate at the inner disk radius. Now, however, this energy must be dissipated in a small region at the star's equator instead of over a disk many AU in extent. This means that the temperature of the radiating region will be far higher. While temperatures in the disk range from 10 K far from the star to about 3000 K near the star, the boundary layer's temperature will be from 7000 to 12000 K (Bertout 1986; Kenyon and Hartmann 1987), and will radiate in the ultraviolet and visible part of the spectrum. As seen below, this is probably the source of the ultraviolet excess and optical veiling observed in many CTTS.

It should, however, be emphasized here that the LBP inner boundary condition maximizes both the boundary layer's luminosity L_{bl} and its temperature. L_{bl} can be reduced in two ways. First, if the star rotates at some angular velocity Ω_*, then the boundary-layer luminosity is reduced to

$$L_{bl} = \frac{GM_* \dot{M}}{2R_*} - \frac{\dot{M} R_*^2 \Omega_*^2}{2}. \tag{5}$$

Second, the boundary-layer size need not be infinitely small. The more general case of a finite boundary layer was recently considered in some detail by Duschl and Tscharnuter (1991), who demonstrated that the fraction of accretion luminosity dissipated in the boundary layer is indeed strongly dependent upon the assumed inner boundary condition. Finally, some of the accretion energy could be released in nonradiative form, e.g., for driving a wind (see Pringle 1989*a*). One should therefore be aware that all comparisons of observed and computed spectral energy distributions done so far assume that half of the accretion luminosity is radiated away in the boundary layer.

B. Comparisons with Observations

Accretion disks and boundary layers in T Tauri systems have been reviewed several times in the last few years (Basri 1987; Kenyon 1987; Hartmann and Kenyon 1987*a*; Bertout 1989; Bertout et al. 1991*a*). Class II infrared sources turn out to be the best examples of "classical" accretion disks. Their infrared spectra are at least as shallow as demanded by the simple flat disk model; most are actually shallower (see Chapter by Beckwith and Sargent). Rydgren and Zak (1987) showed that the average slope of CTTS infrared spectra beyond 5 μm or so is $\propto \lambda^{-3/4}$ rather than $\lambda^{-4/3}$, which probably means that the simple model discussed above needs refinement. Two basic suggestions have been made to explain the fact that the observed infrared spectra are flatter than the LBP disk model predicts.

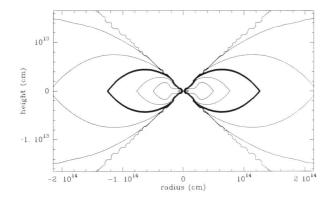

Figure 1.   The optical depth distribution in a typical T Tauri accretion-disk model. The heavy-lined contour corresponds to a Rosseland optical depth of 1, and successive contours differ by 0.5 dex. In the case illustrated here, the optically thick disk has a diameter of about 20 AU.

The first of these assumes that the disk flares up out of the plane at large radii, thus intercepting (and thermally re-emitting) more stellar photons in the outer parts than in a thin disk. Some flaring is expected even in classical accretion disks because the local disk height H determined from hydrostatic equilibrium scales as $R^{9/8}$. Kenyon and Hartmann (1987) showed that such a geometry would partially explain the infrared discrepancy if the disk remains opaque over its full height. However, flared-up disks do not really do an adequate job of explaining the truly flat infrared spectra. Also, the disks need to subtend a large part of the sky as seen from the star in these extreme cases and then one would expect a greater incidence of large extinctions in CTTS. Furthermore, Malbet and Bertout (1991) computed the vertical structure of T Tauri disks in a self-consistent manner and found that the optically thick parts of the disk are confined to regions close to the central plane of the disk even if gas and dust remain mixed together for the required length of time, which is another problem. This result is illustrated by Fig. 1, which displays the optical thickness distribution in a typical T Tauri disk with mass-accretion rate $1 \times 10^{-7} M_\odot$ yr^{-1} as a function of both disk radius and disk height over the midplane. The innermost contour corresponds to Rosseland optical thickness $\log\tau = 1.5$, and successive contours differ by 0.5 dex. The heavy-lined contour corresponds to $\tau = 1$, and thus approximately indicates the physical disk size, here about 20 AU in diameter. In order for the infrared spectrum to steepen up with respect to the usual $\lambda^{-4/3}$ law, the disk should be optically thick all the way out to the outermost contour, which roughly corresponds to $H \sim R^{9/8}$.

The second suggestion, made by Adams et al. (1988), assumes that the temperature distribution in T Tauri disks is flatter than in LBP disks, which implies that the disk's temperature distribution does not result from kinematic viscosity. Following this suggestion, it has become common practice to parameterize the temperature law index and to use different values of this parameter to model infrared distributions of young stellar objects, particularly those with flat infrared spectra (see, e.g., Beckwith et al. 1990). Using the temperature law as a free parameter is equivalent to implicitly assuming that the energy dissipation mechanism in the disk differs from star to star. A possible physical mechanism to accomplish this has been speculated on by Shu et al. (1990), who invoke a lowest-mode gravitational instability which would operate in disks which are more than 0.24 the mass of the star (see Chapter by Adams and Ruden).

At this point, one must conclude that the far-infrared spectral energy distributions of many T Tauri stars do not fully support the hypothesis that T Tauri disks are really classical accretion disks as proposed by LBP, although there are a few stars, such as DF Tau, which fit the classic model very well. A further problem is that the infrared spectrum alone does not allow one to distinguish between passive reprocessing disks and true accretion disks as long as the accretion luminosity is smaller than or comparable to the reprocessed luminosity, i.e., $\dot{M}_{acc} \lesssim 5 \times 10^{-8} M_\odot$ yr^{-1}.

Why then do we believe that accretion disks actually surround most CTTS even when the infrared luminosity excess is not decisive? The decisive argument comes from the blue and ultraviolet spectral ranges. The excess light there has all the properties expected of boundary-layer emission. At the same time, the accretion only occurs near the equator of the star and so avoids the problems that a strong accretion shock over the entire star posed for spherical accretion models. Even more important, the amount of energy available depends solely on the accretion rate and not on the star's resources. This resolves the mystery of how the photospheric spectrum can be so veiled in some cases.

The disk paradigm explains naturally the observed correlation between the respective amounts of infrared and ultraviolet excesses. Even the widespread observations of Balmer continuum emission jumps can easily be explained if the boundary layer is optically thin in the Paschen continuum (Basri and Bertout 1989). Of course, there are still questions regarding the actual geometry and extent of the interface region between disk and star, and to what extent it dissipates energy in nonradiative forms (e.g., material motions). For example, the maximum velocities present in the boundary layer are similar to those seen in the broad emission-line components, suggesting that the broad emission lines are partly formed in the boundary layer and connected regions; this is a topic of current work (see Sec. IV).

There are fairly few free parameters in the simple models that have been analysed so far. Some are associated with the star itself: the stellar mass, effective temperature and radius. In principle, these are all known from the comparison of the position of the star in the Hertzsprung-Russell (HR) diagram, with stellar pre-main-sequence evolutionary tracks. In practice, only the stellar temperature is known with reasonable accuracy. Even that is somewhat uncertain because some temperature estimates are based on the optical colors of the system, with no consideration of the boundary layer's effect on them. Other spectral-type estimates based on stellar spectral features should be more reliable. The effect of disk accretion on the luminosity of the system makes it difficult to estimate the stellar radius precisely. An additional constraint on the radius is provided when both the rotation period (through starspots) and the projected rotation velocity (through spectral line broadening) are known. These constrain the radius and inclination jointly. The mass is also uncertain, because the evolutionary tracks themselves are somewhat uncertain (Sec. V), and among the lower mass stars the convective tracks are crowded together. The central star is typically 0.5 to 1.3 $M_\odot$, with a radius from 1.5 to 3.5 $R_\odot$ and luminosity several times that of the Sun.

Another parameter not directly related to the disk is the external extinction to the system. It is difficult to distinguish between circumstellar extinction caused by dust which is near enough to the star to reprocess optical light into the infrared, and true extinction which is due to dust far enough away that the light is fully lost from the beam. A distinction must be made between them, because the latter changes the bolometric flux observed, while the former merely redistributes energy from one wavelength to another. Neither is directly associated with the disk, because a flat disk is unlikely to remove any light coming directly from the star toward the observer. Extinctions are estimated from the reddening of the observed light compared to the expected stellar intrinsic spectral energy distribution. Obviously the presence of the boundary layer makes such determinations uncertain, as its intrinsic spectral energy distribution is now not known *a priori*.

In addition to those, there are several parameters associated with the disk itself. Of these, the mass and size of the disk are not really a concern because it is assumed that the disk is optically thick, and that its size is sufficiently large not to affect the infrared spectrum. Of course, if one were to consider the spectrum down into the sub-millimetric range these parameters become important, along with details of the grain opacity (see Chapter by Beckwith and Sargent). Another parameter in the disk is the value of the α parameter which is a measure of the strength of turbulent viscosity in the disk. Here there is little theoretical guidance. The largest values derived theoretically are of order 10^{-2} from convection (Lin and Papaloizou 1985). Inferred values of α range up to unity, or greater in cataclysmic variables (Lin et al. 1988). Basri and Bertout (1989) found that setting α to unity was acceptable for most cases, although for lower accretion rates it might easily be of order 10^{-1} to 10^{-2}. They adopted the philosophy that it should not be used as a free parameter, but fixed arbitrarily at a certain value. For optically thick disks, this leaves the mass accretion rate as the main free parameter for the disk. The other parameter associated with the disk is the inclination of the disk plane to the line of sight, which acts primarily as a scaling parameter on the observed flux, due to foreshortening.

Bertout et al. (1988) compared quasi-simultaneous sets of data in the ultraviolet/optical and optical/near-infrared ranges to synthetic spectra emitted by models of a T Tauri system made up of a late-type active star, an accretion disk and its boundary layer. They found that typical T Tauri disks are optically thick over most of their surface so long as $\alpha \leq 1$ and that the spectral energy distribution of typical CTTS can be reproduced from about 0.2 to $10\,\mu$m if emission from the (isothermal) boundary layer is confined to an equatorial region with width comparable to the local disk scale height ($\sim$2% of the stellar radius).

Positive aspects of this simple model are its self-consistency and its small number of free disk parameters (essentially the disk mass-accretion rate and view angle), while a major drawback is the assumption of an optically thick boundary layer. Observed Balmer jumps indicate that the Paschen continuum is at least partially optically thin. Basri and Bertout (1989) therefore computed monochromatic gas opacities in the boundary layer, which is again assumed to be isothermal. They made the boundary layer width δ_{BL} an additional free parameter (rather than α) needed to control the optical depth, and computed emergent spectral energy distributions that they compared to observations of the Balmer and Paschen continuum regions. While the head of the Balmer continuum is optically thick in these models, the Paschen continuum is partially optically thin and the Balmer jump consequently appears in emission. A similar analysis by Kenyon and Hartmann (1990) leads to similar results. An example of the observations and fit to them for DF Tau appears in Fig. 2.

Line emission from the Balmer lines with high quantum number appears consistent with optically thick line emission from the boundary layer. The emitting area is at most a few percent of the stellar surface area. There is obviously a more extended region of emission which contributes to the flux in the lowest members of the Balmer series, as these are predicted to have very little emission contrast in the simple boundary-layer model.

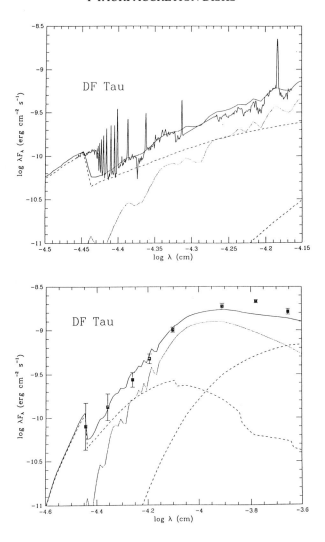

Figure 2. Simple disk model fits for the classical T Tauri star DF Tau. *Upper panel*: The solid lines are spectrophotometry (with a gap in data near $\log \lambda = -4.3$) and the composite model fit in the optical range. The dashed lines are flux from the disk (lower right) and an optically thin boundary layer. The lower faint line is a model stellar spectrum by itself. The parameters discussed in the text have been chosen to give a good fit. *Lower panel*: The same model in the optical and near infrared. The data points are simultaneous photometric fluxes from the star, and the lines are as in the upper panel. Note how the star is not the dominant contributor to the light on either side of the optical range.

C. Derived Disk Parameters

Modeling the spectral energy distributions of T Tauri stars allows one to derive key parameters such as the accretion rate and disk mass. Finding accurate mass-accretion rates from the disk onto the star is important because: (i) if sufficiently high, accretion is expected to affect the evolution of the star in the HR diagram (Kenyon and Hartmann 1990); (ii) one also expects that accretion of disk matter (which possesses high specific angular momentum) will affect the evolution of rotation in CTTS; and (iii) there is growing evidence of a relationship between mass accretion and mass loss (Cohen et al. 1989*b*; Cabrit et al. 1990; Chapter by Edwards et al.). The efficiency of mass gain/loss conversion is important both for understanding the mass-loss mechanism and for issues (i) and (ii) above. By assuming a classical disk model, mass-accretion rates ranging from a few times 10^{-9} to a few times 10^{-7} $M_{\odot}$ yr^{-1} can be determined from models of individual stars (Basri and Bertout 1989). Crucial questions remain about the validity of these values: are they highly dependent on the assumed disk model, and are they more or less unique, or can we find several solutions leading to the same spectrum but with very different mass-accretion rates? These questions are discussed in detail by Hartmann and Kenyon (1987*a*), Basri and Bertout (1989), Bertout and Bouvier (1989) and Bouvier and Bertout (1991).

Bertout and Bouvier (1989) studied the uniqueness problem by constructing maps of the quantity $1/\chi^2$, which measures the goodness of the fit between observed and computed spectral energy distributions, for all possible pairs of parameters. They used 6 computational parameters: the stellar radius R_*, the visual extinction in front of the system A_V, the inclination angle i, the accretion rate $\dot{M}_{\rm acc}$, the viscosity parameter α, and the width δ_{BL} of the emission region associated with the boundary layer. In their analysis of DF Tau, for example, the best solutions (with high $1/\chi^2$) span a small range of mass-accretion rates: $\dot{M}_{\rm acc} \sim 1$ to $2 \times 10^{-7} M_{\odot}$ yr^{-1}. It thus appears that the mass-accretion rate can be estimated within a factor of 2 from these models. This result stems from the fact that the mass-accretion rate primarily reflects the relatively well-determined quantity of integrated excess flux in the near infrared.

The two parameters α and δ_{BL} contain most of the assumed physics for the disk and boundary layer. Given the large range of parameter values that the solution spans, it is reassuring that best fits to the overall spectrum (also including the Balmer jump) were obtained for values $\alpha \sim 1$ and $\delta_{BL}/R_* \sim 0.02$. This appears physically reasonable and gives us some confidence both in the validity of the underlying disk physics and in the α parametrization. Even more reassuring, the best fits produce continuum veiling compatible with the observed amount.

Similar computations were made for 10 CTTS (using new simultaneous data sets to be published by Bouvier, Basri and Bertout spanning the ultraviolet to the near infrared) in the spectral-type range K1–M1. They yield an average mass-accretion rate of $<\dot{M}_{acc}> = (1.4 \pm 1.2) \times 10^{-7} M_\odot$ yr^{-1}. Because these systems are rather variable (due to unsteady accretion?), observations from the ultraviolet, optical, and near infrared must be gathered at the same time to make confident determinations of system parameters.

Approximate estimates of the mass and maximum radius of CTTS disks can be made by modeling the spectral energy distribution in the sub-millimeter and millimeter range, using the sharp turnover of the spectrum that occurs in that spectral range because the outer disk becomes optically thin. Adams et al. (1990) and Beckwith et al. (1990) recently presented such computations, based on the assumption discussed above that the temperature distribution in the disk is a free parameter that can be adjusted to get an overall fit to a given spectral energy distribution. While this *ad hoc* assumption introduces some uncertainty in derived disk masses, the main source of uncertainty stems from lack of knowledge of dust opacities in the millimeter range: depending on the assumed opacity law, mass estimates can vary by up to 1 order of magnitude. One thus finds that masses of CTTS disks may range from $<10^{-2}$ to perhaps as much as 1 $M_\odot$.

IV. SPECTROSCOPIC DIAGNOSTICS OF DISK ACCRETION

It is clear from the last section that further observational constraints on the parameters of the disk models will be very helpful. One clearly relevant accretion diagnostic is the amount of spectral veiling (actually the amount of excess continuum light) in the optical spectral lines. The strong broad optical emission lines are a second such diagnostic; they should in principle reflect the physical properties of the region where the mass loss originates.

A. Spectral Veiling

Veiling is defined by

$$r(\lambda) = \frac{f_{obs}(\lambda)}{f_*(\lambda)} - 1 \tag{6}$$

where f_{obs} is the observed flux and f_* the photospheric flux. It can be estimated by comparing the veiled (observed) spectrum with an appropriate spectral standard. One then finds that all the same lines are present and in the same ratios, but all the line depths are reduced in the CTTS. This was done quantitatively at one wavelength by Hartmann and Kenyon (1990), and over a broad spectral range for one star by Hartigan et al. (1989a) and for an extensive sample of stars by Basri and Batalha (1990) and Hartigan et al. (1991). Their efforts yield excess light as expected from accretion disk models, with wavelength dependence as predicted by hot boundary-layer models. The excess light does not show features unexpected for the optically thin emission that has been postulated.

A measurement of veiling is useful for several reasons. It allows one to correct the observed spectrum for nonstellar emission independent of extinction corrections. This allows in turn a proper estimation of the extinction from the remaining reddened stellar spectrum (Hartigan et al. 1991), and thereby a constraint on the radius of the star (assuming distance is known). It also fixes the level of the boundary-layer emission. Thus, one should demand that disk models produce the observed veiling in detail, constraining the boundary-layer temperature and size along with the extinction, accretion rate and stellar radius. Obviously veiling by itself does not determine all these parameters, but inasmuch as it acts as a further constraint on disk models, it significantly improves the uniqueness of solutions. As veiling is the result of a particular disk model rather than an input parameter, there is no *a priori* guarantee that models which fit the overall spectral energy distribution best will also give the right veiling value.

In their detailed study of DF Tau, Bouvier and Bertout (1991) compare model predictions with empirical average veiling values of 1.6 ± 0.3 at 5000 Å and 1.0 ± 0.3 at 6500 Å (Basri and Batalha 1990). The acceptable solutions that were found for the disk parameters, after the χ^2 minimization procedure mentioned above, produced veiling values ranging from 1.55 to 1.9 at 5000 Å and from 0.6 to 0.75 at 6500 Å , in fair agreement with the observations although they were not obtained simultaneously with the spectral energy distributions. Veiling varies in time by factors which can be as large as 10 in some stars (Basri and Batalha 1990). In order to take full advantage of this additional constraint on the models, it is thus mandatory to measure veiling at the same time as the spectral energy distribution.

Basri and Batalha (1990) and Hartigan et al. (1990*a*) showed that the measured veiling has the relation to infrared excess expected if both are due to accretion and also that Hα emission flux is closely related to veiling. Cabrit et al. (1990) and Bertout et al. (1991*b*) showed that the emission lines in general are correlated with infrared excess, as expected from accretion (and not from a chromospheric origin).

B. Broad Emission Lines

Hα in particular has the breadth and profile expected from a moving extra-stellar region, yet it remains one of the least understood aspects of the T Tauri phenomenon and one of the most fascinating because of the variety of line profiles and their temporal changes. First statistics of line shapes were presented by Kuhi (1978) in the first volume of *Protostars and Planets* (Gehrels 1978) and by Ulrich and Knapp (unpublished catalog). It has recently become possible to monitor emission lines spectroscopically with high resolution and signal to noise, which will hopefully allow clarification of our understanding of the emission lines in the next few years. A first important result of this monitoring is the fact that the Hα line comes in three basic shapes.

1. The most common is broad emission with fairly symmetric far wings extending to 200 to 400 km s^{-1} and a blueshifted absorption feature near 100 km s^{-1} that usually does not go below the continuum. This yields emission peaks with the red usually brighter than the blue.

2. Less common is a more or less flat-topped emission feature, possibly with central absorption that can be unshifted or shifted to either side.
3. Finally there are fairly symmetric triangular (more sharply peaked than gaussian) shaped emission lines, often with little or no absorption.

These shapes can be seen in the other Balmer lines (although for these, the absorption goes more often below the continuum) and in other strong emission lines. The weakest of these seem to have a preference for the triangular shape with little absorption. The large degree of symmetry seen in broad lines argues for a substantial orbital or turbulent component in the velocity broadening. Although the origin of triangular lines is still controversial, such shapes may well arise due to variable turbulent velocities of comparable brightness (Basri 1990). In some cases (see, e.g., Basri 1987), especially in weaker lines like the Ca II infrared triplet or He I lines, one clearly sees both a low-lying broad component and a narrow, undisplaced emission peak. The less active stars tend to show only the narrow component. It is tempting to associate this component with the chromosphere or with filled magnetic loops on the star.

The variability of broad emission lines is puzzling. The same line can have quite a stable appearance in some objects, while in others it undergoes large intensity changes with little change in its profile (e.g., Hα in DF Tau); and still again in others, it changes its shape dramatically. Figure 3 shows an example of Hα line profile variations seen in twice nightly observations of DR Tau and obtained by Basri with the 24″ coudé feed and the Hamilton echelle spectrograph at Lick Observatory. They show that the intensity of the line varies substantially within a few hours, and also that the absorption velocity structure is quite changeable. The appearance of a true P Cygni absorption feature is ephemeral, as is the more usual lower-velocity feature common to T Tauri stars. This means that these features arise in a fairly small region themselves, and the mass outflow is probably very clumpy and changes its structure.

A smooth axisymmetric outflow is probably not a good model, although it may be a useful first approximation. The blue-displaced absorption components do not vary as much as the emission intensity and probably arise in a larger, more distant region. One can look for correlated changes in accretion and outflows in such data, which could establish a direct connection between them. As another example of an extreme case, the same Ca II line in the same star (RW Aur) has shown all three basic shapes discussed above (Bertout et al. 1991b), which may mean that they are compatible with different manifestations of a common underlying structure. The Balmer decrement can be very large, i.e., Hα can be much brighter than the higher Balmer lines relative to the local continuum. This probably means that Hα arises from a substantially larger geometrical area than other Balmer lines. These lines can vary in intensity and shape on time scales down to an hour or less (see, e.g., Mundt and Giampapa 1982; Basri 1990), indicating that they most likely arise in small regions quite near the star.

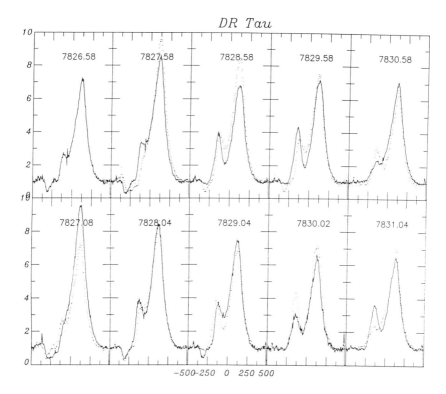

Figure 3.   A time series of Hα profiles from the classical T Tauri star DR Tau. The dotted line in each panel is the preceding profile repeated for comparison. The numbers are the Julian date (without the initial 244), corresponding to 27–31 Oct. 1989 (twice nightly). Each profile is in the same continuum units, and all the abcissas are stellar rest velocities from −500 to 500 km s^{-1}. The profile has both low and high velocity blueshifted absorption for the first half of the run, with the typical simpler T Tauri profile in the latter part. There is a pronounced P Cygni character on 7827.1 JD, and almost no absorption at 7830.6 JD. The red wing is much less variable than the blue side or peak. The total equivalent width is also rather variable, due to changes in both absorption and emission levels.

The boundary layer is an obvious candidate for the line emission region (Basri and Bertout 1989), as it is small, fast moving, turbulent, and energetic. However, inverse P Cygni line profiles seen in YY Orionis stars indicate infall velocities of several hundred km s^{-1} which cannot arise in the boundary layers of classical disk models, where accretion velocities are restricted to the subsonic. It is our intuition that strong stellar magnetic loops may interrupt the flow of the disk within a few stellar radii of the surface, generating the shocks, turbulence and heating that we have been designating as the boundary layer. Flow down such tubes could reach near free-fall velocities, and a modulation of the accretion (emission intensity or veiling) with the stellar rotation period would establish that the emitting region was in the interface between star and disk. There have already been a few reports of "hot spot" modulations (see, e.g., Bertout et al. 1988; Simon et al. 1991) which could be interpreted as accretion columns onto magnetic stellar regions.

It may well be the same thing to say that the lines arise in the boundary layer (now being used in the less restrictive sense of the interface region between star and disk) or that they arise in the base of the wind. In this picture, some material would accrete down through the magnetic loops, while other material would be turned back and flung out in the T Tauri wind. Some relevant theoretical ideas on this still controversial view can be found in the Chapter by Königl and Ruden. The development of this picture, where magnetohydrodynamic processes are probably dominant, will be a major task of the coming decade.

V. IMPLICATIONS OF DISKS AND ACCRETION

One must be careful about what is meant by "classical" T Tauri stars and "weak line" T Tauri stars, because there is really a continuum of stars between the two classes as defined by Hα strength. If there is (i) a substantial near-infrared excess (and this is best measured from 2 to 10 μm) compared to the flux expected from a star alone, (ii) Balmer continuum emission, or (iii) strong Hα emission (say more than 10 Å equivalent width in a late K star), then we conclude from the evidence discussed above that an accretion disk is present. Its presence could serve as the (physical) definition of a CTTS. Note that accretion is the crucial part of this definition. Stars without disks or with disks which do not reach their surfaces will appear as WTTS, while a lack of excess emission at any wavelength is required for the star to truly be a naked (and thus diskless) T Tauri star.

The exact relationship between CTTS and WTTS is unclear at present, and work to clarify it should be encouraged as it is one major key for understanding the formation of low-mass stars. A crucial fact is that WTTS are well intermingled with the CTTS both spatially and on the HR diagram (i.e., temporally). Also, the WTTS have lithium lines similar to the CTTS (Strom et al. 1989e; Basri et al. 1991) after correction for continuum veiling. This result also points toward similar ages in the two populations. Furthermore, the distributions of rotational speeds in WTTS and CTTS are similar (see Feigelson et al. 1991). These findings are curious because if steady-state accretion proceeds during a sizable time, then disk accretion could contribute a significant fraction of the stellar mass in CTTS, with noticeable consequences in the star's evolution, and should also alter its angular momentum history. Königl (1991) has suggested that if the stellar magnetic fields are strong enough and stretch well beyond the corotation radius, the disk will not transfer much angular momentum to the star anyway.

Current estimates based on classical convective-radiative evolutionary tracks indicate that accretion disks last up to a few Myr (Strom et al. 1989d;

Chapter by Strom et al.), and are a relatively common phenomenon among young solar-type stars, with perhaps less than one-half and more than one-fifth of them appearing currently as CTTS. This is cause for optimism if one had been hoping that planetary systems are a relatively common phenomenon, although it remains to be seen how many of these "solar nebulae" actually give rise to planets.

Two reasons make it difficult at this point to estimate the actual amount of mass that is ultimately accreted by the star even if we were optimistic enough to believe that the average mass-accretion rates derived from our disk models were representative of accretion during most of the CTTS phase. First, it is becoming increasingly clear that either the disk or the boundary layer is also driving the strong wind that characterizes the CTTS phase (see Chapter by Edwards et al.). Second, disk lifetime estimates depend crucially on the validity of the convective-radiative evolutionary tracks, which may not be relevant for accreting CTTS. The evolution of mass-accreting pre-main-sequence stars was recently computed by Stringfellow (1989), who indeed finds that their paths in the HR diagram bear little resemblance to classical convective-radiative tracks. Furthermore, even if classical tracks are approximately correct, previous determinations of the position of CTTS in the HR diagram (see, e.g., Cohen and Kuhi 1979) did not take into account the presence of the disk and must, therefore, be regarded as uncertain by a factor of 2 or more. These issues have been addressed most directly by Kenyon and Hartmann (1990). For these reasons, age estimates of T Tauri stars based on the HR diagram appear suspect, as do derived disk lifetimes.

Even though the absolute age of T Tauri stars is doubtful, we assume here the conservative point of view that pre-main-sequence stars sharing the same region of the HR diagram have similar masses and ages, i.e., that young stars do not loop through the HR diagram during pre-main-sequence evolution. The observational evidence reviewed above then implies that accretion has little effect on the global stellar properties, perhaps because most of the disk matter is turned into a wind rather than accreted and/or because disk lifetimes are shorter than estimated on the basis of classical evolutionary tracks. Several different possibilities can then be envisioned for the evolutionary status of CTTS and WTTS:

1. WTTS may simply be CTTS which have stopped accreting. Of course, every CTTS must eventually lose its accretion disk because main-sequence stars are never observed to have active disks. But if all WTTS were once surrounded by disks with comparable physical properties, WTTS should be older than CTTS on the average, which is not the case. Thus, data demand a range in disk masses and lifetimes that may represent a variety of initial conditions for star formation.

2. Alternatively, WTTS may be young stars which never had disks. Some of the WTTS are near the "birthline" for optically visible stars and so may never have been CTTS if one believes in the validity of Stahler's (1983a)

scenario. The existence of Class III radio sources deeply imbedded in the core of the ρ Ophiuchi cloud where the expectation is that they are very young, might also support this possibility (see André et al. 1987).

3. Another possibility is that many T Tauri stars go through CTTS and WTTS phases several times during early pre-main-sequence evolution. In this picture, one could speculate that disk instabilities lead to a large range of recurrent eruptive phenomena from FU Orionis events down to "normal" T Tauri aperiodic variability or even weak-lined phases. Temporal spectroscopic changes seen, e.g., in RY Tau, where the optical spectrum can range from pure absorption to relatively strong emission, perhaps support this hypothesis.

These are probably the three most conservative scenarios for the evolution of CTTS/WTTS. If it turns out that the evolution of T Tauri stars in the HR diagram is very different from what we think today, more radical options follow. Obviously, much remains to be done to test all possibilities. Realistic (magnetohydrodynamic) simulations of T Tauri systems' evolution would of course be extremely useful, but because of the complexity of this approach, less ambitious projects such as investigations of the stability and evolution of protostellar disks and boundary layers are welcome first steps. An observational approach to these problems based on systematic spectroscopic and photometric monitoring of some T Tauri stars is also possible, and will offer some clues about the basic physical mechanisms of aperiodic variability in pre-main-sequence objects.

VI. CONCLUSIONS

Both direct and indirect evidence that visible young stars are sometimes still surrounded by disks is now available. Disks seem to be found around a substantial fraction of newly born low-mass stars, and apparently are the agents responsible for their continuum excesses in the ultraviolet, optical and infrared, for their strong broad emission lines, and indirectly for their strong winds.

Unified steady-state accretion-disk models have been constructed which self-consistently and simultaneously satisfy many of the observational constraints which differentiate CTTS from normal stars. The disk has dimensions comparable to our solar system and contains perhaps 0.01 to 1.0 $M_\odot$ of dust and gas. The overall continuum shape and observed luminosity between 0.1 and 10 μm, as well as the Balmer jump and some emission-line fluxes, can be accounted for by simple disk models. The disk is optically thick at these wavelengths, but the boundary layer is often partially optically thin. The accretion rates lie between 10^{-8} and 5×10^{-7} $M_\odot$ yr^{-1} for the CTTS. The models are both very simplified and parameterized, and do not do a good job of matching the slope of the infrared continuum beyond 10 μm without some modification to either the disk geometry or internal heating mechanisms.

While first results on the stability of the boundary layer and on the detailed structures of both boundary layer and disk are now becoming available, more work is needed on some fundamental aspects of disk theory, particularly on the angular momentum transport mechanism, as well as on details of the interaction between disk and star and on the production of protostellar winds. High-resolution infrared imaging, as well as high-resolution spectroscopy in the ultraviolet, optical and near infrared, will provide major constraints for further work along these lines.

Accretion disks appear responsible in some way (along with the resulting strong mass outflow) for most facets of the "T Tauri phenomena." None of these characteristics hold for the weak emission-line T Tauri stars, and therefore the presence of an *accretion* disk delineates the difference between the two classes of young stars. The underlying stars are probably similar in both cases, and both exhibit very strong magnetic activity. The accretion disks may well be examples of "solar nebulae" that hold the exciting potential for allowing current observations relevant to the planetary-formation process.

ENERGETIC MASS OUTFLOWS FROM YOUNG STARS

SUZAN EDWARDS
Smith College

TOM RAY
Dublin Institute for Advanced Studies

and

REINHARD MUNDT
Max-Planck-Institut für Astronomie

Energetic winds emerge from young stars in two seemingly distinct stages of evolution: the youngest known stars still deeply embedded in their molecular cloud cores and their progeny, the optically visible T Tauri and Herbig Ae/Be stars. In the first section, we focus on winds from low-mass optically visible T Tauri stars, and show that (1) the presence of energetic T Tauri winds is coupled with the presence of solar system-sized circumstellar disks which are optically thick in their inner regions and that (2) it is likely that these disks are in a state of mass accretion and that the source for the T Tauri winds probably derives from the innermost region of the accretion disk. We conclude this section by noting that energetic T Tauri winds are unlikely to be the primary agents in the final clearing of gas and other debris in forming planetary systems, as the cessation of energetic winds apparently coincides with the cessation of inner disk accretion. We then compare the wind energetics of the optically visible classical T Tauri stars and their presumed precursors, low-luminosity embedded infrared sources with molecular outflows. It is a common misconception that T Tauri winds are fundamentally different from, and less energetic than, those from embedded infrared sources, where fast neutral winds are thought to drive molecular outflows. While many of the outward manifestations of the outflows from embedded infrared sources and classical T Tauri stars differ, it is shown that they possess remarkably similar wind velocities, and wind energy and momentum fluxes when objects of the same bolometric luminosity are compared. In addition, optical outflow diagnostics, such as collimated jets and Herbig Haro objects, are found associated with both types of objects. Both also possess spectral energy distributions which require a contribution from an optically thick circumstellar disk. These similarities in outflow and disk properties suggest that energetic winds from both embedded sources and T Tauri stars share a common origin. Finally we examine one of the outflow diagnostics shared by the optically visible classical T Tauri stars and the embedded infrared sources—the spatially extended optical outflows characterized by collimated jets and Herbig Haro objects. While these optical phenomena do not dominate the wind energetics, their collimated structures, extending up to a parsec from their source, raise important issues in the quest to understand energetic mass outflows from young stars. We review the current observations of jets and Herbig Haro objects emerging from young stellar objects, including their morphology, kinematics, formation and propagation, and discuss their relation to other outflow phenomena.

I. INTRODUCTION

The ubiquity of energetic winds from the youngest known stars, deeply embedded in their molecular cloud cores and believed to be in a state of vigorous mass accretion, suggests that the energetic winds are as intrinsic a part of the star formation process as is gravitational collapse. In fact, it has been suggested that a star cannot form without them: they may provide the only means for a collapsing and rotating object to shed sufficent angular momentum to grow to stellar dimensions (see Chapter by Shu et al.; Hartmann and MacGregor 1982). An additional impact which these energetic winds may have on forming stars is reversing the infall of core material and thereby determining the final mass of the star (Shu et al. 1987). Energetic winds also affect the circumstellar environs by dispersing remnant core material and providing pressure support to parent clouds.

In addition to the embedded sources, energetic winds are also observed from optically visible pre-main-sequence stars of ages on the order of a few million years, including both those low-mass stars which exhibit strong excess emission at all wavelengths, known as classical T Tauri stars (Chapter by Basri and Bertout) and their higher-mass analogs, the Herbig Ae/Be stars (see Catala 1989). T Tauri winds are widely invoked, in the case of our own youthful Sun, as a means of clearing debris from the solar nebula during the early stages of planet formation (Horedt 1978; Elmegreen 1978b).

In this review we discuss energetic winds from these two seemingly distinct phases in the evolution of young stars, focusing primarily on the more thoroughly studied low-luminosity (≤ 100 L$_\odot$) sources. Although the wind signatures from low-luminosity embedded infrared sources and optically visible classical T Tauri stars (CTTS) can be very different, they possess remarkably similar wind velocities, mechanical energies and momentum fluxes. Collimated jets and Herbig Haro (HH) objects are found associated with both types of object. Both possess spectral energy distributions that require a contribution from a solar system-sized optically thick circumstellar disk. Evidence is beginning to suggest that the powering mechanism for the energetic winds from both the embedded infrared sources and the CTTS derives from the same source, namely energy released in the innermost regions of a circumstellar accretion disk.

In Sec. II, we address the issue of the powering mechanism for energetic winds from young stars by focusing first on T Tauri winds, because these optically visible stars have spectral energy distributions which are most easily separated into a stellar and a disk contribution. By examining both wind and disk diagnostics for the T Tauri stars we show (1) that energetic winds are *only* present in those systems with optically thick circumstellar disks; (2)

the energy source for the T Tauri winds is non-stellar in origin and probably derives from the innermost regions of an accretion disk; and (3) energetic T Tauri winds are unlikely to be the final clearing agent for gas and other debris in forming planetary systems, as the cessation of energetic winds in T Tauri stars apparently coincides with the cessation of disk accretion.

In Sec. III, we compare the wind energetics from low-luminosity embedded infrared sources, where fast neutral winds drive molecular outflows, with those from the optically visible CTTS. We show that the winds from these sources are comparably energetic when objects of similar bolometric luminosity are examined. Winds from both types of objects are likely to derive from an energy source related to disk accretion.

In Sec. IV we discuss the spatially resolved optical outflows from young stars: jets and HH objects. These outflow signatures, found both from embedded sources and from the youngest optically visible CTTS and Herbig Ae/Be stars, are highly collimated, enigmatic structures which delineate regions where the energetic wind is being shocked, often along the axis of the larger-scale bipolar molecular outflow. Their morphology, kinematics, formation and propagation is discussed, and their relation to other outflow phenomena examined.

II. THE T TAURI STARS

A. Classical and Weak Emission T Tauri Stars

Optically visible low-mass pre-main-sequence stars with ages $\leq 10^7$ yr show spectral energy distributions of modestly reddened, late-type photospheres overlaid with varying degrees of excess X-ray, ultraviolet, optical, infrared and radio emission. Those with the strongest emission excesses were the first low-mass pre-main-sequence population to be identified, largely on the basis of strong Hα emission detected in objective prism surveys, and are known today as the "classical" T Tauri stars (CTTS; Joy 1945; Herbig 1962b; Bertout 1989; Appenzeller and Mundt 1989; Chapter by Basri and Bertout). For decades, the source of the excess emission from the CTTS remained elusive, although the possibility that it was fundamentally stellar in origin, deriving from a youthful analog of a solar-type magnetic dynamo seemed very plausible (Herbig 1970; Bertout 1989). However, repeated attempts to account rigorously for the emission excesses with active chromospheric models fell short in the more extreme members of the CTTS class, leaving nagging doubts regarding the general applicability of this picture (Calvet et al. 1984; Calvet and Albarran 1984). Recently, the discovery of a significant population of additional low-mass pre-main-sequence stars with masses and ages overlapping those of the CTTS but with much weaker emission excesses (the "weak" emission T Tauri stars, WTTS) allowed the CTTS to be viewed in a new perspective (Walter et al. 1988; Strom et al. 1989d; Montemerle and André 1989).

Although these two groups share similar masses and ages and appear nearly identical in terms of their stellar properties (see Sec. II.D), the CTTS possess energetic winds, diagnosed by "P Cygni-like" profiles at Hα, Ca II and Na D (Kuhi 1964; Hartmann 1982; Mundt 1984) and more recently, by broad blueshifted low excitation metallic forbidden lines of [O I] and [S II] (Appenzeller et al. 1984; Edwards et al. 1987). Some also show spatially extended optical outflow diagnostics, as is discussed in Sec. IV. None of these wind signatures are found in their less demonstrative siblings.

Additionally, the CTTS possess other emission excesses not shared by the WTTS. These excesses include: (1) infrared emission from $\lambda = 1\ \mu$m to 1 mm that can only be explained as originating in a flattened optically thick solar system-sized circumstellar disk (Rucinski 1985; Adams et al. 1987; Kenyon and Hartmann 1987; Bertout and Bouvier 1989; Chapter by Beckwith and Sargent); (2) optical and ultraviolet continuum "veiling" emission which has been attributed to boundary layer emission arising as material from the inner regions of a Keplerian accretion disk falls onto the slowly spinning star (Lynden-Bell and Pringle 1974; Basri and Bertout 1989, also see their Chapter; Hartigan et al. 1989b,1990,1991).

In the following sections we first examine the wind diagnostics and the determination of the mass loss rates for the CTTS, then present the evidence that energetic T Tauri winds are found only in those stars with optically thick circumstellar disks, and finally present evidence that the T Tauri winds likely derive from a source related to disk accretion.

B. Wind Diagnostics and Mass Loss Rates from T Tauri Winds

Strong winds from CTTS have long been inferred on the basis of "P Cygni-like" profiles in optically thick lines of hydrogen, Na D, Ca II and Mg II (Kuhi 1964; Hartmann 1982; Mundt 1984; Imhoff and Appenzeller 1987). However, determination of mass loss rates from these optically thick lines presumably formed in the dense inner wind ($r < 5R_\star$), has been plagued by the sensitivity of these lines to the thermal and dynamic properties of the wind (DeCampli 1981; Hartmann 1986). Modeling efforts have also been hampered by the wide variety of line profile shapes found at Hα, with between 10 to 20% showing classic type I P Cygni structure.

The most common Hα profile is the type III, with broad wings and a blueshifted reversal, at a velocity less than that of the blue emission wing. A representative type III line profile from the CTTS, DO Tau is shown in the upper left panel of Fig. 1. The line is strong, with broad symmetric wings, and is cut by a deep reversal with a centroid velocity of about -100 km s^{-1}. While the depth and velocity of the central reversal vary among the observed stars, one characteristic common to Hα in all CTTS is the broad, symmetric emission wings. It has been suggested that a large turbulent component in the inner wind, as would occur if MHD waves were dominant in this region, might account for this symmetry (DeCampli 1981; Hartmann et al. 1982).

Spherical wind models have met with some success in accounting for

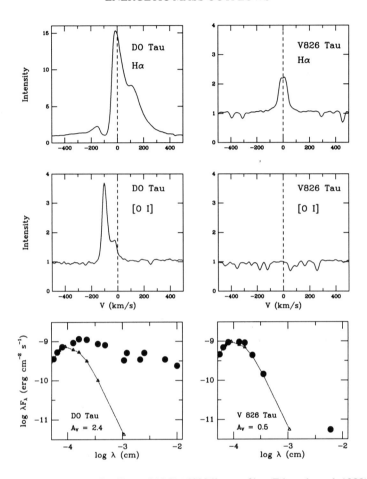

Figure 1. Representative Hα and [O I] λ6300 line profiles (Edwards et al. 1989) and de-reddened spectral energy distributions (K. Strom, unpublished) for the CTTS DO Tau and the WTTS V826 Tau. The line profiles are plotted as relative intensity against velocity, measured with respect to the stellar rest velocity. The open triangles connected by dotted lines in the spectral energy distributions represent the expected stellar contribution to the luminosity.

the fluxes in the hydrogen and low excitation metallic lines (Hartmann et al. 1982,1990; Natta et al. 1988; Natta and Giovanardi 1990). In these models, the atomic rate equations and radiation transfer equations for optically thick lines of several different elements are solved in an expanding spherical wind. Those stars with the most reliably determined mass loss rates are those with deep blueshifted Na D absorption features, representing ∼10% to 20% of the CTTS, and requiring $\dot{M}_w$ on the order of 10^{-7} $M_\odot$ yr^{-1} (Natta and Giovanardi 1990). Lower-mass loss rates, on the order of 10^{-8} $M_\odot$ yr^{-1}, seem to be more representative of the class (Hartmann et al. 1990).

A nagging problem with the spherical wind models is their difficulty in reproducing the line profile structure in the optically thick lines such as Hα. While the models of Hartmann and collaborators have been able to generate a type III P Cygni profile in a spherical wind characterized by both an expansion and a turbulent velocity, the blueshifted absorption is too broad and deep to be representative of most CTTS spectra, especially in the higher Balmer lines. Recent investigations by Calvet et al. (1991b) calculate hydrogen line profiles in an optically thick cone wind, as might be expected for a wind arising in the inner part of an accretion disk or a boundary layer. The strong aspect dependence of the resultant line profiles provides a simple means of producing a variety of profile types, many of which are qualitatively similar to those observed.

In addition to the optically thick hydrogen and permitted metallic lines with wind signatures, the low excitation forbidden lines common in CTTS spectra are also important wind diagnostics. These optically thin lines are formed in low-density, shock-ionized outflowing gas far from the star ($r >> 1$ AU), as determined from the density-sensitive [S II] lines and the forbidden line emission measure (Appenzeller et al. 1984; Edwards et al. 1987). A representative [O I] λ6300 Å line for the CTTS DO Tau is shown in the middle left panel of Fig. 1. While the forbidden lines are also broad, their width is considerably less than the Hα lines, indicating that they are formed well beyond the turbulent inner wind region, after the wind has reached terminal speed. Evidence supporting this is provided by the correspondence in velocity of the maximum in the blueshifted forbidden line emission with the absorption reversal at Hα. Particularly noteworthy is the lack of redshifted emission, characteristic of nearly all T Tauri star forbidden lines, which is attributed to occultation of the receding hemisphere of the outer regions of the outflow by a highly flattened, opaque circumstellar disk.

While the forbidden lines, with fluxes presumably proportional to wind mass loss rates, are important diagnostics of CTTS winds and disks, quantitative mass loss rate determinations from these optically thin lines are not feasible due to uncertainty regarding their formation mechanism. Several suggestions have been put forth to account for the re-heating of the outer wind regions to the requisite temperatures for forbidden line emission, including oblique shocks as winds impact outer, raised disk surfaces (Hartmann and Raymond 1989), ambipolar diffusion (Safier and König1 1991, in preparation), or shocks in spatially unresolved collimated jets (Kwan and Tademaru 1988; Edwards et al. 1989; Mundt et al. 1990).

C. Coupling of Winds and Disks in T Tauri Stars

A comparison of wind and disk diagnostics for a representative sample of low-mass pre-main sequence stars in the Tau-Aur star-forming complex reveals the intimate relation between these two aspects of T Tauri activity. Wind/disk relations have been explored by various workers (see, e.g., Strom et al. 1988; Cohen et al. 1989b; Cabrit et al. 1990; Bertout et al. 1991a; Hartmann 1990;

Hartmann and Kenyon 1990). In what follows, the discussion of wind/disk relations is based on an unpublished set of sky subtracted echelle spectra from the KPNO 4 m Mayall telescope and published infrared photometry for 40 stars of comparable age (<10 Myr), spectral type (K7/M0 or cooler), and inferred mass ($\sim$1 $M_\odot$), which cover the full range of emission excesses found among the (classical to weak) T Tauri stars.

For wind diagnostics, we use the equivalent widths of Hα, probably formed, at least in part, in the inner wind region, and [O I] λ 6300 Å, from the outer reheated wind. In Fig. 1, we display representative Hα and [O I] profiles from stars at opposite extremes of the emission excesses characterizing the T Tauri stars. The Hα and [O I] profiles for the CTTS DO Tau, discussed in the previous section, show evidence of energetic winds. In contrast, the same lines in the representative WTTS, V826 Tau, of the same spectral type and apparent age, show no evidence for energetic winds. In the WTTS, the Hα emission, weak, narrow and symmetric, is presumably chromospheric in origin (see Walter et al. [1988] for additional examples of Hα profiles in WTTS). The [O I] λ 6300 Å line is not seen (equivalent width <0.02 Å). Note also the greater depth of the photospheric features in V826 Tau, which are filled in by continuum "veiling" in DO Tau.

Disk diagnostics are provided by the infrared excess emission from these young stars. The lower panels in Fig. 1 contrast the reddening-corrected spectral energy distributions for the representative CTTS and WTTS. The significant infrared excess for the CTTS must originate in a region that covers a wide range of temperatures, from photospheric values down to 100 K, extending close to the stellar surface, yet still providing a relatively clear line of sight to the star (see Chapter by Basri and Bertout). In contrast, essentially photospheric emission out to at least 10 μm is found for the WTTS.

For a simple diagnostic of the inner disk, we use the near infrared $K - L$ color of the stars in our sample (see Kenyon and Hartmann 1990). This color index provides a means of probing two distinct aspects of TTS inner disks: (1) whether the inner disk is optically thin or thick (Chapter by Strom et al.); and (2) the radial temperature gradient in the inner disk. Only the former aspect is discussed here; the latter, which depends on accretion disk physics, is discussed in the following section.

The discrimination between an optically thin or thick inner disk depends on whether $K - L$ is less or greater than $\sim$0.3 to 0.5. For the range of spectral types represented here, the photospheric $K - L$ colors should all be constant, $\sim$0.2$\pm$0.1, and extinction corrections will be small. When inner disk material is present, the relative contribution of disk to photospheric emission will increase between 2.2 μm and 3.4 μm, so that $K - L$ becomes progressively redder as the disk optical depth increases. Thus TTS with optically thin inner disks will have $K - L$ values between those indicative of photospheres and those from flat, optically thick reprocessing disks, which have $K - L$ colors between 0.4 ($i = 80°$) and 0.7 ($i = 0°$), depending on the viewing angle (Kenyon and Hartmann 1987,1990). If the disk is not only optically thick but is also heated by nonsteady accretion, $K - L$ values can be even higher, as discussed in the the next section.

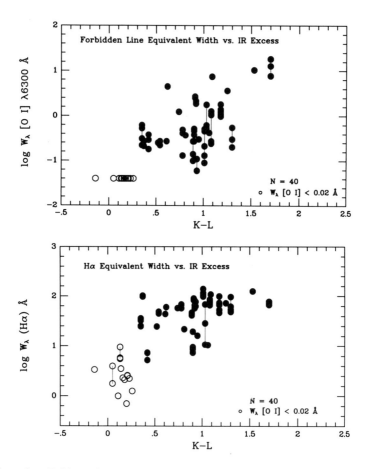

Figure 2.   Evidence for the coupling of energetic T Tauri winds and optically thick T Tauri disks ($K - L > 0.3$) for a representative sample of 40 CTTS and WTTS in Tau-Aur, with spectral types between K7 to M5. In the upper panel (a), a diagnostic of the outer wind, the equivalent width of [O I] $\lambda6300$, is plotted against $K - L$. In the lower panel (b), a diagnostic of the inner wind, the equivalent width of Hα, is plotted against $K - L$. The equivalent widths come from echelle spectra (Edwards, unpublished) and the $K - L$ values are from the literature. Vertical lines connect the same stars observed spectroscopically at different epochs. In both panels, open symbols denote those (WTTS) stars with upper limits on the [O I] equivalent widths, <0.02 Å.

In Fig. 2 the wind and disk diagnostics discussed above are compared for this sample of 40 T Tauri stars in Tau-Aur (see also Hartigan et al. 1990; Hartmann and Kenyon 1990; Cohen and Kuhi 1979; Strom et al. 1973 for related figures). This figure discriminates between optically thin and thick TTS disks, and allows one to distinguish between the presence or absence of an energetic TTS wind. In the upper panel, where the equivalent width of [O I] is plotted against $K-L$, it is seen that those stars with $K-L$ values indicative of optically thick inner disks all exhibit forbidden line emission from shock-heated outflowing gas, while those with $K-L$ values indicative of optically thin or absent inner disk material do not. In the lower panel, where the equivalent width of Hα is plotted against $K-L$, it is seen that those stars with optically thick $K-L$ values and detectable [O I] have Hα equivalent widths ranging from 10 Å to 100 Å while those stars with photospheric $K-L$ values and no [O I] have Hα equivalent widths $\leq$10 Å. An examination of the corresponding Hα profiles shows that, while stars with optically thick inner disks have Hα velocity structure indicative of outflows, the stars with optically thin or absent inner disk material and weaker Hα emission have narrow Hα lines with no evidence for mass loss, as shown in the example in Fig. 1.

This simple comparison of wind and disk diagnostics for a representative sample of CTTS and WTTS of comparable inferred mass and age demonstrates that *the presence of energetic winds is coupled with the presence of optically thick circumstellar disks* in all stars which were examined. In the next section we will examine the arguments that this is a causal relation, where the optically thick disks appear to be in a state of active accretion, which is likely to provide, at least in part, the energy which is the driving force for T Tauri winds.

The ubiquitous coupling of energetic TTS winds and optically thick inner disks has an important implication for the interaction of T Tauri winds with forming planetary systems: energetic T Tauri winds with mass loss rates of 10^{-8} to 10^{-7} M$_\odot$ yr^{-1} cannot be the final debris-clearing agent in forming solar systems, including our own, because once the inner disk has become optically thin, such energetic winds no longer exist. Presumably it is the WTTS, with optically thin or absent inner ($r \leq 1$ AU) disks, which are in the process of scouring clean any planetary systems. These stars are still very active by solar standards, however, and should have enhanced analogs of a solar-type wind, as well as other forms of solar activity, producing strong ultraviolet radiation fields and high-speed particle ejection. Detection of the expected lower-density, higher-temperature solar-type winds ($T \sim 10^5$ K) from WTTS will require high-resolution ultraviolet spectroscopy to search for P Cygni features in transition region lines such as N V or C IV. If present, their mass loss rates will be much lower than those characterizing the energetic CTTS winds.

D. Evidence That Energetic CTTS Winds Are Powered by Accretion

While a comparison of the wind and disk diagnostics of the CTTS and WTTS shows that the presence of energetic winds and optically thick circumstellar disks appear to be coupled in low-mass pre-main-sequence stars, a comparison of the stellar properties of these siblings argues for an external, rather than an internal (stellar) source for the energetic CTTS winds. When WTTS and CTTS of similar inferred mass and age are compared, their apparent underlying stellar structure and surface angular momentum appear similar. Thus their level of internally generated magnetic dynamos would be expected to be similar as well. That this appears to be the case is indicated by examination of the traditional diagnostics of solar/stellar chromospheric and coronal activity (rotational velocities, the presence of dark starspots and the properties of those spots, and the X-ray flux level and its correspondence to the rotational period of the star), which are found to be similar in both the WTTS and the CTTS (see Cabrit et al. 1990; Montemerle and André 1989). Moreover, their stellar activity levels are comparable to those found for evolved stars with similar photospheric temperatures, luminosities, v sin i, and surface gravity, such as the RS CVn stars (Bertout 1989). While both the WTTS and CTTS clearly have significantly enhanced solar-type magnetic dynamo activity, the similar levels in both groups implies that an additional external energy source is required to explain the CTTS winds.

The evidence that CTTS with energetic winds are also surrounded by solar system-sized disks which are optically thick in their inner regions ($r < 1$ AU), suggests that it is the presence of those disks which is somehow responsible for driving the winds. Evidence is growing that the optically thick CTTS disks are in a state of active accretion. A compelling case for accretion is made by the successful modeling of the optical and ultraviolet continuum veiling in CTTS as boundary-layer emission, which is expected at the star/accretion disk interface (Lynden-Bell and Pringle 1974; Bertout et al. 1988; Hartigan et al. 1991; Chapter by Basri and Bertout). Furthermore, the magnitude of this veiling has been found to correlate with the infrared excess from the inner disk (Hartmann and Kenyon 1990). In the remainder of this section, we examine the empirical evidence that links diagnostics of wind energetics with diagnostics of disk accretion, and then briefly discuss some of the theoretical models proposed to account for the dynamical origin of these winds.

If accretion disks, possibly in conjunction with the stellar magnetic field, do provide the energy source for driving CTTS winds, then a correlation of wind mass loss rates $\dot{M}_w$ and disk accretion rates $\dot{M}_{acc}$, might be expected. Because each of these rates is difficult to quantify, appropriate surrogate diagnostics for $\dot{M}_w$ and $\dot{M}_{acc}$ must be selected which will be proportional to the desired physical quantities. As in Cabrit et al. (1990), we adopt the reddening corrected luminosities for lines presumably arising in the inner wind (Hα) and the outer wind ([O I] λ 6300) as diagnostics of $\dot{M}_w$.

Selection of a surrogate diagnostic for $\dot{M}_{acc}$ is less straightforward, especially as the physics of accretion disks is poorly understood at present. In theory, one could use either the excess luminosity in the optical and ultraviolet, arising in the boundary layer, or the excess luminosity in the infrared, arising in the optically thick CTTS circumstellar disks, to compare with the surrogates for $\dot{M}_w$. The former is not well determined for a large sample of stars (see Hartigan et al. 1991) and the latter suffers from the complication that the disk radiation comes from two distinct, but difficult to disentangle, effects, only one of which is $\dot{M}_{acc}$. The component of the infrared excess luminosity due to accretion, L_{acc}, will be proportional to $\dot{M}_{acc} (M_*/R_*)$ and will be comparable to the stellar luminosity for a typical T Tauri star if $\dot{M}_{acc} = 10^{-7}$ $M_\odot$ yr^{-1} (Kenyon and Hartmann 1987). A second contribution to the excess infrared luminosity will arise, however, in a flat dusty circumstellar disk which will "reprocess" stellar photons into the infrared, with resultant infrared luminosities due solely to reprocessing that can be 25% of the stellar luminosity. Moreover, if pure reprocessing disks have flared surfaces, with surface heights scaling as r^z, then the infrared excess luminosity from reprocessing will grow to 50% of the stellar luminosity for $z = 5/4$ (Kenyon and Hartmann 1987). The fact that reprocessing T Tauri star disks with variable morphology and inclination can potentially account for ratios of disk to stellar luminosity up to 0.5 requires that this ratio not be the sole discriminant in searching for evidence for disk accretion (Kenyon and Hartmann 1989).

Given these caveats, it is best to explore as many potential accretion diagnostics as possible. A simple test is to examine the relation between the $\dot{M}_w$ diagnostics, the line luminosities of Hα and [O I], and $K-L$, an observable quantity. As discussed in the previous section, for optically thick reprocessing disks, this color index will range from ~0.4 to 0.7, depending on aspect. In a classic, steady viscous accretion disk, with $T \sim r^{-3/4}$, $K-L$ values range from 0.4 to 0.9, depending on both the inclination and mass accretion rate (Bertout et al. 1988; Hartigan et al. 1990). An accretion disk which departs significantly from the standard model in generating its luminosity may have considerably higher values of $K-L$, which will be determined by the radial temperature gradient (see the Chapter by Adams and Lin).

The comparision of the $\dot{M}_w$ diagnostics with $K-L$ is shown in Fig. 3 for the identical set of 40 T Tauri stars in Tau-Aur shown in Fig. 2. When line luminosities rather than equivalent widths are compared with $K-L$, not only do the T Tauri stars sort between the presence and absence of a wind and the optical depth of the disk, as in Fig. 2, but a correlation of the $\dot{M}_w$ diagnostics is seen with $K-L$ as well. While this approach makes no effort to disentangle reprocessing and accretion contributions in the near infrared, the observed correlation of $\dot{M}_w$ diagnostics with $K-L$ beyond the regime allowed either by reprocessing or standard viscous accretion disks ($K-L>0.9$) rules out reprocessing dominated effects and also suggests that standard accretion disk models are likely to be oversimplifications of the real situation. In the following, we show that the directly observable quantity $K-L$ is also well correlated with two quantities that are likely to be good surrogate diagnostics for $\dot{M}_{acc}$: the mid-infrared excess luminosity and the ratio of the mid-infrared excess luminosity to the stellar luminosity. This, in turn, will support the suggestion that in CTTS, $\dot{M}_w$ is proportional to $\dot{M}_{acc}$.

As discussed in Cabrit et al. (1990), isolating the mid-infrared excess luminosity from the total infrared excess luminosity will provide a diagnostic less sensitive to reprocessing effects, and consequently more sensitive to $\dot{M}_{acc}$.

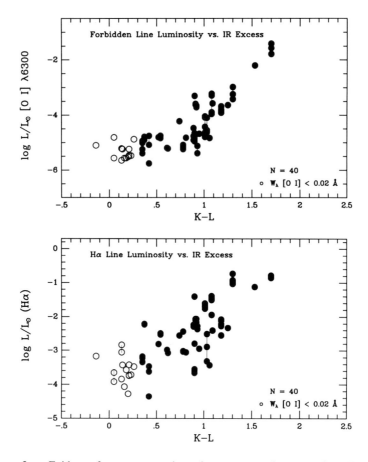

Figure 3.   Evidence for a correspondence between mass loss rates from T Tauri
    winds and the temperature structure in optically thick T Tauri disks. The dataset and
    symbols are identical to those shown in Fig. 2; in this case, however, the luminosities
    of the [O I] and Hα lines are plotted against $K - L$. These $\dot{M}_w$ diagnostics are seen
    to correlate with $K - L$ for the CTTS (filled symbols); the correlation extends to
    $K - L$ beyond the regime explained either by flared reprocessing disks or classic,
    steady viscous accretion disks ($K - L > 0.9$).

This follows because of the geometry of flared reprocessing disks; even in a
very flared disk, the inner disk ($r < 1$ AU; $\lambda = 1$ to $10\,\mu$m) will remain relatively
flat, in contrast to the outer disk regions, which can intercept significantly
more stellar photons from raised surfaces. Thus the large variations in excess
infrared luminosity which can arise due to reprocessing effects alone are most
significant at $\lambda > 10\,\mu$m.

    In a method similar to that of Cabrit et al. (1990), we have calculated
mid-infrared excess luminosities, $L_{\rm mir}$ ($\lambda = 0.9$ to $10.6\,\mu$m) and stellar lu-
minosties, L_*, for the 40 T Tauri stars in Figs. 2 and 3. (In this compilation,

J is the "normalization" wavelength, in which it is assumed that the flux at J is entirely photospheric [see Kenyon and Hartmann (1989); reddenings are from K. Strom].) Figure 4a shows the relation between L_{mir} and $K - L$, and Fig. 4b shows the relation between $L_{mir}/L_\star$ and $K - L$.

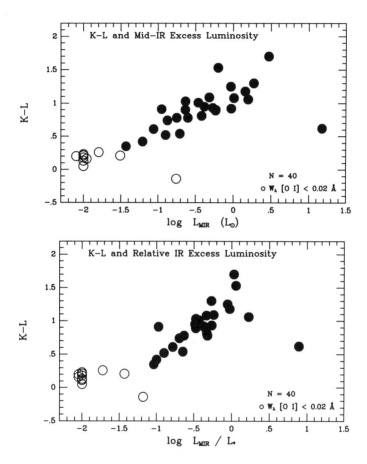

Figure 4.   Evidence for a correlation between $K - L$ and quantities proportional to $\dot{M}_{acc}$. (a) The mid-infrared excess luminosity, L_{mir}, and (b) the ratio of mid-infrared excess luminosity to stellar luminosity, $L_{mir}/L_\star$. The data and symbols are for the same stars shown in Figs. 2 and 3.

In Fig. 4, it is seen that the observed $K - L$ values for these 40 T Tauri stars correlate remarkably well both with L_{mir}, a quantity which will be sensitive to $\dot{M}_{acc}$, and with $L_{mir}/L_\star$ (which should be limited to values of 0.1 to 0.2 for a reprocessing disk [Kenyon, personal communication]). The extension of the $L_{mir}/L_\star$ correlation beyond the regime of reprocessing provides strong evidence that the observed correlations are dominated by effects due to disk accretion. The combination of Figs. 3 and 4 suggests that the surrogate diagnostics of $\dot{M}_w$ are correlated with diagnostics of $\dot{M}_{acc}$.

The arguments presented above implicate accretion disks as the source of energetic CTTS winds. However, the physical mechanism by which an active accretion disk might drive an energetic T Tauri wind is still controversial (see Bertout et al. [1991a] and Chapter by König and Ruden for a discussion of mechanisms under consideration). The terminal wind speeds of several hundred km s^{-1} suggest that the winds originate in the near-stellar regions of the gravitational potential well.

The possibility that the wind derives from hydromagnetic flows in strong toroidal magnetic fields produced by shear in the boundary layer regions has been investigated by Pringle (1989a). Other possibilities under theoretical investigation include magnetized accretion disks that power centrifugally driven outflows (Blandford and Payne 1982; König 1989; Chapter by König and Ruden; Camenzind 1990), and an interaction of the stellar magnetic field with that of the inner disk (Shu et al. 1988; Chapter by Shu et al.). Detailed pursuit of each of these models should soon provide sufficient observational diagnostics to begin discriminating among them.

III. COMPARISON OF WINDS FROM LOW-LUMINOSITY EMBEDDED INFRARED SOURCES AND OPTICALLY VISIBLE T TAURI STARS

A. Fast Neutral Winds and Molecular Outflows

Energetic winds emerge from low-luminosity young stars in two seemingly distinct stages of evolution; in addition to the winds diagnosed from the optically visible CTTS described in the previous section, powerful winds also emerge from objects believed to be T Tauri star predecessors—low-luminosity youthful stars still deep in their cloud cores, known as embedded infrared sources (EIR). Like the CTTS, interpretation of the spectral energy distributions of the EIR suggests a contribution from an optically thick circumstellar disk to explain the large mid-infrared excesses. (Adams et al. 1987; Myers et al. 1987; Chapter by Adams and Lin). The likelihood is that the EIR are in a state of vigorous accretion, with low angular momentum material infalling directly onto the stellar surface and high angular momentum material accreting through the circumstellar disk (Shu et al. 1987a).

Unlike the CTTS, however, the EIR with outflows are observed over a very wide range of bolometric luminosities (0.1 to 10^5 L$_\odot$), indicating that energetic winds emerge from embedded, forming stars of all masses. The progeny of the higher mass EIR are presumably the optically visible Herbig Ae/Be stars, which appear to share many properties in common with the lower mass classical T Tauri stars, including broad and strong Hα lines with velocity structure indicative of mass loss, broad low excitation metallic forbidden emission lines, and large infrared and mm-continuum excesses which are compatible with emission from optically thick circumstellar disks (Finkenzeller and Mundt 1984; Thé et al. 1986; Hu et al. 1989; Dent et al. 1989; Hamann and Persson 1989; Hessman et al. 1991; see Chapter by Strom et al.). Because mass loss rates and disk diagnostics for these higher-mass analogs of the CTTS are not as well determined at the present time (Catala 1988,1989), in this section we will confine our comparison of the wind energetics from embedded and optically visible young stars to those of low luminosity (see also Sec. IV.E).

The energetic outflows from the EIR are primarily traced by swept-up shells of high velocity molecular gas (tens of km s^{-1}), often exhibiting a bipolar morphology, extending to scales of up to several pc (see Chapter by Fukui et al.). In a few cases, the driving wind interior to the molecular shells has been detected as a very high speed, predominantly neutral wind with velocities on the order of hundreds of km s^{-1} (Lizano et al. 1988; Koo 1989; Margulis and Snell 1989). Additional outflow signatures from EIR include jets and HH objects (see next section) and knots of shocked molecular hydrogen (Lane 1989). The determination of the mechanical luminosities and momentum fluxes from molecular outflows is discussed by Lada (1985).

B. Comparison of Energetic Winds

It is a common misconception that CTTS winds are fundamentally different from, and less energetic than, those from EIR, where fast neutral winds drive molecular outflows. While many of the outward manifestations of the outflows from T Tauri stars and EIR differ, a comparision between the two reveals many similarities. Not only are (1) the wind velocities for the T Tauri winds and fast neutral winds from EIR very similar; (2) jets and HH objects found emerging from both types of objects (see next section); and (3) molecular outflows found from some CTTS (Edwards and Snell 1982; Calvet et al. 1983); but also (4) the wind luminosities and momenta from these two types of objects are similar as well, when objects of similar bolometric luminosities are compared.

The comparision of the wind mechanical luminosities L_w and momentum fluxes $\dot{P}_w$ between T Tauri winds and fast neutral winds from EIR is best made by using molecular outflows as a surrogate diagnostic for the fast neutral wind, as few fast neutral winds have been directly detected to date. The comparison is shown in Fig. 5a and 5b, where L_w and $\dot{P}_w$ are shown as a function of the bolometric luminosity of the source, either EIR (data points) or CTTS (locus within dashed box).

As seen in Fig. 5, *the winds from EIR and from CTTS have similar energies when objects of similar luminosity are compared* (see also Edwards and Strom 1987).

Additionally, some evidence suggests that while the CTTS and EIR winds from objects of similar source luminosity share much in common, in contrast to the CTTS, the molecular outflows swept up by the fast neutral winds from EIR have the momentum, but not the energy originally carried by their driving wind:

1. Close inspection of Fig. 5 reveals that while the average $\dot{P}_w$ of T Tauri winds and EIR molecular outflows for sources of comparable luminosity is very similar, the L_w for the CTTS winds with $\dot{M}_w$ between 10^{-7} and $10^{-8} M_\odot$ yr^{-1} appears to be systematically larger than typical values for molecular outflows. If this offset is real, it might imply that the (inferred) fast neutral winds from the EIR are originally comparably energetic to the CTTS winds, but when they impact the wind/molecular shell boundary, momentum is conserved but mechanical energy is not. This conclusion must be viewed with some caution, as the error in determining L_w and $\dot{P}_w$ for the molecular outflows is large (Cabrit 1989).

2. Comparison of the momentum and energy between the fast neutral wind and the swept up molecular shell from L1551 IRS-5 shows that while their momentum fluxes are identical, the energy deposition rate is an order of magnitude greater in the fast wind than in the molecular outflow (Natta 1989). The energy deficit is comparable to that seen radiated away in the far infrared as diffuse emission with a morphology similar to the molecular outflow (Edwards et al. 1986; Clark and Laurejis 1986; Natta 1989). Unfortunately, additional examples of extended far infrared emission from molecular outflows comparable to that found from L1551 IRS5 are not easily studied with the IRAS data, which is too severely constrained in angular resolution and sensitivity (Jarrett et al. 1987; see also Hughes et al. 1989).

3. Comparison of the momentum and energy between directly observed CTTS winds and their accompanying molecular outflows for the two stars in which both are observed (Natta 1989), shows that while the momentum assessed from the optical and molecular line diagnostics for the same star is identical, the wind mechanical luminosities are again more than a magnitude lower in the molecular outflow than in the optically visible T Tauri wind (this follows from the identical $\dot{M}_w$ found for the optical and molecular flows, but wind and outflow velocities differing by an order of magnitude).

This comparision of the EIR and CTTS winds shows no evidence that the winds from CTTS are less energetic than those from their more deeply embedded, and presumably younger, counterparts when objects of the same luminosity are compared. This argues that a similar wind generation mechanism is operating for young stellar objects over a wide range of ages. Further support for similar wind origins from the EIR and the CTTS is given in the next section. The lack of observed molecular outflow diagnostics from the majority of CTTS is likely to be attributed to cleaner circumstellar surroundings, not to less energetic winds.

C. Evolution of Wind and Accretion Disks

The likelihood that the energetic winds from CTTS derive from a process related to disk accretion is based on the correlation of surrogate diagnostics for

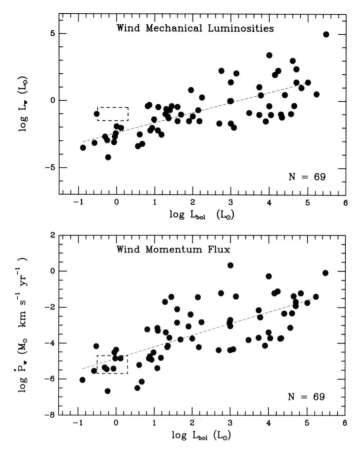

Figure 5.   (a) A comparison of wind mechanical luminosities for embedded sources
with those from optically visible CTTS. The plotted points are data for 69 molecular
outflows taken from the list of Fukui (1989) for which published data are available.
The locus of (L_w, L_{bol}) points for typical CTTS with mass loss rates of 10^{-7} to
10^{-8} $M_\odot$ yr^{-1} and wind speeds of 200 km s^{-1} is enclosed by a dashed box.   (b)
A comparison of wind momentum fluxes for molecular outflows from embedded
sources with those from optically visible CTTS. The plotted points are from the
same molecular outflows from the list of Fukui (1989) shown in Fig. 4a and the
locus of CTTS points for the wind properties in Fig. 4a are shown by the box. In each
panel, the dotted line corresponds to a linear fit to the displayed data, illustrating the
correlations of L_{bol} with L_w and $\dot{P}_w$. The proportionalities are very similar to those
found by Lada (1985) and Levreault (1988*b*).

$\dot{M}_w$ and $\dot{M}_{acc}$ in these stars (Sec. II). The embedded nature of the EIR prevents
the separation of photospheric and excess luminosities. Thus correlations of
$\dot{M}_w$ and $\dot{M}_{acc}$ similar to those presented in Figs. 3 and 4 for the CTTS are not
easily made. It is certainly possible, however, that the observed correlation
of L_w and $\dot{P}_w$ with L_{bol} for the molecular outflows (Lada 1985; Levreault

1988b) shown in Fig. 5 derives at least in part from an $\dot{M}_w/\dot{M}_{acc}$ correlation, provided that a significant fraction of L_{bol} comes from L_{acc}.

Associating the winds from deeply embedded EIR and optically visible CTTS with the presence of an accretion disk does not speak to the evolution of the wind as a given object emerges from its embedded phase to become an optically visible object. For example, if higher disk accretion rates are found in the earlier EIR phase compared to the CTTS phase, then a given object might evolve down the L_w/L_{bol} relation shown in Fig. 5 as it becomes optically visible. This would follow if L_{bol} is dominated by L_{acc} from infall and accretion luminosity in the EIR stage, overwhelming the stellar luminosity produced by the release of gravitational energy from stellar contraction $L_\star$. Even for the optically visible T Tauri stars, where the stellar and disk contributions can be distinguished, many examples are found where the disk contribution to L_{bol} is comparable to or greater than $L_\star$ (Cabrit et al. 1990; Cohen et al. 1989b).

As yet we know little about the evolution of the mass loss rate in the energetic wind (which might in turn trace the evolution of the accretion rate through the disk) as a given source moves from an embedded state to an optically visible, but still accreting, object. Study of the spatially resolved optical outflow diagnostics, the jets and HH objects, which are found both from embedded and optically visible sources, gives some insight into the wind (and disk) evolution, as described in the next section.

IV. OPTICAL OUTFLOWS: JETS AND HH OBJECTS

A. Background

When Herbig (1951) and Haro (1952) discovered the first HH objects in Orion, the nature of these nebulous semi-stellar knots with emission line spectra was far from understood although it was clear from the start that they were associated with star formation. Schwartz (1975) was the first to suggest that HH spectra probably arise in the cooling region of a high-velocity shock, the shock itself being located in the wind from a young stellar object (YSO). At the time this idea was put forward, spherically symmetric YSO winds were assumed but later it became apparent that HH objects, far from tracing spherical outflows, instead delineated high surface brightness knots in or at the leading edge of collimated bipolar optical outflows with velocities of several hundred km s^{-1}. The possibility that at least some HH objects result from the interaction of high-velocity jets with their environment was proposed by Königl (1982) who discussed several possible shock excitation mechanisms. That jets are indeed a relatively frequent phenomenon was first demonstrated by Mundt and Fried (1983) and Mundt et al. (1984) with the help of deep CCD imaging of known YSO and HH objects. Because of their spectral similarity to HH objects, these jets are sometimes referred to as HH jets, but the link between jets and HH objects goes much deeper.

We now know that together they delineate those parts of the energetic outflows from YSO with the highest degree of collimation and velocity. They are frequently found along the outflow axis defined by the swept-up bipolar molecular flows, which have velocities that are an order of magnitude slower and are considerably less collimated. In at least two well-documented cases, the axes of the optical and molecular outflows have been shown to be perpendicular to the equatorial planes of 500 to 1000 AU diameter disks traced by dense molecular gas (Sargent 1989). The realization that the optical outflows traced by the jets and HH objects do not carry the energy and momentum necessary to drive the molecular outflows (Mundt et al. 1987) was instrumental in instigating a search for the dominant, neutral component of the winds from embedded YSO; this was carried out successfully by Lizano et al. (1988). More recently it has been realized that HH objects and jets, like molecular outflows, are associated not only with low-luminosity YSO ($L_{bol} \leq 10^2$ $L_\odot$) but high-luminosity YSO ($L_{bol} \sim 10^3 - 5 \times 10^4$ $L_\odot$) as well (see, e.g., Hartigan et al. 1986; Poetzel et al. 1989; Ray et al. 1990).

The origin of jets and HH objects and in particular their connection with other outflow phenomena is still controversial. Below we discuss these problems and examine jets and HH objects in detail.

B. Morphology

Both jets and HH objects display a wide range of morphologies. In Figs. 6, 7 and 8 we illustrate the morphological variety among YSO jets. Figure 6 shows an image of the HH 34 jet, which is one of the best examples of a straight, knotty jet emanating from a low-mass YSO (Reipurth et al. 1986; Bührke et al. 1988). Like several other known jets, it is pointing towards a bow shock-like HH object (HH 34S). In this case, we can even observe a "counter bow shock" on the opposite side (HH 34N). Radial velocity measurements show that HH 34N is redshifted, while the brighter object HH 34S is blueshifted. In contrast the HH 110 jet (Fig. 7; Reipurth 1989c) shows strong wiggles and its knots are irregularly spaced. A similar knotty but straight jet is associated with the bright bow shock-like HH object RNO 43N (see Fig. 8).

HH objects also come in a variety of morphologies. For example, some sources are linked with only one small and nearly circular HH object such as HH 57 (Reipurth 1989c), while others are associated with extended and highly-structured HH objects consisting of many condensations, e.g., GGD 37 (Lenzen 1988) or HH 2 (Herbig and Jones 1981).

If one studies a large sample of jets and HH objects the following general features are found:

1. All known jets have a knotty structure and in many cases these knots are spaced periodically or quasi-periodically.
2. Bow shock-like HH objects or bright knots often mark the end of a jet. As discussed below these sometimes rather prominent features are almost certainly the "working surface" of the jet where part of its kinetic energy

is transferred into heat via shock waves. Often these bright features were recorded photographically, and noted as HH objects, long before the first deep CCD studies of their surroundings.

3. Approximately 50% of the flows traced by HH objects and jets are bipolar. Of the remainder, it turns out that normally only the blueshifted components are seen, presumably because the redshifted outflow is directed into the parent cloud and obscured.

4. The full projected opening angles of the most highly collimated flows (e.g., the HH 30 jet) are as small as $1°$ with aspect (length-to-diameter) ratios of more than 30. Usually, however the opening angles traced by HH objects and jets range from 5 to $20°$ although in a few cases larger values are known, e.g., V645 Cyg, (Goodrich 1986; Zou and Solf, in preparation).

5. Typically the lengths, velocities and number densities of jets are in the range 0.01 to 1 pc, 200 to 600 km s^{-1}, and 10 to 200 cm^{-3}, respectively. As their diameter can often be resolved, jet mass fluxes in the range 10^{-10} to 3×10^{-6} M$_{\odot}$ yr^{-1} and mechanical luminosities of 10^{-3} to 50 L$_{\odot}$ have been derived.

The rather complex morphologies of jets and the collimated flows traced by HH objects (e.g., HH 84 or HH 85, Reipurth 1989b) and even that of the of the HH objects themselves may arise in part from the inhomogeneous environment with which the flow interacts. However, factors such as temporal variability of the flow and various parameters (e.g., opening angle) are also certainly very important. Finally, it should be kept in mind that HH objects and jets may not trace all of the outflow from a YSO, because they *only* mark those locations where the *normally neutral* (but highly supersonic) flow becomes ionized via shock waves. The reasons why such shock waves arise are discussed below.

C. The Nature of the Emission from HH Objects/Jets and Their Kinematics

The individual spectra of HH objects show a broad range in the degree of their excitation. Those of high-excitation have spectra characterized by strong [O III] λ 5007 emission but comparatively weak [S II] λλ 6716, 6731 lines. The reverse is true for low-excitation HH objects and in many of the latter [O III] is not even detectable. Both high- and low-excitation HH objects show the [O I] λλ 6300, 6363 lines but this doublet is much stronger in the low-excitation case.

Böhm (1956), Haro and Minkowski (1960) and other pioneers in the field of HH object spectroscopy quickly realized that the large number of low-excitation lines (e.g., Ca II, [Fe II], [O I] and [S II]) seen in many HH objects cannot be due to photoionization as in a H II region. In the latter, atoms like oxygen, which have a first ionization potential similar to hydrogen, exist almost exclusively in an ionized state; hence [O I] emission is always

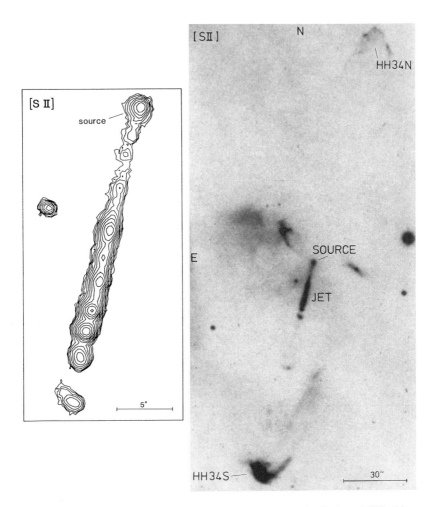

Figure 6.   A [S II] image of the HH 34 jet and the bow shock-shaped HH object
HH 34S from Bührke et al. (1988). Note the extensive wings of HH 34S and its
knotty appearance near the apex. At least 12 quasi-periodically spaced knots are
visible in the jet within 26″ from the source and roughly halfway between the source
and the apex another bow-shaped knot is seen. The latter observation supports the
idea that episodic outbursts are common among optical outflow sources.

relatively weak. The origin of their line emission was to remain a mystery
until Schwartz (1975) noted the resemblance of the spectra of HH objects to
supernova remnants like the Cygnus Loop and this in turn led him to suggest

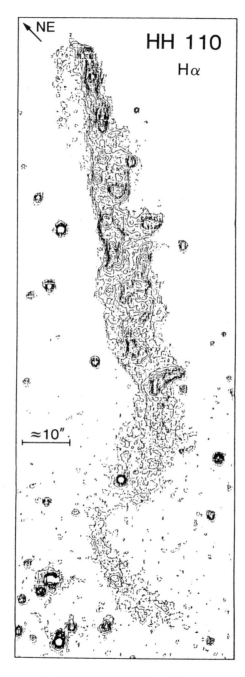

Figure 7.   A contour plot of the HH 110 jet in Hα from Reipurth (1989b). Here the positions of the knots seem random and the jet oscillates markedly from side to side giving the jet an overall turbulent or chaotic appearance.

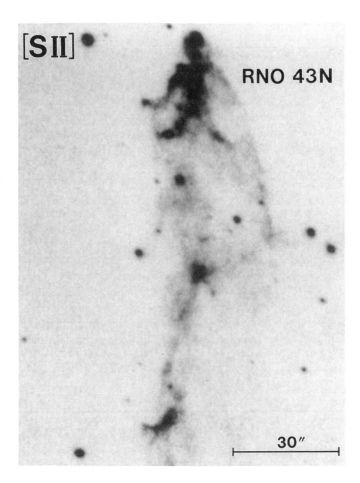

Figure 8.   The RNO 43N jet and bow shock in [S II] which again, as in the case of HH 110, shows a very chaotic jet, but somewhat surprisingly, a well-defined bow shock. The projected size of this system is at least 0.3 pc although its source has not been clearly identified.

that HH emission was also due to radiative shocks. The precise geometrical form of such shocks was unclear and still is in some cases. Early models assumed the shocks to be plane parallel (see, e.g., Raymond 1979; Dopita 1978) but, although qualitatively it was possible to reproduce many of the spectral characteristics of HH objects, it turned out to be extremely difficult to fit the wide range of line profiles and ionization species found.

Given the spectral resemblance of HH objects and jets, it is clear that the emission from jets is also due to shocks. Here, however, the shocks cannot be perpendicular to the direction of the flow but instead must be oblique. Otherwise it is extremely difficult to explain why for example high-excitation lines like [O III]λ5007 have rarely been observed in a jet, (except perhaps at its terminus) despite some jets having velocities of up to 600 km s^{-1}. Instead, only low-excitation lines are seen implying the presence of oblique shocks with normal velocities of around 30 to 70 km s^{-1} (Mundt et al. 1987; Bührke et al. 1988). Such shocks have been seen not only in laboratory jets but in numerically modeled ones as well (see Norman et al. 1982; Norman 1986; Falle et al. 1987 and references therein). Despite the attractiveness of this model, it should be emphasized that, presumably because of seeing limitations, oblique shocks have never been observed in a jet. The jet emission, which comes from the recombination zone behind the shock, originates in a very thin region; typical values are around 10^{15} cm or 0.''4 at the distance of Taurus Auriga (see, e.g., Raymond 1979). The radiative zones of HH objects are of similar size and this accounts for the low filling factors for HH objects (and jets) of around 10^{-2} to 10^{-1} (see, e.g., Böhm 1983). Obviously spatial stratification in the ionized species (Raga 1989) is even harder to observe but it may have been seen in the case of HH 34S (Bührke et al. 1988).

Dyson (1987) has emphasized that if HH emission comes from radiative shocks, then HH objects can be produced in a variety of ways. If this is the case then a unified theory for their origin is not possible. Certainly, HH objects come in several forms. Some, which were first identified photographically, turned out upon deeper inspection with CCDs, to be bright knots within jets. Examples include HH 33/40 (Mundt et al. 1984) and the RNO 43N jet (Ray 1987). Others, such as HH 28/29 (Strom et al. 1974) and GGD 37 (Hartigan et al. 1986; Lenzen 1988), appear to be due to the interaction of a poorly collimated YSO wind with its ambient medium. Perhaps, however, the class of HH objects which have attracted the greatest amount of attention recently from a theoretical standpoint are those resembling bow shocks. This group includes HH 1 (Raga et al. 1988), HH 32 (Raga et al. 1986), HH 34N and HH 34S (Bührke et al. 1988), HH 39 (Mundt et al. 1987), HH 83 (Reipurth 1989b) and RNO 43N (Ray 1987). Other examples of HH objects for which a bow shock model seems to be appropriate include HH 47 (Hartigan 1989) and HH 101 (Hartigan and Graham 1987) although applying such a model may be difficult for HH 101 (Hartigan 1989). All of these objects appear to mark the end of outflows and are presumably the regions where the jet flows into the surrounding medium. The bow shock itself is where ambient material is accelerated in the direction of the jet while at an inner (jet) shock, the jet material is decelerated. Work is done on the ambient medium as the shock system moves outward, so the latter is often referred to as the "working surface."

Spectroscopically bow shock-like HH objects and jets differ in two major respects: first, as mentioned previously, mostly low excitation lines are seen in jets whereas bow shock-shaped HH objects display high-excitation lines, at least at their apex. Second, while the oblique shocks within jets produce very narrow emission lines (FWHM~10 km s^{-1}), much broader lines (with FWHM ~100 to 200 km s^{-1}) are found within bow shock-shaped HH objects.

Both of these differences can be adequately explained by simple bow-shock models (Hartigan et al. 1987; Raga et al. 1986). According to such models, the thermal energy in the post-shock zone is approximately equal to the kinetic energy associated with the component of velocity normal to the shock. Thus the apex of the bow shock is expected to be at a much higher temperature, and hence excitation, than the wings. Moreover it can be easily shown that the maximum width of the emission-line profile from a bow shock has a FWZI (full width zero intensity) $\sim V_{bs}$ where V_{bs} is the velocity of the bow shock (see, e.g., Hartigan et al. 1987). Given the large values of the latter, the broad line widths are not surprising.

Bow shock-shaped HH objects often appear clumpy although the cause of their clumpiness is unknown. It may, for example, be due to thermal instabilities (Innes et al. 1987; Raga and Böhm 1987) or inhomogeneities in the ambient medium. Blondin et al. (1989) have shown that dynamical instabilities could also play a role. According to bow-shock models for HH objects, the highest excitation lines appear near the apex of the shock and this is supported by the observations (Ray and Mundt 1988). Moreover, a property of bow shocks is that, although the shock velocity at the apex may be as high as 250 km s^{-1}, that of the wings is much less. Thus, using lines like the [S II] $\lambda\lambda6716,6731$ doublet, which are excited when $V_{shock} \geq 40$ km s^{-1}, one can trace the actual wings out to large distances from the apex. Aside from the bow shock, the inner jet shock, which is in the form of a Mach disk, should be observable in some cases. Hartigan (1989) has shown that the surface brightness of both shocks may be comparable if the contrast in densities of the jet and surrounding medium is not more than 1 or 2 orders of magnitude. This agrees with the spectroscopic observations by Bührke et al. (1988) who found that the terminal shock in the HH 34 jet may contribute significantly to the emission from HH 34S. Moreover the location of the terminal shock appears to have been identified by Reipurth (1989b).

While optical emission lines characterize the spectra of moderate-to-high-velocity shocks (40 to 300 km s^{-1}), lower-velocity shocks can be detected using the near-infrared lines of molecular hydrogen (see Lane 1989). Historically, the association between shocked molecular H_2 and HH objects has been known for at least a decade (Elias 1980). However, early studies suffered from low resolution (typical beams had apertures ≥ 5 to $10''$) often making it impossible to locate accurately the position of the H_2 emission with respect to the optical. This situation changed with the availability of infrared array detectors making it possible, for the first time, to map large areas very efficiently. These maps showed (Lane 1989) that the shocked H_2 and optical emission are spatially closely related and that the optical and H_2 line intensities are correlated. Moreover it has been established using high resolution spectroscopy that the velocity structure in the H_2 lines is similar to that of the optical lines but the actual widths and velocities are usually much smaller by at least a factor of 2 (Zinnecker et al. 1989).

In the case of bow shocked-shaped HH objects, one expects the H_2

emission to arise far back in the bow-shock wings where the velocities are small enough ($V_{shock} \leq 30$ km s^{-1}) not to dissociate hydrogen. Somewhat higher velocities may be allowed ($V_{shock} \sim 50$ km s^{-1}) if one assumes the shock is not a classical jump (J) shock but instead is a softer MHD continuous (C) shock (Draine 1980; Draine et al. 1983; Smith and Brand 1990). It is then surprising to see, for example, in the case of RNO 43N (Lane 1989), that some of the H_2 emission comes from the vicinity of the bow-shock apex. One explanation for this might be that the jet contains molecular hydrogen (perhaps due to entrainment) and that this hydrogen passes through the jet's terminal shock without significantly being decelerated as is consistent with the high proper motion of RNO 43N. Moreover a possibly related puzzle is that in some outflow sources, e.g., OMC-1 (Brand et al. 1989) and DR21 (Garden 1987), the H_2 line widths reach nearly 140 km s^{-1}. Certainly such broad profiles are difficult to explain using shocks having simple flow geometries without dissociating molecular hydrogen. In principle, however, these broad profiles can be understood in terms of more complex flow patterns with oblique internal shocks. One way around both problems (see Smith et al. 1991) may be C shocks with high Alfvènic velocities (~ 50 km s^{-1}). Whether such values are realistic at large distances from the YSO is a matter of debate; certainly this problem requires further investigation.

H_2 emission from jets has been detected in relatively few cases, e.g., HH 7–11 (Garden et al. 1990), HH 40 (Zinnecker et al. 1989; Lane 1989) and possibly the HH 1/2 VLA 1 jet (Zealey et al. 1989). In addition, a few well collimated H_2 flows have been found with no known optical counterparts (see, e.g., Garden et al. 1990). The origin of the emission, like that of the optical, is presumably due to oblique shocks but whether these might, for example, be due to entrainment (Zinnecker et al. 1989) at the jet boundaries or shocks driven into the ambient medium by the passage of the jet itself (see Cohn 1983) is not known.

As HH objects and jets are at least partially ionized, they are also sources of thermal free-free radio emission. At radio flux, however, scales with the $H\alpha$ surface brightness, and this is very low, the observed fluxes are very small. Pravdo et al. (1985) detected radio continuum emission from HH 1/2 and from the source of these HH objects (HH 1/2 VLA 1). Tentative marginal detections of HH 12 (Snell and Bally 1986) and HH 101 (Brown 1987) have been reported but the faintness of most HH objects make such observations extremely difficult. Recently, Rodríguez (1989) has claimed that the HH objects associated with high-luminosity stars are easier to detect in the radio continuum because of their high intrinsic luminosity and in spite of their large distances. Examples include HH 80/81 (Rodríguez and Reipurth 1989), and the Orion "Streamers" in L1641 (Yusef-Zadeh et al. 1990). However, only in the case of HH 80/81 is the HH nature of these objects beyond question.

The radio continuum images of the sources of HH 1/2 (HH 1/2 VLA 1) and HH 80/81 (IRAS 18162-2048) are interesting in that they display an elongated jet-like morphology with the major axis directed along the outflow direction. DG Tau, L1551 and Z CMa (Bieging and Cohen 1985) also show a similar shape, and as argued, for example, by Natta (1989), the spatial coincidence of this radio emission with the observed optical jets suggests that the jet contributes significantly to the radio emission from these YSO. This hypothesis is also supported by the observation that those T Tauri stars which show evidence for optical outflows are much more likely to be radio sources than ordinary T Tauri stars (Bieging and Cohen 1985). Thus in some cases the observed radio emission from T Tauri stars may be due to spatially unresolved jets or some other shock-excited gas associated with outflow.

Both HH objects and flows are sources of ultraviolet line and continuum emission. All observations in this case have come from the IUE satellite and the reader is referred to Brugel (1989) for an excellent review of this topic; here only some salient points are made. From an examination of the ultraviolet line spectrum, it is seen that the optical spectroscopic division of HH objects into low- and high-excitation classes carries over. Objects which are classified *optically* to be of low excitation have only weak ultraviolet lines although fluorescent H_2 emission may also be present at short wavelengths. By comparison, high-excitation objects display C IV λ 1549, [C III] λ 1909 and [C II] λ 2326 in emission. An interesting feature of the ultraviolet spectra of HH objects is the presence of a "strong blue continuum," the existence of which was already known from optical studies (Brugel et al. 1981) although it was not suspected that it would extend so far into the ultraviolet (Brugel 1989). It has been proposed that its origin is two photon H I emission (Dopita et al. 1982) although this mechanism cannot explain the energy distribution of the continuum at short ultraviolet wavelengths (see, e.g., Böhm et al. 1987; Cameron and Liseau 1990). Here Böhm et al. (1987) suggest the destruction of H_2 may play a role.

D. Proper Motions

Determining proper motions is important for two main reasons. First they can be combined with spectroscopic radial measurements as an aid to establishing the true, as opposed to the projected, outflow direction. Second, in the many cases where there is confusion, one can determine the source of the HH emission by tracing its proper motion vector backwards. Of course, because HH objects were discovered approximately three decades before jets, large amounts of proper motion data is available for HH objects but has only recently become available for jets. Studies (see, e.g., Schwartz 1986) show that most HH objects have relatively small tangential velocities ($\sim$0 to 200 km s^{-1}) but higher velocities are known; for example, the knots within HH 1 reach at least 350 km s^{-1} (Herbig and Jones 1981). Moreover, given the recently discovered very high radial velocities of HH objects and jets associated with high-luminosity YSO (see below), it is clear that much higher tangential velocities remain to be uncovered.

In the case of stellar jets, measurements are confined to determining the proper motions of their bright knots. Although this gives a handle on the jet velocity (when combined with the known radial velocity), it should be emphasized that the tangential velocities of the knots may often not reflect the jet's true tangential velocity. The knots may, for example, be instabilities and hence move with a velocity that is less than the bulk velocity of the jet (see below). Typical proper motions for these knots are in the range (100 to 300 km s^{-1}) and again we expect the upper limit to increase when data becomes available for higher-mass stars. Examples of jets for which proper motions have been measured include the L1551-IRS 5 jet (Neckel and Staude 1987), the HH 34 system (Eislöffel et al. 1989; Reipurth 1989c), and several of the jets of the XZ/HL Tau region (Mundt et al. 1990). As discussed below, such large tangential velocities rule out steady shocks as the origin of the knots.

E. Dependence of Optical Outflow Characteristics on Source Luminosity

Until recently, most known optical outflows were associated with low-luminosity YSO ($L_* \leq 100$ L$_\odot$), including both embedded infrared sources and optically visible CTTS. However, given that molecular outflows are common across a wide range of pre-main-sequence stars, it is not surprizing that HH objects and jets also occur at the higher-mass end of the YSO spectrum as well. Examples where HH objects have been found associated with high-luminosity sources include M42 (Axon and Taylor 1984), Cep A (Hartigan et al. 1986; Lenzen 1988), V645 Cyg (Goodrich 1986; Hamann and Persson 1989; Zou and Solf, in preparation), AFGL 2591 (Poetzel et al., in preparation), HH 80/81 (Rodríguez and Reipurth 1989; Reipurth 1989c) and the Herbig Be stars LkHα 198 and MWC 1080 (Poetzel 1990). Jets have been reported from LkHα 234 (see Fig. 9), AFGL 4029 (Ray et al. 1990) and Z CMa (Poetzel et al. 1989).

While emphasizing that we are dealing with a statistically small sample, several important points can be made:

1. The optical outflows from high-luminosity stars are in general faster than those from low-luminosity stars. Line profiles of the former may extend from 450 to 700 km s^{-1} in the wings. In addition, mass loss rates and mechanical luminosities are, as one would expect, larger, scaling roughly with $L_{bol}^{0.6}$ (see Fig. 10).
2. In some cases, e.g., Cep A and V645 Cyg, the HH emission does not seem to be confined to a well-defined central axis but is instead scattered over a broader range of angles. Such scattered emission could be construed as evidence for a poorly collimated high-velocity wind.

Although several of the high-luminosity YSO with optical outflows are embedded sources, of those that are optically visible (LkHα 198, LkHα 234, MWC 1080, Z CMa and V645 Cyg), virtually all show a Herbig Ae/Be spectrum (Herbig 1960; Finkenzeller and Mundt 1984; Goodrich 1986). Z CMa is the exception as it appears to be an FU Orionis star (Hartmann et al. 1989) deriving most of its luminosity from an accretion disk (see Chapter by Hartmann et al.).

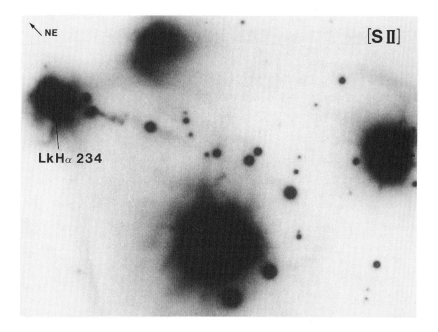

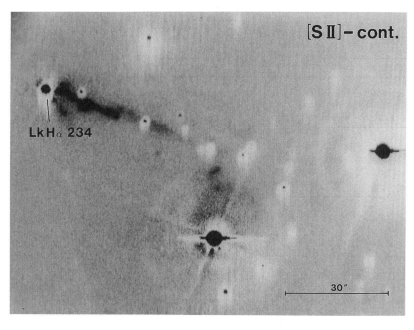

Figure 9. An example of a jet from a high-luminosity pre-main-sequence star LkHα 234. This image of the core of NGC 7129, taken in [S II], also shows the edge of a reflection nebula that has been seen as a cavity in CO. The jet is directed along the axis of this cavity.

Most of the optically visible YSO which do have jets and HH objects are also associated with molecular outflows as well. This is true both for the higher-luminosity Ae/Be stars and the lower-luminosity CTTS (Levreault 1988a; Cantó et al. 1984; Schwartz 1983; Edwards and Snell 1982). In both cases, the fraction of the known population of optically visible sources which display these spatially resolved outflow phenomena is small, indicating that they may be transition objects between the embedded and optically visible phases.

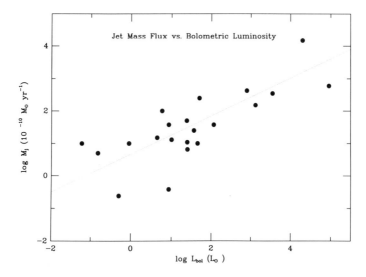

Figure 10.   A log-log plot of jet mass flux vs bolometric luminosity of the parent source. The same power-law dependence on L_{bol} is seen as for the mass flux required to drive molecular outflows shown in Fig. 5.

F. The Formation and Propagation of Jets

Although various models have been proposed as to how jets form, there is no consensus on their origin. Raga and Cantó (1989), for example, basing their scenario on the earlier models of Cantó (1980) and Königl (1982) , suggest that jets form when an initially isotropic YSO wind is collimated by pressure gradients in its surrounding medium. However, we will argue in Sec. IV.G that the jets come directly from the disks and if this is so then the wind will from the start be bipolar (Pudritz 1988) perhaps leaving the innermost regions of the disk in a poorly collimated form. Subsequent focusing by the magnetic field (see, e.g., Uchida and Shibata 1985; Uchida 1989) or the thermal pressure of the ambient medium may then play a role. With regard to the question as to where the jet is focussed the observations of Raga et al. (1991) and Mundt et al. (1991) are interesting. They have mapped, for some jets, the variation in diameter with distance from the source and found that significant collimation is often achieved within several hundred AU. Recollimation may even occur on larger scales of several thousand AU. Initial opening angles appear to be quite wide before being reduced to a few degrees by external forces over the range of 10^2 to 10^3 AU.

Early numerical models for jets (see, e.g., Norman et al. 1982) all assumed the jet to be adiabatic. This is strictly only true for extragalactic jets because in YSO jets radiative losses, especially in the region where they form and at their working surfaces, can be significant. En route to the working surface, however, the adiabatic approximation appears to be a good one, as radiative losses at the oblique shocks scales with $(V_{shock}/V_{jet})^2$ and thus HH jets can propagate over a large portion of their length without significant energy losses. Nonadiabatic propagating (i.e., time-dependent) jets have been studied by Tenorio-Tagle et al. (1988) and Raga (1988), while the detailed behavior of the head of the jet has been explored by Blondin et al. (1989). As pointed out by Raga (1989), far away from the head, and upstream, a pattern of stationary biconical shocks develops and it has been proposed (Falle et al. 1987) that these be identified with the quasi-periodic knots seen in many jets. Obviously if this were the case one could study these shocks in detail by numerically modeling the flow using steady-state equations; this is essentially the approach taken by Falle et al. (1987). However, it appears that such shocks are not the origin of the knots observed, for example, in the HH34 jet. In the first place, such shocks are predicted to be spaced around $M_j D_j$ apart where D_j and M_j are the jet's diameter and Mach number, respectively. As typical Mach numbers and diameters are about 30 and $1''$, the observed spacings are nearly an order of magnitude smaller than predicted by the stationary shock model. Second, as shown, for example, by the proper motion studies of Mundt et al. (1990), Eislöffel et al. (1989) and Reipurth (1989c), knots within YSO jets move outwards with tangential velocities comparable to their observed radial velocities (i.e., a few hundred km s^{-1}). Thus the shocks certainly cannot be regarded as stationary. A third problem (see Raga 1989) concerns the shock velocities predicted by such models. As we are dealing with crossing shocks, the angle they make with the jet's central axis $\theta \sim \sin^{-1}(1/M_j)$. The shock velocities will then be approximately $V_j/M_j = c_s$ where V_j is the jet's velocity and c_s is the sound speed. Assuming a jet temperature of $\sim 10^4$ K, c_s will have a value of 10 to 20 km s^{-1} and the shock velocity is hardly enough to dissociate molecular hydrogen. At best one could use such models to explain very low-excitation jets.

What then is the origin of the quasi-periodic knots seen, for example, in the HH 34 (Bührke et al. 1988), HH 83 and HH 111 jets (Reipurth 1989c)? Bührke et al. (1988) proposed Kelvin-Helmholtz (K–H) pinching instabilities (see, e.g., Ray 1981) as one possible explanation for the knots but to achieve the correct spacing one required a hot tenuous medium outside the jet. As there is no evidence for the latter, an alternative suggestion is an external medium of similar density that travels at a velocity comparable to the jet. As the knot spacing is determined by the *relative velocity* of the jet and its ambient medium, the spacing in this case can be much smaller than that for the steady-shock models. Moreover, a K–H model can not only explain why the knots move (as pinching instabilites) but also can explain the wiggles seen in some jets, e.g., HH 30 (Ray et al. 1991) and HH 83 (Reipurth 1989c), through the growth of the K–H helical mode. An alternative proposal (Uchida 1989) is that such wiggles are due to the helical instability of a twisted magnetic field. Here the twist is due to the rotation of the star itself. Whatever the origin of the wiggles, it seems unlikely that they are due to precession (see Uchida 1989, and references therein).

Aside from pronounced wiggles, bipolar jets have also been seen to bend gradually in an S-shaped pattern (Mundt et al. 1990) centered on their source. One explanation for the bends might be a misalignment between the ambient magnetic field and the direction of the jet (Mundt et al. 1990). For this mechanism to work the jet has to have a magnetic field of its own. Although magnetic fields in jets and HH objects seem likely, if only because of their ubiquitousness, firm evidence for their presence is lacking. However, the recent observations by Yusef-Zadeh et al. (1990) of nonthermal emission from an HH-like object automatically implies the presence of magnetic fields in at least one case.

G. Relationship of Optical Outflows to Other Outflow Phenomena

While a coherent picture of the role of energetic winds in the star formation process is beginning to emerge, the origin of the optical parts of the outflows, the jets and HH objects, remains the most enigmatic. Although they do not dominate the energetics of the winds, their compelling morphology, with regularly spaced knots delineating collimated structures extending up to a parsec from their source, and their alignment along the axes of the bipolar molecular outflows, pose important questions to those seeking to understand energetic winds from young stars. In particular, study of these optical outflows may hold the key to understanding the timing of accretion events in the disks which ultimately power the outflows, and to identifying the mechanism which powers energetic YSO winds.

In spite of the fact that the optical outflows possess only a trace of the energy and momentum carried by the energetic winds, it is clear that they must both ultimately derive from the same energy source. Not only are the optical outflows found along the axes of bipolar outflows (and presumably the driving winds) from YSO of all luminosities, but their energetics are seen to scale with the bolometric luminosity of the source in precisely the same way as found for the driving winds. The correlations between wind mechanical luminosites and momentum flux of the molecular flows with the bolometric luminosity of the parent YSO was shown in Fig. 5. Levreault (1988b) has shown that the mass loss rates in the driving wind scale with the source bolometric luminosity roughly as $L_{bol}^{0.6}$. The same scaling law is also found for the mass loss rates from jets with source luminosity, as shown in Fig. 10 (see also Ray 1991), although the magnitude of the jet mass loss rate for a given L_{bol} is down by a factor of 30 from that for the driving winds. The fact that such a correlation exists at least over the range 0.1 to 2×10^4 $L_\odot$ argues strongly for a universal mechanism for powering jets and molecular outflows. (It is interesting to note that one of the jet sources in Fig. 10, Z CMa, is best modeled as deriving most of its luminosity from an accretion disk [see Chapter by Hartmann et al.], further suggesting that jets originate from processes related to disk accretion.)

A puzzle which must be explained by any model of energetic winds from young stars is why the outflows have a dual nature (see Stocke et al. 1988). The evidence suggests that neutral atomic winds surround shock-ionized jets in all outflows from embedded sources. Although the atomic winds have only been directly detected from a limited number of outflow sources, the requirement for them on energetic grounds is compelling (Sec. III). Furthermore, images of many outflow regions seem to require their presence as well. For instance, jets are often seen to emerge from conical nebulae, e.g., HH 34 (Bührke et al. 1988), 1548C27 (Mundt et al. 1984), Z CMa (Poetzel et al. 1989) and L1551 IRS-5 (Stocke et al. 1988) where the morphology of the nebula suggests it has also been blown by such a wind. These "cavities" appear much too large (e.g., LkHα 234; see Ray et al. 1990) to have been cleared by the jets themselves.

There is preliminary evidence that the energetic winds from optically visible CTTS may have a dual nature as well. Interpretation of the velocity structure of the forbidden-line profiles in CTTS has suggested that many of these stars have sub-arcsecond, and therefore as yet spatially unresolved jets, similar in velocity and excitation to those which have been imaged (Edwards et al. 1989; Solf 1989; Mundt et al. 1990). Additionally, Mundt (1988) has pointed out that strong emission line CTTS often show narrow blueshifted sodium doublet (Na D) absorption with modest velocities (~ 70 km s^{-1}). This, he proposes, is due to a second, less collimated, wind component.

A comparison of the average dynamical time scales $\bar{\tau}_D$ for the optical and molecular outflows provides strong evidence that episodes of mass loss leading to the formation of jets and HH objects may be intermittent during the time interval when a source is in a molecular outflow phase (Mundt et al. 1990). Values for $\bar{\tau}_D$ are derived by simply dividing the length scale of the observed outflows by their characteristic velocities and then averaging. The average dynamical times for molecular outflows $\bar{\tau}_{DM}$ is found to be larger by at least an order of magnitude than the corresponding value $\bar{\tau}_{DO}$ for optical outflows. Rough estimates of $\bar{\tau}_{DO}$ and $\bar{\tau}_{DM}$ are 10^3 and 2–5×10^4 yr, respectively (Mundt et al. 1990). If one accepts that the dynamical time of the molecular outflow is indicative of the total time that the source has been in the outflow phase, then this suggests that optical outflows may be an intermittent phenomenon, perhaps caused by multiple outbursts from the parent YSO, and lasting for about a time $\bar{\tau}_{DO}$.

An alternative approach to this comparison is to measure the number of known optical outflow sources in, for example, Taurus Auriga (Mundt et al. 1987), and to compare this with the total number of known CTTS in this cloud. If the fraction is multiplied by the typical age of CTTS (see Stahler 1983a), i.e., $\sim 3 \times 10^5$ yr, one can obtain a statistically estimated total time scale τ_{SO} during which optical outflows are detectable. Mundt et al. (1987) deduced the latter to be $\sim 2 \times 10^4$ yr which again when compared with $\bar{\tau}_{DO}$ suggests optical outflows switch themselves "on and off" within a period of a few hundred to a thousand years. Note, however, the time scale τ_{SO} may be longer for YSO in a closely knit group (Mundt et al. 1988).

Obviously if jet/HH outflows are intermittent, there should exist a class of CTTS which are potential jet/HH sources but where none are present. Mundt et al. (1990) list several candidate members of this class. Intermittency is also indicated by the observations of Reipurth (1989c) and Hartigan et al. (1990) who describe several objects (e.g., HH 46/46 and HH 111) with suggestions of velocity variations in their outflows on time scales of 10^2 to 10^3 yr. These objects show not only a leading bow-shock but a second bow shock structure closer to the source. Such a secondary working surface is also seen in the HH 34 system (Bührke et al. 1988; Fig. 6).

V. SUMMARY AND FUTURE PROSPECTS

The role that energetic winds play in the star formation process is beginning to be understood. Theoretical arguments suggest that stars of all masses form largely by accretion of material through circumstellar disks and that the forming stars cannot accept this high angular momentum disk material unless angular momentum is shed via winds from the inner disk/near stellar regions (Chapter by Shu et al.). Observational evidence presented here suggests that energetic winds, with mechanical luminosities which are a significant fraction of the bolometric luminosity of the source, emerge from embedded, forming stars of all luminosities. We have argued on empirical grounds that these energetic winds must derive their energy from a process related to disk accretion and must originate in the near-stellar regions of the star/disk system. We see that the phase of disk accretion/energetic winds can persist after a star has become optically visible, when it will appear as a CTTS or a Herbig Ae/Be star.

The duration of the disk accretion/energetic wind phase appears to vary among stars of a given mass. This is best seen in the low-mass optically visible stars of ages $<10^7$ yr, where some objects (CTTS) persist in a disk accretion/energetic wind phase for longer than several times 10^6 yr, while some others (WTTS) have apparently cleared their inner disks, terminating, at least temporarily, the disk accretion/energetic wind phase on time scales as short as 10^6 yr (Strom et al. 1989d). By 10^7 yr, apparently all low-mass stars have finished the phase in which disk-related activity is a dominant energy source (see Chapter by Strom et al. for a full discussion of the time scales for disk clearing). Presumably initial conditions in the molecular cloud core, such as angular momentum, or gravitational interactions with close neighbors in the same star-forming complex are responsible for the observed range of disk-clearing time scales (Pringle 1989b).

If the above scenario is the correct one, and much work remains to be done to verify it, then there are some important implications for star and planetary system formation:

1. The energetic winds may be agents for carrying away the significant angular momentum which resides in the disk. This would apply even during the optically visible CTTS phase, in order to account for the identical distribution of v sin i found for the CTTS, presumed to be accreting material from their inner disks rotating with Keplerian velocities, and for the WTTS, in which disk accretion and energetic winds have halted (Hartmann et al. 1986; Walter et al. 1988).

2. If all stars form largely via accumulation of material through circumstellar accretion disks, they may also have the potential to form planetary systems such as our own. If a planetary system is formed from a circumstellar disk, it is unlikely that energetic winds from a T Tauri star can play a major role in the final debris clearing. This follows from the fact that once the inner disk has become optically thin, energetic winds no longer exist.

3. One must look to the properties of the WTTS to evaluate the impact that a young star will have on forming planets, asteroids and comets. These stars are still very active by the solar/stellar standards of old main-sequence stars, and much work remains to be done to characterize their chromospheric/coronal activity.

The optical outflow phenomena associated with the phase of disk accretion/energetic winds from YSO probably hold important clues regarding the mechanism driving the energetic winds and the timing of accretion events in the disks. These shock-ionized jets and HH objects, while not dominating the wind energetics, can possess remarkable degrees of collimation and extend up to a parsec from their source. In at least a few well-studied cases, these jets emerge perpendicular to the disk planes. The dual nature of energetic winds from YSO is not well understood, but the evidence that shock-ionized jets are surrounded by neutral atomic winds from embedded sources is strong. A similar dual-nature wind may persist during the CTTS phase as well, if the majority of these systems are also characterized by spatially unresolved, but highly collimated jets, as suggested by their forbidden-line structure. The degree to which these neutral winds interact with the circumstellar accretion disks, and thus their direct impact on forming planetary systems before the time of disk clearing, remains to be established.

The dynamical time scale for the optical outflow phenomena and the occurence of sequential, multiple jets and bow shocks from the same source, provide evidence that large, intermittent increases in mass outflow rates, and possibly disk accretion rates as well, may occur. The largest-scale intermittent accretion/outflow events are probably associated with the FU Ori phenomenon, where a disk characterized by extremely high accretion rates appears to dominate the spectrum of a YSO and corresponding very high wind mass loss rates are also inferred (see Chapter by Hartmann et al.). The recent discovery of a very high-velocity and spatially extended jet from an optically visible YSO that appears to be a Fuori object, Z CMa, supports this interpretation (Poetzel et al. 1989).

High spatial resolution imaging of optical outflow phenomena, either from space or with adaptive optics on groundbased telescopes, should clarify some of the uncertainties associated with the formation and propagation of jets. Ascertaining the degree of collimation of jets on spatial scales of several AU will help constrain models for generating energetic winds from YSO and observing the proper motion of fine structure in the jets might, for example, reveal evidence for angular momentum in these outflows.

Acknowledgments. We wish to thank S. Strom for a critical reading of the manuscript, G. Basri and S. Kenyon for valuable comments and suggestions, and C. Andrulis, L. Ghandour and L. Hillenbrand for assistance in preparation of figures. S. Edwards wishes to acknowledge partial support from the NASA Planetary Program. T. Ray acknowledges support from EOLAS, the Irish Science and Technology Agency.

MOLECULAR OUTFLOWS

YASUO FUKUI, TAKAHIRO IWATA, AKIRA MIZUNO
Nagoya University

JOHN BALLY
A. T. and T. Bell Laboratories

and

ADAIR P. LANE
Boston University

We review observations of molecular outflows which mark the earliest phase of stellar evolution. Millimeter-wave observations are compared with optical and near-infrared data, and a general description of molecular outflows is given. In addition, a few sources of both low- and high-mass star formation are discussed in detail. The statistical properties of molecular outflows suggest that this phase corresponds to the main accretion phase of protostellar objects. Models of outflow collimation and acceleration, as well as the contribution of outflows to the mechanical support of molecular clouds against gravitational collapse, are discussed in light of the most recent radio data.

I INTRODUCTION

As stars form inside molecular clouds, the first observational manifestation of the star-forming event is usually the production of an energetic mass outflow. Detectable by its broad mm-wave emission lines, especially CO, the outflowing gas is typically observed to move with velocities of order 10 km s^{-1}, but in some flows, this motion exceeds 100 km s^{-1}. Because the mm-wave emission lines are easily excited by collisions with atoms and molecules at the ambient cloud temperature, these lines can be used to trace the total mass, momentum, and energy contained in cold, accelerated molecular gas. Outflows are frequently associated with other observational signatures of dynamical interaction between high-velocity and ambient gas. Shock-excited optical emission in the form of Herbig-Haro (HH) objects and stellar jets, shock-excited near-infrared emission produced by molecular hydrogen, and mid- and far-infrared cooling line radiation such as that from [O I], [C II], OH, and CO can be used to study gas which has been heated to thousands of degrees Kelvin in outflows and to trace the cooling postshock layers. Since the cooling time for the hot component of molecular outflows is short, typically 1 to 100 yr, the infrared and optical lines trace the present location of shocks and energy dissipation in the flow.

In this chapter, we focus on phenomena which are most likely related to the earliest phase of stellar formation. We use a sample of $\sim$150 molecular outflows to describe the characteristics of the outflow phase of pre-main-sequence evolution. In analyzing these phenomena, we rely heavily on CO data, which do not suffer from the effects of extinction. In addition, we use a sample of $\sim$230 HH objects which have been discovered so far to infer the characteristics of the shocked material in molecular outflows.

In this analysis, we must be cautious about possible differences between the outflows produced by high-mass and low-mass objects. As noted by Shu et al. (1987a), high-mass stars may reach the main sequence while they are still deeply embedded in molecular clouds because of their short evolutionary time scale. This makes it difficult to discern evolutionary status of high-mass protostars precisely from observations. On the other hand, the longer Kelvin-Helmholtz contraction times associated with low-mass protostars imply that they may end their accretion phase without having reached the main sequence, making it easier to recognize their evolutionary status. Although many of the best-known outflows are produced by high-mass objects, it is outflows associated with low-mass, solar-type stars which are most relevant to the study of solar system formation.

We also discuss the current status of our understanding of three outstanding issues concerning molecular outflows: the acceleration and collimation mechanism, the relation of the outflow phase to the evolutionary status of protostars, and the role that outflows play in cloud support, especially in regions of low-mass star formation.

II RADIO OBSERVATIONS OF CARBON MONOXIDE

A. Searches for CO Outflows

Although various signatures of outflow activity from young stellar objects were recognized during the 1960s, it was the discovery of the widespread occurrence of bipolar CO outflows during the late 1970s and early 1980s which led to the realization that outflows are a fundamental feature of the star-formation process. Between 1976 and 1985, searches for broad millimeter CO emission lines were made toward selected optical or near-infrared objects (Bally and Lada 1983; Levreault 1985); HH objects (Snell and Edwards 1981, 1982; Edwards and Snell 1983, 1984); T Tauri stars (Kutner et al. 1982; Edwards and Snell 1982; Calvet et al. 1983); Ae and Be stars (Cantó et al. 1984); and dark clouds (Frerking and Langer 1982; Goldsmith et al. 1984). These works led to the discovery of $\sim$60 outflows as listed by Lada (1985).

Since 1985, unbiased surveys of entire molecular clouds have been carried out by two groups using small millimeter telescopes having relatively large beam sizes (2$'$ to 3$'$). These surveys were conducted with the Nagoya 4 m telescope for Orion, Monoceros, Cepheus and Ophiuchus (Fukui 1988,1989; Fukui et al. 1986,1989,1992, in preparation; Sugitani and Fukui 1988; Sugitani et al. 1989; Nozawa et al. 1991), and with the Texas 4.9 m telescope for Monoceros (Margulis and Lada 1986; Margulis et al. 1988). The Orion-Monoceros region has been most intensively surveyed in an unbiased way. Four clouds have been searched with the Nagoya 4 m telescope, including the Orion A or south cloud (containing L1641 and the Orion Nebula), the Orion B or north cloud (containing L1630 and NGC2023, NGC2024, NGC2068 and NGC2071), Mon R2, and S287, and a fifth cloud, Mon OB1 (containing NGC2264), with the Texas 4.9 m telescope. Figure 1 shows the locations of 36 outflows, including 23 that were discovered by unbiased searches. A new unbiased survey has been made for L1641 with the Bell Laboratories 7 m telescope (Morgan and Bally 1991; Morgan et al. 1991).

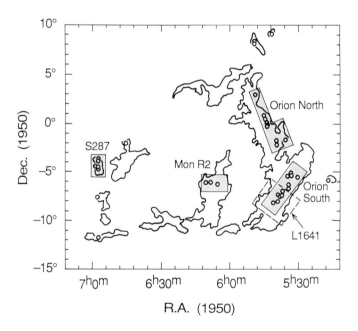

Figure 1.   Locations of 36 molecular outflows shown as circles superposed on a contour map of ^{12}CO ($J=1-0$) total intensity of the Orion-Monoceros region obtained with the Columbia mm-wave telescope (Maddalena et al. 1986). Hatched areas denote the regions where the outflow sources, newly found by the unbiased survey with the Nagoya 4 m telescope (see, e.g., Fukui et al., in preparation), are located. A dashed rectangle denotes the region of the L1641 dark cloud.

The IRAS satellite opened new opportunities for searches for molecular outflows by identifying embedded young stellar objects by their infrared emission. Several surveys of IRAS infrared sources have been carried out (Casoli et al. 1986; Heyer et al. 1987; Snell et al. 1988; P. R. Schwartz et al. 1988; Parker et al. 1988; Fukui 1989; Wouterloot et al. 1989; Snell et al. 1990; Wilking et al. 1990; Fukui et al. 1992, in preparation; Morgan and Bally 1991). These, unbiased and IRAS-based surveys in the late 1980s, have discovered another 90 flows, of which 52 were found with the Nagoya survey as listed by Fukui (1989).

In Table I, we summarized mapped or confirmed molecular outflows. It includes 163 outflows published in the literature until June 1991 and/or discovered in the Nagoya CO survey. Out of the 163 presently known outflows, ∼100 are located within 1 kpc of the Sun. This is about twice the number listed by Lada (1985) to lie within 1 kpc. A typical lifetime of 5×10^4 yr, inferred from Table III in Fukui (1989), implies that the outflow formation frequency is $\sim 6 \times 10^{-4}$ kpc^{-2} yr^{-1}, a value comparable to the formation rate of 1 M$_\odot$ stars (see, e.g., Miller and Scalo 1979). The total number of outflows expected to lie within 1 kpc of the Sun can be estimated from the data shown in Fig. 1, coupled with extensive mapping of the molecular clouds in which they reside. For three clouds, including the Orion south, S287 and Mon OB1 clouds, the ratio of total cloud mass to the number of outflows found in the cloud is 1700, 3000, and 3000 M$_\odot$/outflow, respectively. If we take 2000 as an average value for this ratio, and a total molecular mass of 6×10^5 M$_\odot$ lying within 1 kpc of the Sun, the total number of outflows associated with this gas is estimated to be 300. As ∼100 have been detected so far, 200 remain to be discovered within 1 kpc of the Sun.

B. Physical Properties of CO Outflows

Roughly 85% of CO outflows are bipolar, and almost all are associated with infrared sources. The physical characteristics of the outflows discovered after 1985 are similar to the ones listed by Lada (1985) in his Table III, except that they tend to have lower kinetic energy and lower mechanical luminosity. Of the 52 outflows discovered at Nagoya, 90% have stellar luminosities fainter than 1000 L$_\odot$, while of the sources listed by Lada (1985), ∼60% have $L_* \geq 1000$ L$_\odot$, showing that the recent searches have resulted in greater numbers of outflows from low-luminosity objects.

Various physical parameters for CO outflows have ranges as follows: R_{max} = 0.04 to 4 pc; V_{max} = 3 to 150 km s^{-1}; dynamical age = 1×10^3 to 2×10^5 yr; molecular mass = 0.1 to 170 M$_\odot$; outflow momentum = 0.1 to 1000 M$_\odot$ km s^{-1}; kinetic energy = 10^{43} to 10^{47} erg; and mechanical luminosity = 0.001 to 2600 L$_\odot$. The large outflow mass indicates that most of the accelerated mass has been swept up from the molecular cloud surrounding the central star rather than originating in mass ejected from the protostar.

The bulk of the mass in molecular outflows is cold. Margulis and Lada (1985) and Snell et al. (1984) used observations of the $J = 2-1$ and $J = 1-0$ transitions of ^{12}CO and ^{13}CO to demonstrate that in a sample of 7 typical outflow sources, the excitation temperature of ^{13}CO is in the range 10 to 90 K, and is on average colder than the surrounding quiescent gas. Another estimate of gas temperature by using the NH$_3$(1,1) and (2,2) transitions indicates gas

TABLE I
Catalog of CO Outflow Sources[a]

No.	Object[b]	R.A. (1950)[b]			Decl. (1950)[b]			d (kpc)	Outflow Structure	IRAS No.	Ref.[e]
		00[h]	08[m]	47.9[s]	58°	33'	09"				
1.	LkHα 198	00	08	47.9	58	33	09	1.0	bipolar	00087 + 5833	1
2.	00213+6530	00	21	22.0	65	30	25	0.85	bipolar	00213 + 6530	2,3
3.	00259+6510	00	25	59.8	65	10	12	0.85	bipolar	00259 + 6510	2,3
4.	L1287	00	33	53.3	63	12	32	0.85	bipolar	00338 + 6312	2,4,5
5.	L1293	00	37	57.4	62	48	26	0.85	bipolar	00379 + 6248	2
6.	NGC281-west	00	49	27.8	56	17	28	2.1	bipolar	00494 + 5617	2,4
7.	NGC281-east	00	51	18.0	56	17	07	2.1	bipolar	00512 + 5617	2
8.	01133+6434	01	13	18.2	64	34	50	3.0	bipolar	01133 + 6434	6
9.	S187-IRS	01	19	58	61	33	08	2.0	bipolar	01202 + 6133[d]	1,7
10.	W3-IRS5	02	21	55.4	61	52	34	2.3	bipolar	02219 + 6152	1
11.	IC1805-west	02	25	14.5	61	20	10	2.3	bipolar	02252 + 6120	2
12.	AFGL4029	02	57	35.6	60	17	22	2.2	bipolar	02575 + 6017	8
13.	AFGL437	03	03	33.2	58	19	21	2.0	bipolar	03035 + 5819	1
14.	RNO13	03	22	04.7	30	35	49	0.2	blue	03220 + 3035	1
15.	L1448	03	22	33.5	30	33	34	0.3	bipolar	(no)	9
16.	AFGL490	03	23	39.1	58	36	33	0.9	bipolar	03236 + 5836	1
17.	RNO15 FIR	03	24	34.9	30	02	36	0.35	bipolar[c]	03245 + 3002	1,10
18.	NGC1333 IRAS1	03	25	33.6	31	03	14	0.35	bipolar	03255 + 3103	11
19.	HH7–11 SSV13	03	25	57.9	31	05	50	0.35	bipolar[c]	03259 + 3105	1,12
20.	B5 IRS3	03	43	56.5	32	33	55	0.35	red	03439 + 3233	13
21.	L1489	04	01	40.6	26	10	48	0.14	bipolar	04016 + 2610	14

22.	T Tau	04	19	02.4	19	25	00	0.14	bipolar	04190 + 1924	1
23.	IR04191+1523	04	19	08.5	15	23	16	0.14	bipolar	04191 + 1523	2
24.	Haro 6–10	04	26	21.7	24	26	26	0.14	bipolar	04263 + 2426	2,15
25.	ZZ Tau	04	27	50.6	24	35	24	0.14	red	04278 + 2435	16
26.	L1551-IRS5	04	28	43.8	18	01	51	0.12	bipolar	04287 + 1801	1
27.	HL/XZ Tau	04	28	44.8	18	07	34	0.12	blue	04287 + 1807	1
28.	L1535	04	32	31.6	24	02	08	0.14	red	04325 + 2402	16
29.	L1642	04	32	32.0	–14	19	18	0.1	bipolar	04325 – 1419	17
30.	IC2087	04	36	54.6	25	39	17	0.14	red	04369 + 2539	16
31.	L1634	05	17	21.9	–05	55	05	0.5	bipolar	05173 – 0555	2
32.	AFGL5142	05	27	27.6	33	45	37	1.8	bipolar	05274 + 3345	8
33.	RNO43 S	05	29	32.7	12	47	33	0.5	bipolar^c	05295 + 1247	1,18
34.	Ori A-west	05	30	14.5	–05	37	52	0.5	bipolar	05302-0537	2,19
35.	Ori KL	05	32	47	–05	24	14	0.5	bipolar	(no)	1
36.	OMC-1 FIR4	05	32	45.9	–05	26	06	0.5	bipolar	(no)	20
37.	OMC-2	05	32	59.6	–05	11	32	0.5	bipolar	05329 – 0512^d	21
38.	L1641-north	05	33	52.7	–06	24	02	0.5	bipolar	05338 – 0624	19,22
39.	NGC1999	05	33	59.4	–06	44	45	0.5	red^c	05339 – 0644	1,10
40.	Ori A-east	05	34	11.0	–05	30	03	0.5	red	05341 – 0530	2,19
41.	AFGL5157	05	34	32.6	31	57	40	1.8	bipolar	05345 + 3157	8
42.	Ori-I-2	05	35	33.2	–01	46	50	0.4	bipolar	05355 – 0146	23
43.	S233	05	35	48.8	35	43	41	1.8	bipolar	05358 + 3543	4
44.	L1641-center	05	36	20.9	–07	02	43	0.5	blue	05363 – 0702	2,24
45.	Haro 4–255	05	36	56.4	–07	28	14	0.5	bipolar	05369 – 0728	1
46.	GGD 4	05	37	21.3	23	49	22	1.0	bipolar	05373 + 2349	25
47.	S235 B	05	37	31	35	39	55	1.8	bipolar	05375 + 3540^d	1,26
48.	L1641-south3	05	37	31.1	–07	31	59	0.5	bipolar	05375 – 0731	2
49.	L1641-south	05	38	02.7	–07	28	59	0.5	bipolar	05380 – 0728	19

No.	Name									IRAS	Ref.
50.	L1641-south4	05	38	24.6	−08	08	20	0.5	bipolar	05384 − 0808	2
51.	NGC2024 / Ori-B	05	39	18.0	−01	56	42	0.5	bipolar[c]	05393 − 0156	1,27,28,29
52.	L1641-south2	05	40	23.2	−08	18	26	0.5	bipolar	05403 − 0818	2,24,30
53.	L1594 / B35	05	41	45.3	09	07	40	0.5	bipolar	05417 + 0907	2,14
54.	HH26 IR	05	43	31.1	−00	15	28	0.5	bipolar	05435 − 0015	1
55.	HH24	05	43	34.2	−00	11	08	0.5	bipolar	05435 − 0011	1
56.	NGC2068 H_2O	05	43	58	−00	04	00	0.5	bipolar	05437 − 0001[d]	1
57.	NGC2071	05	44	30.3	00	20	42	0.5	bipolar	05445 + 0020	1
58.	NGC2071-north	05	45	07.8	00	37	41	0.5	bipolar	05451 + 0037	19,31
59.	HH110	05	48	52.8	02	55	25	0.5	bipolar	05487 + 0255	32
60.	S242	05	49	05.2	26	58	52	2.1	bipolar	05490 + 2658	4
61.	L1617 / HH111	05	49	09.1	02	47	48	0.5	bipolar	05491 + 0247	2,32
62.	L1598-NW	05	49	27.9	08	20	48	0.9	bipolar	05494 + 0820	2
63.	L1598	05	49	39.1	08	12	55	0.9	bipolar	05496 + 0812	33
64.	AFGL5173	05	55	20.3	16	31	46	2.5	bipolar	05553 + 1631	4
65.	HD250550	05	59	07	16	13	06	1.0	red	(no)	1
66.	Mon R2	06	05	20.4	−06	22	31	0.8	bipolar	06053 − 0622	1
67.	AFGL6366-S	06	05	40.9	21	31	32	1.5	bipolar	06056 + 2131	8
68.	AFGL5180	06	05	53.9	21	38	57	1.5	blue	06058 + 2138	8
69.	GGD 12–15	06	08	24.5	−06	11	12	1.0	bipolar	06084 − 0611	1
70.	S254–258	06	09	57.9	18	00	12	2.5	bipolar	06099 + 1800	1,34
71.	Mon R2-east	06	10	21.8	−06	12	28	0.95	bipolar	06103 − 0612	2
72.	RNO73	06	30	52.7	04	02	27	1.6	bipolar	06308 + 0402	4
73.	AFGL961	06	31	59.0	04	15	09	1.6	bipolar	06319 + 0415	1
74.	R Mon	06	36	25.6	08	46	57	0.7	bipolar	06364 + 0846	35
75.	Mon OB1-H	06	38	11.9	10	39	41	0.76	bipolar	06381 + 1039	36
76.	Mon OB1-D	06	38	17.8	09	39	03	0.76	bipolar	06382 + 0939	36
77.	Mon OB1-I	06	38	19	10	52	39	0.76	bipolar	(no)	36

78.	NGC2264	06	38	26.2	09	32	25	0.76	red	06384 + 0932	1
79.	Mon OB1-G	06	38	27	09	58	28	0.76	bipolar	06384 + 0958[d]	36
80.	S287-north	06	45	19.2	−02	09	32	2.3	blue	06453 − 0209	2
81.	BFS 56	06	56	45.1	−03	50	41	2.3	bipolar	06567 − 0350	2
82.	BIP 14	06	56	46.5	−03	55	28	2.3	bipolar	06567 − 0355	2
83.	CMa-west	06	56	52.9	−11	54	46	1.1	bipolar	06568 − 1154	2
84.	S287-B	06	57	06.4	−04	41	48	2.3	bipolar	06571 − 0441	2
85.	S287-C	06	57	08.2	−04	36	10	2.3	bipolar	06571 − 0436	2
86.	L1654	06	57	16.8	−07	42	16	1.1	bipolar	06572 − 0742	2
87.	S287-A	06	57	54.5	−04	32	22	2.3	bipolar	06579 − 0432	2
88.	L1660	07	18	00.9	−23	56	42	1.0	bipolar	07180 − 2356	33
89.	HH120	08	07	40.2	−35	56	07	0.43	bipolar	08076 − 3556	37
90.	HH46/47	08	24	16.5	−50	50	44	0.43	bipolar	08242 − 5050	37
91.	HH132	08	33	42.6	−40	28	02	0.95	bipolar	08337 − 4028	38
92.	ε Cha I Center	11	05	28.0	−77	06	32	0.14	bipolar	11054 − 7706	39
93.	ε Cha I North	11	08	22.0	−76	18	09	0.14	bipolar	11083 − 7618	39
94.	GSS30 / ρ Oph A	16	23	24.8	−24	17	46	0.16	bipolar	(no)	40,41
95.	ρ Oph B	16	24	13.3	−24	18	50	0.16	bipolar[c]	16242 − 2422[d]	1,42,43
96.	L1709	16	28	33.5	−23	56	32	0.16	bipolar	16285 − 2356	2
97.	ρ Oph-south	16	28	53.6	−24	50	06	0.16	bipolar	16288 − 2450	2
98.	ρ Oph-east	16	29	20.9	−24	22	13	0.16	bipolar	16293 − 2422	19,44,45
99.	RNO91 / L43	16	31	37.7	−15	40	52	0.16	bipolar	16316 − 1540	1,14
100.	L146	16	54	27.2	−16	04	48	0.16	red	16544 − 1604	2
101.	L100	17	13	03.9	−20	53	39	0.23	bipolar	17130 − 2053	46
102.	NGC6334	17	16	35.1	−35	54	48	1.7	blue	17165 − 3554	1,21
103.	W28-A2	17	57	28.5	−24	03	59	3.0	bipolar	17574 − 2403	47
104.	M8 E	18	01	49.7	−24	26	56	1.5	bipolar	18018 − 2426	1
105.	L483	18	14	50.6	−04	40	49	0.25	bipolar	18148 − 0440	46

No.	Name	RA (h)	(m)	(s)	Dec (°)	(′)	(″)	Size	Type	IRAS	Refs
106.	GGD27	18	16	12.8	−20	48	51	1.7	bipolar	18162 − 2048	48
107.	L379 IRS3	18	26	32.9	−15	17	51	0.2	bipolar	18265 − 1517	49
108.	L379 IRS2	18	27	43.4	−15	16	45	2.0	bipolar	18277 − 1516	33
109.	S68 / Serpens	18	27	50.0	01	11	37	0.5	bipolar	18278 + 0111	1
110.	G35.2−0.74	18	55	41.2	01	36	28	2.0	bipolar	18556 + 0136	11,50
111.	R CrA	18	58	31.6	−37	01	30	0.7	bipolar[c]	18585 − 3701	1,10
112.	W49	19	07	51.7	09	01	11	14	bipolar	19078 + 0901	51
113.	L723	19	15	41.3	19	06	47	0.30	bipolar	19156 + 1906	1
114.	L673	19	18	08	11	14	00	0.3	bipolar	19180 + 1116[d]	42,52
115.	AS353	19	18	08.8	10	56	10	0.20	bipolar	19181 + 1056	1
116.	L778	19	24	26.4	23	52	37	0.25	bipolar	19244 + 2352	14
117.	B335	19	34	35.4	07	27	24	0.40	bipolar	19345 + 0727	1
118.	L810	19	43	21.7	27	43	37	1.5	bipolar	19433 + 2743	53
119.	S87	19	44	13.5	24	28	00	2.7	bipolar	19442 + 2427	1,54
120.	S88-B	19	44	41.4	25	05	17	2.0	bipolar	19446 + 2505	1,11
121.	K3−50	19	59	50.0	33	24	20	9.0	bipolar	19598 + 3324	11
122.	20126+4104	20	12	41.0	41	04	20	1.7	bipolar	20126 + 4104	55
123.	20188+3928	20	18	50.7	39	28	18	4	bipolar	20188 + 3928	56
124.	AFGL2591	20	27	35.2	40	01	09	1.2	bipolar	20275 + 4001	1
125.	W75-N	20	36	51.1	42	27	20	2.0	bipolar	(no)	21
126.	DR21	20	37	13	42	08	50	2.0	blue	(no)	1,21
127.	L1157	20	38	39.6	67	51	33	0.44	bipolar	20386 + 6751	2
128.	PV Cep	20	45	23.6	67	46	36	0.5	bipolar	20453 + 6746	1
129.	V1057 Cyg	20	57	06.2	44	03	47	0.7	blue	(no)	1
130.	L1228	20	58	14.5	77	24	05	0.15	bipolar	20582 + 7724	43,57,58
131.	V1331 Cyg	20	59	32.3	50	09	53	0.7	red	20595 + 5009	1
132.	L988-a	21	00	44.9	49	51	13	0.7	bipolar	21007 + 4951	59
133.	L1172-D	21	01	44.2	67	42	24	0.44	bipolar	21017 + 6742	14

No.	Name	RA (h)	(m)	(s)	Dec (°)	(′)	(″)	Size	Type	IRAS	Ref.
134.	L988-e	21	02	19.6	50	02	40	0.7	bipolar	21023 + 5002	59
135.	L988-f	21	02	24.5	49	55	50	0.7	bipolar	21024 + 4955	59
136.	IC1396-west	21	24	38.7	57	43	14	0.75	bipolar	21246 + 5743	2
137.	V645 Cyg	21	38	11.3	50	00	45	6.0	bipolar	21381 + 5000	1
138.	GN21.38.9	21	38	53.2	56	22	18	0.75	bipolar	21388 + 5622	60
139.	IC1396-north	21	39	10.3	58	02	29	0.75	bipolar	21391 + 5802	23
140.	NGC7129	21	41	52	65	49	50	1.0	bipolar	(no)	1
141.	LkHα 234	21	41	53.2	65	52	42	1.0	red	21418 + 6552	1
142.	IC1396-east	21	44	30.8	57	12	29	0.75	red	21445 + 5712	2
143.	EL 1-12	21	45	26.8	47	18	08	0.9	red	21454 + 4718	1
144.	BD+46°3471	21	50	39.4	46	59	42	0.9	bipolar	21506 + 4659	1
145.	S140	22	17	41.1	63	03	41	0.9	bipolar	22176 + 6303	1
146.	S140-north	22	17	51.1	63	17	50	0.9	bipolar	22178 + 6317	19
147.	L1204-A	22	19	50.7	63	36	33	0.9	bipolar	22198 + 6336	2
148.	L1204-B	22	19	55.7	63	22	12	0.9	blue	22199 + 6322	2
149.	L1221	22	26	37.2	68	45	52	0.2	bipolar	22266 + 6845	2,61
150.	L1203	22	26	46.7	62	44	22	0.9	bipolar	22267 + 6244	2
151.	L1206	22	27	12.2	63	58	21	0.9	blue	22272 + 6358	23
152.	L1251-A	22	34	22.0	75	01	32	0.2	bipolar	22343 + 7501	33,62
153.	L1251-B	22	37	40.8	74	55	50	0.2	bipolar	22376 + 7455	62
154.	L1211	22	45	23.3	61	46	07	0.75	bipolar	22453 + 6146	2
155.	Cep A	22	54	20.2	61	45	55	0.7	bipolar	22543 + 6145	1
156.	Cep E	23	01	10.1	61	26	16	0.75	bipolar	23011 + 6126	2
157.	23032+5937	23	03	16.9	59	37	40	3.5	bipolar	23032 + 5937	30
158.	Cep C	23	03	45.6	62	13	49	0.75	bipolar	23037 + 6123	2
159.	NGC7538 IRS1	23	11	36.9	61	11	57	2.8	bipolar[c]	23116 + 6111	1,63
160.	23139+5939	23	13	57.9	59	39	00	3.5	bipolar	23139 + 5939	30

161.	23151+5912	23	15	08.7	59	12	25	3.5	bipolar	23151 + 5912	30
162.	MWC1080	23	15	14.6	60	34	21	2.5	bipolar	23152 + 6034	1
163.	L1262	23	23	47.9	74	01	03	0.2	bipolar	23238 + 7401	46,64

[a] This table is an extension of that published in Fukui (1989), and includes only mapped outflows published in the literature until June 1991 or discovered in the Nagoya CO survey. 51 of them appear in the list of Lada (1985). The remaining 17 sources in Lada's table (1985) are not included here because they are not confirmed to be outflow by mapping observations. Two sources in Fukui (1989), i.e., IRAS 16442−0930 and L1036, were not confirmed to be outflows and were not included in this table. This catalog was edited by T. Iwata, A. Mizuno and Y. Fukui in June 1991.

[b] Positions of IRAS point sources.

[c] These sources may consist of several outflows.

[d] These are possible candidates for driving sources, and should be confirmed by further observations.

References: (1) Lada 1985, and references therein. (2) Fukui 1989. (3) Yang et al. 1990. (4) Snell et al. 1990. (5) Yang et al. 1991. (6) Arquilla and Kwok 1987. (7) Casoli et al. 1985. (8) Snell et al. 1988. (9) Bachiller et al. 1990a. (10) Levreault 1988a. (11) Phillips et al. 1988. (12) Liseau et al. 1988. (13) Goldsmith et al. 1986. (14) Myers et al. 1988. (15) Takaba and Iwata, in preparation. (16) Heyer et al. 1987. (17) Liljeström et al. 1989. (18) Cabrit et al. 1988. (19) Fukui et al. 1986. (20) Schmid-Burgk et al. 1990. (21) Fischer et al. 1985. (22) Fukui et al. 1988. (23) Sugitani et al. 1989. (24) Fukui 1988. (25) Casoli et al. 1986. (26) Nakano and Yoshida 1986. (27) Sanders and Willner 1985. (28) Richer et al. 1989. (29) Richer 1990. (30) Wouterloot et al. 1989. (31) Iwata et al. 1988. (32) Reipurth and Olberg 1991. (33) P. R. Schwartz et al. 1988. (34) Heyer et al. 1989. (35) Cantó et al. 1981. (36) Margulis et al. 1988. (37) Olberg et al. 1989. (38) Iwata et al. 1992, in preparation. (39) Mattila et al. 1989. (40) Tamura et al. 1990. (41) André et al. 1990a. (42) Loren 1989b. (43) Armstrong 1989. (44) Wootten and Loren 1987. (45) Walker et al. 1988. (46) Parker et al. 1988. (47) Harvey and Forveille 1988. (48) Yamashita et al. 1989. (49) Hilton et al. 1986. (50) Dent et al. 1985. (51) Scoville et al. 1986. (52) Armstrong and Winnewisser 1989. (53) Xie and Goldsmith 1990. (54) Barsony 1989. (55) Wilking et al. 1990. (56) Little et al. 1988. (57) Winnewisser 1988. (58) Haikala and Laureijs 1989. (59) Clark 1986. (60) Duvert et al. 1990. (61) Umemoto et al. 1991. (62) Sato and Fukui 1989. (63) Kameya et al. 1989. (64) Terebey et al. 1989.

temperature of 50 ± 10 K for NH_3 clumps in the NGC2071 outflow (Takano et al. 1985). However, infrared observations of vibration-rotation lines of H_2 and far-infrared/submillimeter observations of [O I], [C II], [Si II], and high-J CO indicate that outflows contain some warm and hot gas at various temperatures up to $\sim$3000 K (Gautier et al. 1976; Watson et al. 1985; Lane and Bally 1986; Jaffe et al. 1987; Jaffe et al. 1989, Boreiko and Betz 1989; Lane et al. 1990). Although the amount of gas traced by these excited species decrease with increasing energy above the ground state, the warm and hot components of outflows may be energetically important because they tend to have larger linewidth than the low-J CO transitions (see, e.g., Persson et al. 1981; Crawford et al. 1986; Stacey et al. 1987). While the warm ($\geq$200 K) component of outflows may constitute from $\sim$10% to >50% of the high-velocity molecular material in at least some sources (e.g., Orion/IRc 2: Storey et al. 1981; Cepheus A: Stacey et al. 1987; DR 21: Jaffe et al. 1989), it seems likely that cold gas is dominant in most molecular outflows.

The high-velocity molecular gas does not fill the volume or projected surface area of outflows. Bally and Lada (1983) and Snell et al. (1984) estimate that the mean projected area filling factor ranges from about 0.01 to 0.8, suggesting that the high-velocity molecular gas is concentrated into clumps, filaments, or sheets. Furthermore, the filling factor appears to decrease with increasing velocity away from the rest velocity of the host cloud.

C. Morphology of Molecular Outflows

1. Degree of Collimation. CO mapping of dozens of outflows with small beams ($10''$ to $30''$ resolution) indicates that the majority are bipolar with a moderate degree of collimation (e.g., L1641-North: Fukui et al. 1988). The ratio of major to minor axis reaches values as high as $\sim$10 for the best collimated CO flows such as the recently discovered ρ Oph A flow (André et al. 1990a), and the L1448 flow (Bachiller et al. 1990b). On average, CO outflows subtend a relatively large opening angle as seen from the source ($\sim$10° to 50°).

2. CO Lobe Symmetry and Structure. Detailed structural information has been derived from mapping of large angular diameter outflows. Most flows exhibit considerable overlap between red- and blue-shifted lobes. Each flow exhibits unique traits, indicating large variations in both the flow properties and in the structure of the cloud into which it is propagating. The prototypical L1551 flow is neatly bipolar with comparable flux and spatial extent in the red and blue lobes. On the other hand, the flow in Mon R2 is asymmetric with large-scale bends and small-scale wiggles. In some sources, multiple outflow axes or flow reversals with alternating pairs of redshifted and blueshifted lobes are seen (e.g., ρ Oph-East and Cep A, see Sec. IV).

In some outflows, high-velocity CO emission is confined to shells or to the walls of evacuated cavities. In L1551, the accelerated CO is concentrated along the walls of a relatively empty cavity and shows evidence of rotation about the outflow axis (Uchida et al. 1987b; Moriarty-Schieven et al. 1987).

Direct evidence for evacuated cavities has been found in optically thin tracers such as ^{13}CO which show holes in the parent cloud in maps of the integrated emission at the position of the high-velocity lobes (Hayashi et al., in preparation). At a velocity a few km s^{-1} different from the cloud rest velocity, the ^{13}CO emission is concentrated into swept-up shells surrounding the cavities; the highest-velocity ^{12}CO is observed to lie just inside the ^{13}CO shells. In other cases, such as Mon R2, the outflow lobes appear to be filled with accelerated gas, while in Cep A, CO emission is confined to cavity rims in one portion of the flow but appears to fill in the flow volume in other regions. In this way, the CO observational data show considerable variety, making it difficult to model the CO emitting region in detail.

III. OPTICAL AND NEAR-INFRARED OBSERVATION OF OUTFLOWS

High-velocity CO emission, HH objects and shock-excited near-infrared lines of H$_2$ are complimentary tracers of outflow activity. As shown below, there exists a relatively high correlation among these three tracers in spite of the effects of extinction in the optical and near-infrared lines. While the CO lines trace the bulk of the accelerated gas, the average velocity of this component is low (1 to 20 km s^{-1}). The H$_2$ lines tend to have linewidths of order 20 to 100 km s^{-1}. The optical lines in HH objects tend to exhibit the highest velocities, with values ranging from less than 100 to over 600 km s^{-1}. Much work still remains to be done in order to understand the relationships among optical, near-infrared, far-infrared and mm-wave emission lines. In the following section, we discuss coincidence of the three tracers on the basis of the most recent catalogs for them, and discuss some aspects relevant to our understanding of molecular outflows.

A. Statistics of Radio, Optical and Near-Infrared Outflow Phenomena

Herbig-Haro (HH) objects are small shock-excited nebulae frequently found in regions of active star formation (see reviews by R. D. Schwartz 1983 and Raga 1989). A catalog of 184 HH objects has recently been presented by von Hippel et al. (1988) and about 45 new HH objects have been reported by Reipurth and Graham (1988), Reipurth (1989a, c), Ogura (1990) and Ogura and Walsh (1991). The majority of HH objects occur in groups, with each group probably associated with a specific outflow source. About 230 individual HH objects are catalogued which can be assigned to about 90 individual outflow sources.

Lists of HH objects and CO outflows are necessarily incomplete, making statistical comparisons of these two manifestations of mass loss difficult. Most of the recently discovered HH objects have not yet been searched for CO outflows and, conversely, most of the outflows listed by Fukui (1989) have not been searched for HH objects. Nonetheless, despite the incompleteness of the searches, the overlap of these tracers is already striking. About 26

CO outflows, out of about 40 that have been searched, are known to contain HH objects within their boundaries. On the other hand, out of 31 HH object complexes which have been observed with high sensitivity (≤ 1 K) in CO, 26 have associated CO outflows. The frequent occurrence of HH objects among the sample of *well-studied* CO outflow regions suggests that HH objects are an important feature of outflow activity.

The nearest CO outflows tend to be the ones associated with HH objects. Good examples include L1551/IRS5 and HL/XZ Tau, the HH 7–11 complex, and the NGC1999 region in Orion. In most cases (although not all), the visible HH objects are located within the blue (approaching) lobe of the CO outflow which usually suffers less extinction than the red (receding) lobe.

Although about 80% of the HH objects have been found in regions of low-mass star formation, this probably reflects a selection effect because low-mass star-forming regions are more numerous, are found closer to the Sun, and tend to be located in smaller cloud cores suffering less extinction than high-mass star-forming regions. Indeed, a number of HH objects (5 to 10% of the known sources) are associated with high-luminosity sources such as the Orion/IRc2 outflow (M42-HHs: Axon and Taylor 1984) and Cepheus A (GGD37 and HH-NE: Lenzen 1988).

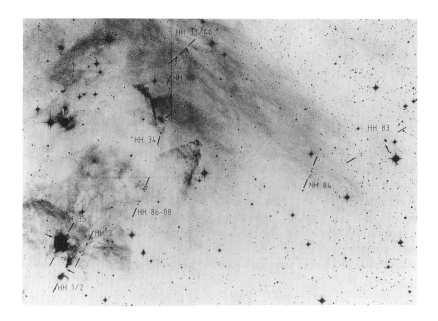

Figure 2.   An ESO Schmidt telescope plate showing a 20′ by 30′ field illustrating the locations and orientation of 9 optical outflows (Herbig-Haro objects) in the northern part of the L1641 cloud. The HH 1/2 region near NGC1999 is near the bottom left of the plate (figure from Reipurth 1989*a*).

About 42 molecular outflows are associated with shock-excited H_2 emission at near-infrared wavelengths. So far, at least 60 regions have been searched for H_2 emission: out of 42 detections, 32 are CO outflows, and 29 are HH objects, with 19 being both HH objects and CO outflows. Several H_2 sources (such as RU Lupi) are not listed as either HH objects or CO outflows.

Nearly 20 of the HH objects associated with low-luminosity young stellar objects (YSO) in the Orion region occur in portions of the L1641 and L1630 clouds which have been mapped in ^{12}CO with the Nagoya and Bell telescopes. Of these HH objects, only a few have been detected as CO outflows. This may be due to another selection effect resulting from the difficulty in distinguishing low-velocity accelerated CO against the turbulent background emission of a typical molecular cloud. Because the CO linewidth of outflows correlates with the luminosity of the driving star, fainter sources having lower-velocity CO flows are harder to distinguish against the background emission. Alternatively, these optical objects may be more evolved than typical CO outflows, having much less molecular gas around them to supply outflow with CO gas. This interpretation is consistent with apparent low luminosities of the driving sources of optical jets, showing no sign of vigorous mass accretion producing thereby large luminosity excess. The ubiquity of optical outflows and HH objects is dramatically illustrated by the recent work of Reipurth (1989a) who shows that at least 11 distinct optical flows from YSO's are present in a $30' \times 30'$ field in the northern part of the L1641 region of Orion (see Fig. 2). Only three CO outflows have so far been detected in this region [HH 1–2 and HH 35 (Levreault 1985), HH 83 (Bally et al. in preparation.)].

B. Models for Shocks in Herbig-Haro Objects

HH objects are the optical manifestation of luminous shocks formed when two fluids collide with a relative velocity in the range 10 to >300 km s^{-1}. These shocks can arise in four general circumstances:

1. Internal shocks in a moving fluid such as a wind or jet (Raga 1989);
2. A shock formed where a wind or jet rams denser quiescent gas in a bow shock (Reipurth et al. 1986);
3. In a bow shock resulting from a wind blowing past an obstacle (Schwartz 1978);
4. In a bow shock formed when a fast-moving knot of dense gas rams into lower-density gas at rest ("an interstellar bullet"; Norman and Silk 1979).

Models (2) and (4) are similar except that the density ratio of the moving and stationary gas is inverted. Although all four kinds of shocks have been seen in HH objects, many HH objects remain difficult to classify.

Internal shocks in a moving fluid (type 1 HH object) are seen in about a dozen stellar jets such as those associated with HH 33/40 (Lane 1989) and HH 34 in Orion (Reipurth et al. 1986). The HH 34 jet emerges from a 0.5 $L_\odot$ source at a velocity of >100 km s^{-1}, yet mostly excites low-excitation species such as [S II]. A bow shock (type 2 HH object) is seen about 90″ south of the jet where it slams into denser gas in the surrounding cloud. Shock-excited H_2 emission has been seen from the HH 33/40 jet (Lane 1989) but not from the HH 34 region, possibly because either the cloud has too low a density, or is mostly atomic. This interpretation is supported by the low average extinction towards this object, and may explain the absence of a CO outflow there, as well.

The objects HH 7–11 may contain examples of bow shocks formed by a wind blowing past stationary obstacles (type 3 HH object). Although the optical velocities are large, mm-wave interferometric observations show stationary knots of dense molecular gas near the positions of the HH objects (Rudolph and Welch 1988). Even without detection of the dense knot, the flow past an obstacle can be identified by the relative spatial position of high- and low-excitation emission. Such an analysis has indicated that HH 43 in Orion also is a type 3 HH object (R. D. Schwartz et al. 1988).

Some HH objects form when a rapidly moving object moves through a stationary medium as an "interstellar bullet." The high proper motion and spatial and velocity structure of regions such as HH 1–2 and GGD37 in Cep A, as well as the relative location of the high- and low-excitation lines, suggest that these may be type 4 HH objects.

C. Near-Infrared Observations of Shock-Excited Molecular Hydrogen from Outflows

Many molecular outflows are associated with bright, shock-excited H_2 emission at near-infrared wavelengths. The most prominent line is usually the $v=1-0$ S(1) line at $2.122\,\mu m$ which is frequently bright in HH objects and outflows located in relatively obscured molecular clouds. HH objects lying in regions of low obscuration or towards which background star fields are readily visible tend not to produce detectable H_2 lines. This property of H_2 emission can be understood in terms of environmental factors: only shocks propagating into relatively dense molecular gas can produce H_2 emission. Those propagating into low-density or mostly atomic gas cannot produce these lines since H_2 is not present in the medium and the H_2 formation time scale in the post-shock medium is longer than the shock dynamic time scale.

Although molecular hydrogen emission is usually found in the same vicinity as the optical line emission which defines HH objects, the H_2 emission region frequently exhibits different morphology (Lane and Bally 1986; Harvey et al. 1986; Lane 1989; Garden et al. 1990). These structural differences can be understood as a consequence of the shock temperature and density structure, and the mechanism of line excitation (Lane and Bally, in preparation).

The most luminous sources of shock-excited H_2 emission tend to be associated with energetic, luminous molecular outflows such as Orion/IRc2, NGC2071, DR21 and Cep A (see Lane 1989, and references therein). The close association between HH objects near low luminosity outflows and shock-excited molecular hydrogen emission gives us some confidence that the excitation conditions in the high luminosity outflow sources are similar in nature to those found in the HH objects. Therefore, the 2 μm lines of H_2 can be used to infer the presence of extensive shock-excited gas in these sources which is similar in nature to HH objects. The highly energetic outflows such as DR21 and Orion/IRc2 clearly contain extensive regions of shock-excited gas. On the other hand, some massive flows such as Mon R2, which are very old as inferred from the dynamic age of the CO emission, have only fluorescent rather than shock-excited H_2 emission (Aspin et al., in preparation). There may be no active radiative shocks in these old flows, and the high-velocity CO may be coasting without active acceleration. In highly obscured sources, it may be possible to use the near-infrared lines to identify knots of H_2 emission as invisible, near-infrared "HH objects."

IV. DETAILED DISCUSSION OF SELECTED OUTFLOWS

The most extensively studied outflows are those associated with the low-luminosity sources L1551/IRS5 and HH 7–11/SVS13 and the high luminosity sources Orion/IRc2 and NGC 2071. Recent references and/or summaries of the extensive work on these outflows may be found in Stocke et al. (1988) and Moriarty-Schieven and Snell (1988) for L1551; Masson et al. (1990) and Liseau et al. (1988) for HH 7–11; Masson et al. (1987), Lane (1989), and Lane and Bally (in preparation) for the Orion outflow; Moriarty-Schieven et al. (1989) and Garden et al. (1990) for NGC 2071. We have selected two sources for detailed discussion for which recent high-resolution observations have revealed particularly interesting features.

A. ρ Oph-East (IRAS 16293–2422): Low-Luminosity Multiple Outflow Lobes

Rho Oph-East (IRAS 16293–2422) is one of the molecular outflows discovered in the Nagoya CO survey of star-formation regions (Fukui et al. 1986), and was independently discovered by Wootten and Loren (1987) and Walker et al. (1988). It is located ~1.5° east of the ρ Oph main cloud at a distance of 160 pc, and is associated with an IRAS point source (IRAS 16293–2422) whose luminosity is ~27 $L_\odot$ (Mundy et al. 1986b). The IRAS source has an unusually cold infrared spectrum, which can be fitted by a single ~40 K blackbody spectrum. Such a very cold dust temperature strongly suggests that IRAS 16293–2422 is one of the best candidates for a solar-type protostar.

The high-velocity ^{12}CO ($J=1-0$) emission has been resolved into four separate compact lobes, suggesting two coincident bipolar outflows, in addition to an extended monopolar blueshifted lobe (Mizuno et al. 1990). Figure 3 shows a map of the two compact bipolar flows obtained with the Nobeyama 45 m telescope. Although the spatial extent of the four compact lobes is smaller than 4' (~0.2 pc), the additional extended blue lobe extends ~10' east from the IRAS source. Physical properties of the four compact lobes are similar to each other but differ from those of the additional extended

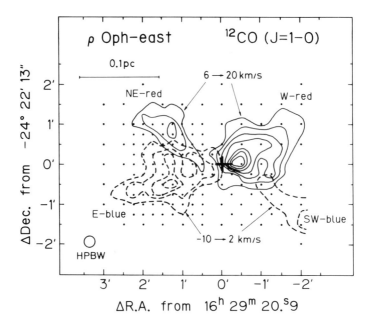

Figure 3.   Distribution of the high-velocity ^{12}CO (J=1−0) emission in ρ Oph-East obtained with the Nobeyama 45 m telescope (Mizuno et al. 1990). The integrated intensity of the redshifted wing is shown by solid contours and that of the blueshifted wing is shown by dashed contours. Contours are every 5.4 K km s^{-1} from 10.8 K km s^{-1}. The position of an IRAS point source (IRAS 16293−2422) is denoted by a cross.

monopolar lobe. The typical velocity of the high-velocity CO gas and the dynamical time scale of the four compact lobes are ∼14 km s^{-1} and ∼1 × 10^4 yr, respectively. The extended blue lobe with lower velocity, ∼6 km s^{-1}, has a dynamical time scale of ∼1 × 10^5 yr, an order of magnitude longer than that of the compact bipolar outflows. We find no candidate for the driving source other than IRAS 16293−2422 within the sensitivity limits of the IRAS point source catalog and POSS (Palomar Observatory Sky Survey) plate in the outflow lobes, suggesting that five outflow lobes are emanating from the common driving source IRAS 16293−2422.

It is important to understand what is the origin of such a multiplicity of outflow lobes. The simplest model is multiple driving sources clustered in a small region which is unresolved by the IRAS beam. Interferometry in the 2.7 mm continuum and in ^{13}CO (J=1−0) by Mundy et al. (1986b) revealed a compact and rapidly rotating dust disk with a projected size of ∼1800 AU by less than 800 AU toward IRAS 16293−2422. VLA maps of 6 cm and 2 cm continuum emission by Wootten (1989) reveal two continuum peaks separated by ∼750 AU in the NW-SE direction of the dust disk, suggesting that IRAS 16293−2422 may contain a binary system.

Two driving engines with stationary axes in the E-W and NE-SW directions, respectively, may explain the four compact lobes. In this case, the extended monopolar outflow and the E-W compact bipolar outflow are both supposed to be driven by a common E-W oriented engine. An order-of-magnitude difference of time scales and mechanical luminosities between the extended monopolar outflow and the compact outflow implies that the outflow activity driven by the E-W oriented engine may have been enhanced drastically in the last $\sim 10^4$ yr.

A third possible model is a precessing jet. If one of the stars in this binary has a disk tilted to the orbital plane, the disk should undergo precession due to the gravitational quadruple interaction with the companion. If the molecular outflow is driven by a circumstellar disk, as predicted in magnetohydrodynamical models (see, e.g., Uchida and Shibata 1985), the outflow can undergo precession. Details of the ^{12}CO map, showing that the four compact lobes appear well separated from each other, however, are not well explained by the precession model.

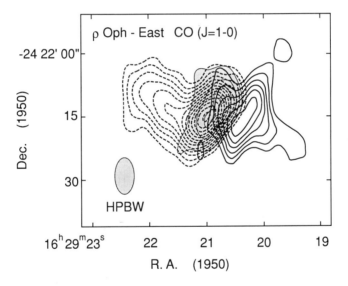

Figure 4. Interferometric ^{12}CO map of ρ Oph-East taken with the Nobeyama Millimeter Array is superposed on the 2.7 mm continuum interferometric map of Mundy et al. (1986*b*). The velocity intervals are -0.4 to -3.6 km s^{-1} (low-velocity blue wing) and 8.5 to 12.5 km s^{-1} (high-velocity red wing). The contours extend from 2σ rms noise with a 1σ step. The 1σ rms noise level is 70 mJy beam^{-1}.

A high resolution ^{12}CO ($J=1-0$) map (Fig. 4) made with the Nobeyama Millimeter Array (Mizuno et al., in preparation) reveals the distribution of the high-velocity gas near the driving source. The blueshifted gas is located on the east side and the redshifted gas on the west side, suggesting that the axis of bipolar outflow is originally oriented in the E-W direction. In the vicinity of the IRAS source, the NE-SW bipolar flow is not seen. Taking this result into consideration, it may be most likely that the NE-SW bipolar flow is formed by some secondary effect such as a dynamical interaction between the high-velocity outflow and the dense ambient cloud. In fact, there is a clear indication for dynamical interaction of the outflow with the nearby NH_3 dense core which is located ~80″ east of the IRAS source (Mizuno et al. 1990). The NH_3 core is located just toward the eastern edge of the eastern blue lobe. At the interface of the outflow and the dense core, the velocity of the NH_3 core is abruptly blueshifted from the rest velocity of the NH_3 cloud. The calculated momentum of the CO outflow is large enough to accelerate the NH_3 core and to cause such an abrupt velocity shift. The outflow pressure is so large that the NH_3 core is not in equilibrium. It is likely that the north-south-elongated structure of the NH_3 core may be formed by the outflow.

Dynamical interaction of outflow with ambient material has been noted in some cases, e.g., L1641-North (Fukui et al. 1988), NGC2071-North (Iwata et al. 1988) and NGC2071 (Takano et al. 1986). Rho Oph-East provides an even clearer example of such interaction. The distribution of the northeastern red lobe shows a remarkable anticorrelation with the NH_3 dense core (Fig. 9 in Mizuno et al. 1990). It suggests that the CO lobe morphology may be influenced by the existence of ambient dense gas; i.e., the outflow gas may be excluded from a volume with higher density than its surroundings. Although this dynamical interaction model can explain the compact lobes' morphology on the east side of the IRAS source, it is still obscure how the extended blue lobe, which is by an order of magnitude older, is formed and how the interaction occurs on the west side. The puzzle of the multiplicity of ρ Oph-East is not resolved completely.

B. Cepheus A: Multiple Outflow Lobes and Time Variable Wind

Cepheus A is a region of massive star formation with a total luminosity of 2×10^4 $L_\odot$. The driving source of a molecular outflow and associated Herbig-Haro objects is buried in a region of high extinction (visual extinction $Av \geq 75$ mag) and is marked by a strong H_2O maser and a cluster of compact radio continuum sources, one of which has the radio spectral index of an optically thick outflow source (Hughes 1988, and references therein). A bright far-infrared source and a near-infrared reflection nebula mark the general vicinity of the driving source, which is invisible even at $2\,\mu$m. The region has been imaged at ~1″ resolution in the 2.122 μm line of H_2 (Lane 1989; Bally and Lane 1990) and mapped with 15″ resolution in the ^{12}CO ($J=1-0$) line (Bally et al. 1991). Figure 5 shows a superposition of the H_2 and high-velocity CO maps, with the location of the outflow driving source marked by a cross.

Cep A contains one of the brightest and highest velocity HH objects known. It is a clear example of a source where CO emission, optical emission, and infrared emission arise from the same outflow region. The GGD37 cluster of HH objects is located ~1′.5 to 2′ west of the driving source and appears

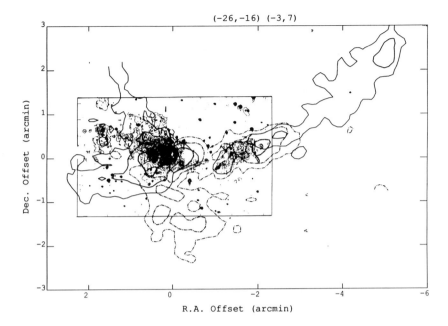

Figure 5. A gray-scale image of the 2.122 μm $v = 1 - 0$ S(1) line of the molecular hydrogen toward Cep A superposed on a contour map of high-velocity ^{12}CO
(J=1−0) emission. Solid contours indicate blueshifted CO, integrated from −26
to −16 km s^{-1}. Dashed contours indicate redshifted CO, integrated from −3 to 7
km s^{-1}. Contour levels are set at 3, 6 and 12 K km s^{-1}. The GGD37 HH object
lies ∼2′ west of the bright infrared reflection nebula, while HH-NE lies ∼1′ to the
northeast. The location of the driving source for the molecular outflow is denoted
by a cross.

as a complex of arcs, bubbles, and twisted filaments in H$_2$ line emission (see
Figs. 2 and 4 in Lane 1989). A second HH object (Cep A HH-NE; Lenzen
1988), also prominent in H$_2$ emission, lies ∼1′ to the northeast of the bright
infrared reflection nebula.

The highest velocity CO extends mostly E-W of the core over a 3′ × 7′
region, corresponding to ∼0.6 × 1.5 pc at a distance of 730 pc. Although
the CO outflow is bipolar within ∼1′ of the driving source (with redshifted
gas to the west and blueshifted gas to the east), the flow structure is more
complex away from the center. To the west, the redshifted lobe is followed
by a blueshifted one starting in the vicinity of GGD37 and continuing with
a slight bend to the north for several arcminutes. A second west lobe of
redshifted gas is also found about 100″ west of the core, indicating that the
CO lines are broader in the vicinity of GGD37.

Only the infrared core contains luminous embedded sources. There are
no bright objects between the extra-blue and redshifted lobes found to the
west. Polarization measurements of the scattered 2 μm continuum radiation

in the east and the west indicate that the illumination for both regions comes from the central infrared source (Lenzen et al. 1984). This suggests that the entire extent of the CO outflow is energized by the activity of young stars embedded in the infrared core.

The optical/infrared reflection nebula in the east and the optical reflection nebula near GGD37 in the west form a bipolar structure in reflected light which confirms the nearly E-W orientation of the high-velocity outflow and flow channel near the source. Optical proper motions of knots of emission in GGD37 and HH-NE are consistent with this flow axis (Lenzen 1988) but suggest that, away from the infrared source, there is a deflection of the outflow towards the north which is also seen in the large-scale CO morphology.

To the east, only blueshifted gas is seen. However, in addition to the (alternating) E-W lobes, a second pair of bipolar lobes having a lower velocity range extends out from the core along a NE-SW axis. The coincidence of the E and NE blue lobes with the edges of the infrared reflection nebula suggests that they mark the walls of a cavity through which a blueshifted wind emerges. The shock-excited HH object, HH-NE, lies squarely between the two blueshifted CO lobes. At low relative velocities, the W and SW redshifted lobes from a second pair of ridges oriented opposite to the blueshifted lobes. However, the reflected light is associated only with the eastern lobe near GGD37. The low-velocity S-W redshifted lobe may be recoiling from the deflection of the higher-velocity western flow to the north. Alternatively, it is possible that this component represents either an unrelated, lower energy flow superimposed on the main flow, or is the debris left over from an earlier epoch when the Cep A wind directed energy in another direction.

There are some interesting correlations between the CO, optical, and infrared properties of Cep A. The CO outflow appears to reverse polarity in the vicinity of the HH object GGD37. Near the shocks, a second blueshifted lobe of emission appears, and extends 3 arcminutes to the west. The transverse width of the optical emission region and the CO lobe are similar, indicating that much of the interior region of the CO lobe is filled with ultra-high-velocity gas having a velocity of order 300 to 600 km s^{-1} as determined from optical proper motions, radio proper motions (Hughes and Moriarty-Schieven 1990) and the highest observed optical radial velocities.

The optical emission from the GGD37 shocks consists of filled cones of emission suggesting that the low-excitation tracers must be excited throughout the volume of the bow-shock structures. Turbulence in the postshock gas lying behind the reverse shock through which the wind passes provides a viable mechanism for the excitation of ionic and atomic lines in this region. The H_2 line emission forms a series of arcs surrounding the $H\alpha$ and [S II] structures. Their morphology and location are consistent with excitation in the bow-shock skirts where pre-existing H_2 can survive passage through the shock (Lane 1989; Bally and Lane 1990). The observed brightness of the S(1) line places a constraint of $n \geq 10^4$ cm^{-3} on the density of the radiating medium.

The CO lines broaden downwind from GGD37 (away from the central source) where the western lobe of blueshifted emission first appears. This suggests that the fast wind responsible for GGD37 may also accelerate CO in the lobes of Cep A by a few km s^{-1}. A possible model for the accelerated CO in this (and other outflows) is that all of the CO has been accelerated by shocks similar in nature to the GGD37 shocks. Like H_2, CO is a robust molecule which can survive accelerations of several tens of km s^{-1} which would be produced in the skirts of a bow shock such as GGD37. In this picture, the radiating shocks seen as HH objects or as near-infrared luminous regions of shock-excited H_2 emission are *the sites of CO acceleration*. The localization of the shocks to a region much smaller than the projected area of the outflow can be interpreted as one indication of time-variable winds (Bally and Lane 1990).

We have discussed our recent knowledge on ρ Oph-East and Cep A as examples of low- and high-mass molecular outflows with multilobes. At present, we do not have a general consensus about the driving mechanism of such multilobe outflows. Further detailed observations at radio, infrared and optical wavelengths are necessary to elucidate the mechanism.

V. ACCELERATION AND COLLIMATION

A. Background

Molecular outflows are collimated and accelerated to velocities of order 3 to 30 km s^{-1} and sometimes to more than 100 km s^{-1}. The molecular component requires some acceleration mechanism other than stellar radiation to provide the observed momentum. The mechanical luminosity of molecular outflows L_{mech} is defined as the total kinetic energy of the high-velocity molecular gas divided by the dynamical time scale of the outflow. In all molecular flows, the ratio L_{mech}/L_* is less than unity, with a value in the range 0.002 to 0.2. However, the average force needed to accelerate outflows, defined as the momentum in the high-velocity gas divided by the dynamical age, is greater than the force of stellar radiation pressure L_*/c. The force required to accelerate molecular flows typically exceeds that produced by radiation pressure by a factor of 100 to 1000, indicating that radiation pressure cannot drive outflows unless the photon scattering optical depth is of order of 100 to 1000 at the wavelength where the radiation field is in thermal equilibrium with the gas temperature (i.e., the mid- to far-infrared). This is the basic reason why some nonradiative acceleration mechanism is needed (Bally and Lada 1983).

A second problem is the physical nature of the collimating agent. Early studies suggested that large-scale "interstellar" ($\sim 10^{17}$ cm) toroids of dense gas constrained the CO outflows (see, e.g., Torrelles et al. 1983). However, recent interferometric imaging has shown that some CO outflows are already collimated on scales much smaller than this (e.g., ρ Oph-East: Mizuno et al., in preparation, see Sec. IV.A). This is consistent with the suggestion that collimation is made very close to the cental star, $\lesssim 10^{16}$ cm, on the basis of observations of the compact dense gas clouds associated with the central stars (Takano et al. 1984; Fukui et al. 1988). On the other hand, optical jets appear to be collimated within a few arcsec of the source (see, e.g., Mundt et al. 1987), indicating that for at least some sources, collimation of the highest velocity ($V \gtrsim 200$ km s^{-1}) outflow occurs closer than 1000 AU, the size scale of a protoplanetary disk.

Models of outflows can be grouped into two categories: (1) magneto-hydrodynamic (MHD) models, in which the wind is primarily accelerated by magnetic fields which thread the protostellar disk; and (2) stellar-wind models, in which the wind is produced in the immediate vicinity of the star.

B. Magnetohydrodynamic Models

The basic idea of magnetohydrodynamic (MHD) models such as discussed by Uchida and Shibata (1985) and by Pudritz and Norman (1986) is that a compact, rotating, circumstellar disk of $\sim 10^{15}$ to 10^{16} cm radius forces corotation of the surrounding interstellar magnetic field by magnetic tension. The distorted magnetic field has a radial component as illustrated in Fig. 6, which accelerates ions frozen to the magnetic field. The ions are coupled to the neutral molecular gas by collisions and accelerate this component. The predicted velocities in these models are 30 to 40 km s^{-1}, depending on the adopted disk parameters. A further prediction of these models is that the wind velocity increases with decreasing radius, which can result in the formation of a high-speed inner jet at a radius of $\sim 10^{11}$ to 10^{12} cm. The inner jet may have velocities up to ~ 300 km s^{-1}. MHD models naturally produce cold flows without any shock heating of the gas and therefore can explain the outflow component observed in the low-J CO transitions having temperatures of 10 to 100 K. This low temperature, however, may not be unique to the MHD models; the cooling time of gas at temperatures of order a few times 10^3 K with densities above a few hundred cm^{-3} is a few years, so even in models in which the entire CO flow is accelerated by shocks, most of the mass is expected to be near the temperature determined by collisional and thermal equilibrium.

The basic observed characteristics of CO outflows, e.g., a typical flow velocity of ~ 10 km s^{-1} and bipolar symmetry, are readily explained by the MHD models. On the other hand, direct evidence for massive and magnetized disks is still lacking, in part due to current limitations of resolution and sensitivity. Future submillimeter and far-infrared line studies and polarimetry (see, e.g., Dragovan 1986; Novak et al. 1989) may provide crucial tests of MHD models by probing magnetic field orientation and grain alignment and by searching for velocity differences between ions and neutral species on arcsecond angular scales.

MHD models make two testable predictions (Uchida and Shibata 1985). The CO lobes are expected to rotate about the symmetry axis, and the molec-ular gas is expected to be accelerated gradually over a linear scale of order

0.1 pc. Uchida et al. (1987*b*) present high angular resolution CO observations of the L1551 outflow with ~20″ resolution obtained with the Nobeyama 45 m telescope. Their data indicate rotation of the lobes about the outflow axis with a velocity of order 1 km s⁻¹. Furthermore, the CO radial velocity vector is seen to increase gradually to its maximum amplitude over a distance of order 0.1 pc, as predicted by the model. Moriarty-Schieven and Snell (1988) confirm these results. Unfortunately, as only the radial velocity component of the flow vector is observed, it is difficult to distinguish true acceleration from projection effects. L1551 is so far the only source with a clear indication of rotation in the CO lobes. Although it is not entirely clear why L1551 IRS5 is unique in this respect, its strong ^{12}CO emission in the outflow and relatively quiescent ambient cloud (^{12}CO linewidth of ~2 km s⁻¹) may favor the detection. Several other sources exhibit radial velocity fields consistent with gradual acceleration over a 0.1 pc scale.

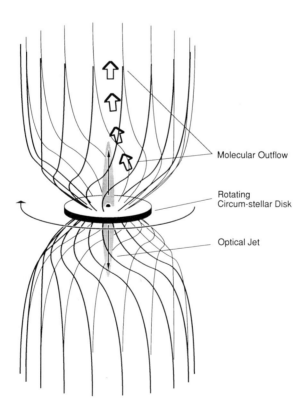

Molecular Outflow

Rotating
Circum-stellar Disk

Optical Jet

Figure 6. A schematic drawing of the magnetohydrodynamical model.

C. Stellar-Wind Models

In this class of models, most of the high-velocity CO has been swept up from the molecular cloud surrounding the YSO by a faster primary wind which is hard to detect directly as it may be mostly atomic. Radio continuum observations (see, e.g., Snell and Bally 1986) have revealed that many outflow-exciting sources have compact components with a spectrum rising as ν^α where α is in the range 0 to 2. Models indicate that $\alpha \approx 0.6$ for a fully ionized wind with an r^{-2} electron-density profile (Wright and Barlow 1975; Panagia and Felli 1975). For a collimated (or partially collimated) flow, the spectral index α can lie in the range 0 to 2 (Reynolds 1986). Under the assumption of steady state and full ionization, the mass-loss rates inferred from the radio continuum parameters typically fail by 2 orders of magnitude to provide the observed momentum in the CO flow. However, the winds may be mostly atomic with only a trace of plasma which is responsible for the radio continuum emission. Several outflows have been found which exhibit broad 21-cm H I lines, consistent with models in which the primary wind is mostly atomic (Bally and Stark 1983; Lizano et al. 1988). Very faint "extremely high-velocity" CO emission with velocities up to $\sim$155 km s^{-1} from line center has been observed towards a few sources (Lizano et al. 1988; Koo 1989; Margulis and Snell 1989). This component of the flow may trace a wind in which hydrogen is in atomic form but carbon has been partially converted into CO (Glassgold et al. 1989).

The primary wind rams into the quiescent cloud surrounding the YSO, accelerating the swept-up gas to the velocity observed in the CO outflow. The primary wind may be accelerated and collimated by a magnetized disk, and in one recent model, the "X-celerator" model of Shu et al. (1988), the wind originates from the inner edge of the disk where gas passes though an accretion shock and decelerates from a Keplerian velocity to the rotation velocity of the young star.

Several outflows (e.g., L1551: Uchida et al. 1987b; Moriarty-Schieven and Snell 1988), exhibit a hollow shell morphology which is consistent with acceleration of the molecular gas in a snowplow shock produced by the ram pressure of the primary wind. High-velocity HH objects are frequently found near low-mass stars, indicating that fast winds occur in these sources. Lane and Bally (in preparation) and Bally and Lane (1990) argue that HH objects and H$_2$ emission regions mark the locations where the primary wind drives shocks into the ambient cloud, accelerating the CO seen in the outflows. Both lobes of the L1551 outflow contain extensive amounts of shock-excited gas in the form of HH objects (Rodrí guez et al. 1989a). The total momentum in the visible shocks appears to be too small to produce the extended massive CO lobes. However, the momentum in the fast-wind may have been greater in the past, and the optical shocks may provide only a lower limit to the fast-wind momentum.

D. Discussion: A Comparison of the Models

The hollow-shell morphology of the CO lobes in L1551 is consistent with both models discussed above. Such a hollow cavity will be created by acceleration of the CO in a snowplow shock produced by the ram pressure of the primary wind, while MHD models equally well explain the hollow distribution, as demonstrated by the numerical simulations by Uchida and Shibata (1985). Flow morphology may not give a clue to discriminate between these models.

The extremely high-velocity (EHV) CO lines seen towards some sources pose constraints on source models. Six outflows are now known to exhibit faint CO components with linewidths ranging from 100 to over 300 km s^{-1}. Most outflows have not been searched with sufficient sensitivity to detect these faint high-velocity features. While three of the EHV CO flows are associated with luminous sources: NGC2071, CRL490 and S140 with $L \approx 10^3$ to 10^4 L$_\odot$ (Margulis and Snell 1989). Three others driven by low-luminosity sources exhibit EHV CO components: the HH 7–11 outflow 90 L$_\odot$ (Koo 1990; Bachiller and Cernicharo 1990; Masson et al. 1990), the flow in L1448 with $L < 11$ L$_\odot$ (Bachiller et al. 1990a), and the flow in L723 (Margulis and Snell 1989). Recent observation (Masson et al. 1990; Bachiller and Cernicharo 1990) show that the EHV gas in HH 7–11 in CO (J=2−1, and J=3−2) is confined to two knots of emission extending over a 60″ region and located in the vicinity of the shocks traced by H$_2$ and the stationary HCO$^+$ clumps observed by Rudolph and Welch (1988). Unlike the classical CO line wings, the EHV CO is seen as high-velocity CO cloudlets with a velocity of about 100 km s^{-1} or greater with respect to the molecular cloud, and linewidths of about 20 km s^{-1}. The spatial extent of the EHV CO is somewhat smaller than the lower-velocity "classical" CO flow as mapped by Liseau et al. (1988) and Bachiller and Cernicharo (1990) who find that the spatial extent of the CO flow is about 30″ by 120″. The velocity of the EHV CO is comparable to that of the optical HH object in this region. The total mass in the EHV CO component is estimated to be about 0.03 M$_\odot$, about 1% of the total outflow mass. Although the momentum in the EHV component is about an order of magnitude less than that in the classical flow (2.2 vs 11 M$_\odot$ km s^{-1}), the mechanical power is about 2 orders of magnitude larger (450 vs 3.3 L$_\odot$). The EHV CO may be part of a fast wind emerging from SVS13, possibly clumped into "bullets," or may represent gas swept from the surrounding cloud and accelerated to the observed velocities by momentum transfer from an even faster wind. In the former model, the CO may form by ion-molecule reactions within a mostly atomic wind, in which H is atomic, but some of the carbon is converted to molecules. CO may have a much lower relative abundance with respect to hydrogen than is typical of molecular clouds (Glassgold et al. 1989), leading to an underestimate of the mass and momentum in the EHV CO flow.

The MHD models cannot produce velocities as high as those observed in the EHV CO component, at least from the outer parts of disks. However, in some versions of the MHD models, the inner portion of the disk give rise to high-velocity jets (Uchida and Shibata 1984; Pudritz and Norman 1986) with velocities similar to the EHV CO. It remains to be seen if the EHV CO is collimated into a jet or if the models can produce the amount of gas seen in this component.

A question recently raised concerning MHD models is the balance between the angular momentum involved in the CO lobe vs that of the driving disk. The study by Moriarty-Schieven and Snell (1988) indicates that the total molecular mass contained in the outflowing gas in L1551 may be a factor of 3 to 10 greater than previously supposed, suggesting that the total molecular mass in the CO lobes may be $\gtrsim 3.6$ $M_\odot$. The rotation speed observed in the CO lobes is ~ 1 km s^{-1}. Assuming that we are looking at the outflow nearly from the direction normal to the outflow axis, the total angular momentum in the flow is as large as ~ 3 $M_\odot$ km s^{-1}. On the other hand, the assumed circumstellar disk has much smaller angular momentum, maybe <0.1 of the above value. However, this may not be a difficulty. First, the disk sitting near the central star should have already released angular momentum to the CO lobes, thus having much less angular momentum at present than before the CO lobe acceleration. Second, it is likely that not all the molecular gas is rotating as indicated by the ^{12}CO velocity maps. The ^{12}CO maps show rotation is obvious only at a distance $\lesssim 0.1$ pc from the center. Third, we should also take account of the angular momentum of the ambient molecular gas. Most molecular mass in outflows is not from the central small volume near the driving source, but from the ambient molecular gas distributed over the interstellar scale. This interstellar gas should have a certain amount of angular momentum before the acceleration. Alfvén waves act to wrap up the ambient gas, thereby bringing it to a smaller radius via the pinching effect, as has been demonstrated by the numerical simulations by Uchida et al. (in preparation). This suggests that there may be additional spin-up from the velocity of the ambient gas, so that the apparent CO lobe rotation may be a combined effect of the angular momentum of the driving disk and that of the ambient gas. In order to account for such an effect properly, theoretical models which incorporate the effect of the ambient material in CO lobe formation are needed.

It is premature to conclude which model is correct. However, the two models share some common characteristics. In particular, the coupling of the magnetic field with rotation of the circumstellar disk or the star itself may be playing an essential role in acceleration and collimation. A point of controversy is the scale of the accelerator. We may accept a general argument that material having velocities ≥ 100 km s^{-1} comes from close to the stellar surface, and that lower-velocity material, with speeds ≤ 30 km s^{-1}, may come from the outer regions as a manifestation of the deepness of the gravitational potential well (see, e.g., Shu et al. 1988). Another question which may be raised is whether the acceleration mechanism is common among stars having different masses. At the present time, the highest-velocity outflows are found mostly toward high-mass stars, suggesting the possibility that high-mass stars may have a more efficient acceleration mechanism than low-mass stars.

VI. THE ORION SOUTH GIANT MOLECULAR CLOUD: TESTS FOR EVOLUTIONARY STATUS AND CLOUD SUPPORT

The giant molecular cloud, Orion south, is one of the most active sites of star formation and has been intensively studied in mm-wave molecular spectra (see, e.g., Fukui et al. 1986,1989; Bally et al. 1987). A new ^{13}CO ($J=1-0$) map of the cloud is shown in Fig. 7 (Fukui et al., in preparation). The total molecular mass estimated from ^{13}CO data is 3×10^4 M$_\odot$. In Fig. 7, the Orion KL region and the H II region M42 are located at $l = 210°$, and the dark cloud L1641 lies at $l = 212°-213°$. The Orion KL region is a site of massive star formation, while the L1641 region is a site of low- to intermediate-mass star formation. In addition to many T Tauri stars, HH objects and reflection nebulae, there are 12 molecular outflows in this cloud. We also note the molecular cloud is highly filamentary showing a trend of smaller widths in latitude toward smaller longitudes. In this section, we use the new molecular data on this cloud to illustrate our current understanding of the evolutionary status of molecular outflows and the role outflows may play in supporting giant molecular clouds against gravitational collapse.

A. Evolutionary Status of Molecular Outflows in L1641

Comparison of the outflow detection rates for visible and embedded objects suggests that molecular outflows correspond to an evolutionary state earlier than T Tauri stars (Lada 1985). A study of molecular outflows in dark cloud cores (Myers et al. 1988) suggested a trend of higher bolometric luminosities in CO outflow sources, although samples of outflows used by these authors were not statistically complete.

Other studies providing some information on the evolutionary status of molecular outflows include optically selected pre-main-sequence objects (Levereault 1988a) and IRAS/H$_2$O maser sources (Wouterloot et al. 1989; Snell et al. 1990). A study of published data on outflows in various star-forming regions (Berilli et al. 1989), as well as the above previous studies, did not find an evolutionary trend of molecular outflows. These studies are not statistically complete, however, being biased toward luminous and perhaps therefore more massive objects.

Most recently, two unbiased surveys have been used to study the evolutionary status of flows. One is the survey of the Mon OB1 cloud by Margulis et al. (1988). The distance to the cloud (760 pc was adopted by the authors), however, sets a lower limit for the infrared luminosity measured by IRAS of ~3 L$_\odot$, too large for studying low mass protostars, and source confusion within the IRAS beam makes it difficult to identify individual sources. The other is an unbiased survey of the L1641 cloud by Fukui et al. (1989); we discuss their results below.

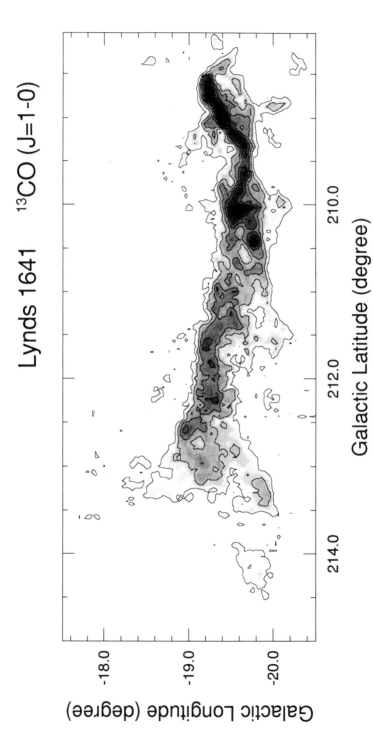

Figure 7. ^{13}CO total intensity map of the Orion south cloud (Fukui et al., in preparation). Contours are every 5 K km s^{-1} from 5 K km s^{-1}.

In order to characterize outflow sources, Fukui et al. (1989) used a complete sample of molecular outflows in L1641 (part of the Orion south cloud). Below declination $\leq 7°$, there are six molecular outflows (see Table 1 in Fukui et al. 1989). All of them have been mapped in CO with three different beam sizes from $2\!.\!7$ to $20''$, and their driving sources are identified in the IRAS point source catalog. The sensitivity limit of the sample corresponds to $L_{mech} \sim 0.001$ $L_\odot$, 2 orders of magnitude smaller than the L1551 outflow. There is a possibility that outflows having much smaller mechanical luminosities than currently known remain undetected in the unbiased survey, and the broad CO emission having a velocity range of ≤ 8 km s^{-1} from the quiescent gas in L1641 may mask outflows having line widths smaller than that. The IRAS point sources in the L1641 region have been studied by Strom et al. (1989c). The authors selected 48 sources detected at 12, 25 and 60 μm from the list of Strom et al. and located them in the ^{13}CO total intensity map (Fig. 8).

The L1641 region contains a sufficiently large number of sources for statistical studies, and its proximity to the Sun (~ 500 pc) allows one to obtain reasonably high sensitivity both for outflows and IRAS sources. In L1641, it is shown that optical emission-line stars are mostly T Tauri stars having masses around 1 $M_\odot$ (Cohen and Kuhi 1979; Herbig and Bell 1988). Thus, solar-type low-mass stars are the main stellar component in the L1641 cloud. A 12/25/60 μm two-color diagram of the 48 sources (Fig. 9) shows that all the outflows are located in the lower left part of the diagram. This indicates that the outflows are characterized by a significantly cooler dust color temperature than T Tauri stars, which are marked by a box in Fig. 9. Most of the other 42 sources without outflows are likely T Tauri stars, as suggested from their color shown in Fig. 9. About 1/5 of the 42 sources are in fact identified as T Tauri stars catalogued by Herbig and Bell (1988). The appreciable increase in the 25 μm flux is the cause of the cooler color temperature of the outflow sources. The outflow sources have ~ 10 times more massive dust envelopes than the T Tauri stars, suggesting that the outflow sources are embedded more heavily in the dust envelope than the T Tauri stars according to a spectral fitting by a two-temperature model (Fukui 1988). This is consistent with the idea that outflow is a signature of an evolutionary stage earlier than pre-main-sequence stars represented by T Tauri stars, i.e., a dynamical phase in which the protostellar matter has not yet established quasi-static equilibrium.

Figure 10 shows a bolometric luminosity-color diagram for the IRAS sources. The detection limit is ~ 1 $L_\odot$. Most of the sources located in or near the T Tauri box in Fig. 9 have luminosities in the range 2 to 20 $L_\odot$. On the other hand, all the outflows have considerably larger luminosities of 20 to 300 $L_\odot$. The molecular outflow sources are distinguished from the majority by their luminosity excess. Fukui et al. (1989) attributed the large luminosity of the outflow sources to gravitational energy released in the dynamical mass accretion onto the protostellar core. The accretion luminosity is calculated by using the equation $L_{acc} = GM_*\dot{M}/R_*$, where $\dot{M}$ is a mass accretion rate

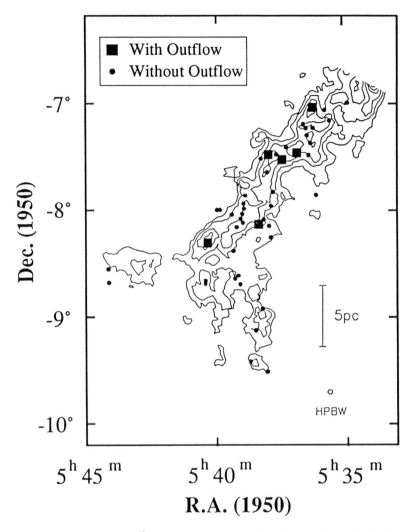

Figure 8.   Contour map of ^{13}CO ($J=1-0$) total intensity of the L1641 molecular cloud at Decl. (1950) $\leq -7°$ obtained with the Nagoya 4 m telescope. IRAS point sources associated with and not associated with molecular outflow are denoted by filled squares and small circles, respectively (Fukui et al. 1989).

(Stahler et al. 1980). For a typical mass accretion rate in a giant molecular cloud of 10^{-5} $M_{\odot}$ yr^{-1}, and a typical radius for a 1 $M_{\odot}$ protostar of ~ 4 $R_{\odot}$, the maximum accretion luminosity of a 1 $M_{\odot}$ star is estimated to be ~ 70 $L_{\odot}$. This value may explain the large luminosity of the outflow sources.

The completeness of the sample allows us to estimate the ratio of the outflow time scale to the T Tauri lifetime to be $\sim 1/7$. The actual number of T Tauri stars may be even larger than that shown in Fig. 10 because the IRAS

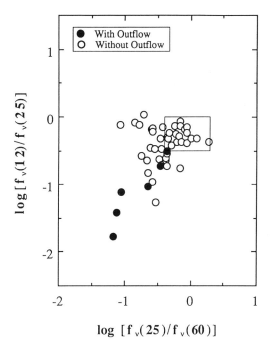

Figure 9.   12/25/60 μm color-color diagram for 48 IRAS point sources from Strom et al. (1989c) in L1641 (Fukui et al. 1989). A rectangular area corresponds to the T Tauri box (Beichman et al. 1986). Filled circles denote IRAS point sources associated with molecular outflow.

detected only T Tauri stars with a dust envelope. The above ratio is then likely <1/10 if we correct for T Tauri stars not detected by the IRAS. This is consistent with the typical dynamical time scale for outflows of $\sim 10^5$ yr, less than 1/10 of a T Tauri star age of $\sim 3 \times 10^6$ yr (Cohen and Kuhi 1979). On the other hand, the accretion time scale is 1×10^5 yr for the above mass accretion rate. This suggests that for low-mass stars in L1641, outflow lasts for almost the entire main accretion phase.

If we accept this interpretation, molecular outflows may correspond to the main accretion phase of solar-type low-mass protostars. It is then likely that accretion takes place inside a flattened circumstellar disk around a protostellar core, and that outflow occurs along the poles (see, e.g., Fig. 11 in Uchida and Shibata 1985). If the outflow is ultimately powered by the energy released by accretion (see, e.g., Strom et al. 1988), the mechanical luminosity of outflow should be smaller than the bolometric luminosity. We find this is actually the case for all known outflows.

The work on L1641 has provided the first observational evidence for the extremely young evolutionary status of molecular outflows. The conclusion

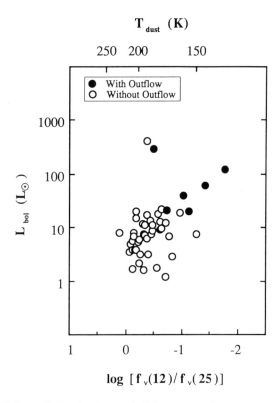

Figure 10. Bolometric luminosity vs 12/25 μm color for 48 IRAS point sources selected from Strom et al. (1989c) in L1641 (Fukui et al. 1989). Upper horizontal scale is the dust color temperature derived from 12 μm and 25 μm flux densities on the assumption of a λ^{-1} emissivity law. Filled circles denote IRAS point sources associated with molecular outflow.

reached by Fukui et al. (1989) is basically consistent with a scenario of protostellar evolution in which molecular outflow coincides with the main accretion phase of protostars (Shu et al. 1987a). We note that the trend of higher luminosities of outflows is also found in other regions studied by unbiased surveys, S287 and Mon R2 (Fukui et al., in preparation).

The time scale of outflows is several times 10^4 yr, nearly comparable to, or a little smaller than, that of the main accretion phase of a protostar. We should learn this more exactly when outflow is ignited and how it grows in time. This probably corresponds to understanding the time evolution of the main accretion phase. In this connection, it is interesting to note a recent discovery of a compact outflow in Ophiuchus (IRAS 16285−2356). IRAS 16285−2356 is detected only in the 100 μm band, is invisible on the POSS plate, and has a total luminosity $\leq 2 \ L_\odot$ (Mizuno et al., in preparation). Its dynamical time scale is $\sim 1 \times 10^4$ yr. If we assume the luminosity is solely supplied by mass accretion and take a typical mass accretion rate of $\sim 5 \times 10^{-6}$ M$_\odot$ yr^{-1} in a dark cloud, the protostellar core mass is estimated to be $\sim 5 \times 10^{-2}$ M$_\odot$. This gives an age of the protostellar core of $\leq 1 \times 10^4$ yr, nearly equal to the outflow age. Then, this outflow may truly be ignited at the beginning of the main accretion phase. Further more-sensitive unbiased surveys for compact outflows should be extremely important in shedding more light on these questions.

Another way to infer the time of ignition of CO outflow may be provided by observations of CO outflows formed in small bright rimmed globules in H II regions. Including the three outflows (IC1396-North, Ori-I-2 and L1206), eight CO outflows have so far been detected and mapped in bright rimmed globules having cloud masses of 3 to 150 M$_\odot$ (Sugitani et al. 1989,1991). Dynamical evolution of such globules has been extensively studied theoretically, and the starting time of cloud collapse is estimated to be $\sim$ a few times 10^4 yr in most cases. This time scale is comparable to the dynamical time scales of CO outflows, suggesting again that the ignition of CO outflows is very early in star formation.

If CO outflow is powered by mass accretion, as is suggested by magneto-hydrodynamical models, termination of outflow must correspond to the end of mass accretion. Dispersal of the protostellar cloud core is likely accomplished both by mass accretion and by outflow, with similar mass flow rates, $\sim 10^{-5}$ to 10^{-6} M$_\odot$ yr^{-1}, leading to the end of mass accretion (Shu et al. 1988). These luminosity excesses are mostly emitted at infrared wavelengths longer than 25 μm. If this far-infrared excess is characteristic of mass accretion from interstellar gas, gas density toward infrared sources should be correlated with the far-infrared excess, or with the luminosity excess. The correlation between luminosity and molecular gas column density $N(H_2)$ derived from ^{13}CO does indicate that this is actually the case for IRAS sources in L1641 (Chen et al. 1991).

B. Cloud Support by Outflows

It has been a long-standing question how a giant molecular cloud is prevented from free-fall collapse. Supersonic turbulent motion in molecular gas may be supporting a cloud against gravity. In fact, molecular linewidths in giant molecular clouds are commonly supersonic, ~ 2 to 3 km s^{-1}, representing random motion needed for cloud support. It is, however, not obvious how such supersonic turbulent motion is generated and maintained because such turbulence must quickly dissipate via interstellar shocks. The highly filamentary and clumpy structure seen in large-scale maps of clouds suggests that the area filling factor of gas is in the range $f = 0.05$ to 0.3, so a cloud fragment may survive for approximately $1/f$ crossing times. For a typical cloud like Orion south, this implies that the total internal turbulent energy, $M_{cloud} V_{turb}^2 / 2 =$ several times 10^{47} erg (for $M_{cloud} =$ several times 10^4 M$_\odot$ and $V_{turb} = 1$ km s^{-1}), would be dissipated over a period of 10 Myr. To maintain this turbulence in a steady state, kinetic energy and momentum must be supplied to the cloud. Molecular clouds must be supported against gravitational collapse as the star formation rate averaged over the Galaxy is about 3 $M_\odot$ yr^{-1} from a total molecular mass of $\sim 3 \times 10^9$ $M_\odot$, implying that the steady-state clouds collapse at a rate $\sim 10^{-3} \tau_{ff}$. Two general mechanisms may inhibit rapid gravitational collapse: a strong magnetic field and a supply of kinetic energy. There is evidence that magnetic fields play an important role in regulating cloud collapse and may contribute to cloud support. However, fields are "slippery" in that gas can readily move along the field lines, permitting collapse of the gas into flattened disks, although Alfvén waves may inhibit this mode. When a cloud becomes sufficiently large that the fractional ionization drops below about 10^{-8}, ambipolar diffusion sets in, making the magnetic field ineffective in supporting the cloud. Even in the presence of B magnetic fields, energy injection is required in order to maintain star formation at the observed rate.

Several distinct mechanisms may inject kinetic energy into molecular clouds, which can be categorized into external and internal energy sources. Large-scale external energy injection includes processes such as cloud-cloud collisions, while on smaller scales, CO outflows and optical and near-infrared shocks may provide an internal source of energy input. In particular, molecular outflows involving large momenta may be a major source of turbulent motion if they are generated nearly constantly and uniformly in a cloud over the cloud lifetime (Bally and Lada 1983; Fukui et al. 1986). The results of the unbiased surveys may well serve as a test of the importance of outflows in cloud support.

Figure 1 shows that molecular outflows are fairly evenly distributed in each molecular cloud of 10 to 40 pc length. In the case of the Orion south and S287 clouds, the mean separation of outflows is roughly 5 pc. Such uniform distribution is favorable for cloud support, as a single molecular outflow can cover a radius not larger than a few pc. A more quantitative test is made by estimating the time scale needed to feed the turbulence momentum by molecular outflows (see, e.g., Margulis et al. 1988).

The number of generations of molecular outflows required may be calculated by dividing $M_{cloud} \times V_{turb}$ by the sum of the outflow momenta observed at the present epoch. The time scale needed to feed the turbulence momentum is then obtained by multiplying the number of generations by the mean lifetime of outflows. If this time scale is smaller than the cloud collapse time scale, molecular outflows are able to maintain the turbulence. From the results of unbiased surveys toward the three giant molecular clouds, Orion south, S287 (Fukui 1989), and Mon OB1 (Margulis et al. 1988), typical total momentum of outflows in a single cloud and a single generation is estimated to be 200 to 300 $M_\odot$ km s^{-1} (Table II). On the other hand, the total cloud mass and turbulent velocity for them are several times 10^4 $M_\odot$ and ~ 1 km s^{-1}, respectively. The 50 to 160 generations of molecular outflows are needed for cloud support, and it takes ~ 3 to 9×10^6 yr to feed the turbulent momentum with the mean outflow lifetime of 3 to 17×10^4 yr. This time scale is similar to the cloud free-fall time scale estimated from typical ^{13}CO linewidths (~ 2 km s^{-1}) and

TABLE II

Momentum Injection by Outflows in Giant Molecular Clouds

		Orion South[e]	S287[e]	Mon OB1[e]
Number of Outflows[a]		12	5	5
Outflow Momentum[b]	($M_\odot$ km s^{-1})	320	270	190
Turbulent Momentum	($M_\odot$ km s^{-1})	2×10^4	1.4×10^4	3×10^4
Generations		60	50	160
τ_{outflow}[c]	(yr)	5×10^4	1.7×10^5	3×10^4
τ_{support}[d]	(yr)	3×10^6	9×10^6	5×10^6

[a] Number of outflows observed at the present epoch; this is assumed as the number of outflows in a single generation.
[b] Total momentum of outflows in a single cloud in a single generation.
[c] Mean outflow lifetime.
[d] Time scale required to feed the turbulent momentum by molecular outflows. References: Orion South (Fukui 1989); S2871 (Iwata et al. 1992b); Mon OB1 (Margulis et al. 1988).

typical cloud sizes (20 pc). It is therefore suggested that molecular outflows play an important role in cloud support.

VII. CONCLUSION

Recent progress in both high-resolution mm-wave observations and infrared imaging has provided a wealth of information on the mass outflow phenomenon from protostars. Systematic searches for molecular outflows have begun to provide the unbiased samples needed for statistical studies of outflow properties. Such studies have shown that molecular outflow sources are characterized by significantly larger luminosity than T Tauri stars in L1641. This suggests that molecular outflows correspond to the main accretion phase of protostellar evolution, with significant luminosity excess resulting from gravitational energy released by mass accretion. There is also increasing evidence for the role of molecular outflows in cloud support, and for dynamical interaction between individual outflows and their ambient gas. It is important to make even more systematic efforts in unbiased surveys, as our coverage of the sky in high-resolution CO surveys is still extremely small. These surveys and high-resolution observations with millimeter interferometers and large-format infrared cameras will provide deeper understanding of the role and evolutionary status of molecular outflows in star formation.

Acknowledgment. We are grateful to Y. Uchida and K. Ogura for valuable discussions, and to all those who made useful comments on the earlier versions of this review. This work was financially supported by Grant-in-Aid for Scientific Research of the Ministry of Education, Science and Culture of Japan.

ORIGIN OF OUTFLOWS AND WINDS

ARIEH KÖNIGL
University of Chicago

and

STEVEN P. RUDEN
University of California, Irvine

A successful interpretation of the energetic bipolar outflows in young stellar objects should account for their ubiquity and for their inferred association with circumstellar accretion disks. The theory also has to explain the apparent presence of two distinct outflow components: an extended, neutral wind which has sufficient momentum to drive the large-scale molecular lobes, and a pair of collimated, ionized jets which emerge along the symmetry axis at high velocities but with a comparatively small momentum discharge. It is argued that the weakly ionized wind most likely represents an MHD outflow driven centrifugally from the disk surfaces or from the boundary between the disk and the star. Specific wind models for each of these alternatives are presented, and it is noted that both provide a natural explanation of the observed correlation between accretion and outflow. The kinematic, thermal and chemical wind properties predicted by these models are described and their observational implications are considered. In particular, it is pointed out that the wind characteristics may be reflected in the observed forbidden line and infrared continuum emission of T Tauri stars and in the measured abundances of various molecular species. The origin of the high-velocity jets is at present less well established. The jets may represent the ionized inner regions of a disk-driven wind or else could correspond to an isotropic, MHD wave-driven stellar wind that has been collimated by the ambient neutral outflow. Yet another possibility is that the jets are powered by the energy released in a shear boundary layer between the disk and the star. Episodic accretion events associated, perhaps, with FU Orionis outbursts might account for the nonsteady mass ejection indicated in certain jets.

The existence of high-velocity winds in T Tauri stars has been known for some thirty years now (see Herbig 1962b). However, only during the last decade, following the discovery of bipolar molecular flows and highly collimated optical jets (see Lada 1985 and Mundt 1985 for reviews), has it become apparent that energetic outflows are a ubiquitous phenomenon in young stellar objects (YSOs) ranging from deeply embedded infrared sources (DEIS) to classical T Tauri stars (CTTS). It is now recognized that the mechanical energy injected by YSOs into their parent molecular clouds could have a strong impact on the evolution of the clouds in that it may provide the required support against gravitational collapse (see, e.g., McKee 1989). It has also been suggested (Shu and Terebey 1984) that these energetic outflows are responsible for terminating the YSO accretion phase by reversing the infall of the surrounding gas toward the star, which would have direct implications to the distribution of YSO masses and to the clearing of protoplanetary disks. Understanding the origin of YSO outflows is thus of fundamental importance to the theory of star and planet formation.

An important recent development has been the emergence of evidence that the outflow activity is strongly linked to the presence of circumstellar accretion disks and that many of the radiative properties of YSOs are direct manifestations of the accretion and outflow processes. However, despite the substantial observational progress, there is still no firm theoretical understanding of the origin of the outflows and of their role in the evolution of YSOs. Ordinary stellar wind mechanisms have proved inadequate to account for the large momentum discharges that are needed to drive the bipolar flows, so other mechanisms have been explored. In particular, the possibilities of tapping the liberated gravitational energy of the accreted material or the rotational kinetic energy of the star have been investigated. One of the most promising ideas has been that of a centrifugally driven wind (CDW), wherein magnetic fields anchored in a rotating disk or star accelerate the outflowing gas by means of the centrifugal force. This mechanism can readily account for the inferred momentum discharges and is consistent with the growing evidence for the presence of strong magnetic fields in YSOs and their environments. CDWs generally also carry large angular momentum discharges and can exert strong braking torques on the driving disks or stars. In fact, such winds could dominate the angular momentum transport in circumstellar disks and may be responsible for the ability of stars that had been spun up to breakup speeds to continue accreting. In this way they could play an essential role in the evolution of YSOs, which might account for their ubiquity.

Our review focuses on the recent developments concerning the accretion-outflow connection and the role of magnetic fields. In Sec. I we summarize the observational evidence for the two main outflow components, namely the ionized jets and the neutral (atomic and molecular) winds, as well as evidence for the association of energetic outflows with circumstellar disks and the dynamical importance of magnetic fields. Section II is devoted to the origin of the recently discovered neutral outflows. We concentrate on the theory of CDWs and discuss their kinematic, thermal, and chemical properties and their observational implications. In Sec. III we consider the formation of the less massive but higher velocity ionized jets. In Sec. IV we discuss a simple model that unifies star formation and bipolar flows. Finally, our conclusions are summarized in Sec. V.

I. OBSERVATIONAL BACKGROUND

A. The Two Outflow Components

Since the original discovery of a molecular bipolar flow source in L1551 IRS 5 by Snell et al. (1980), approximately 150 such sources have been identified (see Chapter by Fukui et al.). The most striking property of these flows is their bipolarity. Although they are often rather poorly collimated (see Lada 1985), these flows nevertheless almost always exhibit a two-lobe morphology, with molecular gas moving in two opposite directions away from a central YSO. The characteristic scales and dynamical ages of the flows are ~ 0.1 to 1 pc and $\gtrsim 10^4$ yr, respectively, and their momentum discharges F_{lobe} are of the order of 10^{-6} to 10^{-2} $M_\odot$ yr^{-1} km s^{-1}. Both the momentum discharge and the kinetic power L_{lobe} appear to be correlated with the bolometric luminosity L_{bol} of the central source, but the force F_{lobe} acting on the lobes is typically 100 to 1000 times greater than L_{bol}/c (where c is the speed of light), indicating that, with the possible exception of high-luminosity YSOs, they are not driven by radiation pressure.

It is now generally accepted that the bipolar molecular lobes, whose velocities V_{lobe} are $\lesssim 25$ km s^{-1}, consist mostly of ambient gas that has been swept up by a much higher velocity wind emanating from the central region. This conclusion is consistent with the comparatively large ($\gtrsim 1$ $M_\odot$) inferred masses of the moving lobes and with the fact that the emission-weighted extent of the flow is often nearly independent of velocity, as expected for an expanding shell (Snell 1987). In the case of the L1551 IRS 5 flow, the presence of a thin, expanding shell has been directly confirmed by high-resolution CO imaging (see, e.g., Moriarty-Schieven et al. 1987). On the assumption that the lobes represent momentum-driven interstellar bubbles (Steigman et al. 1975; see Sec. IV), one can equate the force F_{lobe} on the lobes to the momentum discharge of the wind (i.e., to the product $\dot{M}_w V_w$ of the mass outflow rate and the wind velocity). The wind kinetic powers inferred in this way, $L_w \approx (V_w/V_{lobe}) L_{lobe}$, are generally limited to $\lesssim 0.1$ L_{bol}. In several well-collimated and highly resolved bipolar flow sources, systematic velocity gradients have been observed along the outflow major axis (Snell et al. 1984; Moriarty-Schieven et al. 1987; Fridlund et al. 1989). The line-of-sight velocity $v_\parallel$ is seen to increase linearly with the projected distance $r_\perp$ along a lobe

$$v_\parallel \propto r_\perp \tag{1}$$

for projected distances up to (typically) several tenths of a parsec away from the flow center. This "Hubble" law may arise if the outflow consists of clumps of gas with a range of velocities. As time progresses, the clumps naturally sort themselves in such a way that the highest velocity gas has traveled the farthest from the central source (as with freely flying shrapnel from an explosion). We will return to the origin of this linear Hubble relation in Sec. IV.

Recent surveys of outflows in molecular clouds (Fukui et al. 1986; Margulis et al. 1988; Snell et al. 1988; Wouterloot et al. 1989) clearly indicate that a large fraction (∼50%) of embedded young stellar objects go through the outflow phase some time during their evolution. High-resolution interferometric observations (Terebey et al. 1989) have shown that the outflow phase can occur very early in the evolution of a YSO and that outflow (presumably along the poles) can occur coevally with infall (presumably onto an equatorial disk). In the solar neighborhood, the frequency of occurrence of outflows is sufficiently high when compared to the stellar birth rate that this phase may be common to all stars with masses greater than that of the Sun (Margulis et al. 1988). The momentum and energy in a typical outflow are sufficient to disrupt the natal molecular cloud core surrounding a YSO (Mathieu et al. 1988); furthermore, outflows are predominantly associated with deeply embedded sources but only rarely with optically revealed YSOs (Lada 1988b; Margulis et al. 1989). On the basis of these facts, it seems likely that outflows are responsible for removing circumstellar material and for driving the evolution of YSOs from DEIS to CTTS. By dispersing the infalling molecular cloud core, outflows are also instrumental in determining the final mass of the combined star–disk system (Shu et al. 1987a).

Ionized optical jets are usually associated with DEIS or strong-emission CTTS. In contrast to the molecular flows, they are generally highly collimated (opening angles 3° to 10°), with the collimation lengths inferred to be ≲100 AU (see, e.g., Mundt et al. 1987). Their velocities typically lie in the range 200 to 400 km s^{-1}, although some luminous sources (with $L_{bol} \gtrsim 10^3$ $L_\odot$) have measured radial speeds of 450 to 700 km s^{-1} (see, e.g., Poetzel et al. 1989; Ray et al. 1990). The jet velocities are inferred both from spectroscopy and from proper motion measurements of optical emission knots. For example, the system of knots in the IRS 5 jet in L1551 exhibits maximum radial speeds of 350 km s^{-1} (Stocke et al. 1988) and a transverse velocity of 190 km s^{-1} (Neckel and Staude 1987), which is comparable to the proper motions of 150 and 170 km s^{-1} measured for the associated Herbig-Haro (HH) objects HH 28 and HH 29 (Cudworth and Herbig 1979). The characteristic dynamical ages (∼200 to 3000 yr) are 1 to 2 orders of magnitude lower than those of the molecular flows, and there is growing evidence for velocity variability and for nonsteady outflow episodes (Reipurth 1989b; see also Mitchell et al. 1988a; Bachiller et al. 1990a; Raga et al. 1990; Reipurth and Heathcote 1991).

The mass outflow rates in the jets have been inferred from optical line emission (in particular, the density sensitive [S II] λλ6716, 6731 lines) and radio continuum measurements to be $\dot{M}_j \approx 10^{-9}$–$10^{-8}$ $M_\odot$ yr^{-1}, and for the few cases where both jets and molecular flows are detected in the same source, it was found that their kinetic powers are comparable (Mundt et al. 1987). This raises the possibility that the jets could drive the molecular lobes if the latter represent energy-driven (Weaver et al. 1977), rather than momentum-driven, interstellar bubbles. This possibility could in principle be realized if the cooling time t_{cool} of the shocked jet material that, in this picture, fills the bubble, were longer than the age $t_{lobe} \approx R_{lobe}/V_{lobe}$ of the molecular lobes. We defer discussion of this possibility to Sec. III, where we also reexamine the derivation of $\dot{M}_j$. It should be noted, however, that many molecular lobes are apparently not powered in this way. For example, in the source L1551 IRS 5, measurements of the far infrared emission from the molecular outflow region directly indicate that the expanding molecular shell is not energy driven (see, e.g., Clark et al. 1986). Furthermore, an analysis of the optically luminous Herbig-Haro objects that are associated with the heads of stellar jets suggests that, at least in some cases, the shocked jet material cools within the jet head and contributes to the observed emission (see, e.g., Hartigan 1989).

One of the most significant developments during the last few years has been the discovery in several DEIS of a *neutral* outflow component that apparently involves a sufficiently large momentum discharge to drive the molecular lobes. This component has been detected in the form of high-velocity ($\sim$150 km s^{-1}) atomic (Lizano et al. 1988) and molecular (see, e.g., Mitchell et al. 1988*b*; Koo 1989,1990; Mitchell et al. 1989; Masson et al. 1990; Mitchell and Hasegawa 1991) winds. It is as yet unclear whether the molecular gas, which appears to be clumpy, represents entrained ambient material or is due to the formation of molecules in the high-velocity atomic outflow (see Sec. II.C). In any case, there are indications that even some of the lower-velocity molecular gas in a source like L1551 IRS 5 originates in the vicinity of the YSO and is not part of the swept-up shell (Fridlund and White 1989). The presence of neutral atomic winds in CTTS has previously been inferred from Na I D line measurements (see, e.g., Mundt 1984), and the corresponding momentum discharges also appear to be large enough to drive the associated molecular flows (see, e.g., Natta and Giovanardi 1990). The existence of two distinct outflow components has been deduced in L1551 IRS 5 from an analysis of the optical emission-line spectra (Stocke et al. 1988). This analysis indicates that the CO lobe is pervaded by a $\sim$160 km s^{-1} wind whose momentum discharge is an order of magnitude greater than that of the 440 km s^{-1} jet and could account for the expanding molecular lobes.

B. The Disk-Wind Connection

There is growing evidence that low-mass ($\lesssim$2 M$_\odot$) YSOs, including both DEIS and CTTS, are surrounded by circumstellar disks with estimated masses $\sim$0.01 to 1 M$_\odot$ on scales $\sim$10^2 to 10^3 AU (see Chapters by Beckwith and Sargent and by Basri and Bertout). The most direct evidence has come from molecular line and millimeter continuum interferometric measurements as well as from high-resolution infrared observations. For example, the presence of a rotating, Keplerian disk has been inferred in this way in HL Tau (Sargent and Beckwith 1987) and in T Tau (Weintraub et al. 1989*a*). A compact molecular core (of radius $\sim$45 AU) detected around IRS 5 in L1551 (Keene and Masson 1990) is also a likely candidate for a circumstellar disk. Additional evidence has been provided by analyses of infrared and ultraviolet spectra, optical emission line profiles, and optical polarization maps. Furthermore, there are strong indications that Keplerian disks form in a similar fashion around high-mass YSOs (see, e.g., Vogel et al. 1985; Persson et al.

1988; Plambeck et al. 1990).

Evidence is also mounting for the existence of a strong link between circumstellar disks and energetic outflows in YSOs (see Chapter by Edwards et al.). Perhaps the most compelling argument in favor of such an association in low-mass YSOs derives from a comparison of DEIS and CTTS with weak-line T Tauri stars (WTTS; Walter et al. 1988). WTTS are similar to CTTS in all respects except that they do not exhibit infrared excesses and strong linear polarizations (a signature of circumstellar dust) or ultraviolet excesses (commonly attributed to emission from accreted gas that reaches the stellar surface) and that they lack the strong low-ionization and forbidden line emission as well as the P Cygni line profiles which are the hallmark of energetic outflows in CTTS. The simultaneous absence of evidence for substantial disks and strong winds in WTTS points to a likely connection between the accretion and outflow processes in YSOs. The properties of DEIS and CTTS suggest that the basic relationship lies in the outflows being powered by accretion. In particular, the observed correlation between L_w and L_{bol} is then understood from the fact that the luminosity excesses in YSOs can also be attributed to accretion. In a similar vein, the fact that DEIS and CTTS with comparable bolometric luminosities also have similar outflow parameters indicates that the wind-driving mechanism is the same in both types of objects and that it is tied to the presence of an active accretion disk. Additional support for the disk-wind connection comes from observations of CTTS of the YY Ori type, which show evidence for concurrent mass outflow and inflow (manifested, respectively, in P Cygni and inverse P Cygni line profiles) as well as for the existence of an opaque disk (indicated by the absence of redshifted forbidden line emission). The correlation between disk and wind signatures in DEIS and CTTS is apparently present also in higher-mass stars (Strom et al. 1988), which suggests that the accretion-outflow connection is a basic property of all YSOs.

The fact that the inferred wind kinetic powers are not much smaller than the bolometric luminosities ($L_w \lesssim 0.1\, L_{bol}$) indicates that the bulk of the energy and momentum of the outflows is injected near the central YSO. Whether the wind emanates from the disk, the star, or the boundary between them is still an open question (see Sec. II). However, there are indications that at least part of the outflow is launched at a finite distance from the YSO, presumably from the surface of the associated disk. For example, several CTTS exhibit a narrow, blueshifted (-30 to -100 km s^{-1}) absorption component in the Na I D lines which is sometimes superposed on a classical P Cygni profile and which is plausibly interpreted as forming in a largely neutral wind on a scale of $\gtrsim 10$ stellar radii (Mundt 1988). In the case of neutral outflows detected in high-luminosity DEIS such as GL 2591 (Mitchell et al. 1989), one can argue that the wind is unlikely to originate at the stellar surface because of the high effective temperature of the YSO.

C. The Dynamical Role of Magnetic Fields

Magnetic fields are believed to play a central role in the evolution of molecular clouds and in the initiation of star formation by cloud core collapse (see Chapter by McKee et al.). Their importance in bipolar flows is suggested by the finding that the axes of the molecular lobes and of the associated jets are aligned within 30° of the ambient field direction (as determined by polarization measurements) in over 70% of the surveyed sources (Strom et al. 1986). In those cases where a flattened circumstellar mass distribution has been imaged, its minor axis is found to be parallel to the bipolar flow direction and is thus also aligned with the local magnetic field. Although dense molecular cores are apparently not magnetic pressure-dominated and hence their contraction is not directly controlled by magnetic stresses, this result suggests that the magnetic field nevertheless influences the orientations of the embedded circumstellar disks. This is presumably achieved through magnetic braking (see, e.g., Mouschovias and Paleologou 1980), which removes the angular momentum component normal to the field more efficiently than the parallel component and thus leads to preferential contraction along the field lines and to an alignment of the minor axis of the cloud core with the local magnetic field. Since a wind that originates in the vicinity of a YSO would be naturally directed along the axis of the associated disk, it is then not surprising that the bipolar flows are found to be aligned with the ambient magnetic field. The field is also expected to contribute to the subsequent collimation of the outflows both directly, through its confining transverse stress, and indirectly, through the anisotropic density distribution that it induces in the molecular cloud (Königl 1982; Shu et al. 1991; see Sec. IV).

Although the collapse of molecular cloud cores that form low-mass protostars is likely triggered by the reduction of the magnetic field support of the cloud by ambipolar diffusion (Lizano and Shu 1989), significant magnetic flux may still be carried in by the inflowing matter (see, e.g., Umebayashi and Nakano 1988). The advected magnetic field would not be strong enough to inhibit the accumulation of mass in the center, but it could play an important role in the removal of the excess angular momentum of the accreted matter, as we discuss in Sec. II.A. It is interesting to note in this connection that meteoritic evidence points to the existence of a ~ 1 gauss field at a distance of ~ 3 AU from the center of the protosolar nebula (Levy and Sonett 1978), which is of the right order of magnitude for this effect. The YSO itself could develop a strong magnetic field by dynamo action when it becomes convective (see, e.g., Shu and Terebey 1984). In fact, T Tauri stars exhibit vigorous magnetic activity that implies a surface field of a few hundred gauss (see Chapter by Montmerle et al.), and although direct Zeeman measurements have proven difficult, there are several cases where even higher values ($\gtrsim 10^3$ gauss) are consistent with the data (see, e.g., Johnstone and Penston 1986). Magnetic fields of this magnitude could lead to efficient braking of the stellar rotation (see Sec. III.B). Furthermore, a kilogauss-strength dipole field could disrupt a circumstellar accretion disk at a distance of a few stellar radii and thereby alter the nature of the accretion flow onto the YSO. This situation has already been indicated in several objects (see, e.g., Bertout et al. 1988) and could have important implications to the dynamical and radiative properties of CTTS (Königl 1991).

II. ORIGIN OF NEUTRAL OUTFLOWS

As we pointed out in Sec. I.A, the large inferred momentum discharges in the bipolar flows and the comparatively low bolometric luminosities of most YSOs imply that radiation pressure is generally much too weak to drive the molecular lobes. An alternative possibility, that the high-velocity winds which push on the expanding molecular shells are driven by thermal pressure, can be ruled out on the basis of efficiency considerations. To generate the energetic YSO outflows, the temperatures at the base of the wind would be too high and the radiative losses too severe to be consistent with the observations (DeCampli 1981). The only potentially viable mechanisms proposed to date involve tapping the rotational energies of the YSO or its circumstellar disk through the agency of a magnetic field that threads the rotating mass. In these models, the magnetic field lines are wound up at the base of the wind and accelerate the outflowing gas either centrifugally, like beads on a rotating wire, or by axial pressure gradients, like an uncoiling spring. Similar ideas have been considered in connection with jets in active galactic nuclei, and it may therefore be fruitful to study bipolar flows in the context of the general theory of cosmic jets (Königl 1986). We discuss the basic principles of magnetohydrodynamic (MHD) winds from accretion disks and from the central YSOs in Secs. II.A and B, respectively. The thermal and chemical properties of the winds and their observational consequences are considered in Sec. II.C.

A. MHD Winds from Accretion Disks

1. CDW Models. The basic theory of steady-state, axisymmetric MHD outflows from rotating objects was developed in the context of stellar winds and was originally limited to a two-dimensional (equatorial plane) geometry (Weber and Davis 1967). Blandford and Payne (1982) generalized these ideas to a three-dimensional disk geometry using a self-similar model in the zero temperature limit, and Sakurai (1985,1987) further extended the formalism by re-introducing thermal pressure effects and dropping the self-similarity assumption. The theory was first applied to YSOs and circumstellar disks by Hartmann and MacGregor (1982) and Pudritz and Norman (1983,1986), respectively.

The structure of a perfectly conducting CDW from the surface of a Keplerian accretion disk can be described by referring to the three characteristic surfaces that form the loci of the points where the poloidal (meridional plane) component of the flow velocity is equal to that of the slow magnetosonic

(SMS), the Alfvén, and the fast magnetosonic (FMS) wave velocities, respectively. The wind is initially magnetically dominated, in the sense that the magnetic energy density is larger than the thermal, the gravitational, and the kinetic energy densities. In the limit of a cold outflow (speed of sound C_s much smaller than the Alfvén speed $V_A = B/(4\pi\rho)^{1/2}$, where B is the magnetic field amplitude and ρ is the mass density), the SMS speed reduces to C_s and the SMS surface is located very close to the face of the disk. Below the SMS surface, the poloidal acceleration is essentially thermal. However, above this surface the acceleration is centrifugal and is well approximated by the "bead on a rigid wire" picture (Henriksen and Rayburn 1971). Using this mechanical analogy, one can readily derive the condition for a centrifugal launching of the wind (Blandford and Payne 1982). If the wire intersects the disk at a distance r_0 from the center and makes an angle θ to the disk normal, then the balance of gravitational and centrifugal forces along the wire implies that, in equilibrium, the effective potential $\Phi_{\text{eff}}(y)$ satisfies

$$\frac{\partial \Phi_{\text{eff}}}{\partial y} = \frac{y + \sin\theta}{(1 + 2y\sin\theta + y^2)^{3/2}} - y\sin^2\theta - \sin\theta = 0 \qquad (2)$$

where the dimensionless variable $y \equiv s/r_0$ measures the distance s along the wire. The equilibrium is unstable when $\partial^2\Phi_{\text{eff}}/\partial y^2$ is negative, which at $y = 0$ occurs for $\theta > 30°$. Hence, centrifugal driving sets in if the poloidal component of the field is inclined at an angle of $<60°$ to the outward radius vector.

The magnetic field continues to dominate the outflow up to the Alfvén surface. The inner region of the wind is then effectively force free ($[\nabla \times \mathbf{B}] \times \mathbf{B} = 0$), with the magnetic pressure gradient approximately balancing the magnetic tension. The tension is associated with the bending of the field lines toward the rotation axis (Fig. 1), which provides the initial collimation of the flow. Another consequence of the magnetic field dominance is that the gas along a given magnetic flux tube corotates with the disk material at the foot of the tube (i.e., the azimuthal velocity component satisfies $V_\phi \approx \Omega_K[r_0]\,r$, where $\Omega_K[r_0]$ is the Keplerian angular velocity at r_0 and r is the distance from the rotation axis). The wind can thus exert a substantial back torque on the disk, with the effective lever arm being of the order of the cylindrical radius r_A of the Alfvén surface. CDWs therefore represent a potentially very effective mechanism for removing angular momentum from the accretion flow. In the limit when virtually all the excess angular momentum of the inflowing matter is transported by the wind, the angular momentum conservation equation and the corotation property imply that $\dot{M}_w$ is related to the mass accretion rate $\dot{M}_{\text{in}}$ by

$$\frac{\dot{M}_w}{\dot{M}_{\text{in}}} \approx \left(\frac{r_0}{r_A}\right)^2 \qquad (3)$$

which shows that converting only a small fraction of the inflow into an outgoing wind may suffice for this purpose (see Blandford and Payne 1982).

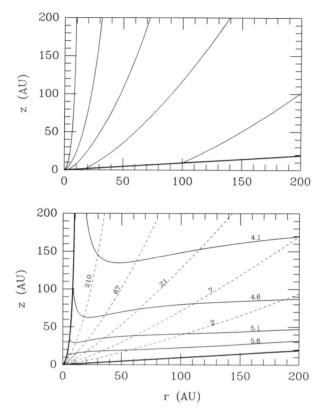

Figure 1.   (*top*) The poloidal magnetic field lines in a self-similar, centrifugally driven
wind from the surface (*heavy* line) of a Keplerian accretion disk. Under the assumed
good coupling conditions, the plotted lines also represent the meridional projections
of the wind flow lines. The model used in this plot corresponds to a central mass
of 0.5 $M_\odot$, a mass accretion rate of 3×10^{-6} $M_\odot$ yr^{-1}, and a mass outflow rate
between 0.1 and 10 AU of 5×10^{-8} $M_\odot$ yr^{-1} (with an equal contribution from each
decade in radius). (*bottom*) Contours of constant poloidal velocity (*dashed* lines,
labeled in km s^{-1}) and particle density (*solid* lines, labeled in log[n/cm^{-3}]) for the
same wind model. The heavy line on the left indicates the flow line that originates
at $r_0 = 0.1$ AU.

The wind inertia becomes important beyond the Alfvén surface, causing
the field lines to become progressively more toroidal as the azimuthal flow
velocity decreases below the corotation value (Fig. 2). The magnetic "hoop"
stresses associated with the toroidal field component act to further collimate
the flow. This field component is induced by a poloidal current, and it
has, in fact, been shown (Heyvaerts and Norman 1989; Eichler 1992) that
the magnetic flux surfaces of any MHD wind that encloses a finite poloidal
current asymptotically assume a cylindrical or a paraboloidal morphology.
The twisting of the field lines also results in the magnetic pressure gradient

replacing the centrifugal force as the main acceleration mechanism in this region.

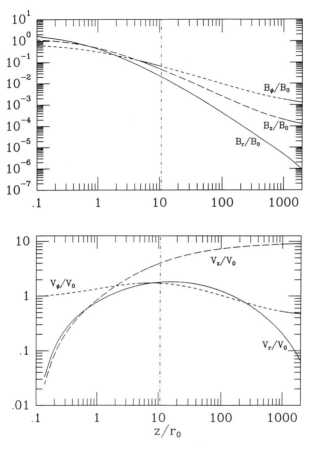

Figure 2.   The distribution of the magnetic field (*top*) and of the velocity (*bottom*) along a flow line for the cold, self-similar, disk-driven wind model shown in Fig. 1. Both *B* and *V* are normalized by their values at the base of the wind. The dash-dotted line marks the position of the critical (Alfvén) point of the flow. Note that V_ϕ decreases above the critical point because of inertial effects but that V_z continues to increase because of the magnetic pressure-gradient force.

The acceleration effectively continues up to the FMS surface, whose location depends on the initial temperature of the wind and on the shape of the field lines (Phinney 1983). In the cold flow limit, the FMS speed approaches V_A and the asymptotic poloidal wind velocity can be approximated by

$$V_w \approx \Omega_K(r_0)\, r_A \qquad (4)$$

(see, e.g., Pudritz and Norman 1986). The asymptotic velocity is thus a factor $\sim r_A/r_0$ larger than the Keplerian velocity at the base of the outflow.

The relations (3) and (4) can be used to estimate the kinetic power of the wind. As we noted in Sec. I.B, the observations suggest that most of the energy injection into the wind occurs near the central YSO, and this is also the theoretical expectation for Keplerian disks (Blandford and Payne 1982). In particular, evaluating the kinetic power

$$L_{\rm w} \approx \dot{M}_{\rm w} V_{\rm w}^2 \approx \dot{M}_{\rm w} \left[\Omega_{\rm K}(r_0)\, r_{\rm A}\right]^2 \approx \dot{M}_{\rm in} \left[\Omega_{\rm K}(r_0)\, r_0\right]^2 \approx G\, M_* \, \dot{M}_{\rm in}/r_0 \quad (5)$$

(where G is the gravitational constant) at the inner edge ($r_{\rm min}$) of the disk and noting that the accretion-dominated bolometric luminosity is

$$L_{\rm bol} \approx \frac{G\, M_* \, \dot{M}_{\rm in}}{R_*} \gtrsim \frac{G M_* \dot{M}_{\rm in}}{r_{\rm min}} \quad (6)$$

(where M_* and R_* are the mass and radius of the YSO), one sees that this model provides a natural explanation for the observed correlation between $L_{\rm w}$ and $L_{\rm bol}$ and for the limit $L_{\rm w} \lesssim 0.1\, L_{\rm bol}$. Most of the momentum injection is also expected to take place near the central star and the predicted momentum discharge can be as high as $\sim L_{\rm bol}/V_{\rm w}(>>L_{\rm bol}/c)$. In the case of L1551 IRS 5, this value is consistent with the inferred magnitude of $F_{\rm lobe}$ and supports the conclusion that the expanding molecular shell is momentum driven. In the original application to bipolar flows (Pudritz and Norman 1983,1986), the molecular lobes were identified with large-scale MHD winds from the surfaces of massive disks and most of the momentum and energy injection were assumed to occur at large radii. This interpretation of the lobes has not been substantiated by subsequent observations (see Sec. I.A) and also faces severe theoretical difficulties in view of the large mass accretion rates that it implies (Shu et al. 1987a).

One can express the effective value of $r_{\rm A}/r_0$ in terms of observable quantities by using the momentum-driven bubble approximation $F_{\rm lobe} \approx \dot{M}_{\rm w} V_{\rm w}$ and the condition $\dot{M}_{\rm in} \gtrsim R_* L_{\rm bol}/G\, M_*$ in Eq. (1), which gives

$$\frac{r_{\rm A}}{r_0} \gtrsim \left(\frac{R_* V_{\rm w} L_{\rm bol}}{G\, M_* \, F_{\rm lobe}}\right)^{1/2}. \quad (7)$$

For self-consistency, the right-hand side of Eq. (7) must be greater than 1; if this condition is not satisfied, then both the accretion and the outflow might be nonsteady (see Sec. III). The magnetic field amplitude B_0 at the base of the wind can be estimated in a similar fashion from the assumption that the outflow carries away the excess angular momentum of the accreted matter (see Eq. [11] below), which implies

$$B_0 \gtrsim \frac{L_{\rm bol}^{1/2}}{(G\, M_* \, r_0^3)^{1/4}} \left(\frac{R_*}{r_0}\right)^{1/2}. \quad (8)$$

For a low-mass YSO like IRS 5 in L1551, this yields a field strength of a few gauss at $r_0 = 1$ AU, which is consistent with the available evidence for the protosolar nebula (Sec. I.C).

The construction of general wind solutions for a perfectly conducting gas with an adiabatic equation of state was discussed by Sakurai (1985). A steady-state, axisymmetric MHD wind possesses five field-line constants: the specific energy e and angular momentum l, which include both material and magnetic contributions, the field angular velocity ω, the ratio k of the mass flux to the magnetic flux, and the magnitude of the coefficient K in the adiabatic relation $P = K\rho^{\Gamma}$ (where P is the pressure and Γ is the adiabatic index). These constants uniquely determine the kinematic and thermodynamic structure of the wind. For a given distribution of angular velocity, pressure, and density at the base of the flow, one can solve for e, l, k, and K in terms of the magnetic stream function Ψ (which determines the poloidal field components) by requiring that the wind pass through the SMS and FMS critical points. The distribution of Ψ is, in turn, evolved from the top of the disk by solving the cross-field force balance equation and imposing the regularity condition at the Alfvén surface. Sakurai (1987) obtained solutions for self-gravitating disks with a split-monopole field configuration. Blandford and Payne (1982) assumed self-similarity in the *spherical* radial coordinate (which for a Keplerian disk implies $B_0 \propto r_0^{-5/4}$) and neglected thermal effects (which is a fair approximation for a wind that is driven mainly by centrifugal forces; cf. Belcher and MacGregor 1976). They further assumed that the flow starts with zero poloidal velocity from the equatorial plane. In this case the solutions only need to pass through one critical point and their topology is completely specified by the dimensionless versions of the parameters k and l. Figures 1 and 2 show a representative solution of this type, slightly generalized to include the effects of finite initial poloidal velocity and height above the midplane.

Several investigators have discussed MHD wind models in which centrifugal forces play a secondary role and the magnetic pressure gradient dominates the acceleration from the outset. An early version of such models was constructed by Draine (1983), who considered a YSO threaded by interstellar magnetic field lines and suggested that the rapid winding of the field by the stellar rotation would drive a strong shock wave into the ambient medium and produce an evacuated magnetic bubble. Uchida and Shibata (1985) performed numerical simulations of disks that are initially rotating at sub-Keplerian velocities and showed that the relaxation of the field-line twist induced by the contracting disk pushes out matter in the polar directions. Both of these models are manifestly time dependent. Lovelace et al. (1991) examined a steady-state version of this idea and proposed that the gradient of the zz component of the magnetic stress, $(B_r^2 + B_\phi^2 - B_z^2)/8\pi$, can drive a wind from the disk surface to escape speeds. In this picture, the wind need not carry a net poloidal current and hence may not be collimated by magnetic hoop stresses. However, it is still assumed to transport the excess angular momentum of the accreted gas. A potential difficulty with these models is that it has not yet been demonstrated that the large z-gradients of B_ϕ that are postulated to exist at the base of the outflow can actually be achieved except as transient events. For one thing, tightly wrapped field configurations (as in Draine's model) are susceptible to field-line reconnection. Furthermore, the azimuthal field generated by differential rotation in a disk that is threaded by open field lines would tend to compress the gas rather than expel it (and similarly for the B_r component induced by the radial inflow), which suggests that the surface boundary conditions envisioned in the Lovelace et al. scenario may be hard to realize. In particular, the transition from inflow to outflow might occur only after the density in the disk had dropped sufficiently to allow the field to control the motion of the gas, in which case the field may be expected to assume the force-free configuration postulated in the CDW model (see Sec. II.A.2 for an illustration of these points).

The preceding discussion of MHD winds has been based on the assumption that the magnetic field is effectively "frozen" into the outflowing gas. In reality, atomic and molecular winds from circumstellar disks would be mostly neutral, so the validity of this assumption must be verified. The charged particles typically consist of free electrons and of positively charged ions. The magnetic flux is transported by the electrons, which are usually well coupled to the positive ions by electrostatic forces. In a weakly ionized medium, it is thus convenient to divide the gas into two components: the neutrals, which provide the dominant contribution to the inertia and thermal pressure; and the ions, which carry the magnetic flux and are subject to the Lorentz force. These two components are coupled by collisions. To an excellent approximation, the steady-state equations of motion of the neutrals and the ions (subscript i) are, respectively,

$$\rho(\mathbf{V} \cdot \nabla)\mathbf{V} + \nabla P + \rho\nabla\Phi = \gamma\rho_i\rho(\mathbf{V}_i - \mathbf{V}) \qquad (9)$$

and

$$\frac{1}{4\pi}(\nabla \times \mathbf{B}) \times \mathbf{B} = \gamma\rho_i\rho(\mathbf{V}_i - \mathbf{V}) \qquad (10)$$

where Φ is the gravitational potential and $\gamma \approx 3.5 \times 10^{13}$ cm^3 g^{-1} s^{-1} (valid for $|\mathbf{V}_i - \mathbf{V}| \lesssim 2 \times 10^6$ cm s^{-1}; Draine et al. 1983) is the collisional coupling coefficient. Equations (9) and (10) show explicitly that the neutrals experience the effect of magnetic stresses only through their collisions with the ions. The wind dynamics can be treated in the single fluid approximation if the magnitude $|\mathbf{V}_i - \mathbf{V}|$ of the ion-neutral drift velocity (calculated from Eq. [10]) is everywhere much smaller than the bulk outflow speed V, which will be the case if the ion-neutral collision time is much shorter than the outflow time. Figure 3 illustrates how the collisional ionization induced by ambipolar diffusion heating (see Sec. II.C) and the radiative ionization of "heavy" atoms (primarily carbon) by the ultraviolet photons from the YSO contribute to the fulfillment of this condition in the wind model presented in Figs. 1 and 2.

Inasmuch as most of the high-energy photons originate in the accretion flow, the radiative ionization represents a feedback mechanism by which the inflow influences the coupling of the magnetic field to the wind and hence the removal of the excess angular momentum of the accreted matter (Pudritz and Norman 1983). It is important to note, however, that even though the outflow can be well represented *dynamically* as a single-component medium, the energy dissipation associated with the ion-neutral drift cannot be neglected in the calculation of the *thermal* properties of the gas (see Sec. II.C).

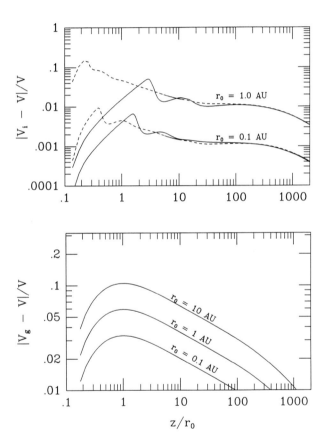

Figure 3.   (*top*) The normalized ion-neutral drift velocity along two flow lines for the model shown in Figs. 1 and 2. The solid lines represent the effect of collisional ionization alone whereas the dashed lines depict the added effect of the ionization of "heavy" atoms (C, O, Si, and Fe) by the stellar radiation field (assuming a photosphere of radius 3 $R_\odot$ and an effective temperature of 4×10^3 K). (*bottom*) The minimum grain-neutral drift velocity that is required for uplifting dust from the disk. Characteristic values of the grain radius (0.1 μm) and density (3 g cm^{-3}) and a normal dust-to-gas ratio have been assumed.

Figure 3 also demonstrates that the neutrals, after being accelerated by neutral-ion collisions, can uplift the dust in the wind formation zone by means of grain-neutral collisions. This has significant implications to the optical and infrared properties of YSOs (Safier and Königl 1992). A direct consequence of the efficient acceleration of grains is that the effective dust photosphere for absorption and reprocessing of stellar radiation is raised above the midplane, which can strongly affect the observed continuum spectrum (see, e.g., Kenyon and Hartmann 1987). In particular, the reprocessed radiation may account for the approximately flat ($\nu F_\nu \approx const$) near- and mid-infrared spectra measured in some of the CTTS that show evidence for circumstellar disks and energetic outflows (see, e.g., Rucinski 1985). A dusty disk-driven wind could also give rise to the $\sim 10^2$ to 10^3 AU cometary reflection nebulae that have been imaged around several YSOs (including IRS 5 in L1551; see, e.g., Strom et al. 1985; Campbell et al. 1988a). Furthermore, such a wind could lead, through multiple scattering of stellar radiation, to the aligned linear polarization pattern exhibited (in both visible and infrared light) by many bipolar outflow sources (see Bastien 1989).

2. *Magnetized Disk Models.* One of the key challenges of CDW models is the construction of self-consistent disk-wind configurations in which the global magnetic field geometry and the mass outflow rate are calculated for given inflow parameters. To account for the shearing (by differential rotation) and radial advection of the magnetic field lines, steady-state models of Keplerian accretion disks must include a field dissipation mechanism: either ambipolar diffusion or (for densities $\gtrsim 10^{11}$ cm^{-3}) Ohmic diffusivity. Königl (1989) studied self-similar models of this type that were matched onto the cold wind solutions of Blandford and Payne (1982). He assumed that the disk is threaded by open field lines whose stresses regulate the angular momentum transport of the inflowing gas and solved explicitly the angular momentum conservation equation

$$\frac{\rho V_r}{r} \frac{\partial}{\partial r} (r V_\phi) = \frac{B_r}{4\pi r} \frac{\partial}{\partial r} (r B_\phi) + \frac{B_z}{4\pi} \frac{\partial B_\phi}{\partial z} \tag{11}$$

(with V_z set equal to zero, appropriate for a geometrically thin disk) and the induction equation, which in the ambipolar diffusion-dominated regime takes the form

$$\nabla \times (\mathbf{V}_i \times \mathbf{B}) = 0 \tag{12}$$

(with $\mathbf{V}_i$ determined from Eq. [1]). The solutions were derived subject to the constraint

$$\nabla \cdot \mathbf{B} = 0 . \tag{13}$$

By imposing the continuity of the magnetic field components and of the field-line constant $\omega = [V_{i\phi} - (V_{ir} B_\phi / B_r)]/r$ at the disk-wind interface (which follow from the continuity of the momentum flux density as well as of the tangential electric field and normal magnetic field components at the boundary), the disk solutions (Fig. 4) were joined smoothly onto the corresponding self-similar wind solutions. The main result of this study was that continuous disk-wind configurations in which the bulk of the excess angular momentum as well as most of the liberated gravitational energy of the accreted matter are carried away by a super-Alfvénic and magnetically collimated CDW can be readily constructed for representative values of the (four) free parameters of the model.

This motivated the suggestion that the ubiquity of energetic outflows in accreting YSOs can be attributed to the fact that the winds are a *necessary* ingredient in the accretion process in that they carry away most of the angular momentum that needs to be removed in order for the accretion to proceed. It was argued that this mechanism should be highly effective because only a small fraction of the accreted mass is typically required to transport away the liberated angular momentum and energy of the inflowing matter (see Eq. [3]), and because this mass can be removed fairly easily on account of the near balance of the gravitational and centrifugal forces acting on it (see also Blandford and Payne 1982). Furthermore, it was noted that the disk may be expected to evolve to a steady state under the action of several feedback mechanisms. For example, an increase in the mass accretion rate over its equilibrium value would lead to a higher density at the base of the wind and hence to a higher outflow rate that would remove the excess energy and angular momentum. Similarly, a sudden increase in the angular momentum inside the disk should result in a larger torque at the base of the wind that would increase the magnitude of B_ϕ and hence the angular momentum carried by the outflow (Blandford and Payne 1982). Finally, any increase in the magnetic-field amplitude in the disk over its equilibrium value should be rapidly smoothed out by the ion-neutral drift, which scales as B^2 (see Eq. [10]).

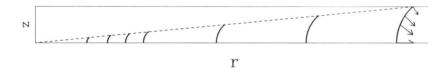

Figure 4.   The poloidal magnetic field geometry in a self-similar, ambipolar diffusion-dominated accretion disk. The heavy lines depict a few of the field lines whereas the dashed line delineates the surface of the disk. The arrows represent the drift velocities of the ions relative to the neutral disk material. Contrast the bending of the field lines away from the symmetry axis with the bending toward the axis in a centrifugally driven wind (top panel of Fig. 1).

The self-similar model employed by Königl did not address the question of whether the derived field configuration can actually be set up in a disk with realistic density and ionization profiles. To make further progress, Wardle and Königl (1992) focused on the vertical structure of a localized region in a thin Keplerian disk and assumed a constant sound speed C_s (with $P = C_s^2 \rho$)

and a uniform ion density ρ_i. (These assumptions may, in fact, apply in real molecular disks in the density regime $\sim 10^8$ to 10^{10} cm^{-3}; see Nakano 1984.) They then solved Eqs. (9), (10), (12), (13), and the mass conservation equation

$$\nabla \cdot (\rho \mathbf{V}) = 0 \qquad (14)$$

neglecting radial derivatives in comparison with z-derivatives whenever possible. An illustrative example of their calculations is presented in Fig. 5. It is seen that most of the disk material is in quasi-hydrostatic equilibrium, with the vertical compression by the combined magnetic and tidal stresses being balanced by the thermal pressure gradient. The neutral gas is in near-Keplerian rotation, and the frictional torque that it exerts on the ions causes the field lines (which are "frozen" into the ionized component) to move around at a slightly lower azimuthal speed. The resulting transfer of angular momentum from the neutrals to the field allows the neutral gas to move toward the center. Under the assumed steady-state conditions there is, however, no inward transport of magnetic flux, so the field lines remain stationary. The neutrals thus move across the field lines, maintaining sub-Keplerian azimuthal speeds on account of the radial neutral-ion drag. The radial drag on the ions in turn causes the field lines to bend away from the rotation axis (see Eq. [10] and Fig. 4), which, as we discussed in Sec. II.A.1, is a necessary condition for launching a CDW.

The angular-momentum conservation equation (the ϕ component of Eq. [9]) together with the constancy of ω along a field line imply that the difference between the angular velocities of the neutrals and the ions decreases as the field lines bend away from the symmetry axis toward regions of lower V_ϕ / r. Eventually a transition point is reached where the ions overtake the neutrals: above this point the field transfers angular momentum back to the matter (a process which in a fully developed CDW continues all the way up to the Alfvén point). Because V_z is much smaller than V_ϕ in a thin disk, the angular-momentum conservation equation also implies that V_r changes sign near this point. In fact, it can be readily shown that the radial drag vanishes and V_ϕ becomes equal to the Keplerian speed V_K at practically the same location (see Fig. 5). Beyond this point the field pushes on the matter and, if the bending angle is large enough (see Sec. II.A.1), a centrifugally driven outflow is initiated. The locus of these transition points can thus be naturally identified with the boundary between the disk and the wind. In a disk where most of the matter transfers angular momentum and energy to the field and only a small fraction of the accreted mass is expelled, the distance of this boundary from the midplane must exceed a density scale height. Under these conditions the base of the wind is effectively force free and the field is usually strong enough to dominate the vertical compression of the disk.

Much more work remains to be done in establishing the validity and applicability of this scenario. The main task is to construct global disk-wind solutions in which the outflow passes through all the relevant critical points. As part of this effort, the self-similar model would have to be generalized to

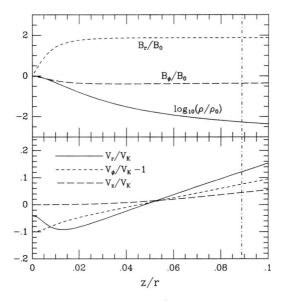

Figure 5.   The distribution of the magnetic field, the density, and the velocity in an ambipolar diffusion-dominated accretion disk. All the quantities are normalized by their midplane values (subscript 0), with V_K denoting the Keplerian speed. The solutions were derived for an isothermal gas with a constant ion density and assuming a geometrically thin disk (so that $B_z = B_0$ = const.). Note the large drop in density below the point where the inflow changes into an outflow ($V_r = 0$) and the change in the sign of the gradient of B_r and B_ϕ above that point. The dash-dotted line marks the location of the sonic point. The parameters used in this model correspond to a gas with temperature $T \approx 10$ K, particle density $n \approx 10^9$ cm^{-3}, electron density $n_e \approx 0.1$ cm^{-3}, and magnetic field amplitude $B \approx 1$ mG at a distance $r_0 \approx 10^2$ AU from a solar-mass star that accretes at a rate $\dot{M}_{in} \approx 10^{-6}$ M$_\odot$ yr^{-1}.

take account of the boundary conditions in the inner and outer radii of the disk. It is also important to investigate the magnetic field configurations that are likely to result from the initial collapse of the parent molecular cloud and the magnitude of the magnetic flux that might be subsequently brought in through the disk. Furthermore, in studying the denser regions of the disk, the effects of Ohmic diffusivity would have to be included. There are also interesting questions of stability and of the global current flow in a magnetically self-collimating wind that would need to be addressed in a complete theory.

B. MHD Winds from YSOs: The X-celerator Mechanism

We summarized the evidence for the presence of disks surrounding YSOs in Sec. I.B. It is generally believed that a large fraction of the final proto-stellar mass must be accreted through a geometrically thin disk (see Chapter by Shu et al.). For this matter to flow inward, it must lose both energy and angular momentum (Lynden-Bell and Pringle 1974). The physical mechanism that governs the loss of angular momentum determines the rate $\dot{M}_{in}$ at which the protostar can gain mass. Processes such as global spiral density waves (Adams et al. 1989), local turbulent convective viscosity (Lin and Papaloizou 1985; Ruden and Lin 1986), and centrifugally driven disk winds (see Sec. II.A) have been invoked as the dominant source of angular momentum loss (see also the chapter by Adams and Lin). No matter which mechanism is operative, as the protostar gains mass from the disk, it also gains angular momentum at the rate $\dot{M}_{in}j_{in}$, where j_{in} is the specific angular momentum at the protostellar equator R_e. During the disk accretion phase, the protostar can easily gain sufficient angular momentum to be spun up to equatorial breakup velocities $V_e \approx (GM_*/R_e)^{1/2}$ (Ruden and Pollack 1991). After this critical state is reached, additional matter cannot flow past the centrifugal potential barrier and be added to the protostar unless either the accreting gas can lose angular momentum or the protostar can spin down below breakup. Two recent papers (Paczyński 1991; Popham and Narayan 1991) have investigated the possibility that a rapidly rotating star can remain below the breakup speed by transferring angular momentum to the surrounding disk via extremely strong viscous torques. In this subsection we discuss an alternate means by which a magnetic protostar can continue to accrete while rotating near breakup, namely, the ejection of a powerful wind that carries away a small amount of mass but a large amount of angular momentum.

Before discussing the physical mechanism behind the protostellar wind, we estimate the mass and angular momentum loss rates that are needed in order for the protostar to continue accreting (Shu et al. 1988; hereafter SLRN). We start with a low-mass protostar on the "deuterium birthline," where nuclear burning of deuterium acts as a thermostat that keeps the central temperature at $\sim 10^6$ K while the star accretes (Stahler 1983a). Such an object is fully convective and can be adequately modeled as an $n = 3/2$ polytrope; it obeys a linear mass–polar radius relation

$$R_* \approx 8\,R_\odot \left(\frac{M_*}{M_\odot}\right) \tag{15}$$

(Shu et al. 1987a). For analytical simplicity, we assume that the protostar is rotating uniformly at a rate Ω_* and has total angular momentum $\mathcal{J}_* = bM_*(GM_*R_e)^{1/2}$. For a critically rotating $n = 3/2$ polytrope, James (1964) has shown that $R_e \approx 1.63\,R_*$, $\Omega_* \approx (GM_*/R_e^3)^{1/2}$, and $b \approx 0.136$. Using these formulae and Eq. (15), it is easy to show that the equatorial rotation velocity V_e is *constant* along the deuterium birthline,

$$V_e = \Omega_* R_e \approx \left(\frac{GM_*}{R_e}\right)^{1/2} \approx 125\ \text{km s}^{-1} \tag{16a}$$

and that the angular momentum is proportional to the square of the stellar mass,

$$\mathcal{J}_* = b\,\frac{GM_*^2}{V_e}. \tag{16b}$$

In the presence of both disk accretion and an outflowing wind (but neglecting viscous transport of angular momentum between star and disk), the protostar gains mass and angular momentum at the *net* rates of

$$\dot{M}_* = \dot{M}_{in} - \dot{M}_w \qquad (17a)$$

and

$$\dot{\mathcal{J}}_* = \dot{M}_{in}j_{in} - \dot{M}_w j_w$$
$$= (\dot{M}_{in} - J\,\dot{M}_w)\,\Omega_* R_e^2 \qquad (17b)$$

respectively, where the streamline-averaged specific angular momentum carried by the wind is $j_w \equiv J\Omega_* R_e^2$. To remain on the deuterium birthline, the protostar must gain angular momentum at a rate given by the time derivative of Eq. (16b),

$$\dot{\mathcal{J}}_* = 2b\,\frac{GM_*}{V_e}\,\dot{M}_*. \qquad (18)$$

Equating Eqs. (17b) and (18) and using equations (16a) and (17a), we find that the mass flux in the wind must be related to the mass flux through the disk by

$$\dot{M}_w = \frac{1 - 2b}{J - 2b}\,\dot{M}_{in} \qquad (19)$$

(SLRN; compare Eq. [3]). For the star to gain more mass from the disk than is lost in the wind, Eqs. (17a) and (19) show that J must exceed unity. In fact, an asymptotic analysis of centrifugally driven winds shows that $J > 3/2$ (Sakurai 1985; SLRN; Heyvaerts and Norman 1989). Furthermore, if the Poynting stress of the magnetic field becomes asymptotically small, J is related to the terminal wind velocity V_∞ by

$$J = \frac{3}{2} + \frac{1}{2}\left(\frac{V_\infty}{V_e}\right)^2. \qquad (20)$$

In the case of the outflow associated with HH 7–11, Lizano et al. (1988) measured the terminal wind velocity to be 170 km s^{-1} ($= 1.36\,V_e$; see Eq. [16a]), so by Eqs. (19) and (20), $\dot{M}_w/\dot{M}_{in} \approx 0.3$. The predicted value of this ratio is in encouragingly good agreement with the value found by dividing the outflow rate $\dot{M}_w \approx 3 \times 10^{-6} M_\odot$ yr^{-1} inferred by these investigators by the disk accretion rate $\dot{M}_{in} \approx 10^{-5}\ M_\odot$ yr^{-1} deduced from infrared spectral modeling (F. C. Adams, personal communication). An important test of this theory will be to verify Eq. (19) for sources other than HH 7–11.

The physics of centrifugally driven mass loss from rapidly rotating magnetic protostars has been analyzed in a variety of contexts by Draine (1983), by Hartmann and MacGregor (1982), and by SLRN. The latter group of authors specifically consider the loss of mass from *cool* stars where the isothermal sound speed of the photospheric layers C_s is much less than the equatorial breakup speed (the virial velocity, Eq. [16a]). In low-mass protostars the

dimensionless ratio $\varepsilon \equiv C_s/V_e$ is typically ~ 0.05. Suppose that a rapidly rotating, *nonmagnetic* protostar attempts to launch a wind using just the thermal
pressure of the photosphere. The outflowing gas, which maintains the same
specific angular momentum as the stellar surface, does not have sufficient
rotation to remain in orbit at a larger radius and, for $\varepsilon \ll 1$, must settle back
onto the star. The presence of a magnetic field is essential for supplying
the additional angular momentum that allows the gas to continue moving out
(recall the "bead on a rigid wire" analogy discussed in Sec. II.A.1). We can
estimate the magnitude of the required magnetic field by demanding that the
Alfvén velocity $V_A = B/(4\pi\rho)^{1/2}$ be comparable to the virial velocity V_e in
order that magnetic stresses can overcome the gravitational potential barrier.
(An analogous estimate of the coronal temperature needed to launch a wind
by purely thermal means would imply $C_s \approx V_e$, i.e., $\varepsilon \approx 1$.) A fiducial measure
of the density in the wind is $\rho \approx \dot{M}_w/4\pi R_*^2 V_e$, which can be substituted into
the criterion $V_A \approx V_e$ to yield

$$B_* \approx \left(\frac{\dot{M}_w V_e}{R_*^2}\right)^{1/2}. \tag{21}$$

Because the bolometric luminosity satisfies $L_{bol} \approx GM_* \dot{M}_{in}/R_*$ and because
$\dot{M}_{in} \approx \dot{M}_w$ by Eq. (19), we can rewrite this last equation as

$$B_* \approx \frac{L_{bol}^{1/2}}{(GM_* R_*^3)^{1/4}} \tag{22}$$

which may be compared with the earlier estimate (Eq. [8]) for B in disk-
driven winds. For typical protostellar parameters ($M_* \approx 0.5\ M_\odot$, $R_* \approx 4\ R_\odot$),
Eq. (21) shows that magnetic fields of order 100 gauss (averaged over the
stellar surface) are required to launch winds with mass-loss rates of order
$10^{-6}\ M_\odot\ yr^{-1}$. Such field strengths are consistent with the values commonly
inferred in YSOs (see Sec. I.C).

Mechanistically, we may think of the wind mass-loss rate as determined
at the sonic point and the wind angular momentum loss rate as determined
at the Alfvén point (Weber and Davis 1967). For outflow velocities and
mass loss rates of the order inferred in HH 7–11, the density at the base of
the wind, where the sonic transition is made, must be high—comparable to
photospheric values. To make analytic calculations tractable, SLRN assumed
that the magnetic field lines at the sonic point rotate at the breakup speed.
Because the field is anchored deep in the protostar, this assumption requires
the protostar to rotate at breakup as well. In this particular case, the wind is
best analyzed in the rotating frame of the star, where an effective gravitational
potential Φ_{eff} that includes the centrifugal effects of rotation can be rigorously
defined. In the rotating frame, ideal MHD requires the magnetic field to be
exactly parallel to the flow velocity, which means that there is no Lorentz force

along the streamlines of the outflowing wind. Bernoulli's equation requires that the outflow velocity satisfy

$$\frac{1}{2} \left| \mathbf{V} - \Omega_* \hat{\mathbf{z}} \times \mathbf{r} \right|^2 + h(\rho) + \Phi_{\mathrm{eff}} = e(\Psi) \tag{23}$$

where $h(\rho)$ is the enthalpy of the gas and $e(\Psi)$ is the (conserved) energy on the streamline labeled by the stream function Ψ. The density and the poloidal velocity are related to the stream function through $\rho \mathbf{V}_{\mathrm{pol}} = \nabla \times [\Psi(r, z)\hat{\phi}]$, which means that the equation of continuity (Eq. [14]) is automatically satisfied. Force balance *across* different streamlines is governed by the analog of the Grad-Shafranov equation from plasma fusion, which describes axisymmetric, steady-state fluid configurations in tokomaks (Shafranov 1966; Heinemann and Olbert 1978). In the outflow problem, the transfield force balance equation takes the form

$$\nabla \cdot (\mathcal{A} \nabla \Psi) = \mathcal{Q} \tag{24}$$

where $\mathcal{A}$ is an Alfvén discriminant that changes sign across the Alfvén surface and $\mathcal{Q}$ is a sum of terms involving the derivative *across* streamlines of each of the five functions that are constant *along* streamlines (energy, angular momentum, angular velocity, ratio of magnetic flux to mass flux, and entropy; see Sec. II.A.1). Equation (24) has the (apparent) form of a steady-state heat conduction equation in which force imbalance across streamlines (detailed in $\mathcal{Q}$) leads to the spreading of the flow streamlines (rather than to the diffusion of heat). The solution of the coupled Eqs. (23) and (24) determines the density and velocity structure in the wind.

At the base of the outflow, pressure forces are required to accelerate the wind to sonic speeds. For a smooth transition from subsonic to supersonic flow to be made (no shocks), SLRN have shown that the sonic transition must occur within a distance $\sim \varepsilon R_*$ from the "X-point" of the effective potential (defined to be where $\nabla \Phi_{\mathrm{eff}} = 0$). For a star rotating at the breakup speed, this "point" is actually a narrow ring at the stellar equator where the centrifugal force balances gravity. Although thermal pressure forces initially are important in launching the wind, well downstream of the sonic point, they are negligibly small compared to magnetic forces, and the wind can be treated as dynamically cold. Once the gas has made the sonic transition, magnetocentrifugal acceleration takes over and enforces approximate corotation of the wind at the angular velocity Ω_* out to the Alfvén radius. If the sonically outflowing gas from the X-point is to pass smoothly through the Alfvén point, the magnetic field strength near the X-point must be $\sim \varepsilon^{-1}$ larger than the estimate above (Eq. [21]) and the gas density must be $\sim \varepsilon^{-2}$ larger than $\dot{M}_{\mathrm{w}}/4\pi R_*^2 V_{\mathrm{e}}$. This large value of the density at the X-point can be understood from the fact that the gas which moves with a speed $C_{\mathrm{s}} = \varepsilon V_{\mathrm{e}}$ through two narrow bands just above and below the stellar equator, each with area $2\pi R_* \cdot \varepsilon R_*$, must carry the full mass discharge $\dot{M}_{\mathrm{w}}$. The concentration of the stellar magnetic field near the locus of the X-points in the equatorial plane is crucial in several ways. First, the magnitude of **B** is sufficiently large to ensure that the gas flowing from the sonic point can be torqued up to super-Alfvénic speeds. Second, the poloidal field component has the correct topology to launch a centrifugally driven wind, i.e., it is inclined by less than 60° to the equatorial plane (see Sec. II.A.1). Third, the high magnetic field strengths mean that the region inside the Alfvén surface is nearly force free. The strong magnetic pressure pushes the streamlines poleward and is responsible for the initial collimation of the outflow. Outside the Alfvén surface, the inertia of the gas dominates the magnetic field, and the "hoop" stress associated with the azimuthal component of **B** focuses the outflow on a global scale (Nerney and Suess 1975; see Sec. II.A.1).

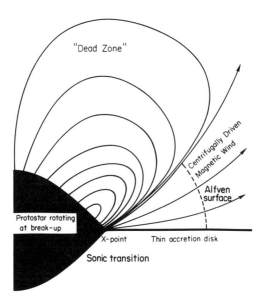

Figure 6.   The geometry of the X-celerator protostellar wind model. Material enters the star from a thin disk, and the wind leaves from two narrow equatorial bands lying above and below the inflowing matter. The gas makes a sonic transition near the X-point of the effective gravitational potential. The primary acceleration of the wind occurs magnetocentrifugally between the X-point and the Alfvén surface. The location and shape of the Alfvén surface depend on the magnitude of the magnetic field and the distribution of mass on field lines. The material in the "dead zone" is on closed field lines and is static. In some protostars, a fraction of these closed field lines may be blown open and may give rise to the ionized jets seen in some CTTS. Any wind that emanates from the "dead zone" is confined and focused by the more massive wind from the X-point.

In Fig. 6 we schematically diagram this outflow mechanism, which SLRN have dubbed the "X-celerator" because of the crucial role played by the X-point (see also Chapter by Shu et al.). The wind begins in a band of height $\sim\varepsilon R_*$ in the high-density photospheric regions at the protostellar equator. The magnetic field, with magnitude $\sim\varepsilon^{-1}(\dot{M}_w\Omega_*/R_*)^{1/2}$, forces the gas to corotate at the protostellar angular velocity Ω_* until the Alfvén transition is made, after which the gas flows out to infinity while undergoing more gradual collimation by the toroidal magnetic field. Our estimate of the ratio $\dot{M}_w/\dot{M}_{in}$ in Eq. (19) was based on the assumption that the protostellar wind can carry off *whatever* angular momentum flux is necessary for the protostar to remain on the deuterium birthline. We expect the X-celerator mechanism to be able to regulate its mass-loss rate to this equilibrium value because of the sensitivity of $\dot{M}_w$ to the value of the mass density at the sonic point. For example, if the wind carries away too little angular momentum ($\dot{M}_w$ smaller than required by Eq. [19]), the outer layers of the star will spin up as they accrete fresh mass and angular momentum from the disk, the increased centrifugal support will cause the stellar surface to rise relative to the location of the X-point, and the density at the base of the wind will increase. The higher-density material making the sonic transition will then tend to restore $\dot{M}_w$ to its equilibrium value. Conversely, if the wind carries away too much angular momentum ($\dot{M}_w$ larger than required by Eq. [19]), the equatorial regions of the star will spin down, the stellar surface will fall relative to the X-point, and the density at the base of the wind will decrease, which will again tend to restore the equilibrium value of $\dot{M}_w$.

For analytical convenience and mathematical tractability, SLRN adopted a magnetic field anchored into a protostar uniformly rotating at exactly the critical rate required for centrifugal balance. The magnetic field topology and the focusing of the gas streamlines in the outflow are then determined from Eqs. (23) and (24). Numerical solutions that smoothly pass through the Alfvén surface (where $\mathcal{A} = 0$) are currently being calculated. For protostars rotating at less than the breakup speed, the analytic development of SLRN does not strictly apply. However, Galli and Shu (in preparation) have recently shown that the equatorial accretion of mass and angular momentum from a disk induces a global circulation pattern in the outermost layers of a protostar that rotates below breakup. They find that newly accreted material flows from the equator towards the poles of the star; in a steady state, a return current at the stellar surface completes the loop by carrying matter from the poles back towards the equator. It is tempting to speculate that this surface circulation could transport stellar magnetic flux into the rapidly rotating and sheared equatorial regions, where a modified form of the X-celerator process would give rise to an outflow. Clearly, much more work remains to be done in analyzing such scenarios that involve more complicated field topologies and possibly time-dependent effects.

C. The Thermal and Chemical Properties of the Winds

The predicted distributions of the kinetic temperature T and degree of ionization $f_e \equiv n_e/n$ (where n_e and n are the electron and total number densities, respectively) of the outflow as well as of the chemical composition of the wind lead to important observational diagnostics which may be used to test the CDW models that we have discussed. In particular, the thermal structure of the winds turns out to depend sensitively on their driving mechanism and on their geometry and could thus be useful for identifying the relevant models. In a similar vein, the freeze-out abundances of various molecular species may provide a unique probe of the physical conditions in the observationally inaccessible wind acceleration region.

The thermal balance of the outflowing gas, treated as a single-phase medium characterized by the temperature T, is governed by the first law of thermodynamics,

$$\frac{3}{2} n k_B \frac{dT}{dt} = k_B T \frac{dn}{dt} + \Gamma - \Lambda \tag{25}$$

where n is the local number density of particles and d/dt is the convective derivative along a flow line. The first term on the right-hand side represents the adiabatic cooling of the gas due to expansion ($dn/dt < 0$). The next two terms represent the total heating and cooling rates per unit volume for the translational degrees of freedom of the gas. Energy sources or sinks due to the interconversion of chemical species (such as dissociation of molecular hydrogen) must also be included in Γ and Λ. In the absence of net sources or sinks of energy ($\Gamma - \Lambda = 0$), Eq. (25) indicates that the gas temperature becomes proportional to $n^{2/3}$. For a constant-velocity wind expanding into a cone with a constant opening angle, the temperature decreases away from the source as $T \propto r^{-4/3}$. This adiabatic-expansion cooling is typically the dominant energy loss mechanism and represents a severe constraint on potential heat sources in the wind.

The dominant heating mechanism of neutral MHD winds in YSOs is the ambipolar diffusion heating induced by the relative drift between ions and neutrals that underlies the wind acceleration process (Ruden et al. 1990; Safier and Königl 1992). This is just the Joule heating that accompanies the electromagnetic driving of the outflow. The net collisional force per unit volume acting on the neutrals (which transfers the magnetic stress from the ions) is

$$\mathbf{f} = \gamma \rho_i \rho (\mathbf{V}_i - \mathbf{V}) \tag{26}$$

(see Eqs. [9] and [10]). The rate at which this force does work on the neutral component is the ambipolar diffusion heating rate (per unit volume)

$$\Gamma = \mathbf{f} \cdot (\mathbf{V}_i - \mathbf{V})$$
$$= \frac{|\mathbf{f}|^2}{\gamma \rho_i \rho} . \tag{27}$$

Equation (27) plays a central role in the ensuing discussion. As noted above, the main energy-loss mechanism is the adiabatic cooling induced by the expansion of the gas, although radiative and photochemical cooling can also play a role.

1. Thermal Structure of Disk-Driven Winds. Safier and Königl (1992) calculated the thermodynamic properties of a self-similar atomic CDW from the surface of a Keplerian disk (see Sec. II.A.1). Using Eq. (10) to substitute for $\mathbf{f}$ in Eq. (27), the ambipolar diffusion heating rate becomes

$$\Gamma = \frac{|(\nabla \times \mathbf{B}) \times \mathbf{B}|^2}{16\pi^2 \gamma m_H^2 n^2 f_e (1 - 2f_e)} \tag{28}$$

where a pure-hydrogen composition has been adopted for simplicity. Because the large-scale structure of the wind is not affected by the thermal pressure, the distribution of the density and magnetic field in the wind is effectively fixed, and the only unknown quantities in Eqs. (25) and (28) are T and f_e. They can be determined with the help of the ionization balance equation

$$\frac{df_e}{dt} = \left\{ I(T)f_e[1 - 2f_e] - R(T)f_e^2 \right\} n \tag{29}$$

where

$$I(T) = 7.8 \times 10^{-11} T^{1/2} \exp(-1.6 \times 10^5 / T) \ \text{cm}^3 \ \text{s}^{-1} \tag{30a}$$

and

$$R(T) = 4.8 \times 10^{-11} T^{-1/2} \ \text{cm}^3 \ \text{s}^{-1} \tag{30b}$$

are the ground-state collisional ionization and recombination coefficients, respectively.

Figure 7 exhibits the behavior of T and f_e along two representative flow lines for the disk–wind model of Figs. 1–3. The striking property of these solutions is the rapid rise of the temperature to a value of $\sim 10^4$ K that is maintained over three decades in distance irrespective of the initial radius, and the correspondingly high values attained by f_e. This behavior is a manifestation of a feedback effect involving T and f_e: a low degree of ionization leads to an increase in the temperature (since the heating rate scales as $\sim f_e^{-1}$), which in turn increases the collisional ionization rate (through the exponential term in $I[T]$) and hence the value of f_e. The coupling between the temperature and the degree of ionization is terminated when the drop in n brought about by the expansion of the flow quenches the growth of f_e (see Eq. [29]). A necessary condition for this behavior is that the adiabatic cooling, which causes the eventual drop in T, should not dominate the heating near the base of the flow. This condition is not fulfilled for the YSO wind model discussed in Sec. II.B, in which the rapid radial expansion reduces

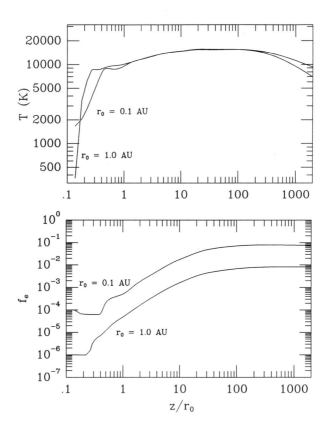

Figure 7.   The distribution of the temperature (*top*) and the degree of ionization (*bottom*) along two flow lines for the self-similar wind model shown in Figs. 1–3. The effects of ambipolar-diffusion heating, adiabatic and radiative cooling, as well as collisional and radiative ionization have been included.

the temperature from its initial photospheric value to $\lesssim 50$ K at a distance of $\lesssim 10^3$ AU from the center (see Fig. 8).

The high values of T and f_e that are obtained in the disk-driven wind model at large distances from the origin provide a natural explanation of some of the characteristic radiative properties of YSOs (Safier and Königl 1992). In particular, they may account for the forbidden line emission (typically [O I]λ6300 and [S II]λ6716, λ6731) that has been measured in a number of CTTS on scales of $\sim 10^2$ AU (see Chapter by Edwards et al.). The inferred temperatures ($\sim 10^4$ K) and electron densities ($\gtrsim 10^4$ cm^{-3}) in the emission region are, in fact, in good agreement with the predicted values, and the double-peaked line profiles often exhibited by these sources are consistent with the latitude-dependent velocity distribution that characterizes the CDW model (Edwards et al. 1987; see Fig. 1). The predicted rapid rise of T above the disk surface could also explain the unique temperature inversion pattern inferred from Ca II triplet line measurements in several YSOs (Hamann and Persson 1989). Furthermore, the high-f_e regions might account for the extended ($\gtrsim 10^2$ AU) thermal radio emission detected in several CTTS (see, e.g., Bieging et al. 1984). However, the radio emission as well as the blueshifted component of the forbidden line emission could, in principle, also originate in the high-velocity ionized jets (see Kwan and Tademaru 1988). High-resolution observations as well as detailed calculations of the expected forbidden line profiles and radio intensity maps would be required to discriminate between these two possibilities.

2. *Chemistry of protostellar winds.* Motivated by the detection of CO and atomic hydrogen in a neutral wind from HH 7–11 (Lizano et al. 1988), Ruden et al. (1990) have analyzed the thermal and chemical structure of winds from cool protostars. They considered the quasi-spherical steady flow of gas with velocity $V_w(r)$ into an effective area $A(r)$; the local gas number density follows from mass conservation as

$$n(r) = \frac{\dot{M}_w}{m V_w(r) A(r)} \tag{31}$$

where m is the mean molecular mass per gas particle. To mimic steady acceleration from the sound speed near the protostellar photosphere to the terminal velocity at large radii, the wind velocity was taken to be $V_w(r) = V_\infty (1 - R_*/r)$. For this velocity profile the wind makes the sonic transition at $r_s \approx R_* (1 + C_s/V_\infty)$, which serves as the radial starting point for their thermal structure calculations. The geometry of the outflow is parameterized by $A(r)$. For spherically symmetric winds $A(r) = 4\pi r^2$, while for the X-celerator mechanism $A(r) \approx 4\pi r^2 (1 - R_*/r)$, reflecting the fact that the flow emanates from a narrow cylindrical ring at the stellar equator rather than from the entire stellar surface. Ruden et al. concentrated on the detailed hydrogen chemistry, taking into account the ground ($n = 1$) and first excited ($n = 2$) states of hydrogen, the negative and positive hydrogen ions H^- and H^+, as well as molecular hydrogen H_2 and its ion H_2^+. In view of its low ionization potential and small photoionization cross section, sodium was included in the models to represent the level of ionization due to singly ionized metals. As sodium remains ionized in all wind models, it sets the minimum electron fraction at $f_e \gtrsim 2.3 \times 10^{-6}$. Because of the strong radiative trapping of Lyman α photons in the wind (line optical depths exceeding 10^{10}), a population of $n = 2$ excited hydrogen that greatly exceeds the LTE level can be built up. Photoionization of hydrogen from this excited state by the Balmer continuum of the protostellar photosphere is the dominant mechanism for increasing the electron fraction above the minimum set by sodium. At the base of the wind, the gas temperature is set equal to the photospheric temperature, and the hydrogen is atomic and predominantly neutral ($f_e \ll 1$). This is in contrast to winds from disks where the hydrogen starts out in molecular form except for the innermost disk regions near the protostar.

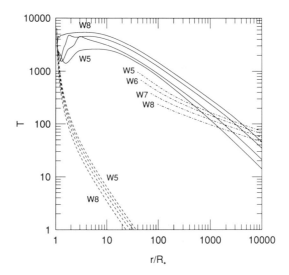

Figure 8.   The thermal structure of protostellar wind models. The physical properties
of the models are listed in Table I. The models are labeled "W" followed by the
(negative) logarithm of the mass-loss rate in $M_\odot$ yr^{-1}. The solid curves show the
gas kinetic temperature, the dashed curves show the gas temperature in the absence
of net sources of heat (adiabatic expansion), and the dash-dotted curves show the
equilibrium dust grain temperature. The latter curves begin where dust grains are
first formed.

The most important source of energy in the wind is ambipolar diffusion
heating. In the case of cool, quasi-spherical outflows, Eq. (9) can be used to
write the ion-neutral frictional force as

$$|\mathbf{f}| \approx \rho \left(V_w \frac{dV_w}{dr} + \frac{GM_*}{r^2} \right) \qquad (32)$$

which can be substituted into Eq. (27) to determine the magnitude of the
ambipolar diffusion heating rate. This expression assumes that the Lorentz
force (Eq. [10]) provides the necessary thrust for producing the postulated
velocity profile. Equation (32) is useful because only the wind velocity $V_w(r)$
but not the exact topology and strength of the magnetic field must be specified
in any given model. Najita (1991, personal communication) has shown that
this simple formula leads to a surprisingly accurate estimate of the ambipolar
diffusion heating rate near the star for the X-celerator wind geometry. Near
the photosphere of the protostar the wind density is high (for large mass-
loss rates), and three-body reactions among hydrogen atoms can produce
molecular hydrogen. The H_2 formation energy (≈ 4.5 eV) is liberated as heat
and can be an additional source of energy for large values of $\dot{M}_w$. As the wind
expands and cools, dust grains can form at $r \gtrsim 10\, R_*$. The grains are heated by
absorption of protostellar radiation and in turn can heat the gas through direct

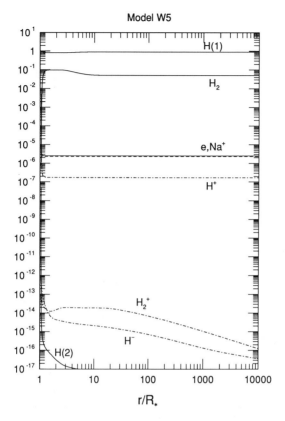

Figure 9. The chemistry in the protostellar wind model W5 depicted in Fig. 8. The ordinate is the number abundance (relative to hydrogen nuclei). Solid curves depict electron and neutral hydrogen species: e, H($n = 1$), H($n = 2$), and H_2; dash-dotted curves depict ionized hydrogen species: H^+, H^-, and H_2^+; and the dashed curve depicts once-ionized sodium: Na^+.

collisions. However, this heating mechanism is of secondary importance in comparison with ambipolar diffusion.

The temperature distribution in the wind is found by integrating Eq. (25) for T using Eqs. (27) and (32) for the ambipolar diffusion heating rate and Eq. (31) for the density. Additional heating and cooling terms are added to describe the energy gain or loss accompanying chemical reactions (such as H_2 formation) or dust formation. These equations are solved simultaneously with ones that describe the production and destruction of each of the species in the chemical network. The ion fraction is of particular importance because of the role it plays in the ambipolar diffusion heating rate (Eq. [27]). Figure 8 shows the thermal structure of a sequence of wind models designed to mimic the simultaneous change of physical parameters in going from high-luminosity

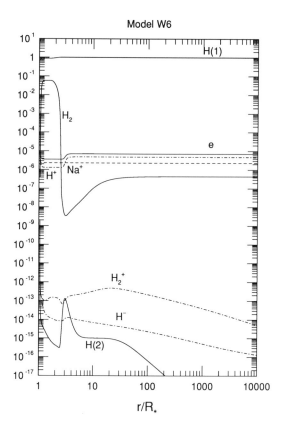

Figure 10.   Same as Fig. 9 but for Model W6.

sources with high mass-loss rates (DEIS) to their low-luminosity, low-mass-loss-rate counterparts (CTTS). Figures 9–12 show the chemical structure along this representative sequence. In all cases, the geometry of the flow was taken to be that of the X-celerator, and the terminal wind speed was $V_\infty = 150$ km s^{-1}. Because the wind kinetic luminosity and force vary roughly linearly with the bolometric luminosity (see Sec. I.A), the protostellar luminosity L_* was chosen in each case so as to make the ratios $\dot{M}_w V_\infty^2/2L_*$ and $\dot{M}_w V_\infty c$ /L_* equal to 0.02 and 80, respectively. The photospheric temperature T_* was lowered from 6500 K to 3500 K as the mass-loss rate decreased from 10^{-5} to 10^{-8} M$_\odot$ yr^{-1} along the sequence. The physical properties of the central protostar are listed in Table I for the cases shown. The calculations have revealed that the thermal structure and wind chemistry are primarily sensitive to the density and the flow crossing time $r/V_w(r)$ rather than to the protostellar luminosity or the photospheric temperature. In all of the cases, there is an adiabatic drop in the gas temperature as the wind expands away from the stellar equator. The high wind densities and cool temperatures are

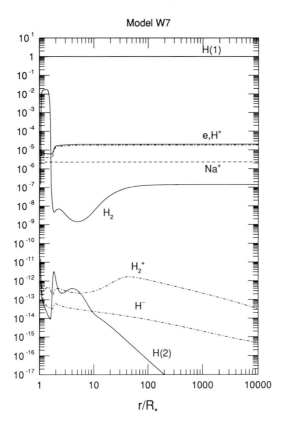

Figure 11.   Same as Fig. 9 but for Model W7.

conducive to the formation of molecular hydrogen, which reheats the gas by liberating the formation energy. The gas temperature reaches a plateau ranging from 2600 K (for $\dot{M}_w = 10^{-5}$ M$_\odot$ yr^{-1}) to 5400 K (for $\dot{M}_w = 10^{-8}$ M$_\odot$ yr^{-1}), which is high enough to dissociate the newly formed H$_2$ (the formation and destruction of H$_2$ near the protostar is clearly evident in Fig. 10). In all the models, ambipolar-diffusion heating dominates in the wind for radii greater than 5 R_*. It can be shown that ambipolar diffusion heating leads asymptotically to a gas temperature profile

$$T(r) \approx 32\pi \, \frac{m^2}{k_B \langle \sigma v \rangle} V_\infty^4 \frac{R_*}{f_e \dot{M}_w} \left(\frac{R_*}{r} \right) \qquad (33)$$

(Ruden et al. 1990), where m is the mean molecular weight of the gas and $\langle \sigma v \rangle \approx 2.4 \times 10^{-9}$ cm^3 s^{-1} is the cross section for collisions between ions and neutral hydrogen atoms (Draine et al. 1983). Equation (33) indicates that the gas temperature falls off inversely with distance from the protostar and is

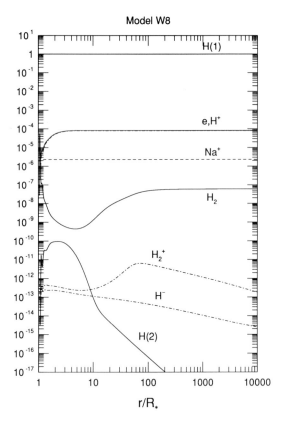

Figure 12.   Same as Fig. 9 but for Model W8.

inversely proportional to the ionization fraction and mass-loss rate (see also Eq. [27]). These findings are in accord with the numerical results shown in Fig. 8. Dust grains may form in the outflowing wind but do not provide a significant heat input; their temperature distribution is also shown in Fig. 8.

For mass-loss rates $\gtrsim 10^{-6}$ $M_\odot$ yr^{-1} the wind density is so high that hydrogen is almost entirely neutral and singly ionized sodium donates the electrons (see Figs. 9–12). For lower values of $\dot{M}_w$, the wind density is lower and hydrogen is the dominant ionized species. In all the cases presented in Figs. 9–12, however, the wind remains predominantly neutral. It is also important to note that, although molecular hydrogen may be formed near the protostar, it is usually destroyed as the wind flows outward and heats up. The exception is model W5, with $\dot{M}_w = 10^{-5}$ $M_\odot$ yr^{-1} (see Fig. 9). In this case, the density is high enough and the temperature low enough for a significant amount of molecular hydrogen to remain in the wind at large radii ($n_{H_2}/n \approx 0.05$). For mass-loss rates greatly exceeding 10^{-5} $M_\odot$ yr^{-1}, the hydrogen in the wind becomes predominantly molecular.

TABLE I

Protostellar Wind Parameters

Model	L_*	T_*	R_*	$\dot{M}_{\mathrm{w}}$
	$[L_\odot]$	[K]	$[R_\odot]$	$[M_\odot \ \mathrm{yr}^{-1}]$
W5	1000	6500	25.0	10^{-5}
W6	100	5500	11.0	10^{-6}
W7	10	4500	5.2	10^{-7}
W8	1	3500	2.7	10^{-8}

Although Ruden et al. (1990) did not specifically calculate the carbon chemistry in their models, Glassgold et al. (1989,1991) have shown that in similar wind environments there is significant production of CO even though the wind remains atomic as a whole, which fits the observations in HH 7–11. Recent high-resolution millimeter-wave observations of this source have indicated that the outflowing gas may be clumpy and that some of the CO may not originate in the wind but rather as entrained ambient cloud material (Koo 1990; Masson et al. 1990). Further observations and theoretical diagnostics are clearly called for to determine how much of the CO is produced in the wind itself, to identify the physical mechanism causing the clumping (e.g., thermal or interfacial shear instabilities), and to differentiate between the chemical signatures of YSO and disk-driven outflows.

III. ORIGIN OF IONIZED JETS

Although several promising mechanisms have been proposed to explain the collimated ionized outflows, only a few quantitative models have appeared in the literature and there is still no consensus on the origin of the jets. The construction of viable models is hindered by the existing uncertainty over the physical parameters of the jets, notably the mass-loss rate $\dot{M}_j$. The values $\sim 10^{-9}$ to 10^{-8} $M_\odot \ \mathrm{yr}^{-1}$ quoted in Sec. I.A were derived on the assumption that the electron densities inferred from the measured [S II] $\lambda 6716/\lambda 6731$ line ratios represent the *postshock* densities behind weak, oblique shocks in the jets, and that the mean (preshock) densities are significantly lower (see, e.g., Mundt et al. 1987). These estimates appear to be consistent with those obtained from radio continuum measurements (see, e.g., Rodríguez et al. 1990) after taking into account the nonspherical geometry of the emission region (Reynolds 1986), and imply that the jets could in principle originate as stellar winds (see, e.g., DeCampli 1981). However, at least part of the optical radiation in certain jets has been attributed to low-level collisional excitation that acts throughout the emission region and is probably associated with the growth of a mixing layer near the boundary of the jet (see, e.g., Solf 1987; Meaburn and Dyson 1987; Cantó and Raga 1991). In this picture, the measured [S II] line ratios are interpreted as corresponding to the actual mean electron densities in the jets. This raises the estimates of $\dot{M}_j$ by over an order of magnitude to values as high as a few times 10^{-7} $M_\odot \ \mathrm{yr}^{-1}$ (see, e.g.,

Hartigan 1989). Even higher values are inferred by noting that the observed low-excitation gas is only partially ionized (with a degree of ionization as low as $\sim 10\%$), in which case the implied momentum discharges in the jets would be large enough to drive the typical molecular lobes (Raga 1991). Such energetic outflows are unlikely to originate as stellar winds and one possible interpretation of the jets is then that they represent the ionized inner regions of disk-driven winds.

An explicit constraint on $\dot{M}_j$ can be obtained in jets whose heads are associated with bow-shaped Herbig-Haro objects. The observed morphology suggests that the leading bow shock (which propagates into the ambient medium) is radiative, which in turn implies that the postshock cooling length

$$d_{\text{cool}} \approx 4.5 \times 10^{16} n_a^{-1} V_{\text{bs},7}^4 \text{ cm} \qquad (34)$$

(where n_a is the ambient particle density and $V_{\text{bs},7}$ is the bowshock speed in units of 10^7 cm s^{-1}) is smaller than the radius $r_h = 10^{16} r_{h,16}$ cm of the jet head (see, e.g., Blondin et al. 1989). The jet (mass) density ρ_j is related to the corresponding ambient density ρ_a through the force balance equation

$$\rho_j (V_j - V_{\text{bs}})^2 \mathcal{A}_j = \rho_a V_{\text{bs}}^2 \mathcal{A}_h \qquad (35)$$

where V_j is the jet speed, $\mathcal{A}_h = \pi r_h^2$ is the transverse area of the jet head, and $\mathcal{A}_j$ is the instantaneous cross-sectional area of the jet (which in principle could be smaller than $\mathcal{A}_h$; see, e.g., Hardee and Norman 1990). Combining the above relations gives a lower limit on $\dot{M}_j = \rho_j V_j \mathcal{A}_j$,

$$\dot{M}_j \gtrsim 4.8 \times 10^{-10} V_{\text{bs},7}^6 V_{j,7}^{-1} (1 - V_{\text{bs}}/V_j)^{-2} r_{h,16} \text{ M}_\odot \text{ yr}^{-1}. \qquad (36)$$

Although the value of V_j is generally not directly measurable, it is safe to assume that it does not exceed ~ 500 km s^{-1} in low-mass YSOs (see, e.g., Hartigan 1989). Equation (36) then leads to a lower bound on $\dot{M}_j$ that effectively depends only on the observable quantities r_h (obtained from high-resolution imaging) and V_{bs} (inferred from proper motion measurements). For example, in the case of the archetypal object HH 1, this equation implies $\dot{M}_j \gtrsim 7 \times 10^{-8} r_{h,16} \text{ M}_\odot \text{ yr}^{-1}$ (with $r_{h,16} \gtrsim 1$) if one identifies the bowshock speed with the mean proper motion (~ 240 km s^{-1}) measured by Herbig and Jones (1981) for the optical emission knots at the jet head, and $\dot{M}_j \gtrsim 2 \times 10^{-6} r_{h,16}$ M$_\odot$ yr^{-1} if V_{bs} is set equal to the measured proper motion (~ 350 km s^{-1}) of knot F at the apex of the bow shock. The latter limit, in particular, is significantly larger than the value $\lesssim 4 \times 10^{-8}$ $(V_{j,7}/5)$ M$_\odot$ yr^{-1} calculated using the low-ρ_j estimates of Mundt et al. (1987) and Rodríguez et al. (1990) for this source. This limit could, however, be reduced if the jet is "heavy" $(\rho_j >> \rho_a)$ and has a speed relative to the jet head that is significantly lower than V_{bs}. In that case, the jet material (which passes through a shock of speed $V_{\text{js}} = V_j - V_{\text{bs}}$) may be able to cool within the jet head even if the cooling length of the shocked ambient gas at the apex exceeds r_h. In this picture, the emission from the apex is dominated by the jet shock, and the bow shock only becomes visible at the wings (where it is oblique and hence weaker). According to this interpretation (which could in principle be tested by future high-resolution imaging), the bound given by Eq. (36) could be lowered by as much as $(V_{js}/V_{bs})^6(\mathcal{A}_j/\mathcal{A}_h)$. For knot F in HH 1 (setting $V_{js} = 150$ km s^{-1} on the basis of shock emission models), this factor reduces the lower limit on $\dot{M}_j$ to $\sim 1 \times 10^{-8} r_{h,16} (\mathcal{A}_j/\mathcal{A}_h)$ M$_\odot$ yr^{-1}, which is consistent with the above-mentioned low-density estimates.

A. Jets as Disk-Driven Winds

The idea that jets with high mass and momentum discharges could be associated with disk-driven winds is supported by the calculated thermal structure of CDWs from Keplerian disks (Safier and Königl 1992; see Fig. 7). These calculations have revealed that ambipolar-diffusion heating in the wind could readily maintain a degree of ionization approaching (or even exceeding) 10% on the scale of the observed jets. An apparent difficulty with this interpretation is that the dynamical ages of jets are significantly shorter than those of bipolar flows (see Sec. I.A). This difference may, however, just be a reflection of the fact that the jets are launched intermittently, with each outflow episode lasting only a fraction $\lesssim 0.1$ of the total YSO outflow phase. This scenario is supported by the fact that the estimated duration of YSO jet visibility ($\gtrsim 10^4$ yr) is comparable to the characteristic dynamical age of bipolar flows (Mundt et al. 1987) and by the direct observational evidence for nonsteady mass ejection in jets (Sec. I.A). The outflow episodes may be associated with FU Orionis outbursts, which apparently represent rapid disk accretion events ($\dot{M}_{in} \gtrsim 10^{-4}$ M$_\odot$ yr^{-1}) accompanied by massive ($\dot{M}_w \approx 10^{-5}$ M$_\odot$ yr^{-1}) outflows (see Chapter by Hartmann et al.). In the context of the magnetized disk model discussed in Sec. II.A, the rapid accretion events may be linked to the episodic release (by gravitational instabilities) of mass that has accumulated in the disk as a result of a drop in the efficiency of magnetic angular momentum transport at small radii (Pudritz and Norman 1986).

The association with accretion disks underlies several other proposals for the origin of the jets. Torbett (1984) suggested that the outflows could be powered by the rotational kinetic energy liberated in the shear boundary layer that develops between a Keplerian disk and a slowly rotating star. In this picture, the layer is sufficiently optically thick for the emitted radiation to contribute to the thermal pressure driving rather than diffuse out. Pringle (1989a) has similarly considered a boundary layer origin for the outflows but assumed that radiative cooling is not inhibited. He suggested that the jets are driven, instead, by strong toroidal magnetic fields that had been amplified by the shear and then risen to the surface of the layer while separating from the matter. Although tapping the substantial free energy of the boundary layer is a promising idea, more detailed calculations (involving two-dimensional numerical simulations; cf. Kley 1989) are needed to confirm the validity of these scenarios, to establish the mass-loss efficiencies, and to determine the amount of outflow collimation. Yet another idea, which combines a rotating magnetic star (as the source of energy) and a disk (as the source of mass), was put forward by Arons et al. (1984; see also Arons 1987). They proposed that mass is loaded (by means of turbulent mixing) onto stellar field lines that sweep past the disk and is then flung out in the form of a CDW. It is not clear, however, that the loaded field lines would assume an open configuration rather than penetrate into the disk (see Lamb 1989). Again, final judgment must be reserved pending the construction of a more quantitative model.

B. Jets as Stellar Winds

We now consider the possibility, indicated by the low mass-discharge estimates for optical jets, that the ionized outflows represent collimated stellar winds. In the case of CTTS with typical (low) rotation rates and a dipole magnetic-field geometry, centrifugal driving is not expected to be efficient (see, e.g., Mestel 1968). However, stellar magnetic fields could in principle drive the observed CTTS winds through the action of MHD waves (see, e.g., DeCampli 1981; Hartmann et al. 1982; Lago 1984). In this picture, Alfvén waves generated in the stellar convection zone deposit their momentum on the way out and can produce winds with mass outflow rates of up to a few times 10^{-8} $M_\odot$ yr^{-1} for typical stellar parameters, which is consistent with the low-$\dot{M}_j$ estimates discussed above. (A similar mechanism could also operate in a convective accretion disk; cf. Galeev et al. 1979b.) The predicted terminal speeds of the winds are of the order of the escape velocity from the stellar surface (a few hundred km s^{-1}) and thus are also compatible with the proposed identification with jets. In order for the waves not to be too rapidly damped, the surface magnetic-field strength must be at least a few hundred gauss, which is consistent with the observational constraints (see Sec. I.C). The wave flux can be estimated from the measured chromospheric and coronal emission. Although the observational implications of these models have been considered in some detail, they are still hampered by the incomplete theory of MHD wave generation and propagation in stellar envelopes.

The collimation of the (quasi-spherical) stellar wind into a pair of directed jets is most likely achieved through a dynamical interaction with the surrounding (magnetically collimated) neutral outflow (see, e.g., SLRN; Kwan and Tademaru 1988). Optical and radio continuum images of the jets imply that at least some jets are already collimated on scales $\lesssim 10^{15}$ cm. One attractive possibility is that the winds are shocked and subsequently collimated by passing through a transonic de Laval nozzle (Königl 1982). However, this mechanism would only be efficient if the shock remained nonradiative (Raga and Cantó 1989). This requires the postshock cooling length (given by Eq. [34] with V_{bs} and n_a replaced by the preshock wind speed and density, respectively) to be *greater* than the shock radius R_s, which, in turn, implies that the jet velocity must exceed

$$V_{j,cool} \approx 470 \left(\frac{\dot{M}_j}{10^{-8} \, M_\odot \, yr^{-1}} \right)^{0.2} \left(\frac{R_s}{10^{14} \, cm} \right)^{-0.2} \quad km \, s^{-1}. \qquad (37)$$

It is interesting to note that jets that are associated with radiative bow shocks at their heads and satisfy the condition given by Eq. (36) would have

$$\frac{d_{cool}(R_s)}{R_s} = 0.2 \left(\frac{R_s}{10^{-2} r_h} \right) \left(\frac{V_{bs}}{0.5 V_j} \right)^{-6} \left(1 - \frac{V_{bs}}{0.5 V_j} \right)^2 \qquad (38)$$

and thus are not likely to be collimated by this mechanism.

A similar model to the one just outlined for CTTS jets can plausibly be expected to apply to the jets in DEIS, as stellar magnetic activity is probably triggered by the onset of convection at an early phase of the YSO evolution (see Chapter by Stahler and Walter). In fact, in the case of the embedded object IRS 5 in L1551, there is evidence from scattered light that a high-velocity wind is present at the source (Mundt et al. 1985). Furthermore, as we noted in Sec. I.A, there is also evidence in this source for a distinct lower-velocity (but higher-momentum-discharge) wind that pervades the space around the jets and that could conceivably contribute to their collimation.

While the expanding molecular lobes around L1551 IRS 5 are apparently driven by a relatively low-velocity wind that is distinct from the ionized jets, the fact that the kinetic powers of the jets and of the molecular flows are comparable in those sources in which they are both detected has motivated the suggestion that some of the bipolar lobes are driven by the "cocoons" of shocked jet material that expand like energy-driven interstellar bubbles (see Sec. I.A). To evaluate this possibility, one can adopt the model of Scheuer (1974) and assume that the expansion along the jet axis is determined by the jet momentum discharge according to Eq. (35) (where we set $A_j \approx A_h \approx \Omega Z^2$, with Z being the distance from the origin and with the solid angle Ω taken to be constant), and that the transverse expansion is driven by the (uniform) thermal pressure of the shocked jet material and is directed (at each point along the boundary) at right angles to the axis. Assuming, furthermore, that the ambient density n_a is uniform and that the expansion speed V_Z along the axis is much smaller than V_j, one finds that the internal energy of the bubble at a given time t after the onset of the outflow is a fraction 6/13 of the energy $\dot{M}_j V_j^2 t$ supplied up to that time by the two jets (with the remainder expended on pushing against the ambient medium), and that the ratio of the major and minor axes of the bubble is given by

$$\frac{Z}{R} \approx 2.8 \left(\frac{\Omega}{2 \times 10^{-2}} \right)^{-\frac{1}{4}} \left(\frac{V_j}{300 \, km \, s^{-1}} \right)^{-\frac{1}{4}} \left(\frac{V_Z}{15 \, km \, s^{-1}} \right)^{\frac{1}{4}} \qquad (39)$$

where we have scaled by representative parameters. This value is consistent with the observed morphologies of bipolar CO lobes (see, e.g., Bally and Lada

1983). The mass loss rate implied by Eq (35) is

$$
\dot{M}_j = \frac{\Omega Z^2 V_Z^2 \rho_a}{V_j}
$$

$$
\approx 10^{-8} \left(\frac{\Omega}{2 \times 10^{-2}} \right) \left(\frac{Z}{0.2 \text{ pc}} \right)^2 \left(\frac{V_Z}{15 \text{ km s}^{-1}} \right)^2 \tag{40}
$$

$$
\times \left(\frac{n_a}{500 \text{ cm}^{-3}} \right) \left(\frac{V_j}{300 \text{ km s}^{-1}} \right)^{-1} M_\odot \text{ yr}^{-1}
$$

which for typical parameters is compatible with the low-jet-density interpretation of the spectroscopic data. The energy-driven-bubble approximation should be valid so long as the cooling time of the bubble interior exceeds the dynamical time $t_{\text{lobe}} = Z/2V_Z$. If the density in the interior is dominated by the mass deposited by the jets, then, using Eq. (34),

$$
\frac{t_{\text{cool}}}{t_{\text{lobe}}} \approx 4 \left(\frac{V_j}{300 \text{ km s}^{-1}} \right)^{\frac{9}{2}} \left(\frac{V_Z}{15 \text{ km s}^{-1}} \right)^{-\frac{1}{2}} \left(\frac{Z}{0.2 \text{ pc}} \right)^{-1}
$$

$$
\times \left(\frac{\Omega}{2 \times 10^{-2}} \right)^{-\frac{1}{2}} \left(\frac{n_a}{500 \text{ cm}^{-3}} \right)^{-1} \tag{41a}
$$

whereas if it is dominated by the mass evaporated from the shell of swept-up ambient gas (see Weaver et al. 1977), this ratio decreases to

$$
\frac{t_{\text{cool}}}{t_{\text{lobe}}} \approx 0.7 \left(\frac{V_j}{300 \text{ km s}^{-1}} \right)^{\frac{3}{14}} \left(\frac{V_Z}{15 \text{ km s}^{-1}} \right)^{\frac{13}{14}} \left(\frac{Z}{0.2 \text{ pc}} \right)^{-\frac{2}{7}}
$$

$$
\times \left(\frac{\Omega}{2 \times 10^{-2}} \right)^{\frac{3}{14}} \left(\frac{n_a}{500 \text{ cm}^{-3}} \right)^{-\frac{2}{7}}. \tag{41b}
$$

Equations (41a) and (41b) indicate that the assumption of an energy-driven bubble can be justified for low-$\dot{M}_j$ sources with characteristic parameters and that it remains adequate even when thermal evaporation effects become important. In fact, as was pointed out by Weaver et al. (1977), a bubble in which evaporation is the main source of mass may be expected to act as a self-regulating thermostat that would keep the internal thermal pressure at a dynamically significant level.

It thus appears that stellar jets expanding into a moderately dense molecular cloud could, in principle, drive the observed bipolar lobes by a combination of ram pressure (along the jet axis) and thermal pressure (in the transverse direction). The heads of these jets would not give rise to visible Herbig-Haro objects because the bow shock would be too slow to emit strongly in the optical and the shocked jet material would be too hot to satisfy $d_{\text{cool}}/r_h < 1$ (see Eq. [34]). We have already noted in Sec. I.A that, at least in the case of L1551 IRS 5, there is direct evidence from infrared measurements that the associated lobes are *not* energy driven. It may also be possible to test this scenario by high-resolution measurements of the molecular line profiles near the boundaries of the lobes (see Lada 1985).

According to the evolutionary scenario developed by SLRN (see also Chapter by Shu et al.), the neutral X-celerator outflow drives the bipolar molecular lobes and ultimately disperses the ambient molecular cloud core. Ionized jets originate from the polar regions of the star as "ordinary" stellar winds (i.e., outflows launched by MHD wave stresses as opposed to the X-celerator) and are collimated by the surrounding dense neutral wind. As the cloud core is dispersed, both the mass-accretion rate through the disk and the neutral-wind mass-outflow rate drop (see Eq. [19]). The "ordinary" wind can no longer be fully confined by the neutral wind and, as the effective lever arm of the outflow (the Alfvén radius r_A) increases, it begins to spin the star down by magnetic braking (Weber and Davis 1967). The mass-loss rate of the X-celerator declines rapidly as the rotation of the star decreases, and the system takes on the characteristics of a CTTS: moderate stellar rotation and moderate mass loss.

We now make a rough estimate of the spindown time by the ordinary stellar wind, which is also the time scale over which the X-celerator shuts off. A star loses angular momentum to the wind at the rate $\dot{J}_* \approx -\dot{M}_w \Omega_* r_A^2$ (Weber and Davis 1967; Mestel 1968; Belcher and MacGregor 1976; Hartmann and MacGregor 1982). For a star with angular momentum $J_* = \beta M_* \Omega_* R_*^2$, where $\beta < 1$ is a dimensionless moment of inertia factor (see Sec. II.B), the spindown time is

$$t_{\text{spin}} = -\frac{J_*}{dJ_*/dt}$$
$$\approx \beta \left(\frac{M_*}{\dot{M}_w}\right)\left(\frac{R_*}{r_A}\right)^2 \tag{42}$$

which is much shorter than the time scale for the star to lose mass because $r_A \gg R_*$. For winds from rapidly rotating stars, the terminal wind speed and the Alfvén radius scale roughly as $V_\infty \approx \Omega_* r_A$ (Belcher and MacGregor 1976; see Eq. [4]). The terminal wind speed may also be related to the stellar magnetic-field strength by $V_\infty \approx (B_*^2 R_*^4 \Omega_*^2 / \dot{M}_w)^{1/3}$, which can be derived by equating the Alfvén speed at the Alfvén radius to the terminal velocity (Michel 1969; Belcher and MacGregor 1976). Substituting these results into Eq. (42) allows us to write

$$t_{\text{spin}} \approx \beta \left(\frac{M_* V_\infty}{B_*^2 R_*^2}\right) \tag{43}$$

which has the advantage of being formally independent of the mass-loss rate. For a typical CTTS ($M_* \approx 0.5$ $M_\odot$, $R_* \approx 4$ $R_\odot$) which has a 100 gauss surface magnetic field and which blows a 200 km s^{-1} wind, the spindown time from near breakup ($\beta \approx 0.4$) is only 3×10^5 yr. Although the actual value of t_{spin} is probably somewhat higher because of the reduction in the angular momentum loss rate brought about by the presence of "dead zones" in the magnetic-field geometry (see Mestel 1968), it is nevertheless evident from this estimate that, in the X-celerator picture, the protostar makes a fairly rapid transition from rotating near breakup to the more moderate rates associated with CTTS.

An alternative explanation of the observed low rotation rates of CTTS has been proposed by Königl (1991) and is based on the Ghosh and Lamb (1979a, b) model of accreting magnetic neutron stars and white dwarfs. In this picture, the star possesses a kilogauss-strength dipole magnetic field which disrupts the accretion disk at a distance of a few stellar radii from the center and channels the inflowing matter into high-latitude accretion columns on the stellar surface (see Sec. I.C). In a steady state, the braking torque transmitted by the field lines that penetrate the disk outside the corotation radius $r_{\rm co}$ (where the Keplerian angular velocity $\Omega_K[r_{\rm co}]$ is equal to Ω_*) exactly balances the spinup torque exerted by the material inside $r_{\rm co}$. A YSO that initially rotates near breakup will have angular momentum $\mathcal{J}_*$ much higher than the equilibrium value and will approach a steady state over a spindown time $t_{\rm spin} = -\mathcal{J}_*/\dot{\mathcal{J}}_*$ that can be estimated by setting $\dot{\mathcal{J}}_* \approx -B_*^2 R_*^6 r_{\rm min}^{-3}$ (see, e.g., Davidson and Ostriker 1973), where the inner radius $r_{\rm min}$ of the disk is given by

$$r_{\rm min} \approx B_*^{4/7} R_*^{12/7} (GM_*)^{-1/7} \dot{M}_{\rm in}^{-2/7} \qquad (44)$$

(determined from the requirement that the magnetic stress be large enough to remove the excess angular momentum of the nearly Keplerian flow over a narrow transition zone). The value of $t_{\rm spin}$ estimated in this way for representative parameters ($M_* \approx 0.5$ M$_\odot$, $R_* \approx 4$R$_\odot$, $B_* \approx 10^3$ G, $\dot{M}_{\rm in} \gtrsim 10^{-7}$ M$_\odot$ yr^{-1}) is comparable to that obtained in the stellar wind model and in reality is likely to be shorter than the age of the youngest visible CTTS. An important test of this scenario will be provided by direct measurements of the surface magnetic field strengths in CTTS. In addition, this model predicts that changes in the mass accretion rate (which would be manifested as changes in the bolometric luminosity) should lead to episodes of spinup or spindown in the stellar rotation rate (corresponding, respectively, to increases or decreases in $\dot{M}_{\rm in}$ relative to the equilibrium value; see Ghosh and Lamb 1979b).

In closing this section we note that, although the question of whether the jets emanate from the disk or from the star cannot yet be answered by direct imaging, it nevertheless appears that some important clues might be provided by high-resolution spectral measurements (see, e.g., Hamann and Persson 1989; Hartmann et al. 1990).

IV. STAR FORMATION AND THE ORIGIN OF BIPOLAR FLOWS

Bipolar outflows play a crucial role in the physics of star formation by dispersing the infalling molecular cloud core that surrounds a protostar. Observational evidence indicates that outflow activity begins early in the evolution of a YSO, is common during the DEIS stage, and is rare once the star is optically revealed (Lada 1988b). Both theoretical calculations (Shu et al.

1987a; see Chapter by Shu et al.) and interferometric observations (Terebey et al. 1989) suggest that bipolar flows occur during an evolutionary phase when infall (along the equatorial regions) and outflow (along the poles) take place simultaneously. At very early times, the ram pressure of the infalling molecular gas is sufficient to contain the wind, whether it comes from the disk (Sec. II.A) or from the protostar (Sec. II.B). With the passage of time, the wind first breaks through in the direction of least resistance along the rotation axis and subsequently reverses the molecular infall within an ever increasing solid angle about each pole. We now discuss a simple dynamical model (Shu et al. 1991) that interprets the bipolar outflow phenomenon in the context of star formation.

We consider an axially symmetric molecular cloud core that has the density distribution of a (flattened) singular isothermal sphere,

$$\rho(r, \theta) = \frac{a^2}{2\pi G r^2} Q(\mu) \tag{45}$$

where a is the effective isothermal sound speed, which includes contributions from all mechanical support processes: thermal, magnetic and turbulent (Shu et al. 1987a). We have adopted spherical coordinates (r, θ, ϕ), with θ being the polar angle measured from the symmetry axis. The flattening of the isodensity contours is due to rotation (Terebey et al. 1984) and to magnetic fields (Lizano and Shu 1989) and is determined by the function $Q(\mu)$ (with $\mu \equiv \cos\theta$), which has been normalized so that

$$\int_0^1 Q(\mu) d\mu = 1. \tag{46}$$

Isodensity contours are given by $r(\mu) \propto \sqrt{Q(\mu)}$, so a 2:1 flattening of the core implies $Q(0)/Q(1) = 4$.

Lizano and Shu (1989) have shown that the r^{-2} density dependence of Eq. (45) arises in a natural manner during the later stages of cloud evolution as the core loses its magnetic and turbulent support. Dynamical collapse of the central regions ensues in an "inside-out" manner that produces a mass infall rate equal to

$$\dot{M}_{in} \approx \frac{a^3}{G} \approx 2 \times 10^{-6} \left(\frac{a}{0.2 \text{ km s}^{-1}}\right)^3 M_\odot \text{ yr}^{-1} \tag{47}$$

(Shu 1977). Accretion of this infalling material onto the disk and the protostar produces a powerful CDW (see Secs. II.A and B) whose outflow rate is related to the infall rate by

$$\dot{M}_w = f \dot{M}_{in} \tag{48}$$

where $f \approx (r_0/r_A)^2$ for a disk-driven wind (see Eq. [3]) and $f \approx 0.3$ for the X-celerator mechanism (see Eq. [19]). The wind momentum discharge per steradian can be written as

$$\frac{\dot{M}_w V_w}{4\pi} P(\mu) \tag{49}$$

where the angular distribution function $P(\mu)$ describes the collimation of the wind by MHD stresses (and, in the case of a wind from a Keplerian disk, also reflects the radial dependence of the disk parameters; see Fig. 1) and is normalized so that

$$\int_0^1 P(\mu)\,d\mu = 1. \tag{50}$$

There is indeed observational evidence for the existence of a latitude-dependent wind from the young star R Mon, whose wind speed varies from $\sim$50 km s^{-1} near the equator ($\mu = 0$) to $\sim$250 km s^{-1} near the pole ($\mu = 1$) (Jones and Herbig 1982; Dutkevitch et al. 1989).

The wind sweeps up a thin shell of matter (see Sec. I.A) which moves outward with speed $V_{lobe} = dR_{lobe}/dt$ that can be a function of μ. To describe the interaction of the wind with the circumstellar medium, Shu et al. (1991) adopted a momentum-conserving snowplow model which assumes that no redistribution of matter occurs in the θ direction and that the outward motion of each segment of the shell is strictly radial so as to satisfy the vector conservation of momentum. In this approximation, the mass $\mathcal{M}$ (per steradian) of swept-up gas in the lobe increases with time as

$$\frac{d\mathcal{M}}{dt} = \frac{a^2}{2\pi G} Q(\mu) V_{lobe} \tag{51}$$

where we have used Eq. (45) for the circumstellar density and have ignored the small amount of mass contributed by the wind. Neglecting the small external pressure of the circumstellar gas, the momentum (per steradian) of the lobe, $\mathcal{M}V_{lobe}$, increases with time as

$$\frac{d}{dt}(\mathcal{M}V_{lobe}) = \frac{\dot{M}_w V_w}{4\pi} P(\mu). \tag{52}$$

Because each piece of the shell is assumed to move outward radially, μ is constant along each trajectory, and the solution to Eqs. (51) and (52) is

$$V_{lobe} = \left(\frac{\dot{M}_w}{2\dot{M}_{in}}\right)^{1/2} (a\,V_w)^{1/2} \left(\frac{P(\mu)}{Q(\mu)}\right)^{1/2} \tag{53a}$$

$$R_{lobe} = V_{lobe}\,t \tag{53b}$$

$$M_{lobe} = 2\pi\mathcal{M} = \dot{M}_{in}\,t\left(\frac{V_{lobe}}{a}\right) Q(\mu) \tag{53c}$$

where we have used Eq. (47) to write a^2/G as $\dot{M}_{in}/a$ and where the time t is measured from the instant the outflow begins to break through the infalling cloud envelope. In this solution, the ram pressure of the incoming circumstellar gas exactly balances the ram pressure of the outflowing wind so that the shell speed (at each μ) is *constant* in time, a result specific to an r^{-2} density profile. Equation (53a) indicates that the net collimation of the outflow is due both to the intrinsic focusing of the driving wind (through P) and to the anisotropic distribution of ambient matter (through Q).

Using Eq. (48), we can write the outflow velocity as

$$V_{lobe} = \left(\frac{f}{2}\frac{P(\mu)}{Q(\mu)}\right)^{1/2} (aV_w)^{1/2} \tag{54}$$

which shows that, apart from the first square-root factor (which is typically of order unity), the characteristic velocity of bipolar flows is the geometric mean of the isothermal sound speed in the molecular core and the wind speed. Because a ranges from a few times 10^{-1} km s^{-1} (for low-mass cores) to a few kilometers per second (for high-mass cores) and the wind speed is typically several hundred kilometers per second, we have an explanation for why the observed values of V_{lobe} are intermediate between these two scales, ranging from a few to $\gtrsim 10$ km s^{-1}. Furthermore, this characteristic velocity is maintained for as long as the outflow propagates through a region with an r^{-2} density profile and is independent of the overall size (or, equivalently, the age) of the molecular lobes.

Equation (53b) indicates that a snapshot of the outflow at any given time will produce a linear "Hubble law" $V_{lobe} \propto R_{lobe}$, which is a natural consequence of a latitude-dependent wind propagating into an anisotropic medium. If the major axis of the lobes is oriented at an angle i to the direction of an external observer, then the line-of-sight velocity $v_{\parallel}$ measured by the observer is related to the projected distance $r_{\perp}$ from the central source by

$$\frac{v_{\parallel}}{r_{\perp}} = \frac{\cot(\theta - i)}{t}. \tag{55}$$

This will result in an apparent linear law, as observed in several well-resolved outflows (see Sec. I.A), provided that the opening angle of the outflow is not too large compared to the inclination angle (i.e., if the source is fairly well collimated).

It is worth pointing out in this connection that an approximate Hubble law may also be expected to apply in the case of molecular lobes that are driven by a combination of the ram pressure of collimated jets and the thermal pressure of the cocoon of shocked jet material (see Sec. III.B) instead of purely by the ram pressure of a latitude-dependent, quasi-spherical wind. This is because an energy-driven bubble that expands into a medium with an r^{-2} density distribution will move with a constant speed just like a momentum-driven bubble (Königl 1982). Using the simple model outlined in Sec. III.B, one can arrive at this conclusion directly from Eq. (39), which shows that the ratio ($\gtrsim 2$) of the major and minor axes of the bubble is independent of the ambient density.

Finally, the total amount of mass accreted by the star–disk system is $M_* + M_{disk} = \dot{M}_{in}t_{tot}$, where t_{tot} is the time elapsed since the onset of the infall. Equation (53c) shows that the total swept-up mass in each lobe is approximately

$$M_{\text{lobe}} \approx M_{\text{tot}} \left(\frac{t}{t_{\text{tot}}} \right) \left(\frac{V_{\text{lobe}}}{a} \right) \qquad (56)$$

which is larger than the star-plus-disk mass by roughly the ratio of the bipolar flow speed to the cloud core sound speed. This relationship would cease to hold if the time since the onset of the flow were significantly shorter than the duration of the infall ($t < < t_{\text{tot}}$). However, because outflows have been detected even in very young objects (Terebey et al. 1989), the duration of the pure infall phase is apparently quite short (i.e., $t \approx t_{\text{tot}}$) and protostars probably spend most of their DEIS phase as combined inflow-outflow sources (see also Chapter by Shu et al.).

V. CONCLUSIONS

Recent observational findings concerning the apparent connection between energetic outflows and circumstellar disks and the presence of massive neutral winds have led to significant advances in our understanding of outflows and winds in young stellar objects. Although we are still far from possessing a complete theory, it seems that at least some of the basic ingredients of a viable model have already been identified and that ongoing research efforts should lead to further progress in the near future. Our discussion of the current state of the theory can be summarized as follows:

1. Magnetic driving is the most promising mechanism for launching the high-momentum neutral winds that give rise to the bipolar molecular lobes. We have focused on centrifugally driven winds in which particles are accelerated along inclined magnetic-field lines like beads on rotating wires. We considered both CDWs from Keplerian disks that are threaded by interstellar magnetic-field lines and winds launched along stellar field lines from the boundary between a disk and a rapidly rotating protostar. In both cases, most of the energy and momentum injection into the wind occurs near the stellar surface. Efficient production of CDWs depends, in the case of disks, on the presence of sufficiently strong magnetic fields and on their adequate coupling to the matter, and, in the case of the X-celerator mechanism, on the equatorial regions of the star rotating near breakup and having a suitable field geometry. Disk-driven winds may regulate the transport of the excess angular momentum of the accreted matter whereas the X-celerator mechanism could enable YSOs to continue accreting even after they had been spun up to breakup speeds. Such winds may thus be an essential ingredient of the accretion process, which could be the reason for the observed correlation between accretion and outflow in YSOs.
2. The kinematic, thermal and chemical properties of CDWs lead to specific predictions that can be tested by observations. In particular, the distribution of uplifted dust in disk-driven winds may have a significant impact

on the infrared continuum spectra and on the optical and infrared polarization properties of YSOs. Furthermore, ambipolar diffusion heating in disk CDWs may give rise to extended ($\gtrsim 100$ AU), high-temperature ($\sim 10^4$ K) and high-ionization ($f_e \lesssim 0.1$) regions that would be detectable through their optical forbidden line and radio continuum emission. In a similar vein, YSO-driven neutral outflows have calculable freeze-out abundances of various molecular species that could be detected at large distances from the origin.

3. Magnetic fields (in the form of Alfvén waves generated in the stellar convection zone) could play a central role also in driving the high-velocity, ionized jets. In this interpretation, the jets arise from a quasi-spherical stellar wind that is collimated by the surrounding neutral outflow. Alternatively, the jets could be powered by accretion, in which case they either represent the ionized inner regions of a disk-driven wind or else are launched by thermal or magnetic processes from a shear boundary layer between the disk and the YSO. In this picture, the nonsteady mass ejection inferred in several jets may be associated with episodic mass accretion events of the FU Orionis type. Despite their small dimensions, the jet formation regions could be probed by high-resolution spectral measurements. Important constraints on the competing models could also be provided by accurate determinations of the jet mass-outflow rates.

Acknowledgments. We are grateful to P. Safier and M. Wardle for their valuable input into this work and for their help in preparing the figures. We are also grateful to N. Hillis for assistance in the preparation of the manuscript. This research was supported in part by grants from the NSF, NASA, and the Illinois Space Institute.

MAGNETIC FIELDS, ACTIVITY AND CIRCUMSTELLAR MATERIAL AROUND YOUNG STELLAR OBJECTS

THIERRY MONTMERLE
Centre d'Etudes de Saclay

ERIC D. FEIGELSON
Pennsylvania State University

JÉRÔME BOUVIER
Canada-France-Hawaii Telescope Corporation

and

PHILIPPE ANDRÉ
Centre d'Etudes de Saclay

Young stellar objects are characterized by a wide variety of activity phenomena and a strong interaction with their environment, on a wide range of spatial scales. We review the solar-like activity in T Tauri stars, and the X-ray and radio [cm] evidence for extended magnetic structures ($\lesssim 10~R_*$). In the case of classical T Tauri stars, the close circumstellar material ($\lesssim 1000$ AU) comprises circumstellar accretion disks and their associated boundary layers, ionized and neutral winds. By contrast, weak-line T Tauri stars are not necessarily deprived of circumstellar material, but may be surrounded by cold matter at a distance, without boundary layer, which can be best detected in the mm range. A tentative unified picture emerges, linking magnetic fields, accretion disks and mass-loss phenomena up to the much larger spatial scales of molecular outflows (several 10^4 AU). The possible relevance of this new picture for the origin of the solar system is indicated.

I. WHAT IS A YOUNG STELLAR OBJECT?

Thanks mainly to the advent of telescopes operating at wavelengths less sensitive to interstellar absorption than the optical (X-rays, infrared, mm, cm), rapid progress has been accomplished in recent years in our understanding of the earliest stages of star formation. Indeed, very young or forming stars are embedded inside molecular clouds, and it is only at the above wavelengths that one is able to see them. However, when these objects are too embedded to be visible in the optical domain, it is difficult to characterize them in the same way as optically visible stars (absence of HR diagram classification in many cases, for instance). On the other hand, the very fact that these stars are embedded in molecular clouds testifies of their youth, typically less than a few Myr. In addition, as we shall see, many stars are surrounded by a complex circumstellar environment: magnetic fields, dust shells or disks, ionized or neutral winds, molecular outflows, etc., and therefore do not appear as "stars" in the conventional sense. At the cost of some vagueness, it has become customary in recent years to call the various sources close to, or embedded in, molecular clouds, by the generic name of young stellar objects (YSOs).

In most cases, YSOs can be shown to have moderate masses ($\leq$ a few $M_\odot$) and temperatures ($\leq$6000 K). Their luminosities, however, can be quite high (up to several 10 $L_\odot$), placing them generally high above the main sequence: in that sense, the expressions "pre-main-sequence objects" (PMS objects) and YSOs are synonymous. However, in practice YSO tends to be applied to objects embedded in their parent molecular clouds, while PMS objects are visible in the optical and can be plotted on the HR diagram. In that sense, they constitute a subclass of the YSO population. As a rule, we will follow this distinction here. Other YSOs have higher masses (10 to 20 $M_\odot$, for instance), and their luminosities may reach several 10^3 $L_\odot$. Being massive, they are characterized by a much shorter evolutionary time scale, and excite compact H II regions. Since we are interested here primarily in solar-type stars, we will discuss only the *low-mass* objects ($\leq$2 $M_\odot$). (For a discussion of high-mass stars in general, see, e.g., Lada [1987a], and for a discussion of Herbig Ae-Be stars, see Catala [1989]).

The most well-known (and historically oldest-known) of the low-mass PMS objects are the T Tauri stars. Very good reviews on this subject have recently appeared (Bertout 1989; Appenzeller and Mundt 1989). For details, we refer the interested reader to these, and to the Chapters by Basri and Bertout and by Hartmann et al., but summarize here their relevant properties for the sake of self-consistency of this chapter.

T Tauri stars were discovered in the vicinity of dark clouds (mainly in Taurus and Orion) thanks to their peculiar emission-line spectra, which made them conspicuous for instance in Hα objective-prism surveys. In the last few years, however, mainly as a result of X-ray observations by the *Einstein* satellite, these stars have been divided into two broad classes: the classical T Tauri stars (CTTS), and the weak-line T Tauri stars (WTTS), also called naked T Tauri stars (NTTS; see Chapter by Stahler and Walter). Although these last two terms have often been used equivalently in the past on the basis of the initial definition of Walter (1986) and Walter et al. (1988), a better understanding of the circumstellar environment of these stars (including in particular of its cold component) probably warrants a further distinction between both terms using refined definitions.

CTTS constitute the vast majority of the originally catalogued TTS (Herbig and Rao 1972; for a revised list, see Herbig and Bell 1988). Their spectra display strong optical and ultraviolet emission lines. The equivalent width of Hα is >5 to 10 Å, and may reach 200 Å or more in extreme TTS. The underlying spectrum is that of a late-type photosphere (K7, typically), with respect to which there is a large ultraviolet and infrared excess. The amount of ultraviolet excess is correlated with the strength of the emission lines, and so does, to a lesser extent, the near-infrared part of the infrared excess. On the contrary, the WTTS show very weak, if any, emission lines (Hα equivalent width <5 to 10 Å, depending on authors, i.e., comparable with solar activity levels), no ultraviolet excess, and little or no infrared excess. Their only conspicuous feature is their large X-ray emission, comparable to that of CTTS, which made their discovery possible. Strictly speaking, the customary definition of the WTTS based only on the strength of the Hα emission line is insufficient, because the true line flux depends on the spectral type: for instance, SU Aur, which has a small Hα equivalent width (3.5 Å), is an active CTTS. In practice, however, because most T Tauri stars have more or less similar spectral types (while SU Aur is a rare G2), the use of the Hα equivalent width yields a clear cut and easy-to-use classification, but it should be kept in mind that there is a good continuity between CTTS and WTTS. By contrast, because the adjective "naked" refers to a physical property (i.e., absence of circumstellar material), and not to a purely observational one, the proper definition of NTTS must include additional ingredients, such as the absence of near-infrared excess, age determination, etc., and other subtleties may have to be introduced on the basis of recent mm observations which trace the cold dust (see Sec. V). In what follows, we will adopt strictly observational criteria to categorize T Tauri stars, and hence refer in most cases to CTTS and WTTS only, including the NTTS in this last class except otherwise specifed.

YSOs too deeply embedded to be visible in the optical and X-ray ($A_v >$ a few magnitudes) generally remain visible at longer wavelengths, and have been characterized by their near- to mid-infrared spectral distribution (Lada 1987b,1988a). They have been broken down in Class I (very strong infrared excess, with a λF_λ spectrum ascending with wavelength), Class II (infrared excess comparable to that of CTTS, i.e., flat or moderately descending spectrum), and Class III (no infrared excess, i.e., pure blackbody spectrum). Current theories link this classification of infrared spectra with evolutionary stages, in terms of decreasing amount of warm dust, from the protostar stage (Class I), in which the central stellar condensation is surrounded by a dust shell, to the dustless, advanced PMS stage (optically visible Class III) (see, e.g., Shu et al. 1987a, b). We shall see that new results in the radio mm and mid-infrared ranges suggest a revision of this scheme.

Perhaps some of the most intriguing characteristics of PMS stars and of embedded YSOs are their intense activity, and a strong interaction with their environment, in the form of mass loss (stellar winds, collimated jets, molecular outflows), accretion from a circumstellar disk, and various magnetic phenomena. In this Chapter, we will be mainly concerned with the question of the "close" circumstellar environment of these objects, i.e., within a radius $\sim$1000 AU. This includes magnetic activity and stellar magnetic fields (Secs. I and II), stellar winds (Sec. III), circumstellar disks (Secs. IV and V), but excludes a detailed discussion of large-scale phenomena such as jets or bipolar flows, which extend up to a fraction of a pc, or of the related Herbig-Haro objects. (The interested reader is referred, respectively, to Lada [1985], Mundt [1988], and Reipurth [1989c] for reviews on these topics; see also the Chapters by Edwards et al. and by Fukui et al.)

II. SOLAR-LIKE ACTIVITY IN LOW-MASS PMS STARS

A. X-Ray Evidence: Flares

The bulk of the data we now have comes from observations using the *Einstein* X-ray Observatory, the first satellite-borne focusing X-ray telescope, which obtained images in the spectral band $\sim$0.4 to $\sim$4 keV (for relevant reviews and additional references to the literature, see Montmerle and André [1988], Bertout [1989], Feigelson et al. [1990], Herbig and Bell [1988]). Most of the observed regions were nearby dark clouds, well known to undergo active star formation and contain many T Tauri stars: ρ Ophiuchi (Montmerle et al. 1983), Taurus-Auriga (Feigelson et al. 1987; Walter et al. 1988), Chameleon (Feigelson and Kriss 1989), and others, all at distances $\sim$160 pc. Images within the Orion star-forming complex at $\sim$450 pc distance also revealed dozens of additional X-ray emitting PMS stars (Ku et al. 1982; Caillault and Zoonematkermani 1989; Strom et al. 1990). Note that some of these regions also contain stars of earlier spectral types which copiously emit X-rays. From these extensive data sets, a variety of PMS stellar X-ray characteristics have been established, which we now summarize.

X-ray emission between 10^{29} and up to several 10^{31} erg s^{-1} in the *Einstein* band is a general property of low-mass PMS stars, and is not confined to any particular subclass. In particular, existing evidence shows that this emission is present independently of the strength of classical T Tauri properties, such as broad emission lines, ultraviolet and infrared excesses. This indicates that, contrary to earlier predictions, PMS X-ray emission is not an indirect consequence of classical T Tauri winds. It also indicates that the X-ray emission is not generally absorbed (or "smothered" as suggested by Walter and Kuhi [1981]) in classical T Tauri winds.

These X-ray levels are much higher than those seen in the main-sequence stars. The average PMS star emits 10^3 times more than old disk stars and 10 times those seen on young main-sequence stars such as the Pleiades. There is some debate over the exact form of the relation between stellar age and X-ray intensity parameters. Walter et al. (1988) choose selected main sequence stars and X-ray-detected PMS stars in Taurus-Auriga to derive a relation for the surface flux F_X of the form $F_X \sim \exp(-t/t_o)$ with $t_o = 4 \times 10^8$ yr (for a thorough discussion, see Walter and Barry [1991]). Feigelson and Kriss (1989) use unbiased X-ray luminosity functions of optically selected samples of main-sequence stars and Chameleon PMS stars (Fig. 1), treating X-ray nondetections as well as detections, to derive the relation $L_X \sim t^{-0.6}$ for the entire range $10^6 \leq t \leq 10^{10}$ yr. It is therefore unclear at present whether

X-ray activity decreases with age throughout the PMS phase, or whether it is relatively constant as the star descends the Hayashi track. X-ray emission appears to be present as soon as convection begins when stars emerge from their accreting envelopes at the "birthline" (Stahler 1988; also Chapter by Stahler and Walter), but it is as yet uncertain whether it is already present during the protostar stage.

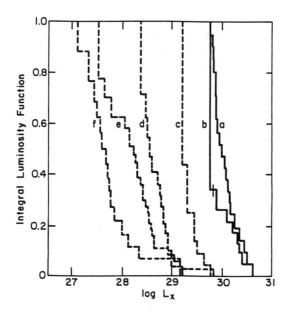

Figure 1.   Integral X-ray luminosity function for the Chameleon PMS stars (solid line), and for main-sequence low-mass stars. (a) Chameleon X-ray sources; (b) Hα emission line stars; (c) Pleiades K stars; (d) Hyades K stars; (e) young disk K stars; (f) old disk K stars (for details, see Feigelson and Kriss 1989). One can clearly see the decrease in X-ray luminosity, which traces solar-type activity, with age.

The X-ray spectra from PMS stars are hard and variable. When a good PMS X-ray spectrum is available, it is consistent with bremsstrahlung emission from a hot ($kT \sim 1$ keV, i.e., $T \sim 10^7$ K) plasma with line-of-sight absorption consistent with optical obscuration (typical $A_v = 1$–2 and $N_H = 10^{21}$ cm^{-2}). Concerning variability, much evidence has accumulated in favor of the existence of large X-ray flares. This feature is the most conspicuous of the X-ray emission, and is apparent as soon as time-dependent observations are available. By comparison, there is little evidence for quiescent emission in time-dependent data. The coronal emission level therefore seems to be weak, at best barely at the level of sensitivity of *Einstein*, i.e., $\gtrsim 10^{30}$ erg s^{-1} at ~ 160 pc. The most extensive data are from the ρ Ophiuchi cloud, for which several dozen X-ray emitting stars are seen blinking on and off giving the impression of an "X-ray Christmas Tree" (Montmerle et al. 1983*a, b*).

The variations are typically factors of 2 to 20 in amplitude and are statistically similar to the distribution of peak X-ray emission in solar flares. These flares have X-ray luminosities up to 10^6 times those seen on the modern Sun, are considerably stronger than flares on dMe flare stars (see discussion in Montmerle et al. 1983), and are quite similar to flares observed on RS CVn close binary systems (see, e.g., Agrawal et al. 1986). However, strong flaring activity is estimated to take place only ∼5% of the time (Gahm 1988).

More precisely, a few individual rapid X-ray flares that fortuitously occurred during *Einstein* and EXOSAT satellite exposures have been analyzed in the context of simple solar flare loop models (Feigelson and DeCampli 1981; Montmerle et al. 1983*a*; Walter and Kuhi 1984; Tagliaferri et al. 1988). These largest X-ray flares are characterized by rise times of minutes to hours, fall times of several hours, and peak luminosities up to 5×10^{31} erg s^{-1}. Application of simple loop models (based on simplifying assumptions which may not strictly hold) suggest that the temperatures ($T \sim 10^7$ K), plasma densities ($n \sim 10^{10}$ cm^{-3}) and magnetic field strengths ($B_* \sim 10^3$ G at the base of the loop) are not remarkably different from solar flares. But the inferred loop sizes ($\ell \sim 10^{11}$ cm) are very large, implying that the size of the X-ray emitting region is in general comparable to $R_\odot$ (i.e., ∼$2R_*$, typically), and sometimes as large as 2 R_*.

In summary, the soft X-ray band has provided extensive evidence for the presence of very powerful solar-like magnetic activity on the surfaces of low-mass PMS stars. Many additional X-ray observations are currently taking place using the ROSAT X-ray telescope, launched in June 1990, which will provide an all-sky survey (expected to identify several hundred PMS X-ray emitters) and pointed exposures an order of magnitude more sensitive than those obtained with the *Einstein* satellite a decade earlier.

B. Consequences for Molecular Clouds and Star Formation

A major by-product of the X-ray surveys was the identification of dozens of previously unrecognized PMS stars with weak T Tauri properties. Estimates of the ratio of WTTS to CTTS populations range from ∼10:1 for Taurus-Auriga (Walter et al. 1988), to ∼3:1 for L1641 in Orion (Strom et al. 1990), to ∼2:1 in Chameleon (Feigelson and Kriss 1989) and Ophiuchus (Bouvier and Appenzeller 1991). The latter two values are more reliable than the first because the *Einstein* image coverage was very incomplete for the Taurus-Auriga complex. The accuracy of a PMS stellar census in any cloud is limited by several factors: Hα line strengths vary in all PM stars, permitting stars to cross the WTTS/CTTS boundary (usually chosen to be 5 to 10 Å equivalent width, as mentioned above); X-ray data have different sensitivities between and within star-forming regions due to different image exposures, distances and obscurations; and the large-amplitude intrinsic X-ray variability causes some stars to be absent in any given X-ray exposure. The possible presence of sub-groups of different ages may also be a source of difficulty (Hartmann et al. 1991). Despite these limitations, the preponderance of evidence shows that the total population of low-mass PMS stars is 3 to 4 times larger than the population of CTTS ($H\alpha > 10$ Å) stars alone, and is about 2 times larger than the PMS stellar population known prior to X-ray surveys (see, e.g., Cohen and Kuhi 1979). Estimates of low-mass star-formation efficiency in these clouds, and in the Galaxy as a whole, should be correspondingly increased.

An important consequence of the increased population of X-ray emitting objects inside molecular clouds is its influence on star formation and on cloud chemistry. The physical effect is basically the increased ionization degree in large volumes around individual stars. Norman and Silk (1983) have shown that it regulates star formation by preventing stars from collapsing in these volumes; on the other hand, Krolik and Kallman (1983) have found that the relative abundance of radicals such as C_2^+, C_2H^+, CH^+, $C_2H_2^+$, or HCN^+, may be changed by factors as large as 10^3 or more. Regretfully, these important papers are generally overlooked in current work on star formation and molecular cloud chemistry.

Because the young Sun was presumably a T Tauri star, we should also take the opportunity to mention here the very important role this intense X-ray flux must have played on the irradiation of the primitive nebula (see, e.g., Feigelson 1982), or on the escape of the Earth's primordial atmosphere (see, e.g., Nakazawa and Nakagawa 1981). These points are central to several issues discussed elsewhere (see Walter and Barry 1991), but have largely been overlooked in the past.

C. Optical and Ultraviolet Evidence: Flares and Starspots

T Tauri stars have long been known to be variable from near-ultraviolet to near-infrared wavelengths, on time scales ranging from a few minutes to a few decades. In the early fifties, they were put in the category of "flare stars" by such pioneers as Haro and Ambartsumian. The flare activity is best observed in the near-ultraviolet band, because the contrast with the essentially red photosphere is enhanced. Analysis of the time structure of these flares (Worden et al. 1981) has shown that they are distributed according to a power law, smaller flares being more frequent than larger ones. The optical flares are solar-like, but enhanced several thousand times with respect to solar flares. According to a recent study, there is some difference in the time distribution of ultraviolet flares between CTTS and WTTS (Gahm 1990).

That some TTS exhibit quasi-cyclic light variations on a time scale of a few days has been known for more than 25 years. Yet, it is only during the past decade that systematic photometric studies on such a time scale have been started and this led to the detection of periodic lightcurves for a number of TTS (CTTS and WTTS alike). The amplitude of the photometric wave, usually of the order of a few tenths of a magnitude, decreases from near-ultraviolet to near-infrared wavelengths and can often be modeled by a single large starspot (or an assembly of smaller starspots) covering typically 10% of the stellar surface, cooler than the stellar photosphere by several 100 K (see, e.g., Bouvier et al. 1986b; Vrba et al. 1986,1989; Bouvier and Bertout 1989).

Therefore, in addition to confirming the existence of solar-like activity, albeit on a much larger fractional area, these observations allow the determination the rotation period of the TTS, a key parameter both for activity-related aspects and for their early evolution (see Sec. IV.C).

The properties of these starspots are very similar to those found on other types of magnetically active stars, such as RS CVn systems and BY Draconis stars. By analogy with the Sun, chromospheric plages are expected to be associated with cool starspots. Spectroscopic evidence for such chromospheric plages has been reported by Herbig and Soderblom (1980), and the co-existence of cool and hot spots at the surface of TTS is sometimes required to model their periodic lightcurves (Vrba et al. 1986). In some cases, only hot spots can account for the large amplitude (up to several magnitudes) of the stellar modulation observed at near-ultraviolet wavelengths. Then, the luminosity of these spots usually amounts to a significant fraction of the star's luminosity, and they have been interpreted as accretion shock regions near the stellar surface, where the accreted material is channelled along magnetic field lines (Bertout et al. 1988; see also Königl 1991). That WTTS only have cool spots while hot spots have only been found on CTTS so far supports this idea.

The lifetime of TTS spots greatly varies from one star to another. Long-term photometric monitoring of the WTTS V410 Tau (Vrba et al. 1988) has shown that two large, cool spots survived for several years at its surface. In other stars, however, the photometric wave has been found to disappear from one season to another. Short spot lifetimes (less than a few days) may explain why rotational modulation was detected in only about one third of the TTS in which it was searched for. In fact, both periodic and nonperiodic variability on a time scale of days usually have comparable amplitudes and similar wavelength dependence. This suggests that modulation of the stellar luminosity by spots is indeed the main mechanism responsible for day-to-day variability in these stars, whether periodic or not (there are some exceptions, however, see, e.g., Gahm et al. [1989])

The optical variability on a time scale of several years is the most conspicuous, with an amplitude that can reach a few magnitudes. Yet, its origin is still a puzzle. Both magnetic and nonmagnetic processes have been invoked to explain it, e.g., a large disruption of the convective zone by magnetic fields (Appenzeller and Dearborn 1984), a variable obscuration of the stellar disk by a dusty condensation (Bellingham and Rossano 1980), or nonsteady disk accretion (Bertout et al. 1988). All of these mechanisms may in fact be at work at one time or another in TTS. Important clues as to whether nonsteady disk accretion is a likely mechanism may be gained by comparing the long-term photometric behavior of WTTS and CTTS, as well as by investigating how spectral variations correlate with photometric ones (see, e.g., Walker 1987; Holtzman et al. 1986).

D. Radio Evidence: Magnetic Fields

Radio observations of YSOs are doubly interesting. First, starting from the solar-flare analogy of the X-ray emission of TTS described in Sec. II.B, it is a logical next step to look for radio emission in the cm (GHz) band, as observed on the Sun; in this case, the emission mechanism is expected to be nonthermal and associated with energetic electrons gyrating in magnetic loops. Second, the ionized component of the strong stellar winds observed around some YSOs (see, e.g., Sec. IV.A) is expected to give rise to free-free (thermal) radio emission.

Taking advantage of the increased sensitivity and imaging capability in the cm range offered by the Very Large Array (VLA) interferometer in New Mexico, a number of workers have made radio observations following two approaches. In the first approach, known YSOs, selected according to a variety of criteria, are the targets of pointed observations: optically visible CTTS or WTTS found in early catalogs (see, e.g., Bieging et al. 1984; O'Neal et al. 1990), sources of molecular outflows (Snell and Bally 1986; Rodríguez et al. 1989), or Herbig-Haro objects (see, e.g., Curiel et al. 1989). In the second approach, extended regions are subject to unbiased surveys, with no prior selection of targets.

The first approach gives evidence that $\sim 10\%$ of stars now known as CTTS or WTTS, as well as the majority of outflow exciting sources, are radio emitters at the level of a flux density sensitivity ~ 0.5 mJy, while almost all the classical, low-luminosity Herbig-Haro objects remain undetected at this level (a major exception is the HH 1–2 system; see below). In the case of CTTS and outflow sources, the radio emission is most likely of thermal origin, in agreement with the fact that some of these sources are resolved by the VLA on a scale of $\sim 0.1''$–$10''$ (i.e., $\sim 10^{14}$–10^{16} cm, at the distance of the nearest molecular clouds). This is consistent with the existence of intense ionized stellar winds around CTTS, already suspected to be present on the basis of some Hα and NaD (see, e.g., Mundt 1984) or [O III] (Edwards et al. 1987) emission-line profiles. As discussed in Sec. IV.A, the size of the emitting region is typically $R \sim 1000 R_*$. However, there are clear examples, like the exciting source of the HH 1–2 system (Rodríguez et al. 1990), where the simple, spherical model does not fit the observed radio spectrum and source shape, suggesting that the wind is rather a bipolar, confined jet (see, e.g., Reynolds 1986). In the case of WTTS, the emission is almost certainly nonthermal, and probably due to gyrosynchrotron radiation occurring in large (several R_*) magnetic loops near the stellar surface. The evidence includes strong variability and detection by the Very Long Baseline Interferometer (VLBI) (see below).

The second approach makes no assumption about the nature of the putative radio emitters, and has revealed the existence of hitherto unknown objects, in ρ Oph (André et al. 1987), CrA (Brown 1987), Orion (Trapezium region: Garay et al. 1987; Churchwell et al. 1987; and L1641: Morgan et al. 1990). Several stellar radio sources had been previously found in X-rays, but many were discovered to be so deeply embedded in the clouds that even the X-rays, if present, would be absorbed. Most of these embedded sources are associated with optically invisible near-infrared sources, and a few of them are "radio-discovered" YSOs, so far undetected even in the infrared (like VLA 1623 in ρ Oph; André et al. 1990a). The number of radio-emitting YSOs found down to a flux density of ~1 mJy is small, on the order of 10% of known near-infrared sources (K-magnitude limit typically 12). However, as the sensitivity increases, the number of radio-emitting objects also increases, as does the fraction of identifications with near-infrared sources, which reaches ~50% in the ρ Oph cloud when the minimum detectable flux density is ~0.1 mJy (Leous et al. 1991).

Understanding the nature of the emission from sources detected through unbiased surveys is difficult, precisely because it turns out that a wide range of objects are represented. Conceptually, one may expect YSOs surrounded by circumstellar disks to be thermal radio emitters by analogy with CTTS, and, conversely, young stars deprived of massive circumstellar material to be nonthermal emitters. As a matter of fact, the majority of the stellar radio sources emitting above ~1 mJy do not exhibit an infrared excess, and are therefore probably nonthermal and of the WTTS type even when they are embedded. (Sources detected in the radio with a lower flux have in most cases been observed at only one wavelength, and their nature is more uncertain at present; see discussion in Leous et al. [1991].) The direct confirmation of the above picture is the subject of active observational work.

One important first step towards understanding the emission mechanism results from a long-term monitoring of the sources. As in X-rays, strong variability was found by way of long or repeated observations: the discovery of the first radio flare in a PMS star came as early as 1985 (on the ρ Oph star DoAr 21; Feigelson and Montmerle 1985), and was followed by other similar findings (see, e.g., Stine et al. 1988). However, the frequency of occurrence and/or duty cycle of PMS radio flares seems much lower than in X-rays (Stine et al. 1988; Bieging and Cohen 1989). For lack of adequate coverage, the exact time scale for variability is unknown, but variations in flux of factors of at least 2 have been found in a few hours, and up to 10 between observations separated by a few months (see, e.g., Cohen and Bieging 1986). Such time scales can only be explained in terms of a nonthermal emission mechanism, because the interpretation in terms of an ionized wind implies emission sizes of at least several 100 R_* and thus long time scales, ~ months or years.

Strictly speaking, only the variability can be considered as evidence for nonthermal, solar-like activity in the form of flares. However, since most of the time the stellar radio flux is not variable, but more or less quiescent, one could then think of generally considering the spectral index α ($S_\nu \propto \nu^\alpha$) to find the emission mechanism, by applying the commonly used criterion stating that if $\alpha > 0$, it is thermal and if $\alpha < 0$, it is nonthermal. This is, however, often a dangerous procedure, because the spectral index reflects only the source *opacity* [resp. optically thick, and optically thin], even if it remains true that $\alpha < -0.1$ is a reliable indicator of a nonthermal mechanism such as synchrotron (see, e.g., André 1987; Dulk 1985). It is then necessary to do polarization and mapping studies (e.g., by VLBI), as discussed in detail in Sec. III below, to demonstrate that the radio emission of several quiescent sources is also nonthermal and associated with large magnetic fields. In this case, the term "solar activity" should be understood in a much broader, looser sense only because modeling the radio emission still relies mostly on loop models devised for the Sun. But beyond these subtleties, the bottom-line result is that radio studies have provided a quasi-direct evidence for the existence of magnetic fields associated with a small, but important class of YSOs.

III. MAGNETIC FIELD STRUCTURE

A. Radio Polarization Data

The most direct signature of the presence of magnetic fields that radio observations can provide is the detection of circular and/or linear polarization. Circularly polarized radio emission generally traces nonrelativistic or mildly relativistic electrons radiating in magnetic fields of ~ 1 to 1000 G through the gyroresonance, gyrosynchrotron, or even free-free processes, while linearly polarized emission generally traces synchrotron-emitting, ultrarelativistic electrons in weaker magnetic fields, $<<0.1$ G (see, e.g., Dulk 1985). The main difficulty of this approach is that detecting polarization at a level of a few percent on sources with a flux density at most a few mJy is often a challenge. However, the power of this technique has recently been illustrated in two different stellar cases.

First, André et al. (1988) detected weak ($\sim 7\%$) circularly polarized radio emission in the very young, embedded B3 star S1 associated with the ρ Ophiuchi cloud core A (see Loren et al. 1990 for core names). The modeling of the emission led to the suggestion that S1 was surrounded by a large-scale (i.e., several stellar radii), axisymmetric (possibly dipolar) magnetosphere, oriented essentially pole-on, and with a surface magnetic field similar to what prevails on the well-known magnetic B stars (1–10 kG). This interpretation was later confirmed by the VLBI (see below), and the level of polarization was repeatedly detected in subsequent observations spread over several years.

Second, Yusef-Zadeh et al. (1990) discovered weak (6–8%) linearly polarized emission at 6 cm near the core of an HH-like object, known as the Orion "streamers," in the L1641 cloud. This detection implies the presence of an organized magnetic field of at least $\sim 5 \times 10^{-4}$ G on a scale of 2500 AU. The authors suggest that this magnetic field is generated by a rapidly rotating, deeply embedded pre-main-sequence star featuring a magnetosphere rather similar to that proposed for S1. In both sources, it is important to stress that the radio emission is essentially non-variable, suggesting a permanent or at least long-lived magnetic field configuration.

B. Very Long Baseline Interferometer Measurements

We have seen above (Sec. II.A) that the X-rays were emitted by bremsstrahlung of a $\sim 10^7$ K optically thin plasma, likely confined in large ($H_X \lesssim 2R_*$)

magnetic loops, as can be shown indirectly from the X-ray emission measures and variability time scales.

More direct measurements of the size of the magnetic structures can be obtained in the radio (cm) range, by means of VLBI observations. Using radio telescopes separated by hundreds or thousands of kilometers, VLBI studies can resolve emission regions on milliarcsecond angular scales, corresponding to linear scales around 10^{12} cm (i.e., a few stellar radii) in nearby star-forming regions. By combining VLBI with simultaneous measurements on arcsecond scales using smaller interferometers like the VLA (Very Large Array), one can determine whether the radio emission is also produced in regions several hundred times larger, such as in ionized winds.

The very young star S1, already mentioned in Sec. III.A, was the first YSO detected by this technique (André et al. 1991). The full width at half maximum (FWHM) size ($\sim$13 stellar radii) and brightness temperature ($\sim$2 × 10^8 K) which were measured at 6 cm are consistent with the main lines of the magnetospheric model proposed on the basis of the circular polarization detection (see above, Sec. III.A). The combination of polarization and size measurements for this source provides good and almost direct constraints on the value of the magnetic field: assuming a dipolar field, a value of $B_* \sim 2$ kG is derived at the stellar surface. Future multi-wavelength VLBI observations of S1 may further improve this value as they should probe the magnetic field at different distances from the star, and thus test the dipolar hypothesis.

Extensive VLBI studies of the "bright" ($S_{6cm} > 1$ mJy) stellar radio sources of Taurus and Ophiuchus (mostly WTTS or class III infrared objects) are now in progress, but already prove extremely successful. The first set of observations has demonstrated that most or all of WTTS radio emission is nonthermal, with brightness temperatures between 4 × 10^8 to >2 × 10^9 K (Phillips et al. 1991). Two radio-bright WTTS, DoAr 21 and HD283447, were mapped in detail and were spatially resolved, with the radio emission coming from regions 5 to 15 stellar diameters in size. In these cases, the size of the emitting region (but not necessarily that of the magnetic structure) also varies on time scales of days or hours. The radio luminosities are of order 10^{16} erg s^{-1} Hz^{-1}, or 10^5 times more powerful than the strongest contemporary solar flares.

These direct imaging results therefore confirm inferences from earlier nonimaging variability characteristics of WTTS radio emission, which, as mentioned in Sec. II.C, suggested radio emitting regions between one and several stellar radii in extent on the basis of models of large solar magnetic loops.

C. Morphology

In spite of these similarities in overall size, it is not clear at present whether the magnetic structures observed around these (late-type) WTTS are of the same nature as the magnetosphere surrounding the earlier-type (B3) star S1. An important difference is that the radio emission appears highly variable

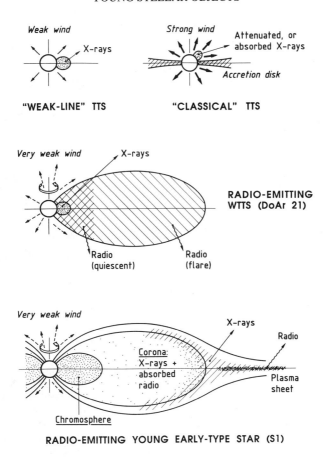

Figure 2.   Sketch (to scale) of the magnetic structures of weak-line and classical
T Tauri stars, as deduced from their X-ray and radio (cm) properties (updated from
Montmerle and André 1988).

on WTTS but nearly stable on S1. The view held by Montmerle and André
(1988) was that the morphology of the magnetic field is different (for instance
a large dipolar loop in the case of WTTS, a dipolar magnetosphere in the
case of S1; see Fig. 2), and that variability in WTTS was due to injection and
acceleration of electrons basically solar-like in origin, while the constancy of
the radio flux of S1 was due both to an axisymmetric magnetosphere and to a
steady injection of electrons from outside, in a fashion similar to that of the
giant planets (see André et al. 1988 for details). More recent work suggests
that the magnetic field configurations may be more similar, based on the
analogy between radio-emitting WTTS and the tidally spun-up, close binaries
known as RS CVn- and Algol-type systems. These active, late-type post-
main-sequence stars, which also emit (variable) nonthermal radio emission

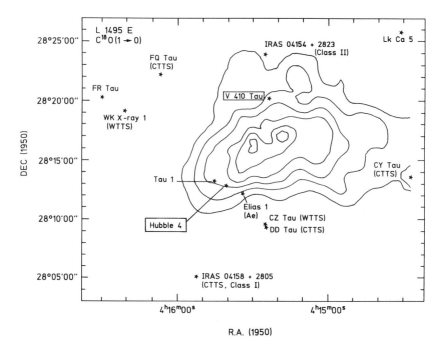

Figure 3.   Map of the gas column density of the molecular concentration in L1495E
in Taurus-Auriga as traced by recent $C^{18}O$ $(1 \rightarrow 0)$ observations made at the NRAO
12-m telescope (André et al., in preparation). The PMS stars and bright infrared
sources known in this region are superimposed. All have been observed at the VLA,
but the only clear detections ($S_{6cm} > 1$ mJy; framed on the figures) are the WTTS
V410 Tau, Hubble 4, HD283447, and Anon 1, all of which are nonthermal emitters
(see O'Neal et al. 1990). ρ Oph cloud core region (André et al. 1987), and must
therefore be very young, in spite of the absence of evidence for the circumstellar
material generally typical of young stellar objects.

of order 10^{16} erg s^{-1} Hz^{-1}, were in fact the first stars detected by VLBI
(see Lestrade 1988 for a review). The most recent model of the numerous
circular-polarization and VLBI measurements made on RS CVns calls for a
stable, large-scale and axisymmetric magnetosphere associated with a single
active component (Morris et al. 1990). It is therefore not excluded that this
model, which presents many similarities with the one proposed for S1, may,
in essence, apply to radio-bright WTTS as well.

In any case, the main conclusion which can be drawn from the recent
VLBI studies described in Sec. III.B is that the extent of the magnetic fields
surrounding nonthermal young radio stars is much larger than on the Sun and
reaches several stellar radii. In particular, this excludes an alternative model,
initially proposed for RS CVn radio flares, and which involves magnetic fields
erupting through the stellar surface rather than large magnetic structures (Mullan 1985). By contrast, the derived strength of the surface magnetic field (B_*

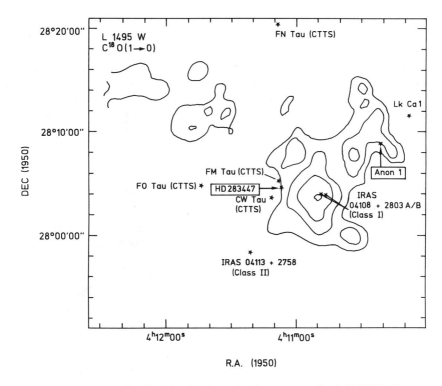

Figure 4.   Same as for Fig. 3 but for the molecular concentration L1495W in Taurus-Auriga.

between a few 100 G and a few kG), which agrees well with the first direct measurement recently obtained by Basri and Marcy (1991) on a WTTS ($<B>$ = 1700 ± 500 G), does not appear to be much higher than the peak values measured on the Sun. There is an important difference with the Sun: the "filling factor" for strong magnetic fields is much larger, a result consistent with that derived from X-ray flares, and optical and ultraviolet starspot observations (see above, Secs. II.C and D). In other words, the magnetic field of young radio stars is not significantly stronger but, at least in the case of the large radio-emitting loops, is probably much more organized than the tangled fields observed in solar active regions.

It seems that only a fraction of WTTS possess these large magnetic structures. Nonthermal radio emission has been confidently detected at levels 3×10^{16} erg s^{-1} Hz^{-1} in only $\sim$10% of WTTS (O'Neal et al. 1990). In the ρ Ophiuchi cloud, up to half of the embedded YSOs emit at levels $\sim$2 $\times$ 10^{15} erg s^{-1} Hz^{-1}, but it is unlikely that the emission is nonthermal in all these cases (Leous et al. 1991). (As for CTTS and related embedded objects, their partially ionized, massive circumstellar material may make their "peristellar" magnetic field inaccessible in the radio; see, e.g., Montmerle and André

1988.) The radio-emitting, WTTS-like objects therefore make up a specific, and apparently quite rare, population of YSOs, precisely distinguished by the large extent of their closed magnetic structures. These nonthermal radio YSOs display another remarkable feature: they are probably younger on average than nondetected WTTS, and perhaps even younger than some CTTS (see, e.g., O'Neal et al. 1990). In particular, both in ρ Ophiuchi (André et al. 1987) and in Taurus (O'Neal et al. 1990), as shown in Figs. 3 and 4, they appear to be concentrated at the edge of the highest column-density regions, in contrast to WTTS and CTTS in general. We shall return to this point in Sec. IV.D.

D. Origin of Magnetism

As on the Sun, the origin of the magnetism is widely thought to be the dynamo mechanism, in which seed magnetic fields are amplified by the combined action of convection and differential rotation (the so-called $\alpha\omega$ mechanism; see, e.g., Gilman 1983). Unfortunately, dynamo theory is not developed well enough in stars to lead to any quantitative prediction. In particular, it is not clear why the dynamo mechanism should operate in fully convective stars such as T Tauri stars: in principle, the amplification of the magnetic field requires it not to escape owing to excessive buoyancy, which implies that the field must be somehow anchored to the bottom of the convective zone—which does not exist in fully convective stars. In the most elementary scenario, one simply expects some correlation between the level of magnetic activity and rotation (itself somehow correlated with differential rotation), both being defined in a broad sense. Because, in the case of embedded YSOs, there is currently no observational access to rotation (such as could be given by high-resolution near-infrared lines), the existing evidence for a dynamo mechanism comes only from observations of optically visible PMS stars.

In these stars, different authors use different parameters to quantify the "magnetic activity": it can be the X-ray luminosity, or chromospheric activity tracers such as the MgII H + K or CaII H + K line fluxes (see, e.g., Rutten and Schrijver 1987; Bouvier 1990). "Rotation," on the other hand, is also ill defined: it can be the projected rotational velocity ($v \sin i$, with an embarassing unknown geometrical factor), the rotational period (which can be directly measured from rotational modulation, see Sec. II.D), or even the so-called "Rossby number" (see, e.g., Mangeney and Praderie 1984; Hartmann and Noyes 1987), which is a mixture of rotation and convection (and equal to the ratio of the observed rotation period and of the computed convective turnover time). As it turns out, however, all combinations tend to yield the same rather comforting trend, namely a decreasing magnetic activity with an increasing rotation period, as expected from dynamo theory (for a more extensive discussion, see e.g., Montmerle 1987). In addition to the properties of dark spots covering the surface of TTS, the most direct evidence for dynamo-generated magnetic fields comes from the correlation between X-ray luminosity and rotational period found for TTS (Bouvier 1990). This correlation is similar to that existing in more evolved, magnetically active stars, such as RS CVn systems, and reinforces the conclusion that surface magnetic fields in TTS, CTTS and WTTS alike, are of the order of a few 10^3 G. Other activity diagnostics, such as the MgII and CaII lines, are more ambiguous. In WTTS, these lines are weak and narrow, and the line flux is consistent with a purely chromospheric origin. The broad emission-line profiles of CTTS such as Hα, however, cannot be reproduced by chromospheric models, and the measured line flux is far too large to be accounted for by a solar-type dynamo mechanism (Calvet and Albarrán 1984; Bouvier 1990). Other nonsolar mechanisms that might be at work in TTS are discussed in the next section.

Considering young stars of earlier spectral types may yield some additional clues on the dynamo mechanism. The PMS counterparts of A stars known as the Herbig Ae-Be stars have masses just above those of T Tauri stars, i.e., $\sim$2 M$_\odot$ and up (see, e.g., Catala 1989). According to the standard models, stars of such masses are already entirely radiative in their PMS phase. Yet the large sample of X-ray-discovered young intermediate mass stars found by Strom et al. (1990) in L1641 and by Caillault and Zoonematkermani (1989) in Orion shows that these stars share many X-ray properties of the T Tauri stars and give a puzzling evidence for magnetic activity. One possible interpretation is that all have invisible, X-ray emitting low-mass companions, as is often found for main-sequence A stars. A perhaps more exciting possibility is suggested by very recent work by Palla and Stahler (1991) on PMS evolution of accreting stars in the $\sim$1 to 8 M$_\odot$ range (see also the Chapter by Stahler and Walter; Palla 1991). For a mass $\geq$2.5 M$_\odot$, deuterium burns in a subsurface shell, and may drive an outer convective zone. It is therefore conceivable that the dynamo mechanism may operate in such particular conditions and give rise to the observed magnetic activity traced by the X-rays.

As for stars of earlier types like S1 (and for the large fraction of all B stars which are magnetic "Bp" stars), the dynamo mechanism probably cannot be invoked, because these stars most likely lack a sizable outer convective zone. The youth of S1 ($\sim$ a few thousand years only, based on the existence of a small H II region associated with it; see André et al. 1988) rather suggests that its dipolar magnetic field is a fossil from its formation stages. At any rate, the S1 case should help understand the origin of magnetic fields of B stars in general, which is unclear at present (see Borra et al. 1982).

IV. NONSOLAR ACTIVITY AND CIRCUMSTELLAR MATERIAL

A. Mass-Loss Phenomena in Pre-Main-Sequence Stars

As mentioned above, there are several indications that strong stellar winds are associated with CTTS. The standard interpretation is that they are high-temperature, fully ionized winds, possibly accelerated by Alfvén waves generated in the outer convective zone (Lago 1984; Hartmann et al. 1982). However, it is very difficult to account in particular for the strength of Hα in such winds (see DeCampli 1981), unless one makes particular assumptions on the expansion velocity field (Hartmann et al. 1990).

A new approach suggests that these winds are in fact *cold* ($T_w \sim$ a few 10^3 K) and only partially ionized (Natta and Giovanardi 1990; see the review by Natta 1989). The actual mass-loss rate would then be the radio rate multiplied by a factor 1/[ionization degree], i.e., by at least an order of magnitude. Although the cold wind model depends on some unmeasured parameters, there is some direct observational support for the existence of "neutral" extended winds with the detection of H I moving at several tens of km s^{-1} near a few well-known embedded objects, like HH 7–11 and L1551 (Lizano et al. 1988; Natta 1989). The associated crossing times (= size/flow velocity) are $\sim$ several 10^3 yr, pointing to very young exciting sources. Additional, indirect support comes from the consideration of the energetics of CO outflows. Indeed, these large-scale phenomena, extending over several 10^4 AU, imply high mass-loss rates (up to several 10^{-6} M$_\odot$ yr^{-1}, but there are large uncertainties on the published figures, Cabrit and Bertout 1990), much higher than those computed for fully ionized stellar winds, (see, e.g., Evans et al. 1987). One can causally link molecular outflows and stellar winds if a momentum-conserving shock transfers the cold material from one to the other (Natta 1989). The driving mechanism for such high mass loss is still controversial; we will return to this question in Sec. VI.

B. Accretion Disks Around Young Stellar Objects

In the case of CTTS, other activity tracers, like the ultraviolet emission lines, and perhaps more conspicuously Hα (which cannot, as we have seen, be easily explained by stellar winds), are in excess by up to 2 orders of magnitude with respect to the other tracers as a function of rotation (Bouvier 1990; see Fig. 5). By contrast, for WTTS, Hα for instance (which has a few Å equivalent width at most), seems to be entirely of solar-like (chromospheric) origin. In other words, there is clear evidence in CTTS that an intense *nonsolar* activity exists. As we shall now discuss, the current paradigm is that the reason for this does not lie in the stars themselves, but in the presence of close circumstellar material, and that there is in fact a continuum across the YSO classes (including TTS).

Let us begin with the YSOs for which the most advanced models are available, the CTTS. Current models (see Bertout et al. 1988; Hartmann and Kenyon 1987*a*, *b*; Chapter by Basri and Bertout) depict a CTTS as a complex object comprising three distinct components: a late-type star; a circumstellar disk; a boundary layer between the star and the disk.

The existence of a Keplerian circumstellar disk of gas and dust has first been inferred from the infrared excess. The idea is that this excess comes from the existence of an additional energy source, which the star itself is unable to provide. The introduction of a disk is natural: it is the logical outcome of the gravitational collapse of rotating structures like the dense "ammonia cores" (Myers et al. 1987), half of which contain infrared sources thought to be protostars. It is worthy of note that this idea has been around since Kant and Laplace in the late 18th century. Laplace (1796) remarked that the coplanar

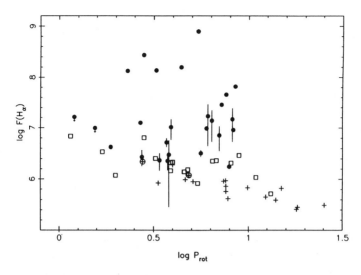

Figure 5.   Relation between the Hα emission line flux $F(\text{H}\alpha)$ (in erg s^{-1} cm^{-2}) and the rotation period P (in days) deduced from the modulation of the stellar light by large starspots for a sample of T Tauri stars, as compared with other magnetically active stars (Bouvier 1990). The excess Hα emission of TTS over the other stars reaches 2 orders of magnitude. Symbols: • = TTS, + = late-type dwarfs, ⊕ = dMe stars, □ = RCVn close binary systems.

orbits of the main bodies of the solar system implied the past existence of a flat, rotating "primitive nebula" out of which planets would form.

To first order, the infrared excess can be fitted by a continuous blackbody temperature distribution as a function of radius, implying in turn a continuous, optically thick matter distribution. This has a fundamental consequence: because the Keplerian rotation velocities are different at r and at $r + dr$, there must be some shear between successive rings, hence viscous stresses and heating. The energy radiated in the process at r must exactly balance the input of potential energy from $r + dr$ to r. But this also means that matter must slowly spiral inwards, and eventually fall onto the star. In other words, any spatially continuous disk losing energy by viscous processes must be an *accretion disk*. Because viscosity is a poorly known process on a microscopic scale (especially under these astrophysical conditions), the temperatures and densities used in modeling the disks are generally taken as power laws as a function of radius, resulting in a power-law for the infrared excess, which in many cases gives a satisfactory fit to the observational data. In spatially thin, optically thick viscous Keplerian disks, for instance, the temperature decreases with radius as $r^{-3/4}$, and the density as r^{-2}; the resulting near-infrared spectrum λF_λ has a slope $-4/3$. The radiated infrared luminosity is $\sim 1/2$ of the available gravitational energy, and is proportional to the accretion rate (see, e.g., Lynden-Bell and Pringle 1974; Pringle 1988). Typical accretion rates computed from such models are $\sim$ a few 10^{-7} $M_\odot$ yr^{-1}. The near-infrared ($\sim$2 μm) traces warm material ($\sim$1500 K) relatively close to the star ($\leq$0.1 AU); longer infrared wavelengths (up to the longest IRAS band, i.e., $\sim$10–100 μm) trace cooler material (a few 10–100 K) and correspond to a few AU. Because these dimensions are small with respect to the overall size of the disk (which possibly extends up to 1000 AU or more; see below, Sec. V.A), and because the disk is optically thick, it is clear that the infrared is essentially a temperature tracer, not a mass tracer; to trace the mass, one has to go to longer wavelengths, where, in addition, there is mounting direct evidence for the existence of extended dust disks (mm range, where they are mostly optically thin: see Sec. V.B).

C. Boundary Layers

Close to the star, an additional, very important phenomenon must occur. At $r \simeq R_*$, the Keplerian velocity in the disk around a $\sim$1 $M_\odot$ star is $\sim$250 km s^{-1}, which is also (by definition) the break-up velocity of this star. But rotation velocity measurements (like the abovementioned starspot analysis; Sec. II.C) show that, contrary to early expectations, T Tauri stars (in general) are slow rotators, with equatorial velocities in the range $\sim$20 km s^{-1} or even less, implying that they have very quickly lost most of their initial angular momentum. As a result, near the equator, the star rotates $\sim$10 times more slowly than the disk. Then a *boundary layer* (a well-known phenomenon in aerodynamics) must develop, to ensure a continuous transition of the fluid velocities between the disk and the star. However, again because viscosity is the central dissipative mechanism and is poorly known, the physical conditions within the boundary layer, its dimensions, etc. are also poorly known and are the subject of active research (see, e.g., Basri and Bertout 1989; Bertout et al. 1991*b*; for basic principles, see, e.g., Frank et al. 1985). The condition of energy balance of the accretion disk shows that the other $\sim$1/2 of the gravitational energy lost during the inward motion of the material must be radiated at the level of the boundary layer. The way in which this energy is radiated will of course depend on the opacity. Several arguments lead us to think that it must be optically thick (Bertout et al. 1988); some of the emission, like the Balmer jump, must however arise from optically thin regions (Basri and Bertout 1989). The temperatures reached are, much like in H II regions (balance between *in situ* heating and cooling), $\sim$8000 K. The bulk of the radiation will therefore belong to the optical and near-ultraviolet domains, and this is why it is believed that the observed emission lines (and in particular Hα) and the ultraviolet excess originate in the boundary layer. No definitive calculations on this point exist yet, because these require a knowledge of the physical conditions within the boundary layer which we do not have at present. But this explains qualitatively nicely the correlation between the infrared and ultraviolet excesses and the strength of the emission lines. In some models (see Bertout 1989) the combined contribution to the overall spectrum of the boundary layer (ultraviolet range) and the disk (infrared range) is so high that it can completely dominate the photospheric contribution (optical range): from the point of view of spectral emission properties, the star light is then completely dominated by the disk.

D. What about Weak-Line T Tauri Stars?

At the other extreme, the implication is that because they have no ultraviolet or infrared excess, and little, if any, Hα emission, the WTTS should not be surrounded by circumstellar material, but simply represent evolved, active, solar-like PMS stars. However, many young stars, although visible in the near-infrared (see, e.g., Wilking et al. 1989a), display no infrared excess: this is the case of the radio-emitting ones found by André et al. (1987) in the ρ Oph cloud, and by O'Neal et al. (1990) in the Taurus-Auriga clouds. In spite of the absence of infrared excess, these stars are probably very young because they are located just at the edge of dense cores (see Fig. 6). Strictly speaking, however, no age can be attributed to most of them because the majority are not seen in the optical and hence cannot be placed on an HR diagram. According to the infrared classification, they are Class III objects. So we are faced with a problem: how is it that very young stars, still embedded in the vicinity of cloud cores, apparently do not show evidence for circumstellar material?

A related problem exists more generally when considering optically visible CTTS and WTTS as a whole. Indeed, when put on an HR diagram, these two classes of PMS stars appear mixed (see discussion in Sec. V.B), whereas, according to "standard" evolutionary models (see, e.g., Adams et al. 1987), CTTS, being surrounded by accretion disks, should be younger, and WTTS, deprived of such disks, should represent a more advanced evolutionary stage.

One probable answer to this problem lies in the fact that the "standard" scenario of early stellar evolution is based on infrared data, which, as mentioned above, traces only warm material comparatively close to the star ($\lesssim 1$ AU). Outer ($\gtrsim 10$ AU), cold ($<< 100$ K) material is visible most conveniently in the mm range, and we shall discuss in the next section how recent work in this range has given new insight into the presence of circumstellar matter around young stars, and into the possible role of magnetic fields (see Montmerle and André 1989). However, other properties, more intrinsic to the stars themselves, may also play a role, like rapid rotation or binarity (see discussion in Montmerle and André 1988). At any rate, an important conclusion is that the near-infrared data alone are not sufficient to qualify the evolutionary state of YSOs.

V. NEW DEVELOPMENTS: MILLIMETER DATA AND EVOLUTIONARY IMPLICATIONS

A. Observational Results

The first evidence for a flattened molecular structure around a YSO has been the detection by Beckwith et al. (1986) of strong emission near 2.6 mm from the extreme TTS HL Tau, both in continuum (dust) and in the ^{12}CO (1 → 0) line. Because these observations were performed with an interferometer (at Owens Valley, California) yielding maps with an angular resolution of $\sim$6″, the fact that the continuum source was unresolved implied a size <1000 AU, whereas the resolved $\sim$30″ size of the elongated ^{12}CO source implied a size $\sim$4000 AU. Later ^{13}CO observations with the same instrument led Sargent and Beckwith (1987) to the conclusion that this structure was consistent with a circumstellar disk of gas and dust in Keplerian rotation. We note, however, that current models linking near-infrared emission with the presence of an accretion disk require only the inner disk (e.g., < 1000 AU) to be Keplerian. The outer gaseous disk may simply be a distinct structure orbiting the star, up to large distances (several 10,000 AU; see, e.g., Rodríguez 1988).

More extensive mm studies of several tens of T Tauri stars have recently been reported, as new, large single-dish telescopes and sensitive receivers (bolometers) came into operation. With the James Clerk Maxwell telescope in Hawaii, operating in atmospheric windows from 0.35 to 1.1 mm, Weintraub et al. (1989) and Adams et al. (1990) selected small samples comprising almost exclusively CTTS, and found a very high detection rate. Beckwith et al. (1990) used the IRAM 30-m radio telescope near Granada, Spain, equipped with the MPIfR (Max-Planck-Institut für Radioastronomie) bolometer operating at 1.3 mm, to study a much larger sample of 86 CTTS and WTTS stars in Taurus-Auriga: 53% of the CTTS, and 29% of the WTTS were detected. André et al. (1990b) have independently undertaken a study similar to that of Beckwith et al. (1990) in the ρ Oph core using the IRAM 30-m telescope, but selecting a sample of embedded sources spanning a broader range of infrared properties than CTTS and WTTS from Class I to Class III, with a very good sensitivity (down to a detection level of $\sim$5 mJy). Out of 18 such objects, 14 were detected, belonging to all classes and spanning a large range in fluxes, from $\sim$30 to $\sim$400 mJy. In addition, the sample included two exceptional objects exciting molecular outflows and believed to be extremely young (a few 10^3 yr), IRAS 16293 (Walker et al. 1988) and VLA 1623 (André et al. 1990a); both are very bright at mm wavelengths (resp. 5 and 0.9 Jy in the 30-m beam).

While the high detection rate of Weintraub et al. (1989) and Adams et al. (1990) was not unexpected, because there was already from infrared data strong independent evidence for circumstellar material around their program stars, the results of Beckwith et al. (1990) are more surprising. Indeed, a significant fraction of the WTTS turn out to be surrounded by cold material, whereas there are indications that no or little warm material is present. Indeed,

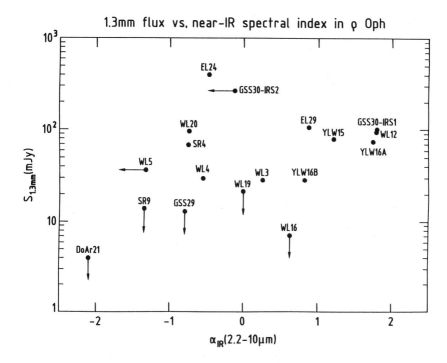

Figure 6.   Scatter diagram of the 1.3 mm flux density vs the near-infrared spectral index (slope of the λF_λ spectrum between 2.2 μm and 10 μm) for a selected sample of sources in the ρ Oph cloud (André et al. 1990b). Class I, Class II and Class III infrared sources lie to the right, center, and left parts of this diagram, respectively. There is no obvious correlation between the mm flux (which traces optically thin, cold dust, at several AU from the central star) and the near-infrared spectral energy distribution (which traces the optically thick, warm material within 1 AU).

André et al. (1990b) find a lack of obvious correlation between the mm emission and near-infrared excess of the objects in the sample. In particular, as shown on Fig. 5, heavily obscured (Class I) objects do not appear brighter at 1.3 mm than CTTS (Class II) in general. On the other hand, there is one case of an object without near-infrared excess (Class III) showing a mm emission. Although the samples of Beckwith et al. (1990) and André et al. (1990b) are qualitatively different, the two results are quite consistent with each other. In the framework of the disk interpretation of the emission of YSOs at long wavelengths, this suggests that, in several cases, these disks are not continuous. More precisely, Beckwith et al. (1990) found that a sizable fraction ($\sim$25%) of the stars classified as WTTS have substantial mm emission ($>3\sigma$ detections, i.e., above $\sim$20 mJy). Now a distinction may be made here between the WTTS and NTTS using the study by Skrutskie et al. (1989) of K (2 μm) and N (10 μm) excesses of various classes of TTS. These authors find that a number of WTTS stars have optically thin near-infrared emission ($\Delta K < 0.2$ and $\Delta N < 0.8$): in the current definition, these stars would constitute the "naked" subsample of the WTTS. In addition, one NTTS (WK X-ray 1) is detected, hence is not, strictly speaking, "naked." Other such stars might be detectable at lower mm fluxes. Even though they are not numerous, these stars are particularly interesting in that the combination of optically thin near-infrared emission and detectable mm emission suggests that they are surrounded by hollow, "45 rpm" disks. Such a picture is consistent with the absence of ultraviolet excess or of significant Hα emission, since this implies the absence of a boundary layer, in other words that there is indeed no physical contact between the star and the disk. The optically thin near-infrared emission of these stars can be attributed to a small amount of warm dust in the inner "hole." On the other hand, CTTS with weak or absent mm emission ($\sim 50\%$) could be surrounded by much less massive dust disks, perhaps already partially condensed in large grains or protoplanets.

The mm flux density gives the total mass (H_2 + dust) of the detected disks, irrespective of their structure. As discussed by André et al. (1990b) and Beckwith et al. (1990), because of uncertainties in the dust emissivity at mm wavelengths, this mass is uncertain. It is found to lie within the range ~ 0.001 to ~ 0.5 $M_\odot$, which, interestingly, encompasses the estimated mass of the primitive solar nebula, ~ 0.005 to 0.1 $M_\odot$ (see, e.g., Cameron 1988). There is no evidence that any of the detected T Tauri stars is resolved within the $\sim 12''$ resolution of the 30-m observations, but there is evidence that embedded objects like VLA 1623 or IRAS 16293 are slightly extended. This implies that in all cases most of the mass of such dust disks must be comprised within radii of 1000 AU at most.

B. Evolutionary Consequences

The consequences of the new picture presented above are important for the evolution of low-mass YSOs. Indeed, it is now well established that WTTS and CTTS co-exist in large regions of the HR diagram, corresponding to ages from $\sim 3 \times 10^5$ yr to $\sim 3 \times 10^7$ yr (Strom et al. 1989; see also Skrutskie et al. 1989; Walter et al. 1988; Chapter by Strom et al.). One may argue that, precisely because of the presence of the disks, the positions of CTTS in the HR diagram are uncertain (Kenyon and Hartmann 1990); still, this cannot seriously affect the fact that the range of ages is very large. As Strom et al. (1989) point out, on average, CTTS tend to be younger than WTTS, but some WTTS are very young ($< 3 \times 10^5$ yr; the nonthermal radio-emitting WTTS or Class III sources of ρ Oph and Taurus-Auriga, as discussed above, are also probably very young), and some CTTS are very old ($> 3 \times 10^7$ yr), as are stars surrounded by conspicuous disks like β Pic (A star, $\geq$ several 10^6 yr; see, e.g., Smith and Terrile 1984; Lagrange-Henri et al. 1989).

Therefore, the transition between CTTS and WTTS occurs at different times from star to star, somewhere between $< 3 \times 10^5$ and $> 3 \times 10^7$ yr. In our framework, this transition is linked not to the disappearance of the disk, but *to the disappearance of the boundary layer.* Its duration is estimated at $<10^6$ yr (perhaps as short as a few 10^5 yr), compatible with the age of the younger CTTS and short compared to the age of the older WTTS (Walter et al. 1988, Skrutskie et al. 1989). The WTTS having detectable mm emission would then be objects undergoing this transition, while the mm-emitting NTTS, having a central hole devoid of dust, would represent the final stage. Based on the frequency of these stars (1 case, i.e., $\sim10\%$ at face value), this last stage must last a short time: a time scale of a few 10^5 yr can be estimated, consistent with the above figures.

This means that the disk evolution of YSOs must be decoupled from the stellar evolution. Comparable stars (that is, identically located on the HR diagram) have different disks: continuous "accretion" disks, or more or less "hollow" ones (WTTS). At least to a first approximation, a star of a given initial mass evolves according to standard laws of stellar structure, but some additional factors make the disks evolve differently for a given star, some quickly, some slowly, with a comparatively abrupt transition between CTTS and WTTS. This is a complex subject, because the problem of disk evolution *per se* is still in its infancy. In addition to the fact that, after all, what we call a "disk" is in fact still very poorly known observationally, probably important factors which must be incorporated in any theory include:

1. Stellar or *in situ* generated magnetic fields;
2. Spatial evolution: appearance of a central cavity, perhaps even of rings, etc. (possibly under the influence of the stellar magnetic field, see below), or disappearance of the outer regions; change in thickness;
3. Dust grain evolution and possible condensation or evaporation phenomena, presumably (but not necessarily) leading to planet formation;
4. Influence of mass loss (likely also from the disk itself; Sec. VI and possible sporadic bipolar flow phases, or intense winds, as is the case for FU Orionis events (in which the mass-loss rate is typically 100 times stronger than in CTTS: Herbig 1989*a*; Kenyon and Hartmann 1988; Chapter by Hartmann et al.);
5. Appropriate initial conditions (in particular initial angular momentum).

Of course, binarity (Zinnecker 1989*a*) may also play an important role on disk evolution as a result of tidal effects, in the case of a sufficiently tight system (separation $\lesssim 100$ AU). Impressive results on finding binary or multiple systems with angular separations down to a milliarcsecond scale have already been obtained (Simon et al. 1987,1991), and will likely also play an important role in the future.

C. Disk Stability and Magnetic Fields

One recent avenue is that of the intrinsic stability of the accretion disks around single stars, and some promising theoretical developments have recently appeared. Adams et al. (1989) and Tagger et al. (1990) have shown that spatially thin disks (scale height $<<$ size) are subject to efficient dynamical instabilities ("swing amplification phenomenon"; see also Chapter by Adams and Lin). Supplementing the initial work of Papaloizou and Pringle (1987) by including magnetic fields (here assumed to be locally perpendicular to the disk) in addition to gravitation, Tagger et al. (1990) have shown that, depending on the respective strength of the magnetic field and the local gravity, several instabilities are possible. (The magnetic field may be of stellar origin, or produced locally by some kind of dynamo resulting from differential rotation.)

If the magnetic field dominates gravity, these instabilities may result in a cavity in the inner regions of the disk and thus explain the existence of hollow disks, although it is premature to say whether they will explain the transition from a regular accretion disk to a hollow disk and the subsequent disappearance of the boundary layer. Support for this situation may be found in the existence of generally large magnetic structures around the nonthermal radio sources; we also note that the widespread X-ray activity of WTTS is indicative of strong surface magnetic fields (see Montmerle and André 1988; Sec. III). Farther out, other instabilities are possible: if the magnetic field is negligible and the self-gravity small, the hydrodynamical instabilities of Papaloizou and Pringle (1987) occur, and if self-gravity is large enough, the spiral instabilities suggested to be present in galactic disks take over (Pellat et al. 1990). Unfortunately, information about the magnetic vector orientation cannot be gained through the interpretation of dust polarization observations (Bastien and Ménard 1990). Work is in progress to determine the time scales over which such instabilities develop in circumstellar disks, but obviously such considerations leave ample room for many possible, time-dependent configurations (including spiral arms in the outer regions).

VI. CONCLUSIONS

There is today no doubt about the presence of warm and cold circumstellar material around YSOs. Its exact morphology is in general not known, but the existing evidence supports, albeit indirectly, the idea that this material should be close to the central star in the form of a flattened, disk-like structure, perhaps as small as a few 100 AU in radius or less. The disk masses are highly uncertain, but still roughly within one order of magnitude both ways of the mass of the primitive solar nebula. This recent image therefore essentially confirms what had been guessed, or speculated, over more than two centuries.

But what is perhaps more surprising is that many YSOs *do not* have continuous disks around them. Would we then be witnessing some fragmentation of the initial, supposedly continuous disks, in other words, the birth of planetary systems? It is certainly premature to answer this question, because we have yet to elucidate many aspects of the stability of circumstellar disks, and in particular the role of magnetic fields. In the above discussion, we have shown ample evidence of their presence on YSOs, and we have mentioned the possible role extended stellar magnetic fields may play.

However, magnetic fields probably also play a key role in another important and related aspect of the circumstellar environment of YSOs, namely the
mass loss associated with CTTS or embedded molecular flow sources. An
assumption common to all existing models is that open magnetic field lines
confine the outflowing material in a bipolar fashion; we have seen that a number of arguments point to the possibility that the stellar wind is in fact mostly
neutral even close to the star, and becomes molecular farther out. Because
the implied mass-loss rate is very high (possibly up to a few 10^{-6} $M_\odot$ yr^{-1}),
it is very difficult to account for it in terms of a purely stellar mechanism.
For instance, Alfvén waves, generated by convection at the stellar surface, act
only on the ionized part of the wind, but likely cannot, even through friction
between ions and neutral atoms, lift the remaining neutral part. This is why
current models draw the mass lost through the wind from the accretion disk
itself, much in the way the disk is also invoked to explain the extra ultraviolet
and infrared luminosity. In other words, we are led to the concept that *accretion drives mass loss.* The general scheme is that accretion is equatorial (disk),
and mass ejection is polar (jets) or, at least, confined to a relatively small angle
to the polar axis of the star. And what about the energetics? In many cases
(see, e.g., Lada 1985), molecular outflows have a kinetic "luminosity" which
approaches, or even exceeds, that of the exciting source. Again, the trick is
to tap this energy from the ultimate reservoir, namely gravity, by using the
rotation of the star, or of the disk.

In one school of thought (Shu et al. 1988), the central star is strongly
magnetized and rotates near break-up; the disk material flows inwards, touches
the stellar equator at the critical, "X point," and then "rebounds" in open
magnetic field lines bending away from the disk (so-called "X-celerator"
mechanism). Because this magnetic field is anchored to the star, the ex-
disk material is dragged along the lines by the centrifugal force and follows
their direction, increasingly parallel to the rotation axis. Obviously, given
the fact that no YSO has yet been observed to rotate close to the break-up
velocity, this model could be applied only to the youngest stellar objects,
before their initial angular momentum is lost. An alternative would be that
the inner disk is kept in corotation with the star by the magnetic field, so that
the X-celerator mechanism takes place farther out, at the corotation radius,
where the Keplerian velocity is much smaller. This, however, requires large
magnetic fields at the surface, like in the case of S1 in ρ Oph (see Sec. II.D).
Deeply embedded sources with outflows having very short dynamical time
scales (like IRAS 16293 and VLA 1623 in ρ Oph) might be in this case,
but none of these sources shows at present evidence for strong or extended
magnetic fields, as would be indicated by nonthermal radio emission.

A second school of thought assumes that the accretion disk is itself
entirely responsible for the mass loss: the disk "evaporates," its material
flowing along magnetic field lines this time anchored to the disk (Uchida and
Shibata 1985; Pudritz and Norman 1986; Königl 1987; Kwan and Tademaru
1988; Camenzind 1991; see Chapter by Königl and Ruden). Here again

the centrifugal force plays a role, but so does the shear and spiraling of the magnetic field lines, following the shear of the disk material. The bending of the field lines is more efficient in this case (they even join at some distance from the star along the polar axis), and it seems that the collimated jets are easier to explain in this framework. In addition, the central star plays no other role than to set up a gravitational potential well holding the accretion disk together, thus removing the stellar rotation velocity constraint. Other arguments, for instance based on forbidden optical lines (see Cabrit et al. 1990), tend to support a mass loss directly from the disk, at least in the case of CTTS.

An interesting consequence is that, in the X-celerator mechanism, a physical contact is required between the star and the disk (which is not a boundary layer, since the star and the disk are assumed to rotate at the same velocity). Thus, irrespective of the stellar rotation, this mechanism could not exist for WTTS surrounded by a hollow disk, whereas the "evaporation" mechanism could conceivably still work: a possible test would be the existence of cold outflows from these stars, for which there is no evidence up to now.

Whatever the (significant) differences between these models, a unified picture of YSOs tends to emerge. At an early stage, the initial protostar, at the center of a rotating spheroidal shell of gas and dust, evolves into a star surrounded by a (possibly, but not necessarily massive) accretion disk. Then, by means of the rotation of either the star or the disk, and because of the presence of stellar and/or disk magnetic fields, an intense, mostly neutral wind is accelerated and collimated. This wind gives rise to bipolar jets and molecular outflows farther away, and disperses what is left of the protostellar shell. Probably because of this intense magnetic field-driven mass loss, the central star slows down quickly, and a boundary layer appears, giving rise to a CTTS if optically visible. Under the influence of these magnetic fields, gravitational instabilities occur, and at some point the disk breaks, forming for instance rings or spiral arms. If only the outer part breaks, we have CTTS with near-infrared excess but no mm emission; if, on the contrary, a central cavity forms, the boundary layer disappears, leading to a WTTS with little, if any, near-infrared excess but surrounded by cold, mm emitting material. This transition may take place at any time (depending on as yet unidentified conditions, as discussed in Sec. V.C) between a few 10^5 yr and a few 10^7 yr, and is rather fast ($<10^6$ yr). Furthermore, the transition is not necessarily unique: if it is confirmed that there is a strong causal link between accretion disks and outflow phenomena, the increasing evidence for structure in bipolar flows, e.g., "blobs" (André et al. 1990a) or "bullets" (Bachiller et al. 1990a) suggests that accretion disks may be in contact with the stars several times, perhaps as a result of instabilities. The lifetime of the outer material is not known, but at some point between $\gtrsim 10^5$ and $\gtrsim 10^7$ yr all the remaining disk material rapidly disappears, leading to a truly naked TTS on its way to the main sequence. One reason for this disappearance may be, of course, the formation of planetary systems.

Admittedly, such a scheme is still qualitative, and even (in part) speculative. But it provides at least a stimulating line for future work, and the advent of many new instruments operating at many wavelengths, both on the ground and in space, as well as of improved theoretical tools (plasma physics, etc.), should help us very soon to give a better answer to the ultimate question: where do the Sun and the solar system come from?

PART IV
Disk Processes and Planetary Matter

TRANSPORT PROCESSES AND THE EVOLUTION OF DISKS

FRED C. ADAMS
University of Michigan

and

D. N. C. LIN
Lick Observatory

In this chapter, we discuss the evolution of circumstellar disks and review the physical mechanisms which lead to the transport of angular momentum and energy. Circumstellar disks arise naturally during the star-formation process. One observational signature of these disks is the infrared and ultraviolet excesses in the spectra of young stellar objects. Analysis of the observed properties of these disks is important for the investigation of star formation in general and for determining the initial conditions for the formation of the solar system (especially formation of the planets). The evolution of circumstellar disks is largely determined by the efficiency of angular momentum transport. During the formation stages, disk dynamics is greatly affected by the addition of material from the infall-collapse flow. In this phase of evolution, the disk mass is likely to be comparable to the stellar mass; as a result, the self gravity of the disk can excite the growth of nonaxisymmetric perturbations, which can induce angular momentum transport and may also lead to the formation of binary companions if the disk is sufficiently massive. As infall gradually ceases, convectively driven turbulence may play an important role in inducing angular momentum transport. In this case, a disk with mass $M_D \approx 0.02$ $M_\odot$ and a physical extent comparable to that of the solar system will evolve viscously on a time scale of 10^5 to 10^6 yr. Transport instabilities may induce modulations in the mass transfer through the disk. Analysis of observed time dependent variations provides valuable information on the mass and temperature distribution as well as the magnitude of the viscosity. When the mass of the disk is sufficiently depleted, surface heating from the central star can stabilize the disk against convection. In the absence of other sources of viscosity, grains will settle towards the midplane of the disk and the disk will become optically thin. Thereafter, the system evolves towards a naked T Tauri phase. The eventual clearing of the protoplanetary disk may be due to photoevaporation.

I. INTRODUCTION

The purpose of this review is to describe the structure and evolution of disks associated with young stellar objects (YSOs). In order to fully understand these disks, we must begin by considering the current paradigm of star formation (see the Chapter by Shu et al.). In this paradigm, stars form within the core of molecular clouds. These core regions, which are small subcondensations within the much larger molecular clouds, evolve in a quasi-static manner through the process of ambipolar diffusion. In this process, magnetic field lines slowly diffuse outward and the central regions of the core become increasingly centrally concentrated (see Shu 1983; Lizano and Shu 1989). The magnetic contribution to the pressure support decreases with time until the thermal pressure alone supports the core against its self gravity (at least in the central regions). At this point, the core is in an unstable quasi-equilibrium state which constitutes the initial conditions for dynamic collapse.

When a core begins to collapse, a small hydrostatic object (i.e., the protostar) forms at the center of the collapse flow and an accompanying circumstellar disk collects around it. This phase of evolution—*the protostellar phase*—is thus characterized by a central star and disk, surrounded by an infalling envelope of dust and gas. In this phase of evolution, the disk is greatly affected by the infall.

As a protostar evolves, both its mass and luminosity increase; the protostar eventually develops a strong stellar wind which breaks through the infall at the rotational poles of the system and creates a bipolar outflow. During much of this *bi-polar outflow phase* of evolution, the outflow will be well collimated (in angular extent) and *infall* will be taking place over most of the solid angle centered on the star. The outflow gradually widens in angular extent and the visual extinction to the central source gradually decreases. The outflow eventually separates the newly formed star/disk system from its parental core and the object enters *the T Tauri phase* of evolution. The newly revealed star then follows a pre-main-sequence track in the HR diagram and evolves toward the main sequence.

Circumstellar disks play an important role in the process of star formation outlined above. In recent years, compelling observational evidence (cf. the reviews of Appenzeller and Mundt 1989; Bertout 1989; Shu et al. 1987a) has established the presence of disks associated with YSOs, although the exact properties of such disks remain controversial. The available observational evidence (see, e.g., Bertout et al. 1988; Adams et al. 1988,1990; Kenyon and Hartmann 1987; see also the Chapters by Strom, Beckwith and Sargent, Basri and Bertout, Hartmann et al.) indicates that YSO disks may produce significant luminosity ($L_D \sim L_* \sim 1$ $L_\odot$) and may have masses comparable to the stellar mass ($M_D \sim M_* \sim 1$ $M_\odot$). The stability, structure and evolution of these disks are determined by mass, angular momentum and radiation transport processes. In this review, we discuss the possible types of transport processes that can occur in these circumstellar disks. We first review the observational evidence and related theory that determine the basic disk properties (see Sec. II). We then consider the growth of *global* gravitational instabilities in star/disk systems (see Sec. III). These instabilities may lead to accretion through the disk and hence to the observed disk luminosities; in the limit of runaway growth of a single mode, these instabilities may also lead to the formation of a binary companion. Next, we discuss the role of convective instabilities (see

Sec. IV) in circumstellar disks. We then discuss the dynamical evolution of these disks (see Sec. V) and conclude with a summary and a look to the future (see Sec. VI).

II. BASIC DISK PROPERTIES

The basic characteristics of circumstellar disks vary with the stage of evolution. For purposes of this review, we will conceptually divide the evolution of disks into two regimes. The first regime corresponds to the protostellar phase of evolution (see the Chapter by Shu et al.; Shu et al. 1987a), i.e., the phase when the central star/disk system is still deeply embedded within an envelope of infalling gas and dust. The second regime corresponds to the T Tauri phase of evolution where the star/disk system is optically revealed, i.e., the system is essentially isolated from its environment.

A. Protostellar Disks

In the protostellar phase of evolution, the central star/disk system is deeply embedded in an infalling envelope of gas and dust. Essentially all of the intrinsic radiation from the star/disk system is absorbed and re-radiated by dust grains in the envelope. As a result, no direct signature can be measured for these disks. However, we can establish the existence of these disks and estimate their properties through indirect methods. One way to view protostellar evolution is through the $L-A_V$ diagram (Adams 1989,1990), which is roughly analogous to the HR diagram for actual stars. In this diagram (see Fig. 1), protostars evolve toward the upper right as they continually gain mass from the interstellar medium.

As the mass increases, the luminosity increases, and the column density (and hence the visual extinction) decreases (see Shu 1977). Any successful theory of star formation must produce $L-A_V$ diagram tracks that are consistent with observations of protostellar candidates. As shown by the dashed curve in Fig. 1, a purely spherical theory (with no rotation and hence no disk) is overluminous; we can thus conclude that disks must be present in protostellar systems. As shown by the dotted curve in Fig. 1, a theory with rotation and no disk accretion is severely underluminous; we can thus conclude that some disk accretion must occur during the protostellar phase. The solid curves in Fig. 1 are calculated on the basis of the current picture of star formation (see the Chapter by Shu et al.); the amount of disk accretion is determined by the gravitational stability of the disk (see Sec. III below; Shu et al. 1990; Adams 1990).

The comparison of theoretical tracks in the $L-A_V$ diagram with observed sources provides a *necessary* but not *sufficient* condition on the theory. The theory must also be able to account for both the spectral energy distribution and the spatial distribution of emission. In the protostellar (embedded) phase, circumstellar disks are required to produce the correct spectral energy distri-

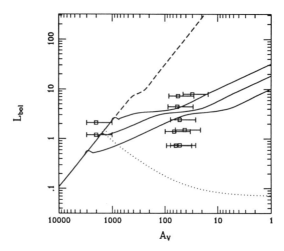

Figure 1. $L-A_V$ diagram for the Taurus molecular cloud. Solid lines show theoretical
tracks for $a = 0.20$ km s^{-1} and $\Omega/(10^{-14}$ rad s$^{-1}) = 1.0, 3.0$ and 10.0. The dashed
curve shows the spherical ($\Omega = 0$) limit; the dotted curve shows an alternate model
with no disk accretion ($\eta_D = 0$). Symbols represent observed protostellar sources in
the cloud.

butions (cf. Fig. 2; Adams et al. 1987; Ladd et al. 1991) and to understand the
observed emission distribution (see Keene and Masson 1990).

The properties of protostellar disks can be summarized as follows. The
disk masses are likely to be substantial because much of the infalling material
falls directly onto the disk rather than onto the star. However, as we discuss
in Sec. III below, the disk masses cannot be arbitrarily large because such
disks would become highly unstable. Disk accretion must occur to account
for the observed protostellar luminosities (again, see Fig. 1) and the intrinsic
disk luminosity L_D is expected to be comparable to the stellar luminosity. We
therefore expect the protostellar disk mass M_D to be a significant fraction of
the total mass M_{total}. The physical size of protostellar disks is determined by
the angular momentum of the collapse (see Sec. V below) and is likely to be
~ 10 to 100 AU for much of the protostellar phase of evolution.

B. T Tauri Disks

For T Tauri disk systems, the intrinsic spectral energy distribution is directly
measurable and we can study more directly the properties of the circumstellar
disks. One of the most striking properties of these systems is the large infrared
excesses (see, e.g., Rucinski 1985; Rydgren and Zak 1987) which indicate the
presence of circumstellar disks. As the observational qualities of T Tauri disk
systems have been extensively discussed elsewhere in this book (see Chapters
by Strom et al., Beckwith and Sargent, Basri and Bertout and Hartmann et
al.), we will only summarize here the basic properties of these disks. The

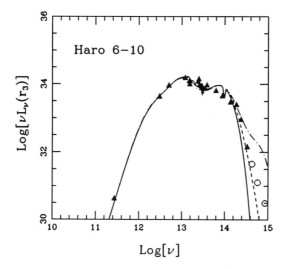

Figure 2. The spectral energy distribution of protostellar candidate Haro 6–10. Curves show the theoretical spectrum calculated on the basis of the current protostellar theory; points show the observed data (figure adapted from Adams et al. 1987).

characteristic radial size is ∼100 AU; this size can be deduced from the characteristics of the spectral energy distributions (Adams et al. 1988,1990), from studies of spectral lines in outflows (Edwards et al. 1987), or from our own solar system (the mean distance of Neptune from the Sun is ∼30 AU). The disks in T Tauri systems can be either *passive* and merely reprocess stellar radiation (see, e.g., Fig. 3a), or *active* and possess additional intrinsic luminosity (see, e.g., Fig. 3b).

For sources with active disks, the intrinsic disk luminosity L_D can be comparable to that of the star, i.e., L_D can be a significant fraction of the total luminosity. In addition, the spectral energy distributions of active disk sources indicate that the disk effective temperature distribution is often much flatter than the expected form $T_D \sim r^{-3/4}$ for a classical steady Keplerian accretion disk (Lynden-Bell and Pringle 1974); many disks require temperature profiles of the form $T_D \sim r^{-1/2}$, which suggests that *nonlocal* processes may be at work. The disk mass is estimated to lie in the range

$$0.01 M_\odot \leq M_D \leq 1.0 M_\odot \tag{1}$$

(see, e.g., Weintraub et al. 1989*b*; Adams et al. 1990; Chapter by Beckwith and Sargent). However, these mass estimates depend directly on the assumed dust opacity at submillimeter wavelengths (and this quantity remains uncertain at present).

Notice that the disk masses and radii (and hence the angular momenta) for the T Tauri disks are consistent with that expected from the protostellar theory

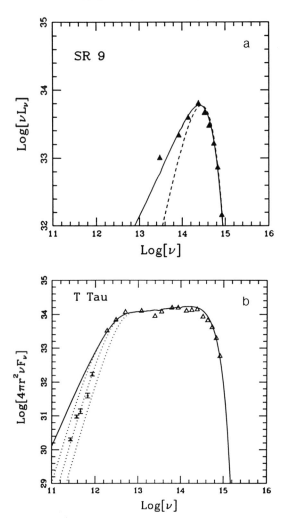

Figure 3. (a) Observed and theoretical spectral energy distribution of passive disk
system of SR 9. Dashed curve shows a blackbody spectrum; solid curve shows the
effects of including a passive disk (figure adapted from Adams et al. 1987). (b)
Observed and theoretical spectral energy distribution of active disk system T Tauri.
Solid curve shows the spectrum in the limit of an optically thick disk ($M_D \rightarrow \infty$);
dotted curves show spectra with finite disk masses of 0.01, 0.1 and 1.0 $M_\odot$ (figure
adapted from Adams et al. 1987,1990).

(see above; Chapter by Shu et al.; Shu et al. 1987a). Note also that both the
protostellar disks and the T Tauri disks must be capable of disk accretion, but
no definitive theory of disk accretion exists at this time. In the following two
sections, we will discuss two important physical mechanisms which may lead
to disk accretion. The first mechanism is self-gravitating disk instabilities,

which require a fairly large disk mass (i.e., M_D comparable to M_*) and are expected to be important during the protostellar phase of evolution. The second mechanism is convective instabilities, which are expected to arise during the T Tauri phase of evolution (when M_D is of order 10^{-2} $M_\odot$).

III. GRAVITATIONAL INSTABILITIES

The basic objective of accretion disk theories is to identify the physical processes which (1) generate the observed spectrum and luminosity, and (2) determine the structure and regulate the evolution of the disk. From accretion disk models, we can delineate the conditions which lead to the formation of stellar companions and planetary systems. One of the central issues of accretion disk theory is the physical mechanism through which angular momentum is redistributed in the disk. In this section, we discuss gravitational instabilities in star/disk systems. During the early phase of protostellar evolution when the mass of the disk is comparable to that of the forming star itself, these gravitational instabilities may be particularly relevant in producing angular momentum transport in the disk.

A. The Initial Unperturbed State

Two general classes of gravitational instabilities can arise: *global gravitational instabilities* in which extended regions of the disk participate in promoting the growth of some initial disturbance and *local gravitational instabilities* in which the characteristic wavelengths are much shorter than the physical dimension of the disk. Global gravitational instabilities may occur when the mass of the disk is comparable to that of the central star. Local gravitational instabilities may occur in disks which are considerably less massive than the star if the disks are sufficiently cool.

In this section, we concentrate our discussion on the global gravitational instabilities which may be responsible for inducing energy dissipation and global energy transport in disks with relatively flat spectra. In particular, we will calculate the growing normal modes of the system for the case of modes with azimuthal wavenumber $m = 1$ (see Adams et al. 1989; Shu et al. 1990). We begin by specifying the basic unperturbed state. The physical system consists of a star and an accompanying gaseous disk. The growth of spiral modes is mainly determined by three elements: self-gravity, pressure and differential rotation. The gravitational forces are determined by the potential of the star and by the disk's surface density distribution $\sigma_0(r)$, which we take to be a simple power law in radius r from the central star (the disk is also assumed to be infinitesimally thin and in centrifugal equilibrium). The pressure is determined by the temperature distribution, or equivalently, the distribution of sound speed in the disk; we take the sound speed $a(r)$ to be a power law in radius. The rotation curve $\Omega(r)$ in the disk is then determined self consistently from the potential of the star, the potential of the disk, and the pressure gradients. Because the potential well of the star dominates that of the disk everywhere except near the disk's outer edge, the rotation curve is nearly Keplerian throughout most of the disk's radial extent (for thin gaseous disks, the pressure gradients are small compared to the gravitational forces and do not significantly affect the rotation curve). Finally, we must specify the radial size of the disk. As discussed in the previous section, both observations and theory indicate that the characteristic disk size is approximately 100 AU, i.e., the disk radius R_D can be up to $\mathcal{O}(10^4)$ times the radius R_* of the star (only the ratio R_D/R_* enters into the calculations of self-gravitating instabilities discussed below).

B. Modes with Azimuthal Wavenumber m = 1

Our discussion concentrates on modes with azimuthal wavenumber $m = 1$, as these modes can be global in extent and may also be the most difficult modes to suppress in unstable gaseous disks. Modes with $m = 1$ correspond to elliptic streamlines (i.e., eccentric particle orbits), a special characteristic of Keplerian potentials. Thus, for a disk with an exact Keplerian rotation curve and no interactions between particles, $m = 1$ disturbances correspond to purely kinematic modes of the system (see Fig. 4); for realistic disks (with pressure), a relatively "small" amount of self gravity is required to overcome the pressure (and shear) and sustain modal growth.

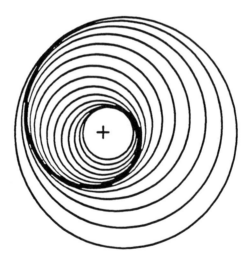

Figure 4. Schematic diagram of $m = 1$ elliptical streamlines oriented to produce a one-armed spiral. The cross denotes the position of the star (figure adapted from Adams et al. 1989).

In addition, modes with $m = 1$ are the most global, i.e., these modes can affect a larger radial extent in the disk than higher m modes (see Fig. 1 of Adams et al. 1989).

One unique and important aspect of $m = 1$ modes is that the center of mass of the perturbation in the disk does *not* lie at the geometrical center of the system; hence, the frame of reference centered on the star is *not* an inertial reference frame. The star is actually in orbit about the center of mass (i.e., the star is accelerating) and creates an effective forcing potential, the "indirect potential," which has the form

$$\psi_{\text{ind}} = \omega^2 R_0 r \exp[i(\omega t - \theta)] \tag{2}$$

where ω is the (complex) eigenvalue of the mode. The quantity R_0 measures the strength of the indirect term and is given by

$$R_0 = \frac{\pi}{M_D + M_*} \int_{R_*}^{R_D} r^2 dr \, \sigma_1(r) \tag{3}$$

where σ_1 is the perturbation of the surface density of the disk. Under some circumstances (see Shu et al. 1990), the indirect potential can have additional contributions from the displacement of the disk edge. Recent results (Adams et al. 1989; Shu et al. 1990) show that this indirect potential is essential for the growth and maintainence of spiral modes with azimuthal wavenumber $m = 1$. In fact, the interaction of this indirect potential with the outer Lindblad resonance in the disk (see below) can be the dominant amplification mechanism for these modes.

C. Wave Physics and Spiral Instabilities in Gaseous Disks

In the simplest description of spiral instabilities, self-excited disturbances (spiral modes) can grow through the feedback and amplification of spiral density waves. In this section, we briefly review the theory of spiral density waves. In the following sections, we describe the feedback cycle and the amplification mechanism for spiral modes, especially those with $m = 1$.

In the asymptotic (WKBJ) limit, the dispersion relation for spiral density waves in a gaseous disk has the form

$$k^2 a^2 - 2\pi G \sigma_0 |k| + \kappa^2 = (\omega - m\Omega)^2 \tag{4}$$

where k is the radial wavenumber, κ is the epicyclic freqeuncy, and where ω is the complex eigenvalue of the system (see, e.g., Lin and Lau 1979). Since the gravitational term is proportional to $|k|$, this dispersion relation has four branches,

$$k = \pm(k_0 \pm k_1) \tag{5}$$

where

$$k_0 \equiv \frac{\pi G \sigma_0}{a^2}, \qquad k_1 \equiv \frac{\pi G \sigma_0}{a^2}[1 - Q^2(1 - v_m^2)]^{1/2},$$

$$\text{and} \qquad v_m \equiv (\omega - m\Omega)/\kappa. \tag{6}$$

Here, the quantity Q determines the stability of the system to *axisymmetric* disturbances ($Q > 1 \Rightarrow$ axisymmetric stability) and is defined by (Toomre 1964):

$$Q \equiv \frac{\kappa a}{\pi G \sigma_0}. \tag{7}$$

The overall sign of k determines whether the waves are *leading* ($|k| > 0$) or *trailing* ($|k| < 0$); the inner sign determines whether the waves are *short* [$k \propto (k_0 + k_1)$] or *long* [$k \propto (k_0 - k_1)$].

The quantity ν_m is a dimensionless frequency of the spiral density waves. The radius in the disk where $\nu_m = 0$ [i.e., where $\Re(\omega) = m\Omega$] is known as the *corotation resonance*; the energy and angular momentum of the perturbation (and the action) are positive outside the corotation radius and negative inside. Note that for $Q > 1$, the wavenumber k_1 becomes imaginary for radii sufficiently close to the corotation resonance, i.e., a classical turning point exists for the density waves. The resulting "forbidden" region surrounding the corotation resonance is known as the Q-barrier. Notice also that for long waves, $k \rightarrow 0$ at any radius where $|\nu_m| = 1$. The radius in the disk where $\nu_m = +1$ is known as the *outer Lindblad resonance* and plays an important role in the physics of $m = 1$ modes. In particular, long waves have a classical turning point at the Lindblad resonances and are thus confined to radii in the disk between the Q-barrier and the Lindblad resonances.

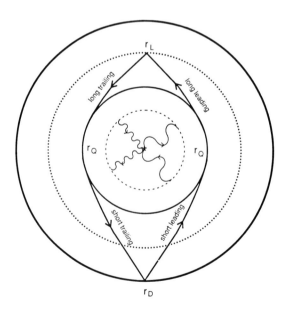

Figure 5. The four-wave feedback cycle for $m = 1$ modes. The propagation of each type of wave is shown by the solid lines with arrows. (See text for further discussion; figure adapted from Shu et al. 1990).

D. Feedback Loop: The Four-Wave Cycle

We now describe the feedback cycle for $m = 1$ modes in gaseous disks (see Shu et al. 1990). One unique aspect of this feedback cycle is that all four types of waves are utilized (see Fig. 5):

1. Begin (somewhat arbitrarily) with the excitation of a long trailing spiral density wave at the outer Linblad resonance by the indirect term. The long trailing wave propagates inward (its group velocity is negative) until it encounters the outer edge of the Q-barrier.
2. At the Q-barrier, the long trailing wave refracts into a short trailing spiral density wave that propagates back outward, through the outer Lindblad resonance to the outer disk edge.
3. The short trailing waves that propagate to the outer disk edge reflect there to become short leading waves. The short leading waves then propagate back to the interior, through the outer Lindblad resonance, until they encounter the outer edge of the Q-barrier, where they refract into long leading spiral density waves that propagate back toward the outer Lindblad resonance.
4. At the outer Lindblad resonance, the long leading waves reflect to become long trailing waves. If the reflected long trailing wave possesses the correct phase relative to the long trailing wave launched from the outer Lindblad resonance by the indirect term in step 1 above, then we have constructive reinforcement of the entire wave cycle, and the basis for the establishment of a *resonant wave cavity.*

Using a WKBJ analysis, this four wave cycle can be used to derive a quantum condition for $m = 1$ spiral modes:

$$\oint k\,dr + \sum \Delta\phi = 2\pi n \tag{8}$$

where the wavenumber k is integrated over the entire four wave cycle described above; the changes in phase $\Delta\phi$ occur due to the reflection and refractions of the waves at their turning points, and where n is an integer. This quantum condition accurately predicts the pattern speeds (i.e., the real part of the eigenfrequencies) for these modes; for strongly growing modes, the analytical results (Shu et al. 1990) and numerical results (Adams et al. 1989) agree to within $\sim 1\%$.

E. SLING Amplification

In this section, we describe the amplification mechanism for eccentric modes. Detailed analysis indicates that the dominant mechanism for amplification arises from the indirect potential, which provides an effective forcing term. The indirect term varies slowly with radius in the disk; as a slowly varying force can only couple to oscillatory disturbances at the disk edges or at the Lindblad resonances (Goldreich and Tremaine 1979b), the main coupling occurs at the outer Lindblad resonance for the modes considered here. Thus, this amplification mechanism differs substantially from the previously studied mechanisms, which utilize the process of *super-reflection* across the corotation resonance (super-reflection can still occur in these disks and is implicitly included in the numerical treatment, but it does not dominate the amplification). In an analytic treatment, the growth rates can be determined for the modes under the assumption that *all* of the amplification arises from this coupling of the indirect term to the outer Lindblad resonance in the disk, and that the indirect term arises mostly near the outer disk edge. In other words, the indirect potential is considered as an external forcing term acting on the disk and the torque exerted on the disk at the outer Lindblad resonance is calculated accordingly. Because the long-range coupling of the star to the outer disk provides the essential forcing, this new instability mechanism is called SLING, Stimulation by the Long-range Interaction of Newtonian Gravity.

The combined numerical and analytical treatments indicate the dependence of the growth rates (i.e., the imaginary part of the eigenfrequencies) on the parameters of the problem. Most importantly, a finite threshold exists for the SLING amplification mechanism. When all other properties of the star/disk system are held fixed, this effect corresponds to a threshold in the ratio of disk mass M_D to the total mass $M_* + M_D$. We find that the growth rates are largest for the case of equal masses $M_D = M_*$ and decrease rapidly with decreasing relative disk mass. In the optimal case, $M_D = M_*$, the growth rates can be comparable to the orbital frequency at the outer disk edge, i.e., the modes can grow on nearly a dynamical time scale. On the other hand, the presence of the finite threshold implies a critical value of the relative disk mass, i.e., the maximum value of $M_D/(M_* + M_D)$ that is stable to $m = 1$ disturbances; for the simplest case of a perfectly Keplerian disk and $Q(R_D) = 1$, this critical ratio has the value

$$\frac{M_D}{(M_* + M_D)} = \frac{3}{4\pi} \approx 0.24. \tag{9}$$

This critical value of the disk mass is known as the maximum solar nebula (Shu et al. 1990).

For SLING-amplified modes, the requirement of conservation of angular momentum luminosity leads to the following behavior for the relative surface density perturbation:

$$S \equiv \left| \frac{\sigma_1}{\sigma_0} \right| \propto r^{-1/2} a_0^{-2} \propto r^{-1/2} T^{-1}. \tag{10}$$

For the expected case of $S = constant$ (see the discussion in Shu et al. 1990), we obtain the temperature profile required for disks with flat infrared spectra: $T(r) \sim r^{-1/2}$ [note that the dynamical calculations ignore the temperature variations in the vertical direction and hence this form for $T(r)$ is that required by the flat-spectrum T Tauri sources (see Sec. II)].

F. Shutting off Gravitational Instabilities

The gravitational instabilities described above can be turned off via several mechanisms. If these instabilities lead to mass accretion through the disk, the fraction of the total mass that resides in the disk must eventually decrease below the critical fraction (see above) and the growth rates for the instabilities will decrease substantially. These instabilities can also be stabilized through excessive heating of the disk. If the Q parameter becomes sufficiently large everywhere in the disk, the growth rates for these instabilities will again decrease. Finally, these modes can be stabilized through the behavior of the outer disk edge. If the disk edge is very efficient at absorbing the (outgoing) short leading waves, the feedback cycle will be broken. However, the SLING amplification mechanism produces a strong amplifier and relatively little feedback is required to sustain the growth of these modes. The disk edge can also distort itself in such a manner as to cancel the effects of the indirect potential (see Shu et al. 1990). However, a fairly wide distribution of disk edge configurations will allow the growth of these modes.

G. Results and Conclusions

Eccentric spiral modes in gaseous disks have been studied using both numerical and analytical treatments of the problem. In both cases the (complex) eigenvalues of the system have been determined and the two approaches are in good agreement. The analysis indicates that the basic modal mechanism involves the four-wave cycle (see Fig. 5; Shu et al. 1990), which provides the feedback loop, and the SLING mechanism, which provides the amplification. The results also indicate that a wide range of YSO disks will be unstable to the growth of these eccentric distortions. When the disk mass is comparable to the stellar mass ($M_D \sim M_*$), these distortions can grow on nearly a dynamical time scale. In addition, these modes can grow when the disk is safely stable to axisymmetric disturbances (i.e., Q substantially > 1). However, the results (e.g., the exact spectrum of unstable modes) are particularly sensitive to the treatment of the outer disk edge (see Shu et al. 1990) and a more complete understanding of these edge effects is a goal of future theoretical studies.

The $m = 1$ modes described here may have important astrophysical applications. In the earliest stage of star formation—the protostellar phase—the mass of the disk is likely to be comparable to that of the star; $m = 1$ modes are thus likely to grow and may lead to mass accretion through the disk and the observed disk luminosities. Perturbations with $m = 1$ prove especially interesting because they force the star to move from the center of mass and thereby transfer angular momentum to the stellar orbit. This effect can be seen in Fig. 6 which shows the lowest-order $m = 1$ mode for a moderately massive disk $M_D = M_*$; note that most of the mass of the perturbation lies in one banana-shaped lump.

These $m = 1$ modes may thus lead to the formation of a binary companion within the disk (or perhaps to the formation of giant planets). Numerical simulations of spherical rotating systems (see, e.g., Monaghan and Lattanzio,

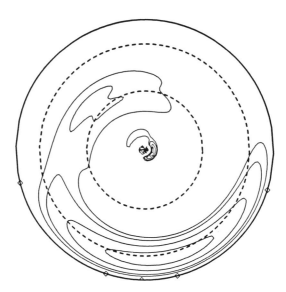

Figure 6. Equidensity contour plot for the lowest order ($n = 0$) SLING amplified
$m = 1$ mode. The two dashed circles show the location of the corotation and outer
Lindblad resonances. The spiral arm trails in the sense of rotation of the material
(figure adapted from Adams et al. 1989).

unpublished calculation) also show the tendency to produce an orbiting pair of
mass condensations. However, further calculations, including both dissipation
and nonlinear effects, are needed to follow up these possibilities.

H. Other Gravitationally Unstable Modes

In addition to the case of $m = 1$ modes discussed above, other gravitationally
unstable modes can arise in circumstellar disks. For example, the condition
for stability against the growth of axisymmetric disturbances in a gaseous
disk does not require the disk mass to be comparable to the central star.
In the cool outer regions of protoplanetary disks, the temperature can be
sufficiently cool so that $Q < 1$ even when $M_D/(M_* + M_D) < 3/4\pi$. In
this limit, local gravitational instabilities may be excited; in this case, the
characteristic wavelength derived from the dispersion relation (see Eq. 4)
would be considerably less than the disk radius. Gravitational instability
against nonaxisymmetric perturbations, when they can be excited, may also
provide an effective mechanism for angular momentum transfer (Paczynski
1978a; Lin and Pringle 1987).

   In the context of protoplanetary disk dynamics, two major theoretical
issues arise concerning nonaxisymmetric gravitational instabilities: the origin
of the instability and the effective magnitude of the torque induced by nonlinear (unstable) modes. Recent progress has been made on both issues (see,
e.g., the reviews of Toomre 1981; Sellwood 1989; Papaloizou and Savonije

1989). The motivation for these investigations came mostly from dynamicists interested in the spiral structure of disk galaxies. For circumstellar disks with $Q \sim 1$, the effect of self gravity can strongly amplify transient disturbances; however, in the absence of large surface density gradients, these disturbances are damped by shear and their amplification does not lead to genuine growth (Goldreich and Lynden-Bell 1965b; Julian and Toomre 1966). For regions with large surface density gradients, Toomre (1981,1989) showed that genuinely unstable "edge" modes can grow strongly. In a recent series of numerical simulations of self-gravitating particle disks, Sellwood and Lin (1989) found that these edge modes are not only readily excited but also promote a recurrent nonaxisymmetric instability. These growing perturbations scatter particles off their Lindblad resonances through tidal torques and induce "grooves" in the phase-space distribution; these grooves, in turn, provide effective edges for the growth of subsequent edge modes (Sellwood and Kahn 1991).

Nonaxisymmetric normal mode analyses have been carried out for narrow rings and disks composed of both particles and an incompressible gas (Papaloizou and Lin 1989). These analyses indicate that the existence and growth rate of nonaxisymmetric instabilities is determined by the distribution of *vortencity*, i.e., the ratio of the surface density to the specific vorticity. There are at least three classes of unstable models: (1) those associated with extrema in vortencity, (2) those promoted by resonant cavities; and (3) those which have corotation points exterior to the disk. The first two types can exist for non-self-gravitating disks (Papaloizou and Pringle 1984,1985,1987) although the effect of self gravity amplifies the growth. The third type can only occur for self-gravitating disks and can promote nonaxisymmetric instabilities even for $Q > 1$.

Nonlinear simulations of self gravitating disks have been performed using a sticky particle hydrodynamics scheme (Anthony and Carlberg 1988) and using a three-dimensional (traditional) hydrodynamics code (Yang et al. 1991); these calculations indicate that nonaxisymmetric perturbations can grow into the nonlinear regime and can indeed transfer angular momentum through their tidal torques (see also Larson 1984). A series of detailed two dimensional hydrodynamic simulations (Savonije and Papaloizou 1989) showed that the amplitude of the perturbation eventually saturates at some finite value. For relatively massive disks, the torque induced by the growing nonaxisymmetric modes transports angular momentum outward and causes a significant amount of matter to spiral inward within a few orbital periods. The growth rate is a sensitive function of Q because the width of the evanescent zone around the corotation resonance (the Q-barrier; see Eqs. [4–7] above) is proportional to $(1 - 1/Q^2)^{1/2}$. When the effect of self gravity is no longer important, propagation through this evanescent region becomes impossible and the disk becomes stable.

For the self-gravitating outer regions of protoplanetary disks with moderate masses, these "local" gravitational instabilities cannot lead to binary star formation and significant global energy redistribution as we have indicated in our discussion above on the $m = 1$ gravitational instabilities. Nevertheless, the effective angular momentum transport induced by these growing modes can affect the *secular* evolution of protoplanetary disks during their formation epochs (Lin and Pringle 1990; see also the discussion in Sec. V).

IV. VISCOUS EVOLUTION AND CONVECTIVE INSTABILITIES

In the later stages of disk evolution, the disk mass will become small enough that gravitational instabilities will shut off, or at least become less important. In the realm of intermediate mass disks, i.e., $M_D = \mathcal{O} (10^{-1}$ to $10^{-2})$ M$_\odot$, convective instabilities (and other sources of viscosity) are likely to play an important role (see, e.g., Lin and Papaloizou 1985; Ruden and Lin 1986). We expect this regime of disk mass to occur after the infall has been terminated, i.e., after the disk no longer has a source for gaining mass.

A. Viscous Evolution of Circumstellar Disks

The evolution of circumstellar disks is governed by the laws of fluid dynamics. In the limit that viscous forces drive the evolution of the disk, the behavior is described by a time-dependent diffusion equation (Lüst 1952; Lynden-Bell and Pringle 1974; Lin and Papaloizou 1985), which can be written in the form

$$\frac{\partial \sigma}{\partial t} - \frac{3}{r} \frac{\partial}{\partial r} \left[r^{1/2} \frac{\partial}{\partial r} (\sigma v r^{1/2}) - \frac{2 S_\sigma (r, t) J (r, t)}{\Omega} \right] - S_\sigma (r, t) = 0 \quad (11)$$

where Ω, σ and v are the angular frequency, surface density and viscosity of the disk material, respectively, $S_\sigma (r, t)$ and $J (r, t)$ are the mass flux and excess angular momentum, respectively, of the infalling material. The evolutionary time scale and pattern depend on both the magnitude of the viscosity and the properties of the infalling material. During the initial formation of the disk and the entire protostellar phase of evolution, the source term $S_\sigma (r, t)$ will be large and may be particularly important in determining the structure of the disk. In the T Tauri (post-infall) phase of evolution [when $S_\sigma (r, t) = 0$], the mass of the disk will eventually become small enough that self-gravitating modes die out and viscous diffusion dominates the evolution of the disk. Note that the viscosity v determines the evolution of the disk; in particular, the radial velocity through the disk is given by $u_r \propto -v/r$. Note also that viscous effects depend on the *local* disk properties, in contrast to the *global* nature of $(m = 1)$ self-gravitating instabilities.

In a differentially rotating, geometrically thin disk, mass flow requires both angular momentum transport and energy dissipation (see, e.g., Lynden-Bell and Pringle 1974). These two processes can operate simultaneously through an effective viscous stress. However, molecular viscosity in accretion disks is generally too small to be of astrophysical interest. In a variety of accretion disks, a "turbulent" viscosity is often assumed to be responsible for both angular momentum transport and viscous dissipation despite the lack of rigorous proof that turbulence may occur intrinsically (Shakura and Sunyaev 1973; Lynden-Bell and Pringle 1974). In such a prescription, an effective viscosity is introduced through the ansatz

$$\nu_{\text{eff}} = \frac{2}{3}\alpha_s a_s H \tag{12}$$

where a_s is the sound speed, H is the pressure scale height in the disk, and α_s is a dimensionless efficiency factor. The evolution of the disk is specified once the viscosity (or, equivalently, the α_s parameter) is given. The goal of any study of viscous evolution in disks is thus to calculate the parameter α_s in an *a priori* manner. In the case of YSO disks, thermal convection can occur in the direction normal to the plane of the disk (the vertical direction) and can produce turbulent flow and hence an effective viscosity (Lin and Papaloizou 1980,1985; Lin 1981).

B. Intrinsic Nature of Convective Instability

In order to show that YSO disks are intrinsically unstable against thermal convection, we analyze the structure of a turbulent-free disk in which rotation prevents gas from migrating in the radial direction. However, gas can contract in the vertical direction. If the disk is not in hydrostatic equilibrium initially, it will rapidly evolve towards such a state. Using a one-dimensional numerical hydrodynamic scheme, Ruden (1986) showed that a disk of cold gas contracts towards the midplane. After the disk has settled into a quasi-hydrostatic equilibrium, slow contraction continues as thermal energy is lost from the disk surface. In the absence of an energy source, surface radiation leads to heat diffusion from the midplane to the surface region. The associated reduction in pressure support leads to a readjustment towards a new hydrostatic equilibrium. In the typical temperature range of 10 to 10^3 K, such adjustment can lead to a *superadiabatic* structure in the vertical direction (see Ruden 1986). This structure is produced because the opacity, which is primarily due to dust grains, is a sufficiently strong increasing function of temperature. In the case of YSO disks, the cooler surface region has a lower opacity and cools more efficiently (Lin and Papaloizou 1980,1985; Lin 1981), although this condition is not generally satisfied in other accretion disks. The standard Schwarzschild criterion for convection can be written

$$\frac{d \log T}{d \log P} \geq \frac{\gamma - 1}{\gamma} \equiv \nabla_{\text{ad}} \tag{13}$$

where γ is the adiabatic exponent of the gas (see, e.g., Schwarzschild 1958). When the temperature gradient in the disk becomes superadiabatic, the disk becomes convectively unstable in the vertical direction.

C. Global Convective Pattern and Angular Momentum Transport

Once convection has become established in a circumstellar disk, a self-consistent cycle of disk accretion can occur. In this cycle, the convective motions lead to turbulence. This turbulence leads to the production of eddies on a wide range of size scales, i.e., a cascade of eddies. On the smallest eddy scale, energy dissipation occurs through an effective viscosity. This viscosity, in turn, leads to energy generation in the disk and hence to disk accretion. The energy generation leads to steep vertical temperature gradients, which can be superadiabatic and therefore can support further convection. Thus, a self consistent cycle of convective instability can be maintained (see Fig. 7).

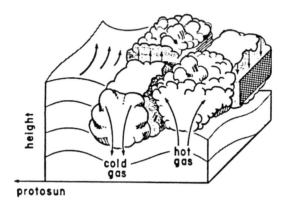

Figure 7. Schematic diagram of convective cell in a circumstellar disk (see text).

The role of convection in a disk is not limited to heat transport in the vertical direction. Convective eddies can induce mixing over a radial extent comparable to their own size. Through this mixing process, angular momentum is transferred. Convection also generates turbulence which causes dissipation of energy stored in differential rotation. Perhaps the simplest treatment for convection is to use the mixing-length prescription in which the eddy viscosity is assumed to be the product of the convective speed and an effective mixing length which is comparable to the size of the eddies (Lin and Papaloizou 1980). From such a treatment, we can build self-consistent models in which convection is responsible for (1) energy dissipation; (2) heat transport in the vertical direction; and (3) angular momentum and mass transport in the radial direction. Thus, energy dissipation is distributed from the midplane to the surface throughout the disk.

Notice, however, that the mixing-length model is based on an *ad hoc* prescription of eddy viscosity. In a convective disk, eddies with a variety of scales are generated. Similar to typical turbulent shear flows, the largest eddies often provide the dominant momentum transfer whereas the smallest eddies provide most of the energy dissipation in the disk. The scale of the largest convective eddies is comparable to the vertical scale of the entire convectively unstable zone which itself extends over a significant fraction of the thickness of the disk. On these large scales, global effects such as rotation and radiative losses are important. Thus, convection must be examined with a global analysis. In an attempt to carry out such a global analysis, Cabot et al. (1987a, b) computed a vertically averaged effective viscosity which is derived from integrating the linear growth rate through various distances above the midplane of the disk. This growth rate varies greatly in the vertical direction and therefore cannot be attributed to any given eddy.

A more appropriate global treatment is to determine a unique growth rate (which is an eigenvalue) and its associated eigenfunction for each characteristic convective mode (Ruden et al. 1988; see Fig. 8).

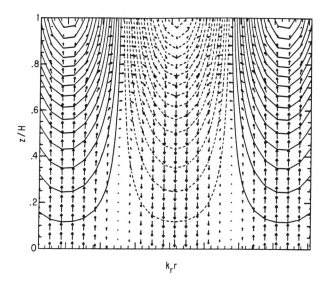

Figure 8. Convective cell of eddies for an unstable convective mode in the protosolar nebula (figure adapted from Ruden et al. 1988).

These eigenfunctions extend over finite radial distances. In the thin-disk limit, the WKB approximation may be used to describe the radial dependence of temperature and density. These global linear stability analyses of axisymmetric perturbations indicate that:

1. Rotation and compressibility tend to reduce the growth rate of the disturbances; however, they cannot suppress the onset of convective instability in protostellar disks.
2. The growth rate is proportional to the square root of the radial wavenumber and is bounded by the maximum value of the Brunt-Väisälä frequency.
3. The maximum radial size of the eddies scales as the square root of the superadiabaticity times the size of the convective region.
4. Due to radiative losses, the short wavelength modes become overstable and only the fundamental and the first harmonic modes can grow effec-

tively (in these modes, the wavelength is comparable to the disk thickness).

5. Both even and odd modes exist, i.e., a single eddy may either be confined to one side of the disk or may thread through the midplane and have a characteristic scale comparable to the thickness of the entire disk.

Convective eddies can thus provide a relatively effective coupling between different parts of the disk and can induce both heat dissipation and angular momentum transport. The magnitude of the effective viscosity can be derived under the assumption that all of the gas within a radial wavelength mixes efficiently during the characteristic growth time scale. This estimate generally agrees well with that derived on the basis of the mixing-length model and yields an effective $\alpha \sim 10^{-2}$. For this value of α, the corresponding evolutionary time scale for the disk is $\sim 10^6$ yr and the mass accretion rate is 10^{-7} to 10^{-8} $M_\odot$ yr^{-1}. The resulting intrinsic luminosity of the disk is fairly low, $L_D \leq 1$ $L_\odot$, and compares well with estimates of L_D from observations of boundary layers (see, e.g., Bertout et al. 1988). This calculation should be extended to include a global analysis of the initial linear growth of nonaxisymmetric disturbances and an analysis of the nonlinear regime where dissipative processes become important. The determination of the torque associated with growing nonaxisymmetric disturbances will provide a more rigorous estimate of the efficiency of angular momentum transport.

D. Suppression of Convection: Infall and Surface Heating

In order for convection to occur, the vertical gradient of temperature in the disk must be sufficiently large (superadiabatic); in particular, the central disk temperature must be sufficiently hotter than the disk surface. This temperature structure can be significantly modified by the boundary condition at the surface of the disk. In fact, any heating of the disk surface will tend to flatten the vertical temperature gradient and thereby suppress convection. In this section, we discuss two mechanisms that can heat the surface of the disk and can suppress convection: (1) disk heating through surface shocks from infalling material; and (2) disk heating by the stellar radiation field.

In the protostellar phase of evolution, infalling material falls onto the disk surface and dissipates its kinetic energy; this dissipation is achieved through a shock on the disk surface (see, e.g., Adams and Shu 1986). This shock heats the disk from its surface and thereby decreases the vertical temperature gradient; if the shock is sufficiently energetic, this heating from above can stabilize convection. If the infall rate onto the disk remains constant, shock dissipation near the surface can dominate the energy loss due to contraction towards the midplane and convection can be suppressed. However, when the infall rate is reduced, the superadiabatic gradient may be established as the surface heating becomes less important (Ruden 1986).

Surface heating due to radiation from the central star can also reduce the temperature gradient in the vertical direction in a manner similar to shock dissipation associated with infall (Watanabe and Nakazawa 1988). In particular, convection may be stabilized for a relatively large stellar radiation field incident onto the surface of a disk with a relatively low surface density. A more general analysis (Bell et al. 1991) indicates that convection is suppressed only when the surface heating is sufficiently large to induce a blackbody temperature comparable to temperature near the midplane. This critical flux increases with the surface density of the disk because the opacity (and consequently the midplane temperature) also increases with the surface density.

In the limit where surface heating exceeds the critical flux required to stabilize convection, the subsequent evolution of the disk remains unclear (see the discussion in Lin 1989). Consider an axisymmetric perturbation in which one region of the disk has a slightly higher temperature. In that region, the disk thickness would increase and additional stellar radiation could be intercepted. Although the disk may be stable against convection, this relatively thick region of the disk would cast shadows on the outer disk and thereby reduce the surface heating for the exterior regions. A decrease in the incident stellar radiation in the shielded region can cause a reduction in the disk temperature. Consequently, the disk opacity is reduced and the cooling efficiency is increased. Although this process may not be thermally unstable, it may generate a large temperature gradient in the *radial* direction across the interface between the exposed inner hot region and the shielded outer cool regions. For perturbations with wavelengths comparable to or shorter than the thickness of the disk, the disk may become convectively unstable in the radial direction. Mixing in the radial direction would limit the magnitude of the radial temperature gradient. One possible consequence of these surface heating effects is the formation of "ripples" on the disk surface.

In the absence of infall and (strong) surface heating by the stellar radiation field, convective instabilities are expected to occur. However, these instabilities lead to accretion through the disk (see above) and reduce the disk mass, and the disk surface density. When the surface density in the disk becomes sufficiently small, the disk becomes optically thin and can no longer maintain vertical gradients of temperature. The disk thus becomes nearly isothermal, in the vertical direction, and convection again becomes suppressed.

E. Other Mechanisms for Angular Momentum Transport

In addition to the mechanisms for angular momentum transport outlined above, other possibilities exist. For example, if the magnetic field strength is large enough, magnetic torque can lead to the transport of angular momentum (see, e.g., Hoyle 1960; Hayashi 1981; Stepinsky and Levy 1990). One way to view magnetic effects is to define an effective magnetic viscosity (Hayashi 1981)

$$\nu_B \equiv \frac{B_r^2 \tau_B}{4\pi\rho} \tag{14}$$

where B_r is the radial component of the magnetic field, τ_B is the decay time of the field, and where ρ is the density. This magnetic viscosity drives disk evolution in a manner which is analogous to the usual viscous evolution described above (see Sec. IV). On the other hand, if the magnetic field is relatively weak, i.e., if the energy density of the field is less than the thermal energy density, then a local axisymmetric shear instability can arise (Balbus and Hawley 1991; Hawley and Balbus 1991). This instability leads to turbulence in the disk and may provide a mechanism for angular momentum transport; further work is necessary to follow up these possiblities.

Magnetic fields may also play an important role in the generation of winds from YSOs (see Shu et al. 1988; Chapter by Königl and Ruden). In many wind models, the outflow is coupled to mass accretion through the disk; in addition, the associated magnetic fields tend to brake the rotation of the star and may expand the disk. The star/disk/wind interface is thus another important issue concerning the evolution of circumstellar disks.

In addition to the particular case of $m = 1$ self-gravitating modes (see Sec. III), other (higher m) modes and spiral density waves may also play a role in transporting angular momentum. Such effects have been studied by Larson (1984,1989) and Spruit (1987); see Chapter by Morfill et al.

Tidal effects can also be important in the disks of YSOs. If planets form in the disk, they can clear out gaps in much the same way as moons of Saturn clear gaps in the rings (see, e.g., Lin and Papaloizou 1986a, b). As most stars are found in binary systems, binary companions should be present in many star/disk systems and can also transport angular momentum through the disk by tidal torques (Chapter by Bodenheimer et al.). These tidal effects are reviewed in the Chapter by Lin and Papaloizou.

V. DYNAMICAL EVOLUTION AND DISCUSSION

In this section, we discuss the various epochs of disk evolution. As the above sections concentrate on the specific topics of $m = 1$ instabilities and convective instabilities, we will also review other calculations related to the evolution of circumstellar disks.

A. Protostellar Disks and Disk Formation

The formation and evolution of protostellar disks are closely associated with the collapse of the protostellar cloud and the formation of the central star itself. The specific angular momentum of a typical protostellar cloud core is 10^{20} to 10^{21} cm^2 s^{-1} (Goldsmith and Arquilla 1985). At the lower end of this range, the specific angular momentum of the cloud is comparable to that in our solar system today, whereas at the upper end, rotational effects would inhibit the collapse of the cloud at a size $\sim 10^{16}$ cm. Recently, numerical calculations have been carried out for both limits.

For a cloud with relatively low angular momentum, collapse calculations have been performed (Morfill et al. 1985) in two dimensions using an *ad hoc* α prescription (Shakura and Sunyaev 1973) for the angular momentum transport. Because the magnitude of viscosity is a function of temperature, the detailed treatment of radiation transfer is important in determining the structure and evolution of the disk. Morfill et al. modeled the radiation transfer process with an Eddington approximation. The outcome of collapse was the formation of a central condensation surrounded by a disk. The energy dissipation associated with the formation and subsequent infall keeps the disk moderately hot. In a more comprehensive analysis, Bodenheimer et al. (1988) adopted a different numerical hydrodynamic scheme and modeled the radiative transport with a diffusion equation. Their results indicate the formation of a rapidly rotating central object which is surrounded by a relatively hot and geometrically thick disk spreading from 1 to 60 AU. The disk is essentially stable against axisymmetric gravitational instabilities.

The collapse of high angular momentum clouds can lead to the formation of more extended disks, which are likely to be optically thin. In such regions, the cooling time scale may be short compared with the dynamical time scale and the disk may be relatively cool. The combination of low temperature and a large distance from the central condensation (the star) provides favorable conditions for the growth of self-gravitating instabilities. Using a simple prescription for the effective torque induced by growing nonaxisymmetric perturbations (Lin and Pringle 1987), we can study the formation and evolution of an extended disk. In these computations, we also include turbulent transport of angular momentum (with an α prescription), but we assume that the turbulent viscosity is relatively small because convection is probably stabilized by the shock dissipation near the disk surface (see above). Radiative transfer for both the optically thin and thick limits are included. Finally, the difference in the specific angular momentum carried by the infalling and disk material is taken into account. In general, we find that nonaxisymmetric instability can induce efficient angular momentum transfer in the outer regions of the disk and regulate mass transfer through the disk; the associated mass accretion rate is $\sim 10^{-6}$ to 10^{-4} $M_\odot$ yr^{-1}. The typical time scale for disk evolution is thus $\sim 10^5$ to 10^6 yr, which is comparable to the time scale of the infall phase. As a result, a substantial amount of material is likely to remain in the disk after the infall has been terminated.

In addition to the numerical results discussed above, analytical calculations have been performed for the collapse of rotating isothermal cloud cores (see, e.g., Terebey et al. 1984). These results have been used extensively as a basis for studying the formation of low-mass stars (see, e.g., Chapter by Shu et al.). In this rotating collapse scenario, the infall is nearly spherical outside a centrifugal radius R_C defined by

$$R_C \equiv \frac{G^3 M^3 \Omega^2}{16 a^8} \tag{15}$$

where M is the mass of the central star/disk system, Ω is the initial rotation rate of the core, and where a is the isothermal sound speed of the core. Inside the radius R_C, the infall becomes highly aspherical. The infalling material with the highest specific angular momentum will encounter a centrifugal barrier at R_C and cannot fall to smaller radii; as a result, the radius R_C defines the outer radius of the disk (in the absence of angular momentum redistribution). Note that R_C grows with time, i.e., with mass M that is accumulating in the center of the collapse flow. The formation of the disk will occur when the centrifugal radius R_C becomes larger than the stellar radius R_*, which is nearly a constant in time (see Stahler et al. 1980). After the initial formation of the disk, most of the infalling material joins onto the disk rather than the star and the relative disk mass M_D/M grows with time (see also Cassen et al. [1985] for a more detailed review on this topic).

The calculations discussed above can be roughly summarized as follows. During the protostellar phase of evolution, the formation of a circumstellar disk is essentially unavoidable because of conservation of angular momentum. These protostellar disks are likely to be relatively massive; in particular, the relative disk mass M_D/M may be sufficiently large to sustain the growth of self-gravitating instabilities, especially the $m = 1$ modes considered in Sec. III above. The radius of the disk generally grows with time and is expected to be comparable to the size of our solar system, i.e., $R_D = 10$ to 100 AU, for much of the protostellar phase of evolution.

B. T Tauri Post-Infall Disks

We now consider the evolutionary phase after infall onto the disk has ended. During the early part of this stage, the disk mass may be large enough to continue to sustain self-gravitating modes; however, the disk mass must eventually become small enough that such modes are stabilized. The dynamics of the disk is then determined by the viscous diffusion process and can be analyzed by applying a simplified prescription for convectively induced turbulent viscosity (see Sec. IV above) in the diffusion equation; keep in mind, however, that other sources of viscosity are possible. In principle, the surface density distribution at the end of the infall stage may be used as the initial condition for the viscous diffusion phase. However, the evolution of the disk can also be computed with arbitrary initial conditions.

In the context of solar system formation, one interesting initial condition is the "minimum mass" nebula model (Cameron 1973; Hayashi 1981). In this model, the surface density distribution is derived by augmenting gas to the present mass distribution in the solar system based on the assumption that the protoplanetary disk had a solar composition and planetary formation is totally efficient at retaining all the heavy elements in the disk. Using the effective viscosity associated with convective viscosity, we find the evolution of the minimum-mass nebula with a physical dimension comparable to that of the present day solar nebula takes place on the time scale of $\sim 10^6$ yr (Lin and Papaloizou 1985). The temperature distribution resembles that deduced from the condensation temperature for various terrestrial planets and satellites (Lewis 1972), i.e., we obtain $T_D \sim r^{-3/4}$. At radii interior to the orbit of Mercury, the midplane temperature exceeds 2000 K, so that dust grains will be mostly evaporated. Notice also that in protostellar models with Sun-like masses, the dust destruction front (i.e., the radius where dust grains are vaporized in the infalling envelope) is approximately 10^{12} cm, which is within the orbit of Mercury (Adams and Shu 1986).

For disks with moderate mass [i.e., $M_D/M_* = \mathcal{O}\ (0.1\text{--}1)$] and a surface density which decreases with radius (as expected), the outer regions of the disk will become optically thin. Observations of T Tauri disk systems show that the associated disks become optically thin for radii $r \geq 100$ AU (see Edwards et al. 1987; Adams et al. 1988,1990). Theoretical models of the solar nebula become optically thin exterior to the orbit of Neptune and attain an isothermal vertical structure. In these optically thin regions, convection cannot occur and dust can settle towards the midplane. Even if some other local instabilities can occur, we do not expect the resulting turbulence to be sustained for an extended period. In the marginally optically thin region, the characteristic time scale for heat loss is comparable to, or shorter than, the orbital time scale of the disk. Energy transfer between eddies with different sizes occurs on the eddy turnover time scale, which is comparable to or longer than the orbital time scale. As thermal energy loss is thus *faster* than the kinetic energy transport of the eddies, the eddy motions become supersonic. Shocks can thus occur during eddy mixing and can lead to the decay of turbulence.

However, the absence of local turbulence may not imply the termination of disk evolution. Accompanying the decay of turbulence is the decline in mass diffusion and energy dissipation so that the disk temperature and optical depth decrease with the surface density. The effects of self gravity also increase with time. Eventually, the stability parameter Q can decrease to order unity and gravitational instability can lead to the growth of nonaxisymmetric disturbances and angular momentum transfer (Lin and Pringle 1990). In this regime, the evolutionary time scale is determined by the cooling time scale, because cooling is essential in maintaining a sufficiently low value for Q. In a relatively extended massive disk (i.e., $M_D \sim 0.1$ M$_\odot$ and $R_D \sim 100$ AU), when the temperature decreases to ~ 10 K, the cooling time scale at the disk edge exceeds 10^6 to 10^7 yr. However, the disk temperature cannot decrease below the temperature (~ 10 K) of the protostellar cloud. Thus, when the disk surface density decreases to a sufficiently small value, the effect of self gravity can no longer be important and the evolution of the disk must stop. A natural outcome for this process is the development of an extended, optically thin, circumstellar disk.

For the minimum-mass solar nebula model, the surface density of the disk is so low that self gravity becomes at best marginally important beyond the orbit of Neptune even when the disk temperature is 10 K. If the outer region is both optically thin and non self gravitating, no disk evolution can occur; this finding may be the reason why no major planet is found beyond the orbit of Neptune. In the inner region, the disk remains opaque and viscous evolution continues. As a result, some fraction of the disk material can be deposited in the outer disk region in order to conserve angular momentum.

C. Optically Thin Disks: The Naked T Tauri Phase

On the evolutionary time scale of typical T Tauri stars ($\tau \sim 10^6$ yr), approximately half of these objects become naked T Tauri stars, i.e., they lose any signature of a circumstellar disk (cf. Walter 1986). For the case of convectively driven turbulent viscosity, we can deduce the time scale for the disk to evolve into an optically thin system, i.e., $\tau \sim 10^8$ to 10^9 yr. Even with the assumption that some hypothetical transonic turbulence may be responsible for angular momentum transfer, the time scale for the disk to evolve into an optically thin state is still $> 10^7$ yr. These arguments imply that the disappearance of disk signatures (infrared and ultraviolet excesses) cannot be entirely due to viscous diffusion.

One possible mechanism for eliminating the ultraviolet and infrared excess from the disk is through dust settling. As discussed above, surface heating can stabilize the disk against convection (Watanabe and Nakazawa 1988) provided that the surface density is sufficiently low (Bell et al. 1991). During the evolution of the disk, the surface density of the disk continually decreases. For a minimum-mass nebula model, the surface density of the outer regions of the disk is reduced to a sufficiently low value such that the outer regions are stabilized after a few million years.

When the disk is stabilized, turbulence decays unless other instabilities can occur. In the absence of turbulence, dust can settle toward the midplane on the relatively short time scale of 10^3 to 10^4 yr (Hayashi et al. 1985). Once dust has settled to the midplane, the grains can either undergo gravitational instability to form kilometer-size planetesimals (Goldreich and Ward 1973) or coagulate rapidly (Weidenschilling 1984). These processes both cause the protoplanetary disk to become transparent so that it can no longer reprocess radiation from the central star. Furthermore, the lack of turbulent viscosity implies that no viscous dissipation can occur to heat the disk or to supply disk mass to the central star. Through this mechanism, radiative flux from the disk is significantly reduced while most of the disk gas is retained. The disk gas may be eventually eliminated on a somewhat longer time scale by (1) stellar wind ablation; (2) wind generated by the dissociation or ionization of disk gas by the incident solar radiation; and (3) disk-protoplanet tidal interaction. If naked T Tauri stars have no observable disk signatures because they are optically thin (for the reasons cited above), a clear prediction exists: these objects still have an appreciable amount of circumstellar *gas* with mass $\sim 10^{-2}$ $M_\odot$. Further observations must be carried out to test this hypothesis.

VI. SUMMARY

In the last decade, a workable paradigm of star formation has emerged. In the paradigm, transport processes in circumstellar disks play an important role. In the earliest stages of evolution, the protostellar stage, the star and disk are still gaining mass from the surrounding infalling envelope. In this stage, the disk can have a mass comparable to that of the central star and gravitational instabilities are likely to occur. Self-gravitating spiral modes with azimuthal wavenumber $m = 1$ are likely to dominate; these modes may lead to mass accretion through the disk and/or to the formation of binary companions within the disk. After the star/disk system becomes optically revealed (in the T Tauri phase of evolution), global self-gravitating instabilities decrease in importance as the disk mass decreases relative to that of the star. Once global gravitational instabilities have effectively shut off, the disk evolves viscously; however, the source of the viscosity remains undetermined. Angular momentum transfer and mass redistribution is induced by tidal torques associated with growing nonaxisymmetric perturbations in the self-gravitating outer regions of the disk. In the opaque inner regions of protoplanetary disks, convective instabilities provide one important mechanism for producing an effective viscosity. The amount of mass accretion through the disk is expected to decrease with time. In the later stages, the amount of luminosity generated in the disk itself becomes small compared to the amount of energy intercepted from the star (and reprocessed into the infrared), i.e., the disk becomes *passive*. Eventually, the disk becomes optically thin as the dust grains accumulate into larger bodies; the spectral appearance of the star/disk system then approaches that of an "ordinary" star, i.e., the infrared and ultraviolet excesses in the spectrum disappear. Finally, the remaining gas in the disk clears away through some sort of photoevaporation process. At this stage, the formation is complete and the star/disk system becomes a star/planetary system.

Although our understanding of circumstellar disks has improved greatly since the production of the *Protostars and Planets II* volume, much work remains to be done. The following areas of study are likely to be fruitful in the coming years:

1. Self-gravitating instabilities in circumstellar disks are likely to be important and may lead to disk accretion and/or the formation of binary companions. However, the present studies must be expanded to include both nonlinear and dissipative effects. In addition, the effects of the outer disk edge must be better understood.

2. Convective instabilities are likely to provide a source of viscosity for circumstellar disks and can therefore greatly affect their evolution. As convection, and hence the viscosity, can be turned off through various mechanisms (see above), future studies must determine the conditions under which convection can occur. Further studies of the global evolution of viscous disks, including nonlinear effects, should also be undertaken. In addition, models with realistic settling of dust grains need to be considered.

3. Other sources of viscosity and turbulence should also be considered. The effects of magnetic fields might be particularly rewarding.

4. Since energetic winds are ubiquitous phenomena and can produce a substantial back-reaction on circumstellar disks, the interface of the star, disk, and wind is another important area of study.

5. The circumstellar disks associated with forming stars provide the background state for the formation of planets. Now that the formation and evolution of circumstellar disks is becoming understood, studies of planet formation should be carried out for more general initial conditions, i.e., more typical circumstellar disks. In particular, planet formation should be studied in disk environments that are substantially more massive than the minimum-mass solar nebula.

ON THE TIDAL INTERACTION BETWEEN PROTOSTELLAR DISKS AND COMPANIONS

D. N. C. LIN
Lick Observatory

and

J. C. B. PAPALOIZOU
Queen Mary and Westfield College

Formation of protoplanets and binary stars in a protostellar disk modifies the structure of the disk. Through tidal interactions, energy and angular momentum are transferred between the disk and protostellar or protoplanetary companion. Here, we summarize recent progress in theoretical investigations of the disk-companion tidal interaction. We show that low-mass protoplanets excite density waves at their Lindblad resonances and that these waves are likely to be dissipated locally. When a protoplanet acquires sufficient mass, its tidal torque induces the formation of a gap in the vicinity of its orbit. Gap formation leads to the termination of protoplanetary growth by accretion. For proto-Jupiter to attain its present mass, we require that (1) the primordial solar nebula is heated by viscous dissipation; (2) the viscous evolution time scale of the nebula is comparable to the age of typical T Tauri stars with circumstellar disks; and (3) the mass distribution in the nebula is comparable to that estimated from a minimum-mass nebula model. Tidal interaction between a disk and a binary star also leads to gap formation in the disk. Observations of binary T Tauri stars in a disk environment may provide important constraints on the origin of binary stars as well as on the nature of the companion-disk interaction.

I. INTRODUCTION

Recent observations indicate that a significant fraction of pre-main-sequence stars are deduced to have disks with masses in the range between 0.01 and 0.1 $M_\odot$ and sizes from 10 to 100 AU (Strom et al. 1989a). Evidence for protoplanetary accretion disks comes from infrared and ultraviolet excesses which are interpreted as radiation from the disk and the accretion disk-star boundary layer, respectively (Adams et al. 1987; Shu et al. 1987a; Kenyon and Hartmann 1987). These observational data indicate that environments similar to those that led to the formation of the solar system may be common (see Chapter by Shu et al.). Theoretical investigation of these disks provides estimates on the physical conditions and evolutionary time scale in the disk and therefore imposes constraints on models of solar-system formation (Cameron 1985,1988). For example, on the time scale ~1 Myr, a large fraction of

T Tauri stars lose any indication of the presence of a circumstellar disk (Walter 1987a). If protostellar disks viscously evolve on this time scale, the magnitude of effective viscosity needs to be much larger than that of molecular viscosity. One possible source of effective transport of angular momentum is convection driven turbulence (Lin and Papaloizou 1980,1985; Chapter by Adams and Lin). From self consistent convective accretion disk models, we can deduce the temperature and density distribution near the present location of the planets (Ruden and Lin 1986).

To apply protoplanetary disk models to cosmogonical theory, it is important to note that protogiant planets must have been formed on a time scale comparable to or shorter than the viscous time scale of the disk. A similar formation time scale is needed for prototerrestrial planets as for protogiant planets as the latter proceed through the formation of solid cores, which have masses comparable to those of the terrestrial planets before acquiring their massive gaseous envelopes (Bodenheimer and Pollack 1986). One important task for a cosmogonical theory is to use this short formation time scale constraint to deduce the physical condition in the protostellar disks. In this chapter, we discuss the tidal interaction between a protoplanet and a protostellar disk and its implications for protogiant planetary formation.

Planetesimals coagulate through cohesive collisions (Safronov 1972). When the planetesimals acquire a few Earth masses, they begin to accrete nebula gas rapidly and become protogiant planets (Bodenheimer 1985). Through its tidal torque on the protostellar disk, a companion, such as a protoplanet or a binary protostar, can excite waves which carry angular momentum and energy. In the regions of the disk interior to the companion's orbit, the waves carry negative angular momentum and energy with respect to the local fluid. When the wave is dissipated, it deposits negative angular momentum and causes the disk material to drift inward. Similarly, in the regions of the disk exterior to the protoplanet, positive angular momentum is deposited and the disk material drifts outward. In Sec. II, we determine the rate of angular momentum exchange between a companion and the disk.

Companion-protostellar disk interaction can lead to modification in the disk structure. If tidally excited waves can propagate deep into the interior of the disk, the tidal influence of the companion is distributed over an extended region, whereas if these waves are dissipated in the close vicinity of the companion's orbit, the tidal effect is localized. In Sec. III, we discuss the response of a gaseous disk to the companion's tide under various physical conditions. Typical protostellar disks have thermally stratified vertical structure, in the sense that the gas temperature decreases with distance from the midplane. Differential propagation speed due to thermal stratification retards wavefronts and refraction transmits waves into the tenuous disk atmosphere. Consequently, wave propagation in the radial and vertical directions are linked. In Sec. IV we illustrate the propagation of density waves in a gaseous protostellar disk.

When the mass of the companion is sufficiently large, its tidal torque may induce the formation of a gap in the vicinity of its orbit. There are two conditions for gap formation: the Roche radius of the companion must be comparable to or larger than the disk thickness and the rate of angular momentum transfer between the companion and the disk must exceed that in the disk due to viscous effects. In Sec. V, we illustrate these conditions, and in Sec. VI discuss the implication of gap formation on the growth of protoplanets in protostellar disks in Sec. VI.

After a protoplanet opens a gap in the vicinity of its orbit, mass growth of the protoplanet is terminated. However, its tidal interaction with the protoplanetary disk continues to operate until the disk material is dissipated. Such interaction causes the orbit of the protoplanet to evolve. If most of the interactions occur at the Lindblad resonance, there may be an imbalance between the exchange rate of energy versus angular momentum which would lead to an excitation of the protoplanet's orbital eccentricity. We determine the rate of change in the companion's orbital eccentricity in Sec. VII.

When gap formation first occurs, the amount of material cleared to either side of the gap is determined by the motion of the protoplanet. The disk responds by modifying the local surface density gradient such that the angular momentum transfer between the protoplanet and the interior regions of the disk is balanced by that between the protoplanet and the exterior regions of the disk. However, the disk interior to the protoplanet would be depleted on a viscous time scale whereas that exterior to the protoplanet retains its mass. This process could lead to substantial orbital migration for the protoplanet. In Sec. VIII, we illustrate the secular evolution of protoplanetary orbit and the disk structure.

Binary stars are very common and they are most likely to have formed in protostellar disks. In contrast to a protoplanet, the tidal effect of a binary with nearly equal mass ratio is widely distributed rather than concentrated near the Lindblad resonances on the disk. In Sec. IX, we determine the disk structure under the tidal influence of a binary star. Finally, in Sec. X, we outline some outstanding issues to be addressed in the future.

II. RATE OF ANGULAR MOMENTUM EXCHANGE FOR EXTREME MASS RATIO

The simplest derivation of the rate of tidally induced angular momentum transfer is based on an impulsive approximation in which the disk is composed of a collection of particles orbiting around a proto-Sun whose mass M is much larger than that of its companion M_p. In this limit, the tidal effect of the companion acts only during close encounters which deflect the orbit of a disk particle, assumed initially circular, by a small angle δ_e during a close passage such that

$$\cot^2(\delta_e/2) = u_e^4 \Delta^2 / G^2 M_p^2 \qquad (1)$$

where u_e is the relative velocity which is counted positive if the relative motion is in the same sense as the binary, and $\Delta = |r_0 - r_c|$, where r_c and r_0 are the radii of the unperturbed circular orbit of the disk particle and companion, respectively. The corresponding change in angular momentum of the particle is

$$\Delta h_e = -u_e(1 - \cos\delta_e)r_c. \tag{2}$$

If the unperturbed particle and the companion circulate around the proto-Sun with angular frequency Ω and ω respectively, $u_e = (\Omega - \omega)r_o$, so that

$$\Delta h_e = -\frac{2G^2M_p^2}{r_c^2(\Omega - \omega)^3\Delta^2}. \tag{3}$$

For the disk that is interior to the companion, $\Omega > \omega$, and Δh_e is negative corresponding to a loss of angular momentum from the disk to the companion. For an external disk, the transfer of angular momentum is from the companion to the disk. In addition to this angular momentum exchange, the particle is given some epicyclic motion so that in the absence of interactions, the subsequent encounter will be such that the particle is in an eccentric orbit. However, if we assume that the epicyclic motion of the particle is damped out by some internal dissipative process between successive encounters, the particle will return to a circular orbit. There will then be a net angular momentum transfer rate between the disk and the companion.

Because the time between encounters is $2\pi/|\Omega - \omega|$, the rate of angular momentum transfer becomes

$$\dot{h}_e = -\frac{G^2M_p^2|\Omega - \omega|}{\pi\Delta^2r_c^2(\Omega - \omega)^3}. \tag{4}$$

Using Taylor's expansion on $\Omega - \omega$ and working correct to first order in Δ, we can deduce the total rate of angular momentum change in an entire disk interior to the companions orbit to be

$$\dot{H}_T = \int_{\Delta_o}^{\infty} 2\pi\Sigma\dot{h}_e r_o d\Delta = -\frac{2G^2M_p^2\Sigma}{r_o(d\Omega/dr)_o^2}\int_{\Delta_o}^{\infty}\frac{d\Delta}{\Delta^4} = -\frac{8q^2\Sigma r_o^4\omega^2}{27}\left(\frac{r_o}{\Delta_o}\right)^3 \tag{5}$$

where $q \equiv M_p/M$ is the mass ratio, Σ is the surface density in the disk, assumed uniform in the interaction region, and Δ_o is the distance between the disk edge and the orbit of the companion. For an external disk, the same formula applies but the sign of $\dot{H}_T$ is reversed.

The above result provides a simple evaluation of $\dot{H}_T$. However, it does not include the effect of the proto-Sun which causes the particle's orbit to curve prior to and after the impulsive encounters. For distant encounters in which Δ is larger than the Roche radius $r_R = (q/3)^{1/3}r_o$, the tidal effect of the proto-Sun is very important. This can be incorporated with a more sophisticated perturbation analysis of the particle orbit around the proto-Sun. Consider a particle which, in the absence of the companion, has a circular orbit with a radius r_c and angular frequency Ω. We define the cylindrical polar coordinates of the particle based on the proto-Sun (r_p, ϕ_p) in an inertial frame to be such that $\phi_p = 0$ is the line joining the proto-Sun, its companion, and the particle on a unperturbed orbit at $t = 0$. The equations of motion for the particle are

$$\frac{d^2 r_p}{dt^2} - r_p \left(\frac{d\phi_p}{dt} \right) = -\frac{\partial(\Psi - \psi_c)}{\partial r_p} \tag{6}$$

and

$$r_p^2 \frac{d^2 \phi_p}{dt^2} + 2 \frac{dr_p}{dt} \frac{d\phi_p}{dt} r_p = -\frac{\partial \psi_c}{\partial \phi_p} \tag{7}$$

where the potential of the proto-Sun and a companion in circular orbit are

$$\Psi = -GM/r_p \tag{8}$$

and

$$\psi_c = -\frac{GM_p}{[r_o^2 + r_p^2 - 2r_o r_p \cos(\phi_p - \omega t)]^{1/2}} + \frac{GM_p r_o \cos(\phi_p - \omega t)}{r_p^2}. \tag{9}$$

We determine the amount of angular momentum transfer from a particle to the companion as $\phi_p - \omega t$ goes from $-\pi/2$ to $\pi/2$. In the limit $M_p << M$, Eqs. (6) and (7) may be linearized in terms of variables $\delta_p = r_p - r_c$ and $\alpha_p = \phi_p - \Omega t$, with $\alpha_p = \dot{\alpha}_p = \delta_p = \dot{\delta}_p = 0$ initially, such that

$$\frac{d^2 \delta_p}{dt^2} + \Omega^2 \delta_p = S \tag{10}$$

and

$$r_c^2 \frac{d\alpha_p}{dt} + 2\Omega r_c \delta_p = \frac{\psi_c}{(\omega - \Omega)} \tag{11}$$

where $S = -(\partial \psi_c / \partial r) + 2\psi_c \Omega / [r_c(\omega - \Omega)]$ is evaluated for $r_p = r_c$ and $\phi_p = \Omega t$. Using a Fourier series expansion

$$S = \sum_{m=0}^{\infty} a_m \cos m(\omega - \Omega)t \tag{12}$$

the solution of Eq. (10) becomes

$$\delta_p = \sum_{m=0}^{\infty} \frac{a_m [\cos m(\omega - \Omega)t - A_m \cos(\Omega t + \epsilon_m)]}{[\Omega^2 - m^2(\omega - \Omega)^2]} \tag{13}$$

where $\epsilon_m = \pi \Omega / 2(\Omega - \omega) + m\pi/2$ and $A_m = 1$ for even m and $A_m = m(\omega - \Omega)/\Omega$ for odd m. In the limit that $|r_0 - r_c| << r_0$ but $m >> 1$ such that $\xi = m|r_0 - r_c|/r_0$ is of order unity,

$$a_m = \frac{2GM_p r_0 \xi}{\pi r_0^2 |r_0 - r_c|} \left[K_1(\xi) + \frac{4}{3\xi} K_0(\xi) \right] \tag{14}$$

where K_0 and K_1 are modified Bessel functions (Goldreich and Tremaine 1982).

The angular momentum transfer to the companion is

$$\int_{-\pi/|2(\Omega-\omega)|}^{\pi/|2(\Omega-\omega)|} \frac{\partial \psi_c}{\partial \phi} dt = -\Delta h_e. \tag{15}$$

This must be evaluated to second order. After some algebra, one obtains

$$\Delta h_e = \frac{-1}{2(\Omega-\omega)} \left[\sum_m \frac{2\Omega a_m A_m \sin \epsilon_m}{[\Omega^2 - m^2(\omega-\Omega)^2]} \right]^2 \tag{16}$$

which differs only by a correction factor of $\frac{1}{2}$ from the corresponding previous expressions (Lin and Papaloizou 1979a).

If the disk is interior to the orbit of the companion, $\Omega > \omega$ and angular momentum is transferred to the companion. For a close encounter the major contribution to the sum comes from a few values of $m \sim m_0 = \Omega/(\Omega-\omega) = 2r_0/3|r_0 - r_c|$, where we can use a first-order Taylor expansion to evaluate $(\Omega-\omega)$. To perform the summation in Eq. (16) in the limit of a close encounter we assume that a_m is slowly varying with m so that it may be approximated as the constant value a_{m_0}. Using the results that for any w

$$\sum_{n=-\infty}^{n=\infty} \frac{(-1)^n}{(w^2 - 4n^2)} = \frac{\pi}{2w \sin(\pi w/2)} \tag{17}$$

and

$$\sum_{n=-\infty}^{n=\infty} \frac{(-1)^n (2n+1)}{w[w^2 - (2n+1)^2]} = \frac{-\pi}{2w \cos(\pi w/2)} \tag{18}$$

we may perform the summation in Eq. (16) with the result that to within a fractional error of order $1/m_0$,

$$\Delta h_e = \frac{-\pi^2}{2(\Omega-\omega)^3} \left(a_{m_0}\right)^2 \tag{19}$$

or equivalently after using Eq. (14)

$$\Delta h_e = -\frac{64G^2 M_p^2 r_0}{243\omega^3 \Delta^5} \left[2K_0\left(\frac{2}{3}\right) + K_1\left(\frac{2}{3}\right) \right]^2 \tag{20}$$

(see also Julian and Toomre 1966; Goldreich and Tremaine 1982).

Equation (20) differs from Eq. (3) by a numerical factor of order unity and so one derives an expression for $\dot{H}_T$, which differs from that given by Eq. (5), by the same factor, namely

$$\dot{H}_T = -\frac{32q^2 \Sigma r_0^4 \omega^2}{243} \left(\frac{r_0}{\Delta_o}\right)^3 \left[2K_0\left(\frac{2}{3}\right) + K_1\left(\frac{2}{3}\right) \right]^2. \tag{21}$$

If the disk is viscous and in Keplerian rotation, there will be an outward flux of angular momentum at radius r given by

$$\dot{H}_v = 3\pi \Sigma v r^2 \Omega \qquad (22)$$

where v is the effective vertically averaged kinematic viscosity (Lynden-Bell and Pringle 1974). The angular momentum exchange between the disk and the companion is always such as to oppose the effect of viscosity. If for $r \sim r_0$, we have $\dot{H}_v < |\dot{H}_T|$, the companion's tidal effect is more effective in transporting angular momentum than the viscous process. In this case, the disk region in the vicinity of the companion's orbit may be depleted and a gap formed (Lin and Papaloizou 1979a). The physical process and implication for gap formation will be discussed in more detail in Secs. V and VI. Note that if a gap is formed as a result of tidal truncation, the width of the gap must be at least comparable to the companion's Roche radius so that $\Delta_o \sim r_R$. Thus, using Eqs. (5) and (21), the critical condition for gap formation becomes

$$q > 81\pi v/(8r_o^2 \omega) \qquad (23)$$

such that the companion must have sufficient mass to induce gap formation in the disk.

The particle impulse approximation used above is simple to derive and can be adequately applied to particle disks such as planetary rings. However, caution in applying these results to gaseous disks has been urged by Lunine and Stevenson (1982). They have stressed the need to consider the detailed nature of the dissipative process, and have noted that the tidally induced disturbance may propagate throughout the disk, delocalizing the tidal torque, when the dissipation is very weak. They also noted that the formation of the gap changes the surface density profile, and, in turn, that could change the position of the Lindblad resonances where the tidal disturbances are launched. Greenberg (1982) has also emphasized the need for a proper consideration of the dissipative process. He suggested that the rate of tidally induced angular momentum transfer may be significantly reduced if the dissipation is weak. In the next section, we examine the linear propagation of a companion's tidal disturbances in a gaseous disk and the importance of dissipation in determining the angular momentum transfer rate.

III. RESPONSE OF A GASEOUS DISK TO THE COMPANION'S TIDE

In typical models of protoplanetary disks, the vertical scale height of the disk H_D is typically 0.05 to 0.1 times the distance to the proto-Sun (Lin and Papaloizou 1985). This scale height is comparable to the Roche radius of companions with mass in the range of 10^{-4} to 10^{-3} $M_\odot$. When low-mass companions, such as protogiant planets, become sufficiently massive to open up a gap, the pressure scale height in the radial direction is approximately

r_R which is comparable to or smaller than that in the vertical direction H_D and cannot be neglected in describing the tidal interaction near the disk edge. The dominant contribution of the pressure effect in the radial direction is to delocalize the companion's tidal torque.

A. Basic Equations

The first step in the analysis of the tidal interaction between a companion and a gaseous disk is to determine the disk's structure. The structure and evolution of a general (not necessarily thin) gaseous disk can be deduced from the equations of fluid dynamics. The rate of change in density ρ is described by the continuity equation

$$\frac{\partial \rho}{\partial t} + \nabla \cdot \rho \mathbf{v} = 0. \tag{24}$$

The evolution of the velocity $\mathbf{v}$ is determined by the momentum equation

$$\frac{D\mathbf{v}}{Dt} = -\frac{1}{\rho}\nabla p + \frac{1}{\rho}\nabla \underline{\underline{\sigma}} - \nabla(\Psi + \psi_c + \psi_d) \tag{25}$$

where p is the pressure. The material derivative D/Dt is composed of both the local time derivative $\partial/\partial t$ and the advective contribution $\mathbf{v} \cdot \nabla$. The stress tensor $\underline{\underline{\sigma}} = \sigma_{ij} = 2\rho v_s(e_{ij} - \Delta_d \delta_{ij}/3) + \rho v_b \Delta_d \delta_{ij}$, where v_s and v_b are the shear and bulk viscosity, the rate of strain tensor $e_{ij} = (\partial v_i/\partial x_j + \partial v_j/\partial x_i)/2$ and $\Delta_d = e_{ii}$. The gravitational potential due to the central object is Ψ, that due to an orbiting companion is ψ_c, and that due to the disk matter is ψ_d. The rate of change in thermal energy can be derived from the energy equation such that

$$C_v \rho \left[\frac{DT}{Dt} - (\Gamma_3 - 1)\frac{T}{\rho}\frac{D\rho}{Dt} \right] = D_v - \nabla \cdot \mathbf{F} \tag{26}$$

where C_v and Γ_3 are the specific heat and adiabatic index and $\mathbf{F}$ is the total heat flux. The local rate of dissipation is given by

$$D_v = 2\rho v_s(e_{ij} - \Delta_d \delta_{ij}/3)(e_{ij} - \Delta_d \delta_{ij}/3) + \rho v_b \Delta_d^2. \tag{27}$$

Together with a viscosity prescription and an equation of motion, these basic equations describe completely the response of a disk due to the interplay between a companion's tidal torque, thermal effects, and viscous evolution.

To deduce the disk response to the tidal perturbation of the companion, it is convenient to adopt a thin disk approximation in which there is no vertical motion, and $\Sigma(\equiv \int_{-\infty}^{\infty} \rho dz)$, and $P (\equiv \int_{-\infty}^{\infty} p dz)$ are the surface density and vertically integrated pressure, respectively. In a nonrotating cylindrical coordinate system (r, ϕ, z) defined with the origin at the proto-Sun and z being the vertical coordinate, the radial component of the equation of motion is

$$\frac{\partial v_r}{\partial t} + v_r \frac{\partial v_r}{\partial r} + \frac{v_\phi}{r}\frac{\partial v_r}{\partial \phi} - \frac{v_\phi^2}{r} = -\frac{1}{\Sigma}\frac{\partial(P + q_v)}{\partial r} - \frac{\partial(\Psi + \psi_c + \psi_d)}{\partial r} \tag{28}$$

where v_r and v_ϕ are the radial and azimuthal components of velocity, and for our numerical work to be presented below, we use a pseudoviscous pressure q_v which enables the propagation of shock waves. The azimuthal component of the equation of motion in the flat disk model is

$$\frac{\partial v_\phi}{\partial t} + \frac{v_\phi}{r}\frac{\partial v_\phi}{\partial \phi} + v_r\frac{\partial v_\phi}{\partial r} + \frac{v_\phi v_r}{r} = -\frac{1}{r\Sigma}\frac{\partial(P + q_v)}{\partial \phi} - \frac{\partial(\psi_c + \psi_d)}{\partial r} + F_v \quad (29)$$

where

$$F_v = \frac{1}{\Sigma r^2}\frac{\partial}{\partial r}\left(v_s \Sigma r^3 \frac{\partial \Omega}{\partial r}\right). \quad (30)$$

The gravitational potential due to the disk material is

$$\psi_d = -G\int\int \frac{\Sigma(r', \phi', t)r'dr'd\phi'}{[r^2 + r'^2 - 2rr'\cos(\phi - \phi')]^{1/2}}. \quad (31)$$

In Eqs. (28) and (29), the detailed treatment of an equation of state is not important. For illustrative purposes and computational convenience, a barotropic equation of state $P = P(\Sigma)$ is usually used and the local sound speed c is defined through $c^2 \equiv dP/d\Sigma$. Finally, the continuity equation becomes

$$\frac{\partial \Sigma}{\partial t} + \frac{1}{r}\frac{\partial}{\partial r}\Sigma r v_r + \frac{1}{r^2}\frac{\partial}{\partial \phi}\Sigma r v_\phi = 0. \quad (32)$$

The main motion in the disk is rotation in the azimuthal direction and its angular frequency can be deduced from Eq. (28) in the limit of negligible v_r, q, ψ_c and ψ_d such that

$$v_\phi^2/r = \Omega^2 r = \frac{1}{\Sigma_o}\frac{dP_o}{dr} + \frac{d\Psi}{dr} \quad (33)$$

where the subscript o denotes an unperturbed axisymmetric value. In the thin disk limit when the sound speed of the disk is very small, $\Omega^2 \sim GM/r^3$ which is the Keplerian rotation law. Using these equations, we calculate the response of the disk to the companion's tidal disturbance and the associated rate of angular momentum transfer.

B. Linear Calculations of Disk Response

In order to calculate the response of a gaseous disk and the propagation of disturbances induced therein by the presence of a companion, it is useful to carry out a linear analysis in which the nonaxisymmetric perturbations to the disk motion are assumed small. We linearize Eqs. (28) and (29) such that

$$\frac{\partial v_r'}{\partial t} + \Omega\frac{\partial v_r'}{\partial \phi} - 2\Omega v_\phi' = -\frac{\partial}{\partial r}\left(\frac{\Sigma' c^2}{\Sigma_o}\right) - \frac{\partial \psi_c'}{\partial r} + F_r' \quad (34)$$

and

$$\frac{\partial v'_\phi}{\partial t} + \Omega \frac{\partial v'_\phi}{\partial \phi} + \frac{v'_r}{r}\frac{d(\Omega r^2)}{dr} = -\frac{1}{r}\frac{\partial}{\partial \phi}\left(\frac{\Sigma' c^2}{\Sigma_o}\right) - \frac{1}{r}\frac{\partial \psi_c}{\partial \phi} + F'_\phi. \tag{35}$$

In the above equations, perturbed quantities are denoted by a prime. In the unperturbed axisymmetric flow, $\Sigma = \Sigma_o$, $v_r = 0$ and $v_\phi = \Omega r$. Contributions arising from the viscous terms are included in F'_r and F'_ϕ. For most cases to be considered here, the contribution from self gravity is neglected. When the companion is in a circular orbit with orbital frequency ω, ψ_c is a function of ϕ and t only through the combination $\phi - \omega t$ and therefore the perturbed quantities are also functions of the same combination. Thus, $\partial/\partial t = -\omega \partial/\partial \phi$ and the equations of motion can be written

$$(\Omega - \omega)\frac{\partial v'_r}{\partial \phi} - 2\Omega v'_\phi = -\frac{\partial}{\partial r}\left(\frac{\Sigma' c^2}{\Sigma_o}\right) - \frac{\partial \psi_c}{\partial r} + F'_r \tag{36}$$

and

$$(\Omega - \omega)\frac{\partial v'_\phi}{\partial \phi} + \frac{v'_r}{r}\frac{d(\Omega r^2)}{dr} = -\frac{1}{r}\frac{\partial}{\partial \phi}\left(\frac{\Sigma' c^2}{\Sigma_o}\right) - \frac{1}{r}\frac{\partial \psi_c}{\partial \phi} + F'_\phi. \tag{37}$$

The perturbed continuity equation is

$$\frac{\partial \Sigma'}{\partial t} + \Omega\frac{\partial \Sigma'}{\partial \phi} = -\frac{1}{r}\left[\frac{\partial}{\partial r}(r v'_r \Sigma_o) + \frac{\partial}{\partial \phi}(\Sigma_o v'_\phi)\right] \tag{38}$$

which becomes

$$(\Omega - \omega)\frac{\partial \Sigma'}{\partial \phi} = -\frac{1}{r}\left[\frac{\partial}{\partial r}(r v'_r \Sigma_o) + \frac{\partial}{\partial \phi}(\Sigma_o v'_\phi)\right]. \tag{39}$$

One important quantity to be determined is the flux of angular momentum through a circle of constant radius

$$F_H = \int_0^{2\pi} \Sigma_o r^2 v'_r v'_\phi d\phi. \tag{40}$$

The lowest-order contribution to F_H is of second order in the perturbed variables. This is because the first-order terms average to zero. In a region where the direct forcing and hence ψ_c may be neglected, after some algebra (Papaloizou and Lin 1984), F_H may be expressed in terms of the radial Lagrangian dispacement, ξ_r, such that

$$F_H = -\int_0^{2\pi} \Sigma' c^2 r \frac{\partial \xi_r}{\partial \phi} d\phi - \int_0^{2\pi} \Sigma_o r^2 \xi_r F'_\phi d\phi. \tag{41}$$

Using Eqs. (36), (37), (39) and (40), we can also derive

$$\frac{dF_H}{dr} = \frac{1}{(\omega - \Omega)} \left\{ \frac{d}{dr} \left[(\Omega - \omega) \int_0^{2\pi} \Sigma_o r^2 \xi_r F_\phi' d\phi \right] \right.$$
$$\left. + \int_0^{2\pi} (F_r' v_r' + F_\phi' v_\phi') \Sigma_o r d\phi \right\}. \tag{42}$$

To illustrate the direction of angular momentum flow, we consider the special case where $F_\phi' = 0$ but $F_r' = -kv_r'$, with k being a positive constant corresponding to a dissipative force (similar conclusions follow in more general cases) so that $dF_H/dr = -\epsilon_D/(\omega - \Omega)$ where $\epsilon_D = -\int_0^{2\pi} F_r' v_r' \Sigma_o r d\phi$ is the rate of energy dissipation per unit length which is positive. If the companion is interior to the disk, $\omega > \Omega$ and F_H decreases with r. Because F_H vanishes at large distances from the companion, F_H is positive close to the companion and the disk receives angular momentum. Similarly, the disk loses angular momentum when the companion is exterior to the disk. In both cases, there is an outward flux of angular momentum with $F_H > 0$.

To determine the magnitude of F_H as a function of r, we need to determine the disk response from the perturbed equation of motion. Note that in the absence of dissipation ($F_r' = F_\phi' = 0$), Eq. (42) implies F_H is constant and there is zero transfer of angular momentum. For computational convenience, we first consider the dissipation-free limit and incorporate the effect of dissipation at the end of this section. In the limit that $M_p << M$, it is convenient to expand ψ_c in Eqs. (28) and (29) into a Fourier series (Brouwer and Clemence 1961) where

$$\psi_c = \text{Re} \sum_{m=0}^{\infty} W_m \exp[im(\phi - \omega t)] \tag{43}$$

where for the case of extreme mass ratio

$$W_m = -\frac{GM_p}{\pi} \int_0^{2\pi} \frac{\cos m\phi \, d\phi}{(r_o^2 + r^2 - 2rr_o \cos \phi)^{1/2}} = -\frac{2GM_p}{\pi r_o} K_0 \left(\frac{m\Delta}{r_o} \right) \tag{44}$$

where $\Delta = r - r_o$, and r_o is the orbital radius of the companion (Goldreich and Tremaine 1982). Equations (36) and (37) become

$$\frac{1}{\Sigma_o r} \frac{d}{dr} (\Sigma_o r \xi_{rm}) - \frac{2\Omega \xi_{rm}}{r(\Omega - \omega)} = \left[\frac{K_m}{r^2(\Omega - \omega)^2} - \frac{K_m}{c^2} \right] + \frac{W_m}{r^2(\Omega - \omega)^2} \tag{45}$$

and

$$\frac{dK_m}{dr} + \frac{2\Omega K_m}{r(\Omega - \omega)} = \xi_{rm}[m^2(\Omega - \omega)^2 - \Omega^2] - \frac{dW_m}{dr} - \frac{2\Omega W_m}{r(\Omega - \omega)}, \tag{46}$$

where $K_m = \Sigma_m' c^2 / \Sigma_o$, Σ_m' and ξ_{rm} are the coefficients in the Fourier expansion of Σ' and ξ_r, respectively.

The two linearized equations of motion may be combined to give a single second-order differential equation for ξ_{rm} from which the disk response to the companion's tide may be determined. For illustration purposes, let us consider the case in which the disk is tidally truncated and Σ_o vanishes near r_o. To avoid singularity, the boundary condition at the disk edge is

$$K_m = -\frac{\xi_{rm}}{\Sigma_o}\frac{dP_o}{dr}. \tag{47}$$

In the asymptotic limit far away from the companion, the general solution of Eqs. (45) and (46) can be obtained using a WKB approximation such that

$$\xi_{rm} = \frac{(\eta_1 e^{i\mu} + \eta_2 e^{-i\mu})}{\{\Sigma_o rc[m^2(\Omega - \omega)^2 - \Omega^2]^{1/2}\}^{1/2}} \tag{48}$$

where η_1 and η_2 are complex constants,

$$\mu = \int_{r_1}^{r} \frac{[m^2(\Omega - \omega)^2 - \Omega^2]^{1/2}}{c}\,dr \tag{49}$$

and r_1 is an arbitrary constant radial location chosen such that the WKB approximation is valid there. In addition r must be such that the WKB approximation is valid in the interval $[r_1, r]$. Equation (48) implies that waves are propagating away from and towards the companion with amplitudes η_1 and η_2 respectively. Inserting this solution into the expression (41), the flux of angular momentum carried by the waves is

$$F_H = \pi m(|\eta_1|^2 - |\eta_2|^2). \tag{50}$$

The limiting case $\eta_2 = 0$ corresponds to pure outward transport of angular momentum while $\eta_1 = \eta_2$ corresponds to zero transport. In general η_2 may be finite if there is some wave reflection at large distances from the location where the waves are launched. The ratio η_1/η_2 has to be determined from the boundary condition at large distances from the companion.

Equations (45) and (46) can be combined into a single second-order ordinary differential equation which can be solved by the method of variation of parameters. It turns out that F_H can be determined if the solution of the unforced equations ($W_m = 0$) that satisfies the correct boundary condition (47) is known (Papaloizou and Lin 1984). We denote ξ_{rm} for this solution by X_m. At large distances from the companion, say for $|r - r_o| > r_1$, this takes on the WKB form

$$X_m = (Ce^{i\mu} + C^*e^{-i\mu})/\{\Sigma_o rc|[m^2(\Omega - \omega)^2 - \Omega^2]^{1/2}|\}^{1/2} \tag{51}$$

where C is a complex constant which satisfies

$$|C|^2 = \frac{\Sigma_o rc^3}{4}\left[\frac{(dX_m/dr)^2}{|[m^2(\Omega - \omega)^2 - \Omega^2]^{1/2}|} + \frac{X_m^2|[m^2(\Omega - \omega)^2 - \Omega^2]^{1/2}|}{c^2}\right]. \tag{52}$$

We then find that

$$F_H = \frac{\pi nf}{4|C|^2} \left| \int_{r_o}^{r_1} \Sigma_o r S_m X_m \mathrm{d}r \right|^2 \tag{53}$$

where

$$f = \frac{(1 - y^2)}{(1 + y^2 - 2y\cos\chi)} \tag{54}$$

$$y = |\eta_2|/|\eta_1| \tag{55}$$

$$\chi = 2\arg(C) + \arg(\eta_2) - \arg(\eta_1) \tag{56}$$

$$S_m = \frac{\mathrm{d}}{\mathrm{d}r}\left(\frac{W_m\alpha}{c^2}\right) + \frac{2\Omega W_m\alpha}{r(\Omega - \omega)c^2} \tag{57}$$

and

$$\alpha = c^2 r^2 (\Omega - \omega)^2 / [c^2 - r^2(\Omega - \omega)^2]. \tag{58}$$

The effect of distant boundary conditions is contained in the factor f through the phase argument χ. Equation (58) indicates that the asymptotic solution of X_m oscillates on a very small scale so that only a very slight change in the condition at large Δ can produce a large change in χ. The factor y is a measure of the strength of dissipation and we suppose $y<1$. If $y = 0$, only outward transport of angular momentum is present which means that energy is dissipated before it can reach any distant boundary and be reflected. If $y = 1$, no dissipation takes place. In this case, f becomes infinite if cos $\chi = 1$. This corresponds to a resonance and means that the orbital rotation frequency is equal to some free normal mode rotation frequency. The response is singular because there is no dissipation. In principle, if nonlinear effects are neglected, the magnitude of y and χ are determined by the size of the viscosity and details of very distant behavior. In practice, χ is so sensitive to small changes in distant conditions that we can take an average over 2π and replace f with $\int_0^{2\pi} f\mathrm{d}\chi/2\pi$. This corresponds to acknowledging that the disk and the companion are only slowly evolving, and calculating an average angular momentum transport over a range of disk-defining parameters or orbital frequencies. Physically, one requires that, as the system evolves, the full tidal response can be established, i.e., the time required to evolve χ through 2π be at least equal to the dissipation time scale associated with the tide. If we are considering a disk changing through viscous evolution, with viscosity as the dissipative agent acting on the tide, the above condition should be met even if the viscosity is small. Now

$$\frac{1}{2\pi} \int_0^{2\pi} f\mathrm{d}\chi = 1, \tag{59}$$

independent of the size of y so that f in Eq. (53) can be replaced by unity. Thus, the net angular momentum flux carried by waves associated with a particular m value is

$$F_H = \frac{\pi m}{4|C|^2} \left| \int_{r_o}^{r_1} \Sigma_o r S_m X_m dr \right|^2 \equiv \frac{\pi m}{4|C|^2} \left| \int_{r_o}^{r_1} \frac{\Sigma_o r W_m K_m}{c^2} dr \right|^2 . \quad (60)$$

Here the second equivalent form of the integral may be derived from Eqs. (45) and (46) (with $W_m \equiv 0$ and $\xi_{rm} \equiv X_m$) by integrating by parts under the assumption that the effect of the perturbing potential is negligible at the boundary most distant from the companion. To get the total angular momentum flux we must sum over m. We remark that Eq. (60) holds for any mass ratio and does not require the presence of a Lindblad resonance directly in the disk.

C. A Companion with Extreme Mass Ratio

To calculate the rate of angular momentum flux in a particular case, a specific equation of state and Σ_o distribution needs to be specified in order to evaluate X_m. For illustrative purposes, we adopt a polytropic equation of state

$$P = K \Sigma_o^{1+1/n} \quad (61)$$

and a disk model with a surface density profile which cuts off to avoid the companion

$$\Sigma = \Sigma_\infty (\tanh(\Delta - \Delta_o)/H_r)^n . \quad (62)$$

The characteristic radial scale length can be parameterized in terms of $H_r = (\frac{n}{\beta})^{1/2} c_\infty/\omega$ where c_∞ is the asymptotic sound speed and β is a constant. From Eq. (33) we obtain, for the case when the companion orbits inside the disk,

$$\frac{\Omega}{\omega} - 1 = \frac{H_r}{2r_o} \left(-3x + \frac{\beta}{\cosh^2(x - \delta)} \right) \quad (63)$$

with $x = \Delta/H_r$ and $\delta = \Delta_o/H_r$. Consequently the epicycle frequency becomes

$$\kappa^2 = \frac{2\Omega}{r} \frac{d(\Omega r^2)}{dr} = \omega^2 \left(1 - \frac{2\beta \tanh(x - \delta)}{\cosh^2(x - \delta)} \right) \quad (64)$$

such that the Rayleigh's stability criterion (which requires $\kappa^2 > 0$) is satisfied if the dimensionless constant $\beta < 3^{3/2}/4$ or equivalently,

$$H_r > 2n^{1/2}/3^{3/4} H_\infty \simeq H_\infty \quad (65)$$

where $H_\infty = c_\infty/\omega$ is the disk thickness; i.e., the characteristic density scale length in the radial direction cannot be less than the vertical disk scale height.

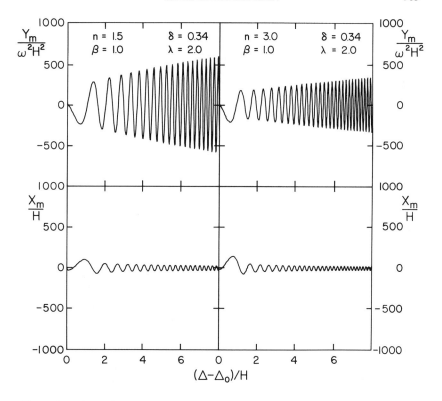

Figure 1.   Excitation of waves in a gaseous disk. The variables X_m and Y_m are the values of ξ_{rm} (the perturbed displacement) and K_m (the perturbed surface density), respectively, in the homogeneous solutions of Eqs. (45) and (46) in the limit $W_m = 0$. These examples correspond to a perturber at approximately one vertical disk scale height away from the disk edge, where the surface density tails off on a similar scale length, ($H \equiv H_r$, $\lambda = mH/r_0$).

A similar conclusion may be obtained when the companion orbits exterior to the disk.

With the magnitude of n, β, and Δ_o/H_r specified, Eqs. (45) and (46) can be solved numerically and the results obtained therefrom indicate that the WKB approximation becomes reasonably good $2H_r$ into the disk (Fig. 1). The total angular momentum exchange rate can be deduced by summing F_H over m. Because, in the case of extreme mass ratio, most of the contribution comes from large m, m can be treated as a continuous variable so that

$$\dot{H}_T = \int_0^\infty F_H \, dm. \qquad (66)$$

For illustration purposes, we adopt a particular case with $n = 1.5$ and $\beta = 1$ which corresponds to the disk thickness comparable to the gap width. The discussion in Sec. IV show that $\beta \simeq 1$ during the epoch that the conditions for gap formation are marginally satisfied. Using the asymptotic value of X_m calculated as a function of m from Eqs. (45) and (46), and F_H from Eq. (60), Eq. (66) yields a maximum value for the total outward angular momentum flow given by

$$\dot{H}_T = 0.23q^2 \Sigma_\infty r_o^4 \omega^2 \left(\frac{r_o}{H_\infty}\right)^3. \tag{67}$$

This value applies independently of whether the disk orbits externally or internally to the disk and is very similar to the $\dot{H}_T$ derived from the impulsive approximation given by Eq. (5), provided H_∞ and Δ_o are identified with each other.

The cold disk limit corresponds to the case where $\Delta_o/H_\infty \gg 1$. In this case, the torque decreases with Δ_o much as predicted by the impulse approximation in Eq. (5). The maximum value of the torque is attained when Δ_o is comparable to H_∞. The latter in turn must satisfy $H_\infty \geq r_0 q^{1/3}$ for the disk to lie exterior to the companions Roche lobe. Taking the lower limit, the viscous condition for gap formation derived from Eqs. (67) and (22) is

$$q > \frac{40\nu}{\omega r_o^2}. \tag{68}$$

This is very similar to the condition given by Eq. (23) derived from consideration of particle orbits. Note that when a gap is marginally formed, the maximum torque occurs when the edge of the disk around the protostar approaches the Roche lobe of the companion. In this case, from the reasoning that led to Eq. (65), a necessary condition to form a gap and avoid the violation of the Rayleigh stability criterion becomes $r_R \geq H_\infty$ or equivalently

$$q > (c_\infty/\omega r)^3. \tag{69}$$

Although the total rate of angular momentum transfer is independent of viscosity whether or not nonlinear effects operate, the extent of wave propagation is determined by the dissipation processes. From the dissipation-free results, we can estimate the condition for the tidally induced waves to reach a distance r_o before they are damped by some effective dissipation processes. We introduce an effective viscosity ν to parameterize the magnitude of viscosity. A wave with wavenumber $k \sim m\omega/c_\infty$ would decay by a factor $e^{-\nu k^2 t}$ after it has traveled for a time interval t. At a distance $\sim r_o$ from the companion, from the dispersion relation (to be given by Eq. 78 below) we may approximate the group velocity of the waves to be the sound speed. The appropriate value of $t \sim r_o/c_\infty$ so that the decay factor is $\sim e^{-\nu m^2 \omega^2 r_0/c_\infty^3}$. To reach r_o without large decay implies

$$\nu < \frac{c_\infty^3}{m^2 \omega^2 r_o}. \tag{70}$$

For important values of $m \sim r_o/H_\infty$ the above condition becomes

$$\nu/H_\infty^2 \omega < (H_\infty/r_o)^3.$$

In terms of the standard α-prescription (Shakura and Sunyaev 1973) in which $\nu = \alpha_D H_\infty^2 \omega$, waves are dissipated near the protoplanet if

$$\alpha_D > (H_\infty/r_o)^3. \tag{71}$$

Recent models of protostellar disks indicate that convection-driven turbulence may provide an effective viscosity with an $\alpha_D \sim 10^{-2}$ (Lin and Papaloizou 1985; Ruden et al. 1988; Chapter by Adams and Lin, and references therein) which suggests that tidally induced waves may be viscously dissipated in the vicinity where they are launched. In the following section, we show that even in the absence of any significant turbulent viscosity, tidally induced density waves cannot propagate over extended regions in the disk.

IV. PROPAGATION OF DENSITY WAVES EXCITED AT RESONANCES

When the companion's mass is very small compared with the central protostar, low order Lindblad resonances fall interior to the Roche radius of the protostar. If the disk is sufficiently extensive, so that it contains one or more of these resonances, density waves may be excited in their vicinity (Shu 1984). Density waves induced by such resonant tidal interactions have been observed by the voyager spacecraft in planetary rings (Cuzzi et al. 1984). The wave excited at these resonances carries both energy and angular momentum (Lynden-Bell and Kalnajs 1974). When the wave is dissipated, angular momentum is deposited in the fluid causing the disk material to evolve. If the waves can propagate deep into the interior region of the disk, the tidal influence of the satellite is distributed over an extended region whereas if these waves are dissipated in the close vicinity of the satellite's orbit, the tidal effect is localized. Thus, the propagation and dissipation of the waves determines the disk response to the companion's tidal field. In subsection IV.B, we examine how far the tidally induced waves can propagate in a protostellar disk.

A. Angular Momentum Flux at the Companion's Lindblad Resonances

Not all the resonances can be found in a disk. Consider the case where the companion is exterior to the disk. In the limit of a low mass companion and a geometrically thin disk, the disk's main motion is Keplerian rotation, $\Omega = (r_o/r)^{3/2}\omega$. In order for a low-order Lindblad resonance with $\omega/\Omega = (m-1)/m$ to fall inside the Roche lobe of the proto-Sun, we must have $q < 3\{1 - [(m-1)/m]^{2/3}\}^3$. For companions with $q > 0.3$, no low-order Lindblad resonances can be found inside the Roche radius of the proto-Sun.

Near the resonant radius r_s, the accumulated tidal effect can lead to an efficient angular momentum transfer. Several idealized cases have been analyzed in a variety of astrophysical contexts: gaseous disks with pressure only

(Goldreich and Tremaine 1978; Donner 1979), gaseous disks with viscosity only (Lin and Papaloizou 1979a), and particle disks with a low collision frequency (Franklin et al. 1980). In all cases excepting the last, an identical angular momentum flux is derived. Pressure and self gravity promote the propagation of the tidal disturbances whereas viscosity and nonlinear damping determine where the disturbance is damped and dissipated. To illustrate the derivation of the angular momentum flux, we present the analysis of the case of a disk with pressure but no viscosity which links with the more general calculations discussed above.

As the resonance is interior to the disk, Σ_o and c are taken to be constant. In addition, we suppose the response varies on a small scale so that $d/dr \gg 1/r$, and $r^2(\Omega - \omega)^2 \gg c^2$. Then Eqs. (45) and (46) yield the following second-order ordinary differential equation.

$$-c^2\frac{d^2\xi_{rm}}{dr^2} = \xi_{rm}[m^2(\Omega - \omega)^2 - \Omega^2] - \frac{dW_m}{dr} - \frac{2\Omega W_m}{r(\Omega - \omega)}. \qquad (72)$$

To simplify further, we make a local expansion such that

$$[m^2(\Omega - \omega)^2 - \Omega^2] = -Dx \qquad (73)$$

where $x = r - r_s$ and

$$D = [3m^2(\Omega - \omega)\omega/r]_{r=r_s} \qquad (74)$$

may be approximated as constant.

In order to apply Eq. (60), we need to evaluate X_m. In this case, this is the solution of Eq. (72) (with $W_m = 0$) which is wavelike interior to an inner Lindblad resonance if appropriate, or exterior to an outer Lindblad resonance if that is appropriate, but which then decays exponentially on the other side of the appropriate resonance. This can be written in the form

$$X_m = \int_{-\infty}^{\infty} \exp i\left(kx + \frac{c^2k^3}{3D}\right)\frac{dk}{|D|}. \qquad (75)$$

This integral can be expanded asymptotically using the well known method of steepest descents and then by comparison with Eq. (51), one finds

$$|C| = \left(\frac{\pi\Sigma_o r_s}{|D|}\right)^{1/2}. \qquad (76)$$

If all quantities apart from X_m in the integral in Eq. (60) are taken to be constant and equal to their values at $r = r_s$, and the range extended to become $[-\infty, \infty]$, the integral may be evaluated analytically with the result

$$F_H = \left[\frac{\pi^2\Sigma_0}{3\Omega\omega}\left(r\frac{dW_m}{dr} + 2\frac{\Omega W_m}{(\Omega - \omega)}\right)^2\right]_{r=r_s}. \qquad (77)$$

This result is completely independent of viscosity provided the tidal disturbance induced by the resonant torque is completely dissipated. It gives the angular momentum flux as a result of resonance.

However, it is not strictly necessary for a resonance to be interior to the disk for it to produce an effect. Equation (75) shows that the scale of variation in the neighborhood of resonance is $\delta w \sim r_s [c/(r\Omega)]^{2/3}$. Therefore we would expect a strong virtual effect from a resonance as long as it was within a distance δw of the disk edge. In this way we would expect to account for the two-arm spiral features associated with the 2:1 inner Lindblad resonance often seen in the numerical calculations of disks in cataclysmic binaries, even though the resonance is outside the disk (see, e.g., Lin and Papaloizou 1979a, Whitehurst 1988a, b; Hirose and Osaki 1989). To calculate F_H in such cases one must solve for X_m numerically and then evaluate F_H using Eq. (60).

B. Waves Propagation and Dissipation in the Disk

We now examine the dissipation of the density waves excited at the Lindblad resonances. As these waves propagate away from the resonances, the tidal effect of the companion vanishes rapidly and the WKBJ approximation adequately describes the nature of wave propagation. In the absence of viscosity, these waves obey the standard dispersion equation (Lin and Shu 1964,1968) such that

$$c^2 k^2 - 2\pi G \Sigma_o |k| + \kappa^2 - (m\Omega - m\omega)^2 = 0, \qquad (78)$$

where k is the radial wavenumber. The effect of self gravity, which is contained in the second term of Eq. (78), allows the propagation of long waves between the Lindblad resonances (Lin and Lau 1979). In a non-self gravitating inviscid disk, for any wave to exist, we require that $m^2(\Omega - \omega)^2$ be bigger than $\kappa^2(= \Omega^2)$; i.e., waves exist only interior to the inner Lindblad resonance for which $m(\Omega - \omega) = \Omega$ and exterior to the outer Lindblad resonance for which $m(\Omega - \omega) = -\Omega$.

The dispersion relation in (78) indicates that the wavelength decreases as a wave propagates inward from the inner Lindblad resonance. Linear solutions of the momentum Eqs. (45) and (46) indicate that the contrast between the waves' peaks and troughs also increases as waves propagate away from the locations where they are launched (Fig. 1 in Sec. III). From Eq. (48) one can show that Σ' increases as the fourth root of the distance from the resonance. This increase in Σ' with distance from the resonance is due to the conservation of wave action or the constancy of F_H with distance expressing conservation of angular momentum (Toomre 1969; Shu 1970). In the presence of viscous dissipation, the rate of dissipation may also increase as k increases. In the strongly viscous limit, propagation over a distance $\sim r_s/m$ from the Lindblad resonances occurs with a group velocity significantly less than the sound speed (in contrast to the situation considered in Sec. III.C). With a constant effective kinematic viscosity ν, the amplitude of the waves is significantly damped over a distance

$$\Delta x \sim \left(\frac{c^3 r_s^{1/2}}{3 v m^{1/2} \Omega^2} \right)^{2/3} \tag{79}$$

from the resonance (Goldreich and Tremaine 1978,1980). In the weakly viscous limit (see Sec. III.C) when the propagation region extends to $\sim r_s$, the group velocity may be approximated as the sound speed and

$$\Delta x \sim \sqrt{\frac{c^3 r_s}{3 v m^2 \Omega \omega}}. \tag{80}$$

In the absence of viscosity, the waves will eventually become nonlinear when $|\Sigma'| \sim |\Sigma_o|$ (Cuzzi et al. 1981). Shock dissipation would lead to effective wave dissipation and angular momentum deposition (Lubow and Shu 1975; Shu 1976; Donner 1979; Shu et al. 1985b; Borderies et al. 1984b,1986; Spruit 1987). Because density waves carry negative angular momentum in the disk interior to the companion, their dissipation would induce an inward migration of the disk gas. Similarly, nonlinear wave dissipation in the disk exterior to the companion would induce an outward migration.

If nonlinear waves propagate over a large region of the disk, the tidal effect of the companion could regulate the evolution of the entire disk (Larson 1989). To discuss this possibility, a self-similar, steady-state, but nonaxisymmetric disk model has been constructed (Spruit 1987,1989b). This involves a nonlinear wave with zero pattern speed giving dissipation and angular momentum flow at shock discontinuities. The companion's tidal torque is not explicitly considered but must be thought of as ultimately responsible for the angular momentum which flows out of the system. For steady inviscid flow with no perturbing potential due to a companion, Eqs. (28) and (29) reduce to

$$v_r \frac{\partial v_r}{\partial r} + \frac{v_\phi}{r} \frac{\partial v_r}{\partial \phi} - \frac{v_\phi^2}{r} = -\frac{1}{\Sigma} \frac{\partial P}{\partial r} - \frac{GM}{r^2} \tag{81}$$

and

$$v_r \frac{\partial v_\phi}{\partial r} + \frac{v_\phi}{r} \frac{\partial v_\phi}{\partial \phi} + \frac{v_r v_\phi}{r} = -\frac{1}{r \Sigma} \frac{\partial P}{\partial \phi}. \tag{82}$$

The energy Eq. (27) reduces to

$$\Sigma T \left(v_r \frac{\partial S}{\partial r} + \frac{v_\phi}{r} \frac{\partial S}{\partial \phi} \right) + 2 F_z = 0 \tag{83}$$

where S is the entropy at the midplane and F_z is the radiative flux in the vertical direction. Using a thin disk approximation, $F_z = 8 \sigma T^4 / (3 \Sigma \kappa)$ and ideal equation of state and opacity $\kappa \propto \Sigma^{-1}$, Eqs. (81) to (83) have solutions of the form $v_r = r^{-1/2} f_r(\varphi)$, $v_\phi = r^{-1/2} f_\phi(\varphi)$, $P = r^{-5/2} f_p(\varphi)$, and $\Sigma = r^{-3/2} f_\rho(\varphi)$ where the similarity variable is $\varphi = \phi + B \ln r$ and

B is a constant which defines a logarithmic spiral where the streamlines converge and there is shock dissipation. Energy dissipation occurs across the shock front and the change in the flow variables across the shock front can be computed with the standard jump conditions (Shu et al. 1973; Larson 1990b). Two-dimensional numerical simulations for disk models in which the characteristic sound speed is a significant fraction of the Keplerian speed, lead to results which seem compatible with these semi-analytic calculations (Sawada et al. 1987; Spruit et al. 1987; Różyczka and Spruit 1989; Matsuda et al. 1989; Spruit 1989b).

The self similar solutions indicate the possibility that the companion's tidal disturbance may propagate and be dissipated over extended regions of the disk. However, the stability of the self-similar solutions remains uncertain. To distribute nonlinear dissipation and loss of wave action over the entire disk, a special temperature and density profile is needed which may not be generally realized. It is possible that when dissipation occurs, it would be localized in the region where the waves first become nonlinear (see Sec. V below).

C. Wave Refraction

The nonlinear development of density waves requires little or no dissipative damping of wave action as the waves undergo linear propagation in the disk. For waves propagating interior to the inner Lindblad resonance, $|k|$ increases rapidly as r decreases because Ω increases rapidly, and we have approximately

$$k^2 = (m^2 - 1)\Omega^2/c^2. \tag{84}$$

For sufficiently large values of m, the characteristic radial wavelength, $2\pi/|k|$, is comparable to or smaller than the disk's density scale height in the vertical direction, $H = 2^{1/2}c/\Omega$. In this case, wave propagation in the vertical direction can no longer be ignored. This effect is particularly important in a thermally stratified protostellar disk in which the propagation speed of the waves is a function of z because refraction of waves may turn the direction of propagation towards the vertical and cause an effective loss of wave action.

In order to examine the propagation of waves in a thermally stratified disk, a three-dimensional analysis is needed. This problem can be analyzed numerically with a linear numerical scheme (Lin et al. 1990a). In such a scheme, the equilibrium model may be arbitrarily chosen for computational convenience such that the flow velocity is $v = (0, \Omega r, 0)$ with $\Omega = (GM/r^3)^{1/2}$. In the equilibrium state, the z-component of the equation of motion reduces to

$$\frac{1}{\rho}\frac{\partial p}{\partial z} = -\Omega^2 z. \tag{85}$$

We are interested in studying wave propagation in a vertically stratified geometrically thin disk where there is the possibility of propagation through an extended atmosphere. Such a structure may be obtained from a barotropic equation of state in which

$$p = c_\infty^2 \rho \left[1 + \left(\frac{An}{1+n} \right) \left(\frac{\rho}{\rho_c} \right)^{1/n} \right] \tag{86}$$

where the quantities A, n, and c_∞ are constants. When the density at the midplane ρ_c is taken to be constant, the Keplerian rotation law is satisfied at the midplane. Equation (86) also implies

$$c^2 = \frac{dp}{d\rho} = c_\infty^2 \left[1 + A \left(\frac{\rho}{\rho_c} \right)^{1/n} \right]. \tag{87}$$

so that from Eq. (85) we have

$$A_n \left[1 - \left(\frac{\rho}{\rho_c} \right)^{1/n} \right] - \log \left(\frac{\rho}{\rho_c} \right) = \frac{z^2}{H^2} \tag{88}$$

where $H^2 \equiv 2c^2/\Omega^2$. Near the midplane, the gas is nearly polytropic with $c = (1 + A)^{1/2}c_\infty$. Above H, the disk is isothermal with $c = c_\infty$. Provided the perturbation is modest in strength, wave propagation can be examined by using linearized fluid Eqs. (24) and (25) which, in a frame rotating with angular velocity Ω_p, are such that

$$\frac{\partial v_r'}{\partial t} + (\Omega - \Omega_p) \frac{\partial v_r'}{\partial \phi} - 2\Omega v_\phi' = -\frac{\partial W}{\partial r} + T_r \tag{89}$$

$$\frac{\partial v_\phi'}{\partial t} + (\Omega - \Omega_p) \frac{\partial v_\phi'}{\partial \phi} + \frac{\Omega v_r'}{2} = -\frac{1}{r} \frac{\partial W}{\partial \phi} + T_\phi \tag{90}$$

$$\frac{\partial v_z'}{\partial t} + (\Omega - \Omega_p) \frac{\partial v_z'}{\partial \phi} = -\frac{\partial W}{\partial z} + T_z \tag{91}$$

where

$$W \equiv \frac{p'}{\rho} = \frac{\rho'}{\rho} c^2,$$

T_r, T_ϕ and T_z are the external tidal forces per unit mass in the radial, azimuthal, and z directions, respectively. The linearized continuity equation gives

$$\frac{\partial W}{\partial t} + (\Omega - \Omega_p) \frac{\partial W}{\partial \phi} = -\frac{c^2}{\rho} \left[\frac{1}{r} \frac{\partial}{\partial r}(r\rho v_r) + \frac{\rho}{r} \frac{\partial v_\phi}{\partial \phi} + \frac{\partial}{\partial z}(\rho v_z) \right]. \tag{92}$$

Before looking into detailed numerical solutions of the linearized equations of motion (89) to (92), we demonstrate that in the absence of any external force and dissipation, the angular momentum of the propagating wave is conserved. Consider wave-like disturbances which contain ϕ and t in the form

$$\phi - (\omega - \Omega_p)t \tag{93}$$

so that ω is the pattern speed measured in an inertial frame. It is convenient to introduce the Lagrangian displacement $\xi = (\xi_r, \xi_\phi, \xi_z)$ in terms of which the equations of motion (89–92) may be written as

$$(\Omega - \omega)^2 \frac{\partial^2 \xi_r}{\partial \phi^2} - 2\Omega(\Omega - \omega)\frac{\partial \xi_\phi}{\partial \phi} + 2r\Omega\frac{d\Omega}{dr}\xi_r = -\frac{\partial W}{\partial r} \tag{94}$$

$$(\Omega - \omega)^2 \frac{\partial^2 \xi_\phi}{\partial \phi^2} + 2\Omega(\Omega - \omega)\frac{\partial \xi_r}{\partial \phi} = -\frac{1}{r}\frac{\partial W}{\partial \phi} \tag{95}$$

and

$$(\Omega - \omega)^2 \frac{\partial^2 \xi_z}{\partial \phi^2} = -\frac{\partial W}{\partial z}. \tag{96}$$

For any ring-like internal region of volume v, when no external force acts, the rate of gain of angular momentum is given by

$$\dot{J} = -\int_v \rho'\frac{\partial W}{\partial \phi}dv = 0 \tag{97}$$

and this of course must be identically zero.
From the continuity equation, we have

$$\rho' = -\nabla \cdot (\rho\xi) \tag{98}$$

so that

$$\dot{J} = \int_v \nabla \cdot \left(\rho\xi\frac{\partial W}{\partial \phi}\right)dv + \int_v \left(\rho\frac{\partial \xi}{\partial \phi} \cdot \nabla W\right)dv = 0. \tag{99}$$

It is readily seen after using Eqs. (94–96) that the latter term vanishes on integration with respect to ϕ. Using the fact that

$$v_r = (\Omega - \omega)\frac{\partial \xi_r}{\partial \phi} \tag{100}$$

it follows from Eq. (90)

$$W = -r(\Omega - \omega)\left(v_\phi + \frac{1}{2}\Omega\xi_r\right). \tag{101}$$

Equation (99) may be recast, after integration with respect to ϕ, in the form of conservation law that

$$\dot{J} = \int_v \left(\frac{1}{r}\frac{\partial}{\partial r}rF_r + \frac{\partial}{\partial z}F_z\right)r\,dr\,dz = 0 \tag{102}$$

where

$$F_r = \int_0^{2\pi} r \rho v_r v_\phi \, d\phi \tag{103}$$

and

$$F_z = \int_0^{2\pi} \rho v_z r (v_\phi + \frac{1}{2} \Omega \xi_r) \, d\phi \tag{104}$$

are the components of the angular momentum flux in the r and z direction.

Equation (102) indicates that if a wave component is transmitted to regions above the disk without reflection, then the component of the angular momentum flux in the radial direction should decrease as a disturbance propagates radially. One key issue is how far waves can propagate in the vertical direction. If we consider a disturbance moving vertically under the assumption that $\rho = \rho(z)$ and $W = W(z) \exp(im\phi)$ in Eqs. (94–96) (these assumptions can only be consistent if Ω is a constant so that which follows is only valid locally in r), we obtain the following equation for W

$$\frac{\partial}{\partial z} \left(\rho \frac{\partial W}{\partial z} \right) + \frac{m^2 W \rho \bar\sigma^2}{r^2(\Omega^2 - \bar\sigma^2)} = - \left(\frac{\bar\sigma^2}{c^2} \rho W \right) \tag{105}$$

where $\bar\sigma = m(\Omega - \omega)$. If we set $W \equiv Q/\rho^{1/2}$, we find

$$\frac{\partial^2 Q}{\partial z^2} + Q \left[-\frac{\partial^2 \rho / \partial z^2}{2\rho} + \left(\frac{\partial \rho / \partial z}{2\rho} \right)^2 + \frac{\bar\sigma^2}{c^2} + \frac{m^2 \bar\sigma^2}{r^2(\Omega^2 - \bar\sigma^2)} \right] = 0. \tag{106}$$

In the upper regions of the isothermal atmosphere of the disk, this takes the form of

$$\frac{\partial^2 Q}{\partial z^2} + Q \left[-\frac{\Omega^2}{2c_\infty^2} - \frac{\Omega^4 z^2}{4c_\infty^4} + \frac{\bar\sigma^2}{c_\infty^2} + \frac{m^2 \bar\sigma^2}{r^2(\Omega^2 - \bar\sigma^2)} \right] = 0. \tag{107}$$

Equation (107) suggests that waves propagate vertically upwards until they are reflected at a level given, when $\Omega \gg \omega$, by

$$z_c^2 = \frac{4c_\infty^2}{\Omega^2} \left(m^2 - \frac{1}{2} \right) - \frac{4c_\infty^4 m^4}{\Omega^4 r^2 (m^2 - 1)}. \tag{108}$$

From this we see that for $m \sim r\Omega/c_\infty \sim r/H$, $z_c \sim r$. Thus for large m, we expect the wave energy to be readily lost through vertical propagation. At a sufficient distance above the midplane where the density is low, the wave amplitude becomes nonlinear and shock dissipation can lead to the deposition of energy and angular momentum into the disk (Murray and Lin 1991). However, for smaller m there may be a reflection before a sufficiently low density for shock formation to be attained. What happens in this case will be more sensitive to the details of the vertical structure. We present some numerical results to illustrate the refraction of waves as they propagate through a thermally stratified disk.

In these numerical results, we set

$$T_r(r) = -\frac{m}{r_t}\left(\frac{r}{r_t}\right)^{(m-1)} \exp\{-i\left[(m-1)\Omega(r_t) - m\Omega_p\right]t + im\phi\}. \quad (109)$$

$$T_\phi = iT_r \quad \text{and} \quad T_z = 0. \quad (110)$$

In Eq. (109), the forcing frequency $m\omega$ (as seen in the inertial frame), is replaced by $(m-1)\Omega(r_t)$ and therefore $r = r_t$ corresponds to the inner Lindblad resonance of the companion (see Sec. IV.A). Both T_r and T_ϕ are independent of z and no forcing is imposed in the vertical direction so that the front of the excited wave should be initially perpendicular to the midplane and the propagation should therefore be in the radial direction near the location where the waves are launched. A useful quantity for analyzing the numerical results is

$$T(r) \equiv \int_{r_i}^r \pi r^2 dr \int_0^\infty \frac{\rho}{c^2} \text{Re}\left(W^* T_\phi\right) dz \quad (111)$$

which is the total rate of increase of angular momentum, due to external tidal forces, between r and r_i in that region of the disk above the midplane. Note that this quantity is negative for the calculations presented here. Because the companion orbits slower than the disk, the disk material loses angular momentum.

At a radius r, the angular momentum flowing in the radial direction is given by

$$J_r \equiv \text{Re} \int_0^\infty \pi r^2 \rho v_r v_\phi^* dz \quad (112)$$

and the angular momentum flowing normal to the surface given by $z = z(r)$, between r and r_i is given by

$$J_z(r) \equiv \text{Re} \int_{r_i}^r \pi r^2 \rho \left[v_z U^* - v_r v_\phi^* \left(\frac{dz(r)}{dr}\right)\right] dr \quad (113)$$

with

$$U \equiv v_\phi - \frac{i\Omega v_r}{2m(\Omega - \omega)} \quad (114)$$

where r_i is the radius of the inner boundary. From these two quantities, we can measure the efficiency of refraction.

For the numerical model presented here, dimensionless units for which $GM = 1$, $\Omega_p = 1$, $\rho_c = 1$, $c_\infty = 10^{-1.5}$, $A = 3$, and $n = 1.5$ are used. With these structural parameters, the sound speed near the midplane is twice that at the disk surface. Such a vertical structure corresponds to an optical depth of the order a few hundred which is typical of the expected optical depth for the protostellar disks. The inner and outer boundaries are set at

$r = r_i = 0.8$ and $r = r_o = 1.6$, respectively with radiative conditions for the inner boundary in the radial direction and surface boundary in the vertical direction. A reflection condition is used at the outer boundary in the radial direction. Reflection symmetry in the midplane is used so that W is an even function of z. For the perturbed quantities, $m = 5$ and $r_t = 1.55$, so that the inner Lindblad resonance is just inside the outer boundary of the main flow domain. To construct transient-free solutions, the external torque is switched on slowly and the asymptotic steady-state solution is presented below.

For the steady-state solutions, we plot the radial and vertical distribution of the real part of W, v_r, v_ϕ, and v_z (Fig. 2). In the steady-state solutions, all variables vary sinusoidally in azimuth so that the form of the radial and vertical distribution plots is independent of the azimuthal phase. The excited waves essentially propagate in the radial direction near the radius r_t where they are launched. As the waves propagate away from r_t, the retardation of the wave front at the larger values of z becomes more apparent. The sound speed decreases with z, and the differential speed of propagation induces a curvature in the wave front which leads to an upward propagation towards the disk surface. The refraction of the wave is also well demonstrated by the $v_z(r, z)$ plot. Near the midplane, v_z is relatively small and the wave propagates approximately radially. Near the disk surface, however, the amplitude of v_z increases considerably.

The characteristic radial wavelength near the inner Lindblad resonance, $r_t = 1.55$, is comparable to the disk radius and at $r \sim 0.8$ to 0.9, it is reduced to one tenth the radius near the midplane and about half that near the disk surface. The midplane result is consistent with the expected wavelength from the dispersion relation. The fact that the wavelength is longer near the midplane compared to the surface, implies relatively fewer wave crests near the midplane.

We also plot the azimuthal variation of W at the midplane and near the disk surface as functions of r and ϕ (Fig. 3). The amplitude of W is not plotted to the same scale in these plots. The azimuthal extent in Fig. 3 is $2\pi/5$, W being periodic in ϕ with this period. The wavelength in the radial direction is smaller near the disk surface than at the midplane due to the lower sound speed. Near the midplane, the amplitude of the wave decreases with distance from the r_t where the wave is launched.

In Fig. 4, we also plot the angular momentum flow integral J_r with a dashed line, J_z with a dotted line, and the torque integral T with a solid line. Working outward from the inner boundary of the disk, J_z increases linearly with radius which is an indication that a nearly constant angular momentum flow rate per unit radius is being diverted vertically. Consequently, the angular momentum flux in the radial direction decreases linearly with distance from r_t. At the inner boundary there is a small amount of residual angular momentum flux which is transmitted through the inner edge in this particular case.

The torque integral $T(r)$ provides an indicator to the distribution of the angular momentum deposition rate in the disk. Our results indicate that most

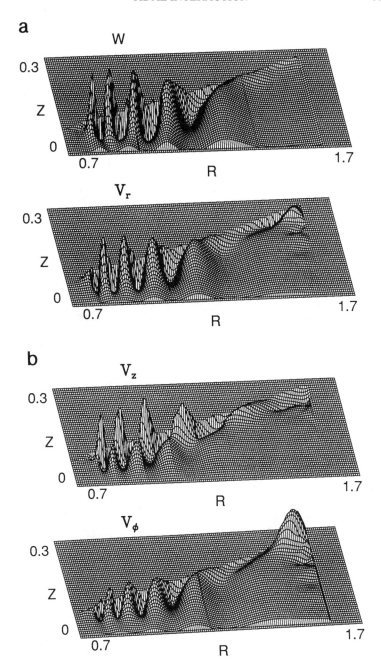

Figure 2.   Wave propagation pattern in the steady-state solution for a disk model in which $c_\infty = 10^{-1.5}$, $A = 3$ and $n = 1.5$. The real parts of perturbed quantities are plotted here. A resonant disturbance is imposed at $r = 1.55$. The quantities W, v_r, v_z and v_ϕ correspond to the perturbed surface density, radial, vertical, and azimuthal velocities.

W

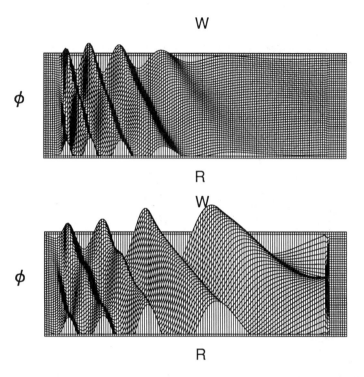

Figure 3.   Radial and azimuthal distribution of the perturbed surface density in the midplane (lower panel) and at the disk surface (upper panel). The azimuthal extent is $2\pi/5$ and the radial extent is identical to that in Fig. 2.

of the angular momentum is deposited into the disk near the inner Lindblad resonance where the characteristic wavelength is a significant fraction of the radius. From conservation of angular momentum, we expect

$$T(r_o) = J_z(r_o) - J_r(r_i) + J_r(r_o) \tag{115}$$

which is satisfied within 10% accuracy.

It is of interest to compare the total angular momentum transfer rate $|T(r_o)|$ with that expected from a two-dimensional cold disk. For a Keplerian disk, Eq. (77) gives for this case

$$|T(r_o)| = \frac{\pi^2 r^2 \Sigma m}{3\Omega^2(m-1)} \left(\frac{dW_m}{dr} + \frac{2mW_m}{r} \right)^2 \tag{116}$$

where in the above, only the surface density above the mid-plane should be used and the right-hand side is to be evaluated at the inner Lindblad resonance, with $|T_r| = dW_m/dr = mW_m/r$. A straightforward application of this result gives a value for $|T(r_o)|$ a factor of ~ 8 larger than that actually found in the

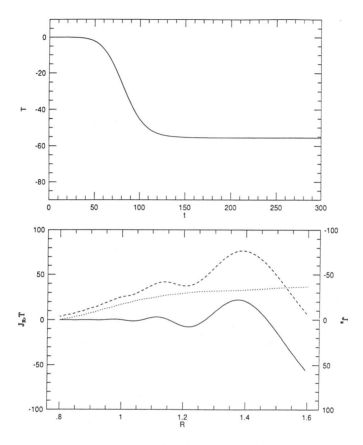

Figure 4.   Evolution of total torque integral (upper panel) and the steady-state radial
distribution of angular momentum flux $J_R(r)$ (dashed line) and $J_z(r)$ (dotted line)
and the torque integral $T(r)$ (solid line) curves (lower panel). Note that the radial
flux of angular momentum vanishes as waves propagate away from the location
($r = 1.55$) where a resonant disturbance is imposed.

calculation. However, we believe that this discrepancy can be explained by
the fact that the resonance is very close to the outer boundary and the fact
that the sound speed is finite. The latter effect causes the active region around
resonance to have a width of order $H^{2/3}r^{1/3} \sim 0.1r$. Thus, the resonance must
be regarded as being at the outer boundary. A simple calculation of the type
done in Sec. III for the case when the resonance is precisely on the rigid outer
boundary, shows that $|T(r_o)|$ should be reduced by a factor of ~ 3. We further
note that because of the finite resonance width, there is some uncertainty as
to where the formula should be applied. If it is applied 10% farther in, the
torque is reduced by a factor of 4. The formula is thus very sensitive to the
radius of application. When several values of r_o and r_t are used, numerical

results show that $T(r_o)$ indeed increases with the distance between r_t and r_o as expected.

Numerical models with a variety of sound speeds, vertical thermal stratifications, and external torques have been computed (Lin et al. 1990b). These results indicate that the gradient of the sound speed in the vertical direction can induce tilting of the wave front as an excited wave propagates away from the region where it is launched. This leads to refraction which induces a propagation component in the direction normal to the plane of the disk. The wave carries energy and angular momentum as it propagates and so a flux of angular momentum is transmitted in both the radial and vertical directions.

In most astrophysical contexts such as interacting binary stars, the disk is optically thick. Then the temperature near the midplane is expected to be several times larger than that near the disk surface. Furthermore, the sound speed at the midplane is a small fraction of the local Keplerian velocity. Under these conditions, thermal stratification promotes efficient refraction as a wave propagates over a radial distance comparable to the vertical scale height from the region where it is tidally induced. Thus, in a vertically thin disk, tidally induced waves are localized in the region where they are launched and angular momentum transport by wave propagation is unlikely to be effective over extended regions of the disk.

The effect of refraction may be suppressed if the disk has an isothermal vertical structure or if the sound speed at the midplane is a significant fraction of the local Keplerian velocity. These conditions may be attained in protostellar disks when they are exposed to intense surface heating (Watanabe and Nagazawa 1988) or when they become optically thin or when there is infall onto the disk surface (Ruden 1986). In these cases, tidally induced waves may propagate over an extended region in the disk and angular momentum transport by wave propagation may be efficient. If the companion has a comparable mass to the proto-Sun so that Lindblad resonances are excluded from the Roche lobe of the proto-Sun, a strong wave-like response is possible only if the characteristic wavelength of the generated waves is comparable to the disk radius. But the dispersion relation indicates that the wavelength of a propagating wave is $\sim c/m\Omega$, and there is generally a wavelength mismatch. A large response is possible only if the sound speed is sufficiently large such that the characteristic scale height in the vertical direction is comparable to the disk radius (Sawada et al. 1986,1987).

V. GAP FORMATION AND DISK STRUCTURE

The results in Sec. III indicate that the disk structure is regulated by the efficiency of density wave propagation and dissipation. Energy and angular momentum transferred from the companion is deposited into the disk through either viscous or nonlinear dissipation. Although the magnitude of the sound speed and hence the effective viscosity in protostellar disks remains uncertain, the results in Sec. IV indicate that even in the inviscid limit, refractory effects inhibit significant radial wave propagation and induce effective dissipation through vertical losses. In this section, we present self-consistent two-dimensional numerical calculations to illustrate the nonlinear disk response to the companion's tide.

A. Numerical Methods

Several numerical schemes have been introduced to study companion-disk interactions. These include the sticky particle scheme (Lin and Pringle 1976; Lin and Papaloizou 1979a, b; Cameron 1979; Whitehurst 1988a, b,1989; Hirose and Osaki 1989), smooth particle hydrodynamics (Miyama et al. 1984; Artymowicz et al. 1991), and finite difference methods (Prendergast and Burbidge 1968; Flannery 1975; Miki 1982; Matsuda et al. 1987; Różyczka and Spruit 1989). Although these schemes provide important information on global features of the disk such as total angular momentum flux and gap formation, they generally lack sufficient resolution for the investigation of resonant tidal interactions, linear wave propagation, and nonlinear wave dissipation especially when the companion has a low mass. To determine the fine structure of the disk, two versatile methods have been employed: a two-dimensional Lagrangian scheme (Lin and Papaloizou 1986a) and a two-dimensional second-order Lax-Wendroff scheme. The two-dimensional non-axisymmetric Lagrangian scheme was developed in order to calculate the disk structure in a steady state. In this scheme, the disk has one fixed rigid boundary at a disk radius which is far enough from the companion so that tidal effects are unimportant. At this location, the flow is assumed to be Keplerian. This condition introduces a viscous stress acting at the boundary such that it induces transfer angular momentum to the disk if the disk lies outside the boundary and remove angular momentum from the disk if it lies inside. The other boundary is assumed to be completely free so that it may move to a location close enough to the companion for its tidal effect to become important. An external companion removes angular momentum from the disk while an internal companion does the opposite. For a companion with a circular orbit, a steady-state solution may be attained in which the disk appears stationary in a frame corotating with the companion. In the inertial frame, the steady-state solutions are functions of ϕ and t through the combination of $\phi - \omega t$. To describe solutions of Eqs. (28) to (32), we introduce variable $\xi = \theta(\phi - \omega t)$ where $\theta = 1 \ (-1)$ for a disk located interior (exterior) to the companion's orbit. The equations of motion then become

$$(\Omega - \omega)\theta \frac{\partial v_r}{\partial \xi} + v_r \frac{\partial v_r}{\partial r} - \frac{v_\phi^2}{r} = -\frac{1}{\Sigma} \frac{\partial (P + q_v)}{\partial r} - \frac{\partial \psi}{\partial r} \quad (117)$$

$$(\Omega - \omega)\theta \frac{\partial v_\phi}{\partial \xi} + v_r \frac{\partial v_\phi}{\partial r} + \frac{v_\phi v_r}{r} = -\frac{\theta}{\Sigma r} \frac{\partial (P + q_v)}{\partial \xi} - \frac{\theta}{r} \frac{\partial \psi}{\partial \xi} + F_v. \quad (118)$$

Only steady-state solutions of Eqs. (117) and (118) have physical meaning and they correspond to solutions that are periodic in ξ with period 2π.

To study the formation of gaps, an accurate treatment of the free boundary is essential. Because the location of the edge cannot be specified *a priori*, but must be determined in the course of solution, it is convenient to use a Lagrangian rather than a Eulerian description of the flow. Another advantage of the Lagrangian scheme is that by adjusting the initial mesh spacing, the spatial resolution may be improved greatly. In this scheme, the Lagrangian coordinate r is a function of ξ, which we treat as a time coordinate, and r_f the value of r at $\xi = 0$. The radial velocity is then given by

$$\left(\frac{\partial r}{\partial \xi} \right)_{r_f} = \frac{\theta v_r}{\Omega - \omega} \tag{119}$$

and the advective derivative becomes

$$\left(\frac{\partial}{\partial \xi} \right)_r + \frac{\theta v_r}{\Omega - \omega} \left(\frac{\partial}{\partial r} \right)_\xi \equiv \left(\frac{\partial}{\partial \xi} \right)_{r_f}. \tag{120}$$

The continuity Eq. (32) can be written

$$\left[\frac{\partial}{\partial \xi} \left(r \frac{\partial r}{\partial r_f} (\omega - \Omega) \Sigma \right) \right]_{r_f} = 0 \tag{121}$$

which implies the existence of a conserved Lagrangian mass element $\mu(r_f) = r(\omega - \Omega) \Sigma \partial r / \partial r_f$. After some algebra, Eqs. (117) and (118) become

$$\frac{\partial v_r}{\partial \xi} = \frac{\theta}{\Omega - \omega} \left[\frac{v_\phi^2}{r} - \left(\frac{\partial r_f}{\partial r} \right)^{-1} \left(\frac{1}{\Sigma} \frac{\partial (P + q_v)}{\partial r_f} + \frac{\partial \psi}{\partial r_f} \right) \right] \tag{122}$$

and

$$\frac{\partial}{\partial \xi} \left(\frac{v_\phi^2}{2} + \frac{v_r^2}{2} - r \omega v_\phi + \psi + H_\Sigma + \frac{q_v}{\Sigma} \right) = \theta r F_v - q_v \Sigma^{-2} \frac{\partial \Sigma}{\partial \xi} \tag{123}$$

respectively where

$$H_\Sigma(\Sigma) \equiv \int_0^\Sigma \frac{1}{\Sigma} \frac{dP}{d\Sigma} d\Sigma. \tag{124}$$

The four dependent variables r, Σ, v_r, and v_ϕ are obtained from the solutions of Eqs. (119) and (121–123). The numerical analysis is considered as posing an initial value problem with ξ playing the role of time, using standard finite difference techniques (Lin and Papaloizou 1986a). The solution is started from initial conditions chosen for convenience. It is then followed until it becomes periodic in ξ indicating that a physically meaningful solution has been obtained. For computational convenience, we adopt an equation of state

$$P(\Sigma) = c_o^2 \Sigma (\Sigma / \Sigma_o)^2 \tag{125}$$

where c_o and Σ_o are constants so that the sound speed $c = 3^{1/2} c_o \Sigma / \Sigma_o$. The constant Σ_o sets the surface density scale and may be taken, without loss of generality, to be unity. The maximum value of the surface density in the calculated profiles presented here is always close to Σ_o or eqivalently unity. For shear viscosity we adopt $v_s = v_{so} \Sigma / (\Sigma + \Sigma_o)$, where v_{so} is constant. The pseudoviscous pressure may be decomposed into two parts $q_v = q_1 + q_2$ where

$$q_1 = -\frac{v_{bo} \Sigma^2}{(\Sigma + \Sigma_o)} \frac{\partial v_r}{\partial r} \qquad (126)$$

and

$$q_2 = -a^2 (\Delta r)^2 \Sigma \left(\frac{\partial v_r}{\partial r} \right) \left(\left| \frac{\partial v_r}{\partial r} \right| - \frac{\partial v_r}{\partial r} \right). \qquad (127)$$

The variable q_2 is introduced to handle strong shocks, Δr is the radial separation of two adjacent grid points and a^2 is a constant normally taken to be 2. These equations can be solved using a finite difference scheme with the time step restricted by the usual requirements for stability (Richtmeyer and Morton 1967). The initial surface density and width of each zone may be chosen to provide good resolution. These calculations are computed until the solution becomes periodic to within one part in 10^4 which, for typical models, occurs in a few hundred to a few thousand periods. The steady-state solution does not depend on the arbitrary choice of the initial surface-density distribution.

Although the periodic Lagrangian method is convenient to use, it is useful only in cases where periodic solutions exist. In order to verify that there is a tendency for the disk to evolve towards an equilibrium, we also computed the flow pattern with a two-step Lax-Wendroff scheme which is a second-order accuracy method that minimizes large numerical diffusion. This method is useful for computing-time-dependent behavior such as (1) evolution towards equilibrium disk structure; (2) disk response to a companion on an eccentric orbit; and (3) the evolution of a ring between two satellites.

B. Dynamical Equilibrium

The existence of periodic solution can be established with the fully time-dependent Lax-Wendroff method. For illustrative purposes, we adopt a model 1 in which $q = 10^{-2}$, $c_o = 0.04 \omega r_o$, and $v_{so} = v_{bo} = 10^{-5} \omega r_o^2$. This model is computed for more than 100 orbital periods when an asymptotic periodic solution is attained. The Σ distribution of model 1 (Fig. 5) is compared with that obtained from the periodic solution using the Lagrangian scheme in model 2 (Fig. 6). The model parameters have identical values in models 1 and 2. There are general agreement and many similarities between these two models: (1) strong density waves are excited near the 2:1 resonance (at $r \sim 0.6 \, r_o$); (2) the disk is tidally truncated just exterior to the 2:1 resonance; (3) density waves propagate well into small disk radii; and (4) the characteristic wavelengths in models 1 and 2 have similar radial dependence and magnitude. There are also some minor differences: (1) the amplitude of the waves in

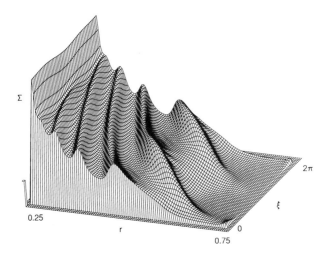

Figure 5.   Steady-state surface density distribution in model 1. The Lax-Wendroff scheme is used in this model. In this and subsequent figures, the position of the companion is located at $r = 1$ and $\xi = 0$ and the maximum value of the surface density plotted is close to unity in terms of the dimensionless units used in the text.

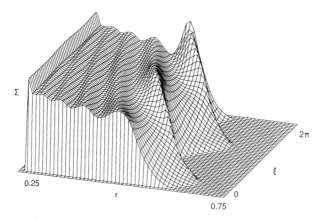

Figure 6.   Steady-state surface density distribution in model 2. A periodic Lagrangian scheme is used in this and most of the subsequent models. The values of the physical model parameters used here are identical to those in model 1. The 2:1 (inner Lindblad) resonance is located at $r \simeq 0.6$ where an $m = 2$ density wave is excited.

model 2 is less than in model 1. This difference is due to a slight difference in the equation of state; (2) the disk is slightly more extended in model 1 than in model 2 which may be attributed to the small numerical diffusion in the Lax-Wendroff scheme.

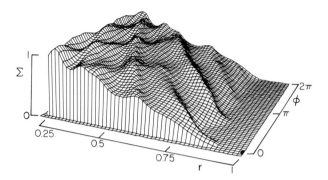

Figure 7.   Steady-state surface density distribution in model 3. Wave propagation is enhanced by the large sound speed. There is wave reflection at the inner boundary. An $m = 4$ density wave is excited at the 4:3 resonance. Similarly, an $m = 3$ and an $m = 2$ wave is excited at the 3:2 and 2:1 resonances, respectively. In this and several subsequent figures, the azimuthal coordinate ϕ is identical to ξ in the text and units are adopted such that $\omega = r_o = 1$.

The results in Figs. 5 and 6 are also in good agreement with the analytic and numerical results presented in Sec. II–IV. With the parameters in models 1 and 2, both conditions [Eqs. (68) and (69)] for gap formation are satisfied. Indeed, the results in Figs. 5 and 6 show a tidally truncated disk structure. The model parameters in models 1 and 2 correspond to the weakly viscous limit. In agreement with criterion (70), the density wave propagates over an extended radial range as shown in Figs. 5 and 6. Applying the model parameters to Eq. (79), we find the damping length scale to be comparable to r_o as is found in models 1 and 2. The radial dependence of wavelength is also in good agreement with predictions made using the dispersion relation (78).

These comparisons indicate that when periodic solutions exist, the Lagrangian scheme can provide reliable results. Although the Lax-Wendroff method is a truly two-dimensional time-dependent scheme, it is very time consuming for numerical computation. Therefore, the Lagrangian scheme is used in the analysis of disk-companion tidal interactions presented below.

C. The Standard Model

In this and subsequent sections, we focus our discussion on cases where the companion's mass is much smaller than that of the central star. This calculation is particularly relevant to the formation of protogiant planets. To illustrate disk response to the companion's tide, a standard model (model 3) is adopted in which $q = 10^{-3}$, $c_o = 0.1 \, \omega r_o$, and $v_{so} = v_{bo} = 1.5 \times 10^{-5} \omega r_o^2$. According to Eqs. (68) and (69), the mass of this companion is sufficiently large to marginally open a gap. These parameters are particularly relevant to the formation of Jupiter in the solar nebula.

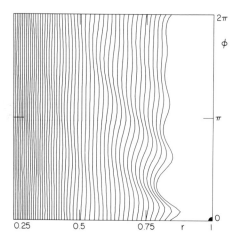

Figure 8.   Steady-state streamlines in model 3. The companion's perturbation is most intense at $\phi = 0$.

Surface Density Distribution. The equilibrium Σ distribution is plotted in Fig. 7. The outer edge of the disk approaches to within 2 Roche radii ($r_R \sim 0.1$) from the companion. A 4-arm spiral pattern is clearly visible near the outer edge of the disk. This pattern is excited by the 4:3 Lindblad resonance. Similarly 3- and 2-arm spiral waves are evident near the 3:2 and 2:1 Lindblad resonance. Although sound waves are excited at these resonances, that they can propagate throughout the disk is indicated by the fact that some reflection near the inner boundary is observed. Application of these model parameters to the discussion in Sec. IV indicate that waves can indeed propagate over a large radial extent. In particular, Eq. (78) indicates that deep inside the disk, the expected wavenumber is $k \sim m\omega/c_0$ and the condition for waves to propagate to the inner boundary is approximately that $[c_0/(r_0\omega) > m^{2/3}(\nu_{s0}/r_0^2\omega)^{1/3}]$. In the standard model, $\nu_{s0} = 10^{-5}\omega r_0^2$, this requires $c_0 > 0.02\ \omega r_0$ which is satisfied for $m = 3$ or 4.

Streamlines and Shock Dissipation. The streamlines and non-Keplerian velocity vectors are plotted in Figs. 8 and 9. At the outer edge of the disk, the protoplanet's tide strongly perturbs the disk flow near $\xi = 0$. At later orbital phases, the amplitude of radial motion is damped by both viscosity and nonlinear shock dissipation. The streamlines depart strongly from sinusoidal form. In Fig. 10, the dimensionless shear gradient, $\Omega^* \equiv (r/\Omega)(\partial\Omega/\partial r)$ is plotted as a function of ξ for several streamlines. Note that although the shear is modified near $\xi = 0$, the protoplanet's perturbations are not sufficient to induce shear reversal. Nevertheless, $\kappa^2 = \Omega^2(4 + 2\Omega^*)$ is modified near the disk edge and the resonances become delocalized. Wave crests in the Σ distribution correspond to locations where the streamlines converge and shock. These shocks correspond to maxima in the dissipation due to the pseudoviscous pressure, $D_r = (q_\nu/\Sigma^2)(\partial\Sigma/\partial\xi)(\Omega - \omega)$ (see Fig. 11). At relatively small disk radii, tidal disturbances result from the effects of low-

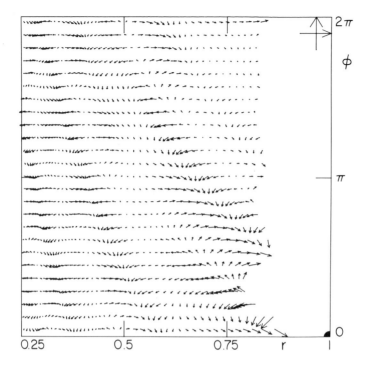

Figure 9.   Perturbed velocity field in model 3. The reference vector at the upper right
     corner represents 0.2 ωr_o where ωr_o is the orbital velocity of the companion.

order Lindblad resonances and do not have strong phase dependence. The
streamlines there thus take on a sinusoidal form.

D. Model Parameters

In a protoplanetary disk, a wide range of physical conditions may be possible.
The three main model parameters are c_o, v_{so}, and q. Using a few models, we
show below the disk response as a function of these parameters.

  Sound Speed. The discussion in Sec. III indicates that gap formation
in the disk is only possible if the Roche radius of the protoplanet is larger
than the disk scale height H. Models with the same values of q and v_{so} but
larger c_o do not lead to gap formation (see also the discussion in Sec. VI). In
model 4, a cold disk is presented. We adopt $c_o = 10^{-3} r_o \omega$, and take all other
parameters to be identical to those in model 3. The streamlines in Fig. 12
clearly indicate the presence of 2- and 3-arm spiral patterns near the 2:1 and
3:2 Lindblad resonances.

  However, the Σ distribution (Fig. 13) shows that the tidally induced

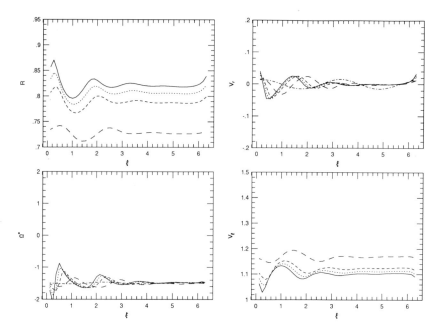

Figure 10.   Perturbed velocities, radii and shear ($\Omega^* \equiv \partial \ln \Omega / \partial \ln r$) at different orbital phases of four representative Lagrangian fluid elements near the outer edge of the disk.

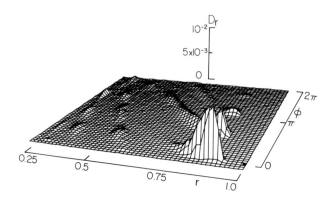

Figure 11.   Distribution of dissipation rate D_4 in model 3. The bulk dissipation rate attains local maxima at regions where the streamlines converge (compare to Fig. 9).

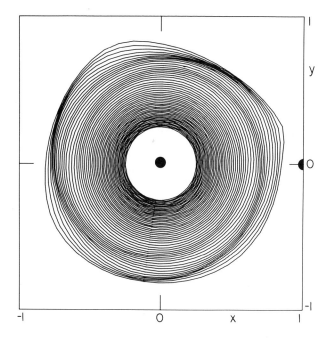

Figure 12.   Steady-state streamlines in the corotating frame (counter-clockwise) of a
companion which is located at $x = 1$ and $y = 0$.

density waves cannot propagate significantly from the location where they are
launched.

Viscosity. In Sec. II, we showed that in the limit of large ν_{so}, the rate
of viscous transport of angular momentum dH_ν/dt exceeds that due to the
protoplanet's tide dH_T/dt so that gap formation is prevented. In Sec. VI, we
deduce the critical value of ν_{so} for gap formation. In the limit of small ν_{so},
dH_ν/dt may be small compared to the rate of angular momentum transfer that
would occur if the disk includes Lindblad resonances. In this case, the disk
may be tidally truncated at a Lindblad resonance. From Eqs. (22) and (60), an
approximate condition for tidal truncation at a low-order Lindblad resonance
is

$$2\pi\nu_s < q^2 r_o^2 \omega. \tag{128}$$

The physical parameters for model 5 are identical to those for model 4 except
that $\nu_{so} = 10^{-7} r_0^2 \omega$ so that the condition for tidal truncation at the 2:1 Lindblad
resonance is marginally satisfied. The Σ distribution (Fig. 14) indeed has a
relatively sharp edge near the 2:1 Lindblad resonance.

Mass Ratio. In Sec. IX, we present disk structure in nearly equal-mass
binaries. For protoplanets with a small q and thus a small r_R, the disk may
include Lindblad resonances with values of m up to $(2/3)(r_o/r_R) \sim (2/3)q^{-1/3}$.
In model 6, $q = 10^{-4}$, $c_o = 10^{-2} r_o \omega$, and $\nu_{so} = 10^{-7} r_0^2 \omega$ so that structures
manifesting m as large as 10 may be produced in principle. The actual Σ

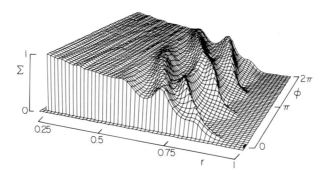

Figure 13.   Steady-state surface density distribution in model 4. There is little wave
    propagation because the sound speed is low. Density waves are dissipated at the
    resonant locations where they are excited.

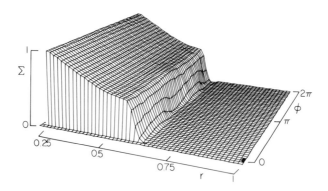

Figure 14.   Steady-state surface density distribution in model 5. The disk is truncated
    at the 2:1 resonance because the viscosity is small. Models with even lower viscosity
    are also truncated at the 2:1 resonances.

distribution (Fig. 15) shows a $m = 8$ pattern. The smaller values of m seen are
probably due to the fact that there is not enough material in the outer regions
for the higher m resonances to be excited. At the disk edge, the streamlines
depart significantly from a sinusoidal pattern. Viscous dissipation introduces
an azimuthal dependence in the wave amplitude. There is no clear separation
between the density waves which are excited at the several resonances within
the disk. In fact, density waves propagate between resonances throughout the
disk. This is due to a modest pressure effect which promotes wave propagation
and slightly modifies the resonant locations.   Despite the interference of
density waves excited by different resonances, the wave pattern is dominated
by local effects. At the inner edge of the disk ($r = 0.55$), there is a partial
reflection of the density wave which is excited at the 2:1 resonance ($r \simeq 0.63$).
    The small q models are useful to demonstrate that the gap width is also a

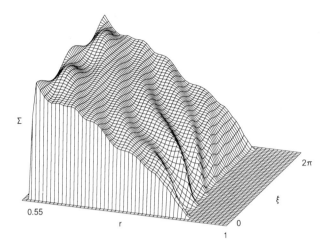

Figure 15.   Steady-state surface density distribution in model 6 where $q = 10^{-4}$. The disk extends closer to the companion than in models where q is larger. Consequently high-m Lindblad resonances are excited.

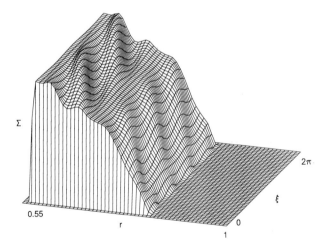

Figure 16.   Steady-state surface density distribution in model 7. The resonant zones are well separated due to the low sound speed. Waves are dissipated where they are excited.

function of c_o and v_{so} as well as q. Model 7 has identical parameters to model 6 but $c_o = 10^{-3} r_o \omega$. The size of the gap is a factor of 2 larger for the lower-c_o model 7 (Fig. 16) so that only $m = 4$ waves are seen in the disk. The waves are confined to the vicinity of the resonant locations where they are excited so different low-order resonances are clearly separated. Model 7 can be classified as a strongly viscous case as Eq. (79) implies that the dissipation length scale is less than the separation of the resonances. We compute another model with identical parameters to model 7 but $v_{so} = 10^{-9} r_o^2 \omega$ so that $2\pi v_s < q^2 \omega r_o^2$. The disk is truncated at the 2:1 Lindblad resonance. In the absence of overlapping resonances and wave propagation, the disk edge is very sharp.

E. Circumcompanion Disks

The disk structure exterior to the protoplanet's orbit is essentially similar to that interior to it since, because of the symmetry that applies in these extreme mass ratio cases, the hydrodynamics of the disk flow is essentially identical. In model 8, $q = 10^{-3}$, $c_o = 5 \times 10^{-2} r_o \omega$, and $v_{so} = 10^{-5} r_0^2 \omega$. Near the inner edge of the outer disk, 2- and 1-arm spiral patterns are excited by the 3:2 and 2:1 outer Lindblad resonances (Fig. 17). In this relatively high sound speed case, the waves do not propagate very far from the gap edge. This occurs because the wave amplitude decreases as the wavelength decreases outward; this is also indicated by the dispersion relation (78). The gap attains a modest size which is about 2 to 3 r_R from the companion. Models with smaller viscosity attain larger gap because the viscous dissipation is distributed over extended regions of the disk.

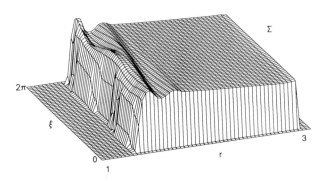

Figure 17.   Steady-state surface density distribution in circumbinary disk in model 8 where $q = 10^{-3}$.

F. Proto-Sun's Tidal Effects on Protoplanetary Disks

It is also of interest to consider the disk structure around the protoplanet where the "companion's" mass is much larger than that at the center of the disk (Lissauer and Cuzzi 1985). In model 9, $q = 10^3$, $v_{so} = 2 \times 10^{-5}$, and $c_o = 0.1$ which corresponds to a gaseous disk around proto-Jupiter. Streamlines in the outer region of the disk are very elongated (Fig. 18). The region near the protostar extends to ~ 0.5 r_R whereas on the opposite side the disk is strongly compressed. Density waves are not excited in the disk because the characteristic length scale for the protostar's tidal disturbance is comparable to the disk radius and much larger than the characteristic wavelength so that waves cannot be excited except near Lindblad resonances. In the disks around low-mass companions, the low-order Lindblad resonances are excluded.

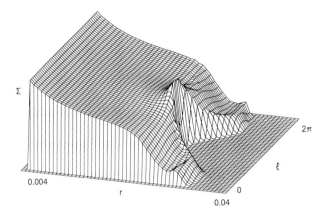

Figure 18.   Steady-state surface density profile in model 9. In this case, the companion is the proto-Sun whose mass is 10^3 larger than that of the protoplanet at the center of the disk.

In this model, the streamlines have eccentricities approaching ~ 0.15 (Fig. 19). The magnitude of v_r and v_ϕ are in units of $(GM_p/r_o)^{1/2}$ which is $[M_p/(M_p+M)]^{1/2}$ smaller than proto-Jupiter's orbital velocity. Nevertheless, they are sufficient to induce large compression in the streamlines. At $\xi \sim \pi$, streamlines converge with the material on them carrying differing specific angular momenta. In particular, material on outer streamlines has more specific angular momentum than material on inner streamlines. But material on outer streamlines undergoes a larger inward radial excursion than that on inner streamlines in order to produce convergence at periapse. As specific angular momentum is approximately conserved during the excursion, the increase in v_ϕ for material on the outer streamlines is larger than that for material on the inner streamlines. Consequently, the local shear is significantly reduced and, for a limited range in ξ, even reversed. However, this flow pattern does not modify the outward direction of angular momentum flux when it is averaged over all phases. In models where the sound speed is much less than that adopted here, the compressional region becomes highly nonlinear and the shear reversal is more pronounced.

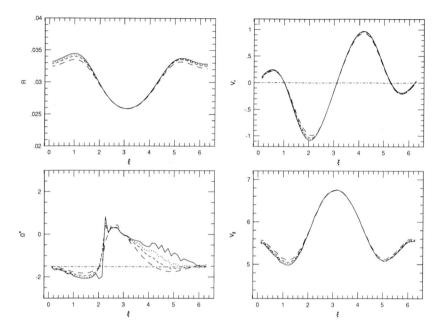

Figure 19. Perturbed velocities, radii and shear along five Lagrangian fluid elements in the outer regions of the circumplanetary disk. Note that at $\xi \sim \pi$ where the streamlines are near their perigee, the shear decreases and even changes sign. The eccentricity of the outermost streamlines is nearly 0.15.

VI. IMPLICATIONS FOR PROTOPLANETARY GROWTH

A. Necessary Conditions for Gap Formation

Protostellar disk-protoplanet interactions are very important in determining the final mass of the protoplanet. When the mass of the protoplanet is relatively small, its tidal influence is unlikely to perturb the disk structure strongly. The protoplanetary mass increases as it accretes gas from the disk. In Secs. II and III, we indicated that when a protogiant planet has acquired a sufficiently large mass, tidal effects will induce the formation of a gap in the vicinity of its orbit. Thereafter, mass growth of the protoplanet is essentially terminated. In this section, we examine the gap formation process and its implications for protoplanetary formation. These results are verified by a series of numerical computations based on the periodic Lagrangian scheme (see Fig. 20). In this series, $c_o = 10^{-2} \omega r_o$, and $10^{-4} \omega r_o^2 > \nu_{so} = \nu_{bo} > 10^{-7} \omega r_o^2$. In each case, radius and mass scale with the binary separation and the central star's mass, respectively, in such a way that the sound speed and viscosity are in units of ωr_o and ωr_o^2, respectively. In most regions of the disk, $\Sigma \sim 1$. Models in which a gap is formed are marked with open circles and models in which a gap is overrun by viscous spreading are marked with filled dots. The results in Fig. 20

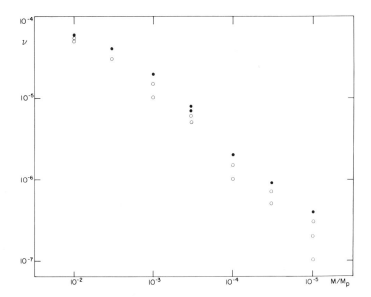

Figure 20.   Viscous condition for gap formation. The 22 models have various values of q and $v_{so} = v_{bo}(= v)$. The sound speed for all models is $c_o = 10^{-2}$. The open circles represent models with gaps whereas filled circles represent models where gap formation is prevented. The critical relation for gap formation is $q \sim 40 v_{so}$.

clearly delineate a critical line at $q \sim 40 v_{so}$, which is in good agreement with the analytic approximation (68). A necessary condition for gap formation is determined by the gas pressure which induces wave propagation and modifies the epicyclic frequency κ. From the results in Sec. III, it is apparent that for a gap to exist, we need the radial scale length H_r at the edge of the gap to be greater than or comparable to the disk thickness H_∞. If this condition is not satisfied, the radial pressure gradient at the boundary of the disk near the protoplanet would become sufficiently large to cause angular momentum to decrease with radius and κ to become imaginary. In this case, the disk edge near the gap would be dynamically unstable such that the disk flow would overrun the gap. The optimum torque occurs when the vertical and radial scale heights are equal to each other and to the distance between the disk edge and protoplanet. When a gap is first formed, the width of the gap must be larger than the Roche radius of the protoplanet; a necessary condition for gap formation is thus $r_R > H$ or $q > 3(H/r_o)^3$ as indicated by Eq. (69).

The analytic approximation to this necessary condition is also verified by numerical computations using the Lagrangian scheme for determining periodic solutions. The results for three mass ratios, $q = 10^{-2}$, 10^{-3}, and 10^{-4} are plotted in Figs. 21, 22, and 23, respectively. In each case, models with

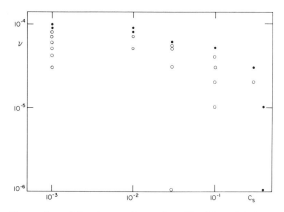

Figure 21.   Thermal condition for gap formation. The 25 models have various values
of c_o and ν_{s0}. The mass of the companion is $q = 10^{-2}$ in all cases. Note that the
violation of the thermal condition prevents the gap formation in models with $c_o \geq 0.4$.

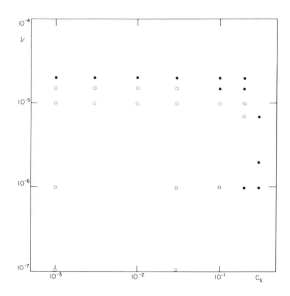

Figure 22.   Thermal condition for gap formation for $q = 10^{-3}$. For $c_o \geq 0.3$, gap
formation is no longer possible.

several values of c_o and ν_{so} are used. These results indicate that gap formation
is attainable only for $c_o < q^{1/3}$ which is in good agreement with the analytic
approximation. For large values of c_o, gap formation is prevented for cases
in which the condition for truncation based on torque comparisons is satisfied
(i.e., cases with negligible ν_{so}). In the low-c_o limit, this condition regulates
gap formation independent of the value of c_o.

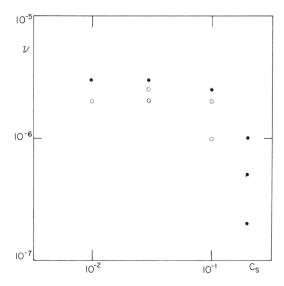

Figure 23.   Thermal condition for gap formation for $q = 10^{-4}$. For $c_o \geq 0.2$, gap formation is prevented. The thermal condition for gap formation is that the Roche radius of the companion must exceed the disk thickness.

B. Implications for the Disk Structure

The above results indicate that gap formation is determined by the magnitude of viscosity as well as by the disk thickness, both of which are uncertain quantities. Let us consider the scenario in which the protoplanetary disk is not turbulent. In this case, viscosity is due to molecular processes and is negligible that only the necessary condition (69) is relevant. To avoid gap formation before a protoplanet has acquired a mass comparable to that of Jupiter, the disk scale height must be $\gtrsim (q/3)^{1/3} r_o = 0.1\, r_o$.

In the absence of viscous dissipation, an isothermal vertical hydrostatic equilibrium may be maintained in the disk by surface heating from the central star such that $\rho = \rho_o \exp(-z^2/H^2)$ where $H = 2^{1/2} c/\omega = (2R_g T/\mu)^{1/2}/\omega$. Here ρ_0 denotes the midplane density, T the temperature, R_g the gas constant, and μ the mean molecular weight. In the limit that the entire disk is optically thin, radiation from the central star with luminosity L, would heat dust in the disk to a temperature $T = 280(L/L_\odot)^{1/4}(r_o/\, 1\mathrm{AU})^{-1/2}$ so that $H/r_o = 0.047(L/L_\odot)^{1/8}(r_o/\, 1\mathrm{AU})^{1/4}$ (Hayashi et al. 1985). In reality, the temperature in the outer regions of the disk may be considerably lower than this value if the inner regions of the disk near the star are opaque. Therefore, unless the central star is much more luminous than typical T Tauri stars, the vertical scale height maintained by stellar radiation alone would not be able to prevent gap formation for protoplanets even with mass significantly less than the present mass of Jupiter. Thus, additional sources of heating are required to maintain a large enough H/r_o that gap formation will be delayed until a protoplanet has acquired a mass comparable to that of Jupiter.

Viscous dissipation in the disk can provide a sufficiently high temperature but it requires an effective viscosity much larger than the molecular viscosity which is inconsistent with the assumption that the disk flow is laminar. To deduce the required magnitude of ν_s, we deduce the midplane disk temperature corresponding to conditions under which a gap could just form:

$$T_c \sim (q/3)^{2/3} \mu \omega^2 r_o^2 / R_g. \tag{129}$$

For Jupiter, this temperature is ~ 200 K which is comparable to the melting temperature of ice grains in a low-pressure environment. If this temperature is maintained by viscous dissipation in the disk, the standard thin-disk approximation (Shakura and Sunyaev 1973) indicates that the surface effective temperature would be

$$T_e^4 = 8\, T_c^4 / (3 \kappa_c \Sigma). \tag{130}$$

Here the opacity in the midplane $\kappa_c = \kappa_o T_c^2$, where in cgs units $\kappa_o = 2 \times 10^{-4}$ is primarily due to ice grains. When a viscous protoplanetary disk is in thermal equilibrium such that the dissipation rate is balanced by the surface radiated flux,

$$2.25 \, \nu_s \Sigma \omega^2 = 2 \, \sigma T_e^4. \tag{131}$$

From the above equations, the magnitude of effective viscosity needed is found to be

$$\nu_s \sim \frac{64 \, \sigma \mu^2 \omega^2 r_o^4}{27 \, \kappa_o \Sigma^2 R_g^2} \left(\frac{q}{3}\right)^{\frac{4}{3}}. \tag{132}$$

We note that a decrease in the magnitude of κ_o, possibly due to grain coagulation (Pollack et al. 1985), would increase the required viscosity. During its evolution, most regions of the disk attain a quasi steady state (Lin and Bodenheimer 1982; Ruden and Lin 1986) such that the mass transfer rate is

$$\dot{m} \simeq 3\pi \, \Sigma \nu_s. \tag{133}$$

Taking $q = 10^{-3}$, at Jupiter's current location we obtain from Eq. (132) in cgs units

$$\nu_s \sim 0.1 \, \omega r_o^2 / \Sigma^2 \tag{134}$$

and then

$$\dot{m} \sim \left(10^{-6}/\Sigma\right) \, M_\odot \, \text{yr}^{-1}. \tag{135}$$

The corresponding diffusion time scale is

$$\tau_d \sim r_o^2 / \nu_s \sim 10 \, \Sigma^2 \text{yr}. \tag{136}$$

Adopting the value of Σ from the minimum-mass nebula model (Cameron 1973; Hayashi 1981), we find from Eqs. (134) to (136) $\nu_s / \omega r_o^2 \sim 10^{-5}$, $\dot{m} \sim 10^{-8}$

$M_\odot$ yr^{-1}, and $\tau \equiv \pi \Sigma r_o^2/\dot{m} \sim 10^{5-6}$ yr. These results imply that $q \sim 40\ v_s/\omega r_o^2$ and thus both the conditions (68) and (69) for gap formation are marginally satisfied in a minimum-mass nebula. According to these estimates, the diffusion time scale for a minimum mass nebula is relatively short and protoplanetary formation must proceed on a time scale $\sim 10^{5-6}$ yr. This time scale is comparable to the typical age of classical T Tauri stars with an observed signature of protostellar disks (Walter 1987a; Strom et al. 1989b).

The magnitude of effective viscosity estimated with Eq. (134) is much larger than the molecular viscosity. In typical astrophysical accretion disks, turbulent viscosity is usually invoked to account for efficient angular momentum transfer and energy dissipation. An often used prescription for turbulent viscosity is the *ad hoc* α-model (Pringle 1981), in which it is assumed that $v_s \sim \alpha c H$ despite the lack of rigorous demonstration of the presence of turbulence (Shakura and Sunyaev 1973). In terms of this prescription, the minimum-mass model implies $\alpha \sim 0.01$. Theoretical investigation has shown that protostellar disks are unstable against thermal convection in the vertical direction (see Chapter by Adams and Lin). Turbulence generated from this thermal convection may provide an effective viscosity with $\alpha \sim 10^{-2}$ (Ruden et al. 1988) which is consistent with the required viscosity (134) for the minimum-mass nebula. For an accretion rate of $\sim 10^{-8}$ $M_\odot$ yr^{-1}, self-consistent convective accretion-disk models yield $H/r_o \sim 0.1$ at the present position of Jupiter (Lin and Papaloizou 1985). Using these models, the derived disk-temperature distribution is consistent with that inferred from the condensation temperature for various planets and satellites (Lewis 1972,1974).

Despite the success of theoretical nebula models in satisfying the gap-formation conditions, there exist uncertainties in both the minimum mass nebula model and the self consistent convection model. For example, if the fractionation of heavy elements of the solar nebula and their subsequent accretion onto the protoplanets are relatively inefficient processes, the value of Σ inferred from the present mass of the planets would be larger than that deduced for the minimum mass nebula model. The corresponding viscosity is also smaller. The effective viscosity may also be less than that deduced from theoretical analyses if the viscous stress provided by the convective eddies is less than that estimated from the linear normal mode analysis and mixing-length model. In this case, the viscous condition (68) would be satisfied prior to the condition (69). However, gap formation cannot occur until $q > (H_\infty/r_o)^3$. We refer to gap formation in this case as weakly viscous disk truncation.

Strongly viscous disk truncation can occur if the thermal condition (69) is satisfied prior to the viscous condition (68). For example, in the self-gravitating outer regions (Toomre 1964) of protostellar disks where $\pi G \Sigma/(\omega c) \sim 1$ or $\Sigma \sim (M/\pi r_o^2)(H/r_o)$, the condition (69) implies a critical $\Sigma \sim q^{1/3}$ $M_\odot/\pi r_o^2$ which corresponds to a disk mass ~ 0.1 $M_\odot$ for proto-Jupiter. When this condition is satisfied, angular momentum transfer by gravitational torques due to growing nonaxisymmetric perturbations (Larson 1984; Lin and Pringle

1987; Papaloizou and Lin 1989; Papaloizou and Savonije 1991) may be more efficient than effects arising from protoplanets such that no gap is formed until the protoplanet's mass becomes substantially larger than that of Jupiter. In the context of the solar nebula, the gravitational instability scenario seems unlikely unless planets in the solar system only acquired a small fraction of the total mass that resided in the primordial solar nebula (Cameron 1978a; Lin 1981a).

C. Growth of Protogiant Planets

The short formation time scale inferred from the gap formation conditions places strong constraints on models of protoplanetary formation. The most probable scenario for protoplanetary formation is accretion onto solid cores (Bodenheimer 1985). The first stage of this process requires the formation of planetesimals (Safronov 1972; Cameron 1962,1978a; Hayashi 1981). Surface density and temperature in the protoplanetary disk both decrease with time (Lin and Papaloizou 1985; Ruden and Lin 1986) leading to condensation and growth of dust particles. Dust particles segregate from the gaseous nebula and coagulate (Wasserberg 1985). Eventually, kilometer-size planetesimals form, either through gravitational instability (Goldreich and Ward 1973), or through continuing cohesive collisions (Weidenschilling 1987; also see Chapter by Weidenschilling and Cuzzi). Subsequently, the collisional evolution of low mass planetesimals results in the formation of Earth-mass prototerrestrial planets and protogiant planetary cores (see Chapter by Lissauer and Stewart). The present estimates of the growth time scales for Earth-size planetesimals (Safronov and Ruzmainkina 1985; Hayashi et al. 1985) near Jupiter's orbit are $\sim$10 to 100 Myr which is somewhat longer than the $\sim$0.1 to 1 Myr protogiant planet formation time scales deduced above. Detailed analysis of the expected kinematic properties of planetesimals (Wetherill and Stewart 1989) has shown that a relatively short growth time scale for Earth-sized planetesimals is possible if runaway coagulation occurs in which a few massive planetesimals grow much faster than the rest of the population.

When a protoplanetary core has attained $\sim$1 $M_\oplus$, it begins to accrete a low-mass quasi-hydrostatic gaseous envelope (Mizuno et al. 1978; Mizuno and Nakazawa 1988). As the core continues to grow to 10 to 20 Earth masses, the hydrostatic equilibrium can no longer be maintained and the envelope collapses (Sekiya et al. 1987,1988). We then have a situation approximating spherically symmetric accretion in which the gas accretion rate is

$$\dot{M}_p = 4\pi\rho c r_C^2 \qquad (137)$$

where r_C is the radius at the sonic point in the flow and ρ_C is the density there. For the isothermal Bondi solution,

$$r_C = GM_p/(2c^2) \qquad (138)$$

where M_p is the protoplanet mass. The accretion rate then becomes

$$\dot{M}_p = \pi \exp(3/2)G^2M_p^2\rho_\infty/c^3 \qquad (139)$$

where ρ_∞ is the ambient density. For a typical solar nebular model, the characteristic growth time scale near the present location of Jupiter is $\tau_g = M_p/\dot{M}_p \sim 10^5 (M_\oplus/M_p)$ yr. Provided that the coagulation time scale for Earth-mass cores is sufficiently rapid, a protoplanet could then accrete enough gas to become a giant planet on a time scale comparable to the viscous diffusion time scale of the protoplanetary disk.

During this early gas accretion phase, infall toward the protoplanet is essentially spherical because $r_C < r_R$. However, as a protogiant planet grows, r_C increases until it exceeds r_R or H. It turns out that as the protoplanet grows, the condition that $r_C = r_R$ is reached approximately simultaneously with the condition that $r_C = H$. The critical mass ratio for $r_C \sim r_R$ is $q \simeq (H/r_o)^3$. This is also the minimum mass ratio needed for gap formation. Subsequent protoplanetary growth and disk evolution depend on the viscosity in the disk. In the strongly viscous limit, gas accretion will continue until the viscous condition for gap formation is established when $q \sim 40v/\omega r_o^2$. During this final stage of gas accretion, $H < r_C$ implies that gas accretion proceeds primarily in the plane and $r_R < r_C$ indicates that the tidal effect of the central star regulates mass flow onto the protoplanet. Several numerical simulations of accretion in this strongly viscous limit have been carried out (Cameron 1979; Miki 1982; Sekiya and Miyama 1988a; Sekiya et al. 1988). These calculations have shown that as gas flows across the Roche lobe, a directly rotating circumplanetary disk is formed. This disk extends to a significant fraction of the protoplanet's Roche radius (see Fig. 18 in Sec. V.F) and is much larger than the final radius of the planet. Thus, the newly accreted material must lose more than 90 % of its original spin angular momentum if the giant planets and their satellite systems are to end up with their present spin and orbital angular momenta. Such angular momentum losses cannot be induced by the tidal field of the proto-Sun because such tides are ineffective near Jupiter's final radius which is $\sim 10^{-3} r_R$.

A more probable scenario for gap formation is that disk truncation occurs in the marginal or weakly viscous limits because little angular momentum loss may be needed. In this case, gap formation occurs while $r_C \leq r_R$ so that throughout the protoplanet's growth, gas accretion is essentially spherical. The details of gap formation in the disk and the final stages of protoplanetary accretion must be analyzed taking three-dimensional effects fully into account. Such analysis has yet to be carried out.

D. Termination of Protoplanetary Growth and Disk Clearing

After a protoplanet opens up a gap in the vicinity of its orbit, its mass growth is essentially terminated though energy and angular momentum continue to be exchanged between it and the disk. The orbital evolution of the protoplanet is thus locked to that of the disk. In the absence of continuous infall, Σ, c and v decrease with time. The results in Sec. V showed that the width of the gap increases as c and v decrease. For $q > 10^{-4}$, gas in the region between the inner and outer 2:1 Lindblad resonances may be expelled for modest values of c and v. Beyond the inner and outer 2:1 Lindblad resonances, there are no additional low-order orbital resonances and therefore the disk feels only a weak tidal effect from the protoplanet. Disk matter can accumulate in the neighborhood of the outer 2:1 Lindblad resonance and provide a favorable formation site for another protoplanet (see Sec. VIII).

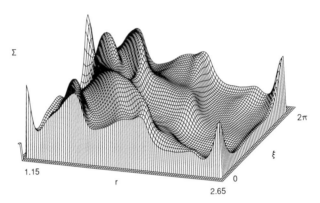

Figure 24.   Surface density distribution for a gaseous ring with two shepherding companions (model 10). The mass of both companions is $3 \times 10^{-3} M_\odot$. In the disk, $c_o = 0.04 \omega r_o$ and $v_{so} = 10^{-5}$. The companions have orbital radii 1 and 3, respectively. The plot is in the corotating frame of the outer companion at an epoch when both companions are at $\xi = 0$.

In a protoplanetary disk, several planets may be formed while there is still residual gas in the disk. Gaseous rings may be trapped between protoplanets in the same way that planetary rings are shepherded by guardian satellites (Goldreich and Tremaine 1979a). As the protoplanets' masses increase and Σ and c decrease, gaps around the protoplanets widen and the inner and outer edges of the trapped ring retreat toward the outermost Lindblad resonance of the inner protoplanet and the innermost Lindblad resonance of the outer planet. In the present solar system, the outermost Lindblad resonance of any given planet is exterior to the innermost Lindblad resonance of its neighboring outer planet. Consequently, for sufficiently small c and v, the trapped ring may be squeezed.

To demonstrate the tidal interaction between a ring and two shepherding companions, we need to use the Lax-Wendroff scheme as a set of truly time-dependent, two-dimensional solutions is needed. In order to save computing time, models with massive protoplanets are computed. In each case, a gas ring is initially placed between two shepherding protoplanets with identical masses. In each model, $c_o = 0.04 \omega r_o$ and $v_{so} = v_{bo} = 10^{-5} \omega r_o^2$. The inner companion is placed at r_o and the outer planet is placed at $3r_o$ and $2r_o$ in models 10 and 11, respectively. In both models 10 and 11, $q = 3 \times 10^{-3}$. In model 10, the outermost Lindblad resonance of the inner planet is located at

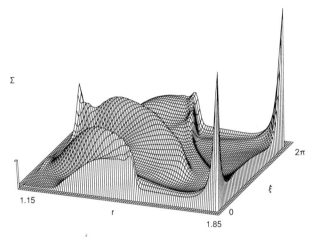

Figure 25.   Surface density distribution in model 11. All the parameters here are identical to model 10 (Fig. 24) except the orbital radius of the outer companion is 2. The spike at $\xi = 0$ near the outer edge of the disk is due to accretion onto the Roche volume of the outer companion. Its amplitude does not change with time.

a similar location as the innermost Lindblad resonance of the outer planet. In model 11, the outermost Lindblad resonance of the inner planet is outside the innermost Lindblad resonance of the outer planet. More massive protoplanets ($q = 0.01$) are examined in model 12 with the outer planet at $3r_o$. In each model, equilibrium is established in about a hundred inner planet's orbital periods. In Fig. 24, we plot the Σ distribution for model 10 in a frame corotating with the outer planet for the epoch at which both planets are located $\xi = 0$, where ξ here represents the azimuthal angle in the rotating frame. The analogous plot for model 11 is illustrated in Fig. 25. In this corotating frame, the azimuthal phase of the inner planet ξ_s changes on a synodic period and the disk structure in the inner part of the ring also evolves. In Fig. 26, the radial distribution of Σ at $\xi = 0$ ($\equiv \Sigma_o$) is plotted as a function of phase (ξ_s). The phase-dependent variation of Σ is much larger at small r than at large r because the outer region is predominantly under the influence of the outer planet with which the frame is corotating. The inner region of the ring is most strongly influenced by the inner planet and the variations there are due to the orbital phase change in this frame. This is best illustrated in Fig. 27 where the phase distribution of Σ at the inner boundary of the ring (Σ_i) is plotted against ξ_s.

The results in Sec. V show that the density waves excited at low-order Lindblad resonances which are in close vicinity to the protoplanet's orbit. In the low sound-speed limit, these waves are dissipated locally. The above results are in agreement with these qualitative expectations as the local ring structure is most affected by the nearest protoplanet. In comparison to

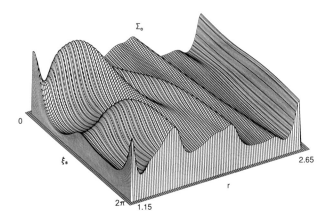

Figure 26.   Surface density distribution at $\xi = 0$ as a function of the synodic phase ξ_s of the inner companion in model 10. Trailing density waves are excited near the inner companion.

model 10, the protoplanets are less separated in model 11 and the entire ring is consequently under strong tidal perturbation from both planets. The spike at the outer edge of the disk near $\xi = 0$ indicates accretion of gas into the Roche radius of the outer planet. The ring is modestly depleted and contains strongly nonlinear structures. Because of the strong perturbations, rings are unstable and Σ is rapidly depleted in models with smaller radial separation between the two shepherding protoplanets. In model 12 where the shepherding protoplanets are more massive, the ring is also strongly squeezed and it becomes unstable and is depleted on a relatively short time scale.

VII. ORBITAL ECCENTRICITIES

In addition to the above dynamical constraints, important clues on the structure and evolution of the solar nebula can be deduced from companion-disk interactions in the post gap formation epoch. In this section, we first show that tidally induced energy and angular momentum exchange between the disk and the companion can excite the growth of the companion's eccentricity. The small eccentricities of the planets today imply that Σ cannot much exceed the estimates in the minimum mass nebula model. The corresponding effective viscosity would imply a viscous evolution time scale ~ 1 Myr for Σ to be significantly depleted after a gap is formed.

When a companion with an initially circular orbit interacts with a disk, it is sometimes possible that an initially small eccentricity may be amplified through the tidal interaction. In order to discuss this issue, we consider the tidal interactions due to a companion that has an orbit with small eccentricity. We first consider the nature of the perturbing potential acting on an axisymmetric

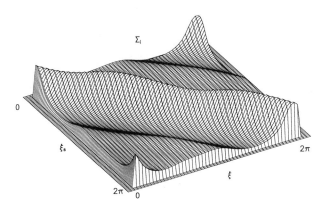

Figure 27. Azimuthal distribution of surface density at the disk's inner boundary in model 10 ($r = 1.15$) as a function of ξ_s. The ridge marks the disk response to the inner companion's tidal disturbance.

disk centered on the primary protostar. We show that, to first order in the eccentricity, the tides can be considered to be produced by three independent companions in circular orbits with different frequencies. The tidal response can then be found by applying the theory developed above for systems with circular orbits to each component in turn and then superposing the results; this procedure will be valid in the linear regime. The rate of growth of orbital eccentricity can then be found by consideration of the orbital energy and angular momentum exchange rates associated with each tide.

The perturbing potential due to a companion in a nonrotating frame centered on the primary is

$$\psi_c = -\frac{GM_p}{[r^2 + r_p^2 - 2rr_p\cos(\phi - \Phi)]^{1/2}} + \frac{GM_p r\cos(\phi - \Phi)}{r_p^2}. \quad (140)$$

Here, r_p and Φ are the radius and angular coordinate of the companion, respectively. For an orbit with small eccentricity e, these are given to an adequate approximation by

$$r_p = r_o + er_o\cos(\omega t) \quad (141)$$

and

$$\Phi = \omega t - 2e\sin(\omega t) \quad (142)$$

where r_o is now the semimajor axis. We now expand ψ_c to first order in e and perform a Fourier analysis in ϕ with the result that

$$\psi_c = \sum_{m=0}^{\infty} W_m \cos[m(\phi - \omega t)] + \sum_{m=0}^{\infty} W_{1m} \cos[m(\phi - \omega_1 t)]$$

$$+ \sum_{m=0}^{\infty} W_{2m} \cos[m(\phi - \omega_2 t)] \tag{143}$$

where for a disk interior to the orbit

$$W_m = C_m(\epsilon)/r_o \tag{144}$$

$$W_{1m} = \frac{e}{2r_o} \left[-\epsilon C_m'(\epsilon) + (2m - 1)C_m(\epsilon) \right] \tag{145}$$

and

$$W_{2m} = \frac{e}{2r_o} \left[-\epsilon C_m'(\epsilon) - (2m + 1)C_m(\epsilon) \right]. \tag{146}$$

Here $\epsilon \equiv r/r_o$, $\omega_1 = \frac{(m+1)}{m}\omega$, and $\omega_2 = \frac{(m-1)}{m}\omega$, and the $C_m(\epsilon)$ are related to the Laplace coefficients.

For a disk exterior to the orbit the same formal expression for the potential holds but now

$$W_m = C_m(\epsilon^{-1})/r_o \tag{147}$$

$$W_{1m} = \frac{e}{2r_o} \left[\epsilon^{-1} C_m'(\epsilon^{-1}) + 2m C_m(\epsilon^{-1}) \right] \tag{148}$$

and

$$W_{2m} = \frac{e}{2r_o} \left[\epsilon^{-1} C_m'(\epsilon^{-1}) - 2m C_m(\epsilon^{-1}) \right]. \tag{149}$$

Sometimes when discussing an external disk it is convenient to use a coordinate system based on the center of mass of the binary system rather than that of the primary. If this is done, an expansion of the same formal type as that given above is obtained and the discussion given below also holds in this case.

Thus we see that for any $m>0$ the tidal forcing can be considered to be produced by three companions in circular orbits with frequencies ω, ω_1 and ω_2. The frequencies (ω_1) one obtains that are larger than ω are $2\omega, 3\omega/2, 4\omega/3, \dots$ for $m = 1, 2, 3\dots$, respectively, while the frequencies (ω_2) smaller than ω are $\omega/2, 2\omega/3, 3\omega/4, \dots$ for $m = 2, 3, 4\dots$, respectively. Axisymmetric forcing from terms with $m = 0$ is always weak and never resonant; in what follows it will be neglected. Both sets of frequencies (ω_1, ω_2) occur for internal and external disks. Because frequencies different from ω occur, in some cases, it is possible that disk material corotates with one of the tidal perturbations. One must then consider the effect of the corotation resonance on the response. This requires disk material orbiting with frequencies between 2ω and $\omega/2$ which is possible only for a fairly extreme mass ratio. Therefore effects arising from corotation resonances can generally be ignored for the binary star problem where the mass ratio is not extreme. However, they should be considered in the protoplanet problem. Goldreich and Tremaine (1979b,1981) found that the effects of corotation resonances can dampen the eccentricity in a weakly perturbed disk. They also point out that a strong tidal interaction at the Lindblad resonances could lead to gap formation such that the region around the corotation resonances is depleted as in the circular orbit case (see Sec. V). Consequently, eccentricity damping by the corotation resonances is rendered ineffective (also see Ward 1988). If there are gaps in the disk at corotation resonances, we should in principle consider the tidal response of more than one section for each m and this will be considered in what follows. However, in most cases one only gets a significant response from the section containing the appropriate Lindblad resonance.

We now consider the post gap formation epoch in which there are no corotation resonances, so that the theory given in Sec. III may be applied directly. This is correct in the binary star case and in the protoplanetary case if material is depleted at the Lindblad resonances appropriate to the circular orbit tide. In order to discuss the evolution of the eccentricity, we first note that for each m the terms in the tidal potential (140) proportional to e are the most important. The other term is of course the term that appears in the circular orbit case and this by itself cannot excite the eccentricity. We then evaluate the angular momentum exchange rate for each of the terms proportional to e independently, using the theory developed in Sec. III. We may, in fact, apply Eq. (53) directly, so obtaining an angular momentum flux for each m

$$F_{H1} = \frac{\pi m}{4|C_1|^2} \left| \int_{r_o}^{r_1} \Sigma_o r S_{1m} X_{1m} dr \right|^2 \tag{150}$$

and

$$F_{H2} = \frac{\pi m}{4|C_2|^2} \left| \int_{r_o}^{r_1} \Sigma_o r S_{2m} X_{2m} dr \right|^2. \tag{151}$$

Here F_{Hi}, C_i, S_{im}, and X_{im} for $i = 1, 2$ have the same meanings as F_H, C, S_m, and X_m, respectively as used in Sec. III but they apply to the situation where the forcing term is

$$\psi_c = W_{im} \cos[m(\phi - \omega_i t)]. \tag{152}$$

In order to evaluate the effect on the companion's orbit, we use the fact that a tidal wave in the disk proportional to $\cos[m(\phi - \omega_i t)]$ causes the binary orbit to evolve such that changes to its energy E and angular momentum J are related by

$$\frac{dE}{dt} = \omega_i \frac{dJ}{dt}. \tag{153}$$

Also we use the fact that correct to second order in eccentricity,

$$\frac{dE}{dt} - \omega \frac{dJ}{dt} = \omega \frac{G^{1/2}MM_p}{2\sqrt{(M+M_p)}} \frac{d(r_o^{1/2}e^2)}{dt}. \tag{154}$$

Only the effect of tidal perturbations proportional to e contribute to the left-hand side of the above equation so that we have for an interior disk

$$\frac{dE}{dt} - \omega \frac{dJ}{dt} = \omega \frac{G^{1/2}MM_p}{2\sqrt{(M+M_p)}} \frac{d(r_o^{1/2}e^2)}{dt} = \sum_m [(\omega_1 - \omega)F_{H1}]$$

$$+(\omega_2 - \omega)F_{H2}]. \tag{155}$$

For an external disk the signs of the terms on the right-hand side are reversed. Considering the effects of both an internal and external disk, we obtain after substituting the values of the ω_i

$$\frac{G^{1/2}MM_p}{2\sqrt{(M+M_p)}}\frac{d(r_o^{1/2}e^2)}{dt} = \sum_m \frac{1}{m}[F_{H1} - F_{H2}]_I + \sum_m \frac{1}{m}[F_{H2} - F_{H1}]_E \tag{156}$$

where the subscript I denotes the internal disk if it exists and the subscript E denotes the external disk if that exists. We now argue that the contributions from both the internal and external disk to the right-hand side of Eq. (156) are always positive.

In the protoplanetary case, when there is a gap, the F_{Hi} are only nonzero when appropriate Lindblad resonances occur in the disk. For high frequencies ω_1, Lindblad resonances occur only in the internal disk, while for low frequencies ω_2, Lindblad resonances occur only in the external disk. Thus the F_{H2} are zero in the inner disk and the F_{H1} are zero in the outer disk. It is then clear that

$$\frac{G^{1/2}MM_p}{2\sqrt{(M+M_p)}}\frac{d(r_o^{1/2}e^2)}{dt} > 0. \tag{157}$$

Therefore, for a slowly evolving disk and gap for which dr_o/dt may be neglected, we conclude that $de^2/dt > 0$. Thus in the absence of corotation resonances the eccentricity of the protoplanet grows. Similar conclusions can be obtained for tidal interaction between a binary system and an accretion disk (Artymowicz et al. 1991; also see Sec. IX).

The rate of eccentricity growth depends on the disk structure. In the high sound speed and marginal truncation limit, the disk is extended to the vicinity of the companion's radius. From Eq. (156) we find very approximately that for a gap with a width of ~ 2 Roche radii

$$\frac{1}{e^2}\frac{de^2}{dt} \sim \frac{q^{-1/3}}{t_d}\frac{M_N}{M_\odot} \tag{158}$$

where $M_N = \pi \Sigma r_o^2$ is the characteristic disk mass and $t_d = 2\pi/\omega$ is the orbital period of the companion. Using a minimum mass nebula model, $de^2/dt \sim 10^{-4}$ yr^{-1} for $e = 0.1$ at Jupiter's current location. Thus, on the evolution time scale of the disk, the orbits of protoplanets with mass comparable to Jupiter could become highly eccentric. A much slower rate of eccentricity increase may be attained in the weakly viscous and relatively cool limit where the disk is truncated near the 2:1 resonances (see numerical results in Sec. V). In this case, Eqs. (77) and (156) give very approximately

$$\frac{1}{e^2}\frac{de^2}{dt} \sim \frac{q}{t_d}\frac{M_N}{M_\odot} \tag{159}$$

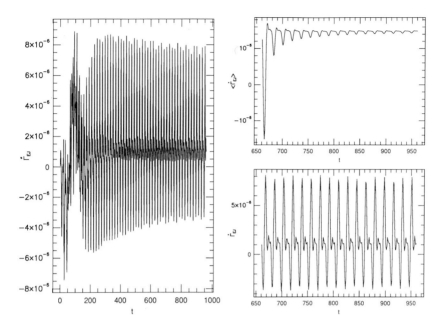

Figure 28.   Growth of companion's eccentricity in model 13. On the left panel, the instantaneous values of $\dot{\Gamma}_{EJ}$ are obtained from Eq. (160). When an equilibrium is achieved, the asymptotic values of $\dot{\Gamma}_{EJ}$ are plotted in the lower right panel. The time averaged value of $\langle\dot{\Gamma}_{EJ}\rangle$ (upper right panel) is derived from Eq. (161). Positive values of $\langle\dot{\Gamma}_{EJ}\rangle$ indicate eccentricity growth. The surface density distribution in model 13 is identical to that in model 1 (see Fig. 5).

so that proto-Jupiter's eccentricity would not increase significantly beyond its present value in <1 Myr.

These results are verified with numerical computations. Because the potential of a companion in an eccentric orbit is time dependent, we use the Lax-Wendroff scheme. Other than the companion's eccentricity (0.1), model 13 has identical model parameters to model 1 (i.e., $q = 0.01$). The Σ distribution of the disk is essentially identical to that in Fig. 5, i.e., the disk is tidally truncated at the inner 2:1 resonance of the protoplanet. From the disk response, the value of

$$\dot{\Gamma}_{EJ} \equiv \frac{dE}{dt} - \omega\frac{dJ}{dt} \qquad (160)$$

is computed and plotted as a function of time (Fig. 28). This result indicates that an equilibrium is essentially established at $t \sim 200$. Although the value of $\dot{\Gamma}_{EJ}$ has a phase dependence, when it is averaged over a sufficiently long time interval ΔT such that for some initial t_0,

$$\langle \dot{\Gamma}_{EJ} \rangle \equiv \frac{1}{\Delta T}\int_{t_o}^{t_o+\Delta T}\dot{\Gamma}_{EJ}dt \qquad (161)$$

a steady-state value independent of t_0 can be found. In the computations, Σ is of order unity so that Eqs. (155) and (156) would lead us to expect $<\dot{\Gamma}_{EJ}>\sim10^{-2}\,q^2\omega^3r_o^4\sim10^{-6}\omega^3r_o^4$. This is in adequate agreement with the results shown in Fig. 28 which adopted units such that $r_o = \omega = 1$.

When tidal truncation occurs in the weakly viscous limit, the gap is relatively wide. However, in order to deplete a significant fraction of the disk mass in <1 Myr so that a small eccentricity for proto-Jupiter may be sustained, a modest value ($\alpha\sim10^{-2}$) of viscosity is needed. In addition, tidal interaction can also induce the growth of the inclination angle of protoplanets' orbits (Borderies et al. 1984a). The inclined protoplanets also excite density waves (Shu et al. 1983). The dynamical constraints that can be inferred from the present orbital inclinations of the planets are yet to be determined.

The mass deduced for protoplanetary disks around several T Tauri stars ranges from 10^{-2} to $10^{-1}M_\odot$ (Sargent and Beckwith 1987; Weintraub et al. 1989b; Adams et al. 1990; Beckwith et al. 1990; also see Chapters by Beckwith and Sargent and by Hartmann et al.). The above discussion indicates that giant planets formed in the relatively massive protostellar disks could have large orbital eccentricities. An interesting issue to investigate is that of the long-term stability of a system of planets with modest orbital eccentricities (see Chapter by Duncan and Quinn).

VIII. ORBITAL MIGRATION OF PROTOPLANETS

In the previous sections, we derived the tidal interaction between a proto-planet and the disk on the assumption of a constant r_o. However, angular momentum exchange can cause secular orbital evolution of the protoplanet's orbit (Goldreich and Tremaine 1980; Ward 1984). This orbital migration may also result in additional modification to the local disk structure.

A. Differential Torque

The discussion in Secs. II–IV indicates that angular momentum is transferred to a protoplanet from the disk interior to its orbit. Similarly, angular momentum is transferred from a protoplanet to the disk exterior to its orbit. Orbital migration of a protoplanet is induced by an imbalance in the losses and gains of angular momentum. For protoplanets with circular orbits, angular momentum flux is produced at the Lindblad resonances. The imbalance in the angular momentum exchange rate can be expressed as the sum of the differences between the torques, F_H^+ and F_H^-, at the outer and inner resonances, respectively, for each m

$$\Delta F_H = \left(F_H^+ - F_H^-\right). \qquad (162)$$

The net rate of angular momentum gain of the protoplanet is then given by

$$\frac{dH}{dt} = -\sum_m \Delta F_H. \qquad (163)$$

There are several cases where we would expect a nonzero ΔF_H. In a cold disk where the flow velocity is essentially Keplerian, a gradient in Σ according to Ward (1986) leads to

$$\Delta F_H = \frac{16}{9} m \Sigma q^2 \omega^2 r^4 \left[K_1 \left(\frac{2}{3} \right) + 2K_o \left(\frac{2}{3} \right) \right]^2 \frac{d \ln \Sigma}{d \ln r}. \qquad (164)$$

This differential torque calculation is valid only in the limit $\xi_c \equiv mc/r\Omega \ll 1$ and near the protoplanet where the relevant m is large this condition breaks down. For $\xi_c > 1$, the spacing between resonances becomes smaller than the disk thickness, the effects of the disk's vertical structure and its small deviation from Keplerian rotation must be taken into account. In addition coupling between different resonances may be induced by sound waves so that individual resonances can no longer be separated. In principle, a three-dimensional nonlinear analysis would be appropriate. However, some of the effects may be taken into account by multiplying each resonant contribution by a correction factor, $f_Q(\xi_c)$ which attains unity for $\xi_1 \ll 1$ and vanishes for some critical value $\xi_c \sim 1$ (Goldreich and Tremaine 1980; Ward 1989a). The total differential torque now becomes

$$\frac{dH}{dt} = -\sum_m f_Q(mc/r\omega) \Delta F_H. \qquad (165)$$

A Σ gradient then leads to

$$\frac{dH}{dt} \simeq -5.6 \xi_1^2 q^2 \Sigma \Omega^2 r^4 \left(\frac{r\Omega}{c} \right)^2 \frac{d\ln\Sigma}{d\ln r}. \qquad (166)$$

A pressure gradient in the disk can also contribute to ΔF_H. Equation (33) indicates that a modest sound speed in the disk changes the resonance locations. In an isothermal thin disk, the locations of high-m inner and outer Lindblad resonances become

$$r_s = r_o \left(1 \mp \frac{1}{m} + \frac{1}{2} \left(\frac{c}{r\Omega} \right)^2 \frac{d\ln\Sigma}{d\ln r} \right)^{2/3}. \qquad (167)$$

The shift in the resonance locations introduces a difference between the protoplanet's distance to the inner and outer Lindblad resonances which then produces a torque imbalance leading to the angular momentum flux

$$\frac{dH}{dt} \simeq 5.19 \, \xi_1^4 q^2 \Sigma \Omega^2 r^4 \left(\frac{r\Omega}{c} \right)^2 \frac{d\ln\Sigma}{d\ln r}. \qquad (168)$$

The results in Eqs. (166) and (168) should adequately approximate the magnitude of the differential torque induced by a protoplanet with a mass

large enough to have induced a gap. But there may be additional modifications due to asymmetries arising through the finite thickness of the disk
and the correction functions. When there is no gap, the effects of material
corotating with the protoplanet have to be taken into account. This requires a
calculation which takes account of the disk's vertical structure which has yet
to be performed.

B. Disk Response and Orbital Migration

The results in Eqs. (166) and (168) are obtained on the assumption that the Σ
and pressure distribution are unperturbed by the protoplanet. In reality, Σ is
actively modified by the protoplanet's tide and viscous stress. In the limit that
the protoplanet does not strongly perturb the Keplerian flow, we integrate, in
the azimuthal direction, Eq. (30) such that

$$\Sigma v_r \frac{d(r^2\Omega)}{dr} = \frac{1}{r}\frac{\partial}{\partial r}\left(v_s \Sigma r^3 \frac{d\Omega}{dr}\right) + \Sigma\Lambda, \tag{169}$$

where Λ is the local injection rate of angular momentum per unit mass into
the disk gas by the protoplanet's tidal torque. The radial dependence of Λ
will be determined by the location where the density waves are dissipated
rather than where they are excited. Conservation of total angular momentum
implies that

$$\sum_m \Delta F_H = 2\pi \int_{r_{in}}^{r_{out}} \Lambda r \Sigma dr. \tag{170}$$

A protoplanet undergoes orbital migration such that

$$M_p \omega r_o \dot{r}_o/2 = -\sum_m \Delta F_H. \tag{171}$$

The protoplanet's orbital evolution and disk response may be analyzed
self consistently by choosing a prescription for Λ. The discussion in Secs. III
and IV indicates that if v_s is relatively large, density waves are dissipated
near the location where they are launched. If v_s is relatively small, nonlinear
shocks and refraction effects may also lead to effective dissipation near the
protoplanet. Thus, it is probably adequate to utilize the results derived from
both the impulsive approximation and the resonant interaction calculation in
Sec. II. Then

$$\Lambda = \text{sign}(r - r_o)\frac{1}{2\pi r_0 \Sigma}\frac{d\dot{H}_T}{d\Delta_0} \tag{172}$$

where we can use Eq. (5) to evaluate $d\dot{H}_T/d\Delta_0$, with the result that

$$\Lambda = \text{sign}(r - r_o)f q^2 \omega^2 r_0^2 (r_0/|\Delta_0|)^4 \tag{173}$$

where $f = 4/9\pi$ is a constant of order unity. Equations (169), (171) and
(173) coupled with the continuity equation

$$\frac{\partial \Sigma}{\partial t} + \frac{1}{r}\frac{\partial (r\Sigma v_r)}{\partial r} = 0 \tag{174}$$

and a viscosity prescription comprise a complete set of self-consistent equations from which the protoplanet's orbital migration and the disk's Σ evolution may be computed. Within the approximation scheme associated with the impulse approximation, we may replace r_0 by r and ω by Ω. To be consistent with the notation in some of our other work, we relabel $|\Delta_0|$, the distance to the protoplanet as $|\Delta_p|$.

Using these equations, Hourigan and Ward (1984) postulate that orbital migration may inhibit gap formation. Equations (169) and (173) indicate that the time scale for modifying an initially smooth Σ gradient is

$$\tau_{\text{grad}} \simeq \left| \frac{\partial \Sigma}{\partial r} \Big/ \frac{\partial^2 \Sigma}{\partial r \partial t} \right| \simeq \Omega \Delta_p^2 / \Lambda \simeq \left(\frac{M}{M_p} \right)^2 \left(\frac{\Delta_p}{r_o} \right)^6 \frac{1}{\omega}. \tag{175}$$

Equation (171) implies that the time scale for migrating across the gap, assuming a smooth background surface density, is

$$\tau_\Delta \simeq |\Delta_p / \dot{r}_o| \simeq \frac{M_p \omega}{\Lambda \Sigma} \frac{r_o}{|\Delta_p|} \simeq (M/M_p)(M/\Sigma r_o^2)(|\Delta_p|/r_o)^3/\omega. \tag{176}$$

For $\tau_{\text{grad}} > \tau_\Delta$, the protoplanet drifts too fast to allow the disk to respond. Thus, there is a critical mass $M_c \sim \Sigma r_o^2 (\Delta_p/r_o)^3$ below which gap formation is inhibited by radial drift (Ward and Hourigan 1989). According to the results in Sec. III, the Rayleigh stability criterion requires $\Delta_p > H$ so that $M_c < (H/r_o)^3 \Sigma r_o^2 < (H/r_o)^3 M_\odot$. Thus, the prohibition of gap formation by tidal drift can only be realized in cases where the protoplanet's mass is already too small to satisfy the necessary condition for tidal truncation.

C. Self-Consistent Calculations

Because the disk does not have time to respond, Eqs. (166), (168) and (171) can be used to compute the evolution of protoplanets with $M_p < M_c$ once it is known how to make the modifications needed to account for neglected three-dimensional effects near the corotation radius. Simple estimates indicate that the characteristic drift time scale for resonant interaction between the disk and low-mass protoplanets is considerably shorter than that expected from hydrodynamic drag. But the drift time scale may be comparable to or longer than the viscous diffusion time scale of the disk gas so that low-mass protoplanets probably do not migrate over substantial radial distances prior to gas depletion in the disk.

For protoplanets with $M_p > M_c$, Eqs. (169), (171) and (173) are relevant. To apply these equations in accordance with the torque cutoff factor, $|\Delta_p|$ is taken to be the maximum of the disk thickness, c/Ω, or $|r - r_o|$. Note that this prescription does not prevent a discontinuous change of sign for Λ at r_o. Equations (169) and (171) can be solved for an arbitrary viscosity prescription. For example, according to both the simple self consistent convective model

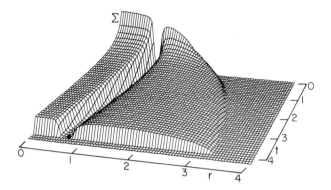

Figure 29.   Evolution of disk surface density and orbital migration of embedded
protoplanet in model 14 with $r_o = 0.6$ initially. The time scale t is in terms of
the diffusion time scale. Gap formation occurs early during the evolution and the
companion migrates inward monotonically. Note also that the disk is spreading as
a result of viscous diffusion.

and the α prescription $v_s = v_o(\Sigma/\Sigma_0)^2$ in a solar nebula (Lin and Papaloizou
1985).

    For computational convenience, we can choose the unit of time to be the
viscous diffusion time scale r_o^2/v_0, where the unit of radius r_0 is some suitably
large radius, and the unit of Σ is Σ_o. Each model is then defined by two
parameters, $A \equiv \omega r_o^2 q^2/3v_o$ which measures the ratio of viscous and orbital
time scales, and $B \equiv 3\pi \Sigma_o r_o^2/M_p$ which measures the ratio of disk mass to
protoplanet mass. The magnitudes of A and B can be easily estimated for
any solar nebula model. For example, by adopting the assumption that the
protostellar disk had a solar composition and that most of the volatile material
in the disk was not accreted by the planets, we can determine a lower limit
on the mass of the nebula to be $\sim 0.02\ M_\odot$ (Cameron 1962). For a protogiant
planet with a Jupiter mass, $B \sim 10$. According to self-consistent convective
disk models (Lin and Papaloizou 1980), $v_s \sim 10^{-5}\ \omega r_o^2$ so that $A \sim 10^{-3}$. Based
on these estimates, we computed a series of models with $10^{-4} > A > 10^{-2}$ and
$10^2 > B > 0.1$ (Lin and Papaloizou 1986b).

    It is of interest to examine the protoplanet's orbital migration under a
variety of initial conditions. For illustrative purpose, we show in Figs. 29–
31 the evolution for an arbitrarily chosen model 14 with $A = 10^{-3}$ and
$B = 20$. At $t = 0$, $\Sigma = 1$ at $r < 1$ and 0 at $r > 1$ in the disk. Three initial
protoplanetary positions 0.6, 0.7 and 0.8 are chosen. These results indicate
that the protoplanet opens up a gap in the disk shortly after the onset of the
calculation. After an initial transitory stage, Σ, near the protoplanet's orbital
position, adjusts continually so that the angular momentum transferred from

the inner disk region to the protoplanet is approximately balanced by that transferred from the protoplanet to the exterior disk region. This has the effect that the protoplanet's orbital evolution ultimately occurs on the disk's viscous diffusion time scale.

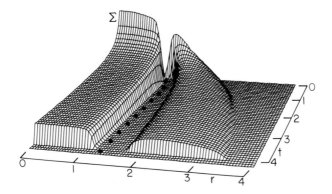

Figure 30. Evolution of disk surface density and orbital evolution of embedded protoplanet in model 14 with $r_o = 0.7$ initially. During the early stage of the evolution when the disk mass inside is greater than that outside to the protoplanet's orbit, the protoplanet migrates outward. At later stages, the surface density in the inner region is depleted due to accretion. Angular momentum removal from the protoplanet by the external disk becomes more efficient than the angular momentum supply from the interior disk so that the protoplanet migrates inward.

A comparison of different models also indicates that the protoplanet's orbital evolution is sensitive to the location of the protoplanet with respect to the mass distribution of the disk. If a protoplanet is located in the inner regions of the disk, it evolves towards the center on a local viscous evolution time scale. A protoplanet may be induced to evolve into the Sun if its mass is much less than that of the disk exterior to its orbit. If a protoplanet is located in the outer regions of the disk, it generally migrates outward on the viscous diffusion time scale of the outer disk region which is generally longer than that of the inner region. However, a protoplanet cannot double its distance from the Sun unless the disk mass interior to the protoplanet far exceeds that exterior to it and that of the protoplanet. In the intermediate region, the protoplanet first moves outward and then inward following a similar pattern to a local fluid element. The initial outward migration is induced by the absorption of angular momentum from the inner parts of the disk. As the inner region is depleted, the location of maximum viscous stress moves outward (Lynden-Bell and Pringle 1974). When the protoplanet's loss of angular momentum to the outer region cannot be replenished, it migrates inward. Throughout its evolution, the protoplanet maintains a strong tidal coupling with the disk. This is well illustrated by the relative position of the protoplanet in the gap and the Σ distribution. Note that the protoplanet is slightly closer to the inner edge of the gap (Fig. 30) because Σ adjacent to the gap is slightly larger in the inner region. Despite the lack of mass transfer through the gap, the two regions of the disk are dynamically connected by the protoplanet's tidal torque. Note that in the exterior regions of the disk, Σ attains a local maximum value at the outer Lindblad resonance of the protoplanet. This disk structure provides a favorable location for the formation of a subsequent protoplanet which is consistent with the presently observed orbital distribution of the planets.

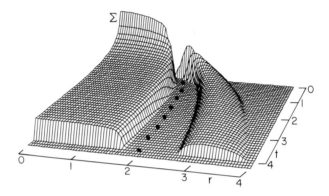

Figure 31.   Evolution of disk surface density and orbital evolution of embedded protoplanet in model 14 with $r_o = 0.8$ initially. Throughout the calculation, the rate of angular momentum transfer to the protoplanet from the disk interior to its orbit exceeds that from the protoplanet to the disk exterior to its orbit. Consequently, the companion migrates outward monotonically.

In Fig. 32, the protoplanet's position $r_p(= r_o)$ is plotted as a function of time for models 15 and 16 which have the same value of A but $B = 5$ and 80, respectively. These values of B represent a relatively low-mass model and a modestly massive solar-nebula model, respectively. Model 15 showed that in the low-disk-mass limit, the protoplanet regulates the structure of the disk but does not undergo significant orbital migration. In model 16 where the disk mass is relatively high, the protoplanet's orbital evolution is regulated by the disk's viscous evolution. From these results, we speculate that the disk mass exterior to proto-Jupiter's orbit did not exceed ~ 0.1 $M_\odot$, otherwise Jupiter's orbit would have evolved into the Sun as the inner regions of the disk were depleted in <0.1 to 1 Myr. It is also unlikely that the disk mass interior to Jupiter's orbit exceeded that exterior to its orbit by ~ 0.1 $M_\odot$, otherwise Jupiter's outward motion would have led to resonant capture of protoplanets and asteroids at its outer Lindblad resonance. The results in Sec. VII provide stronger limits on the disk mass as it is shown that modest radial migration would led to a significant orbital eccentricity for the protoplanets. The small orbital eccentricity of the planets today implies that the mass of the nebula is

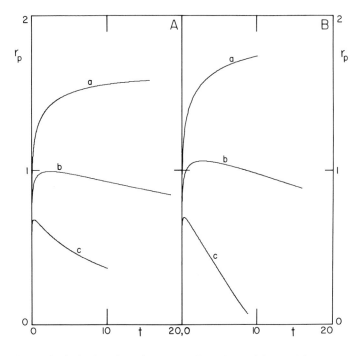

Figure 32.   Orbital migration of the protoplanet in models 15 (left panel) and 16 (right panel) where $A = 10^{-3}$ and $B = 5$ and 80, respectively. The labels a, b, and c represent three initial values for $r_o = 0.6, 0.7, 0.8$, respectively.

probably comparable to that deduced from the minimum mass nebula model.

IX. BINARY STAR-DISK INTERACTION

In addition to protoplanet-disk interaction, the above discussion can also be applied to binary star-disk interaction. Such interaction is most likely to occur during the formation of binary stars. In the past few years, several binary T Tauri stars have been observed (see Chapter by Bodenheimer et al.). These binary stars probably formed in a disk environment (Adams et al. 1989; Shu et al. 1990). In some binary T Tauri stars, infrared excesses have been observed and interpreted as evidence for protostellar disks (Mathieu et al. 1991). It is therefore timely to investigate tidal interactions between a binary star and an accretion disk. Such interactions are commonly observed in interacting binary stars in the post-main-sequence phase such as cataclysmic variables (Warner 1976; Robinson 1976; Wade and Ward 1985).

Unlike the star-planet configuration, low-order Lindblad resonances are excluded from the Roche lobe of both components of a binary system with modest mass ratio. Nevertheless, low order resonances may have a virtual effect on disk structure because of their finite width. One task here is to determine the range of mass ratios for which coherent density waves are excited. Because the tidal perturbation of the companion is strong, the disk response is expected to be nonlinear. We will also evaluate the validity of extrapolating results based on linear approximations. The same numerical techniques used in Secs. V–VIII are ideally suited to these problems.

A. The Standard Model

To illustrate the disk response to a binary star, a standard model (model 16) is adopted in which $q = 1$, $c_0 = 10^{-2}$, and $v_{s0} = 10^{-5}$. Most of the following discussion will be centered on the response of an interior disk the inner boundary of which is set at $r = 0.25$. We refer to the star with the disk as the primary star and its companion as the secondary star.

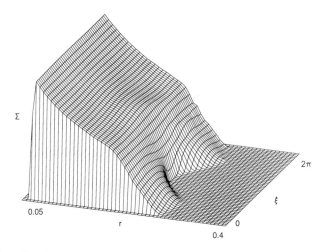

Figure 33.   Steady-state surface density distribution in model 16 in which $q = 1$, $c_o = 10^{-2}$ and $v_{so} = 10^{-5}$. Although the 2:1 resonance is excluded from the disk, the effect of virtual resonance regulates the $m = 2$ structure in the outer region of the disk.

Surface Density Distribution. The equilibrium Σ distribution is plotted in Fig. 33. The outer edge of the disk is an off-center ellipse with the major axis perpendicular to the line joining the two stars. The apoapse of the outer edge is ~ 0.3 which is about half of the Roche radius of the primary. Although no Lindblad resonances are inside the disk, some influence of the lowest-order 2:1 resonances is implied by the presence of a low amplitude $m = 2$ response in the outer regions of the disk. In contrast to the disk response to a protoplanet's tidal torque (e.g., Fig. 6), Σ decreases monotonically with radius without any clear evidence of waves being excited. This is because the characteristic wavelength of the companion's tidal disturbance is comparable with the disk radius which is much larger than the disk thickness, which produces an "impedance mismatch" in the absence of resonance. This is in contrast to some numerical computations (Matsuda et al. 1987) for thick disks where the sound speed is a large fraction of the Keplerian speed and waves may be excited.

Streamline Eccentricity and Shear Reversal. The streamlines of the flow depart significantly from those of a Keplerian pattern. The outer radius of the disk forms an asymmetric $m = 2$ pattern corotating with the binary star (left panel Fig. 38 below). The disk radius attains a maximum at $\xi \sim \pm\pi/2$ and a minimum at $\xi \sim \pi$. In the outer regions of the disk, the streamline eccentricity is approximately 0.1. The outermost streamlines extend slightly beyond the maximum nonintercepting periodic orbits (Paczynski 1977). Those orbits are computed for noninteracting particles whereas gas pressure prevents the crossing of streamlines.

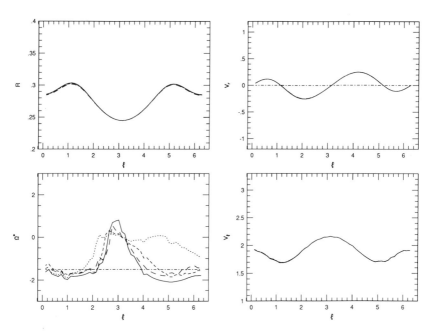

Figure 34.   Perturbed velocities and shear of five Lagrangian fluid elements in model 16. The local shear vanishes at $\xi = \pi$ where the streamlines approach to their perigee. The eccentricity of the streamlines exceeds 0.1.

In Fig. 34, we plot the orbital radius, v_r, v_ϕ and $\Omega^* \equiv (r/\Omega)(\partial\Omega/\partial r)$ as functions of ξ. Near $\xi = \pm\pi/2$, the streamlines in the outer regions of the disk are most radially extended and have essentially uniform specific angular momentum ($\Omega^* \sim -2$). Although the epicyclic frequency κ essentially vanishes near $\xi = 0$, Rayleigh's stability criterion is not violated. The variations in r and V_ϕ indicate that the specific angular momentum on each streamline is essentially conserved. As streamlines converge toward their perigee at $\xi \sim \pi$, the local shear vanishes (i.e., $\Omega^* \sim 0$) and reverses as is shown in Fig. 19 (Sec. V.F) where $q = 10^3$. Nevertheless, the total angular momentum luminosity remains in the outward direction when the flux is integrated over all phases (Borderies et al. 1982,1983). The radial velocity in the outer regions of the disk varies with azimuth approximately sinusoidally with an amplitude ~ 0.1 of the expected local Keplerian velocity. The azimuthal velocity varies similarly but with half the amplitude. These large amplitude variations imply that caution must be applied when using Doppler imaging techniques to map the disk structure and deduce the binary orbital properties (Marsh and Horne 1990a, b; Marsh et al. 1988; Hessman and Hopp 1990). In eclipsing interactive binaries, these velocity variations may be directly observed from the evolution of emission line profiles (Young et al. 1981; Marsh et al. 1987).

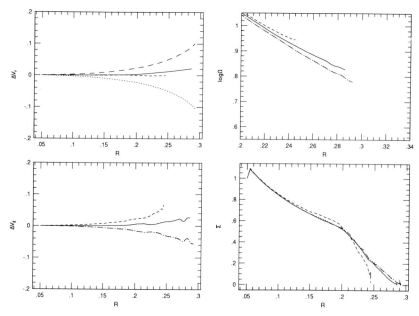

Figure 35.   Amplitude of the perturbed velocities and surface density distribution at four orbital phases near $\xi = 0$, $\pi/2$, and $3\pi/2$. The values of the perturbed velocities are normalized to the local Keplerian velocity. These amplitudes exceed 10%.

The velocity variations also strongly modify the surface density profile in such ways that Σ decreases more rapidly with radius during the compression phase ($\phi \simeq \pi$) than during the rarifaction phase ($\phi \simeq \pm \pi/2$) (Fig. 35). At $\phi \sim 0$, the characteristic surface density scale height is $\sim 10^{-2}$ and the pressure gradient provides an outward force per unit mass such that $|(1/\Sigma)(\partial\Sigma c^2/\partial r)| \sim 10^{-2}GM/r^2$. This is insufficient to change Ω significantly, but can strongly modify κ in this case. Also plotted in Fig. 35 are $\Delta v_r \equiv v_r(r/GM)^{1/2}$ and $\Delta v_\phi \equiv v_\phi(r/GM)^{1/2} - 1$. During some phases, Δv_r exceeds 0.1.

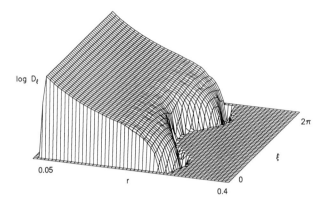

Figure 36.   The distribution of viscous shear dissipation in model 16. Tidal disturbances do not introduce significant departure from the Keplerian values.

Dissipation Pattern. The modest departure from Keplerian flow in the outer regions of the disk slightly modifies the dissipation rate per unit area $D_\xi = \Sigma v_s r^2(d\Omega/dr)^2$ from that of an isolated accretion disk where $D_K = 9\Sigma v_s\Omega^2/4$. These changes are brought about by both the variation in the shear and the Σ dependence of the viscosity. With the prescription adopted here, the reduction in the shear in the compressed regions (at $\xi \simeq \pi$) is compensated by a slight increase in the viscosity. Consequently, the D_ξ distribution differs little from D_K apart from the absence of dissipation in the gap (Fig. 36). The relatively small departure of D_ξ from D_K is because the tidal torque of the companion is distributed throughout the disk. In the absence of Lindblad resonances, the characteristic length scale of the tidal forcing is comparable to the disk radius. In addition, wave propagation may further diffuse the tidal disturbances. Azimuthal variations would be larger if v increases more rapidly with Σ. We note that other than a dip at the relevant wavelength, the spectrum of a disk which is truncated by a binary is essentially identical to that of a disk which is unperturbed. This expectation is in agreement with observations of the binary T Tauri system GW Ori (Mathieu et al. 1991).

B. Model Parameters

Disk Temperature and Sound Speed. The size and structure of the outer disk is relatively insensitive to changes in the magnitude of the sound speed. In model 17, $c_0 = 0.08$ and all other parameters are identical to those in model 16. The Σ distribution in this case (Fig. 37) is essentially the same as that in model 16. Similarly, the stream lines (right panel, Fig. 38) do not undergo significant changes. Despite a relatively large sound speed, the characteristic wavelength is an order of magnitude less than the disk radius so no coherent response to the companion's tidal effect is found. The results in Sec. IV indicate that a large sound speed also promotes the propagation of

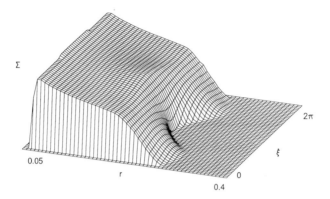

Figure 37.   Steady-state surface density distribution in model 17 where $c_o = 0.08$. Despite the large sound speed, density waves are not excited and the disk structure in this case resembles that in model 16 (Fig. 33).

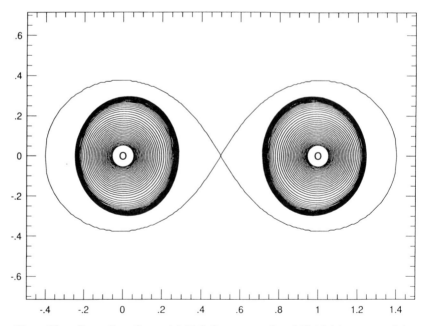

Figure 38.   Streamlines for model 16 (left component) and 17 (right component) in a corotating frame of the binary.

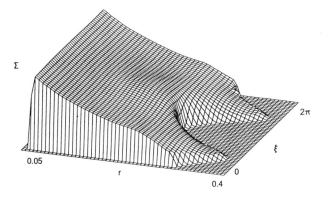

Figure 39.   Steady-state surface density distribution for model 18 where the viscosity
is high and sound speed is modest. The apogee of the streamlines reaches $0.39\ r_o$
and the eccentricity is large (≥ 0.15).

tidal disturbances throughout the disk. With a much larger sound speed, it may
also become possible for the disk to undergo a coherent response (Matsuda et
al. 1989). However, in this limit, it becomes inappropriate to use the standard
vertically averaged approximation for the equation of motion.

   Viscosity. In the limit of low sound speed and high viscosity, most of
the tidal disturbance is dissipated close to the location where it is excited.
Thus at each radius, angular momentum flux induced by the companion's
tidal torque is balanced by that due to viscous stress. With an increase in the
magnitude of viscosity, the viscous transport of angular momentum becomes
more efficient and the disk expands in radius. In model 18, we adopt the
same parameters as in model 16 except $\nu_{s0} = 5 \times 10^{-5}$, a value 5 times
bigger than that in model 16. The disk size increases by $\sim 30\%$ (Fig. 39). For
some phases, the streamlines expand beyond the Roche radius (right-hand
panel, Fig. 43 below). At the outermost regions of the disk, the eccentricity
of the streamlines becomes ~ 0.2 and the shear is significantly reduced from
its Keplerian value (Fig. 40). Although shear reversal occurs during some
phases, the average value of the shear continues to correspond to an outward
viscous transport of angular momentum.

   The values of effective viscosity we have adopted here are much larger
than what can be accounted for by molecular process; the physical process
usually invoked to explain this excess is turbulence. In a turbulent disk, a
relatively large viscosity may require a relatively large sound speed. The
above results are not strongly altered if the sound speed is only a factor of 2
larger than in model 18. However, if the value of α in the α-prescription of
turbulent viscosity is small and a large sound speed is thus needed for even a
modest magnitude of viscosity, large gas pressure may efficiently induce wave
propagation and redistribution of tidal disturbances over extended regions of

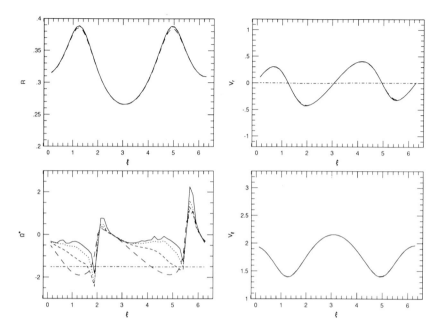

Figure 40.   Perturbed velocities and shear of five Lagrangian fluid elements in the outer regions of the disk in model 18. Over most orbital phases, the local shear is significantly reduced from the Keplerian value.

the disk. In model 19, we adopt $\nu_{s0} = 5 \times 10^{-5}$ and $c_0 = 0.08$. In this case, the Σ distribution (Fig. 41) is similar to that of model 17 despite the fact that the viscosity is a factor of five larger. The size of the disk is smaller than that of model 18 (left-hand panel, Fig. 43), but the outermost streamlines in this smaller disk have eccentricity ~ 0.1 (Fig. 42). The dissipation pattern is much more smoothly distributed over the entire disk.

These theoretical results can be tested with observational data in the context of post-main-sequence cataclysmic variable stars. These systems are semi-detached close binaries in which the companion fills its Roche volume and transfers mass onto the accretion disk around the primary star (Warner 1976). In some cases, the orbital plane of the binary is essentially in the line of sight and the disk is eclipsed periodically. In these systems, the duration of the disk eclipse increases significantly during outbursts (Smak 1984). The most likely cause for these outbursts is thermal instability in the disk (Meyer and Meyer-Hofmeister 1981; Faulkner et al. 1983; Lin et al. 1985). During the onset of thermal instability, both c_s and ν increase. Because viscous transport of angular momentum increases with ν, the disk expands. As the disk fills the Roche volume, tidal transport increases rapidly with disk size (Papaloizou and Pringle 1977). A new equilibrium may thus be established with a larger disk (Smak 1984). Such a scenario is consistent with the results in models

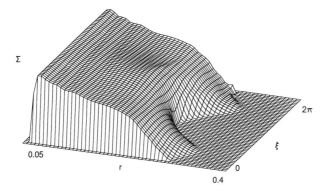

Figure 41.   Surface density distribution in model 19 which has the same viscosity
as model 18 but much larger sound speed. Tidal disturbances are transmitted
throughout the disk although no coherent waves are excited. Consequently, they
are easily dissipated near the inner disk region.

16–19 where $\alpha \sim 0.1 - 1$.

Mass Ratio. In models 20–23, a range of mass ratios are considered
where $q = 0.1, 0.3, 3$ and 10. The sound speed and viscosity in these models
are identical to those in model 16. In all these models, the disk extends to a
significant fraction of the Roche lobe such that the streamlines in the outermost
regions of the disk have eccentricity >0.1. The Σ distribution of models 20
and 21 shows a $m = 2$ response with the elongated axis perpendicular to
the line joining the centers of the binary. Such a disk pattern indicates that
the flow is under the influence of the inner Lindblad resonance. Such a
pattern is observed even for $q = 1$. Although, for $q > 0.2$, the inner Lindblad
resonance is not contained in the primary's Roche volume, it appears that
because the resonance can be thought of as having a finite width, it still has
a strong influence on the disk flow (Figs. 44 and 45). However, for $q > 1$,
the Lindblad resonances are sufficiently remote from the disk that their finite
width is insufficient for them to regulate the flow at the outer regions of the
disk (Figs. 46–48). The disk structure in these cases are very similar to that for
model 9 in which the companion's mass is 10^3 that of the gravitating object
at the center of the disk.

C. External Disks and Width of Gap

The above models indicate that tidal truncation is a natural consequence of
binary-star formation in a disk. This phenomenon can be observed as a dip in
the power-law continuum spectrum from the disk. The range of wavelength
for the dip is determined by the width of the gap which can be determined
from the structure of both the inner and outer disks. Calculations based

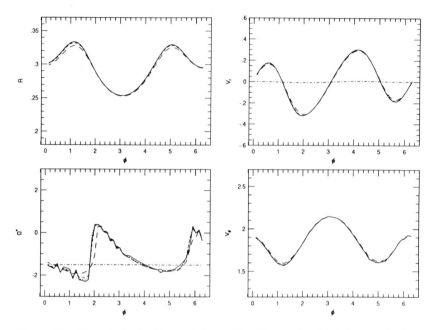

Figure 42.   Perturbed velocities and shear of five Lagrangian fluid elements in the outer regions of the disk in model 19. The flow pattern is different to both model 17 and 18. The large viscosity and sound speed remove the azimuthal symmetry about $\xi = \pi$.

on sticky particle hydrodynamics (Lin and Papaloizou 1979b) and smoothed particle hydrodynamics (Artymowicz 1990) show that the inner edge of a circumbinary disk is typically a 2-binary separation from the center of mass. The models presented in this chapter show that the width of the gap may be dependent on the thermal and viscous structure of the disk. In model 24, we use the Lagrangian scheme to determine the structure of a circumbinary disk which has the identical structure property as model 16, i.e., $\nu_{s0} = 10^{-5}$ and $c_0 = 10^{-2}$. In this case, the inner edge of the disk extends to a 2-binary separation from the center of mass. For the viscosity prescription, the inner edge of the disk retreats to larger values for disks with higher sound speed. However, other parameters being fixed, the inner edge moves to smaller values for disks with larger viscosity. These properties are consistent with similar behavior for the inner disk.

D. The Companion's Orbital Eccentricity

In Sec. VII, we showed that the orbital eccentricity of a protoplanetary companion grows in the post gap formation epoch. This eccentricity growth is driven by energy and angular momentum transfer at the Lindblad resonances. In binary stars with comparable masses, for Lindblad resonances associated

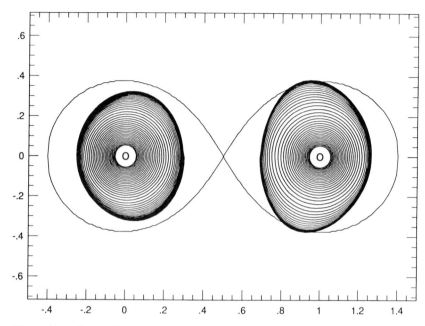

Figure 43. Streamlines for models 18 (right panel) and 19 (left panel) in a rotating frame. Note that the disk in model 18 is extended to the Roche surface. The disk size determined from eclipsed observation for model 18 is considerably larger than for models 16, 17 and 19. Although viscosity in models 18 and 19 are identical, the disk size is smaller for model 19 because the large sound speed promotes the distribution of tidal disturbance well into the inner regions of the disk.

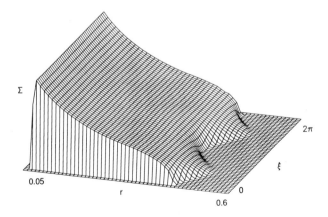

Figure 44. Steady-state surface density distribution in model 20 where $q = 0.1$.

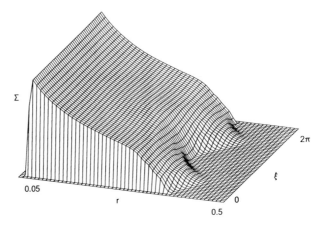

Figure 45.   Steady-state surface density distribution in model 21 where $q = 0.3$.

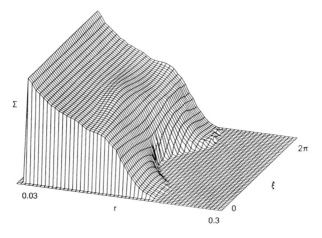

Figure 46.   Steady-state surface density distribution in model 22 where $q = 3$. Note
that the flow in the outer regions of the disk has an $m = 1$ pattern which is different
from the $m = 2$ response for disks in systems with $q \leq 1$.

with the ω_1 terms to occur in the inner disk, we require disk material with
$\Omega \geq 3\omega$ (see discussion in Sec. VII). For Lindblad resonances associated with
the ω_2 terms to occur in the outer disk, there needs to be finite amount of disk
material with $\Omega \leq \omega/3$. Equality here corresponds to 3:1 resonances and such
a resonance can exist inside the Roche lobe of the primary only for $q \leq 0.3$.
Therefore, we would expect the tidal response to the terms proportional to the
eccentricity to be dominated by either the presence in the disk of, or the virtual
effect through the finite width of the 3:1 resonance even for moderately high
mass ratios.

    More generally, because of an impedance mismatch, an inner disk re-

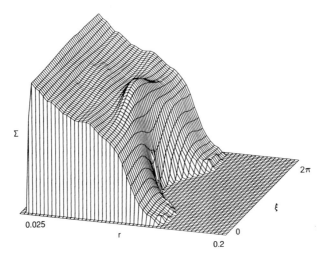

Figure 47.   Steady-state surface density distribution in model 23 where $q = 10$. The flow pattern is similar to model 9 where $q = 10^3$ (Fig. 18). The local surface density maximum at $\xi = \pi$ is due to the pressure gradient as streamlines converge near their perigee.

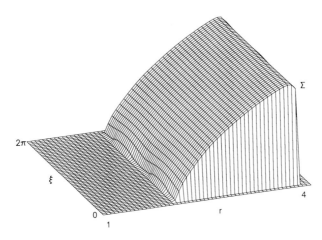

Figure 48.   Steady-state surface density distribution of a circumbinary disk in model 24 where $q = 1$, $c_o = 10^{-2}$ and $v_{so} = 10^{-5}$. For binaries with identical q, the size of the gap increases with c_o and decreases with v_{so}.

sponds poorly to the low ω_2 frequencies as compared to the higher ω_1 frequencies. Similarly, an outer disk responds poorly to the higher-frequency ω_1 terms as compared to the lower-frequency ω_2 terms. Thus just as for the protoplanet case, we expect

$$\frac{G^{1/2}MM_p}{2\sqrt{(M+M_p)}} \frac{d(r_0^{1/2}e^2)}{dt} > 0. \tag{177}$$

Therefore, if either $dr_0/dt < 0$, which would occur if there was an external disk alone, or r_0 is very slowly changing as would occur either in a disk-gap situation or in an interacting binary system, we expect $de^2/dt > 0$. In all of these cases, because the F_{Hi} are proportional to e^2, the eccentricity grows exponentially on a time scale roughly equal to the ratio of the orbital angular momentum to the tidally induced angular momentum flow rate, calculated as if e were of order unity. This time scale is independent of, and can be much smaller than, the viscous evolution time of the disk-gap configuration so that eccentricities can grow within the lifetime of the system. However, we should stress the limitations of the above discussion. Because in the linear approximation, we have neglected the interaction of the tides induced by terms proportional to e with the circular orbit tides and have assumed that material was depleted at the Lindblad resonances associated with the circular orbit tides. In order to relax these assumptions it is necessary to do nonlinear time-dependent calculations of disk structure with an eccentric binary orbit. We have performed such calculations with the Lax-Wendroff scheme. The results obtained for these cases are similar to that in Fig. 26 (Sec. VI.D). These results are also in qualitative agreement with numerical simulations of binary star-disk interaction using a smoothed particle hydrodynamics scheme (Artymowicz et al. 1991).

E. Eccentric Disks

It has been suggested that the accretion disks associated with the SU Ursa Majoris subgroup of cataclysmic binaries might be eccentric (Whitehurst 1989; Hirose and Osaki 1989). Such behavior may result from possible excitation of parametric instabilities associated with 3:1 resonances in the disk. The disk is then made to appear eccentric through the presence of an $m = 1$ mode. A similar process may operate in planetary rings (Goldreich and Tremaine 1980).

To understand the proposed instability mechanism we consider a simplified model of a disk which is initially stationary when viewed from a reference frame corotating with the binary with a circular orbit which has period $2\pi/\omega$. We treat the disk as an axisymmetric and initially inviscid object being forced tidally by the nonaxisymmetric potential ψ_c. This tidal potential is supposed to have a negligible effect on the initial steady state which is approximated as axisymmetric, but is retained when linear perturbations are considered. We consider such a perturbation associated with a Lagrangian displacement $\xi \exp(i\sigma t)$. This can be found from an equation of the general form

$$\mathcal{O}(\sigma)\xi = -(\xi \cdot \nabla)\nabla\psi_c \tag{178}$$

which may be considered as giving the response to the perturbing potential given on the right-hand side exactly in the same way as in Sec. III.B. Here $\mathcal{O}$ is a generalised linear operator, the detailed form of which is not needed to illustrate the instability.

Parametric instability occurs when two normal modes with different m in the axisymmetric disk have nearly the same eigenfrequency. These are uncoupled in the axisymmetric limit when ψ_c is neglected but become coupled and may interact unstably when it is included. To show this we look for a solution when $|\psi_c|$ is small of the form in cylindrical coordinates

$$\xi = a_1 \xi_1(r) \exp(im_1\phi) + a_2 \xi_2(r) \exp(im_2\phi) \tag{179}$$

where a_1 and a_2 are constants and ξ_1 and ξ_2 give the radial parts of the eigenfunctions for the axisymmetric disk associated with azimuthal mode numbers m_1 and m_2, respectively. In order to give an eccentric disk, we require m_1 or m_2 to be 1. To obtain an equation for σ, we substitute the above specified form of ξ into Eq. (178) noting that differentiating with respect to ϕ is equivalent to multiplying by an appropriate value of m. We then multiply by $\Sigma \xi_k^* \exp(-im_k\phi)$, $(k = 1, 2)$ and integrate over the disk so obtaining two linear equations for a_k, $(k = 1, 2)$. Setting the determinant of the coefficients to zero then gives an equation for σ in the form

$$\left| \mathcal{O}_{i,j} + \mathcal{P}_{i,j} \right| = 0 \tag{180}$$

where

$$\mathcal{O}_{ij} = \delta_{ij} \int_{\text{disk}} \Sigma \xi_i^* \cdot \mathcal{O}(\sigma) \xi_j r \, dr \, d\phi \tag{181}$$

with δ_{ij} being the Kronnecker delta, and

$$\mathcal{P}_{i,j} = \int_{\text{disk}} \exp\left[i \left(m_j - m_i \right) \phi \right] \Sigma \xi_i^* \cdot \left(\xi_j \cdot \nabla \right) \nabla \psi_c r \, dr \, d\phi. \tag{182}$$

The cross terms $\mathcal{O}_{i,j}$, $i \neq j$, are zero because of the orthogonality of modes with different m regardless of axially symmetric weight. Without loss of generality, we may assume that when ψ_c is Fourier analyzed in azimuth, there is no axisymmetric ($m = 0$) component. Then the diagonal elements $\mathcal{P}_{ii}$ will be zero. In addition, we always have $\mathcal{P}_{ij} = \mathcal{P}_{ji}^*$. We suppose that the ξ_i, $i = 1, 2$ are eigenmodes of the axisymmetric disk with the same eigenvalue σ_0, (but note that the assumption of strict equality here is not essential to the analysis), so that $\mathcal{O}_{ii}(\sigma_0) = 0$. For σ close to σ_0, we adopt the first order Taylor expansion

$$\mathcal{O}_{i,i}(\sigma) = B_{ii}(\sigma - \sigma_0) \tag{183}$$

where

$$B_{ii} = \left[\frac{\partial}{\partial \sigma} \left(\mathcal{O}_{ii}(\sigma) \right) \right]_{\sigma = \sigma_0}. \tag{184}$$

Evaluating the determinant to find σ then gives

$$B_{11}B_{22}(\sigma - \sigma_0)^2 = |\mathcal{P}_{12}|^2. \tag{185}$$

From this we see that instability occurs if B_{11} and B_{22} are of opposite sign. This means that the modes have opposite signs of energy and angular momentum as measured in the rotating frame. If friction were applied to reduce the Lagrangian displacement to zero in this frame the positive energy mode would decay and the negative energy mode would grow. The parametric instability then occurs because of the interaction of positive and negative energy modes. To ensure that the modes have opposite signs of energy, m_1 and m_2 must be of opposite sign so that the modes which have the same eigenfrequency, propagate in opposite directions when viewed from the rotating frame. Also, we require that the modes are in fact coupled through the tidal potential which therefore must contain a nonzero Fourier component with $m = m_1 - m_2$.

The resonance condition that both modes must have $\sigma = \sigma_0$ in the axisymmetric disk can be evaluated by noting that in a thin disk that if there is to be significant spatial overlap of the modes, their inner/outer Lindblad resonances must coincide. Applying the dispersion relation (78) with $\sigma_0 \equiv -m\omega$, $k = 0$ and m set equal to m_1 and m_2, respectively we get

$$[\sigma_0 + m_i(\Omega - \omega)] = \pm\kappa, i = 1, 2 \tag{186}$$

where ω has been subtracted from the gas angular velocity because we are in a rotating frame. If we take m_1 to be positive with the $+$ alternative and m_2 to be negative with the $-$ alternative, elimination of σ_0 leads to the condition that at the Lindblad resonance

$$\Omega - \frac{2\kappa}{(|m_2| + m_1)} = \omega. \tag{187}$$

For an eccentric disk, we must take $m_1 = 1$, then the smallest possible value of $|m_2|$ is 2. Then in a Keplerian disk, we require that at the Lindblad resonance, $\Omega = 3\omega$. Thus there needs to be a 3:1 resonance in the disk. In a slender planetary ring, larger values of $|m_2|$ are needed to satisfy the parametric resonance condition inside it.

The instability as described here requires two coherent modes in the unperturbed axisymmetric disk. However, because the instability occurs as the result of the interaction of positive and negative energy modes it is likely that only the $m = 1$ mode needs to have reasonable coherence. The excited $m_2 = -2$ disturbance may not need to be in the form of a normal mode. It could be a wave taking away energy and angular momentum, the excitation of which causes the $m_1 = 1$ mode to grow. We remark that in the inertial frame the $m_1 = 1$ mode rotates with a prograde pattern speed equal to $\Omega - \kappa$ evaluated at the 3:1 resonance.

It is unclear whether realistic models of disks in close binary systems allow a parametric instability of the type described above to occur. There is some evidence for the effect in particle simulations (Whitehurst 1989; Hirose and Osaki 1989). However, attempts to find the effect using the finite difference codes described here have so far been unsuccessful. It may be that conditions for the instability to occur are sensitive to the disk model.

X. SUMMARY

We briefly recapitulate the main conclusions and discuss some outstanding uncertainties.

A. Disk Response

A companion around a protostar exerts tidal torque on the disk. For a low-mass companion such as a protoplanet, density waves are excited at Lindblad resonances. These waves carry an energy and an angular momentum flux which are determined entirely by the gravitational force of the companion; they are independent of physical conditions in the disk. For a disk under the tidal influence of a companion with either a comparable mass (such as a binary T Tauri system) or a much larger mass (such as disk around a protoplanet), the Lindblad resonances are excluded from both the disk and the Roche lobe which contains the disk. In this case, unless the disk is geometrically thick, an impedance mismatch between the natural length scale of the tidal field and the characteristic wavelength in the disk prevents the excitation of density waves.

The propagation and dissipation of tidal disturbances is determined primarily by the thermal and viscous properties of the disk. In a non-self-gravitating disk, tidal disturbances propagate away from the companion at essentially the sound speed. In a vertically thermally stratified disk, refraction induces wavefront retardation and wave propagation into the disk atmosphere. Nonlinear dissipation in the tenuous disk atmosphere then leads to effective loss of wave action, a local deposition of the wave's angular momentum and reduction in the thermal stratification of the vertical structure. Thus, refraction may be a self-limited process (Larson 1990b). A natural next step of theoretical investigation is to carry out a self-consistent nonlinear analysis of wave propagation in a disk with a modest thickness. In addition to sound waves, internal and Alfvén waves may also be excited (Vishniac et al. 1991). The propagation speeds of these waves increase with the distances from the midplane and thus they may be focused toward the midplane. At large radii, the disk's self gravity may become important such that the characteristic wavelength becomes large compared with the disk thickness and the vertical thermal stratification may only negligibly impede wave propagation. The propagation and nonlinear dissipation of these waves needs to be further investigated.

Surface heating of the disk by nonlinear dissipation and irradiation can effectively eliminate the vertical temperature gradient in the disk. In an isothermal disk where the refraction effect is ineffective, density waves can propagate with little or no loss of wave action. As waves propagate inward from the inner Lindblad resonances, their wavelength decreases and amplitude increases. At large distances from the Lindblad resonances, these waves may either be dissipated viscously or become nonlinear leading to a local deposition of angular momentum. Energy dissipation then modifies the temperature distribution in the disk which may further modify the wave propagation speed and the condition for nonlinear dissipation. Self-consistent investigation of this issue is underway.

B. Implications for Protoplanetary Formation

Tidal interaction between a protoplanet and the protoplanetary disk is particularly important during the late stages of protoplanetary formation. The flux of tidally induced angular momentum transfer is determined by the mass of the protoplanet. When the protoplanet's mass becomes sufficiently large, it can induce gap formation in the vicinity of its orbit. Thereafter, gas in the disk can no longer flow through its orbit; a protoplanet's growth by accretion is terminated. There are two conditions for gap formation: (1) the mass ratio between the protostar and the protoplanet must exceed about 10^{-2} the effective Reynolds number so that the angular momentum flux induced by the companion's tidal torque exceeds that due to viscous stress; and (2) the Roche radius of the the protoplanet must not exceed the vertical density scale height in the disk so that the flow near the truncated edges of the disk is stable. Discussions in Sec. VI imply that if gap formation terminates protogiant planetary growth, a protoplanet can only acquire the present mass of Jupiter (1) if the disk is heated by turbulent viscosity; (2) if the mass of the disk is comparable to that estimated from the minimum-mass nebula scenario; and (3) if the viscous evolution time scale of the disk is $\sim$0.1 to 1 Myr. With milliarcsecond resolution, we should be able to resolve the spiral dissipation pattern induced by this protoplanet-disk interaction as far away as the Orion star-forming regions.

The actual gap-formation process determines not only the mass but also the spin angular momentum of a protoplanet. The remarkably low spin angular momentum of Jupiter and its satellites suggest that most of proto-Jupiter's mass is acquired through nearly spherical accretion rather than through mass transfer from the protoplanetary disk through Lagrangian points. This scenario is feasible if the gap formation is regulated by the second criterion. Such a condition requires the disk temperature near Jupiter to be relatively high ($H/r_o \sim 0.1$) and viscosity to be relatively low ($\alpha < 0.1$). The implication here is that a three-dimensional scheme must be used to analyze the secular evolution of the disk during the final stages of protoplanetary growth when the protoplanet acquires most of its final mass. Such a calculation is a natural next step in studying protoplanetary growth.

A lower limit for α may be obtained from the orbital properties of the protoplanet. After a gap is formed, gas is expelled from regions around the Lindblad resonances associated with the circular orbit tide, which are also located at corotation resonances when the orbit has a slight eccentricity. The disk and the protoplanet then continue their interaction only through the Lindblad resonances. For disk regions interior to a protoplanet's slightly eccentric orbit, energy is transferred from the disk to the protoplanet at a faster rate than that required to keep it in a circular orbit consistent with the amount of angular momentum transferred. This results in the growth of the protoplanet's orbital eccentricity. Interaction between a protoplanet and an external disk also amplifies the protoplanet's orbital eccentricity. The rate of eccentricity growth is determined by the disk's surface density. In order to restrain the protoplanetary orbits from attaining eccentricities greater than their present observed value, a solar nebula with a mass comparable to that in the minimum-mass-nebula model must have had an evolutionary time scale comparable to or shorter than ~ 1 Myr. This implies that $\alpha \geq 10^{-2}$. For a relatively low-mass protoplanet, the orbital eccentricity will grow until its orbit undergoes an excursion which takes it beyond the gap region. Subsequent nonlinear interactions remain uncertain and require further investigation. Another important issue to be investigated is the growth of the orbital inclination of protogiant planets.

C. The Companion's Orbital Evolution

An important element in deducing physical conditions in the solar nebula is the assumption that both the terrestrial and giant planets are formed in the vicinity of their present orbits. However, tidal interaction between the disk and protoplanets can lead to orbital migration. For protogiant planets, after their mass becomes large enough to induce gap formation, their orbital migration is locked to the viscous evolution of the disk. If the disk is very massive, a protoplanet can migrate over a large radial extent. Such orbital migration would, however, lead to very large orbital eccentricities. The small observed eccentricity of the giant planets today implies that they are unlikely to have participated in large-scale radial migration and the mass of the nebula must thus be relatively low. These constraints, however, do not apply to prototerrestrial planets and protogiant planetary cores. The Roche radius of these bodies is smaller than the vertical scale height of the disk and gap formation will not occur. Nevertheless, these bodies participate in tidal interactions with the disk. Surface density gradients and velocity differences between the protoplanets and the gas can lead to nonvanishing differential tidal torques and thus to radial orbital migration. However, analyses of these processes remain uncertain because they are carried out with a thin disk approximation. A more appropriate analysis requires the investigation of wave excitation and propagation in three dimensions.

D. Binary Star-Disk Interaction

A large fraction of stars in the Galaxy are members of binary or multiple-star systems (Abt 1983). One of the most exciting recent observational results has been the discovery of binary T Tauri stars with circumbinary disks (Mathieu et al. 1991; Chapter by Bodenheimer et al.). Binary star-disk tidal interactions are similar to protoplanet-disk interactions in that they induce energy and angular momentum transfer. In binaries with nearly equal masses, density waves may not be excited because the Lindblad resonances are excluded from both the disk and the Roche lobe which contains the disk. Nevertheless, the finite width of Lindblad resonances permits the virtual effect of the resonances to be present. These effects promote eccentricity growth. Consequently, unless the residual disk mass in the vicinity of the binary is relatively small, binary T Tauri stars should acquire large orbital eccentricities on a time scale comparable to or less than the typical age of T Tauri stars (Artymowicz et al. 1991).

The rate of eccentricity growth may be reduced if the binary induces a sufficiently large gap near its orbit. Because tidal effects are large, both conditions for gap formation will generally be satisfied in circumbinary disks unless the disk's mass in the vicinity of the binary is comparable to the mass of the binary. The width of the gap is determined by the local sound speed and viscosity in the disk. Hot and inviscid disks have wide gaps whereas relatively cool and viscous disks have narrow gaps. A relatively wide gap would significantly reduce the effect of virtual resonances. The width of the gap should be directly observable as a dip in the disk spectrum (Mathieu et al. 1991).

The disk response to the companion's tidal disturbance remains a controversial issue. The results of some calculations show the excitation of self-similar density waves (Matsuda et al. 1989). Other calculations based on sticky particle schemes suggest the possible excitation of parametric resonances at 3:1 resonances (Whitehurst 1989; Hirose and Osaki 1989). Despite our efforts to reproduce these results, neither response is found with either the Lax-Wendroff scheme or the Lagrangian periodic scheme for disks with appropriate sound speeds and viscosities. Another interesting feature of these calculations is the strong nonlinear behavior of the streamlines near the outer edge of the disk. For eclipsing cataclysmic variable stars, the flow pattern of these streamlines can be observed directly using Doppler imaging techniques (Marsh and Horne 1990a, b). During the quiescent state when the disk is optically thin, the existence of the self-similar nonlinear wave dissipation may also be tested directly from observed temperature distribution in the disk (Horne 1985; Wood et al. 1986,1989a). Thus, observations of post-main-sequence interacting binary systems may thus provide valuable clues to the formation of binaries in a disk environment.

Acknowledgments. We thank P. Bodenheimer, J. Cuzzi, P. Goldreich,

D. Korycansky, T. Matsuda, S. Miyama, G. Savonije, F. Shu, H. Spruit, S. Tremaine and W. Ward for useful conversations, collaborations and constructive interaction over the past few years, and R. Bell for suggestions and assistances on manuscript preparation. This work is supported in part by NSF and a U. S.-U. K. collaborative research grant by NASA through an Astrophysical Theory Program on accretion process in astrophysics, and an Origin of Solar System Program on observational, theoretical, and experimental cosmogony. We also thank SERC and NATO for a collaborative research grant. Part of this work has been conducted under the auspices of a special NASA Astrophysics Theory Program that supports a Joint Center for Star Formation Studies, at NASA-Ames Research Center, University of California, Berkeley, and University of California, Santa Cruz. Support from Non Linear Science Institute, University of California, Santa Cruz, is also acknowledged.

EVOLUTIONARY TIME SCALES FOR CIRCUMSTELLAR DISKS ASSOCIATED WITH INTERMEDIATE- AND SOLAR-TYPE STARS

STEPHEN E. STROM, SUZAN EDWARDS and
MICHAEL F. SKRUTSKIE
University of Massachusetts

Sensitive infrared and millimeter-continuum measurements provide the basis for diagnosing the frequency with which disks are formed around solar-type and intermediate-mass stars, and the time scales for disk evolution. Optically thick disks in regions sampled by excess emission at $\lambda \leq 10 \, \mu$m are found around 30% to 50% of young ($t < 3$ Myr) stars of all masses ($M < 3 \, M_\odot$); such disks are also found around young, higher-mass stars but with unknown frequency. Nearly all stars with optically thick disks have disk masses comparable to or greater than the minimum-mass solar nebula. Disk sizes range from $\simeq 10$ to $\simeq 1000$ AU. Hence, disks of solar system size, and of masses at least that of a minimum-mass nebula appear to form around a significant fraction of stars with $M < 3 \, M_\odot$. Disks survive as massive optically thick structures extending nearly to the stellar surface for times ranging from $t << 3$ Myr to $t \simeq 10$ Myr around solar-type pre-main-sequence stars; no optically thick disks have been identified around solar-type stars with ages $t \geq 30$ Myr. The survival times for disks surrounding more massive stars may be shorter. Evidence of infrared excess emission consistent with emission arising in optically thin disks is found for stars of all masses. These disks are presumably the descendents of objects originally surrounded by massive, optically thick disks. The presence of micron-size dust grains in optically thin disks requires that such grains be continuously supplied to the disk, possibly from (1) massive outer disk regions; (2) collisions among larger grains or planetesimals; or (3) evaporation of cometesimals. Disks with apparent "inner holes" have been discovered. These are most likely structures in transition between massive, optically thick disks that extend inward to the stellar surface, and disks that are optically thin throughout. The presence of an inner hole probably requires that the inner regions of a disk be isolated from the outer regions, and may indicate the formation of a Jupiter mass planet. The discussion of disk properties summarized in this review rests entirely on observations of micron-size dust embedded in circumstellar disks. Possible observations aimed at detecting the dominant component of primordial solar nebulae, the gas, are reviewed and discussed briefly.

I. INTRODUCTION

The formation of circumstellar disks appears to be a natural, if not inevitable, consequence of the process which gives birth to stars. Observations of young, optically visible pre-main-sequence (PMS) stars show that many solar-type

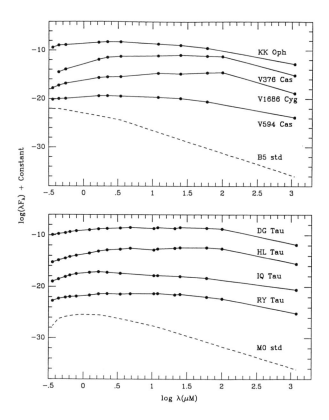

Figure 1.   Spectral energy distributions for a selection of Herbig Ae/Be stars (top) and
T Tauri stars (bottom). Superposed on this figure are spectral energy distributions
for unreddened B5 (top) and K0 stars (bottom). The millimeter-continuum points
plotted are from Beckwith et al. (1990) for the T Tauri stars, and from unpublished
data obtained at the Cal Tech Submillimeter Observatory by Edwards, Strom and
Keene.   Note that the shapes of the excess emission above photospheric levels
at $\lambda > 2.2\,\mu$m are similar for both young intermediate-mass and solar-type pre-
main-sequence stars. We conclude that massive, optically thick disks of similar
characteristics surround young solar-type and intermediate-mass stars.

$(0.2 < M/M_\odot < 1.5)$ and intermediate-mass $(1.5 < M/M_\odot < 10)$ stars are surrounded by disks of solar-system dimension with masses comparable to or
greater than the "minimum-mass solar nebula"[a] (Adams et al. 1989,1990;
Edwards et al. 1987; Strom et al. 1989d; Strom et al. 1989; Skrutskie et al.
1990; Beckwith et al. 1990) as evidenced by:

1. Broad forbidden line profiles, which diagnose outflowing gas driven by

---

[a] The minimum-mass for the solar nebulae, $M_{mm} \simeq 0.01\ M_\odot$, is computed from the
total mass of the planets, adjusting for the difference in chemical composition between
the planets and the interstellar medium.

PMS stars and show only *blueshifted* components; no redshifted emission is observed. The receding gas in the outflow is presumed to be occulted by an *optically thick disk* of dimension about 10 AU to hundreds of AU (see, e.g., Appenzeller et al. 1984; Edwards et al. 1987; Hamann and Persson 1989; Cabrit et al. 1990).

2. Infrared, submillimeter, and millimeter-continuum radiation in excess of photospheric levels which finds most straightforward interpretation in terms of emission arising in circumstellar disks (Myers et al. 1987; Adams et al. 1987; Kenyon and Hartmann 1987; Chapters by Edwards et al. and by Hartmann et al.). Such excesses (see Fig. 1) arise both from photospheric radiation absorbed and re-radiated by circumstellar dust grains ("reprocessing"), and from heating of circumstellar gas and dust via accretion (see Lynden-Bell and Pringle 1974; Kenyon and Hartmann 1987; Adams et al. 1987; Bertout et al. 1988; Bertout 1989).

For most solar-type PMS stars surrounded by optically thick disks, the masses deduced from modeling infrared and mm spectra lie in the range $0.01 < M_{disk}/M_\odot < 0.1$ (Beckwith et al. 1990), while sizes range from 10 AU to 1000 AU (Edwards et al. 1987; Adams et al. 1990). The range of masses and sizes for disks surrounding young intermediate mass stars is not as well established because systematic surveys of infrared and mm-continuum spectral energy distributions and forbidden line profiles are currently lacking.

No main-sequence stars exhibit strong infrared excess emission characteristic of the massive, optically thick disks which surround many stars during earlier evolutionary phases. Hence, the mass of disk material in the form of small dust grains must decrease with time. The observation of infrared and millimeter-continuum emission arising in massive, optically thick disks surrounding young PMS stars of all masses, combined with the absence of such signatures among their older stellar counterparts provides the basis for establishing time scales for disk evolution from studies of intermediate-age stars. Such time scales provide an astronomical constraint on the time available for building large grains and possibly planetesimals.

The goals of this contribution are the following:

1. To determine the frequency with which massive, optically thick disks (with mass in the form of *distributed* micron-size dust and orbiting gas greater than that of the minimum-mass solar nebula, M_{mm}) form around solar-type and intermediate mass stars. Our strategy will be to develop criteria for diagnosing massive, optically thick disks, and to determine the fraction of young, optically visible stars surrounded by such disks. Because some disks may evolve on time scales short compared to the time required for stars to emerge from optically opaque protostellar cores, this fraction will represent a *lower limit* to the fraction of all stars which form disks massive enough to make planetary systems similar to our own.

2. To determine the range of *survival times* for massive, optically thick

disks by determining the fraction of stars surrounded by such disks as a function of stellar age. Our discussion of the frequency of disk occurence and the time scales for disk evolution will rest exclusively on observations of emission from distributed circumstellar *dust*. Establishing time scales for the evolution of the *gas* component of disks is essential, but technically far more difficult at present.

3. To establish the existence of the probable descendents of massive, optically thick disks: *optically thin* dust disks, perhaps analogous to the structures surrounding Vega and β Pictoris (Aumann et al. 1984; Gillett 1986; Backman and Gillett 1987; Chapter by Backman and Paresce). The mass in *distributed* material (as opposed to dust and gas which have been assembled into larger grains, planetesimals, or planets) will be small ($M_d << M_{mm}$). Such disks are best detected at present from high precision infrared observations capable of measuring small excess emission above photospheric levels. We shall argue that optically thin disks must represent secondary structures, continuously replenished by small grains produced either *in situ* via collisions between large grains or planetesimals (see Gillett 1986; Backman and Gillettx 1987; Chapter by Backman and Paresce), or supplied from material (e.g., small grains and cometesimals) stored in outer regions of the disk and introduced into the inner disk via Poynting-Robertson drag or gravitational scattering. If planetesimal collisions are the dominant source of dust replenishment, then measurement of optical depths for disks surrounding stars sampling a wide range of ages will provide a measure of the rate of dust production via collisions as a function of stellar age, and thus an astronomical constraint on the time scale(s) for assembling planetesimals into larger bodies.

4. To establish the existence of structures in transition between those which are optically thick from the stellar surface outward, and those which are optically thin throughout. Observational practicalities force us to search for transition structures in which the inner disk regions have cleared and are optically thin (the absence of warm dust in the inner regions will result in small near-infrared excesses), while the outer disk regions remain optically thick (the presence of cool dust in the outer regions will produce large far-infrared excesses). We shall argue that this type of transition structure may require that the inner and outer disks be isolated, and that the formation of a giant planet represents a natural mechanism for ensuring that material from the outer disk is prevented from entering the inner disk. So interpreted, the ages of these systems would provide upper limits to the time scale for building a giant planet. Their number, relative to the number of stars surrounded by optically thick disks, would provide an estimate of the time required following giant planet formation for gas and small dust grains in the initially optically thick inner disk regions to be accreted by the star.

5. To review current efforts and future prospects for measuring the gas

content of disks and discuss their implications for constraining the time
scale for building gas-rich planets.

II. OBSERVATIONAL CRITERIA FOR DIAGNOSING
DISK PROPERTIES

Direct determinations of disk masses depend on modeling of far-infrared, sub-
millimeter and millimeter-continuum excesses arising in circumstellar disks
(Beckwith et al. 1990; Adams et al. 1990). Submillimeter and millimeter-
continuum measurements are particularly critical because at wavelengths
$\lambda \leq 100 \, \mu$m, disks are usually optically opaque (see below); flux measurements below $100 \, \mu$m are therefore insensitive to the disk mass. Unfortunately, mass determinations are thus far available for a relatively small and
not completely representative sample of PMS stars (Beckwith et al. 1990;
Chapter by Beckwith and Sargent). However, examination of the frequency
distribution of disk masses derived for solar-type stars by Beckwith et al.
(1990) reveals that virtually all disks with masses $M \geq M_{mm}$ have optical
depths $\tau \geq 1$ as inferred from measurements of excess infrared emission at
wavelengths $\lambda \leq 10 \, \mu$m. This result is not surprising because if a disk has
(1) a mass $M_d = 0.01 \, M_\odot = M_{mm}$; (2) a radius of 50 AU; (3) a gas/dust
ratio and a grain size distribution characteristic of the interstellar medium;
and (4) dust grains which are distributed uniformly throughout the disk, then
the disk will have an optical depth, $\tau(\lambda) \sim 1$ at $100 \, \mu$m, and will be optically
thick at all wavelengths $\lambda \leq 100 \, \mu$m (for an assumed grain emissivity $\simeq \lambda^{-1}$
for $\lambda > 1 \, \mu$m. Disks with $M_d < M_{mm}$ will have correspondingly smaller optical
depths at all wavelengths. Hence, it is possible to use estimates of disk optical
depth inferred from near- and mid-infrared excesses as a surrogate means
of identifying disks having masses in distributed material $M_d \geq M_{mm}$. Such
measurements are available for large samples of stars covering a wide range
in age and mass.

Different wavelength regimes probe the optical depth in different regions
of the disk, as the equilibrium dust temperature falls with increasing distance
from the star. For example, in an optically thick reprocessing disk surrounding
a star with luminosity $L_* = 1 \, L_\odot$, the dominant contribution to the observed
fluxes at $100 \, \mu$m, $10 \, \mu$m and $2 \, \mu$m arises from dust located at radial distances
of ~ 4 AU, 0.2 AU and 0.04 AU, respectively. Given observations of sufficient
sensitivity, the observed infrared excess relative to the photospheric flux can
determine whether the disk is optically thick or thin at a radius corresponding
to the wavelength of observation. Figure 2a shows the computed spectral
energy distribution for a "typical" solar-type PMS star of spectral type M0 in
Taurus-Auriga, and the spectral energy distribution for such a star surrounded
by a perfectly flat, optically thick reprocessing disk viewed at inclination angles, $i = 0°$ (pole-on) and $i = 80°$; Fig. 2b shows the corresponding stellar,
and star-plus-disk spectral energy distributions for a main-sequence A0 star
at the same distance. Accretion and disk "flaring" can increase the observed

infrared spectral energy distribution above the solid line (Kenyon and Hart-mann 1987); inclination effects reduce the observed disk contribution from its maximum for pole-on stars, by a factor $\sim\cos(i)$.

In the following discussion, we will consider stars showing excess emission falling below the expected emission for a flat, optically thick disk viewed at $i = 80°$ at a given wavelength λ to be *candidates* for objects surrounded by disks that are *optically thin* at those radii which dominate contributions to the observed flux at λ. Among a large sample of such candidates, 1/6 may in fact be optically thick disks viewed at $i > 80°$. We shall consider stars showing excess emission falling above this level as *candidates* for objects surrounded by *optically thick* disks, recognizing that among a large sample of such candidates, we could include some optically thin disks with $0.2 < \tau < 1$ (if such disks exist).

By using the observed infrared excess radiation to infer the radial variation of disk optical depths, we can define 3 disk classes:

1. Optically thick disks: if $\tau(\lambda) \geq 1$ for all wavelengths $\lambda \leq 100\ \mu$m, the radial surface density distribution of dust is *consistent* with that of a nebula with $M_{disk} \geq M_{mm}$ extending to the stellar photosphere.

2. Optically thin disks: in practice, observations will reveal two categories of stars: (1) those in which there are no infrared excesses above photospheric levels to within observational uncertainties; and (2) those that show small excesses consistent with $\tau \ll 1$. Stars in category (1) either were never surrounded by disks or were surrounded by disks which were quickly ($t \ll 1$ Myr) disrupted or have evolved over longer times to a point where the mass of distributed material is too small to detect. We will argue (see below) that disks in category (2) must represent structures in which small dust grains are continuously supplied to the inner disk regions probed by infrared observations at $\lambda \leq 100\ \mu$m.

3. Transition disks: if for $\lambda < 100\ \mu$m, $\tau(\lambda) \geq 1$ at wavelengths λ longer than some wavelength, λ_0, and $\tau(\lambda) \ll 1$ for $\lambda \leq \lambda_0$, then we conclude that such a disk lacks "warm" distributed dust (diagnosed by small excess emission at $\lambda < \lambda_0$) in its inner regions, but is surrounded by an optically thick outer disk (inferred from the observed infrared excess at $\lambda > \lambda_0$). The "hole" in the inner disk regions could result from: (1) more rapid assembly of small grains into larger bodies in the inner disk, where the shorter dynamical time scales may hasten the agglomeration process; or (2) accretion of material in the inner disk following the formation of a giant planet which isolates the inner and outer disk regions (S. Ruden, personal communication). If disk optical depths in the inner regions are high, while those in the outer regions are low, we would conclude that either such disks initially had small radii, or that assembly of dust grains into larger bodies may have begun first in the outer disk regions. In practice (see below) the current sensitivity of long-wavelength ($\lambda \geq 25\ \mu$m) infrared measurements precludes detection of this class of transition structure.

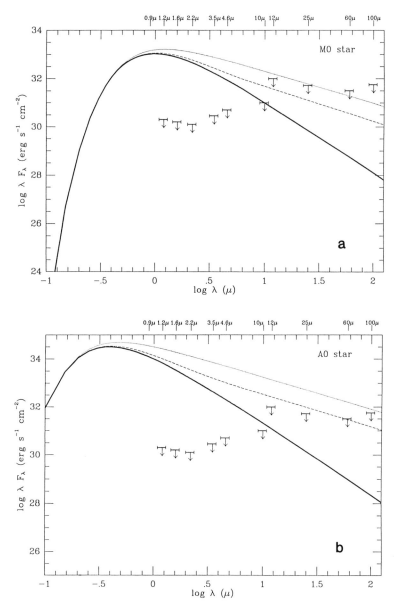

Figure 2.   (a) A plot of the spectral energy distribution for a "diskless" M0 star (dark solid line) of radius 3 $R_\odot$, and an M0 star surrounded by a flat reprocessing disk viewed at inclinations $i = 0°$ (pole-on; light dotted line) and $i = 80°$ (light dashed line). Superposed on this plot are the 5 σ limiting fluxes which can be achieved in an integration time of 1 hr on the 3-m diameter IRTF at wavelengths $\lambda \leq 10\,\mu$m, along with the limits reached by the IRAS for $\lambda \geq 12\,\mu$m for such a star observed at the distance of the nearby ($d = 150$ pc) Taurus-Auriga complex. (b) Identical to (a) except for an A0 main-sequence star.

Classification of disks into these categories depends critically upon the ability of infrared measurements to distinguish between optically thin and optically thick disks. We indicate in Fig. 2 the 5σ sensitivity limits for groundbased IRTF ($1.6\,\mu m \leq \lambda \leq 10\,\mu m$; assuming 1 hr integration) and IRAS measurements for a "typical" solar-type PMS star in Taurus-Auriga ($d = 150$ pc). The IRAS sensitivity limits lie near or *above* the level of excess radiation expected from a flat reprocessing disk, and well above the limit required to detect stellar photospheric emission. Hence IRAS measurements cannot, in general, reveal weak infrared excesses arising in optically thin disks surrounding solar-type PMS stars in Taurus-Auriga. Diagnosis of optically thin regions in disks surrounding PMS stars in nearby star-forming regions must therefore depend primarily on groundbased measurements at wavelengths $\lambda \leq 10\,\mu m$. IRAS measurements are however, capable of detecting excess radiation at $\lambda \geq 12\,\mu m$ from many optically thick disks surrounding known solar-type PMS stars in this region. For stars with photospheric luminosities $L_* > 100\,L_\odot$ and located within ~ 200 pc, IRAS measurements are sufficiently sensitive at $12\,\mu m$ and $25\,\mu m$ to permit detection of emission from all optically thick, and some optically thin disks.

Let us define a quantitative measure of the infrared excess at a given wavelength X as $\Delta X = \log_{10}[F(X)_0/F(X)_*]$. Here, $F(X)_0$ is the observed flux for a PMS star corrected for interstellar reddening, while $F(X)_*$ is the photospheric flux predicted for a disk-free star of spectral type identical to the PMS star. Thus ΔK and ΔN, for example, represent the logarithm of the ratio of the observed flux to the flux expected from the PMS star photosphere at $2.2\,\mu m$ and $10\,\mu m$, respectively (see Strom et al. 1989d). For the discussion that follows, we choose the excess infrared radiation at $2.2\,\mu m$ and $10\,\mu m$ as diagnostic of the state of the inner disk, because $10\,\mu m$ is the longest wavelength where achievable sensitivity permits the detection of typical stellar photospheres of solar-type PMS stars in Taurus-Auriga, and $2.2\,\mu m$ is the shortest wavelength where an optically thick reprocessing disk contributes more than 30% of the total flux in such stars.

Candidate *optically thick* disks will be those with ΔK and ΔN values which exceed the critical values ΔK_c and ΔN_c, listed in Table I for a flat, optically thick reprocessing disk viewed at an inclination $i = 80°$; candidate *optically thin* disks will be those having ΔK and ΔN values *smaller* than these critical values. For a central star of luminosity $L_* = 1\,L_\odot$, the $2.2\,\mu m$ and $10\,\mu m$ radiation will arise from small grains located in disk regions centered at distances $r < 1$ AU, and thus diagnoses the optical depth in the inner terrestrial planet region; these distances will scale as $L_*^{1/2}$.

III. FREQUENCY OF OPTICALLY THICK DISK OCCURENCE AMONG YOUNG STARS

A. Solar-Type Stars

Strom et al. (1989*d*) and Skrutskie et al. (1990) have recently compiled infrared fluxes for a sample of 83 solar-type PMS stars in Taurus-Auriga; most stars in this sample are spectral types between K0 and M5. We make use of these flux values along with their estimates of the expected level of photospheric radiation to compute ΔK and ΔN values based either on $10\,\mu$m fluxes measured at the NASA IRTF, or on IRAS $12\,\mu$m fluxes. In Fig. 3, we plot the frequency distribution of ages among the stars in the Strom et al. sample. Ages derived from comparing the location of these stars in the HR diagram with computed pre-main-sequence evolutionary tracks; the uncertainties in such ages may exceed ± 0.5 dex in some cases (see Kenyon and Hartmann 1990; Hartmann and Kenyon 1990). We also indicate the locations of stars with $\Delta K \geq 0.2$ dex (the critical value, ΔK_c, for a F0-type star, separating candidate optically thick from optically thin disks at $2.2\,\mu$m; see Table I) and note that 22/48 stars with ages $t < 3$ Myr are surrounded by optically thick disks as judged from ΔK; the corresponding fraction as judged from $\Delta N \geq 0.8$ dex is 28/37 for the smaller subset of stars having adequate N-band or IRAS measurements.

TABLE I
Predicted Excess Emission From Flat Reprocessing Disks
Viewed at Inclinations $i = 80°$ and $0°$

Spectral Type	ΔK_c (dex)	ΔN_c (dex)	ΔK_c (dex)	ΔN_c (dex)
	$i = 80°$		$i = 0°$	
B0	0.78	1.83	1.48	2.59
B3	0.49	1.45	1.11	2.20
B5	0.38	1.27	1.95	2.01
A0	0.25	1.05	0.74	1.78
F0	0.16	1.84	0.55	1.55
G0	0.12	0.73	0.46	1.43
K0	0.10	0.67	0.40	1.35
M0	0.06	0.52	0.28	1.16

We thus conclude that approximately 50% of all young ($t < 3$ Myr) solar-type stars are currently surrounded by disks that are optically thick at $r < 1$ AU. Although the mass contained within 1 AU represents a small fraction of the total mass in a minimum-mass nebula, nearly all PMS stars in our sample with optically thick *inner* disks appear to have masses in the range $0.01 < M/M_\odot < 0.1$ as judged from the recent millimeter-continuum measurements of Beckwith et al. (1990). [We note that the sample of solar-type PMS stars in Taurus-Auriga is incompletely known, particularly among "weak

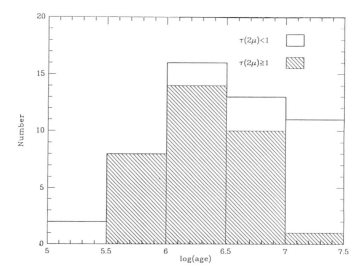

Figure 3. The observed frequency distribution of derived ages for the sample of stars
in the Taurus-Auriga complex compiled by Strom et al. (1989a). We separate the
sample into two categories: (1) stars with $\Delta K \geq 0.2$ dex (the condition for a candidate
optically thick disk surrounding a G0 star in those regions contributing to the 2.2 μm
flux; see Table I), the locations of which are indicated by the hatched region; and
(2) stars with $\Delta K < 0.2$ dex. We draw two conclusions from this figure: (i) $\sim$ half
of all *known* solar-type stars in Taurus-Auriga with ages $t < 3$ Myr are surrounded
by optically thick disks; (ii) as judged from ΔK, the survival time of optically thick
disks ranges from $t < < 3$ Myr to $t \sim 10$ Myr; few disks survive as infrared-detectable
entities at ages in excess of 10 Myr.

emission" T Tauri stars (WTTS) heretofore detected primarily by X-ray observations. These stars, in general, do not show signatures of optically thick
disk emission (Walter et al. 1988; Strom et al. 1989d; Chapter by Montmerle
et al.).]

    To ensure that in most cases ΔK and ΔN provide consistent diagnostics
of disk optical depth, we plot ΔN against ΔK (Fig. 4); the uncertainty in
ΔN and ΔK is estimated to be ± 0.10 dex. This figure reveals that for a
sample of 51 stars having both ΔK and ΔN values, the large majority, 39/51,
are selected as candidate optically thick or optically thin disks based on *both*
ΔN and ΔK. Figure 4 also reveals that the sample stars fall in two clearly
separated regions: the domain occupied by optically thick disks ($\Delta N > 0.8$
dex), and the domain occupied by candidate optically thin disks. The group
of objects with $\Delta N > 0.8$ dex is dominated by "classical" T Tauri stars. Their
location in the ΔN-ΔK plane finds ready explanation if all such T Tauri stars
are surrounded by optically thick disks heated by a combination of radiation
from the central star and accretion. Stars in our sample with $\Delta N < 0.8$ dex
and $\Delta K < 0.2$ dex are dominated by weak emission T Tauri stars (WTTS; see
Walter et al. 1988; Strom et al. 1989a; Chapter by Montmerle et al.). From the

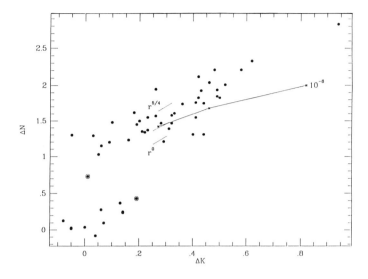

Figure 4.   A plot of ΔN against ΔK, the logarithmic ratios of observed to photo-spheric fluxes at $10\,\mu$m and $2.2\,\mu$m, respectively, for all stars in the Strom et al. (1989) and Skrutskie et al. (1990) samples for which adequate ($>3\sigma$) estimates of these quantities are available. Note that systematic errors in determining ΔK and ΔN are estimated to be $\sim\pm0.1$ dex. The sample stars fall in two domains separated by a gap. We superpose on this figure the loci defined by (1) flat reprocessing disks viewed at inclination angles ranging from $i = 0°$ (pole-on) to $i = 60°$; (2) two flared reprocessing disks viewed over the same range of inclination angles; the disk scale height h varies with radial distance r, as $r^{9/8}$ (middle dotted line), and $r^{5/4}$ (top dotted line); and (3) slightly flared ($h \sim r^{9/8}$ accretion disks with mass accretion rates ranging from 10^{-8} (leftmost star symbol), to 10^{-7} (middle star symbol), to 10^{-7} M$_\odot$yr^{-1}. These model disks are assumed to surround a star of type M0, typical of the solar-type PMS population in Taurus-Auriga. The two circled points denote the locations of the candidate "transition" cases V819 Tau and DI Tau.

above arguments, these stars must be surrounded by optically thin inner disks. The paucity of stars with ΔN values between 0.3 and 0.8 argues strongly that (1) the majority of stars with $\Delta N < 0.2$ dex represent optically thin disks rather than the extrema (disks viewed nearly equator-on) in a continuum of optically thick disks ($\Delta N > 0.8$ dex) viewed at differing inclinations; and (2) the transition between disks with $\tau \geq 1$ and $\tau \ll 1$ must be rapid as there are few disks with infrared excesses consistent with disks having intermediate ($0.2 < \tau < 1$) optical depths (see the Chapter by Edwards et al.). This result provides strong *ex post facto* confirmation of the practical utility of our criteria for selecting optically thick and optically thin disks.

It is also noteworthy that nearly all solar-type PMS stars selected to have massive, optically thick disks (with both ΔK and ΔN exceeding ΔK_c and ΔN_c, respectively) also show evidence of (1) boundary layer emission diagnosed by excess ultraviolet and optical continuum emission, and indicative of accretion of disk material onto the surface of the star; and (2) broad Hα

emission and forbidden line profiles (Chapter by Edwards et al.; Hartigan et al. 1990*a*), indicative of energetic winds. Disks with smaller excesses generally show neither measurable optical veiling, nor energetic wind signatures.

B. Intermediate-Mass Stars

Observations of a group of stars known as Herbig Ae/Be stars provide convincing evidence that some intermediate-mass stars with $M > 1.5$ $M_\odot$ are surrounded by massive, optically thick disks during early evolutionary phases.[b] These stars were selected (Herbig 1960) on the basis of (a) spectral type (type A or earlier); (b) strong Hα emission (a characteristic shared with solar-type PMS stars of the T Tauri class); and (c) close association with molecular cloud material as evidenced by their illumination of reflection nebulae. Strom et al. (1972) and Finkenzeller and Mundt (1984) located these stars in the HR diagram and found them to be PMS stars with ages in the range 0.2 to 1 Myr, and with masses $1.5 < M/M_\odot < 10$. Values of ΔK and ΔN for Herbig Ae/Be stars derived from the observations of Strom et al. (1972), Lorenzetti et al. (1983) and from the IRAS *Point Source Catalog*, are consistent in a large majority of cases with emission arising from massive, optically thick disks. Unfortunately, the fraction of the population of young ($t < 1$ Myr) stars with $M > 1.5$ $M_\odot$ represented by stars selected with Herbig's criteria is unknown. Hence, the infrared properties of Herbig Ae/Be stars establishes the *existence*, but not the *frequency* with which massive, optically thick disks form around stars in this mass range.

The disk occurence frequency for intermediate stars must rest at present on estimates of disk optical depth in the inner disk provided by the 2.2 μm excess, ΔK, in two well-studied young clusters: (a) the Orion Nebula cluster (Walker 1969; Penston 1973; McNamara 1976); and (b) NGC 2264 (Walker 1956; Warner et al. 1977); both clusters contain stars with ages ranging from 1 to 10 Myr. There are as yet no surveys which permit optical depth estimates based on ΔN for representative samples of young, intermediate-mass stars.

We estimate ΔK values for intermediate-mass stars in Orion and NGC 2264 from the reddening-corrected $(V - K)$ color index $\equiv 2.5 \log[F(2.2\,\mu m)/F(0.55\,\mu m)]$. Reference to Fig. 2b shows that for an optically thick reprocessing disk, the flux at V (0.55 μm) is dominated by the stellar photosphere, while at K (2.2 μm) the flux is dominated by disk emission. Hence, the difference between the reddening-corrected $(V - K)$ index and

---

[b] Perhaps the most persuasive arguments are provided by (1) the similarity of Ae/Be star infrared spectral energy distributions to those of solar-type PMS stars believed to be surrounded by disks (see Fig. 1); and (2) the recent millimeter-continuum observations obtained at the Caltech Submillimeter Observatory by Strom et al. (1990, unpublished). The observed millimeter-continuum fluxes lead to mass estimates of $\sim$0.1 to 1 $M_\odot$ for the objects included in their sample. Were the gas and dust corresponding to this mass distributed in a spherical envelope of radius 500 AU, the optical depth at 0.55 μm would be $\sim$100 if M(env) $\sim$0.1 $M_\odot$. By confining the material to a relatively flattened disk, one can produce the same mm-continuum fluxes while preserving an optically thin line of sight to the star at most viewing angles.

the $(V-K)$ color for a standard star of identical spectral type, will yield $\Delta K \sim [(V-K)_0 - (V-K)_{std}]/2.5$. Stars with $\Delta K \geq \Delta K_c$ (see Table I) appropriate to their spectral type are considered candidates for systems surrounded by optically thick disks.

Ages for individual stars are estimated from their location in the HR diagram relative to computed pre-main-sequence evolutionary tracks (Iben and Talbot 1966). In practice, such determinations are possible only for objects that have yet to reach the main sequence: stars with mass $M \leq 3$ $M_\odot$ in these clusters. For proper motion members (probability $\geq 80\%$) of these clusters in the mass range $1.5 \leq M/M_\odot \leq 3$, having ages $t < 3$ Myr, we find that in Orion, the fraction of stars with optically thick disks as judged from ΔK is 9/23 (Penston [1973] sample), and 10/19 (McNamara [1976] sample); in NGC 2264, the corresponding fraction is 7/18 (Warner et al. [1977] sample). Hence the frequency with which massive, optically thick disks are found among young stars of mass $1.5 < M/M_\odot < 3$ may be comparable to that characterizing solar-type stars of similar age.

Both clusters contain PMS stars with masses $M < 3$ $M_\odot$, and ages as small as $t \sim 1$ Myr as judged by their location in the HR diagram relative to computed evolutionary tracks. While age determinations are impossible for more massive stars (because they have reached the main sequence), it is likely that at least some of these stars have ages comparable to those of the youngest PMS stars ($t \sim 1$ Myr). Nevertheless none of the 11 stars in NGC 2264 with $M > 3$ $M_\odot$ and nominal ages $t \geq 1$ Myr have ΔK values signifying the presence of optically thick disks (neither the McNamara nor Penston sample in Orion include a significant number of main-sequence B stars). However, the presumed precursors of these stars, Herbig Be stars with masses $M > 3$ $M_\odot$ and ages $t < 1$ Myr, show evidence for massive, optically thick disks. Thus, the *survival time* for massive optically thick disks may be short for stars in this mass range. Observations of a much larger sample of B stars in young clusters is clearly desirable in order to confirm this result.

IV. SURVIVAL TIME SCALES FOR OPTICALLY THICK DISKS

The range of survival times for massive, optically thick disks is derived from the range of ages over which such disks are detected around young stars. The presence of optically thick disks is diagnosed from the magnitude of the observed infrared excesses (Fig. 2; Table I). Ages are estimated from the observed location of such stars in the HR diagram relative to computed PMS evolutionary tracks. Such ages, particularly for classical T Tauri stars, should be treated with caution, because the effects of infall and accretion have not been incorporated into currently available pre-main-sequence evolutionary tracks (see Hartmann and Kenyon 1990; Kenyon and Hartmann 1990); ages for weak line T Tauri stars in which accretion has ceased may be more reliable.

A. Solar-Type Stars

The information required to estimate the range in survival times for optically thick disks is also provided in Fig. 3. We interpret this histogram as follows: if *all* solar-type PMS stars are surrounded initially by optically thick disks, then by an age $t \sim 3$ Myr, $\sim 50\%$ of these disks have been accreted, disrupted or have begun to form larger bodies. Candidate optically thick disks (as judged from ΔK) are found around only 1/11 stars in our sample older than 10 Myr. Hence, we conclude that few disks survive as optically thick structures for ages much in excess of 10 Myr.

Strong confirmation of this limit on survival times is provided by unpublished near-infrared observations (J. Stauffer and R. Joyce, personal communication) of 15 main-sequence G stars in the α Perseii cluster (age ~ 30 to 50 Myr). None of these stars exhibits a near-infrared excess ΔK, consistent with emission arising in an optically thick disk.

Note that our discussion of disk survival time relies exclusively upon ΔK estimates; in principle, similar estimates could be made from ΔN and mm-continuum excesses. Unfortunately, while the sample of stars with $10\,\mu$m measurements of precision adequate to distinguish between excess emission arising in optically thick and optically thin disks includes a large fraction (90%) of known solar-type PMS stars in Taurus-Auriga with ages $t < 3$Myr, more than 55% of those stars with ages $t > 3$ Myr lack such measurements. A similar under-representation of older PMS stars is found among the Beckwith et al. (1990) sample of millimeter-continuum measurements. Hence, at present we can only estimate the range in times over which disks remain optically thick in their innermost regions ($r \ll 1$ AU). We emphasize as well that infrared and millimeter-continuum measurements probe only emission arising from circumstellar dust; *gas* survival times should not be inferred from these observations.

B. Intermediate-Mass Stars

At present, no comprehensive studies yielding ΔK, ΔN, or mm-continuum excesses are available for intermediate mass stars having a wide range of ages. Quantitative estimates of disk survival times must thus await observations of infrared spectral energy distributions for B, A and F stars in clusters and associations sampling a broad range of ages: for example, NGC 6611 (age ~ 1 to 3 Myr; distance ~ 2 kpc), Sco-Cen (age ~ 7 to 10 Myr; $d = 150$ pc) and α Per (age ~ 30 to 50 Myr; $d = 140$ pc), in addition to the Orion and NGC 2264 clusters already discussed, would be good candidates for study.

Qualitatively, however, the observation of massive, optically thick disks among Herbig Be stars with ages $t < 1$ Myr and masses $M > 3$ M$_\odot$, combined with the apparent lack of massive disks surrounding B stars in NGC 2264 (with ages ~ 1 to 3 Myr) suggests that disk survival times for stars with $M > 3$ M$_\odot$ may be considerably shorter ($t < 1$ Myr) than the values characterizing solar-type stars (~ 3 to 10 Myr).

V. OPTICALLY THIN DISKS

Candidate optiacally thin disks will show infrared emission which lies below the excess emission produced by a flat, optically thick reprocessing disk, but above photospheric emission (see Fig. 2; Table I). Recent observations provide persuasive evidence that such disks surround both solar-type and intermediate-mass stars.

A. Young Solar-Type Stars

Dutkevitch et al. (in preparation) carried out a series of observations of solar-type PMS stars in Taurus-Auriga aimed at determining infrared excesses as small as 5% above photospheric levels. Their technique involves *simultaneous* measurement of stellar fluxes over the wavelength range 1.2 to 4.6 μm, and thus avoids the uncertainties ($\sim\pm0.1$ dex) inherent in extant estimates of excess disk radiation based on more heterogeneous data sets obtained at different times for these photometrically variable objects (see Strom et al. 1989d; Skrutskie et al. 1990). Reference to Fig. 2 shows that for a typical solar type PMS star, the flux at 1.2 μm is dominated by photospheric emission, while the flux at $\lambda \geq 2.2\,\mu$m has a significant (2.2 μm) or dominant (4.6 μm) contribution from the disk. Hence, accurate estimates of the quantities ΔK (2.2 μm), $\Delta L(3.5\,\mu$m), $\Delta M(4.6\,\mu$m) can be derived from reddening-corrected color excesses $(J-X)_0$, as follows: $\Delta X \sim [(J-X)_0 - (J-X)_{std}]/2.5$, where $(J-X)_{std}$ is the color index for a standard star of identical spectral type.

The Dutkevitch et al. sample is selected from the set of stars in Skrutskie et al. (1990) with tabulated ΔK and ΔN values which fall near the critical values ΔK_c and/or ΔN_c (see Table I); the nominal age range represented in their sample is 0.3 to 20 Myr. In Fig. 5, we plot the reddening-corrected index $(J-L)_0$ against spectral type for the stars observed by Dutkevitch et al. Also plotted in this figure are (a) the color-spectral type relationship for dwarf standard stars obtained on the same night with the same instrument; and (b) the expected values of $(J-L)$ for optically thick disks viewed at inclination angles $i = 0°$, 60° and 80°. Note that 10 of the 15 stars in this sample show $(J-L)$ values consistent with their identification as candidate optically thin disks with discernible infrared excesses; 4 show colors consistent with photospheric colors, while one lies in the region occupied by candidate optically thick disks. For the number of stars in the Taurus-Auriga sample surrounded by optically thick disks (33), we expect to observe $\sim$5 optically thick disks at inclination $i > 80°$, and thus to show infrared excesses that fall below the critical values for this view angle, but above photospheric levels. Such stars would be expected to show the accretion and wind signatures found to characterize nearly all objects surrounded by optically thick disks (as judged from $\Delta K \geq \Delta K_c$) in the Cabrit et al. (1990) sample; indeed, 5 stars in the Dutkevitch et al. sample have Hα luminosities which are typical of stars surrounded by optically thick disks. The remaining 5 stars with intermediate $(J-L)_0$ values may be objects surrounded by disks which are optically thin in their inner regions.

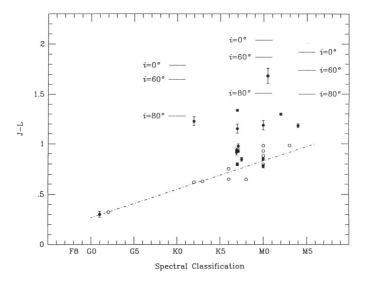

Figure 5.   A plot of reddening-corrected color index, $(J-L)_0$, against spectral type for solar-type PMS stars in Taurus-Auriga selected to have $\Delta N < 1.2$ and $\Delta K < 0.2$ dex (filled circles), and dwarf standard stars (open circles) observed on the same system. The dashed line is the color-spectral type relationship which defines the photospheric colors expected for stars in this spectral-type range. The horizontal lines labeled $i = 0°$, $60°$ and $80°$ represent the colors expected for a star of the indicated spectral type surrounded by a slightly flared ($h \sim r^{9/8}$) reprocessing disk. Solar-type PMS stars with colors which fall below the $i = 80°$ line and above the expected photospheric colors either (1) are surrounded by optically thick disks viewed at $i > 80°$, or (2) are surrounded by disks which are optically thin in the regions giving rise to the observed (small) $(J-L)$ excesses.

B. Older Solar-Type Stars

Evidence of emission arising from optically thin disks surrounding older solar-type stars comes from observation of significant infrared emission at $\lambda \geq 12 \, \mu m$ for nearby G and K main-sequence stars (Walker and Wolstencroft 1988; Backman and Gillett 1987). In all cases, the reported fluxes exceed expected photospheric emission, but fall below the critical values predicted for flat, optically thick disks. Because there is no evidence of emission arising from optically thick disks among main-sequence G and K stars, inclination effects cannot account for small infrared excesses; hence, if these excesses arise from disks, the disks must be optically thin. As the *minimum* ages for these stars exceed 50 Myr (the PMS evolution time for a G star), we must conclude that if this emission arises in optically thin disks, such disks must persist as observable entities for at least this long.

It would be extremely valuable to carry out high-precision, near-infrared observations of older solar-type stars both in order to diagnose the inner-disk structure in objects which show mid- and far-infrared excesses, and perhaps more importantly, to examine a much larger sample of objects spanning a wide range of (known) ages by virtue of the higher sensitivity of current groundbased measurements to small excess emission.

C. Young Intermediate-Mass Stars

Examination of the IRAS *Point Source Catalog* (Waters 1986; Waters et al. 1987; Cote 1987) reveals significant infrared excesses among main-sequence B stars. In some cases, this excess emission can be attributed to free-free emission in gaseous envelopes surrounding rapidly rotating Be stars (not to be confused with Herbig Be stars where excess emission is believed to arise in disks; Strom et al. 1972). However, in many cases, *normal* main-sequence B stars show significant excesses, typically of a size consistent with that expected from optically thin disks. The ages represented by stars in these samples cannot exceed their main-sequence lifetimes (20 Myr at B0; 500 Myr at A0; Iben 1967), and could be as short as their PMS evolution times (0.3 Myr at B0; 3 Myr at A0)

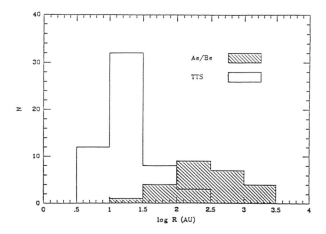

Figure 6.   A histogram depicting the frequency distribution of $12\,\mu$m excesses $(\log_{10}[F(\text{observed})/F(\text{photosphere})])$ derived for main-sequence B stars in the young ($t\sim$7 to 10 Myr) Scorpio-Centaurus Association; only objects with detections at a level 3σ or greater are included. From Table I, we note that candidate optically thick disks surrounding a B5 star would have an excess $\Delta N>1.5$ dex.

An investigation currently underway (S. E. Strom and G. Condon) yields a preliminary estimate of the fraction of *young* B stars surrounded by optically thin disks. In Fig. 6, we present the frequency distribution of $12\,\mu$m excesses for a sample of 42 stars in the Sco-Cen OB Association (age $\sim$7 to 10 Myr). Values of $\Delta(12\,\mu$m) were estimated from $\Delta\,(12\,\mu$m) =

$[(V-12\,\mu m)_0 - (V-12\,\mu m)_{std})]/2.5$.   Here $(V-12\,\mu m)_0$ is the reddening-corrected $(V-12\,\mu m)$ color index derived from the co-added IRAS survey, and from extant visible photometry and spectral types; $(V-12\,\mu m)_{std}$ is the color index appropriate to a standard star of identical spectral type.   The $\Delta(12\,\mu m)$ value corresponding to a flat, optically thick disk surrounding a star of type B5 is 1.5 dex (if viewed at an inclination $i = 80°$). The typical 12 μm flux uncertainty is about 20% for the stars in their sample. We conclude from Fig. 6, that $\sim$17/39 main-sequence B stars have $0.15 < \Delta(12\,\mu m)$ < 1.5 dex and may thus be surrounded by optically thin disks. We caution that the IRAS beam is relatively large ($\sim$1′ at 12 μm) and may include emission from either nearby red companions or infrared cirrus. Careful study of each object (ultimately involving imaging) will be necessary before unambiguous attribution of excess 12 μm flux to disk emission can be made.

D. Older Intermediate-Mass Stars

One of the most dramatic revelations of the IRAS survey was the "Vega" phenomenon: the discovery of extended infrared emission associated with α Lyr and several other nearby A stars.   Aumann et al. (1984), Gillett et al. (1984), and Aumann (1985) argue that this emission very likely arises in optically thin circumstellar disks.   Persuasive evidence that this excess emission indeed arises from disks (as opposed to circumstellar shells) is provided by optical observations of light scattered by small dust grains in the nearly edge-on ($i > 76°$) disk surrounding β Pictoris (Smith and Terrile 1984; Paresce and Burrows 1987; Artymowicz et al. 1989). The β Pic disk has a radius, $r \sim 1000$ AU, a mean visual optical depth $\tau \sim 3 \times 10^{-3}$ and a mass of 10^{-7} M$_\odot$ in micron-size grains.

Subsequent analysis of IRAS survey data reveals that a significant number of nearby main-sequence A and F stars exhibit infrared excesses consistent with emission arising from optically thin disks (Backman and Gillett 1987; Chapter by Backman and Paresce), though none comes within an order of magnitude of the excess emission observed for β Pic. In the sample of main-sequence A and F stars studied by Aumann (1985), disk optical depths range from $-5 < \log \tau$ (0.55 μm) < -2.5, and estimated disk sizes range from 3 to 100 AU. The ages of these stars range from 100 to 1000 Myr.

From the range of excess emission observed around nearby stars, combined with the known limiting fluxes of the IRAS survey, Backman and Gillett (1987) estimate that 23% of all A stars, and $\geq 10\%$ of all F and G stars are surrounded by disks with emission optical depths $\tau > 10^{-5}$; by comparison the zodiacal cloud has an optical depth $\tau \sim 10^{-7}$. The true number of optically thin disks may be even greater. IRAS sensitivity levels are sufficient to locate disks with optical depths as small as 10^{-6} among luminous nearby A stars (though not around less luminous F and G stars). At this level, the fraction of A stars exhibiting excess emission increases from the 23% with excesses consistent with $\tau \geq 10^{-5}$, to 45% consistent with disks with $\tau \geq 10^{-6}$. This result suggests that the detected fraction of F and G stars surrounded by disks would likely increase were higher sensitivity observations available.

E. The Origin and Implications of Optically Thin Disks

Although considerably more work is needed to establish the frequency with which optically thin disks are found around stars of differing ages, it seems clear that some main-sequence stars of all types exhibit small infrared excesses above photospheric levels, and may thus be surrounded by optically thin disks.

Accretion in Optically Thin Disks. All PMS stars which lack infrared excesses above photospheric levels (no disk, or undetectable optically thin disk), and most PMS stars (with small excesses consistent with identification as *candidate* optically thin disks, show no evidence of either (1) the strong boundary layer emission; or (2) the energetic winds which diagnose disk accretion. In contrast, nearly all PMS stars surrounded by optically thick disks extending inward to the stellar surface show (a) the excess ultraviolet and blue continuum emission produced in a boundary layer; (b) the strong, broad $H\alpha$ emission arising in the inner regions of energetic winds; and (c) large infrared excess luminosity, some of which must be produced by self-luminous accretion disks (Bertout 1989; Cabrit et al. 1990; Hartigan et al. 1989a,1990a; Chapter by Edwards et al.). We thus conclude that when disks become optically thin, they can no longer sustain accretion, at least at observable rates ($\dot{M}_{acc} \geq 10^{-8}$ M$_\odot$ yr^{-1}).

Replenishment of Optically Thin Disks. The observed excesses in optically thin disks are produced by small ($a \sim 1$ μm diameter) radiating grains. If these grains are located in a gas-free region, then Poynting-Robertson drag will cause them to spiral into the central PMS star from distances r on time scales $t_{PR} < 400$ $(r/\mathrm{AU})^2$ yr (Burns et al. 1979). If the grains are suspended within a residual quiescent gas layer, sedimentation and coagulation of small grains into $a > 1$ centimeter-sized objects can occur in ~ 1000 $(r/\mathrm{AU})^{3/2}$ yr (Weidenschilling 1980). Aerodynamic drag on these larger bodies forces them to spiral into the protostar in $< 10^4$ yr (Weidenschilling 1977). Hence, the detection of near-infrared emission from small grains surrounding stars with $t >> 1$ Myr means that their inner disks must be re-supplied continuously with fresh grain material (see Gillett 1986; Backman and Gillett 1987).

Fresh grains could be introduced into the inner disk regions via Poynting-Robertson drag operating on micron-size grains located in outer disks at $r >> 4$ AU; such cold grains would be detectable only from sensitive millimeter-continuum measurements. The efficacy of this mechanism depends on the total momentum which can be transferred to the grains, and therefore on the geometry and optical depth of the outer disk. The magnitude of the observed near- and mid-infrared excesses produced by small grains introduced into the inner disk will reflect a balance between the rate of injection of fresh grains from the outer disk, and the rate of depletion by Poynting-Robertson or aerodynamic drag in the inner disk. Grains can also be produced *in situ* as a by-product of collisions which fragment pieces of larger parent bodies (e.g., large grains; planetesimals), or as a result of evaporation of cometesimals gravitationally scattered into the inner disk by one or more massive planets. The mass in small ($a \sim 1 \, \mu$m) grains which must be resupplied over a Poynting-Robertson or aerodynamic drag time is $\sim 10^{-10}$ $M_\odot$ for a star of luminosity $L_* = 1 \, L_\odot$ which exhibits $\Delta N = 0.2$ dex.

It is currently believed that Earth-mass planets were built by inelastic collisions of planetesimals over time scales $t > 100$ Myr; micron-size dust grains are by-products of such collisions. If the presence of cold, outer disk regions containing a significant store of small grains can be excluded, then measurement of infrared excesses diagnosing optically thin inner disks might establish the rate at which micron-size dust is replenished via planetesimal collisions: a decrease in the infrared excess surrounding older stars would indicate a decrease in the rate of planetesimal collisions. Determining the time scale for near-infrared excess emission to become undetectable would thus provide an important astronomical constraint on the time scale required to build terrestrial planets from planetesimals. It is thus critical (1) to determine disk optical depths around main-sequence stars having ages ranging from 10 Myr to 1000 Myr; (2) to establish from millimeter-continuum observations whether stars that show evidence of infrared emission arising from optically thin disks, are surrounded by massive outer disks comprised in part of cold micron-size dust which might "feed" small grains into the inner disk regions; and (3) to ascertain the dominant source of grain replenishment.

There is some evidence that the infrared excess (and by inference, the disk optical depth) decreases with age. The bulk of the evidence thus far comes from the work of Backman and Paresce (see their Chapter) who examine the excess infrared emission associated with main-sequence A stars in three nearby ($d \le 150$ pc) clusters: α Per (age $\sim$30–50 Myr); the Pleiades (age $\sim$70–100 Myr); and the Ursa Major cluster (age $\sim$200–300 Myr). They find that the typical excesses among the α Per and Pleiades A stars are significantly larger than those characterizing the A stars in the older Ursa Major cluster. Systematic examination of mid- and far-infrared excesses for lower-mass stars in these clusters is impossible from the IRAS survey, because these stars have photospheric (and thus optically thin disk) luminosities that are too faint to admit reliable detection at IRAS sensitivity limits.

Groundbased near- and mid-infrared ($1.2 \, \mu$m $\le \lambda \le 10 \, \mu$m) observations have the sensitivities required to establish the frequency with which optically thin disks surround main-sequence stars of all types, and to derive characteristic time scales for their evolution from observation of main-sequence stars located in clusters of differing age. Systematic studies aimed at measuring infrared excesses for large and representative samples in clusters sampling the age range 10 to 1000 Myr are currently underway (Backman, Witteborn and Stauffer, unpublished; Strom, Edwards, Skrutskie and Dutkevitch, unpublished).

Large millimeter-antennae located at good sites (IRAM, CSO, JCMT, respectively) are capable of detecting millimeter-continuum excesses arising from disks containing dust masses as small as 10^{-4} to 10^{-5} $M_\odot$ around main-sequence G stars in nearby clusters. Strom, Edwards and Skrutskie are currently carrying out a program on the CSO aimed at detecting or placing strong limits on millimeter-continuum excesses for stars in the α Perseii, Pleiades and Ursa Major clusters, and thus constraining the mass of cold dust grains located in outer disk regions.

VI. TRANSITION DISKS

The arguments presented in Secs. III and V establish the presence of both (1) massive, optically thick disks surrounding young stars ($t < 10$ Myr) of all masses, and (2) optically thin disks surrounding older stars of all masses. If optically thin disks represent the descendents of massive, optically thick structures, it may be possible to identify disks *in transition* between these two states, and to define a characteristic time scale for the transition phase. Currently available observational techniques force us to search for a particular type of transition structure: one in which the inner disk regions are optically thin, while the outer disk regions are optically thick. Such disks can be diagnosed from infrared spectral energy distributions in the manner described in Sec. II.

To select candidate transition cases, we make use of the infrared excesses tabulated by Strom et al. (1989a) and Skrutskie et al. (1990) for a representative sample of 83 solar-type PMS stars in Taurus-Auriga. We examine the spectral energy distributions of all stars in their sample in order to determine whether any have optically thin *inner* but optically thick *outer* disks. Three candidate cases (DI Tau, V819 Tau, SAO76411A) exhibit the small near- and mid-infrared, and large far-infrared excesses which diagnose such structures. As noted in Sec. II, current limits on the sensitivity of mid- and far-infrared measurements preclude detection of disks which might first become optically thin in their outer regions.

In Fig. 7, we plot the spectral energy distribution of V819 Tau along with that of a "typical" T Tauri star, AA Tau. Superposed in each case is a line indicating the location of the composite star plus disk spectrum (for a flat reprocessing disk viewed at $i = 60°$). Note that the observed spectrum of AA Tau lies above that of the predicted composite spectrum at all infrared wavelengths $\lambda > 1.2\,\mu$m, and is thus consistent with that of an optically thick disk that extends inward to within a few stellar radii of the PMS star surface. In contrast, the observed excesses for V819 Tau at $\lambda \leq 10\,\mu$m must be produced in a disk with an "inner hole" ($\tau << 1$ in regions $r < 0.2$ AU). At radii $r > 0.5$ AU, the disk appears to be optically thick, as the observed infrared fluxes at $\lambda > 25\,\mu$m fall above the flat disk spectrum. V819 Tau is also detected as a millimeter-continuum source (Beckwith et al. 1990); the amount of disk material at $r >> 0.5$ AU is ~ 0.03 M$_\odot$, or comparable to that of a minimum-mass solar nebula.

The ratio of the number of transition cases (3) to the number of PMS stars surrounded by optically thick disks in the Strom et al. (1989a) and Skrutskie

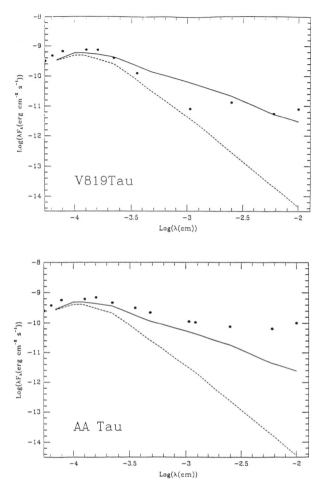

Figure 7.   A plot of the spectral energy distribution observed for V819 Tau (upper panel) and AA Tau (lower panel). Superposed on these plots in each case, is a solid line indicating the location of the composite star plus optically thick, flat reprocessing disk ($i = 60°$); the dashed line corresponds to the emission arising from the stellar photosphere. The observed spectrum for AA Tau lies above that of the predicted thick disk + star spectrum at all wavelengths $\lambda \geq 1.2\,\mu$m. The observed spectrum for V819 Tau lies *below* the thick disk + star spectrum at wavelengths $2.2\,\mu$m $\leq \lambda \leq 25\,\mu$m and above at longer wavelengths. These data suggest that the *inner* disk of V819 Tau is optically thin, while the outer disk is optically thick. It is noteworthy that V819 Tau is readily detected at millimeter wavelengths.

et al. (1990) samples (33 as judged from ΔK), yields a *transition time* of $3/33 \times 3 \times 10^6$ yr or ~ 0.3 Myr, assuming a mean age of 3 Myr for PMS stars surrounded by optically thick disks.

A. Possible Implications of Inner Holes

The End of the Disk Accretion Phase. The development of a "transition" disk with an optically thin inner region appears to signal the cessation of disk accretion. None of the three stars (DI Tau, V819 Tau, SAO 76411A) apparently surrounded by transition disks shows evidence of either (1) the strong boundary-layer emission, or (2) the energetic winds which diagnose disk accretion. As noted in Secs. III and V, their presumed precursors, young stars surrounded by optically thick disks extending inward to the stellar surface, show evidence of (correlated) accretion and wind signatures; their presumed descendents, disks which are optically thin throughout or show no evidence of excess emission, also lack accretion signatures (see also the Chapter by Edwards et al.).

Possible Evidence for Giant Planet Formation. Creation of an optically thin region in the interior regions of circumstellar disks may reflect the shorter dynamical times and consequently shorter grain agglomeration times in such regions: the agglomeration of small grains into larger bodies will decrease the optical depth in the inner disk regions. However, the presence of an optically thin region interior to an optically thick outer disk may be indicative of more profound effects. Such structures appear to require a mechanism to keep the "inner hole" isolated from the outer regions of the disk. In the absence of an isolating mechanism, Poynting-Robertson drag will cause small grains located at the inner boundary of the outer disk to spiral in toward star on a time scale $t < 6000$ yr (see above), or much shorter than the estimated "transition time" ($t \sim 0.3$ Myr). The mass in small grains introduced into the inner disk (and thus the infrared excess produced) depends critically on the geometry and transparency of the outer disk; more physically extended, optically thinner outer disks can introduce more grains into the inner disk, and produce larger near- and mid-infrared excesses.

S. Ruden (1989, personal communication) and G. Morfill (1989, personal communication) suggest that the formation of a Jupiter mass protoplanet can isolate regions of a disk interior to its orbit from those outside its path as a result of tidal torques generated by the protoplanet. Preliminary calculations carried out by Ruden indicate that once a Jupiter-mass body forms in the disk, it creates a "gap." Small dust particles and gas interior to the gap accrete onto the parent star on a time scale $\sim 5 \times 10^5$ yr (for a Jupiter-mass planet initially located at a distance 5 AU from a solar mass star surrounded by a minimum-mass nebula; the time scale is set by the orbital time scales and the viscous coupling efficiency in the inner disk). Material exterior to the gap can no longer enter the inner disk. The resulting structure (an optically thick outer disk with an inner hole) will produce an infrared signature in qualitative agreement with those of the transition structures discussed above, and on a time scale comparable to our estimated transition time, $t \sim 0.3$ Myr. In the context of the hypothesis that giant planets produce and sustain inner holes, the ages of our three candidate transition cases, $t \sim 0.3$ to 1 Myr (DI Tau), and

$t\sim3$ to 10 Myr (V819 Tau and SAO 76411A), provide upper limits to the time required to agglomerate a Jupiter mass body.

By isolating the inner disk, the formation of a giant planet may permit the initial clearing of the inner disk via accretion. Although the inner disk will thus be depleted of gas and small dust grains, larger bodies (e.g., planetesimals) will remain behind, presumably to continue agglomeration into planet-sized bodies. High-precision infrared observations may permit indirect diagnosis of the presence of planetesimals or cometesimals located in or scattered into the inner regions of transition disks. Because the giant planet keeps the outer reservoir of gas and dust dammed up, and because Poynting-Robertson drag sweeps the optically thin inner disk clean on short time scales, dust responsible for excess near-infrared radiation must be continuously resupplied to the inner disk regions.

VII. SEARCHES FOR CIRCUMSTELLAR GAS

The evidence regarding the frequency of disk formation, as well as estimates of time scales for disk evolution rests entirely on observations of excess infrared and millimeter-continuum emission produced by circumstellar dust. What is the fate of the dominant constituent of circumstellar disks—the gas? Determining the survival time for the gas component is essential to understanding the evolution of nebular disks, and is particularly crucial for establishing the time available for building the giant, gas rich outer planets.

Thus far, the only direct evidence regarding the gas content of circumstellar disks derives from high angular resolution CO observations carried out with the Owens Valley Radio Observatory millimeter-wave interferometer. CO emission has been detected for HL Tau and L1551 IRS 5, two extremely young ($t<1$ Myr) solar-type pre-main-sequence stars (see, e.g., Sargent and Beckwith 1987; Sargent et al. 1988). The gas mass in these disks is ~0.1 to 1.0 $M_\odot$, while the optical depth in the ^{12}CO (1–0) transition is $\tau \sim$ several thousand. Presumably, as the disks evolve and assemble material into planets, the gas content should decrease from this initial value. Ultimately, the gas will either be accreted by rocky or icy cores, removed by winds, or stripped by encounters between solar-type stars and other molecular clouds. However, ^{12}CO will remain optically thick until the disk mass decreases to $\sim10^{-4}$ $M_\odot$ (for a disk of diameter ~100 AU).

In order to predict the required sensitivity for detection of remnant gaseous disks, we rely upon a purely geometrical argument, assuming the CO emission is optically thick out to an outer disk radius of 100 AU, with T_K (CO) = 20 K. Such a disk surrounding a star located at $d\sim25$ pc will fill $\sim4\%$ of the Five College Radio Astronomy Observatory (FCRAO) 14-m telescope antenna beam; with the IRAM 30-m diameter telescope, disks surrounding stars located at $d\sim100$ pc will have similar beam-filling fractions. For a beam dilution of 4%, an assumed beam efficiency of 0.5, we predict a peak antenna

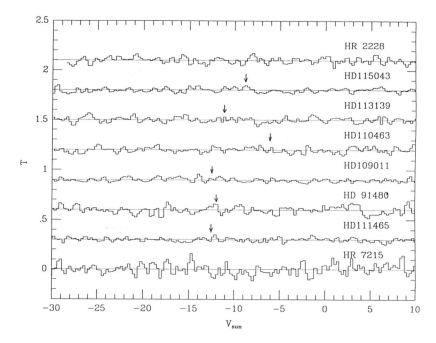

Figure 8.   A plot of antenna temperature against heliocentric velocity for 8 stars in the Ursa Major Stream observed in the ^{12}CO (1–0) transition at the FCRAO 14-m telescope by Skrutskie et al. (in preparation). The arrows locate the heliocentric velocities of the stars where known. The typical 1σ detection limit is estimated as 25 mK/100 kHz channel. No star shows evidence for ^{12}CO emission.

temperature $T_A = 40$ mK for a disk viewed at an inclination $i = 60°$, and 400 mK for a disk viewed pole-on.

Recently, Skrutskie et al. (1991) have carried out preliminary searches at FCRAO for ^{12}CO (1–0) emission arising from disks surrounding main-sequence F and G stars in the nearby ($d = 25$ pc) Ursa Major cluster (age $\sim$200 to 300 Myr). Typical integration times were 2 to 3 hr on source, and similar times off source. The resulting position-switched difference spectra for the sources examined to date are reproduced in Fig. 8; the typical noise level achieved was 25 mK/100 kHz channel (100 kHz corresponds to 0.26 km s^{-1}).

None of these objects was detected. Hence, either (1) the disks have radii $r << 100$ AU; (2) the CO/H$_2$ ratio is much smaller than the ratio which obtains in molecular clouds (as might be the case if the CO in an unshielded, dust-free disk is dissociated); or (3) the ^{12}CO emission is optically thin, and the disk mass is $\leq 10^{-4}$ M$_\odot$ (or 0.01 times the mass of the "minimum-mass" solar nebula). If the disk mass is indeed small, our observations imply that the gas survival time in disks cannot exceed $t \sim 200$ to 300 Myr. It is essential to extend such studies to solar-type stars located in younger clusters. Because such clusters are more distant than the Ursa Major cluster, the power of larger millimeter-wave telescopes (e.g., the IRAM antenna) is required in order to detect disk gas or to place significant upper limits on the disk gas content.

VIII. SUMMARY AND FUTURE DIRECTIONS

Efforts to date allow us to draw the following conclusions regarding the frequency of disk occurence and the time scales for disk evolution as follows:

A. Optically thick disks ($\tau > 1$ in regions sampled by excess emission $\lambda \leq 10$ μm) are found around 30 to 50% of young ($t < 3$ Myr) stars of all masses $M < 3$ M$_\odot$. Optically thick disks are found around PMS stars of higher mass, but with unknown frequency. For the relatively limited sample of objects for which sub-millimeter and millimeter-continuum measurements are available, nearly all stars with $\tau > 1$ show disk masses $M_d > 0.01$ M$_\odot$ ($\equiv M_{mm}$). Disk sizes range from ~10 to ~1000 AU. Hence disks of solar system sizes, and of masses comparable to or greater than the minimum-mass solar nebula appear to form around a large fraction of stars with $M < 3$ M$_\odot$.

Three major uncertainties associated with these conclusions are:

1. The inference of disks associated with intermediate-mass stars. Current arguments rest primarily on the observation of excess infrared and millimeter-continuum emission similar to that observed for solar-type PMS stars (T Tauri stars) surrounded by optically thick disks (see Fig. 1), and economy of hypotheses. Blueshifted forbidden line emission has been observed for only a few intermediate-mass stars. Moreover, the links between infrared excess emission arising in optically thick disks, energetic winds diagnosed by Balmer line and forbidden line emission, and ultraviolet and optical excess emission arising in a boundary layer are not nearly so well established for intermediate-mass stars. Efforts to solidify the arguments supporting the existence of disks associated with these stars are badly needed.

2. The completeness of our census of newly born stars. While the proper motion studies in NGC 2264 and Orion provide comprehensive membership lists for stars $M \geq 1.5$ M$_\odot$, no comparably complete lists of young solar-type stars are available either in these clusters or in nearby star-forming complexes such as Taurus-Auriga. Present samples of solar-type stars are amalgams of samples selected on the basis of strong Hα emission (classical T Tauri stars or CTTS) or X-ray emission (objects which include CTTS and weak emission T Tauri stars, or WTTS), and may not be representative of either the true proportion of CTTS and WTTS, or the total population of young solar-type stars.

3. The mass determinations. While sub-millimeter and millimeter-continuum excesses measure emission arising from optically thin dust and thus provide the most sensitive probe of disk masses, the emissivity of the dust is not well known. Hence, *absolute* mass estimates may be highly uncertain (perhaps by a factor of 10 or more).

B. Nearly all optically thick disks show evidence of (1) boundary layer emission diagnosed by "spectral veiling" at optical wavelengths, and indicative of accretion of material onto the stellar surface; and (2) strong, broad Hα emission and forbidden line emission arising in energetic winds whose mechanical luminosity is proportional to the disk accretion luminosity (Cabrit et al. 1990; Chapter by Edwards et al.). We conclude that optically thick disks are accretion disks with typical accretion rates $\dot{M}_{acc} > 10^{-8}$ M$_\odot$ yr^{-1}.

C. Disks survive as massive, optically thick structures extending nearly to the stellar surface for times ranging from $t << 3$ Myr to $t \sim 10$ Myr around solar-type PMS stars. No optically thick disks have been identified around solar-type stars with ages $t \geq 30$ Myr. The survival times for disks around more massive stars are not well established, particularly for stars $M > 3$ M$_\odot$. However, the limited data available suggests that the disk survival times for such massive stars may be considerably shorter ($t < 1$ Myr) than those for solar-type stars.

D. The descendents of massive, optically thick disks are disks containing masses of distributed micron-size dust grains $M < 10^{-4}$ M$_\odot$; such disks are optically thin, and perhaps analogous to the structures observed to surround the intermediate-mass stars Vega and β Pic. Evidence of infrared excess emission consistent with emission arising in optically thin disks is found for stars of all masses. Stars surrounded by optically thin disks show none of the signatures associated with disk accretion (strong boundary layer emission; emission from spectral features arising in energetic winds). Hence, when disks become optically thin, disk accretion ceases.

E. The presence of micron-size grains in optically thin circumstellar disks at distances $r < 1$ AU from the surface of solar-type stars, requires that such grains be continuously supplied to these disk regions; Poynting-Robertson or aerodynamic drag would otherwise deplete the grains on time scales short compared to the age of stars surrounded by optically thin disks. Small grains can be supplied either by collisions among larger grains or planetesimals; by evaporation of cometesimals gravitationally scattered into the inner disk regions; or by grains spiraling in from cold, outer disk regions ($r >> 1$ AU). Eventually, millimeter-continuum observations can place stringent limits on the amount of distributed cold dust.

F. Older stars of intermediate mass surrounded by optically thin disks appear on average to show smaller mid- and far-infrared excesses than do younger stars surrounded by such disks. This result, based on results from the IRAS survey for nearby clusters, motivates sensitive near- and mid-infrared measurements aimed at establishing characteristic time scales for systematic changes in disk optical depth (a measure of the mass in micron-size dust). Establishing such time scales may provide an astronomical constraint on the rate of planetesimal collisions or cometesimal evaporation as a function of time during the planet-building epoch.

G. Disks with apparent "inner holes"($r \gtrsim 0.2$ AU) have been discovered. These are most likely structures in transition between disks that extend inward to the stellar surface and are optically thick throughout, and those that are optically thin. The presence of an inner hole may require that the inner regions of a disk be isolated from the exterior, optically thick disk regions. If so, the formation of a giant planet provides an attractive mechanism for tidally isolating the outer and inner disk regions. In this picture, the formation of a giant planet creates a gap in the disk. Gas and small grains located inward of the gap accrete onto the parent star; small grains, and possibly gas exterior to the gap survive for times $t > 0.1$ Myr.

H. Our discussion of disk properties rests entirely on observations of emission arising from the dust component. Interferometric observations reveal CO millimeter-line emission arising in disk gas surrounding two extremely young solar-type PMS stars (HL Tau and L1551, IRS 5). The estimated gas mass for these systems lies in the range 0.1 to 1 $M_\odot$. Systematic searches for disk gas associated with stars of differing ages and masses are critical to determining characteristic times for the evolution of the gas component. However, no such searches are yet available, primarily because the sensitivity required demands extensive time on large millimeter-wave telescopes. We report preliminary results of a search for CO emission associated with G and K main-sequence stars in the nearby Ursa Major cluster ($d = 25$ pc; $t \sim 200$ to 300 Myr). We find no evidence of CO emission associated with these stars, despite the fact that our measurements are sensitive enough to permit detection of $M \sim 10^{-4}$ $M_\odot$ of gaseous material with a CO/H_2 ratio comparable to interstellar values. If CO survives in the circumstellar environment, then these observations suggest a gas survival time $t < 200$ to 300 Myr.

The presence and properties of disks associated with solar-type PMS stars have thus far been inferred from indirect observations. *Direct detection* of a low surface brightness region of dust-scattered light from a disk at an angular distance $\theta < 0\rlap{.}''5$ ($r < 70$ AU for PMS stars in Taurus-Auriga and other nearby star-forming complexes) from a bright central star, represents a daunting task for groundbased telescopes. Despite the application of speckle imaging and resolution enhancement techniques, the great range of light intensity over the small angular scale of the star/disk system has precluded detection of candidate PMS disks except in a few cases. Over the next few years, the Hubble Space Telescope (HST), equipped with the second generation Wide

Field Planetary Camera, should permit direct detection and imaging of such disks. HST, with a diffraction-limited beam size of $\sim0\rlap{.}''05$ at 5000 Å and a stable point spread function is uniquely capable of:

1. Placing decisive limits on the occurence frequency of disks, by virtue of its ability to image dust-scattered light from low-mass, low optical depth, low-surface-brightness disks. With HST, a 600 second exposure with Wide Field/Planetary Camera II can image a disk with a dust mass as low as 10^{-9} $M_\odot$ (comparable to the dust mass in the disk surrounding the nearby main sequence A star, β Pictoris) distributed uniformly within a 50-AU region surrounding a typical PMS star in Taurus-Auriga. The ability of HST to image disks around PMS stars will thus allow us to learn for the first time from direct observations whether our planetary system is likely to represent a common and expected result of star formation, or whether formation of our solar system required a fortuitous combination of circumstances reproduced only rarely in nature.

2. Establishing characteristic time scales for disk evolution by imaging disks during a wide range of evolutionary phases: from the earliest stages when the disk mass and surface brightness may be high, to the epoch of dust settling and planetesimal building when the mass of *distributed* material and the disk surface brightness is low. HST will be able to obtain images of disks to limiting surface brightnesses $V \sim 21$ mag arcsec^{-2} around a typical PMS star in Taurus-Auriga, and should be capable of establising the presence of circumstellar "holes," possibly resulting from planet-building episodes in the disk interior to ~ 10 AU.

Sensitive near- and mid-infrared and millimeter-continuum measurements should permit indirect detection of circumstellar material in disk regions within $r \sim 10$ AU of the surfaces of typical solar-type and intermediate-mass PMS stars in nearby star-forming regions ($d < 200$ pc), and main-sequence stars in nearby clusters. By increasing the size of PMS star samples, and by studying stars with ages ranging from 10 to 1000 Myr, we should be able to place more convincing limits on characteristic disk evolutionary time scales (1) from a more complete census of optically thick disks, leading to better estimates of their frequency and survival times; (2) from identification of candidate optically thin disks surrounding stars sampling a wide range in age, providing the basis for determining the rate at which the optical depth in such disks decreases with time; and (3) from identification of additional transition cases analogous to the 3 objects discussed here. It will be of particular interest to locate stars which lack near- and mid-infrared excesses, but which show strong millimeter-continuum fluxes. The age of systems which exhibit large holes ($r > 5$ AU), may provide an astronomical constraint on the time scale over which giant planet building takes place.

Finally, CO observations (carried out with large millimeter-wave telescopes) of large samples of stars covering a wide range of ages and masses are required in order to measure the gas content of circumstellar disks, to determine the range in disk gas mass associated with young stars, and gas survival times.

Acknowledgments. We wish to extend our deepest thanks to K. M. Strom, D. Dutkevitch, L. Hillenbrand, G. Condon, and L. Gauvin for their many helpful contributions. This work was supported by grants from the National Science Foundation, the NASA Planetary Program, and the NASA Astrophysics Data Program.

EXTINCT RADIONUCLIDES AND EVOLUTIONARY TIME SCALES

TIMOTHY D. SWINDLE
University of Arizona

Extinct radionuclides, radioactive nuclides with half-lives of 10^5 to 10^8 yr, potentially provide sensitive clocks for dating events that occur within a few half-lives of their nucleosynthesis. Strong evidence for at least seven of these has been found in meteorites. In principle, these can constrain several time scales, including the length of time between the last episode of nucleosynthesis and the formation of solids in the solar system, the duration of the nebula, the length of time from the formation of solids to the formation of bodies large enough to differentiate, and the length of time from the formation of solids to the formation of terrestrial planets.

I. INTRODUCTION

The study of the primitive solar nebula is essentially a historical study. Although we can perform experiments and make observations today to provide constraints, our ultimate goal is to reconstruct events that happened long ago. As in any historical study, chronology is an important facet; to fully understand what happened, we must know the order and duration of events.

One of the most effective ways to study early solar system chronology turns out to be the study of the abundances of extinct radionuclides in meteorites. These are radioactive isotopes with half-lives long enough for them to have survived from their nucleosynthesis until incorporation into solar system solids, but not long enough to survive until the present. In practice, this translates into half-lives of roughly 10^5 to 10^8 yr. Their abundances are frequently measured with accuracies of a few percent (sometimes even greater), so each extinct radionuclide potentially provides chronological information with an accuracy of a fraction of its half-life.

A number of reviews of extinct radionuclides are available (see, e.g., Podosek 1978; Begemann 1980; Wasserburg and Papanastassiou 1982; Wasserburg 1985; Podosek and Swindle 1988). As many of these contain detailed descriptions of various searches for extinct radionuclides, the present chapter contains only a general description of such searches. Instead, it emphasizes the chronological information that extinct radionuclides can provide. The time scales addressed include: (1) the length of time from nucleosynthesis to incorporation into solar system material; (2) the duration of the solar nebula;

(3) the length of time between the formation of early solids and the formation of bodies big enough to differentiate; and (4) the length of time required to form a planet-sized body (Earth).

II. EXTINCT RADIONUCLIDES

A. Detection

Extinct radionuclides are, by definition, no longer present in meteorites. Therefore, the only way to determine that they were once present is to find indirect evidence of their existence. That evidence is usually an excess of the daughter isotope of the extinct radionuclide that correlates with the presence of another isotope of the parent element. An example is given in Fig. 1, from the work that established the presence of ^{26}Al (Lee et al. 1977). The data points represent the measured Mg isotopic composition and Al/Mg ratios for mineral separates from an inclusion in the Allende meteorite. The ^{26}Mg/^{24}Mg ratio varies by 10%, far more than the experimental uncertainties or the normal variations in isotopic ratios. Furthermore, the samples with the highest Al/Mg ratios have the highest ^{26}Mg/^{24}Mg ratios. Because ^{26}Al decays to ^{26}Mg, the simplest explanation for the correlation of excess ^{26}Mg with Al is that the inclusion formed with isotopically normal Mg but that the Al contained some ^{26}Al, which subsequently decayed. The slope of the correlation line gives the initial ^{26}Al/^{27}Al ratio, 5×10^{-5} in this case. Analogous plots have been used for the identification of all other extinct radionuclides presently known except for ^{244}Pu, which was detected by the presence of a xenon isotopic pattern matching the ^{244}Pu fission yield spectrum.

   In order to detect a given extinct radionuclide in a sample, two conditions must be met: 1) the sample must have contained that extinct radionuclide, and 2) there must be some parts of that sample where the ratio of the parent to daughter element is high enough that the decay-produced excesses are measurable. The first condition is frequently met by using samples from Allende, a large meteorite with inclusions that formed early enough to contain several extinct radionuclides. The second condition is frequently a showstopper. The fractional excess produced by an extinct radionuclide can be written as

$$(D/S_D)_{ex} = (E/S_E) \times (S_E/S_D) \qquad (1)$$

where E is the extinct radionuclide (e.g., ^{26}Al), D is the daughter isotope (^{26}Mg), and S_E and S_D are stable isotopes of the same elements as the extinct radionuclide and its daughter (^{27}Al and ^{24}Mg). Because most extinct radionuclides that have been observed are no more than 10^{-5} to 10^{-3} as abundant as stable isotopes of the same element (E/S_E), that means that elemental ratios (S_E/S_D) of 10 to 1000 are necessary to produce an excess of 1%. Thus it is not surprising that the first two extinct radionuclides confirmed (^{129}I and ^{244}Pu) were the two with noble gas daughter products, and hence large typical

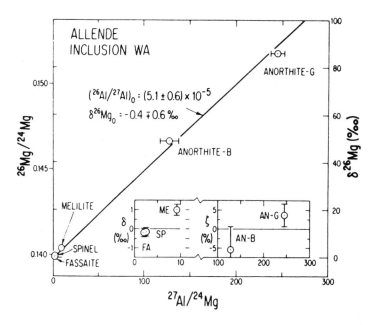

Figure 1. Correlation of $^{26}Mg/^{24}Mg$ with $^{27}Al/^{24}Mg$ in Allende inclusion WA (figure from Lee et al. 1977).

elemental fractionation. Further detections occur as mass spectrometric precision and sensitivity increase, as more highly fractionated samples are found, or both.

B. The Roster

Table I lists the six extinct radionuclides whose presence has been confirmed, along with their half-lives and their daughter products. Wasserburg (1985) and Podosek and Swindle (1988) have reviewed the detection and implications of each of these, so those discussions will not be repeated here.

Databases for two of the six, ^{53}Mn and ^{146}Sm, have expanded considerably since the previous reviews. The first evidence for ^{53}Mn was a correlation of ^{53}Cr excesses (and deficits) in Allende inclusions with higher (and lower) than normal Mn/Cr ratios (Birck and Allègre 1985). Since then, evidence for extinct ^{53}Mn has been found in about a dozen other meteorites by at least four different groups (Birck and Allègre 1988; Birck et al. 1990; Davis and Olsen 1990; Hutcheon and Olsen 1991; Nyquist et al. 1991). Although some evidence for the presence of ^{146}Sm has been available for more than a decade (Lugmair and Marti 1977), early results were of marginal statistical significance, and measurements by different laboratories gave different results. Recently, three different laboratories have reported new evidence for the existence of ^{146}Sm in several different samples (Prinzhofer et al. 1989; Nyquist et al. 1990,1991; Lugmair and Galer 1989), although it is still not

TABLE I
Extinct Radionuclides[a]

Radio-nuclide	$T_{1/2}$ (Myr)	Daughter	Abundance	Comments
			Confirmed	
^{26}Al	0.7	^{26}Mg	^{26}Al/^{27}Al $= 5 \times 10^{-5}$	Shortest $T^{1/2}$; possible heat source
^{53}Mn	3.7	^{53}Cr	^{53}Mn/^{55}Mn $= 4 \times 10^{-5}$	Most recent discovery
^{107}Pd	6.5	^{107}Ag	^{107}Pd/^{108}Pd $= 2 \times 10^{-5}$	Only found in metal
^{129}I	16	^{129}Xe	^{129}I/^{127}I $= 1 \times 10^{-4}$	First discovered; found in >75 meteorites
^{146}Sm	103	^{142}Nd	^{146}Sm/^{144}Sm $= .005 - .015$	Uncertain abundance
^{244}Pu	82	Fission Xe	^{244}Pu/^{238}U $= .004 - .007$	Also α-decays, leaving damage tracks
			Strong evidence	
^{92}Nb	36	^{92}Zr	^{92}Nb/^{93}Nb $= (2 \pm 1) \times 10^{-5}$	Based on single Nb-rich grain
			Hints	
^{60}Fe	1	^{60}Ni	^{60}Fe/^{56}Fe $\leq 2 \times 10^{-6}$	Excess ^{60}Ni observed; could be nucleosynthetic effect
^{135}Cs	2.3	^{135}Ba	^{135}Cs/^{133}Cs $\leq 2 \times 10^{-4}$	Deficit in ^{135}Ba observed; could be nucleosynthetic
^{182}Hf	9	^{182}W	^{182}Hf $= 1-10\times10^{-5}$	Deficit in ^{182}W observed in Hf-depleted metal
			Interesting upper limits	
^{41}Ca	0.13	^{41}K	^{41}Ca/^{40}Ca $< 1 \times 10^{-8}$	
^{247}Cm	16	^{235}U	^{247}Cm/^{235}U $< .004$	Must be cogenetic with ^{244}Pu

[a] From compilation of Podosek and Swindle (1988), with additional data from Prinzhofer et al. (1989), Birck and Lugmair (1988), McCulloch and Wasserburg (1978), and Harper et al. (1991a, b).

possible to reconcile the data from all the laboratories with a single simple scenario.

In addition, Harper and colleagues have recently reported strong evidence for another extinct radionuclide, ^{92}Nb. The evidence consists of the measurement of a clear excess (13σ) of ^{92}Zr in an Nb-rutile sample from the Toluca type IAB iron meteorite (Harper et al. 1991b). While it seems likely that the excess ^{92}Zr was produced by ^{92}Nb decay in this grain, there is no guarantee that a single grain is representative of the bulk solar system.

The number of extinct radionuclides that were actually present in the solar nebula is probably far larger than the seven for which strong evidence exists. Wasserburg (1985) lists another two dozen isotopes that have half-lives comparable to those seven. Searches have established upper limits for many of those. Two of the more interesting upper limits, which are used later in this chapter, are given in Table I. It is likely that more extinct radionuclides will be discovered as a result of improvement in analytical techniques or simply of finding the right samples. Three candidates for which hints already exist are ^{60}Fe, ^{135}Cs and ^{182}Hf. Variations in their daughter isotopes, ^{60}Ni, ^{135}Ba and ^{182}W, have been observed (Birck and Lugmair 1988; McCulloch and Wasserburg 1978; Harper et al. 1991a). For the first two, it has not been possible to determine whether these variations correlate with the parent element. For ^{182}Hf, a deficit of ^{182}W occurs in the (Hf-free) metal phase of the Toluca iron meteorite. In this case, the argument would be that ^{182}Hf decay increased the bulk solar system ^{182}W/^{184}W ratio after the separation of the Toluca metal. Further work will be needed to confirm this story.

III. TIME SCALES

A. Duration from Stellar Nucleosynthesis to Collapse of the Solar Nebula

The first time scale that can be addressed, and the one where the results are most difficult to reconcile with theoretical models, is the length of time from nucleosynthesis to the formation of surviving solids. In principle, an upper limit can be determined from the difference between the nucleosynthetic production ratio and the observed early solar system ratio for any element pair. In practice, there are a number of stumbling blocks along the path of this simple calculation, which are addressed below.

(1) We must know the nucleosynthetic production ratio. Of course, to know that, we must know the production mechanism. Table II lists the stellar sites and/or processes that may be responsible for each of the observed extinct radionuclides. Although the mechanism and production rate seem to be well understood for some nuclides (most notably ^{244}Pu), they are not for others. The identities of the sources of ^{26}Al and ^{146}Sm are particularly controversial.

(2) Even if we can determine a time scale, if the nucleosynthesis occurred locally rather than in a stellar environment, the meaning of that time scale is quite different. At least one of the extinct radionuclides, ^{244}Pu, must have been produced in a stellar environment, as its production requires extremely

TABLE II
Nucleosynthesis of Extinct Radionuclides[a]

Nuclide	Process	Site[b]
^{26}Al	nonexplosive H-burning	MMLS
	explosive H-burning	N
	C- or Ne-burning	SN Type II
	production in	AGB
	spallation	Neb
^{53}Mn	Si-burning or NSE	SN Type I or II
	spallation	Neb
^{92}Nb	p-process	SN
	γ-process	SN
	proton irradiation	Neb
^{107}Pd	s-process	MMLS, AGB
	proton irradiation	Neb
	r-process	SN[c]
^{146}Sm	p-process	SN
	γ-process	SN
^{129}I	r-process	SN[c]
^{244}Pu	r-process	SN[c]

[a] References: Wasserburg and Arnould (1987), Rayet (1987), Kaiser and Wasserburg (1983), Dearborn and Blake (1988), Matthews and Ward (1985), Harper et al. (1991b).

[b] MMLS: Massive mass-losing star (Wolf-Rayet); N: nova; SN: supernova; AGB: Asymptotic Giant Branch; Neb: solar nebula.

[c] Other possible r-process sites are reviewed by Matthews and Ward (1985).

high doses of neutrons that seem impossible to obtain in the solar system. However, many of the others might be produced with high-energy particle fluxes that could exist in the solar system. Shortly after the discovery of the first extinct radionuclide, ^{129}I, Fowler et al. (1961) suggested that it might have been synthesized in the solar system. Although their specific model predicts effects that are not observed (Reynolds 1967), the idea of a local irradiation is still attractive. Wasserburg (1985) has pointed out that ^{26}Al/^{27}Al, ^{53}Mn/^{55}Mn, ^{107}Pd/^{108}Pd and ^{129}I/^{127}I all have observed ratios of $\sim 10^{-4}$, despite half-lives that differ by a factor of 25. Synthesis of 10^{-4} of the nebular material in the nebula itself, perhaps in local irradiations, would be a tidy explanation. For example, Wasserburg and Arnould (1987) have suggested that spallation reactions caused by solar energetic particles might produce both ^{26}Al and

^{53}Mn. This model would then predict that these two extinct radionuclides would correlate with each other, a prediction that will be tested as more data is acquired for ^{53}Mn. The discovery of a γ-ray line due to ^{26}Al in the interstellar medium (Mahoney et al. 1984) suggests that there are stellar sources, but these results do not preclude the possibility of local production.

(3) The surviving grains that we see must have been produced in the solar nebula. If, for example, we are merely seeing grains that condensed around a supernova and then were transported to the solar nebula and incorporated intact into meteorites, the chronological information does not apply to the solar nebula. For most of the confirmed entries in Table I, there is at least one case where the evidence for the existence of the extinct radionuclide comes from a differentiated object with a formation age close to the age of the solar system, and because differentiation normally leads to isotopic equilibration, decay almost certainly occurred in the solar nebula. The only nuclide which has not been found in a differentiated object is ^{26}Al (but see discussion in Sec. III.C), the one with the shortest half-life.

Several models have been developed, in particular by Clayton et al., that would remove ^{26}Al decay from the solar nebula (Clayton and Leising 1987). In one model, the correlation of excess ^{26}Mg with Al would occur because Al is more refractory than Mg. If objects condensed with extremely high ^{26}Al/^{27}Al ratios (0.1), were transported to the solar nebula over some length of time much longer than the ^{26}Al half-life, then heated to temperatures high enough to drive off most of the Mg but little of the Al, and finally mixed with isotopically normal Mg, objects containing random amounts of presolar grains would define a mixing line that would mimic an isochron in Fig. 1. However, this model would predict Mg isotopic fractionation (produced during distillation) that is not observed. Furthermore, it is not clear whether it is physically possible to add normal Mg without equilibrating Mg isotopes, particularly as many of the objects in question have been molten at some point in their histories. Less complicated models fare less well. For example, could macroscopic samples (e.g., Ca-Al-rich inclusions) be presolar? There is no evidence for the spallogenic rare gases that would be produced by interactions with cosmic rays during the interstellar passage, nor is there any strong evidence from other isotopic systems for formation ages that vary by more than a few Myr (see, e.g., Chen and Wasserburg 1981b). While it might be possible to select a temperature high enough to volatilize and/or re-equilibrate noble gases without equilibrating Mg isotopes, it seems unlikely that all the chronometers based on elements other than the noble gases could be reset without affecting Mg.

(4) We must know what fraction of the material in the solar nebula arrived with the extinct radionuclide. For example, for a production ratio of 1 for ^{26}Al/^{27}Al the observed abundance ratio of 5×10^{-5} can occur either because 5×10^{-5} of the ^{27}Al was synthesized with the ^{26}Al and then incorporated in meteorites within a fraction of a half-life of ^{26}Al, or because all the ^{27}Al and ^{26}Al were synthesized in a single event $\sim$10 Myr before incorporation into

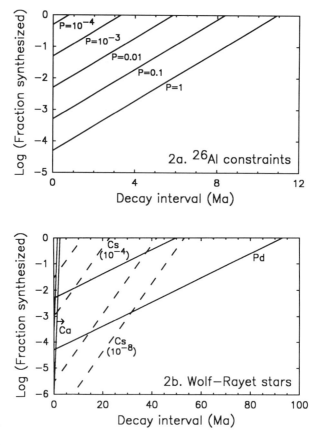

Figure 2. Combined constraint on the decay interval (the time between nucleosynthesis and incorporation into solids) and the fraction of the element synthesized in
the last event. (a) ^{26}Al. The solid lines give the constraints for an early solar system
^{26}Al/^{27}Al ratio of 5×10^{-5} and relative ^{26}Al/^{27}Al production rates P as indicated.
(b) Constraints based on calculations for Wolf-Rayet star nucleosynthesis of ^{41}Ca,
^{135}Cs, and ^{107}Pd (Dearborn and Blake 1988). In each case, the lower line represents
the maximum production rate calculated, and the upper line represents the average
production rate for a 50 $M_\odot$ star. Early solar system ratios of ^{41}Ca/^{40}Ca $= 1 \times 10^{-9}$
(the upper limit), ^{107}Pd/^{108}Pd $= 2 \times 10^{-5}$, and ^{135}Cs/^{133}Cs $= 1 \times 10^{-4}$ and 1×10^{-8}
are plotted. Note that the fraction synthesized does not have to be the same for all
three elements.

solids. The truth is likely to lie somewhere in between (Fig. 2a).

With these caveats in mind, we can evaluate the time scales provided.
Because the extinct radionuclides were probably not all produced at the same
time and place (Table II), the constraints provided by different extinct radionuclides probably apply to different nucleosynthetic events. Thus, as long as
the list of extinct radionuclides does not contain a single pair that has to have

been produced together, it is difficult to use the time scale of one to constrain that of another. However, there are already some useful upper limits, and there are candidate extinct radionuclides that could be extremely useful.

The best constraints are on the actinide-producing r-process. Although ^{244}Pu is the only extinct radionuclide whose production requires such a high neutron flux, the upper limit on shorter-lived ^{247}Cm suggests a decay interval of > 70 Myr between the last significant actinide production and the formation of the solar nebula (Chen and Wasserburg 1981b), while the abundance ratios of longer-lived ^{238}U, ^{235}U and ^{232}Th indicate that actinide production must have occurred over a period of time much longer than the half-life of ^{244}Pu (Schramm 1982).

Similarly, Dearborn and Blake (1988) calculated that nucleosynthesis in a Wolf-Rayet star, followed by 2 to 80 Myr of decay, could satisfy the constraints on both ^{107}Pd and ^{41}Ca (Fig. 2b). These calculations predict the presence of ^{135}Cs, but indicate that Wolf-Rayet stars could not be the source of ^{129}I. Note that there are other possible sites for production of ^{107}Pd.

The mechanism and rate of production of proton-rich nuclides are still uncertain (Rayet 1987). If the detection of ^{92}Nb can be confirmed, and the initial ^{92}Nb/^{93}Nb and ^{92}Nb/^{146}Sm ratios pinned down, this pair of proton-rich nuclides could constrain both the timing and the mechanism of the p- (or γ-) process (Harper et al. 1991b). Local production of ^{92}Nb is a possible complication, as ^{92}Nb could be produced within the solar nebula with proton fluxes that would contribute only 2% of the ^{146}Sm (Harper et al. 1991b). Even without considering ^{92}Nb, the early solar system abundance of ^{146}Sm has led to a reconsideration of γ-process production rate calculations (Woosley and Howard 1990).

The most interesting constraint is provided by ^{26}Al, whose presence seems to demand that some nuclides were synthesized within 10 Myr of their incorporation into the meteoritic objects in which they are now found (Fig. 2a). It is possible that the source of the ^{26}Al γ-ray line, which is still not known with certainty, is also the source of the meteoritic ^{26}Al, although the interstellar ^{26}Al/^{27}Al ratio appears to be lower than the meteoritic one (Clayton and Leising 1987). It also is possible that the time scale has nothing to do with interstellar transport, if either the nucleosynthesis was nebular or if the ^{26}Mg excesses were fossil. It is worth noting that some of the strongest suggestions that the chemistry be reinterpreted come from an astrophysical perspective (see, e.g., Clayton and Leising 1987), while some of the strongest suggestions that the astrophysics be reinterpreted come from a chemical perspective (see, e.g., Wasserburg 1985).

Models in which 10^{-3} to 10^{-4} solar masses of newly synthesized material is added to the solar system in a single event shortly before formation of meteorites can account for the common 10^{-4} ratio of radionuclides to their stable neighbors (Birck and Allègre 1988), but it does not seem astrophysically reasonable to have all extinct radionuclides synthesized by the same star. In his chapter, Cameron shows that it is possible to construct a series of individually plausible astrophysical events that can reproduce the solar system abundances of the extinct radionuclides. Whether his specific scenario is, as a whole, plausible remains to be seen, but there is now sufficient information available that better and better models can be constructed and tested.

B. Duration of the Solar Nebula

In principle, we could set a lower limit to the duration of the solar nebula if we could determine the elapsed time between any two nebular events. Since that time is probably a few Myr or less, the most promising chronometers are extinct radionuclides, particularly ^{129}I and ^{26}Al, for which the most data exist. In practice, we cannot set any limits, both because of the difficulty of identifying isotopic systems that have been affected only by nebular events and because of uncertainties about the extent of isotopic inhomogeneity.

The condensation of CAIs and formation of chondrules are both high-temperature nebular events that could be expected to equilibrate the Xe and Mg isotopic systems (i.e., reset the clocks). For the I-Xe system, the difference in apparent age between the oldest (CAI-rich) carbonaceous chondrite, Vigarano, and the oldest chondrules from unequilibrated ordinary chondrites is ∼5 Myr (Swindle and Podosek 1988). However, the I-Xe system is based on two volatile elements, so processes such as aqueous alteration, mild thermal metamorphism and shock all might be capable of resetting the clock. Furthermore, although there are cases where correlations with elemental and isotopic parameters that indicate the I-Xe system is varying because of decay rather than inhomogeneity, this is not one of them.

The Al-Mg system is less likely to be susceptible to such secondary processes, but is demonstrably susceptible to Al isotopic heterogeneity. This was suspected based on the existence of petrographically similar inclusions with grossly different $^{26}Al/^{27}Al$ ratios (Wasserburg and Papanastassiou 1982). The strongest confirmation of heterogeneity comes from an inclusion in the carbonaceous chondrite Efremovka that has a $^{26}Al/^{27}Al$ ratio that is an order of magnitude lower than that of the rim, which surrounds the inclusion and must post-date it (Fahey et al. 1987). Thus, although differences of three orders of magnitude in $^{26}Al/^{27}Al$ ratios (which would take several Myr of decay) are observed among nebular objects, they may not say anything about the duration of the nebula.

C. Time from First Solids to First Differentiation

A time scale that is easier to address is that from the formation of the first solids to the accretion of bodies large enough to differentiate. In this case, we can set a lower limit by comparing an age set by a nebular event with one set by a differentiation event. Comparisons based on four different radionuclide systems are summarized in Table III. The time scale seems to have been less than 10 Myr, perhaps as short as 2 Myr.

Dating techniques based on long-lived radionuclides such as U isotopes can provide some information. Tilton (1988) has reviewed the subject, con-

TABLE III

Time from Early Solids to Differentiation

Technique	Primitive Sample	Differentiated Sample	ΔT (Myr)
U-Pb ages	Allende (4559 Myr)	Angra dos Reis (4551 Myr)	8
$^{53}Mn/^{55}Mn$	Allende (4.4×10^{-5})	Springwater (1.4×10^{-5})	6
$^{129}I/^{127}I$	Vigarano (1.60×10^{-4})	Happy Canyon (1.16×10^{-4})	7
$^{26}Al/^{27}Al$	Allende (5×10^{-5})	Semarkona chondrule (8×10^{-6})	2

cluding that comparison of Pb-Pb ages of Angra dos Reis (a differentiated meteorite) with Allende (a primitive meteorite) provides the best data. The range in apparent ages for Allende ($\sim$15 Myr) is larger than can be explained by statistical uncertainties, and may mean that sampling is important (i.e., that Allende contains objects with a range of ages). The ages of Angra dos Reis are in better agreement. Even taking the most extreme values, the difference in age between the two meteorites is <20 Myr. Tilton (1988) argues that 8 Myr is the best estimate.

Extinct radionuclides provide more precise apparent ages. The ^{53}Mn/^{55}Mn ratio in the (differentiated) pallasite Springwater (1.4×10^{-5}; Hutcheon and Olsen 1991), when compared to the ratio in Allende (4.4×10^{-5}; Birck and Allègre 1988), gives an apparent age difference of 6 Myr. The presence of ^{107}Pd in iron meteorites (Chen and Wasserburg 1990) also suggests a time scale of a few times 10 Myr or less. However, without knowing the ^{107}Pd/^{108}Pd ratio in any undifferentiated object (see Chen and Wasserburg 1991), it is impossible to be more quantitative than that.

For ^{129}I, Swindle and Podosek (1988) have suggested that the oldest well-defined apparent age comes from the meteorite Vigarano. Vigarano belongs to the same class, CV3, as the canonical "oldest" meteorite Allende (most I-Xe studies of Allende objects have produced variable younger apparent ages, probably dating secondary alteration involving halogens rather than nebular condensation). Several differentiated meteorites contain ^{129}I-derived Xe, with the highest ^{129}I/^{127}I ratios found in the enstatite achondrites Happy Canyon and Shallowater. These correspond to apparent ages $\sim$7 Myr later than Vigarano.

The most precise apparent ages can be provided by the shortest-lived extinct radionuclide, a ^{26}Al. However, there is only one case, an anorthite-bearing chondrule in the ordinary chondrite Semarkona, where the presence of ^{26}Al has been detected in what might be a differentiated object. Hutcheon and Hutchison (1989) found a ratio of ^{26}Al to ^{27}Al of $\sim$8 $\times$ 10^{-6}, which would correspond to an apparent age of 1.9 Myr after typical Allende inclusions (5×10^{-5}). However, this chondrule (and similar objects in Allende) could have been products of nebular processes, rather than differentiation (see, e.g., Kring and Boynton 1990), so it may not be telling us anything at all about the time scale for differentiation. Searches in unambiguously differentiated meteorites have so far failed to find any evidence for live ^{26}Al (see, e.g., Bernius et al. 1991).

The possibility of isotopic heterogeneity should not be forgotten when trying to infer ages from the presence of extinct radionuclides, and ages based on any single radionuclide are certainly suspect. However, unless natural processes have conspired to have lower intrinsic abundances of ^{26}Al, ^{53}Mn and ^{129}I in the region of formation of the carbonaceous chondrites than in the regions of formation of the ordinary chondrites, pallasites and enstatite achondrites, respectively, a 10 Myr upper limit on the time between formation of early solids and differentiation is valid.

D. Time from First Solids to Planets

There are only two planet-sized solar system bodies, the Earth and its moon, for which we have any chronologically significant isotopic data. In each case, ages of 50 to 100 Myr after the formation of primitive meteorites are suggested (Swindle et al. 1986a; Carlson and Lugmair 1988).

   Two different approaches have been applied to the problem of the age of the Earth. The first approach, pioneered by Patterson (1956), involves identifying times of early U-Pb differentiation. Such calculations give ages ranging from contemporaneous with meteorites to ~100 Myr later. However, this first differentiation could be within the planetesimals from which the Earth formed, rather than the time of formation of the Earth as a planet. The second makes use of the apparent presence of ^{129}I-derived Xe (and, in some models, ^{244}Pu-derived Xe) in the Earth's atmosphere, and dates the time of the last loss of Xe from the Earth's atmosphere or the material that was destined to become the source of the atmosphere. Determination of the abundance of the decay products in the atmosphere is dependent on models of primordial Xe isotopic composition, and the age calculations are extremely sensitive to the poorly constrained I and actinide abundances of the atmospheric source. As in the case of the U-Pb system, ages of ~100 Myr after meteorite formation are typical, but contemporaneous ages cannot be excluded (Wetherill 1975b).

For the Moon, the U-Pb and I-Pu-Xe systems can be used in slightly different fashions, and there are some rocks that may be close to the age of the Moon itself. Analyses of the U-Pb system suggest differentiation events at 4420 Myr and perhaps 4590 Myr (Wasserburg et al. 1977; Tera and Wasserburg 1976). The I-Pu-Xe chronometer can, in principle, be applied to rocks which have ^{129}I- and ^{244}Pu-derived Xe, unaccompanied by the ^{129}I or actinides that would be expected for *in situ* decay. This Xe, which is found only near the surfaces of grains, has apparently been produced by decay within the Moon and then transported to the lunar surface where it was somehow implanted into the rocks where it is now found. Swindle et al. (1986a) have suggested that the ratio of ^{129}I- to ^{244}Pu-derived Xe in some samples suggests that the decay occurred in the Moon no more than 50 to 100 Myr after the time of formation of meteorites. Finally, there are some very old lunar rocks. In particular, there are two rocks (76535 and 60025), both of types expected to be created in the earliest lunar crust, for which two different dating methods give an age of 50 to 120 Myr after the time of formation of meteorites (Table IV). However, in each case, the ages derived by different methods differ by more than the statistical errors, leaving the results open to interpretation (Carlson and Lugmair 1988).

Thus for both the Earth and Moon, formation intervals of ~50 to 100 Myr after the formation of primitive meteorites are suggested, in agreement with accretion calculations (Wetherill 1980a). However, there are large enough uncertainties in the calculation of these formation intervals that the agreement

TABLE IV

Ages of the Oldest Lunar Rocks

Sample	Age (Myr)	Technique	Reference
76535	4340 ± 80	K-Ar	Bogard et al. 1975
	$4080 - 4230$	K-Ar	Huneke and Wasserburg 1975
	4260 ± 20	K-Ar	Husain and Schaeffer 1975
	4260 ± 60	Sm-Nd	Lugmair et al. 1976
	4530 ± 70	Rb-Sr	Papanastassiou and Wasserburg 1976
	4510 ± 80	Pu-REE-Xe	Caffee et al. 1981
60025	4510 ± 10	U-Pb	Hanan and Tilton 1987
	4440 ± 20	Sm-Nd	Carlson and Lugmair 1988

with calculated time scales should not be taken as proof that the calculated time scales are correct.

IV. SUMMARY

From the study of extinct radionuclides, we calculate the following time scales:

1. The time between nucleosynthesis and formation of surviving meteoritic grains was <10 Myr for at least 10^{-4} of the Al in some meteorites. Unless virtually all the Al was synthesized in a single late event, the time was probably no more than a few Myr. Although suggestions of local nucleosynthesis and of incorporation of non solar system grains have been made, it seems likely that these time scales apply to the transport of freshly synthesized material from another star to the solar nebula. Other extinct radionuclides set different, less stringent, limits, but as production probably occurred in different settings, these need not be dating the same event.
2. We cannot set any firm constraints on the duration of the nebula. Major confounding factors are the possibility of isotopic inhomogeneities in the solar nebula (particularly for ^{26}Al) and a lack of understanding of what nonnebular events could reset isotopic clocks (particularly for ^{129}I).
3. The time from the formation of surviving solids to the formation of bodies large enough to undergo differentiation is almost certainly <10 Myr, and may be as short as 2 Myr.
4. The time from the formation of solids to the formation of the Earth and Moon seems to be 50 to 100 Myr. Most of these calculations are quite uncertain, but the results are comfortingly close to theoretical predictions.

Future work that is likely to occur includes:

(a) Identification of the mechanisms and sites of production of extinct radionuclides. It is particularly important to understand whether some of these could have been synthesized within the solar nebula. If not, then the challenge is to understand how ^{26}Al was transported into the solar nebula so quickly (see Cameron's Chapter).

(b) Determination of the distribution of extinct radionuclides within the solar nebula. Demonstrations of isotopic homogeneity, if possible to achieve, might make it possible to address some of the early time scales with more confidence, while demonstrations of inhomogeneity could be used to constrain the amount of mixing within the nebula. The distribution of ^{26}Al is particularly important, not just because of its chronological significance, but because it is a potential heat source in early solar system bodies. The search for clear evidence of ^{26}Al in clearly differentiated objects is certain to continue.

(c) Identification of more extinct radionuclides. This is more than just an exercise in expanding Table I. A single extinct radionuclide, even if its production rate is well known, can only constrain the combination of (1) time since nucleosynthesis, and (2) fraction of material synthesized. Adding a second radionuclide that was produced at the same time and place can resolve that ambiguity. The current list does not contain a single pair of extinct radionuclides that *must* have been synthesized together, although there are several pairs that *may* have been synthesized together. Although there are some upper limits for the abundance of some potential extinct radionuclides that can be paired with confirmed extinct radionuclides to provide limits on transport times (Fig. 2), actual values will be even more useful.

PHYSICAL AND ISOTOPIC PROPERTIES OF SURVIVING INTERSTELLAR CARBON PHASES

U. OTT

Max-Planck-Institut für Chemie

Primitive meteorites contain grains that clearly represent surviving interstellar material. These grains are carbonaceous in the three cases that are well documented and are discussed here: tiny (size $\sim0.003\,\mu m$) diamonds, graphitic carbon and silicon carbide. Unambiguous proof for the interstellar origin of these grains is their isotopic composition, which not only for the trace elements they carry (notably the noble gases) is unusual, but also for the structural elements (C, N, Si) is both unusual and (at least in the case of the graphite and SiC grains) variable on a grain-to-grain basis, pointing to an origin from specific nuclear sources. The survival of these grains as well as their ability to retain the volatile noble gases that we find in them puts constraints on the temperature and/or the chemical environment to which they have been exposed. It remains to be shown, however, that they were part of the solar system during its formative stage if, from their survival, one wants to draw conclusions regarding conditions during solar system formation.

I. INTRODUCTION

It has become clear during the past 2 decades, through the study of the isotopic composition of meteorites, that the matter from which the solar system formed cannot have been totally homogenized at any one stage. Rather, isotopic heterogeneity on a small scale (refractory Ca-Al rich inclusions in carbonaceous meteorites; see, e.g., Papanastassiou and Brigham 1989; Ireland 1990) and on a large scale (variations in the oxygen isotopic composition among different meteorite parent bodies; Clayton et al. 1976) appears to have been the rule despite evidence that, in establishing the *elemental* composition of these objects, high-temperature events before or during their formation must have played a crucial role (cf. Ireland 1990; Chapter by Palme and Boynton).

In addition, work performed during the past 3 yr (i.e., after *Protostars and Planets II*) has identified refractory carbon phases as the carriers of isotopically anomalous noble gases which testifies to the almost intact survival of whole grains of interstellar material. The properties of such refractory carbon phases will be described in this review. Unless specified otherwise, properties of the phases of interest such as their abundances etc. refer to the observations on the C2M meteorites Murray and Murchison, where their study has been most complete. For another recent survey see Anders (1988).

II. CONDITIONS FOR SURVIVAL OF INTERSTELLAR MATTER AND ITS IDENTIFICATION

Before a detailed discussion of specific phases of surviving interstellar matter, we shall address the question of what one might expect to find, that is, the likelihood of interstellar matter surviving and being identified as such.

A. Identification

The only generally accepted way to prove an interstellar origin of matter that we find in meteorites today (and, hopefully, in the future on other primitive bodies such as comets) is by means of its isotopic composition. Only if it differs from the common, solar-system composition in a way not explicable by prosaic processes such as mass fractionation, radioactive decay or cosmic ray interaction, can we be reasonably sure about its interstellar origin. Because in the region of space where the solar system formed the average composition of interstellar matter 4.6 Gyr ago must have been (for the nuclides not affected by these prosaic processes) just that of the solar system now, our ability for identifying interstellar matter as such is surely restricted. In addition, we have to distinguish between matter formed *in* the solar system from insufficiently mixed presolar material and truly surviving presolar matter. To identify the latter without any doubt may be possible only in the case of direct stellar condensates, which still carry the isotopic signature of a single nucleosynthetic source.

B. Survival

Two properties appear useful. First, the matter of interest should be refractory (i.e., temperature-resistant). And second, it should be chemically resistant (inert) under the conditions prevailing in the interstellar medium, during formation of the solar system and during evolution of the meteorite parent body, which has been the host until its delivery to Earth via a meteorite. (Chemical resistivity in the laboratory, i.e., resistance to acids, has also been essential in the isolation of the presolar carbon phases discussed here.) Based on these requirements, one can expect basically two forms of matter to have survived and to be identifiable as interstellar: refractory oxides and refractory carbon compounds, which are the chemically stable forms for different C/O ratios. For a solar system-like C/O ratio (0.42; Anders and Grevesse 1989) the oxides, not the carbon compounds, are the stable phases. However, it is for the carbon phases, not the oxides, that the case for survival of presolar matter is strongest. Isotopic anomalies in refractory oxides have been found in the high-temperature Ca-Al-rich inclusions (CAI's) in carbonaceous meteorites (see, e.g., Papanastassiou and Brigham 1989; Ireland 1988,1990); but, although the isotopic anomalies there appear not to have been fully wiped out, the CAI's and their constituents do *not* represent unaltered, presolar material, and may constitute a case of formation *in* the solar system from insufficiently mixed presolar material (MacPherson et al. 1988; Ireland 1990). In this review, therefore, we concentrate on the C phases.

In addition to the other isotopic properties that will be discussed below in detail, one of the strongest arguments that sets aside the refractory carbon phases as almost certainly presolar against these oxides is the level in which extinct radioactive ^{26}Al ($T_{1/2} = 7 \times 10^5$ yr; now showing up as an excess of stable ^{26}Mg) occurs in them (Zinner et al. 1991). Ratios ^{26}Al/^{27}Al at the time of formation of these phases (or introduction of Al into them) of up to 0.2 have been inferred from the measurements on graphite and silicon carbide; this is close to the expected production ratio and requires almost immediate formation of these phases (or introduction of the Al) after the nucleosynthetic event that produced this Al. In contrast, the corresponding value of typically 5×10^{-5} for the oxides in the refractory inclusions is 4000 times lower, leaving some extra 10 Myr time between formation and fixation of the observed ratio.

Besides the refractory carbon phases which are likely to be of circumstellar origin (Anders 1988), there is evidence for the presence of (also carbon-bearing, but more volatile) material for which an origin in interstellar *clouds* is indicated (Zinner 1988), notably because of its high D/H ratio. In addition, there are at least two cases of clearly strongly processed material possibly carrying extreme isotopic anomalies: Ag in iron meteorites (Kaiser and Wasserburg 1983) and nitrogen in the stony-iron meteorites Bencubbin and Weatherford (Prombo and Clayton 1985; Franchi et al. 1986). However, a prosaic origin, as defined above, of the large deviations from the isotopic normals has not yet been ruled out in these cases.

III. NOBLE GAS COMPONENTS AND THEIR ISOLATION

The identification of carbonaceous interstellar matter has been largely the result of the quest for the host phases of isotopically anomalous trapped (i.e., not *in situ* produced) noble gases. Noble gases are easier to separate, study and characterize than other elements (Anders 1988), and, especially important, are "diluted" to a much smaller extent by "normal" matter because even the most primitive bulk meteorites contain only a small fraction ($\sim 10^{-4}$ for Xe to $\sim 10^{-9}$ for He) of their solar abundances. For that very reason, at least in bulk matter isotopic variations in noble gases are quite often orders of magnitudes larger than the isotopic variations in the rock-forming elements (Clayton et al. 1988; Lee 1988) which are usually measured in per-mill (10^{-3}) or ϵ-units (10^{-4}) deviations from the normal. As discussed below in more detail, the anomalous noble gases reside in a small amount of very resistant carbonaceous carrier material. As such they have served as the guide in the isolation of these phases, in which isotopic anomalies found in other elements are actually of the *same* order of magnitude as those of the noble gases. Along the way, more phases with distinct C isotopic compositions have been found, but no connection to other elements has been established yet (see Ash et al. 1990). As only the noble-gas-carrying phases have been identified so far, we restrict the discussion to these. Contributions from other local sources (^{4}He and ^{40}Ar of radiogenic origin, ^{3}He and Ne of cosmogenic origin in the case of the noble gases) are usually only a minor complication in the isotopic study of these phases.

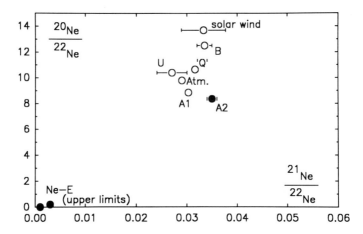

Figure 1.   Composition of important Ne components in the solar system in a 3-isotope $^{20}Ne/^{22}Ne$ vs $^{21}Ne/^{22}Ne$ diagram. The presolar phases discussed here carry the Ne components marked by filled circles: Ne-A2 (carried by diamonds) and Ne-E (carried by graphite and SiC). Compositions as summarized in Swindle (1988) except for Q-Ne (Wieler et al. 1989) and Ne-A1 (Tang and Anders 1988c). See also Table I.

Most prominent are the isotopic variations in Ne and Xe. Although, formally, for an element with 3 isotopes, any isotopic composition can be described by a mixture of 3 extreme components, there is compelling evidence for the existence of more than 3 distinct components of trapped Ne the relation between which is not well understood (see, e.g., Swindle 1988). Figure 1 shows the position of major components found in solar system materials in a 3-isotope diagram of $^{20}Ne/^{22}Ne$ vs $^{21}Ne/^{22}Ne$. Note that in this representation with a common denominator mixtures of 2 components lie along a straight line connecting the end-member compositions. Most spectacular is Ne-E, which is almost pure ^{22}Ne (Table I). Based on their occurrence in different host phases (see below) two varieties, Ne-E(L) and Ne-E(H) have been distinguished; there appears to be also a slight difference in isotopic composition, with Ne-E(H) containing finite amounts of ^{21}Ne (Lewis et al. 1990).

The case of Xe with its large number of isotopes (9) is most complex; for a detailed discussion see the review by Swindle (1988). Figure 2 shows the composition of the 2 isotopically anomalous Xe components which are most interesting in the context of this chapter. Plotted is also AVCC-Xe, which is not a pure component but the composition measured for total Xe in carbonaceous meteorites of type 1 and 2. Shown are the deviations in $°/_{oo}$ of the ratios $^{i}Xe/^{130}Xe$ from the same ratios in solar wind xenon as identified in lunar soil. ^{129}Xe has been omitted because of possible radiogenic contribu-

TABLE I

Planetary Ne and Xe Components in Carbonaceous
Chondrites and Their Host Phases[a]

Ne Comp.	^{20}Ne/^{22}Ne	Xe Comp.	^{136}Xe/^{132}Xe	Host Phase
Q-Ne	10.65	Q-Xe	0.31	Q
Ne-A1	8.86	—	0.31	Cξ
Ne-A2	8.37	Xe-HL	0.65	Cδ, diamond
Ne-E(L)	0.01	—	—	Cα, graphitic
Ne-E(H)	< 0.28	—	—	Cε, SiC
—	—	Xe-S	0	Cβ, SiC

[a] Source: Anders (1988) except for isotopic composition of Q-Ne (Wieler et al. 1989) and Ne-Al (Tang and Anders 1988c).

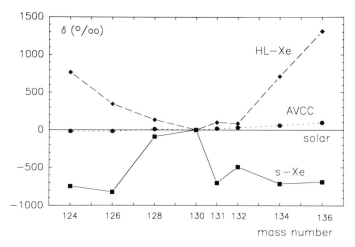

Figure 2. Exotic Xe components Xe-HL (carrier: diamond) and s-process Xe (carrier: SiC) compared with solar Xe as inferred from measurements of solar wind Xe implanted in lunar soil (Swindle 1988). Shown are the deviations in °/$_{oo}$ in the ^{130}Xe-normalized isotopic ratios from the solar composition. Note that not pure inferred endmember compositions are shown but highly extreme *measured* compositions that are still diluted to some extent with "normal" Xe: Xe in A-II-D/HClO$_4$ (Ott et al. 1981) for Xe-HL and Murchison CFOc/1080°C (Ott et al. 1988b) for Xe-S. Also shown for comparison is the composition of AVCC-Xe, not a pure component, but what is measured in bulk analyses of C1 and C2 carbonaceous chondrites.

tions from now-extinct ^{129}I with a 16 Myr half-life. Note that what is plotted for the exotic components Xe-HL and Xe-S are the *measured* ratios for samples extremely enriched in these components; the inferred compositions of the pure components would be even more extreme. Xe-HL is characterized, relative to ^{130}Xe and the solar wind composition, by an overabundance of the light and heavy Xe isotopes by up to a factor >2. Xe-S shows approximately, though not in detail, opposite deviations from solar Xe. Relative to ^{128}Xe and ^{130}Xe, which are produced in the s-process only (with minor p-process contributions), all other isotopes are strongly depleted. The pattern is consistent with the addition to isotopically normal xenon of Xe as it is produced in the slow neutron capture process (s-process) of nucleosynthesis.

A major breakthrough in the search for the host phases of both exotic and isotopically ordinary noble gases occurred when their resistance to acids was noted by the Chicago group of Anders and coworkers (Lewis et al. 1975) in their work on the carbonaceous chondrite (C3V) Allende. [A similar, earlier, result for the carrier phase of the trapped noble gases in the ureilite Haverö has been obtained by Weber et al. (1971), but ureilite meteorites do not contain isotopically anomalous noble gases.] Later work (Alaerts et al. 1980; Lewis et al. 1987; Tang and Anders 1988a) showed that almost all the phases of interest were resistant to extremely harsh chemical treatments (with the notable exception of the host phase of Ne-E(L); see Tables I and II).

The procedures leading to their isolation can be summarized as consisting of 3 basic steps:

1. Dissolution of ~99% of meteoritic material (the exact amount depending on meteorite type) in HF/HCl;
2. Use of oxidizing agents that remove (the carriers of) the normal type trapped noble gases (Q; Table I);
3. Separation of the exotic components from each other according to physical properties such as grain size and density.

Most of the work described here has been done by the Chicago group of Anders and coworkers (isolation of carrier phases, noble gas work) in collaboration with the group of Pillinger at Milton Keynes, U. K. (formerly Cambridge, U. K.; C and N isotopes) and E. Zinner and coworkers at St. Louis (ion probe analyses of C, N, Si, Mg isotopes). The author's own group has been involved in the isotopic characterization of the heavy noble gases Kr and Xe and other so-called heavy elements.

In some way, results of the work since 1975 are reminiscent of nested Russian dolls, with the noble gas of interest always contained in another subfraction of the isolated material. Table II gives a summary of our current understanding of the compositions of the Ne and Xe components and the host phases assigned. Their physical properties are summarized in Table II. Below, each is discussed in detail. Of these phases, diamond (Cδ) and SiC (Cβ, Cϵ) appear to be fairly well characterized by now (Anders 1988; Lewis et al. 1989; Tang et al. 1989), with progress being currently made in the case of Cα (Amari et al. 1990a). But even within each of these phases there still exist isotopic variations, as has been shown by ion probe work (Zinner et al. 1989,1990), which may not be surprising given the fact that there may be a number of sources (stars) for these phases (see below). In addition, as pointed out above, more phases distinct in the isotopic composition of C, but

TABLE II

Physical Properties of Host Phases[a]

Host Phase	Gas Comp.	Grain Size μm	Release-T[b] °C	Combust.-T °C	$\delta^{13}C_{PDB}$ ⁰/₀₀	Abundance μg g^{-1}
Cα	Ne-E(L)	1–10	700	600	see text	~ 1, graphite
Cβ	Xe-S	0.1	1400	1000	see text	~ 3, SiC
Cδ	Xe-HL	0.003	1000	500	−32	~ 400, diamond
Cε	Ne-E(H)	0.1–1	1200	1000	see text	~ 2, Sic

[a] Source: Anders (1988) except for abundance of Cα (Amari et al. 1990a) and the carbon isotopic composition of Cδ (Ash et al. 1990).

[b] Release-T refers to the temperature where major release of noble gases occurs during laboratory analyses (heating in vacuum). Ion probe measurements have shown the C isotopic composition of SiC and Cα graphite to vary from grain to grain. No comparable information is available for the Cδ diamonds because of their small size.

with unknown relation to other elements, have been identified by stepwise combustion experiments (see summary by Ash et al. 1990).

IV. DIAMOND (Cδ)

Identification of the Cδ phase (Swart et al. 1983) that carries Xe-HL and Ne-A2 as being a form of diamond had to wait until 1987 (Lewis et al. 1987), 12 yr after the discovery of its acid resistance (Lewis et al. 1975). The principal reason for the surprisingly long time span was probably its extremely small grain size of only ~2.6 nm on the average (Lewis et al. 1989). The log-normal size distribution contrasts with the inferred power-law distribution for interstellar dust (Mathis et al. 1977).

Diamond is by far the most abundant of the anomalous gas-carrying phases. In carbonaceous meteorites of type C2, it amounts to something like $400 \, \mu g \, g^{-1}$ of bulk meteorite. It has also been observed in the less primitive type 3 meteorites (carbonaceous as well as ordinary and enstatite types), although in lower abundance (Schelhaas 1987; Huss 1990; Schelhaas et al. 1990). Under laboratory conditions, noble gas release occurs at around 1000°C; in an oxygen atmosphere it combusts at around 500°C. The somewhat lower combustion temperature as compared to terrestrial diamond is probably caused by its small grain size (Ash et al. 1987), but possibly also its H-rich composition (Bernatowicz et al. 1990) plays a role. In the case of the other 2 carrier phases discussed below, graphite and SiC, grain-to-grain variations in the isotopic composition of the major elements (C, N, Si) have been found. Similar information on the diamonds cannot be obtained with currently available techniques because of their small grain size. Their inferred carbon isotopic composition, which therefore could be only a mean value, is within the range observed in terrestrial materials and (by itself) does not provide a basis for identification as interstellar material. $\delta^{13}C_{PDB}$ is $-32 \, ^{o}/_{oo}$ (Ash et al. 1990), i.e., $^{13}C/^{12}C$ is $32 \, ^{o}/_{oo}$ lower than in the terrestrial PDB standard or $^{12}C/^{13}C = 92$. Nitrogen carried by the diamonds, however, is clearly isotopically unusual. The deviation of $^{15}N/^{14}N$ from the AIR value, i.e., $\delta^{15}N$, is $-330 \, ^{o}/_{oo}$ (Lewis et al. 1983a; Anders 1988), which corresponds to $^{14}N/^{15}N = 406$ rather than the normal 272.

A. Compositional Properties

The compositional properties of the meteoritic microdiamonds have been extensively described in a review by Lewis et al. (1989). In aqueous solution, the diamonds behave like a weak acid, forming colloids at high pH values and coagulating at lower pH. The presence at the surface of weakly acidic groups such as $-COOH$ has been suggested as an explanation. The presence of $-COOH$ may also be responsible (in part) for the O-H peak at 3402 cm^{-1} and the C=O feature at 1777 cm^{-1} in the infrared spectrum (Lewis et al. 1989). There are also other infrared features that can be attributed to C-H and nitrogen (or C-O).

Other spectral data indicate an amorphous character of (part of) the diamonds (Blake et al. 1988; Bernatowicz et al. 1990). Electron energy loss spectra (EELS), in particular, suggest that the diamond residues actually consist of a mixture of diamond proper and some form of amorphous carbon, probably "diamond-like hydrocarbon," in roughly equal proportions (Bernatowicz et al. 1989,1990).

B. Formation of Diamond, Cδ

In principle, 2 mechanisms appear possible: formation from graphitic precursor material under high pressure (static or shock) or direct condensation from the vapor phase (chemical vapor deposition; CVD). Static high pressure can be safely ruled out. Formation by interstellar shock has been suggested by Tielens et al. (1987), and comparison was made by Greiner et al. (1988) to diamonds produced as solid detonation products. Formation by CVD as suggested by the discoverers (Lewis et al. 1987) appears more likely, however. The available evidence is discussed by Anders (1988) and Lewis et al. (1989). Their conclusions are supported by the result of recent calculations by Badziag et al. (1990) which indicates that CVD diamond may be not the metastable condensation product as previously thought but rather the thermodynamically stable product. According to these authors, for grain sizes ≤ 3 nm, the formation of diamonds is energetically favored over that of polycyclic aromatics (thought to be graphite precursors). Such a formation process may also imply a closer connection between the isotopic compositions of the trapped noble gases and that of their host phase.

C. Origin of Xe-HL

First suggestions for the origin of Xe-HL (formerly also called CCF-Xe, CCFX, Xe-X and DME-Xe) were based on the observed enrichments in the heavy Xe isotopes and called for a fission origin of these. Because the fission spectrum did not match any of the known fission spectra, *in-situ* fission of a superheavy element was suggested (Srinivasan et al. 1969; Anders and Heymann 1969; Dakowski 1969; Anders et al. 1975; Anders 1981). This interpretation, however, has been discarded meanwhile. Following Manuel et al. (1972), who pointed out the need to account for the excesses not only of the heavy, but also of the light (not fission-produced) Xe isotopes, a nucleosynthetic origin is generally accepted since at least 1983. Crucial observations in that year were (a) the detection of a nitrogen isotopic anomaly in the Xe-HL carrier phase (Lewis et al. 1983*a*) and (b) the negative result of Lewis et al. (1983*b*) in their search for CCF-Ba associated with CCF-Xe (i.e., Xe-HL). Such a result could not be explained by the *in situ* fission model, where comparable amounts of fission-Ba and fission-Xe were expected to be present. Lewis et al. obtained an upper limit of 0.03 for the ratio HL-^{135}Ba/HL-^{136}Xe, but in retrospect it appears that in their work (which was performed *before* the identification of Cδ as diamond) the diamonds proper could not have been analyzed for their Ba. New data on diamonds from the Allende meteorite give only an upper limit of $\sim$1 (U. Ott, unpublished), but it seems likely that a limit lower than this can be set after more complete separation of extraneous Ba from the diamonds.

Formation of Xe-HL during nucleosynthesis appears to require the action of a supernova. The most extensive studies have been performed by Heymann and Dziczkaniec (1979,1980). Other, later work (see, e.g., Clayton 1989) has mostly focused on the H part (heavy isotopes) of Xe-HL only. Heymann and Dzieczkaniec have investigated explosive nucleosynthesis in a massive star of 25 $M_\odot$. According to their results, Xe-H can be produced in the C shell by neutron capture reactions, a "mini-r-process" intermediate between the slow s and rapid r neutron capture processes of nucleosynthesis that phenomenologically describe the synthesis of the bulk of the heavy nuclides in their solar system abundances. Xe-L could have been produced in the same star in deeper zones (O-shell) at higher temperatures (2.7×10^9 K) by photodisintegration of a Xe/Ba seed.

D. Trapping Process

Lewis and Anders (1981) have pointed out that there is no evidence for a significant ^{129}Xe enrichment in Xe-HL. Such an enrichment would be expected if present-day HL-^{129}Xe had been largely trapped in the form of its progenitor ^{129}I, and if the trapping process were condensation or adsorption as these processes would favor chemically active elements over the noble gases (Anders 1988). This reasoning is also supported by the lack of enrichment of Ba-HL relative to Xe-HL. Anders (1988) suggests trapping by ion implantation in a scenario proposed by Clayton (1981), where the hot plasma of a supernova, containing Xe-HL, overtakes a dust shell previously expelled during the star's red giant or planetary nebula stage. However, according to Jørgensen (1988) Xe-HL and the Cδ diamonds that carry it cannot have been produced by the same star. He suggests that both were produced in the two stars of a close binary system, the diamonds in the smaller of the two companions near the end of its evolution, Xe-HL in the compact remnant (white dwarf) of the bigger star, after it accreted enough mass from its companion to explode as a type I supernova. However, clearly more work on the nucleosynthetic processes and possible sites needs to be done. Only then will we be able, by building on the observed properties of the surviving interstellar diamonds in meteorites, to identify also the trapping process.

V. GRAPHITIC CARBON (Cα)

Being less resistant to chemical attack than diamond or silicon carbide and, at the same time, being much rarer ($<1\ \mu$g g^{-1} bulk C2 meteorite), Cα, the host phase of Ne-E(L) has proved to be the most difficult one to characterize, but progress is currently being made. First results indicated Cα to be a graphitic phase, being destroyed by HClO$_4$ treatment at 200°C. In the laboratory, a combustion temperature for Cα of ~650°C (Ash et al. 1990) and a Ne-E(L) release temperature, upon heating in vacuum, of ~700°C were inferred (Anders 1988). Stepwise combustion analyses indicated both carbon and nitrogen to be isotopically unusual. The inferred compositions (Table II) are $\delta^{13}C_{PDB} \sim +340\,^o/_{oo}$ and $\delta^{15}N_{AIR} > +250\,^o/_{oo}$, corresponding to $^{12}C/^{13}C \sim 66$ and $^{14}N/^{15}N < 220$ (Tang et al. 1988).

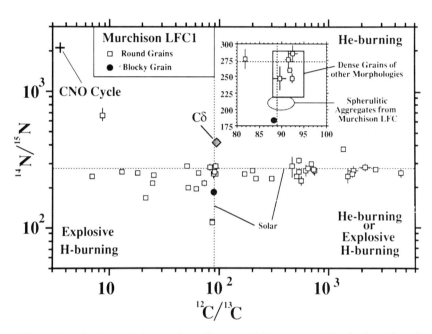

Figure 3. Carbon and nitrogen isotopic compositions measured in single grains of separate Murchison LFC1 (Zinner et al. 1990) which is enriched in the carrier phase of Ne-E(L) (Cα). Note the logarithmic scales on both axes. The "normal" solar compositions are indicated by the dashed lines. Especially large are the variations in $^{12}C/^{13}C$. The composition of Cδ diamond is also shown for comparison (figure provided by E. Zinner).

However, the latest ion probe and noble gas work (Amari et al. 1990a, b, c) shows growing complexity. Ion probe analyses reveal that the C and N isotopic compositions for individual grains are variable (Fig. 3; note the logarithmic scales) so that the $\delta^{13}C$ of $+340\,^o/_{oo}$ must be a mean value for the earlier prepared Cα-rich samples. They do not support a link with heavy nitrogen (Amari et al. 1990b; E. Zinner, personal communication). According to this latest work, round grains with 1 to 6 μm diameter appear to be those that carry Ne-E(L). The composition of the individual Cα grains correlates with morphology, with $^{12}C/^{13}C$ for the round grains ranging from 0.05 to 50 times the solar value, but C in the other forms close to normal (Amari et al. 1990b, c).

Noble gas work on a new set of Ne-E rich samples (Amari et al. 1990a) indicates the existence of two Ne-E(L) carrier phases, Cα1 and Cα2, both of grain size $>1\,\mu$m, but differing in release temperature of the anomalous Ne ($\sim$700°C vs $\sim$900°C) and density. The apparent host phase of the low-temperature Ne-E(L),Cα1, consists of heavier carbon of lower density with generally higher contents of other elements, whereas Cα2 is isotopically light and consists of well-crystallized graphite with low impurity content.

A. Nuclear Source of Ne-E

There appears to be a variety of possible sources that can produce Ne-E. In thermonuclear reactions under nonexplosive conditions neon with ^{20}Ne/^{22}Ne <0.01 can be produced at low density and high temperatures (<10 g cm^{-3}, 4×10^{8} K). This neon, however, would be accompanied by large amounts of ^{4}He (Arnould and Nørgaard 1978) from which it has to be separated. The extreme isotopic purity of Ne-E(L) favors production via the precursor ^{22}Na ($T_{1/2} = 2.6$ Gyr), which would allow easy separation from the noble gases. Partly based on the apparent enrichment of ^{15}N, novae have been favored as a source of Ne-E (Clayton and Hoyle 1976; Arnould and Nørgaard 1981), but this will have to be re-assessed with the new ion probe N isotope data for Cα (Amari et al. 1990b). On the other hand, the composition of Ne-E(H), which is isotopically less pure ^{22}Ne (Table I) and is carried by silicon carbide (see below) is consistent with an origin from a low-mass AGB (asymptotic giant branch) star, as are the other anomalies detected in SiC.

VI. SILICON CARBIDE

Silicon carbide (Fig. 4) occurs in C2 meteorites in abundances of a few ppm (Tang et al. 1989; Zinner et al. 1989). According to these authors (and references cited therein) SiC occurs in two varieties, Cβ and Cϵ, that differ in grain size and associated noble gas components. Cβ with a grain size between 0.03 and 0.2 μm and, like the Cδ diamonds, a log-normal size distribution (Tang and Anders 1988a), is defined as the carrier of s-process Xe, while Cϵ (grain size $>0.1\,\mu$m) is thought to carry Ne-E(H). There is some doubt, however, about this division into two types of SiC. For one thing, the latest noble gas data for SiC by Lewis et al. (1990) indicate nearly uniform concentrations of ^{82}Kr-S in various size ranges, with Xe-S declining and Ne-E(H) increasing in abundance with grain size. [Note that this behavior is different from the trend in Table1 of Zinner et al. (1989), because these authors assumed constant Ne-E(H) content of SiC in order to estimate the SiC content of their samples.] In addition, as already pointed out by Ott et al. (1988a) and confirmed by these newer data, changes in the Ne-E(H)/Xe-S (as well as Kr-S/Xe-S) ratio are accompanied by changes in the isotopic composition of s-process Kr, with higher Ne-E/Xe-S ratios accompanying krypton that has been produced under more neutron-rich conditions. Such a correlation is not a natural consequence of mixing two SiC carrier types, one carrying

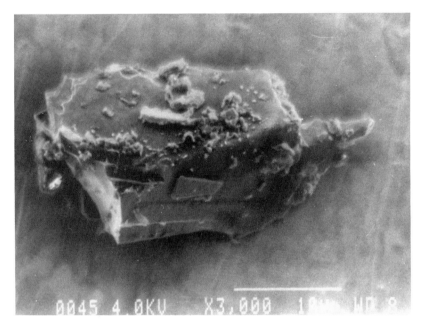

Figure 4.   SEM photograph of one of the largest SiC crystals (figure courtesy of
E. Zinner). Scale bar is 10 μm.

N-E(H), the other carrying s-process material.   Independent of whether a
two-component mixture or a continuous spectrum is a better description of
reality, it appears that the s-process Kr and Xe found in meteoritic SiC may
be naturally accompanied be Ne-E(H) as both can be produced in low-mass
thermally pulsing asymptotic giant branch (TP-AGB) stars (Gallino et al.
1990). To complicate matters further, there are more recent noble gas results
of Amari et al. (1990a) obtained in conjunction with their work on graphite
that suggest the existence of a "super-carrier" of the s-process gases (an even
smaller Russian doll), that possibly makes up only a small subfraction of the
SiC; but even the identification of SiC as the (sole) carrier of the s-process
anomalies may have to be questioned based on their results.

   Variations in the isotopic composition of the structural elements of SiC
are extreme (Table III). $^{12}C/^{13}C$ varies by a factor >350, $^{30}Si/^{28}Si$ by a factor
of almost 3, and, of course, the variations in nitrogen ($^{14}N/^{15}N$ varies by
>300 times) and the noble gases that appear associated with SiC are huge as
well. As an example, ion probe data for Si obtained by Zinner and coworkers
for individual grains (Wopenka et al. 1989) are shown in Fig. 5 (see also
the results obtained for a morphological subclass of SiC grains by Stone et
al. [1990]). From the fact that (in a previous set of samples) single grains
occupied isotopically distinct places, Zinner et al. (1989) have inferred the
existence of at least 6 isotopically independent Si components.   Since the

C and N data (see below) indicated the action of processes that could have altered the Si isotopic composition in only 2 of these cases (but cf. Gallino et al. 1990), they have argued that they were dealing with SiC contributions from at least 4 individual stars.

TABLE III
Isotopic Ratios in Meteoritic SiC[a]

Ratio	Solar System	Silicon Carbide	
		Absolute	Normalized
$^{12}C/^{13}C$	89.0	$3.0 - 1135$	$0.03 - 13$
$^{14}N/^{15}N$	272.2	$18 - 6200$	$0.07 - 23$
$^{29}Si/^{28}Si$	0.0506	$0.031 - 0.064$	$0.62 - 1.26$
$^{30}Si/^{28}Si$	0.0336	$0.014 - 0.038$	$0.41 - 1.14$
$^{20}Ne/^{22}Ne$	13.7	0.23	0.017
$^{130}Xe/^{132}Xe$	0.55	1.38	2.51

[a] Source: Zinner et al. (1989) and Zinner (1990, personal communication). See also Fig. 6. The last row shows the measured ratios normalized to the solar system values. Cα (graphite) has a range in the C isotope ratio that extends to even higher values (Fig. 3).

Carbon and nitrogen isotopic data for the same samples as in Fig. 5 are shown in Fig. 6, where the composition of "normal" carbon and nitrogen is indicated by the horizontal and vertical lines. (Note the logarithmic scale for both axes.) Most of the data points plot in the upper left, indicating dominant contributions from H burning via the CNO-cycle, although only in one case is the CNO-cycle equilibrium value (cross in Fig. 6) approached. A rightward shift (higher $^{12}C/^{13}C$) may be due to contributions from He-burning or ^{12}C dredged up during the AGB phase. Contributions from these processes are consistent with a red giant origin for the SiC which had been proposed early on after the discovery of s-process Xe (Srinivasan and Anders 1978) as red giants are considered as the place where the s-process occurs.

A downward shift (to lower $^{15}N/^{14}N$), on the other hand, could possibly result from He or explosive H-burning. There is no obvious regular relationship between the Si and C, N isotopic compositions in Figs. 5 and 6. Again, this is not unexpected for a red giant origin, because, due to dredge up, red giant atmospheres keep changing in C and N composition probably without much affecting Si.

A. Implications for the s-Process

Enhanced contributions from the process of slow neutron capture (s-process) in trace elements other than the noble gases have recently been reported for Ba and Sr in acid-resistant residues from Murchison by Ott and Begemann (1990a, b). While they have not actually demonstrated the presence of SiC in their samples, it seems likely, after the chemical treatments to which their samples were subjected, that SiC is the carrier phase of Ba-S and Sr-S in their

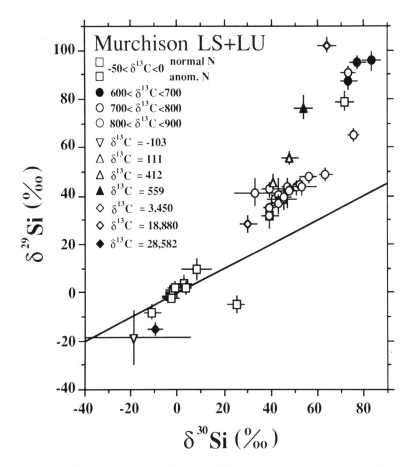

Figure 5.   Si isotopic variations in single SiC grains of Murchison LS and LU (ion probe data by Wopenka et al. 1989; figure courtesy of E. Zinner). The straight line is the terrestrial mass fractionation line. Several independent components are necessary in order to describe the isotopic variations.

samples as well. Combined with the noble gas data, these results can be used to constrain the conditions prevailing during the s-process. Of course, this can be done with a much higher degree of confidence from such data than from average solar system abundances because more or less pure s-process matter reveals its isotopic signature much clearer than does average solar system matter where the contributions from all relevant nucleosynthetic processes are always inseparably intermingled.

In Ba, an enrichment of s-only [134]Ba and [136]Ba relative to the p-only isotopes [130]Ba and [132]Ba of ~50% has been found (Ott and Begemann 1990a; Fig. 7). The inferred [138]Ba/[136]Ba ratio of the s-process component indicates that the conditions for the production of this s-process component were dif-

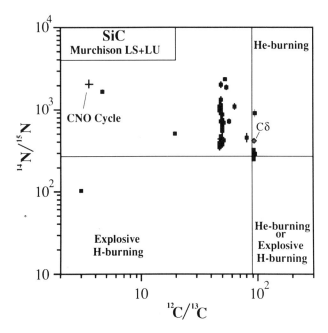

Figure 6.   Isotopic variations of carbon and nitrogen in single SiC grains from
   Murchison LS and LU. Ion probe data by Wopenka et al. (1989) provided by
   E. Zinner. The solar system ratios are indicated by the lines. Note logarithmic
   scales on both axes. For comparison also the composition of the Cδ diamonds is
   shown. See text for discussion.

ferent (lower neutron *dose*) from those for the solar system "main" *s*-process
component that phenomenologically describes the solar system abundances
in the Ba region of the nuclides of *s*-process origin (Käppeler et al. 1989).
Data for krypton first reported by Ott et al. (1988*b*) indicate that $(^{80}Kr/^{82}Kr)_s$
is lower and $(^{86}Kr/^{82}Kr)_s$ higher for Murchison SiC than for the average
s-process contribution to the solar system, implying higher neutron *density* and lower *temperature* (Fig. 8). Beer and Macklin (1989) have shown
that such krypton can be produced in intermediate mass AGB stars with the
$^{22}Ne(\alpha, n)^{25}Mg$ reaction acting as the neutron source but that the same models cannot reproduce the composition of the average solar-system *s*-process
contribution. In addition, the data of Ott et al. (1988*b*) showed variations
of $(^{86}Kr/^{82}Kr)_s$ and $(^{80}Kr/^{82}Kr)_s$ with combustion temperature. From the
most recent data by Lewis et al. (1990), it appears that these variations were
caused by SiC of different grain sizes combusting at different temperatures.
Their new, high-precision, data extend the observed range of variations (as
a function of SiC grain size). Calculations reported in a companion paper
by Gallino et al. (1990) show that the data for Kr as well as those for the
other noble gases can be described by mixing *s*-process material produced

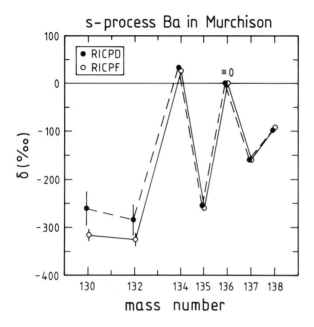

Figure 7. Composition of Ba in Murchison R1CPD and R1CPF (Ott and Begemann 1990*a*). Relative to the *p*-process-only isotopes ^{130}Ba and ^{132}Ba the *s*-only isotopes ^{134}Ba and ^{136}Ba are enriched by ~50% when compared to the "normal" composition.

during thermal pulses in low mass TP-AGB stars with envelope matter from the same star. At present, it is an open question, however, how important or how typical the kind of nuclear contributions we see in the silicon carbide grains from meteorites are with regard to the solar system as a whole.

B. Trapping Process

In contrast to the case of the diamonds that carry Xe-HL, trapping by SiC of the *s*-process nuclides seems to have occurred under conditions that resulted in elemental fractionation. Ott and Begemann (1990*a*) report a fractionation factor between *s*-process Ba and *s*-process Xe of ~1800 favoring Ba. As an explanation they suggest that trapping occurred by implantation of ions with the fractionation factor reflecting the different degrees of ionization of the elements involved, i.e., ambipolar diffusion as has already been suggested for trapping of noble gases in other cases (Jokipii 1964; Göbel et al. 1978). From the Ba/Xe fractionation, a relevant temperature around 10,000 K has been implied (Ott and Begemann 1990*a*), quite similar to the temperatures implied from the elemental abundance pattern of the isotopically normal ureilite noble gases. Interestingly, there have been recent identifications of former red giants, post-AGB stars, having high effective temperatures in that range, strong, rapid winds, and many circumstellar grains (Buss et al.

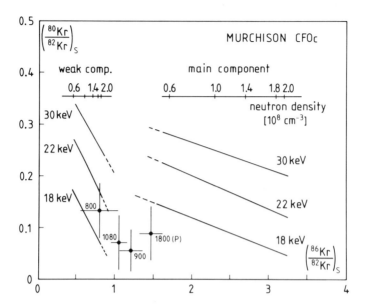

Figure 8.   Composition of s-process Kr released in different analysis steps from
Murchison CFOc (Ott et al. 1988b). The data show variations in the composition of
s-process Kr; the individual compositions cannot be produced under the conditions
of temperature and neutron density as commonly assumed in the phenomenological
description of the s-process contributions to the solar abundances. See also Beer
and Macklin (1989) and the new, more precise data by Lewis et al. (1990) and
calculations by Gallino et al. (1990).

1989). Noting the strong fractionation between Ba and Xe on the one hand,
but comparably little fractionation between the noble gas elements, when
compared to the TP-AGB calculations, on the other hand, Lewis et al. (1990)
have suggested that SiC grains were impregnated with s-process material
by *two* types of stellar wind: a high-temperature, highly ionized one, that
accounts for the noble gases and the small fractionation among them and a
cool wind that shows strong elemental fractionation and which could account
for the high abundance of Ba relative to Xe. Still, these ideas must be
regarded as speculative at this time. The high abundance of Al relative to
Mg (Zinner et al. 1991) as well as the relative lack of Rb-S relative to Sr-S
(U. Ott, unpublished) certainly indicate that volatility must have played a role
in establishing the elemental abundance pattern.

VII. SUMMARY AND IMPLICATIONS

The survival of interstellar materials such as diamond, graphite and silicon
carbide puts significant constraints on the thermal and environmental history
of the medium/media they have been stored in from their birth until their

arrival at Earth. As discussed in Sec. II, these phases belong to those that can (under nonoxidizing conditions) be expected to be among the most resistant. Nevertheless, we know that under laboratory conditions (oxygen pressure ~10 mbar, time scale ~1 hr) diamond/graphite and SiC combust at ~600°C and 1000°C, respectively, and it is clear that the surviving material cannot have experienced high-temperature episodes under oxidizing conditions, for any extended period of time. And even for the case of nonoxidizing conditions, there are severe limits to the temperature they can have experienced which are given by the temperature at which the associated noble gases would be lost (Table II). Provided diamond, Cα graphite and SiC have been part of the solar nebula during solar system formation, it follows that there must have existed niches where the corresponding limits of temperature and/or oxygen fugacity have not been exceeded. The question then is how large a fraction of interstellar material did survive in those niches.

The case of SiC has been discussed by Zinner et al. (1989). In their view there must have been a tremendous loss/destruction of the SiC phase (of at least some 99.9%). Their conclusion rests on the differences in the fraction of Si that occurs in the form of SiC between what is thought to be produced at the source and what is observed in the meteorites. On the one hand, there are the carbon stars, which are thought to contribute to the interstellar medium ~1/3 of the total matter ejected by stars, and which appear to eject at least 20% (and possibly 100%) of their Si in the form of SiC; on the other hand, even in the most primitive C2 meteorites, only 4×10^{-5} of Si is present in that form. Zinner et al. (1989) favor that most of the destruction did not occur in the solar nebula, but earlier, perhaps by supernova shocks or some other selective interstellar process(es). According to their estimates, the survival rate of that fraction of SiC that entered the solar system must have been on the order of 50%. However, with Cδ diamond both much more abundant and (under laboratory conditions as they are employed for isotopic analysis by conventional mass spectrometry of C, N, and the noble gases) more susceptible to oxidation and noble gas loss, the survival of this phase and its noble gases may put more severe limitations on early solar system history. From these laboratory experiences it seems that 700°C, the temperature at which noticeable noble gas loss starts, cannot have been exceeded for any extended period of time. As in these experiments oxygen fugacity is not well controlled, dedicated investigations that try to reproduce relevant conditions in the solar nebula are desirable.

In discussions of these observations, it has been generally assumed that the presolar phases took part in the processes during formation of the solar system (see, e.g., Huss and Lewis 1990; Huss 1990). In fact, this is something that remains to be shown and to do so is not a trivial task. At present, we cannot exclude that the interstellar material we are detecting today entered the solar system *after* some of the critical steps in its formation. We do not even know whether these phases predate formation of the solar system. Carrying none of the elements that can be used for age dating via their radioactive decay (K, Rb, U, etc.) to any significant degree and, in addition, coming from exotic nucleosynthetic sources for which we do not know with certainty the underlying composition of the elements that would be changed by the decay process, there is no way to obtain a definite absolute age for them. The only thing safe to say is that their introduction into the solar system must have happened before compaction of the meteorites because of their fine dissemination throughout the meteorite matrices; SiC at least appears to exist as single grains there with little, if any, connection to other minerals (Alexander et al. 1990). Even their presolar age based on early cosmic ray interaction with the grains before incorporation into the meteorite parent bodies (Tang and Anders 1988*b*; Zinner et al. 1989) is hard to obtain in the case of SiC because of the uncertainty of the amount of recoil losses in such small particles, possibility of other losses (Lewis et al. 1990) and the possibility (Amari et al. 1990*a*) that only part of the SiC is the phase of interest; it is all but impossible in the case of diamonds with a size of only ~30 Å.

In addition, as pointed out in Sec. VI, the *s*-process trace elements carried by the SiC do not have the isotopic composition of the typical *s*-process contribution to the solar nebula. And, similarly in the case of the diamonds, Xe-HL is quite different from the typical *r*-process contribution to the solar abundance of that element. In other words, the fractions of those phases that survived may have been quite atypical of the matter that formed the solar system and may be even unlike the bulk of C and SiC that was present in the early solar nebula.

In conclusion, the analyses of surviving interstellar materials allow us to have a glimpse of various nuclear sources of solar system materials, and *provided* they were part of the solar system during its formative stage, their existence implies that there must have been niches that allowed survival of some, possibly tiny, fraction of such material. Conditions in the niches must have been characterized by comparatively low temperature and/or oxygen fugacity. What remains to be shown is that these materials were really part of the solar system during its formation.

Acknowledgments. I thank E. Zinner for both generously providing Figs. 3–6 including some unpublished data and a constructive review, and F. Begemann for discussions and a critical reading of this manuscript that contributed to its improvement.

ENERGETIC PARTICLE ENVIRONMENT IN THE EARLY SOLAR SYSTEM: EXTREMELY LONG PRE-COMPACTION METEORITIC AGES OR AN ENHANCED EARLY PARTICLE FLUX

DOROTHY S. WOOLUM
California State University, Fullerton

and

CHARLES HOHENBERG
Washington University, St. Louis

Individual mineral grains from meteorites have been exposed to energetic particles prior to inclusion in the host meteorite matrix, an event which occurred quite early in solar system history. Spallation-produced noble gases in such grains are observed to be orders of magnitude greater in abundance than could be produced after meteorite formation and thus provide a record of pre-compaction irradiation by energetic particles. If attributed to the contemporary particle environment (current galactic and solar cosmic rays) a minimum exposure time for these grains in the CM parent-body regolith of 150 Myr is obtained, with more realistic models suggesting an active parent body regolith for 300 Myr. Constraints on when meteorite compaction occurred and current models for meteorite evolution suggest that this is unreasonably long, that the time available for exposure to energetic particles on the surface of the parent-body regolith is much less than this. Pre-compaction exposure, compressed into a shorter span of time by constraints on the compaction times of carbonaceous meteorites, would require energetic particle fluxes in excess of those in the contemporary solar system and point toward an active early (T Tauri) Sun as its source.

I. INTRODUCTION

We are fortunate to have pristine samples from our own early solar system. These can provide a measure of "ground truth" by which we are able to assess our understanding of conditions and processes important in the formation and subsequent evolution of nebulae, protostars and planetary bodies. When compared to all known natural solar-system materials, a class of meteorites called chondrites, is known to: (1) have the oldest formation ages (Patterson 1956; Tilton 1988, and references therein); (2) have a bulk chemical composition closest to that of the Sun, which represents 99.9% of the present solar system's mass, presumably reflecting the average composition of the original solar nebula (Suess and Urey 1956; Anders and Grevesse 1989); and (3) contain the products of short-lived radionuclides that are long since extinct but that appear to have been "alive" at the time they were incorporated in their present host materials (see, e.g., Podosek and Swindle 1988; Chapter by Swindle). Recent detailed assessment of nuclide abundance smoothness for certain (type CI) carbonaceous chondritic meteorites strengthens the case for associating the composition of these particular chondrites with average solar-system abundances (see, e.g., Woolum 1988; Burnett et al. 1989,1990). Furthermore, convincing cases have been made for the preservation of pristine materials of solar nebula origin, as well as those of presolar origin (see, e.g., Kerridge and Matthews, eds. 1988; Chapter by Ott).

Studies of chondritic meteorites have provided insight and constraints for models of the formation and early evolution of our solar system and of the nucleosynthesis of the elements, and the status of these efforts as of 1987 is well summarized in Kerridge and Matthews (1988). Timely updates in select areas appear in this book, provided by Swindle, by Morfill et al., by Palme and Boynton, and by Prinn.

In this chapter, we focus on the meteorite record regarding the energetic particle environment in the early solar system. We review the evidence for the exposure of individual meteoritic grains to energetic particles prior to their ultimate incorporation in the host meteorite, and we discuss the interpretations of these data, concluding that the observed pre-compaction exposures can only be explained by either extremely long pre-compaction meteorite time intervals or by exposure to a primitive Sun much more active than the present Sun. We include a discussion of the astrophysical implications of both of these alternatives, and compare these with information available from astronomical observations. We conclude with a look toward the future and the need for additional studies.

II. PRE-COMPACTION EXPOSURES OF INDIVIDUAL METEORITE GRAINS TO ENERGETIC PARTICLES

Goswami et al. (1984), Caffee et al. (1988), Wieler et al. (1989) and Hohenberg et al. (1990) give recent overviews of the diverse irradiation records recorded in meteorites. Of interest here are the records produced by solar-flare and galactic cosmic-ray (GCR) particles. Both sources contain protons and heavy ions with broad energy spectra, but solar-flare particles have energies peaking between 1 and 100 MeV/nucleon, and they penetrate in geologic materials to depths of hundreds of microns and the order of 1 cm for heavy ions and protons, respectively. GCR particles have a broader energy spectrum and energies are typically well in excess of 100 Mev/nucleon (more like a few GeV/nucleon), and they penetrate centimeters and meters for heavy ions and protons, respectively.

Energetic heavy ions (in the solar [SCR] or galactic [GCR] cosmic rays) near the end of their range can produce permanent radiation damage in the lattice which can be revealed with an appropriate chemical etchant. The etchant preferentially removes the damaged material and produces a conical hole that can be viewed in an optical or an electron microscope. These radiation-damage scars in the lattice are called nuclear particle tracks, or, more simply, tracks.

Energetic particles in the SCR and GCR are capable of inducing nuclear reactions in geologic materials, producing stable and radioactive nuclides relatively close in mass to the target nuclei; these are called spallation reactions. The depth dependence of these effects are strongly dependent on the range (and hence the energy and charge) of the specific cosmic ray particle involved. Secondary neutrons and primary protons and alpha particles, being much more prevalent and having a lower Coulomb barrier to surmount, are most efficient in inducing nuclear reactions in planetary materials. The most important reaction for this work is nuclear spallation in which a small fragment is ejected, leaving a large residual nucleus.

^{21}Ne is one of the more important cosmogenic nuclides produced in nuclear reactions. It is very low in natural abundance so ^{21}Ne is very prominent as a nuclear reaction product. Furthermore, it is produced in copious amount by energetic protons and secondary neutrons in spallation reactions on common rock-forming elements like Mg, Al, Si and Fe. Thus spallogenic ^{21}Ne is both abundant and easy to resolve from other sources of neon. Because ^{21}Ne is stable, the measured accumulation of spallogenic ^{21}Ne provides a measure of the total integrated exposure to energetic particles. This includes both the recent, well-studied, cosmic-ray exposure just before terrestrial impact when the meteorite was a small object, penetrable by energetic particles. It also can include any exposure that may have occurred before the meteoritic material was consolidated into a large solid object.

It is important here to point out how these two exposures can be distinguished. Adjacent mineral grains within a meteorite must have received comparable exposure during the recent cosmic-ray exposure because the object is a rigid solid. However, as we shall show, some grains have received energetic particle doses far in excess of that received by a similar grain residing nearby. The difference in exposure must, therefore, have occurred before the meteorite was consolidated, when different grains could have had different energetic particle exposures. Differential exposure simply could not have happened after the grains were fixed by the rigid geometry of the host meteorite.

The vast majority of meteoritic grains were exposed only during the recent cosmic-ray exposure age, indicating that during the pre-compaction era only a small fraction of the grains could have spent significant time near the surface of the parent body. Consequently, the spallogenic ^{21}Ne measured in bulk meteoritic material provides only a record of the time during which the meteorite existed as a meter-sized object in its transit to the Earth (referred to as the pre-atmospheric meteorite body). Until grain-by-grain studies were done in the manner presented here, little was known about the energetic particle environment of the pre-compaction era although the conventional cosmic ray exposure ages were well documented.

In 1983, Caffee et al. reported cosmogenic ^{21}Ne results for sets of in-dividually selected large grains from the Murchison carbonaceous chondrite, (analyzed in groups of ~10 grains each). They discovered that some of the sets of grains contained amounts of cosmogenic ^{21}Ne exceeding (by more than an order of magnitude) that ^{21}Ne which accumulated during the conven-tional cosmic ray exposure age of the meteorite, i.e., the time the meteorite existed in space as an object small enough to be exposed to energetic particles from galactic cosmic rays. As all grains reside, embedded side by side, in a fine-grained matrix in the meteorite, Caffee et al. concluded that the excess cosmogenic ^{21}Ne must have been produced prior to the time that the indi-vidual grains were finally incorporated in the meteorite. Further, they found that the excess ^{21}Ne was correlated with the occurrence of solar-flare particle tracks at the grain surfaces, and in no case was excess ^{21}Ne found in the sets of grains selected by the absence of solar-flare tracks. Thus, the excess ^{21}Ne was correlated with an effect that indicated an exposure to solar radiation, either in free space, or at the very surface of the meteorite parent body. Additionally, the pre-compaction spallation effects observed by Caffee et al. were not only large, but there was a significant variation among the different sets of solar-flare irradiated grains analyzed. They therefore expected that if they were capable of analyzing individual grains, they would find even greater excesses than were evident as the ensemble averages observed for the sets of grains.

Since then, Hohenberg et al. (1990) have succeeded in analyzing indi-vidual grains ranging in mass from a few tens of μg down to somewhat less than one μg. Briefly, the meteorite samples are disaggregated and the grains are extracted, mounted in epoxy, polished and etched to determine whether they have solar-flare particle tracks; this is an important selection criterion, because in carbonaceous chondrites only about 2% of the extracted grains are solar flare irradiated. The elemental compositions of the irradiated grains and a representative subset of unirradiated grains are determined using energy dis-persive X-ray analysis on a scanning electron microscope. The grains are then extracted from the epoxy mount, individually weighed, and transferred to in-dividual wells in a target holder which is connected to the mass spectrometer. With the aid of optical microscopic viewing, the grains are then individually volatilized using a 70-watt CW Nd-Yag laser (a 20-watt CW argon-ion laser has been used also) and, in sequence, the extracted noble gases for each grain are purified, separated and isotopically analyzed.

Large olivines were analyzed from Murchison, Murray and Cold Bok-keveld, all CM carbonaceous chondrites. In all cases, cosmogenic ^{21}Ne dominated the signal at mass 21 and corrections made for solar neon and molecular interferences were not significant. Based on the determined grain chemistry and calculated production rates (Reedy et al. 1979; Hohenberg et al. 1978), it is possible to calculate apparent exposure ages from the excess spallogenic neon assuming it was produced by GCR.

Figure 1 shows the exposure ages determined for the track-free grains, which show no evidence of direct exposure to solar radiation. The exposure ages in this case are calculated assuming that only the minimally required (4-π geometry) exposure of the meter-sized meteorite to GCR in transit to Earth is relevant. The peaks in the histograms of data from individual grains (at 1.93 Myr for Murchison, 6.08 Myr for Murray, and 340×10^3 yr for Cold Bokkeveld) occur at ages that agree well with the known conventional cosmic-ray exposure ages of these meteorites (Goswami et al. 1984), particularly considering uncertainties in the pre-atmospheric meteorite size and uncertainties in the depth of the grain in the pre-atmospheric meteorite object.

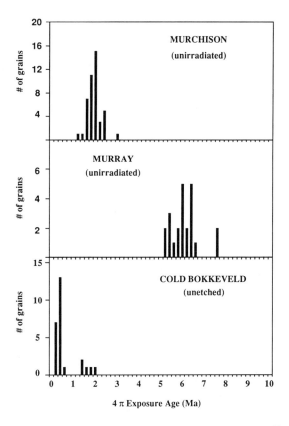

Figure 1.   A histogram of 4-π exposure ages calculated from excess ^{21}Ne contents of unirradiated Murchison and Murray olivines and unetched Cold Bokkeveld olivines. The peaks correspond well with the known conventional cosmic-ray exposure ages of these meteorites. Deviation of grains to the right of the prominent peaks possibly indicate some pre-compaction irradiation.

The data for the track-rich grains show large excesses of spallation-produced ^{21}Ne, much more than could have been produced during the cosmic-ray exposure age. Figure 2 shows the measured quantity of excess ^{21}Ne in the track-rich grains, converted to apparent exposure ages, this time assuming that the grains were irradiated by GCR in a parent-body regolith, therefore with 2-π exposure geometry.

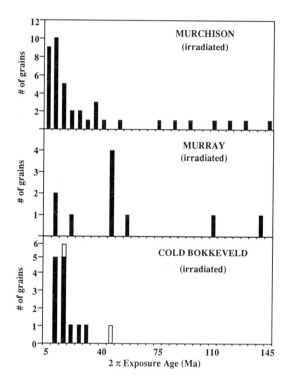

Figure 2.   A histogram of model 2-π exposure ages for irradiated Murchison and Murray olivines and irradiated (closed boxes) and unetched (open boxes) Cold Bokkeveld olivines. Maximum ^{21}Ne production rates are assumed, so these are minimum GCR exposure ages.

GCR are responsible for essentially all spallation effects currently observable in meteorites. The production rates are depth dependent, and the production rate assumed in the calculation of these exposure ages is the *maximum* modern-day production rate, as if all of the exposure occurred at the optimum depth. Thus, these apparent exposure ages, which for Murchison and Murray are about 150 Myr, represent the *minimum* regolith exposure time for these grains. This is a strict lower limit, because it requires that the grains spend their entire regolith history at the depth of maximum spallation production; we are confident that cannot be true because we know that, because the grains contains solar-flare tracks, they had to have spent some time at the surface as well, where the GCR production rate is lower. Similar minimum pre-compaction regolith exposure times are obtained if maximum SCR (energetic solar flares) production rates at 3 AU (i.e., at the asteroid belt, the presumed source of these meteorites) are assumed, although it requires a contrived regolith history to achieve a dominance of SCR production over GCR production with current fluxes (Caffee et al. 1983; Hohenberg et al. 1990).

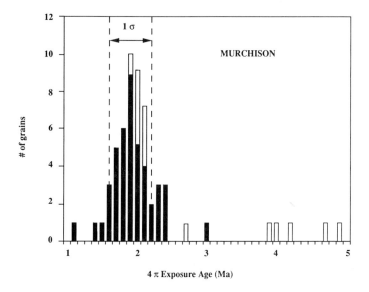

Figure 3.   A histogram of 4-π exposure ages of unirradiated (closed boxes) and irradiated (open boxes) Murchison olivines having <10 Myr 2-π exposure ages. There is little overlap between the irradiated and unirradiated sets.

Figure 3 shows a superposition of solar-flare irradiated and unirradiated grains from Murchison, which demonstrates the strong correlation between the presence of solar flare tracks and pre-compaction spallation effects. This is rather surprising, given the difference in the range of GCR and solar-flare ions. If due to GCR irradiation, such a correlation appears to require a two component regolith: one very mature with virtually all grains exposed to GCR in the upper few meters cycled to the surface to register the solar-flare tracks, as well; and one very immature, having neither tracks nor pre-compaction spallation effects.

The correlation could be more easily understood without this requirement if spallation effects were due to solar-flare protons. However, at the current level of solar activity GCR spallation effects dominate over SCR effects in most materials. Only fluxes of solar-flare protons orders of magnitude greater than present fluxes could readily reconcile the spallation neon–solar heavy ion effects with a simple exposure history. Figure 4 demonstrates the fact that while there is a good correlation between the presence of solar-flare tracks and pre-compaction spallation neon, there is no quantitative correlation between solar-flare track density and spallation neon. Such a separation of the two effects is not unreasonable given the range difference between solar-flare heavy ions and SCR or GCR protons.

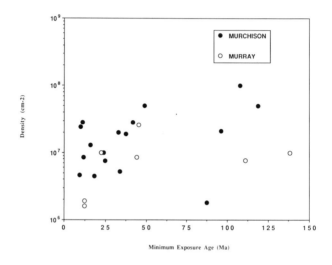

Figure 4.   Track density is plotted versus minimum exposure age for Murchison and Murray grains. Although there is no quantitative correlation between the track density and the quantity of pre-compaction spallogenic ^{21}Ne, a good correlation exists between the presence of the two effects.

III. IMPLICATIONS OF THE PRE-COMPACTION EXPOSURES

It is possible to assess the ^{21}Ne results and attempt to model a more realistic exposure history by comparison with lunar results. The Moon is, albeit, a larger body, but it is atmosphereless; its regolith histories are relatively well constrained, providing us with some basis for interpreting the meteoritic record in terms of regolith processing. Data from the Heidelberg group (Kirsten et al. 1972), discussed by Langevin and Maurette (1976), provide ^{21}Ne age data from lunar regolith grains extracted from regoliths of known age. Comparison of the age distributions for the meteoritic and lunar cases yields a "model" meteorite regolith age of about twice the minimum age inferred from the most heavily irradiated meteorite grains. If the pre-compaction exposure effects in the CM meteorites are due to the contemporary particle environment, the regolith exposure ages required are certainly greater than

the 150 Myr minimum inferred from the most heavily irradiated grains; the
"model" regolith age for the CM parent body is about twice that, or 300 Myr.

Regolith exposure ages of hundreds of Myr appear very long when compared to current models for present-day asteroidal regoliths (Housen and
Wilkening 1982, and references therein), where exposure ages range largely
between 1 and 10 Myr. If the regolith exposures relevant to the meteoritic
data are ancient, such long exposures may not be unreasonable. It is difficult to assess the relevant parameters needed in the models. However, based
on the lunar comparison, where exposures on the order of a few 100 Myr
are obtained, meteorites appear to represent far less mature regoliths. Table I compares regolith maturity factors like volume-percent impact glass and
abundances of track-rich grains, glassy spherules and micrometeorite craters
for lunar soils and meteorites. In the lunar case, maturity indicators are much
more prevalent, which, taken at face value, would indicate much shorter regolith exposures for the meteorites. However, more complex two-component
models, with 1 to 10% admixtures of mature, lunar-type regolith with 90 to
99% immature regolith are also consistent with the data.

TABLE I[a]
Abundance of Maturity Features

Feature	Gas Rich Meteorites	Lunar Soils and Breccias
Impact Glass	rarely > 1 vol%	up to 50%
Glassy Spherules	rare	up to ~ 10%
Micrometeorite Craters	rare	pervasive
^{4}He (cm^3 STP/g)	10^4–10^5	$> 10^7$
Track-Rich Grains	1–10%	20–100%

[a] Table adapted from Housen et al. (1979) and Crozaz (1980).

Based on our present understanding of regolith histories and modeling
and the short evolutionary histories inferred for carbonaceous meteorites, the
long regolith exposures implied by conventional interpretation of the ^{21}Ne
data do seem unlikely. A more plausible alternative explanation for the data
may be offered by the possibility for exposure of the grains to an enhanced
particle flux. It is unlikely that such an enhanced flux would be due to an
increased GCR activity as the Galaxy 4.6 Gyr ago was probably much as
it is today. The Sun, however, was contracting toward the main sequence
and the existence of an early active Sun is astrophysically reasonable. It is
widely agreed among stellar evolution theorists that the Sun and stars similar
to the Sun evolve through especially active phases (so-called T Tauri phases)
just prior to settling down to the main sequence. As is clear from a number
of chapters in this book, astrophysical observations of T Tauri stars in the
last decade have placed important empirical constraints on the nature and
evolution of these young stars. An active Sun could easily have flare activity

three (or more) orders of magnitude greater than the contemporary Sun.

Despite the fact that it is reasonable to presume that all solar-type stars go through a T Tauri phase, there is no well-documented unambiguous evidence of this portion of solar-system history in the record provided by previous studies of solar-system materials; now the ^{21}Ne data provided by the St. Louis group presents the first serious suggestion of this possibility. It may be that the T Tauri phase was too violent an epoch for most solids to survive. It may be that most of the matter in the solar accretion disk was well shielded from the stellar particles, and irradiation of solids was possible only after removal of most of the nebula gas (during the so-called naked T Tauri phase), after accretion, in which case only a small fraction (the skins) of solid bodies would have been exposed. Because grains of interstellar origin have been identified in primitive meteorites, there is encouragement that there may be preserved grains which were exposed to the early active Sun, too. Perhaps we have not looked in the right place(s), or in the right way(s) yet.

If, in fact, exposure to a T Tauri Sun is the explanation for the pre-compaction ^{21}Ne spallation excesses, one thing is certain: the compaction of the CM meteorites is required to be very early. Time scales estimated for the T Tauri phases are on the order of from 10^5 to somewhat more than 10^7 yr.

IV. COMPACTION AGES OF THE CARBONACEOUS METEORITES

In attempting to resolve the question as to whether very long regolith exposures or exposure to an early active Sun was responsible for producing the pre-compaction ^{21}Ne excesses, an attempt to date the time at which these meteorites were compacted could prove critical. An unambiguous compaction age would place a lower limit on the time, relative to the present, at which the pre-compaction exposure could have occurred. If these irradiations occurred very close to 4.55 to 4.6 Gyr ago, the age inferred for the major solar-system differentiation events, and are constrained to a short duration by the compaction ages, then the pre-compaction record provided would have to be most reasonably attributed to a very active early Sun, as this would indicate that there were high fluxes and insufficient time to allow for very lengthy regolith residences.

In 1976, Macdougall and Kothari reported compaction ages for five meteorites, two of which are represented in the Hohenberg group data. The ages were based on the nuclear particle track densities on large euhedral olivine crystal surfaces in contact with the matrix of these meteorites. These are model ages for the times at which the matrix (containing all the elements in roughly cosmic proportions) and the large olivines (serving as essentially actinide free detectors) became associated. The model is illustrated in Fig. 5. Briefly, the normal sources of the nuclear particle track densities at the olivine surfaces would be expected to be GCR and SCR and the energetic charged products of the fissioning of uranium (both in the olivine and in the matrix).

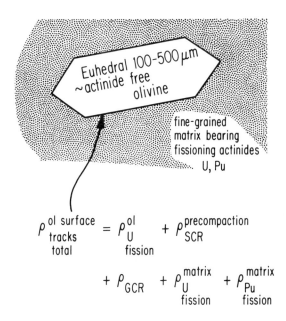

Figure 5.   A cartoon showing the sources of possible contributions to the surface track density of the isolated olivine grains. Olivine is actinide poor, and pre-compaction SCR exposure is readily recognizable. Further, the GCR contribution is generally small. Thus, grains selected for study have surface track densities dominated by actinide fission contributions from the matrix. Correction for the determined U contribution, allows for the calculation of the Pu fission surface track density, which can be related to a contact age, assuming an initial Pu/U (or Pu/LREE) ratio at 4.56 Gyr.

If the olivine came into contact with the matrix early enough, an additional contribution due to the fissioning of extinct ^{244}Pu (72 Myr half-life) could be expected. Of these contributions to the total olivine surface track density the olivine U fission contribution was negligible, based on the measured U content for the olivines. The SCR contribution could be identified by the steep depth dependence of the tracks in the grains and, thus, eliminated. The GCR track density would be essentially uniform throughout the grain and was found to be negligible, or at most require a small correction to the total track density. Contributions from matrix U fission were significant based on the measured matrix U content, but the dominant contribution was actually from matrix Pu fission. This already implies that the matrix and olivine were in contact very early, at least 4.2 Gyr ago, or before, as the ^{244}Pu half-life is so very short. The excess track density associated with the fission of Pu in the matrix was determined for each grain, and contact ages of ~4.4 Gyr were calculated for those meteorites identified as having received a heavy pre-compaction exposure. This was done assuming an initial (4.56 Gyr ago) ^{244}Pu/^{238}U ratio in the matrix materials of 0.0154, the chondritic ratio from Podosek (1970). Currently, the accepted value of the same initial Pu/U isotopic ratio is 0.007 (Hudson et al. 1989), and so the revised model compaction ages now range from 4.48 to 4.52 Gyr for these meteorites (Fig. 6). Compaction this early would appear to preclude the extremely long regolith exposures necessary to explain the excess spallation ^{21}Ne in terms of GCR. However, nothing is quite so straight forward, and there are problems with accepting these compaction ages at face value.

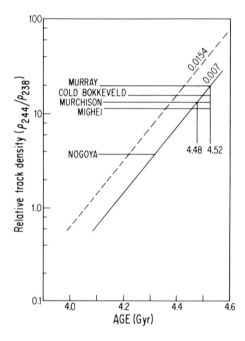

Figure 6.   The relative surface track densities expected for the isolated olivine grains versus contact age, the time before present at which the grain came into contact with the actinide-rich matrix. The dotted line shows the theoretical expectation, assuming the initial (at 4.56 Gyr) Pu/U ratio of 0.0154, the previously accepted value (Podosek 1970). The current value of this ratio is 0.007 (Hudson et al. 1988), and this yields higher compaction ages based on the relative track densities determined in the meteorite (shown by the labelled horizontal line in the figure). For example, Murray and Murchison model compaction ages are about 4.53 and 4.48, respectively. This figure is adapted from Macdougall and Kothari (1976).

    The primary problem is that the ages are calculated using the average olivine crystal surface track density calculated for each meteorite, but the track densities for individual grains vary significantly, by factors ranging from about 50 to more than 2 orders of magnitude in the meteorites studies, excluding the grains with obvious solar-flare track gradients. This variation is well beyond uncertainties due to counting statistics. Because the measured matrix U distribution in all the meteorites appeared homogeneous on a 100 μm scale (the typical grain surface dimensions), the variations could imply a varying Pu (and thus Pu/U ratio) in the matrices of the meteorites. The simplest interpretation of the Macdougall-Kothari (1976) work is that Pu (and Pu/U) is variable, but that the mean value is interpretable in terms of a compaction age. On the other hand, it may also be that the dispersion is telling us that the story is more complicated, involving multiple compactions, with different grains registering different contact ages. These may not provide the most satisfactory estimates of compaction ages for CM meteorites, but they are all we have at present.

There are more confident constraints on CI carbonaceous chondrite compaction ages, and CI meteorites are thought to represent our best samples of unfractionated solar-system materials (Anders and Grevesse 1989; Burnett and Woolum 1989). In CI meteorites, there are veins of calcium carbonates and sulfates due to aqueous activity on the parent body (DuFresne and Anders 1962; Richardson 1978; Kerridge and Bunch 1979). Simple physical examination of these meteorites shows that the emplacement of these veins into the host matrix surely cannot predate the compaction of the meteorite. Thus, if such veins could be dated they would provide confident lower limits to the compaction ages. Richardson (1978) recognized three generations of mineralization: first the carbonates, dolomite $CaMg(CO_3)_2$ and breunnerite Fe, $Mg(Mn)CO_3$—these occur as single crystals and sometimes aggregates which are thought to be residual vein fragments; second, Ca-sulphates; and third, Mg-sulphates. Because these phases are expected on geochemical grounds to contain Sr (a $+2$ anion, as are Ca and Mg) but virtually no Rb (a $+1$ anion) and because the CI meteorites have high bulk Rb/Sr ratios, these phases should be sensitive indicators of the time of their formation. This reasoning was used by Macdougall et al. (1984) and Macdougall and Lugmair (1989) to obtain a compaction age for the CI meteorite Orgueil illustrated in Fig. 7.

Starting with a primitive $^{87}Sr/^{86}Sr$ ratio of 0.699 4.55 Gyr ago, indicated by a star in the figure, the bulk meteorite would evolve with time along the line labeled bulk Orgueil as ^{87}Rb beta decayed to ^{87}Sr. If at some time later, Δt, aqueous activity resulted in the production of Rb-poor phases from the bulk material, the Sr evolution would be halted, locking in the 87/86 Sr ratio at the time of the vein emplacement. Isotopic ratios can be very accurately determined making the time resolution achieved very accurate.

Macdougall et al. (1984) and Macdougall and Lugmair (1989) prepared mineral separates of vein materials. Their separates were not pure, but samples with no detectable Rb gave initial Sr ratios which must be upper limits to the true carbonate formation age values, considering the possibility of contamination with the high Rb ground mass. These initial Sr values are equivalent to the lowest yet observed in solar-system materials (Allende CAI inclusions; Gray et al. 1973; Minster et al. 1982) and imply that carbonate deposition occurred *contemporaneously with parent body formation*, or shortly

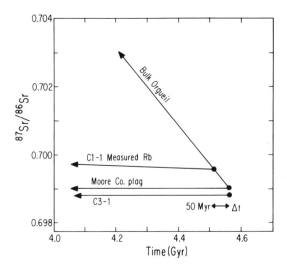

Figure 7.   An illustration of the Sr evolution method for determining the time of
formation of Rb-poor phase in the CI meteorite Orgueil. The curve labeled bulk
Orgueil shows the expected evolution of the Rb 87/86 isotopic ratio of the meteorite
with time, as the beta decay of [87]Rb to [87]Sr alters the Sr 87/86 ratio. The arrow on the
line indicates the evolution, starting from the most primitive initial Sr 87/86 ratio
preserved, represented by the Moore County plagioclase value. Two carbonates
(Rb-poor mineral) from Orgueil yielded Sr isotopic ratios indicating that they were
formed and isolated from the bulk meteorite (thus, preserving a very primitive,
nonradiogenic Sr 87/86 ratio) at formation intervals (Δt) that were short: from 0 to
50 Myr. This figure was adapted from Macdougall et al. (1984) and used data from
Macdougall and Lugmair (1989).

afterwards—certainly within 50 Myr, but probably even less. If the CI meteorite studied were the CM meteorite Murchison or Murray, this study would
effectively preclude a 150 to 300 Myr pre-compaction regolith residence time
for that meteorite and would thus compel us to conclude that the [21]Ne data
reflected the exposure to an active early Sun.

    In summary, then, while constraints on compaction ages do exist for
the relevant meteorites (CM), the compaction ages are model dependent and
rest on assumptions that have not been adequately tested. More rigorous
constraints exist for related meteorites (CI), but the applicability of these
compaction ages for the case of the CM meteorites is not clear. Future work
is necessary to address this issue.

V. ASTROPHYSICAL IMPLICATIONS OF THE EARLY SOLAR
EXPOSURE MODEL

If additional studies support the idea that the [21]Ne data reflect a pre-compaction
exposure to an active early Sun, this would have important consequences for

nebular modeling. Because the source region of the CM meteorites is thought to be the asteroid belt, this would imply that the irradiation occurred at a time when the nebula was already well cleared, assuming that the irradiation occurred near the plane of the ecliptic. T Tauri protons might be expected only to penetrate a few 10 g cm^{-2} of matter, which would imply average nebula densities of only about 10^{-12} g cm^{-3} within the inner 3 AU radius of the solar nebula. This might mean that the exposure occurred not during the initial T Tauri phase but later, in the weak-line T Tauri phase (the so-called naked T Tauri phase, which can last hundreds of Myr [Micela et al. 1985; Feigelson and Kriss 1989]), when the circumstellar disk is believed to be substantially cleared or in the process of clearing. However, the solar-flare track densities observed in the track-rich grains are rather low in comparison with solar-flare track densities in lunar grains (Goswami et al. 1976), and lower than might otherwise be expected from the amount of spallation-produced ^{21}Ne observed in the grains. If we accept that enhanced solar-flare activity is responsible for both effects some residual shielding by nebular matter may be implied, suggesting, perhaps, irradiation during the transition. Our quantitative understanding of the detailed astrophysical conditions is somewhat fuzzy, but based on estimates of enhancements of 3 orders of magnitude in particle fluxes in the weak-line T Tauri phase (Feigelson, personal communication) and considering that the activity during this phase declines relatively slowly, it appears that a reasonable number of solar protons are available to explain the neon data.

VI. FUTURE WORK

Looking to the future, one obvious study would be to extend the ^{21}Ne studies to CI meteorites, where compaction ages are more confidently accepted. The problem in this is that the isolated olivines in these meteorites are very small and very scarce, making not only the track work, but also the ^{21}Ne determinations, very difficult.

It also is important to attempt to re-evaluate the CM compaction ages. Detailed elemental mappings of U and the light rare earth elements (LREE) should be made at the matrix-olivine grain contacts on fine spatial scale (order of 10 μm) in order to assess the homogeneity of these elements specifically in the matrix areas relevant to the olivine surface track data (fission fragments have only about 15 μm ranges in the matrix). Under varied REDOX conditions Pu could be expected to behave cosmochemically in different ways (see, e.g., Burnett et al. 1982). Under reducing conditions Pu can be trivalent and would have an ionic radius between that of Ce and Pr, and so in terms of crystal chemistry, it would be expected to behave like a LREE. On the other hand, in more oxidizing conditions, it could be expected to act like U, which is tetravalent (Boynton 1978a; Benjamin et al. 1978). Homogeneity of the LREE and U, if documented, would lend support to the assumption that Pu homogeneity could be expected at the same level; there is no direct way to check for this homogeneity, as Pu is extinct. Further, it would be important to check that LREE/U is chondritic ("cosmic"); then it is reasonable to assume the chondritic $^{244}Pu/^{238}U$ ratio (Hudson et al. 1989) or the chondritic Pu/Nd ratio for the matrix material in calculating an olivine-matrix contact compaction age. Finally, it would be very important to determine contact ages for individual olivine grains if at all possible. Fortuitous associations of the olivine grains and actinide-rich phases in the matrix might make this feasible.

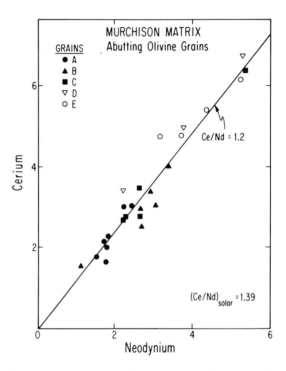

Figure 8.   Cerium counts versus neodymium counts for matrix abutting isolated olivine grains in Murchison. These preliminary data have not been corrected for relative efficiencies, but the differences are expected to be small compared to the observed variations. Cd/Nd values are well correlated and the elemental ratio agrees well with the chondritic (solar) ratio. Variations up to about a factor of 2 are observed, but this is small compared to the 1 to 2 orders of magnitude observed in the olivine surface track densities attributable to ^{244}Pu. Additional determinations are necessary, but if Pu behaves cosmochemically like the LREE, as is observed in some meteorites, this may imply a complex, multi-compaction history for this meteorite. Alternatively, the track data may be indicating that Pu was heterogeneously distributed in the meteorite matrix.

We (Woolum et al., unpublished data) have recently made preliminary determinations of Nd and Ce variations in the Murchison matrix adjacent to large euhedral olivine grains. The matrix was analyzed adjacent to 5 isolated olivine grains using an ion probe with a $\sim$30 μm beam spot size. The total variation in both the Nd and Ce abundances was a factor of $\pm$2. Further, Ce and Nd abundance variations were correlated. These results are illustrated in Fig. 8, where Ce vs Nd results are plotted. The data are uncorrected for relative counting efficiencies, but we estimate that the Ce and Nd counting efficiencies might vary only by up to 20 to 30% for Murchison matrix. The line in the figure corresponds to the average Ce/Nd value of 1.22 ($\pm$0.15, one sigma), which is in good agreement with the CI Ce/Nd ratio of 1.39, given the present uncertainties.

Assuming that the Macdougall-Kothari (1976) U results are applicable for our Murchison sample, these limited data imply that for Murchison one could expect to have obtained a Pu/U ratio that varied in the matrix associated with large euhedral olivines by a factor of $\sim$2. This does not explain the 1 to 2 orders of magnitude variations in the track densities observed by Macdougall and Kothari (1976) and calls into question the validity of basing compaction ages on the average olivine track densities. These preliminary results may imply that the variable track densities reflect (at least to some extent) actual time differences and, thus, multiple compaction events. However, this is only one possible explanation, and considerably more work is necessary before any confident interpretation is possible. Extending the neon studies to other CM meteorites might shed light on the irradiation features ascribable to a common parent-body history. And the discovery of even more heavily irradiated grains could reduce further the plausibility of the long pre-compaction regolith residence scenario.

Acknowledgments. The authors gratefully acknowledge the partial support of this work by several NASA grants.

FORMATION OF THE PROTOSOLAR NEBULA

W. M. TSCHARNUTER
Universität Heidelberg

and

A. P. BOSS
Carnegie Institution of Washington

In this review we discuss theoretical models of the collapse of a dense molecular cloud core to form the protosolar nebula that produced the Sun and planets. The theoretical models use the equations of hydrodynamics, gravitation and radiative transfer to follow the time evolution of a cloud collapsing under its own self-gravity; magnetic fields are not thought to be important during the collapse phase. Both semi-analytical and fully numerical solutions (in two and three spatial dimensions) have been calculated by several workers. Single protostars appear to result from the collapse of clouds that are initially very slowly rotating, close to thermal equilibrium, or strongly centrally condensed. Given one set of these initial conditions, the challenge is to calculate the collapse as far as possible, ideally far enough to make contact with theoretical models of planetary formation. This goal has not yet been fully achieved, largely because of numerical difficulties and because the resulting protosolar nebula must undergo significant dynamical evolution in order to transfer mass inward to form the Sun, and to transfer angular momentum outward. The agent of this large-scale, angular momentum-mass differentiation is problematical, though several promising candidates have been identified: turbulent viscosity, gravitational torques and magnetic fields. Detailed results are given for a two-dimensional model with turbulent viscosity for redistributing angular momentum, and for three-dimensional models investigating the strength of gravitational torques associated with nonaxisymmetry produced during the collapse phase. These models imply that the early protosolar nebula may have been fairly hot ($\sim$1500 K) in the inner few AU. As the numerical models are further refined in the future, we shall be able to improve our predictions of crucial quantities such as surface densities and midplane temperatures in the solar nebula, at a time when the Sun has largely formed and the process of planetesimal accumulation is well underway.

I. INTRODUCTION

This chapter is devoted to the formation of the protosolar nebula through the collapse of a dense molecular cloud core. The protosolar nebula is the rotationally flattened cloud of gas and dust from which the Sun and the planets are thought to have formed, as was first hypothesized in the eighteenth century by Kant and Laplace. In the intervening two hundred years, most of the work on protosolar nebula formation has necessarily been theoretical in nature (recently reviewed by Cameron [1988], Cassen and Boss [1988], Boss et al. [1989] and Larson [1989]), because of the severe difficulties associated with observations of collapsing protostellar clouds. While direct observational evidence for collapse in low-mass protostellar clouds is still lacking, millimeter-wave interferometers are now beginning to provide measurements of physical conditions in low-mass protostellar disks that might well be analogs of the protosolar nebula (see the Chapter by Beckwith and Sargent), and optical and infrared observations are yielding constraints on the amount of gas and dust in the circumstellar environments of young stellar objects and T Tauri stars (see the Chapters by Basri and Bertout and by Strom et al.). Our review will summarize progress made in theoretical modeling of the formation of the protosolar nebula, with the emphasis being on studies finished after *Protostars and Planets II* (Black and Matthews, eds. 1985).

Protosolar nebula formation models can be conveniently divided into two categories, depending on the symmetry assumed in the model. Axisymmetric models assume symmetry about the rotation axis of the nebula; examples are the hydrodynamical models of rotating cloud collapse of Terebey et al. (1984), assuming isothermal collapse, and of Tscharnuter (1978) and Boss (1984a), both of which allowed compressional heating to occur. All three of these studies assumed initially slowly rotating clouds, but with the Terebey et al. model starting from a centrally condensed configuration with $\rho_i \propto r^{-2}$ (Shu 1977), and with the latter two starting from uniform density initial conditions. Cassen and Moosman (1981), Cassen and Summers (1983) and Ruzmaikina and Maeva (1986) constructed axisymmetric models of viscous disks formed by accretion of envelope gas, and followed the evolution as the disks grew through addition of mass and through the effects of viscous torques. The other category is nonaxisymmetric models, which allow for arbitrary deformation of the cloud about the rotational axis. Boss (1982) considered the possibility of protosolar nebula formation as a member of a three-body system, while Boss (1985) studied the formation of bar-like nebulae from the collapse of slowly rotating clouds.

Although approximate analytical investigations are very useful for getting deeper insight into collapse dynamics, they cannot compete with detailed numerical studies. Analytical results (if available at all) are always based on very stringent assumptions, e.g., on the existence of asymptotic solutions for (isothermal) collapse flows, or on expansions which make use of the fact that there are only small deviations from spherical symmetry (cf. Terebey et al. 1984). Analytical results pertaining to the early, isothermal phases have been found to be in reasonable agreement with the numerical models with respect to the spatial dependencies of the state variables. As a matter of fact, asymptotic solutions became attractive (Shu 1977) only after detailed collapse calculations had been carried out (Larson 1969). The formation of protostellar cores, the details of the ensuing accumulation process resulting from effective redistribution of angular momentum, and the radiative energy transport necessary for realistic models of nonisothermal protostellar nebulae are well beyond the scope of any serious analytical treatment.

We restrict ourselves here to consideration of models of nonmagnetic collapse, in large part because much of the progress on protosolar nebula formation has been made with this simplification (however, see the Chapter by McKee et al., and the reviews by Mouschovias [1987a] and Nakano and Umebayashi [1988]). More importantly, magnetic fields decrease in dynamical importance during contraction as a result of ambipolar diffusion (Mouschovias 1987a) or ohmic dissipation (Nakano and Umebayashi 1988). Lizano and Shu (1989) and Tomisaka et al. (1990) have shown that even in an initially magnetically supported, quasi-equilibrium cloud (a subcritical cloud; Shu et al. 1987a), ambipolar diffusion coupled with cloud contraction eventually leads to a situation where dynamical collapse ensues, and while the magnetic field is carried along and amplified by this collapse, the magnetic field should not dominate the collapse dynamics. Nonmagnetic calculations thus should be appropriate for this phase of subcritical cloud evolution, as well as for clouds that are not initially supported primarily by magnetic fields (supercritical clouds). In the case of the latter type of cloud, collapse may be externally initiated by compression, such as that due to passage of a shock wave derived from a nearby massive star (see, e.g., Herbst and Rajan 1980); the presence of short-lived radioactivities in meteoritical inclusions (see the Chapter by Swindle) limits the amount of time that the protosolar cloud could have spent in any quasi-equilibrium phase to $\sim$1 Myr.

In this brief review we do not have space to describe the observational implications of protostellar collapse models. Suffice it to say that the collapse models are generally consistent with observations of embedded protostellar objects (see Shu et al. [1987a], Boss and Yorke [1990] and Bodenheimer et al. [1990] for more details and references).

II. BOUNDARY AND INITIAL CONDITIONS

In mathematical terms, protostellar collapse models are solutions of a complex initial-boundary value problem for a nonlinear system of partial differential equations. The equations express the basic conservation laws of physics, i.e., conservation of mass, momentum and energy, and relate the gravitational potential to the density distribution. In addition, there are important transport phenomena, such as redistribution of energy and momentum by radiation and viscosity, respectively, and mixing processes associated with turbulence. The structure equations also contain functions that define the physical properties of the material. These "constitutive" relations include the equations of state for the pressure and internal energy, and relations for the opacity, viscosity coefficient(s), and thermal and electrical conductivities.

Gravitational collapse of a single cloud fragment, which takes place in the cores of dense molecular clouds and leads eventually to a star surrounded by a dusty, gaseous, disk-like nebula (i.e., to structures like the solar nebula that formed some 4.5 Gyr ago), can be treated as a localized phenomenon, restricted to a finite domain of the interstellar space. Hence the set of structure equations and constitutive relations has to be completed by certain boundary conditions which connect the interior solution (to be determined) with the "known" exterior. For example, one can prescribe the radiation field impinging on the collapsing protostellar fragment from the outside, say, by choosing an "equivalent" (radiation) temperature T_{ex} as the simplest approximation— this determines the net energy flow across the boundary—and fix the external pressure P_{ex}, or keep the volume within which the collapse takes place unchanged with time; the gravitational field and, if taken into consideration, the magnetic field are determined by matching the interior solution and (given) exterior configuration at the outer boundary in a consistent way. One can also simulate the loss and gain of mass (cf. Wuchterl 1990b, c) and/or the transfer rate of angular momentum across the boundary of the fragment. Thus, boundary conditions serve as the necessary completion of the physical assumptions and approximations made for the modeling, and coupled with a particular set of initial data, these choices determine the resulting evolutionary sequence. The boundary conditions and initial conditions should be suggested either by a more general theory or by observations.

It is an attractive feature of protostellar collapse models that the general features and results do not depend critically on the boundary conditions, unless very peculiar and rather artificial constraints are adopted. The opposite is true for the initial configuration from which the collapse starts out. This is easily seen by looking at the characteristic time scale t_{ff} of gravitational collapse (free-fall time) which is inversely proportional to the square-root of the mean density $\bar{\rho}$ of the collapsing fragment, independent of the total mass. For a pressureless spherical collapse $t_{ff} = \sqrt{3\pi/32G\bar{\rho}}$ (G is the gravitational constant), which is also the typical time scale for the realistic case. According to Jeans' criterion for the onset of collapse, $\bar{\rho}$ has to be greater than a critical density ρ_{crit} for a given temperature (typically 10 K for dense cores of molecular clouds), if a thermally supported cloud fragment of 1 M$_\odot$ (say) is to become gravitationally unstable. Then $\rho_{crit} \approx 10^{-19}$ g cm^{-3} and the corresponding collapse time scale $t_{ff} \approx 2 \times 10^5$ yr. But the critical Jeans density ρ_{crit} is just a lower limit, because self-gravitation has to overcome not only gas pressure but also the magnetic and the turbulent pressure. Therefore, the actual mean density $\bar{\rho}_i$ at which dynamical collapse is able to start could in principle be orders of magnitude higher than ρ_{crit}, which would correspond to a much smaller evolutionary time scale t_{ff}.

If we assume that magnetic fields dominate the structure of some cloud cores during their slow quasi-magnetohydrostatic pre-collapse evolution, e.g., in the subcritical regime (cf. Lizano and Shu 1989), then the density at which collapse actually commences is related to the leakage of magnetic flux due to ambipolar diffusion. Thereafter the neutrals can slip across the field lines, while the field strength remains almost unchanged. The transition from the quiescent, flux-conserving regime to the collapse phase should in principle be observable by measuring the magnetic field strengths in cloud cores at still higher number densities ($> 10^5$ cm^{-3}, Fiebig 1990, personal communication) and much better angular resolution (by 1 to 2 orders of magnitude at least) than is feasible today. It is expected that above some characteristic density ρ_m, the exponent κ in the (theoretical) relation between density ρ and the field strength $B \propto \rho^\kappa$, $1/3 \leq \kappa \leq 1/2$ (cf. Mouschovias 1978) will tend to vanish, unless there is some (not yet understood) coherent physical link to the very strong magnetic fields measured in OH and H_2O masers at 10^8 and 10^{10} cm^{-3}, for which $\kappa = 0.5 \pm 0.1$ has been derived (Fiebig and Güsten 1989). Mouschovias (1987a) estimates that ambipolar diffusion should lead to magnetic field decoupling above densities on the order of 10^4 to 10^6 cm^{-3}, while Umebayashi and Nakano (1990) maintain that ambipolar diffusion is ineffective and that ohmic diffusion leads to field decoupling only above $\sim 10^{11}$ cm^{-3}. For supercritical fragments there are no such theoretical constraints on the initial density distribution.

To summarize, we can state that, on the basis of the observational findings, a wide range of densities between 10^{-19} and 10^{-14} g cm^{-3} and a variety of different density distributions are admitted as possible initial values for modeling protostellar collapse, depending on the history of magnetic field support and the mechanism that induces collapse. Accordingly, the time scales for collapse can vary from a few 10^5 yr down to somewhat less than 10^3 yr; most workers expect that the former time scale is likely to be applicable.

III. NUMERICAL METHODS

The structure equations and constitutive relations mentioned in the preceding section form a set of nonlinear partial differential and algebraic equations. There is no hope of finding a purely analytical solution of the complete collapse problem. It is even hard to invent an algorithm which is robust enough to yield reliable numerical model sequences pertaining to protostellar collapse and stellar formation for the simplest case of spherical symmetry. For the present state of art in spherical symmetry and more details, see Balluch (1991b, c). Needless to say, the computational difficulties increase with two dimensional (axisymmetric) and three dimensional models.

Basically, there are two quite distinct methods for representing continuous, nonstationary, hydrodynamical flows on a discrete numerical grid: (i) "explicit" schemes (forward time differences); and (ii) "implicit" schemes (backward time differences). Both methods have their advantages and shortcomings. For example, it is relatively easy and straightforward to write an explicit hydrodynamics code, but the disadvantage is the fact that the time step and the spacing of the numerical grid, i.e., the spatial resolution, are coupled through the famous Courant-Friedrich-Lewy (CFL) condition for numerical stability. To be more specific, a numerical scheme of type (i), with which the physical quantities at time level n are essentially determined from the (known) quantities at level $n - 1$ by simple forward extrapolation, is CFL-stable, only if the time step fulfills the inequality

$$\Delta t < \min_{grid} \left\{ \frac{\Delta x}{|\mathbf{u}| + c} \right\} \tag{1}$$

where Δx indicates the grid spacing and $|\mathbf{u}|$ and c denote the absolute value of the flow velocity (with respect to the grid) and the (local) speed of sound, respectively. Implicit schemes based on methods of type (ii) are always CFL-stable. In this case, the only restriction on the time step is that it must be small enough to resolve any important dynamical changes that might occur in the flow. However, for each time step a huge system of nonlinear algebraic equations has to be solved iteratively in some way or another. Particularly for the axisymmetric collapse, this very often raises the tedious question of numerical convergence, and requires almost unlimited availability of a supercomputer.

A quick inspection of a typical protostellar collapse and accretion flow gives a rough estimate of the variation of the physical variables in space and with time. The compression of the cloud to the final star spans almost 20 orders of magnitude in density, while the temperature variation is ~5 to 6 orders of magnitude. As a consequence, the CFL-time step drops to extremely low values as compared to the time needed for accumulation of the protostellar envelope which is at least of the order of the initial free-fall time t_{ff}. Typically more than 10^9 time steps would then be required in order to cover the main accretion phase. These simple estimates lead to the conclusion that complete and consistent models of star formation, e.g., the accumulation of the Sun together with the evolution of the protosolar nebula, are not likely to be obtained with explicit methods.

The numerical difficulties of explicit methods are considerably lessened if the protostellar accumulation process is simulated in a more qualitative way (by introducing an artificial inner boundary) so that only the global bookkeeping for the net amount of mass and angular momentum flowing toward this central "hole" is considered, but not the detailed structure of the stellar core (Boss and Black 1982). This procedure may seem to be an oversimplification of the problem, but it has the advantage that the evolution of the disk can be followed up to a much more advanced state. This was done by Morfill et al. (1985) who showed that the collapse of a rotating turbulent cloud leads to the formation of a central star surrounded by an accretion disk that is continuously fed by material still infalling from the much larger envelope. Similar techniques are used by Bodenheimer et al. (1990) for axisymmetric models of the solar nebula and by Boss (1989b) for nonaxisymmetric models (see Sec. V).

The partial derivatives appearing in the structure equations are replaced by finite differences in such a way that the global conservation of mass and (angular) momentum is numerically guaranteed. The conservation form is also adopted for the balance equation of the internal energy (first law of thermodynamics), but a similar equation for the total energy generally does not exist and so total energy conservation cannot be guaranteed. It appears that there might be an accuracy problem for the overall energy balance of newly formed stellar embryos, which could trigger oscillations of very large amplitudes (Tscharnuter 1989). According to spherically symmetric calculations recently made by Balluch (1991c) to investigate this particular point, the number of gridpoints has to be drastically increased (at least by a factor of 10, from about a few 10^2 to a few 10^3) in order to arrive at an accuracy level that is high enough to guarantee total energy conservation to $<1\%$ (measured relative to the gravitational energy). Such high accuracy is necessary to remain on the safe side energetically, particularly during the second ("stellar") collapse triggered by the onset of H_2-dissociation in the first, optically thick, core. During this period the density increases by ~ 6 orders of magnitude, which corresponds to ~ 2 orders of magnitude decrease in the radius of a sphere containing a fixed amount of mass (typically a few 10^{-2} $M_\odot$). Hence, both the gravitational energy W and the sum of the internal and kinetic energy E rise by a factor of $\sim 10^2$ (while $E + W \approx$ const), because the second collapse is adiabatic to a high degree of approximation. Thus, if one does not solve the total energy equation, we have a numerical problem. But in trying to enforce total energy conservation, one could run into trouble again, for instance, within free-falling regions of the protostellar envelope, where the kinetic and potential energies almost cancel so that the internal energy and hence the temperature is not well defined numerically. It seems as if the detailed energy balance of stellar core formation is one of the major bottle-necks in the numerical modeling of interstellar cloud collapse, even in the simplest case of spherical symmetry. This is also the reason why the complete axisymmetric model sequences, including both the formation of the disk and the detailed structure of the central star, cannot yet be conducted much beyond the end of the second collapse phase (see Sec. IV below).

IV. AXISYMMETRIC MODELS

A first step toward a realistic numerical approach to a theory of protosolar nebula formation is to study the collapse of a rotating cloud with axial symmetry. As mentioned in Sec. III, the calculations require extremely large amounts of computer time, and the complexity of the numerical methods and solutions often obscures any deeper insight into the physical connections and interrelations we want to understand first and foremost. To this end, Lin and Pringle (1990) have developed a semi-analytical approach to protosolar nebula formation, utilizing the ballistic collapse approximation of Cassen and Moosman (1981) to model the collapse, and using a simplified physical model to model the effects of turbulent viscosity and gravitational torques on the growing nebula. Lin and Pringle (1990) find that, using typical values for observed rapidly rotating molecular clouds, the resulting nebulae should be

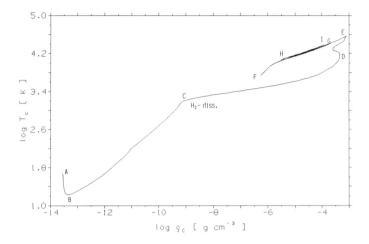

Figure 1.   Axisymmetric protostellar core evolution: log central density vs log central temperature. Labels A-I indicate salient evolutionary stages. A-B is the initial cooling phase (<100 yr). B-C (≈ 500 yr) is the first collapse and C-D (≈ 10 yr) is the second collapse, triggered by dissociation of H_2. During D-E (≈ 15 yr) there is a substantial loss of angular momentum from the core region caused by turbulent viscosity stresses. At epoch E the core starts to oscillate quasi-adiabatically. E-F-G and G-H-I are the first two cycles with periods of ~ 3.5 and 2 yr, respectively.

relatively large (~ 1000 AU), massive (disk mass comparable to stellar mass), and long-lived (several to ~ 10 free-fall times).

In a recent numerical study, Bodenheimer et al. (1990) calculated the formation of a protostellar disk starting from a centrally condensed sphere of radius 5×10^{15} cm, a mean density of 4×10^{-15} g cm^{-3}, a total mass of 1 $M_\odot$ and a total angular momentum of 10^{53} g cm^2 s^{-1}. Because of the aforementioned numerical difficulties with resolving stellar cores (see Sec. III), only the region between 1 and 60 AU is resolved. At the end of the calculation a relatively thick, warm (several 10^2 to 10^3 K) disk formed, close to hydrostatic equilibrium. At the same time the (nonresolved) central region contained a rapidly rotating stellar core of 0.6 $M_\odot$ which could be unstable to nonaxisymmetric perturbations. In order to get an idea of what such an object would look like for an external observer, frequency-dependent radiative transfer calculations have been performed as well. Even if the precise initial conditions should turn out to be inappropriate, such calculations provide a rigorous analysis of the thermodynamics associated with disk formation and at least include the phases of collapse where by far the bulk of the potential energy is released.

Tscharnuter (1987a) investigated the problem of the joint disk-core for-

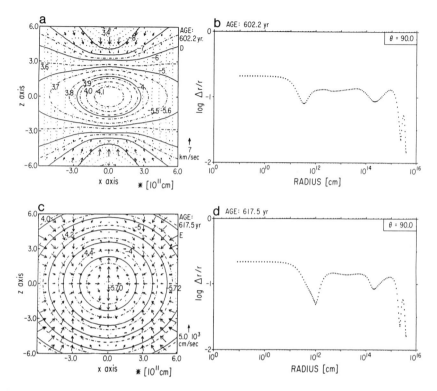

Figure 2. Axisymmetric model of Fig. 1. Panels (a) and (c) illustrate the spin down of the core region, while (b) and (d) show the distribution of the radial gridpoints at the epochs D and E, respectively. Evidently, the stellar core is well resolved by automatically adjusting the numerical grid. The dependency on the polar angle is expressed by a Legendre expansion up to order 14. In panels (a,c), contours of constant density (full lines), temperature (dash-dotted lines) and angular velocity (dotted lines) are plotted in a meridional cross section. The projected velocity field is indicated by arrows; the length of the single arrow on the right side gives the scale of typical velocities. Numbers in the first, second and fourth quadrant are logarithms of the density, temperature and angular velocity, respectively. The core contains ~ 0.06 $M_\odot$.

mation in a rotating turbulent protostellar cloud fragment. Turbulence is taken into account by introducing an α-type "turbulent" viscosity with $\alpha = 0.1$. The initial conditions are very similar to those given above by Bodenheimer et al. (1990), but the total angular momentum is three times larger. Redistribution of angular momentum due to the amount of viscosity chosen is so efficient that the collapse shows qualitatively the same features that have long been known for the spherical case. This is best illustrated by the central density, the central temperature diagram (Fig. 1). The cloud starts with a cooling phase (A-B), since the Kelvin-Helmholtz time t_{KH} of the fragment is shorter than the free-fall time $t_{ff} \approx 720$ yr. After the core has become optically thick (densities greater than $\sim 10^{-13}$ g cm^{-3}), the collapse proceeds quasi-adiabatically (B-C). Angular momentum is continuously removed from the central regions to such an extent that the core is able to contract quickly further down to a rather flattened object at point D (see Fig. 2a). Between D and E a substantial loss of the excess angular momentum ensues and the core becomes closer and closer to a spherical shape.

The evolution of the collapse and the formation of the first, optically thick core and of the outer disk are displayed in Fig. 3a–c and Fig. 3d–f on length scales of 6×10^{14} cm (about the diameter of the solar system), and 2×10^{14} cm, respectively, for epochs C, D and E. Notice the anisotropic nature of the second collapse (Fig. 3f), which proceeds parallel to the axis of rotation and gives rise to the formation of an inner disk around the stellar core. Fig. 4a-d shows two disk regions of $\sim 10^{15}$ cm and 10^{13} cm linear scale, respectively, at epoch E. The (inner) accretion shock has the shape of two "polar caps," but at lower latitudes the flow is subsonic and fairly smooth. After most of the material of the outer disk will have fallen onto the inner disk, there will emerge a single flattened structure that is on the way to becoming an ordinary α-accretion disk, which could then be justly referred to as the protoplanetary nebula.

Although an implicit numerical Newton-Raphson-Henyey technique was adopted for these axisymmetric models, for which no CFL-limitations of the time step have to be observed, the onset of core oscillations (see Fig. 1, labels E through I) on a short time scale has impeded further progress, as a prohibitively large amount of computer time would be required to follow the oscillations. According to the results obtained by Tscharnuter (1987b) and Balluch (1991b, c) for spherical protostellar collapse and by Wuchterl (1990a) for giant gaseous planet formation, quasi-stationary accretion flows are basically unstable, and could conceivably account for the observational inability to detect accretion in protostars. The quasi-hydrostatic parts of such flows are found to be vibrationally and also dynamically (e.g., the second collapse) unstable; the amplitudes of the oscillations may become quite large because the (mass-weighted) adiabatic exponent Γ_1 is always around the critical value 4/3, due to the fact that ionization of H and/or dissociation of H_2 is not yet complete in newly formed (dense and relatively cool) regions of the protostar.

V. NONAXISYMMETRIC MODELS

In a nonaxisymmetric model of protosolar nebula formation, the possibility exists that the cloud could break up during the collapse phase. If the cloud fragments into two or more nearly equal mass objects, the end result is likely to be a binary or multiple protostellar system, rather than the single star and protoplanetary nebula desired in this chapter. There appear to be three criteria (see the Chapter by Bodenheimer et al.) for enhancing the probability of single solar-type star formation: (a) slowly rotating clouds with specific

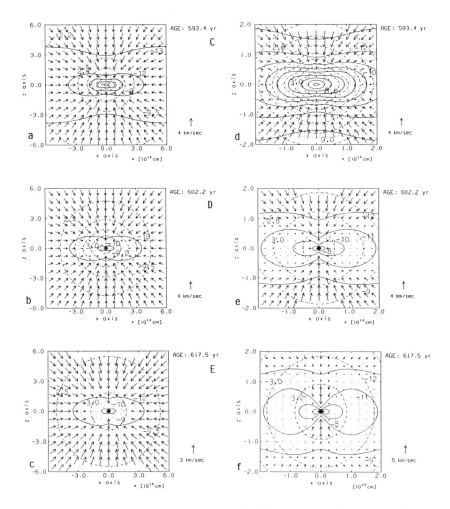

Figure 3.   Axisymmetric model of Fig. 1. Meridional cross sections at two different
    length scales show the formation of the outer disk (a–c), the onset of the second
    collapse and the appearance of an inner disk (d–f) during the period C-E. Contour
    lines and numbers have the same meaning as in Fig. 2.

angular momentum $J/M < 10^{20}$ cm^{2}s^{-1} (Safronov and Ruzmaikina 1978;
Boss 1985); (b) centrally condensed clouds with power-law initial density
profiles ($\rho_i \propto r^{-n}$), shown to be stable against fragmentation numerically for
$n = 1$ by Boss (1987) and analytically for $n = 2$ (the singular isothermal
sphere of Shu [1977]) by Tsai and Bertschinger (1989); and possibly (c)
initially uniform density clouds that begin their isothermal collapse very close
to Jeans stability (Miyama et al. 1984), though the evolution of such clouds
in the nonisothermal regime is largely unknown.

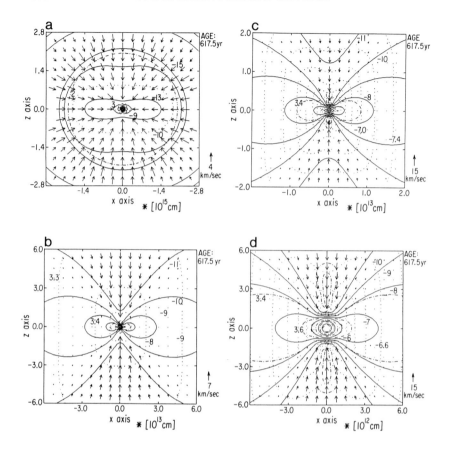

Figure 4.   Axisymmetric models of Fig. 1. Panels (a–d) display meridional cross sections at various length scales at epoch E (maximum central density). Contour lines and numbers have the same meaning as in Fig. 2.

Starting from initial conditions obeying at least one of these three criteria, the challenge is straightforward: calculate the nonaxisymmetric collapse of a dense molecular cloud. However, given the formidable difficulties encountered in calculating even axisymmetric models of protosolar nebula formation (see previous sections), it is premature for definitive models of fully three-dimensional, nonaxisymmetric collapse to have emerged. Nevertheless, considering the apparent (though limited) success of nonaxisymmetric models of binary star formation through fragmentation (see the Chapter by Bodenheimer et al.), preliminary nonaxisymmetric models of protosolar nebula formation can be generated and are discussed in this section. A new three-dimensional code is being developed to study this general problem (Myhill and Kaula 1990), but at present only one three-dimensional code exists that is capable of studying the coupled dynamical and thermodynamical evolution of protostellar nebulae (Boss 1989a, b).

A. Angular Momentum Transport

As advanced by Larson (1984), the protosolar nebula can be expected to become nonaxisymmetric to some extent, and hence the possibility arises of performing at least some of the large-scale angular momentum transport necessary for nebula evolution through gravitational torques (see the Chapter by Adams and Lin). Cassen et al. (1981) studied the growth of nonaxisymmetry in pre-existing, isothermal, thin disks of particles simulating a fluid, and found that marginally unstable disks produced strong trailing spiral arms (see, e.g., Cassen et al. 1981, their Fig. 5); they noted in passing that such models would lead to gravitational torques, and that the nebula might evolve along a marginally unstable path. Boss (1984b) used a simple analytical model to estimate the efficiency of angular momentum transport by gravitational torques, and found that $d(J/M)/dt \sim G\rho R^2$, where G is the gravitational constant and R is a characteristic length. Coupled with typical nebula densities and sizes, this yields a time scale for angular momentum transport $\tau_J = (J/M)/[d(J/M)/dt] \sim 10$ Myr. Cassen and Tomley (1988), improving on the calculations of Cassen et al. (1981) by including radiative cooling, found $\tau_J \sim 100$ yr for the evolution through spiral density waves of the very innermost (<0.1 AU) regions of their nebula model. Miyama (1989) also found that gravitational torques are likely to be effective during the nebula formation phase. Anthony and Carlberg (1988) estimated the viscosity associated with spiral waves excited in self-gravitating accretion disks by carrying out extensive numerical n-body experiments.

The main motivation for the Boss (1989b) study was to try to improve upon previous estimates of the efficiency of angular momentum transport by gravitational torques. Boss (1989a, b) calculated a suite of low-mass protostellar disk models with wide variations in the initial parameters, such as the initial mass of the cloud (0.1 to 1.0 $M_\odot$), the initial mass of the central protostar (0.0 to 1.0 $M_\odot$, the detailed structure of which was not considered), and the initial specific angular momentum ($J/M = 2 \times 10^{18}$ to 6.9×10^{19} cm^{2}s^{-1}). The initial density profile was varied from uniform to strongly centrally condensed ($\rho_i \propto r^{-1}$); the initial angular velocity profiles were also varied from solid body rotation to $\Omega_i \propto R$, where R is cylindrical radius. In this section, we give detailed results (Fig. 5) for a model starting from $\rho_i \propto r^{-1}$, roughly consistent with the possibility that the solar system was derived from a subcritical, magnetically supported dense cloud core, and with the tendency of such clouds to evolve toward a power-law density profile prior to the onset of dynamic collapse (Tomisaka et al. 1989; Lizano and Shu 1989). While such subcritical cores are thought to be common in the Taurus molecular cloud, the situation may be different in other star-forming regions, such as Ophiuchus and Orion, and we cannot be sure in which type of environment the solar system formed. As it turns out, the results of the model shown in Fig. 5 are largely independent of the initial density profile; similar results obtain even for uniform initial densities (Boss 1989b).

Because of the time-step limitation on explicit hydrodynamics codes (see Sec. III), the three-dimensional models were started from abnormally high initial densities, typically $\rho_i \sim 10^{-13}$ to 10^{-12} g cm^{-3}, leading to very high effective mass accretion rates $\dot{M} \sim 10^{-3}$ to 10^{-2} M$_\odot$ yr^{-1}, compared to the usually assumed values of 10^{-6} to 10^{-4} M$_\odot$ yr^{-1}. In the absence of refined calculations, the effect of this assumption on the models is uncertain. However, densities this high are consistent with models of magnetic fields that do not become dynamically insignificant until similar densities are reached (Nakano and Umebayshi 1988; see Sec. II), and they also sidestep the problem of protostellar core instabilities (Sec. IV). Most importantly, compared to pre-existing nebula models that effectively are created instantaneously (yielding $\dot{M} = \infty$), a large but finite value of $\dot{M}$ does allow the effect of collapse on nebula dynamics to be addressed in at least a preliminary fashion.

Figure 5(a) shows that an initially nearly axisymmetric cloud will become significantly nonaxisymmetric during its formation (note that numerical dissipation should tend to damp nonaxisymmetry, so the models probably underestimate the amplitude of the density asymmetries). This nonaxisymmetry results from a combination of nonlinear coupling to the large supersonic infall velocities (Boss 1989b) and rotational instability of the nebula $(T/|W| = \beta = E_{\rm rot}/|E_{\rm grav}| > 0.274)$. The model is stable to axisymmetric perturbations, according to Toomre's (1964) Q stability criterion (cf. the unstable models of Cassen et al. [1981], which were Q unstable).

While the density contours in Fig. 5(c) do not reveal a trailing spiral arm pattern, this pattern is evident in the innermost nebula in the phase pattern of Fig. 5(b). The trailing spiral arm region (out to $\sim$14 AU) is largely losing angular momentum, at a peak rate of dJ/dt of $\sim 10^{39}$ to 10^{40} g cm^2 s^{-2}, while the outer regions are gaining angular momentum. If this rate of transfer of angular momentum could be sustained (the calculations generally are limited to fractions of a rotation period), the time scale for angular momentum evolution by gravitational torques in this model would be $\tau_J \sim 10^5$ to 10^6 yr. Smaller time scales can result in models that come closer to undergoing fragmentation during their formation (Boss 1989b). However, the former time scales are already comparable to those (10^5 to 10^7 yr) inferred for the bulk of nebula dissipation from observations of weak-lined T Tauri stars (see Chapter by Strom et al.) and naked T Tauri stars (Walter et al. 1988).

For computational economy, the Boss (1989b) models were limited to even-m modes; recent linearized models of gravitational instability in preexisting, thin disks by Adams et al. (1989) and Shu et al. (1990) have shown that the $m = 1$ mode might be more important than the even-m modes, because the $m = 1$ mode, corresponding to eccentric particle orbits, may be the most unstable mode of all. The eccentric mode can grow even in a nebula that is

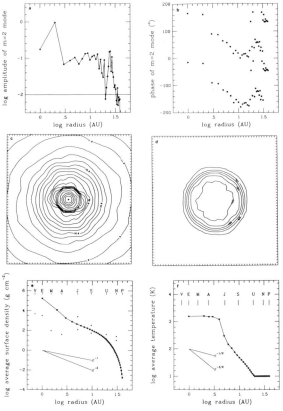

Figure 5. Three-dimensional model of a protostellar nebula at a time when the disk mass is 0.191 $M_\odot$ and the central protostellar mass is 0.837 $M_\odot$ (Boss 1989a). The precollapse configuration consisted of a 0.010 $M_\odot$ protostar surrounded by a 1.018 $M_\odot$ spherical cloud, with a power-law density profile ($\rho_i \propto r^{-1}$), and with a uniform angular velocity such that the specific angular momentum $J/M = 1.86 \times 10^{19}$ cm^{2}s^{-1}. (a) Amplitude of the $m = 2$ mode of the density as a function of radius in the equatorial plane of the disk. Horizontal line denotes the initial density perturbation, which is approximately equal to the level of nonaxisymmetric noise generated by the code. Significant growth of the $m = 2$ mode is evident. (b) Phase of the $m = 2$ mode of the density as a function of radius in the equatorial plane of the disk. Two phases are plotted for each radius, one for each end of the bar mode. A trailing spiral pattern occurs in the inner 10 AU, and a leading spiral pattern outside 10 AU. (c) Equatorial density contours for a region 10 AU in radius. Each contour represents a factor of 2 change in density; the innermost 0.5 AU is not resolved in this calculation. Contours are labeled by densities in g cm^{-3}. (d) Equatorial temperature contours for a region 10 AU in radius. Each contour represents a change of 250 K in temperature. Contours are labeled by temperatures in K. (e) Azimuthally averaged surface density as a function of radius in the disk. Locations of planets are noted. Plus signs refer to reconstituted solar nebula of Weidenschilling (1977b). (f) Azimuthally averaged midplane temperature as a function of radius in the disk. The temperature levels off at ∼1500 K in the inner nebula because of the thermostatic effect of the evaporation of iron grains (the dominant source of opacity) around that temperature.

stable with respect to axisymmetric disturbances (Toomre 1964). However, the nonlinear behavior of this mode is uncertain, and even in the linear regime, growth of the eccentric mode requires the presence of a "sharp" outer edge to the disk; note that the relatively sharp outer edge evident in Fig. 5(e) is an artifact of the boundary conditions, not a real effect. The degree to which sharp edges exist in realistic disks is also uncertain. Regardless of the nonlinear behavior of the eccentric mode in a realistic nebula, the inclusion of odd modes should lead to decreased estimates of τ_J.

B. Surface Density and Temperature Profiles

The three-dimensional models also produced surface density and temperature profiles that are important links between nebula models and planetary formation models. The models showed no strong tendency even for massive nebulae to break up directly into giant gaseous protoplanets. Surface densities in the inner nebula were always adequate to account for the formation of the terrestrial planets (Fig. 5e), even in a minimum-mass (~ 0.05 $M_\odot$) nebula. However, surface densities in the outer nebula always appear to be inadequate to account for rapid formation of the ~ 10 $M_\oplus$ cores needed for giant planet formation (see the Chapters by Lissauer and Stewart and by Podolak et al.), unless the nebula is substantially more massive than the minimum-mass nebula (Boss 1989a, b). This result is consistent with two-dimensional models that map clouds into thin, equilibrium disks (Stemwedel et al. 1990); the implication is that if enhanced surface densities are to occur in the outer regions of a minimum-mass nebula, the enhancement must be achieved through nebula evolution.

Midplane temperatures typically reach 1500 K in the inner nebula, but remain at 150 K or lower in the outer nebula (Fig. 5f). This implies melting of many solids in the innermost regions of the early nebula, subsequently leading to condensation (Grossman 1972) as the nebula cooled. Because the time scale for this global cooling is $\sim 10^5$ to 10^6 yr (based on a crude optical depth argument), globally high temperatures cannot account for the formation of meteoritical chondrules, whose textures require very rapid cooling (Hewins 1988). However, a hot inner nebula seems necessary in order to account for the volatile metal depletions seen in the terrestrial planets (see the Chapter by Palme and Boynton). Figure 5(f) shows that it is possible to have a hot inner nebula, while still having a nebula cool enough at 5 AU for ice condensation and cool enough at 20 AU and beyond for comet formation (see the Chapter by Mumma et al.). In addition, some nebula models are strongly nonaxisymmetric (Boss 1989b), opening up the possibility of having both high- and low-temperature regions at the same radius, possibly allowing for the survival of relict grains that remain closely coupled to their gas parcel until accumulation.

Because the three-dimensional calculations start from initial densities that are $\sim 10^6$ times those of dense cloud cores, the collapse times are 1000 times shorter than usual (~ 100 yr), possibly leading to higher temperatures than would be the case for collapse starting from lower densities. However, if the cooling time for a dust-enshrouded nebula is $\sim 10^5$ to 10^6 yr (as estimated by Boss 1989b), then the amount of nebula cooling during collapse may not be significant. Counterbalancing this effect is the fact that the three-dimensional calculations of Boss (1989b) ignore the heating caused by the central proto-star; two-dimensional calculations by Tscharnuter (1987a) and Bodenheimer et al. (1990), which include protostar heating and start from lower initial densities, produce comparable or even higher midplane temperatures in the inner nebula (see Sec. IV). Refined models will be needed to improve upon these preliminary estimates of nebula thermal structure.

IV. CONCLUSIONS

It is quite obvious that, in spite of the progress made in the past five years, the calculation of completely rigorous model sequences pertaining to the formation of the solar nebula is far beyond the reach of our present computational power. Because there is a stringent limitation on the time step given by the CFL-condition 1 (Sec. III) for explicit schemes, we are able to study only the formation of the disk in some detail, whereas the accumulation of the central proto-Sun cannot be adequately treated in three dimensional models. The most important result of nonaxisymmetric model sequences carried out so far is that angular momentum can be effectively redistributed by means of gravitational torques exerted by nonaxisymmetric (bar-like or spiral-like) structures in the disk.

Major progress in multidimensional models may not be possible until still more powerful implicit methods (which are insensitive to the CFL-condition) are available. To date, variants of the classical Newton-Raphson-Henyey technique, which implies the repeated inversion of large matrices (typically $10^4 \times 10^4$-matrices of a 10×10^2 block-diagonal structure for the axisymmetric problem), have been used very successfully. However, it turned out that the condition of these huge matrices becomes worse if the number of the numerical zones is increased. This unpleasant situation is further amplified if a self-adaptive grid is used, with the possibility of a strong local refinement in the radial direction. Current investigations are attempting to overcome this technical difficulty.

It would also be desirable to develop a technique for handling fully adaptive (both spatial directions) two-dimensional grids, but such an enterprise does not seem feasible in the near future. Likewise, there is also not much hope to attack the three-dimensional collapse problem by means of an implicit method, because an unrealistically large amount of computer time would be needed to arrive at a sufficiently accurate and reliable numerical solution. Interim progress on the multidimensional collapse problem may have to continue to rely on numerical tricks such as the use of "sink cells" (Boss and Black 1982; Boss 1989b) and "multiple collapses" (Bodenheimer et al. 1990).

In axisymmetric models turbulent friction is an efficient mechanism of angular momentum transport. Unfortunately, no generally accepted, consistent theory is available for the onset, maintenance and decay of turbulence, nor is it possible to determine the viscosity coefficient from first principles with complete generality (for some promising steps in this direction, however, see Cabot et al. [1987b] and Canuto [1989]). Hence, the α-viscosity, widely used in accretion disk theory, must still be viewed as somewhat of a free parameter that can be adjusted both to astrophysical observations and cosmochemical findings.

The role of magnetic fields during protostellar collapse and disk formation is not quite clear. Although the dynamics of the collapse are most probably not affected by magnetic forces—that is exactly why collapse is able to commence—there might be an influence on the long-term evolution of the disk; magnetic flux may be regenerated when the temperatures have become high enough and dynamo processes in the disk and/or in the central star may become active. Magnetic forces should therefore be taken into account in the next phase of modeling protostellar collapse through detailed numerical experiments. With the caveats above pertaining to the prospects of the availability of more powerful numerical tools in mind, this step could, in principle, be made immediately with axisymmetric field geometries (axis of rotation parallel to the magnetic axis), but it is our opinion that the solution of the nonmagnetic collapse and accretion problem should have a firm basis consisting of reliable and sufficiently accurate numerical results, before still more complexity is introduced in the models.

Acknowledgments. We thank P. Cassen for a thorough review of the manuscript, an anonymous reviewer for much-appreciated moral support, and J. Dunlap for assistance in manuscript preparation.

PHYSICAL PROCESSES AND CONDITIONS ASSOCIATED WITH THE FORMATION OF PROTOPLANETARY DISKS

G. MORFILL
Max-Planck-Institut für extraterrestrische Physik

H. SPRUIT
Max-Planck-Institut für Physik und Astrophysik

and

E. H. LEVY
The University of Arizona

Stars and planetary systems are thought to develop more or less contemporaneously from extended disk-shaped nebulae. Because the dynamical states of such disks are far from their final equilibrium configurations, the nebulae can dissipate large amounts of energy, as matter accumulates in the center and angular momentum moves to the peripheries. Experience with cosmical systems that are far from equilibrium indicates that such rapid dissipative evolution frequently occurs through a variety of collective behaviors, sometimes producing phenomena of surprising violence. This chapter reviews the range of types of collective processes that may occur in protostellar disks and play significant roles in speeding the evolution of the disks, as well as affecting the physical state of the disk and altering the state of protoplanetary matter. The processes considered here include collective angular momentum transport processes, electrostatic lightning and magnetic flares.

I. INTRODUCTION

During approximately the past decade a great deal of attention has been given to developing a physically realistic picture of planet-forming regions around protostars and young pre-main-sequence (PMS) stars. This recent work complements the detailed chemical, isotopic and mineralogical analyses of primitive meteorites, which have been carried out during the past several decades, and which have allowed important inferences to be made about the conditions that attended the birth of our own planetary system. Among the principal theoretical developments have been numerical collapse calculations (see, e.g., Shu 1977; Morfill 1985; Morfill et al. 1985; Cassen and Boss 1988; Boss 1981; Tscharnuter 1987) and accretion disk models (see, e.g., Lin 1981; Cabot et al. 1987a, b; Ruden and Lin 1986; Wood and Morfill 1988; Sec. III).

The starting point of the collapse calculations can be identified with observations of dense cores in interstellar clouds. The starting point for accretion disk models is a PMS star, surrounded by remnants of the cloud from which it formed, orbiting in an equatorially flattened, disk configuration. Although collapse calculations suggest evolution toward a PMS star accretion disk system, the assumption nevertheless was that star formation occurs sufficiently rapidly, and disk dispersal sufficiently slowly, that this scenario was viable. Until recently, no observational evidence existed to support this assumption.

During these stages of development of our theoretical understanding it was natural to treat quiescent accretion disks first, to establish the basic mathematical framework, to determine the relative importance of physical processes and assess them quantitatively (e.g., the nature of the disk viscosity, density waves, opacity due to dust, coagulation and transport, etc.) and to relate these models to observable properties of our solar system. (For references see, e.g., Larson 1984; Bodenheimer et al. 1980; Pollack 1985; Levy et al. 1991; Morfill and Völk 1984; Morfill 1983,1988.) Although many of the questions which arose have not been answered yet, it has become increasingly clear that a number of the observed phenomena will not find their explanation in quiescent accretion disk models. Prominent examples are the formation of chondrules, the CAIs and possibly the isotopic anomalies.

This chapter describes active, dynamic, even violent, accretion disk processes and relates them to measurements where possible. We shall start with a brief review of the recently obtained observational evidence of accretion disks around PMS stars in order to provide a quantitative background. After that, we discuss angular momentum transport, which, together with radiation transfer, is the most important globally occuring process in accretion disks. We then turn our attention to more localized processes that might occur, and whose signatures may have been observed in our solar system: electrostatic discharges (lightning) and magnetic field flares.

II. ACCRETION DISKS AROUND PMS-STARS

The photometric identification of stars (or protostars) with protoplanetary disks is very difficult because of the small size of these objects, typically 100 AU, and the need to observe them at infrared, sub-mm and radio wavelengths. The best established example is HL Tau, which has a massive disk extending to about 1000 AU and has been resolved in radio-line emission from CO molecules (Fig. 1). Of course, observation of one (still) poorly resolved object is not sufficient to establish the character of the PMS-star/disk relationship, let alone provide quantitative disk properties. Fortunately, spectroscopic measurements have yielded a wealth of additional information so that a number of features have become clearly established during the past few years. These are the existence of disks (or disk-like objects) surrounding the central stars, the occurrence of molecular outflows and winds, the emission from the disk-star boundary layer, disk gaps, etc. The most recent comprehensive survey of a large number of PMS stars and associated circumstellar matter is due to Beckwith et al. (1990).

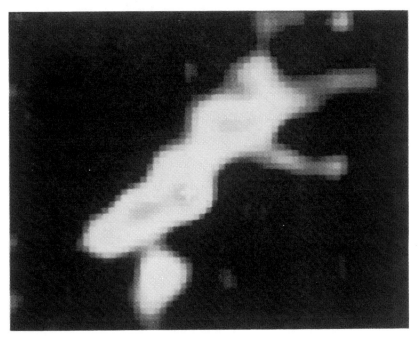

Figure 1.   Flattened disk of dust and gas orbiting the PMS star HL Tauri (located at the white cross). The figure shows the radio line emission from CO molecules. Figure taken from a review by Black (1991).

Of a sample of 86 PMS stars, evidence for circumstellar disks was found in 37, or some 43%. If planets are formed in all such circumstellar disks, planetary systems would appear to be quite common around solar-type stars. Recent studies have attempted to ascertain the systematic character of this class of objects, from which the following conclusions seem to emerge (Beckwith et al. 1990):

1. Disk masses typically fall in the range: 2×10^{-3} to 10^{-1} $M_\odot$.
2. There is no systematic correlation of disk mass with stellar age between 10^5 and 10^7 yr.
3. No disks have been detected for stars older than about 10^7 yr.
4. Disk temperatures, at 1 AU, fall in the range from 50 K to 300 K.
5. Disk temperatures vary as $T \propto r^{-q}$, temperature index $0.5 \leq q \leq 0.75$.
6. The hotter disks have $q \approx 0.5$.
7. There is a tendency for hotter disks to be found around younger PMSs.

8. Some disks have holes or gaps in their centers, ranging from 0.1 to 0.3 AU in radius.
9. Disks typically extend to about 100 AU.

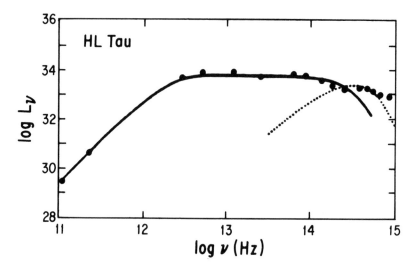

Figure 2.   Spectral energy distribution of the emission from HL Tauri. The dotted line is the Planck function at the stellar effective temperatures fitted to the short wavelength data; the solid line is a theoretical fit to the disk emission.

In Fig. 2 we show the measured spectrum for HL Tau, the star which has a photometrically resolved disk (see Fig. 1), as an example of the kind of signature that can be obtained and analyzed. In Fig. 3, we show a histogram of the computed temperature power-law index, q, (from $T \propto r^{-q}$) for the Beckwith et al. sample.

Theoretically, the surface temperature of a simple, self-luminous disk (i.e., one which is not seen in reflected light from the central star) should follow the relationship

$$T \propto r^{-3/4} \tag{1}$$

that is, with a temperature index $q = 0.75$. The measurements show that disks typically have lower temperature indices, with 0.75 being the upper limit. This discrepancy may not seem startling at first glance, until it is remembered how the theoretical result is derived.

The release of gravitational energy, as matter evolves inwards under the influence of viscous interactions, is (per unit mass of disk material):

$$\dot{\epsilon} = v_r \frac{\partial}{\partial r} \left(\frac{GM_c}{r} \right) \tag{2}$$

where M_c is the central mass (the PMS star or the protostar, whichever the case may be), G is the gravitational constant, v_r the radial flow velocity and

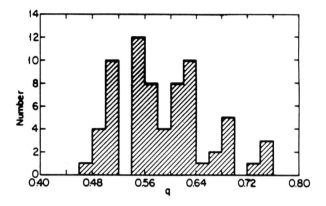

Figure 3.   Distribution of temperature power-law indices g derived from a sample of
37 PMS stars with spectroscopic signatures of accretion disks.

r the distance from the center. The radial flow velocity arises from viscous
interactions and need not be specified. Utilizing the continuity condition for
the mass flow,

$$\dot{M} = 2\pi r \Sigma v_r \qquad (3)$$

with Σ the disk surface density, we obtain the energy production in the disk
in a radius interval dr, from

$$\dot{E} = 2\pi r \Sigma \dot{\epsilon} dr. \qquad (4)$$

The heat generated in this way is radiated into space from the disk surface.
Radiative losses from the same radius interval are

$$L = 4\pi r \sigma T_s^{\,4} dr \qquad (5)$$

where blackbody radiation has been assumed at a characteristic surface temperature, T_s. Stefans constant is denoted by σ. In equilibrium, putting $L \equiv \dot{E}$,
we get

$$T_s^4 = \frac{3GM_c\dot{M}}{8\pi\sigma r^3}. \qquad (6)$$

This result is quite general for self-luminous steady disks. The magnitude of
the temperature at a given radial distance depends only on the central mass,
M_c, and the rate of mass flow through the disk, i.e., the accretion rate $\dot{M}$.

Departures from expression (6) imply departures from the assumptions
that went into deriving this result. One of these assumptions was axial
symmetry. It is conceivable that, e.g., density waves, macroscopic instabilities
or spiral shocks may modify Eq. (6). The reason is that small localized
variations in the surface temperature are enhanced by the T^4 dependence
and may simulate overall a different radial gradient. In Figs. 4a and 4b, we

demonstrate this effect. In Fig. 4a, we have plotted two examples of the radial temperature variation, two power laws with index $q = 0.5$ (short dashed) and 0.75 (long dashed). Figure 4b shows the computed spectrum from such a disk. The spectrum corresponding to the temperature index of 0.5 is flat over a large frequency range, the one due to $q = 0.75$ increases with frequency. In fact, theory gives $L_\nu \propto \nu^{4-2/q}$. Departures from this power law are due to the inner and outer disk edges, the spectra can therefore give information on the size of these objects. The curves also demonstrate that measurement uncertainties in the observed spectra should not be responsible for the discrepancy with the theoretical result derived in Eq. (6).

In Fig. 4a, we also show a radial temperature profile of a perturbed disk (solid line). The perturbation has a radially growing wavelength simulating a macroscopically unstable structure or a density wave type fluctuation (although our synthetic model is axially symmetric). The spectrum obtained from this perturbed disk is also shown in Fig. 4b. As can be seen, the spectrum is relatively flat and would be interpreted as having a temperature index $q \approx 0.5$.

Of course, this interpretation does not cover all of the possibilities. It is interesting to see how such perturbations may simulate flatter temperature gradients, because disk instabilities, nonaxisymmetric structures, etc. have received increasing attention in the past few years. Alternative processes which may also yield flatter spectra will have to be explored. Among other possible spectrum-flattening processes are:

1. Effects of "viewing geometry," taking account of radial thickness variations of the disk. Spectral changes are due to different weighting of high (inner) and low (outer) temperature regimes.
2. Radiation transfer, taking account of the fact that the penetration depths of different wavelengths are not the same.
3. Dust properties, taking account of the possibly fractal geometry of aggregates and the resulting modification of absorption and emission characteristics.
4. Dust size distribution, taking account of the possible dynamical and radial evolution of dust coagulation and the associated size spectra, effects of major condensation/sublimation zones (e.g., ice) on the size distribution.
5. Dust sedimentation, leading to a height-dependent size distribution.
6. Reflection of light emitted from the central star, if the disk geometry is flared.
7. Disk perturbations, e.g., macroscopic instabilities, density waves, spiral shocks, pile-up due to viscosity variations as function of radius.
8. Time dependence in the disk evolution.

Summarizing this section, it is widely accepted now that protoplanetary disks are a common feature amongst PMS stars. These disks appear to have lifetimes of about 10^7 yr, which constrains the planet formation rates. The

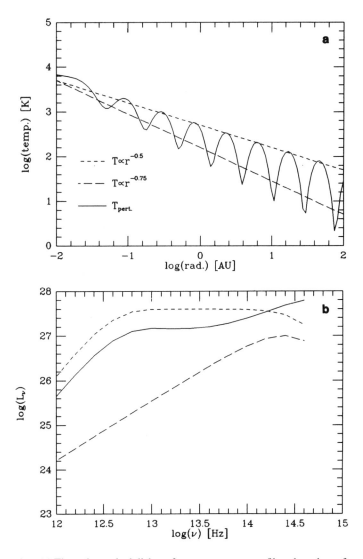

Figure 4. (a) Three theoretical disk surface temperature profiles plotted as a function of disk radius. (b) Corresponding spectral energy distribution. Note that the perturbed disk has a relatively flat spectrum. Figure taken from Morfill and Sterzik (1990).

disks are cool, with temperatures not higher than a few hundred degrees at 1 AU, and evidence is mounting that they are not quiescent objects, with instabilities and nonaxisymmetric perturbations playing important roles. We want to emphasize the important advance which these disk observations and the determination of their properties, have brought to theoretical modeling of both star and planet formation. Before these disks were observed and measured, the prime observational information came from protostellar clouds, believed to be the dense cores found in the more diffuse interstellar clouds. Measured properties of such cores, including mass, size, density and angular momentum, provided the starting point for theoretical collapse calculations and the studies of star formation (see, e.g., Boss et al. 1989). Two-dimensional hydrodynamic calculations by, e.g., Tscharnuter (1987) yielded disks with quite interesting properties, surrounding and feeding the central protostar.

From the point of view of planet formation these early disks, surrounding the growing protostar, are not especially relevant. Most of the mass in these disks will end up in the star and hence the conditions in and properties of such disks cannot be directly related to the protoplanetary disks from which the planets ultimately derive. Disk modeling, for the purposes of studying the protoplanetary environment, had to assume that the early disks computed in the context of collapse calculations had properties similar to the later disks in which planets may form (without ending up inside the star). For a detailed description see Morfill (1985).

With the disk properties around young PMS stars now a subject of observational study, modeling of planet formation has been provided with an informative new starting point, a basis from which to proceed with much greater confidence than was possible before. Already there are some puzzles—observation that some disks apparently have central holes, and the temperature gradient, which we discussed above, being just two of these. The solutions of such puzzles will help us to understand the environment in which the planets of our solar system were born and to learn how the solar system was formed, provided it was not created under very special circumstances—differing greatly from the PMS disks which we can study now.

III. ANGULAR MOMENTUM TRANSPORT PROCESSES

We briefly review here several specific angular momentum transport processes. Other processes and further details are given elsewhere in this book.

A. Shear Turbulence

Historically, one of the first processes proposed to allow mass in a disk to lose angular momentum and thereby accrete onto the central object is *shear turbulence* (see, e.g., Zeldovich 1981; see also von Weizsäcker 1948). The shear in the fluid flow between neighboring orbits is similar to that in a rotating Couette flow experiment (apart from the compressibility in the astrophysical case, the equations are nearly identical). For experimental parameters (rotation rates of the inner and outer cylinders) that correspond closest to the astrophysical case, turbulence has not been observed in the laboratory, as was pointed out by Kippenhahn and Thomas (1981). Because shear flows in the laboratory always seem to lead to turbulence at sufficiently high Reynolds numbers, it is often considered obvious (see, e.g., Di Prima and

Swinney 1981) that this will also happen in the rotating Couette experiment, for all parameter values. In the astrophysical community, the question is usually considered unresolved; the effect of rotation (not present in most laboratory shear flows) is often believed to be stabilizing (Paczyński 1978b; cf. Pringle 1981). Numerical simulations of (parts of) thin accretion disks (Siregar and Léorat 1989; Kaisig 1989a, b and in preparation) have not found the expected turbulence, though instabilities operating on a dynamical time scale have been found in thick disks (Drury 1979; Papaloizou and Pringle 1985; Zurek and Benz 1986), in thin tori (Goodman et al. 1987; Hawley 1989), and at the edges of disks (Blaes and Hawley 1988). If thin disks are in fact unstable to shear turbulence, it will probably be found only in three-dimensional calculations as it is known (see, e.g., Orszag and Kells 1980; Bayly et al. 1988) that the third dimension is often essential for the instability mechanism.

Because the instability would be a dynamical one, the resulting viscosity would be high (with α—the ratio of the "turbulent" gas speed to the sound speed—of the order unity), and it would operate in the same way in *all* disks, at least those where a fluid approximation is justified. It seems that observations may well be able to shed light on the question. For example, very low values of α in some disks, or strong variations in α between objects would be hard to understand in the shear turbulence picture. Do protostellar disks give us any clues? If taken at face value, the observed typical sizes (100 AU), temperatures ($T = 300r^{-1/2}$ K; Hayashi et al. 1985), and time scales (10^6 yr or longer) imply α's less than 10^{-2}. Even a modest improvement in the reliability of these numbers would be enough to justify doubt about viscosities due to shear turbulence.

Inasmuch as the observations of protostellar disks do not require very high α's (compared with the case of cataclysmic variables [Pringle et al. 1986]), several proposed mechanisms are viable, especially for the inner parts of the disks. We discuss here two of these, namely angular momentum loss due to a wind from the disk, and spiral waves generated by self-gravitating processes or planets embedded in the disk. Convective instabilities, another possible turbulence-generating mechanism, has been reviewed elsewhere (see, e.g., Lin 1981; Cabot et al. 1987a, b) and will not be discussed here.

B. Angular Momentum Loss Due to a Magnetic Wind

The idea that a magnetized disk could lose angular momentum and thereby allow accretion to take place via a wind was proposed by Bisnovatyi and Ruzmaikin (1976), Blandford (1976) and Lovelace (1976). The idea relies on the existence of a fairly organized vertical (perpendicular to the disk) magnetic field component of uniform sign. Such a field would evolve naturally in the inner parts of an extended disk from a random field captured by accretion in the outer parts. Any nonaxisymmetric component of this field would quickly be sheared to small length scales and dissipate away (Zeldovich et al. 1983) leaving an axisymmetric part that is concentrated inward by the accretion flow. [In addition to this field, there could be an internal field, possibly stronger, generated by dynamo action (Levy 1978). To the extent that this would fluctuate in sign on a small scale, its contribution to the acceleration of a wind would be minor (cf. Blandford and Payne 1982).] In the protostellar case, it is not even necessary to start from a random field at the disk edge, because the observations indicate that the disks are usually perpendicular to the (strong and fairly orderly) ambient cloud field (presumably because of the very strong influence of the magnetic field on the collapse of the cloud; see Chapter by McKee et al. The resulting poloidal field, kept in place by the high gas density in the disk, can accelerate a cool wind in the same way as in magnetic stellar winds (see Sakurai 1985, and references therein). Full solutions of such winds have been obtained (Blandford and Payne 1982; Sakurai 1987). In these calculations, the mass loss from the disk surface is specified as a function of radius. If no dissipative processes (such as Ohmic diffusion, ambipolar diffusion or viscosity) are present in the disk, the disk can accrete with *all* the angular momentum being carried away by the wind. In practice, angular momentum transport by the wind may be competing with, say, a viscous process in the disk. In this case, the magnetic wind will dominate only if the magnetic field accelerating the wind, and the mass-loss rate are sufficiently high. The terminal speed of the wind can (if the mass-loss rate is sufficiently low) be much higher than the escape speed from the point on the disk at which the wind started. This makes the process attractive for explaining jets, especially in cases (active galactic nuclei, SS433) in which high speeds are seen. In protostellar disks, ambipolar diffusion is likely to be effective in the outer parts of the disk. Königl (1989), and Wardle and Königl (1990; see also Chapter by Königl and Ruden) have obtained detailed solutions for the field structure and flow field in the disk, taking into account ambipolar or Ohmic diffusion.

In the present state of the theory, the main uncertainty lies in the mass flux in the wind. As in the case of stellar magnetic winds, the actual amount of mass lost from the disk is sensitive to details of the heating and cooling processes near the disk surface. As the mass loss is increased, the resulting wind becomes slower and less collimated, but it carries more angular momentum away from the disk.

C. Angular Momentum Transport by Spiral Waves

Nonaxisymmetric waves in accretion disks have a spiral shape, due to the differential rotation (axisymmetric waves are also possible but less interesting because they cannot transport angular momentum). Angular momentum exchange by such waves is possible in a number of ways. A companion, if present, exerts gravitational torques on the nonaxisymmetric density distribution in the spirals, adding or extracting angular momentum from the disk. If the disk is massive enough for self-gravity to be important, the perturbations in the gravitational field due to the spirals produce torques on the fluid. Trailing spirals lead to an outward transport of angular momentum (Toomre 1964;

Larson 1984). But even in the absence of self-gravity or companions, angular momentum transport is possible because spiral waves effectively carry angular momentum. That is to say, the total angular momentum of a disk with waves is different from that of the same system without waves. As long as the waves are linear, no interaction takes place with the fluid, and the angular momenta of fluid and waves are conserved separately (for a discussion of wave angular momentum in disks see Narayan et al. [1987]). With dissipation, for example in the form of shock waves, angular momentum is transferred from the waves to the fluid. *Trailing* waves carry a *negative* angular momentum (in the sense described). Where they dissipate, they *reduce* the angular momentum of the fluid, allowing accretion to take place. Waves generated at the outer edge of a disk by a stationary or a corotating perturber are such trailing waves. As they propagate inward, their negative angular momentum provides a sink of angular momentum for the fluid that is available as far as the wave can travel inward.

It may sound counterintuitive at first that a wave source at the edge of a disk can cause matter to accrete in the inner parts of the disk, because the specific angular momentum of the fluid is large at the edge. It can be made more understandable by describing the waves in terms of wave packets or "phonons." Assume that at the outer edge of the disk we have a corotating source of phonons, emitted isotropically. The average momentum of these particles in the corotating frame is therefore zero. A phonon traveling inward meets an increasingly higher rotational velocity. Assuming for a moment that the phonon energy in an *inertial* frame is conserved, its energy measured in the corotating frame increases due to the Doppler effect. At the same time, in a comoving frame the particles have an average momentum *opposite* to the direction of the fluid. Interaction between the particles and the fluid therefore will brake the fluid. The dissipation of the waves thus communicates to the fluid the angular *velocity* of the outer edge, rather than the angular *momentum* there. The energy released in the process comes not from the initial wave energy of the emitted phonons but from the "Doppler-boosted" energy of the waves in the corotating frame, and therefore from the orbital motion of the fluid. The orbiting fluid feels a "friction" that slows it down with respect to the field of phonons emitted at the edge. The continous inward increase of the orbital speed allows the phonons to continue losing energy by interaction with the fluid while gaining energy by Doppler effect, so that a steady state (in an inertial frame) with a net accretion is possible.

The same process takes place if waves are generated at the inner edge of a disk, for example, by nonaxisymmetries on a a slowly rotating central star. This has been proposed by Michel (1984). Such waves could propagate outward, and where they dissipate they would communicate the slow rotation of the central object to the fluid. In this case however, the energy of the waves measured in the comoving frame *decreases* outward. Dissipation of the waves therefore limits their range, and the torque exerted on the disk is limited to the neighborhood of the central object. Numerical simulations of this process were done by Kaisig (1989b).

The wave-particle duality that we appeal to here can be given an exact basis by noting (Whitham 1970; see Lighthill 1978) that wave packets moving in a shearing flow conserve their wave action $s = e/\omega$, where e and ω are the wave energy and wave frequency measured in the fluid frame. One verifies that this conservation property corresponds to the assumed conservation of phonon energy in an inertial frame, if the phonons are treated as particles of zero rest mass. We note that the analogy holds independent of the precise nature of the waves (acoustic or internal gravity waves, for example) as long as one can meaningfully build localized wave packets from them.

D. Self-Gravitating Spiral Waves

In disks of sufficiently high surface density or sufficiently low temperature, clumps can form by Jeans instability if the growth rate is large enough to offset the disruption by the shear in the Kepler flow. In terms of the Q parameter, $Q = c_s \Omega/(G\Sigma)$ this means $Q<1$, where c_s is the sound speed and Σ the surface mass density. In the absence of (external or internal) heat sources, a disk would cool by radiation and eventually reach instability. The instability will lead to internal dissipation (extracted from the shear), heating the disk again. Paczyński (1978a) has proposed that at least certain disks might reach a quasi-equilibrium in this way, in which the heating by the instability balances cooling by radiation. Such disks would have a Q just below unity. Koslowski et al. (1979) and Lin and Pringle (1987,1990; see also Chapter by Adams and Lin) have calculated models of disks in which the instability is taken to act as a source of effective viscosity, which can be modeled in a way similar to the α prescription. Numerical simulations have been done by Anthony and Carlberg (1988). The angular momentum transport by such instabilities can also be understood in terms of the wave packet description given above. The instability generates waves which propagate some distance away from their origin. Because $Q<1$, the gravitational torques between the density fluctuations in the waves and the fluid are important and produce an effective coupling between the wave packets and the fluid. At radii inward of their source, the waves extract angular momentum from the fluid in the same way as discussed above; outward of their source, they add angular momentum through this interaction. The net effect is an outwardly directed angular momentum flux, similar to that produced by a viscosity.

E. Angular Momentum Transport by Spiral Shocks

Angular momentum transport by global spiral-shaped shock waves has been proposed by Donner (1979; see also Lynden-Bell 1974), Michel (1984), Sawada et al. (1986), and Spruit (1987). Shu (1970) suggested that in binaries such shocks could be generated by the impact of the mass-accumulation stream on the disk. Suppose that, at the outer edge of a disk, low-amplitude sound waves are generated. As these propagate inward, their energy density increases, as described above, and eventually they form shock waves. The dissipation in these shocks limits the increase and at the same time transfers the negative angular momentum of the waves to the fluid, so that accretion can take place.

Numerical simulations of mass transfer in a Roche geometry by Sawada et al. (1986), Matsuda et al. (1987,1990) showed that this process can lead to a steady (in a corotating frame) two-armed shock pattern, and a corresponding steady accretion. In the outer parts of disks formed in this way, rather strong shocks are maintained by tidal forces (at least if a sufficiently massive companion is present). As these waves propagate inward, their strength decreases. In the inner parts of the disks, the tidal effects are small, and the strength of the shocks is determined entirely by the balance between shock dissipation and the inward increase in energy density of the wave. The angular momentum transport in these inner regions is therefore *independent* of the strength of the wave excited at the outer edge (Spruit et al. 1987; Spruit 1989*a, b*). Low-amplitude waves amplify, high-amplitude waves dissipate, propagating inward until they reach the strength at which amplification and dissipation balance. Spruit (1987) described nonlinear solutions of the hydrodynamic equations that apply to this situation. The efficiency of the accretion process can be measured by comparing the radial drift speed with the standard α description for viscous disks.

The effective α of the process, in regions where there is a balance between amplification and dissipation of the shocks, is of the order $0.01(c_s/\Omega r)^{3/2}$, where c_s is the isothermal sound speed. This dependence was also derived by Larson (1989). It must be stressed that this "effective" viscosity does not imply a real viscous behavior but only measures the accretion rate. In particular, time-dependent processes in spiral-shocked disks may well be very different from those in viscous ones. The main importance of shock wave accretion lies in the fact that it provides a reliable means of angular momentum transport in non-self-gravitating disks in places that are far from the inner and outer edges of the disk. For a moderately "hot" protoplanetary disk ($r = 5$ AU, $T = 800$ K) one gets $\alpha = 5 \times 10^{-4}$, corresponding to an accretion time scale of 10^6 yr. Though this is not a very high value, we nevertheless see that spiral shock accretion could be relevant for angular momentum transport in the giant planet region of a protostellar disk. Figure 5 shows a numerical solution of a self-similar, spiral-shocked flow.

F. Planet-Generated Shock Waves

Embedded objects in disks (protoplanets, for example) are a plausible source of shock waves. The interaction of protoplanets with a disk has been considered by Lin and Papaloizou (1979*b*) and Goldreich and Tremaine (1980), both having assumed a linear response of the disk to the tidal force. These authors derived expressions for the tidal torque exerted on the disk, and showed that a protoplanet will in general (excepting the case of a very high disk viscosity) clear a gap between its orbit and the disk's outer edge. The position of this edge is determined by a balance between the tidal forces which extract angular

Figure 5.   Self similar spiral shocked flow. The image shows the density distribution
(on a logarithmic scale).

momentum from the edge, and the outward diffusion of angular momentum
by viscosity. When the viscosity is small, shocks form at the disk edge,
invalidating the original assumption of linearity. These shocks have been
studied in some detail by Lin and Papaloizou (1986a), though the viscosity
used was so high that the shocks did not propagate very far into the disk. If
we are interested in the angular momentum transport by shocks, and therefore neglect viscosity altogether, the problem of finding the edge of the disk
becomes somewhat ill posed. Shocks at the edge of the disk extract angular
momentum from it and return it to the orbit of the companion by gravitational interaction. In addition, Lindblad resonances (Goldreich and Tremaine
1980) between the disk and the companion extract angular momentum in a
manner that is independent of viscosity (at low, but finite, viscosity). In the
absence of an additional "spreading agent," such as viscosity, the disk would
continue to shrink until the tidal forces became negligible. One may wonder
if this behavior is consistent with the known behavior of disks in binaries.
Observational information on the sizes of disks, compared with the orbits of
companions is available for cataclysmic variables. The disks in these cases
have radii of about 0.3 times the orbital separation, for systems with mass
ratio of the order 0.1, in quiescence (Wood et al. 1986,1989a, b; Zola 1988);
during outburst the disks are slightly larger. At this size, the tidal forces

produce velocity variations at the edge that are of the order of the sound speed. If this is correct, it suggests that the main factor limiting the disk size is angular momentum extraction through shock waves at the outer edge. Only weak shocks would then propagate inward, and the scenario described for accretion by spiral shocks would give, for the protoplanetary case, the value $\sim 5 \times 10^{-4}$ quoted above. On the other hand, during their outbursts, the disks in cataclysmic variables (CV) definitely increase in size (see, e.g., Zola 1988) on a fairly short time scale. It is not clear at present whether this can result from tidally generated shocks alone; perhaps these observations indicate a truly "viscous" behavior.

Another estimate of the maximum size of a disk in the presence of a companion is that of Paczyński (1978a). He calculates the largest orbit that can fit inside a Roche lobe without intersecting other orbits. Orbits larger than this would be impossible because of the formation of strong shocks. The maximum sizes obtained are larger than the quiescent CV disks, but perhaps compatible with the disk sizes (Warner and O'Donoghue 1988; Horne and Cook 1985) at outburst. On the other hand, it is clear that this estimate is an upper limit, as shocks may form well before orbits intersect. This could happen as soon as the tidal perturbations are of the order of the sound speed.

Numerical simulations of disks in binaries should be useful in answering these questions. Current simulations (Schwarzenberg-Czerny and Różycka 1988; Sawada et al. 1986; Matsuda et al. 1987,1990; Whitehurst 1988a; Hirose and Osaki 1990; Różycka and Spruit 1989) do address them, but the interpretation of the results is not entirely unambiguous. Hydrodynamical simulations without viscosity have been done for a gas with constant ratio of specific heats, ignoring radiation losses (Matsuda et al. 1990). Although these calculations start with fairly cool disks, the dissipation in the shocks quickly heats the gas up to a value near the virial temperature. The resulting disks are stationary only in a rather average sense, and do not have well-defined edges. The accretion time scale in these simulations is roughly inversely proportional to the companion mass. It remains to be seen how inviscid disks with more realistic low temperatures behave. Hydrodynamic simulations of a viscous disk have been done by Schwarzenberg-Czerny and Różycka (1988). Particle simulations of disks in binaries have been done by Whitehurst (1988a) and Osaki and Hirose (1990). The way in which particle interactions model the hydrodynamic forces in these simulations prevents one from assigning a sound speed to the calculations but it is believed that they represent a highly supersonic, viscous flow. The disks in these simulations are somewhat larger than the observed quiescent disks in CV's, probably as a result of the high effective viscosity.

G. One Armed Spirals, Eccentric Instability

Kato (1983,1989) and Okazaki and Kato (1985) have pointed out the existence of a class of low-frequency one-armed spiral modes whose frequency (in an inertial frame) is proportional to the disk temperature. The fundamental mode of oscillation is equivalent to a slow precession of an eccentric disk. In the limit of vanishing temperature (pressureless disk), a disk in this mode is a collection of nested elliptic Kepler orbits. These modes can become unstable in the presence of a companion. This instability so far has been studied in the zero-pressure limit by orbital mechanics methods (Piotrowski and Ziołkowski 1970; Paczyński 1977; Osaki 1989; Osaki and Hirose 1990). The instability is associated with the 3:1 resonance between a disk orbit and the orbit of the companion. For small companion mass, the range of unstable orbits near the resonance shrinks and the growth rate vanishes asymptotically (Hirose and Osaki 1990). Thus, though the instability is very important for cataclysmic variables (Whitehurst 1988a), it is not certain that it is very relevant for the protoplanetary case.

The global $m = 1$ modes studied by Kato can also be destabilized by self-gravity of the disk. Such instabilities have been described by Adams et al. (1989). An important ingredient in the instability mechanism is the motion of both the central object and the disk with respect to the common center of mass. They are discussed in more detail elsewhere in this book. The disk mass required for instability is roughly proportional to the stability parameter Q. ($M_D/M_{tot} \approx 0.24 \, Q$; Shu et al. 1990). For disk masses less than 0.24, $Q < 1$ is needed for instability. For $Q < 1$ the disk is also unstable to local self-gravitating instabilities discussed by Paczyński et al. and Carlberg (1988) (see above). For hot ($Q > 1$) massive ($M_D/M > 0.24$) disks the global instability would be important, but for cool ($Q < 1$) less massive disks the local self-gravitating instabilities would be more important.

IV. EVIDENCE FOR TRANSIENT HEATING EVENTS

A. Observational Evidence

Astronomical evidence for heating events exists (see, e.g., Gahm et al. 1989), but it is not clear whether these events are related to stellar or disk activity. We shall not discuss astronomical evidence here, instead we restrict ourselves to information derived from measurements in our solar system.

There are basically three types of evidence for short-term heating events, which are obtained from the study of meteorites. These are chondrules, Calcium-Aluminum-rich-Inclusions (CAIs) and oxygen isotope anomalies. A discussion of the energy problem associated with the chondrule observations is given by Levy (1988). Here we summarize major characteristics of chondrules relevant to understanding the role of rapid heating events in the protosolar nebula.

1. Chondrules are spherical molten droplets. They have a narrow size distribution—typically of the order of a millimeter with masses of the order of a milligram.

2. Chondrules make up a significant fraction of the total mass of all primitive meteorites. Whilst they are a negligible component in the CI meteorites,

they can account for up to 80% by mass in other chondrites. The rest of the meteorite consists of fine-grained (micron to sub-micron-sized) matrix material. In some primitive meteorites another class of objects is found, the so-called "Calcium-Aluminum-rich Inclusions" (CAIs), which we discuss separately later.

3. Chondrules are clearly a nebular product. This is underscored by age determinations and their chemical and textural properties.

4. Dynamical crystallization experiments show that the cooling times of these chondrules are of the order of hours or less (Hewins 1991).

5. By implication, the time scale for heating the prechondrular dust accumulations was shorter, it could have been of the order minutes or seconds; however, this is not clear at present. During the heating events there appears to have been a "stalling" of the temperature at $\sim$1800 K. This is inferred from the chemical composition, although it is probably fair to say that this evidence is not conclusive and still subject to debate.

6. The energy requirement for raising the temperature of a chondrule to $\sim$1800 K and then melting the material (silicates, e.g., spinel, mellilite, olivine) is $\sim$2000 J g^{-1}. This translates into a time integrated heat flux of at least 400 J cm^{-2}. Losses of heat by radiation and by evaporation increases this value and has consequences for the duration of the heating event.

7. Chondrules are often surrounded by an "accretion mantle" of fine grained material. This mantle was accumulated in the time between formation and incorporation into the parent body.

Additional evidence of rapid nebular heating events seen in CAIs can be summarized as follows:

1. CAIs are aggregates of refractory-rich material, with sizes reaching the centimeter range.

2. In the Allende meteorite (a type CV3 carbonaceous chondrite) they account for about 5% by mass.

3. The cooling times for whole CAIs, as inferred from dynamic crystallisation experiments, are of the order days (cooling rates are 0.5 to 50 K hr^{-1}). These time scales are similar to nebular thermal-convection time scales, so that these objects could be produced in the inner (hot) portions of an accretion disk and then become convectively transported out into cooler regions (Morfill 1983).

4. Practically all CAIs have rims, which have the characteristics of being molten during some intense heating event. The rims are micron-sized; the total mass of a rim is typically about a milligram. High-temperature rims have been reported around chondrules as well (Kring 1991a).

5. Heat conduction and radiative loss considerations give the result that the rim-producing heating events are not longer than a few seconds and subsequent cooling times are also very short.

6. The minimum energy needed to make rims is about 300 J cm^{-2}. The previous statements about cooling and evaporation losses apply here as well, of course.

The evidence from isotopic anomalies is more indirect, but potentially important. During evaporation and condensation, which is a natural consequence of the heating events discussed here, mass dependent fractionation is to be expected purely from, e.g., Rayleigh distillation (see, e.g., Morfill and Clayton 1986a, b). In addition, it was shown by Thiemens and Heidenreich (1983) that, in the presence of strong radiation fields, non-mass-dependent oxygen isotope fractionation may be produced. The full physics and chemistry for this effect is not completely understood but seems to be related to the formation of ozone (O_3). Isotopic anomalies of these types are observed in CAI's (see e.g. the reviews by Begemann 1980 and Clayton et al. 1985) and could be further evidence of short-term intense heating events. However, a great deal more experimental and theoretical work is required in order to make this a quantitative connection.

The evidence for short-duration intense heating events is overwhelming. Furthermore, quantitatively it appears as if practically every CAI was exposed to such an event at least once, and a sizeable fraction of chondrule precurser material had the same fate. Based on this observational input, we can estimate the efficiency with which the available energy—basically the gravitational energy of the disk in the field of the central PMS star (or protostar)—is channeled into this particular form of dissipation (see also Levy 1988).

B. Energetics and Frequency of Short-Term Heating Events

Energetically it is immediately obvious that short-term heating events also have to be localized. Let us assume then that these events, irrespective of their possible physical origin, are cylindrical in shape, with radius R and length d. Let us further assume that these events are localized in an "activity zone" in the disk at the radius r, from the center and occupy an annulus of size Δr. The probability that a given dust grain should encounter one of these "events" while it traverses this activity zone is then given by

$$P = \frac{v_\phi}{v_r} \frac{\Delta r}{\lambda} \tag{7}$$

where v_ϕ is essentially the Kepler velocity and

$$\lambda \equiv \frac{1}{N2Rd} \tag{8}$$

is the "mean free path" between encounters of "heating events." The density of these localized hot gas regions in the disk is

$$N = \dot{N} t_L / V \tag{9}$$

where $\dot{N}$ is their production rate (s^{-1}), t_L their lifetime and V the portion of the disk volume where these events occur ($V = 2\pi r \Delta r h$, with h the disk thickness).

The lifetime of such events is determined by the cooling of the hot gas (initially at temperature T_H). This in turn is determined by the adiabatic expansion, and is approximately

$$t_L = \frac{R}{c_s(T)} \left[\left(\frac{T_H}{T_M}\right)^{1/2(\gamma-1)} - 1 \right] \tag{10}$$

where c_s is the sound speed in the cold (temperature T) external disk gas and $T_M \approx 1800$ K is the temperature below which melting essentially stops. γ is the adiabatic index of the gas. The production rate of hot gas regions can be estimated from energy arguments. It is equal to $\dot{E}/E_H$, the total rate of energy release of the disk in the activity zone, divided by the thermal energy content of a single hot region; $E_H = \pi R^2 d \frac{3}{2} k T_H n_H$. There has to be an efficiency factor η for converting energy into this nonlinear localized hot phase, so that

$$\dot{E} \equiv \eta \frac{GM}{2r^2} v_r \rho V. \tag{11}$$

Substituting all these expressions in Eq. 8 yields finally

$$P = \frac{\eta}{\pi} \left[\frac{v_\phi}{c_s(T)} \right]^3 \left(\frac{\Delta r}{r}\right) \frac{T}{T_H} \left(\frac{T_H}{T_M}\right)^{1/(\gamma-1)} \tag{12}$$

where we have made use of the fact that $T_H >> T_M$. Note that this expression is independent of all "hot spot" parameters with the exception of the initial temperature of the gas T_H. All parameters entering this expression are therefore determined quite well. Putting $\Delta r/r \equiv 0.5$, $v_\phi/c_s = 10$, $T_H/T_M = 100$ and $T_M/T = 3$, i.e., an ambient disk temperature of 600 K, yields

$$P \approx 500 \, \eta. \tag{13}$$

The meteorite data suggest that the probability of an encounter with a localized rapid heating event is $0.3 \leq P \leq 1$. This implies that the efficiency η with which the available energy dissipation is converted into such events, has to be of the order 0.1 to 1%. This is a high efficiency considering the nonlinear and strongly inhomogeneous nature such processes must possess.

V. DISK ELECTROSTATIC DISCHARGES

A. Charge Production and Separation

In terrestrial thunderclouds lightning is a surprisingly efficient process, in which between 1 and 10% of the kinetic energy of the cloud convective motion is released. In addition, lightning discharges have all the short duration, localized (etc.) characteristics required by the meteorite observations, hence the speculation, first voiced by Whipple 1966, that electrostatic discharges may be responsible for the production of chondrules, should be investigated seriously. (For comprehensive reading about terrestrial lightning, the reader is referred to a special issue of the *Journal of Geophys. Res.*, Vol. 94, 1989.)

The problem of generating electrostatic discharges in atmospheres can be subdivided into three parts:

1. Generating charges.
2. Separating charges and building up a large-scale electric field.
3. Maintaining a sufficiently low conductivity so that eventually the breakdown potential is obtained.

At some stage there will be a local breakdown somewhere; electrons will be accelerated over one mean free path to energies above the ionization potential, there will be further collisional ionization leading to an avalanche. In the case of H_2 gas, there are additional losses not only due to ionization, but also due to excitation, charge exchange and dissociation. As shown by Cravens et al. (1975), the required energy to which the electrons need to be accelerated (before colliding with the molecules) is ~60 eV—otherwise the efficiency of the collisional ionization is too low and the cascade may simply fizzle out.

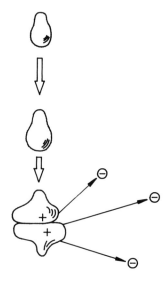

Figure 6.   Illustration of the collisional Lenard effect.

In order that the condition holds that a great deal of charge be separated to build up a sufficiently strong electric field, while at the same time not having the charges so mobile to short circuit the system, it is clear that the charge has to be concentrated on slowly moving dust grains and not on highly mobile ions and electrons. In thunderclouds this role is played by water droplets and/or hail.

There are three major processes that have been discussed in the literature as being relevant for terrestrial lightning. These processes presumably also play a role in Venusian lightning and Jovian lightning—we see no reason, in principle, why these processes should not be expected to operate in protoplanetary disks. The first is the Lenard effect (1904), and is illustrated in Fig. 6. It is found that when a water drop breaks up, the large main residue carries a positive charge whereas the fine spray that has been shed carries a corresponding amount of negative charge. This process has been tested in the laboratory for different conditions such as break-up (ablation) due to drag forces, break-up due to collisions and electrostatic disruption (see Richards and Dawson 1971; Crabb and Latham 1974; Taylor 1964; Latham and Myers 1970).

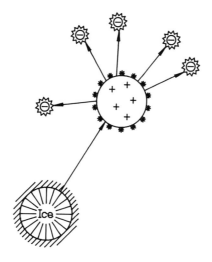

Figure 7.   Illustration of the Findeisen effect.

The second process is the Findeisen effect (1943), illustrated in Fig. 7. This is associated with the freezing of water drops. Initially, a thin surface ice layer forms when a water drop is cooled below the freezing point. Subsequently, as freezing continues, the interior also turns to ice and expands (because the density of ice is less than that of water). This leads to cracking of the surface layer and flaking off of little "chips." As in the previous case, the large particle is found to carry a positive charge and the small "chips" a corresponding negative charge.

There is no mention in the available literature whether, for instance, collisional break-up of conglomerate grains also leads to charge production, but based on the above evidence this seems likely. Some experimental work is clearly required here.

The third process is the Elster-Geitel effect (1885), which is illustrated in Fig. 8. This process enhances the charge separation, once it has started, in the following way: consider charge separation by gravitational sedimentation. Large grains sediment faster than small grains because the gas drag, which slows the descent, is proportional to the particle's cross section, whereas gravity acts on its mass. From the previous discussion this naturally produces a large-scale charge separation and the associated electric field. Large dust grains exposed to this electric field will become polarized, the resultant charge distribution is as shown in Fig. 8, with a preponderance of negative charges on the downward-facing side. As these polarized particles sediment, they may overtake and collide with smaller grains. If the collision is elastic, charge transfer during contact may then leave an enhanced charge difference between large and small grains. This electrical induction process can produce large charge separations. It has been found that in an electric field of 1 volt per meter charge separation of 10^5 to 10^6 electron charges per cubic centimeter can be obtained in this way.

If the collision results in coagulation, no net charge separation occurs. On the other hand, if the collision is disruptive, the previously discussed effects will even be enhanced because of the existence of negative surface charges at the impact site.

For the protoplanetary disk there appears to be no problem in generating charges on grains by essentially the same processes that play a role in the terrestrial atmosphere. We must qualify this statement by the constraint that this applies particularly well to the ice transition zone, where central portions of the disk are hotter than the melting (or sublimation) temperature of ice ($\sim$150 K at the ambient pressures) and the regions closer to the surface are cooler. Ice is of course a major constituent and disk models show that the transition zone is quite large, $\Delta r/r \approx 0.5$ (see, e.g., Morfill 1988). It would be interesting to test the charge generation mechanisms discussed above for other abundant constituents of cosmic dust particles and other gas pressures, to see whether charging might be possible throughout the protoplanetary disk. Here again is a need for experimental effort.

We have seen that the charge production processes on grains automatically lead to charge separation by vertical or radial size sorting. Hence the most obvious charge separation process to be considered for protoplanetary disks is essentially the same as in terrestrial lightning: precipitation. There is a difference, however. In terrestrial lightning we can basically consider a constant gravitational acceleration g and a constant gas density, everything happening in the vertical (z) direction. (This is not true for convective up-

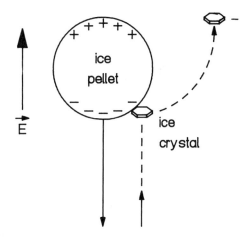

Figure 8.   Illustration of the Elster-Geitel effect. This inductive process is able to enhance considerably the charge density—all charges residing on dust particles.

welling, of course, but our meaning will become obvious soon.)

In a protoplanetary disk, particulate material sediments vertically (in the z direction) toward the disk midplane; the gravitational acceleration, almost exclusively due to the mass of the central PMS star (or protostar), is given by

$$g = -\frac{GM}{r^3} z. \qquad (14)$$

In other words g is strongly z-dependent, vanishing as z becomes small. The gas density also varies with z.

In addition, there is a radial size sorting due to the effect of gas drag on the particles. For the small grains we consider here, the radial drift increases with particle size, in other words, the positively charged large grains move towards the inner parts of the disk faster than the negatively charged small grains.

These two effects combine to give a charge-separation electric field which has a vertical (z) component and a radial (r) component. Thus, even in the presence of efficient vertical mixing, e.g., by strong turbulence, a net radial charge separation and associated electric field would be present in any case.

A steady-state electrical field strength can be calculated by considering current continuity. Although the radial electric field may be stronger than the vertical component, we illustrate the calculations by the latter in this chapter. A full discussion is given elsewhere.

We consider the general case of free electrons, positive and negative ions, small charged "dust" grains (index S) and large charged "dust" grains (or ice particles, index L) based on the preceeding discussions. Free charges are produced by cosmic-ray interactions or radioactive decay. The equation governing the current continuity is

$$n_L Q_L e \tau_{fL} \left[\Omega^2 z - \frac{eEQ_L}{m_L} \right] = n_S Q_S e \tau_{fS} \left[\Omega^2 z + \frac{eEQ_S}{m_S} \right]$$
$$+ \frac{n_e e^2 E}{m_e \nu_{eo}} + \frac{n_- e^2 E}{M_- \nu_{-o}} - \frac{n_+ e^2 E}{m_+ \nu_{+o}}. \tag{15}$$

The term on the left-hand side of Eq. (16) is the current due to large positively charged dust grains, the four terms on the right are the current due to small negatively charged dust particles, electrons, negative ions and positive ions, respectively. The number densities of the various species are denoted n and the masses are m with the appropriate subscripts added. The resulting electric field is E, the ions are assumed to be singly charged, the dust grains carry charges Q_L and Q_S in units of the electron charge, e, respectively. The conductivity of the free charge species is dominated by collisions with the neutral gas particles, collision frequency is ν, again with appropriate subscripts. Finally, the mobility of the dust grains depends on the frictional coupling to the gas—the appropriate time scales are denoted τ_{fL} and τ_{fS}, respectively. The quantity Ω is the angular Kepler frequency.

In order to build up an electric field of sufficient strength so that breakdown may occur, the conductivity of the medium has to be sufficiently small, because large-scale electric fields cannot be maintained otherwise. In practice this means that the dust grains have to be the dominant charge carriers. From Eq. (16) we can then derive the requirements for the free charge densities of ions and electrons. Based on the production mechanisms (cosmic rays, radioactivity), the speed of ion-molecule reactions and on the recombination processes (gas phase and dust surface recombination) we can then estimate requirements on the disk properties, such as gas density and surface density. We will not discuss such details here, however.

We see from Eq. (16) that the equilibrium solution for the electric field E depends on radial distance (through Ω) and on the vertical distance z. The vertical electric field goes to zero, of course, when z goes to zero—i.e., in the disk midplane, the radial electric field does not. Note that here we are only discussing the average field, not the fluctuating component introduced by, e.g., turbulence or compressive waves. For a discussion of the latter see Pilipp et al. (1992).

We may summarize our results in the following way:

1. For a height z above the midplane, which is between 0.1 and 1 of the disk scale height H, the required electric field strengths of about 1% of the breakdown value can be obtained.
2. This requires two particle populations.
3. Nonlinear charging processes must operate to produce surface potentials on the larger dust (or ice) grains of the order 1 volt.

4. Free charge must be negligible. This may mean significant shielding of even minimum ionizing cosmic rays, i.e., disk surface density in excess of 100 g cm^{-2}. Other nonthermal ionization sources—e.g., from the decay of radioactive nuclides—must also produce negligibly small amounts of free electric charge.

In order to get a breakdown in the gas which carries sufficient power, enough current must flow through the discharge channel. One way of achieving this is that the discharge connects two reservoirs of mobile charges located at different potentials. (Note, however, that on Earth, cloud to cloud lightning occurs as frequently as ground lightning without there being an obvious reservoir of mobile charges.) Clearly there is a surface layer of the disk, being irradiated by cosmic rays, ultraviolet photons, etc. which is such a reservoir. A second reservoir can be obtained in the vicinity of the disk midplane where the dust material (e.g., ices) sublimes and in this way "frees" the dust charges in the form of molecular ions. Thus we envisage breakdown to occur perhaps preferentially, but not exclusively, between these two reservoirs, and expect a tendency for this process to cluster around major sublimation zones of abundant chemical components, particularly ice.

Another requirement is the availability of sufficient energy. This increases with decreasing distance from the central protostar—the implication is that electrostatic discharges should be less frequent in the outer regions of the protoplanetary disk. An interesting consequence is that comets, if they formed sufficiently far outside should not exhibit signs of flash heating—provided electrostatic discharges are responsible for these signatures. On the other hand, if comets formed in the inner solar system and were then scattered out by proto-Jupiter, they might well have sampled major condensation zones and their component particles might therefore show signs of flash heating by electrostatic discharges.

It is widely believed that Jupiter was located at the "ice zone" when it formed (at around 150 K). The evidence comes from disk models as well as planet and satellite composition measurements (see, e.g., Morfill 1985; Stevenson and Lunine 1988).

B. Heating of Dust Grains in the Discharge Channel

Assuming that the energetics and the processes discussed earlier conspire to produce electrostatic discharges, it is then reasonable to ask the question: how does this affect dust grains in the discharge channel and can we reproduce the observed signatures?

We mentioned earlier that a minimum of 300 to 400 J cm^{-2} are required for the time-integrated heat flux. Including losses, e.g., by radiation from the heated grains, latent heat of vaporization, kinetic energy taken away by the evaporated parts of the embedded dust particle, etc., the energy flux has to be raised considerably.

The full physics of dust in a discharge channel is too involved to describe here. It will be discussed elsewhere. Here it is sufficient to mention some of the stages of development. These are, in terms of increasing time scales: charging of the dust to a negative potential in the hot discharge plasma, heating of dust particles by ion and electron impacts taking account of the effective cross section for each species, electron-ion thermalization, associated change in the grain potential and effective cross section, expansion of the discharge channel, adiabatic cooling, recombination of the plasma in the gas phase, collisional ionization, heating of dust by impact of neutrals as well as ions and electrons, radiative loss, latent heat effects, losses by evaporation, etc.

One of the key results is that the time integrated energy flux turns out to be a constant at around 10^4 J cm^{-2}. Without going into all the details of a full calculation, which leads to this result, we can describe an approximate approach, which illustrates the physics quite well.

The energy flux onto a grain due to collisions with plasma particles of type i is

$$F = \sum_i n_i A_i c_i E_i \tag{16}$$

where n_i is the plasma density, A_i the cross section, c_i the velocity of the particles and E_i their energy at impact. Because of Coulomb effects, the cross sections of different species will be different, too. Integrated over the duration of the heat pulse and normalized to the grain cross section πa^2, the energy deposition per cm^2 per heat pulse is

$$E_a = \frac{1}{\pi a^2} \int F \, dt. \tag{17}$$

The discharge channel is assumed to be fully ionized, with initially (after thermalization of electrons and ions) a plasma temperature of about 30 eV. In such a plasma, a dust grain gets charged to a potential $e\phi \approx -2.5kT$. The cross section for ion bombardment is enhanced by the Coulomb forces to a value

$$A = \pi a^2 (1 + e|\phi|/kT) \tag{18}$$

and the energy of the plasma ions, when hitting the dust grains is $|\phi|$, because they are accelerated in the electrostatic field of the grain.

Utilizing our earlier model of an expanding channel, we may write Eq. (17) as

$$E_a = \frac{e|\phi|}{kT_i}\left(1 + \frac{e|\phi|}{kT_i}\right) \int n_i c_i \frac{3}{2} kT_i \, dt \tag{19}$$

where we only need consider heating by ion impacts. With respect to the initial channel radius R_o, and an initial temperature $T_H (\approx 30$ eV) we then get on changing $dt = dR/c_s$, with c_s the sound speed in the unperturbed disk:

$$E_a = 13.1 \, n \, c_H \, k \, T_H \, \frac{R_o}{c_s} \int_1^{X_M} \left(\frac{1}{x}\right)^2 \cdot \left(\frac{1}{x}\right)^{3(\gamma-1)} dx. \tag{20}$$

We have defined $x \equiv R/R_o$ and

$$X_M \equiv \left(\frac{T_H}{T_M}\right)^{1/2(\gamma-1)}. \tag{21}$$

The solution is:

$$E_a = 4.38 \, n \, c_H \, k \, T_H \, \frac{R_o}{c_s}. \tag{22}$$

For a discharge to take place, collisional ionization must be possible, which means that $T_H \approx 30$ eV. It cannot be much lower than this value, nor should it be much higher.

The size of the discharge channel should be proportional to the mean free path of the electrons in the gas; this means $R_o \sim 1/n$. Scaling this relationship with lightning in the Earth's atmosphere, and correcting for the cross sections (H_2 as opposed to N_2 and O_2) yields

$$E_a = 4.5 \times 10^3 \frac{c_H}{c_s} T_H. \tag{23}$$

We have used the observed thickness of lightning channels of ~ 1 mm in the terrestrial atmosphere with a density of 2×10^{19} cm^{-3} for scaling. This yields finally

$$E_a = 5.36 \times 10^3 \frac{T_{30}^{3/2}}{T_{300}^{1/2}} \text{ J cm}^{-2}. \tag{24}$$

T_{30} is the temperature corresponding to 30 eV and T_{300} is the ambient disk temperature taken to be 300 K for this quantitative example. The greatest uncertainty in Eq. (24) is the scaling relation used from terrestrial lightning.

The interesting consequence of Eq. (24) is that the result does not depend on the local gas density, only on the temperature (this controls the channel expansion). In hotter regions of the disk, the energy available per cm^2 is less than in cooler regions of the disk. This may be significant in the context of rim formation for CAIs, which is conjectured to occur in the hot inner portions of the disk, as compared to chondrule formation, which is thought to occur in the ice transition zone at greater radial distances.

The time scale for the heat pulse imparted to the grain depends inversely on the gas density. In the Earth's atmosphere it is of the order milliseconds, in the protosolar nebula at a representative gas density of 10^{14} cm^{-3} it is of the order 10 to 100 s. The integrated heat flux is the same, however.

From these considerations and the quantitative estimates associated with them it appears plausible that chondrules may be an observable consequence of nebular lightning. At the same time, it seems acceptable that the rims found on CAI's could also be formed by such a process, albeit at a different location—further in towards the protostar—than the chondrules. The density increases with decreasing distance, which in turn implies that the duration of the heat pulse decreases. The CAI rims seem to have been produced by a more rapid heat flash than that required for the chondrules.

VI. DISK-CORONAL MAGNETIC FLARES

A. Magnetic Fields in the Protosolar Nebula

Many cosmical systems have magnetic fields that are strong enough to influence evolutionary and dynamical processes. These magnetic influences arise because a magnetic field exerts force on the gas within which it is embedded, through the Lorentz force:

$$\mathbf{F}_L = \frac{1}{c}\mathbf{j} \times \mathbf{B} = \frac{1}{4\pi}(\nabla \times \mathbf{B}) \times \mathbf{B}. \tag{25}$$

Generally speaking, the Lorentz stress has two main effects that are of interest in understanding the behavior and evolution of protoplanetary disks. Because the Lorentz force can exert strong torques and therefore efficiently transmit angular momentum, magnetic stresses can play an important role in the evolution and radial accretion of disks. In addition, magnetic fields distorted away from their equilibrium geometries can store large amounts of energy, which are sometimes released in rapid explosions or flares. Such explosive energy releases are observed to occur in magnetized bodies in the solar system: solar flares and geomagnetic storms are among the most familiar of these manifestations; magnetic flaring is thought to be responsible for observed outbursts also in many distant astrophysical systems.

The empirical evidence for strong magnetic fields in the protosolar nebula is found in primitive meteorites (for reviews, see Levy and Sonett [1978] and Cisowski and Hood [1991], as well as references therein). Remnant magnetism (frequently called "remanence") is found in a wide variety of meteorites, spanning the different meteorite chemical classes as well as the varying degrees of meteorite metamorphism. For the present discussion, the most primitive meteorites—the carbonaceous chondrites—are most pertinent. While other classes of evolved meteorites have been altered by a variety of metamorphic processes on parent bodies, and in some cases may record parent-body magnetic fields, the carbonaceous chondrites appear to be relatively unevolved products of nebular accretion, and their remanence probably records nebular magnetic fields. The remanence of carbonaceous chondrites is very complicated, occurring in the rock at a variety of strengths over a variety of spatial scales. Although significant ambiguities and uncertainties remain in the interpretation of carbonaceous chondrites, some characteristics seem to emerge from the data. Generally speaking, the inferred magnetizing-field intensities are higher for the smaller rock components than for large pieces of the rock. Inferred magnetizing intensities for whole-rock pieces typically range from a tenth of a Gauss to approximately a Gauss. Inferred magnetizing intensities for small components have ranged as high as a few Gauss, and in isolated cases to as much as ten Gauss or higher, though questions remain about the reliability of the very highest of these intensities, owing to the possibility of magnetic contamination of the samples.

The magnetizing-field intensities inferred from measured meteorite remanence are "model" intensities, the interpretation of which still involves several uncertainties. Most of our understanding of rock magnetism is derived from the much larger body of experience measuring and interpreting paleomagnetic intensities of terrestrial rocks. First, detailed differences between terrestrial rocks and meteorites remain to be fully studied and understood, especially with respect to the physical processes by which meteorities acquired their remanence, whether through thermal, physical, or chemical changes in the presence of magnetic fields. Second, it is important to realize that the remanence does not record the actual magnetic field at some point in space at some time. Rather, what is recorded is the magnetic field in the frame of the magnetized rock, averaged over a time period as long as the time scale of remanence acquisition. Thus, for example, a rotating object acquiring its magnetization over a time scale longer than the rotation period will record only the static component of the magnetic field in the rotating frame, that is to say the projection of the ambient magnetic field along the axis of rotation. Similar considerations apply to the fact that meteorites may accumulate and acquire their magnetization over periods of time longer than the orbit period and even over periods of time comparable to or longer than the oscillation periods of magnetic dynamo modes. These considerations suggest that the most reliable indicators of actual nebular magnetic field strengths should be the highest of the inferred magnetizing fields recorded in the smallest objects, if, of course, both the measurements and the assumed relationship between recorded remanence and magnetizing-field intensity are reliable, and assuming that the magnetizing fields were of nebular origin. Altogether then, leaving aside the most extreme of the inferred magnetizing fields, the meteorite remanence record suggests the presence of magnetizing fields as intense as a few Gauss existing in the nebula at a distance of several AU from the center, at least episodically.

B. The Influence of Magnetic Stresses

The possibility of nebular magnetic fields with intensities falling in the range of a tenth of a Gauss to a few (and possibly, though the evidence is weak only suggestive, ten or more) Gauss (Cisowski and Hood 1991) is very provocative in terms of disk characteristics and disk evolution. This can easily be seen by considering the rate of angular momentum transport induced by such a magnetic field. Consider the angular momentum $\mathcal{L}$ contained in an annulus in Keplerian motion and having radius R, width ΔR, thickness $2h$, and density ρ:

$$\mathcal{L} \sim 4\pi \Delta R h \rho \sqrt{GMR^3}. \tag{26}$$

A magnetic field having toroidal and poloidal intensities B_ϕ and B_p, respectively will exert a torque $\mathcal{T}$ on the annulus given by

$$\mathcal{T} \sim \langle B_\phi B_p \rangle R^2 h. \tag{27}$$

Combining Eqs. (26) and (27), the time scale τ for angular momentum transport is

$$\tau \sim \frac{\mathcal{L}}{\mathcal{T}} \sim \frac{4\pi \Delta R \rho}{\langle B_\phi B_p \rangle} \sqrt{\frac{GM}{R}}. \tag{28}$$

Taking $\rho \sim 10^{-9}$ g s^{-1}, $R \sim 3$ AU, $\Delta R \sim 1$ AU, and $M \sim M_\odot$, we find $\tau \sim 10^4 \langle B_\phi B_p \rangle$ yr. Thus for $\langle B_\phi B_p \rangle^{1/2}$ ranging from a nominal value of about 0.1 Gauss to the limit of plausibility in the vicinity of 10 Gauss, we find τ in the range 10^6 yr to 10^2 yr. Thus the magnetic-torque-induced angular momentum transport times associated with nebular magnetic field strengths inferred from meteorite remanence correspond with the time scales thought to typify protostellar-nebula evolution. Moreover, these magnetic field strengths fall in the range that could plausibly be produced by a dynamo generation process acting in such a nebula (Levy 1978; Levy and Sonett 1978; Levy et al. 1991; Stepinski and Levy 1990).

Now consider the effect of the magnetic pressure. For a nebula with gas density $\sim 10^{-9}$ g cm^{-3} and temperature ~ 300 K, the pressure is about 10 dynes cm^{-2}. In comparison, a five-Gauss magnetic field—near upper end of plausibility for the protosolar nebula—exerts a magnetic pressure of about one dyne cm^{-2}, some 10% of the gas pressure; even a one Gauss magnetic field exerts a pressure approaching 1% of the gas pressure.

Because of the magnetic pressure, a magnetic field tends to induce buoyancy in the gas through which it threads. The origin of this buoyancy is easy to see. For simplicity, imagine a uniform magnetic field of intensity B confined within a straight cylinder, and suppose that both the interior of the cylinder and the surrounding space is filled with an ideal gas having temperature T and exterior number density n_0. Then pressure balance between the interior of the cylinder and the surrounding gas requires that

$$n_0 kT = n_i kT + \frac{B^2}{8\pi} \tag{29}$$

where n_i is the number density of the gas within the cylinder. Obviously, from Eq. (29) $n_0 > n_i$. By virtue of the lower gas density within, the magnetized cylinder is buoyant in a gravitational field, and, all things being equal, it will rise. Of course, the real situation is more complicated than this simple analysis. The buoyant magnetized gas might be produced in a stratified equilibrium, which can be unstable to breaking up in a kind of magnetically driven Rayleigh- Taylor instability, which has been analyzed in a number of different contexts (see, e.g., Parker 1955, 1966,1979; Galeev et al. 1979b). This buoyancy is a manifestation of a more general physical property of magnetic fields. It can be seen from a generalized virial theorem (Parker 1954) that a magnetic field always produces net expansive stress in a system, which, in equilibrium, must be confined by other stresses; in cosmical systems, the confining stress is normally gravity. Because the force of gravity acting on a magnetic field itself is negligible, a gravitating system, partially supported by magnetic stress, can always lower its overall energy by releasing the magnetic field and contracting. Generally speaking, unless thermal effects produce a significant adverse temperature contrast that can negate the buoyancy of the magnetic field (Vainshtein and Levy 1991), we expect that the magnetic flux will bubble buoyantly out of the disk. The strongly magnetized solar corona is one of the most familiar and best studied manifestations of a magnetic field's propensity to bubble out of gravitating systems.

Before passing on to the main topic of this section, it is appropriate to consider briefly the magnetic field's coupling to the nebular gas. Indeed, this is among the most significant and puzzling questions in understanding possible magnetohydrodynamic and plasma effects associated with protoplanetary disks. The primary question involved here is the ionization level (and the resulting electrical conductivity) of the gas. There are several principal sources of ionization in such a nebula: thermal ionization, cosmic-ray ionization, and ionization produced by nonthermal processes intrinsic to the disk gas. In the inner parts of the disk, where the temperature exceeds about 10^3 K, thermal ionization dominates and the magnetic field is inevitably closely coupled to the gas. Farther from the center, however, well beyond an astronomical unit, the temperature is normally so low that thermal ionization plays no significant role. In these regions, both cosmic rays and live radioactive nuclides can, at least in principle, produce enough ionization to couple the gas and the magnetic field (Consolmagno and Jokipii 1978; Levy 1978; Levy et al. 1991; Umebayashi and Nakano 1988). Briefly, two physical questions need to be considered. First it is necessary that the electrical conductivity be sufficiently high that the time scale for Joule dissipation be longer than the time scales characteristic of the fluid motion; this is a purely kinematical constraint, and is necessary in order for any magnetic field to be coupled to the fluid. Second, it is necessary that the time scale for the ions to drift a significant distance through the neutral gas—driven by the magnetic field stress, a phenomenon frequently referred to as ambipolar diffusion—is not shorter than the fluid's relevant dynamical time scales. This latter constraint is a dynamical one, in that it depends on the strength of the magnetic field. Generally speaking, disk magnetic fields of the order of a Gauss in intensity, and consistent with the meteorite paleomagnetic record, seem not be ruled out by either the electrical-conductivity or the ambipolar-diffusion constraints (Levy et al. 1991); however, pinning down the electrical conductivity of protostellar nebulae, as a function of space and time, remains a pressing challenge to understanding the physical processes in these systems.

C. Coronal Flares

If strong magnetic fields occur in protostellar disks—as seems to be indicated by the meteorite remanence—then it is likely that the disks have magnetized coronae in the tenuous regions above their midplanes. Experience with similar systems suggests that such coronae are likely to be the sites of explosive magnetic flaring. Flares of this kind occur when the magnetic field connects two highly electrically conducting regions of space, in one of which (the fluid body, say) the motion of field is controlled by the fluid, while in the other (the "corona") the field itself provides the dominating force and controls the fluid motion. If the corona were vacuum, then the magnetic field there would be a potential field, completely determined by the boundary conditions imposed on the surrounding surfaces. As the magnetic field in the fluid body was dragged around by the fluid motion, this boundary magnetic field would change, producing a changing magnetic structure in the corona. The coronal field structure would evolve rapidly, always adopting the unique vacuum structure consistent with the boundary conditions, which structure is also the lowest energy state of the field consistent with the boundary conditions. Changes in the field structure would propagate through the corona at the speed of light.

Now consider the situation when the corona is not vacuum, but rather is filled with tenuous, electrically conducting gas; two differences result in this case. First, changes in the magnetic structure no longer propagate from the boundary at the speed of light; rather they propagate at the Alfvén speed. But more importantly, the coronal magnetic field is no longer able freely to assume a vacuum, potential-field structure in response to changing conditions on the boundary. Under these conditions, the magnetic field lines can be regarded as continuous and ponderable, and as maintaining their topological structures under distortion. Because of this, the field lines cannot freely pass through one another and the coronal magnetic field is prevented from rearranging itself into a potential-field structure matching the boundary conditions imposed from the fluid body below. As a consequence, the magnetic field lines become tangled and distorted away from equilibrium, and the resulting magnetic-field structure stores large amounts of energy in its highly disequilibrated state.

In a protostellar disk, the coronal magnetic field will become distorted both by the radially shearing motions associated with the nearly Keplerian overall motion of the disk as well as by the convective and turbulent movement of the gas within the disk. Inasmuch as flaring release of energy associated with rapid relaxation of distorted magnetic structures seems to be a common manifestation in cosmical systems, it is instructive to explore the possibilities in a protostellar disk.

A variety of specific mechanisms have been explored to account for magnetic flares (see, e.g., the review and references in Spicer et al. 1986), and much remains to be understood about the details. For the purpose of this discussion, we will focus on the possibility of rapid magnetic reconnection as a flare mechanism; this will capture the essential physics of the process and illustrate the magnitudes of energies that can be released. For specificity, consider the classical magnetic reconnection geometry sketched in Fig. 9. Regions of oppositely directed magnetic field are separated by a "neutral layer." The magnetic field collapses from the sides (in this diagram) toward the neutral layer; magnetic energy is converted and escapes from the system, largely in the form of fast, hot plasma and energetic particles. The energy flowing into the system is largely the energy of the magnetic field, $B^2/8\pi$ erg cm^{-3}. If u is the "merging" speed of the inflow, and if L is the characteristic transverse scale of the system, then the overall rate of energy release is

$$\dot{E} \sim uL^2\frac{B^2}{8\pi} \text{ erg s}^{-1}. \tag{30}$$

Observations of energy release rates in solar system objects (such as the solar corona and the terrestrial magnetosphere) indicate that the magnetic merging speed u must be of the order of a tenth to several tenths of the Alfvén speed, $B/\sqrt{4\pi\rho}$. Theoretical analyses (Petschek 1964; Sonnerup 1970; Yeh and Axford 1970; Yeh 1976) indicated that such high rates of energy release may occur as the magnetic tension in the reconnected field lines ejects gas from the central layer, allowing rapid collapse of the surrounding magnetic field and gas, though the physical details remain obscure.

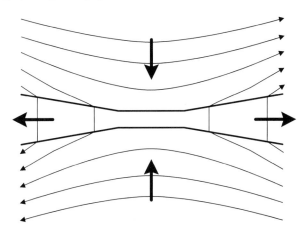

Figure 9.   Idealized sketch of a magnetic field line reconnection region. Magnetically driven gas flow (heavy vertical arrows), in from the sides, carries the magnetic field lines (light lines). Magnetic reconnection in the central region relieves topological constraints on the field's structure and the resulting relaxation of the stressed magnetic structure releases stored magnetic energy in the ejected fluid (heavy horizontal arrows). The energy conversion essentially takes place in the central reconnection region bounded by the medium-weight lines in the figure.

Here we are concerned with two properties of such magnetic flares: the energy *flux* emerging from the flare and the total energy emitted in a typical event. Using a simple model for reconnection flares (Petschek 1964), which, while it is not likely to represent an exact model for real flares, probably captures the main features of fast reconnection. Levy and Araki (1989) estimated that the energy flux F emerging from the flare is

$$F \sim \frac{B^3}{8\pi\sqrt{4\pi\rho}} \; \text{erg cm}^{-2} \, \text{s}^{-1}. \tag{31}$$

Interestingly, the outgoing energy flux corresponds to the magnetic energy density in the incoming fluid, moving at the Alfvén speed. At least within the context of this mechanism, the energy *flux* is independent of the merging speed u. The total rate of energy release still depends on u in that the fixed flux given in Eq. (31) emerges over a larger area, and is in fact proportional to u.

Inasmuch as magnetic flares are so commonly associated in energetic outbursts in cosmical systems, it is natural to inquire whether such flares could provide the transient heating mechanism that melted chondrules. Imagine that milligram-sized nebular dust accumulations are heated by the magnetic-flare energy flux given in Eq. (31), and that the cooling is by blackbody radiation from the heated droplet, with emissivity ϵ. Crudely, the steady-state temperature to which the chondrule precursor can be heated is

$$T_{eq} \sim \left(\frac{B^3}{16\pi^{5/2}\sigma\epsilon\sqrt{\rho}} \right)^{1/4} \tag{32}$$

where we have neglected the much lower external temperature into which the heated material is radiating. The maximum temperature to which the chondrule precursor can be heated depends only very weakly on the structure of the particle—both the heating and the cooling rates are proportional to the surface area, and in Eq. (32) we have made the most conservative assumption that the particle is heated from only one side, while it cools into 4π steradians of cold space. In addition, for energy fluxes capable of heating matter into the range of interest for making chondrules, the heating time is very short, of the order of a second or less (Levy and Araki 1989).

Now we turn to consider the conditions that must be satisfied if chondrules are to be the result of magnetic-flare-driven heating events. Taking 1400°C as the temperature needed for incipient melting, and taking the high-plausible end of the magnetic field intensities inferred from meteorite remanence of 5 Gauss, we find from Eq. (32) that the flares would have to occur in an ambient gas density of about 10^{-18} g cm^{-3}. (For this estimate, we have made the conservative assumption that, the emissivity $\epsilon = 1$.) Interestingly, this density is characteristic of the cloud cores from which stars are believed to form (see Boss et al. 1989, and references therein); however, such cores are some 3 orders of magnitude larger in spatial scale than the far more compact nebular disks considered in the present discussion. More relevant perhaps is the variation of gas density with altitude above the midplane of a disk. With $\rho(z) = \rho(0)\exp[-z^2/\Lambda^2]$, $\Lambda^2 = 2kT\,r^3/GM_\odot m$, and taking $\rho(0) = 10^{-9}$ g cm^{-3} and using a disk temperature of 200 K, we find $\rho \sim 10^{-18}$ g cm^{-3} at about 1 AU above the midplane, when $r \sim 3$ AU. Thus, if energy released in magnetic flares is to make chondrules, then the flares must occur at least 1 AU above the disk midplane. And, indeed, it is problematical whether a real disk corona (with its possibly higher temperature and highly dynamical state) will have a number density as low as 10^6 cm^{-3} at that altitude.

D. The Locale of Chondrule Formation

It is important to realize that the low-ambient-density requirement severely constrains the environments in which flares could occur with energy fluxes sufficient to melt chondrules—at least with magnetic-field intensities limited by even the highest values plausible for a nebula. Although there is a range of combinations of field strength and ambient density that can produce the requisite energy fluxes, the restriction on the strength of plausible nebular magnetic fields essentially limits the occurrence of possible chondrule melting flares to very low density regions. The maximum temperature scales roughly as the eighth rcot of the density. [For high densities and low temperatures, a correction is needed in Eq. (32) to account for the temperature of the background into which the matter is radiating.] A consequence of this dependence on density is that flares embedded in the disk itself are wholly incapable of raising temperatures significantly: at $\rho \sim 10^{-9}$ g cm^{-3}, T_{eq} given by Eq. (32) is only slightly higher than 100 K, colder than the prevailing ambient temperature.

The form in which energy emerges also constrains the conditions under which magnetic flares could melt chondrules. To estimate the form in which the bulk of the energy emerges from such a flare, consider that the magnetic energy density $B^2/8\pi$ partitioned among the particles with number density $\sim 10^6$ cm^{-3} yields approximately 1 MeV particles for a magnetic field of 5 Gauss. Or looked at another way, with the same parameters, the characteristic Alfvén speed is about 1.4×10^9 cm s^{-1}, which again corresponds to ~ 1 MeV per particle. Thus is it is reasonable to expect much of the energy to emerge from these events in the form of 1-MeV protons. These energetic particles will be channeled along the magnetic field lines, down toward the higher-density layers of the disk, where they encounter the bulk of the matter, possibly melting dust assemblages to make chondrules and inducing radiation—analogous to a solar flare or to a terrestrial aurora—which might be an observable manifestation of the flare.

The path length of 1-MeV protons through matter is very short, about 10 mg per cm^{-2}. Thus, although the actual melting of material may occur some distance below the altitude of the flares, there cannot be more than ~ 10 mg cm^{-2} of intervening matter. At the ambient densities required, and in the context of a simple nebular model described above, the chondrules melting could happen at about a scale height below the minimum height of the flares. Altogether then, in a strongly magnetized, low-density nebular corona, chondrule-melting flares could occur at about four scale heights above the disk. Energy in the form of energetic particles could be channeled downward to about one scale height below, where the actual chondrule formation would take place, assuming the presence there of the requisite milligram assemblages of silicate dust. Implicit in the discussion of disk coronal flares is the assumption that the coronal magnetic field strength (at an altitude of about 1 AU above the midplane) is comparable to the stronger of the magnetic fields inferred from the most straightforward interpretation of the meteorite magnetization record. This question is problematical, and requires detailed analysis of the possible structures of disk-corona magnetic fields as they are tangled by motions in the disk and wound up by the disk's differential rotation.

In order for chondrules to have formed in this manner, it would have been necessary for milligram assemblages of precursor silicate dust to have been present at about 3 scale heights above the disk midplane. There are two possibilities. Dust assemblages in the form of very fluffy fractal aggregates could, in principle, be lofted to the necessary altitude. Alternatively, dust assemblages could have been heated during their infall from the precursor cloud core, assuming that such massive assemblages could have grown there in the first place.

Here we will consider the possibility that the material was lofted to high altitude, inasmuch as that seems to offer the greatest freedom in terms of when, during the development of the nebula, chondrule formation might have taken place. In the extreme case, consider assemblages of fractal dimensionality ~ 2, so that the mass grows as a^2, where a is diameter of the assemblage. In that case, an assemblage made from $10\,\mu$m grains presents about a square centimeter of cross-sectional area to the gas blowing by it. Because the particle and aggregate size is much smaller than the ambient molecular mean-free path, the gas drag is of the order of $a^2 n_g m c_s v$, where n_g is the ambient gas density, m is the mean molecular weight of the gas, c_s is the sound speed, and v is the velocity of the gas blowing by the assemblage. Taking $c_s v \sim 10^{10}$ cm s^{-1}, and $m \approx 3 \times 10^{-24}$ g, we find that a grain can be lofted against the force of gravity so long as

$$n_g \gtrsim \sigma_m g_v / m c_s v \tag{33}$$

where σ_m is the projected surface density of the precursor dust assemblage. At 3 AU, the vertical component of gravity, g_v is about 10^{-2} cm s^{-2}, and with $\sigma_m \sim 10$ mg cm^{-2} and $m \sim 10^{-24}$ g, we find that the precursor assemblages can be lofted against gravity where $n_g \gtrsim 10^9$ cm^{-3}.

In principle then, very fluffy fractal aggregate precursors of chondrules could be lofted by gas motions to altitudes at which they could be melted by energetic particles accelerated in disk coronal flares. However, for this result to be meaningful, it is necessary that the precursors be lofted to such altitudes in sufficient numbers to account for the apparently large fraction of some meteorite material that is found in the form of chondrules—some tens of percent in certain classes of meteorites. To see the magnitude of this problem, consider that a stringent limitation on the altitude of chondrule formation by disk flares comes from the requirement that the flare particles be able to propagate to the site of chondrule formation without losing their energy. As noted earlier, this requires that the intervening material not exceed more than about 10 mg cm^{-2}. Consider a disk with projected mass density above the midplane of 500 g cm^{-2}. In that case, in order for a dust aggregate to be exposed to a chondrule-melting flare, it must rise to an altitude such that the layer intervening between it and the flare contains only one part in 50,000 of the disk mass. Making the crude assumption that, so long as Eq. (33) is satisfied, the chondrule precursor are more or less tied to the gas, then at any time, only a very small fraction (a part in 50,000) of the potential chondrule material would be at risk of being heated. However, suppose that the convective motions are a few $\times 10^4$ cm s^{-1}, then the convective turnover time (for the largest convective eddies) could be of the order of 10 yr. Roughly speaking, with reasonably effective mixing, it would take some 500,000 yr to mix a significant fraction of the potential chondrule precursors to the required altitude. This rate of mixing is sufficient to expose a large fraction of nebular silicates to chondrule melting, assuming that they are assembled into sufficiently large and fluffy assemblages.

Perhaps the more severe constraint involves the formation and persistence of such fluffy, fractal aggregates in the nebula. Numerical simulations suggest that fractal aggregates of dimensionality $\sim$2 can be formed as a consequence of Brownian motion dominated accumulation of smaller dust clusters. However, macroscopic motions, such as those associated with turbulence, may prevent the formation of large, low-dimensionality fractal dust accumulations (J. Blum, personal communication). In this case, it may be that dust balls lofted from below cannot provide the raw material for chondrules melted by disk coronal flares. It is possible that dust accumulations entering the disk during accretion from the precursor cloud could provide the raw material that is melted to form chondrules, a suggestion that has been made in a different context by Wood (1984).

E. The Energy Production of Chondrule-Making Flares

Chondrules heated by disk-coronal flares described here have characteristic heating and intrinsic cooling times of the order of a second—short in comparison with the rate of temperature change thought to characterize chondrules as they cool through the solidus. Assuming that the cooling occurs into a cool background, this suggests that the cooling rates were controlled by variations in the heat source. This assumption allows us to estimate the total energy evolved in such flares. Taking τ_f to be the flare duration during the MHD collapse phase, the total evolved energy is (Levy and Araki 1989),

$$E_f \sim \frac{B^5}{64\pi^{5/2}\rho^{3/2}}\tau_f^3.$$
(34)

With $B \sim 5$ Gauss and $\rho \sim 10^{-18}$ g cm^{-3}, we find for the energy of a flare:

$$E_f \sim 3 \times 10^{33}\tau_{f1000}^3 \text{ erg}$$
(35)

where $\tau_{f_{1000}}$ is the duration of a flare measured in units of a thousand seconds. Thus the total energy output of such a flare falls in the range of flares observed to be associated with T Tauri stars (Feigelson 1982).

VII. SUMMARY

In this Chapter we have discussed several nonlinear dynamical behaviors of the protoplanetary nebula that may provide mechanisms capable of explaining some of the major mysteries presented by observations of disks and solar system matter surviving from the time of planet formation. We have concentrated on only a few processes; others are discussed elsewhere in this book. We have focused on the possible influences of nonlinear collective behaviors on angular momentum transport in disks, and on the possible manifestations of electrostatic and electromagnetic effects. Our purpose has been to provide a tentative physical context for a variety of observations, both solar-system derived and astronomical in origin, that is firmly within the confines of known physical phenomena, satisfying the essential constraints on energy, momentum and continuity.

It has become clear that "quiescent" or quasi-linear disk models are unable to account for a number of important observed phenomena—hence the growing interest in nonlinear processes. For example, the rapid temporal evolution of disks is inconsistent with the time scales that could be driven by classical molecular processes. In a related fashion, the observed temperatures of disks point strongly toward mechanisms that tap the disk's underlying gravitational energy source more rapidly than can classical molecular processes. Therefore, even though the details still remain obscure, it is clear that global disk evolution is driven by collective processes. There are a number of candidate mechanisms, including turbulent transport, collective gravitational effects and magnetic torques.

Among the most provocative observations is the co-existence of chondrules—relics of short-lived hot events—and matrix material—which does not appear to have been similarly heated—in the same meteorites. This observation points strongly toward extremely spatially and temporally heterogeneous thermal structures occurring in the protoplanetary disk. Observations of other protoplanetary disks around young pre-main-sequence stars indicate a cool environment with temperatures less than a few hundred degrees Kelvin. These apparently conflicting observations, together with the assumption that our protoplanetary disk was not in any way special, represent a major challenge in cosmogony.

The observational evidence for short-term heating events is especially provocative. As we have discussed, a protoplanetary nebula has more than sufficient energy to drive the high-temperature episodic events inferred from the meteorite record. With reasonable estimates of the efficiency, the observed abundance of meteorite manifestations (such as chondrules) can be accommodated. However, as we have discussed, a rich selection of physical phenomena may be needed to account for these things, requiring detailed consideration of the relevant dynamics. A variety of possible chondrule-forming mechanisms have been proposed (see, e.g., Hood and Horanyi [1991] and the review given in Levy [1988]); in this chapter, we have focused on the possibility that chondrules could result from disk lightning or disk coronal flares.

Large electrostatic discharges (lightning) may occur in nebulae with power, frequency and distribution sufficient to account for chondrule formation. However, it should be borne in mind that, so far at least, arguments for the effects of nebular lightning are plausible rather than compelling. On the other hand, it should also be borne in mind that lightning, as an important form of energy dissipation in our own atmosphere, would not have been predicted to occur had it not been for the well-known empirical evidence of its existence. Moreover, evidence for lightning has been found in other planetary atmospheres (Jupiter and Venus, for example), suggesting that it may be a widespread and common phenomenon. In this sense, "nebular lightning" is at least a plausible candidate mechanism for localized, episodic heating events and might furnish an explanation for a significant portion of the thermal complexity observed in meteorites, though considerably more analysis is needed to elucidate fully the mechanisms and effects of possible nebular lightning.

Experience with a wide range of cosmic systems teaches us that magnetic fields are frequently important both for dynamical evolution and for large, transient energetic events. Such "flares" are known to occur in a variety of objects ranging from planetary magnetospheres to stellar coronae, as well as high-energy astrophysical sources. Evidence from primitive meteorites suggests the presence of a strong magnetic field in the protoplanetary nebula. Flares produced in a disk corona during the evolution of this magnetic field provide another possible mechanism that could account for the transient heating events that seem to be recorded in the primitive meteorites. However, even if disk coronal flares are prevalent, chondrule production by such flares presents problems. Two problems stand out. The presence of sufficient quantities of sufficiently massive (~ 1 mg) precursor aggregates at high enough altitudes to be heated by coronal flares is one such problem. The other is the question of the actual magnetic field strengths prevailing in the nebula, especially at several scale heights above the disk midplane.

Altogether, evidence points to the occurrence, in the protoplanetary nebula, of unexpected transient events that released large amounts of energy in localized volumes of space over short periods of time. Such events would, *a priori*, not be expected to occur. However, other cosmic systems with large amounts of free energy to expend are known to exhibit such unexpected and complex behaviors. Provoked by meteoritic and other evidence, we confront the likelihood that nonlinear collective phenomena occurred in the protoplanetary nebula and had important consequences. Our comprehension of these phenomena remains in an early and incomplete state. While final answers are not yet in hand, it seems likely that a complete understanding of the conditions and processes associated with the formation of stars in general, and of our own solar system in particular, will have to encompass physical processes as complex as those that we confront in such better known systems as the Earth's atmosphere and magnetosphere, as well as the Sun's corona.

METEORITIC CONSTRAINTS ON CONDITIONS IN THE SOLAR NEBULA

H. PALME
Max-Planck-Institut für Chemie

and

W. V. BOYNTON
University of Arizona

Conditions in the early solar nebula may be inferred from the bulk chemical composition of undifferentiated, or chondritic, meteorites as well as from the bulk composition of the planets. Differentiated meteorites, those processed in the interior of planets, may also be useful if reasonable estimates of the chemical composition of their parent bodies can be made. From the bulk chemical compositions of meteorite parent bodies and planets, representative chemical trends can be deduced and some constraints for large-scale conditions in the inner solar nebula at the time of formation of solid matter can be inferred. In contrast, clasts, inclusions and mineral fragments in undifferentiated meteorites provide detailed information about conditions in local environments. All chemical fractionation trends observed in bulk chondritic meteorites require formation temperatures at around the condensation temperature of metallic iron or magnesian olivine (1200–1440 K). The pronounced depletion of moderately volatile elements (Mn, Na, Ga, Zn, S etc.) is a characteristic signature of most chondritic meteorites. Several possibilities for establishing this depletion pattern are discussed, and it is concluded that they are most likely due to fractionation during condensation requiring temperatures of 1200 to 1400 K. Evidence for processing bulk meteorite material at higher temperatures is lacking. Reconstruction of the original composition of planets and parent bodies of differentiated meteorites leads to fractionation trends similar to those observed for chondritic meteorites. It is therefore concluded that the inner solar system, including most of the asteroidal belt, had at one point in the evolution of the solar system experienced temperatures of 1200 to 1400 K. Some Ca, Al-rich inclusions in carbonaceous chondrites require formation temperatures as high as 1600 K. Estimates from trace-element fractionation trends and simple thermodynamic calculations for the growth of refractory metal nuggets in carbonaceous chondrites indicate that these temperatures must have prevailed, at least locally, for tens of years. Refractory-rich rims on Ca, Al-rich inclusions suggest that much more intense heating also occurred, but for times only on the order of one second. Evidence is presented that requires some objects to have formed in rather oxidizing conditions at elevated temperatures. However, it is not clear what fraction of the solar nebula was affected by these processes, in particular when considering that the high-temperature indicators in carbonaceous chondrites only represent a small fraction of the bulk meteorites.

I. INTRODUCTION

Ideally, for meteorites to provide constraints on conditions in the solar nebula, little should have happened to them since they formed. Meteorites, however, have only spent a comparatively short timespan of several Myr as individual bodies in space. Most of the time, for more than 4 Gyr, they were buried in the interior of km-sized planetesimals. The pressure, temperature and oxygen fugacity recorded in the mineralogy of a meteorite is dependent on the type of parent body, i.e., its composition, and the position of the meteorite inside the planetesimal. Meteorites may thus reflect in their chemical and mineralogical compositions a variety of conditions, from complete melting or minor heating to low-temperature alteration by fluid phases (e.g., H_2O).

Clearly, we can only hope to identify relics of nebular processes in those meteorites that were not exposed to high temperatures in the interior of planetesimals. Either the planetesimals from which these meteorites were derived were not heated to sufficiently high temperatures, or meteorites from them were buried at shallow depths where temperatures were low. Meteorites which have never been exposed to temperatures sufficient for melting are called chondritic meteorites, or chondrites. They may still preserve the original texture and mineralogy that they had acquired in the solar nebula, or they may have a texture and mineralogy adjusted to the new "planetary environment," according to metamorphic conditions or low-temperature alteration. If the low-temperature alteration was isochemical, as often found in carbonaceous chondrites, their bulk chemical composition will still record the nebular origin.

Chondritic meteorites contain all nonvolatile elements in the proportions in which they occur in the Sun, i.e., they have the average solar system composition. This is only true to a first approximation; on a finer scale, compositional variations are found among chondritic meteorites such as variable contents of volatile elements, variations in total iron, variable Mg/Si ratios, etc. In addition, chondritic meteorites also display a wide variety of the degree of oxidation, from water-containing CI chondrites to the completely reduced enstatite chondrites with Si-containing FeNi. These variations (oxygen fugacity and composition) must reflect different environments during condensation or aggregation of solid matter in the solar nebula. For example, different environments could be characterized by small deviations from solar abundances in the local environment (nebula and/or average grain composition) in which they formed. In addition, gas-solid fractionation processes, such as gravitational settling of early-formed grains to the midplane, could have been important. Chondritic meteorites, particularly those of low petrologic type (indicating very little heating inside a larger body) contain components (inclusions, mineral fragments, etc.) that record nebular processes and thus allow one to infer composition, temperature and perhaps pressure of the nebular environment.

In meteorites derived from molten planetesimals (differentiated meteorites), as well as in samples from larger planets (Earth, Moon), mineralogical or textural evidence of solar nebular processes is completely erased. There is, nevertheless, some information contained in the chemical composition of individual samples. For example, the composition of basalts (terrestrial, lunar or meteoritic) allows certain conclusions regarding the chemical and mineralogical composition of their source region, and, in some cases, of the bulk chemical composition of their parent planet. In some cases, ratios of incompatible elements, for example, the K/U ratio (see, e.g., Taylor 1988), are constant during igneous fractionations and thus reflect the bulk planet ratio. The composition of a basaltic rock sample from the surface of a planet is also a sensitive indicator of the mineralogy of the source region in the interior of the planet. It is possible by these means to obtain a reasonable estimate of the bulk composition of planets (see, e.g., Dreibus and Wänke 1980,1990).

This approach of using the bulk composition allows one to compare the compositions of the Earth, the Moon and Mars with the composition of a wide range of meteorite parent bodies (differentiated and undifferentiated). Most of the planets and the differentiated planetesimals appear to have compositions that differ slightly from average solar system composition, in particular with respect to volatile elements. These compositional variations are in the same range as those within the group of chondritic meteorites and may also reflect a similar variability in formation conditions in the nebular environment.The advantage of considering the bulk composition of differentiated bodies is that it provides a more representative sampling of the solar nebula. By comparing simultaneously the compositions of the inner planets, meteorite parent bodies, comets (e.g., Halley) and interplanetary dust (results of analyses of interplanetary dust particles, IDPs), it may be possible to identify processes that operated in the inner part of the solar system during formation of solid matter and which are ultimately responsible for the variations in elemental abundances and conditions of formation of planets and planetesimals.

The evidence provided by individual components in meteorites, such as exotic mineral grains, may be more definitive in terms of identifiable solar nebula processes, but the significance of these processes for the bulk matter of the inner solar system may be difficult to evaluate. For example, deducing high temperatures for the conditions of formation of some refractory mineral grains recovered from the Allende meteorite is straightforward, but the extent in space and time of this high-temperature zone is unclear. Was the zone a local phenomenon, or was it a major part of the inner solar system?

In the first part of this chapter (Sec. II) we describe evidence derived from the bulk composition of meteorites, planetesimals and planets and discuss their significance for the evolution of the inner part of the solar nebula. In the second part (Sec. III) we consider evidence for solar nebula processes recorded in individual meteoritic components, such as chondrules, inclusions or mineral grains.

II. BULK CHEMICAL COMPOSITION

A. Significance of Chondritic Meteorites

In cosmochemistry it is useful to arrange elements according to their condensation temperatures (see review by Larimer 1988). Such temperatures are calculated assuming cooling of a hot gas of solar composition at a given pressure under equilibrium conditions. The sequence of elements condensing from this gas is reasonably well known, except for trace elements where, in many cases, a lack of relevant thermodynamic data prevents precise calculations. Condensation temperatures are a convenient measure for volatility, and they provide a useful framework for classification of chondritic meteorites, considering that variations in abundances of volatile elements account for the most pronounced differences among chondrites. According to their condensation temperatures, elements can be divided into four groups:

1. The refractory component (Ca, Al, Ti, Zr, REE, Ir, Os, etc.) which makes up about 5% of the total condensible matter.
2. Mg-silicates (forsterite Mg_2SiO_4, enstatite $MgSiO_3$) and metal (FeNi) which represent the major fraction of condensible matter.
3. Moderately volatile elements (Na, K, Cu, Zn, etc.) with condensation temperatures below Mg-silicates and FeNi and above (but including) S (as FeS).
4. Highly volatile elements (In, Cd, Pb, etc.) with condensation temperatures below FeS.

Compositional variations of some representative elements in various groups of chondritic meteorites are shown in Fig. 1. Abundances are divided by Si and normalized to those in CI chondrites (see below). Meteorite groups are arranged in the order of decreasing oxygen content, reflecting increasingly reducing conditions from carbonaceous to enstatite chondrites.

The solar abundances as recently compiled by Anders and Grevesse (1989) are plotted on the left end of the meteorite data in Fig. 1. It is obvious that the best agreement between solar and meteorite data is obtained with CI abundances. This is true for Ca, representing the group of refractory elements, Mg and volatile elements (Na, Zn, S). The solar Fe abundance is an exception being some 50% above the CI abundance. The reason for this discrepancy is not known. It is certainly not due to an overabundance of metal, because the solar abundances of Ni and other siderophile elements agree with the CI abundances (see Anders and Grevesse [1989], for detailed discussion). In a very recent redetermination of the solar iron abundance, Holweger et al. (1991) found agreement between solar and meteorite Fe abundance, suggesting that older solar Fe abundances were in error. Rare gases and some light elements, including C, N and O, are deficient in CI chondrites compared to solar abundances due to their high volatility. The CI-oxygen abundance, for example, is a factor of 3 lower than the solar abundance (Fig. 1), reflecting incomplete water condensation, incomplete formation of

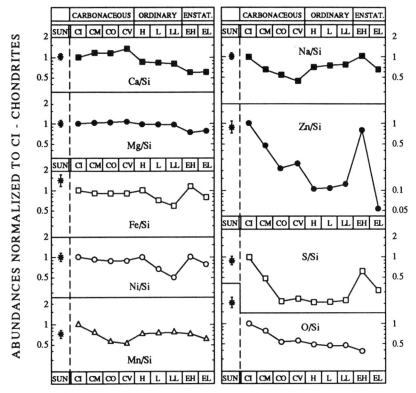

Figure 1.   Selected element/Si ratios for all groups of chondritic meteorites. Meteorite groups are arranged in order of decreasing oxygen content (lower right). Solar
ratios are indicated on the right side. CI meteorites agree best with solar abundances,
e.g., refractory elements (Ca) and volatiles (Zn,S). Oxidized (CI) and reduced meteorites (EH) are both rich in volatiles. Sources of data: Mason 1979; Kallemeyn
and Wasson 1981; Anders and Grevesse 1989, unpublished data Cosmochemistry,
Max-Planck-Institut für Chemie, Mainz.

hydrous silicates or saturation of silicates with water.

The refractory element Ca (Fig. 1), and the other refractory elements
(Al, Sc, Ir, etc.), vary by about a factor of 2 within the group of chondritic
meteorites (see difference in Ca between CV and E chondrites in Fig. 1).
Ca, Al-rich inclusions in carbonaceous chondrites have, on the average, 20
times higher contents of refractory elements than CI chondrites. This factor of 20 would be expected from complete condensation of all refractory
elements, corresponding to 5% of condensible matter (see Grossman and
Larimer 1974). These inclusions were, therefore, thought to represent high-
temperature condensates. Formation of Ca, Al-rich inclusions as a condensate
requires their isolation from equilibrium with the nebula above the condensation temperatures of the more volatile, major components, metallic iron

and the Mg-silicates, i.e., from 1225 K at 10^{-6} bar to 1450 K at 10^{-3} bar. Calculated condensation temperatures of refractory elements at a pressure of 10^{-4} bar are shown in Fig. 2. Methods for calculations are described in Palme and Wlotzka (1976) and Kornacki and Fegley (1986). Major-element condensation temperatures were deduced from data in Kornacki and Fegley (1984). In some refractory inclusions and minerals in carbonaceous chondrites refractory elements with different volatilities are fractionated. For example, highly variable ratios of the very refractory Lu to the less refractory Yb or, similarly, variable ratios of the strongly refractory Zr to the much less refractory Nb are observed. These variations can be explained by incomplete condensation or evaporation of refractory elements as discussed in Sec. III.B.1. Here we wish to emphasize that such fractionations are practically absent in bulk meteorites, i.e., refractory elements always occur in the same relative proportions, despite some variation in absolute contents. This is shown for the Zr/Y and Zr/Nb ratios in Fig. 3. The difference in condensation temperature between Zr and Nb is more than 200 K, while Zr and Y have similar volatilities (Fig. 2). The observation of phases with fractionated ratios in carbonaceous chondrites implies either closed system behavior or a very small proportion of material with fractionated refractory elements. In the first case, high Zr/Nb phases would be balanced by low Zr/Nb phases. Evidence presented later suggests the second possibility, a very small contribution of material processed at high temperatures.

Chondritic meteorites have variable Si/Mg ratios, with maximum variations of about 30% (Fig. 1). These variations may be ascribed to fractionation of olivine, as suggested by Larimer (1979). Forsterite (Mg olivine) has an atomic Mg/Si ratio of 2, as compared to the solar ratio of $\sim$1. Therefore, addition or removal of forsterite has a large influence on the Mg/Si ratio. Isolation of forsterite after condensation requires similarly high temperatures as addition or removal of refractory components.

Parallel variations in Fe and Ni within the different groups of chondritic meteorites are indicated in Fig. 1. Although fractionation of iron as oxide is possible, it is difficult to fractionate Ni as oxide. Simultaneous variations in Fe and Ni contents may therefore be ascribed to addition or removal of metal components established by processes in the solar nebula. Again, similar temperatures as those required for olivine and refractory element fractionations are indicated, because in a gas of solar composition, metal and forsterite have similar condensation temperatures (Grossman and Larimer 1974). Details of the fractionation processes remain unclear, however, as there is no parallel variation of the Mg/Si ratios (reflecting olivine fractionation) with the contents of metallic iron (Kerridge 1979). The evidence for high temperatures is, therefore, indirect. The essential point to emphasize is that variations in Mg/Si ratios and variable metal contents in chondritic meteorites were not established by melting processes, as chondritic meteorites were never molten. Therefore, a nebular formation history with temperatures of 1250 to 1450 K is suggested, although the mechanical separation of metal and

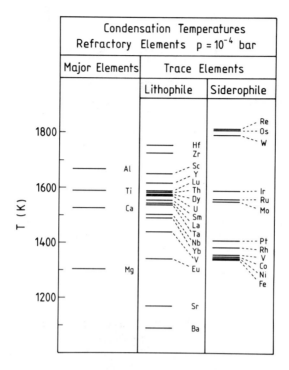

Figure 2.   Calculated condensation temperatures of refractory elements at 10^{-4} atm. Hf, Zr, Sc and Y would condense before a major phase condenses.   All other elements were assumed to condense in solid solution with perovskite. Ideal solid solution was in all cases assumed. This leads to unrealistically low condensation temperatures for Sr and Ba. Both elements obviously require a different host phase. Elemental fractionations according to differences of condensation temperatures are found in components of meteorites but never in bulk meteorites. Calculations were performed according to programs described by Palme and Wlotzka (1976), Fegley and Palme (1985) and Kornacki and Fegley (1986).   Major elements are from Kornacki and Fegley (1984).

forsterite could have occurred at lower temperatures.

The systematics and the significance of abundance trends of moderately volatile elements in chondritic meteorites are discussed in some detail by Palme et al. (1988). Here we review the most important aspects. Abundance trends for several moderately volatile elements are shown in Fig. 1 (Mn, Na, Zn, S). Volatility is increasing, i.e., condensation temperatures are decreasing from Mn through Na to Zn and S, reflecting a tendency for increasing variability in abundances with increasing volatility. In Fig. 4, concentrations of moderately volatile elements in CV3 carbonaceous chondrites (e.g., Allende, Vigarano) relative to CI chondrites and normalized to Si are shown. Elements are arranged according to condensation temperatures. The regular decrease in abundances, independent of the geochemical character of the el-

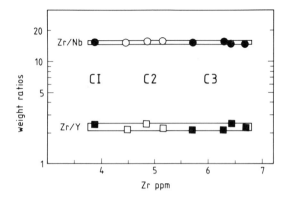

Figure 3.   Constant ratios among refractory elements in the major types of carbonaceous chondrites. Despite large differences in condensation temperatures of Zr and
Nb (Fig. 2), the ratio is constant. Data from Jochum et al. (1986).

ements (siderophile, chalcophile or lithophile), suggests a volatility-related
process, i.e., condensation or evaporation, as being responsible for this trend.
All chondritic meteorites, except CI chondrites with solar abundances, have a
similar type of abundance pattern (i.e., a trend of decreasing abundances with
increasing volatility) for the moderately volatile elements (Palme et al. 1988).

Under equilibrium conditions condensation and evaporation would lead
to the same results. However, evaporation of silicates that occurred after
partial or complete dissipation of the nebular gas (equivalent to an enrichment
of O_2-rich dust over H_2-rich gas) will lead to more oxidizing conditions,
primarily due to the absence of H_2 and through decomposition of oxides and
evaporation of water. In most cases, therefore, evaporation should occur
under more oxidizing conditions than condensation.

Figures 5 and 6 and show some results of heating experiments on Allende
chips at various oxygen fugacities (Wulf and Palme 1991). The reducing conditions of the solar nebula correspond to an oxygen fugacity indicated at the
left of the oxygen fugacity scale of Figs. 5 and 6, at about 10^{-20} bar. On the
right, CI-normalized abundances for Allende are indicated, allowing a comparison of the Allende pattern depleted in volatiles to the pattern produced by
the volatilization experiments. If the Allende pattern is produced by heating
from a reservoir with solar abundances, e.g., Orgueil, then the Allende trace-
element pattern should display a similar trend as the pattern produced by the
heating experiments. Figures 5 and 6 demonstrate quite clearly that the Allende pattern cannot be produced by evaporation under oxidizing conditions.
At oxidizing conditions, the high losses of Au and As and the small losses of
Zn and the alkalis preclude any similarity to the observed depletion pattern
in Allende. The sequence fits better at low oxygen fugacities, although the
absence of Mn depletion in the experiments is not compatible with the ob-

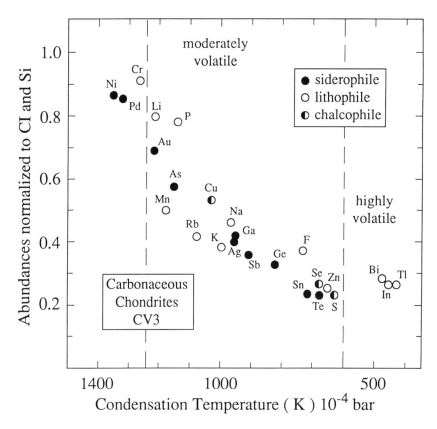

Figure 4.   Moderately volatile element abundances in bulk CV chondrites decrease more or less continually with condensation temperatures. The depletion sequence is independent of the geochemical character of the elements, suggesting that a volatility-related process is responsible for the depletion, either evaporation or incomplete condensation. For details and for sources of data see Palme et al. (1988).

served low Mn content. In addition, Se depletions are higher than observed in Allende (in the heating experiments only upper limits could be determined). Heating experiments by Bart et al. (1980) on Allende show that there is no loss of Se at 1000°C, while significant losses are observed at 1050 similar to what was found in the experiments by Wulf and Palme (1991). This and the Bart et al. data at higher temperatures suggest a sudden increase of the rate of evaporation of S between 1000°C and 1050°C, implying some kind of a threshold temperature. Other elements appear to behave similarly; once the threshold temperature for losing volatiles is reached, a small temperature increase is sufficient to cause extensive losses. Obviously loss of Mn does not occur in the experiments because the threshold is not reached. At the same time, however, Se is almost completely lost, contrary to the observed pattern.

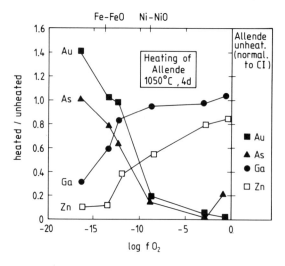

Figure 5.   Loss of volatile elements by heating of Allende (1050°C, 4 days) is strongly dependent on oxygen fugacity (Wulf and Palme 1991; Wulf 1990). The oxygen fugacity of the Fe-FeO buffer would give the best fit to the Allende pattern (normalized to CI). Highly oxidizing environments which would be produced by heating after dissipation of the H_2-rich (reducing) gas are excluded. The high Au value for the most reducing sample is probably due to contamination.

An additional problem arises for the highly volatile element abundances (Tl, Pb, Cd, etc.). These elements have similar abundance levels in carbonaceous chondrites as S, Se and Zn (Anders et al. 1976). However, in heating experiments they are much more readily lost than the moderate volatiles (Bart et al. 1980). Partial loss of Na or Zn during thermal metamorphism would therefore imply complete loss of the highly volatile elements. These elements would have to be brought in again after thermal metamorphism. The same trends in volatile element depletions observed in heating experiments on Allende were also found in experiments with Murchison (Wulf 1990). Considering the different texture and mineralogy of this meteorite, it may be concluded that the observed depletions reflect a general trend and are not a result of the specific Allende mineralogy. The results of the heating experiments thus imply that the loss of volatiles occurred in the early solar nebula at a time when the hydrogen-rich, reducing gas was still present, evaporation in the absence of an H_2 atmosphere would have a very different trace-element pattern than that observed.

Anders (1964) and Larimer and Anders (1967) have suggested that moderately volatile and volatile elements are lost during chondrule formation. These authors assume that, for example, Allende parent material initially (before chondrule formation) had the same composition as CI-chondrites, i.e., the full complement of volatiles. By converting three-quarters of this mate-

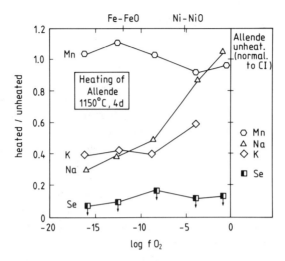

Figure 6. Heating experiments of Allende (Wulf and Palme 1991; Wulf 1990). In these experiements no loss of Mn is found, although Allende is depleted by nearly 50% (see also Fig. 1). Selenium and S (not shown here) are completely lost in all experiments (below detection limit). Both elements are present in Allende at a level of 0.25 (relative to Si and CI). The pattern of K and Na are, similar to Fig. 5, more compatible with reducing conditions.

rial to chondrules (through the chondrule-forming process), a corresponding fraction of highly volatile elements would be completely lost from this fraction, while moderately volatile elements would only be partially lost, thus producing the abundance pattern of Fig. 4. However, Allende chondrules are, on the average, enriched, not depleted in Na and K (Rubin and Wasson 1987; Spettel et al. 1989), indicating that heating during chondrule formation is not sufficient for losing alkalis. Also, fine-grained matrix material of Allende is, except for a volumetrically insignificant fraction (with grain size $<1\,\mu$m), not enriched in volatile elements as expected from this model but has similar abundances in bulk Allende (Ireland et al. 1990a). Large compositional variations in chondrules suggest that chondrules are made of coarse-grained precursor materials and not of fine-grained matrix (Grossman 1988). There is, therefore, little evidence that the depletion of volatiles in Allende and other chondrites is associated with chondrule formation.

Alternate explanations for the depletion of moderately volatile elements involve condensation. However, incomplete condensation alone is not sufficient to explain the observed pattern. Removal of a condensate from the solar nebula at a given temperature would lead to a similar pattern as envisioned for evaporation in a hydrogen-rich environment. Wasson and Chou (1974) have, therefore, suggested dissipation of the solar gas during condensation. The abundances of volatiles in the gas, i.e., their partial pressures, would decrease during condensation as gas would be continually removed, leading to the observed smooth depletion sequence of moderately volatile elements (Fig. 4). The abundance level of the highly volatile elements would then reflect the total fraction of gas lost.

Whatever the origin of the depleted-element pattern, there is little doubt that temperatures up to 1127 K at 10^{-6} bar or 1284 K at 10^{-4} bar, the 50% condensation temperatures of Au (Wai and Wasson 1977), are required. The parent material of most chondritic meteorites must have been processed at this temperature in a solar nebula environment, although it should be emphasized that the duration of the heating event may have been quite short as indicated in Figs. 5 and 6 where significant losses of volatiles are shown to occur within a few days (the time scale for incomplete condensation would depend on kinetics and is thus difficult to predict).

That the loss or incomplete condensation of moderately volatile elements and volatiles occurred very early in the evolution of the solar system is apparent from Sr and Pb isotopic data. At the time of formation of the solar system, chondritic meteorites had $^{87}Sr/^{86}Sr$ ratios not very different from that of the lowest measured ratio in Ca, Al inclusions from Allende and slightly lower than the lowest ratio determined in lunar rocks and basaltic achondrites (see review by Tilton 1988). Such data indicate widespread homogenization of Sr isotopes at the time of formation of the solar system. The time of separation of the volatile Rb from the refractory Sr in chondrites cannot be determined precisely because of uncertainties in the Rb decay constant (Minster et al. 1982). However, as data of H, LL and E chondrites fall on a single isochron, it is very likely that the Rb-Sr fractionation is a global event that reflects the general fractionation of volatile elements from a solar reservoir, i.e., condensation of chondritic meteorites. The determination of the isotopic composition of Pb is the most precise dating method for the early separation of volatile (Pb) from refractory (U, Th) elements. Ca, Al inclusions in Allende, for example, have ages from 4.565 to 4.575 Gyr (Manhes et al. 1987). Phosphates in primitive (type 3) chondrites are about 4.55 Gyr old (Göpel et al. 1990). A similar age of 4.551 ± 0.004 Gyr was reported for the differentiated meteorite Angra dos Reis (Chen and Wasserburg 1981b). These data indicate a spread of 25 Myr for the separation of Pb from U and the subsequent undisturbed growth of radiogenic Pb. The Pb separation must have occurred very early during the formation of solid material in the early solar system and cannot have been produced in a late reheating event.

B. Differentiated Meteorites and Planets

The depletion of moderately volatile elements is, however, not confined to chondritic meteorites. There is ample evidence that meteoritic basalts (eucrites), lunar and terrestrial basalts are all deficient in moderately volatile elements, as for example, recorded in their K/U ratios (see, e.g., Taylor 1988). More sophisticated models for the bulk compositions of planets lead to similar conclusions: volatile elements when compared to the solar system average are depleted in the Earth, the Moon, the eucrite parent body and the Shergotty parent body (perhaps Mars) (Dreibus and Wänke 1980,1990).

It is not very likely that the low contents of volatile elements in planets and planetesimals were established during accretion. The extent of the depletions does not depend on the size of the planet. The Moon and the eucrite parent body have lower contents of volatiles than the Earth and the Shergotty parent body, presumably Mars. Volatile element depletions in the Moon may be the result of special conditions during its formation. However, this argument cannot be applied to the much smaller eucrite parent body (*Basaltic Volcanism Study Project* 1981).

The depletion of volatile and moderately volatile elements is also visible in the trace-element pattern of iron meteorites. In their siderophile element abundances, these pieces of planetary cores have retained the signature of the composition of the original planet, i.e., material that accreted to form the planet. A particularly strong depletion pattern is observed in the IVB iron meteorites (see Fig. 7.5.10 in Palme et al. 1988). This is, for example, reflected in the high Pd/Ag ratio (see, e.g., Kaiser and Wasserburg 1983), with a difference in condensation temperatures between Pd and the more volatile Ag of some 400 K (Wasson 1985). Results of the analysis of Ag isotope ratios indicate that the separation of Ag from Pd occurred very early in the history of the solar system. It is, therefore, most likely that the parent body of these meteorites had a fractionated (high) Pd/Ag ratio, i.e., the trace-element pattern may have been established by condensation processes in the solar nebula.

This example also demonstrates the lack of a correlation between heliocentric distance and chemical composition of meteorites and planets. There is no obvious trend for enrichment of volatiles with heliocentric distance. The Earth is richer in moderately volatile elements than the eucrite parent body in the asteroid belt. The IVB iron meteorite parent body, another object of the asteroid belt, is very deficient in volatiles (Kelly and Larimer 1977). Thus, CI chondrites with their full complement of moderately volatile and volatile elements and the volatile poor eucrites and IVB irons are all derived from the asteroid belt. Also, there is no correlation between the amount of volatiles and the degree of oxidation among parent bodies of meteorites. Such a correlation would be expected if volatiles were removed from condensed matter by heating, because water would be removed at low temperatures and oxides would be decomposed at elevated temperatures. Figure 1 demonstrates high Zn and S contents in both, the most oxidized (CI) and in the most reduced (EH) meteorites. The absence of a clear trend of the degree of volatile element depletion with distance from the Sun within different groups of meteorites and between meteorites and planets does not necessarily reflect the original structure of the solar system. Meteorite parent bodies may have formed at very different places than their present location would indicate. However, there is always the difficulty to understand the formation of the volatile-rich and completely reduced enstatite chondrites within the framework of the equilibrium condensation model. Additional assumptions such as an enhanced C/O ratio are required (Larimer and Bartholomy 1979).

It appears that each group of chondritic meteorites was formed in its own separate nebular reservoir that was slightly different from the average composition of the solar system, i.e., the Sun, producing a characteristic chemical signature for each group.

Depletion of volatiles, however, has occurred in almost all matter of the inner solar system. Consequently, conditions prevailing in the inner solar system through at least part of the asteroidal belt, included a H_2-rich gas with temperatures on the order of 1100°C. The CI meteorites, in particular Orgueil, are exceptional; either this material is derived from comets, or it has escaped the high-temperature event, or it was added to the asteroid belt after the high-temperature event was over. Mixing of unheated material to the inner solar system is required by recent findings of unprocessed interstellar material (Anders 1988; Zinner 1988). However, at least some interplanetary dust particles (IDPs) may have the full complement of volatile elements, even including carbon (Schramm et al. 1989). Those particles may be derived from comets. The high contents of volatile elements in Comet Halley supports this assumption (Jessberger et al. 1988).

Other chemical fractionation trends in chondritic meteorites discussed earlier, such as variations in Mg/Si ratios or in the content of metal (NiFe) also seem to be present in planets. These fractionations would also require temperatures of around 1200 K. However, because the estimation of bulk planet compositions are not always unambiguous, and because variations in these ratios are small, they will not be discussed further.

Although there is widespread evidence for loss of volatiles, there is very little evidence for gain of volatiles as pointed out by Palme et al. (1988). Except for a few volatile-rich clasts in chondrites, there is no meteorite class with abundances of moderately volatile elements or volatiles clearly in excess of the solar photospheric level. This may be an important clue in identifying the process that caused the volatile element depletion. If it occurred during condensation, the uncondensed volatiles may have been blown out of the inner solar system. However, as mentioned before, they did not recondense in the asteroid belt, at least not in that part that is sampled by meteorites present in our collections.

An upper limit to the temperature where the bulk of protoplanetary material has been processed may be deduced from a lack of fractionation among refractory elements. Bulk planet ratios of Zr/Nb and Hf/Ta are essentially the same as ratios in primitive chondritic meteorites as shown by Jochum et al. (1986), despite large differences in the volatilities of Zr and Nb and Hf and Ta, respectively (Fig. 2). Such fractionations are, however, observed in high-temperature components of carbonaceous chondrites as is discussed in the next section.

III. CONDITIONS IN THE SOLAR NEBULA DERIVED FROM COMPONENTS IN METEORITES

A. Chondritic Meteorites are Complex Mixtures of Components Formed Under Highly Variable Conditions

As techniques for chemical and isotopic analyses have improved over the years, a continuous decrease in the amount of material required for precise chemical and isotopic analysis of meteorites and their constituents allows one to study meteorites in great detail. A fascinating body of data on minerals and inclusions of unequilibrated meteorites, particularly from carbonaceous chondrites, has been obtained. Sophisticated techniques of chemical dissolution of meteorites have led to the identification of extremely small amounts of pre-solar material in these meteorites (see Chapter by Ott). On the other hand, isolation of refractory materials in carbonaceous chondrites has provided abundant evidence for processing material at temperatures of 1600 to 1800 K (at 10^{-4} bar; see Fig. 2). These meteorites contain a bewildering variety of components that require very different conditions of formation, from low temperature phases such as carbonates to mineral grains that record gas-solid equilibria at rather high temperatures. In many cases, processing of these materials must have occurred in distinctly nonsolar environments, characterized by high H_2O/H_2 ratios, i.e., excess oxygen.

The multitude of components assembled in these meteorites reflect the absence of equilibration processes in a parent body. This mixture of high- and low-temperature materials formed under both extremely reducing and highly oxidizing conditions, comprising mineral grains and clasts which are in apparent disequilibrium, has nevertheless a well-defined chemical composition, which is basically solar for nonvolatile elements and shows systematic trends for volatile elements (see Fig. 4). This apparent paradox may be resolved by assuming a nearly closed system with respect to chemical composition that provides a variety of local environments but allows the addition of only a very small fraction of chemically unusual material. Small amounts of isotopically unusual material found in meteorites, presumably of pre-solar origin, does not show up in bulk analyses. On the other hand, chondrules and matrix, both produced by very different processes, are significant components in the Murchison CM2 meteorite needed to account for the "primitive" bulk composition (Wood 1985), suggesting the nearly closed system behavior.

An example of this mixture of materials formed under different processes can be seen in Fig. 7, which shows a thin section of the Murchison meteorite. In the lower half of the figure, there are two granular olivine chondrules. Extremely fine-grained matrix, rich in FeO, appears to have accreted with these FeO poor chondrules. In the upper right, there is a large olivine crystal, some of these olivine crystals are almost free of FeO and may have formed by condensation from the solar nebula. Thus the Murchison meteorite is a breccia containing a variety of lithic and mineral fragments that were mechanically mixed on the surface of a parent body. The components of this meteorite formed at different temperatures and under very different oxygen fugacities. Low- and high-temperature components (e.g., calcite vs forsterite) display large differences in the composition of the oxygen isotopes, excluding a common single source of the oxygen in these minerals (Clayton and Mayeda 1984). Despite the chaotic nature of this meteorite and the apparently random mixing of various components, the chemical composition of Murchison is well defined, resembling average solar system composition.

Figure 7. Ultra-thin section of Murchison, a CM meteorite. The brecciated character of this meteorite is obvious. Several olivine crystals are clearly visible (e.g.,
upper right). Chondrules appear to be surrounded by fine-grained rims probably
formed during accretion in space. There is no thermodynamic equilibrium among
the various components of Murchison. They formed under different oxygen fugacity (Si-bearing metal and H_2O-containing phyllosilicates), and they show large
variations in oxygen iotopic composition (Clayton and Mayeda 1984). The bulk
chemical composition of this meteorite is, however, well defined. Long side 2.4 mm.
Source: R. Beauchamp, Battelle-Institut, NW.

Recently Kring (1988) and Metzler et al. (1992) have shown that chondrules and fragments in CM meteorites are often surrounded by fine-grained
matrix material compositionally similar to CM matrix. The authors suggest
that these "dust mantles" formed by accretion in space and not by processes

Figure 8.   A Type B Ca, Al-rich inclusion from the Allende meteorite, consisting primarily of melilite (bright) and fassite (dark). The dark area in the lower middle inside the inclusion is matrix. Diameter of the inclusion is 1 cm.

in the regolith of a parent body. This supports the view that CM components formed independently in space and that there is very little processing after assembling of the meteorite parent body.

We will now give some examples to illustrate the variety of local environments that are required to explain the diversity of components identified in carbonaceous chondrites.

B. High-Temperature Components

1. Ca, Al-Rich Inclusions (CAIs) and Their Components. There exists a large body of literature on these inclusions, and the reader is referred to review papers by Grossman and Larimer (1974) and by MacPherson et al. (1988). Here we will restrict the discussion to a few aspects that bear on the question of temperature, pressure, oxygen fugacity and time scale of the environment in which they formed.

One of our best sources of information on high-temperature processes are the Ca, Al-rich inclusions found in carbonaceous chondrites. A large Ca, Al-rich inclusion from the Allende meteorite is shown in Fig. 8. The high content of refractory elements, including low-vapor-pressure metals, such as W, Ir, Pt, etc. (see Fig. 2), and the absence of igneous trends in the trace-element pattern leave no doubt that these objects are composed of material formed at high temperatures, either by condensation from a gas of solar composition or by vaporization of chondritic material. Figure 9, for example, demonstrates the uniform enrichment of refractory elements in Allende inclusion A4. Often, however, more complicated patterns are found, such as the pattern of A10 in Fig. 9. These puzzling patterns were interpreted by Boynton (1975) as condensates from a gas that has suffered previous removal of a first very refractory condensate. This first condensate is expected to be rich in Hf, Zr, Y, Sc, Lu and Dy, according to condensation temperatures (see Fig. 2). Removal of such a condensate from equilibrium with the solar gas would leave the residual gas depleted in Hf, Zr, and other refractory elements in low abundance in A10. Any component formed by subsequent condensation would lack these elements just as it is observed in the A10 inclusion in Fig. 9. This picture may seem complicated and somewhat contrived; any other explanation, however, is even more complex. These so-called group II patterns are common in certain types of fine-grained inclusions in Allende, but the complementary "superrefractory pattern" (enriched in Hf, Zr, etc.) indicative of the first condensate was never found in Allende.

Inclusions with this superrefractory pattern have been found in other meteorites, however. Boynton et al. (1980) found such a pattern in a spinel-hibonite inclusion from Murchison, and Palme et al. (1982) isolated a tiny grain (about 1 μg) from the Ornans meteorite that had an even more extreme super-refractory pattern (Fig. 10). Condensation of only the most refractory elements, for example in perovskite ($CaTiO_3$) or hibonite ($CaAl_{12}O_{19}$) as suggested by Boynton (1975) requires thermodynamic equilibrium between gas and solid at high temperatures. The rare-earth element, Lu, should practically completely condense, while the six-times more abundant and geochemically similar rare-earth element Sm, with a condensation temperature of only 50 K lower, would remain in the gas (Fig. 10). Effects due to supersaturation must have been absent or were not important. Perhaps pre-solar, tiny, unvaporized grains acted as condensation nuclei.

Removal of most of Lu from a given volume requires time in the order of days to years as a simple calculation demonstrates. Given grains of perovskite with 1 μm diameter, the number of encounters per second of Lu with these grains is given by:

$$Z = n_{Lu} n_{gr} \pi (d_{Lu} + d_{gr})^2 \sqrt{v_{Lu}^2 + V_{gr}^2}. \tag{1}$$

n_{Lu} and n_{gr} are number densities of Lu atoms and perovskite grains, d_{Lu} and d_{gr} are the sizes of the Lu atom and the perovskite grain the v's are the

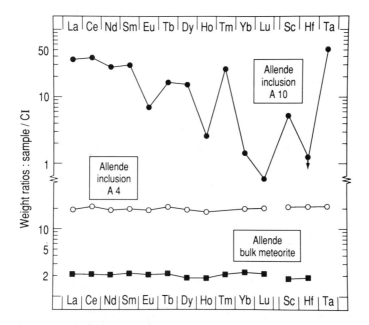

Figure 9.   Refractory elements in bulk Allende, a Type B inclusion with uniform enrichment factors (about 20), and a fine-grained inclusion with a Group II pattern. These patterns reflect solid-gas equilibrium after removal of a super-refractory component (see Fig. 2 for volatilities of refractory elements). Sources of data: Cosmochemistry, Max-Planck-Institut für Chemie, Mainz (unpublished).

corresponding mean velocities, given by

$$v = \sqrt{8\frac{kT}{\pi m}}. \qquad (2)$$

The time between two encounters of Lu atoms with perovskite grains is $1/Z$ and the lower limit in time for accumulating as many Lu atoms as would correspond to the Ti atoms in the perovskite grain, assuming a chondritic Ti/Lu ratio, is $(1/Z)cl_{Ti}(a_{Lu}/a_{Ti})$ where a_{Lu} and a_{Ti} are the cosmic abundances of both elements and cl_{Ti} is the number of Ti atoms in a 1 μm grain of perovskite. With some simplification, such as unity sticking coefficients and no consideration of depletion of the gas in Lu during condensation, a time of 4 days for Lu condensation on perovskite is obtained for a temperature of 1600 K and a pressure of 10^{-4} bar. Because of the assumptions made, this should be considered a lower limit only. After condensation of Lu and other trace elements, additional time is required for the perovskite grains to have been removed, perhaps by settling to the midplane, so that the other refractory elements could condense on other phases and produce the peculiar Group II pattern.

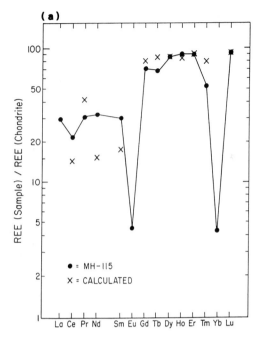

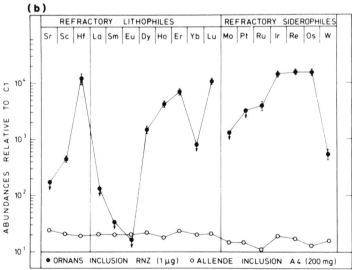

Figure 10.   (a) Super-refractory pattern in a spinel-hibonite inclusion from Murchison. (b) Super-refractory pattern of a tiny (1 μg) grain in Ornans. These patterns are complementary to the Group II pattern in Fig. 9 and are thought to represent the most refractory material that was isolated from the gas before Group II inclusions began to condense.

Ireland et al. (1988) have shown that the Group II pattern is commonly found in very refractory mineral grains, such as hibonite and perovskite from the Murchison meteorite. The time significance for formation of components carrying the Group II pattern may be similarly constrained because similar times are needed to grow the Group II grains yet leave in the gas the less refractory elements, such as Eu and Yb. The equilibrium distribution of refractory trace elements between gas and solids thus requires a rather slow cooling from high temperatures (days to perhaps years).

It is therefore clear that rapid heating, provided by lightning for example, is not a suitable mechanism for explaining these gas-solid equilibria. In some of the refractory grains analyzed by Ireland (1990), isotope anomalies were found, suggesting that interstellar grains may have acted as condensation nuclei. Ireland et al. (1990b) analyzed the isotopic composition of lead in perovskite grains and found ages of around 4.56 Gyr, indicating that these are solar nebula products, at least heated at that time to temperatures required to completely vaporize lead.

There is good evidence that refractory metals (Re, Os, Ir, etc.) condensed as alloys in the solar nebula. Tiny refractory metal grains either contain all refractory metals in chondritic relative proportions or they have a volatility-dependent pattern of refractory metals, suggesting a condensation origin. An origin as evaporative residue is less likely. Under moderately oxidizing conditions W and Mo will be lost from the alloy, because both elements form volatile oxides. Conditions during evaporation are likely to be more oxidizing than a solar gas as stated before. There are inclusion and metal grains with depletions in W and Mo. Fegley and Palme (1985) have, however, argued that even these grains formed by condensation, but under more oxidizing conditions than assumed for the solar nebula.

A similar, simplified calculation as the one indicated before was made for the time required to condense these alloys (see Palme and Wlotzka [1976] for details). For example, for the growth of an Os cluster with a diameter of $10 \mu m$ by collisions of Os atoms with the growing cluster a time of 1500 yr is calculated for a pressure of 10^{-4} bar and a temperature of 1600 K. The corresponding time for growing $1 \mu m$ and $0.1 \mu m$ grains are 150 and 15 yr, respectively. These times are longer than those calculated for condensation of Lu, simply because in the Lu case we have assumed that existing perovskite grains capture the Lu atoms, while in the Os case the metal-clusters are built up by addition of single Os atoms. As there is evidence that refractory metals condense as alloys, Os is used in the calculations as representing all refractory metals. In a few cases, the isotopic composition of some of the refractory metals has been determined (Hutcheon et al. 1987). No anomalies larger than 1% were found, suggesting that the alloys are solar nebula products.

2. Isolated Olivine Crystals. There are other coarse-grained components in carbonaceous chondrites that may be linked to condensation. Forsterite is the first major Mg silicate to condense from a cooling gas of solar composition. Isolated FeO-poor olivine grains found in CM, CV and CO chondrites have indeed morphologies and trace-element contents that strongly suggest a condensation origin (Steele 1988, and references therein). Recent data on the oxygen isotopic composition of forsteritic cores showed that they are rich in ^{16}O and different from chondrules (Weinbruch et al. 1989), excluding formation of forsteritic olivine grains inside chondrules, thus supporting a condensation origin. Without making detailed calculations, it is clear that time scales of at least days are needed to produce these 100 to 1000 μm objects (Fig. 11).

3. Oxidizing Conditions in the Solar Nebula. The FeO contents of even the most Mg-rich olivines, however, are still too high for condensates from a gas of solar composition (Wood 1985; Kring 1991b; Palme and Fegley 1990). These olivines must have condensed from a gas considerably enriched in oxygen compared to a solar gas. Still higher oxygen partial pressures are required for the formation of broad (10 μm) rims of FeO-rich olivine commonly observed in Allende. There is now a large body of evidence suggesting that these rims formed by condensation in an environment far more oxidizing than generally assumed for the solar nebula (Peck and Wood 1987; Kring 1988; Hua et al. 1988; Weinbruch et al. 1990). Condensation of the FeO-rich rims must have occurred rapidly to retain the sharp boundary (steep compositional gradient) between FeO-rich and FeO-poor lithologies (Fig. 11). The FeO-poor olivine grains were either transported into a region of the nebula that had higher oxygen partial pressure or a gas more enriched in oxygen arrived at the location of these mineral grains. A high-temperature formation at oxidizing conditions for FeO-rich matrix grains was earlier postulated by Kornacki and Wood (1984).

Additional evidence for oxidizing conditions at high temperatures is recorded in Ca, Al-rich inclusions from type 2 and 3 carbonaceous chondrites (see Rubin et al. [1988] for a comprehensive discussion):

a. Some inclusions have a negative Ce anomaly. Lack of condensation of Ce can be explained by gas-solid equilibrium at high temperatures and oxidizing conditions (Boynton 1978b; Davis et al. 1982).

b. Among the first phases to condense in a gas of solar composition is a refractory metal alloy containing W, Os, Ir, Ru, Mo, etc. Such alloys were observed in Ca, Al-rich inclusions (see, e.g., Palme and Wlotzka 1976), but many of these inclusions have lower W and Mo contents than expected. The deficit of Mo and W can be explained by high-temperature condensation in an oxidized gas (Fegley and Palme 1985).

c. Similarly, oxidation of existing metal alloys led to formation of tungstate and molybdate. There are good reasons to assume that this process occurred at high temperatures (Bischoff and Palme 1987).

These examples should demonstrate that the high-temperature stage experienced by mineral grains and metals was not caused by a local fast-heating event, such as impacts or lightning. In these cases, it is unlikely that the

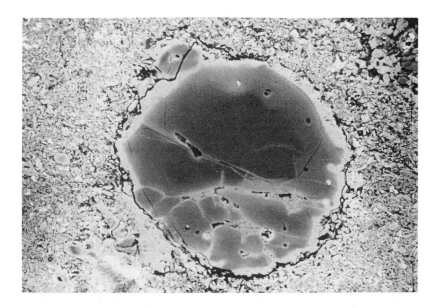

Figure 11.   Single olivine grain (∼0.3 mm diameter) from the Allende meteorite. The sharp transition from the FeO-poor interior (dark) to the FeO-rich rim (bright) reflects late condensation of FeO-rich olivine under oxidizing conditions (see text). Several earlier episodes of oxidation are reflected in the variable brightness of the interior, reflecting variable FeO- contents (SEM-image, S. Weinbruch).

temperature would have stayed high sufficiently long to enable equilibrium distribution of trace elements between gas and solid, and furthermore, there is not enough time to form solid condensates that were either removed later or are still present as refractory metal alloys.

Direct evidence for evaporation processes is provided by the isotopic fractionation of Si and Mg in some inclusions from the Allende meteorite (Clayton et al. 1985). The enrichment of heavy isotopes of Mg and Si in these inclusions is most likely the result of preferential evaporation of the light isotopes. This process most likely requires liquid residual phases where isotopic equilibrium is possible, as demonstrated for olivine by Davis et al. (1990). Therefore, temperatures near or above the liquidus temperatures are necessary. These temperatures may be as high as 1700° C. Also the time for evaporation must be comparatively slow to allow isotopic re-equilibration in the residual liquid. In many cases complex processes of evaporation and condensation are required to produce the observed fractionations of Si, Mg and Ca (Clayton et al. 1985).

At this point, it must be emphasized that, contrary to general belief, only a small fraction (20–30%) of the refractory elements in the type 3 carbonaceous chondrites are residing in Ca, Al-rich inclusions or minerals. The major fraction of refractories is sited in a very fine-grained component, intimately associated with olivine, as chemical data on grain-size fractions from the Allende meteorite have shown (Ireland et al. 1990a). The fine-grained mixture of olivine, chromite, sulfide and refractory phases may have formed by rapid condensation, reflecting fast cooling, quite different from conditions experienced by large mineral grains and coarse inclusions discussed before. Because the oxygen isotopic composition of the coarse-grained high-temperature materials and the fine-grained matrix material are also different, two separate reservoirs are required for the formation of both components, at least with respect to oxygen.

4. *Chondrule Formation.* As most of the processes discussed above are restricted to Ca, Al-rich inclusions in carbonaceous chondrites, a relatively small fraction of the bulk meteorites is actually involved. Another high-temperature event, the formation of chondrules, has affected major fractions of most chondritic meteorites. Chondrule formation may, however, be different from processes discussed above in that heating and cooling may be much faster than estimated for gas-solid equilibria. The extent of Si isotopic variations in Allende chondrules is, for example, six times smaller than that observed in Ca, Al-rich inclusions (Clayton et al. 1985), suggesting a shorter heating period for chondrules. Estimates by Hewins (1988) indicate chondrule formation temperatures of up to 1600 K and cooling rates from 100 K to 2000 K hr^{-1}, too slow for cooling by radiation into space but faster than the time scale of Ca, Al-rich inclusions discussed above. It is therefore likely that the chondrule-forming process records a different event than the high-temperature gas-solid equilibria.

In addition, there is good evidence that the formation of chondrules affected pre-existing compositionally heterogeneous matter, suggesting that the formation of chondrules occurred later, after solid solar system material was formed by condensation processes (see, e.g., Grossman 1988). This suggestion is supported by the rather high contents of alkali elements in Allende chondrules. Na and K must have condensed in order to be associated with solid chondrule precursors, and most models for chondrule formation assume solid coarse-grained precursor materials. Chondrule formation, therefore, is likely to be caused by a transient heat source which locally transformed assemblages of mineral grains into melt spherules, and thus chondrule formation must have occurred some time after condensation of moderately volatile elements. Attempts to date chondrule formation, relative to the formation of other meteorite components, by the I-Xe method have not been successful. It can only be stated that components of unequilibrated meteorites formed within about 10 Myr of each other (Swindle and Podosek 1988).

5. *Formation of Refractory Rims on CAIs.* A much more rapid heating event may have also affected many Ca, Al-rich inclusions from Allende. This event is the flash heating that formed rims on most of the coarse-grained CAIs.

Like the chondrule formation process, this process also acted on pre-existing matter, but it is not clear what relationship, if any, this heating event has to the formation of the CAIs.

Wark and Lovering (1977) showed that most coarse-grained CAIs are surrounded by thin $\sim$50 μm rims that are made up of several layers of different minerals. Many speculations were advanced on the origin of these rims including several episodes of condensation (Wark and Lovering 1977), crystallization from a thin layer of melt (Bunch and Chang 1980a; Wark 1981; Korina et al. 1982), metasomatic alteration (MacPherson et al. 1981; Nord et al. 1982), and nonequilibrium condensation (Korina et al. 1982; Wark 1983), but each of these mechanisms has difficulties in explaining the data.

Boynton and Wark (1987) showed that the rims were enriched in refractory elements by uniform factors of 3 to 6 relative to the interior of the CAIs. An important observation was that the chondrite-normalized patterns of the refractory element patterns in the rim were always identical (except for the uniform enrichment ratio) to that of the underlying CAI. This was true even for CAIs in which the underlying pattern was formed as a Group II condensate from a refractory-depleted nebular gas or was formed as an igneous fractionation product. Because the rims reflected the elemental abundances in the original CAIs, it was clear that they formed from the interior, and because they were more refractory, it was clear they formed by heating with the loss of the more volatile elements such as Mg and Si. Finally, because the rims were thin, and the rest of the CAI seemed unaffected by the event, it must have been brief—on the order of a few seconds at most.

Murrell and Burnett (1987) considered six different possibilities for the origin of rims on CAIs but did not choose a preferred mechanism. They did not endorse the flash-heating hypothesis of Boynton and Wark (1987) because they felt it would not account for the Mg isotopic data of Fahey et al. (1987), who showed that the ^{26}Mg/^{24}Mg ratio was higher in the rim than in the interior of a CAI from Efremovka—in the opposite direction to that expected from volatilization during the flash heating. An important part of the Mg data not considered by these groups was the fact that the ^{26}Mg/^{24}Mg ratio in the rim had the normal solar system value, but that the interior was lower than normal. The flash-heating hypothesis, however, predicts that most of the Mg was lost from the proto-rim during the vaporization event, and that a later metasomatic diffusion, which formed the different layers now observed in the rim, was the source of the rim-sited Mg. The normal Mg isotopic composition of the rim is thus expected if the metasomatic reservoir had normal Mg isotopic composition.

Boynton (1988) calculated that the time scale was, indeed, about one second or less and showed that the energy required to vaporize enough mass to account for the observed enrichments of refractory elements was about 300 J cm^{-2}. He also noted the difficulty in providing this heat source in an astrophysical environment. Heating by photons, electromagnetic discharge, and atmospheric entry were considered and dismissed. Recently, Morfill et al. (see their Chapter) showed that lightning, if it can occur, has both the appropriate time scale and energy flux to account for the rim observations.

IV. DISCUSSION AND SUMMARY

The inner part of the solar system, including most of the asteroid belt, was heated to temperatures of about 1300 K at the beginning of the formation of the solar system. The evidence is derived from the low volatile-element contents of meteorites and planets, possibly caused by dissipation of gas during condensation. There is no evidence for systematic variations in volatile-element contents with heliocentric distance. Variations in Fe-metal and in Mg/Si ratios among meteorites and planets, require separation of components that formed at high temperatures, similar to those necessary to produce the volatile-element variations.

Midplane temperatures obtained from theoretical modeling of planetary formation by Tscharnuter and Boss (see their chapter) are about 1500 K in the inner nebula with a flat profile and a steep gradient at a distance of 4 AU where temperatures drop to 150 K. These results were calculated for a time when the central protosolar mass had reached 0.84 $M_\odot$. The upper temperature limit in this model is determined by the stability of metallic iron through a buffering effect caused by reduction in opacity on the evaporation of silicates and iron grains.

It is obvious that there is some agreement between constraints derived from meteoritic evidence and the calculated temperature profiles. It may be mere coincidence. However, both temperatures are associated with the stability of the major phase forsterite and metallic iron and the only elements affected are those with condensation temperatures below the condensation temperatures of the major phases. During the high-temperature stage and later cooling, uncondensed nebular gas could be blown out of the inner part of the nebula, producing the reduction in partial pressure required by the low contents of volatile elements.

In contrast to this global picture, temperatures and oxygen partial pressures inferred from individual components in meteorites, particularly in the unequilibrated carbonaceous chondrites, reflect local environmental conditions. Fractionations among refractory elements indicate temperatures up to 1800 K. Time scales required to establish the gas-solid equilibria range from days to years.

Similarly high temperatures but much faster cooling is registered by the melting of chondrules. Even faster and more sporadic events are recorded in sudden melting of rims of Ca, Al-rich inclusions. These violent events may have been caused by lightning (Chapter by Morfill et al.).

CHEMISTRY AND EVOLUTION OF GASEOUS CIRCUMSTELLAR DISKS

RONALD G. PRINN
Massachusetts Institute of Technology

The chemical and physical processes which determine the composition and evolution of gas-rich circumstellar disks are discussed. The composition of accreted interstellar material is an important determinant of subsequent chemical evolution. Elemental composition (e.g., C and O abundances) and molecular composition (e.g., N_2 vs NH_3 and hydrous vs anhydrous silicates) are both significant in this respect. Major disk chemical processes are thermochemistry (not necessarily equilibrium) and shock chemistry (during accretion and by lightning). Strong mixing in a thermoclinic environment like an accretion disk leads to thermochemical disequilibration due to "kinetic inhibition" induced by chemical time constants becoming longer than outward mixing (or equivalently cooling) time constants. In this case, species thermodynamically stable at high temperatures but not at low temperatures (e.g., CO, N_2) dominate at all temperatures in the disk. Nonaxisymmetric accretion of material at hypersonic speeds is a major forcing mechanism for mixing in the disk and can produce eddy speeds of 1% of the sound speed. Typical resulting mixing times for trace species are about 10^{10} s compared to a typical disk lifetime of 10^{13} s. Nonlinear processes yield much longer mixing times for momentum than for trace species. Kinetic inhibition in the carbon, nitrogen and anhydrous/hydrous silicate families has major implications for the compositions of the terrestrial planets, giant planets, ice-rich satellites, Pluto, comets, meteorites and asteroids which are discussed. Ice-rich bodies in the outer solar system have rock-to-ice ratios which can be explained by formation in either the CO- and hydrocarbon-rich circumsolar disk or a CH_4- and CO-rich circumJovian planetary disk with some post-formation ice loss being required in some cases. The CO, CH_4, NH_3 and H_2O abundances in comet Halley can be explained by its formation in the circumsolar disk. Observations of key species like CO, CH_4, NH_3 and H_2O in gaseous circumstellar disks could provide important information about these disks (and thus about the original circumsolar gaseous disk).

I. INTRODUCTION

The advancement of our understanding of the chemistry of the gaseous and dusty circumsolar nebular disk in which the planets formed has been guided traditionally by a combination of observations of the composition of the planets, planetary satellites, meteorites and comets together with development of *a priori* models of the circumsolar disk. More recently, astronomical observations of circumstellar disks have begun to provide an important new

[1005]

source of information on disk structure and constraints on both circumsolar and other circumstellar disk models.

In view of the interdisciplinary goals of this book, this chapter will focus more on the general principles determining the composition and chemistry of gas-rich circumstellar disks. A more specific (and very detailed) discussion of the chemistry of the circumsolar nebula already exists (Prinn and Fegley 1989) and these details will not be repeated here. However, the conclusions in the latter paper will be updated where necessary, based on more recent results.

As a preface to the chapter it is useful to paraphrase the principal conclusions from Prinn and Fegley (1989):

1. The starting materials for forming the circumsolar disk were gases and grains in interstellar clouds.
2. As the circumsolar gaseous disk formed and evolved, these gases and grains were at least partially reprocessed with the greatest reprocessing occurring near the proto-Sun.
3. Thermochemical reactions (using thermal energy and shock heating) constitute the major chemical processing mechanism.
4. The gas-rich and later dust-rich circumsolar disk was highly opaque limiting severely the influence of ultraviolet radiation from the proto-Sun in chemically processing the body of the disk.
5. The reduction of N_2 and CO to NH_3 and CH_4 was kinetically inhibited in the circumsolar disk but not in higher pressure disks like circumJovian planetary disks. Thus the circumsolar disk was rich in N_2 and CO (and organics formed from it) while the circumJovian planetary disks were rich in NH_3 and CH_4.
6. Vapor-phase hydration of silicates is kinetically inhibited in the circumsolar disk but not in circumJovian planetary disks.
7. Regular satellites can form in circumJovian planetary disks with distinctly different compositions than bodies formed in the circumsolar disk.
8. Thus certain volatile ratios (e.g., CO/CH_4, N_2/NH_3, H_2O ice/silicate) in ice-rich bodies are diagnostic of their origin.
9. After dissipation of the gaseous circumsolar disk, collisions between objects formed in the circumsolar disk and those formed in circumJovian planetary disks imbedded within it are inevitable and can lead to hybrid heterogeneous objects.

This chapter attempts to generalize the above ideas about the circumsolar nebula disk to other circumstellar nebula disks by addressing the underlying chemical and physical processes. There are five sections devoted respectively to: chemical theories; chemical-dynamical theories (kinetic inhibition); comparison between theory and planetary, satellite observations; comparison between theory and observations of comets, meteorites and asteroids; and finally key new astrophysical observations to test and extend current theories.

II. CHEMICAL PROCESSES

A. Initial Composition and Its Relevance

The *elemental* composition of a circumstellar disk (or equivalently its pre-cursor interstellar cloud) is obviously important in determining the chemical state of the disk. Observations of the gaseous component of dense interstellar clouds show a complex mixture of hydrogen-, oxygen-, carbon-, nitrogen-, and sulfur-containing compounds while observations of the dust component suggest the presence of various silicates, magnetite, carbon-containing compounds (including graphite), and water ice. As reviewed by Irvine and Knacke (1989), hydrogen and helium are generally dominant in interstellar clouds with oxygen, carbon, neon and nitrogen being of the order of 10^{-4} to 10^{-5} by number of the H abundance and silicon, magnesium, sulfur and iron being of the order of 10^{-6} of the H abundance. Two estimates of the elemental abundances in a solar composition medium are given in Table I. In this chapter we will denote the number density of an element or compound i by $[i]$.

TABLE I[a]

Ratios Relative to H_2 of the Ten Most Abundant
Elements in Solar Composition Material

	Anders and Grevesse	Cameron
H_2	1.0	1.0
He	0.195	0.135
O	1.17×10^{-3}	1.38×10^{-3}
C	7.24×10^{-4}	8.35×10^{-4}
Ne	2.47×10^{-4}	1.95×10^{-4}
N	2.24×10^{-4}	1.74×10^{-4}
Mg	7.70×10^{-5}	7.97×10^{-5}
Si	7.17×10^{-5}	7.52×10^{-5}
Fe	6.45×10^{-5}	6.77×10^{-5}
S	3.69×10^{-5}	3.76×10^{-5}

[a] Table uses the abundance estimates of Anders and
Grevesse (1989) and Cameron (1982).

Certain elemental ratios are extremely important in defining the disk chemistry. In particular, if carbon equals or exceeds oxygen, the majority of the oxygen can end up as CO leading to: relatively very low H_2O (vapor or ice) abundances (Lewis and Prinn 1980); an excess of carbon leading to abundant graphite and hydrocarbons; and very reducing conditions leading to formation in thermochemical equilibrium of CaS, Fe_3C, MgS, SiC and TiN (Larimer 1975). Conversely, if oxygen significantly exceeds carbon (as it did on the average for the circumsolar nebula), H_2O (vapor or ice) is abundant even if essentially all the carbon is in CO (Lewis and Prinn 1980) and the oxidation state of the nebula is much higher making minerals such as SiC thermochemically unstable if [O] exceeds [C] by more than 10% (Larimer 1975).

The [C]/[O] ratio also potentially affects the density of ice-rich condensates in circumstellar disks through its control of the H_2O to silicate ratio. For the circumsolar nebula disk the precise [C]/[O] ratio is controversial with Anders and Grevesse (1989) and Stone (1989) reporting the solar [C]/[O] ratio ≈0.42 and Cameron (1982) recommending [C]/[O] ≈ 0.60. If CO is the dominant C compound ([CO]/[C] ≈ 1) and if we model anhydrous rock as SiO_2, MgO, FeO and FeS, then the ratio of the oxygen abundances in CO, H_2O and anhydrous silicates (i.e. [CO]:[H_2O]:2[Si] + [Mg] + [Fe] − [S]) is 3.2:1.1:1.0 for [C]/[O] = 0.60 and 2.9:3.0:1.0 for [C]/[O] = 0.42. Thus there is 3.0/1.1 = 2.7 times more water available for forming ice-rich bodies when [C]/[O] = 0.42 than when [C]/[O] = 0.60. Water availability is also sensitive to the oxidation state of carbon because if [CO]/[C] ≈ 0 (i.e., CH_4 and/or organics are the major forms of carbon), then [CO]:[H_2O]:2[Si] + [Mg] + [Fe] − [S] is 0.0:4.3:1.0 for [C]/[O] = 0.60 and 0.0:5.9:1.0 for C/O = 0.42. Thus there is 4.3/1.1 = 3.9 ([C]/[O] = 0.6) or 5.9/3.0 = 2.0 ([C]/[O] = 0.42) times more water available for forming ice in a CH_4-rich or organic-rich region than in a CO-rich region.

The *molecular* composition and *physical state* of interstellar material collapsing onto the disk are also important in determining the susceptibility of this material to chemical processing during accretion and later processing within the disk. Highly refractory solid material (e.g., Al_2O_3, $CaTiO_3$, $Ca_2MgSi_2O_7$, Fe-Ni alloy, Mg_2SiO_4) is more likely to survive than less refractory solid material (e.g., graphite, hydrated silicates, polycyclic aromatic compounds, FeS) and both are much more likely to survive than volatile solids (H_2O ice, $NH_3 \cdot H_2O$, $CH_4 \cdot 7H_2O$, $CO \cdot 6H_2O$). Survival of solids will also depend on grain size, and on whether compounds are in crystalline or amorphous states. In the gas phase, strongly bound molecules like N_2 are more likely to survive processing than more weakly bound species like NH_3. Some of the organic material in some meteorites may be unaltered interstellar material and one certainly expects a greater probability for survival of such unaltered material in bodies in the least explored outermost regions of the solar system, particularly comets.

B. Energy for Chemical Reactions

Potential sources of energy to drive chemical reactions include the internal thermal energy of the disk, shock heating due to impact of hypersonic interstellar material accreting onto the disk, lightning discharges and thundershocks associated with convection in the disk, ultraviolet photons from the central protostar and the surrounding proximal stars, and very high energy photons and particles from decay of radioactive elements (especially ^{26}Al) in the disk. As discussed by Prinn and Fegley (1989), it is useful to compare the usable energy fluxes ϕ available from the various sources to drive chemical reactions to the net outward flux $\phi_T = sT_e^4$ in the disk where s is Stefan's constant and

$T_e(R)$ is the effective temperature of the disk at radius R. For conditions at $R = 1$ AU in the circumsolar nebula disk they compute

$$\phi(\text{thermal})/\phi_T \approx \exp(-E/NkT) = 4 \times 10^{-5} \tag{1}$$

$$\phi(\text{thundershock})/\phi_T \approx 4 \times 10^{-6} \tag{2}$$

$$\phi(\text{solar photons})/\phi_T \approx 10^4 \times F\exp(-\tau)/\phi_T \approx 0 \tag{3}$$

$$\phi(\text{stellar photons})/\phi_T \approx 10^{-9} \tag{4}$$

$$\phi(\text{radioactivity})/\phi_T \approx 3 \times 10^{-12} \tag{5}$$

where $E = 5 \times 10^4$ J/mole is a typical activation energy of a chemical reaction quenching at $R = 1$ AU, k is Boltzmann's constant, N is Avogadro's number, τ is the optical depth for solar ultraviolet photons between the proto-Sun and radius R, and F is the present-day solar ultraviolet energy flux at $R = 1$ AU. Evidently for these five energy sources, thermal energy dominates at 1 AU except for reactions with very large E (for which thundershocks could be significant). Solar photons (due to the very large τ value resulting from abundant H_2O, dust, etc., in the disk) and radioactivity (due to the low abundance of ^{26}Al) are negligible at 1 AU. As we approach the proto-Sun ($R<<1$ AU) thermal energy dominates even more while thundershocks become relatively more important further from the proto-Sun ($R>>1$ AU). Stellar photons are unimportant for $R\leq1$ AU but could be relatively significant for $R\geq10$ AU. Very close to the protoSun ($\tau\approx0$, $R\approx0$), protosolar ultraviolet photon fluxes will be very large but due to the large opacity, the photons are absorbed wholly in the very hot (temperature $>>2500$ K) relatively dense innermost part of the gaseous disk. Here, large thermochemical reaction rates involving abundant thermally produced atoms, free radicals, electrons and ions will generally overwhelm the dissociative and ionizing effects of the ultraviolet photons. For this purpose, efficient thermal ionization requires temperatures about 1000 K greater than thermal dissociation. We note parenthetically that protosolar photons emitted in directions other than into the disk (e.g., in polar directions) may suffer little or no absorption and therefore be remotely observable. Thus observations of ultraviolet emissions from T Tauri stars do not provide evidence for photochemistry driven by photons from the central star occurring other than in the innermost regions of a protosolar-type gaseous disk.

The sixth energy source mentioned above, namely shock heating during accretion onto the disk, was not considered quantitatively by Prinn and Fegley (1989). To obtain a measure of the intensity of the shock heating which can occur, we use here the Hugoniot shock relation in an ideal polytropic gas (see Courant and Friedrichs 1976) to obtain the temperature T_s of the shocked disk gas relative to the temperature T_o of the pre-shocked gas

$$T_s \approx T_o[\mu^2(1 - M^{-2}) + 1][\mu^2(M^2 - 1) + 1]$$
$$\approx 1031 \text{ K}(\mu^2 = 0.18, T_o = 60 \text{ K}, R = 10AU)$$
$$\approx 1847 \text{ K}(\mu^2 = 0.17, T_o = 120 \text{ K}, R = 5AU) \qquad (6)$$
$$\approx 8224 \text{ K}(\mu^2 = 0.16, T_o = 600 \text{ K}, R = 1AU)$$

where μ^2 is the average over the interval between T_o and T_s of the gas constant divided by the sum of the heat capacities at constant volume and constant pressure, $M \approx 22$ is the Mach number of the shock front which we equate to the Mach number of the free falling impacting interstellar material, and T_o is a representative solar nebula temperature profile allowing ice condensation in the Jovian formation region. Note that because the free fall speed varies as $R^{-0.5}$, the sound speed varies as $T_o^{0.5}$ and T_o varies approximately as R^{-1}, then the latter Mach number is independent approximately of distance R. Evidently, the disk gas (and to a lesser extent the lower-temperature-impacting interstellar gas) is intensely shock heated for $R \leq 1$ AU but only mildly shock heated for $R \geq 10$ AU. Thus, quite apart from any subsequent processing within the disk, the disk accretion process itself probably led to significant thermochemical reprocessing of disk (and to a lesser extent impacting interstellar) material inside $R = 10$ AU. Depending on the relative magnitudes of the mass per unit area (in the direction of impact) for the accreting gas mass and the nebula disk, the accreting material will be decelerated and force strong mixing near the surface of the disk or throughout its thickness. Chemical reprocessing will occur in both the region of deceleration and beyond this region through propagating shock waves.

C. Equilibrium Products of Thermochemical Reactions

The *potential* products of thermochemical reactions are conveniently summarized by addressing the predicted composition in *thermochemical equilibrium*. As discussed later, such equilibrium is possible only if the relevant cooling time (dynamical or radiative) for the gas exceeds the relevant chemical reaction times and it is thus most likely to occur at high temperatures in the disk where reactions are fast (e.g., near the protostar or in a region shock heated by an impact or lightning). Figures 1, 2 and 3 illustrate how three key ratios CO/CH_4, N_2/NH_3 and CO_2/CO vary in thermochemical equilibrium as a function of pressure and temperature in a gas with solar H, O, C and N ratios. Sample disk adiabatic pressure-temperature profiles are shown for the circumsolar nebula disk (Lewis and Prinn 1980) and the circumJovian planetary nebula disk (Prinn and Fegley 1981). Pressure-temperature profiles for any arbitrary near-solar-composition circumstellar disk may also be drawn on these diagrams to illustrate the composition of these disks if thermochemical equilibrium is attained. Similarly, shock heating of the disk gas can be illustrated on these diagrams by imagining a temporary excursion of

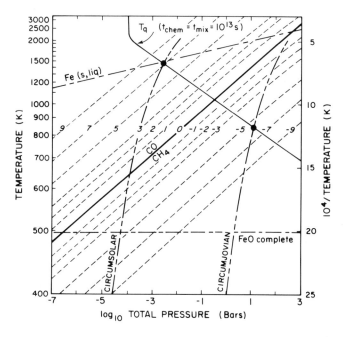

Figure 1.   Thermochemical equilibrium $\log_{10}([CO]/[CH_4])$ values in a solar composition medium with $[C]/[O] = 0.6$. Also shown are illustrative temperature-pressure profiles for the circumsolar and circumJovian disks and the quench temperature T_q for homogeneous (gas-phase) conversion of CO to CH_4 assuming an upper limit for $t_{mix} = 10^{13}$ s. Heterogeneous (iron-catalyzed) conversion is possible between the lines designating the iron condensation temperature labeled "Fe(s,liq)" and the temperature where oxidation of Fe to FeO is complete, labeled "FeO complete."

the shocked gas up an adiabat (as shown) or up a shock-Hugoniot (which would be closer to vertical than the adiabats shown) to higher temperatures and pressures. Evidently, for all these near-adiabatic temperature-pressure profiles the transition to higher temperatures and pressures is accompanied by a shift of the three major equilibria

$$CH_4 + H_2O = CO + 3H_2 \tag{7}$$

$$NH_3 + NH_3 = N_2 + 3H_2 \tag{8}$$

$$CO_2 + H_2 = CO + H_2O \tag{9}$$

to the right to yield CO, N_2 and CO_2 until temperatures become high enough (≥ 3000 K) for atomization and ionization to begin. In addition, the equilibria

$$2CH_4 + N_2 = 2HCN + 3H_2 \tag{10}$$

$$CH_4 + NH_3 = HCN + 3H_2 \tag{11}$$

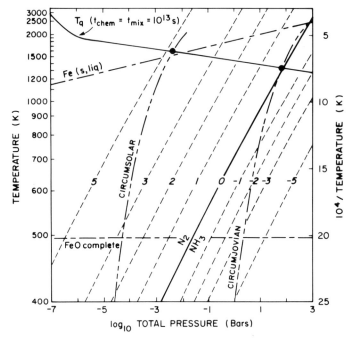

Figure 2.   As in Fig. 1 but for the $[N_2]/[NH_3]$ ratio and conversion of N_2 to NH_3.

$$2CO + N_2 + 3H_2 = 2HCN + 2H_2O \tag{12}$$

$$2CH_4 = C_2H_2 + 3H_2 \tag{13}$$

$$2CO + 3H_2 = C_2H_2 + 2H_2O \tag{14}$$

$$2CH_4 = C_2H_6 + H_2 \tag{15}$$

$$2CO + 5H_2 = C_2H_6 + 2H_2O \tag{16}$$

also shift to the right for $T \leq 3000$ K yielding trace amounts of HCN, C_2H_2 and C_2H_6 (see, e.g., Prinn and Fegley 1989, Figs. 5, 6, 9, 10). Figures 1–3 use a [C]/[O] ratio of 0.6 and are quantitatively but not qualitatively altered for [C]/[O] = 0.42.

The equilibrium composition of solid materials under circumstellar disk conditions is also of considerable interest (see Fig. 4). At sufficiently high temperatures, silicates exist in anhydrous forms (olivine $(Fe,Mg)_2SiO_4$; pyroxene $(Fe,Mg)SiO_3$; etc.) and Fe exists as the metal or (as FeO) in olivines and pyroxenes. As the temperature is lowered, Fe oxidation to FeO which is incorporated into olivines and pyroxenes and to FeS which forms troilite depletes this metal, specifically

$$Fe + H_2O = FeO + H_2 \tag{17}$$

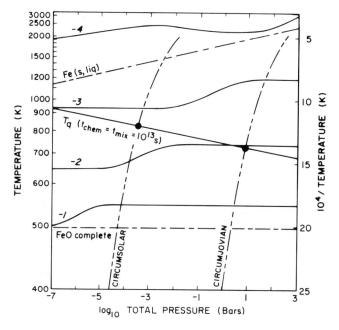

Figure 3.   As in Fig. 1 but for the $[CO_2]/[CO]$ ratio and conversion of CO to CO_2.

$$Fe + H_2S = FeS + H_2. \tag{18}$$

Also, as the temperature is lowered further, hydration of olivines and pyroxenes to form hydrated silicates like serpentine and talc and hydroxides like brucite occur. For example,

$$2Mg_2SiO_4 + 3H_2O = Mg_3Si_2O_5(OH)_4 + Mg(OH)_2 \tag{19}$$

$$4MgSiO_3 + 2H_2O = Mg_3Si_4O_{10}(OH)_2 + Mg(OH)_2. \tag{20}$$

Because these latter hydrated silicates and hydroxides are 8 to 16% by weight H_2O they lower the density of the rock material in equilibrium from about 4 g cm^{-3} (anhydrous) to 3 g cm^{-3} (hydrated).

At even lower temperatures (Fig. 4), the following transitions occur in order: H_2O condenses as ice; NH_3 (the major form of N in equilibrium) condenses as the hydrate $NH_3 \cdot H_2O$; CH_4 (the major form of C in equilibrium) condenses as the clathrate $CH_4 \cdot 6H_2O$; and finally, excess CH_4 condenses as the pure solid.

For both the gas and condensed phases, we emphasize again that the rates of the above conversions must be fast relative to changes in temperature and pressure for the thermochemical equilibrium products to be relevant. As discussed in the next section, this condition is not always met.

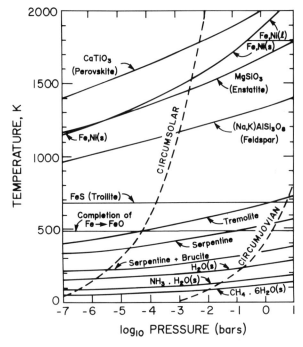

Figure 4.   Thermochemical equilibrium stability fields for condensed solid material
in a solar composition medium. Also shown are illustrative temperature-pressure
profiles for the circumsolar and circumJovian disks.

III. DYNAMICAL-CHEMICAL PROCESSES

Chemistry circumstellar disks is profoundly influenced by sufficiently strong
circulation and mixing if it occurs in the disk. Mixing processes in disks are
often discussed in the context of a standard model (see, e.g., Pringle 1981;
Lin and Papaloizou 1985; Morfill et al. 1985). This model parameterizes the
radial angular momentum flux which is fundamentally quadratically nonlinear
in velocity solely by a gradient eddy or viscous diffusion expression which is
linear in velocity. This simplification (or oversimplification) has enabled conclusions to be reached by simply choosing a viscosity which ensues evolution
of the disk on the observed time scale (10^5 to 10^6 yr). A particular difficulty
arises when this viscosity is considered applicable also to the transport of
trace materials in the disk.

There are arguments (albeit not universally accepted) favoring strong
trace material mixing in the circumsolar disk and by analogy probable strong
mixing in other circumstellar disks. First, there is considerable observational
evidence for extensive inner circumsolar nebular mixing and reprocessing
contained in chondritic meteorites. Specifically, these meteorites are an inhomogeneous mixture of materials (chondrules, refractory inclusions, matrix)

which formed in distinctly different nebula environments in which they underwent varying degrees of condensation, vaporization and recondensation (see, e.g., review by Wood 1988). Formation of this mixture requires strong mixing of these materials with different histories (although the timing and spatial scale of the mixing is ill defined in part because the separation of these nebula environments is unknown).

Second, the time scale for mixing can be much shorter than the time scale for disk evolution because basic nonlinear acceleration processes enable the shear in an accretion disk to be maintained against viscous dissipation (plus gravitational torques if applicable), so that trace constituents in these disks can be well mixed without momentum being well mixed (Prinn 1990). Stevenson (1990) presents an opposing viewpoint. He concludes that momentum and trace constituents are approximately equally (and in this context both weakly) mixed. These arguments are based on an acceptance of the standard model which as discussed above neglects the nonlinear (in velocity) nature of momentum transport compared to the linear (in velocity) nature of trace material transport. Prinn (1990) argues that the poor trace constituent mixing predicted in accretion disk models which omit nonlinear processes is an artifact of this omission. This is almost certainly true when the system is externally forced by the disk accretion process but the Stevenson (1990) case cannot be ruled out when accretion ceases. An important question to be answered is, therefore, whether the material which formed the present solar system was defined while accretion onto the disk was still proceeding or afterwards.

Third, as we discuss in Sec. III.B, the disk accretion process itself provides an important and perhaps dominant forcing mechanism for strong mixing.

A. Kinetic Inhibition Models

The major effect of strong mixing in the disk is to cause chemical disequilibration. The so-called "kinetic inhibition" theory for thermochemical disequilibration induced by rapid mixing in a thermoclinic environment was first proposed by Prinn and Barshay (1977) to explain the unexpected discovery of CO in the CH_4-rich Jovian atmosphere. It was later applied also to the circumsolar nebula disk (Lewis and Prinn 1980) and to circumplanetary disks around the Jovian planets (Prinn and Fegley 1981). The disequilibration results specifically when the cooling time due to transport in a medium with a temperature gradient becomes shorter than the relevant chemical reaction time for maintaining equilibrium.

In a hot thermoclinic environment like a gaseous circumstellar disk one can define for each chemical species a quench radius (R_q) inside which the temperatures are high enough (and thus thermochemical reactions fast enough) to ensure that the species concentration [i] is essentially equal to that in thermochemical equilibrium (see Fig. 5). To a sufficient approximation R_q is the radius at which the chemical destruction time $t_{chem} = -[i]/(d[i]/dt)$ for the species equals the radial mixing time t_{mix}. Also, since temperature, pressure and hence [i]/t_{chem} generally decrease with altitude z above the disk plane,

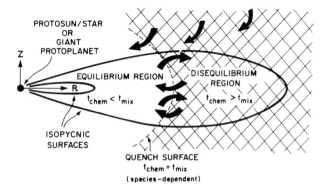

Figure 5.   Illustration of the species-dependent quench surface separating thermochemical equilibrium and transport-induced-disequilibrium regions in gaseous
circumsolar or circumJovian planetary disks.

the equilibrium region bounded by the quench surface is convex as shown in
Fig. 5. We use the mixing ratio form of the species continuity equation in
cylindrical coordinates in the region $R > R_q$ with the assumption that the gas
is quenched chemically by moving outward a distance l_{chem} from R_q which
is much less than the characteristic length L of eddy circulations in the disk.
This leads to the solution that the mixing ratio f in the disequilibrium region
$(R > R_q)$ equals approximately the equilibrium value of f at the quench radius
R_q (see Eqs. (13)–(18) of Prinn and Fegley [1989] for details). This latter
solution is the one appropriate to reactions with large activation energies
such as the conversions of CO to CH_4 and N_2 to NH_3 in the hydrogen-rich
circumsolar nebula disk. Referring back to the thermochemical equilibrium
calculations in Figs. 1 to 3, the general approach is to compute t_{chem} values for
the relevant conversion using equilibrium concentrations and then to compare
them to estimates of t_{mix} (see next section). Then we define the quench radius
to be where $t_{chem} = t_{mix}$. The temperature at the quench radius is referred to
as the quench temperature T_q and it is often more convenient to define the
quench position in the disk by T_q rather than R_q. Sample T_q values defined by
considering only homogeneous gas-phase conversion reactions to define t_{chem}
and equating t_{mix} to the approximate lifetime of the circumsolar disk (10^{13} s)
are shown in Figs. 1 to 3. The actual t_{chem} values are shorter (lowering T_q)
than the homogeneous ones due to heterogeneous catalysis reactions on Fe
particles but this is countered by the t_{mix} values also being less than 10^{13} s
(raising T_q). The importance of knowing t_{mix} is clear and in the next section
we deduce t_{mix} values for a disk mixed by eddies forced by nonaxisymmetric
accretion of interstellar material.

B. Accretionally Forced Turbulence

The forcing of turbulence in circumstellar accretion disks by the infalling material itself has been recognized by Cameron (1978a) and Prinn (1990), but it has not been quantified and indeed has received remarkably little attention. As summarized by Prinn (1990), the thermodynamics of accretion involves the conversion of the initial gravitational potential energy into kinetic energy (mostly orbital but some radial). Next, a large fraction of the kinetic energy is converted into internal (thermal) energy by compression or frictional dissipation as the accreting material collides and merges with the gaseous disk. Some of the latter thermal energy is then lost by radiation to space and some is converted back to kinetic energy by various internal instabilities. Convective instability has received some attention but there are many other instabilities (baroclinic, barotropic, etc.) yet to be considered. However, these internal instabilities cannot reduce the temperature gradient or the absolute temperature to zero and they are capable therefore of converting only a very small fraction of the internal energy back to kinetic energy. Hence the production (through internal instabilities) of disk kinetic energy from disk internal energy is expected to be considerably less than the production of disk kinetic energy (through accretion) from initial gravitational potential energy (Prinn 1990). An approximate estimate of the speed in eddy circulations forced in the disk by accretion is provided here by considering the eddy angular momentum equation in cylindrical polar coordinates for the disk (Prinn 1990, Eq. 4):

$$\frac{1}{R}\frac{\partial}{\partial R}(R^2\sigma\langle V'U'\rangle)+R\sigma\langle V\rangle\frac{1}{R}\frac{\partial}{\partial R}(R\langle U\rangle) = R(\langle U_a\rangle - \langle U\rangle)\left(\frac{d\sigma}{dt}\right)_a \quad (21)$$

where R is the radius, σ is the surface density, $(d\sigma/dt)_a$ is the the rate of change of σ due to accretion, V and U are the radial and azimuthal components of velocity, respectively, U_a is the azimuthal velocity of infalling material, t is time and $\langle i\rangle$ and i' denote, respectively, the azimuthal average and deviation from this azimuthal average of the quantity i. Assuming the radial advective flux is small relative to the radial eddy flux of azimuthal (angular) momentum, equating $\langle U_a\rangle - \langle U\rangle$ to the product of the free-fall velocity $(2GM/R)^{1/2}$ of accreting material and a constant α of order unity, and neglecting any small radial dependence for σ, we have:

$$\frac{2}{R}\langle V'U'\rangle + \frac{\partial}{\partial R}\langle V'U'\rangle = \frac{\alpha}{\tau}\left(\frac{2GM}{R}\right)^{1/2} \quad (22)$$

where $\tau = (d\ln\sigma/dt)^{-1}$ is the evolution time for mass in the disk. Assuming in Eq. (22) the scaling relations $\langle V'U'\rangle \approx v^2$ and $\partial\langle V'U'\rangle/\partial R \approx v^2/L$ where v and L are the characteristic speed and length scale of the eddy circulations, we obtain the following simple expression for the eddy speed

$$v = \left[\frac{GMR\alpha^2}{2\tau^2(1+R/2L)^2}\right]^{1/4}$$
$$= 18\left(\frac{M}{M_s}\right)^{1/4}\left(\frac{R}{R_0}\right)^{1/4}\left(\frac{\tau_0}{\tau}\right)^{1/2}\alpha^{1/2}(1+R/2L)^{-1/2} \quad (23)$$

in m s^{-1} where M_s is the mass of the Sun, $R_o = 1$ AU, and $\tau_o = 10^{13}$ s is the approximate lifetime of the gaseous circumsolar disk. The nominal eddy speed (v_o) of $18/(1 + R/2L)^{1/2}$ m s^{-1} is $0.95\%/(1 + R/2L)^{1/2}$ of the sound speed in H$_2$ (1900 m s^{-1} at 600 K). For all L exceeding about $5 \times 10^{-3}R$ this speed equals or exceeds the maximum speed expected from free convection in the disk (Cabot et al. 1987b). This v_o also implies nominal radial eddy mixing times $t_{mix} = R_o/v_o$ of $6 \times 10^9/(1 + R/2L)^{1/2}$ s. Thus for all L exceeding about $4 \times 10^{-7}R$, $t_{mix} < \tau_o$ so that the disk is well mixed. Finally, we note parenthetically that the advective velocity $\langle V \rangle$ estimated by equating the right-hand side of Eq. (21) to the second (rather than the first) term on the left-hand side and using the Keplerian velocity for $\langle U \rangle$ is

$$\langle V \rangle = 2^{3/2} \frac{\alpha R}{\tau}$$
$$= 0.04 \frac{R \tau_o}{R_o \tau} \alpha \text{ m s}^{-1} \tag{24}$$

which is evidently much less than the eddy speed v.

C. Kinetic Inhibition Results

The quench points for the conditions in Figs. 1 to 3 are located at the intersections of the disk temperature profiles and the quench temperature curves. Remembering that the species mixing ratios at the quench points determine the ratios at all points exterior to them (i.e., at all $T < T_q$ or all $R > R_q$), we conclude from Figs. 1 and 2 that [CO] $\gg$ [CH$_4$] and [N$_2$] $\gg$ [NH$_3$] for all T in the circumsolar nebula disk. Conversely, [CH$_4$] $\gg$ [CO] and [NH$_3$] $\gg$ [N$_2$] in the circumJovian planetary nebula disk. From Fig. 3, [CO$_2$] is 0.2 to 1% of [CO] in both disks but as [CO] is negligible in the circumJovian planetary disk, it is only in the circumsolar disk that CO$_2$ becomes an important carbon-containing compound. These conclusions from the above simplified approach using Figs. 1 to 3 are qualitatively similar to the conclusions from the more detailed approaches which take into account both homogeneous and heterogeneous (catalyzed) reactions and a wide range of feasible mixing times (Lewis and Prinn 1980; Prinn and Fegley 1981; Prinn and Barshay 1977; Prinn and Olaguer 1981; Prinn and Fegley 1989).

The chemical lifetimes and quench temperatures for formation of hydrated silicates (e.g., serpentine, tremolite in Fig. 4) in the solar nebula have been considered by Prinn and Fegley (1989). They conclude that even the simple hydration (reaction 19) of 0.1 μm radius forsterite (Mg$_2$SiO$_4$) grains has $t_{chem} \geq 10^{18}$ s (which certainly exceeds t_{mix} and even exceeds the lifetime of the solar system) at the 225 K temperature at which its hydration to serpentine and brucite becomes thermodynamically possible in the circumsolar nebula disk. Formation of tremolite from hydration of enstatite plus diopside at around 500 K and of serpentine from hydration of enstatite plus forsterite at around 350 K requires solid-solid diffusion leading to even longer t_{chem}

times. Tremolite is in any case a minor predicted mineral in the nebula and rare in meteorites. The anhydrous silicates accreted onto the circumsolar disk or produced from thermal reprocessing of accreted interstellar hydrated silicates are expected therefore to be very stable in the disk. In contrast, Fegley and Prinn (1989) conclude that the 100 K higher temperatures (325 K) and 10^5 times greater H_2O concentrations at the serpentine formation point in the circumJovian disk mean that $t_{chem} \approx 10^9$ s which may feasibly be faster than the mixing time due to convection in this disk. Thus unlike in the circumsolar disk, the formation of hydrated silicates may not be kinetically inhibited in the circumJovian disk.

Fegley and Prinn (1989) discussed for the solar nebula the conversion of Fe to FeS (in troilite) and FeO (in olivine and pyroxene) and concluded that FeS formation is not kinetically inhibited at the FeS stability point ($\approx$690 K) whereas the oxidation of Fe and its incorporation into magnesium silicates was almost certainly kinetically inhibited. In this case, the excess Fe not converted to FeS could thermodynamically begin to react with H_2O vapor at about 380 K to form magnetite (Fe_3O_4) but laboratory data are not available to determine if this reaction is not also inhibited.

The kinetic inhibition of the $N_2 \rightarrow NH_3$, $CO \rightarrow CH_4$ and anhydrous $\rightarrow$ hydrous silicate transitions in the circumsolar disk but not in the circumJovian planetary disks means that low-temperature condensation of ice-rich phases in the two disks are distinctly different (Lewis and Prinn 1980; Prinn and Fegley 1981). In the solar disk, we expect at successively lower temperatures formation of anhydrous rock, H_2O ice, NH_4HCO_3 salt, NH_4COONH_2 salt, CO_2 (dry) ice, $CO \cdot 6H_2O$ clathrate, $N_2 \cdot 6H_2O$ clathrate, CO solid and N_2 solid. Conversion of some of the CO through Fischer-Tropsch-type reactions to involatile carbonaceous compounds (condensing at higher temperatures than ice) and light hydrocarbons and hydrocarbon clathrates (condensing coincident with or at lower temperatures than ice) is also possible particularly if the metallic Fe catalyst required is not completely oxidized or sulfurized as discussed above or de-activated by organic coatings. If volatilization of interstellar involatile carbonaceous compounds was kinetically inhibited during accretion shock heating and within the circumsolar nebula disk then this would add to the involatile carbon component.

In the circumJovian planetary disks in contrast we have the condensation sequence: hydrous rock, H_2O ice, $NH_3 \cdot H_2O$ hydrate, $CH_4 \cdot 6H_2O$ clathrate, CO solid and N_2 solid. This looks superficially like the condensation sequence if chemical equilibrium prevailed in the circumsolar nebula but on closer examination it is clearly recognized as different due to the presence of some C and N as CO and N_2. The significant differences between the circumsolar and circumJovian planetary condensation sequences which are summarized in Table II lead to the conclusion that low-temperature condensates in ice-rich bodies may be diagnostic of the type of nebula disk in which they formed (Prinn and Fegley 1981).

Certain caveats concerning the formation of the clathrates in the above

TABLE II[a]

Condensation Sequences Predicted in the Circumsolar Gas-Dust Disk and CircumJovian Planet Gas/Dust Disk[b]

Circumsolar Disk		CircumJovian Planet disk CH_4-Rich
CO-Rich	CO + Hydrocarbons	
Largely anhydrous rock[c]	Largely anhydrous rock[c] Involatile carbonaceous compounds[e,f]	Largely hydrous rock[d]
H_2O (150 K)	H_2O	H_2O (235 K)
NH_4HCO_3/NH_4COONH_2 (150, 130 K)[e]	NH_4HCO_3/NH_4COONH_2[e]	$NH_3 \cdot H_2O$ (160 K)
	Light hydrocarbons and hydrocarbon clathrates[g]	$CH_4 \cdot 6H_2O$ (94 K)
CO_2 (70 K)[e]	CO_2[e]	CH_4 (40 K)
$CO \cdot 6H_2O$ (60 K)[h]	$CO \cdot 6H_2O$[h]	CO (20 K)[e]
$N_2 \cdot 6H_2O$ (55 K)[h]	$N_2 \cdot 6H_2O$[h]	N_2 (20 K)[e]
CO (20 K)	CO (20 K)	
N_2 (20 K)	N_2 (20 K)	

[a] Prinn and Fegley 1989.

[b] Condensation temperatures in models (Lewis and Prinn 1980; Prinn and Fegley 1981) are given for each condensate where available. For the circumsolar disk the CO-plus-hydrocarbons sequence is preferred with the condensation temperatures being slightly below those in the CO-rich case (the precise temperatures will depend on the degree of CO to hydrocarbons conversion).

[c] Olivine, pyroxene, troilite, unoxidized Fe, feldspar, etc.

[d] Serpentine, talc, magnetite, troilite, tremolite, etc.

[e] These compounds contain $\leq 10^{-2}$ of the total C or N.

[f] Exemplified by organic material in chondrites, and formed by nebular Fischer-Tropsch-type reactions and/or of interstellar origin.

[g] Exemplified by the light (C_1–C_5) hydrocarbons formed as initial products in high H_2: CO environments (see Prinn and Fegley 1989). Precise condensation or enclathration temperature depends on the chain length and hydrocarbon abundance.

[h] There is insufficient H_2O to allow condensation of all C and N as CO and N_2 clathrates.

sequences are in order. For the Anders and Grevesse [C]/[O] ratio and even more so for the Cameron [C]/[O] ratio, there is insufficient H_2O to enclathrate the majority of the CO let alone the N_2 in the circumsolar disk. Lunine and Stevenson (1985) give theoretical arguments and Davidson et al. (1987) give experimental evidence that $CO \cdot 6H_2O$ is more stable than $N_2 \cdot 6H_2O$ so the latter clathrate may be very rare indeed. Similar difficulties exist for CO and N_2 clathrates in circumJovian planetary disks where there is insufficient water to enclathrate even the majority of the CH_4. Fegley and Prinn (1989) also point out possible kinetic inhibition of clathrate formation in the circumsolar (but not circumJovian planetary) disk unless the activation energy for diffusion of gas into the ice is very small. Lunine (1989b) argues in reply that this activation energy is indeed small. However, if for CO one equates it (arguably) to the recently reported binding energy for CO on H_2O ice (14.5 kJ mole^{-1}; Sandford and Allamandola 1990b) this is indeed large enough to cause $CO \cdot 6H_2O$ formation to be kinetically inhibited in the circumsolar nebula. Specifically, we compute $t_{chem} = 1.6 \times 10^{17}$ s for conversion of ice particles (2 μm dimension) to $CO \cdot 6H_2O$ at its formation temperature of about 60 K in the circumsolar nebula (Table II). Conversely, this large activation energy would lead to efficient *adsorption* of CO on ice provided the ice is sufficiently *amorphous* to expose a large enough surface area to enable $[CO]/[H_2O]$ stoichiometries like the 1/6 value in the clathrate to be realized. This may be the way in which CO was incorporated into ice to explain its presence in comets. Note also that the salts ammonium bicarbonate (NH_4HCO_3) and ammonium carbamate (NH_4COONH_2) will enable up to 1% of the solar abundances of C and N to be incorporated into ice.

IV. PLANETS AND SATELLITES

The amounts of H, O, C and N in the terrestrial planets, giant planets, satellites of giant planets, Pluto/Charon, and comets provide crucial tests of theories of their origin. This subject is thoroughly discussed by Prinn (1982), Lewis and Prinn (1984), Prinn and Fegley (1987,1989) and Lunine (1989a, c) so here we present only a brief summary and update (the latter focusing on recent observations of ice-rich satellites and comets).

A. Terrestrial Planets

It has been understood ever since Urey (1952) proposed the idea (based on evidence such as the observed depletion of noble gases like Ne relative to reactive elements like N) that the volatile elements on the terrestrial planets originated not from accretion of disk gas but from accretion of disk solids (before and after disk gas dissipation). Fegley and Prinn (1989) and Prinn and Fegley (1987) review the evidence that the major source of carbon and some hydrogen in the terrestrial planets was accretion of material rich in organic compounds (e.g., of the type produced from CO by Fischer-Tropsch reactions) with thermochemical equilibrium sources like carbon dissolved in Fe-Ni alloy being secondary. This argues in favor of the CO plus hydrocarbons sequence in Table II. The small amount of nitrogen in these planets is easily provided by the same organic compounds. As noted earlier, the kinetic inhibition of hydrous silicate formation in the circumsolar nebula makes the most likely source of H_2O for the terrestrial planets to be accretion of ice-rich material in the post-gas-rich phase of the circumsolar disk (this material could also include hydrous minerals formed in asteroid-sized bodies once these evolved).

B. Giant Planets

It is also well understood that the giant planets are composed of circumsolar disk gases plus additional solid material either contained in a precursor core or accreted later during both the gas-rich and post-gas-rich epochs (see the review by Pollack and Bodenheimer [1989]). Prinn and Fegley (1989) conclude that the solid material must have been 9 to 30% (Jupiter) and 3 to 12% (Saturn) carbon by mass to explain the bulk properties and atmospheric composition of these two planets. This also argues in favor of the CO plus hydrocarbons sequence in Table II because the CO-rich sequence has insufficient carbon at the formation temperatures for Jupiter and Saturn which certainly exceeded 60 K. Similar conclusions for large carbon inputs pertain to Uranus and Neptune with the organic carbon content of the added solid material being at least comparable to the carbon in (solid or clathrate) carbon monoxide (Pollack et al. 1986; Fegley and Prinn 1986; Simonelli et al. 1989).

C. Ice-Rich Satellites and Pluto

The ice-rich satellites of the outer planets and Pluto provide a wealth of significant information on the circumsolar nebula disk. The information refers not only to the outer regions of the nebula disk but also to inner regions as kinetic inhibition of CO $\rightarrow$ CH_4, N_2 $\rightarrow$ NH_3, and anhydrous rock $\rightarrow$ hydrous rock, and catalytic conversion of CO to light or heavy hydrocarbons are inner nebula processes impacting the composition of outer nebula objects (Lewis and Prinn 1980; Prinn and Fegley 1981,1989; Lunine 1989a, c).

Using observations and models for satellite interiors, Johnson et al. (1987), Johnson (1990), McKinnon and Mueller (1988,1989), and Simonelli et al. (1989) have computed the ratio r = (rock mass)/(rock mass + ice mass) in various ice-rich bodies in the outer solar system and these are shown in Fig. 6. To compare with these "observed" r values, we have computed models for a variety of kinetic inhibition scenarios and for the C and O abundances A_C and A_O (measured relative to $A_{Si} = 10^6$) of both Anders and Grevesse (1989) and Cameron (1982). We model the rock as SiO_2+ MgO + FeO + FeS (i.e., anhydrous) and noting the similarity of the Si, Mg, Fe and S abundances of Anders and Grevesse (1989) and Cameron (1982), we use the average of these two abundance estimates for these latter 4 elements. We consider carbon in three forms: CO (noncondensing), light hydrocarbons including CH_4 (noncondensing), and heavy hydrocarbons (condensing). We model the heavy condensing hydrocarbons as CH_2 and add them to the ice mass in the

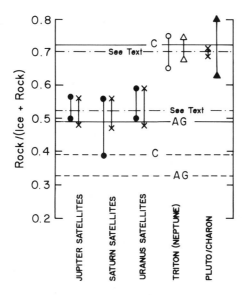

Figure 6.   Ranges of r = (rock mass)/(ice plus rock mass) for ice-rich bodies in the solar system deduced from observations by Johnson et al. (1987; filled circles), Johnson (1990; open circles), Simonelli et al. (1989; crosses), McKinnon and Mueller (1988; filled triangles) and McKinnon and Mueller (1989; open triangles). Also shown are predicted r values for condensates in a CO-rich solar-composition environment (solid horizontal lines) and a CH_4-rich solar-composition environment (dashed horizontal lines). Predictions are shown for solar C and O abundances from Cameron (1982; labeled "C") and from Anders and Grevesse (1989; labeled AG). The dot-dashed horizontal lines at $r = 0.52$ and $r = 0.70$ are predictions from a variety of models discussed in the text.

definition of r. We define the mole fractions of total carbon in CO, light hydrocarbons, and heavy hydrocarbons as α, $1 - \alpha - \beta$, and β, respectively. Thus the CO-rich nebula of Lewis and Prinn (1980) has $\alpha \approx 1 - \beta$, $1 - \alpha - \beta \approx 0$, and $\beta < \alpha$ (allowing for some conversion of CO to heavy hydrocarbons by Fischer-Tropsch-type reactions). Conversely the CH_4-rich nebula of Prinn and Fegley (1981) has $\alpha \approx 0$, $1 - \alpha - \beta \approx 1$, and $\beta \approx 0$, respectively.

With the above definitions, we obtain

$$r = 9.755 \times 10^6 [A_O - \left(\alpha - \frac{14}{18}\beta\right) A_C + 6.296 \times 10^6]^{-1}. \quad (25)$$

Note that r does not depend on the [C]/[O] ratio A_C/A_O alone but on the *separate values* of A_O and A_C.

In Fig. 6, we see distinctly lower values of r for the satellites of Jupiter, Saturn and Uranus ($r \approx 0.52$) than for Triton, Pluto and Charon ($r \approx 0.7$). For A_C and A_O from Anders and Grevesse (1989), we could approximately explain the Jovian, Saturnian and Uranian satellites as objects formed in a CO-rich ($\alpha = 1$, $\beta = 0$) solar nebula disk but must then rely on *ad hoc* explanations (e.g., massive post-formation water loss) to explain Triton, Pluto and Charon. Conversely for A_C and A_O from Cameron (1982), we could easily explain Triton, Pluto and Charon as objects formed in a CO-rich ($\alpha = 1$, $\beta = 0$) solar nebula disk but must then rely on *ad hoc* explanations (e.g., massive post-accretional water addition) to explain the Jovian, Saturnian and Uranian satellites.

To address this failure of the simplest models to fit the data, we can use Eq. (25) to deduce the values of $\alpha - (14/18)\beta$ which will fit the apparent r values (0.52, 0.7) of the above two classes of objects. We thus obtain for the Anders and Grevesse (1989) abundances

$$\alpha - \frac{14}{18}\beta = 1.104 \text{ (Jovian, Saturnian, Uranian)}$$
$$= 1.586 \text{ (Triton, Pluto, Charon)} \tag{26}$$

and for the Cameron (1982) abundances

$$\alpha - \frac{14}{18}\beta = 0.349 \text{ (Jovian, Saturnian, Uranian)}$$
$$= 0.831 \text{ (Triton, Pluto, Charon)} \tag{27}$$

As $\alpha+\beta\leq1$, we see that for the Anders and Grevesse (1989) abundances no values of α and β will suffice (consistent with both the Anders and Grevesse values being less than 0.52 in Fig. 6).

The almost ubiquitous presence of some dark presumably carbonaceous material in ice-rich bodies in the outer solar system argues either that $\beta \neq 0$ or that the bodies contained some CH_4 which was photochemically converted to carbonaceous material (i.e., $1 - \alpha - \beta \neq 0$). Also, the presence of CH_4 on Titan, Triton (Spencer et al. 1990; Broadfoot et al. 1989), and Pluto argues either for $1 - \alpha - \beta \neq 0$ or for thermochemical conversion of organic material in object interiors to form CH_4 (i.e., $\beta \neq 0$). Therefore we can fit the observations in Fig. 6 with the Cameron (1982) abundances if we make the following reasonable assumptions: (a) the Jovian, Saturnian and Uranian satellites formed in a CH_4-rich circumplanetary disk with the carbon being 65% CH_4, 35% CO, and only trace amounts of heavy hydrocarbons (i.e., $\alpha = 0.349$, $1-\alpha-\beta = 0.651$, $\beta\approx0$), and (b) Triton, Pluto and Charon formed in a CO-rich circumsolar disk with the carbon being 90% CO, 9% heavy hydrocarbons and 1% CH_4 (i.e., $\alpha = 0.9$, $\beta\approx0.0887$, $1-\alpha-\beta = 0.0113$).

Alternatively, we can assume that the C and O abundances are unknowns and ask what values of A_C and A_O in Eq. (25), would allow the Jovian, Saturnian and Uranian satellites to form in very CH_4-rich circumplanetary nebulae ($\alpha = 0$, $\beta = 0$) while Triton, Pluto and Charon formed in the CO-rich environment discussed in the previous paragraph ($\alpha=0.9$, $\beta=0.0887$, $1-\alpha-\beta = 0.0113$). This requires $A_o = 1.25\times10^7$ and $A_c = 0.59\times10^7$ which are each significantly less than either the Cameron (1982) or Anders and Grevesse (1989) values. This explanation would therefore require the C and O abundances in the circumsolar disk to be much less than the current solar values due to some unforeseen evolutionary process which seems very unlikely.

Finally, we can model the data by postulating post-formation water loss. Using Anders and Grevesse (1989) abundances, the Jovian, Saturnian and Uranian satellites could have formed in a CH_4-rich ($\alpha=0$, $\beta=0$) environment if they subsequently suffered 56% loss of their original ice content while Triton, Pluto and Charon could be explained as being formed in a CO-rich ($\alpha = 1$, $\beta = 0$) environment if they subsequently lost 59% of their ice content. For the Cameron (1982) abundances, formation of the Jovian, Saturnian and Uranian satellites in the same CH_4-rich environment requires 41% post-formation loss of ice while formation of Triton, Pluto and Charon in the same CO-rich environment requires essentially no post-formation water loss.

In summary, a variety of explanations can be put forward to explain the r values of ice-rich bodies. To differentiate between them, more knowledge is needed about current A_O and A_C values, the possibility, if any, of evolution of A_O and A_C over time, and mechanisms for post-formation evolution of r.

V. COMETS, METEORITES AND ASTEROIDS

A. Comets

Comets also contain important information on the circumsolar nebula disk. Prinn and Fegley (1989) concluded that the analyses of Halley observations available at that time which indicated the presence of several percent CO, a few percent CH_4, and up to 1% NH_3 relative to H_2O could be explained by this comet being a heterogeneous mixture. Specifically, they proposed a mixture of a small amount of condensate containing CH_4 and NH_3 from a circumJovian planetary disk (Prinn and Fegley 1981) and a large amount of condensate containing CO from either the processed circumsolar disk (Lewis and Prinn 1980) or from unprocessed interstellar material. Engel et al. (1990) proposed an alternative model in which CH_4 was provided by partial conversion of CO to CH_4 in the inner circumsolar disk followed by outward mixing and the NH_3 was of interstellar origin. The Prinn and Fegley (1989) proposal involves ejection of circumJovian planetary disk material out into the solar system through (post-gas-phase) collisions but the process has not been quantified. The Engel et al. (1990) proposal requires catalysis of the CO $\rightarrow$ CH_4 conversion down to temperatures of 600 to 800 K which is unlikely (Fegley and Prinn 1989), and also survival by interstellar NH_3 condensate of accretional shock heating.

Recent Halley analyses indicate that the two above proposed mechanisms for producing heterogeneous mixtures may not be necessary. Boice et al. (1990) have specifically reanalyzed the Giotto mass spectrometer data for Halley and concluded that abundant CH_4 is not required because the hydromonocarbon ions observed can be explained by the presence of heavier organic compounds (e.g., polymerized formaldehyde) on Halley. Specifically, they conclude that CH_4 is <0.5% of CO which is consistent with its origin exclusively in the CO-rich circumsolar nebula disk (Lewis and Prinn 1980). As noted earlier, up to 1% of the solar C and N is predicted to condense as NH_4COONH_2 and NH_4HCO_3 in a CO-rich circumsolar disk so that provided there are no kinetic barriers to their formation these species are potential sources of NH_3, CO_2 and HCN. Also Boice et al. (1990) note that organics could also be the sources of the HCN and inferred NH_3 in the Halley coma. Thus sources of NH_3 beyond those expected in circumsolar disk condensates may also not be required.

Finally, on the issue of interstellar material being present in comets Tokunaga and Brooke (1990) have compared infrared spectra of comets and interstellar dust and gas in molecular clouds and found significant differences in both the silicate and organic band profiles and in the relative abundances of some molecules (CH_4, CH_3OH). They conclude that comets did not therefore form from unaltered interstellar grains in agreement with the Boice et al. (1990) conclusion.

B. Chondritic Meteorites

Chondritic meteorites provide the rare luxury of detailed laboratory chemical, physical and isotopic analyses of ancient material in the solar system. A detailed review of current knowledge is provided in a recent Space Science Series book (Kerridge and Matthews, eds. 1988). We focus here only on a few results relevant to the preceding discussion. The reader should be aware, however, that there are many detailed aspects of meteorites not easily explained by current models (e.g., origin of chondrules).

One idea which can be addressed through meteorites is whether the CO-rich nebula predicted by kinetic inhibition theory is conducive to organic synthesis. As reviewed by Hayatsu and Anders (1981), Fischer-Tropsch-type reactions in the circumsolar disk could have led to some of the organic material in these meteorites. For example, CI1 and CM2 chondrites contain organic material light in ^{13}C and carbonates heavy in ^{13}C and this pattern has been reproduced in organics and CO_2 synthesized by Fischer-Tropsch reactions from CO and H_2 (note that this has not yet been done in laboratory environments appropriate to the circumsolar disk). On the other hand, Yang and Epstein (1983) and Pillinger (1984) note that the D/H ratio in chondritic organic matter can be up to 50 times that in solar composition requiring equilibration of organic material with the major D reservoir (HD) at very low temperatures (e.g., <100 K). Such low-temperature equilibration is undoubtedly kinetically inhibited in the circumsolar nebula but may proceed over much longer time scales in (ionized) interstellar clouds. How much of the organic material in meteorites is circumsolar and how much is interstellar remains an open question.

Petrographic evidence (see, e.g., Barber 1985) favors an origin for the hydrous silicates found in CI1 and CM2 meteorites in environments which are expected in "asteroid-like" meteorite parent bodies but not expected in the circumsolar disk. This is in accordance with the prediction discussed earlier that the hydration of silicates in the circumsolar disk is kinetically inhibited. Concerning the origin of the water needed to carry out hydration on the parent bodies, Bunch and Chang (1980b) presented water ice as a possibility. Prinn and Fegley (1989) concurred noting that the D/H ratio in the meteoritic hydrated silicates (Yang and Epstein 1983; Pillinger 1984) was consistent with (but did not demand) the source of this water being simply ice condensed at about 200 K.

C. Asteroids

Asteroids have so far been investigated only by remote sensing but these studies have already provided insight into circumsolar nebula processes. Jones et al. (1990) have recently presented results of laboratory and telescopic observations of the low-albedo outer-belt asteroids (the C, P and D asteroids between 2.4 and 5.2 AU). Some 66% of the C asteroids have hydrated silicate surfaces. In contrast, the P and D asteroids appear to have anhydrous silicate surfaces. They conclude that their data is consistent with the initial asteroidal composition being anhydrous silicates, water ice and complex organics (i.e., the expected asteroidal composition in the CO-plus-hydrocarbons sequence of Prinn and Fegley [1989] as summarized in Table II), with subsequent evolution forced by solar wind induction heating leading to selective silicate hydration reactions. The latter reactions would maximize in the nearest C asteroids and minimize in the most distant P and D asteroids thus explaining the observed variations in the degree of surface silicate hydration.

VI. CONCLUSIONS

Observations of solar system objects have already provided much significant information on the gaseous circumsolar disk environment in which they originated. Current observations of many volatile-rich bodies, though impressive in some cases, are still however insufficient to totally constrain current models. The need is great ultimately to sample and analyze unperturbed material from comets, asteroids, and ice-rich satellites as well as to continue work on remote sensing of these bodies and on meteorites. A great deal of information is contained in the trace compounds as well as the dominant compounds in these bodies. Such information may be necessary, for example, to differentiate definitively between pristine interstellar material and reprocessed disk material in these solar system objects.

Future observations of certain key chemical species in circumstellar disks obtained with sufficient sensitivity and spectral and spatial resolution could yield very significant new information about the chemistry and physics of these disks. Answers to the following key questions would be especially revealing:

1. Is the $[CO]/[H_2]$ ratio constant over the disk or does it decrease with radius due to conversion of CO to light (e.g., CH_4) or heavy hydrocarbons?
2. Is NH_3 detectable and if so is the $[NH_3]/[H_2]$ ratio constant over the disk or does it increase with radius due to conversion of N_2 to NH_3?
3. Are silicates in the disk in anhydrous forms (olivines, pyroxenes, feldspar, etc.) as expected in the circumsolar disk or hydrous forms (serpentine, talc, etc.)? Are there changes with radius?
4. Is iron present as the metal, as FeO in olivines and pyroxenes, or as magnetite? Are there changes with radius?
5. Is there evidence for a sudden large drop in the $[H_2O]/[H_2]$ ratio at a particular radius indicating condensation of ice? Is there evidence for the expected increase in the ratio of condensate mass to gas mass at this radius?
6. What is the $[CO]/[H_2O]$ ratio at temperatures above the H_2O condensation point in the disk? If $[CO]/[H_2O] >> 1$ indicating reducing conditions, one would expect to find also abundant graphite and hydrocarbons and exotic minerals like SiC.

Finally, we still lack quantitative laboratory studies of the kinetics and mechanisms of a wide range of homogeneous and heterogeneous reactions (e.g., Fischer-Tropsch-type reactions) under conditions relevant to circumsolar, circumJovian planetary, and circumstellar gas-rich disks. This knowledge is essential to further development toward more realistic models of circumstellar disk chemistry.

Acknowledgments. I thank T. Johnson and W. McKinnon for discussions concerning their work and D. Sykes for manuscript preparation. J. Lewis and particularly an anonymous reviewer provided constructive comments on the manuscript. This research was supported by the Planetary Materials and Geochemistry Program of the National Aeronautics and Space Administration and by the Atmospheric Chemistry Program of the National Science Foundation.

PART V
Planetesimals and Planets

FORMATION OF PLANETESIMALS IN THE SOLAR NEBULA

S. J. WEIDENSCHILLING
Planetary Science Institute

and

JEFFREY N. CUZZI
NASA Ames Research Center

This chapter describes the evolution of solid particles in the solar nebula (or other circumstellar disk). Motions of bodies $\lesssim$km in size were dominated by gas drag rather than gravity. An original population of microscopic grains had to produce > km-sized planetesimals before gravitational accretion of planets could begin. Planetesimals probably formed by coagulation of grain aggregates that collided due to differential settling, turbulence, and drag-induced orbital decay. Growth of such aggregates depended on sticking mechanisms and their mechanical properties, which are poorly understood. Their growth was aided by concentration of larger bodies toward the central plane of the disk. The nebula could remain optically thick during this process. It is unlikely that a particle layer formed by settling would undergo gravitational instability, as a small amount of turbulence (e.g., $\alpha \sim 10^{-4}$ in a convective disk) would keep the particle layer from reaching the critical density. This conclusion is independent of the particle size, as even large bodies do not effectively decouple from the gas. Even in a laminar disk, shear in the particle layer would generate enough turbulence to keep it stirred up. This shear-induced turbulence produces complex flow patterns that could result in radial transport and size sorting of particles.

I. INTRODUCTION

The most widely accepted theory of the origin of planets is that they formed by accretion from an initial population of small solid bodies, or planetesimals, in orbit about the Sun. When their orbits intersected, collisions and binding by self-gravity resulted in growth of larger bodies, which eventually reached planetary size. The usual assumption, used as the starting a point for most calculations of planetary accretion (cf. Chapter by Lissauer and Stewart), is that planetesimals had initial sizes in the range 1 to 10 km. This is large enough so that they moved in Keplerian orbits, relatively unaffected by forces other than gravity. In a dynamical sense, there was little difference between a km-sized planetesimal and a planet some 10^4 times larger. In contrast, smaller bodies were dominated by their interactions with nebular gas, while it was present. Planetesimals must have formed from smaller bodies, i.e., surviving presolar grains and/or condensates from the gas of the solar nebula. With presently available observational techniques, dust is the dominant, perhaps only, detectable solid matter in protostellar disks and surrounding newly formed stars (see Chapter by Strom et al.). The dust is the principal source of opacity in optically thick disks. The presence of larger bodies, either planetesimals or planets, must be inferred indirectly. We cannot tell whether dust means that large bodies have not yet formed, or that they are merely hidden within a dusty cocoon.

At one time, it was generally accepted that planetesimals formed by gravitational instability. In that model, dust grains settled to the central plane of the gaseous disk, forming a layer of enhanced concentration. This dust layer became thinner until its density reached a critical value about equal to the Roche density:

$$\rho_c \simeq 3M_\odot / 2\pi r^3 \qquad (1)$$

where $M_\odot$ is the solar mass, and r the distance from the Sun. At this density, the layer would become unstable to perturbations by its own self-gravity, and develop condensations with a characteristic size

$$\lambda_c \simeq 4\pi G\sigma_s / \Omega^2 \qquad (2)$$

where G is the gravitational constant, σ_s the surface density of the dust layer, and $\Omega = (GM_\odot / r^3)^{\frac{1}{2}}$ the Kepler frequency. The mass of a condensation would be $\sim \sigma_s \lambda_c^2$; plausible values of σ_s imply that the resulting planetesimals would have sizes ~ 1 to 10 km. The quantitative aspects of this model were developed independently by Safronov (1972) and Goldreich and Ward (1973).

The simplicity of this model is appealing, but there are problems that render it untenable, at least in the extreme form of direct conversion of μm-sized dust grains to km-sized bodies by gravity alone, without any other sticking mechanism. Wetherill (1980a) warned that ". . . it would be a mistake to conclude that the solar system *must* have developed dust-layer instabilities simply because this does not require specification of sticking processes that are poorly understood, but that quite possibly may have occurred anyway." Weidenschilling (1988) pointed out that gravitational instability depends on quiescent nebula. An extremely small amount of turbulence in the gas would prevent the dust layer from reaching the critical density; in order to do so, its thickness must be $\lesssim 10^{-6}r$. This condition implies turbulent velocities $\lesssim$ one particle diameter s^{-1}, which are implausible for dust grains and unlikely even for macroscopic bodies. Over the last two decades both theory and observations have led away from the simple picture of the solar nebula as a passive reservoir of raw material for making planets. Rather, it was a dynamic, evolving, and sometimes violent place. Small solid bodies were more strongly affected by drag of nebular gas than by gravitational forces, and gas was probably turbulent. Formation of planetesimals must have involved some collisional sticking, at least to form bodies large enough to settle toward the central plane and begin to decouple from the gas. In the follow sections we describe briefly some of the effects of particle/gas interaction in the solar nebula (for a fuller account, see Adachi et al. [1976], Weidenschilling [1977a,1980,1984,1988] and Nakagawa et al. [1981,1986]).

II. STRUCTURE OF THE NEBULAR DISK

A zero-order description of the solar nebula is a flattened disk of gas in orbit about the proto-Sun. The process of formation of planetesimals was dominated by the first-order corrections to this picture, i.e., the disk had a finite thickness; its mean rotation was not strictly Keplerian, and may have had turbulent motions; and most importantly, some small fraction of its mass consisted of solid matter. Present observational methods do not constrain the structures of protostellar disks at this level of detail; of necessity, the "solar nebula" described here is essentially schematic.

For simplicity we assume that the surface density Σ has a power law distribution:

$$\Sigma(r) = \Sigma_0(r/r_0)^{-n} \qquad (3)$$

where Σ_0 is the value at some arbitrary radius r_0. The masses and orbital radii of the planets are roughly consistent with $n = 3/2$ (Weidenschilling 1977b). We assume the disk's structure is in equilibrium between gravitational, centrifugal, and pressure forces. If Σ is low enough (disk mass ≤ 0.1 M$_\odot$), the disk's gravity can be neglected, and the force normal to the plane of the disk is simply the vertical component of solar gravity:

$$g_z = \frac{GM_\odot}{r^2} \left[\frac{z}{r} \right] = \Omega^2 z \qquad (4)$$

where z is the distance from the midplane. If the local temperature T is independent of z, then the gas pressure at the midplane is (Safronov 1972)

$$P_c = \Omega \Sigma c / 4 \qquad (5)$$

where c is the mean thermal velocity of the gas molecules (nearly equal to sound velocity), and at other values of z is

$$P(z) = P_c \exp(-z^2/H^2) \qquad (6)$$

where $H = \sqrt{\pi} c / 2\Omega$ is the characteristic half-thickness of the disk. Other plausible temperature profiles, e.g., adiabatic gradient in the z direction, yield similar values. Typical model disks have $H \leq 0.1r$.

Equation (5) implies a radial pressure gradient. Most disk models have both Σ and T (hence c) decreasing with r, while $\Omega \propto r^{-3/2}$. If we assume that $T(r) = T_0(r/r_0)^{-k}$, then

$$P_c(r) \propto r^{-(n+k/2+3/2)} \qquad (7)$$

and $\partial P / \partial r < 0$, unless Σ actually increases with r more rapidly than $r^{(3+k)/2}$. The pressure gradient partially supports the gaseous disk against the Sun's gravity; for equilibrium, its rotation must be slightly slower than the Kepler velocity V_k. The fractional deviation is (Weidenschilling 1977a)

$$\frac{\Delta V}{V_k} = \frac{-(n + k/2 + 3/2)}{2} \frac{\Re T / \mu}{(GM_\odot / r)} \tag{8}$$

where $\Re$ is the gas constant and μ the mean molecular weight. Thus, $\Delta V / V_k$ is of order $c^2 / V_k{}^2$, or the ratio of the thermal energy of a gas molecule to its orbital kinetic energy. This quantity is typically only a few times 10^{-3}, but even this small deviation from Keplerian rotation has important consequences for solid bodies within the disk.

A. Turbulence

The solar nebula must have been turbulent during its formation from the collapsing protostar, because of velocity discontinuities as the infalling matter struck the disk (Cassen and Moosman 1981). The infall probably did not stop suddenly, but decayed, not necessarily monotonically, over some interval. Another possible source of turbulence was a massive outflow from the proto-Sun, impinging on the surface of the disk (Elmegreen 1978b).

Some researchers (ter Haar 1950; Safronov 1972; Hayashi et al. 1985) have concluded that in the absence of such extrinsic stirring mechanisms any residual turbulence would have decayed quickly. This conclusion was based on the stability of a rotating disk against convection in the radial direction This stability would allow a quiescent period of indefinite length, during which planetesimals, or even planets, could form. However, the concept of a laminar nebula was challenged by Lin and co-workers (Lin and Papaloizou 1980; Lin and Bodenheimer 1982; Ruden and Lin 1986), who developed a self-consistent model for turbulence driven by the differential rotation of the disk. If convection occurs normal to the plane of the disk, eddies cause the effective viscosity to be large. Shear in the viscous, differentially rotating disk causes dissipation of energy, and the resulting heating drives the convection. Lin and Papaloizou showed that the disk is unstable to convection in the z direction if it is optically thick and the opacity increases sufficiently rapidly with temperature; both conditions are met if the opacity is due to small grains suspended in the gas. The ultimate energy source for the turbulence is the disk's potential energy in the solar gravity well (viscous spreading causes a net inward flow of gas).

Published models of turbulent accretion disks generally define the turbulent viscosity $\nu_t = \alpha H c$, where α is a coefficient chosen to make this expression correct (cf. Weidenschilling 1988a). This expression is appropriate for turbulence acting on the full vertical scale (H) of the nebular gas. If H is taken to be the length scale of the largest turbulent eddies, then the maximum random velocities of gas in those eddies is of order αc, but this should be regarded as no more than a crude estimate. Lin's early models suggested $\alpha \simeq 1/3$, with convective velocities a significant fraction of the sound speed. A more elaborate analysis of convective instability modes in a rotating disk led Cabot et al. (1987a, b) to conclude that likely values of α were in the range 10^{-4} to 10^{-2}, or $\nu_t \sim 10^{13}$ to 10^{15} cm^2 s^{-1}. The corresponding turbulent velocities V_t, are $\lesssim 0.01$ c, or less than a few tens of meters per second in the inner part of the disk, and less in the outer regions.

The size of the largest eddies is set by the dimensions of the disk. The energy in those eddies cascades through a spectrum of smaller eddies, down to a size where they are damped by molecular viscosity. Dimensional arguments (Tennekes and Lumley 1972) imply that the rate of dissipation per unit mass is of order

$$\epsilon \sim V_t^3/L \qquad (9)$$

where V_t, and L are the velocity and length scales of the largest eddies. The smallest eddies have the so-called "inner scales" of velocity, length and time given by

$$u_i \sim (\nu\epsilon)^{\frac{1}{4}}$$

$$\ell_i \sim (\nu^3/\epsilon)^{\frac{1}{4}} \qquad (10)$$

$$t_i \sim (\nu/\epsilon)^{\frac{1}{2}}$$

where ν is the molecular viscosity. On smaller scales, the flow is locally laminar, though varying with time. For example, if $V_t = 10$ m s^{-1} and $L = H \simeq 0.05$ AU, then u_i, ℓ_i, t_i are respectively a few cm s^{-1}, a few hundred meters, and a few hours.

III. AERODYNAMICS OF SOLID BODIES IN THE NEBULA

The behavior of an individual solid body in the nebular gas (laminar or turbulent) is largely determined by a single parameter, its response time to the force exerted by gas drag:

$$t_e = mV/F_d \qquad (11)$$

where m is the particle's mass and V its velocity relative to the gas. The drag force F_d depends on V and two dimensionless parameters, the Reynolds and Knudsen numbers. The Reynolds number Re is the ratio of inertial to viscous forces; for a particle of diameter d, $Re = Vd/\nu$. The Knudsen number Kn is defined as λ/d, where λ is the mean free path of the gas molecules. At the low gas densities in the solar nebula, λ is typically >1 cm, so Kn is large for typical grains and even aggregates of many grains. In this regime, t_e has the simple form

$$t_e = d\rho_s/2\rho c \qquad (12)$$

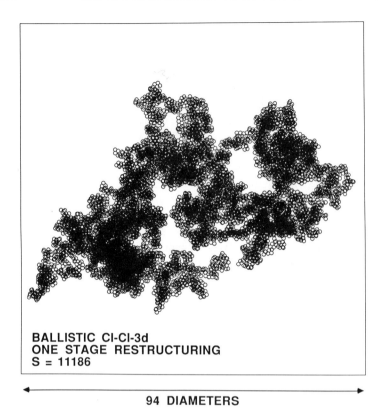

BALLISTIC Cl-Cl-3d
ONE STAGE RESTRUCTURING
S = 11186

← 94 DIAMETERS →

Figure 1.   View of a fractal aggregate of dimension $D \simeq 2.1$, generated by a computer model of P. Meakin (cf. Weidenschilling et al. 1989). Fluffy structures resembling this aggregate are found in laboratory studies of particle coagulation (Meakin and Donn 1988), and should have been produced in the solar nebula.

where ρ is the gas density and ρ_s is the particle's density. Expressions for t_e in other in other regimes are given by Weidenschilling (1977a).

Equation (12) refers to a spherical particle of diameter d. However, coagulation of solid grains does not produce spherical particles, but rather aggregates with highly irregular shapes (Fig. 1). Such structures have fractal-like properties, i.e., their bulk densities decrease with increasing size (Meakin and Donn 1988), with $\rho_s \propto d^{D-3}$, where D is the fractal dimension of the aggregate. Particle aggregates typically have $D \simeq 2$, so density varies inversely with size. In the regime of large Kn, t_e is proportional to the ratio of particle mass to its projected area. Computer models by P. Meakin (cf. Weidenschilling et al. 1989) give the mass/area ratio for fractal aggregates with $D \simeq 2$ shown in Fig. 2. As one would expect, a low-density aggregate of irregular shape is more strongly coupled to the gas (has smaller t_e) than a dense, compact body with the same mass. The larger cross section also means that the opacity due

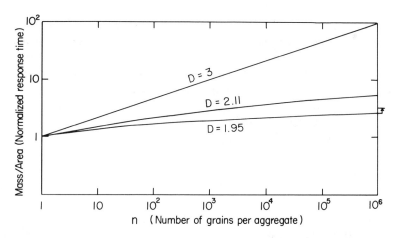

Figure 2.   Mass/area ratio (or response time t_e) for aggregates of different fractal dimension D, vs number of grains (or mass). $D = 3$ corresponds to a spherical particle of constant density. Aggregates with $D \simeq 2$ have larger cross sections and smaller t_e, i.e., are more closely coupled to the gas by drag forces than are compact particles of the same mass.

to grains remains relatively high, despite their coagulation.

Solar gravity affects the particle's motion on a time scale of $1/\Omega$, the inverse of the Kepler frequency. If $t_e \ll 1/\Omega$, drag force dominates and the particle moves with the angular velocity of the gas. A solid body is not supported by the pressure gradient in the gas, and so there is a residual gravitational force in the radial direction. The particle drifts radially inward (relative to the gas) at a terminal velocity

$$V_r = -2\Omega t_e \Delta V. \qquad (13)$$

Similarly, the vertical component of solar gravity (Eq. 4) causes a particle at distance z from the central plane to settle vertically at a rate

$$V_z = -\Omega^2 z t_e. \qquad (14)$$

A sufficiently large body has $t_e \gg 1/\Omega$; its motion is dominated by gravitational forces. Such a body follows a Keplerian orbit, moving faster than the gas. This "headwind" of magnitude ΔV causes the orbit to decay. A circular orbit of semimajor axis a decays at a rate

$$V_r = da/dt = -2\Delta V/(\Omega t_e). \qquad (15)$$

The radial velocity has a maximum value of ΔV when $t_e = 1/\Omega$ (Weidenschilling 1977a). Bodies of different sizes (or t_e) have relative velocities that are readily found from Eqs. (13–15). Effects of drag on eccentric and inclined Keplerian orbits are discussed by Adachi et al. (1976).

In addition to these systematic motions induced by the nebula's non-Keplerian rotation, particles may have other motions driven by turbulence in the gas. The relevant parameters controlling their behavior are the response time t_e, the inner time scale of the turbulence t_i, and the turnover time of the largest eddies, t_0. If $t_e < t_i$, a particle is effectively coupled to all motions of the gas, down to the smallest eddies. If $t_e > t_0$, it is decoupled from the turbulence at all scales, and can cross the largest eddies. For intermediate values of t_e, the particle responds to eddies larger than some size between the smallest and largest scales. The smallest scales are given by Eqs. (10), while in a rotating system it is generally a good assumption that $t_0 \sim 1/\Omega$.

Diffusive transport of particles relative to the mean (nonturbulent) component of gas flow is accomplished mainly by the largest eddies. The viscosity ν_t due to gas turbulence can be expressed as $\overline{v^2}t_0$, where $\overline{v^2}$ is the mean square of the fluctuating part of the gas velocity, and t_0 is the mean eddy mixing time. A particle acquires its random velocity v_p by drag interactions with fluid eddies, described by the equation $\dot{v}_p = (v - v_p)/t_e$. Because $\dot{v} \simeq v/t_0$, v_p/v should depend on the time scale ratio t_e/t_0, known as the Stokes number (Crowe et al. 1985,1988). The Schmidt number is defined as $Sc \equiv \overline{v^2}/\overline{v_p^2}$ and, by analogy with the fluid viscosity, a particle diffusion coefficient $D = \nu_t/Sc$ may be written $\overline{v_p^2}$, where $\overline{v_p^2}$ is the mean square fluctuation of the particle velocity (see Appendix below). Safronov (1972) assumed that $t_0 \simeq 1/\Omega$, and also argued that a particle would build up a typical velocity $v_p = V_t/(1 + t_e/t_0)$, implying $Sc = (1 + t_e/t_0)^2$. A more careful analysis, including averaging over a size spectrum of rotating eddies, indicates that the Schmidt number becomes simply (Völk et al. 1980)

$$Sc = 1 + t_e/t_0$$

or

$$v_p = V_t/(1 + t_e/t_0)^{\frac{1}{2}}. \tag{16}$$

For small Stokes numbers, Sc is near unity since the particles respond to the eddies almost immediately, like the fluid molecules themselves. The Schmidt number becomes large at high Stokes numbers because of the reduced coupling of the particles to the eddies.

The relative velocity between two particles embedded in turbulent gas is more complex, and depends on t_0, t_i, and the value of t_e for each particle. These relative velocities are needed to study collisions between particles and accretion that may result (Sec. V). A first solution by Völk et al. (1980) neglected the inner cutoff scale of the turbulence spectrum. An improved solution by Mizuno et al. (1988) takes this cutoff into account; its effect is to decrease the relative velocities of small grains for which $t_e < t_i$. Two bodies with the same value of t_e have no systematic relative motion, but will have nonzero relative velocities due to turbulence (small particles may have significant thermal motion as well). Relative velocities as a function of

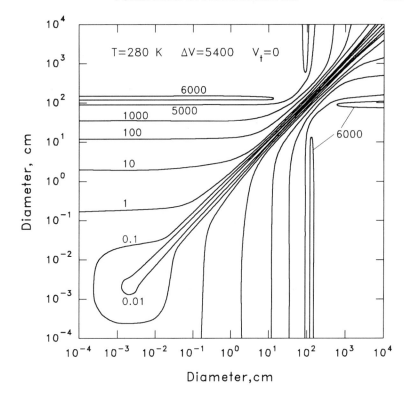

Figure 3. Relative velocities between two particles of unit density vs size in a laminar nebula. The nebular parameters are those of the model of Hayashi et al. (1985) at $r = 1$ AU, $z = 0$: $\rho = 1.4 \times 10^{-9}$ g cm^{-3}, $T = 280$ K, $\Delta V = 5400$ cm s^{-1}. Relative motions are due to thermal motion and radial and transverse velocities induced by non-Keplerian rotation of the gas. Thermal motion dominates for $d \lesssim 0.01$ cm; relative velocities are low for equal-sized bodies.

particle size are shown in Figs. 3 and 4 for one set of nebular parameters, with the gas assumed to be either laminar or turbulent.

IV. COUPLED MOTIONS IN A PARTICLE-GAS LAYER

Goldreich and Ward (1973) recognized that if a particle layer reached a density greater than that of the gas, the particles would dominate. Their velocities relative to the gas would not be given by Eqs. (13–15); rather, the particle layer would tend toward Keplerian motion, and the gas within that layer would be dragged with it. Weidenschilling (1980) showed by dimensional arguments that such shear would be unstable, and should produce turbulence in the particle layer. Any density gradient in the z direction is too small to prevent the dust layer from becoming turbulent. This localized shear-

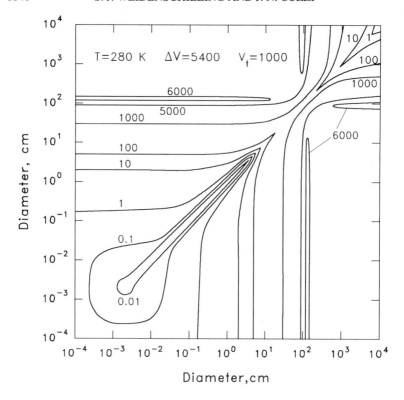

Figure 4.   Same as Fig. 3, but with turbulence added; $V_t = 1000$ cm s^{-1}. Bodies with $d \simeq 100$ cm have $t_e \simeq 1/\Omega$; in this size range relative velocities are $\simeq V_t$ even for equal-sized bodies.

induced turbulence would occur even if the gaseous disk had no other source of turbulence, such as convection.

Nakagawa et al. (1986) developed an analytic solution for the coupled equations of motion for a two-phase inviscid system of particles and gas. The solutions are expressed as functions of the mass loading, i.e., the ratio of densities of the particle layer and the gas. To a good approximation, if the particles are small, the deviation from Keplerian motion within the layer is given by

$$\Delta V' = \Delta V / (1 + \rho_p / \rho) \tag{17}$$

where ρ_p and ρ are the space densities of the particles and gas, respectively. Nakagawa et al. assumed purely laminar motions; their solution is inappropriate if such a layer became turbulent. The presence of sufficiently large ($\gtrsim 1$ cm) particles could cause some damping of turbulence, but it is not clear whether it is possible to have particles that are simultaneously large enough and abundant enough to damp the turbulence effectively. This question is clearly a fruitful subject for future work.

The shear due to collective motion of the particle/gas layer becomes significant for mass loadings of order unity. Weidenschilling (1988b) speculated that this effect could lead to a quasi-stable, long-lived state. Formation of a layer with solids/gas mass ratio of order unity might induce enough turbulence to prevent further settling. Any decrease (increase) in layer's thickness would increase (decrease) the intensity of shear and turbulence, tending to restore its original state. It appears that the only escape from such a state would be a change in the particles themselves, i.e., coagulation into bodies large enough to decouple from the gas and settle in spite of the turbulence. This scenario needs to be tested by numerical modeling; preliminary results of such an effort are described in Sec. VI.B.

A. Viscosity in the Boundary Layer

The particle layer interacts with the surrounding gas primarily in a region of limited thickness, or boundary layer, in which the shear velocity and particle concentration vary most strongly. In a boundary layer of thickness δ surrounding the particle layer, the orbital velocity of the nebular gas changes by an amount ΔV. Shear flows with velocity scale ΔV and length scale δ develop turbulence when their Reynolds number $Re = \Delta V \delta / \nu$ exceeds a critical value $Re^* \sim 50$ to 500. The difference in orbital velocity is typically tens of meters per second and $\delta \sim 10^3$ km. For these values, the boundary layer is indeed turbulent. Thereafter the turbulent viscosity in the boundary layer becomes

$$\nu_t \sim (\Delta V \delta / Re^*). \tag{18}$$

In a system rotating at angular velocity Ω, the boundary layer thickness itself depends on the viscosity as

$$\delta \sim (\nu_t / \Omega)^{\frac{1}{2}} \tag{19}$$

leading to a crude mixing length estimate of the boundary layer viscosity:

$$\nu_t \sim \Delta V^2 / \Omega Re^{*2}. \tag{20}$$

For eddies with length scale δ, this implies turbulent velocities of order $\Delta V / Re^*$ in the boundary layer.

Shear flow has been studied in laboratories for decades, and the thickness of the boundary layer and the properties of the turbulence have been well characterized. In fact, numerical techniques now exist for modeling the structure of individual eddies and coherent structures within the turbulence. Canuto and Battaglia (1988) have derived expressions for turbulent viscosity and particle velocities that take into account anisotropic turbulence in a rotating system. However, for the purposes of modeling the nebula, homogeneous, isotropic turbulence in the shear-generated boundary layer is not a bad assumption. Several parameterizations exist for obtaining its magnitude. Champney and

Cuzzi (1990) have used two such parameterizations to study the turbulence induced by the particle layer in an otherwise laminar nebula. A simple model due to Prandtl, which is still an improvement over Eq. (20), is

$$v_t = C\delta^2 (\nabla \times \mathbf{V}_g) \qquad (21)$$

where $C \sim 0.1$ is a constant determined in the laboratory and $\mathbf{V}_g$ is the (vector) gas velocity. In more sophisticated models in computational fluid dynamics, a "two-equation" model is often used, in which the turbulence at a point is a balance between ongoing creation and dissipation processes, and is transported as is any other quantity. This latter model may be adapted to a two-phase environment in which the particles can damp turbulence, as well as generate it by maintaining the velocity gradient. Champney and Cuzzi (1990) have made preliminary studies of the merits of these two models, and further work is underway. The overall conclusion, at present, is that in the boundary layer $v_t \sim 10^{10}$ cm^2 s^{-1}. The turbulent viscosity calculated in this way for the boundary layer alone is much less than the values $\sim 10^{13}$ to 10^{15} cm^2 s^{-1} for turbulence through the entire thickness of the disk with $\alpha \sim 10^{-4}$ to 10^{-2}. The viscosity is confined to a boundary layer surrounding the particle layer, with thickness rather well estimated by Eq. (19).

Given the viscosity in the boundary layer, one may calculate the shear torque between the particle layer as a whole and the surrounding, more slowly rotating pressure-supported nebular gas. Goldreich and Ward (1973) assumed the particle layer acted as a rigid, impermeable rotating disk embedded within the gas. This situation, the Ekman problem, has been thoroughly studied in theoretical and laboratory work, and inspection of the analytical solutions leads to considerable insight into the more complex nebula problem (see, e.g., Batchelor 1967). The shear torque exerted between the disk and the fluid is

$$S = \rho v_t \Delta V / \delta = \rho \Delta V^2 / Re^*. \qquad (22)$$

The result of this torque is a loss of angular momentum by the particle layer, which causes it to drift inwards at a rate that depends on the boundary layer's turbulent viscosity v_t. Goldreich and Ward considered the orbital decay of a dust layer due to shear between it and the more slowly moving gas on either side. They did not discuss the effects on the gas, possibly thinking of it as merely providing a vast reservoir of angular momentum and not significantly affected by the particle layer. Nevertheless, interesting dynamical structure is present in both the particles and the gas in this important region of interaction, driven primarily by the turbulence that the particle layer itself generates. For example, the angular momentum lost by the particle layer is transferred to the gas in the boundary layer, which must then move outward (Weidenschilling 1980). Study of these motions in more detail requires a numerical model. Such a model is described in Sec. VI.B.

V. COAGULATION AND SETTLING OF PARTICLES

For an assumed structure of the nebular disk, the expressions in Sec. III yield the velocities of particles as functions of their sizes and location. In principle, one can use these values to compute the evolution of their spatial distribution in the disk. Also, by assuming mechanical properties (i.e., whether particles stick together in collisions), the evolution of the size distribution from some assumed initial state can be computed. Of course, the spatial and size distributions are coupled, and their evolution must be calculated simultaneously. In practice, all such efforts are subject to simplifying assumptions to render the problem tractable for analytic or numerical solution.

A. Coagulation in a Laminar Nebula

Safronov (1972) derived the settling velocity of a particle (Eq. 14). Because $V_z \propto z$, the distance from the central plane decreases exponentially with an e-folding time of $(\Omega^2 t_e)^{-1}$. If there is no turbulence and the particles do not coagulate, the entire dust layer becomes thinner and denser uniformly, in homologous fashion. Safronov considered settling with coagulation, in which a single large particle falls through a field of small grains and grows by sweeping them up. If only vertical motion of the particle is considered, it can grow to a maximum size $d_{max} = \sigma_s/4\rho_s$, where σ_s is the surface density of the grain component of the disk. Plausible values of σ_s and ρ_s give d_{max} of the order of a few cm. The initial growth rate of the particle is exponential with a time constant $\lesssim 4\rho c/\sigma_s \Omega^2$. It reaches d_{max} (and $z \to 0$) in a few thousand orbital periods. Goldreich and Ward (1973) independently derived a similar result. These analytic models refer only to the first (and largest) particles to "rain out" to the central plane of the disk. Modeling the subsequent buildup of a particle layer and depletion of the smaller particles at large values of z requires numerical simulation. Nakagawa et al. (1986) described the settling and coagulation in a model that included collective effects of the particles upon the gas; i.e., ΔV varied with particle concentration. In order to make this problem analytically tractable, they had to make two mutually inconsistent assumptions: that growth was by the sweeping up of the small particles by the largest ones, and that settling was uniform, as if all particles had the same size.

Numerical models were developed independently by Weidenschilling (1980) and Nakagawa et al. (1981). These were one-dimensional models that computed the vertical distribution of particles at a given radius in the disk. The evolution of the size distribution due to coagulation was calculated in a series of levels, with mass transported between layers by settling (both assumed a laminar nebula, with transport only toward the central plane). Their results were very similar; assuming perfect sticking, a dense particle layer forms in the central plane within a few times 10^3 yr at $r = 1$ AU. The largest bodies grow to 10^2 to 10^3 cm in size, much larger than Safronov's value of d_{max}, because the radial motion of the particles allows more growth. A modest (but still significant) fraction of the mass of solids remains as small particles suspended in the gas to much later times (10^4–10^5 yr), due to their slow settling velocities. Thus, the disk could remain optically thick long after planetesimals formed. Both Weidenschilling and Nakagawa et al. assumed that settling of the large aggregate bodies would eventually result in gravitational instability of the particle layer, but neither model had fine enough spatial resolution to show this result directly. Also, they neglected the possibility of turbulence in the disk.

B. Turbulent Coagulation

Although it is widely accepted that the solar nebula passed through a stage as a turbulent accretion disk, there has been relatively little study of the consequences of turbulence for planetesimal formation. Most cosmogonical scenarios assume that planetesimals either formed despite turbulence (even values of α as large as $1/3$), or else that any turbulence eventually decayed completely, and they formed after the disk became perfectly laminar. Particle coagulation in a turbulent disk has been considered by Wieneke and Clayton (1983), Morfill (1983,1988), Mizuno et al. (1988) and Mizuno (1989). These studies concerned radial transport of solid matter in the disk, while averaging the distribution and properties of particles vertically through the thickness of the disk. Only Mizuno computed actual size distributions of particles, rather than some effective mean size. The formation of planetesimals, whether by gravitational instability of a particle layer or direct collisional growth, appears to be intimately linked with concentration of particles toward the central plane of the disk. Thus, averaging through the disk thickness can cast little light on planetesimal formation.

Weidenschilling (1984) modeled the vertical distribution of particles at a given radius in a turbulent disk. The numerical method was based on the one-dimensional model described above (Weidenschilling 1980), but particles were allowed to settle toward the central plane only if they were large enough so that their settling velocities exceeded their turbulent diffusion velocities. The nebular model used had high turbulent velocities ($\alpha \sim 1/3$), as suggested by the convective disk models of Lin and Bodenheimer (1982). The vigorous turbulence resulted in high collision rates, and if perfect sticking was assumed, large particles would accrete much more rapidly than for a laminar disk. However, this assumption of perfect sticking seems unrealistic. If the particles were assigned plausible impact strengths, in the range 10^4 to 10^6 erg cm^{-3}, accretion came to a halt after a few centuries, with the largest aggregates (<1 cm in size), being destroyed in collisions as rapidly as they formed. Battaglia (1987) performed a similar calculation for the low-α disk model of Cabot et al. (1987a, b). He concluded that even for the lower turbulent velocities in that model nebula, unrealistically high strengths were required to allow aggregates to grow large enough to decouple from the gas and settle to the central plane.

The results of Weidenschilling and Battaglia can be readily understood in terms of their assumptions about the collisional behavior of the particles. They assumed that for a given material there exists a critical energy density E_c, such that the target is shattered if an impact yields a higher value. For lesser impact energies, "cratering" erosion excavates and removes some amount of mass that is proportional to the impact energy; the excavated mass is then C_{ex} times the impact energy, where C_{ex} is the excavation coefficient. Then the excavated mass exceeds the projectile mass, i.e., there is net erosion, if the impact velocity V_i, is $>(2/C_{ex})^{\frac{1}{2}}$, and net mass gain occurs at lower V_i. E_c and C_{ex} are related by the condition that there is a largest cratering impact that the target can sustain without shattering. If the largest crater contains a fraction f of the target's mass, then $C_{ex} = f\rho_s/E_c$; typically, $f \simeq 0.1$ (Weidenschilling 1984).

Solid particles embedded in turbulent gas have relative velocities that increase with size, up to a maximum value roughly equal to the eddy velocity V_t for bodies with $t_e \simeq t_0$. It is significant that a body of this size has velocity of $\simeq V_t$ relative to *all* smaller bodies, up to those of its own size. This means that the target body will be eroded by collisions with smaller particles, losing mass rather than gaining, if V_t is $\gtrsim (2/C_{ex})^{\frac{1}{2}} = (2 E_c/f\rho_s)^{\frac{1}{2}}$, or unless E_c exceeds the critical value $f\rho_s V_t^2/2$. If E_c is smaller than this value, solid bodies can never grow large enough to decouple from the turbulence ($t_e > t_0$). If E_c is $\gtrsim \rho_s V_t^2/2$, even collisions of equal-sized bodies will not disrupt them. For $f\rho_s V_t^2/2 < E_c < \rho_s V_t^2/2$, a body will grow due to impacts of much smaller bodies, but may be shattered by large impacts; its fate will depend on the size distribution of those it encounters. Taking $\rho_s = 2$ g cm^{-3}, $f = 0.1$, we see that, e.g., $V_t = 10^3$ cm s^{-1} requires $E_c \gtrsim 10^5$ erg cm^{-3} for collisional growth to occur. This simple model explains why Weidenschilling (1984) and Battaglia (1987) required high values of E_c in order to form large bodies by collisional coagulation. For any value of E_c we can find the maximum value of turbulent velocity that allows coagulation (although lower values of V_t do not necessarily assure it, unless the particles also stick when they collide).

Similar results can be obtained for the systematic velocities due to non-Keplerian rotation of the disk. However, bodies of comparable size (or t_e) have low relative velocities. The maximum relative velocity between bodies of very different sizes is $\simeq \Delta V$, so collisions always result in mass gain if $E_c \gtrsim f\rho_s \Delta V^2/2$. For typical values of ΔV ($\sim 10^4$ cm s^{-1} at 1 AU), growth is assured if $E_c \gtrsim 10^7$ erg cm^{-3}. This value is quite high, of the order of values measured for solid rock (Fujiwara et al. 1989). However, the model of impact behavior was extrapolated from experiments involving solid projectiles and targets; even for these, there is some indication that impact strength does not correlate with more conventional properties such as compressive strength (Fujiwara et al. 1989). It is possible that C_{ex} is much smaller than assumed by Weidenschilling, or even zero, i.e., that fluffy aggregates stick rather than "crater" each other below some critical velocity or energy density, as in the model of Donn (1990), discussed below.

C. The Problem of Particle Sticking

The major unsolved problem is the degree to which particles stick together in collisions, and the mechanism (or mechanisms) by which they adhere. It is common experience under a wide range of laboratory condition that very small ($\lesssim \mu$m-sized) particles stick together into larger aggregates. The dominant sticking mechanism in this size range is van der Waals bonding; this force is relatively insensitive to the compositions of the particles. Electrostatic forces may also play a role in coagulation. Weidenschilling (1980) argued that van der Waals forces alone would suffice to produce ~cm-sized aggregates of grains in a laminar nebula, where relative velocities are due to thermal motion and differential settling. This conclusion should be valid for a nebula with moderate turbulence ($\alpha \lesssim 10^{-2}$); particles in this size range are strongly coupled to the gas, so their turbulent motions are correlated and have little effect on their relative velocities.

The growth of larger bodies poses more of a problem. Relative velocities increase with size and differences in size (cf. Fig. 3), and there is a real possibility of outcomes other than sticking: rebound, erosion with net loss of mass, and even disruption of the colliding bodies. There are few relevant experimental data on the collisional behavior of weakly bonded aggregates. Weidenschilling (1988c) performed drop tests of unconsolidated pumice dust into a dust target and concluded that net mass loss would occur for impact velocities approximately $> 10^3$ cm s^{-1}. Pinter et al. (1989) and Blum (1989) have conducted experimental collisions of sub-cm-sized fluffy aggregates at relative velocities up to a few meters per second. They found coagulation occurred with sticking probabilities of a few tens of percent. Their estimate that fragmentation would occur at relative velocities $\gtrsim 600$ cm s^{-1} implies impact strengths of order 10^5 erg cm^{-3}. The aggregates consisted of glass spheres bonded by a coating of hydrocarbons; their degree of similarity to actual nebular material is uncertain.

Donn (1990) argues that collisions of porous bodies would result primarily in their interpenetration, with essentially complete agglomeration up to relative velocities of $\simeq 10^3$ cm s^{-1}, and net mass gain to $\simeq 5 \times 10^3$ cm s^{-1}. His analysis is based mainly on the behavior of porous materials under static compression, and needs to be verified by collisional experiments. Donn assumes very high porosities, making the aggregate bodies very compressible. We note that in most solar nebula models, peak velocities relative to most other particles are reached for roughly meter-sized bodies. Thus, bodies of this size are likely to be fairly well compacted, and may not coagulate as effectively as Donn's model implies.

As seen in Figs. 3 and 4, relative velocities decrease for bodies with sizes larger than ~10^2 cm, so if meter-sized bodies can form by collisional coagulation, then there is no obstacle to the formation of km-sized planetesimals. The crucial gap is in the centimeter-to-meter range. The collisional strengths that seem necessary for growth in this size (or velocity) range are higher than expected for aggregates composed of small grains. There are alternatives to assuming that the aggregates are extremely strong. The relative velocities are caused mainly by differential motion due to non-Keplerian rotation of the gas. The maximum velocities are $\simeq \Delta V$, so there is less of a problem in the outer part of the disk where velocities are lower (Eq. 8). It is possible that the nebula's temperature and radial structure were such that ΔV was lower than implied by most nebular models. Less plausibly, accretion might have produced a narrow size distribution so that relative velocities of particles comprising most of the mass were less than ΔV. The most probable explanation is that a concentration of small (cm-sized?) aggregates decreased ΔV in a layer near the central plane (Eq. 17, cf. Sec. VI.B), allowing growth to proceed with lower relative velocities and minimizing collisional breakup.

VI. RECENT RESULTS OF NUMERICAL MODELS

The formation of planetesimals involves complex feedbacks between all of the processes described above, and no current model describes the full situation. Recent work has approached the problem from complementary perspectives, by emphasizing or suppressing different aspects of the problem. Below, we describe two such efforts which together paint a fairly realistic picture of, at least, the scope of possible situations. As a natural outgrowth of the discussion above, we first describe a model of particle growth by coagulation, which neglects coupling of the mean flow regimes and treats the gas eddy velocity in an *ad hoc* fashion. We then describe a complementary model of the fully coupled gas-particle flow dynamics near the midplane, which neglects particle growth and treats only a single typical particle size. The results of the two approaches are instructive and illustrate the variety of possible states in which a protoplanetary nebula might be found (possibly all, at different times). In the coagulation models, it is seen that (depending on the assumed turbulent gas velocity) a flattened state in which the particle mass density significantly exceeds that of the gas may or may not occur. In the coupled flow models, it is seen that even in a globally laminar nebula, relatively flat (thousands to tens of thousands of km vertical thickness) particle layers of mass density which exceeds that of the gas by 1 or even 2 orders of magnitude are stable against gravitational settling for particles smaller than tens of meters or so, due to shear-induced turbulence.

A. Coagulation and Settling with Turbulence

Weidenschilling has recently developed an improved code for modeling coagulation and settling in a turbulent disk. As in the earlier models (Weidenschilling 1980,1984) this is a one-dimensional computation, with the nebula divided into a series of layers at a given radial distance. The size distribution of particles in each layer is computed as it evolves due to collisions; sources of relative velocity include thermal motion, systematic settling, and turbulence.

In contrast to the earlier models, transport between layers by turbulent diffusion is computed explicitly. Particles can diffuse upward if their turbulent diffusion velocity (cf. Eq. 16) exceeds the settling velocity (Eq. 14). The spatial resolution (number and thickness of layers) is improved over earlier work, as is the resolution of the size distribution. A complete description of the model and results is published elsewhere (Weidenschilling 1993).

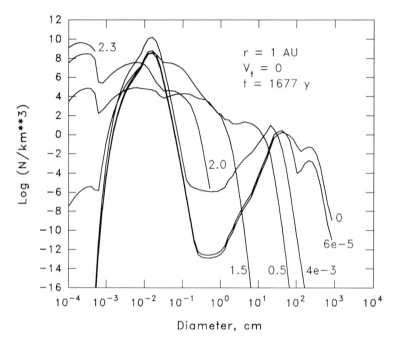

Figure 5.   Size distribution at different levels for a laminar nebula, at $r = 1$ AU, at the time when the solids/gas mass ratio reaches unity at $z = 0$. Numbers give values of z in units of scale height H. At the highest level, aggregates reach a maximum size $\simeq 10^{-3}$ cm; the low density (4×10^{-3} times that at $z = 0$) allows rapid settling of larger particles. At $z = 0$, most of the mass is in bodies $\sim 10^2$ to 10^3 cm in size.

Because of uncertainties in the mechanisms for producing turbulence and its consequent strength, calculations are performed for "generic" turbulence. The largest eddies are assumed to have an arbitrary velocity V_t and size H, with a Kolmogorov spectrum of smaller eddies down to the appropriate inner scale (Eqs. 9 and 10). In the examples shown here, aggregate particles are assumed to have fractal properties with dimension 2.11 (Meakin and Donn 1988; Weidenschilling et al. 1989) in the size range 10^{-4} to 10^{-2} cm. At sizes >1 cm, their density begins to increase again to a limiting value of 1 g cm^{-3} at sizes $>10^2$ cm. The qualitative justification for this variation is that while grain assemblages formed by low-velocity collisions will be fluffy, higher-velocity collisions of larger aggregates will cause compaction (the qualitative results are not affected by these assumptions about particle density).

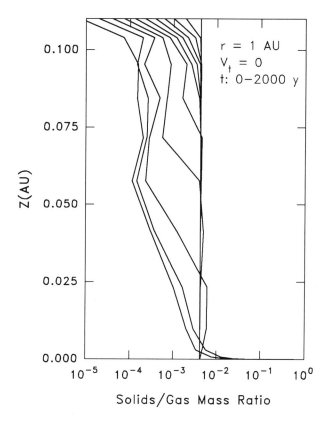

Figure 6.   Vertical distribution of solids/gas mass ratio with time. At $t = 0$, the ratio is assumed to be 4.2×10^{-3}, independent of z. Contour interval is 200 yr.

In these calculations and those in Sec. VI.B, the nebular parameters are taken from the model of Hayashi et al. (1985). In the two examples shown here, an initial population of single grains, all of size 10^{-4} cm, is initially uniformly mixed with the gas. Perfect sticking is assumed in order to set a lower limit on the evolution time. The calculations are carried out until the particle/gas mass ratio reaches unity at $z = 0$, at which time collective effects become important, and a different method must be used (Sec. VI.B).

Figures 5, 6 and 7 illustrate the results for a purely laminar, nonturbulent disk. Figure 5 shows the size distribution at the various levels when the solids/gas ratio reaches unity, after a model time of 1677 yr (the much longer settling time reported by Weidenschilling et al. [1989] appears to be an artifact of the coarse resolution of their size distribution). Several peaks in the

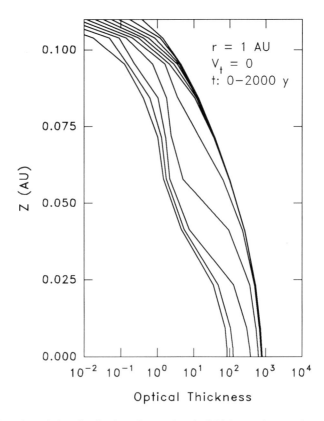

Figure 7.   Cumulative distribution of normal optical thickness above a given value of
z, vs time. Rosseland mean opacities from Pollack et al. (1985) are used. Opacities
due to particles smaller than the peak emission wavelength are assumed proportional
to the mass loading, with geometric extinction by (opaque) particles of larger size.
The optical half-thickness drops from the initial value of $\simeq 800$ to $\simeq 100$ after 2000 yr,
for perfect sticking of particles.

distribution correspond to sizes at which thermal motion (10^{-4} to 10^{-3} cm),
vertical settling ($\sim 10^{-2}$ to 10^{-1} cm), and radial motion (~ 10 cm) dominate
the collision rate. A fourth peak at 10^2 to 10^3 cm, appears at $z = 0$, where the
largest bodies have settled. The changing vertical distribution of the dust/gas
ratio is shown in Fig. 6, through most of the disk; the solids are depleted by
$\simeq 1$ order of magnitude in this time. Figure 7 shows the distribution of the
optical thickness.

Turbulence with velocity of 10^3 cm s^{-1} has very little effect on relative
velocities and coagulation rates of small particles, because their motions
are highly correlated. However, the vigorous vertical diffusion of small
particles affects the overall rate of growth and settling. There is very little
variation of the size distribution with z (Fig. 8). The time to reach a high
mass concentration at the central plane is much longer than for the laminar

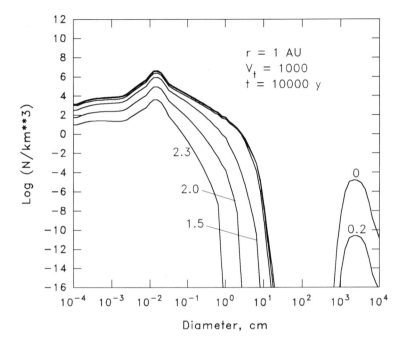

Figure 8.   Same as Fig. 5, but for turbulence with velocity 10^3 cm s^{-1} through the thickness of the nebula. The size distribution has less variation with z, due to diffusion of particles between levels. Most of the mass is in large bodies $\sim 10^3$ to 10^4 cm near the central plane. The computation was halted after 10^4 yr, when solids/gas $\simeq 0.4$ at $z = 0$ (see text).

case, partly because particles must grow larger to settle appreciably, and also because the high-density layer near $z = 0$ is thicker due to stirring by the turbulence, and requires more total mass to reach a given solids/gas ratio. After $\gtrsim 10^4$ yr, most of the mass is in bodies $\sim 10^3$ to 10^4 cm in size. Bodies $\sim 10^2$ cm are strongly depleted because of the peak in radial velocity at that size, which causes them to grow rapidly through this size range, or to be swept up by the larger bodies.

From Fig. 9 it is seen that the dust/gas ratio varies only slightly with z through most of the disk's thickness, but it decreases steadily with time as small aggregates diffuse downward and are accreted by the large bodies near the central plane. (Figure 10 shows the evolution of the optical thickness.) The thickness of the densest layer is a few times 10^{-4} AU (5×10^4 km). The calculation was halted after 10^4 yr because the solids/gas mass ratio reached a peak value $\simeq 0.4$, and then began to *decrease*. This decrease is due to the fact that the settling rate for large bodies (actually, the rate of damping inclinations for Keplerian orbits by gas drag) is proportional to t_e^{-1} (Adachi et al. 1976), while the random velocity induced by turbulence (Eq. 16) varies as $t_e^{-1/2}$. Therefore, sufficiently large bodies ($t_e \gg 1/\Omega$) are stirred more effectively than they are damped. In this simulation, the solid bodies may continue to grow by collisions to much larger sizes, but will never form a layer that is sufficiently thin and dense to become gravitationally unstable.

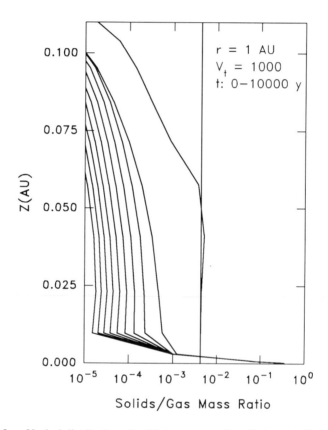

Figure 9.   Vertical distribution of solids/gas mass ratio with time for $V_t = 10^3$ cm s^{-1}. Contour interval is 1000 yr.

These simulations assumed perfect sticking of particles and thus give only a lower limit to the time scales for growth and settling. If an arbitrary sticking efficiency β (≤ 1) is assumed, then the results are very similar, but with all time scales increased by a factor of $1/\beta$. The outcomes of such simulations may well be different if fragmentation is allowed, as the sticking efficiency would be expected to vary with relative velocity (or size). Also, the long time scale associated with the turbulent case implies that radial transport of solids would be significant. The small grains that are coupled to the gas will diffuse both inward and outward over a distance of order $(V_t H t)^{\frac{1}{2}}$, in this case more than 0.5 AU. In addition, orbital decay of the largest bodies exceeds 1 AU, meaning that they would be lost into the Sun (presumably replaced by others spiraling inward from larger distances). Thus, the one-dimensional model is incomplete at best, and its results should be interpreted cautiously. Further progress will require a two-dimensional model with both vertical and radial resolution.

B. Fluid-Dynamical Modeling of a Dense Particle-Gas Layer

Simulations of settling with coagulation generally produce a significant concentration of large (meter-sized?) bodies toward the central plane of the disk. Whether the nebula is assumed to be laminar or turbulent, such a layer can attain a solids/gas mass ratio of order unity or greater. At such densities, coupling of the particles and gas, as described in Sec. III, becomes significant. The complex flow patterns that result must be modeled numerically by computational fluid dynamical methods. Current models are limited to a single particle size, without coagulation; in effect, particles interact with each other only by their mutual coupling with the gas (Champney and Cuzzi 1990; Cuzzi et al. 1993). The numerical methods used are described in the chapter Appendix.

This model is sufficiently stable and robust to allow modeling of particle layers with mass density exceeding that of the gas by more than 2 orders of magnitude. Due to the present neglect of particle collisional viscosity, these results are of questionable quantitative validity for particle mass densities much larger than that of the gas; nevertheless, major new qualitative aspects of the entire family of viscous, two-phase solutions may be readily observed in cases with a particle density of 1 to 10 times that of gas, where particle viscosity does not play a dominant role.

For example, Figs. 11 and 12 show the gas and particle velocities relative to their assumed unperturbed values (pressure supported orbital motion, $\Delta V = 5.4 \times 10^3$ cm s^{-1}, with zero radial and vertical velocity for the gas, Keplerian orbital motion with zero radial velocity and simple vertical settling for the particles). In this simulation, there is no turbulence in the disk, except that which is generated locally by shear near or within the particle layer. In both figures, initial and final particle density profiles are shown, with mass density along the top axis (the gas density at this location, chosen at 1 AU, is 1.4×10^{-9} g cm^{-3}). The fundamental property is the large vertical gradient in the gas orbital velocity; this leads to turbulence which diffuses the particles into a much thicker layer than that in which they were assumed to lie initially. In cases where a thicker initial layer was assumed, the layer flattens into the identical final state, which thus represents a steady-state balance between gravitational settling and vertical diffusion for these nebula and particle properties.

In model calculations to date, using the Schmidt number model describe in Sec. III (Cuzzi et al. 1993), it appears that particles of unit density must grow to at least 10^5 to 10^6 g (depending on location) before being able to settle out from the nebula into a gravitationally unstable state. If they are more "fractal" or fluffy than solid in nature, they need to become more massive by about

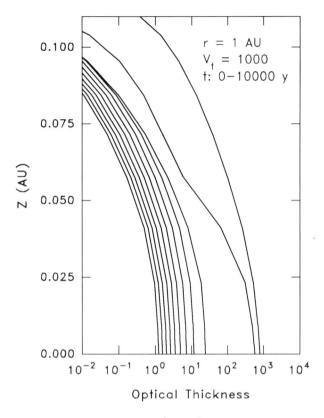

Figure 10.   Same as 7, but with $V_t = 10^3$ cm s^{-1}. The nebula becomes optically thin at $r = 1$ AU on a time scale of 10^4 yr.

an order of magnitude. Initially flattened particle layers of mass density 10 to 100 times that of the gas are seen to "puff up" to vertical thicknesses of several thousand km, as long as particle diameters are $\lesssim 100$ cm. For minimum mass solar nebulae, this keeps the particle layer mass density ρ_p well below the critical value for gravitational instability, confirming the conclusions of Weidenschilling (1980,1988) that collisional coagulation is a fundamental process in the formation of planetesimals.

The mean velocity profiles are seen to relate to the location of the particle layer. Well above the layer, when the particles exert no significant influence on the gas, the solutions of Weidenschilling (1977) and Nakagawa et al. (1986) are appropriate, i.e., particles experience a strong headwind in the pressure supported gas, their orbital velocity falls below Keplerian, and they drift radially inwards. At deeper levels within the layer, the particles are increasingly shielded from the headwind as the surrounding gas is speeded up more closely to the Keplerian velocity (several thousand cm s^{-1} faster

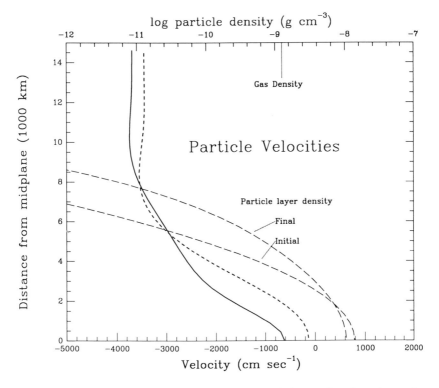

Figure 11.   Particle velocities relative to planar Keplerian motion, for a layer of
midplane mass density initially 10 times that of the gas at 1 AU. The particle orbital
velocity profile is shown by the short-dashed curve, and the particle radial velocity
profile is shown by the solid curve. The particles have $d = 100$ cm and $\rho_s = 0.1$
g cm^{-3}. The values shown are steady state, and have converged after $\sim$100,000
Cray YMP timesteps, or $\sim$2 yr. The initial and final profiles of mass density in
the particle layer are shown by the long-dashed curves; the initial profile is flatter,
and the layer "puffs up" over the course of the run due to the diffusive effects of
shear-driven gas turbulence. The steady-state density of the center of the layer is
$\simeq$7 times the gas density, while the critical density for gravitational instability of
the particle layer at this location is $\sim$10^{-7} g cm^{-3}, or $\sim$10 times larger.

than its pressure-supported value). The lack of a headwind causes the inward
drift of the particles to diminish. Simultaneously, as the particles begin to
drive the gas to orbital velocities which exceed its pressure-balanced value,
the gas acquires an *outward* radial drift in the boundary layer consistent with
Ekman layer analogy (Weidenschilling 1980). Another way to understand
this is as the result of transfer of angular momentum between the more rapidly
rotating particle layer and the more slowly rotating surrounding gas. The
effect is confined to a region of thickness comparable to the boundary layer,
because that is where the velocity gradients are sufficiently large to generate
a significant viscosity and shear stress. The particle drift rates in the upper

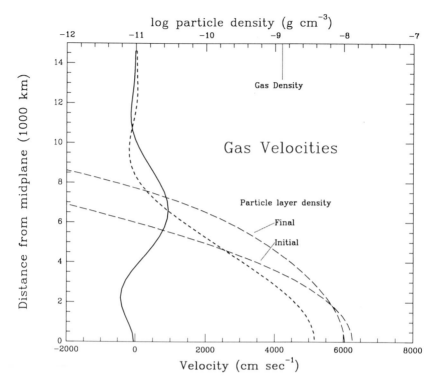

Figure 12.   Gas velocities relative to pressure supported motion, for the final profile of mass density in the particle layer. The gas orbital velocity profile is shown by the short-dashed curve, and the gas radial velocity profile is shown by the solid curve. These are mean flow velocities; in addition, there is turbulence with fluctuations of comparable magnitude within the layer.

reaches of the layer are in fairly good agreement with the "Ekman" estimates discussed earlier, but deeper within the layer the solid material is less affected. Naturally, this assumes there is no source of turbulence other than the local shear, and that the gas viscosity at this stage and in this region is considerably less than the "alpha model" values of 10^{13} to 10^{15} cm^2 s^{-1}. Within the particle layer, the gas drifts inward along with the p rticles. In some cases, the vertical shear in the *radial* velocities can compete with the vertical shear in the *orbital* velocities in contributing to the viscosity, which emphasizes that this is a highly coupled problem.

In detail, the drift rates of gas and particles will probably change once viscosity due to particle collisions is included. For example, one simple estimate of the particle viscosity may be obtained from the expression widely used in studies of planetary rings (Goldreich and Tremaine 1978), using the expressions relating $\bar{v}_p^2$ to $\bar{v}^2$ from Eq. (16) (Dobrovolskis 1991, personal communication);

$$v_p \sim \frac{v_p^2}{2\Omega}\left[\frac{\tau_0}{1+\tau_0^2}\right] \sim \frac{v_t}{2Sc}\left[\frac{\tau_0}{1+\tau_0^2}\right] \tag{23}$$

where τ_0 is the dynamical optical depth of the layer (i.e., that due to particles of sizes comparable to the mean mass). Substituting, we obtain

$$\tau_0 \sim \frac{\sigma_s}{\rho_s d} \sim \frac{\rho_p h}{\rho_s d} \sim 10^{-1} \tag{24}$$

where $\rho_p \sim 10^{-8}$ g cm^{-3} is the volume mass density in the particle layer, $d \sim 60$ to 100 cm is the typical particle diameter, and $\rho_s \sim 1$ g cm^{-3} is the density of an individual particle. The particle viscosity is comparable to the gas viscosity for the models presented above. When included, this effect will probably increase the inward radial drift of the densest part of the layer, causing a simultaneous increase in the outward flow of the surrounding gas.

VII. CONCLUSIONS

The preceding sections sketch out parts of the possible course of growth of particles from μm to km sizes in the context of current ideas about the solar nebula. It seems likely that planetesimal formation involved at least an early stage of collisional sticking and coagulation of particles. This process depended on poorly constrained properties of the nebula (e.g., temperature, pressure gradients, turbulence) and of the particles themselves (stickiness, mechanical strength, fractal dimension). Many variables need to be examined in more detail, especially collisional behavior of particle aggregates, for which relevant experimental data are sorely lacking.

It is apparent that if grains have a reasonable probability of sticking in collisions, then their size distribution and the vertical distribution of solids in the disk evolve rapidly compared with the expected astrophysical lifetime of the disk ($\sim 10^6$ yr). The assumption used by some observers, that size distribution is similar to that of interstellar grains, is not likely to be valid. Moreover, the actual size distribution may not resemble a simple power law, and may vary with location (radial and vertical) in the disk.

In the later stages of accumulation, collective effects (a dense particle layer dragging the gas) could have important consequences. Within the particle layer, ΔV is decreased, which may promote accretion of fragile aggregates; however, there is increased turbulence generated by the shear associated with this layer's motion, which can have the opposite effect. Complex flow patterns can occur, with radial gas motion inward at the central plane, and outward in the boundary layer. In addition to vertical sorting (larger bodies nearer the center of the layer), there may be radial segregation as well, if smaller particles are entrained in the boundary layer. If this stage is long-lived, there could be significant mixing of solids formed at different distances from the Sun. It is not clear that any such effects would leave visible traces in the meteoritic record, but the possibility should be considered. The computational fluid dynamical model described here is still in the early stages of development; the obvious next step will be to include a spectrum of particle sizes and coagulation within the particle layer.

The later stage ($\sim$ m to km) of planetesimal formation has not yet been examined in detail. Two (at least) possible outcomes can be inferred. If the bodies in the turbulent, shearing layer near the central plane can grow large enough by mutual collisions, they may decouple from the gas sufficiently to allow further settling. In that case, there might be gravitational instability in a layer of > meter-sized bodies, rather than one made of dust grains. On the other hand, a plausible amount of global turbulence, e.g., due to convection in the disk, could prevent gravitational instability no matter how large the solid bodies grow. In that case, collisional coagulation will be the only mechanism of growth at all sizes. These outcomes are model dependent; indeed, both could have occurred in different parts of the disk, or at different times.

Regardless of the actual mechanism by which planetesimals formed, that process was not so efficient as to deplete the dust in the solar nebula and render the disk optically thin. Detection of dust in a circumstellar disk does not rule out the presence of planetesimals, or even planets, within it.

The parts of this chapter do not make a seamless and self-consistent scenario for the origin of planetesimals. One should remember that the accretion of planets from planetesimals involves growth in size by a factor 10^4, while the formation of km-sized bodies from μm-sized grains covers 9 orders of magnitude. There is still room for plenty of work within that range.

APPENDIX: MODELING OF A PARTICLE-GAS LAYER

Here we present for reference a self-consistent formulation for computational fluid dynamical simulation of a viscous two-phase (particles plus gas) layer near the central plane of the solar nebula. A more detailed description of this model is given by Champney and Cuzzi (1990) and Cuzzi et al. (1993).

A. Solution of the Mass and Momentum Equations

The full equations of motion in cylindrical coordinates are:

$$\frac{\partial \rho}{\partial t} + w \frac{\partial \rho}{\partial z} + \rho \frac{\partial w}{\partial z} + u \frac{\partial \rho}{\partial r} + \frac{\rho \partial u}{\partial r} + \frac{\rho u}{r} = 0 \qquad (A.1)$$

$$\frac{\partial w}{\partial t} + w \frac{\partial w}{\partial z} + v \frac{\partial w}{\partial r} + \frac{1}{\rho} \frac{\partial P}{\partial z} =$$

$$-\frac{GM}{R^3} z - A\rho_p (w - w_p) + \frac{1}{\rho} \frac{\partial \tau_{zz}}{\partial z} + \frac{1}{\rho r} \frac{\partial (r \tau_{zr})}{\partial r} \qquad (A.2)$$

$$\frac{\partial u}{\partial t} + w\frac{\partial u}{\partial z} + u\frac{\partial u}{\partial r} + \frac{1}{\rho}\frac{\partial \rho}{\partial r} =$$

$$\frac{v^2}{r} - \frac{GM}{R^3}r - A\rho_p(u - u_p) + \frac{1}{\rho}\frac{\partial \tau_{zr}}{\partial z} + \frac{1}{\rho r}\frac{\partial (r\tau_{rr})}{\partial r} - \frac{\tau_{\theta\theta}}{\rho r} \qquad (A.3)$$

$$\frac{\partial v}{\partial t} + w\frac{\partial v}{\partial z} + u\frac{\partial v}{\partial r} =$$

$$\frac{-uv}{r} - A\rho_p(v - v_p) + \frac{1}{\rho}\frac{\partial \tau_{z\theta}}{\partial z} + \frac{1}{\rho r}\frac{\partial (r\tau_{r\theta})}{\partial r} + \frac{\tau_{r\theta}}{\rho r} \qquad (A.4)$$

$$\frac{\partial \rho_p}{\partial t} + w_p\frac{\partial \rho_p}{\partial z} + \rho_p\frac{\partial w_p}{\partial z} + u_p\frac{\partial \rho_p}{\partial r} + \rho_p\frac{\partial u_p}{\partial r} + \frac{\rho_p u_p}{r} =$$

$$\frac{\partial}{\partial z}\left[\frac{v_t}{Sc}\frac{\partial \rho_p}{\partial z}\right] + \frac{\partial}{\partial r}\left[\frac{v_t}{Sc}\frac{\partial \rho_p}{\partial r}\right] \qquad (A.5)$$

$$\frac{\partial w_p}{\partial t} + w_p\frac{\partial w_p}{\partial z} + u_p\frac{\partial w_p}{\partial r} = \frac{-GM}{R^3}z - A\rho(w_p - w) \qquad (A.6)$$

$$\frac{\partial u_p}{\partial t} + w_p\frac{\partial u_p}{\partial z} + u_p\frac{\partial u_p}{\partial r} = \frac{v_p^2}{r} - \frac{GM}{R^3}r - A\rho(u_p - u) \qquad (A.7)$$

$$\frac{\partial v_p}{\partial t} + w_p\frac{\partial v_p}{\partial z} + u_p\frac{\partial v_p}{\partial r} = -\frac{u_p v_p}{r} - A\rho(v_p - v) \qquad (A.8)$$

where u, v, w are the mean velocity components in the radial (r), transverse (θ), and vertical z directions, and ρ is the spatial density. Quantities without subscripts refer to the gas, and the subscript p refers to the particles. A is a drag coefficient (equal to $1/\rho t_e$), $R^2 = r^2 + z^2$, v_t is the turbulent kinematic viscosity of the gas phase, and Sc is a dimensionless parameter known as the Schmidt number, which is discussed in Sec. III. The terms τ_{ij} are the components of the viscous stress tensor for the gas phase. In the above equations, no similar terms are included for particle viscosity. Initially in the nebula, the particle volume density is sufficiently low that this is an acceptable approximation. However, as increasingly dense layers are studied, this does become a concern.

Champney and Cuzzi (1990) and Cuzzi et al. (1993) present a numerical model which includes all of the viscous and nonlinear terms, and overcomes the significant numerical challenges by using a perturbation technique in which an analytical solution is first subtracted from the problem. The analytical, or unperturbed, solution consists of Keplerian orbital motion for the particles, pressure supported orbital motion for the gas, and hydrostatic equilibrium in the vertical direction for the gas. In the unperturbed solution, the vertical particle velocity is merely the terminal settling velocity. These solutions are subtracted from the exact equations, and the remaining terms are solved numerically.

In Secs. III and IV we discussed solutions for the mean gas and particle velocities, and for the gas turbulent (eddy) viscosity. The effects of the turbulent boundary layer on dispersion of the particles is modeled as a diffusion term in the particle mass conservation equation (Eq. A.5) where the term v_t/Sc is the particle diffusion coefficient D. Certain subtleties are associated with derivation of the diffusion term from averaged versions of the exact equations (see Champney and Cuzzi 1990), but models of this form are extensively used in two-phase fluid engineering applications.

GROWTH OF PLANETS FROM PLANETESIMALS

JACK J. LISSAUER
State University of New York at Stony Brook

and

GLEN R. STEWART
University of Colorado

The formation of terrestrial planets and the cores of Jovian planets is reviewed in the framework of the planetesimal hypothesis, wherein planets are assumed to grow via the pairwise accumulation of small solid bodies. The rate of (proto)planetary growth is determined by the size and mass of the protoplanet, the surface density of planetesimals, and the distribution of planetesimal velocities relative to the protoplanet. Planetesimal velocities are modified by mutual gravitational interactions and collisions, which convert energy present in the ordered relative motions of orbiting particles (Keplerian shear) into random motions and tend to reduce the velocities of the largest bodies in the swarm relative to those of smaller bodies, as well as by gas drag, which damps eccentricities and inclinations. The evolution of the planetesimal size distribution is determined by the gravitationally enhanced collision cross-section, which favors collisions between planetesimals with smaller velocities. Deviations from the 2-body approximation for the collision cross-section are caused by the central star's tidal influence; this limits the growth rates of protoplanets. Runaway growth of the largest planetesimal in each accretion zone appears to be a likely outcome. The subsequent accumulation of the resulting protoplanets leads to a large degree of radial mixing in the terrestrial planet region, and giant impacts are probable. Gravitational perturbations by Jupiter probably were responsible for preventing runaway accretion in the asteroid belt, but detailed models of this process need to be developed. In particular, the method of removal of most of the condensed matter (expected in a nebula of slowly varying surface density) from the asteroid region and the resulting degree of radial mixing in the asteroid belt have yet to be adequately modeled. Accumulation of Jupiter's core before the dispersal of the solar nebula may require more condensable material at 5 AU than predicted by standard minimum-mass solar-nebula models.

I. INTRODUCTION

The nearly circular and coplanar orbits of the planets argue for planetary formation in a flattened disk orbiting the Sun (Kant 1755; Laplace 1796). Astrophysical evidence suggests that such disks are the natural byproducts of the collapse of molecular cloud cores leading to star formation (Chapters by Beckwith and Sargent and by Basri and Bertout). The most highly developed theory for explaining planetary growth within such a circumstellar disk is accretion (aggregation) of solid planetesimals via binary collisions, followed in the case of Jovian type planets by accretion of gas onto solid cores (Safronov 1972; Hayashi et al. 1985).

In this chapter, we review the dynamics of the accretion process from kilometer sized planetesimals to terrestrial planets and the cores of giant planets. The formulas presented herein are valid for single stars of any mass; however, the complexities of planetary accretion in multiple star systems are not treated. We assume as our initial conditions the presence of a disk of stellar composition in orbit about the star. Moreover, our calculations begin when the bulk of the condensed material in the nebula has settled out and agglomerated into bodies at least ~ 1 km in size. The microphysics of the growth of the sub-centimeter grains is very different than the dynamical processes important to later stages of planetary accretion, and growth in the intermediate-size range is believed to be very rapid (Chapter by Weidenschilling and Cuzzi). Our review also neglects the accretion of the gaseous envelopes of the giant planets, except insofar as an atmosphere may enhance the planetesimal accretion cross-section of a protoplanet. This final stage of giant planet growth is reviewed in the Chapter by Podolak et al.

Strictly speaking, the planet formation process is unlikely to be as purely sequential as we have lain out. Grain growth and even the accretion of large planetesimals described in detail herein may well begin during the epoch when a protoplanetary disk is still accreting and redistributing matter (cf. Chapters by Shu and by Adams and Lin). Given that the theories of each of these epochs are still rather primitive, a sequential study of each stage is probably adequate. However, to the extent that planetary growth depends on, e.g., the initial size distribution of planetesimals, we must recognize that various processes currently being treated as separate events occur simultaneously and affect each other.

The star's gravity is the dominant force upon planetesimals, and other forces are generally included as perturbations on the planetesimal's Keplerian orbit. The dominant perturbations to a planetesimal's heliocentric trajectory are usually due to gravitational attraction of other planetesimals and proto-planets. (We use the term protoplanets to refer to exceptionally large plan-etesimals, not the giant gaseous protoplanets once popularized by Cameron [1962].) When a single protoplanet is the dominant perturber in a given region of the protoplanetary disk, it is convenient to treat its perturbations separately. The dominant nongravitational forces upon planetesimals are mutual inelastic collisions (which may lead to accretion and/or fragmentation), and gas drag. Whole body magnetic forces are believed to be negligible for the dynamics of bodies of kilometer size and larger. However, if electromagnetic induction heating (Herbert 1989) is sufficient to melt planetesimal interiors, then the outcome of physical collisions could be altered.

The distribution of planetesimal velocities is one of the key factors which control the rate of planetary growth. In Sec. II, we review the physical factors important to determining the equilibrium velocity distribution of a swarm of planetesimals of various sizes when accretion is neglected. The growth rate of a protoplanet in a uniform surface density disk of planetesimals with known velocity dispersion is discussed in Sec. III. In Sec. IV, we follow the simultaneous evolution of planetesimal masses and velocities and show that under a wide variety of initial conditions the largest body in any given accretion zone grows very rapidly and "runs away" from the mass distribution of other accreting bodies in its region of the solar system. The limits to such runaway accretion are quantified in Sec. V. In Sec. VI, we review models of the final stages of planetary accretion, with emphasis on the growth times of planets and the possibility of giant impacts. We conclude with a summary of the major results of planetesimal dynamics and a list of outstanding questions in Sec. VII.

II. PLANETESIMAL VELOCITIES FOR A STATIC MASS DISTRIBUTION

The simplest analytic approach for calculating the evolution of planetesimal velocities uses a "particle-in-a-box" approximation in which the evolution of the mean square planetesimal velocities are calculated via the methods of the kinetic theory of gases. Originally, the particle-in-a-box calculations were developed by Safronov (1972) using relaxation time arguments similar to those used by Chandrasekhar (1942) in stellar dynamics. More recently, these calculations have been refined by employing modern kinetic theory methods (Hornung et al. 1985; Stewart and Wetherill 1988; Barge and Pellat 1990; Ida 1990). A kinetic theory approach appears to be the only viable method for treating the initial stages of planetesimal accumulation because the number of initial planetesimals is extremely large. During the final stages of planetesimal accumulation, the number of planetesimals eventually becomes small enough that a more direct treatment of individual planetesimal orbits is feasible. With a modest number of planetesimals, the most straightforward approach would be a numerical n-body integration of the planetesimal orbits, but the exceedingly long time scales required for the final stages of planetary accumulation (e.g., between 10 and 100 Myr in the inner solar system) severely limit the usefulness of this method. An alternative approach that has enjoyed considerable success is to assume that the planetesimals follow slowly precessing elliptic orbits that are occasionally altered by rare close encounters with other planetesimals. A Monte Carlo procedure is then used to choose successive pairs of planetesimals that interact according to the two-body gravitational scattering formula. This Monte Carlo approach has been extensively developed by Wetherill (1980b,1985,1986,1988,1990b).

A. Random Velocity Distribution

In the particle-in-a-box approximation, one ignores the details of individual planetesimal orbits and uses a probability density to describe the distribution of orbital elements in the planetesimal population. Specifically, the orbital

perihelia and longitudes of the ascending node are assumed to be random
and the orbital eccentricities e and inclinations i are assumed to be Rayleigh
distributed,

$$f(e, i) = 4 \frac{\sigma}{m} \frac{ei}{\langle e^2 \rangle \langle i^2 \rangle} \exp \left[-\frac{e^2}{\langle e^2 \rangle} - \frac{i^2}{\langle i^2 \rangle} \right] \qquad (1)$$

where m is the mass of the individual planetesimals and σ is the surface
mass density of planetesimals with a particular semimajor axis; $\langle e^2 \rangle$ and
$\langle i^2 \rangle$ are the mean square eccentricity and inclination. Although the form of
Eq. (1) is difficult to justify rigorously, the randomization of orbits caused
by planetesimal interactions in n-body simulations has been found to yield
orbital distributions similar to Rayleigh distributions (Wetherill 1980b; Ida
and Makino 1992a).

Another important property of planetesimal orbits is the fact that planetesimals with different semimajor axes orbit the star with different mean
velocities. This Keplerian shear is usually introduced into a local velocity
distribution by postulating a local mean velocity that reproduces the differential rotation between coplanar circular orbits. Because the orbital phase angles
are averaged out in this approximation, the deviations from coplanar circular
orbits are reduced to a distribution of random velocities relative to the local
mean velocity of a circular orbit. The distribution of random velocities that is
locally equivalent to Eq. (1) is a triaxial Gaussian distribution in cylindrical
coordinates,

$$f_o(z, \mathbf{v}) = \frac{\Omega \sigma}{2\pi^2 c_r{}^2 c_z{}^2 m} \exp \left[-\frac{v_r{}^2 + 4v_\theta{}^2}{2c_r{}^2} - \frac{v_z{}^2 + \Omega^2 z^2}{2c_z{}^2} \right] \qquad (2)$$

where $2c_r{}^2 = \langle e^2 \rangle v_K{}^2$, $2c_z{}^2 = \langle i^2 \rangle v_K{}^2$, v_θ is the azimuthal component of
velocity relative to the local circular Keplerian velocity, $v_K = (GM_\star/r)^{1/2}$,
and $\Omega = v_K/r$ is the orbit frequency. The integral of f_o over velocity space
yields the local number density,

$$n = \int d^3 v f_o = \frac{\Omega \sigma}{\sqrt{2\pi} c_z m} \exp \frac{-\Omega^2 z^2}{2c_z^2}. \qquad (3)$$

B. Velocity Evolution

Given the local random velocity distribution in Eq. (2), one can write down
a kinetic equation for the time evolution of the random velocity distribution
caused by mutual planetesimal interactions:

$$\frac{\partial f}{\partial t} + \mathbf{v} \cdot \nabla_r f - \left[\frac{GM_\star \mathbf{r}}{r^3} + (\mathbf{v} + \mathbf{v}_K) \cdot \nabla_r \mathbf{v}_K \right] \cdot \nabla_v f = \frac{\delta f}{\delta t} |_{\text{coll}} + \frac{\delta f}{\delta t} |_{\text{grav}}. \qquad (4)$$

The two terms on the right-hand side of the kinetic equation denote the two
kinds of planetesimal interactions: (1) physical collisions which dissipate

some or all of the relative kinetic energy of the colliding bodies; and (2) gravitational scattering which conserves the relative kinetic energy but results in a rotation of the relative velocity vector. Physical collisions are usually modeled with a Boltzmann collision operator for hard spheres that has been modified to allow for inelastic collisions (see Trulsen 1971; Hornung et al. 1985). Most published calculations of planetesimal accumulation assume collisions are completely inelastic because physically plausible rebound velocities rarely exceed the mutual escape velocity except for the very smallest planetesimals in the distribution (cf. Sec. III). Violent collisions that produce a size distribution of collision fragments are more likely to occur once sizeable protoplanets are formed. In order to account for the gravitational enhancement of the collision cross-section, a correction factor must be applied to the hard sphere collision operator. An accurate determination of this enhancement factor is rather difficult in general, owing to the tidal influence of the star; a detailed discussion of its calculation is presented in Sec. III.

Gravitational scattering between planetesimals can be modeled with a Fokker-Planck operator similar to that used in stellar dynamics (cf. Binney and Tremaine 1987). The use of a Fokker-Planck operator requires two assumptions. First, the result of a close encounter must be well approximated by the 2-body Rutherford scattering formula, which ignores the gravitational influence of the star for the duration of the encounter. Between successive encounters, the star's gravitational influence is properly taken into account by the left-hand side of Eq. (4). Numerical integrations of the 3-body problem indicate that the average perturbations given by the 2-body approximation are valid within a factor of 2, so long as the random velocities are greater than 0.07 times the surface escape velocity of the two bodies at contact (Wetherill and Cox 1984). More extensive numerical investigations of the 3-body problem support the accuracy of the Fokker-Planck relaxation and energy exchange rates provided that the encounter velocities do not become too small (Ida 1990). Second, the relative velocity between planetesimals must be primarily determined by their random velocities, rather than by Keplerian shear. This assumption is valid for the early stages of planetesimal accumulation, when a very large number of planetesimals can be found within a spherical volume of radius equal to the scale height c_z / Ω of the planetesimal disk.

The essential reason why the Fokker-Planck operator cannot correctly model a gravitational encounter more distant than a scale height is that the Fokker-Planck operator implicitly assumes that gravitational encounters are local events. In particular, it assumes that the two interacting planetesimals have the same local mean velocity, so that their relative velocity is entirely determined by the random velocity distribution (Eq. 2). This assumption is violated for distant encounters, where the relative velocity between planetesimals is largely determined by the difference in their semimajor axes (i.e., by Keplerian shear). Although it is possible to write down formal expressions for the velocity evolution due to distant encounters (see Chapter by Ohtsuki et al.), an explicit analytic formula is generally not available to replace the Rutherford scattering law, except in a few limiting cases.

One such limiting case occurs when the separation in semimajor axes is large compared to both the orbital eccentricities and the Hill sphere radius h (see Sec. III.C), and is also small compared to both semimajor axes (Hénon and Petit 1986; Hasegawa and Nakazawa 1990). Weidenschilling (1989) evaluated the contribution of distant gravitational encounters in this limit and concluded that they may safely be neglected compared to close gravitational encounters except when a few protoplanets are so large that their surface escape velocities exceed the local random velocity by a factor of ≥ 100. For these reasons, the maximum encounter distance is set equal to the disk scale height in the Fokker-Planck operator.

Once the interaction terms have been specified, an approximate solution to Eq. (4) can be obtained by linearizing the kinetic equation about the Gaussian velocity distribution f_o stated in Eq. (2):

$$\left[\frac{\partial}{\partial t} + 2\Omega v_\theta \frac{\partial}{\partial v_r} - \left(\frac{\Omega v_r}{2} \right) \frac{\partial}{\partial v_\theta} \right] f_1 = \frac{\delta f_o}{\delta t}\Big|_{\text{coll}} + \frac{\delta f_o}{\delta t}\Big|_{\text{grav}} \qquad (5)$$

where $|f_1| << |f_o|$. Note that f_o does not appear on the left-hand side of Eq. (5) because it is an integral of the free orbital motion. Linearization is an excellent approximation because the orbit frequency greatly exceeds both the collision rate and relaxation rate due to gravitational scattering. Evolution equations for the mean square random velocities are obtained by taking second-order velocity moments of the linearized kinetic equation. The rate of change of the mean square random velocity is given by

$$\frac{\partial \langle nv^2 \rangle}{\partial t} = \int d^3v \left(\frac{\partial f_1}{\partial t} \right) v^2. \qquad (6)$$

In general, one obtains two coupled equations for the two components of the random velocity associated with the orbital eccentricity and inclination (Hornung et al. 1985). However, during the early stages of planetesimal accumulation the ratio of rms inclination to rms eccentricity is likely to be nearly constant due to an approximate equipartition of energy between the planar and vertical orbital motions. Barge and Pellat (1990) have derived a value of this ratio of approximately 0.6 by simultaneously solving coupled equations for the "thermal" motions in the vertical and horizontal directions. This value may be compared with values close to 0.5 that were obtained by Wetherill (1980b) using a Monte Carlo simulation, and by Ida and Makino (1992a) using direct n-body calculations. Calculations of steady-state planetesimal velocities indicate that the difference between 0.5 and 0.6 is unimportant compared to the other approximations described above.

Thus, setting $c_z/c_r = 0.5$ in Eq. (2), a single equation is derived for the velocity evolution of a test body of mass m_i interacting with a swarm of field bodies of mass m_k. Following Stewart and Wetherill (1988), we split the equation into four separate contributions and add a fifth term that models the velocity evolution caused by gas drag in the solar nebula. The rate of change in v_i, the rms velocity of bodies of mass m_i, is given by:

$$\frac{dv_i}{dt} = A + B + C + D + E \tag{7}$$

where the five terms on the right-hand side are:

1. *"Viscous stirring" resulting from gravitational scattering*:

$$A = \frac{3}{4}\frac{\sqrt{\pi}G^2}{v_i V_{ik}^3}\rho_k \ln\Lambda [(9L-12\sqrt{3})(m_i+m_k)v_i^2+(5L-4\sqrt{3})(m_k v_k^2-m_i v_i^2)] \tag{8}$$

where ρ_k is the density (specific gravity) of the planetesimals of mass m_k, $V_{ik}^2 \equiv v_i^2+v_k^2$, $L \equiv \ln[(2+\sqrt{3})/(2-\sqrt{3})] \approx 2.634$, and $\Lambda \equiv \sin(\Psi_{max}/2)/\sin(\Psi_{min}/2)$. The angle Ψ is related to the impact parameter b according to the Rutherford scattering formula,

$$\sin\left(\frac{\Psi}{2}\right) = \left(1 + \frac{b^2}{b_o^2}\right)^{-1/2} \tag{9}$$

where $b_o = G(m_i + m_k)V_{ik}^{-2}$. The minimum deflection angle, Ψ_{min}, is calculated with an impact parameter $b = b_{max} = \max(v_i/\Omega, v_k/\Omega)$. The maximum deflection angle, Ψ_{max}, is calculated using the impact parameter corresponding to the 2-body gravitational capture cross section,

$$b = b_{min} = R_g = (R_i + R_k)\left[1 + \frac{2b_o}{(R_i + R_k)}\right]^{1/2}. \tag{10}$$

2. *Viscous stirring caused by inelastic collisions*:

$$B = \frac{\sqrt{\pi}}{8}\left(\sqrt{3} - \frac{5L}{12}\right)\frac{V_{ik}}{v_i}\rho_k R_g^2\frac{m_k(v_i^2 - v_k^2) + 2m_i v_i^2}{(m_i + m_k)^2}. \tag{11}$$

3. *Velocity damping due to energy dissipated by inelastic collisions*:

$$C = -\sqrt{\pi}\left(\frac{11\sqrt{3}}{18} + \frac{L}{24}\right)\frac{V_{ik}}{v_i}\rho_k R_g^2\frac{m_k(v_i^2 - v_k^2) + 2m_i v_i^2}{(m_i + m_k)^2}. \tag{12}$$

4. *Energy transfer from large bodies to small ones via dynamical friction*:

$$D = \frac{4\sqrt{\pi}LG^2}{v_i V_{ik}^3}\rho_k \ln\Lambda(m_k v_k^2 - m_i v_i^2). \tag{13}$$

5. *Energy damping caused by gas drag*:

$$E = -\pi \left(\frac{C_D}{2m_i} \right) \rho_g R_i^2 v_i (v_i + \eta) \tag{14}$$

where C_D is the drag coefficient, ρ_g is the gas density, and η is the velocity of a body moving in a circular Keplerian orbit relative to the that of the nebular gas, which orbits at a slower velocity because it is partially supported by thermal pressure (cf. Adachi et al. 1976; Weidenschilling 1977*a*).

The viscous stirring terms are so denoted because their sum is proportional to the product of the shear stress and the rate of strain generated by the differential rotation of the local mean velocity. The physical source of the energy for the viscous stirring terms is the kinetic energy contained in the sheared mean flow. Mutual planetesimal interactions transform this energy into the random motion associated with orbital eccentricities and inclinations. In a uniformly rotating disk, these two terms would vanish. The dynamical friction term tends to drive the system toward an equipartition of random kinetic energy. In a broad distribution of planetesimal sizes, energy equipartition is never actually achieved because the viscous stirring terms tend to drive the system away from the equipartitioned state towards a state in which velocities are independent of mass. For the special case of $m_i = m_k$ and $v_i = v_k$, these expressions are similar to those found by Safronov (1972) and Kaula (1979*a*) using relaxation time estimates. The equation for gas drag is adapted from the theory of Adachi et al. (1976). In this formula, the drag coefficient C_D appropriate for the large Reynolds number flows occurring around planetesimals in the circumstellar nebulae is ~ 0.5 (Whipple 1972).

The steady-state velocities predicted by Eqs. (7–14) are displayed in Fig. 1 for several different mass distributions. The mass distributions are labeled by the exponent q in the differential power law $dn/dm \propto m^{-q}$; the planetesimal masses range from 10^{18} to 9.766×10^{24} g in each case. In all the cases plotted, the largest planetesimals have the smallest velocities as a result of the transfer of energy from large bodies to small bodies via dynamical friction. This result is distinctly different from what one obtains from Safronov's relaxation time theory, which omitted dynamical friction. The power law with exponent $q = 2$ is a transitional case because equal amounts of mass are distributed in each logarithmic mass interval in that case. Mass distributions with $q > 2$ have most of their mass in the smaller bodies and are therefore more efficient at draining away the energy from the larger bodies. For large values of q, the velocity distribution achieves its maximum value at an intermediate mass planetesimal where the viscous stirring by gravitational scattering is strong relative to both inelastic collisions, which dominate for small planetesimals, and dynamical friction, which slows the largest bodies. Somewhat surprisingly, omitting the gas drag term from Eq. (14) would hardly change the velocity curves in Fig. 1 at all. Apparently, the absence of gas drag is mostly compensated by a larger energy loss due to inelastic collisions if the steady-state velocities

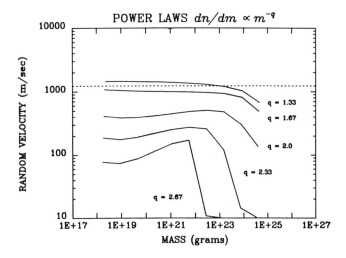

Figure 1.   Steady-state random velocities calculated from Eqs. (7–14) for five different power law mass distributions. The curves are labeled by the exponent q in the differential power law $n(m)dm \propto m^{-q}dm$. The horizontal dotted line gives the escape velocity of the largest body. In all cases, the surface mass density of solids was 20 g/cm^2, the masses ranged from 10^{18} to 9.766×10^{24} g and the gas density was 1.18×10^9 g/cm^3.

increase by only a few percent. We suspect that gas drag would have a more significant affect if planetesimals less massive than 10^{18} g were included in the simulations.

C. Relationship Between Random Velocities and Planetesimal Orbital Elements

Planetary accretion occurs within disks of planetesimals on nearly Keplerian orbits about a central star. The particle trajectories can be described using Keplerian orbital elements. Particle-in-a-box calculations approximate the relative orbital motions of planetesimals with a random (or relative) velocity. The relationship between random velocity and orbital elements is by no means straightforward; several different relationships have been used by various authors, causing a great deal of confusion in the field (especially because all of these different quantities are denoted by the symbol v). We attempt to clarify the situation below.

The appropriate conversion between orbital elements and random velocities depends on the specific aspect of the accretion problem being examined. Four different conversions have been used, each of which assumes the epicyclic approximation ($e, i \ll 1$):

1. The velocity of a planetesimal relative to the mean circular orbit in the disk midplane with the same semimajor axis as that of the planetesimal,

i.e., its epicyclic velocity plus a contribution due to inclination:

$$v_{ep} = (e^2 + i^2)^{1/2} v_K. \tag{15}$$

2. The velocity of a planetesimal relative to the *local* mean circular orbit, averaged over an epicycle:

$$v_{lc} = \left(\frac{5e^2}{8} + \frac{i^2}{2} \right)^{1/2} v_K. \tag{16}$$

This is the local rms velocity one would calculate from the velocity distribution equations (cf. to transform between Eqs. [1] and [2]) and is also the velocity which appears in Eqs. (7–14).

3. The velocity of a planetesimal relative to other planetesimals in the swarm, averaged over an epicycle *and* over a vertical oscillation:

$$v_{sw} = \left(\frac{5e^2}{4} + i^2 \right)^{1/2} v_K. \tag{17}$$

This is the local rms relative velocity that one would calculate from the product of two velocity distributions:

$$v_{sw}{}^2 = \int d^3 v_1 d^3 v_2 f_1(\mathbf{v}_1) f_2(\mathbf{v}_2)(\mathbf{v}_1 - \mathbf{v}_2)^2 / n^2.$$

4. The weighted averaged approach velocity of a planetesimal to a protoplanet on a circular orbit averaged over an epicycle. The averaging is weighted by a factor of $1/v^2$ in order to produce the appropriate 2-body approximation to the gravitational enhancement in cross sections for the calculation of accretion rates (cf. Greenzweig and Lissauer 1990):

$$v_{cs} = (e^2 + i^2)^{1/2} \left(\frac{E(k)}{K(k)} \right)^{1/2} v_K \tag{18}$$

where $k \equiv (4(I^2 + 1)/3)^{-1/2}$, $I \equiv \sin i/e$, and where $K(k) \equiv \int_0^{\frac{\pi}{2}} (1 - k^2 \sin^2 \theta)^{-1/2} d\theta$ and $E(k) \equiv \int_0^{\frac{\pi}{2}} (1 - k^2 \sin^2 \theta)^{1/2} d\theta$ are complete elliptic integrals of the first and second kinds.

III. COLLISION CROSS-SECTIONS AND GROWTH RATES

The size distribution of planetesimals evolves principally due to physical collisions among its members. Stresses caused by tidal forces during close encounters between planetesimals may also fragment very weak bodies, but such disruptive encounters require special circumstances (Boss et al. 1991), and thus can be neglected or folded into the formalism of fragmentation via physical collisions. The evolution of the size distribution of planetesimals can be studied using coagulation theory (cf. Sec. IV). The inputs required in the coagulation calculations are the collision frequency and physical assumptions regarding the outcome of collisions.

The velocity at which two bodies of radii R_1 and R_2 and masses m_1 and m_2 collide is given by:

$$v_c = (v^2 + v_e^2)^{1/2} \tag{19}$$

where v is the relative velocity of the two bodies far from encounter and v_e is the escape velocity from the point of contact:

$$v_e = \left(2G\frac{m_1 + m_2}{R_1 + R_2}\right)^{1/2}. \tag{20}$$

The rebound velocity is equal to ϵv_c, where $\epsilon \leq 1$ is the coefficient of restitution. If $\epsilon v_c \leq v_e$, then the two bodies remain bound gravitationally and soon re-collide and accrete. Net disruption requires both fragmentation, which depends on the internal strength of the bodies, and post-rebound velocities greater than the escape speed. As we saw in Sec. II, relative velocities of planetesimals are generally less than the escape velocity from the largest typical bodies in the swarm. Thus, unless ϵ is very close to unity, the largest members of the swarm are likely to accrete the overwhelming bulk of material with which they collide. Fragmentation is likely to be most important for very small planetesimals.

In this section, we shall give formulas for the calculation of the collision rates between planetesimals. For the case of the largest bodies in the swarms, which eventually are to grow to planetary size, the rate of collision with material is essentially identical to the accretion rate.

A. The Particle-in-a-Box Approximation

We wish to compute the accretion rate of a given body, which we shall refer to as the protoplanet, embedded within a uniform surface density and velocity dispersion disk of bodies which we shall call planetesimals. The simplest model for computing the collision rate of planetesimals ignores their motion about the central star entirely. This problem can be treated using methods of the kinetic theory of gases (Safronov 1972; Wetherill 1980a). Collisions occur when the separation between the centers of two particles becomes less than the sum of their radii, R_s. The accretion rate of a body in a swarm of planetesimals of density ρ_{sw} (not to be confused with the density or specific gravity of the individual bodies, ρ) at relative velocity v ($= v_{cs}$) is given by:

$$\dot{M}_{PiB} = \rho_{sw} v \pi R_s^2 \left[1 + \left(\frac{v_e}{v}\right)^2\right] = \rho_{sw} v \pi R_s^2 (1 + 2\theta) \tag{21}$$

where the second term in the parentheses represents the gravitational enhancement of the accretion cross section. The Safronov number, $\theta \equiv (v_e/v)^2/2$, is frequently used to quantify the effects of gravitational focusing.

Note that in a disk of given surface mass density σ, the volume density of the planetesimal swarm is inversely proportional to v, assuming the mean ratio of horizontal to vertical motions (eccentricities to inclinations) remains fixed:

$$\rho_{sw} \approx \frac{\sigma}{2H} = \frac{\sigma}{2a\sin i} \approx \frac{\sigma\Omega}{2v_z} \approx \frac{\sqrt{3}}{2}\frac{\sigma\Omega}{v} \tag{22}$$

where $H = a\sin i$ is the half-thickness ($\approx$ scale height) of the disk and the factor of $\sqrt{3}$ assumes the velocity dispersion is isotropic. Substituting Eq. (22) into Eq. (21), we find that the mass accretion rate of a protoplanet depends on v only through the gravitational focusing factor:

$$\dot{M}_{PiB} = \frac{\sqrt{3}}{2}\sigma\Omega\pi R_s^2[1 + \left(\frac{v_e}{v}\right)^2]. \tag{23}$$

The exact value of the constant in front of the right-hand side of Eq. (23) depends on the velocity distribution; thus, various values have been quoted in the literature and used in estimating growth times for planets.

B. Formulas for the 2-Body Approximation Including Kepler Shear

The particle-in-a-box approximation discussed above works well *locally* as long as encounters are rapid compared to an orbital period, i.e., that relative velocities are high. Actual planetary encounters can be much more complicated than the simple particle-in-a-box model. In addition to the gravitational forces between the colliding bodies, the gravity of the star needs to be taken into account and the 3-body problem must be solved in order to calculate accretion rates. For sufficiently rapid relative velocities, the encounter is brief enough that the tidal effect of stellar gravity can be neglected during the interval in which the gravitational forces between the secondaries is important. Stellar gravity must, however, still be incorporated to determine the flux and velocities of approaching bodies. As the problem is broken into a set of 2-body calculations (which, unlike the 3-body problem, may be solved analytically), this method is known as the 2-body approximation.

The dynamics of planetesimal-protoplanet encounters are the same in the 2-body approximation as in the particle-in-a-box case. The additional problem is finding the average density of planetesimals which a protoplanet encounters and the appropriately weighted encounter velocities "at infinity" in terms of the orbital elements of the bodies. Intermediate steps in the derivation of the accretion rate are omitted in the discussion below, but may be found in Greenzweig and Lissauer (1990, henceforth GL90).

Several simplifying assumptions are required in order to produce a simple analytic expression for the 2-body accretion rate analogous to Eq. (23). First, we assume that orbital eccentricities and inclinations are sufficiently small that terms quadratic in e and $\sin i$ may be neglected (i.e., we make the epicyclic approximation); this assumption is valid for all but possibly the very final stages of accretion. On the other hand, the radial and vertical oscillations of the planetesimals during one orbital period, ae and $a \sin i$, are assumed to be large compared to the particle-in-a-box accretion radius of the protoplanet, $R_s(1 + v_e^2/v^2)^{1/2}$; accretion rates for the planer case (i.e., where $i = 0$) for both zero and nonzero eccentricities are given by GL90. We also assume that the pre-encounter inclinations and eccentricities of the planetesimal orbits relative to that of the protoplanet are the same for all planetesimals. (If the protoplanet is on a circular orbit, as is often the case [cf. Sec. II], then the relative eccentricity is identical to the eccentricity of the planetesimals; see Eq. [10] of GL90 for the general definition of relative eccentricity.) Distributions in e, i and R_s may be accounted for by integrating the results presented below over these distributions (cf. Sec. III.D).

Under the assumptions stated above, the accretion rate of a protoplanet is given by (cf. Eqs. 30–32 of GL90):

$$\dot{M}_{2B} = P_{2B}(e, i)\sigma h^2 \frac{2\pi}{T} = \frac{\sigma R_s^2}{T} F(I) F_{2B} \qquad (24)$$

where $P_{2B}(e, i)$ is the normalized 2-body collision probability as defined by Nakazawa et al. (1989), T is the protoplanet's orbital period, and

$$F(I) \equiv 4 \frac{\sqrt{(1 + I^2)}}{I} \mathbf{E}(k) \qquad (25)$$

where I is defined by Eq. (18) above. The 2-body gravitational enhancement factor, F_{2B}, is the ratio of a protoplanet's 2-body collision rate to that of a nongravitating protoplanet of identical size:

$$F_{2B} = 1 + \frac{v_e^2}{v^2} = 1 + 2\theta \qquad (26)$$

where $v \equiv v_{cs}$.

C. The 3-Body Gravitational Enhancement Factor

The 3-body problem can be solved analytically only for a few special equilibrium cases; thus, numerical integrations are necessary to calculate the 3-body gravitational enhancement factor F_g. The general problem of computing F_g for a variety of planetesimal velocity dispersions over the entire range of different protoplanet masses, radii and orbital locations plausible during the planetary accretion epoch would be intractable were it not for a very useful set of scaling laws based on Hill's (1878) equations. A desirable feature of Hill's equations is their lack of dependence on the ratio of the protoplanet's mass to that of the star as long as $m_1/M_\star \ll 1$. The radius of a protoplanet's Hill sphere is

$$h = \left(\frac{m_1}{3M_\star}\right)^{1/3} a. \qquad (27).$$

An encounter with a planetesimal whose mass is not negligible compared to that of the protoplanet is best described by replacing m_1 in Eq. (27) by $m_1 + m_2$. It is also useful to define the Hill eccentricity e_H, Hill inclination i_H, and relative Hill semimajor axis b_H as:

$$e_H \equiv \frac{ea}{h}, \quad i_H \equiv \frac{ia}{h}, \quad b_H \equiv \frac{a_2 - a_1}{h}. \tag{28}$$

and the scaled accretion radius of a protoplanet as:

$$r_H \equiv \frac{R_s}{h}. \tag{29}$$

Three-body integrations of planetesimal-protoplanet encounters can be scaled to protoplanets of different masses, sizes and separations from stars of any mass if e_H, i_H and r_H remain fixed (Nishida 1983; GL90). Note that r_H is the same for all protoplanets of a given density at a fixed distance from any particular star. A detailed analytic development of the Hill scaling of the accretion problem is presented by GL90.

Following Ida and Nakazawa (1989), we define the collision probability, $P(e_H, i_H, r_H)$, as the ratio of the rate at which bodies hit a protoplanet in a uniform surface density disk to the flux of bodies with semimajor axes in the range $-1 < b_H < 1$ which would pass the planets if their orbits about the star were unperturbed. The value of $P(e_H, i_H, r_H)$ may be calculated from numerical experiments (Ida and Nakazawa 1989; GL90). The mass accretion rate may be obtained from P using the relationship:

$$\dot{M}_{3B} = P(e_H, i_H, r_H) \sigma h^2 \frac{2\pi}{T} \tag{30}$$

and F_g may be determined by dividing $\dot{M}_{3B}$ by the accretion rate for a non-gravitating planet of the same size.

Numerical results for $P(e_H, i_H, 0.005)$, which represents, e.g., a protoplanet of density $\rho = 3.4$ g cm^{-3} orbiting 1 AU from a 1 M$_\odot$ star, are presented for a variety of different values of e_H and i_H by Ida and Nakazawa (1989) and GL90. Figure 2 shows the dependence of F_g on random velocities for $r_H = 0.005$ and $e_H = 2i_H$. Similar plots for other values of r_H are presented in GL90. Note that the 2-body approximation is valid for $v/v_e \gtrsim 0.1$, 3-body accretion rates exceed those given by the 2-body formula by up to a factor of 2 when $0.02 \lesssim v/v_e \lesssim 0.1$, and F_g rises less steeply than the 2-body formula at smaller v/v_e, approaching the value of $\sim 1.7 \times 10^4$ as $v \to 0$.

D. Cross Sections for a Gaussian Velocity Distribution

The formulas quoted above are valid for uniform surface density disks in which all planetesimals either approach the protoplanet at the same speed (particle-in-a-box case) or have the same unperturbed values of e and i (2-body and 3-body cases). While these homogeneous cases provide insight

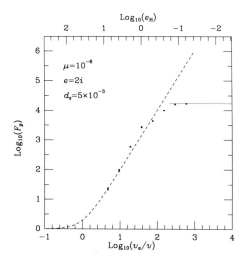

Figure 2.   The 3-body gravitational enhancement of the accretion cross-section of a protoplanet on a circular orbit plotted as a function of planetesimal random velocities (v_{cs}). The radius of the protoplanet is $r_H = 0.005$, which represents, e.g., a protoplanet of density $\rho = 3.40$ g cm^{-3} orbiting 1 AU from a star of mass 1 M$_\odot$. The ratio of the mass of the protoplanet to that of the star used in the numerical calculations is $\mu = 10^{-6}$; however, the results should be valid for any $\mu \ll 1$. Individual points correspond to results for separate runs; in each run all of the planetesimals have identical eccentricities and inclinations with $e_H = 2i_H$, and are distributed in semimajor axis as a uniform surface density swarm. Error bars represent statistical ($\sqrt{n}$) uncertainties in the numerical experiments. The dotted line represents the limiting case of planar circular orbits. The dashed curve is the 2-body approximation given by Eq. (24). Figure from GL90.

into the dynamics of the orbits leading to collisions, they do not represent a realistic disk of planetesimals. Planetesimal motions are much better approximated by a Rayleigh distribution in e and i (Eq. 1) or, equivalently, a triaxial Gaussian distribution in local velocities (Eq. 2). Using the particle-in-a-box approximation, Vityazev and Pechernikova (1981) found that when $v \ll v_e$, the average eccentricity of *colliding* planetesimals is $\sim 3^{-1/2} \langle e^2 \rangle^{1/2}$; this implies a factor of ~ 3 enhancement in accretion rates. Greenzweig and Lissauer (1992) compute formulas for the accretion rate of a protoplanet embedded in a uniform surface density disk of planetesimals with such a velocity distribution. They find that in the 2-body approximation, accretion rates when $v \ll v_e$ are enhanced compared to the case where all planetesimals have e and i equal to the rms of the Rayleigh distribution by a factor which depends weakly on $I\,(\equiv \sin i/e)$ and always exceeds 2.6. Numerical 3-body integrations show comparable enhancements, except when $\langle e_H{}^2 \rangle^{1/2} < 1$ (Greenzweig and Lissauer 1992); at very low eccentricities, the enhancement disappears due to the flattening of the accretion rate as $e \to 0$ (cf. Fig. 2). Qualitatively similar results have been reported by Ida and Nakazawa (1988) and Ohtsuki and Ida (1990).

IV. EARLY STAGES OF PLANETESIMAL ACCUMULATION

The velocity dependence of the collision cross section described above leads to a strong coupling between the evolution of the velocity and size distributions of planetesimals. The study of planetary accumulation therefore requires a simultaneous calculation of the velocity evolution and size evolution of the planetesimal swarm. The early stages of planetesimal evolution has been modeled by several authors (Greenberg et al. 1978,1984; Nakagawa et al. 1983; Ohtsuki et al. 1988; Wetherill and Stewart 1989). All of these papers use some variant of the particle-in-a-box approximation, but they differ in their treatment of the velocity evolution and their method of simulating the planetesimal size evolution. The most important result that has emerged from this work is that the evolving planetesimal size distribution can follow two qualitatively distinct paths that are characterized by very different time scales. The slower evolutionary path exhibits an orderly growth of the entire size distribution so that all the planetesimals remain tied to the continuous size distribution throughout the early stages of planetesimal accumulation. The term *runaway accretion* refers to a different evolutionary path where the largest planetesimal in the local region grows much more rapidly than the remainder of the population and therefore becomes detached from the continuous size distribution. Recent modeling efforts have focused on the task of delineating the circumstances under which *runaway* accretion may or may not occur (Ohtsuki et al. 1990; Ohtsuki and Ida 1990; Wetherill 1989,1990*a*). In this section we review the causes of runaway accretion and discuss the relevance of this process to planet formation.

A. Solutions of the Coagulation Equation

The origin of runaway accretion can be best understood in the context of various solutions to the discrete form of the coagulation equation,

$$\frac{dn_k}{dt} = \frac{1}{2} \sum_{i+j=k} A_{ij} n_i n_j - n_k \sum_i A_{ik} n_i \tag{31}$$

which describes the time evolution of the number of bodies n_k with mass m_k. The merger of smaller bodies increases n_k, whereas the incorporation of bodies of mass m_k into larger bodies causes n_k to decrease. Analytical solutions to Eq. (31) are known for a few simple forms of the collision probability A_{ik} (Safronov 1972; Trubnikov 1971; Wetherill 1990*a*). Although none of these special cases accurately represent the velocity-dependent collision probabilities described in Sec. III, the few analytical solutions to the coagulation equation provide a rigorous test for the accuracy of any numerical algorithm used to solve the coagulation equation. Two such solutions which exhibit orderly growth rather than runaway accretion are the cases when A_{ij} equals a constant and when A_{ij} is proportional to the sum of the masses, $m_i + m_j$. Ohtsuki et al. (1990) used these two solutions to show how numerical algorithms that employ fixed mass-coordinate divisions to represent the mass distribution can produce erroneous results. They conclude that numerical calculations of planetesimal accumulation which use a mass ratio between neighboring mass coordinates that exceeds $\sqrt{2}$ can substantially overestimate the rate of planetary growth.

In contrast to the fixed mass-coordinate algorithms used by most previous workers, Wetherill and Stewart (1989) represented the size distribution with a number of moving "batches," each containing a large number of bodies of the same mass. In this Lagrangian-type scheme, the mass value characterizing each batch is allowed to grow by the sweep-up of smaller bodies as well as by merger with other bodies in the same batch. Wetherill (1990a) has presented test calculations using this method which show better agreement with the two analytical solutions described above than was obtained by Ohtsuki et al.

A more stringent test of any numerical procedure for solving the coagulation equation is the ability to reproduce the solution when A_{ij} is proportional to the product of the masses $m_i \, m_j$ because this is a case known to exhibit runaway growth of the largest body. As pointed out by Wetherill (1990a), the analytical solution for this case is best understood by replacing Eq. (31) with a set of equations which explicitly separate out the evolution of the largest body in the distribution:

$$\frac{dn_k}{dt} = \frac{\gamma}{2} \sum_{i+j=k} ij n_i n_j - \gamma k n_k \sum_i i n_i - \gamma k_R k n_k \tag{32}$$

$$\frac{dk_R}{dt} = \gamma k_R \sum_k k^2 n_k \tag{33}$$

where the masses have been normalized such that $A_{ij} = \gamma ij$ and k_R is the normalized mass of the largest body. The last term on the right-hand side of Eq. (32) represents the mass lost from bin "k" as a result of collisions with the largest body. These equations are consistent with a time independent total mass n_o where

$$n_o = k_R + \sum_k k n_k. \tag{34}$$

The solution of Eq. (32) was found by Trubnikov (1971) to be

$$n_k(\eta) = \frac{n_o (k\eta)^{k-1} e^{-k\eta}}{k! \, k} \tag{35}$$

where $\eta = n_o \gamma t$ is a dimensionless time variable. The time evolution of the runaway body is obtained by substituting the solution (35) into Eq. (34) and solving for k_R.

Wetherill (1990*a*) has utilized the above solution to verify the accuracy of the numerical procedure used by Wetherill and Stewart (1989) in their calculation of planetesimal evolution. A crucial attribute of Wetherill and Stewart's calculation is the probabilistic procedure used to simulate abrupt changes in the mass of the largest body. During a time step of length Δt, the number of merging collisions between bodies in batch "i" and bodies in batch "j," v_{ij}, is given by the formula

$$v_{ij} = n_i n_j A_{ij} \Delta t. \tag{36}$$

When "i" and "j" represent very large bodies, v_{ij} often falls between 0 and 1. It is physically reasonable to interpret a fractional value of v_{ij} as a collision probability per time step because collisions between very large bodies do not really occur during every time step. The procedure used by Wetherill and Stewart is therefore to create a new batch containing one body only when $m_i + m_j$ exceeds the mass of the largest body by a factor δ (where 1.07 $<\delta<1.15$) *and* v_{ij} exceeds a random number between 0 and 1. Numerical calculations using this procedure are able to reproduce the runaway growth predicted by Eqs. (34) and (35) quite accurately, aside from a small time lag that depends on the value chosen for δ (Wetherill 1990*a*). Alternative numerical procedures which fail to simulate the abrupt changes in the size of the largest body can artificially suppress runaway growth. Taken together, these studies of the coagulation equation provide an important lesson: it is always advisable to test one's numerical procedures against known solutions of the coagulation equation before drawing any conclusions about the physical causes of runaway accretion of the planets.

B. The Causes of Runaway Accretion

The three analytic solutions of the coagulation equation described above indicate a bifurcation between solutions which display orderly growth and solutions which display runaway growth of the largest body. The decisive factor that determines which branch the solution will take during planetesimal accumulation is the velocity dependence of the collision rate described in Sec. III. In general, a velocity distribution which exhibits smaller velocities for the larger bodies (as is shown in Fig. 1) will increase the effective mass dependence of the collision rate and will therefore increase the likelihood of runaway accretion. Wetherill and Stewart (1989) presented a series of calculations which serve to delineate the conditions required for runaway accretion to occur. In those calculations which omitted the energy equipartition terms proportional to $(m_k v_k^2 - m_i v_i^2)$ in Eqs. (7–14), orderly growth of the planetesimal size distribution was found. These results show qualitative agreement with the earlier results of Safronov (1972) and Nakagawa et al. (1983), who also neglected the energy equipartitioning caused by gravitational scattering.

When Wetherill and Stewart (1989) included the energy equipartition terms in their velocity evolution equations, they found runaway growth of the largest planetesimal to occur. Complementary calculations by Ohtsuki and Ida (1990) confirm the result that runaway accretion occurs when the random velocities decrease with increasing planetesimal mass. Another mechanism which may decrease the velocities of the largest bodies is the enhanced gas drag that would result from gravitational concentration of nebular gas around massive planetesimals (Takeda et al. 1985). Ohtsuki et al. (1988) present calculations including this enhancement of gas drag which also display runaway accretion. These recent investigations were motivated to large degree by the pioneering work of Greenberg et al. (1978) that first reported runaway accretion. However, Wetherill and Stewart were unable to reproduce runaway growth when using the same physical assumptions that were stated by Greenberg et al., and the validity of those early results remains controversial (cf. Spaute et al. 1991; Kolvoord and Greenberg 1992).

As energy equipartition via gravitational scattering plays a pivotal role in the bifurcation between orderly and runaway growth, is important to establish the limits of validity of the velocity evolution equations discussed in Sec. II. Ida (1990) has presented an extensive series of 3-body orbit integrations in order to determine directly the ability of a protoplanet on a circular orbit to alter the velocity distribution of the surrounding planetesimal swarm as well as the ability of a planetesimal swarm to damp the eccentricity and inclination of a protoplanet via dynamical friction. Ida finds that the energy equipartition rate agrees well with the 2-body results at high velocities. However, when e_H of the smaller planetesimals drops below 2, the rate of equipartition stops increasing with decreasing velocity (unlike the 2-body case, where the rate increases indefinitely). Ida's work therefore implies that dynamical friction continues to operate at low velocities, but the energy equipartition rate attains a maximum asymptotic value that is given by setting $e_H = 2$ in the 2-body formula. When the velocity evolution Eqs. (7–14) are modified to include the limiting equipartition rate, one finds that the steady-state velocities of the largest planetesimals are somewhat greater than shown in Fig. 1. Nevertheless, the velocities of the largest bodies are still found to be substantially smaller than $e_H = 2$ in size distributions where most of the mass is contained in the smaller planetesimals. In light of this result, it appears likely that runaway growth of the largest planetesimals will occur in spite of the limiting equipartition rate found by Ida.

Although the causes of runaway accretion are fairly well established, the precise rate of runaway growth remains somewhat uncertain because the statistical arguments used to derive the velocity evolution equations begin to break down once runaway growth has begun. The Fokker-Planck operator used to derive Eqs. (7–14) implicitly assumes that successive gravitational encounters are uncorrelated, so that random velocities evolve in a random-walk fashion. However, once a protoplanet grows much more massive than any other body in the local accretion zone, each successive gravitational encounter with this protoplanet tends to be correlated with the preceding encounter. In the limit that one can neglect mutual interactions among the smaller planetesimals, the smaller planetesimals will always approach the protoplanet with the same velocity owing to the conservation of their Jacobi constant with respect to the protoplanet. Wetherill and Stewart (1989) placed limits on the importance of this effect by presenting an accretion calculation which omitted perturbations by the largest body in the zone. The rate of runaway accretion was accelerated in that case, due to the reduced random velocities.

When a single protoplanet is the dominant perturber in a zone, the velocity dispersion ceases to be isotropic. Because of the coupling between azimuthal and radial motions, the protoplanet is able to excite eccentricities in planetesimals initially on circular orbits, but due to the symmetry of the equations of motion about the plane of the protoplanet's orbit, the protoplanet cannot excite inclinations in coplanar planetesimals. Numerical studies of planetesimals with initially small e_H and i_H imply much more rapid growth of eccentricity than inclination (Ida 1990; GL90). The rate of planetary growth can be greatly accelerated in such hot, flat disks. (For details, see the discussion surrounding Eq. [50] in GL90.)

Realistically, the precise rate of runaway growth is determined by the relative frequency of encounters with the protoplanet compared to the rate of velocity evolution due to gas drag and mutual interactions among the smaller planetesimals. Hayashi et al. (1977) have stressed the importance of calculating the evolution of both semimajor axes and eccentricities in order to determine the long-term evolution of the Jacobi "constant" when all of these processes are active. More detailed simulations of this problem are needed to better constrain the maximum accretion rate during runaway growth.

C. The Problem of the Asteroid Belt

The absence of a large terrestrial planet in the region between the orbits of Mars and Jupiter provides a strong constraint for models of solar system formation. Numerical simulations of planetesimal accumulation in the primordial asteroid belt yield runaway growth of a protoplanet more massive than 10^{27} g on a time scale of several times 10^5 yr if the influence of Jupiter is neglected (Wetherill 1989). Gravitational perturbations by Jupiter could have curtailed runaway in the asteroid belt only if Jupiter's core had also formed on a time scale of several times 10^5 yr. The early runaway growth of Jupiter's core has therefore emerged as a desirable feature in scenarios of planet formation.

The accumulation of Jupiter's core on a time scale of <1 Myr appears to be possible if the surface density of condensable materials at Jupiter's orbit was ≥ 20 g cm^{-2} (Lissauer 1987; Wetherill 1989). This large a surface density is not predicted by standard minimum-mass solar-nebula models. (A minimum-mass solar nebula is a model circumsolar disk which contains only the amount of condensable material currently present in the planets, augmented with volatiles to solar composition and spread out to give a smooth density distribution in radius. The total mass of the protoplanetary disk in such models is 0.01 to 0.02 $M_\odot$.) One possibility is that the surface density of solids may have been enhanced at Jupiter's orbit due to water ice condensation and by diffusive transport of condensable water vapor from the inner solar system (Stevenson and Lunine 1988). However, we believe that the more likely explanation is that the surface mass density of the protoplanetary disk in the region where the giant planets accreted was a factor of several greater than that predicted by minimum-mass models. This "excess" mass not only could account for the rapid formation of the cores of the giant planets (Sec. V; cf. Lissauer 1987), but also offers a source of condensable material large enough to have produced the $\sim$50 $M_{\oplus}$ of comets believed to exist in the Oort cloud at the present epoch, even after losses over the age of the solar system have been accounted for (Fernandez and Ip 1984; Duncan et al. 1987; Weissman 1990). Moreover, disk stability analysis, which suggests a preferred mass of protoplanetary disks equal to $\sim$1/3 that of the central star (Chapter by Adams and Lin; cf. Shu et al. 1990), and observations of relatively massive disks around many young stars (Chapter by Beckwith and Sargent), reinforce our opinion that minimum-mass models of the solar nebula are headed towards the dustbin of history.

Although Jovian perturbations are widely invoked to explain the asteroid belt, the precise mechanism that halted planet formation is still a subject of some dispute. The only way to stop planet growth is to increase the planetesimal velocities to a value substantially greater than the surface escape velocity of the largest bodies in a local region. The difficulty with invoking gravitational perturbations by Jupiter for this purpose is that the resultant eccentricity pumping is only appreciable at narrow resonance locations. Several authors have discussed how the dissipation of the nebular gas could have shifted the resonance locations, thereby causing Jupiter's resonances to sweep through the entire asteroid belt (Heppenheimer 1980; Torbett and Smoluchowski 1980; Ward 1981). Detailed models which predict the time scale for removal of the gas are required to evaluate this suggestion properly. Changes in Jupiter's semimajor axis due to accretion of gas and/or ejection of planetesimals from the solar system could also have caused Jovian resonances to have swept across the asteroid belt (Safronov and Gusseinov 1989).

Another possible way to pump up velocities in the asteroid belt is to postulate a population of Jupiter-zone planetesimals that are scattered into the asteroid belt once Jupiter becomes sufficiently massive. Recent Monte Carlo calculations of this scenario show that the necessary large velocities are produced in the asteroid belt if the Jupiter zone planetesimals are as massive as the Earth and if one of these bodies becomes trapped in the asteroid belt for an extended period of time (Ip 1987; Wetherill 1989). This result is somewhat unsatisfying, because one must then require this Earth-size body to end up near a strong Jupiter resonance in order to provide a mechanism for its removal. It is also not clear that such a population of Jupiter-zone planetesimals would be created during the rapid runaway growth of Jupiter's core.

One variant of the previous model is that a planet-sized body did form at $\sim$3 AU from the Sun. This body could then have excited the eccentricities of the current asteroids prior to being ejected from the solar system by resonant perturbations from Jupiter. In this case, asteroid belts might not be a common feature among planetary systems otherwise much like our own. More detailed simulations of planetesimal accumulation which accurately model Jupiter perturbations should help narrow the possibilities (cf. Wetherill 1991c).

V. LIMITS TO RUNAWAY GROWTH

Runaway accretion with high F_g requires low random velocities, and thus small radial excursions, $2ae$, of planetesimals. This implies that a proto-planet's feeding zone is limited to the annulus of planetesimals which it can gravitationally perturb into intersecting orbits. Thus, rapid runaway growth must cease when a protoplanet has consumed most of the planetesimals within its gravitational reach (Lissauer 1987).

For the case of a protoplanet on a circular orbit, the standard theory of the restricted 3-body problem places an upper bound on the initial semimajor axis, b_H, that may lead to collision. Neglecting gas drag and interactions with other planetesimals, a planetesimal whose orbital elements satisfy the inequality:

$$\frac{3}{4}b_H{}^2 - e_H{}^2 - i_H{}^2 \geq 9 \tag{37}$$

cannot enter the protoplanet's Hill sphere (see, e.g., Artymowicz 1987). For example, a particle which initially has $e_H = i_H = 0$ and $b_H > 2\sqrt{3} \approx 3.5$ remains in superior orbit to the protoplanet, although it's path may be perturbed (this perturbation preserves the left-hand side of Eq. (37), which is a version of the Jacobi constant). For a single encounter (one synodic period), GL90 find that planetesimals with initial $e_H = i_H = 0$ and $b_H > 2.6$ do not approach closer than $0.1 h$ from the protoplanet; however, Kary et al. (1993) show that most planetesimals with initial $e_H = i_H = 0$ and $1.3 < b_H < 3.2$ come within $0.1 h$ of the protoplanet during 20 synodic periods. For nonzero "initial" eccentricity and/or inclination, the accretion zone expands slightly, but for e and i low enough for F_g to be large, planetesimals with $|b_H| > 4$ cannot be accreted. Thus, the accretion zone of a protoplanet embedded in a disk of low random-velocity planetesimals extends over the region:

$$|b_H| < B \tag{38}$$

where B depends on the magnitude of other perturbations on the planetesimals, and typically is ~ 3.5 to 4 in a quiescent disk.

The size of a protoplanet's Hill sphere expands as it accretes matter. The mass of a protoplanet which has accreted all of the planetesimals within an annulus of width $2\Delta r$ is:

$$m = \int_{r-\Delta r}^{r+\Delta r} 2\pi r' \sigma(r') dr' \approx 4\pi r \Delta r \sigma(r). \tag{39}$$

Setting $\Delta r = Bh = Br(m/3M_\star)^{1/3}$, we obtain the isolation mass to which a protoplanet orbiting at a distance r from a star of mass $M_\star$ may grow:

$$m = \frac{(4\pi Br^2\sigma)^{3/2}}{(3M_\star)^{1/2}} = 2.10 \times 10^{-3} \left(\frac{Br^2\sigma}{2\sqrt{3}}\right)^{3/2} \left(\frac{M_\odot}{M_\star}\right)^{1/2} M_\oplus \quad (40)$$

where the mass of the Earth $M_\oplus = 5.98 \times 10^{27}$ g, r is expressed in AU and σ in g cm^{-2} (Lissauer 1987). For example, assuming $B = 2\sqrt{3}$, a minimum-mass solar nebula with $\sigma = 10$ g cm^{-2} at 1 AU implies protoplanet isolation at 0.066 $M_\oplus$; whereas $\sigma = 3$ g cm^{-2} at 5 AU implies protoplanet isolation at 1.36 $M_\oplus$.

Runaway growth can persist beyond the isolation mass given by Eq. (40) only if additional mass can diffuse into the protoplanet's accretion zone. Three plausible mechanisms for such diffusion are scattering between planetesimals, perturbations by protoplanets in neighboring accretion zones and gas drag. The process of radial drift due to scattering within the vicinity of a protoplanet has not yet been analyzed quantitatively and remains a major open question.

Drift due to gas drag has been modeled in more detail. Weidenschilling and Davis (1985) suggested that a protoplanet can enhance the effects of gas drag on material orbiting just outside its accretion zone in the following manner: As planetesimals drift slowly inwards due to gas drag, they eventually encounter and are trapped into small integer commensurabilities (resonances) with the protoplanet. The eccentricities induced at such resonances lead to high-velocity collisions which grind planetesimals into small debris. Small planetesimals are very strongly affected by gas drag; they cannot be stopped by protoplanet resonances, and thus rapidly drift into the protoplanet's accretion zone.

Alternatively, radial motion of the protoplanet may bring it into zones not depleted of planetesimals. Gravitational torques due to excitation of spiral density waves in the gaseous component of the protoplanetary disk have the potential of inducing rapid radial migration of protoplanets; however, as the migration rate depends on the difference between comparable positive and negative torques (due to excitation of waves at resonances interior and exterior to the protoplanet's orbit, respectively), the magnitude of this effect is extremely difficult to quantify (Goldreich and Tremaine 1980; Ward 1986). Recently it has been suggested that gravitational focusing of gas could vastly increase the rate of inward drift of protoplanets due to gas drag (Takeda et al. 1985; Ohtsuki et al. 1988). This result would have profound implications for models of planetary accretion. However, the fluid calculations assumed low Reynolds number, which may be appropriate for a turbulent protoplanetary disk, but not for a laminar one. The effects of the stellar gravitational field upon the flow pattern also must be examined.

Radial drift of planetesimals relative to a protoplanet increases the protoplanet's isolation mass only if the protoplanet is able to efficiently accrete those planetesimals which approach its orbit. The case of planetesimals decaying inwards (due to gas drag) towards a planet on a circular orbit has been studied by Kary et al. (1993). They find that unless the planet's accretion radius is $\gtrsim 0.01$ h (which would imply a distended thick atmosphere), a majority of the planetesimals miss the planet, and continue to drift inwards towards the star. Thus, radial drift is not as promising a mechanism to increase a planet's isolation mass as was previously believed.

VI. FINAL STAGES OF PLANETESIMAL ACCUMULATION

The self-limiting nature of runaway growth strongly implies that massive protoplanets form at regular intervals in semimajor axis throughout the inner solar system. The mutual accumulation of these protoplanets into a small number of widely spaced planets necessarily requires a stage characterized by large orbital eccentricities, significant radial mixing, and giant impacts. Mutual gravitational scattering can pump up the relative velocities of the protoplanets to values comparable to the surface escape velocity of the largest protoplanet, which is sufficient to ensure their mutual accumulation into planets. The large velocities imply small collision cross sections and hence long accretion times. In the outer solar system, the limits of runaway growth are less severe; it is feasible that runaway growth of Jupiter's core continued until it attained the necessary mass to rapidly capture its massive gas envelope (Lissauer 1987).

A. Simulations of Terrestrial Planet Formation

The transition from runaway growth in isolated accretion zones to the mutual accumulation of protoplanets in large eccentricity orbits has so far only been studied qualitatively. The limiting runaway mass given by Eq. (40) suggests that runaway growth in the inner solar system will yield protoplanets of mass $\sim 10^{26}$ g, with their semimajor axes spaced 0.01 to 0.02 AU apart. At this stage, most of the original mass will be contained in the large protoplanets, so their random velocities will no longer be strongly damped by energy equipartition with the smaller planetesimals. Even if the protoplanets form in circular orbits, mutual gravitational perturbations among several bodies can eventually induce eccentricities of order 0.01 ($e_H \approx 5$), which is sufficient to enable their orbits to cross so the protoplanets can suffer close gravitational encounters. Stagnation of protoplanets in isolated orbits is unlikely because the width of the accretion zone at the end of runaway growth roughly coincides with the maximum orbit separation from which neighboring pairs of protoplanets can significantly perturb each other.

Once the protoplanets have perturbed one another into crossing orbits, their subsequent orbital evolution is governed by close gravitational encounters and violent, highly inelastic collisions. Wetherill (1980a,1985,1986,1988, 1990b) has described numerous simulations of this final stage of accretion, exploring a wide range of initial conditions. Surprisingly, the results of these simulations are not strongly affected by the earlier runaway growth stage. Regardless of whether the simulations start from a swarm of 500 bodies of mass 10^{25} g or just 30 protoplanets of mass several times 10^{26} g, the end result is the formation of 2 to 5 terrestrial planets on a time scale of about 100 Myr. An important feature of Wetherill's simulations is that planetesimal orbits execute a random walk in semimajor axis due to successive gravitational encounters. The resulting widespread mixing of material throughout the terrestrial planet region greatly diminishes any chemical gradients that may have existed during the early stages of planetesimal formation. Although, some correlations between the final heliocentric distance of a planet and the region where most of its constituents originated are preserved in the simulations (Wetherill 1988). Nevertheless, Mercury's high iron abundance is therefore less likely to arise from chemical fractionation in the solar nebula and more likely to be caused by a catastrophic giant impact during the final stages of accretion (see Sec. VI.B below).

Although radial mixing does occur, the total mass and orbital angular momentum of the swarm is nearly conserved during its accumulation into planets. The energy lost as heat in collisions, however, amounts to a few percent of the total orbital energy, resulting in a significant spreading of the disk. Thus, in order to end up with the same angular momentum distribution as is observed for Mercury, Venus, Earth and Mars, one must confine the initial swarm of protoplanets to a narrow annulus with most of the mass between 0.7 and 1.1 AU (Wetherill 1988). More plausible initial mass distributions which vary smoothly with heliocentric distance are incapable of reproducing the observed angular momentum distribution of the inner planets unless the planetesimal swarm can be supplied with extra "free" energy from an external source, e.g., Jupiter. By free energy, we mean energy in excess of that required for a circular orbit at a given semimajor axis:

$$E_f \equiv E - \Omega L \qquad (41)$$

where E is orbital energy and L is the magnitude of the orbital angular momentum. Thus, the free energy may be increased by the removal of energy and angular momentum, provided $\Delta E / \Delta L < \Omega$.

The problem of removing excess angular momentum from the terrestrial planets is intimately tied to the larger problem of explaining the extreme depletion of mass between the orbits of Mars and Jupiter. A large radial redistribution of mass is apparently required, but the candidate mechanisms proposed to accomplish this redistribution are poorly constrained at present. Whatever process pumped up the velocities of the asteroids must have yielded a vast quantity of small collision fragments that would have been susceptible to gas drag. These collision fragments would tend to spiral in toward the Sun, thereby enhancing the density of solids in the terrestrial planet zone. Large protoplanets may also experience significant orbital decay by exciting spiral density waves in the gaseous disk (Ward 1986,1988,1989b). Ward calculates that protoplanets more massive than 10^{27} g suffer significant orbital evolution inwards towards the protostar on a 1 to 10 Myr time scale via this process. Numerical simulations of planetesimal accumulation which include orbital migration due to density wave torques also show significant redistribution of mass and angular momentum in the terrestrial planet zone (Wetherill 1990*b*,1991*b*). More detailed studies are needed to determine if density wave torques can fully resolve the angular momentum problem in the inner solar system.

B. Giant Impacts

The mutual accumulation of numerous protoplanets into a small number of planets must have entailed many collisions between protoplanets of comparable size. As, in the post-runaway era, the random velocities keep pace with the escape velocity of the larger protoplanets, collisions between smaller bodies result in disruption instead of aggregation. The largest bodies are resistant to collisional disruption because their gravitational binding energy exceeds the kinetic energy of the collision. When the largest protoplanet reached a mass comparable to the Earth's mass, bodies near the size of Mercury became marginal cases, with collisions just as likely to strip off material as to add to the final mass of the planet. This result led Wetherill (1988) and Vityazev et al. (1988) to suggest that Mercury's silicate mantle was stripped off in a giant impact, leaving behind an iron-rich core (cf. Benz et al. 1988). Wetherill's simulations also lend support to the giant impact hypothesis for the origin of the Earth's Moon (e.g., Stevenson 1987); during the final stage of accumulation, an Earth-size planet is typically found to collide with several objects as large as the Moon and frequently one body as massive as Mars. The obliquities of the rotation axes of the planets provide independent evidence of the occurrence of giant impacts during the accretionary epoch (Safronov 1966, Lissauer and Safronov 1991).

C. Accretion Time Scales

Growth of planet-sized bodies from kilometer-sized planetesimals involves an increase in radius of 4 to 5 orders of magnitude. The planetary accretion rates given by Eqs. (19–26) all vary as the surface area of the planet R_s^2. Thus, for all other parameters constant, a planet's radius grows at a (statistically) uniform rate, and most of the growth time is spent in the last decade of radial expansion. The late phases of planetary growth are therefore crucial to determining the overall length of the accretionary epoch. Thus, a stage of runaway growth which ends in isolated protoplanets which must pump up velocities in order to collide and continue their agglomeration to planetary size, leads to accretion time scales similar to those models in which runaway never occurs.

Of course, protoplanet radius is not the only parameter in the equations which varies with time. Runaway accretion starts slowly (small F_g) and accelerates with time; thus it is possible that the "initial" size distribution of planetesimals significantly influenced the growth times of the planets. For example, a "protoplanet" initially of radius 10 km, at 5.2 AU from a 1 $M_\odot$ star, in a disk of surface mass density $\sigma = 15$ g cm^{-2} whose velocity dispersion is controlled by planetesimals 5 km in radius, would require $\sim 2 \times 10^5$ yr to grow large enough that $F_g = 3000$ (at which point $e_H = 2$ and protoplanet perturbations become an important factor in exciting noncircular motions of planetesimals; cf. GL90). This protoplanet would take a subsequent $\sim 6 \times 10^5$ yr to attain a mass of 15 $M_\oplus$ (Lissauer 1987). A protoplanet in an equivalently skewed initial distribution of larger planetesimals would take much longer to reach $F_g = 3000$, whereas the runaway would proceed much faster in a swarm of smaller planetesimals, because the radius doubling time for the protoplanet would be shorter.

VII. CONCLUSIONS

The planetesimal hypothesis provides a viable theory of the growth of the terrestrial planets, the cores of the giant planets and the smaller bodies present in the solar system. The formation of solid bodies of planetary size should be a common event, at least around stars which do not have binary companions orbiting at planetary distances. The formation of giant planets, which contain significant fractions of H_2 and He, requires rapid growth of planetary cores, so that gravitational trapping of gas can occur prior to the dispersal of the gas from the protoplanetary disk. According to the scenario which we have outlined, the largest bodies in any given zone are the most efficient accreters, in the sense that they double in mass the fastest. Such runaway accretion of a few large solid protoplanets can lead to core formation in ~ 1 Myr, provided disk masses are a few times as large as those given by minimum-mass models of the solar nebula. Thus, it appears possible that giant planets may also be common, although this conclusion must be regarded as much more tentative.

The ultimate sizes and spacings of solid planets are determined by their ability to gravitationally perturb each other into crossing orbits. Such perturbations often are due to weak resonant forcing, and occur on time scales much longer than the bulk of planetesimal interactions discussed herein. These interactions are not yet fully understood, although a great deal of progress has been made in recent years (Wisdom 1983; Chapter by Duncan and Quinn). Although quantitative formulas for scaling planetary sizes and spacings will require a better grasp of these processes, a few qualitative remarks can be made. First, a more massive protoplanetary disk will probably produce larger but fewer planets. Second, stochastic processes are important in planetary accretion, so nearly identical initial conditions could produce quite different outcomes, e.g., the fact that there are 4 terrestrial planets in our solar system as opposed to 3 or 5 or 6 is probably just the luck of the draw (cf. Wetherill 1988). Third, migration of some planetesimals over significant distances within the protoplanetary disk probably occurred, leading to a radial mixing of material which condensed in differing regions of the solar nebula. Fourth, although the spacing of giant planets is probably determined by similar processes to that of solid planets, their ultimate sizes may depend more on such factors as how fast they grew relative to the dispersal of the nebula and their potential ability to halt their own growth by gravitationally truncating the gaseous disk (Lin and Papaloizou 1979*a*).

Many aspects of planetary growth remain poorly understood. The asteroid belt currently contains much less than a planetary mass of material, and that material is spread over countless bodies which move at high velocities relative to each other. Models of the protoplanetary disk suggest this is unlikely to be due to an initial absence of condensed material. Perturbations by Jupiter and/or Jupiter-scattered protoplanets have been invoked for preventing planetary growth in the asteroid zone by increasing relative velocities of planetesimals, and for clearing material from that region; however, many problems remain with these scenarios (cf. Sec. IV.C). Planetesimal dynamics in the Uranus-Neptune region are complicated by the fact that bodies in this region are not tightly bound to the solar system: The difference between circular orbit velocity and escape speed from the solar system at 30 AU is less than the escape speed from the Moon. Using a planetary accumulation model analogous to those which Wetherill has successfully applied to the late stages of growth of the terrestrial planets (cf. Ipatov 1987), Ipatov (1989) finds that the mass of material ejected into hyperbolic orbits from the Uranus-Neptune region during the accretionary epoch exceeds the amount of solid matter incorporated into these planets by approximately an order of magnitude. Moreover, much of the material remaining in heliocentric orbit has spread to tens of AU beyond the original outer boundary of the disk. However, Ipatov's calculations start with planetesimals of identical size and neglect dynamical friction; these assumptions suppress runaway growth and thus lengthen the accretion time scales to unreasonably large values. Further studies of planetesimal accumulation in the outer regions of protoplanetary disks are needed to answer the many interesting questions raised by this study (Lissauer et al., in press).

Acknowledgments. We thank Y. Greenzweig, D. Kary, K. Ohtsuki, V. Safronov, A. Vityazev and G. Wetherill for their valuable comments. This work was supported in part by the NASA Planetary Geology and Geophysics Program under grants at the State University of New York at Stony Brook and at the University of Colorado. JJL is an Alfred P. Sloan Research Fellow.

PLANETARY ACCRETION IN THE SOLAR GRAVITATIONAL FIELD

KEIJI OHTSUKI, SHIGERU IDA, YOSHITSUGU NAKAGAWA
University of Tokyo

and

KIYOSHI NAKAZAWA
Tokyo Institute of Technology

We describe accumulation of planetesimals based on current studies on collisions and gravitational scattering of planetesimals in the solar gravitational field. First we examine the condition of runaway growth of protoplanets based on the collision rate between planetesimals and express it in terms of relative velocity between protoplanet and planetesimals. We find that random velocity (i.e., eccentricity and inclination of Kepler orbit) sensitively controls runaway growth of protoplanets. Next we describe the evolution of random velocity due to gravitational scattering and inelastic collisions, both in the solar gravitational field, and show that these processes drive the planetesimals toward equipartition of energy of random motion. Using these results, we estimate planetesimal velocities in each stage of accumulation, and examine the sequence of planet formation from planetesimals. Runaway growth of protoplanets is expected in an intermediate stage of accumulation, when the growth time scale is considerably shorter, while in the subsequent stage with high relative velocity, their growth will be decelerated. Finally, we discuss an appropriate treatment in numerical simulation of planetesimal accumulation.

Among the unsolved questions in the origin of the solar system, the most serious one is that no theory or numerical simulation has so far succeeded in explaining the formation of Uranus or Neptune within the age of the solar system without any *ad hoc* assumptions. The formation of Jupiter and Saturn before the dispersal of the solar nebula also remains to be explained. Although runaway growth of protoplanets has been recently discussed as a possible mechanism to considerably shorten the time scale of planetary formation (Wetherill and Stewart 1989), it has not yet been clarified how runaway growth proceeds in the late stage of accretion which determines the whole growth time of planets. In order to understand the whole planetary accretion process, various approaches have so far been taken to clarify individual fundamental processes. Safronov (1969) first studied them quantitatively by analytical methods. Subsequently, they have been studied in more detail mainly through computer simulation incorporating various effects such as gas drag of the solar nebula, gravitational interaction between planetesimals, or collisional disruption (see, e.g., Greenberg et al. 1978; Nakagawa et al. 1983; Ohtsuki et al. 1988; Wetherill and Stewart 1989; see Chapter by Lissauer and Stewart). However, as yet the effects of the solar gravitational field on the velocity change caused by gravitational scattering and collisions have not been taken into account in these studies. We cannot obtain a reliable time scale of planetary formation until we clarify all these fundamental processes in the solar gravitational field.

In recent years, the kinetic behavior of planetesimals (i.e., collisions and gravitational scattering) in the solar gravitational field has been extensively studied through orbital calculations in the three-body problem or direct N-body simulations. In this chapter, we review recent studies of fundamental processes and, using these results, we examine the sequence of planetary growth from planetesimals. First, we briefly describe the Hill's approximations, which provide us with useful tools to study the interaction between planetesimals (Sec. I). In Sec. II, we examine the condition of runaway growth of a protoplanet, using the collision rate in the solar gravitational field. We express the condition in terms of relative velocity between protoplanet and planetesimals. Subsequently, we discuss the velocity evolution caused by gravitational scattering (Sec. III) and by collisional accretion (Sec. IV), and find that both processes drive planetesimals toward energy equipartition for random motion. The fundamental processes determining the random velocity in each stage of planetesimal accumulation are summarized in Sec. V, where we find that runaway growth of protoplanets is expected in an intermediate stage of planetary growth. Finally, in Sec. VI, we discuss an appropriate numerical treatment of planetesimal accumulation.

I. HILL'S APPROXIMATIONS IN THE THREE-BODY PROBLEM

A planetesimal moves along an elliptic orbit around the Sun. When it occasionally comes very close to another planetesimal, the orbit is disturbed through gravitational scattering or collisions. Thus, we may focus on a small region when we describe such interactions between planetesimals. The coordinates of planetesimals in this region are referred to a reference point that moves on a circular orbit with semimajor axis a_{ref} at the Keplerian angular velocity $\Omega = (GM_\odot/a_{ref}^3)^{1/2}$. In this case, equations for the motion of the planetesimal relative to the point of reference may be linearized. These linearized equations are called Hill's equations (Hill 1878; Hénon and Petit 1986; Nakazawa and Ida 1988).

In general, if particles move under a potential force linear to their position vectors and the interaction term is a function of the relative distance between the particles alone, the motion of particles can be separated into the relative motion and the center of mass motion. Such separation is impossible in the general three-body problem, because the potential force $GM_\odot r_i/r^3$ in this case is nonlinear to the position vectors. In Hill's equations, on the other hand, the solar gravity is expressed in a linear form, and the motions are separable. This is a striking advantage of the use of Hill's equations.

Hénon and Petit (1986) and Nakazawa and Ida (1988) introduced orbital elements defined by

$$\begin{cases} \mathbf{e}_i \equiv (e_i \cos \tau_i, e_i \sin \tau_i) \\ \mathbf{i}_i \equiv (i_i \cos \omega_i, i_i \sin \omega_i) \end{cases} \tag{1}$$

where τ_i and ω_i are the longitudes of the perihelion and the ascending nodes. We can confirm that the solutions to Hill's equations are expressed by linear combinations of the components of $\mathbf{e}_i$ and $\mathbf{i}_i$ (Nakazawa and Ida 1988). This means that the orbital elements $\mathbf{e}_i$ and $\mathbf{i}_i$ are also separated into those for the relative motion and for the center of mass motion: the orbital elements for the center of mass motion are given by

$$\begin{cases} \mathbf{e}_R \equiv (e_R \cos \tau_R, e_R \sin \tau_R) = m_1' \mathbf{e}_1 + m_2' \mathbf{e}_2 \\ \mathbf{i}_R \equiv (i_R \cos \omega_R, i_R \sin \omega_R) = m_1' \mathbf{i}_1 + m_2' \mathbf{i}_2 \end{cases} \tag{2}$$

with

$$m_i' = \frac{m_i}{m_1 + m_2} \tag{3}$$

and those for the relative motion are

$$\begin{cases} \mathbf{e}_{21} \equiv (e_{21} \cos \tau, e_{21} \sin \tau) = \mathbf{e}_2 - \mathbf{e}_1 \\ \mathbf{i}_{21} \equiv (i_{21} \cos \omega, i_{21} \sin \omega) = \mathbf{i}_2 - \mathbf{i}_1 \end{cases} . \tag{4}$$

Because the center of mass motion is unaffected by gravitational scattering or collisions, the orbital elements $\mathbf{e}_R$ and $\mathbf{i}_R$ of the center of mass motions before and after encounter are the same, while those of the relative motion are affected by these interactions.

The above characteristics of Hill's approximations enable us to study the behavior of planetesimals directly in terms of the orbital elements, without any ambiguity which might be caused when we define "random velocity of planetesimals" in several different ways (Chapter by Lissauer and Stewart). Kinetic behavior of planetesimals discussed in this chapter is essentially described in terms of their eccentricity and inclination, which are more fundamental quantities than "random velocity." However, we need to define "relative random velocity v_{21}" when we compare some quantities (e.g., collision rate) of planetesimals in the solar gravitational field with those in free space. In this case, we use a simple form defined by

$$v_{21} = (e_{21}^2 + i_{21}^2)^{1/2} v_K \tag{5}$$

with $v_K (= a_{\mathrm{ref}} \Omega)$ being circular Keplerian velocity.

Another remarkable advantage in Hill's approximations is that we can rewrite the equations of relative motion into a nondimensional form in which the planetesimal masses m_1 and m_2 or heliocentric distance a_{ref} never appear explicitly. We scale time by Ω^{-1} and length by ha_{ref}, where

$$h = \left(\frac{m_1 + m_2}{3M_\odot}\right)^{1/3}. \tag{6}$$

The size of radial excursion $e_{21}a_{ref}$ and the amplitude of vertical oscillation $i_{21}a_{ref}$ for the relative motion are then scaled respectively as

$$\begin{cases} \tilde{e}_{21} = e_{21}/h \\ \tilde{i}_{21} = i_{21}/h \end{cases} \tag{7}$$

($\tilde{e}_{21}$ and $\tilde{i}_{21}$ correspond to "Hill eccentricity e_H " and "Hill inclination i_H" defined by Eq. (28) of the Chapter by Lissauer and Stewart). Thus once we obtain the solutions to the nondimensional equations, we may apply them to planetesimals with arbitrary masses and arbitrary a_{ref} by use of the relations (6) and (7).

II. COLLISION RATE AND RUNAWAY GROWTH

A. Condition of Runaway Growth

Here we examine the condition of runaway growth of a protoplanet. We consider a simple situation wherein test protoplanets are embedded in a swarm of small field planetesimals with mass m_2 and the protoplanets grow through the collision with these planetesimals. The growth rate of the test protoplanet with mass m_1 ($>m_2$) is

$$\frac{dm_1}{dt} = m_2 n_s <P> \tag{8}$$

where n_s is the number of field planetesimals in a unit surface of $(ha_{ref})^2$ [$h \simeq (m_1/3M_\odot)^{1/3}$; see Eq. (6)], and $<P>$ is the collision rate per unit surface number density averaged over the velocity distribution of the planetesimals.

Let the right-hand side of Eq. (8) be proportional to m_1^α. The growth time scale of the protoplanet defined by

$$T_{grow} = \left(\frac{1}{m_1}\frac{dm_1}{dt}\right)^{-1} \tag{9}$$

is then proportional to $m_1^{1-\alpha}$. It means that for $\alpha > 1$ larger protoplanets grow more rapidly than smaller protoplanets (or planetesimals) and runaway growth occurs, while for $\alpha < 1$ larger ones grow more slowly and the growth should be orderly (Wetherill and Cox 1985). From its definition, α is given by

$$\alpha = \frac{d\ln(n_s <P>)}{d\ln m_1} \tag{10}$$

$$= \frac{2}{3} + \frac{d\ln\tilde{v}_{21}}{d\ln m_1}\frac{d\ln <P>}{d\ln\tilde{v}_{21}}$$

where we used the relation $n_s \propto (h a_{ref})^2 \propto m_1^{2/3}$, and $\tilde{v}_{21}$ is the "relative random velocity" in a nondimensional form, defined by $\tilde{v}_{21} = v_{21}/hv_K = (\tilde{e}_{21}^2 + \tilde{i}_{21}^2)^{1/2}$ [see Eqs. (5) and (7)]. Equation (10) shows whether runaway growth occurs or not depends on the $\tilde{v}_{21}$-dependence of $<P>$ and the m_1-dependence of $\tilde{v}_{21}$ (Ohtsuki and Ida 1990). We examine the above dependence in the followings, and describe the condition of runaway growth in terms of m_1-dependence of $\tilde{v}_{21}$.

B. $\tilde{v}_{21}$-Dependence of P

First, we examine $\tilde{v}_{21}$-dependence of the collision rate. The collision rate in the solar gravitational field has been numerically obtained by calculating the orbits in the three-body problem with various initial conditions and counting the collision orbits (Nishida 1983; Wetherill and Cox 1985; Ida and Nakazawa 1989; Greenzweig and Lissauer 1990). These results are described in detail in the Chapter by Lissauer and Stewart.

In Fig. 1, we plot collision rate P (without velocity distribution of planetesimals) as a function of $\tilde{v}_{21}$. We assume $\tilde{e}_{21}:\tilde{i}_{21} = 2:1$ because this ratio is expected for relatively large random velocities as a result of gravitational scatterings (Ida 1990; Ida and Makino 1992a). In this figure, P_{2B} is the collision rate with the two-body approximation and P is numerically obtained by orbital calculation; P for $\tilde{v}_{21} < 20$ is drawn using the numerical results for 12 discrete sets of $(\tilde{e}_{21}, \tilde{i}_{21})$ with $\tilde{e}_{21} = 2\tilde{i}_{21}$ (data points obtained by numerical calculation are shown by triangles), while that for $\tilde{v}_{21} > 20$ is drawn by extrapolation of P_{2B} (for detailed definitions of P and P_{2B}, see Chapter by Lissauer and Stewart). As first noted by Wetherill and Cox (1985), a discrepancy between P and P_{2B} exists in the low-velocity region, where P takes constant value for $\tilde{v}_{21} < 0.1$. In such a low-velocity case, the approach velocity is almost determined by the Keplerian shear velocity rather than random velocity due to $\tilde{e}_{21}$ and $\tilde{i}_{21}$ (see, e.g., Binney and Tremaine 1987; Greenberg et al. 1988; Ida 1990), thus P is independent of $\tilde{e}_{21}$ or $\tilde{i}_{21}$. In the two-body approximation, on the other hand, such systematic relative motion as the Keplerian shear is not taken into account, which gives rise to the above discrepancy. Collision rate in this shear-dominated regime is discussed in detail by Greenberg et al. (1991). On the other hand, P is well approximated by P_{2B} for $\tilde{v}_{21} > 5$. This is because the random motion is large enough to dominate the relative velocity in such cases, and the two-body approximation in gravitational encounters is valid. We will call this velocity regime the dispersion-dominated regime.

We also need take into account the velocity distribution of planetesimals in evaluating collision rate. Ida and Makino (1992a) confirmed by N-body simulation (three-dimensional, 400 nonaccreting equal-mass planetesimals) that the distributions of $\tilde{e}_2$ and $\tilde{i}_2$ become Gaussian through mutual gravitational scatterings. The collision rates $<P>$ averaged with the Gaussian velocity distribution is presented by Ida and Nakazawa (1988), Ohtsuki and Ida (1990), and Greenzweig and Lissauer (1992). They find that the velocity dependence of the averaged collision rates is almost the same as P, and $<P>$

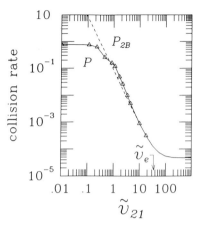

Figure 1. Collision rate P (in the solar gravitational field) and P_{2B} (in free space) per unit surface number density as a function of $\tilde{v}_{21}$ in the case of $\tilde{e}_{21}:\tilde{i}_{21} = 2:1$. The value of the escape velocity ($\tilde{v} \simeq 35$ at 1 AU) is also indicated. The data points obtained by our calculations are shown by triangles and these values are consistent with Greenzweig and Lissauer (1990). The original values obtained by Ida and Nakazawa (1989), which were also used in Ohtsuki and Ida (1990), are slightly smaller than these values for $\tilde{v}_{21}>2$, as pointed out by Greenberg et al. (1991).

is enhanced by a few times compared to P; the velocity dependence of $<P>$ is given by

$$\frac{d\ln <P>}{d\ln\tilde{v}_{21}} \simeq \begin{cases} 0 & \text{for } \tilde{v}_{21} < 0.1 \\ -1 & \text{for } 0.1 < \tilde{v}_{21} < 1 \\ -2 & \text{for } 1 < \tilde{v}_{21} < 10 \\ -1.5 & \text{for } \tilde{v}_{21} \simeq 20 \\ -1 & \text{for } \tilde{v}_{21} \simeq 40 \end{cases} \tag{11}$$

and, for $\tilde{v}_{21}>40$, d ln$<P>$/d ln $\tilde{v}_{21}$ gradually increases to zero with increasing $\tilde{v}_{21}$. Hence, the condition of runaway growth, $\alpha>1$, is equivalent to the condition

$$\frac{d\ln\tilde{v}_{21}}{d\ln m_1} < \begin{cases} -\infty & \text{for } \tilde{v}_{21} < 0.1 \\ -1/3 & \text{for } 0.1 < \tilde{v}_{21} < 1 \\ -1/6 & \text{for } 1 < \tilde{v}_{21} < 10 \\ -2/9 & \text{for } \tilde{v}_{21} \simeq 20 \\ -1/3 & \text{for } \tilde{v}_{21} \simeq 40. \end{cases} \tag{12}$$

C. m_1-Dependence of $\tilde{v}_{21}$

Next, we reduce the condition of runaway growth to the condition of mass-dependence of random velocity. We introduce the following assumptions: (i) each field planetesimal has the equal mass m_2; (ii) the mass m_2 and velocity $v_2 = [(e_2^2 + i_2^2)^{1/2}v_K]$ of field planetesimals are constant; (iii) the random velocity is proportional to m^q. In this case, the velocity of a test protoplanet, $v_1 = (e_1^2 + i_1^2)^{1/2}v_K$, is given by

$$v_1(m_1) = \left(\frac{m_1}{m_2}\right)^q v_2. \tag{13}$$

Because $h \propto m_1^{1/3}$ for $m_1 >> m_2$, the scaled velocity of the protoplanet, $\tilde{v}_1$ $(= v_1/hv_K)$, and that of the field planetesimal, $\tilde{v}_2$, have the relations $\tilde{v}_1 \propto m_1^{q-1/3}$ and $\tilde{v}_2 \propto m_1^{-1/3}$, respectively. From Eqs. (4) and (5), we then have

$$\tilde{v}_{21}^2 \simeq \tilde{v}_1^2 + \tilde{v}_2^2 \simeq \begin{cases} \tilde{v}_1^2 & \text{for } q > 0 \ (\tilde{v}_1^2 >> \tilde{v}_2^2) \\ \tilde{v}_2^2 & \text{for } q < 0 \ (\tilde{v}_1^2 << \tilde{v}_2^2). \end{cases} \tag{14}$$

Using Eqs. (13) and (14), we obtain

$$\frac{d \ln\tilde{v}_{21}}{d \ln m_1} = \begin{cases} q - 1/3 & \text{for } q > 0 \\ -1/3 & \text{for } q < 0. \end{cases} \tag{15}$$

We find that $d \ln\tilde{v}_{21}/d \ln m_1$ cannot be smaller than $-1/3$. From Eqs. (12) and (15), we find that runaway growth is possible only for $1 < \tilde{v}_{21} < 40$. In an intermediate stage of planetary accretion, $\tilde{v}_{21}$ takes the values nearly equal to or smaller than 20 (Sec. V). Thus, for $\tilde{v}_{21} < 20$, we can write the condition of runaway growth as

$$\frac{d \ln\tilde{v}_{21}}{d \ln m_1} < -\frac{2}{9} \tag{16}$$

which is equivalent to the condition

$$q < \frac{1}{9}. \tag{17}$$

Assuming that the steady state in random velocity is achieved through the balance between the enhancement due to gravitational scattering and damping due to inelastic collisions, Safronov (1969) obtained $q = 1/3$. Nakagawa et al. (1983) considered that the damping is due to gas drag of the solar nebula and obtained $q = 1/9$. These authors did not find runaway growth. On the other hand, Stewart and Wetherill (1988) and Wetherill and Stewart (1989) considered the effects of dynamical friction caused by gravitational scatter-ing, which lead to energy equipartition between planetesimals with different masses. They show that the random velocity of large planetesimals is effec-tively suppressed to lead to negative q (if energy equipartition is completely realized, we have $q = -1/2$) and runaway growth of a small number of protoplanets was found in their simulation. Greenberg et al. (1978) also obtained negative q and found runaway growth, although the accuracy of their numerical code has been questioned in several respects (Wetherill and Stewart 1989; Ohtsuki et al. 1990; Spaute et al. 1991; see also Kolvoord and Greenberg [1992] for refutations). Thus, whether runaway growth occurs or not is determined by q, the mass dependence of the random velocity. The random velocity is governed by mutual gravitational scattering, inelastic collisions, and gas drag. In all the above studies, gravitational encounters are treated by the two-body approximation. Wetherill and Cox (1984) demonstrated that the two-body approximation for gravitational scattering is valid for $\tilde{v}_{21} > 2$, but it is not valid for $\tilde{v}_{21} < 2$. We need to determine the velocity of planetesimals including the effects of solar gravitational field for such low velocity cases. In the next two sections, we review the recent studies of gravitational scattering and inelastic collisions in the solar gravitational field.

III. GRAVITATIONAL SCATTERING

Ida (1990) considered a bimodal population of planetesimals with arbitrary mass ratio and examined the change in the mean random velocity of the Population 1 with mass m_1 caused by gravitational encounters with the Population 2 with mass m_2. He derived the equations for the change rates of eccentricity as

$$\frac{d < \tilde{e}_1^2 >}{dt} \simeq m_2' < \frac{m_2' \tilde{e}_{21}^2}{T_{VSe}} + \frac{m_2' \tilde{e}_2^2 - m_1' \tilde{e}_1^2}{T_{DFe}} > \qquad (18)$$

with

$$\begin{cases} T_{VSe}(\tilde{e}_{21}, \tilde{i}_{21}) = \tilde{e}_{21}^2 / n_s(m_2)[\Delta \tilde{e}_{21}^2] \\ \\ T_{DFe}(\tilde{e}_{21}, \tilde{i}_{21}) = -\tilde{e}_{21} / 2 n_s(m_2)[\Delta \tilde{e}_{21\parallel}] \end{cases} \qquad (19)$$

where $\Delta \tilde{e}_{21}^2$ and $\Delta \tilde{e}_{21}$ are the changes in $\tilde{e}_{21}^2$ and $\tilde{e}_{21}$ during an encounter, and $\Delta \tilde{e}_{21\parallel}$ is the component of $\Delta \tilde{e}_{21}$ parallel to $\tilde{e}_{21}$. " $<>$ " in Eq. (18) denotes average over the Gaussian velocity distributions of Populations 1 and 2, and "[ ]" in Eq. (19) denotes ensemble average taking into account the synodic period of interacting planetesimals (for detail, see Ida 1990). The first term of the right-hand side of Eq. (18) represents the transformation of the solar gravitational potential energy to energy of random motion, and is called "viscous stirring" term. The second term represents the energy exchange toward energy equipartition (i.e., $m_1 \tilde{e}_1^2 = m_2 \tilde{e}_2^2$) between the planetesimals with different masses and is called "dynamical friction" term. T_{VSe} and T_{DFe} are the time scales of viscous stirring and dynamical friction. Similar equations for inclination are also obtained, and the time scales of viscous stirring and dynamical friction for inclination are denoted by T_{VSi} and T_{DFi}, respectively.

Although the orbital changes in general cases cannot be obtained analytically, those for distant encounters, where the minimum distance during
encounter is so large that the changes are quite small, can be obtained by a
perturbation method. In such cases, Goldreich and Tremaine (1982) derived
$\Delta \tilde{e}_{21}$ for coplanar and initially circular ($\tilde{e}_{21}, \tilde{i}_{21} = 0$) orbits, and Hasegawa and
Nakazawa (1990) derived $\Delta \tilde{e}_{21}$ and $\Delta \tilde{i}_{21}$ for general cases with $\tilde{e}_{21}, \tilde{i}_{21} \neq 0$.
Weidenschilling (1989) extended the result of Goldreich and Tremaine to gen-
eral cases in an approximate manner and calculated the viscous stirring rate
due to distant encounters. He found that the stirring rate does not include the
logarithmically divergent term which appears in the two-body relaxation rate
in free space [cf. Eq. (22) below].

Ida (1990), on the other hand, evaluated the above time scales through
orbital calculations. He evaluated the contribution from close encounters by
numerical integration of about 10^8 orbits and that from distant encounters by
using the results of Hasegawa and Nakazawa (1990). He found that distant
encounters in the solar gravitational field hardly contribute to the viscous
stirring and dynamical friction. He also found that T_{DFe} and T_{DFi} take
nearly equal positive values, so we simply denote T_{DFe} and T_{DFi} by T_{DF};
these positive values ensure that mutual gravitational scattering works in the
direction of energy equipartition in both planar and vertical motions. We
define the time scales of total viscous stirring T_{VS} by

$$\frac{1}{T_{VS}} = \frac{1}{T_{VSe}} + \frac{1}{T_{VSi}}. \tag{20}$$

In the dispersion-dominated regime, $T_{VSi} \simeq T_{VSe} (\simeq T_{VS})$. On the other hand,
in the shear-dominated regime where relative motion is almost horizontal,
viscous stirring in the horizontal direction is much more effective than that in
the vertical direction, and we have $T_{VSi} >> T_{VSe} (\simeq T_{VS})$.

We plot T_{VS} and T_{DF} in Fig. 2. Using T_{VS} and T_{DF}, the evolution of
$<\tilde{v}_1^2>$ is written as [cf. Eq. (18) above]

$$\frac{d < \tilde{v}_1^2 >}{dt} \simeq m_2' < \frac{m_2' \tilde{v}_{21}^2}{T_{VS}} + \frac{m_2' \tilde{v}_2^2 - m_1' \tilde{v}_1'^2}{T_{DF}} > . \tag{21}$$

In most previous studies of planetesimal accumulation, dynamical friction
was neglected and the time scale of viscous stirring was simply assumed
to be equal to the two-body relaxation time T_{2B} in free space obtained by
Chandrasekhar (1942); T_{2B} is given by

$$T_{2B} \simeq \frac{1}{\pi \tilde{r}_G^2 n \tilde{v}_{21} \ln \Lambda} \simeq \frac{\tilde{v}_{21}^4}{10^2 n_s \ln \Lambda} \tag{22}$$

where $n = n_s / 2\tilde{i}_{21}$ is the spatial number density, $\ln \Lambda$ comes from the contribution of distant encounters ($\Lambda = v_{21}^2 / G(m_1 + m_2) n^{1/3}$ and $\ln \Lambda \sim 10$), and
$\tilde{r}_G$ is the scaled gravitational radius given by

$$\tilde{r}_G = \frac{G(m_1 + m_2)}{v_{21}^2 h a_{\text{ref}}} = \frac{3}{\tilde{v}_{21}^2}. \tag{23}$$

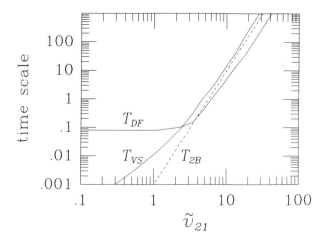

Figure 2.   Normalized time scales in units of Ω_K^{-1}, as a function of $\tilde{v}_{21}$ ($n_s = 1$). In the dispersion-dominated regime ($\tilde{v}_{21} > 5$), T_{VS} and T_{DF} are well approximated by ordinary two-body relaxation time T_{2B}. On the other hand, they deviate from T_{2B} in the shear-dominated regime ($\tilde{v}_{21} < 2$) where the effects of the solar gravity are important.

We also plot T_{2B} in Fig. 2. Stewart and Wetherill (1988) pointed out the importance of the dynamical friction and calculated its rate as well as viscous stirring rate by solving the Boltzmann equation following the formalism of Hornung et al. (1985) (Chapter by Lissauer and Stewart). In their study, they treated gravitational encounters with the two-body approximation neglecting the solar gravitational field and, hence, the time scales of the viscous stirring and the dynamical friction are given by T_{2B}. It can been seen in Fig. 2 that T_{VS} and T_{DF} obtained by the three-body calculation are in good agreement with T_{2B} in the dipersion-dominated regime, which means that previous treatment with the two-body approximation is valid in these cases. However, shear-dominated encounters become important in the runaway accretion stage (Barge and Pellat 1991; see also Sec. V). In this case, T_{VS} and T_{DF} are much greater than T_{2B} and, hence, the two-body approximation breaks down.

With the time scales T_{VS} and T_{DF} given in Fig. 2, we examine the relative importance of dynamical friction to viscous stirring. We assume $m_1 >> m_2$ ($m_1' \sim 1$, $m_2' << 1$). First, we consider the random velocity change in the dispersion-dominated regime. If $\tilde{v}_1^2 \geq \tilde{v}_2^2$ (i.e., $\tilde{v}_{21}^2 \simeq \tilde{v}_1^2$), Eq. (21) reduces to

$$\frac{d < \tilde{v}_1^2 >}{dt} \simeq m_2' \tilde{v}_1^2 < \frac{m_2'}{T_{VS}} - \frac{1}{T_{DF}} > \simeq -m_2' \tilde{v}_1^2 < \frac{1}{T_{DF}} > . \qquad (24)$$

Hence, dynamical friction is dominant and $\tilde{v}_1$ will decrease rapidly. Ida and Makino (1992b) examine the effects of dynamical friction on a protoplanet embedded in a swarm of small planetesimals by N-body simulation. They confirm that eccentricity and inclination of the protoplanet decay in a time scale of T_{DF}, as shown in Eq. (24). Thus, we may consider that $\tilde{v}_1$ is reduced so that $m_2\tilde{v}_2^2$ and $m_1\tilde{v}_1^2$ are of the same order. In this case, $\tilde{v}_{21}^2 \simeq \tilde{v}_2^2$, and Eq. (21) reduces to

$$\frac{d < \tilde{v}_1^2 >}{dt} \sim m_2'^2 \tilde{v}_2^2 < \frac{1}{T_{VS}} - \frac{1}{T_{DF}} > \tag{25}$$

where we find that both viscous stirring and dynamical friction contribute to velocity evolution. We conclude that the high random velocity of large protoplanets is effectively suppressed through dynamical friction. It gives negative q in Eq. (13), which should lead to runaway growth and, consequently, the growth time of planets is substantially shortened (see Sec. V). However, if the runaway continues, $\tilde{v}_{21}$ should be reduced to the low velocity regime where $T_{VS} << T_{DF}$, and viscous stirring becomes more effective than dynamical friction. The protoplanet would then have a larger velocity than is expected from energy equipartition (Ida and Makino 1992b; Barge and Pellat 1991).

IV. INELASTIC COLLISIONS AND ACCRETION

Colliding planetesimals accrete or rebound from one another depending on the relative impact velocity and restitution coefficient. The magnitude of relative impact velocity v_{imp} is approximately given by $v_{imp}^2 \simeq v_e^2 + v_{21}^2$, where v_e is the escape velocity and v_{21} is relative random velocity. If the collision is sufficiently dissipative that the relative velocity just after the impact is smaller than the escape velocity, then the colliding planetesimal should ultimately accrete. In some previous numerical simulations of planetesimal accumulation, the velocity damping caused by collisions was taken into account according to the data of laboratory impact experiments (see, e.g., Hartmann 1978; Greenberg et al. 1978; Wetherill and Stewart 1989). However, these authors evaluated the random velocity with the two-body approximation in free space.

Recently, Ohtsuki (1992a) examined the probability of gravitational sticking for inelastic collisions between planetesimals in heliocentric orbits, using three-body trajectory integrations. He found that most collisions lead to accretion if $v_{21} \lesssim v_e$ and the restitution coefficient is less than 0.7. It can be shown that the random velocity of planetesimals does not considerably exceed the escape velocity, because gas drag of the solar nebula is effective (Sec. V). Thus, as long as the restitution coefficient is small enough, it can be considered that most collisions between planetesimals may lead to accretion. (In the final stage of accumulation where random velocity of small planetesimals is enhanced by gravitational scattering by protoplanets, collisions between small planetesimals may result in fragmentation.) From this point of view, Ohtsuki (1992b) derived the equation for the random velocity evolution due to accretion by coupling the evolutions of both the random velocity and the mass distribution. The evolution of mean eccentricity of Population 1 with mass m_1 is written as

$$n_s(m_1)\frac{\mathrm{d}}{\mathrm{d}t} < \tilde{e}_1^2 = -n_s(m_1)\int_0^\infty n_s(m) < P(m_1, m) >$$

$$\left\{ <\tilde{e}_1^2>_{\mathrm{col}} - <\tilde{e}_1^2> \right\}\mathrm{d}m + \frac{1}{2}\int_0^{m_1} n_s(m_2)n_s(m_3) \quad (26)$$

$$< P(m_2, m_3) > [< m_2'^2\tilde{e}_2^2 + m_3'^2\tilde{e}_3^2 >_{\mathrm{col}} -$$

$$< \tilde{e}_1^2 >]\mathrm{d}m_2$$

with

$$m_3 = m_1 - m_2 \text{ and } m_i' = m_i/m_1 \quad (27)$$

where $P(m_i, m_j)$ is the collision rate between m_i and m_j, $<>$ denotes the average over the velocity distributions, and $<>_{\mathrm{col}}$ denotes the average over the velocity distributions of colliding pairs weighted with their collision rate. In the above, eccentricities are scaled by $h = (m_2 + m_3/3\mathrm{M}_\odot)^{1/3} = (m_1/3\mathrm{M}_\odot)^{1/3}$. The first term on the right-hand side of Eq. (26) represents the collision between planetesimals with mass m_1 and those with other masses, and the second term expresses the merger of m_2 and m_3 to form a planetesimal with mass m_1. We obtain a similar equation for the evolution of inclination.

The first term of Eq. (26) has the following consequences: if the average energy $<\tilde{e}_1^2>_{\mathrm{col}}$ leaving from the original assemblage 1 as a result of a collision with other planetesimals is larger than the average energy $<\tilde{e}_1^2>$ of the original assemblage, this term works so as to decrease $<\tilde{e}_1^2>$, and vice versa. The contribution of the first term depends on the form of the velocity distribution function and the collision rate. Because the collision rate of planetesimals with lower velocity is larger (Fig. 1), planetesimals with lower velocity leave from the assemblage more frequently than those with high velocity. We then expect that this term should work so as to increase the average random velocity as long as we consider the velocity change caused by accretion alone.

The terms in the braces of the second term on the right-hand side of Eq. (26) are written as

$$< m_2'^2\tilde{e}_2^2 + m_3'^2\tilde{e}_3^2 >_{\mathrm{col}} - <\tilde{e}_1^2> = (<\overline{E}_c>_{\mathrm{col}} - <E_1>)/m_1 \quad (28)$$

where $E_i = m_i\tilde{e}_i^2$, and $\overline{E}_c = m_2'E_2 + m_3'E_3$ is the weighted mean of random energy of colliding planetesimals. The right-hand side of Eq. (28) may be interpreted as follows; if $<\overline{E}_c>_{\mathrm{col}}$ is larger than $<E_1>$, a newly formed planetesimal with mass m_1 contributes to the increase in the average random energy of Population 1. If $<\overline{E}_c>_{\mathrm{col}}$ is smaller than $<E_1>$, on the other hand, it contributes to the decrease in the average random energy. As a result, accretion proceeds so that a planetesimal should keep the same random energy on average, although its mass continues to increase due to accretion. We then expect that accretion should drive the system toward energy equipartition between planetesimals with different masses. If the energy equipartition $E_i = E_{eq}$ (= const.) is already achieved through the above accretion or gravitational scattering described in Sec. III, it is not violated by the subsequent accretion, as in Eq. (28) we have

$$m_2'^2 \tilde{e}_2^2 + m_3'^2 \tilde{e}_3^2 = (m_2' + m_3') E_{eq}/m_1 = E_{eq}/m_1 \quad (29)$$

which shows that the weighted mean of random energy of colliding planetesimals is the same as that before collision.

V. SUMMARY

We have described the velocity evolution caused by gravitational scattering and accretion in the last two sections. Another important process, which we have not discussed in this chapter and still remains to be solved, is the gas drag force on a planetesimal moving in the solar nebula. The gas drag force on a Keplerian particle with small size ($m < 10^{24}$g) was studied by Adachi et al. (1976) and Weidenschilling (1977a), and that on a massive planetesimal ($m \geq 10^{25}$g) with its gravitational effect in uniform flow with relatively low Reynolds number ($\simeq 20$) was numerically studied by Takeda et al. (1985). Applying the drag law of Takeda et al. to planetesimals, Ohtsuki et al. (1988) simulated accumulation of planetesimals to the full-size Earth. They found that the drag force on a protoplanet at the final stage becomes so large that its orbit should greatly decay toward the Sun during accumulation. Ohtsuki et al. then concluded that the drag force obtained by Takeda et al. in uniform flow with low Reynolds number must be overestimated for planetesimals in Keplerian motion which move in shear flow of the solar nebula.

In Table I we summarize fundamental processes which determine the random velocity in each stage of planetesimal accumulation.

A. Early Stage

First, we consider an early stage where the typical mass of planetesimal m is about $10^{18} \sim 10^{20}$g. We define collision time T_{coll} by

$$T_{\text{coll}} = \frac{1}{n_s P}. \quad (30)$$

Using the gas drag law obtained by Adachi et al. (1976), we define the time scale of velocity damping due to gas drag by

$$T_{\text{gas}} = \frac{m v_m}{\frac{1}{2} C_D \pi r_m^2 \rho_{\text{gas}} v_m^2} \sim 5 \times 10^6 \left(\frac{a}{1 \text{ AU}}\right)^{13/4} \tilde{v}_m^{-1} \text{yr} \quad (31)$$

where m, r_m, and v_m are the mass, radius, and random velocity of a planetesimal, C_D is drag coefficient (we put $C_D = 1$), and ρ_{gas} is the mass density of the

TABLE I

Fundamental Processes Determining the Random Velocities of Planetesimals in Each Accretion Stage[a]

Accretion Stage	Typical Mass at 1 AU	Processes Determining Random Velocity
Early stage	$m = 10^{18-20}$ g	VS Inelastic collisions and accretion Gas drag
Intermediate stage	$M \leq 10^{25}$ g	$v_m \begin{cases} \text{VS by small planetesimals} \\ \text{Gas drag} \end{cases}$
Runaway era: Protoplanet formation	$(M << M_t)$	v_M: DF by small planetesimals $(Mv_M^2 \sim mv_m^2)$
Final stage	$M \geq 10^{25}$ g	$v_m \begin{cases} \text{Gravitational scattering by protoplanets} \\ \text{Gas drag, if nebular gas exists} \end{cases}$ $v_M \begin{cases} \text{DF by small planetesimals or VS by protoplanets} \\ \text{Gas drag, if nebular gas exists} \end{cases}$

[a] M and m are the typical masses of a protoplanet and a small planetesimal, and v_M and v_m are their typical random velocities. Total mass of planetesimals in a feeding zone of the protoplanet is denoted by M_t. VS and DF represent viscous stirring and dynamical friction due to gravitational scattering, respectively.

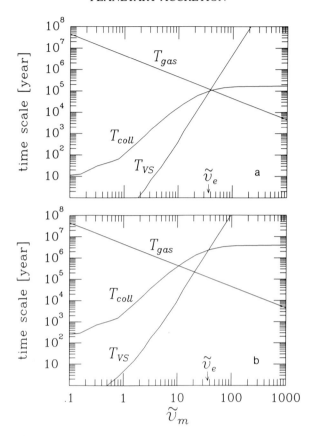

Figure 3. Time scales T_{VS}, T_{gas}, and T_{coll} (in units of year) at 1 AU with $\Sigma = 7$ g cm^{-2}: (a) in an early stage for $m = 10^{20}$g, and (b) in an intermediate stage for $m = 10^{24}$g.

nebular gas [we adopted $\rho_{gas} = 1.4 \times 10^{-9}$ $(a/1\text{AU})^{-11/4}$g cm^{-3}, following Hayashi (1981)]. Scaled velocity $\tilde{v}_m$ is given by

$$\tilde{v}_m = v_m/h_m v_K \text{ with } h_m = \left(\frac{2m}{3\text{M}_\odot}\right)^{1/3}. \tag{32}$$

Because the difference in sizes of planetesimals is not appreciable in this stage, the relative velocity $\tilde{v}_{mm}$ for planetesimal-planetesimal encounter is approximately given by $\tilde{v}_{mm} \simeq \sqrt{2}\tilde{v}_m$. In Fig. 3a, T_{coll} and T_{gas}, together with T_{VS} defined by Eq. (20) in this stage ($m = 10^{20}$g) at 1 AU are plotted as a function of $\tilde{v}_m$. In evaluating T_{coll} and T_{VS}, we adopted $\Sigma = 7$g cm^{-2} for the surface mass density of planetesimals, following Hayashi (1981). If we consider that the random velocity of the planetesimal is determined by the balance between the enhancement by gravitational scattering and damping by gas drag, the crossing point of the two lines for T_{VS} and T_{gas} gives the value of the equilibrium velocity. It can be seen in Fig. 3a that an equilibrium state is expected at $\tilde{v}_m \simeq 40$. We also find that in the neighborhood of this equilibrium velocity, T_{coll} is also comparable to T_{gas} and T_{VS}, thus inelastic collision and accretion discussed in Sec. IV are also important in this stage. That is, in addition to gravitational scattering, accretion may make a large contribution to the velocity evolution toward energy equipartition as discussed in Secs. III and IV.

We find in Fig. 3a that the equilibrium velocity is slightly larger than the escape velocity $\tilde{v}_e$ ($\tilde{v}_e \simeq 35$ at 1 AU), which lies quite close to the upper boundary of the runaway growth condition given in Sec. II. Thus, it is not clear whether runaway growth occurs in this stage or not.

B. Intermediate Stage

Next, we consider an intermediate stage, where the typical mass of planetesimals is 10^{22-24}g. The time scales T_{coll}, T_{gas}, and T_{VS} for $m = 10^{24}$g at $a = 1$ AU are plotted in Fig. 3b as a function of $\tilde{v}_m$. From the value of $\tilde{v}_m$ at the crossing point of T_{VS} and T_{gas}, we expect that an equilibrium state is achieved at

$$\tilde{v}_m \simeq 20 \left(\frac{m}{10^{24}\text{g}} \right)^{-1/15} \tag{33}$$

where we again used Eq. (31) and adopted $\Sigma = 7$g cm^{-2}, and also used a relation $T_{VS} \propto \tilde{v}_m^4$, which is valid for $\tilde{v}_m > 5$ (Fig. 2). This equilibrium velocity is slightly smaller than the escape velocity. In this stage, T_{coll} is much larger than T_{gas} or T_{VS} at the above equilibrium velocity. It shows that an equilibrium state between viscous stirring by gravitational scattering and damping by gas drag is achieved and inelastic collision is not important in determining the velocity. In such a high-velocity regime as is given by Eq. (33), $T_{DF} \simeq T_{VS}$ and dynamical friction is also effective in reducing the random velocity of the large planetesimal and we have negative q, as mentioned in Sec. III. The condition (16) is then satisfied and several planetesimals start runaway growth in this stage. Let the mass of the largest planetesimal formed in this stage be M ($>m$) and its random velocity be v_M. The scaled relative velocity in an encounter between planetesimals with masses M and m is given by

$$\tilde{v}_{Mm} \simeq \frac{(v_M^2 + v_m^2)^{1/2}}{h_M v_K} \simeq 20 \left(\frac{m}{10^{24}\text{g}} \right)^{-1/15} \left(\frac{2m}{M} \right)^{1/3} \tag{34}$$

where

$$h_M = \left(\frac{M+m}{3M_\odot} \right)^{1/3} \simeq \left(\frac{M}{3M_\odot} \right)^{1/3}. \tag{35}$$

From Eqs. (34) and (35) we find that once a large planetesimal is formed, $\tilde{v}_{Mm}$ decreases as M grows and the condition of runaway growth continues to be satisfied. As a result, the large planetesimal should grow much more rapidly than its neighbors. We will call it a protoplanet. When M increases to be larger than $10^3 m$, we have $\tilde{v}_{Mm} \lesssim 2$, where we cannot use the two-body approximation for collision rate or gravitational scattering: runaway growth would lead to the situation where the three-body effects are important. It should be noted that runaway growth should terminate if $\tilde{v}_{21}$ becomes so small that the condition (12) is violated.

If we do not take into account the effects of dynamical friction on a protoplanet, the planetary growth should be orderly as demonstrated by Wetherill and Stewart (1989). In this case, the encounter velocity between largest planetesimals is given by

$$\tilde{v}_{MM} \simeq \frac{(v_M^2 + v_M^2)^{1/2}}{h_M v_K} \simeq \sqrt{2} \times 20 \left(\frac{M}{10^{24} \text{g}} \right)^{-1/15} \tag{36}$$

which is larger than $\tilde{v}_{Mm}$ given by Eqs. (34) and (35) at least by a factor of $(M/m)^{1/3}$. As shown in Eq. (11), the collision rate is almost inversely proportional to the square of the scaled encounter velocity for $1 < \tilde{v}_{21} < 20$. Hence, we expect that the growth time of a protoplanet in the runaway accretion stage is substantially shorter than that in the orderly accretion case: we have a relation

$$T_{\text{grow}}^{\text{runaway}} \simeq (\frac{m}{M})^{2/3} T_{\text{grow}}^{\text{orderly}}. \tag{37}$$

C. Final Stage

In a final accretion stage, runaway protoplanets become so large that their masses cannot be neglected compared to the total mass of small planetesimals. In this case, the aspects of gravitational scatterings change drastically from those in the intermediate stage. This would have the following two consequences: first, the process which determines the random velocity v_m of small planetesimals would change. In the final stage, the velocity v_m would be largely influenced by gravitational scattering caused by protoplanets, while in the preceding stages v_m is determined by mutual scattering between small planetesimals. Consequently, $\tilde{v}_{Mm}$ would become much larger than the value given by Eq. (34). Second, the process which determines the random velocity v_M of protoplanets would also change. As Lissauer and Stewart state in their chapter, mutual scattering between protoplanets would become more important in determining v_M, rather than dynamical friction caused by the small planetesimals. In this case, $\tilde{v}_M$ would be much larger than the value in the case where dynamical friction is effective.

If the first aspect is dominant, even planetesimals in relatively distant orbits will be able to come close to protoplanets and feeding zones of the protoplanets would gradually expand. As a result, runaway growth of a protoplanet will occur from several protoplanets in the expanded feeding zone, which will lead to rather calm final accretion. If the second aspect is dominant, on the other hand, orbits of the protoplanets will be largely perturbed and may cross one another, which will lead to the final accretion with giant impacts as Lissauer and Stewart state. However, we do not yet know which aspect is dominant.

It should be noted that the growth time of protoplanets in this high velocity stage becomes much larger than that in the preceding runaway stage with low relative velocity; it should be of the same order of $T_{grow}^{orderly}$ rather than $T_{grow}^{runaway}$ given by Eq. (37). In view of the giant planet formation scenario through gravitational instability of its surrounding atmosphere, the protoplanet in the Jovian region should grow large enough (say, ten times the Earth's mass; see Mizuno 1980) before the dispersal of the solar nebula gas to cause the instability. Hence, it is not so easy to solve the problem of the formation time scale of the Jovian planets even in the runaway growth scenario.

VI. DISCUSSION

In recent years, collisions and gravitational scattering between planetesimals in the solar gravitational field have been appreciably clarified through both analytic studies and numerical integration of a large number of orbits. Through these studies we have made considerable progress in understanding the elementary processes governing the planetary growth.

After the fundamental physical processes determining the random velocity of planetesimals have been clarified, we can study the planetary accretion in detail by numerically solving the so-called coagulation equation which, of course, includes the effects of the spatial change of the mass distribution caused by radial diffusion due to gravitational scattering and inward drift due to gas drag force (Nakagawa et al. 1983; Ohtsuki et al. 1988; Spaute et al. 1991). When we use a numerical code to solve the coagulation equation, accuracy of the code must be tested by, for example, comparing numerical and analytic solutions (Ohtsuki et al. 1990; Wetherill 1990a). However, a more serious problem in using the coagulation equation is that if a small number of protoplanets start runaway growth, it becomes difficult to describe the growth of both a small number of runaway protoplanets and many small planetesimals by the coagulation equation alone (Wetherill 1990a). Recently, Tanaka and Nakazawa (1992) derived a new coagulation equation for a stochastic distribution function, which can exactly describe the accretion even at the final stage of formation of a single body. They also showed that the ordinary coagulation equation, which has been used in the previous studies of planetesimal accumulation (see, e.g., Safronov 1969; Greenberg et al. 1978; Nakagawa et al. 1983; Ohtsuki et al. 1988), cannot describe the coagulation process when an appreciable number of particles with mass greater than $(M_t m_0)^{1/2}$ are formed (M_t is the total mass of the system and m_0 is the mass of a smallest particle). For example, if we consider a system of planetesimals with $M_t = 10^{28}$g and $m_0 = 10^{18}$g, we cannot correctly simulate the evolution of the number of planetesimals with mass greater than 10^{24}g by the use of the ordinary coagulation equation. The stochastic coagulation equation derived by Tanaka and Nakazawa (1992) is, however, difficult to solve for general forms of coalescence rate, even numerically. Thus, we have to combine the statistical treatment with a mass-distribution function for small planetesimals and direct N-body simulation for large protoplanets to simulate the final stage of accretion. Wetherill and Stewart (1989) and Spaute et al. (1991) introduced a stochastic component into their simulation and developed a numerical code which can follow the evolution of both the size distribution of a planetesimal swarm and the size of a small number of the largest bodies treated as discrete objects. These codes need to be further extended to include the radial motion of planetesimals caused by gas drag or gravitational scattering.

With the elucidation of these elemental processes and an appropriate treatment in numerical simulation of the planetary growth, we will be able to simulate the planetary accretion in a more realistic manner for a given initial distribution of planetesimals.

Acknowledgments. We thank E. H. Levy for encouraging us to write this chapter. We are also grateful to S. Watanabe, D. R. Davis, S. J. Weidenschilling, J. J. Lissauer, and G. R. Stewart for fruitful discussions and helpful comments, and to K. Kudo for cooperation in preparing the manuscript.

GASEOUS ACCRETION AND THE FORMATION
OF GIANT PLANETS

MORRIS PODOLAK
Tel Aviv University

WILLIAM B. HUBBARD
University of Arizona

and

JAMES B. POLLACK
NASA Ames Reseach Center

This chapter presents a review of the structure and composition of the giant planets and the theory of their formation and growth. All of the giant planets have heavy-element cores of 5 to 15 $M_\oplus$, and have envelopes which contain large amounts of high-Z material in addition to hydrogen and helium. The planets most probably formed through the core instability mechanism. This is a much more complex mechanism than was previously thought, depending, as it does, on several time-dependent parameters. We present the results of new, more detailed, simulations. Towards the end of accretion, the transfer of angular momentum to the outer layers of the contracting protoplanet should lead to the formation of a disk. This disk may be the site of satellite formation. Some recent simulation results are shown.

I. INTRODUCTION

The events that formed our solar system are not, in themselves, accessible to observation, but the various members of the solar system do provide us with a number of clues to their origin. The outer planets are a particularly promising area to search for such clues. In the first place, if we neglect the Sun itself, almost all of the remaining mass of the solar system resides in the outer planets (the precise fraction depends on the mass of the Oort cloud comets). In addition, unless the Oort cloud mass is very large indeed, almost all of the angular momentum of the system (including the Sun) resides in these planets.

A second way in which the outer planets can provide us with clues to the origin of the solar system is through their compositions. In their present state, mass loss from these planets is very low. Using the standard Jeans escape formula (Chamberlain and Hunten 1987) and taking the Jovian exospheric temperature to be 1000 K, we find that even for H^+, which has an effective mass of only 8.3×10^{-25} g, only several thousand ions are lost in the lifetime of the solar system. For Uranus, with an exospheric temperature of 700 K, and a much lower gravitational acceleration, the escape rate for H^+ is much higher, $\sim 2 \times 10^{17}$ g in the lifetime of the solar system, but this too is only a small fraction of the hydrogen reservoir of that planet. In other words, the outer planets are expected to have maintained their primordial composition.

A third way in which the outer planets provide clues to the origin of the solar system is through their satellite systems. The present state of these satellites, both dynamical and compositional, is the direct result of the evolutionary history of their parent planets. We should therefore expect them to contain clues to that history.

In Sec. II, we review the procedure for computing model planetary interiors, and present the results of such models. Section III summarizes the current view of how the outer planets formed, through the core instability mechanism. Section IV presents the results of evolutionary models for the outer planets. We present the main conclusions in Sec. V.

II. GIANT PLANETS TODAY

A. Modeling Procedure

The procedure for constructing model interiors is straightforward in principle, and has been reviewed in the past (see, e.g., Podolak and Reynolds 1985; Podolak et al. 1991). For the sake of completeness, we briefly review the relevant equations here. Planetary rotation causes a noticeable departure from sphericity. Most modeling efforts deal with the problem by defining a "level surface" as the surface on which the sum of the gravitational and rotational forces is constant. A particular level surface is denoted by:

$$s(r, \theta) = r \left[1 + \sum_{n=0}^{\infty} \epsilon_{2n}(r) P_{2n}(\cos \theta) \right]. \tag{1}$$

Here θ is the colatitude, the P_{2n} are Legendre polynomials, the ϵ_{2n} define the shape of the surface, and r is the radius of a sphere with a volume equal to the volume contained within the level surface. With this definition of r, conservation of mass is given by

$$\frac{dM(r)}{dr} = 4\pi r^2 \rho(r) \tag{2}$$

where $M(r)$ is the mass contained within a sphere of radius r, and $\rho(r)$ is the density at r. The assumption of hydrostatic equilibrium for a rotating fluid, including a latitude–averaged centrifugal force gives

$$\frac{dP}{dr} = -\left[\frac{GM(r)}{r^2} - \frac{2\omega^2 r}{3} \right] \rho(r). \tag{3}$$

Here G is Newton's constant of gravitation, P is the pressure, and ω is the angular velocity.

The departure from sphericity induced by the rotation appears in the gravitational field as a departure from a $1/r$ potential field.

$$V(s,\theta) = -\frac{GM}{s}\left[1 - \sum_{n=1}^{\infty}\left(\frac{R_e}{s}\right)^{2n} J_{2n}P_{2n}(\cos\theta)\right] \qquad (4)$$

where M is the mass of the planet, R_e is the equatorial radius, and the J_{2n} are the multipole moments of the gravitational field. The measured values of the J_{2n} can be compared to those computed from the internal density distribution via

$$J_n = -\frac{1}{MR_e{}^n}\int \rho(r)r^n P_n(\cos\theta)\mathrm{d}\tau \qquad (5)$$

where $\mathrm{d}\tau$ is an element of volume. Further details on the level surface approach can be found in the book by Zharkov and Trubitsyn (1978). An alternative approach to evaluating the effects of rotation is discussed by Hubbard et al. (1975).

Perhaps the point of greatest uncertainty in computing interior models is the equation of state

$$P = P(\rho, T) \qquad (6)$$

where T is the temperature. Even before one can worry about the problem of computing an equation of state, one must decide which materials the equation of state will be computed for. Jupiter and Saturn have densities so low that hydrogen must be the major constituent. Helium should be present because of its high cosmic abundance. These two species are so volatile that they can only enter the planet in the gas phase. Additional materials can be incorporated into the planet as solids if they are sufficiently refractory, or as either solid or gas (depending on temperature) if they are of intermediate volatility (Podolak and Cameron 1974). Models of the outer planets require, in general, equations of state for three classes of material: gas, ice and rock. It must be understood that these names are generic, and refer to the volatility of the material, not the phase in which it is found in the planet. "Gas" refers to hydrogen and helium in solar proportions, "ice" is generally taken to be a solar mix of H_2O, CH_4, and NH_3, and "rock" is usually taken to be a mix of SiO_2, MgO, FeS, and FeO. In some models Fe and Ni are used in place of FeS and FeO. Hydrogen exists in two forms: a low-pressure molecular phase, and a high-pressure metallic phase. The pressure at which a transition from metallic to molecular hydrogen occurs is not well determined, but is probably in the 3 to 5 Mbar range (Marley and Hubbard 1988). This is the phase transition that is most important for the determination of the density profile. There are other transitions that could be of importance for modeling planetary interiors. In shock experiments water becomes an ionic melt, $H_3O^+OH^-$ (Mitchell and Nellis 1982), and Smoluchowski (1975) has suggested that molecular hydrogen undergoes a transition to a more conducting (but still molecular) phase before it becomes a metal. These transitions and others do not substantially affect the computed density profile, but will be important for determining the conductivity of the material, and hence for the computation of magnetic fields. There are various techniques used for computing the equations of state of the relevant materials. Details of these techniques, their relative merits and accuracies have been summarized by Podolak and Reynolds (1985) and Podolak et al. (1991).

One other piece of information that is needed for computing the equation of state is the temperature profile in the planet. For the case of Jupiter, the profile can be computed in a straightforward manner. Jupiter's intrinsic luminosity is considerably higher than the value expected in equilibrium with the Sun (Hanel et al. 1981). It is possible to estimate the thermal conductivity of metallic hydrogen, and compute the temperature gradient required to provide the observed thermal flux. The result is that over most of the planet the thermal gradient must be much higher than the adiabatic gradient, and that most of the interior must therefore be convecting (Hubbard 1968; Bishop and DeMarcus 1970). This implies that the temperature gradient throughout most of the interior is very close to adiabatic, provided that there are no major phase transitions (Stevenson and Salpeter 1976). Saturn, with a luminosity about one third that of Jupiter's and a greater fraction of its hydrogen in the low-conductivity molecular state, must also have an adiabatic gradient throughout a large fraction of its interior. A similar argument can be given for Neptune, and has been applied to models of Uranus as well. Thus all current model of the outer planets assume that the temperature gradient is given by

$$\frac{dT}{dr} = \frac{dP}{dr}\left(\frac{dT}{dP}\right)_{adiabatic}. \tag{7}$$

It is not completely clear that the adiabatic approximation applies to Uranus, however. Unlike the other planets, Uranus has a very small internal heat source, so the argument regarding the efficiency of heat transport fails. It is true that the cooling time for both Uranus and Neptune is extremely long, so that almost any reasonable initial temperature profile (due to the release of accretional energy, for example) would result in a planet that is convecting today. The problem is that such profiles result in planets that are much hotter than Uranus and Neptune are today, indicating that the cooling of these planets is not yet sufficiently well understood. This problem will be discussed in more detail in Sec. IV. Finally, Uranus does have a magnetic field (Connerney et al. 1987) and this indicates that at least part of the interior is convecting, although there may be large regions of the planet where other mechanisms are sufficient to transport the small flux (see Podolak et al. [1991] for a more complete discussion of the relation between the low heat flux and the magnetic field). For our purposes it is sufficient to be aware that the arguments for an adiabatic temperature profile in the outer planets are not as compelling for Uranus.

TABLE I

Observational Data

Property	Jupiter	Saturn	Uranus	Neptune
Mass (10^{29} g)	18.99	5.68	0.8682	1.0243
Equatorial Radius (10^9 cm)	7.1492	6.0268	2.5559	2.4764
J_2 ($\times 10^6$)	14736 ± 1	16331 ± 18	3516 ± 3	3538 ± 9
J_4 ($\times 10^6$)	-587 ± 5	-914 ± 61	-31.9 ± 0.5	-38.0 ± 1
J_6 ($\times 10^6$)	31 ± 20	108 ± 5		
Rotation Period	$9^h 55^m 29^s.7$	$10^h 39^m 24^s$	$17^h 14^m 24^s$	$16^h 6^m 36^s$
Atmospheric Parameters				
T_{1bar} (K)	165 ± 5	135 ± 5	76 ± 2	69 ± 2
$E_{emitted}/E_{abs}$	1.67 ± 0.09	1.78 ± 0.09	a	2.7 ± 0.3
He mass fraction	0.18 ± 0.04	0.06 ± 0.05	0.262 ± 0.048	0.32 ± 0.05

a The observed value is consistent with 1 to the precision of the measurements.

1. Jupiter. The flybys of Jupiter by Pioneers 10 and 11, and by Voyagers 1 and 2 have allowed rather precise determinations of the parameters required for interior models. Although refinements are continually being made in these parameters, their values have been well determined since about 1975, and all models published after that date can, in some sense, be considered "current." As for the best modern values for the radius, mass, and gravitational field parameters, they are shown in Table I. The equatorial radius of Jupiter is that determined by Lindal et al. (1981) at the 0.1 bar level, and recomputed at the 1 bar level (Hubbard and Horedt 1983). The most recent analysis of the gravity field as determined by all four of the spacecraft flybys is given by Campbell and Synnott (1985). The values for the multipole moments are all normalized to a radius of 7.1398×10^9 cm. J_3 $[= (1.4 \pm 5) \times 10^{-6}]$ and the sectorial harmonics are small, $[C_{22} = (-3 \pm 15) \times 10^{-8}, S_{22} = (-0.7 \pm 15) \times 10^{-8}]$ and consistent with zero, as expected for a body in hydrostatic equilibrium. The rotation period of the main body of the planet is taken to be equal to that of the magnetic field (Seidelmann and Divine 1977). The atmospheric properties are also shown in Table I. Particular attention should be paid to the helium mass fraction, $Y = 0.18 \pm 0.04$. Many of the earlier models used a value considerably larger than this, so while they give a density distribution that matches the gravitational moments, the composition represented by that distribution may need to be reinterpreted.

Shortly after the Pioneer flybys of Jupiter, a meeting was held in Tucson to review the latest results, and how they affected our view of the planet. At this meeting three new sets of models were presented. Each of these sets was representative of a different approach to the problem of planetary modeling. The work of Stevenson and Salpeter (1976) concentrated mostly on improving the physics that went into the models. Thus they examined the miscibility of the hydrogen–helium mixture, the strength of the assumption that the volume of a mixture of species is simply the sum of the volumes occupied by the individual species (the "additive volume law"), and the extent of superadiabaticity in the models. Of necessity, they limited their investigation to envelopes composed of only hydrogen and helium. They found that the observed parameters could be fit by models in which this hydrogen–helium envelope surrounded a dense core (rock) of some 9–11 $M_\oplus$. It should be noted, that they chose $Y = 0.25$–0.32, values that were then consistent with the expected abundance on Jupiter. Since we now know that Y is considerably smaller, those models may now be reinterpreted to indicate that the envelope is enriched in some high-Z material above the usual solar composition.

A second set of models was presented by Zharkov and Trubitsyn (1976). Their approach was to allow for a variable composition envelope, and to attempt to assess the abundance of high-Z material in the envelope. This, of course, could only be done at the expense of a less rigorous treatment of the equations of state. Thus, the densities of materials other than hydrogen were simply taken to be proportional to the density of hydrogen. They found that if they took the envelope to be in solar composition, then a rock core of 10–13

$M_\oplus$ was needed to match the gravitational moments, quite similar to the result of Stevenson and Salpeter (1976). This is especially so, since Zharkov and Trubitsyn chose a similar value of the helium abundance ($Y = 0.26$). Here too, these models must be reinterpreted to show that the envelope is indeed enriched in high-Z material. A second set of models was computed that had 30–40 $M_\oplus$ of high-Z material in the envelope, and no rock core. It is interesting that the observational data do not allow us to decide whether the high-Z material is in the envelope or in the core. This is due to the fact that, as can be seen from Eq. (5), J_n is proportional to the integral of the density weighted by the radius to the nth power. Because the core is confined to small radii, it does not contribute significantly to J_n. Instead, the J_n determine the run of density in the outer part of the envelope, and this density distribution accounts for most of the mass of the planet. The core provides any additional mass that is required. The core is thus a second-order effect, and can be present or absent, depending on the details of the envelope density. This is true for models of all the outer planets.

The third modeling paper (Hubbard and Slattery 1976) suggested still another approach: concentrate on finding the class of density profiles that fit the observed mass, radius and gravitational moments. The composition represented by such a profile can be determined afterwards. For Jupiter, Hubbard and Slattery pointed out that the polytrope $P = 1.9545\rho^{2.05263}$ provides an excellent fit to the mass, radius and J_2 of Jupiter (although J_4 is a bit high). Models of this type are important for defining the limits placed on the allowed density distributions by the observed gravitational field. This particular density distribution corresponds to a planet-wide enhancement of high-Z material of some 30 $M_\oplus$. Once again, since Hubbard and Slattery used a helium abundance corresponding to $Y \simeq 0.25$, this enhancement must be revised upwards, but this will be offset by the fact that J_4 was too high (in absolute value) indicating that the density in the outer envelope must be reduced. This model has no dense core. These three approaches have formed the basis for all later modeling attempts.

In the intervening years a number of other Jupiter models have been computed (Podolak 1977; Slattery 1977; Hubbard et al. 1980; Grossman et al. 1980; Hubbard and Horedt 1983) and these were summarized in the last Protostars and Planets volume (Podolak and Reynolds 1985). Since that book, there have been two new sets of Jupiter models. The first (Gudkova et al. 1989) follows the second approach outlined above, and assumes $Y = 0.18$, in agreement with observation. The model consists of four layers: an outer layer of molecular hydrogen, helium, and "ice," with the ice comprising 2% by mass, a second layer with both rock and ice mixed in with the gas, a third layer which is identical to the second, except that the hydrogen here is metallic, and has a rock-ice core of 4.99 $M_\oplus$. The ratio of ice to rock in the planet is 3, in agreement with solar abundances, The total mass of ice in the planet is 50.08 $M_\oplus$, while the total mass of rock is 16.55 $M_\oplus$. When allowance is made for the lower helium abundance, this is consistent with earlier models.

The second of the most recent models (Hubbard and Marley 1989) follows the third approach and concentrates on finding the density distribution that optimizes the fit to the observed parameters within the framework of a model that has a dense core surrounded by a hydrogen-helium envelope. They find that if the core is composed of rock, then it has a mass of 8.3 $M_\oplus$, and the envelope has $Y = 0.25$. Assuming that the ice that complements the rock in solar composition ($\sim$25 $M_\oplus$) is mixed into the envelope, then the actual helium abundance is reduced to $Y = 0.21$, which is consistent with observation. The mass fraction of ice in the envelope is then about 0.08, which is a bit high, but of the order of what is observed. If the core is taken to consist of a mixture of rock and ice, then because of the lower density of such a mixture, its mass must be increased to 12 $M_\oplus$. In this case, however, the value of $Y = 0.25$ must be explained as an additional enrichment of high-Z material above solar composition.

In conclusion, Jupiter seems to consist of a high-density core of some 5 to 15 $M_\oplus$ surrounded by an envelope which is itself substantially enriched in high-Z material over solar composition. The most recent models put the high-Z enrichment at about 50 $M_\oplus$ of ice and rock, although the models cannot, at present, differentiate between these two materials.

2. Saturn. For Saturn, the model input parameters are again shown in Table I. The equatorial radius was determined by Lindal et al. (1985) at the 1-bar level. The mass, determined from satellite orbits and spacecraft tracking is from Null et al. 1981. The gravitational moments are taken from Nicholson and Porco (1988), but are normalized to the 1-bar radius. The rotation period of the main body of the planet is taken to be equal to that inferred from periodicities in kilometer-wavelength radio storms (Desch and Kaiser 1981). The 1-bar temperature is from Tyler et al. 1982 and the internal heat flux is from Hanel et al. 1983. Finally, the helium mass fraction in the atmosphere (Conrath et al. 1984) is well below the solar value assumed by all but the most recent Saturn models. This is the reason that most of these earlier models had difficulties in matching the observed J_4 (Podolak and Reynolds 1985).

Here too there have been two recent sets of Saturn models. That by Gudkova et al. (1989) consists of five layers. The outer layer consists of molecular hydrogen, helium, and ice, with the ice comprising 2% by mass, and $Y = 0.064$. Below it is a second layer with both rock and ice mixed in with the gas. This is followed by a third layer which is identical to the second, except that the hydrogen here is metallic, a fourth layer composed of helium, rock and ice (to allow for the excess helium that has settled out of the outer layers), and a rock-ice core of 5.7 $M_\oplus$. Once again the ratio of ice to rock in the planet is 3, in agreement with solar abundances. The total mass of ice in the planet is 21.42 $M_\oplus$, while the total mass of rock is 6.98 $M_\oplus$. It is interesting that the 20 $M_\oplus$ cores that were determined by earlier models are here replaced with a helium-rock-ice mixture which also has a mass of about 20 $M_\oplus$, while the nearly solar composition envelope of the earlier is replaced by one that is significantly enriched in high-Z material, but depleted in helium. The density profile remains relatively unchanged, but its interpretation in terms of composition is very different.

The models of Hubbard and Marley (1989) are not as detailed in their choice of compositional layers, but because of this they can be used to explore the space of allowable density distributions more widely. They find that the core mass must be between 9 and 20 $M_\oplus$, depending on its composition. This is surrounded by a metallic hydrogen envelope which is heavily enriched in helium or some other dense component. Surrounding this is a molecular hydrogen envelope with $Y = 0.06$, but with an enrichment of some other high-Z material. If this other material is assumed to be H_2O ice, then it comprises some 20 $M_\oplus$. This is quite similar to the Gudkova et al. result.

3. Uranus. Uranus' vital statistics are given in Table I. The equatorial radius is from Lindal et al. (1987) at the 1-bar level. The mass is from Anderson et al. 1987, and the gravitational moments, from French et al. (1988), but normalized to the 1-bar radius. The rotation period of the main body is taken to be equal to the magnetospheric rotation rate determined by Warwick et al. (1986). The 1-bar temperature is from Lindal et al. (1987), and the helium mass fraction in the atmosphere is from Conrath et al. (1987). In the case of Uranus, the helium abundance in the upper atmosphere is consistent with solar composition.

The rotation period and gravitational moments are now known with much greater accuracy than before the Voyager flyby, so that only those models computed after 1986 can be said to be adequate representations of the planet. Such models have been computed by Podolak and Reynolds (1987), Gudkova et al. (1988), Hubbard and Marley (1989), and Stevenson (1989). They are reviewed in Podolak et al. (1991). The models all have very similar density distributions, characterized by a small or vanishing core surrounded by a region where the density slowly decreases from about 4.5 g cm^{-3} to about 2 g cm^{-3} at 70% of the planetary radius. At this point there is a rapid decrease to zero density at the edge of the planet. Although the models essentially agree on the form of the density distribution, there is an ambiguity with regard to the composition represented by that distribution. This comes about because many different combinations of materials give very similar densities. For example, a mixture, by mass, of 0.75 rock and 0.25 gas will give a density very similar to ice for a large range of pressure (Podolak et al. 1991), so that an arbitrary mixture consisting of 75% rock and 25% gas as one component and ice as the second component will always have the density of pure ice. For this reason, it is not possible to derive the ratio of ice to rock in Uranus from interior models alone. This same ambiguity applies to models of the other planets as well, but in Jupiter and Saturn these high-Z materials comprise only a small fraction of the mass. Because of this, their contribution to the gravitational field is small, and it is at present impossible to resolve rock from ice in these planets even when they are not mixed. The solar ratio ice to rock is assumed, but it can not be verified. In Uranus and Neptune, where ice and rock make up the bulk of the planet, such a resolution is possible if the materials are not mixed, and could have interesting cosmogonic consequences (Podolak and Reynolds 1984). The ambiguity introduced by the possibility of mixtures, however, complicates the problem.

4. Neptune. The most recent determination of Neptune's parameters was made by the Voyager 2 flyby (see Table I). Tyler et al. (1989) measured the equatorial radius, at the 1-bar level and the mass. The gravitational moments are from Owen et al. (1991) but normalized to the 1-bar radius. The 1-bar temperature and the ratio of absorbed solar power to thermally emitted power are from Conrath et al. (1989). The helium mass fraction in the atmosphere seems to be similar to that for Uranus (Conrath et al. 1991).

Of the above parameters, the rotation period, J_2, and J_4 have been revised substantially as a result of the recent Voyager flyby. As these parameters are very important for interior models, all of the pre-Voyager models lose their relevance. There are two recent sets of calculations done using the revised parameters, those of Podolak et al. (1990), and those of Hubbard et al. (1991). The models of Podolak et al. consisted of three layers, a rock core, an ice layer, and an envelope consisting of hydrogen, helium, H_2O, NH_3, and CH_4. The hydrogen to helium ratio is solar, in agreement with the observations of both Uranus' and Neptune's atmospheres, and the H_2O, NH_3, and CH_4 are in the solar ratio to each other, but may be enhanced by some factor A over their solar ratio to hydrogen. They find a rock core of 1.2 $M_\oplus$, and an envelope of solar composition (i.e., $A = 1$). Their model also gives a ratio of ice to rock for the planet of 13. This is about four times the solar value. Once again, due to the ambiguity described above, we can replace the ice layer in the model by a mixture of rock, gas, and ice, so the ice to rock ratio can be brought down to the solar value, or less. They showed that Uranus and Neptune may have very similar compositions (in the sense of the ice to rock ratio) in spite of their differences in density.

The work of Hubbard et al. (1991) investigated the effects of differential rotation. Although there are strong latitudinal gradients in the atmospheric wind profile, Hubbard et al. showed that these differences do not extend very deeply into the planet. As a result, the density profiles computed for a differentially rotating planet are similar to those computed for a uniformly rotating planet. Their models are very similar to those of Podolak et al. and they too conclude that the compositions of Uranus and Neptune are probably very similar.

B. Atmospheric Composition

The models described above give considerable information about the density distribution in the planetary interior. They cannot, however, set anything more than limits on the compositional details a particular density distribution may represent. Observations of the upper atmosphere of these planets, on the other hand, give precise information on the composition, but are, of necessity, limited to the very outer layers of the planet. Nonetheless, some interesting conclusions may be drawn. The different H/He ratios in the outer planets have already been mentioned above. Because both H_2 and He are thought to have been accreted directly from the nebula in the gas phase, it is difficult to see how the various planets could have accreted different relative amounts of these two gases. The different ratios observed are believed to be relevant only to the outer layers of these planets. Theoretical studies indicate that at high pressures helium may not be miscible in hydrogen (Stevenson, unpublished; Hubbard and MacFarlane 1985; Hubbard and DeWitt 1985). It is possible that the low helium abundance observed in Saturn is due to this imiscibility, and the subsequent rain-out of helium in the outer layers of the planet. The He/H ratio on Jupiter is larger than that on Saturn, but is lower than that of Uranus, Neptune, and the Sun (Gautier and Owen 1989). It is possible that helium rain-out has begun in Jupiter as well. The total (volume integrated) He/H ratios in all of the outer planets is expected to be solar, however.

The deuterium/hydrogen (D/H) ratio observed in the atmospheres of the giant planets provides important clues to the origin of these bodies. In Jupiter, in which ~90% of the mass consists of H and He, the atmospheric D/H should reflect the average value in the solar nebula at the time of formation. The observed Jovian value deduced from abundances of monodeuterated methane is $D/H = (2.0 \pm 0.6) \times 10^{-5}$ (Lutz et al. 1983), in agreement with the presumed interstellar D/H value at the time of origin of the solar system. It is also consistent with the fact that the lower mass limit for deuterium burning is about 20 Jupiter masses.

About 70% of the mass of Saturn consists of H and He, and again one expects the atmospheric D/H to reflect the primordial value. The observed value from monodeuterated methane, consistent with the Jovian value, is $D/H = (1.6^{+1.6}_{-1.0}) \times 10^{-5}$ (de Bergh et al. 1986,1988).

The situation is quite different in Uranus and Neptune. In these planets, the H and He component is less than 10% of the total planetary mass, so that exchange of deuterium with an icy reservoir could significantly affect the atmospheric ratio. The error bars on deuterium enhancement are still quite large. For Uranus, $D/H = (7.2^{+7.2}_{-3.6}) \times 10^{-5}$, and for Neptune, $D/H = (12^{+12}_{-8}) \times 10^{-5}$ (de Bergh et al. 1990). If we assume that the initial nebular D/H was 2×10^{-5} and that D/H in the initial icy planetesimals which were accumulated to form Uranus and Neptune was 10 times larger due to isotopic fractionation effects, as indicated in Halley's comet (de Bergh et al. 1990), then a mass balance calculation leads to limits on the ratio of icy material to rocky material in Uranus and Neptune. The lower limits on D/H suggest that the mass of nebular hydrogen and helium was about equal to the mass of icy material (i.e., the enhanced icy D/H was diluted about twofold with nebular D/H), while the upper limits suggest that the atmospheric D/H mainly represents the enhanced value in the icy material, so that the mass of hydrogen brought in with the icy component considerably exceeds the hydrogen captured directly from the nebula. Quantitatively, this means that

$$M_{H-He}/M_{ice} < 0.8 \qquad (8)$$

where M_{H-He}/M_{ice} is the ratio of the mass of nebular H-He to the mass of ice in Uranus or Neptune. The interior models discussed above suggest that $M_{H-He}/M_{ice} \sim 0.1$. Unfortunately, the error bars and our ignorance of the initial deuterium enrichment factor for the icy material preclude more definitive limits for now.

In addition to hydrogen and helium, there are also observations of other elements. Of particular interest is the C/H ratio. Although some of the atmospheric carbon is locked up in such molecules as C_2H_2, and C_2H_6, these are nonequilibrium species produced by photolysis and lightning. Most carbon is in the form of CH_4, and can be readily observed in this form in Jupiter and Saturn. Although CH_4 does condense out in the tropospheres of Uranus and Neptune, it can still be observed in the deeper atmospheres of these planets. There is a definite increase in the C/H ratio as one goes farther out in the solar system. Thus the values of C/H in Jupiter, Saturn, Uranus, and Neptune are 2.3, 2–6, $\sim$25, and $\sim$35 times the solar ratio, respectively (Gautier and Owen 1989). This trend can be understood in the context of the core instability model (see Sec. III.B).

The N/H ratios on Jupiter and Saturn are similar to the C/H ratios (Gautier and Owen 1989), but the values for Uranus and Neptune appear to be even lower than the solar ratio (de Pater and Massie 1985). No completely satisfactory explanation has yet been given for this (see de Pater et al. 1989), but these observations must somehow be incorporated into any theory of outer planet formation.

C. Satellites of the Giant Planets

In certain ways, the giant planet systems are miniature solar systems that contain not only the giant planet itself, but an entire retinue of satellites and a magnificent ring system located close to the planet. Here, we summarize the basic characteristics of the satellites, with emphasis on properties that constrain models of their formation. We lump the rings with the satellites, since it is now widely believed that the micron to meter sized particles that constitute the rings of the outer planets are derived by meteoroid impact of larger parent objects, i.e., small satellites in the region of the rings (see Esposito et al. 1991).

All four giant planets have large numbers of satellites surrounding them and all four have ring systems. This abundance of surrounding objects may be contrasted with the comparative paucity of satellites around the terrestrial planets, that collectively possess only three moons and that have no rings. Thus, some aspect of the formation of the giant planets, perhaps connected with their having gaseous envelopes or forming when the solar nebula was still present, promoted the development of a satellite system around them;

whereas, such a situation did not readily occur for the terrestrial planets. Indeed, all three moons of the terrestrial planets are thought to have gone into orbit around their planets either as a result of capture processes (Phobos and Deimos) or as a result of a giant impact event (the Earth's moon), events that did not result in many objects at a variety of distances.

The orbital properties of the satellites of the giant planet permit them to be divided into two major categories: regular and irregular satellites. Regular satellites have prograde orbits of low eccentricity and low obliquity (with respect to the planet's equatorial plane), whereas irregular satellites have high obliquities, often possess high eccentricities (Triton is an exception, probably because of tidal evolution), and travel in both retrograde and prograde directions. All four giant planets contain numerous regular satellites. They are situated from a fraction of the planet's radius above its atmosphere to several tens of planetary radii away. In all four systems, the largest regular satellites, having diameters up to about 5000 km (Titan, Ganymede, and Callisto), are found farthest from the planet, with their sizes tending to more or less systematically decrease to values on the order of tens of kilometers at the closest distances. Their orbital properties and propinquity to their associated planets suggest that the regular satellites formed in the same region of space where they are currently located and that they therefore formed as part of the same process that led to the assembly and/or early evolution of their planet. Possible source materials for them include the primordial solar nebula and the outer envelope of the giant planet.

Irregular satellites are known to be present around all the giant planets, except for Uranus. They are always found at distances greater than that of the regular satellites. Typically, they are found at distances of several hundred planetary radii, although Triton is located at 15 planetary radii (probably due to significant tidal orbital evolution following capture). We note that distances of several hundred plantary radii correspond to positions close (within a factor of several) to the outermost position at which a satellite can be maintained in a stable orbit against solar and other planetary perturbations over the age of the solar system. Thus, the irregular satellites, including Triton, were presumably formed in the outer solar nebula and later captured into loosely bound orbits about their planet. This means that solid objects as large as several thousand km in diameter (Triton's diameter is 2700 km) were formed in the outer solar system. (Pluto is probably another example of such a sized body.) More typically, irregular satellites have sizes that range from several tens to several hundreds of kilometers.

Table II summarizes some of the salient properties of the regular satellites, irregular satellites, and rings of the giant planets. In this table, we have not tried to enumerate the poorly known properties of the ring moons and so have separated the regular satellites outside of the major rings from those inside them. The variables M_{sat}/M_{pl} and J_{sat}/J_{pl} in this table are the ratios of the total mass and orbital angular momenta of the regular satellites to the mass and spin angular momenta of their parent planets, respectively. Both ratios

TABLE II

Comparison of Satellite Systems

Property[a]	Jupiter	Saturn	Uranus	Neptune
Regular Satellites				
Number	8	17	15	6
Location (R_{pl})	$1.8 - 27$	$2.3 - 59$	$2.0 - 23$	$1.9 - 4.8$
M_{sat}/M_{pl}	2.1×10^{-4}	2.4×10^{-4}	1.1×10^{-4}	2.9×10^{-4}
J_{sat}/J_{pl}	6.5×10^{-3}	6.6×10^{-3}	6.5×10^{-3}	2.1×10^{-2}
Irregular satellites				
Number	8	1	0	1
Location (R_{pl})	$156 - 333$	216		227
Rings				
Location (R_{pl})	$1.3 - 1.8$	$1.1 - 8^{b}$	$1.6 - 2$	$1.7 - 2.5$
M_{ring}/M_{pl}	5×10^{-5}	6×10^{-8}	6×10^{-11}	
Max τ		> 1.5	> 1.5	0.1

[a] The symbols R, M, and J refer to radius, mass, and angular momentum, respectively; subscripts sat, pl, and $ring$ refer to satellite, planet, and rings, respectively; and max τ is the maximum value of the normal optical depth at visible wavelengths.

[b] The major rings of Saturn, the A, B, and C rings, are located between 1.2 and 2.3 R_{pl}.

have very similar values for the Jupiter, Saturn, and Uranus systems. These similarities, in conjunction with the similarity in the architecture of these three systems, suggest that they formed through some common process. The lower values of these ratios for the Neptune system may be a result of a disruption of the outer part of the original regular satellite system that followed the capture and orbital evolution of Triton (Goldreich et al. 1989; McKinnon and Leith 1990).

The ratios of J_{sat}/J_{pl} in Table II show that there is a very important difference between the regular satellite systems of the outer planets and the planetary system itself. More than 99% of the angular momentum of the solar system is contained in the orbital angular momenta of the planets, whereas a very sizeable fraction of the angular momenta of the outer planet systems is contained in the spin angular momenta of the planets. Even augmenting the orbital angular momenta of the regular satellite systems by the gases that may have been present in the disks within which they formed would at best result in orbital angular momenta comparable to that of their associated planets. Put another way, the regular satellites formed much closer to their parent planets (in units of the planet's size) than did the planets with respect to the Sun. Moreover, the occurrence of low obliquity, prograde orbits in both types of systems strongly suggest that the regular satellites formed within flattened disks that contained both gases and solid matter.

The bulk composition of the satellites of the outer solar system provides an important constraint on the composition of the nebular disk within which they formed, and, by implication, on the source of this nebula. Estimates of the bulk composition of the larger satellites can be obtained from the measured mean densities and assumptions about their major constituents (see, e.g., Johnson et al. 1987). The Voyager spacecraft flybys of all four giant planet systems have provided accurate mean densities for the larger satellites in these systems. If one assumes that water ice (ice) and silicates (rock) constitute the two major constituents of their interiors, then the mass fractions of ice and rock may be derived by fitting models of their interior structure to the observed mean density and size of these bodies. Useful limiting cases are provided by alternatively assuming that the ice and rock are homogeneously distributed throughout their interior (homogeneous models) and that the rock and ice are fully segregated from one another, with the heavier rock forming a central core (differentiated models). For objects of the size of the largest satellites (i.e., Titan, and the Galilean satellites), the differentiated model is probably closer to the truth, whereas for bodies having a size on the order of a few hundred kilometers or smaller, the homogeneous model is probably closer to the truth (Johnson et al. 1987).

Figure 1 summarizes the silicate mass fractions inferred for the larger satellites of the giant planet systems (Smith et al. 1989). The horizontal lines labeled "CO-rich" and "CH_4-rich" refer to the silicate mass fractions expected for nebulae having a solar abundance of elements that are cold enough for water to be present almost entirely in the solid phase, rather than the vapor

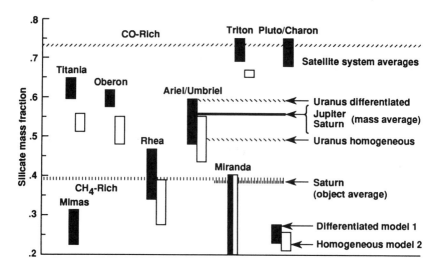

Figure 1.   Comparison of model silicate mass fractions for several outer planet
satellites and the Pluto/Charon system. Two types of satellite interior structure
are given, a fully differentiated body with a silicate core and an ice mantle, and a
homogeneous undifferentiated mix of water ice and silicates. Radiogenic heating
is taken into account in the current thermal structure of the interiors. Also shown
are approximate values for silicate mass fraction for bodies formed in CO-rich and
CH_4-rich nebular conditions and the system averages from several satellites.

phase. These lines are displaced by a significant amount from one another
because C has an abundance that is about half of that of O. Thus, if CO is the
dominant form of C in the nebula, there is a lot less O left to form H_2O. In
both of these cases, about 15% of the O is tied up in silicates. When the most
current values of the solar abundances of C, O, and Si are used (Anders and
Grevesse 1989; Grevesse et al. 1991), we find that the CO-rich line in Fig. 1
shifts considerably down to 0.53, while the CH_4-rich line decreases a bit to
0.33. Results shown for the mass-weighted Jupiter and Saturn system refer to
the average silicate mass fraction of the fully icy Galilean satellites Ganymede
and Callisto and to a value heavily dominated by Titan, respectively.

    There are several important trends to note from Fig. 1. First, the silicate
mass fraction inferred for any of the regular satellites shown in this figure is
distinctly smaller than the silicate mass fractions characterizing Triton and
Pluto/Charon. Since the latter presumably formed in the outer solar nebula
(see above), this difference implies that the nebular disks within which the
regular satellites formed were compositionally different from that of the outer
solar nebula. While the silicate mass fractions of the regular satellites span
the full range of values expected for nebulae having oxidation states ranging
from fully reduced to fully oxidized, the silicate mass fractions inferred for
Triton and Pluto/Charon are distinctly larger than those expected even for a

fully reduced nebula. The latter puzzling circumstance may imply that there are still nontrivial errors in the solar abundances, that the capture of Triton by Neptune and its subsequent orbital evolution and the formation of the Pluto/Charon system, perhaps by a giant impact, resulted in a preferential loss of water, or that our compositional assumptions are in error.

Second, the silicate mass fractions of the larger Uranian satellites (Ariel, Umbriel, Titania, and Oberon) are very similar to those of the largest icy satellites of the Jupiter and Saturn systems. Since the former are sufficiently small that the gravitational energy of their accretion is much less than the latent heat needed to sublimate all their water (these two energies are comparable for the latter), not much water could have been lost in the assembly of the largest satellites, contrary to the suggestion made by Stevenson et al. (1986). One could attempt to rescue this suggestion by Stevenson et al. (1986) by postulating that high velocity impacts by comets resulted in the preferential loss of water from Ganymede, Callisto, Titan, and the Uranian satellites. However, it would then be puzzling that the smaller regular satellites of Saturn that are situated deeper in Saturn's gravitational well (except for Iapetus) are more ice rich than is Titan. Third, the silicate mass fraction of Titan (Saturn mass average) is significantly larger than those for some of the other, smaller Saturnian satellites (Mimas, Rhea, and Miranda). This difference could be attributed alternatively to the vaporization and loss of water accompanying the formation of Titan (Stevenson et al. 1986; but see above) or to a time evolution in the properties of the Saturnian nebula (Pollack et al. 1991).

Figure 1 does not include the two inner Galilean satellites, Io and Europa. These objects have a much higher mean density than do Ganymede and Callisto and hence have silicate mass fractions close to unity. This difference is thought to reflect a trend towards warmer temperatures at closer distances in the Jovian nebula, which limited the amount of water vapor that condensed at the distances where Io and Europa formed (Pollack and Reynolds 1974). One could alternatively attribute the systematic variation of the density of the Galilean satellites deeper in Jupiter's gravitational well by comets. However, the lack of similar trends among the Saturnian and Uranian satellites casts doubt on this latter hypothesis. The dominant presence of water ice in the rings of Saturn and the occurrence of low-albedo material in the rings of Uranus (perhaps derived from C-containing ices) suggest a progressive decrease in temperature from the Jovian to the Saturnian to the Uranian nebula (Pollack et al. 1991).

Information on the composition of the irregular satellites of the giant planets, with the notable exception of Triton, can best be inferred from their spectral reflectivity properties. Strictly speaking, composition inferred in this fashion refers only to the satellites' surface. However, since the objects of interest are sufficiently small that they probably have not differentiated and since their surfaces are excavated to considerable depths by impacts, the inferred composition is likely to apply to the body as a whole. The larger irregular satellites of Jupiter and Saturn's one known irregular satellite, Phoebe, have low albedos, comparable to those of dark asteroids and carbonaceous chondrites (Degewij et al. 1980). By contrast, the small regular satellites of both systems have much higher albedos. This difference again suggests a difference in the compositions of the satellite-forming nebulae and the outer solar nebula.

III. FORMATION AND GROWTH OF THE GIANT PLANETS

A. Gas Instability Model

There have been two scenarios proposed in the literature for the formation of the giant planets. We will refer to them as the gas instability model and the core instability model. The gas instability model postulates that the primitive solar nebula was sufficiently massive that large portions of it could become unstable against gravitational collapse. The end products of such a collapse, giant gaseous protoplanets (GGP's), are supposed to evolve into the giant planets we see today. This hypothesis has the advantage that it allows naturally for the accretion of large masses of solar composition material on a short (dynamic) time scale. The most detailed studies of this model were those of Cameron and his collegues (DeCampli and Cameron 1979; Cameron et al. 1982; Cameron 1985).

This hypothesis has a number of disadvantages, however. First, as we have seen above, the outer planets all contain substantial components of high-Z material. GGP scenarios require special mechanisms to supply this additional nonsolar component. Second, as solids are accreted by the GGP, they will be dissolved and mixed with the gas at the high temperatures and pressures in the GGP interior. It is hard to see how a core would form under such conditions (Stevenson 1982a). Third, such a scenario is especially difficult to justify for the case of Uranus and Neptune, which are composed mostly of high-Z material, with relatively little hydrogen and helium. Fourth, as Cameron himself pointed out (Cameron 1989), an instability of this type requires a nebular mass of the order of a solar mass or more. This is higher than the value currently suggested for the solar nebula (Strom et al. 1989a). A final argument can be brought from the theory of brown dwarfs.

The largest giant planet, Jupiter, has many physical properties in common with the so-called brown dwarfs, which are objects of solar composition having masses less than about 0.08 $M_\odot$. This similarity (electron-degenerate convective interior, low intrinsic luminosity primarily derived from depletion of thermal energy) suggests that a continuum of such objects exists, extending from objects with mass $M \sim 0.001$ $M_\odot$ (Jupiter) up to the hydrogen-burning limit at $M \sim 0.08$ $M_\odot$ (Burrows et al. 1989). However, Boss (1986b) has shown that a minimum mass exists for collapse of a fragment of an interstellar H-He cloud into a condensed object of the brown-dwarf or Jovian type, and this minimum mass is ~ 0.02 $M_\odot$, or about 20 Jovian masses. Thus giant planets must apparently form by a two-stage nucleation process, in a solar nebula where solid particle densities are sufficiently high for trigger nuclei to form before the gas dissipates. The formation of brown dwarfs may be more analogous to the formation of giant gaseous protoplanets. The end products of either formation path are quite similar, differing principally in the existence or nonexistence of a dense rocky core.

B. Core-Instability Model

The core-instability model of the giant planets postulates that they initially grew through the accretion of planetsimals, i.e., in the same fashion that the terrestrial planets formed. However, in contrast to the terrestrial planets, the giant planets grew quickly enough and to large enough masses so as to gravitationally capture progressively larger amounts of gas from the surrounding solar nebula, prior to its dissipation. Eventually, a point was reached in the accretional growth of Jupiter and Saturn that gas accretion became a runaway process and these planets ended up with more gas mass than planetesimal mass. Uranus and Neptune may not have quite reached the point of having runaway gas accretion before the solar nebula dissipated. The term core refers to the planetesimal derived mass, which, at least initially, was concentrated in the central region of the protoplanet. The term instability refers to the eventual occurrence of a gas runaway phase, i.e., a period when gas was added extremely rapidly. A recent suggestion for forming Jupiter's core quickly has been made by Stevenson and Lunine (1988), but it is not clear how this mechanism can be applied to the other giant planets.

C. Critical Core Mass

These authors assumed that the planetesimal mass (high-Z material) was totally segregated in a central core from the gas (low-Z material) and that the gaseous envelope was in convective equilibrium throughout its extent and hence had an adiabatic temperature profile. They constructed a series of hydrostatic models, in which the core mass was varied. Above a core mass of about 100 $M_{\oplus}$ for a particular choice of solar nebula conditions, they were unable to construct a static model and suggested that larger-sized cores promoted a hydrodynamical instability in the gas of the surrounding solar nebula. Furthermore, the core mass depended sensitively on the choice of the nebular boundary conditions. Because of this sensitivity and the high core mass required for instability for the nominal choice of parameters, this model did not seem to readily explain the near constancy of the core masses of the giant planets nor the absolute values of these masses.

In the late 1970s, Mizuno and his colleagues revisited the core instability model. Rather than assuming a convective envelope, they calculated the radiative temperature gradient and only in places where it was superadiabatic was it reset to the adiabatic value. In their classic paper, Mizuno et al. (1978) found that large radiatively stable regions existed within the envelopes of their static models. As a result, the critical core mass above which it was no longer possible to construct a static model was found to be about 10 $M_{\oplus}$. Furthermore, the value of the critical core mass was shown to be very insensitive to the temperature/density values in the surrounding solar nebula and to the amount of grains postulated to be present in the envelopes. Such a result appeared to offer a natural explanation for the great similarity (within a factor of several) of the high-Z masses of the giant planets, with these masses ranging from a little more than 10 $M_\oplus$ for Uranus and Neptune to several tens of $M_\oplus$ for Jupiter and Saturn. Some addition of planetesimals following the onset of runaway gas accretion could readily account for the modest differences in the high-Z masses of the giant planets.

At the time Mizuno was performing his calculations, Harris (1978) was also examining this class of models and estimating the factors that controlled the critical core mass. He pointed out that the value of the critical core mass might depend sensitively on the rate of core accretion, an insight that has turned out to be quite important in some subsequent studies of this problem. Stevenson (1982a) was able to derive analytically Mizuno's major results. He found that the critical core mass decreased significantly when the opacity in the envelope decreased, the accretion time scale increased, or the mean molecular weight of the envelope increased. Studies of the core instability in the case of an isothermal atmosphere were done by Sasaki (1989).

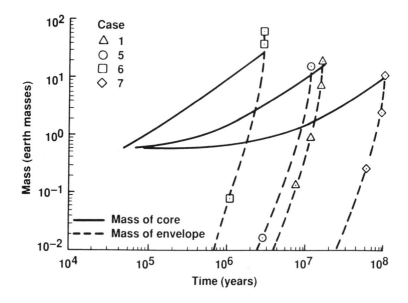

Figure 2.   Mass of core (solid curve) and envelope (dashed curve) as a function of time. Cases 1, 6, and 7 are for accretion rates of 10^{-6}, 10^{-5}, and 10^{-7} $M_\oplus$ yr^{-1}, respectively. The full solar abundance of small grains is assumed. Case 5 is the same as case 1, but the grain abundance has been reduced by a factor of 50.

The first evolutionary calculations of the core instabilty model were performed by Bodenheimer and Pollack (1986). These workers used a code, similar to ones used to model the pre-main-sequence evolution of stars, to follow the time history of accretion. They did this by constructing a sequence of static models that were connected in time by a prescribed, time-independent accretion rate of planestimals and a calculated rate of gas addition from the surrounding solar nebula. The latter was based on the amount of contraction that the envelope underwent in a given time step, with the amount of gas added depending on the gas density in the solar nebula and the size of the outer zone temporally opened up by contraction. Figure 2 illustrates the temporal evolution of the core (solid curves) and envelope (dashed curves) masses for several models of Bodenheimer and Pollack (1986). Cases 1, 6, and 7 correspond to planetesimal accretion rates of 10^{-6}, 10^{-5}, and 10^{-7} $M_\oplus$ yr^{-1}, respectively, with there being a full solar abundance of small grains in all these models. Case 5 is the same as case 1, except that the grain abundance has been decreased by a factor of 50. According to this figure, during the early growth of the protoplanet, when its core mass is on the order of a few $M_\oplus$, the mass of the envelope represents only a tiny fraction of the total mass. However, as the core mass approaches a critical value of about 15 $M_\oplus$, the envelope mass rapidly increases and becomes comparable to the core mass. Beyond this point, the planet's mass rapidly increases due almost entirely to the addition of gas to the envelope. Although the gas accretion rate becomes progressively more rapid as the critical core mass is approached and exceeded, the gas accretion is never hydrodynamic as long as the solar nebula can supply gas fast enough. For the purposes of the current discussion, we will define the critical core mass as being the value at which the envelope mass equals the core mass (note that in these evolutionary calculations, the rate of gas accretion varies smoothly with time); and the runaway gas accretion phase as commencing when a critical core mass is reached (while the gas accretion rate varies smoothly with time, it exponentially increases).

Figure 3 illustrates the temporal variation of the luminosity of the protoplanets of Fig. 2. When the core mass lies below the critical value, the luminosity is provided chiefly by part of the gravitational energy released by the accreting planetesimals. The rest of this energy goes into heating the envelope. However, as the critical core mass is approached and exceeded, the luminosity is generated chiefly by part of the gravitational energy released by the envelope's contraction. At this stage, a very strong positive feedback exists between envelope contraction and gas accretion in the sense that the more massive the envelope is the more rapidly it contracts and the more rapidly it contracts the greater the rate of gas addition. Thus, the critical core mass reflects the transition point between when the planetesimal accretion rate is large enough to keep the envelope puffed up and its rate of contraction moderate to when these conditions are no longer met.

Bodenheimer and Pollack (1986) found that the critical core mass depended most sensitively on the rate of planetesimal accretion. As can be

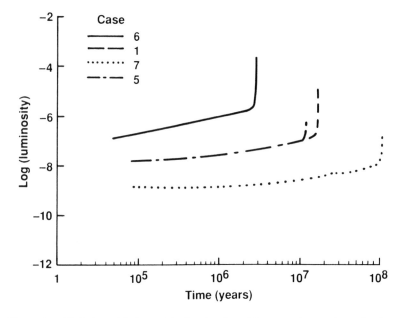

Figure 3.   Variation with time of the luminosity of the protoplanets shown in Fig. 2.

anticipated from the above discussion, the critical core mass decreased when the planetesimal accretion rate was lowered. In particular, the critical core mass equalled 11.2, 16.8, and 28.9 $M_\oplus$ for models 7, 1, and 6, which had planetesimal accretion rates of 10^{-7}, 10^{-6} and 10^{-5} $M_\oplus$ yr^{-1}, respectively. The critical core mass changed very little when the solar nebula boundary conditions were varied, in accord with Mizuno's results, and when the grain abundance was decreased by a factor of 50. The latter result does not appear to agree with Stevenson's (1982a) results. The reason for this may be that molecular opacity (e.g., water vapor) has a more important influence on the overall structure of the envelope than does grain opacity: the critical core mass did change significantly when both the grain and water abundances were decreased by sizable amounts.

   All the calculations discussed so far have been based on a fairly naive specification of the planetesimal accretion rate, namely that it is constant in time. However, significant variations in this rate can be expected. As the planet's mass increases, its gravitational influence extends further out, and the size of its feeding zone increases. In addition, its envelope becomes massive enough to stop planetesimals passing through it, hence increasing the capture cross section. Both of these processes tend to increase the accretion rate. On the other hand, as the feeding zone becomes depleted of planetesimals, the accretion rate will decrease. Pollack et al. (1990,1991) are carrying out the first set of calculations in which both the planetesimal and gas accretion rates are being calculated in a self-consistent fashion. The planetesimal accretion rate is being determined for situations in which their random velocities are determined by a combination of gravitational interactions with the protoplanet (this yields a velocity comparable to the escape velocity at its tidal radius) and with one another (this yields a velocity comparable to the escape velocity at their surfaces). The three-body calculations (planetesimal, planet, Sun) of Greenzweig and Lissauer (1990, 1992) are being used to evaluate the capture cross section and hence accretion rate, with allowance being made for capture occurring within the envelope. The planetesimal accretion rate calculated in this fashion (it is now time variable) is used in the protoplanet model of Bodenheimer and Pollack (1986) to evaluate the gas accretion rate.

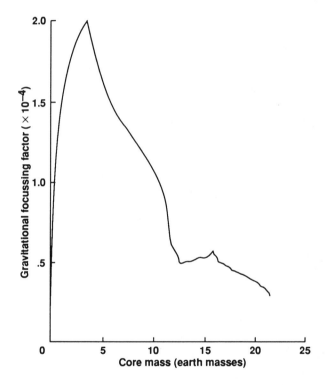

Figure 4.   Results of new simulations for a protoplanet at Jupiter's distance from the Sun. Ratio of computed accretion cross section to geometrical cross section as a function of core mass.

Figures 4–8 show typical results from these new simulations, for a model at Jupiter's distance from the Sun that begins with a surface density of planetesimals of about 10 g cm^{-2}, i.e., a factor of several times larger than the so-called minimum-mass solar nebula (just enough mass to make all the planets). All the planetesimals were assumed to be 100 km in radius. Figure 4 shows the ratio of the calculated accretion cross section of the protoplanet to

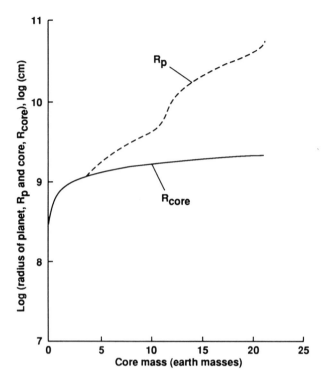

Figure 5.   Results of new simulations for a protoplanet at Jupiter's distance from the
    Sun. Radius of core (solid curve) and effective radius of capture (dashed curve) as
    a function of core mass.

its geometrical cross section as a function of core mass. Throughout the entire
accretion period shown, this ratio is on the order of 10^4 due to the dominant
effect of gravitational focussing: the random velocities of the planetesimals
are small compared to the escape velocity at the effective capture radius of
the protoplanet. Thus, accretion is fast.

    Figure 5 compares the radius of the core (solid curve) to the effective
capture radius (dashed curve). When the core mass is less than a few $M_\oplus$, the
envelope has too little mass to stop incoming planetesimals and hence they
are stopped at the surface of the core. For larger core masses, however, the
envelope is massive enough to stop incoming planetesimals well above the
core interface due to a slowing down of the planetesimals by gas drag and
their fragmentation by gas dynamical pressure. This larger "stopping radius"
leads to a larger three-body capture cross section (Greenzweig and Lissauer
1990). As a result, the accretion rate is significantly augmented.

    Figure 6 shows the envelope and total masses as a function of core mass.
Just as in the older calculations of Bodenheimer and Pollack (1986), a critical
core mass is achieved when the core mass equals about 16 $M_\oplus$. The reasons
are, however, very different. As shown in Fig. 7, the surface density of

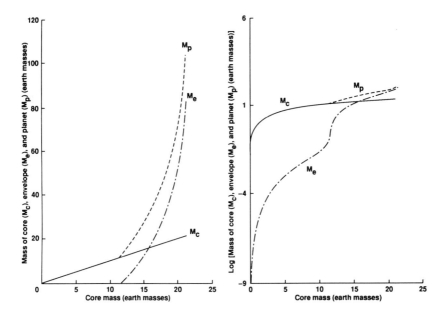

Figure 6.   Results of new simulations for a protoplanet at Jupiter's distance from the Sun. Envelope mass and total mass as a function of core mass.

planetesimals in the planet's feeding zone is severely reduced from its initial value when the core mass exceeds about 12 $M_\oplus$. This causes a very large drop in the rate of planetesimal accretion (Fig. 8), which results in the onset of a rapid contraction of the envelope and eventually runaway gas accretion. Prior to the depletion of planetesimals in the feeding zone, the rate of planetesimal accretion increases with time due to a combination of the envelope being able to stop planetesimals at increasing distances from the center and a growth in the size of the feeding zone (it varies as the cube root of the planet's total mass). Thus, it is more and more difficult to reach a critical core mass until the feeding zone is severely depleted. These results suggest that the critical core mass is *not* a universal constant, but rather depends sensitively on the surface density of planetesimals. Conversely, it may be possible to constrain the surface density of planetesimals from the joint constraints of the estimated high-Z masses of the giant planets (the surface density cannot be too large) and the lifetime of the solar nebula (the surface density cannot be too small). For the model shown in Figs. 4–8, a critical core mass is reached in about 7 Myr. Since the lifetime of the solar nebula was probably not much longer than this, the surface density at Jupiter's distance was probably not much less than 10 g cm^{-2}. Similarly, the surface density could not have been much more than a factor of several times larger than this number or Jupiter would have accreted too much high-Z mass.

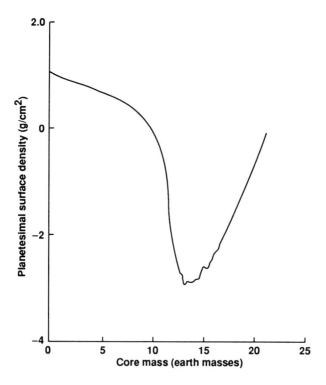

Figure 7.   Results of new simulations for a protoplanet at Jupiter's distance from the
Sun. Surface density of planetesimals as a function of core mass.

D. End of Accretion

We have followed the accretion of the giant planets to the point at which they
have accreted about 15 $M_\oplus$ of high-Z material, their envelope masses have
become comparable to that of the core, and they are approaching or have
entered into a runaway gas accretion stage. Here, we consider the factors that
may have led to a termination of gas and planetesimal accretion. At this point,
we note the important difference between Jupiter and Saturn on the one hand
and Uranus and Neptune on the other hand: most of Jupiter's and Saturn's
masses are contained in their low-Z components (H and He). Thus, they
experienced a runaway gas accretion phase, according to the core instability
model. However, the opposite is true for Uranus and Neptune. Therefore,
they did not experience such a phase. By considering factors that limited the
accretion of both components, we may gain some insight into the reasons for
this essential difference.

    We first consider the potential amount of gas mass that Jupiter might have
accreted. When fully grown, Jupiter's tidal radius equalled about 0.36 AU.
Jupiter's gravity severely perturbed the orbits of both solids and gases whose

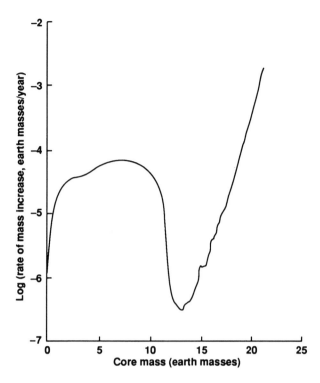

Figure 8.   Results of new simulations for a protoplanet at Jupiter's distance from the Sun. Planetesimal accretion rate as a function of core mass.

orbital radii lay within about 3 tidal radii on either side of its orbit (Greenzweig and Lissauer 1990). Augmenting the planetesimal surface density estimated above by a factor of 100 to allow for the surface mass density of gas, we find that Jupiter could have readily had access to about 2500 $M_\oplus$ of gas when it was fully grown. This may be compared to its current gas mass of about 270 $M_\oplus$. At earlier times, when its mass was less than its current value, Jupiter would have access to as much mass as it needed, since the tidal radius scales as the cube root of its mass. Thus, Jupiter (and Saturn too) should not have been limited in the amount of gas mass they accreted by the need to transport gas from zones outside of the neighboring region of the nebula that its gravity strongly affected. Indeed, it might seem puzzling at this point that Jupiter and Saturn did not acquire more gas.

One obvious limitation on the amount of gas mass accreted by the giant planets was set by the dissipation of the solar nebula. If a giant planet took long enough to reach a critical core mass, the amount of gas available in the solar nebula may have been substantially reduced from its earlier values by this dissipation. Since there was probably a fair spread in the time scales over which the four giant planets approached a critical core mass, with Jupiter probably reaching this point first and Saturn second, the dissipation of the solar nebula probably had its biggest effect on the amounts of gas that Uranus and Neptune accreted.

Another limitation on the amount of gas mass that the giant planets attained may have been set by their gravitational interaction with the gas in the surrounding solar nebula. Just like a small satellite in the region of planetary rings can open up a gap in the rings around them, so too can a protoplanet clear a gap around itself in the solar nebula when it reaches a large enough mass (Lin and Papaloizou 1979a). The condition for just being able to clear a gap is that the tidal gravitational torque that the planet exerts on the surrounding solar nebula equals the viscous torque by which the nebula tries to fill in vacated regions. The former torque is proportional to the planet's mass, while the latter torque depends on the effective (probably turbulent) viscosity coefficient. Tidal truncation could have occurred when Jupiter reached its current mass for plausible values of the turbulent viscosity coefficient (Lin and Papaloizou 1979a,1985). One might speculate that tidal truncation limited Saturn's gas accretion and perhaps even prematurely shut off gas accretion for Uranus and Neptune before they achieved critical core masses. Alternatively, the gas masses of Uranus and Neptune could have been limited by the solar nebula dissipating before their planetesimal feeding zones were severely depleted.

A planet's gravitational torque that can produce a gap results from a combination of its perturbation of the orbits of individual entities (e.g., gas molecules or small volumes of gas) and the mutual interaction (gravitational, viscous) of these entities. Both processes alter the affected orbits. Thus, gap clearing could have readily occurred for the gas component of the solar nebula, but may have been ineffective for planetesimals (they interact too infrequently). Furthermore, Jupiter at its present mass is quite effective at altering the orbits of planetesimals situated within a few AU of its orbit into ones that cross its orbit (Gladman et al. 1990). A similar statement can be made for virtually the entire region of the outer solar system once all four giant planets had formed (*ibid*). Thus, within a time scale much shorter than the age of the solar system, virtually any planetesimal located in the outer solar system (from a little beyond Neptune's orbit to somewhat inside Jupiter's orbit) would have had its orbit perturbed into one that crossed that of a giant planet. The fate of the planetesimal depended on the size of the planet at the time their orbits crossed. If the giant planet's envelope filled a fair fraction of its Hill sphere, as it did during all or most of the later stages of gas accretion, then the planetesimal would have been accreted. If the planet had shrunk to a size comparable to its current dimensions, then the escape velocity at its surface would have been large enough to scatter the planetesimal gravitationally to other parts of the solar system (Uranus and Neptune) or entirely out of the solar system (Jupiter and Saturn). In this case, a large fraction of the planetesimals would have either been lost from the solar system or sent out to the Oort cloud. Such a situation must have prevailed for the vast majority of the planetesimals in the outer solar system or the giant planets would have ended up with too large a high-Z mass. This conclusion may be readily seen from the following logic. The surface density of planetesimals had to be large enough so that the giant planets reached or came close to a critical core mass before the solar nebula dissipated. In that case, it took a range of radial distances of only a few tenths of an AU to supply Jupiter, for example, with about 15 $M_\oplus$ of planetesimals. However, Jupiter's high-Z mass is no more than several times larger than this value and it could have accreted planetesimals that were as much as a factor of 30 times further away than the outer limits of this early feeding zone. Similar conclusions hold for the other giant planets.

E. Planetesimal Dissolution

In discussing the core instability model for the formation of the giant planets, we may have given the impression that their two major building blocks— planetesimals and gases—remained spatially separated from one another, with the high-Z material located exclusively in a segregated core. Certainly, this picture is a good approximation to the situation during the early phase of formation, when the core mass was not more than a few $M_\oplus$. However, as the envelope grew in mass, it was increasingly difficult for planetesimals to reach the core before dissolving in the envelope (e.g., see Fig. 5) (Pollack et al. 1986; Podolak et al. 1988). Here, we briefly review the physics of the interaction of planetesimals with the envelopes of the forming giant planets and the implications for the interior structure and atmospheric abundances.

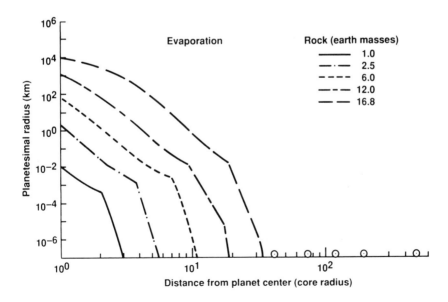

Figure 9.   Distance from planet center to which a planetesimal penetrates for core masses of 1.0, 2.5, 6.0, 12.0, and 16.8 $M_\oplus$ for rock planetesimals.

As a planetesimal passed through a giant planet's envelope, it was slowed down by gas drag; it was heated and vaporized by radiation from the shock in front of it (when its speed was supersonic) and from the warm ambient gases; and it was subjected to compressional stresses by gas dynamic pressure, that broke it into small "digestible" pieces when this pressure exceeded its compressional strength. Figures 9 and 10 show the distances from the planet's center (in units of the core's radius) at which rocky and icy planetesimals, respectively, would have been totally vaporized as a function of the size of the planetesimal (Pollack et al. 1986). The five curves in each figure correspond to 5 different stages in the growth of a giant planet, as denoted by the value of its core mass, for case 1 of Bodenheimer and Pollack (1986) (see Fig. 2). According to these results, which are representative of other cases of Bodenheimer and Pollack (1986), all but the smallest-sized planetesimal (less than a km) can penetrate to the planet's core when its core mass is less than several $M_\oplus$. However, only planetesimals larger than the size of the Earth are capable of penetrating this far when the core mass approaches its critical value. Thus, a significant fraction of the planetesimals that formed the giant planets, especially the later arriving ones, were dissolved in their envelopes.

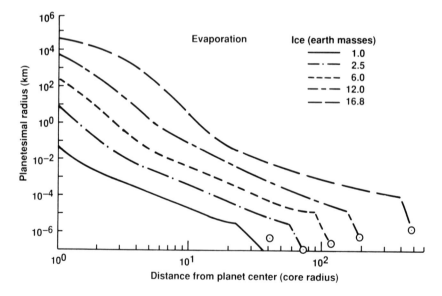

Figure 10.   Distance from planet center to which a planetesimal penetrates for core masses of 1.0, 2.5, 6.0, 12.0, and 16.8 $M_\oplus$ for ice planetesimals.

This result has several interesting implications. First, some nontrivial amount of mixing between the low- and high-Z components of the giant planets characterizes their present interiors, in accord with recent model results.

Second, there could exist significant molecular weight gradients within the envelopes of the giant planets, which could significantly affect the transport of heat from the planet's interior to its photosphere. This possibility may be important for understanding the current excess luminosities of the giant planets, as has been pointed out by Stevenson. Third, those dissolved high-Z elements that were mixed up to the observable atmosphere could lead to marked enhancements in the abundances of certain gas species. For example, the abundances of CH_4 in the present atmospheres of the giant planets imply C/H ratios that are factors of several (Jupiter) to several tens (Uranus and Neptune) larger values than that expected from solar elemental abundances. This result can be attributed to a combination of the giant planets forming preferentially from planetesimals as opposed to gas, and the dissolution of a significant fraction of the planetesimals in the envelopes of the forming giant planets (Pollack et al. 1986; Simonelli et al. 1989). Indeed detailed modeling of the data suggest that 10% or more of the carbon in the outer solar nebula was present in the solid component of the nebula (Simonelli et al. 1989). Finally, materials formed from the dissolved planetesimals, if mixed towards the top of the envelopes, served as the source material for making satellites, according to one model of their formation, which is discussed below.

F. Origin of Satellites

The origin of both the irregular and regular satellites of the giant planets is tied in intimate ways to the early history of their parent planets. Here, we briefly summarize some of the more popular hypotheses on the origin of each of these classes of satellites, and relate them to the early evolution of the giant planets. A more detailed discussion of some of these matters may be found in review articles by Stevenson et al. (1986) and Pollack et al. (1991).

The irregular satellites formed within the outer solar nebula, as was discussed above, and were subsequently captured into planet circling orbits. Table II summarizes some of their key properties. To affect capture, the relative velocities of the bodies needed to be reduced to below the escape velocity from the planet's gravitational field. This may have been accomplished either through the gas drag they experienced in passing through the outer envelope of the planet or its associated nebular disk (Pollack et al. 1979; McKinnon and Leith 1990) or through collision with a regular satellite or another planetesimal (Columbo and Franklin 1971; Harris and Kaula 1975; Goldreich et al. 1989).

First consider the gas drag capture hypothesis. For the velocity of a planetesimal to be substantially reduced, it needed to encounter a gas mass comparable to its own mass along its trajectory (Pollack et al. 1979). This requirement can be met by models of the outer envelopes of the giant planets when they filled a fair fraction of their Hill spheres (Pollack et al. 1989) or by models of the disks within which the regular satellites formed (McKinnon 1989). In both cases, the gases needed to extend out to several hundred planetary radii to be consistent with the present locations of the irregular satellites (note that McKinnon's [1989] gas drag model of Triton's capture supposes that its semimajor axis just after capture was much larger than its present value and that it tidally evolved to its present location).

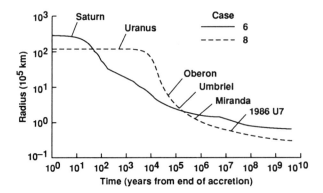

Figure 11.   Tidal radius of Saturn and Uranus as a function of time. Also shown are the positions of some of Uranus' satellites.

Following the capture of a planetesimal, gas drag continued to act on it, causing a reduction in its semimajor axis. If this occurred for too long (tens to thousands of orbital periods, depending on the initial capture conditions), the planetesimal would have evolved into the planet's deep interior and therefore would have been lost from the satellite system. Relatively rapid removal of gas from the position of the captured body is needed to prevent its loss. This requirement can be met by some plausible scenarios of the early history of the giant planets. For example, following tidal truncation of the solar nebula, the outer radius of the envelopes of the giant planets may have shrunk very rapidly, as illustrated in Fig. 11 (but, see the discussion below). Finally, the occurrence of "families" of the irregular satellites of Jupiter (irregular satellites that have orbits with similar semimajor axes and inclinations) can be accounted for by postulating that gas dynamical pressure experienced during the initial capture exceeded the compressional strength of the original object, causing it to fragment into several large pieces (Pollack et al. 1979).

Alternatively, the irregular satellites may have been captured by colliding with another planetesimal passing through the Hill sphere of the giant planet (Harris and Kaula 1975) or with a regular satellite (Columbo and Franklin 1971; Goldreich 1989). In this case, capture occurred after the giant planet evolved to a small size and/or after its regular satellites had formed. Given the large number of planetesimals that were accreted to form each of the giant planets, most of the time with the capture radius being much smaller than the tidal radius, it seems almost inevitable that pairs of planetesimals would have collided within the planet's gravitational sphere of influence, but outside of its envelope. Furthermore, such collisions would almost certainly fragment the original bodies, thus accounting for the occurrence of two families of irregular satellites around Jupiter. It is not clear, however, why Saturn's Phoebe and Neptune's Nereid do not apparently have companions. If the irregular satellites were captured by collision with a regular satellite, then the regular satellites were initially present to much larger distances than they currently are (cf. Table II). It is then a bit puzzling why there is not a greater overlap in the locations of the irregular and regular satellite systems.

In contrast to the irregular satellites of the giant planets, the regular satellites formed within circumplanetary disks that were situated in the equatorial planes of their parent (true rather than adopted) planets. These disks could have formed in at least four different ways (Pollack et al. 1991). The disks may have formed directly from gases and solids in the surrounding solar nebula that flowed into the Hill spheres around the giant planets (accretion-disk model); from the outermost portions of the envelopes of the giant planets that rotated rapidly and so were left behind as the the envelope contracted (spin-out model); from the blowout of part of the planet's envelope that accompanied the impact of a massive planetesimal (blow-out model); and from the collision of a pair of planetesimals (co-accretion model). The nature of these models and their pros and cons are extensively discussed in Pollack et al. (1991). Here, we focus on the two most plausible models: the accretion-disk and spin-out models and relate them to aspects of the early histories of the giant planets.

Consider first the accretion-disk model. This model postulates that near the end of the gas accretion phase, a giant planet shrank to a size not much larger than its present size and the inflowing solar nebula material formed a disk around the planet, rather than being directly accreted. If the solar nebula failed to supply gas quickly enough to fill the region vacated by the contracting envelope, then the planet would have shrunk on a very short time scale from dimensions comparable to its tidal radius to dimensions not much smaller than its present size (Bodenheimer, personal communication). In that case, the angular momentum of gas flowing into the Hill sphere may have been enough for it to have formed a disk around the planet. Although Kepler shear caused the gas outside the Hill sphere to have retrograde angular momentum, it would have been changed to prograde angular momentum due to three-body gravitational effects (Sun/planet/gas) as it flowed into the Hill sphere (Coradini et al. 1989). The solids that flowed in with the gas (e.g., small grains) as well as planetesimals that were captured by gas drag in passing through the disk could have provided the raw materials from which the regular satellites formed before the gas component of the disk was dissipated (e.g., by viscous processes that caused most of it to flow into the planet).

Alternatively, the satellite-forming disks of the giant planets were derived from the outer envelopes of the giant planets. According to this model, the spin-out disk model, the outermost portions of the envelope were rotating sufficiently rapidly when the planet filled a sizeable fraction of its Hill sphere, that they reached centrifugal force balance as the planet contracted following the end of gas accretion. Korycansky et al. (1990) have investigated the viability of this hypothesis by simulating the transfer of angular momentum within the envelopes of the giant planets during their accretion and subsequent contraction phases. They assumed that convective regions of the envelope tended towards a state of uniform angular rotation on a turbulent viscous time scale, as estimated from mixing length theory; and they derived a parameterization for the time scale and final state for angular momentum transfer in radiatively stable zones when there is an unstable gradient of specific angular momentum, j (j increasing with decreasing radial distance). In the latter case, the radiative zone tends towards a state of constant j.

Figure 12 illustrates the distribution of j within the envelope of a forming giant planet at three stages in its accretion, as marked by the value of the planet's total mass. For this simulation, the accreted gas was assumed to have a time-independent value of specific angular momentum. Thus, the large gradients in j shown in this figure are strictly the result of a very efficient outward tranfer of angular momentum. In this figure, convective regions are places where j increases steeply outward, while radiative zones are locations of approximately constant j. By the time the planet reached a mass comparable to that of Saturn, the outermost envelope had a j that is about 4 orders of magnitude larger than that of the innermost region. In fact at this point, a large fraction of the envelope's total angular momentum is concentrated in the outer few percent of the envelope's mass. Very similar results were obtained for other choices of the specific angular momentum associated with the accreted gas.

When gas accretion ceased, the model planet began a sustained contraction phase. During the early phases of this contraction, there was an inner and an outer convection zone, with a radiative zone sandwiched in between. The planet soon reached a stage where the outermost zones had rotation rates comparable to those needed for centrifugal force balance and hence were left behind as contraction proceeded. When the evolution reached the point where the two convection zones coalesced together, a net inward transfer of angular momentum occurred due to the nonhomologous nature of the contraction. This prevented further shedding from occurring.

Figure 13 illustrates the nature of the disk shedding that accompanied the early part of the contraction of a model for which the accreted gas had a j that increased somewhat as the mass of the planet increased. The subpanels in this figure show the planet's radius and rate of mass shedding as a function of time after the end of accretion and the initial surface mass density of the resulting nebula. For this particular model, the disk's mass is about 1 $M_\oplus$; the total angular momentum of the disk is about twice that of a minimum-mass satellite-forming disk around Saturn; the disk ends up with about 80 % of the system's angular momentum; and the planet retains a total angular momentum that is about 1/3 that of current Saturn. In this regard, the simulated properties of the disk and its planet are comparable to those of the Saturn system. However, the initial disk extends out to several hundred planetary radii, about

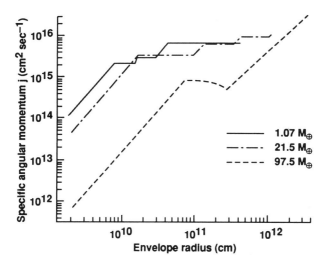

Figure 12.   Distribution of specific angular momentum within the envelope of a forming giant planet for core masses of 1.07, 21.5 and 97.5 $M_{\oplus}$.

an order of magnitude larger than the dimensions of the current system of regular satellites.

Thus, these first detailed simulations of the spin out disk model yield results that bear some resemblance to an actual giant planet system. Perhaps, more to the point, they indicate that outward transfer of angular momentum could have been a very effective process during the formation and early contraction of the giant planets and therefore the spin out model offers a possibly viable means for forming satellite-making disks around the giant planets. However, there are many issues that must be resolved before this model can be considered highly successful. These issues include obtaining a more quantitative match to the observed properties of the giant planets and their satellite systems, determining the degree to which the shedded portions of the envelopes contain an adequate amount of high-Z material to form satellites (especially whether they have enough silicates), and having an improved understanding of the transition from gas accretion to contraction (e.g., is there a rapid contraction before gas accretion halts?).

IV. Giant Planet Evolution

It is instructive to estimate the highest temperature that could be attained within a giant planet following a process of rapid accumulation. If a spherical body is assembled from cold mass fragments which are initially at rest at

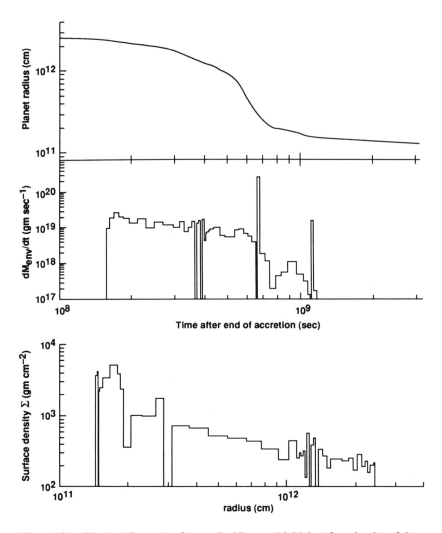

Figure 13.   Planet radius, rate of mass shedding, and initial surface density of the
    resulting subnebula as a function of time after the end of accretion.

infinite separation, the resulting object has a gravitational binding energy

$$\Omega = -q\frac{GM^2}{a} \qquad (9)$$

where a is the radius of the body, and the dimensionless coefficient q depends
on the distribution of mass within the body. For a body like Jupiter, theoretical
models give $q \sim 3/4$. Since the initial energy of the fragments was zero, an
amount of energy equal to $-\Omega$ has been dissipated. If we assume that the
accumulation was so rapid and efficient that none of this energy was radiated

into space, then the energy must show up as heat in the accumulated body, whose resulting mean temperature T_{max} we may estimate according to

$$T_{\max} \sim -\Omega/MC \qquad (10)$$

where C is a mean heat capacity per unit mass, which we may estimate as $C \sim k_B/Am$. Here k_B is Boltzmann's constant, m is the atomic mass unit, and A is the mean atomic mass of the accumulated material, assumed to be ~ 1 for Jupiter and Saturn, and ~ 6 for Uranus and Neptune.

It turns out that T_{max} is quite similar for all four giant planets, and is $\sim 10^5$ K. Furthermore, this value of T_{max} is coincidentally quite close to the degeneracy temperature of these bodies, which is the temperature above which the pressure equation of state would approach that of an ideal gas, and below which the pressure equation of state would become insensitive to the temperature.

Theoretical calculations of the thermal evolution of the giant planets can provide important information about the initial accumulation of these bodies. The problem is (in its simplest version) a well-posed one: at time $t = 0$ we form a giant planet at temperature T_{max}, and we follow the thermal evolution of the body until $t = 4.5$ Gyr, at which point the theoretical value of the effective temperature T_e (the temperature of a blackbody with the same bolometric power as the planet) is evaluated and compared with the observed value $T_{e,o}$. If $T_e \sim T_{e,o}$, the model of rapid and efficient collapse of the protoplanet is validated. A result that $T_e > T_{e,o}$ suggests that the initial accumulation was a slow or inefficient process, or that the current radiation of stored heat of accumulation is impeded by a process which is not considered by the model. A result that $T_e < T_{e,o}$ implies that the planet possesses energy sources in addition to stored heat of accumulation.

A. Model Results for Jupiter and Saturn

The simplest thermal evolution model for Jupiter or Saturn assumes that there are no phase boundaries in the interior, and that the temperature distribution in the interior follows an adiabat in the form

$$T = f(g, T_e)\rho^\gamma \qquad (11)$$

where ρ is the mass density, γ is the Grüneisen parameter, and $f(g, T_e)$ is a coefficient which depends on the planet's surface gravity g and effective temperature T_e. Such a model gives the result

$$t \sim \frac{\alpha}{2.757} T_e^{-2.757} \qquad (12)$$

where α is a constant which depends on the planet's mass and composition (Hubbard and Stevenson 1984). This model has no free parameters, and requires only that the effective temperature be much larger than the present value when t is much smaller than the present value. For Jupiter, the model gives $T_e \sim T_{e,o}$ at the present t, but for Saturn, it gives $T_e \sim T_{e,o}$ at $t = 3.9$ Gyr, so that $T_e < T_{e,o}$ at present. The most commonly accepted explanation of this discrepancy is that Saturn derives a significant component of its present luminosity from unmixing of helium from hydrogen due to immiscibility at high pressures (Stevenson 1975).

Thus, both Jupiter and Saturn have present interior heat flows which are consistent with an initial high-temperature, high-luminosity state, and with initial interior temperatures approximately one order of magnitude higher than present values.

B. Model Results for Uranus and Neptune

A simple thermal evolution model for Uranus or Neptune gives results similar to Eq. (12), but with a smaller value of α (Hubbard and MacFarlane 1980a). However, $T_e > T_{e,o}$ for both planets. For Uranus, the most recent determination from Voyager 2 measurements gives $T_{e,o} = 59.1 \pm 0.3$ K, which is not significantly different from the effective temperature that the planet would have solely from equilibrium reradiation of absorbed sunlight, $T_{eq} = 58.2 \pm 1.0$ K (Pearl et al. 1990). Neptune has an almost identical effective temperature, $T_{e,o} = 59.3 \pm 1$ K, but a lower value of T_{eq} because of its greater distance from the Sun, $T_{eq} = 46$ K (Conrath et al. 1989; Lowenstein et al. 1977).

Hubbard and MacFarlane supposed that these results imply slow, inefficient accumulation of both planets, such that initial interior temperatures were much smaller than T_{max}, and were in fact only slightly higher than present values. Thus the accumulation of Uranus and Neptune may have been a much different process than that of Jupiter and Saturn, with the higher value of T_{eq} for Uranus compared with Neptune playing a crucial role in the planet's subsequent thermal evolution. On the other hand, Podolak et al. (1991) argue that the accumulation of Uranus and Neptune must have been rapid enough to cause substantial heating in any case, and that the present low heat flow from Uranus is a result of poor transport of internal heat, with convection substantially impeded by chemical gradients.

Current interior models of Uranus and Neptune posit an outermost layer of nebular hydrogen and helium, extending to pressures ~ 100 kbar. The adiabatic variation of temperature with density in hydrogen is quite rapid, and if it is followed throughout this region, one infers temperatures of several thousand K at the highest pressures. The temperature variation in deeper ice-rich layers is much more modest, so that present-day interior temperatures are primarily established by the outermost $\sim 10\%$ of the planetary mass in adiabatic models. At present there is little direct evidence that such a rapid temperature rise actually exists in the deep atmosphere of either Uranus or Neptune. More work is needed on these planets' thermal state.

V. Conclusions

Modeling the interiors of the outer planets has shown that these planets all have departed significantly from solar composition. In all cases these planets have cores, composed primarily of rock and/or ice of between 5 and 20 $M_\oplus$. Surrounding these high density regions are regions of much lower density, composed primarily of hydrogen and helium. The observed differences in the He/H ratio among the outer planets is probably the result of a rearrangement of helium in the planetary interior, and does not reflect the actual volume integrated abundance. The C/H ratio shows a steady increase with increasing distance from the Sun. This does seem to reflect the actual abundance of the high-Z material in these bodies. The core-instability model provides a means for understanding both the structure and composition of the outer planets (the low observed abundance of N on Uranus and Neptune is still a problem, however). In addition, it seems to be capable of explaining the observed properties of the satellite systems of these bodies. For the most part, these explanations are still qualitative, but it is hoped that when the next Protostars and Planets volume is written, some stronger results will be available.

ORIGINS OF OUTER-PLANET SATELLITES

JONATHAN I. LUNINE and WILLIAM C. TITTEMORE
University of Arizona

The natural satellites of the planets represent a collection of physically and chemically diverse objects which originated in a number of distinct ways. In the context of the present book, which concerns the formation of planetary systems as a general phenomenon of galactic evolution and as a specific event in the case of our solar system, we present a highly selective treatment of the physics and chemistry of outer solar system satellites. Models for the formation of the regular satellite systems of the giant planets generally invoke a disk of material, but the origin of the disk, and its genetic relationship to the solar nebula and the forming giant planet are vigorously debated. In the case of Neptune, the only remains of a regular satellite system appears to be the small inner satellites discovered by Voyager 2; Triton was probably captured from solar orbit. Much of what we see today in terms of satellite surface geology and recorded tectonic activity may in principle be a result of tidal heating induced by resonant interactions between satellites as their orbits tidally evolved early in the history of the solar system. The volatile budgets of the large icy satellites and Pluto/Charon provide an interesting set of compositional constraints on the relationship between interstellar gas and grains, and the material which formed planetesimals in the outer solar system. The ratio of CO to CH_4 on both Titan and Triton suggest significant chemical evolution beyond the original interstellar composition, and beyond the mixture seen in comet Halley. The abundance of deuterated methane on Titan and elsewhere in the outer solar system has been used to argue that there is a component of relatively unaltered interstellar material preserved today in Titan; however, other mechanisms may have operated after formation of Titan to enhance its atmospheric deuterium abundance.

The natural satellite systems of the planets constitute a field of study in their own right, (Burns and Matthews 1986, eds.). Ideas regarding the origins of satellites are diverse and numerous, due not merely to uncertainty but also to the intrinsic variety of the systems themselves. Recent review articles dealing with satellite origins include Stevenson et al. (1986), who discuss the general issue of satellite formation, and reviews specialized to the outer solar system (Coradini et al. 1989), and the regular satellite systems of Saturn (Pollack and Consolmagno 1984) and Uranus (Pollack et al. 1991). Earlier reviews of Jovian satellite formation exist but should be regarded as out of date. For the inner solar system, the Martian satellites are covered by Burns (1992) and the Earth's Moon by an entire book (Hartmann et al. 1986). The present review, then, is intended to be highly selective and addresses those aspects of satellite origin and evolution which bear on the formation of the solar system itself. We have further chosen to specialize the chapter toward the chemical evidence contained in satellites concerning the modification of interstellar material which went to form the solar system. This is an area which has seen the acquisition of qualitatively new data in the past five years both for satellite and comet composition, and for abundances in interstellar clouds (see Chapters by Mumma et al. and Van Dishoeck et al.). In addition, we cover new results on the tidal evolution of outer-planet satellite systems, as this process is significantly more important than previously thought in determining satellite properties.

The chapter is organized as follows. In Sec. I we briefly outline the gross physical and chemical properties of the satellite systems of the outer planets, largely deferring to previous reviews for older information which remains up to date. Section II then outlines models for regular satellite formation, as understanding the physical and chemical processes occurring during formation is a prerequisite to relating satellite properties to the larger issue of solar system formation. The subsequent evolution of the satellites, with particular emphasis on tidal evolution, is covered in Sec. III, including a discussion of the dynamical origin of Triton, an important cosmochemical tiepoint. Section IV then concerns the volatile budgets of Titan, Triton and Pluto/Charon, and presents models for the origin of these budgets which tie them to the larger set of processes associated with solar system formation. Finally, we summarize in Sec. V with a list of some outstanding issues in this field.

I. PROPERTIES OF THE SATELLITE SYSTEMS

A. Physical Properties

The basic physical properties of the satellite systems of the giant planets are summarized in Table II in the Chapter by Podolak et al., updated from Pollack et al. (1991). The interesting systematics and regularities provide a compelling argument for a common origin for at least the Jupiter, Saturn and Uranus systems (Pollack et al. 1991). The Neptune system is distinguished from the other three by having fewer satellites, a truncated spatial extent of regular satellites, and a much higher satellite system angular momentum (relative to the planet). All these features support models in which Triton is captured from solar orbit and evolves to its present state while collisionally consuming, or interrupting the formation of, a regular satellite system (McKinnon 1984; Goldreich et al. 1989). One might ask why the mass of Triton normalized by that of Neptune comes so close to the value for the satellite systems of the other giant planets. If this is simply a coincidence, then one turns to the broader issue of why Pluto/Charon and Triton exist at all, what determined their sizes, and how many other such objects are in solar orbit (Stern 1991).

The irregular satellites appear to have no systematic properties from planet to planet. Those in the Jupiter system are best explained as the result of a head-on collision of circum-Jovian material with a solar orbiting body, thereby producing families of pro- and retrograde satellites (Colombo and Franklin 1971); although fragmentation due to gas drag has also been invoked

(Pollack et al. 1979). Insofar as this type of collision is now invoked as a way of locking Triton into a retrograde Neptunian orbit (Goldreich et al. 1989), the formation of the Jovian irregular satellites has become of increased interest.

B. Chemical Properties

Rock-to-Ice Ratios. The satellite systems contain bodies with densities consistent with compositions ranging from pure silicate to essentially pure water ice. The validity of this statement is predicated on the notion that elemental abundances (Anders and Grevesse 1989) are such that the two most abundant condensibles are silicates ("rocks") and water ice. In fact, rock should include iron, nickel and other metals; water ice, depending on the temperature, may include ammonia hydrate, carbon dioxide ice, and condensed or trapped methane, carbon monoxide, noble gases, etc. Involatile organic species may be hard to assign to either category. Unless an object has a density, corrected for the effects of internal compression, which falls outside the range encompassed by silicates and ice (between 3.5 and 0.92 g cm^{-3}), one cannot make a more precise statement without applying detailed chemical models.

One can, however, compare the inferred rock-to-ice ratio in the satellites with predicted values for the environment of the solar nebula and circumplanetary formation regions (which are described in Sec. II). Simonelli et al. (1989) and Simonelli and Reynolds (1989) have most recently done this for the outer solar system, and their results are to be preferred over earlier similar computations. The conceptual argument goes as follows: the amount of oxygen available to form water is primarily determined by the solar photospheric C/O ratio. For quite some time, this was thought to be 0.60 (Anders and Ebihara 1982), but was revised downward to 0.43 by Anders and Grevesse (1989). A new determination puts the number at 0.47 (Grevesse et al. 1991). Examination of molecular chemistry in the H_2-rich gas of the solar nebula indicates that oxygen is taken up in the following order: refractory (silicate) minerals, carbon monoxide, water (Prinn and Fegley 1989). Other molecules like CO_2 will also remove oxygen, but not to an extent sufficient to affect the water abundance. When one computes the resulting rock-to-ice ratio for CO being the dominant carbon-bearing species (Chapter by Prinn), it corresponds to a rock mass fraction of 0.51 to 0.71. (This rock mass fraction is relative to rock plus ice, and corresponds to C/O ranging from 0.42 to 0.60 according to calculations from Simonelli and Reynolds [1989].) On the other hand, if a body formed in a disk of material in which most of the carbon was in the form of methane (as we will argue for some of the satellites), much more oxygen is available for making water molecules, and the resulting rock mass fraction is between 0.34 and 0.39. Note that, for a C/O value of 0.60, the rock mass fraction is a very sensitive function of the oxidation state of the carbon, ranging from 0.39 to 0.71. With the new, lower C/O value of 0.43, the range is much smaller: 0.34 to 0.51 from CH_4-rich to CO-rich. A small additional effect comes from the state of hydration of the silicates, as shown by Prinn and Fegley (1989). The magnitude of the uncertainty introduced is much less than that caused by the uncertainty in the elemental C/O ratio, and is therefore ignored in the present chapter. A superb summary table of rock and ice mixing ratios is given in Zharkov and Gudkova (1991).

The resulting rock mass fractions can be tied to the satellite densities and masses by constructing interior models which take account of the varying compressibilities of the materials. This has been done by many groups over the years (an example of the early work is Lupo and Lewis 1979). As better information on physical properties has become available through spacecraft encounters (and for Pluto/Charon, mutual eclipses), the calculations have been revised. Current results are presented in Fig. 1 for objects of interest to us in this chapter, along with the expected rock mass fractions for objects formed in the solar nebula versus highly reducing protosatellite nebulae. Values for C/O = 0.43 and C/O = 0.60 are presented.

When interpreting the numbers in this figure, it is crucial to bear in mind that we are assuming that the rock-to-ice ratio in the feeding zones for various bodies reflects cosmic elemental abundances (as defined by the solar photosphere and meteorites). This may not be the case for objects forming in disks around giant planets, because physical processes in such disks and in the atmospheres of the proto-giant planets may alter the rock-to-ice ratio, as described in Pollack et al. (1991) and reviewed below. Even in the case of the solar nebula, one can envision processes in the disk which may alter the rock-to-ice ratio from that inferred from cosmic elemental composition (see, e.g., Stevenson and Lunine 1988). Thus, the results one might obtain from Fig. 1 are only a part of the story.

The icy Galilean satellites Ganymede and Callisto, as well as Titan, contain less ice than one would predict for a high-pressure, fully reduced disk around Jupiter or Saturn (Prinn and Fegley 1981). The simplest explanation is that these objects were large enough to lose water during accretion (Stevenson et al. 1986), which is energetically possible because the gravitational potential energy, per unit mass, for these bodies is comparable to or greater than the latent heat of vaporization. Alternative explanations, that Titan and the Galilean satellites formed from an admixture of solar nebula material, or that the Jovian and Saturnian nebulae were more rock-rich than the models predict, do not seem to be consistent with the lower densities of the other Saturnian satellites.

The Uranian satellites represent the interesting case of bodies which are clearly part of a regular satellite system, are too small to have undergone significant water loss during accretion, but have proportionately less ice that their comparably sized Saturnian counterparts. The extreme obliquity of Uranus has led to the suggestion that the satellite disk was spun out as a result of a giant impact, with consequent shock chemistry producing a water-depleted nebula (Stevenson 1984). On the other hand, disks formed in the absence of a giant impact may have a silicate enrichment relative to cosmic abundance (Pollack et al. 1991), and the very similar normalized masses and specific angular momenta of the Saturnian and Uranian systems suggest like

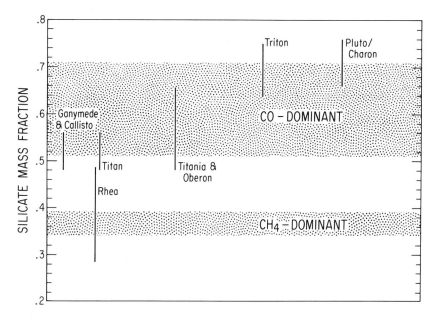

Figure 1.   Rock (silicate) mass fraction, relative to total rock plus ice, in outer solar system objects, compared to predictions for nebular disks in which CO or CH_4 is the dominant carbon species. The objects are assumed to be purely silicate and water ice. The upper-end of each field corresponds to the elemental C/O determination of Anders and Ebihara (1982); the lower end to Anders and Grevesse (1989). The fields are from Simonelli and Reynolds (1989). Object silicate mass fractions are shown as vertical lines. In each case, the range represents uncertainties in densities, and assumptions regarding the interior models: differentiated vs undifferentiated, and (anhydrous) silicate densities of 3.36 to 3.66 g cm^{-3} (see discussion in Simonelli et al. 1989). The Pluto/Charon model is based on a differentiated Pluto model and the bulk density determined for Pluto and Charon combined. For other references for object mass fractions see Johnson et al. (1987), and Smith et al. (1989), with some adjustments.

origins. In this instance, the chemical difference expressed by the rock-to-ice ratios goes unexplained, but may be found in the disparate sizes of the parent planets and/or random effects.

The Neptunian satellite Triton and the Pluto/Charon system have remarkably similar densities (Binzel 1989; Tyler et al. 1989), very close to that predicted for rock-ice objects formed from a CO-rich solar nebula for C/O of 0.60. However, the notion that these are "classical" solar nebula objects runs into three problems: (1) for the Anders and Grevesse C/O value of 0.43, these objects are significantly overdense compared to the predicted value for the solar nebula; (2) we do not know yet Pluto's density independent of Charon's; (3) there is some expectation that significant water loss may have occurred on Pluto during formation of the binary (McKinnon 1989), and some water loss could have occurred during the capture of Triton or subsequent tidal evolution of the orbit (McKinnon 1984; Goldreich et al. 1989). The difficulty lies in arranging the amount of water loss (per total mass) to be the same for the two bodies, which is a peculiar requirement given that their histories are thought to be unrelated.

Volatile Species. The most volatile components of the satellite systems, as well as refractory organic phases, cannot be determined uniquely from density information, but must be directly detected on surfaces and in atmospheres. Material more volatile than water ice has been detected on Titan, Triton, Pluto and comets. Relatively involatile organic phases have been detected in meteorites and inferred to be present on the rings and satellite surfaces which exhibit a very dark, reddish component. It is unclear whether such material is primordial or processed from methane by energetic chemistry after formation, but the latter hypothesis has been generally favored for the rings and satellites (see, e.g., Thompson et al. 1987*a*). Inference has been made of the presence of ammonia or other compounds which produce low melting point aqueous liquids based on the existence of resurfaced areas on small Saturnian and Uranian satellites (Stevenson 1982*b*; Croft et al. 1988; Jankowski and Squyres 1988). Any ammonia present on the surface of a body exposed to magnetospheric or charged particle irradiation is likely to have been converted to molecular nitrogen or perhaps intermediate products, and have escaped the satellite or migrated to polar regions.

Io, Titan and Triton are the only satellites which have an extensive surficial inventory of materials more volatile than water ice. Pluto is in a class with these objects as a small icy planet with surficial volatiles and an atmosphere. On Io, sulfur dioxide and possibly hydrogen sulfide (Nash and Howell 1989; but see Lellouch et al. 1990) appear as atmospheric constituents and surface frost deposits. Both of these volatiles are likely the product of an involved chemical chain involving sulfur, silicates, and possibly water at some time in the past. The extensive tidal heating of Io (Peale et al. 1979) has ensured significant chemical alteration of this satellite's volatile budget. Lewis (1982) argued that metal-free C3V or C2M chondrites would evolve into buoyant, sulfur-rich liquids and SO_2 upon heating at depth in Io. Whether these are plausible starting materials for Io cannot be evaluated at present; nor can alternative models involving leaching of sulfur during silicate dehydration (Fanale et al. 1974). Further progress in these areas may come from remote sensing of Galilean satellite mineralogy from the Galileo Near Infrared Mapping Spectrometer (NIMS). A thorough review of Io's atmosphere is given by Johnson and Matson (1989); we shall not say more about Io in this chapter.

Titan, Triton and Pluto are known to contain methane in their atmospheres (Kuiper 1944; Fink et al. 1980; Broadfoot et al. 1989). Methane surface frost is known or strongly suspected to be present on Triton and Pluto (Cruikshank and Apt 1984; Cruikshank and Brown 1986), and a surface reservoir containing liquid methane appears to be required on Titan to sustain the prodigious photochemical destruction of that molecule (Lunine et al. 1983).

The atmospheres of Titan and Triton are both dominated by molecular nitrogen (Lindal et al. 1983; Broadfoot et al. 1981,1989). In each case, the predominance of N_2 over CH_4 is a volatility effect, and does not directly correspond to the surface plus atmosphere ratio of these species, much less the original, bulk-interior value. In the case of Triton, the atmospheric nitrogen is almost assuredly in equilibrium with surface nitrogen ice, discovered spectroscopically from Earth (Cruikshank et al. 1984); if a deep methane-ethane ocean exists on Titan, then the atmosphere supports a significant dissolved nitrogen component at the surface of that satellite (Lunine et al. 1983). The 1988 stellar occultation by Pluto provided indirect evidence that its atmosphere contains N_2 and/or CO, though deciding how much of each is not possible at present (Yelle and Lunine 1989). The nitrogen or carbon monoxide in Pluto's atmosphere, curiously enough, cannot be in equilibrium with surface frosts if the IRAS-inferred temperature (Sykes et al. 1987) is correct (Hubbard et al. 1990b).

Other constituents in Titan's atmosphere include carbon monoxide, carbon dioxide and a suite of hydrocarbons and nitriles, all of which abundances are controlled by photochemical processes. Even the CO and CO_2, which on the basis of cosmochemistry would be present in the primordial satellite, appear to be controlled by photochemistry. The abundance of argon is very poorly constrained, and the other noble gases are completely unknown. In the case of Triton, an atmospheric upper limit on the mixing ratio CO to N_2 of 0.01 is obtained from Voyager ultraviolet spectrometer data (Broadfoot et al. 1989). Both CO and CO_2 frost have been detected on the surface from groundbased spectroscopy (Cruikshank et al. 1991), although abundances relative to N_2 or CH_4 are not yet determined. As noted above, the CO/N_2 value for Pluto's atmosphere is undetermined. The argon abundance is unknown on both Triton and Pluto.

Spectroscopic evidence strongly suggests that Charon's surface is composed of water ice with no methane (Marcialis et al. 1987); some weak evidence for an atmosphere exists but no quantitative limits are available (Elliot et al. 1991). Finally, observation of a coma development around the solar orbiting object Chiron has been interpreted as driven by "CO or a gas of similar volatility" (Meech and Belton 1990); we point out that N_2 has a vapor pressure which is similar. Comet composition is a complex subject unto itself; the best inventory to date is that provided by the extensive observations of Comet Halley. Though still controversial, the currently understood inventory is given in Table V of the Chapter by Mumma et al. and Table X of the Chapter by Van Dishoeck et al.

Table I summarizes the abundance ratios of key cosmochemical species in the atmospheres and on the surfaces of the volatile-rich bodies Titan, Triton and Pluto. Where appropriate, total mass estimates for individual species are also given. Figure 1 and Table I offer the basis for the treatment of satellite origin from the cosmochemical point of view in Sec. IV. We now set the stage by discussing physical environments of satellite formation.

TABLE I
Volatile Budgets of Selected Objects

Object	Total CH_4, g	Total N_2, g	Total CO, g
Titan			
Atmosphere	3×10^{20}	9×10^{21}	10^{18}
Dissolved in ocean	1×10^{22}	3×10^{21}	$10^{17} - 10^{18}$
In ocean & sediment as			
higher hydrocarbons	6×10^{22}		
Total ocean plus atmosphere	7×10^{22}	1×10^{22}	$1-2 \times 10^{18}$
Triton			
Atmosphere	3×10^{12}	5×10^{16}	$<5 \times 10^{14}$
Surface[a]	$>10^{19}$	10^{20}	$\leq 10^{19}$
Bulk primordial[b]	$1-20 \times 10^{22}$	2×10^{21}	$1 - 7 \times 10^{23}$
Pluto			
Atmosphere	$2-5 \times 10^{15}$	$\longrightarrow 3-6 \times 10^{15} \longleftarrow$[c]	

[a] Nitrogen and carbon monoxide numbers assume of order 1 meter thickness of nitrogen ice over one hemisphere (this is the maximum amount which can be moved from the summer to winter hemisphere over a major season). Methane is based on the rate of photolysis (Broadfoot et al. 1989), and thus includes methane already photolyzed and present as a surface deposit of higher hydrocarbons. Carbon monoxide number is based on atmospheric upper limit (Broadfoot et al. 1989) as discussed in text; CO frost has been detected spectroscopically (Cruikshank et al. 1991) but no estimated abundance is available.

[b] Estimated from cometary abundances given in the Mumma et al. Chapter.

[c] Approximately one-half the atmosphere by number is carbon monoxide and/or nitrogen, but the relative proportions cannot be distinguished (Hubbard et al. 1990b).

II. MODELS FOR THE FORMATION OF REGULAR SATELLITE SYSTEMS

As detailed in Pollack et al. (1991), the normalized masses and specific angular momenta of the regular satellite systems of Jupiter, Saturn and Uranus strongly suggest a common mechanism for formation. Angular momentum constraints dictate that a disk of material must be involved, but whether this disk was gaseous or particulate, derived from solar nebula material or the atmosphere of the proto-giant planet is at issue. Four distinct disk models can be enumerated (Pollack et al. 1991):

a. *Accretion disk:* By analogy with the solar nebula, material falling into the giant planet's gravitational well is redirected into an equatorial disk by virtue of its angular momentum. The disk material consists of grains, plus gas of solar-nebula composition. Further chemical evolution of the gas in the high-temperature, high-pressure portion of the disk will cause alteration of the trace molecular composition toward a more reduced state.

b. *Spin-out disk:* As the distended envelopes of the forming giant planets contract, material is left behind in the form of a disk, again by virtue of its angular momentum. The gas is a mixture of solar nebula gas and sublimated solid material, chemically reprocessed in the protoplanetary envelope. Additionally, some solid material may have survived entry into the envelope without sublimating, and was then spun out with little reprocessing. This disk could have an elemental and molecular composition very different from the solar nebula.

c. *Blow-out disk:* A giant impact with the protoplanetary atmosphere may spin out a disk of material. The impact generates high temperatures enabling shock chemistry and subsequent quenching to alter the molecular composition of the gas from that of the protoplanetary atmosphere (Stevenson 1984).

d. *Co-accretional disk:* Collision of planetesimals within the protoplanetary sphere of influence will generate a disk of particles, which may reaccrete to form satellites. Such disks consist of a mixture of rock and ice dictated by the original planetesimal composition, altered possibly by the thermal effects of the collision.

As discussed in Pollack et al. (1991), disk model (d) is implausible because it predicts a satellite system whose size is slightly less than the Hill sphere radius of the planet, which is orders of magnitude larger than the observed extent of the regular satellites. Model (a) requires certain constraints on the timing of giant planet envelope contraction relative to the lifetime of the solar nebula: for the solar nebula gas to be the direct source of the satellite systems, the parent planet envelopes must have contracted to a point inward of roughly $2 R_p$, where R_p is the current planetary radius. Model (c) is compelling for Uranus, given the planet's axial tilt, but the similarities between the Uranian satellite system and those of the other giant planets (see Table II of the Chapter by Podolak et al.) are not predicted by the impact model.

We now focus on the implications of conditions within satellite forming disks for the current composition of the satellites. We dispense quickly with the co-accretional disk because it appears not to explain the extent of regular satellite systems, but may be relevant to the irregular satellites, for which composition and density data are essentially nonexistent. The composition of satellites derived from such a disk depends upon the rock-to-ice ratio and volatile budget of the original impactors, and therefore on the original gaseous nebulae (solar or circum-planetary) from which they formed. The details of the collision process, and final particle size distribution, will tend to increase the refractory component fraction, as vaporized constituents may not recondense and in fact could be swept out of the system by magnetospheric processes (Stevenson et al. 1986).

Disk model (a), in which material is captured from the solar nebula, will have a total elemental gas and condensate composition identical to that of the solar nebula. However, processes associated with the infall and subsequent steady-state thermodynamic condition of the disk will alter the molecular composition of the material, including the rock-to-ice ratio. Such processes include (i) sublimation and chemical reactions associated with gas dynamic heating at the surface of the circumplanetary disk, and (ii) thermochemistry in gas-phase and gas-grain reactions. Although these effects have been examined for the solar nebula (Lewis and Prinn 1980; Hollenbach and Neufeld 1990; Lunine et al. 1991), only (ii) has been quantified for giant planet disks (Prinn and Fegley 1981). Process (i) is primarily of interest as a means of subliming water ice and more volatile constituents; the importance of this effect can be crudely estimated by comparing the latent heat of sublimation to the kinetic energy dissipated by gas drag as the grains go into Keplerian orbit. For infalling water ice achieving Saturn orbit at Titan's radial position, this ratio is roughly 0.4, indicating that substantial quantities of water may sublime off of grains (the actual mass fraction sublimed depends on particle size, and hence emissivity, as well as the rock fraction of the original grain). For Uranus, at the orbit of Oberon, the ratio is approximately unity, indicating less potential for water ice sublimation. In both systems, a portion of the water ice and other volatiles may be in the gas phase with the remainder in grains, complicating the chemical history of satellite-forming material.

Process (ii) acts upon molecules which may be transported inward by radial mixing either in the gas phase or on grains swept inward by gas drag and then sublimated. The outer solar nebula is expected to have $CO/CH_4 > 1$, and perhaps $>> 1$ if radial mixing is efficient (Chapter by Prinn). CO_2 is never more than 1% of the total carbon. The predicted outer solar nebula NH_3/N_2 value is less certain, since it depends very strongly on whether material was mixed toward the inner portion of the solar nebula where ammonia is effectively destroyed, and on the poorly-determined ratio in interstellar material (Chapter by Van Dishoeck et al.). When solar nebula material is mixed into the inner parts of circumplanetary disks, thermochemistry will act to increase the CH_4 abundance, to the point where it dominates over CO (Prinn and Fegley 1981); consequently the rock mass fraction drops from >0.5 to as low as 0.34 (Fig. 1). CO_2 continues to be a minor species, but may become competitive in abundance with CO. The NH_3 abundance in the circumplanetary disk equals or exceeds that of N_2. In the simplest view, then, we expect satellites formed in an accretion disk to have densities consistent with most of the oxygen locked up in water, abundant methane and ammonia, and little carbon monoxide and nitrogen.

Complicating the story is the realization that radial mixing of material from outer portions of the disk to the inner, higher temperature parts may not have been efficient enough to process most of the material available to make satellites. Although we make this statement by analogy with the solar nebula (Stevenson 1990; Prinn 1990), its validity is dependent upon detailed models of the transport of angular momentum and mass in circumplanetary nebulae, which have not been constructed. Also, although temperature profiles have been constructed for accretion disks around Jupiter and Saturn (Prinn and Fegley 1981; Lunine and Stevenson 1982), the same is not true for Uranus and Neptune. It is conceivable that around these smaller planets, the high temperature zones were sufficiently close in and limited in extent that chemical processing of most of the disk material is implausible. Pollack et al. (1991) point out that this type of disk may be decoupled from the planetary envelope, so that high temperatures may not be accessible at all in the Uranus and Neptune disks, and one might even argue in favor of this possibility for Jupiter and Saturn.

If we consider disk models (b) and (c), the chemical predictions become even more ambiguous. Model (c) postulates a shock event, for which the temperature and pressure history can be formulated and used to compute molecular abundances after cooling and consequent quenching (Stevenson 1984); such models predict an enhancement of CO, and increase in the rock mass fraction, relative to the accretion disk chemistry described above. However, as the source material for the disk is now the shocked planetary envelope itself, the total elemental inventory and specifically the abundances of rocky and icy materials are poorly constrained. There is no compelling argument that the inventory need be solar, because much of the rocky planetesimals (if large enough) could have survived to greater depths in the envelope without sublimating than could the water ice. Such compositional gradients may have survived in the face of convection (Stevenson 1985; Pollack et al. 1991). Alternatively, the impactor may have gouged deeply enough such that the source material for the blow-out disk contained a near-solar mix of species.

Model (b) suffers from similar uncertainties, as again the disk material is derived from the planetary envelope, in this case contracting from a highly distended configuration. If the dissolution of rocky and icy planetesimals in the envelope proceeded to differing degrees, the gaseous disk would have a nonsolar elemental abundance which altered the relationship between the rock mass fraction (observed through the satellite bulk density) and the oxidation state of the carbon-bearing species in the nebula (Simonelli et al. 1989). The fact that the intermediate-sized Saturnian satellites, which likely did not lose water during accretion (Stevenson et al. 1986), have rock-to-ice ratios consistent with a highly reducing nebula, is weak evidence that the rock-to-ice ratio of the Saturn disk was unaffected by the dissolution process. However, the Uranian satellite system implies just the opposite: if formed via model (b), the densities of those satellites suggest an enhancement in the rocky component of the disk.

The conclusion to be drawn is that the predicted oxidation state of satellite-forming material is highly uncertain, and even the satellite densities may reflect a long suite of processes which modified the basic elemental abundances of rock- and ice-forming materials from their solar values. One additional effect which must be considered is loss of water during accretion. Comparison of the virialized energy of collapse to the latent heat of vaporization of water shows that, energetically, the Galilean satellites and Titan could have lost substantial quantities of water during accretion, and plausibly enhanced their rock-to-ice ratios above that of the planetesimals from which they formed. Both Triton and Pluto/Charon suffered significant dynamical effects during their early evolution to the current state, and jetting of water during the formation of the Pluto/Charon binary has been proposed by McKinnon (1989). The similar observed densities of Triton and the Pluto/Charon system raise the interesting issue of whether their separate dynamical histories could plausibly produce similar final densities. Nonetheless, if the Grevesse et al. (1991) elemental abundances are correct, such water loss is required to explain the current densities if these bodies began with a solar mix of oxygen, carbon, silicon, etc., even for a CO-dominated nebula (Fig. 1; Simonelli and Reynolds 1989).

With regard to the deuterium-to-hydrogen ratio in the satellite disks, little work has been done. It is impossible, given the lifetime of protosatellite disks ($\sim$0.1 Myr; Stevenson et al. 1986) to create significant enhancement in deuterated methane and other heavy species, relative to their "solar" (essentially Jovian; Lutz et al. 1990) abundance. In fact, if the methane which we see in Titan today is assumed to have been produced thermochemically in the Saturn circumplanetary nebula, the required temperatures would have produced an enhancement factor of deuterated methane of only 1 to 2, less than the observed enrichment (Lutz et al. 1990). Either Titan's methane is of interstellar origin, unprocessed in the solar nebula or circumplanetary nebulae, and retentive of an enhanced deuterium signature, or processes occurred after Titan's formation to produce an enhancement (Pinto et al. 1986). We will return to this important point in Sec. IV where the volatile budgets of individual objects are examined more closely. Beyond Titan, and excepting comets, deuterium has not been detected in any icy body. Clearly this is a key measurement to be attempted in the coming years.

In this section, we have illustrated the complexity of the molecular composition of gases and grains which went to form satellites and other icy bodies of the outer solar system. Although there is no evidence for gross deviations from solar in the rock-to-ice ratios in the outer planet satellites, modest variations mask the chemical state of the gaseous disk from which these objects formed. As yet, we cannot decide between the three gaseous disk models [(a)–(c)], on chemical grounds, though dynamical considerations associated with giant planet formation (Chapter by Podolak et al.) weakly favor (b) over (a), and there is no compelling reason to argue that the Uranian satellites had to form in the blow-out disk rather than the spin-out disk (Pollack et al. 1991). The densities of the Saturnian satellites suggests that the disk carbon was sequestered in CH_4, with Titan having lost water during accretion. The Uranian satellites have a slightly higher density than the intermediate-sized satellites of Saturn, implying perhaps a much less massive satellite disk in which thermochemical conversion of CH_4 was incomplete, or alternatively the signature of differential planetesimal dissolution in the Uranian envelope. Before examining in greater detail the origin of the volatile budgets of Titan, Triton and Pluto/Charon, we consider the history of satellite systems after accretion.

III. SUBSEQUENT PHYSICAL EVOLUTION OF SATELLITES

Subsequent to the formation of the outer-planet satellite systems, a number of important physical processes have modified their dynamical and geophysical properties. Tidal interactions between the satellites and their primaries have altered their rotational and orbital states. Thermal histories have been affected by radiogenic, and in some cases tidal, heating. In this section, we consider the effect of these processes on the evolution of satellite systems.

A. Tidal Evolution of Satellite Spins and Orbits

Neighboring objects tidally distort one another, because of the dependence of the gravitational force on distance. If the tidal distortion of a planet or satellite varies with time, frictional dissipation of energy occurs in the interior of the body. This energy dissipation may have profound effects on the rotational and orbital states of the satellites as well as their thermal histories. By decreasing the dynamical energy of the system, tidal evolution tends to drive a planet and its satellite towards a stable state where both rotate synchronously and the tidal distortion on all objects is invariant with time, unless the satellite(s) either escape or collide with the planet (see, e.g., Counselman 1973). The Pluto-Charon system is an example of a system that appears to have reached the stable state (Peale 1986); all of the other outer planet satellite systems are currently tidally evolving. Because the differential tidal force depends inversely on the cube of the distance between objects, inner satellites are affected by tides much more than the outer ones.

The rotation states of most planetary satellites are highly evolved, with the exception of the small, irregular outer satellites of Jupiter, Saturn and Neptune. Therefore, the current rotational states of most satellites do not well constrain their initial states. Extensive reviews of the processes affecting rotation may be found in Peale (1977,1986); here we briefly summarize these processes.

The rotation of a solid body results in a slight oblate distortion of its figure. If the rotation axis is not coincident with one of the principal axes, motion of the rotational bulge with respect to the figure of the object dissipates energy due to internal friction. For a nearly spherical body, the rotational energy is approximately

$$E = \frac{L^2}{2I} \tag{1}$$

hence if rotational angular momentum L is conserved, the minimum energy state corresponds to rotation about the principal axis corresponding to the maximum moment of inertia I. Any initial rotation state of a body will be rapidly driven towards this state by internal friction, on a time scale (Peale 1977)

$$T_{\text{wobble}} = \frac{3GIQ}{k_2 R^5 \omega^3} \tag{2}$$

where G is the gravitational constant, Q is the specific dissipation function, k_2 is the Love number, R is the equatorial radius, and ω is the angular rate of rotation. For planetary satellites, this time scale is much shorter than the age of the solar system (Peale 1977).

Differential gravitational forces cause a slight prolate distortion of a satellite's figure along the direction to the planet. If the rotation period of a satellite is different from its orbital period, the body of the satellite moves with respect to the tidal distortion potential. Because of internal friction, there is a time lag in the response of the body to the tidal potential, which results in a slight displacement between the satellite's tidal bulge and the satellite-planet direction. The planet therefore exerts a net torque on the satellite, driving the latter's rotation rate towards a state where the net torque is zero on a time scale (Peale 1977)

$$T_{\text{despin}} = \frac{2\Delta\omega m a^6 Q}{15 k_2 GM^2 R^5} \tag{3}$$

where $\Delta\omega$ is the change in spin rate, m and M are the masses of the satellite and planet, respectively, and a is the orbital semimajor axis. Except for the small irregular outer satellites of Jupiter, Saturn and Neptune, the despinning time scale is shorter than the age of the solar system (Peale 1977).

For a satellite without a permanent asymmetric mass distribution and which moves in an eccentric orbit, the final spin rate is slightly larger than the orbital mean motion (Goldreich 1966). However, if certain conditions are met, the planetary torque on any permanent asymmetry in the satellite may counteract the torque on the displaced tidal bulge, preventing the satellite from completely despinning. These conditions may be met at spin-orbit resonances, which occur when the rotational and orbital periods form near-integer ratios (Goldreich and Peale 1966). Many such resonances may be encountered by a tidally despinning satellite. Exact spin-orbit resonance occurs when the axis of largest principal moment of inertia coincides with the planet-satellite line at periapse, which ensures that the torque on the permanent asymmetry is zero when the differential gravitational force is strongest. For nearly spherical satellites with low orbital eccentricities, the dynamical behavior at spin-orbit resonances can be described using a pendulum-like model, and the probability of being captured into such a state can be determined analytically (Goldreich and Peale 1966; Counselman and Shapiro 1970). For satellites with very irregular shapes and/or relatively large orbital eccentricities, the dynamical behavior is more complex, capture into resonance occurring via a chaotic zone (Wisdom et al. 1984; Wisdom 1987a).

The strongest spin-orbit resonance, with the largest probability of capture, is the synchronous rotation state. Almost all tidally evolved bodies in the solar system rotate synchronously. The exceptions are Mercury, which was captured into a 3:2 spin-orbit resonance (Goldreich and Peale 1966), and Hyperion (Wisdom et al. 1984). Because of Hyperion's large orbital eccentricity and very irregular shape, the synchronous spin-orbit resonance is unstable, and Hyperion probably tumbles chaotically.

If a satellite has a free obliquity, i.e., its rotational axis is not coincident with the orbit normal, the tidal bulge is pulled alternately above and below the rotational equator as the satellite revolves about the planet. This also dissipates energy through internal friction, driving the obliquity towards zero on a time scale similar to the despinning time scale. However, if the satellite orbit is inclined and precessing, the obliquity is forced due to occupancy of a Cassini state (Peale 1969,1973,1977; Jankowski et al. 1989). In this case, the rotational and orbital motions are coupled, and the damping of the satellite's obliquity due to tidal dissipation leads to a decrease in the orbital inclination. The damping occurs on a much longer time scale than the damping of the free obliquity (see, e.g., Yoder and Peale 1981). For most outer-planet satellites, the forced obliquity is small.

A synchronously rotating satellite on an eccentric orbit is subject to a varying tidal potential (see, e.g., Kaula 1964; Peale and Cassen 1978). As the distance to the planet varies during an orbital period, the magnitude of the tidal distortion changes. More importantly, the orbital angular speed varies while the rotational angular speed remains nearly constant, which results in an oscillation of the tidal bulge relative to the satellite-planet direction (Yoder 1979b). The resulting torques on the tidal bulge rock it back and forth relative to the principal axes of the bodies. These variations in the tidal potential result in the dissipation of energy due to internal friction, which may significantly heat the interiors of some satellites. For a locked rotation state, the dissipated energy must come from the orbit, and the orbital semimajor axis tends to decrease as the orbital eccentricity damps at a rate (see, e.g., Goldreich 1963)

$$\frac{de}{dt} = -\frac{21}{2} k_2 n \frac{M}{m} \left(\frac{R}{a}\right)^5 \frac{e}{Q} \qquad (4)$$

where n is the orbital mean motion. The eccentricity decay time scales are much shorter than the age of the solar system for the inner large satellites of Jupiter (Yoder and Peale 1981), Saturn (Peale et al. 1980), Uranus (Squyres et al. 1985), and Neptune (Chyba et al. 1989a). We would therefore expect these satellites to have negligible eccentricities currently. However, some of the inner satellites of Saturn (Peale et al. 1980) and Uranus (Squyres et al. 1985) have anomalously large orbital eccentricities. The orbital eccentricities of some of the satellites of Jupiter and Saturn are forced due to orbital resonances, and the anomalously large orbital eccentricities of the inner large Uranian satellites are most likely remnants of past orbital resonances (see below, and also Tittemore and Wisdom 1988,1990).

Orbital eccentricity decay may have played a major role in the orbital and thermal history of Triton (McKinnon 1984; Goldreich et al. 1989). If

Triton was captured from heliocentric orbit into an elongated orbit around Neptune, tidal dissipation in the satellite would have led to circularization of the orbit into its present configuration on a time scale of order 100 Myr. Any pre-existing satellites beyond about 5 R_N would have been cannibalized by the massive Triton as it periodically crossed their orbits, and Nereid may have been perturbed into its current very eccentric orbit (Goldreich et al. 1989).

   A satellite raises a tidal bulge on the planet (Darwin 1880; MacDonald 1964). If the planet rotates at a different rate than the satellite revolves, friction in the planetary interior results in a time lag in the response of the planet to the tidal potential, and therefore a phase lag between the orientation of the tidal bulge and the planet-satellite direction. This results in a net torque between the tidal bulge of the planet and the satellite. The effects of this torque depend on the relative rates of planet rotation and satellite revolution. If the satellite motion is prograde with an orbital period longer than the planetary rotation period, the torque tends to decelerate the planet's rotation and accelerate the orbital motion of the satellite, tending to increase the semimajor axis. The orbits of most outer planet satellites are tidally evolving in this manner. If the satellite motion is prograde with an orbital period shorter than the rotation period of the planet, the torque accelerates the planetary rotation rate and decelerates the orbital motion of the satellite, tending to decrease the semimajor axis. For example, Phobos is spiralling in towards Mars, and may collide with the surface in about 50 Myr (Burns 1986). Likewise, if the satellite motion is retrograde, the planetary rotation rate increases and the satellite spirals in; Triton is currently undergoing this process, although it is in no danger of crashing into Neptune anytime soon (Chyba et al. 1989a). The rate of tidal evolution of the orbital semimajor axis of a satellite is given by (Darwin 1880):

$$\frac{1}{a}\frac{da}{dt} = 3k_{2p}n\frac{M}{m}\left(\frac{R_p}{a}\right)^5\frac{1}{Q_p}\text{sign}(\omega_p - n) \qquad (5)$$

where k_{2p}, Q_p and ω_p are, respectively, the Love number, specific dissipation function, and frequency of rotation of the planet. Planetary tides also result in slow changes in the orbital eccentricity (Kaula 1964)

$$\frac{1}{e}\frac{de}{dt} = \frac{19}{8}\frac{1}{a}\frac{da}{dt} \qquad (6)$$

and inclination (Darwin 1880)

$$\frac{1}{\iota}\frac{d\iota}{dt} = -\frac{1}{4}\frac{1}{a}\frac{da}{dt} \qquad (7)$$

of the satellite.

   As discussed above, tidal forces evolve the system towards a configuration where the tidal distortion of objects is invariant with time. For a planet with a single satellite, these conditions would be met if the satellite is on a circular, equatorial orbit and both planet and satellite rotate synchronously (this ignores effects of the Sun, other planets, and the Galaxy). The Pluto-Charon system has apparently reached this end state (Peale 1986). For the giant planets, however, the effect on the planetary rotation rate is negligible; the most important effect of planetary tides is to increase the semimajor axes of satellite orbits.

The regular satellite systems of Jupiter and Saturn are involved in many mean motion resonances, which occur when the ratio between the orbital periods of two satellites is near a small integer. The inner three Galilean satellites, Io, Europa and Ganymede, are currently involved in the Laplace resonance, in which the orbital period of Io is approximately half that of Europa, which is in turn approximately half the orbital period of Ganymede. In the Saturnian system, Enceladus occupies a 2:1 resonance with Dione, Mimas and Tethys occupy a 4:2 orbital resonance, and Titan and Hyperion a 3:2 resonance. Goldreich (1965) showed that the tidal expansion of satellite orbits could lead to the establishment of some of these resonances. In contrast, the Uranian satellite system is currently devoid of mean-motion resonances, although the satellite orbits may have undergone significant tidal evolution (Goldreich and Soter 1966; Tittemore and Wisdom 1989,1990). The reasons for this may be understood by considering the nature of the dynamical interactions between satellites at a mean motion resonance.

The gravitational attraction between a pair of satellites is strongest near conjunctions. At a mean-motion resonance, conjunctions occur with the satellites in nearly the same configuration. Under these conditions, the cumulative effects of many conjunctions can significantly perturb the shapes of the orbits. A mean motion resonance is exact when the configuration of the satellites is the same at successive conjunctions, i.e., conjunctions are stationary in a coordinate frame referred to one or both orbits, both of which precess due to planetary oblateness. At, for example, a 3:1 mean motion resonance, the most important resonances affecting the orbital eccentricities involve the following combinations of angles:

$$3l_O - l_I - 2\varpi_I$$

$$3l_O - l_I - \varpi_I - \varpi_O$$

$$3l_O - l_I - 2\varpi_O \tag{8}$$

where l is the mean longitude and I and O refer to the inner and outer satellite, respectively. Near a 3:1 mean motion resonance, these angles may either circulate or oscillate: each may be described by a pendulum-like Hamiltonian function in the absence of other perturbations (see, e.g., Peale 1976). The longitudes of the pericenter ϖ and ascending node (Ω) circulate at rates

$$\frac{d\varpi}{dt} \approx \frac{3}{2} n J_2 \left(\frac{R_p}{a}\right)^2 \approx -\frac{d\Omega}{dt} \tag{9}$$

where J_2 is the second order spherical harmonic of the planet. These rates are sensitive functions of the orbital distance, and may be significantly different for two satellites. Therefore, the frequency of each of the above resonance angles is slightly different. If the planetary oblateness is large, then at a given semimajor axis ratio the effects of one resonant term may dominate the effects of the others, in which case the motion may be described analytically, and the probability that the resonant combination of angles is captured into oscillation can be determined (Yoder 1979a; Henrard 1982; Henrard and Lemaitre 1983,1984; Borderies and Goldreich 1984). In this case, if the satellites are captured into a resonance, this may lead to an increase of one or both orbital eccentricities and/or inclinations, until the system reaches an equilibrium state determined by the rate of tidal energy dissipation in the satellite(s). However, if the planetary oblateness is small, the resonant terms may interact strongly, resulting in more complicated behavior.

Dermott (1984) suggested that mean-motion resonances among the Uranian satellites might be chaotic and therefore unstable, because of the relatively small oblateness of that planet. In addition, because of the relatively large satellite masses, there are strong secular interactions between the Uranian satellites (Greenberg 1975; Laskar 1986; Dermott and Nicholson 1986), resulting in variations of the orbital precession rate, which further complicate their motions. Modern methods in nonlinear dynamics have enabled us to determine the nature of the behavior at resonances among the Uranian satellites. Specifically, the motion is "chaotic" (Tittemore and Wisdom 1988,1989,1990; Tittemore 1990b; Dermott et al. 1988; Malhotra and Dermott 1990), i.e., it is characterized by irregularities and extreme sensitivity to initial conditions. At the mean-motion resonances that may have been encountered by the Uranian satellites, one or more resonant angles may be temporarily captured into oscillation; however, the resonance ultimately becomes unstable and is disrupted. During the temporary phase of resonant behavior, the orbital eccentricities and/or inclinations may undergo significant increases. For example, Miranda's anomalously large orbital inclination has been explained as a consequence of passage through the 3:1 mean-motion commensurability with Umbriel (Tittemore and Wisdom 1989). Miranda's orbital eccentricity may also have increased considerably during evolution through this commensurability (Tittemore and Wisdom 1990; Malhotra and Dermott 1990); the large orbital inclination had a significant effect on the evolution through the eccentricity resonances (Tittemore and Wisdom 1990). Ariel, too, may have undergone a large increase in orbital eccentricity during passage through a 4:1 mean-motion resonance with Titania, as shown in Fig. 2 (Tittemore 1990b). By determining which resonances lead to unstable, chaotic behavior, Tittemore and Wisdom have strongly constrained the rate of internal tidal energy dissipation in Uranus: $11,000 \lesssim Q_U \lesssim 39,000$ (Tittemore and Wisdom 1989,1990).

Chaotic motion may also have played an important role in the Jovian satellite system. Prior to the establishment of the current Laplace resonance

Ariel-Titania 4:1 Resonance

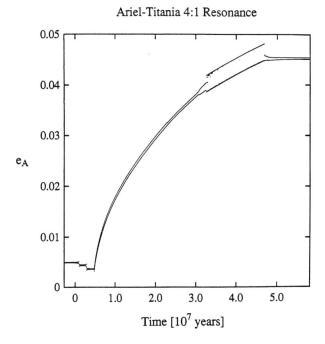

Figure 2.   The orbital eccentricity of Ariel increasing dramatically during evolution
through the 4:1 mean-motion commensurability with Titania (Tittemore 1990b).
Shown are the maximum and minimum eccentricities as a function of time. Initially,
the eccentricity increases in a regular manner and can be analytically modeled;
however, at $e_M \sim 0.04$, the orbits become chaotic, and eventually escape from the
resonance. During evolution through the 4:1 resonance, and afterwards as the
eccentricity damps, the interior of Ariel may have been warmed by up to 20 K,
perhaps triggering the geologic activity that led to its late resurfacing.

among the Galilean satellites, Europa and Ganymede may have been involved
in a chaotic 3:1 mean-motion resonance, leading to large increases in both
orbital eccentricities (Tittemore 1990a; Malhotra 1991). Although the J_2 of
Jupiter is much larger than that of Uranus, the relatively large orbital distances
of Europa and Ganymede result in orbital precession rates comparable to those
among the inner Uranian satellites, and the satellite-to-planet mass ratios
are similar, resulting in strong interactions between resonances and chaotic
behavior at the 3:1 commensurability.

B. Thermal Evolution of Satellites by Tidal Heating

Satellites may be heated by a number of sources (Schubert et al. 1986); here we
focus on tidal effects, which over the past few years have assumed increased
importance in the literature. Tidal flexing of a satellite frictionally dissipates
energy, heating the interior of a satellite. For a synchronously rotating satellite
with a small obliquity θ and orbital eccentricity, the rate of tidal heating is

approximately (Peale and Cassen 1978)

$$\frac{dE}{dt} = -\frac{k_2 M n^3 R^5}{a^3 Q}\left(\frac{21}{2}e^2 + \frac{3}{2}\theta^2\right).$$
(10)

For most outer-planet satellites, the eccentricity term dominates because the forced obliquities are very small, and free obliquities are damped on a very short time scale.

Tidal heating due to the decay of a free eccentricity is probably not important to the thermal history, because of the short time scale (see, e.g., Cassen et al. 1982). A notable exception to this rule may be Triton. As discussed above, if Triton was captured from a heliocentric orbit, its initial orbit about Neptune may have been very eccentric, and estimates of the tidal heating rate indicate that the water ice component of Triton could have melted and remained largely liquid as the eccentricity decreased (McKinnon 1984; Goldreich et al. 1989).

Satellites involved in mean motion resonances may have significant forced eccentricities. Io, the innermost Galilean satellite, moves in an eccentric orbit forced by its interactions with Europa and Ganymede in the Laplace resonance. Peale et al. (1979) predicted that the tidal heating of Io was intense enough to cause melting in its interior, possibly with observable surface activity; a week later, the Voyager 1 spacecraft discovered that this satellite is indeed volcanically active (Morabito et al. 1979). Europa's orbital eccentricity is also forced by the resonant interaction, and tidal heating may be sufficient to maintain a liquid layer beneath the satellite's smooth icy crust (Cassen et al. 1979,1980,1982; Squyres et al. 1983a; Ross and Schubert 1987; Ojakangas and Stevenson 1989), although the heating is not currently large enough to initiate melting of water ice in a completely frozen ice mantle. Currently, tidal heating of Ganymede in the Laplace resonance is negligible (Cassen et al. 1982). However, both Ganymede and Europa may have been strongly tidally heated in the past, during passage through a 3:1 resonance (Tittemore 1990a; Malhotra 1991). These satellites may have reached large orbital eccentricities during passage through a chaotic zone, followed by escape from the resonance and subsequent evolution into the Laplace resonance. Tidal distortion may have fractured the rigid lithospheres of both satellites, and melting of water ice in the mantles may have occurred leading to episodes of resurfacing. Because this resonance provided Ganymede with a large heat source that Callisto lacked, this could explain the dichotomy between these satellites, which have similar bulk properties but apparently very different thermal histories.

Tidal heating has most likely played a role in the thermal history of Enceladus, which is relatively small but has had a very active geologic history (Smith et al. 1982). However, its current forced orbital eccentricity due to the 2:1 resonance with Dione is only 0.0044. According to elastic tidal heating models, this value is about a factor of 20 too small to initiate melting of water ice, and 5 to 7 times too small to maintain a molten interior (Squyres et al. 1983b); however, consideration of viscoelastic effects suggest that the tidal heating rate in fact may be much larger (Ross and Schubert 1989).

The orbital resonances encountered by the Uranian satellites may have had an important effect on the thermal histories of Ariel and Miranda. During evolution through the 4:1 resonance with Titania, about 3.8 Gyr ago or earlier, the orbital eccentricity of Ariel may have reached a value up to about 0.045 before the satellite escaped from the resonance (Tittemore 1990b). During resonance passage, and afterwards as the eccentricity decayed, the internal temperature of Ariel may have increased by up to 20 K. Before the resonance was encountered, Ariel's internal temperature was probably around 220 K due to accretional and radiogenic heating. The increase of internal temperature to about 240 K conceivably may have triggered the geologic activity that led to the resurfacing of Ariel, by thinning the rigid lithosphere and thus increasing thermal stresses on it. Surface fracturing may have resulted, providing conduits to the surface for relatively buoyant, mobile materials such as ammonia dihydrate. During evolution through the 3:1 resonance with Umbriel, the internal temperature of Miranda may have exceeded the minimum melting point · of ammonia dihydrate (approximately 176 K), however, resonance passage most likely did not result in significant melting of water ice (Tittemore and Wisdom 1990; Malhotra and Dermott 1990).

If a satellite is not rotating sychronously, e.g., during the initial spindown or a phase of chaotic rotation, the rate of tidal heating may be enhanced over the rate for synchronous rotation by a factor of order $1/e^2$, with a corresponding decrease in the eccentricity decay time scale (Wisdom 1987a). This may result in a very intense but brief heating episode, which has been suggested as a mechanism for thermally altering the surface of Miranda (Marcialis and Greenberg 1987; however, see Wisdom 1987a; Tittemore and Wisdom 1990). Hyperion, despite having both a chaotic rotation state and a significant forced orbital eccentricity, is too small and too far from Saturn for tidal heating to be important in its thermal history.

IV. ORIGIN OF THE VOLATILE BUDGETS OF OUTER SOLAR SYSTEM BODIES

The accumulation of data on the volatile component of solid bodies in the outer solar system provides another approach to understanding the environments which produced regular satellites and the double planet system of Pluto/Charon. In this section, we explore the following particular assertions:

1. The volatile budgets of Pluto and Triton reflect partial processing of interstellar material in the outer solar nebula. In particular, sublimation of volatiles from infalling interstellar grains was followed by selective retrapping onto grains, which preferentially excluded carbon monoxide and molecular nitrogen. The molecular nitrogen seen on Triton was con-

verted from ammonia, by photochemical or shock-chemical processing after Triton's formation.

2. Methane in Titan's atmosphere-surface system is the result of thermo-chemical processing in the circum-Saturnian disk from which it formed. The deuterium enhancement seen in Titan's atmosphere is not necessarily a sign of interstellar origin of methane, but may be related to processes occurring after formation.

Table I contains a partial reconstruction of the total surface and atmo-sphere volatile budgets for Titan and Triton. The Titan reconstruction (Lunine 1985*a*) is based on the ethane ocean model for resupply of methane to Titan's atmosphere (Lunine et al. 1983), and uses a photolysis rate for methane from Yung et al. (1984) and the thermodynamics discussed in Lunine and Steven-son (1985). The values for Triton are lower limits computed assuming the minimum amount of nitrogen required to provide year-round coverage of each hemisphere with a layer of nitrogen (which corresponds to of order 1 meter thickness), and with the areal methane fraction estimated from groundbased spectroscopy. For Pluto, where the atmospheric and surface properties are more poorly known, this exercise is less well constrained than for Triton, and we therefore do not attempt an estimate.

A. Origin of the Volatile Budgets of Triton and Pluto

In Sec. II we argued that the bulk densities of Triton and the Pluto/Charon system are to some degree consistent with formation in a relatively water-poor environment, plausibly due to sequestration of oxygen by carbon monoxide as has been proposed for the solar nebula by Lewis and Prinn (1980). The atmospheric value of $CO/N_2 < 0.01$ (Broadfoot et al. 1989) limits the ratio of surface-exposed CO to N_2 to <0.15, using vapor pressure data from Brown and Ziegler (1979) and Raoult's law for an ideal solid solution

$$x_{CO} = \frac{e_{N_2}}{e_{CO}} y_{CO} \qquad (11)$$

where x_{CO}, y_{CO} are the mole fractions of CO, relative to N_2, in the surface ice and atmosphere, respectively, and e_i is the vapor pressure of component i. (The reader is cautioned that a numerical error exists in the coefficients for nitrogen latent heats in Brown and Ziegler [1979]. However, the vapor pressure coefficients are correct, and can be used with the Clausius-Clapeyron relationship to derive latent heats if these are desired.)

The liquid solution of CO and N_2 is nearly ideal, meaning it obeys Eq. (11) (Verschoyle 1931), and the solid solution is expected to be likewise, particularly if the presence of α-CO stabilizes α-N_2 a few degrees above the latter's α-β transition temperature of 35 K (see the phase diagram in Scott [1976]). Therefore, the estimate afforded by Eq. (11) is probably good to better than, say, a factor of 2.

As Triton's atmospheric CH_4 mixing ratio (again relative to N_2) is close to the saturated value of 10^{-4} at 38 K (Herbert and Sandel 1991), we cannot use the atmospheric CO/CH_4 as an indicator of their relative surface abundances. In Table I, we use the rate of photochemical destruction of CH_4 (Broadfoot et al. 1989), along with the argument that we do not view Triton at a special time in its history, to argue for a minimum reservoir of CH_4 at or near the surface which will not be depleted by photochemistry for at least hundreds of Myr. We would then conclude from Table I that $CO/CH_4 < 1$.

We are therefore presented with the possibility that Triton's surface carbon budget consists at best of comparable quantities of CH_4 and CO, or even more extreme, that CO is a minor species. Perhaps CO was not outgassed with the same efficiency as CH_4, or was preferentially destroyed by atmospheric chemistry over the age of the solar system. Neither of these is particularly plausible, based on analogous studies of Titan (Samuelson et al. 1983; Lunine and Stevenson 1985).

Table I also gives an estimated primordial volatile budget for Triton, based on cometary abundances from the Chapter by Mumma et al. Most striking in comparing surface and primordial abundances is the loss of CO relative to N_2 and CH_4. The loss relative to CH_4 could be explained by fractionation effects associated with sublimation and recondensation of infalling precursor interstellar grains (Lunine et al. 1991), provided such a process did not similarly alter comet grains. This requires cometary grains to have entered the nebula at a somewhat larger radial distance than those which went to form Triton. Alternatively, the higher volatility of CO compared to CH_4 could have resulted in preferential escape of the former after formation. In either case, most of the N_2 should have been lost as well. The problem therefore reduces to explaining the present predominance of nitrogen.

The most sensible way out of the nitrogen dilemma is to assume that Triton's nitrogen was brought in as ammonia, NH_3. This molecule is fairly abundant in comet Halley and interstellar clouds. Ammonia's low volatility compared to carbon monoxide, and strong propensity for hydrogen bonding with itself and water ice, allowed it to be retained or recondensed during most energetic processes associated with nebular infall of grains (Lunine et al. 1991). The resulting composition of Triton-forming planetesimals was rich in ammonia and methane, and poor in carbon monoxide.

Subsequent to the formation of Triton, conversion of ammonia to molecular nitrogen may have occurred photochemically (Atreya 1978) or through thermochemistry generated by shock-heating of the atmosphere or surface by a projectile (Jones and Lewis 1987; McKay et al. 1988). Photochemical conversion requires warm conditions (in excess of 150 K) to avoid freezeout of the intermediate product hydrazine (Atreya et al. 1978). If Triton was captured from solar orbit, such a warm environment, in the form of a massive "greenhouse" atmosphere, may have been possible early on, as tidal evolution circularized Triton's initially eccentric capture orbit around Neptune (Lunine and Nolan 1992). With ammonia present in the warm atmosphere, and scaling the photochemical yield of N_2 from Titan (the limiting factor being solar flux), we find it is possible to produce a layer of nitrogen hundreds of meters thick, which may be an underestimate if shock chemistry was also operative (McKay et al. 1988).

A key test of this hypothesis, that NH_3 is the source of Triton's present-day N_2, is measurement of the atmospheric ratio of argon to nitrogen. As described for Titan in Lunine et al. (1989), because argon is excluded from the planetesimal ice to nearly the same degree as CO and N_2, the Ar/N_2 ratio in the atmosphere of Triton should be orders of magnitude less than the solar elemental value of 0.1, if the present-day N_2 had been manufactured from ammonia. A ratio close to 0.1 would be more consistent with the N_2 being primordial, and brought into Triton (along with argon) trapped in water ice. Unfortunately, the Voyager UVS instrument was not sufficiently sensitive to provide a useful upper limit on argon.

With regard to Pluto, although methane is the only species spectroscopically detected there, evidence from a stellar occultation observed in 1988 indicates the presence of a heavier gas, most plausibly N_2 or CO (Yelle and Lunine 1989). The fact that the surface temperature and atmosphere pressure are not consistent with the existence of either species as a surface ice leads to a dilemma, because the estimated escape rate of the atmosphere corresponds to a 3 kilometer layer of CO or N_2 ice over the age of the solar system (Hubbard et al. 1990b). Alternative surface models include slow outgassing of the heavier species, from subsurface clathrates or deeper interior reservoirs. Until a surface model consistent with the data and loss processes is developed, it will be impossible to estimate Pluto's volatile budget and hence test the hypothesis presented above for Triton. The expectation that N_2 should dominate over CO is only valid if the NH_3-N_2 conversion was operative on Pluto as it was on Triton. The dynamical histories of these bodies are so different that such an assumption is not a safe one. If conversion of ammonia to nitrogen did not occur, CH_4 should be very much overabundant compared to N_2 as well as CO in Pluto's bulk volatile budget, CO should be more abundant than N_2, and neither species may be sufficiently plentiful to produce significant deposits of ice (in contrast to CH_4). Other speculations on the volatile composition of Pluto may be found in Simonelli et al. (1989) and McKinnon and Mueller (1988).

B. Origin of Methane on Titan

Prinn and Fegley (1981) established that methane was an important, if not dominant, carbon-bearing molecule in the disk of material from which the Saturn satellites formed. This conclusion is robust provided that the disk is of elemental solar composition, gaseous, optically thick and continues sufficiently close in to Saturn that high (>600 K) temperatures could be obtained. Such a disk is consistent with the accretion disk model of Lunine and Stevenson (1982), or the spin-out disk concept described in Pollack et al. (1991). Given that CH_4 dominated the carbon budget, it would also have sequestered most of the adsorption or cage sites in the water ice planetesimals which formed Titan (Lunine and Stevenson (1985). Consequently, little CO or N_2 could have been incorporated in Titan (Lunine and Stevenson 1987), and the observed nitrogen is likely a product of ammonia photolysis or shock chemistry (Atreya et al. 1978; Jones and Lewis 1987; McKay et al. 1988).

This picture, while consistent with Titan's bulk density and other details of satellite formation, has been challenged by the observation of a deuterium enrichment in atmospheric methane on Titan (Lutz et al. 1990). Adopting a $D/H_{primordial} = 3 \times 10^{-5}$ (Anders and Grevesse 1989), the D/H value of 0.4 to 1.5×10^{-4}, $\pm 50\%$, (de Bergh et al. 1988) in Titan's atmosphere corresponds to a factor of 3 to 10 enrichment over the primordial value. If this reflects Titan's initial, bulk deuterated methane abundance, it is straightforward to show that Titan's methane could not have been manufactured by high-temperature thermochemical processes in the solar nebula or a circumplanetary disk. According to Prinn and Fegley (1989), the minimum temperature at which the CO-CH_4 interchange reactions could have proceeded, over the lifetime of the solar nebula, was 600 K. This is also close to the quench temperature for deuterium fractionation between HD and heavier molecules (Grinspoon and Lewis 1987), and would yield a D/H enrichment in methane of 1.4 to 1.5. Clearly, if the observed enrichment is primordial, then Titan's methane must contain a significant component of interstellar origin, which has not been subjected to high temperatures.

Alternatively, methane in Titan may have started with only modest deuterium enrichment which was then enhanced through physical and chemical processing. Pinto et al. (1986) examined various possibilities, including cloud formation, partitioning between hydrocarbon ocean and atmosphere, and photochemical evolution of the atmosphere. The latter process proved to be most promising. Methane photolysis occurs mainly through a catalytic cycle involving the key reaction $C_2H + CH_4 \longrightarrow C_2H_2 + CH_3$, followed by photolysis of the acetylene to reform the catalyst, the C_2H radical; direct photolysis of methane is much slower. Pinto et al. (1986) proposed that the analogous reaction for deuterated methane was slower, because of the higher binding energy of the deuterium atom, and estimated rates 3/4 to 7/8 of that for the nondeuterated methane. The enrichment of deuterium then depends on the ratio of the total initial methane reservoir to the total present reservoir.

The result is plotted in Fig. 3, adapted from Pinto et al. (1986). Models of a surface ocean on Titan (Lunine et al. 1983; Dubouloz et al. 1989) posit an ethane-methane mix of 15 to 80% methane. The ocean's ethane is regarded as the primary product of methane photolysis, with solid acetylene (at the ocean base) the secondary product at 20% of the ethane abundance (Yung et al. 1984). Hence the ocean ethane to methane ratio, with the ethane augmented by 20% to account for acetylene, constitutes the ratio of total-initial to total-present CH_4 reservoir. The maximum ratio then, is roughly 8, which corresponds to an atmospheric enrichment of deuterium of no more than 1.7. If we allow for slight vapor-pressure differences between methane and its deuterated form, there is an additional ocean-atmosphere partitioning which gives an enrichment of at most 1.3. Combined with the allowable

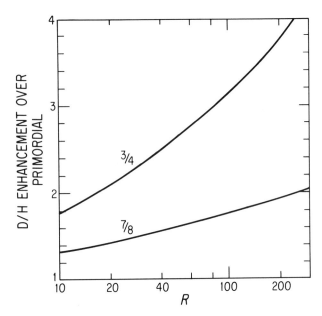

Figure 3.   Deuterium enrichment from photochemical destruction of methane in
Titan's atmosphere, redrawn from Pinto et al. (1986). The D/H enrichment over
the primordial value is plotted as a function of R, the ratio between the total initial
reservoir of methane and the present reservoir. The two curves represent different
assumptions about the rate of photolysis of CH_3D, either 3/4 or 7/8 the rate of CH_4
photolysis. The calculation assumes that photosensitized dissociation of methane is
the dominant destruction mechanism, as described in Yung et al. (1984).

enrichment in the Saturn nebula of 1.4, the total enrichment from these two
processes is ~3, marginally consistent with that derived from the observations
of de Bergh et al. (1988).

    Although consistent with the deuterium data, the case would be more
convincing if additional enhancement mechanisms could be found. There is
compelling evidence for an early period of high ultraviolet flux from the Sun
(Zahnle and Walker 1982), with consequent increased rate of photolysis. This,
plus other processes such as shock chemistry due to impacts might increase
the amount of methane processed, and hence the atmospheric deuterium enrichment. This would have had to occur, however, after the departure of the
massive hydrogen envelope of the Saturn nebula, to prevent re-equilibration
of deuterated species with H_2. Such a requirement is plausible, and consistent
with the fact that the primitive Titan atmosphere only contained molecular
hydrogen as a minor constituent soon after formation (Lunine 1985b).

    More restrictive, however, is the requirement that the primordial photo-
chemical products not interact thermodynamically with the present-day liquid
methane-ethane reservoir. If, for example, ethane or propane were the primary
product of the early enhanced photolysis, they would be present in the current
ocean, whose composition determines the measured atmospheric methane

abundance. By this argument, the Voyager data on the composition of Titan's lower atmosphere constrain the total amount of surficial ethane, propane and other soluble compounds which have been produced since the formation of Titan. Acetylene, however, is a solid, and it as well as other heavy organics have only limited thermodynamic contact with the liquid. If primordial photochemical products exist today as acetylene and other heavy organics, rather than ethane, one could select a larger x-axis value in Fig. 3, without violating the Voyager constraints on the present-day composition of the ocean. Ongoing processing of ethane to solid organics by surface chemistry (driven by impacts, perhaps) would achieve this end.

To produce a deuterium enhancement of 3, solely by photochemistry, would require depositing 10 km of primordial photochemical detritus on the surface (Pinto et al. 1986). If this material were solid and sufficiently porous ($> 10\%$), the proposed ocean of Lunine et al. (1983) could largely be confined to pore spaces within the sediment. Interestingly, such a configuration might simultaneously satisfy the requirements of the ethane ocean model and explain recent radar data, which appear to have detected a solid surface (Muhleman et al. 1990).

V. SUMMARY OF OUTSTANDING ISSUES

Our current picture of the formation of regular satellites and the Pluto/Charon system has benefitted greatly from the era of outer planet exploration, and from the spacecraft and groundbased investigations of comet Halley. The chemical information is at the stage wherein useful comparisons can be made between observed volatile abundances (and ice/rock ratios, through the bulk densities) on the one hand, and models of chemical and associated physical processes in the solar nebula, protosatellite disks, and interstellar clouds on the other. At the same time, models of giant-planet formation have become sufficiently detailed that they now make rather specific predictions concerning the nature of protosatellite disk systems. A complete, detailed list of currently outstanding issues would be lengthy; we close this chapter with an eclectic choice of the most interesting:

1. The origin and chemical nature of protosatellite disks must be clarified, particularly with regard to the ice-to-rock ratio and the accessibility of high-temperature, chemically active, inner zones. The first *in-situ* measurements of the Jovian atmosphere from Galileo may help in constraining giant-planet formation, which will certainly bear on the properties of the disks.

2. Elemental abundances, particularly C/O, in primitive solar system material remain controversial but are crucial to interpreting bulk densities of satellites in terms of origins scenarios.

3. The abundances of volatile molecules in molecular clouds, particularly methane, ammonia and carbon monoxide, must continue to be deter-

mined to higher spatial resolution and increasing sensitivity. Also, our understanding of the parent molecule composition of comets is poor and it is of high value to improve these numbers.

4. The natures of the surfaces of Titan, Triton and Pluto are still incompletely understood and hinder our ability to construct volatile budgets for these objects. The origin of methane, nitrogen and carbon monoxide on these three bodies must be tied together, but these relationships are still only dimly perceived.

Acknowledgments. Careful reviews by C. F. Chyba and D. P. Simonelli improved the content and writing. Conversations with D. J. Stevenson, W. B. McKinnon, A. G. W. Cameron, J. F. Kerridge, D. M. Hunten, J. B. Pollack and E. Van Dishoeck were stimulating and helpful. Preparation of this chapter, and some of the work herein, was supported by grants from NASA's Planetary Atmospheres and Geology and Geophysics Programs.

COMETS AND THE ORIGIN OF THE SOLAR SYSTEM: READING THE ROSETTA STONE

MICHAEL J. MUMMA
NASA Goddard Space Flight Center

PAUL R. WEISSMAN
Jet Propulsion Laboratory

and

S. ALAN STERN
Southwest Research Institute

Comets are the most primitive bodies in the solar system and preserve a cosmochemical record of the primordial solar nebula at the time of their formation. Spacecraft missions to, and intensive groundbased studies of comet Halley (and other recent comets) have strongly reinforced this view, and given us the first detailed insights into the nature of the refractory solids and frozen volatiles contained in cometary nuclei. The volatile record reflects a composition similar to that observed in dense molecular cloud cores, with little evidence of reprocessing in the solar nebula or in giant planet subnebulae. From the measured volatile abundances, it appears likely that comets formed at temperatures near or below ~60 K and possibly as low as ~25 K. Grains in comet Halley were found to be of two types: silicates and organics. The organic grains consist of the elements carbon, hydrogen, oxygen and nitrogen (CHON) and appear to act as an extended source of molecules in the cometary coma. Isotopic evidence shows that comet Halley formed from material with the same compositional mix as the rest of the solar system, and is consistent with comets having been a major contributor to the volatile reservoirs on the terrestrial planets. A variety of processes have been shown to modify and reprocess the outer layers of comets both during their long residence time in the Oort cloud and following their entry back into the planetary system. These include: irradiation by energetic photons and particles, heating by passing stars and nearby supernovae, competing accretion and erosion by interstellar gas and grains, regolith gardening by debris impacts, and solar heating during perihelion passage. The last process results in sublimation of volatiles to form the visible cometary coma, and also causes the transformation of amorphous water ice to crystalline ice, thermal stressing of the nuclear surface and structure, and the likely formation of a lag-deposit crust of larger silicate and organic grains on the nucleus surface. However, the pristine nature of cometary materials at depth within the nucleus appears to be well preserved. The most likely formation site for comets is in the Uranus-Neptune zone or just beyond, with dynamical ejection by the growing proto-planets to distant orbits to form the Oort cloud. A substantial flux of interstellar comets was likely created by the same process, and may be detectable if cometary formation is common in planetary systems around other stars.

I. INTRODUCTION

Comets are the most pristine bodies in our solar system, and their compositions and structure encode seminal information on processes that affected their
formation and subsequent evolution. For this reason, the nature of the cometary nucleus is the central problem in cometary science. Prior to 1986, studies
of the composition of cometary nuclei were entirely remote, being based on
spectroscopy and radiometry of the dust and of (mainly) dissociation products of the volatile ices. Production rates of dust were measured, grain sizes
were estimated, and silicate components were identified. The volatile parents
of observed radicals and ions were identified in a few cases, notably water
and carbon dioxide (the parent of CO_2^+), and water production rates were
obtained for many comets. Direct quantitative detections of candidate parent
volatiles, however, were limited to CO in two comets (West 1976 VI and
Bradfield 1979 X), along with a few tentative detections of other species at
radio wavelengths.

In recent years, however, our perspective of comets has changed dramatically. The combination of improved understanding of spectral line formation
in the coma and advances in instrumentation has enabled direct spectroscopic
study of parent volatiles in many comets. These produced quantitative detections of H_2O, CO_2, H_2CO, CH_3OH, H_2S, HCN, and CO, and provided a
provisional detection of CH_4 in a new comet while setting important upper
limits for it in two other comets. The detection of crystalline silicates and
of the so-called organic-grain feature in comet P/Halley (1986 III) (and in
every bright comet since then) introduced a similar revolution in the remote
study of cometary dust. By far the most significant development, however,
was the enormous effort expended on the study of comet Halley, involving
a fleet of spacecraft and virtually every significant astronomical observatory
on the ground, in the air, and in space. Two of the most stunning advances
were the discovery of a previously unknown population of organic grains, and
the finding that fully reduced species are underrepresented among cometary
volatiles. The results returned by these and subsequent investigations revo-
lutionized our understanding of cometary compositions, and of the messages
carried by comets on their origin and subsequent evolution.

Cometary material may have experienced various kinds of processing
while enroute from the dense molecular cloud core that predated our solar
system to the final cometary nucleus. Understanding the nature and degree
of that processing is a principal objective of cometary studies. Interstellar
ices are subject to substantial processing in the natal cloud core (e.g., by
ultraviolet photons; see Chapter by van Dishoeck et al.), and may be vaporized
while falling into the solar nebula. If the latter occurs, volatile species may
be chemically modified before recondensing as cometary ices. Even after
aggregation into the nucleus, cometary material is not necessarily free from
further change. Cometary nuclei are exposed to a variety of processes while
in the Oort cloud, and after injection into the inner solar system, and so have

been modified to varying degrees. Energetic particles and photons, acting over the long residence time in the Oort cloud, chemically and physically modify the surface layers of cometary nuclei. These nuclei are also subjected to heating by passing stars and nearby supernovae, to collisional gardening by Oort cloud debris, and to episodic erosion by impact of interstellar grains during encounters of the solar system with molecular clouds. Following their orbital evolution into the inner solar system, the nuclei of short-period comets experience insolation-induced heating that may modify the structure of their ices (from amorphous to crystalline). This modification may occur throughout the entire body in some cases, and may mobilize highly volatile molecules within the ice-organic-silicate matrix. Although the plausibility of these effects is supported by detailed modeling and by laboratory experiments, many key cometary and environmental parameters are highly uncertain and thus the significance of the predicted effects can be fully assessed only after making measurements on real cometary nuclei.

Each process imposes its characteristic signature on cometary material, and this is preserved as a record in the form of chemical, elemental, isotopic and isomeric compositions, and in the morphology and physical structure of the material within the nucleus. New capabilities in remote sensing and the possibility of detailed investigations of cometary material and of cometary nuclei by future spacecraft (see Sec. V), promise major advances in our ability to read this record, the "Rosetta stone" of cometary and solar system origin.

A. The Problem of Cometary Origins

The origin of cometary nuclei is still unclear. Several possibilities have been advanced, and are presented here in the context of defining key tests for them. Most associate the formative region with the solar nebula, but Cameron (1973) suggested that comets formed in associated satellite-nebulae (fragments of the pre-solar nebula), while Clube and Napier (1982) proposed formation in giant molecular clouds with subsequent capture when the solar system traverses such clouds. A schematic representation of processes relating comets to the formation and evolution of the solar system is shown in Fig. 1.

The solar-nebula hypotheses are distinguished primarily by the degree of reprocessing experienced by interstellar material prior to formation of cometary nuclei (see Lunine [1989b], Prinn and Fegley [1989], and Fegley and Prinn [1989] for recent reviews of chemistry in the solar nebula). The present models may be regarded as single-component and dual-component accretional scenarios. In one single-component model, comets accreted from unmodified ice- and organic-mantled interstellar grains (cf. Greenberg 1982; Greenberg and Hage 1990) at distances greater than about 20 AU from the proto-Sun, and at temperatures < 100 K. Alternatively, accretion could have occurred after the volatiles in icy mantles were vaporized, then were chemically processed in the main solar nebula, and later were recondensed onto the grains which finally aggregated into macroscopic cometesimals. The dual-component models represent admixtures of interstellar and processed fractions. For example,

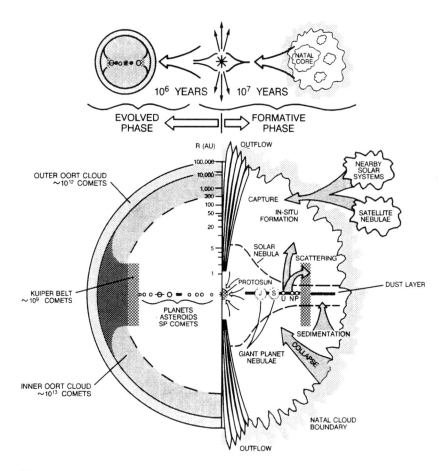

Figure 1.   Schematic representation of proposed theories for the origins of comets,
and their relationship to the formation of planetary systems. The formation of a
planetary system from a dense cloud core is illustrated schematically at top, with
time evolving from right to left. Processes in the solar nebula are schematically
illustrated at bottom (right), and the present state of planets and comets is illustrated
on the left. See text for discussion of theories for cometary origins, and evolution
of the Oort cloud and Kuiper belt.

Engel et al. (1990) argued that volatiles in icy mantles on most interstellar
grains would vaporize and be dissociated in the accretion shock as they fall
into the solar nebula. Thus these interstellar volatiles could be reprocessed
before being incorporated into cometary nuclei. It also seems unlikely that
chain molecules such as polyoxymethylene (POM), and the CHON organic
grains (composed of carbon, hydrogen, oxygen and nitrogen) could survive
this infall process without disruption. These materials are fairly fragile and
were easily destroyed even in the relatively benign environment of the coma of

comet Halley. The various accretion models can thus be tested by comparing cometary compositions with model predictions.

Temperatures in Cameron's accretion-disk model of the solar nebula range from about 180 K at the orbit of Jupiter to about 18 K at 100 AU. The range is 130 K to 25 K in Hayashi's radiative steady-state model (Cameron 1978a; Hayashi 1981). These temperature profiles are compared in Fig. 2, and the distances at which various volatiles would be stable as ices are indicated (Yamamoto 1985) (see also Table I). Lewis and Prinn (1980) examined chemical conversion in this region of the solar nebula, and demonstrated that reduction of CO and N_2 could not be achieved in the nebular collapse time. Thus, the conversion was said to be "quenched." In this scenario, comets would contain volatile carbon and nitrogen mainly as CO and N_2, with very little in the reduced forms (CH_4, NH_3), and would contain almost no easily reacted or destroyed interstellar material such as H_2CO and S_2.

Early reports of significant amounts of fully reduced volatile carbon (CH_4) and nitrogen (NH_3) in comet Halley stimulated work on ways to bring chemically equilibrated material into the trans-Uranian region of the solar nebula, where comets presumably formed. Proposed mechanisms include processing in the giant planet sub-nebulae (Fegley and Prinn 1989b), and radial diffusion of inner solar-system material (Lunine 1989b; Engel et al. 1990). Chemical reduction of oxidized species (e.g., CO to CH_4, N_2 to NH_3, etc.) can proceed efficiently in regions of relatively high temperatures and densities, such as the giant planet sub-nebulae. Comets formed from this material would be rich in reduced species, but extremely poor in "interstellar" chemicals such as CO, N_2, H_2CO and S_2. The high temperatures and densities predicted by models of giant planet sub-nebulae would also seem to be inconsistent with the survival of CHON particles and POM, which vaporizes at about 400 K. Further difficulties lie in providing a mechanism for removing this material (or alternatively, icy planetesimals) from the local gravitational wells of the giant planets. The radial diffusion models differ primarily by transporting equilibrated material from the inner solar nebula outward to the comet-forming zone, thus avoiding the need for an escape mechanism from the gravitational wells of the giant planet sub-nebulae. However, the abundance estimates of reduced volatiles in comet Halley have been revised downward to the point where it may not be necessary to invoke dual-component models at all.

In this chapter, we begin by reviewing the evolution of the Oort cloud from its formation to the present time, and the processes which may modify cometary nuclei while in the Oort cloud and after re-injection into the inner solar system. We then review key results obtained from detailed investigations of comet Halley, and of several comets since then, and we compare them with current models of cometary formation and evolution. Finally, we review present and expected future capabilities for reading the information encoded in a significant number of comets. By doing so, it should be possible to establish the statistical properties of groups of cometary nuclei and the significance of individual comets, within the context of the origin of our solar system.

TABLE I

Sublimation Temperatures of Observed Cometary Volatiles and Corresponding Distances in the Solar Nebula, and a Comparison of Cometary and Interstellar Abundances

Species	Temperature[a] (K)	Distance[a] (AU)	Comets	Relative Abundance[b]	
				ISM-gas	ISM-ice
H_2O	152	3.4–5.4	100	≤100	100
CH_3OH	99		1–5[c]	0.01–0.1	7–40
HCN	95		0.02–0.1[d]	0.01–0.1	4
SO_2	83		<0.002	0.01–0.1	?
			<0.01[e]		
NH_3	78	13.3–14.6	0.1–0.3[d]	0.2–2	< 5
CO_2	72		3	<10	?
H_2CO	64		0–5[d]	0.1–0.3	< 0.2
			0.1–0.04[f]		
H_2S	57		0.2	<0.01	0.3
(CO)	(50)	22–25	(if co-deposited with water)		
CH_4	31	60–50	0.2–1.2	1	< 1
			<0.2[g]		
CO	25	84–63	~7	500–2000	0–5
			0–20[h]		
N_2	22	112–79	0.02	100	?
S_2	20		0.025[i]	?	?

[a] Temperature and distance of sublimation/condensation in models of the solar nebula. The first distance given represents a radiative equilibrium model (Hayashi 1981) while the second represents an accretion disk model (Cameron 1978a). Gas density of $n = 10^{13}$ cm^{-3} and mixing ratio $H_2O/H_2 = 10^{-4}$ are assumed. After Yamamoto (1985). [b] Relative abundances are for comet Halley, except as noted (see also Table V). For the interstellar medium, we follow van Dishoeck et al. (see their Chapter) for values in low-mass star-forming regions (see their Tables VI and IX), while for interstellar ice we adopt the values given in Table VI of this Chapter. [c] 1% in comet Levy (1990 XX), 5% in comet Austin (1990 V). See Hoban et al. (1991a) and Bockelée-Morvan et al. (1991). [d] Both NH_3 and H_2CO were temporally variable in comet Halley. It is assumed that NH_3 is the sole parent of NH_2. See text. [e] Typical of several comets. See Kim and A'Hearn (1991). [f] In comet Austin (1990 V), assuming a parent distribution. See Colom et al. (1992). [g] In comet Levy (1990 XX). See Brooke et al. (1991b). [h] Range exhibited by recent comets. See text, and Feldman (1983). [i] In comet IRAS-Araki-Alcock (1983 VII), but in no other comets.

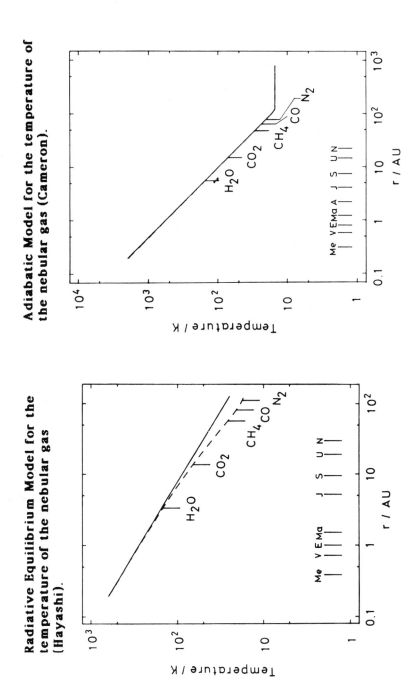

Figure 2. Temperature profiles in the solar nebula and sublimation distances for volatile ices. (a) Radiative equilibrium model (Hayashi 1981) for the temperature of the nebular gas. (b) Adiabatic model (Cameron 1978a) (figure after Yamamoto 1985).

II. COMETARY RESERVOIRS AND ORBITAL DYNAMICS

A. Comets and the Oort Cloud

Cometary orbits are classified as either long or short period, depending on whether their periods are greater than or less than 200 yr, respectively. The orbits of long-period (LP) comets are randomly oriented on the celestial sphere, whereas the short-period (SP) comets are generally confined to direct orbits with inclinations less than ~30 deg. Most of the known SP orbits have periods between 5 and 20 yr, while the LP orbits range up to 10 Myr.

Approximately one third of all long-period comets observed passing through the planetary region are on weakly hyperbolic orbits. However, it is only the osculating (instantaneous) orbits that appear hyperbolic. Integration of the orbits backward in time to points outside the planetary region, and conversion from a heliocentric to a barycentric coordinate system (Bilo and van de Hulst 1960), showed that those comets in fact had highly eccentric but still gravitationally bound orbits. Planetary perturbations (primarily by Jupiter) scatter the LP comets in orbital energy, either ejecting them on hyperbolic orbits or capturing them to more tightly bound ellipses.

The distribution of orbital energies for the observed LP comets is shown in Fig. 3 (Marsden 1989). The numbers of comets are plotted versus $1/a_o$, the inverse original semimajor axis of the orbit (prior to entry into the planetary region), which is proportional to orbital energy. Positive $1/a_o$ values indicate bound orbits, while negative values denote hyperbolic ones. The distribution is characterized by a sharp spike of comets at nearly zero (but bound) energies, and a low continuous distribution of comets in more tightly bound (less eccentric) orbits.

Oort (1950) showed that this unique distribution could be explained by a vast spherical cloud of comets surrounding the planetary system and extending halfway to the nearest stars. Oort demonstrated that comets in the cloud are repeatedly scattered by perturbations from random passing stars. The orbits diffuse in velocity phase space and occasionally are perturbed to perihelia within the planetary region.

However, once comets enter the planetary region they are scattered in $1/a_o$ primarily by Jupiter, with a typical $\Delta(1/a)$ of $\pm 630 \times 10^{-6}$ AU^{-1} on each perihelion passage (van Woerkom 1948; Everhart 1968). This is more than six times the width of the Oort cloud spike in Fig. 3. Thus, the comets rapidly diffuse in $1/a_o$. A typical LP comet from the Oort cloud makes an average of 5 returns with a mean time of 6×10^5 yr between its first and last perihelion passage. Approximately 65% of LP comets are hyperbolically ejected to interstellar space, 27% are randomly disrupted (they break up for reasons that are not well understood), and the rest are lost to a variety of mechanisms such as perturbation to a Sun-impacting orbit (Weissman 1979). A very small fraction of the long-period comets, on the order of 10^{-3} to 10^{-4}, evolve to short-period orbits.

The few hyperbolic comets in Fig. 3 are believed to be the result of small

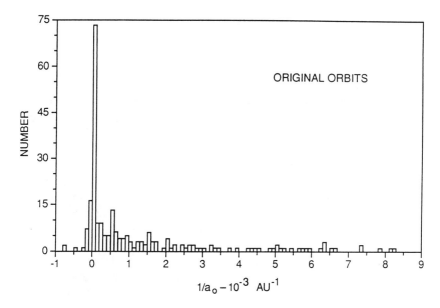

Figure 3.   The distribution of original inverse semimajor axes for 264 long-period comets as found by Marsden (1989). The large spike at very small positive values of $1/a_o$ corresponds to comets in very long-period orbits, extending to interstellar distances. These are the "dynamically new" comets from the Oort cloud. The low continuous distribution with $1/a_o > 1 \times 10^{-4}$ AU^{-1} are returning comets which have been scattered in $1/a_o$ by planetary perturbations, primarily by Jupiter.

errors in their orbit determinations and/or nongravitational forces (i.e., jetting of volatiles from the nuclear surface), which cause the orbits to appear more eccentric than they actually are. True interstellar comets would be expected to have $1/a_o$ values on the order of -0.50 AU^{-1}, for hyperbolic encounter velocities of 20 km s^{-1} (the average velocity of the Sun relative to nearby stars), more than 600 times greater than the largest value observed.

The dynamical evolution of LP comets is controlled by the combined action of planetary and stellar perturbations, by rare encounters between the solar system and giant molecular clouds (GMCs), by the tidal field of the galactic disk (and to a lesser extent the galactic nucleus), and by nongravitational forces from jetting during perihelion passage. Monte Carlo numerical simulations of the comets' dynamical evolution in the Oort cloud (Weissman 1982,1985a; Fernandez 1982; Heisler and Tremaine 1986; Heisler 1990), can be used to estimate the cloud population by comparing the predicted and observed flux of LP comets through the planetary region, after correction for observational selection effects (Everhart 1967). The current best estimate for the classical Oort cloud (Fig. 1) is $\sim 10^{12}$ comets (Weissman 1990).

Studies of the integrated perturbations on comets in the classical Oort cloud have shown that their typical dynamical lifetime is only about half the age of the solar system (Hut and Tremaine 1985). Thus, it has been suggested that the cloud needs to be replenished. The two possible sources are an unseen inner Oort cloud which is not dynamically sampled except by the largest perturbations (i.e., those due to penetrating stellar passages or encounters with GMCs), and the capture of comets from interstellar space. Capture has been shown to be highly improbable at typical stellar encounter velocities (Valtonen and Innanen 1982; Valtonen 1983). On the other hand, dynamical models (Duncan et al. 1987; Shoemaker and Wolfe 1984) have shown that an inner Oort cloud with a probable population 5 to 10 times greater than that of the outer cloud is a natural consequence of the dynamical ejection of comets from the planetary region (Fig. 1). The inner cloud is capable of replenishing the outer Oort cloud as comets there are stripped away by stellar and GMC perturbations. The estimated population of the Oort cloud (inner and outer) is then 6 to 11×10^{12} comets. Assuming an average nuclear mass of 3.8×10^{16} g (Weissman 1991a), the total mass of comets in the Oort cloud is between 38 and 70 Earth masses ($M_\oplus$). Because estimates of cometary masses are highly uncertain, some Oort cloud mass estimates are much larger, up to $\sim$500 $M_\oplus$ (Bailey 1990). A more complete review of Oort cloud dynamics is given in Weissman (1990).

B. Formation of the Oort Cloud

Hypotheses on the origin of the Oort cloud can generally be divided into two groups: (a) that comets are of primordial origin, and were formed coincident with the Sun and planetary system, or (b) that comets were formed and/or were captured episodically, either once or many times over the history of the solar system.

The many different proposals for cometary origin are reviewed in detail in *Protostars and Planets II* (see Weissman 1985b). The major primordial origin hypotheses include:

1. Formation as icy planetesimals in and/or just beyond the outer planets zone, followed by dynamical scattering to distant orbits in the Oort cloud by the growing proto-planets (Oort 1950; Kuiper 1951; Safronov 1972; Cameron 1978b;

2. *In-situ* accretion in the solar nebula at large solar distances (Biermann and Michel 1978; Hills 1982; Bailey 1987);

3. Formation in satellite nebulae of the solar nebula (Cameron 1973);

4. Formation of the Sun in a closely spaced cluster of stars such that comets formed around each star were dynamically mixed, populating each of their respective Oort clouds (Donn 1976).

The episodic hypotheses include:

5. Gravitational focusing of material in the solar wake after solar system passage through interstellar clouds (Lyttleton 1948);

6. Eruption from the giant planets and/or their satellites (Vsekhsvyatskii 1967);
7. Formation in compressed interstellar clouds at galactic spiral-arm shocks (McCrea 1975);
8. Formation in GMCs, followed by capture during encounters with the solar system (Clube and Napier 1982).

All the episodic hypotheses have serious difficulties and will not be discussed further here (see discussion of problems and weaknesses in Weissman [1985b]). Of the "primordial origin" hypotheses, (1) and (2) are the leading candidates. A cometary origin among the outer planets provides for formation in a relatively dense but still cold region of the solar nebula, with the problem that scattering to Oort cloud distances is dynamically inefficient. Conversely, formation at Oort cloud distances is dynamically efficient but is difficult to understand given the very low nebular densities expected at 10^3 to 10^4 AU from the proto-Sun. The first four hypotheses are illustrated schematically in Fig. 1.

Oort (1950) suggested that comets were dynamically ejected from the asteroid belt by Jupiter and the other giant planets. Kuiper (1951) pointed out that the proposed icy composition of comets required that they be formed farther from the Sun, at the orbit of Jupiter or beyond. Safronov (1972) subsequently showed that gravitational scattering by Jupiter and Saturn tended to eject most planetesimals from their zones into hyperbolic orbits, rather than to bound orbits in the Oort cloud. However, Safronov also showed that the lower masses and larger heliocentric distances of Uranus and Neptune would allow them to provide considerable dynamical scattering, with far fewer ejections. This work has served as the basis for the most widely accepted and widely studied theory of cometary origin.

The present ideas about planetesimal formation are reviewed elsewhere in this book (see Chapter by Weidenschilling and Cuzzi). Icy planetesimals, formed in the outer solar system, are thought to have been the building blocks of the cores of the giant planets. They probably also provided a substantial fraction of the mass of Uranus and Neptune, which apparently formed after much of the nebular gas had dispersed. The composition of icy planetesimals undoubtedly varied with temperature (and hence, with heliocentric distance) in the solar nebula, though dynamical scattering by the growing protoplanets may have blurred those differences somewhat.

Safronov (1972) showed that if the current masses of Uranus and Neptune were dispersed into small planetesimals in their zones, their accretion times would be $\sim 10^{11}$ yr, longer than the age of the solar system. To surmount this difficulty, Safronov suggested that the original mass of planetesimals in the Uranus-Neptune zone was 10 times larger than the current masses of the two planets combined. According to Safronov, most of this material was ejected to hyperbolic orbits but about 1 to 2% of it reached distant elliptical orbits to form the Oort cloud.

Numerical simulation of this dynamical scattering (Fernandez and Ip 1981,1983) confirmed many of Safronov's ideas, but also found some important differences. Fernandez and Ip showed that material was placed in the Oort cloud most efficiently by Neptune, with 72% of all ejecta ending up in the Oort cloud, the rest escaping to interstellar space or being sent to short-period orbits among the inner planets. The comparable Oort cloud fractions are 57, 14 and 3% for Uranus, Saturn and Jupiter, respectively. These efficiencies are considerably greater than the values of 2, 1.2, 0.5 and 0.2% for Neptune, Uranus, Saturn and Jupiter found by Safronov (1972). A portion of the disagreement can be attributed to Safronov's use of a much narrower capture range in orbital energy for the Oort cloud. Scaling Safronov's work to the same capture zone would lead to probabilities of 5.4, 3.2, 1.3, and 0.5%, still a factor of 6 to 17 less than those of Fernandez and Ip. The remaining disagreement may be due to other differences in the models. For example, Fernandez and Ip considered capture to short-period orbit to be an endstate, whereas in reality those comets continue to circulate and eventually will most likely be dynamically ejected.

Greenberg et al. (1984) constructed another model simulating accretion, and studied the problem of forming Neptune while simultaneously populating the Oort cloud. They found that the planet and comet cloud had to be built from a primordial population of relatively small bodies, $\sim$4 to 8 km in radius. Larger initial sizes on the order of 100 km diameter did not lead to a sufficient number of comets in the Oort cloud. It is interesting to compare the resulting size distribution of icy planetesimals in Fig. 4 with the mass distribution for cometary nuclei found by Weissman(1991b), based on Everhart's intrinsic distribution of cometary magnitudes, corrected for observational selection effects. The two distributions have similar bimodal forms and slopes, with the size break occurring at nearly the same radius ($\sim$7–8 km). It is notable that these two relatively independent approaches arrive at such similar mass distributions for cometary nuclei. Note that a substantial number of large comets are accreted, some up to 10^3 km in diameter or more. Stern (1991) also argued that a substantial number of icy bodies similar in size to Triton and Pluto may have accreted, and may be resident in the Oort cloud or may have been ejected to interstellar space. The accretional and collisional evolution of this population of large comets in the Uranus-Neptune zone, prior to their ejection to the Oort cloud, may provide for some degree of physical processing of the proto-comets. Fernandez (1980) also suggested that large comets, $\sim$$10^3$ km diameter, were required to perturb comets out of the hypothetical ring of comets beyond the orbit of Neptune (see Kuiper belt discussion below).

Ejection of comets to large aphelion distances is not, by itself, sufficient to create the Oort cloud, because the comets will return to small perihelion distances where they can again be perturbed by the giant planets. However, Duncan et al. (1987) showed that comets with aphelia of $\sim$5 to 6×10^3 AU would be sufficiently perturbed by galactic tides during a single orbit to raise their perihelia out of the planetary region. In this manner, a large inner Oort cloud could be populated. Cometary orbits would continue to evolve in the cloud under the combined influence of stellar, GMC, and galactic tidal perturbations. The net effect of those perturbations is to pump angular momentum into the cloud, randomizing the inclinations of the comet orbits and further raising their perihelia. Duncan et al. showed that the Oort cloud would be essentially randomized at semimajor axes $>8 \times 10^3$ AU. At semimajor axes $>2 \times 10^4$ AU, typical perturbations by random passing stars and the galactic tide are sufficient to throw some comets back into the planetary region where they are observed as "dynamically new" comets.

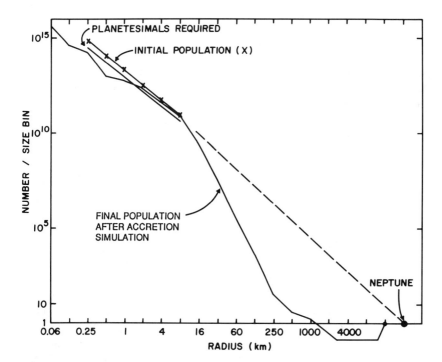

Figure 4.   Size distribution of accreted and disrupted planetesimals in the Uranus-Neptune zone after a simulated evolution of 1.4×10^5 yr, as modeled by Greenberg et al. (1984). The initial planetesimal size distribution, shown at upper left, was chosen to match the number and size distribution of observed comets. A runaway accretion results in the formation of Neptune, but also leaves a bimodal double-sloped spectrum of smaller-sized bodies which are the protocomets.

The dynamical problems associated with populating the Oort cloud from the outer planet region are eliminated if comets formed at Oort cloud distances, $\sim 10^3$ to 10^4 AU. One possible description of this process is provided by Hills (1982) and Hills and Sandford (1983a,b). They suggested that infalling, low opacity dust clumps in the outer solar nebula experienced a net external radiation pressure which forced them to collapse, forming cometary nuclei. An important assumption in their study is that the Sun was a member of a star-forming cluster and that the dust clump experienced a roughly isotropic radiation field, forcing material in toward the center of the clump. Hills (1982) estimated that an initial clump with a radius of ~ 0.02 AU, a dust density of 4×10^{-20} g cm^{-3}, and a gas density of 2×10^{-18} g cm^{-3}, would collapse in about 400 yr, considerably less than the 5×10^3 yr needed for the clump to fall into the proto-Sun. The resulting comet nucleus would have a radius of 1 km and a mass of 4×10^{15} g. Hills and Sandford (1983a, b) showed that the radiation field need not be perfectly isotropic for collapse to proceed, but also that the radiation field from only two protostars would not focus material towards the center of the clump sufficiently for proto-comets to form.

Although this is an intriguing and novel idea that could explain the origin of some comets, several aspects must be explored before it can be regarded as robust. Unanswered questions include: what is the effect of turbulence in the nebula on the formation of irregular dust clumps? How efficient is the coupling between radiation and grains in this irregular field of objects, i.e., do clumps shadow one another? What is the total number of comets that could be formed by this mechanism and what is their subsequent interaction with the inner nebula? Late infalling clumps may be delayed by their own angular momentum and this may also prevent them from falling directly into the proto-Sun, but is it sufficient to prevent them from interacting gravitationally with the growing proto-planets? If not, they could be scattered to interstellar space just like the icy planetesimals formed in the inner nebular disk.

A somewhat different mechanism was proposed by Whipple and Lecar (1976) and developed by Bailey (1987). Whipple and Lecar suggested that a strong protosolar wind, akin to observed T Tauri winds, would produce a turbulent boundary layer between infalling nebular and outflowing protosolar material, and that material in this region would be driven into a thin circum-stellar shell. Densities may be sufficient to allow rapid accretion of dust grains which are then sufficiently massive to de-couple from the gas motion. Local concentrations of dust grains may then collect and gravitationally collapse into cometary-sized bodies. In this manner, it may be possible to produce a large number of cometary bodies, either bound to the proto-Sun or circulating in nearby interstellar space. The problems with this scenario are similar to those of the previous one; the effects of turbulence, heating at shock boundaries, and subsequent dynamical evolution are not well understood or quantified.

If comets are formed in nearly circular orbits about the proto-Sun, then those in the region between roughly 10^2 and 2×10^3 AU experience no dynamical mechanism for being perturbed either into the planetary region or far enough outward to fall under the influence of perturbations by stellar and galactic tides. Thus, this population of comets would circulate indefinitely without ever producing visible comets. One possible means for increasing the semimajor axes of such orbits would be rapid loss of mass by the central protostar and nebula (Cameron 1978*b*). In particular, if the mass loss is slightly less than 50% of the total central mass, nearly circular orbits are transformed into nearly parabolic ones, thereby populating the Oort cloud. However, the range of values for the mass loss that would produce this transformation is very narrow, and thus relatively unlikely. Dermott and Gold (1978) proposed a solution to this problem by suggesting that the loss of central mass in a number of smaller, repeated steps could pump up comets a bit at a time to Oort cloud distances. An important requirement is that the mass loss must be accomplished on a time scale long compared with the orbital periods of the planets, but short compared with the cometary periods. The required scenarios are probably too constraining to be physically plausible. Although mass ejection rates up to 10^{-5} $M_\odot$ yr^{-1} are observed in some T Tauri stars, the total integrated mass loss for solar-mass stars probably does not exceed 0.5 $M_\odot$, or about one-third of the initial nebular mass.

The discussion so far has emphasized orbital dynamics as tests of theories of cometary origins, but these theories also have implications for the orbital histories of the major planets. Marochnik et al. (1988) estimated that the angular momentum of the Oort cloud is 0.5 to 2×10^{53} g cm^2 s^{-1}, 2 to 3 orders of magnitude greater than the total angular momentum of the current planetary system. They argued that had this angular momentum been ejected from the planetary system, the giant planets would have spiraled in toward the Sun. Thus, they suggested that comets must have formed *in situ* at Oort cloud distances. However, Weissman (1991*b*) showed that most of the angular momentum in the present-day Oort cloud resulted from the action of external perturbers over the history of the solar system. In addition, some Oort cloud parameters used by Marochnik et al. tend to be higher than current best estimates. Weissman showed that the total angular momentum of the current Oort cloud is probably in the range 0.6 to 1.1×10^{51} g cm^2 s^{-1}, and that the original angular momentum was perhaps a factor of 5 smaller. These more modest estimates are consistent with comets having been ejected from the Uranus-Neptune zone.

This point was also examined by Fernandez and Ip (1984) who modeled the exchange of angular momentum between the giant planets and the evolving protocomets. They showed that the giant planets scatter material both inward and outward, resulting in little net movement in the semimajor axes of the protoplanets. Jupiter is responsible for most of the hyperbolic ejection because it has no larger planet to which to pass the evolving comets, but because of its great mass, Jupiter only moves inward a few tenths of an AU.

C. Extra-Solar Oort Clouds and Interstellar Comets

Presumably, the same processes that led to the formation of our solar system's Oort cloud also occur around other forming stars. Because the expected dimensions of Oort clouds are so large, it may be possible to detect and resolve such comet clouds. One approach is to search for thermal radiation from dust grains, created by collisions and sputtering in the cloud. Stern et al. (1991) analyzed IRAS 60 and 100 μm sky flux plates for 17 nearby stars with known infrared excesses (e.g., β Pictoris, ϵ Eridani). They searched for infrared excess emission in summed circular annuli at distances of several thousand AU from each star; however, no detections were made. They also showed that the expected brightness of Oort clouds with populations similar to our own was below the IRAS detection limits, because dust at Oort cloud distances will be very cold, and radiation pressure and collisions with interstellar dust and gas should rapidly sweep fine dust from the comet clouds (Stern 1990). Stern et al. (1990) also proposed searching for Oort clouds around red giant stars. These stars are sufficiently luminous that comets in their Kuiper belts (see short-period comet discussion below), if present, would be actively sublimating at rates comparable to the gas production rates of comets at 1 AU in our own solar system. Stern et al. suggested that observed OH/IR stars may be examples of this phenomena.

It is interesting to speculate on the fate of the many comets ejected to interstellar space in forming the Oort cloud, and over its history. Dynamical ejection is the most common loss mechanism for comets in the cloud, either due to close stellar and GMC encounters or as a result of Jovian perturbations during passes through the planetary system.

If other planetary systems exist and have generated Oort clouds in a similar fashion, then there should be a substantial population of comets in interstellar space, and our solar system should occasionally encounter these interstellar wanderers. However, no comet has ever been observed passing through the planetary system on a clearly interstellar trajectory. Sekanina (1976) showed that this fact sets an upper limit on the space density of interstellar comets of 6×10^{-4} $M_\odot$ pc^{-3} ($\sim 4 \times 10^{12}$ comets pc^{-3}, using Sekanina's mean nuclear mass of 3×10^{17} g). For comparison, this is ~ 300 times smaller than the density of material in the solar neighborhood (~ 0.185 $M_\odot$ pc^{-3}; Bahcall 1984), so interstellar comets cannot contribute significantly to the "missing mass" problem in the Galaxy. It is about half the space density of comets in the outer Oort cloud, assuming a population of 10^{12} comets in a sphere of radius 10^5 AU centered on the Sun. Thus, the limit is not very strict.

The same problem was studied by McGlynn and Chapman (1989) who suggested that at least 6 interstellar comets should have been observed passing within 2 AU of the Sun in the past 150 yr, even after accounting for the fact that only 7% of all long-period comets passing within 2 AU are expected to be discovered. Their estimate was based on an average ejected population of 10^{14} comets per star, or $\sim 8 \times 10^{12}$ comets pc^{-3}. This is twice Sekanina's upper limit of 4×10^{12} comets pc^{-3}. The reason for this difference is not clear, though it may not be significant.

The above limits can be compared with the estimated space density of interstellar comets, derived by assuming that all stars produce cometary clouds. Dynamical models estimate that between 3 (Fernandez and Ip 1981) and 50 (Safronov 1972) times as many comets are ejected by the protoplanets as are placed in the Oort cloud (though, as discussed above, Safronov's estimate is extreme because it assumes a too narrow range of semimajor axes for capture to the Oort cloud). Of the original population placed in the Oort cloud, one-half to two-thirds have been ejected over the history of the solar system. Thus, taking a nominal current Oort cloud population of 7×10^{12} comets, the solar system has ejected $\sim 4 \times 10^{13}$ to $\sim 10^{15}$ comets to interstellar space. Taking a mean volume per star in the solar neighborhood of ~ 12 pc^3 (Allen 1976) and assuming that all stars produce comet clouds similar to our Oort cloud, the predicted space density is 3×10^{12} to 9×10^{13} comets pc^{-3}. This is 0.8 to 23 times the upper limit estimated from the absence of observed comets on interstellar trajectories (Sekanina 1976).

Because half of all stars form in multiple systems, and that process may prevent the formation of a protoplanetary disk leading to icy planetesimals (though that conjecture has not been demonstrated), the factor of 0.8 excess is probably not a problem. However, a factor of 23 excess is not consistent with the Oort cloud models and population estimates presented here, and thus represents a difference that clearly needs to be resolved. As noted above, at least a factor of 2.7 (and probably twice that) reduction can be obtained by assuming a wider range of Oort cloud semimajor axes than in Safronov's work. The remaining difference clearly merits further study.

D. Origin of Short-Period Comets

It has generally been thought that the SP comets are LP comets which have random-walked to orbits with small semimajor axes as a result of perturbations by Jupiter and the other planets. Capture of comets to short-period orbits in a single planetary encounter is highly unlikely (Newton 1893) and could not provide the observed number of SP comets. Planetary perturbations are larger for direct, low-inclination orbits and it was suggested that this acted as a selection mechanism for producing the direct, low-inclination SP comet population. However, evolution from an Oort cloud orbit to a short-period one would be expected to take approximately 400 returns (Weissman 1979), and it is most likely that a comet would be ejected over that time, or destroyed by one of several poorly understood physical mechanisms (e.g., disruption, sublimation). A variation which results in improved capture efficiency was proposed by Everhart (1972) who suggested that comets with perihelia among the outer planets random walk in $1/a$ without ever coming close to the Sun, and then are dumped into small perihelia orbits late in their dynamical evolution. In this manner, the SP comets can be spared some of the physical loss associated with solar heating or passage through the more densely populated regions of the planetary system. This also increases the planetary system's cross-section for capturing LP comets to SP orbits, thus more easily supplying the observed number of SP comets.

Estimates of the number of short-period comets produced by planetary perturbation from the Oort cloud have varied considerably (Joss 1973; Delsemme 1973), with some suggestion that the above mechanisms can not produce the observed number of SP comets. An alternative suggested source of SP comets is a belt or ring of comets beyond the orbit of Neptune (cf. Fig. 1) (Kuiper 1951; Whipple 1964; Fernandez 1980). Dynamical injection from this source may be up to 300 times more efficient than repeated perturbation of LP comets from the Oort cloud. Fernandez (1980) pointed out that some massive comets, $\sim 10^2$ to 10^3 km in diameter, would be required in the distant comet ring, slowly perturbing other comets back into the planetary region and providing the SP comet flux.

Duncan et al. (1988) compared the two proposed sources and found that LP comets tended to preserve their inclinations as they evolved inward to SP orbits. Thus, if SP comets were initially LP comets from the Oort cloud, far more high inclination SP comet orbits would be expected. Duncan et al. (1988) showed that the observed distribution of inclinations was more consistent with a low-inclination trans-Neptunian population of comets as the source of the SP comets. They called this population the Kuiper belt in honor of Gerard Kuiper who first suggested its existence in 1951. Subsequently, Torbett (1989) showed that planetary perturbations would lead to chaotic motion in a disk of comets beyond Neptune, throwing those with perihelia near 30 AU into Neptune-crossing orbits in only 10^7 yr. However, Torbett and Smoluchowski (1990) showed that the motion would not be chaotic for Kuiper belt orbits with perihelia >45 AU. Levison (1991) confirmed this by studying the evolution of orbits with perihelia >30 AU and aphelia <100 AU, and showed that many can persist over the age of the solar system, in particular those with initially low-eccentricity orbits beyond 45 AU.

Stagg and Bailey (1989) and Bailey and Stagg (1990) raised a number of questions with regard to the Kuiper belt scenario. First, physical loss mechanisms might remove high inclination LP comets during their longer evolution inward from the Oort cloud to SP comet orbits (longer because of the smaller average perturbations for high inclination orbits), leading to a predominantly low-inclination population of observed SP comets. This would remove the objection raised by Duncan et al. (1988) to this source for the SP comets. Second, they suggested that the combination of LP comet evolution from the Oort cloud plus Kuiper belt might produce too many SP comets, and that this possibly implied a lower than estimated population for the Oort cloud or the Kuiper belt, or both. Finally, they noted that Duncan et al. (1988) had increased the planetary masses by a factor of 40 to speed the numerical integrations, a common technique in simulations of celestial mechanics but one that could lead to erroneous results. However, Quinn et al. (1990) repeated the integrations with the mass enhancement factor reduced to 10 and obtained identical results for the Kuiper belt comet evolution. In addition, Wetherill (1991) was able to recreate the same results with no mass enhancement, using a simpler Öpik-type integrator for simulating the evolution of cometary orbits.

Duncan et al. (1988) estimated that 0.02 $M_{\oplus}$ (1.2 × 10^{26} g) of comets are required in the Kuiper belt to maintain the current population of SP comets. Duncan et al. used an average cometary mass of 3.2 × 10^{17} g, implying a population of 3.8×10^8 objects. However, using the average mass of 3.8×10^{16} g noted above (Weissman 1991a), the same number of Kuiper belt comets would have a total mass of ~ 0.0025 $M_\oplus$, a factor of eight smaller.

Observational searches for Kuiper belt comets have so far been negative. Based on a search of 4.9 square degrees of sky near the ecliptic, Levison and Duncan (1990) set an upper limit of less than one object brighter than magnitude $V = 22.5$ per square degree. Magnitude $V = 22.5$ corresponds to a 60 km radius cometary nucleus with albedo of 0.05 at 50 AU from the Sun. This limit translates to $N \leq 10^{10}$ objects (total mass $\sim 3.8 \times 10^{26}$ g, or ~ 0.06 $M_\oplus$) if the Kuiper belt comets are in orbits between 35 and 60 AU from the Sun. Yeomans (1986) set an upper limit of ~ 1 $M_\oplus$ on the unknown mass just beyond the orbit of Neptune, based on the absence of observed gravitational perturbations on the orbit of comet Halley. An earlier limit of 5 $M_\oplus$ was found by Anderson and Standish (1986) based on tracking of the Pioneer 10 spacecraft. Future tracking of the Voyager and Pioneer spacecraft can be expected to further refine these limits.

III. EVOLUTION OF COMETARY NUCLEI

A. Evolution of Comets in the Oort Cloud

Because the Oort cloud is cold, remote, and relatively sparsely populated, it was long believed to act as a perfect storage environment for comets. However, over the past decade several physical processes have been shown to modify the outer layers of cometary nuclei while they are in the cloud. These are: irradiation by high-energy photons and charged particles, heating by luminous passing stars and supernovae, erosive interactions with grains in the interstellar medium, and gardening by debris impacts within the Oort cloud. We review the effects of each process individually, and then comment on the net effects of these acting in concert.

Radiation Damage. Radiation bombardment of icy surfaces falls into two categories: photon bombardment and charged particle bombardment. We consider photon bombardment first.

Ultraviolet photons (i.e., $h\nu > 3$ eV) provide the energy necessary to initiate substantial chemical changes in cometary surfaces. In a classic series of laboratory experiments and theoretical studies, J. M. Greenberg showed that ultraviolet irradiation would produce significant alteration of the composition, color and volatility of icy mantles on silicate grains in the interstellar medium prior to their aggregation into cometary nuclei (cf. Greenberg 1982). Further work on the spectroscopic effects of ultraviolet irradiation of volatile ices was performed by Thompson et al. (1987), who demonstrated that ultraviolet photons induce chemical changes accompanied by surface darkening and color change (i.e., reddening). Such optical effects were shown to become progressively more severe with dose. The expected interaction depth of ultraviolet photons in a cometary surface is just a few optical depths, so the expected processing damage is likely to be confined to the outermost layer, only a few tens to hundred atoms thick (1–10 nm).

More energetic X-ray photons from the galactic and cosmic background, and from nearby supernovae (Stern and Shull 1988) penetrate to greater depths, perhaps as deeply as 10 to 100 μm below the cometary surface. Ionizing radiation deposits $\sim$30 to 35 eV for each ion electron pair produced, and much of this energy is dissipated thermally. For this reason, X-ray interactions could cause more extensive thermal and structural changes in the ice than ultraviolet photons. Because of the relatively small depths affected, photoprocessing is most important for pre-cometary ices, e.g., icy mantles on grains in dense molecular cloud cores or in the solar nebula, prior to accretion of cometary nuclei. A detailed discussion of such processing exceeds the scope of this chapter (see the Chapter by van Dishoeck et al.; see also Greenberg 1982; d'Hendecourt et al. 1986; Allamandola et al. 1988).

Charged particle radiation is capable of breaking bonds, inducing chemical reactions, and re-ordering or mechanically disturbing the ice matrix to much greater depths than ultraviolet and X-ray radiation (Strazzulla and Johnson 1991). The column density (g cm^{-2}) to which charged particles penetrate depends strongly on their energy, and to a lesser extent on the specific composition of the material. Low-energy solar protons of 1 to 300 keV penetrate only about 10^{-3} g cm^{-2} before being stopped. Cosmic-ray protons with typical energies of $\sim$100 MeV penetrate about 1 g cm^{-2}; while those with energies of 1 GeV penetrate 100 to 200 g cm^{-2} (Johnson et al. 1987; Strazzulla and Johnson 1991). Figure 5 illustrates the variation in energy dose with depth for the expected cosmic-ray fluence on comets (integrated over the age of the solar system). Because penetration depth is inversely proportional to the density of surface material, lower-density ices suffer damage to proportionately greater depth. If cometary surfaces are indeed low in density, then the radiation damaged crust may extend to many meters in depth.

Laboratory experiments simulating high-energy electron, proton and ion bombardment of cometary surfaces have been performed for over a decade now. In such experiments, keV to MeV accelerator beams have been used to irradiate various pure and aggregate mixtures of H_2O, CO_2, NH_3, SO_2, S_2 and CH_4 frosts to fluences of 10^{12} to 10^{19} ions cm^{-2}. Such experiments (Moore et al. 1983; Johnson et al. 1984,1985,1987; Lanzerotti et al. 1987) have confirmed that cosmic-ray bombardment introduces important structural and chemical changes to these ices. The effects resulting from bombardment by higher-energy (GeV) cosmic rays are estimated by theoretical scaling, because the limited energy range of the accelerators used precludes direct measurement. In a recent detailed review of these effects, Strazzulla and Johnson (1991) report simultaneous increases in the porosity and density of proton- and ion-bombarded ices. Such bombardment also causes preferential breaking of hydrogen bonds, leading to the formation of C—C, C—N, C—S, C—O, S—O and S—S bonds. Free H atoms created by the bond-breaking irradiation migrate in the ice matrix to form substantial amounts of

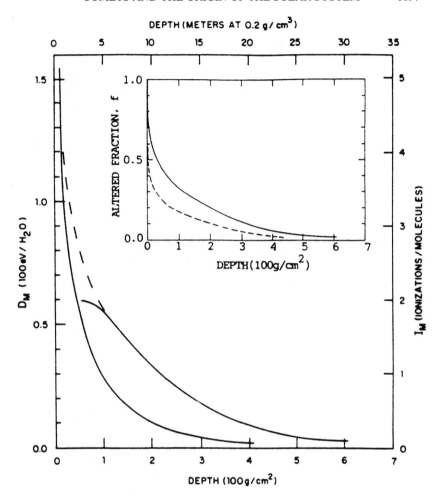

Figure 5.   Cosmic-ray energy deposition in H_2O vs depth for an exposure age of 4.5 Gyr. Upper curve: extracted from atmospheric data (see Strazzulla and Johnson 1991). Lower curve: sum of dose rates given by Moore (1981) and Ryan and Draganic (1986) (note difference from that in Johnson et al. [1987], based on Ryan and Draganic data which were plotted incorrectly). Dashed: extrapolation. The inset panel shows the fraction of chemical bonds altered by radiation processing vs depth (after Johnson 1989).

the extremely volatile H_2 molecule, and may initiate spin conversion in water and other symmetric molecules through exchange reactions (Johnson 1991). In addition, reactive molecular radicals are generated in the resulting matrix. The resulting surface layer loses its original chemical identity as new volatiles (e.g., S_2 from S_8, and CO from CO_2) and organic solids (with many CH_2,

CH_3, C=O and C=C groups) are created. The similarity of some end products of these processes to exotic cometary material (e.g., S_2, CHON grains and long chain hydrocarbons; see Sec. IV) has been noted.

Radiation processing experiments relevant to comets have also been conducted on refractory solids, including graphite, organics, amino acids and a number of commercial polymers (G. Strazzulla, personal communication; Caffee et al. 1988). Such materials have been subjected to both ion and ultraviolet irradiation; apparently, the effects of galactic cosmic rays on such materials have not been studied. The effects of such bombardment are that with increasing fluence, organic and refractory materials exhibit increasingly disordered structures, increased C/H, C/N and C/O ratios, and the loss or conversion of weakly bonded molecular volatiles such as NH_3, H_2 and CO_2 (cf. Strazzulla, personal communication 1991; Foti et al. 1990). These changes in turn result in sample compaction, densification, spectral reddening, darkening, and then flattening in the 400 to 800 nm regime (Thompson et al. 1987; Andronico et al. 1987). As a result of these studies, it is clear that the nonvolatile component of cometary surfaces also undergoes substantial radiation-induced chemical and structural change in the Oort cloud. However, the depth of penetration of ion and ultraviolet radiation is small (millimeters, micrometers, respectively), and the evidence of such irradiation may not survive when comets are close enough to the Sun for study. As with ices, energetic cosmic rays should penetrate refractory solids 10 to 100 times more deeply than either ions or ultraviolet photons, and should induce substantial damage and chemical conversion in the refractory material.

For sufficiently cold environments, such as in the Oort cloud, the time required for radicals and most newly created volatiles to diffuse to the surface is so long as to be essentially infinite (with the exception of H_2). Therefore, the newly created species (excepting H_2) remain trapped in the ice-dust matrix until the comet is heated or the surface becomes otherwise exposed e.g., by erosional effects. The irreversible conversion of the original, water-dominated ice matrix to this more complex "crust" may lead to darkening and reddening of the optically active ice surface, by forming long-chain hydrocarbons. A nonvolatile surface crust may form, beginning the process of sealing off the nuclear surface against sublimation even before the onset of crustal development during perihelion passage (Brin and Mendis 1979). Further, the radicals and volatiles produced by radiation bombardment may be responsible for observed activity at large solar distances (i.e., 5–10 AU) during the approach of dynamically new comets to perihelion. However, other processes may mitigate the observable consequences of radiation damage.

Heating by Passing Stars and Supernovae. A second process which acts to modify cometary surfaces in the Oort cloud is heating by luminous stars and supernovae. It has long been recognized that passing stars regularly penetrate the Oort cloud (cf. Smoluchowski et al. 1986). This process results in the diffusion of comets from the inner cloud to the outer cloud, and the dynamical randomization of orbits in the outer cloud. Stern and Shull (1988) modeled the implications of such encounters for the heating of comets in the cloud. They found that although several thousand encounters between stars and the Oort cloud should have occurred, the vast majority involved dwarf stars which are too faint to heat a significant fraction of the comets. Further, most comets heated by encounters with low-mass stars are gravitationally ejected from the cloud during the heating event. However, the rare encounters with luminous O and supergiant stars can heat the *entire* Oort cloud to interesting temperatures from distances so great that they do not substantially perturb orbits in the cloud. For example, a typical O star with luminosity $5 \times 10^5 \, L_\odot$ can heat the Oort cloud to 20 K from about a parsec (pc) away, but such a star would have to pass within a few thousand AU to eject comets from the cloud. Because vapor pressures vary exponentially with temperature, this thermal regime contrasts sharply with the ambient 5 to 6 K environment in the cloud. By combining the encounter rate of various stellar types with the influence of each stellar type on the Oort cloud, Stern and Shull showed that luminous passing stars have probably caused all comets to undergo at least one long (i.e., $\sim 3 \times 10^4$ yr) heating episode to 19 to 22 K.

Stern and Shull also estimated the effects of nearby supernovae on Oort cloud heating. Because supernovae are so much more luminous than even the brightest main-sequence stars, a typical supernova exploding at 19 pc can raise surface temperatures on comets to 30 K. Using supernova luminosity and formation-rate statistics (Tammann 1982; Narayan 1987), it was found that ~ 30 supernova heating events to 30 K should have occurred since the formation of the Oort cloud, that there is a high probability that the Oort cloud should have experienced one supernova heating event to 50 K, and that there is a 50% probability of a supernova heating event to 60 K over the age of the solar system. Figure 6 summarizes these results. Temperatures in the range 20 to 60 K can induce potentially important effects on cometary surfaces in the Oort cloud; these include mechanical stresses, porosity/densification changes, and the sublimation of highly volatile trace constituents (e.g., S_2, Ne, Ar, CO, N_2 and CH_4) from the surface. Table II gives additional information on important thermal milestones in this temperature regime.

The depth of penetration of the thermal wave resulting from a stellar or supernova encounter is determined from the diffusion equation, and is based on the diffusivity of the surface ice, the relative orientation of the cometary rotation pole to the encounter plane, and the duration of the encounter. The major uncertainty here is the actual diffusivity of the surface ice, which depends on its composition, porosity and crystal state (i.e., amorphous or crystalline). Adopting a range of surface diffusivities consistent with conductivity measurements on water ice (Klinger 1985), Stern and Shull estimated that the thermal wave of a typical, O star encounter (lasting 3×10^4 yr) could penetrate as far as 150 m into the comet, but is more likely to lie in the range 7 to 50 m. Because their model shows that all comets have unit probability of being heated by passing stars to temperatures ~ 19 to 22 K, one expects the outer layers of all Oort cloud comets to have been heat soaked at these

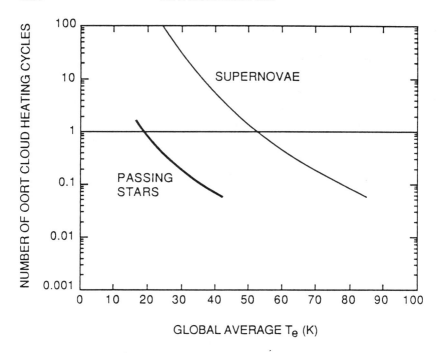

Figure 6.   The number of times a typical comet in the Oort cloud was warmed to a
global average temperature (T_e), by supernovae and by stellar transients (cf. Stern
and Shull 1988).   A cometary surface albedo of 0.04 and emissivity of 0.9 was
assumed. Heating by stellar transients is dominated by rare, O-star events. Though
less severe in thermal magnitude than supernovae events, the stellar transients last
much longer ($\sim 10^4$ yr vs ~ 0.1 yr for supernovae), causing the thermal wave to
penetrate several hundred times deeper (after Stern 1989).

temperatures for many centuries.   Supernova encounters, while generating
a stronger thermal pulse, are much briefer ($\sim 10^7$ s), therefore their thermal
effects are estimated to propagate only 0.1 to 2 m into the surface.   These
heating effects may play an important role in the behavior of "new" comets
as they approach the Sun.

Interactions with the Interstellar Medium.   The first study of the interaction between comets and the interstellar medium was made by O'Dell (1973),
who investigated the accretion of interstellar gas onto cometary surfaces. He
found that a layer 10 to 100 mm thick should be accumulated over an interval
of 4.5 Gyr. Whipple (1977) speculated that this layer of interstellar volatiles
accounted for the anomalous brightness of dynamically new comets on their
first pass through the planetary region. However, it now appears that O'Dell's
analysis was too simple, leaving out the important effects of impacts by interstellar grains on cometary surfaces. Stern (1986) investigated the competition

TABLE II

Critical Temperatures and Processes in Laboratory
Analogs of Cometary Ices

Temperature (K)	Process
5–6	Ambient Oort cloud temperatures.
6–20	Maximum temperature at 10^4 AU in solar nebula (Cassen and and Boss 1988).
16	Neon sublimes (Greenberg 1982).
17–26	Irradiated ices display luminescence (Moore et al. 1983).
25	All neon trapped in water clathrates is expelled.
27	"Yellow stuff" explodes on rapid heating (seconds), or slow warming as a chain reaction in radical chemistry (Greenberg 1986).
31	Argon sublimes (Bar-Nun et al. 1985).
40	Heavy cracking, whitening of mixed water ice (Bar-Nun et al. 1985).
40	Formation of formaldehyde (H_2CO) and formamide (NH_2HCO) (Greenberg et al. 1980).
57	H_2S sublimes.
60	Temperature in Uranus-Neptune zone of solar nebula.
70	40 m s^{-1} grain impact produces spikes to this temperature (Greenberg 1986; d'Hendecourt 1980).
125–150	Exothermic phase transition (to crystalline form) occurs in amorphous water ice. Excess trapped gases are released.
160–170	Water ice sublimation rates become significant.

between grain-driven erosion and gas-driven accretion. Based on laboratory studies of micro-cratering in ices and snows with porosities ranging from 0.01 to 1 (cf. Cintala 1981; Croft 1982b; Lange and Ahrens 1987), Stern demonstrated that grain-driven erosion on icy surfaces is some 700 to 1000 times more efficient than gas accretion, causing comets to actually lose material due to interstellar medium interactions. The primary reason for this large erosion-to-sticking ratio is that gas sticking is inefficient (typical sticking ratios of gases being a few percent), while energetic impacts of grains (at typical relative velocities of 20 km s^{-1}) cause much more material to be lost from the cometary surface than the original mass of the impacting grain. Lissauer and Griffith (1989) found similar results in a study of the source mechanisms of fine dust in β Pictoris type stellar disks.

   A more recent and comprehensive study by Stern (1990) included four interaction processes: grain erosion, accretion due to molecular sticking, gas sputtering, and thermal evaporation due to grain impacts. This study employed

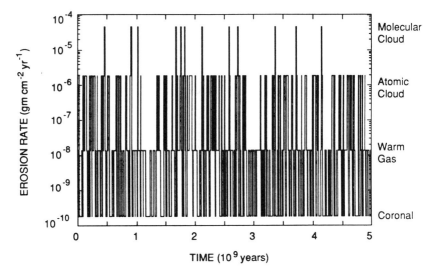

Figure 7.   The modeled mass-loss rate from surfaces in the Oort cloud. The model
assumes a random sequence of encounters with the four interstellar medium phases.
The total amount of time spent in each phase is proportional to their galactic filling
factors. Atomic and molecular cloud encounters typically last 2 to 10×10^5 yr. The
depicted encounters have been exaggerated in width by a factor of 10 for readability.
Cloud encounters actually occupy only ~3% of the total elapsed time.

a detailed model of the various phases of the interstellar medium, as well as the
evolution of the Sun's velocity relative to that medium over time. The results
of one such simulation, which assumed erosion on H_2O ice and a constant solar
velocity (20 km s^{-1}), are shown in Fig. 7. Owing primarily to strong density
gradients between cloud and inter-cloud phases in the interstellar medium,
interstellar medium interactions act in a highly time-dependent way. Their
cumulative effect causes comets in the Oort cloud to lose 60 to 600 g cm^{-2}
of surface material over 4.5 Gyr (Fig. 8). Due to correlations in the location
of coronal-phase parcels, the Sun will spend long periods in this rarefied,
essentially dust-free environment. During these periods, the erosion rate is
reduced to very low levels. Stern's analysis of the typical ejection velocity of
eroded material indicates that debris lost from comets in the Kuiper belt does
not directly escape the solar system. However, most debris from comets in the
inner and outer Oort clouds is injected directly into the interstellar medium,
thereby opening up an interesting (if minor) mass and chemical feedback
between Oort clouds and the interstellar medium (Stern and Shull 1990).

   Collisions in the Oort Cloud. Collisional interaction of comets with small
debris in the Oort cloud was first identified as an important process by Stern
(1988). A simple particle-in-a-box model was used, and both number density
gradients in the cloud and debris feedback to the population were ignored.
Power-law distributions with both constant area and constant mass per bin

OORT CLOUD INTEGRATED ISM EROSION

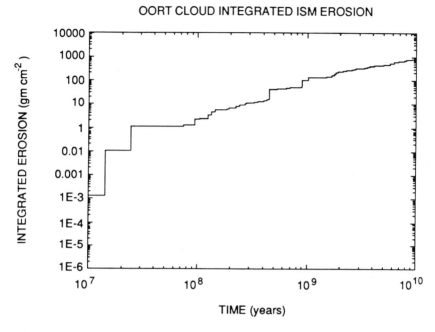

Figure 8.   The time-integrated mass loss experienced by cometary nuclei and other debris in the Oort cloud, from the model of Fig. 7 (after Stern 1990).

were found to produce frequent collisions between comets and small debris bodies (up to 10 m in size) in the inner Oort cloud. The rate of collisions in the outer Oort cloud was found to be much lower. Comet-on-comet collisions were found to be rare throughout the cloud, with an upper limit of perhaps one comet in 10^4 having suffered a catastrophic collision over the age of the solar system. The amount of ejecta produced by comet-on-comet collisions over 4.5 Gyr was evaluated in simulations which started with no objects smaller than ∼3 km. Such collisions cannot generate enough debris to create either constant area or constant mass per bin population distributions. Stern also found that debris created from collisions in the Oort cloud will not be energetic enough to escape directly to interstellar space but will instead enhance the population of small-sized bodies in the cloud.

Based on these results, Stern (1989) constructed a more complete time-dependent feedback model, with radial gradients for the Oort cloud in both density and velocity. This model included the weathering destruction of small (centimeter to meter class) debris by interstellar medium erosion, the loss of dust from the Oort cloud via interstellar medium-induced gas drag, and the feedback of collisional debris to the Oort cloud population distribution. Dust loss is important because it cleanses the Oort cloud of grains created through high-velocity impacts. However, radial transport of ejecta within the cloud was neglected, i.e., each macroscopic ejected body was assumed trapped in the (several thousand AU-wide) distance bin in which it had been created.

This model was evaluated for Oort cloud masses ranging from 10 to 10^3 M$_\oplus$, and for initial population distributions featuring constant area per bin, constant mass per bin, and collisional equilibrium. It was found that collisions do provide an important feedback mechanism for the production of small debris in the cloud, but erosion and drag by the interstellar medium prevent the population of small ejecta from growing without bound. Indeed, by rapidly removing the very numerous centimeter-to-meter-sized bodies that may have initially populated the cloud, erosion and drag act to severely truncate the collision rate within a few hundred Myr after the formation of the cloud. This in turn implies that present-day impact rates throughout the cloud are very low. Modeled estimates of the surface cratering fraction on comets indicate that regolith development can take place in the inner Oort cloud only if: (1) the inner reservoir is massive ($>10^2$ M$_\oplus$); and (2) the number of objects of <30 m size was initially large compared to the comet population itself. The predicted gardened depths are several centimeters to tens-of-centimeters. However, in the denser, inner portions of the Oort cloud, and in the Kuiper belt, the potential exists for significant cometary regolith development over time. This suggests that short- and long-period comets may experience differing degrees of collisional modification while in storage, if they do originate from different parts of the comet cloud.

B. Cometary Processing in the Planetary Region

When cometary nuclei return to the planetary region, their physical evolution is dominated by the heating they receive from direct solar radiation. Other processes such as irradiation by solar-wind protons and impacts by interplanetary dust particles will also intensify, but these are much less significant than solar heating in terms of both mass removal and depth of penetration. Meech (1991) has recently reviewed observational evidence for physical aging in comets.

The effects of heating will manifest themselves in several ways. On a comet's first approach to the Sun, the first effect to occur will be conversion of amorphous water ice to the crystalline form. The amorphous-crystalline phase transition occurs at ~140 K. Prialnik and Bar-Nun (1987) showed that the slow heating of the surface of an amorphous ice nucleus, presumably formed at low temperature, <100 K, would cause a transition to crystalline ice at about 5 AU inbound on the first perihelion passage. Because this reaction is exothermic, an additional heat pulse would propagate inward, converting a layer 10 to 15 m thick to crystalline ice. Prialnik and Bar-Nun showed that a chain reaction converting the entire nucleus to crystalline ice does not occur, because the pulse is eventually dissipated as it reaches colder ice layers at greater depths, and by warming of nonvolatile dust mixed with the ice. The amorphous-crystalline ice transition may supply sufficient energy to blow off pieces of the primitive irradiated crust, resulting in the anomalously bright behavior at large solar distance often displayed by dynamically new comets.

After the first perihelion passage inside 5 AU, the amorphous-crystalline transition is predicted not to repeat for several orbits, until a sufficiently thick layer of the overlying crystalline ice has sublimated and the orbital heat pulse can penetrate to the buried amorphous core. Using a more detailed model, which included the effect of a nonvolatile dust crust on the nuclear surface (see below), Prialnik and Bar-Nun (1988) found that the thickness of the crystalline ice layer overlying the amorphous ice was at least ∼15 m, and was often much thicker, ∼25 to 40 m.

The second insolation-related process is the sublimation of volatile ices at the nuclear surface, which results in the development of the extended cometary atmosphere, the coma. The evolving gases will entrain solid grains of dust and ice, creating the dust coma and (perhaps) an ice/hydrocarbon halo around the nucleus. Larger grains of nonvolatile materials not entrained by the escaping gas will begin to form a lag deposit on the nuclear surface, accumulating to form a crust that will begin to seal the nucleus against further mass loss and to insulate thermally the layers below it (Brin and Mendis 1979; Prialnik and Bar-Nun 1988).

There are thus two processes likely to contribute to formation of a nonvolatile crust. The comet may retain some or all of its original cosmic-ray-irradiated crust, and this may serve as the foundation for additional crustal growth. Or, debris gardening and the energy release from the amorphous-crystalline ice phase transition may blow away all of the primitive crust, and the comet may then grow a new lag deposit of heavy nonvolatile grains, possibly "glued" together by complex organics. A variety of models have been developed to study crustal growth on cometary nuclei. Calculations have shown that a nonvolatile layer only 1 or 2 cm thick would probably be sufficient to insulate the ices below and to reduce greatly the sublimation rate (Brin and Mendis 1979; Fanale and Salvail 1984; Horanyi et al. 1984). Estimates of the number of perihelion returns required to form the crust range from ∼1 to 20. Various workers have proposed that the crust is either porous, allowing continued gas diffusion through it to the surface, or that it is sealed, resulting in a buildup of pressure beneath it that might cause violent rupture events.

Violent rupture of the crust due to the pressure of evolving gases below it may result in visible outbursts or even a disruption of the nucleus. Further heating of amorphous ice within the nucleus will result in sporadic transitions to crystalline ice, also possibly resulting in visible outbursts. Thermal stresses on the nucleus caused by substantial temperature gradients within the ice may also result in cracking and exposure of "fresh" ices, and possibly outburst or disruption phenomena. The outburst of comet Halley near 14.3 AU post-perihelion (West et al. 1991) may be related to such processes. More subtle effects will include the migration of highly volatile molecules, both outward through the still frozen water ice matrix, and inward towards cooler regions of the nucleus where they will recondense (cf. Espinasse et al. 1991).

As mass is lost from the rotating nucleus, its moments of inertia will change and the nucleus will precess, changing the orientation of the rotation pole and the balance of insolation across the nuclear surface, with perhaps additional interesting implications. For example, the changing axis orientation will change the rotational stresses, perhaps causing the break off of weakly bonded nuclear fragments and their appearance as secondary nuclei. However, the reasons for cometary splitting are still largely unknown (see Sekanina 1982).

The most important factor in modifying cometary material at depth is the heating of the cometary nucleus. The thermal skin depth δ is defined as

$$\delta = (KP/\pi\rho C)^{1/2} \tag{1}$$

where K is the conductivity, P is the period (rotation period for diurnal skin depth, orbital period for orbital skin depth), ρ is the density, and C is the specific heat. The thermal skin depth is the distance over which a temperature perturbation at the surface will decrease by a factor of $1/e$, upon diffusing into the nucleus. For conductivities typical of solid, crystalline water ice, $\delta = 0.2$ m for a rotation period of 24 hr, or 9.2 m for an orbital period of 6 yr. However, conductivity decreases sharply for porous, low-density structures such as are suspected for cometary nuclei, and actual values of δ may be considerably smaller. Most measured values of surface thermal conductivity in the solar system, including the icy Galilean satellites, are extremely low (though these are usually due to gravitationally bound regoliths which may not be present on comets). If comets are similar, then a more realistic estimate of the orbital thermal skin depth may be only a few centimeters. The detailed thermal/physical properties of the cometary nucleus constitute an important area of future research.

The internal temperature profiles for cometary nuclei are complex functions of their orbital parameters, the obliquities of their rotation poles, and the thermal properties of their surface materials. For example, the variation of temperature with depth for a 1 km radius comet nucleus in an orbit like that of P/Encke, ignoring the diurnal temperature cycle and assuming solid crystalline ice, is shown in Fig. 9 for four points around the comet's orbit (Herman and Weissman 1987). Although the surface layer undergoes extreme temperature variations, the temperature is virtually constant at depths greater than several orbital thermal skin depths. Below this depth, the nucleus is heated to its average orbital temperature, but not appreciably higher.

Temperature profiles for real nuclei may not be as smooth as shown in Fig. 9. Because of the low thermal conductivities expected for cometary nuclei, many hundreds of returns (or more) will be required for a comet to be warmed to its equilibrium internal temperature. The comet's orbit is likely to change significantly over that same period of time and, as the changes are essentially random, some will result in additional internal heating, while others will result in a net cooling of the cometary nucleus. The final result might be a fairly complex interweaving of warm and cooler layers. Nevertheless, the

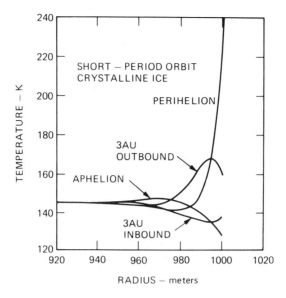

Figure 9.   Equilibrium temperature profiles in the near surface layers of a hypothetical (1 km radius) crystalline ice nucleus in a short-period orbit, for four points around the orbit (after Herman and Weissman 1987).

equilibrium internal temperature of cometary nuclei can be approximated to within $\pm10\%$ by the mean temperature (T_m) for a nucleus in a circular orbit with the same semimajor axis. T_m is given by

$$T_m = 280 \, (1 - A)^{1/4} a^{-1/2} \varepsilon^{-1/4} K \qquad (2)$$

where A is the Bond albedo of the nuclear surface, a is the semimajor axis in AU, and ε is the surface emissivity. For comets in noncircular orbits the problem cannot be solved analytically, so numerical techniques must be employed. The intense solar heating near perihelion and increased time for cooling near aphelion, as well as the temperature dependence of the thermal conductivity, lead to a complex behavior of the central temperature as a function of semimajor axis and eccentricity. That behavior is illustrated in Fig. 10 for the case of a (1 km radius) crystalline ice nucleus (Herman and Weissman 1987) where the "normalized temperature" is the calculated central temperature divided by the mean temperature from Eq. (2).

The final central temperature reached by the nucleus is also a function of the thermal conductivity, but is not a function of the nuclear radius (Herman and Weissman 1987). Lower values of thermal conductivity lead to slower heating, and suppress the variations in the central temperatures with orbital eccentricity shown in Fig. 10, giving values even closer to the mean temperature calculated with Eq. (2). Weissman (1987) estimated that the thermal

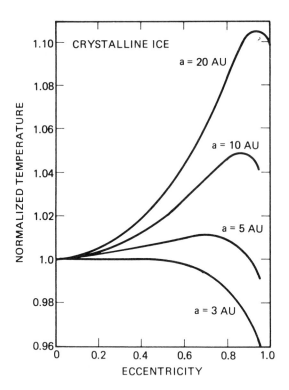

Figure 10.   Variation of central temperature for short-period comets as a function of orbital semimajor axis and eccentricity. The normalized temperature is that for a comet in a circular orbit given by Eq. (2) (after Herman and Weissman 1987).

inertia of the surface ices in comet Halley (1986 III) was about a factor of 10 less than that for solid water ice, based on the turn-on of the Halley coma at 5.8 AU inbound.

The thermal inertia I is given by: $I = (K\rho C)^{1/2}$, where K is the thermal conductivity, ρ is the bulk density, and C is the specific heat. If the cometary surface is a low-density, porous ice structure then low thermal conductivities are expected. Weissman (1987) found $I = 3 \times 10^{-3}$ cal cm^{-2} s$^{-1/2}$ K^{-1}, somewhat larger than that for the Moon but not very different from that found for the surfaces of the icy Galilean satellites (Morrison 1977). Low values of the conductivity and/or large nuclear radii lead to very slow internal heating that may take hundreds or thousands of orbits to reach equilibrium, much longer than the time scale for major dynamical changes in the cometary orbits.

As a result of this internal warming, volatile ices with sublimation points below the central temperature could diffuse through the icy matrix and possibly escape the nucleus, unless they are trapped (e.g., as clathrate hydrates).

TABLE III

Estimated Central Temperatures for Several Short-Period Comets[a]

Comet	q (AU)	P (yr)	T_c (K)	T'_c (K)
Encke	0.341	3.30	171.0	183.4
Honda-Mrkos- Pajdusakova	0.581	5.28	149.8	154.9
Tempel 2	1.381	5.29	150.8	156.9
Wild 2	1.491	6.17	144.7	148.9
D'Arrest	1.291	6.38	141.9	146.1
Kopff	1.572	6.43	143.4	147.0
Churyumov- Gerasimenko	1.298	6.59	140.8	145.2
Halley	0.587	76.10	67.8^b	68.7^b

[a] Table from Herman and Weissman (1987). T_c includes water ice sublimation; T'_c neglects sublimation of water ice.
[b] P/Halley is the only comet (of those tabulated) whose central temperature
likely remains low enough to retain amorphous ice in its interior.

The near-surface layers may well be depleted in many of the more volatile ices. As a result, gas production rates (relative to water) for some volatile species may show a temporal dependence: as the comet approaches perihelion, the more volatile ices may not appear until the thermal wave has penetrated deeply enough into the nucleus to reach undepleted layers. In addition, some deeper layers in the nucleus may be enriched in certain volatiles which were sublimated from the warm near-surface layers, diffused into the cooler interior, and recondensed there. Espinasse et al. (1991) have discussed these and other aspects in some detail.

Calculated central temperatures for 8 well-known SP comets, most of which are possible targets for comet rendezvous and/or sample return missions, are given in Table III. The quantity T_c is the equilibrium central temperature for the comet's present orbit, assuming normal water ice sublimation over the entire nuclear surface. T'_c is the central temperature assuming no sublimation, a condition the comet would evolve to as it develops a nonvolatile crust covering the nuclear surface. With the exception of P/Halley, all the short-period comets in Table III have a central temperature above 140 K, the transition temperature for amorphous to crystalline water ice. Thus, it is highly unlikely that any sample of nuclear ice returned from such a comet will be in the amorphous form. If so, this certainly would represent a major departure from the state in which the comet formed. For a low-albedo nucleus, Eq. (2) gives a central temperature of 140 K at a semimajor axis of 4.0 AU, or an orbital period of 8.0 yr. Of 155 known periodic comets in the most recent catalog (Marsden 1989), 91 have periods less than that value. Thus, those comets can be expected to have completely converted to crystalline ice in their interiors.

For comets with longer orbital periods, and for those comets recently arrived in the inner planets region, a sizeable fraction of the nuclear interior can still be expected to be amorphous ice (provided that the comet has not been significantly closer to the Sun in the past). However, according to Prialnik and Bar-Nun (1988), the amorphous ice is located at depths greater than 15 to 40 m beneath the surface.

The results of the spacecraft flybys of comet Halley have produced strong evidence for the existence of an insulating crust on the nuclear surface (Sagdeev et al. 1986), but at the same time have presented us with several apparent paradoxes. The fraction of active area on the nucleus of the comet has been estimated at 20 to 30% of the sunlit surface, based on spacecraft imaging and measurements of the gas production rate (Weissman 1987; Keller 1990). However, it is not clear why such a small fraction of the Halley nucleus is active. Virtually all predictions for Halley were that the nuclear surface would be crust-free, the crust having been blown away by the high gas flux.

Given that 70 to 80% of the nuclear surface was crusted over (cf. Keller 1990), why was it not 100%? Did the few active areas serve as pressure release points for the entire nucleus, implying a highly porous nuclear structure? Did the active areas change as the comet moved along its orbit, or from one perihelion passage to another? Earth-based observations of the comet suggested that the active areas did turn on and off irregularly, and Giotto imaging suggested at least one surface structure that may be a crusted over, formerly active area. On the other hand, some observers alleged a link between active areas observed in 1985–86 and locations of active areas derived from observations in 1910 (Sekanina and Larson 1986), a very tenuous possibility which is exceedingly difficult to prove.

A further question is whether or not the allegedly inactive areas on the Halley nucleus were really inactive. Clearly, the sources of the dust jets appear to be confined to relatively small areas. Is it possible that gas from subliming ices diffuses through the dark porous crust in other areas, while being unable to carry entrained dust with it?

Examination of the total mass lost from Halley on its recent apparition gives another interesting result. Feldman et al. (1987) estimated a total loss of water of 3×10^{14} g, based on IUE observations during the 1986 perihelion passage. Assuming that H_2O represented 80% of the volatiles by mass, and taking a dust:gas ratio of 2, the total mass loss was 1.1×10^{15} g. Given a total surface area of 400 km^2 (Keller et al. 1987; Keller 1990) with 30% of it active, and a mean density of 1.0 g cm^{-3}, then the excavated depth at each active area is $\sim$10 m. The IUE-based estimate for the total water lost is almost twice Weissman and Kieffer's (1981) prediction of 1.7×10^{14} g for the total H_2O that could be sublimed from the Halley nucleus (scaled to the Halley dimensions and albedo obtained from the spacecraft encounters). Scaling this estimate in a fashion similar to the IUE results, the predicted total mass loss would be $\sim$6.4 $\times 10^{14}$ g, and the excavated depth would be $\sim$5 m. The difference may best reflect uncertainties in both the observational measurements and the theoretical model.

Giotto imaging of the Halley nucleus did not resolve the active areas, suggesting that they are <100 m in size. But if the dimensions of the active areas are comparable to the depths above, then shadowing by the crater walls will quickly reduce the total sublimation. How then does the sublimation process not become self-limiting, particularly over repeated perihelion passages?

Another piece of this complex puzzle comes from estimates of nongravitational forces on the motion of comet Halley. The highly elongated nucleus seen in the spacecraft imaging, with its irregular and asymmetrically distributed active areas, would be expected to precess rapidly. But the fitted nongravitational forces on the comet have been virtually constant over the past 2200 yr (Yeomans and Kiang 1981). How then can the nongravitational forces be so constant when the nucleus appears to be so dynamic and ever changing?

A final area of interest here is that of thermo/mechanical stresses on cometary nuclei (Kührt 1984; Green 1986; Tauber and Kührt 1987). The dust jets in the Halley spacecraft images appear to be highly collimated, and appear to occur along linear features (Sagdeev et al. 1987). This suggests deep crevices opened up by stresses on the nucleus, or perhaps freshly exposed voids in a fractal structure. These crevices may penetrate sufficiently deeply into the nucleus to reach relatively unmodified volatile-rich layers, several thermal skin depths below the crusted surface. They also may be the best sites for obtaining samples of the cometary nucleus. However, little is known yet about thermal stresses and researchers in this area cannot even agree on whether the stresses are due to tension or compression. More work in this area is clearly needed.

IV. MEASUREMENTS OF COMETARY COMPOSITION

Theories of cometary origin and evolution can only be tested and refined if the composition and structure of cometary nuclei are measured. Having done so, we may obtain improved insight into processes affecting formation in the early solar system. Until recently, most of our knowledge of cometary composition was obtained indirectly, from studies of decomposition products of the directly sublimed volatiles, and of the dust (cf. Delsemme 1982). Of the 20-some proposed parent volatiles, only two (CO and S_2) had been detected directly before 1985. Even these were problematic, since CO is also a daughter product (e.g., of CO_2) and S_2 may not be original cometary material.

Beginning in 1985, however, a revolution occurred in our knowledge of cometary composition, driven primarily by new capabilities for remote sensing at infrared and millimeter wavelengths, and by *in-situ* investigations carried out with the fleet of spacecraft that visited comet Halley. Together, these provided direct compositional measurements of parent volatiles and dust, a macroscopic view of nuclear properties, and improved knowledge of the physical parameters of the cometary coma. Substantial new insights were gained into the properties of the nucleus itself and the asymmetry of gas production and outflow from it, and into such coma properties as distributed sources for "parent molecules," the ionization state, bulk outflow and thermal velocities, and rotational and kinetic temperatures. The combined results provided the best picture to date of the true nature of the cometary nucleus. In this section, we review insights gained through *in situ* investigations of comet Halley, and through remote observations of P/Halley and other recent comets, and we discuss important aspects of their nuclear compositions and structure.

A. Cometary Dust: Organics and Silicates

Considering the many pre-solar organic and silicate compounds that could be present in comets (cf. Turner 1989c) and their possible role in initiating pre-biotic organic chemistry on the early Earth (cf. Marcus and Olsen 1991), it is vital to identify those that are actually present. Two of the major surprises in the exploration of comet Halley were the discovery of a new population of grains, composed of carbon, hydrogen, oxygen and nitrogen (called CHON), and the discovery of a new cometary emission feature extending from 3.2 to 3.5 μm. The C-H stretching bands of many organic compounds fall within this wavelength range, and an immediate qualitative connection was made between these two discoveries, extending even to labeling the latter the "organic grain" feature. Both condensed and gaseous hydrocarbon species could contribute to the emission feature, however, and the nature of the progenitors is still uncertain. This spectral region has received intense scrutiny in recent comets, and it is now recognized that while some of the feature is probably contributed by thermal emission from grains, a significant part is due to emission from gaseous species, including methanol (see later discussion). We shall use the term "cometary organic feature" to refer to this new emission, because it originates from both condensed- and gas-phase material.

The composition of cometary dust was addressed by the Particle Impact Analyzer instruments, called PIA on Giotto, and PUMA on Vega 1 and 2. The principal findings have been reviewed recently by Jessberger and Kissel (1991) (also see McDonnell et al. 1991). The PIA/PUMA instruments measured elemental compositions by permitting dust particles to hit a clean target surface (Ag, or Ag-doped Pt), then conducting mass analysis by time-of-flight spectroscopy (see Kissel 1986). Two different chemical end-types were found: an organic phase (CHON) and a Mg-rich silicate phase. The CHON material contributed ~30% of the total mass of the measured particles, which were individually in the range 10^{-16} to 10^{-11} g. About one-third of the measured grains contained no significant organic component, and the remainder were mixtures of the two phases. The pure silicate particles were more heavily represented in the lower mass ranges.

From energy discrimination measurements on the mixed-phase particles, it was found that the CHON elements were produced with significantly higher kinetic energy than were the silicate elements. Krueger and Kissel (1987) inferred a grain structure in which the organic component coats a silicate core. The CHON ions from the refractory organic would leave the impact site with higher initial energies (up to 150 eV) on average than would ions of the silicate cores (Jessberger et al. 1987). This would seem to confirm Greenberg's model of core/mantle interstellar grains for cometary dust (Greenberg 1982). However, this model is clearly not valid for all cometary grains, because some contained no significant organic material (Jessberger and Kissel 1991). Brownlee and Kissel (1990) have commented that "to state that Halley is actually composed of interstellar grains, all having core/mantle structure, would be a gross over-interpretation." Nevertheless, the discovery of bi-modal grain types demonstrates that the refractory component of comets is disequilibrated. Indeed, the pristine nature of the dust in comet Halley is demonstrated by a lack of equilibrium among the grains, as revealed by their diverse chemical, isotopic, and elemental compositions.

The retrieved densities were $\sim$2.5 g cm^{-3} for the silicates, and $\sim$1 g cm^{-3} for the mixed particles (Maas et al. 1990). These densities are in the approximate ratio of their constituent (atomic) masses, suggesting that the two grain types are comparably compact. However, if the entire intensity of the cometary organic infrared emission feature was assigned to completely filled CHON grains, the implied dust fluences would exceed the measured value by orders of magnitude (see discussion in McDonnell et al. 1991). Greenberg and Hage (1990) argue that this discrepancy can be resolved if the dust particles are fluffy, with a porosity of $p = 0.97$. Effectively this increases the surface area, while keeping the particle mass unchanged. However, if volatiles contribute significantly to the cometary organics feature, as now seems to be so, this porosity factor would need downward revision. This is in the direction needed to resolve the apparent discrepancy in density implied by the infrared and the PIA/PUMA results.

Elemental abundances in the dust were retrieved by Jessberger et al. (1988) and revised abundances are reviewed by Jessberger and Kissel (1991) (Table IV). The dust is far more enriched in H, C and N than even the most primitive meteoritic material, the CI-chondrites. Thus the dust in comet Halley is the most primitive material ever sampled. The O/C ratio in the gas is $\sim$5.9, while in the dust it is $\sim$1.1, and if the total C/Mg and O/Mg are forced to fit the solar value, then an estimate of the dust/gas ratio can be obtained. The required ratio is 1.7 (Jessberger and Kissel 1991). In deriving this estimate, the average composition of the small grains sampled by PIA/PUMA was assumed to be representative of the mean dust in comet Halley. In this case, H and N are depleted in Halley, by factors of 650 and 3 relative to solar abundances, while all other elements are present in solar abundance, to within a factor of 2 (Table IV). A third possible exception is the Fe/Si ratio, which appears to be low in comet Halley. The discovery that a major portion of cometary carbon is locked in the dust has resolved the "missing carbon" problem, discussed by Delsemme (1982).

A note of caution should be introduced, though. The PIA/PUMA instruments sampled the smallest grains acquired during the spacecraft encounters. The sampled grains, if spherical (and of zero porosity), would have diameters

TABLE IV

Average Elemental Abundances in Comet Halley[a]

Element	Halley		Solar System[b]	CI-Chondrites[b]
	Dust	Dust and Ice		
H	2,025	4,062	2.6×10^6	492
C	814	1,010	940	70.5
N	42	95	291	5.6
O	890	2,040	2, 216	712
Na	10	10	5.34	5.34
Mg	100	100	100	100
Al	6.8	6.8	7.91	7.91
Si	185	185	93.1	93.1
S	72	72	46.9	47.9
K	0.2	0.2	0.35	0.35
Ca	6.3	6.3	5.69	5.69
Ti	0.4	0.4	0.223	0.223
Cr	0.9	0.9	1.26	1.26
Mn	0.5	0.5	0.89	0.89
Fe	52	52	83.8	83.8
Co	0.3	0.3	0.21	0.21
Ni	4.1	4.1	4.59	4.59

[a] Table after Jessberger and Kissel 1991.
[b] All abundances are by number, not by mass. The solar photospheric abundances of the listed elements are practically indistinguishable from the solar-system abundances, with the exception of Fe, which has a photospheric abundance ratio of 123 (Anders and Grevesse 1989).

in the range 0.06 to 3 μm. Grains in this size range represent only a tiny fraction of the total dust mass, however, given the overall mass distribution function (cf. McDonnell et al. 1991). Grains as massive as $\sim 10^{-4}$ g were measured by DIDSY (dusty impact detection system), and perturbations in the Giotto spin axis (as recorded by the on-board HMC camera) were interpreted as resulting from the impact of grains as large as $\sim 2 \times 10^{-2}$ g. A fully filled mixed-phase grain with this mass would have a diameter of several millimeters. Millimeter-sized grains contribute most of the infrared flux seen in the comet trails discovered by the IRAS satellite (Sykes et al. 1986,1990), and so large grains are apparently common in comets. From the mass in these dust trails and in the measured gas production rates, Sykes and Walker (1992) estimate the dust/gas ratio in comets to be ~ 3 (with a range of 1–5), somewhat larger than the estimate of 1.7 in Halley mentioned above (Jessberger and Kissel 1991). McDonnell et al. (1991) estimate the dust/gas ratio in Halley to be ~ 2.

The mass-12 and -13 peaks in the PIA/PUMA spectra have been interpreted as ^{12}C and ^{13}C, and the retrieved values of ^{12}C/^{13}C range from 1 to

5000, compared with the solar value of 89.91. Similar ranges in $^{12}C/^{13}C$ ratios are seen in microscopic analyses of carbonaceous chondrite grains. $^{12}CH^+$ could also contribute to the signal at mass 13, however, leading to smaller apparent ratios of $^{12}C/^{13}C$. While the majority of grains show $^{12}C/^{13}C$ abundance ratios to be consistent with solar, some others differ substantially. No secure information can be obtained regarding enrichment of ^{13}C, but some grains showed major depletion. Depletions of ten fold were often found and, in the most extreme case, $^{12}C/^{13}C$ was ~5000, probably representing a remnant condensed in stellar outflow and not subjected to later modification (it may have formed in a He-burning or explosive H-burning stellar process, see discussion in Jessberger and Kissel 1991).

Although the composition of CHON material is superficially similar to the "tar balls" found in some interplanetary dust particles collected at stratospheric altitudes, the CHON particles are fragile bodies that showed evidence of fragmentation during outflow in the coma of P/Halley. If CHON particles are the antecedents for tar balls, the collected interplanetary dust particles have been severely devolatilized and possibly metamorphosed during their interplanetary period prior to collection in the stratosphere. With respect to other indicators, the distributions in both the Fe/(Fe + Mg) ratio and in the system Mg-Si-Fe are quite different in comet Halley than in carbonaceous chondrite matrix and lattice layer silicate interplanetary dust particles. However, they agree much more closely with those found for anhydrous interplanetary dust particles (see discussion in Grün and Jessberger 1990; Jessberger and Kissel 1991).

The cometary organic feature was discovered in IKS (infrakrasnoi spectrometre) spectra of comet Halley acquired on the Vega 1 spacecraft (Combes et al. 1986), and confirmed later by groundbased observers (Wickramasinghe and Allen 1986; Baas et al. 1986; Danks et al. 1987; Knacke et al. 1986). It has also been seen in every suitably bright comet since comet Halley. Possible contributors include thermal emission from CHON grains, emission from small super-hot grains, and resonance fluorescence of volatiles. The relative contributions of condensed phase and gaseous matter are as yet unknown (Encrenaz et al. 1988; Encrenaz and Knacke 1991; Chyba et al. 1989*b*; Brooke et al. 1991*a*). Chyba et al. showed that the shape of this feature was reproduced reasonably well by synthetic spectra based on optical constants of organic laboratory analogs. However, its heliocentric intensity evolution was not well reproduced except by small hot grains. Brooke et al. (1991*a*) found that the heliocentric dependence in several comets varied as expected for a volatile precursor, and methanol has been identified as a contributor in recent comets (see below).

Brooke et al. (1991*a*) compared the shape of this feature in five recent comets: comets Halley (1986 III), Wilson (1987 VII), Bradfield (1987 XXIX), Okazaki-Levy-Rudenko (1989 XIX) and P/Brorsen-Metcalf (1989 X). Their spectra are shown in Fig. 11. While the shapes resemble one another they are not identical. Brooke et al. argue that, as a group, the cometary spectra exhibit greater similarity to one another, than do laboratory spectra of synthesized organic compounds. They suggest that the material incorporated in these five comets was more closely related than the synthesized compounds were. This is perhaps noteworthy because two are short-period comets (Halley and Brorsen-Metcalf), while two were dynamically new comets, and one (Bradfield) had an intermediate period. Brooke et al. (1991*a*; Tokunaga and Brooke 1990) suggest that their similarity results from common origins.

Recent work suggests that both volatiles and refractories contribute to spectral features in the 3.2 to 3.5 μm region. On the basis of their identification of the ν_3 band of methanol at 3.52 μm, Hoban et al. (1991*a*) commented that four-fold stronger emissions were expected from the ν_1 and ν_9 bands near 3.35 μm, and that these should contribute significantly to the cometary organic feature. Reuter (1992) developed a fluorescence model for these new bands and found that methanol could have contributed 10 to 40% of the total intensity of the cometary organic feature in recent comets.

B. Direct Detection of Parent Volatiles

Prior to 1985, the only cometary parent volatile detected directly was CO, achieved at ultraviolet wavelengths in two comets (West 1976 VI and Bradfield 1979 X; cf. Feldman 1983). S_2 was detected in comet IRAS-Araki-Alcock (1983 VII), but not in ten other comets (cf. A'Hearn et al. 1983; Budzien and Feldman 1992). However, it is not yet clear whether S_2 is a parent volatile.

Through the use of ultraviolet spectroscopy it has been possible to determine water production rates and their dependence on heliocentric distance (through measurements of the OH radical), to measure several key elemental production rates (e.g., O, C, S), to characterize other products such as H, CS and $CO_2{}^+$, and to explore coma aeronomy in detail. Recently, sensitive upper limits to the production rates of SO and SO_2 were also obtained (see later discussion of sulfur chemistry). Selected details are discussed below, and comprehensive reviews are given by Festou (1990*a*) and Feldman (1991*a*; see also Feldman 1983, 1991*b*). A few early infrared searches produced only nonrestrictive upper limits, and are not discussed further in this chapter. The 18-cm radio lines of OH were first detected in comet Kohoutek (1973 XII) (Biraud et al. 1974; Turner 1974), and are now routinely used for productive studies of cometary activity and coma kinematics (see the review by Crovisier and Schloerb 1991). Radio searches for parent volatiles before 1985 were less successful, and provided only tentative molecular detections (e.g., HCN, NH_3, H_2O and CH_3CN) or upper limits (see the reviews by Snyder [1982] and by Crovisier and Schloerb [1991]).

The direct detection of parent molecules at infrared wavelengths became tractable with the development of theoretical models for fluorescence in their vibrational bands (cf. Crovisier and Encrenaz 1983; Weaver and Mumma 1984), and of suitable instrumentation for observing them. The first successful application of this approach was to the water molecule in comet Halley (Mumma et al. 1986), and it provided the first detection of water vapor in

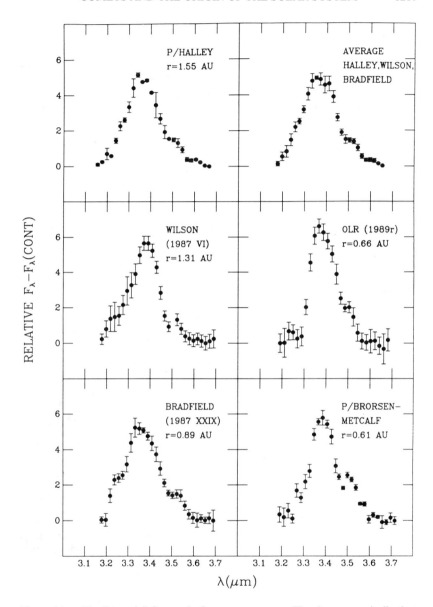

RELATIVE $F_\lambda - F_\lambda(CONT)$

P/HALLEY
r=1.55 AU

AVERAGE
HALLEY,WILSON,
BRADFIELD

WILSON
(1987 VI)
r=1.31 AU

OLR (1989r)
r=0.66 AU

BRADFIELD
(1987 XXIX)
r=0.89 AU

P/BRORSEN-
METCALF
r=0.61 AU

$\lambda(\mu m)$

Figure 11.   The "organic" feature in five recent comets. The shapes are similar but
not identical (after Brooke et al. 1991*a*).

comets (Fig. 12). Subsequently, spectra of comet Halley were acquired by an
infrared spectrometer (IKS) carried on the Vega 1 spacecraft (cf. Combes et
al. 1986). A spectrum measured during the flyby on 6 March 1986 is shown in
Fig. 13. The IKS featured low spectral resolution but wide spectral coverage

(2.5 to 12 μm), and it achieved the first detections of CO_2 and of the cometary organic feature, as well as the detections of water; formaldehyde (H_2CO), and possibly CO (cf. Moroz et al. 1987, Combes et al. 1988; Mumma and Reuter 1989). Infrared spectroscopy of recent comets has led to detections of CH_3OH, and possibly of CO, and to sensitive upper limits for CH_4, H_2CO and OCS (Hoban et al. 1991a; DiSanti et al. 1990,1992; Brooke et al. 1991b; Reuter et al. 1992).

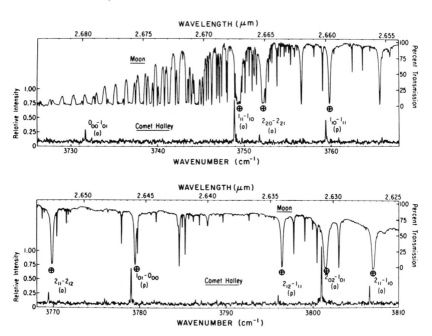

Figure 12.   Part of the infrared spectrum of comet Halley obtained on 24 December 1985, from the Kuiper Airborne Observatory. The lower trace in each panel shows the emission spectrum from comet Halley, the upper trace the lunar reflectance spectrum. The cometary geocentric velocity causes the H_2O lines to be Doppler shifted by about $\sim$0.43 cm^{-1} from the terrestrial absorption lines seen in the lunar spectrum. The lines are labeled with quantum numbers and the ortho or para designation (after Mumma et al. 1986).

The first secure detections of parent volatiles at radio wavelengths occurred in comet Halley (see discussion in Crovisier and Schloerb 1991). Velocity-resolved intensity profiles of HCN lines were measured over a wide range of heliocentric distance. From these profiles, production rates were obtained along with their diurnal and seasonal variability and the physics of outflow in the cometary coma was determined (Despois et al. 1986; Schloerb et al. 1986,1987; Winnberg et al. 1987). The advent of large aperture millimeter telescopes enabled the secure identification of H_2CO, CH_3OH, and H_2S in several recent comets, while sensitive upper limits were reported for

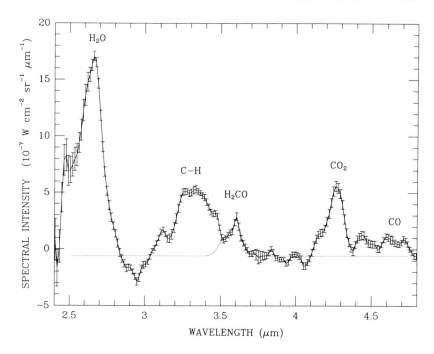

Figure 13.   A representation of the spectrum of comet Halley acquired on 6 March 1986, with the IKS spectrometer on the Vega 1 spacecraft. This representation emphasizes the portion of the detected flux that is contributed by directly sublimed species. The low instrumental resolution causes each molecular band to appear without rotational structure; however, detections of water, carbon dioxide and formaldehyde are evident. The broad feature spanning 3.2 to 3.5 μm corresponds to the C-H stretch in many solid and volatile species, and both probably contribute to this emission. A comparison of the modeled band shape for H_2CO ν_1 and ν_5 bands at rotational temperature $T = 50$ K, is also shown (Mumma and Reuter 1989). The retrieved production rate (4.5% relative to H_2O) represents a direct nuclear source for H_2CO, to which must be added any extended source that may be present (e.g., POM). Although the shape of the CO (1−0) band is not obvious in this spectrum, the dependence of integrated intensity on distance from the nucleus supports its presence. IKS data from Combes et al. (1988).

other species, e.g., for OCS, SO_2, H_2CS and HC_3N in comet Levy (1990 XX) (Bockelée-Morvan et al. 1990,1991; Colom et al. 1992).

Production rates and spatial distributions within the coma of comet Halley were also obtained for many parent volatiles by the neutral and ion mass spectrometers on the Giotto spacecraft, including H_2O, CO, CO_2, H_2CO, CH_3OH, HCN and H_2S. A summary of the parent molecular abundances found in recent comets is given in Table V.

We now examine several key volatile species and their interrelationships in greater detail.

TABLE V
Volatile Abundances in Recent Comets

Molecule	Relative Abundance (by number)	Comments
Comet P/Halley		
H_2O	100	Remote and *in-situ* detections[1]
CO	~7	Direct (native) source[2]
	~8	Distributed source[2]
H_2CO	0–5	Variable[3]
CO_2	3	Infrared (Vega 1 IKS)
CH_4	<0.2–1.2	Groundbased infrared[4]
	0–2	Giotto IMS[5]
NH_3	0.1–0.3	Variable[6]; based on NH_2
	1–2	Giotto IMS[7]
HCN	0.1	Variable; groundbased radio
	<0.02	Giotto IMS
N_2	~0.02	Groundbased N_2^+ emission
SO_2	<0.002	Ultraviolet (IUE)
H_2S	—	Giotto IMS[8]
CH_3OH	~1	Giotto NMS and IMS[9]
Other Comets		
CO	20	West (1976 VI)
	2	Bradfield (1979 X)
	1–3	Austin (1990 V)
CH_4	<0.2	Levy (1990 XX)[10]
	1.5–4.5	Wilson (1987 VII)[11]
CH_3OH	1–5	Variable[12]
H_2CO	0.1–0.04	If a parent species[13]
HCN	0.03–0.2	Several comets[14]
H_2S	0.2	Austin (1990 V); Levy (1990 XX)[15]
S_2	0.025	IRAS-Araki-Alcock (1983 VII)[16]

[1] Water was detected directly by infrared spectroscopy (Mumma et al. 1986; Combes et al. 1986) and by mass spectroscopy (cf. Krankowsky et al. 1986).

[2] CO was detected directly at ultraviolet wavelengths (Feldman et al. 1987; Woods et al. 1986,1987) and in neutral mass spectra (Eberhardt et al. 1987b). A tentative detection at infrared wavelengths (Combes et al. 1988) provided production rates in agreement with the native source.

[3] The H_2CO abundance is variable, relative to water. The largest value found for comet Halley was 4.5%±0.5%, measured by both IKS and Giotto NMS (also IMS, see text), but at other times the production rates were 10 times smaller (cf. Mumma and Reuter 1989). The values retrieved for comets Austin and Levy were much smaller than the values found in comet Halley.

[4] Retrieved from a single spectral line of CH_4. The range reflects the uncertainty in rotational temperature for cometary CH_4 (50–200 K). The extrapolation from a single line to the ensemble production rate is therefore highly uncertain. See Kawara et al. (1988).

[5] The production rate retrieved from the ion mass spectra on Giotto is highly model dependent, and could be zero in comet Halley (Allen et al. 1987; Boice et al. 1990). Brooke et al. (1991b) recently found $CH_4 < 0.2\%$ in comet Levy (1990 XX).

[6] Assuming that NH_2 is produced solely from NH_3 (Magee-Sauer et al. 1989; Mumma et al. 1990; Krasnopolsky and Tkachuk 1991; Wyckoff et al. 1991a).

[7] Allen et al. (1987). Boice et al. (1990) re-analyzed the IMS spectra, retrieving 1%, while Ip et al. (1990) retrieved 0.5% from the Giotto IMS data.

[8] Marconi et al. (1990). An incorrect lifetime was used in deriving the abundance of H_2S (see Crovisier et al. (1991).

[9] See Geiss et al. (1991) and Eberhardt et al. (1991).

[10] Based on groundbased infrared spectroscopy (Brooke et al. 1991b).

[11] Based on airborne infrared spectroscopy (Larson et al. 1989).

[12] The value in comet Levy was $\sim 1\%$, but in comet Austin it was about 5% (Bockelée-Morvan et al. 1991; Bockelée-Morvan, personal communication); Hoban et al. 1991a).

[13] 0.1% in comet Austin (1990 V), and 0.04% in comet Levy (1990 XX). The production rate would be about ten-fold larger if formaldehyde were a daughter product (Colom et al. 1992).

[14] In P/Brorsen-Metcalf (1989 X), Austin (1990 V), Levy (1990 XX) (compare Bockelée-Morvan et al. 1990).

[15] Crovisier et al. (1991).

[16] cf. Kim et al. (1990).

C. Water

Water was long thought to be the dominant volatile species in the cometary nucleus (based on strong indirect evidence), but it was first detected more than 35 years after Whipple (1950) introduced his icy conglomerate model for the cometary nucleus. Infrared spectroscopy of comet Halley on UT 21–24 December 1985 from the Kuiper Airborne Observatory provided a detection of 10 lines of the ν_3 band in emission (Mumma et al. 1986), as predicted by models for solar infrared fluorescence from rotationally relaxed cometary water (Crovisier 1984; Weaver and Mumma 1984). Water was detected on 6 March 1986 in IKS spectra acquired on the Vega 1 spacecraft (cf. Combes et al. 1986,1988), and in-depth studies were conducted on 20–26 March 1986 from the Kuiper Airborne Observatory (Weaver et al. 1986). Absolute production rates were obtained that, when compared with those for other volatiles, confirmed that water was indeed the dominant volatile species in the cometary nucleus. The production rates were found to vary from day-to-day, in concert with the visible lightcurves. A detailed analysis of the relative line intensities and shapes permitted retrieval of the rotational temperature and outflow velocity in the intermediate coma, and confirmed the predicted transition from collisionally dominated to fluorescence dominated regimes in the outer coma (Larson et al. 1987; Weaver et al. 1987; Bockelée-Morvan and Crovisier 1987; Bockelée-Morvan 1987). The spin temperatures of cometary water were measured in comets Halley and Wilson (1987 VII) with significantly different results (see later discussion in Sec. IV.J).

D. Carbon Dioxide

CO_2 was detected for the first time in comets in infrared spectra acquired on Vega 1, and in mass spectrometric measurements on Giotto. The Giotto neutral mass spectra (NMS) showed a strong peak at 44 amu/e⁻ (Krankowsky et al. 1986; Krankowsky 1991). CO_2, CS, C_3H_8 and several other species are potential contributors to this feature, but it was argued from the measured mass distribution that propane (C_3H_8), acetaldehyde (CH_3CHO), and ethylene oxide (($CH_3)_2O$) could contribute no more than one-third of the measured peak intensity. The spatial profile showed no evidence of parent decay over the entire measurement interval from the nucleus (2500–6300 km). However, this is not very restrictive because the largest distance is still much smaller than the scale length of CO_2, while the smallest is much larger than the scale length of the presumed parent of CS (i.e., CS_2). Thus, the measured spatial distribution is consistent with either CS or CO_2. By assigning the entire peak to CO_2, Krankowsky et al. derived an upper limit to its production rate of 3.5% relative to water. CO_2 was also directly detected in infrared spectra of comet Halley (by IKS on Vega 1), through its ν_3 band near 4.3 μm (Combes et al. 1986). A production rate of 2.7% relative to water was obtained (Combes et al. 1988). CO_2^+ is ubiquitous in ultraviolet (see, e.g., Festou et al. 1982) and blue (groundbased) spectra of comets, and is potentially a useful tracer for CO_2. However, direct observations of the parent species are most desired, at least until the ionization rates and processes are verified and the fluorescence efficiencies are well in hand. Unfortunately, the strong infrared bands of CO_2 are observable only from space or from balloon altitudes.

E. Methane and Carbon Monoxide

The abundance of CH_4 relative to CO may provide a key test of the manner in which cometary ices formed. Interstellar methane has recently been measured in both the solid and gaseous forms (Lacy et al. 1991) in several dense cloud cores (e.g., W33A) and star-forming regions (e.g., NGC 7538 IRS 1 and IRS 9). The abundance ratio CH_4/CO in the condensed phase was 2.4 in W33A and in the range 3.2 to 5.6 in IRS 1, but only 0.13 in IRS 9. The ratios in the gas phase were 0.001–0.011, 0.0015, and 0.001–0.003, respectively, in agreement with earlier upper limits (Knacke et al. 1985), and the ratios for total CH_4/CO were 0.044, 0.01–0.017, and 0.013, respectively. The ranges correspond to uncertainties in the rotational temperatures for CH_4 and CO. Lacy et al. (1991) suggest that the unusually high abundance ratio in the solid phase is consistent with production of methane on the grains (see Chapter by van Dishoeck et al.). Whether or not this is so, comets that accreted from such grains would exhibit large CH_4/CO ratios (of order unity) while those that accreted at very low temperatures (such that all of the gaseous CO was condensed) would show CH_4/CO~0.01. The variability in the ratio CH_4/CO is much larger for the condensed phase in these three sources than it is for the total abundances, and this may indicate that the ratio CH_4/CO in cometary ice will depend on the exact temperature of condensation, thermal history, and formation site.

Cometary CO was first detected at ultraviolet wavelengths in comet West (1976 VI), and later in comets Bradfield (1979 X), Halley (1986 III), Austin (1990 V) and Levy (1990 XX) (Feldman and Brune 1976; Feldman et al. 1987; Woods et al. 1986,1987; Budzien et al. 1990; Sahnow et al. 1990; Feldman et al. 1991). It was searched for, but not detected in several dozen other comets (Festou and Feldman 1987; Festou 1990b). The reported production rates (relative to water) varied more than ten fold for the five comets in which CO was detected (cf. Feldman 1983). The rocket detections (West, Halley) yielded production rates $\sim$15 to 20% relative to water, while the IUE results (Halley, Austin, Levy) were in the range of 1 to 7%. However, the direct comparison of production rates measured with large (e.g., rocket-UV) and small (e.g., IUE) beam sizes is problematical if CO is a product of both direct and distributed sources. The small beam size emphasizes the direct source at the expense of the distributed one, while the large beam samples CO produced from both sources. Part of the spread in measured values may be due to this effect (see Feldman 1983; Weaver 1989).

Eberhardt et al. (1987b) retrieved the production rate of CO from the 28 amu/e$^-$ peak in the NMS spectra on Giotto. The production rate was $\sim$15% relative to water, in agreement with rocket-ultraviolet results ($\sim$15 to 20%; Woods et al. 1986). Eberhardt et al. found that the spatial profile of CO was not consistent with a direct nuclear source alone, but required a second distributed source whose parent scale length was about 1×10^4 km. The nuclear source contributed $\sim$7%, consistent with the value retrieved by IUE ($\sim$7%; Feldman et al. 1987). The destruction scale length found for H_2CO by Krankowsky (1991) was comparable to that found for the extended source of CO. The production rate for CO from formaldehyde in the combined polymeric and monomeric forms would exceed that from the monomer alone (which was 4.5%, relative to water), so it is plausible that H_2CO was responsible for a significant fraction of the distributed source of CO.

The $(1-0)$ band of CO was tentatively identified in IKS spectra (see Combes et al. 1988, Fig. 8d) near 4.7 μm, and the P3 line was also detected during an outburst in comet Austin (DiSanti et al. 1990,1992). Earlier groundbased infrared searches were carried out on comet IRAS-Araki-Alcock (1983 VII) (Weaver et al. 1983, as reported in Chin and Weaver 1984), but the instrumental approach suffered from limited sensitivity. The search in comet Austin was the first to use a modern high-resolution cryogenic echelle spectrometer (IRSHELL; Lacy et al. 1989a), however the detector was optimized for longer wavelengths and improved sensitivity is expected with other detectors.

Early attempts to detect CH_4 in comets were unsuccessful (see, e.g., Roche et al. 1975). The upper limits obtained for comet Kohoutek (1973 XII) did not impose useful contraints on the formation models ($CH_4/CO<50$, $CH_4/H_2O<5$). A search was made in comet Halley from the KAO and an upper limit of 4% relative to water was obtained (Drapatz et al. 1987). A more restrictive upper limit (0.2–1.2%, the range representing the uncertainty in rotational temperature) was obtained for Halley by Kawara et al. (1988) from groundbased infrared spectroscopy. Taken with the direct source of CO (~7%) measured by the IMS on Giotto (Krankowsky et al. 1986), this suggests a ratio $CH_4/CO<0.03$ to 0.17. Modeling of the Giotto High Intensity Spectrometer data (Balsiger et al. 1986) provided an abundance of 2% for CH_4 relative to water (Allen et al. 1987; Wegmann et al. 1987). However Boice et al. (1990) obtained satisfactory fits to these mass spectra by including the fractionation fragments of polymeric formaldehyde (POM), with no CH_4, and also reproduced the spatial distribution found for the extended source of CO by the NMS. They did not fit the CH peak (amu = 13) well, however, and so invoked an unidentified source (grains were suggested).

Methane has been detected only in comet Wilson (1987 VII), a dynamically new comet. Larson et al. (1989) found emission at the 5σ level at the expected Doppler-shifted frequencies, and reported it as a possible 3σ detection above the noise level. The rotational temperature was not measured, so the retrieved production rate (1.4–4.5%, relative to water) was presented for two temperatures. The lower value assumes the kinetic temperature (50 K) derived from rotational populations measured for polar molecules (H_2O and HCN), while the higher value represents an upper limit to the kinetic temperature (300 K) for neutral molecules obtained from the NMS spectrometer on Giotto (Lämmerzahl et al. 1987). The production rate (3.3%) was also reported for 215 K. Owing to the curious manner in which these results were reported, all production rates refer to the 3σ level. They would be slightly higher (2.3–7.5%) had the measured level (5σ) been used. IUE observations provided $CO/H_2O<0.1$ (Roettger 1991), so that $CH_4/CO>0.2$ for this new comet.

This lower limit for comet Wilson exceeds the ratio found for ices in the star-forming region, NGC 7538 IRS 9 (Lacy et al. 1991). Recent observations of another comet (Levy 1990 XX) support a low ratio for CH_4/CO, (<0.1, at the 3σ confidence limit) (Brooke et al. 1991b). While it is not certain whether comet Levy was dynamically new or was a returning long-period comet (B. Marsden 1991, personal communication), its chemical abundances (e.g., of CH_3OH and CH_4) and the CH_4/CO ratio appear to be consistent with those in P/Halley, but not with comets Austin or Wilson. Still, the upper limits for CH_4/CO in comets P/Halley and Levy are smaller than the ratio found for solid phase CH_4/CO around embedded stars (Lacy et al. 1991).

Bar-Nun and Kleinfeld (1989) studied the condensation properties of laboratory analogs of cometary ices, and compared their findings with the reported (NMS) measurement of $CH_4/H_2O\sim0.02$ and $CO/H_2O\sim0.07$ in comet Halley (see Table V). Beginning with gaseous mixtures of H_2O and CO, they found that the ratio of trapped CO in H_2O ice reached 0.07 at a condensation temperature of ~50 K, and they argued that the ice in Halley's comet formed at this temperature in the solar nebula. They further argued that a ratio of

$CH_4/CO \sim 0.3$ in comet Halley required a gaseous ratio $CH_4/CO \sim 0.01$ in the solar nebula, if both ices were condensed from the gas phase at 50 K. This value is consistent with the range of total abundance ratios for CH_4/CO found by Lacy et al. (1991). However, the abundance of CH_4 in comet Halley is now thought to be substantially less than the value of 0.02 reported by Allen et al. (1987).

Prinn and Fegley (1989) and Fegley and Prinn (1989) examined disequilibrium chemistry in the solar nebula and found that the long time scales needed for chemical conversion prevent efficient reduction of CO to CH_4. Thus, they predict $CH_4/CO < 10^{-7}$ for disequilibrium condensation, far lower than the values reported (~ 0.03–0.17) for comet Halley. They therefore suggested the giant planet sub-nebulae as a source of pre-cometary material enriched in methane. Chemical reduction is efficient in the giant planet sub-nebulae, owing to higher temperatures and densities, and thus material in them is CH_4-rich. Lunine (1989b) and Engel et al. (1990) examined radial transport of processed material from the inner solar nebula to the comet-forming zone, and concluded that although this could produce CH_4 with 1 to 4% relative abundance in Halley, an admixture of NH_3-rich material was needed (however, the NH_3 abundance may be much smaller than originally claimed; see later discussion). They invoked interstellar material as its source, and this of course admits the possibility of delivering the required methane as well. The high ratio of CH_4/CO found in the solid phase in the interstellar medium (Lacy et al. 1991) avoids the need to mix reduced material from the solar nebula into comets, if comets accreted from such grains. In any event, the abundance of reduced material (both carbon and nitrogen) in the nucleus of P/Halley and other short-period comets has been revised downward substantially from the high levels reported initially.

F. Formaldehyde and Methanol

The presence of formaldehyde is expected if the cometary nucleus contains unmodified interstellar material, as formaldehyde is known to be ubiquitous in interstellar clouds. The abundance in the solid phase around the embedded infrared source W33A is $\sim 0.2\%$ (relative to water ice; Table VI). Four independent lines of evidence support its presence in comet Halley at much higher concentrations, but its distribution within the nucleus may be inhomogeneous. Mitchell et al. (1987) and Huebner (1987) interpreted certain ion mass spectra, acquired by the PICCA ion mass spectrometer on the Giotto spacecraft (Korth et al. 1986), as cracking fractions of polymeric formaldehyde (polyoxymethylene or POM). The dissociation of POM also could explain the observed extended source for CO. However, this interpretation may not be unique because temporal variability or other sources of CO could also be significant. Further, Mitchell et al. (1989) challenged the identification of POM as progenitor of the mass spectra, by showing that a blend of compounds of CH_2, NH, O and N connected to complex hydrocarbons could also produce the observed PICCA mass spectra. Moore and Tanabe (1990) reproduced the PICCA spectrum by ion sputtering of para formaldehyde, but reported the yield to be 10^3 too small to account for the PICCA results quantitatively. Formaldehyde was also detected by the NMS on Giotto at the production rate of 4.5% (relative to water), and its spatial profile seemed to require a distributed source (Krankowsky et al. 1991; Krankowsky 1991).

TABLE VI
Molecular Abundances in Interstellar Icy Grain Mantles[a]

Species	Band Position		Relative Abundance	
	λ (μm)	ν (cm^{-1})	Observed	Calculated[b]
H_2O	3.08	3250	—	—
CO	4.68	2135	0–5	0.4
CH_3OH	3.5	2830	7	(40)[c]
X-CN	4.61	2167	(4)[d]	—
OCS	4.90	2040	(0.05)[e]	0.4
NH_3	2.95	3375	< 5	1
H_2S	3.94	2540	0.3	0.03
CH_4	7.70	1300	< 1	0.007
H_2CO	3.53	2835	< 0.2	(0)[c]
CO_2	4.28	2337	—	9

[a] Relative abundances refer to number, not mass. Most abundances from Tielens (1989); CH_3OH from Grim et al. (1991); CH_4 from Lacy et al. (1991).
[b] Tielens and Hagen (1982).
[c] Assuming all H_2CO is converted to CH_3OH by grain surface reactions.
[d] Thought to result from ultraviolet photolysis. Abundance estimated from laboratory studies (d'Hendecourt et al. 1986).
[e] If the feature is due to OCS.

Snyder et al. (1989) reported a spectral line of monomeric H_2CO in comet Halley (the 110–111 rotational line at 4.8 GHz, detected with S/N~2.5), and found a relative production rate of ~1.5% for late January, 1986. They required an extended source of formaldehyde in the coma to explain their spectrum. However, based on a new model for the rotational spectrum of cometary H_2CO, Bockelée-Morvan and Crovisier (1992) argue that the radio detection in P/Halley must be spurious.

Combes et al. (1988) claimed a tentative detection of the ν_1 and ν_5 bands of formaldehyde, from the 3.6 μm feature detected by IKS. Their retrievals were based on incomplete cometary models and the band shape was not well reproduced by their synthetic models, leading them to label the three channels at band center as spurious. Their synthetic spectrum assumed a rotational temperature of 300 K. They analyzed only the flux which exhibited a cometocentric dependence consistent with that of a parent molecule, and retrieved a production rate of ~4.5% relative to water. Reuter et al. (1989) developed a new fluorescence model for these bands at low rotational temperatures, and Mumma and Reuter (1989) applied it to the IKS spectra, achieving an excellent fit to the band shape and accounting for all spectral channels satisfactorily (Fig. 13). They obtained a production rate of 4.5%±0.5% relative to water, and found that the formaldehyde was rotationally relaxed ($T \sim 50$ K), in accord with other estimates for the temperature in the intermediate coma.

Mumma and Reuter (1989) also applied their model to the entire set of infrared spectra obtained from March through May 1986 by various observers. This interval spans eight cycles of the 7.37 day period found in the lightcurve for March and April (Millis and Schleicher 1986; Schleicher et al. 1990). When arranged against rotational phase, the production rates relative to water were found to correlate strongly with the lightcurves for C_2, CN, and other species. The detections occurred over a narrow range of phases in the lightcurve, corresponding to one of the three maxima in activity. This phase range was experienced during both the Giotto and Vega 1 encounters and also for the sole groundbased spectrum in which formaldehyde was marginally present. Upper limits for the production rates at other phases were 5 to 10 times smaller, including phases corresponding to the remaining two maxima in the lightcurves and to the Vega 2 encounter. Reuter and Mumma (1989) suggested that formaldehyde was heterogeneously distributed in the Halley nucleus, relative to water. Robust detections of formaldehyde rotational lines were achieved in two recent comets (Colom et al. 1992; Schloerb and Weiguo 1992). Colom et al. (1992) found H_2CO in comet Austin (1990 V) and comet Levy (1990 XX) at levels of 0.1% and 0.04% relative to water, respectively, assuming that it was produced solely as a parent volatile. Upper limits (3σ) were obtained from infrared spectra of these two comets ($<0.4\%$ and $<0.2\%$, respectively), again assuming a parent distribution (Reuter et al. 1992), however, these were not as sensitive as the millimeter results. The ratios in both cases would be $\sim$ ten times greater, were formaldehyde produced solely from an extended source. Colom et al. (1992) found that an extended source of H_2CO was required in comet Austin, similar to that reported for comet Halley.

Although only one of the groundbased spectra of comet Halley showed (marginally), the 3.6 μm feature, a feature was nearly always present near 3.52 μm. In their analysis of the IKS spectrum, Mumma and Reuter (1989) found that the entire flux near 3.6 μm could be attributed to formaldehyde, without requiring any additional flux from the 3.52 μm feature. The distribution of the two species may therefore be anti-correlated within the nucleus. Brooke et al. (1989) investigated the 3.25 to 3.65 μm region in comet Wilson at moderate resolving power, and found a high degree of correlation between the spectral intensity profile of the 3.45 to 3.65 μm region in that comet and the spectrum of comet Halley (Baas et al. 1986). The spectrum could not be explained simply by fluorescence from formaldehyde, however, and following earlier authors (Knacke et al. 1986) they suggested methanol as another possibility.

Cometary methanol was first identified through infrared and millimeter spectroscopy of comet Austin (Bockelée-Morvan et al. 1991; Hoban et al. 1991a). The discovery spectra are shown in Fig. 14. Hoban et al.

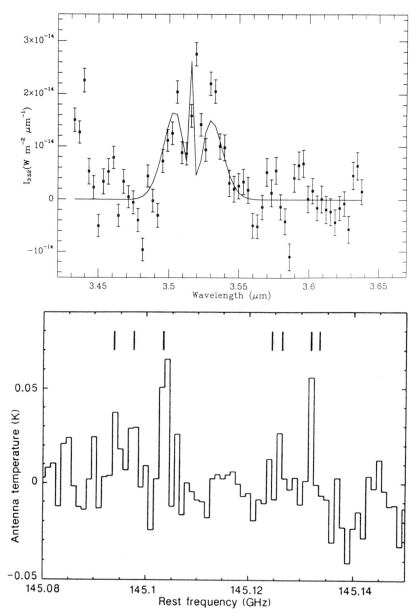

Figure 14. Discovery of methanol in comet Austin (1990 V). Top: vibrational band emission (ν_3) near 3.52 μm wavelength on UT May 5–7, 1989 (after Hoban et al. 1991a). Bottom: rotational emission lines near 2.07 mm wavelength on UT 25 May 1990 (after Bockelée-Morvan et al. 1991). Both data sets indicate low rotational temperatures. The synthetic fit for the infrared band is shown for a model with rotational temperature $T = 70$ K. The line-by-line mm intensities also indicate a low rotational temperature. The production rates retrieved from these spectra are comparable (~4% relative to water; see text).

(1991a) studied the spectral shape and spatial profile of the 3.52 μm feature in comets Austin, Brorsen-Metcalf, and Levy (1990 XX) and tentatively identified methanol as its progenitor, based on a new fluorescence model for that molecule. The production rate (relative to water) was 4% in comet Austin on UT 5–7 May 1990. Bockelée-Morvan et al. (1991) detected methanol on UT 24–25 May 1990 at millimeter wavelengths and reported a production rate of 1% relative to water. However, this was subsequently revised upward as a result of improved knowledge of the telescope tracking, and of the application of non-LTE excitation models. The revised value agrees with the infrared result (D. Bockelée-Morvan, personal communication). The infrared production rate for CH_3OH in Austin is not sensitive to rotational excitation models, because the spectral range includes all rotational levels in the molecular band. The integrated band intensity is easily related to the production rate through a fluorescence model. Infrared and millimeter retrievals of methanol in comet Levy (1990 XX) (on the same day) were both $\sim$1% relative to water. In comet Halley, methanol has recently been proposed as a progenitor of the mass 32 and 33 (CH_3OH^+ and $CH_3OH_2^+$) peaks in the Giotto IMS data, with a relative production rate of 0.3 to 1.5% (Geiss et al. 1991). A relative production rate of 1% was also retrieved from the mass 33 peak in the NMS data (Eberhardt et al. 1991).

The methanol abundance in comet Austin was close to the value in W33A. Recent detections of CH_3OH ice in W33A show it to be present at the level 7% relative to water (Grim et al. 1991), while formaldehyde ice is present at 0.2% abundance (Tielens 1989), so that $CH_3OH/H_2CO \sim 35$. In comet Austin, the ratio was >40 (taking H_2CO to be a parent species) comparable with the interstellar value. The ratio would be $\sim$10 times lower (i.e., 4) were H_2CO produced from an extended source. In comet Halley the ratio was in the range 0.3 to 0.06, significantly lower than the interstellar value. It is interesting that the ratios CH_3OH/H_2O and CH_3OH/H_2CO in comet Austin are comparable to those in W33A, as methanol is easily converted to formaldehyde by photo-processing (Allamandola et al. 1988).

It is unknown whether cosmic-ray processing also has this effect, but this line of inquiry seems warranted since Austin was a dynamically new comet and it is not known whether the radiation processed layer had been completely lost by the time of observation (see Sec. III). These results hint that such effects are identifiable from remote observations.

G. Volatile Nitrogen: NH_3, HCN and N_2

Nitrogen has proven particularly difficult to measure remotely, because neither the atomic nor diatomic forms are easily detected, and the fraction of total N locked in the grains cannot be studied remotely. Of the potential reservoirs of volatile N, only HCN has been detected directly, although the abundances of N_2 and NH_3 have been inferred from measurements of N_2^+, NH and NH_2. Wyckoff et al. (1991b) have reviewed the abundance of total nitrogen in comet Halley, retrieved from mass spectrometry of the gas and dust on Giotto and Vega, and from groundbased spectroscopy of NH, NH_2 and HCN. They argue that total nitrogen is depleted by about two-to-six fold, relative to solar abundance, in agreement with the depletion deduced by Jessberger and Kissel (1991) (see Table IV).

It must be noted that NH_3 has itself not yet been detected in comets. Models based on the ratio of intensities measured for species with mass 18 and mass 19 (taken to be NH_4^+ and H_3O^+, respectively) in the Giotto ion mass spectra indicated a relative production rate (NH_3/H_2O) of $\sim$1 to 2% (Allen et al. 1987; Wegmann et al. 1987). However, Ip et al. (1990) pointed out that reactions of H_3O^+ with HCN influenced the retrieved values for NH_3. They re-analyzed the spectrum, and obtained a value of $\sim$0.5% for NH_3. Boice et al. (1990) developed a model which included the cracking pattern of POM, and reproduced the measured mass spectrum with no CH_4 and only 1% NH_3. Furthermore, the NH_3 production rate (0.1 to 0.3%) retrieved from groundbased spectra of NH and NH_2, taken during the Giotto encounter, was 10 times smaller than the IMS result (Tegler and Wyckoff 1989; Wyckoff et al. 1991a, b; Allen et al. 1989; Magee-Sauer et al. 1989). Krasnopolsky and Tkachuk (1991) found an NH_3 production rate of 0.15% from the TKS spectra of NH and NH_2 on Vega 2. The spatial profiles of NH_2 require a distributed source whose lifetime is consistent with that of NH_3, but the progenitor could be either a grain or a volatile species so long as it has the required destruction scale length. Some fraction of the NH_2 could originate from a source other than NH_3. It is interesting to note that Kim et al. (1989) have argued that NH_3 was not the principal parent of NH in comet Halley. Given the difficulties in extracting unique production rates for NH_3 from mass spectra, greater weight must be extended to the retrievals from NH_2 at this time. The latter provide the most restrictive values for the production rate of NH_3.

When arranged against rotational phase, Mumma et al. (1990) found that the NH_2 production rate was strongly correlated with the lightcurves in C_2 and other radical species (Millis and Schleicher 1986), suggesting that the progenitors of these species are homogeneously mixed within the active regions of the nucleus. Wyckoff et al. (1991a) also argued that NH_3 was uniformly mixed in the nucleus of comet Halley, because its production rate (relative to water) did not change significantly over a span of many months.

Other forms of volatile nitrogen detected in comet Halley include HCN and N_2. Gaseous HCN is a familiar interstellar molecule, and the abundance of X-CN in interstellar ice is $\sim$4% relative to water ice (Table VI). While not all of the X-CN is likely to be HCN, the abundance of HCN predicted by chemical equilibrium models of the outer solar nebula is extremely small $(HCN/H_2O \sim 10^{-55}$ (sic) at $T=100$ K; Lewis 1972). Prinn and Fegley (1989) showed that kinetic inhibition would limit chemical conversion of HCN (to N_2) in the outer solar nebula $(HCN/H_2O \sim 10^{-6.8})$ and of HCN to NH_3 in the giant planet sub-nebulae $(HCN/H_2O \sim 10^{-6.4}$ at $T \sim 1220$ K). Lightning induced shock chemistry $(T=3000-4000$ K) could produce significant HCN, and isotopic equilibration at solar values (e.g., of $^{12}C/^{13}C$) would also be expected. However, only a small dilution factor of shocked gas by unshocked gas could be tolerated to produce the observed value (HCN/H$_2$O~1 × 10^{-3}) in comet Halley (see Fegley 1992). The abundance of HCN (and the ratios HCN/NH$_3$; HCN/N$_2$) may prove to be a very significant indicator of the degree of processing experienced by pre-cometary material. HCN was tentatively detected in comet Kohoutek (1973 XII) (Huebner et al. 1974), but it was first securely detected in comet Halley (Despois et al. 1986; Schloerb et al. 1986,1987; Winnberg et al. 1987). HCN has often been suggested as a likely progenitor of the CN seen at optical wavelengths, however, HCN production rates were found to be significantly smaller than production rates for CN in IRAS-Araki-Alcock (1983 VII). Thus, HCN could not have been the principal volatile source of CN in that older long-period comet (Bockelée-Morvan and Crovisier 1985). Production rates of HCN in comet Halley (HCN/H$_2$O~1 × 10^{-3}; Schloerb et al. 1986; Bockelée-Morvan et al. 1987) are in approximate agreement with production rates for CN (~1 × 10^{-3}) given by Catalano et al. (1986) and Wyckoff et al. (1988). However, other CN production rates (see Schleicher et al. 1986; Osip et al. 1992) are several times larger. As the retrieved production rates are dependent on assumed model parameters, Bockelée-Morvan et al. (1987) argue that a definitive conclusion on this question will require a reconsideration of the entire data set, accounting for temporal variability and also involving a critical analysis of assumptions in the cometary models. Using a coma model based on ion-molecule chemistry, Ip et al. (1990) reanalyzed the Giotto IMS results for ions with 28 and 29 amu/e^{-}, identifying them with H$_2$CN^{+}, and concluded that the production rate ratio HCN/H$_2$O was <2 × 10^{-4}.

Disregarding model sensitivities to coma physical parameters, the IMS and millimeter results can only be reconciled if the production rate for HCN varies by a factor of 5, relative to water. Millimeter production rates for HCN did show significant day-to-day variability, and CN emission showed both jets and shells (Schloerb et al. 1986; A'Hearn et al. 1986a; Cosmovici et al. 1988; Schlosser et al. 1986).

A'Hearn et al. (1986a, b) argued that ~20 to 50% of the CN was produced from dust, and this is consistent with the retrieved HCN and CN production rates at the accuracy of the retrievals. While it is clear that HCN is present in the nucleus at ~0.1%, relative to water, it seems that CN is produced from both gaseous and refractory sources; for an alternative view, see Combi (1987). The millimeter and optical observations represent many observations on a wide range of dates (hence rotational phases), while the IMS retrieval represents the rotational phase at the time of the Giotto encounter. It is not yet clear whether the measurements of Giotto are representative of other rotational phases or not.

N$_2$ itself has not been detectable spectroscopically in comets, having no allowed vibrational or rotational transitions, and with its first allowed electronic transition (b^1Π_u-X^1Σ_g^+) at 99.5–85.5 nm, below the short wavelength cut-off of IUE and the High Resolution Spectrograph and Faint Object Spectrograph instruments of the Hubble Space Telescope (HST). The production rate ratio (N_2/H_2O) was found to be <0.15, based on the mass 14 amu/e$^-$ peak in the IMS spectra on Giotto (Balsiger et al. 1986), and was <0.1 from the mass 28 peak in the NMS (Eberhardt et al. 1987b). N_2 was also retrieved from observations of N_2^+ (Wyckoff et al. 1991b). The ratio N_2^+/CO^+ was measured in the tail of comet Halley, and the production rate of N_2 was retrieved from the CO production rate using a model in which photo-ionization was the dominant production mechanism for both ions, while the main loss mechanism was dissociative recombination (Wyckoff et al. 1991b). The abundance ratio $CO/H_2O = 0.1$ was used to obtain the N_2/H_2O production rate ratio $\sim 2 \times 10^{-4}$. This approach is the most direct one available for N_2; however, its accuracy is difficult to assess because the rates for dissociative recombination are sensitive to electron temperatures, and these are not well known.

H. Sulfur Chemistry

Sulfur has relatively high cosmic abundance (S/O ~ 0.02) and is well represented in dense clouds in both oxidized and reduced forms, but investigation of its chemistry in the cometary nucleus has only recently become tractable. Production rates for S and CS have been retrieved from ultraviolet spectra of several dozen comets since 1976 (see below). S_2 was observed for the first time in any astrophysical source during an outburst in comet IRAS-Araki-Alcock (1983 VII), and its short decay length (~ 300 km) suggested that it sublimed directly from the cometary nucleus (A'Hearn et al. 1983). Its production rate has recently been revised downward to 2.5×10^{-4}, relative to water, based on an improved fluorescence model (Kim et al. 1990).

A'Hearn and Feldman (1985) argued that the S_2 may have been produced by irradiation of ices on interstellar grains, then incorporated into the cometary nucleus. While laboratory simulation experiments for sulfur compounds in an oxygen-free matrix do produce S_2, those conducted with an oxygen rich matrix (e.g., H_2O ice) often produce more SO than S_2 (Grim and Greenberg 1987b). Moore et al. (1988) demonstrated that synthesis of S_2 in oxygen-rich ices also produced significant SO_2. Kim and A'Hearn (1991) established upper limits for SO and SO_2 in recent comets, including IRAS-Araki-Alcock (1983 VII), for which they found $SO/S_2 < 2 \times 10^{-2}$ (<0.6), if SO is treated as a parent (daughter), and $SO_2/H_2O < 8 \times 10^{-7}$. For other comets, SO_2/H_2O is typically $< 10^{-4}$ and was $< 2 \times 10^{-5}$ in P/Halley. The absence of SO in IRAS-Araki-Alcock appears to rule out models for production of S_2 by irradiation of sulfur-bearing icy mantles on pre-cometary interstellar grains (Kim and A'Hearn 1991). S_2 has not been observed in any other comet, either before or after Iras-Araki-Alcock (the detection in P/Halley reported by Wallis and Krishna-Swamy [1987] has been challenged by Feldman 1991a). Russell et al. (1987) suggested that the detection of S_2 in comet I-A-A may have been related to an unusual, nearly radial alignment of the interplanetary magnetic field, permitting solar-wind particles to penetrate much deeper than usual into the coma. If so, additional sources and excitation mechanisms might be created. The origin and significance of S_2 in comets is still highly uncertain at present.

Cometary H_2S was first detected in comets Austin (1990 V) and Levy (1990 XX) and its abundance was found to be 0.2% relative to water (Crovisier et al. 1991). The discovery spectrum and a partial energy level diagram are shown in Fig. 15. The production rate for H_2S in comet Halley had been obtained from the Giotto Positive Ion Cluster Composition Analyzer (PICCA) ion spectra and was found to be impossibly large (Marconi et al. 1990); however Crovisier et al. (1991) found that an incorrect lifetime was used. With the correct lifetime of 4000 s, the revised abundance is consistent with the values found for Austin and Levy. The abundance of H_2S is about ten times less than the abundance of total volatile sulfur, estimated to be $S/H_2O \sim 0.02$ (Azoulay and Festou 1986). However, the retrieval of total sulfur from IUE observations of atomic sulfur depends on the assumed parent lifetime. Roettger (1991) has established an upper limit (5×10^{-3}) for the sum of measured (H_2S, OCS) and inferred (CS, S_2 upper limit, CS_2, SO_2) parent species. Because it is not clear that all S-bearing species were counted, the abundance of volatile sulfur is still open to question. Upper limits have been established at millimeter wavelengths for OCS and H_2CS at 0.2% and 0.1% in comet Levy, thus they remain possible significant contributors to the total sulfur budget. CS_2 is a prime candidate, and is likely to be present at the 0.1% level if it is the principal parent of CS (Jackson et al. 1982). However, being linear and symmetric, CS_2 has no allowed pure rotational transitions, so is undetectable at millimeter and far-infrared wavelengths. It is detectable at infrared wavelengths but only from airborne altitudes or higher, as its strong ν_3 fundamental overlaps the ν_2 band of water (near 6.5 μm). The ν_2 band occurs near 25 μm and might be accessible at favorable sites but the fluorescence efficiency is very small.

Sulfur was detected in grains in comet Halley by the PIA and PUMA instruments on Vega 1 and Giotto with $S/O \sim 0.08$. The bulk ratio in dust and ice in comet Halley was estimated to be ~ 0.035 (Table IV), somewhat higher than the solar abundance ratio (0.02). Jessberger and Kissel (1991) did not include any volatile sulfur compounds in arriving at the bulk S/O ratio, so the S/O discrepancy will be made worse when compounds such as H_2S and CS_2 are included (see Table IV). There is at present no clear approach for detecting sulfur in condensed phase matter in comets, other than by direct sampling.

I. Noble Gases

Owing to their low polarizability, the frosts of noble gases are extremely volatile. They are therefore sensitive "thermometers" of the formation temperature and subsequent thermal history of comets. Although other low temperature thermometers exist (e.g., CO, CH_4, H_2CO), noble gas abundances are easier to interpret because they are less affected by chemistry. In principle, the maximum temperature experienced by cometary ices can be determined

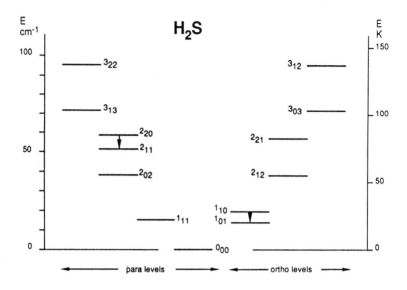

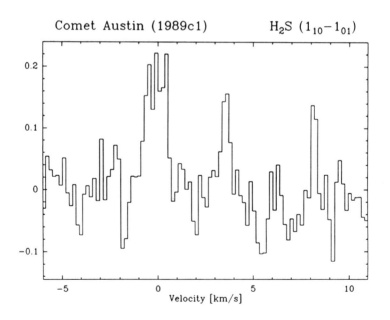

Figure 15.   Discovery of H_2S in comets. Top: Partial energy level diagram for
H_2S showing ortho and para levels, and the two rotational transitions observed in
comets Austin (1990 V) and Levy (1990 XX). Bottom: Spectra of the 110–101 line
measured in comet Austin on UT 24–25 May 1990 (after Crovisier et al. 1991).

from the relative abundances of these temperature sensitive gases. However, the inferred temperature depends also on poorly known physical factors, such as the trapping efficiency of gases in the ice matrix (Bar-Nun and Kleinfeld 1989). The change in relative abundances of noble gases in ices formed at different temperatures has been described (with applications to comets) by Owen et al. (1991).

Noble gases were not identified from *in-situ* measurements of comet Halley. However, searches for noble gases in recent comets were conducted using EUV spectroscopy. Stern et al. (1992) observed the dynamically new comet Austin (1990 V) with a sounding rocket payload in April 1990. Based on the upper limit for the He I resonance line at 58.4 nm, they concluded that the He/O ratio in this comet was a factor of 1.5×10^4 smaller than the solar value. Feldman et al. (1991) observed comet Levy (1990 XX) with the ASTRO-1 Spacelab observatory during December 1990. Unfortunately, the ASTRO-1 observations took place in daylight, and geocoronal emission prevented detection of cometary He I. Also, the instrumental spectral resolution was too low to separate the Ne I 73.6 nm line (observed in second order) from the S I 147 nm line. Weak upper limits were obtained for the cometary Ar and Ne abundances (<5–10 times their solar abundances), using O as a standard. As noted by both reports, the interpretation of these results remains open given the ease with which noble gases sublime from the surface and (possibly) diffuse from the interior when the nucleus is warmed, for example, either during its residence time in the Oort cloud or after entering the planetary region (see earlier discussion). The uncertainty in the predicted de-volatilization is due in part to our poor knowledge of the thermal conductivity and microscopic pore structure of the cometary nucleus.

Future abundance determinations for temperature sensitive noble gases (e.g., Ne, Xe, Ar; cf. Table II) may provide useful constraints on the condensation temperature and processing history of cometary ices, and may assist in interpreting the abundances of species that are strongly affected by both chemistry and by temperature dependent fractionation during condensation (e.g., S_2, N_2, CH_4, CO, NH_3, H_2CO, CH_3OH; cf. Table I).

J. Nuclear Spin Temperatures

Infrared spectroscopic studies of water vapor in comets Halley (1986 III) and Wilson (1987 VII), were conducted with sufficient spectral resolution to reveal the intensities of individual ro-vibrational lines, and even to measure their velocity profiles (Mumma et al. 1986; Larson et al. 1987; Weaver et al. 1987; Larson et al. 1989). An especially surprising finding was that cosmogonic information may have been preserved in the relative abundances of the nuclear spin species of cometary water (Mumma et al. 1987,1988*a*, *b*; Mumma 1989).

The energy levels for water are organized into two ladders, according to the total value of the nuclear spin ($I = 1$ or 0), corresponding to the ortho- and para-H_2O nuclear spin species (note: "nuclear" in this context refers

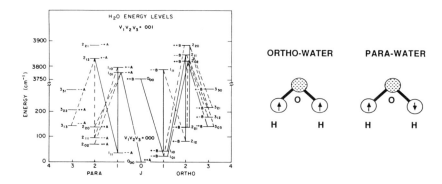

Figure 16.   Right: Water consists of two noncombining nuclear spin species, ortho and para, with the value of the total nuclear spin $I = 1$ or 0, respectively. Left: Partial energy level diagram of the water molecule. Ro-vibrational transitions measured in comet Halley are shown by solid lines, while those which must have been present but which were obscured by terrestrial CO_2 are shown as dashed lines (after Mumma et al. 1986).

to sub-atomic structure, not the cometary nucleus). Ortho- and para-water may be regarded as independent and noninteracting species, under normal conditions (see below). The physical orientation of the proton spins is shown schematically in Fig. 16. The measured water lines (Fig. 12) represent the major part of the (001–000) ν_3 water band in comets, but a few additional lines are obscured by CO_2 atmospheric extinction. Transitions observed in comet Halley are represented by solid lines and the inferred transitions by broken ones in Fig. 16. The latter intensities can be inferred from the measured lines using known branching ratios, if the coma is assumed to be optically thin. Infrared optical trapping is not too important for the KAO observations, owing to the large beam diameters used (34,000 km and 20,000 km in Dec. 1985 and March 1986, respectively) (cf. Mumma et al. 1986; Bockelée-Morvan and Crovisier 1987). Some 81% (70% for March 1986) of the sampled molecules lie more than 5000 km from the nucleus, and fewer than 4% (10% for March 1986) lie within the zone where infrared optical depths are significant ($R < 1000$ km).

The abundance ratio of the ortho and para spin species (the OPR) is readily modeled for the equilibrium case, using the defining relation for rotational distribution (Mumma et al. 1987). An ortho-para ratio may be obtained from the measured cometary line intensities by summing the total line intensities for each spin species separately. The ratio of summed intensities is the observed ortho-para ratio, and the nuclear spin temperature may be determined by comparison with the equilibrium case (cf. Mumma et al. 1988a, b). Summing over the intensities is equivalent to summing over the rotational population if all levels are sampled and all lines are measured. Mumma et al. (1987)

summed only over levels sampled directly in the KAO spectra, while Mumma et al. (1988b) corrected for several levels whose lines were obscured by terrestrial extinction, by assuming that a single temperature characterized the rotational distribution. The ortho-para ratio and rotational temperature were then retrieved from a nonlinear least-squares analysis of the measured line intensities. A discussion of the approach and its uncertainties is given by Mumma et al. (1988b).

The results for comet P/Halley clustered near OPR of 2.5±0.1, equivalent to a spin temperature of 29 K, even though the conditions of observation were very different and the measurements were separated in time by four months (W. E. Blass and M. J. Mumma, personal communication). Over this time interval, active regions of the nucleus likely eroded by ∼5 to 10 m (assuming 30% active surface area), and thus the measured values are characteristic of the ensemble average spin temperature for a range of depths and surface areas of Halley's nucleus. Comet Wilson, however, showed an OPR of ∼3.2 ± 0.2, which is consistent with thermal equilibrium at temperatures >50 K (Mumma et al. 1988b).

Bockelée-Morvan and Crovisier (1990) applied a much more sophisticated model for formation of cometary water lines, and were able to fit the KAO line intensities for P/Halley with OPR of 3 when opacity effects were included, indicating thermal equilibrium at >50 K. However, the same model applied to comet Wilson returned OPR of 3.9. The meaning of the latter result is not clear, because there is no known mechanism for producing a ratio >3 (technically, this would be a spin "maser"). The retrieved ortho-para ratios for both comets were substantially lower when opacity effects were neglected, and the value for Halley post-perihelion (2.5) then agreed with the result of Mumma et al. (1988b), and the line-by-line intensities were satisfactorily reproduced. However, although the pre-perihelion result (OPR = 2.5) also agreed with Mumma et al. (1988a), the retrieved rotational temperature was lower than that demonstrated by the KAO spectra and the modeled line-by-line intensities were not in good agreement with the KAO spectra. The ortho-para ratio for Wilson (3.65) still substantially exceeded that of Halley. As the water production rates for comet Wilson were comparable to those for comet Halley pre-perihelion, and both were much smaller than for comet Halley post-perihelion, it is unlikely that opacity effects can explain the large difference found for the ortho-para ratios, which is evident even in the individual spectra.

Several aspects of the optically thick model may need further development; for example, the use of the escape-to-space approximation has not yet been verified for cometary comae, and electron collisions should be included. Xie and Mumma (1992) showed that electron collisions dominate rotational excitation of water in the intermediate coma of comet Halley (the region sampled most efficiently by the KAO spectra), increasing the distance for which the rotational states are in equilibrium and lending additional strength to the LTE assumption. In any event, both groups find that the ortho-para ratio for comet Wilson differs significantly from that of comet Halley, and the fact that they differ is not dependent on model assumptions. A possible interpretation of this difference is discussed below.

Conversion between the spin species is an exceedingly unlikely process, requiring a strong nonuniform magnetic field, such as might be experienced by collisions with a paramagnetic species. Although spin species conversion is poorly studied for water, the data that do exist suggest very long conversion times (see Mumma et al. 1988b). Better data exist for molecular hydrogen, and we may employ this as an analog. Theoretical studies of gaseous molecular hydrogen demonstrate a radiative conversion time ($\sim 10^{20}$ s) exceeding the age of the universe, and laboratory studies confirm a collisional half-life of 3 yr at STP ($\sim 2 \times 10^{17}$ collisions) (Dodelson 1986; Farkas 1935). The conversion time is reduced significantly if a paramagnetic collision partner is introduced (e.g., O_2, NO, Pt), and this approach has been used to prepare laboratory samples of pure para-hydrogen. Once prepared, however, the para-hydrogen may be stored separately, and will retain its relaxed spin temperature for an indefinite period. In laboratory ices, spin conversion proceeds slowly for H_2 (even at temperatures as low as 10 K) because migration of ortho-H_2 enables ortho-ortho collisions, thus providing the required asymmetric magnetic field (cf. Silvera 1980). Amorphous water ice is weakly hydrogen bonded, however, and this will likely prevent migration of ortho-water molecules, thus inhibiting spin conversion in low temperature cometary ice.

Spin relaxation has now been found in astrophysical systems for several phases related to formation of planetary systems. In regions of bi-polar outflow from young stellar objects, the spin temperature for H_2 is typically ~ 100 K (corresponding to formation on grains at a temperature ~ 65 K (cf. Takayanagi et al. 1987; Hoban et al. 1991b). Minh et al. (1991) found relaxed ortho-para ratios for H_2CS in several dense cloud cores (e.g., TMC-1, where $T_{spin} \sim 15$ K, comparable to the kinetic temperature), while Madden (1990) found cyclopropenylidene (C_3H_2) to be relaxed in Orion KL. However, ketene (CH_2:CO) is not relaxed (W. M. Irvine, personal communication). The mechanism responsible for relaxation is not yet certain, and the different behavior exhibited by ketene is not understood.

Mumma et al. (1988b) have suggested that the spin temperatures found for Halley and Wilson may be cosmogonic. The water now released by the active regions in comet Halley may have been last processed as an ice on grains at a temperature of ~ 29 K, then incorporated into cometesimals and buried deep within the original nucleus, where the overburden protected it from cosmic-ray damage. The outer layers of the nucleus experience significant processing while in the Oort cloud, as discussed earlier in this chapter, and this (particularly radiation processing) would reset the ortho-para ratio to the high temperature limit. Thus, a dynamically new comet (e.g., Wilson) might be expected to show a statistically equilibrated ortho-para ratio (i.e., 3 for H_2O). Johnson (1991) has modeled radiation damage for the outer layer of new comets, and estimates that more than 30% of the chemical bonds have been altered in this way. He argues that hydrogen atom exchange could be an efficient mechanism for spin-species alteration in new comets.

K. Elemental and Isotopic Abundances

Elemental abundances can be deduced by summing the contributions identified in gaseous and refractory progenitors. For example, Wyckoff et al. (1991b) examined the abundance of volatile nitrogen and argued that it is deficient in Halley, relative to solar abundance. The heavy elements in grains were discussed by Jessberger et al. (1988) and Jessberger and Kissel (1991). With the exception of H (deficient by $\sim$650 fold), N (deficient by $\sim$3 fold), and Fe (deficient by $\sim$2 fold), all total elemental abundances appear to be solar, to within model and measurement accuracy (Table IV, see Jessberger and Kissel 1991; Wyckoff et al. 1991b; cf. the discussion of total sulfur in Sec. IV.H).

However, extrapolations of cometary ensemble averages are subject to uncertainties, such as: (1) the use of unverified models to retrieve a parent abundance from the observed species (e.g., N_2 from N_2^+, CO from CO^+); (2) the assumption that all significant reservoirs have been tallied; and (3) the inability to measure directly production rates for certain key atoms (e.g., the resonance line of N near 120 nm is difficult to separate from H Lyman α at 121.6 nm with moderate resolution spectrometers, such as that on IUE). Further discussion exceeds the scope of this chapter, and the reader is referred to the excellent reviews which already exist on this topic (cf. Delsemme 1991) and to the sources already cited.

Along with the gross chemical identities of the cometary volatiles, the abundances of their isomeric and isotopic variations should provide key information on the conditions of volatile formation and subsequent processing. Isomeric and isotopomeric studies are well advanced for interstellar molecular clouds, and recent reviews of cloud observations and cloud chemistry have been provided by Irvine and Knacke (1989) and Turner (1989c) (see also Chapter by van Dishoeck et al.). Considering that detections of most cometary parent volatile species post-date 1985, it is perhaps not surprising that their isomers (e.g., HNC) have yet to be detected. Swade et al. (1986) reported HNC/HCN <0.3 in comet Halley. As detection capabilities improve, this may well become a fruitful line of investigation.

Isotopic ratios have been regarded as relatively less subject to interpretational difficulties, but measurements require both very high sensitivity and very high resolution. The only ratio measured in comets prior to comet P/Halley was $^{12}C/^{13}C$ in C_2. The derived cometary ratios were consistent with a solar system value of 89, except for comet West (1976 VI), for which Lambert and Danks (1983) found somewhat lower values (50$\pm$15, 60$\pm$15). The value of $^{12}C/^{13}C$ was measured in CN and was found to be 89$\pm$17 in comet Halley (Jaworski and Tatum 1991). The lower value for Halley reported originally by Wyckoff et al. (1989) (65$\pm$9) has been revised recently (100$\pm$15) as a result of improved analytical procedures (S. Wyckoff 1991, personal communication). Thus, $^{12}C/^{13}C$ in CN is consistent with solar abundance.

The interpretation of isotopic ratios in molecular fragments like CN and C_2 must be approached with care. About 20 to 50% of the CN in the coma was produced in jets (A'Hearn et al. 1986a). These jets are not congruent with the jets seen in the optical continuum (A'Hearn et al. 1986b), and are possibly associated with the CHON carbonaceous grains discovered in comet Halley (Krueger and Kissel 1987; Kissel and Krueger 1987). The CHON grains show a wide range of $^{12}C/^{13}C$ ratios, although some uncertainty in the ^{13}C retrieval is introduced by the coincidence with CH at mass 13 (Jessberger et al. 1988). Wyckoff et al. (1989) discounted the refractory source of CN and suggested that their earlier value (65±9) was inconsistent with formation of cometary nuclei in the Uranus-Neptune region. However, Krankowsky and Eberhardt (1991) demonstrated that the claimed enrichment (note: since revised, see above) in ^{13}C in CN could originate entirely from the refractory source. A solar $^{12}CN/^{13}CN$ ratio can be accommodated by reducing the fraction of ^{13}CN contributed by grains, however, this only emphasizes the lack of uniqueness in the interpretation. Thus, even with a solar ratio for $^{12}CN/^{13}CN$, the isotopic ratio in the volatile source of CN is unresolved at present. As the volatile parent of CN is a minor constituent (e.g., HCN is <0.1%, relative to water), the interpretation of $^{12}C/^{13}C$ in volatile carbon in comets also requires measurements on the major carriers (CO, CO_2, CH_3OH, H_2CO, etc.).

More definitive results were obtained for D/H, in the form of the following: H_2DO^+/H_3O^+. Eberhardt et al. (1987b) found the number ratio for D/H to be within the range 0.6 to 4.8×10^{-4}, which exceeds the local interstellar value significantly but agrees with the values found in molecular cloud "hot cores," as discussed in Sec. V (Fig. 17). It is larger than the values for the proto-Sun, Jupiter and Saturn (all bodies that received their hydrogen in the form of nebular H_2), while it is comparable to values for Uranus, Neptune, Titan and the Earth (seawater is 1.6×10^{-4}), bodies that are believed to have acquired their hydrogen in the form of water ice and/or hydrated silicates (for the Earth). The D/H ratio found in comet Halley thus supports the idea that Earth's ocean water may have been contributed by cometary impact. The cometary D/H value also suggests that the water in comet Halley did not re-equilibrate with H_2, the principal reservoir of hydrogen in the solar nebula. Grinspoon and Lewis (1987) argued that the time scale for such equilibration exceeds the estimated lifetime of the solar nebula. It would therefore appear that the D/H ratio in cometary water was established in the natal dense cloud core (see later discussion). The value retrieved for isotopic oxygen was $^{18}O/^{16}O = 0.0023\pm0.0006$, in agreement with the terrestrial value (0.00205). The ratio $^{15}N/^{14}N$ in CN in comet Halley is also typical of solar system values (Wyckoff et al. 1989). A good review of isotopic compositions appears in Vanysek (1991).

At this writing, the isotopic measurements support a common origin for

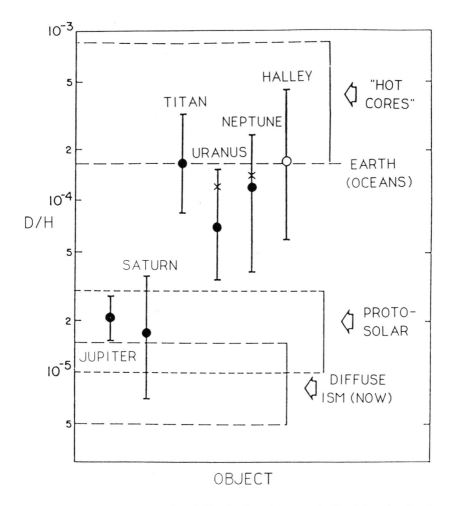

Figure 17.   A comparison of the D/H ratios for solar system bodies. The value found for comet Halley significantly exceeds values in the diffuse interstellar medium (ISM) and solar nebula, but is consistent with values found for water in molecular cloud "hot cores" (see Sec. V), and with values measured for solar system bodies that acquired their volatile inventory in the form of ices. The x's superimposed on the Uranus and Neptune results are the predictions of an accretion model in which pre-planetary ices are assumed to be enriched in deuterium to the SMOW level by condensation fractionation, and nebular H_2 is disproportionately captured by the growing protoplanets (Hubbard and MacFarlane 1980b) (adapted from de Bergh et al. 1990).

the ices in comet Halley, and for solar system bodies that acquired their hydrogen in the form of ices.

V. SYNTHESIS

From a chemical perspective, there are three important questions regarding the composition of cometary nuclei. First, are the mean chemical abundances consistent with interstellar composition, or do they require some degree of processing (e.g., reduction or isotopic exchange) either in the solar nebula or after incorporation into the final cometary nucleus? Second, is the nucleus chemically homogeneous or heterogeneous, and if the latter, is that heterogeneity natal or derived? Third, are there important differences between comets that result from their mode and site of origin? We have reviewed the chemistry of cometary nuclei, and have assembled a typical cometary composition in Table I (see also Tables IV, V). However, an impressive array of evidence is accumulating in support of the idea that comets exhibit both intra- and inter-nuclear heterogeneity. The ratio of total production rates for dust and gas varies strongly from comet to comet, as do the production rates of some trace volatiles relative to water. The variability in the dust/gas ratio may also be related to surface processing and to received insolation, however, and must be evaluated at a uniform heliocentric distance to remove temperature dependence. These effects must be considered along with cosmogonic differences. CO may vary strongly from comet to comet, and measurements of H_2CO and of CH_3OH in recent comets suggest that they also exhibit intra- and inter-nuclear variability. The former (CO) may be related to variation in the condensation temperature of pre-cometary ices or its thermal history after formation, while the latter (H_2CO, CH_3OH) may reflect varying degrees of processing prior to incorporation into the nucleus.

Several lines of evidence indicate the presence of macroscopic internal heterogeneity in comet Halley (Table VII). The relative production rate for formaldehyde varied more than ten fold in March and April 1986, and the positive detections corresponded with only one of the three peaks revealed in the lightcurves for the species CN, C_2 and NH. This region was active during both the Vega 1 and Giotto encounters (but not during Vega 2), suggesting that the spacecraft data should be re-examined from the perspective of revealing macroscopic chemical heterogeneity in the nucleus. Other notable evidence includes the CO_2-rich outburst observed by Feldman et al. (1986), and the unusual outburst of March 20 observed in H_2O and other species (cf. Larson et al. 1990). The latter outburst occurred during direct measurements of water at infrared wavelengths, and it more than doubled the water content of the inner coma. Larson et al. (1990) suggested that the latent heat released by conversion of amorphous ice to the crystalline form may have provided the required energy. This is consistent with the view that the ices in comet Halley formed at low temperatures, perhaps in the outer reaches of the solar nebula (beyond the orbit of Jupiter) or in the interstellar medium, and that

TABLE VII
Evidence Supporting Macroscopic Heterogeneity
in the Nucleus of Comet Halley

Short-Term Variability	
Diurnal Activity	
H_2CO	Mumma and Reuter 1989
Outbursts	
Dust	e.g., Russell et al. 1986
H_2O (Amorphous?)	Larson et al. 1990
CO_2	Feldman et al. 1986
Pre-Post Perihelion Asymmetry	
Organics Feature	e.g., Danks et al. 1987
Possible Explanations:	
Cosmogony:	
Chemically distinct icy planetesimals, (some) 50 to 70 m diameter	
Formed in different regions of solar nebula or interstellar medium	
Gravitational diffusion in solar nebula	
Agglomeration into final nucleus	
Post-Accretional Processing:	
Thermal migration of chemical species within nucleus	

the pre-cometary ices in comet Halley were not later processed at elevated temperatures during infall in the solar nebula. Further, comet Halley could not have spent significant time in an orbit with insolation high enough to trigger amorphous-to-crystalline conversion throughout the nucleus. The exothermic conversion of amorphous ice to the crystalline form may expel excess trapped gas, whose accumulation in pockets could drive some outbursts (Prialnik and Bar-Nun 1987), but comet Wilson did not show such outbursts. However, the repetitive production of formaldehyde from only one of several active regions on the nucleus of comet Halley would appear to require macroscopic heterogeneity.

The "mean" chemical abundances may also be examined in the context of either a solar nebula or interstellar origin for cometary material. Table VIII summarizes certain key indicators, discussed in detail earlier. The evidence at present supports the presence of pre-solar volatiles, as found in dense molecular cloud cores, and of low-temperature refractory organics in cometary nuclei. Thus, it appears that any successful model for cometary formation must account for the absence of significant processing of pre-cometary material in the

TABLE VIII

Some Key Indicators of the Nature of Cometary Material

CO:	Variable abundance consistent with condensation near 25–50 K.
H_2O:	Enriched D/H consistent with unmodified interstellar material (UIM)[a]. Low spin temperature in Halley consistent with UIM. High spin temperature in Wilson consistent with cosmic ray damage.
HCN:	Abundance much higher than expected for inner solar nebula or giant planet sub-nebulae; consistent with UIM.
H_2CO:	Abundance much higher than expected for inner solar nebula or giant planet sub-nebulae, but near that for UIM. Polymer not expected for solar nebula origin.
$^{12}C/^{13}C$:	Value in CN is consistent with solar abundance ratio. The ratio in CHON grains is highly variable, as in primitive meteorites, and demonstrates that the grains are chemically disequilibrated. Consistent with stellar and interstellar origin for the grains. It is not known whether the solar nebula could produce similar effects.
NH_3:	Low abundance (0.2% or less) consistent with UIM.
CH_4:	Low abundance (0.2% or less) consistent with UIM.
CH_3OH:	Abundance consistent with UIM, but not with processed material from the solar nebula.
N_2/NH_3:	Ratio consistent with UIM.
CO/CH_4:	Ratio consistent with UIM.
S_2:	Inconclusive. Detected in only one comet, but not certain whether S_2 is native to the nucleus or is a secondary product.
CHON:	Easily destroyed, unique chemistry favors UIM.
Outbursts:	Significance not yet clear. Could represent either heterogeneous accretion, or thermal processing of nucleus.

[a] By "unmodified interstellar material" (UIM) we mean that the observed quantity is consistent with values found in dense molecular clouds, implying little or no subsequent processing in the solar nebula prior to incorporation into the cometary nucleus. For example, an abundance would be similar to those found in interstellar ice, and a temperature would be similar to those found in dense cloud cores.

solar nebula and giant planet sub-nebulae. Several key indicators point to an interstellar composition, unmodified by processing in the solar nebula. These include evidence for the low nuclear spin temperature for water in Halley, the presence of easily destroyed CHON particles and of HCN, the presence of both H_2CO and CH_3OH, the presence of substantial CO but of virtually no CH_4, and the virtual absence of NH_3. However, the heterogeneity (in H_2CO) revealed by comet Halley may indicate the presence of chemically distinct bodies within the nucleus, and this could be the signature of heterogeneous aggregation. Gravitational scattering by the giant protoplanets will cause radial diffusion of cometesimals formed in different regions of the solar nebula, possibly resulting in aggregation of chemically distinct icy bodies into a single final cometary nucleus.

The D/H ratio (in H_2DO^+/H_3O^+) is distinctly higher in comet Halley than it is in solar system bodies that acquired their hydrogen mainly in the form of nebular H_2 (e.g., Jupiter, Saturn, the Sun), and it significantly exceeds the value in the local interstellar medium. However, D/H is often enriched in trace species in dense molecular cloud cores (Irvine and Knacke 1989), and so the value in comet Halley could reflect fractionation through ion molecule processes in the natal cloud. The high D/H value in Halley demonstrates that its water was never re-equilibrated with the principal reservoir of hydrogen (H_2, HD) in the solar nebula. Re-equilibration with H_2 in the solar nebula is kinetically inhibited though, so the D/H ratio does not test the question of vaporization of icy mantles from infalling grains.

The gas phase abundances in "hot cores" of giant molecular clouds are quite different from those found in cold dense clouds (where volatile species reside in both vapor and condensed phases), and are thought to reflect evaporation of icy mantles from interstellar grains (cf. Brown and Charnley 1990; Chapter by van Dishoeck et al.). Abundances in hot cores may thus be directly comparable to cometary ices. Jacq et al. (1990) measured the HDO/H_2O ratio in the hot cores of several giant molecular clouds. Using the standard rotation diagram approach for estimating column abundances in four clouds having the best quality data, they obtained values (HDO/H_2O) of 3 to 6×10^{-4}. However, it is not certain that the measured lines are optically thin, and so Jacq et al. also presented abundance ratios obtained from lines having similar excitation energies and frequencies. In this case, the abundance ratios (HDO/H_2O) were larger by factors of 2 to 3.

The D/H ratios in hot cores, as retrieved from the first method (1.5 to 3×10^{-4}) are consistent with the value found in comet Halley (0.6 to 4.8×10^{-4}) and are about ten times larger than the values found for the local interstellar medium. A rigorous comparison must await more accurate analysis of the Halley NMS data (in progress, P. Eberhardt, personal communication) and further evaluation of the question of spectral line formation in hot molecular cloud cores. We cannot exclude the possibility that the D/H ratio in comet Halley may be a typical value for condensed phase water in dense molecular cloud cores, and may not be unique to our solar system. Given that, we cannot absolutely conclude that Halley is of "solar system" origin, as compared with the capture scenario, though capture remains a very low probability event and cannot yield the population of the Oort cloud. However, all isotopic evidence (including D/H) in comet Halley is consistent with an origin as a member of our solar system.

The presence of H_2S and CO in cometary nuclei is consistent with formation of cometary ices at temperatures likely <60 K (cf. Table I). The low spin temperature for water in comet Halley (if Mumma et al. [1988b] are interpreting it correctly) suggests that water was last processed at ~29 K, and this low temperature would also be consistent with direct condensation of CO (~25 K). However, CO can also be trapped at ~7% relative abundance by condensation from a vapor-phase mixture of CO and water at ~50 K. The presence of N_2, the presumed parent of N_2^+, also supports a low condensation temperature (~22 K) if it condensed directly. However, it is not known whether N_2 can be trapped at higher temperatures, as CO is. Thus, the present evidence supports the formation of cometary ices at temperatures below ~60 K, and possibly as low as 20 to 30 K. If these represent physical temperatures in the solar nebula, they imply that cometary ices formed at distances greater than ~20 AU from the Sun.

Our understanding of the formation of comets in the Uranus-Neptune zone is complicated by our poor knowledge of the formation of icy planetesimals in that region, and poor understanding of the formation of Uranus and Neptune themselves. Solar-system formation studies have tended to focus on the formation of the terrestrial planets and the problem of forming Jupiter and Saturn prior to dispersal of the nebular gas. However, if we allow that icy planetesimals of cometary dimensions did form in the Uranus-Neptune zone, and that some accreted to form those two outer planets, then the ejection of comets to the Oort cloud follows as a natural dynamical consequence of this process. Furthermore, several researchers (see, e.g., Safronov 1972) have argued that an initially much larger mass of icy planetesimals was necessary in the Uranus-Neptune zone to provide sufficient initial surface density to decrease the formation times for those planets to a value less than the age of the solar system. The extra material was dynamically ejected, either to the Oort cloud or to interstellar space. Estimates of the total mass and angular momentum of comets in the Oort cloud are fully consistent with ejection from the outer planet zone. The failure to observe interstellar comets passing through the solar system raises questions about the ubiquity of cometary formation around other stars, but the uncertainties in the calculations are too great to make any definitive conclusions in this regard.

If the long-period comets originated in the Uranus-Neptune zone, and if the short-period comets also originated there, or in the Kuiper belt region just beyond, then what conclusions can be drawn about processes in that region of the solar nebula? Given the close identity of cometary materials with their expected pre-solar state in dense molecular cloud cores, it would seem that very little reprocessing went on in the solar nebula at distances of ~20 AU or greater. Infalling material was apparently decelerated gently, perhaps across only a weak shock, with little heating, and icy mantles on interstellar grains may have been preserved. Temperatures may have been even lower than predicted by Hayashi (1981) or Cameron (1978a) (Fig. 2). Compared with most models for the inner nebula, the solar nebula at 20 AU was a fairly benign environment.

Formation of comets at much larger distances in the solar nebula, on the order of several thousand AU or more, is also consistent with the compositional evidence presented in this chapter. However, the detailed dynamics of that formation are so poorly defined that it is difficult to give much credence to such hypotheses. If detailed examination of cometary materials continues to argue for an unprocessed pre-solar composition, then this may provide sufficient impetus to better quantify the details of possible cometary formation at larger heliocentric distances.

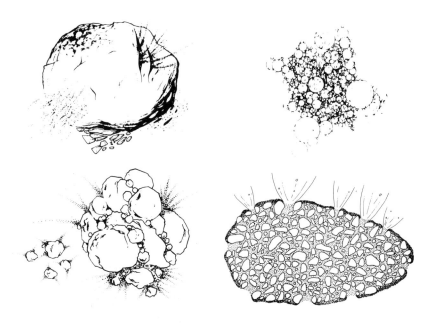

Figure 18.   Four suggested models for the structure of cometary nuclei. Top left: the icy conglomerate model (Whipple 1950; drawing from Weissman and Kieffer 1981); top right: the fractal model (Donn et al. 1985); bottom left: the primordial rubble pile (Weissman 1986a); and bottom right: the icy-glue model (Gombosi and Houpis 1986). All but the icy-glue model were suggested prior to the Halley spacecraft encounters in 1986.

The internal structure of cometary nuclei is very uncertain (see reviews by Keller 1990 and Rickman 1991). A variety of models have been suggested (Whipple 1950; Donn et al. 1985; Weissman 1986a; Gombosi and Houpis 1986) and several of these are illustrated in Fig. 18 (from Weissman 1986b). The current consensus is that nuclei are weakly bonded, fractal assemblages of smaller icy-conglomerate planetesimals, possibly "welded" into a single nucleus by thermal processing and sintering. Dynamical scattering of planetesimals by the outer planets may have resulted in mixing of bodies formed at different heliocentric distances in the solar nebula, and hence, in different temperature regimes. These different regimes may be reflected in compositional differences between the fragments making up each cometary nucleus.

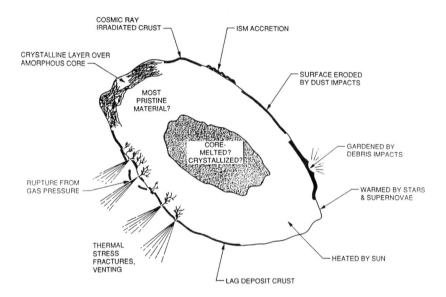

Figure 19.   A schematic view of the various processes that cause comets to evolve. Effects shown on the left are driven by thermal- and photo-processing during perihelion passages in the planetary region. Effects shown on the right are due to processes which take place in the Oort cloud (after McSween and Weissman 1989).

Several processes are thought to cause cometary evolution in the Oort cloud (Fig. 19). How do these various evolutionary processes interrelate? Clearly, collisions with Oort cloud debris and erosion by interstellar grains compete with irradiation and heating damage. One expects irradiation and impact processes to make cometary surfaces more dense, in turn enhancing the efficiency of erosion and thermal wave penetration. One also expects the most severely heat- and radiation-damaged crust to be removed by interstellar grains. The erosion mechanism could be tested by obtaining cosmic-ray exposure ages for surface material from a new comet. Because radiation damage is a continual process while erosion is episodic, an anomalously young exposure age is expected. The record of earlier supernova damage should also be missing, unless the most recent nearby supernova heating pulse came since the last erosional encounter with a dense galactic cloud (i.e., within the past 300 Myr). However, because the heat wave from encounters with passing O-type stars would likely penetrate to much greater depths than can be removed by interstellar grain erosion, some thermal damage is expected in all comets (unless the actual surface diffusivity is so low as to limit the heating to a very thin layer near the surface).

No detailed, quantitative model has been attempted for a cometary surface undergoing the full suite of Oort cloud processing mechanisms. Although such a model would be instructive, the present lack of knowledge about the physical properties and conditions on and under cometary surfaces prevents such a detailed description from being profitably carried out. Perhaps a more useful route would be to perform laboratory experiments simulating irradiation, interstellar grain bombardment, and stellar heating on the same cometary surface analog. Among the most interesting results anticipated from such an experiment would be the degree of interactive effects between the various processes. One wishes to know, for example, if hypervelocity impacts of interstellar grains result in a net *erosion*, or instead in a net *accretion* of material on the nuclear surface. Similarly, one wishes to know the penetration depth of a simulated supernova or O-star thermal wave after irradiation and interstellar grain bombardment has reworked the surface. Although we now know that comets are evolved from their formation state, it is important to stress the comparative subtlety and low rate of evolution experienced by objects in the Oort cloud. By obtaining an understanding of the processes which serve to modify cometary surfaces in the Oort cloud, we are better equipped to interpret both cometary behavior and cometary samples. Still, comets are almost certainly the best preserved samples from the epoch of planetary formation. In comparison to evolutionary time scales on planetary surfaces, cometary modification is slow and modest, indeed. Owing to the cold and rarefied environment of the Oort cloud, the cometary reservoir is a very quiet (though not silent) place. At least while the Sun remains on the main sequence, the Oort cloud will likely remain the most benign evolutionary environment in the solar system.

Future remote sensing of comets can be employed both to determine the degree of processing by radiation and by heating, and to refine model predictions. For example, CO, N_2, CH_4, Ar, S_2, and Ne should be depleted in the outer layers of new comets, due to thermal processing in the Oort cloud, while the ratios of less volatile species (e.g., CH_3OH, H_2CO) may be influenced more strongly by radiation processing than by thermal effects. From abundance ratios, it should be possible to estimate the maximum processing temperatures experienced. However, the provisional detection of CH_4 in the dynamically new comet Wilson, coupled with much lower abundances in one short-period comet (P/Halley) and one returning long-period comet (Levy 1990 XX), is consistent with enhancement of CH_4 in the outer layers of Oort cloud comets.

The spin temperatures of ices in the surface layer of new comets may also be related to radiation processing in the Oort cloud. The results for Halley and Wilson are most easily explained by the reset of spin temperatures in radiation-processed material in new comets while their interiors remain protected, preserving the spin temperature at which the ice was last processed (prior to accretion into the nucleus). The low spin temperatures in P/Halley are consistent with temperatures in dense cloud cores, suggesting that the H_2O ice in comets was last processed in the interstellar medium. The hypothesis of the ortho-para ratio as a cosmogonic indicator is testable. Measurements are needed on other comets, and observations from space would permit access to the entire ν_3 band, thereby reducing the sensitivity of the retrieved ortho-para ratio to modeling uncertainties. Extension to other molecules with nuclear symmetry (e.g., H_2S, CH_4, NH_3, H_2CO and H_2CS) is an obvious objective.

Our capabilities for remote characterization of cometary composition are still expanding rapidly. Infrared searches of comets Austin and Levy in 1990 were carried out with cryogenic grating and Fabry-Perot spectrometers and these provided detections of CH_3OH and possibly CO, and sensitive upper limits for H_2CO and CH_4. Instrumental sensitivities are improving rapidly, and detections should become routine now that higher-resolution echelle spectrometers equipped with large format array detectors have become available. Similar advances at millimeter wavelengths led to the detections of HCN, H_2S, H_2CO and CH_3OH, and to sensitive upper limits for OCS, SO_2, H_2CS and HC_3N. Detections of additional molecular species are likely.

The two approaches provide complementary information. Infrared observations permit spatial mapping at high spatial and temporal resolution, of both polar and nonpolar molecules. Their relatively broad spectral grasp permits measurement of the entire ro-vibrational band simultaneously, thereby retrieving the rotational distribution. This complements the intensity measurements of individual spectral lines obtained at millimeter wavelengths (albeit with larger beam sizes). The millimeter range is useful for polar species, and is less subject to spectral confusion for C-H bearing compounds as compared with infrared measurements. Ultra-high velocity resolution is easily achieved at millimeter wavelengths, permitting measurement of intensity profiles of individual spectral lines and thus providing direct measurements of the kinematics of the coma. These velocity fields are crucial to the retrieval of production rates from observations made with small beams in other spectral domains.

Similar major advances are expected when the sub-millimeter wavelength range becomes available for cometary studies, and indeed that field was recently opened with the detection of H_2CO in comet Levy at sub-millimeter wavelengths. Already the most sensitive probe of SO and of noble gases, ultraviolet observations can be expected to play an increasingly important role in direct detection of parent volatiles, as higher spectral resolution (e.g., with HST) and observations at extreme ultraviolet wavelengths become routine. Future remote sensing should provide new insights into the composition and processing history of individual comets, and will permit the development of a taxonomy for comets.

However, many key questions can only be addressed definitively by measurements made *in situ*. For the cometary nucleus, these include fundamental properties such as the porosity, tortuosity, and mean density; the surface thermal conductivity and its variation with depth; the degree of compositional differentiation with depth and with lateral position; the nature of the inactive surface material; and the detailed physical, thermal, chemical and mineralogical structure of the nucleus. Once these properties are known, the nature of the cometary accretion process and of subsequent evolutionary processes can then be inferred more clearly.

Microscopic investigations of individual cometary grains are required for many key investigations and these can best be accomplished from spacecraft near the cometary nucleus, or in terrestrial laboratories after return of cometary samples to the Earth. These studies would provide the relative abundances of pristine interstellar dust and of refractories processed in the inner solar nebula, key indicators of the history of cometary material prior to accretion of the cometary nucleus. Such direct studies could also establish the identities and abundances of pre-biotic chemicals, including (possibly) the simple amino acids, which are critical for understanding the role of comets in delivering these materials to the early Earth. In comets, most of these organics are locked in the grains, and cannot be identified remotely. The need for direct sampling and analysis can be appreciated by noting that more than 500 distinct organic compounds have been isolated from the Murchison meteorite alone. Direct sampling and analysis of cometary volatiles are also needed. Production rates for some of the more abundant volatile species can be obtained from remote sensing, but several of the more important ones are inaccessible (e.g., N_2) and many trace species will be missed entirely because of inadequate sensitivity of Earth-based instruments. While Earth-based studies may identify macroscopic chemical heterogeneities of the nucleus, their detailed spatial distribution on the nucleus can only be accomplished from a close-orbiting spacecraft which can sample and analyze the outflows from individual activity sites, and can supplement these with maps of surface composition.

There is also considerable interest in the physical processes that affect cometary material as it flows outward in the coma. The complex processes that occur there are still poorly understood. Direct measurements of parent molecules and their destruction products will help to clarify our understanding of the chemical kinetics and hydrodynamics of the coma, while measurements of plasma properties will lead to improved understanding of the interaction between the coma and the solar wind. This understanding can then be extended to remote observations of comets, leading to an improved ability to interpret the cosmochemical record encoded in them.

In-situ analyses will provide a major advance for both refractory and volatile material, but ultimately cometary material must be returned to Earth and subjected to a full suite of diagnostic investigations in terrestrial laboratories. These will provide an unbiased inventory of all significant constituents and will permit direct investigations of minor volatile and refractory components, including abundance measurements for isotopomers. Other key parameters will then be open to study, too. For example, radiation exposure ages of bulk samples may be compared with formation ages of processed refractories (obtained by rhenium-osmium dating, for example). In this way, it may be possible to date the epoch of nucleus formation, and to constrain accretion times for cometary nuclei during the formative phase of the solar system.

Several *in-situ* investigations of comets are planned. The Giotto spacecraft (Reinhard 1987) is scheduled to fly by a second short-period comet Grigg-Skjellerup on 10 July 1992. The surviving instrument complement (which does not include the multicolor camera) will emphasize characterization of the solar wind interaction with the cometary coma. The Comet Rendezvous Asteroid Flyby (CRAF) mission would place a spacecraft in orbit about a short-period comet and would conduct detailed investigations of the nucleus and coma over a period of three years, from its dormant phase near aphelion throughout its most active phase near perihelion (Weissman and Neugebauer 1992). CRAF was approved for a New Start by NASA in 1989, but is was terminated in 1992 for budgetary reasons. Given the potential benefits to studies of solar system origin, the loss of the cometary science results from the CRAF mission is a significant setback. Attention now is focusing on the Rosetta mission, in which an intact sample of cometary material would be returned to Earth for detailed analysis (Schwehm and Langevin 1991). Rosetta will seek to obtain a core sample from a cometary nucleus, along with samples of crustal material and perhaps a pristine sample from deep within the nucleus. Rosetta is now being studied jointly by NASA and ESA.

Future physical and dynamical studies will continue to test the various hypotheses of cometary formation. At present, the weight of evidence and physical plausibility favors cometary formation from relatively unmodified interstellar material, and the low temperatures inferred would place the region of accumulation in or just beyond the Uranus-Neptune zone of the solar nebula. However, it would be imprudent to extrapolate the findings obtained for a few comets to sweeping conclusions on the general cometary population. Remote compositional studies of many comets, combined with detailed direct examination (both *in situ* and through sample return) of cometary materials from a few, are needed to clarify the true formation site(s) and mechanisms of cometary nuclei, and to identify the evolutionary processes that affected this material before and after its incorporation into the cometary nucleus.

Acknowledgments. This work was supported by the Solar System Exploration Division, National Aeronautics and Space Administration, under grants from the Planetary Astronomy Program to M. J. Mumma and from the Planetary Geology and Geophysics Program to P. Weissman. A portion of this work was performed at the Jet Propulsion Laboratory under contract to NASA. We thank D. Bockelée-Morvan, J. Crovisier, M. A. DiSanti, S. Hoban, R. E. Johnson, D. Reuter, E. Roettger and H. A. Weaver for useful comments. We thank D. Brownlee and another referee for helpful comments.

MAIN-SEQUENCE STARS WITH CIRCUMSTELLAR SOLID MATERIAL: THE VEGA PHENOMENON

DANA E. BACKMAN
NASA Ames Research Center

and

FRANCESCO PARESCE
Space Telescope Science Institute

Solid grains with temperatures of 50 to 125 K and fractional bolometric luminosities ($L_{grains}/L_{\star}$) in the range 10^{-5} to 10^{-3} were found early in the IRAS mission around three nearby A main-sequence stars, α Lyrae (Vega), α Piscis Austrinus (Fomalhaut), and β Pictoris. Spatial resolution of the emission indicates that: (1) the grains are larger than interstellar grains; (2) the material probably lies in disks in the stellar equatorial planes; (3) the disks extend to distances of 100 to 1000 AU from the stars; and (4) zones a few tens of AU in radius around the central stars are relatively empty. The total mass included in the small grains in each case is only 10^{-3} to 10^{-2} $M_{\oplus}$ although more mass could be in undetectable larger bodies. Subsequent surveys of IRAS data reveal more than 100 main-sequence stars of all spectral classes having unresolved excesses with similar temperatures and fractional luminosities to the three prototypes. Some stars with excesses have estimated ages of 1 to 5 Gyr. Thus, main-sequence far-infrared excesses appear to be widespread and are present in systems old enough to be probably past the stage of active planet formation. The sizes of these disks correspond most closely to the hypothetical Kuiper disk reservoir of short-period comets that is supposed to lie in the ecliptic plane just outside the planetary region of our solar system.

I. INTRODUCTION

Expecting to obtain routine calibration observations of α Lyrae during the first months of 1983, IRAS science team members instead found a powerful source of far-infrared radiation an order of magnitude brighter than the emission from the star's photosphere at wavelengths of 60 and 100 μm. Turning to other stars, they discovered several more with similarly strong excesses. Initial studies of the infrared sources around α Lyrae (Vega) (Aumann et al. 1984), α Piscis Austrinus (Fomalhaut), and β Pictoris (Gillett 1986) resulted in the exciting realization that these objects were the first main-sequence stars apart from our Sun known to have attendant solid material in the absence of significant mass loss. Those studies made it clear that the emission was caused by thermal radiation from grains warmed by the respective stars to temperatures of about 50 to 125 K.

As a result of the IRAS announcements Smith and Terrile (1984) observed β Pic with a coronagraph also constructed for investigations of planetary rings and satellites (Vilas and Smith 1987). Their images revealed that the grains lie in an edge-on disk, considered strong evidence of a relationship between this object and a planetary system. It was initially thought that the material detected by IRAS might represent ongoing planet formation because these three stars are in spectral class A and have lifetimes of order 10^9 years or less, roughly corresponding to the era of planet formation and subsequent heavy bombardment in our solar system. However, investigators continuing to survey the IRAS data have found more than 100 additional examples of similar far-infrared excesses around main-sequence stars of all types and ages.

This review will cover relevant physics and astronomy background information in Sec. II, observations and models of the three spatially resolved examples in Sec. III, search methods and general properties revealed by surveys in Sec. IV, time scales and evolution in Sec. V, and plausibility of a connection between these objects and our solar system in Sec. VI. The latter section includes a comparison between the characteristics of the Vega phenomenon and possible properties of the Kuiper disk.

Several notes of caution are necessary:

1. Almost all the information about these systems comes from IRAS, supplemented in a few cases by groundbased infrared and sub-millimeter observations and in only one case (β Pic) by visual wavelength images. Thus, the available data are limited.

2. It is not clear whether the sources of far-infrared excess in the 100+ candidate systems are the same as in the three resolved prototypes despite similar excess color temperatures and fractional luminosities ($L_e/L_\star$). Until the candidates are resolved by present or future telescopes such as HST, ISO, SOFIA, or SIRTF, the physical mechanism remains uncertain. However, for the purposes of this review these numerous candidates will be considered as examples of the Vega phenomenon.

3. The properties of the three resolved examples which suggest planetary systems are inferred with varying degrees of certainty but are not directly observed. Those properties are: (a) grains larger than interstellar grains; (b) material orbiting rather than falling into or flowing away from the stars; (c) material arranged in disks in the stellar equatorial planes rather than, e.g., in spherical clouds; (d) central regions similar in size to the planetary region of our solar system which are relatively lacking in small grains; (e) a necessary reservoir of larger undetected parent bodies to resupply the presently observed small grain population against various rapid removal processes; (f) transient spectral absorption line events attributed to comets impacting β Pic. These inferences will be examined in later sections.

II. INFRARED EMISSION FROM STARS AND GRAINS

A. Expectations for Stellar Photospheres in the Infrared

The peak emission of even the coolest stars, $T_{eff} \sim 2500$ K, occurs at wavelengths shorter than 2.5 μm. The continuum flux densities of stars follow a Rayleigh-Jeans (RJ) slope ($F_\nu \propto \lambda^{-2}$) to about 10% accuracy from wavelengths of 3 to 30 μm and greater, with real stellar infrared continua generally being slightly steeper than RJ (see, e.g., Dreiling and Bell 1980; Engelke 1990 and references therein). Thus, it is relatively easy to estimate the far-infrared flux from a star given its flux at shorter wavelengths and to distinguish excess flux which is not photospheric. An excess with a given luminosity which itself has a spectrum like a Planck function will be more easily detected the cooler it is with respect to the star.

Determination of far-infrared photospheric and excess flux requires a choice of wavelengths at which a star's photosphere can be confidently considered the sole source of significant flux. Figures 1 and 2 compare the observed spectral energy distributions of four systems having strong far-infrared excesses with those of four stars which have little or no far-infrared excess but similar spectral types and similar near-infrared (1–5 μm) colors (see, e.g., Johnson 1966; Koornneef 1983a, b).

B. Thermal Emission from Grains

Temperatures of spherical grains warmed by stellar radiation can be calculated from equilibrium between absorbed and emitted energy:

$$g_1 \pi a^2 \left(\frac{R_\star}{r}\right)^2 \int_0^\infty \pi \epsilon_\nu B_\nu(T_\star) d\nu = 4 g_2 \pi a^2 \int_0^\infty \pi \epsilon_\nu B_\nu(T_g) d\nu \quad (1)$$

where a is the grain radius, r is the distance of the grain from the illuminating star, g_1 and g_2 are geometric factors depending on grain shape, conductivity, and rotation, and ϵ is the grain radiative efficiency. The spectrum of the star is assumed to be a Planck distribution with $T_\star$ equal to the effective temperature T_{eff}. If a grain is small or thermally conductive or rapidly rotating such that its entire surface is at the temperature T_g, then $g_1 = g_2$.

The efficiency parameter ϵ is crucial to consideration of the thermal equilibrium of solid material. Particles efficiently absorb and emit radiation with wavelengths smaller than the grain size, but not with wavelengths much larger than the grain size. For a given grain size a and ignoring spectral features of the grain material, ϵ can be considered roughly constant, $\epsilon \sim (1-\text{albedo}) \sim 1$, for absorption and emission of radiation of wavelengths shorter than a critical wavelength $\lambda_o \sim a$. Efficiency ϵ decreases for interaction with radiation of wavelength longer than λ_o.

The ratio between critical wavelength λ_o and grain radius, $\xi \equiv \lambda_o/a$, can vary widely depending on the grain optical properties. For example, $\xi \sim 2\pi$ for strongly absorbing material, while $\xi \sim 1/2\pi$ for weakly absorbing

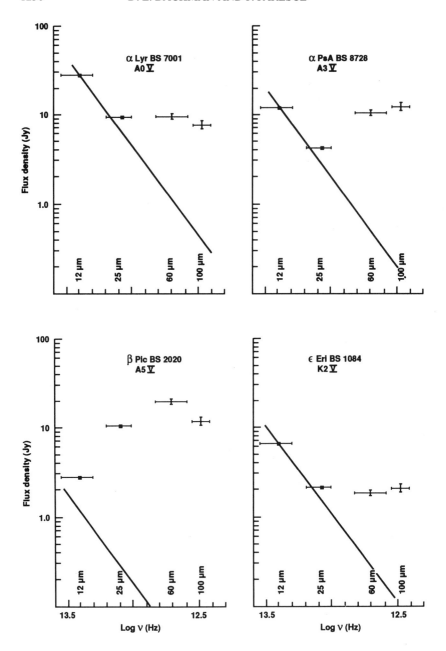

Figure 1.   Spectral energy distribution of four nearby main-sequence stars with strong far-infrared excesses detected by IRAS. The diagonal lines show Rayleigh-Jeans extrapolations of the 3.5 μm (L-band) photospheric flux densities. The four excesses have 25–100 μm color temperatures which imply flux densities at 3.5 μm of less than 10^{-10} of the photosphere. The flux density uncertainties include uncertainties in the absolute calibration (figure courtesy of F. C. Gillett 1986).

material. A value of $\xi \sim 1$ may correspond roughly to moderately absorbing dielectrics like "dirty ice" (see, e.g., Greenberg 1978).

The decrease in radiative efficiency for $\lambda > \lambda_o$ can be described by a power law in λ with exponent controlled by the grain properties. A slope $\epsilon \sim (\lambda_o/\lambda)^1$ is appropriate for amorphous materials whereas a steeper decrease $\epsilon \sim (\lambda_o/\lambda)^2$ corresponds to crystalline dielectrics and metals (see, e.g., review by Witt 1989). Grains in the interstellar medium (ISM) may be characterized by an exponent value of about 1.5 (see, e.g., review by Helou 1989).

Assumption of power law absorptive and emissive efficiencies allows an analytic solution to Eq. (1). Specifically, if, ϵ (absorp) is $\epsilon_o(\lambda_o/\lambda)^p$ and ϵ (emiss) is $\epsilon_o(\lambda_o/\lambda)^q$, then for simple geometry ($g_1 = g_2$) the grain temperature is:

$$T_g = \left[\frac{\gamma(4+p) \sum_{n=1}^{\infty} n^{-(4+p)}}{\gamma(4+q) \sum_{n=1}^{\infty} n^{-(4+p)}} \left(\frac{hc}{k\lambda_o} \right)^{q-p} \frac{L_\star}{16\pi\sigma_{SB}r^2} T_\star^p \right]^{\frac{1}{4+q}} \qquad (2)$$

where σ_{SB} is the Stefan-Boltzmann constant.

Grain size has a strong effect on the grain temperature. Arbitrarily large (blackbody) grains with efficient emission at all wavelengths will have the lowest temperature at a given location, while very small grains (like interstellar grains) with inefficient emission at all relevant wavelengths will have the highest temperature.

The most important consequences of this regarding the "Vega phenomenon" are: (1) infrared emission can be proven to be due to grains illuminated by the star in question if the temperature and spatial scale of the emission are consistent, and (2) an estimate of the typical grain size can be made from measurements of the emission temperature and scale. For example, if the grains are much larger than the peak wavelengths of both the absorbed and emitted spectra (blackbody grains), then $p=q=0$ and Eq. (2) reduces to:

$$T_g = 278 \, L_\star^{\frac{1}{4}} \, r_{AU}^{-\frac{1}{2}} \text{ K} \qquad (3)$$

where $L_\star$ is in units of $L_\odot$, 3.9×10^{33} erg s^{-1}. The temperature can also be calculated using the source *angular* scale and the *apparent* brightness of the central star. For example, Eq. (3) for blackbody grains can be rewritten as:

$$T_g = 88 \, \theta^{-\frac{1}{2}} \, 10^{0.10(4.75-m_v-B.C.)} \text{ K} \qquad (4)$$

where θ is the grain distance from the star in arcsec, m_v is the star's visual magnitude, and B.C. is the bolometric correction in magnitudes for the star's type (see, e.g., Allen 1976).

A grain much larger than the peak wavelength of the incoming stellar radiation but much smaller than the peak wavelength of the grain thermal emission would have efficient absorption ($p = 0$) but inefficient emission ($q > 0$). If $p = 0$ and $q = 1$, one obtains:

$$T_g = 468 \, L_\star^{\frac{1}{5}} \, r_{AU}^{-\frac{2}{5}} \, \lambda_o^{-\frac{1}{5}} \text{K} = 186 \, \theta^{-\frac{2}{5}} 10^{0.08(4.75-m_v-B.C.)} \, \lambda_o^{-\frac{1}{5}} \text{ K} \qquad (5)$$

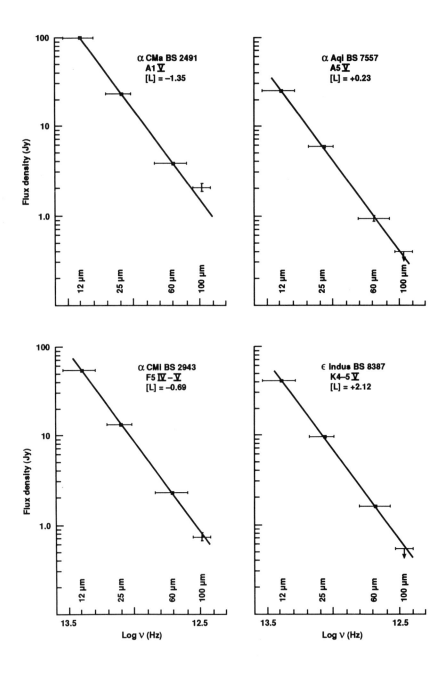

Figure 2. Comparison of spectral energy distributions of four stars of similar spectral types to those in Fig. 1, but with little or no far-infrared excess emission.

where λ_o is in μm.

ISM grains are small enough that they both absorb and emit stellar radiation inefficiently as well as emit inefficiently. Assuming $p=q=1.5$, one obtains:

$$T_g = 636 \, L_\star^{\frac{2}{11}} \, r_{\mathrm{AU}}^{-\frac{4}{11}} \left(\frac{T_\star}{T_\odot}\right)^{\frac{3}{11}} \mathrm{K}$$

$$= 275 \, \theta^{-\frac{4}{11}} \, 10^{(0.80/11)(4.75-m_v-B.C.)} \left(\frac{T_\star}{T_\odot}\right)^{\frac{3}{11}} \mathrm{K} \qquad (6)$$

where $T_\odot$ is 5770 K.

The temperatures described in Eqs. (1)–(6) are physical temperatures of the grain material. The received grain emission spectrum will be affected by the radiative efficiency for $\lambda > \lambda_o$ such that:

$$F_\nu = \epsilon_\nu B_\nu \left[T_g(r,a)\right] \Omega = \left(\frac{\xi a \nu}{c}\right)^q B_\nu \left[T_g(r,a)\right] \Omega \qquad (7)$$

where ξa represents the critical wavelength λ_o and Ω is the solid angle subtended by the total of the geometric cross sections of the grains.

Assuming an optically thin disk and single-size grains, the flux density received at the Earth from a system of grains is:

$$F_\nu = 2\pi \times 10^{23} \int_{r_1}^{r_2} \sigma_o \left(\frac{r}{r_o}\right)^\gamma \left(\frac{\xi a \nu}{c}\right)^q B_\nu \left[T_g(r,a)\right] \frac{r \, dr}{D^2} \, \mathrm{Jy} \qquad (8)$$

in Janskys [$= 10^{-26}$ W m^{-2} (Hz bandwidth)$^{-1}$], where $B_\nu[T_g(r,a)]$ is in cgs units, D is the distance to the system, and the amount of material in the disk is expressed as a face-on fractional geometric surface density $\sigma(r)$, cm^2/cm^2, following a power law in radius with exponent γ. The face-on optical depth for such a disk would be $\tau_\perp(r,\nu) = \sigma(r)(\xi a \nu/c)^q$. The volume density distribution producing a spectrum would be $n(r) \propto r^{\gamma-1}$.

Equation (8) can be converted to an expression which allows integration over temperature rather than radius, concentrating most of the "modeling power" in those regions where the temperature and thus the thermal emission is largest (Artymowicz et al. 1989):

$$F_\nu \propto \int_{T_{\min}}^{T_{\max}} \epsilon_\nu B_\nu(T_g) y(T) T^{-\kappa} \frac{dT}{D^2} \qquad (9)$$

where $\kappa = 4$ for blackbody grains (Eq. 3) or $\kappa = 3$ for "intermediate" sized grains (Eq. 5), and $y(T(r)) \propto r\sigma(r)$.

A spectrum falling more steeply than an RJ slope at long wavelengths may indicate the grain spatial distribution rather than a limit on grain size. Conversely, for some spatial distributions an RJ slope can be produced by an ensemble of grains substantially smaller than the wavelengths of observation.

Combining Eq. (3) for blackbody grains with Eq. (8), assuming $r_1 = 0$ and $r_2 = \infty$ but also assuming that total optical depth along the disk plane is small, there are exact solutions for the spectral energy distribution if $\gamma > -2$. These have the form:

$$F_\nu \propto \lambda^{2\gamma+1}. \tag{10}$$

A spectrum declining faster than an RJ slope results from a disk spatial distribution with $\gamma < -1.5$ even if the grains are assumed arbitrarily large. Note that this is an idealization since total grain area, mass, and optical depth actually diverge for an infinite distribution with $\gamma > -2$. A realistic spatial distribution would have finite inner and outer radii (the inner radius perhaps fixed by the temperature of grain evaporation) and would therefore emit a spectrum falling below the simple power law at both short and long wavelengths.

III. RESOLVED EXAMPLES: α LYR, α PSA, AND β PIC

Examples of the "Vega phenomenon" which have been definitely spatially resolved are the regions around the three stars α Lyr, α PsA and β Pic, although Aumann (1991) may have found a measurable extent for ϵ Eri as well. The three resolved systems deserve special attention because resolution allows structural models and discussion of origins and evolution. This section will discuss observational data, models and interpretations of these systems.

Characteristics of the stars are shown in Table I. All three systems are well within the range of accurate trigonometric parallax distance determinations. A comparison of the visual/near-infrared colors of these stars ("obs": Koornneef 1983a; Campins et al. 1985) to the average colors for main sequence stars of their spectral types ("std": Koornneef 1983b) in columns 7 and 8 demonstrates that these three have no significant near-infrared excesses. The expected main sequence lifetimes in the last column corresponding to the estimated masses are interpolated from Iben (1967).

A. General Analyses of the "Big 3"

1. IRAS Discovery. Table II presents the photospheric and excess flux densities in the IRAS bands from "pointed" observations (Gillett 1986). The uncertainties include estimates of absolute calibration uncertainty; the relative photometric uncertainties are only a few percent. The 12 μm IRAS measurements are close to photospheric for α Lyr and α PsA but β Pic has a significant excess even at that wavelength.

The possibility that the observed excesses include substantial line emission can be rejected. The infrared continuum of α Lyr is smooth across the range of wavelengths (≤ 13 μm) over which it has been carefully compared to standards with no far-infrared excess (Witteborn and Cohen, personal communication). Reports of weak visual wavelength emission lines and variability of α Lyr (Johnson and Wisniewski 1978, and references therein) have not been confirmed and are likely due to difficulties in observing such a bright object

TABLE I
Stellar Characteristics

	BS #	Spectral Type	T_e	m_v	B.C.	obs $V-L$	std $V-L$	dist	$L_\star$	$M_\star$	Expected Lifetime
α Lyr	7001	A0V	9700 K	+0.02	−0.40	0.00	0.00	8.1pc	60L$_\odot$	2.5M$_\odot$	4×10^8 yr
α PsA	8728	A3V	8800	+1.15	−0.23	+0.15	+0.21	7.0	13	2.0	7×10^8
β Pic	2020	A5V	8200	+3.85	−0.12	+0.38	+0.36	16.4	6	1.5	2×10^9

TABLE II
Photometric Characteristics

	IRAS Flux Densities, Janskys				Single temperature Fit to Spectrum	
	12 μm excess (photosphere)	25 μm excess (photosphere)	60 μm excess (photosphere)	100 μm excess (photosphere)	$T_{B.B.}$	Bolometric $f \equiv L_e/L_\star$
α Lyr	0.0 ± 0.8 (28.0)	2.75 ± 0.5 (6.45)	8.2 ± 0.5 (1.1)	7.1 ± 0.8 (0.4)	89 K	2×10^{-5}
α PsA	0.3 ± 0.8 (11.3)	1.4 ± 0.2 (2.6)	9.35 ± 0.5 (0.45)	11.1 ± 1.1 (0.2)	72	8×10^{-5}
β Pic	1.6 ± 0.1 (1.2)	10.1 ± 0.5 (0.3)	18.8 ± 0.9 (0.05)	11.2 ± 1.0 (0.02)	108	3×10^{-3}

optically and finding suitable reference standards. Recent 1-μm-resolution spectrophotometry of β Pic near 10 μm shows that a small amount of silicate line emission is present (Telesco and Knacke 1991).

The excess flux densities at 25 to 100 μm can be fit fairly well in all three cases by Planck distributions with single temperatures listed in column 6 of Table II. A single-temperature fit implying a single radial location for the material is of course only schematic. As discussed previously, many combinations of radial density and grain size distributions can produce a spectrum which resembles portions of a Planck distribution. In the case of β Pic, accounting for all of the 12 μm excess (excluding the weak line emission) in the simplest way requires addition of a second hotter component.

Fractional luminosities f of the excesses relative to the stars are presented in the last column of Table II. The optical depth of grains along a radial line from the star to infinity is $\tau \sim f/\sin i$ if the grains are in a "wedge" or "flaring" disk with thickness proportional to radius and opening angle $2i$. The range of fractional luminosities for these three stars is surprising: f for β Pic is more than 1 order of magnitude larger than for α PsA and 2 orders larger than for α Lyr. Note that the fractional luminosity of the material around β Pic is 4 orders of magnitude larger than for the zodiacal dust in our solar system (Sec. VI).

IRAS profiles of these three sources show widths that are significantly wider than profiles of point sources. Figure 3 compares the average of 12 scans at 60 μm across α PsA, the largest of the three resolved sources, with a point source profile which is a combination of 24 scans of α Boo and β Gru.

Table III presents intrinsic 60 μm source sizes for the three resolved cases estimated from simple deconvolution (Gillett 1986). Note that these are characteristic scales of the strongly emitting regions but are not limiting outer diameters of the sources; substantial amounts of cooler material may lie at larger distances. In the cases of α Lyr and α PsA the sizes given are the largest of values measured along several scan position angles. These can be used to estimate diameters of circular figures seen in projection as ellipses (Sec. III.A.6). In the case of β Pic the infrared source sizes along two orthogonal directions have been de-projected using the disk orientation angle seen in optical images (Sec. III.B.1).

2. α Lyr as Prototype: Large Grains In Orbit. Explanations for infrared excesses other than an optically thin collection of orbiting grains can be eliminated by arguments from the first α Lyr IRAS paper (Aumann et al. 1984), *which also should be considered for all candidate analogs to α Lyr found in IRAS data.* Note that points (c), (d) and (e) depend on successful spatial resolution of the infrared emitting region.

(a) Plasma (free-free) emission, e.g., from a hot wind: the excess is well fit by a Planck function and rises in flux density between 12 and 60 μm, inconsistent with a thin or partially thick free-free spectrum which would have $f_\nu \propto \lambda^{-2}$ to λ^0;

(b) Chance alignment with background interstellar cirrus: the source

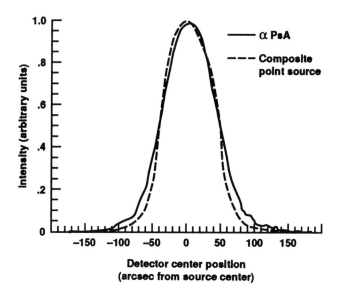

Figure 3.   Comparison of IRAS 60 μm scan profiles of α PsA, the most extended of the "Big 3," with a composite of several point sources. The scans were made at 1/2 the all-sky-survey scan rate with a sampling interval of 3.6 arcsec. Scan widths represent a combination of detector FOV, detector response time, telescope diffraction and intrinsic source width; the FWHM is equal to the 90 arcsec detector width parallel to the scan direction ("in-scan"). The profiles are normalized to 1 at their peaks, which makes the broader source appear narrower above the FWHM points.   Irregularities in the α PsA profile wings indicate the noise level (data provided courtesy of F. Low, F. Gillett, G. Neugebauer, J. Good and H. Aumann).

temperature of 89 K is much warmer than even unusually warm cirrus knots (Low et al. 1984);

(c) Chance alignment with an 89 K infrared object: the 60 μm source scale of 29 arcsec is quite consistent with heating of grains many μm in size by the star (Eqs. 4 and 5) and the centroid of the far-infrared source is found to be within 2 to 3 arcsec of the star's position, strongly indicating that the emission must be from grains around α Lyr;

(d) A single optically thick blackbody, either a cool self-luminous companion or a single object heated by Vega: its diameter from the flux and temperature would be 0.2 arcsec, much smaller than the observed diameter of the source;

(e) Emission from ordinary circumstellar/ISM grains around α Lyr with efficiencies approximately proportional to $\lambda^{-1.5}$: the diameter of such a shell with the observed color temperature would be almost 2000 arcsec (Eq. 6);

(f) Emission from ordinary grains in a shell around a low-luminosity companion close to the line of sight to the primary star: based on normal ISM grain efficiencies, the shell optical depth would be 4×10^{-3} and the luminosity

TABLE III

Source Sizes, Estimated Grain Sizes, Total Area, and Mass

	Source Diameters, Arcsec			Single-Size Grain Models			
	Observed 60 μm	Hypothetical B.B. Grains	ISM Grains	Grain Size, λ_o, μm	Total Grain Area, cm^2	Minimum Mass, $M_\oplus$	"Maximum" Mass, $M_\oplus$
α Lyr	29	21	1800	80	1×10^{27}	2×10^{-3}	9×10^2
α PsA	36	17	1300	27	3×10^{27}	2×10^{-3}	2×10^3
β Pic	26	1	180	1	3×10^{29}	7×10^{-3}	4×10^4

of the companion would be 0.4 $L_\odot$. Such a companion would be detectable as a close optical double, but none is observed (see, e.g., Hanbury Brown et al. 1974);

(g) Large grains condensing in a stellar wind flowing from Vega: a mass-loss limit of $\dot{M} < 3.4 \times 10^{-10}$ $M_\odot$ yr^{-1} was obtained for α Lyr from radio and ultraviolet measurements (Hollis et al. 1985, and references therein). Even with optimistic assumptions about the carrying capacity of a wind with α Lyr's metallicity, i.e., all the elements except H and He condensed into grains and a wind speed just above escape velocity, this upper limit on mass loss precludes enough mass in a continuous spherically symmetric wind to be carrying the minimum grain mass of 10^{-8} $M_\odot$ (Table III) across the indicated range of radii. The crucial conclusion that the grains must be in orbit around the star rests on (1) the impossibility of interstellar grains being found at the observed radii and temperatures, and (2) inconsistency of mass loss limits with the amount of solid material observed.

Regarding mass loss from the other prototypes: β Pic's shell is like typical A–F shells in showing no overt signs of mass loss like asymmetric or violet-displaced lines (Slettebak and Carpenter 1983), but precise evaluation of mass-loss limits for β Pic is complicated by transient absorption line events (Sec. III.B.5). α PsA has ultraviolet absorption lines which may be from circumstellar gas (Kondo and Bruhweiler 1985) although an analysis of optical Na I and Ca II lines (Hobbs et al. 1985) indicates otherwise. Assuming for a limit that all of the column density tabulated in Bruhweiler and Kondo (1982) represents material in a spherically symmetric wind with solar abundance, fractional ionization $X(Mg^+) = 0.01$, v_∞ of 15 km s^{-1}, and purely gravitational deceleration, then the mass-loss rate from α PsA is less than 10^{-9} $M_\odot$ yr^{-1}.

3. Grain Sizes and Total Grain Mass. Columns 3 and 4 in Table III show the source sizes expected from the source temperatures if the grains are blackbody grains (Eq. 4) or ISM grains (Eq. 6). The actual grains in all three cases can be judged to be smaller in size than 100 μm but also clearly larger than ISM grains because the observed source sizes lie between these extremes. The grains around α Lyr appear to be the largest, approaching blackbody size, while the grains around β Pic appear to be relatively small.

Typical grain sizes can be estimated by assuming the grains are inter-mediate-sized ($\epsilon < 1$ across the IRAS bands) and calculating a best fit to the observed source size and 25 to 100 μm flux densities. Resulting grain size parameters ($\lambda_o \sim a$) are presented in column 5 of Table III. The total grain areas in column 6 of Table III are geometric cross sections. The values of grain size and total area depend on the assumed value of emissivity exponent q but the relative ranking of the three systems is preserved if the same q is used for each. Note that the amount of material is anti-correlated with the mass and luminosity of the central stars.

The quality of the fits of these single-size grain models to the observed spectra are not great. It is more likely that the grains have a range of locations and sizes, but this exercise gives some indication of effective or typical grain sizes. The calculated grain size for α Lyr for $q = 1$ appears to be nearly large enough to violate the assumption implicit in Eq. (5) that most of the emission takes place at wavelengths larger than λ_o. A solution for α Lyr with blackbody grains yields nearly the same grain area.

The mass associated with the circumstellar grains is of primary importance and unfortunately is also uncertain. The grain size distribution slope and maximum sizes are not constrained by the IRAS observations. Minimum total masses derived with the assumptions that all the grains have size $a = \lambda_o$ and density 1 g cm^{-3} are presented in column 7 of Table III (1 M$_\oplus$ = 3 × 10^{-6} M$_\odot$ = 6 × 10^{27} g). These masses are quite small; note especially in the case of α Lyr that 2 × 10^{-3} M$_\oplus$ of 80 μm grains can produce a far-infrared signal which overpowers the star's flux and is easily detectable at stellar distances. In the case of β Pic the lower limit to the grain mass is approximately consistent with solar composition and an upper limit on neutral gas mass (Sec. III.B.4).

A maximum grain mass presented in the last column of Table III can be estimated by assuming: (1) collision fragmentation equilibrium size distribution $n(a) \propto a^{-3.5}$ (Dohnanyi 1969); (2) minimum size $a_{\min} = \lambda_o$; (3) maximum size $a_{\max} = 1000$ km estimated from the gravitational instability in a thin dust disk at 100 AU (Goldreich and Ward 1973; Greenberg et al. 1984); and (4) density of 5 g cm^{-3}. These values are larger than the total mass of the planets in our solar system (450 M$_\oplus$) and scale with the assumed maximum size as $a_{\max}^{1/2}$. The grain upper mass limit for β Pic is 0.1 M$_\odot$. The grain population that contains most of the surface area and is most easily detected does not represent the population containing most of the mass in this distribution: half of the total area is contained in grains smaller than 4 times the minimum size, while half the mass is contained in grains larger than 1/4 the maximum size.

Note that the deduced maximum mass is quite sensitive to the size distribution exponent: assuming a power law a^{-4} rather than $a^{-3.5}$ produces "maximum" mass estimates smaller by a factor of 10^4. Grain size distributions caused by mechanisms differing from collisions, such as sublimation of icy bodies, seem to give size distributions differing from $a^{-3.5}$ in our solar system (see, e.g., Kyte and Wasson 1986).

4. *Inner and Outer Boundaries.* An important fact provided by the spectral energy distributions of these objects is that they have relatively empty central regions (Gillett 1986). This conclusion is based on consideration of maximum grain temperature determined by minimum wavelength of excess flux. If in Eq. (8) the outer boundary radius $r_2 \to \infty$, then the entire disk emission can be characterized by the value of the inner boundary radius r_1. Table IV presents estimates of inner boundary radii for each of the three systems constrained by the hottest IRAS color temperature and assuming grain sizes from Table III. The solutions differ significantly depending on whether a flat or steep spatial distribution is assumed. Only in the case of β Pic is there independent information about the actual spatial gradient ($\gamma \sim -1.7$ at r greater than about 65 AU; Sec. III.B.1).

TABLE IV
Inner Boundary Radii

	γ	r_1	$T_g(\text{max})$
α Lyr	0.0	—	—
	−1.7	26	119
α PsA	0.0	30	104
	−1.7	67	75
β Pic	0.0	20	199
	−1.7	38	154

These calculations show that the grain number density in these three systems cannot increase monotonically toward the stars or even be constant with radius. Instead, there must be central regions of relatively low density. These central "voids" have sizes similar to each other and also similar to the planetary region of our solar system. Note that this analysis finds that the amount of material at $r < r_1$ was undetectable with IRAS but it may not be zero, and there is not necessarily a sharp boundary at r_1. One conclusion of Aumann et al. (1984) was that a hypothetical hot grain component around α Lyr with $T = 500$ K could have up to 10^{-3} of the grain area in the 90 K component and not violate the limit on excess at 12 μm. Because PR radiation drag (Sec. V.B.2) would act quickly to fill in the voids, grain removal mechanisms such as ice sublimation, radiation pressure expulsion of small collision fragments, or gravitational perturbations are probably required to prevent a warm dust signature. Better photometry at infrared wavelengths much shorter than the emission peaks would further define the inner structure of these systems and the processes controlling the location of the grains.

The ability of infinite spatial distributions to yield positive or negative spectral slopes has been discussed. The IRAS data therefore do not allow specification of *outer* boundary radii for these systems. IRAS sensitivity to outlying material would have been limited by detector field of view and grain temperature. The IRAS field of view at 100 μm corresponds to roughly 1500 by 2500 AU at the distance of α Lyr and 3000 by 5000 AU at β Pic. Material colder than about 30 K has an emission peak beyond the long wavelength cutoff of the IRAS 100 μm band and would be difficult to detect without substantial optical depth. That temperature corresponds to distances of about 1000 AU from α Lyr and 2300 AU from β Pic for the model grain characteristics in Table III.

5. *Is the β Pic System Really a Disk?* We are likely to be viewing the β Pic system from near the star's equatorial plane because the star's projected rotation velocity is above the mean for main-sequence stars of its spectral type (Table V). Is it possible that the linear structure seen in the coronagraph images is not a disk of orbiting grains? Could it instead

TABLE V

Stellar Rotation and IRAS Source Orientation

	Mean v sin i for Spectral Type, km s^{-1}	Observed v sin i, km s^{-1}	Apparent Grain Disk Inclination, i, degrees
α Lyr	145	15	30^{+15}_{-30}
α PsA	135	100	>70
β Pic	125	139	>80

represent equatorial mass loss or even bi-polar jets emanating from a post-main-sequence star as Herbig (1989b) speculated based on the disk optical appearance and Slettebak's (1975) classification of the central star as a sub-giant? The argument that the observed system must actually be a disk of orbiting material is provided by Paresce and Artymowicz (1989), which we summarize:

(a) An equatorial flow capable of reaching the great distances of the observed disk can be rejected because the escape velocity from the β Pic atmosphere is about 500 km s^{-1} but the observed rotation rate is 139 km s^{-1};

(b) The bipolar jet hypothesis can be rejected because there is little resemblance between bipolar nebulae and the β Pic disk in terms of scale, mass, grain temperature, expansion velocity, lifetime, or molecular content;

(c) The central "void" would require that the grains condense at much colder temperatures than in all other known examples of grain condensation in stellar winds;

(d) Such a jet would need to be directed by another substantial disk that would necessarily also lie close to our line of sight to the star, yet no such disk is observed;

(e) There are no obvious spectroscopic signs of mass loss (Slettebak and Carpenter 1983);

(f) The surface brightness declines rapidly as r^{-4} or steeper, but material on an escape trajectory decelerating gravitationally and presenting a "wedge" in projection (Sec. III.B.1) would instead have scattering surface brightness following $r^{-2.5}$ (r^1 increasing viewing path length $\times$ $r^{-1.5}$ volume density decrease $\times$ r^{-2} decrease of input photons).

6. *Source Morphologies and Stellar Rotation.* Previously described analyses by Gillett (1986) of IRAS 60 μm in-scan and cross-scan profiles of α Lyr yields minimum and maximum source scales of 25 and 29 arcsec, interpreted as roughly FWHM of the surface brightness distribution. Because these scans were made along a number of position angles, it appears that this source is nearly circular in projection. The same analysis for α PsA yields a scale of 36$\pm$2 arcsec along position angle 29° W of N, but less than 13 arcsec perpendicular to that direction. In this case there is only one scan axis and therefore size estimates only along two position angles. The actual source shape for α PsA could be between two extremes: (1) elliptical, with axial

ratio less than 13/36 and major axis along the IRAS scan direction, or (2) a one-dimensional source lying along a position angle 10° E of N.

The IRAS spatial information on β Pic is of lower quality but is consistent with a one-dimensional source oriented like the optical disk. A careful analysis of the scattered light isophotes (Artymowicz et al. 1989) shows that the disk inclination to the plane of the sky is greater than 80° (Sec. III.B.1).

These source shapes provide crucial information about the true location of the grain material relative to the rotation axes of these stars. Table V compares the observed (projected) stellar rotation velocities (Uesugi and Fukuda 1970) with the average (projected) rotation velocities for stars of the same spectral types (Schmidt-Kaler 1982, and references therein). This is statistical evidence that the star α Lyr is nearly pole-on to us, that β Pic is nearly equator-on to us, and that α PsA lies between those extremes. There is also independent evidence from detailed spectral line analysis (Gray 1986) that what we see in the case of α Lyr is the pole of a rapidly rotating star.

The inclination angles of disks which would result in elliptical projections with the observed aspect ratios are in the last column of Table V. This comparison of the rotational orientation of the three stars with the projected shapes of the regions containing the grains is the basis for concluding that the material is probably arranged as disks in the stellar equatorial planes.

7. IRAS Followup. A number of investigators have extended the original IRAS results on source sizes and spectral energy distributions for the three resolved systems. Harvey et al. (1984) confirmed the α Lyr excess from the Kuiper Airborne Observatory (KAO) and measured a source size at 47 and 95 μm consistent with, but marginally larger than, the IRAS values. Lester et al.'s (1990) KAO 100 μm characteristic size for α PsA is in agreement with Gillett (1986) along the major axis, and they find evidence for some emission at least 240 AU (35 arcsec) from the star. Aumann (1991) re-analyzed IRAS slow-scan observations of α Lyr, α PsA, β Pic, and other stars assuming Gaussian intrinsic source profiles and found sizes like Gillett's (1986) except an upper limit twice as large as Gillett's for α PsA's minor axis. Early KAO photometry by Harper et al. (1984) showed flux density from α Lyr at 193 μm well below an extrapolation of the IRAS 25 to 100 μm color temperature. Becklin and Zuckerman (1990) found a similar result for α Lyr and β Pic (but not α PsA) at 450 and 800 μm using the 15 m James Clerk Maxwell Telescope (JCMT). They also noted that the excess flux from α PsA was stronger at an offset of 16 arcsec (one beamwidth) from the star than in a beam centered on the star, which they interpreted as direct confirmation of the central "void" inferred from the spectrum. Becklin and Zuckerman modeled the emission from the three systems as due to emission from material at one temperature (i.e., no spatial distribution) and thus interpreted their and Harper et al.'s flux-density shortfalls relative to IRAS extrapolations as evidence of: (1) smaller grain sizes ($a < < \lambda_{obs}$); and (2) lower total dust masses, than previously published. However, rapidly declining long-wavelength spectra can also be produced by some spatial distributions (Sec. II.B). Chini et al. (1990) used the

30 m Instituto de Radioastronomia Millimetrica (IRAM) telescope and found excesses relative to the stellar photospheres which decrease as $\lambda \rightarrow 1$ mm, but claim an inconsistency exists between their beamsize/flux density/source size information and that of Becklin and Zuckerman (1990). Chini et al. find their results in agreement with the IRAS source sizes.

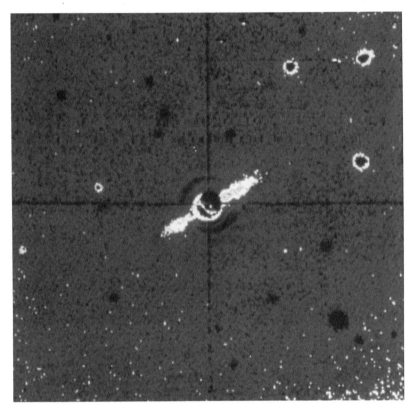

Figure 4.   Red (0.89 μm) CCD coronagraph image of the β Pic disk from the 100-inch telescope at Las Campanas. The extent of each wing of the disk is roughly 25 arcsec or 400 AU. Processing involved use of comparison images of α Pic to remove the profile of light scattered in the atmosphere and instrument (figure from Smith and Terrile 1984).

B. Detailed Analyses of β Pic

It has already been noted that β Pic possesses the warmest and proportionately most luminous infrared excess of the "Big 3". Following the IRAS discovery Smith and Terrile (1984) used a CCD coronagraph to detect 0.89 μm (I-band) light scattered by grains in a nearly edge-on disk around the star (Fig. 4). β Pic is the only example which has been so far detected optically because it is by a wide margin the densest main-sequence grain disk within 20 pc (Sec. IV.A.4) and is fortuitously inclined so as to enhance the surface brightness by roughly an extra factor of 10. β Pic is also nearly unique among these systems in that it exhibits substantial excess emission at wavelengths as short as 10 to 20 μm (Aumann and Probst 1991). β Pic is therefore the object in this class about which most is known and considerable discussion of its properties is warranted.

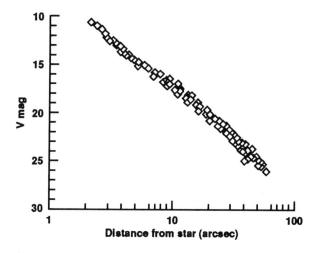

Figure 5.   Radial surface brightness distribution in V band of the β Pic disk midplane, both wings plotted together. (Smith and Terrile, unpublished data.)

1. Disk Structure: Optical. The disk was visible beyond the edge of Smith and Terrile's coronagraph mask at $r \geq 6$ arcsec (100 AU) extending about 25 arcsec (450 AU) to the north-east and 20 arcsec to the southwest along a position angle of 30° (Fig. 4). The disk is resolved along its short dimension, with a seeing-deconvolved thickness of 3 arcsec (50 AU) at $r = 350$ AU. The I-band surface brightness is 16th magnitude per square arcsec at 100 AU, decreasing approximately as $r^{-4.3}$.

Subsequent images (Smith and Terrile 1987) show the disk apparently extending inward to radii of a few arcsec and outward to 60 arcsec (1100 AU). The most complete surface brightness profile combining the two "wings" is shown in Fig. 5. The surface brightness fades below present detectability beyond 1100 AU but the disk may well extend farther.

Smith and Terrile's (1984) Fig. 2 implies that there is an asymmetry of the disk involving truncation of the southwest wing and increasing thickness of that wing near its outer terminus. A brown dwarf companion for β Pic was suggested (Whitmire et al. 1988) as a possible cause for such a disturbance of the disk. However, Paresce and Burrows (1987) found no systematic asymmetry in the disk structure in their optical coronagraph images. In fact, asymmetries with both the same *and opposite* sense of that originally proposed

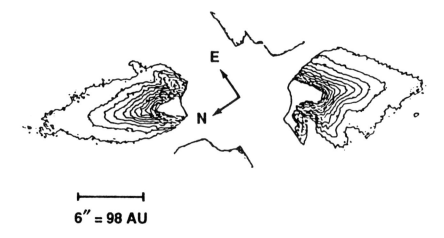

6″ = 98 AU

Figure 6.   Contour plot of the β Pic disk surface brightness at 0.79 μm. The isophotes are most easily modeled as a "wedge" disk (thickness increasing with radius) convolved with atmospheric seeing (figure from Artymowicz et al. 1989).

can be easily produced by small registration errors in the nontrivial process of reducing coronagraph images using reference star images.

Model pertaining to the specific radii range of 6–15 arcsec (100–250 AU) show that the likeliest structure capable of producing the observed elliptical isophotes (Fig. 6) is a "wedge" disk having an opening angle of about 14° and a presentation angle (versus exactly edge-on) of 10° or less, convolved with terrestrial seeing (Artymowicz et al. 1989). Those model calculations show that the isophotes for a constant thickness disk would be quite different and easily distinguished from a wedge and that the symmetry of the surface brightness distribution is consistent with a scattering phase function similar to that for zodiacal dust (see especially Artymowicz et al. Figs. 1 and 3–7). It is more difficult to distinguish between model isophotes of a wedge versus a very thin, inclined disk. However, the latter hypothesis is rejected because it would require an unlikely scattering phase function more isotropic than Lambert's law.

Given a slightly inclined wedge geometry, the optical depth *perpendicular to the disk plane* is deduced to follow approximately $r^{-1.7}$ for $r \geq 65$–100 AU (Artymowicz et al. 1989,1990). This corresponds to a mid-plane grain volume number density declining approximately as $r^{-2.7}$.

2. Models Combining Optical and Infrared Data. The area corresponding to the infrared luminosity of the entire disk is within a factor of two of the scattering area observed outside coronagraph masks (Artymowicz et al. 1989; Backman et al. 1992). The total grain surface cross-sectional area is related to the scattering area σ_s, the infrared (absorption/emission) area σ_i, and the albedo $\mathcal{A}$ via:

$$\sigma_{total} = \sigma_s + \sigma_i = \sigma_s/\mathcal{A}. \tag{11}$$

Plausible albedos imply there cannot be much more grain area hiding within the regions obscured in coronagraph images.

The grain temperature in 4 and 8 arcsec fields of view centered on the star is 180 K from observations at 10 and 20 μm if radiative efficiency is assumed to be $\epsilon \propto \lambda^{-1}$ (Telesco et al. 1988; Backman et al. 1992). Because the temperature on these scales is not growing rapidly with decreasing aperture size, one can estimate that the maximum temperature of detectable material is $T_{max} \sim 200$ K, which corresponds quite well to the limits in Table IV based on IRAS data.

Models attempting to match both infrared data and results of analyses of optical observations have been calculated by Diner and Appleby (1986), Nakano (1988), Artymowicz et al. (1989) and Backman et al. (1992). All combined optical-infrared models agree on a relatively flat spatial distribution for the grains producing the bulk of the grain thermal luminosity. These models in general show:

1. An outer component corresponding to the disk visible in coronagraph images, extending from $r \leq 100$ AU to $r \geq 1000$ AU with mid-plane grain number density decreasing steeply as $n \sim r^{-3}$; this component emits most of the radiation detected at $\lambda > 20 \mu$m;
2. A transition region extending approximately from $r = 10$–100 AU as yet hidden from easy view in coronagraph images, with a lower density and either smaller grains or a less steep spatial gradient than the outer component; most of the radiation at $\lambda < 20 \mu$m originates from this region;
3. An innermost region with grain area below present detection limits.

Details of the structure in the region hidden by coronagraph masks such as the sizes and properties of the transition region and innermost "void" are not uniquely determined by the available infrared observations (see, e.g., Diner and Appleby 1986). The grain cross-section area in these regions is as little as 1% of the amount there would be if the outer component gradient continued to $r = 0$. The face-on optical depth in the inner component is of order 10^{-4} in contrast to a maximum in the outer disk of a few times 10^{-3}. The physical mechanisms that may contribute to the relatively low density of material in the inner regions of the β Pic disk are discussed in Secs. V.B and C.

The midplane visual extinction Δm based on assumption of a "wedge" structure and following the definitions in equation (8) is:

$$\Delta m_\lambda \sim 1.1\tau_\lambda \sim 1.1 \int_{r_1}^{\infty} \frac{\sigma_o}{\theta r} \left(\frac{r}{r_o}\right)^\gamma \, dr = \frac{1.1\sigma_o}{-\gamma\theta} \left(\frac{r_1}{r_o}\right)^\gamma \text{ mag} \tag{12}$$

where r_o is a reference radius of 100 AU, r_1 is the radius of the disk inner edge, and θ is the wedge angle ~ 0.1 rad (Artymowicz et al. 1989). Models with small grains (Sec. III.B.3) predict the correct infrared spectrum and source size if $r_1 \sim 30$ to 100 AU for the main outer disk component containing most of the optical depth and if $\sigma_o \sim 1$–5×10^{-3}, yielding $\Delta m \leq 0.04$ mag. This agrees with careful analysis (Paresce 1991; see Sec. V.A below) of β Pic photometry versus new metallicity indices and isochrones showing that the star lies on the zero age main sequence (ZAMS) within the uncertainties involved. Thus, our line of sight through the disk does not suffer significant extinction. The conclusion by Smith and Terrile (1984) that there must be a central cavity of roughly 30 AU radius is therefore only fortuitously correct because their argument was based on an assumption that substantial gray extinction is observed toward β Pic.

  3. *Grain Properties.* The composition and size of the grains around β Pic are significant in comparing this system to our solar system. Grain sizes are necessary input to calculations of grain lifetimes against removal processes discussed in Sec. V.A such as direct radiation pressure expulsion, mutual collisions, collisions with interstellar grains, PR radiation drag, and ice sublimation. Grain lifetimes in turn are crucial to hypotheses of the source and evolution of the grains.

  One complication is that the grains are unlikely to have a single size. A steady-state size distribution $n(a) \propto a^{-3.5}$ has been derived from theoretical studies of inelastic collisions and fragmentation processes and compared to observations of asteroids and micrometeoroids (Dohnanyi 1969). This distribution has been adopted in some discussions of the β Pic system (see, e.g., Smith and Terrile 1984; Matese et al. 1987). Model calculations based on observations usually yield a single size parameter, but the relation between such a size and the properties of a realistic size distribution are quite uncertain (see, e.g., Backman et al. 1992).

  Scattering properties of the grains will be set in part by their size. Smith and Terrile (1987) and Paresce and Burrows (1987) find that the color of the scattered light is the same as the color of the star within observational uncertainty across the range 0.4 to 0.9 μm. This would indicate grains 1 μm or larger in size. Gradie et al. (1987) have found very red colors for the disk, $V - I \sim 0.6$ magnitudes redder than the the star, also implying a grain size larger than 1 μm. Gledhill et al.'s (1991) optical polarization measurements yield a (model-dependent) size of about 0.25 μm.

  As previously discussed, the grain size can be characterized by a comparison of the run of temperatures represented by the thermal spectrum to measurements of the size of the infrared emitting region. Telesco et al. (1988) determined the size of the warmest grains radiating at 10 μm to be 0.1 to 0.4 μm; Backman et al. (1992) find typical sizes of 2 to 20 μm or minimum sizes of 0.05 to 3 μm for the infrared emitters depending on choice of emissivity exponent and spatial gradient.

  Artymowicz (1988) has calculated the ratio of radiation pressure forces to gravitational forces acting on small grains in the β Pic system, similar to the work of Burns et al. (1979) for grains of various types in our solar system (Sec.

V.B.1). The deduced minimum stable size for a wide variety of compositions is of order 1 μm, and could be larger if the grains are porous, delicate, or icy. There may also be a size regime (perhaps smaller than ≤ 0.1 μm, depending on composition) below which grains would couple inefficiently to the β Pic radiation field and also be stable. Grains in the range 0.1 to 1 μm should be subject to rapid ejection from the system.

Various groups have thus reported substantially different grain size results, and the disagreements are difficult to interpret. Comparison of the radiation pressure calculation with the optical and infrared observations indicates that the β Pic grain size distribution may not be continuous but may consist of one population smaller and one larger than the size range excluded by direct radiation pressure.

Artymowicz et al. (1989) concluded that the β Pic disk has 9×10^{28} cm^2 of optical scattering cross-section area in the region 100 AU$<r<$500 AU, yielding an albedo of $\mathcal{A} = 0.6$ (Artymowicz et al. 1989) or 0.35 (Backman et al. 1992) for the grains at $r>100$ AU depending on the amount of thermal radiating area attributed to the same region. These albedos are very high compared to small grains found in various parts of our solar system such as zodiacal dust grains ($\mathcal{A} \leq 0.1$; Hauser and Houck 1986) or grains in comet comae ($\mathcal{A} \sim 0.2$; Ney 1982). Even material which is nearly pure ice by mass with carbon added at solar abundance has albedo generally below 0.1 (Clark and Lucey 1984).

4. Spectroscopy of the Classical Shell Gas Component. β Pic has been classified as a shell star because of the presence of strong narrow circumstellar absorption features superimposed on the cores of some of the broad photospheric lines at ultraviolet wavelengths (Slettebak and Carpenter 1983; Kondo and Bruhweiler 1985; Lagrange et al. 1987; Lagrange-Henri et al. 1988) and at visible wavelengths (Slettebak 1975; Hobbs et al. 1985; Vidal-Madjar et al. 1986; Ferlet et al. 1987; Hobbs et al. 1988). The lines involved are listed in Table VI and are transitions from resonance (= ground) or metastable states, signifying gas with low density and temperature relative to the stellar atmosphere. 'I' signifies a neutral species, 'II' singly ionized, and 'III' doubly ionized.

The shell absorption lines in the β Pic spectrum themselves have complex multi-component structures. In general this structure can be decomposed into a strong stable component at the star's 21 km s^{-1} heliocentric radial velocity, plus additional components especially noticeable in the Al III and Mg II lines which are extremely variable in strength and velocity on time scales of days and months and are always redshifted with respect to the stable component (Sec. III.B.5). Both the stable and variable components were recently observed with the Goddard High-Resolution Spectrograph on the Hubble Space Telescope (Boggess et al. 1991).

The stable component may correspond to a "classical" A–F star shell. Detectability of the Ca II 3d and Fe II metastable transitions indicates that the radiation dilution (solid angle subtended by the star) must be $W>10^{-5}$ in

TABLE VI
Shell Lines in the β Pic Spectrum

Line	λ, Å
UV	
Al III resonance	1854
Fe II resonance multiplet	~2600
Fe II metastable	2750
Mg II resonance doublet	2795, 2802
Mg I resonance	2852
VISUAL/FAR-RED	
Ca II K resonance	3934
Na I D2 resonance doublet	5890, 5896
Ca II 3d metastable	8542

order that the rate of depopulation of the lower levels via spontaneous decay be negligible compared to radiative absorption (see, e.g., Viotti 1976). This in turn indicates that the stable lines from ionized gas originate in material at $r \leq 1$ AU. Thus, the presence of β Pic's "classical" shell may be unconnected to the infrared/optical grain disk lying at roughly 10 to 1000 AU. There is no correlation between strength of far-infrared excess and strength of spectroscopic A–F shell indicators (Hobbs 1986; Jaschek et al. 1986), and contrary to Jaschek et al. (1986), the frequency of far-infrared excess in A shell stars is not anomalously high compared to ordinary A stars (Sec. IV.A.3).

Velocity dispersion parameters of $b = 1.8$ km s^{-1} for Na I versus 3.7 km s^{-1} for Ca II K are evidence that the material producing the neutral sodium absorption is connected to the grain disk because it seems to lie farther from the star than the ionized material in the shell. Most of the gas mass around β Pic would be expected to lie in this outer component considering the relative volumes and densities, and would distinguish β Pic from "classical" A–F shell stars. Estimates of the total gas mass (including hydrogen) indicate that it is less than 2 M$_\oplus$, (<0.5% of the mass in our planetary system) (Vidal-Madjar et al. 1986; Hobbs et al. 1988). Model results indicate that the gas near the star is roughly solar-composition but the line strengths imply that the region where the Na is located must be extremely depleted in Ca relative to Na, possibly because the Ca is trapped in grains (Vidal-Madjar et al. 1986). A recent search (Lagrange-Henri et al. 1990) for spectroscopic analogs to β Pic is discussed in Sec. IV.B.

5. *Transient Spectroscopic Events: Comet Impacts?* Although spectral variability is also found in other shell stars, the unique occasional highly redshifted absorption events at β Pic have been claimed as evidence of impact of comet-like bodies onto the star (see, e.g., Ferlet et al. 1987; Lagrange-Henri et al. 1988; Beust et al. 1990; Beust et al. 1991). The most dramatic changes occur on time scales as short as an hour for the Al III line and 6 to 8 hours for

the Ca II line. Individual components of the Al III and Mg II lines can reach equivalent widths of 400 and 900 mÅ and velocities of 350 and 300 km s^{-1}, respectively. These velocities are comparable to free-fall speeds at 2–3 $R_\star$. In contrast, the Ca II K variable components never exceed about 40 km s^{-1} red shift. Estimated frequencies of redshifted events for the Ca II K line are 10 to 100 yr^{-1} and 100 to 200 yr^{-1} for the Al III line depending on red shift. Typical velocities and frequencies appear to vary somewhat over the course of years.

A highly developed but not unique explanation for these phenomena is described by Beust et al. (1990,1991). In their model the transient events in the metallic lines are attributed to the passage through the observer's line of sight of infalling asteroids or comets producing clouds of dust grains that are sublimating or evaporating as they approach the star. The amount of absorbing material is equivalent to the complete vaporization of cometary objects several kilometers in diameter. The observed multi-component structure of the time-variable features can be plausibly explained in this scenario as resulting from objects breaking into pieces due to tidal disruption or extreme heating. The ratio of radiation pressure to gravity is a factor of 10 larger for Ca II ions than for Al III and Mg II ions, which is crucial to explaining the large differences in their observed behavior. Their model also includes a planetary mass in the β Pic system that sends the infalling bodies on the inferred trajectories. A complete dynamical simulation described in Beust et al. (1991) is able to fit the spectroscopic data well with only a few free parameters.

A cometary hypothesis is not as far-fetched as it might seem since small Sun-grazing comets that are totally disrupted in the ensuing collision are known to occur in our own solar system as often as 10 times per year (Michels et al. 1982). It is therefore an interesting possibility that the younger and perhaps more crowded β Pic system is host to many active comets.

IV. SEARCHES FOR VEGA-LIKE SYSTEMS: SURVEYS AND INTERPRETATION

A. Infrared Searches

The statistical properties of the IRAS Point Source Catalog (PSC) including detectability and characteristic far-infrared properties of various types of objects are discussed by Chester (1985). The IRAS Faint Source Survey (FSS) released in September 1990 consists of coadded survey data and is thus 2 to 3 times as sensitive as the PSC, but avoids the galactic plane and does not contain sources detected only at 100 μm.

1. IRAS Limitations in Observations of Stars. Normal main-sequence photospheres are difficult targets of study in IRAS data for the following reasons:

(a) Stellar photospheres of different types all have approximately RJ spectra and are nearly indistinguishable in the IRAS data, with [12]–[25], etc.

colors close to 0.0 mag. As a corollary, cool main-sequence *physical* companion stars are essentially undetectable with IRAS—the maximum possible color excess E(V-[12]) due to a main sequence binary companion to any main sequence primary is roughly 0.2 magnitudes.

(b) IRAS was only slightly more sensitive than the human eye to sources with stellar temperatures; the PSC 12 μm completeness limit of ~0.4 Jy corresponds to visual magnitude +5.0 for A and +7.5 for G main-sequence stars ($d < 30$ pc!).

(c) Most stars cooler than the Sun detected by IRAS are giants; the dividing line between detection of mostly dwarfs versus mostly giants is at $B-V$ ~0.75 (approximately spectral type G5) (Waters et al. 1987).

In the absence of circumstellar material one would therefore expect that only a few nearby and bright main-sequence stars would be found in the IRAS data and that even these would not be detected beyond 12 μm.

Unlike binary companions, *background* stars are capable of producing substantial excesses in IRAS observations of nearby stars in measures of V-[12], V-[25], etc. In a groundbased 2.2 μm photometric study of the IRAS fields of view around approximately 60 candidate stars with substantial V-[12] excesses, Aumann and Probst (1991) found that only 2 (β Pic and ζ Lep) had true intrinsic 12 μm excess and the rest were caused by red background stars. This implies that the phenomenon of far-infrared excesses around main-sequence stars is usually limited to temperatures below about 200 K.

IRAS discovered that the sky is covered with low-temperature emission from the general ISM, now called infrared "cirrus," which is concentrated toward the galactic plane. The usually extended cirrus emission includes concentrations which sometimes mimicked the characteristics of a point source in IRAS data processing. One of the warmest such sources found at high galactic latitude and discussed by Low et al. (1984) has a color temperature of 34 K. A source with color temperature less than 40 K is most significantly detected in the 100 μm band in IRAS data. Thus, sources associated by position with nearby stars but having $T_c < 40$ K, and also excesses detected only in the 100 μm band, are possibly due to background cirrus. This is more likely statistically at low galactic latitudes. One way to eliminate candidate Vega-like sources caused by background cirrus would be improved spatial resolution in future infrared instruments which might show a cold source to be an extension of a larger structure or not centered on the star in question.

The far-infrared color typical for galaxies and infrared galactic components (HII regions, planetary nebulae, parts of molecular clouds and star forming regions) of $f_\nu(25) < f_\nu(60)$ (Chester 1985) also describes the prototype main-sequence circumstellar sources. However, infrared galactic components are strongly concentrated toward the galactic plane and thus would likely produce only a few spurious candidates. External galaxies, especially starburst galaxies, can have spectra and fluxes similar to the Vega-like sources (Helou and Beichman 1991) and far-infrared sizes of order 10 arcsec, so it is at least possible that some Vega-like candidates could be produced by a galaxy lying in the background of a star, although the number density of sufficiently bright galaxies versus IRAS 60/100 μm detector field of views of 7 to 15 square arcmin makes for agreement with subject. Again, future comparison of source centroids with star positions can help eliminate this possibility.

Finally, reflection nebulae have been found to emit in all 4 IRAS bands, generally with $f_\nu(12) < f_\nu(25) < f_\nu(60) < f_\nu(100)$ after removal of direct flux from the illuminating star (Sellgren et al. 1990). Some of the 60/100 μm ratios are not far from 1 and thus are similar to the circumstellar disk prototypes. This similarity should not be surprising because the β Pic disk is in a sense a reflection nebula. Characteristics which might distinguish orbiting material from ISM grains warmed by a luminous star might be: symmetric versus asymmetric source shape and/or angular scale too small for emission involving small ISM grains. Once again, spatial resolution superior to that available with IRAS may be needed to distinguish normal reflection nebulae from Vega-like systems.

2. IRAS Data on Typical Main-Sequence Stars. Waters et al. (1987) and Cohen et al. (1987) summarize PSC data on objects in the Bright Star Catalog (BSC) (Hoffleit and Jaschek 1982). Both papers present typical V-[12], V-[25], etc. colors for a range of spectral types and luminosity classes and confirm the expectation that there are no significant flux *deficits* relative to expectations of RJ or slightly steeper slopes; disturbances to photospheric colors are exclusively *excesses*. An independent check by Graps (personal communication) confined to luminosity class V found V-[12] colors similar to Waters et al.'s means for all spectral types. These colors are about 0.1 mag redder than predicted by stellar atmosphere models.

Figure 7 presents the median IRAS 12/25 μm flux density ratios versus spectral type for the 670 BSC stars of luminosity classes IV, IV–V, and V detected in both bands (Graps, personal communication). These continuum slopes are quite similar to those found for Gliese catalog (nearby) stars (Backman and Gillett 1987) and for SAO stars (Stencel and Backman 1991).

Among A and early F stars, even typical stars have modest 25 μm excesses with median T_c cooler than T_{eff} by several thousand degrees. Note that the ratio for α Lyr is exactly at the median for types B9–A1, yet approximately 25% of the 25 μm flux density in that case is known to be excess from grains. For these types, 50th percentile falls closer to the blue end of the distribution, because the population includes only a few stars with substantial excesses. Median (50th percentile) late F, G and early K stars fall near the average of the range, and the range is consistent with IRAS photometric uncertainty, implying that few of these stars have detectable excess at 25 μm compared to 12 μm. Typical B IV–V stars have substantially redder colors than stars of later types and 50th percentile falls nearer the red end of the ranges, approaching the colors of a free-free plasma emission spectrum. This is consistent with the typical B star having a free-free excess but with a blue tail in the population toward stars with more nearly photospheric colors.

3. Published Surveys and Present "Master List." A number of searches

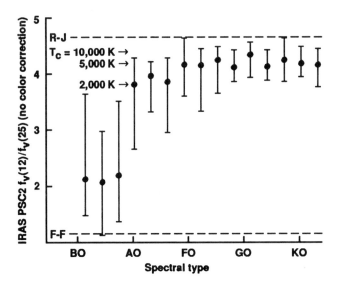

Figure 7.   Median IRAS 12/25 μm flux density ratios (raw, i.e., without bandwidth
color corrections) vs spectral type for main-sequence stars in the Yale Bright Star
Catalog detected in both bands. Stars with giant or supergiant companions were
excluded. There were few enough main-sequence stars later than K4 that they were
also excluded. The type bins represent combinations of groups of 3 subtypes, i.e.,
the median labelled A0 is for B9, A0, and A1 stars considered together. The error
bars represent the range from 25th percentile to 75th percentile. The R-J line at the
top represents the color of a blackbody of infinite temperature. The F-F line at the
bottom represents the color of pure free-free (plasma) emission.

have been conducted for examples of stellar far-infrared excesses. Aumann
(1985) examined PSC data on stars with $d < 25$ pc and found 12 stars with
galactic latitude $|b_{II}| > 10$ deg having [12]–[60] > 1.0 mag, defined there as
Vega-like, including the three resolved prototypes. The fact that the ones with
excess are mostly A–F stars was recognized as probably due to a luminosity
selection effect. Sadakane and Nishida (1986) examined PSC data on BSC
members and found 12 more Vega-like stars using criteria identical with
Aumann's (1985), again mostly A-F stars with *no lack of excesses in binary
systems,* which may be very significant information about the mechanisms
for forming these grain clouds. One of the stars, λ Boo, is known to be
metal deficient, which is interesting in light of the low metallicity of β Pic
(Sec. V.A), and raises the issue of composition of the circumstellar grains.
Coté (1987) examined PSC data on BSC B and A stars and found 24 examples
of infrared excess, some of which were also noted in the previous studies.
Twenty-one of these cases are attributed by the author to dust emission, 3 to
free-free emission.
    Odenwald (1986*a, b*) examined PSC data on G stars of all luminosity

TABLE VII
Compendium of Published Lists of Main-Sequence Stars
with Far-Infrared Excess

	HR	Type	V	References[a]
49 Cet	451	A3V	5.6	SN, WW
DM+19 279	493	K1V	5.2	B2
τ Cet	509	G8V	3.5	B1, B2
γ Tri	664	A1Vnn	4.0	Co
τ^1 Eri	818	F6V	4.5	Au, B2
τ^3 Eri	919	A4V	4.1	B2
α For	963	F8V	3.9	WW
ϵ Eri	1084	K2V	3.7	Au, B1, B2
π^1Ori	1570	A0V	4.7	SN
HD 31648		A2ps	7.7	Ja
κ Lep	1705	B9V	4.4	SN
DM+79 169	1686	F6V	5.0	Au, Ho, WW
HD 34700		G0	8.7	WW
HD 35187		A2	8.2	WW
ζ Lep	1998	A3Vn	3.6	B2, Co
δ Dor	2015	A7V	4.4	Co
β Pic	2020	A5V	3.8	Au, B2, Co, Ja
HD 41511	2148	A2s	4.9	Ja
ψ^5 Aur	2483	G0V	5.3	B2
HD 53143		K0V	6.8	WW
B Car	3220	F5V	4.8	B2
HD 233517		K2	8.7	WW
	3314	A0V	3.9	Co
δ Vel	3485	A1V	2.0	B2, Co, WW
β UMa	4295	A1V	2.4	Au, Co
HD 98800		K5V	8.6	WW
Ross 128		M5V	11.1	B1
β Leo	4534	A3V	2.1	Au, B2, Co WW
HD 121384	5236	G6IV-V	6.0	WW
λ Boo	5351	A0p	4.2	SN
σ Boo	5447	F2V	4.5	B2
HD 135344		A0V	8.7	WW
γ Tra	5671	A1V	2.9	Co
α CrB	5793	A0V	2.2	Au, Co, Ho
HD 139614		A7V	8.0	WW
HD 139664	5825	F5IV-V	4.6	WW
HD 142666		A3	8.2	WW
HD 144432		A9+F0 V	8.4	WW
σ Her	6168	B9V	4.2	SN, Co
	6297	A5IV-V	5.7	Co

TABLE VII (cont.)

	HR	Type	V	References[a]
HD 155826	6398	F7+G2 V	6.0	WW
DM-24 13337	6486	A3m	4.2	B2
μ Ara	6585	G3IV-V	5.2	B2
γ Oph	6629	A0V	3.8	SN, Co
	6670	F3IV-V	5.8	SN
HD 169142		B9V	8.3	WW
α Lyr	7001	A0V	0.0	Au, B2, Ho, Co
	7012	A5IV-V	4.8	Co
61 Cyg	8085	K2+K5 V	5.2	B1, B2
δ Equ	8123	F5+G0 V	4.7	B2
DM-47 13928	8323	G2V	5.6	Au, WW
α PsA	8728	A3V	1.2	Au, B2, Co
	8799	A5V	6.0	SN
HD 221354		K0V	6.7	WW

[a] References: Au = Aumann 1985 (Gliese+Woolley catalogs; PSC+coadds);
B1 = Backman et al. 1986 ($d < 5$ pc stars; survey coadds + pointed obs.);
B2 = Backman and Gillett 1987 (Gliese subset; survey coadds); Co = Cote
1987 (BS catalog B+A stars; PSC); Ho = Hobbs 1986 (shell stars; PSC); Ja
= Jaschek et al. 1986 (shell stars; PSC); SN = Sadakane and Nishida 1986
PASP 86, 685 (BS catalog; PSC); WW = Walker and Wolstencroft 1988 (HD
catalog; PSC-extended sources).

classes and found strong infrared excesses mostly from dusty shells around
supergiants, but also at least one case of a main sequence excess. Johnson
(1986) found warm excesses around some K dwarfs, most of which are
classified as T Tauri stars. Jaschek et al. (1986) examined PSC data on 19
Ae/A shell stars comparable in optical spectral qualities to β Pic and found 8
with strong far-infrared excess. This is similar to the frequency of excesses
among normal A stars in samples of nearby stars (Sec. IV.A.4). There appears
to be no clear correlation between the optical and infrared properties of shell
and emission line stars.

Walker and Wolstencroft (1988) selected HD stars ($m_v > 10$) with PSC
60/100 μm flux density ratios similar to the prototypes. In addition, they
examined evidence of the extent of the emission via the IRAS data "confusion"
flag, which was set when a source was seen by 3 detectors in one waveband,
nominally requiring a size of at least 20 arcsec. The data were then analyzable
in terms of grain size and total mass. Some of the objects discovered in this
survey have stronger far-infrared fluxes than the prototypes, and since these
HD stars are in general dim the fractional luminosities (optical depths) in
some cases approach 1, substantially different from the prototypes. A search
for CO emission (Walker and Butner 1991) from some of these stars did not
detect any molecular material other than probable background emission.

Backman et al. (1986) combined groundbased near-infrared photometry with IRAS pointed observation ("AO") data on stars within 5 pc and found several examples of weak excess which had been missed at the sensitivity level of the PSC. One of those stars, Ross 128, is the only M dwarf so far known to have a Vega-like excess. α CMa (Sirius) was found to have a weak 100 μm excess, but the characteristics seem to exclude emission from orbiting grains, as confirmed by Chini et al. (1990) with mm-wave measurements.

Table VII is a "master list", a union of lists of main-sequence (luminosity class IV, IV–V, and V) stars found to have far-infrared excesses. Because the selection criteria and procedures were not uniform and the IRAS data used had various sensitivities, this list is of somewhat mixed quality.

Backman and Gillett (1987) and Aumann (1988) showed by examining coadded survey photometry of nearby stars that the phenomenon of far-infrared excesses (a) may be the rule rather than the exception for main-sequence stars, (b) appears in all spectral types, and (c) is not restricted to stars of young age. The implication that our solar system might also have a cool excess when viewed from outside is considered in Sec. VI.D.3.

4. Bright Star Catalog and Gliese Catalog Surveys. Table VIII presents a complete list of BSC main-sequence stars found to have Vega-like far-infrared excesses, as defined below. There is some overlap between this list and the "master list" in Table VII. The original sample consisted of all BSC stars in luminosity class V with detected flux at 12 μm and having no higher-luminosity class companion. In case of dimmer companions, the system is listed by the primary's spectral type. IRAS PSC2 data were used and all infrared sources within 60 arcsec of the star position corrected for proper motion to 1983.5 were accepted.

It was found in a plot of $f_\nu(25)/f_\nu(60)$ versus $f_\nu(12)/f_\nu(25)$ (Fig. 8) that the three prototypes α Lyr, α PsA, and β Pic spanned a broad stretch of the diagram, and a number of objects were found to lie in the same stretch. Stars of spectral class O or Be were found to lie often in a region with $f_\nu(25)/f_\nu(60)>1$ and $f_\nu(12)/f_\nu(25)<2.5$ which contains the locus of a purely free-free spectrum and were rejected as having excesses likely due to plasma emission. Stars with detected flux at 100 μm but with $f_\nu(60)/f_\nu(100)<0.2$, i.e., $T_c < 30$ K, were excluded as being likely due to background cirrus. Finally, stars with $f_\nu(25)/f_\nu(60)>4.0$, i.e., $T_c>500$ K, were excluded; over that wavelength range, those colors do not differ significantly from photospheric colors. These criteria in combination are essentially equivalent to a search for excesses most prominent at 60 μm.

Note the two trajectories in Fig. 8 which show the loci of points corresponding to: (a) a system like β Pic if the disk luminosity is decreased relative to the star from its actual value of 2×10^{-3} toward 1×10^{-5}, approaching the upper right of the diagram where the colors of a pure stellar photosphere lie; and (b) a system like α PsA if the disk luminosity is increased relative to the star from its actual value of 8×10^{-5} toward 1×10^{-2}, approaching the locus of β Pic. This indicates that the different loci of β Pic and α PsA in this color-color plot simply represent different fractional grain luminosities but

TABLE VIII
Bright Star Catalog Main-Sequence Stars with
Vega-like Far-Infrared Excesses

HR	Type	V	IRAS fqual	F_ν, Jy 12μm	Excesses, Jy 25μm	60μm	100μm
123	B8Vn	4.73	3133	0.43	0.00	0.96	2.44
189	B5V	5.67	1333	0.28	0.41	1.01	2.03
241	B9.5V	6.21	3332	0.28	0.21	3.87	11.18
287	A3V	6.47	1133	0.25	0.00	0.76	2.70
333	A3V	5.62	3131	0.26	0.00	0.34	0.00
451	A3V	5.63	3233	0.33	0.33	1.99	1.90
506	F8V	5.52	3331	0.82	0.14	0.81	0.00
533	B1.5V	5.52	2332	0.26	0.76	2.29	3.74
664	A1Vnn	4.01	3332	1.07	0.21	0.81	0.83
811	B7V	4.25	3333	0.66	0.32	1.32	1.76
818	F6V	4.47	3322	1.88	0.20	1.55	4.98
890	B7V+B9V	5.03	3223	0.53	0.56	3.10	14.07
963	F8V	3.87	3321	4.11	−0.02	0.11	0.00
1074	B1V	5.90	1321	0.25	0.41	1.07	0.00
1082	A3V	6.38	1131	0.26	0.00	0.58	0.00
1084	K2V	3.73	3333	9.52	0.28	1.23	1.73
1144	B8V	5.64	1231	0.47	0.48	2.97	0.00
1151	B8V	5.76	3333	0.42	1.04	4.66	10.90
1307	B8Vn	6.23	3333	1.15	6.16	26.27	37.19
1399	B7V	5.53	1333	0.46	1.07	5.77	16.13
1415	B3V	5.55	3333	0.35	0.98	2.60	3.48
1448	A2Vs	5.68	1131	0.39	0.00	0.54	0.00
1570	A0V	4.65	3121	0.73	0.00	0.42	0.00
1686	F6V	5.05	3221	1.14	−0.04	0.23	0.00
1719	B5V	6.13	1132	0.30	0.00	1.02	2.73
1798	B2Vn	6.25	1331	0.26	0.61	1.54	0.00
1808	B5V	5.42	1233	0.43	0.41	1.57	3.59
1839	B5V	4.20	3121	0.56	0.00	0.55	0.00
1868	B1V	5.34	3323	0.38	1.86	7.50	6.16
2015	A7V	4.35	3332	1.39	0.04	0.45	1.11
2020	A5V	3.85	3333	3.46	8.22	19.75	11.23
2124	A2V	4.12	3333	1.50	0.63	3.20	2.75
2161	B3V	6.66	1131	0.25	0.00	0.57	0.00
2522	B6V	5.39	3333	0.43	1.52	4.58	5.67
2806	O9V	6.43	1331	0.25	0.56	1.43	0.00
3415	B3V+B3Vn	5.26	1131	0.51	0.00	0.51	0.00
3720	A1V	5.29	3233	0.47	0.07	1.26	2.74
3927	A0V	5.72	1131	0.29	0.00	0.68	0.00
4295	A1V	2.37	3331	4.80	0.24	0.43	0.00

4534	A3V	2.14	3331	6.97	0.44	0.88	0.00
4732	B3Vn	4.82	3131	0.27	0.00	0.79	0.00
5250	B8VpShell	5.15	2123	0.24	0.00	1.07	2.16
5336	B4V	5.06	3332	0.31	0.63	6.04	10.04
5793	A0V	2.23	3331	5.92	0.31	0.50	0.00
5902	B2.5V	5.03	3231	0.56	0.26	0.63	0.00
5933	F6V	3.85	3331	3.85	−0.08	0.24	0.00
6168	B9V	4.20	3321	0.96	0.02	0.22	0.00
6211	B8V	6.46	1131	0.52	0.00	1.06	0.00
6532	B9.5V	6.42	1132	0.25	0.00	0.69	1.28
6533	A1V	5.62	3133	0.30	0.00	0.46	1.23
6629	A0V	3.75	3331	1.41	0.17	1.23	0.00
7001	A0Va	0.03	3333	41.56	1.08	7.75	7.12
7030	B8V	6.41	1133	0.25	0.00	0.83	1.43
7035	B5.V	5.83	1133	0.66	0.00	3.33	8.49
7329	A0Vn	5.05	3331	0.54	0.35	0.51	0.00
7474	B3V+B3V	5.17	2332	0.68	1.37	5.08	7.70
8323	G0V	5.58	3221	0.81	0.02	0.21	0.00
8549	B2V	6.46	1133	0.25	0.00	1.01	1.33
8728	A3V	1.16	3333	18.21	0.45	8.25	10.91
8854	B0Vn	6.53	3321	0.60	0.79	13.68	0.00

not fundamental differences in disk structure or the mechanism of far-infrared emission. The region of the diagram these trajectories cross is interpreted as the region of Vega-like excesses.

A study made by Backman and Gillett (1987) examined coadded IRAS photometry of 134 stellar systems from the Gliese Catalog (Gliese 1969; Gliese and Jahreiss 1979). The sample was limited to stars of spectral types A–K, luminosity classes V and IV–V, trigonometric parallax $\pi_t \geq 0.045$ ($d < 22$ pc), and color temperatures above 35 K. This study contrasts with the aforementioned Bright Star survey in that the sample was defined by a flux density limit at $12 \, \mu$m, with the result that a greater proportion of later-type stars were included.

Those stars with significant excesses at 25, 60, or $100 \, \mu$m with color temperatures above 35 K are listed in Table X. Significance was defined as greater than 3 times the rms noise in the off-source scan baseline. The summed luminosity of components listed by Gliese (1969) was used in calculating the fractional cloud luminosities in column 11. Table IX lists as a function of spectral type an estimate of the sample distance limit (reciprocal of 25th percentile trigonometric parallax), the number of stars in the sample, the fraction of sample stars with significant excesses, and the fraction of sample stars with significant excesses stronger than a fractional luminosity of 2×10^{-5}, approximately equivalent to α Lyr.

The possible meaning of the statistics in Table IX is that there is no strong dependence on spectral type when frequencies of far-infrared excesses

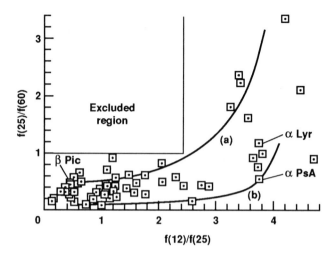

Figure 8.   IRAS color-color diagram for main-sequence stars in the Yale Bright Star Catalog. The colors of the three prototype objects are indicated. The two trajectories respectively show the range of colors which the β Pic and α PsA disks would have if all other disk parameters were kept constant while the optical depth was varied. The "excluded" region contains the color of pure free-free emission.

TABLE IX

Far-Infrared Excesses in Gliese Catalog Subsets

Spectral Type	N	Excess $>3\sigma$	Excess $\tau \geq 2 \times 10^{-5}$	Approx. limit (pc)
A	22	41%	18%	20
F	51	10%	10%	18
G	39	10%	8%	15
K	22	14%	14%	7

are compared by fractional luminosity rather than by significance of IRAS detection. Similar results are reported by Aumann (1988). The frequency of far-infrared excesses around A stars at this sensitivity limit is near 50%; it may be that a more sensitive instrument than IRAS will find that fraction approaching 100%. Note also in column 12 of Table X that there are several stars with ages approaching the age of our Sun determined from photometric isochrones (Varsik 1987), lithium abundances (Simon et al. 1985), or emission-line strengths (Duncan 1981). This implies that the Vega phenomenon is not limited to proto-stellar or proto-planetary contexts.

B. Optical Spectroscopic and Coronagraph Searches

Lagrange-Henri et al. (1990) have undertaken a comprehensive spectroscopic survey for β Pic analogs. Since β Pic has both an infrared excess and a

TABLE X

Gliese Catalog Stars with Far-Infrared Excesses

Gliese Number		Spectral Type	25 μm Excess (Jy)	e/σ_e	60 μm Excess (Jy)	e/σ_e	100 μm Excess (Jy)	e/σ_e	T_c	τ	Age (Gyr)
68.0	DM+19 279	K1	0.10	1.7	0.11	3.4	0.38	2.9	85	7.1e-5	
71.0	τ Cet	G8	0.06	0.7	0.08	2.1	0.42	6.0	76	1.2e-5	3.5
111.0	τ^1 Eri	F6	0.17	3.9	0.89	35.6	3.65	22.7	62	2.2e-4	2.2 ± 1.6
121.0	τ^3 Eri	A4	0.02	0.8	0.04	1.6	0.15	3.4	75	7.5e-6	
144.0	ε Eri	K2	0.29	3.9	1.33	41.7	2.27	41.6	74	9.7e-5	
167.1	γ Dor	F0	0.05	2.2	0.21	9.3	0.21	4.1	83	3.0e-5	
217.1	ζ Lep	A3	0.68	15.4	0.40	12.9	<0.11	—	139	7.1e-5	
219.0	β Pic	A5	9.05	194.1	20.44	363.4	13.15	115.3	106	2.4e-3	
245.0	ψ^5 Aur	G0	0.11	2.6	0.43	8.9	0.53	2.4	81	1.7e-4	3.8
248.0	α Pic	A5	0.13	3.5	0.00	0.1	<0.19	—	79	9.2e-6	
297.1	B Car	F5	0.04	2.0	1.76	51.8	3.05	16.9	52	3.5e-4	4.5 ± 1.1
321.3	δ Vel	A0	0.03	0.4	0.29	6.9	<0.02	—	63	2.8e-6	
364.0	DM-23 8646	G0	0.12	2.8	0.13	5.4	0.24	3.0	98	7.2e-5	
448.0	β Leo	A3	0.41	5.5	0.77	14.9	0.63	7.0	109	1.9e-5	2.6
557.0	σ Boo	F2	0.06	2.2	0.09	3.5	<0.06	—	83	2.3e-5	
673.1	DM-24 13337	A9	0.15	1.3	1.02	2.9	11.48	5.9	52	2.9e-4	2.0 ± 1.5
691.0	μ Ara	G5	0.22	7.3	0.08	1.1	<0.96	—	67	1.0e-4	
721.0	α Lyr	A0	1.14	3.4	7.85	114.7	8.89	48.1	74	1.5e-5	
820.0	61 Cyg	K5	0.38	5.1	0.57	5.0	2.38	1.9	77	1.5e-4	
822.0	δ Eql	F8	0.15	3.8	0.07	2.2	0.48	4.9	88	5.4e-5	6.4 ± 1.0
881.0	α PsA	A3	0.34	2.3	8.66	157.2	11.95	109.3	58	5.0e-5	

shell spectrum, they searched 57 candidates that had at least one of those two attributes for the telltale signature of narrow, mostly redshifted and possibly variable circumstellar absorption lines of Ca II and Na I. Although a few (especially HR 10, 2174, 8519 and 9043) had some similarities or were remarkable in some way, none in the sample were found to resemble β Pic in detail.

Coronagraph surveys based on lists of IRAS excesses and/or shell spectra have been made separately by Brahic, Smith and Terrile, and Burrows and Paresce. The latter team examined 20 of the objects in Lagrange-Henri's list mentioned above, and found no disk detectable at radii of 6 arcsec from the stars to an I-band surface brightness limit of about 18 mag arcsec^{-2}. This limit is roughly 2 mag fainter than the β Pic disk, but null results are not surprising considering that β Pic's grain optical depth and orientation are so exceptional.

V. TIME SCALES, DYNAMICS, AND EVOLUTION

A. Ages of The "Big 3"

The metallicities of α Lyr and α PsA are close to solar. Their ages on the Green et al. (1987) revised Yale isochrones are 4×10^8 and 2×10^8 yr, respectively, with uncertainties of roughly 30%. The derived age for α Lyr is near the end of the stage of core hydrogen burning for its mass, and in fact that star is a factor of 3 to 4 times brighter than its ZAMS luminosity.

The indicated metal abundance for β Pic from a new calibration of Geneva photometry (Kobi and North 1990) is $\log_{10}[\text{M/H}]=-0.6\pm0.3$, or 1/4 of solar, rather depleted with respect to normal nearby stars of early spectral class. Fundamental parameters of $R_\star = 1.2\ R_\odot$ and $M_\star = 1.5\ M_\odot$ are found, both slightly lower than typical A5 dwarfs but within a plausible range. The ZAMS bolometric magnitude from the Green et al. isochrones for $Z = 0.004$, $Y = 0.2$, $T_e = 8200$ K is $M_{\text{bol}} = 2.80\pm0.15$ mag (1 σ) versus an observed value from its parallax, V magnitude, and bolometric correction of 2.63. β Pic's age can be estimated to be about 100 Myr, above but close to the ZAMS. The most evolved position on the Yale isochrones allowable (considering the range in possible helium and metal abundances) corresponds to an age of 200 Myr, about 10% of the main sequence lifetime of a 1.5 $M_\odot$ star (Paresce 1991).

B. Intrinsic Grain Removal Processes

Radiation pressure, particle collisions, and ice sublimation are processes which can remove or destroy grains orbiting stars. Their time scales offer clues to the source and evolution of the grain material. Individual processes will be discussed separately in the next sections, followed by discussion and comparison of their rates at inner and outer radii in the three prototype systems.

1. Radiation Pressure Expulsion. A simple derivation of the overpressure ratio δ of radiation to gravitational forces acting on a grain near a star assuming spherical black ($\mathcal{A} = 0$) grains of radius a is:

$$\delta \equiv \frac{F_{\text{rad}}}{F_{\text{grav}}} = \left(\frac{3}{16\pi c G}\right)\left(\frac{L_\star}{M_\star}\right)\left(\frac{1}{a\rho}\right) = \left(\frac{0.57}{a_{\mu m}\rho}\right)\left(\frac{L_\star}{L_\odot}\right)\left(\frac{M_\odot}{M_\star}\right). \quad (13)$$

The value of δ is inversely proportional to grain size in this simple treatment. For a given value of grain density there will be one size for which $\delta = 1$, defining a minimum stable or "blowout" size for grains released at rest which depends only on the central star's luminosity and mass. Both forces depend on r^{-2} so their ratio δ has the same value at all locations for a given grain and all regions will be cleared equally well. Grains are propelled beyond radii of 10^3 AU on free-fall time scales of order 10^4 yr by this means. Column 2 of Table XI presents the "blowout" size for each of the three systems. It is interesting that this size is comparable to the typical or abundant grain sizes estimated in Table III, as if this process provides lower limits to grain size distributions weighted toward the smallest grains.

A more careful treatment of radiation pressure effects (Burns et al. 1979) including grain albedo and radiative efficiency shows that there is a maximum value in our solar system of $\delta \sim 0.5$ at a grain size of about 0.5 μm (close to the peak wavelength of the Sun's radiation) for most realistic grain materials. Grains smaller and larger than that size are more stable against ejection, meaning there is no unstable size for grains released at rest in our system. One can approximately scale Burns et al.'s results to other stellar systems with an increase of overpressure by a factor of $(L_\star/L_\odot)(M_\odot/M_\star)$ and a shift of the grain size of maximum overpressure to smaller sizes by a factor of $T_\star/T_\odot$. Artymowicz (1988) employed a similar analysis for the case of β Pic. The general expectation for stars with the luminosities of the three A stars is an excluded grain size range between about 1/30 and 1/2 times the simply calculated "blowout" size.

2. Poynting-Robertson Radiation Drag. Particles stable against radiation pressure ejection are subject to the PR effect in which incoming radiation in the moving frame of the grain has a component that always opposes the grain velocity. This causes grains orbiting in a radiation field to spiral toward their orbit center even if the orbit center and the radiation source are not the same. The orbit decay time due to PR drag (see, e.g., Burns et al. 1979) is:

$$t_{PR} = \left(\frac{4\pi a\rho}{3}\right)\left(\frac{c^2 r^2}{L_\star}\right) \text{cgs} = 7.1 \times 10^2 a_{\mu m}\rho r_{\text{AU}}^2\left(\frac{L_\odot}{L_\star}\right) \text{yr}. \quad (14)$$

An important characteristic of PR drag is that it is guaranteed to apply to all stably orbiting grains of any composition in any optically thin situation whereas other processes such as mutual collisions or ice sublimation may not apply in some cases.

3. Grain Collisions. The mutual collision time scale is approximately:

$$t_{\text{coll}} \sim \frac{t_{\text{orb}}}{8\sigma(r)} \sim \frac{r_{\text{AU}}^{\frac{3}{2}}}{8\sigma(r)}\left(\frac{M_\odot}{M_\star}\right)^{\frac{1}{2}} \text{yr} \quad (15)$$

where $\sigma(r)$ is the fractional surface density as in Eq. (8). This expression assumes that each orbiting grain encounters the full surface density roughly twice per orbit as it oscillates perpendicular to the disk plane and that the collision cross section is 4 times the cross-sectional area of individual grains to include tangential collisions.

Collisions can be assumed to result in destruction rather than accretion because even tangential collisions at relative speeds greater than 0.1 km s^{-1} (10^8 erg g^{-1}) should result in catastrophic fragmentation (see, e.g., references in Lissauer and Griffith 1989). The relative velocity of grains due to their motion perpendicular to the plane of a disk is approximately $V_z \sim iV_{\text{orb}}$, where $2i$ is the opening angle of the disk subtended at the star. For example, for β Pic with an opening angle of about 0.1 rad and $M_\star = 1.5$ M$_\odot$, $V_z > 0.1$ km s^{-1} at $r < 1000$ AU. Most fragments from such high-speed collisions involving small grains would likely be smaller than the "blowout" size and be rapidly ejected from the system.

Grains of various sizes will move at different orbital speeds for the same orbit because the radiation force reduces the effective potential. The range of orbital speeds resulting from this effect for grains between 1 and 10 times the "blowout" size would also imply relative speeds greater than the collisional destruction threshold of 0.1 km s^{-1} throughout the relevant radii ranges even if motions perpendicular to the disks are unexpectedly small.

4. Ice Sublimation. The time scale for sublimation of pure water ice grains can be estimated from expressions in Isobe (1970):

$$t_{\text{subl}} = 1.5 \times 10^{-12} a\rho \frac{10^{2480/T_g}}{T_g^{7/2}} \text{yr} \qquad (16)$$

where T_g is the grain temperature. This rate is an extremely powerful function of temperature; i.e., for temperatures of 90 to 120 K, Eq. (16) can be approximated by:

$$t_{\text{subl}} \sim 10^6 a_{\mu m} \left(\frac{T_g}{100\text{K}}\right)^{-55} \text{yr.} \qquad (17)$$

A more complete treatment can be found in Lien (1990), including the result that "dirty ice" grains have a much shorter sublimation time scale than pure ice at a given location.

5. Comparison of Grain Removal Time Scales, and Mass-Loss Rates. Columns 4–6 of Table XI contain the time scales for each process at inner boundaries and outer locations in the models of the three systems using grain sizes from Table III and assuming grain density of 1 g cm^{-3}. The PR decay time would increase linearly with increased density. The inner boundary radii are for the $\gamma = -1.6$ case. The outer positions considered correspond roughly to the optically traced extent of the β Pic disk. The face-on optical depths $\sigma(r_1)$ for Eq. (15) can be approximated by the fractional luminosity f in Table III. Time scales labelled ∞ are much greater than 10^{10} yr.

TABLE XI

Grain Removal Processes

	Blowout Size, μm	r_1, AU r_2, AU	Removal Time Scales, yr		
			t_{PR}	t_{coll}	t_{subl}
α Lyr	14	26	2×10^5	5×10^5	9×10^5
		1000	3×10^8	5×10^{10}	∞
α PsA	4	67	6×10^6	6×10^5	∞
		1000	1×10^9	3×10^9	∞
β Pic	2	38	2×10^5	8×10^3	4×10^{-4}
		1000	1×10^8	2×10^8	∞

A number of conclusions can be drawn from this information and a comparison with the stellar lifetimes and ages in Table I and Sec. V.A:

1. The grain lifetimes are shorter than the stellar main sequence lifetimes throughout the systems to $r \sim 1000$ AU.
2. The grain lifetimes are so short at the inner boundaries that the grains must be either continually replenished from some reservoir or else the inner boundaries move rapidly compared to stellar time scales.
3. Icy grains could persist for the lifetimes of the stars throughout the α PsA system and in the outer regions of the α Lyr and β Pic systems.
4. The ice sublimation time is only a few hours in the inner zone of the β Pic disk, so the grains there must be refractory, apparently confirmed by the detection of silicate emission close to the star (Telesco and Knacke 1991).
5. The maximum temperatures in the three systems are so different in terms of implied ice lifetimes that it seems unlikely the inner boundaries are all defined by a phase transition from icy to refractory grains unless the grain composition differs significantly from system to system.
6. Within uncertainties of grain density and radial structure the PR process is competitive with the collision process throughout these systems, although collisions appear to be generally dominant. If PR drag were in fact dominant, the existence of inner voids would require some additional mechanism to prevent grains from moving toward the stars and erasing the voids.

The time scales in Table XI clearly indicate that most of the small grains detected by IRAS are short-lived and must be replenished somehow. Detection statistics for analogs to the prototypes show that these systems are too common to represent chance detection of events as brief as the grain lifetimes.

Collisions can be a mechanism for production as well as destruction of small grains via destruction of larger bodies. Collisional or thermal fragmentation of undetected parent bodies to balance loss of the detectable grains were originally discussed by Weissman (1984), Harper et al. (1984), and Matese et al. (1987).

The grain lifetimes do not seem to offer a general explanation for the inner boundary locations, temperatures, and surface densities of the three prototypes. It is possible that the inner boundaries reflect original structural features such as orbits of major planets. Planets orbiting the stars at the inner boundaries of the infrared-emitting regions and consuming or perturbing grains inbound under the influence of the PR effect could explain the central voids. On the other hand, in analogy with our solar system a large planet might result in an asteroid belt interior to its orbit which would be a source of extra dust, especially in a young planetary system. Surface photometry of the disks at $\lambda = 10$ to $20\,\mu$m with spatial resolution of 1 arcsec or better would be extremely valuable in constraining the sizes of the central voids and determining which processes are responsible for their creation and maintenance.

Minimum mass loss rates from collisions alone can be estimated by integrating Eq. (15) over radii, assuming that all the surface area is in grains with sizes as in Table III and density 1 g cm^{-3}. If this rate were to remain constant over the main sequence lifetime of the stars, a mass in small grains of at least 1×10^{-2} M$_\oplus$ for α Lyr and 1×10^{1} M$_\oplus$ for β Pic would be destroyed.

C. Speculation on History and Evolution

1. Protostellar and Protoplanetary Stages. The broad context of the study of protostellar and protoplanetary evolution tends towards elaboration of a basic planetesimal theory. This proceeds from the gravitational collapse of a rotating molecular cloud core to the formation of an optically thick circumstellar gas and dust accretion disk, which could be the precursor to optically thin grain disks like the ones under consideration here.

The choice between minimum-mass (see, e.g., Cassen and Summers 1983) and massive (see, e.g., Lin and Pringle 1990) protoplanetary disk scenarios is presently a basic controversy in the field of planetary formation. The relevance to this review is the question of origin of material at radii of 10^3 AU like the grains in the disk around β Pic, and by inference the other main-sequence far-infrared excess stars. In the minimum-mass scenario, material could reach that position via (1) the process of angular momentum distribution which dumps much of the initial disk into the star but greatly expands the scale of the remainder, and/or (2) orbital perturbation from the inner solar system; for example, many Uranus-Neptune zone planetesimals could be tossed outward by encounters with the nearly completed planets and found at intermediate semimajor axes of 10^2–10^3 AU, falling short of the Oort cloud. In contrast, in the massive-disk scenario planetesimals can form *in situ* at radii of 10^3 AU.

Calculations (Lin and Pringle 1990) of the evolution of massive disks show that the operation of some viscosity mechanisms makes any original surface distribution produced by infall eventually approach $\sigma_m(r) \propto r^{-1.5}$. This may indicate that the β Pic optical disk with $\sigma \propto r^{-1.7}$ from 100 AU to 1000 AU is a fossil remnant of an original disk rather than material sorted into that region by planetary perturbations.

At later development stages there is ample observational evidence that a significant fraction of young main-sequence stars are surrounded by optically thick circumstellar disks of solar-system size with masses $\sim$0.01 $M_\odot$ (Strom et al. 1989d; Beckwith et al. 1990) which corresponds to the entire disk in a low-mass scenario or the central portions of a more massive disk. Recent near- and mid-infrared (Skrutskie et al. 1990) and mm-continuum (Beckwith et al. 1990) studies of a large sample of pre-main-sequence stars in the Taurus-Auriga complex suggest that half of the known very young ($t < 3$ Myr) solar-type stars have excess emission arising from heated dust embedded within such optically thick disks (see also Walter et al. 1988). At ages $t \geq 10$ Myr, fewer than 10% of the Taurus-Auriga sample show the infrared signatures of optically thick disks. Hence, disks apparently evolve from optically thick to optically thin structures. If all solar-type pre-main-sequence stars are initially surrounded by disks, then roughly half have either fully accreted their initial disk material or have begun to assemble grains into planetesimals by an age of 3 Myr and few optically thick disks survive for more than 10 Myr.

In a series of papers Nakano (1987a, b,1988) has attempted to extend a detailed theory of planetesimal and planet formation to explain the observations of β Pic and α Lyr in a self-contained way. This ambitious model is at odds in some ways with other planetesimal scenarios for the formation of our planetary system in having low initial disk mass and long time scales. Nakano (1987a) "finishes" Neptune in 3 Gyr, more than 1/2 the age of our solar system, whereas models of "runaway" accretion which begin with disks an order of magnitude more massive than the minimum mass produce planets much faster (Lissauer 1987) and put leftover material into substantial inner and outer Oort clouds. The most troubling disagreement between Nakano's model and observations of β Pic is that the model implies that we are viewing the star nearly in the disk midplane through significant optical depth ($\tau \sim 0.5$), but as discussed in Sec. V.A, the metallicity of β Pic explains the star's apparent low luminosity without extinction. Despite this and other problems, the value of these papers lies in the profound importance of the general issues raised by connecting the β Pic and α Lyr systems with theories of planet formation.

Nakano (1988) models the scattered light from the β Pic disk as due to small grains released by planetesimal collisions. He predicts an edge-on surface brightness of the disk following $r^{-4.375}$, quite close to the observed value $r^{-4.3}$, and also an approximately constant disk opening angle in the outer parts of the disk, agreeing with analyses of coronagraph observations. Nakano's model in addition predicts that there should be three regions in the β Pic disk, $r < 4.7$ AU in which the planets are finished, $4.7 < r < 23$ AU in which the planetesimals have had at least one encounter and an equilibrium thickness has been reached, and 23 AU $< r < \infty$, a realm of colliding ice planetesimals which corresponds to the region visible in coronagraph images. The innermost region in the model represents a striking theoretical prediction of a central zone of the appropriate size which might be empty of small grains to the IRAS sensitivity limit. It is interesting that the outer region would not be full of small grains but instead because of low planetesimal collision rates would contain discontinuous regions of relatively high density resulting from individual recent collisions, analogous to the asteroidal dust bands discovered in our solar system in IRAS data (Low et al. 1984; Sykes and Greenberg 1986). If this is true the surface brightness of a face-on disk would not be uniform but instead would be composed of myriad discrete and incomplete bands and streaks.

Planetesimal models of planet formation in general predict a progression of the planet formation process with time to greater radii, reducing the number of planetesimals and therefore the infrared luminosity from small grains. There would also be a decrease in the color temperature of the emission as the boundary between regions containing finished planets (no planetesimals) and unfinished planets (plus planetesimals and plentiful small grains) moves to greater radii. Nakano's models imply that a disk originally of β Pic's particle density and fractional luminosity would be more than an order of magnitude less prominent by an age of 5 Gyr.

2. Encounters with the Interstellar Medium. Lissauer and Griffith (1989) pointed out that erosion by dust grains in the ISM could be a significant destruction process for small grains orbiting far from a star like β Pic. This is true only if ISM grains can reach the circumstellar grains in the face of repulsion by the stellar luminosity; the work of Burns et al. (1979) and Artymowicz (1988) imply that very small grains may be relatively unaffected by radiation forces. The steady erosion of grains around stars with lifetimes of several times 10^8 yr or more would be mostly due to encounters with ISM grains in atomic clouds. The fragments would likely leave the circumstellar disk due to ISM gas drag, or radiation pressure, or ejection velocity above escape velocity.

Over a 100 Myr period in which no molecular clouds are encountered, erosion during the nominal 3% of the time spent in atomic clouds with $n_H \sim$ 20 cm^{-3} would decrease a grain's radius by an average of $\triangle a = 50$ (v/10 km s^{-1})^3 μm. Note the dependence on v^3; that plus the fact that stellar random velocities are larger than cloud random velocities means that case-by-case stellar velocity will be important. The main point of the Lissauer and Griffith paper is that β Pic has a surprisingly low velocity relative to the "local standard of rest" (LSR), defined as the galactic circular orbital velocity vector for the solar neighborhood. The average motion of interstellar Ca II is within 3 km s^{-1} of the LSR (Mihalas and Binney 1981), indicating the LSR is a reasonable choice for the velocity frame of atomic clouds.

Lissauer and Griffith's conclusion is that β Pic's grains should be suffering markedly less ISM erosion than grains around other stars. This may provide an explanation as to why β Pic's disk density is so high and also why the typical grain size is so small compared to other examples. As further evidence, they find a weak correlation between grain luminosity L_g/L_* and stellar velocity for 10 nearby stars which have Vega/β Pic analog far-infrared excesses.

Whitmire et al. (1992) propose instead that passage through a dense region of the ISM would act to temporarily *increase* the total grain area around stars via multiple-impact fragmentation of large grains in pre-existing disks. The grains in our inner solar system which have been produced by asteroid collisions or released from active comets have typical sizes of $100 \, \mu$m (Grün et al. 1985). This is roughly the same size as the grains in the α Lyr disk. Whitmire et al. argue that multiple collisions during passage through an atomic cloud could result in destructive fragmentation even of large grains. In this scenario most circumstellar disks are in quiescent states of low detectability during which large grains are produced by parent body collisions or sputtering. This alternates with brief periods during which the large grains suffer fragmentation so that detectability is greatly enhanced until the various internal removal mechanisms clear out the small grains.

Whitmire et al. (1992) show that five A stars with the most luminous circumstellar disks or candidate disks (α Lyr, α PsA, β Pic, β Leo and ζ Lep) have space motions which are highly correlated. The trajectories of all 5 of these stars passed through or near the Lupus-Centaurus (also called Sco-Cen) ISM concentration ($d = 100$–200 pc) between 5 and 10 Myr ago. Although groups of stars with small dispersion velocities in at least two components are usually young stars of common origin which can often be associated with a known concentration, this is unlikely to be the explanation for this group of 5 stars because their mean velocity is not close to that of objects in the Lupus-Cen concentration or any other known concentrations. Also, evidence in Sec. V.A indicates that the ages at least of α Lyr and β Pic are discordant, and both stars are much too old to have formed in Lupus-Cen. Another explanation would be a selection effect whereby stars which independently passed through the nearest concentrations of the ISM call attention to themselves by their newly enhanced small-grain far-infrared excesses. Such stars which are currently in our vicinity would necessarily have similar velocity vectors.

The essential differences between the Lissauer and Griffith and the Whitmire et al. scenarios is that the former group imagines an initial population of grains of a range of sizes down to $1 \, \mu$m which is not replenished but is eroded progressively except in special cases of low stellar velocity. The latter group imagines continual replacement of large grains which are episodically converted into smaller grains.

3. Post-Main-Sequence Evolution. Matese et al. (1989) investigated the question of the fate of circumstellar disk material when the parent star leaves the main sequence and progresses through red giant and asymptotic giant branch (AGB) stages. The luminosity of 1 to 2 $M_{\odot}$ stars increase by factors of 10^{3}–10^{4} during these stages over periods of 10^{7}–10^{8} yr, with corresponding increases of circumstellar grain temperatures by factors up to 10. For example, small water ice grains located anywhere within 10^{4} AU of β Pic would be heated to above sublimation temperature and be rapidly destroyed. Larger icy parent bodies would become active and release gas and grains as comets do in our inner solar system.

Matese et al. conclude that the grain release rate from activated comets is probably not sufficient to supply the grain mass observed in AGB star winds usually interpreted as condensing directly from the wind material. However, small grains from the comets could provide the necessary cores on which the AGB stellar mass loss would condense into larger grains. The source of those nucleation cores has been somewhat of a puzzle in ISM grain and AGB star research especially given the fact that other types of stars (e.g., hot supergiants and most Wolf-Rayet stars) have significant mass loss but negligible grain condensation. Their model of the environment of AGB stars yields testable predictions of grain condensation location and morphology of the grain flow (flat rather than spherical) which differ significantly from the predictions of classical nucleation theory. Jura (1990) (see also Judge et al. 1987) finds that less than 5% of G–K *giants* have far-infrared excesses in IRAS PSC data. G giants are mostly descendants of A–F main-sequence stars, do not generally show signs of mass loss, and precede the AGB evolutionary stage. If giants had disks with optical depths like those seen around at least 20% of nearby main-sequence stars, they would have had detectable excesses (a G giant has about 3 times the luminosity of its A main-sequence progenitor). Jura concluded that either main-sequence circumstellar grains are destroyed by PR drag without resupply, or the grains are icy such that grains originally near 100 K are quickly destroyed by small temperature increases at the beginning of post-main-sequence evolution.

VI. RELATIONSHIP TO OUR SOLAR SYSTEM

The crucial question of the relation of these Vega/β Pic systems to our solar system and to planetary systems in general needs to be addressed. If the examples of cool debris around main-sequence stars detailed in Sec. IV are each sign of a planetary system, then the Drake equation factor expressing the fraction of stars with planets f_p can be estimated to be greater than 0.15 based on these IRAS investigations.

A. Planetary Systems Around The "Big 3" Prototypes?

As previously discussed, there is evidence that the infrared radiating material in all three prototypes probably lies in a disk in the stellar equatorial planes. Central clearings of solar system scale have been found in these examples because flux densities are high enough to allow a plausible determination at IRAS sensitivity of a maximum grain temperature. This has been interpreted as evidence that planets have accreted and removed an original small grain population from detectability (Diner and Appleby 1986). On the other hand, the depleted zones could also be regions cleared of ice grains by sublimation, or could signify a failure of any material to accumulate in those areas when these systems formed.

Weissman (1984) and Harper et al. (1984) have suggested that the emission around α Lyr is from small grains generated by sublimation or collisions in a cloud of comet-like bodies similar to that thought to have condensed in the outer solar system. Weissman pointed out that it was improbable that planetary-sized bodies are present in the far-infrared emitting region itself because the accretion time for large bodies at these distances from the star are longer than the age of the system.

In the β Pic system, there may be a planet-mass perturber propelling dust particles close to the star to produce the transient absorption line events (Beust et al. 1991). The grain evaporation products may accumulate in a coma of ions observed in the stable heliocentric component of the Ca ion representing the massive "classical" shell.

B. Infrared Detectability of Our Solar System from "Outside"

One can consider how our solar system would appear to an instrument with IRAS' capabilities if viewed from β Pic ($d \sim 16$ pc).

1. The Sun itself would be detectable only at 12 and 25 μm. The diameter of the planetary region (Pluto's orbit) would subtend 5 arcsec. The sum of the infrared flux from the planets, moons, and large asteroids would be a factor of 10^4 below IRAS' best sensitivity limit in "pointed" observations.

2. The bolometric luminosity of the zodiacal dust in the inner solar system is about 8×10^{-8} $L_\odot$, with characteristic temperature of 230 K and typical grain sizes of 100 μm (Good et al. 1986; Grün et al. 1985) The presence of that material would produce at $\lambda \sim 20\,\mu$m (best contrast) a tiny excess of 2×10^{-4} Jy, less than 10^{-2} of IRAS' limit, on top of a solar photospheric flux density of 0.2 Jy. The emission would be contained within a diameter of less than an arcsec. Not only would the signal be too small in an absolute sense, but without high spatial resolution even a sufficiently sensitive infrared instrument would need to be coupled with extremely precise knowledge of the Sun's photospheric infrared spectrum to identify that amount of nonphotospheric emission.

This illustrates once again the principle that direct detection of optical or infrared radiation from extra-solar planetary material is sensitive to area rather than mass; the zodiacal dust contains less than 10^{-10} of the total mass of the planets, yet has an infrared luminosity 10^2 times larger. It is possible that other planetary systems might have more prominent zodiacal clouds. A zodiacal cloud around α Centauri with roughly twice the particle area of our own would be barely within the best sensitivity limit of IRAS in an ideal sense, but that star is seen against the background of a dense stretch of the galactic plane which frustrates attempts at sensitive searches.

3. The classical Oort cloud from which long-period comets are believed to come (see, e.g., review by Weissman 1990), contains of order 100 $M_\oplus$ in comet nuclei (typical radii of order 10 km). These bodies orbit at least 2×10^4 AU (0.1 pc) from the Sun out to an outer limit of roughly 1.5×10^5 AU (0.75 pc) set by encounters with other stars and the galactic tide (Heisler and Tremaine 1986). Objects in this range of radii are loosely enough bound to the Sun that encounters with passing stars can perturb them into orbits that pass into the planetary region. Our Oort cloud would subtend a diameter of 5°

when viewed from β Pictoris. This would allow easy spatial resolution, but the material would be undetectable to IRAS if the mass were contained solely in comet-sized bodies because the total surface area and thus the luminosity would be extremely small (Stern 1990). Also, the temperature would be less than 5 K and its emission would be difficult to find in contrast with galactic material and cosmological background radiation. Therefore, if the classical Oort cloud consists only of larger bodies, it would not be seen by IRAS from a distance of 16 pc. Stern (1990) noted that erosion by ISM grains should result in the production of small grains and greatly enhanced surface area and detectability, but a search of several nearby stars resulted in no detections and an appreciation for how dirty the Galaxy is at the relevant wavelengths.

4. Many authors have speculated that the process of forming the Oort cloud could have left as much as 5 to 100 times more mass in a so-called "inner Oort cloud" or "Hills cloud" than in the classical Oort cloud (see, e.g., Tremaine 1990; Weissman 1990). If there are in fact objects closer to the Sun than about 2×10^4 AU, their orbits are more tightly bound than orbits in the classical Oort cloud and thus are not usually perturbed into the planetary region. The inner Oort cloud is imagined as a plausible central concentration of the known Oort cloud which might approximate a disk in its inner regions ($r \leq 5000$ AU; Tremaine 1990) but would tend towards sphericity in transition to the outer observed Oort cloud due to perturbations from passing stars.

Note that the "maximum" mass estimates for the resolved circumstellar disks in Table III are in the same range as estimates of the mass of the inner Oort cloud, which more than doubles the minimum gas mass needed in the original protoplanetary disk. However, even this large amount of material would have too little surface area for the IRAS sensitivity limits if it consisted only of large bodies. Also, material at $r > 100$ AU around the Sun would be colder than 30 K which is the approximate limiting temperature for good contrast to background material at low galactic latitudes. From β Pic the Sun appears near the boundary between Draco and Hercules, at $b_{II} \sim 30°$, so the limiting temperature in this specific hypothetical case might be cooler.

In summary, given the known characteristics of our system, IRAS would detect nothing from $d = 16$ pc at temperatures corresponding to the planetary region, $T > 50$ K, $R < 30$ AU, and the amount of material known or hypothesized to exist at $T < 30$ K, $R > 100$ AU is insufficient for IRAS sensitivity. The region of potential IRAS detection by process of elimination is therefore the region roughly between 30 and 100 AU from the Sun lying just outside the planetary region which may contain the Kuiper belt.

C. The Kuiper Belt

The Kuiper belt is imagined alternatively as (1) an extension of the proto-planetary disk in which low density of planetesimals and long encounter time scales prevented the formation of planets, or (2) a region into which Uranus-Neptune planetesimals were propelled during the formation of those planets. An excellent review of limits on "dark" (undiscovered) matter in the solar system, including speculation regarding the properties of the Kuiper belt, has been written by Tremaine (1990). The simplest reason to believe there may be such a structure is the low inclination to the ecliptic of the orbits of short period ($P < 20$ yr) comets, also known as "Jupiter-family" comets. There is no simple way to produce this population dynamically unless their reservoir is a flattened system, in contrast to the Oort cloud (long-period) comets which have random inclinations and thus probably a spherical reservoir. Figure 9 shows that the surface density of the planetary region follows a power law which when extrapolated predicts about 5 $M_\oplus$ in the region between 30 and 100 AU, below the present limits of detectability placed by measurement of planetary and spacecraft motions.

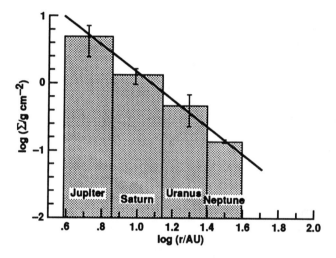

Figure 9.   Mass surface density profile of the solar system if each Jovian planet's mass were spread half-way (in a logarithmic sense) to the adjacent planets. This can be imagined to represent the pre-solar nebula's surface density at some stage during the formation of the solar system. It is possible that the mass distribution continues beyond the known planets, allowing roughly 1 to 10 Earth masses to reside between 30 and 100 AU from the Sun in some form.

If one imagined that the nearby stars all have planets, zodiacal clouds, Kuiper belts, and Oort clouds with approximately the characteristics of the components of our planetary system, an instrument like IRAS would be sensitive only to the Kuiper belt component temperatures and scales. Table XII summarizes the detectability of the components of our solar system and makes comparison to the general properties of the Vega/β Pic systems. The radius in this table characterizing the main-sequence grain disks is not the outer radius, but rather is the far-infrared size of the three resolved systems and the peak surface density in models of the β Pic disk. The best analog in scale to the main sequence disks is the Kuiper belt. However, if only large bodies (comet nuclei) exist there, the solid surface area is insufficient for infrared detection.

TABLE XII

Circumstellar Dust Properties

Location	R (AU)	T (K)	Log Area, AU^2	Log $L_g/L_\star$	Morphology
Zodiacal Dust	3	200	−5	−7	flattened
Kuiper Belt (1 M$_\oplus$,10 km objects)	> 50	40	−5	−9	flattened
Kuiper Belt (1 M$_\oplus$, 10 μm–10 km size distr.)	> 50	40	0	−5	flattened
Typical M.S. Grain Envelopes	100	30–150	0–2	−5 to −3	flattened ?
Oort Cloud (100 M$_\oplus$, 10 km objects)	10^4	3	−5	−12	spherical

An order-of-magnitude check on the plausibility of our solar system having a collection of small grain fragments from Kuiper-belt comet collisions shows that such a disk would have a grain area very comparable to Vega/β Pic disks. If 1 M$_\oplus$ of 10 km bodies with material density 1 g cm^{-3} are imagined confined between 30 and 100 AU from the Sun (figure from Tremaine 1990), the collision time per parent body is about 10^{11} yr so that most of them would be intact at the present age of the solar system. The number of collisions per year throughout the region would be approximately 10^{-1}.

If the fragments from each collision are assumed to have an $a^{-3.5}$ size distribution running from a minimum size somewhere in the range of 1 to 100 μm (like dust in comet comae, the zodiacal cloud, and the prototype main-sequence far-infrared sources) to a maximum size of 1 km (1/10 the size of the parents), then the PR destruction time scale for the smallest grains (those dominating the surface area and therefore the most detectable) would be in the range 10^6 to 10^9 yr depending on original orbit radius. The total amount of new small grain fragments produced by parent collisions during a PR time scale would yield a small-grain optical depth perpendicular to the ecliptic plane at $r = 30$–100 AU in the range 1 Myr to 1 Gyr strikingly similar to the lower range of extrasolar examples. The small-grain mutual collision time scale in this case would be less than but within an order of magnitude of the PR time scale, just like in the 3 prototype sources. In the case of our Sun, the distances would make the characteristic temperature of the emission 50 K or colder.

D. Steps Toward a Connection

Is it true, then, that what we are observing with IRAS around nearby main-sequence stars corresponds to their "Kuiper belts"? There are at least three ways to investigate the connection. *First,* are there planetary perturbations in any of the IRAS-detected systems, and where do the massive objects lie with respect to the IRAS material? *Second,* can systems be found with material at warm temperatures/radii corresponding to the planetary zone along with cooler Vega-type material? In other words, are the phenomena of clear zones in the "Big 3" resolved systems, and also the general lack of detection of material warmer than roughly 200 K around nearby stars (Aumann and Probst 1991), signs of *failed* planetary systems or *completed* planetary systems? *Third,* does our solar system have a cool grain disk outside the planetary zone?

1. Solar-Type Stars With Planetary Perturbations. Campbell et al. (1988*b*) investigated 12 slowly rotating, single, bright F-K dwarfs for radial velocity variations in the first search sensitive enough to find Jupiter-mass planets. Their technique depends on passing starlight through a cell of hydrogen fluoride which provides sharp-reference velocity lines insensitive to effects like flexure in the spectrograph. Their resulting mean external error is about 13 m s^{-1}.

Seven stars in the sample have marginal detections of velocity trends over 8 years' time, some with concavity or convexity implying periods of order a decade, although no system has been seen to go through a full cycle yet. There is no correlation between the radial velocity variations and stellar activity indicators such as equivalent line widths of Ca II H and K line emission. If the perturbations are real, the implied sizes of the companions are of order 1 to 10 Jupiter masses.

The significance of these results in the context of this review is that one of the systems with possible Jovian companions, ϵ Eri, is an IRAS far-infrared excess system. The emission temperature of 70 K could indicate material within or beyond the region of planetary companions, depending on grain size. The other six Campbell candidates do not have far-infrared excesses at IRAS sensitivity and thus would be prime targets for observation with more sensitive future instruments such as ISO or SIRTF.

2. Young Stars with Hot Debris. Calculations of collision dynamics and orbit evolution of planetesimals in our solar system yield a particle number density 1000 times larger than the present value as late as 500 Myr after the Sun reached the main sequence, with about a 100 Myr half-life for subsequent decline (Grinspoon 1988; Grinspoon and Sagan 1991). Witteborn et al. (1982) calculated the amount of infrared radiation which would be emitted from an $a^{-3.5}$ size distribution of grains and planetesimals orbiting main sequence stars assuming a spatial distribution like that of the planetary masses in our solar system. Their results are that 80% of the infrared excess will come from warm material in the "terrestrial" region despite the 100 times larger mass in the Jovian region, and that 1% of the solid mass of the planets is easily found at 10 to 20 μm if the photosphere of the parent star is detectable. This mass corresponds to the particle number density calculated for roughly 200 to

500 Myr after the formation of the solar system. Backman et al. (1990) made a study of coadded IRAS survey data on main-sequence A stars in open clusters younger than 1 Gyr. IRAS photometry was not accurate enough to discern individual stars in most cases because of the distances to the clusters, but statistical statements can be made about the presence or absence of infrared excesses in the cluster stars considered as groups. There appears to be greater emission from A stars at 12 μm in the younger two clusters than in the older two. The trend with age may represent progressive destruction or accretion of grains.

Although the Pleiades are passing through a molecular cloud that is responsible for the spectacular optical reflection nebulae (Breger 1987), and calculations of radiation pressure efficiency imply that very small grains like ISM grains might be able to come close to luminous stars, the 12 to 25 μm ratios suggest that ambient cloud grains are not the cause of the observed excesses. Assuming that a thin undisturbed distribution of grains around an unrelated star would be roughly $n(r) \sim r^0$, the expected thermal emission spectrum from ISM-size grains would rise steeply toward longer wavelengths, $F_v \sim \lambda^2$, because of the increase of grain number with distance (cf. Eq. 10).

These results need to be confirmed via groundbased observations especially in light of Aumann and Probst's (1991) studies of false 12 μm excesses around nearby field stars. Also, sub-millimeter searches for this effect in the same clusters have returned null results (Zuckerman, personal cummunication) although that technique would be somewhat less sensitive to material at warm temperatures. Unfortunately these clusters are so distant that IRAS data could not be used to check for complementary cool excesses with properties like the Vega/β Pic systems, so a direct comparison of location of material around the cluster stars versus generally older nearby field stars will have to await future instruments such as ISO or SIRTF.

3. Does Our Solar System Have a Cool Grain Disk? If our solar system had a grain disk outside the planetary region in the region corresponding to the Kuiper belt, it is surprising but true that grain surface area and temperatures like the examples turned up in IRAS surveys would be difficult to detect from *inside* because of the powerful interference viewed from Earth of bright warm foreground zodiacal dust. Backman and Gillett (1987) and Aumann and Good (1990) found that the difference between IRAS scans at 60 and 100 μm of the zodiacal dust emission in the ecliptic plane, and models of the expected emission, could represent some emission from material beyond the planetary region. It is possible that all the emission observed is actually in the inner solar system and will be explained in the future by more sophisticated models of the zodiacal cloud. However, the uncertainties at this point in our knowledge of the zodiacal system *allow* a disk at temperatures below 40 K, located at $r > 50$ AU, with total optical depth somewhat smaller than the material around α Lyr.

The upper limit on optical depth for such a cloud is larger than the apparent mean amount of material observed in ensemble for nearby G dwarfs.

Aumann and Good (1990) argue that, since the typical G dwarf appears to have some cool circumstellar material, it would be peculiar if the Sun *did not*. If there is such material in our solar system, it cannot be concentrated too strongly towards the ecliptic plane, or its surface brightness would be too large. "Wedge" angles ι greater than 5° are consistent with the observations.

Tremaine (1990) points out that upper limits on number density of comet-nuclei-sized bodies in the Kuiper belt from the best present optical searches are just beginning to approach values corresponding to upper mass limits from dynamical calculations. Cruikshank et al. (1990), however, find that these bodies might be detectable with SIRTF.

VII. SUMMARY AND FUTURE DIRECTIONS

A detailed understanding of the genesis of planets is one of the major goals for the astronomical community, equal in import to investigation of the origin of the universe as a whole. The IRAS mission resulted in the startling discovery that normal main-sequence stars often host cool orbiting grains with likely but not certain connection to planets. The state of our knowledge about these systems after the IRAS mission can be divided into three categories: (a) a few systems close enough/luminous enough that hints of shape and structure have been discerned; (b) one hundred or more stars with cool far-infrared excesses but without IRAS spatial resolution or sufficiently good photometry to demonstrate conclusively that the emission is due to orbiting grains rather than to possible alternate mechanisms; and (c) most stars, up to 80% of the nearby A-K dwarfs and all but one M dwarf, with no circumstellar material apparent at IRAS sensitivity.

Future infrared observations should aim to: (1) explore the three proto-type disks in detail, addressing questions of radial and azimuthal symmetry and precise nature of boundaries; also, perform reflectance mineralogic spectroscopy where possible; (2) spatially resolve IRAS candidates, checking whether the sources are centered on the stars (i.e., not background) and are disks in the stellar equatorial planes, estimating grain sizes from angular scale vs temperature, and performing sufficiently precise photometry that maximum temperatures (inner boundary locations) can be deduced; and (3) search for grains around stars with IRAS nondetections; a hundred-fold better sensitivity would make detection of systems as sparse as our solar system's zodiacal dust cloud possible in some circumstances.

The story behind the origin, maintenance and destruction of such disks will gradually become clearer as main-sequence circumstellar disks come to be understood in the context of stellar evolution. For example, we can ask what type of circumstellar disks normally remain after star formation? This could be answered by further examination of classical and "naked" (weak) T Tauri stars and open cluster stars, with ages of order 10^7 to 10^8 yr, in order to investigate inner disk clearing time scales and persistence of outer disks. Also, what happens to these disks when the stars evolve off the main sequence? This calls for further observations of evolved stars in search of evidence of increased sublimation and possible augmented dust nucleation occurring when remnant disks are exposed to post-main-sequence luminosity increases and mass loss. Finally, do such disks represent success or failure modes in planet building? We need to continue searches of the distant regions of our own solar system's ecliptic plane for evidence of planetesimal-sized parent bodies and collision fragments, and to compare expanded lists of infrared excess stars with lists of suspected planetary-mass astrometric and spectroscopic binary companions. We can hope for eventual firm knowledge of the frequency with which normal stars possess material left over from the process of planet formation.

Acknowledgment. D. B. acknowledges a National Research Council post-doctoral fellowship, a NASA Origins Initiative grant, the NASA-Ames SETI Institute, and Franklin and Marshall College for support. We thank F. Witteborn, M. Werner, P. Cassen, J. Lissauer, P. Artymowicz, and the referee M. Skrutskie for many helpful comments. We also thank A. Graps for computer searches relating to stellar surveys and N. Jennerjohn for assistance in preparing the tables and references.

PLANETARY NOBLE GASES

KEVIN ZAHNLE
NASA Ames Research Center

Noble gas elemental and isotopic abundance patterns should in principle provide
constraints on the formation and early evolution of the terrestrial planets and their
atmospheres, but at present these patterns are poorly understood. The seemingly
reasonable suppositions that noble gases would behave like other volatiles (only more
so), and that noble gases on planets derive from noble gases in meteorites have both
proved more wrong than right. It is now fairly clear that the so-called "planetary" rare
gases are not found on planets. The unexpected discovery that neon and argon are vastly
more abundant on Venus than on Earth points to the solar wind rather than condensation
as the fundamental process for placing noble gases in the atmospheres of the terrestrial
planets; however, solar wind implantation may not be able to fully reproduce the
observed gradient, nor does it obviously account for similar planetary Ne/Ar ratios
and dissimilar planetary Ar/Kr ratios. More recent studies have emphasized escape
rather than accretion. Hydrodynamic escape, which is fractionating, readily accounts
for the difference between atmospheric neon and isotopically light mantle neon. It can
also with more difficulty account for isotopically heavy Martian argon. Atmospheric
cratering, which is nearly nonfractionating, can account for the extreme scarcity of
nonradiogenic noble gases (and other volatiles) on Mars. Escape is less satisfying
when used to force Ar and Ne on Venus and Earth into accord. Outstanding problems
remain, namely, the different Ar/Kr ratios of Venus and Earth and the fractionated and
underabundant Xe found on both Earth and Mars. At least two sources seem to be
required.

For reasons which are becoming increasingly hard to recall, it was once
thought that noble gases, being the most volatile elements and the simplest
chemically, would tell a relatively straightforward story of volatile evolution
in the solar system. The noble gases are chemically inert, or nearly so. Among
real substances they are the closest approximations to that ideal atmospheric
species, the obligate atmophile, that exists only in air. To the extent that
they approach the ideal, the history of the noble gases is the history of the
atmosphere. Of course, with the possible exception of a direct solar nebular
source, the very existence of noble gases in planetary atmospheres demands
that they fall somewhat short of the ideal. They are found in the mantle, small
amounts are dissolved in seawater, and xenon at least can be adsorbed in crustal
rocks. Noble gases are also found trapped in meteorites or driven into exposed
surfaces by the solar wind. Where temperatures are low enough they might
even be expected to freeze out. Because none of these means of incorporating
noble gases into planetary atmospheres is very efficient they are rare, and so

they are also called the "rare gases." Noble gases are also interesting for their radiogenic isotopes, which can be relatively prominent in view of the extreme scarcity of their nonradiogenic siblings. These are particularly useful as chronometers of planetary outgassing and early atmospheric evolution.

I. A BRIEF, TRUNCATED HISTORY

Apart from helium, noble gases do not readily escape from modern planetary atmospheres. Their low abundance in air implies that if Earth ever had a primary atmosphere of solar composition, it must have either been very thin or very efficiently lost. It was therefore early concluded that Earth's was a "secondary" atmosphere, i.e., one degassed from solid materials (see Walker 1977). It was assumed that the noble gases had degassed from the mantle, probably early and catastrophically (Fanale 1971). The mantle, of course, is just a stop along the way, and the original source materials were naturally sought among meteorites.

Nonradiogenic noble gases in meteorites were divided into two distinct classes based on relative elemental abundances (see Ozima and Podosek 1983; Swindle 1988). These were suggestively named "solar" and "planetary." The solar pattern, characterized by roughly solar abundances, appears to have a genetic relationship with the solar wind. The "planetary" pattern is greatly depleted in the lighter noble gases. With the obvious exception of xenon and the more subtle exception of neon, it is elementally and isotopically like air (see Fig. 1).

Of the two meteoritic components, the "planetary" component better matches air (hence the unfortunate name), and therefore it was for some time presumed that the noble gases actually found on planets descend directly from "planetary" noble gases in meteorites. Moreover, the "planetary" pattern qualitatively resembles the elemental pattern produced experimentally when solar noble gases are adsorbed on grains (Fanale and Cannon 1972; Ozima and Podosek 1983). Quantitatively, terrestrial argon and krypton could be accounted for either by accumulating the bulk of the planet from meteorites like the H-type ordinary chondrites, or by accreting a relatively thin veneer of meteorites like the carbonaceous chondrites, which have higher intrinsic abundances of volatiles and rare gases (Anders and Owen 1977). A basically simple picture seemed to emerge. There were some unresolved problems—in particular, the discord at xenon was large, and could only be resolved if 96% of Earth's xenon were hidden somewhere other than the atmosphere and the isotopes were ignored—but these seemed surmountable.

Then we began to gather data from other planets. At first things seemed to go well enough, if not entirely according to plan. Viking's Mars resembled nothing so much as Earth depleted in volatiles across the board by a factor of about 30 (Anders and Owen 1977). Such apparent volatile poverty was distressing, both intellectually (prevailing theory had predicted that volatiles would be more abundant at greater distance from the Sun) and emotionally

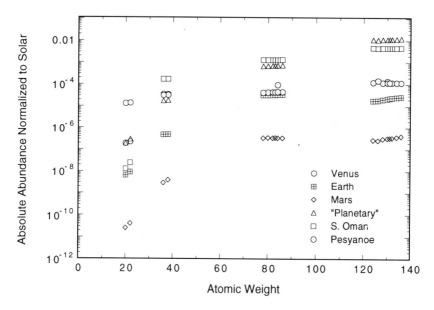

Figure 1.   Abundances of the nonradiogenic noble gas isotopes with respect to solar abundances [g/g] for a few solar-system objects of interest. Sources are listed in Tables I and II. The "planetary" meteoritic pattern plotted here is the AVCC (average carbonaceous chondritic) pattern. South Oman is an enstatite chondrite with high argon. It is an example of the "subsolar" pattern. Pesyanoe illustrates the "solar" meteoritic pattern. The abundance of ^{36}Ar on Venus is quantitatively similar to that in many meteorites.

("one of the great disappointments of the space age"; Anders and Owen 1977), but could be accomodated within a veneer model simply by assigning Mars a thinner veneer. The veneer became identified with carbonaceous chondrites. Again there was a problem with "missing" xenon, but, as with Earth, it was thought likely to be hidden, adsorbed in some crustal rock (Fanale et al. 1978).

Venus and Earth have comparable surficial carbon and nitrogen inventories, and so it was predicted that Venus and Earth would have comparable inventories of the nonradiogenic rare gases, with the exception of xenon, which could not reasonably be adsorbed in crustal materials at Venusian surface temperatures (Anders and Owen 1977). What Pioneer Venus found at Venus was something completely different. Nonradiogenic neon and argon are, respectively, 30 and 70 times more abundant than on Earth (Donahue and Pollack 1983), and the Ar/Kr ratio is more "solar" than "planetary" (Donahue et al. 1981). Venusian xenon is extremely uncertain, but the Kr/Xe ratio reported by Donahue (1986) is much closer to solar than to "planetary" meteorites. The absolute abundance of ^{36}Ar on Venus is strikingly large (Fig. 1). Of known solar system materials, it is matched only by a few meteorites and lunar soil samples heavily exposed to the solar wind. Venus not only killed the hypothesis of a thin meteoritic veneer as the source of planetary rare gases but also the presumed linkage between the inventories of condensible volatiles and noble gases that gave rise to it.

Since then it has become apparent that the SNC meteorites come from Mars (Bogard et al. 1984; McSween 1985; Pepin and Carr 1992). Atmospheric gases trapped in one of these meteorites have greatly increased our knowledge of the isotopic composition of the Martian atmosphere, which for all four elements differ significantly from planetary. These data reinforce a conclusion that had become obvious after Pioneer Venus: "planetary" rare gases are not found on planets (Wetherill 1981). There remain some likenesses, particularly the Ne/Ar ratio, but there are at least as many differences: for both Earth and Mars, the "planetary" Xe/Kr ratio is some 30 times too high and the xenon isotopic patterns are grossly discordant, and neon is isotopically too light; Martian nonradiogenic argon is strikingly heavy; and the data for Venus, such as they are, more closely resemble the solar pattern than the "planetary" pattern (Fig. 2). It now appears that noble gases on planets do not look all that much like either the "planetary" or solar meteoritic components, and the resemblances that are perceived in the elemental pattern tend to vanish when one looks carefully at the isotopes.

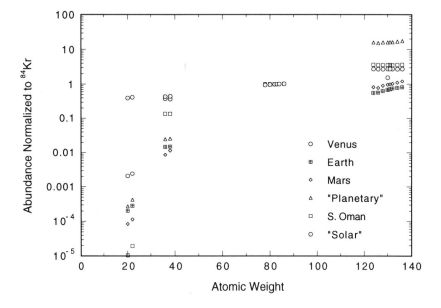

Figure 2. The same as Fig. 1, but with all reservoirs also normalized to ^{84}Kr. This better illustrates the qualitative differences between reservoirs. Xenon fractionations on Earth and Mars are clearly visible even on this scale, as is a general tendency for the less abundant elements to be more isotopically fractionated.

In particular, the discrepancies with xenon never went away. When only Earth was known, it was suggested that the bulk of terrestrial xenon is hidden, adsorbed inside crustal rocks, particularly shales (Canalas et al. 1968; Fanale and Cannon 1971); a similar explanation (without the shales) was offered for Mars when Viking showed that it too was missing xenon (Fanale et al. 1978). Buried xenon has been vigorously sought on Earth, but it has not been found (Podosek et al. 1981; Ozima and Podosek 1983; Wacker and Anders 1984; Bernatowicz et al. 1984,1985; Matsuda and Matsubara 1989). It now appears that crustal materials harbor no more xenon than does the atmosphere, and no attempt has ever been made to account for the isotopes. If some terrestrial hiding place is eventually found that can store 96% of the xenon while leaving the atmospheric residuum highly mass fractionated, then one can again seriously entertain the "planetary" hypothesis.

II. THE DATA

There is voluminous literature on noble gases in planetology. Ozima and Podosek (1983) survey the entire field, from nucleogenesis to Pioneer Venus. Donahue and Pollack (1983) review the wreckage of noble gas comparative planetology after Pioneer Venus. The last serious attempt to explain atmospheric rare gases by adding together different sources was made by Bogard (1987). Escape processes began to receive serious consideration with Sekiya et al. (1980b) and Donahue (1986), and have dominated the literature since Hunten et al. (1987) and Sasaki and Nakazawa (1988) pointed out that mass fractionation patterns similar to those observed on Earth and Mars are a predictable consequence of hydrodynamic escape of hydrogen-rich early atmospheres. By far the most ambitious study of this kind is Pepin's (1991). Ozima and Igarashi (1989) and Hiyagon and Sasaki (1988) have recently reviewed terrestrial rare gases; the latter review is mainly concerned with xenon. This chapter will emphasize escape.

Detailed reviews of the data are given by Ozima and Podosek (1983) and Pepin (1989,1991). Swindle (1988) reviews meteorites; Donahue (1986) is the most recent source for Venus. Terrestrial MORB and Loihi data are presented by Allègre et al. (1987), Ozima and Igarashi (1989), and Honda et al. (1991). Some of these data are presented in Tables I and II and shown graphically in Figs. 1–4. Error bars have been omitted; the interested reader is urged to consult the sources noted above for detailed discussion.

Nonradiogenic rare gases on planets are neither "solar" nor "planetary," but something in between. The key observations for comparative planetology are that (i) atmospheric inventories of argon and neon fall off precipitously with increasing distance from the Sun, such that ^{36}Ar, e.g., drops as 11,000:160:1 going from Venus to Earth to Mars. (ii) The Ar/Kr ratio on Venus is nearly solar while the Ar/Kr ratio on Earth and Mars is "planetary"; as a result, unlike argon and neon, krypton abundances on Earth and Venus are not very different. (iii) Xenon on both Earth and Mars is isotopically strongly

TABLE I

Elemental Abundances

	^{36}Ar [g/g]	$\dfrac{^{20}\text{Ne}}{^{36}\text{Ar}}$	$\dfrac{^{36}\text{Ar}}{^{84}\text{Kr}}$	$\dfrac{^{84}\text{Kr}}{^{130}\text{Xe}}$
Solar[a]	7.6×10^{-5}	40	3300	125
Solar wind[b]				
Pesyanoe	2.2×10^{-9}	16	2270	46
Venus[c]	2.5×10^{-9}	0.2	1200	13
S. Oman[b]	1.2×10^{-8}	0.003	447	36
Earth				
air[b]	3.5×10^{-11}	0.52	48	183
MORB[d]	1.2×10^{-13}	2.3	23	46
Loihi[d]	1.4×10^{-12}	0.67	43	126
Mars[e]	2.2×10^{-13}	0.36	28	130
"Planetary"				
AVCC[e]	1.3×10^{-9}	0.4	84	8

[a] Anders and Grevesse 1989.
[b] Pepin 1991; AVCC = average carbonaceous chondrite.
[c] Pollack and Black 1982; Donahue 1986.
[d] Allègre et al. 1987; MORB = mid-ocean rift basalt.
[e] Pepin 1989; AVCC = average carbonaceous chondrite.

mass fractionated and elementally underabundant with respect to krypton. (iv) The absolute abundance of ^{36}Ar on Venus is similar to that of the most argon-rich meteorites. (v) The Ne/Ar elemental ratio is roughly the same on Earth, Mars, Venus and "planetary" meteorites; however, (vi) neon on planets is isotopically between solar wind and "planetary" neon, and argon on Mars is isotopically heavier than either.

A. Neon

The ^{20}Ne/^{22}Ne ratio is quite variable in the solar system (Fig. 2, Table II). Meteoritic neon is notable for two oddities: the existence of a pure ^{22}Ne component, Ne-E, that apparently derives from the decay of ^{22}Na in the presolar grains; and solar-wind ^{20}Ne/^{22}Ne is $\sim$13.7, but the solar flare ratio is only $\sim$10 (Mewaldt et al. 1984). The arguments given for accepting the solar wind ratio as truly solar seem stronger than those favoring the flare ratio, but the subject is not yet closed. The "planetary" meteoritic ratio is $\sim$8.9 (Pepin 1991). Planetary ratios are intermediate: air is 9.8; Mars (SNC) is 10.1$\pm$0.7 (Swindle et al. 1986b; Wiens et al. 1986); and Venus is reported as 11.7$\pm$0.7 (Donahue 1986; Donahue et al. 1981). Terrestrial mantle and diamond ratios tend to be isotopically light (Ozima and Igarashi 1989).

B. Argon

Argon has stable isotopes of atomic weight 36, 38 and 40 (Table II). Argon in

TABLE II

Isotope Ratios

	$\dfrac{^{20}\text{Ne}}{^{22}\text{Ne}}$	$\dfrac{^{36}\text{Ar}}{^{38}\text{Ar}}$	$\dfrac{^{40}\text{Ar}}{^{36}\text{Ar}}$	$\dfrac{^{129}\text{Xe}}{^{130}\text{Xe}}$	$\dfrac{^{136}\text{Xe}}{^{130}\text{Xe}}$
Solar Wind[a]	13.7	5.3–5.8			
Lunar soils[a]	12.9	5.33		6.35	1.82
Pesyanoe[a]				6.31	1.79
U-Xe[b]				6.05	1.66
Venus[c]	11.7	5.6	1.0		
S. Oman[a]	7.4[b]	5.46	8	8.28[b]	1.87
Earth					
air[d]	9.8	5.32	295.5	6.48	2.17
MORB[d]	10–13	5.32	17000	6.95	2.38
Loihi[d]	10–11[e]	5.32	390	6.48	2.19
Mars[f]	10.1	4.1	2260	16.4	2.29
"Planetary"					
AVCC[b]	8.9	5.3		6.35–6.71	1.98

[a] Swindle 1988.
[b] Pepin 1991; AVCC = average carbonaceous chondrite.
[c] Donahue 1986; Pollack and Black 1982.
[d] Allègre et al. 1987; MORB = mid-ocean rift basalt.
[e] Honda et al. 1991.
[f] Wiens et al. 1986.

air is mostly radiogenic ^{40}Ar, formed by the decay of ^{40}K (half-life 1.25 Gyr). The accumulation of ^{40}Ar in the atmosphere is the primary measure for the degassing of the mantle over geologic time. Dreibus and Wänke (1987) estimate that 53% of Earth's ^{40}Ar has been degassed; other models for Earth's K content differ somewhat. Because the nonradiogenic abundance of ^{40}Ar is negligible, the ratio of ^{40}Ar to ^{36}Ar in mantle samples is an excellent indicator of the extent to which the mantle has been degassed (Ozima and Podosek 1983).

Earth and most meteorites have an $^{36}\text{Ar}/^{38}\text{Ar}$ ratio of $\sim$5.3. The solar wind value is also usually given as $\sim$5.3. The actual solar value is probably higher; estimated values range from 5.4 to 5.8 (Swindle 1988; Pepin 1989,1991). The Venusian ratio has been reported as 5.6±0.5 (Donahue 1986). The SNC meteorites yield a Martian ratio of 4.1±0.3 (Wiens et al. 1986; Swindle et al. 1986b) that is clearly quite distinctive.

C. Krypton

To first order, all four known (Earth, Mars, solar wind and planetary) krypton inventories are isotopically similar. Modest fractionation patterns have been identified. It was once thought that terrestrial krypton was isotopically lighter than solar-wind krypton (Ozima and Podosek 1983), but newer data have reversed this (Pepin 1989). SNC krypton appears to be slightly lighter than terrestrial krypton, while "planetary" krypton is very slightly heavier than Earth's. In all cases, the reported krypton fractionations are quantitatively small, no more than 20% that of xenon. Krypton fractionations are small enough to admit many explanations. They are within the observed variance for adsorption (see adsorption data in Ozima and Podosek [1983]) and, as the difficulty defining solar Kr implies, the same goes for solar-wind implantation.

D. Xenon

Xenon has nine stable isotopes, several of which have been affected by the decay of extinct radionuclides and the apparent preservation of presolar isotopic anomalies (Ozima and Podosek 1983; Swindle 1988). Xenon isotopes with substantial radiogenic sources are ^{129}Xe, formed by decay of extinct ^{129}I (half-life 17 Myr) and $^{131-136}$Xe, formed occasionally by the spontaneous fission of ^{238}U (half-life 4.47 Gyr) or the (no longer) extinct ^{244}Pu (half-life 82 Myr). About 6.7% of Earth's ^{129}Xe is radiogenic, and about 60% of Martian ^{129}Xe is radiogenic (Table II). Fission xenon is harder to measure because it is spread over several isotopes, and therefore obscured by the general trend of fractionation of terrestrial xenon and the uncertain composition of primordial xenon. Pepin and Phinney (1978) find that 4.6% of the ^{136}Xe in air is plutogenic, while Igarashi and Ozima (1988) get 3.1%. Modern terrestrial Xe is then the sum of mass fractionated U-Xe and ^{244}Pu fission Xe. Fission xenon is not evident on Mars (Swindle et al. 1986b).

The known xenon components are all quite distinct. "Planetary" and solar wind (SUCOR) xenon differ in that the former has a large excess of the heavy isotopes ^{134}Xe and ^{136}Xe; the excess, which is not consistent with known fission products, is called Xe-H (Ozima and Podosek 1983). Both terrestrial and Martian xenon are profoundly mass fractionated. Within measurement error, Martian xenon looks like SUCOR Xe fractionated by about 3.8% per amu or like planetary Xe fractionated by about 2.5% per amu (Fig. 3). Terrestrial Xe is fractionated to a similar extent (Fig. 4), but it is not yet clear what it has been fractionated from.

One of the problems with xenon is that the composition of solar xenon is uncertain. Pepin and Phinney (1978) argue that solar wind xenon has also been polluted with Xe-H. They suggest that the original terrestrial Xe is the true ancestral solar xenon, which they call U-Xe. If it is assumed that the heavier Xe isotopes in air have been supplemented by ^{244}Pu fission products after mass fractionation of nonradiogenic Xe took place, and that the extent of mass fractionation is that deduced for the light isotopes, then the known composition of plutonium fission products can be used to determine both the amount of plutogenic xenon and the composition of U-Xe.

E. Internal Rare Gases

Earth's mantle may harbor considerable quantities of rare gases, both radiogenic and nonradiogenic. There are basically two kind of mantle sam-

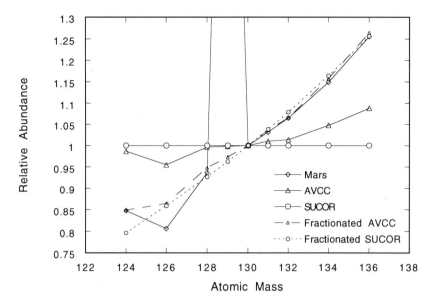

Figure 3. Xenon isotopes on Mars can be produced by mass fractionation of AVCC (average carbonaceous chondritic) xenon at 2.5% per amu or by mass fractionation of solar-wind (SUCOR) xenon at 3.8% per amu. It is not matched by mass fractionation of U-Xe.

ples: mid-ocean ridge basalts (MORB) that sample the highly degassed upper mantle, and hot spot volcanism that may sample a deeper, more primitive, possibly undegassed lower mantle.

Gases trapped in the glassy margins of MORB sample the depleted mantle (the depleted mantle is that part of the mantle from which the continents have been extracted). MORB rare gases are especially rare, and marked by a relatively high abundance of radiogenic isotopes. With the exceptions of helium and neon, the nonradiogenic rare gases are isotopically like air. Neon appears to be isotopically somewhat lighter than air; this may imply preferential neon loss from the atmosphere (Allègre et al. 1987; Staudacher 1987; Ozima and Igarashi 1989; Zahnle et al. 1990a). Neon in terrestrial diamonds is also often light (Honda et al. 1987; Ozima and Igarashi 1989). Very high $^{40}Ar/^{36}Ar$, greater than $\sim 10^4$, when combined with the low K abundance in depleted mantle, indicates that ^{36}Ar has been degassed to better than 99.6% (Ozima and Podosek 1983; Allègre et al. 1987). The xenon in these samples is also marked by small excesses relative to air of the radiogenic isotopes, both ^{129}Xe and fissogenic (Allègre et al. 1987; Ozima and Igarashi 1989). It is the presence of excess ^{129}Xe, in particular, that implies the upper mantle was outgassed quickly, before all the ^{129}I decayed (see, e.g., Thomsen 1980; Ozima and Podosek 1983; Allègre et al. 1987). Typical estimates are

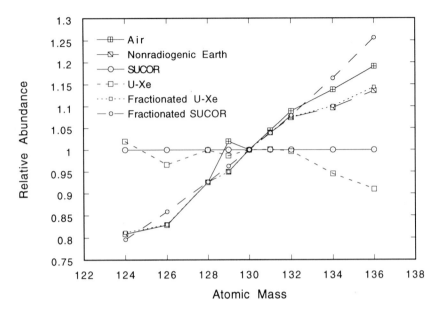

Figure 4.   Nonradiogenic xenon isotopes on Earth can be produced by mass fractionation of U-Xe at 3.8% per amu. U-Xe is the hypothetical composition of primordial solar system xenon (Pepin 1991). Air is the sum of fractionated U-Xe and ^{129}Xe from decay of extinct ^{129}I and ^{131}Xe-^{136}Xe from spontantaneous fission of extinct ^{244}Pu.

that mantle degassing took place within 50 Myr (Allègre et al. 1987; Zhang and Zindler 1989).

   Despite evidence for rapid, near total degassing of the upper mantle, there is also evidence that can be interpreted as implying that a considerable fraction of the mantle remains to this day practically undegassed (Allègre et al. 1987; Zhang and Zindler 1989). For example, Allègre et al. (1987), using a model in which they assume that the atmosphere was generated by degassing part of the mantle, estimate that the amount of nonradiogenic rare gases still held within the lower mantle exceeds the amount in the atmosphere. Primordial ^{3}He, with a present outgassing rate of order 4 atoms cm^{-2} s^{-1}, is the most celebrated product of the undegassed mantle (Ozima and Podosek 1983). The most intriguing samples come from Loihi, an undersea volcano that is destined to be the next island in the Hawaiian chain. Loihi neon is isotopically light, and there is growing evidence that primitive neon is isotopically solar (Craig and Lupton 1976; Honda et al. 1991). Honda et al. (1991) also infer that ^{3}He/Ne is roughly solar, and so raise the possibility that Earth's internal rare gases are all solar. However, as the He/Ne ratio in both the solar and the "planetary" meteoritic rare gases is roughly solar (Ozima and Podosek 1983), it would have been more surprising if Earth's were not. Loihi krypton and xenon both look like air (Allègre et al. 1987), which, taken at face value, would imply a strong genetic relationship between interior and atmospheric rare gases. Of course, with such an observation one fears contamination, here by atmospheric rare gases dissolved in seawater or by subducted air (Patterson et al. 1990)—a suggestion touching off a lively debate (Staudacher et al. 1991; Patterson et al. 1991). The reported Loihi $^{40}Ar/^{36}Ar$ ratio of 390 (Allègre etal. 1987) is only modestly larger than the atmospheric value of 296. If the Loihi samples really are snapshots of the lower mantle, one must conclude that the lower mantle is practically undegassed. Loihi argon would then also support the premise that the atmospheric rare gases are indeed the outgassed complement of the highly degassed upper mantle, and so would rule out a very late veneer as a plausible source of atmospheric rare gases. On the other hand, if the Loihi gases are contaminants, no such sweeping conclusions can be drawn.

Ott (1988) has reported evidence for nonatmospheric noble gases in Shergotty, Nakhla and Chassigny. These are strikingly unlike the Martian atmosphere, and they are also strikingly unlike either terrestrial mantle source. They are characterized by very low abundances of the radiogenic isotopes ^{40}Ar and ^{129}Xe, low Ar/Xe and Kr/Xe, and an unfractionated solar-like xenon isotope pattern. If these are representative of internal Martian noble gases, they imply an unearthly relationship between the interior and the atmosphere.

III. SOURCES

The very chemical simplicity that makes noble gases such hopeful tracers of volatile evolution also poses a problem: it is hard to get noble gases out of the solar nebula and into a planetary atmosphere. If anything, the problem is not that the planets have too little noble gas; the problem is that they have so much (Podosek 1991). The key here is Venus. Among meteorites only South Oman, an E-chondrite, has significantly more ^{36}Ar on a gram per gram basis than does Venus. There are four potential sources for atmospheric rare gases:

1. Direct gravitational attraction of a solar nebula atmosphere by very large planetesimals or full-grown planets (the Kyoto school, including Hayashi et al. 1979; Mizuno et al. 1980; Sekiya et al. 1980a,1981; Mizuno and Wetherill 1984; Sasaki and Nakazawa 1988; Pepin 1989; Sasaki and Nakazawa 1990).

2. Adsorption and trapping in grains and planetesimals, which accreted to form planets (see, e.g., Fanale and Cannon 1972; Frick et al. 1979; Honda et al. 1979; Yang and Anders 1982a, b; Pollack and Black 1979,1982; Wacker et al. 1985; Zadnik et al. 1985; Hunten et al. 1988; Wacker 1989; Pepin 1991).

3. Solar-wind implantation, mainly into smaller bodies that accreted to form planets (Wetherill 1981; McElroy and Prather 1981; Hostetler 1981; Bogard 1988; Sasaki 1991).

4. Condensation at very low temperature, either as clathrates, adsorbate, or

even as ices, in distant parts of the solar system, and subsequent delivery
to planets by cometary impact (Bar-Nun et al. 1985; Owen 1985; Owen
et al. 1991; Pepin 1991).

There are also some hybrid schemes in which solar nebular gases are
gravitationally attracted by planetesimals, but in which adsorption also plays
a role, either on grains in the atmosphere, or on the surface or inside the
planetesimal.
    Solar-wind implantation and grain accretion are alike in that noble gases
are assumed to accrete homogeneously, i.e., all materials destined for a particular planet are more or less equally endowed with noble gases. Gravitationally
captured atmospheres may occur early or late, given the presence of the nebula; cold condensates are more likely to arrive late. Gravitationally captured
rare gases are necessarily of solar composition, both elementally and isotopically. Solar-wind implants are similar, but are somewhat mass fractionated in
favor of the heavier gases. Adsorption tends to produce a pattern rather like
the "planetary" pattern, but there is much variation. The compositions of cold
condensates are not well known.

A. Nebular Atmospheres

A planet or large protoplanet that accretes within the solar nebula (i.e., before
the nebular gases dispersed) attracts to itself a substantial primary atmosphere
("kyotosphere") of nebular composition. This is the only model I will discuss
that has the problem of predicting larger rare-gas inventories than are observed.
Typically Earth's primary atmosphere is calculated to be of order 10^{26} g, but
this estimate is highly model dependent; an extremely dusty atmosphere can
be as small as 10^{24} g (Mizuno et al. 1980; Mizuno and Wetherill 1984;
Sasaki and Nakazawa 1990). The small amount of neon in air demands
that most of the primary atmosphere has escaped, but in principle some
atmospheric gases could be fractionated remnants of the primary atmosphere;
a detailed discussion will be given in Sec. IV.B on hydrodynamic escape.
The greenhouse effect of the kyotosphere is so enormous that the mantle is
melted to considerable depth. Substantial quantities of the noble gases enter
the planet in proportion to their atmospheric abundance and their solubilities
in silicate melt. These models predict that 2 to 200 times the neon presently
in air would have entered the mantle (Mizuno et al. 1980), in proportion to
the mass of the primary atmosphere. That the larger amounts of neon are not
evident in Earth has been taken as an argument against the Kyoto model. In
any event, there is no obvious evidence that favors it.
    Gravitationally concentrated nebular gases at the surfaces of protoplanets
offers a closely related way to put rare gases into planetary materials. This
is more plausible than the Kyoto model, mainly because protoplanets can
grow very quickly by runaway accretion, so that growth within the nebula is
plausible (Wetherill 1990b). The higher nebular gas densities at the surface
can greatly increase the efficiency at which nebular gases are trapped. The

mass of the primary atmosphere is very sensitive to the mass of the protoplanet, so that harvesting the "right" amount of noble gases is in principle achievable. Zahnle et al. (1990*b*) make some use of this to trap xenon in planetesimals; Pepin (1991) employs it to place interesting quantities of adsorbed nebular gases into planetary interiors.

B. Adsorption and Physical Trapping

Adsorption involves a variable range of elemental fractionations, but isotopes are generally not greatly fractionated (Frick et al. 1979; Ozima and Podosek 1983; Bernatowicz and Fahey 1986; Bernatowicz and Podosek 1986). The heavier elements are usually, but not always, more prone to adsorption (Bernatowicz and Podosek 1986; Pepin 1991). In particular, Xe is usually more readily adsorbed than Kr, which is an attractive feature for explaining "planetary" rare gases, but a less attractive one for explaining rare gases on planets. There is another difficulty with adsorption. Laboratory adsorption experiments usually fall many orders of magnitude short of being able to match the quantity of noble gases actually found in meteorites (Zadnik et al. 1985; Wacker 1989; Pepin 1991). Quantitative agreement can be reached at very low temperatures, but even then, the comparison may not be apt: Fanale and Cannon (1972) were able to adsorb almost enough Xe onto ground Allende powder at 110 K given a maximum mass solar nebula, but the adsorbed Xe was very loosely bound. Some other process, perhaps collisional shock (Honda et al. 1979; Wiens and Pepin 1988), is needed to trap the adsorbate. The problem is actually much worse than might be deduced by simply comparing the noble gases of meteorites on a gram per gram basis with the laboratory data, because most of the "planetary" noble gases are localized in the carbonaceous "Q" carrier, which is by mass a tiny fraction of the meteorite.

In the grain-accretion model, noble gases are adsorbed onto grains in the solar nebula (Pollack and Black 1979,1982). The primary purpose of the model was to explain the sunward gradient of planetary Ar while retaining the "planetary" Ne/Ar ratio. The gradient was attributed to the pressure gradient in the solar nebula. Grains accreted locally with little radial mixing, preserving the gradient in the completed planets. Because the quantity of adsorbed gas is expected to be linearly proportional to pressure but exponentially sensitive to temperature when adsorption is efficient enough to be of interest, Pollack and Black had to postulate that adsorption occurred at the same temperature throughout the inner solar system. This could plausibly be argued only by assuming that the prime adsorber formed and prospered within a very narrow temperature range. In essence, grain accretion was a post-Pioneer Venus version of the "planetary" hypothesis, and therefore it suffers from the same discrepancies as Xe. Nor does the model realistically account for the Venusian Ar/Kr ratio. The model could of course be broadened to other carriers than Q, with different adsorptive affinities (e.g., South Oman?). Grain accretion suffers from the quantity gap that characterizes other adsorption models. Finally, as Wetherill (1981) emphasized, it is very difficult to maintain a steep radial gradient between Venus and Earth during accretion.

A different approach to fractionating Xe is to assume gravitational settling of nebular gas inside porous planetesimals (Ozima and Nakazawa 1980;
Igarashi and Ozima 1988; Zahnle et al. 1990b). The planetesimals are presumed to be immersed in the solar nebula. Within the planetesimal each
isotope assumes its own scale height, so that, at depth, concentrations of the
heavier isotopes are much enhanced over their concentrations in the nebula.
Like hydrodynamic escape (see Sec. IV.B), this model has the potential to produce mass fractionated xenon. Large planetesimals (>1000 km) are needed
to achieve terrestrial xenon fractionation.

Ozima and Nakazawa (1980) and Igarashi and Ozima (1988) assume that
the planetesimals are porous at all depths, regardless of size. Gases are not
trapped, so that accretion of planets is implicitly assumed to occur in the presence of the nebula. Igarashi and Ozima (1988) attempt to produce the whole
suite of terrestrial rare gases by summing a distribution of these planetesimals.
They regard their model as a failure because it incorporates three orders of
magnitude too little fractionated xenon to supply Earth. However, they do not
allow for adsorption, which can greatly increase the yield at low temperature.

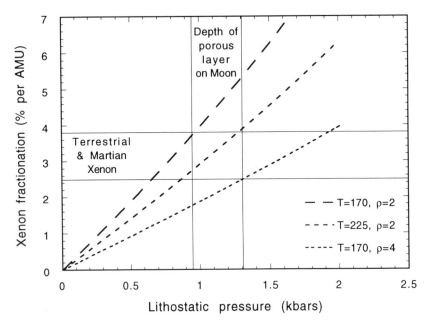

Figure 5. Xenon fractionation produced by gravitational settling of nebular gases
    inside porous planetesimals. Fractionated Xe is trapped when pores are closed off
    by lithostatic pressure. The extent of Xe fractionation is determined by planetesimal
    density ρ [g cm^{-3}], temperature T [K], and the characteristic pressure at which pores
    are closed (Zahnle et al. 1990b).

Zahnle et al. (1990*b*) replace the assumption that large planetesimals remain porous at all depths with the more realisitic assumption that gases are trapped when pores are closed by high lithostatic pressure. This relates the degree of fractionation to the strength of rock, a material property, and thus the degree of fractionation is predicted and is the same in every planetesimal large enough to reach the pore-closing pressure (Fig. 5). Lunar seismic observations imply that the pore-closing horizon in the regolith occurs at a lithostatic pressure of ∼1 to 1.2 kbar. The model's best point (i) is that at reasonable temperatures the predicted degree of Xe fractionation is that observed on Earth and Mars. It has other positives. (ii) It can explain why Mars and Earth have the same degree of Xe fractionation. (iii) Because the xenon is physically trapped, accretion of the planetesimals to form planets can occur after nebular dissipation. (iv) With adsorptive amplification included, the model planetesimals can provide enough xenon to supply Earth and Venus. (v) The corresponding amount of fractionated Kr is negligible. (vi) It is physically realistic. An analogous process is known to occur in the Antarctic ice cap (Craig et al. 1988), where at −30°C and 70 m depth porous snow transforms to impermeable ice. The trapped atmospheric gases are observed to be isotopically fractionated to just the extent predicted by application of the barometric law.

The planetesimal models can plausibly produce fractionated xenon, but at present they are not too useful otherwise. Planetesimal xenon could be used to replace hydrodynamically fractionated xenon on Earth and Mars. But, like hydrodynamic escape, it does not account for the high Kr/Xe ratios found on these planets. The difficulty is the same one that dogs hydrodynamic escape: what is the source of fresh Kr, and why is it not accompanied by fresh Xe?

C. Solar-Wind Implantation

Wetherill (1981) and McElroy and Prather (1981) suggested solar-wind implantation as a way to supply Venus with its observed high quantity of ^{36}Ar, and account for the steep sunward gradient of ^{36}Ar, abundances on Venus, Earth and Mars. (Hostetler [1981] proposes a variant in which differential ionization and separation of ions from neutrals in a magnetic solar wind affects elemental ratios.) The model assumes that the solar nebula cleared of gas before the terrestrial planets accreted, so that a high surface area of dust, fragments, and planetesimals were exposed to an intense early solar wind. The dust and small fragments are envisioned as constantly being formed by collisions and re-accreted into planetesimals. Most of this material is eventually incorporated into planets. Because much of the planetary material becomes exposed to the solar wind, noble gases accrete early and more or less homogenously.

Solar-wind implantation has no difficulty supplying enough ^{36}Ar for Venus, which may be its greatest strength. T Tauri winds are very high, typically 10^6 times higher than today, but are probably associated with the accretion of gas onto the star, and thus predate the gas-free debris disk. Later winds are no doubt less severe, but it would be rash to imagine that they fell abruptly to the modern level. A sensible compromise would be to assume solar-wind enhancements comparable to solar EUV enhancements, as both are ultimately related to solar activity. Wetherill achieves Venusian levels of argon over 10^7 yr with solar-wind enhancements of order 50 to 500, and this estimate may be too pessimistic: using somewhat more liberal assumptions, McElroy and Prather get there in 2×10^7 yr with the modern solar wind.

However, solar wind implantation does less well with the sunward gradient than one might initially suppose. A steep sunward gradient in noble gas abundance can be produced early if the debris disk is optically thick in the radial direction, i.e, if the inner edge of the disk absorbs all the incident solar wind, shadowing the interior. To the extent that planets form from local material in isolated feeding zones, the sunward gradient will be manifested in the completed planets. But, as Wetherill emphasizes, Venus and Earth very likely did not accumulate in isolation, so that it is difficult to maintain a steep gradient between them. Venus and Earth, as co-accumulating planets, necessarily scattered material between them. Wetherill finds that it is difficult to get a gradient larger than 10:1, which falls roughly an order of magnitude short of what is needed (note that this restriction applies to the grain accretion model, as well). Radial mixing is not the only reason why solar-wind implantation might not produce a steep inner solar-system rare-gas gradient. The debris disk may not be optically thick when the gas clears, for one. Also, popular disks tend to be flared owing to weaker solar gravity at greater distances from the Sun. Hence the surface of the disk may be obliquely exposed to the solar wind at all distances, permitting solar-wind implantation to occur with reasonable efficiency at asteroidal distances (Sasaki 1991). A possible patch to the model is to put essentially all the noble gas into the innermost planetesimal, and have it be incorporated into Venus (Wetherill 1981).

The solar-wind implantation model has two serious difficulties. The model predicts more implanted neon than argon, and to the extent that the solar-wind implantation model fails to solve the gradient problem, it may be required that Earth lose much of its original argon and neon, apparently without comparable losses of krypton and xenon.

The importance of producing the observed inner solar-system planetary argon gradient by accretional processes fades if escape is accepted as important. Certainly this must be so for Mars; no one seriously entertains the notion that the present Martian atmosphere is the atmosphere Mars was formed with. The present Martian inventory of Xe would be doubled by accretion of just 50 m of average carbonaceous chondrites—what is more, the Xe isotopes are so discordant that even 10 m of the stuff would be apparent. Granted, carbonaceous chondrites may have little to do with planets in bulk, but this is still a very thin veneer for a material that is so evident among modern impactors. Small bodies are particularly susceptible to impact erosion—the expulsion of atmospheric gases by impact—especially towards the end of accretion, when high-velocity collisions with widely scattered stray bodies are most likely (Melosh and Vickery 1989). Escape can easily account for a thin Martian atmosphere (Melosh and Vickery 1989). Whether it can reconcile Venus and Earth is another question.

Solar-wind-implanted noble gases, such as those in lunar soils, have roughly solar elemental and isotopic abundances, colored by a modest elemental fractionation favoring retention of the heavier elements (Swindle 1988). The typical Ne/Ar ratio of $\sim$10 in implanted gases is less than the solar $\sim$40, but clearly far removed from the Venusian ratio of $\sim$0.2. Thus solar-wind implantation can account for nearly solar Ar/Kr on Venus, but fails to account for low Ne/Ar. Highly selective neon loss is required, which is not observed in some heated lunar soils (Pollack and Black 1982). However, there are classes of meteorites with very high argon content and very low Ne/Ar ratios. Examples are ureilites and, most intriguingly, the E-chondrites. South Oman, the most argon-rich meteorite, has Ne/Ar of just 0.003 (Crabb and Anders 1981). This noble gas pattern differs from either the "planetary" or the solar pattern, and has been called "subsolar." As emphasized by Wetherill (1981) and Bogard (1987), the very existence of subsolar rare gases demonstrates that an otherwise apparently solar noble-gas reservoir can lose its neon without losing its argon. Hostetler's (1981) model is constructed to account for low Ne/Ar ratios by differential thermal ionization in a cool, dense solar wind. The more highly ionized Ar is presumed to get preferentially implanted.

South Oman is an enstatite chondrite, a highly reduced class of meteorites that is so fascinating chemically that it is extremely tempting to try to assign them an important role in the evolution of the solar system (see, e.g. Crabb and Anders 1981; Prinn and Fegley 1989; Pepin 1991). Its rare gases are somewhat similar to Venus', but even using South Oman, the most Ar-rich member of the class, Venus would need to accrete 20% of its mass as enstatite chondrites to supply enough [36]Ar. Like all other veneering candidates, enstatite chondrites carry far too much carbon and nitrogen for this to work easily; the vast majority of carbon and nitrogen would need to be hidden elsewhere.

It is interesting to note that the subsolar noble gas carrier in South Oman is not the ubiquitous carbonaceous Q carrier but appears instead to be the enstatite itself (Crabb and Anders 1982). This may indicate that the nebular adsorption onto optimal carriers is not the cause of high noble-gas abundances in meteorites. The noble gases may have been incorporated into solids by some other process—solar-wind implantation is a plausible source—and then were later released to move about inside planetesimals, migrating to preferred mineral phases like Q, as less-mobile elements do. This might explain the similar bulk concentrations of noble gases in meteorites and Venus. Such a scheme could also account for the high intrinsic noble-gas partial pressures needed to reconcile laboratory adsorption experiments with the extremely high noble-gas contents of the Q carrier. This is also the kind of process that might allow a solar-wind implantation model to yield a Ne/Ar ratio characteristic of adsorption.

D. Cold Comets

Low-temperature condensates are essentially unknown. Clathrates are expected to favor the heavier noble gases, while direct freezing of ices might be expected to follow the solar pattern for the heavier elements, with an abrupt cut-off at the uncondensed element. Neither seems likely to discriminate among isotopes of an element. Adsorption on ice may be more variable, with some suggestion that Kr may be more readily adsorbed than Xe at very low temperatures (Owen et al. 1991). Also at low temperatures, very high adsorption efficiencies might be expected.

Comets offer the only possibility for a simple, single source explanation of nonradiogenic noble gases. In a cometary model noble gases would have arrived in a late veneer. Venus, with its high native noble-gas abundance and lighter-elemental pattern, would have to be regarded as more or less reflecting the rare-gas composition of the comets. There are nonetheless some severe constraints that can be placed on the composition of the hypothetical comets. If a cometary source is to supply Earth and Mars—why should it not?—it would need to contain intrinsically mass-fractionated xenon and a slightly supersolar Kr/Xe ratio. This is a general requirement of any single-source model. The distinct differences between terrestrial and Martian xenon further complicates this demand. It may be possible for Xe to be mass fractionated in adsorption at very low temperatures, but if so, it must be postulated that a roughly solar Ar/Kr/Xe pattern can occur when none of these is harvested quantitatively. Such a model is still left with the need to reduce both terrestrial and Martian Ne and Ar in such a way that on both planets the Ne/Ar/Kr ratios come out essentially the same, but with Martian ^{36}Ar/^{38}Ar coming out quite different. This may be possible with judicious application of hydrodynamic (or other) escape, but in doing so, much of the simplicity that makes comets attractive is lost.

Another argument that can be raised against a late cometary veneer is that if the noble gases in Loihi basalts truly sample an undegassed mantle (and they may not; see Sec. II.E), ^{36}Ar in air is quantitatively consistent with it being the outgassed complement of the degassed upper mantle. If icy planetesimals are to be important, they must arrive early enough in accretion that they can be mixed throughout the mantle. This may rule out very distant sources.

To some extent these difficulties can be ameliorated by chance differences among the impactors on Earth and Venus. In general, it must be expected that a few large impactors dominated the delivery of volatiles. Venus' largest contributing comet could have had a "colder" composition than Earth's. Or it may be that chance availed Venus through a large comet striking at an improbably low impact velocity, such that its volatile inventory was retained with unusual efficiency.

IV. SINKS

Jeans escape is thermal escape at low flux. In Jeans escape the atmosphere effectively has a top, the exobase, above which atoms are as free as satellites. Hydrodynamic escape refers to a planetary wind, analogous to the solar wind, that occurs when thermal escape fluxes are high. The escaping gas remains collisional and behaves as a fluid at all heights; i.e., it obeys the equations of hydrodynamics. Hydrodynamic escape is a plausible process in the early solar system, and it is inherently mass fractionating. Impact erosion, which refers to the expulsion of atmospheric gases by impact, is also plausible, especially towards the end of accretion when impact velocities are expected to be high (Walker 1986; Ahrens and O'Keefe 1987; Hunten et al. 1989; Ahrens et al. 1989; Melosh and Vickery 1989). It is expected to be almost nonfractionating, and so shrinks an atmosphere without much altering its composition. Impact erosion offers a straightforward explanation for the extreme scarcity of Martian rare gases. Giant impacts, such as the hypothetical Moon-forming event, may behave a little differently in this respect. For example, Cameron (1983) suggests that escape may occur when the atmosphere climbs up a massive disk of Earth-orbiting debris. By prolonging escape over an extended period, such a scenario could lead to significant mass fractionation.

A. Jeans Escape

Donahue (1986) considers moderately large planetesimals that accumulated after the dispersal of the nebula. Isotopically solar noble gases that had been adsorbed on grains at an earlier stage of nebular evolution are released to form a noble-gas atmosphere. Other less volatile volatiles are retained in the planetesimal interiors. The noble gases are then fractionated by Jeans escape. Because Jeans escape is extremely sensitive to mass, it produces extreme fractionations too easily; the gentler fractionating powers of hydrodynamic escape would probably work better. The model at present fails to produce Venusian argon; and it does not address Xe or Kr. Planetesimal atmospheres are also problematic. They may be difficult to form, difficult to retain in the face of numerous erosive processes, and easily lost during collisions with other planetesimals.

B. Hydrodynamic Escape

For hydrodynamic escape to occur from the terrestrial planets, it is probably necessary that hydrogen be a major atmospheric constituent, at least at the high altitudes from which escape begins. The likeliest energy source driving escape would have been solar EUV radiation ($\lambda < 100$ nm), i.e., radiation of short enough wavelength that it can be directly absorbed by hydrogen (Sekiya et al. 1980a; Watson et al. 1981). Hydrodynamic escape is a reasonable process to expect in the early solar system because hydrogen is for many reasons a likely component of early atmospheres, and EUV fluxes from the young Sun are likely to have been very large. Hydrodynamic escape from Venus is a natural consequence of the runaway greenhouse (Watson et al. 1981; Kasting and Pollack 1983). It has also been invoked for the removal of massive primordial hydrogen atmospheres predicted by the Kyoto school of planetary accretion (Sekiya et al. 1980*a, b*), and as the means for removing the enormous amount of hydrogen formed by reaction of water and iron in other planetary accretion models (Dreibus and Wänke 1987,1989; Ringwood 1979).

A reasonable estimate for the H_2 escape flux at high solar EUV fluxes is obtained by balancing the absorbed EUV energy with the energy needed to lift the escaping atmosphere out of the planet's potential well (Watson et al. 1981),

$$F = \frac{R_p \epsilon}{GM_p m} \approx 5 \times 10^{11} \epsilon \quad cm^{-2}s^{-1} \tag{1}$$

where R_p and M_p are the radius and mass of the planet; m and F are the molecular mass and the escape flux; and ϵ is the globally averaged incident solar EUV flux. The numerical values refer to an H_2 atmosphere escaping from Earth. For the modern quiet Sun, $\epsilon \sim 0.3$ erg cm^{-2} s^{-1} ($\sim 10^{-6}$ L$_\odot$); the Sun is quite variable at these wavelengths. Observations of young Sun-like stars imply that the young Sun would have emitted some 100 to 1000 times more EUV energy, with EUV fluxes declining roughly inversely with the Sun's age. Zahnle and Walker (1982) estimated that

$$\epsilon \approx 0.3 \left(\frac{10^{10} yr}{t} \right) \quad erg \ cm^{-2}s^{-1} \tag{2}$$

while Pepin (1991), using more recent X-ray data and excluding the more extreme T Tauri stars (see Fig. 6), estimates that

$$\epsilon \approx 300 \exp \left(\frac{-t}{9 \times 10^7 yr} \right) erg \ cm^{-2}s^{-1}. \tag{3}$$

Thus Eq. (1) implies that hydrogen escape fluxes of order 10^{13} to 10^{14} cm^{-2} s^{-1} could reasonably be expected from the terrestrial planets ca. 10 to 100 Myr, provided that hydrogen were available to escape. At the higher rate it would take just 3 Myr to lose the hydrogen present in the Earth's oceans. It need not have taken long to turn an Earth into a Venus.

Even higher escape rates would be possible if more abundant photons at longer wavelengths could be utilized. For example, Sekiya et al. (1981) have suggested using far-ultraviolet radiations with $\lambda < 185$ nm that can be absorbed by H_2O. This allows them to rid Earth of its hypothesized 10^{26}g kyotosphere in a reasonable time. However, in this study, Sekiya et al. neglected radiative cooling, which for H_2O at >500 K is probably considerable. It is far from obvious that any of this far-ultraviolet energy is actually available for escape. This illustrates the need for a quantitative, self-consistent hydrodynamic escape model that includes both radiative heating and cooling, as well as the necessary chemistry and transport affecting the abundance of possible coolants.

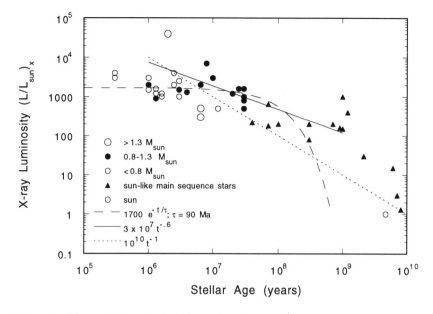

Figure 6. Observed X-ray luminosities $L_x/L_\odot$ of young stars relative to the modern Sun vs stellar age. The three curves represent different approximations to solar X-ray evolution (figure adapted from Pepin 1991).

Mass Fractionation by Hydrodynamic Escape.   Mass fractionation occurs in hydrodynamic escape when the escape flux is high enough that heavier molecules are dragged along with the flow. In a two-gas atmosphere, the transition occurs when the hydrogen escape flux exceeds the "diffusion-limited" flux defined by Hunten (1973). The escape flux of a trace constituent j in a hydrogen escape flux F_1 is approximately (Hunten et al. 1987)

$$F_j = F_1 \frac{N_j}{N_1} \left(1 - \frac{(m_j - m_1)gb_{1j}}{F_1 kT} \right); \quad F_j \geq 0 \qquad (4)$$

where N_1 and N_j are the column densities of the major and minor constituents; F_1 and F_j their escape fluxes; b_{1j} is the appropriate binary diffusion coefficient (Marrero and Mason 1972); and k is Boltzmann's constant, T temperature, and g gravity. Hunten et al. (1987) and Pepin (1991) present their analyses in terms of a "crossover mass," defined as

$$m_{cj} \equiv m_1 + \frac{F_1 kT}{g b_{1j}}. \qquad (5)$$

The crossover mass is the heaviest mass that can be dragged to space by the escaping hydrogen. The crossover mass is different for different elements, because binary diffusion coefficients vary. Zahnle and Kasting (1986) consider a more general case, and verify that Eq. (4) is a good approximation for escape of a trace constituent provided that

$$\frac{m_{cj} - m_j}{m_{cj} - m_1} > \frac{m_1}{m_j}. \tag{6}$$

A similar argument is derived by Sasaki and Nakazawa (1988). For noble gases escaping in a hydrogen-dominated atmosphere, Eq. (4) is good for Ne, Ar, Kr and Xe; it is less good for He. With $b_{1j} = 2.7 \times 10^{17} T^{0.712}$ for diffusion for Xe and H_2 (Marrero and Mason 1972), hydrogen escape fluxes from Earth must exceed

$$F_1 > 7 \times 10^{13} \left(\frac{T}{500K}\right)^{-0.288} cm^{-2}s^{-1} \tag{7}$$

to fractionate xenon. The needed escape flux is the same order of magnitude as that expected, and therefore mass fractionation in hydrodynamic escape must be considered seriously.

A simple case to treat analytically is that for which the hydrogen escape flux is high enough that all the elements (isotopes) of interest escape; i.e., if

$$F_1 > \frac{(m_j - m_1)gb_{1j}}{kT} \tag{8}$$

for all j of interest. Then the fractionation of two species i and j produced by hydrodynamic escape is just

$$\log\left(\frac{N_i(t)}{N_j(t)} \div \frac{N_i(0)}{N_j(0)}\right) = [(m_i - m_1)b_{1i} \\ - (m_j - m_1)b_{1j}]\frac{g}{kT}\int_0^t \frac{dt}{N_1(t)}. \tag{9}$$

This expression shows that, provided F_1 is large enough to effect escape of the relevent isotopes, the mass fractionation pattern produced is independent of F_1. If $b_{1j} \sim b_{1i}$, as it is for isotopes of a noble-gas element (other than helium) in H_2, then

$$\log\left(\frac{N_i(t)}{N_j(t)} \div \frac{N_i(0)}{N_j(0)}\right) \sim (m_i - m_j)\frac{g}{kT}\int_0^t \frac{dt}{N_1(t)}. \tag{10}$$

Because both the solar EUV flux and the hydrogen supply are expected to decline with time, albeit fitfully, the lighter noble-gas elements are more depleted and relatively more fractionated than the heavier elements by real hydrodynamic escape. It is possible that something like this pattern is seen in "planetary" rare gases (Pepin 1991), but the predicted pattern is most certainly not seen on Earth and Mars, the only planets for which we have adequate data. In particular, xenon is predicted to be more abundant and less fractionated isotopically than krypton, relative to the composition of the source. Precisely the opposite is seen on Earth and Mars. Xenon is profoundly mass fractionated on both planets, while krypton is not; on Mars the Kr/Xe ratio is roughly solar, and on Earth it's even higher.

Total Hydrogen Loss During Hydrodynamic Escape. The assumption that hydrodynamic hydrogen escape proceeds at the maximum rate allowed by the solar EUV flux necessarily implies an enormous hydrogen loss and therefore an enormous hydrogen source. To illustrate, assume that the solar EUV flux history is given by Eq. (3). The total amount of hydrogen lost is then obtained by integrating Eq. (1):

$$\Delta M = \frac{4\pi R_p^3}{GM_p} \int \epsilon(t) \, dt \sim 7 \times 10^{24} \text{ g} \tag{11}$$

from Earth. A similar, somewhat smaller estimate of $\sim 2 \times 10^{24}$ g is obtained if the solar EUV flux history is obtained from Eq. (2). This is the hydrogen content of 15 to 50 oceans. The larger estimate is equivalent to accreting Earth from material that is 1% water by mass, a water content consistent with the geochemically driven terrestrial accretion models proposed by Ringwood (1979) and Dreibus and Wänke (1987). In such a model, most of the hydrogen is produced by the reaction of water with iron during accretion (Lange and Ahrens 1984). Evidently this magnitude of hydrogen escape is not beyond the bounds of reasonable expectation. Nevertheless, these enormous hydrogen losses are not demanded by the hydrodynamic escape model; they are simply a consequence of assuming that there is always enough hydrogen available for escape to proceed at the maximum possible rate. This need not be so. For example, the planet could simply run out of hydrogen, as Venus has. If Venus had had only an ocean's worth of hydrogen, it could have been lost in as little as 3 Myr, well before the solar EUV flux declined. Or escape may occur only during brief intervals. Consider an impact-dominated environment. The planet may only have had hydrogen to lose for short intervals following major impacts. Such, for example, may be the case for Venus today (Grinspoon and Lewis 1988). Or consider Earth; after a large impact the oceans are vaporized, and for a little while hydrodynamic hydrogen escape would be expected. But the oceans quickly condense and rain out again, so that, even with many impacts, most time is spent with the ocean condensed and therefore at most times little hydrogen escapes.

None of these hydrogen-poor scenarios would necessarily produce a different fractionation pattern than does the hydrogen-rich scenario. As noted earlier in Eq. (9), the magnitude of mass fractionation produced by hydrodynamic escape is proportional to $\int N_1^{-1}(t) dt$ where $N_1(t)$ is the hydrogen column density during episodes of escape. The same degree of fractionation is produced in a tenth the time if N_1 is on average a tenth the size. Pepin (1991) treats N_1 as a free parameter, so clearly his model can adapt to low hydrogen regimes. In summary, mass fractionation by hydrodynamic escape does not necessarily imply enormous hydrogen losses, although obviously it

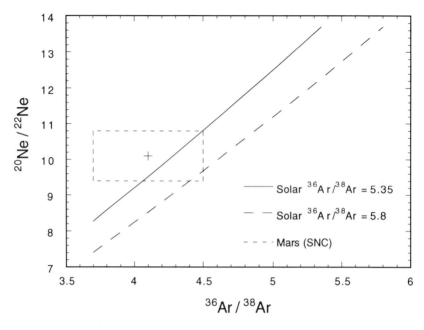

Figure 7.   The minimum ^{20}Ne/^{22}Ne ratio produced while fractionating argon by hydrodynamic escape from Mars. The initial ^{20}Ne/^{22}Ne ratio is assumed to be solar 13.7. The presence of an abundant impurity (like CO_2) or any extended period in which the hydrogen escape flux was large enough for neon to escape but not large enough for argon to escape would lower the predicted ^{20}Ne/^{22}Ne below the indicated curves.

can accomodate them. It does imply extremely vigorous escape during those intervals during which hydrogen loss occurs.

Some Applications. Mass fractionation in hydrodynamic escape may have contributed to the observed high D/H ratio on Mars (Owen et al. 1989). It has been invoked to explain the fractionation pattern of the isotopes of xenon on Earth (Hunten et al. 1987; Sasaki and Nakazawa 1988; Pepin 1991) and Mars (Pepin 1991), the high Martian ^{38}Ar/^{36}Ar ratio (Pepin 1989; Zahnle et al. 1990*a*; Pepin 1991), and isotopic differences between atmospheric and mantle neon (Sasaki and Nakazawa 1988; Zahnle et al. 1990*a*; Pepin 1991). Mass fractionation by hydrodynamic escape could also account for the observed modest elemental fractionation of the Martian noble gases with respect to Earth (Hunten et al. 1987); however, it cannot easily explain both the elemental and the isotopic fractionations, because the relatively strong isotopic fractionations are incompatible with the relatively weak elemental fractionation.

Martian Argon. Martian argon, with a $^{36}Ar/^{38}Ar$ ratio of 4.1±0.3, is strikingly heavy (Wiens et al. 1986; Swindle et al. 1986*b*). By comparison the terrestrial ratio is 5.32 and "planetary" ~5.3. The solar-wind value is usually given as ~5.3, but may actually be as high as ~5.8 (Swindle 1988; Pepin 1991). The only component as low as Mars' is solar-flare argon (see Swindle 1988), but pure solar flares are an unlikely source of Martian argon. Fractionation acting on Mars itself seems to be required. Hydrodynamic escape can account for fractionation of Martian argon (Zahnle et al. 1990*a*; Pepin 1991). This is marginally consistent with the composition of Martian neon, $^{20}Ne/^{22}Ne$ = 10.1±0.7 (Swindle et al. 1986*b*; Wiens et al. 1986). If the original Martian ratios of $^{36}Ar/^{38}Ar$ and $^{20}Ne/^{22}Ne$ were 5.35 and 13.7, respectively, hydrodynamic escape predicts a final $^{20}Ne/^{22}Ne$ ratio no greater than 9.5±1.3 (see Fig. 7). This is barely consistent with observation. To produce neon this light requires hydrogen escape fluxes high enough that both argon and neon escape at all times, and the escaping atmosphere must be mainly H_2. The presence of abundant heavy atmospheric pollutants like CO_2 or any extended period in which the hydrogen escape flux was large enough for neon to escape but not large enough for argon to escape would lower the predicted $^{20}Ne/^{22}Ne$ below the range permitted by observation (Zahnle et al. 1990*a*).

Terrestrial Neon. Atmospheric neon is isotopically heavier than mantle neon (Craig and Lupton 1976; Staudacher 1987; Ozima and Igarashi 1989; Honda et al. 1991). The evidence points to an initially solar $^{20}Ne/^{22}Ne$ ratio. By contrast, mantle and atmospheric reservoirs of the nonradiogenic isotopes of argon, krypton and xenon appear to be isotopically the same (Allègre et al. 1987; Ozima and Igarashi 1989). Heavy atmospheric neon can be explained by the preferential loss of the lighter neon isotope from Earth's atmosphere, as has been noted by Craig and Lupton (1976), Allègre et al. (1987), and Staudacher (1987). Allègre et al. (1987) estimate that 28% of degassed neon has escaped.

Zahnle et al. (1990*a*) extended the theory of mass fractionation in hydrodynamic escape to atmospheres in which hydrogen was not the only major constituent. We were particularly interested in cases where an abundant heavy constituent did not escape. Under such circumstances hydrogen escape would be regulated by the diffusion-limited flux (Hunten 1973). This could be the case if the heavy constituent was an effective radiative coolant, and would surely be the case if the heavy constituent were overwhelmingly abundant. There is no lack of plausible candidates; N_2, CO and CO_2 could all suffice.

Steam atmospheres are a likely, if intermittent, feature of the accreting Earth (Matsui and Abe 1986; Zahnle et al. 1988). In our calculations, steam atmospheres would have lasted some 10 to 50 Myr. They occur because, on average, the energy of accretion places Earth above the runaway-greenhouse threshold, so that liquid water is not stable at the surface. Zahnle et al. (1990*a*) showed that neon fractionation would have been a natural byproduct of hydrogen escape in diffusion-limiting flux through a steam atmosphere polluted with significant amounts of CO_2, N_2, or CO. The reason neon can escape and argon cannot is that neon is less massive than any of the likely pollutants. Assuming that the initial $^{20}Ne/^{22}Ne$ ratio was solar, we found that it would have taken some 10 Myr to effect the observed neon fractionation with a 30-bar steam atmosphere and 10 bar of CO (see Fig. 8). Thicker atmospheres would have taken longer; less CO, shorter. This mechanism for fractionating neon has about the right level of efficiency. Because the lighter isotope escapes much more readily, total neon loss is fairly minimal; less than half of the initial neon endowment escapes.

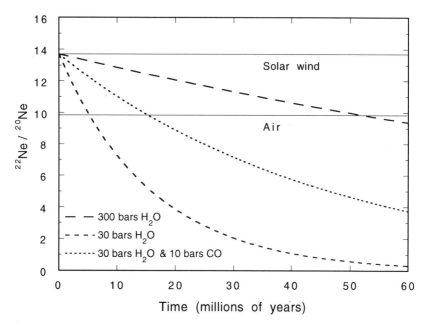

Figure 8. Neon fractionation produced by vigorous hydrogen escape from steam atmospheres during accretion is shown as a function of the lifetime of the steam atmosphere. Three cases are shown. An initial solar $^{20}Ne/^{22}Ne$ ratio is consistent with reports of isotopically light neon emanating from the modern undegassed mantle (Ozima and Igarashi 1989; Honda et al. 1991). The time scales for neon fractionation are consistent with the expected lifetimes of steam atmospheres during accretion (Matsui and Abe 1986; Zahnle et al. 1988; figure adapted from Zahnle et al. 1990a).

Xenon. Xenon fractionation on Earth has been a long-standing problem (see Ozima and Podosek 1983). Hydrodynamic escape can fractionate xenon (Hunten et al. 1987; Sasaki and Nakazawa 1988). Indeed, hydrodynamic escape can produce terrestrial xenon whether one starts with U-Xe, SUCOR xenon, or "planetary" xenon as the raw material (Hunten et al. 1987). With U-Xe the fit is straightforward, which is not surprising, because the composition of U-Xe is in part derived on the assumption that terrestrial nonradiogenic xenon is mass fractionated U-Xe (Pepin 1991). Terrestrial nonradiogenic xenon can be derived from "planetary" or SUCOR xenon if the three heavier xenon isotopes are subject to much less escape than are the lighter isotopes; i.e., much of the escape must have taken place with a xenon crossover mass fixed at 132. A problem specific to this explanation is that the hydrogen escape flux cannot have varied by more than ~1%; it must have stayed tuned to a specific crossover mass for a period of time comparable to the time spent at all higher fluxes. As there is no obvious reason why the hydrodynamic escape flux should be so tuned, this explanation is not really satisfactory.

There are three general difficulties with producing terrestrial xenon fractionation by hydrodynamic escape. (i) The basic problem is that any escape episode that fractionated xenon must have fractionated krypton more greatly. Because Kr is both relatively abundant and relatively unfractionated, xenon fractionation must have preceded the arrival of atmospheric krypton. Unless most of the "atmospheric" xenon is buried (adsorbed) in yet undiscovered crustal reservoirs, the fresh krypton could not have been accompanied by the expected quantity of fresh xenon, lest the latter overwhelm the hydrodynamic signature. Thus at a minimum there must have been two atmospheres, an earlier one of which fractionated xenon is the only clear remnant, and a later one that was more or less xenon free. (ii) Martian xenon imposes precisely the same constraints on the evolution of the Martian atmosphere. This poses something of a puzzle, because Mars and Earth are not so much alike as planets that parallel evolution should be expected. (iii) Because xenon is one of the heaviest atmospheric gases, large-scale xenon escape demands that all other atmospheric gases must also escape. It is not obvious that such vigorous escape could actually occur from a planet as large as Earth. Wholesale escape of potentially abundant heavy atmospheric constituents, like H_2O, CO_2, CO and N_2, might place unreasonable demands on the energy source fueling escape. This problem would be exacerbated if radiative cooling by these or other infrared-active species were important.

Fractionating the Kyotosphere. Sasaki and Nakazawa (1988) also found that hydrodynamic escape could fractionate terrestrial xenon. The main difference is that they approach the problem from the perspective of the Kyoto school: they assume that Earth accreted in the presence of the solar nebula and thereby acquired an immense ($\sim 10^{26}$ g) primordial atmosphere which, of course, had to be removed. Sekiya et al. (1980a) showed that it might be possible to do this with solar EUV-fueled hydrodynamic escape. Sasaki and Nakazawa (1988) consider terrestrial Xe and Ne as possible remnants of the lost kyotosphere. They account for terrestrial nonradiogenic Xe by mixing a small amount of highly fractionated U-Xe, produced by hydrodynamic fractionation of the kyotosphere, with the fresh xenon that accompanied Kr and Ar outgassing. Kyotospheric Ar and Kr are completely lost, but some highly fractionated Ne remains, which is supplemented with outgassed isotopically solar Ne to give air.

Because they presume the validity of the Kyoto model, Sasaki and Nakazawa make some rather specific predictions regarding the fate of the original 10^{26} g primary atmosphere. At the end of Ne loss, only about 10^{18} g (or less) of the primary atmosphere remain, i.e., 200 μbar. This hardly seems realistic, but neither is it realistic to expect all the assumptions of the Kyoto model to be rigorously obeyed. Because Mars is not massive enough to have attracted a thick primordial atmosphere, the model does not attempt to explain why Martian xenon should also be fractionated.

A Comprehensive Model. Pepin (1991) proposes a comprehensive model for the present distribution of noble gases in the inner solar system. The actual model is rather complicated, but the underlying premise is straightforward. Pepin assumes that the primary source of the noble gases was the solar nebula and that the processes fixing noble gases into atmospheres, chiefly adsorption, are not isotopically fractionating. Thus the initial noble gas inventories are isotopically, but not necessarily elementally, solar. Subsequent elemental fractionations and all isotopic fractionations are caused by hydrodynamic escape. The main reason the model is complicated is because, as noted above, noble gas isotopes cannot be explained by hydrodynamic escape of an isotopically solar atmosphere. Xenon precludes this on Earth and Mars.

Pepin assumes the following:

1. Hydrogen escapes at the maximum rate allowed by solar EUV. The latter is chosen to be consistent with emissions of young Sun-like stars (as discussed above, this assumption is not absolutely necessary to the model).
2. The mass fractionation produced is adequately described by equations derived for the escape of trace constituents in an atmosphere composed mainly of hydrogen.
3. There are two important initial reservoirs of planetary rare gases. One is an external primary atmosphere, possibly supplied by comets or carbonaceous chondrites. The other is internal, and dates back to the planet's youth as a protoplanet immersed in the solar nebula.
4. On Earth and Venus the external source is mainly cometary; carbonaceous chondrites are relatively more important on Mars. The internal and external rare gases may have different elemental compositions, but both are isotopically solar.
5. The external atmospheres escape to varying degrees from the three planets. On Earth and Mars escape is severe and fractionated Xe is the only clear remnant. On Venus much of the external atmosphere remains, and is only modestly fractionated.
6. On Earth and Mars internal Ne, Ar and Kr are outgassed after the escape of the external atmosphere. These replacement atmospheres are also subject to hydrodynamic losses, but the losses are less severe. Outgassing on

Venus may also occur but against the background of the massive remnant of the primary atmosphere it is not large enough to be noticed.

7. Internal xenon is not outgassed, but is instead incorporated into the cores; i.e., it is assumed that xenon becomes a siderophile at high pressure.

With continuing hydrodynamic escape acting upon the replacement atmosphere, the model can be made to match the terrestrial data quite well. It can also accomodate subtle effects like isotopic fractionation of krypton and argon.

Pepin suggests that the internal noble gases were incorporated into the growing protoplanet by gravitational attraction of nebular gases and subsequent adsorption at the surface. Large gravitational enhancements are necessary because noble gases are too scarce in the nebula to be effectively captured in the needed quantity. Rather large protoplanets are required, at least Moon-sized and probably Mars-sized for Earth. Thus the internal rare gases are identified with the "cold core," the original solid, undegassed protoplanet predicted to lie at the heart of a growing planet until it is ultimately dislodged by core formation (Stevenson 1981). This also puts internal xenon in a good position to go into the core.

Venus' present atmospheric composition is almost independent of Earth's. Venus's high Ar/Kr and "planetary" Ne/Ar ratio are essentially properties of the external source material, most likely comets. An earlier attempt to identify the Venusian atmosphere with a kyotosphere failed because it left too much neon behind (Pepin, personal communication). Rare gases are mass fractionated from cold cometary composition. Therefore Venus defines the comets. Pepin predicts that Venusian xenon will be isotopically much less fractionated than terrestrial or Martian xenon.

Earth's original external atmosphere is also presumed to be cometary, and so would have had the same composition as Venus' external atmosphere. Essentially all the Ne, Ar and Kr from the external atmosphere must escape. In Pepin's model, hydrodynamic escape alone does not suffice to remove enough Ne and Ar. Additional selective escape of Ne and Ar is assumed to take place in the aftermath of the giant, Moon-forming impact. Present atmospheric Ne, Ar and Kr largely reflect the composition of the internal source, although they too have been subject to some late mass-fractionating escape.

The story for Mars is very much like the story for Earth, absent the giant impact. However, for Mars the original atmosphere is identified with "planetary noble gases, because Martian xenon is consistent with mass-fractionated "planetary" xenon. The dominance of "planetary" over cometary rare gases is consistent with the high carbonaceous chondritic composition of Mars that characterizes recent geochemical models (see, e.g., Dreibus and Wänke 1987). Like Earth, xenon is the only clear remnant of the primary atmosphere and, like Earth, the xenon in the outgassed component was scavenged by the core. Hydrodynamic escape acting on the replacement atmosphere fractionates argon. Because Pepin uses a very light solar argon ($^{36}Ar/^{38}Ar =$

5.8), neon becomes too heavy (^{20}Ne)/^{22}Ne = 8.5) during argon fractionation. Consequently neon must be replenished a third time, here by later addition of solar-wind-impregnated cosmic dust. It is stark testimony to the thinness of Mars's atmosphere that the modern flux of cosmic dust is large enough to perceptibly alter its composition.

I have not here done full justice to the model. Pepin (1991) also tracks carbon and nitrogen, and he develops a relatively specific and interesting hydrodynamical escape explanation for isotopic fractionation of "planetary" rare gases. These matters may be pursued by the reader.

A key observation is that Pepin's (1991) model does not succeed in generating Venus, Earth and Mars from a single universal rare-gas reservoir. Instead, the rare gases on each planet are derived from different initial reservoirs. The exception is that Earth and Mars derive their Ne, Ar and Kr mainly from similar internal sources (although Martian Ne is supplemented by a late solar-wind source). Earth and Mars fractionate different xenons: Earth's is from cold comets, and Mars's from carbonaceous chondrites. Venus generates its rare gases from cold comets. Thus many of the perceived elemental and isotopic patterns that hint so intriguingly of a big picture are either reduced to coincidence or pushed back a frame into a darker past. Presumably many of the general similarities and particular dissimilarities are ultimately due to the consistencies and vagaries of adsorption, the physical process at the beginning of every thread.

How much of the model is likely to be right? It is unsettling that Venus, the one planet for which we have good reason to believe that a lot of hydrogen has selectively escaped, is the planet that shows the least evidence for selective hydrodynamic escape in its rare gases. The model's appeal to the giant impact as an agent of selective escape is also problematic, but in my opinion, the giant impact is not really germane to the issues at hand. In principle hydrodynamic escape can selectively remove any neon and argon; Pepin's argument that it cannot is rooted in the specific model he adopts for the EUV flux history of the ancient Sun. A potential inconsistency with the model arises from the controversial claim that internal rare gases in Earth's undegassed lower mantle are elementally and isotopically like air (Allègre et al. 1987; for a dissenting opinion see Patterson et al. [1990]). If verified, this near match of internal and external rare gases argues against different histories for xenon and the other rare gases, nor does it support the notion that xenon becomes siderophile at lower mantle pressures. Still, the assumption that xenon is siderophile at Martian core-mantle pressures is testable in the laboratory; it would be a serious blow to the model if it were to prove otherwise.

V. RADIOGENIC NOBLE GASES

Because they are born in rock and accumulate in the atmosphere, radiogenic noble gases serve as chronometers of planetary degassing and atmospheric evolution. Radiogenic isotopes include ^{4}He (from α decay of uranium and thorium), ^{40}Ar, ^{129}Xe, and fissogenic $^{131-136}$Xe. The short half-lives of ^{129}I and ^{244}Pu make the xenon isotopes useful for early events; ^{40}Ar is useful for determining the long term degassing history of the planets. About 6.7% of atmospheric ^{129}Xe is radiogenic (Pepin and Phinney 1978). Pepin and Phinney (1978) find that 4.6% of the ^{136}Xe in air is plutogenic, while Igarashi and Ozima (1988) get 3.1%.

Outgassing models assume that the atmosphere was generated by the degassing of part of the mantle, often identified with the upper mantle. Such models usually imply that the degassed mantle was degassed very thoroughly very early (see, e.g., Ozima and Igarashi 1989). For example, Allègre et al. (1987) conclude that about half the mantle was degassed, and that degassing was more than 99% efficient and took place within 50 Myr. Early catastrophic degassing is consistent with degassing taking place during accretion, and so can be consistent with degassing occuring promptly on impact.

A. Xenon

Using meteoritic estimates for the initial ^{129}I/^{127}I and ^{244}Pu/^{238}U ratios (^{127}I is the only stable iodine isotope, and uranium must serve as proxy for plutonium), it has been deduced that the atmosphere contains about 0.5% of the radiogenic ^{129}Xe expected from decay of the original solar system complement of ^{129}I, while it contains some 20 to 40% of the fission xenon expected from decay of primordial plutonium (Ozima and Podosek 1983). The first point to note is that the amount of ^{129}Xe in the atmosphere is very small. This cannot be due to low outgassing efficiency; daughters of longer-lived parents (^{4}He, ^{40}Ar and ^{136}Xe) have all been rather efficiently degassed (Ozima and Podosek 1983), and so has ^{129}Xe itself, judging by the very small quantities in MORB samples. The small amount of ^{129}Xe in the atmosphere must either be due to crustal burial (which, as discussed above, is not really a viable option), or to escape, either from Earth or else from the grains, rocks, planetesimals, and protoplanets that accreted to form Earth. At 17 Myr, the half-life of ^{129}I is long enough that most ^{129}Xe escape probably had to take place from large objects. (This should remind the reader that planetesimal atmospheres are a plausible site for noble gas fractionation.) A second point to note is that, if xenon in the lower mantle looks like air (Allègre et al. 1987), most ^{129}Xe was lost before nonradiogenic xenon was placed in the mantle.

If plutonium xenon has been correctly identified in the atmosphere, xenon in the atmosphere can be dated to 120 Myr after the origin of the solar system from isotope ratios alone (Wetherill 1975a). Other authors have repeated the exercise with somewhat different assumptions and similar results (Pepin and Phinney 1976; Ozima and Podosek 1983; Hiyagon and Sasaki 1988). Again, the age of ~120 Myr is most simply interpreted as implying substantial loss of any earlier terrestrial atmosphere, although other interpretations are possible (see, e.g., Hiyagon and Sasaki 1988).

While Earth's mantle contains detectable and correlated excesses of ^{129}Xe and uranium fission xenon, there is no trace of plutonium fission xenon (Staudacher 1987; Hiyagon and Sasaki 1988). Because ^{244}Pu has a longer half-life than ^{129}I, the plutonium model for atmospheric xenon plainly predicts that plutonium xenon should be more evident in the mantle than either uranium or iodine xenon. The absence of plutonium fission xenon in MORB must be explained by postulating a special hiding place to which Earth's plutonium was removed immediately after catastrophic degassing took place. Alternatively, plutonium abundance in the early solar system has been badly overestimated, and fissogenic xenon in Earth's atmosphere has been misidentified.

The question of missing plutonium recurs on Mars. Compared to its known reservoirs of other volatiles, Mars has relatively large amounts of radiogenic ^{40}Ar and ^{129}Xe. How much would be expected? The release factor for ^{244}Pu fission xenon should be between that for ^{129}Xe and ^{40}Ar. The amount of ^{40}Ar in the present Martian atmosphere is 2.6% of what has been generated internally by ^{40}K decay (Dreibus and Wänke 1987), or about 5% that of Earth. On a gram per gram basis, Mars has about 1/3 as much radiogenic ^{129}Xe as Earth. According to Dreibus and Wänke (1987,1989), Mars was endowed with 2.4 times more iodine than was Earth, so that the equivalent scaling factor for ^{129}Xe would be ∼14%. Thus one would expect that between 15 and 50% of Martian ^{136}Xe should be plutonium fission xenon. However, although Mars has a clear excess of heavy xenon isotopes with respect to Earth, it has no discernable plutonium xenon (Swindle et al. 1986*b*).

B. Argon

In contrast to nonradiogenic argon, Venus has about 25% as much atmospheric ^{40}Ar as Earth (Pollack and Black 1982). If Venus has the same amount of potassium as Earth, this observation would imply that Venus is about 25% as degassed as Earth, corresponding to a release factor of about 10 to 15%. Most of this difference is probably timing. Venusian atmospheric ^{40}Ar is consistent with outgassing having effectively ceased ca 3.5 Gyr (Pollack and Black 1982). Martian atmospheric ^{40}Ar corresponds to a release factor of only 2.6% (Dreibus and Wänke 1987). Some of this may be due to escape, but mostly this too should be ascribed mainly to an effective early cessation of outgassing, probably well before 4 Gyr.

VI. CONCLUSIONS

Two issues that have dominated discussion of planetary rare gases were first a general resemblance between the so-called "planetary" rare gases found in meteorites and those in air, and later the discovery that atmospheric inventories of argon and neon fall off precipitously with increasing distance from the Sun. The former encouraged identifying atmospheric rare gases with certain classes of meteorites, but it is now fairly clear that "planetary" rare gases are not found on planets. In particular, the postulated missing xenon has not been found. The apparent and very unexpected heliotropism of the noble gases points to the solar wind rather than condensation as the fundamental process for placing noble gases in the atmospheres of the terrestrial planets; however, extant models are unable to fully reproduce the observed gradient, nor do they account for similar planetary Ne/Ar ratios and dissimilar planetary Ar/Kr ratios.

More recent studies have emphasized escape. Hydrodynamic escape, which is fractionating, readily accounts for the difference between atmospheric neon and isotopically light mantle neon. It can also with more difficulty account for isotopically heavy Martian argon. Atmospheric cratering, which is nearly nonfractionating, can account for the extreme scarcity of nonradiogenic noble gases (and other volatiles) on Mars. Escape is less satisfying when used to force argon and neon on Venus and Earth into accord. Because the two planets appear to have similar amounts of krypton, carbon, and nitrogen, catastrophic volatile loss from Earth provoked by the famous Moon-forming impact is not the easy solution that it is sometimes claimed to be. If escape from Earth were nonfractionating, it would have preserved the Ar/Kr ratio. If escape were fractionating enough to produce the typically planetary ratios from an originally Venusian Ar/Kr ratio, it would also have greatly fractionated argon and neon isotopically, which is not observed.

In my opinion, some combination of solar-wind implantation, adsorption, and escape is probably required to begin explaining planetary noble gases. Solar-wind implantation helps to explain the sunward gradient, and it is a proven means of getting enough noble gas into the building blocks of Venus. Subsequent mobilization and adsorption in planetesimals might then account for the roughly adsorptive pattern common to planets and "planetary" meteorites. The high xenon abundance of "planetary" rare gases would be a peculiarity of the Q carrier. Where the Q carrier was not available, other carriers would have to suffice. In enstatite chondrites it is the enstatite itself that carries the subsolar rare gases (Crabb and Anders 1982).

There remain two outstanding problems: (i) Xe fractionation on Earth and Mars, and (ii) the different Ar/Kr ratios on Venus and Earth. Neither is close to being understood. The former imposes severe constraints on the composition of hypothetical source materials or on the evolution of the atmosphere, or both. It is not especially difficult to imagine environments in which xenon can be mass fractionated, but it is very difficult to fractionate xenon while not fractionating krypton. Thus one is forced to invoke at least two noble gas sources, one that consists mainly of fractionated xenon, and another for the other noble gases that is essentially Xe-free. An example of a two-source model is Pepin's (1991). Fractionated xenon is the only significant remnant of an early escaped atmosphere. Neon, argon and krypton are later outgassed from the mantle, but xenon is not outgassed; it is incorporated into the core.

The only possibility for a single-source explanation is a cometary veneer, if the comets are assigned the desired composition: a high Kr/Xe ratio and naturally mass fractionated xenon. Despite the freedom of working with a material of which we know next to nothing, even with prodigious escape from Earth, comets have great difficulty solving the second problem. Another potential difficulty, unique to late-accreting sources, is that Earth's degassed mantle probably degassed within ~50 Myr, and the noble gases in Earth's undegassed mantle may be quantitatively consistent with the rare gases in air being the degassed complement of the undegassed mantle. If correct, these observations would leave no place for a late veneer in the atmosphere.

Acknowledgments. I would like to thank R. O. Pepin for many detailed discussions of nobility in general and this manuscript in particular. I also thank R. Miller for TEXnical aid.

IMPACTS AND THE EARLY ENVIRONMENT AND EVOLUTION OF THE TERRESTRIAL PLANETS

H. J. MELOSH, A. M. VICKERY, and W. B. TONKS
University of Arizona

Early models for the accretion of the planets generally assumed that the planets grew gently by the steady addition of small grains swept up by the growing planetary embryos from the solar nebula. More recent models of the formation of the solar system, however, show that the population of impactors bombarding the growing planets increased in size and relative velocity with time, so that accretionary impacts tended to become progressively more violent until the last of the giant planetesimals were destroyed in collisions with planets. Accretion then tapered off during the period of heavy bombardment and continues at a much diminished rate today. The recognition of the importance of large individual impacts as well as of the cumulative effects of smaller impacts has led to a revolution in our approach to the modeling of the processes of planetary formation and differentiation, including the origin of the Moon and the origin and evolution of atmospheres. Once the planetary embryos reached a critical size, impact velocities became high enough to liberate volatiles due to shock heating, and proto-atmospheres could begin to form. Larger impacts resulted in planetary heating as the time between impacts became smaller than the time for cooling by radiation, an effect that would have been enhanced by the presence of an atmosphere. Very large impacts may have melted significant fractions of the planets, leading to rapid core formation. If the Moon was created in a giant impact, as is now widely believed, it is likely that melting would be widespread, at least in the hemisphere of the Earth in which the impact occurred and possibly throughout the entire planet. A magma ocean would form on the Earth, but because of the high gravity, convection would be sufficiently vigorous to prevent significant fractional crystallization, as is inferred to have occurred on the Moon. The effects of such an impact on a pre-existing atmosphere, and its implications for atmospheric evolution, have not yet been explored. Impacts during heavy bombardment probably resulted in some net atmospheric loss from the Earth and Venus, but the effect would have been much more pronounced on Mars, consistent with its very thin atmosphere. Such impacts may fractionate a planet's volatile inventory by sweeping away atmospheric gases but leaving condensible phases (such as water) little changed.

I. INTRODUCTION

Over the last two decades knowledge about the physics of high-speed impact processes has grown enormously. At the same time the role of impact and related processes in the formation and early evolution of the terrestrial planets has been increasingly well-defined. The most recent developments in this field are the recognition of the importance of "giant impacts"—collisions between embryo planets and bodies of about 10% of their mass—in the last stages of accretion, and the realization that impact processes may have played a major role in the growth and retention of planetary atmospheres. In this chapter we review the nature and implications of these new ideas as they currently stand, although the reader should be warned that these are both areas of very active research at the present time, so that new developments may make this review obsolete in a short time.

The recognition of the importance of giant impacts grew mainly out of recent research on the Moon's origin. The idea that the Moon was born in a gigantic collision between the proto-Earth and a Mars-size protoplanet has continued to gain adherents since it was first proposed by Hartmann and Davis (1975) and by Cameron and Ward (1976). This theory received a great deal of attention at the 1984 Conference on the Origin of the Moon in Kona, Hawaii (Hartmann et al. 1986), and has subsequently been the subject of a number of reviews (Boss 1986c; Stevenson 1987; Newsom and Taylor 1989). It is probably not an exaggeration to claim that it has become the current consensus theory of the Moon's origin. Although a great deal of work has been, and is being, done on the details of this scenario, most work up until recently concentrated on the Moon, with little consideration given to the more general effects of such large impacts on the Earth or other planets. It is clear that such an event would have profound effects on a growing planet: by the time a planet has reached the size of Mars, impacts with objects half its diameter (i.e., about 10% of its mass) are capable of melting either the entire planet or at least one hemisphere, as the shock wave from the impact traverses the planet. In this scenario, core formation is quick and inevitable; metallic iron particles separate rapidly from the silicate melt, forming a pool of iron at the bottom of the molten volume. This pool may be large enough to displace even strong underlying silicate rocks and form a proto-core (Tonks and Melosh 1991). Subsequent large impact and melting events will add more material to the core in a batchwise fashion. Vigorous convection in the melted regions (magma oceans) causes the global magma sea to cool rapidly (cooling to 50% crystallization requires only a few thousands to tens of thousands of years) and prevents crystal-liquid density segregation when the Rayleigh number is high enough to ensure turbulent convection.

At a later stage of planetary evolution, the smaller impacts during late heavy bombardment may have played an important role in bringing volatiles to the planet, in stripping away the original gaseous atmospheres of the planets and in segregating condensible substances, such as water, from volatile ones, such as CO_2. A large meteorite or comet that strikes the surface of a planet possessing an atmosphere can interact with the atmosphere in three ways: during its initial passage through it, by means of high-speed ejecta thrown out from the growing crater, and by the expanding high-energy ejecta plume that is produced at sufficiently high impact velocities. The leading edge of this plume may have a velocity many times greater than the impact velocity. In sufficiently large impacts on Earth (meteorites >300 m in diameter), the

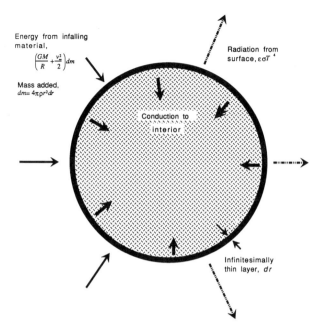

Figure 1.   The "physicist's model" of planetary accretion. This model, which was
current between about 1950 and 1975, assumes that the gravitational energy of
infalling material is balanced primarily by thermal radiation from the surface with
a small additional contribution from conduction. This model is valid only if the
planet grows from planetesimals smaller than about 10 km in diameter.

plume may have enough momentum to blow aside the atmosphere and vent
vaporized rock and other debris directly into space. In still larger impacts
(>10 km diameter), the overlying atmopheric gases may be accelerated to
greater than escape velocity so that atmospheric erosion occurs. Mars may
have lost most of its primordial atmosphere in this way during the era of heavy
bombardment.

II. ENERGY DEPOSITION BY IMPACTS ON GROWING PLANETS

A. Accretion of Infinitesimally Small Impactors

The history of accretion models of the Earth and planets mirrors a gradually
increasing appreciation for the importance of large impacts. Early models
(Hanks and Anderson 1969; Mizutani et al. 1972; Urey 1952) of the Earth's
accretion assumed that material is added by the deposition of infinitesimally
thin, laterally homogeneous layers. The accreted material adds a net energy
per unit mass consisting of gravitational binding energy and initial kinetic

energy of the mass, $GM/R + v_\infty^2/2$, where G is Newton's gravitational constant, M is the planet's mass when it has grown to radius R, and v_∞ is the velocity of encounter between the growing planet and the infalling material at great distance from the planet. Some of this energy is lost by radiation, and the rest goes into heating the planet—initially into heating the surface layer, then gradually into heating the interior of the planet by conduction. The basic equation describing the equilibrium between infall energy, radiation and conduction is

$$\rho \left(\frac{GM(t)}{R(t)} + \frac{v_\infty^2}{2} \right) \frac{dR(t)}{dt} = \epsilon\sigma[T^4(R,t) - T_a^4]$$

$$+\rho c_P T(R,t) \frac{dR(t)}{dt} + k\left(\frac{\partial T}{\partial r} \right)_{r=R} \tag{1}$$

where $T(R,t)$ is the temperature at the surface of the growing planet, T_a is the effective temperature of the atmosphere, σ is the Stefan-Boltzman constant, ϵ is the emissivity (generally set equal to 1), c_P is heat capacity and k is the thermal conductivity. Radius r is the position *within* the growing planet of total radius $R(t)$. The second term on the right side accounts for the heat content of the hot added material and should include latent heat if melting occurs. The thermal state of a growing planet is determined by Eq. (1) as soon as the growth rate, $dR(t)/dt$, is specified.

The studies based on Eq. (1) had great difficulty explaining why any of the planets should be differentiated. Thermal radiation is so efficient at removing energy that the planets would have had to grow extremely rapidly for temperatures to reach the melting point anywhere in their interiors. Thus, Hanks and Anderson (1969) require the Earth to grow in 10^5 to 10^6 yr for melting to occur, while Mizutani et al. (1972) find the Moon must accrete within 1,000 yr if melting is to occur in its outer portions. These times are far shorter than the 10^7 to 10^8 yr time scale derived from standard planetesimal accretion models (Wetherill 1980a). This implies that the Earth and Moon accreted cold, and the Earth later differentiated when enough radiogenic heat had accumulated to cause internal melting. This model also implies that the Moon should not have differentiated, a prediction that was quickly proved wrong when the Apollo missions returned Moon rocks to Earth for analysis.

B. Accretion of Small Planetesimals

A solution to this quandary was first suggested by Safronov (1972,1978) and more recently was elaborated by Kaula (1979b,1980). They realized that the idealization of the gradual addition of infinitesimally thin global layers is not a valid representation of the actual process of accretion, which involves the impact of individual planetesimals onto the planet's surface. Each impact heats the target rocks directly beneath the impact site and causes the deposition of an ejecta blanket of finite thickness and localized extent, as shown in Fig. 2. Under these conditions radiation is able to cool only a thin layer on the top of the ejecta blanket before yet another ejecta sheet is deposited on top of it. A large fraction of the initial energy of the infalling material is thus retained in the overlapping ejecta blankets.

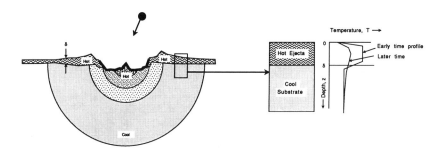

Figure 2.   Heat deposition in the vicinity of a large impact. Shock waves deposit heat directly in the rocks beneath the crater, the amount of heating decreasing with increasing distance from the impact site, illustrated by the schematic temperature contours. The heavy black line inside the crater indicates a thick layer of melted target rocks.   Such layers are observed inside large fresh lunar craters such as Copernicus and Tycho. The ejecta also contain a considerable amount of heat and thickly blanket the terrain to a distance of about 1 crater diameter from the rim. The temperature in the ejecta blanket (inset) declines as a result of thermal radiation from the surface and thermal conduction into the cooler substrate.

It is easy to estimate the thickness δ of an ejecta sheet for which heat retention outweighs radiative loss. If the planet grows at a rate dR/dt, the average time interval Δt between deposition of ejecta sheets of thickness δ at any given site is simply $\delta/(dR/dt)$. But the conductive cooling time of an ejecta sheet is of order δ^2/k, where k is the ejecta's thermal diffusivity. This cooling time equals the average time between deposition events for $\delta_0 = k/(dR/dt)$. Layers thinner than δ_0 radiate all of their heat before the next ejecta deposition event, whereas thicker layers do not have time to cool between events. Making the conservative assumption that the Earth grew over a period of about 100 Myr and that $\kappa \approx 10^{-6}$ m^2 s^{-1}, yields a crossover thickness δ_0 of about 3 km. The maximum ejecta thickness of a fresh crater is about 0.02 of the crater diameter, which itself is about 10 times larger than the projectile diameter (Melosh 1989), so the crossover condition is met by planetesimals about 15 km in diameter. Moreover, much of a projectile's energy is deposited more deeply beneath the crater floor than in the ejecta blanket, so that it is plausible that most of the heat added by 10 km or larger diameter planetesimals is retained by the growing planet rather than lost by radiation.

The details of the overall heat deposition process are complex and depend sensitively on impact mechanics. Thus, of the projectile's initial kinetic energy, about 30% is initially partitioned into the kinetic energy of the ejecta and most of the remainder is directly deposited as heat in the target rocks (Melosh 1989). The kinetic energy of the ejecta is converted into heat when it comes to rest on the surface after mixing with a variable amount of preexisting surface material, resulting in a mixed sheet of hot rock debris thickly covering the surface within one or two crater diameters of the crater's original rim. The rocks beneath the crater (and the expelled ejecta) are directly heated by the shock from the impact which, depending upon the impact velocity, may melt or vaporize large quantities of material. The amount of material thus heated depends mainly upon the projectile's size and the square of its velocity (Melosh 1989). On the other hand, a large impact cools the target planet by raising deeply buried materials closer to the surface, where their heat may be more readily conducted to the surface and ultimately lost by radiation. Some of the initial heat of the ejecta may be lost by radiation during its ballistic flight from the impact crater to its site of deposition, while at high impact velocities a portion of the ejecta (especially that in the vapor plume) may travel at velocities greater than escape velocity and thus leave the planet entirely. The net thermal effect of an impact is thus a sum of gains and losses which tend to offset one another and hence make accurate estimation of the net heat deposition difficult. Crude consideration of these processes indicates that large impacts are more efficient at depositing energy than small impacts, and that for kilometer-size planetesimals the radiation term in Eq. (1) is almost entirely negligible.

Kaula's (1980) solution to the uncertainty in estimating the net heat deposition of impacts is to lump all the poorly known processes into a single numerical factor h and write Eq. (1) in a form that neglects both radiation and conduction. Thus, rearranging Eq. (1),

$$T(R, t) = \frac{h}{c_P} \left(\frac{GM(t)}{R(t)} + \frac{v_\infty^2}{2} \right). \tag{2}$$

Note that with the neglect of radiation and conductive heat loss the accretion rate $dR(t)/dt$ drops out, so that the interiors of planets growing over the 10^8 yr accretion time scale may easily reach temperatures high enough to initiate melting and differentiation. The unknown dimensionless factor h must lie somewhere between 0 (no net burial of heat) and 1 (all kinetic energy of infalling matter is retained as heat). In the absence of better information h is generally assumed to be about 0.5 for kilometer-size or larger planetesimals.

In Safronov's (1972) theory, the random velocity component v_∞ of approaching planetesimals at any time t is proportional to the escape velocity $v_{esc} = (2GM/R)^{1/2}$ of the planet growing in their neighborhood, so the entire right-hand side of Eq. (2) is proportional to GM/R times factors of the order of unity. Because M is proportional to $R^3(t)$, the surface temperature T of a growing planet at time t is proportional to $R^2(t)$. This temperature is "locked in" at radius $r = R(t)$ as more material accumulates on the former surface of the planet, establishing an internal temperature distribution of the form

$T(r) \sim r^2$. This relation holds until T approaches the melting temperature in the outer portion of the planet, at which time convection begins and the mantle temperature remains near the melting point (Kaula 1980). Melting in the Earth begins when it has reached about 10% of its final mass, or about half its final diameter. A core is presumed to form at about this time, further heating the mantle and stirring it so that the original $T \sim r^2$ thermal structure is wiped out in a manner similar to that described by Stevenson (1981).

Subsequent to core formation the mantle temperature must have been closer to the solidus than the liquidus, because convection in a completely liquid mantle is so vigorous that its cooling time is only about 10^4 yr (Tonks and Melosh 1990), far shorter than the time scale for Earth's growth from planetesimals. Once more than about 50% of the mantle material crystallizes, however, convection is regulated by the high viscosity of solid-state creep in the crystalline fraction, thus greatly lowering the cooling rates. Craters formed by impacts subsequent to core formation would therefore have excavated either the conductive boundary layer or, if sufficiently large, the semimolten convecting mantle. The extent to which the deposition of impact energy is altered in a planet with a hot mantle has not yet been thoroughly investigated (although Minear [1980] studied the disrupting effect of impacts on the boundary layer of a cooling lunar magma ocean, concluding that impacts decrease the cooling time), nor has the related problem of convection in the presence of rapid accumulation of thick ejecta sheets been studied.

Although Eq. (2) is more realistic than Eq. (1), it is nevertheless limited by the implicit assumption that the impacting planetesimals are small in comparison to the growing planetary embryo. Impact energy is still assumed to be added in thin (although not infinitesimally thin) shells that are, on average, uniformly distributed over the planet's surface (Fig. 3). Thermal radiation losses can be neglected because most of the infalling planetesimals' energy is buried below the surface, but the overall pattern of energy deposition is much the same.

C. Accretion of Large Planetesimals

In recent years, however, it has seemed increasingly likely that the distribution of planetesimal sizes follows a rough power law extending from the smallest sizes up to objects half the size of the largest planetary embryo (Hartmann and Davis 1975; Wetherill 1985), at least during the later stages of accretion. This power distribution is expected to be of the form $N_{cum}(D) = CD^{-b}$, where N_{cum} is the number of planetesimals with diameters greater than or equal to D, C is a constant, and the power b is frequently observed to be close to 2 in numerical simulations of accretion processes (Greenberg et al. 1978), impact fragmentation experiments (Fujiwara 1986) and in the size distribution of comet nuclei (Delsemme 1987). The $b = 2$ distribution has the special property that the total surface area of planetesimals in successive logarithmically increasing size intervals is constant. That is, the surface area of planetesimals with diameters between, say, D_1 and $2D_1$ is the same as

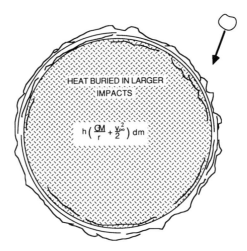

Figure 3.   A more realistic model of planetary accretion. In this model, a large
fraction of the planetesimal's energy is trapped in the growing planet as heat.
Melting temperatures are reached by the time the Earth has grown to about 10% of
its present mass. In this model the planetesimals are assumed to be much smaller
than the diameter of the planet so that growth is gradual; no provision is made for
the effects of very large impacts that deposit their heat catastrophically through a
substantial depth of the planetary embryo's mantle.

the surface area of planetesimals with diameters between $2D_1$ and $4D_1$ (see
Fig. 4). It seems natural that a distribution of this kind should arise from
processes such as impact or coagulation that depend upon cross-sectional area
(Chapman and Morrison 1989).

While the largest numbers of planetesimals in a $b = 2$ distribution are
concentrated at the small sizes, most of the mass (and therefore energy) resides
in the largest objects (see Fig. 5). Thus, although a growing planetary embryo
would be constantly battered by small planetesimals, most of the overall mass
and heat transfer would occur in large, rare events that bury their heat deep
within the embryo's interior. If this catastrophic mode of growth was indeed
important, then the assumption that the infalling planetesimals were much
smaller than the growing planet fails for the most significant events, and a
revised thermal analysis must take account of impacts between objects of
comparable size. In the following section, we examine the effects of a single
such giant impact between the proto-Earth and an object half its size.

III. EFFECT OF A GIANT IMPACT ON EARTH'S THERMAL STATE

The collision between a Mars-size protoplanet and the proto-Earth adds a
truly prodigious amount of energy to the Earth over a time interval measured

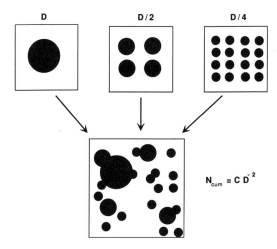

Figure 4.   Schematic illustration of the distribution of planetesimal sizes in a population described by a $b = 2$ cumulative number distribution. In this distribution, the projected area of planetesimals in logarithmically decreasing size intervals is constant.

in hours. The mass m, velocity v, and impact parameter b of the projectile are constrained only by the total angular momentum L of the Earth-Moon system, 3.49×10^{34} kg m^2 s^{-1}. Because the angular momentum is a product of all three terms, $L = mvb$, the value of each individual quantity is uncertain within broad limits. The impact parameter is bounded between zero and the sum of the Earth's and the projectile's radii. The impact velocity must be at least as large as the proto-Earth's escape velocity. The overall geochemical similarity of the Earth and Moon implies that the projectile had an initial orbit close to that of Earth's, which in turn implies that the impact velocity was probably not as much as twice the escape velocity. These constraints point to an impact by a body with a mass about 10% of the Earth's mass, hence a diameter approximately half of the Earth's—about the size of Mars.

The energy released in such a collision is the sum of gravitational and kinetic energies, and was probably within the range 2×10^{31} to 5×10^{31} J. Averaged over the Earth's mass, this is about 7.5×10^6 J kg^{-1}. Several authors have used this energy in conjunction with a single value of silicate heat capacity to infer rather high average temperatures for the Earth. Thus, a silicate heat capacity of 10^3 J kg^{-1} K yields an average temperature of 7500 K for the Earth. However, such estimates neglect the relatively small latent heat of melting, $\sim 4 \times 10^5$ J kg^{-1}, the large latent heat of vaporization, $\sim 5 \times 10^6$ J kg^{-1}, and the rapid increase of heat capacity with temperature, ranging from zero at 0 K and reaching a constant maximum at the Debye temperature (about 676 K for dunite, the main constituent of the Earth's mantle). According to the ANEOS equation of state for dunite (Benz et al. 1989), an internal energy

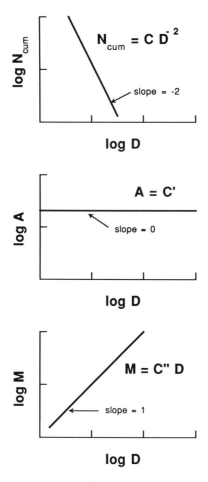

Figure 5.   This figure shows dependence of various quantities on planetesimal diameter D in a population where the cumulative number N_{cum} is a function of D^{-2}. In such a population the number distribution is dominated by the smallest objects, the area (either total surface or cross sectional) is uniformly distributed, and the mass distribution is dominated by the largest objects.

of 7.5×10^6 J kg^{-1} corresponds to a temperature rise of about 3200 K for dunite near the Earth's surface and about 7000 K deep in the mantle, where vaporization does not limit the temperature rise. Nevertheless, it is clear that melting, even vaporization, should be widespread in a giant collision.

Impacts at velocities on the order of 10 km s^{-1} in dunite generate pressures on the order of 300 GPa and will thus melt and partially vaporize a few projectile masses of the target, but such melting is confined to regions within about 2 projectile radii of the impact site. More distant regions of the planet will be warmed by impact heating, but will not necessarily melt unless they are close to melting already (see Fig. 6). Large impacts thus deposit their energy deeper than small impacts, and impacts with objects half the size of the proto-Earth can be expected to melt at least one hemisphere of the Earth's mantle right down to the core. The melt pool following a large impact will undergo further change in shape after it forms, because the melt will have a different (generally smaller) density than surrounding rocks at the same depth. Because the surrounding rocks were likely to be hot, and therefore relatively fluid, subsolidus viscous deformation subsequent to the melting event should close the initial melt-solid crater, producing a global magma ocean of nearly uniform depth overlying hot, more dense, solid mantle material. Although the time scale of this relaxation is difficult to estimate, is was probably not longer than the time scale of post-glacial relaxation in the present Earth; i.e., a few thousand years.

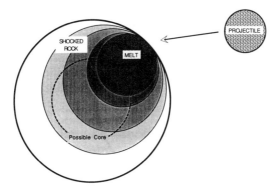

Figure 6.   Schematic illustration of the pattern of heat deposition in a planet struck by a projectile of comparable size (1/4 its diameter in this figure). Adjacent to a melted region roughly twice the projectile's diameter, the shock level, and thus temperature, falls off steeply with increasing distance from the impact site.

Although the total energy available from the collision of a Mars-size projectile with the proto-Earth is impressive, the distribution of the energy within the Earth is equally important. If, as has been suggested by Stevenson (1987), this energy is mainly expended in vaporizing the projectile, the Earth may acquire a transient silicate vapor atmosphere without strongly heating the deeper mantle. This is much more likely to occur, however, if the impactor were much less dense than the target, i.e., a comet. The similarity between the compositions of the Earth and Moon, however, strongly suggest that the impactor was a silicate body instead of a comet. For this case, the simple considerations of shock wave geometry discussed above indicate that only partial vaporization occurs (although jetting [Melosh and Sonett 1986] may enhance the local production of hot vapor), and that deep melting should be widespread in at least the hemisphere that the impact occurs.

An additional factor not previously considered here (but see also Benz and Cameron [1990]) is that the projectile might have an iron core which will sink through the Earth's mantle shortly after the collision and merge with the Earth's core, releasing its gravitational potential energy. Assuming a core equal to 30% of the projectile's mass, the energy released by sinking 3000 km through Earth's mantle is of order 3×10^{30} J, which itself will cause strong heating of the mantle through which it sinks and of the Earth's core when it arrives. After the cores have merged, this heat is applied to the bottom of the mantle, so that any portion of the mantle that escaped melting by the direct shock wave will likely be melted by this means.

To address the temperature rise in the Earth during a Moon-forming collision more exactly, H. J. Melosh performed a series of 3-dimensional numerical hydrocode computations in conjunction with M. E. Kipp of Sandia National Laboratory. These computations were designed to simulate the impact between the proto-Earth and a Mars-size protoplanet. They used the code CTH, implemented on the Cray X/MP supercomputers of the Sandia National Laboratory. This computation uses the ANEOS equations of state for dunite in the mantles and iron in the cores of the two colliding planets. The Earth has a central gravitational field, and is adjusted so that its initial temperature profile is similar to that of the present-day Earth. These models thus start out relatively cold, with mantle temperatures well below the solidus of dunite. The computations were performed at a variety of initial velocities and impact parameters, including pairs that give the Earth-Moon system its present angular momentum.

At the lowest velocity ($v_\infty = 0$), for impact parameters b of 0.88 (Fig. 7a) and 1.25 times the Earth's radius R_e, the strongest heating upon impact is confined to the hemisphere on which the projectile strikes. Shock-induced temperature rises are typically 2000 to 3000 K between the site of the impact and the Earth's core. A crater forms that extends most of the way down to the core. The gravitational energy of this excavation itself is of order 10^{30} J, which appears as heat within an hour or two as mantle material flows inward to fill the crater cavity. Unfortunately, in the high impact parameter runs the computations do not extend to long enough times for the entire projectile to merge with the Earth. This limitation is a result of the finite grid size: by the time fallback should have occurred a substantial amount of material had left the grid and could not be taken into account properly. These computations were thus stopped while some of the projectile was still falling on portions of the Earth more distant from the impact site. In these cases, more than half the mantle will be strongly heated, so the quoted results are a lower limit. Figure 7b illustrates the temperature contours for the $b = 0.88R_e$ computation 1800 s after the impact (this impact parameter and velocity correspond to the angular momentum of the present Earth-Moon system). Most of the Earth lies within contour D, that is, is at temperatures greater than 1800 K, and so is probably melted. Note that in this computation a very hot (>4200 K) low-velocity vapor plume is expelled backwards from the impact site. This plume eventually spreads over the entire Earth, producing a transient silicate vapor atmosphere.

The results for the higher impact velocity $v_\infty = 7.8 \text{ km s}^{-1}$ are more spectacular. For the impact parameters studied (0.59 and $1.25R_e$) the hemisphere near the impact was heated nearly uniformly by 1000 to 3000 K. The projectile's core was almost entirely vaporized and a much larger crater formed in the proto-Earth. A fast, hot vapor plume also carries several lunar masses of material out along trajectories that eventually take up elliptical orbits about the Earth. Figure 8b illustrates temperature contours for $b = 0.59R_e$, corresponding to the angular momentum of the present Earth-Moon system, at 1200 s after the impact. Again, a hot, low-velocity backwards vapor plume is formed that will eventually cover the Earth's surface.

IV. IMPLICATIONS OF GLOBAL MELTING

The three-dimensional hydrocode computations, in conjunction with the more general considerations described above, indicate that a Moon-forming impact would have had a profound effect on the Earth's thermal state. The shock produced by the impact would have heated the Earth to great depths, raising at least the hemisphere adjacent to the impact above the melting temperature. Later phenomena, such as the merger of the projectile's and proto-Earth's cores and the collapse of the mantle-deep crater created by the impact would have added comparable amounts of energy to the Earth. There seems to be no way to avoid the conclusion that a large Moon-forming impact is inevitably accompanied by widespread melting of most or all of the Earth's mantle.

Although the idea of molten planets is not new, having arisen most recently in the context of the lunar "magma ocean" (Wood 1970), modern geochemists have a difficult time accepting that the Earth once had a similarly extensive molten mantle. Kato et al. (1988) argue that fractionation of a small amount of perovskite and/or majorite garnet from a terrestrial magma ocean would drive the nearly chondritic ratios of most refractory lithophile elements substantially away from their observed chondritic values. Fractionation would also result in a primitive crust rich in Ba, K, Rb and Cs and relatively depleted in U, Th, Pb, Sr, Zr and Hf. In a similar vein, McFarlane and Drake (1990) argue that fractionation of olivine by flotation at high pressure would yield ratios of Sc/Sm, Ni/Co, and Ir/Au more than 10% above their observed chondritic values.

Ringwood (1990) argues that Kato et al.'s (1988) results can be used as a test of the giant impact hypothesis and Safronov accretion theory. Because the giant impact leads to global melting of the Earth, he infers that such an impact could not have occurred because the oldest ($\sim$4.2 Gyr) zircons and most ancient ($\sim$3.9 Gyr) crustal rocks would show no signature of the element fractionations noted above. Ringwood thus proposes that the Earth accreted

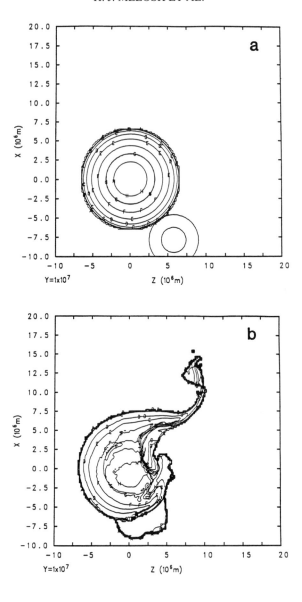

Figure 7.   Temperature contours in the collision between the proto-Earth and a
protoplanet half its diameter. This computation is for $v_\infty = 0$ km s^{-1} (8 km s^{-1} at
contact) and an impact parameter of 0.88 R_e at contact. Plot (a) shows the initial
configuration before impact in which the projectile is traveling upward (positive
x-direction) and the proto-Earth is at rest. Plot (b) shows the configuration 1802
s after contact. The contour values are $A = 300$ K, $B = 600$ K, $C = 1200$ K, $D =$
1800 K, $E = 2400$ K, $F = 3000$ K, $G = 3600$ K and $H = 4200$ K. Figures 7 and 8
were computed by M. E. Kipp at Sandia National Laboratory, Albuquerque, NM
using the three-dimensional hydrocode CTH. These plots are in the symmetry plane
of the two colliding spheres.

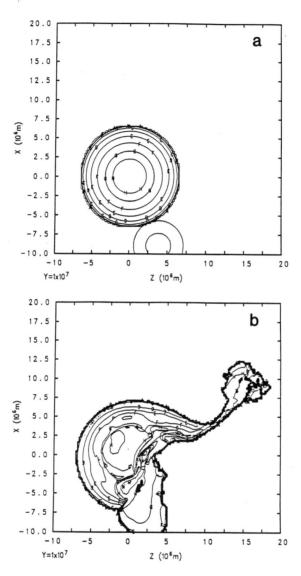

Figure 8.   (a) Initial configuration and (b) temperature contours 1193 s after the impact
between the proto-Earth and a protoplanet half its diameter moving at $v_\infty = 7.8$ km
sec^{-1} (12 km s^{-1} at contact) and an impact parameter of 0.59 R_e at contact. Contour
values are the same as in Fig. 7.

without widescale melting and suggests several conditions under which this
might occur.

    All of these objections make the implicit assumption that a magma ocean
necessarily undergoes fractional crystallization, that is, that the crystals that

grow within the cooling magma are removed from the magmatic system by either settling or floating (because of density differences) and do not subsequently interact with the magma. However, it seems obvious that if convection is sufficiently vigorous, separation may be either inhibited or even prohibited. Thus, the reconciliation of the theoretical arguments for widespread melting in the Earth and the geochemical arguments prohibiting it may be rooted in the mechanical behavior of crystals in a convecting magma ocean.

The mode of convection in such a magma ocean, or any other naturally convecting system, is approximately characterized by the Rayleigh number Ra and Prandtl number σ (Chandrasekhar 1961), given by

$$Ra = \frac{\alpha g \Delta T D^3}{\kappa \nu} \tag{3}$$

$$\sigma = \frac{\nu}{\kappa} \tag{4}$$

where α is the thermal expansion coefficient, g the acceleration of gravity, D the depth of the convecting fluid, κ is the fluid's thermal diffusivity, ν its viscosity, and ΔT is the superadiabatic temperature gradient. Although these two parameters must be supplemented by others describing the temperature dependence of viscosity and the change of viscosity with depth in an exact model of magma ocean convection, the success of parameterized convection models suggests that the Rayleigh and Prandtl numbers define the major convective regimes to first order.

On the basis of the values of the Rayleigh and Prandtl numbers, magma ocean convection is utterly different from that of the more familiar sub-solidus convection now occurring in the planets, shown in Fig. 9. Magma ocean convection and crystal settling has more affinity to aeolian transport in a deep planetary atmosphere than to traditional studies of convection. In this range of conditions, a vigorously convecting system can be divided into three distinct regions (Kraichnan 1962): an upper conductive boundary layer, a sublayer dominated by viscous forces, and a volume of turbulent fluid whose flow is controlled by inertial forces (see Fig. 10). This inertial flow zone, which does not exist in the subsolidus convection familiar to most geophysicists, is dominated by large scale turbulent eddies that are analogous to gusty winds in a planetary atmospheric convective system. The inertial flow zone appears in the central part of a convecting fluid when the following inequality between the Rayleigh and Prandtl numbers is satisfied:

$$Ra^{1/3} \geq 35\sigma^{1/2}. \tag{5}$$

Convection in a magma ocean is expected to remain vigorous so long as the liquid can lose its heat near the free outer surface. Although a thin conductive skin may form at the immediate surface, differential convective

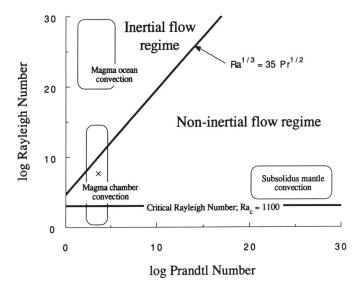

Figure 9.   Rayleigh number vs Prandtl number, showing the convective regimes of common geologic processes and of planetary magma oceans. The inertial flow zone exists above and to the left of the inertial flow zone disappearance boundary and disappears to the right of the boundary. See the text and Fig. 10 for further explanation. The x marks where an experimental study (Martin and Nokes 1988) showed the onset of crystal suspension.

velocities are generally high enough to overcome the skin's strength and keep the heat loss near the theoretical value. This situation is quite different from that of a magma chamber, where a thick roof may strongly inhibit convection (Brandeis and Marsh 1989). The presence of an atmosphere may also have an important effect in limiting the rate at which heat is lost from a magma ocean. By blocking or limiting the rate of heat loss to space by infared radiation, an atmosphere can effectively blanket the top of the magma ocean, becoming a partial substitute for the thin upper conductive skin and controlling the cooling time scale of the planet's interior. A very thick initial atmosphere may even lead to the formation of a magma ocean without the intervention of a giant impact (Abe and Matsui 1985). The role of such atmospheric blanketing in controlling the surface temperature and rate of heat loss from a planet with a magma ocean is not well understood, and is an obvious target for future research.

A criterion for crystal suspension near the upper and lower margins of a convecting magma ocean can be derived from empirical studies of sediment transport. The Rouse number is defined as the ratio between the terminal

Free Surface

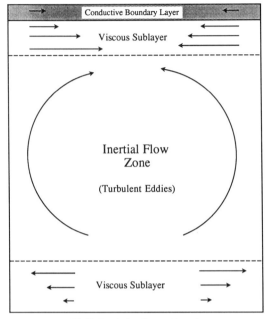

Rigid, Insulating Lower Boundary

Figure 10.   Diagram showing convection breaking into three distinct regions of behavior at very high Rayleigh number. The topmost region is a conductive boundary
layer underlain by a viscous sublayer. In the third region, flow is dominated by
inertial forces which produce large-scale turbulent eddies (the inertial flow zone).

velocity of the crystals in the fluid and the turbulent friction velocity. Settling
(or flotation) takes place only for Rouse numbers greater than about 1 (Vanoni
1975). A study of the evolution of an ultramafic magma ocean with realistic
magma rheology (Tonks and Melosh 1990) found that on both the Earth and
Moon crystals up to 1 to 2 cm in diameter can be suspended from the onset
of cooling. In the later stages of cooling, as viscosity increases, crystals of
10 cm to more than 1 m in diameter may easily be suspended. A surprising
feature of the analysis is that the ability of a magma ocean to suspend crystals
is only weakly dependent on gravity and depth: viscosity is the major factor
determining suspension. This result, although gratifying for the Earth, raises
a serious problem for the *lunar* magma ocean hypothesis, because the differentiation of an anorthositic crust on the Moon strongly suggests that crystal
fractionation did occur there.

     One explanation for this apparent paradox appeals to the low lunar grav-
ity. Because pressure in the Moon increases slowly with depth compared to
the Earth, the solidus and liquidus temperatures are nearly independent of

depth. In a convecting magma ocean on the Moon, adiabats lie between the solidus and liquidus at all depths, so that growing crystals circulate through the ocean for long periods of time without dissolving. Under these circumstances crystals may grow large enough to settle out, producing the observed differentiation. In contrast, on the Earth the liquidus and solidus profiles are steeper than an adiabat so crystallization only takes place over a restricted depth range. Convective velocities are so high that crystals nucleating in this zone do not have time to grow larger than a few microns and are thus incapable of settling out. In this way, even a totally molten initial Earth may have failed to differentiate by fractional crystallization.

Although magma ocean convection may be sufficiently vigorous to prevent fractionation between silicate melt and silicate crystals (whose density differs only a few hundred $kg\ m^{-3}$ from that of the melt), the density difference between the silicate melt and droplets of molten iron (typically 2000–3000 $kg\ m^{-3}$) is great enough to allow iron droplets to rain out of the magma ocean, especially in the early stages when melt viscosities are low, and accumulate in a dense pool at the base of the magma ocean. Such dense pools of molten iron may, if sufficiently large, descend to the center of the growing planet, forming (or adding to) its protocore (Tonks and Melosh 1991).

V. ORIGIN AND EVOLUTION OF ATMOSPHERES

The origin and evolution of planetary atmospheres is a complex and poorly understood process. Elemental and isotopic compositions of planetary atmospheres differ from one another and from the composition of any single (or simple combination) of plausible volatile reservoirs in the early solar system (Pepin 1989). These reservoirs include solar nebular gas, solar wind, comets, and volatile-rich meteorites. Accretionary impacts have two competing effects on the development of planetary atmospheres: the release of volatiles due to shock heating and/or vaporization of the impactor and part of the target in relatively low-energy events may add to the planet's volatile inventory (Arrhenius et al. 1974; Benlow and Meadows 1977; Chyba 1990), and sufficiently large impacts may cause the loss of part of the existing volatile inventory.

During most of the accretionary process, impact velocities generally differ little from the escape velocity of the growing protoplanet because most of the collisions are between bodies in nearly matching orbits. At some stage in planetary growth, the impact velocities become high enough that the shock pressures resulting from the impact are high enough to devolatilize the impactor (and possibly part of the target). For carbonaceous chondrites, partial devolatilization can begin on the Earth and Venus when the planetary embryos reach approximately 12% of their current size, and complete devolatilization begins when the embryos are approximately 30% of their current size (Ahrens et al. 1989; Lange and Ahrens 1982). The corresponding values for Mars are roughly 23% for partial devolatilization and 56% for complete devolatilization. Devolatilization of comets presumably occurs at lower planetary radii because comets are more volatile than silicates and because their lower density with respect to the target means that a larger fraction of the impact energy is partitioned into heating the projectile. Abe and Matsui (1985) consider thermal models of accretion in which CO_2, H_2O and fine impact ejecta are produced in accretionary impacts. They conclude that the proto-atmosphere forms a thermal blanket that traps the internal energy added by impacts at or near the planet's surface. The result of this thermal trapping is that the surface temperature becomes high enough to induce devolatilization regardless of the shock pressure produced by an impact, once the Earth (or Venus) achieves $\sim$30% of its final radius. This modeling thus strongly suggests that substantial atmospheres/hydrospheres could develop this way.

VI. ATMOSPHERIC EROSION

Toward the end of accretion, however, collisions are rarer but much more energetic, involving large planetesimals and higher impact velocities, as discussed above. Such impacts may cause a net loss of atmosphere from a planet, so that the cumulative effect of impacts during the period of heavy bombardment might have dramatically depleted the planets' original atmospheres (Cameron 1983). The basic concept is that the fraction of the projectile's momentum that is transferred to the atmosphere may be sufficient to accelerate a portion of the atmosphere to velocities greater than the escape velocity of the planet.

A meteorite or comet may transfer momentum to the atmosphere of the target planet in three ways: first, during the initial passage of the impactor through the atmosphere, there is a direct transfer of momentum as the impactor penetrates the atmosphere, compressing and accelerating the gas in front of it. A small meteorite or comet may be broken up by, or significantly ablated during, its passage through the atmosphere, and thus deposit most or all of its energy directly in the atmosphere, mainly as shock waves and heat. This is apparently the explanation for the famous Tunguska event of 1908, in which shock waves from a meteoroid that did not quite reach the ground leveled about two thousand square kilometers of dense Siberian forest, snapping off meter-diameter trees like matchsticks. Large impactors, however, are not significantly slowed by the atmosphere: O'Keefe and Ahrens (1982) showed that in this case, the impactor delivers only a small fraction of its momentum directly to the atmosphere, and Walker (1986) showed that this momentum is distributed in such a way that no significant amount of atmosphere escapes from a planet with an escape velocity ≥ 10 km s^{-1}. This mechanism is thus negligible for the Earth and Venus, but it may have contributed to atmospheric erosion on Mars. Second, solid ejecta thrown out of the growing crater can transfer momentum to the atmosphere in a similar fashion, but again this has been shown to result in negligible atmospheric loss even for large, high-speed impacts (Melosh and Vickery 1988), even on Mars. Third, for a sufficiently energetic impact, a great deal of very highly shocked impactor (and possibly target) material expands upward and outward at high velocities, driving the overlying atmosphere ahead of it (Melosh and Vickery 1989; Vickery and Melosh 1990; Watkins 1983). This last mechanism is by far the most important for atmospheric erosion, and we discuss it in more detail below.

A. Minimum Projectile Size

In order that a high-energy ejecta plume should form, the most basic requirement is that the projectile should be large enough to strike the target surface at high velocity and remain intact. The minimum diameter L_{min} of a projectile that reaches the surface with about half of its initial (pre-atmospheric) velocity, neglecting ablation and crushing effects, is

$$L_{min} = P_a/(\rho_0 g \sin \theta) \tag{6}$$

where P_a is the surface atmospheric pressure, g is the acceleration of gravity at the planet's surface, ρ_0 is the projectile's density and θ is the angle of entry, measured from the horizontal. Thus, a vertically incident ($\theta = 90°$) stony projectile on Earth with its current atmospheric pressure must be larger than about 3.4 m in diameter to strike with half or more of its initial velocity, while on Venus such a projectile must exceed a diameter of 350 m. If atmospheric pressures were greater than their current values at early times in the history of a given planet, this minimum size would have been correspondingly greater. In actuality, these limits are probably much too small because they neglect ablation and atmospheric breakup, both of which tend to increase the minimum size of a projectile that can reach the surface with high velocity. Ablation is most important for meter-sized projectiles, so that a limit placed by the process of atmospheric breakup is probably more realistic.

Projectiles entering a planetary atmosphere encounter aerodynamic stresses of order $\rho(z)v_e^2$, where $\rho(z)$ is the atmospheric density at altitude z and v_e is the entry velocity. These stresses may reach many kilobars in the lower atmospheres of the Earth and Venus, and so the entering projectiles are nearly certain to be crushed. This effect is independent of the diameter of the projectile. Although the broken fragments of smaller projectiles may be scattered by these stresses, creating the familiar km-wide elliptical strewn fields, the fragments of larger projectiles do not have time to separate significantly. These fragments thus fall in nearly the same location and produce a crater almost indistinguishable from one produced by a single intact projectile. For a vertically incident projectile, the diameter L_c above which breakup is not important is given by (Melosh 1981)

$$L_c = 2(\rho_a/\rho_0)^{1/2}H \tag{7}$$

where ρ_a is the density of the atmosphere at the surface and H is the scale height. Because the surfaces of both planets are quite young, we use current values for the atmospheric parameters to calculate L_c, then use scaling relations to predict minimum crater sizes, and compare these to observations. On

Earth, L_c is about 300 m for stony projectiles, while on Venus it is 4500 m. This implies that the minimum-size hypervelocity impact crater that can form on Earth is a few km diameter (for iron projectiles, the minimum is about a kilometer—about the size of Meteor Crater, Arizona, which was clearly formed by an iron projectile that at least partially fragmented in the upper atmosphere), while the minimum size crater that can form on Venus is about 30 km in diameter, in good agreement with the observations of the Magellan spacecraft (Phillips et al. 1991). We believe therefore that this model gives a reasonably accurate estimate of the minimum size projectile that can survive atmospheric passage as a function of the density structure, and so may be applied to early (hypothetically) denser atmospheres.

B. Melt and Vapor Plume Formation

When a projectile strikes a planet's surface at high speed, a rapid but orderly sequence of events is initiated that ultimately results in a crater. The principal topic here, however, is not the crater itself but the plume of hot melt and vapor that expands into the atmosphere above it. The ultimate origin of this plume lies in the thermodynamics of shock compression and release. When the projectile first strikes the surface, a region of high pressure develops at the interface that spreads into both the projectile and surface rocks as shock waves expand away from the contact area. These waves quickly reach the sides and rear free surface of the projectile and are reflected from there as rarefaction waves that then propagate back downward, releasing the compressed projectile to low pressure. In the surface rocks the shock waves travel outward and downward, engulfing more material and weakening as they proceed.

The resulting pattern of maximum shock pressure is shown schematically in Fig. 11. The projectile and a roughly equal mass of target (surface) rocks are shocked to the highest pressure P_o in a volume comparable to that of the projectile itself (strong shock compression decreases the initial volume of most materials by a factor of about 2). For a target and projectile of the same composition, P_o can be simply estimated using the Hugoniot equations and impedence matching to give

$$P_o = \rho_o \left(C + S \frac{v_i}{2} \right) \frac{v_i}{2} \tag{8}$$

where v_i is the impact velocity, ρ_o is the density of the uncompressed material and C and S are material constants. For a typical crustal rock (granite), $\rho_o = 2630$ kg m^{-3}, $C = 3.68$ km s^{-1} and $S = 1.24$ (Kieffer and Simonds 1980). At an impact velocity of 20 km s^{-1}, this yields a maximum pressure P_o of about 425 GPa, somewhat higher than the pressure at the Earth's center. For the more complex case where the target and projectile differ in composition see Melosh (1989).

The maximum shock pressure outside this "isobaric core" (Croft 1982a) falls approximately as a power of distance away from the impact site

$$P = P_o \left(\frac{a}{r} \right)^n \tag{9}$$

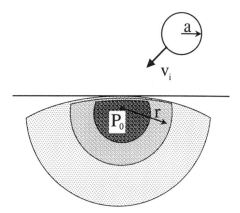

Figure 11.   Schematic illustration of shock contours beneath the site of an impact.
The projectile of radius a strikes the surface at a velocity v_i. A region of nearly
uniform high pressure P_0 develops with a volume comparable to that of the projectile
itself. Outside this region the pressure falls off with increasing distance r from the
impact site. The free surface is constrained to zero pressure, so the pressure contours
fall off rapidly in a narrow near-surface zone.

where a is the radius of the projectile and the power n ranges between 3 for
strong shocks, declining to 2 at shock pressures lower than 100 GPa. The
shock wave thus establishes a pattern of highly shocked rocks near the impact
site surrounded by roughly concentric zones of successively less shocked
material. The near-surface zone is a special case, where a very thin layer
of target material is protected from high shock pressure by the boundary
condition at the free surface. The small amount of material within this zone
experiences only low shock pressures, but the pressure gradients are high, and
the material is ejected at high speed. The concentric shock pressure structure
is established long before the excavation flow has time to open the crater.
The pattern is subsequently altered by the cratering flow, and may not be
discernible except in the deformed rocks directly beneath the final crater.

The heating of shocked materials is the result of a thermodynamic cycle.
Shock compression is a thermodynamically irreversible process. At a shock
front, mass, momentum and energy are conserved, but not entropy. On a
P-T plot, the state of the shocked material jumps discontinuously from its
initial state to a final state lying on the Hugoniot curve, a locus of points each
of which results from different degree of shock compression, i.e., this curve
does *not* define a thermodynamic path traversed by the shocked material. The
shocked material then decompresses along an adiabat, which does define a
continuous path of P-T states attained by the material.

The standard P-T plot, however, is not very convenient for illustrating phase transformations induced by shock, as the adiabats are complicated
curves whose paths give little insight into the ultimate state upon decompression, especially near the critical point. In this chapter, we introduce a P-S

plot (S is entropy) which, although perhaps unfamiliar, contains far more information about the ultimate fate of shocked material. One major advantage of this plot is that release adiabats are vertical lines (constant entropy), and the state of the material at 1 bar (10^{-4} GPa) is generally well known. Thus, if the entropy along the Hugoniot can be computed, the phases present upon decompression to 1 bar, at least, are well determined.

Figure 12 shows an example of a Hugoniot and the phase structure of forsterite on a P-S plot. The Hugoniot and liquid-vapor phase curve were computed from the ANEOS equation of state (Benz et al. 1989), modified slightly to correct a small error in the input parameters. The thermodynamic cycle of shock compression followed by adiabatic release does work on the material, and upon release to low pressure the shocked material may be in the form of a hot solid (for weak shocks), a liquid or a gas (at the highest pressures). Figure 12 indicates that at shock pressures less than about 500 GPa (with a corresponding particle velocity of 10 km s^{-1}), the entropy of the shocked material is less than that required to reach the critical point, and the release curve crosses the liquid/vapor phase curve from the liquid side. Thus, for most asteroidal impact velocities, silicates never vaporize completely (for this to occur the release adiabat must pass to the right of the critical point in Fig. 12), but instead come to a boil, rapidly releasing vapor bubbles and breaking the melt apart into discrete clumps as the pressure declines (Melosh and Vickery 1991). Once the material reaches the liquid/vapor phase curve, its temperature and pressure are linked by the phase relation P (GPa) $\approx$ 220 exp($-48,900/T$). The relative masses of liquid and vapor (or solid and vapor below the triple point at about 10^{-9} GPa) can be determined from this diagram by the lever rule. When considering the problem of atmospheric erosion, it is not critical whether the material expands into the liquid field and breaks up into droplets or expands into the vapor field and condenses into droplets: in both the cases the plume expands adiabatically, engulfing and accelerating the ambient atmosphere. The difference may be important, however, for the problem of tektite and microspherule formation.

The specific internal energy initially deposited in the shocked material is approximately equal to one-half of the particle velocity squared. Because the particle velocity itself is about one-half of the impact velocity (for projectile and target composed of the same material), the internal energy of the shocked material per unit mass is given by $E \approx v_i^2/8$. When the impact velocity is high enough that melting or vaporization occur, this energy is available for acceleration of the material out of the growing crater. The average final velocity attained by the isentropic expansion of a fluid with initial internal energy E is given by $v_\infty = (2E)^{1/2}$, or in this case $v_\infty = v_i/2$. Thus, the plume of melt and vapor expanding away from the site of an impact eventually achieves a mean speed of about half the impact velocity. Furthermore, detailed numerical computations with the ANEOS equation of state for dunite show that the edge of the expanding plume typically moves at twice this speed, or at nearly the impact velocity itself.

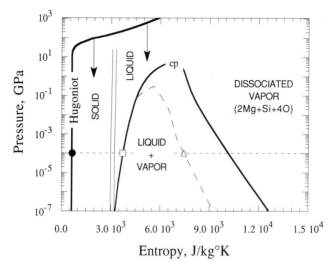

Figure 12.   The thermodynamic path of shocked dunite in P-S (pressure-entropy) space. The Hugoniot curve and solid liquid-vapor phase curve are based on the ANEOS equation of state for dunite (Benz et al. 1989), with the slightly corrected parameters $ZB(11) = 2.23 \times 10^{11}$, $COT(1) = 0.571$, $COT(2) = 0.143$, $COT(3) = 0.286$. The entropy of the 1 bar (10^{-4} GPa) reference state (from the JANAF tables) is shown as a heavy black circle, the near-vertical light lines denoting incipient and complete melting are constrained at 1 bar by JANAF data, and the points for incipient (square) and complete (triangle) vaporization are from Ahrens and O'Keefe (1972). Note that although the ANEOS phase curve agrees well with the point for incipient vaporization, it lies at much too high an entropy at complete vaporization. Similarly, the critical point is at about a factor of 10 higher pressure than that derived from the hard sphere formalism (Ahrens and O'Keefe 1972). These displacements are a result of ANEOS's implicit assumption that the gas phase is a monatomic mixture of species, whereas vaporized dunite is likely to consist mainly of a mixture of MgO, SiO and O_2. The long-dashed curve is an interpolation through the hard-sphere critical point indicating the probable true nature of this phase boundary. Release adiabats are shown schematically as downward pointing arrows. Note that even for the hard-sphere critical point at $S = 5600$ J kg^{-1} K^{-1} the shock pressure must exceed about 500 GPa (with corresponding particle velocity of 10 km s^{-1}) to reach the liquid-vapor curve from the vapor side.

C. Atmospheric Erosion by Melt and Vapor Plumes

The structure of this expanding plume of hot melt and vapor is initially rather complex, as revealed by detailed numerical computations (O'Keefe and Ahrens 1982), but as time goes on, it quickly attains a roughly hemispherical shape (Basilevsky et al. 1983). The mean density of the plume declines below the melt density by the time it has expanded to about twice the projectile radius, indicating that by this time the plume is composed of droplets of melt in local thermodynamic equilibrium with vapor. The gas and melt droplets are strongly coupled, however, until much later in the expansion. As this fast-moving dispersion of liquid droplets and vapor meets the ambient atmosphere, shock waves form at the interface that decelerate the plume while accelerating the surrounding air.

In the case of a small impact, the mass of the atmosphere may be sufficiently great that the plume's expansion is entirely halted. If this occurs over a distance small compared to the atmospheric scale height H, the hot melt and vapor form a hemispherical bubble that rises into the stratosphere under the influence of its own buoyancy, like a hot air balloon. This is the classic fireball scenario, familiar to everyone as the mushroom cloud emblematic of nuclear explosions. An upper limit on the projectile size for this scenario can be roughly derived by equating the radius of a vapor "fireball" to the atmosphere's scale height H (see Melosh 1989; Jones and Kodis 1982). Assume that the projectile and an equal mass of target begin expanding from an initial pressure given by the second Hugoniot equation, $P_i = \rho_p(v_i/2)[C + S(v_i/2)]$, where v_i is the impact velocity on the surface (the initial particle velocity is $v_i/2$ if projectile and target materials are similar), C and S are material constants in the linear shock-particle velocity equation of state, and ρ_p is the projectile density. At high velocities, only the S term is important: $P_i \approx \rho_p S v_i^2/4$. Then the radius of the smallest projectile that can generate an atmosphere-piercing plume is:

$$a_p = H \left(\frac{P_a}{S\rho_p v_i^2}\right)^{1/3\gamma} \tag{10}$$

where P_a is the surface atmospheric pressure and γ is the ratio of specific heats ($\gamma = 1.4$ for air). As $S \approx 1.3$ for most silicates (Melosh 1989, Table AII.2), $H = 10$ km on Earth, and taking $v_i = 20$ km s^{-1} this equation yields $a_p \approx 200$ m, corresponding to a transient crater about 6 km in diameter. Larger projectiles than this produce vapor plumes that cannot be stopped by the atmosphere.

Because the smallest mass of atmosphere is vertically above the crater (Fig. 13), it is expected that projectiles somewhat larger than the limit (Eq. 10) will create strongly focused vertical plumes above the impact site. The maximum velocities in these plumes may approach the impact velocity, and thus some material (mostly derived from the projectile itself) may be ejected back into interplanetary space. A simple method of estimating whether the atmospheric gases will be ejected is to compute the momentum of the expanding melt and vapor plume in a small element of solid angle extending outward from the impact site. The mass of the atmospheric gases in the same solid angle is then computed and added to the mass of the crater ejecta. Momentum conservation then determines the mean velocity of the mixture. If the mean velocity is greater than escape velocity, then the atmosphere in this element of solid angle is considered to be ejected. A minimum requirement is that the mean velocity of the melt and vapor plume must exceed escape velocity, so that the initial impact velocity must be twice the escape velocity, or about 20 km s^{-1} on the Earth. Similarly, the mass of the plume must be sufficient to accelerate the atmosphere along with it. This puts a lower limit on the projectile size for atmospheric loss to occur, which we have found to be close to 10^{13} kg or roughly 2 km in diameter (Vickery and Melosh 1990) for silicate projectiles and Earth's current atmosphere. This lower limit would be greater for Venus and less for Mars at the present time, because it increases with increasing atmospheric mass, and decreases with increasing H/R, the ratio of the atmospheric scale height to planetary radius.

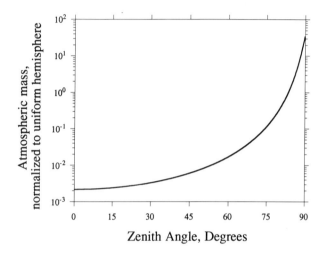

Figure 13.   Mass in the Earth's atmosphere as a function of zenith angle ($0°$ is vertical, $90°$ is horizontal), normalized by the mass in a uniform hemisphere of the same total mass. This gives the "coupling factor" between an expanding hemisphere of gas and the ambient atmosphere. The mass in the Earth's atmosphere is small directly overhead and strongly concentrated toward the horizon. The mass in an element of solid angle is the same as a uniform hemisphere at a zenith angle of $83°2$ (based on the U. S. Standard Atmosphere).

As the size of the projectile increases, more and more of the atmosphere is liable to be stripped away from the planet by the expanding plume of vapor mixed with melt droplets. The maximum that can be stripped is given approximately by the mass lying above the plane tangent to the planet at the impact point, $m_{tp} = 2\pi P_a H R_P/g$, where R_P is the radius of the planet, P_a is the surface atmospheric pressure and g is the surface acceleration of gravity (Melosh and Vickery 1989). This amounts to about 0.1% of the total mass of the atmosphere (the fraction is H/R_E), or 3×10^{15} kg on the Earth. Detailed numerical computations (Vickery and Melosh 1990) have shown that at impact velocities of >20 km s^{-1}, the atmospheric mass ejected is not a strong function of impact velocity and that projectiles with masses in the range of 10^{17} kg and above are capable of removing essentially all of the atmosphere above the tangent plane on Earth. These computations are difficult, however, because most of the atmosphere's mass is concentrated at low elevations from the impact point (see Fig. 13), whereas the expanding plume of melt and vapor is more or less isotropic. The two thus do not couple well together. The coupling may be enhanced, however, by the concentration of the plume's mass at low angles, as is expected for oblique impacts (Schultz and Crawford 1987). Such low-angle processes as jetting (Melosh 1989) or ballistic curvature of lower-speed ejecta may also enhance the efficiency of atmospheric loss. To achieve accurate estimates of the process of atmospheric erosion by impacts, it is thus necessary to perform full scale numerical computations on the fast, low-angle ejecta from an impact, and to consider the radius of curvature of the target planet. Such computations have not yet been done with sufficient accuracy: this is clearly a direction that future work on atmospheric erosion by impacts must take.

Although an exact treatment of the projectile mass needed to eject the atmosphere above the tangent plane is still lacking, the computations discussed above (Vickery and Melosh 1990) show that a projectile mass m_* between 1 and 10 times m_{tp} is needed to strip the atmosphere above the tangent plane. Using this estimated threshold along with a postulated cratering flux permits the rate of atmospheric erosion by impacts to be roughly computed. The present cumulative flux (number/sec/m^2) of projectiles with mass greater than or equal to m is parameterized by $N_{\text{cum}}(m) = am^{-b}$, where a and b are constants. Unfortunately, neither the over-all flux a nor the slope of the distribution b are well known. A pair of constants that predicts the correct present lunar crater distribution via the revised Schmidt-Holsapple scaling law (Schmidt and Housen 1987) is $a = 1.55 \times 10^{-23}(MKS), b = 0.47$. The slope of the crater distribution on the Martian plains is the same as the lunar distribution (Neukum and Wise 1976; Strom 1984), while the overall cratering rate on Mars at present is estimated to between 1 and 4 times the lunar rate, with a preferred mean of about 2 (BVSP 1981).

Using a constant cratering rate, the rate of mass loss from a planet's atmosphere dM_{atm}/dt is given by the flux $N_{\text{cum}}(m_*)$ of projectiles large enough to remove the atmosphere above the tangent plane, times the planet's surface area $4\pi R^2$, times the mass above the tangent plane m_{tp}. As m_* itself depends upon M_{atm}, a simple differential equation for M_{atm} (or, equivalently, P) results whose solution, converted to a convenient and universal form, is:

$$\frac{M_{\text{atm}}(t)}{M_0} = \frac{P(t)}{P_0} = \left(1 - \frac{t}{t_*}\right)^{1/b} \tag{11}$$

where the M_0 is the present $(t = 0)$ atmospheric mass, P_0 is the present surface atmospheric pressure, and t_* is the length of time required for impacts to reduce the atmospheric pressure to zero. In terms of previously defined quantities,

$$t_* = \frac{f_{\text{crit}}^{\,b}}{2\pi ab(RH)^{1-b}} \left(\frac{4\pi P_0}{g}\right)^b \tag{12}$$

where f_{crit} is the ratio of the critical projectile mass m_* to the mass above the tangent plane m_{tp}. This ratio is assumed to be a constant. As $b = 0.47$, the characteristic time t_* is not very sensitive to uncertainty in f_{crit}.

Note that the above equation for the time dependence of atmospheric mass or pressure can be used for times before the present ($t < 0$), so that the atmospheric pressure at any previous era can also be computed. It is interesting to note that the atmospheric pressure declines rigorously to zero at time t_*—it does not simply fall exponentially towards zero. This interesting fact is a unique characteristic of the impact erosion mechanism. As the atmospheric pressure declines, smaller projectiles are capable of removing the atmosphere above the tangent plane. But by the distribution law, there are more smaller projectiles than larger ones, so a greater fraction of the atmosphere is removed in each unit time interval. The net result is the complete stripping of the atmosphere after time t_*, barring volcanic or other sources of replenishment.

The post-late-heavy-bombardment flux, along with other parameters and the assumption $f_{crit} \approx 1$, gives a value $t_* \approx 60$ Gyr for Mars at the present time (note that t_* for Earth and Venus is longer by several orders of magnitude, so that impact erosion of their atmospheres is entirely negligible). This value for t_* predicts an atmospheric pressure on Mars at the end of late heavy bombardment (3.2 Gyr) only 1.1 times the present low pressure. This would not explain the geomorphic evidence (Baker and Partridge 1986) for running water on Mars' surface early in its geologic history. However, it is widely known (*BVSP* 1981) that the impact flux on the Moon in the first Gyr of its history was many orders of magnitude larger than at present. It is presumed (Neukum and Wise 1976) that Mars, too, shared in this era of heavy bombardment, so that the enhanced impact fluxes of this era may have produced an early period of rapid atmospheric pressure change. The heavy bombardment flux can be adequately fit by the expression

$$N_{cum}(m, t) = a \left(1 + Be^{-\lambda(t+4.6)}\right) m^{-b} \tag{13}$$

where a and b are the same as before, and $B = 1.17 \times 10^4$ and $\lambda = 6.94$ Gyr^{-1} (Chyba 1990).

The differential equation for the rate of atmospheric mass loss implied by Eq. (13) can be readily integrated, giving a simple analytic expression similar to Eq. (11) for the evolution of atmospheric pressure as a function of time t from the present:

$$\frac{M_{atm}(t)}{M_0} = \frac{P(t)}{P_0} = \left(1 - \frac{t}{t_*} - \frac{Be^{-4.6\lambda}}{\lambda t_*}[1 - e^{\lambda t}]\right)^{1/b}. \tag{14}$$

The atmospheric pressure of Mars predicted by this expression is shown in Fig. 14. It is clear that early in Mars' history its atmospheric pressure may have been as much as 100 times its present value, or approximately 1 bar. It is interesting to note that this pressure is near that considered necessary for Mars to have supported free water on its surface (Cess et al. 1980; Kasting and Toon 1989). Of course, other processes besides impact erosion probably played major roles in determining Mars' atmospheric pressure, so this computation should be regarded merely as an indication of the magnitude of the effect and time interval over which impact erosion may have been important.

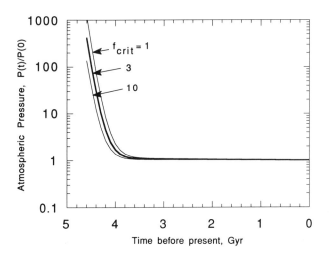

Figure 14.   The history of atmospheric pressure on Mars in a model where impact erosion alone can affect the mass of the atmosphere (although this is an unrealistic assumption, it serves to highlight the effect of this process). The different curves reflect different assumptions about the ratio f_{crit} between projectile mass at the threshold of atmospheric erosion and the mass above the tangent plane. The atmospheric pressure declines rapidly through the period of heavy bombardment.

In the above computation, we neglected the flux of volatiles delivered by impacts (as well as those from mantle degassing, etc.). Whether a given impact will add or erode volatiles is a function of its impact velocity as well as its size, so that the spectrum of velocities as well as the compositions of projectiles striking a planet's surface must be known (or assumed from a plausible model). A beginning on this problem has been made by (Chyba 1990), although it is clear that a great deal of additional work must be done before this topic is fully understood.

One additional consequence of the impact erosion of atmospheres, however, is worth noting. Volatile constituents in the atmosphere above an impact site are vulnerable to stripping by the expanding plume of vapor. The mass that may be removed is $m_{tp} = 2\pi P_a H R_P/g$, which is a fraction $H/2R_P$ of the total atmospheric mass. However, if the volatile is condensed on the surface of the planet, the maximum amount expelled with the expanding vapors is roughly equal that intersected by the projectile, equal to the area of the projectile projected onto the surface of the planet times the thickness t of the condensed volatile, $\pi a^2 t$, where a is the projectile radius. As the total amount of this material on the planet is $4\pi R_p^2 t$ (assuming a uniform distribution) the fraction of the consensed volatile removed is thus $a^2/4R_p^2$. At the threshold for removal by atmospheric erosion $m_\star = f_{crit} m_{tp}$, so at radius $a_\star = (3m_\star/4\pi\rho_0)^{1/3}$ we can show

$$\frac{\text{Condensibles Ejected}}{\text{Total Inventory}} = \left\{\left[\frac{3}{4}f_{crit}\left(\frac{\rho_a}{\rho_0}\right)\right]^{2/3}\left(\frac{H}{2R_P}\right)^{1/3}\right\}\frac{H}{2R_P}. \quad (15)$$

Because this ratio is generally much smaller than $H/2R_P$, it is clear that surface condensates are much less vulnerable to impact erosion than atmospheric gases. In this way, condensible volatiles may have been fractionated from noncondensible ones during accretion or heavy bombardment. For example, Mars may have acquired a substantial inventory of water (which condenses or freezes on or near the surface) while only retaining a small amount of the noncondensible gas nitrogen. The fate of CO_2 is more complex, as it can both form carbonates on the surface as well as remain uncondensed in the atmosphere.

VII. SUMMARY

The effects of "giant" impacts (i.e., impacts by bodies roughly half the diameter of the primary) on the thermal state of a growing planetary embryo is a more general problem than that of the hypothetical Moon-forming impact on the proto-Earth. If the cumulative spectrum of planetesimal sizes is close to a power law of slope -2, then most of the mass and energy added to a growing planet will be deposited by such "giant" impacts. In this case a simple pattern of temperature versus radius may never develop, as the thermal state at any given era will depend upon the time, velocity and obliquity of the last large collision. The process of core formation and differentiation in a catastrophically growing body may be qualitatively different than that suggested by current gradualist models of planetesimal growth. Global melting and the formation of "magma oceans" may have occurred repeatedly in the history of all planetary-size bodies. The implications of such catastrophic growth have yet to be worked out, but it is clear that much more work needs to be done on the effects of large impacts on protoplanets of all sizes, and the chemical and mechanical evolution of planetary-scale magma bodies.

Impacts also contribute in a fundamental way to the origin and evolution of atmospheres. Devolatilization of impactors (and of surface rocks, if they have a volatile-rich composition) during accretion may lead to the formation of relatively dense proto-atmospheres. Giant impacts, such as that which is believed to have formed the Moon, influence atmospheric development in ways that are presently unclear: although a number of authors declare that giant impacts will simply strip away any pre-existing atmosphere, the loss mechanism is sufficiently obscure that this conclusion is open to doubt.

The cumulative effect of the smaller, but much more numerous, impacts during heavy bombardment was to strip away a fraction of the primordial atmospheres; this effect would have been much more pronounced for Mars than for the Earth or Venus. Impact erosion is much more effective for atmospheric gases than for condensed liquids or ices on the surface, and may thus have helped to segregate noncondensible from condensible substances.

THE LONG-TERM DYNAMICAL EVOLUTION AND STABILITY OF THE SOLAR SYSTEM

MARTIN J. DUNCAN
Queen's University

and

THOMAS QUINN
Oxford University

The question of the long-term stability of the solar system is one of the oldest unsolved problems in Newtonian physics. Although the general solution remains elusive, several recent theoretical and numerical advances suggest that the simple "clockwork" model of the solar system envisaged by early workers must be replaced by one incorporating the deterministic chaos recently found in many nonlinear dynamical systems. We begin with a brief review of research on the gravitational N-body problem, with an emphasis on recent developments which pertain to the long-term evolution of orbits. We then describe how this work may be used to understand the results of computer simulations of (1) the structure of the asteroid belt, (2) the formation and evolution of the Oort comet cloud and the origin of short-period comets, (3) the gravitational sweeping of minor bodies between the giant planets, and (4) the possible existence of a belt of comets just beyond Neptune. We continue with a summary of the status of current research into the long-term dynamical evolution and stability of the solar system and conclude with a discussion of promising areas for future investigation.

I. INTRODUCTION

Most of the chapters in this book review current theories of the formation of the solar system up to the time several billion years ago when the planets acquired most of their material and settled into nearly circular, nearly coplanar orbits. In this chapter, we describe efforts to study the subsequent dynamical evolution of such a system. The issue of the long-term stability of the solar system is of course one of the oldest unsolved problems in Newtonian physics, but recent (largely numerical) work has provided some insight into the problem. In particular, several lines of investigation suggest that the solar system is subject to the deterministic chaos recently found in many nonlinear Hamiltonian systems. This inherent unpredictability has profound implications for the dynamics of the solar system, not the least of which is the demise of the "clockwork" picture favored by many earlier workers in the field.

We return at the end of this review to the results of simulations of the evolution of the planets. However, many aspects of the dynamical evolution of the orbits of "test" bodies (such as comets and asteroids) under the perturbing influence of one or more planets (1) are of considerable intrinsic interest, (2) can be numerically more tractable, and (3) may offer considerable insight into the more general dynamical problem. Therefore, in the next section we summarize progress in the past decade in the gravitational N-body problem, using the restricted 3-body problem to illustrate the distinction between quasi-periodic and chaotic orbits and the useful features of area-preserving mappings, Poincaré surfaces of section and Lyapunov exponents. With this mathematical machinery as a basis, we then describe in Sec. III an example of chaos in action—the origins of the 3:1 gap in the asteroid belt. We also discuss the extent to which it is possible that the structure in the outer asteroid belt has been shaped by the gravitational influences of the giant planets (particularly Jupiter).

In Sec. IV, we describe recent simulations of the formation of the Oort comet cloud and of the origins of short-period comets. The latter simulations provide indirect evidence for an as yet undiscovered belt of icy planetesimals beyond Neptune (the Kuiper belt) and we will briefly describe observational searches for the members of this belt. In Sec. V, we discuss numerical simulations designed to test the dynamical stability of test particles between the giant planets and beyond Neptune. In Sec. VI, we return to the larger question of the long-term stability of the planets themselves. We conclude (Sec. VII) with a summary of the results of the previous sections and briefly discuss promising areas for future research.

II. RECENT DEVELOPMENTS IN THE GRAVITATIONAL N-BODY PROBLEM

Insofar as the Sun, planets and minor bodies can be approximated as point masses interacting solely via their mutual gravitational forces, the solar system can be viewed as a nonlinear Hamiltonian system. It is therefore not surprising that many of the mathematical techniques used in the study of other nonlinear dynamical systems can be applied to the problem at hand. In the remainder of this section we present a very nonrigorous summary of concepts which are useful in understanding the numerical results presented in subsequent sections. Interested readers will find more information in the superb reviews by Hénon (1983) and Berry (1978) on Hamiltonian systems in general and by Wisdom (1987b) on chaotic dynamics in the solar system in particular.

Consider then a self-gravitating system of N particles, for which one can write down the Hamiltonian in the standard way. Each particle in the system will in general have three degrees of freedom and with each degree of freedom one can associate a generalized coordinate and its conjugate momentum. The state of the system at any instant in time may then be represented as a point in a $6N$-dimensional phase space and the evolution of the system is then a trajectory (line) in this space which begins at some point determined by the initial spatial coordinates and momenta for each particle.

In some cases, there exist one or more independent single-valued functions of the spatial coordinates and momenta which are conserved along each trajectory of the system. Each of these functions is called an integral of the motion (sometimes called an isolating integral; cf. Binney and Tremaine 1987). If the functions are mutually independent (technically if they are all "in involution"; see, e.g., Berry 1978) then each integral restricts a given trajectory of the system to a submanifold of one less dimension of the $6N$-dimensional phase space. If there exists an independent integral for every degree of freedom then the system is said to be integrable and it is always possible via canonical transformations to the phase space coordinates to cast the Hamiltonian in a form in which there is no explicit dependence on the spatial coordinates. The relevant coordinates are known as action-angle variables and Hamilton's equations can be used to show that each trajectory is then confined to a manifold which is topologically a torus of dimension equal to one half of the dimension of the phase space. One can then envision a given trajectory winding its way around the multi-dimensional torus upon which it is confined. For each degree of freedom and set of initial conditions, there exists a unique angular frequency (given by the partial derivative of the Hamiltonian with respect to the associated action variable) so that unless two of the frequencies are commensurable (i.e., are in the ratio m/n where m and n are integers), a given trajectory will densely cover its confining torus. These trajectories are called quasi-periodic and are the general case for an integrable Hamiltonian. In some cases (such as the Kepler problem) two or more of the angular frequencies are commensurate, in which case there is said to be a resonance and the trajectory is restricted to a lower-dimensional submanifold on the torus. For example, in most integrable systems with two degrees of freedom, a typical trajectory covers the two-dimensional surface of a "donut," but if there exists a commensurability between the two relevant angular frequencies, then the trajectory is a closed line on the surface of a donut. Resonances play a key role both in the theory of Hamiltonian systems and in the actual solar system, as we shall see.

It is clear that for systems with only one degree of freedom the trajectories under time-dependent Hamiltonians are always integrable, because the Hamiltonian itself is an integral. For systems with two or more degrees of freedom, the question which must now be answered is: do most Hamiltonians possess an integral for every degree of freedom—i.e., is integrability exceptional or generic? Furthermore, what are the consequences of nonintegrability? Historically, researchers have attacked the problem by beginning with an integrable system (such as the Kepler problem) and introducing a small perturbation (such as the influence of a third object of small mass). Most of classical perturbation theory assumed that all problems are integrable and their perturbation expansions typically pushed the nonintegrable parts of the Hamiltonian to successively higher orders in the perturbation parameter.

However, Poincaré (1892) showed that these perturbation series are in general divergent and have validity only over finite time spans. We return to this issue in Sec. VI.

Significant progress on the mathematical issue was provided by the work of Kolmogorov, Arnold and Moser (see, e.g., Hénon 1983 for a review and references). This work (which is now termed the KAM theorem) showed that the quasi-periodic trajectories of an integrable problem usually remain quasi-periodic under the influence of a "sufficiently small" perturbation to the Hamiltonian. The exceptions are trajectories where the ratios of characteristic frequencies of the original problem are sufficiently well approximated by rational numbers—i.e., the theorem failed near resonances. As rational numbers are inextricably mixed with irrationals along the real number line, regions of quasi-periodic and chaotic behavior are similarly intertwined (see, e.g., Berry 1978 for a good discussion). Furthermore, the tori which are "destroyed" form a finite set which grows with the strength of the perturbation. The KAM theorem is thus fundamental in showing the persistence of tori under very small perturbations, but much of our understanding of the structure of phase space in more realistic Hamiltonians comes from numerical investigations.

Numerical integration of orbits (as exemplified in the classic paper of Hénon and Heiles [1964] in their investigation of the existence of another integral in two-dimensional potentials) has generally shown results which are consistent with the KAM theorem—quasi-periodic and chaotic orbits are inextricably intermixed in phase space. For systems with two degrees of freedom, the chaotic regions are bounded by tori corresponding to quasi-periodic trajectories. A useful tool for visualizing the phase space in such cases was employed by Poincaré and is sometimes called a Poincaré surface of section. For systems with two degrees of freedom, the trajectory in four-dimensional phase space is in general a 3-surface because the Hamiltonian itself is an integral of the motion and hence constrains the motion. If we now consider a slice through this three-dimensional volume (for example by plotting one coordinate and its conjugate momentum whenever the other coordinate is zero), the resulting two-dimensional plot is the locus of points created as the trajectory "pierces" the surface. If there exists a second integral, the resultant plot will be a one-dimensional curve; if the orbit is chaotic, an area on the plane will be populated (and in a seemingly random manner). This approach can be illustrated with a particularly relevant example. A classic case of orbits in a system with two degrees of freedom is the planar restricted circular 3-body problem, in which a test particle (taken to have zero mass) moves under the gravitational influence of two massive objects moving in circular orbits about their mutual center of mass. In the appropriate rotating frame, the massive objects are fixed and the resultant time-independent Hamiltonian is known as the Jacobi integral (cf. Binney and Tremaine 1987). If one integrates the trajectory of the test particle, surfaces of section such as those shown in Fig. 1 ensue. The particular example is for the case of two equal massive bodies and is discussed more fully in Hénon (1966a, b,1983). The x-axis is along the line joining the two masses and the section is obtained by plotting the velocity in the x-direction versus x whenever the particle goes through $y = 0$ with positive y-velocity. For the particular value of the Jacobi integral depicted here ($C = 0.39$), but for otherwise different conditions, there is a zone of quasi-periodic orbits centered near $x = 0.74$ (we show three orbits here) and a region where orbits are chaotic, as exemplified by the dotted area. Note that the unconnected dots are produced by the motion of *one* particle.

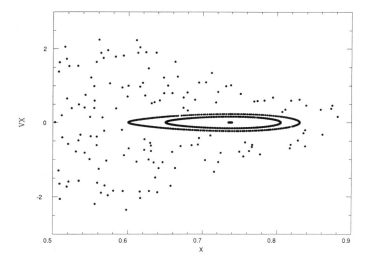

Figure 1.   Surfaces of section for the circular, planar, restricted 3-body problem. Shown are the trajectories of test particles started with the same Jacobi constant but otherwise different initial conditions. See the text for further discussions of the initial conditions. Three regular orbits are shown, which are centered on a period orbit near $x = 0.74$. Note that the unconnected dots are produced by the motion of *one* particle.

A particularly appealing aspect of the use of surfaces of section is that in some cases it is possible to devise what is known as an area-preserving map which carries one point on the surface to the next algebraically (i.e., without having to integrate the particle's trajectory in between passages through the surface). This approximation (if it can be found) typically speeds up the time necessary to explore the phase space by a factor of 100 to 1000.

For systems with more than two degrees of freedom, a similar analysis would require multidimensional hypersurfaces of section. However, regardless of the number of degrees of freedom, there is a key feature of the irregular orbits which we will use here as a definition of "chaos": two trajectories which begin arbitrarily close in phase space in a chaotic region will typically diverge exponentially in time. Ironically, the time scale for that divergence in a given chaotic region does not typically depend on the initial conditions. Thus, if one computes the distance in six-dimensional phase space $d(t)$ between two particles having an initially small separation it can be shown that for quasi-periodic orbits, $d(t) - d(t_0)$ grows as a power of time t (typically linearly) whereas for irregular orbits $d(t)$ grows exponentially as $d(t_0)e^{\gamma(t-t_0)}$ where γ is conventionally called the Lyapunov exponent. Thus, chaotic orbits show such a sensitive dependence on initial conditions that the detailed long-term behavior of the orbits is lost within several Lyapunov time scales.

In Fig. 2, we illustrate the distinction between adjacent trajectories which are regular versus those which are quasi-periodic as characterized by the Lyapunov exponent described above. The quantity $\gamma \equiv \ln[d(t)/d(0)]/t$, for chaotic trajectories eventually levels off at a value which is the Lyapunov value in that region (the inverse of the characteristic time scale for divergence) whereas γ continues to decline for quasi-periodic orbits when plotted as a function of time. The examples shown correspond to orbits within and without a chaotic region of the asteroid belt near the 3:1 mean-motion resonance with Jupiter (see, e.g., Wisdom 1983, and references therein).

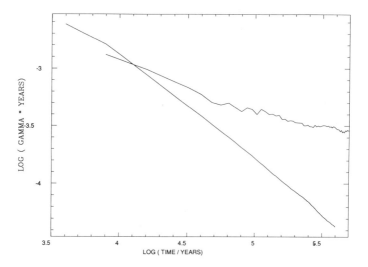

Figure 2.   Distinction between regular (lower curve) and chaotic (upper curve) trajectories as characterized by the Lyapunov exponent discussed in the text. For chaotic trajectories, a plot of $\log(\gamma)$ vs $\log(t)$ eventually levels off at a value of γ which is the inverse of the characteristic time scale for the divergence of initially adjacent trajectories.

For systems of more than 2 degrees of freedom, the chaotic regions of phase space are no longer bounded by quasi-periodic tori so that the chaotic regions can communicate. The process by which a particle evolves from one chaotic region to another is termed "Arnold diffusion." The time scale for this process is very important for our understanding of solar system stability, as we shall see.

III. CHAOS IN ACTION: STRUCTURE IN THE ASTEROID BELT

The number density of the asteroids in the asteroid belt as a function of semimajor axis (or mean motion) shows an intricate structure. There are several influences that could affect this structure, but we will limit the discussion to the gravitational hypothesis. (Other explanations for gaps which we will not discuss include the statistical hypothesis, the collisional hypothesis, and the cosmogonic hypothesis; see Greenberg and Scholl [1979] for a review of these.)

The gravitational hypothesis states that virtually all of the structure in the asteroid belt can be attributed to gravitational interactions with the planets which clear asteroids from certain regions over the age of the solar system. This is supported by the coincidence of depletion and enhancement bands with mean motion resonances of Jupiter, the main gravitational perturber in the asteroid region. As well as gaps corresponding to mean motion resonances, there are gaps corresponding to secular resonances, i.e., regions where the precession rate of an asteroid's longitude of perihelion or line of nodes corresponds to one of the eigenfrequencies of the solar system. The existence of secular resonances was noted as early as the 19th century (Le Verrier 1856), and it has long been evident that the inner edge of the asteroid belt near 2 AU is near such a resonance. Williams (1969,1971) introduced a powerful approach to analyzing secular resonances, and showed that asteroids with $a < 2.6$ AU are severely depleted at the locations of the three strongest resonances. Further applications of this approach and numerical studies have accounted for much of the observed depletion in the inner regions of the asteroid belt (see Scholl et al. 1989 for a recent review). Knezevic et al. (1991) have extended such studies to the outer planetary region.

An obvious mean-motion resonance is at the location of the Trojan asteroids, a 1:1 resonance with Jupiter. These asteroids librate about the points $60°$ behind or ahead of Jupiter and therefore, never suffer a close approach to Jupiter. Another example of a protection mechanism provided by a resonance is the Hilda group at a mean motion 3:2 resonance. These asteroids have a libration about $0°$ of their critical argument, $\sigma \equiv 2\lambda' - \lambda - \varpi$, where λ' is Jupiter's longitude, λ is the asteroid's longitude, and ϖ is the asteroid's longitude of perihelion (Schubart 1968). In this way, whenever the asteroid is in conjunction with Jupiter ($\lambda = \lambda'$), the asteroid is close to perihelion, ($\lambda' \sim \varpi$) and well away from Jupiter.

Using resonances to explain the gaps in the outer asteroid belt and the general depletion of the outer belt proves to be more difficult. A feature subject to much investigation has been the gap at the 3:1 mean motion resonance. Scholl and Froeschlé (1974) investigated this commensurability using the averaged planar elliptic restricted 3-body problem. They found that most orbits starting at small eccentricity were regular and showed very little variation in eccentricity or semimajor axis over time scales of 5×10^4 yr.

Investigation of more realistic models was limited because of the amount of computer time needed to follow orbits over long periods of time. A breakthrough occurred when Wisdom (1982) devised an algebraic mapping of phase space onto itself with a resonant structure similar to the 3:1 commensurability. His surprising result was that an orbit near the resonance could maintain a low ($e < 0.1$) eccentricity for nearly a million years and then have a sudden increase in eccentricity to over 0.3. Wisdom also showed that this behavior could not have been seen in the averaged planar eccentric restricted 3-body problem because the averaged Hamiltonian admits a quasi-integral which confines the orbits to low eccentricities.

Any doubts about this apparent chaotic behavior were later dispelled by numerical integrations and the measurement of a non-zero maximum Lyapanov exponent (Wisdom 1983) and by a semianalytic perturbative theory which explained many of the features found in the mappings (Wisdom 1985). Figure 3 plots the eccentricity as a function of time for a typical chaotic trajectory near the 3:1 resonance in Wisdom's perturbative theory. Note that the time is measured in Myr, so that the particle can remain in a low-eccentricity state for many tens of thousands of orbits before relatively rapidly entering a high-eccentricity phase. Approximate surfaces of section explaining this behavior may be found in Wisdom (1985). There it can be seen that the particle spends some time in low-eccentricity islands near the origin, while being free to explore a fairly large chaotic zone of higher eccentricity.

Wisdom (1987*b*) gives other (albeit rarer) examples of trajectories which exhibit long periods of low-eccentricity punctuated by bursts of short-lived but highly eccentric behavior. The relevant surfaces of section clearly demonstrate that although a wide chaotic zone surrounds the origin, there also exists a narrow branch that extends to high eccentricity. An orbit wandering in the band near the origin could appear to be constrained to low eccentricities, and suddenly suffer a large jump in eccentricity as it went down the branch.

The outer boundaries of the chaotic zone as determined by Wisdom's work has been shown to coincide well with the boundaries of the Kirkwood gap as shown in the numbered minor planets and the Palomar-Leiden survey. Since orbits which begin on near-circular orbits in the gap acquire sufficient eccentricities to cross the orbit of Mars, the perturbative effects of Mars are believed capable of clearing out the 3:1 gap over the age of the solar system.

The depletion of the outer asteroid belt (3.6 AU to 5.2 AU) has long been assumed to be due to the gravitational influence of Jupiter (see, e.g., Nobili 1989 for a review). This seems plausible because of the large number of strong resonances that occur in the outer belt. These resonances overlap at moderate eccentricities which corresponds to a sharp cutoff in the asteroid distribution (Dermott and Murray 1983). The argument for depletion by Jupiter's perturbations was first checked by a numerical integration by Lecar and Franklin (1973) who numerically integrated orbits in the planar elliptic

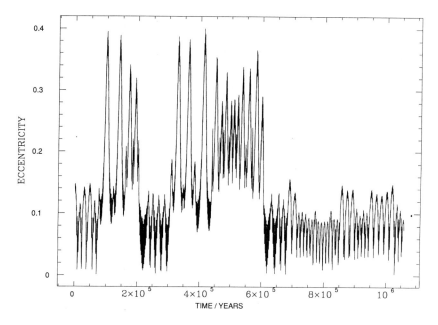

Figure 3.   Eccentricity as a function of time of a typical chaotic trajectory near the 3:1 resonance in the elliptic restricted 3-body problem with parameters appropriate to the Sun-Jupiter-asteroid case. Note that the particle can remain in a low-eccentricity state for many tens of thousands of orbits before relatively rapidly entering a high-eccentricity phase.

restricted 3-body problem for a period of a few thousand years. They found that the region outside of 4.0 AU could be depleted in this time span, but the depletion inside 4.0 AU was very small. Froeschlé and Scholl (1979) extended this by integrating orbits in the three-dimensional Sun-Jupiter-Saturn model for 10^5 yr. For small eccentricities, they again found very little depletion inside 4.0 AU. By carefully selecting their initial conditions, Milani and Nobili (1985) were able to find orbits that crossed Jupiter very quickly. However, for an initial semimajor axis less than 4.0 AU, only orbits with initially high eccentricities ($e>0.15$) were found to be Jupiter crossers in their relatively short integrations.

Gladman and Duncan (1990) have performed integrations for 12 Myr of the equations of motion of the Sun, Jupiter, Saturn and 80 test particles which began on circular orbits at various radii distributed between 3.1 and 3.9 AU. Particles which evolved into orbits that crossed that of Mars or which brought them to a close approach with Jupiter or Saturn were removed from the integrations. Gladman and Duncan found depletions at some of the stronger mean motion resonances, in particular the 2:1 resonance, but most of their orbits were stable over this time scale. It appears likely that longer direct integrations will show considerable depletion over Gyr time scales in the region beyond the 2:1 resonance, but this remains to be shown. Indeed, the study of the orbits near resonances other than the 3:1 remains an active area of research which is beyond the scope of this review, but it is quite possible that cosmogonical effects have played a role in shaping some of the gaps (see, e.g., Yoshikawa 1991).

IV. ORIGINS OF LONG- AND SHORT-PERIOD COMETS

Four decades have passed since Oort (1950) deduced the existence of the cloud of comets which bears his name and proposed a mechanism for delivering a steady-state flux out of that reservoir into orbits which brings one every few years close enough to the Sun at pericenter to become visible as a new long-period comet. Although many of the basic ingredients of Oort's classic picture remain, there has been considerable refinement and some modification of the theory (see Weissman 1990). Chapters by Mumma et al. and by Van Dishoeck et al. describe in more detail the present orbital properties of current comets, their chemical compositions and theories of their formation. In keeping with the theme of this chapter, however, we discuss herein our current understanding of the dynamical processes which (1) may have formed and thereafter sculpted the Oort cloud, and (2) may be responsible for producing the current population of short-period comets (i.e., those with periods shorter than 200 yr, hereafter called SP comets).

The current version of Oort's original theory is that comets formed as icy planetesimals on nearly circular orbits in the outer planetary system. Some of these accreted to form Uranus and Neptune, while the remainder were repeatedly scattered by the growing planets until they reached semimajor axes large enough for galactic tidal fields and stellar perturbations to remove their perihelia from the planetary region, after which they are relatively immune to planetary perturbations. The bodies which reach "safety" in this way comprise the comet cloud which is present today. Tidal torquing by the smooth distribution in the galactic disk has only recently been recognized as being more important (by a factor of 2 to 3) than stellar perturbations (Heisler and Tremaine 1986; Morris and Muller 1986; Smoluchowski and Torbett 1984; Torbett 1986) although stellar perturbations are still important in randomizing the orbits in the cloud (see, e.g., Weissman 1990) and occasional close stellar encounters can produce an unusually large influx of comets called comet showers (Hills 1981; Heisler 1990). Several competing theories of comet formation have been proposed (for example, that comets form *in situ* in the outskirts of an extended solar nebula), but they are not considered here (see Fernández 1985 for a review).

It has not yet been possible of course to follow the detailed evolution of the planetesimal swarm as the outer planets grew to their final masses. However, Duncan et al. (1987) have "picked up the story" at the stage when the outer planets have attained their final masses and have gravitationally scattered a large fraction of the unaccreted planetesimals to semimajor axes of a few hundred AU (although the perihelia of the comets remained in the planetary region). Details of the approach used and approximations made in the numerical simulations may be found in the original reference. (See also a simulation using an Öpik approximation by Shoemaker and Wolfe [1984].) The simulations showed that planetary perturbations acting on moderately eccentric orbits caused a form of random walk in energy with little change in pericentric distance. Thus the formation of the comet cloud was driven by the interaction between planetary perturbations which drove diffusion in semimajor axis a at constant pericentric distance q, and galactic tidal torques which changed q at fixed a, thereby removing cometary perihelia from the planetary region for sufficiently large a. Consequently, a typical comet evolved more or less in the ecliptic plane with pericenter near to its birthplace until it was either ejected or torqued by the galactic tidal field into a roughly spherical cloud with a more nearly isotropic velocity distribution. An inner edge to the cloud was found at $\sim$3000 AU—roughly the radius at which the time scales for the two effects are equal for comets formed in the Uranus-Neptune region. The density profile between 3000 and 50,000 AU is roughly a power law proportional to $r^{-3.5}$. The inner cloud ($a < 2 \times 10^4$ AU) thus contains roughly five times more mass than the classical Oort cloud.

Figure 4 depicts the evolution of the simulated comet swarm as a function of time. Note that the galactic plane in each snapshot is a horizontal line, so that the ecliptic is inclined at an angle of $\sim$60° in these diagrams. The dotted circle in each snapshot is at a radius of 20,000 AU indicating the inner edge of the classical Oort cloud. It is evident from Fig. 4 that the distribution after 100 Myr was still biased toward the ecliptic for orbits with $a \lesssim 10^4$ AU. However, by 1 Gyr the distribution was isotropic for $a \lesssim 2000$ AU, and roughly 20% of the survivors were in the Oort cloud, with the remainder populating the inner cloud (except for the $\sim$4% that became SP comets).

Between 1 and 4.5 Gyr, the total number of survivors decreased by a factor of $\sim$2, but the relative populations of the two clouds remained essentially unchanged. It is important to note, however, that the simulations did not include encounters with passing molecular clouds which may, over Gyr time scales, have a perturbing influence on the comets which is comparable to that of stars (Biermann and Lüst 1978; Biermann 1978; Bailey 1983a,1986a). This is particularly true for comets with $a \gtrsim 2.5 \times 10^4$ AU which are not protected by adiabatic invariance (i.e., those with orbital periods longer than the duration of a typical encounter). However, the influence of molecular clouds is difficult to estimate because their physical parameters are so uncertain. Indeed, the large observed fractions of wide binary stars with separations of order 20,000 AU suggests that the Oort cloud has probably not been disrupted by giant molecular clouds (Hut and Tremaine 1985). Nonetheless, a complete treatment of the long-term dynamical evolution of the Oort cloud will probably require the inclusion of molecular clouds (cf. Heisler 1990).

Having discussed the formation and evolution of the reservoir which

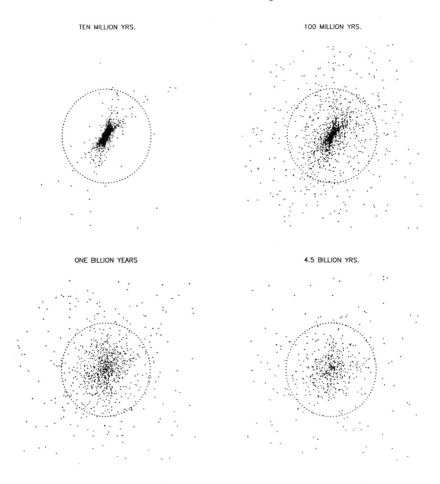

Figure 4.   The formation of the Oort comet cloud in the simulation of Duncan et al.
(1987). The galactic plane in each snapshot is a horizontal line. The dotted circle
denotes a radius of 20,000 AU, indicating the inner edge of the classical Oort Cloud.

supplies the current population of long-period comets, let us now turn to
the origins of the SP population. There are several striking features in the
distribution of orbital elements of SP comets: (1) the distribution is strongly
peaked towards periods $\lesssim 15$ yr; there are 100 known comets with orbital
period $P \leq 15$ yr but only 21 with 15 yr $< P < 200$ yr (Marsden 1983); (2)
the SP comets are mostly on low-inclination prograde orbits; out of 121
known SP comets, only four are on retrograde orbits; the inclinations i satisfy
$\langle \cos i \rangle = 0.88$ ($\langle \cos i \rangle = 0$ for isotropic orbits and $= 1$ for prograde orbits in
the ecliptic); (3) the arguments of perihelion ω (the angle between perihelion
and ascending node) of the SP comets are strongly peaked near 0 and 180°.

Until recently, it has been believed that the SP comets originate in the Oort comet cloud. In an influential paper, Everhart (1972) argued from an extensive set of orbital integrations that repeated interactions with Jupiter can produce SP orbits from near-parabolic orbits, so long as the initial inclination is small and the initial perihelion distance is near the orbit of Jupiter. He showed that the distribution of orbital elements for SP comets formed by this mechanism agreed well with observations. There was immediate concern, however, that Everhart's mechanism might not produce the correct number of SP comets. Joss (1973) has argued that the efficiency of this process is too low (by a factor of order 10^3 to 10^4) to produce the observed number of SP comets from the known flux of near-parabolic comets, although Delsemme (1973) argued that the discrepancy could be removed by the proper inclusion of all return passages of comets initially from the Oort cloud. (See Stagg and Bailey [1989] and Yabushita and Tsujii [1991] for further references.)

In an ingenious Monte Carlo simulation, Everhart (1977) subsequently showed that a fraction of near-parabolic orbits with perihelion as large as Neptune's orbit will be gravitationally scattered by the outer planets into orbits that are Jupiter crossing and that a fraction of these will eventually become SP comets. In conventional models of the Oort cloud, the efficiency of this process is too low to remove the flux discrepancy noted by Joss. However, Bailey (1986b) has suggested that if a very massive inner Oort cloud (inner radius $\simeq$5000 AU, mass $\simeq$700 Earth masses) is present, the flux of near-parabolic comets into Neptune-crossing orbits may be sufficient to supply the SP comets. (Note however that this inner cloud is much more massive than that found in the simulations of Duncan et al. 1987.)

An alternative theory proposes that SP comets originate in a belt of low-inclination comets just beyond the orbit of Neptune, between about 35 and 50 AU (e.g., Fernández 1980; Fernández and Ip 1983). The belt could be a natural remnant of the outermost parts of the solar nebula (Kuiper 1951; Whipple 1964), possibly producing a component of the infrared background at $100\,\mu$m detected by IRAS (Low et al. 1984). IRAS data also indicates the presence of flattened dust shells around other stars, extending from 20 to 100 AU (see Weissman 1986c for a review). If some of the belt comets can be perturbed into Neptune-crossing orbits, subsequent scattering by the outer planets converts some of these into observable SP comets, in the manner described by Everhart (1977).

In order to test these two hypotheses, Duncan, Quinn and Tremaine (1988: hereafter called DQT88) performed an extensive series of numerical integrations of a representative sample of comet orbits in the field of the Sun and the giant planets. Comets which began with perihelia either near Jupiter (q between 4 and 6 AU) or in the outer planetary region (q between 20 and 30 AU) were integrated until the comet was either ejected or became visible to an Earth-based observer, which the authors assumed occurred when $q <$ 1.5 AU. A numerical obstacle that the authors had to overcome was that orbital evolution can be a slow process: evolution from a Neptune-crossing orbit to a visible orbit typically takes millions of orbits. However, as gravitational scattering is a diffusion process, the authors argued that multiplying the mass of all the giant planets by a fixed factor μ should change the rate of evolution but not the statistical properties of the final distribution of orbits. (We shall return to this point below.)

The results showed that the inclination distribution of comets with large perihelion ($q \lesssim 30$ AU) that evolve to observable comets (i.e., those with $q \lesssim 1.5$ AU) is approximately preserved. Thus, the short-period (SP) comets, which are mostly in prograde, low-inclination orbits, cannot arise from gravitational scattering of any spherical population of comets (such as the Oort cloud). However, the distribution of orbital elements of SP comets arising from a population of low-inclination Neptune-crossing comets is in excellent agreement with observations. The authors concluded that the SP comets arise from a cometary belt in the outer solar system (now called the Kuiper belt). An example of the agreement between the Kuiper belt simulations and the observed short-period comet orbital elements is shown in Fig. 5.

Stagg and Bailey (1989) have argued that a massive inner Oort cloud may still be a viable source for the SP comets if the mass enhancement factor used by DQT88 badly underestimated the effects of close encounters and if there exists a large population of unobserved extinct high-inclination comets. Quinn et al. (1990) have replied with a more extensive series of experiments, taking into account a variety of nongravitational and selection effects, and conclude that those SP comets with periods less than 20 yr (sometimes called the Jupiter family) cannot arise from an isotropic distribution. They argue that the results from simulations with $\mu = 40$ do not differ substantially from those with $\mu = 10$. Furthermore, Wetherill (1991a) used an Öpik (1951) approximation using Arnold's (1965a, b) method and $\mu = 1$ and obtained results very similar to DQT88. A definitive resolution to the problem requires an extremely CPU-intensive direct integration with $\mu = 1$, with modeling for observational selection effects and the finite lifetimes of comets.

If the proposed Kuiper belt indeed exists, two questions naturally arise: (1) as comets on Neptune-crossing orbits typically are ejected or evolve to SP comets on time scales which are much shorter than the age of the solar system, what mechanism injects comets from the Kuiper belt into planet-crossing orbits, and (2) can one detect the larger members of this putative belt? We shall return to a possible answer to the first question in the next section. Let us now briefly consider the second.

It is possible that the minor body Chiron (Kowal 1979) originated in the Kuiper belt. Chiron, which is a roughly 100-km-sized object, is on a Saturn-crossing orbit which is unstable on a time scale of 10^5 to 10^6 yr (Oikawa and Everhart 1979). It has recently been shown to exhibit cometary behavior (Luu and Jewitt 1990) such as the formation of a resolved coma (Meech and Belton 1990). Because of the short lifetime of its current orbit, it seems likely that it is representative of a much larger population of similar objects which currently reside in the Kuiper belt. Such objects might be detected by their

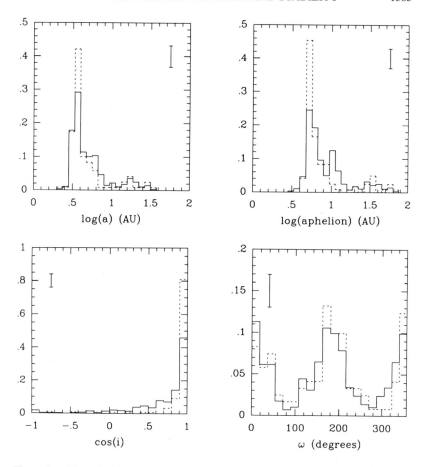

Figure 5. The solid histograms show the distributions of semimajor axis a, inclination i, argument of perihelion ω and aphelion for comets that evolve to short-period orbits with $q < 1.5$ AU in the simulations of Duncan et al. (1988). The initial inclinations are uniformly distributed in $\cos(i)$ for i between 0 and 18 deg, the initial perihelia are uniformly distributed between 20 and 30 AU and the initial semimajor axis is $a = 50$ AU. The dashed histograms show the distributions in the same four orbital elements for the comets in Marsden's (1983) catalog. Some typical statistical error bars are plotted. The histograms are based on 281 simulated comets and 121 observed comets.

retrograde motion as seen at opposition from the Earth, which amounts to roughly $150/r$ arcsec per hr, where r is the heliocentric distance of the object in AU. Figure 6 demonstrates the lower limit to the diameter of objects which could be detected at a given heliocentric distance for two assumed limiting magnitudes (assuming an albedo of 0.1). The proper motion corresponding to the distance is also shown. Note that Chiron itself would be fainter than $V = 22$ if it were beyond 40 AU.

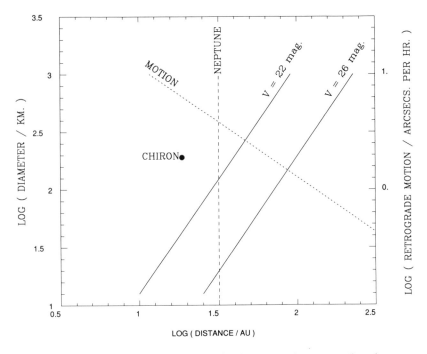

Figure 6.   Detectability of planetesimals in the outer solar system (based on an
unpublished figure of W. Baum). The dashed curve (calibrated by the vertical scale
on the right) shows the retrograde motion as seen from the Earth (in seconds of
arc per hour) exhibited by an object in a low-eccentricity orbit at the indicated
distance from the Sun. The two solid lines indicate the distance from the Sun out
to which objects of a given diameter (calibrated by the vertical scale on the left)
can be detected above a given limiting magnitude. The limiting magnitudes shown
are $V = 22$ and 26, respectively, assuming an albedo of 0.1 for the object. Shown
for reference are the approximate diameter and semimajor axis of Chiron and the
location of Neptune.

There have been four major proper motion surveys designed to search
for objects in the outer solar system, and with the exception of Chiron, the
surveys have found no objects. Figure 7 shows the upper limits to the surface
density of objects in the sky as a function of the surveys' limiting magnitudes.
However, as has been discussed in Levison and Duncan (1990), the lack
of detections cannot be used to place severe constraints on the mass of the
Kuiper belt because the size distribution of the belt members is unknown.
Indeed, if all of the proto-comets had sizes less than 40 km, then no objects
beyond 25 AU would have been detectable in Levison and Duncan's survey.
Furthermore, we shall see in the next section that there are now theoretical
reasons to believe that the region from Jupiter's orbit to somewhere beyond
Neptune's orbit has been gravitationally swept clean of planetesimals. Thus,

the inner edge of the putative Kuiper belt is probably beyond 40 AU, making the detection of even Chiron-sized objects difficult (but not impossible).

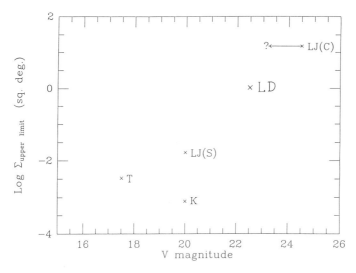

Figure 7.   Upper limits to the surface density of objects in the sky Σ, as a function of various surveys' limiting magnitudes in V. The symbols are: T = Tombaugh (1961), K = Kowal (1989), LJ(C) = Luu and Jewitt (1988) CCD work, LJ(S) = Luu and Jewitt (1988) Schmidt telescope, and LD = Levison and Duncan (1990) CCD survey.

V. CHAOTIC ORBITS IN THE OUTER PLANETARY REGIONS

In a sense, the solar system seems to be remarkably devoid of objects. In the generally favored model for the formation of the solar system discussed in other chapters, the planets accumulate through the accretion of vast numbers of planetesimals that formed on nearly circular orbits in the early solar system. As it is unlikely that all of these objects would have been incorporated into the present planets, a question that immediately arises is: can any of these objects have survived to the present day? Because of the physical complications present in the early solar system (e.g., gas drag, nebular dispersal, etc.) one can ask the following simplified question: are there any regions in the *current* solar system where test particles (negligible mass) placed on initially circular orbits are stable against having a future close approach with one of the giant planets? Having a close approach is not in itself a guarantee of ejection (although over sufficiently long time scales it usually is) but it usually means that the orbit is chaotic.

   A complete answer to this question would of course require the direct integration over Gyr time scales of all of the planets together with many test particles distributed between them. This is not technically feasible at this time (see Sec. VII). More modest investigations over shorter time spans have been performed for the case of test particles between Jupiter and Saturn by Lecar and Franklin (1973) and Franklin et al. (1989). In the first of these papers, a band of semimajor axis was discovered in which initially low eccentricity orbits were found to be stable over the time scales studied (500 Jupiter orbits). In the second paper, the stable region (between 1.30 and 1.55 Jupiter semimajor axes) was studied more intensely, and the authors concluded that essentially no particles between Jupiter and Saturn were stable against becoming planet crossers for longer than 10 Myr.

Wisdom (1980) showed for the planar, circular restricted 3-body case that there exists a band in semimajor axis centered on each planet's semimajor axis within which virtually all test particle trajectories are chaotic due to resonance overlap. These chaotic zones are shown for each planet as the shaded regions in Fig. 8.

Gladman and Duncan (1990) have performed the longest and most accurate integrations to date of the evolution of a swarm of test particles on initially circular orbits ranging from the outer asteroid belt to the Kuiper belt. Using a very accurate symplectic integration scheme (Gladman et al. 1991), the orbits of roughly one thousand test particles were followed for up to 22 Myr. The test particles were placed on initially circular orbits about the Sun and felt the gravitational influence of the Sun and four (or in some cases two) of the giant planets. The initial conditions of the planets were obtained from their current orbital elements and their mutual gravitational interactions were fully included. Test particles that underwent a close approach to a planet were removed from the integration.

The dynamical clearing of gaps found near some of the resonances with Jupiter in the asteroid belt has been discussed in Sec. III. Exterior to Neptune there appeared to be a dynamical erosion of the inner edge of the Kuiper belt, although, as we shall see, much longer integrations are needed in this region. The majority of the test particles *between* the giant planets were perturbed to a close approach to a planet on time scales of millions of years. These results suggest that there are very few initially circular orbits between the giant planets that are stable against a close approach to a planet over the lifetime of the solar system. This may explain the apparent absence of a large number of minor bodies between the giant planets discussed in the last section.

The results of the simulations are summarized in Fig. 8. As stated above, with the exception of regions in the asteroid belt and the outer Kuiper belt, most of the orbits become planet crossing on surprisingly short time scales. This of course raises some rather disturbing questions about the stability of the planets themselves (see the next section) and may have implications in the field of planetary formation (see Gladman and Duncan 1990).

An approximate mapping technique described in Duncan et al. (1989) suggested that most near-circular orbits with semimajor axes >33 AU (i.e.,

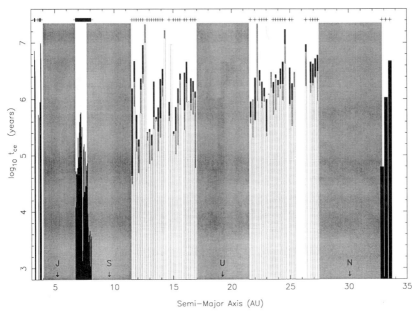

Stability of Test Particles in the Outer Solar System (Composite)

Figure 8. Composite diagram of removal times for all of the simulations of Gladman and Duncan 1990. The single-planet crossing zones (see text) of the giant planets are shaded. A cross at the upper portion of the diagram indicates that all three of the test particles placed on an initially circular orbit at that semimajor axis were removed. The top of a histogram bar indicates the removal time of a test particle originally at that value of semimajor axis. The simulation ran 22.5 Myr (indicated by the horizontal line). See the text for additional explanations.

in the proposed Kuiper belt) were stable against becoming Neptune crossers over the lifetime of the solar system. However, Torbett (1989) performed direct numerical integration of test particles in this region including the perturbative effects of the four giant planets, although the latter were taken to be on fixed Keplerian orbits. He found evidence for chaotic motion with an inverse Lyapunov exponent on the order of Myr for moderately eccentric, moderately inclined orbits with perihelia between 30 and 45 AU (a "scattered disk"). Torbett and Smoluchowski (1990) extended this work and suggested that even particles with initial eccentricities as low as 0.02 are typically on chaotic trajectories if their semimajor axes are <45 AU. Except in a few cases, however, the authors were unable to follow the orbits long enough to establish whether or not most chaotic trajectories in this group led to Neptune crossing. Levison (1991) treated the same question as a diffusion problem and used direct integrations over short time scales to calculate diffusion coefficients on a grid of perihelion vs aphelion. Using the theory of Markov chains he found using this approach that objects tend to diffuse through a surprisingly large region of 70 AU in extent over the lifetime of the solar system. He argued that even objects that formed near Neptune have a significant chance of evolving to orbits with $a > 100$ AU and can subsequently diffuse back on Gyr time scales to become Neptune crossers and potentially short-period comets. Levison's approach has the potentially serious drawback that the "diffusion coefficients" may only measure the linear part of long-term sinusoidal variations in the orbital elements which may average out over time. Nonetheless these provocative results make it clear that longer direct integrations are needed in order for us to understand the long-term behavior of the Kuiper belt.

Having considered the somewhat surprising results for the stability of *test* particles in the solar system, and having discussed promising areas for future work, let us now turn to the question of the stability of the planets themselves.

VI. LONG-TERM STABILITY OF THE PLANETS

We return now to one of the oldest problems in dynamical astronomy: whether the planets will continue indefinitely in nearly circular, nearly co-planar orbits. As we have described in Sec. III, if a Hamiltonian system has an integral of motion for each degree of freedom, then the system will be quasi-periodic, as can be shown by expressing the Hamiltonian in terms of action-angle variables. The orbits will be confined to a multi-dimensional torus and the orbital elements should be describable by a sum of periodic terms in the sense that the Fourier transform of the time evolution of any planet's coordinates will involve only integer combinations of the fundamental frequencies (one per degree of freedom). Arnold (1961) has shown that such stable orbits would describe the solar system if the masses, eccentricities and inclinations were sufficiently small. The real solar system, however, does not satisfy Arnold's requirements, so the question of its stability is unresolved. Laplace and Lagrange showed that, if the mutual planetary perturbations were calculated to first order in the masses, inclinations and eccentricities, the orbits could indeed be described by a sum of periodic terms, indicating stability. Successive work by Brouwer and van Woerkom (1950), Bretagnon (1974) and others has shown that this is still the case if the perturbations are expanded to higher orders. However, the work of Poincaré (1892) casts doubt on the long-term convergence of the various perturbation schemes. The problem with the perturbation expansion is that although the expansion is done in powers of small parameters, the existence of resonances between the planets will introduce small divisors into the expansion terms. Such small divisors can make high-order terms in the power series unexpectedly large and destroy the convergence of the series.

In particular, the method of Laplace involves considering the disturbing function averaged over the mean motions of the planets, known as the secular part of the disturbing function. If the disturbing function is further limited to terms of lowest order, the equations of motion of the orbital elements of the planets can be expressed as a coupled set of first-order linear differential equations. This system can then be diagonalized to find the proper modes, which are sinusoids. The evolution of a given planet's orbital elements is, therefore, a sum of the proper modes. With the addition of higher-order terms, the equations are no longer linear; however, it is sometimes possible to find a solution of a form similar to the linear solution, except with shifted proper mode frequencies and terms involving combinations of the proper mode frequencies (Bretagnon 1974).

There are two separate points in the construction of the secular system at which resonances can cause nonconvergence of the expansion. The first is in averaging over mean motions. Mean-motion resonances between the planets can introduce small divisors leading to divergences when forming the secular disturbing function. Secondly, there can be resonances between the proper mode frequencies leading to problems trying to solve the secular system using an expansion approach.

Laskar (1989) performed a critical test of the quasi-periodic hypothesis by numerically integrating the perturbations calculated to second order in mass and fifth order in eccentricities and inclinations. Such an expansion consists of about 150,000 polynomial terms. By numerically integrating the secular system, he avoids the small divisor problem caused by resonances between proper modes. The Fourier analysis of this 200-Myr integration showed that it was not possible to describe the solution as a sum of periodic terms. Laskar also estimated the maximum Lyaponov exponent by the divergence of nearby orbits. The Lyaponov exponent should be zero for quasi-periodic orbits. Instead, he found the surprisingly high value of ($1/5$ Myr $^{-1}$). In a subsequent paper, Laskar (1990) argued that the exponential divergence is due to the transition from libration to circulation of the critical argument of a secular resonance related to the motions of perihelia and nodes of the Earth and Mars. He argued from his results that the chaotic nature of the inner solar system is robust against small variations in the initial conditions or in the model. These very important conclusions must clearly be checked by direct numerical calculations (see below).

The analytical complexity of the perturbation techniques and the development of ever faster computers has led others to the investigation of stability by purely numerical models. The first numerical computation of planetary orbits was by Eckert et al. (1951) who did a simulation of the outer planets for 350 yr. This was extended by Cohen and Hubbard (1965) and Cohen et al. (1973) to 120,000 yr and 1 Myr, respectively. These integrations compared well with the perturbation calculations of Brouwer and van Woerkom, showing quasi-periodic behavior for the four major outer planets. Pluto's behavior, however, was sufficiently different to inspire further study. Williams and Benson (1971) performed a 4.5 Myr integration of the secular motion of Pluto under the influence of the four Jovian planets. The motion of the Jovian planets were determined from Brouwer and van Woerkom's (1950) analytic solution. They found that the angle $3\lambda - 2\lambda_N - \varpi$ was in libration with a period of 20,000 yr, where λ and λ_N are the mean longitudes of Pluto and Neptune, respectively, and ϖ is the longitude of perihelion of Pluto. As well, they found that the argument of perihelion of Pluto librates with a period of 4 Myr and that the angles $\Omega - \Omega_N$ and $\varpi - \varpi_N$ seemed to be in resonance with this libration. All these resonances acted to prevent close encounters of Pluto with Neptune and hence protect the orbit of Pluto.

The longest numerical integration done on general purpose computers was the LONGSTOP 1B integration done by Nobili et al. (1989) which ran for 100 Myr. This integration followed the mutual interactions of the 5 outer planets, but also included the secular effects of the inner planets and the effects of general relativity. They have shown that these latter effects significantly affect the orbits over the time span of their integrations. Spectral analysis of their results suggested but did not prove that the orbits are chaotic. The evidence for this are regions of the spectrum where many lines of comparable amplitude accumulate, indicating that the series of Fourier terms is not converging.

Integrations of the outer planets for period up to 845 Myr or 20% of the age of the solar system have been done with a special purpose machine: the Digital Orrery (Applegate et al. 1986; Sussman and Wisdom 1988). This machine consists of one CPU per planet, with all the CPUs arranged in a ring. A force evaluation for all planets is accomplished in $\mathcal{O}(N)$ time instead of the usual $\mathcal{O}(N(N-1)/2)$, and the integration is performed in $\mathcal{O}(1)$ time. A disadvantage of using such a machine is that the nature of the force calculations are "hard wired" and the inclusion of effects such as general relativity could not be accomplished with out a rebuild of the machine. The dynamical system investigated by the Orrery is therefore slightly different than that investigated by Nobili et al. The longest of the integrations performed on the Orrery show that Pluto's orbit is not quasi-periodic. There is evidence for the existence of very long period changes in Pluto's orbital elements and Sussman and Wisdom (1988) calculate a Lyapunov exponent of (1/20 Myr^{-1}). This has recently been confirmed using a symplectic integrator by Wisdom and Holman (1991). A subsequent re-examination of the LONGSTOP 100 Myr data led Milani et al. (1989) to suggest that Pluto is locked in a complicated system of three resonances and that the value of the Lyapunov exponent for its motion could be sensitive to the assumed initial conditions and planetary masses. Thus, a detailed understanding of Pluto's behavior is likely to be obtained only with the next generation of simulations (see Sec. VII).

The detection of such large Lyapunov exponents indicates chaotic behavior. However, the apparent regularity of the motion of the Earth and Pluto and indeed the fact that the solar system has survived for 4.5 Gyr implies that the chaotic regions must be narrow. What the chaotic motion does mean (if confirmed) is that there is a horizon of predictability for the detailed motions of the planets. Thus, the exponential divergence of orbits with a 5 Myr time scale claimed by Laskar means that an error as small as 10^{-10} in the initial conditions will lead to a 100% discrepancy in 100 Myr.

Until very recently, the length of direct integrations was limited, not by CPU time, but by round-off error. Milani and Nobili (1988) concluded that it was impossible to integrate reliably the orbits of the outer planets for a period of 10^9 yr or more with then current computer hardware and software. The main problems they found were limited machine precision, and the empirical evidence that the longitude error after n steps was proportional to n^2 instead of $n^{3/2}$, as expected if the round-off errors were uncorrelated. However, Quinn and Tremaine (1990) have proposed several corrections to the integration algorithms which considerably reduce the round-off error, and Quinlan and Tremaine (1990) have proposed a high-order symmetric scheme. Taken together, these two techniques appear to maintain a longitude error growth which is linear in the number of steps and which may therefore allow for accurate integrations of the planets for tens of Gyr. Another promising avenue for certain solar system integrations are the symplectic schemes designed specifically to maintain the Hamiltonian structure of such systems of equations (see, e.g., Yoshida 1990; Gladman et al. 1991; Forest 1991; Wisdom and Holman 1991).

The longest accurate integration of the entire solar system as of this writing is an integration of the nine planets and the Earth's spin axis for 3.05 Myr into the past (Quinn et al. 1991). Previously, Richardson and Walker (1987,1989) performed a 2 Myr integration of all nine planets, but they neglected the rather important effects (for the inner planets) of General Relativity and the finite size of the Earth-Moon system. Quinn et al. (1991) have computed the long-term variations (periods >2000 yr) of the orbital elements of all the planets and the Earth's spin direction. These can be used to check or replace the results of secular perturbation theory and as input into geophysical models that test the Milankovich hypothesis that climate variations are caused by changes in the Earth's orbit. All of the planetary orbits appear to be regular over the 3 Myr integration span. This integration was not long enough to detect the chaotic motion found by Laskar, but the two simulations are in remarkable agreement over their common range. In particular, the numerical integrations demonstrated the same secular resonance which Laskar claims is responsible for the chaos seen in his longer simulation.

VII. SUMMARY AND PROSPECTS FOR THE FUTURE

We have shown how the results from research in nonlinear Hamiltonian systems can be used to help us understand the newly emerging picture of the long-term dynamical evolution of orbits in the solar system. Although algebraic mappings, analytic approaches and perturbation theory have greatly aided our intuition in such nonlinear problems, much of our understanding usually comes initially from numerical simulations. It is straightforward to estimate that with an efficient and sufficiently accurate integration scheme it will require roughly 10^{16} floating point operations to integrate the equations of motion of a model solar system for a simulation time of 5 Gyr. This corresponds to a total of roughly 100 Mflop-years, where 1 Mflop is a million floating point operations per second. Although the floating point speeds of the fastest workstations are currently 20 Mflops, solar system simulations are ideally suited to the sorts of modest parallelism (4–64 processors) that can boost system performance by an order of magnitude, as has been demonstrated by the Digital Orrery. As a result, the ability to perform simulations of the evolution of model solar systems over interesting time scales are now becoming feasible. It is also clear from what has been discussed here and elsewhere (see, e.g., the Chapter by Lissauer and Stewart; Wetherill 1991c) that long-range gravitational interactions must have influenced planet formation itself. It is now possible to undertake more realistic simulations of the late stages of planet formation, including the long-range mutual interactions of the planetary embryos. We are thus at the dawn of a new era—that of the numerical exploration of the solar system.

Acknowledgments. The authors acknowledge helpful discussions with B. Gladman, H. Levison, G. Quinlan, S. Tremaine and P. Weissman. MJD is grateful for continuing financial support from the Natural Sciences and Engineering Research Council of Canada.

COLOR SECTION

Plate 1. A true-color near-infrared (1–2.5 μm) image of the IC 348-IR region, covering ~3.2×3.2 arcmin (~0.3×0.3 pc at 300 pc). Images at three different near-infrared wavelengths (1.2, 1.65, and 2.1 μm) were superimposed to create this composite, the first as blue, the second as green, the last as red, giving a true indication of the near-infrared colors of sources. For example, objects that are generally hot and/or relatively unobscured by dust appear as bluish, while cooler objects and/or those hidden by dust at the shorter wavelengths appear orange and red. The true-color images shown in Plates 2, 4, 5, and 6 were made with the same technique. In this image of IC 348-IR, a previously unknown jet-like feature is revealed, possibly a Herbig-Haro object. The associated driving source, which must be a very young and highly obscured star, has not yet been identified. The data were obtained using a 256×256 pixel HgCdTe array on the University of Hawaii 2.2-m telescope. (See the chapter by Zinnecker et al.)

Plate 2. A true-color near-infrared (1–2.5 μm) mosaic covering the central 5×5 arcmin (∼0.65 × 0.65 pc at 450 pc) of the Trapezium Cluster. Some 500 stars are seen to a completeness limit of K∼15.5 mag. The data were obtained using a 62×58 pixel InSb array on the United Kingdom 3.8-m Infrared Telescope (McCaughrean et al. 1992). (See the Chapter by Zinnecker et al.)

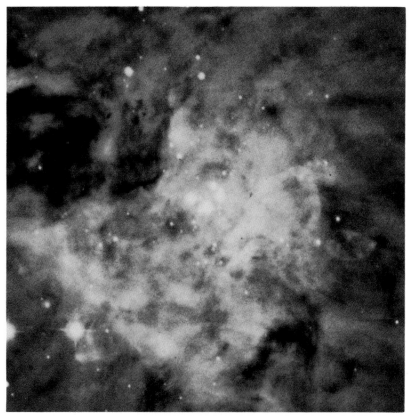

Plate 3.   For comparison with Plate 2, an optical photograph of the same region is shown (photograph taken at the Anglo-Australian Telescope, and kindly provided by D. Malin). (See the Chapter by Zinnecker et al.)

Plate 4.   A true-color near-infrared (1–2.5 μm) mosaic of the H II region S 106, covering ~10×10 arcmin (~1.75×1.75 pc at 600 pc).   The data were obtained using a 256×256 pixel HgCdTe infrared array at the University of Hawaii 2.2-m (Rayner 1992).   Some 200 stars are probable members of a cluster centered near S 106 IRS 4. (See the Chapter by Zinnecker et al.)

Plate 5.   A true-color near-infrared (1–2.5 μm) mosaic of M 17 covering ∼9×9 arcmin (∼5.8×5.8 pc at 2.2 pc). Approximately 100 stars (B9 or earlier) towards the center of the image have been identified as cluster members. The data were obtained using a 256×256 pixel PtSi array on the KPNO 2.1-m telescope (courtesy National Optical Astronomy Observatories). (See the Chapter by Zinnecker et al.)

Plate 6. A true-color near-infrared (1–2.5 μm) mosaic of the S 255/S 257/S 255-IR region covering 8×5.2 arcmin (~5.8×3.8 pc at 2.5 kpc). At left center and right center are the two B0 stars which excite the H II regions S 255 and S 257: note that they have no cospatial clusters. Between the two H II regions, the infrared cluster associated with the source S 255-IR is seen. The data were obtained using a 128×128 HgCdTe infrared array at the Steward Observatory 2.3-m telescope (McCaughrean et al. 1991). (See the Chapter by Zinnecker et al.)

Plate 7. A Hubble Space Telescope image of a portion of the Orion Nebula west of
the bright star θ^2A Ori. Red indicates [N II] emission, green Hα, and blue [O III].
Orion is a well known region of recent star formation, including stars of less than
one million years age. This image shows the main ionization front as a bright bar.
Extending above the bar is a shock front with the Herbig-Haro object M42-HH3 as
the brightest portion. Such objects are believed to be formed by the interaction of
a jet of high-velocity material originating from a young star with ambient nebular
gas. Recent observations show evidence for protoplanetary disks around many of
the young stars. These disks are ionized from the outside by the hot, massive stars
in the nebula and the resulting outflow of gas is distorted by the stellar wind from
the same stars. One of these disks is seen at the upper right, just below the bright
bar and the HH3 plume. Filters isolating continuum emission show a young star
near the center of this object (courtesy of C. R. O'Dell, Rich University).

Glossary

GLOSSARY*

AGB star

asymptotic giant branch star; a large red giant, 100–1000 solar radii, with a hot carbon-oxygen white dwarf at its center. Such stars are precursors of planetary nebulae, and get their name from the path they take on a H-R diagram (plot of luminosity versus effective temperature).

Alfvén surface

geometric surface in a magnetized wind where the poloidal components of the flow speed and of the local Alfvén speed are equal.

Alfvén velocity (v_A)

the speed at which hydromagnetic waves are propagated along a magnetic field: $v_A = B/(4\pi\rho)^{1/2}$, where B is the magnetic field strength and ρ is the density.

Allende meteorite

A C3 carbonaceous chondrite which fell in northern Mexico on 1969 February 8. It is one of the largest recorded stony meteorite falls and one of the most primitive examples in the solar system. Almost 2 tons of material were recovered, with the largest stone having a mass of 15 kg. The meteorite was later found to have many isotopic anomalies—e.g., an excess of ^{26}Mg which can most plausibly be attributed to *in situ* decay of ^{26}Al (half-life 7×10^5 yr). The oxygen ratios in the different inclusions are different from each other and also different from the rest of the solar system. Variations in Ca, Ba, and Nd also exist.

---

* We have used some definitions from *Glossary of Astronomy and Astrophysics* by J. Hopkins (by permission of the University of Chicago Press, copyright 1980 by the University of Chicago), from *Astrophysical Quantities* by C. W. Allen (London: Athlone Press, 1973), and from *The Planetary System* by David Morrison and Tobias Owen (Reading, Mass.: Addison-Wesley Publishing Co., 1988). We also acknowledge definitions and helpful comments from various chapter authors.

"alpha model"   a heuristic model for turbulence in which the angular
                       momentum transport is parameterized in terms of an
                       eddy (or turbulent) viscosity. This viscosity in turn
                       is proportional to the turbulent velocity and the char-
                       acteristic length scale in the problem, the constant of
                       proportionality being traditionally labeled "alpha."

ANEOS equation   a semi analytic equation of state that links pressure,
                       temperature and density for materials of interest in im-
                       pact computations.

Arrhenius rela-   the rate of a chemical reaction is inversely proportional
    tionship   to the exponential of the inverse temperature, times a
                       constant.

AU   the mean distance between Earth and Sun (1.496×10^{13}
                       cm $\simeq 500$ light seconds).

Barnett effect   paramagnetic relaxation associated with the internal
(in interstellar   magnetic field produced directly by grain rotation.
   grains)

BN-KL   the Becklin-Neugebauer Object/Kleinman-Low Neb-
                       ula region, a small dense region in the Orion Molecular
                       Cloud that contains luminous IR sources, interstellar
                       and circumstellar masers, hot gas, and young stars. It
                       is the nearest location of active star formation.

bolometric   referring to electromagnetic radiation received from an
                       object at all wavelengths of the spectrum.

brown dwarf   a stellar-like object which is not massive enough to sus-
                       tain hydrogen fusion as its main source of luminosity,
                       but instead radiates by virtue of the energy of collapse
                       of the original gas from which it formed.

Brownian motion   random motion of microscopic particles suspended in
                       a gas due to collisions with gas molecules.

CAI   Ca-, Al-rich inclusions; mm- to cm-size components
                       of chondritic meteorites. Compositions are greatly de-
                       pleted from cosmic in the relatively volatile elements,
                       hence enriched in Ca, Al, and other involatile elements.
                       They are thought to have formed as dispersed objects
                       in the solar nebula, at high temperatures.

carbonaceous chondrites types 1,2,3
a rare type of chondrite (about thirty are known) characterized by the presence of carbon compounds. Carbonaceous chondrites are the most primitive samples of matter known in the solar system. They resemble the solid material expected when a gas cloud of solar composition cools to a temperature of about 300 K at low pressure. Wiik (1956) classification: Type C1 are the least dense and are strongly magnetic. They are the most primitive and show only minimal chemical alteration. Type C2 are weakly magnetic or nonmagnetic. Type C3 are the densest, are water-poor, and usually are largely olivine.

Cassini state
particular obliquity states of tidally locked satellites for which the spin direction, the normal to the orbit plane and the normal to the invariable plane (roughly the planet's equatorial plane) are all coplanar.

CDW
centrifugally driven wind.

CHON
an acronym for the organic grains found in comet Halley, and composed of Carbon, Hydrogen, Oxygen, and Nitrogen.

Class I and II sources
a class I source is a deeply embedded, usually invisible, young stellar object with a spectral energy distribution that rises with increasing wavelength of 2.2 microns. Class I sources are believed to be extremely young, and, at least for relatively low luminosity objects (i.e., $L < 100$ solar luminosities), protostellar in nature. A Class II source is a young stellar object whose spectral energy distribution decreases with increasing wavelength longward of 2.2 microns. Such objects are often visible and have energy distributions which are power-like in form in the near to mid infrared. Classical T Tauri stars are typically Class II sources. The energy distribution of a Class II source is often well modelled by emission from a circumstellar disk.

classical novae
explosions that result from thermonuclear runaways on the surfaces of white dwarfs accreting matter in close binary systems.

Clausius-Clapey-ron relation	equation describing the dependence of pressure on temperature at the boundary between two phases of a substance in equilibrium.
cometesimal	an icy planetesimal.
coronagraph	device used to block light from a bright object to allow examination of surroundings like the corona of the Sun or the dust disk around beta Pictoris. A complete coronograph consists of a focal-plane or sky-plane mask to eliminate direct light plus a pupil-plane mask to reduce light scattered in the telescope and instrument.
Coulomb barrier	the potential energy "hill" due to the repulsive electrostatic force between like-charged particles. For nuclear reactions, the Coulomb barrier acts to bar low energy (energies less than the height of the barrier) positively charged particles from approaching the positively charges atomic nucleus to the point where a nuclear reaction can proceed.
Drake equation	an empirical formula which relates the number of possible intelligent civilizations in the galaxy to various well-known and poorly known physical parameters such as star and planetary system formation rates, etc.
DEIS	deeply embedded infrared source.
Eddington limit L_{EDD}	in essence, for the stability of a star, radiation pressure must not exceed gravitational pressure. It is the luminosity limit beyond which the radiation force on hydrogen atoms is greater than the gravitational force binding them to the star.
Einstein coefficient	an emission (or absorption) coefficient. A_{ji} is the coefficient of spontaneous emission; B_{ji} is the coefficient of stimulated emission; B_{ij} is the coefficient of absorption, where i is the lower level and j is the upper level.
Ekman layer	upper boundary layer within which the amplitude of a wave changes exponentially.

fake Zeeman splitting	circular polarization of a spectral line that mimics Zeeman splitting, but that is in fact a result of linear birefringence.
Faraday rotation	rotation of the position angle of linear polarization in a magnetized plasma.
Fischer-Tropsch reactions	production of organic molecules by the hydrogenation of carbon monoxide in the presence of a suitable catalyst.
FMS	fast magnetosonic wave speed.
force-free field	in MHD, a magnetic field configuration that is characterized by a vanishing net stress and arises when the magnetic terms dominate all other terms in the equation of motion.
FWHM	full width at half maximum.
Gould's belt	the local system of stars and gas within about 300 pc of the Sun. It is an expanding belt ≤ 1 kpc in diameter and inclined about $20°$ to the galactic plane in which OB stars, young clusters and associations, and dark clouds are concentrated. Discovered by B. Gould in 1879. It is a source of high-energy γ-rays.
GRO	Gamma Ray Observatory, now operational and called the Compton Gamma Ray Observatory, one of NASA's Earth-orbiting "Great Observatories," designed to observe astrophysical gamma ray sources at high spatial and spectral resolution.
Hamiltonian systems	systems for which the dynamical evolution from a given star initial state is governed by Hamilton's equations of classical mechanics.
Hayashi phase	the period of pre-main-sequence stellar evolution during which a star has negligible nuclear energy production and has sufficiently low internal temperatures so that convective energy transport dominates through most or all of the interior. The star evolves in the H-R diagram with decreasing luminosity and nearly constant effective temperature.

Herbig-Harbo
objects (HH
objects)

small, faint patches of nebulosity seen on the surfaces
of many dark clouds, believed to be a very early phase
of stellar evolution. All known Herbig-Haro objects
have been found within the boundaries of dark clouds.
They are strong infrared sources and are characterized
by mass loss.

High-Z material

in stellar and planetary models, that component composed of atoms with an atomic number higher than
2.

Hill's approxima-
tion

an approximation for the equations of nearly circular and planar motion of bodies in a central gravitational field, under which these equations can be linearized. This is a special case of so-called "epicyclic
approximation."

HR diagram

Hertzsprung-Russell diagram; a plot of luminosity versus spectral type. More loosely, a plot of magnitude
versus color. In present usage, a plot of bolometric
magnitude against effective temperature for a population of stars. Related plots are the color-magnitude plot
(absolute or apparent visual magnitude against color
index) and the spectrum-magnitude plot (visual magnitude versus spectral type, the original form of the HR
diagram).

HST

Hubble Space Telescope, the first of NASA's orbiting
"Great Observatories," designed to provide high spatial
and spectral resolution observations of objects in the
ultraviolet, visible and (eventually) near-infrared.

Hubble law

also called law of redshifts, the distance of galaxies
from us is linearly related to their redshift. More generally, the relative velocity (resulting from large-scale
cosmic expansion) between particles that are *locally* at
rest is proportional to their separation. Discovered by
Hubble in 1929.

Hubble time
(H_0^{-1})

the characteristic age of the universe since the big bang
($19.4\pm1.6\times10^9$ yr for $H_0 = 50$ km per second per
megaparsec and a constant expansion rate).

Hugoniot curve/ equation	the relation between pressure and volume of material subjected to sudden shock compression, also known as the Hugoniot shock relation. These equations express the conservation of mass, energy and momentum across the shock.
IKS	InfraKrasnoi Spectrometre (an infrared spectrometer flown on the Vega 1 and Vega 2 spacecraft).
Initial mass function (IMF)	the distribution of stellar masses at birth; relative number of stars formed per unit mass.
IRAS	Infrared Astronomical Satellite; joint UK-USA-Dutch satellite which provided the first surveys of the sky at thermal infrared wavelengths with moderate spatial and spectral resolution.
ISO	Infrared Space Observatory, a European infrared space telescope to be launched in (approximately) 1993.
JANAF tables	numerical tables of thermochemical data on a wide variety of substances, published by the U.S. Department of Commerce.
jansky (Jy)	unit of flux density adopted by the IAU in 1973. $1\,\text{Jy} = 10^{-26}\ \text{W m}^{-2}\ \text{Hz}^{-1}$; named after Karl Jansky, who discovered galactic radio waves in 1931.
Jeans escape formula	a formula giving the rate of thermal excape of molecules from an atmosphere.
Jeans instability	gravitational instability in an idealized, infinite, homogeneous medium.
Jeans length	the critical wavelength at which the oscillations in an infinite, homogeneous medium become gravitationally unstable. Any disturbance whose characteristic dimension is greater than the Jeans length will decouple by self-gravitation from the rest of the medium to become a stable, bound system.
Jeans mass	the mass necessary for protostellar collapse $M_j \geq 1.795(RT/G)^{3/2}\rho^{-1/2}$, where R is the gas constant, T the cloud temperature, G the gravitational constant, and ρ the cloud density.

Jeans rate

the characteristic rate of collapse or evolution of a gas at density ρ, equal to about $(4\pi G\rho)^{1/2}$, that results from self-gravitational attraction.

Julian Date (JD)

the Julian day number followed by the fraction of the day elapsed since the preceding noon. The Julian day number is the number of days that have elapsed since Greenwich noon on January 1, 4713 B. C. The Julian period was worked out by Joseph Scaliger and applied in 1582, the year the Gregorian calendar was adopted. The Julian period contains exactly 7980 Julian yr of 365 1/4 days each.

Kelvin-Helm-
holtz instability

the tendency of waves to grow on a shear boundary between two regions in relative motion parallel to the boundary.

Kelvin-Helm-
holtz time

the time over which a self-gravitating star radiates away enough thermal energy to contract by a significant fraction of its initial radius. It is assumed that the star is not being supported by internal nuclear fusion.

Keplerian shear

the velocity gradient that results from the differential rotation of particles whose orbits about a massive primary obey Kepler's Laws.

Kirchhoff's laws

(1) To each chemical species there corresponds a characteristic spectrum. (2) Every element is capable of absorbing the radiation which it is able to emit; it gives rise to the phenomenon of the reversal of the lines. A precise statement of the second law is that the absorption coefficient (κ_v) of material in thermal equilibrium at a temperature T is related to the emissivity (j_v) by $\kappa_v = j_v B_v(T)$, where $B_v(T)$ is the Planck function for the temperature T.

Knudsen number

the ratio of the mean free path length of the molecules in a fluid to a characteristic length of the structure in the fluid stream.

Kolmogorov
spectrum

in a homogeneous and isotropic turbulent medium, energy is continually transfering between turbulent eddies of different sizes. The distribution of energy cascading down the different scales is referred to as the Kolmogorov spectrum.

Kolmogorov turbulence	turbulence, as in a laboratory, that appears to satisfy the scaling relation proposed by Kolmogorov (1941), namely, relative velocity proportional to the cube root of separation.
Kramers-Kronig dispersion relation	relates the real and imaginary parts of the dielectric constant of a medium.
Kuiper disk or belt	hypothetical component of our planetary system extending in the ecliptic plane beyond the planetary region out to the Oort cloud. Proposed as a dynamically plausible reservoir for short-period comets.
Langevin theory	a theory which explains diamagnetic and paramagnetic properties of materials in terms of a classical picture (Langevin, P. 1905. *Ann. Chim. Phys.* 5:245).
Langrangian points	the five equilibrium points in the restricted three-body problem.
Lindblad resonance	a resonance of first order in the eccentricity of a perturbed particle. This terminology comes from galactic dynamics; in the 1920s Lindblad invoked such resonances to explain spiral arms.
Lindblad's ring	a ring of neutral gas, discovered by Lindblad et al. (1973), that surrounds the Sun at a distance of several hundred parsecs and expands away from the local standard of rest in all directions at a speed of several kilometers per second (from Lindblad, P. O., Grape, K., Sandqvist, Å., and Schober, J. 1973). On the kinematics of a local component of the interstellar hydrogen gas possibly related to Gould's belt. *Astron. Astrophys.* 24:309–312.
linear birefringence	circular polarization of a spectral line caused by linear dichroism when the direction of linear polarization is not parallel to either optical axis.
linear dichroism	linear polarization of a spectral line caused by differing populations of the relevant magnetic substates.

Lyapunov expo-   the time for the distance between two chaotic orbits to
nent   increase by e (2.718...), when the initial conditions are
                   altered infinitesimally.

MHD   magnetohydrodynamics; sometimes called hydromag-
                   netics. The study of the collective motions of charged
                   particles in a magnetic field.

Mie-scattering   the scattering of light by particles with size compa-
theory   rable to a wavelength of the light. Exact albedo and
                   phase functions are highly oscillatory with direction
                   and wavelength; however, slight irregularities and dis-
                   tributions of size smooth out the features, leaving pre-
                   dominantly forward scattering behavior.

minimum mass   the minimum mass for the solar nebula ($\sim$0.01 solar
solar nebula   masses) is computed from the total mass of the planets,
                   adjusting for the difference in chemical composition
                   between the planets and the interstellar medium.

MRN grain size   distribution of grain sizes introduced by Mathis et al.
distribution   (1977) to account for observations of interstellar ex-
                   tinction from 0.11 μm to 1.0 μm. The distribution has
                   the form $N(a)da \propto a^{-3.5}da$, where a is the grain ra-
                   dius. It extends from 5 nm to about 1 μm for graphite
                   and over a narrower range for other materials.

NMS   Neutral Mass Spectrometer, an instrument flown on
                   Giotto.

OMC-1   Orion Molecular Cloud 1, a very dense north-south
                   ridge about 1 pc in length near the Orion nebula. It
                   contains the active BN-KL region.

Oort constants   parameters that characterize the differential rotation of
(A and B)   our Galaxy at the Sun's distance form the center: A =
                   15 km s^{-1} kpc^{-1}; B = -10 km s^{-1} kpc^{-1}.

Oort cycle   a cycle of cloud formation and destruction proposed
                   by Oort (1954) in which small clouds randomly collide
                   and build up larger clouds, and then stars form in the
                   larger clouds, breaking them into small clouds again.

Orion-A molecular ridge	a giant molecular cloud near the Orion nebula within which the Orion Molecular Cloud 1 (OMC-1) and the BN-KL regions are located.
oxygen-neon-magnesium (O-Ne-Mg) white dwarfs	massive (≤ 1.2 $M_\odot$) white dwarfs that are the degenerate remains of the cores of intermediate mass (≈ 10 $M_\odot$) stars. Their enhanced abundances of O, Ne, and Mg result from an extended helium burning phase in the progenitor star.
PAH	polycyclic aromatic hydrocarbon: a class of organic compounds thought to be important in interstellar and possibly proto-planetary chemistry.
P-Cygni line profile	spectral line profile characterized by asymmetric emission and absorption components indicative of massive winds around a star. Named for a particular B-type supergiant star with a massive wind.
Parker instability	instability caused by energetic particles in the magnetic field interacting with the interstellar gas.
pc	parsec; 1 parsec = the distance where 1 AU subtends 1 arcsec = 206,265 AU = 3.26 lightyear = 3.086×10^{18} cm.
PIA	Particle Impact Analyzer, an instrument for measuring dust composition, flown on Giotto.
PICCA	Positive Ion Cluster Composition Analyzer, an instrument flown on Giotto.
Planck Black-body function	the dependence of emitted radiance on wavelength for an object which behaves as a perfect blackbody, i.e., is capable of absorbing all radiation incident upon it.
poloidal	refers to the components of a vector, such as the magnetic field, that lie in the meridional (constant longitude) plane (compare toroidal).
POM	polyoxymethylene, a chemical found in comet Halley.

Populations I and II	two classes of stars introduced by Baade in 1944. In general, Population I (now sometimes called arm population) are young stars with relatively high abundances of metals, and are found in the disk of a galaxy, especially the spiral arms, in dense regions of interstellar gas. Population II (now sometimes called halo populations) are older stars with relatively low abundances of metals, and are typically found in the nuclear bulge of a galaxy or in globular clusters. The Sun is a rather old Population I star; old Population I is sometimes called the disk population.
position angle	angle in the sky, for example, of the direction of linear polarization. It is normally denoted by θ and is measured with $\theta = 0°$ toward the north and θ increasing towards the east.
p-process	the name of the hypothetical nucleosynthetic process thought to be responsible for the synthesis of the rare heavy proton-rich nuclei which are bypassed by the r- and s-processes. It is manifestly less efficient (and therefore rarer) than the s- or r-process because the protons must overcome the Coulomb barrier, and may in fact work as a secondary process on the r- and s-process nuclei. It seems to involve primarily (p, γ) reactions above cerium (where neutron separation energies are low). The p-process is assumed to occur in supernova envelopes at a temperature $\geq 10^9$ K and at densities $\leq 10^4$ g cm^{-3}.
PUMA	a dust composition analyzer flown on Vega 1 and 2. It is identical to PIA.
Raoult's law	the mole fraction in solution of a solute times its saturation vapor pressure is equal to its partial pressure over the solution. The law is an idealization but holds approximately for solutions involving chemically-similar species.
Rayleigh-Jeans side	the low frequency portion of the Planck blackbody emission law, for which $h\nu/kT << 1$.
Rayleigh-Taylor instability	a type of hydrodynamic instability for static fluids (e.g., cold dense gas above hot rarefied gas).

Reynolds number a dimensionless number ($R = Lv/\nu$, where L is a typical dimension of the system, v is a measure of the velocities that prevail, and ν is the kinematic viscosity) that governs the conditions for the occurrence of turbulence in fluids.

Roche lobe the largest closed zero-velocity potential surface surrounding a secondary body (satellite) in orbit about a primary. This surface passes through the Lagrange L_2 point which lies between the primary and secondary bodies.

Roche radius the limiting distance between two bodies below which tidal disruption of one of the objects would occur.

r-process the capture of neutrons on a very rapid time scale (i.e., one in which a nucleus can absorb neutrons in rapid succession, so that regions of great nuclear instability are bridged), a theory advanced to account for the existence of all elements heavier than bismuth as well as the neutron-rich isotopes heavier than iron. The essential feature of the r-process is the release of great numbers of neutrons in a very short time (less than 100 s). The presumed source for such a large flux of neutrons is a supernova, at the boundary between the neutron star and the ejected material.

Schmidt-Holsapple scaling law a widely used relation that links impact crater diameter or volume to the size density and velocity of the projectile that created it.

self-absorption an atomic or molecular spectral line seen in absorption against an emission feature produced by the line itself arising in a background cloud.

SIRTF Space Infrared Telescope Facility; a large cryogenically cooled telescope with state-of-the-art infrared detectors proposed by NASA for launch to high earth orbit around the year 2000.

SMS slow magnetosonic wave speed.

SOFIA Stratospheric Observatory for Infrared Astronomy; a proposed NASA infrared telescope.

s-process | a process in which heavy, stable, neutron-rich nuclei are synthesized from iron-peak elements by successive captures of free neutrons in a weak neutron flux, so there is time for β decay before another neutron is captured (c.f. r-process). This is a slow process of nucleosynthesis which is assumed to take place in the intershell regions during the red-giant phase of evolution, at densities up to 10^5 g cm $^{-3}$ and temperatures of about 3×10^8 K (neutron densities assumed are 10^{10} cm^{-3}).

Stokes number | dimensionless number associated with dynamics of a particle in a fluid. It is the product of the fluid dynamic viscosity times the characteristic vibration time of the particle, divided by a length scale and the density of the fluid.

Stromgren sphere | a more or less spherical H II region surrounding a hot star.

Supernovae Type Ia | supernovae which have a spectrum characterized by detailed structure without a well-defined continuum, and appear to be deficient in hydrogen. They occur in the halo and in old disk populations (Type II stellar population) in galaxies.

Supernovae Type II | supernovae which have a spectrum characterized by a continuum with superimposed hydrogen lines and some metal lines. They occur in the arms of spiral galaxies, implying that their progenitors are massive stars (Type I stellar population).

Thomson (electron) scattering | the non-relativistic scattering process involving the radiation from a free electron accelerated by an incident electromagnetic wave. It is the dominant scattering process in low-density, high-temperature plasmas.

toroidal | refers to the component of a vector, such as the magnetic field, that points in the azimuthal direction (compare poloidal).

van der Waals forces | the relatively weak attractive forces operative between neutral atoms and molecules.

virial theorem	for a bound gravitational system the long-term average of the kinetic energy is one-half of the potential energy.
VLA	Very Large Array; world's largest radio telescope facility, composed of strings of individual telescopes arrayed together near Socorro, New Mexico.
white dwarf (wd or D)	a star of high surface temperature (typically on the order of 3×10^4 K), low luminosity, and high density (10^5–10^8 g cm^{-3}) with roughly the mass of the Sun (average mass about 0.7 solar mass) and the radius of the Earth, which has exhausted most or all of its nuclear fuel, believed to be a star in its final stage of evolution. DA white dwarfs are hydrogen-rich; DB white dwarfs are helium-rich; DC are pure continuum; DF are calcium-rich; DP are magnetic stars (some have magnetic fields as high as 10^8 gauss).
Wien side	the high-frequency portion of the Planck blackbody emission law, for which $h\nu/kT >> 1$.
WKB (Wentzel-Kramers-Brillouin) method	a method for obtaining an approximate solution to Schrödinger's equation.
Wolf-Rayet star	one of a class of very luminous, very hot (as high as 50,000 K) stars whose spectra have broad emission lines (mainly He I and He II), which are presumed to originate from material continually ejected from the star at very high (about 2000 km s^{-1}) velocities by stellar winds. They may be the exposed helium cores of stars that were at one time on the H-burning main sequence. Some Wolf-Rayet spectra show dominantly emission lines from ions of carbon (WC stars); others show dominantly emission lines from the ions of nitrogen (WN stars).
WSRT	Westerbork Synthesis Radio Telescope; a very large radio telescope facility in the Netherlands.
X-point	the saddlepoint in the effective gravitational potential where the centrifugal force balances gravity.

YSO   young stellar object: general term for stars which are
                             in the several phases of evolution before reaching the
                             main sequence.

ZAMS   zero-age main sequence.

Zeeman effect   line broadening due to the influence of magnetic fields.
                             A multiplet of lines is produced, with distinct polariza-
                             tion characteristics. In stellar spectroscopy the Zeeman
                             effect is used to determine longitudinal magnetic fields
                             by measuring the difference between right-hand and
                             left-hand circular polarization across a spectral line.

Zeeman splitting   the removal of degeneracy and small energy shift of
                             atomic or molecular energy levels caused by an external
                             magnetic field.

Bibliography

BIBLIOGRAPHY

Compiled by Mary Guerrieri

Abe, Y., and Matsui, T. 1985. The formation of an impact-generated H_2O atmosphere and its implications for the early thermal history of the Earth. *Proc. Lunar Planet. Sci. Conf.* 16, *J. Geophys. Res. Suppl.* 90:C545–C559.

Abt, H. A. 1983. Normal and abnormal binary frequencies. *Ann. Rev. Astron. Astrophys.* 21:343–372.

Abt, H. A. 1987. Implications for solar-type statistics of the Morbey-Griffin improved binary analysis. *Astrophys. J.* 317:353–354.

Abt, H. A., and Levy, S. G. 1976. Multiplicity among solar-type stars. *Astrophys. J. Suppl.* 30:273–306.

Adachi, I., Hayashi, C., and Nakazawa, K. 1976. The gas drag effect on the elliptical motion of a solid body in the primordial solar nebula. *Prog. Theor. Phys.* 56:1756–1771.

Adams, F. C. 1989. The infrared spectral signature of star formation. In *Infrared Spectroscopy in Astronomy*, ed. B. H. Kaldeich, ESA SP-290, pp. 233–243.

Adams, F. C. 1990. The $L–A_V$ diagram for protostars. *Astrophys. J.* 363:578–588.

Adams, F. C., and Shu, F. H. 1986. Infrared spectra of rotating protostars. *Astrophys. J.* 308:836–853.

Adams, F. C., Lada, C. J., and Shu, F. H. 1987. Spectral evolution of young stellar objects. *Astrophys. J.* 312:788–806.

Adams, F. C., Lada, C. J., and Shu, F. H. 1988. The disks of T Tauri stars with flat infrared spectra. *Astrophys. J.* 326:865–883.

Adams, F. C., Ruden, S. P., and Shu, F. H. 1989. Eccentric gravitational instabilities in nearly Keplerian disks. *Astrophys. J.* 347:959–975.

Adams, F. C., Emerson, J. P., and Fuller, G. A. 1990. Submillimeter photometry and disk masses of T Tauri disk systems. *Astrophys. J.* 357:606–620.

Adams, M. T., Strom, S. E., and Strom, K. M. 1983. The star-forming history of the young cluster NGC2264. *Astrophys. J.* 53:893–936.

Adams, N. G., and Smith, D. 1988. Laboratory studies of dissociative recombination and mutual neutralization and their relevance to interstellar chemistry. In *Rate Coefficients in Astrochemistry*, eds. T. J. Millar and D. A. Williams (Dordrecht: Kluwer), pp. 173–192.

Adamson, A. J., Whittet, D. C. B., and Duley, W. W. 1990. The 3.4 μm interstellar absorption feature in Cyg OB2 no. 12. *Mon. Not. Roy. Astron. Soc.* 243:400–404.

Adler, D. S. 1988. The Distribution of Molecular Clouds in Spiral Galaxies. Ph. D. Thesis, Univ. of Virginia.

Adler, D. S., and Roberts, W. W., Jr. 1988. The mass spectrum of molecular clouds in computational studies of spiral galaxies. In *Molecular Clouds in the Milky Way and External Galaxies*, eds. R. L. Dickman, R. L. Snell and J. S. Young (Berlin: Springer-Verlag), pp. 291.

Adler, D. S., and Roberts, W. W. 1992. Ambiguities in the identification of giant molecular cloud complexes from longitude-velocity diagrams. *Astrophys. J.* 389:95–104.

Agrawal, A., Rao, A., and Riegler, G. 1986. Detection of an X-ray flare in the RS CVn binary σ Borealis. *Mon. Not. Roy. Astron. Soc.* 219:777–784.

A'Hearn, M. F., and Feldman, P. D. 1985. S_2: A clue to the origin of cometary ice? In *Ices in the Solar System*, eds. J. Klinger, D. Benest, A. Dollfus and R. Smoluchowski (Dordrecht: D. Reidel), pp. 463–471.

A'Hearn, M. F., Feldman, P. D., and Schleicher, D. G. 1983. The discovery of S_2 in comet IRAS-Araki-Alcock 1983d. *Astrophys. J. Lett.* 274:99–103.

A'Hearn, M. F., Hoban, S., Birch, P. V., Bowers, C., Martin, R., and Klinglesmith, D. A.,III. 1986a. Cyanogen jets in comet Halley. *Nature* 324:649–651.

A'Hearn, M. F., Hoban, S., Birch, P. V., Bowers, C., Martin, R., and Klinglesmith, D. A.,III. 1986b. Gaseous jets in comet P/Halley. In *20th ESLAB Symposium on the Exploration of Halley's Comet*, vol. 1, eds. B. Battrick, E. J. Rolfe and R. Reinhard, ESA SP-250, pp. 483–486.

Ahrens, T. J., and O'Keefe, J. D. 1972. Shock melting and vaporization of lunar rocks and minerals. *The Moon* 4:214–249.

Ahrens, T. J., and O'Keefe, J. D. 1987. Impact on the Earth, ocean, and atmosphere. *J. Impact Eng.* 5:13–32.

Ahrens, T. J., O'Keefe, J. D., and Lange, M. A. 1989. Formation of atmospheres during accretion of the terrestrial planets. In *Origin and Evolution of Planetary and Satellite Atmospheres*, eds. S. K. Atreya, J. B. Pollack and M. S. Matthews (Tucson: Univ. of Arizona Press), pp. 328–385.

Aitken, D. K., Bailey, J. A., Roche, P. F., and Hough, J. M. 1985. Infrared spectropolarimetric observations of BN-KL: The grain alignment mechanism. *Mon. Not. Roy. Astron. Soc.* 215:815–831.

Alaerts, L., Lewis, R. S., Matsuda, J., and Anders, E. 1980. Isotopic anomalies of noble gases in meteorites and their origins.—VI. Presolar components in the Murchison C2 chondrite. *Geochim. Cosmochim. Acta* 44:189–209.

Albinson, J. S., and Evans, A. 1987. Possible role of the white dwarf in grain formation in cataclysmic variable systems. *Astrophys. Space Sci.* 131:443–447.

Alexander, C. M. O.'D., Swan, P., and Walker, R. M. 1990. In situ measurement of interstellar silicon carbide in two CM chondrite meteorites. *Nature* 348:715–717.

Alfvèn, H., and Arrhenius, G. 1976. *Evolution of the Solar System*, NASA SP-345.

Allamandola, L. J. 1984. In *Galactic and Extragalactic Infrared Spectroscopy*, eds. M. F. Kessler and J. P. Phillips (Dordrecht: D. Reidel), pp. 5–35.

Allamandola, L. J. 1989. The infrared emission features and polycyclic aromatic hydrocarbons. In *Interstellar Dust: Proc. IAU Symp. 135*, eds. L. J. Allamandola and A. G. G. M. Tielens (Dordrecht: Kluwer), pp. 129–139.

Allamandola, L. J., Tielens, A. G. G. M., and Barker, J. R. 1987. Infrared absorption and emission characteristics of interstellar PHAs. In *Interstellar Processes*, eds. D. J. Hollenbach and H. A. Thronson, Jr. (Dordrecht: D. Reidel), pp. 471–489.

Allamandola, L. J., Sandford, S. A., and Valero, G. J. 1988. Photochemical and thermal evolution of interstellar/pre-cometary ice analogs. *Icarus* 76:225–252.

Allégre, C. J., Staudacher, T., and Sarda, P. 1987. Rare gas systematics: formation of the atmosphere, evolution and structure of the Earth's mantle. *Earth Planet. Sci. Lett.* 81:127–150.

Allen, C. W. 1976. *Astrophysical Quantities*, 3rd ed. (London: Athlone Press).

Allen, R. J., Atherton, P. D., and Tilanus, R. P. J. 1986. Large scale dissociation of molecular gas in galaxies by newly formed stars. *Nature* 319:296–298.

Allen, M., Delitsky, M., Huntress, W., Yung, Y., Ip, W.-H., Schwenn, R., Rosenbauer, H., Shelley, E., Balziger, H., and Geiss, J. 1987. Evidence for methane and ammonia in the coma of comet P/Halley. *Astron. Astrophys.* 187:502–512.

Allen, M., Bochner, B., van Dishoeck, E., Tegler, S., and Wyckoff, S. 1989. A photochemical model simulation of comet Halley NH_2 and CH observations.

Bull. Amer. Astron. Soc. 21:934 (abstract).

Alonso-Costa, J. L., and Kwan, J. 1989. Hydrogen line and continuum emission in young stellar objects. II. Theoretical results and observational constraints. *Astrophys. J.* 338:403–416.

Altenhoff, W. J., Downes, D., Goad, L., Maxwell, A., and Rinehart, R. 1970. Survey of the galactic plane at 1.414, 2.695, and 5.000 GHz. *Astron. Astrophys. Suppl.* 1:319–355.

Altenhoff, W. J., Downes, D., Pauls, T., and Schraml, J. 1978. Survey of the galactic plane at 4.875 GHz. *Astron. Astrophys. Suppl.* 35:23–54.

Alvarez, H., May, J., and Bronfman, L. 1990. The rotation of the galaxy within the solar circle. *Astrophys. J.* 348:495–502.

Amari, S., Lewis, R. S., and Anders, E. 1990a. Interstellar graphite in meteorites: Growing complexity, implied by its noble-gas components. *Lunar Planet. Sci.* XXI:19–20 (abstract).

Amari, S., Anders, E., Virag, A., and Zinner, E. 1990b. Interstellar graphite in meteorites. *Nature* 345:238–240.

Amari, S., Zinner, E., and Lewis, R. S. 1990c. Two types of interstellar carbon grains in the Murchison carbonaceous chondrite. *Meteoritics* 25:348–349.

Ambartsumian, V. A. 1980. On some trends in the development of astrophysics. *Ann. Rev. Astron. Astrophys.* 18:1–13.

Ambartsumian, V. A., Iskudarian, S. G., Shahbazian, R. K., and Sahakian, K. A. 1963. *Bull. Byurakan Obs.* 33:1.

Anders, E. 1964. Origin, age, and composition of meteorites. *Space Sci. Rev.* 3:583–714.

Anders, E. 1981. Noble gases in meteorites: Evidence for presolar matter and superheavy elements. *Proc. Roy. Soc. London* A374:207–238.

Anders, E. 1988. Circumstellar material in meteorites: Noble gases, carbon and nitrogen. In *Meteorites and the Early Solar System*, eds. J. F. Kerridge and M. S. Matthews (Tucson: Univ. of Arizona Press), pp. 927–955.

Anders, E., and Ebihara, M. 1982. Solar system abundances of the elements. *Geochim. Cosmochim. Acta* 46:2363–2380.

Anders, E., and Grevesse, N. 1989. Abundances of the elements: Meteoritic and solar. *Geochim. Cosmochim. Acta* 53:197–214.

Anders, E., and Heymann, D. 1969. Elements 112 to 119: Were they present in meteorites? *Science* 164:821–823.

Anders, E., and Owen, T. 1977. Mars and Earth: Origin and abundance of volatiles. *Science* 198:453–465.

Anders, E., Higuchi, H., Gros, J., Takahashi, H., and Morgan, J. W. 1975. Extinct superheavy element in the Allende meteorite. *Science* 190:1262–1271.

Anders, E., Higuchi, H., Ganapathy, R., and Morgan, J. W. 1976. Chemical fractionations in meteorites—IX. C3 chondrites. *Geochim. Cosmochim. Acta* 40:1131–1139.

Andersen, J., Lindgren, H., Hazel, M. L., and Mayor, M. 1989. The pre-main-sequence binary system AK Scorpii. *Astron. Astrophys.* 219:142–150.

Anderson, J. D., and Standish, E. M., Jr. 1986. Dynamical evidence for Planet X. In *The Galaxy and the Solar System*, eds. R. Smoluchowski, J. N. Bahcall and M. S. Matthews (Tucson: Univ. of Arizona Press), pp. 286–296.

Anderson, J. D., Campbell, J. K., Jacobson, R. A., Sweetnam, D. N., Taylor, A. H., Prentice, A. J. R., and Tyler, G. L. 1987. Radio science with Voyager 2 at Uranus: Results on masses and densities of the planet and five principal satellites. *J. Geophys. Res.* 92:14877–14883.

Andersson, B.-G., Wannier, P. G., and Morris, M. 1991. Warm neutral halos around molecular clouds. II. Interpretation of H I and CO $J = 1 - 0$ data. *Astrophys. J.*

366:464–473.

André, Ph. 1987. Radio emission from young stellar objects. In *Protostars and Molecular Clouds*, eds. T. Montmerle and C. Bertout (Gif-sur-Yvette: C. E. A. Doc.), pp. 143–187.

André, Ph., Montmerle, T., and Feigelson, E. D. 1987. A VLA survey of radio-emitting young stars in the ρ Ophiuchi dark cloud. *Astron. J.* 93:1182–1198.

André, Ph., Montmerle, T., Stine, P. C., Feigelson, E. D., and Klein, K. L. 1988. A young radio-emitting magnetic B star in the ρ Ophiuchi cloud. *Astrophys. J.* 335:940–952.

André, Ph., Martín-Pintado, J., Despois, D., and Montmerle, T. 1990a. Discover of a remarkable bipolar flow and exciting source in the ρ Ophiuchi cloud core. *Astron. Astrophys.* 236:180–192.

André, Ph., Montmerle, T., Feigelson, E. D., and Steppe, H. 1990b. Cold dust around young stellar objects in the ρ Ophiuchi cloud core. *Astron. Astrophys.* 240:321–330.

André, Ph., Phillips, R. B., Lestrade, J.-F., and Klein, K. L. 1991. Direct detection of the magnetosphere surrounding the young star S1 in ρ Ophiuchi. *Astrophys. J.*, in press.

Andronico, G., Baratta, G. A., Spinella, F., and Strazzulla, G. 1987. Optical evolution of laboratory-produced organics: applications to Phoebe, Iapetus, outer belt asteroids and cometary nuclei. *Astron. Astrophys.* 184:333–336.

Anthony, D. M., and Carlberg, R. G. 1988. Spiral wave viscosity in self-gravitating accretion disks. *Astrophys. J.* 332:637–645.

Anthony-Twarog, B. J. 1982. The $H\beta$ distance scale for B stars: The Orion association. *Astron. J.* 87:1213–1222.

Appenzeller, I. 1983. Recent advances in the theoretical interpretation of T Tauri stars. *Rev. Mexicana Astron. Astrophys.* 7:151–168.

Appenzeller, I., and Dearborn, D. S. P. 1984. Brightness variations caused by surface magnetic fields in pre-main sequence stars. *Astrophys. J.* 278:689–694.

Appenzeller, I., and Mundt, R. 1989. T Tauri stars. *Astron. Astrophys. Rev.* 1:291–234.

Appenzeller, I., Jankovics, I., and Krautter, J. 1983. Spectroscopy and infrared photometry of southern T Tauri stars. *Astron. Astrophys. Suppl.* 53:291–309.

Appenzeller, I., Jankovics, I., and Östreicher, R. 1984. Forbidden-line profiles of T Tauri stars. *Astron. Astrophys.* 141:108–115

Applegate, J. H., Douglas, M. R., Gürsel, Y., Sussman, G. J., and Wisdom, J. 1986. The outer solar system for 210 million years. *Astron. J.* 92:176–194.

Armstrong, J. T. 1989. Molecular outflows. In *The Physics and Chemistry of Interstellar Molecular Clouds*, eds. G. Winnewisser and J. T. Armstrong (Berlin: Springer-Verlag) pp. 143–151.

Armstrong, J. T., and Winnewisser, G. 1989. An extended outflow in L673. *Astron. Astrophys.* 210:373–377.

Arnett, W. D., and Truran, J. W. 1969. Carbon-burning nucleosynthesis at constant temperature. *Astrophys. J.* 157:339–365.

Arnett, W. D., and Wefel, J. P. 1978. Aluminum-26 production from a stellar evolutionary sequence. *Astrophys. J. Lett.* 224:139–142.

Arnold, J. R. 1965a. The origin of meteorites as small bodies. II. The model. *Astrophys. J.* 141:1536–1547.

Arnold, J. R. 1965b. The origin of meteorites as small bodies. III. General considerations. *Astrophys. J.* 141:1548–1556.

Arnold, V. I. 1961. Small denominators and the problem of stability in classical and celestial mechanics. In *Report to the IVth All-Union Mathematical Congress*, Leningrad, pp. 85–191.

Arnould, M., and Nørgaard, H. 1975. The explosive thermonuclear formation of ^{7}Li and ^{11}B. *Astron. Astrophys.* 42:55–70.

Arnould, M., and N orgaard, H. 1978. Thermonuclear origin of Ne-E. *Astron. Astrophys.* 64:195–213.

Arnould, M., and N orgaard, H. 1981. Isotopic anomalies in meteorites and cosmic rays, and the heavy neon puzzle. *Comments Astrophys.* 9:145–154.

Arnould, M., N orgaard, H., Thielemann, F.-K., and Hillebrandt, W. 1980. Synthesis of ^{26}Al in explosive hydrogen burning. *Astrophys. J.* 237:931–950.

Arons, J. 1987. Accretion onto magnetized neutron stars: polar cap flow and centrifugally driven winds. In *The Origin and Evolution of Neutron Stars: Proc. IAU Symp. 125*, eds. D. J. Helfand and J.-H. Huang (Dordrecht: D. Reidel), pp. 207–225.

Arons, J., and Max, C. E. 1975. Hydromagnetic waves in molecular clouds. *Astrophys. J. Lett.* 196:77–82.

Arons, J., Burnard, D. J., Klein, R. I., McKee, C. F., Pudritz, R. E., and Lea, S. M. 1984. Accretion onto magnetized neutron stars: magnetospheric structure and stability. In *High Energy Transients in Astrophysics: AIP Conf. Proc. 115*, ed. S. E. Woosley (New York: American Inst. of Physics), pp. 215–234.

Arp, H. C. 1956. Novae in the Andromeda nebula. *Astron. J.* 61:15–34.

Arquilla, R. A., and Goldsmith, P. F. 1986. A detailed examination of the kinematics of rotating dark clouds. *Astrophys. J.* 303:356–374.

Arquilla, R., and Kwok, S. 1987. CO observations of IRAS Circular No. 9 sources 19520+2759 and 01133+6434: Regions of star formation. *Astron. Astrophys.* 173:271–278.

Arrhenius, G., De, B. R., and Alfvén, H. 1974. Origin of the ocean. In *The Sea*, ed. E. D. Goldberg (New York: Wiley), pp. 839–861.

Artymowicz, P. 1987. Self regulating protoplanet growth. *Icarus* 70:303–318.

Artymowicz, P. 1988. Radiation pressure forces on particles in the Beta Pictoris system. *Astrophys. J. Lett.* 335:79–82.

Artymowicz, P. 1990. Ph. D. Thesis, Warsaw Univ.

Artymowicz, P., Burrows, C., and Paresce, F. 1989. The structure of the Beta Pictoris circumstellar disk from combined IRAS and coronagraphic observations from combined IRAS and coronagraphic observations. *Astrophys. J.* 337:494–513.

Artymowicz, P., Paresce, F., and Burrows, C. 1990. The structure of the Beta Pictoris disk and the properties of its particles. *Adv. Space Res.* 10:81–84.

Artymowicz, P., Clarke, C. J., Lubow, S. H., and Pringle, J. E. 1991. The effect of an external disk on the orbital elements of a central binary. *Astrophys. J. Lett.* 370:35–38.

Ash, R. D., Wright, I. P., Grady, M. M., Pillinger, C. T., Lewis, R. S., and Anders, E. 1987. An investigation of carbon and nitrogen isotopes in Cδ and the effects of grain size upon combustion temperature. *Meteoritics* 22:319 (abstract).

Ash, R. D., Arden, J. W., Grady, M. M., Wright, I. P., and Pillinger, C. T. 1990. Recondite interstellar carbon components in the Allende meteorite revealed by preparative precombustion. *Geochim. Cosmochim. Acta* 54:455–468.

Aspin, C., and Walther, D. M. 1991. IR imaging and polarimetry of Mon R2 IRS. In *Astrophysics with Infrared Arrays*, ed. R. Elston (San Francisco: Astron. Soc. of the Pacific) pp. 288–290.

Assendorp, R., Wesselius, P. R., Whittet, D. C. B., and Prusti, T. 1990. A study of the Chamaeleon I dark cloud and T-association. II. High-resolution *IRAS* maps around HD 97048 and 97300. *Mon. Not. Roy. Astron. Soc.* 247:624–631.

Atreya, S. K., Donahue, T. M., and Kuhn, W. R. 1978. Evolution of a nitrogen atmosphere on Titan. *Science* 201:611–613.

Aumann, H. H. 1985. IRAS observations of matter around nearby stars. *Publ. Astron. Soc. Pacific* 97:885–891.

Aumann, H. H. 1988. Spectral class distribution of circumstellar material in main-sequence stars. *Astron. J.* 96:1415–1419.

Aumann, H. H. 1991. Circumstellar material in main-sequence stars. In *The Infrared Spectral Region of Stars*, eds. C. Jaschek and Y. Andrillat (Cambridge: Cambridge Univ. Press), pp. 363–379.

Aumann, H. H., and Good, J. C. 1990. IRAS constraints on a cold cloud around the solar system. *Astrophys. J.* 350:408–412.

Aumann, H. H., and Probst, R. G. 1991. Search for Vega-like nearby stars with 12 micron excess. *Astrophs. J.* 368:264–271.

Aumann, H. H., Gillett, F. C., Beichman, C. A., de Jong, T., Houck, J., Low, F. J., Neugebauer, G., Walker, R., and Wesselius, P. R. 1984. Discovery of a shell around Alpha Lyrae. *Astrophys. J. Lett.* 278:23–27.

Axon, D. J., and Taylor, K. 1984. Discovery of a family of Herbig-Haro objects in M42: Implications for the geometry of the high velocity molecular flow? *Mon. Not. Roy. Astron. Soc.* 207:241–261

Avedisova, V. S. 1974. Formation of structure in the interstellar gas behind a shock front. *Soviet Astron.* 18:283–288.

Avery, L. W. 1987. Radio and millimetre observations of larger molecules. In *Astrochemistry: Proc. IAU Symp. 120*, eds. M. S. Vardya and S. P. Tarafdar (Dordrecht: D. Reidel), pp. 187–197.

Azoulay, G., and Festou, M. C. 1986. The abundance of sulphur in comets. In *Asteroids, Comets, Meteors II*, eds. C. I. Lagerkvist, B. A. Lindblad, Lundstedt and H. Rickman (Uppsala: Reprocentralen HSC), pp. 273–277.

Baade, W., and Minkowski, R. 1937. The Trapezium cluster of the Orion Nuebula. *Astrophys. J.* 86:119–122.

Baas, F., Geballe, T. R., and Walther, D. M. 1986. Spectroscopy of the 3.4 μm emission feature in comet Halley. *Astrophys. J. Lett.* 311:97–101.

Baas, F., Grim, R. J. A., Geballe, T. R., Schutte, W., and Greenberg, J. M. 1988. The detection of solid methanol in W33A. In *Dust in the Universe*, eds. M. E. Bailey and D. A. Williams (Cambridge: Cambridge Univ. Press), pp. 55–60.

Bachiller, R., and Cernicharo, J. 1984. Molecular observations of B1: A dense globule in Perseus. *Astron. Astrophys.* 140:414–420.

Bachiller, R., and Cernicharo, J. 1990. Extremely high-velocity emission from molecular jets in NGC 6334I and NGC 1333 (HH 7–11). *Astron. Astrophys.* 239:276–286.

Bachiller, R., Cernicharo, J., Martín-Pintado, J., Tafalla, M., and Lazareff, B. 1990*a*. High-velocity molecular bullets in a fast bipolar outflow near L1448/IRS3. *Astron. Astrophys.* 231:174–186.

Bachiller, R., Menten, K. M., and del Río-Alvarez, S. 1990*b*. Anatomy of a dark cloud: A multimolecular study of Barnard 1. *Astron. Astrophys.* 236:461–471.

Backman, D. E., and Gillett, F. C. 1987. Exploiting the infrared: IRAS observations of the main sequence. In *Cool Stars, Stellar Systems and the Sun*, eds. J. Linsky and R. E. Stencel (Berlin: Springer-Verlag), pp. 340–350.

Backman, D. E., Gillett, F. C., and Low, F. J. 1986. IRAS observations of nearby main sequence stars and models of excess emission. *Adv. Space Res.* 6:43–46.

Backman, D. E., Stauffer, J. R., and Witteborn, F. C. 1990. Planetary accretion debris and IR excesses in open stellar clusters. Protostars and Planets III, March 5–9, Tucson, Arizona.

Backman, D. E., Gillett, F. C., and Witteborn, F. C. 1992. Infrared observations and thermal models of the β Pictoris disk. *Astrophys. J.* 385:670–679.

Badziag, P., Verwoerd, W. S., Ellis, W. P., and Greiner, N. R. 1990. Nanometre-sized diamonds are more stable than graphite. *Nature* 343:244–245.

Bahcall, J. N. 1984. Self-consistent determination of the total amount of matter near the Sun. *Astrophys. J.* 276:169–181.

Bahcall, J. N. 1986. Brown dwarfs: conference summary. In *Astrophysics of Brown Dwarfs*, eds. M. C. Kafatos, R. S. Harrington and S. P. Maran (Cambridge: Cambridge Univ. Press), pp. 233–237.

Baier, G., Bastian, U., Keller, E., Mundt, R., and Weigelt, G. 1985. Speckle interferometry of T Tauri stars and related objects. *Astron. Astrophys.* 153:278–280.

Bailey, M. E. 1983*a*. The structure and evolution of the Solar System comet cloud. *Mon. Not. Roy. Astron. Soc.* 204:347–368.

Bailey, M. E. 1983*b*. Comets, Planet X and the orbit of Neptune. *Nature* 302: 399–400.

Bailey, M. E. 1986*a*. The near-parabolic flux and the origin of short-period comets. *Nature* 324:350–352.

Bailey, M. E. 1986*b*. The mean energy transfer rate of comets in the Oort cloud and implications for cometary origins. *Mon. Not. Roy. Astron. Soc.* 218:1–30.

Bailey, M. E. 1987. The formation of comets in wind-driven shells around protostars. *Icarus* 69:70–82.

Bailey, M. E. 1990. Cometary masses. In *Baryonic Dark Matter*, eds. D. Lynden-Bell and G. Gilmore (Dordrecht: Kluwer), pp. 7–35.

Bailey, M. E., and Stagg, C. R. 1990. The origin of short-period comets. *Icarus* 86:2–8.

Bak, P., Tang, C., and Wiesenfeld, K. 1988. Self-organized criticality. *Phys. Rev. Ann.* 38:364–374.

Baker, V. R., and Partridge, J. B. 1986. Small Martian valleys: pristine and degraded morphology. *J. Geophys. Res.* 91:3561–3572.

Balbus, S. A. 1988. Local interstellar gasdynamical stability and substructure in spiral arms. *Astrophys. J.* 324:60–70.

Balbus, S. A. 1990. Large scale interstellar gas dynamics in disk galaxies. In *The Interstellar Medium in Galaxies*, eds. H. A. Thronson, Jr. and J. M. Shull (Dordrecht: Kluwer), pp. 305–321.

Balbus, S. A., and Cowie, L. L. 1985. On the gravitational stability of the interstellar medium in spiral arms. *Astrophys. J.* 297:61–75

Balbus, S. A., and Hawley, J. F. 1991. A powerful local shear instability in weakly magnetized disks. I. Linear analysis. *Astrophys. J.* 376:214–222.

Baldwin, J. A., Ferland, G. J., Martin, P. G., Corbin, M. R., Cota, S. A., Peterson, B. M., and Slettebak, A. 1991. Physical conditions in the Orion nebula and an assessment of its helium abundance. *Astrophys. J.* 374:580–609.

Balick, B., Gammon, R. H., and Hjellming, R. M. 1974. The structure of the Orion nebula. *Publ. Astron. Soc. Pacific* 86:616–634.

Balluch, M. 1991*a*. Structure and stability of steady protostellar accretion flows. I. The flow structure. *Astron. Astrophys.* 243:168–186.

Balluch, M. 1991*b*. Structure and stability of steady protostellar accretion flows. II. Linear stability analysis. *Astron. Astrophys.* 243:187–204.

Balluch, M. 1991*c*. Structure and stability of steady protostellar accretion flows. III. Non-linear instabilities. *Astron. Astrophys.* 243:205–218.

Bally, J. 1987. Massive bipolar outflows around young stars. *Irish Astron. J.* 17:270–279.

Bally, J. 1989. The structure and kinematics of star forming clouds. In *ESO Workshop on Low Mass Star Formation and Pre-Main Sequence Objects*, ed. B. Reipurth (Garching: European Southern Obs.), pp. 1–32.

Bally, J., and Lada, C. J. 1983. The high-velocity molecular flows near young stellar objects. *Astrophys. J.* 265:824–847.

Bally, J., and Lane, A. P. 1990. Shocked molecular hydrogen associated with Herbig-Haro objects and molecular outflows: the Cepheus-A flow. In *Astrophysics with*

Infrared Arrays, ed. R. Elston (San Francisco: Astron. Soc. of the Pacific), p. 273.

Bally, J., and Stark, A. A. 1983. Atomic hydrogen associated with the high-velocity flow in NGC 2071. *Astrophys. J. Lett.* 266:61–64.

Bally, J., Snell, R. J., and Predmore, R. 1983. Radio images of the bipolar H II region S 106. *Astrophys. J.* 272:154–162.

Bally, J., Langer, W. D., Stark, A. A., and Wilson, R. W. 1987. Filamentary structure in the Orion molecular cloud. *Astrophys. J. Lett.* 312:45–49.

Bally, J., Langer, W. D., Wilson, R. W., Stark, A. A., and Pound, M. W. 1992. *Astrophys. J.*, in press.

Balsiger, H., Altwegg, K., Bühler, F., Geiss, J., Ghielmetti, A. G., Goldstein, B. E., Goldstein, R., Huntress, W. T., Ip, W.-H., Lazarus, A. J., Meier, A., Neugebauer, M., Rettenmund, U., Rosenbauer, H., Schwenn, R., Sharp, R. D., Shelley, E. G., Ungstrup, E., and Young, D. T. 1986. The ion composition and dynamics at comet Halley. *Nature* 321:330–336.

Bania, T. M., and Lyon, J. G. 1980. OB Stars and the structure of the interstellar medium: Cloud formation and effects of different equations of state. *Astrophys. J.* 239:173–192.

Baran, G. P. 1982. Ph. D. Thesis, Columbia Univ.

Barber, D. 1985. Phyllosilicates and other layer-structured minerals in stony meteorites. *Clay Min.* 20:415–454.

Barge, P., and Pellat, R. 1990. Self-consistent velocity dispersions and mass spectra in the protoplanetary cloud. *Icarus* 85:481–498.

Barge, P., and Pellat, R. 1991. Mass spectrum and velocity dispersions during planetesimal accumulation. I. Accretion. *Icarus* 93:270–287.

Barker, D. M., and Mestel, L. 1990. Disk-like magneto-gravitational equilibria—II. *Mon. Not. Roy. Astron. Soc.* 245:147–153.

Barnard, E. E. 1919. On the dark markings of the sky with a catalogue of 182 objects. *Astrophys. J.* 49:1–23.

Barnard, E. E. 1927. In *A Photographic Atlas of Selected Regions of the Milky Way*, eds. E. B. Frost and M. R. Calvert (Washington: Carnegie Inst. of Washington).

Barnes, P. J., and Crutcher, R. M. 1990. Orion B (NGC 2024). II. Hat Creek synthesis observations of the molecular core in the $J = 1 - 0$ line of HCO^+. *Astrophys. J.* 351:176–188.

Barnes, P. J., Crutcher, R. M., Bieging, J. H., Storey, J. W. V., and Willner, S. P. 1989. Orion B (NGC 2024). I. VLA and IR observations of the H II region. *Astrophys. J.* 342:883–907.

Bar-Nun, A., and Kleinfeld, I. 1989. On the temperature and gas composition in the region of comet formation. *Icarus* 80:243–253.

Bar-Nun, A., Herman, G., Laufer, D., and Rappaport, M. L. 1985. Trapping and release of gases by water ice and implications for icy bodies. *Icarus* 63:317–332.

Bar-Nun, A., Dror, J., Kochavi, E., and Laufer, D. 1987. Amorphous water ice and its ability to trap gases. *Phys. Rev.* B35:2427–2435.

Bar-Nun, A., Kleinfeld, I., and Kochavi, E. 1988. Trapping of gas mixtures by amorphous water ice. *Phys. Rev.* B38:7749–7754.

Barrett, A. H., Meeks, M. L., and Weinreb, S. 1964. High resolution microwave spectra of H and OH absorption lines of Cassiopeia A. *Nature* 202:475–480.

Barsony, M. 1989. A close-up view of the S87 molecular outflow. *Astrophys. J.* 345:268–281.

Barsony, M., Burton, M. G., Russell, A. P. G., Carlstrom, J. E., and Garden, R. 1989a. Discovery of new 2 micron sources in ρ Ophiuchi. *Astrophys. J. Lett.* 346:93–96.

Barsony, M., Scoville, N. Z., Bally, J., and Claussen, M. J. 1989b. No Molecular gas disk in S106. *Astrophys. J.* 343:212–221.

Barsony, M., Schombert, J. M., and Kis-Halas, K. 1991. The LkHα101 infrared cluster. *Astrophys. J.* 379:221–231.

Bart, G., Ikramuddin, M., and Lipschutz, M. 1980. Thermal metamorphism of primitive meteorites. IX. On the mechanism of trace element loss from Allende heated up to 1400°C. *Geochim. Cosmochim. Acta* 44:719–730.

Barvainis, R., Clemens, D. P., and Leach, R. 1988. Polarimetry at 1.3 mm using Millipol: methods and preliminary results for Orion. *Astron. J.* 95:510–515.

Basaltic Volcanism Study Project. 1981. *Basaltic Volcanism on the Terrestrial Planets* (New York: Pergamon), esp. p. 649.

Bash, F. N., Green, E. M.,, and Peters, W. L., III. 1977. The galactic density wave, molecular clouds, and star formation. *Astrophys. J.* 217:464–472.

Basilevsky, A. T., Ivanov, B. A., Florensky, K. P., Yakolev, O. I., Fel'dman, V. I., Granovsky, L. V., and Sandovisky, M. A. 1983. *Impact Craters on the Moon and Planets* (Moscow: Nauka).

Basri, G. 1987. The T Tauri stars. In *Fifth Cambridge Workshop on Cool Stars, Stellar Systems, and the Sun*, eds. J. L. Linsky and R. E. Stencel (Berlin: Springer-Verlag), pp. 411–420.

Basri, G. 1990. Strong emission line profiles from T Tauri stars. In *High Resolution Spectroscopy in Astrophysics*. Special issue of *Mem. Soc. Astron. Italiana*, vol. 61, no. 3, ed. R. Pallavicini, pp. 707–721.

Basri, G., and Batalha, C. 1990. Hamilton echelle spectra of young stars. I. Optical veiling. *Astrophys. J.* 363:654–669.

Basri, G., and Bertout, C. M. 1989. Accretion disks around T Tauri stars. II. Balmer emission. *Astrophys. J.* 341:340–358.

Basri, G., and Marcy, G. 1991. Limits on the magnetic flux on a pre-main sequence star. In *The Sun and Cool Stars: Activity, Magnetism, Dynamos: Proc. IAU Coll. 130*, ed. I. Tuominen, D. Moss and G. Rüdiger (Berlin: Springer-Verlag), pp. 401–410.

Basri, G., Martin, E., and Bertout, C. 1991. The lithium resonance line in T Tauri stars. *Astron. Astrophys.* 252:625–638.

Bastian, U., and Mundt, R. 1985. FU Orionis star winds. *Astron. Astrophys.* 144:57–63.

Bastien, P. 1982. A linear polarization survey of T Tauri stars. *Astron. Astrophys. Suppl.* 48:153–164.

Bastien, P. 1983. Gravitational collapse and fragmentation of isothermal, nonrotating cylindrical clouds. *Astron. Astrophys.* 119:109–116.

Bastien, P. 1989. Polarization properties of T Tauri stars and other pre-main-sequence objects. In *Polarized Radiation of Circumstellar Origins*, eds. G. V. Coyne, A. F. J. Moffat, S. Tapia, A. M. Magalhães, R. E. Schulte-Ladbeck and D. T. Wickramasinghe (Vatican City: Vatican Press), pp. 541–582.

Bastien, P., and Menard, F. 1988. On the interpretation of polarization maps of young stellar objects. *Astrophys. J.* 326:334–338.

Bastien, P., and Ménard, F. 1990. Parameters of disks around young stellar objects from polarization observations. *Astrophys. J.* 364:232–241.

Bastien, P., Carmelle, R., and Nadeau, R. 1989. Circular polarization in T Tauri stars. II. New observations and evidence for multiple scattering. *Astrophys. J.* 339:1089–1092.

Bastien, P., Arcoragi, J.-P., Benz, W., Bonnell, I., and Martel, H. 1991. Fragmentation of elongated cylindrical clouds. I. Isothermal clouds. *Astrophys. J.* 378:255–265.

Batchelor, G. K. 1967. *The Theory of Homogeneous Turbulence* (Cambridge: Cambridge Univ. Press).

Bates, D. R., and Herbst, E. 1988a. Radiative association. In *Rate Coefficients in Astrochemistry*, eds. T. J. Millar and D. A. Williams (Dordrecht: Kluwer), pp.

17–40.

Bates, D. R., and Herbst, E. 1988*b*. Dissociative recombination: Polyatomic positive ion reactions with electrons and negative ions. In *Rate Coefficients in Astrochemistry*, eds. T. J. Millar and D. A. Williams (Dordrecht: Kluwer), pp. 41–48.

Batrla, W., and Menten, K. M. 1985. A rotating gas disk around L1551 IRS 5? *Astrophys. J. Lett.* 298:19–20.

Battaglia, A. 1987. Growth and Sedimentation of Dust Grains in the Primitive Solar Nebula. Ph. D. Thesis, City Univ. of New York.

Bayly, B. J., Orszag, S. A., and Herbert, T. 1988. Instability mechanisms in shear-F low transition. *Ann. Rev. Fluid Mech.* 20:359–391.

Bazan, G. 1991. The Galactic Evolution of s-Process Elements via AGB Stars. Ph. D. Thesis, Univ. of Illinois.

Bazell, D., and Desert, F. X. 1988. Fractal structure in interstellar cirrus. *Astrophys. J.* 333:353–358.

Beck, R. 1991. Magnetic fields in spiral galaxies. In *The Interpretation of Modern Synthesis Observations of Spiral Galaxies*, eds. N. Duric and P. C. Crane (San Francisco: Astron. Soc. of the Pacific).

Beck, R., Loiseau, N., Hummel, E., Berkhuijsen, E. M., Grave, R., and Wielebinski, R. 1989. High resolution polarization observations of M31: I. Structure of the magnetic field in the southwestern arm. *Astron. Astrophys.* 222:58–68.

Becklin, E. E., and Zuckerman, B. 1988. A low-temperature companion to a white dwarf star. *Nature* 336:656–658.

Becklin, E. E., and Zuckerman, B. 1990. Submillimeter emission from small dust grains orbiting nearby stars. In *Submillimeter Astronomy*, eds. G. D. Watt and A. S. Webster (Netherlands: Kluwer), pp. 147–153.

Beckwith, S. V. W., and Sargent, A. I. 1991. Particle emissivities in circumstellar disks. *Astrophys. J.* 381:250–258.

Beckwith, S. V. W., Zuckerman, B., Skrutskie, M. F., and Dyck, H. M. 1985. Discovery of solar system-sized halos around young stars. *Astrophys. J.* 287:793–800.

Beckwith, S. V. W., Sargent, A. I., Scoville, N. Z., Masson, C. R., Zuckerman, B., and Phillips, T. G. 1986. Small-scale structure of the circumstellar gas of HL Tauri and R Monocerotis. *Astrophys. J.* 309:755–761.

Beckwith, S. V. W., Sargent, A. I., Chini, R. S., and Güsten, R. 1990. A survey for circumstellar disks around young stars. *Astron. J.* 99:924–945.

Beech, M. 1987. Are Lynds dark clouds fractals? *Astrophys. Space Sci.* 133:193–195.

Beer, H., and Macklin, R. L. 1989. Measurement of the ^{85}Rb and ^{87}Rb capture cross sections for s-process studies. *Astrophys. J.* 339:962–977.

Beer, R., and Taylor, F. W. 1973. The abundance of CH_3D and the D/H ratio in Jupiter. *Astrophys. J.* 179:309–327.

Beetz, M., Elsässer, H., Poulakos, C., and Weinberger, R. 1976. Several H II regions in the near infrared. *Astron. Astrophys.* 50:41–46.

Begelman, M. C. 1990. Thermal phases of the interstellar medium in galaxies. In *The Interstellar Medium in Galaxies*, ed. H. A. Thronson, Jr., and J. M. Shull (Dordrecht: Kluwer), pp. 287–354.

Begemann, F. 1980. Isotopic anomalies in meteorites. *Rept. Prog. Phys.* 43: 1309–1356.

Beichman, C. A. 1987. The *IRAS* view of the galaxy and the solar system. *Ann. Rev. Astron. Astrophys.* 25:521–563.

Beichman, C. A., Becklin, E. E., and Wynn-Williams, C. G. 1979. New multiple systems in molecular clouds. *Astrophys. J. Lett.* 232:47–51.

Beichman, C. A., Neugebauer, G., Habing, H. J., Clegg, P. E., and Chester, T. J., eds. 1984. *IRAS Catalogs and Atlases, Explanatory Supplement* (Washington, D. C.: U. S. Government Printing Office).

Beichman, C. A., Myers, P. C., Emerson, J. P., Harris, S., Mathieu, R., Benson, P. J., and Jennings, R. E. 1986. Candidate solar-type protostars in nearby molecular clouds cores. *Astrophys. J.* 307:337–349.

Bel, N., and Leroy, B. 1989. Zeeman splitting in interstellar molecules. In *Galactic and Extragalactic Magnetic Fields: Proc. IAU Coll. 140*, eds. R. Beck, P. P. Kronberg and R. Wielebinski (Dordrecht: Kluwer), pp. 304.

Belcher, J. W., and MacGregor, K. B. 1976. Magnetic acceleration of winds from solar-type stars. *Astrophys. J.* 210:498–507.

Bell, K. R., Lin, D. N. C., and Ruden, S. 1991. Nonlinear evolution of accretion disks induced by radiative feedback processes. *Astrophys. J.* 372:633–645.

Bellingham, J. G., and Rossano, G. S. 1980. Long-period variations in R CrA, S CrA, T CrA, and R Mon. *Astron. J.* 85:555–559.

Benjamin, B., and Dinerstein, H. L. 1990. Near-infrared spectroscopy of classical novae in the coronal phase. *Astron. J.* 100:1588–1600.

Benjamin, T. M., Heuser, W. R., and Burnett, D. S. 1978. Laboratory studies of actinide partitioning relevant to Pu^{244} chronometry. *Proc. Lunar Planet. Sci. Conf.* 9: 1393–1406.

Benlow, A., and Meadows, A. J. 1977. The formation of the atmospheres of the terrestrial planets by impact. *Astrophys. Space Sci.* 46:293–300.

Benson, P. J., and Myers, P. C. 1989. A survey for dense cores in dark clouds. *Astrophys. J. Suppl.* 71:89–108.

Benz, W., and Cameron, A. G. W. 1990. Terrestrial effects of the giant impact. In *Origin of the Earth*, eds. H. E. Newsom and J. H. Jones (New York: Oxford Univ. Press), pp. 61–68.

Benz, W., Slattery, W. L., and Cameron, A. G. W. 1988. Collisional stripping of Mercury's mantle. *Icarus* 74:516–528.

Benz, W., Cameron, A. G. W., and Melosh, H. J. 1989. The origin of the Moon and the single impact hypothesis III. *Icarus* 81:113–131.

Bernatowicz, T. J., and Fahey, A. J. 1985. Xe isotopic fractionation in a cathodeless glow discharge. *Geochim. Cosmochim. Acta* 50:445–452.

Bernatowicz, T. J., and Podosek, F. A. 1986. Adsorption and isotopic fractionation of Xe. *Geochim. Cosmochim. Acta* 50:1503–1507.

Bernatowicz, T. J., Podosek, F. A., Honda, H., and Kramer, F. E. 1984. The atmospheric inventory of xenon and noble gases in shales: the plastic bag experiment. *J. Geophys. Res.* 89:4597–4611.

Bernatowicz, T. J., Kennedy, B. M., and Podosek, F. A. 1985. Xe in glacial ice and the atmospheric inventory of noble gases. *Geochim. Cosmochim. Acta* 49:2561–2564.

Bernatowicz, T., Gibbons, P., and Lewis, R. 1989. Meteoritic diamonds: Nature of the amorphous component. *Lunar Planet. Sci.* XX:65–66 (abstract).

Bernatowicz, T., Gibbons, P., and Lewis, R. S. 1990. Electron energy loss spectrometry of interstellar diamonds. *Astrophys. J.* 359:246–255.

Bernius, M. T., Hutcheon, I. D., and Wasserburg, G. J. 1991. Search for evidence of ^{26}Al in meteorites that are planetary differentiates. In *Lunar Planet. Sci.* XXII:93–94 (abstract).

Berrilli, F., Ceccarelli, C., Liseau, R., Lorenzetti, D., Saraceno, P., and Spinoglio, L. 1989. The evolutionary status of young stellar mass loss driving sources as derived from IRAS observations. *Mon. Not. Roy. Astron. Soc.* 237:1–15.

Berriman, G., and Reid, N. 1987. Observations of M dwarfs beyond $2.2\mu m$. *Mon. Not. Roy. Astron. Soc.* 227:315–329.

Berry, M. V. 1978. Regular and irregular motion. In *Topics in Nonlinear Dynamics*, eds. S. Jorna and E. C. Bullard (New York: American Inst. of Physics), pp. 16–146.

Bertiau, F. C. 1958. Absolute magnitudes of stars in the Scorpio-Centaurus association. *Astrophys. J.* 128:533–561.

Bertoldi, F. 1989. The photoevaporation of interstellar clouds. I. Radiation driven implosion. *Astrophys. J.* 346:735–755.

Bertoldi, F., and McKee, C. F. 1992. Pressure-confined clumps in magnetized molecular clouds. *Astrophys. J.* 395:140.

Bertoldi, F., McKee, C. F., and Klein, R. I. 1991. Induced star formation in photoevaporating molecular clouds. In preparation.

Bertout, C. 1986. Circumstellar matter of young low-mass stars: Observations versus theory. In *Circumstellar Matter: Proc. IAU Symp. 122*, eds. I. Appenzeller and C. Jordan (Dordrecht: D. Reidel), pp. 23–38.

Bertout, C. 1989. T Tauri stars: Wild as dust. *Ann. Rev. Astron. Astrophys.* 27:351–395.

Bertout, C., and Bouvier, J. 1989. T Tauri disk models. In *ESO Workshop on Low Mass Star Formation and Pre-Main Sequence Objects*, ed. B. Reipurth (Garching: European Southern Obs.), pp. 215–232.

Bertout, C., and Yorke, H. W. 1978. The spectral appearance of solar-type collapsing protostellar clouds. In *Protostars and Planets*, ed. T. Gehrels (Tucson: Univ. of Arizona Press), pp. 648–689.

Bertout, C., Basri, G., and Bouvier, J. 1988. Accretion disks around T Tauri stars. *Astrophys. J.* 330:350–373.

Bertout, C., Basri, G., and Cabrit, S. 1991a. The classical T Tauri stars—Future solar systems? In *The Sun in Time*, eds. C. P. Sonett, M. S. Giampapa and M. S. Matthews (Tucson: Univ. of Arizona Press), pp. 682–709.

Bertout, C., Collin-Souffrin, S., Lasota, J.-P., and Tran Thanh Van, J., eds. 1991b. *Structure and Emission Properties of Accretion Disks: Proc. IAU Coll. 129* (Gif-sur-Yvette: Editions Frontières).

Beust, H., Lagrange-Henri, A. M., Vidal-Madjar, A., and Ferlet, R. 1990. The Beta Pictoris circumstellar disk. X. Numerical simulations of infalling evaporating bodies. *Astron. Astrophys.* 236:202–216.

Beust, H., Vidal-Madjar, A., Lagrange-Henri, A. M., and Ferlet, R. 1991. The Beta Pictoris circumstellar disk. XI. New Ca II absorption features reproduced numerically. *Astron. Astrophys.* 241:488–492.

Bica, E., and Alloin, D. 1986a. A base of star clusters for stellar population synthesis. *Astron. Astrophys.* 162:21–31.

Bica, E., and Alloin, D. 1986b. A grid of star cluster properties for stellar population synthesis. *Astron. Astrophys. Suppl.* 66:171–179.

Bieging, J. H., and Cohen, M. 1985. Multifrequency radio images of L 1551 IRS 5. *Astrophys. J. Lett.* 289:5–8.

Bieging, J. H., Cohen, M., and Schwartz, P. R. 1984. VLA observations of T Tauri stars. II. A luminosity-limited survey of Taurus-Auriga. *Astrophys. J.* 282:699–708.

Biermann, L. 1978. Dense interstellar clouds In *Astronomical Papers Dedicated to Bengt Strömgren*, eds. A. Reiz and T. Anderson (Copenhagen: Copenhagen Observatory), pp. 327–337.

Biermann, L. and Lüst, R. 1978. *Sitz. ber. Bayer. Akad. Wiss. Mat.-Naturw. Kl.*

Biermann, L., and Michel, K. W. 1978. On the origin of cometary nuclei in the pre-solar nebula. *Moon and Planets* 18:447–464.

Bigeleisen, J. 1961. Statistical mechanics of isotope effects on the thermodynamic properties of condensed systems. *J. Chem. Phys.* 34:1485–1493. Erratum 35:2246.

Biglari, H., and Diamond, P. H. 1988. Cascade and intermittency model for turbulent compressible self-gravitating matter and self-binding phase space density fluctuations. *Phys. Rev. Lett.* 61:1716–1719.

Bilo, E. H., and van de Hulst, H. C. 1960. Methods for computing the original orbits of comets. *Bull. Astron. Inst. Netherlands* 15:119–127.

Binney, J., and Tremaine, S. D. 1987. *Galactic Dynamics* (Princeton: Princeton Univ. Press).

Binzel, R. P. 1989. Pluto-Charon mutual events. *Geophys. Res. Lett.* 16:1205–1208.

Biraud, F., Bourgois, G., Crovisier, J., Fillit, R., Gerard, E., and Kazes, I. 1974. OH observations of comet Kohoutek (1973f) at 18 cm wavelength. *Astron. Astrophys.* 34:163–166.

Birck, J.-L., and Allègre, C. J. 1985. Evidence for the presence of ^{53}Mn in the early solar system. *Geophys. Res. Lett.* 12:745–748.

Birck, J.-L., and Allègre, C. J. 1987. ^{53}Cr isotopic anomalies related to ^{53}Mn decay in old solar system matter. *Meteoritics* 22:325–326 (abstract).

Birck, J.-L., and Allègre, C. J. 1988. Manganese-chromium isotope systematics and the development of the early solar system. *Nature* 331:579–584.

Birck, J.-L., and Lugmair, G. W. 1988. Nickel and chromium isotopes in Allende inclusions. *Earth Planet. Sci. Lett.* 90:131–143.

Birck, J.-L., Rotaru, M., and Allègre, C. J. 1990. ^{53}Mn in carbonaceous chondrites. *Meteoritics* 25:349–350 (abstract).

Bischoff, A., and Palme, H. 1987. Composition and mineralogy of refractory-metal-rich assemblages from a Ca, Al-rich inclusion in the Allende meteorite. *Geochim. Cosmochim. Acta* 51:2733–2748.

Bishop, E. V., and DeMarcus, W. C. 1970. Thermal histories of Jupiter models. *Icarus* 12:317–330.

Bisnovatyi Kogan, G., and Ruzmaikin, A. A. 1976. The accretion of matter by a collapsing star in the presence of a magnetic field. II. Self-consistent stationary picture. *Astrophys. Space Sci.* 42:401–424.

Blaauw, A. 1952. The age and evolution of the ζ Persei group of O- and B-type stars. *Bull. Astron. Inst. Netherlands* 11:405–413.

Blaauw, A. 1956. On the luminosities, motions, and space distribution of the nearer northern O-B5 stars. *Astrophys. J.* 123:408–439.

Blaauw, A. 1964. The O associations in the solar neighborhood. *Ann. Rev. Astron. Astrophys.* 2:213–247.

Blaauw, A. 1984. Some remarks on the OB associations in our neighborhood. *Irish Astron. J.* 16:141–147.

Blaauw, A. 1985. Star formation in the Orion arm. In *The Milky Way Galaxy*, eds. H. van Woerden, R. J. Allen and W. B. Burton (Dordrecht: D. Reidel), pp. 335–392.

Blaauw, A. 1991. OB associations and the fossil record of star formation. In *The Physics of Star Formation and Early Stellar Evolution*, eds. C. J. Lada and N. D. Kylafis (Dordrecht: Kluwer), pp. 125–154.

Black, D. C. 1991. Worlds around other stars. *Sci. American* 264:76–82.

Black, D. C., and Matthews, M. S., eds. 1985. *Protostars and Planets II* (Tucson: Univ. of Arizona Press).

Black, D. C., and Scott, E. H. 1982. A numerical study of the effects of ambipolar diffusion on the collapse of magnetic gas clouds. *Astrophys. J.* 263:696–715.

Black, J. H., and Smith, P. L. 1984. Interstellar O_2. I. Abundance, excitation and prospects for detection of $^{16}O^{18}O$ at radio frequencies. *Astrophys. J.* 277:562–568.

Black, J. H., van Dishoeck, E. F., Willner, S. P., and Woods, R. C. 1990. Interstellar absorption lines toward NGC 2264 and AFGL 2591: Abundances of H_2, H_3^+, and CO. *Astrophys. J.* 358:459–467.

Blaes, O. M., and Hawley, J. F. 1988. Nonaxisymmetric disk instabilities: A linear and nonlinear synthesis. *Astrophys. J.* 326:277–291.

Blake, D. F., Freund, F., Krishnan, K. F. M., Echer, C. J., Shipp, R., Bunch, T. E., Tielens, A. G., Lipari, R. J., Hetherington, C. J. D., and Chang, S. 1988. The nature and origin of interstellar diamond. *Nature* 332:611–613.

Blake, G. A., Sutton, E. C., Masson, C. R., and Phillips, T. G. 1986. The rotational emission line spectrum of Orion A between 247 and 263 GHz. *Astrophys. J. Suppl.* 60:357–374.

Blake, G. A., Sutton, E. C., Masson, C. R., and Phillips, T. G. 1987. Molecular abundances in OMC-1: The chemical composition of interstellar molecular clouds and the influence of massive star formation. *Astrophys. J.* 315:621–645.

Blake, G. A., van Dishoeck, E. F., and Sargent, A. I. 1992. Chemistry in circumstellar disks: CS towards HL Tauri. *Astrophys. J. Lett.* 391:99–103.

Blake, G. A., Groesbeck, T., van Dishoeck, E. F., and Jansen, D., and Mundy, L. G. 1993. Millimeter and spectral line surveys of *IRAS* 16293–2422. *Astrophys. J.*, in preparation.

Blandford, R. D. 1976. Accretion disk electrodynamics—A model for double radio sources. *Mon. Not. Roy. Astron. Soc.* 176:465–481.

Blandford, R. D., and Payne, D. G. 1982. Hydromagnetic flows from accretion disks and the production of radio jets. *Mon. Not. Roy. Astron. Soc.* 199:883–903.

Blitz, L. 1978. A Study of the Molecular Complexes Accompanying Mon OB1, Mon OB2, and CMa OB1. Ph. D. Thesis, Columbia Univ.

Blitz, L. 1980. Star forming molecular clouds towards the galactic anticentre. In *Giant Molecular Clouds in the Galaxy*, eds. P. M. Soloman and M. G. Edmunds (Oxford: Pergamon), pp. 211–229.

Blitz, L. 1987a. Large molecular-cloud complexes. In *Millimetre and Submillimetre Astronomy*, eds. R. D. Wolstencroft and W. B. Burton (Dordrecht: Kluwer), pp. 269–280.

Blitz, L. 1987b. The structure of molecular clouds. In *Physical Processes in Interstellar Clouds*, eds. G. E. Morfill and M. Scholer (Dordrecht: D. Reidel), pp. 35–58.

Blitz, L. 1990. The evolution of galactic giant molecular clouds. In *The Evolution of the Interstellar Medium*, ed. L. Blitz (San Francisco: Astron. Soc. of the Pacific), pp. 273–289.

Blitz, L., and Shu, F. H. 1980. The origin and lifetime of molecular cloud complexes. *Astrophys. J.* 238:148–157.

Blitz, L., and Stark, A. A. 1986. Detection of clump and interclump gas in the Rosette molecular cloud complex. *Astrophys. J. Lett.* 300:89–93.

Blitz, L., and Thaddeus, P. 1980. Giant molecular complexes and OB associations. I. The Rosette molecular complex. *Astrophys. J.* 241:676–696.

Blitz, L., Bazell, D., and Desert, F. X. 1990. Molecular clouds without detectable CO. *Astrophys. J. Lett.* 352:13–16.

Bloemen, J. B. G. M., Caraveo, P. A., Herson, W., Lebrun, F., Madoalena, R. J., Strong, A. W., and Thaddeus, P. 1984. Gamma rays from atomic and molecular gas in the large complex of clouds in Orion and Monoceros. *Astron. Astrophys.* 139:37–42.

Bloemen, J. B. G. M., Strong, A. W., Blitz, L., Cohen, R. S., Dame, T. M., Grabelsky, D. A., Hersen, W., Lebrun, F., Mayer-Hasselwander, H.A., and Thaddeus, P. 1986. The radial distribution of galactic gamma rays. III. The distribution of cosmic rays in the galaxy and the CO-H_2 calibration. *Astron. Astrophys.* 154:25–41.

Bloemen, J. B. G. M. 1987. On stable hydrostatic equilibrium configurations of the galaxy and implications for its halo. *Astrophys. J.* 322:694–705.

Blondin, J. M., Königl, A., and Fryxell, B. A. 1989. Herbig-Haro objects as the heads of radiative jets. *Astrophys. J. Lett.* 337:37–40

Blum, J. 1989. Coagulation of protoplanetary dust. *Bull. Amer. Astron. Soc.* 22:1082 (abstract).

Bockelée-Morvan, D. 1987. A model for the excitation of water in comets. *Astron. Astrophys.* 181:169–181.

Bockelée-Morvan, D., and Crovisier, J. 1985. Possible parents for the cometary CN radical: photochemistry and excitation conditions. *Astron. Astrophys.* 151:90–100.

Bockelée-Morvan, D., and Crovisier, J. 1987. The 2.7 μm water band of comet Halley: interpretation of observations by an excitation model. *Astron. Astrophys.* 187:425–430.

Bockelée-Morvan, D., and Crovisier, J. 1990. The ortho-para ratio of water in comets: can reliable values be derived from the observations of the 2.7μm band by the KAO? In *Asteroids, Comets, Meteors 3*, eds. C. I. Lagerkvist, H. Rickman, B. A. Lindblad and M. Lindgren (Uppsala: Reprocentrelen-HSC), pp. 263–265.

Bockelée-Morvan, D., and Crovisier, J. 1992. Formaldehyde in comets: excitation of the rotational lines. *Astron. Astrophys.*, submitted.

Bockelée-Morvan, D., Crovisier, J., Despois, D., Forville, T., Gerard, E., Schraml, J., and Thum, C. 1987. Molecular observations of comets P/Giacobini-Zinner 1984e and P/Halley 1982i at millimetre wavelengths. *Astron. Astrophys.* 180:253–262.

Bockelée-Morvan, D., Crovisier, J., Colom, P., Despois, D., and Paubert, G. 1990. Observations of parent molecules in comets P/Borsen-Metcalf (1989o), Austin (1989cl) and Levy (1990c) at millimetre wavelengths: HCN, H_2S, H_2CO and CH_3OH. In *Formation of Stars and Planets, and the Evolution of the Solar System*, ed. B. Battrick, ESA SP-315, pp. 143–148.

Bockelée-Morvan, D., Colom, P., Crovisier, J., Despois, D., and Paubert, G. 1991. Microwave detection of hydrogen sulphide and methanol in comet Austin (1989c1). *Nature* 350:318–320.

Bode, M. F. 1988. Observations and modelling of circumstellar dust. In *Dust in the Universe*, eds. M. E. Bailey and D. A. Williams (Cambridge: Cambridge Univ. Press), pp. 73–102.

Bode, M. F. 1989. Infrared spectroscopy of novae and related objects. In *Infrared Spectroscopy in Astronomy*, ed. B. H. Kaldeich, ESA SP-290, pp. 317–327.

Bode, M. F., and Evans, A. 1989. Infrared observations of novae. In *Classical Novae*, eds. M. F. Bode and A. Evans (London: Wiley), pp. 163–186.

Bodenheimer, P. 1965. Depletion of deuterium and beryllium during pre-main-sequence evolution. *Astrophys. J.* 144:701–722.

Bodenheimer, P. 1966. Studies in stellar evolution. IV. The influence of initial conditions on pre-main-sequence calculations. *Astrophys. J.* 144:709–722.

Bodenheimer, P. 1978. Evolution of rotating interstellar clouds. III. On the formation of multiple star systems. *Astrophys. J.* 224:488–496.

Bodenheimer, P. 1985. Evolution of the giant planets. In *Protostars and Planets II*, eds. D. C. Black and M. S. Matthews (Tucson: Univ. of Arizona Press), pp. 873–894.

Bodenheimer, P., and Boss, A. P. 1981. Fragmentation in a rotating protostar: a re-examination of comparison calculations. *Mon. Not. Roy. Astron. Soc.* 197:477–485.

Bodenheimer, P., and Ostriker, J. P. 1970. Rapidly rotating stars. VI. Pre-main-sequence evolution of massive stars. *Astrophys. J.* 161:1101–1113.

Bodenheimer, P., and Pollack, J. B. 1986. Calculations of the accretion and evolution of giant planets: The effects of solid cores. *Icarus* 67:391–408.

Bodenheimer, P., and Sweigert, A. 1968. The dynamical collapse of the isothermal sphere. *Astrophys. J.* 152:515–522.

Bodenheimer, P., Tohline, J. E., and Black, D. C. 1980. Criteria for fragmentation in a collapsing rotating cloud. *Astrophys. J.* 242:209–218.

Bodenheimer, P., Yorke, H. W., Różyczka, M., and Tohline, J. E. 1988. Collapse of a rotating protostellar cloud. In *Formation and Evolution of Low-Mass Stars*, eds. A. K. Dupree and M. T. V. T. Lago (Dordrecht: Kluwer), pp. 139–151.

Bodenheimer, P., Yorke, H. W., Różyczka, M., and Tohline, J. E. 1990. The formation phase of the solar nebula. *Astrophys. J.* 355:651–660.

Boesgaard, A., and Tripico, M. 1986. Lithium in the Hyades Cluster. *Astrophys. J. Lett.* 302:49–53.

Bogard, D. D. 1987. On the origin of Venus' atmosphere: possible contribution from single component mixtures and fractionated solar wind. *Icarus* 74:3–20.

Bogard, D. D., Nyquist, L. E., Bansal, B. M., Wiesmann, H., and Shih, C. Y. 1975. 76535: An old lunar rock. *Earth Planet. Sci. Lett.* 26:69–80.

Bogard, D. D., Nyquist, L. E., and Johnson, P. 1984. Noble gas contents of shergottites and implications for the Martian origin of SNC meteorites. *Geochim. Cosmochim. Acta* 48:1723–1739.

Boggess, A., Bruhweiler, F. C., Grady, C. A., Ebbets, D. C., Kondo, Y., Trafton, L. M., Brandt, J. C., and Heap, S. R. 1991. First results from the Goddard high-resolution spectrograph: resolved velocity and density structures in the Beta Pictoris circumstellar disk. *Astrophys. J. Lett.* 377:49–52.

Bohlin, R. C., Savage, B. D., and Drake, J. F. 1978. A survey of interstellar HI from $L\alpha$ absorption measurements. II. *Astrophys. J.* 224:132-142.

Böhm, K.-H. 1956. A spectrophotometric analysis of the brightest Herbig-Haro object. *Astrophys. J.* 123:379–391

Böhm, K.-H. 1983. Optical and UV observations of Herbig-Haro objects. *Rev. Mexicana Astron. Astrof.* 7:55–70.

Böhm, K.-H., Bührke, T., Raga, A. C., Brugel, E. W., Witt, A. N., and Mundt, R. 1987. Ultraviolet spectra of HH 1 and HH 2: Spatial variations and the continuum problem. *Astrophys. J.* 316:349–359.

Boice, D. C., Huebner, W. F., Sablik, M. J., and Konno, I. 1990. Distributed coma sources and the CH_4/CO ratio in comet Halley. *Geophys. Res. Lett.* 17:1813–1816.

Boland, W., and de Jong, T. 1982. Carbon depletion in turbulent molecular cloud cores. *Astrophys. J.* 261:110-114.

Bonazzola, S., Falgarone, E., Hayvaerts, J. Perault, M., and Puget, J.L. 1987. Jeans collapse in a turbulent medium. *Astron. Astrophys.* 172:293–298.

Bonnell, I., and Bastien, P. 1991. The collapse of cylindrical isothermal and polytropic clouds with rotation. *Astrophys. J.* 374:610–622.

Bonnell, I., Martel, H., Bastien, P., Arcoragi, J. P., and Benz, W. 1991. Fragmentation of elongated cylindrical clouds. III. Formation of binary and multiple systems. *Astrophys. J.* 377:553–558.

Bonnor, W. B. 1956. Boyle's law and gravitational instability. *Mon. Not. Roy. Astron. Soc.* 116:351–359.

Borderies, N., and Goldreich, P. 1984. A simple derivation of capture probabilities for the J+1:J and J+2:J orbit-orbit resonance problems. *Celest. Mech.* 32:127–136.

Borderies, N., Goldreich, P., and Tremaine, S. 1982. Sharp edges of planetary rings. *Nature* 299:209–211.

Borderies, N., Goldreich, P., and Tremaine, S. 1983. Perturbed particle disks. *Icarus* 55:124–132.

Borderies, N., Goldreich, P., and Tremaine, S. 1984a. Excitation of inclinations in ring-satellite systems. *Astron. J.* 284:429–434.

Borderies, N., Goldreich, P., and Tremaine, S. 1984b. Unsolved problems in planetary ring dynamics. In *Planetary Rings*, eds. R. Greenberg and A. Brahic (Tucson: Univ. of Arizona Press), pp. 713–734.

Borderies, N., Goldreich, P., and Tremaine, S. 1986. Nonlinear density waves in planetary rings. *Icarus* 68:522–533.

Boreiko, R. T., and Betz, A. L. 1989. Heterodyne spectroscopy of the J = 22 − 21 CO line in Orion. *Astrophys. J. Lett.* 346:97–100.

Borra, E. F., Landstreet, J. D., and Mestel, L. 1982. Magnetic stars. *Ann. Rev. Astron. Astrophys.* 20:191–220.

Boss, A. P. 1980. Protostellar formation in rotating interstellar clouds. III. Nonaxisymmetric collapse. *Astrophys. J.* 237:866–876.

Boss, A. P. 1981. Collapse and fragmentation of rotating adiabatic clouds. *Astrophys. J.* 250:636–644.

Boss, A. P. 1982. Hydrodynamical models of presolar nebula formation. *Icarus* 51:623–632.

Boss, A. P. 1984*a*. Protostellar formation in rotating interstellar clouds. IV. Non-isothermal collapse. *Astrophys. J.* 277:768–782.

Boss, A. P. 1984*b*. Angular momentum transfer by gravitational torques and the evolution of binary protostars. *Mon. Not. Roy. Astron. Soc.* 209:543–567.

Boss, A. P. 1985. Three dimensional calculations of the formation of the presolar nebula from a slowly rotating cloud. *Icarus* 61:3–9.

Boss, A. P. 1986*a*. Protostellar formation in rotating interstellar clouds. V. Nonisothermal collapse and fragmentation. *Astrophys. J. Suppl.* 62:519–552.

Boss, A. P. 1986*b*. Theoretical determination of the minimum protostellar mass. In *Astrophysics of Brown Dwarfs*, eds. M. C. Kafatos, R. S. Harrington and S. P. Maran (Cambridge: Cambridge Univ. Press), pp. 206–211.

Boss, A. P. 1986*c*. The origin of the Moon. *Science* 231:341–345.

Boss, A. P. 1987. Protostellar formation in rotating interstellar clouds. VI. Nonuniform initial conditions. *Astrophys. J.* 319:149–161.

Boss, A. P. 1988. Binary stars: formation by fragmentation. *Comment. Astrophys.* 12:169–190.

Boss, A. P. 1989. Cloud collapse and fragmentation. In *Highlights of Astronomy*, vol. 8, ed. D. McNally (Dordrecht: Kluwer), pp. 123–126.

Boss, A. P. 1989*a*. Surface density and temperature profiles in the early solar nebula. *Lunar Planet. Sci.* V:99–100 (abstract).

Boss, A. P. 1989*b*. Evolution of the solar nebula. I. Nonaxisymmetric structure during nebula formation. *Astrophys. J.* 345:554–571.

Boss, A. P. 1990. Fragmentation of collapsing molecular cloud cores and formation of unequal mass binaries. Paper presented at Protostars and Planets III, Tucson, Ariz., 5–6 March.

Boss, A. P., and Black, D. C. 1982. Collapse of accreting, rotating, isothermal clouds. *Astrophys. J.* 258:270–279.

Boss, A. P., and Yorke, H. W. 1990. Spectral and isophotal appearance of three dimensional protostellar models. *Astrophys. J.* 353:236–244.

Boss, A. P., Morfill, G. E., and Tscharnuter, W. M. 1989. Models of the formation and evolution of the solar nebula. In *Origin and Evolution of Planetary and Satellite Atmospheres*, eds. S. K. Atreya, J. B. Pollack and M. S. Matthews (Tucson: Univ. of Arizona Press), pp. 35–77.

Boss, A. P., Cameron, A. G. W., and Benz, W. 1991. Tidal disruption of inviscid protoplanets. *Icarus* 92:165–178.

Bossi, M., Gaspani, A., Scardia, M., and Tadini, M. 1989. θ^1 Orionis A: a pre-main sequence low Q binary system? *Astron. Astrophys.* 222:117–120.

Bottinelli, L., and Gouguenheim, L. 1964. *Ann. Astrophys.* 27:685

Boulanger, F., and Perault, M. 1988. Diffuse infrared emission from the galaxy. I. Solar neighborhood. *Astrophys. J.* 330:964–985.

Bouvier, J. 1990. Rotation in T Tauri stars. II. Clues for magnetic activity. *Astron. J.* 99:946–964.

Bouvier, J. 1991. Rotation in pre-main-sequence stars: Properties and evolution. In *Angular Momentum Evolution of Young Stars*, eds. S. Catalano and J. R. Stauffer (Dordrecht: Kluwer), pp. 41–62.

Bouvier, J., and Appenzeller, I. 1991. A magnitude-limited spectroscopic and photometric survey of ρ Ophiuchi X-ray sources. *Astron. Astrophys. Suppl.* 92:481–516.

Bouvier, J., and Bertout, C. 1989. Spots on T Tauri stars. *Astron. Astrophys.* 211:99–114.

Bouvier, J., and Bertout, C. 1991. Accretion Disks Around T Tauri Stars. III. A χ^2 Analysis of Disk Parameters. Preprint.

Bouvier, J., Bertout, C., Benz, W., and Mayor, M. 1986a. Rotation in T Tauri stars. I. Observations and immediate analysis. *Astron. Astrophys.* 165:110–119.

Bouvier, J., Bertout, C., and Bouchet, P. 1986b. DN Tauri: A spotted T Tauri star. *Astron. Astrophys.* 158:149–157.

Boynton, W. V. 1975. Fractionation in the solar nebula: condensation of yttrium and the rare earth elements. *Geochim. Cosmochim. Acta* 39:569–584.

Boynton, W. V. 1978a. Fractionation in the solar nebula. II. Condensation of Th, U, Pu, and Cm. *Earth Planet. Sci. Lett.* 40:63–70.

Boynton, W. V. 1978b. Rare-earth elements as indicators of supernova condensation. *Lunar Planet. Sci.* IX:120–122(abstract).

Boynton, W. V. 1988. Nebular processes associated with CAI rim formation. *Meteoritics* 23:259.

Boynton, W. V., and Wark, D. A. 1987. Origin of CAI rims—I: the evidence from the rare earth elements. *Lunar Planet. Sci.* XVIII:117–118 (abstract).

Boynton, W. V., Frazier, R. M., and Macdougall, J. D. 1980. Identification of an ultra-refractory component in the Murchison meteorite. *Lunar and Planet. Sci.* XI:103–105.

Brand, P. W. J. L., and Zealey, W. J. 1975. Cloud structure in the galactic plane. A cosmic bubble bath? *Astron. Astrophys.* 38:363–371.

Brand, P. W. J. L., Toner, M. P., Geballe, T. R., and Webster, A. S. 1989. The velocity profile of the $1 - 0$ S(1) line of molecular hydrogen at peak 1 in Orion. *Mon. Not. Roy. Astron. Soc.* 237:1009–1018

Brandeis, G., and Marsh, B. D. 1989. The convective liquidus in solidifying magma chamber: a fluid dynamic investigation. *Nature* 339:613–616.

Breger, M. 1987. The Pleiades cluster. IV. The visit of a molecular CO cloud. *Astrophys. J.* 319:754–755.

Bregman, J. N. 1980. The galactic fountain of high velocity clouds. *Astrophys. J.* 36:577–579.

Bretagnon, P. 1974. Termes à longues périodes dans le systme solaire. *Astron. Astrophys.* 30:141–154.

Breukers, R. 1991. Thermal and Chemical Processes in the Evolution of Interstellar Dust and Gas. Ph. D. Thesis, University of Leiden.

Brin, G. D., and Mendis, D. A. 1979. Dust release and mantle development in comets. *Astrophys. J.* 229:402–408.

Brinks, E., and Bajaja, E. 1986. A high resolution hydrogen line survey of Messier 31. *Astron. Astrophys.* 169:14–42.

Broadfoot, A. L., Sandel, B. R., Shemansky, D. E., Holberg, J. B., Smith, G. R., Strobel, D. F., McConnell, J. C., Kumar, S., Hunten, D. M., Atreya, S. K., Donahue, T. M., Moos, H. W., Bertaux, J. L., Blamont, J. E., Pomphrey, R. B., and Linick, S. 1981. Extreme ultraviolet observations from Voyager 1 encounter with Saturn. *Science* 212:207–211.

Broadfoot, A. L., Atreya, S. K., Bertaux, J. L., Blamont, J. E., Dessler, A. J., Donahue, T. M., Forrester, W. T., Hall, D. T., Herbert, F., Holberg, J. B., Hunten, D. M.,

Krasnopolsky, V. A., Linick, S., Lunine, J. I., McConnell, J. C., Moos, H. W., Sandel, B. R., Schneider, N. M., Shemansky, D. E., Smith, G. R., Strobel, D. F., and Yelle, R. V. 1989. Ultraviolet spectrometer observations of Neptune and Triton. *Science* 246:1459–1466.

Brooke, T. Y., Knacke, R. F., Owen, T. C., and Tokunaga, A. T. 1989. Spectroscopy of emission features near 3 m in comet Wilson (1986l). *Astrophys. J.* 336:971–978.

Brooke, T. Y., Tokunaga, A. T., and Knacke, R. F. 1991a. Detection of the 3.4μm emission feature in comets P/Brorsen-Metcalf and Okazaki-Levy-Rudenko (1989r) and an observational summary. *Astron. J.* 101:268–278.

Brooke, T. Y., Tokunaga, A. T., Weaver, H. A., Chin, G., and Geballe, T. R. 1991b. A sensitive upper limit on the methane abundance in comet Levy (1990c). *Astrophys. J. Lett.* 372:113–116.

Brouwer, D., and Clemence, G. M. 1961. *Methods of Celestial Mechanics* (New York: Academic Press).

Brouwer, D., and van Woerkom, A. J. J. 1950. The secular variations of the orbital elements of the principal planets. *Astron. Papers Amercan Ephem. and Nautical Almanac*, vol. 13, pt. 2 (Washington: Nautical Almanac Office, U. S. Govt. Printing Office).

Brown, A. 1987. Radio emission from pre-main-sequence stars in corona Australis. *Astrophys. J. Lett.* 322:31–34

Brown, A., Mundt, R., and Drake, S. A. 1985. Radio continuum emission from pre-main sequence stars and associated structures. In *Radio Stars*, eds. R. M. Hjellming and D. M. Gibson (Dordrecht: D. Reidel), pp. 105–110.

Brown, A., Walter, F., Carpenter, K., Jordan, C., and Judge, P. 1985. Emission line variability of RY Tau, DR Tau, and SU Aur. *Bull. Amer. Astron. Soc.* 17:556 (abstract).

Brown, G. N., Jr., and Ziegler, W. T. 1979. Vapor pressure and heats of sublimation of liquids and solids of interest in cryogenics below 1-atm. pressure. In *Advances in Cryogenic Engineering*, eds. K. Timmerhaus and H. A. Snyder (New York: Plenum Press), pp. 662–670.

Brown, P. D., and Charnley, S. B. 1990. Chemical models of interstellar gas-grain processes—I. Modelling and the effect of accretion on gas abundances and mantle composition in dense clouds. *Mon. Not. Roy. Astron. Soc.* 244:432–443.

Brown, P. D. and Millar, T. J. 1989a. Models of the gas-grain interaction—deuterium chemistry. *Mon. Not. Roy. Astron. Soc.* 237:661–671.

Brown, P. D. and Millar, T. J. 1989b. Grain-surface formation of multi-deuterated molecules. *Mon. Not. Roy. Astron. Soc.* 240:25P-29P.

Brown, P. D., Charnley, S. B., and Millar, T. J. 1988. A model of the chemistry in hot molecular cores. *Mon. Not. Roy. Astron. Soc.* 231:409–417.

Brownlee, D. E., and Kissel, J. 1990. The composition of dust particles in the environment of comet Halley. In *Comet Halley 1986: World-Wide Investigations, Results, and Interpretations* (West Sussex, England: Ellis Horwood Ltd.), pp. 89–98.

Brugel, E. W. 1989. Ultraviolet spectroscopic observations of Herbig-Haro objects. In *ESO Workshop on Low Mass Star Formation and Pre-main Sequence Objects*, ed. B. Reipurth (Garching: European Southern Obs.), pp. 311–332.

Brugel, E. W., Böhm, K.-H., and Mannery, E. 1981. The blue continua of Herbig-Haro objects. *Astrophys. J.* 243:874–882.

Bruhweiler, F. C., and Kondo, Y. 1982. The detection of interstellar CI in the immediate vicinity of the Sun. *Astrophys. J. Lett.* 260:91–94.

Budzien, S. A., and Feldman, P. D. 1992. Upper limits to the S_2 abundance in several comets observed with the International Ultraviolet Explorer. *Icarus*, in press.

Budzien, S. A., Feldman, P. D., Roettger, E. E., A'Hearn, M. F., and Festou, M. C.

1990. IUE observations of comet Austin (1989c1). In *Workshop on Observations of Recent Comets*, Albuquerque N. M., June 15–16, eds. W. F. Huebner, P. A. Wehinger, J. Rahe and I. Konno, pp. 64–68.

Bührke, T., Mundt, R., and Ray, T. P. 1988. A detailed study of HH 34 and its associated jet. *Astron. Astrophys.* 200:99–119.

Bujarrabal, V., Guélin, M., Morris, M., and Thaddeus, P. 1981. The abundance and excitation of carbon chains in interstellar molecular clouds. *Astron. Astrophys.* 99:239–247.

Bunch, T. E., and Chang, S. 1980*a*. An alternative origin for Allende CAI inclusions rims, or a correlation between the early solar system and a British steel furnace. *Meteoritics* 15:270.

Bunch, T. E., and Chang, S. 1980*b*. Carbonaceous chondrite phyllosilicates and light element geochemistry as indicators of parent body processes and surface conditions. *Geochim. Cosmochim. Acta* 44:1543–1577.

Burdyuzha, V. V., and Varshalovich, D. A. 1972. Spin alignment of OH molecules by infrared radiation. *Astron. Zh.* 49:727.

Burnett, D. S., and Woolum, D. S. 1990. The interpretation of solar system abundances at the $N = 50$ neutron shell. *Astron. Astrophys.* 228:253–259.

Burnett, D. S., Woolum, D., Benjamin, T., Rogers, P., Duffy, C., and Maggiore, C. 1989. A test of the smoothness of the elemental abundances of carbonaceous chondrites. *Geochim. Cosmochim. Acta* 53:471–481.

Burns, J. A. 1986. The evolution of satellite orbits. In *Satellites*, eds. J. A. Burns and M. S. Matthews (Tucson: Univ. of Arizona Press), pp. 117–158.

Burns, J. A. 1992. Contradictory clues to the origin of the Martian satellites. In *Mars*, eds. H. H. Kieffer, B. M. Jakosky, C. W. Snyder and M. S. Matthews (Tucson: Univ. of Arizona Press), pp. 1283–1301.

Burns, J. A., and Matthews, M. S., eds. 1986. *Satellites* (Tucson: Univ. of Arizona Press).

Burns, J. A., Lamy, P. L., and Soter, S. 1979. Radiation forces on small particles in the solar system. *Icarus* 40:1–48.

Burrows, A. S., Hubbard, W. B., and Lunine, J. I. 1989. Theoretical models of very low mass stars and brown dwarfs. *Astrophys. J.* 345:939–958.

Burstein, P., Borken, R. J., Kraushaar, W. L., and Sanders, W. T. 1976. Three-band observations of the soft X-ray background and some implications of thermal emission models. *Astrophys. J.* 213:405–420.

Burton, M. G., Hollenbach, D. J., and Tielens, A. G. G. M. 1990. Line emission from clumpy photodissociation regions. *Astrophys. J.* 365:620–639.

Burton, M. G., Minchin, N. R., Hough, J. H., Aspin, C., Axon, D. J., and Bailey, J. A. 1991. Molecular hydrogen polarization images of OMC-1. *Astrophys. J.* 375:611–617.

Buss, R. H., Jr., Lamers, H. J. G. L. M., and Snow, T. P., Jr. 1989. Grain ultraviolet extinction properties of recently discovered post asymptotic giant branch stars. *Astrophys. J.* 347:977–988.

Butchart, I., McFadzean, A. D., Whittet, D. C. B., Geballe, T. R., and Greenberg, J. M.1986. Three micron spectroscopy of the galactic centre source IRS7. *Astron. Astrophys. Lett.* 154:5–7.

Butner, H. M., Evans, N. J., II, Lester, D. F., Levreault, R. M., and Strom, S. E. 1991. Testing models of low mass star formation: high resolution far-infrared observations of L 1551 IRS 5. *Astrophys. J.* 376:636–653.

Butner, H. M., Lada, E. A., and Loren, R. B. 1992. The physical properties of dense cores: DCO$^+$ observations. *Astrophys. J.*, submitted.

Byrne, P. B. 1986. HD319139. *Irish Astron. J.* 17:294–300.

Cabot, W., Canuto, V. M., Hubickyj, O., and Pollack, J. B. 1987*a*. The role of turbulent convection in the primitive solar nebula. I. Theory. *Icarus* 69:387–422.

Cabot, W., Canuto, V. M., Hubickyj, O., and Pollack, J. B. 1987*b*. The role of turbulent convection in the primitive solar nebula. II. Results. *Icarus* 69:423–457.

Cabrit, S. 1989. Interpretation of CO emission from molecular flows. In *ESO Workshop on Low Mass Star Formation and Pre-Main Sequence Objects*, ed. B. Reipurth (Garching: European Southern Obs.), pp. 119–139.

Cabrit, S., and Bertout, C. 1990. CO line formation in bipolar flows. II. Decelerated outflow case and summar of results. *Astrophys. J.* 348:530–541.

Cabrit, S., Goldsmith, P. F., and Snell, R. L. 1988. Identification of RNO 43 and B335 as two highly collimated bipolar flows oriented nearly in the plane of the sky. *Astrophys. J.* 334:196–208.

Cabrit, S., Edwards, S., Strom, S. E., and Strom, K. M. 1990. Forbidden-line emission and infrared excess in T Tauri stars: Evidence for accretion-driven mass loss? *Astrophys. J.* 354:687–700.

Caffee, M. W., Hohenberg, C. M., and Hudson, B. 1981. Troctolite 76535: A study in the preservation of early isotopic records. *Proc. Lunar Planet. Sci. Conf.* 12, *J. Geophys. Res. Suppl.* 12B:99–115.

Caffee, M. W., Goswami, J. N., Hohenberg, C. M., and Swindle, T. D. 1983. Cosmogenic neon from pre-compaction irradiation of Kapoeta and Murchison. *Proc. Lunar Planet. Sci. Conf.* 14, *J. Geophys. Res. Suppl.* 88B:267–273.

Caffee, M. W., Goswami, J. N., Hohenberg, C. M., Marti, K., and Reedy, R. C. 1988. Irradiation records in meteorites. In *Meteorites in the Early Solar System*, eds. J. F. Kerridge and M. S. Matthews (Tucson: Univ. of Arizona Press), pp. 205–245.

Caillault, J.-P., and Zoonematkermani, S. 1989. Detection of a dozen X-ray-emitting main-sequence B6-A3 stars in Orion. *Astrophys. J. Lett.* 338:57–60.

Calvet, N., and Albarrán, J. 1984. Energy sources of T Tauri stars. *Rev. Mexicana Astron. Astrofis.* 9:35–47.

Calvet, N., Cantó, J., and Rodríguez, L. F. 1983. Stellar winds and molecular clouds: T Tauri stars. *Astrophys. J.* 268:739–752.

Calvet, N., Basri, G., and Kuhi, L. V. 1984. The chromospheric hypothesis for the T Tauri phenomenon. *Astrophys. J.* 277:725–737.

Calvet, N., Patino, A., Magris, G., and D'Alessio, P. 1991*a*. Irradiation of accretion disks around young stars I. Near-infrared CO bands. *Astrophys. J.* 380:617–630.

Calvet, N., Hartmann, L. and Hewitt, R. 1991*b*. Winds from T Tauri stars. II. Balmer line profiles for inner disk winds. *Astrophys. J.*, submitted.

Camenzind, M. 1990. Magnetized disk winds and the origin of bipolar outflows. *Rev. Modern Astron.* 3:234–265.

Cameron, A. G. W. 1962. The formation of the Sun and planets. *Icarus* 1:13–69.

Cameron, A. G. W. 1973. Accumulation processes in the primitive solar nebula. *Icarus* 18:407–450.

Cameron, A. G. W. 1978*a*. Physics of the primitive solar accretion disc. *Moon and Planets* 18:5–40.

Cameron, A. G. W. 1978*b*. The primitive solar accretion disk and the formation of the planets. In *The Origin of the Solar System*, ed. S. F. Dermott (New York: Wiley and Sons), pp. 49–75

Cameron, A. G. W. 1979. The interaction between giant gaseous protoplanets and the primitive solar nebula. *Moon and Planets* 21:173–183.

Cameron, A. G. W. 1982. Elemental and nuclidic abundances in the solar system. In *Essays in Nuclear Astrophysics*, eds. C. A. Barnes, D. D. Clayton and D. N. Schramm (Cambridge: Cambridge Univ. Press), pp. 23–43.

Cameron, A. G. W. 1983. Origin of the atmospheres of the terrestrial planets. *Icarus* 56:195–201.

Cameron, A. G. W. 1984. Star formation and extinct radioactivities. *Icarus* 60:416–

427.

Cameron, A. G. W. 1985. Formation and evolution of the primitive solar nebula. In *Protostars and Planets II*, D. C. Black and M. S. Matthews, eds. (Tucson: Univ. of Arizona Press), pp. 1073–1099.

Cameron, A. G. W. 1988. Origin of the solar system. *Ann. Rev. Astron. Astrophys.* 26:441-472.

Cameron, A. G. W. 1989. Comment following "The formation of the solar system: Consensus, alternatives, and missing factors" by G. W. Wetherill. In *The Formation and Evolution of Planetary Systems*, eds. H. A. Weaver and L. Danly (Cambridge: Univ. of Cambridge Press), p. 29.

Cameron, A. G. W., and Truran, J. W. 1977. The supernova trigger for the formation of the solar system. *Icarus* 30:447–461.

Cameron, A. G. W., and Ward, W. R. 1976. Origin of the Moon. In *Lunar Science* (Houston: Lunar Planetary Inst.).

Cameron, A. G. W., DeCampli, W. M., and Bodenheimer, P. 1982. Evolution of giant gaseous protoplanets embedded in the primitive solar nebula. *Icarus* 49:298–312.

Cameron, A. G. W., Cowan, J. J., Klapdor, H. V., Metzinger, J., Oda, T., and Truran, J. W. 1983. Steady flow approximations to the helium r-process. *Astrophys. Space Sci.* 91:221–234.

Cameron, M., and Liseau, R. 1990. UV observations of Herbig-Haro objects associated with bipolar molecular outflows: HH 7, HH 11 and HH 29. *Astron. Astrophys.* 240:409–428.

Campbell, B. 1989. A search for planetary-mass companions to nearby stars. In *Highlights of Astronomy*, vol. 8, ed. D. McNally (Dordrecht: Kluwer), pp. 109–110.

Campbell, B., Persson, S. E., Strom, S. E., and Grasdalen, G. L. 1988*a*. Images of star-forming regions. II. The circumstellar environment of L1551 IRS 5. *Astron. J.* 95:1173–1184.

Campbell, B., Walker, G. A. H., and Yang, S. 1988*b*. A search from substellar companions to solar-type stars. *Astrophys. J.* 331:902–921.

Campbell, J. K., and Synnott, S. P. 1985. Gravity field of the Jovian system from Pioneer and Voyager tracking data. *Astron. J.* 90:364–372.

Campbell, M. F., Lester, D. F., Harvey, P. M., and Joy, M. 1989. High spatial resolution far-infrared scans of W3(OH). *Astrophys. J.* 345:298–305.

Campins, H., Rieke, G. H., and Lebofsky, M. J. 1985. Absolute calibration of photometry at 1 through 5 μm. *Astron. J.* 90:896–899.

Canalas, R. E., Mayeda, T. K., and Manuel, O. 1968. Terrestrial abundances of noble gases. *J. Geophys. Res.* 73:3331–3334.

Cantó, J. 1980. A stellar wind model for Herbig-Haro objects. *Astron. Astrophys.* 86:327–338.

Cantó, J., and Raga, A. C. 1991. Mixing layers in stellar outflows. *Astrophys. J.* 372:646–658.

Cantó, J., Rodríguez, L. F., Barral, J. F., and Carral, P. 1981. Carbon monoxide observations of R Monocerotis, NGC 2261, and Herbig-Haro 39: the interstellar nozzle. *Astrophys. J.* 244:102–114.

Cantó, J., Rodríguez, L. F., Calvet. N., and Levreault, R. M. 1984. Stellar winds and molecular clouds: Herbig Be and Ae type stars. *Astrophys. J.* 282:631–640.

Cantó, J., Raga, A.C., and Binette, L. 1989. On the structure of steady stellar jets: An analytical model. *Rev. Mexicana Astron. Astrof.* 17:65–74.

Cantó, J., Raga, A.C., Binette, L., and Calvet, N. 1990. Stellar Jets with Intrinsically Variable Sources. Preprint.

Canuto, V. M. 1989. AMLT: Anisotropic mixing length theory. *Astron. Astrophys.* 217:333–343.

Canuto, V. M., and Battaglia, A. 1988. Turbulent diffusivity. *Astron. Astrophys.* 193:313–326.

Capaccioli, M., Della Valle, M., D'Onofrio, M., and Rosino, L. 1989. *Astron. J.* 97:1622–1633.

Carballo, R., Wesselius, P. R., and Whittet, D. C. B. 1991. Identification of IRAS point sources in Scorpius-Centaurus-Lupus. *Astron. Astrophys.*, submitted.

Cardelli, J. A., Clayton, G. C., and Mathis, J. S. 1989. The relationship between infrared, optical, and ultraviolet extinction. *Astrophys. J.* 345:245-256.

Cardelli, J. A., Suntzeff, N. B., Edgar, R. J., Savage, B. D. 1990. Molecules toward HD62542: A high density peculiar extinction sight line in the Gum Nebula complex. Astrophys. J. 362:551–562.

Carlberg, R. G., and Pudritz, R. E. 1991. Magnetic support and fragmention of molecular clouds. *Mon. Not. Roy. Astron. Soc.* 247:353–366.

Carlson, R. W., and Lugmair, G. W. 1988. The age of ferroan anorthosite 60025: Oldest crust on a young Moon? *Earth Planet. Sci. Lett.* 90:119–130.

Carpenter, J. M., Snell, R. L., and Schloerb, F. P. 1991. The stellar population in massive star forming regions. In *Astrophysics with Infrared Arrays*, ed. R. Elston (San Franciso: Astron. Soc. of the Pacific), pp. 241–243.

Carr, J. S. 1987. A study of clumping in the Cepheus OB3 molecular cloud. *Astrophys. J.* 323:170–178.

Carr, J. S. 1989. Near infrared CO emission in young stellar objects. *Astrophys. J.* 345:522–535.

Carr, J. S., Harvey, P. M., and Lester, D. F. 1987. The two micron spectrum of L1551 IRS5. *Astrophys. J. Lett.* 321:71–74.

Casoli, F., and Combes, F. 1982. Can giant molecular clouds form in spiral arms? *Astron. Astrophys.* 110:287–294.

Casoli, F., Combes, F., and Gérin, M. 1985. CO observations of high velocity gas around S187. In *Nearby Molecular Clouds,* ed. G. Serra (Berlin: Springer-Verlag), pp. 136–139.

Casoli, F., Dupraz, C., Gérin, M., Combes, F., and Boulanger, F. 1986. ^{13}CO and ^{12}CO observations of cold IRAS unidentified point sources in the Galaxy. *Astron. Astrophys.* 169:281–297.

Casoli, F., Clausset, F., Viallefond, F., Combes, F., and Boulanger, F. 1990. $^{12}CO(1-0)$ and $^{13}CO(2-1)$ mapping of gas complexes in the spiral galaxy NGC 6946: Comparison with HI and HII regions. *Astron. Astrophys.* 233:357–371.

Cassen, P. M., and Boss, A. P. 1988. Protostellar collapse, dust grains, and solar-system formation. In *Meteorites and the Early Solar System*, eds. J. F. Kerridge and M. S. Matthews (Tucson: Univ. of Arizona Press), pp. 304–328.

Cassen, P. M., and Moosman, A. 1981. On the formation of protostellar disks. *Icarus* 48:353–376.

Cassen, P. M., and Summers, A. L. 1983. Models of the formation of the solar nebula. *Icarus* 53:26–40.

Cassen, P. M., and Tomley, L. 1988. The dynamical behavior of gravitationally unstable solar nebula models. *Bull. Amer. Astron. Soc.* 20:815 (abstract).

Cassen, P. M., Reynolds, R. T., and Peale, S. J. 1979. Is there liquid water on Europa? *Geophys. Res. Lett.* 6:731–734.

Cassen, P. M., Peale, S. J., and Reynolds, R. T. 1980. Tidal dissipation in Europa: a correction. *Geophys. Res. Lett.* 7:987–988.

Cassen, P. M., Smith, B. F., Miller, R. H., and Reynolds, R. T. 1981. Numerical experiments on the stability of preplanetary disks. *Icarus* 48:377–392.

Cassen, P. M., Peale, S. J., and Reynolds, R. T. 1982. Structure and thermal evolution of the Galilean satellites. In *Satellites of Jupiter*, ed. D. Morrison (Tucson: Univ. of Arizona Press), pp. 93–122.

Cassen, P. M., Shu, F. H., and Terebey, S. 1985. Protostellar disks and star formation: An overview. In *Protostars and Planets II*, eds. D. C. Black and M. S. Matthews (Tucson: Univ. of Arizona Press), pp. 448–482.

Castelaz, M. W., Grasdalen, G. L., Hackwell, J. A., Capps, R. W., and Thompson, D. 1985. GL 961-W: A pre-main-sequence object. *Astron. J.* 90:1113–1116.

Catala, C. 1988. Line formation in the winds of Herbig Ae/Be stars. The CIV resonance lines. *Astron. Astrophys.* 193:222, 228.

Catala, C. 1989. Herbig Ae and Be stars. In *ESO Workshop on Low Mass Star Formation and Pre-Main Sequence Objects*, ed. B. Reipurth (Garching: European Southern Obs.), pp. 471–490.

Catalano, F. A., Baratta, G. A., Lo Presti, C., and Strazzulla, G. 1986. Pre-perihelion photometry of P/Halley (1982i) at Catania (Italy) Observatory. *Astron. Astrophys.* 168:341–345.

Centurión, M., and Vladilo, G. 1991. Redetermination of the interstellar $^{12}C/^{13}C$ ratio in the solar vicinity. *Astron. Astrophys.* 251:242–252.

Cernicharo, J., Guélin, M., and Askne, J. 1984. TMC1–like cloudlets in HCL2. *Astron. Astrophys.* 138:371–379.

Cernicharo, J., Bachiller, R., and Duvert, G. 1985. The Taurus-Auriga-Perseus complex of dark clouds. I. Density structure. *Astron. Astrophys. Suppl.* 149:273–282.

Cernicharo, J., Gottlieb, C. A., Guélin, M., Thaddeus, P., and Vrtilek, J. M. 1989. Astronomical and laboratory detection of the SiC radical. *Astrophys. J. Lett.* 341:25-28.

Cernicharo, J., Thum, C., Hein, H., John, D., Garcia, P., and Mattioco, F. 1990. Detection of 183 GHz water vapor maser emission from interstellar and circumstellar sources. *Astron. Astrophys. Lett.* 231:15–18.

Cernicharo, J., Gottlieb, C. A., Guélin, M., Killian, T. C., Paubert, G., Thaddeus, P., and Vrtilek, J. M. 1991*a*. Astronomical detection of H_2CCC. *Astrophys. J. Lett.* 368:39–42.

Cernicharo, J., Gottlieb, C. A., Guélin, M., Killian, T. C., Thaddeus, P., and Vrtilek, J. M. 1991*b*. Astronomical detection of H_2CCCC. *Astrophys. J. Lett.* 368:43-45.

Cess, R. D., Ramanathan, V., and Owen, T. 1980. The Martian paleoclimate and enhanced atmospheric carbon dioxide. *Icarus* 41:159–165.

Chaboyer, B., and Henriksen, R.N. 1990. Turbulent magnetic fields. I. *Astron. Astrophys.* 236:275–288.

Chamberlain, J. W., and Hunten, D. M. 1987. *Theory of Planetary Atmospheres* (New York: Academic Press).

Champagne, A. E., Howard, A. J., and Parker, P. D. 1984. Nucleosynthesis of ^{26}Al at low stellar temperatures. *Astrophys. J.* 269:686–689.

Champney, J., and Cuzzi, J. 1990. A Turbulent Two-Phase Flow Model for Nebula Flows. Paper presented at the 28th Aerospace Sciences Meeting, Reno Nev., Jan. 1990, AIAA 90–0211.

Chandrasekhar, S. 1942. *Principles of Stellar Dynamics* (Chicago: Univ. of Chicago Press).

Chandrasekhar, S. 1951*a*. The fluctuations of density in isotropic turbulence. *Proc. Roy. Soc.* A210:18–25.

Chandrasekhar, S. 1951*b*. The gravitational instability of an infinite homogeneous turbulent medium. *Proc. Roy. Soc.* A210:26–29.

Chandrasekhar, S. 1960. The stability of non-dissipative Couette flow in hydromagnetics. *Proc. Nat. Acad. Sci.* 46:253–257.

Chandrasekhar, S. 1961. *Hydrodynamic and Hydromagnetic Stability* (Oxford: Oxford Univ. Press).

Chandrasekhar, S. 1965. *Hydrodynamic and Hydromagnetic Stability* (Oxford: Clarendon Press), pp. 384–389.

Chandrasekhar, S., and Fermi, E. 1953. Problems of gravitational instability in the presence of a magnetic field. *Astrophys. J.* 118:116–141.

Chapman, C. R., and Morrison, D. 1989. *Cosmic Catastrophes* (New York: Plenum Press).

Charnley, S. B., Dyson, J. E., Hartquist, T. W., and Williams, D. A. 1988. Chemical limit cycles for models of a region of low-mass star formation. *Mon. Not. Roy. Astron. Soc.* 235:1257–1271.

Charnley, S. B., Dyson, J. E., Hartquist, T. W., and Williams, D. A. 1990. Chemical evolution in molecular clump-stellar wind interfaces. *Mon. Not. Roy. Astron. Soc.* 243:405–412.

Charnley, S. B., Whittet, D. C. B., and Williams, D. A. 1990. Ice mantles in Barnard 5 IRS 1. *Mon. Not. Roy. Astron. Soc.* 245:161–163.

Chelli, A., Zinnecker, H., Carrasco, L., Cruz-Gonzalez, I., and Perrier, C. 1988. Infrared companions to T Tauri stars. *Astron. Astrophys.* 207:46–54.

Chen, H., Fukui, Y., and Yang, J. 1992. ^{13}CO and HCO$^+$ observations in IRAS sources in L1641. *Astrophys. J.* 398:544–551.

Chen, J. H., and Wasserburg, G. J. 1981*a*. Isotopic determination of uranium in picomole and subpicomole quantities. *Anal. Chem.* 53:2060–2067.

Chen, J. H., and Wasserburg, G. J. 1981*b*. The isotopic composition of uranium and lead in Allende inclusions and meteorite phosphates. *Earth Planet. Sci. Lett.* 52:1–15.

Chen, J. H., and Wasserburg, G. J. 1990. The isotopic composition of Ag in meteorites and the presence of ^{107}Pd in protoplanets. *Geochim. Cosmochim. Acta* 54:1729–1743.

Chen, J. H., and Wasserburg, G. J. 1991. The Pd-Ag systematics in chondrites and mesosiderites. *Lunar Planet. Sci.* XXII:199–200 (abstract).

Chen, W. P. 1990. Near-Infrared Milliaresecond Observations by the Lunar Occultation Technique. Ph. D. Thesis, State Univ. of New York, Stony Brook.

Chen, W. P., Simon, M., Longmore, A. J., Howell, R. R., and Benson, J. A. 1990. Discovery of five pre-main-sequence binaries in Taurus. *Astrophys. J.* 357:224–230.

Chernoff, D. F., Hollenbach, D. J., and McKee, C. F. 1982. Molecular shock waves in the BN-KL region of orion. *Astrophys. J. Lett.* 259:97–102.

Chester, T. J. 1985. A statistical analysis and overview of the IRAS point source catalog. In *Light on Dark Matter*, ed. F. P. Israel (Dordrecht: D. Reidel), pp. 3–22.

Chiang, W. H., and Prendergast, K. 1985. Numerical study of a two fluid hydrodynamic model of the interstellar medium and population I stars. *Astrophys. J.* 297:507–530.

Chin, G., and Weaver, H. A. 1984. Vibrational and rotational excitation of CO in comets. *Astrophys. J.* 285:858–869.

Chini, R. 1981. Multicolour photometry of stars in the Ophiuchus dark cloud region. *Astron. Astrophys.* 99:346–350.

Chini, R., Elsässer, H., Hefele, H., and Weinberger, R. 1977. On the infrared sources in the Ophiuchus dark cloud region. *Astron. Astrophys.* 56:323–325.

Chini, R., Elsässer, H., and Neckel, Th. 1980. Multicolour UBVRI photometry of stars in M17. *Astron. Astrophys.* 91:186–193.

Chini, R., Krügel, E., and Kreysa, E. 1990. Large dust particles around main sequence stars. *Astron. Astrophys. Lett.* 227:5–8.

Chiosi, C., and Maeder, A. 1986. The evolution of massive stars with mass loss. *Ann. Rev. Astron. Astrophys.* 24:329–375.

Chromey, F. R., Elmegreen, B. G., and Elmegreen, D. M. 1989. Atomic hydrogen in the Orion star-forming region. *Astron. J.* 98:2203–2209.

Churchwell, E., Felli, M., Wood, D. O. S., and Massi, M. 1987. Solar-system sized condensations in the Orion Nebula. *Astrophys. J.* 321:516–529.

Chyba, C. F. 1990. Impact delivery and erosion of planetary oceans in the early inner solar system. *Nature* 343:129–133.

Chyba, C. F., Jankowski, D. G., and Nicholson, P. D. 1989a. Tidal evolution in the Neptune-Triton system. *Astron. Astrophys. Lett.* 219:23–26.

Chyba, C. F., Sagan, C., and Mumma, M. J. 1989b. The heliocentric evolution of cometary infrared spectra: results from an organic grain model. *Icarus* 79:362–381.

Ciardullo, R., Ford, H. C., Williams, R. E., Tamblyn, P., and Jacoby, G. H. 1990. The nova rate in the elliptical component of NGC 5128. *Astron. J.* 99:1079–1087.

Cintala, M. J. 1981. Meteoroid impact into short-period comet nuclei. *Nature* 291:134–136.

Cisowski, S. M., and Hood, L. L. 1991. The relict magnetism of meteorites. In *The Sun in Time*, eds. C. P. Sonett, M. S. Giampapa and M. S. Matthews (Tucson: Univ. of Arizona Press), pp. 761–784.

Clark, B. G. 1965. An interferometer investigation of the 21-centimeter hydrogen line absorption. *Astrophys. J.* 142:1398–1422.

Clark, F. O. 1986. The pincushion cloud: the bipolar flows in L988. *Astron. Astrophys. Lett.* 164:19–21.

Clark, F. O., and Johnson, D. R. 1974. Magnetic fields in the orion molecular cloud from the zeeman effect in SO. *Astrophys. J. Lett.* 191:87–91.

Clark, F. O., and Laureijs, R. J. 1986. IRAS observations of the L 1551 bipolar outflow. *Astron. Astrophys.* 154:L26–L29.

Clark, F. O., Laureijs, R. J., Chlewicki, G., Zhang, C. Y., van Oosterom, W., and Kester, D. 1986. The extended infra-red radiation of the L1551 bipolar flow, $L_0 > 19 \, L_\odot$. *Astron. Astrophys.* 168:L1–L4.

Clark, R. N., and Lucey, P. 1984. Spectral properties of ice-particulate mixtures and implications for remote sensing. I. Intimate mixtures. *J. Geophys. Res.* 89:6341–6348.

Clarke, C., and Pringle, J. E. 1991a. Star-disc interactions and binary star formation. *Mon. Not. Roy. Astron. Soc.* 249:584–587.

Clarke, C., and Pringle, J. E. 1991b. The role of disks in the formation of binary and multiple star systems. *Mon. Not. Roy. Astron. Soc.* 249:588–595.

Clarke, C. J., Lin, D. N. C., and Papaloizou, J. C. B. 1989. Accretion disc flows around FU Orionis stars. *Mon. Not. Roy. Astron. Soc.* 236:495–503.

Clarke, C. J., Lin, D. N. C., and Pringle, J. E 1990. Pre-conditions for disc-generated FU Orionis outbursts. *Mon. Not. Roy. Astron. Soc.* 242:439–446.

Clayton, D. D. 1968. *Principles of Stellar Evolution and Nucleosynthesis* (New York: McGraw-Hill).

Clayton, D. D. 1981. Some key issues in isotopic anomalies: Astrophysical history and aggregation. *Proc. Lunar Planet. Sci.* 12:1781–1802.

Clayton, D. D. 1982. Cosmic chemical memory: A new astronomy. *Quart. J. Roy. Astron. Soc.* 23:174–212.

Clayton, D. D. 1989. Origin of heavy xenon in meteoritic diamonds. *Astrophys. J.* 340:613–619.

Clayton, D. D., and Hoyle, F. 1976. Grains of anomalous isotopic composition from novae. *Astrophys. J.* 203:490–496.

Clayton, D. D., and Leising, M. D. 1987. [26]Al in the interstellar medium. *Physics Rept.* 144:1–50.

Clayton, R. N., and Mayeda, T. K. 1984. The oxygen isotope record in Murchison and other carbonaceous chondrites. *Earth Plant. Sci. Lett.* 67:151–161.

Clayton, R. N., Onuma, N., and Mayeda, T. K. 1976. A classification of meteorites based on oxygen isotopes. *Earth Planet. Sci. Lett.* 30:10–18.

Clayton, R. N., Mayeda, T. K., and Molini-Velsko, C. A. 1985. Isotopic variations in solar system material: Evaporation and condensation of silicates. In *Protostars and Planets II*, eds. D. C. Black and M. S. Matthews (Tucson: Univ. of Arizona Press), pp. 755–771.

Clayton, R. N., Hinton, R. W., and Davis, A. M. 1988. Isotopic variations in the rock-forming elements in meteorites. *Phil. Trans. Roy. Soc. London* 325A:483–501.

Clemens, D. P. 1985. Massachusetts-Stoney Brook galactic plane CO survey: The galactic disk rotation curve. *Astrophys. J.* 295:422–436.

Clemens, D. P., and Barvanis, R. 1988. A catalog of small, optically selected molecular clouds: optical, infrared, and millimeter properties. *Astrophys. J. Suppl.* 68:257–286.

Clifford, P., and Elmegreen, B. G. 1983. A collision cross section for magnetic diffuse clouds. *Mon. Not. Roy. Astron. Soc.* 202:629–646.

Clube, S. V. M., and Napier, W. M. 1982. Spiral arms, comets, and terrestrial catastrophism. *Quart. J. Roy. Astron. Soc.* 23:45–66.

Cohen, C. J., and Hubbard, E. C. 1965. Liberation of the close approaches of Pluto to Neptune. *Astron. J.* 70:10–13.

Cohen, C. J., Hubbard, E. C., and Oesterwinter, C. 1973. *Astron. Papers American Ephem.* vol. 22, part I.

Cohen, M. 1984. The T Tauri stars. *Phys. Rept.* 116:173–249.

Cohen, M., and Bieging, J. H. 1986. Radio variability and structure of T Tauri stars. *Astron. J.* 92:1396–1402.

Cohen, M., and Kuhi, L. V. 1979. Observational studies of pre-main-sequence evolution. *Astrophys. J. Suppl.* 41:743–843.

Cohen, M., and Schwartz, R. D. 1983. The exciting stars of Herbig-Haro objects. *Astrophys. J.* 265:877–900.

Cohen, M., and Schwartz, R. D. 1987. IRAS observations of the exciting stars of Herbig-Haro objects. *Astrophys. J.* 316:311–322.

Cohen, M., Schwartz, D. E., Chokshi, A., and Walker, R. G. 1987. IRAS colors of normal stars. *Astron. J.* 93:1199–1219.

Cohen, M., Tielens, A. G. G. M., and Bregman, J. D. 1989*a*. Mid-infrared spectra of WC 9 stars: The composition of circumstellar and interstellar dust. *Astrophys. J. Lett.* 344:13–16.

Cohen, M., Emerson, J. P., and Beichman, C. A. 1989*b*. A reexamination of luminosity sources in T Tauri stars. I. Taurus-Auriga. *Astrophys. J.* 339:455–473.

Cohen, R. H., and Kulsrud, R. M. 1974. Nonlinear evolution of parallel propagating hydromagnetic waves. *Phys. Fluids* 17:2215–2225.

Cohen, R. J., Rowland, P. R., and Blair, M. M. 1984. The source of the bipolar outflow in Cepheus A. *Mon. Not. Roy. Astron. Soc.* 210:425–438.

Cohen, R. S., Cong, H.-I., Dame, T. M., and Thaddeus, P. 1980. Molecular clouds and galactic spiral structure. *Astrophys. J. Lett.* 239:53–56.

Cohn, H. 1983. The stability of a magnetically confined radio jet. *Astrophys. J.* 269:500–512.

Colom, P., Crovisier, J., Bockelée-Morvan, D., Despois, D., and Paubert, G. 1992. Formaldehyde in comets: microwave observations of P/Brorsen-Metcalf (1989 X), Austin (1989C1), and Levy (1990c). *Astron. Astrophys.*, in press.

Columbo, G., and Franklin, F. A. 1971. On the formation of the outer satellite groups of Jupiter. *Icarus* 15:186–191.

Combes, F., and Gerin, M 1985. Spiral structure of molecular clouds in response to bar forcing: a particle simulation. *Astron. Astrophys.* 150:327–338.

Combes, M., Moroz, V. I., Crifo, J. F., Lamarre, J. M., Charra, J., Sanko, N. F., Soufflot, A., Bibring, J. P., Cazes, S., Coron, N., Crovisier, J., Emerich, C., Encrenaz, T.,

Gispert, R., Grigoryev, A. V., Guyot, G., Krasnopolsky, V. A., Nikolsky, Yu. V., and Rocard, F. 1986. Infrared sounding of comet Halley from Vega 1. *Nature* 321:266–268.

Combes, M., Moroz, V. I., Crovisier, J., Encrenaz, T., Bibring, J. -P., Grigoriev, A. V., Sanko, N. F., Coron, N., Crifo, J. F., Gispert, R., Bockelée-Morvan, D., Nikolsky, Yu. V., Krasnopolsky, V. A., Owen, T., Emerich, C., Lamarre, J. M., and Rocard, F. 1988. The 2.5 – 12μm spectrum of comet Halley from the IKS-Vega experiment. *Icarus* 76:404–436.

Combi, M. R. 1987. Sources of cometary radicals and their jets: gases or grains? *Icarus* 71:178–191.

Connerney, J. E. P., Acuña, M. H., and Ness, N. F. 1987. The magnetic field of Uranus. *J. Geophys. Res.* 92:15329–15336.

Conrath, B. J., Gautier, D., Hanel, R., and Hornstein, J. S. 1984. The helium abundance of Saturn from Voyager measurements. *Astrophys. J.* 282:807–815.

Conrath, B. J., Gautier, D., Hanel, R., Lindal, G., and Marten, A. 1987. The helium abundance of Uranus from Voyager measurements. *J. Geophys. Res.* 92:15003–15010.

Conrath, B. J., Flasar, F. M., Hanel, R., Kunde, V., Maguire, W., Pearl, J., Pirraglia, J., Samuelson, R., Gierasch, P., Weir, A., Bezard, B., Gautier, D., Cruikshank, D., Horn, L., Springer, R., and Shaffer, W. 1989. Infrared observations of the Neptunian System. *Science* 246:1454–1459.

Conrath, B. J., Gautier, D., Lindal, G. F., Samuelson, R. E., and Schaeffer. 1992. The helium abundance of Neptune from Voyager measurements. *J. Geophys. Res.* 96:18907–18919.

Consolmagno, G. J., and Jokipii, J. R. 1978. ^{26}Al and the partial ionization of the solar nebula. *Moon and Planets* 19:253–259.

Coradini, A., Cerroni, P., Magni, G., and Federico, C. 1989. Formation of the satellites of the outer solar system: Sources of their atmospheres. In *Origin and Evolution of Planetary and Satellite Atmospheres*, eds. S. K. Atreya, J. B. Pollack and M. S. Matthews (Tucson: Univ. of Arizona Press), pp. 723–762.

Cosmovici, C. B., Schwarz, W., Ip, W.-H., and Mack, P. 1988. Gas and dust jets in the inner coma of comet Halley. *Nature* 332:705–709.

Coté, J. 1987. B and A type stars with unexpectedly large color excesses at IRAS wavelengths. *Astron. Astrophys.* 181:77–84.

Counselman, C. C., III. 1973. Outcomes of tidal evolution. *Astrophys. J.* 180:307–314.

Counselman, C. C., III., and Shapiro, I. I. 1970. Spin-orbit resonance of Mercury. *Symp. Mathematica* 3:121–169.

Courant, R., and Friedrichs, K. 1976. *Supersonic Flow and Shock Waves* (New York: Springer-Verlag).

Covino, E., Terranegra, A., Vittone, A., and Russo, G. 1984. Spectroscopic and photometric observations of the Herbig Be star Z Canis Majoris. *Astron. J.* 89:1868–1875.

Cowan, J. J., Thielemann, F.-K., and Truran, J. W. 1991. The *r*-Process and Nucleochronology. *Phys. Rept.* 208:267–394.

Cowie, L. L. 1981. Cloud fluid compression and softening in spiral arms and the formation of giant molecular cloud complexes. *Astrophys. J.* 245:66–71.

Cox, D. P. 1988. Overview of the interstellar medium: supernova related issues. In *Supernova Remnants and the Interstellar Medium*, eds. R. S. Roger and T. L. Landecker, pp. 73–90.

Cox, D. P. 1990. The diffuse interstellar medium. In *The Interstellar Medium in Galaxies*, eds. H. A. Thronson, Jr. and J. M. Shull (Dordrecht: Kluwer), pp. 181–200.

Cox, D. P., and Smith, B. W. 1974. Large scale effects of supernova remnants in the galaxy: Generation and maintenance of a hot network of tunnels. *Astrophys. J. Lett.* 189:105–108.

Cox, J. P. and Giuli, R. T. 1968. *Principles of Stellar Structure, Vol. II* (New York: Gordon and Breach).

Cox, P. 1989. The line of sight towards AFGL 961: Detection of the librational band of water ice at 13.6 micron. *Astron. Astrophys. Lett.* 225:1–4.

Cox, P., Walmsley, C. M., and Gusten, R. 1989. C_3H_2 observations in dense dark clouds. *Astron. Astrophys.* 209:382–390.

Cox, P., Deharveng, L., and Leene, A. 1990. IRAS observations of the Rosette nebula complex. *Astron. Astrophys.* 230:181–192.

Crabb, J., and Anders, E. 1981. Noble gases in E-chondrites. *Geochim. Cosmochim. Acta* 45:2443–2464.

Crabb, J., and Anders, E. 1982. On the siting of noble gases in E-chondrites. *Geochim. Cosmochim. Acta* 46:2351–2361.

Crabb, J. A., and Latham, J. 1974. Corona from colliding drops as a possible mechanism for the triggering of lightning. *Quart. J. Roy. Meteorol. Soc.* 100:191–202.

Craig, H., and Lupton, J. 1976. Primordial neon, helium, and hydrogen in oceanic basalts. *Earth Planet. Sci. Lett.* 31:369–385.

Craig, H., Horibe, Y., and Sowers, T. 1988. Gravitational separation of gases in polar ice caps. *Science* 243:1675–1678.

Cram, L. 1979. The atmospheres of T Tauri stars: The photosphere and low chromosphere. *Astrophys. J.* 234:949–957.

Crane, P., Hegyi, D. J., and Lambert, D. L. 1991. Interstellar $^{12}C/^{13}C$ ratio revisited. *Astrophys. J.* 378:181–185.

Cravens, T. E., Victor, G. A., and Dalgarno, A. 1975. The absorption of energetic electrons by molecular hydrogen gas. *Planet. Space Sci.* 23:1059–1070.

Crawford, M. K., Lugten, J. B., Fitelson, W., Genzel, R., and Melnick, G. 1986. Observations of far-infrared line profiles in the Orion-KL region. *Astrophys. J. Lett.* 303:61–65.

Croft, S. K. 1982a. A first-order estimate of shock heating and vaporization in oceanic impacts. In *Geological Implications of Impacts of Large Asteroids and Comet on the Earth*, eds. L. T. Silver and P. H. Schultz, Geol. Soc. America Special Paper 190, pp. 143–152.

Croft, S. K. 1982b. Impacts on ice and snow: implications for crater scaling on icy satellites. *Lunar Planet. Sci.* XIII:135–136. (abstract)

Croft, S. K., Lunine, J. I., and Kargel, J. S. 1988. Equation of state of ammonia-water liquid: derivation and planetological application. *Icarus* 73:279–293.

Croswell, K., Hartmann, L., and Avrett, E. H. 1987. Mass loss from FU Orionis objects. *Astrophys. J.* 312:227–242.

Crovisier, J. 1984. The water molecule in comets: fluorescence mechanisms and thermodynamics of the inner coma. *Astron. Astrophys.* 130:361–372.

Crovisier, J., and Encrenaz, T. 1983. Infrared fluorescence of molecules in comets: the general synthetic spectrum. *Astron. Astrophys.* 126:170–182.

Crovisier, J., and Schloerb, F. P. 1991. The study of comets at radio wavelengths. In *Comets in the Post-Halley Era*, eds. R. Newburn, M. Neugebauer and J. Rahe (Dordrecht: Kluwer), pp. 149–173.

Crovisier, J., Despois, D., Bockelée-Morvan, D., Colom, P., and Paubert, G. 1991. Microwave observations of hydrogen sulfide and searches for other sulphur compounds in comets Austin (1989c1) and Levy (1990c). *Icarus* 93:246–258.

Crowe C., Gore, R., and Troutt, T. 1985. Particle dispersion by coherent structures in free shear flows. *Particulate Sci. Tech.* 3:149–158.

Crowe, C., Chung, J., and Troutt, T. 1988. Particle mixing in free shear flows. *Prog. Energy Combust. Sci.* 14:171–194.

Crozaz, G. 1980. Solar flare and galactic cosmic ray tracks in lunar samples and meteorites: What they tell us about the ancient sun. In *The Ancient Sun: Fossil Record in the Earth, Moon and Meteorites*, eds. R. O. Pepin, J. A. Eddy and R. B. Merrill (New York: Pergamon), pp. 331–346.

Cruikshank, D. P., and Apt, J. 1984. Methane on Triton: physical state and distribution. *Icarus* 58:306–311.

Cruikshank, D. P., and Brown, R. H. 1986. Satellites of Uranus and Neptune, and the Pluto-Charon system. In *Satellites*, eds. J. A. Burns and M. S. Matthews (Tucson: Univ. of Arizona Press), pp. 836–873.

Cruikshank, D. P., Brown, R. H., and Clark, R. N. 1984. Nitrogen on Triton. *Icarus* 58:293–305.

Cruikshank, D. P., Werner, M. W., and Backman, D. E. 1990. SIRTF: capabilities for planetary science. *Adv. Space Res.* XX.

Cruikshank, D. P., Owen, T. C., Geballe, T. R., Schmitt, B., DeBergh, C., Maillard, J.-P., Lutz, B. L., and Brown, R. H. 1991. Tentative detection of CO and CO_2 ices on Triton. *Bull. Amer. Astron. Soc.* 23:1208 (abstract).

Crutcher, R. M. 1977. Excitation of OH toward interstellar dust clouds. *Astophys. J.* 216:308–319.

Crutcher, R. M. 1979. Nonthermal OH main lines and the abundance of OH in interstellar dust clouds. *Astrophys. J.* 234:881–890.

Crutcher, R. M. 1988. OH zeeman effect studies of magnetic fields in molecular clouds. In *Molecular Clouds in the Milky Way and External Galaxies*, eds. R. H. Dickman, R. Snell and J. Young (Berlin: Springer-Verlag), pp. 105–117.

Crutcher, R. M., Hartkopf, W. I., and Giguere, P. T. 1978. The NGC 2264 molecular cloud: CO observations. *Astrophys. J.* 226:839–850.

Cudworth, K. M., and Herbig, G. H. 1979. Two large proper-motion Herbig-Haro objects. *Astron. J.* 84:548–551.

Curiel, S., Rodríguez, L. F., Cantó, J., and Torrelles, J. M. 1989. A search for radio sources near double Herbig-Haro objects. *Rev. Mexicana Astron. Astrof.* 17:137–141.

Cuzzi, J. N., Lissauer, J. J., and Shu, F. H. 1981. Density waves in Saturn's rings. *Nature* 292:703–707.

Cuzzi, J. N., Lissauer, J. J., Esposito, L. W., Holberg, J. B., Marouf, E. A., Tyler, G. L., and Boichot, A. 1984. Saturn's rings: properties and processes. In *Planetary Rings*, eds. R. Greenberg and A. Brahic (Tucson: Univ. of Arizona Press), pp. 73–199.

Cuzzi, J. N., Champney, J., and Dobrovolskis, A. 1993. Coupled particle-gas dynamics in the protoplanetary nebula. In preparation.

Dakowski, M. 1969. The possiblity of extinct superheavy elements occurring in meteorites. *Earth Planet. Sci. Lett.* 6:152–154.

Dalgarno, A. 1984. The chemistry of shocked regions of the interstellar gas. In *Molecular Astrophysics*, eds. G. H. F. Diercksen, W. F. Huebner and P. W. Langhoff (Dordrecht: D. Reidel), pp. 281–293.

Dalgarno, A. 1986. Is interstellar chemistry useful? *Quart. J. Roy. Astron. Soc.* 27:83–89.

Dalgarno, A. 1987. Chemical processes in the interstellar gas. In *Physical Processes in Interstellar Clouds*, eds. G. Morfill and M. S. Scholer (Dordrecht: D. Reidel), pp. 219–239.

Dalgarno, A. 1991. Interstellar chemistry. In *Chemistry in Space*, eds. J. M. Greenberg and V. Pirronello (Dordrecht: Kluwer), pp. 71–87.

Dalgarno, A., and Lepp, S. 1984. Deuterium fractionation mechanisms in interstellar clouds. *Astrophys. J. Lett.* 287:47–50.

Dame, T. M., Elmegreen, B. G., Cohen, R. S., and Thaddeus, P. 1986. The largest molecular cloud complexes in the first galactic quadrant. *Astrophys. J.* 305:892–908.

Dame, T. M., Ungerechts, H., Cohen, R., DeGeus, E., Grenier, I., May, J., Murphy, D., Nyman, L.-A., and Thaddeus, P. 1987. A composite CO survey of the entire Milky Way. *Astrophys. J.* 322:702–706.

Danks, A. C., Encrenaz, Th., Bouchet, P., LeBertre, T., and Chalabaev, A. 1987. The spectrum of comet P/Halley from 3.0 to 4.0 μm. *Astron. Astrophys.* 184:329–332.

D'Antona, F. 1986. Very low mass stars vs. brown dwarfs: a common approach. In *Astrophysics of Brown Dwarfs*, eds. M. C. Kafatos, R. S. Harrington and S. P. Maran (Cambridge: Cambridge Univ. Press), pp. 148–159.

D'Antona, F., and Mazzitelli, I. 1985. Evolution of very low mass stars and brown dwarfs. I. The minimum main-sequence mass and luminosity. *Astrophys. J.* 296:502–513.

Darwin, G. H. 1880. On the secular changes in the elements of the orbit of a satellite revoloving about a tidally distorted planet. *Phil. Trans. Roy. Soc. London* 17:713–891.

Davidson, D., Desando, M., Gough, S., Handa, Y., Ratcliffe, C., Ripmeester, J., and Tse, J. 1987. A clathrate hydrate of carbon monoxide. *Nature* 328:418–419.

Davidson, K., and Ostriker, J. P. 1973. Neutron star accretion in a stellar wind: model for a pulsed X-ray source. *Astrophys. J.* 179:585–598.

Davies, R. D. 1974. Magnetic fields in OH maser clouds. In *Galactic Radio Astronomy*, eds. F. J. Kerr and S. C. Simonson, III, IAU Symp. No. 60 (Dordrecht: D. Reidel), pp. 275–292.

Davis, A. M., and Olsen, E. J. 1990. Phosphates in the El Sampal IIIA iron meteorite have excess ^{53}Cr and primordial lead. *Lunar Planet. Sci.* XXI:258–259 (abstract).

Davis, A. M., Tanaka, T., Grossman, L., Lee, T., and Wasserburg, G. J. 1982. Chemical composition of HAL, an isotopically unusual Allende inclusion. *Geochim. Cosmochim. Acta* 46:1627–1651.

Davis, A. M., Hashimoto, A., Clayton, R. N., and Mayeda, T. K. 1990. Isotope mass fractionation during evaporation of Mg_2SiO_4. *Nature* 347:655–658.

Davis, L., and Greenstein, J. L. 1951. The polarization of starlight by aligned dust grains. *Astrophys. J.* 114:206–240.

Dearborn, D. S. P., and Blake, J. B. 1988. Possible contributions by Wolf-Rayet stars to the protosolar nebula: Extinct radioactivities, or grains of truth from Wolf-Rayet stars. *Astrophys. J.* 332:305–312.

de Bergh, C., Lutz, B. L., Owen, T., Brault, J., and Chauville, J. 1986. Monodeuterated methane in the outer solar system. II. Its detection on Uranus at 1.6 microns. *Astrophys. J.* 311:501–510.

de Bergh, C., Lutz, B. L., Owen, T. and Chauville, J. 1988. Monodeuterated methane in the outer solar system. III. Its abundance on Titan. *Astrophys. J.* 329:951–955.

de Bergh, C., Lutz, B. L., Owen, T., and Maillard, J. P. 1990. Monodeuterated methane in the outer solar system. IV. Its detection and abundance on Neptune. *Astrophys. J.* 355:661–666.

DeCampli, W. M. 1981. T Tauri winds. *Astrophys. J.* 244:124–146.

DeCampli, W. M., and Cameron, A. G. W. 1979. Structure and evolution of isolated giant gaseous protoplanets. *Icarus* 38:367–391.

de Geus, E. 1988. Stars and Interstellar Matter in Scorpio-Centaurus. Ph. D. Thesis, Sterrewacht Leiden.

de Geus, E., de Zeeuw, P. T., and Lub, J. 1989. Physical parameters of stars in the Scorpius-Centaurus OB association. *Astron. Astrophys.* 216:44–61.

de Geus, E. J., Bronfman, L., and Thaddeus, P. 1990. A CO survey of the dark clouds in Ophiuchus. *Astron. Astrophys.* 231:137–150.

Degewij, J., Cruikshank, D. P., and Hartmann, W. K. 1980. Near-infrared colorimetry of J6 Himalia and S9 Phoebe: A summary of 0.3 to 2.2 μm reflectances. *Icarus* 44:541–547.

Deguchi, S., and Watson, W. D. 1984. Linear polarization of molecular lines at radio frequencies. *Astrophys. J.* 285:126–133.

Deguchi, S., and Watson, W. D. 1985. Circular polarization of interstellar absorption lines at radio frequencies. *Astrophys. J.* 289:621–629.

Deguchi, S., and Watson, W. D. 1990. Linear polarized radiation from astrophysical masers due to magnetic fields when the rate for stimulated emissions exceeds the zeeman frequency. *Astrophys. J.* 354:649–659.

de la Reza, R., Quast, G., Torres, C. A. D., Mayor, M., Meylan, G., and Llorente de Andres, F. 1986. Simultaneous UV-optical observations of isolated T Tauri stars: the V4046 Sgr case. In *New Insights in Astrophysics*, ESA SP-263, pp. 107–111.

Delsemme, A. H. 1973. Origin of the short-period comets. *Astron. Astrophys.* 29:377–381.

Delsemme, A. H. 1982. Chemical composition of nuclei. In *Comets*, ed. L. L. Wilkening (Tucson: Univ. of Arizona Press), pp. 85–130.

Delsemme, A. H. 1987. Diversity and similarity of comets. In *Proc. Symp. on the Diversity and Similarity of Comets*, eds. M. Nicolet, E. Rolfe and B. Battric, ESA SP-278, pp. 19–30.

Delsemme, A. H. 1991. Nature and history of the organic compounds in comets: an astrophysical view. In *Comets in the Post-Halley Era*, eds. R. Newburn, M. Neugebauer and J. Rahe (Dorbrecht: Kluwer), pp. 327–428.

Dent, W. R. F., Little, L. T., Kaifu, N., Ohishi, M., and Suzuki, S. 1985. A plan view of the bipolar molecular outflow source G35.2 N. *Astron. Astrophys.* 146:375–380.

Dent, W. R. F., Sandell, G., Duncan, W. D., and Robson, E. I. 1989. The structure of dust discs around G35.2N, NGC 2071 and LkHα 234. *Mon. Not. Roy. Astron. Soc.* 238:1497–1512.

de Pater, I., and Massie, S. T. 1985. Models of the millimeter-centimeter spectra of the giant planets. *Icarus* 62:143–171.

de Pater, I., Romani, P. N., and Atreya, S. K. 1989. Uranus deep atmosphere revealed. *Icarus* 82:288–313.

DePoy, D. L., Lada, E. A., Gatley, I., and Probst, R. 1990. The luminosity function in NGC 2023. *Astrophys. J. Lett.* 356:55–58.

Dermott, S. F. 1984. Origin and evolution of the Uranian and Neptunian satellites: some dynamical considerations. In *Uranus and Neptune*, ed. J. Bergstralh, NASA CP-2330, pp. 377–404.

Dermott, S. F., and Gold, T. 1978. On the origin of the Oort cloud. *Astron. J.* 83:449–450.

Dermott, S. F., and Murray, C. D. 1983. Nature of the Kirkwood gaps in the asteroid belt. *Nature* 301:201–205.

Dermott, S. F., and Nicholson, P. D. 1986. Masses of the satellites of Uranus. *Nature* 319:115–120.

Dermott, S. F., Malhotra, R., and Murray, C. D. 1988. Dynamics of the Uranian and Saturnian satellites: a chaotic route to melting Miranda? *Icarus* 76:295–334.

Desch, M. D., and Kaiser, M. L. 1981. Voyager measurement of the rotation period of Saturn's magnetic field. *Geophys. Res. Let.* 8:253–256.

Désert, F.-X., Boulanger, F., and Puget, J.L. 1990. Interstellar dust models for extinction and emission. *Astron. Astrophys.* 237:215–236.

Despois, D., Crovisier, J., Bockelée-Morvan, D., Schraml, J., Forveille, T., and Gerard, E. 1986. Observations of hydrogen cyanide in comet Halley. *Astron. Astrophys. Lett.* 160:11–12.

d'Hendecourt, L. B. 1980. Laboratory Evidence for Molecule Ejection at Low Temperatures. Internal Report, Laboratory for Astrophysics, Univ. of Leiden.

d'Hendecourt, L. B., and Allamandola, L. J. 1986. Time dependent chemistry in dense molecular clouds. III. Infrared band cross sections of molecules in the solid state at 10 K. *Astron. Astrophys. Suppl.* 64:453–467.

d'Hendecourt, L. B., and de Muizon, M. J. 1989. The discovery of interstellar carbon dioxide. *Astron. Astrophys. Lett.* 223:5–8.

d'Hendecourt, L. B., Allamandola, L. J., Baas, F., and Greenberg, J. M. 1982. Interstellar grain explosions: Molecule cycling between gas and dust. *Astron. Astrophys. Lett.* 109:12–14.

d'Hendecourt, L. B., Allamandola, L. J., and Greenberg, J. M. 1985. Time dependent chemistry in dense molecular clouds. I. Grain surface reactions, gas/grain interactions and infrared spectroscopy. *Astron. Astrophys.* 152:130–150.

d'Hendecourt, L. B., Allamandola, L. J., Grim, R. J. A., and Greenberg, J. M. 1986. Time dependent chemistry in dense molecular clouds. II. Ultraviolet processing and infrared spectroscopy of grain mantles. *Astron. Astrophys.* 158:119–134.

Deul, E. R., and Burton, W. B. 1990. Unravelling the kinematic structure of the infrared cirrus. *Astron. Astrophys.* 230:153–171.

Deul, E. R., and Hartog, R. H. 1990. Small scale structure in the HI distribution in M33. *Astron. Astrophys.* 229:362–377.

Dickel, H. R., and Goss, W. M. 1987. VLA observations of the 6 cm and 2 cm lines of H_2CO in the direction of W3(OH). *Astron. Astrophys.* 185:271–282.

Dickey, J.M., and Garwood, R.W. 1989. The mass spectrum of interstellar clouds. *Astrophys. J.* 341:201–207.

Dickman, R. L. 1985. Turbulence in molecular clouds. In *Protostars and Planets II*, eds. D. C. Black and M. S. Matthews (Tucson: Univ. of Arizona Press), pp. 150–174.

Dickman, R. L., Horvath, M. A., Margulis, M. 1990. A search for scale-dependent morphology in five molecular cloud complexes. *Astrophys. J.* 365:586–601.

Digel, S., Bally, J., and Thaddeus, P. 1990. Giant molecular clouds in the outer arm of the galaxy. *Astrophys. J. Lett.* 357:29–33.

Diner, D. J., and Appleby, J. F. 1986. Prospecting for planets in circumstellar dust: sifting the evidence from β Pictoris. *Nature* 322:436–438.

Di Prima, R. C., and Swinney, H. L. 1981. Instabilities and transition in flow between concentric rotation cylinders. In *Hydrodynamic Instabilities and the Transition to Turbulence*, eds. H. L. Swinney and J. P. Gollub (New York: Springer-Verlag), pp. 140–180

DiSanti, M., Mumma, M., Hoban, S., Reuter, D., Espenak, F., Lacy, J., and Parmar, R. 1990. A search for CO emission in comet Austin (1990c1). *Bull. Amer. Astron. Soc.* 22:1094 (abstract).

DiSanti, M., Mumma, M., Lacy, J., and Parmar, R. 1992. A possible detection of infrared emission from CO in comet Austin (1989c1). *Icarus* 96:151–160.

Dischler, B., Bubenzer, A., and Koidl, P. 1983. Hard carbon coatings with low optical absorption. *Appl. Phys. Lett.* 42:636–638.

Dodelson, S. 1986. Relativistic treatment of ortho-para H_2 transitions. *J. Phys. Atom. Molec. Phys.* B19:2871–2879.

Dohnanyi, J. 1969. Collisional model of asteroids and their debris. *J. Geophys. Res.* 74:2531–2554.

Dolidze, M. V., and Arakelyan, M. A. 1959. The T-association near ρ Ophiuchi. *Soviet Astron. J.* 3:434–438.

Donahue, T. M. 1986. Fractionation of noble gases by thermal escape from accreting planetesimals. *Icarus* 66:195–210.

Donahue, T. M., and Pollack, J. B. 1983. Origin and evolution of the atmosphere of Venus. In *Venus*, eds. D. Hunten, L. Colin, T. Donahue, and V. Moroz (Tucson: Univ. of Arizona Press), pp. 1003–1036.

Donahue, T. M., Hoffman, J. H., and Hodges, R. R. 1981. Krypton and xenon in the atmosphere of Venus. *Geophys. Res. Lett.* 8:513–516.

Donn, B. 1976. Comets, interstellar clouds, and star clusters. In *The Study of Comets*, eds. B. Donn, M. J. Mumma, W. Jackson, M. A'Hearn, and R. Harrington, NASA SP-393, pp. 663–672.

Donn, B. 1990. The formation and structure of fluffy cometary nuclei from random accumulation of grains. *Astron. Astrophys.* 235:441–446.

Donn, B., Daniels, P. A., and Hughes, D. W. 1985. On the structure of the cometary nucleus. *Bull. Amer. Astron. Soc.* 17:520 (abstract).

Donner, K. J. 1979. On the Dynamics of Nonaxisymmetric discs. Ph.D. Thesis, Cambridge Univ.

Dopita, M. A., 1978. Optical emission from shocks. IV. The Herbig-Haro objects. *Astrophys. J. Suppl.* 37:117–144.

Dopita, M. A., Binette, L., and Schwartz, R. D. 1982. The two-photon continuum in Herbig-Haro objects. *Astrophys. J.* 261:183–194.

Dorfi, E. 1982. 3D models for self-gravitating, rotating magnetic interstellar clouds. *Astron. Astrophys.* 114:151–164.

Dorfi, E. 1990. Numerical studies on magnetic braking of interstellar clouds. *Astron. Astrophys.* 225:507–516.

Dragovan, M. 1986. Submillimeter polarization in the Orion nebula. *Astrophys. J.* 308:270–280.

Draine, B. T. 1980. Interstellar shock waves with magnetic precursors. *Astrophys. J.* 241:1021–1038.

Draine, B. T. 1983. Magnetic bubbles and high velocity outflows in molecular clouds. *Astrophys. J.* 270:519–536.

Draine, B. T. 1985a. Grain evolution in dark clouds. In *Protostars and Planets II*, eds. D. C. Black and M. S. Matthews (Tucson: Univ. of Arizona), pp. 621–640.

Draine, B. T. 1985b. Tabulated optical properties of graphite and silicate grains. *Astrophys. J. Suppl.* 87:587–594.

Draine, B. T. 1989a. Interstellar extinction in the infrared. In *Infrared Spectroscopy in Astronomy*, eds. B. H. Kaldeich (Noordwijk: ESA), pp. 93–98.

Draine, B.T. 1989b. On the interpretation of the λ 2175 Å feature. In *Interstellar Dust: Proc. IAU Symp. 135*, eds. L. J. Allamandola and A. G. G. M. Tielens (Dordrecht: Kluwer), pp. 313–327.

Draine, B. T. 1990. Mass determinations from far-infrared observations. In *The Interstellar Medium in Galaxies*, eds. H. A. Thronson, Jr. and J. M. Shull (Dordrecht: Kluwer), pp. 483–492.

Draine, B. T., and Anderson, N. 1985. Temperature fluctuations and infrared emission from interstellar grains. *Astrophys. J.* 292:494–499.

Draine, B. T., and Katz, N. 1986. Magnetohydrodynamic shocks in diffuse clouds. II. production of CH^+, OH, CH, and other species. *Astrophys. J.* 310:392–407.

Draine, B. T. and Lee, H. M. 1984. Optical properties of interstellar graphite and silicate grains. *Astrophys. J.* 285:89–108.

Draine, B. T., and Roberge, W. G. 1982. A model for the intense molecular line emission from OMC-1. *Astrophys. J. Lett.* 259:91–96.

Draine, B. T., and Sutin, B. 1987. Collisional charging of interstellar grains. *Astrophys. J.* 320:803–817.

Draine, B. T., Roberge, W. G., and Dalgarno, A. 1983. Magnetohydrodynamic shock waves in molecular clouds. *Astrophys. J.* 264:485–507.

Drapatz, S., Haser, L., Hofmann, R., Oda, N., and Iyengar, K.V.K. 1983. Far-infrared spectrophotometry of the Orion molecular cloud 1 ridge. *Astron. Astrophys.* 128:207–211.

Drapatz, S., Larson, H. P., and Davis, D. S. 1987. Search for methane in comet Halley.

Astron. Astrophys. 187:497–501.

Dreher, J. W., and Welch, W. J. 1981. Discovery of shell structure in ultracompact H II region W3(OH). *Astrophys. J.* 245:857–865.

Dreibus, G., and Wänke, H. 1980. The bulk composition of the eucrite parent asteroid and its bearing on planetary evolution. *Z. Naturforsch.* 35:204–216.

Dreibus, G., and Wänke, H. 1987. Volatiles on Earth and Mars: a comparison. *Icarus* 71:225–240.

Dreibus, G., and Wänke, H. 1989. Supply and loss of volatile constituents during the accretion of the terrestrial planets. In *Origin and Evolution of Planetary and Satellite Atmospheres*, eds. S. K. Atreya, J. B. Pollack and M. S. Matthews (Tucson: Univ. of Arizona Press), pp. 268–288.

Dreibus, G., and Wänke, H. 1990. Comparison of the chemistry of Moon and Mars. *Adv. Space Res.* 10:3–4.

Dreiling, L. A., and Bell, R. A. 1980. The chemical composition, gravity, and temperature of Vega. *Astrophys. J.* 241:736–758.

Drukier, G. A., Fahlman, G. G., Richer, H. B., and VandenBerg, D. A. 1988. A deep luminosity function of the globular cluster M13. *Astron. J.* 95:1415–1421.

Drury, L. O'C. 1979. Ph. D. Thesis, Cambridge Univ.

Dubouloz, N., Raulin, F., Lellouch, E., and Gautier, D. 1989. Titan's hypothesized ocean properties: the influence of surface temperature and atmospheric composition uncertainties. *Icarus* 82:81–96.

Duerbeck, H. W. 1987. A reference catalogue and atlas of galactic novae. *Space Sci. Rev.* 45:1–212; republished as *A Reference Catalog and Atlas of Galactic Novae.* 1987 (Dordrecht: D. Reidel).

Duerr, R., Imhoff, C. L., and Lada, C. J. 1982. Star formation in the λ Orion regions. I. The distribution of young objects. *Astrophys. J.* 261:135–150.

DuFresne, E. R., and Anders, E. 1962. On the chemical evolution of the carbonaceous chondrites. *Geochim. Cosmochim. Acta* 26:1085–1114.

Duley, W.W., and Williams, D.A. 1984. *Interstellar Chemistry* (New York: Academic Press).

Duley, W. W., Jones, A. P., and Williams, D. A. 1989. Hydrogenated amorphous carbon-coated silicate particles as a source of interstellar extinction. *Mon. Not. Roy. Astron. Soc.* 236:709–725.

Duley, W. W., Jones, A. P., Whittet, D. C. B., and Williams, D. A. 1989. Mantle desorption from amorphous grains. *Mon. Not. Roy. Astron. Soc.* 241:697–705.

Dulk, G. A. 1985. Radio emission from the Sun and stars. *Ann. Rev. Astron. Astrophys.* 23:169–224.

Duncan, D. K. 1981. Lithium abundances, K line emission and ages of nearby solar type stars. *Astrophys. J.* 248:651–669.

Duncan, M., Quinn, T., and Tremaine, S. 1987. The formation and extent of the solar system comet cloud. *Astron. J.* 94:1330–1338.

Duncan, M., Quinn, T., and Tremaine, S. 1988. The origin of short-period comets. *Astrophys. J. Lett.* 328:69–73.

Duncan, M., Quinn, T., and Tremaine, S. 1989. The long-term evolution of orbits in the solar system: a mapping approach. *Icarus* 82:402–418.

Duquennoy, A., and Mayor, M. 1991. Multiplicity among solar-type stars in the solar neighborhood. II. Distributions of the orbital elements in an unbiased sample. *Astron. Astrophys.* 246–289.

Durisen, R. H., Gingold, R. A., Tohline, J. E., and Boss, A. P. 1986. Dynamic fission instabilities in rapidly rotating n=3/2 polytropes: a comparison of results from finite-difference and smoothed particle hydrodynamics codes. *Astrophys. J.* 305:281–308.

Durisen, R. H., Yang, S., and Grabhorn, R. 1989*a*. Numerical simulations of fission.

In *Highlights of Astronomy*, vol. 8, eds. D. McNally (Dordrecht: Kluwer), pp. 133–135.

Durisen, R. H., Yang, S., Cassen, P., and Stahler, S. W. 1989*b*. Numerical models of rotating protostars. *Astrophys. J.* 345:959–971.

Duschl, W., and Tscharnuter, W. M. 1991. On the inner boundary condition of thin keplerian accretion disks. *Astron. Astrophys.* 241:153–158.

Dutkevitch, D., Edwards, S., and Strom, S. E. 1989. The angular dependence of the wind from R Mon. *Bull. Amer. Astron. Soc.* 20:1093 (abstract).

Duvert, G., Cernicharo, J., Bachiller, R., and Gómez-González, J. 1990. Star formation in a small globule in IC1396. *Astron. Astrophys.* 233:190–196.

Dyck, H. M., and Howell, R. R. 1982. Speckle interferometry of molecular cloud sources at 4.8 μm. *Astron. J.* 87:400–403.

Dyck, H. M., and Staude, H. J. 1982. Near-infrared slit scans of molecular cloud sources. II. *Astron. Astrophys.* 109:320–325.

Dyck, H. M., Simon, T., and Zuckerman, B. 1982. Discovery of an infrared companion to T Tauri. *Astrophys. J. Lett.* 255:103–106.

Dyson, J. 1987. Theoretical models of Herbig-Haro objects. In *Circumstellar Matter: Proc. IAU Symp. 22*, eds. I. Appenzeller and C. Jordan (Dordrecht: D. Reidel), pp. 159–172.

Dziembowski, W. A., Goode, P. R., and Libbrecht, K. G. 1989. The radial gradient in the Sun's rotation. *Astrophys. J. Lett.* 337:53–59.

Eberhardt, P., Dolder, U., Schulte, W., Krankowsky, D., Lämmerzahl, P., Hoffman, J. H., Hodges, R. R., Berthelier, J. J., and Illiano, J. M. 1987*a*. The D/H ratio in water from comet P/Halley. *Astron. Astrophys.* 187:435–437.

Eberhardt, P., Krankowsky, D., Schulte, W., Dolder, U., Lämmerzahl, P., Berthelier, J. J., Woweries, J., Stubbemann, U., Hodges, R. R., Hoffmann, J. H., and Illiano, J. M. 1987*b*. The CO and N_2 abundance in comet P/Halley. *Astron. Astrophys.* 187:481–484.

Eberhardt, P., Meier, R., Krankowsky, D., and Hodges, R. R. 1991. Methanol abundances in comet P/Halley from in-situ measurements. *Bull. Amer. Astron. Soc.* 23:1161 (abstract).

Ebert, R. 1955. Temperatur des interstellaren Gases bei großen Dichten. *Z. Astrophys.* 37:222–229.

Ebert, R., Hoerner, S. von, and Temesvary, S. 1960. *Die Enstehung von Sternen durch Kondensation diffuser Materie* (Berlin: Springer-Verlag), pp. 184–324.

Eckart, A., Downes, D., Genzel, R., Harris, A. I., Jaffee, D. T., and Wild, W. 1990. Warm gas and spatial variations of molecular excitation in the nuclear regions of IC 342. *Astrophys. J.* 348:434–447.

Eckert, W. J., Brouwer, D., and Clemence, G. 1951. Coordinates of the five outer planets. *Astron. Papers American Ephem. and Nautical Almanac* (Washington: Nautical Almanac Office, U.S. Gov. Printing Office) vol. 12

Edwards, S., and Snell, R. L. 1982. A search for high velocity molecular gas around T Tauri stars. *Astrophys. J.* 261:151–160.

Edwards, S., and Snell, R. L. 1983. A survey of high-velocity molecular gas in the vicinity of Herbig-Haro objects. I. *Astrophys. J.* 270:605–619.

Edwards, S., and Snell, R. L. 1984. A survey of high-velocity molecular gas near Herbig-Haro objects. II. *Astrophys. J.* 281:237–249.

Edwards, S., and Strom, S. E. 1987. Energetic winds from low mass young stellar objects. In *Proc. Fifth Cambridge Workshop on Cool Stars, Stellar Systems and the Sun*, eds. J. L. Linsky and R. Stencel (Berlin: Springer-Verlag), pp. 443–545.

Edwards, S., Strom, S. E., Snell, R. L., Jarrett, T. H., Beichman, C. A., and Strom, K. M. 1986. Extended far-infrared emission associated with mass outflow from young stars: L 1551 IRS 5. *Astrophys. J. Lett.* 307:65–68.

Edwards, S., Cabrit, S., Strom, S., Heyer, I., Strom, K., and Anderson, E. 1987. Forbidden line and Hα profiles in T Tauri spectra: a probe of anisotropic mass outflows and circumstellar disks. *Astrophys. J.* 321:473–495.

Edwards, S., Cabrit, S., Ghandour, L., and Strom, S. E. 1989. Forbidden lines in T Tauri star spectra: A clue to the origin of T Tauri winds. In *ESO Workshop on Low Mass Star Formation and Pre-Main Sequence Objects*, ed. B. Reipurth (Garching: European Southern Obs.), pp. 385–398

Efremov, Yu. N. 1989. *Stellar Complexes* (London: Harwood).

Efremov, Yu. N., and Sitnik, T. G. 1988. Young galactic star gas complexes. *Soviet Astron. Lett.* 14:347–352.

Eggen, O. J. 1976. Is star formation bimodal? The early main sequence. *Quart. J. Roy. Astron. Soc.* 17:472–487.

Eichler, D. 1992. Magnetic Confinement of Jets. Preprint.

Eiroa, C., and Casali, M. M. 1989. The Serpens sources SVS4 and FIRS 1: new results from infrared images. *Astron. Astrophys. Lett.* 223:17–19.

Eiroa, C., and Leinert, Ch. 1987. Speckle observations of the ice feature in the young double source Serpens SVS20. *Astron. Astrophys.* 188:46–48.

Eiroa, C., Lenzen, R., Leinert, Ch., and Hodapp, K.-W. 1987. Serpens-SVS20: a new young infrared double source. *Astron. Astrophys.* 179:171–175.

Eislöffel, J., Mundt, R., and Ray, T. P. 1989. Proper motion measurements of jets from young stars: First results. *Astron. Ges. Abstr.*, Ser. 3, p. 35.

Eislöffel, J., Hessman, F. V., and Mundt, R. 1990. High resolution spectroscopy of the new FU Orionis object BBW 76. *Astron. Astrophys.* 232:70–74

Elias, J. H. 1978*a*. An infrared study of the Ophiuchus dark cloud. *Astrophys. J.* 224:453–472

Elias, J. H. 1978*b*. A study of the IC 5146 dark cloud complex. *Astrophys. J.* 223:859–875.

Elias, J. H. 1978*c*. A study of the Taurus dark cloud complex. *Astrophys. J.* 224:857–872.

Elias, J. H. 1980. H_2 emission from Herbig-Haro objects. *Astrophys. J.* 241:728–735.

Elitzur, M., Hollenbach, D. J., and McKee, C. 1989. H_2O masers in star-forming regions. *Astrophys. J.* 346:983–990.

Elliot, J. L., and Young, Y. A. 1991. Limits on the radius and possible atmosphere of Charon from its 1988 stellar occultation. *Icarus* 89:244–254.

Elmegreen, B. G. 1978*a*. On the determination of magnetic fields in dense cloud complexes by the observation of Zeeman splitting. *Astrophys. J. Lett.* 225:85–88.

Elmegreen, B. G. 1978*b*. On the interaction between a strong stellar wind and a surrounding disk nebula. *Moon and Planets* 19:261–277.

Elmegreen, B. G. 1979*a*. Gravitational collapse in dust lanes and the appearance of spiral structure in galaxies. *Astrophys. J.* 231:372–383.

Elmegreen, B. G. 1979*b*. Magnetic diffusion and ionization fractions in dense molecular clouds: the role of charged grains. *Astrophys. J.* 232:729–739.

Elmegreen, B. G. 1979*c*. On the disruption of a protoplanetary disk nebula by a T Tauri-like solar wind. *Astron. Astrophys.* 80:77–78.

Elmegreen, B. G. 1981*a*. Magnetic coupling between various phases of interstellar matter. In *The Phases of the Interstellar Medium*, ed. J. Dickey (Green Bank, W. V.: NRAO), pp. 25–34.

Elmegreen, B. G. 1981*b*. The role of magnetic fields in constraining the translational motions of giant cloud complexes. *Astrophys. J.* 243:512–525.

Elmegreen, B. G. 1982*a*. The formation of giant cloud complexes. In *Submillimeter Wave Astronomy*, eds. J. E. Beckman and J. P. Phillips (Cambridge: Cambridge Univ. Press), pp. 3–14.

Elmegreen, B. G. 1982*b*. The formation of giant cloud complexes by the Parker-Jeans instability. *Astrophys. J.* 253:655–665.

Elmegreen, B. G. 1985*a*. Molecular clouds and star formation: an overview. In *Protostars and Planets II*, eds. D. C. Black and M. S. Matthews (Tucson: Univ. of Arizona Press), pp. 33–58.

Elmegreen, B. G. 1985*b*. Energy dissipation in clumpy magnetic clouds. *Astrophys. J.* 299:196–210.

Elmegreen, B. G. 1985*c*. The initial mass function and implications for cluster formation. In *Birth and Infancy of Stars*, eds. R. Lucas, A. Ormont and R. Stora (Amsterdam: North-Holland), pp. 257–277.

Elmegreen, B. G. 1987*a*. Cloud formation and destruction. In *Interstellar Processes*, eds. D. J. Hollenbach and H. A. Thronson, Jr. (Dordrecht: D. Reidel), pp. 259–280.

Elmegreen, B. G. 1987*b*. Large-scale star formation: Density waves, superassociations and propagation. In *Star Forming Regions: Proc. IAU Symp. 115*, eds. M. Peimbert and J. Jugaku (Dordrecht: D. Reidel), pp. 457–481.

Elmegreen, B. G. 1987*c*. Supercloud formation by gravitational instabilities in sheared magnetic galaxy disks. *Astrophys. J.* 312:626–634.

Elmegreen, B. G. 1987*d*. Formation and evolution of the largest cloud complexes in spiral galaxies. In *Physical Processes in Interstellar Clouds*, eds. G. E. Morfill and M. Scholer (Dordrecht: D. Reidel), pp. 1–12.

Elmegreen, B. G. 1988*a*. Magnetic cloud collision fronts. *Astrophys. J.* 326:616–638.

Elmegreen, B.G. 1988*b*. Structure and motions in molecular clouds. In *Interstellar Matter*, eds. P. C. Myers, J. Moran and P. Ho (New York: Gordon and Breach), pp. 55–64.

Elmegreen, B. G. 1989*a*. A pressure and metallicity dependence for molecular cloud correlations and the calibration of mass. *Astrophys. J.* 338:178–196.

Elmegreen, B. G. 1989*b*. On the gravitational collapse of decelerating shocked layers in OB associations. *Astrophys. J.* 340:786–811.

Elmegreen, B. G. 1989*c*. Gravitational instabilities in shearing, magnetic galaxies with a cloudy interstellar gas. *Astrophys. J. Lett.* 342:67–70.

Elmegreen, B. G. 1989*d*. Molecular cloud formation by gravitational instabilities in a clumpy interstellar medium. *Astrophys. J.* 344:306–310.

Elmegreen, B. G. 1989*e*. A revised mass spectrum for the random collisional build-up model of molecular cloud formation. *Astrophys. J.* 347:859–862.

Elmegreen, B. G. 1990*a*. Theories of molecular cloud formation. In *The Evolution of the Interstellar Medium*, ed. L. Blitz (San Francisco: Astron. Soc. of the Pacific), pp. 247.

Elmegreen, B. G. 1990*b*. A wavelike origin for clumpy structure and broad line wings in molecular clouds. *Astrophys. J. Lett.* 361:77–80.

Elmegreen, B. G. 1991*a*. Cloud formation by combined instabilities in galactic gas layers: evidence for a Q threshold in the fragmentation of shearing wavelets. *Astrophys. J.* 378:139–156.

Elmegreen, B. G. 1991*b*. The origin and evolution of giant molecular clouds. In *Physics of Star Formation and Early Stellar Evolution*, eds. C. J. Lada and N. Kalafis (Dordrecht: Kluwer), p. 35.

Elmegreen, B. G., and Clemens, C. 1985. On the formation rate of galactic clusters in clouds of various masses. *Astrophys. J.* 294:523–532.

Elmegreen, D. M., and Elmegreen, B. G. 1980. The location of star-forming regions in barred Magellanic-type systems. *Astron. J.* 85:1325–1327.

Elmegreen, B. G., and Elmegreen, D. M. 1983. Regularly spaced H II regions and superclouds in spiral galaxies: Clues to the origins of cloudy structure. *Mon. Not. Roy. Astron. Soc.* 203:31–45.

Elmegreen, B. G., and Elmegreen, D. M. 1987. HI superclouds in the inner galaxy. *Astrophys. J.* 320:182–198.

Elmegreen, B. G., and Elmegreen, D. M. 1989. The arms of spiral galaxies. In *Evolutionary Phenomena in Galaxies*, eds. J. Beckman and B. Pagel (Cambridge: Cambridge Univ. Press), pp. 83–99.

Elmegreen, B. G., and Lada, C. J. 1977. Sequential formation of subgroups in OB associations. *Astrophys. J.* 214:725–741.

Elmegreen, B. G., Dickinson, D. F., and Lada, C. J. 1978. Heat sources for bright-rimmed molecular clouds: CO observation of NGC 7822. *Astrophys. J.* 220:853–863.

Elmegreen, B. G., Lada, C. J., and Dickinson, D. F. 1979. The structure and extent of the giant molecular cloud near M17. *Astrophys. J.* 230:415–427.

Elmegreen, B. G., Elmegreen, D. M., and Morris, M. 1980. On the abundance of carbon monoxide in galaxies: A comparison between spiral and magellanic irregular galaxies. *Astrophys. J.* 240:455–463.

Elmegreen, D. M., Phillips, J., Beck, K., Thomas, H., and Howard, J. 1988. A search for near-infrared counterparts of *IRAS* embedded sources in the M 17 SW giant molecular cloud. *Astrophys. J.* 335:803–813.

Elphick, C., Regev, O., and Spiegel, E. A. 1991. Complexity from thermal instability. *Mon. Not. Roy. Astron. Soc.* 250:617–628.

Elsässer, H., and Staude, H. J. 1978. On the polarization of young stellar objects. *Astron. Astrophys. Lett.* 70:3–6.

Elson, R. A. W., Fall, S. M., and Freeman, K. C. 1989. The stellar content of rich young clusters in the Large Magellanic Cloud. *Astrophys. J.* 336:734–751.

Elster, J., and Geitel, H. 1885. Über den elektrischen Vorgang in den Gewitterwolken. *Wiedemann's Ann. Phys.* 25:116.

Encrenaz, T., and Knacke, R. 1991. Carbonaceous compounds in comets: infrared observations. In *Comets in the Post-Halley Era*, eds. R. Newburn, M. Neugebauer and J. Rahe (Dordrecht: Kluwer), pp. 107–138.

Encrenaz, T., d'Hendecourt, L., and Puget, J. L. 1988. The interpretation of the 3.2–3.5 μm emission feature in the spectrum of comet P/Halley: abundances in the comet and in interstellar matter. *Astron. Astrophys.* 207:162–173.

Endal, A.S. and Sofia, S. 1982. Rotation in solar-type stars. I. Evolutionary models for the spin-down of the Sun. *Astrophys. J.* 243:625–640.

Engel, S., Lunine, J. I. and Lewis, J. S. 1990. Solar nebula origin for volatile gases in Halley's comet. *Icarus* 85:380–393.

Engelke, C. W. 1990. LWIR Stellar Calibration: Infrared Spectral Curves for 30 Standard Stars. Lincoln Lab. Project Rept. SDP-327.

Erickson, E.F., Knacke, R.F., Tokunaga, A.T., and Haas, M. R. 1981. The 45 micron H_2O ice band in the Kleinmann-Low nebula. *Astrophys. J.* 245:148–153.

Erickson, N. R., Goldsmith, P. J., Snell, R. L., Berson, R. L., Hugunenin, G. R., Ulich, B. L., and Lada, C. 1982. Detection of bipolar CO outflow in Orion. *Astrophys. J. Lett.* 261:103–107.

Espinasse, S., Klinger, J., Ritz, C., and Schmitt, B. 1991. Modeling of the thermal behavior and of the chemical differentiation of cometary nuclei. *Icarus* 92:350–365.

Esposito, L. W., Brahic, A., Burns, J. A., and Marouf, E. A. 1991. Particle properties and processes in Uranus' rings. In *Uranus*, eds. J. Bergstralh, E. Miner and M. S. Matthews (Tucson: Univ. of Arizona Press), pp. 410–465.

Evans, N. J., II. 1985. Star formation: An overview. In *Protostars and Planets II*, eds. D. C. Black and M. S. Matthews (Tucson: Univ. of Arizona Press), pp. 175–187.

Evans, N. J., II. 1991. Star formation—an observational view. In *Frontiers of Stellar Evolution*, ed. D. L. Lambert (San Francisco: Astron. Soc. of the Pacific), pp.

45–96.

Evans, N. J., II, Blair, G. N., and Beckwith, S. 1977. The energetics of molecular clouds. I. Methods of analysis and application to the S255 molecular cloud. *Astrophys. J.* 217:448–463.

Evans, N. J., II, Levreault, R. M., Beckwith, S., and Skrutskie, M. 1987. Observations of infrared emission lines and radio continuum emission from pre-main-sequence objects. *Astrophys. J.* 320:364–375.

Evans, N. J., II, Lacy, J. H., and Carr, J. S. 1991. Infrared molecular spectroscopy toward the Orion IRc2 and IRc7 sources: A new probe of physical conditions and abundances in molecular clouds. *Astrophys. J.* 383:674–692.

Everhart, E. 1967. Intrinsic distributions of cometary perihelia and magnitudes. *Astron. J.* 72:1002–1011.

Everhart, E. 1968. Changes in total energy of comets passing through the solar system. *Astron. J.* 73:1039–1052.

Everhart, E. 1972. The origin of short-period comets. *Astrophys. J. Lett.* 10:131–135.

Everhart, E. 1977. In *Comets, Asteroids, and Meteorites*, ed. A. H. Delsemme (Ohio: Univ. of Toledo), pp. 99.

Ewald, R., Imhoff, C. L., and Giampapa, M. S. 1986. IUE Observations of the Eruptive Pre-Main Sequence Object FU Orionis. In *New Insights in Astrophysics*, ed. E. J. Rolfe, ESA SP-263, pp. 205–207.

Ezer, D., and Cameron, A. G. W. 1967. A study of solar evolution. *Canadian J. Phys.* 43:1497–1517.

Fahey, A. J., Zinner, E., Crozaz, G., and Kornacki, A. S. 1987. Microdistributions of Mg isotopes and REE abundances in a Type A calcium-aluminum-rich inclusion from Efremovka. *Geochim. Cosmochim. Acta* 51:3215–3229.

Fahlman, G. G., Richer, H. B., Searle, L., and Thompson, I. B. 1990. Faint star counts in NGC 6397. *Astrophys. J. Lett.* 343:49–51.

Falgarone, E., and Lequeux, J. 1973. A discussion of the distribution of interstellar matter close to the Sun. *Astron. Astrophys.* 25:253–260.

Falgarone, E., and Pérault, M. 1987. Structure and physics of cool giant molecular complexes. In *Physical Processes in Interstellar Clouds*, eds. G. Morfill and M. S. Scholer (Dordrecht: D. Reidel), pp. 59–73.

Falgarone, E., and Pérault, M. 1988. Structure at the 0.02 pc scale in molecular gas of low H_2 column density. *Astron. Astrophys.* 205:L1–L4.

Falgarone, E., and Phillips, T. 1990. A signature of the intermittency of interstellar turbulence: The wings of molecular line profiles. *Astrophys. J.* 359:344–354.

Falgarone, E., and Phillips, T. 1991. Signatures of turbulence in the dense interstellar medium. In *Fragmentation of Molecular Clouds and Star Formation*, eds. E. Falgarone, F. Boulanger and G. Duvert (Dordrecht: Kluwer), pp. 119–136.

Falgarone, E., and Puget, J. L. 1986. Model of clumped molecular clouds. II. Physics and evolution of the hierarchical structure. *Astron. Astrophys.* 162:235–247.

Falgarone, E., Phillips, T., and Walker, C. K. 1991. The edges of molecular clouds: Fractal boundaries and density structure. *Astrophys. J.* 378:186–201.

Falle, S. A. E. G., Innes, D. E., and Wilson, M. J. 1987. Steady stellar jets. *Mon. Not. Roy. Astron. Soc.* 225:741–759.

Fanale, F. P. 1971. A case for catastrophic early degassing of the Earth. *Chem. Geol.* 8:79–105.

Fanale, F. P., and Cannon, W. A. 1971. Physical adsorption of rare gas on terrigenous sediments. *Earth and Planet. Sci. Lett.* 11:362–268.

Fanale, F. P., and Cannon, W. A. 1972. Origin of planetary rare gas: the possible role of adsorption. *Geochim. Cosmochim. Acta* 36:319–328.

Fanale, F. P., and Salvail, J. R. 1984. An idealized short-period comet flyby. *Astron. Astrophys.* 187:835–838.

Fanale, F. P., Johnson, T. V., and Matson, D. L. 1974. Io: a surface evaporite deposit? *Science* 186:922–925.

Fanale, F. P., Cannon, W. A., and Owen, T. 1978. Mars: regolith adsorption and the relative concentrations of atmospheric rare gases. *Geophys. Res. Lett.* 5:77–80.

Farkas, A. 1935. *Orthohydrogen, Parahydrogen and Heavy Hydrogen* (New York: Cambridge Univ. Press).

Faulkner, J., Lin, D. N. C., and Papaloizou, J. 1983. On the evolution of accretion disc flow in cataclysmic variables. I. The prospect of a limit cycle in dwarf nova systems. *Mon. Not. Roy. Astron. Soc.* 205:359–375.

Federman, S. R., and Willson, R. F. 1982. Diffuse interstellar clouds associated with dark clouds. *Astrophys. J.* 260:124–127.

Federman, S. R., Glassgold, A. E., and Kwan, J. 1979. Atomic to molecular hydrogen transition in interstellar clouds. *Astrophys. J.* 227:466–473.

Federman, S. R., Huntress, W. T., Jr., and Prasad, S. S. 1990. Modelling the chemistry of dense interstellar clouds. I. Observational constraints for the chemistry. *Astrophys. J.* 354:504–512.

Fegley, B., Jr. 1992. Disequilibrium chemistry in the solar nebula and early solar system: implications for the chemistry of comets. In *Proc. Comet Sample Return Workshop*, ed. S. Chang, in press.

Fegley, B., Jr., and Palme, H. 1985. Evidence for oxidizing conditions in the solar nebula from Mo and W depletions in refractory inclusions in carbonaceous chondrites. *Earth Planet. Sci. Lett.* 72:311–326.

Fegley, B., Jr., and Prinn, R. 1986. Chemical models of the deep atmosphere of Uranus. *Astrophys. J.* 307:852–865.

Fegley, B., Jr., and Prinn, R. G. 1988. The predicted abundances of deuterium-bearing gases in the atmospheres of Jupiter and Saturn. *Astrophys. J.* 326:490–508.

Fegley, B., Jr., and Prinn, R. G. 1989. Solar nebula chemistry: Implications for volatiles in the solar system. In *The Formation and Evolution of Planetary Systems*, eds. H. A. Weaver and L. Danly (Cambridge: Cambridge Univ. Press), pp. 171–211.

Feigelson, E. D. 1982. X-ray emission from young stars and implications for the early solar system. *Icarus* 51:155–163.

Feigelson, E. D., and DeCampli, W. M. 1981. Observations of X-ray emission from T Tauri stars. *Astrophys. J. Lett.* 243:89–93.

Feigelson, E. D., and Kriss, G. A. 1989. Soft X-ray observations of pre-main sequence stars in the Chamaeleon dark cloud. *Astrophys. J.* 338:262–276

Feigelson, E. D., and Montmerle, T. 1985. An extremely variable radio star in the ρ Ophiuchi cloud. *Astrophys. J. Lett.* 289:19–23.

Feigelson, E. D., Jackson, J. M., Mathieu, R. D., Myers, P. C., and Walter, F. M. 1987. An X-ray survey for pre-main sequence stars in the Taurus-Auriga and Perseus molecular cloud complexes. *Astron. J.* 94:1251–1259.

Feigelson, E. D., Giampapa, M. S., and Vrba, F. J. 1990. Magnetic activity in pre-main-sequence stars. In *The Sun in Time*, eds. C. P. Sonett, M. S. Giampapa and M. S. Matthews (Tucson: Univ. of Arizona Press), pp. 658–681.

Feldman, P. D. 1983. Ultraviolet spectroscopy and the composition of cometary ice. *Science* 219:347–354.

Feldman, P. D. 1991a. The volatile composition of comets deduced from ultraviolet spectroscopy. In *Chemistry in Space*, eds. J. M. Greenberg and V. Pironello (Dordrecht: Kluwer), pp. 339–361.

Feldman, P. D. 1991b. Ultraviolet spectroscopy of cometary comae. In *Comets in the Post-Halley Era*, eds. R. Newburn, M. Neugebauer and J. Rahe (Dordrecht: Kluwer), pp. 139–148.

Feldman, P. D., and Brune, W. H. 1976. Carbon production in comet West (1975n).

Astrophys. J. Lett. 209:45–48.

Feldman, P. D., A'Hearn, M. F., Festou, M. C., McFadden, L. A., Weaver, H. A., and Woods, T. N. 1986. Is CO_2 responsible for the outbursts in comet Halley? *Nature* 324:433–436.

Feldman, P. D., Festou, M. C., A'Hearn, M. F., Arpigny, C., Butterworth, P. S., Cosmovici, C. B., Danks, A. C., Gilmozzi, R., Jackson, W. M., McFadden, L. A., Patriarchi, P., Schleicher, D. G., Tozzi, G. P., Wallis, M. E., Weaver, H. A., and Woods, T. N. 1987. IUE observations of comet P/Halley: evolution of the ultraviolet spectrum between September 1985 and July 1986. *Astron. Astrophys.* 187:325–328.

Feldman, P. D., Davidsen, A. F., Blair, W. P., Bowers, C. W., Dixon, W. V., Durrance, S. T., Ferguson, H. C., Henry, R. C., Kimble, R. A., Kriss, G. A., Kruk, J. W., Long, K. S., Moos, H. W., Vancura, O., and Gull, T. R. 1991. Observations of comet Levy (1990c) with the Hopkins ultraviolet telescope. *Astrophys. J. Lett.* 379:37–40.

Felenbok, P., Praderie, F., and Talavera, A. 1983. The Herbig Ae stars AB Aur: Absorption along line of sight and chromospheric emission. *Astron. Astrophys.* 128:74–83.

Ferland, G. J. 1979. Helium abundance in ejecta from CP Lacertae and V446 Herculis. *Astrophys. J.* 231:781–788.

Ferland, G. J., and Shields, G. A. 1978a. Fine structure lines and the 10 micron excess of nova Cygni 1975. *Astrophys. J. Lett.* 224:15–18.

Ferland, G. J., and Shields, G. A. 1978b. Heavy element abundances of nova Cygni 1975. *Astrophys. J.* 226:172–185.

Ferlet, R., Hobbs, L. M., and Vidal-Madjar, A. 1987. The Beta Pictoris circumstellar disk. *Astron. Astrophys.* 185:267–270.

Fernández, J. A. 1980. On the existence of a comet belt beyond Neptune. *Mon. Not. Roy. Astron. Soc.* 192:481–491.

Fernández, J. A. 1982. Dynamical aspects of the origin of comets. *Astron. J.* 87:1318–1332.

Fernández, J. A. 1985. The formation and dynamical survival of the comet cloud. In *Dynamics of Comets: Their Origin and Evolution*, eds. A. Carusi and G. B. Valsecchi (Dordrecht: D. Reidel), pp. 45–70.

Fernández, J. A., and Ip, W.-H. 1981. Dynamical evolution of a cometary swarm in the outer planetary region. *Icarus* 47:470–479.

Fernández, J. A., and Ip, W.-H. 1983. On the time evolution of the cometary influx in the region of the terrestrial planets. *Icarus* 54:377–387.

Fernández, J. A., and Ip, W.-H. 1984. Some dynamical aspects of the accretion of Uranus and Neptune: The exchange of angular momentum with planetesimals. *Icarus* 58:109–120.

Ferriere, K. M., Zweibel, E. G., and Shull, J. M. 1988. Hydromagnetic wave heating of the low density interstellar medium. *Astrophys. J.* 332:984–994.

Ferriere, K. M., MacLow, M. M., and Zweibel, E. G. 1991. Expansion of a superbubble in a uniform magnetic field. *Astrophys. J.* 375:239–253.

Festou, M. C. 1990a. Comparative cometology from IUE observations. In *Evolution in Astrophysics*, ESA SP-310, pp. 3–10.

Festou, M. C. 1990b. *Comets: International Ultraviolet Explorer. Uniform Low Dispersion Archive. IUE-ULDA Access Guide No. 2*, ESA SP-1134.

Festou, M. C., and Feldman, P. D. 1987. Comets. In *Exploring the Universe with the IUE Satellite*, ed. Y. Kondo (Dordrecht: D. Reidel), pp. 101–118.

Festou, M. C., Feldman, P. D., and Weaver, H. A. 1982. The ultraviolet bands of the $CO_2{}^+$ ion in comets. *Astrophys. J.* 256:331–338.

Fiebig, D., and Güsten, R. 1989. Strong magnetic fields in interstellar H_2O maser clumps. *Astron. Astrophys.* 214:333–338.

Field, G. B. 1965. Thermal instability. *Astrophys. J.* 142:531–567.

Field, G. B., and Saslaw, W. C. 1965. A statistical model of the formation of stars and interstellar clouds. *Astrophys. J.* 142:568–583.

Field, G. B., Goldsmith, D. W., and Habing, H. J. 1969. Cosmic ray heating of the interstellar gas. *Astrophys. J. Lett.* 155:149–154.

Findeisen, W. 1943. Untersuchung über die eissplitterung an reifschichten. Meteor. Zeit. 60:145–154.

Fink, U., Smith, B. A., Benner, D. C., Johnson, J. R., Reitsema, H. J., and Westphal, J. A. 1980. Detection of a CH_4 atmosphere on Pluto. *Icarus* 44:62–71.

Finkenzeller, U., and Basri, G. 1987. The atmospheres of T Tauri stars. I. High-resolution calibrated observations of moderately active stars. *Astrophys. J.* 318:823–843.

Finkenzeller, U., and Jankovics, I. 1984. Line profiles and radial velocities of Herbig Ae/Be stars. *Astron. Astrophys. Suppl.* 57:285–326.

Finkenzeller, U., and Mundt, R. 1984. The Herbig Ae/Be stars associated with nebulosity. *Astron. Astrophys. Suppl.* 55:109–141.

Fischer, J., Sanders, D. B., Simon, M., and Solomon, P. M. 1985. High-velocity gas flows associated with H_2 emission regions: how are they related and what powers them? *Astrophys. J.* 293:508–521.

Fitzpatrick, E. L., and Massa, D. 1986. An analysis of the shapes of ultraviolet extinction curves. I. The 2175 Å bump. *Astrophys. J.* 307:286–294.

Flannery, B. P. 1975. Gas flow in cataclysmic variable stars. *Astrophys. J.* 201:661–694.

Fleck, R. C., Jr. 1983. A note on compressibility and energy cascade in turbulent molecular clouds. *Astrophys. J. Lett.* 272:45–48.

Fleck, R. C., Jr. 1984. The Kelvin-Helmoltz interface instability in the interstellar environment. I. The morphology of the Corona Australis dark cloud complex. *Astron. J.* 89:506–508.

Fleck, R. C., Jr. 1989. The Kelvin-Helmholtz instability in the interstellar Environment. II. Interstellar cloud rotation. *Astron. J.* 97:783–785.

Fleck, R. C., Jr. 1990. Comment on 'Magnetic reconnection flares in the protoplanetary nebula and the possible origin of meteorite chondrules'. *Icarus* 87:241–243.

Flett, A. M., and Murray, A. G. 1991. First results from a submillimetre polarimeter on the James Clerk Maxwell telescope. *Mon. Not. Roy. Astron. Soc.* 249:4P–6P.

Forest, E. 1991. Sixth-Order Lie Group Integrators. Lawrence Berkeley Lab. Preprint.

Forrest, W. J., Ninkov, Z., Garnett, J. D., Skrutskie, M. F., and Shure, M. 1989. Discovery of low mass objects in Taurus. In *Third Infrared Detector Technology Workshop*, ed. C. R. McCreight, NASA TM-102209, pp. 221–230.

Foti, A. M., Baratta, G. A., Leto, G., and Strazzulla, G. 1991. Molecular alteration and carbonization of glycine by ion irradiation. *Europhys. Lett.* 16:201–204.

Fowler, A. M., Gillett, F. C., Gregory, B., Joyce, R. R., Probst, R. G., and Smith, R. 1987. The NOAO infrared imagers: description and performance. In *Infrared Astronomy with Arrays*, eds. C. G. Wynn-Williams and E. E. Becklin (Honolulu: Univ. of Hawaii), pp. 197–203.

Fowler, W. A., Greenstein, J. L., and Hoyle, F. 1961. Nucleosynthesis during the early history of the solar system. *Geophys. J.* 6:148–220.

Franchi, I. A., Wright, I. P., and Pillinger, C. T. 1986. Heavy nitrogen in Bencubbin—a light-element isotopic anomaly in a stony-iron meteorite. *Nature* 323:138–140.

Franco, J. 1990. Scenarios for large scale star formation. In *Chemical and Dynamical Evolution of Galaxies*, eds. F. Ferrini, J. Franco and F. Matteucci (Pisa-Lugano: Giardini), in press.

Franco, J., Tenorio-Tagle, G., Bodenheimer, P., Rozyczka, M., and Mirabel, I. F.

1988. On the origin of the Orion and Monoceros molecular cloud complexes. *Astrophys. J.* 333:826–839.

Frank, J., King, A. R., and Raine, D. J. 1985. *Accretion Power in Astrophysics* (Cambridge: Cambridge Univ. Press).

Franklin, F., Lecar, M., Lin, D. N. C., and Papaloizou, J. 1980. Tidal torque on infrequently colliding particle disks in binary systems and the truncation of the asteroid belt. *Icarus* 42:272–280.

Franklin, F., Lecar, M., and Soper, P. 1989. On the original distribution of asteroids. II. Do stable orbits exist between Jupiter and Saturn? *Icarus* 79:223–227.

French, R. G., Elliot, J. L., French, L. M., Kangas, J. A., Meech, K. J., Ressler, M. E., Buie, M. W., Frogel, J. A., Holberg, J. B., Fuensalida, J. J., and Joy, M. 1988. Uranian ring orbits from Earth-based and Voyager occultation observations. *Icarus* 73:349–378.

Frerking, M. A., and Langer, W. D. 1982. Detection of pedestal features in dark clouds evidence for formation of low mass stars. *Astrophys. J.* 256:523–529.

Friberg, P., and Hjalmarson, Å. 1990. Molecular clouds in the Milky Way. In *Molecular Astrophysics*, ed. T. W. Hartquist (Cambridge: Cambridge Univ. Press), pp. 3–34.

Friberg, P., Madden, S.C., Hjalmarson, Å., and Irvine, W.M. 1988. Methanol in dark clouds. *Astron. Astrophys.* 195:281–289.

Frick, U., Mack, R., and Chang, S. 1985. Noble gas trapping and fractionation during synthesis of carbonaceous matter. *Proc. Lunar Planet. Sci. Conf.* 10:1961–1973.

Fricke, K. 1969. Stability of rotating stars. II. The influence of toroidal and poloidal magnetic fields. *Astron. Astrophys.* 1:388–398.

Fridlund, C. V. M., and White, G. J. 1989. High signal/noise ^{13}CO observations of the bipolar outflow in L1551. *Astron. Astrophys. Lett.* 223:13–16.

Fridlund, C. V. M., Sandqvist, A., Nordh, H. L., and Olofsson, G. 1989. The L1551 IRS5 CO bipolar outflow: acceleration and origin. *Astron. Astrophys.* 213:310–322.

Froeschlé, C., and Scholl, H. 1979. New numerical experiments to deplete the outer part of the asteroidal belt. *Astron. Astrophys.* 72:246–255.

Fujiwara, A. 1986. Results obtained by laboratory simulations of catastrophic impact. *Mem. Soc. Astron. Italiana* 57:47–64.

Fujiwara, A., Cerroni, P., Davis, D., Ryan, E., DiMartino, M., Holsapple, and Housen, K. 1989. Experiments and scaling laws on catastrophic collisions. In *Asteroids II*, eds. R. P. Binzel, T. Gehrels and M. S. Matthews (Tucson: Univ. of Arizona Press), pp. 240–265.

Fukui, Y. 1988. An unbiased survey of star formation regions with the Nagoya 4m radio telescope. *Vistas in Astronomy* 31:217–226.

Fukui, Y. 1989. Molecular outflows: their implication on protostellar evolution. In *ESO Workshop on Low Mass Star Formation and Pre-Main Sequence Objects*, ed. B. Reipurth (Garching: European Southern Obs.), pp. 95–117.

Fukui, Y., and Mizuno, A. 1991. A comparative study of star formation efficiencies in nearby molecular cloud complexes. In *Fragmentation of Molecular Clouds and Star Formation*, eds. E. Falgarone, F. Boulanger and G. Duvert (Dordrecht: Kluwer), pp. 275–288.

Fukui, Y., Sugitani, K., Takaba, H., Iwata, T., Mizuno, A., Ogawa, H., and Kawabata, K. 1986. Discovery of seven bipolar outflows by an unbiased survey. *Astrophys. J. Lett.* 311:85–88.

Fukui, Y., Takaba, H., Iwata, T., and Mizuno, A. 1988. A bipolar outflow: L1641-North and its ambient dense cloud. *Astrophys. J. Lett.* 325:13–15.

Fukui, Y., Iwata, T., Takaba, H., Mizuno, A., Ogawa, H., Kawabata, K., and Sugitani, K. 1989. Molecular outflows in protostellar evolution. *Nature* 342:161–163.

Fuller, G. A. 1989. Molecular Studies of Dense Cores. Ph. D. Thesis, Univ. of California, Berkeley.

Fuller, G. A., and Myers, P. C. 1987. Dense cores in dark clouds. In *Physical Processes in Interstellar Clouds*, eds. G. Morfill and M. Scholer (Dordrecht: D. Reidel), pp. 137–160.

Fuller, G. A., Myers, P. C., Welch, W. J., Goldsmith, P. F., Langer, W. D., Campbell, B. G., Guilloteau, S., and Wilson, R. W. 1991. Anatomy of the Barnard 5 core. *Astrophys. J.* 376:135–149.

Furth, H. P., Killeen, J., and Rosenbluth, M. N. 1963. Finite resistivity instabilities of a sheet pinch. *Phys. Fluids* 6:459

Gahm, G. F. 1988. Some aspects of T Tauri variability. In *Formation and Evolution of Low Mass Stars*, eds. A. K. Dupree and M. T V T. Lago (Dordrecht: Kluwer), pp. 295–304.

Gahm, G. F. 1990. Flares on T Tauri stars. In *Flare Stars in Star Clusters, Associations, and the Solar Vicinity: Proc. IAU Symp. 137*, eds. L. V. Mirzoyan, B. K. Pettersen and M. K. Tsvetkov (Dordrecht: Kluwer), pp. 193–207.

Gahm, G. F., Fischerström, C., Liseau, R., and Lindroos, K. P. 1989. Long- and short-term variability of the T Tauri Star RY Lupi. *Astron. Astrophys.* 211:115–130

Galeev, A. A. Rosner, R., Serio, S. and Vaiana, G. S. 1979*a*. Dynamics of coronal structures: magnetic-field related heating and loop energy balance. *Astrophys. J.* 243:301–308.

Galeev, A. A., Rosner, R., and Vaiana, G. S. 1979*b*. Structured coronae of accretion disks. *Astrophys. J.* 229:318–326.

Gallagher, J. S., III, and Hunter, D. A. 1984. Structure and evolution of irregular galaxies. *Ann. Rev. Astron. Astrophys.* 22:37–74.

Gallagher, J. S., and Starrfield, S. G. 1978. Theory and observations of classical novae. *Ann. Rev. Astron. Astrophys.* 16:171–214.

Gallagher, J. S., Hege, E. K., Kopriva, D. A., Williams, R. E., and Butcher, H. R. 1980. *Astrophys. J.* 237:55–60.

Galli, D. 1990. Formazione di Stelle Galattica ed Extragalattica: Problematiche Teoriche. Ph. D. Thesis, Univ. of Florence.

Gallino, R., Busso, M., Picchio, G., and Raiteri, C. M. 1990. On the astrophysical interpretation of isotope anomalies in meteoritic SiC grains. *Nature* 348:298–302.

Garay, G., Moran, J. M., and Reid, M. J. 1987. Compact continuum radio sources in the Orion Nebula. *Astrophys. J.* 314:535–550.

Garay, G., Moran, J. M., and Haschick, A. D. 1989. The orion-KL super water maser. *Astrophys. J.* 338:244–261.

Garden, R. P. 1987. An extremely luminous H_2 flow in the DR21 star forming region. In *Star Forming Regions: Proc. IAU Symp. 115*, eds. M. Peimbert and J. Jugaku (Dordrecht: D. Reidel) pp. 325–328.

Garden, R. P., Russell, A. P. G., and Burton, M. G. 1990. Images of shock-excited molecular hydrogen in young stellar outflows. *Astrophys. J.* 354:232–241.

Garrison, L. M., Jr. 1979. Observational studies of the Herbig Ae/Be stars. III. Spectrophotometry. *Astrophys. J.* 224:535–545.

Gatley, I., Becklin, E. E., Matthews, K., Neugebauer, G., Penston, M. V., and Scoville, N. 1974. A new infrared complex and molecular cloud in Orion. *Astrophys. J. Lett.* 191:121–125.

Gatley, I., Becklin, E. E., Sellgren, K., and Werner, M. W. 1979. Far-infrared observations of M17: the interaction of an HII region with a molecular cloud. *Astrophys. J.* 233:575–583.

Gatley, I., Merrill, K. M., Fowler, A. M., and Tamura, M. 1991. The luminosity function in regions of massive star formation. In *Astrophysics with Infrared Arrays*, ed. R. Elston, pp. 230–237.

Gaustad, J. E. 1963. The opacity of diffuse cosmic matter and the early stages of star formation. *Astrophys. J.* 138:1050–1073.

Gautier, D., and Owen, T. 1989. The composition of outer planet atmospheres. In *Origin and Evolution of Planetary and Satellite Atmospheres*, eds. S. K. Atreya, J. B. Pollack and M. S. Matthews (Tucson: Univ. of Arizona Press), pp. 487–512.

Gautier, T. N., III, Fink, U., Treffers, R. R., and Larson, H. P. 1976. Detection of molecular hydrogen quadrupole emission in the Orion nebula. *Astrophys. J. Lett.* 207:129–133.

Gauvin, L. S., and Strom, K. M. 1991. A Study of the Stellar Population in the Chamaeleon Dark Clouds. Preprint.

Geballe, T. R. 1986. Absorption by solid and gaseous CO towards obscured infrared objects. *Astron. Astrophys.* 162:248–252.

Geballe, T. R., Baas, F., Greenberg, J. M., and Schutte, W. 1985. New infrared absorption features due to solid phase molecules containing sulfur in W33A. *Astron. Astrophys.* 146:L6–L8.

Gehrz, R. D. 1988. The infrared temporal development of classical novae. *Ann. Rev. Astron. Astrophys.* 26:377–412.

Gehrz, R. D. 1989*a*. Sources of star dust in the galaxy. In *Interstellar Dust: Proc. IAU Symp. 135*, eds. L. Allamandola and A. G. G. M. Tielens (Dordrecht: Kluwer), pp. 445–453.

Gehrz, R. D. 1989*b*. In *Proc. of the NAS/ASUSSR Workshop on Planetary Sciences* (Washington, D. C.: National Academy Press), in press.

Gehrz, R. D. 1990. New infrared results for classical novae. In *Physics of Classical Novae: Proc. IAU Coll. 122*, eds. A. Cassatella and R. Viotti (Berlin: Springer-Verlag), pp. 138–147.

Gehrz, R. D., and Hackwell, J. A. 1976. A search for anonymous AFCRL infrared sources. *Astron. J.* 206:L161–L164.

Gehrz, R. D., Grasdalen, G. L., Hackwell, J. A., and Ney, E. P. 1980*a*. The evolution of the dust shell of noval Serpentis 1978. *Astrophys. J.* 237:855–865.

Gehrz, R. D., Hackwell, J. H., Grasdalen, J. A., Ney, E. P., Neugebauer, G., and Sellgren, K. 1980*b*. The optically thin dust shell of nova Cygni 1978. *Astrophys. J.* 239:570–580.

Gehrz, R. D., Ney, E. P., Grasdalen, G. L., Hackwell, J. A., and Thronson, H. A. 1984. The mysterious 10 micron emission feature in the spectrum of nova Aquilae 1982. *Astrophys. J.* 281:303–312.

Gehrz, R. D., Grasdalen, G. L., and Hackwell, J. A. 1985. A neonova: Discovery of a remarkable 12–8 micron [ne II] emission line in noval Vulpeculae 1984 number 2. *Astrophys. J. Lett.* 298:47–50. Erratum *Astrophys. J. Lett.* 306:49.

Gehrz, R. D., Grasdalen, G. L., Greenhouse, M. A., Hackwell, J. A., Hayward, T., and Bentley, A. F. 1986. The neon nova II. Condensation of silicate grains in ejects of nova Vulpeculae 1984 number 2. *Astrophys. J. Lett.* 308:63–66.

Geiss, J., Altwegg, K., Anders, E., Balsiger, H., Ip, W.-H., Meier, A., Neugebauer, M., Rosenbauer, H., and Shelley, E. G. 1991. Interpretation of the ion mass spectra in the mass per charge range 25–35amu/e⁻ obtained in the inner coma of Halley's comet by the HIS-sensor of the Giotto IMS experiment. *Astron. Astrophys.* 247:226–234.

Genzel, R. 1991*a*. Molecular clouds in regions of massive star formation. In *Chemistry in Space*, eds. J. M. Greenberg and V. Pirronello (Dordrecht: Kluwer), pp. 123–170.

Genzel, R. 1991*b*. Structure and energy balance of molecular clouds. In *Molecular Clouds*, eds. R. James and T. Millar (Cambridge: Cambridge Univ. Press), pp. 75–96.

Genzel, R., and Downes, D. 1977. H₂O in the galaxy: sites of newly formed OB stars.

Astron. Astrophys. J. Suppl. 30:145–168.

Genzel, R., and Stutzki, J. 1989. The Orion molecular cloud and star-forming region. *Ann. Rev. Astron. Astrophys.* 27:41–86.

Genzel, R., Reid, M. J., Moran, J. M., and Downes, D. 1981. Proper motions and distances of H₂O master sources. I. The outflow in orion-KL. *Astrophys. J.* 244:844–902.

Genzel, R., Downes, D., Ho, P.T.P., and Bieging, J. 1982. NH₃ in Orion-KL: A new interpretation. *Astrophys. J. Lett.* 259:103–107.

Gerin, M., Combes, F., Wlodarczak, G., Encrenaz, P., and Laurent, C. 1992*a*. Interstellar detection of deuterated methyl acetylene. *Astron. Astrophys.* 253:L29–L32.

Gerin, M., Combes, F., Wlodarczak, G., Jacq, T., Guélin, M., Encrenaz, P., and Laurent, C. 1992*b*. Interstellar detection of deuterated methyl cyanide. *Astron. Astrophys.* 259:L35–L38.

Gerlich, D., and Kaefer, G. 1989. Ion trap studies of association processes in collisions of CH₃⁺ and CD₃⁺ with *n*-H₂, *p*-H₂, D₂ and He at 80 K. *Astrophys. J.* 347:849–854.

Ghez, A., Gorham, P., Hariff, C., Kulkarni, S., Neugebauer, G., Matthews, K., Soifer, T., Beckwith, S., and Koresko, C. 1990. The components of T Tauri. *Bull. Amer. Astron. Soc.* 22:1254.

Ghosh, P., and Lamb, F. K. 1979*a*. Accretion by rotating magnetic neutron stars. II. Radial and vertical structure of the transition zone in disk accretion. *Astrophys. J.* 232:259–276.

Ghosh, P., and Lamb, F. K. 1979*b*. Accretion by rotating magnetic neutron stars. III. Accretion torques and period changes in pulsating X-ray sources. *Astrophys. J.* 234:296–316.

Gilden, D. L. 1984. Thermal instability in molecular clouds. *Astrophys. J.* 283:679–686.

Gillett, F. C. 1986. IRAS observations of cool excess around main sequence stars. In *Light on Dark Matter*, ed. F. P. Israel (Dordrecht: D. Reidel), pp. 61–69.

Gillis, J., Mestel, L., and Paris, R. 1974. Magnetic braking during star formation—I. *Astrophys. Space Sci.* 27:167–194.

Gillis, J., Mestel, L., and Paris, R. 1979. Magnetic braking during star formation—II. *Mon. Not. Roy. Astron. Soc.* 187:311–335.

Gilman, P. A. 1983. Dynamos of the Sun and stars, and associated convection zone dynamics. In *Solar and Stellar Magnetic Fields: Origin and Coronal Effects: Proc. IAU Symp. 102*, ed. J. O. Stenflo (Dordrecht: D. Reidel), pp. 247–270.

Gilman, R. C. 1974*a*. Planck mean cross-sections for four grain materials. *Astrophys. J. Suppl.* 28:397–403.

Gilman, R. C. 1974*b*. Free-free and free-bound emission in low-surface-gravity stars. *Astron. J.* 188:87–94.

Gladman, B., and Duncan, M. J. 1990. On the fates of minor bodies in the outer solar system. *Astron. J.* 100:1669–1675.

Gladman, B., Duncan, M., and Candy, J. 1991. Symplectic integrators for long-term integrations in celestial mechanics. *Celest. Mech.*, in press.

Glasby, J. S. 1974. *The Nebular Variables* (Oxford: Pergamon).

Glassgold, A. E., Mamon, G. A., and Huggins, P. J. 1989. Molecule formation in fast neutral winds from protostars. *Astrophys. J. Lett.* 336:29–31.

Glassgold, A. E., Mamon, G. A., and Huggins, P. J. 1991. The formation of molecules in protostellar winds. *Astrophys. J.* 373:254–265.

Gledhill, T. M., Scarrott, S. M., and Wolstencroft, R. D. 1991. Optical polarization in the disc around β Pictoris. *Mon. Not. Roy. Astron. Soc.* 252:50P–54P.

Gliese, W. 1969. Catalogue of nearby stars. *Veröffentl. Astron. Rechen-Instituts*, Nr. 22.

Gliese, W., and Jahreiss, H. 1979. Nearby star data published 1968-1978. *Astron.*

Astrophys. Suppl. 38:423–448.

Göbel, R., Ott, U., and Begemann, F. 1978. On trapped noble gases in ureilites. *Proc. Lunar Planet. Sci. Conf.*, 9, *J. Geophys. Res.* 83:855–867.

Goldader, J. D., and Wynn-Williams, C. G. 1991. Sub-arcsecond resolution $2 \mu m$ imaging of W51. *Bull. Amer. Astron. Assoc.* 23:977.

Goldman, I., and Mazeh, T. 1991. On the orbital circularization of close binaries. *Astrophys. J.* 376:260–272.

Goldreich, P. 1963. On the eccentricity of satellite orbits in the solar system. *Mon. Not. Roy. Astron. Soc.* 126:257–268.

Goldreich, P. 1965. An explanation of the frequent occurrence of commensurable mean motions in the solar system. *Mon. Not. Roy. Astron. Soc.* 130:159–181.

Goldreich, P. 1966. Final spin states of planets and satellites. *Astron. J.* 71:1–7.

Goldreich, P., and Lynden-Bell, D. 1965a. I. Gravitational instability of uniformly rotating disks. *Mon. Not. Roy. Astron. Soc.* 130:97–124.

Goldreich, P., and Lynden-Bell, D. 1965b. II. Spiral arms as sheared gravitational instabilities. *Mon. Not. Roy. Astron. Soc.* 130:125–158.

Goldreich, P., and Peale, S. J. 1966. Spin-orbit coupling in the solar system. *Astron. J.* 71:425–438.

Goldreich, P., and Porco, C. C. 1987. Shepherding of the Uranian rings: II. Dynamics. *Astron. J.* 93:730–737.

Goldreich, P., and Soter, S. 1966. Q in the solar system. *Icarus* 5:375–389.

Goldreich, P., and Tremaine, S. 1978. The velocity dispersion in Saturn's rings. *Icarus* 34:227–239.

Goldreich, P., and Tremaine, S. 1979a. Towards a theory for the Uranian rings. *Nature* 277:97–99.

Goldreich, P., and Tremaine, S. 1979b. The excitation of density waves at the Lindblad and corotation resonances of an external potential. *Astrophys. J.* 233:857–871.

Goldreich, P., and Tremaine, S. 1980. Disk–satellite interactions. *Astrophys. J.* 241:425–441.

Goldreich, P., and Tremaine, S. 1981. The origin of the eccentricities of the rings of Uranus. *Astrophys. J.* 243:1062–1075.

Goldreich, P., and Tremaine, S. 1982. The dynamics of planetary rings. *Ann. Rev. Astron. Astrophys.* 20:249–283.

Goldreich, P., and Ward, W. R. 1973. The formation of planetesimals. *Astrophys. J.* 183:1051–1061.

Goldreich, P., Keeley, D. A., and Kwan, J. Y. 1973. Astrophysical masers II. Polarization properties. *Astrophys. J.* 179:111–134.

Goldreich, P., Murray, N., Longaretti, P. Y., and Banfield, D. 1989. Neptune's story. *Science* 245:500–504.

Goldsmith, P. F. 1987. Molecular clouds: an overview. In *Interstellar Processes*, eds. D. Hollenbach and H. A. Thronson (Dordrecht: D. Reidel), pp. 51–70.

Goldsmith, P. F., and Arquilla, R. 1985, Rotation in dark clouds. In *Protostars and Planets II*, eds. D. C. Black and M. S. Matthews (Tucson: Univ. of Arizona Press), pp. 137–149.

Goldsmith, P. F., and Langer, W. D. 1978. Molecular cooling and thermal balance of dense interstellar clouds. *Astrophys. J.* 222:881–895.

Goldsmith, P. F., Snell, R. L., Heyer, M. H., and Langer, W. D. 1984. Bipolar outflows in dark clouds. *Astrophys. J.* 286:599–608.

Goldsmith, P. F., Snell, R. L., Erickson, N. R., Dickman, R. L., Schloerb, F. P., and Irvine, W. M. 1985. Search for molecular oxygen in dense interstellar clouds. *Astrophys. J.* 289:613–617.

Goldsmith, P. F., Langer, W. D., and Wilson, R. W. 1986. Molecular outflows, gas density distribution, and the effects of star formation in the dark cloud Barnard 5.

Astrophys. J. Lett. 303:11–15.

Goldsmith, P. F., Snell, R. L., Hasegawa, T., and Ukita, N. 1987. Small-scale structure and chemical differentiation in the central regions of the Sagittarius B2 molecular cloud. *Astrophys. J.* 314:525–534.

Gombosi, T. I., and Houpis, H. L. 1986. The icy-glue model of the cometary nucleus. *Nature* 324:43–44.

Gonatas, D. P., Hildebrand, R. H., Platt, S. R., Wu, X. D., Davidson, J. A., Novak, G., Aitken, D. K., and Smith, C. 1990. The far-infrared polarization of the orion nebula. *Astrophys. J.* 357:132–137.

Good, J. C., Hauser, M. G., and Gautier, T. N. 1986. IRAS observations of the zodiacal background. *Adv. Space Res.* 6(7):83–86.

Goodman, A. A. 1989. Interstellar Magnetic Fields: An Observational Perspective. Ph. D. Thesis, Harvard Univ.

Goodman, A. A. 1991 Magnetic fields: A photo essay. In *Atoms, Ions, and Molecules: New Results in Spectral Line Astrophysics*, eds. A. D. Haschick and P. T. P. Ho (San Francisco: Astron. Soc. of the Pacific), pp. 333–348.

Goodman, A. A., and Myers, P. C. 1991. A study of alignment among the features of interstellar clouds. In preparation.

Goodman, A. A., Crutcher, R. M., Heiles, C., Myers, P. C., and Troland, T. H. 1989. Measurements of magnetic field strengths in the dark cloud Barnard 1. *Astrophys. J. Lett.* 338:61–64.

Goodman, A. A., Bastien, P., Myers, P.C., Menard, F. 1990. Optical polarization maps of star-forming regions in Perseus, Taurus, and Ophiuchus. *Astrophys. J.* 359:363–377.

Goodman, A. A., Myers, P. C., Fuller, G. A., and Benson, P. J. 1991*a*. Dense cores in dark clouds VI: Shapes. *Astrophys. J.*. 376:561–572.

Goodman, A. A., Benson, P. J., Fuller, G. A., and Myers, P. C. 1991*b*. Dense cores in dark clouds VII: Velocity gradients. In preparation.

Goodman, J., Narayan, R., and Goldreich, P. 1987. The stability of accretion Tori—II. Non-linear evolution to discrete planets. *Mon. Not. Roy. Astron. Soc.* 225:695–711.

Goodrich, R. W. 1986. V645 Cygni and the Duck Nebula. *Astrophys. J.* 311:882–894.

Göpel, C., Manhes, G., and Allegre, C. J. 1990. U-Pb study of phosphates in chondrites. *Meteoritics* 25:367–368 (abstract).

Goswami, J. N., Hutcheon, I. D., and Macdougall, J. D. 1976. Microcraters and solar flare tracks in crystals from carbonaceous chondrites and lunar breccias. *Proc. Lunar Planet. Sci. Conf.* 7:543–562.

Goswami, J. N., Lal, D., and Wilkening, L. L. 1984. Gas-rich meteorites: Probes for particle environment and dynamical processes in the inner solar system. *Space Sci. Rev.* 37:111–159.

Gottlieb, C. A., Gottlieb, E. W., Litvak, M. M., Ball, J. A., and Penfield, H. 1978. Observations of interstellar sulfur monoxide. *Astrophys. J.* 219:77–94.

Gottlieb, C. A., Gottlieb, E. W., Thaddeus, P., and Kawamura, H. 1983*a*. Laboratory detection of the C_3N and C_4H free radicals. *Astrophys. J.* 275:916–921.

Gottlieb, C. A., Gottlieb, E. W., and Thaddeus, P. 1983*b*. Laboratory and astronomical measurement of the millimeter wave spectrum of the ethynyl radical CCH. *Astrophys. J.* 264:740–795.

Grabelsky, D. A., Cohen, R. S., Bronfman, L., and Thaddeus, P. 1987. Molecular clouds in the Carina Arm: Large scale properties of molecular gas and comparison with HI. *Astrophys. J.* 315:122–141.

Gradie, J., Hayashi, J., Zuckerman, B., Epps, H., and Howell, R. 1987. *Lunar Planet. Sci. Conf.* XVIII:351 (abstract).

Graff, M. M., and Dalgarno, A. 1987. Oxygen chemistry in shocked interstellar clouds.

II. Effects of non-thermal internal energy on chemical evolution. *Astrophys. J.* 317:432–441.

Graham, J. A., and Frogel, J. A. 1985. An Fu Orionis star associated with Herbig-Haro object 57. *Astrophys. J.* 289:331–341.

Grasdalen, G. L. 1974. An infrared study of NGC 2024. *Astrophys. J.* 193:373–383.

Grasdalen, G. L., and Joyce, R. R. 1976. Coronal lines in near infrared spectrum of nova Cygni 1975. *Nature* 259:187–189.

Grasdalen, G. L., Strom, K. M., and Strom, S. E. 1973. A 2-micron map of the Ophiuchus dark-cloud region. *Astrophys. J. Lett.* 184:53–57.

Gray, C. M., Papanastassiou, D. A., and Wasserburg, G. J. 1973. The identification of early condensates from the solar nebula. *Icarus* 20:213–219.

Gray, R. 1986. Ph. D. Thesis, Univ. of Toronto.

Gredel, R. 1987. The $^{12}C_2/^{12}C^{13}C$ Abundance Ratio in Comet Halley. Ph. D. Thesis, Univ. of Heidelberg.

Gredel, R., Lepp, S., Dalgarno, A., and Herbst, E. 1989. Cosmic-ray-induced photodissociation and photoionization rates of interstellar molecules. *Astrophys. J.* 347:289–293.

Green, E. M., Demarque, P., and King, C. R. 1987. *The Revised Yale Isochrones and Luminosity Functions* (New Haven: Yale Univ. Obs.).

Green, J. R. 1986. Stress, fracture, and outburst in cometary nuclei. *Bull. Amer. Astron. Soc.* 18:800 (abstract).

Greenberg, J. M. 1978. Physics and astrophysics of interstellar dust. In *Infrared Astronomy*, eds. G. Setti and G. Fazio (Dordrecht: D. Reidel), pp. 51–95.

Greenberg, J. M. 1982. What are comets made of? A model based on interstellar grains. In *Comets*, ed. L. L. Wilkening (Tucson: Univ. of Arizona Press), pp. 131–163.

Greenberg, J. M. 1986. Predicting that comet Halley is dark. *Nature* 321:385–387.

Greenberg, J. M. 1989a. The core-mantle model of interstellar grains and the cosmic dust connection. In *Interstellar Dust: Proc. of IAU Symp. 135*, eds. L. Allamandola and A. G. G. M. Tielens (Dordrecht: Kluwer), pp. 345–355.

Greenberg, J. M. 1989b. Interstellar dust: An overview of physical and chemical evolution. In *Evolution of Interstellar Dust and Related Topics*, eds. A. Bonnetti, J. M. Greenberg and S. Aiello (Amsterdam: North-Holland Press), pp. 7–52.

Greenberg, J. M., and Hage, J. I. 1990. From interstellar dust to comets: a unification of observational constraints. *Astrophys. J.* 361:260–274.

Greenberg, J. M., Allamandola, L. J., Hagen, W., van de Bult, C. E. P. M., and Bass, R. 1980. Laboratory and theoretical results on interstellar molecule production by grains in molecular clouds. In *Interstellar Molecules*, ed. B. H. Andrew (Dordrecht: D. Reidel), pp. 353–363.

Greenberg, R. 1975. The dynamics of Uranus' satellites. *Icarus* 24:325–332.

Greenberg, R. 1983. The role of dissipation in shepherding of ring particles. *Icarus* 53:207–218.

Greenberg, R., and Scholl, H. 1979. Resonances in the asteroid belt. In *Asteroids*, ed. T. Gehrels (Tucson: Univ. of Arizona Press), pp. 310–333.

Greenberg, R., Wacker, J. F., Hartmann, W. L., and Chapman, C. R. 1978. Planetesimals to planets: Numerical simulations of collisional evolution. *Icarus* 35:1–26.

Greenberg, R., Weidenschilling, S. J., Chapman, C. R., and Davis, D. R. 1984. From icy planetesimals to outer planets and comets. *Icarus* 59:87–113.

Greenberg, R., Carusi, A., and Valsecchi, G. B. 1988. Outcomes of planetary close encounters: a systematic comparison of methodologies. *Icarus* 75:1–29.

Greenberg, R., Bottke, W. F., Carusi, A., and Valsecchi, G. B. 1991. Planetary accretion rates: analytical derivation. *Icarus* 94:98–111.

Greene, T. P. 1991. Infrared Studies of Star Formation in the Rho Ophiuchi Dark

Cloud. Ph. D. Thesis, Univ. of Arizona.

Greene, T. P., and Young, E. T. 1989. *IRAS* observations of dust heating and energy balance in Rho Ophiuchi dark cloud. *Astrophys. J.* 339:258–267.

Greenhouse, M. A., Grasdalen, G. L., Hayward, T. L., Gehrz, R. D., and Jones, T. J. 1988. *Astron. J.* 95:172–177.

Greenhouse, M. A., Grasdalen, G. L., Woodward, C. E., Benson, J., Gehrz, R. D., Rosenthal, E., and Skrutskie, M. F. 1990. The infrared coronal lines of recent novae. *Astrophys. J.* 352:307–317.

Greenzweig, Y., and Lissauer, J. J. 1990. Accretion rates of protoplanets. *Icarus* 87:40–77.

Greenzweig, Y., and Lissauer, J. 1992. Accretion rates of protoplanets II. Gaussian distribution of planetesimal velocities. In preparation.

Greiner, N. R., Phillips, D. S., Johnson, J. D., and Volk, F. 1988. Diamonds in detonation soot. *Nature* 333:440–442.

Greve, A., and Pauls, T. 1980. On the Zeeman splitting of high *n* recombination lines. *Astron. Astrophys.* 82:388.

Grevesse, N., Lambert, D. L., Sauval, A. J., van Dishoeck, E. F., Farmer, C. B., and Norton, R. H. 1991. Vibration-rotation bands of CH in the solar infrared spectrum and the solar carbon abundance. *Astron. Astrophys.* 242:488–495.

Grim, R. J., and Greenberg, J. M. 1987*a*. Ions in grain mantles: The 4.62 micron absorption by OCN^- in W33A. *Astrophys. J. Lett.* 321:91–96.

Grim, R. J., and Greenberg, J. M. 1987*b*. Photo-processing of H_2S in interstellar grain mantles as an explanation for S_2 in comets. *Astron. Astrophys.* 181:155–168.

Grim, R. J., Greenberg, J. M., Schutte, W., and Schmitt, B. 1989. Ions in grain mantles: A new explanation for the 6.86 micron absorption in W33A. *Astrophys. J. Lett.* 341:87–90.

Grim, R. J., Baas, F., Geballe, T. R., Greenberg, J. M., and Schutte, W. 1991. Detection of solid methanol in W33A. *Astron. Astrophys.* 243:473–477.

Grinspoon, D. H. 1989. Large Impact Events and Atmospheric Evolution on the Terrestrial Planets. Ph. D. Thesis, Univ. of Arizona.

Grinspoon, D. H., and Lewis, J. S. 1987. Deuterium fractionation in the presolar nebula: kinetic limitations on surface catalysis. *Icarus* 72:430–436.

Grinspoon, D. H., and Lewis, J. S. 1988. Cometary water on Venus: implications of stochastic impacts. *Icarus* 74:21–35.

Grinspoon, D. H., and Sagan, C. 1991. Impact dust and climate on primordial Earth. *Icarus*, submitted.

Groesbeck, T., Phillips, T. G., and Blake, G. A. 1993. The integrated molecular line emission to dust continuum ratio of Orion/KL at 345 GHz. *Astrophys. J.*, in preparation.

Grossman, A. S., Hays, D., and Graboske, H. C. 1974. The theoretical low-mass main sequence. *Astron. Astrophys.* 30:95–103.

Grossman, A. S., Pollack, J. B., Reynolds, R. T., Summers, A. L., and Graboske, H. C., Jr. 1980. The effect of cores on the structure and evolution of Jupiter and Saturn. *Icarus* 42:358–379.

Grossman, J. N. 1988. Formation of chondrules. In *Meteorites and the Early Solar System*, eds. J. F. Kerridge and M. S. Matthews (Tucson: Univ. of Arizona Press), pp. 680–696.

Grossman, L. 1972. Condensation in the primitive solar nebula. *Geochim. Cosmochim. Acta* 36:597–619.

Grossman, L., and Larimer, J. W. 1974. Early chemical history of the solar system. *Rev. Geophys. Space Phys.* 12:71–101.

Grün, E., and Jessberger, E. K. 1990. Dust. In *Physics and Chemistry of Comets*, ed. W. F. Huebner (Berlin: Springer-Verlag), pp. 113–176.

Grün, E., Zook, H. A., Fechtig, H., and Giese, R. H. 1985. Collisional balance of the meteoritic complex. *Icarus* 62:244–272.

Gudkova, T. V., Zharkov, V. N., and Leont'ev, V. V. 1988. Models of Uranus and Neptune with partially mixed envelopes. *Astron. Vestnik* 22:23–40.

Gudkova, T. V., Zharkov, V. N., and Leont'ev, V. V. 1989. Models of Jupiter and Saturn having a two–layer molecular envelope. *Solar System Res.* 22:159–166.

Guélin, M. 1988. Organic and exotic molecules in space. In *Molecules in Physics, Chemistry, and Biology*, vol. II, ed. J. Maruani (Dordrecht: Kluwer), pp. 175–187.

Guélin, M., and Cernicharo, J. 1988. Mass distribution in the Taurus complex. In *Molecular Clouds in the Milky Way and External Galaxies*, eds. R. L. Dickman, R. L. Snell and J. S. Young (Berlin: Springer-Verlag), pp. 81–90.

Guélin, M., Langer, W. D., and Wilson, R. W. 1982. The state of ionization in dense molecular clouds. *Astron. Astrophys.* 107:107–127.

Güsten, R., and Fiebig, D. 1990. Magnetic fields in dark cloud cores and H_2O masers. In *Galactic and Extragalactic Magnetic Fields: Proc. IAU Symp. 140*, eds. R. Beck, P. P. Kronberg and R. Wielebinski (Dordrecht: Kluwer), pp. 305–308.

Güsten, R., and Mezger, P. G. 1982. Star formation and abundance gradients in the galaxy. *Vistas in Astron.* 26:159–224.

Haas, M. R., Leinert, Ch., and Zinnecker, H. 1990. XZ Tau resolved as double infrared source. *Astron. Astrophys.* 230:L1–L4.

Haas, M. R., Hollenbach, D., and Erickson, E. E. 1991. Observations of [Si II] (35 μm) and [S I](25 μm) in Orion: evidence for a windshock near IRc2. *Astrophys. J.* 379:555–563.

Hachisu, I., Tohline, J. E., and Eriguchi, Y. 1987. Fragmentation of rapidly rotating gas clouds. I. A universal criterion for fragmentation. *Astrophys. J.* 323:592–613.

Hachisu, I., Tohline, J. E., and Eriguchi, Y. 1988. Fragmentation of rapidly rotating gas clouds. II. Polytropes—clues to the outcome of adiabatic collapse. *Astrophys. J. Suppl.* 66:315–342.

Hackwell, J. A. 1971. Ph. D. Thesis, Univ. College, London.

Hackwell, J. A. 1972. Long wavelength spectrometry and photometry of M, S and C-stars. *Astron. Astrophys.* 21:239–248.

Hageman, R., Nief, R. G., and Roth, E. 1970. Absolute isotopic scale for deuterium analysis of natural waters. *Tellus* 22:712–715.

Haikala, L. K., and Laureijs, R. J. 1989. CO and IR in L1228: extended molecular outflow and strongly self-absorbed ^{12}CO emission. *Astron. Astrophys.* 223:287–292.

Halbwachs, J.-L. 1987. Distribution of mass ratios in spectroscopic binaries. *Astron. Astrophys.* 183:234–240.

Hamann, F., and Persson, S. E. 1989. High-resolution spectra of the luminous young stellar object V645 Cygni. *Astrophys. J.* 339:1078–1088.

Hambly, N. C., and Jameson, R. F. 1991. The luminosity and mass functions of the Pleiades: low mass stars and brown dwarfs. *Mon. Not. Roy. Astron. Soc.* 249:137–144.

Hanan, B. B., and Tilton, G. R. 1987. 60025: Relict of primitive lunar crust? *Earth Planet. Sci. Lett.* 84:15–21.

Hanbury Brown, R., Davis, J., and Allen, L. R. 1974. The angular diameter of 32 stars. *Mon. Not. Roy. Astron. Soc.* 167:121–136.

Hanel, R. A., Conrath, B. J., Hearth, L. W., Kunde, V. G., Pirraglia, J. A. 1981. Albedo, internal heat, and energy balance of Jupiter: Preliminary results of the Voyager infrared investigation. *J. Geophys. Res.* 86:8705–8712.

Hanel, R. A., Conrath, B. J., Kunde, V. G., Pearl, J. C., and Pirraglia, J. A. 1983. Albedo, internal heat flux, and energy balance of Saturn. *Icarus* 53:262–285.

Hanks, T. C., and Anderson, D. L. 1969. The early thermal history of the Earth. *Phys.*

Earth Planet. Int. 2:19–29.

Hardee, P. E., and Norman, M. L. 1990. Asymmetric morphology of the propagating jet. *Astrophys. J.* 365:134–158.

Harju, J., Walmsley, C. M., and Wouterloot, J. G. A. 1991. Young ammonia clumps in the Orion molecular cloud. *Astron. Astrophys.* 245:643–647.

Haro, G. 1949. Faint spectra showing Hα in emission in the obscuring clouds of Ophiuchus and Scorpius and in a region in Sagittarius. *Astron. J.* 54:188–189.

Harper, C. L., Völkening, J., Heumann, K. G., Shih, C. Y., and Wiesmann, H. 1991*a*. [182]Hf-[182]W: New cosmochronometric constraints on terrestrial accretion, core formation, the astrophysical site of the r-process, and the origin of the solar system. *Lunar Planet. Sci.* XXII:515–516 (abstract).

Harper, C. L., Wiesmann, H., Nyquist, L. E., Howard, W. M., Meyer, B., Yokoyama, Y., Rayet, M., Arnould, M., Palme, H., Spettel, B., and Jochum, K. P. 1991*b*. [92]Nb/[93]Nb and [92]Nb/[146]Sm ratios of the early solar system: Observations and comparison of p-process and spallation models. *Lunar Planet. Sci.* XXII:519–520 (abstract).

Harper, C. L., Wiesmann, H., and Nyquist, L. E. 1991*c*. [135]Cs-[135]Ba: An application of very high precision mass spectrometry to identifying the astrophysical site of the origin of the solar system. Abstract of talk presented at the Alfred O. Nier Symp. on Inorganic Mass Spectrometry, Durango, Colo.

Harper, C. L., Wiesmann, H., and Nyquist, L. E. 1991*d*. [135]Cs-[135]Ba: A new cosmochronometric constraint on the origin of the Earth and the astrophysical site of the origin of the solar system. *Meteoritics* 26:341.

Harper, D. A., Loewenstein, R. F., and Davidson, J. A. 1984. On the nature of the material surrounding Vega. *Astrophys. J.* 285:808–812.

Haro, G. 1952. Herbig's nebulous objects near NGC 1999. *Astrophys. J.* 115:572–573.

Haro, G., and Minkowski, R. 1960. The Herbig-Haro objects near NGC 1999. *Astron. J.* 65:490–491.

Harris, A. H. 1978. The formation of the outer planets. *Lunar Planet. Sci.* IX:459–461 (absract).

Harris, A. H., and Kaula, W. M. 1975. A co-accretional model of satellite formation. *Icarus* 24:516–524.

Harris, M. J., Fowler, W. A., Caughlan, G. R., and Zimmerman, B. A. 1983. Thermonuclear Reaction Rates, III. *Ann. Rev. Astron. Astrophys.* 21:165–176.

Harris, A., Townes, C. H., Matsakis, D. N., and Palmer, P. 1983. Small rotating clouds of stellar mass in Orion Molecular Cloud I. *Astrophys. J. Lett.* 265:63–66.

Hartigan, P. 1989. The visibility of the Mach disk and the bow shock of a stellar jet. *Astrophys. J.* 339:987–999.

Hartigan, P., and Graham, J. A. 1987. Outflows in the star-formation region near R CrA. *Astron. J.* 93:913–919.

Hartigan, P., Mundt, R., and Stocke, J. 1986. A detailed study of HH 32 and the highly collimated outflow from the T Tauri Star AS 353A. *Astron. J.* 91:1357–1371.

Hartigan, P., Lada, C. J., Stocke, J., and Tapia, S. 1986. The extraordinary HH objects near the star-formation region Cepheus A. *Astron. J.* 92:1155–1161.

Hartigan, P., Raymond, J., and Hartmann, L. 1987. Radiative bow shock models of Herbig-Haro objects. *Astrophys. J.* 316:323–348.

Hartigan, P., Hartmann, L., Kenyon, S., Hewett, R., and Stauffer, J. 1989*a*. How to unveil a T Tauri star. *Astrophys. J. Suppl.* 70:899–914.

Hartigan, P., Curiel, S., and Raymond, J. 1989*b*. Molecular hydrogen and optical images of HH 7–11. *Astrophys. J. Lett.* 347:31–34.

Hartigan, P., Hartmann, L., Kenyon, S. J., Strom, S. E., and Skrutskie, M. E. 1990*a*. Correlations of optical and infrared excesses in T Tauri stars. *Astrophys. J. Lett.* 354:25–28.

Hartigan, P., Raymond, J., and Meaburn, J. 1990b. Observations and shock models of the jet and Herbig-Haro objects HH 46/47. *Astrophys. J.* 362:624–633.

Hartigan, P., Kenyon, S. J., Hartmann, L., Strom, S. E., Edwards, S., Welty, A. D., and Stauffer, J. 1991. Optical excess emission in T Tauri stars. *Astrophys. J.* 382:617–635.

Hartley, M., Manchester, R. N., Smith, R. M., Tritton, S. B., and Goss, W. M. 1986. A catalogue of southern dark clouds. *Astron. Astrophys. Suppl.* 63:27–48.

Hartmann, L. 1982. Line profiles of T Tauri stars: Clues to the nature of the mass flow. *Astrophys. J. Suppl.* 48:109–126.

Hartmann, L. 1986. Theories of mass loss from T Tauri stars. *Fund. Cosmic Phys.* 11:279–310.

Hartmann, L. 1990. Emission lines and winds from T Tauri stars. In *Cool Stars, Stellar Systems, and the Sun*, ed. G. Wallerstein (San Francisco: Astron. Soc. of the Pacific), pp. 289–300.

Hartmann, L., and Kenyon, S. J. 1985. On the nature of the FU Orionis objects. *Astrophys. J.* 299:462–478.

Hartmann, L., and Kenyon, S. J. 1987a. Further evidence for disk accretion in FU Orionis objects. *Astrophys. J.* 312:243–253.

Hartmann, L., and Kenyon, S. J. 1987b. High spectral resolution infrared observations of V1057 Cygni. *Astrophys. J.* 322:393–398.

Hartmann, L., and Kenyon, S. J. 1988. Accretion disks around young stars. In *Formation and Evolution of Low Mass Stars*, eds. A. K. Dupree and M. T. V. T. Lago (Dordrecht: Kluwer), pp. 163–179.

Hartmann, L., and Kenyon, S. 1990. Optical veiling, disk accretion, and the evolution T Tauri stars. *Astrophys. J.* 349:190–196.

Hartmann, L., and MacGregor, K. B. 1982. Protostellar mass and angular momentum loss. *Astrophys. J.* 259:180–192.

Hartmann, L., and Noyes, R. W. 1987. Rotation and magnetic activity in main sequence stars. *Ann. Rev. Astron. Astrophys.* 25:271–301.

Hartmann, L., and Raymond, J. C. 1989. Wind-disk shocks around T Tauri stars. *Astrophys. J.* 337:903–916.

Hartmann, L., and Stauffer, J. 1989. Additional measurements of pre-main-sequence stellar rotation. *Astron. J.* 97:873–880.

Hartmann, L., Edwards, S., and Avrett, A, 1982. Wave-driven winds from cool stars. II. Models for T Tauri stars. *Astrophys. J.* 261:279–292.

Hartmann, L., Hewett, R., Stahler, S., and Mathieu, R. D. 1986. Rotational and radial velocities of T Tauri stars. *Astrophys. J.* 309:275–293.

Hartmann, L., Soderblom, D. R., and Stauffer, J. R. 1987. Rotation and kinematics of the pre-main-sequence stars in Taurus-Auriga with C II emission. *Astron. J.* 93:907–912.

Hartmann, L., Kenyon, S. J., Hewett, R., Edwards, S., Strom, K. M., Strom, S. E., and Stauffer, J. R. 1989. Pre-main-sequence disk accretion in Z Canis Majoris. *Astrophys. J.* 338:1001–1010.

Hartmann, L., Calvet, N., Avrett, E. H., and Loeser, R. K. 1990. Winds from T Tauri stars. I. Spherically symmetric models. *Astrophys. J.* 349:168–189.

Hartmann, L. W., Jones, B. F., Stauffer, J. R., and Kenyon, S. J. 1991. A proper motion survey for pre-main sequence stars in Taurus-Auriga. *Astron. J.* 101:1050–1062.

Hartmann, W. K. 1978. Planet formation: mechanism of early growth. *Icarus* 33:50–61.

Hartmann, W. K., and Davis, D. R. 1975. Satellite-sized planetesimals and lunar origin. *Icarus* 24:504–515.

Hartmann, W. K., Phillips, R. J., and Taylor, G. J., eds. 1986. *Origin of the Moon* (Houston: Lunar and Planetary Inst.).

Hartquist, T. W., ed. 1990. *Molecular Astrophysics* (Cambridge: Cambridge Univ. Press).

Hartquist, T. W., Oppenheimer, M., and Dalgarno, A. 1980. Molecular diagnostics of interstellar shocks. *Astrophys. J.* 236:182–188.

Harvey, P. M. 1985. Observational evidence for disks around young stars. In *Protostars and Planets II*, eds. D. C. Black and M. S. Matthews (Tucson: Univ. of Arizona Press), pp. 484–492.

Harvey, P. M., and Forveille, T. 1988. A remarkable molecular outflow in W28. *Astron. Astrophys.* 197:L19–L21.

Harvey, P. M., Wilking, B. A., and Joy, M. 1984. On the far-infrared excess of Vega. *Nature* 307:441–442.

Harvey, P. M., Joy, M., Lester, D. F., and Wilking, B. A. 1986. Infrared studies of the Herbig-Haro object 1–2 region. *Astrophys. J.* 301:346–354.

Hasegawa, M., and Nakazawa, K. 1990. Distant encounter between Keplerian particles. *Astron. Astrophys.* 227:619–627.

Hassall, B. J. M., Snijders, M. A. J., Harris, A. W., Cassatella, A., Dennefeld, M., Friedjung, M., Bode, M., Whittet, D., Whitelock, P., Menzies, J., Evans, T. L., and Bath, G. T. 1990. Measurements of outburst characteristics, temperatures, densities and abundances in the ejecta of nova Muscae 1983. In *Physics of Classical Novae: Proc. IAU Coll. 122*, eds. A. Cassetalla and R. Viotti (Berlin: Springer-Verlag), pp. 202–203.

Hauser, M. G., and Houck, J. R. 1986. The zodiacal background in the IRAS data. In *Light on Dark Matter*, ed. F. P. Israel (Dordrecht: D. Reidel), pp. 39–44.

Hausman, M. A. 1981. Collisional mergers and fragmentation of interstellar clouds. *Astrophys. J.* 245:72–91.

Hawkins, I., and Jura, M. 1987. The $^{12}C/^{13}C$ isotope ratio of the interstellar medium in the neighborhood of the Sun. *Astrophys. J.* 317:926–950.

Hawley, J. F. 1989. Simulations of three-dimensional slender tori. In *Theory of Accretion Disks*, eds. F. Meyer, W. J. Duschl, J. Frank, E. Meyer-Hofmeister (Dordrecht: Kluwer), pp. 259–291.

Hawley, J. F., and Balbus, S. A. 1991. A powerful local shear instability in weakly magnetized disks. II. Nonlinear evolution. *Astrophys. J.* 376:223–233.

Hayashi, C. 1961. Stellar evolution in early phases of gravitational contraction. *Publ. Astron. Soc. Japan* 13:450–452.

Hayashi, C. 1966. Evolution of protostars. *Ann. Rev. Astron. Astrophys.* 4:171–192.

Hayashi, C. 1981. Structure of the solar nebula, growth and decay of magnetic fields and effects of magnetic and turbulent viscosities on the nebula. *Prog. Theor. Phys. Suppl.* 70:35–53.

Hayashi, C., Hoshi, R., and Sugimoto, D. 1962. Evolution of the stars. *Prog. Theor. Phys. Suppl.* 22:1–183.

Hayashi, C., Nakazawa, K., and Adachi, I. 1977. Long term behavior of planetesimals and the formation of the planets. Publ. Astron. Soc. Japan 29:163–196.

Hayashi, C., Nakazawa, K., and Mizuno, H. 1979. Earth's melting due to the blanketing effect of the primordial dense atmosphere. *Earth Planet. Sci. Lett.* 43:22–28.

Hayashi, C., Nakazawa, K., and Nakagawa, Y. 1985. Formation of the solar system. In *Protostars and Planets II*, eds. D. C. Black and M. S. Matthews (Tucson: Univ. of Arizona Press), pp. 1100–1153.

Hayashi, M., Kobayashi, H., and Hasegawa, T. 1989. Velocity field in the fragmented molecular cloud core of W3: Evidence for large-scale turbulence. *Astrophys. J.* 340:298–306.

Hayashi, S. S., Hasegawa, T., Tanaka, M., Hayashi, M., Aspin, C., McLean, I. S., Brand, P. W. J. L., and Gatley, I. 1990. Infrared images of ionized and molecular hydrogen emission in S 106. *Astrophys. J.* 354:242–246.

Hayatsu, R., and Anders, E. 1981. Organic compounds in meteorites and their origins. In *Topics in Current Chemistry, Cosmochemistry and Geochemistry*, vol. 99 (Berlin: Springer-Verlag), pp. 1–39.

Heaton, B. D., Little, L. T., and Bishop, I. S. 1989. The "ultracompact hot core" of G 34.4 + 0.15: arcsec resolution ammonia observations. *Astron. Astrophys.* 213:148–154.

Heiles, C. 1979. H I shells and supershells. *Astrophys. J.* 229:533–544.

Heiles, C. 1988. L204: A gravitationally confined dark cloud in a strong magnetic environment. *Astrophys. J.* 324:321–330.

Heiles, C. 1989. Magnetic fields, pressure and thermally unstable gas in prominent HI shells. *Astrophys. J.* 336:808–821.

Heiles, C., and Stevens, M. 1986. Zeeman splitting of 18 centimeter OH lines toward Cassiopeia A and other sources. *Astrophys. J.* 301:331–338.

Heiles, C., and Troland, T. H. 1982. Measurements of magnetic field strengths in the vicinity of Orion. *Astrophys. J. Lett.* 260:23–26.

Heinemann, M., and Olbert, S. 1978. Axisymmetric ideal MHD stellar wind flow. *J. Geophys. Res.* 83:2457–2460.

Heintz, W. D. 1969. A statistical study of binary stars. *J. Roy. Astron. Soc. Canada* 63:275–298.

Heisler, J. 1990. Monte Carlo simulations of the Oort comet cloud. *Icarus* 88:104–121.

Heisler, J., and Tremaine, S. 1986. The influence of the galactic tidal field on the Oort comet cloud. *Icarus* 65:13–26.

Helou, G. 1989. Far infrared emission from galactic and extra galactic dust. In *Interstellar Dust: Proc. of IAU Symp. 135*, eds. L. Allamandola and A. G. G. M. Tielens (Dordrecht: Kluwer), pp. 285–301.

Helou, G., and Beichman, C. A. 1991. The confusion limits to the sensitivity of submillimetre telescopes. In *From Ground-Based to Space-Borne Sub-Millimeter Astronomy*, Proc. 29th Liège Intl. Astrophys. Coll., pp. 117–123.

Henderson, A. P., Jackson, P. D., and Kerr, F. J. 1982. The distribution of neutral atomic hydrogen in our galaxy beyond the solar circle. *Astrophys. J.* 263:116–122.

Henkel, C., Wilson, T. L., and Mauersberger, R. 1987. A multilevel study of ammonia in star forming regions. *Astron. Astrophys.* 182:137–142.

Hénon, M. 1966a. *Bull. Astron. Paris* 1, fax. 1, pp. 57.

Hénon, M. 1966b. *Bull. Astron. Paris* 1, fax. 2, pp. 49.

Hénon, M. 1983. Numerical exploration of Hamiltonian systems. In *Chaotic Behaviour of Deterministic Systems*, eds. G. Iooss, R. H. G. Helleman and R. Stora (Amsterdam: North-Holland), pp. 54–169.

Hénon, M., and Heiles, C. 1964. The applicability of the third integral of motion: some numerical experiments. *Astron. J.* 69:73–79.

Hénon, M., and Petit, J. M. 1986. Series expansions for encounter-type solutions of Hill's problem. *Celest. Mech.* 38:67–100.

Henrard, J. 1982. Capture into resonance: an extension of the use of adiabatic invariants. *Celest. Mech.* 27:3–22.

Henrard, J., and Lemaitre, A. 1983. A second fundamental model for resonance. *Celest. Mech.* 30:197–218.

Henriksen, R. N., and Raybrun, D. E. 1971. Relativistic stellar wind theory: 'near' zone solutions. *Mon. Not. Roy. Astron. Soc.* 152:323–332.

Henry, T. J. 1991. A Systematic Search for Low Mass Companions Orbiting Nearby Stars and the Calibration of the End of the Stellar Main Sequence. Ph. D. Thesis, Univ. of Arizona.

Henry, T. J., and McCarthy, D. W. 1990. A systematic search for brown dwarfs orbiting nearby stars. *Astrophys. J.* 350:334–347.

Henyey, L.G., LeLevier, R., and Levee, R. D. 1955. The early phases of stellar evolution. *Publ. Astron. Soc. Pacific* 67:154–160.

Heppenheimer, T. A. 1980. Secular resonances and the orgin of eccentricities of Mars and the asteroids. *Icarus* 41:76–88.

Herbert, F. 1989. Primordial electrical induction heating of asteroids. *Icarus* 78:402–410.

Herbert, F., and Sandel, B. R. 1991. CH$_4$ and haze in Triton's lower atmosphere. *J. Geophys. Res.* 96:19241–19252.

Herbig, G. H. 1951. The spectra of two nebulous objects near NGC 1999. *Astrophys. J.* 113:697–699.

Herbig, G. H. 1960. The spectra of Be and Ae-Type stars associated with nebulosity. *Astrophys. J. Suppl.* 4:337–368.

Herbig, G. H. 1962a. Spectral classification of faint members of the Hyades and Pleiades and the dating problem in galactic clusters. *Astrophys. J.* 135:736–747.

Herbig, G. H. 1962b. The properties and problems of T Tauri stars and related objects. *Adv. Astron. Astrophys.* 1:47–103.

Herbig, G. H. 1970. *Mem. Roy. Sci. Liège* Ser. 5, 9:13.

Herbig, G. H. 1974. On the nature of the small dark globules in the Rosette nebula. *Publ. Astron. Soc. Pacific* 86:604–608.

Herbig, G. H. 1975. The diffuse interstellar bands. IV. The region 4400–6850 Å. *Astrophys. J.* 196:129–160.

Herbig, G. H. 1977. Eruptive phenomena in early stellar evolution. *Astrophys. J.* 217:693–715.

Herbig, G. H. 1978. The post T Tauri stars. In *Problems of Physics and Evolution of the Universe*, ed. L. Mirzoyan (Yervan: Academy of Sciences of the Armenian S. S. R.), pp. 171–183.

Herbig, G. H. 1982. Stars of low to intermediate mass in the Orion Nebula. In *Symp. on the Orion Nebula to Honor Henry Draper*, eds. A. E. Glassgold, P. J. Huggins and E. L. Schucking (New York: New York Academy of Science), pp. 64–78.

Herbig, G. H. 1989a. FU Orionis eruptions. In *ESO Workshop on Low-Mass Star Formation and Pre-Main Sequence Objects*, ed. B. Reipurth (Garching: European Southern Obs.), pp. 233–246.

Herbig, G. H. 1989b. Summarizing remarks on the astronomical evidence for circumstellar disks. In *The Formation and Evolution of Planetary Systems*, eds. H. A. Weaver and L. Danly (Cambridge: Cambridge Univ. Press), pp. 296–304.

Herbig, G. H. 1990. The unusual pre-main sequence star VY Tauri. *Astrophys. J.* 360:639–649.

Herbig, G. H., and Bell, K. R. 1988. Third catalog of emission-line stars of the Orion population. *Lick Observatory Bull. No. 1111* (Univ. of California).

Herbig, G. H., and Goodrich, R. W. 1986. Near-simultaneous ultraviolet and optical spectrophotometry of T Tauri stars. *Astrophys. J.* 309:294–305.

Herbig, G. H., and Jones, B. E. 1981. Large proper motions of the Herbig-Haro objects HH 1 and HH 2. *Astron. J.* 86:1232–1244.

Herbig, G. H., and Rao, N. K. 1972. Second catalog of emission-line stars of the Orion Population. *Astrophys. J.* 174:401–423.

Herbig, G. H., and Soderblom, D. R. 1980. Observations and interpretation of the near-infrared line spectra of T Tauri stars. *Astrophys. J.* 242:628–637.

Herbig, G. H., and Terndrup, D. M. 1986. The trapezium cluster of the Orion Nebula. *Astrophys. J* 307:609–618.

Herbig, G. H., Vrba, F. J., and Rydgren, A. E. 1986. A spectroscopic survey of the Taurus-Auriga dark clouds for pre-main-sequence stars having Ca II H, K emission. *Astron. J.* 91:575–582.

Herbst, E. 1985. Gas-phase carbon chemistry in dense clouds. In *Protostars and Planets II*, eds. D. C. Black and M. S. Matthews (Tucson: Univ. of Arizona Press), pp. 668–684.

Herbst, E., and Klemperer, W. 1973. The formation and depletion of molecules in dense interstellar clouds. *Astrophys. J.* 185:505–533.

Herbst, E., and Leung, C. M. 1986. Effects of large rate coefficients for ion-polar neutral reactions on chemical models of dense interstellar clouds. *Astrophys. J.* 310:378–382.

Herbst, E., and Leung, C. M. 1989. Gas-phase production of complex hydrocarbons, cyanopolyynes, and related compounds in dense interstellar clouds. *Astrophys. J. Suppl.* 69:271–300.

Herbst, E., and Rajan, R. S. 1980. On the role of a supernovae in the formation of the solar system. *Icarus* 42:35–42.

Herbst, E., and Winnewisser, G. 1987. Organic molecules in space. In *Topics in Current Chemistry*, vol. 139, pp. 121–172.

Herd, C. R., Adams, N. G., and Smith, D. 1990. OH production in the dissociative recombination of H_3O^+, HCO_2^+, N_2OH^+: Comparison with theory and interstellar implications. *Astrophys. J.* 349:388–392.

Herman, G., and Weissman, P. R. 1987. Internal temperatures of cometary nuclei. *Icarus* 69:314–328.

Hessman, F. V., and Hopp, U. 1990. The massive, nearly face-on cataclysmic variable GD 552. *Astron. Astrophys.* 228:387–398.

Hessman, F. V., Eislöffel, J., Mundt, R., Hartmann, L., Herbst, W. and Krautter, J. 1991. The high state of the FU Ori variable Z CMa. *Astrophys. J.* 370:384–395.

Hewins, R. H. 1988. Experimental studies of chondrules. In *Meteorites and the Early Solar System*, eds. J. F. Kerridge and M. S. Matthews (Tucson: Univ. of Arizona Press), pp. 660–679.

Hewins, R. H. 1991. Retention of sodium during chondrule melting. *Geochim. Cosmochim. Acta* 55:935–42.

Hewitt, D. W., Francis, G. E., and Max, C. E. 1989. New regimes of magnetic reconnection in collisionless plasmas. *Phys. Rev. Lett.* 61:893–896.

Heyer, M. H. 1988. The magnetic evolution of the Taurus molecular clouds. II. A reduced role of the magnetic field in dense core regions. *Astrophys. J.* 324:311–320.

Heyer, M. H., and Graham, J. A. 1989. Newborn stars and stellar winds in Barnard 228. *Publ. Astron. Soc. Pacific* 101:816–831.

Heyer, M. H., Snell, R. L., Goldsmith, P. F., Strom, S. E., and Strom, K. M. 1986. A study of the morphology and kinematics of the dense gas associated with star-forming regions. *Astrophys. J.* 308:134–143.

Heyer, M. H., Vrba, F. J., Snell, R. L., Schloerb, F. P., Strom, S. E., Goldsmith, P. F., and Strom, K. M. 1987. The magnetic evolution of the Taurus molecular clouds. I. Large scale properties. *Astrophys. J.* 321:855–876.

Heyer, M. H., Snell, R. L., Goldsmith, P. F., and Myers, P. C. 1987. A survey of IRAS point sources in Taurus for high-velocity molecular gas. *Astrophys. J.* 321:370–382.

Heyer, M. H., Strom, S. E., and Strom, K. M. 1987. The magnetic field geometry in the vicinity of HH 7–11/HH 12 and HH 33/HH 40. *Astron. J.* 94:1653–1656.

Heyer, M. H., Snell, R. L., Morgan, J., and Schloerb, F. P. 1989. A CO and far-infrared study of the S254-S258 region. *Astrophys. J.* 346:220–231.

Heymann, D., and Dziczkaniec, M. 1979. Xenon from intermediate zones of super-novae. *Proc. Lunar Planet. Sci. Conf.* 10:1943–1959.

Heymann, D., and Dziczkaniec, M. 1980. A process of stellar nucleosynthesis which mimicks mass fractionation in p-xenon. *Meteoritics* 15:15–24.

Heyvaerts, J., and Norman, C. 1989. The collimation of magnetized winds. *Astrophys. J.* 347:1055–1081.

Higdon, J. C., and Fowler, W. A. 1987. Gamma-ray constraints on ^{22}Na yields in nova explosions. *Astrophys. J.* 317:710–716.

Higdon, J. C., and Fowler, W. A. 1989. Angular distribution of 1.809 MeV gamma rays generated in the decay of ^{26}Al produced by galactic novae. *Astrophys. J.* 339:956–961.

Hildebrand, R. H. 1983. The determination of cloud masses and dust characteristics from submillimeter thermal emission. *Quart. J. Roy. Astron. Soc.* 24:267–282.

Hildebrand, R. H. 1988. Magnetic fields and stardust. *Quart. J. Roy. Astron. Soc.* 29:327–351.

Hill, G. W. 1878. Researches in the lunar history. *American J. Math.* 1:5–26; 129–147; 245–260.

Hill, R. M., and Gordy, W. 1954. Zeeman effect and line breadth studies of the microwave lines of oxygen. *Phys. Rev.* 93:1019–1022.

Hillebrandt, W., and Theilemann, F.-K. 1982. Nucleosynthesis in novae: A source of Ne-E and ^{26}Al? *Astrophys. J.* 255:617–623.

Hills, J. G. 1972. An explanation of the cloudy structure of the interstellar medium. *Astron. Astrophys.* 17:155–160.

Hills, J. G. 1981. Comet showers and the steady-state infall of comets from the Oort cloud. *Astron. J.* 86:1730–1740.

Hills, J. G. 1982. The formation of comets by radiation pressure in the outer protosun. *Astron. J.* 87:906–910.

Hills, J. G., and Sandford, M. T., II. 1983*a*. The formation of comets by radiation pressure in the outer protosun. II. Dependence on the radiation-grain coupling. *Astron. J.* 88:1519–1521.

Hills, J. G., and Sandford, M. T., II. 1983*b*. The formation of comets by radiation pressure in the outer protosun. III. Dependence on the anisotropy of the radiation field. *Astron. J.* 88:1522–1530.

Hilton, J., White, G. J., Cronin, N. J., and Rainey, R. 1986. Lynds 379: a new source of bipolar molecular outflow. *Astron. Astrophys.* 154:274–278.

Hirose, M., and Osaki, Y. 1989. Hydrodynamic simulation of accretion disks in cataclysmic variables. In *Theory of Accretion Disks*, eds. F. Meyer, W. J. Duschl, J. Frank and E. Meyer-Hofmeister (Dordrecht: Kluwer), pp. 207–212.

Hirose, M., and Osaki, Y. 1990. Hydrodynamic simulations of accretion disks in cataclysmic variables: Superhump phenomenon in SU Vma stars. *Publ. Astron. Soc. Japan* 42:135–163.

Hiyagon, H., and Sasaki, S. 1988. Noble gas constraints on the early history of the Earth. *Prog. Theor. Phys. Suppl.* 96:1–15.

Ho, P. C. T., and Haschick, A. D. 1986. Formation of OB clusters: radiation-driven implosion? *Astrophys. J.* 305:714–720.

Hoban, S., Mumma, M. J., Reuter, D. C., DiSanti, M., Joyce, R. R., and Storrs, A. 1991*a*. A tentative identification of methanol as the progenitor of the 3.52 μm feature in several comets. *Icarus* 93:122–134.

Hoban, S., Reuter, D. C., Mumma, M. J., and Storrs, A. 1991*b*. Molecular hydrogen in the vicinity of NGC 7538 IRS 1 and IRS 2: temperature and ortho-para ratio. *Astrophys. J.* 370:228–236.

Hobbs, L. M. 1986. Observations of gaseous circumstellar disks. III. *Astrophys. J.* 308:854–858.

Hobbs, L. M., Vidal-Madjar, A., Ferlet, R., Albert, C. E., and Gry, C. 1985. The gaseous component of the disk around Beta Pictoris. *Astrophys. J. Lett.* 293:29–33.

Hobbs, L. M., Lagrange-Henri, A. M., Ferlet, R., Vidal-Madjar, A., and Welty, D. E. 1988. The location of Ca II ions in the Beta Pictoris disk. *Astrophys. J. Lett.* 334:41–44.

Hodapp, K. W. 1984. Infrared polarization of sources with bipolar mass outflow. *Astron. Astrophys.* 141:255–262.

Hodapp, K.-W. 1987. The magnetic field in star-forming large globules. *Astrophys. J.* 319:842–849.

Hodapp, K.-W. 1990. Magnetic field and spatial structure of bipolar outflow sources. *Astrophys. J.* 352:184–191.

Hodapp, K.-W., and Rayner, J. T. 1991. The S 106 star-forming region. *Astron. J.* 102:1108–1117.

Hoffleit, D., and Jaschek, C. 1982. *The Bright Star Catalog* (New Haven: Yale Univ. Obs.).

Hohenberg, C. M., Marti, K., Podosek, F. A., Reedy, R. C., and Shirck, J. R. 1978. Comparisons between observed and predicted cosmogenic noble gases in lunar samples. *Proc. Lunar Planet. Sci. Conf.* 9:2311–2344.

Hohenberg, C. M., Nichols, R. H., Jr., Olinger, C. T., and Goswami, J. N. 1990. Cosmogenic neon from individual grains of CM meteorites: Extremely long pre-compaction exposure histories or an enhanced early particle flux. *Geochim. Cosmochim. Acta* 54:2133–2140.

Hollenbach, D. J., and McKee, C. F. 1979. Molecule formation and infrared emission in fast molecular shocks. I. Physical processes. *Astrophys. J. Suppl.* 41:555–592.

Hollenbach, D. J., and McKee, C. F. 1989. Molecule formation and infrared emission in fast molecular shocks. III. Results for J shocks in molecular clouds. *Astrophys. J.* 342:306–336.

Hollenbach, D. J., and Neufeld, D. A. 1990. Accretion shocks at protostellar disks. Protostars and Planets III, Tucson, Arizona, 5–9 March, p. 58, Abstract book.

Hollenbach, D. J., and Salpeter, E. E. 1971. Surface recombination of hydrogen molecules. *Astrophys. J.* 163:155–164.

Hollenbach, D. J., Chernoff, D. F., and McKee, C. F. 1989. Infrared diagnostics of interstellar shocks. In *Infrared Spectroscopy in Astronomy*, ed. B. H. Kaldeich, ESA SP-290, pp. 245–258.

Hollis, J. M., Chin, G., and Brown, R. L. 1985. An attempt to detect mass loss from α Lyrae with the VLA. *Astrophys. J.* 294:646–648.

Hollowell, D., and Iben, I., Jr. 1989. Neutron production and neutron-capture nucleosynthesis in a low-mass, low metallicity asymptotic red giant star. *Astrophys. J.* 340:986–984.

Hollowell, D., and Iben, I., Jr. 1990. Algorithms for neutron density and s-process nucleosynthesis in a low-mass asymptotic giant branch star. *Astrophys. J.* 349:208–221.

Holtzman, J. A., Herbst, W., and Booth, J. F. 1986. Photometric variations of Orion Population stars. IV. Coordinated spectroscopy in 1984–1985 with some success for RY Tau. *Astron. J.* 92:1387–1395.

Holweger, H., Bard, A., Kock, A., and Kock, M. 1991. A redetermination of the solar iron abundance based on new FeI oscillator strengths. *Astron. Astrophys.* 249:545–549.

Honda, M., Ozima, M., Nakada, Y., and Onaka, T. 1979. Trapping of rare gases during the condensation of solids. *Earth and Planet. Sci. Lett.* 43:197–200.

Honda, M., Reynolds, J. H., Roedder, E., and Epstein, S. 1987. Noble gases in diamonds: occurrences of solar-like helium and neon. *J. Geophys. Res.* 92:12507–12521.

Honda, M., McDougall, I., Patterson, D. B., Doulgeris, A., and Clague, D. A. 1991. Possible solar noble-gas component in Hawaiian basalts. *Nature* 349:149–151.

Hong, S. S., and Greenberg, J. M. 1980. A unified model of interstellar grains: a connection between alignment efficiency, grain model size, and cosmic abundance. *Astron. Astrophys.* 88:194–202.

Hood, L. L., and Horanyi, M. 1991. Gas dynamic heating of chondrule precursor grains in the solar nebula. *Icarus* 93:259–269.

Hopper, P. B., and Disney, M. J. 1974. The alignment of interstellar dust clouds. *Mon. Not. Roy. Astron. Soc.* 168:639–650.

Horanyi, M., Gombosi, T. I., Cravens, T. E., Korosmezey, A., Kecskemety, K., Nagy, A., and Szego, K. 1984. The friable sponge model of a cometary nucleus. *Astrophys. J.* 278:449–455.

Horedt, G. P. 1978. Blow-off of the proto-planetary cloud by a T Tauri-like solar wind. *Astron. Astrophys.* 64:173–178.

Horne, K. 1985. Images of accretion discs—I. The eclipse mapping method. *Mon. Not. Roy. Astron. Soc.* 213:129–141.

Horne, K., and Cook, M. C. 1985. UBV images of the Z CHA accretion disk in outburst. *Mon. Not. Roy. Astron. Soc.* 214:307–317.

Hornung, P., Pellat, R., and Barge, P. 1985. Thermal velocity equilibrium in the protoplanetary cloud. *Icarus* 64:295–307.

Hostetler, C. J. 1981. A possible common origin for the rare gases on Venus, Earth, and Mars. *Proc. Lunar Planet. Sci. Conf.* 12:1387–1393.

Hourigan, K., and Ward, W. R. 1984. Radical migration of preplanetary material: Implications for the accretion time scale problem. *Icarus* 60:29–39.

Housen, K. R., and Wilkening, L. L. 1982. Regoliths on small bodies in the solar system. *Ann. Rev. Earth Planet. Sci.* 20:355–376.

Housen, K. R., Wilkening, L. L., Chapman, C. R., and Greenberg, R. 1979. Asteroidal regoliths. *Icarus* 39:317–351.

Howard, W. M., Meyer, B. S., and Woosley, S. E. 1991. A new site for the astrophysical gamma-process. In *Nuclei in the Cosmos*, eds. H. Oberhummer and J. H. Applegate (Berlin: Springer-Verlag).

Hoyle, F. 1953. On the fragmentation of gas clouds into galaxies and stars. *Astrophys. J.* 118:513–528.

Hoyle, F. 1960. On the origin of the solar system. *Quart. J. Roy. Astron. Soc.* 1:28–55.

Hu, J. Y., The, P. S., and de Winter, D. 1989. Photometric and spectroscopic study of 3 candidate Herbig Ae/Be stars. *Astron. Astrophys.* 64:173–178.

Hua, X., Adam, J., Palme, H., and El Goresy, A. 1988. Fayalite-rich rims, veins, and halos around and in forsteritic olivines in CAIs and chondrules in carbonaceous chondrites: types, compositional profiles and constraints of their formation. *Geochim. Cosmochim. Acta* 52:1389–1408.

Hubbard, W. B. 1968. Thermal structure of Jupiter. *Astrophys. J.* 152:745–754.

Hubbard, W. B., and Horedt, G. P. 1983. Computation of Jupiter interior models from gravitational inversion theory. *Icarus* 54:456–465.

Hubbard, W. B., and DeWitt, H. E. 1985. Statistical mechanics of elements at high pressure. VII. A perturbative free energy for arbitrary mixtures of H and He. *Astrophys. J.* 290:388–393.

Hubbard, W. B., and MacFarlane, J. J. 1980a. Structure and evolution of Uranus and Neptune. *J. Geophys. Res.* 85:225–234.

Hubbard, W. B., and MacFarlane, J. J. 1980b. Theoretical predictions of deuterium abundances in the Jovian planets. *Icarus* 44:676–682.

Hubbard, W. B., and MacFarlane, J. J. 1985. Statistical mechanics of elements at high pressure. VIII. Thomas-Fermi-Dirac theory for binary mixtures of H with He, C, and O. *Astrophys. J.* 297:133–144.

Hubbard, W. B., and Marley, M. S. 1989. Optimized Jupiter, Saturn, and Uranus interior models. *Icarus* 78:102–118.

Hubbard, W. B., and Slattery, W. L. 1976. Interior structure of Jupiter: theory of gravity sounding. In *Jupiter*, ed. T. Gehrels (Tucson: Univ. of Arizona Press), pp. 176–194.

Hubbard, W. B., and Stevenson, D. J. 1984. Interior structure of Saturn. In *Saturn*, eds. T. Gehrels and M. S. Matthews (Tucson: Univ. of Arizona Press), pp. 47–87.

Hubbard, W. B., Slattery, W. L., and De Vito, C. 1975. High zonal harmonics of rapidly rotating planets. *Astrophys. J.* 199:504–516.

Hubbard, W. B., MacFarlane, J. J., Anderson, J. D., Null, G. W., and Biller, E. D. 1980. Interior structure of Saturn inferred from Pioneer 11 gravity data. *J. Geophys. Res.* 85:5909–5916.

Hubbard, W. B., Burrows, A., and Lunine, J. I. 1990*a*. The initial mass function for very low mass stars in the Hyades. *Astrophys. J. Lett.* 358:53–55.

Hubbard, W. B., Yelle, R. V., and Lunine, J. I. 1990*b*. Non-isothermal Pluto atmosphere models. *Icarus* 84:1–11.

Hubbard, W. B., Nellis, W. J., Mitchell, A. C., Holmes, N. C., Limaye, S. S., and McCandless, P. C., 1991. Interior structure of Neptune: comparison with Uranus. *Science* 253:648–651.

Hudson, G. B., Kennedy, B. M., Podosek, F. A., and Hohenberg, C. M. 1989. The early solar system abundance of ^{244}Pu as inferred from the St. Severin chondrite. *Proc. Lunar Planet. Sci. Conf.* 19:547–557.

Heubner, W. F. 1987. First polymer in space identified in comet Halley. *Science* 237:628–630.

Huebner, W. F., Snyder, L. E., and Buhl, D. 1974. HCN radio emission from comet Kohoutek (1973f). *Icarus* 23:580–585.

Hughes, J. D. 1989. Ph. D. Thesis, Queen Mary College, Univ. of London.

Hughes, J. D., Emerson, J. P., Zinnecker, H., and Whitelock, P. A. 1989. IRAS 12496-7650: An Ae Star with outflow? *Mon. Not. Roy. Astron. Soc.* 236:117–127.

Hughes, V. A. 1988. Radio observations of Cepheus A. I. The evolving pre-main-sequence stars in Cepheus A east? *Astrophys. J.* 333:788–800.

Hughes V. A., and Moriarty-Schieven, G. H. 1990. A comparison of the radio continuum and optical properties of the HH objects in Cepheus A West (GGD37). *Astrophys. J.* 360:215–220.

Huneke, J. C., and Wasserburg, G. J. 1975. Trapped ^{40}Ar in troctolite 76535 and evidence for enhanced ^{40}Ar-^{39}Ar age plateaus. *Lunar Sci.* VI:417–419 (abstract).

Hunten, D. M. 1973. The escape of light gases from planetary atmospheres. *J. Atmos. Sci.* 30:1481–1494.

Hunten, D. M., Pepin, R. O., and Walker, J. C. G. 1987. Mass fractionation in hydrodynamic escape. *Icarus* 69:532–549.

Hunten, D. M., Pepin, R. O., and Owen, T. C. 1988. Planetary atmospheres. In *Meteorites and the Early Solar System*, eds. J. F. Kerridge and M. S. Matthews (Tucson: Univ. of Arizona Press), pp. 565–591.

Hunten, D. M., Donahue, T. M., Walker, J. C. G., and Kasting, J. F. 1989. Escape of atmospheres and loss of water. In *Origin and Evolution of Planetary and Satellite Atmospheres*, eds. S. K. Atreya, J. B. Pollack and M. S. Matthews (Tucson: Univ. of Arizona Press), pp. 386–422.

Hunter, D. A., and Massey, P. 1990. Small galactic H II regions. I. Spectral classifications of massive stars. *Astron. J.* 99:846–856.

Hunter, J. H., Jr., and Whitaker, R. W. 1989. Anisentropic, magnetic, Kelvin-Helmholtz and related instabilities in the interstellar medium. *Astrophys. J. Suppl.* 71:777–798.

Husain, L., and Schaeffer, O. A. 1975. Lunar evolution: The first 600 million years. *Geophys. Res. Lett.* 2:29–32.

Huss, G. R. 1990. Ubiquitous interstellar diamonds and SiC in primitive chondrites.

Nature 347:159–161.

Huss, G. R., and Lewis, R. S. 1989. Interstellar diamonds and SiC from type 3 ordinary chondrites. *Meteoritics* 24:278–279 (abstract).

Huss, G. R., and Lewis, R. S. 1990. Interstellar diamonds and silicon carbide in enstatite chondrites. *Lunar Planet. Sci.* XXI:542–543 (abstract).

Hut, P., and Tremaine, S. 1985. Have interstellar clouds disrupted the Oort comet cloud? *Astron. J.* 90:1548–1557.

Hutcheon, I. D., and Hutchison, R. 1989. Evidence from the Semarkona ordinary chondrite for ^{26}Al heating of small planets. *Nature* 337:238-241.

Hutcheon, I. D., and Olsen, E. 1991. Cr isotopic composition of differentiated meteorites: A search for ^{53}Mn. *Lunar Planet. Sci.* XXII:605–606 (abstract).

Hutcheon, I. D., Armstrong, J. T., and Wasserburg, G. J. 1984. Excess ^{41}K in Allende CAZ: Confirmation of a hint. *Lunar Planet. Sci.* XV:387–388 (abstract).

Hutcheon, I. D., Armstrong, J. T., and Wasserburg, G. J. 1987. Isotopic studies of Mg,Fe,Mo,Ru and W in Fremdling from Allende refractory inclusions. *Geochim. Cosmochim. Acta* 51:3175–3192.

Hyland, A. R., and Jones, T. J. 1991. Candidate protostars in the vicinity of 30 Doradus. In *The Magellanic Clouds*, eds. R. Haynes and D. Milne (Dordrecht: Kluwer), pp. 202–204.

Hyland, A. R., and MacGregor, P. J. 1989. In *Interstellar Dust: Contributed Papers: Proc. IAU Symp. 135*, NASA CP-3036, pp. 101–106.

Hyland, A. R., Jones, T. J., and Mitchell, R. M. 1982. A study of the Chamaeleon dark cloud complex: survey, structure, and embedded sources. *Mon. Not. Roy. Astron. Soc.* 201:1095–1117.

Hyland, A. R., Allen, D. A., Barnes, P. J., and Ward, M. J. 1984. The distribution and nature of the 2 μm radiation of the inner Orion Nebula. *Mon. Not. Roy. Astron. Soc.* 206:465–474.

Iben, I. 1965. Stellar evolution. I. The approach to the main sequence. *Astrophys. J.* 141:993–1018.

Iben, I. 1967. Stellar evolution within and off the main sequence. *Ann. Rev. Astron. Astrophys.* 5:571–626.

Iben, I., and Talbot, R. J. 1966. Stellar formation rates in young clusters. *Astrophys. J.* 144:968–977.

Ichikawa, T., and Nishida, M. 1989. *IRAS* point sources in the Ophiuchus molecular cloud complex: optical identification. *Astron. J.* 97:1074–1088.

Ida, S. 1990. Stirring and dynamical friction rates of planetesimals in the solar gravitational field. *Icarus* 88:129–145.

Ida, S., and Makino, J. 1992a. N-body simulation of gravitational interaction between planetesimals and a protoplanet: I. Velocity distribution of planetesimals. *Icarus* 96:107–120.

Ida, S., and Makino, J. 1992b. N-body simulation of gravitational interaction between planetesimals and a protoplanet: II. Dynamical friction. *Icarus*, in press.

Ida, S., and Nakazawa, K. 1988. Collision rate between planetesimals and the growth of planets. *Prog. Theor. Phys. Suppl.* 96:211–227.

Ida, S., and Nakazawa, K. 1989. Collisional probability of planetesimals revolving in the solar gravitational field, III. *Astron. Astrophys.* 224:303–315.

Igarashi, G., and Ozima, M. 1988. Origin of isotopic fractionation of terrestrial xenon. *Proc. NIPR Symp. Antarctic Meteorites* 1:315–320.

Iglesias, E. 1977. The chemical evolution of molecular clouds. *Astrophys. J.* 218:697–715.

Igumentshchev, I. V., Shustov, B. M., and Tutukov, A. V. 1990. Dynamics of supershells: Blow out. *Astron. Astrophys.* 234:396–402.

Imhoff, C. L., and Appenzeller, I. 1987. Pre-main sequence stars. In *Scientific*

Accomplishments of the I. U. E., ed. Y. Kondo (Dordrecht: D. Reidel), pp. 295–319.

Innes, D. E., Giddings, J., and Falle, S. A. E. G. 1987. Dynamical models of radiative shocks. II. Unsteady shocks. *Mon. Not. Roy. Astron. Soc.* 236:117–127.

Ip, W.-H. 1987. Gravitational stirring of the asteroid belt by Jupiter zone bodies. *Gerlands. Beitrag. Geophysik* 96:44–51.

Ip, W.-H., Balsiger, H., Geiss, J., Goldstein, B. E., Kettman, G., Lazarus, A. J., Meier, A., Rosenbauer, H., Schwenn, R., and Shelley, E. 1990. Giotto IMS measurements of the production rate of hydrogen cyanide in the coma of comet Halley. *Ann. Geophys.* 8:319–326.

Ipatov, S. I. 1987. Solid body accumulation of terrestrial planets. *Solar System Res.* 21:129–135.

Ipatov, S. I. 1989. Evolution of the orbital eccentricities of planetesimals during the formation of the giant planets. *Solar System Res.* 23:119–125.

Ireland, T. R. 1988. Correlated morphological, chemical, and isotopic characteristics of hibonites from the Murchison carbonaceous chondrite. *Geochim. Cosmochim. Acta* 52:2827–2839.

Ireland, T. R. 1990. Presolar isotopic and chemical signatures in hibonite-bearing refractory inclusions from the Murchison carbonaceous chondrite. *Geochim. Cosmochim. Acta* 54:3219–3237.

Ireland, T. R. 1991. The abundance of ^{182}Hf in the early solar system. *Lunar Planet. Sci.* XXII:609–610 (abstract).

Ireland, T. R., Fahey, A. J., and Zinner, E. K. 1988. Trace-element abundances in hibonites from the Murchison carbonaceous chondrite: constraints on high-temperature processes in the solar nebula. *Geochim. Cosmochim. Acta* 52:2841–2854.

Ireland, T. R., Palme, H., and Spettel, B. 1990*a*. Trace-element inventory of the Allende (CV3) meteorite. *Lunar Planet. Sci.* XXI:546–547(abstract).

Ireland, T. R., Compston, W., Williams, I. S., and Wendt, I. 1990*b*. U-Th-Pb systematics of individual perovskite grains from the Allende and Murchison carbonaceous chondrites. *Earth Planet. Sci. Lett.* 101:379–387.

Irvine, W. M., and Knacke, R. F. 1989. The chemistry of interstellar gas and grains. In *Origin and Evolution of Planetary and Satellite Atmospheres*, eds. S. K. Atreya, J. B. Pollack and M. S. Matthews (Tucson: Univ. of Arizona Press), pp. 3–34.

Irvine, W. M., Schloerb, F. P., Hjalmarson, Å., and Herbst, E. 1985. The chemical state of dense interstellar clouds: an overview. In *Protostars and Planets II*, eds. D. C. Black and M. S. Matthews (Tucson: Univ. of Arizona Press), pp. 579–620.

Irvine, W. M., Goldsmith, P. F., and Hjalmarson, Å. 1987. Chemical abundances in molecular clouds. In *Interstellar Processes*, eds. D. Hollenbach and H. A. Thronson (Dordrecht: D. Reidel), pp. 561–610.

Irvine, W. M., Ohishi, M., and Kaifu, N. 1991. Chemical abundances in cold, dark interstellar clouds. *Icarus* 91:2–6.

Isobe, S. 1970. Evaporation of dirty ice particles surrounding early type stars. I. The Orion nebula. *Publ. Astron. Soc. Japan* 22:429–445.

Israel, F. P., de Graauw, T., van de Stadt, H., and de Vries, C. P. 1986. Carbon monoxide in the magellanic clouds. *Astrophys. J.* 303:186–197.

Israel, F. P., Hawarden, T. G., Geballe, T. R., and Wade, R. 1990. The molecular hydrogen content of NGC 604 and other M33 HII region complexes. *Mon. Not. Roy. Astron. Soc.* 242:471–977.

Issa, M., MacLaren, I., and Wolfendale, A. W. 1990. The size-linewidth relation and the mass of molecular hydrogen. *Astrophys. J.* 352:132–138.

Iwata, T., Fukui, Y., and Ogawa, H. 1988. A molecular line study of a bipolar outflow object NGC2071-North in L1630. *Astrophys. J.* 325:372–381.

Jackson, W. M., Halpern, J. B., Feldman, P. D., and Rahe, J. 1982. Production of CS and S in comet Bradfield (1979 X). *Astron. Astrophys.* 107:385–389.

Jacq, T., Jewell, P. R., Henkel, C., Walmsley, C. M., and Baudry, A. 1988. $H_2^{18}O$ in hot dense molecular cloud cores. *Astron. Astrophys.* 199:L5–L8.

Jacq, T., Walmsley, C. M., Henkel, C., Baudry, A., Mauersberger, R., and Jewell, P. R. 1990. Deuterated water and ammonia in hot cores. *Astron. Astrophys.* 228:447–470.

Jaffe, D. T., Harris, A. L., and Genzel, R. 1987. Warm dense gas in luminous protostellar regions: A submillimeter and far-infrared CO line study. *Astrophys. J.* 316:231–241.

Jaffe, D. T., Genzel, R., Harris, A. I., Lugten, J. B., Stacey, G. J., and Stutzki, J. 1989. Strong, spatially extended CO 7-6 emission from luminous cloud cores: W51 and DR21. *Astrophys. J.* 344:265–276.

James, R. A. 1964. The structure and stability of rotating gas masses. *Astrophys. J.* 140:552–582.

Jameson, R. F., and Skillen, I. 1989. A search for low-mass stars and brown dwarfs in the Pleiades. *Mon. Not. Roy. Astron. Soc.* 239:247–253.

Jankowski, D. G., and Squyres, S. W. 1988. Solid-state volcanism on the satellites of Uranus. *Science* 241:1322–1325.

Jankowski, D. G., Chyba, C. F., and Nicholson, P. D. 1989. On the obliquity and tidal heating of Triton. *Icarus* 80:211–219.

Jarrett, T. H., Edwards, S., Strom, S., and Snell, R. L. 1987. The L1551 far-infared phenomenon. *Bull. Amer. Astron. Soc.* 19:197 (abstract).

Jarrett, T. H., Dickman, R. L., and Herbst, W. 1989. Far-infrared emission in the ρ Ophiuchi region: a comparison with molecular gas emission and visual extinction. *Astrophys. J.* 345:881–893.

Jaschek, M., Jaschek, C., and Egret, D. 1986. A-type shell stars and infrared sources. *Astron. Astrophys.* 158:325–328.

Jasniewicz, G., and Mayor, M. 1988. Radial velocity measurements of a sample of northern metal-deficient stars. *Astron. Astrophys.* 203:329–340.

Jaworski, W. A., and Tatum, J. B. 1991. Analysis of the Swings effect and Greenstein effect in comet P/Halley. *Astrophys. J.* 377:306–317.

Jen, C. K. 1948. The Zeeman effect in microwave molecular spectra. *Phys. Rev.* 74:1396–1406

Jenkins, E. B., and Meloy, D. A. 1974. A survey with Copernicus of interstellar OVI absorption. *Astrophys. J. Lett.* 193:121–125

Jessberger, E. K., and Kissel, J. 1991. Chemical properties of cometary dust and a note on carbon isotopes. In *Comets in the Post-Halley Era*, eds. R. Newburn, M. Neugebauer and J. Rahe (Dordrecht: Kluwer), pp. 1075–1092.

Jessberger, E. K., Kissel, J., Fechtig, H., and Krueger, F. R. 1987. On the average chemical composition of cometary dust. In *Physical Processes in Comets, Stars, and Active Galaxies*, eds. W. Hillenbrandt, E. Meyer-Hofmeister and H. C. Thomas (Heidelberg: Springer-Verlag), pp. 26–33.

Jessberger, E. K., Christoforidis, A., and Kissel, J. 1988. Aspects of the major element composition of Halley's dust. *Nature* 332:691–695.

Jewell, P. R., Hollis, J. M., Lovas, F. J., and Snyder, L. E. 1989. Millimeter- and submillimeter-wave survey of Orion A emission lines in the range 200.7–202.3, 203.7–205.3, and 330–360 GHz. *Astrophys. J. Suppl.* 70:833–864.

Joblin, C., Maillard, J. P., d'Hendecourt, L., and Leger, A. 1990. Detection of diffuse interstellar bands in the infrared. *Nature* 346:729–731.

Jochum, K. P., Seufert, H. M., Spettel, B., and Palme, H. 1986. The solar system abundances of Nb, Ta and Y, and the relative abundances of refractory lithophile elements in differentiated planetary bodies. *Geochim. Cosmochim. Acta* 50:1173–

1183.

Johansson, L. E. B., Andersson, C., Ellder, J., Friberg, P., Hjalmarson, Å, Höglund, B., Irvine, W. M., Olofsson, H., and Rydbeck, G. 1984. Spectral line survey of Orion A and IRC+10216 from 72 to 91 GHz. *Astron. Astrophys.* 130:227–256.

Johns, T. C., and Nelson, A. H. 1986. Global simulations of gas flow in disc galaxies— I. Response to a spiral density wave. *Mon. Not. Roy. Astron. Soc.* 220:165–183

Johnson, H. L. 1966. Astronomical measurements in the infrared. *Ann. Rev. Astron. Astrophys.* 4:193–206.

Johnson, H. L., and Wisniewski, W. Z. 1978. Emission lines in the spectrum of Vega. *Publ. Astron. Soc. Pacific* 90:139–206.

Johnson, H. M. 1986. Far-infrared characteristics of K dwarfs with H α emission. *Astrophys. J.* 300:401–405.

Johnson, J. J., Gehrz, R. D., Jones, T. J., Hackwell, J. A., and Grasdalen, G. L. 1990. An infrared study of Orion Molecular Cloud-2 (OMC-2). *Astron. J.* 100:518–529.

Johnson, R. E. 1989. Effect of irradiation on the surface of Pluto. *Geophys. Res. Lett.* 16:1233–1236.

Johnson, R. E. 1991. Irradiation effects in a comet's outer layer. *J. Geophys. Res.* 96:17553–17557.

Johnson, R. E., Lanzerotti, L. J., and Brown, W. L. 1984. Sputtering processes: erosion and chemical changes. *Adv. Space Res.* 4:41–51.

Johnson, R. E., Barton, L. A., Boring, J. W., Jesser, W. A., Brown, W. L., and Lanzerotti, L. J. 1985. Charged particle modification of ices in the Jovian and Saturnian systems. In *Ices in the Solar System*, eds. J. Klinger, D. Benest, A. Dollfus and R. Smoluchowski (Dordrecht: D. Reidel), pp. 301–316.

Johnson, R. E., Cooper, J. F., Lanzerotti, L. J., and Strazzulla, G. 1987. Radiation formation of a nonvolatile comet crust. *Astron. Astrophys.* 187:889–892.

Johnson, T. V. 1990. Paper presented at the COSPAR XVIII Plenary Meeting, The Hague, The Netherlands, July, 1990.

Johnson, T. V., and Matson, D. L. 1989. Io's tenuous atmosphere. In *Origin and Evolution of Planetary and Satellite Atmospheres*, eds. S. K. Atreya, J. B. Pollack and M. S. Matthews (Tucson: Univ. of Arizona Press), pp. 666–681.

Johnson, T. V., Brown, R. H., and Pollack, J. B. 1987. Uranian satellites: Densities and composition. *J. Geophys. Res.* 92:14884–14894.

Johnston, K. J., Migenes, V., and Norris, R. P. 1989. The spatial distribution of the OH masers in Orion-KL. *Astrophys. J.* 341:847–856.

Johnstone, R. M., and Penston, M. Y. 1986. A search for magnetic fields in the T Tauri stars GW Ori, CoD-34 7151, and RU Lup. *Mon. Not. Roy. Astron. Soc.* 219:927–941.

Jokipii, J. R. 1964. The distribution of gases in the protoplanetary nebula. *Icarus* 3:248–252.

Jones, A. P., and Williams, D. A. 1984. The 3 μm ice band in Taurus: Implications for interstellar chemistry. *Mon. Not. Roy. Astron. Soc.* 209:955–960.

Jones, B. F., and Herbig, G. H. 1979. Proper motions of T Tauri variables and other stars associated with the Taurus-Auriga dark clouds. *Astron. J.* 84:1872–1889.

Jones, B. F., and Herbig, G. H. 1982. Proper motions of Herbig-Haro objects. II. The relationship of HH-39 to R Monocerotis and NGC 2261. *Astron. J.* 87:1223–1232.

Jones, B. F., and Walker, M. F. 1988. Proper motion and variabilities of stars near the Orion Nebula. *Astron. J.* 95:1755–1782.

Jones, E. M., and Kodis, J. W. 1982. Atmospheric effects of large body impacts: the first few minutes. In *Geological Implications of Impacts of Large Asteroids and Comets on the Earth*, eds. L. T. Silver and P. H. Schultz, Geol. Soc. America Special Paper 190, pp. 175–186.

Jones, T. D., and Lewis, J. S. 1987. Estimated impact shock production of N_2 and organic compounds on early Titan. *Icarus* 72:381–393.

Jones, T. D., Lebofsky, L., Lewis, J., and Marley, M. 1990. The composition and origin of the C, P, and D asteroids: Water as a tracer of thermal evolution in the outer belt. *Icarus* 88:172–192.

Jones, T. J. 1989. Infrared polarimetry and the interstellar magnetic field. *Astrophys. J.* 346:728–734.

Jones, T. J., Hyland, A. R., and Bailey, J. 1984. The inner core of a Bok globule. *Astrophys. J.* 282:675–682.

Jones, T. J., Hyland, A. R., Harvey, P. M., Wilking, B. A., and Joy, M. 1985. The Chamaeleon dark cloud complex. II. A deep survey around HD 97330. *Astron. J.* 90:1191–1195.

Jørgensen, U. G. 1988. Formation of Xe-HL-enriched diamond grains in stellar environments. *Nature* 332:702–705.

Joss, P. C. 1973. On the origin of short-period comets. *Astron. Astrophys.* 25:271–273.

Joy, A. H. 1945. T Tauri variable stars. *Astrophys. J.* 102:168–195.

Joy, A. H. 1949. Bright-line stars among the Taurus dark clouds. *Astrophys. J.* 110:424–437.

Joy, A. H., and van Biesbroeck, G. 1944. Five new double stars among variables of the T Tauri class. *Publ. Astron. Soc. Pacific* 56:123–124.

Judge, P. G., Jordan, C., and Rowan-Robinson, M. 1987. Delta Andromeda (K3 III): an IRAS source with an unusual ultraviolet spectrum. *Mon. Not. Roy. Astron. Soc.* 224:93–106.

Julian, W. H., and Toomre, A. 1966. Non axisymmetric responses of differentially rotating disks of stars. *Astrophys. J.* 146:810–830.

Jura, M. 1974. Formation and destruction rates of interstellar H_2. *Astrophys. J.* 191:375–379.

Jura, M. 1975. Interstellar clouds containing optically thin H_2. *Astrophys. J.* 197:575–580.

Jura, M. 1980. Origin of large interstellar grains toward ρ Ophiuchi. *Astrophys. J.* 253:63–65.

Jura, M. 1990. The absence of circumstellar dust debris around G giants. *Astrophys. J.* 365:317–320.

Kahane, C., Frerking, M. A., Langer, W. D., Encrenaz, P., and Lucas, R. 1984. Measurement of the formaldehyde ortho to para ratio in three molecular clouds. *Astron. Astrophys.* 137:211–222.

Kahn, F. D., and Breitschwerdt, D. 1990. Stellar winds in H II regions: II. Turbulent mixing between an H II layer and a stellar wind bubble in the presence of a magnetic field. *Mon. Not. Roy. Astron. Soc.* 242:209–214.

Kaifu, N., Suzuki, S., Hasegawa, T., Morimoto, M., Inatani, J., Nagane, K., Miyazawa, K., Chikada, Y., Kanzawa, T., and Akabane, K. 1984. Rotating gas disk around L1551 IRS-5. *Astron. Astrophys.* 134:7–12.

Kaiser, T., and Wasserburg, G. J. 1983. The isotopic composition and concentration of Ag in iron meteorites and the origin of exotic silver. *Geochim. Cosmochim. Acta* 47:43–58.

Kaisig, M. 1989a. Numerical simulation to acoustic instabilities in thin accretion disks. *Astron. Astrophys.* 218:89–101.

Kaisig, M. 1989b. Numerical simulation of the formation of shock wave in thin accretion disks and the resulting angular momentum transport. *Astron. Astrophys.* 218:102–110.

Kallemeyn, G. W., and Wasson, J. T. 1981. The composition and classification of meteorites. I. The carbonaceous chondrite groups. *Geochim. Cosmochim. Acta* 45:1217–1230.

Kameya, O., Hasegawa, T. I., Hirano, N., Takakubo, K., and Seki, M. 1989. High-velocity flows in the NGC7538 molecular cloud. *Astrophys. J.* 339:222–230.

Kant, I. 1755. *Allegmeine Naturgeschichte und Theorie des Himmels.*

Käppeler, F., Beer, H., and Wisshak, K. 1989. S-process nucleosynthesis—nuclear physics and the classical model. *Rept. Prog. Phys.* 52:945–1013.

Kary, D. M., Lissauer, J. J., and Greenzweig, Y. 1993. Collision probabilities in the presence of nebular gas drag. In preparation.

Kasting, J. F., and Toon, O. B. 1989. Climate evolution on the terrestrial planets. In *Origin and Evolution of Planetary and Satellite Atmospheres*, eds. S. K. Atreya, J. B. Pollack and M. S. Matthews (Tucson: Univ. of Arizona Press), pp. 423–449.

Kato, S. 1983. Low-frequency one-armed oscillations of Keplerian gaseous disks. *Publ. Astron. Soc. Japan* 35:249–261.

Kato, S. 1989. One-armed oscillations of disks and their application. In *Theory of Accretion Disks*, eds. F. Meyer, W. Duschl, J. Frank, E. Meyer-Hofmeister, (Dordrecht: Kluwer), pp. 173–181.

Kato, T., Ringwood, A. E., and Irifune, T. 1988. Experimental determination of element partitioning between silicate perovskites, garnets and liquids: constraints on the early differentiation of the mantle. *Earth Planet. Sci. Lett.* 89:123–145.

Kaufman, M., Elmegreen, D. M., and Bash, F. N. 1989. The dust lanes in M81: Tracers of molecular gas? *Astrophys. J.* 345:697–706.

Kaula, W. M. 1964. Tidal dissipation by solid friction and the resulting orbital evolution. *Rev. Geophys.* 2:661–685.

Kaula, W. M. 1979a. Equilibrium velocities of a planetesimal population. *Icarus* 40:262–275.

Kaula, W. M. 1979b. Thermal evolution of the Earth and Moon growing by planetes-imal impacts. *J. Geophys. Res.* 84:999–1008.

Kaula, W. M. 1980. The beginning of the Earth's thermal evolution. In *The Continental Crust and its Mineral Deposits*, ed. D. W. Strangway, Geol. Assoc. of Canada Special Paper 20, Waterloo, Ontario, pp. 25–34.

Kawara, K., Gregory, B., Yamamoto, T., and Shibai, H. 1988. Infrared spectroscopic observation of methane in comet P/Halley. *Astron. Astrophys.* 207:174–181.

Kazès, I., and Crutcher, R. M. 1986. Measurements of magnetic-field strengths in molecular clouds: detection of OH-line Zeeman splitting. *Astron. Astrophys. J.* 164:328–336.

Keene, J., and Masson, C. 1990. Detection of a 45 AU radius source around L1551-IRS5—A possible accretion disk. *Astrophys. J.* 355:635–644.

Keene, J., Blake, G. A., Phillips, T. G., Huggins, P. J., and Beichman, C. A. 1985. The abundance of atomic carbon near the ionization fronts in M17 and S140. *Astrophys. J.* 299:967–980.

Kegel, W. H. 1989. The interpretation of correlations between observed parameters of molecular clouds. *Astron. Astrophys.* 225:517–520.

Keller, H. U. 1990. The nucleus. In *Physics and Chemistry of Comets*, ed. W. F. Huebner (Berlin: Springer-Verlag), pp. 13–68.

Keller, H. U., Delamere, W. A., Huebner, W. F., Reitsema, H. J., Schmidt, K., Schmidt, H. U., Whipple, F. L., Wilhelm, K., Curdt, W., Kramm, R., Thomas, N., Arpigny, C., Barbieri, C., Bonnet, R. M., Cazes, S., Coradini, M., Cosmovici, C. B., Hughes, D. W., Jamar, C., Malaise, D., Schmidt, W. K. H., and Seige, P. 1987. Comet P/Halley's nucleus and its activity. *Astron. Astrophys.* 187:807–823.

Kelly, W. R., and Larimer, J. W. 1977. Chemical fractionations in meteorites. VIII. Iron meteorites and the cosmochemical history of the metal phase. *Geochim. Cosmochim. Acta* 41:93–111.

Kennicutt, R. C., Jr. 1984. Constraints on the masses of supernova progenitors. *Astrophys. J.* 277:361–366.

Kennicutt, R. C., Jr. 1989. The star formation law in galactic disks. *Astrophys. J.* 344:685–703.

Kenyon, S. J. 1987. Accretion as an energy source of pre-main sequence stars. In *Fifth Cambridge Workshop on Cool Stars, Stellar Systems and the Sun*, eds. J. L. Linsky and R. E. Stencel (Berlin: Springer-Verlag), pp. 431–441.

Kenyon, S. J., and Hartmann, L. 1987. Spectral energy distributions of T Tauri stars: disk flaring and limits on accretion. *Astrophys. J.* 323:714–733.

Kenyon, S. J., and Hartmann, L. 1988. The FU Orionis variables: Accretion and mass loss. In *Pulsation and Mass Loss in Stars*, eds. R. Stalio and L. A. Willson (Dordrecht: Kluwer), pp. 133–154.

Kenyon, S. J., and Hartmann, L. W. 1990. On the apparent positions of T Tauri stars in the H-R diagram. *Astrophys. J.* 349:197–207.

Kenyon, S. J., and Hartmann, L. 1991. The dusty envelopes of FU Orionis variables. *Astrophys. J.* 383:664–673.

Kenyon, S. J., Hartmann, L., and Hewett, R. 1988. Accretion disk models for FU Orionis and V1057 Cygni: Detailed comparisons between observations and theory. *Astrophys. J.* 325:231–251.

Kenyon, S. J., Hartmann, L., Imhoff, C. L., and Cassatella, A. 1989. Ultraviolet spectroscopy of pre-main sequence accretion disks. *Astrophys. J.* 344:925–931.

Kenyon, S. J., Hartmann, L. W., Strom, K. M., and Strom, S. E. 1990. An *IRAS* survey of the Taurus-Auriga molecular cloud. *Astron. J.* 99:869–887.

Kerr, F. J., and Lynden-Bell, D. 1986. Review of galactic constants. *Mon. Not. Roy. Astron. Soc.* 221:1023–1038.

Kerridge, J. F. 1979. Fractionation of refractory elements in chondritic meteorites. *Lunar Planet. Sci. Conf.* X:655–657 (abstract).

Kerridge, J. F., and Bunch, T. E. 1979. Aqueous alteration on asteroids: Evidence from carbonaceous meteorites. In *Asteroids*, ed. T. Gehrels (Tucson: Univ. of Arizona Press), pp. 745–764.

Kerridge, J. F., and Matthews, M. S., eds. 1988. *Meteorites and the Early Solar System* (Tucson: Univ. of Arizona Press).

Keto, E. R., and Lattanzio, J. C. 1989. Collisions between high latitude molecular clouds: Theory meets observations. *Astrophys. J.* 346:184–192.

Keto, E. R., Ho, P. T. P., and Haschick, A. D. 1987. Temperature and density structure of the collapsing core. *Astrophys. J.* 318:712–728.

Kieffer, S. W., and Simonds, C. H. 1980. The role of volatiles and lithology in the impact cratering process. *Rev. Geophys. Space Phys.* 18:143–181.

Kiguchi, M., Narita, S., Miyama, S. M., and Hayashi, C. 1987. The equilibria of rotating isothermal clouds. *Astrophys. J.* 317:830–845.

Kim, S. J., and A'Hearn, M. F. 1991. Upperlimits of SO and SO_2 in comets. *Icarus* 90:79–95.

Kim, S. J., A'Hearn, M. F., and Cochran, W. D. 1989. NH emissions in comets: fluorescence vs. collisions. *Icarus* 77:98–108.

Kim, S. J., A'Hearn, M. F., and Larson, S. M. 1990. Multi-cycle fluorescence: application to S_2 in comets IRAS-Araki-Alcock 1983d. *Icarus* 87:440–451.

Kimura, T., and Tosa, M. 1988. Formation of a gas condensation in a perturbed shock. *Mon. Not. Roy. Astron. Soc.* 234:51–65.

King, A. R. 1989. Mass transfer in cataclysmic binary systems. In *Classical Novae*, eds. M. F. Bode and A. Evans (London: Wiley), pp. 17–37.

Kippenhahn, R., and Thomas, H.-C. 1981. Rotation and stellar evolution. In *Fundamental Problems in the Theory of Stellar Evolution*, eds. D. Sugimoto, D. Lamb and D. N. Schramm (Dordrecht: D. Reidel), pp. 237–256.

Kirsten, T., Deubner, J., Horn, P., Kaneoka, I., Kiko, J., Shaffer, O. A., and Thio, S. K. 1972. The rare gas record of Apollo 14 and 15 samples. *Proc. Lunar Planet. Sci.*

Conf. 3:1865–1889.

Kissel, J. 1986. The Giotto particulate impact analyzer. *The Giotto Mission—Its Scientific Investigations*, ESA SP-1077, pp. 67–68.

Kissel, J., and Krueger, F. R. 1987. The organic component in dust from comet Halley as measured by the PUMA mass spectrometer on board Vega 1. *Nature* 326:755–760.

Kitta, K., and Kratschmer, W. 1983. Status of laboratory experiments on ice mixtures and on the 12 μm H_2O ice feature. *Astron. Astrophys.* 122:105–110.

Klebe, D. I., and Jones, T. J. 1990. Infrared polarimetry of Bok globules. *Astron. J.* 99:638–649.

Klein, R. I., Sandford, M. T., II, and Whittaker, R. W. 1980. Two-dimensional radiation hydrodynamics calculations on the formation of OB associations in dense clouds. *Space Sci. Rev.* 27:275–281.

Klein, R. I., Sandford, M. T., II, and Whitaker, R. W. 1983. Star formation with OB subgroups: implosion by multiple sources. *Astrophys. J. Lett.* 271:69–73.

Klein, R. I., Sandford, M. T., II, and Whittaker, R. W. 1985. Processes and problems in secondary star formation. In *Protostars and Planets II*, eds. D. C. Black and M. S. Matthews (Tucson: Univ. of Arizona Press), pp. 340–367.

Kleiner, S. C., and Dickman, R. L. 1984. Large-scale structure of the Taurus molecular complex. I. Density fluctuations—a fossil jeans length? *Astrophys. J.* 286:255–262.

Kleiner, S. C., and Dickman, R. L. 1985. Large-scale structure of the Taurus molecular complex. II. *Astrophys. J.* 295:466–478.

Kleiner, S. C., and Dickman, R. L. 1987. Small-scale structure of the Taurus molecular clouds: Turbulence in Heiles' cloud 2. *Astrophys. J.* 312:837–847.

Kley, W. 1989. Radiation hydrodynamics of the boundary layer in accretion disks. *Astron. Astrophys.* 208:98–110.

Klinger, J. 1985. Composition and structure of the comet nucleus and its evolution on a periodic orbit. In *Ices in the Solar System*, eds. J. Klinger, D. Benest, A. Dollfus and R. Smoluchowski (Dordrecht: D. Reidel), pp. 407–417.

Knacke, R. F., and McCorkle, S. M. 1987. Spectroscopy of the Kleinmann-Low nebula: Scattering in a solid absorption band. *Astron. J.* 94:972–976.

Knacke, R. F., and Larson, H. P. 1991. Water vapor in the Orion molecular cloud. *Astrophys. J.* 367:162–167.

Knacke, R. F., McCorkle, S., Puetter, R. C., Erikson, E. F., and Kratschmer, W. 1982. Observation of interstellar ammonia ice. *Astrophys. J.* 260:141–146.

Knacke, R. F., Geballe, T. R., Noll, K. S., and Tokunaga, A. T. 1985. Search for interstellar methane. *Astrophys. J. Lett.* 298:67–69.

Knacke, R. F., Brooke, T. Y., and Joyce, R. R. 1986. Observations of the 3.2–3.6 μm emission features in comet Halley. *Astrophys. J. Lett.* 310:49–53.

Knacke, R. F., Kim, Y. H., Noll, K. S., and Geballe, T. R. 1988a. Search for interstellar methane. In *Molecular Clouds in the Milky Way and External Galaxies*, eds. R. L. Dickman, R. L. Snell and J. S. Young (Münich: Springer-Verlag), pp. 180–181.

Knacke, R. F., Larson, H. P., and Noll, K. S. 1988b. Evidence for interstellar H_2O in the Orion molecular cloud. *Astrophys. J. Lett.* 335:27–30.

Knapp, G. R. 1974. Observations of H I in dense interstellar dust clouds: I. A survey of 88 clouds. *Astron. J.* 79:527–540.

Knapp, G. R., Kuiper, T. B., Knapp, S. L., and Brown, R. L. 1976. CO observations of NGC 1579 (5222) 5239. *Astrophys. J.* 206:443–451.

Knežević, Z., Milani, A., Farinella, P., Froeschlé, Ch., and Froeschlé, Cl. 1991. Secular resonances from 2 to 50 AU. *Icarus* 93:316–330.

Knowles, S. H., and Batchelor, R. A. 1978. Linear polarization of 22-GH_z water-vapor line emissions in southern sources. *Mon. Not. Roy. Astron. Soc.* 184:107–117.

Kobi, D., and North, P. 1990. A new calibration of the Geneva photometry in terms of T_e, Log g, and mass for main sequence A4 to G5 stars. *Astron. Astrophys. Suppl.* 85:999–1014.

Koch, R. H., and Hrivnak, B. J. 1981. On Zahn's theory of tidal friction for cool main-sequence close binaries. *Astron. J.* 86:438–441.

Kogure, T., Yoshida, S., Wiramihardja, S. D., Nakano, M., Iwata, T., and Ogura, K. 1989. Survey observations of emission-line stars in the Orion region. II. The Kiso area A-0903. *Publ. Astron. Soc. Japan* 41:1195–1213.

Kolmogorov, A. N. 1941. The local structure of turbulence in an incompressible viscous fluid for very large Reynolds numbers. *Compt. Rend. Acad. Sci. URSS* 30:301–305. Reprinted in Friedlander, S. K., and Topper, L., eds. 1961. *Turbulence: Classic Papers on Statistical Theory* (New York: Interscience).

Kolotilov, E. A., and Petrov, P. P. 1985. Study of the FU Orionis stars. III. The photometric observations of FU Ori in 1978–1985. *Pis'ma Astron. Zh.* 11:846–854.

Kolvoord, R. A., and Greenberg, R. 1992. A critical reanalysis of planetary accretion models. *Icarus* 98:2–19.

Kondo, Y., and Bruhweiler, F. C. 1985. IUE observations of Beta Pictoris: an IRAS candidate for a proto-planetary system. *Astrophys. J. Lett.* 291:1–5.

Königl, A. 1982. On the nature of bipolar sources in dense molecular clouds. *Astrophys. J.* 261:115–134.

Königl, A. 1986. From molecular clouds to active galactic nuclei: the universality of the jet phenomenon. *Annals New York Acad. Sci. Vol. 12: Proc. Texas Symp. on Relativistic Astrophysics* 470:88–107.

Königl, A. 1987. Magnetic braking in weakly ionized media. *Astrophys. J.* 320:726–740.

Königl, A. 1989. Self-similar models of magnetized accretion disks. *Astrophys. J.* 342:208–223.

Königl, A. 1991. Disk accretion onto magnetic T Tauri stars. *Astrophys. J. Lett.* 370:39–43.

Koo, B.-C. 1989. Extremely high-velocity molecular flows in young stellar objects. *Astrophys. J.* 337:318–331.

Koo, B.-C. 1990. Molecular line observations of the extremely high-velocity molecular flow near HH 7–11. *Astrophys. J.* 361:145–149.

Koornneef, J. 1983a. Near-infrared photometry. I. Homogenization of near-infrared data from southern bright stars. *Astron. Astrophys. Suppl.* 51:489–503.

Koornneef, J. 1983b. Near-infrared photometry. II. Intrinsic colours and the absolute calibration from one to five micron. *Astron. Astrophys.* 128:84–93.

Koresko, C. D., Beckwith, S. V. B., and Sargent, A. I. 1989. Diffraction-limited infrared observations of the young star Z CMa. *Astron. J.* 98:1394–1397.

Korina, M. I., Nazarov, M. S., and Ulyanov, A. A. 1982. Efremovka CAI's: composition and origin of rims. *Lunar Planet. Sci.* XIII:399–400 (abstract).

Kornacki, A. S., and Fegley, B., Jr. 1984. Origin of spinel-rich chondrules and inclusions in carbonaceous and ordinary chondrites. *Proc. Lunar Planet. Sci. Conf.* 14:588–596.

Kornacki, A. S., and Fegley, B., Jr. 1986. The abundance and relative volatility of refractory trace elements in Allende Ca, Al-rich inclusions: implications for chemical and physical processes in the solar nebula. *Earth Planet. Sci. Lett.* 79:217–234.

Kornacki, A. S., and Wood, J. A. 1984. The mineral chemistry and origin of inclusion matrix and meteorite matrix in the Allende CV3 chondrite. *Geochim. Cosmochim. Acta* 48:1663–1676.

Korth, A., Richter, A. K., Loidl, A., Anderson, K. A., Carlson, C. W., Curtis, D. W.,

Lin, R. P., Réme, H., Sauvaud, J. A., d'Uston, C., Cotin, F., Cros, A., and Mendis, D. A. 1986. Mass spectra of heavy ions near comet Halley. *Nature* 321:335–336.

Korycansky, D. G., Bodenheimer, P., Cassen, P., and Pollack, J. B. 1990. One-dimensional calculations of a large impact on Uranus. *Icarus* 84:528–541.

Koslowski, M., Wiita, P. J., and Paczyński, B. 1979. Self-gravitating accretion disk models with realistic equations of state and opacities. *Acta Astron.* 29:157–176.

Kowal, C. T. 1979. Chiron. In *Asteroids*, ed. T. Gehrels (Tucson: Univ. of Arizona Press), pp. 436–439.

Kraft, R. P. 1970. Stellar rotation. In *Spectroscopic Astrophysics*, ed. G. H. Herbig (Berkeley: Univ. of California Press), pp. 385–422.

Kraichnan, R. H. 1962. Turbulent thermal convection at arbitrary Plandtl number. *Phys. Fluids* 5:1374–1389.

Krankowsky, D. 1991. The composition of comets. In *Comets in the Post-Halley Era*, eds. R. L. Newburn, J. Rahe and M. Neugebauer (Dordrecht: Kluwer), pp. 855–877.

Krankowsky, D., and Eberhardt, P. 1991. Evidence for the composition of the ices in the nucleus of comet Halley. In *Comet Halley—Investigations, Results, and Interpretations*, ed. J. Mason (Chichester: Ellis Horwood Ltd.), pp. 273–289.

Krankowsky, D., Lämmerzahl, P., Herrwerth, I., Woweries, J., Eberhardt, P., Dolder, U., Herrmann, U., Schulte, W., Berthelier, J. J., Illiano, J. M., Hodges, R. R., and Hoffmann, J. H. 1986. In situ gas and ion measurements at comet Halley. *Nature* 321:326–330.

Krankowsky, D., Eberhardt, P., Meier, R., Schulte, W., Lämmerzahl, P., and Hodges, R. R. 1991. Formaldehyde in Halley derived from the Giotto NMS measurements. *Astron. Astrophys.*, in preparation.

Krasnopolsky, V. A., and Tkachuk, A. Yu. 1991. TKS-Vega experiment: NH and NH_2 bands in comet Halley. *Astron. J.* 101:1915–1919.

Krautter, J. 1986. Th28: a new bipolar Herbig-Haro jet. *Astron. Astrophys.* 161:195–200.

Krautter, J. 1991. The star forming regions in Lupus. In *Low Mass Star Formation in Southern Molecular Clouds ESO Sci. Report No. 11*, ed. B. Reipurth, 127–148.

Krautter, J., and Kelemen, J. 1987. The T Tauri associations Chamaeleon and Lupus. *Mitt. Astron. Ges.* 70:397–398.

Kring, D. A. 1988. The Petrology of Meteoritic Chondrules: Evidence for Fluctuating Conditions in the Solar Nebula. Ph. D. Thesis, Harvard Univ.

Kring, D. A. 1991*a*. High temperature rims around chondrules in primitive chondrites: evidence for fluctuating conditions in the presolar nebula. *Earth Planet. Sci. Lett.* 105:65–80.

Kring, D. A. 1991*b*. Redox conditions in the solar nebula gas in the zones of CM2, CR2, CO3, CV3, and UOC chondrule formation: implications for O/H, O/C, and dust/gas fractionation. *Geochim. Cosmochim. Acta* 55:1737–1742.

Kring, D. A., and Boynton, W. V. 1990. Chemical memory in chondrules of precursor dust with fractionated compositions. *Meteoritics* 24:289–290 (abstract).

Krolik, J. H., and Kallman, T. R. 1983. X-ray ionization and the Orion molecular cloud. *Astrophys. J.* 267:610–624.

Kroupa, P., Tout, C. A., and Gilmore, G. 1990. The low-luminosity stellar mass function. *Mon. Not. Roy. Astron. Soc.* 244:76–85.

Kroupa, P., Tout, C. A., and Gilmore, G. 1991. The effects of unresolved binary stars on the determination of the stellar mass function. *Mon. Not. Roy. Astron. Soc.* 251:293–302.

Krueger, F. R., and Kissel, J. 1987. The chemical compostion of the dust of comet P/Halley as measured by "PUMA" on board Vega 1. *Naturwissenschaften* 74:312–316.

Ku, W. H.-M., and Chanan, G.A. 1979. EINSTEIN observations of the Orion nebula. *Astrophys. J. Lett.* 234:59–63.

Ku, W. H.-M., Righini-Cohen, G., and Simon, M. 1982. High-resolution X-ray observations of the Orion nebula. *Science* 215:61–64.

Kuhi, L. V. 1964. Mass loss from T Tauri stars. *Astrophys. J.* 140:1409–1433.

Kuhi, L. V. 1974. Spectral energy distributions of T Tauri stars. *Astron. Astrophys. Suppl.* 15:47–89.

Kuhi, L. V. 1978. Spectral characteristics of T Tauri stars. In *Protostars and Planets*, ed. T. Gehrels (Tucson: Univ. of Arizona Press), pp. 708–717.

Kührt, E. 1984. Temperature profiles and thermal stress on cometary nuclei. *Icarus* 60:512–521.

Kuiper, G. P. 1944. Titan: a satellite with an atmosphere. *Astrophys. J.* 100:378–383.

Kuiper, G. P. 1951. On the origin of the solar system. In *Astrophysics: A Topical Symposium*, ed. J. A. Hynek (New York: McGraw-Hill), pp. 357–424.

Kulkarni, S. R., and Heiles, C. 1988. Neutral hydrogen and the diffuse interstellar medium. In *Galactic and Extragalactic Radio Astronomy*, eds. G. L. Verschuur and K. I. Kellermann (New York: Springer-Verlag), p. 95.

Kulsrud, R. M. 1971. Rotational deceleration of magnetized stars. *Astrophys. J.* 163:567–576.

Kulsrud, R. M., and Pearce, W. P. 1969. The effect of wave-particle interactions on the propagation of cosmic rays. *Astrophys. J.* 156:445–469.

Kumar, S. S. 1963. The structure of stars of very low mass. *Astrophys. J.* 137:1121-1125.

Kuo, A. Y., and Corrsin, S. 1971. Experiments on internal intermittency and fine-structure distribution functions in a fully turbulent fluid. *J. Fluid Mech.* 50:285–319.

Kusaka, T., Nakano, T., and Hayashi, C. 1970. Growth of solid particles in the primordial solar nebula. *Prog. Theor. Phys.* 44:1580–1596.

Kutner, M. L., Tucker, K. D., Chin, G., and Thaddeus, P. 1977. The molecular complexes in Orion. *Astrophys. J.* 215:521–528.

Kutner, M. L., Leung, C. M., Machnik, D. E., and Mead, K. N. 1982. Broad carbon monoxide line wing near T Tauri stars. *Astrophys. J. Lett.* 259:35–39.

Kwan, J. 1979. The mass spectrum of interstellar clouds. *Astrophys. J.* 229:567–577.

Kwan, J. 1988. The life cycle of interstellar clouds. In *Molecular Clouds in the Milky Way and External Galaxies*, eds. R. L. Dickman, R. L. Snell and J. S. Young (Berlin: Springer-Verlag), pp. 281–288.

Kwan, J., and Sanders, D. B. 1986. The physical and kinematic structures of molecular clouds. *Astrophys. J.* 309:783–803.

Kwan, J., and Tademaru, E. 1988. Jets from T Tauri stars: Spectroscopic evidence and collimation mechanism. *Astrophys. J. Lett.* 332:41–44.

Kwan, J., and Valdes, F. 1983. Spiral gravitational potentials and the mass growth of molecular clouds. *Astrophys. J.* 271:604–610.

Kylafis, N. D. 1983a. Polarization of interstellar radio-frequency lines and magnetic field direction. *Astrophys. J.* 267:137–150.

Kylafis, N. D. 1983b. Linear polarization of interstellar radio-frequency absorption lines and magnetic field direction. *Astrophys. J.* 275:135–144.

Kylafis, N. D., and Shapiro, P. R. 1983. Polarization of interstellar molecular radio frequency absorption lines. *Astrophys. J. Lett.* 272:35–39.

Kyte, F. T., and Wasson, J. T. 1986. Accretion rate of extraterrestrial matter: iridium deposited 33 to 67 million years ago. *Science* 232:1225–1229.

Lacey, C. G., and Fall, S. M. 1985. Chemical evolution of the galactic disk with radial gas flows. *Astrophys. J.* 290:154–170.

Lacy, J. H., Baas, F., Allamandola, F. J., Persson, S. E., McGregor, P. J., Lonsdale, C. J.,

Geballe, T. R., and van de Bult, C. E. P. 1984. 4.6 micron absorption features due to solid phase CO and cyano group molecules toward compact infrared sources. *Astrophys. J.* 276:533–543.

Lacy, J. H., Achtermann, J. M., Bruce, D. E., Lester, D. F., Arens, J. F., Peck, M. C., and Gaalema, S. D. 1989*a*. IR-SHELL: a mid-infrared cryogenic echelle spectrograph. *Publ. Astron. Soc. Pacific* 101:1166–1175.

Lacy, J. H., Evans, N. J., Achtermann, J. M., Bruce, D. E., Arens, J. F., and Carr, J. S. 1989*b*. Discovery of interstellar acetylene. *Astrophys. J. Lett.* 342:43–46.

Lacy, J. H., Carr, J. S., Evans, N. J., Baas, F., Achtermann, J. M., and Arens, J. F. 1991. Discovery of interstellar methane: observations of gaseous and solid CH_4 absorption toward young stars in molecular clouds. *Astrophys. J.* 376:556–560.

Lada, C. J. 1985. Cold outflows, energetic winds, and enigmatic jets around young stellar objects. *Ann. Rev. Astron. Astrophys.* 23:267–317.

Lada, C. J. 1987. Star formation: from OB associations to protostars. *Star Forming Regions: Proc. IAU Symp. 115*, eds. M. Peimbert and J. Jugaku (Dordrecht: D. Reidel), pp. 1–18.

Lada, C. J. 1988*a*. Infrared energy distributions and the nature of young stellar objects. In *Formation and Evolution of Low Mass Stars*, eds. A. K. Dupree and M. T. V. T. Lago (Dordrecht: Kluwer), pp. 1–18.

Lada, C. J. 1988*b*. On the importance of outflows for molecular clouds and star formation. In *Galactic and Extragalactic Star Formation*, eds. R. E. Pudritz and M. Fich (Dordrecht: Kluwer), pp. 5–24.

Lada, C. J., and Lada, E. A. 1991. The nature, origin, and evolution of embedded star clusters. In *The Formation and Evolution of Star Clusters*, ed. K. A. Janes (San Francisco: Astron. Soc. of the Pacific), pp. 3–22.

Lada, C. J., and Shu, F. H. 1990. The formation of sunlike stars. *Science* 248:564–572.

Lada, C. J., and Wilking, B. A. 1984. The nature of the embedded population in the ρ Ophiuchi dark cloud: Mid-infrared observations. *Astrophys. J.* 287:610–621.

Lada, C. J., Blitz, L., and Elmegreen, B. G. 1978*a*. Star formation in OB associations. In *Protostars and Planets*, ed. T. Gehrels (Tucson: Univ. of Arizona Press), pp. 341–367.

Lada, C. J., Elmegreen, B. G., Cong, H.-I., and Thaddeus, P. 1978*b*. Molecular clouds in the vicinity of W3, W4, and W5. *Astrophys. J. Lett.* 226:39–42

Lada, C. J., Margulis, M., and Dearborn, D. 1984. The formation and early dynamical evolution of bound stellar systems. *Astrophys. J.* 285:141–152.

Lada, C. J., Margulis, M., Sofue, Y., Nakai, N., and Handa, T. 1988*a*. Observation of molecular and atomic clouds in M31. refit Astrophys. J. 328:143–160.

Lada, C. J., Margulis, M., Sofue, Y., Nakai, N., and Handa, T. 1988*b*. Observations of molecules around young stellar objects. *Ann. Rev. Astron. Astrophys.* 23:267–317

Lada, C. J., DePoy, D., Merrill, K. M., and Gatley, I. 1991. Infrared images of M17. *Astrophys. J.* 374:533–539.

Lada, E. A. 1990. Global Star Formation in the L1630 Molecular Cloud. Ph. D. Thesis, Univ. of Texas, Austin.

Lada, E. A. 1992. Global star formation in the L1630 molecular cloud. *Astrophys. J. Lett.* 393:25–28.

Lada, E. A., and Blitz, L. 1988. Two populations of diffuse molecualr clouds. *Astrophys. J. Lett.* 326:69–73.

Lada, E. A., Bally, J., and Stark, A. A. 1991*a*. An unbiased survey for dense cores in the Lynds 1630 molecular cloud. *Astrophys. J.* 368:432–444.

Lada, E. A., DePoy, D. L., Evans, N. J., and Gatley, I. 1991*b*. A 2.2 micron survey in the L1630 molecular cloud. *Astrophys. J.* 371:171–182.

Ladd, E. F., Adams, F. C., Casey, S., Davidson, J. A., Fuller, G. A., Harper, D. A., Myers, P. C., and Padman, R. 1991. Far infrared and submillimeter wavelength observations of star forming dense cores. I. Spectra. *Astrophys. J.* 366:203–220.

Lago, M. T. V. T. 1984. A new investigation of the T Tauri star RU Lupi-III. The wind model. *Mon. Not. Roy. Astron. Soc.* 210:323–340.

Lagrange, A.-M., Ferlet, R., and Vidal-Madjar, A. 1987. The Beta Pictoris circumstellar disk. IV. Redshifted UV lines. *Astron. Astrophys.* 173:289–292.

Lagrange-Henri, A.-M., Vidal-Madjar, A., and Ferlet, R. 1988. The β Pictoris circumstellar disk. VI. Evidence for material falling onto the star. *Astron. Astrophys.* 190:275–282.

Lagrange-Henri, A.-M., Beust, H., Ferlet, R., and Vidal-Madjar, A. 1989. The circumstellar gas around β Pictoris. VIII. Evidence for a clumpy structure of the infalling gas. *Astron. Astrophys.* 215:L5–L8.

Lagrange-Henri, A.-M., Ferlet, R., Vidal-Madjar, A., Beust, H., Gry, C., and Lallement, R. 1990. Search for β Pictoris-like stars. *Astron. Astrophys. Suppl.* 85:1089–1100.

Lamb, F. K. 1989. Accretion by magnetic neutron stars. In *Timing Neutron Stars*, eds. H. Ögelman and E. P. J. van den Heuvel (Dordrecht: Kluwer), pp. 649–722.

Lambert, D. L., and Danks, A. C. 1983. High resolution spectra of C_2 Swan bands from comet West 1976 VI. *Astrophys. J.* 268:428–446.

Lämmerzahl, P., Krankowsky, D., Hodges, R. R., Stubbemann, U., Woweries, J., Herrwerth, I., Berthelier, J. J., Illiano, J. M., Eberhardt, P., Dolder, U., Schulte, W., and Hoffman, J. H. 1987. Expansion velocity and temperatures of gas and ions measured in the coma of comet P/Halley. *Astron. Astrophys.* 187:169–171.

Lane, A. P. 1989. Near-infrared imaging of H_2 emission from Herbig-Haro objects and bipolar flows. In *ESO Workshop on Low Mass Star Formation and Pre-Main Sequence Objects*, ed. B. Reipurth (Garching: European Southern Obs.), pp. 331–348.

Lane, A. P., and Bally, J. 1986. Shocked molecular hydrogen and jets in star-forming clouds. II. *Astrophys. J.* 310:820–831.

Lane, A. P., Haas, M. R., Hollenbach, D. J., and Erickson, E. F. 1990. Far-infrared spectroscopy of the DR21 star formation region. *Astrophys. J.* 361:132–144.

Lange, M. A., and Ahrens, T. J. 1982. The evolution of an impact-generated atmosphere. *Icarus* 51:96–120.

Lange, M. A., and Ahrens, T. J. 1984. FeO and H_2O and the homogeneous accretion of the Earth. *Earth Planet. Sci. Lett.* 71:111–119.

Lange, M. A., and Ahrens, T. J. 1987. Impact experiments in low-temperature ice. *Icarus* 69:506–518.

Langer, W. D. 1978. The stability of interstellar clouds containing magnetic fields. *Astrophys. J.* 225:95–106.

Langer, W. D., and Glassgold, A. E. 1990. Silicon chemistry in interstellar clouds. *Astrophys. J.* 352:123–131.

Langer, W. D., and Graedel, T. E. 1989. Ion-molecule chemistry of dense interstellar clouds: Nitrogen-, oxygen-, and carbon-bearing molecule abundances and isotopic ratios. *Astrophys. J. Suppl.* 69:241–269.

Langer, W. D., and Penzias, A. A. 1992. $^{12}C/^{13}C$ isotope ratio in the local interstellar medium from observations fo $^{13}C^{10}O$ in molecular clouds. *Astrophys. J.*, in press.

Langer, W. D., Frerking, M. A., and Wilson, R. W. 1986. Multiple star formation and the dynamical evolution of B335. *Astrophys. J. Lett.* 306:29–32.

Langevin, Y., and Maurette, M. 1976. A Monte Carlo simulation of galactic cosmic ray effects in the lunar regolith. *Proc. Lunar Planet. Sci. Conf.* 7:75–91.

Lanzerotti, L. J., Brown, W. L., and Marcantonio, K. J. 1987. Experimental study of erosion of methane ice by energetic ions and some consideration for astrophysics. *Astrophys. J.* 131:910–919.

Laplace, P. S. 1796. *Exposition du Système du Monde* (Paris). English translation: H. H. Harte. 1830. *The System of the World* (Dublin: University Press).

Larimer, J. W. 1975. The effect of C/O ratio on the condensation of planetary material. *Geochim. Cosmichim. Acta* 39:389–392.

Larimer, J. W. 1979. The condensation and fractionation of refractory lithophile elements. *Icarus* 40:446–454.

Larimer, J. W. 1988. The cosmochemical classification of the elements. In *Meteorites and the Early Solar System*, eds. J. F. Kerridge and M. S. Matthews (Tucson: Univ. of Arizona Press), pp. 375–388.

Larimer, J. W., and Anders, E. 1967. Chemical fractionations in meteorites. II. Abundance patterns and their interpretations. *Geochim. Cosmochim. Acta* 31:1239–1270.

Larimer, J. W., and Bartholomy, M. 1979. The role of carbon and oxygen in cosmic gases: some applications to the chemistry and mineralogy of enstatite chondrites. *Geochim. Cosmochim. Acta* 43:1455–1466.

LaRosa, T. N. 1983. Radiatively induced star formation. *Astrophys. J.* 274:815–821.

Larson, H. P., Mumma, M. J., and Weaver, H. A. 1987. Kinematic properties of the neutral gas outflow from comet P/Halley. *Astron. Astrophys.* 187:391–397.

Larson, H. P., Weaver, H. A., Mumma, M. J., and Drapatz, S. 1989. Airborne infrared spectroscopy of comet Wilson (1986l) and comparisons with comet Halley. *Astrophys. J.* 338:1106–1114.

Larson, H. P., Hu, H. Y., Mumma, M. J., and Weaver, H. A. 1990. Outbursts of water in comet P/Halley. *Icarus* 86:129–151.

Larson, R. B. 1969. Numerical calculations of the dynamics of a collapsing protostar. *Mon. Not. Roy. Astron. Soc.* 145:271–295.

Larson, R. B. 1972. The collapse of a rotating cloud. *Mon. Not. Roy. Astron. Soc.* 156:437–458.

Larson, R. B. 1978. Calculations of three-dimensional collapse and fragmentation. *Mon. Not. Roy. Astron. Soc.* 184:69–85.

Larson, R. B. 1981. Turbulence and star formation in molecular clouds. *Mon. Not. Roy. Astron. Soc.* 194:809–826.

Larson, R. B. 1982. Mass spectra of young stars. *Mon. Not. Roy. Astron. Soc.* 200:159–174.

Larson, R. B. 1984. Gravitational torques and star formation. *Mon. Not. Roy. Astron. Soc.* 206:197–207.

Larson, R. B. 1985. Cloud fragmentation and stellar masses. *Mon. Not. Roy. Astron. Soc.* 214:379–398.

Larson, R. B. 1986. Bimodal star formation and remnant-dominated galactic models. *Mon. Not. Roy. Astron. Soc.* 218:409–428.

Larson, R. B. 1988. Large-scale aspects of star formation and galactic evolution. In *Galactic and Extragalactic Star Formation*, eds. R. E. Pudritz and M. Fich (Dordrecht: Kluwer), pp. 459–474.

Larson, R. B. 1989. The evolution of protostellar disks. In *The Formation and Evolution of Planetary Systems*, eds. H. A. Weaver and L. Danly (Cambridge: Cambridge Univ. Press), pp. 31–54.

Larson, R. B. 1990a. Formation of star clusters. In *Physical Processes in Fragmentation and Star Formation*, vol. 8, eds. R. Capuzzo-Dolcetta, C. Chiosi and A. Di Fazio (Dordrecht: Kluwer), pp. 389–400.

Larson, R. B. 1990b. Non-linear acoustic waves in discs. *Mon. Not. Roy. Astron. Soc.* 243:588–592.

Larson, R. B. 1991. Some physical processes influencing the stellar initial mass function. In *Fragmentation of Molecular Clouds and Star Formation*, eds. E. Falgarone, F. Boulanger and G. Duvet (Dordrecht: Kluwer), pp. 261–273.

Laskar, J. 1986. A general theory for the Uranian satellites. *Astron. Astrophys.* 166:349–358.

Laskar, J. 1989. A numerical experiment on the chaotic behaviour of the solar system. *Nature* 338:237–238.

Laskar, J. 1990. The chaotic motion of the solar system: a numerical estimate of the size of the chaotic zones. *Icarus* 88:266–291.

Latham, D. W. 1989. New observational clues on binary formation in the galaxy. In *Highlights of Astronomy*, vol. 8, ed. D. McNally (Dordrecht: Kluwer), pp. 103–107.

Latham, D. W., Mazch, T., Carney, B. W., McCrosky, R. E., Stefanik, R. P., and Davis, R. J. 1988. A survey of proper motion stars. VI. Orbits for 40 spectroscopic binaries. *Astron. J.* 96:567–587.

Latham, J., and Myers, V. 1970. Loss of charge and mass from raindrops falling in intense electric fields. *J. Geophys. Res.* 75:515–520.

Lattanzio, J. C., and Elmegreen, B. G. 1991. Numerical simulations of fragmenting collisions between self-gravitating clouds. In preparation.

Lebovitz, N. R. 1989. Mathematical status of the fission theory. In *Highlights of Astronomy*, vol. 8, ed. D. McNally (Dordrecht: Kluwer), pp. 129–131.

Lecar, M., and Franklin, F. 1973. On the original distribution of the asteroids. *Icarus* 20:422–436.

Lee, H. M., and Draine, B. T. 1985. Infrared extinction and polarization due to partially aligned spheriodal grains: models for the dust toward the BN object. *Astrophys. J.* 290:211–228.

Lee, T. 1988. Implications of isotopic anomalies for nucleosynthesis. In *Meteorites and the Early Solar System*, eds. J. F. Kerridge and M. S. Matthews (Tucson: Univ. of Arizona Press), pp. 1063–1089.

Lee, T., Papanastassiou, D. A., and Wasserburg, G. J. 1977. ^{26}Al in the early solar system: Fossil or fuel? *Astrophys. J. Lett.* 211:107–110.

Lee, Y., Snell, R. L., Dickman, R. L. 1990. Analysis of ^{12}CO and ^{13}CO emission in a 3 square degree region of the galactic plane between $l = 23$deg and 25deg. *Astrophys. J.* 355:536–545.

Léger, A., and Puget, J. L. 1984. Identification of the unidentified IR emission features of interstellar dust? *Astron. Astrophys. Lett.* 146:5–8.

Léger, A., Jura, M., and Omont, A. 1985. Desorption from interstellar grains. *Astron. Astrophys.* 144:147–160.

Léger, A., Verstraete, L, d'Hendecourt, L., Defourneau, D., Dutuit, O., Schmidt, W., and Lauer, J. C. 1989. The PAH hypothesis and the extinction curve. In *Interstellar Dust: Proc. IAU Symp. 135*, eds. L. J. Allamandola and A. G. G. M. Tielens (Dordrecht: Kluwer), pp. 173–180.

Leggett, S. K., and Hawkins, M. R. S. 1988. The infrared luminosity function for low-mass stars. *Mon. Not. Roy. Astron. Soc.* 234:1065–1090.

Leggett, S. K., and Hawkins, M. R. S. 1989. Low mass stars in the region of the Hyades cluster. *Mon. Not. Roy. Astron. Soc.* 238:145–153.

Leinert, C., and Haas, M. 1989. Detection of an infrared companion to Haro 6–10. *Astrophys. J. Lett.* 342:39–42.

Leisawitz, D. 1990. A CO survey of regions around 34 open clusters II. Physical properties of catalogues molecular clouds. *Astrophys. J.* 359:319–343.

Leisawitz, D., Bash, F. N., and Thaddeus, P. 1989. A CO survey of regions around 34 open clusters. *Astrophys. J. Suppl.* 70:731–812.

Lellouch, E., Belton, M., de Pater, I., Gulkis, S., and Encrenaz, T. 1990. Io's atmosphere from microwave detection of SO$_2$. *Nature* 346:639–641.

Lemaitre, A. 1984. High-order resonances in the restricted three-body problem. *Celest. Mech.* 32:109–126.

Lenard, P. 1904. Über regen. *Meteor. Zeitschrift* 21:249.

Lenzen, R. 1988. High velocity Herbig-Haro objects near Cep A. *Astron. Astrophys.* 190:269–274.

Lenzen, R., Hodapp, K. W., and Solf, J. 1984. Optical and infrared observations of Cep-A/GGD37. *Astron. Astrophys.* 137:202–210.

Leorat, J., Passot, T., and Pouquet, A. 1990. Influence of supersonic turbulence on self-gravitating flows. *Mon. Not. Roy. Astron. Soc.* 243:293–311.

Leous, J. A., Feigelson, E. D., André, P., and Montmerle, T. 1991. A rich cluster of radio stars in the ρ Ophiuchi cloud cores. *Astrophys. J.*, in press.

Lepp, S., and Dalgarno, A. 1988. Polycyclic aromatic hydrocarbons in interstellar chemistry. *Astrophys. J.* 324:553–556.

Lepp, S., McCray, R., Shull, J. M., Woods, D. T., and Kallman, T. 1985. Thermal phases of interstellar and quasar gas. *Astrophys. J.* 288:58–64.

Lepp, S., Dalgarno, A., and Sternberg, A. 1987. The abundance of H_3 ions in dense interstellar clouds. *Astrophys. J.* 321:383–385.

Lester, D. F., Harvey, P. M., Joy, M., and Ellis, H. B., Jr. 1986. Far-infrared image restoration analysis of the protostellar cluster in S 140. *Astrophys. J.* 309:80–89.

Lester, D., Harvey, P., Smith, B., Colome, C., and Low, F. 1990. The far-IR size of the dust cloud around Fomalhaut. Presented at the 175th Meeting of the Amer. Astron. Soc., Washington, D. C.

Lestrade, J.-F. 1988. VLBI observations of radio stars. In *The Impact of VLBI on Astrophysics and Geophysics: Proc. IAU Symp. 129*, eds. J. Moran and M. Reid (Dordrecht: D. Reidel), pp. 265–274.

Leung, C. M., Herbst, E., and Huebner, W. F. 1984. Synthesis of complex molecules in dense interstellar clouds via gas–phase chemistry: A pseudo time–dependent calculation. *Astrophys. J. Suppl.* 56:231–256.

Leventhal, M., MacCallum, C., and Watts, A. 1977. A search for gamma-ray lines from Nova Cygni 1975, Nova Serpentis 1970, and the Crab Nebula. *Astrophys. J.* 216:491–502.

Le Verrier, U.-J. 1856. *Ann. Obs. Paris* 2:165.

Levison, H. F. 1991. The long-term dynamical behaviour of small bodies in the Kuiper belt. *Astron. J.* 102:787–795.

Levison, H. F., and Duncan, M. J. 1990. A search for proto-comets in the outer regions of the solar system. *Astron. J.* 100:1669–1675.

Levreault, R. M. 1983. Interactions between pre-main sequence objects and molecular clouds. I. Elias 1–12. *Astrophys. J.* 265:855–863.

Levreault, R. M. 1985. Molecular Outflows and Mass Loss in Pre-Main-Sequence Stars. Ph. D. Thesis, Univ. of Texas.

Levreault, R. M. 1988a. A search for molecular outflows toward pre-main-sequence objects. *Astrophys. J. Suppl.* 67:283–371.

Levreault, R. M. 1988b. Molecular outflows and mass loss in pre-main-sequence stars. *Astrophys. J.* 330:897–910.

Levy, E. H. 1978. Magnetic field in the early solar system. *Nature* 276:481.

Levy, E. H. 1988. Energetics of chondrule formation. In *Meteorites and the Early Solar System*, eds. J. F. Kerridge and M. S. Matthews (Tucson: Univ. of Arizona Press), pp. 697–711.

Levy, E. H. and Araki, S. 1989. Magnetic reconnection flares in the protoplanetary nebula and the possible origin of meteorite chondrules. *Icarus* 81:74–91.

Levy, E. H., and Sonnett, C. P. 1978. Meteorite magnetism and early solar system magnetic fields. In *Protostars and Planets*, eds. T. Gehrels (Tucson: Univ. of Arizona Press), pp. 516–532.

Levy, E. H., Ruzmaikin, A. A., and Ruzmaikina, T. V. 1991. Magnetic history of the Sun. In *The Sun in Time*, eds. C. P. Sonnett, M. S. Giampapa and M. S. Matthews (Tucson: Univ. of Arizona Press), pp. 589–632.

Lewis, J. S. 1972. Low temperature condensation from the solar nebula. *Icarus* 16:241–252.

Lewis, J. S. 1974. The temperature gradient in the solar nebula. *Science* 186:440–443.

Lewis, J. S. 1982. Io: geochemistry of sulfur. *Icarus* 50:103–114.

Lewis, J. S., and Ney, E. P. 1979. Iron and the formation of astrophysical dust grains. *Astrophys. J.* 234:154–157.

Lewis, J. S. and Prinn, R. G. 1980. Kinetic inhibition of CO and N_2 reduction in the solar nebula. *Astrophys. J.* 238:357–364.

Lewis, J. S., and Prinn, R. 1984. *Planets and their Atmospheres: Origin and Evolution* (New York: Academic Press).

Lewis, R. S., and Anders, E. 1981. Isotopically anomalous xenon in meteorites: A new clue to its origin. *Astrophys. J.* 247:1122–1124.

Lewis, R. S., Srinivasan, B., and Anders, E. 1975. Host phase of a strange xenon component in Allende. *Science* 190:1251–1262.

Lewis, R. S., Anders, E., Wright, I. P., Norris, S. J., and Pillinger, C. T. 1983*a*. Isotcpically anomalous nitrogen in primitive meteorites. *Nature* 305:767–771.

Lewis, R. S., Anders, E., Shimamura, T., and Lugmair, G. W. 1983*b*. Barium isotopes in Allende meteorite: Evidence against an extinct superheavy element. *Science* 222:1013–1015.

Lewis, R. S., Tang, M., Wacker, J. F., Anders, E., and Steel, E. 1987. Interstellar diamonds in meteorites. *Nature* 326:160–162.

Lewis, R. S., Anders, E., and Draine, B. T. 1989. Properties, detectability and origin of interstellar diamonds in meteorites. *Nature* 339:117–121.

Lewis, R. S., Amari, S., and Anders, E. 1990. Meteoritic silicon carbide: Pristine material from carbon stars. *Nature* 348:293–298.

Liebert, J., and Probst, R. G. 1987. Very low mass stars. *Ann. Rev. Astron. Astrophys.* 25:473–519.

Lien, D. J. 1990. Dust in comets. I. Thermal properties of homogenous and heterogenous grains. *Astrophys. J.* 355:680–692.

Lighthill, M. J. 1978. *Waves in Fluids* (Cambridge: Cambridge Univ. Press)

Liljeström, T., Mattila, K., and Friberg, P. 1989. CO outflow and properties of the molecular gas around the far-infrared point source IRAS+4325–1419 in Lynds 1642. *Astron. Astrophys.* 210:337–344.

Lilley, A. E., and Palmer, P. 1968. Tables of radio-frequency recombination lines. *Astrophys. J. Supp.* 16:143–173.

Lin, C. C., and Lau, Y. Y. 1979. Density wave theory of spiral structure of galaxies. *Studies in Applied Math.* 60:97–163.

Lin, C. C., and Shu, F. H. 1964. On the spiral structure of disk galaxies. *Astrophys. J.* 140:646–655.

Lin, C. C., and Shu, F. H. 1968. Theory of spiral structure. In *Galactic Astronomy*, eds. H. Y. Chiu and A. Muriel (New York: Gordon & Breach), pp. 1–93.

Lin, D. N. C. 1981. Convective accretion disk model for the primordial solar nebula. *Astrophys. J.* 246:972–984.

Lin, D. N. C. 1989. Dynamical properties of protoplanetary disks. In *The Formation and Evolution of Planetary Systems*, eds. H. A. Weaver and L. Danly (Cambridge: Cambridge Univ. Press), pp. 314–321.

Lin D. N. C., and Bodenheimer, P. 1982. On the evolution of convective accretion disk models of the primordial solar nebula. *Astrophys. J.* 262:768–779.

Lin, D. N. C., and Papaloizou, J. C. B. 1979*a*. Tidal torques on accretion disks in binary systems with extreme mass ratios. *Mon. Not. Roy. Astron. Soc.* 186:799–812.

Lin, D. N. C., and Papaloizou, J. C. B. 1979*b*. On the evolution of a circumbinary accretion disk and the tidal evolution of commensurable satellites. *Mon. Not. Roy. Astron. Soc.* 188:191–201.

Lin, D. N. C., and Papaloizou, J. C. B. 1980. On the structure and evolution of the primordial solar nebula. *Mon. Not. Roy. Astron. Soc.* 191:37–48.

Lin, D. N. C., and Papaloizou, J. C. B. 1985. On the dynamical origin of the solar system. In *Protostars and Planets II*, eds. D. C. Black and M. S. Mathews (Tucson: Univ. of Arizona Press), pp. 981–1072.

Lin, D. N. C., and Papaloizou, J. C. B. 1986*a*. On the tidal interaction between proto-planets and the primordial solar nebula II. Self-consistent non-linear interaction. *Astrophys. J.* 307:395–409.

Lin, D. N. C., and Papaloizou, J. C. B. 1986*b*. On the tidal interaction between pro-toplanets and the primordial solar nebula. III. Orbital migration of protoplanets. *Astrophys. J.* 309:846–857.

Lin, D. N. C., and Pringle, J. E. 1976. Numerical simulation of mass transfer and accretion disc flow in binary systems. In *Structure and Evolution of Close Binary Systems: Proc. IAU Symp. 73*, eds. P. P. Eggleton, S. Mitton and J. A. J. Whelan (Dordrecht: D. Reidel), pp. 237–252.

Lin, D. N. C., and Pringle, J. E. 1987. A viscosity prescription for a self-gravitating accretion disc. *Mon. Not. Roy. Astron. Soc.* 225:607–613.

Lin, D. N. C., and Pringle, J. E. 1990. The formation and initial evolution of protostellar disks. *Astrophys. J.* 358:515–524.

Lin, D. N. C., Papaloizou, J. C. B., and Faulkner, J. 1985. On the evolution of accretion disk flow in cataclysmic variables–III. Outburst properties of constant and uniform α model disks. *Mon. Not. Roy. Astron. Soc.* 212:105–149.

Lin, D. N. C., Williams, R. E., and Stover, R. J. 1988. Stark broadened emission lines in the accretion disks of cataclysmic variables. *Astrophys. J.* 327:234–247.

Lin, D. N. C., Papaloizou, J., and Savonije, G. 1990*a*. Wave propagation in gaseous accretion disks. *Astrophys. J.* 364:326–334.

Lin, D. N. C., Papaloizou, J., and Savonije, G. 1990*b*. Propagation of tidal disturbance in gaseous accretion disks. *Astrophys. J.* 365:748–756.

Lind, K. R., Payne, D. G., Meier, D. L., and Blandford, R. D. 1989. Numerical simulations of magnetized jets. *Astrophys. J.* 344:89–103.

Lindal, G. F., Wood, G. E., Levy, G. S., Anderson, J. D., Sweetnam, D. N., Hotz, H. B., Buckles, B. J., Holms, D. P., Doms, P. E., Eshleman, V. R., Tyler, G. L., and Croft, T. A. 1981. The atmosphere of Jupiter: An analysis of the Voyager radio occultation measurements. *J. Geophys. Res.* 86:8721–8727.

Lindal, G. F., Wood, G. E., Hotz, H. B., Sweetnam, D. N., Eshleman, V. R., and Tyler, G. L. 1983. The atmosphere of Titan: an analysis of the Voyager 1 radio occultation measurements. *Icarus* 53:348–363.

Lindal, G. F., Sweetnam, D. N., and Eshleman, V. R. 1985. The atmosphere of Saturn: an analysis of the Voyager radio occultation measurements. *Astron. J.* 90:1136–1146.

Lindal, G. F., Lyons, J. R., Sweetnam, D. N., Eshleman, V. R., Hinson, D. P., and Tyler, G. L. 1987. The atmosphere of Uranus: results of radio occultation measurements with Voyager 2. *J. Geophys. Res.* 92:14987–15001.

Lioure, A., and Chieze, J. P. 1990. Interstellar gas cycling powered by star formation. *Astron. Astrophys.* 235:379–386.

Lis, D. C., Goldsmith, P. F., Dickman, R. L., Predmore, C. P., Omont, A., and Cernicharo, J. 1988. Linear polarization of millimeter wavelength emission lines in clouds without large velocity gradients. *Astrophys. J.* 328:304–314.

Liseau, R., Sandell, G., and Knee, L. B. G. 1988. The structure of the molecular outflow near SSV13 and HH7–11 in the NGC1333 region. *Astron. Astrophys.*

192:153–164.

Lissauer, J. J. 1987. Time scales for planetary accretion and the structure of the protoplanetary disk. *Icarus* 69:249–265.

Lissauer, J. J., and Cuzzi, J. N. 1985. Rings and moons: clues to understanding the solar nebula. In *Protostars and Planets II*, eds. D. C. Black and M. S. Matthews (Tucson: Univ. of Arizona Press), pp. 920–956.

Lissauer, J. J., and Griffith, C. A. 1989. Erosion of circumstellar particle disks by interstellar dust. *Astrophys. J.* 340:468–471.

Lissauer, J. J., and Safronov, V. S. 1991. The random component of planetary rotation. *Icarus* 93:288–297.

Liszt, H. S., and Burton, W. B. 1983. ^{13}CO in the galactic plane: the cloud-to-cloud velocity dispersion in the inner galaxy. In *Kinematics, Dynamics, and Structure of the Milky Way*, ed. W. L. Shuter (Dordrecht: D. Reidel), pp. 135–142.

Liszt, H. S., and Vanden Bout, P. A. 1985. Upper limits on the O_2/CO ratio in two dense interstellar clouds. *Astrophys. J.* 291:178–182.

Liszt, H. S., Burton, W. B., and Xiang, D. 1981. ^{13}CO in the inner galactic plane. *Astron. Astrophys.* 140:303–313.

Little, L. T., MacDonald, G. H., Riley, P. W., and Matheson, D. N. 1979. The relative distribution of ammonia and cyanobutadiyne emission in Heiles 2 dust cloud. *Mon. Not. Roy. Astron. Soc.* 189:539–550.

Little, L. T., Bergman, P., Cunningham, C. T., Heaton, B. D., Knee, L. B. G., MacDonald, G. H., Richards, P. J., and Toriseva, M. 1988. IRAS20188+3928: a molecular cloud with a very dense bipolar outflow. *Astron. Astrophys.* 205:129–134.

Livio, M., and Truran, J. W. 1990. Elemental mixing in classical nova systems. In *Nonlinear Astrophysical Fluid Dynamics*, eds. J. R. Buchler and S. T. Gottesman (New York: New York Academy of Sci.), pp. 126–137.

Livio, M., Ögelman, H., and Truran, J. W. 1990. Preprint.

Lizano, S., and Shu, F. H. 1987. Formation and heating of molecular cloud cores. In *Physical Processes in Interstellar Clouds*, eds. G. Morfill and M. Schöler (Dordrecht: Reidel), pp. 173–193.

Lizano, S., and Shu, F. H. 1989. Molecular cloud cores and bimodal star formation. *Astrophys. J.* 342:834–854.

Lizano, S., Heiles, C. Rodríguez, L. F., Koo, B.-C., Shu, F. H., Hasegawa, T., Hayashi, S. S., and Mirabel, I. F. 1988. Neutral stellar winds that drive bipolar outflows in low-mass protostars. *Astrophys. J.* 328:763–776.

Lo, K. Y., Ball, R., Masson, C. R., Phillips, T. G., Scott, S., and Woody, D. P. 1987. Molecular spiral structure in M51. *Astrophys. J. Lett.* 317:63.

Lockman, J. 1990. Recombination lines and galactic structure. In i *Radio Recombination Lines: 25 Years of Investigation*, eds. Gordon and Sorochenko (Dordrecht: Kluwer), pp. 225–236.

Lohsen, E. 1976. A tentative spectroscopic orbit of θ^1 OriA. *Info. Bull. Var. Stars* 1211.

Lord, S. D., and Kenney, J. D. P. 1991. A molecular gas ridge offset from the dust lane in a spiral arm of M83. *Astrophys. J.* 381:130–136.

Loren, R. B. 1989a. The cobwebs of Ophiuchus: I. Strands of ^{13}CO: the mass distribution. *Astrophys. J.* 338:902–924.

Loren, R. B. 1989b. The cobwebs of Ophiuchus: II. ^{13}CO filament kinematics. *Astrophys. J.* 338:925–944.

Loren, R. B., and Wootten, A. 1986. A massive prestellar molecular core and adjacent compression front in the ρ Ophiuchi cloud. *Astrophys. J.* 306:142–159.

Loren, R. B., Sandqvist, A., and Wootten, A. 1983. Molecular clouds on the threshold of star formation: the radial density profile of the cores of the Rho Ophiuchi and R Coronae Australis clouds. *Astrophys. J.* 270:620–640.

Loren, R. B., Wootten, A., and Wilking, B. A. 1990. Cold DCO^+ cores and protostars in the warm ρ Ophiuchi cloud. *Astrophys. J.* 365:269–286.

Lorenzetti, D., Sarceno, P., and Strafella, F. 1983. The near-infrared spectrum of the Herbig Ae-Be stars. *Astrophys. J.* 264:554–559.

Loushin, R., Crutcher, R. M., and Bieging, J. H. 1990. Observations of an expanding molecular ring in S 106. *Astrophys. J. Lett.* 362:67–69.

Lovas, F. J., Johnson, D. R., Buhl, D., and Snyder, L. E. 1976. Millimeter emission lines in Orion A. *Astrophys. J.* 209:770–777.

Lovas, F. J., Snyder, L. E., and Johnson, D. R. 1979. Recommended rest frequencies for observed interstellar molecular transitions. *Astrophys. J. Suppl.* 41:451–480.

Lovejoy, S. 1982. Area-perimeter relation for rain and cloud areas. *Science* 216:185–187.

Lovelace, R. V. E. 1976. Dynamo model of double radio sources. *Nature* 262:649–652.

Lovelace, R. V. E., Berk, H. L., and Contopoulos, J. 1991. Magnetically driven jets and winds. *Astrophys. J.* 379:696–705.

Low, C., and Lynden-Bell, D. 1976. The minimum Jeans mass or when fragmentation must stop. *Mon. Not. Roy. Astron. Soc.* 176:367–390.

Low, F. J., Bientema, D. A., Gautier, T. N., Gillett, F. C., Beichman, C. A., Neugebauer, G., Young, E., Aumann, H. H., Boggess, N., Emerson, J. P., Habing, H. J., Hauser, M. G., Houck, J. R., Rowan-Robinson, M., Soifer, B. T., Walker, R. G., and Wesselius, P. R. 1984. Infrared cirrus: New components of the extended infrared emission. *Astrophys. J. Lett.* 278:19–23.

Lowenstein, R. F., Harper, D. A., and Moseley, H. 1977. The effective temperature of Neptune. *Astrophys. J. Lett.* 218:145–146.

Lubow, S. H., and Shu, F. H. 1975. Gas dynamics of semidetached binaries. *Astrophys. J.* 198:383–405.

Lucy, L. B. 1977. A numerical approach to the testing of the fission hypothsis. *Astron. J.* 82:1013–1024.

Lucy, L. B. 1981. The formation of binary stars. In *Fundamental Problems in the Theory of Stellar Evolution: Proc. IAU Symp. 93*, eds. D. Sugimoto, D. Q. Lamb and D. N. Schramm (Dordrecht: D. Reidel), pp. 75–83.

Lugmair, G. W., and Galer, S. J. G. 1989. Isotopic evolution and age of angrite LEW 86010. *Meteoritics* 24:140 (abstract).

Lugmair, G. W., and Marti, K. 1977. Sm-Nd-Pu timepieces in the Angra dos Reis meteorite. *Earth Planet. Sci. Lett.* 35:273–284.

Lugmair, G. W., Marti, K., Kurtz, J. P., and Scheinin, N. B. 1976. History and genesis of lunar troctolite 76535 or: How old is old? *Proc. Lunar Sci. Conf.* 7:2009–2033.

Lunine, J. I. 1985a. Titan's surface: implications for Cassini. In *The Atmospheres of Saturn and Titan*, eds. E. Rolfe and B. Battrick, ESP SP-241, pp. 83–88.

Lunine, J. I. 1985b. Volatiles in the Outer Solar System. Ph. D. Thesis, California Inst. of Technoloy.

Lunine, J. I. 1989a. Origin and evolution of outer solar system atmospheres. *Science* 245:141–147.

Lunine, J. I. 1989b. Primitive bodies: molecular abundances in comet Halley as probes of cometary formation environments. In *The Formation and Evolution of Planetary Systems*, eds. H. A. Weaver and L. Danly (Cambridge: Cambridge Univ. Press), pp. 213–242.

Lunine, J. I. 1989c. The Urey Prize lecture. Volatile processes in the outer solar system. *Icarus* 81:1–13.

Lunine, J. I., and Nolan, M. 1992. A massive early atmosphere on Titan. *Icarus* 100:221–234.

Lunine, J. I. and Stevenson, D. J. 1982. Formation of the Galilean satellites in a gaseous nebula. *Icarus* 52:14–39.

Lunine, J. I. and Stevenson, D. J. 1985. Thermodynamics of clathrate hydrate at low and high pressures with application to the outer solar system. *Astrophys. J. Suppl.* 58:493–531.

Lunine, J. I., and Stevenson, D. J. 1987. Clathrate and ammonia hydrates at high pressure: application to the origin of methane on Titan. *Icarus* 70:61–77.

Lunine, J. I., Stevenson, D. J., and Yung, Y. L. 1983. Ethane ocean on Titan. *Science* 222:1229–1230.

Lunine, J. I., Atreya, S. K., and Pollack, J. B. 1989. Present state and chemical evolution of the atmospheres of Titan, Triton and Pluto. In *Origin and Evolution of Planetary and Satellite Atmospheres*, eds. S. K. Atreya, J. B. Pollack and M. S. Matthews (Tucson: Univ. of Arizona Press), pp. 605–665.

Lunine, J. I., Engel, S., Rizk, B., and Horanyi, M. 1991. Sublimation and reformation of icy grains in the primitive solar nebula. *Icarus* 94:333–344.

Lupo, M. J., and Lewis, J. S. 1979. Mass-radius relationships in icy satellites. *Icarus* 40:125–135.

Lüst, R. 1952. Die entwicklung einer um einen zeutralkorper rotierenden gasmasse. I. Loesungen de hydrodynamischen gleichungenmit mit turulenter reibung. *Z. Naturforschung* 7a:87–98.

Lüst, R., and Schlüter, A. 1955. Drehimpulstransport durch Magnetfelder und die abbremsung rotierender sterne. *Zeits. Astrophys.* 38:190–211.

Lutz, B. L., de Bergh, C., and Maillard, J. P. 1983. Monodeuterated methane in the outer solar system. I. Spectroscopic analysis of the bands at 1.55 and 1.95 microns. *Astophys. J.* 273:397–409.

Lutz, B. L., Owen, T., and de Bergh, C. 1990. Deuterium enrichment in the primitive ices of the protosolar nebula. *Icarus* 86:329–335.

Luu, J., and Jewitt, D. 1988. A two-part search for slow-moving objects. *Astron. J.* 95:1256–1262.

Luu, J., and Jewitt, D. 1990. Cometary activity in 2060 Chiron. *Astron. J.* 100:913–932.

Lynden-Bell, D. 1974. In *Galaxies and Relativistic Astrophysics*, eds. B. Barbanis and J. D. Hadjidemetriou (Berlin: Springer-Verlag), pp. 224.

Lynden-Bell, D., and Kalnajs, A. 1974. On the generating mechanism of spiral structure. *Mon. Not. Roy. Astron. Soc.* 157:1–30.

Lynden-Bell, D., and Pringle, J. E. 1974. The evolution of viscous disks and the origin of the nebular variables. *Mon. Not. Roy. Astron. Soc.* 168:603–637.

Lynds, B. T. 1962. Catalogue of dark nebulae. *Astrophys. J. Suppl.* 7:1–52.

Lyttleton, R. A. 1948. On the origin of comets. *Mon. Not. Roy. Astron. Soc.* 108:465–475.

Ma, S.-K. 1976. *Modern Theory of Critical Phenomena* (London:Benjamin).

Maas, D., Krueger, F. R., and Kissel, J. 1990. Mass and density of silicate- and CHON-type dust particles released by comet P/Halley. In *Asteroids, Comets, Meteors III*, eds. C. I. Lagerkvist, H. Rickman, B. A. Lindblad and M. Lindgren (Uppsala: Reprocentralen HSC), pp. 389–392.

Maaswinkel, F., Bortoletto, F., Buananno, R., Buzzoni, B., D'Odorico, S., Gulli, B., Huster, G., and Nees, W. 1988. Image stabilization with DISCO: results of first observing run. In *Very Large Telescopes and Their Instrumentation*, ed. M.-H. Ulrich (Garching: ESO), pp. 751–760.

MacDonald, G. J. F. 1964. Tidal friction. *Rev. Geophys.* 2:467–541.

Macdougall, J. D., and Kothari, B. K. 1976. Formation chronology for C2 meteorites. *Earth Planet. Sci. Lett.* 33:36–44.

Macdougall, J. D., and Lugmair, G. W. 1989. Chronology of chemical change in the Orgueil CI chondrite based on Sr isotope systematics. *Meteoritics* 24:297

(abstract).

Macdougall, J. D., Lugmair, G. W., and Kerridge, J. F. 1984. Early solar system aqueous activity: Sr isotope evidence from the Orgueil CI meteorites. *Nature* 307:249–251.

MacLow, M. M., McCray, R., and Norman, M. L. 1989. Superbubble blowout dynamics. *Astrophys. J.* 337:141–154.

MacPherson, G. J., Grossman, L., Allen, J. M., and Beckett, J. R. 1981. Origin of rims on coarse-grained inclusions in the Allende meteorite. *Lunar Planet. Sci.* XII:1079–1091 (abstract).

MacPherson, G. J., Wark, D. A., and Armstrong, J. T. 1988. Primitive material surviving in chondrites: refractory inclusions. In *Meteorites and the Early Solar System*, eds. J. F. Kerridge and M. S. Matthews (Tucson: Univ. of Arizona Press), pp. 746–807.

Maddelena, R. J., and Morris, M. 1987. An expanding system of molecular clouds surrounding λ Orionis. *Astrophys. J.* 323:179–192.

Maddalena, R. J., and Thaddeus, P. 1985. A large cold and unusual moecular cloud in Monoceros. *Astrophys. J.* 294:231–237.

Maddalena, R. J., Morris, M., Moscowitz, J., and Thaddeus, P. 1986. The large system of molecular clouds in Orion and Monoceros. *Astrophys. J.* 303:375–391.

Madden, S. C. 1990. A Multi-Transition Study of the Cyclic Molecule Cyclopropenylidene (C_3H_2) in the Galaxy. Ph. D. Thesis, Univ. of Massachusetts.

Madden, S. C., Irvine, W. M., Matthews, H. E., Friberg, P., and Swade, D. A. 1989. A survey of cyclopropenylidene (C_3H_2) in galactic sources. *Astron. J.* 97:1403–1422.

Magee-Sauer, K., Scherb, F., Roesler, F. L., and Harlander, J. 1989. Fabry-Perot observations of NH_2 emission from comet Halley. *Icarus* 82:50–60.

Magnani, L. 1987. Molecular Clouds at High Galactic Latitudes. Ph.D. Thesis, Univ. of Maryland.

Magnani, L., Blitz, L., and Mundy, L. 1985. Molecular gas at high galactic latitudes. *Astrophys. J.* 295:402–421.

Magnani, L., Lada, E. A., Sandell, G., and Blitz, L. 1989. CH observations of diffuse molecular clouds. *Astrophys. J.* 339:244–257.

Magnani, L., Carpenter, J. M., Blitz, L., Kassin, N. E., and Nath, B. B. 1990*a*. Structure in small molecular clouds: Pedestals and clumping. *Astrophys. J. Suppl.* 73:747–768.

Magnani, L., Caillault, J.-P., and Armus, L. 1990*b*. A search for T Tauri stars in high latitude molecular clouds. I. IRAS sources and CCD imaging. *Astrophys. J.* 357:602–605.

Maihara, T., and Kataza, H. 1991. A study of a spatially resolved T Tau system. *Astron. Astrophys.* 249:392–396.

Mahoney, W. A., Ling, J. C., Jacobson, A. S., and Ligenfelter, R. E. 1982. Diffuse galactic gamma-ray line emission from nucleosynthetic ^{60}Fe, ^{26}Al, and ^{22}Na: Preliminary limits from HEAO 3. *Astrophys. J.* 262:742–748.

Mahoney, W. A., Ling, J. C., Wheaton, W. A., and Jacobson, A. S. 1984. HEAO3 discovery of ^{26}Al in the interstellar medium. *Astrophys. J.* 286:578–585.

Makalkin, A. B. 1987. Thermal conditions of the protoplanetary disk. *Astron. Vestn.* 21:324–327.

Malbet, F., and Bertout, C. 1991. The vertical structure of T Tauri accretion disks. Preprint.

Malhotra, R. 1991. Tidal origin of the Laplace resonance and the resurfacing of Ganymede. *Icarus* 94:399–412.

Malhotra, R., and Dermott, S. F. 1990. The role of secondary resonances in the orbital history of Miranda. *Icarus* 85:444–480.

Maloney, P. 1988. The turbulent interstellar medium and pressure-bounded molecular clouds. *Astrophys. J.* 334:761–770.

Maloney, P. 1990*a*. Are molecular clouds in virial equilibrium? *Astrophys. J. Lett.* 348:9–12

Maloney, P. 1990*b*. Mass determinations from CO observations. In *The Interstellar Medium in Galaxies*, eds. H. A. Thronson, Jr. and J. M. Shull (Dordrecht: Kluwer), pp. 493–523.

Mangeney, A., and Praderie, F. 1984. The influence of convection and rotation on X-ray emission in main sequence stars. *Astron. Astrophys.* 130:143–150.

Mangum, J. G., Wootten, A., Loren, R. B., and Wadiak, E. J. 1990. Observations of the formaldehyde emission in Orion-KL: Abundances, distribution, and kinematics of the dense gas in the Orion molecular ridge. *Astrophys. J.* 348:542–556.

Mangum, J. G., Plambeck, R. L., and Wootten, A. 1991. Fossil DCN in Orion-KL. *Astrophys. J.* 396:169–174.

Manhes, G., Göpel, C., and Allègre, C. J. 1987. High resolution chronology of the early solar system based on lead isotopes. *Meteoritics* 22:453–454.

Manuel, O. K., Hennecke, E. W, and Sabu, D. D. 1972. Xenon in carbonaceous chondrites. *Nature Phys. Sci.* 240:99–101.

Marcialis, R. L., and Greenberg, R. 1987. Warming of Miranda during chaotic rotation. *Nature* 328:227–229.

Marcialis, R. L., Rieke, G. H., and Lebofsky, L. A. 1987. The surface composition of Charon: tentative identification of water ice. *Science* 237:1349–1351.

Marconi, M. L., Mendis, D. A., Korth, A., Pin, R. P., Mitchell, D. L., and Reme, H. 1990. The identification of H_3S^+ with the ion of mass per charge (m/q) 35 observed in the coma of comet Halley. *Astrophys. J. Lett.* 352:17–20.

Marcus, J. N., and Olsen, M. A. 1991. Biological implications of organic compounds in comets. In *Comets in the Post-Halley Era*, eds. R. Newburn, M. Neugebauer and J. Rahe (Dordrecht: Kluwer), pp. 439–462.

Marcy, G. W., and Benitz, K. J. 1989. A search for substellar companions to low-mass stars. *Astrophys. J.* 344:441–453.

Margulis, M., and Lada, C. J. 1985. Masses and energetics of high-velocity molecular outflows. *Astrophys. J.* 299:925–938.

Margulis, M., and Lada, C. J. 1986. An unbiased survey for high-velocity gas in the Monoceros OB1 molecular cloud. *Astrophys. J. Lett.* 309:87–90.

Margulis, M., and Lada, C. J. 1988. Molecular outflows in the Monoceros OB1 molecular cloud. *Astrophys. J.* 333:316–331.

Margulis, M. and Snell, R. L. 1989. Very high velocity emission from molecular outflows. *Astrophys. J.* 343:779–784.

Margulis, M., Lada, C. J., and Snell, R. 1988. Molecular outflows in the Monoceros OB1 molecular cloud. *Astrophys. J.* 333:316–331.

Margulis, M., Lada, C. J., and Young, E. T. 1989. Young stellar objects in the Monoceros OB1 molecular cloud. *Astrophys. J.* 345:906–917.

Marley, M., and Hubbard, W. B. 1988. Thermodynamics of dense molecular hydrogen-helium mixtures at high pressure. *Icarus* 73:536–544.

Marochnik, L. S., Mukhin, L. M., and Sagdeev, R. Z. 1988. Estimates of mass and angular momentum in the Oort cloud. *Science* 242:547–550.

Marraco, H. G., and Rydgren, A. E. 1981. On the distance and membership of the R Cr A T association. *Astron. J.* 86:62–68.

Marrero, T. R., and Mason, E. A. 1972. Gaseous diffusion coefficients. *J. Phys. Chem. Ref. Data* 1:3–118.

Marschall, L., and Mathieu, R. D. 1988. Parenago 1540: A pre-main-sequence double-lined spectroscopic binary near the Orion Trapezium. *Astron. J.* 96:1956–1964.

Marsden, B. G. 1989. *Catalogue of Cometary Orbits*, 6th ed. (Cambridge, Mass.:

Smithsonian Astrophys. Obs.).

Marsh, T. R., and Horne, K. 1990*a*. Emission-line mapping of the dwarf Nova Ip Pegasi I outburst and quiescence. *Astrophys. J.* 349:593–607.

Marsh, T. R., and Horne, K. 1990*b*. Images of accretion disks—II. Doppler tomography *Mon. Not. Roy. Astron. Soc.* 235:269–286.

Marsh, T. R., Horne, K., and Shipman, H. L. 1987. A spectophotometric study of the emission lines in the quiescent dwarf nova Z Chamaeleontis. *Mon. Not. Roy. Astron. Soc.* 225:551–580.

Marsh, T. R., Horne, K., Schlegel, E. M., Honeycutt, K., and Kaitchuck, R. H. 1990. Doppler imaging of the dwarf nova U Geminorun. *Astrophys. J.* 364:637–646.

Martin, D., and Nokes, R. 1988. Crystal settling in a vigorously convecting magma chamber. *Nature* 332:534–536.

Martin, P. G. 1989. Photoinonization models of the evolution of nova D Q Her 1934. In *Classical Novae*, eds. M. F. Bode and A. Evans (London: Wiley), pp. 113–141.

Martín-Pintado, J., Wilson, T. L., Gardner, F. F., and Henkel, C. 1983. High density molecular gas in the ρ Ophiuchi cloud. *Astron. Astrophys.* 117:145–148.

Martín-Pintado, J., Bachiller, R., and Thum, C. 1990. Radio recombination line maser emission in MWC349. In *Radio Recombination Lines: 25 Years of Investigation*, eds. M. A. Gordon and R. L. Sorochenko (Dordrecht: Kluwer), pp. 161–167.

Martín-Pintado, J., Bachiller, R., and Fuente, A. 1992. SiO emission as a tracer of shocked gas in molecular outflows. *Astron. Astrophys.* 254:315–326.

Marsden, B. 1983. *Catalog of Comet Orbits* (Hillside: Enslaw).

Mason, B. 1979. Cosmochemistry Part 1. Meteorites. In *Data of Geochemistry*, ed. M. Fleischer. U. S. G. S. Prof. Paper 440-B-1 (Washington, D. C.: U. S. Government Printing Office).

Masson, C. R., and Mundy, L. G. 1988. The hot core of Orion: $4''$ maps of HC_3N emission. *Astrophys. J.* 324:538–543.

Masson, C. R., Lo, K. Y., Phillips, T. G., Sargent, A. I., Scoville, N. Z., and Woody, D. P. 1987. CO maps of the OMC-1 outflow. *Astrophys. J.* 319:446–455.

Masson, C. R., Mundy, L. G., and Keene, J. 1990. The extremely high velocity CO flow in HH 7–11. *Astrophys. J. Lett.* 357:25–28.

Mateo, M. 1988. Main-sequence luminosity and initial mass functions of six Magellanic Cloud star clusters ranging in age from 10 megayears to 2.5 gigayears. *Astrophys. J.* 331:261–293.

Mateo, M. 1990. The initial mass functions of Magellanic Cloud star cluster. In *Physical Processes in Fragmentation and Star Formation*, eds. R. Capuzzo-Dolcetta, C. Chiosi and A. Di Fazio (Dordrecht: Kluwer), pp. 401–414.

Matese, J. L., Whitmire, D. P., Lafleur, L. D., Reynolds, R. T., and Cassen, P. M. 1987. Modeling the circumstellar matter around Beta Pictoris and other nearby main sequence stars. *Bull. Amer. Astron. Soc.* 19:830.

Matese, J. J., Whitmire, D. P., and Reynolds, R. T. 1989. Dust clouds around red giant stars: evidence of sublimating comet disks. *Icarus* 81:24–30.

Mathews, G. J., and Cowan, J. J. 1990. New insights into the astrophysical r-process. *Nature* 245:491–494.

Mathews, G. J., and Ward, R. A. 1985. Neutron capture processes in astrophysics. *Rept. Prog. Phys.* 48:1371-1418.

Mathewson, D. S., and Ford, V. L. 1970. Polarization observations of 1800 stars. *Mem. Roy. Astron. Soc.* 74:139–182.

Mathieu, R. D. 1986. The dynamical evolution of young clusters and associations. In *Highlights of Astronomy*, vol. 7, ed. J.-P. Swings (Dordrecht: Kluwer), pp. 481–488.

Mathieu, R. D. 1989. Spectroscopic binaries among low-mass pre-main-sequence stars. In *Highlights of Astronomy*, vol. 8, ed. D. McNally (Dordrecht: Kluwer), pp. 111–115.

Mathieu, R. D., and Mazch, T. 1988. The circularized binaries in open clusters: a new clock for age determination. *Astrophys. J.* 326:256–264.

Mathieu, R. D., Benson, P. J., Fuller, G. A., Myers, P. C., and Schild, R. E. 1988. L43: an example of interaction between molecular outflows and dense cores. *Astrophys. J.* 330:385–398.

Mathieu, R. D., Walter, F. M., and Myers, P. C. 1989. The discovery of six pre-main-sequence spectroscopic binaries. *Astron. J.* 98:987–1001.

Mathieu, R. D., Adams, F., and Latham, D. W. 1991. The T Tauri spectroscopic binary GW Orionis. *Astron. J.* 101:2184–2198.

Mathis, J. S. 1990. Interstellar Dust and Extinction. *Ann. Rev. Astron. Astrophys.* 28:37–70.

Mathis, J. S., Rumpl, W., and Nordsieck, K. H. 1977. The size distribution of interstellar grains. *Astrophys. J.* 217:425–433.

Matsuda, J., and Matsubara, K. 1989. Noble gases in silica and their implication for the terrestrial "missing" Xe. *Geophys. Res. Lett.* 16:81–84.

Matsuda, T., Inoue, M., Sawada, K., Shima, E., and Wakamatsu, K. 1987. A reinvestigation of gas response to an ovally deformed gravitational potential. *Mon. Not. Roy. Astron. Soc.* 229:295–314.

Matsuda, T., Sekino, N., Shima, E., Sawada, K., and Spruit, H. C. 1989. Mass transfer by tidally induced spiral shocks in an accretion disk. In *Theory of Accretion Disks*, eds. F. Meyer, W. J. Duschl, J. Frank and E. Meyer-Hofmeister (Dordrecht: Kluwer), pp. 355–371.

Matsuda, T., Sekino, N., Shima, E., Sawada, K., and Spruit, H. C. 1990. Mass transfer by tidally induced spiral shocks in an accretion disk. *Astron. Astrophys.* 235:211–218.

Matsui, T., and Abe, Y. 1986. Impact-induced oceans on Earth and Venus. *Nature* 322:526–528.

Matsumoto, R., Horiuchi, T., Shibata, K., and Hanawa, T. 1988. Parker instability in non-uniform gravitational fields. II. Nonlinear time evolution. *Publ. Astron. Soc. Japan* 40:171–195.

Matthews, H. E., and Sears, T. J. 1983. Detection of the $J = 1$ to 0 transition of CH_3CN. *Astrophys. J. Lett.* 267:53–57.

Mattila, K., Liljeström, T., and Toriseva, M. 1989. Recent observations with *SEST* of the star forming cloud ϵ Cha I. In *ESO Workshop on Low Mass Star Formation and Pre-Main Sequence Objects*, ed. B. Reipurth (Garching: European Southern Obs.), pp. 153–171.

Mauersberger, R., Henkel, C., Jacq, T., and Walmsley, C. M. 1988. Deuterated methanol in Orion. *Astron. Astrophys.* 194:L1–L4.

Mauersberger, R., Wilson, T. L., Mezger, P. G., Gaume, R., and Johnston, K. J. 1992. The internal structure of molecular clouds III. Evidence for molecular depletion in the NGC 2024 condensations. *Astron. Astrophys.* 256:640–6512.

Mayer, E., and Pletzer, R. 1986. Astrophysical implications of amorphous ice: A microporous solid. *Nature* 319:298–301.

Mayle, R. W., and Wilson, J. R. 1988. Nucleosynthesis and supernovae produced by late time neutrino heating. In *Origin and Distribution of the Elements*, ed. G. J. Mathews (Singapore: World Scientific), pp. 433–443.

Mayor, M., and Mermilliod, J.-C. 1984. Orbit circularization time in binary stellar systems. In *Observational Tests of Stellar Evolution Theory: Proc. IAU Symp. 105*, eds. A. Maeder and A. Renzini (Dordrecht: D. Reidel), pp. 411–414.

Mazch, T., Latham, D. W., Mathieu, R. D., and Carney, B. W. 1990. On the orbital circularization of close binaries. In *Active Close Binaries*, ed. C. Ibanoglu (Dordrecht: Kluwer), pp. 145–154.

Mazzitelli, I., and Moretti, M. 1980. Early pre-main-sequence evolution with deuterium burning. *Astrophys. J.* 235:955–959.

McBreen, B., Fazio, G. G., Stier, M., and Wright, E. L. 1979. Evidence for a variable far-infrared source in NGC 6334. *Astrophys. J. Lett.* 232:183–187.

McCarthy, D. W. 1982. Triple structure of infrared source 3 in the Monoceros R2 molecular cloud. *Astrophys. J. Lett.* 257:93–97.

McCaughrean, M. J. 1988. The Astronomical Application of Infrared Array Detectors. Ph. D. Thesis, Univ. of Edinburgh.

McCaughrean, M. J., and Gezari, D. Y. 1991. 12 μm dust shells and cavities in the Trapezium/Ney-Allen nebula. In *Astrophysics with Infrared Arrays*, ed. R. Elston (San Francisco: Astron. Soc. of the Pacific), pp. 301–303.

McCaughrean, M. J., Aspin, C., McLean, I. S., and Zinnecker, H. 1989. In *ESO Workshop on Low Mass Star Formation and Pre-Main-Sequence Objects*, ed. B. Reipurth (Garching: European Southern Obs.).

McCaughrean, M. J., Zinnecker, H., Aspin, C., and McLean, I. S. 1991. Low mass pre-main sequence clusters in regions of massive star formation. In *Astrophysics with Infrared Arrays*, ed. R. Elston (San Francisco: Astron. Soc. of the Pacific), pp. 238–240.

McCaughrean, M. J., Zinnecker, H., Aspin, C., and McLean, I. S. 1992. Infrared imaging photometry of the Trapezium Cluster. In preparation.

McClure, R. D., VandenBerg, D. A., Smith, G. H., Fahlman, G. G., Richer, H. B., Hesser, J. E., Harris, W. E., Stetson, P. B., and Bell, R. A. 1986. Mass functions for globular cluster main sequences based on CCD photometry and stellar models. *Astrophys. J. Lett.* 307:49–53.

McCray, R., and Kafatos, M. 1987. Supershells and propagating star formation. *Astrophys. J.* 317:190–196.

McCray, R., and Stein, R. F. 1975. Thermal instability in supernova shells. *Astrophys. J.* 196:565–570.

McCrea, W. H. 1957. The formation of population I stars. Part I. Gravitational contraction. *Mon. Not. Roy. Astron. Soc.* 117:562–578.

McCrea, W. H. 1975. Solar system as space probe. *Observatory* 95:239–255.

McCulloch, M. T., and Wasserburg, G. J. 1978. Ba and Nd isotopic anomalies in the Allende meteorite. *Astrophys J. Lett.* 220:15-19.

McCutcheon, W. H., Vrba, F. J., Dickman, R L., and Clemens, D. P. 1986. The Lynds 204 Complex: magnetic field controlled evolution? *Astrophys. J.* 309:619–627.

McDonnell, J. A. M., Lamy, P. L., and Pankiewicz, G. S. 1991. Physical properties of cometary dust. In *Comets in the Post-Halley Era*, eds. R. Newburn, M. Neugebauer and J. Rahe (Dordrecht: Kluwer), pp. 1043–1073.

McElroy, M. B., and Prather, M. J. 1981. Noble gases in the terrestrial planets. *Nature* 293:535–539.

McFadzean, A. D., Whittet, D. C. B., Longmore, A. J., Bode, M. F., and Adamson, A. J. 1989. Infrared studies of dust and gas towards the Galactic centre: 3–5 μm spectroscopy. *Mon. Not. Roy. Astron. Soc.* 241:873–882.

McFarlane, E. A., and Drake, M. J. 1990. Element partitioning and the early thermal history of the Earth. In *Origin of the Earth*, eds. H. E. Newsom and J. J. Jones (New York: Oxford Univ. Press), pp. 135–150.

McGee, R. X., and Milton, J. A. 1964. A sky survey of neutral hydrogen at 21 cm. III. Gas at higher radial velocities. *Australian J. Phys.* 17:125–157.

McGlynn, T. A., and Chapman, R. D. 1989. On the nondetection of extrasolar comets. *Astrophys. J. Lett.* 346:105–108.

McGonagle, D., Ziurys, L. M., Irvine, W. M., and Minh, Y. C. 1990. Detection of nitric oxide in the dark cloud L134N. *Astrophys. J.* 359:121–124.

McGregor, P. J., and Hyland, A. R. 1981. Infrared studies of the two stellar populations in 30 Doradus. *Astrophys. J.* 250:116–134.

McKay, C. P., Scattergood, T. W., Pollack, J. B., Borucki, W. J., and Van Ghysegahm, H. T. 1988. High temperature shock formation of N_2 and organics on primordial Titan. *Nature* 332:520–522.

McKee, C. F. 1989. Photoionization-regulated star formation and the structure of molecular clouds. *Astrophys. J.* 345:782–801.

McKee, C. F., and Cowie, L. L. 1975. The interaction between the blast wave of a supernova remnant and interstellar clouds. *Astrophys. J.* 195:715–725.

McKee, C. F., and Lin, J.-Y. 1988. Star formation. In *Origin, Structure, and Evolution of Galaxies*, ed. F. L. Zhi (Singapore: World Scientific), pp. 47–59.

McKee, C. F., and Ostriker, J. P. 1977. A theory of the interstellar medium: Three components regulated by supernova explosions in a inhomogeneous substrate. *Astrophys. J.* 218:148–169.

McKee, C. F., and Zweibel, E. G. 1992. On the virial theorem for turbulent molecular clouds. *Astrophys. J.*, in press.

McKinnon, W. B. 1984. On the origin of Triton and Pluto. *Nature* 311:355–358.

McKinnon, W. B. 1989. Impact jetting of water ice with application to the accretion of icy planetesimals and Pluto. *Geophys. Res. Lett.* 16:1237–1240.

McKinnon, W. B., and Leith, A. C. 1990. Gas drag and the orbital evolution of a captured Triton. *Icarus*, in press.

McKinnon, W. B., and Mueller, S. 1988. Pluto's structure and composition suggest origin in the solar and not a planetary nebula. *Nature* 335:240–243.

McMullin, J. P., Mundy, L. G., and Blake, G. A. 1992. Structure and chemistry of Orion-S. *Astrophys. J.*, in press.

McNamara, B. J. 1976. Pre-main-sequence masses and the age spread in the Orion cluster. *Astron. J.* 81:845–854.

McNamara, B. J., Hack, W. J., Olson, R. W., and Mathieu, R. D. 1989. A proper-motion membership analysis of stars in the vicinity of the Orion Nebula. *Astron. J.* 97:1427–1439.

McPherson, G. J., Wark, D. A., and Armstrong, J. T. 1988. Primitive material surviving in chondrites: Refractory inclusions. In *Meteorites and the Early Solar System*, eds. J. F. Kerridge and M. S. Matthews (Tucson: Univ. of Arizona Press), pp. 746–807.

McSween, H. Y. 1985. SNC meteorites: clues to martian petrologic evolution. *Rev. Geophys.* 23:391–416.

McSween, H. Y., and Weissman, P. R. 1989. Cosmochemical implications of the chemical processing of cometary nuclei. *Geochim. Cosmochim. Acta* 53:3263–3271.

Meaburn, J., and Dyson, J. E. 1987. The dynamics of Herbig-Haro objects HH 46 and 47 A and their remarkable connecting filament HH47B. *Mon. Not. Roy. Astron. Soc.* 225:863–872.

Mead, K. M., and Kutner, M. L. 1988. Molecular clouds in the outer galaxy. III. CO studies of individual clouds. *Astrophys. J.* 330:399–414.

Meakin, P. and Donn, B. 1988. Aerodynamic properties of fractal grains: implications for the primordial solar nebula. *Astrophys. J. Lett.* 329:39–42.

Mebold, U. 1989. High latitude molecular clouds and IR Cirrus. In *The Physics and Chemistry of Interstellar Molecular Clouds*, eds. G. Winnewisser and J. T. Armstrong (Berlin: Springer-Verlag), pp. 45–52.

Meech, K. J. 1991. Physical aging in comets. In *Comets in the Post-Halley Era*, eds. R. Newburn, M. Neugebauer and J. Rahe (Dordrecht: Kluwer), pp. 629–669.

Meech, K. J., and Belton, M. J. S. 1990. The atmosphere of 2060 Chiron. *Astron. J.* 100:1323–1338.

Melnick, J. 1985. The 30 Doradus nebula. I. Spectral classification of 69 stars in the central cluster. *Astron. Astrophys.* 153:235–244.

Melnick, J. 1989. High-mass star formation. *The Messenger* 57:4–6.

Melosh, H. J. 1981. Atmospheric breakup of terrestrial impactors. In *Multi-ring Basins*, eds. P. H. Schultz and R. B. Merrill (New York: Pergamon), pp. 29–35.

Melosh, H. J. 1989. *Impact Cratering: A Geologic Process* (New York: Oxford Univ. Press).

Melosh, H. J., and Sonnett, C. P. 1986. When worlds collide: jetted vapor plumes and the Moon's origin. In *Origin of the Moon*, eds. W. K. Hartmann, R. J. Phillips and G. J. Taylor (Houston: Lunar and Planet. Inst.), pp. 621–642.

Melosh, H. J., and Vickery, A. M. 1988. Atmospheric erosion by high speed impact ejecta. *EOS: Tans. AGU* 69:388 (abstract).

Melosh, H. J., and Vickery, A. M. 1989. Impact erosion of the primordial atmosphere of Mars. *Nature* 338:487–489.

Melosh, H. J., and Vickery, A. M. 1991. Melt drop formation in energetic impacts. *Nature* 350:487–489.

Mendoza, E. E. 1966. Infrared photometry of T Tauri stars and related objects. *Astrophys. J.* 143:1010–1014.

Mendoza, E. E. 1968. Infrared excess in T Tauri stars and related objects. *Astrophys. J.* 151:977–989.

Menten, K. M., and Walmsley, C. M. 1985. Ammonia observations of L1551. *Astron. Astrophys.* 146:369–374.

Menten, K. M., Serabyn, E., Gusten, R., and Wilson, T. L. 1987. Physical conditions in the *IRAS* 16293–2422 parent cloud. *Astron. Astrophys.* 177:L57–L60.

Menten, K. M., Walmsley, C. M., Henkel, C., and Wilson, T. L. 1988. Methanol in the Orion region. I. Millimeter-wave observations. *Astron. Astrophys.* 198:253–266.

Menten, K. M., Harju, J., Olano, C. A., and Walmsley, C. M. 1989. The high density molecular cores near L1551–IRS5 and B335FIR. *Astron. Astrophys.* 223:258–266.

Menten, K. M., Melnick, G. J., Phillips, T. G., and Neufeld, D. A. 1990. A new submillimeter water maser transition at 325 GHz. *Astrophys. J. Lett.* 363:27–31.

Mercer-Smith, J. A., Cameron, A. G. W., and Epstein, R. I. 1984. On the formation of stars from disk accretion. *Astrophys. J.* 279:363–366.

Mestel, L. 1965a. Problems of star formation—I. *Quart. J. Roy. Astron. Soc.* 6:161–198.

Mestel, L. 1965b. Problems of star formation—II. *Quart. J. Roy. Astron. Soc.* 6:265–298.

Mestel, L. 1966a. The magnetic field of contracting gas cloud. I. Strict flux freezing. *Mon. Not. Roy. Astron. Soc.* 133:265–284.

Mestel, L. 1966b. A note on the spin of sub-condensations forming in a differentially rotating medium. *Mon. Not. Roy. Astron. Soc.* 131:307–310.

Mestel, L. 1968. Magnetic braking by a stellar wind—I. *Mon. Not. Roy. Astron. Soc.* 138:359–391.

Mestel, L. 1985. Magnetic fields. In *Protostars and Planets II*, eds. D. C. Black and M. S. Matthews (Tucson: Univ. of Arizona Press), pp. 320–339.

Mestel, L., and Paris, R. B. 1979. Magnetic braking during star formation—III. *Mon. Not. Roy. Astron. Soc.* 187:337–356.

Mestel, L., and Paris, R. B. 1984. Star formation and the galactic magnetic field. *Astron. Astrophys.* 136:98–120.

Mestel, L., and Ray, T. P. 1985. Disk-like magneto-gravitational equilibria. *Mon. Not. Roy. Astron. Soc.* 212:275–300.

Mestel, L., and Spitzer, L., Jr. 1956. Star formation in magnetic dust clouds. *Mon. Not. Roy. Astron. Soc.* 116:505–514.

Mestel, L., and Strittmatter, P. A. 1967. The magnetic field of contracting gas cloud: II. Finite diffusion effect—an illustrative example. *Mon. Not. Roy. Astron. Soc.* 137:95–105.

Metzler, K., Bischoff, A., and Stöffler, D. 1992. Accretionary dust mantles in Cm chondrites: Evidence for solar nebula processes. *Geochim. Cosmochim. Acta* 56:2873–2897.

Mewaldt, R. A., Spalding, J. D., and Stone, E. C. 1984. A high resolution study of solar flare nuclei. *Astrophys. J.* 280:892–901.

Meyer, F., and Meyer-Hofmeister, E. 1981. On the elusive case of cataclysmic variable outbursts. *Astron. Astrophys.* 104:L10–L12.

Meyerdierks, H., Brouillet, N., and Mebold, U. 1990. High molecular abundances in galactic cirrus clouds. *Astron. Astrophys. 230:172–180.*

Mezger, P. G., and Smith, L. F. 1977. Radio observations related to star formation. In *Star Formation*, eds. T. de Jong and A. Maeder (Dordrecht: D. Reidel), pp. 133–177.

Mezger, P. G., Chini, R., Kreysa, E., Wink, J. E., and Salter, C. J. 1988. Dust emission at submillimeter wavelengths from cloud cores and protostellar condensations in NGC 2024 and S 255 IR. *Astron. Astrophys.* 191:44–56.

Mezger, P. G., Wink, J. E., and Zylka, R. 1990. λ 1.3mm dust emission from the star-forming cloud cores OMC 1 and 2. *Astron. Astrophys.* 288:95–107.

Mezger, P. G., Sievers, A. W., Haslam, C. G. T.., Kreysa, E., Lemke, R., Mauersberger, R., and Wilson, T. L. 1992. Dust emission from star forming regions. II. The NGC 2024 cloud core—revisted. *Astron. Astrophys.* 256:631–639.

Micela, G., Sciortinos, S., and Serio, S. 1985. Relationship between X-ray luminosity and Rossby number for a sample of late-type stars. In *Intl. Symp. X-Ray Astronomy*, pp. 43–36.

Michel, F. C. 1969. Relativistic stellar-wind torques. *Astrophys. J.* 158:727–738.

Michel, F. C. 1984. Hydraulic jumps in "viscous" accretion disks. *Astrophys. J.* 279:807–813.

Michels, D. J., Sheeley, N. R., Howard, R. A., and Kooman, R. J. 1982. Observations of a comet on collision course with the Sun. *Science* 215:1097–1102.

Migenes, V., Johnston, K. J., Pauls, T. A., and Wilson, T. L. 1989. The distribution and kinematics of ammonia in the Orion-KL nebula: high sensitivity *VLA* maps of the $NH_3(3,2)$ line. *Astrophys. J.* 347:294–301.

Mihalas, D. 1978. *Stellar Atmospheres* (San Francisco: Freeman).

Mihalas, D., and Binney, J. 1981. *Galactic Astronomy: Structure and Kinematics* (San Francisco: Freeman).

Miki, S. 1982. The gaseous flow around a protoplanet in the primordial solar nebula. *Prog. Theor. Phys.* 67:1053–1067.

Milani, A., and Nobili, A. M. 1985. The depletion of the outer asteroid belt. *Astron. Astrophys.* 144:261–274.

Milani, A., and Nobili, A. M. 1988. Resonant structure of the outer asteroid belt. *Celes. Mech.* 34:343–355.

Milani, A., Nobili, A. M., and Carpino, M. 1989. Dynamics of Pluto. *Icarus* 82:200–217.

Millar, T. J. 1990. Chemical modelling of quiescent dense interstellar clouds. In *Molecular Astrophysics*, ed. T. W. Hartquist (Cambridge: Cambridge Univ. Press), pp. 115–131.

Millar, T. J., and Nejad, L. A. M. 1985. Chemical modeling of molecular sources—IV. Time-dependent chemistry of dark clouds. *Mon. Not. Roy. Astron. Soc.* 217:507–522.

Millar, T. J., Leung, C. M., and Herbst, E. 1987. How abundant are complex interstellar molecules? *Astron. Astrophys.* 183:109–117.

Millar, T. J., Bennett, A., and Herbst, E. 1989. Deuterium fractionation in dense interstellar clouds. *Astrophys. J.* 340:906–920.

Millar, T. J., Herbst, E., and Charnley, S. B. 1991*a*. The formation of oxygen-containing organic molecules in the Orion compact ridge. *Astrophys. J.* 369:147–156.

Millar, T. J., Rawlings, J. N. C., Bennett, A., Brown, P. D., and Charnley, S. B. 1991*b*. Gas phase reactions and rate coefficients for use in astrochemistry. The UMIST rate file. *Astron. Astrophys. Suppl.* 87:585–619.

Miller, G. E., and Scalo, J. M. 1978. On the birthplaces of stars. *Publ. Astron. Soc. Pacific* 90:506–513.

Miller, G. E., and Scalo, J. M. 1979. The initial mass function and stellar birthrate in the solar neighborhood. *Astrophys. J. Suppl.* 41:513–547.

Miller, S. L. 1961. The occurrence of gas hydrates in the solar system. *Proc. Natl. Acad. Sci. U. S.* 47:1798–1808.

Millis, R. L., and Schleicher, D. G. 1986. Rotational period of comet Halley. *Nature* 324:646–649.

Minear, J. W. 1980. The lunar magma ocean: a transient lunar phenomenon? *Proc. Lunar Planet. Sci. Conf.* II:1941–1955.

Minh, Y. C., Irvine, W. M., and Ziurys, L. M. 1988. Observations of interstellar $HOCO^+$: Abundance enhancements toward the galactic center. *Astrophys. J.* 334:175–181.

Minh, Y. C., Irvine, W. M., and Ziurys, L. M. 1989. Detection of interstellar hydrogen sulfide in cold, dark clouds. *Astrophys. J. Lett.* 345:63–66.

Minh, Y. C., Ziurys, L. M., Irvine, W. M., and McGonagle, D. 1990. Observations of H_2S toward OMC-1. *Astrophys. J.* 360:136–141.

Minh, Y. C., Irvine, W. M., and Brewer, M. K. 1991. H_2CS abundances and ortho-to-para ratios in interstellar clouds. *Astron. Astrophys.* 244:181–189.

Minster, J.-F., Birck, J.-L., and Allègre, C. J. 1982. Absolute age of formation of chondrites studied by the Rb-87 = Sr-87 method. *Nature* 300:414–419.

Mitchell, A. K., and Nellis, W. J. 1982. Equation of state and electrical conductivity of water and ammonia shocked to 100 GPa (1 Mbar) pressure. *J. Chem. Phys.* 76:6273–6281.

Mitchell, D. L., Lin, R. P., Anderson, K. A., Carlson, C. W., Curtis, D. W., Korth, A., Réme, H., Sauvaud, J. A., d'Uston, C., and Mendis, D. A. 1987. Evidence for chain molecules enriched in carbon, hydrogen, and oxygen in comet Halley. *Science* 237:626–628.

Mitchell, D. L., Lin, R. P., Anderson, K. A., Carlson, C. W., Curtis, D. W., Korth, A., Réme, H., Sauvaud, J. A., d'Uston, C., and Mendis, D. A. 1989. Complex organic ions in the atmosphere of comet Halley. *Adv. Space Res.* 9(2):35–39.

Mitchell, G. F. 1984. Effects of shocks on the molecular composition of a dense interstellar cloud. *Astrophys. J.* 54:81–101.

Mitchell, G. F., and Hasegawa, T. I. 1991. An extremely high velocity CO outflow from NGC 7358 IRS 9. *Astrophys. J. Lett.* 371:33–36.

Mitchell, G. F., and Watt, G. D. 1985. Molecular abundances in shocked diffuse clouds: results from time-dependent modelling and comparison with diffuse cloud abundances. *Astron. Astrophys.* 151:121–130.

Mitchell, G. F., Allen, M. and Maillard, J.-P. 1988. The ratio of solid to gas-phase CO in the line of sight to W33A. *Astrophys. J. Lett.* 333:55–58.

Mitchell, G. F., Allen, M., Beer, R., Dekany, R., Huntress, W., and Maillard, J.-P. 1988*a*. The detection of a discrete outflow from the young stellar object GL 490. *Astron. Astrophys.* 201:L16–L18.

Mitchell, G. F., Allen, M., Beer, R., Dekany, R., Huntress, W., and Maillard, J.-P. 1988*b*. The detection of high-velocity outflows from M8E-IR. *Astrophys. J. Lett.*

327:17–21.

Mitchell, G. F., Curry, C., Maillard, J.-P., and Allen, M. 1989. The gas environment of the young stellar object GL 2591 studied by infrared spectroscopy. *Astrophys. J.* 341:1020–1034.

Mitchell, G. F., Maillard, J.-P., Allen, M., Beer, R., and Belcourt, K. 1990. Hot and cold gas toward young stellar objects. *Astrophys. J.* 363:554–573.

Mitchell, J. B. A. 1990. The dissociative recombination of molecular ions. *Phys. Repts.* 186:215–248.

Miyama, S. M. 1989*a*. Criteria for collapse and fragmentation of rotating clouds. In *Highlights of Astronomy*, vol. 8, ed. D. McNally (Dordrecht: Kluwer), pp. 127–128.

Miyama, S. M. 1989*b*. Formation of the proto-sun and the evolution of the solar nebula mechanism of angular momentum transfer. In *The Formation and Evolution of Planetary Systems*, eds. H. A. Weaver and L. Danly (Cambridge: Cambridge Univ. Press), pp. 284–290.

Miyama, S. M., Hayashi, C., and Narita, S. 1984. Criteria for collapse and fragmentation of rotating, isothermal clouds. *Astrophys. J.* 279:621–632.

Miyawaki, R., Hasegawa, T., Hayashi, M. 1988. Diffuse molecular gas toward W49A. *Publ. Astron. Soc. Japan* 40:69–78.

Mizuno, H. 1980. Formation of the giant planets. *Prog. Theor. Phys.* 64:544–557.

Mizuno, H. 1989. Grain growth in the turbulent accretion disk solar nebula. *Icarus* 80:189–201.

Mizuno, H., and Nakazawa, K. 1988. *Prog. Theor. Phys.* 96:266.

Mizuno, H., and Wetherill, G. W. 1984. Grain abundance in the primordial atmosphere of the Earth. *Icarus* 59:74–86.

Mizuno, H., Nakazawa, K., and Hayashi, C. 1978. Instability of gaseous envelope surrounding a planetary core and formation of giant planets. *Prog. Theor. Phys.* 60:699–710.

Mizuno, H., Nakazawa, K., and Hayashi, C. 1980. Dissolution of the primordial rare gases into the molten Earth's material. *Earth Planet. Sci. Lett.* 50:202–210.

Mizuno, H., Markiewicz, W., and Völk, H. 1988. Grain growth in turbulent photo-planetary accretion disks. *Astron. Astrophys.* 195:183–192.

Mizuno, A., Fukui, Y., Iwata, T., Nozawa, S., and Takano, T. 1990. A remarkable multilobe molecular outflow: ρ Ophiuchi East, associated with *IRAS* 16293–2422. *Astrophys. J.* 356:184–194.

Mizutani, H., Matsui, T., and Takeuchi, H. 1972. Accretion process of the Moon. *Moon* 4:476–489.

Moffatt, K. H. 1978. *Magnetic Field Generation in Electrically Conducting Fluids* (Cambridge: Cambridge Univ. Press).

Monaghan, J. J., and Lattanzio, J. C. 1991. A simulation of the collapse and fragmentation of cooling molecular clouds. *Astrophys. J.* 375:177–189.

Moneti, A., and Zinnecker, H. 1991. Infrared imaging photometry of binary T Tauri stars. *Astron. Astrophys.* 242:428–432.

Moneti, A., Pipher, J. L., Helfer, H. L., McMillan, R. S., and Perry, M. L. 1984. Magnetic field structure in the Taurus dark cloud. *Astrophys. J.* 282:508–515.

Monin, A. S., and Yaglom, A. M. 1981. *Statistical Fluid Mechanics* (Cambridge, Mass.: MIT Press).

Monin, J.-L., Pudritz, R. E., and Rouan, D., and Lacombe, F. 1990. Infrared images of HL Tauri: Scattering from an inclined, flaring disk. *Astron. Astrophys.* 215:L1–L14.

Montmerle, T. 1987. Stellar vs. solar activity: the case of pre-main sequence stars. In *Solar and Stellar Physics*, eds. E. H. Schröter and M. Schüssler, *Lecture Notes in Physics*, vol. 292 (Berlin: Springer-Verlag), pp. 117–138.

Montmerle, T., and André, P. 1988. X-rays, radio emission, and magnetism in low mass young stars. In *Formation and Evolution of Low-Mass Stars*, eds. A. K. Dupree and M. T. V. T. Lago (Dordrecht: Kluwer), pp. 225–246.

Montmerle, T., and André, P. 1989. The evolutionary status of weak-line T Tauri stars. In *ESO Workshop on Low Mass Star Formation and Pre-Main Sequence Objects*, ed. B. Reipurth (Garching: European Southern Obs.), pp. 407–422.

Montmerle, T., Koch-Miramond, L., Falgarone, E., and Grindlay, J. E. 1983*a*. *Einstein* observations of the Rho Ophiuchi dark cloud: An X-ray Christmas tree. *Astrophys. J.* 269:182–201.

Montmerle, T., Koch-Miramond, L., Falgarone, E., and Grindlay, J. E. 1983*b*. X-ray variability of pre-main sequence objects associated with the Rho Oph dark cloud. *Physica Scripta* T7:59–61.

Mooney, T. J., and Solomon, P. M. 1988. Star formation rates and the far-infrared luminosity of giant molecular clouds. *Astrophys. J. Lett.* 334:51–54.

Moore, M. H. 1981. Studies of Proton-Irradiated Cometary-Type Ice Mixtures. Ph. D. Thesis, Univ. of Maryland.

Moore, M. H., and Tanabe, T. 1990. Mass spectra of sputtered polyoxymethylene: implications for comets. *Astrophys. J. Lett.* 365:39–42.

Moore, M. H., Donn, B., Khanna, R., and A'Hearn, M. 1983. Studies of proton-irradiated ice mixtures. *Icarus* 54:388–392.

Moore, M. H., Donn, B., and Hudson, R. L. 1988. Vaporization of ices containing S_2: implications for comets. *Icarus* 74:399–412.

Moore, T. J. T., Mountain, C. M., Yamashita, T., and McLean, I. S. 1991. High-resolution $1-2\mu$m imaging polarimetry of W 75 N. *Mon. Not. Roy. Astron. Soc.* 248:377–388.

Morabito, L. A., Synnott, S. P., Kupferman, P. N., and Collins, S. A. 1979. Discovery of currently active extraterrestrial volcanism. *Science* 204:972.

Morbey, C. L., and Griffin, R. F. 1987. On the reality of certain spectroscopic orbits. *Astrophys. J.* 317:343–352.

Morfill, G. E. 1983. Some cosmochemical consequences of a turbulent protoplanetary cloud. *Icarus* 53:41–54.

Morfill, G. E. 1985. Physics and chemistry in the primitive solar nebula. In *Birth and Infancy of Stars* (Amsterdam: North-Holland Press), pp. 693–792.

Morfill, G. E. 1988. Protoplanetary accretion disks with coagulation and evaporation. *Icarus* 75:371–379.

Morfill, G. E., and Clayton, R. N. 1986*a*. Oxygen isotope fractionation in the primitive solar nebula: theory. *Lunar Planet. Sci.* XVII:569–570 (abstract).

Morfill, G. E. and Clayton, R. N. 1986*b*. Oxygen isotope fractionation in the primitive solar nebula: bulk properties. *Lunar and Planet. Sci.* XVII:567–568 (abstract).

Morfill, G. E., and Sterzik, M. 1990. Protoplanetary disks: Observations and physical processes. In *Formation of Stars and Planets and the Evolution of the Solar System*, ed. B. Battrick, ESA SP-315, pp. 219–227.

Morfill, G. E. and Völk, H. J. 1984. Transport of dust and vapor and chemical fractionation in the early protosolar cloud. *Astrophys. J.* 287:371–395

Morfill, G. E., Tscharnuter, W., and Völk, H. J. 1985. Dynamical and chemical evolution of the protoplanetary nebula. In *Protostars and Planets II*, eds. D. C. Black and M. S. Matthews (Tucson: Univ. of Arizona Press), pp. 493–533.

Morgan, J. A., and Bally, J. 1991. Molecular outflows in the L1641 region of Orion. *Astrophys. J.* 372:505–517.

Morgan, J. A., Snell, R. L., and Strom, K. M. 1990. Radio continuum emission from young stellar objects in L1641. *Astrophys. J.* 362:274–283.

Morgan, J. A., Schloerb, F. P., Snell, R. L., and Bally, J. 1991. Molecular outflows associated with young stellar objects in the L1641 region of Orion. *Astrophys. J.*

376:618.

Morgan, W. W., Hiltner, W. A., Neff, J. S., Garrison, R., and Osterbrock, D. E. 1965. Studies in spectral classification. III. The H-R diagrams of NGC 2244 and NGC 2264. *Astrophys. J.* 142:974–978.

Moriarty-Schieven, G. H., and Snell, R. L. 1988. High-resolution images of the L1551 molecular outflow. II. Structure and kinematics. *Astrophys. J.* 332:364–378.

Moriarty-Schieven, G. H., and Wannier, P. G. 1991. A second outflow from L1551/-IRS-5? *Astrophys. J. Lett.* 373:23–26.

Moriarty-Schieven, G. H., Snell, R. L., Strom, S. E., Schloerb, F. P., Strom, K. M., and Grasdalen, G. L. 1987. High-resolution images of the L1551 bipolar outflow: evidence for an expanding, accelerated shell. *Astrophys. J.* 319:742–753.

Moriarty-Schieven, G. H., Snell, R. L., and Hughes, V. A. 1989. CO $J = 2 - 1$ observations of the NGC 2071 molecular outflow: A wind-driven shell. *Astrophys. J.* 347:358–364.

Moroz, V. I., Combes, M., Bibring, J. P., Coron, N., Crovisier, J., Encrenaz, T., Crifo, J. F., Sanko, N., Grigoreyev, A. V., Bockelée-Morvan, D., Gispert, R., Nikolsky, Y. V., Emerich, C., Lamarre, J. M., Rocard, F., Krasnopolsky, V. A., and Owen, T. 1987. Detection of parent molecules in comet P/Halley from the IKS-Vega experiment. *Astron. Astrophys.* 187:513–518.

Morris, D. E., and Muller, R. A. 1986. Tidal gravitational forces: the infall of new comets and comet showers. *Icarus* 65:1–12.

Morris, D. H., Mutel, R. L., and Su, B. 1990. A magnetospheric model for radio emission from active, late-type binary stars. *Astrophys. J.* 362:299–307.

Morrison, D. 1977. Radiometry of satellites and the rings of Saturn. In *Planetary Satellites*, ed. J. A. Burns (Tucson: Univ. of Arizona Press), pp. 269–301.

Moss, D. L. 1973. Models for rapidly rotating pre-main-sequence stars. *Mon. Not. Roy. Astron. Soc.* 161:225–237.

Mottmann, J. 1977. Origin of the late heavy bombardment. *Icarus* 31:412–413.

Mouschovias, T. 1976a. Nonhomologous contraction and equilibria of self gravitating, magnetic interstellar clouds embedded in an intercloud medium: star formation. I. Formulation of the problem and method of solution. *Astrophys. J.* 206:753–767.

Mouschovias, T. 1976b. Nonhomologous contraction and equilibria of self gravitating, magnetic interstellar clouds embedded in an intercloud medium: star formation. II. Results. *Astrophys. J.* 207:141–158.

Mouschovias, T. 1977. A connection between the rate of rotation of interstellar clouds, magnetic fields, ambipolar diffusion, and the periods of binary stars. *Astrophys. J.* 211:147–151.

Mouschovias, T. 1978. Formation of stars and planetary systems in magnetic interstellar clouds. In *Protostars and Planets*, ed. T. Gehrels (Tucson: Univ. of Arizona Press), pp. 209–242.

Mouschovias, T. 1979. Magnetic braking of self-gravitating, oblate interstellar clouds. *Astrophys. J.* 228:159–162.

Mouschovias, T. 1987a. Star formation in magnetic interstellar clouds: I. Interplay between theory and observation. In *Physical Processes in Interstellar Clouds*, eds. G. E. Morfill and M. Scholer (Dordrecht: D. Reidel), pp. 453–489.

Mouschovias, T. 1987b. Star formation in magnetic interstellar clouds: II. Basic theory. In *Physical Processes in Interstellar Clouds*, eds. G. E. Morfill and M. Scholer (Dordrecht: D. Reidel), pp. 491–552.

Mouschovias, T. 1989. Magnetic fields in molecular clouds: regulators of star formation. In *The Physics and Chemistry of Interstellar Molecular Clouds*, eds. G. Winnewisser and J. T. Armstrong (Berlin: Springer-Verlag), pp. 297–312.

Mouschovias, T. 1991a. Single-stage fragmentation and a modern theory of star formation. In *The Physics of Star Formation and Early Stellar Evolution*, eds.

C. J. Lada and N. D. Kylafis, (Dordrecht: Kluwer), pp. 449–468.

Mouschovias, T. 1991*b*. Magnetic braking, ambipolar diffusion, cloud cores, and star formation: natural scale lengths and protostellar masses. *Astrophys. J.* 373:169–186.

Mouschovias, T., and Morton, S. A. 1991. Ambipolar diffusion, cloud cores, and star formation: two-dimensional, cylindrically symmetric contraction. I. The issues, formulation of the problem, and method of solution. *Astrophys. J.* 371:296–316.

Mouschovias, T., and Paleologou, E. V. 1979. The angular momentum problem and magnetic braking: an exact, time-dependent solution. *Astrophys. J.* 230:204–222.

Mouschovias, T., and Paleologou, E. V. 1980. Magnetic braking of an aligned rotator during star formation: an exact, time-dependent solution. *Astrophys. J.* 237:877–899.

Mouschovias, T., and Paleologou, E. V. 1981. Ambipolar diffusion in interstellar clouds: time-dependent solutions in one spatial dimension. *Astrophys. J.* 246:48–64.

Mouschovias, T., and Paleologou, E. V. 1986. The effect of ambipolar diffusion on magnetic braking of molecular cloud cores: an exact, time-dependent solution. *Astrophys. J.* 308:781–790.

Mouschovias, T., and Spitzer, L. 1976. Note on the collapse of magnetic interstellar clouds. *Astrophys. J.* 210:326–327.

Mouschovias, T., Shu, F. H., and Woodward, P. R. 1974. On the formation of interstellar cloud complexes, OB associations and giant H II regions. *Astron. Astrophys.* 33:73–77.

Mouschovias, T., Paleologou, E. V., and Fiedler, R., A. 1985. The magnetic flux problem and ambipolar diffusion during star formaton: one-dimensional collapse. II. Results. *Astrophys. J.* 291:772–797.

Mufson, S. L. 1975. The structure and stability of shock waves in a multiphase interstellar medium. II. The effect of a magnetic field, thermal conduction, and viscosity. *Astrophys. J.* 202:372–388.

Muhleman, D. O., Grossman, A. W., and Butler, B. J. 1990. Radar reflectivity of Titan. *Science* 248:975–980.

Mullan, D. J. 1985. Radio outbursts in RS Canum Venaticorum stars: Coronal heating and electron runaway. *Astrophys. J.* 295:628–633.

Mumma, M. J. 1989. Probing solar system objects at infrared wavelengths. *Infrared Physics* 29:167–174.

Mumma, M. J. and Reuter, D. C. 1989. On the identification of formaldehyde in Halley's comet. *Astrophys. J.* 344:940–948.

Mumma, M. J., Weaver, H. A., Larson, H. P., Davis, D. S., and Williams, M. 1986. Detection of water vapor in Halley's comet. *Science* 232:1523–1528.

Mumma, M. J., Weaver, H. A., and Larson, H. P. 1987. The ortho-para ratio of water in comet P/Halley. *Astron. Astrophys.* 187:419–424.

Mumma, M. J., Blass, W. E., Weaver, H. A., and Larson, H. P. 1988*a*. Measurements of the ortho-para ratio and nuclear spin temperature of water vapor in comets Halley and Wilson (1986l) and implications for their origin and evolution. *Bull. Amer. Astron. Soc.* 20:826 (abstract).

Mumma, M. J., Blass, W. E., Weaver, H. A., and Larson, H. P. 1988*b*. Measurements of the ortho-para ratio and nuclear spin temperature of water vapor in comets Halley and Wilson (1986l) and implications for their origin and evolution. In *Proc. Conf. on Formation and Evolution of Planetary Systems* (Baltimore: Space Telescope Science Inst.)

Mumma, M. J., Reuter, D. C., and Magee-Sauer, K. 1990. Heterogeneity of the nucleus of comet Halley. *Bull. Amer. Astron. Soc.* 22:1088 (abstract).

Münch, G. 1957. Interstellar absorption lines in distant stars I. Northern Milky Way.

Astrophys. J. 125:42–65.

Münch, G. and Zirin, H. 1961. Interstellar matter at large distances form the galactic plane. *Astrophys. J.* 133:11–28.

Mundt, R. 1984. Mass loss in T Tauri stars: Observational studies of the cool parts of their stellar winds and expanding shells. *Astrophys. J.* 280:749–770.

Mundt, R. 1985. Highly collimated mass outflows from young stars. In *Protostars and Planets II*, eds. D. C. Black and M. S. Matthews (Tucson: Univ. of Arizona Press), pp. 414–433.

Mundt, R. 1988. Flows and jets from young stars. In *Formation and Evolution of Low Mass Stars*, eds. A. K. Dupree and M. T. V. T. Lago (Dordrecht: Kluwer), pp. 257–279.

Mundt, R., and Fried, J. W. 1983. Jets from young stars. *Astrophys. J. Lett.* 274:83–86.

Mundt, R., and Giampapa, M. S. 1982. Observations of rapid line profile variability in the spectra of T Tauri stars. *Astrophys. J.* 256:156–167.

Mundt, R., Walter, F. M., Feigelson, E. D., Finkenzeller, U., Herbig, G. H. and Odell, A. P. 1983. Observations of suspected low-mass post-T Tauri stars and their evolutionary status. *Astrophys. J.* 269:229–238.

Mundt, R., Bührke, T., Fried, J. W., Neckel, T., Sarcander, M., and Stocke, J. 1984. Jets from young stars III. The case of Haro 6–5 B, HH 33/40, HH 19 and 1548C27. *Astron. Astrophys.* 140:17–23.

Mundt, R., Stocke, J., Strom, S. E., Strom, K. M., and Anderson, E. R. 1985. The Optical Spectrum of L1551 IRS5. *Astrophys. J. Lett.* 297:41–45.

Mundt, R., Brugel, E. W., and Bührke, T. 1987. Jets from young stars: CCD imaging, long-slit spectroscopy, and interpretation of existing data. *Astrophys. J.* 319:275–303.

Mundt, R., Ray, T.P. and Bührke, T. 1988. A close association of five jet and outflow sources in the HL Tau region. *Astrophys. J. Lett.* 333:69–72.

Mundt, R., Ray, T. P., Bührke, T., Raga, A., and Solf, J. 1990. Optical jets and outflows in the HL Tau region. *Astron. Astrophys.* 232:37–61.

Mundt, R., Ray, T., and Raga, A. 1991. Collimation of stellar jets. Constraints from the observed spatial structure. II. Observational results. *Astron. Astrophys.* 252:740–761.

Mundy, L. G., Scoville, N. Z., Baath, L. B., Masson, C. R., and Woody, D. P. 1986a. Protostellar condensations within the Orion ridge. *Astrophys. J. Lett.* 304:51–55.

Mundy, L. G., Wilking, B. A., and Myers, S. T. 1986b. Resolution of structure in the protostellar source IRAS 16293-2422. *Astrophys. J. Lett.* 311:75–80.

Mundy, L. G., Evans, N. J., Snell, R. L., and Goldsmith, P. F. 1987. Models of molecular cloud cores. III. A multitransition study of H_2CO. *Astrophys. J.* 318:392–409.

Mundy, L. G., Cornwell, T. J., Masson, C. R., Scoville, N. Z., Baath, L. B., and Johansson, L. E. B. 1988. High resolution images of the Orion molecular ridge in the CS $J = 2$ to 1 transition. *Astrophys. J.* 325:382–388.

Mundy, L. G., Wootten, H. A., and Wilking, B. A. 1990. The circumstellar structure of IRAS 16293-2422: $C^{18}O$, NH_3, and CO observations. *Astrophys. J.* 352:159–166.

Mundy, L. G., Wootten, H. A., Wilking, B. A., Blake, G. A., and Sargent, A. I. 1992. IRAS 16293-2422: A very young binary system? *Astrophys. J.* 385:306–313.

Murray, S., and Lin, D. N. C. 1991. Thermal instabilities in proto-globular clusters resulting form time-dependent potentials. *Astrophys. J.* 367:149–154.

Myers, P. C. 1978. A compilation of interstellar gas properties. *Astrophys. J.* 225:380-389.

Myers, P. C. 1983. Dense cores in dark clouds. III. Subsonic turbulence. *Astrophys. J.* 270:105–118.

Myers, P. C. 1985. Molecular cloud cores. In *Protostars and Planets II*, eds. D. C.

Black and M. S. Matthews (Tucson: Univ. of Arizona), pp. 81–103.

Myers, P. C. 1989. Physical conditions in dark clouds. In *The Physics and Chemistry of Interstellar Molecular Clouds: Millimeter and Sub-millimeter Observations*, eds. G. Winnewisser and J. T. Armstrong (Berlin: Springer-Verlag), pp. 38–44.

Myers, P. C. 1990. Molecular cloud structure, motions, and evolution. In *Molecular Astrophysics*, ed. T. W. Hartquist (Cambridge: Cambridge Univ. Press), pp. 328–342.

Myers, P. C. 1991. Clouds, cores, and stars in the nearest molecular complexes. In *Fragmentation of Molecular Clouds and Star Formation*, eds. E. Falgarone, F. Boulanger and G. Duvert (Dordrecht: Kluwer), pp. 221–228.

Myers, P. C., and Benson, P. J. 1983. Dense cores in dark clouds: II. NH_3 observations and star formation. *Astrophys. J.* 266:309–320.

Myers, P. C., and Goodman, A. A. 1988*a*. Evidence for magnetic and virial equilibrium in molecular clouds. *Astrophys. J. Lett.* 326:27–30.

Myers, P. C., and Goodman, A. A. 1988*b*. Magnetic molecular clouds: indirect evidence for magnetic support and ambipolar diffusion. *Astrophys. J.* 329:392–405.

Myers, P. C., and Goodman, A. A. 1991. On the dispersion in direction of interstellar polarization. *Astrophys. J.* 373:509–524.

Myers, P. C., Ho, P. T. P., Schneps, M. H., Chin, B., Pankonin, V., and Winnberg, A. 1978. Atomic and molecular observations of the ρ Ophiuchi dark cloud. *Astrophys. J.* 220:864.

Myers, P. C., Linke, R. A., and Benson, P. J. 1983. Dense cores in dark clouds. I. CO observations and column densities of high-extinction regions. *Astrophys. J.* 264:517–537.

Myers, P. C., Dame, T. M., Thaddeus, P., Cohen, R. S., Silverberg, R. F., Dwek, E., and Hauser, M. G. 1986. Molecular clouds and star formation in the inner galaxy: a comparison of CO, H II, and far-infrared surveys. *Astrophys. J.* 301:398–442.

Myers, P. C., Fuller, G. A., Mathieu, R. D., Beichman, C. A., Benson, P. J., Schild, R. E., and Emerson, J. P. 1987. Near-infrared and optical observations of *IRAS* sources in and near dense cores. *Astrophys. J.* 319:340–357.

Myers, P. C., Heyer, M., Snell, R. L., and Goldsmith, P. F. 1988. Dense cores in dark clouds. V. CO Outflow. *Astrophys. J.* 324:907–919.

Myers, P. C., Fuller, G. A., Goodman, A. A., and Benson, P. J. 1991*a*. Dense cores in dark clouds VI. Shapes. *Astrophys. J.* 376:561–572.

Myers, P. C., Ladd, E. F., and Fuller, G. A. 1991*b*. Thermal and nonthermal motions in dense cores. *Astrophys. J. Lett.* 372:95–98.

Myhill, E. A., and Kaula, W. M. 1990. Protostellar collapse and nebula formation: A second order model including radiative transfer. Protostars and Planets III, 5–9 March, Tucson, Arizona, Abstract Book, p. 34.

Nakagawa, Y., Nakazawa, K., and Hayashi, C. 1981. Growth and sedimentation of dust grains in the primordial solar nebula. *Icarus* 45:517–528.

Nakagawa, Y., Hayashi, C., and Nakazawa, K. 1983. Accumulation of planetesimals in the solar nebula. *Icarus* 54:361–376.

Nakagawa, Y., Sekiya, M., and Hayashi, C. 1986. Settling and growth of dust particles in a laminar phase of a low-mass solar nebula. *Icarus* 67:355–390.

Nakajima, T., Nagata, T., Nishida, M., Sato, S., and Kawara, K. 1986. A near-infrared survey of the L 1641 dark cloud. *Mon. Not. Roy. Astron. Soc.* 221:483–498.

Nakano, T. 1979. Quasistatic contraction of magnetic protostars due to magnetic flux leakage. I. Formulation and an example. *Publ. Astron. Soc. Japan* 31:697–712.

Nakano, T. 1982. Quasistatic contraction of magnetic clouds due to plasma drift. II.

The effect of grain friction. *Publ. Astron. Soc. Japan* 34:337–350.

Nakano, T. 1984. Contraction of magnetic interstellar clouds. *Fund. Cosmic Phys.* 9:139–232.

Nakano, T. 1987*a*. Formation of planets around stars of various masses. I. Formulation and a star of one solar mass. *Mon. Not. Roy. Astron. Soc.* 224:107–130.

Nakano, T. 1987*b*. The formation of planets around stars of various masses and the origin and the evolution of circumstellar dust clouds. In *Star Forming Regions: Proc. IAU Symp. 115*, eds. M. Peimbert and J. Jugaku (Dordrecht: D. Reidel), pp. 301–313.

Nakano, T. 1988. Formation of planets around stars of various masses. II. Stars of two and three solar masses and the origin and evolution of circumstellar dust clouds. *Mon. Not. Roy. Astron. Soc.* 230:551–571.

Nakano, T. 1989. Conditions for the formation of massive stars through nonspherical accretion. *Astrophys. J.* 345:464–471.

Nakano, T., and Umebayashi, T. 1986*a*. Dissipation of magnetic fields in very dense interstellar clouds—I. Formulation and conditions for efficient dissipation. *Mon. Not. Roy. Astron. Soc.* 218:663–684.

Nakano, T., and Umebayashi, T. 1986*b*. Dissipation of magnetic fields in very dense interstellar clouds—II. Final phases of star formation and the magnetic flux of a newborn star. *Mon. Not. Roy. Astron. Soc.* 221:319–338.

Nakano, T., and Umebayashi, T. 1988. Effects of magnetic fields on star formation. *Prog. Theor. Phys. Suppl.* 96:73–84.

Nakano, M., and Yoshida, S. 1986. Molecular line observations of the S235B region. *Publ. Astron. Soc. Japan* 38:531–545.

Nakazawa, K., and Ida, S. 1988. Hill's approximation in the three-body problems. *Prog. Theor. Phys. Suppl.* 96:167–174.

Nakazawa, K., and Nakagawa, Y. 1982. Origin of the solar system: Planetary growth in the gaseous nebula. *Suppl. Prog. Theor. Phys.* 70:11–34.

Nakazawa, K., Ida, S., and Nakagawa, Y. 1989. Collision probability of planetesimals revolving in the solar gravitational field. I. Basic formulation. *Astron. Astrophys.* 220:293–300.

Narayan, R. 1987. The birthrate and spin period of single radio-pulsars. *Astrophys. J.* 319:162–179.

Narayan, R., Goldreich, P., and Goodman, J. 1987. Physics of modes in a differentially rotating system-analysis of the shearing sheet. *Mon. Not. Roy. Astron. Soc.* 228:1–41.

Nash, D. B., and Howell, R. R. 1989. Hydrogen sulfide on Io: evidence from telescopic and laboratory infrared spectra. *Science* 244:454–457.

Natta, A. 1989. T Tauri winds and their relationship to molecular outflows. In *ESO Workshop on Low Mass Star Formation and Pre-Main Sequence Objects*, eds. B. Reipurth (Garching: European Southern Obs.), pp. 365–384.

Natta, A., and Giovanardi, C. 1990. Sodium lines in T Tauri stars: Diagnostics of pre-main sequence winds. *Astrophys. J.* 356:646–661.

Natta, A., Giovanardi, C., and Palla, F. 1988. Ionizaion structure and emission of winds from low-luminosity pre-main-sequence stars. *Astrophys. J.* 332:921–939.

Nazakura, T. 1990. In homogeneous distribution of interstellar clouds induced by collisional fragmentations. *Mon. Not. Roy. Astron. Soc.* 243:543–552.

Neckel, T., and Staude, J. 1987. Temporal changes of the IRS 5 jet in L 1551. *Astrophys. J. Lett.* 322:27–30.

Nedoluha, G. E., and Watson, W. D. 1990*a*. Spectra of circularly polarized radiation from astrophysical OH masers. *Astrophys. J.* 361:653–662.

Nedoluha, G. E., and Watson, W. D. 1990*b*. Modifications to the relationship between the magnetic field and weak zeeman features in the spectra of astrophysical masers. *Astrophys. J. Lett.* 361:53–55.

Nedoluha, G. E., and Watson, W. D. 1990c. Linearly polarized radiation from astrophysical masers due to fields of intermediate strength. *Astrophys. J.* 354:660–675.

Nedoluha, G. E., and Watson, W. D. 1991. Spectral line profiles and luminosities of astrophysical water masers. *Astrophys. J. Lett.* 367:63–67.

Nerney, S. F., and Suess, S. T. 1975. Restricted three-dimensional stellar wind modelling. I. Polytropic case. *Astrophys. J.* 196:837–847.

Neufeld, D. A., and Dalgarno, A. 1989. Fast molecular shocks. I. Reformation of molecules behind a dissociative shock. *Astrophys. J.* 340:869–893.

Neukum, G., and Wise, D. U. 1976. Mars: a standard crater curve and possible new time scale. *Science* 194:1381–1387.

Newsom, H. E., and Taylor, S. R. 1989. Geochemical implications of the formation of the Moon by a single giant impact. *Nature* 338:29–34.

Newton, H. A. 1893. On the capture of comets by planets, especially their capture by Jupiter. *Mem. Natl. Acad. Sci.* 6:7–23.

Ney, E. P. 1982. Optical and infrared observations of bright comets in the range 0.5 μm to 20 μm. In *Comets*, ed. L. L. Wilkening (Tucson: Univ. of Arizona Press), pp. 323–340.

Nicholson, P. D., and Porco, C. C. 1988. A new constraint on Saturn's zonal gravity harmonics from Voyager observations of an eccentric ringlet. *J. Geophys. Res.* 93:10209–10224.

Nisenson, P., Stachnik, R. V., Karovska, M., and Noyes, R. 1985. A new optical source associated with T Tauri. *Astrophys. J. Lett.* 297:17–20.

Nishi, R., Nakano, T., and Umebayashi, T. 1991. Magnetic flux loss from interstellar clouds with various grain-size distributions. *Astrophys. J.* 368:181–194.

Nishida, S. 1983. Collisional processes of planetesimals with a protoplanet under the gravity of the proto-sun. *Prog. Theor. Phys.* 70:93–105.

Nobili, A. M. 1989. Dynamics of the outer asteroid belt. In *Asteroids II*, eds. R. P. Binzel, T. Gehrels and M. S. Matthews (Tucson: Univ. of Arizona Press), pp. 862–879.

Nobili, A. M., Milani, A., and Carpino, M. 1989. *Astron. Astrophys.* 210:313–336.

Nofar, I., Shaviv, G., and Starrfield, S. G. 1991. The formation of ^{26}Al in nova explosions. *Astrophys. J.* 369:440–450.

Nord, G. L., Jr., Huebner, J. S., and McGee, J. J. 1982. Thermal history of a Type A Allende inclusion. *EOS: Trans. Amer. Geophys. Union* 63:462 (abstract).

Norgaard, H. 1980. ^{26}Al from red giants. *Astrophys. J.* 236:895–898.

Norman, C. A., and Ikeuchi, S. 1989. The disk-halo interaction: Superbubbles and the structure of the interstellar medium. *Astrophys. J.* 345:372–383.

Norman, C. A., and Silk, J. 1979. Interstellar bullets: H_2O masers and Herbig-Haro objects. *Astrophys. J.* 228:197–205.

Norman, C. A., and Silk, J. 1980a. Clumpy molecular clouds: A dynamic model self-consistently regulated by T Tauri star formation. *Astrophys. J.* 238:158–174.

Norman, C. A., and Silk, J. 1980b. The evolution of giant molecular clouds. In *Interstellar Molecules: Proc. IAU Symp. 87*, ed. B. H. Andrew (Dordrecht: D. Reidel), pp. 137–149.

Norman, M. L. 1986. Interpretation of extragalactic jets. In *Radiation Hydrodynamics in Stars and Compact Objects*, eds. D. Mihalas and K.-H. A. Winkler (Berlin: Springer-Verlag), pp. 425–437.

Norman, M. L., and Wilson, J. R. 1978. The fragmentation of isothermal rings and star formation. *Astrophys. J.* 224:497–511.

Norman, M. L., Smarr, L., Winkler, K.-H. A., and Smith, M. D. 1982. Structure and dynamics of supersonic jets. *Astron. Astrophys.* 113:285–302.

Norman, M. L., Smarr, L., and Winkler, K.-H. A. 1985. Fluid dynamical mechanisms for knots in astrophysical jets. In *Numerical Astrophysics*, eds. J..Centrella, J. LeBlanc and R. Bowers (Boston: Jones and Bartlett) pp. 88–125.

Norris, R. P. 1984. MERLIN observations of OH masers outflows in Orion-KL. *Mon. Not. Roy. Astron. Soc.* 207:127–138.

Novak, G., Gonatas, D. P., Hildebrand, R. H., Platt, S. R., and Dragovan, M. 1989. Polarization of far-infrared radiation from molecular clouds. *Astrophys. J.* 345:802–810.

Novak, G., Predmore, C. R., and Goldsmith, P. F. 1990. Polarization of the $\lambda = 1.3$ millimeter continum radiation from the Kleinmann-Low nebula. *Astrophys. J.* 355:166–171.

Nozakura, T. 1990. Inhomogeneous distribution of interstellar clouds induced by collisional fragmentation. *Mon. Not. Roy. Astron. Soc.* 243:543–552.

Nozawa, S., Mizuno, A., Teshima, Y., Ogawa, H., and Fukui, Y. 1991. A study of ^{13}CO cores in Ophiuchus. *Astrophys. J. Suppl.* 77:647–675.

Null, G. W., Lau, E. L., Biller, E. D., and Anderson, J. D. 1981. Saturn gravity results obtained from Pioneer 11 tracking data and Earth-based Saturn satellite data. *Astron. J.* 86:456–468.

Nuth, J. A., and Moore, M. H. 1988. SiH and the unidentified 4.6 micron feature. *Astrophys. J. Lett.* 329:113–116.

Nyquist, L. E., Bogard, D. D., Wiesmann, H., Bansal, B. M., Shih, C.-Y., and Morris, R. V. 1990. Age of a eucrite clast from the Bholghati howardite. *Geochim. Cosmochim. Acta* 54:2195–2206.

Nyquist, L. E., Wiesmann, H., Bansal, B., Shih, C.-Y., and Harper, C. L. 1991. ^{53}Mn and ^{146}Sm: Alive and well in an angrite magma. *Lunar Planet. Sci.* XXII:989–990 (abstract).

O'Dell, C. R. 1973. A new model for cometary nuclei. *Icarus* 19:137–146.

Odenwald, S. F. 1986a. A search for infrared excesses in G-type stars. In *Light on Dark Matter*, ed. F. P. Israel (Dordrecht: D. Reidel), pp. 75–76.

Odenwald, S. F. 1986b. An *IRAS* survey of IR excesses in G-type stars. *Astrophys. J.* 307:711–722.

Ögelman, H., Krautter, J., and Beuermann, K. 1987. EXOSTAT observations of X-rays from classical novae during the outburst stage. *Astron. Astrophys.* 177:110–116.

Ogura, K. 1990. Two Herbig-Haro objects discovered by narrow-band CCD imagery. *Publ. Astron. Soc. Pacific* 102:1366–1371.

Ogura, K., and Walsh, J. R. 1991. Five new Herbig-Haro objects in the Orion region. *Astron. J.* 101:185–195.

Ohashi, N., Kawabe, R., Hayashi, M., and Ishiguro, H. 1991. Observations of 11 protostellar sources in Taurus with Nobeyama Millimeter Array: growth of circumstellar disks. *Astron. J.* 102:2054–2065.

Ohishi, M., Kaifu, N., Kawaguchi, K., Murakami, A., Saito, S., Yamamoto, S., Isitikawa, S., Fujita, Y., Sitiratori, Y., and Irvine, W. 1989. Detection of a new circumstellar carbon chain molecule, C_4Si. *Astrophys. J. Lett.* 345:83–86.

Ohtsuki, K. 1992a. Capture probability of colliding planetesimals: constraints on the formation of planets and satellites by accretion. *Icarus*, submitted.

Ohtsuki, K. 1992b. Evolution of random velocities of planetesimals in the course of accretion. *Icarus* 98:20–27.

Ohtsuki, K., and Ida, S. 1990. Runaway planetary growth with collision rate in the solar gravitational field. *Icarus* 85:499–511.

Ohtsuki, K., and Nakagawa, Y. 1988. Dissipation of the solar nebula. *Prog. Theor. Phys.* 96:161–165.

Ohtsuki, K., Nakagawa, Y., and Nakazawa, K. 1988. Growth of the Earth in nebular gas. *Icarus* 75:552–565.

Ohtsuki, K., Nakagawa, Y., and Nakazawa, K. 1990. Artificial acceleration in accumulation due to coarse mass-coordinate divisions in numerical simulation. *Icarus* 83:205–215.

Oikawa, S., and Everhart, E. 1979. Past and future orbit of 1977 UB, object Chiron. *Astron. J.* 84:134–139.

Ojakangas, G. W., and Stevenson, D. J. 1989. Thermal state of an ice shell on Europa. *Icarus* 81:220–241.

Okazaki, A.T. and Kato, S. 1985. Adiabatic oscillations of a nonself-gravitating polytropic disk. I. A disk with constant thickness. *Publ. Astron. Soc. Japan* 37:683–696.

O'Keefe, J. D., and Ahrens, T. J. 1982. The interaction of the Cretaceous/Tertiary extinction bolide with the atmosphere, ocean, and solid Earth. In *Geological Implications of Impacts of Large Asteroids and Comets on the Earth*, eds. L. T. Silver and P. H. Schultz, Geol. Soc. Amer. Special Paper 190, pp. 103–120.

Olano, C. A. 1982. On a model of local gas related to Gould's belt. *Astron. Astrophys.* 112:195–208.

Olano, C. A., and Pöppel, W. G. L. 1987. Kinematic origin of the dark clouds in Taurus and of some nearby galactic clusters. *Astron. Astrophys.* 179:202–218.

Olano, C. A., Walmsley, C. M., and Wilson, T. L. 1988. The relative distribution of NH_3, HC_7N and C_4H in the Taurus Molecular Cloud 1 (TMC 1). *Astron. Astrophys.* 196:194–200.

Olberg, M., Reipurth, B., and Booth, R. S. 1989. First results from observations of southern star forming regions with the Swedish ESO submillimetre telescope. In *The Physics and Chemistry of Interstellar Molecular Clouds*, eds. G. Winnewisser and J. T. Armstrong (Heidelberg: Springer-Verlag), pp. 120–123.

Omont, A. 1986. Physics and chemistry of interstellar polycyclic aromatic molecules. *Astron. Astrophys.* 164:159–178.

O'Neal, D. B., Feigelson, E. D., Myers, P. C., and Mathieu, R. D. 1990. A VLA survey of weak T Tauri stars in Taurus-Auriga. *Astron. J.* 100:1610–1617.

Oort, J. H. 1950. The structure of the cloud of comets surrounding the solar system and a hypothesis concerning its structure. *Bull. Astron. Inst. Netherlands* 11:91–110.

Oort, J. H. 1954. Outline of a theory on the origin and acceleration of interstellar clouds and O associations. *Bull. Astron. Inst. Netherlands* 12:177–186.

Öpik, E. J. 1951. Collision probabilities with the planets and the distribution of interplanetary matter. *Proc. Roy. Irish Acad.* 54A:165–199.

Öpik, E. J. 1953. Stellar associations and supernovae. *Irish Astron. J.* 2:219–233.

Orszag, S. A., and Kells, L. C. 1980. Transition to turbulence in plane Poiseuille and plane Couette flow. *J. Fluid Mech.* 96:159–205.

Osaki, Y. 1989. A model for the superoutburst phenomenon of SU Ursae Majoris stars. *Publ. Astron. Soc. Japan* 41:1005–1033.

Osip, D. J., Millis, R. L., and Schleicher, D. G. 1992. Comets: Ground-based observations of spacecraft mission candidates. *Icarus* 98:115–124.

Ostriker, E., Shu, F. H., and Adams, F. C. 1991. Near-resonant excitation of eccentric density waves by external forcing. *Astrophys. J.*, in preparation.

Ott, U. 1988. Noble gases in SNC meteorites: Shergotty, Nakhla, Chassigny. *Geochim. Cosmochim. Acta* 52:1937–1948.

Ott, U., and Begemann, F. 1990a. Discovery of s-process barium in the Murchison meteorite. *Astrophys. J. Lett.* 353:57–60.

Ott, U., and Begemann, F. 1990b. S-process material in Murchison: Sr and more on Ba. *Lunar Planet. Sci.* XXI:920–921 (abstract).

Ott, U., Mack, R., and Chang, S. 1981. Noble-gas-rich separates from the Allende meteorite. *Geochim. Cosmochim. Acta* 45:1751–1788.

Ott, U., Begemann, F., Yang, J., and Epstein, S. 1988a. S-process Kr and ^{22}Ne in Murchison: A correlation. *Lunar Planet. Sci.* XIX:895–896 (abstract).

Ott, U., Begemann, F., Yang, J., and Epstein, S. 1988b. S-process krypton of variable isotopic composition in the Murchison meteorite. *Nature* 332:700–702.

Owen, T. 1985. The atmospheres of icy bodies. In *Ice in the Solar System*, ed. J. Klinger, D. Benest, A. Dollfus and R. Smolchowski (Dordrecht: D. Reidel), pp. 731–740.

Owen, T., Maillard, J. P., DeBergh, C., and Lutz, B. L. 1989. Deuterium on Mars: the abundance of HDO and the value of D/H. *Science* 240:1767–1770.

Owen, T., Bar-Nun, A., and Kleinfeld, I. 1991. Noble gases in terrestrial planets: evidence for cometary impacts? In *Comets in the Post Halley Era*, eds. R. Newburn, M. Neugebauer, and J. Rahe (Dordrecht: Kluwer), pp. 429–438.

Owen, T., Geballe, T., de Bergh, C., Young, L., Elliot, J., Cruikshank, D., Roush, T., Schmitt, B., Brown, R. H., and Green, J. 1992. Detection of nitrogen and carbon monoxide on the surface of Pluto. *Bull. Amer. Astron. Soc.* 24, in press.

Owen W. W., Jr., Vaughan, R. M., and Synnott, S. P. 1991. Orbits of the six new satellites of Neptune. *Astron. J.* 101:1511–1515.

Ozima, M., and Igarashi, G. 1989. Terrestrial noble gases: constraints and implications on atmospheric evolution. In *Origin and Evolution of Planetary Atmospheres*, eds. S. K. Atreya, J. B. Pollack and M. S. Matthews (Tucson: Univ. of Arizona Press), pp. 306–327.

Ozima, M., and Nakazawa, K. 1980. Origin of rare gases on the Earth. *Nature* 284:313–316.

Ozima, M., and Podosek, F. A. 1983. *Noble Gas Geochemistry* (Cambridge: Cambridge Univ. Press).

Paczyński, B. 1977. A model of accretion disks in close binaries. *Astrophys. J.* 216:822–826.

Paczyński, B. 1978a. A model of a self gravitating accretion disk. *Acta Astron.* 28:91–109.

Paczyński, B. 1978b. Ion viscosity in hot accretion disks. *Acta Astron.* 28:253–274.

Paczyński, B. 1991. A polytropic model of an accretion disk, a boundary layer, and a star. *Astrophys. J.* 370:597–603.

Padin, S., Sargent, A. I., Mundy, L. G., Scoville, N. Z., Woody, D. P., Leighton, R. B., Scott, S. L., Seling, T. V., Stapelfeldt, K. R., and Terebey, S. 1989. Interferometer $C^{18}O$ observations of DR 21(OH) and L1551 IRS 5 at $\lambda = 1.4$ millimeters. *Astrophys. J. Lett.* 337:45–48.

Palla, F. 1991. Theoretical and observational aspects of young stars of intermediate mass. In *Fragmentation of Molecular Clouds and Star Formation: Proc. IAU Symp. 147*, ed. E. Falgarone (Dordrecht: Kluwer), in press.

Palla, F., and Stahler, S. W. 1990. The birthline for intermediate-mass stars. *Astrophys. J. Lett.* 360:47–50.

Palla, F., and Stahler, S. W. 1991. The evolution of intermediate-mass protostars. I. Basic results. *Astrophys. J.* 375:288–299.

Palme, H., and Fegley, B., Jr. 1990. High temperature condensation of iron-rich olivine in the solar nebula. *Earth Planet. Sci. Lett.* 101:180–195.

Palme, H., and Wlotzka, F. 1976. A metal particle from a Ca,Al-rich inclusion from the meteorite Allende and the condensation of refractory siderophile elements. *Earth Planet. Sci. Lett.* 33:45–60.

Palme, H., Wlotzka, F., Nagel, K., and El Goresy, A. 1982. An ultra-refractory inclusion from the Ornans carbonaceous chondrite. *Earth Planet. Sci. Lett.* 61:1–12.

Palme, H., Larimer, J. S., and Lipschutz, M. E. 1988. Moderately volatile elements. In *Meteorites and the Early Solar System*, eds. J. F. Kerridge and M. S. Matthews (Tucson: Univ. of Arizona Press), pp. 436–458.

Palouš, J. 1986. The local velocity field in the last billion years. In *The Galaxy and the*

Solar System, eds. R. Smoluchowski, J. N. Bahcall and M. S. Matthews (Tucson: Univ. of Arizona Press), pp. 47–57.

Palouš, J. 1987. The Local kinematics of young objects. In *Evolution of Galaxies*, ed. L. Perek (Prague: Astronomical Inst. of Czechoslovakia), pp. 209–216.

Palouš, J., and Houck, B. 1986. The Sirius supercluster. *Astron. Astrophys.* 162:54–61.

Palouš, J., Franco, J., and Tenorio-Tagle, G. 1990. The evolution of superstructures expanding in differentially rotating disks. *Astron. Astrophys. 227:175–182.*

Panagia, N., and Felli, M. 1975. The spectrum of the free-free radiation from extended envelopes. *Astron. Astrophys.* 39:1–5.

Pandey, A. K., Paliwal, D. C., and Mahra, H. S. 1990. Star formation efficiency in clouds of various mass. *Astrophys. J.* 362:165–167.

Papaloizou, J. C. B., and Lin, D. N. C. 1984. On the tidal interaction between protoplanets and the primordial solar nebula. I. Linear calculation of the role of angular momentum exchange. *Astrophys. J.* 285:818–834.

Papaloizou, J. C. B., and Lin, D. N. C. 1989. Nonaxisymmetric instabilities in thin self-gravitating rings and disks. *Astrophys. J.* 344:645–668.

Papaloizou, J. C. B., and Pringle, J. E. 1977. Tidal torques on accretion disks in close binary systems. *Mon. Not. Roy. Astron. Soc.* 181:441–454.

Papaloizou, J. C. B., and Pringle, J. E. 1984. The dynamical stability of differentially rotating discs with constant specific angular momentum. *Mon. Not. Roy. Astron. Soc.* 208:721–750.

Papaloizou, J. C. B., and Pringle, J. E. 1985. The dynamical stability of differentially rotating discs. II. *Mon. Not. Roy. Astron. Soc.* 213:799–820.

Papaloizou, J. C. B., and Pringle, J. E. 1987. The dynamical stability of differentially rotating discs. III. *Mon. Not. Roy. Astron. Soc.* 225:267–283.

Papaloizou, J. C. B., and Savonije, G. J. 1989. Non-axisymmetric instabilities in thin self-gravitating differentially rotating gaseous discs. In *Dynamics of Astrophysical Discs*, ed. J. A. Sellwood (Cambridge: Cambridge Univ. Press), pp. 103–114.

Papaloizou, J. C. B., and Savonije, G. J. 1991. Instabilities in gaseous self-gravitating discs. *Mon. Not. Roy. Astron. Soc.* 248:353–369.

Papanastassiou, D. A., and Brigham, C. A. 1989. The identification of meteorite inclusions with isotope anomalies. *Astrophys. J. Lett.* 338:37–40.

Papanastassiou, D. A., and Wasserburg, G. J. 1976. Rb-Sr age of troctolite 76535. *Proc. Lunar Sci. Conf.* 7:2035–2054.

Papoular, R. 1981. On the ice content of the KL nebula in Orion. *Astron. Astrophys. Lett.* 104:1–3.

Parenago, P. P. 1954. *Studies of Stars in the Region of the Orion Nebula* (Moscow: Moscow State Univ.).

Paresce, F. 1991. On the evolutionary status of beta Pictoris. *Astron. Astrophys.* 247:L25.

Paresce, F., and Artymowicz, P. 1989. On the nature of the Beta Pictoris circumstellar nebula. In *Structure and Dynamics of the ISM*, eds. G. Tenorio-Tagle, M. Moles and J. Melnick (Berlin: Springer-Verlag), pp. 221–226.

Paresce, F., and Burrows, C. 1987. Broad-band imaging of Beta Pictoris circumstellar disk. *Astrophys. J. Lett.* 319:23–25.

Park, C. 1990. Large N-body simulation of a universe dominated by cold dark matter. *Mon. Not. Roy. Astron. Soc.* 242:59p–61p.

Parker, D. A. 1973. The equilibrium of an interstellar magnetic gas cloud. *Mon. Not. Roy. Astron. Soc.* 163:41–65.

Parker, D. A. 1974. The equilibrium of an interstellar magnetic disk. *Mon. Not. Roy. Astron. Soc.* 168:331–344.

Parker, E. N. 1954. Tensor virial equations. *Phys. Rev.* 96:1686–1689.

Parker, E. N. 1955. The formation of sunspots from the solar toroidal field. *Astrophys. J.* 121:491–507.

Parker, E. N. 1966. The dynamical state of the interstellar gas and field. *Astrophys. J.* 145:811–833.

Parker, E. N. 1968. The dynamical state of the interstellar gas and fields. VII. Disruptive forces. *Astrophys. J.* 154:875–879.

Parker, E. N. 1979. *Cosmical Magnetic Fields* (Oxford: Oxford Univ. Press).

Parker, N. D., Padman, R., Scott, P. F., and Hills, R. E. 1988. New bipolar outflows in dark molecular clouds. *Mon. Not. Roy. Astron. Soc.* 234:67–72.

Parravano, A. 1987. Condensation of small spherical non-gravitationally bound cool clouds. *Astron. Astrophys.* 172:280–292.

Parravano, A. 1988. Self-regulating star formation in isolated galaxies: Thermal instabilities in the interstellar medium. *Astron. Astrophys.* 205:71–76.

Parravano, A. 1989. A self-regulated star formation rate as a function of global galactic parameters. *Astrophys. J.* 347:812–816.

Parravano, A., Rosenzweig, P., and Teran, M. 1990. Galactic evolution with self-regulated star formation: Stability of a simple one zone model. *Astrophys. J.* 356:100–109.

Parsamian, E. S., and Chavira, E. 1982. Catalog of Hα emission stars in the region of the Orion Nebula. *Bol. Inst. Tonantzintla* 3:69–96.

Passot, T., Pouquet, A., and Woodward, P. 1988. The plausibility of Kolmogorov-type spectra in molecular Clouds. *Astron. Astrophys.* 197:228–234.

Patterson, C. C. 1956. Age of meteorites and the Earth. *Geochim. Cosmochim. Acta* 10:230–237.

Patterson, D. B., Honda, M., and McDougall, I. 1990. Atmospheric contamination: a possible source for heavy noble gases in basalts from Loihi Seamount, Hawaii. *Geophys. Res. Lett.* 17:705–708.

Patterson, D. B, Honda, M., and McDougall, I. 1991. Contamination of Loihi magmas with atmosphere derived noble gases: a reply to comments by T. Staudacher, P. Sarda, and C. Allègre. *Geophys. Res. Lett.* 18:749–752.

Peale, S. J. 1969. Generalized Cassini's laws. *Astron. J.* 74:483–489.

Peale, S. J. 1973. Rotation of solid bodies in the solar system. *Rev. Geophys. Space Phys.* 11:767–793.

Peale, S. J. 1976. Orbital resonances in the solar system. *Ann. Rev. Astron. Astrophys.* 14:215–245.

Peale, S. J. 1977. Rotation histories of the natural satellites. In *Planetary Satellites*, ed. J. Burns (Tucson: Univ. of Arizona Press), pp. 87–112.

Peale, S. J. 1986. Orbital resonances, unusual configurations, and exotic rotation states among planetary satellites. In *Satellites*, eds. J. A. Burns and M. S. Matthews (Tucson: Univ. of Arizona Press), pp. 159–223.

Peale, S. J., and Cassen, P. 1978. Contribution of tidal dissipation to lunar thermal history. *Icarus* 36:245–269.

Peale, S. J., Cassen, P., and Reynolds, R. T. 1979. Melting of Io by tidal dissipation. *Science* 203:892–894.

Peale, S. J., Cassen, P., and Reynolds, R. T. 1980. Tidal dissipation, orbital evolution, and the nature of Saturn's inner satellites. *Icarus* 43:65–72.

Pearl, J. C., Conrath, B. J., Hanel, R. A., Pirraglia, J. A., and Coustenis, A. 1990. The albedo, effective temperature, and energy balance of Uranus, as determined from Voyager IRIS data. *Icarus* 84:12–28.

Peck, J. A., and Wood, J. A. 1987. The origin of ferrous zoning in Allende chondrule olivine. *Geochim. Cosmochim. Acta* 51:1503–1510.

Pehlemann, E., Hofmann, K. H., and Weigelt, G. 1992. Photon bias compensation in the bispectrum and speckle masking observation of R 136. In *High Resolution Imaging by Interferometry II*, eds. J. Beckers and F. Merkle (Garching: European Southern Obs.), in press.

Pellat, R., Tagger, M., and Sygnet, J. F. 1990. Swing amplification of spiral waves in low-mass disks. *Astron. Astrophys.* 231:347–353.

Pendleton, T., Werner, M. W., Capps, R., and Lester, D. 1986. Infrared reflection nebulae in Orion Molecular Cloud 2. *Astrophys. J.* 311:360–370.

Penston, M. V. 1973. Multicolor observations of stars in the vicinity of the Orion nebula. *Astrophys. J.* 183:505–534.

Pepin, R. O. 1989. Atmospheric composition: key similarities and differences. In *Origin and Evolution of Planetary Atmospheres*, eds. S. K. Atreya, J. B. Pollack and M. S. Matthews (Tucson: Univ. of Arizona Press) pp. 291–305.

Pepin, R. O. 1991. On the origin and early evolution of terrestrial planet atmospheres and meteoritic volatiles. *Icarus* 92:2–79.

Pepin, R. O., and Carr, M. H. 1992. 4. Major issues and outstanding questions. In *Mars*, eds. H. H. Keiffer, B. M. Jakosky, C. W. Snyder and M. S. Matthews (Tucson: Univ. of Arizona Press) pp. 120–143.

Pepin, R. O., and Phinney, D. 1976. Formation interval of the Earth. *Lunar Sci.* VII:682–684 (abstract).

Pepin, R. O., and Phinney, D. 1978. Components Oxenon in the Solar System. Unpublished monograph.

Perault, M., Falgarone, E., and Puget, J. L. 1985. ^{13}CO observations of cold giant molecular clouds. *Astron. Astrophys.* 152:371–386.

Perault, M., Falgarone, E., and Puget, J. L. 1986. Fragmented molecular clouds: Statistical analysis of the ^{13}CO ($J = 1 - 0$) emission distribution. *Astron. Astrophys.* 157:139–147.

Pérez, M., Thé, P. S., and Westerlund, B. E. 1987. On the distances to the young open clusters NGC 2244 and NGC 2264. *Publ. Astron. Soc. Pacific* 99:1050–1066.

Perri, F., and Cameron, A. G. W. 1974. Hydrodynamic instability of the solar nebula in the presence of a planetary core. *Icarus* 22:416–425.

Persi, P., Ferrari-Toniolo, M., Busso, M., Origlia, L., Roberto, M., Scaltriti, F., and Silvestro, G. 1990. A search for young stellar objects in southern dark clouds. *Astron. J.* 99:303–313.

Persson, S. E., Geballe, T. R., Simon, T., Lonsdale, C. J., and Baas, F. 1981. High velocity H_2 line emission in the NGC 2071 region. *Astrophys. J. Lett.* 251:85–89.

Persson, S. E., McGregor, P. J., and Campbell, B. 1988. High spatial and spectral resolution observations of the optical counterparts of GL 490 and S106/IRS 3. *Astrophys. J.* 326:339–355.

Petschek, H. E. 1964. Magnetic field annihilation. In *AAS-NASA Symposium on the Physics of Solar Flares*, ed. W. N. Hess, NASA SP-50, pp. 425–439.

Phillips, J. P., White, G. J., Rainey, R., Avery, L. W., Richardson, K. J., Griffin, M. J., Cronin, N. J., Monteiro, T., and Hilton, J. 1988. CO $J = 3$–2 and $J = 2$–1 spectroscopy and mapping of ten high velocity molecular outflow sources. *Astron. Astrophys.* 190:289–319.

Phillips, R. B., Lonsdale, C. J., and Feigelson, E. D. 1991. Milliarcsecond structure of weak-line T Tauri stars. *Astrophys. J.*, in press.

Phillips, R. J., Arvidson, R. E., Boyce, J. M., Campbell, D. B., Guest, J. E., Schaber, G. G., and Soderblom, L. A. 1991. Impact craters on Venus: initial analysis from Magellan. *Science* 252:288–297.

Phillips, T. G., and Huggins, P. J. 1981. Abundance of atomic carbon (C I) in dense interstellar clouds. *Astrophys. J.* 251:533–540.

Phillips, T. G., Scoville, N. Z., Kwan, J., Huggins, P. J., and Wannier, P. G. 1978. Detection of $H_2^{18}O$ and an abundance estimate for interstellar water. *Astrophys.*

J. Lett. 222:59–62.

Phillips, T. G., Kwan, J., and Huggins, P. J. 1980. Detection of submillimeter lines of CO (0.65 mm) and H_2O (0.79 mm). In *Interstellar Molecules: Proc. IAU Symp. 87*, eds. B. H. Andrew (Dordrecht: D. Reidel), pp. 21–24.

Phillips, T. G., van Dishoeck, E. F., and Keene, J. 1992. Interstellar H_3^+ and its relation to the O_2 and H_2O abundances. *Astrophys. J.*, in press.

Phinney, E. S. 1983. A Theory of Radio Sources. Ph. D. Thesis. Univ. of Cambridge.

Pilipp, W., Hartquist, T. W., and Morfill, G. E. 1992. Large electrical fields in acoustic waves and simulations of lightning discharges. *Astrophys. J.* 387:364–371.

Pillinter, C. 1984. Light element stable isotopes in meteorites—from grams to picograms. *Geochim. Cosmochim. Acta* 48:2739–2766.

Pinsonneault, M. H., Kawaler, S. D., Sofia, S., and Demarque, P. 1989. Evolutionary models of the rotating Sun. *Astrophys. J.* 338:424–452.

Pinter, S., Blum, J., and Grün 1989. Mechanical properties of "fluffy" agglomerates consisting of core-mantle particles. In *Proc. Intl. Workshop on the Physics and Mechanics of Cometary Materials*, eds. J. Hunt and T.D. Guyenne, ESA SP-302, pp. 215–219.

Pinto, J. P., Lunine, J. I., Kim, S.-J. and Yung, Y. L. 1986. D to H ratio and the origin and evolution of Titan's atmosphere. *Nature* 319:388–390.

Piotrowski, S. L., and Ziolkowski, K. 1970. Two dimensions of gaseous rings in close binary systems. *Astrophys. Space Sci.* 8:66–73.

Piskunov, A. E., and Malkov, O. Yu. 1991. Unresolved binaries and the stellar luminosity function. *Astron. Astrophys.* 247:87–90.

Plambeck, R. L., and Wright, M. C. H. 1987. Aperture synthesis maps of HDO emission in Orion-KL. *Astrophys. J. Lett.* 317:101–105.

Plambeck, R. L., and Wright, M. C. H. 1988. Aperture synthesis maps of molecular lines toward Orion-KL: Evidence for chemical inhomogeneities. In *Molecular Clouds in the Milky Way and External Galaxies*, eds. R. L. Dickman, R. L. Snell and J. S. Young (Münich: Springer-Verlag), pp. 182–184.

Plambeck, R. L., Wright, M. C. H., Welch, W. J., Bieging, J. H., Baud, B., Ho, P. T. P., and Vogel, S. N. 1982. Kinematics of Orion-KL: aperture synthesis maps of 86 GHz SO emission. *Astrophys. J.* 259:617–624.

Plambeck, R. L., Vogel, S. N., Wright, M. C. H., Bieging, J. H., and Welch, W. J. 1985. Millimeter-wavelength interferometry of Orion-KL. In *Proc. of the Intl. Symp. on Millimeter and Submillimeter Wave Radio Astronomy*, ed. J. Gomez-Gonzales (Granada: URSI/IRAM), pp. 235–245.

Plambeck, R. L., Wright, M. C. H., and Carlstrom, J. E. 1990. Velocity structure of the Orion-IRc2 SiO maser: evidence for an 80 AU diameter circumstellar disk. *Astrophys. J. Lett.* 348:65–68.

Podolak, M. 1977. The abundance of water and rock in Jupiter as derived from interior models. *Icarus* 30:155–162.

Podolak, M., and Cameron, A. G. W. 1974. Models of the giant planets. *Icarus* 22:123–148.

Podolak, M., and Reynolds, R. T. 1984. Consistency tests of cosmogonic theories from models of Uranus and Neptune *Icarus* 57:102–111.

Podolak, M., and Reynolds, R. T. 1985. What have we learned from modeling giant planet interiors? In *Protostars and Planets II*, eds. D. C. Black and M. S. Matthews (Tucson: Univ. of Arizona Press), pp. 847–872.

Podolak, M., and Reynolds, R. T. 1987. The rotation rate of Uranus, its internal structure, and the process of planetary accretion. *Icarus* 70:31–36.

Podolak, M., Pollack, J. B., and Reynolds, R. T. 1988. Interactions of planetesimals with protoplanetary atmospheres. *Icarus* 73:163–179.

Podolak, M., Reynolds, R. T., and Young, R. 1990. Post Voyager comparisons of the interiors of Uranus and Neptune. *Geophys. Res. Let.* 17:1737–1740.

Podolak, M., Hubbard, W. B., and Stevenson, D. J. 1991. Models of Uranus' interior and magnetic field. In *Uranus*, eds. J. Bergstrahl, E. Miner and M. S. Matthews (Tucson: Univ. of Arizona Press), pp. 29–61.

Podosek, F. A. 1970. Dating of meteorites by the high-temperature release of iodine-correlated Xe-129. *Geochim. Cosmochim. Acta* 34:341–365.

Podosek, F. A. 1978. Isotopic structures in solar system materials. *Ann. Rev. Astron. Astrophys.* 16:293–334.

Podosek, F. A. 1991. Solar gases in the Earth? *Nature* 349:106–107.

Podosek, F. A., and Swindle, T. D. 1988. Extinct radionuclides. In *Meteorites and the Early Solar System*, eds. J. F. Kerridge and M. S. Matthews (Tucson: Univ. of Arizona Press), pp. 1093–1113.

Podosek, F. A., Bernatowicz, T. J., and Kramer, F. E. 1981. Adsorption of xenon and krypton on shales. *Geochim. Cosmochim. Acta* 45:2401–2415.

Poetzel, R. 1990. Optical Investigations of Outflows from Luminous Young Stars. Ph. D. Thesis, Univ. of Heidelberg.

Poetzel, R., Mundt, R., and Ray, T. P. 1989. Z CMa: A large-scale high velocity bipolar outflow traced by Herbig-Haro objects and a jet. *Astron. Astrophys.* 224:L13–L16.

Poincaré, H. 1892. *Les Méthodes Nouvelles de la Méchanique Céleste* (Paris: Gauthiers-Villars).

Politano, M., Livio, M., Truran, J. W., and Webbink, R. F. 1990. The theoretical frequency of classical nova outbursts as a function of white dwarf mass. In *Physics of Classical Novae*, eds. A. Cassatella and R. Viotti (Berlin: Springer-Verlag), pp. 368–389.

Polk, K. S., Knapp, J. G., Stark, A. A., and Wilson, R. W. 1988. Molecules in galaxies VI: Diffuse and dense cloud contributions to the large scale CO emission in the galaxy. *Astrophys. J.* 332:432–438.

Pollack, J. B., and Black, D. C. 1979. Implications of the gas compositional measurements of Pioneer Venus for the origin of planetary atmospheres. *Science* 205:56–59.

Pollack, J. B., and Black, D. C. 1982. Noble gases in planetary atmospheres: Implications for the origin and evolution of atmospheres. *Icarus* 51:169–198.

Pollack, J. B., and Bodenheimer, P. 1989. Theories of the origin and evolution of the protoplanetary nebula. In *Protostars and Planets II*, eds. D. C. Black and M. S. Matthews (Tucson: Univ. of Arizona Press), pp. 493–533.

Pollack, J. B., and Consolmagno, G. 1984. Origin and evolution of the Saturn system. In *Saturn*, eds. T. Gehrels and M. S. Matthews (Tucson: Univ. of Arizona Press), pp. 811–866.

Pollack, J. B. and Reynolds, R. T. 1974. Implications of Jupiter's early contraction history for the composition of the Galilean satellites. *Icarus* 21:248–253.

Pollack, J. B., Burns, J. A., and Tauber, M. E. 1979. Gas drag in primordial circumplanetary envelopes: A mechanism for satellite capture. *Icarus* 37:587–611.

Pollack, J. B., McKay, C., and Christofferson B. 1985. A calculation of the Rosseland mean opacity of dust grains in primordial solar system nebulae. *Icarus* 64:471–492.

Pollack, J. B., Podolak, M., Bodenheimer, P., and Christofferson, B. 1986. Planetesimal dissolution in the envelopes of the forming, giant planets. *Icarus* 67:409–443.

Pollack, J. B., Podolak, M., Hubickyj, O., Bodenheimer, P., Lissauer, J., and Greenzweig, Y. 1990. Simulations of the accretion of the giant planets. *Bull. Amer. Astron. Soc.* 22:1081 (abstract).

Pollack, J. B., Lunine, J. I., and Tittemore, W. C. 1991. Origin of the Uranian satellites. In *Uranus*, eds. J. T. Bergstralh, E. D. Miner and M. S. Matthews (Tucson: Univ.

of Arizona Press), pp. 469–512.

Popham, R., and Narayan, R. 1991. Does accretion cease when a star approaches break-up? *Astrophys. J.* 370:604–614.

Popper, D. M. 1987. A pre-main-sequence star in the detached binary EK Cephei. *Astrophys. J. Lett.* 313:81–83.

Popper, D. M., and Plavec, M. 1976. BM Orionis: the enigmatic eclipsing binary in the Trapezium. *Astrophys. J.* 205:462–471.

Pound, M. W., Bania, T. M., and Wilson, R. W. 1990. Subparsec clumping in the nearby molecular cloud MBM 12. *Astrophys. J.* 351:165–175.

Poynter, R. L., and Kakar, R. K. 1975. The microwave frequencies, line parameters, and spectral constants for $^{14}NH_3$. *Astrophys. J. Suppl.* 29:87–96.

Praderie, F., Talavera, A., Felenbok, P., Czarny, J, and Boesgaard, A. M. 1982. The chromosphere and wind of the Herbig Ae star AB Aurigae. *Astrophys. J.* 254:658–662.

Prasad, S. S., and Tarafdar, S. P. 1983. UV radiation field inside dense clouds: Its possible existence and chemical implications. *Astrophys. J.* 267:603–609.

Prasad, S. S., Tarafdar, S. P., Villere, K. R., and Huntress, W. J., Jr. 1987. Chemical evolution of molecular clouds. In *Interstellar Processes*, eds. D. Hollenbach and H. A. Thronson (Dordrecht: D. Reidel), pp. 631–666.

Pratap, P., Batrla, W., and Snyder, L.E. 1990. High resolution molecular observations of NGC 7538 IRS1. *Astrophys. J.* 351:530–537.

Pravdo, S. H., Rodríguez, L. F., Curiel, S., Cantó, J., Torrelles, J. M., Becker, R. H., and Sellgren, K. M. 1985. Detection of radio continuum emission from Herbig-Haro objects 1 and 2 and from their central exciting source. *Astrophys. J. Lett.* 293:35–38.

Prendergast, K. H., and Burbidge, G. R. 1968. On the nature of some Galactic x-ray sources. *Astrophys. J. Lett.* 151:83–88.

Prialnik, D., and Bar-Nun, A. 1987. On the evolution and activity of cometary nuclei. *Astrophys. J.* 313:893–905.

Prialnik, D., and Bar-Nun, A. 1988. The formation of a permanent dust mantle and its effect on cometary activity. *Icarus* 74:272–283.

Pringle, J. E. 1981. Accretion discs in astrophysics. *Ann. Rev. Astron. Astrophys.* 19:137–162.

Pringle, J. E. 1988. Accretion discs. In *Formation and Evolution of Low Mass Stars*, eds. A. K. Dupree and M. T. V. T. Lago (Dordrecht: Kluwer), pp. 153–162.

Pringle, J. E. 1989a. A boundary layer origin for bipolar flows. *Mon. Not. Roy. Astron. Soc.* 236:107–115.

Pringle, J. E. 1989b. Angular momentum loss in star formation. In *ESO Workshop on Low Mass Star Formation and Pre-Main Sequence Objects*, ed. B. Reipurth (Garching: European Southern Obs.), pp. 89–94.

Pringle, J. E. 1989c. On the formation of binary stars. *Mon. Not. Roy. Astron. Soc.* 239:361–370.

Pringle, J. E. 1991. Binary star formation. In *The Physics of Star Formation and Early Stellar Evolution*, eds. N. Kylafis and C. J. Lada (Dordrecht: Kluwer), pp. 437–447.

Pringle, J. E., Verbunt, F., and Wade, R. A. 1986. Dwarf novae in outburst: modelling the observations. *Mon. Not. Roy. Astron. Soc.* 221:169–194.

Prinn, R. G. 1982. Origin and evolution of planetary atmospheres: An introduction to the problem. *Planet. Space Sci.* 30:741–753.

Prinn, R. G. 1990. On neglect of non-linear momentum terms in solar nebula accretion disk models. *Astrophys. J.* 348:725–729.

Prinn, R. G., and Barshay, S. 1977. Carbon monoxide on Jupiter and implications for atmospheric convection. *Science* 198:1031–1034.

Prinn, R. G., and Fegley, B. 1981. Kinetic inhibition of CO and N_2 reduction in circumplanetary nebulae: Implications for satellite formation. *Astrophys. J.* 249:308–317.

Prinn, R. G., and Fegley, B. 1987. The atmospheres of Venus, Earth, and Mars: A critical comparison. *Ann. Rev. Earth Planet. Sci.* 15:171–212.

Prinn, R. G., and Fegley, B., Jr. 1989. Solar nebula chemistry: origin of planetary, satellite and cometary volatiles. In *Origin and Evolution of Planetary and Satellite Atmospheres*, eds. S. K. Atreya, J. B. Pollack and M. S. Matthews (Tucson: Univ. of Arizona Press), pp. 78–136.

Prinn, R. G., and Olaguer, E. 1981. Nitrogen on Jupiter: A deep atmospheric source. *J. Geophys. Res.* 86:9895–9899.

Prinzhofer, A., Papanastassiou, D. A., and Wasserburg, G. J. 1989. The presence of ^{146}Sm in the early solar system and implications for its nucleosynthesis. *Astrophys. J. Lett.* 344:81–84.

Prombo, C. A., and Clayton, R. N. 1985. A striking nitrogen isotope anomaly in the Bencubbin and Weatherford meteorites. *Science* 230:935–937.

Prusti, T., Clark, F. O., Whittet, D. C. B., Laureijs, R. J., and Zhang, C. Y. 1991. A study of the Chamaeleon I dark cloud and T-association. IV. Infrared objects near Cederblad 110. *Mon. Not. Roy. Astron. Soc.* 251:303–309.

Prusti, T., Whittet, D. C. B., and Wesselius, P. R. 1992. A study of the Chamaeleon I dark cloud and T-association. V. Luminosity function for member. *Mon. Not. Roy. Astron. Soc.* 254:361–368.

Puchalsky, R., Blitz, L., and Bania, T.M. 1991. In preparation.

Pudritz, R. E. 1988. The origin of bipolar outflows. In *Galactic and Extragalactic Star Formation*, eds. R. E. Pudritz and M. Fich (Dordrecht: Kluwer), pp. 135–158.

Pudritz, R. E. 1990. The stability of molecular clouds. *Astrophys. J.* 350:195–208.

Pudritz, R. E., and Norman, C. A. 1983. Centrifugally driven winds from contracting molecular disks. *Astrophys. J.* 274:677–697.

Pudritz, R. E., and Norman, C. A. 1986. Bipolar hydromagnetic winds from disks around protostellar objects. *Astrophys. J.* 301:571–586.

Purcell, E. M. 1979. Suprathermal rotation of interstellar grains. *Astrophys. J.* 231:404–416.

Quinlan, G., and Tremaine, S. 1990. Symmetric multistep methods for the numerical integration of planetary orbits. *Astron. J.* 100:1694–1700.

Quinn, T. R., and Tremaine, S. 1990. Roundoff error in long-term planetary orbit integrations. *Astron. J.* 99:1065–1070.

Quinn, T. R., Tremaine, S., and Duncan, M. J. 1990. Planetary perturbations and the origin of short-period comets. *Astrophys. J.* 355:667–679.

Quinn, T. R., Tremaine, S., and Duncan, M. J. 1991. A 3 million year integration of the Earth's orbit. *Astron. J.* 101:2287–2305.

Quirk, W. J. 1972. On the gas content of galaxies. *Astrophys. J. Lett.* 176:9–14.

Radhakrishnan, V., and Goss, W. M. 1972. The Park's survey of 21 centimeter absorption in discrete-source spectra V. Note on the statistics of absorbing HI concentrations in the galactic disk. *Astrophys. J. Suppl.* 24:161–166.

Raga, A. C. 1988. Intensity maps predicted from nonadiabatic stellar jet models. *Astrophys. J.* 335:820–828.

Raga, A. C. 1989. Shock models of Herbig-Haro objects. In *ESO Workshop on Low Mass Star Formation and Pre-Main Sequence Objects*, ed. B. Reipurth (Garching: European Southern Obs.), pp. 281–310.

Raga, A. C. 1991. A new analysis of the momentum and mass-loss rates of stellar jets. *Astron. J.* 101:1472–1475.

Raga, A. C., and Cantó, J. 1989. Collimation of stellar winds by nonadiabatic de Laval nozzles. *Astrophys. J.* 344:404–412.

Raga, A. C., Böhm, K.-H., and Solf, J. 1986. A new test of bow-shock models of Herbig-Haro objects. *Astron. J.* 92:119–124.

Raga, A. C., Mateo, M., Böhm, K.-H., and Solf, J. 1988. An interpretation of observations of HH 1 in terms of a time-dependent bow-shock model. *Astron. J.* 95:1783–1793.

Raga, A. C., Cantó, J., Binette, L., and Calvet, N. 1990. Stellar jets with intrinsically variable sources. *Astrophys. J.* 364:601–610.

Raga, A. C., Mundt, R., and Ray, T. P. 1991. Collimation of stellar jets; constraints from the observed spatial structures. I. Data analysis methods. *Astron. Astrophys.* 252:733–739.

Rand, R. J., and Kulkarni, S. R. 1989. The local galactic magnetic field. *Astrophys. J.* 343:760–772.

Rand, R. J., and Kulkarni, S. 1990. M51: Molecular spiral arms, giant molecular associations, and superclouds. *Astrophys. J. Lett.* 349:43–46.

Rawlings, J. M. C. 1988. Chemistry in the ejecta of novae. *Mon. Not. Roy. Astron. Soc.* 232:507–524.

Rawlings, J. M. C., Hartquist, T. W., Menten, K. M., and Williams, D. A. 1992. Direct diagnosis of infall in collapsing protostars—I. The theoretical identification of molecular species with broad velocity distributions. *Mon. Not. Roy. Astron. Soc.* 255:471–485.

Ray, T. P. 1981. Kelvin Helmholtz instabilities in radio jets. *Mon. Not. Roy. Astron. Soc.* 196:195–207.

Ray, T. P. 1987. CCD observations of jets from young stars. *Astron. Astrophys.* 171:145–151.

Ray, T. P., and Mundt, R. 1988. Bow shock structures near young stellar objects. In *Mass Outflows from Stars and Galactic Nuclei*, eds. L. Bianchi and R. Gilmozzi (Dordrecht: Kluwer), pp. 293–294.

Ray, T. P., Poetzel, R., Solf, J., and Mundt, R. 1990. Optical jets from the high luminosity young stars LKHα 234 and AFGL 4029. *Astrophys. J. Lett.* 357:45–48.

Ray, T. P., Eislöffel, J., and Mundt, R. 1991. The origin of wiggles in YSO jets. In preparation.

Rayet, M. 1987. On the synthesis of the proton-rich nuclei. In *Nuclear Astrophysics*, eds. W. Hillebrandt, R. Kuhfua, E. Müller, and J. W. Truran (New York: Springer-Verlag), pp. 210–221.

Raymond, J. 1979. Shock waves in the interstellar medium. *Astrophys. J. Suppl.* 39:1–27.

Rayner, J. T. 1992. Infrared imaging photometry of the S106 embedded cluster. In preparation.

Rayner, J. T., McLean, I. S., McCaughrean, M. J., and Aspin, C. 1989. Near-infrared imaging and imaging polarimetry of OMC-2. *Mon. Not. Roy. Astron. Soc.* 241:469–494.

Rayner, J. T., Hodapp, K.W., and Zinnecker, H. 1991*a*. NIR imaging of embedded OB clusters. In *Astrophysics with Infrared Arrays*, ed. R. Elston, pp. 264–266.

Rayner, J. T., McCaughrean, M. J., and Zinnecker, H. 1991*b*. An infrared image of NGC 7538. *Sky and Telescope* 82:113

Reedy, R. C., Herzog, G. F., and Jessberger, E. K. 1979. The reaction Mg(n, alpha) Ne at 14.1 and 14.7 MeV: Cross sections and implications for meteorites. *Earth Planet. Sci. Lett.* 44:341–348.

Rees, M. J. 1976. Opacity-limited hierarchical fragmentation and the masses of protostars. *Mon. Not. Roy. Astron. Soc.* 176:483–486.

Reeves, H. 1978. The "Big Bang" theory of the origin of the solar system. In *Protostars and Planets*, ed. T. Gehrels (Tucson: Univ. of Arizona Press), pp.

399–426.

Reid, M. J., and Moran, J. M. 1981. Masers. *Ann. Rev. Astron. Astrophys.* 19:231–276.

Reid, M. J., and Silverstein, E. M. 1990. OH masers and the galactic magnetic field. *Astrophys. J.* 361:483–486.

Reid, M. J., Haschick, A. D., Burke, B. F., Moran, J. M., Johnston, K. J., and Swenson, G. W., Jr. 1980. The structure of interstellar hydroxyl masers: *VLBI* synthesis observations. *Astrophys. J.* 239:89–111.

Reid, M. J., Myers, P. C., and Bieging, J. H. 1987. The circumstellar envelope of W3(OH): NH_3 observations. *Astrophys. J.* 312:830–836.

Reid, N. 1991. Unresolved binaries and the stellar luminosity function. *Astron. J.* 102:1428–1438.

Reinhard, R. 1987. The Giotto mission to Halley's comet. *Astron. Astrophys.* 187:949–955.

Reipurth, B. 1985. Optical/infrared observations of low-mass star formation regions. In *Proc. ESO-IRAM-Onsala Workshop on Sub-Millimeter Astronomy*, eds. P. A. Shaver and K. Kjar (Germany: ESO) pp. 459–472.

Reipurth, B. 1988. Pre-main-sequence binaries. In *Formation and Evolution of Low Mass Stars*, eds. A. K. Dupree and M. T. V. T. Lago (Dordrecht: Kluwer), pp. 305–318.

Reipurth, B. 1989*a*. Herbig-Haro objects in flows from young stars in Orion. *Astron. Astrophys.* 220:249–268.

Reipurth, B. 1989*b*. The HH 111 jet: Multiple outflow episodes from a young star. *Nature* 340:42–44.

Reipurth, B. 1989*c*. Observations of Herbig-Haro objects. In *ESO Workshop on Low Mass Star Formation and Pre-Main Sequence Objects*, ed. B. Reipurth (Garching: European Southern Obs.), pp. 247–280.

Reipurth, B., and Gee, G. 1986. Star formation in Bok globules and low-mass clouds. III. Barnard 62. *Astron. Astrophys.* 166:148–156.

Reipurth, B., and Graham, J. A. 1988. New Herbig-Haro objects in star-forming regions. *Astron. Astrophys.* 202:219–239.

Reipurth, B., and Heathcote, S. 1991. The jet and energy source of HH 46/47. *Astron. Astrophys.* 246:511-534.

Reipurth, B., and Olberg, M. 1991. Herbig-Haro jets and molecular outflows in L1617. *Astron. Astrophys.* 246:535–550.

Reipurth, B., Bally, J., Graham, J. A., Lane, A. P., and Zealey, W. J. 1986. The jet and energy source of HH34. *Astron. Astrophys.* 164:51–66.

Reipurth, B., Lindgren, H., Nordstrom, B., and Mayor, M. 1990. Spectroscopic pre-main-sequence binaries I. Improved orbital elements of V826 Tauri. *Astron. Astrophys.* 235:197–204.

Reuter, D. C. 1992. The contribution of methanol to the $3.4 \mu m$ emission feature in comets. *Astrophys. J.* 386:330–335.

Reuter, D. C., and Mumma, M. J. 1989. Temporal variability of the H_2CO production rates in comet Halley. *Bull. Amer. Astron. Soc.* 21:937 (abstract).

Reuter, D. C., Mumma, M. J., and Nadler, S. 1989. Infrared fluorescence efficiencies for the v_1 and v_5 bands of formaldehyde in the solar radiation field. *Astrophys. J.* 341:1045–1058.

Reuter, D. C., Hoban, S., and Mumma, M. J. 1992. An infrared search for formaldehyde in several comets. *Icarus* 95:329–332.

Reynolds, J. H. 1967. Isotopic abundance anomalies in the solar system. *Ann. Rev. Nuclear Sci.* 17:253–316.

Reynolds, R. J., and Ogden, P. M. 1979. Optical evidence for a very large expanding shell associated with the I Orion OB association, Barnard's Loop, and the high galactic latitude H α filaments in Eridanus. *Astrophys. J.* 229:942–953.

Reynolds, R. J., Roesler, F. L., and Scherb, F. 1974. The intensity distribution of diffuse H alpha emission. *Astrophys. J. Lett.* 192:53–56.

Reynolds, S. P. 1986. Continuum spectra of collimated, ionized stellar winds. *Astrophys. J.* 304:713–720.

Richards, C. N., and Dawson, G. A. 1971. The hydrodynamic instability of water drops falling and colliding in an electric field. *Arch. f. Met. Geophys. Bioklim.* A21:299–306.

Richardson, D. L., and Walker, C. F. 1987. Multivalue integration of the planetary equations over the last one-million years. In *Astrodynamics 1987*, eds. J. K. Soldner, A. K. Misra, R. E. Lindberg and W. Williamson (San Diego: Univelt), pp. 1473–1495.

Richardson, D. L., and Walker, C. F. 1989. Numerical simulation of the nine-body planetary system spanning two million years. *Astronaut. Sci.* 37:159–182.

Richardson, K. J., Sandell, G., White, G. J., Duncan, W. D., and Krisciunas, K. 1989. A high resolution millimetre and submillimetre study of W3. *Astron. Astrophys.* 221:95–99.

Richardson, S. M. 1978. Vein formation in the CI carbonaceous chondrites. *Meteoritics* 13:141–159.

Richer, H. B., Fahlman, G. G., Buananno, R., and Pecci, F. F. 1990. Low-luminosity stellar mass functions in globular clusters. *Astrophys. J. Lett.* 359:11–14.

Richer, H. B., Fahlman, G. G., Buonanno, R., Pecci, F. F., Searle, L., and Thompson, I. B. 1991. Globular cluster mass functions. *Astrophys. J.* 381:147–159.

Richer, J. S. 1990. A compact CO bipolar outflow from NGC2024 FIR6. *Mon. Not. Roy. Astron. Soc.* 245:24p–27p.

Richer, J. S., Hills, R. E., Padman, R., and Russell, A. P. G. 1989. High-resolution molecular line observations of the core and outflow in Orion B. *Mon. Not. Roy. Astron. Soc.* 241:231–246.

Richmeyer, R. D., and Morton, K. W. 1967. *Difference Methods for Initial Value Problems* (New York: Wiley Interscience).

Rickman, H. 1991. The thermal history and structure of cometary nuclei. In *Comets in the Post-Halley Era*, eds. R. L. Newburn, M. Neugebauer and J. Rahe (Dordrecht: Kluwer), pp. 733–760.

Rieke, G. H., and Lebofsky, M. J. 1985. The interstellar extinction law from 1 to 13 microns. *Astrophys. J.* 288:618–621.

Rieke, G. H., and Rieke, M. J. 1990. Possible substellar objects in the ρ Ophiuchus cloud. *Astrophys. J. Lett.* 362:21–24.

Rieke, G. H., Ashok, N. M., and Boyle, R. P. 1989. The initial mass function in the Rho Ophiuchi cluster. *Astrophys. J. Lett.* 339:71–74.

Ringwood, A. E. 1979. *Origin of Earth and Moon* (New York: Sprinter-Verlag).

Ringwood, A. E. 1990. Earliest history of the Earth-Moon system. In *Origin of the Earth*, eds. H. E. Newsom and J. J. Jones (New York: Oxford Univ. Press), pp. 101–134.

Roberts, M. S. 1957. The numbers of early-type stars in the galaxy and their relation to galactic clusters and associations. *Publ. Astron. Soc. Pacific* 69:59–64.

Roberts, W. W., and Steward, G. R. 1987. The role of orbital dynamics and cloud-cloud collisions in the formation of giant molecular clouds in global spiral structures. *Astrophys. J.* 314:10–32.

Robinson, E. L. 1976. The structure of cataclysmic variables. *Ann. Rev. Astron. Astrophys.* 14:119–142.

Roche, A. E., Cosmovici, C. B., Drapatz, S., Michel, K. W., and Wells, W. C. 1975. Search for methane in comet Kohoutek (1973f). *Icarus* 24:120–127.

Roche, P. F., and Aitken, D. K. 1984. An investigation of the interstellar extinction—I. Towards dusty WC Wolf-Rayet stars. *Mon. Not. Roy. Astron. Soc.* 208:481–492.

Roche, P. F., and Aitken, D. K. 1985. An investigation of the interstellar extinction—II. Towards the mid-infrared sources in the Galactic centre. *Mon. Not. Roy. Astron. Soc.* 215:425–435.

Rodríguez, L. F. 1988. Interstellar and circumstellar toroids. In *Galactic and Extragalactic Star Formation*, eds. M. Pudritz and M. Fich (Dordrecht: Kluwer), pp. 97–109.

Rodríguez, L. F. 1989. Radio observations of Herbig-Haro objects. *Rev. Mexicana Astron. Astrof.* 18:45–54.

Rodríguez, L. F. 1990. Bipolar outflows: evolutionary and global consideration. In *The Evolution of the Interstellar Medium*, ed. L. Blitz (San Francisco: Astron. Soc. of the Pacific), pp. 183–189.

Rodríguez, L. F., and Reipurth, B. 1989. Detection of radio continuum emission from Herbig-Haro objects 80 and 81 and their suspected energy source. *Rev. Mexicana Astron. Astrof.* 17:59–63.

Rodríguez, L. F., Cantó, J., Moreno, M. A., and López, J. A. 1989*a*. New nebular objects in the L1551 region. *Rev. Mexicana Astron. Astrof.* 17:111–114.

Rodríguez, L. F., Myers, P. C., Cruz-González, I., and Terebey, S. 1989*b*. Radio continuum observations of IRAS sources associated with dense cores. *Astrophys. J.* 347:461–467.

Rodríguez, L. F., Hartmann, L. W., and Chavira, E. 1990. Radio continuum from FU Orionis stars. *Publ. Astron. Soc. Pacific* 102:1413–1417.

Rodríguez, L. F., Ho, P. T. P., Torrelles, J. M., Curiel, S., and Cantó, J. 1990. VLA observations of the Herbig-Haro 1-2 system. *Astrophys. J.* 352:645–653.

Roettger, E. E. 1991. Comparison of Cometary Comae Using Ultraviolet Spectroscopy: Composition and Variation. Ph. D. Thesis, Johns Hopkins Univ.

Ross, M. N., and Schubert, G. 1987. Tidal heating in an internal ocean model of Europa. *Nature* 325:133–134.

Ross, M. N., and Schubert, G. 1989. Viscoelastic models of tidal heating in Enceladus. *Icarus* 78:90–101.

Roulcau, F., and Bastien, P. 1990. Collapse and fragmentation of isothermal and polytropic cylindrical clouds. *Astrophys. J.* 355:172–181.

Routley, P. M., and Spitzer, L., 1952. A comparison of the components in interstellar sodium and calcium. *Astrophys. J.* 115:227–243.

Rowe, B. R. 1988. Studies of ion-molecule reactions at $T < 80$ K. In *Rate Coefficients in Astrochemistry*, eds. T. J. Millar and D. A. Williams (Dordrecht: Kluwer), pp. 135–152.

Różyczka, M., and Spruit, H. C. 1989. In *Theory of Accretion Disks*, eds. F. Meyer, W. J. Duschl, J. Frank and E. Meyer-Hofmeister (Dordrecht: Kluwer), pp. 341–354

Rubin, A. E., and Wasson, J. T. 1987. Chondrules, matrix and coarse-grained chondrule rims in the Allende meteorite: origin, interrelationships and possible precursor components. *Geochim. Cosmochim. Acta* 51:1923–1937.

Rubin, A. E., Fegley, B., and Brett, R. 1988. Oxidation state in chondrites. In *Meteorites and the Early Solar System*, eds. J. F. Kerridge and M. S. Matthews (Tucson: Univ. of Arizona Press), pp. 488–511.

Rucinski, S. M. 1985. IRAS observations of T Tauri and post-T Tauri stars. *Astron. J.* 90:2321–2330.

Ruden, S. P. 1986. The Structure and Evolution of the Solar Nebula. Ph. D. Thesis, Univ. of California, Santa Cruz.

Ruden, S. P., and Lin, D. N. C. 1986. The global evolution of the primordial solar nebula. *Astrophys. J.* 308:883–901.

Ruden, S. P., and Pollack, J. B. 1991. The dynamical evolution of the protosolar nebula. *Astrophys. J.* 375:740–760.

Ruden, S. P., Papaloizou, J. C. B., and Lin, D. N. C. 1988. Axisymmetric perturbations of thin gaseous disks. I. Unstable convective modes and their consequences for the solar nebula. *Astrophys. J.* 329:739–763.

Ruden, S. P., Glassgold, A. E., and Shu, F. H. 1990. Thermal structure of neutral winds from young stellar objects. *Astrophys. J.* 361:546–569.

Rudolph, A., and Welch, W. J. 1988. Herbig-Haro objects as shocked ambient cloudlets—High resolution radio observations of HH 7–11. *Astrophys. J. Lett.* 326:31–34.

Rudolph, A., Welch, W. J., Palmer, P., and Dubrulle, B. 1990. Dynamical collapse of the W-51 star-forming region. *Astrophys. J.* 363:528–546.

Ruiz, A., Alonso, J. L., and Mirabel, F. 1991. Detection of high velocity neutral wind in T Tau. Preprint.

Russell, C. T., Luhmann, J. G., and Baker, D. N. 1987. An examination of possible solar wind sources for a sudden brightening of comet IRAS-Araki-Alcock. *Geophys. Res. Lett.* 14:991–994.

Russell, R. W., Lynch, D. K., Rudy, R. J., Rossano, G. S., Hackwell, J. A., and Campins, H. C. 1986. Multiple aperture airborne infrared measurements of comet Halley. In *20th ESLAB Symposium on the Exploration of Halley's Comet*, vol. 2, eds. B. Battrick, E. J. Rolfe and R. Reinhard, ESA SP-250, pp. 125–128.

Rutten, R. G. M., and Schrijver, C. J. 1987. Magnetic structures in cool stars. XIII. Appropriate units for the rotation-activity relation. *Astron. Astrophys.* 177:155–162.

Ruzmaikina, T. V. 1981a. On the role of the magnetic field and turbulence in the evolution of the pre-solar nebula. *Adv. Space Res.* 1:49–53.

Ruzmaikina, T. V. 1981b. The angular momentum of protostars creating the proto-planetary disks. *Pis'ma Astron. J.* 7:188–192.

Ruzmaikina, T. V. 1982. Ursprung des Sonnensystems. *Mitt. Astron. Ges.* 57:49–54.

Ruzmaikina, T. V. 1985. Magnetic field of the collapsing solar nebula. *Astron. Vestnik* 19:101–112.

Ruzmaikina, T. V. 1986. Origin of the angular momentum of the presolar nebula. *Astron. Zirk.* 1439:1–3.

Ruzmaikina, T. V. 1988. Distribution of planetary systems. In *Bioastronomy—The Next Steps: Proc. IAU Coll. 99*, ed. G. Marx (Dordrecht: Kluwer), pp. 41–47.

Ruzmaikina, T. V. 1989. Origin and evolution of planetary systems. *Acta Astronautica* 19:859–862.

Ruzmaikina, T. V., and Maeva, S. V. 1986. Process of formation of the protoplanetary disk. *Astron. Vestnik* 20:212–226.

Ryan, M. P., and Draganic, I. G. 1986. An estimate of the contribution of high energy cosmic-ray protons to the absorbed dose inventory of a cometary nucleus. *Astrophys. Space Sci.* 125:49–67.

Rydgren, A. E., and Cohen, M. 1985. Young stellar objects and their circumstellar dust: an overview. In *Protostars and Planets II*, eds. D. C. Black and M. S. Matthews (Tucson: Univ. of Arizona Press), pp. 371–385.

Rydgren, A. E., and Vrba, F. J. 1983. Periodic light variations in four pre-main-sequence K stars. *Astrophys. J.* 267:191–198.

Rydgren, A. E., and Zak, D. S. 1987. On the spectral form of the infrared excess component in T Tauri systems. *Publ. Astron. Soc. Pacific* 99:141–145.

Sadakane, K., and Nishida, M. 1986. Twelve additional 'Vega-like' stars. *Publ. Astron. Soc. Pacific* 98:685–689.

Safier, P. N., and Königl, A. 1992. A unified model of the radiative properties of accreting T Tauri stars. *Astrophys. J. Lett.*, submitted.

Safronov, V. S. 1960. On the gravitational instability in flattened systems with axial symmetry and non-uniform rotation. *Ann. Astrophys.* 23:901–904.

Safronov, V. S. 1966. Sizes of the largest bodies falling onto the planets during their formation. Soviet Astron. J. 9:987–991.

Safronov, V. S. 1972. In *Evolution of the Protoplanetary Cloud and Formation of the Earth and Planets* (Moscow: Nauka Press); also NASA-TT-F-677 (1972).

Safronov, V. S. 1978. The heating of the Earth during its formation. *Icarus* 33:1–12.

Safronov, V. S., and Gusseinov, K. M. 1989. Possible role of resonances in the formation of the asteroid belt. *Circ. Astrophys. Obs. Shemakha* 87:6–7

Safronov, V. S., and Ruzmaikina, T. V. 1978. On angular momentum transfer and accumulation of solid bodies in the solar nebula. In *Protostars and Planets*, ed. T. Gehrels (Tucson: Univ. of Arizona Press), pp. 545–564.

Safronov, V. S., and Ruzmaikina, T. V. 1985. Formation of the solar nebula and the planets. In *Protostars and Planets II*, eds. D. C. Black and M. S. Matthews (Tucson: Univ. of Arizona Press), pp. 959–980.

Sagar, R., and Richtler, T. 1991. Mass functions of five young Large Magellanic Cloud star clusters. *Astron. Astrophys.* 250:324–339.

Sagar, R., Myakutin, V. I., Piskunov, A. E., and Dluzhnevskaya, O. B. 1988. A study of the spatial stellar mass distribution in some open clusters. *Mon. Not. Roy. Astron. Soc.* 234:831–845.

Sagdeev, R. Z., Szabo, F., Avanesov, G. A., Cruvellier, P., Szabo, L., Szego, K., Abergel, A., Balazs, A., Barinov, I. V., Bertaux, J.-L., Blamont, J., Detaille, M., Demarelis, E., Dul'nev, G. N., Endroczy, G., Gardos, M., Kanyo, M., Kostenko, V. I., Krasikov, V. A., Nguyen-Trong, T., Nyitrai, Z., Reny, I., Rusznyak, P., Shamis, V. A., Smith, B., Sukhanov, K. G., Szabo, F., Szalai, S., Tarnopolsky, V. I., Toth, I., Tsukanova, G., Valnicek, B. I., Varhalmi, L., Zaiko, Yu. K., Zatsepin, S. I., Ziman, Ya. L., Zsenei, M., and Zhukov, B. S. 1986. Television observations of comet Halley from Vega spacecraft. *Nature* 321:262–266.

Sagdeev, R. Z., Smith, B., Szego, K., Larson, S., Toth, I., Merenyi, E., Avanesov, G. A., Krasikov, V. A., Shamis, V. A., and Tarnapolski, V. I. 1987. The spatial distribution of dust jets seen during the Vega 2 flyby. *Astron. Astrophys.* 187:835–838.

Sahvow, D. J., Feldman, P. D., McCandliss, S. R., and Martinez, M. E. 1990. Rocket observations of the ultraviolet spectrum of comet Austin (1989c1). *Bull. Amer. Astron. Soc.* 22:1090.

Saito, S., Kawaguchi, K., Yamamoto, S., Ohishi, M., Suzuki, H., and Kaifu, N. 1987. Laboratory detection and astronomical identification of a new free radical, CCS ($^3\Sigma^-$). *Astrophys. J. Lett.* 317:115–119.

Sakurai, T. 1985. Magnetic stellar winds: a 2-D generalization of the Weber-Davis model. *Astron. Astrophys.* 152:121–129.

Sakurai, T. 1987. Magnetically collimated winds from accretion disks. *Publ. Astron. Soc. Japan* 39:821–835.

Salpeter, E. E. 1953. Reactions of light nuclei and young contracting stars. *Mem. Soc. Roy. Sci. Liège* 14:116–121.

Salpeter, E. E. 1955. The luminosity function and stellar evolution. *Astrophys. J.* 121:161–167.

Salpeter, E. E. 1976. Planetary nebulae, supernova remnants, and the interstellar medium. *Astrophys. J.* 206:673–678.

Salzar, P., Starrfield, S., Ferland, G. J., Wagner, R. M., Truran, J. W., Kenyon, S. J., Sparks, W. M., and Williams, R. E. 1990. *Astrophys. J.*, submitted.

Samuelson, R. E., Maguire, W. C., Hanel, R. A., Kunde, V. G., Jennings, D. E., Yung, Y. L., and Aiken, A. C. 1983. CO_2 on Titan. *J. Geophys. Res.* 8:8709–8715.

Sancisi, R., Goss, W. M., Anderson, C., Johansson, L. E. B., and Winnberg, A. 1974. OH and HI observations of the Perseus OB2 dust cloud. *Astron. Astrophys.* 35:445–458.

Sandell, G., Reipurth, B., and Gahm, G. 1987. Low-mass star formation in the high galactic latitude dark cloud L1642. *Astron. Astrophys.* 181:283–288.

Sanders, D. B., and Willner, S. P. 1985. The Orion B jet. *Astrophys. J. Lett.* 293:39–43.

Sanders, D. B., Solomon, P. M., and Scoville, N. Z. 1984. Giant molecular clouds in the galaxy I. The axisymmetric distribution of H_2 *Astrophys. J.* 276:182–203.

Sanders, D. B., Scoville, N. Z., and Solomon, P. M. 1985. Giant molecular clouds in the galaxy II. Characteristics of discrete features. *Astrophys. J.* 289:373–387.

Sandford, S. A., and Allamandola, L. J. 1990*a*. The physical and infrared spectral properties of CO_2 in astrophysical ice analogs. *Astrophys. J.* 355:357–372.

Sandford, S. A., and Allamandola, L. J. 1990*b*. The volume and surface binding energies of ice systems containing CO, CO_2, and H_2O. *Icarus* 87:188–192.

Sandford, S. A., Allamandola, L. J., Tielens, A. G. G. M., and Valero, G. J. 1988. Laboratory studies of the infrared spectral properties of CO in astrophysical ices. *Astrophys. J.* 329:498–510.

Sandford, S. A., Allamandola, L. J., Tielens, A. G. G. M., Sellgren, K., Tapia, M., and Pendleton, Y. 1991. The interstellar C-H stretching material in the diffuse interstellar medium. *Astrophys. J.* 371:607–620.

Sargent, A. I. 1977. Molecular clouds and star formation. I. Observations of the Cepheus OB3 molecular cloud. *Astrophys. J.* 218:736–748.

Sargent, A. I. 1989. Molecular disks and their link to planetary systems. In *The Formation and Evolution of Planetary Systems* eds. H. A. Weaver and L. Danly (Cambridge: Cambridge Univ. Press), pp. 111–129.

Sargent, A. I., and Beckwith, S. 1987. Kinematics of the circumstellar gas of HL Tauri and R Monocerotis. *Astrophys. J.* 323:294–305.

Sargent, A. I., and Beckwith, S. V. W. 1989. Molecular disks around young stars. In *Structure and Dynamics of the Interstellar Medium: Proc. IAU Coll. 120*, eds. G. Tenorio-Tagle, J. Melnick and M. Moles (Berlin: Springer-Verlag), pp. 215–220.

Sargent, A. I., and Beckwith, S. V. W. 1991. The molecular structure around HL Tauri. *Astrophys. J. Lett.* 382:31–35.

Sargent, A. I., Beckwith, S., Keene, J., and Masson, C. R. 1988. Small-scale structure of the circumstellar gas around L1551 IRS 5. *Astrophys. J.* 333:936–942.

Sasaki, S. 1989. Minimum planetary size for forming outer Jovian-type planets: stability of an isothermal atmosphere surrounding a protoplanet. *Astron. Astrophys.* 215:177–180.

Sasaki, S. 1991. Off-disk penetration of ancient solar wind. *Icarus* 91:29–38.

Sasaki, S., and Nakazawa, K. 1988. Origin of isotopic fractionation of terrestrial Xe: hydrodynamic fractionation during escape of the primordial H_2-He atmosphere. *Earth Planet. Sci. Lett.* 89:323–334.

Sasaki, S., and Nakazawa, K. 1990. Did a primary solar-type atmosphere exist around the proto-Earth? *Icarus* 85:21–42.

Sasselov, D. D., and Rucinski, S. M. 1990. Formaldehyde mapping of ρ Ophiuchi B1: The densest cold prestellar core. *Astrophys. J.* 351:578–582.

Sato, F., and Fukui, Y. 1989. Two molecular outflows in L1251. *Astrophys. J.* 343:773–778.

Savage, B. D., Bohlin, R. C., Drake, J. F., and Budich, W. 1977. A survey of interstellar molecular Hydrogen I. *Astrophys. J.* 216:291–307.

Savonije, G. J., and Papaloizou, J. C. B. 1989. Non-linear evolution of non-axisymmetric perturbations in thin self-gravitating gaseous discs. In *Dynamics of Astrophysical Discs*, ed. J. A. Sellwood (Cambridge: Cambridge University Press), pp. 115–118.

Sawada, K., Matsuda, T., and Hachisu, I. 1986. Spiral shocks on a Roche Lobe overflow in a semi detached binary system. *Mon. Not. Roy. Astron. Soc.* 219:75–88.

Sawada, K., Matsuda, T., Inoue, M., and Hachisu, I. 1987. Is the standard accretion disk model invulnerable? *Mon. Not. Roy. Astron. Soc.* 224:307–322.

Scalo, J. M. 1977. Heating of dense interstellar clouds by magnetic ion slip: a constraint on cloud field strengths. *Astrophys. J.* 213:705–711.

Scalo, J. M. 1978. The stellar mass spectrum. In *Protostars and Planets*, ed. T. Gehrels (Tucson: Univ. of Arizona Press), pp. 265–287.

Scalo, J. M. 1984. Turbulent velocity structure in interstellar clouds. *Astrophys. J.* 277:556–561.

Scalo, J. M., 1985. Fragmentation and hierarchical structure in the interstellar medium. In *Protostars and Planets II*, eds. D. C. Black and M. S. Matthews (Tucson: Univ. of Arizona), pp. 201–296.

Scalo, J. M. 1986. The stellar initial mass function. *Fund. Cosmic Phys.* 11:1–278.

Scalo, J. M. 1987. Theoretical approaches to interstellar turbulence. In *Interstellar Processes*, eds. D. J. Hollenbach and H. A. Thronson, Jr. (Dordrecht: D. Reidel), pp. 349–392.

Scalo, J. M. 1988. Theories and implications of hierarchical fragmentation. In *Molecular Clouds in the Milky Way and External Galaxies*, eds. R. L. Dickman, R. Snell, and J. Young (Berlin: Springer-Verlag), pp. 201–213.

Scalo, J. 1990. Perception of interstellar structure: Facing complexity. In *Physical Processes in Fragmentation and Star Formation*, eds. R. Capuzzo-Dolcetta, C. Chiosi and A. Di Fazio (Dordrecht: Kluwer), pp. 151–177.

Scalo, J. M., and Pumphrey, W. A. 1982. Dissipation of supersonic turbulence in interstellar clouds. *Astrophys. J. Lett.* 258:29–33.

Schelhaas, N. 1987. Massenspektrometrische Analyse von Edelgasen in gewöhnlichen Chondriten. Ph. D. Thesis, Mainz.

Schelhaas, N., Ott, U., and Begemann, F. 1990. Trapped noble gases in unequilibrated ordinary chondrites. *Geochim. Cosmochim. Acta* 54:2869–2882.

Scheuer, P. A. G. 1974. Models of extragalactic radio sources with a continuous energy supply from a central object. *Mon. Not. Roy. Astron. Soc.* 166:513–528.

Schleicher, D. G., Millis, R. L., Tholen, D., Lark, N., Birch, P. V., Martin, R., and A'Hearn, M. F. 1986. The variability of Halley's comet during the Vega, Planet-A, and Giotto encounters. In *Proc. 20th ESLAB Symposium on the Exploration of Halley's Comet*, vol. 1, eds. J. Battrick, E. J. Rolfe and R. Reinhard, ESA SP-250, pp. 565–567.

Schleicher, D. G., Millis, R. L., Thompson, D. T., Birch, P. V., Martin, R., Tholen, D. J., Piscitelli, J. R., Lark, N. L., and Hammel, H. B. 1990. Periodic variations in the activity of comet P/Halley during the 1985/1986 apparition. *Astron. J.* 100:896–912.

Schloerb, F. P., and Weiguo, W. 1992. Submillimeter molecular line observations of comet Levy (1990c). In *Asteroids, Comets, and Meteors, IV*, in press.

Schloerb, F. P., Kinzel, W. M., Swade, D. A., and Irvine, W. M. 1986. HCN production from comet Halley. *Astrophys. J. Lett.* 310:55–60.

Schloerb, F. P., Kinzel, W. M., Swade, D. A., and Irvine, W. M. 1987. Observations of HCN in comet Halley. *Astron. Astrophys.* 187:475–480.

Schlosser, W. R., Schulz, R., and Koczet, P. 1986. The cyan shells of comet P/Halley. In *20th ESLAB Symposium on the Exploration of Halley's Comet*, vol. 3, eds. B. Battrick, E. J. Rolfe and R. Reinhard, ESA SP-250, pp. 495–501.

Schmid-Burgk, J., Güsten, R., Mauersberger, R., Schultz, A., and Wilson, T. L. 1990. A highly collimated outflow in core of OMC-1. *Astrophys. J. Lett.* 362:25–28.

Schmidt, R. M., and Housen, K. R. 1987. Some recent advances in the scaling impact and explosion cratering. *Int. J. Impact Eng.* 5:543–560.

Schmidt-Kaler, Th. 1982. Physical parameters of the stars. In *Numerical Data and Functional Relationships in Science and Technology*, eds. K. Schaifers and H. H.

Voigt (Berlin: Springer-Verlag), pp. 1–34.

Schneps, M. H., Ho, P. T. P., and Barrett, A. H. 1980. The formation of elephant-trunk globules in the Rosette nebula: CO observations. *Astrophys. J.* 240:84–98.

Scholl, H., and Froeschlé, Ch. 1974. Asteroidal motion at the 3/1 commensurability. *Astron. Astrophys.* 33:455–458.

Scholl, H., Froeschlé, Ch., Kinoshita, H., Yoshikawa, M., and Williams, J. G. 1989. Secular Resonances. In *Asteroids II*, eds. R. P. Binzel, T. Gehrels and M. S. Matthews (Tucson: Univ. of Arizona Press), pp. 845–861.

Schramm, D. N. 1982. The r-process and nucleocomsochronology. In *Essays in Nuclear Astrophysics*, eds. C. A. Barnes, D. D. Clayton and D. N. Schramm (Cambridge: Cambridge Univ. Press), pp. 325–353.

Schramm, L. S., Brownlee, D. E., and Whellock, M. M. 1989. Major element composition of stratospheric micrometeorites. *Meteoritics* 24:99–112.

Schroeder, M. C., and Comins, N. F. 1988. Star formation in very young galactic clusters. *Astrophys. J.* 326:756–760.

Schubart, J. 1968. Long-period effects in the motion of Hilda-type planets. *Astron. J.* 73:99–103.

Schubert, G., Spohn, T., and Reynolds, R. T. 1986. Thermal histories, compositions, and internal structures of the moons of the solar system. In *Satellites*, eds. J. Burns and M. S. Matthews (Tucson: Univ. of Arizona Press), pp. 224–292.

Schultz, P. H., and Crawford, D. 1987. Impact vaporization by low-angle impacts. *Lunar Planet. Sci.* XVIII:888–889 (abstract).

Schutte, W. A., Tielens, A. G. G. M., and Sandford, S. A. 1991. 10 micron spectra of protostars and the solid methanol abundance. *Astrophys. J.* 382:523–529.

Schwartz, P. R., Gee, G., and Huang, Y.-L. 1988. CO pedestal features from IRAS sources in dark clouds. *Astrophys. J.* 327:350–355.

Schwartz, P. R., Simon, T., Zuckerman, B., and Howell, R. R. 1984. The T Tauri radio source. *Astrophys. J. Lett.* 280:23–26.

Schwartz, R. D. 1975. T Tauri nebulae and Herbig-Haro nebulae: Evidence for excitation by a strong stellar wind. *Astrophys. J.* 195:631–642.

Schwartz, R. D. 1977. A survey of southern dark clouds for Herbig-Haro objects and H-alpha stars. *Astrophys. J. Suppl.* 35:161–170.

Schwartz, R. D. 1978. A shocked cloudlet model for Herbig-Haro objects. *Astrophys. J.* 223:884–900.

Schwartz, R. D. 1983. Herbig-Haro objects. *Ann. Rev. Astron. Astrophys.* 21:209–237.

Schwartz, R. D. 1986. The dynamics of Herbig-Haro objects. *Canadian J. Phys.* 64:414–420.

Schwartz, R. D. 1991. The Chamaeleon dark clouds and T-associations. In *ESO Workshop on Low Mass Star Formation in Southern Molecular Clouds*, ed. B. Reipurth, ESO Sci. Rept. No. 11, pp. 93–112.

Schwartz, R. D., Williams, P. M., Cohen, M., and Jennings, D. G. 1988. High-resolution infrared molecular hydrogen images and optical images of Herbig-Haro object 43. *Astrophys. J. Lett.* 334:99–102.

Schwartzschild, M. 1958. *Structure and Evolution of the Stars* (Princeton, N.J.: Princeton Univ. Press).

Schwarz, K. W. 1990. Evidence for organized small-scale structure in fully developed turbulence. *Phys. Rev. Lett.* 64:415–418.

Schwarz, U. J., Troland, T. H., Albinson, J. S., Bregman, J. D., Goss, W. M., and Heiles, C. 1986. Apperture synthesis observations of the 21-centimeter zeeman effect toward Cassiopeia A. *Astrophys. J.* 301:320–330.

Schwarzenberg-Czerny, A., and Różyczka, M. 1988. On tidal effects in accretion disks. *Acta Astron.* 38:189–205.

Schwehm, G. H., and Langevin, Y. 1991. In *Rosetta: A Comet Nucleus Sample Return*

Mission, ESA SP-1125, pp. 189.

Scott, T. A. 1976. Solid and liquid nitrogen. *Phys. Reports (Phys. Lett. C)* 27:89–157.

Scoville, N. Z., Kleinmann, S. G., Hall, D. N. B., and Ridgway, S. T. 1983. The circumstellar and nebular environment of the Becklin-Neugebauer object: $\lambda =2$–5 micron spectroscopy. *Astrophys. J.* 275:201–224.

Scoville, N. Z., Sargent, A. I., Sanders, D. B., Claussen, M. J., Masson, C. R., Lo, K. Y., and Phillips, T. G. 1986. High-resolution mapping of molecular outflows in NGC2071, W49, and NGC7538. *Astrophys. J.* 303:416–432.

Scoville, N. Z., Min, S. Y., Clemens, D. P., Sanders, D. B., and Waller, W. H. 1987. Molecular clouds and cloud cores in the inner galaxy. *Astrophys. J. Suppl.* 63:821–915.

Seab, C.G. 1987. Grain destruction, formation and evolution. In *Interstellar Processes*, eds. D. Hollenbach and H. A. Thronson (Dordrecht: D. Reidel), pp. 491–512.

Seidelmann, P. K. and Divine, N. 1977. Evaluation of Jupiter longitudes in System III (1965). *Geophys. Res. Lett.* 4:65–67.

Sekanina, Z. 1976. A probability of encounters with interstellar comets and the likelihood of their exsistence. *Icarus* 27:123–133.

Sekanina, Z. 1982. The problem of split comets in review. In *Comets*, ed. L. Wilkining (Tucson: Univ. of Arizona Press), pp. 251–287.

Sekanina, Z., and Larson, S. M. 1986. Dust jets in comet Halley observed by Giotto and from the ground. *Nature* 321:357–361.

Sekiya, M., and Miyama, S. 1988*a*. The stability of a differentially rotating cylinder of an incompressible perfect fluid. *Mon. Not. Roy. Astron. Soc.* 234:107–114.

Sekiya, M., and Miyama, S. 1988*b*. Stability of the primordial solar nebula. *Prog. Theor. Phys. Supp.* 96:85–94.

Sekiya, M., Nakazawa, K., and Hayashi, C. 1980*a*. Dissipation of the primordial terrestrial atmosphere due to irradiation of the solar EUV. *Prog. Theor. Phys.* 64:1968–1985.

Sekiya, M., Nakazawa, K., and Hayashi, C. 1980*b*. Dissipation of the rare gases contained in the primordial Earth's atmosphere. *Earth Planet. Sci. Lett.* 50:197–201.

Sekiya, M., Hayashi, C., and Nakazawa, K. 1981. Dissipation of the primordial terrestrial atmosphere due to irradiation of the solar far-UV during the T Tauri stage. *Prog. Theor. Phys.* 66:1301–1316.

Sekiya, M., Miyama, S., and Hayashi, C. 1987. Gas flow in the solar nebula leading to the formation of Jupiter. *Earth, Moon, and Planets* 39:1–16.

Sekiya, M., Miyama, S., and Hayashi, C. 1988. Gas capture by proto-Jupiter and proto-Saturn. *Prog. Theor. Phys. Suppl.* 96:274–280.

Sellgren, K. 1983. Properties of young clusters near reflection nebulae. *Astron. J.* 88:985–997.

Sellgren, K., Luan, L., and Werner, M. W. 1990. The excitation of 12 micron emission from very small particles. *Astrophys. J.* 359:384–391.

Sellwood, J. A. 1989. Spiral instabilities in N-body simulations. In *Dynamics of Astrophysical Discs*, ed. J. A. Sellwood (Cambridge: Cambridge Univ. Press), pp. 155–171.

Sellwood, J. A., and Kahn, F. D. 1991. Spiral modes driven by narrow features in angular-momentum density. *Mon. Not. Roy. Astron. Soc.* 250:278–299.

Sellwood, J. A., and Lin, D. N. C. 1989. A recurrent spiral instability cycle in self-gravitating particle disks. *Mon. Not. Roy. Astron. Soc.* 240:991–1007.

Shafranov, V. D. 1966. Plasma equilibrium in a magnetic field. *Rev. Plasma Phys.* 2:103–151.

Shakura, N. I., and Sunyaev, R. A. 1973. Black holes in binary systems: Observational appearance. *Astron. Astrophys.* 24:337–355.

Shapiro, S. L., and Teukolsky, S. A. 1983. *Black Holes, White Dwarfs, and Neutron Stars* (New York: Wiley).

Shevchenko, V. S. 1979. The structure of star formation regions. I. Population categories and the evolution of molecular clouds. *Sov. Astron.* 23:163–173.

Shibata, K., Tajima, T., Matsumoto, R., Horiuchi, T., Hanawa, T., Rosner, R., and Ushida, Y. 1989. Nonlinear Parker instability of isolated magnetic flux in a plasma. *Astrophys. J.* 338:471–492.

Shock, E. L., and McKinnon, W. B. 1992. Hydrothermal processing of cometary volatiles—Applications to Triton. *Icarus*, submitted.

Shoemaker, E. M., and Wolfe, R. F. 1984. Evolution of the Uranus-Neptune planetesimal swarm. *Lunar Planet. Sci.* XXV:780–781 (abstract).

Showalter, M. 1991. The Visual Detection of 1981S13 and Its Role in the Encke Gap. Preprint.

Shu, F. H. 1970. On the density-wave theory of galactic spirals. II. The propagation of the density of wave action. *Astrophys. J.* 160:99–112.

Shu, F. H. 1976. Mass transfer in semi-detached binaries. In *Structure and Evolution of Close Binary Systems*, eds. P. P. Eggleton, S. Mitton and J. A. J. Whelan (Dordrecht: D. Reidel), pp. 253–264.

Shu, F. H. 1977. Self-similar collapse of isothermal spheres and star formation. *Astrophys. J.* 214:488–497.

Shu, F. H. 1983. Ambipolar diffusion in self-gravitating isothermal layers. *Astrophys. J.* 273:202–213.

Shu, F. H. 1984. Waves in planetary rings. In *Planetary Rings*, eds. R. Greenberg and A. Brahic (Tucson: Univ. of Arizona Press), pp. 513–561.

Shu, F. H. 1985. Star formation in molecular clouds. In *The Milky Way Galaxy: Proc. IAU Symp. 106*, eds. H. van Woerden, R. J. Allen and W. B. Burton (Dordrecht: D. Reidel), pp. 561–565.

Shu, F. H. 1987. Summary of symposium: low luminosity sources. In *Star Formation in Galaxies*, ed. C. J. L. Persson, NASA CP. 2466, pp. 743–752.

Shu, F. H. 1991. Star formation—A theoretician's view. In *Frontiers of Stellar Evolution*, ed. D. L. Lambert (San Francisco: Astron. Soc. of the Pacific), pp. 23–44.

Shu, F. H., and Terebey, S. 1984. The formation of cool stars from cloud cores. In *Cool Stars, Stellar Systems and the Sun*, eds. S. Baliunas and L. Hartmann (Berlin: Springer Verlag), pp. 78–89.

Shu, F. H., Milione, V., and Roberts, W. W. 1973. Nonlinear gaseous density waves and galactic shocks. *Astrophys. J.* 183:819–841.

Shu, F. H., Cuzzi, J. N., and Lissauer, J. J. 1983. Bending waves in Saturn's rings. *Icarus* 53:185–206.

Shu, F. H., Dones, L., Lissauer, J. J., Yuan, C., and Cuzzi, J. N. 1985a. Nonlinear spiral density waves: viscous damping. *Astrophys. J.* 299:542–573.

Shu, F. H., Yuan, C., and Lissauer, J. J. 1985b. Nonlinear spiral density waves: An inviscid theory. *Astrophys. J.* 291:356–376.

Shu, F. H., Adams, F. C., and Lizano, S. 1987a. Star formation in molecular clouds: Observation and theory. *Ann. Rev. Astron. Astrophys.* 25:23–81.

Shu, F. H., Lizano, S., and Adams, F. C. 1987b. Star formation in molecular cloud cores. In *Star Forming Regions: Proc. IAU Symp. 115*, eds. M. Peimbert and J. Jugaku (Dordrecht: D. Reidel), pp. 417–434.

Shu, F. H., Lizano, S., Ruden, S. P., and Najita, J. 1988. Mass loss from rapidly rotating magnetic protostars. *Astrophys. J. Lett.* 328:19–23.

Shu, F. H., Tremaine, S., Adams, F. C., and Ruden, S. P. 1990. SLING amplification and eccentric gravitational instabilities in gaseous disks. *Astrophys. J.* 358:495–

514.

Shu, F. H., Ruden, S. P., Lada, C. J., and Lizano, S. 1991. Star formation and the nature of bipolar outflows. *Astrophys. J. Lett.* 370:31–34.

Shull, J. M., and Draine, B. T. 1987. The physics of interstellar shock waves. In *Interstellar Processes*, eds. D. Hollenbach and H. A. Thronson (Dordrecht: D. Reidel), pp. 283–320.

Shuter, W. L. H., Dickman, R. L., and Klatt, C. 1987. Velocity waves in 21 cm self-absorption toward the Taurus molecular complex. *Astrophys. J. Lett.* 322:103–108.

Silk, J. 1978. Fragmentation of molecular clouds. In *Protostars and Planets*, ed. T. Gehrels (Tucson: Univ. of Arizona Press), pp. 172–188.

Siluk, R. S., and Silk, J. 1974. On the velocity dependence of the NaI/CaIII ratio. *Astrophys. J.* 192:51–57.

Silvera, I. F. 1980. The solid molecular hydrogens in the condensed phase: fundamentals and static properties. *Rev. Mod. Phys.* 52:393–452.

Simon, M., Howell, R. R., Longmore, A. J., Wilking, B. A., Peterson, D. M., and Chen, W. P. 1987. Milliarcsecond resolution infrared observation of young stars in Taurus and Ophiuchus. *Astrophys. J.* 320:344–355.

Simon, M., Chen, W.-P., Howell, R. R., Benson, J. A., and Slowik, D. 1992. Multiplicity among the young stars in Taurus. *Astrophys. J.*, in press.

Simon, T., and Joyce, R. R. 1988. Infrared photometry of V1057 Cygni (1971–1987). *Publ. Astron. Soc. Pacific* 100:1549–1554.

Simon, T., Herbig, G., and Boesgaard, A. M. 1985. The evolution of chromospheric activity and the spin-down of solar-type stars. *Astrophys. J.* 293:551–574.

Simon, T., Vrba, F. J., and Herbst, W. 1991. The UV and visible light variability of BP Tau: possible clues for the origin of T Tauri star acitivity. *Astron. J.* 100:1957–1967.

Simonelli, D. P., and Reynolds, R. T. 1989. The interiors of Pluto and Charon: structure, composition and implications. *Geophys. Res. Lett.* 16:1209–1212.

Simonelli, D. P., Pollack, J. B., McKay, C. P., Reynolds, R. T., and Summers, A. L. 1989. The carbon budget in the outer solar nebula. *Icarus* 82:1–35.

Simonetti, J. H., and Cordes, J. M. 1986. Rotation measures of extragalactic radio sources viewed through the molecular clouds Cepheus A and L1551. *Astrophys. J.* 303:659–666.

Simons, D. A., and Becklin, E. E. 1991a. A near infrared search for brown dwarfs in the Pleiades. In *Astrophysics with Infrared Arrays*, ed. R. Elston (San Francisco: Astron. Soc. of the Pacific) pp. 310–312.

Simons, D. A., and Becklin, E. E. 1991b. A near infrared search for brown dwarfs in the Pleiades. *Astrophys. J.*, submitted.

Singh, P. D., ed. 1992. *The Astrochemistry of Cosmic Phenomena: Proc. IAU Symp. 150* (Dordrecht: Kluwer).

Siregar, E., and Léorat, J. 1989. Accretion in numerical simulations of two-dimensional flows. *Astron. Astrophys.* 193:131–140.

Skillman, E. D. 1987. Neutral hydrogen and star formation in irregular galaxies. In *Star Formation in Galaxies*, ed. C. J. Lonsdale, NASA CP-2466, pp. 263–266.

Skrutskie, M. F., Forrest, W. J., and Shure, M. A. 1986. An infrared search for low-mass companions of stars within 12 parsecs of the Sun. In *Astrophysics of Brown Dwarfs*, eds. M. C. Kafatos, R. S. Harrington and S. P. Maran (Cambridge: Cambridge Univ. Press), pp. 82–86.

Skrutskie, M. F., Dutkevitch, D., Strom, S. E., Edwards, S., Strom, K. M., and Shure, M. A. 1990. A sensitive 10 micron search for emission arising from circumstellar dust associated with solar-type pre-main-sequence stars. *Astron. J.* 99:1187–1195.

Slattery, W. L. 1977. The structure of the planets Jupiter and Saturn. *Icarus* 32:58–72.

Slettebak, A. 1975. Some interesting bright stars of early type. *Astrophys. J.* 197:137–138.

Slettebak, A., and Carpenter, K. 1983. Ultraviolet spectroscopic observations of some Be stars of later type and A-F type shell stars. *Astrophys. J. Suppl.* 53:869–892.

Smak, J. 1984. Outbursts of dwarf novae. *Publ. Astron. Soc. Pacific* 96:5–18.

Smith, B. A., and Terrile, R. J. 1984. A circumstellar disk around β Pictoris. *Science* 226:1421–1424.

Smith, B. A., and Terrile, R. J. 1987. The Beta Pictoris disk: recent optical observations. *Bull. Amer. Astron. Soc.* 19:829 (abstract).

Smith, B. A., Soderblom, L., Batson, R., Bridges, P., Inge, J., Masursky, H., Shoemaker, E., Beebe, R., Boyce, J., Briggs, G., Bunker, A., Collins, S., Hansen, C., Johnson, T., Mitchell, J., Terrile, R., Cook, A. F., III, Cuzzi, J., Pollack, J. B., Danielson, G. E., Morrison, D., Owen, T., Sagan, C., Veverka, J., Strom, R., and Suomi, V. E. 1982. A new look at the Saturn system: the Voyager 2 images. *Science* 215:504–537.

Smith, B. A., Soderblom, L. A., Banfield, D., Barnet, C., Basilevsky, A. T., Beebe, R. F., Bollinger, K., Boyce, J. M., Brahic, A., Briggs, G. A., Brown, R. H., Chyba, C., Collins, S. A., Colvin, T., Cook, A. F., III, Crisp, D., Croft, S. K., Cruickshank, D., Cuzzi, J. N., Danielson, G. E., Davies, M. E., De Jong, E., Dones, L., Godfrey, D., Goguen, J., Grenier, I., Haemmerle, V. R., Hammel, H., Hansen, C. J., Helfenstein, C. P., Howell, C., Hunt, G. E., Ingersoll, A. P., Johnson, T. V., Kargel, J., Kirk, R., Kuehn, D. I., Limaye, S., Masursky, H., McEwen, A., Morrison, D., Owen, T., Owen, W., Pollack, J. B., Porco, C. C., Rages, K., Showalter, M., Sicardy, B., Simonelli, D., Spencer, J., Sromovsky, B., Stoker, C., Strom, R. G., Suomi, V. E., Synott, S. P., Terrile, R. J., Thomas, P., Thompson, W. R., Verbinger, A. and Veverka, J. 1989. Voyager 2 at Neptune: Imaging science results. *Science* 246:1422–1449.

Smith, M. D. 1989. Atomic and molecular cloud formation: Structural changes during collapse. *Mon. Not. Roy. Astron. Soc.* 238:835–849.

Smith, M. D., and Brand, P. W. J. L. 1990. Cool C-shocks and high-velocity flows in molecular clouds. *Mon. Not. Roy. Astron. Soc.* 242:494–504.

Smith, M. D., Brand, P. W. J. L., and Moorhouse, A. 1991. Bow shocks in molecular clouds: H2 line strengths. *Mon. Not. Roy. Astron. Soc.* 248:451–456.

Smith, R. G. 1991. A search for solid H_2S in dense clouds. *Mon. Not. Roy. Astron. Soc.* 249:172–176.

Smith, R. G., Sellgren, K., and Tokunaga, A. T. 1988. A study of H_2O ice in the 3 micron spectrum of OH 231.8+4.2 (OH 0739–14). *Astrophys. J.* 334:209–219.

Smith, R. G., Sellgren, K., and Tokunaga, A. T. 1989. Absorption features in the 3 micron spectra of protostars. *Astrophys. J.* 344:413–426.

Smoluchowski, R. 1975. Jupiter's molecular hydrogen layer and the magnetic field. *Astrophys. J. Lett.* 200:119–121.

Smoluchowski, R., and Torbett, M. 1984. The boundary of the solar system. *Nature* 311:38–39.

Smoluchowski, R., Bahcall, J. N., and Matthews, M. S., eds. 1986. *The Galaxy and the Solar System* (Tucson: Univ. of Arizona Press).

Snell, R. L. 1987. Bipolar outflows and stellar jets. In *Star Forming Regions: Proc. IAU Symp. 115*, eds. M. Peimbert and J. Jugaku (Dordrecht: D. Reidel), pp. 213–237.

Snell, R. L., and Bally, J. 1986. Compact radio sources associated with molecular outflows. *Astrophys. J.* 303:683–701.

Snell, R. L., and Edwards, S. 1981. High velocity molecular gas near Herbig-Haro objects HH7–11. *Astrophys. J.* 251:103–107.

Snell, R. L., and Edwards, S. 1982. Observations of high-velocity molecular gas near Herbig-Haro objects: HH 24–27 and HH 1–2. *Astrophys. J.* 259:668–676.

Snell, R. L., and Schloerb, F. B. 1985. Structure and physical properties of the bipolar outflow in L1551. *Astrophys. J.* 295:490–500.

Snell, R. L., Loren, R. B., and Plambeck, R. L. 1980. Observations of CO in L1551: evidence for stellar wind driven shocks. *Astrophys. J. Lett.* 239:17–22.

Snell, R. L., Scoville, N. Z., Sanders, D. B., and Erickson, N. R. 1984. High-velocity molecular jets. *Astrophys. J.* 284:176–193.

Snell, R. L., Huang, Y.-L., Dickman, R. L., and Claussen, M. J. 1988. Molecular outflows associated with bright far-infrared sources. *Astrophys. J.* 325:853–863.

Snell, R. L., Heyer, M. H., and Schloerb, F. P. 1989. Comparison of the far-infrared and carbon monoxide emission in Heiles' Cloud 2 and B 18. *Astrophys. J.* 337:739–753.

Snell, R. L., Dickman, R. L., and Huang, Y.-L. 1990. Molecular outflows associated with a flux-limited sample of bright far-infrared sources. *Astrophys. J.* 352:139–148.

Snijders, M. A. J. 1990. Physical properties and abundances of novae in the nebular phase. In *Physics of Classical Novae: Proc. IAU Coll. 22*, eds. A. Cassatella and R. Viotti (Berlin: Springer-Verlag), pp. 188–194.

Snijders, M. A. J., Batt, T. J., Roche, P. F., Seaton, M. J., Morton, D. C., Spoelstra, T. A. T., and Blades, J. C. 1987. Nova Aquilae 1982. *Mon. Not. Roy. Astron. Soc.* 228:329–376.

Snyder, L. E. 1982. A review of radio observations of comets. *Icarus* 51:1–24.

Snyder, L. E., Palmer, P., and de Pater, I. 1989. Radio detection of formaldehyde emission from comet Halley. *Astron. J.* 97:246–253.

Sodrowski, T. J., Dwek, E., Hauser, M. G., and Kerr, F. J. 1987. Large-scale galactic dust morphology and physical conditions from IRAS observations. *Astrophys. J.* 322:101–112.

Solf, J. 1987. The kinematic structure of the HH 24 complex derived from high-resolution spectroscopy. *Astron. Astrophys.* 184:322–328.

Solf, J. 1989. High resolution spectral imaging of T Tauri stars. In *ESO Workshop on Low Mass Star Formation and Pre-Main Sequence Objects*, ed. B. Reipurth (Garching: European Southern Obs.), pp. 399–406.

Solf, J., Böhm, K.-H., and Raga, A. 1986. Kinematical and Hydrodynamical Study of the HH 32 Complex. *Astrophys. J.* 305:795–804.

Solomon, P. M., and Rivolo, A. R. 1989. A face-on view of the first galactic quadrant in molecular clouds. *Astrophys. J.* 339:919–925.

Solomon, P. M., and Sanders, D. B. 1980. Giant molecular clouds as the dominant component of interstellar matter in the galaxy. In *Giant Molecular Clouds in the Galaxy*, eds. P. M. Solomon, and M. G. Edmunds (Oxford: Pergammon), pp. 41–63.

Solomon, P. M., Sanders, D. B., and Rivolo, A. R. 1985. The Massachusetts-Stony Brook galactic plane CO survey: disk and spiral arm molecular cloud populations. *Astrophys. J. Lett.* 292:19–24.

Solomon, P. M., Rivolo, A. R., Barrett, J., and Yahil, A. 1987. Mass, luminosity and line width relations of galactic molecular clouds. *Astrophys. J.* 319:730–741.

Sonnerup, B. U. Ö. 1970. Magnetic-field re-connection in a highly conducting incompressible fluid. *J. Plasma Phys.* 4:161–174.

Sorochenko, R. L., Tolmachev, A. M., and Winnewisser, G. 1986. High resolution measurments of cyanocetylene in dark clouds. *Astron. Astrophys.* 155:237–241.

Spaute, D., Weidenschilling, S. J., Davis, D. R., and Marzari, F. 1991. Accretional evolution of a planetesimal swarm: 1. A new simulation. *Icarus* 92:147–164.

Spangler, S. R., and Gwinn, C. R. 1990. Evidence for an inner scale to the density turbulence in the interstellar medium. *Astrophys. J. Lett.* 353:29–32.

Spencer, J., Buie, M., and Bjoraker, G. 1990. Solid methane on Triton and Pluto from 3 to 4 μm spectrophotometry. *Icarus* 82:1–35.

Spettel, B., Palme, H., and Kurat, G. 1989. Have different parts of Allende sampled compositionally different chondrules? *Meteoritics* 4:326–327 (abstract).

Spicer, D. S., Mariska, J. T., and Boris, J. P. 1986. Magnetic energy storage and conversion in the solar atmosphere. In *Physics of the Sun*, eds. P. A. Sturrock, T. E. Holzer, D. M. Mihalas and R. K. Ulrich (Dordrecht: D. Reidel), pp. 181–248.

Spitzer, L., Jr. 1968*a*. *Diffuse Matter in Space* (New York: Interscience).

Spitzer, L., Jr., 1968*b*. Dynamics of interstellar matter and the formation of stars. In *Stars and Stellar Systems*, vol. 7, eds. B. M. Midddlehurst and L. H. Aller (Chicago: Univ. of Chicago Press), pp. 1–58.

Spitzer, L. 1978. *Physical Processes in the Interstellar Medium* (New York: Wiley).

Spruit, H. C. 1987. Stationary shocks in accretion disks. *Astron. Astrophys.* 184:173–184.

Spruit, H. C. 1989*a*. In *Magnetic Fields and Accretion Disks in Astrophysics*, ed. G. Belvedere (Dordrecht: Kluwer), pp. 59.

Spruit, H. C. 1989*b*. Physics of accretion by spiral shock waves. In *Theory of Accretion Disks*, eds. F. Meyer, W. Duschl, J. Frank and E. Meyer-Hofmeister (Dordrecht: Kluwer), pp. 325–340.

Spruit, H. C., Matsuda, T., Inoue, M., and Sawada, K. 1987. Spiral shocks and accretion discs. *Mon. Not. Roy. Astron. Soc.* 229:517–527.

Squyres, S. W., Reynolds, R. T., Cassen, P. M., and Peale, S. J. 1983*a*. Liquid water and active resurfacing on Europa. *Nature* 301:225–226.

Squyres, S. W., Reynolds, R. T., Cassen, P. M., and Peale, S. J. 1983*b*. The evolution of Enceladus. *Icarus* 53:319–331.

Squyres, S. W., Reynolds, R. T., and Lissauer, J. J. 1985. The enigma of the Uranian satellites' orbital eccentricitites. *Icarus* 61:218–233.

Srinivasan, B., and Anders, E. 1978. Noble gases in the Murchison meteorite: Possible relics of s-process nucleosynthesis. *Science* 201:51–56.

Srinivasan, B., Alexander, E. C., Jr., Manuel, O. K., and Troutner, D. E. 1969. Xenon and krypton from the spontaneous fission of Californium-252. *Phys. Rev.* 179:1166–1169.

Stacey, J. G., Jaffe, D. T., Lugten, J. B., Genzel, R., and Townes, C. H. 1987. Far-infrared observations of hot molecular gas in the Cepheus A flow. *Bull. Amer. Astron. Soc.* 19:1016 (abstract).

Stacey, J. G., Benson, P. J., Myers, P. C., and Goodman, A. A. 1988. Dense cores associated with Herbig Ae/Be stars. In *Interstellar Matter*, eds. J. M. Moran and P. T. P. Ho (New York: Gordon and Breach), pp. 179–182.

Stacy, J. G., Myers, P. C., and de Vries, H. W. 1989. Dense cores associated with diffuse, high-latitude molecular clouds. In *The Physics and Chemistry of Interstellar Molecular Clouds*, eds. G. Winnewisser and J. T. Armstrong (New York: Springer-Verlag), pp. 117–119.

Stagg, C. R., and Bailey, M. E. 1989. Stochastic capture of short-period comets. *Mon. Not. Roy. Astron. Soc.* 241:507–541.

Stahl, O., Wilson, T. L., Henkel, C., and Appenzeller, I. 1989. The $^{12}CH^+/^{13}CH^+$ ratio toward ζ Ophiuchi. *Astron. Astrophys.* 221:321–325.

Stahler, S. W. 1983*a*. The birthline of low-mass stars. *Astrophys. J.* 274:822–829.

Stahler, S. W. 1983*b*. The equilibria of rotating, isothermal clouds. I. Structure and dynamical stability. *Astrophys. J.* 268:155–164.

Stahler, S. W. 1985. The star-formation history of very young clusters. *Astrophys. J.* 293:207–215.

Stahler, S. W. 1988. Deuterium and the stellar birthline. *Astrophys. J.* 332:804–825.

Stahler, S. W. 1989. Luminosity jumps in pre-main sequence stars. *Astrophys. J.* 347:950–958.

Stahler, S. W., Shu, F. H., and Taam, R. E. 1980. The evolution of protostars. I. Global formulation and results. *Astrophys. J.* 241:637–654.

Stark, A. A. 1983. Kinematics of molecular clouds: evidence for agglomeration in spiral arms. In *Kinematics, Dynamics, and Structure of the Milky Way*, ed. W. L. Shuter (Dordrecht: D. Reidel), pp. 127–133.

Stark, A. A. 1984. Kinematics of molecular clouds. I. Velocity dispersion in the solar neighborhood. *Astrophys. J.* 281:624–633.

Stark, A. A., and Blitz, L. 1978. On the masses of giant molecular cloud complexes. *Astrophys. J. Lett.* 225:15–19.

Stark, A. A., and Brand, J. 1989. Kinematics of molecular clouds II. New data on nearby giant molecular clouds. *Astrophys. J.* 339:763–771.

Stark, A. A., Elmegreen, B. G., and Chance, D. 1987. Molecules in galaxies V. CO observations of flocculent and grand-design spirals. *Astrophys. J.* 322:64–73.

Starrfield, S. G. 1989*a*. In *Multiwavelength Studies in Astrophysics*, ed. F. Cordova (Cambridge: Cambridge Univ. Press), in press.

Starrfield, S. G. 1989*b*. Thermonuclear processes and the classical nova outburst. In *Classical Novae*, eds N. Evans and M. Bode (New York: Wiley), pp. 39–60.

Starrfield, S., Truran, J. W., Sparks, W. M., and Arnould, M. 1978. On ^{7}Li production in nova explosions. *Astrophys. J.* 222:600–603.

Starrfield, S., Truran, J. W., Sparks, W. M., and Krautter, J. 1986. Hydrodynamic models for novae with ejecta rich in oxygen, neon, and magnesium. *Astrophys. J. Lett.* 303:5–9.

Staudacher, T. 1987. Upper mantle origin for Harding County well gases. *Nature* 325:605–607.

Staudacher, T., Sarda, P., and Allègre, C. 1991. Comment on "Atmospheric contamination: a possible source for heavy noble gases in basalts from Loihi Seamount, Hawaii" by D. B. Patterson, M. Honda and I. McDougall *Geophys. Res. Lett.* 18:745–748.

Staude, H. J., and Neckel, Th. 1991. RNO 1B—A new FUor in Cassiopeia. *Astron. Astrophys. Lett.* 224:L13–L16.

Staude, H. J., Lenzen, R., Dyck, H. M., and Schmidt, G. D. 1982. The bipolar nebula S 106: photometric, polarimetric, and spectropolarimetric observations. *Astrophys. J.* 255:95–102.

Stauffer, J. R., Hartmann, L. W., Soderblom, D. R., and Burnham, N. 1984. Rotation velocities of low-mass stars in the Pleiades. *Astrophys. J.* 280:202–212.

Stauffer, J. R., Hamilton, D., Probst, R., Rieke, G., and Mateo, M. 1989. Possible Pleiades members with $M \simeq 0.07 M_\odot$: identification of brown dwarf candidates of known age, distance, and metallicity. *Astrophys. J. Lett.* 344:21–24.

Stauffer, J. R., Herter, T., Hamilton, D., Rieke, G. H., Rieke, M. J., Probst, R., and Forrest, W. J. 1991. Spectroscopy of Taurus cloud brown dwarf candidates. *Astrophys. J. Lett.* 367:23–26.

Steel, T. M. and Duley, W. W. 1987. A 217.5 nanometer absorption feature in the spectrum of small silicate particles. *Astrophys. J.* 315:337–339.

Steele, I. M. 1988. Primitive material surviving in chondrites: mineral grains. In *Meteorites and the Early Solar System*, eds. J. F. Kerridge and M. S. Matthews (Tucson: Univ. of Arizona Press), pp. 808–818.

Steigman, G., Strittmatter, P. A., and Williams, R. E. 1975. The Copernicus observations: interstellar or circumstellar material? *Astrophys. J.* 198:575–582.

Stella, L., and Rosner, R. 1984. Magnetic field instabilities in accretion disks. *Astrophys. J.* 277:312–321.

Stemwedel, S. W., Yuan, C., and Cassen, P. 1990. Equilibrium models for self-gravitating inviscid disks resulting from the collapse of rotating clouds. *Astrophys. J.* 351:206-221.

Stencel, R. E., and Backman, D. E. 1991. A survey for infrared excesses among high galactic latitude SAO stars. *Astrophys. J. Suppl.* 75:905–924.

Stenholm, L. G. 1990. Molecular cloud fluctuations. II. Methods of analysis and cloud maps. *Astron. Astrophys.* 232:495–509.

Stepinski, T. F., and Levy, E. H. 1988. Generation of dynamo magnetic fields in protoplanetary and other astrophysical accretion disks. *Astrophys. J.* 331:416–434.

Stepinski, T. F., and Levy, E. H. 1990. Dynamo magnetic field-induced angular momentum transport in protostellar nebulae: The 'minimum mass protosolar nebula'. *Astrophys. J.* 350:819–826.

Stern, S. A. 1986. The effects of mechanical interaction between the interstellar medium and comets. *Icarus* 68:276–283.

Stern, S. A. 1988. Collisions in the Oort cloud. *Icarus* 73:499–507.

Stern, S. A. 1989. The Evolution of Comets and the Detectability of Extra-Solar Oort Clouds. Ph. D. Thesis, Univ. of Colorado.

Stern, S. A. 1990. ISM-induced erosion and gas-dynamical drag in the Oort cloud. *Icarus* 84:447–466.

Stern, S. A. 1991. On the number of planets in the outer solar system: evidence of a substantial population of 1000-km bodies. *Icarus* 90:271–281.

Stern, S. A., and Shull, J. M. 1988. The influence of supernovae and passing stars on comets in the Oort cloud. *Nature* 332:407–411.

Stern S. A., and Shull, J. M. 1990. The influence of Oort clouds on the mass and chemical balance of the interstellar medium. *Astrophys. J.* 359:506–511.

Stern, S. A., Shull, J. M., and Brandt, J. C. 1990. The evolution and detectability of comet clouds during post main sequence stellar evolution. *Nature* 345:305–308.

Stern, S. A., Stocke, J., and Weissman, P. R. 1991. An IRAS search for extra-solar Oort clouds. *Icarus* 91:65–75.

Stern, S. A., Green, J. C., Cash, C., and Cook, T. A. 1992. Helium and argon abundance constraints and the thermal evolution of comet Austin (1989c1). *Icarus* 95:157–161.

Stevenson, D. J. 1975. Thermodynamics and phase separation of dense, fully-ionized hydrogen-helium fluid mixtures. *Phys. Rev.* 12B:3999–4007.

Stevenson, D. J. 1979. Solubility of helium in metallic hydrogen. *J. Phys. Fluids (Metal Phys.)* 9:791–801.

Stevenson, D. J. 1981. Models of the Earth's core. *Science* 214:611–619.

Stevenson, D. J. 1982*a*. Formation of the giant planets. *Planet. Space Sci.* 30:755–764.

Stevenson, D. J. 1982*b*. Volcanism and igneous processes in small icy satellites. *Nature* 298:142–144.

Stevenson, D. J. 1984. Composition, structure, and evolution of Uranian and Neptunian satellites. In *Uranus and Neptune*, ed. J. Bergstralh, NASA CP-2330, pp. 405–423.

Stevenson, D. J. 1985. Cosmochemistry and structure of the giant planets and their satellites. *Icarus* 62:4–15.

Stevenson, D. J. 1987. Orgin of the Moon—the collision hypothesis. *Ann. Rev. Earth Planet. Sci.* 15:271–315.

Stevenson, D. J. 1990. Chemical heterogeneity and imperfect mixing in the solar nebula. *Astrophys. J.* 348:730–737.

Stevenson, D. J. 1991. The search for brown dwarfs. *Ann. Rev. Astron. Astrophys.* 29:163–193.

Stevenson, D. J., and Lunine, J. I. 1988. Rapid formation of Jupiter by diffusive redistribution of water vapor in the solar nebula. *Icarus* 75:146–155.

Stevenson, D. J., and Salpeter, E. E. 1976. Interior models of Jupiter. In *Jupiter*, ed. T. Gehrels (Tucson: Univ. of Arizona Press), pp. 85–112.

Stevenson, D. J., Harris, A. W., and Lunine, J. I. 1986. Origins of satellites. In *Satellites*, eds. J. A. Burns and M. S. Matthews (Tucson: Univ. of Arizona Press), pp. 39–88.

Stewart, G. R., and Wetherill, G. W. 1988. Evolution of planetesimal velocities. *Icarus* 74:542–553.

Stewart, R. W., and Townsend, A. A. 1951. Similarity and self-preservation in isotropic turbulence. *Phil. Trans. Roy. Soc. London* A243:359–386.

Stickland, D. J., Penn, C. J., Seaton, M. J., Snijders, M. A. J., and Storey, P. J. 1981. Nova Cygni 1978—The nebular phase. *Mon. Not. Roy. Astron. Soc.* 197:107–138.

Stine, P. C., Feigelson, E. D., André, Ph., and Montmerle, T. 1988. Multi-epoch radio continuum surveys of the ρ Ophiuchi dark cloud. *Astron. J.* 96:1394–1406.

Stocke, J. T., Hartigan, P. M., Strom, S. E., Strom, K. M., Anderson, E. R., Hartmann, L., and Kenyon, S. J. 1988. A detailed study of the Lynds L1551 star formation region. *Astrophys. J. Suppl.* 68:229–255.

Stone, E. 1989. Solar abundances as derived from solar energetic particles. In *Proc. Symp. on Cosmic Abundances of Matter*, ed. C. J. Waddington (New York: American Inst. of Physics), pp. 72–90.

Stone, J., Hutcheon, I. D., Epstein, S., and Wasserburg, G. J. 1990. Si isotopes in SiC from carbonaceous and enstatite chondrites. *Lunar Planet. Sci.* XXI:1212–1213 (abstract).

Stoney, G. J. 1867. See Aitken, R. G. 1935. *The Binary Stars* (New York: McGraw-Hill).

Storey, J. W. V., Watson, D. M., Townes, C. H., Haller, E. E., and Hansen, W. L., 1981. Far-infrared observations of shocked CO in Orion. *Astrophys. J.* 247:136–143.

Stothers, R., and Frogel, J. A. 1974. The local complex of O and B stars. I. Distribution of stars and interstellar dust. *Astron. J.* 79:456–471.

Strauss, F. M., Poppel, W. G. L., and Vieira, E. R. 1979. The structure of Gould's belt. *Astron. Astrophys.* 71:319–325.

Straw, S. M., and Hyland, A. R. 1989. Global aspects of the NGC 6334 star formation complex: an infrared survey. *Astrophys. J.* 340:318–343.

Straw, S. M., Hyland, A. R., and McGregor, P. J. 1989. The centers of star formation in NGC 6334 and their stellar mass distributions. *Astrophys. J. Suppl.* 69:99–140.

Strazzulla, G., and Johnson, R. E. 1991. Irradiation effects on comets and cometary debris. In *Comets in the Post-Halley Era*, eds. R. Newburn, M. Neugebauer and J. Rahe (Dordrecht: Kluwer), pp. 243–275.

Stringfellow, G. S. 1989. Evolutionary Scenarios for Low-Mass Stars and Substellar Brown Dwarfs. Ph. D. Thesis, Univ. of California, Santa Cruz.

Stringfellow, G. S. 1991. Brown dwarfs in young stellar cluster. *Astrophys. J. Lett.* 375:21–25.

Strittmatter, P. A. 1966. Gravitational collapse in the presence of a magnetic field. *Mon. Not. Roy. Astron. Soc.* 132:359–378.

Strom, K. M., Strom, S. E., and Vrba, F. J. 1976. Infrared surveys of dark-cloud complexes. IV. The Lynds 1517 and Lynds 1551 clouds. *Astron. J.* 81:320–322.

Strom, K. M., Strom, S. E., Wolff, S. C., Morgan, J., and Wenz, M. 1986. Optical manifestations of mass outflows from young stars: an atlas of CCD images of Herbig-Haro objects. *Astrophys. J. Suppl.* 62:39–80.

Strom, K. M., Strom, S. E., Kenyon, S. J., and Hartmann, L. 1988. Luminosity excesses in low-mass young stellar objects: A statistical survey. *Astron. J.* 95:534–542.

Strom, K. M., Margulis, M., and Strom, S. E. 1989a. A study of the stellar population in the Lynds 1641 dark cloud: deep near-infrared imaging. *Astrophys. J. Lett.* 345:79–82.

Strom, K. M., Margulis, M., and Strom, S. E. 1989*b*. A study of the stellar population in the Lynds 1641 dark cloud: a possible dense cluster associated with IRAS 05338–0624. *Astrophys. J. Lett.* 346:33–35.

Strom, K. M., Newton, G., Strom, S. E., Seaman, R. L., Carrasco, L., Cruz-Gonzalez, I., Serrano, A., and Grasdalen, G. L. 1989*c*. A study of the stellar population in the Lynds 1641 dark cloud. I. The *IRAS* catalog sources. *Astrophys. J. Suppl.* 71:183–217.

Strom, K. M., Strom, S. E. Edwards, S., Cabrit, S., and Skrutskie, M. F. 1989*d*. Circumstellar material associated with solar-type pre-main sequence stars: a possible constraint on the timescale for planet building. *Astron. J.* 97:1451–1470.

Strom, K. M., Wilkin, F. P., Strom S. E., and Seaman, R. L. 1989*e*. Lithium abundances among solar-type pre-main-sequence stars. *Astron. J.* 98:1444–1450.

Strom, K. M., Cabrit, S., Edwards, S., Skrutskie, M. F., and Strom, S. E. 1989*f*. The spectral energy distributions of classical and naked T Tauri stars. *Astron. J.* 97:1451–1470.

Strom, K. M., Strom, S. E., Wilkin, F. P., Carrasco, L., Cruz-Gonzalez, I., Recillas, E., Serrano, A., Seaman, R. L., Stauffer, J. R., Dai, D., and Sottile, J. 1990. A study of the stellar population in the Lynds 1641 dark cloud. IV. The *Einstein* X-ray sources. *Astrophys. J.* 362:168–190.

Strom, R. G. 1984. Mercury. In *The Geology of the Terrestrial Planets*, ed. M. H. Carr, NASA SP-469, pp. 13–55.

Strom, S. E. 1985. Protostars and planets: overview from an astronomical perspective. In *Protostars and Planets II*, eds. D. C. Black and M. S. Matthews (Tucson: Univ. of Arizona Press), pp. 17–29.

Strom, S. E., Strom, K.M., Yost, J., Carrasco, L., and Grasdalen, G. 1972. The nature of the Herbig Ae and Be-type stars associated with nebulosity. *Astrophys. J.* 173:353–366.

Strom, S. E., Strom, K. M., and Grasdalen, G. L. 1973. Young stellar objects and dark interstellar clouds. *Ann. Rev. Astron. Astrophys.* 13:187–216.

Strom, S. E., Grasdalen, G. L., and Strom, K. M. 1974. Infrared and optical observations of Herbig-Haro objects. *Astron. Astrophys.* 191:111–142.

Strom, S. E., Strom, K. E., Grasdalen, G. L., Capps, R. W., and Thompson, D. 1985. High-spatial-resolution studies of young stellar objects. II. A thick disk surrounding Lynds 1551, IRS 5. *Astron. J.* 90:2575–2580.

Strom, S. E., Strom, K. M., and Edwards, S. 1988. Energetic winds and circumstellar disks associated with low mass young stellar objects. In *Galactic and Extragalactic Star Formation*, eds. R. E. Pudritz and M. Fich (Dordecht: Kluwer), pp. 53–88.

Strom, S. E., Edwards, S., and Strom, K. M. 1989*a*. Constraints on the properities and environment of primitive stellar nebulae from the astrophysical record provided by young stellar objects. In *The Formation and Evolution of Planetary Systems*, eds. H. A. Weaver and L. Danly (Cambridge: Univ. of Cambridge), pp. 91–106.

Strom, S. E. Strom, K. M., Edwards, S., Cabrit, S., and Skrutskie, M. F. 1989*b*. Circumstellar material associated with solar-type pre-main-sequence stars: A possible constraint on the timescale for planet building. *Astron. J.* 97:1451–1470.

Strömgren, B. 1939. The physical state of interstellar hydrogen. *Astrophys. J.* 89:526–547.

Struck-Marcell, C., and Scalo, J.M. 1984. Continuum models for gas in disturbed galaxies. II. Stability of simplified model systems. *Astrophys. J.* 277:132–148.

Stutski, J., and Güsten, R. 1990. High spatial resolution isotopic CO and CS observations of M17SW: the clumpy structure of the molecular cloud core. *Astrophys. J.* 356:513–533.

Stutzki, J., Stacey, G. J., Genzel, R., Harris, A. I., Jaffe, D. T., and Lugten, J. B. 1988. Submillimeter and far-infrared line observations of M17 SW: A clumpy molecular cloud core penetrated by ultraviolet radiation. *Astrophys. J.* 332:379–399.

Suess, H. E., and Urey, H. C. 1956. Abundances of the elements. *Rev. Mod. Phys.* 28:53–74.

Suess, S. T. and Nerney, S. F. 1973. Meridional flow and the validity of the two-dimensional approximation in stellar-wind modeling. *Astrophys. J.* 184:17–25.

Sugitani, K., and Fukui, Y. 1988. Molecular cloud cores and cold *IRAS* point sources in Cepheus. *Vistas in Astron.* 31:507–511.

Sugitani, K., Fukui, Y., Mizuno, A., and Ohashi, N. 1989. Star formation in bright-rimmed globules: evidence for radiation-driven implosion. *Astrophys. J. Lett.* 342:87–90.

Sugitani, K., Fukui, Y., and Ogura, K. 1991. A catalog of bright-rimmed clouds with IRAS point sources: candidates for star formation by radiation-driven implosion. I. Northern hemisphere.*Astrophys. J. Suppl.* 77:59–66.

Sussman, G. J., and Wisdom, J. 1988. Numerical evidence that the motion of Pluto is chaotic. *Science* 241:433–437.

Sutton, E. C., Blake, G. A., Masson, C. R., and Phillips, T. G. 1985. Molecular line survey of Orion A from 215 to 247 GHz. *Astrophys. J. Suppl.* 58:341–378.

Sutton, E. C., Jaminet, P. A., Danchi, W. C., and Blake, G. A. 1991. Molecular line survey of Sagittarius B2(M) from 330 to 355 GHz and comparison with Sagittarius B2(N). *Astrophys. J. Suppl.* 77:255–285.

Suzuki, H. 1983. Synthesis of chain molecules in regions with partially ionized carbon. *Astrophys. J.* 272:579–590.

Swade, D. A. 1989*a*. Radio wavelength observations of the L134N molecular core. *Astrophys. J. Suppl.* 71:219–244.

Swade, D. A. 1989*b*. The physics and chemistry of the L134N molecular core. *Astrophys. J.* 345:828–852.

Swade, D. A., and Schloerb, F. P. 1992. A source model for the L134N molecular cloud. *Astrophys. J.* 392:543–550.

Swade, D. A., Schloerb, F. P., Irvine, W. M., and Kinzel, W. M. 1986. Search for molecules in comet Halley at millimeter wavelengths. In *Cometary Radio Astronomy*, eds. W. M. Irvine, F. P. Schloerb and Tacconi-Garman (Greenbank, W. Va.: NRAO), pp. 79–83.

Swart, P. K., Grady, M. M., Pillinger, C. T., Lewis, R. S., and Anders, E. 1983. Interstellar carbon in meteorites. *Science* 220:406–410.

Swindle, T. D. 1988. Trapped noble gases in meteorites. In *Meteorites and the Early Solar System*, eds. J. F. Kerridge and M. S. Matthews (Tucson: Univ. of Arizona Press), pp. 535–564.

Swindle, T. D., and Podosek, F. A. 1988. Iodine-xenon dating. In *Meteorites and the Early Solar System*, eds. J. F. Kerridge and M. S. Matthews (Tucson: Univ. of Arizona Press), pp. 1127–1146.

Swindle, T. D., Caffee, M. W., Hohenberg, C. M. and Taylor, S. R. 1986*a*. I-Pu-Xe dating and the relative ages of the Earth and Moon. In *The Origin of the Moon*, eds. W. K. Hartmann, R. J. Phillips and G. J. Taylor (Houston: Lunar and Planetary Inst.), pp. 331–358.

Swindle, T. D., Caffee, M. W., and Hohenberg, C. M. 1986*b*. Xenon and other noble gases in shergottites. *Geochim. Cosmochim. Acta* 50:1001–1015.

Sykes, M. V., and Greenberg, R. 1986. The formation and origin of the IRAS zodiacal dust bands as a consequence of single collisions between asteroids. *Icarus* 65:51–

69.

Sykes, M. V., and Walker, R. G. 1992. Cometary dust trails. I. Survey. *Icarus* 95:180–210.

Sykes, M. V., Lebofsky, L. A., Hunten, D. M., and Low, F. J. 1986. The discovery of dust trails in the orbits of periodic comets. *Science* 232:1115–1117.

Sykes, M. V., Cutri, R. M., Lebofsky, L. A., and Binzel, R. P. 1987. IRAS serendipitous survey observations of Pluto and Charon. *Science* 237:1336–1340.

Sykes, M. V., Lien, D. J., and Walker, R. G. 1990. The Tempel 2 dust trail. *Icarus* 86:236–247.

Tagger, M., Henriksen, R. N., Sygnet, J. F., and Pellat, R. 1990. Spiral waves and instability in magnetized astrophysical disks. *Astrophys. J.* 353:654–657.

Tagliaferri, G., Giommi, P., Angelini, L., Osborne, J. P., and Pallavicini, R. 1988. An X-ray flare from a B9 + post-T Tauri star system in the field of the Seyfert Galaxy III Zw 2. *Astrophys. J. Lett.* 331:113–116.

Takano, T., Fukui, Y., Ogawa, H., Takaba, H., Kawabe, R., Fujimoto, Y., Sugitani, K., and Kawabata, K. 1984. High angular resolution CS($J = 1$–0) observations of the bipolar flow source near NGC2071: can the CS compact cloud collimate the flow? *Astrophys. J. Lett.* 282:69–72.

Takano, T., Stutzki, J., Winnewisser, G., and Fukui, Y. 1985. Detection of high velocity emission in NH_3 (1,1) and (2,2) spectra towards NGC2071. *Astron. Astrophys.* 144:L20–L22.

Takano, T., Stutzki, J., Fukui, Y., and Winnewisser, G. 1986. High-angular-resolution NH_3 observations of the bipolar flow source near NGC2071. *Astron. Astrophys.* 167:333–340.

Takayanagi, K., Sakimoto, K., and Onda, K. 1987. Para/ortho abundance ratio of molecular hydrogen in NGC 2023. *Astrophys. J. Lett.* 318:81–84.

Takeda, H., Matsuda, T., Sawada, K., and Hayashi, C. 1985. Drag on a gravitating sphere moving through gas. *Prog. Theor. Phys.* 74:272–287.

Tammann, G. A. 1982. Supernova statistics and related problems. In *Supernovae: A Survey of Current Research*, eds. M. J. Rees and R. J. Stoneham (Dordrecht: D. Reidel), pp. 371–403.

Tamura, M., and Sato, S. 1989. A two micron polarization survey of T Tauri stars. *Astron. J.* 98:1368–1381.

Tamura, M., Nagata, T., Sato, S., and Tanaka, M. 1987. Infrared polarimetry of dark clouds—I. magnetic field structure in Heiles cloud 2. *Mon. Not. Roy. Astron. Soc.* 224:413–423.

Tamura, M., Yamashita, T., Sato, S., Nagata, T., and Gatley, I. 1988. Infrared polarimetry of dark clouds—III. The relationship between the magnetic field and star formation in the NGC 1333 region. *Mon. Not. Roy. Astron. Soc.* 231:445–453.

Tamura, M., Sato, S., Suzuki, H., Kaifu, N., and Hough, J. H. 1990. CO outflow and infrared reflection nebula of GSS30 in the Rho Ophiuchi core. *Astrophys. J.* 350:728–731.

Tamura, M., Gatley, I., Joyce, R. R., Ueno, M., Suto, H., and Sekiguchi, M. 1991. Infrared polarization images of star-forming regions. I. The ubiquity of bipolar structure. *Astrophys. J.* 378:611–627.

Tanaka, H., and Nakazawa, K. 1992. Stochastic coagulation equation. In preparation.

Tang, M., and Anders, E. 1988*a*. Isotopic anomalies of Ne, Xe, and C in meteorites. II. Interstellar diamond and SiC: Carriers of exotic noble gases. *Geochim. Cosmochim. Acta* 52:1235–1244.

Tang, M., and Anders, E. 1988*b*. Interstellar silicon carbide: How much older than the solar system? *Astrophys. J. Lett.* 335:31–34.

Tang, M., and Anders, E. 1988*c*. Isotopic anomalies of Ne, Xe, and C in meteorites.

III. Local and exotic noble gas components and their interrelations. *Geochim. Cosmochim. Acta* 52:1245–1254.

Tang, M., Lewis, R. S., Anders, E., Grady, M. M., Wright, I. P., and Pillinger, C. T. 1988. Isotopic anomalies of Ne, Xe, and C in meteorites. I. Separation of carriers by density and chemical resistance. *Geochim. Cosmochim. Acta* 52:1221–1234.

Tang, M., Anders, E., Hoppe, P., and Zinner, E. 1989. Meteoritic silicon carbide and its stellar sources; implications for galactic chemical evolution. *Nature* 339:351–354.

Tapia, M., Lopez, J. A., Roth, M., Persi, P., and Ferrari-Toniolo, M. 1991. High resolution images of the embedded cluster associated with GM 24. In *Astrophysics with Infrared Arrays*, ed. R. Elston (San Francisco: Astron. Soc. of the Pacific), pp. 252–254.

Tarafdar, S. P., Prasad, S. S., Huntress, W. T., Jr., Villere, K. R., and Black, D. C. 1985. Chemistry in dynamically evolving clouds. *Astrophys. J.* 289:220–237.

Tassoul, J.-L. 1978. *Theory of Rotating Stars* (Princeton: Princeton Univ. Press).

Tassoul, J.-L. 1987. On synchronization in carly-type binaries. *Astrophys. J.* 322:856–861.

Tassoul, J.-L. 1988. On orbital circularization in detached close binaries. *Astrophys. J. Lett.* 324:71–73.

Tassoul, J.-L. 1990. On orbital circularization in massive close binaries. *Astrophys. J.* 358:196–198.

Tauber, F., and Kührt, E. 1987. Thermal stresses in cometary nuclei. *Icarus* 69:83–90.

Tauber, J. A., and Goldsmith, P. F. 1990. A model for clumpy giant molecular clouds with extended UV heating. *Astrophys. J. Lett.* 356:63–66.

Taylor, G. I. 1964. Disintegration of water drops in an electric field. *Proc. Roy. Soc. London* A280:383–397.

Taylor, K. N. R., and Storey, J. W. V. 1984. The Coronet, an obscured cluster adjacent to R Corona Austrina. *Mon. Not. Roy. Astron. Soc.* 209:5P–10P.

Taylor, S. R. 1988. Planetary compositions. In *Meteorites and the Early Solar System*, eds. J. F. Kerridge and M. S. Matthews (Tucson: Univ. of Arizona Press), pp. 512–534.

Tegler, S. C., and Wyckoff, S. 1989. NH_2 fluorescence efficiencies and the NH_2 abundance in comet Halley. *Astrophys. J.* 343:445–449.

Telesco, C. M., and Knacke, R. F. 1991. Detection of silicates in the beta Pictoris disk. *Astrophys. J. Lett.* 372:29–31.

Telesco, C. M., Becklin, E. E., Wolstencroft, R. D., and Decker, E. 1988. Resolution of the circumstellar disk of β Pictoris at 10 and 20 μm. *Nature* 335:51–53.

Tennekes, H., and Lumley, J. 1972. *A First Course in Turbulence* (Cambridge, Mass.: MIT Press).

Tenorio-Tagle, G. 1981. The collision of clouds with a galactic disk. *Astron. Astrophys.* 94:338–344.

Tenorio-Tagle, G., and Bodenheimer, P. 1989. Large scale expanding superstructures in galaxies. *Ann. Rev. Astron. Astrophys.* 26:145–197.

Tenorio-Tagle, G., and Palous, J. 1987. Giant scale supernova remnants. The role of differential galactic rotation and the formation of interstellar clouds. *Astron. Astrophys.* 186:287

Tenorio-Tagle, G., Cantö, J., and Różyczka, M. 1988. The formation of interstellar jets. *Astron. Astrophys.* 202:256–266.

Tenorio-Tagle, G., Różyczka, M., and Bodenheimer, P. 1990. The hydrodynamics of superstructures produced by multiple supernova explosions. *Astron. Astrophys.* 237:207–214.

Tera, F., and Wasserburg, G. J. 1976. Lunar ball games and other sports. *Lunar Sci.* VII:858–860 (abstract).

Terebey, S., Shu, F. H., and Cassen, P. 1984. The collapse of the cores of slowly rotating isothermal clouds. *Astrophys. J.* 286:529–551.

Terebey, S., Vogel, S. N., and Myers, P. C. 1989. High-resolution CO observations of young low-mass stars. *Astrophys. J.* 340:472–478.

Ter Haar, D. 1950. Further studies on the origin of the solar system. *Astrophys. J.* 111:179–190.

Thaddeus, P., Cummins, S. E., and Linke, R. A. 1984. Identification of the SiCC radical toward IRC+10216: the first molecular ring in an astronomical source. *Astrophys. J. Lett.* 283:45–48.

The, P. S., Wesselius, P. R., Tjin, A., Djie, H. R. E., and Steenman, H. 1986. Studies of the Chameleon star-forming region. II. The pre-main sequence stars HD97048 and HD97300. *Astron. Astrophys.* 155:347–355.

Thielemann, F.-K., Nomoto, K., and Hashimoto, M. 1992. Nucleosynthesis in supernovae. In *Supernovae*, eds. J. Audouze, S. Bludman, R. Mochkovitch and J. Zinn-Justin (Amsterdam: Elsevier), in press.

Thiemens, M. H., and Heidenreich, J. E. 1983. The mass independent fractioning of oxygen. *Science* 219:1073–1075.

Thompson, W. R., Murray, B. G. J. P. T., Khare, B. N., and Sagan, C. 1987. Coloration and darkening of methane clathrate and other ices by charged particle radiation: applications to the outer solar system. *J. Geophys. Res.* 92:14933–14947.

Thomsen, L. 1980. $^{1}29Xe$ on the outgassing of the atmosphere. *J. Geophys. Res.* 85:4374–4378.

Thronson, H. A. 1988. Molecular gas in disk galaxies without spirals: Irregulars and SOs. In *Molecular Clouds in the Milky Way and External Galaxies*, eds. R. L. Dickman, R. L. Snell and J. S. Young (Berlin: Springer-Verlag), pp. 413–420.

Thronson, H. A., and Erickson, E. F., 1984. *Airborne Astronomy Symposium*, NASA CP-2353.

Thronson, H. A., Lada, C. J., and Hewagama, T. 1985. The W3 molecular cloud. *Astrophys. J.* 297:662–676.

Tielens, A. G. G. M. 1983. Surface chemistry of deuterated molecules. *Astron. Astrophys.* 119:177–184.

Tielens, A. G. G. M. 1989. Dust in dense clouds. In *Interstellar Dust: Proc. IAU Symp. 135*, eds. L. J. Allamandola and A. G. G. M. Tielens (Dordrecht: Kluwer), pp. 239–262.

Tielens, A. G. G. M., and Allamandola, L. J. 1987*a*. Evolution of interstellar dust. In *Physical Processes in Interstellar Clouds*, eds. G. Morfill and M. S. Scholer (Dordrecht: D. Reidel), pp. 333–376.

Tielens, A. G. G. M., and Allamandola, L. J. 1987*b*. Composition, structure and chemistry of interstellar dust. In *Interstellar Processes*, eds. D. Hollenbach and H. A. Thronson (Dordrecht: D. Reidel), pp. 379–470.

Tielens, A. G. G. M., and Hagen, W. 1982. Model calculations of the molecular composition of interstellar grain mantles. *Astron. Astrophys.* 114:245–260.

Tielens, A. G. G. M., and Hollenbach, D. 1985. Photodissociation regions. I. Basic model. *Astrophys. J.* 291:722–746.

Tielens, A. G. G. M., Seab, C. G., Hollenbach, D. J., and McKee, C. F. 1987. Shock processing of interstellar dust: Diamonds in the sky. *Astrophys. J. Lett.* 319:109–113.

Tielens, A. G. G. M., Tokunaga, A. T., Geballe, T. R., and Baas, F. 1991. Interstellar solid CO: polar and nonpolar interstellar ices. *Astrophys. J.* 321:181–199.

Tilanus, R. P. J., and Allen, R. J. 1989. Spiral structure of M51: Displacement of the HI from the nonthermal radio arms. *Astrophys. J. Lett.* 339:57–61.

Tilton, G. R. 1988. Age of the solar system. In *Meteorites and the Early Solar System*, eds. J. F. Kerridge and M. S. Matthews (Tucson: Univ. of Arizona Press), pp.

259–275.

Tittemore, W. C. 1990*a*. Chaotic motion of Europa and Ganymede and the Ganymede-Callisto dichotomy. *Science* 250:263–267.

Tittemore, W. C. 1990*b*. Tidal heating of Ariel. *Icarus* 87:110–139.

Tittemore, W. C., and Wisdom, J. 1988. Tidal evolution of the Uranian satellites. I. Passage of Ariel and Umbriel through the 5:3 mean-motion commensurability. *Icarus* 74:172–230.

Tittemore, W. C., and Wisdom, J. 1989. Tidal evolution of the Uranian satellites. II. An explanantion of the anomalously high orbital inclination of Miranda. *Icarus* 77:63–89.

Tittemore, W. C., and Wisdom, J. 1990. Tidal evolution of the Uranian satellites. III. Evolution through the Miranda-Umbriel 3:1, Miranda-Ariel 5:3, and Ariel-Umbriel 2:1 mean-motion commensurabilities. *Icarus* 85:394–443.

Tohline, J. E. 1981. The collapse to equilibrium of rotating adiabatic spheroids. I. Protostars. *Astrophys. J.* 248:717–726.

Tohline, J. E., and Hachisu, I. 1990. The breakup of self-gravitating rings, tori, and thick accretion disks. *Astrophys. J.* 361:394–407.

Tokunaga, A., and Brooke, T. 1990. Did comets form from unaltered interstellar dust and ices? The evidence from infrared spectroscopy. *Icarus* 86:208–219.

Tölle, F., Ungerechts, H., Walmsley, C. M., Winnewisser, G., and Churchwell, E. 1981. A molecular line study of the elongated dark dust cloud TMC 1. *Astron. Astrophys.* 95:143–155.

Tombaugh, C. 1961. In *Planets and Satellites*, eds. G. Kuiper and B. Middlehurst (Chicago: Univ. of Chicago Press), pp. 12

Tomisaka, K. 1984. Coagulation of interstellar clouds in spiral gravitational potential and formation of giant molecular clouds. *Publ. Astron. Soc. Japan* 36:457–475.

Tomisaka, K. 1986. Formation of giant molecular clouds by coagulation of small clouds and spiral structure. *Publ. Astron. Soc. Japan* 38:95–109.

Tomisaka, K. 1987. Galactic shock in cloud fluid and its gravitational instability. *Publ. Astron. Soc. Japan* 39:109–133.

Tomisaka, K. 1990. Blowout of superbubble in galactic magnetic field. *Astrophys. J. Lett.* 361:5–8.

Tomisaka, K. 1991. The equilibria and evolution of magnetized, rotating, isothermal clouds. V. The effect of the toroidal field. *Astrophys. J.* 376:190–198.

Tomisaka, K., Ikeuchi, S., and Nakamura, T. 1988*a*. The equilibria and evolutions of magnetized, rotating, isothermal clouds. I. Basic equations and numerical methods. *Astrophys. J.* 326:208–222.

Tomisaka, K., Ikeuchi, S., and Nakamura, T. 1988*b*. The equilibria and evolutions of magnetized, rotating, isothermal clouds. II. The extreme case: nonrotating clouds. *Astrophys. J.* 335:239–262.

Tomisaka, K., Ikeuchi, S., and Nakamura, T. 1989. The equilibria and evolutions of magnetized, rotating, isothermal clouds. III. Critical mass. *Astrophys. J.* 341:220–237.

Tomisaka, K., Ikeuchi, S., and Nakamura, T. 1990. The equilibria and evolutions of magnetized, rotating, isothermal clouds. IV. Quasistatic evolution. *Astrophys. J.* 362:202–214.

Tomkin, J. 1983. Secondaries of eclipsing binaries. V. EK Cephei. *Astrophys. J.* 271:717–724.

Tonks, W. B., and Melosh, H. J. 1990. The physics of crystal settling and suspension in a turbulent magma ocean. In *Origin of the Earth*, eds. J. H. Jones and H. E. Newsom (New York: Oxford Univ. Press), pp. 151–174.

Tonks, W. B., and Melosh, H. J. 1991. Core formation by giant impacts. *Lunar Planet. Sci.* XXII:1405–1406 (abstract).

Toomre, A. 1964. On the gravitational stability of a disk of stars. *Astrophys. J.* 139:1217–1238.

Toomre, A. 1969. Group velocity of spiral waves in galactic disks. *Astrophys. J.* 158:899–913.

Toomre, A. 1981. What amplifies the spirals? In *The Structure and Evolution of Normal Galaxies*, eds. S. M. Fall and D. Lynden-Bell (Cambridge: Cambridge Univ. Press), pp. 111–136.

Toomre, A. 1989. Non-axisymmetric disturbances in galactic discs. In *Dynamics of Astrophysical Discs*, ed. J. A. Sellwood (Cambridge: Cambridge Univ. Press), pp. 153–154.

Torbett, M. V. 1984. Hydrodynamic ejection of bipolar flows from objects undergoing disk accretion: T Tauri stars, massive pre-main-sequence objects, and cataclysmic variables. *Astrophys. J.* 278:318–325.

Torbett, M. V. 1986. Dynamical influence of galactic tides and molecular clouds on the Oort cloud of comets. In *The Galaxy and the Solary System*, eds. R. Smoluchowski, J. Bahcall and M. S. Matthews (Tucson: Univ. of Arizona Press), pp. 147–172.

Torbett, M. V. 1989. Chaotic motion in a comet disk beyond Neptune: the delivery of short-period comets. *Astron. J.* 98:1477–1481.

Torbett, M. V., and Smoluchowski, R. 1980. Sweeping of the Jovian resonances and the evolution of the asteroids. *Icarus* 44:722–729.

Torbett, M. V., and Smoluchowski, R. 1990. Chaotic motion in a primordial comet disk beyond Neptune and comet influx. *Nature* 345:49–51.

Torrelles, J. M., Rodríguez, L. F., Cantó, J., Carral, P., Marcaide, J., Moran, J. M., and Ho, P. T. P. 1983. Are interstellar toroids the focusing agent of the bipolar molecular outflows? *Astrophys. J.* 274:214–230.

Tremaine, S. 1990. Dark matter in the solar system. In *Baryonic Dark Matter*, eds. D. Lynden-Bell and G. Gilmore (Netherlands:Kluwer Academic Press Pub.) pp. 37–65.

Tremaine, S. 1991. On the origin of the obliquities of the outer planets. *Icarus* 89:85–92.

Troland, T. H. 1990. Observational aspects of magnetic fields in molecular clouds. In *Galactic and Extragalactic Magnetic Fields*, eds. R. Beck, P. P. Kronberg and R. Wielebinski (Dordrecht: Kluwer), pp. 293

Troland, T. H., and Heiles, C. 1986. Interstellar magnetic field strengths and gas densities: observational and theoretical perspectives. *Astrophys. J.* 301:339–345.

Troland, T. H., Crutcher, R. M., and Kazès, I. 1986. Detection of the OH zeeman effect toward Orion A. *Astrophys. J. Lett.* 304:57–60.

Troland, T. H., Heiles, C., and Goss, W. M. 1989. Aperture synthesis observations of the 21 centimeter zeeman effect toward Orion A. *Astrophys. J.* 337:342–354.

Troland, T. H., Crutcher, R. M., Goss, W. M., and Heiles, C. 1989. Structure of the magnetic field in the W3 core. *Astrophys. J. Lett.* 347:89–92.

Troland, T. H., Heiles, C., Goodman, A., Crutcher, R. M., and Myers, P. C. 1991. The magnetic field in the ρ Ophiuchus dark cloud. In preparation.

Trubnikov, B. A. 1971. Solution of the coagulation equations in the case of a bilinear coefficient of adhesion of particles. *Soviet Physics-Doklady* 16:124–126.

Trulsen, J. 1971. Towards a theory of jet streams. *Astrophys. Space Sci.* 12:329–348.

Trumpler, R. J. 1931. The distance of the Orion Nebula. *Publ. Astron. Soc. Pacific* 43:255–260.

Truran, J. W. 1982. Nuclear theory of novae. In *Essays in Nuclear Astrophysics*, eds. C. A. Barns, D. D. Clayton and D. N. Schramm (Cambridge: Cambridge Univ. Press), pp. 467–493.

Truran, J. W. 1985. Nucleosynthesis in novae. In *Nucleosynthesis: Challenges and New Developments*, eds. W. D. Arnett and J. W. Truran (Chicago: Univ. of Chicago Press), pp. 292–306.

Truran, J. W. 1990. Theoretical implications of nova abundances. In *Physics of Classical Novae*, eds. A. Cassatella and R. Viotti (Berlin: Springer-Verlag), pp. 373–385.

Truran, J. W., and Cameron, A. G. W. 1978. ^{26}Al production in explosive carbon burning. *Astrophys. J.* 219:226–229.

Truran, J. W., and Hillebrandt, W. 1986. A gamma-ray impased constraint on the mode of origin of mass 22 in astrophysical environments. *Astron. Astrophys.* 168:384–385.

Truran, J. W., and Livio, M. 1986. On the frequency of occurence of oxygen-neon-magnesium white dwarfs in classical nova systems. *Astrophys. J.* 308:721–727.

Truran, J. W., and Livio, M. 1989. On the masses of white dwarfs in classical nova systems. In *White Dwarfs*, ed. G. Wegner (Berlin:Springer-Verlag), pp. 498–506.

Tsai, J. C., and Bertschinger, E. 1989. Stability of the expansion wave solution of collapsing isothermal spheres. *Bull. Amer. Astron. Soc.* 21:1089 (abstract).

Tscharnuter, W. M. 1978. Collapse of the presolar nebula. *Moon and Planets* 19:229–236.

Tscharnuter, W. M. 1987*a*. A collapse model of the turbulent presolar nebula. *Astron. Astrophys.* 188:55–73.

Tscharnuter, W. M. 1987*b*. Models of star formation. In *Physical Processes in Comets, Stars, and Active Galaxies*, eds. E. Meyer-Hofmeister, H. C. Thomas and W. Hillebrandt (Berlin: Springer-Verlag), pp. 96–104.

Tscharnuter, W. M. 1989. Formation of viscous protostellar accretion disks. In *The Theory of Accretion Disks*, eds. F. Myer, W. Duschl, J. Frank and E. Meyer-Hofmeister (Dordrecht: Kluwer), pp. 113–123.

Tsvetkova, K. P. 1982. Photographic photometry of V1515 Cygni. *I. B. V. S.* No. 2236.

Turner, B. E. 1974. Detection of OH at 18-centimeter wavelength in comet Kohoutek (1973f). *Astrophys. J. Lett.* 189:137–139.

Turner, B. E. 1988. Molecules as probes of the interstellar medium and of star formation. In *Galactic and Extragalactic Radio Astronomy*, eds. G. L. Verschuur and K. I. Kellerman (Berlin: Springer-Verlag), pp. 154–199.

Turner, B. E. 1989*a*. Detection of interstellar C_4D: Implications for ion-molecule chemistry. *Astrophys. J. Lett.* 347:39–42.

Turner, B. E. 1989*b*. Molecular line survey of Sagittarius B2 and Orion-KL from 70 to 115 GHz. I. The observational data. *Astrophys. J. Suppl.* 70:539–622.

Turner, B. E. 1989*c*. Recent progress in astrochemistry. *Space Sci. Rev.* 51:235–337.

Turner, B.E. 1990. Detection of doubly deuterated interstellar formaldehyde (D_2CO): An indication of active grain surface chemistry. *Astrophys. J. Lett.* 362:29–33.

Turner, B.E. 1991. Observations and chemistry of interstellar refractory elements. *Astrophys. J.* 376:573–598.

Turner, B. E., and Ziurys, L. M. 1988. Interstellar molecules and astrochemistry. In *Galactic and Extragalactic Radio Astronomy*, eds. G. L. Verschuur and K. I. Kellerman (Berlin: Springer-Verlag), pp. 200–254.

Turner, D. G. 1976. The value of R in monoceros. *Astrophys. J.* 210:65–75.

Turner, J. L., and Welch, W. J. 1984. Discovery of a young stellar object near water masers in W 3(OH). *Astrophys. J. Lett.* 287:81–84.

Turnshek, D. A., Turnshek, D. E., and Craine, E. 1980. Spectroscopic and polarimetric observations of NGC 1333 and the surrounding dark cloud complex. *Astron. J.* 85:1638–1643.

Tylenda, R. 1978. Photoionization models of the envelope of Nova Delphini 1967 in the nebular stage II. The nova in June–July 1969. *Acta Astron.* 28:333–361.

Tyler, G. L., Eshleman, V. R., Anderson, J. D., Levy, G. S., Lindal, G. F., Wood, G. E., and Croft, T. A. 1982. Radio science with Voyager 2 at Saturn: Atmosphere and ionosphere and the masses of Mimas, Tethys, and Iapetus. *Science* 215:553–558.

Tyler, G. L., Sweetnam, D. N., Anderson, J. D., Borutzki, S. E., Campbell, J. K., Eshleman, V. R., Gresh, D. L., Gurrola, E. M., Hinson, D. P., Kawashima, N., Kursinski, E. R., Levy, G. S., Lindal, G. F., Lyons, J. R., Marouf, E. A., Rosen, P. A., Simpson, R. A., and Wood, G. E. 1989. Voyager radio science observations of Neptune and Triton. *Science* 246:1466–1473.

Uchida, Y. 1989. Gravo-magnetodynamical flows around forming stars—molecular bipolar flows, optical jets, and streamers. In *ESO Workshop on Low Mass Star Formation and Pre-Main Sequence Objects*, ed. B. Reipurth (Garching: European Southern Obs.), pp. 141–152.

Uchida, Y., and Shibata, K. 1984. Magnetically buffered accretion to a young star and the formation of bipolar flows. *Publ. Astron. Soc. Japan* 36:105–118.

Uchida, Y., and Shibata, K. 1985. Magnetodynamical acceleration of CO and optical bipolar flows from the region of star formation. *Publ. Astron. Soc. Japan* 37:515–535.

Uchida, Y., Kaifu, N., Shibata, K., Hayashi, S. S., and Hasegawa, T. 1987a. Hollow cylindrical lobes with a helical velocity field of the L1551 bipolar flow. In *Star-Forming Regions: Proc. IAU Symp. 115*, eds. M. Peimbert and J. Jugaku (Dordrecht: D. Reidel), pp. 287–300.

Uchida, Y., Kaifu, N., Shibata, K., Hayashi, S. S., Hasegawa, T., and Hamatake, H. 1987b. Observations of the detailed structure and velocity field in the CO bipolar flows associated with L1551 IRS-5. *Publ. Astron. Soc. Japan* 39:907–924.

Uchida, Y., Mizuno, A., Nozawa, S., and Fukui, Y. 1990. Velocity split along the Rho Ophiuchi streamer (North): spinning streamer as "angular momentum drain" from massive cloud of active star formation. *Publ. Astron. Soc Japan* 42:69–83.

Uchida, Y., Fukui, Y., Monishima, Y., Mizuno, A., Iwata, T., and Takaba, H. 1991. Evidence for a rotating filament in L1641, part of the Orion cloud complex. *Nature* 349:140–142.

Uesugi, A., and Fukuda, I. 1970. *Mem. Faculty of Sci., Kyoto Univ.* No. 33, Art. 5.

Umebayashi, T., and Nakano, T. 1988. Ionization state and magnetic fields in the solar nebula. *Prog. Theor. Phys. Suppl.* 96:151–160.

Umebayashi, T., and Nakano, T. 1990. Magnetic flux loss from interstellar clouds. *Mon. Not. Roy. Astron. Soc.* 243:103–113.

Umemoto, T., Hirano, N., Kameya, O., Fukui, Y., Kuno, N., and Takakubo, K. 1991. U-shaped outflow in the L1221 dark clouds: an example of interaction of outflows with ambient clouds. *Astrophys. J.* 377:510–518.

Ungerechts, H., and Thaddeus, P. 1987. A CO survey of the dark nebulae in Perseus, Taurus and Auriga. *Astrophys. J. Suppl.* 63:645–660.

Ungerechts, H., Walmsley, C. M., and Winnewasser, G. 1982. Ammonia observations of cold cloud cores. *Astron. Astrophys.* 111:339–345.

Urey, H. 1952. *The Planets: Their Origin and Development* (New Haven: Yale Univ. Press).

Vainshtein, S., and Levy, E. H. 1991. Dynamical behavior of strong magnetic fields in the solar convection zone. *Solar Phys.* 135:261–274.

Valtonen, M. J. 1983. On the capture of comets into the inner solar system. *Observatory* 103:1–4.

Valtonen, M. J., and Innanen, K. A. 1982. The capture of interstellar comets. *Astrophys. J.* 255:307–315.

van Albada, T. S. 1968a. Numerical integration of the N-body problem. *Bull. Astron. Inst. Netherlands* 19:479–499.

van Albada, T. S. 1968*b*. The evolution of small stellar systems and the implications for the formation of double stars. *Bull. Astron. Inst. Netherlands* 20:57–68.

van Altena, W. F., Lee, J.-F., Lee, J. T., Lu, P. K., and Upgren, A. R. 1988. The velocity dispersion of the Orion Nebula cluster. *Astron. J.* 95:1744–1754.

Vandenberg, D. A., Hartwick, F. O. A., Dawson, P., and Alexander, D. R. 1983. Studies of late-type dwarfs. V. Theoretical models for lower main-sequence stars. *Astrophys. J.* 266:747–754.

van der Kruit, P. C. 1987. Comparison of the galaxy with external spiral galaxies. In *The Galaxy*, eds. G. Gilmore and B. Carswell (Dordrecht: D. Reidel), pp. 27–50.

Vandervoort, P. O. 1963. The stability of ionization fronts and the evolution of HII regions. *Astrophys. J.* 138:599–601.

van der Werf, P. P., and Goss, W. M. 1990. Subparsec size HI cloudlets associated with Orion A. *Astrophys. J.* 364:157–163.

van der Werf, P. P., and Goss, W. M. 1990. High resolution H I observations of H II regions. II. H I and the magnetic field near W3. *Astron. Astrophys.* 238:296–314.

van Dishoeck, E. F. 1988*a*. Molecular cloud chemistry. In *Millimetre and Submillimetre Astronomy*, eds. R. D. Wolstencroft and W. B. Burton (Dordrecht: Kluwer), pp. 117–164.

van Dishoeck, E. F. 1988*b*. Photodissociation and photoionization processes. In *Rate Coefficients in Astrochemistry*, eds. T. J. Millar and D. A. Williams (Dordrecht: Kluwer), pp. 49–72.

van Dishoeck, E. F. 1990. The chemical evolution of the diffuse interstellar gas. In *The Evolution of the Interstellar Medium*, ed. L. Blitz, pp. 207–228.

van Dishoeck, E. F., and Black, J. H. 1986. Comprehensive models of diffuse interstellar clouds: Physical conditions and molecular abundances. *Astrophys. J. Supp.* 62:109–145.

van Dishoeck, E. F., and Black, J. H. 1987. The abundance of interstellar CO. In *Physical Processes in Interstellar Clouds*, eds. G. Morfill and M. S. Scholer (Dordrecht: Kluwer), pp. 241–274.

van Dishoeck, E. F., and Black, J. H. 1988*a*. Diffuse cloud chemistry. In *Rate Coefficients in Astrochemistry*, eds. T. J. Millar and D. A. Williams (Dordrecht: Kluwer), pp. 209–238.

van Dishoeck, E. F., and Black, J. H. 1988*b*. The photodissociation and chemistry of interstellar CO. *Astrophys. J.* 334:771–802.

van Dishoeck, E.F., and Black, J.H. 1989. Interstellar C_2, CH and CN in translucent molecular clouds. *Astrophys. J.* 340:273–297.

van Dishoeck, E. F., Phillips, T. G., Keene, J., and Blake, G. A. 1992. Ground-based searches for interstellar H_2D^+. *Astron. Astrophys. Lett.* 26:13–16.

van Leeuwen, F. 1980. Mass and luminosity function of the Pleiades. In *Star Clusters*, ed. J. E. Hesser (Dordrecht: D. Reidel), pp. 157–163.

van Leeuwen, F. 1983. The Pleiades: An Astrometric and Photometric Study of an Open Cluster. Ph. D. Thesis, Univ. of Leiden.

van Leeuwen, F., Alphenaar, P., and Meys, J. J. M. 1983. VBLUW observations of Pleiades G and K Stars. *Astron. Astrophys. Suppl.* 67:483–506.

Vannice, M.A. 1975. The catalytic synthesis of hydrocarbons from H_2/CO mixtures over the Group VIII metals. *J. Catalysis* 37:449–461.

Vanoni, V. A. 1975. *Sedimentation Engineering* (New York: American Soc. of Civil Eng.).

van Woerkom, A. F. F. 1948. On the origin of comets. *Bull. Astron. Inst. Netherlands* 10:445–472.

Vanysek, V. 1991. Isotopic ratios in comets. In *Comets in the Post-Halley Era*, eds. R. Newburn, M. Neugebauer and J. Rahe (Dordrecht: Kluwer), pp. 879–895.

Vardya, M. S., and Tarafdar, S. P., eds. 1987. *Astrochemistry: Proc. IAU Symp. 120*

(Dordrecht: Kluwer).

Varsik, J. 1987. Properties of Stellar Activity in F Stars. Ph. D. Thesis, Univ. of Hawaii.

Vazquez, E. C., and Scalo, J. M. 1989. Evolution of the star formation rate in galaxies with increasing densities. *Astrophys. J.* 343:644–658.

Verbunt, F. 1989. The formation and evolution of binaries in globular clusters. In *Highlights of Astronomy*, vol. 8, ed. D. McNally (Dordrecht: Kluwer), pp. 139–142.

Verschoyle, T. T. H. 1931. The ternary system carbon monoxide-nitrogen-hydrogen and the component binary systems between temperatures of $-185°$ and $-215°$, and between pressures of 0 and 225 atmospheres. *Trans. Roy. Soc.* 230A:189–220.

Verschueren, W., and David, M. 1989. The effect of gas removal on the dynamical evolution of young stellar clusters. *Astron. Astrophys.* 219:105–120.

Verschuur, G.L. 1990. An association between HI concentrations within high velocity clouds A and C and nearby molecular clouds. *Astrophys. J.* 361:497–510.

Viallefond, F. 1988. Star formation in blue compact and irregular galaxies. In *Galactic and Extragalactic Star Formation*, eds. R. E. Pudritz and M. Fich (Dordrecht: Kluwer), pp. 439–458.

Viallefond, F., and Goss, W. M. 1986. HII regions in M33 III. Physical properties. *Astron. Astrophys.* 154:357–369.

Viallefond, F., Goss, W. M., and Allen, R. J. 1982. The giant spiral galaxy M101, VIII: Star formation in HI–HII associations. *Astron. Astrophys.* 115:373–387.

Vickery, A. M., and Melosh, H. J. 1990. Atmospheric erosion and impactor retention in large impacts: application to mass extinctions. In *Global Catastrophes in Earth History*, eds. B. Sharpton and P. Ward, Geological Soc. of America SP-247, pp. 289–300.

Vidal-Madjar, A., Hobbs, L. M., Ferlet, R., Gry, C., and Albert, C. 1986. The circumstellar gas cloud around Beta Pictoris. II. *Astron. Astrophys.* 167:325–332.

Vilas, F., and Smith, B. A. 1987. Coronagraph for astronomical imaging and spectrophotometry. *Applied Optics* 26:664–668.

Viotti, R. 1976. Forbidden and permitted emission lines of singly ionized iron as a diagnostic in the investigation of stellar emission-line spectra. *Astrophys. J.* 204:293–300.

Vishniac, E. T, Jin, L., and Diamond, P. 1990. Dynamo action by internal waves in accretion disks. *Astrophys. J.* 365:648–659.

Vityazev, A. V., and Pechernikova, G. V. 1981. Solution of the problem of the rotation of the planets within the framework of the statistical theory of accumulation. *Soviet Astron.* 25:494–499.

Vityazev, A. V., Perchernikova, G. V., and Safronov, V. S. 1988. Formation of Mercury and removal of its silicate shell. In *Mercury*, eds. F. Vilas, C. R. Chapman and M. S. Matthews (Tucson: Univ. of Arizona Press), pp. 667–669.

Vogel, S. N., and Kuhi, L. V. 1981. Rotational velocities of pre-main-sequence stars. *Astrophys. J.* 245:960–976.

Vogel, S. N., Wright, M. C. H., Plambeck, R. L., and Welch, W. J. 1984. Interaction of the outflow and quiescent gas in Orion: HCO$^+$ aperture synthesis maps. *Astrophys. J.* 283:655–667.

Vogel, S. N., Bieging, J. H., Plambeck, R. L., Welch, W. J., and Wright, M. C. H. 1985. Differential rotation near the Orion Kleinmann-Low region: Aperture synthesis observations of HCN emission. *Astrophys. J.* 296:600–605.

Vogt, S. 1981. A method for unambiguous determination of starspot temperatures and areas: application to II Pegasi, BY Draconis, and HD 209813. *Astrophys. J.*

250:327–340.

Vogt, S. S. 1983. Spots, spot-cycles, and magnetic fields of late-type dwarfs. In *Activity in Red Dwarf Stars*, eds. P. B. Byrne and M. Rodonò (Dordrecht: D. Reidel), pp. 137–156.

Völk, H., Jones, F., Morfill, G. and Röser, S. 1980. Collisions between grains in a turbulent gas. *Astron. Astrophys.* 85:316–325.

von Hippel, T., Burnell, S. J. B., and Williams, P. M. 1988. A working catalogue of Herbig-Haro objects. *Astron. Astrophys. Suppl.* 74:431–442.

Von Sengbusch, K. 1968. Sternentwicklung IX. Die Erste Hydrostatische Kontraktionsphase für eienen Stern von 1 $M_\odot$. *Zeit. Astrophys.* 69:79–111.

von Weizsäcker, C. F. 1948. Die rotation kosmischyer Gasmassen. *Z. Naturforschung* 3a:524–539.

Vrba, F. J. 1977. Role of magnetic fields in the evolution of five dark cloud complexes. *Astron. J.* 82:198–200.

Vrba, F. J., Strom, K. M., Strom, S. E., and Grasdalen, G. L. 1975. Further study of the stellar cluster embedded in the Ophiuchus dark cloud complex. *Astrophys. J.* 197:77–84.

Vrba, F. J., Strom, S. E., and Strom, K. M. 1976. Magnetic field structure in the vicinity of five dark cloud complexes. *Astron. J.* 81:958–969.

Vrba, F. J., Schmidt, G. D., and Hintzen, P. M. 1979. Observations and evaluation of the polarization in Herbig Ae/Be Stars. *Astrophys. J.* 227:185–196.

Vrba, F. J., Coyne, G. V., and Tapia, S. 1981. Observations of grain alignment and magnetic field properties of the R Corona Australis dark cloud. *Astrophys. J.* 243:489–511.

Vrba, F. J., Rydgren, A. E., Chugainov, P. F., Shakovskaya, N. I., and Zak, D. S. 1986*a*. Further evidence for rotational modulation of the light from T Tauri stars. *Astrophys. J.* 306:199–214.

Vrba, F. J., Luginbuhl, C. B., Strom, S. E., Strom, K. M., and Heyer, M. H. 1986*b*. An optical imaging and polarimetric study of the Lynds 723 and Barnard 335 molecular outflow regions. *Astron. J.* 92:633–636.

Vrba, F. J., Strom, S. E., and Strom, K. M. 1988*a*. Optical polarization measurements in the vicinity of nearby star-forming complexes. I. The Lynds L1641 giant molecular cloud. *Astron. J.* 96:680–694.

Vrba, F. J., Herbst, W., and Booth, J. F. 1988*b*. Spot evolution on the T Tauri star V410 Tau. *Astron. J.* 96:1032–1039.

Vrba, F. J., Rydgren, A. E., Chugainov, P. F., Shakovskaya, N. I., and Weaver, W. B. 1989. Additional UBVRI photometric searches for periodic light variability from T Tauri stars. *Astron. J.* 97:483–498.

Vsekhsvyatskii, S. K. 1967. *The Nature and Origin of Comets and Meteors* (Moscow: Prosveschcheniye Press).

Wacker, J. F. 1989. Laboratory simulation of meteoritic noble gases. III. Sorption of neon, argon, krypton, and xenon on carbon: elemental fractionation. *Geochim. Cosmochim. Acta* 53:1421–1433.

Wacker, J. F., and Anders, E. 1984. Trapping of xenon in ice: implications for the origin of the Earth's noble gases. *Geochim. Cosmochim. Acta* 48:2373–2380.

Wacker, J. F., Zadnik, M. G., and Anders, E. 1985. Laboratory simulation of meteoritic noble gases. I. Sorption of xenon on carbon: trapping experiments. *Geochim. Cosmochim. Acta* 49:1035–1048.

Wade, R. A., and Ward, M. J. 1985. Cataclysmic variables: observational overview. In *Interacting Binary Stars*, eds. J. E. Pringle and R. A. Wade (Cambridge: Cambridge Univ. Press), pp. 129–176.

Wadiak, E. J., Wilson, T. L., Rood, R. T., and Johnston, K., J. 1985. VLA observations of formaldehyde emission from Rho Ophiuchi B. *Astrophys. J. Lett.* 295:43–46.

Wai, C. M., and Wasson, J. T. 1977. Nebular condensation of moderately volatile elements and their abundances in ordinary chondrites. *Earth Planet. Sci. Lett.* 36:1–13.

Walborn, N. R. 1990. 30 Doradus, starburst Rosetta. In *Massive Stars in Starbursts*, eds. C. Leitherer, N. R. Walborn, T. M. Heckman and C. A. Norman (Cambridge: Camridge Univ. Press), pp. 145–155.

Walborn, N. R., and Blades, J. C. 1987. Recent massive star formation in 30 Doradus. *Astrophys. J. Lett.* 323:65–67.

Walker, C., Adams, F. C., and Lada, C. J. 1990. 1.3 millimeter continuum observations of cold molecular cloud cores. *Astrophys. J.* 349:515–528.

Walker, C. K., Lada, C. J., Young, E. T., Maloney, P. R., and Wilking, B. A. 1986. Spectroscopic evidence for infall around an extraordinary *IRAS* source in Ophiuchus. *Astrophys. J. Lett.* 309:47–51.

Walker, C. K., Lada, C. J., Young, E. T., and Margulis, M. 1988. An unusual outflow around IRAS 16293–2422. *Astrophys. J.* 332:335–345.

Walker, C. K., Carlstrom, J. E., Bieging, J. H., Lada, C. J., and Young, E. T. 1990. Observations of the dense gas in the *IRAS* 16293–2422 outflow system. *Astrophys. J.* 364:173–177.

Walker, H. J., and Butner, H. M. 1991. CO observations of Vega-like candidates detected by IRAS. In preparation.

Walker, H. J., and Wolstencroft, R. D. 1988. Cool circumstellar matter around nearby main sequence stars. *Publ. Astron. Soc. Pacific* 100:1509–1521.

Walker, J. C. G. 1977. *Evolution of the Atmosphere* (New York: MacMillan).

Walker, J. C. G. 1986. Impact erosion of planetary atmospheres. *Icarus* 68:87–98.

Walker, J. C. G., Turekian, K. K. and Hunten, D. M. 1970. An estimate of the present-day deep-mantle degassing rate from data on the atmosphere of Venus. *J. Geophys. Res.* 75:3558–3561.

Walker, M. F. 1956. Studies of extremely young clusters. I. NGC2264. *Astrophys. J. Suppl.* 2:365–388.

Walker, M. F. 1969. Studies of extremely young clusters. V. Stars in the vicinity of the Orion nebula. *Astrophys. J.* 155:447–468.

Walker, M. F. 1972. Studies of extremely young clusters. VI. Spectroscopic observations of the ultraviolet excess stars in the Orion nebula cluster and NGC 2264. *Astrophys. J.* 175:89–116.

Walker, M. F. 1987. Simultaneous spectroscopic and photometric observations of four T Tauri-type variables. *Publ. Astron. Soc. Pacific* 99:392–406.

Wall, W. F., and Jaffe, D. T. 1990. Detection of ^{13}CO $J = 3$–2 in IC 342: Warm clumpy molecular gas. *Astrophys. J. Lett.* 361:45–48.

Waller, W. H., and Hodge, P. W. 1991. Dynamical constraints on massive star formation. In *Dynamics of Galaxies and Molecular Clouds Distribution*, eds. F. Combes and F. Casoli (Dordrecht: Kluwer), in press.

Wallerstein, G., Herbig, G. H., and Conti, P. 1965. Observations of the lithium content of main-sequence stars in the Hyades. *Astrophys. J.* 141:610–616.

Wallis, M. K., and Krishna-Swamy, K. S. 1987. Some diatomic molecules from comet P/Halley's UV spectra near spacecraft flybys. *Astron. Astrophys.* 187:329–332.

Walmsley, C. M. 1991. Physical and chemical parameters in dense cores. In *Fragmentation of Molecular Clouds and Star Formation: Proc. IAU Symp. 147*, eds. E. Falgarone, F. Boulanger and G. Duvert (Dordrecht: Kluwer), pp. 161–175.

Walmsley, C. M. 1992. Exotic chemistry in star-forming regions. In *Chemistry and Spectroscopy of Interstellar Molecules*, eds. D. K. Bohme, E. Herbst, N. Kaifu and S. Saito (Tokyo: Univ. of Tokyo Press).

Walmsley, C. M., Hermsen, W., Henkel, C., Mauersberger, R., and Wilson, T. L. 1987. Deuterated ammonia in the Orion hot core. *Astron. Astrophys.* 172:311–315.

Walter, F. M. 1986. X-ray sources in regions of star formation. I. The naked T Tauri stars. *Astrophys. J.* 306:573–586.

Walter, F. M. 1987*a*. The naked T Tauri stars: the low-mass pre-main sequence unveiled. *Publ. Astron. Soc. Pacific* 99:31–37.

Walter, F. M. 1987*b*. Naked T Tauri Stars in II Sco and Ori OBIC. In *Cool Stars, Stellar Systems and the Sun*, eds. J. L. Linsky and R. E. Stencel (Berlin: Springer-Verlag) pp. 422–430.

Walter, F. M., and Barry, D. C. 1991. Pre- and main-sequence evolution of solar activity. In *The Sun in Time*, eds. E. Levy, M. S. Giampapa and M. S. Matthews (Tucson: Univ. of Arizona Press), pp. 633–657.

Walter, F. M., and Boyd, W. T. 1991. Star formation in Taurus-Auriga: the high mass stars. *Astrophys. J.* 370:318–323.

Walter, F. M., and Kuhi, L. V. 1981. The smothered coronae of T Tauri stars. *Astrophys. J.* 250:254–261.

Walter, F. M. and Kuhi, L. V. 1984. X-ray photometry and spectroscopy of T Tauri stars. *Astrophys. J.* 284:194–201.

Walter, F. M., Brown, A., Mathieu, R. D., Myers, P. C., and Vrba, F. J. 1988. X-ray sources in regions of star formation. III. Naked T Tauri stars associated with the Taurus-Auriga complex. *Astron. J.* 96:297–325.

Walter, F. M., Brown, A., Vrba, F. J., Mathieu, R. D., and Myers, P. C. 1989. How naked are the naked T Tauri stars? *Bull. Amer. Astron. Soc.* 20:1092.

Wannier, P. G., Scoville, N. Z., and Barvainis, R. 1983. The polarization of millimeter-wave emission lines in dense interstellar clouds. *Astrophys. J.* 267:126–136.

Wannier, P. G., Pagani, L., Kuiper, T. B. H., Frerking, M. A., Gulkis, S., Encrenaz, P., Pickett, H. M., Lecacheux, A., and Wilson, W. J. 1991. Water in dense molecular clouds. *Astrophys. J.* 377:171–186.

Ward, W. R. 1981. Solar nebula dispersal and the stability of the planetary system. I. Scanning secular resonance theory. *Icarus* 47:234–264.

Ward, W. R. 1986. Density waves in the solar nebula: differential Lindblad torque. *Icarus* 67:164–180.

Ward, W. R. 1988. On disk-planet interactions and orbital eccentricities. *Icarus* 73:330–348.

Ward, W. R. 1989*a*. Corotation torques in the solar nebula: the cutoff function. *Astrophys. J.* 336:526–538.

Ward, W. R. 1989*b*. On the rapid formation of giant planet cores. *Astrophys. J. Lett.* 345:99–102.

Ward, W. R., and Hourigan, K. 1989. Orbital migration of protoplanets: the inertial limit. *Astrophys. J.* 347:490–495.

Wardle, M., and Königl, A. 1990. A model for the magnetic field in the molecular disk at the galactic center. *Astrophys. J.* 362:120–134.

Wardle, M., and Königl, A. 1992. The structure of protostellar accretion disks and the origin of bipolar flows. *Astrophys. J.*, submitted.

Wark, D. A. 1981. Alteration and metasomatism of Allende Ca-Al-rich materials. *Lunar Planet. Sci.* XII:1145–1147 (abstract).

Wark, D. A. 1983. The Allende Meteorite: Information from Ca-Al-Rich Inclusions on the Formation and Early Evolution of the Solar System. Ph. D. Thesis, Univ. of Melbourne.

Wark, D. A., and Lovering, J. F. 1977. Marker events in the early evolution of the solar system: evidence from rims on calcium-aluminum-rich inclusions in carbonaceous chondrites. *Lunar Sci.* VIII:95–112 (abstract).

Warner, B. 1976. Observations of dwarf novae. In *Structure and Evolution of Close Binary Systems*, eds. P. P. Eggleton, S. Mitton and J. A. J. Whelan (Dordrecht: D. Reidel), pp. 85–140.

Warner, B. 1989. Properties of novae: an overview. In *Classical Novae*, eds. M. F. Bode and A. Evans (London: Wiley), pp. 1–16.

Warner, B., and O'Donoghue, D. 1988. High-speed photometry of Z Chamaeleontis during outbursts. *Mon. Not. Roy. Astron. Soc.* 223:705–738.

Warner, J. W., Strom, S. E., and Strom, K. M. 1977. Circumstellar shells in NGC2264: A reevaluation. *Astrophys. J.* 213:427–437.

Warren, W. H., Jr., and Hesser, J. E. 1977. A photometric study of the Orion OB 1 association. I. Observational data. *Astrophys. J. Suppl.* 34:115–206.

Warren, W. H., Jr., and Hesser, J. E. 1978. A photometric study of the Orion OB 1 association. III. Subgroup analyses. *Astrophys. J. Suppl.* 36:497–572.

Warwick, J. W., Evans, D. R., Romig, J. H., Sawyer, C. B., Desch, M. D., Kaiser, M. L., Alexander, J. K., Carr, T. D., Staelin, D. H., Gulkis, S., Poynter, R. L., Aubier, M., Boischot, A., Leblanc, Y., Lecacheux, A., Pedersen, B. M., and Zarka, P. 1986. Voyager 2 radio observations of Uranus. *Science* 223:102–106.

Wasserburg, G. J. 1985. Short-lived nuclei in the early solar system. In *Protostars and Planets II*, eds. D. C. Black and M. S. Matthews (Tucson: Univ. of Arizona Press), pp. 703–737.

Wasserburg, G. J., and Arnould, M. 1987. A possible relationship between extinct ^{26}Al and ^{53}Mn in meteorites and early solar activity. In *Nuclear Astrophysics*, eds. W. Hillebrandt, R. Kuhfua, E. Müller and J. W. Truran (New York: Springer-Verlag), pp. 262– 276.

Wasserburg, G. J., and Papanastassiou, D. A. 1982. Some short-lived nuclides in the early solar system—A connection with the placental ISM. In *Essays in Nuclear Astrophysics*, eds. C. A. Barnes, D. D. Clayton and D. N. Schramm (Cambridge: Cambridge Univ. Press), pp. 77–140.

Wasserburg, G. J., Papanastassiou, D. A., Tera, F., and Huneke, J. C. 1977. Outline of a lunar chronology. *Phil. Trans. Roy. Soc. London* A285:7–22.

Wasson, J. T. 1985. *Meteorites* (New York: W. H. Freeman).

Wasson, J. T., and Chou, C.-L. 1974. Fractionation of moderately volatile elements in ordinary chondrites. *Meteoritics* 1:69–84.

Watanabe, S., and Nakagawa, Y. 1988. Cooling of the solar nebula. *Prog. Theor. Physics Suppl.* 96:130–140.

Waters, J. W., Gustincic, J. J., Kakar, R. K., Kuiper, T. B. H., Roscoe, H. K., Swanson, P. N., Rodriguez Kuiper, E. N., Kerr, A. R., and Thaddeus, P. 1980. Observations of interstellar H_2O emission at 183 Gigahertz. *Astrophys. J.* 235:57–62.

Waters, L. B. F. M. 1986. The correlation between rotation and IR colour excess for B-type dwarfs. *Astron. Astrophys.* 159:L1–L4.

Waters, L. B. F. M., Coté, J., and Aumann, H. H. 1987. IRAS far-infrared colours of normal stars. *Astron. Astrophys.* 172:225–234.

Watkins, G. H. 1983. The Consequences of Cometary and Asteroidal Impacts on the Volatile Inventories of the Terrestrial Planets. Ph. D. Thesis, Massachussetts Inst. of Technology.

Watson, A. J., Donahue, T. M., and Walker, J. C. G. 1981. The dynamics of rapidly escaping atmosphere: applications to the evolution of Earth and Venus. *Icarus* 48:150–166.

Watson, D. M., Genzel, R., Townes, C. H., and Storey, J. W. V. 1985. Far-infrared emission lines of CO and OH in the Orion-KL molecular shock. *Astrophys. J.* 298:316–327.

Weaver, H. A. 1989. The volatile composition of comets. In *Highlights of Astronomy*, vol. 8, ed. D. McNally (Dordrecht: Kluwer), pp. 387–393.

Weaver, H. A., and Mumma, M. J. 1984. Infrared molecular emissions from comets. *Astrophys. J.* 276:782–797. Erratum: *Astrophys. J.* 285:872–873.

Weaver, H. A., Chin, G., Mumma, M. J., Espenak, F., Bally, J., Stark, A., and Hinkle,

K. 1983. A search for vibrational and rotational emissions from CO in comet IRAS-Araki-Alcock (1983d). *Bull. Amer. Astron. Soc.* 15:802 (abstract).

Weaver, H. A., Mumma, M. J., Larson, H. P., and Davis, D. S. 1986. Post-perihelion modelling observations of water in comet Halley. *Nature* 324:441–444.

Weaver, H. A., Mumma, M. J., and Larson, H. P. 1987. Infrared investigation of water in comet P/Halley. *Astron. Astrophys.* 187:411–418.

Weaver, R., McCray, R., Castor, J., Shapiro, P., and Moore, R. 1977. Interstellar bubbles. II. Structure and evolution. *Astrophys. J.* 218:377–395.

Weber, E. J. and Davis, L. 1967. The angular momentum of the solar wind. *Astrophys. J.* 148:217–227.

Weber, H. W., Hintenberger, H., and Begemann, F. 1971. Noble gases in the Haverö ureilite. *Earth Planet. Sci. Lett.* 13:205–209.

Wegmann, R., Schmidt, H. U., Huebner, W. F., and Boice, D. C. 1987. Cometary MHD and chemistry. *Astron. Astrophys.* 187:339–350.

Weidenschilling, S. J. 1977a. Aerodynamics of solid bodies in the solar nebula. *Mon. Not. Roy. Astron. Soc.* 180:57–70.

Weidenschilling, S. J. 1977b. The distribution of mass in the planetary system and solar nebula. *Astrophys. Space Sci.* 51:153–158.

Weidenschilling, S. J. 1980. Dust to planetesimals: Settling and coagulation in the solar nebula. *Icarus* 44:172–189.

Weidenschilling, S. J. 1984. Evolution of grains in a turbulent solar nebula. *Icarus* 60:555–567.

Weidenschilling, S. J. 1987. Accumulation of solid bodies in the solar nebula. *Gerlands. Beitr. Geophys.* 96:21–33.

Weidenschilling, S. J. 1988a. Comparisons of solar nebula models. In *Workshop on the Origins of Solar Systems*, eds. J. Nuth and P. Sylvester, LPI Tech Rept. 88-04, pp. 31–37.

Weidenschilling, S. J. 1988b. Formation processes and timescales for meteorite parent bodies. In *Meteorites and the Early Solar System*, eds. J. F. Kerridge and M. S. Matthews (Tucson: Univ. of Arizona Press), pp. 348–371.

Weidenschilling, S. J. 1988c. Dust to dust: Low-velocity impacts of fragile projectiles. *Lunar Planet. Sci.* XIX:1253-1254 (abstract).

Weidenschilling, S. J. 1989. Stirring of a planetesimal swarm: the role of distant encounters. *Icarus* 80:179–188.

Weidenschilling, S. J. 1993. Formation of planetesimals in a solar nebula with "generic" turbulence. In preparation.

Weidenschilling, S. J., and Davis, D. R. 1985. Orbital resonances in the solar nebula: implications for planetary accretion. *Icarus* 62:16–29.

Weidenschilling, S. J., Donn, B., and Meakin, P. 1989. The physics of planetesimal formation. In *The Formation and Evolution of Planetary Systems*, eds. H. Weaver, F. Paresce and L. Danly (Cambridge: Cambridge Univ. Press), pp. 131–150.

Weigelt, G., and Baier, G. 1985. R136a in the 30 Doradus nebula resolved by holographic speckle interferometry. *Astron. Astrophys.* 150:L18–L20.

Weigelt, G., Albrecht, R., Barbieri, C., Blades, J. C., Boksenberg, A., Crane, P., Deharveng, J. M., Disney, M. J., Jakobsen, P., Kamperman, T. M., King, I. R., Macchetto, F., MacKay, C. D., Paresce, F., Baxter, D., Greenfield, P., Jedrzejewski, R., Nota, A., and Sparks, W. B. 1991. First results from the Faint Object Camera: high-resolution observations of the central object R 136 in the 30 Doradus nebula. *Astrophys. J. Lett.* 378:21–23.

Weinbruch, S., Zinner, E. K., Steele, I. M., and Palme, H. 1989. Oxygen-isotopic composition of Allende olivines. *Meteoritics* 9:339 (abstract).

Weinbruch, S., Palme, H., Müller, W. F., and El Goresy, A. 1990. FeO-rich rims and veins in Allende forsterite: evidence for high temperature condensation at oxidizing conditions. *Meteoritics* 25:115–125.

Weintraub, D. A., Masson, C. R., and Zuckerman, B. 1989*a*. Measurements of Keplerian rotation of the gas in the circumbinary disk around T Tauri. *Astrophys. J.* 344:915–924.

Weintraub, D. A., Sandell, G., and Duncan, W. D. 1989*b*. Submillimeter measurements of T Tauri and FU Orionis stars. *Astrophys. J. Lett.* 340:69–72.

Weintraub, D. A., Shelton, C., and Zucherman, B. 1989*c*. Circumstellar disks and infrared binaries: T Tau and other newly resolved T Tauri stars. *Bull. Amer. Astron. Soc.* 21:715–716.

Weischer, M., Gorres, J., Theilemann, F.-K., and Ritter, H. 1986. Explosive hydrogen burning in novae. *Astron. Astrophys.* 160:56–72.

Weiss, A. and Truran, J. W. 1990. ^{22}Na and ^{26}Al production and nucleosynthesis in novae explosions. *Astron. Astrophys.* 238:178–186.

Weissman, P. R. 1979. Physical and dynamical evolution of long-period comets. In *Dynamics of the Solar System*, ed. R. L. Duncombe (Dordrecht: D. Reidel), pp. 277–282.

Weissman, P. R. 1982. Dynamical history of the Oort cloud. In *Comets*, ed. L. L. Wilkening (Tucson: Univ. of Arizona Press), pp. 637–658.

Weissman, P. R. 1984. The Vega particulate shell: comets or asteroids. *Science* 224:987–989.

Weissman, P. R. 1985*a*. Dynamical evolution of the Oort cloud. In *Dynamics of Comets: Their Origin and Evolution*, eds. A. Carusi and G. B. Valsecchi (Dordrecht: D. Reidel), pp. 87–96.

Weissman, P. R. 1985*b*. The origin of comets: implications for planetary formation. In *Protostars and Planets II*, eds. D. C. Black and M. S. Matthews (Tucson: Univ. of Arizona Press), pp. 895–919.

Weissman, P. R. 1986*a*. Are cometary nuclei primordial rubble piles? *Nature* 320:242–244.

Weissman, P. R. 1986*b*. How pristine are cometary nuclei? In *The Comet Nucleus Sample Return Mission*, ed. O. Meliter, ESA SP-249, pp. 15–25.

Weissman, P. R. 1986*c*. The Oort cloud and the galaxy: dynamical interactions. In *The Galaxy and the Solar System*, eds. R. Smoluchowski, J. N. Bahcall and M. S. Matthews (Tucson: Univ. of Arizona Press), pp. 204–237.

Weissman, P. R. 1987. Post-perihelion brightening of Halley's comet: springtime for Halley. *Astron. Astrophys.* 187:873–878.

Weissman, P. R. 1990. The Oort cloud. *Nature* 344:825–830.

Weissman, P. R. 1991*a*. The cometary impactor flux at the Earth. In *Global Catastrophes in Earth History*, eds. V. Sharpton and P. Ward, Geological Soc. of America SP-247, pp. 171–180.

Weissman, P. R. 1991*b*. The angular momentum of the Oort cloud. *Icarus* 89:190–193.

Weissman, P. R., and Kieffer, H. H. 1981. Thermal modelling of cometary nuclei. *Icarus* 47:302–311.

Welch, W. J., Vogel, S. N., Plambeck, R. L., Wright, M. C. H., and Bieging, J. H. 1985. Gas jets associated with star formation. *Science* 228:1389–1395.

Welch, W. J., Dreher, J. W., Jackson, J. M., Terebey, S., and Vogel, S. N. 1987. Star formation in W49A: gravitational collapse of the molecular cloud core toward a ring of massive stars. *Science* 238:1550–1555.

Welty, A.D. 1991. A Spectroscopic Study of Three FU Orionis Objects: Accretion Disks and Wind Signatures. Ph. D. Thesis, Univ. of Massachussetts.

Welty, A. D., Strom, S. E., Strom, K. M., Hartmann, L. W., Kenyon, S. J., Grasdalen, G. L., and Stauffer, J. R. 1990. Further evidence for differential rotation in V1057 Cyg. *Astrophys. J.* 349:328–334.

Werner, M. W., Dinerstein, H. L., and Capps, R. W. 1983. The polarization of the infrared cluster in Orion: the spatial distribution of the 3.8 micron polarization. *Astrophys. J. Lett.* 265:13–17.

Werner, M. W., Crawford, M. K., Genzel, R., Hollenbach, D. J., Townes, C. H., and Watson, D. M. 1984. Detection of shocked atomic gas in the Kleinmann-Low nebula. *Astrophys. J. Lett.* 282:81–84.

West, R. M., Hainaut, O., and Smette, A. 1991. Post-perihelion observations of P/Halley. III. An outburst at $r = 14.3$ AU. *Astron. Astrophys.* 246:L77–L80.

Western, L. R., and Watson, W. D. 1984. Linear polarization of astronomical masers and magnetic fields. *Astrophys. J.* 285:158–173.

Wetherill, G. W. 1975a. Late heavy bombardment of the moon and terrestrial planets. *Proc. Lunar Sci. Conf.* 6:1539–1561.

Wetherill, G. W. 1975b. Radiometric chronology of the early solar system. *Ann. Rev. Nucl. Sci.* 25:283–328.

Wetherill, G. W. 1980a. Formation of the terrestrial planets. *Ann. Rev. Astron. Astrophys.* 18:77–113.

Wetherill, G. W. 1980b. Numerical calculations relevant to the accumulation of the terrestrial planets. In *The Continental Crust and Its Mineral Deposits*, eds. D. W. Strangway, Geological Assoc. of Canada SP-20, pp. 3–24.

Wetherill, G. W. 1981. Solar wind origin of ^{36}Ar on Venus. *Icarus* 46:70–80.

Wetherill, G. W. 1985. Occurence of giant impacts during the growth of the terrestrial planets. *Science* 228:877–879.

Wetherill, G. W. 1986. Accumulation of the terrestrial planets and implications concerning lunar origin. In *Origin of the Moon*, eds. W. K. Hartmann, R. J. Phillips and G. J. Taylor (Houston: Lunar and Planetary Science Inst.), pp. 519–550.

Wetherill, G. W. 1988. Accumulation of Mercury from planetesimals. In *Mercury*, eds. F. Vilas, C. R. Chapman and M. S. Matthews (Tucson: Univ. of Arizona Press), pp. 670–691.

Wetherill, G. W. 1989. Origin of the asteroid belt. In *Asteroids II*, eds. R. P. Binzel, T. Gehrels and M. S. Matthews (Tucson: Univ. of Arizona Press), pp. 661–680.

Wetherill, G. W. 1990a. Comparison of analytical and physical modeling of planetesimal accumulation. *Icarus* 88:336–354.

Wetherill, G. W. 1990b. Formation of the Earth. *Ann. Rev. Earth Planet Sci.* 18:205–256.

Wetherill, G. W. 1991a. End products of cometary evolution: cometary origin of Earth-crossing bodies of asteriodal appearances. In *Comets in the Post-Halley Era*, eds. R. L. Newburn, J. Rahe and M. Neugebauer (Amsterdam: Kluwer), pp. 537–556.

Wetherill, G. W. 1991b. Formation of the terrestrial planets from planetesimals. In *Planetary Sciences, American and Soviet Research*, ed. T. Donahue (Washington, D. C.: Natl. Academy of Science Press), pp. 98–115.

Wetherill, G. W. 1991c. Occurrence of Earth-like bodies in planetary systems. *Science* 253:535–538.

Wetherill, G. W., and Cox, L. P. 1984. The range of validity of the two-body approximation in models of terrestrial planet accumulation in models of terrestrial planet accumulation. I. Gravitational perturbations. *Icarus* 60:40–55.

Wetherill, G. W., and Cox, L. P. 1985. The range of validity of the two-body approximation in models of terrestrial planet accumulation. II. Gravitational cross sections and runaway accretion. *Icarus* 63:290–303.

Wetherill, G. W., and Stewart, G. R. 1989. Accumulation of a swarm of small planetesimals. *Icarus* 77:330–357.

Wexler, A. S. 1967. Integrated intensities of absorption bands in infrared spectroscopy. *Appl. Spectroscopy Rev.* 1:29–98.

Whipple, F. L. 1950. A comet model I: the acceleration of comet Encke. *Astrophys.*

J. 111:375–394.

Whipple, F. L. 1964. Evidence for a comet belt beyond Neptune. *Proc. Natl. Acad. Sci. U. S.* 51:711–718.

Whipple, F. L. 1966. Chondrules: suggestion concerning the origin. *Nature* 153:54–56.

Whipple, F. L. 1972. On certain aerodynamic processes for asteroids and comets. *From Plasma to Planet: Proc. Nobel Symp. No. 21*, ed. A. Elvius (New York: Wiley).

Whipple, F. L. 1977. The constitution of cometary nuclei. In *Comets, Asteroids, Meteorites*, ed. A. H. Delsemme (Toledo: Univ. of Toledo Press), pp. 25–32.

Whipple, F. L., and Lecar, M. 1976. Comet formation induced by solar wind. In *The Study of Comets*, NASA SP-393, pp. 600–662.

White, G. J., and Padman, R. 1991. Images of atomic carbon in the interstellar medium. *Nature* 354:511–513.

White, R. E. 1977. Microturbulence, systematic motions, and line formation in molecular clouds. *Astrophys. J.* 211:744–753.

Whitehurst, R. 1988*a*. Numerical simulations of accretion discs—I. Superhumps: a tidal phenomenon of accretion disks. *Mon. Not. Roy. Astron. Soc.* 232:35–51.

Whitehurst, R. 1988*b*. Numerical simulations of accretion discs—II. Design and implementation of a new numerical method. *Mon. Not. Roy. Astron. Soc.* 233:529–551.

Whitehurst, R. 1989. Simulations of accretion flow in close binary stars. In *Theory of Accretion Disks*, eds. F. Meyer, W. J. Duschl, J. Frank and E. Meyer-Hofmeister (Dordrecht: Kluwer), pp. 213–220.

Whitham, G. B. 1970. Two timing, variational principles and waves. *J. Fluid Mech.* 44:373–395.

Whitham, G. B. 1974. *Linear and Nonlinear Waves* (New York: Wiley).

Whitmire, D. P., Matese, J. J., and Tomley, L. J. 1988. A brown dwarf companion as an explanation of the asymmetry in the Beta Pictoris disk. *Astron. Astrophys.* 203:L13–L15.

Whitmire, D. P., Matese, J. J., and Whitman, P. G. 1992. Velocity streaming of *IRAS* main-sequence stars and the episodic enhancement of particulate disk by interstellar clouds. *Astrophys. J.* 388:190–195

Whitney, B. A., and Clayton, G. C. 1989. The slow Nova V 1819 Cygni (Nova Cygni 1986) *Astron. J.* 98:297–310.

Whittet, D. C. B. 1974. The ratio of total to selective absorption in the Rho Ophiuchi cloud. *Mon. Not. Roy. Astron. Soc.* 168:371–378.

Whittet, D. C. B. 1988. The observed properties of interstellar dust in the infrared. In *Dust in the Universe*, eds. M. E. Bailey and D. A. Williams (Cambridge: Cambridge Univ. Press), pp. 25–53.

Whittet, D. C. B. 1992. Observations of molecular ices. In *Dust and Chemistry in Astronomy*, eds. T. J. Millar and D. A. Williams (Dordrecht: Kluwer).

Whittet, D. C. B., Kirrane, T. M., Kilkenny, D., Oates, A. P., Watson, F. G., and King, D. J. 1987. A study of the Chamaeleon dark cloud and T-association. I. Extinction, distance and membership. *Mon. Not. Roy. Astron. Soc.* 224:497–512.

Whittet, D.C.B., Bode, M. F., Longmore, A. J., Adamson, A. J., McFadzean, A. D., Aitken, D. K., and Roche, P. F. 1988. Infrared spectroscopy of dust in the Taurus dark cloud: Ice and silicates. *Mon. Not. Roy. Astron. Soc.* 233: 321–336.

Whittet, D.C.B., Adamson, A. J., Duley, W. W., Geballe, T. R., and McFadzean, A. D. 1989. Infrared spectroscopy of dust in the Taurus dark cloud: Solid carbon monoxide. *Mon. Not. Roy. Astron. Soc.* 241:707–720.

Whittet, D. C. B., Assendorp, R., Prusti, T., Roth, M., and Wesselius, P. R. 1991*a*. An infrared study of pre-main-sequence stars in the Chamaeleon II association.

Astron. Astrophys. 251:524–430.

Whittet, D. C. B., Prusti, T., and Wesselius, P. R. 1991*b*. A study of the Chamaeleon I dark cloud and T-association. III. Near-infrared photometry of *IRAS*-selected field stars. *Mon. Not. Roy. Astron. Soc.* 249:319–326.

Wickramasinghe, D. T., and Allen, D. A. 1986. Discovery of organic grains in comet Halley. *Nature* 323:44–46.

Wieler, R., Baur, H., Signer, P., Lewis, R. S., and Anders, E. 1989. Planetary noble gases in "Phase Q" of Allende: Direct determination by closed system etching. *Lunar Planet. Sci.* XX:1201–1202 (abstract).

Wieneke, B., and Clayton, D. 1983. Aggregation of grains in a turbulent pre-solar disk. In *Chondrules and Their Origins*, ed. E. A. King (Houston: Lunar and Planetary Inst.), pp. 284–295.

Wiens, R. C., and Pepin, R. O. 1988. Laboratory shock emplacement of noble gases, nitrogen, and carbon dioxide into basalt, and implications for trapped gases in shergottite EETA 79001. *Geochim. Cosmochim. Acta* 52:295–307.

Wiens, R. C., Becker, R. H., and Pepin, R. O. 1986. The case for martian origin of the shergottites. II. Trapped and indigenous gas components in EETA 79001 glass. *Earth and Plan. Sci. Lett.* 77:149–158.

Wiklind, T., Rydbeck, G., Hjalmarson, Å., and Bergman, P. 1990. Arm and interarm molecular clouds in M83. *Astron. Astrophys.* 232:L11–L14.

Wilking, B. A. 1989. The formation of low-mass stars. *Publ. Astron. Soc. Pacific* 101:229–243.

Wilking, B. A. 1992. Star formation in the Ophiuchus molecular cloud complex. In *Low Mass Star Formation in Southern Molecular Clouds*, ed. B. Reipurth, ESO Scientific Rept. No. 11, pp. 159–196.

Wilking, B. A., and Claussen, M. J. 1987. Water masers associated with low-mass stars: a survey of the Rho Ophiuchi infrared cluster. *Astrophys. J. Lett.* 320:133–137.

Wilking, B. A., and Lada, C. J. 1983. The discovery of new embedded sources in the centrally condensed core of the Rho Ophiuchi dark cloud: the formation of a bound cluster? *Astrophys. J.* 274:698–716.

Wilking, B. A., and Lada, C. J. 1985. The formation of bound stellar clusters. In *Protostars and Planets II*, eds. D. C. Black and M. S. Matthews (Tucson: Univ. of Arizona Press), pp. 297–319.

Wilking, B. A., Lebofsky, M. J., Rieke, G. H., and Kemp, J. C. 1979. Infrared polarimetry in the ρ Ophiuchus dark cloud. *Astron. J.* 84:199–203.

Wilking, B. A., Taylor, K. N. R., and Storey, J. W. V. 1986. The nature of the infrared cluster in the R Corona Australis cloud core. *Astron. J.* 92:103–110.

Wilking, B. A., Schwartz, R. D., and Blackwell, J. H. 1987. An Hα emission-line survey of the ρ Ophiuchi dark cloud complex. *Astron. J.* 94:106–110.

Wilking, B. A., Lada, C. J., and Young, E. T. 1989*a*. *IRAS* observations of the ρ Ophiuchi infrared cluster: spectral energy distributions and luminosity functions. *Astrophys. J.* 340:823–852.

Wilking, B. A., Mundy, L. G., Blackwell, J. H., and Howe, J. E. 1989*b*. A millimeter-wave spectral line and continuum survey of cold *IRAS* sources. *Astrophys. J.* 345:257–264.

Wilking, B. A., Blackwell, J. H., and Mundy, L. G. 1990. High-velocity molecular gas associated with cold IRAS sources. *Astron. J.* 100:758–770.

Wilking, B. A., Lada, C. J., Young, E. T., Greene, T. P., and Meyer, M. R. 1991. *IRAS* observations of young stellar objects in the R Corona Australis dark cloud. *Astrophys. J.*, In preparation.

Williams, H. A., and Tohline, J. E. 1988. Circumstellar ring formation in rapidly rotating protostars. *Astrophys. J.* 334:449–464.

Williams, J. G. 1969. Secular Perturbations in the Solar System. Ph. D. Thesis, Univ. of California, Los Angeles.

Williams, J. G. 1971. Proper elements, families, and belt boundaries. In *Physical Studies of Minor Planets*, ed. T. Gehrels, NASA SP-267, pp. 177–181.

Williams, J. G., and Benson, G. S. 1971. Resonances in the Neptune-Pluto system. *Astron. J.* 76:167–177.

Williams, R. E. 1977. Element abundance analysis of novae. In *The Interaction of Variable Stars with Their Environment: Proc. IAU Coll. 42*, eds. R. Kippenhahn, J. Rahe, and W. Strohmeier (Bamberg: Remeis-Sternwaarte), pp. 242–271.

Williams, R. E. 1985. CNO abundances in ejecta. In *Production and Distribution of CNO Elements*, ed. I. J. Danziger (Garching: European Southern Obs.), pp. 225–232.

Williams, R. E. 1991. The ionization of nova ejecta. In *Physics of Classical Novae: Proc. IAU Coll. 122*, eds. A. Cassetalla and R. Viotti (Berlin: Springer-Verlag), pp. 215–227.

Williams, R. E., and Gallagher, J. S. 1979. Spectrophotometry of filaments surrounding Nova RR Pictoris 1925. *Astrophys. J.* 228:482–490.

Williams, R. E., Woolf, N. J., Hege, E. K., Moore, R. L., and Kopriva, D. A. 1978. The shell around Nova DQ Herculis 1934. *Astrophys. J.* 224:171–181.

Williams, R. E., Ney, E. P., Sparks, W. M., Starrfield, S. G., Wyckoff, S., and Truran, J. W. 1985. Ultraviolet spectral evolution and heavy element abundances in nova Coronae Austrinae 1981. *Mon. Not. Roy. Astron. Soc.* 212:753–766.

Willner, S. P., Gillett, F. C., Herter, T. L., Jones, B., Krassner, J., Merrill, K. M., Pipher, J. L., Puetter, R. C., Rudy, R. J., Russell, R. W., and Soifer, B. T. 1982. Infrared spectra of protostars: composition of the dust shells. *Astrophys. J.* 253:174–187.

Wilson, T. L., and Walmsley, C. M. 1989. Small-scale clumping in molecular clouds. *Astron. Astrophys. Rev.* 1:141–176.

Wilson, T. L., Johnston, K. J., Henkel, C., and Menten, K. M. 1989. The distribution of hot thermal methanol in Orion-KL. *Astron. Astrophys.* 214:321–326.

Winkler, K.-H. and Newman, M. J. 1980. Formation of solar-type stars in spherical symmetry. I. The key role of the accretion shock. *Astrophys. J.* 236:201–211.

Winnberg, A., Ekelund, L., and Ekelund, A. 1987. Detection of HCN in comet P/Halley. *Astron. Astrophys.* 172:335–341.

Winnewisser, G. 1988. Molecular observations: I. Molecular chains and rings II. The Cologne 3-m radio telescope on Gornergrat. *Astrophys. Lett. Commun.* 26:227–237.

Wiramihardja, S. D., Kogure, T., Yoshida, S., Ogura, K., and Nakano, M. 1989. Survey observations of emission-line stars in the Orion region. I. The Kiso area A-0904. *Publ. Astron. Soc. Japan* 41:155–174.

Wisdom, J. 1980. The resonance overlap criterion and the onset of stochastic behavior in the restricted three-body problem. *Astron. J.* 85:1122–1133.

Wisdom, J. 1982. The origin of the Kirkwood gaps: a mapping for asteroidal motion near the 3/1 commensurability. *Astron. J.* 87:577–593.

Wisdom, J. 1983. Chaotic behavior and the orgin of the 3/1 Kirkwood gap. *Icarus* 56:51–74.

Wisdom, J. 1985. A perturbative treatment of motion near the 3/1 commensurability. *Icarus* 63:272–279.

Wisdom, J. 1987*a*. Rotational dynamics of irregularly shaped satellites. *Astron. J.* 94:1350–1360.

Wisdom, J. 1987*b*. Urey prize lecture: chaotic dynamics in the solar system. *Icarus* 72:241–275.

Wisdom, J., and Holman, M. 1991. Symplectic maps for the N-body problem. *Astron. J.* 102:1528–1538.

Wisdom, J., Peale, S. J., and Mignard, F. 1984. The chaotic rotation of Hyperion. *Icarus* 58:137–152.

Witt, A. N. 1989. Visible/UV scattering by interstellar dust. In *Interstellar Dust: Proc. IAU Symp. 135*, eds. L. J. Allamandola and A. G. G. M. Tielens (Dordrecht: Kluwer), pp. 87–100.

Witteborn, F. C., Bregman, J. D., Lester, D. F., and Rank, D. M. 1982. A search for fragmentation debris near Ursa major stream stars. *Icarus* 50:63–71.

Wolfendale, A. W. 1991. The mass of molecular gas in the galaxy. In *Molecular Clouds*, eds. R. A. James and T. J. Millar (Cambridge: Cambridge Univ. Press), pp. 41–48.

Wolfire, M. G., and Cassinelli, J. P. 1986. The temperature structure in accretion flows onto massive protostars. *Astrophys. J.* 310:207–221.

Wolfire, M. G., and Cassinelli, J. P. 1987. Conditions for the formation of massive stars. *Astrophys. J.* 319:850–867.

Wolfire, M. G., and Königl, A. 1991. Molecular line emission models of Herbig-Haro objects. I. H_2 emission. *Astrophys. J.* 383:205–225.

Womack, M., Ziurys, L. M., and Wyckoff, S. 1992. Estimates of N_2 abundances in dense molecular clouds. *Astrophys. J.* 393:188–192.

Wood, J. A. 1970. Petrology of the lunar soil and geophysical implications. *J. Geophys. Res.* 32:6497–6513.

Wood, J. A. 1984. On the formation of meteorite chondrules by aerodynamic drag heating in the solar nebula. *Earth Planet. Sci. Lett.* 70:11–26.

Wood, J. A. 1985. Meteoritic constraints on processes in the solar nebula. In *Protostars and Planets II*, eds. D. C. Black and M. S. Matthews (Tucson: Univ. of Arizona Press), pp. 687–702.

Wood, J. A. 1988. Chondritic meteorites and the solar nebula. *Ann. Rev. Earth Planet. Sci.* 16:53–72.

Wood, J. A., and Morfill, G. E. 1988. A review of solar nebula models. In *Meteorites and the Early Solar System*, eds. J. F. Kerridge and M. S. Matthews (Tucson: Univ. of Arizona Press), pp. 329–347.

Wood, J. A., Horne, K., Berriman, G., Wade, R. A., O'Donoghue, D., and Warner, B. 1986. High speed photometry of the dwarf nova Z chain quiescence. *Mon. Not. Roy. Astron. Soc.* 219:629–655.

Wood, J. A., Horne, K., Berriman, G., and Wade, R. A. 1989*a*. Eclipse studies of the dwarf nova O Y Carinae in quiescence. *Astrophys. J.* 341:974–996.

Wood, J. A., Marsh, T. R., Robinson, E. L., Steining, R. R., Horne, K., Stover, R. J., Schoembs, R., Allen, S. L., Bond, H. E., Jones, D. H. P., Grauer, A. D., and Ciardullo, R. 1989*b*. The ephemeris and variations of the accretion disk radius in IP Pegasi. *Mon. Not. Roy. Astron. Soc.* 239:809–824.

Woods, T. N., Feldman, P. D., Dymond, K. F., and Sahnow, D. J. 1986. Rocket ultraviolet spectroscopy of comet Halley and abundance of carbon monoxide and carbon. *Nature* 324:436–438.

Woods, T. N., Feldman, P. D., and Dymond, K. F. 1987. The atomic carbon distribution in the coma of comet P/Halley. *Astron. Astrophys.* 187:380–384.

Woody, D. P., Scott, S. L., Scoville, N. Z., Mundy, L. G., Sargent, A. I., Padin, S., Tinney, C. G., and Wilson, C. D. 1989. Interferometric observations of 1.4 millimeter continuum sources. *Astrophys. J. Lett.* 337:41–44.

Woolum, D. S. 1988. Solar-system abundances and processes of nucleosynthesis. In *Meteorites and the Early Solar System*, eds. J. F. Kerridge and M. S. Matthews (Tucson: Univ. of Arizona Press), pp. 995–1020.

Woosley, S. E., and Hoffman, R. D. 1986. Tables of reaction rates for nucleosynthesis and for charged particle and weak interactions. Unpublished.

Woosley, S. E., and Howard, W. M. 1978. The p-process in supernovae. *Astrophys. J.*

Suppl. 36:285–304.

Woosley, S. E., and Howard, W. M. 1990. [146]Sm production by the gamma-process. *Astrophys. J. Lett.* 354:21–24.

Woosley, S. E., and Weaver, T. A. 1980. Explosive neon burning and [26]Al gamma-ray astronomy. *Astrophys. J.* 238:1017–1025.

Woosley, S. E., Arnett, W. D., and Clayton, D. D. 1973. The explosive burning of oxygen and silicon. *Astrophys. J. Suppl.* 26:231–312.

Wootten, A. 1989. The duplicity of *IRAS* 16293–2422: A protobinary star? *Astrophys. J.* 337:858–864.

Wootten, A., and Loren, R. B. 1987. L1689N: Misalignment between a bipolar outflow and a magnetic field. *Astrophys. J.* 317:220–230.

Wootten, A., and Loren, R. B. 1988. Sulfur oxides in ρ Ophiuchi cloud cores: symptoms of star formation? In *Molecular Clouds in the Milky Way and External Galaxies*, eds. R. L. Dickman, R. L. Snell and J. S. Young (Münich: Springer-Verlag), pp. 178–181.

Wootten, A., Mangum, J. G., Turner, B. E., Bogey, M., Boulanger, F., Combes, F., Encrenaz, P. J., and Gerin, M. 1991. Detection of interstellar H_3^+: a confirming line. *Astrophys. J. Lett.* 380:79–83.

Wopenka, B., Virag, A., Zinner, E., Amari, S., Lewis, R. S., and Anders, E. 1989. Isotopic and optical properties of large individual SiC crystals from the Murchison meteorite. *Meteoritics* 24:342 (abstract).

Worden, S. P., Schneeberger, T. J., Kuhn, J. R., and Africano, J. L. 1981. Flare activity on T Tauri stars. *Astrophys. J.* 244:520–527.

Wouterloot, J. G. A., and Habing, H. J. 1985. OH observations of cloud complexes in Taurus. *Astron. Astrophys. Suppl.* 60:43–59.

Wouterloot, J. G. A., Walmsley, C. M., and Henkel, C. 1988. Star formation in the outer galaxy. *Astron. Astrophys.* 191:323–340.

Wouterloot, J. G. A., Henkel, C., and Walmsley, C. M. 1989. CO observations of IRAS sources in Orion and Cepheus. *Astron. Astrophys.* 215:131–146.

Wright, A. E., and Barlow, M. J. 1975. The radio and infrared spectrum of early-type stars undergoing mass loss. *Mon. Not. Roy. Astron. Soc.* 170:41–51.

Wright, E. L. 1987. Long wavelength absorption by fractal dust grains. *Astrophys. J.* 320:818–824.

Wright, M. C. H., Plambeck, R. L., Vogel, S. N., Ho, P. T. P., and Welch, W. J. 1983. Source of the high-velocity molecular flow in Orion. *Astrophys. J. Lett.* 267:41–45.

Wright, M. C. H., Dickel, H. R., and Ho, P. T. P. 1984. An aperture synthesis map of HCN emission close to W3 IRS 4. *Astrophys. J. Lett.* 281:71–74.

Wuchterl, G. 1990*a*. Hydrodynamics of giant planet formation. I. Overviewing the κ-mechanism. *Astron. Astrophys.* 238:83–94.

Wuchterl, G. 1990*b*. Hydrodynamics of giant planet formation. II. Model equations and critical mass. *Icarus*, submitted.

Wuchterl, G. 1990*c*. Hydrodynamics of giant planet formation. III. Jupiter's nucleated instability. *Icarus* 91:53–64.

Wulf, A. V. 1990. Das Verhalten flüchtiger Elemente bei Aufheizprozessen unter variabler Sauerstofffugazität. Ph. D. Thesis, Univ. Mainz.

Wulf, A. V., and Palme, H. 1991. Origin of moderately volatile elements in primitive meteorites. *Lunar Planet. Sci.* XXII:1527–1528 (abstract).

Wyckoff, S., Tegler, S., Wehinger, P. A., Spinrad, H. and Belton, M. J. S. 1988. Abundances in Comet Halley at the time of the spacecraft encounters. *Astrophys. J.* 325:927–938.

Wyckoff, S., Lindholm, E., Wehinger, P. A., Peterson, B. A., Zucconi, J.-M. and Festou, M. C. 1989. The $^{12}C/^{13}C$ abundance ratio in Comet Halley. *Astrophys. J.*

339:488–500.

Wyckoff, S., Tegler, S. C., and Engel, L. 1991a. Ammonia abundances in four comets. *Astrophys. J.* 368:279–286.

Wyckoff, S., Tegler, S. C., and Engel, L. 1991b. Nitrogen abundance in comet Halley. *Astrophys. J.* 367:641–648.

Wynn-Williams, C. G., Becklin, E. E., and Neugebauer, G. 1972. Infrared sources in the H II region W3. *Mon. Not. Roy. Astron. Soc.* 160:1–14.

Xie, T., and Goldsmith, P. F. 1990. A bipolar outflow in the globule Lynds 810. *Astrophys. J.* 359:378–383.

Xie, X., and Mumma, M. J. 1992. The effect of electron collisions on rotational populations of cometary water. *Astrophys. J.* 386:720–728.

Yabushita, S., and Tsujii, T. 1991. Near-parabolic cometary flux in the outer solar system— II. *Mon. Not. Roy. Astron. Soc.* 252:151–155.

Yamamoto, T. 1985. Formation environment of cometary nuclei in the primordial solar nebula. *Astron. Astrophys.* 142:31–36.

Yamashita, T., Hayashi, S. S., Kaifu, N., Kameya, O., Ukita, N., and Hasegawa, T. 1989. The inner, dense part of the protostellar disk: A CS $J = 7$–6 observation around NGC 2071 IRS. *Astrophys. J. Lett.* 347:85–88.

Yamashita, T., Suzuki, H., Kaifu, N., Tamura, M., Mountain, C. M., and Moore, T. J. T. 1989. A new bipolar flow and dense disk system associated with the infrared reflection nebula GGD27 IRS. *Astrophys. J.* 347:894–900.

Yang, J., and Anders, E. 1982a. Sorption of noble gases by solids, with reference to meteorites. I. Chromite and carbon. *Geochim. Cosmochim. Acta* 46:861–875.

Yang, J., and Anders, E. 1982b. Sorption of noble gases by solids, with reference to meteorites. II. Sulfides, spinels, and other substances; on the origin of planetary gases. *Geochim. Cosmochim. Acta* 46:877–892.

Yang, J. and Epstein, S. 1983. Interstellar organic matter in meteorites. *Geochim. Cosmochim. Acta* 47:2199–2216.

Yang, J., Fukui, Y., Umemoto, T., Ogawa, H., and Chen, H. 1990. A newly discovered molecular cloud in Cepheus OB4. *Astrophys. J.* 362:538–544.

Yang, J., Umemoto, T., Iwata, T., and Fukui, Y. 1991. A millimeter-wave line study of L1287: a case of induced star formation by stellar wind compression? *Astrophys. J.* 373:137–145.

Yang, S., Durisen, R. H., Cohl, H. S., Imamura, J. N., and Toman, J. 1991. Dynamic instabilities in rotating, low-mass protostars during early disk formation, *Icarus* 91:14–28.

Yeh, T. 1976. Reconnection of magnetic field lines in viscous conducting fluids. *J. Geophys. Res.* 81:4524–4530.

Yeh, T., and Axford, W. I. 1970. On the re-connection of magnetic field lines in conducting fluids. *J. Plasma Phys.* 4:207–229.

Yelle, R. V., and Lunine, J. I. 1989. Evidence for a molecule heavier than methane in the atmosphere of Pluto. *Nature* 339:288–290.

Yeomans, D. K. 1986. Physical interpretations from the motions of comets Halley and Giacobini-Zinner. In *20th ESLAB Symposium on the Exploration of Halley's Comet*, vol. 2, eds. B. Battrick, E. J. Rolfe and R. Reinhard, ESA SP-250, pp. 419–425.

Yeomans, D. K., and Kiang, T. 1981. The long term motion of comet Halley. *Mon. Not. Roy. Astron. Soc.* 197:633–643.

Yoder, C. F. 1979a. Diagrammatic theory of transition of pendulum-like systems. *Celest. Mech.* 19:3–29.

Yoder, C. F. 1979b. How tidal heating in Io drives the Galilean orbital resonance locks. *Nature* 279:767–770.

Yoder, C. F., and Peale, S. J. 1981. The tides of Io. *Icarus* 47:1–35.

York, D. G. and Rogerson, J. B. 1976. The abundance of deuterium relative to hydrogen in interstellar space. *Astrophys. J.* 203:378–385.

Yorke, H. W. 1980. The evolution of protostellar envelopes of masses 3 $M_\odot$ and 10 $M_\odot$. II. Radiation transfer and spectral appearance. *Astron. Astrophys.* 85:215–220.

Yorke, H. W., and Shustov, B. M. 1981. The spectral appearance of dusty protostellar envelopes. *Astron. Astrophys.* 98:125–132.

Yoshida, H. 1990. Construction of higher order symplectic integrators. *Phys. Lett. A.* 150:262–268.

Yoshii, Y., and Sabano, Y. 1980. Fragmentation of cosmic gas clouds due to thermal instability. *Publ. Astron. Soc. Japan* 32:229–245.

Yoshikawa, M. 1991. Motions of asteroids at the Kirkwood Gaps. III. On the 5:2, 7:3, and 2:1 resonances with Jupiter. *Icarus* 92:94–117.

Young, E. T., Lada, C. J., and Wilking, B. A. 1986. High resolution *IRAS* observations of the Rho Ophiuchi cloud core. *Astrophys. J. Lett.* 304:45–49.

Young, P., Schneider, D. P., and Schectman, S. A. 1981. The voracious vortex in HT Cassiopeiae. *Astrophys. J.* 245:1035–1042.

Yuan, C. 1969. Application of the density-wave theory to the spiral structure of the Milky Way system. II. Migration of stars. *Astrophys. J.* 158:889–898.

Yuan, C., and Cassen, P. 1985. Protostellar angular momentum transport by spiral density waves. *Icarus* 64:435–447.

Yun, J. L., and Clemens, D. P. 1990. Star formation in small globules: Bart Bok was correct! *Astrophys. J. Lett.* 365:73–76.

Yung, Y. L., Allen, M., and Pinto, J. P. 1984. Photochemistry of the atmosphere of Titan: comparison between models and observations. *Astrophys. J. Suppl.* 55:465–506.

Yung, Y. L., Friedl, R. R., Pinto, J. P., Bayes, K. D. and Wen, J.-S. 1988. Kinetic isotope fractionation and the origin of HDO and CH_3D in the solar system. *Icarus* 74:121–132.

Yusef-Zadeh, F., Cornwell, T. J., Reipurth, B., and Roth, M. 1990. Detection of synchrotron emission from a unique HH-like object in Orion. *Astrophys. J. Lett.* 348:61–64.

Zadnik, M. G., Wacker, J. F., and Lewis, R. S. 1985. Laboratory simulation of meteoritic noble gases. II. Sorption of xenon on carbon: etching and heating experiments. *Geochim. Cosmochim. Acta* 49:1049–1059.

Zahn, J. P. 1977. Tidal friction in close binary stars. *Astron. Astrophys.* 57:383–394.

Zahn, J. P. 1989. Tidal evolution of close binary stars. I. Revisiting the theory of the equilibrium tide. *Astron. Astrophys.* 220:112–116.

Zahn, J. P., and Bouchet, L. 1989. Tidal evolution of close binary stars. II. Orbital circularization of late-type binaries. *Astron. Astrophys.* 223:112–118.

Zahnle, K. J., and Kasting, J. F. 1986. Mass fractionation during transonic escape and implications for loss of water from Mars and Venus. *Icarus* 68:462–480.

Zahnle, K. J., and Walker, J. C. G. 1982. Evolution of solar ultraviolet luminosity. *Rev. Geophys. Space Phys.* 20:280–292.

Zahnle, K. J., Kasting, J. F., and Pollack, J. B. 1988. Evolution of a stream atmosphere during Earth's accretion. *Icarus* 74:62–97.

Zahnle, K. J., Kasting, J. F., and Pollack, J. B. 1990*a*. Mass fractionation of noble gases in diffusion-limited hydrodynamic hydrogen escape. *Icarus* 84:502–527.

Zahnle, K. J., Pollack, J. B., and Kasting, J. F. 1990*b*. Xenon fractionation in porous planetesimals. *Geochim. Cosmochim. Acta* 54:2577–2586.

Zealey, W., Mundt, R., Ray, T. P., Sandell, G., Geballe, T., Taylor, K. N. R., Williams, P. M., and Zinnecker, H. 1989. Stellar outflows and jets. *Proc. Astron. Soc. Australia* 8:62–67.

Zel'dovich, Ya. B. 1970. Gravitational instability: an approximate theory for large density perturbations. *Astron. Astrophys.* 5:84–89.

Zel'dovich, Ya. B. 1981. On the friction of fluids between rotating cylinders. *Proc. Roy. Soc. London* A374:299–312.

Zel'dovich, Ya. B., Ruzmaikin, A. A., and Sokoloff, D. D. 1983. *Magnetic Fields in Astrophysics* (New York: Gordon & Breach).

Zeng, Q., Batrla, W., and Wilson, T. L. 1984. A high density molecular fragment in the ρ Oph cloud. *Astron. Astrophys.* 141:127–130.

Zhang, Y., and Zindler, A. 1989. Noble gas constraints on the evolution of the Earth's atmosphere. *J. Geophys. Res.* 94:13719–13737.

Zharkov, V. N., and Gudkova, T. V. 1991. Models of giant planets with a variable ratio of ice to rock. *Ann. Geophysicae* 9:357–366.

Zharkov, V. N., and Trubitsyn, V. P. 1976. Structure, composition and gravitational field of Jupiter. In *Jupiter*, ed. T. Gehrels (Tucson: Univ. of Arizona Press), pp. 133–175.

Zharkov, V. N., and Trubitsyn, V. P. 1978. *Physics of Planetary Interiors*, ed. and trans. W. B. Hubbard (Tucson: Pachart Press).

Zhou, S., Wu, Y., Evans, N. J., Fuller, G. A., and Myers, P. C. 1989. A CS survey of low mass cores and comparison with NH_3 observations. *Astrophys. J.* 346:168–179.

Zhou, S., Evans, N. J., and Mundy, L. G. 1990a. An NH_3 ring around the infrared sources in NGC 2071. *Astrophys. J.* 355:159–165.

Zhou, S., Evans, N. J., Butner, H. M., Kutner, M. L., Leung, C. M., and Mundy, L. G. 1990b. Testing star formation theories: VLA observations of H_2CO in the Bok globule B335. *Astrophys. J.* 363:168–179.

Zinnecker, H. 1982. Prediction of the protostellar mass spectrum in the Orion near-infrared cluster. In *Symposium on the Orion Nebula to Honor Henry Draper*, eds. A. E. Glassgold, P. J. Huggins and E. L. Schucking (New York: New York Academy of Sci.), pp. 226–235.

Zinnecker, H. 1984a. Binary statistics and star formation. *Astrophys. Space Sci.* 99:41–70.

Zinnecker, H. 1984b. Star formation from hierarchical cloud fragmentation: a statistical theory of the log-normal initial mass function. *Mon. Not. Roy. Astron. Soc.* 210:43–56.

Zinnecker, H. 1985. Star formation as a random multiplicative process. In *Birth and Infancy of Stars*, eds. R. Lucas, A. Ormont and R. Stora (Amsterdam: North-Holland), pp. 473–475.

Zinnecker, H. 1986a. The initial mass function in young star clusters. In *Highlights of Astronomy*, vol. 7, ed. J.-P. Swings (Dordrecht: D. Reidel), pp. 489–499.

Zinnecker, H. 1986b. Population II brown dwarfs and dark halos. In *Astrophysics of Brown Dwarfs*, eds. M. C. Kafatos, R. S. Harrington and S. P. Maran (Cambridge: Cambridge Univ. Press), pp. 212–217.

Zinnecker, H. 1989a. Pre-main sequence binaries. In *ESO Workshop on Low Mass Star Formation and Pre-Main Sequence Objects*, ed. B. Reipurth (Garching: European Southern Obs.), pp. 447–469.

Zinnecker, H. 1989b. Towards a theory of star formation. In *Evolutionary Phenomena in Galaxies*, eds. J. E. Beckman and B. E. J. Pagel (Cambridge: Cambridge Univ. Press), pp. 113–127.

Zinnecker, H. 1990. Observations of fragmentation. In *Physical Processes in Fragmentation and Star Formation*, eds. R. Capuzzo-Dolcetta, C. Chiosi and A. Di Fazio (Dordrecht: Kluwer), pp. 201–209.

Zinnecker, H., Drapatz, S., and Cowsik, R. 1980. A Fokker-Planck model for the initial mass function. *Mitt. Astron. Ges.* 50:64–68.

Zinnecker, H., Mundt, R., Geballe, T. R., and Zealey, W. J. 1985. High spectral resolution observations of the H_2 2.12 micron line in Herbig-Haro objects. *Astrophys.*

J. 342:337–344.

Zinner, E. 1988. Interstellar cloud material in meteorites. In *Meteorites and the Early Solar System*, eds. J. F. Kerridge and M. S. Matthews (Tucson: Univ. of Arizona Press), pp. 956–983.

Zinner, E., Tang, M., and Anders, E. 1989. Interstellar SiC in the Murchison and Murray meteorites: Isotopic composition of Ne, Xe, Si, C, and N. *Geochim. Cosmochim. Acta* 53:3273–3290.

Zinner, E., Wopenka, B., Amari, S., and Anders, E. 1990. Interstellar graphite and other carbonaceous grains from the Murchison meteorite: Structure, composition and isotopes of C, N, and Ne. *Lunar Planet. Sci.* XXI:1379–1380 (abstract).

Zinner, E., Amari, S., Anders, E., and Lewis, R. 1991*a*. Large amounts of extinct ^{26}Al in interstellar grains from the Murchison meteorite. *Nature* 349:51–53.

Zinner, E., Amari, S., and Lewis, R. 1991*b*. Silicon carbide from a supernova? *Meteoritics* 26:413.

Ziurys, L. M. 1990. Millimeter and submillimeter studies of interstellar high temperature chemistry. In *Submillimetre Astronomy*, eds. G. D. Watt and A. S. Webster (Dordrecht: Kluwer), pp. 109–110.

Ziurys, L. M., and Turner, B. E. 1985. Detection of interstellar rotationally excited CH. *Astrophys. J. Lett.* 292:25–29.

Ziurys, L. M., Friberg, P., and Irvine, W. M. 1989. Interstellar SiO as a tracer of high-temperature chemistry. *Astrophys. J.* 343:201–207.

Zola, S. 1988. Disc radius versus time relations in Z Cha. *Acta Astron.* 39:45–49.

Zuckerman, B. 1989. Brown dwarfs in binary systems. In *Highlights of Astronomy*, vol. 8, ed. D. McNally (Dordrecht: Kluwer), pp. 119–122.

Zuckerman, B., and Becklin, E. E. 1987. A search for brown dwarfs and late M dwarfs in the Hyades and the Pleiades. *Astrophys. J. Lett.* 319:99–102.

Zuckerman, B., and Evans, N. J. 1974. Models of massive molecular clouds. *Astrophys. J. Lett.* 192:149–152.

Zuckerman, B., and Palmer, P. 1974. Radio radiation from interstellar molecules. *Ann. Rev. Astron. Astrophys.* 12:279–313.

Zurek, W. H., and Benz, W. 1986. Redistribution of angular momentum by nonaxisymmetric instabilities in a thick accretion disk. *Astrophys. J.* 308:123–133.

Zweibel, E. G. 1990*a*. Magnetic field line tangling and polarization measurements in clumpy molecular clouds. *Astrophys. J.* 362:545–550.

Zweibel, E. G. 1990*b*. Virial theorem analysis of the structure and stability of magnetized clouds. *Astrophys. J.* 348:186–197.

Zweibel, E. G., and Josafatsson, K. 1983. Hydromagnetic wave dissipation in molecular clouds. *Astrophys. J.* 270:511–518.

Acknowledgments

ACKNOWLEDGMENTS

The editors ackowledge NASA Grant NAGW-2160 and The University of Arizona for support in the preparation of this book. They wish to thank J. E. Frecker, who volunteered as one of the proofreaders of this book. The following authors wish to acknowledge specific funds involved in supporting the preparation of their chapters.

Adams, F.: NASA Grant NAGW-2802
Basri, G.: NSF Grant INT-8815464 and CNRS matching grant
Bertout, C.: NSF Grant INT-8815464 and CNRS matching grant
Blitz, L.: NSF Grant AST-8918912
Bodenheimer, P.: NSF Grant 89-14173
Cameron, A. G. W.: NSF Grant AST-8612647 *and* NASA Grants NAGW-1598, NAGW-1963 *and* NAGW-2277
Fukui, Y.: Grant-in-Aid for Specially Promoted Research 01065002, Grant-in-Aid for International Scientific Research 02044064 *and* Grant-in-Aid for General Scientific Research (A) 62420002,63420003
Goodman, A. A.: President's Fellowship at the University of California, Berkeley
Hartmann, L.: NASA Grant NAGW-511 *and* the Scholarly Studies Program of the Smithsonian Institution
Heiles, C.: NSF Grant 443836-21705
Hohenberg, C.: NASA Grants NAG 9-7 *and* NGT-50194
Ida, S.: Ministry of Education, Science, and Culture of Japan Grant-in-Aid for Scientific Research 02452062
Iwata, T.: Grant-in-Aid for Specially Promoted Research 01065002, Grant-in-Aid for International Scientific Research 02044064 *and* Grant-in-Aid for General Scientific Research (A) 62420002,63420003
Kenyon, S. J.: NASA Grant NAGW-511 *and* the Scholarly Studies Program of the Smithsonian Institution
Königl, A.: NASA Grant NAGW-2379
Lin, D. N. C.: NSF grant AST 89-14173
Lissauer, J. J.: NASA Planetary Geology and Geophysics Program Grant NAGW-1107
Lunine, J. I.: NASA Grant NAGW-1039
McKee, C. F.: NSF Grant AST-8918573 *and* a NASA grant to the Center for Star Formation Studies
Mizuno, A.: Grant-in-Aid for Specially Promoted Research 01065002, Grant-in-Aid for International Scientific Research 02044064 *and* Grant-in-Aid for General Scientific Research (A) 62420002,63420003
Mumma, M. J.: NASA RTOP 196-41-54
Nakagawa, Y.: Ministry of Education, Science, and Culture of Japan Grant-in-Aid for Scientific Research 03640391
Nakazawa, K.: Ministry of Education, Science, and Culture of Japan Grant-in-Aid for Scientific Research 02452062

Ohtsuki, K.: Ministry of Education, Science, and Culture of Japan Grant-in-Aid for Encouragement of Young Scientists 02952094

Stahler, S. W.: NSF Grant AST-9014479

Stewart, G. R.: NASA Planetary Geology and Geophysics Program Grant NAGW-769

Swindle, T. D.: NASA Grant NAG 9-240

Truran, J. W.: NSF grant AST 89-17442

van Dishoeck, E. F.: Netherlands Organization for Scientific Research (NWO) and Leiden Observatory

Weidenschilling, S. J.: NASA Contract NASW-4618

Wilking, B. A.: NASA ADP Grant NAG5-1670

Woolum, D. S.: NASA Grant NAG 9-57

Zinnecker, H.: Deutsche Forschungsgemeinschaft Grant Yo 5/7-1

Zweibel, E. G.: NSF Grant ATM85-60032 and the Institute for Theoretical Physics-NSF Grant PHY89-04035

Index

INDEX

luminosities, 2
occurrence of disks, 529–530
optically thin disks: naked T Tauri
 phase, 26. *See also under* T
 Tauri stars
other mechanisms for angular momentum transport, 21–22. *See also*
 Angular momentum
protostellar disks, 3–4, 22–24. *See
 also* Protostellar disks
protostellar phase, 2, 3
star/disk systems, 7
suppression of convection, 20–21
T Tauri phase, 2–7
T Tauri post-infall disks, 24–26. *See
 also under* T Tauri stars
transport processes, 2
viscous evolution, 16–17
YSOs, 2

FU Orionis outbursts objects, 6, 500,
 507–509, 511–513, 641
disk model, 501–506, 515
magnetic field, 514
mass loss, 514, 515
surface temperature, 514
variation with spectral type and wavelength, 497
winds and bipolar flow, 513–518

Galaxy
spiral arms, 106
age, 57
H I emission regions, 106
Ganymede, 1165, 1168
Gaseous disks
boundary layer, 510, 641, 708, 749
chemical and physical properties,
 processes, 1005, 1006–1021
composition, 1006, 1008
energy for chemical reactions, 1008–
 1010
evolution of circumstellar disks, 533–
 535, 837–865, 1005. *See also*
 Evolution of disks
forced turbulence, 1016–1018
growth, 1088
inner holes, 859
lightning, 1005
mixing, mixing ratios, 1005, 1015,
 1018
models, 1015–1016
observations, surveys, 523–525, 841–
 844, 1005
optically thick disks, 845–850
optically thin disks: origin, accretion,
 851, 860
radioactivity, 1009
searches for interstellar gas, 860
shell absorption lines, 1275, 1297

shock heating, 1005, 1006, 1009
solar and stellar photons, 1009
spectroscopic evidence for disks, 549
survival times for disks, 849
thermochemical reactions, 1005, 1006,
 1009, 1010–1013, 1015
time scales for disk evolution, 749,
 837–865
transition between optically thick and
 optically thin disks, 841–844,
 857–860
Giant molecular clouds
age, 131, 145
angular momentum, 137–143
cloud support by outflows, 637–639
clumps, 127, 133, 141, 148–159
CO emission studies, observations,
 surveys, 126, 131, 147, 150
dense cores, 108
formation and evolution, evolutionary
 stages, 126, 134, 142, 143, 145,
 159–161
luminosity, 133
mass, 133
O and OB formation, 130, 135, 145
Orion south, 631–639
properties, 126–134
relation to atomic hydrogen, 147–148
rotation, 127, 138–139, 141
star formation, 125, 134–137. *See
 also* Star formation
structure, 127, 151
temperature, 133
velocity, 138, 141
Giant planets (Jovian planets, outer planets), 1022
angular momentum, 1109
atmospheres, 1118–1120
composition, 1109
core heavy elements, 1109
core instability models, simulations,
 1109, 1127, 1137
critical core mass, 1127–1133
evolution, 1143–1146
formation and growth theories, 859,
 1109, 1126–1143
gaseous accretion, 1109
growth of core, 1087
mass loss, 1109
models, planetary interiors, 1110–
 1118, 1126
structure, 1109
Grains
collisions, 1289
composition, mineral composition, 90
ice sublimation, 1290
interstellar, 76, 883, 1253, 1254, 1265
large grains in orbit, 1262
mass, mass loss, 1266, 1290